FEEDS & NUTRITION DIGEST

(Formerly, FEEDS & NUTRITION—abridged)

by

M. E. ENSMINGER, PH.D.
J. E. OLDFIELD, PH.D.
W. W. HEINEMANN, PH.D.

SECOND EDITION

Copyright © 1990.

by

The Ensminger Publishing Company
648 West Sierra Avenue
P.O. Box 429
Clovis, California 93612
Phone #: 209/299-2263
U.S.A.

All rights reserved under International and Pan-American Copyright Conventions. No part of this publication may be reproduced, stored in a retrieval system or transmitted in any form or by any means (be it electronic, mechanical, photo-copying, recording or any other means currently known or yet to be invented) without prior written permission from the copyright holder.

LIBRARY OF CONGRESS CATALOG CARD NO. 89-083516
ISBN — 0-941218-08-2

Feeds & Nutrition is in several languages

Lest we forget!

Dedicated to the memory of two immortals

W. A. Henry
(1850–1932)

F. B. Morrison
(1887–1958)

authors of

Feeds and Feeding

1898, 1st edition—1956, 22nd edition

(From 1898 to 1910, Henry was sole author;
from 1910 to 1929, Henry and Morrison were coauthors;
from 1929 to 1956, Morrison was sole author.)

ABOUT THE AUTHORS

The three authors are shown seated in the library of the senior author, Dr. M. E. Ensminger. *Left to right:* Dr. James E. Oldfield, Dr. M. E. Ensminger, and Dr. Wilton W. Heinemann.

• **About Dr. M. E. Ensminger (Center)**—Dr. M. E. Ensminger is President of Agriservices Foundation, Clovis, California, a nonprofit foundation serving world agriculture in the area of World Food, Hunger and Malnutrition. Also, he is Adjunct Professor, California State University—Fresno; Adjunct Professor, The University of Arizona—Tucson; Distinguished Professor, University of Wisconsin—River Falls; Honorary Professor, Huazhong Agricultural College—Wuhan, People's Republic of China; and collaborator, U.S. Department of Agriculture.

Dr. Ensminger grew up on a Missouri Farm; completed B.S. and M.S. degrees at the University of Missouri, and the Ph.D. at the University of Minnesota; served on the staffs of the University of Massachusetts, the University of Minnesota, and Washington State University; served as Consultant, General Electric Company, Nucleonics Department; and served as the first President of the American Society of Agricultural Consultants.

Among Dr. Ensminger's honors and awards are: Distinguished Teacher Award, American Society of Animal Science; Washington State University named and dedicated the *Ensminger Beef Cattle Research Center,* in recognition of his contributions to the University; and an oil portrait of him was placed in the 300-year-old gallery of the famed Saddle and Sirloin Club, which is recognized as the highest honor that can be bestowed on anyone in the livestock industry.

Dr. Ensminger founded the International Stockmen's School, which he directed for 40 years. He has lectured and/or conducted seminars in more than 60 countries, including giving 5 invitational lectures before the Chinese Academy of Science and conducting the largest Seminar in China since the revolution. Dr. Ensminger is the author of more than 500 scientific and popular articles, bulletins and columns; and the author or co-author of 19 books, which are in several languages and used all over the world. The whole world is his classroom!

• **About Dr. James E. Oldfield (Left)**—Dr. James E. Oldfield is Director, Nutrition Research Institute, and Professor Emeritus of Animal Science, Oregon State University—Corvallis.

Dr. Oldfield was born in Victoria, B.C., Canada. He received the B.S.A. and M.S.A. degrees from the University of British Columbia; and the Ph.D. from Oregon State University. During World War II, he served in the Canadian Army, and was decorated with the Military Cross.

Dr. Oldfield has conducted nutrition research and/or been involved in nutrition problems pertaining to all animal species, including fur and laboratory animals. He has served in numerous scientific capacities, including: member of the Committee on Animal Nutrition, National Research Council; member, National Technical Advisory Committee on Agricultural Uses of Water, U.S. Department of Interior; member, Animal Nutrition Research Council; Director, Council of Agricultural Science and Technology (CAST); member, Nutrition Study Section, Division of Research Grants, National Institutes of Health; and Consultant to the Office of Economic Cooperation and Development, Ankara, Turkey. Dr. Oldfield is a world-renowned authority on selenium, on which subject he has given invitational lectures, participated in symposia, and/or served on committees throughout the United States, and in Canada, New Zealand, Australia, and China.

Among Dr. Oldfield's numerous professional recognitions are the Morrison Award, American Society of Animal Science; Fulbright Research Scholar, Massey University, New Zealand; and President, American Society of Animal Science.

Dr. Oldfield has published 142 technical papers and journal articles, and 92 reports and popular articles.

• **About Dr. Wilton W. Heinemann (Right)**—Dr. Wilton W. Heinemann, whose expertise is animal nutrition and pasture/range management, is Professor Emeritus, Washington State University, where he had a long and distinguished career. He completed B.S. and M.S. degrees at Washington State University, and the Ph.D. degree at Oregon State University. Dr. Heinemann is a Fellow in both the American Association for the Advancement of Science and the American Society of Animal Science.

Dr. Heinemann has served as Consultant/Nutritionist to the Bureau of Fisheries and Wildlife, U.S. Department of the Interior; the National Hay Association; the U.S. Feed Grains Council; Battelle Northwest; and to farmers, ranchers and agribusinesses, worldwide.

Dr. Heinemann has presented invitational papers in the U.K., Brazil, Finland, Australia, and the USSR; participated in the Nordic Congress of Agricultural Scientists, Helsinki, and in the Symposium on Ruminant Nutrition, Cambridge University, England; and has twice given invitational lectures before the Polish Academy of Science.

Dr. Heinemann has conducted research and/or been involved in problems pertaining to all animal species, including fish and wildlife. He is the author of 151 technical papers and journal articles, 158 popular articles and reports, and the co-author of a textbook. Dr. Heinemann is an international authority on animal nutrition, and on pasture and range management.

PREFACE TO THE SECOND EDITION
Feeds & Nutrition Digest

Genetic wizardry by gene splicing is giving rise to a major scientific revolution called biotechnology and spawning many new developments exceeding our fondest dreams. Biotechnology will involve every facet of animal production from breeding and feeding to the finished product, including the genetic makeup of animals and the feeds they eat; the digestion, physiology, stress tolerance, disease resistance, and efficiency of production of animals; the composition, quality, and quantity of products produced; along with the production of large quantities of drugs and chemicals. While some aspects of biotechnology are decades away from commercial production, others are near, and still others are here now.

But advanced technology calls for advanced animal adaptation, welfare, and environmental control. We need to breed and select animals adapted to an artificially-made environment—animals that not only survive, but thrive, under the conditions in which they are kept. We need to heed the warnings of endangered animals, endangered people, and an endangered planet—presaged by increased pollution, the greenhouse effect, acid rain, depletion of the ozone layer, and destruction of rain forests.

As the senior author of this book, I wish to acknowledge the great contribution of my two co-authors, Dr. James E. Oldfield and Dr. Wilton W. Heinemann, along with their patience with, and understanding of, my idiosyncracies. Audrey Ensminger shepherded the manuscript from beginning to end; Joan Wright deciphered my hieroglyphics and typed the manuscript; Lynn Wright set the type and paged up the copy; Ran Guang Liang prepared the cover design; and Margo Williams did the traditional art work and prepared the camera-ready copy. Also, I am grateful to Verla Rape, Program Analyst, Office of the Administrator, Economic Research Service, U.S. Department of Agriculture, who gave liberally of her time and talents in providing much of the recent source material for updating. Additionally, at appropriate places in the book, due acknowledgment and appreciation is expressed to all those who reviewed portions of the manuscript or responded so liberally to my call for illustrations and information. Without the help of all these fine folks, the task could not have been completed.

All authors are dreamers and doers. They visualize a need, then set out to fill it through the written word. These were the motivating forces back of *Feeds & Nutrition*. The need: to bring together in one book both the art and the science of livestock feeding; to narrow the gap between nutrition research and application—to speed the process; and to assure more and better animals in the future, followed by conversion of more feeds to palatable and nutritious foods for human consumption. If invoking my nocturnal habit and spending night after night in my lexicon garden alone doing my thing—writing this book—makes these dreams come true faster and more abundantly, throughout the world, I shall feel amply rewarded.

Clovis, California *M. E. Ensminger*
1990

FEEDS & NUTRITION DIGEST

CONTENTS

	Page
SECTION I — NUTRITION	vi
1. Food and Animals—a global perspective	1
2. Principles of Nutrition	11
3. Digestion/Absorption	27
4. Nutrients/Metabolism	41
5. Nutritional Disorders/Toxins	63
SECTION II — FEEDS	114
6. Types and Roles of Feedstuffs	115
7. Pasture and Range Forages	127
8. Hay	147
9. Silage/Haylage/High-Moisture Grain	169
10. Grains/High-Energy Feeds	187
11. Protein Supplements	203
12. By-product Feeds/Crop Residues	225
13. Feed Supplements/Additives/Implants	247
14. Feed Processing	265
15. Feed Analysis/Feed Evaluation	283
16. Buying Feeds/Commercial Feeds/Feed Laws	297
SECTION III — FEEDING	306
17. Animal Behavior/Environment	307
18. Feeding Standards/Ration Formulation	335
19. Feeding Beef Cattle	347
20. Feeding Dairy Cattle	425
21. Feeding Sheep	465
22. Feeding Goats	491
23. Feeding Swine	507
24. Feeding Poultry	549
25. Feeding Horses	591
26. Feeding Rabbits	623
27. Feeding Mink	641
28. Feeding Fish	657
SECTION IV — GLOSSARY	674
29. Glossary of Nutrition Terms	675
30. Glossary of Feedstuffs	695
SECTION V — COMPOSITION OF FEEDS	706
SECTION VI — APPENDIX	752
Weights and Measures	753
Animal Units	762
Poison Information Centers	762
Careers in Animal Industries	763

This is an original painting by the noted artist, Tom Phillips (3333 17th Street, San Francisco, California 94110), prepared especially for this book. It portrays the artist's conception of what is in all five chapters of Section I, *Nutrition*.

SECTION I

NUTRITION

Feeding had its beginning as an art, the foundations of which were animal instinct and a blend of the caretaker's fads, foibles, and trade secrets. Then came science, founded on chemistry, physics, physiology, and bacteriology.

For many years, the keepers of the herds and flocks were responsible for the very considerable progress made in the art of feeding. They were intensely practical, never overlooking the utility value or the market requirements. No animal met with their favor unless it was earned by meat upon the back, milk in the pail, weight and quality of wool, pounds gained for pounds of feed consumed, draft or speed ability, or some other performance of practical value. In time, scientists teamed up with caretakers to improve the feeding of animals, slowly, but surely, evolving with the science called *nutrition*.

The successful merger of the art and the science of feeding—the joining of feeds and nutrition—ushered in a new era in animal agriculture. In the process, it also stimulated increased interest in human nutrition, where the requirements are similar. This gave rise to the statement that, "We are gradually learning to feed our children as well as our animals." Many maladies which had long plagued both humankind and beast were traced to dietary deficiencies, imbalances, and toxicities. Rectifying these nutritional problems improved the health and performance of all creatures, including humans.

But the past is prologue! The final chapter of the 20th century, which is now being written, may well be the most revolutionary and dramatic of all in animal nutrition.

Section I covers the fundamentals of nutrition, with separate chapters devoted to each of the following:

Chapter 1, Food and Animals—a global perspective
Chapter 2, Principles of Nutrition
Chapter 3, Digestion/Absorption
Chapter 4, Nutrients/Metabolism
Chapter 5, Nutritional Disorders/Toxins

FOOD AND ANIMALS— a global perspective

Original painting by Tom Phillips

Contents	Page
All Flesh Is Grass!	1
Photosynthesis and Ruminants	1
Conserve Energy	2
Food for the 21st Century	4
Who Shall Eat?	4
Favoring Direct Human Grain Consumption	4
Favoring an Animal Agriculture	6
World Without End—with Animals	8
Questions for Study and Discussion	9

Back of animals are feeds; and back of the feeds are soil resources, spring rains, and the energy of the sun. With the aid of science, technology, and animals, farmers and ranchers combine these to produce a tasty platter of meat and eggs for the table, cream for the peaches, butter for the biscuits, and cheese for the macaroni—all derived from the sun through the process known as photosynthesis.

Fig. 1-1. Without photosynthesis, there would be no oxygen, no plants, no food, no animals, and no people.

But animal products are far more than just very tempting and delicious foods! From a nutrition standpoint, foods of animal origin contribute certain essentials to the American diet; they supply 35% of the energy, 70% of the protein—along with the essential amino acids, 80% of the calcium, 60% of the phosphorus, and significant amounts of the other minerals and vitamins needed in the human diet. It is noteworthy, too, that animal products contain vitamin B-12, which does not occur in plant foods, and that they are a rich source of iron, the availability of which is twice as high as in plants.

ALL FLESH IS GRASS!

Life on earth is dependent upon photosynthesis. Without it, there would be no oxygen, no plants, no feed, no food, no animals, and no people.

As fossil fuels (coal, oil, shale, and petroleum)—the stored photosynthates of previous millennia—become exhausted, the biblical statement, "all flesh is grass" (Isaiah 40:6), comes alive again. The focus is on photosynthesis. Plants, using solar energy, are by far the most important, and the only renewable, energy-producing method,[1] the only basic food-manufacturing process in the world; and the only major source of oxygen in the earth's atmosphere. Even the chemical and electrical energy used in the brain cells of man is the product of sunlight and the chlorophyll of green plants. Thus, in an era of world food shortages, it is inevitable that the entrapment of solar energy through photosynthesis will, in the long run, prove more valuable than all the underground fossil fuels—for when the latter are gone, they are gone forever.

PHOTOSYNTHESIS AND RUMINANTS

Photosynthesis is the process by which the chlorophyll-containing cells in green plants capture the energy of the sun and convert it into chemical energy; it's the process through which plants synthesize and store organic compounds, especially carbohydrates, from inorganic compounds—carbon dioxide, water, and minerals, with the simultaneous release of oxygen.

Ruminants, which include cattle, sheep, goats, and water buffalo, are even-toed, hoofed animals that ruminate (regurgitate and chew a cud) and have a complex four-compartment stomach characterized by much storage space and microbial fermentation and adapted to the effective use of high-fiber feeds and the manufacture of B-complex vitamins and essential amino acids.

Photosynthesis and ruminants team up to provide every ounce of food that we eat, every breath of oxygen that we inhale, and a very large portion of all the B-complex vitamins and all the essential amino acids that we require. Without photosynthesis and ruminants, there would be no plant and animal life on earth—and no human race.

---

[1]Certain types of microorganisms, termed chemoautotrophs, get their energy from inorganic compounds, but aside from this minor exception, the energy that runs the life support systems of the biosphere comes from photosynthesis.

Photosynthesis is dependent upon the presence of chlorophyll, a green pigment which develops in plants soon after they emerge from the soil. Chlorophyll is a chemical catalyst—it stimulates and makes possible certain chemical reactions without becoming involved in the reaction itself. By drawing upon the energy of the sun, it can convert inorganic molecules, carbon dioxide (CO_2) and water (H_2O), into an energy-rich organic molecule such as glucose ($C_6H_{12}O_6$), and at the same time release free oxygen (O_2). It transforms solar energy into a form that can be used by plants, animals, and humans. Because of this capability, chlorophyll has been referred to as the link between nonliving and living matter, or the pathway through which nonliving elements may become part of living matter.

Through the photosynthetic process, it is estimated that more than a billion tons of carbon per day are converted from inorganic carbon dioxide (CO_2) to organic sugars ($C_6H_{12}O_6$— glucose), which can then be converted into other carbohydrates, fats, and proteins—the three main groups of organic materials of living matter.

Photosynthesis is a series of many complex chemical reactions, involving the following two stages (see Fig. 1-2):

Stage 1—The water molecule (H_2O) is split into hydrogen (H) and oxygen (O); and oxygen, the necessary gas for breathing of animals, is released into the atmosphere. Hydrogen is combined with certain organic compounds to keep it available for use in the second step of photosynthesis. Chlorophyll and light are involved in this stage.

Stage 2—Carbon dioxide (CO_2) combines with released hydrogen to form the simple sugar (glucose) and water. This reaction is energized (powered) by ATP (adenosine triphosphate), a stored source of energy. Neither chlorophyll nor light is involved in this stage.

The chemical reactions through which chlorophyll converts the energy of solar light to energy in organic compounds is one of nature's best-kept secrets. Scientists have not been able to unlock it, as they have so many of life's processes. Moreover, photosynthesis is limited to plants; animals store energy in their products—meat, milk, and eggs—but they must depend upon plants to manufacture it.

Although photosynthesis is vital to life itself, it is very inefficient in capturing the potentially available energy. Of energy that leaves the sun in a path toward the earth, only about half reaches the ground. The other half is absorbed or reflected in the atmosphere. Most of that which reaches the ground is dissipated immediately as heat or is used to evaporate water in another important process for making life possible. Only about 2% of the earthbound energy from the sun actually reaches green plants, and only half of this amount (1%) is transformed by photosynthesis to energy storage in organic compounds. Moreover, only 5% of this plant-captured energy is fixed in a form suitable as food for people.

With such a small portion of the potentially useful solar energy actually being used to form plant tissue, it would appear that some better understanding of the action of chlorophyll should make it possible to increase the effectiveness of the process. Three approaches are suggested: (1) increasing the amount of photosynthesis on earth; (2) manipulating plants for increased efficiency of solar energy conversion; and (3) converting a greater percentage of total energy fixed as chemical energy in plants (the other 95%) into a form available to humans. Ruminants are the solution to the latter approach; they can convert energy from such humanly inedible plant materials as grass, cornstalks, and straw into food for humans. (See Fig. 1-3.)

CONSERVE ENERGY

In addition to production, as such, there are two other important steps in the feed-food line as it moves from the producer to the consumer; namely, processing and marketing, both of which require higher energy inputs than to produce the food on the farm. (See Table 1-1.)

Table 1-1 points up the increasing drain that modern food production is putting on the energy supply. In 1980, U.S. farms put in 2.8 calories of fuel per calorie of food grown, 3.1 times more than the on-farm energy input in 1940.

Table 1-1 also shows that, in the United States in 1980, a total of 12.1 calories were used in the production, food processing, and marketing-cooking for every calorie of food consumed, with a percentage distribution of the total cost of energy at each step from producer to consumer as follows: on the farm, 23%; food processing, 39%; and marketing and home cooking, 38%. In 1940, it took only 5.2 cal-

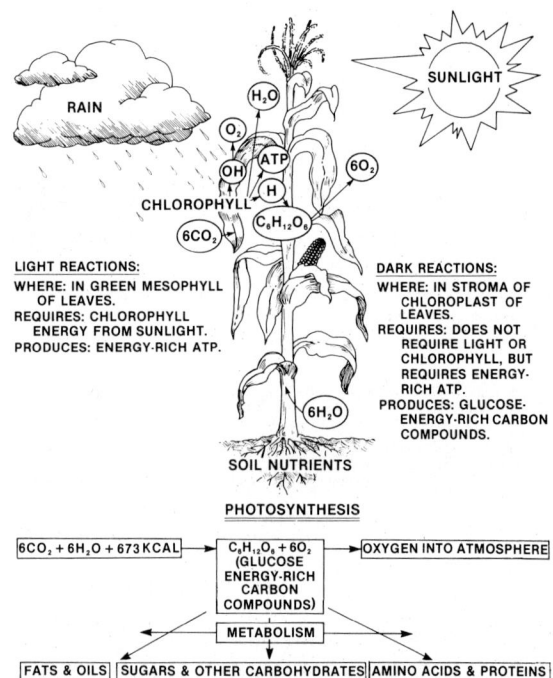

Fig. 1-2. Photosynthesis fixes energy. Diagrammatic summary of (1) photosynthesis, and (2) the metabolic formation of organic compounds from the simple sugars. This diagram shows the following:

1. Carbon dioxide gas from the air enters the green mesophyll cells of plant leaves.
2. Plants take up oxygen from the air for some of their metabolic processes and release oxygen back to the air from other metabolic processes.
3. Plants take up water and essential elements from the soil.
4. The energy essential to photosynthesis is absorbed by chlorophyll and supplied by sunlight.
5. For a net input of 6 molecules of carbon dioxide and 6 molecules of water, there is a net output of 1 molecule of sugar and 6 molecules of oxygen.
6. The process is divided into light and dark reactions, with the light reactions building up the energy-rich ATP required for the dark reactions.
7. In the process, 673 Calories (kcal) of energy are used.
8. The sugar (glucose) manufactured in photosynthesis may be converted into fats and oils, sugars and other carbohydrates, and amino acids and proteins.

Food and Animals—a global perspective

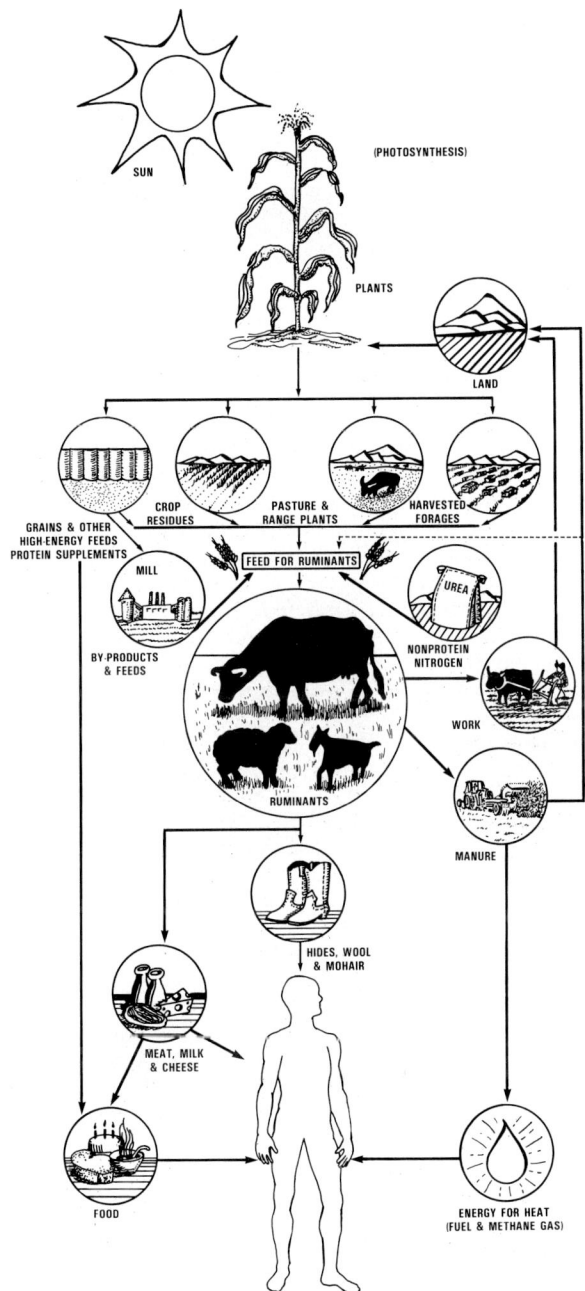

Fig. 1-3. The sun-plant-land-ruminant-human relationship. Ruminants step up energy and manufacture B-vitamins and essential amino acids. Their feed comes from plants which have their tops in the sun and their roots in the soil. Hence, we have the nutrition cycle as a whole—from the sun and the soil, through the plant, thence to the ruminant (and human) and back to the soil again.

ories—slightly less than half the 1980 figure—to get 1 calorie of food on the table. It's noteworthy, too, that more energy is required for food processing and marketing/home cooking than for growing the product; and that, from 1940 to 1980, the on-the-farm energy requirement increased by 3.1 times, in comparison with an increase of 2.1 and 2.2 times for each of the other steps—processing and marketing/home cooking.

Table 1-1
MODERN FOOD PRODUCTION IS INEFFICIENT IN ENERGY UTILIZATION—THE STORY FROM PRODUCER TO CONSUMER[1]

Year	On the Farm	Food Processing	Marketing and Home Cooking	Total/Person /Year
1940[2]				
Million kcal	0.9	2.2	2.1	5.2
Percent	18.0	42.0	40.0	100.0
1980[3]				
Million kcal	2.8	4.7	4.6	12.1[4]
Percent	23.0	39.0	38.0	100.0
Increase, times, 1940-1980	3.1	2.1	2.2	2.3

[1]Energy in million kcal per capita to produce one million kcal of food in the U.S.
[2]Values from Borgstrom, G., "The Price of a Tractor," *Ceres,* FAO of the U.N., Rome, Italy, Nov.-Dec., 1974, p. 18, Table 3.
[3]Authors' estimate based on several reports detailing trends in energy usage.
[4]This means that in 1980 it required 12.1 million kcal to produce 1 million kcal of food for each person, a daily consumption of 2,740 kcal (1,000,000 ÷ 365 = 2,740).

Modern intensive farming has markedly increased crop yields per acre and per man-hour—by as much as 50- to 100-fold. But this has been done at the cost of large inputs of fuel. Today, for a surprising number of cropping systems, a 10- to 50-fold increase in the energy output merely doubles or triples the food energy. Thus, the law of diminishing returns prevails.

Scarce and high-priced fossil fuels have spurred a search for conserving stored energy and for increased energy production through photosynthesis. Higher productivity of the agriculture of tomorrow must be achieved through ingenious approaches in order to reverse the present lopsided energy balance. In obtaining increased feed and food yields, we must consider how many calories of energy are required to produce each calorie of feed or food. We must remember that photosynthesis does not deplete fossil fuels. We must remember, too, that grazing animals do not require fuel outside of their own body use to harvest the energy and other nutrients of grass (solar energy converted into chemical energy by grass), a renewable source. It follows

Fig. 1-4. An Oriental wet rice peasant, using animal power (water buffalo), expends only 1 calorie of energy to produce each 50 calories of food. By comparison, the average U.S. farmer, using mechanical power (tractors), expends 2.5 calories of fuel energy to produce 1 calorie of food. (Courtesy, International Bank for Reconstruction and Development, Washington, D.C.)

that ruminants, which utilize grazing land, offer the best means of stepping up and storing energy for humans.

Energy may also be conserved by lessening waste. Pests cause an estimated 30% annual crop loss in the worldwide potential production of crops, livestock, and forests.[2] Every part of our feed, food, and fiber supply is vulnerable to pest attack, including marine life, wild and domestic animals, field crops, horticultural crops, and wild plants. Obviously, reducing these losses would conserve energy and increase the supply of feed, food, and fiber.

FOOD FOR THE 21st CENTURY

As the ghost of hunger, foretold by the English clergyman Thomas Robert Malthus in 1798, stalks the world, the focus is on animals. During periods of food scarcity, it is inevitable that some will suggest that grain be diverted from livestock and poultry feeding—that they will challenge the efficiency of animals in converting feed to food and the place of animals in the economical production of human food. Animal agriculture will be on trial. Increasingly, the charge will be made that much of the world goes hungry because of the substitution of meat, milk, and eggs for direct grain consumption. A response to this accusation requires that animal agriculturalists substitute knowledge for moral indignation. To this end, the important sections that follow are presented.

Who Shall Eat?

An appalling 500 million people in the world suffer from hunger and malnutrition. More shocking yet, it is estimated that the troubled 21st century will open with 1.3 billion malnourished people. For these starving millions around the globe, life is little more than a heartrending journey to an end which refuses to arrive soon enough to stop their suffering. For the most part, the world's hungry and malnourished are grain eaters.

Cereal grain is the most important single component of the world's food supply, accounting for 50% of the food produced in all the globe. It is the major source of food for many of the world's poorest people, supplying 58% of the total calories in the developing countries (see Fig. 1-5).

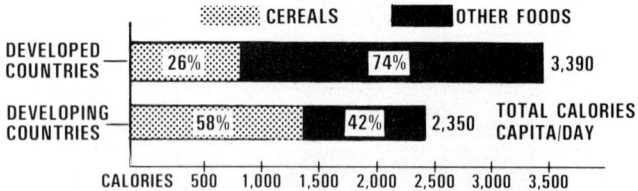

Fig. 1-5. Calories per person per day from cereals vs other foods. In the developed countries, only 26% of the calories comes from direct consumption of cereals, compared with about 58% in developing countries. (Source: The Fifth World Food Survey, FAO, United Nations, Rome, Italy, 1985)

However, in many developed countries, more grain is fed to animals than is consumed directly by humans. Under such circumstances, sporadic food shortages and famine in different parts of the world give rise to the following recurring questions:

1. Who should eat grain—people or animals? Shall we have food or feed?
2. Can we have both food and feed?

In attempting to answer these complex questions, those favoring people going on a grain diet often substitute moral indignation for knowledge. The authors' answers are given in the sections that follow.

Favoring Direct Human Grain Consumption

Historically, the people of new and sparsely populated countries have been meat eaters, whereas the population of the older and more densely populated areas have been vegetarians. The latter group has been forced to eliminate most animals and consume plants and grains directly in an effort to avoid famine.

Among the arguments sometimes advanced by those who favor bread alone—the direct human consumption of grain—are the following:

1. **More people can be fed.** More hunger can be alleviated with a given quantity of grain by completely eliminating animals. About 2,000 lb of concentrates (mostly grain) must be supplied to livestock in order to produce enough meat and other livestock products to support a person for a year, whereas 400 lb of grain (corn, wheat, rice, soybeans, etc.) eaten directly will support a person for the same period of time. Thus, a given quantity of grain eaten directly will feed 5 times as many people as it will if it is first fed to livestock and is eaten indirectly by humans in the form of livestock products.

2. **On a feed, calorie, or protein conversion basis, it's not efficient to feed grain to animals and then to consume the livestock products.** This fact is pointed up in Fig. 1-6 and Table 1-2.[3]

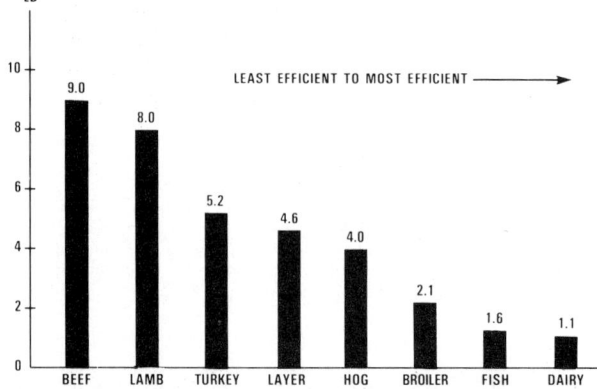

Fig. 1-6. Pounds of feed required to produce 1 lb of product. This shows that it takes 9 lb of feed to produce 1 lb of on-foot beef, whereas it takes only 1.11 lb of feed to produce 1 lb of milk. (Source: Table 1-2 of this chapter)

---

[2]Ennis, Jr., W. B., W. M. Dowler, and W. Klassen, "Crop Production to Increase Food Supplies," *Science*, Vol. 188, No. 4188, May 9, 1975, pp. 593–598.

[3]See **NOTE WELL** at the bottom of Table 1-2.

Food and Animals—a global perspective

TABLE 1-2
FEED TO FOOD EFFICIENCY RATING BY SPECIES OF ANIMALS, RANKED BY PROTEIN CONVERSION EFFICIENCY
(Based on Energy as TDN or DE and Crude Protein in Feed Eaten by Various Kinds of Animals Converted into Calories and Protein Content of Ready-to-Eat Human Food)

Species	Unit of Production (on foot)	Feed Required to Produce One Production Unit				Dressing Yield		Ready-to-Eat; Yield of Edible Product (meat & fish deboned & after cooking)				Feed Efficiency[4]		Efficiency Rating			
		Pounds	TDN[1]	DE[2]	Protein	Percent	Net Left	As % of Raw Product (carcass)	Amount Remaining from One Unit of Production	Cal-orie[3]	Protein[3]	(lb feed to produce one lb product)		Calorie Efficiency[5]		Protein Efficiency[6]	
		(lb)	(lb)	(kcal)	(lb)	(%)	(lb)	(%)	(lb)	(kcal)	(lb)	(%)	(ratio)	(%)	(ratio)	(%)	(ratio)
Broiler	1 lb chicken	2.1[7]	1.7[8]	3,400	0.21[8]	72[13]	0.72	54[14]	0.39	274	0.11	47.6	2.1:1	8.1	12.4:1	52.4	1.9:1
Fish	1 lb fish	1.6[9]	0.98	1,960	0.57	65[10]	0.65	57[11]	0.37	285	0.27	62.5	1.6:1	14.5	6.9:1	47.6	2.1:1
Dairy cow	1 lb milk	1.11[7]	0.9[8]	1,800	0.1[8]	100	1.0	100	1.0	309	0.037	90.0	1.11:1	17.2	5.8:1	37.0	2.7:1
Turkey	1 lb turkey	5.2[7]	4.21[8]	8,420	0.46[8]	79.7[13]	0.797	57[15]	0.45	446	0.146	19.2	5.2:1	5.3	18.9:1	31.7	3.2:1
Layer	1 lb eggs (8 eggs)	4.6[7]	3.73[8]	7,460	0.41[8]	100	1.0	100[12]	1.0[12]	616	0.106	21.8	4.6:1	8.3	12.1:1	25.9	3.9:1
Hog (birth to market weight)	1 lb pork	4.0[18]	3.2	6,400	0.36	70[16]	0.70	44[17]	0.31	341	0.088	0.25	4.0:1	5.3	18.8:1	24.4	4.1:1
Rabbit	1 lb fryer	3.0[19]	2.20	4,400	0.48	55[19]	0.55	79[19]	0.43	301	0.08	35.7	2.8:1	6.8	14.6:1	16.7	6.0:1
Beef steer (yearling finishing period in feedlot)	1 lb beef	9.0[8]	5.85	11,700	0.90	58[8]	0.58	49[17]	0.28	342	0.085	11.1	9.0:1	2.9	34.2:1	9.4	10.6:1
Lamb (finishing period in feedlot)	1 lb lamb	8.0[18]	4.96	9,920	0.86	47[18]	0.47	40[17]	0.19	225	0.052	12.5	8.0:1	2.3	44.1:1	6.0	16.5:1

[1]TDN pounds computed by multiplying pounds feed (column to left) times percent TDN in normal rations. Normal ration percent TDN taken from M. E. Ensminger's books and rations, except for the following: dairy cow, layer, broiler, and turkey from *Agricultural Statistics 1974*, p. 358, Table 518. Fish based on averages recommended by Michigan and Minnesota Stations and U.S. Fish and Wildlife Service.

[2]Digestible Energy (DE) in this column given in kcal, which is 1 Calorie (written with a capital C), or 1,000 calories (written with a small c). Kilocalories computed from TDN values in column to immediate left as follows: 1 lb TDN = 2,000 kcal.

[3]From *Lessons on Meat*, National Live Stock and Meat Board, 1965.

[4]Feed efficiency as used herein is based on pounds of feed required to produce 1 lb of product. Given in both percent and ratio.

[5]Kilocalories in ready-to-eat food = kilocalories in feed consumed, converted to percentage. Loss = kcal in feed ÷ kcal in product.

[6]Protein in ready-to-eat food = protein in feed consumed, converted to percentage. Loss = pounds protein in feed ÷ pounds protein in product.

[7]*Agricultural Statistics 1974*, p. 358, Table 518. Pounds feed per unit of production is expressed in equivalent feeding value of corn.

[8]Pounds feed (column No. 2) per unit of production (column No. 1) is expressed in equivalent feeding value of corn. Therefore, the values for corn were used in arriving at these computations. No. 2 corn values are TDN, 81%; protein, 8.9%. Hence, for the dairy cow 81% × 1.11 = 0.9 lb TDN; and 8.9% × 1.11 = 0.1 lb protein.

[9]Data from report by Dr. Phillip J. Schaible, Michigan State University, *Feedstuffs*, April 15, 1967.

[10]*Industrial Fishery Technology*, edited by Maurice E. Stansby, Reinhold Pub. Corp., 1963, Ch. 26, Table 26-1.

[11]*Ibid.* Reports that "Dressed fish averages about 73% flesh, 21% bone, and 6% skin." In limited experiments conducted by A. Ensminger, it was found that there was a 22% cooking loss on filet of sole. Hence, these values—73% flesh from dressed fish, minus 22% cooking losses—give 57% yield of edible fish after cooking, as a percent of the raw, dressed product.

[12]Calories and protein computed basis per egg; hence, the values herein are 100% and 1.0 lb, respectively.

[13]*Marketing Poultry Products*, 5th Ed., by E. W. Benjamin et al., John Wiley & Sons, 1960, p. 147.

[14]*Factors Affecting Poultry Meat Yields*, University of Minnesota Sta. Bull. 476, 1964, p. 29, Table 11 (fricassee).

[15]*Ibid.* Page 28, Table 10.

[16]Ensminger, M. E., *The Stockman's Handbook*, 6th Ed., Sec. XII.

[17]Allowance made for both cutting and cooking losses following dressing. Thus, values are on a cooked, ready-to-eat basis of lean and marbled meat, exclusive of bone, gristle, and fat. Values provided by National Live Stock and Meat Board (personal communication of June 5, 1967, from Dr. Wm. C. Sherman, Director, Nutrition Research, to the senior author), and based on data from *The Nutritive Value of Cooked Meat*, by Ruth M. Leverton and George V. Odell, Misc. Pub. MP-49, Appendix C, March 1958.

[18]Estimates by the authors.

[19]Based on information in *Commercial Rabbit Raising*, Ag. Hdbk. No. 309, USDA, 1966, and *A Handbook on Rabbit Raising*, by H. M. Butterfield, Washington State University Ext. Bull. No. 411.

NOTE WELL: It could be argued that Table 1-2 makes no provision for the feed used by the sires and dams of these animals—the animals that gave birth to these producers. Others may be critical of using a yearling steer without making provision to get him to the feedlot stage. Finally, it may be contended that any such comparison should be between animals of like age; for example, between broilers and veal calves. Having raised these questions, the authors submit Table 1-2, which in their judgment is as fair a rating on feed to food efficiency as can be made.

Favoring An Animal Agriculture

Among some social reformists, the charge persists that much of the world goes hungry because of the substitution of meat, milk, and eggs for direct grain consumption. A response to this accusation requires far more than a simple denial.

The following facts are presented in favor of sharing grain with animals, then consuming the animal products:

1. **Animals provide needed power.** In the developing nations, cattle, water buffalo, and horses still provide much of the agricultural power. In this capacity, they contribute to human food supply from plant sources.

2. **Animals provide needed nutrients.** It is estimated that the average American gets the percentages of food nutrients shown in Fig. 1-7 from animal products.

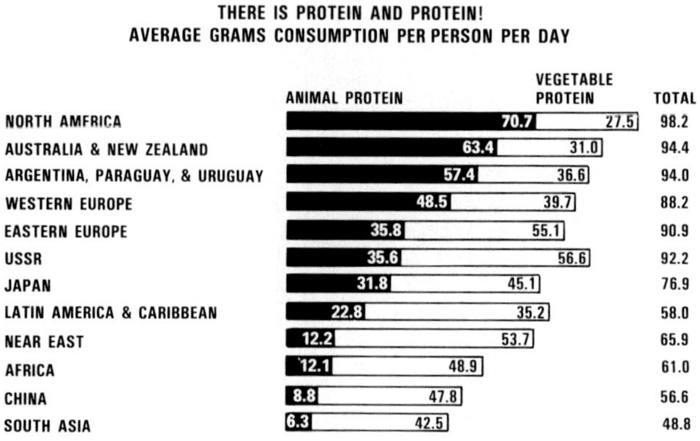

THERE IS PROTEIN AND PROTEIN!
AVERAGE GRAMS CONSUMPTION PER PERSON PER DAY

	ANIMAL PROTEIN	VEGETABLE PROTEIN	TOTAL
NORTH AMERICA	70.7	27.5	98.2
AUSTRALIA & NEW ZEALAND	63.4	31.0	94.4
ARGENTINA, PARAGUAY, & URUGUAY	57.4	36.6	94.0
WESTERN EUROPE	48.5	39.7	88.2
EASTERN EUROPE	35.8	55.1	90.9
USSR	35.6	56.6	92.2
JAPAN	31.8	45.1	76.9
LATIN AMERICA & CARIBBEAN	22.8	35.2	58.0
NEAR EAST	12.2	53.7	65.9
AFRICA	12.1	48.9	61.0
CHINA	8.8	47.8	56.6
SOUTH ASIA	6.3	42.5	48.8

Fig. 1-8. Average grams protein consumption per person per day, with a breakdown into animal and vegetable protein, by geographic areas and countries. (From: *Ceres*, FAO/UN, Vol. 8, No. 3)

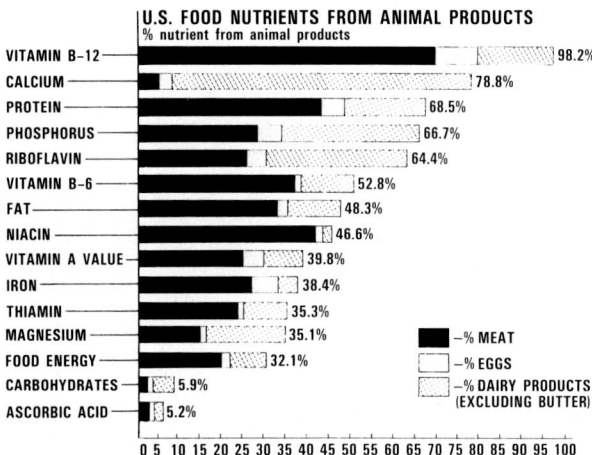

Fig. 1-7. Percentage of food nutrients contributed by animal products of the total nutrient supply in the U.S. (From: *Agricultural Statistics*, USDA)

Foods of animal origin (meat, milk, eggs, and their various by-products) are especially important in the American diet; they provide ⅔ of the total protein, ⅓ of the total energy, ⅘ of the calcium, ⅔ of the phosphorus, and significant amounts of the other minerals, and vitamins needed in the human diet. Note, too, that animal products provide practically all of the Vitamin B-12, which does not occur in plant foods—only in animal sources, and fermentation products. Also, it is noteworthy that the availability of iron in meat is twice as high as in plants.

About ⅔ of the world's protein supply is provided from plant sources, ⅓ from animal sources. (See Fig. 1-8.) The Food and Agriculture Organization of the United Nations reports that the world's diet needs animal protein in amounts equivalent to ⅓ of the total protein requirements. Thus, there should be ample animal protein, *provided* it is equally distributed. But it isn't.

The most important role of animal protein is to correct the amino acid deficiencies of the cereal proteins, which supply about ⅔ of the total protein intake, and which are notably deficient in the amino acid, lysine. The latter deficiency can also be filled by soybean meal, fish, protein concentrates and isolates, synthetic lysine, or high-lysine corn. But such products have neither the natural balance in amino acids nor the appetite appeal of animal protein.

3. **Animals produce protein of higher value than plants.** Proteins from animal sources (meat, milk, and eggs) have a higher value than proteins from plant sources because they have every amino acid needed for growth, including lysine, tryptophan, and methionine, which are deficient in vegetable sources.

4. **Ruminants convert nonprotein nitrogen to protein.** Ruminant animals, by their ability to fix nitrogen through bacterial action, can use nonprotein nitrogen, like urea, to produce protein for humans in the form of meat and milk (see Fig. 1-9).

5. **Animals step up the protein content and quality of foods.** Grains, such as corn, are much lower in protein content in cereal form than after conversion into meat, milk, and eggs. On a dry basis, the protein contents of selected products are corn, 10.45%; beef (Choice grade, total edible, trimmed to retail level, raw), 30.7%; milk, 26.4%; and eggs, 47.0%.[4] Also, animals increase the quality (*i.e.*, biological value) of the protein.

6. **Animals provide products that meet consumer preferences.** Most people who can afford to do so eat a portion of their food in the form of livestock products simply because of preference—because they like them.

7. **Much of the world's land is not cultivatable.** Vast acreages throughout the world—including arid and semiarid grazing lands; and brush, forest, cutover, and swamplands—are unsuited to the production of bread grains or any other type of farming; their highest and best use is, and will remain, for grazing and forest.

---

[4]*Composition of Foods*, Ag. Hdbk. No. 8, Agricultural Research Service, USDA.

Food and Animals—a global perspective

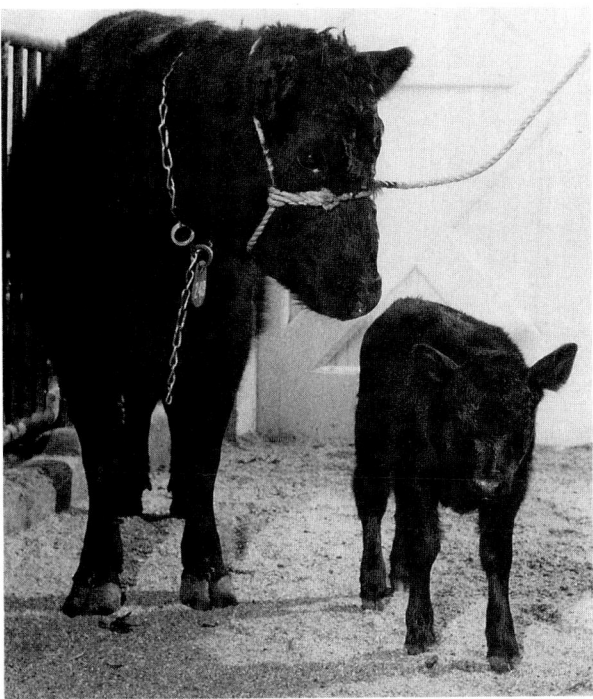

Fig. 1-9. This cow is believed to be the first in the world to have grown, conceived, and given birth to a healthy calf when fed since weaning (7 months) on a protein-free diet that contained urea as the only source of nitrogen. The cow weighed 930 lb and the calf 61 lb at the time of birth. (Courtesy, Robert R. Oltjen, U.S. Meat Animal Research Center, Clay Center, Nebr.)

8. **Forages provide most of the feed for livestock.** Pastures and other roughages—feeds not suitable for human consumption—provide most of the feed for livestock, especially for ruminants, throughout the world. In the 48 states of the U.S., grassland, pasture and range constitute a major land use (see Fig. 1-10).

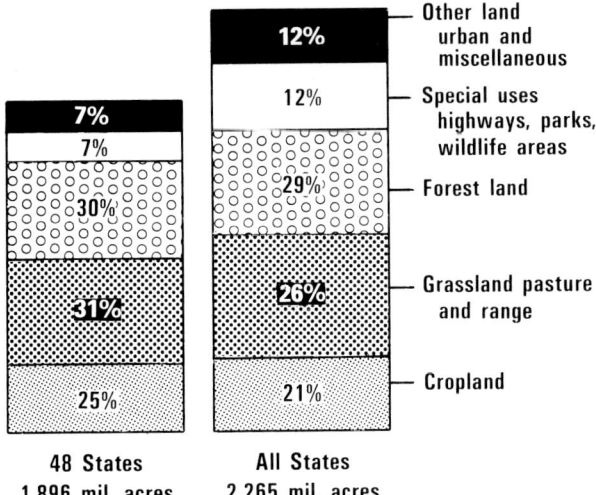

Fig. 1-10. Major uses of land. (From: Ag. Hdbk. No. 652, USDA, p. 16, Chart 41)

Despite grains being relatively plentiful in the United States, forages provide the bulk of animal feeds; pastures and other roughages account for 93.8% of the total feed of sheep and goats, 84.5% of the feed of beef cattle, 58.7% of the feed of dairy cattle, and 61.7% of the feed of all livestock.[5]

9. **Food and feed grains are not synonymous.** Animals do not compete to any appreciable extent with the hungry people of the world for food grains, such as rice or wheat. Instead, they eat feed grains and by-product feeds—such as field corn, grain sorghum, barley, oats, milling by-products, distillery wastes, and fruit and vegetable wastes—for which there is little or no demand for human use in most countries, plus forages and grasses—fibrous stuff that people cannot eat.

10. **Ruminants utilize low-quality roughages.** Cattle, sheep, and goats efficiently utilize large quantities of coarse, high-cellulose roughages, including crop residues, straw, and coarse low-grade hays. Such products are indigestible by humans, but from 30 to 80% of the cellulose material is digested by ruminants.

Fig. 1-11. Cattle can utilize efficiently large quantities of coarse, humanly inedible roughages, like cornstalks. This shows cows feeding on corn residue which had been harvested by mechanical means. (Courtesy, Iowa State University, Ames)

11. **Animals utilize by-products.** Animals provide a practical outlet for a host of by-product feeds derived from plants and animals, which are not suited for human consumption. Among such by-products are corncobs, cottonseed hulls, gin trash, oilseed meals, beet pulp, citrus pulp, molasses (cane, beet, citrus, and wood), wood by-products, rice bran and hulls, wheat milling by-products, and fruit, nut, and vegetable refuse.

12. **Animals provide elasticity and stability to grain production.** Livestock feeding provides a large and flexible outlet for the year-to-year changes in grain supplies. When there is a large production of grain, more can be fed to livestock, with the animals carried to heavier weights and higher finish. On the other hand, when grain supplies are low, herds and flocks can be maintained by reducing the grain that is fed and by increasing the grasses and roughages in the ration.

---

[5]Unpublished data provided to the authors by Commodity Economics Division, Economic Research Service, USDA.

13. **Animals provide medicinal and other products.** Animals are not processed for meat alone. They are the source of hundreds of important by-products, including some 100 medicines such as insulin, epinephrine, thyroxin, estrogen, cortisone, and ACTH, along with a multitude of products used in making everything from candles to cosmetics; without which the health and life-style of many people would be altered.

14. **Animals are an effective method of food storage.** In many countries, there are no facilities for storing or transporting crops. Animals may be fed crops in productive years, store food nutrients until needed, and transport themselves to market.

15. **Animals maintain soil fertility.** Animals provide manure for the fields, a fact which was often forgotten during the era when chemical fertilizers were relatively abundant and cheap.

WORLD WITHOUT END—WITH ANIMALS

World population, which is increasing by 76 million per year, is outrunning acres of cropland (see Fig. 1-12). So, food supplies have become more dependent upon increased productivity. But just as Middle East oil is being depleted, so too are soils. Reversing this alarming trend calls for an increased animal agriculture worldwide.

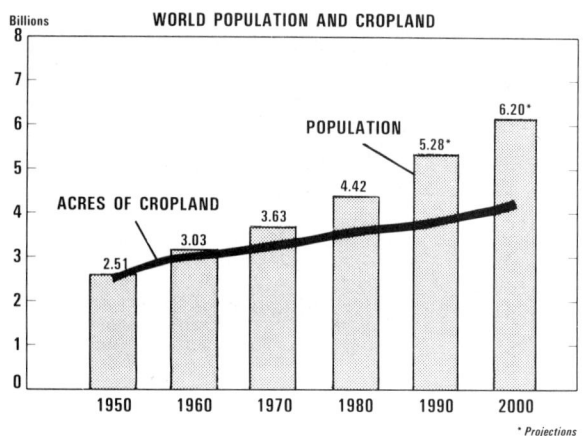

Fig. 1-12. World population is outrunning available cropland. This shows the cropland/population relationship since 1950, along with projections for the year 2000. (World population figures from Worldwatch Institute. World cropland figures from *World Agriculture Outlook and Situation Report*, USDA, Economic Research Service, WAS-36, cover page, June 1984)

World population reached 5 billion in 1987. Population growth during the 1990s is expected to be below the rate of the previous decades—around 1.6% per year, compared to 1.7% in the 1980s, and 1.9% in the 1950s and 1960s. In the year 2000, world population is expected to reach 6.2 billion. (See Fig. 1-12.)

World food consumption is determined by population and the amount eaten per person. It is expected to double over the next three decades, led by greater per capita consumption linked to rising incomes, changing tastes—preference for animal products, and improved food supplies in the developing countries.

Practicality dictates that in the years ahead a hungry world will meet its increased food needs through having plants and animals play complementary roles—and with animal products complementing the deficiencies of plant products. The virtue, even necessity, of using the plant-animal relationship is illustrated in Fig. 1-13. As shown, crops vary in their return of captured solar energy per unit of cultural energy input. Grazing land is highly efficient in the capture of solar energy—requiring little energy for a high return. Hay and silage rank second in energy return, followed by feed grains and oil crops. For the most part, however, these efficient capturers of solar energy do not store the energy in a form available to humans. It follows that ruminants which can utilize grazing land, hay, and silage (not suitable for human consumption) and convert them into meat and milk, are essential.

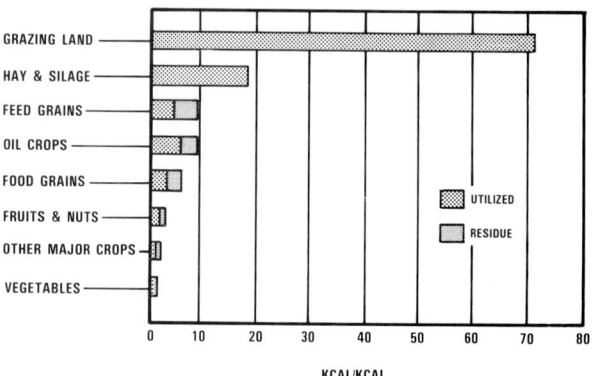

Fig. 1-13. Energy output per unit of cultural energy input (kcal/kcal) for production of food, feed, and fiber crops. (Adapted by the authors from *American Society of Agricultural Engineers*, St. Joseph, Mich., paper No. 75-7505, p. 10, Fig. 5, prepared by L. F. Nelson, W. C. Burrows, and F. C. Stickler, Deere & Company, Moline, Ill.)

For a world without end, the developing countries need a massive infusion of research, technology, and education—self-help programs, with emphasis on a plant-animal relationship. Other approaches serve only to prolong and aggravate the current disparities.

Specific methods for improving the world food situation include the following:
1. Curb population growth.
2. Increase farm prices and profits.
3. Bring more arable land under cultivation.
4. Develop more irrigation.
5. Increase crop yields.
6. Improve pastures and ranges.
7. Feed more roughage and less grain.
8. Produce leaner meats.
9. Develop more efficient animals.
10. Control diseases and parasites.
11. Improve and increase protein sources.
12. Tap the sea for more food.
13. Increase fish farming.
14. Conserve energy.
15. Control pollution.
16. Establish grain reserves.
17. Lessen food waste caused by pests.
18. Increase scientific exchange between countries.
19. Increase research, education, and extension.

QUESTIONS FOR STUDY AND DISCUSSION

1. List the essential nutrients, and give the percentage of each, contributed to the American diet by animal products.

2. Why is it said that "all flesh is grass?"

3. How do animal and plant food products compare in B-12 and iron content?

4. Define photosynthesis, and explain the process.

5. Why is photosynthesis classed as the most vital of all chemical reactions on earth?

6. Discuss the efficiency of photosynthesis from the standpoint of capturing the potentially available energy of the sun.

7. What is the most practical approach for converting a greater proportion of the total energy of plants (the other 95%) into a form available to humans?

8. Discuss the potential for solving the world food problems of the future through manipulating plants for increased solar energy conversion.

9. What is a ruminant?

10. Why and how are ruminants so important in lessening world food shortages, hunger, and malnutrition?

11. Why and how has modern food production become so inefficient in energy utilization?

12. How may we conserve energy?

13. Who was Thomas Robert Malthus? What, and when, did he prophesy relative to world population and food?

14. Why is it important that livestock producers answer by more than simple denial the charge that the world goes hungry because of animals?

15. In the developed countries, little more than one-third of the calories come from direct consumption of cereals, compared with 62% in the developing countries. Why this difference?

16. List and discuss the factors favoring direct human grain consumption.

17. Table 1-2 shows that on a calorie or protein conversion basis it is not efficient to feed grain to animals and then consume the livestock products. Evaluate this table.

18. Beef and lamb are the least efficient of all animals in feed conversion in pounds of feed required to produce one pound of product. Why not eliminate them entirely?

19. List and discuss the factors favoring an animal agriculture sharing grain with animals.

20. Why retain so much animal power in the developing countries, rather than go entirely to mechanical power?

21. What nutrients can best be obtained from animal proteins, rather than from plant proteins?

22. Fig. 1-9 pictures a cow and a calf that were produced on a protein-free diet. What's unique and significant about this?

23. World population is outrunning world cropland. What can be done to check or lessen this situation?

24. How and why should plants and animals play complementary roles in lessening world food shortages, hunger and malnutrition in the years ahead?

25. Rank crops in their return of captured solar energy per unit of cultural energy input.

26. How can world food shortages be lessened through increased research, technology, and education?

27. Will world food production be adequate to meet world demand in the 21st century?

28. What solution do you propose for the world food problem?

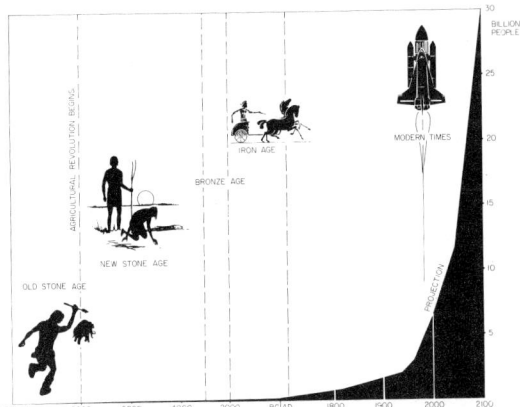

Fig. 1-14. Human population growth, the main determinant of demand for food. This shows the population growth of the world over the past 1½ million years. **NOTE:** If the Old Stone Age were in scale, its base line would extend 8 feet to the left.

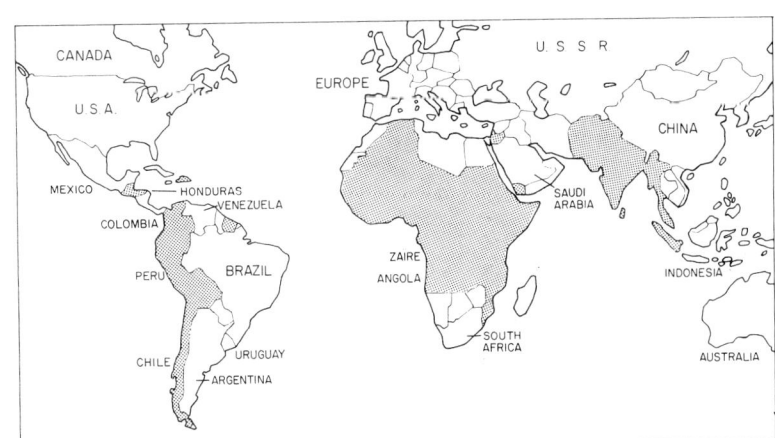

Fig. 1-15. World geography of food problems. The shaded areas are countries where undernourished population exceeds 15%.

In Quest Of Food

Fig. 1-16. Neolithic (New Stone Age) Man was a hunter. (Courtesy, American Museum of Natural History, New York)

Fig. 1-17. The Crest of Lord Bacon, noted English viscount, lawyer, statesman, politician (1561-1626). (Courtesy, Picture Post Library, London, England)

Fig. 1-18. The Knighting of Sir Loin. According to legend, King Charles II was so impressed with a platter of beef that was served at one of his feasts that he ceremoniously arose, touched his sword to the steaming platter and proclaimed: "A noble joint. It shall have a title. Loin, I dub thee Knight—henceforth thou shall be Sir Loin." (Courtesy, Picture Post Library, London, England)

Fig. 1-19. Live animals were taken with an expeditionary force to Persia and India in 1670, because there was no meat preservation. (Courtesy, The Bettmann Archive)

Fig. 1-20. African woman milking a cow. (Courtesy, FAO, Rome, Italy).

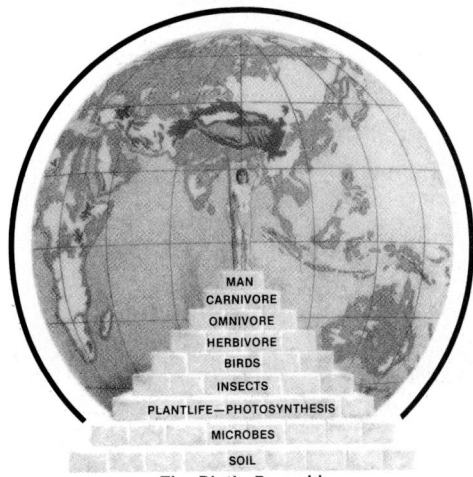

The Biotic Pyramid

Original painting by Tom Phillips

2

PRINCIPLES OF NUTRITION

Contents	Page
Perspective of Nutrition	11
Milestones in Nutrition	12
Naturalistic Era	12
Chemical-Energy Era	12
Mineral Era	12
Vitamin-Biological Era	13
Feeding Standard Era	13
Amino Acid Era	14
Feed Additive and Implant Era	14
Use Additives and Implants Safely	14
Biotechnology Era	14
Summary of Nutritional Milestones	15
The Search Goes On	15
Body Composition	16
Classification of Nutrients	17
Functions of Nutrients	17
Maintenance	18
Factors Affecting Maintenance Requirements	18
Exercise	18
Weather	18
Stress	19
Health	19
Body Size	20
Temperament	20
Individual Variation	20
Level of Production	20
Lactation	20
Other Factors Affecting Maintenance Requirements	20
Growth	20
Measures of Growth	21
Factors Affecting Nutritive Needs for Growth	21
Age	21
Breed	22
Sex	22
Rate of Growth	22
Compensatory Growth	22
Health	22
Reproduction	22
Nutritional Factors Affecting Reproduction	23
Liberal Feeding for Early Sexual Maturity	23
Flushing	23
Nutritional Reproductive Failure	23
Egg Production	24
Lactation	24
Finishing/Fitting	25
Wool and Mohair	25
Work/Running	25
Questions for Study and Discussion	26

A massive purebred bull standing belly deep in straw with a manger full of feed in front of him is the result of two forces—heredity and environment. If turned out on the range, an identical twin to the placid bull would present an entirely different appearance. By the same token, optimum nutrition could never make a champion out of a bull with scrub ancestry. But it might well be added that "fat and hair will cover up a multitude of sins." Thus, animals inherit certain genetic possibilites, but how well these potentialities develop depends upon the environment to which they are subjected; and the most important influence in the environment is the nutrition. In turn, all feed comes directly or indirectly from plants which have their tops in the sun and their roots in the soil. Hence, we have the nutrition cycle as a whole—from the sun and soil, through the plant, thence to the animal, and back to the soil again.

From the above, it may be concluded that the terms *nutrition* and *nutritionist* involve more than just feeding. *Nutrition is the science of feeds as they relate to the requirements, production, and health of animals. Nutritionist is a person who is trained in and able to apply knowledge of feeds to the requirements, production, and health of animals.* Nutrition begins with a knowledge of the fertility of the soil and the composition of plants; and it includes the ingestion of feed, the liberation of energy, the elimination of wastes, and all the syntheses essential for maintenance, growth, reproduction, egg production, lactation, fattening (fitting), wool and mohair production, and/or work (running).

A good understanding of nutrition is important because animals and humans are dependent upon food nutrients for the processes of life. This chapter will be limited to an elucidation of the principles of nutrition, including an historical perspective, body composition, classification of nutrients, and functions of nutrients. Separate chapters are devoted to digestion and absorption; nutrients—metabolism; and nutritional diseases—toxins.

PERSPECTIVE OF NUTRITION

The primary purpose of keeping animals is to transform feeds into substances usable by humans—meat, milk, eggs, wool, mohair, and work (running). But the conversion of feed to these uses must be done efficiently and economically. To do this, the principles of nutrition must be applied; and they must be augmented by superior breeding, good health, and competent management.

Like other sciences, nutrition does not stand alone. It draws heavily on the basic findings of chemistry, biochemistry, physics, microbiology, physiology, medicine, genetics, mathematics, endocrinology, and, most recently, animal behavior and cellular biology. In turn, it also contributes richly to each of these fields of scientific investigation.

The sections that follow are devoted to the historical. In nutrition, more than in any other science, history is most important. In the final analysis, animals and food are not only inseparable from history—they're part of it. Without them, there would be no history—no humankind.

MILESTONES IN NUTRITION

Obviously, all the significant milestones in a subject as broad as nutrition cannot be presented in this chapter because of space limitations. Obviously, too, the milestones are a progressive series of developments, rather than unrelated individual discoveries. In reality, they are overlapping eras. The intention in the sections that follow is to reveal (1) the general development of nutrition as a field of scientific study, and (2) the primary research emphasis that characterized each era. To this end, the history of nutrition has been aptly divided into eight eras: (1) naturalistic era, (2) chemical-energy era, (3) mineral era, (4) vitamin-biological era, (5) feeding standard era, (6) amino acid era, (7) feed additive and implant era, and (8) biotechnology era.

Naturalistic Era

The naturalistic, or prescientific, era of nutrition was characterized by a fascinating maze of sage philosophy, taboos, bizarre superstitions, and religious precepts.

Prior to the 18th century, little of truly scientific nature was accomplished in the field of nutrition. Although the ancient Greek philosophers were interested in science, logical reasoning—rather than experimentation—was the Greek way.

Hippocrates (460 to 357 B.C.), known as the *Father of Medicine*, was the first great physician to indicate an interest in nutrition. The following are among his famous aphorisms: *"Children produce more heat and need more food than adults"*; and *"Persons who are naturally very fat are apt to die earlier than those who are slender."*

Chemical-Energy Era

The great French chemist Antoine Laurent Lavoisier (1743 to 1794) is credited with being the founder of the science of nutrition. Experimenting with guinea pigs enclosed in a chamber that he had constructed, and using the thermometer and balance, Lavoisier measured body heat loss, oxygen consumed, and carbon dioxide expired. He concluded that respiration is a combustion process similar to that occurring when substances are burned outside the body. Further, he was able to show that heat production in the animal body is directly related to oxygen consumption. Lavoisier compared animal heat to that produced by a lamp or candle.

Fig. 2-1. Lavoisier and his beauteous wife. (Courtesy, the Rockefeller Institute)

But the achievements that led to the modern era of nutrition did not bring honor to Lavoisier during his lifetime. Outside his laboratory, he was a public tax collector and landed gentry. During the French Revolution, he was convicted of collecting taxes on water contained in tobacco—a criminal offense, punishable by death. Lavoisier was guillotined at age 51. His close friend, Lagrange, who witnessed the execution, said of him, "It took but a second to cut off his head; yet, a hundred years will not suffice to produce one like it."

Despite some technical inaccuracies in Lavoisier's work and in his interpretation of the results, subsequent refinement in instrumentation and in scientific thought has added little to the basic concepts derived from his experiments. Chemistry and physiology, in which Lavoisier pioneered, served as the foundation upon which the science of nutrition was established. But almost 100 years elapsed following Lavoisier's classical studies before carbohydrates, fats, and proteins were identified as the source of energy for the animal body.

Mineral Era

The occurrence of goiter in both animals and people is as old as antiquity. In 1822, Boussingault, while traveling in South America, observed that the villagers who used salt containing iodine were free from goiter, but that those who used plain salt were afflicted. Thereupon, he advocated the use of iodine at low levels to prevent the malady.

In 1838, Berzelius, noted Swedish chemist, concluded that the iron in hemoglobin made it possible for the blood to absorb much oxygen. As a result of a series of studies initiated in 1925, Hart and his associates, at the University of Wisconsin, reported that white rats developed anemia if fed a milk diet, and that the anemia was not cured by iron salts alone. However, the ash of lettuce, which contained copper, was an effective cure. Mere traces of copper added to pure iron salts were found to be adequate for the cure of iron-deficiency anemia.

The mineral structure of bones and teeth was recognized by early workers. In 1840, Charles J. Chossat, a physician in Geneva, Switzerland, won a prize for showing that a diet of wheat and water must be supplemented with calcium for the bones of a growing pigeon to develop. In 1843, the importance of calcium in the diet was further attested by J. B. Boussingault, a Frenchman, who performed calcium-balance studies with animals.

The general belief at the beginning of the 20th century was that sodium chloride, calcium, phosphorus, and probably iron were the important minerals in animal nutrition. Little attention was given to other elements known to be present in tissues in relatively small amounts. The importance of trace elements in animal nutrition was more readily accepted following the discovery of vitamins.

Vitamin-Biological Era

Until early in the 1900s, if a ration contained proteins, fats, carbohydrates, and minerals—together with a certain amount of fiber—it was considered to be a complete diet. True enough, the disease known as beriberi, having been known to the Chinese as early as 2600 B.C., made its appearance in the rice-eating districts of the orient when milling machinery was introduced from the West; and scurvy was long known to occur among sailors fed salt meat and biscuits. However, for centuries these diseases were thought to be due to toxic substances in the digestive tract caused by pathogenic organisms, rather than food deficiencies; and more time elapsed before the discovery of vitamins.

Largely through the trial-and-error method, it was discovered that specific foods were helpful in the treatment of certain of these maladies. In 1747, Lind, a British naval surgeon, showed that the juice of citrus fruits was a cure for scurvy. In 1881, Lunin had come to the conclusion that certain foods, such as milk, contain small quantities of unknown substances essential to life, in addition to their principal ingredients. In 1897, Eijkman, a Dutch physician working in the East Indies, produced and cured at will the disease beriberi (polyneuritis) in hens, simply by changing the diet of unpolished rice to milled rice, or the reverse. At a very early date, the Chinese used a concoction rich in vitamin A as a remedy for night blindness. Also, cod-liver oil was used in treating or preventing rickets long before anything was known about the cause of the disease.

The significance of these observations relative to diet, however, was not fully appreciated until scientists found it desirable in many types of investigations to use the biological approach, with purified diets to supplement chemical analyses in measuring the value of feeds. These rations were made up of relatively pure nutrients—proteins, carbohydrates, fats, and minerals—from which the unidentified substances were largely excluded. With these purified rations, all investigators shared a common experience—the animals limited to such diets not only failed to thrive, but they even failed to survive if the investigations were continued for any length of time. At first, many investigators explained such failures on the basis of unpalatability and monotony of the rations. Finally, it was realized that these purified rations were lacking in certain substances, minute in amount, and the identity of which was unknown to science. These substances were essential for the maintenance of health and life itself and the efficient utilization of the main ingredients of the food. With these findings, a new era of science was ushered in. The modern approach to nutrition was born.

Casimir Funk, a Polish scientist working in London, discovered evidence of one such substance in the outer hulls of rice, in 1912. He called it a "vital amine" because he (1) believed it to be necessary for life, and (2) found that it was an amine, chemically. This substance, later identified as thiamin, or vitamin B-1, gave rise to the general terminology *vitamine,* later shortened to *vitamin* as we know it today.

Acceptance of the vitamin theory was not immediate, however. It remained for the independent discovery, in 1913, of vitamin A by McCollum and Davis, of the University of Wisconsin, and Osborn and Mendel, of Yale University. These researchers observed that animals which received diets containing lard instead of butterfat would cease to grow and eventually develop an inflammation of the eye. When either butterfat or cod-liver oil was used, the condition was corrected. These discoveries led to the recognition of *fat-soluble A* and *water-soluble B,* as the factors were then designated. But it was not until the decade 1930 to 1940 that the majority of the vitamins were identified, isolated from feeds, and synthesized in the laboratory.

The known existence of vitamins, therefore, dates only to 1912. Since that time, the growth of the vitamin family, the isolation and determination of many vitamins, the partial solution of the puzzle of vitamin functions in the body, the discovery of the amazing therapeutic value of minute quantities of these vitamins in the cure of deficiency diseases, the numerous determinations of feed (food) composition with respect to vitamins, and the synthesis of most of them, have had a profound effect in nutrition.

Feeding Standard Era

Feeding standards are tables showing the amounts of one or more nutrients needed by different species of animals for different purposes. They serve as guides in balancing rations and feeding practices. Most feeding standards are expressed in (1) quantities of nutrients required per day, and/or (2) percent of the ration; the first type is used where animals are provided a given amount of feed during a 24-hour period, and the second is used where animals are provided a ration without limitation on the time in which it is consumed.

The first feeding standard was developed in 1810, by Thaer, a German scientist. He took meadow hay as his standard, compared the extractable nutrients of other feeds to it, then assigned them *hay values.* In 1859, Grouven, another German scientist, made use of analyses of protein, fat, and carbohydrate to formulate the first feeding standard for farm animals. In 1864, the great German scientist, Wolff, devised a standard based on digestible nutrients obtained from feeding trials. In 1897, Lehmann, of Germany, modified Wolff's standards. Other systems followed; among them, the total digestible nutrient system by Henry, of the University of Wisconsin, published in the first edition of his book, *Feeds and Feeding,* in 1898; the starch values of Kellner, a German scientist, in 1907; the Scandinavian feed unit system (Woll, 1912); the dairy cow standards of Haecker, of Minnesota, in 1914; the net energy values of Armsby, of Pennsylvania, in 1915; the productive units developed by Mollgaard, of Denmark, in 1939; and the productive energy values computed by Fraps of the Texas Station, in 1937 and 1941.

Today, the most widely used feeding standards in the United States are those published by the National Academy of Sciences. In England, similar standards are issued by the Agricultural Research Council (ARC). Other countries have similar bodies which make recommendations on the nutritive requirements of animals.

In the United States, the TDN system is gradually giving way to other energy evaluation systems, particularly net energy. England uses metabolizable energy (ME), adjusted according to the efficiency with which a feedstuff or diet is used for a particular purpose. Other European standards are based on starch equivalents, Scandinavian feed units, and other methods of evaluation.

Amino Acid Era

Among the more recent outstanding achievements in nutrition was the identification of the essential amino acids by William C. Rose and co-workers of the University of Illinois. In 1930, they initiated a brilliant series of studies, using a new technique, out of which evolved specific information relative to the amino acids that must be present in feed. By the use of diets otherwise designed to be adequate for the normal growth of rats, in which the sole source of nitrogen was supplied by amino acids, the effect of the addition or deletion of each of the amino acids was studied. Simultaneously, the doctoral students conducting these investigations confirmed the results on themselves. They tested each amino acid at different levels and determined the amount needed for optimal utilization of dietary protein. As a result of these investigations, the Illinois workers were able to classify 10 amino acids as essential dietary constituents and the others as nonessential.

Feed Additive and Implant Era

In the 1950s and 1960s, a new era of livestock production was ushered in when it was found that antibiotics and stilbestrol would increase both rate and efficiency of gain of animals. They were hailed as the "wonder drugs" of our time. Soon the race was on, and other feed additives and implants followed.

In 1949, Jukes, of Lederle Laboratories, found that antibiotics were something new to be added to livestock feeds—that they would improve growth rate and feed efficiency.

In 1952, Burroughs, of Iowa State University, reported that the feeding of diethylstilbestrol (DES) would lower feed usage and increase weight gains of finishing cattle. In 1954, stilbestrol was approved by the Food and Drug Administration (FDA) for use in cattle finishing rations; and in 1956, FDA approved the use of stilbestrol implants for finishing steers. (**NOTE:** In 1948, Andrews of Purdue University reported on the effect of implanted stilbestrol, but no practical application of the finding was made at the time.)

Today, more than 1,000 drugs are approved by FDA for use by livestock and poultry producers, and 80% of all animals raised for food receive some animal drugs during their lifetime. When used properly, these drugs enable livestock producers to provide safe and wholesome meat, eggs, and milk to consumers at lower costs than would otherwise be possible.

Among the newer additives and implants now being tested or used are various antimicrobials (including antibiotics, antibacterial agents, and antifungal agents), hormones or hormonelike substances, enzymes, and ionophores. Additional products are evolving, some of them made commercially possible by recombinant DNA (gene splicing) technology.

(See Chapter 13, Feed Supplements/Additives/Implants for definitions of, and further information relative to, additives and implants.)

USE ADDITIVES AND IMPLANTS SAFELY

Since the beginning of recorded history, people have been concerned about the purity and safety of their food and drink. In 1202, King John of England proclaimed the first English food law, the Assize of Bread, which prohibited adulteration of bread with such ingredients as ground peas or beans. Regulation of food in the United States dates from 1784. Federal controls over the drug supply started in 1848.

As the list of additives and implants grows, so also do the concerns of consumers, augmented by government regulatory bodies and laws. In 1958, the food additive amendment, better known as the Delaney Clause, was passed by the U.S. Congress. This clause gave rise to the policy of *zero tolerance*—that no substance can be used as an additive or implant if it has been shown to produce cancer in either humans or animals.

The key to using animal drugs safely is to read and follow the label directions. Everything that a livestock producer needs to know about approved species, dosage, methods of administration, and withdrawal time is on the label.

More than half of the FDA-approved drug products require pre-slaughter withdrawal times or milk-discard periods. The FDA has found that approximately 90% of residue violations are caused by producers failing to withdraw drugs from animals soon enough. Any illegal residues found in market livestock, milk, or eggs can result in marketing delays and even lead to condemnation of the shipment.

Biotechnology Era

The development of gene-splicing (also known as recombinant DNA) ushered in a new era. On May 23, 1977, scientists at the University of California-San Francisco reported a major breakthrough as a result of altering genes—turning ordinary bacteria into factories capable of producing insulin, a valuable hormone previously extracted at slaughter from pigs, sheep, and cattle, so essential to the survival of diabetics. The feat gave rise to a major scientific revolution, called *biotechnology*.

Biotechnology is the aspect of technology concerned with the application of an array of biological and engineering tools, from gene splicing to manipulating cells, tissues, and genes so as to control certain characteristics. It involves every facet of animal production, from breeding and feeding to finished products, including the genetic makeup of animals; the feeds they eat; their digestion, physiology, stress tolerance, disease resistance, and efficiency of production; the composition, quality, and quantity of products produced; along with the production of large quantities of drugs and chemicals. Potential uses number into the hundreds, perhaps thousands—they're endless. While some aspects are decades away from commercial production, others are near, and still others are here now.

The entire food team—producers, processors, and consumers—stands to benefit from biotechnology, from the greater abundance of high-quality products produced more efficiently. However, progress will be slowed by regulatory and social problems; and some producers and processors will falter and fail along the way. Nevertheless, with the opening of the 21st century, it is predicted that the magnitude of change wrought by biotechnology will rival that of mechanization, which doubled total American agricultural production and increased output per man hour 20-fold.

Summary of Nutritional Milestones

Obviously, all of the significant milestones in a subject as broad as nutrition cannot be summarized in this section because of space limitations. Obviously, too, the milestones were a progressive series of developments, rather than unrelated individual discoveries. All researchers contributed to the kaleidoscope of knowledge that we have today, whether it was to prove or disprove a theory, or to add just one more piece to the puzzle. Neither can the dates given always be pinpointed because some research projects spanned several years; others were completed on a certain date but not reported or published until later, and still others led to the wrong conclusion, which, with further study, was changed.

The Search Goes On

Modern nutritionists move ahead in their search. Food groups—carbohydrates, fats, proteins, minerals, vitamins, and water—do not go far enough in any sophisticated study of nutrient requirements. Corn, alfalfa, soybean meal, and bone meal never get to the cells and tissues. But the nutrient chemicals—more than 40 of them, including amino acids, minerals, vitamins—do reach the body cells and tissues and are essential to their function. These are the ABCs of modern nutrition. Today, there is increasing research emphasis on the microscopic and subcellular components of the living system, and there is more and more fragmentation and specialization. First we had the animal nutritionists. Next came the ruminant and nonruminant nutritionists. Now these are being replaced by species specialists, but with further breakdown and study of a class of nutrients, or even of an individual nutrient. So, the scientific domain of the individual nutritionist has become increasingly narrow. Too often, this has been done at the expense of understanding the animal as a whole. Too often, there has been little attention to the comparative aspects with other animal species. Too often, modern nutritionists may be likened to an automotive engineer who knows all about the production of energy by combustion of fossil fuels; he even knows about each of the car systems—the fuel system, the transmission system, the brake system, etc.—but who cannot put the automobile together and drive it. Too often, in this day of specialization, the nutritionist has neglected the team approach, with scientists of many disciplines working together.

Despite the many advances that typify nutrition, the following are among the unknowns in nutrition, or among the nutrition-related areas that need improvement, which merit the best efforts of nutritionists:

1. **We need to know more about the nutrient requirements of animals.** We need to know if we have identified all of the required nutrients, and we need to know more about the availability of the nutrients and the way that they interact with each other. We need to know more about the unidentified factors, and we need to know if more of the minerals in the periodic table should be listed as requirements. We need to recognize that nutrient availability may be more important than composition.

2. **We need to know more about the net energy system.** As currently measured, net energy represents an animal's predicted response to a given mixture of feedstuffs under an incompletely defined set of environmental conditions. Further refinement is needed.

3. **We need to feed animals more forages and less grain in the future.** As population increases, the demand for grain increases. This, coupled with the increasing cost of fossil fuels, will elevate grain prices to a point where a transition in animal feeding practices will be dictated largely by economics. Shifts to greater reliance on forage in livestock rations and less grain will take place in the future. Ruminants can make this transition easily. Increasingly, they will be "roughage burners," transforming huge quantities of fibrous materials and wastes into high-quality protein found in meat and milk.

4. **We need to know more about animal behavior and environment.** As the art of husbandry evolved into science, people's fluency with the language of their beasts declined; and human-made environments became increasingly artificial until animals were viewed primarily as economic, biological conversion mechanisms. Behavior received little attention; the push was on for quantity and quality of meat, milk, eggs, fiber, and power of animals. But many abnormal animal behaviors evolved to plague those who raise them, including cannibalism, loss of appetite, stereotyped movements, poor mothering, overaggressiveness, dullness, degenerate sexual behavior, tail biting, cribbing, and a host of other behavior disorders. Today, the situation is being righted. Attention is being given to animal space requirements, heat and vapor production, and environmental control. The modern nutritionist must (a) formulate rations for different environments; and (b) develop nutritional programs for periods of stress, such as weaning, shipping, fatigue, illness, and abrupt temperature and weather changes. The development of nutritional programs for periods of stress should receive as much attention as the medical treatment of ills.

5. **We need to conserve energy.** Population growth and feed-food production technology are creating energy stresses of unprecedented scale and urgency—threatening human existence. In an era of world food shortages, the nutritionist is a key person in the entrapment of solar energy through photosynthesis and in the conservation of energy.

6. **We need to use agricultural chemicals and drugs with discretion.** Sometimes choices must be made; for example, between malaria-carrying mosquitoes and some fish, or between hordes of grasshoppers and the crops that they devour, or between the use of antibiotics and hormones as feed additives and higher consumer costs. In the era ahead, food shortages will increasingly concern all people—producers and consumers alike. (See section headed "Use Additives and Implants Safely.")

7. **We need to control pollution.** Today, there is worldwide awakening to the problem of pollution of the environment (air, water, and soil) and its effects on human health and other forms of life. Much of this pollution was ushered in with industry. But some of this concern stems from the sudden increase of animals in confinement and the disposal of waste. Certainly, there have been abuses of the environment; and there is no argument that such neglect should be rectified in a sound, orderly manner. Nutritionists can, and must, contribute mightily to the control of pollution, and they must do so with a minimum of disruption of the economy and without lowering the standard of living.

BODY COMPOSITION

Nutrition encompasses the various chemical and physiological reactions which change feed elements into body elements. It follows that knowledge of body composition is useful in understanding animal response to nutrition.

The figures given in Tables 2–1 and 2–2 have been assembled to provide a picture of the gross composition of animals of different species at different ages and in different nutritional states.

TABLE 2–1
RANGE IN CHEMICAL COMPOSITION OF BODIES OF CATTLE, SHEEP, AND PIGS[1]

Species	Number in Group	Range in Body Composition of Ingesta-Free (empty) Body			
		Water	Fat	Protein	Ash
		%	%	%	%
Cattle....	256	39.8–77.6	1.8–44.6	12.4–20.6	3.0–6.1
Sheep....	221	39.6–73.8	4.9–46.6	10.7–19.5	1.7–5.8
Pigs.....	714	30.7–80.8	1.1–61.5	8.3–19.6	1.3–5.6

[1]Reid, J. T., et al., "Some Peculiarities in the Body Composition of Animals," *Body Composition in Animals and Man*, National Academy of Sciences, 1968, p. 20, Table 1. Pertinent information about the data follows:
  Cattle: Of the 256 animals, 139 were beef type and 117 were dairy type. Seven purebreds and 5 crossbred combinations were represented. Ages ranged from 1 to 4,860 days (13.3 yr). Sex was designated on 249 animals, of which 135 were males and 114 were females.
  Sheep: The study included 4 purebred breeds and 2 crossbred populations, representing a considerable range in body conformation, maturing rate, and mature size. All animals were male castrates ranging from 90 to 895 days (2.5 yr) of age and from 26.4 lb to 147.4 lb in weight.
  Pigs: The pigs represented 8 distinct breeds and 1 crossbred group, ranging in age from 1 to 923 days (2.5 yr). Of the 714 pigs, sex was identified for 248, of which 153 were male castrates and 95 were females.

TABLE 2–2
BODY COMPOSITION OF ANIMALS, AT DIFFERENT WEIGHTS AND AGES, INGESTA–FREE (EMPTY) BASIS[1]

Species	Age or Status	Weight		Water	Fat	Protein	Ash
		(lb)	(kg)	(%)	(%)	(%)	(%)
Cattle:							
Calf.....	Newborn	70	31.8	74.4	2.5	19.0	4.1
Calf.....	Weanling	450	204.1	69.0	9.0	18.0	4.0
Steer....	Feeder	650	294.8	60.3	18.0	17.2	4.5
Steer....	Choice grade	1,050	476.3	53.5	26.0	17.0	3.5
Steer....	Very fat	1,500	680.4	40.0	41.0	16.0	3.0
Cow.....	Breeding condition	1,100	499.0	60.0	18.0	17.5	4.5
Sheep:							
Lamb....	Newborn	9	4.1	72.8	2.0	20.2	5.0
Lamb....	Feeder	65	29.5	63.9	17.0	15.7	3.4
Lamb....	Fat	100	45.4	53.2	29.0	15.0	2.8
Lamb....	Very fat	125	56.7	39.0	44.0	14.4	2.6
Pig:							
Pig......	Newborn	3	1.4	74.0	2.0	19.0	5.0
Pig......	Weanling	30	13.6	70.0	9.0	17.5	3.5
Pig......	Growing-finishing	100	45.4	66.8	16.2	14.9	3.1
Pig......	Market	220	100.0	50.0	34.4	13.0	2.6
Pig......	Very fat	300	136.1	42.5	43.5	12.0	2.0
Poultry:							
Chick....	Newly hatched	0.090	0.04	78.8	4.0	15.3	1.9
Broiler...	Market	3.5	1.6	65.7	12.2	18.4	3.7
Hen.....	Layer	4.5	2.0	59.6	20.0	17.0	3.4
Turkey...	8 weeks	5.1	2.3	70.1	4.5	20.9	4.5
Turkey...	Market	18.0	8.2	59.4	18.4	19.1	3.1
Rabbit:							
Rabbit...	Market	8	3.6	69.0	3.5	9.0	5.0
Horse:							
Foal.....	Newborn	110	49.9	73.0	2.0	20.0	5.0
Foal.....	Weanling	400	181.4	69.0	9.0	18.0	4.0
Horse....	Mature	1,050	476.3	62.0	17.0	17.0	4.0
Man:							
Baby....	Newborn	8	3.6	76.0	9.2	12.0	2.8
Man.....	Mature	150	68.0	59.0	18.0	18.0	5.0

[1]Prepared by the authors from numerous sources.

Tables 2–1 and 2–2 show that there is a wide range in the body composition of animals according to age and nutritional state (degree of fatness). Yet, based on these tables, together with other studies, the following conclusions relative to body composition may be drawn:

1. **Water.** On a percentage basis, the water content shows a marked decrease with advancing age, maturity, and fatness.

2. **Fat.** The percentage of fat normally increases with growth and fattening.

3. **Fat and water.** As the percentage of fat increases, the percentage of water decreases.

4. **Protein.** The percentage of protein remains rather constant during growth, but decreases as the animal fattens.

On the average, there are 3 to 4 lb of water per 1 lb of protein in the body.

5. **Ash.** The percentage of ash shows the least change. However, it decreases as animals fatten because fat tissue contains less minerals than lean tissue.

6. **Composition of body gain.** The data presented in Tables 2–1 and 2–2 clearly indicate that gain in weight tells nothing about the composition of gain.

Principles of Nutrition

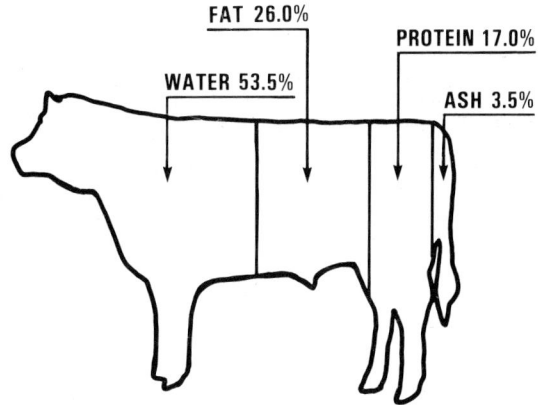

Fig. 2-2. Composition of a 1,050-lb steer, grading Choice.

Composition of gain is affected by feed intake and rate of gain. When feed intake is limited, the rate of gain is depressed and the composition of this gain is primarily protein along with the concomitant deposition of water. As feed (energy) intake is increased, daily weight gain increases correspondingly, but the proportion of protein in this gain decreases and the proportion of fat increases.

From the above, the following deductions and practical applications relative to composition of gain may be made:

a. Rate of gain is determined primarily by total feed energy intake. But it may be depressed by protein deficiency.

b. In young growing cattle, energy intake above needs for daily protein deposition will result in fat deposition. In older cattle, once the maximum protein content (lean) has been reached, most of the energy intake is used for fattening.

c. In young cattle, maximal protein gain can be achieved without excess fat deposition by providing rations of low energy density, which is usually achieved by feeding high roughage grower rations. As cattle get heavier, they are switched to finishing (high energy density) rations.

d. Feeding high energy density (finishing) rations during the growing phase will result in cattle that reach the desired finish quite early in relation to their mature weight, without the total potential protein (lean) gain being expressed. Further feeding of such rations will result in maximum carcass protein deposition, but the carcass will also be overly finished.

The feeding of grower type, or lower energy density rations such as corn silage, will result in the maximum total carcass protein gain, but a longer feeding period is required to obtain the desired amount of finish.

e. Biologically, it would appear desirable to provide for a rate of gain during the growing period that maximizes carcass lean deposition without concomitant excess fattening, followed by providing extra feed energy to encourage rapid finishing.

7. **Variation between organs and tissues.** The chemical composition of the body varies widely between organs and tissues and is more or less localized according to function. Thus, water is an essential of every part of the body, but the percentage composition varies greatly in different body parts; blood plasma contains 90 to 92% water, muscle 72 to 78%, bone 45%, and the enamel of the teeth only 5%. Proteins are the principal constituents, other than water, of muscles, tendons, and connective tissues. Most of the fat is localized under the skin, near the kidneys, and around the intestines. But it is also present in the muscles (known as marbling in a carcass), bones, and elsewhere.

8. **Species comparison.** The following species differences are noteworthy:

a. The bodies of very fat pigs contain more fat and less water than cattle and sheep in comparable condition.

b. Because of their smaller skeletons, the bodies of pigs contain less ash than those of cattle or sheep.

c. Cattle are higher in protein than pigs of comparable age, weight, and finish. At normal market stage, broilers are higher in protein than four-footed animals.

CLASSIFICATION OF NUTRIENTS

Animals do not utilize feeds as such. Rather, they use those portions of feeds called *nutrients* that are released by digestion, then absorbed into the body fluids and tissues.

Nutrients are those substances, usually obtained from feeds, which can be used by the animal when made available in a suitable form to its cells, organs, and tissues. They include carbohydrates, fats, proteins, minerals, vitamins, and water. More correctly speaking, the term *nutrients* refers to the more than 40 nutrient chemicals, including amino acids, minerals, and vitamins.

Knowledge of the basic functions of the nutrients in the animal body, and of the interrelationships between various nutrients and other metabolites within the cells of the animal, is necessary before one can make practical scientific use of the principles of nutrition. To this end, the rest of this chapter is appropriately devoted to a discussion of the "Functions of Nutrients."

FUNCTIONS OF NUTRIENTS

Of the feed consumed, a portion is digested and absorbed for use by the animal. The remaining undigested portion is excreted and constitutes the major portion of the feces. Nutrients from the digested feed are used for a number of different body processes, the exact usage varying with the species, class, age, and productivity of the animal. All animals use a portion of their absorbed nutrients to carry on essential functions, such as body metabolism and maintaining body temperature and the replacement and repair of body cells and tissues. These uses of nutrients are referred to as *maintenance*. That portion of digested feed used for growth, fattening, or the production of milk, eggs, wool, and work is known as *production requirements*. Another portion of the nutrients is used for the development of the fetus and is referred to as *reproduction requirements*.

Based on the quantity of nutrients needed daily for different purposes, nutrient demands may be classed as high, low, variable, or intermediate.

Requirements for milk and egg production are considered *high-demand uses*, whereas wool is a *low-demand use*. Work, which may be strenuous for limited periods of time (as in racing), and the last stages of pregnancy have *variable requirements*. Growth and fattening may be classed as intermediate in nutrient demands. Each of these needs will be discussed in more detail.

Maintenance

Maintenance requirements may be defined as the combination of nutrients which are needed by the animal to keep its body functioning without any gain or loss in body weight or any productive activity. Although these requirements are relatively simple, they are essential for life itself. A mature animal must have (1) heat to maintain body temperature, (2) sufficient energy to keep vital body processes functional, (3) energy for minimal movement, and (4) the necessary nutrients to repair damaged cells and tissues and to replace those which have become nonfunctional. Thus, energy is the primary nutritive need for maintenance. Even though the quantity of other nutrients required for maintenance is relatively small, it is necessary to have a balance of the essential proteins, minerals, and vitamins.

No matter how quietly an animal may be lying in a stall or in a pasture, it requires a certain amount of fuel and other nutrients. The least amount on which it can exist is called its *basal maintenance requirement*. With the exception of horses, most animals require about 9% more fuel (calories) when standing than when lying, and even more is needed when they walk or run. This explains why it is desirable, for economic reasons, that finishing animals eat, then lie down as much as possible.

There are only a few times in the normal life of an animal when only the maintenance requirement needs to be met. Such a status is closely approached by mature males not in service; by mature, dry, nonpregnant females; and by idle horses. Nevertheless, maintenance is the standard bench mark or reference point for evaluating nutritional needs.

Although nutrient needs are minimal during maintenance, it is noteworthy that ⅓ to ½ of the feed consumed by animals as a whole is used to meet the maintenance requirement (see Fig. 2–3). Of course, on an individual basis, the higher the production, the smaller the proportion of nutrients needed for maintenance.

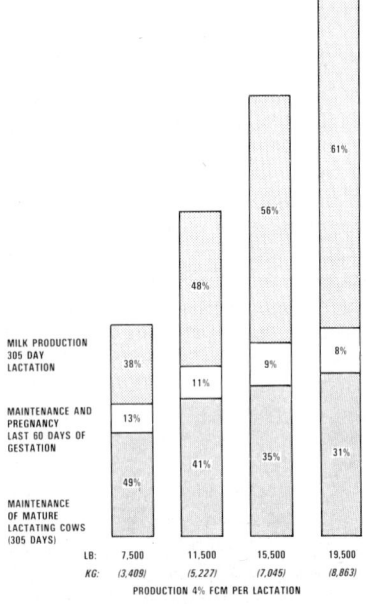

Fig. 2–3. Relative proportions of feed used by a 1,430-lb (650 kg) cow for (1) maintenance, (2) maintenance and pregnancy, and (3) milk production, at levels of 7,500, 11,500, 15,500, and 19,500 lb 4% FCM milk. Note that the percentage of feed used for maintenance decreases as production increases.

FACTORS AFFECTING MAINTENANCE REQUIREMENTS

Even though maintenance requirements might be considered an expression of the nonproduction needs of an animal, there are many factors which affect the amount of nutrients necessary for this vital function; among them, (1) exercise, (2) weather, (3) stress, (4) health, (5) body size, (6) temperament, (7) individual variation, (8) level of production, and (9) lactation. The first four are *external factors*—they are subject to control to some degree through management and facilities. The others are *internal factors*—they are part of the animal itself. Both external and internal factors influence requirements according to their intensity. For example, the colder or hotter it gets from the most comfortable (optimum) temperature, the greater will be the maintenance requirements.

Exercise

Movement of animals might be classed as work, except that in the normal use of the term maintenance, all activities which are not considered productive are grouped under maintenance requirements. With this concept, animals in a confined or restricted lot have a smaller maintenance requirement than those in a pasture or on the range.

Fig. 2–4. Rough terrain and sparse vegetation make for a higher maintenance requirement.

Weather

Weather affects the maintenance requirements of animals. For example, under ideal October weather conditions in Missouri, a horse may require 14 lb of a 60% TDN ration daily, whereas in the same area and doing the same work, the same horse may require 16 lb daily of the same feed in July and August and 20 lb in the winter. The good caretaker senses this situation, and changes the feed allowance accordingly.

The maintenance requirements of animals increase as temperature, humidity, and air movements depart from the comfort zone. Likewise, the heat loss from animals is affected by these three items.

The critical temperature is that temperature at which the heat created by digestion and body metabolism just equals that which the animal dissipates by convection, evaporation, radiation, and conduction. The comfort zone is the range in temperature within which the animal may perform with little or no discomfort. At temperatures below the comfort zone, additional nutrients need to be converted to heat to keep the body warm; and at temperatures above the comfort zone, nutrients are needed to help keep the animal cool. *The optimum temperature is the temperature at which the animal responds most favorably, as determined or measured by maximum rate of gain or production, feed efficiency, and/or reproduction.*

The critical temperature, comfort zone, and optimum temperature vary with different species, ages, breeds, and the physiological and productive status of animals. The species differences result primarily from the kinds of thermoregulatory mechanism provided by nature, such as type of coat (hair, wool, feathers), sweat glands, etc. Thus, hogs, which have a light coat of hair, are very sensitive to extremes of heat and cold. On the other hand, nature gave cattle an assist through growing more hair for winter and shedding hair for summer, with the result that they can withstand higher and lower temperatures than hogs.

The critical temperature varies according to age, too. For example, the comfort zone of newborn lambs is 75° to 80°F, whereas the comfort zone of mature sheep is 45° to 75°F.

There are also breed differences, which make it possible to select animals well adapted to specific environments. For example, haired sheep (devoid of wool) are adapted to the desert, fat-tailed sheep do well in arid zones, *Bos indicus* (Zebu) types of cattle thrive in tropical areas, and *Bos taurus* cattle (European breeds) do well in temperate zones. The Santa Gertrudis breed of cattle, which evolved from a Brahman X Shorthorn cross, is intermediate between its parent breeds in heat tolerance. The Jersey and Brown Swiss breeds of dairy cattle will maintain their milk production in hot summer temperatures better than Holsteins or other dairy breeds.

Animals that consume large quantities of roughage produce more heat during digestion; hence, they have a different critical temperature than the same animals on a high-concentrate ration. Because of this, experienced cattle feeders decrease the roughage and increase the concentrate of finishing cattle during the hot summer months.

Stresses at both high and low temperatures are increased with high humidity. The cooling effect of evaporating sweat is minimized, and the respired air has less cooling effect. As humidity of the air increases, discomfort at any temperature increases, and nutrient utilization decreases.

Air movement (wind) results in body heat being removed at a more rapid rate than when there is no wind. In warm weather, air movement may make the animal more comfortable; but in cold weather, it adds to the stress of temperature. At low temperatures, the nutrients required to maintain body temperature are increased as the wind velocity increases. In addition to the wind, a drafty condition where the wind passes through small openings directly onto some portion or all of the animal body will usually be more detrimental to comfort and nutrient utilization than the wind itself.

While stock raisers have always been concerned with the effect of weather on animals, this concern has taken on a new dimension with rising costs of feed, labor, land, and financing. There was need to minimize maintenance requirements and maximize the nutrients available for production. This prompted interest in building design. Naturally ventilated buildings, consisting mainly of shells to protect animals

Fig. 2-5. Environmentally controlled dairy barn. (Courtesy, Babson Bros. Co., Oak Brook, Ill.)

from rain and snow, are still used for most animal housing. But environmentally controlled buildings (in which the air temperature, relative humidity, and air velocity are regulated) are rather common in poultry and swine housing, and on the increase for other classes of livestock—especially dairy cattle. They make for the ultimate in feed efficiency, along with a saving in labor and land.

In hot climates, increased use is being made of shades for the purpose of enhancing animal comfort and minimizing the maintenance requirements. Also, studies with lactating dairy cows reveal that putting only the head in an air-conditioned chamber, with the rest of the body left exposed to the heat, will increase production and feed efficiency.

Stress

Stress of any kind increases maintenance requirements. Stress is affected by temperament, excitement, presence of strangers, fatigue, number of animals together, changing corral and corral mates, previous training, previous nutrition, breed, age, and management.

Animals can be prepared in such manner as to reduce stress. For example, if calves are properly preconditioned (started on feed, vaccinated, treated for parasites, etc.) prior to weaning, the stress of subsequent weaning and movement to a feedlot will be minimized.

Health

Although animal deaths take a tremendous toll, even greater economic losses—hidden losses—result from poor feed efficiency due to diseases and parasites. It is difficult to assess accurately the effect of animal ill health in terms of the added nutrients necessary for maintenance, for it is affected by the kind and severity of the health problem, the nutritional state of the animal, and the stamina of the individual. Nevertheless, animals that are emaciated give ample evidence of wasted feed.

Body Size

Body size was once used as the basis for determining the maintenance requirements of animals. However, it proved to be inaccurate for such a determination, because it did not represent the relative metabolic activity. Body surface was found to be a better reference point for metabolism than body weight. It was formerly expressed as ⅔ power of body weight ($W^{.67}$). More precise studies, however, have shown that the most accurate measure of metabolic size of an animal is $W^{.75}$, a factor which is generally used by most scientists. In practical terms, this means that a smaller animal has a higher rate of metabolism per pound of body weight than a larger animal. For example, a 1,200-lb animal has less than twice the maintenance requirements of an animal weighing half that amount (600 lb).

Fig. 2-6. On a per pound body weight basis, the small Shetland Pony has a higher rate of metabolism than the big Shire. (Courtesy, Iowa State University, Ames)

Temperament

Nervous animals require more nutrients for maintenance than docile animals. Some animals are naturally nervous; others may be nervous because of the way in which they have been handled. Estrus (heat) is a natural condition; nevertheless, it usually induces excitement and increases nervousness. The presence of strangers and the changing of corrals and corral mates result in induced nervousness. Either natural or induced nervousness has much the same effect on nutrient utilization—both increase it.

Individual Variation

The maintenance requirement varies among individual animals, just as it does in people. Some animals utilize their feed more efficiently than others. A "hard keeper" will require considerably more feed than an "easy keeper." This, of course, is the basis for production testing animals as a means of selecting those which utilize their feed most efficiently.

Level of Production

Animals with accelerated rates of production tend to have higher maintenance requirements than those with lower levels of production. For convenience reasons, most feeding standards make provision for this increased maintenance as part of the production requirements; it simplifies calculations to use a single maintenance value for all production levels.

Lactation

The maintenance requirements of lactating females of all species are higher than those of dry, nonpregnant females. For example, the maintenance requirement of lactating cows is approximately 10% higher than that of dry, nonpregnant cows. This is attributed, in part, to (1) the increased secretion of thyroxin in lactating cows, which is reflected in a faster heartbeat—increased basal metabolic rate; and (2) the approximately 400 lb of blood which must be pumped through the udder to supply the raw material for 1 lb of milk.[1] (Hence, a cow giving 80 lb of milk in a day must pump 32,000 lb, or 16 tons, of blood through her udder daily.)

Other Factors Affecting Maintenance Requirements

Other factors that affect the maintenance requirements follow:

1. **Shearing sheep and goats.** In cool weather, the maintenance requirements increase at shearing due to decreased insulation. In hot weather, however, shearing may decrease maintenance requirements.
2. **Gestation.** Bearing in mind that the maintenance requirements of gestation do not include fetal growth, the lowered activity of pregnant females generally leads to a lower maintenance requirement.
3. **Mature size of breed.** Larger breeds grow more rapidly than smaller breeds; hence, they have a higher maintenance requirement.
4. **Sex.** Young males gain more rapidly and have a higher maintenance requirement than young females.

Growth

Growth may be defined as *the increase in size of bones, muscles, internal organs, and other parts of the body*. It is the normal process before birth, and after birth until the animal reaches its full, mature size. Growth is influenced primarily by nutrient intake. The nutritive requirements become increasingly acute when young animals are under forced production, such as when heifers are bred to calve as 2-year-olds and horses are raced as twos.

Growth is the very foundation of animal production. Young cattle, sheep, swine, poultry, rabbits, and other types

---

[1]Van Sant, W. R., *A Milker's Manual*, The University of Arizona, Bull. A-37, 1965, p. 15.

of meat animals will not make the most economical finishing gains unless they have been raised to be thrifty and vigorous. Likewise, breeding females may have their reproductive ability seriously impaired if they have been improperly grown. Nor can one expect the most satisfactory yields of milk from dairy cows or production of eggs from laying hens unless they were well developed during their growing period. Horses cannot perform the maximum amount of work, and running horses do not possess the desired speed and endurance, if their growth was stunted or if their skeletons were improperly formed by inadequate rations during growth.

Generally speaking, organs vital for the maintenance of life—e.g., the brain, which coordinates body activities, and the gut, upon which the rest of the postnatal growth depends—are early developing; and the commercially more valuable parts, such as muscle, fat, and udder, develop later. However, not all gut is early developing; for example, the growth and functioning of the ruminant stomach is delayed.

MEASURES OF GROWTH

Various methods of measuring growth have evolved, the most common of which follow:
1. **Body weight.** This is the most widely used measure of growth from birth to maturity. It may be measured by (a) weight gain per unit of time (pounds gain per day, per month, or per year); (b) weight per day of age; (c) percent of birth weight, mature weight, or weight during prior period; or (d) cumulative weight gain to any given age.
2. **Body measurements.** Numerous body measurements may be, and are, used. Heart girth measurements are commonly made on replacement dairy heifers; sometimes, height at withers is included, also. Normal horse measurements are height at withers and circumference of girth and bone. Other body measurements that are sometimes used for different species are length from nose to tailhead, length from crown to rump, and width of hips.
3. **Combination body weight and body measurement.** In practical livestock husbandry, a combination of body weight plus a measurement such as height at withers is often of greater value as a guide to feed allowance than either measurement alone.
4. **Feed efficiency.** This value is usually expressed by the conversion factor—pounds of feed eaten per pound of gain in body weight.

FACTORS AFFECTING NUTRITIVE NEEDS FOR GROWTH

The nutritive needs for growth vary with age, breed, sex, rate of growth, and health.

Age

In comparison with older animals, young animals generally (1) consume more feed per unit of body weight; (2) utilize feed more efficiently, in pounds of feed eaten per pound of body gain; (3) have a higher requirement for protein, energy, vitamins, and minerals per unit of body weight; (4) require a more concentrated and more easily digested diet; and (5) are more subject to nutritional deficiencies.

After an initial adjustment period (which may be prolonged if the feed, environment, or disease level are unfavorable), the rate of gain of young animals and birds is very rapid when measured as a percentage of body weight. Table 2-3 shows growth rate as measured by the days needed to double birth weight and the number of months needed to reach 50% of mature body weight. As shown, the long juvenile period of humans is unique.

TABLE 2-3
DAYS NEEDED TO DOUBLE BIRTH WEIGHT AND MONTHS NEEDED TO REACH 50% OF MATURE BODY WEIGHT, FOR SEVERAL SPECIES OF ANIMALS

Species	Days to Double Birth Weight	Months to Reach 50% Mature Weight
Human	150	115–145
Cattle	47–70	12–22
Horse	60	8–9
Goat	22	4–6
Sheep	15	3–5
Swine	14	11
Rabbit	6	2–5
Rat	6	1½–3
Chicken	5	2
Turkey	5	4
Duck	4	1½

Typical growth patterns in body weight gains of four breeds of dairy heifers are shown in Fig. 2-7.

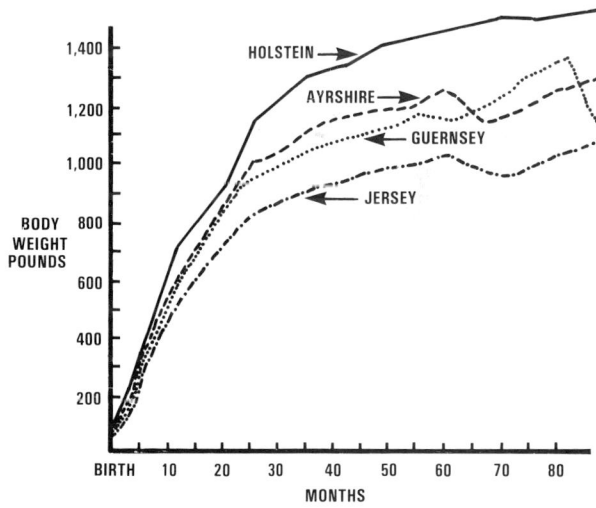

Fig. 2-7. Body weight of four breeds of dairy females from birth to 84 months of age. (Adapted by the authors from Neb. Ag. Exp. Sta. Bull. 179, by H. P. Davis and I. L. Hathaway)

Fig. 2-7, along with studies by Brody (*Bioenergetics and Growth*, by S. Brody) and others, show that growth in dairy heifers, as measured by gain in body weight, is most rapid to about 2 years of age, followed by a more gradual rate of gain. A similar pattern of rapid weight gains early in life occurs in all species.

The typical growth pattern of horses in height at withers and weight is shown in Fig. 2-8.

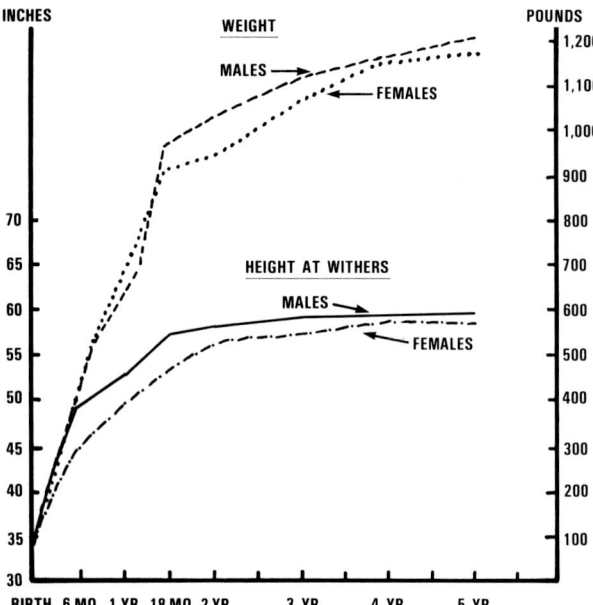

Fig. 2-8. Growth curve (height at withers and weight) of Quarter Horse males and females from birth to maturity. (Based on data from *A Study of Growth and Development in the Quarter Horse,* Bull. No. 546, Louisiana State University, Baton Rouge)

Fig. 2-8 shows that Quarter Horses grow rapidly in weight until about 1 year of age and in height to about 18 months of age, followed by more gradual growth to 5 years of age. Also, it is noteworthy that females reached maturity (tapered off in both weight and height) faster than males. In a study (*Bioenergetics and Growth,* by S. Brody) involving 4 breeds of dairy cows, Brody found a similar pattern of rapid growth in height at the withers to about 2 years of age, followed by a period of slow growth to about 5 years of age. A like growth pattern applies to all species and breeds.

Breed

Larger breeds of all species grow more rapidly than smaller breeds and have a higher nutrient requirement.

Sex

Growth studies involving young animals of both sexes, and of all species, reveal the following: (1) Males gain more rapidly than females and have a higher feed requirement; (2) uncastrated males use feed more efficiently for body weight gains than females, because of the higher water and protein content and the lower fat content of the increased body weight; (3) mature average size is larger in males than in females; and (4) females reach maturity faster than males (see Fig. 2-8).

Rate of Growth

In recent years, the accent in meat production has been on forced production and marketing at an early age; and in breeding animals, it has been on early reproduction. Achieving each of these goals has involved improved nutrition, and, generally speaking, rapid gains and profits have been on the same side of the ledger. Today, broilers may reach market weight in 8 weeks, with notable efficiencies in feed utilization as low as 1 lb of feed per 1 lb of bird. Heifers now calve at 24 months of age, or less, and more and more ewe lambs are being bred to lamb at about 12 months of age.

Rapid gains call for more nutrients. In turn, this necessitates high-energy, palatable, well-balanced rations. For the most part, fast gains are efficient gains; when animals grow at maximum rates, they require fewer nutrients and fewer pounds of feed per pound of gain.

Despite the above factors favoring rapid gains, the following cautions are noteworthy:
1. Fast gains may be fat gains.
2. Fleshy feeder cattle may not be desirable.
3. Rapid gains may be uneconomical.
4. Rapid gains may impair reproduction in females.

COMPENSATORY GROWTH

Compensatory growth is making up for a bad start in life. It is common practice for stocker cattle to be "roughed through" the winter as cheaply as possible, with limited daily gains. Then, in the spring, the animals are turned to lush spring pasture or put in a feedlot on a high-energy ration. Animals so managed exhibit the phenomenon of compensatory growth; that is, on a high-energy diet they gain faster and more efficiently than similar cattle that were fed more liberally during the wintering period. Feedlot operators were quick to sense this phenomenon, and to take advantage of it. This is the chief reason for the popularity of "Okie-type" cattle—animals whose growth has been held back to less than their genetic potential. When fed more liberally, they exhibit a surge in growth rate and feed efficiency. Large compensatory growth usually indicates that someone (the stocker operator) has lost money while someone else (the feeder) has made money. It is noteworthy that Holsteins and the larger exotics should never be handled so as to exhibit compensatory gains. If they're held back in the winter, they're too heavy when they finish.

Health

Ill health—diseases and parasites—results in lack of thrift and poor development in young stock. When the causative factor is severe, growth may be stunted. Feed is always too costly to waste. Besides, the full productive potential of animals is needed.

Reproduction

Being born (or hatched) and born alive are the first and most important requisites of livestock production, for if animals fail to reproduce, the breeder is soon out of business. A "mating of the gods," involving the greatest genes in the world, is of no value unless these genes result in (1) the successful joining of the sperm and egg, and (2) the birth of live offspring. Stock raisers acknowledge that young crop percentage is the biggest single factor affecting profit. Despite this undeniable fact, it has been estimated that 20 to 50% of all matings are infertile, that 25% of all cows culled from dairy herds are removed because of reproductive inefficiency, that the overall average U.S. calf crop of all cattle (beef and dairy combined) is only 88%—the other 12% abort or are sterile, that 5% of all ewes are sterile, that 15% of all sows bred fail to produce litters, and that only 50% of all mares bred actually produce foals.

Even more nutritional problems with developing embryos are encountered in chickens and turkeys. This may be related to maternal turnover. For example, cattle and horses produce birth-weights each year which are from 7% to 10% of maternal liveweight, whereas a value of about 700% is realized with chickens. Although it could be argued that mammals (such as cattle and horses) subsequently provide milk as a maternal output, we have much greater control over this process, and, in problem situations, we can always supplement the milk or wean the offspring. With chickens and turkeys, however, the "package" (egg) is delivered at laying, and as yet we have no way of controlling embryo nutrition during incubation. Considering the very large maternal turnover of chickens and turkeys, and the lack of control of embryo nutrition during incubation, it is not too surprising that nutritional problems with developing embryos sometimes occur.

Certainly, there are many causes of reproductive failure, but scientists are agreed that nutritional inadequacies play a major role.

NUTRITIONAL FACTORS AFFECTING REPRODUCTION

Many factors affect reproduction. For convenience, herein they have been grouped under the following headings: (1) liberal feeding for early sexual maturity, (2) flushing, and (3) nutritional reproductive failure.

Liberal Feeding for Early Sexual Maturity

Rising costs make it imperative, for economic reasons, that breeders get females in reproduction as early in life as possible. This calls for calving 2-year-old heifers, instead of threes; for breeding ewes to lamb as yearlings, and for the early breeding of females of all other species.

It is now known that liberal feeding makes for early sexual maturity. This was confirmed in a longtime experiment conducted at Cornell University, in which 3 groups of Holstein heifers were raised from birth to first calving on 3 different nutritive levels—62% (low), 100% (medium), and 146% (high)—of the standard amount of total digestible nutrients.[2] An extremely important finding in the Cornell study was that, regardless of the plane of nutrition, the heifers (Holsteins) came into heat at about 600 lb body weight, showing that size and weight, not age, determine the time of sexual maturity (see Fig. 2-9). This points up the importance of liberal feeding for early breeding.

For heifers of both beef and dairy breeds to breed at 15 months of age and calve at 24 months of age, following weaning the larger size breeds should gain from 1.4 to 1.8 lb per day, and the smaller breeds should gain 1.0 to 1.4 lb per day.

---
[2]Reid, J. T., et al., *Causes and Prevention of Reproductive Failures in Dairy Cattle: IV.* "Effect of Plane of Nutrition During Early Life on Growth, Reproduction, Production, Health, and Longevity of Holstein Cows"; I. "Birth to Fifth Calving," Feb. 1964.

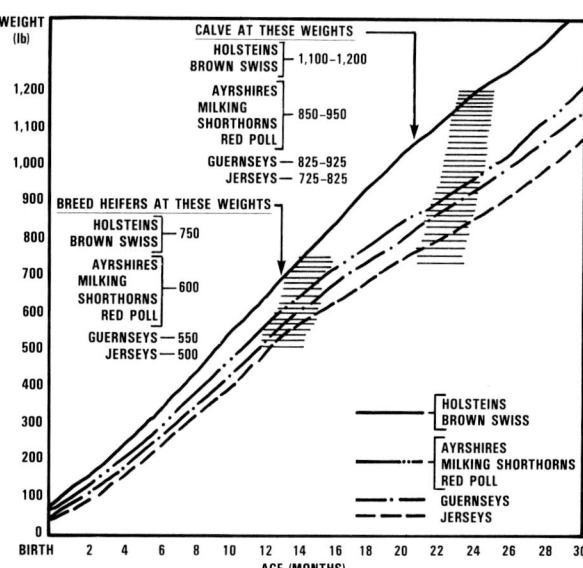

Fig. 2-9. Heifers that are well grown can be bred at the size and age shown, so as to calve at the size and age shown. (Adapted by the authors from Cornell University Ag. Exp. Sta. Bull. 987, Feb. 1964)

Flushing

Flushing is the practice of having females gain in weight just prior to breeding. The purpose of flushing is to increase the number of ova shed during estrus. Although it is not likely that all the benefits ascribed to flushing will be fully realized under all conditions, the general feeling persists that the practice will result in a 15 to 30% increase in lamb and pig crops, and that females of all species will breed both earlier and more nearly at the same time. Hence, it follows that the offspring will be earlier and more uniform in age and size.

Mature females appear to respond better to flushing than virgin females. Also, flushing may be more beneficial early and late in the breeding season than during the peak when the ovulation rate is highest.

Fat females will not respond to flushing. Instead, they should be conditioned for breeding by stepping up the exercise.

Flushing is accomplished by feeding females more liberally 2 to 3 weeks prior to breeding and continuing into the breeding season, so that they are gaining in weight at the time of mating. It may be accomplished by increasing the grain allowance; or in the case of ruminants and mares, by turning them to a fresh, luxuriant pasture.

Soon after breeding, females bred for the first time and dry females should be put back on limited rations. Continuation of a high level of nutrition after breeding will result in a higher embryo mortality.

Nutritional Reproductive Failure

Because livestock producers largely determine their own destiny when it comes to feeding, it is important that they know the causes of reproductive failure and how to rectify them.

A review of the literature clearly points to three reproductive difficulties: (1) a small number of females in heat and bred early in the breeding season, (2) the low conception rate at first service, and (3) the excessive losses at birth or within the first two weeks of age.

Research gives ample evidence that the real cause of most reproductive failure is a deficiency of one or more nutrients just before or immediately following parturition—nutritive deficiencies during the critical period when life begins—a deficiency of energy, protein, minerals, and/or vitamins. Based on an extensive review of the voluminous literature, the authors evolved with the following summary of the nutritional causes of reproductive failures in mammals:

1. **Overfeeding or underfeeding.** Overfeeding, accompanied by extremely high condition, or underfeeding, accompanied by emaciated and run-down condition, usually result in temporary sterility. Overfat females often experience birth difficulties. Excessive thinness results in low birth weights and weak young.

2. **Energy.** A low level of energy during the last third of pregnancy and immediately following parturition will have a marked effect on rebreeding—fewer females will come in heat at the beginning of the breeding season, and fewer will conceive.

3. **Protein.** A low level of protein during gestation results in lowered reproduction, lighter birth weights, and delayed heat following parturition.

4. **Phosphorus.** Low phosphorus will markedly decrease the number of young born.

5. **Iodine.** A deficiency of iodine willl cause impaired reproduction; weak or dead offspring at birth; big-necked (goiterous) calves, lambs, and foals; and hairless pigs.

6. **Vitamin A.** Low vitamin A will result in the birth of weak, malformed, partially blind, or dead young.

With all mammalian species, most of the growth of the fetus occurs during the last third of pregnancy. Additionally, the female must store body reserves during pregnancy, for the demands for milk production are generally greater than can be supplied by the ration fed during early lactation. Hence, the nutrient requirements are very critical during this period, especially for young pregnant females.

It is also known that the ration exerts a powerful effect on sperm production and semen quality. Too fat a condition can lead to temporary or permanent sterility. Moreover, there is abundant evidence that greater fertility of herd sires exists under conditions where a well-balanced ration is provided.

Egg Production

Egg production involves feeding for number of eggs, egg quality, hatchability, and control of molt and broodiness.

The nutritive needs for commercial egg production include those for maintenance of the birds, growth of pullet layers, and the formation of eggs. The nutritive requirements are greater for birds with an inherited capacity for high egg production than for those that lay only a few eggs. The standard-weight egg contains about 95 calories of gross energy, 7.5 grams of crude protein, and 2 grams of calcium.

With poultry, the development and hatching of a fertile egg constitute reproduction. As with mammals, the nutritive requirements of poultry breeders (including chickens, turkeys, ducks, geese, etc.) are more rigorous than those for commerical laying. For high hatchability and good development of young, breeders require greater amounts of vitamins A, D, E, B-12, riboflavin, pantothenic acid, niacin, and of the mineral manganese. Birds intended as breeders should be started on special breeder rations at least a month before hatching eggs are to be saved.

Lactation

Simply stated, milk production is a by-product of the reproductive process.

The nutrient needs for lactation depend on the amount and composition of milk secreted. Some idea of the enormity of these requirements, as well as the efficiency of milk secretion, becomes apparent when it is realized that a dairy cow that produces 14,500 lb of milk during 1 year manufactures 523 lb of milk fat, 674 lb of milk sugar, 477 lb of milk protein, and 109 lb of minerals and vitamins, or a total of over 1,783 lb of food. That's equivalent to the carcass weight produced by 2½ steers in 18 months time. Also, it is noteworthy that the cow remains alive and can repeat the production again and again, whereas the steers must be slaughtered or "spent." It is noteworthy, too, that the cow needs additional nutrients for body maintenance (which are about the same as those of one steer), for development of the unborn calf if she is pregnant, and for growth if she is a young cow.

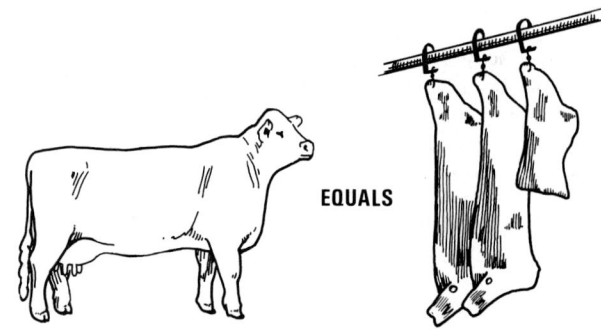

Fig. 2–10. It takes 2½ steers 18 months' time (for each steer) to produce as much carcass weight as one cow produces in milk in 1 year. And the cow remains alive to do it all over again!

Although high-producing cows require more total feed than low producers, they utilize proportionately more nutrients for milk production (Fig. 2–3), and generally they return more net income over feed cost.

One of the most drastic changes in the life cycle of mammals is that which occurs at freshening, when a female suddenly makes the transition from nonlactating to lactating. In high-producing dairy cows, clinical milk fever is an indication of a very sudden drop in the level of calcium in the blood serum and of putting calcium into the milk. If

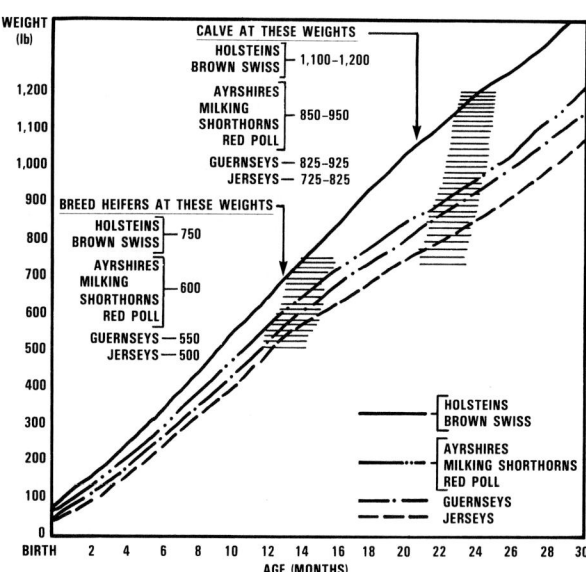

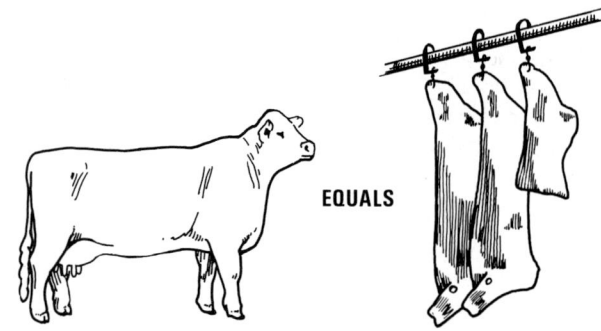

the reversal process is too drastic, the animal may be thrown into a blood calcium deficiency (hypocalcemia), followed by slowing of the heart and eventual coma. Although this problem is most common among high-producing dairy cows, it occasionally occurs in sheep and in other mammals.

Finishing/Fitting

Finishing is what the name implies—the laying on of fat, especially in the tissues of the abdominal cavity and in the connective tissues just under the skin and between the muscles. It is the normal feeding practice followed prior to slaughter, for the purpose of improving the flavor, tenderness, and quality of meat, better to meet consumer demands. Generally speaking, the higher the degree of finish, the higher the dressing percentage and the lower the protein (red meat) content. Also, it takes more nutrients to produce a pound of fat than a pound of lean; hence, excess finish is wasteful and undesirable.

Fattening is usually achieved through the use of high-energy feeds, carbohydrates, and fats—a liberal allowance of grains. However, due to world food shortages, the long-time trend is to incorporate more roughages in finishing rations for ruminants. Such rations are lower in net energy and produce smaller gains than high-concentrate rations, but they may make for more net returns when feed grains are scarce and high in price.

The objective of the livestock producer is to finish animals to the degree of fleshing and carcass weight desired by the consumer, at a maximum of profit for his efforts. Computers have given large operators a big assist in meeting these goals.

Fitting is the conditioning of animals, usually for show or sale, through careful feeding, grooming, and exercising, to enhance their bloom and attractiveness. Fitting animals for show or sale involves the application of similar principles and practices to those followed in fattening (finishing) livestock for market. Animals intended for show or sale should be fed so as to achieve a certain amount of finish or bloom, but they should not be too fat. In general, most fitting rations are similar to the rations used in commercial fattening operations for animals of like species and comparable ages, except that they are usually higher in protein content; experienced caretakers feel that they get more bloom by use of high-protein rations. Also, it is common practice to feed a palatable milk replacer to young animals that are being fitted for show or sale.

Wool and Mohair

Wool and mohair are high-protein products. Also, it is noteworthy that they are especially rich in the sulfur-containing amino acid, cystine. But the latter requirement is usually amply met by the cystine of feeds or by methionine—another amino acid which is also rather widely distributed in feeds, as well as being derived from rumen synthesis.

A lack of energy in the ration of sheep will result in lighter fleeces and lower quality of wool, including breaks, or tender spots, in the fiber. A protein deficiency will also make for lighter fleeces.

Copper-deficient sheep produce "steely" wool, lacking in crimp, tensile strength, affinity for dyes, and elasticity. With a severe deficiency, the wool of black sheep is depigmented.

Work/Running

Fig. 2-11. Work being performed by a hunter. The nutritive requirements of such equine athletes are very exacting. (Courtesy, the American Morgan Horse Association, West Moreland, N.Y.)

In the United States, the function of work or running is limited to horses. But, in certain parts of the world, oxen, water buffalo, camels, reindeer, dogs, and other animals are the chief sources of power.

Racehorses, and other horses used for such purposes as hunting, jumping, cutting, or roping, are equine athletes, whose nutritive requirements are very exacting. And the younger the animal, the more intense the use, and the greater the stress, the higher the level of nutrition needed in order to develop and maintain sound legs and build a strong frame and body. It would appear that their nutritive requirements are not unlike human athletes—for example, college football teams and participants in the Olympics—who are required to eat at special training tables, supervised by expert nutritionists; and who are fed generous quantities of meats and other high-protein foods, along with diets high in energy, minerals, and vitamins. Further credence to this analogy is lent by the estimate that, when racing, the energy requirements of a horse are 100 times greater than at rest. Thus, rations for racehorses, and other horses in heavy use, should be rich in available energy, high in protein, and fortified with minerals and vitamins—with all nutrients in proper balance.

For draft horses not in reproduction, the energy needs are met primarily by carbohydrates and fats, nutrients that can be provided in the form of grain. Adequate protein is necessary to maintain, repair, and replace the muscles used in work. The mineral and vitamin requirements for work horses are practically the same as for comparable idle horses —except for the greater need for Na and Cl to offset the salt losses that accompany increased perspiration.

QUESTIONS FOR STUDY AND DISCUSSION

1. Which is the most important in animal production, heredity or nutrition?

2. Define the terms *nutrition* and *nutritionist*.

3. What is the primary reason for keeping animals?

4. Discuss the leading nutrition developments in each of the following eras, along with the impact of each:
   a. Naturalistic era
   b. Chemical-energy era
   c. Mineral era
   d. Vitamin-biological era
   e. Feeding standard era
   f. Amino acid era
   g. Feed additive and implant era
   h. Biotechnology era

5. What's the purpose of pre-slaughter withdrawal times or milk-discard periods for animals receiving certain drugs? How may livestock producers be certain that they have used drugs safely?

6. How will each member of the food team—producers, processors, and consumers—benefit from biotechnology?

7. As the scientific domain of the nutritionist has become increasingly narrow, frequently such individuals (a) lack understanding of animals as a whole, and (b) pay little attention to the comparative aspects with other animal species. Is this good or bad? If it is bad, how may it be rectified?

8. How, and why, should we improve our store of knowledge relative to the following nutrition-related areas:
   a. Nutrient requirements
   b. Net energy system
   c. Proportion of forage to grain
   d. Behavior and environment of animals
   e. Conservation of energy
   f. Pollution control

9. Why is knowledge of body composition important to the nutritionist? What is the effect of each of the following on body composition:
   a. Age
   b. Feed intake
   c. Rate of gain
   d. Grower type rations
   e. Species

Generally speaking, experienced cattle feeders feed grower rations during the growing (stocker) phase, then, as cattle get heavier, they switch to high-energy finishing rations. Discuss the practicality of such feeding programs.

10. On the average, how many pounds of water per pound of protein are found in the animal body?

11. What is meant by the "functions of nutrients"?

12. Discuss the nutrient needs for each of the following body functions, with appropriate application to different species:
    a. Maintenance
    b. Growth
    c. Reproduction
    d. Egg production
    e. Lactation
    f. Fattening/fitting
    g. Wool and mohair
    h. Work/running

13. List and discuss the factors affecting the nutritional requirements for each of the following:
    a. Maintenance
    b. Growth
    c. Reproduction

14. What is compensatory growth? How does this phenomenon affect (a) the cow-calf producer, and (b) the cattle feeder?

15. Compare the reproductive problems encountered in mammals and poultry. Discuss which is the easiest to improve.

16. What is the justification for liberal feeding of heifers for early sexual maturity?

17. What is flushing? How may females of each species be flushed?

18. Give and discuss some examples of nutritional reproductive failure.

19. What is the difference(s) between (a) finishing animals for market, and (b) fitting animals for show or sale?

20. Why are the nutritive requirements of a 2-year-old racehorse so critical?

Vitamin C Made The Difference!

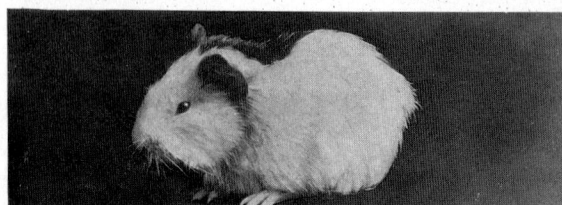
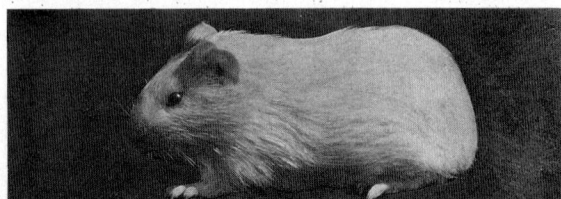

Fig. 2-12. Two guinea pigs of the same age. *Left:* This guinea pig had no vitamin C and developed scurvy; note the crouched position due to sore joints. *Right:* This guinea pig had plenty of vitamin C; note that it is healthy and alert, and that its fur is sleek and fine. (Adapted from USDA sources)

3

DIGESTION/ABSORPTION

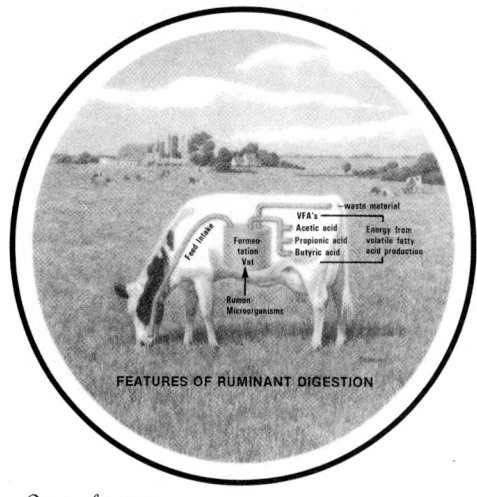

Original painting by Tom Phillips

Contents	Page
Hunger and Appetite	27
Hypothalamic Control of Appetite	28
Gastric Influences on Appetite Regulation	28
Types of Feeding Behavior	28
Anatomy of the Digestive System	28
Nonruminant Digestive System	28
Monogastric (Simple-Stomached) Digestive System	29
Functional Cecum Digestive System (Non-ruminant Herbivore)	29
Ruminant Digestive System	29
Esophageal Groove (Reticular Groove)	30
Avian Digestive System	30
Capacity of the Digestive Tract	31
Process of Digestion	31
Process of Absorption	31
Nutrient Carriers	31
Lymph	31
Blood	32
Physiology of Digestion	32
Oral Region	32
Teeth	32
Tongue	32
Salivary Glands	32
Pharyngeal and Esophageal Region	33
Gastric Region	33
Nonruminant Animals	33
Ruminant Animals	34
Reticulum and Rumen	34
Omasum	35
Abomasum	35
Newborn Ruminants	35
Avian Gastric Digestion	35
Pancreatic Region	35
Hepatic Region	35
Intestinal and Cecum-Colon Region	36
Digestion and Absorption in the Small Intestine	36
Carbohydrates	36
Lipids	36
Proteins	36
Minerals and Vitamins	37
Cecum and Colon	37
Gastrointestinal Hormones	37
Neurological Control of the Gastrointestinal Tract	37
Factors Affecting Digestion and Absorption	38
Influences of Feed Composition and Preparation	38
Rate of Passage of Feed	38
Feed Processing	38
Digestion and Absorption	38
Level of Feed Intake	38
Composition of Feed	39

Contents	Page
Chelates	39
Phytic Acid	39
Oxalates	39
Dysfunctions of the Digestive Tract	39
Reticulitis	39
Ruminal Acidosis	39
Acute Tympany (Bloat)	39
Parakeratosis of the Rumen	39
Colic	40
Diarrhea (Scours)	40
Questions for Study and Discussion	40

In nature, the primary purpose of grains and other seeds is to provide a means of propagating plants—not to supply nutrients to animals that consume them. Likewise, eggs—a universally used food and feed—provide a medium from which the developing embryo can draw vital nutrients. As a result, numerous differences can be observed in the anatomy and physiology of digestion in various livestock species which enable them to utilize feedstuffs efficiently. These differences can be observed through two modes of adaptation: (1) behavioral, and (2) anatomic. Thus, animals that consume large amounts of fibrous feeds, such as hays, differ behaviorally and anatomically from animals that consume meat and other easily digested feeds. Therefore, in order to maximize the use of feed, the caretaker and nutritionist must have an understanding of these differences in digestion and absorption.

When a feed is evaluated by chemical analysis alone, there is a good possibility that this evaluation is not a true indicator of its value to livestock. No matter how accurate the chemical analysis, it tells us little with regard to how the feed will react when subjected to the physiological phenomena occurring in the animal. Nor does it tell us anything about palatability. If a feed is extremely high in nutrients but is refused by animals, its value is nil. Likewise, once the feed is ingested, the animal must be able to digest and absorb the nutrients efficiently.

HUNGER AND APPETITE

Hunger is the physiological desire for feed following a period of fasting.

Appetite is a learned or habitual response to the presence of feed.

An animal that is extremely hungry may not have an appetite for a type of feed that it deems undesirable. Conversely, if the feed is of a desirable nature, an animal may have an appetite for it in spite of the fact that it is not hungry.

Hypothalamic Control of Appetite

The hypothalamus (derived from the terms *hypo* meaning below, and *thalamus,* a region of the brain)—a structure in the ventral region of the diencephalon—has been implicated as one of the major control centers of appetite regulation. Within the hypothalamus, certain areas can be differentiated. Two of these are of particular importance in the regulation of appetite. The first area is that of the lateral hypothalamus. It is commonly called the *feeding center,* because, upon stimulation of this region, the animal commences to eat whether or not it is hungry. If this area is damaged, the animal loses all desire to eat and will eventually starve. The ventro-medial area of the hypothalamus functions as the *satiety center.* Stimulation of this region will depress appetite. If the ventro-medial nuclei are destroyed, there is no inhibition of feed intake, and the animal will have an uncontrollable appetite. It is believed that there is a chronic activity in the lateral hypothalamus which is kept in check by the inhibitory influence of the ventro-medial area.

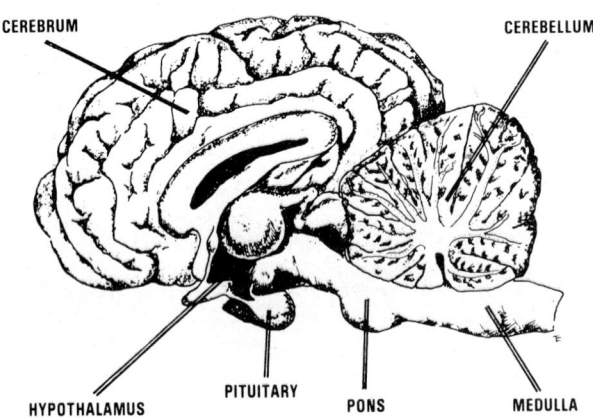

Fig. 3-1. Brain of the horse showing location of important parts, including hypothalamus.

Several theories have been advanced as to the exact physiological mechanism which triggers the hypothalamus to tell the animal when to eat. While each theory has its merits, there is no conclusive proof in favor of any one of them. In the long run, it seems probable that a combination of a number of factors will provide the answer.

The two theories concerning the hypothalamic control of appetite that have received the most attention are (1) the chemostatic hypothesis, and (2) the thermostatic hypothesis. The chemostatic hypothesis reasons that the hypothalamus is sensitive to circulating blood nutrient levels, such as sugar or lipid. When these levels become too low, the hypothalamus sends signals to begin feeding. Once the blood nutrient level is elevated, stimuli from the feeding center are inhibited and the animal feels full. The second theory of appetite control, the thermostatic hypothesis, theorizes that the hypothalamus plays an important role in heat regulation within the body, and that a decrease in hypothalamic temperature will induce feeding.

Gastric Influences on Appetite Regulation

Appetite in ruminants has been shown to be sensitive to volatile fatty acid levels in the rumen. When acetate is injected into the dorsal rumen, feed consumption is reduced. Researchers believe there are receptors in this area which are sensitive to the acetate levels produced in the rumen, thereby influencing feed intake to a limited degree. Rumen propionate levels affect feed intake, but this receptor system may be distinctly different from the one involving acetate. When the gastrointestinal tract is distended, there is normally a cessation of feeding. Thus, the actual physical limit of the digestive system has a direct influence on appetite. If an animal eats bulky or succulent feed, it may become satiated, even though it does not fulfill its energy requirements.

TYPES OF FEEDING BEHAVIOR

In general, the feeding behavior of animals is correlated with the various anatomical adaptions of the gastrointestinal tract. It is logical to assume that animals which consume feed that can be easily digested and absorbed would have gastrointestinal tracts that are smaller and simpler in structure than those animals which utilize feeds that are complex in chemical composition.

Based on the kind of feed eaten, animals are classified as follows:
1. **Carnivores.** These are the flesh eaters.
2. **Herbivores.** These are vegetarians.
3. **Omnivores.** These consume both flesh and plants.

ANATOMY OF THE DIGESTIVE SYSTEM

The digestive system, or alimentary canal, consists of a tube which courses internally from the lips to the anus. It is specialized in regions called the mouth, esophagus, stomach, small intestine, cecum, large intestine, rectum, and anus. Attached to the tube are the liver and the pancreas, which provide essential secretory products for digestion.

There are major differences in the anatomy and physiology of the organs of the digestive tract of the various animal species. These differences are of great nutritional significance, as they affect the nature of the digestive processes—hence, the kind of feed that can be utilized. Based on the anatomy of the digestive tract, animals may be grouped as (1) nonruminants, (2) ruminants, and (3) avian. An understanding of the types of digestive systems and the kind of feed eaten is essential to the intelligent feeding of each class of animal.

Nonruminant Digestive System

These animals possess stomachs that are relatively simple in structure. This group can be subdivided with respect to the functionality of the cecum and colon.

MONOGASTRIC (SIMPLE-STOMACHED) DIGESTIVE SYSTEM

These animals have the simplest of all digestive systems. It consists of the mouth and associated glands, esophagus, stomach, small intestine, large intestine, pancreas, and liver. This is the type of gastrointestinal tract found in the pig, dog, mink, fish, monkey, and humans. It is characterized by limited capacity and limited microbial action and fiber digestion. It follows that these animals are better adapted to the use of concentrated feeds, such as grains and meat products, than to the use of large quantities of roughages.

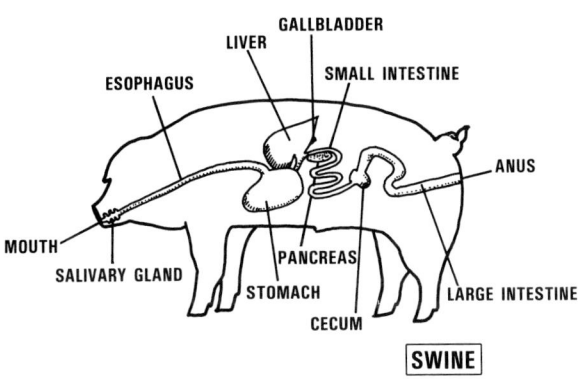

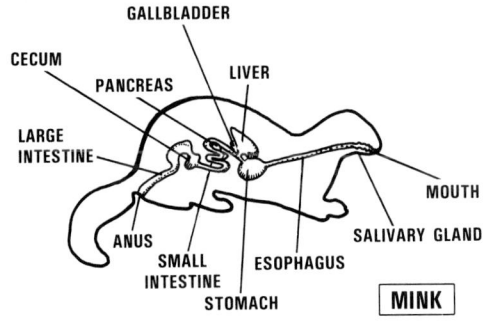

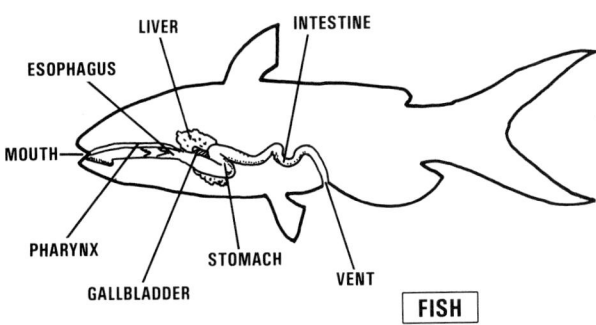

Fig. 3-2. The digestive systems of representative monogastrics with nonfunctional cecums: swine, mink, and catfish.

FUNCTIONAL CECUM DIGESTIVE SYSTEM (Nonruminant Herbivore)

In this type of digestive system—as represented by the horse, rabbit, guinea pig, and hamster—the cecum and colon are extremely large and contain a large population of microorganisms which are capable of digesting fiber as well as synthesizing a number of vitamins. Thus, from a practical feeding standpoint, these animals are between (1) the monogastric or simple-stomached animals, and (2) the ruminants.

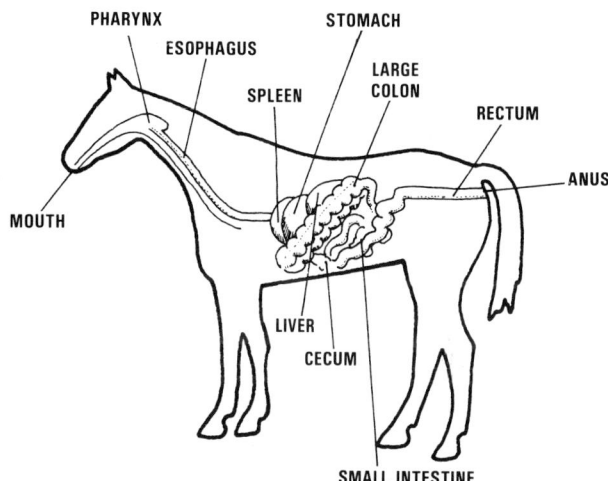

Fig. 3-3. The digestive system of the horse—a nonruminant herbivore with a functional cecum and colon.

Ruminant Digestive System

Cattle, sheep, and goats are ruminants. They differ from monogastric animals in the following important ways:

1. **Mouth.** Ruminants have no upper incisor or canine teeth. Thus, they depend on the upper dental pad and lower incisors, along with the lips and tongue, for the prehension of feed.

2. **Four stomach compartments.** Ruminants possess four stomach compartments—rumen, reticulum, omasum, and abomasum (true stomach)—whereas monogastrics have one. Such a digestive system makes for two primary nutritional differences between ruminants and simple-stomached animals:

   a. **More space.** They have the necessary space for processing large quantities of bulky forages to provide their nutrients. The cow, for example, when compared to the human on a proportion-to-weight basis, has about nine times the digestive tract capacity.

   b. **More microorganisms.** The rumen provides a desirable environment for an enormous population of microorganisms. Typical counts of rumen bacteria range from 25 to 80 billion/ml, and typical counts of protozoa range from 200,000 to 500,000/ml. The number of rumen bacteria varies according to the nature of the diet, feeding regimen, time of sampling after feeding, species differences, individual animal differences, season, availability of green feed, and the presence or absence of ciliate protozoa.

Rumen microorganisms serve two important functions:

(1) They make it possible for ruminants to utilize roughage—to digest the fiber therein. They break down the cellulose and pentosans of feeds into usable organic acids, chiefly acetic, propionic, and butyric acid—commonly called the volatile fatty acids (VFA). These VFAs are largely absorbed through the rumen wall and provide the ruminant 60 to 80% of its energy needs. Microbial digestion is of great practical importance in the nutrition of ruminants; it is the fundamental reason why they can be maintained chiefly on roughages.

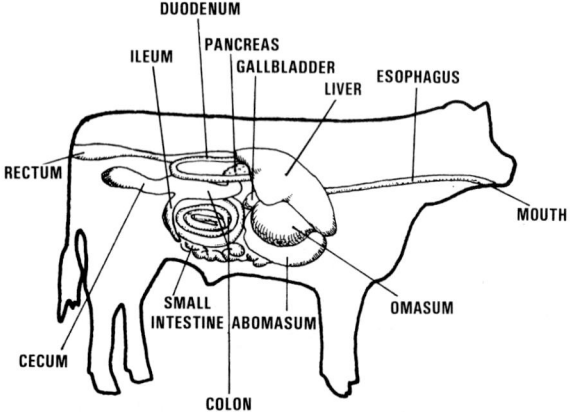

Fig. 3-4. The digestive system of the bovine—a ruminant.

(2) In exchange for their rumen-housing privileges, the microbes synthesize nutrients for their host, in a true symbiotic relationship. Rumen microbes synthesize, or manufacture, all the B-complex vitamins and all the essential amino acids. The latter can even be made from nonprotein nitrogen compounds (NPN), such as urea or ammoniated products, or from proteins that are deficient in one or more of the amino acids. Finally, the microorganisms give their life to their host in payment for food and shelter, being digested farther along in the gastrointestinal tract.

NOTE WELL: Although rumen microbes synthesize all the B-complex vitamins and all the essential amino acids, it has been shown that they do not supply them in adequate amounts for the maximum growth or milk production of ruminants.

3. **Rumination.** A placid cow lying under a tree slowly chewing her cud conveys a special sense of contentment symbolic of the tranquility of the countryside. But this activity, or phenomenon, which is peculiar to ruminants, is of great practical significance.

During rumination, the animal regurgitates and rechews a soft mass of coarse feed particles, called a *bolus*. Each bolus is chewed for about a minute, then swallowed again. Ruminants may spend 8 hours or more per day in rumination, the amount of time varying according to the nature of the diet. Coarse, fibrous diets result in more time ruminating. Rechewing does not improve digestibility. Rather, rumination has an important bearing on the amount of feed the animal can eat and utilize. Feed particle size must be reduced to allow passage of the material from the rumen. Because high-quality forages contain less fiber than low-quality forages, they require much less rechewing and pass out of the rumen at a faster rate; hence, they allow the cow to eat more. The rumination process also stimulates the salivary glands in the mouth. These produce large amounts of saliva which facilitate food passage through the digestive tract. Saliva also contains buffer salts which resist changes in pH of the rumen contents and protect against acidosis.

4. **Eructation (belching of gas).** Substantially more gas is produced in digestion by ruminants than by simple-stomached animals. The microbial fermentation in the rumen results in the production of large amounts of gases (primarily CO_2 and methane) which must be eliminated; otherwise, bloat results. Normally, these gases are expelled quite freely by eructation (belching) and, to a lesser extent, by absorption into the blood draining from the rumen, from which they are eliminated through exhaled air from the lungs.

(Also see Chapter 11, section on "Protein and Amino Acids For Ruminants.")

ESOPHAGEAL GROOVE (RETICULAR GROOVE)

The anatomical peculiarities of the newborn ruminant are: (1) The reticulum, rumen, and omasum are relatively underdeveloped; (2) the esophageal groove, which is formed by two heavy muscular folds, or lips, and which extends from the lower end of the esophagus into the omasum, can convey material from the esophagus directly to the abomasum; and (3) nursing stimulates reflex closure of the esophageal groove and causes the milk consumed to bypass the rumen and reticulum and most of the omasum, thereby escaping bacterial fermentation, and to pass directly into the last compartment of the gastric region—the abomasum.

As the young ruminant grows older, it ingests increasingly larger quantities of solid feed which, in turn, stimulates the growth and development of the reticulum and rumen. Also, through its contact with the environment and other ruminants, the young ruminant becomes naturally inoculated with microorganisms which benefit digestion. At this stage, reflex to milk diminishes and the animal functions as a mature ruminant. Age and type of feed are the main factors associated with loss of response to suckled milk. Normally, most of this transition occurs in the young ruminant by 2 months of age. Thus, at birth the first three compartments (reticulum, rumen, and omasum) of the new born calf represent less than 30% of the total somach capacity; by 2 months of age, they represent 70% of the total stomach capacity; and at maturity, they account for 93% of the total stomach capacity. The greatest development is in the rumen, which accounts for 80% of the total stomach capacity of the mature cow. (Also see subsequent section in this chapter on "Newborn Ruminants.")

Avian Digestive System

The digestive system of poultry differs considerably from that of other monogastric animals. Birds have no teeth; hence, there is no chewing. The esophagus empties directly into the crop, where the feed is stored and soaked. From the crop, the feed passes to the proventriculus (or glandular stomach), the thick-walled organ immediately in front of the gizzard. Here it is stored temporarily while digestive juices are copiously secreted and mixed with it. Thence, it passes to the gizzard, a very muscular organ, which normally contains stones or grit, where it is crushed and ground. Then the feed moves through the small intestine, the ceca, and the large intestine to the cloaca.

Digestion in the fowl is rapid. It requires only about 2½ hours in the laying hen, and 8 to 12 hours in the nonlaying hen, for the feed to pass from the mouth to the cloaca.

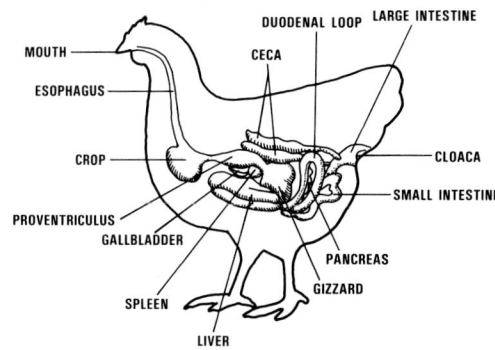

Fig. 3-5. The digestive system of the hen.

CAPACITY OF THE DIGESTIVE TRACT

Due to the anatomic adaptations of the various livestock species, the relative importance of the various digestive organs is reflected in their respective capacities. In ruminants, the stomach has the largest capacity, while in the pig the small intestine and the stomach have about the same capacity. In the horse, the cecum and colon are the largest segments of the gastrointestinal tract. (See Table 3-1.)

PROCESS OF DIGESTION

Digestion, taken in a narrow sense of the word, *can be defined as the process whereby proteins, fats, and complex carbohydrates are broken down into units that are of small enough size to be absorbed through the gut wall, into the animal body, proper.* This process is accomplished by both chemical and physical processes.

Enzymes are organic catalysts produced by certain cells within the body which speed biochemical reactions at ordinary body temperatures without being used up in the process. Enzymatic activity is responsible for most of the chemical changes occurring in feeds as they move through the digestive tract.

Many of the digestive enzymes are stored in an inactive form. When they are in the inactive form, they are called *zymogens* or *proenzymes*. Once secreted into a favorable environment for digestion, generally governed by pH, these inactive enzymes "turn on" and perform their specific digestive function.

PROCESS OF ABSORPTION

Once the various nutrients have been adequately digested, several modes of absorption can occur. These modes are dependent on the chemical nature of the nutrient and the site of absorption. Virtually no absorption takes place before the feed enters the stomach. In the ruminant, the microflora digest a large proportion of the feed, and considerable absorption takes place in the rumen. In the nonruminant species, very little absorption occurs in the stomach. Rather, the primary site for the absorption of most nutrients is in the small intestine. In nonruminants with a functional cecum and colon, some proteins and amino acids may be

TABLE 3-1
PARTS AND AVERAGE CAPACITIES OF DIGESTIVE TRACTS OF SELECTED ANIMALS[1]

	Animal Species									
	Cattle		Sheep or Goat		Horse		Pig		Humans	
	(gal)	*(liter)*	(gal)	*(liter)*	(gal)	*(liter)*	(gal)	*(liter)*	(gal)	*(liter)*
Gastric compartment:										
Rumen (paunch)	53.4	*202.4*	6.2	*23.5*						
Reticulum (honeycomb)	2.0	*7.6*	0.5	*1.9*						
Omasum (manyplies)	5.0	*18.9*	0.2	*0.8*						
Abomasum (true stomach)	6.1	*23.1*	0.9	*3.4*	4.8	*18.2*	2.1	*7.9*	0.3	*1.1*
Subtotal	66.5	*252.0*	7.8	*29.5*	4.8	*18.2*	2.1	*7.9*	0.3	*1.1*
Small intestine	17.4	*65.9*	2.4	*9.1*	16.9	*64.0*	2.4	*9.1*	1.0	*3.8*
Cecum	2.6	*9.8*	0.3	*1.1*	8.9	*33.7*	0.4	*1.5*		
Large intestine	7.4	*28.0*	1.2	*4.5*	25.4	*96.1*	2.3	*8.7*	0.3	*1.1*
Total	93.9	*355.9*	11.7	*44.2*	56.0	*212.0*	7.2	*27.2*	1.6	*6.0*

[1]Adapted by the authors from Swenson, M. J. (ed): *Dukes' Physiology of Domestic Animals,* 9th ed., 1977, by Cornell University, by permission of Cornell University Press.

absorbed in the colon. For the most part, absorption in the large intestine is restricted to water and electrolytes.

Nutrient Carriers

When the nutrients have been digested and absorbed, they must be transported to tissues that either have an immediate demand for them or can store them for later use.

Lymph and blood are the primary transport media for the nutrients which have been absorbed.

LYMPH

Within the mucosal membrane of the intestinal tract, there is a capillary network of lymph vessels. Cholesterol, water, long-chain fatty acids, and some proteins are picked up by this system and transported through a series of larger vessels which ultimately empty into the venous system anterior to the heart.

The immune system of the newborn animal is not well developed. Therefore, it is essential that it receive colostrum from its mother as soon as possible. Through this intake of colostrum, antibodies are passed from the mother to the newborn, imparting a certain degree of immunity to stress and disease which will last for the first critical days of the young animal's life. Many of these antibodies are absorbed intact in the newborn and transported via the lymphatic system.

BLOOD

Most of the low molecular weight products of digestion are absorbed and transported by the blood. These nutrients include water, salts, glycerol, amino acids, short-chain fatty acids, monosaccharides, and certain vitamins. These materials are absorbed into the capillary system of the intestine. The capillary network drains into the venous system, eventually entering the portal vein of the liver. From the liver, the nutrients then travel through the hepatic veins which, in turn, enter the main systemic vein—the vena cava.

PHYSIOLOGY OF DIGESTION

Although there are numerous anatomical differences among the domestic livestock species, the principles of digestion and absorption are quite similar on a physiological basis. In order to compare these processes in the various animals, the discussion of the physiology of digestion will be divided into sections dealing with the regions of digestion instead of individual organs. These regions are the oral region, pharyngeal and esophageal region, gastric region, pancreatic region, hepatic region, and intestinal and cecum-colon region.

Oral Region

Three physical processes occur in the oral region of animals—prehension, mastication, and the initiation of deglutition.

Prehension can be defined as the act of bringing food into the mouth. Numerous modes of prehension can be found in animals. Animals, such as the raccoon and humans, use their forelimbs to bring the food to the oral cavity, while many other types of animals rely on structures of the mouth, such as the tongue, lips, and teeth.

Mastication is the act of chewing food. Most animals chew their food immediately following prehension. One notable exception is the fowl which swallows its food whole. Mastication involves the physical grinding and tearing of the food in addition to the admixture of saliva which lubricates the food as well as initiates a limited amount of enzymatic digestion. *Food that has been masticated and formed into a small compact ball for passage down the digestive tract is called a bolus.*

Deglutition is the act of swallowing. This process involves both voluntary and involuntary reflexes. Upon completion of mastication, the bolus is lifted by the tongue and moved to the back of the mouth. The bolus passes through the pharynx, causing a temporary inhibition of respiration by the reflex closure of the larynx, and finally down the esophagus to the gastric region.

TEETH

The teeth serve primarily as a mechanical aid for mastication. By tearing and grinding the food, they provide a means whereby a large surface area is created which can be exposed effectively to the digestive fluids of the tract. There are four types of teeth, each serving a specialized function: incisors, canines, premolars, and molars. The teeth on the front of the jaw are called incisors and are used for the tearing and slicing of food. Moving progressively to the back of the jaw, the next teeth are called canines. These teeth—sometimes called the eye teeth, or tusks—are also used for tearing. Ruminants do not have these teeth. Following the canines are the premolars and molars, both types of which are used for grinding. Generally grinding can occur on only one side of the jaw at a time.

Poultry do not have teeth. Rather, they swallow their food whole, and the mechanical breakdown of food takes place in the gizzard, aided by the presence of stones or grit.

TONGUE

In many species of domestic animals, the tongue is the primary structure for prehension. In the cow, the tongue is elongated and covered with rough papillae, making it adapted to wrapping around grass and other forages. The cow then brings the forage into the oral cavity where it is sheared by the movement of the incisors against the dental pad.

Throughout the process of mastication, the tongue serves a threefold purpose. First, movement of the tongue transports the feed to the various areas of the mouth to be torn and ground. While doing this, the tongue is also mixing the feed with the various secretions of the mouth, ultimately forming a bolus. Secondly, the presence of taste buds on the tongue provides a neurological control for feed selection and intake. If the feed is bitter or unpalatable, impulses from the taste buds signal the animal to stop eating. Conversely, a desirable taste stimulates appetite. In the cow, the tip of the tongue contains a large number of taste buds while the middle portion has very few. The back portion of the tongue contains the highest density of taste buds. In the chicken, there are very few taste buds; and what few there are lie on the back area of the tongue. Finally, the tongue initiates the process of deglutition. When the bolus has been adequately prepared, the tongue moves it to the back of the mouth where neural receptors are stimulated, and swallowing commences.

SALIVARY GLANDS

The salivary glands represent a network of accessory structures which are essential to digestion. Three pairs of salivary glands are of primary importance—parotid, submaxillary, and sublingual. Fig. 3-6 illustrates the location of these glands.

Saliva, the secretion from these glands, is highly variable in chemical composition. Two basic types of saliva are produced. The first type is extremely thick, being rich in the glycoprotein *mucin*. In this type of saliva, there is very little enzymatic activity. The second type of saliva is serous in

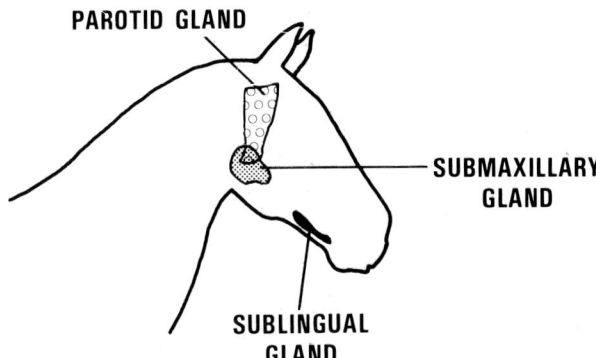

Fig. 3-6. Location of the main salivary glands.

composition; that is, it is watery and thin, containing various proteins and enzymes but little mucin. Depending on the particular species of animal, the salivary glands secrete different mixtures of these two kinds of saliva. Saliva from the parotid glands tends to be serous in composition, while the submaxillary glands normally secrete a mixture of the two. In the horse, cow, pig, and dog, the sublingual glands secrete a mixture of the two types; but in humans and rodents, this pair of glands secretes saliva that is primarily mucous in composition.

The uses of saliva in digestion are manyfold, including the following:

1. **Lubricant.** These secretions act as aids in mastication, the formation of the bolus, and swallowing. Without this moisture, swallowing would be extremely difficult in most animals.

2. **Enzymatic activity.** The enzyme alpha-amylase (ptyalin) is found in serous saliva of nonruminants. It acts to break alpha 1, 4 glucosidic linkages in starch and glycogen.

3. **Buffering capacity.** A large quantity of bicarbonate is secreted in saliva, thus serving as a buffer in the ingesta.

4. **Nutrients for rumen microorganisms.** Saliva contains considerable amounts of urea, mucin, phosphorus, magnesium, and chloride—all of which can be readily utilized by the bacteria and protozoa in the rumen.

5. **Prevention of frothing.** Gas can accumulate in the rumen and cause serious bloating if the eructation process is impaired. Saliva—acting as a surfactant—helps to prevent these problems.

6. **Taste.** Saliva solubilizes a number of the chemicals in the feed which, once in solution, can be detected by the taste buds.

7. **Protection.** The membranes within the mouth must be kept moist in order to remain viable. Saliva provides one means by which this is accomplished.

Pharyngeal and Esophageal Region

The pharynx is the structure which controls the passage of air and feed. In this organ, the openings of the mouth, esophagus, posterior nares, Eustachian tubes, and larynx come together. During the act of swallowing, the opening into the larynx is reflexly closed by the arytenoid cartilages; and the epiglottis is passively folded over the opening of the larynx. This forces food into the esophagus, thus preventing it from passing into the respiratory tract.

The esophagus is a muscular tube extending from the pharynx to the cardia of the stomach. The musculature and innervation of the esophagus are such that peristalic waves move the bolus. *Peristalsis is the coordinated contraction and relaxation of smooth muscles creating an unidirectional movement which pushes the bolus through the digestive tract.* In nonruminant animals, peristalsis normally moves from the mouth to the stomach. Belching and vomiting, reverse peristalsis, are usually dysfunctions of normal digestion. Reverse peristalsis is a normal process in ruminants as eructation and rumination are essential for digestion. Vomiting is rare in horses; if it does occur, most of the material passes out through the nostrils.

In fish, the esophagus is an extremely distensible organ. Generally, if the fish can get an object in its mouth, the esophagus will accommodate its passage down the digestive tract. Herbivorous fish have no differentiated structure resembling the stomach; essentially, the esophagus extends all the way to the small intestine.

At the junction of the cervical segment and the thoracic segment of the esophagus in birds, there is a differentiated outpouching of the esophagus called the crop. If the bird has been starved, feed will bypass this structure and go directly to the proventriculus and gizzard. As feeding progresses, the crop begins to fill and acts as a storage organ. In the crop, there is limited digestion due to the presence of salivary amylase mixed in the food and a small amount of fermentation. Limited absorption of glucose and volatile fatty acids in the crop has been demonstrated in some birds.

Gastric Region

The primary differences in digestion among domestic livestock can be traced to the specialized development of the gastric region. Variations in the structure of the stomach are reflected in the nonruminant, ruminant, and avian species.

NONRUMINANT ANIMALS

In nonruminant animals, the shape of the distended stomach, being relatively simple, resembles that of a kidney bean. Fig. 3-7 illustrates the various external areas of the stomach.

Two specialized types of cells in the stomach provide the gastric secretions needed for the initial stages of digestion. The *parietal cells,* located in the fundic region, secrete hydro-

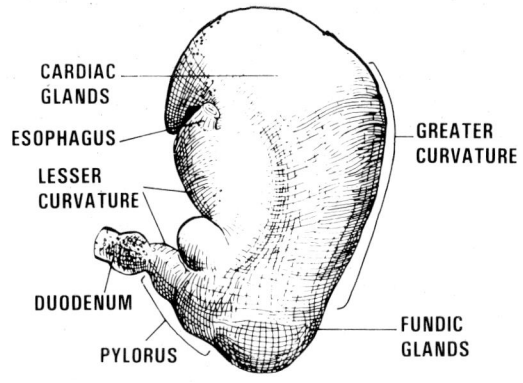

Fig. 3-7. Anatomy of the nonruminant stomach.

chloric acid. Hydrochloric acid hydrolyzes a limited amount of protein, but its main function is to establish an acid environment conducive to the activity of certain hormones and enzymes. The second cell type is called the *chief cell* or *peptic cell*. These cells secrete the enzyme pepsinogen. When pepsinogen is secreted in an acid environment (pH of 1.6 to 3.2), the proenzyme is activated forming pepsin—an enzyme that hydrolyzes certain peptide bonds.

Mucus is secreted by cells in the stomach mucosa. This secretion provides protection for the lining of the stomach. If there is a malfunction in the secretion of mucus, the stomach can digest itself—resulting in ulceration.

Two types of motility have been observed in the stomach. The first type is that of *peristalsis* whereby food is moved toward the duodenum of the small intestine. *Tonic contractions*, the second type of motility, churn and knead the ingesta to ensure thorough mixing, but do not propel the food from one end of the stomach to the other.

Feed ingested by carnivores is passed through the stomach at an extremely fast rate (on the order of a couple of hours), while ingesta of herbivores and omnivores tend to remain in the stomach for relatively long periods, sometimes in excess of 24 hours. The rate of passage of feed depends largely on the nutrient composition of the ration. Carbohydrates pass through the stomach faster than either proteins or fats, with proteins being intermediate in rate of passage. Water can pass directly through to the small intestine, spending very little time in the stomach.

In addition to pepsin, gelatinase and rennin are two enzymes secreted in the stomach. Gelatinase liquifies gelatin, and rennin coagulates milk. The latter enzyme is not found in adult humans but is extremely important in the nutrition of all young mammals.

Intrinsic factor, a protein necessary for proper absorption of vitamin B–12, is produced in the parietal cells of the stomach of humans, pigs, rats, and guinea pigs; but it is not produced by ruminants or dogs. If there is a malfunction in the production of this protein, a condition called pernicious anemia results. Quite often, people suffering from this condition ingest amounts of dietary vitamin B–12 that are normally considered adequate, but the vitamin has no means of being absorbed—thus, pernicious anemia develops.

RUMINANT ANIMALS

The ruminant has been described as having four stomachs. In reality, the ruminant possesses a complex stomach consisting of four morphologically distinct compartments. These compartments are reticulum, rumen, omasum, and abomasum.

Scientists have developed highly advanced methods of fermentation technology in which microorganisms are used to produce food and health products. However, the natural fermentation engineering found in the stomach of the ruminant is unexcelled as a culture system for bacteria, protozoa, and fungi.

Reticulum and Rumen

The reticulum and rumen are closely related as to physiological function and are often discussed together. The esophagus empties into the atrium ventriculi, a convex area formed by both the rumen and reticulum. In the adult ruminant, the rumen is an extremely large compartment lined with a large number of papillae that increase the surface area for the churning of digested material and absorption. The reticulum, a structure which has an interior that very much resembles a honeycomb, acts as a collection compartment for foreign objects as well as an organ for digestion.

Ingested feed passes into these two compartments and is digested thoroughly through the action of various microorganisms (bacteria and protozoa) present in the rumen. The rumen, in effect, is a large physiological fermentation vat.

The microbes of the rumen digest carbohydrates to produce carbon dioxide and volatile fatty acids. Although a number of volatile fatty acids are produced, the vast majority of these end products are acetate, propionate, and butyrate. These products are then absorbed from the rumen and supply much of the energy required by the animal. Quite often, when high-concentrate rations are used, large quantities of lactic acid are produced and the pH of the rumen falls. Since most of the bacteria in the rumen are pH sensitive, any dramatic shift in pH will alter the proportions of the various types of microorganisms. When the ruminal pH drops too low, the animal goes off feed—a symptom of acute digestive problems.

Lipids are degraded by the ruminal microbes to fatty acids and glycerol. Glycerol is then primarily converted to propionate, with the long-chain fatty acids passing down to the small intestine for absorption.

Very few dietary proteins escape the degradation process of the rumen. The degree to which dietary protein is degraded is dependent on its solubility. A highly soluble protein will be rapidly degraded while a highly insoluble protein will probably leave the rumen relatively intact. Most dietary proteins are metabolized by the microorganisms and are incorporated as microbial protein. The microbes can also use simpler forms of nonprotein nitrogen, such as urea, to make protein. The microbes are then passed down the tract, which, through their own degradation, provide protein for the animal. With the degradation of the various

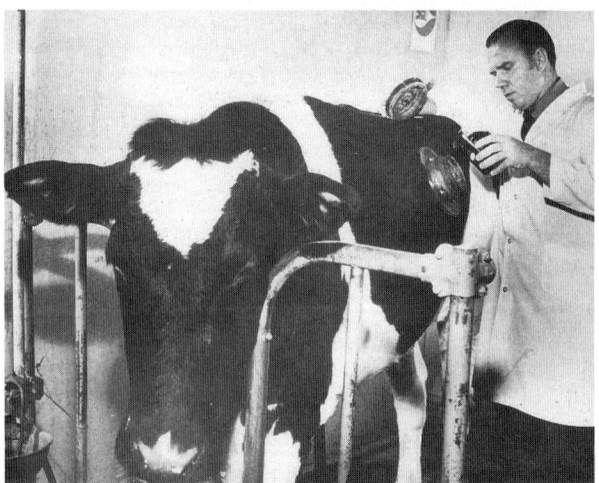

Fig. 3–8. A rumen fistula, a port (opening) that can be opened and closed at will. The feedstuff(s) being studied is placed in porous polyester bags, which are inserted in the rumen. The bags are held by a fishline, suspended in the rumen fluids, so that the microbes digest the contents. The undigested portion of the feedstuff remains in the bag for washing and analysis. The difference in analysis of the feed before and after digestion represents the portion digested. (Courtesy, *Feed Management*, Watt Publishing Co., Mount Morris, Ill.)

dietary proteins, ammonia is produced in the rumen which can then either be absorbed through the rumen wall or provide nitrogenous precursors for the synthesis of bacterial protein. If the ration is high in sugars and starches, ammonia concentration is depressed.

Vitamin K and the B-complex vitamins are all synthesized by the ruminal microbes. Therefore, no supplementation of these vitamins is necessary in the adult ruminants. The young ruminant must obtain these vitamins from exogenous sources, but milk generally supplies the young animal's needs. Vitamin C is synthesized on the tissue level in ruminants as well as most nonruminants. Humans are one of the notable exceptions.

Throughout the fermentation process in the rumen, large quantities of various gases are produced and expelled through eructation (belching). Methane and carbon dioxide are the two gases produced most abundantly in the rumen. It has been estimated that methane constitutes 30 to 40% of the total rumen gas volume, and carbon dioxide 20 to 65%.

Omasum

The omasum, or manyplies, is the next compartment for digestion. It contains numerous laminae (tissue leaves) that help grind ingesta. The exact physiological function of this compartment has not been fully elucidated, but many researchers feel that it serves to absorb water in addition to its function of grinding ingesta.

Abomasum

This compartment is analogous to the stomach of the nonruminant. It is the only compartment of the gastric region of the ruminant containing digestive glands. Digestive processes of this compartment are very similar to those of the stomach in the nonruminant.

Newborn Ruminants

When a calf is born, the rumen is small and the fourth stomach is by far the largest of the compartments. Thus, digestion in the young calf is more like that of a simple-stomached animal than that of a ruminant. The milk which the calf normally consumes bypasses the first two compartments by way of the esophageal groove and goes almost directly to the fourth stomach in which the rennin and other compounds for the digestion of milk are produced. If the calf gulps too rapidly, or gorges itself, the milk may go into the rumen where it is not digested properly and may upset the calf's digestive system. As the calf nibbles at hay, small amounts of material get into the rumen. When certain bacteria become established, the rumen develops and the calf gradually becomes a full-fledged ruminant. (Also see earlier section in this chapter on "Esophageal Groove [Reticular Groove].")

AVIAN GASTRIC DIGESTION

Birds have no teeth; hence, there is no chewing. The esophagus empties directly into the crop, where the feed is stored and soaked.

Gastric digestion in birds occurs in two separate and distinct organs—the proventriculus and the gizzard.

The proventriculus is a small organ, through which ingested feed passes rapidly. Its main function is that of gastric fluid secretion. The fluids secreted by the proventriculus are very similar to those in the stomach of the nonruminant, containing both pepsin and hydrochloric acid. Very little churning and mixing of feed occur in this organ.

The function of the gizzard is the mechanical action of mixing and grinding the feed. Because the bird has no teeth and swallows its feed whole, this muscular organ—sometimes called the "hen's teeth"—acts primarily as an organ for *mastication*. Here, fluids secreted by the proventriculus are mixed in the ingesta during grinding. Grit, such as small pieces of granite, is often added to poultry rations to increase the digestibility of the whole grains or grains with a minimal amount of processing. Grit stimulates motility in the gizzard and provides additional surface for grinding. When feed is provided in mash form, the benefits of grit are minimal.

Pancreatic Region

The pancreatic region involves the pancreas and the pancreatic duct—a duct leading from the pancreas to the small intestine.

The pancreas, an accessory organ of digestion, is a glandular structure that plays an essential role in the digestive physiology of animals. The pancreas—being both an endocrine and exocrine gland—serves two physiologically distinct functions. The endocrine function is that of the secretion of the hormones, insulin and glucagon. The exocrine function deals with the production and secretion of fluids that are necessary for digestion within the small intestine.

The digestive fluid produced by the pancreas is clear and alkaline and consists of two phases—an aqueous phase and an organic phase. Being rich in bicarbonate, the aqueous phase serves primarily to neutralize the highly acid chyme produced in the stomach and passed on to the small intestine. In the organic phase, enzymes produced in the acinar cells of the pancreas are transported to the duodenum. These enzymes are stored in granules in the pancreas and are secreted from the cells through the process of emeiocytosis (cell vomiting). This process is sometimes called reverse pinocytosis because the granule fuses with the cell membrane, followed by a breakdown of the membrane and the evacuation of the granule.

Many of the pancreatic enzymes are stored and secreted in an inactive form to be activated at the site of digestion. Trypsinogen is a proteolytic enzyme that is activated in the small intestine by the enterokinase, an enzyme secreted from the intestinal mucosa. When activated, trypsinogen becomes trypsin. Trypsin, in turn, can then activate chymotrypsinogen to chymotrypsin.

The nucleases, lipases, and pancreatic amylase are secreted in their active form. Many of the enzymes require a specific environment before they will function. For example, amylase requires a pH of about 6.9 and the presence of inorganic ions before it will digest complex carbohydrates.

Hepatic Region

The hepatic region incorporates the liver, gallbladder, and bile duct.

In addition to the pancreas and salivary glands, the liver is an indispensable accessory organ of the gastrointestinal tract. From the stomach and small intestine, most of the absorbed nutrients travel through the portal vein to the liver—the largest gland in the body. The liver not only plays an important part in nutrient metabolism and storage, but also forms bile, a fluid essential for lipid absorption in the small intestine. The numerous physiological functions of the liver follow:

1. Secretion of bile.
2. Detoxification of harmful compounds.
3. Metabolism of proteins, carbohydrates, and lipids.
4. Storage of vitamins.
5. Storage of carbohydrates.
6. Destruction of red blood cells.
7. Formation of plasma proteins.
8. Inactivation of polypeptide hormones.
9. Urea formation.

The primary role of the liver in digestion and absorption is the production of bile. Bile facilitates the solubilization and absorption of dietary fats and also aids in the excretion of certain waste products such as cholesterol and by-products of hemoglobin degradation. The greenish color of bile is due to the end products of red blood cell destruction—biliverdin and bilirubin. Bile contains a number of salts resulting from the combination of sodium and potassium with bile acids.

The volume of bile production is highly variable. An animal that has been starved produces little bile. Conversely, an animal that is fed a high-fat ration will produce substantial quantities in order to keep up with absorptive requirements. Generally, the volume of bile is dependent on (1) blood flow, (2) nutritive state of the animal, (3) type of ration being fed, and (4) the enterohepatic bile salt circulation.

In many animals, the gallbladder is the storage site for bile. Several species of livestock and animals, however, do not have gallbladders; among them, horses, rats, gophers, deer, elk, moose, giraffes, camels, elephants, pigeons, and doves.

Intestinal and Cecum-Colon Region

The small intestine is divided anatomically into three sections—duodenum, jejunum, and ileum. The first segment, the duodenum, originates at the pyloric sphincter of the stomach and is closely attached to the body wall by a short mesentery. Both bile and pancreatic fluids are emptied into this segment. The next section is the jejunum. There is no clear demarcation between the jejunum and the ileum, but it is arbitrarily defined as the free border of the ileocecal fold.

Throughout the luminal surface of the small intestine lies an extensive network of fingerlike projections called villi. In the human, there are about 20 to 40 of these projections per square millimeter of intestine, each one being from 0.5 to 1.0 mm long. Each villus contains a lymph vessel called a lacteal and a series of capillary vessels. On the surface of the villi are a great number of microvilli which provide further surface area for absorption.

Three types of motility can be observed in the small intestine. The first type is called *pendular motion*. These waves do not advance down the intestine. Rather, they are merely a localized shortening and lengthening of the intestine which produces a mixing action. *Segmentation contractions* are the second type of intestinal motility. These intestinal movements are ringlike contractions at regular intervals which periodically relax, whereupon the area that had been previously relaxed contracts. This type of motility provides a means of mixing in addition to the pendular contractions. *Peristalsis*, a form of motility that has been previously discussed, is the third type of intestinal motility, providing a means for movement of chyme (intestinal contents) down the tract.

DIGESTION AND ABSORPTION IN THE SMALL INTESTINE

In nonruminant animals, the small intestine is the primary site of both digestion and absorption. While considerable digestion takes place in the gastric region of the ruminant, the small intestine is still of paramount importance for digestion and absorption, especially for lipids and proteins.

Carbohydrates

The digestion and absorption of most carbohydrates, in nonruminant animals, occurs in the small intestine. Here, such enzymes as amylase, sucrase, maltase, and lactase split carbohydrates into monosaccharides, whereupon absorption takes place. The region of the greatest absorption of sugars is in the jejunum. Glucose and galactose are absorbed through an active transport mechanism. Sodium ion concentration within the intestinal contents has been shown to be critical in this mechanism. A high Na^+ concentration will facilitate rapid absorption of these sugars while a low Na^+ concentration will reduce the rate of absorption. Some pentoses and hexoses are absorbed through diffusion—a process considerably slower than that of active transport.

Lipids

Lipids are digested and absorbed primarily in the upper part of the small intestine, but considerable absorption can take place as far down as the ileum. When lipids, emulsified by bile salts, come into contact with the various lipases that are found in the duodenum, they are broken down into monoglycerides and fatty acids. Short-chain fatty acids are then absorbed directly into the mucosa of the small intestine and are transported to the portal circulation. Monoglycerides and insoluble fatty acids are emulsified by bile salts, forming micelles. By attaching to the surface of epithelial cells, the micelles enable these components to be absorbed into the mucosal cells. Once inside these cells, the long-chain fatty acids are reesterified to form triglycerides. Triglycerides then combine with cholesterol, lipoproteins, and phospholipids to form chylomicrons—minute fat droplets. The chylomicrons are then passed into the lymphatic circulatory system.

Proteins

While protein digestion is initiated in the stomach of nonruminant animals, most digestion and absorption occur in the small intestine. Numerous pancreatic and intestinal enzymes split proteins into their constituent amino acids, which are subsequently absorbed. In humans, it has been estimated that 50% of the digested protein comes from the diet, 25% from the proteins in the digestive fluids, and the remaining 25% from sloughed cells of the gastrointestinal tract. The rate of turnover of mucosal intestinal cells is extremely rapid—1 to 3 days—thereby giving an excellent source of recyclable protein. In ruminants, it has been estimated that up to 80% of the ingested protein is converted to microbial protein.

Amino acid absorption is not clearly understood; but an active transport mechanism involving Na^+, similar to that of glucose absorption, has been implicated. Amino acids are rapidly absorbed in the duodenal and jejunal segments, but are poorly absorbed in the ileum. A limited amount of absorption of small polypeptides can occur, especially in newborn animals. This mechanism of absorption, pinocytotic in nature, facilitates the passage of antibodies from the colostrum of the mother to her young.

Minerals and Vitamins

Mineral absorption occurs throughout the small and large intestine, with the rate of absorption depending on a number of factors; among them, pH and carriers. Numerous mechanisms of mineral absorption have been elucidated. Many minerals, for example iron and sodium, require active transport systems. Others, such as calcium, utilize both carrier proteins and diffusion mechanisms. Additional information on specific minerals can be found in Chapter 4, Nutrients/Metabolism.

Most of the vitamins are absorbed in the upper portion of the intestine, with the exception of vitamin B-12 which is absorbed in the ileum. Water-soluble vitamins are rapidly absorbed, but the absorption of fat-soluble vitamins relies heavily on the fat absorption mechanisms which are generally slow.

CECUM AND COLON

The small intestine terminates at the ileocecal valve—a sphincter that controls the flow of ingesta from the small intestine into the cecum and large intestine. This structure prevents the backflow of ingesta into the small intestine.

The cecum and colon in mammals are composed of several layers of muscle. There is a circular layer of muscle that forms the basic tube of the colon and facilitates movement. In addition to this layer of muscle, there are three strips of longitudinal muscle which form the *taenia coli*. These strips form a series of pouches or sacculations throughout the colon which are called *haustrae*. Ingesta are held in these saclike structures to facilitate the removal of water. Subsequently, the feces generally take on the shape of the haustrae. Numerous mucous-secreting goblet cells can be found in the colon, but villi, such as the type that are found in the small intestine, are absent.

Three types of motility can be observed in the colon: (1) haustral contractions, (2) massive peristalsis, and (3) reverse peristalsis.

1. **Haustral contractions.** This type of motility creates a mixing action of the ingesta; hence, the absorption of water from the material is facilitated. These contractions are localized in the various portions of the colon with no coordinated wave movement traveling along the organ.

2. **Massive peristalsis.** These waves are slow, strong movements that propel the digesta down the colon.

3. **Reverse peristalsis.** Reverse peristalsis is a type of movement that aids in the mixing of the digesta as well as the absorption of nutrients. By moving material proximally, the colon has another opportunity to absorb what was missed the first time.

At the proximal end of the colon is a blind sac called the cecum. In carnivores, the cecum is of little importance because most of the ingested material has been digested prior to entering the colon. Water and electrolytes are absorbed in the colon as well as a number of water-soluble vitamins and vitamin K, which are produced in limited amounts by the microflora of the colon.

The cecum and colon are rather large and well-developed in nonruminant herbivores, such as the rabbit and the horse. In these animals, considerable fermentation can take place in the cecum and colon, which, together, act somewhat like a rumen. Fermentation products are synthesized and absorbed. When one considers that these are nonruminant animals, considerable fiber and cellulose digestion can be achieved. Physiologically, this anatomic adaptation can never be as efficient as the ruminant process of digestion. Ruminal contents are passed into the small intestine—the primary site of digestion and absorption, whereas products which are broken down in the cecum of a nonruminant herbivore are transported down the colon where absorption is severely limited. Rabbits and rats can compensate for this fact through the behavioral adaptation of coprophagy. *Coprophagy is the ingestion of fecal material.* Through this feeding practice, the rabbit gives the digestive tract another chance for additional digestion and absorption. Despite this arrangement, the rabbit is not as efficient in digestion as the ruminant.

The cecum of the ruminant is not very well developed and plays a rather insignificant role in digestion. There is some absorption of volatile fatty acids in the ruminant cecum, and considerable amounts of water and electrolytes are absorbed in the colon. It has been estimated that in sheep 70% of the water in the digesta is absorbed between the mouth and the duodenum. In the colon alone, about 19% of the water is absorbed.

There are two blind sacs (ceca) in the fowl, in which a limited amount of bacterial activity and subsequent nutrient absorption have been observed. The large intestine is extremely short and is very similar in structure to the small intestine. It is generally believed that the large intestine in the fowl does not play any significant role in digestion.

GASTROINTESTINAL HORMONES

A hormone can be defined as a chemical released by a specific area of the body that is transported to another region within the animal where it elicits a physiological response. A number of hormones have been isolated and characterized from the gastrointestinal tract. Gastrointestinal endocrinology is an extremely recent area of study; and new hormones are being found and identified chemically.

NEUROLOGICAL CONTROL OF THE GASTROINTESTINAL TRACT

The nervous system can be divided into two anatomical systems—the somatic nervous system, and the autonomic nervous system. The somatic nervous system enables the body to adapt to stimuli from the external environment. Various stimuli, such as touch, are perceived by specialized receptors within this system, and the body responds accordingly. The autonomic system involves the maintenance of homeostasis—the internal environment of the body. This is the system that controls the gastrointestinal tract.

The autonomic system can be further divided into the *sympathetic autonomic nervous system* and the *parasympathetic autonomic nervous system*. The sympathetic system is generally associated with the traditional "fight or flight" response, and the parasympathetic system is usually associated with routine integration of normal activity.

When the sympathetic system is stimulated, there is a need for large amounts of blood in peripheral tissues, such as skeletal muscle. In order to accommodate this need, blood is shunted from the gastrointestinal tract, resulting in reduced digestive activity. For the most part, salivation ceases, and the mouth becomes dry. Secretions from the digestive glands are inhibited as well as peristalsis throughout the tract. The various sphincters of the gastrointestinal tract contract in response to sympathetic stimulation.

Stimulation of the parasympathetic system induces increased gastrointestinal activity. Generally, the parasympathetic system is stimulatory to the gastrointestinal system during rest and normal activity.

With the knowledge of the action of the sympathetic and parasympathetic autonomic systems, one can understand the action of certain drugs. When acute diarrhea is encountered, sympathetic-type drugs or parasympathetic depressant drugs are often used. Drugs that act as parasympathetic stimulators are frequently used as laxatives.

FACTORS AFFECTING DIGESTION AND ABSORPTION

In all animals, both extrinsic and intrinsic influences alter the efficiency of digestion and absorption. Feed given to a particular animal in one form may not be as digestible as the same feed in another form. Numerous factors concerning the feed itself will influence digestion and absorption. The autoregulation of the digestive system can influence digestion, especially when there are severe dysfunctions.

Influences of Feed Composition and Preparation

When planning a nutrition program for livestock, the producer must understand the digestive and absorptive processes of the particular species being fed. If these processes are understood, a feeding program can be planned so as to maximize the digestibility of the feed.

RATE OF PASSAGE OF FEED

If a certain type of feed is known to pass quickly through an animal, it is likely that digestion will not be very efficient because the feed will not have adequate exposure to the digestive enzymes. Likewise, if nonruminant animals having little storage area for ingested feed are fed only once or twice a day, they will ingest large quantities of feed at a single feeding. This also limits the relative exposure of the feed to the digestive enzymes; hence, a nonruminant fed on a free-choice regimen would utilize feed more efficiently.

FEED PROCESSING

Processing of grains, such as grinding, does not markedly increase the digestibility of the feed in animals that masticate their feed thoroughly. On the other hand, processing can offer great advantages when feeding grains to animals that swallow their feed without much chewing. Unbroken seed coats are not easily digested, and unprocessed grains can pass through the tract undigested. By removing or cracking this seed coat, digestive enzymes can readily degrade the feed.

Grinding is a useful process in the feeding of very young or very old animals. In the young animals, dentition is rudimentary and the digestive tract has not been well developed. Old animals frequently have worn or missing teeth, and grinding substitutes for mastication.

Cooking feeds has been shown, in certain cases, to increase digestibility. But the process involves additional labor and expense; so, one must weigh the economic advantages and disadvantages carefully before processing. In some cases, such as with soybeans, cooking is necessary before the feed can be fed to poultry and other species, due to the presence of metabolic inhibitors. More information regarding the use of cooking in processing can be found in Chapter 14, Feed Processing.

DIGESTION AND ABSORPTION

It is possible to process certain protein sources in such a way that they escape microbial destruction in the rumen. Then, the protein passes down to the small intestine where it is digested and absorbed. This preserves the integrity of the amino acid composition of high-quality protein. Such protein is known as protected or by-pass protein.

By definition, protected or by-pass protein is feed protein which escapes digestion in the rumen, but is digested in the small intestine. Among the feed processing methods used to prepare protected protein are: (1) *natural protection* (corn gluten meal, distillers dried grains, dehydrated alfalfa meal, and animal by-products), (2) heat denaturation, (3) aldehyde treatment, (4) combined heat and aldehyde treatment, (5) lipid coating—treatment with certain tannins, and (6) complexing with bentonite clay. Even though protein may be protected, performance is not necessarily guaranteed. For improved performance, the by-pass protein must contain amino acids that complement those of the microbe-synthesized protein, with the mixture absorbed by the animal balanced. But, eventually, by-pass products will be developed that will either cut cost or increase production—or both.

LEVEL OF FEED INTAKE

As feed is offered to animals in excessive amounts, digestibility decreases. It has been observed in ruminants that the offering of feed at twice the maintenance level can reduce dry matter digestibility by 1 to 2 percentage units. In swine, this trend is also observed, but the magnitude of decreased digestibility is not as great.

COMPOSITION OF FEED

Feeds vary considerably in composition and digestibility. One good estimate of digestibility in nonruminant animals can be obtained by the content of crude fiber. Fiber is extremely undigestible in nonruminants due to the absence of the enzymes necessary to break down the complex cell walls of the feed. Therefore, a feed high in fiber is probably highly undigestible. Roughages are high in fiber; hence, their use is limited in rations for nonruminants. Conversely, the microorganisms of the rumen synthesize enzymes that digest the components of plant cell walls, and roughages are well utilized by ruminants. For additional information on the effects of nutrient composition on digestibility, see Chapter 4, Nutrients/Metabolism.

CHELATES

Chelates are formed when certain organic molecules combine with metallic ions to form cyclic compounds. These chelated complexes possess different solubility characteristics than the unbound metallic ion. By this binding, certain minerals may be more or less readily absorbed in the gastrointestinal tract. Several naturally occurring chelating agents are chlorophylls, cytochromes, hemoglobin, ascorbic acid, vitamin B-12, and some amino acids. The most commonly used synthetic chelating agent is EDTA (ethylenediamine tetraacetic acid)—a compound used in some poultry rations to improve the availability of certain minerals.

PHYTIC ACID

Phytic acid is a hexaphosphoric acid ester of inositol. When the acid form is combined with a cation to form a salt, the compound is referred to as phytin. More than 50% of the phosphorus in mature seeds is in the form of phytin. Numerous studies have shown that animals vary greatly in their ability to absorb phytins. Cattle and sheep have little trouble breaking down phytins and absorbing the phosphorus since their rumen flora produce phytase. In the dog and humans, phytic acid may combine with calcium, thus making the calcium less available for absorption.

OXALATES

Oxalic acid, a compound present in certain leafy plants, may interfere with calcium absorption. The acid combines with calcium to form insoluble calcium oxalate which is less available for absorption. Spinach has a high oxalic acid content, which may subsequently tie up substantial portions of its calcium. The magnitude of this problem is debatable as certain species of livestock have been shown to metabolize oxalic acid in such a way as to render it harmless.

Dysfunctions of the Digestive Tract

The digestive tract, though relatively simple in structure, is complex in function. In order for proper digestion and absorption to take place, various humoral and neural mechanisms coordinate the movement of ingesta throughout the tract. If anything goes wrong either physically or chemically in the digestive organs, the entire integrated process can break down, causing malabsorption syndromes or physical changes in the tract.

RETICULITIS

The condition when the collection of foreign objects irritates or punctures the reticulum is called reticulitis, or hardware disease. Ruminants are grazers by nature and are sometimes indiscriminant in their selection of feed. Often they will consume nails, pieces of wire, and other foreign objects. Due to the motility patterns of the gastric region, this paraphernalia tends to accumulate in the reticulum; and the presence of these sharp objects can pose serious problems, especially if the reticulum should be punctured. Magnets may be given ruminants to hold the metal objects and prevent them from harming the animal.

RUMINAL ACIDOSIS

When ruminants are switched from a high-roughage ration to a high-grain ration too rapidly, ruminal acidosis and atony may occur. The microflora of the rumen do not have time to adapt to the new ration, resulting in serious digestive disturbances. The pH of the rumen contents falls, and lactic acid production increases dramatically. Many of the microorganisms cannot live in this environment resulting in radical changes in the bacterial and protozoal populations. Rumen motility, then, becomes static. Afflicted animals show signs of weakness, diarrhea, and abdominal discomfort, and in many cases die.

ACUTE TYMPANY (Bloat)

Quite often, when ruminants are first introduced into the feedlot or lush, legume pastures, bloat problems arise. Normally, the eructation process allows for the expulsion of gases that are produced in the rumen. In cases of bloat, frothing (the trapping of gas in the ingesta) prevents this process and intraruminal pressure builds, thereupon distending the left side of the abdomen until the animal is thrown off feed and goes down due to pain and the buildup of toxic metabolites. A number of theories of the causes of bloat have been postulated, and several treatments for bloat have proven effective. Additional information on bloat may be found in Chapter 5, Nutritional Disorders/Toxins.

PARAKERATOSIS OF THE RUMEN

When rations are fed in a highly processed form—for example, pelleted—morphological changes can occur in the epithelium of the rumen. The papillae of the rumen become enlarged and hardened. Lesions can also be found. This condition is known as parakeratosis.

COLIC

Improper feeding, working, or watering of horses can cause colic—a condition whereby the horse has excruciating abdominal pain. Depending on the type of colic, the symptoms include distended abdomen, increased intestinal rumbling, violent rolling and kicking, profuse sweating, constipation, and refusal of feed and water. Additional information on the treatment and prevention of colic can be found in Chapter 5, Nutritional Disorders/Toxins.

DIARRHEA (Scours)

Diarrhea is a common digestive ailment, especially in young animals. Severe dehydration and acidosis due to the loss of sodium, potassium, bicarbonate, and water can result in death. Even in mild cases of scours, costly setbacks in the growth of animals result. There are numerous causes of scours. Several of these are overfeeding, proliferation of pathogenic microorganisms, lack of bulk in the ration, abrupt changes in the rations, the feeding of spoiled hay or silage, and the inclusion of excessive levels of laxative types of feeds.

QUESTIONS FOR STUDY AND DISCUSSION

1. What is the difference between hunger and appetite?

2. What happens to an animal that is stimulated in the ventro-medial hypothalamus?

3. Discuss the following theories concerning appetite control:
    a. The chemostatic hypothesis.
    b. The thermostatic hypothesis.

4. Define: (a) carnivore; (b) herbivore; (c) omnivore.

5. Discuss the differences in the digestive systems of the monogastric and the ruminant.

6. What is the function of rumination?

7. Explain how the esophageal groove functions in the young ruminant, and tell of its importance from a nutrition standpoint.

8. How does the digestive system of poultry differ from that of other monogastrics?

9. Why is knowledge of the capacities of the digestive tracts of importance?

10. Define digestion.

11. What are the primary sites for absorption in (a) ruminants and (b) nonruminants?

12. Discuss the role of (a) the lymph and (b) the blood in the digestive process.

13. Define: (a) prehension; (b) mastication; (c) deglutition.

14. Discuss the method and the location of the mechanical breakdown of food in (a) four-footed animals and (b) poultry.

15. List the various functions of the tongue.

16. What are the three main salivary glands and where are they located?

17. Discuss the role of saliva in digestion.

18. What type of gastric cell secretes hydrochloric acid? What is the function of hydrochloric acid in digestion?

19. What is peristalsis?

20. Discuss the role of the intrinsic factor.

21. Discuss the anatomy and function of each of the four stomach compartments of the ruminant: reticulum, rumen, omasum, and abomasum.

22. What are the two main gases produced in the rumen?

23. Discuss the changes that take place in the stomach of a newborn ruminant.

24. What is the function of the crop in birds?

25. Discuss the functions of the avian gastric organs.

26. Why is grit sometimes fed to birds?

27. List the physiological functions of the liver.

28. On what factors is the volume of bile dependent?

29. Describe the various types of motility in the small intestine.

30. How are fats digested and absorbed in the small intestine?

31. What are haustrae?

32. Of what benefit is the practice of coprophagy in rabbits?

33. Why are sympathetic-type drugs used in the treatment of diarrhea (scours)?

34. Discuss how each of the following factors may affect digestion and/or absorption: rate of passage of feed, feed processing, protected or by-pass protein, level of feed intake, composition of feed, chelates, phytic acid, and oxalates.

35. Discuss each of the following disfunctions of the digestive tract: reticulitis, ruminal acidosis, bloat, parakeratosis of the rumen, colic, and scours.

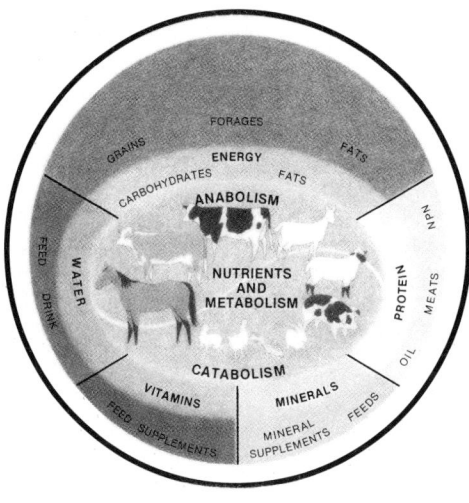

Original painting by Tom Phillips

NUTRIENTS/ METABOLISM

Contents	Page
Cells—Functional Units of Nutrition and Metabolism	41
Nucleic Acids (DNA and RNA)	42
Synthesis of DNA and RNA	43
Genetic Engineering (Recombinant DNA)	43
Biotechnology	44
Nutrients and Their Metabolism	44
Energy (Carbohydrates and Lipids)	44
Carbohydrates	45
Carbohydrate Components Within Feeds	45
Availability of Fiber	45
Fats and Other Lipids	45
Characteristics of Fats	46
Fatty Acids	46
Essential Fatty Acids	47
Other Lipids	47
Metabolism of Fats	47
Measuring and Expressing Energy Value of Feedstuffs	48
Energy Definitions and Conversions	48
Energy Systems	48
Total Digestible Nutrients (TDN)	48
Calorie System	49
Nitrogenous Feeds	50
Nonprotein Nitrogen	50
Urea and Ammoniated Feeds	50
Proteins	51
Amino Acids	51
Synthesis of Amino Acids	53
Biological Value of Protein	53
Protein Sources	54
Minerals	55
History/Discovery of Minerals	55
Definitions/Classification/Functions	56
Vitamins	57
History/Discovery of Vitamins	57
Definitions/Classification/Functions	58
Water	59
Water Balance	59
Water Sources	60
Drinking Water	60
Water in Feeds	60
Metabolic Water	61
Water Excretion	61
Questions for Study and Discussion	61

Corn, alfalfa, soybean meal, and grass never get to the cells and tissues of animals. But their nutrient chemicals—more than 40 of them, including amino acids, minerals, and vitamins—do reach the body cells and tissues and are essential to their life.

Once feeds are eaten and digested, the nutrients are absorbed into the blood and distributed to the cells of the body. Still more chemical changes are required before the nutrients can be put to work in the body, by transforming them into energy or structural material. Thus, nutrients—carbohydrates, fats, proteins, minerals, vitamins, and water—are subjected to various chemical reactions. These chemical reations occur on the cellular and subcellular level. The sum of all these chemical reactions is termed *metabolism*. It has two phases: *catabolism* and *anabolism*. Pertinent definitions follow.

Nutrients are the chemical substances found in feed materials that can be used, and are necessary, for the maintenance, production, and health of animals.

Metabolism is the process by which physiological changes occur in the body through chemical reactions.

Catabolism is the oxidative breakdown of nutrients, liberating energy (exergonic reaction) which is used to fulfill the body's immediate demands.

Anabolism is the process by which nutrient molecules are used as building blocks for the synthesis of complex molecules. Anabolic reactions are endergonic—that is, they require the input of energy into the system.

Knowledge of what nutrients are needed by animals, and how they are used (metabolized) is the basis of modern nutrition.

CELLS—FUNCTIONAL UNITS OF NUTRITION AND METABOLISM

Nutrition is achieved in cells; they are the functional units of nutrition and metabolism. In the cells, the metabolism of carbohydrates, fats, and amino acids (proteins) takes place. In the cells, compounds are built for use in the body (anabolism), and compounds are broken down into simpler units (catabolism) for new uses or for excretion if not useful. The release of energy by some substances and the acceptance of it by others occurs in the cells, providing energy for use whenever and wherever it may be needed.

Each of the many cells is specialized in some way to carry out the functions of the organism (body) of which it is a part. Grouped together, cells serving the same general function form a tissue; thus, there are muscle, nerve, epithelial, and connective tissues. Structural units, made up of two or more tissues, serving a specific function or functions, are organs; for example, the heart, kidneys, and lungs.

Just as the body has organs, the cells of the body contain organelles (little organs) which are also involved in nutrition and metabolism. The two major parts of the cell are the nucleus and the protoplasm surrounding it, called *cytoplasm* (see Fig. 4-1). The nucleus serves as the center of the cell, controlling its functions; the cytoplasm performs the cell's metabolic activities.

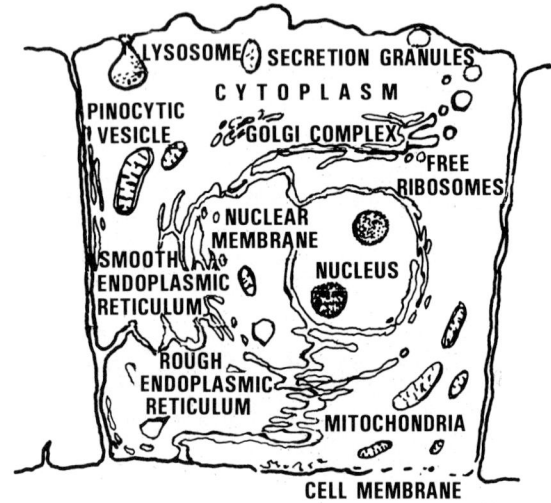

Fig. 4-1. A typical cell showing organelles (little organs) associated with cell nutrition and metabolism.

In the nucleus, the pattern of each of the different proteins is present as deoxyribonucleic acid (DNA). The information stored in DNA is put into action by transcribing it into ribonucleic acid (RNA), which directs the actual protein synthesis in the ribosomes. This process of transcription is the key to nutrition.

Membranes subdivide the cell into compartments and regulate the passage of substances into, out of, and within the cell. The membranes include those of the endoplasmic reticulum, the mitochondria, the golgi body, the lysosomes, the nucleus, and the cell itself. The cell wall separates the cell from the external environment and selectively controls the rate of movement of nutrient and waste material into and out of the cell.

All components for the formation of nutrients in cells come from feeds. Genes determine which ones the body can synthesize and which it cannot.

Nucleic Acids (DNA and RNA)

Nucleic acids were so named because they were originally isolated from cell nuclei. They are the carriers and mediators of genetic information, of which there are 2 types: *deoxyribonucleic acid (DNA)* and *ribonucleic acid (RNA)*. The 2 types of nucleic acids, DNA and RNA, differ in that DNA has 1 less oxygen molecule (on carbon atom number 2) in its component sugar ribose, and is a double strand.

Every cell of the body contains the same amount of DNA with the exception of the sperm and egg cells. It is the DNA of the chromosomes in the nuclei of cells that carries the coded master plans for all of the inherited characteristics—size, shape, and orderly development from conception to birth to death. DNA is different for each species, even for each individual within a species. These differences consist of minor rearrangements of sequences among the nitrogenous bases, which constitute a code containing all the information on the heritable characteristics of cells, tissues, organs, and individuals.

The messages carried by DNA are put into action in the cells by the other nucleic acid, RNA. To do this, DNA serves as a template (as the pattern or guide) for the formation of RNA. The genetic message is coded by the sequence of purine and pyrimidine bases attached to the *backbone* of the DNA structure—a long chain of the sugar deoxyribose and phosphoric acid. Purine bases in the DNA include adenine and guanine, while pyrimidine bases include cytosine and thymine. One molecule of DNA may contain 500 million bases. The *backbone* of RNA is also a sugar, the sugar ribose, plus phosphoric acid. However, in RNA the pyrimidine base thymine is replaced by uracil, another pyrimidine. RNA molecules are considerably smaller than DNA, containing from less than a hundred to hundreds of bases—not millions.

In the 1860s, the Austrian monk, Gregor Mendel, experimenting with garden peas in his monastery garden, showed that heredity could be explained in terms of simple mathematical ratios. Although Mendel was not able to explain how hereditary characteristics are transmitted to offspring, he was observing nucleic acids in action. An explanation of Mendel's work was not generally available until 1916, when Thomas Hunt Morgan and his group at Columbia University, using the fruit fly, demonstrated that genes on the chromosomes carried hereditary information from parent to offspring. In 1933, Morgan won a Nobel Prize for his work. Before 1950, George W. Beadle and Edward L. Tatum demonstrated that genes were long-chain polymers in which proteins are linked with nucleic acids. They received a Nobel Prize in 1958. Next, Alfred E. Mirsky and V. G. Allfrey of the Rockefeller Institute showed that DNA was necessary for the manufacture of RNA in the cell and for the buildup of proteins in the nucleus. Then, Erwin Chargoff noted the important relationship between the bases—that adenine and thymine occur in equal amounts, as do cytosine and guanine. Armed with this information and the x-ray diffraction studies of Maurice Wilkins of King's College in London, James Watson, an American biochemist, and Francis Crick, an English physical chemist, set out to determine the actual structure of DNA. Working together, they explored many possible structures, out of which they finally concluded that DNA was composed of two polynucleotide chains joined by bonds between the bases and wrapped around each other in the form of a double helix (see Fig. 4-2). In 1962, those three scientists—Wilkins, Watson, and Crick—shared the Nobel Prize for Physiology and Medicine for their work. But the story did not end! Building upon the present knowledge and understanding of the nucleic acids, DNA and RNA, scientists are making such concepts as genetic engineering and cloning into realities. Thus, humanity is coming closer and closer to understanding the very essence of life—DNA.

SYNTHESIS OF DNA AND RNA

Each time a cell divides, a new DNA is made, while within most cells RNA is continually being synthesized and broken down into its components. When the purines and pyrimidines are combined with the sugars ribose or deoxyribose, the resultant compound is referred to as a *nucleoside*. *Nucleotides* are nucleosides that are esterified (acid and alcohol combination) with phosphoric acid. When a series of nucleotides are joined together, nucleic acids are formed. The secret of the code is a process called base pairing. During the formation of new DNA or RNA, the same bases always pair off—guanine with cytosine, adenine with thymine, and adenine with uracil in RNA. This ensures that the code is duplicated time and time again. During synthesis, the double helix splits, pulling the base pairs apart. New bases line up with the proper partner and form another sugar-phosphoric acid *backbone* (see Fig. 4-2).

Purines, pyrimidines, ribose, and deoxyribose may be synthesized in the body from other compounds, or they may be recycled. Phosphorus, for the formation of phosphoric acid, is a required dietary mineral, and it is employed in a variety of other compounds besides DNA and RNA.

GENETIC ENGINEERING (RECOMBINANT DNA)

Genetic engineering has been going on for thousands of years, ever since the human race began raising crops and domestic livestock. From that remote day forward, the most productive plants and animals were held back for breeding while the least productive were eaten. That's a rudimentary form of genetic engineering!

But the development of gene-splicing (also known as recombinant DNA [see Fig. 4-3]) ushered in a new era of genetic engineering. On May 23, 1977, scientists at the University of California-San Francisco reported a major breakthrough as a result of altering genes—turning ordinary bacteria into factories capable of producing insulin, a valuable hormone previously extracted at slaughter from pigs, sheep, and cattle, so essential to the survival of diabetics. This feat opened the door to further genetic engineering or splicing. Already, this genetic wizardry has been used in transplanting into bacteria (and recently into yeast cells) genes responsible for the synthesis of many critical biochemicals in addition to insulin.

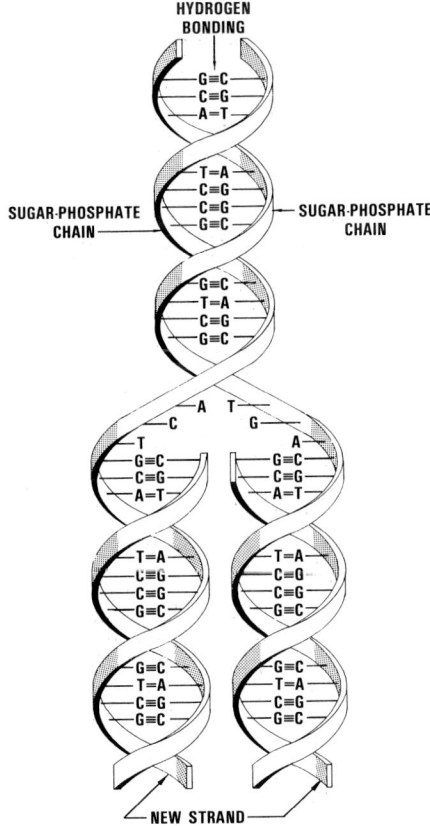

Fig. 4-2. The spiral structure of deoxyribonucleic acid, or DNA—the basic building block of life on earth. It's a double helix (a double spiral structure), with the sugar (deoxyribose)-phosphate (phosphoric acid) *backbone* represented by the 2 spiral ribbons. Connecting the *backbone* are 4 nitrogenous bases (a base is the nonacid part of a salt): adenine (A) paired with thymine (T), and guanine (G) paired with cytosine (C); with the parallel spiral ribbons held together by hydrogen bonding between these base pairs.

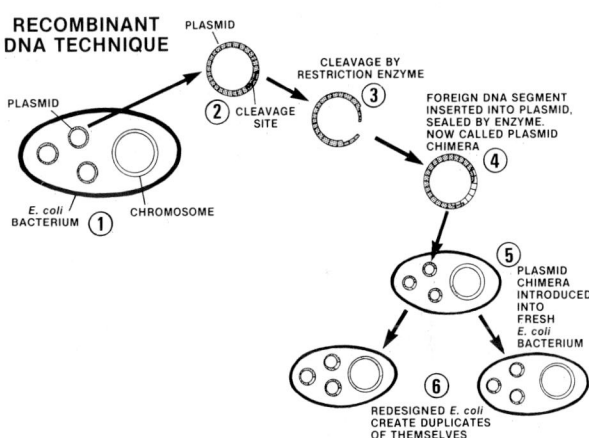

Fig. 4-3. Redesigning *E. coli*, common bacteria of animal intestines. The steps:

1. The scientist places the bacterium in a test tube with a detergent. This dissolves the microbe's outer membrane, causing its DNA strands to spill out.
2. The plasmids (the closed loops), which have only a few genes, are separated from the chromosomal DNA in a centrifuge.
3. The plasmids are placed in a solution with a chemical catalyst called a restriction enzyme, which cuts through the plasmids' DNA strips at specific points.
4. The opened plasmid loops are then mixed in a solution with genes—also removed by the use of restriction enzymes—from the DNA of a plant, animal, bacterium, or virus. In the solution is another enzyme called a DNA ligase, which cements the foreign gene into place in the opening of the plasmids. These new loops of DNA are called plasmid chimeras because, like the chimera—the mythical lion-goat-serpent after which they are named—they contain the components of more than one organism.
5. The chimeras are placed in a cold solution of calcium chloride containing normal *E. coli* bacteria. Then the solution is suddenly heated, at which time the membranes of the *E. coli* become permeable, allowing the plasmid chimeras to pass through and become a part of the microbe's new genetic structure.
6. When the redesigned *E. coli* reproduce, they create duplicates of themselves, new plasmids—and DNA sequences—and all.

BIOTECHNOLOGY

Genetic engineering by gene splicing gave rise to a major scientific revolution, called *biotechnology*.

By definition, *biotechnology is concerned with the application of an array of biological and engineering tools, from gene splicing, to manipulating cells, tissues, and genes so as to control certain characteristics.* It involves every facet of animal production, from breeding and feeding to finished products, including the genetic makeup of animals; the feeds they eat; their digestion, physiology, stress tolerance, disease resistance, and efficiency of production; the composition, quality, and quantity of products produced; along with the production of large quantities of drugs and chemicals. Potential uses are endless. While some aspects of biotechnology are decades away from commercial production, others are near, and still others are here now. Among the major biotechnology breakthroughs visualized, being researched, or reality, are the following:

1. Redesigning animals by gene manipulation and cloning, so that they are more stress tolerant and feed efficient.

2. Modifying ruminant microorganisms to increase their efficiency in digesting forages, from the current approximately 50% to a future 80%.

3. Changing the growth pattern and production of animals by using somatotropin (growth hormone) (a) to increase the milk production of dairy cows by 20 to 40%; and (b) to produce pigs that gain 15 to 18% faster, require 30% less feed per 100 lb gain, and are leaner.

4. Producing safer and more effective vaccines; already, new vaccines have been developed for use against foot-and-mouth disease and pseudorabies, and others will follow.

5. Altering the composition of meat, milk, and eggs.

6. Producing super crops that have built-in defenses to diseases and insects; that are nitrogen-fixing and environmentally adapted; and that yield higher compositions of essential amino acids, starches, low-cholesterol oils, minerals, and vitamins.

7. Developing accurate and simple techniques for the determination of sex, pregnancy, and estrus.

8. Producing multiple births at will.

9. Processing foods that are more attractive, more nutritious, more flavorsome, and longer lasting.

10. Converting wastes into ethanol and high-protein animal feeds.

The entire food team—producers, processors, and consumers—stands to benefit from biotechnology, from the greater abundance of high-quality products produced more efficiently. However, progress will be slowed by regulatory and social problems; and some producers and processors will falter and fail along the way. Nevertheless, with the opening of the 21st century, it is predicted that the magnitude of change wrought by biotechnology will rival that of mechanization, which doubled total American agricultural production and increased output per man hour 20-fold. *Biotechnology will favor those who are there first with the most of the best.*

NUTRIENTS AND THEIR METABOLISM

The diet contains carbohydrates, fats, and proteins. Although each of these nutrients has specific functions in maintaining a normal body, all of them can be used to provide the number one requirement—energy. Hence, the discussion of their metabolism centers first, on their catabolism, and second, on their anabolism. Although they are not used for energy, minerals, vitamins, and water are essential to the metabolic processes.

The sections that follow pertain to the metabolism of nutrients—to all the chemical changes that take place after the end products of the digestion of carbohydrates, fats, and proteins, along with the minerals, vitamins, and water, are absorbed into the body. (See Fig. 4–4.)

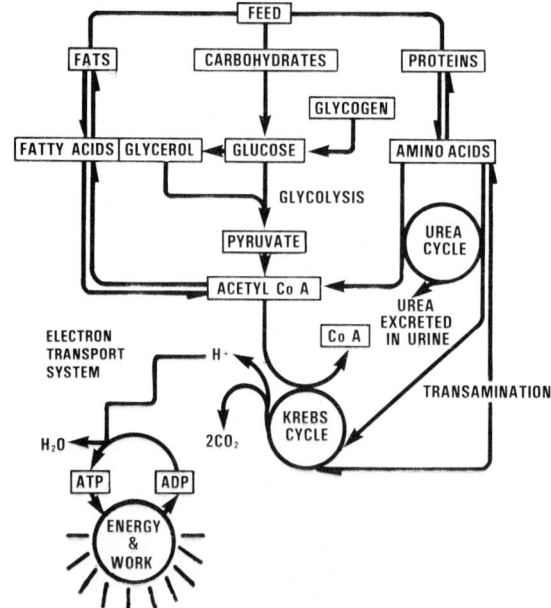

Fig. 4–4. Schematic diagram of the metabolism of nutrients. Note the metabolic interrelationships of carbohydrates, fats, and proteins.

ENERGY (CARBOHYDRATES AND LIPIDS)

Energy is required for practically all life processes—for the action of the heart, maintenance of blood pressure and muscle tone, transmission of nerve impulses, ion transport across membranes, reabsorption in the kidneys, protein and fat synthesis, the secretion of milk, and the production of eggs, wool, and power.

A deficiency of energy is manifested by slow or stunted growth, body tissue losses, and/or lowered production of meat, milk, eggs, fiber, or power, rather than by specific signs such as those which characterize many mineral and vitamin deficiencies. For this reason, energy deficiencies often go undetected and unrectified for extended periods of time.

It is common knowledge that a ration must contain carbohydrates, fats, and proteins. Although each of these has specific functions in maintaining a normal body, all of them can be used to provide energy for maintenance, for work, or for finishing. From the standpoint of supplying the normal energy needs of animals, however, the carbohydrates are by far the most important, more of them being consumed than any other compound, whereas the fats are next in importance for energy purposes. Carbohydrates are usually more abundant and cheaper, and most of them are very easily digested, absorbed, and transformed into body fat. Also, carbohydrate feeds may be more easily stored than fats in warm weather and for longer periods of time. Feeds high in fat content are likely to become rancid, and rancid feed is unpalatable, if not actually injurious in some instances.

Carbohydrates

The carbohydrates are organic compounds composed of carbon, hydrogen, and oxygen. This group includes the sugars, starch, cellulose, gums, and related substances. Although few carbohydrates occur in animal tissues (glucose and glycogen are exceptions), they form the largest part of the feed supply of animals. Carbohydrates make up 75% of the dry weight of the plant world, upon which animal life primarily depends.

The carbohydrates in plants are produced by photosynthesis, the most important chemical reaction in nature. The radiant energy of the sun is captured by the chlorophyll of plants and changed to chemical energy, which, in turn, supports the formation of glucose from carbon dioxide and water—an endergonic process. (See Fig. 4–5.)

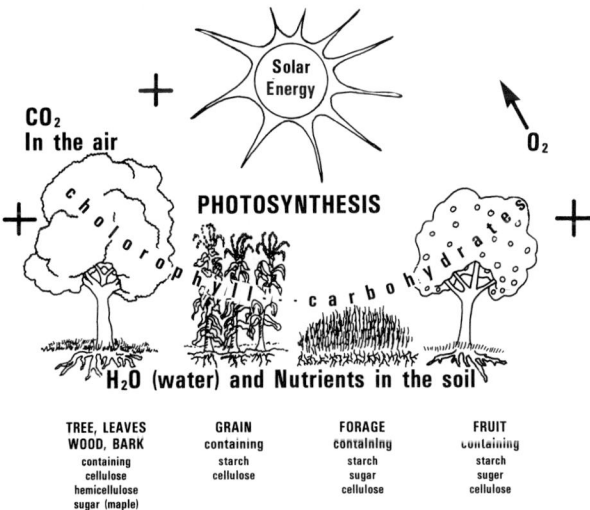

Fig. 4–5. The production of carbohydrates in plants by photosynthesis.

There are many intermediary reactions, but the overall process of glucose formation is as follows:

$6CO_2 + 6H_2O$ + energy from sun = $C_6H_{12}O_6$ (glucose) + $6O_2$

Carbohydrates form the woody framework of plants as well as the chief reserve food stored in seeds, roots, and tubers. For animals, carbohydrates serve as a source of heat, energy, or bulk; and any excess of them is stored in the body as fat. But there is no specific requirement for any individual carbohydrate compound.

CARBOHYDRATE COMPONENTS WITHIN FEEDS

From a feeding standpoint, carbohydrates are components of the nitrogen-free extract (NFE) and the crude fiber fractions of proximate analysis. The nitrogen-free extract includes the more soluble, and, therefore, the more digestible, carbohydrates—such as the starches, sugars, hemicelluloses, and some cellulose and pentosan. Also, NFE contains a limited amount of lignin. Crude fiber is the woody portion of plants (or feeds) which is not dissolved by weak acids and alkalies. Therefore, crude fiber—including cellulose, hemicellulose, and lignin—is digested less easily.

Availability of Fiber

The ability of animals to utilize roughages—to digest the fiber therein—depends chiefly on microbial action. This is confined largely to the first three compartments of the stomach of ruminants, to the cecum and colon of the horse, and to a lesser extent the large intestine of all animals. This bacterial digestion breaks down the cellulose and pentosans of feeds into usable organic acids (chiefly acetic, propionic, and butyric acid). These volatile fatty acids are absorbed directly through the ruminal wall, and furnish much of the energy required for maintenance. The fiber of growing pasture grass, fresh or dried, is more digestible than the fiber of most hay. Likewise, the fiber of early-cut hay is more digestible than that of hay cut in the late-bloom or seed stages. The difference is due to both chemical and physical structure, especially to the presence of certain encrusting substances (notably lignin) which are deposited in the cell wall with age. Young stock of all classes, finishing steers and lambs, high-producing dairy cows, swine, poultry, and horses must have rations in which a large part of the carbohydrate content of the ration is low in fiber but high in the form of nitrogen-free extract. On the other hand, a considerable amount of fiber or bulk in the ration is considered desirable for mature breeding animals of all classes of livestock, especially when high condition is not desired. Likewise, with young animals being developed for breeding purposes, the increased fiber will tend to develop more growth and not so much fat.

To promote good muscle tone and activity in the gastrointestinal tract, one must feed a certain amount of coarse roughage to all classes of farm animals, except to swine and poultry.

Fats and Other Lipids

Lipids is an all-embracing term referring to compounds that are soluble in chloroform, benzene, petroleum, or ether. It includes fats, oils, waxes, sterols, and complex compounds such as phospholipids and sphingolipids.

Lipids, like carbohydrates, contain the three elements—carbon, hydrogen, and oxygen.

Not all lipids are fats, but all fats are lipids; so, the two terms are used interchangeably. Fats and oils are differentiated on the basis of melting points; fats are solid at room temperature, while oils are liquid. As livestock feeds, fats and oils function much like carbohydrates in that they serve as a source of heat and energy and for the formation of fat. Because of the larger proportion of carbon and hydrogen, however, fats and oils liberate more energy than carbohydrates when digested, furnishing on oxidation approximately 2.25 times as much heat or energy per pound as do the carbohydrates. A smaller quantity of fat is required, therefore, to serve the same function. Common belief to the contrary, animals can tolerate a rather high-fat content in the ration. As evidence of this, sucklings normally handle a relatively large amount of fat, for milk contains 25 to 40% of this nutrient on a dry matter basis. Also, except for the soft pork problem, no apparent difficulty is encountered in feeding hogs a rather high-fat ration, such as results when large quantities of peanuts or soybeans are fed.

A small amount of fats in the ration is desirable, as these fats are the carriers of the fat-soluble vitamins. Fortunately, normal farm rations contain ample quantities of these nutrients.

CHARACTERISTICS OF FATS

Triglycerides—the combination of 3 fatty acid molecules and a glycerol molecule—account for about 98% of the fats in feed and over 90% of the fat in the animal body. The remainder of the diet and body fats is comprised primarily of phospholipids and cholesterol. Since triglycerides and their component fatty acids are so abundant in the diet and body, the discussion which follows centers around fatty acids and triglycerides.

Fatty Acids

Fatty acids are key components of lipids. They are called acids because of the carboxyl organic acid group (COOH) which they contain. Their degree of saturation and the length of their carbon chain determine many of the physical characteristics of lipids. Numerous triglycerides exist due to the variety of fatty acids which may bind with glycerol. The properties of fatty acids depend on their chemical characteristics, which follow.

• **Saturation**—This refers to the ratio of hydrogen atoms to carbon atoms. The backbone of the fatty acid consists of a chain of carbon atoms joined by chemical bonds. When a single bond joins each pair of carbon atoms, carbon atoms within the chain have 2 hydrogen atoms joined to them and carbons at the end of the chain have 3 hydrogens. When carbon atoms are joined by double bonds, the carbon atoms within the chain are able to have only 1 hydrogen bound to them. Therefore, saturated fatty acids contain all possible hydrogen and no double bond between carbon atoms. Unsaturated fatty acids contain at least one double bond within the carbon chain (monounsaturated) or 2 or more double bonds within the carbon chain (polyunsaturated). Therefore, unsaturated fatty acids contain the same number of carbon atoms, but fewer hydrogen atoms than their saturated counterparts. Fig. 4-6 illustrates the concept of saturated and unsaturated.

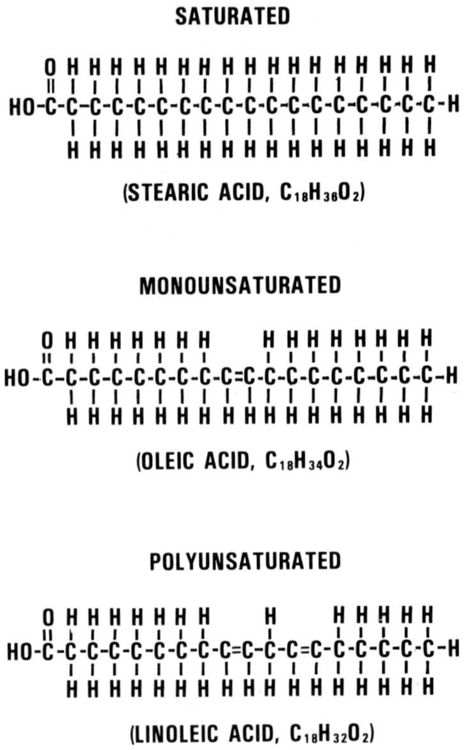

SATURATED

(STEARIC ACID, $C_{18}H_{36}O_2$)

MONOUNSATURATED

(OLEIC ACID, $C_{18}H_{34}O_2$)

POLYUNSATURATED

(LINOLEIC ACID, $C_{18}H_{32}O_2$)

Fig. 4-6. Three fatty acids all composed of 18 carbons but different degrees of saturation or unsaturation. The = indicates a double bond and C stands for carbon, H for hydrogen, and O for oxygen.

Most fatty acids in nature contain an even number of carbon atoms. The nomenclature is such that the following suffixes are used to describe the degree of unsaturation:
1. Anoic—no double bond.
2. Enoic—one double bond.
3. Dienoic—two double bonds.
4. Trienoic—three double bonds.
5. Tetraenoic—four double bonds.
6. Pentaenoic—five double bonds.

• **Iodine number**—Unsaturated fat readily unites with iodine; 2 atoms of this element will add to each double bond. Thus, in experimental work, the number of grams of iodine absorbed by 100 g of fat—the iodine number—is an excellent criterion of the degree of unsaturation. In the past, the iodine test was commonly used when studying the soft pork problem—a problem caused when pigs are fattened on feeds rich in unsaturated fats, such as peanuts or soybeans. At the present time, the chief measure used in such determinations is the refractive index, as determined by a refractometer.

Fatty acids that are unsaturated have the ability to take up oxygen or certain other chemicals. This presents both advantages and disadvantages. The value of linseed oil and varnish is due to their high content of unsaturated fatty acids, by virtue of which oxygen is absorbed when they are exposed to air, resulting in a tough, resistant coating. On the other hand, because of their unsaturation, these fats often become rancid through oxidation, resulting in disagreeable flavors and odors which lessen their desirablility as feeds. Moreover, oxidative rancidity in fats results in formation of unstable compounds called peroxides, which can destroy certain essential nutrients in the diet.

- **Rancidity**—This is the oxidation (decomposition) primarily of unsaturated fatty acids resulting in disagreeable flavors and odors in fats and oils. This process occurs slowly and spontaneously, and may be accelerated by light, heat, and certain minerals. Rancidity may be prevented through proper storage and/or the addition of antioxidants such as BHA (butylated hydroxyanisole). Some fats are naturally protected from oxidation due to the presence of vitamin E. Hydrogenation of fats (adding hydrogen to unsaturated fatty acids) increases their hardness and also lessens the threat of rancidity. This process has been used to improve the keeping qualities of vegetable shortenings and lard.
- **Hydrogenation (hardening)**—This process adds hydrogen to the double bonds of unsaturated fatty acids. It may be accomplished with hydrogen gas in the presence of a nickel catalyst. It also occurs in the rumen as a result of microbial activity. The result of hydrogenation is a harder fat because adding hydrogen increases the melting temperature. It may be used on animal or vegetable fats to produce fats with a desired hardness. Many vegetable oils are converted into a solid or semisolid form for use in shortenings and margarines. Hydrogenation is also known as hardening.

Hydrogenation has a drawback in that it converts the naturally-occurring *cis* fatty acids to *trans* fatty acids. The prefixes *cis* and *trans* refer to the orientation of the atoms around the double bond. The *trans* form of essential fatty acids does *not* function as an essential fatty acid in the body. Also, some researchers have found that (1) *trans* fatty acids are not as effective as their *cis* analogs in lowering blood cholesterol, and (2) fats rich in *trans* fatty acids appear to promote atherosclerosis.

The content of *trans* fatty acids generally increases with the extent to which a vegetable oil has been hydrogenated. For example, hard sticks of vegetable oil margarines may contain from 24 to 35% of *trans* acids, whereas lightly hydrogenated liquid oils usually contain 5% or less.

- **Carbon chain length**—Another variable factor in the makeup of fatty acid molecules is the number of carbon atoms. Fatty acids are designated as having (1) short chains when the number of carbon atoms is 6 or less, (2) medium chains where there are 8 or 10 carbon atoms, and (3) long chains when there are 12 or more carbon atoms.

Together, the degree of saturation and the length of the carbon chain influence the melting point of fats. Short chain fatty acids tend to be more volatile, and acetic, propionic, and butyric acids are collectively called volatile fatty acids (VFA). There is a steady rise in melting point as the chain lengths increase. However, as the number of double bonds increases, the melting point decreases.

- **Saponification**—The combination of a fatty acid with an alkali, such as potassium or sodium hydroxide, forms soap. This reaction is called saponification. Besides forming soap, it is a method of evaluating the average length of the carbon chain in the fatty acids which constitute a fat. The test is performed by reacting fats with potassium hydroxide. The saponification number, or value, is the number of milligrams of potassium hydroxide required for the complete saponification of 1 g of the fat. A high saponification value signifies a short chain length and vice versa.

Saponification may also occur in the alkaline medium of the intestine. For example, calcium may combine with free fatty acids.

- **Emulsification**—Fats (oils) and water do not stay mixed, but often it is desirable for them to do so. Therefore, fats are often emulsified. Minute droplets of fats or oils are evenly distributed throughout a water-based solution. Emulsions are essential for the digestion, absorption, and transport of fats in the body. Emulsifying agents used to create emulsions include some fatlike and fat-derived substances such as monoglycerides (glycerol with one fatty acid), diglycerides (glycerol with two fatty acids), lecithin, and the bile salts.

ESSENTIAL FATTY ACIDS

There is evidence that linoleic, linolenic, and arachidonic acids are dietary essentials. Deficiency symptoms of these fatty acids have been observed in mice, poultry, swine, dogs, guinea pigs, and human infants. Depending on the type of animal, numerous manifestations of these deficiencies are seen; among them, dermatitis, reduced growth, increased water consumption and retention, impaired reproduction, and increases in metabolic rate. Two functions of the essential fatty acids have been postulated: (1) precursors of prostaglandins, and (2) structural components of cells. The evidence for the first theory is conclusive. The second theory has substantial support inasmuch as the essential fatty acids are in highest concentrations in phospholipids, a type of lipid that plays an important role in the structural integrity of the cell.

OTHER LIPIDS

Other important lipids include the phospholipids, lipoproteins, and cholesterol. All cells contain phospholipids. They are structural compounds found in cell membranes and in the blood. The brain, nerves, and liver contain particularly high levels. Lecithin is one of the most abundant phospholipids in the diet and the body. Phospholipids are powerful emulsifying agents. Lipoproteins are the primary vehicle for lipid transport in the blood. There are four main types: chylomicrons; very low density lipoproteins (VLDL); low density lipoproteins (LDL); and high density lipoproteins (HDL). Cholesterol is derived from the diet or synthesized in the body. It is necessary for the formation of hormones, bile salts, and vitamin D.

Metabolism of Fats

In monogastrics, fats are digested in the small intestine, primarily as a result of action (1) of bile which emulsifies the fat, thus greatly increasing the surface area, and (2) of pancreatic lipase, an enzyme which hydrolyzes fatty acids from the glycerol molecule. Some diglycerides are absorbed, but most of the absorption is as monoglycerides and fatty acids. Most of the longer chain fatty acids are absorbed by lacteals into the lymph system which enters the blood stream just before the vena cava vein enters the heart.

In the rumen, many of the nutrients are transformed by the microorganisms into volatile fatty acids—primarily acetate, propionate, and butyrate. These products are then absorbed through the rumen wall and used as energy sources. Acetate is converted to acetyl CoA, whereupon it enters the Krebs cycle. Since 2 ATP are needed in this transformation, the net energy yield is 10 ATP. Propionate is not only a product of rumen fermentation, but it is also the final product of the β-oxidation of odd-carbon fatty acids. Upon absorption, it is converted to propionyl CoA. Propionyl CoA is then converted to succinyl CoA through a series of reactions. Succinyl CoA then enters the Krebs cycle and a net of 18 ATP is produced from propionate. Butyric acid, a 4-carbon fatty acid, undergoes β-oxidation resulting eventually in the formation of 27 ATP.

After absorption as a fatty acid or monoglyceride, triglycerides are resynthesized in the mucosal tissues of the gut. The fats are then transported to the various tissues, particularly the liver, where they are (1) used in synthesis of various compounds required by the body, (2) metabolized as a source of energy, or (3) stored in the tissues (as fat deposits).

Measuring and Expressing Energy Value of Feedstuffs

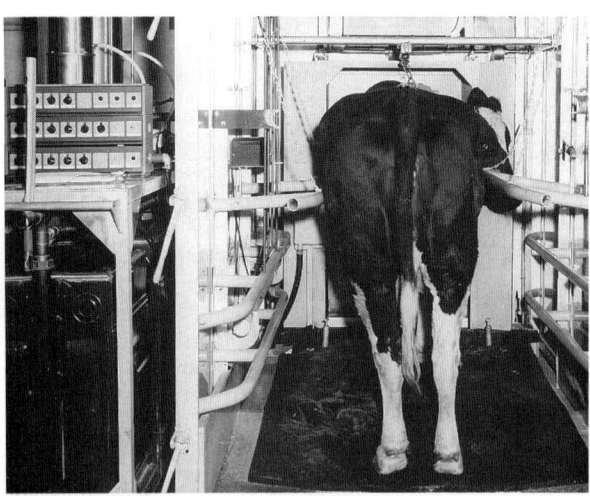

Fig. 4-7. A Holstein cow in an open circuit respiration chamber, used for measuring the energy value of a feedstuff. The unit to the left of the chamber is used to measure the respiratory exchange of the cow (air flow, temperature, relative humidity, gas samples, and animal activity). (Courtesy, Energy Metabolism Lab., USDA, Beltsville, Md.)

One nutrient cannot be considered as more important than another, because all essential nutrients must be present in adequate amounts if efficient production is to be maintained. Yet, historically, feedstuffs have been compared or evaluated primarily on their ability to supply energy to animals. This is understandable because (1) energy is required in larger amounts than anything else, (2) energy is most often the limiting factor in livestock production, and (3) energy is the major cost associated with feeding animals.

Cereal grains are higher in available energy than roughages. Although grains usually cost more on a weight basis than roughages, they are often a cheaper source of energy.

Our understanding of energy metabolism has increased through the years. With this added knowledge, changes have come in both the methods and terms used to express the energy value of feeds.

(Also see Chapter 15, Feed Analysis/Feed Evaluation, Section headed "Types of Feed Evaluation Trials.")

ENERGY DEFINITIONS AND CONVERSIONS

Some pertinent definitions and conversions of energy terms follow:

• **Calorie (cal)**—*The amount of energy as heat required to raise the temperature of 1 g of water 1°C* (precisely from 14.5°C to 15.5°C). It is equivalent to 4.184 joules. Although not preferred, it is also called a *small calorie* and so designated by being spelled with a lower case "c." **NOTE WELL**: In popular writings, the term calorie is frequently used erroneously for the kilocalorie (1,000 small calories).

• **Kilocalorie (kcal)**—*the amount of energy as heat required to raise the temperature of 1 kg of water 1°C* (from 14.5°C to 15.5°C). Equivalent to 1,000 calories. In human nutrition, it is referred to as a kilogram calorie or as a *large Calorie* and is so designated by being spelled with a capital "C" to distinguish it from the *small calorie.*

• **Megacalorie (Mcal)**—*Equivalent to 1,000 kilocalories or 1,000,000 calories.* Also, referred to as a *therm*, but the term megacalorie is preferred.

• **British Thermal Unit (Btu)**—*The amount of energy as heat required to raise 1 lb of water 1°F;* equivalent to 252 calories. This term is seldom used in animal nutrition.

• **Joule**—*A proposed international unit* (4.184J = 1 calorie) *for expressing mechanical, chemical, or electrical energy, as well as the concept of heat.* In the future, energy requirements and feed values will likely be expressed by this unit.

• **Converting TDN to Mcal**—One pound of TDN = 2.0 Mcal or 2,000 kcal. It is recognized, however, that the roughage component in a ration affects its energy value. Thus, when converting all-roughage rations from TDN to calories, some scientists figure that 1 lb of TDN = 1,500 kcal, instead of 2,000.

• **Hay equivalent (HE)**—*This is the energy equivalent of 1 ton of hay which, on the average, contains 800 Mcal of net energy.* With an Animal Unit Month (AUM) being equivalent to 320 Mcal of net energy, 2.5 AUM are required to furnish the same amount of energy as 1 ton of hay.

ENERGY SYSTEMS

Broadly speaking, two methods of measuring energy are employed in this country—the total digestible nutrient system (TDN), and the calorie system. Each system has its advantages and advocates. But, more and more feedstuffs are being evaluated in calories.

Total Digestible Nutrients (TDN)

Total digestible nutrients (TDN) is the sum of the digestible protein, digestible fiber, digestible nitrogen-free extract, and digestible fat × 2.25. It has been the most extensively used measure for energy in the United States.

Back of TDN values are the following steps:

1. **Digestibility.** The digestibility of a particular feed for a specific species is determined by a digestion trial.

2. **Computation of digestible nutrients.** Digestible nutrients are computed by multiplying the percentage of each nutrient in the feed (protein, fiber, nitrogen-free extract [NFE], and fat) by its digestion coefficient (percentage digest-

ibility). The result is expressed as digestible protein, digestible fiber, digestible NFE, and digestible fat. For example, if No. 2 corn contains 8.9% protein of which 77% is digestible, the percentage of digestible protein is 6.9.

3. **Computation of total digestible nutrients (TDN).** The TDN is computed by use of the following formula:

% TDN = % DCP + % DCF + % DNFE + (% DEE × 2.25)

where DCP = digestible crude protein; DCF = digestible crude fiber; DNFE = digestible nitrogen-free extract; and DEE = digestible ether extract.

TDN is ordinarily expressed as a percent of the ration or in units of weight (lb or kg), not as a caloric figure.

The main **advantage** of the TDN system is that it has been used for a very long time and many people are acquainted with it.

The main **disadvantages** of the TDN system are:

1. It is really a misnomer, because TDN is not an actual total of the digestible nutrients in a feed. It does not include the digestible mineral matter (such as salt, limestone, and defluorinated phosphate—all of which are digestible); and the digestible fat is multiplied by the factor 2.25 before being included in the TDN figure, because its energy value is higher than carbohydrates and protein. As a result of multiplying fat by the factor 2.25, feeds high in fat will sometimes exceed 100 in percentage TDN (a pure fat with a coefficient of digestibility of 100% would have a theoretical TDN value of 225% [100% × 2.25]).

2. It is an empirical formula based upon chemical determinations that are not related to actual metabolism of the animal.

3. It is expressed as a percent or in weight (lb or kg), whereas energy is expressed in calories.

4. It takes into consideration only digestive losses; it does not take into account other important losses, such as losses in the urine, gases, and increased heat production (heat increment).

5. It overevaluates roughages in relation to concentrates when fed for high rates of production, due to the higher heat loss per pound of TDN in high-fiber feeds.

Because of these several limitations, in the United States the TDN system is gradually being replaced by other energy evaluation systems, particularly net energy for ruminants and metabolizable energy for nonruminants. However, due to the voluminous TDN data on many feeds and long-standing tradition, it will continue to be used by many people for a long time to come.

Calorie System

Calories are used to express the energy value of feedstuffs. *One calorie (always written with a small "c") is the amount of heat required to raise the temperature of 1 g of water 1°C* (precisely from 14.5°C to 15.5°C).

To measure this heat energy, an instrument known as the bomb calorimeter is used, in which the feed (or other substance) to be tested is placed and burned in the presence of oxygen.

Through various digestive and metabolic processes, much of the energy in feed is dissipated as it passes through the animal's digestive system. About 60% of the total combustible energy in grain and about 80% of the total combustible energy in roughage is lost as feces, urine, gases, and heat. These losses are illustrated in Figs. 4-8 and 4-9.

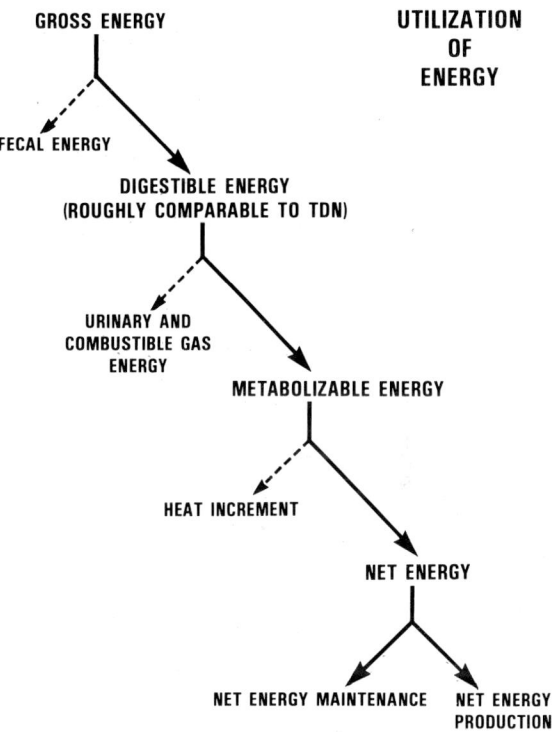

Fig. 4–8. Utilization of energy.

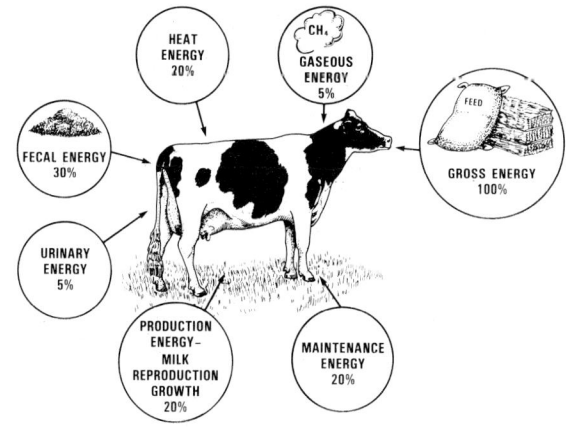

Fig. 4-9. Energy utilization by a lactating cow showing average partition of feed energy by the animal.

As shown in Figs. 4–8 and 4–9, energy losses occur in the digestion and metabolism of feed. Measures that are used to express animal requirements and the energy content of feeds differ primarily in the digestive and metabolic losses that are included in their determination. Thus, the following terms are used to express the energy value of feeds:

• **Gross energy (GE)**—*Gross energy represents the total combustible energy in a feedstuff.* It does not differ greatly among feeds, except for those high in fat. For example, 1 lb of corncobs contains about the same amount of GE as 1 lb of shelled corn. Therefore, GE does little to describe the useful energy in feeds for finishing animals.

• **Digestible energy (DE)**—*Digestible energy is that portion of the GE in a feed that is not excreted in the feces.*

• **Metabolizable energy (ME)**—*Metabolizable energy represents that portion of the GE that is not lost in the feces, urine, and gas.* Although ME more accurately describes the useful energy in the feed than does GE or DE, it does not take into account the energy lost as heat.

• **Net energy (NE)**—*Net energy represents the energy fraction in a feed that is left after the fecal, urinary, gas, and heat losses are deducted from the GE.* The net energy, because of its greater accuracy, is being used increasingly in ration formulations, especially in computerized formulations for large operations.

Although net energy is a more precise measure of the real value of the feed than other energy values, it is much more difficult to determine.

Two systems of net energy evaluation are presently being used. Lofgreen and Garrett [1] developed a system whereby the net energy requirements are listed as dictated by physiological functions—for example, net energy for maintenance (NE_m) and net energy for gain (NE_g). Also, Moe and Flatt [2] developed a net energy system that compares the physiological function to that of lactation through the use of regression analysis. This value, $NE_{lactation}$, is applicable for all physiological functions.

NITROGENOUS FEEDS

The nitrogenous fraction of feeds is generally broken down into two parts—protein and nonprotein. Compounds such as amines, free amino acids, purines, pyrimidines, and urea are included in the nonprotein part.

Nonprotein Nitrogen

When feeds are analyzed by the proximate analysis method, crude protein is determined by multiplying the nitrogen content of the feed by 6.25. This factor is used because protein is approximately 16% nitrogen (100% ÷ 16% = 6.25). This assumes that all of the nitrogen in feed is in the form of protein, but in many cases some of the nitrogen is derived from purines and pyrimidines and other nonprotein nitrogen sources. Ruminal microorganisms have the ability to convert certain nonprotein nitrogen feeds into microbial protein which can subsequently be used by the host animal. Hence, feeds such as urea and ammoniated by-products are commonly incorporated in ruminant rations.

(Also see Chapter 11, section on "Nonprotein Nitrogen (NPN) Feedstuffs.")

UREA AND AMMONIATED FEEDS

It has long been recognized that animals in which there is ruminal fermentation can subsist very well on protein of extremely low quality and that these animals can incorporate nonprotein nitrogen into microbial protein. When dietary protein and nonprotein nitrogen are metabolized by the ruminal microorganisms, ammonia is produced which is, in turn, combined with carbon chains to form the amino acids which are incorporated into microbial protein. In the rumen, there is a constant turnover of microorganisms. Many of these microorganisms are passed down the digestive tract along with ingesta and are subsequently digested and absorbed in the small intestine. When compared with casein, microbial protein has nearly the same biological value, although it is slightly lower in methionine. More recent studies on the need for sulfur and cobalt by rumen microorganisms have accounted for some of the differences in microbial protein utilization.

Urea is a form of nonprotein nitrogen which can be used in the synthesis of protein via microbial fermentation. It is usually the least costly of the nonprotein nitrogen compounds. Its chief disadvantage is that it should be fed along with some readily available carbohydrate and, even then, occasional incidences of toxicity may occur.

The common methods of feeding urea are: urea mixed with concentrates; urea liquid supplements; urea salt blocks; urea mixed with silages; urea added to dry roughages; and slow-release urea products.

In monogastric species, the only microorganisms that can convert urea to protein are found in the lower intestinal tract at a point where absorption of amino acids, peptides, and proteins is rather low or nonexistent. Research with pigs, poultry, horses, and other species indicates only slight utilization of nitrogen from urea.

Nonprotein nitrogen can also be supplied through the ammoniation of certain feedstuffs. Ammoniated molasses represents a logical combination of a nitrogen source and a carbohydrate source. Beet pulp, citrus pulp, rice hulls, and cottonseed hulls are protein sources when ammoniated.

Anhydrous ammonia (NH_3) may be added to forages. Since 1970, the ammoniation of low-quality forages has increased dramatically throughout North America, with the anhydrous ammonia added to silages, and to dry forages in bales, stacks and barns. The ammonia provides an economical source of nitrogen (protein equivalent) which can be utilized by ruminants.

---

[1] Lofgreen, G. P., and W. N. Garrett, "A System for Expressing Net Energy Requirements and Feed Values for Growing and Finishing Beef Cattle," *Journal of Animal Science*, Vol. 27, 1968, p. 793.

[2] Moe, P. W., and W. P. Flatt, "Net Energy of Feedstuffs for Lactation," *Journal of Dairy Science*, Vol. 52, 1969, p. 928.

When nonprotein nitrogen sources are fed to ruminants, the rate of ammonia release must be carefully monitored in order to prevent toxicities. For this reason, only restricted levels of NPN are used in ruminant rations. The following toxicity preventive/control measures are recommended: (1) Do not treat hay with more than 1.0 to 1.5% ammonia by weight, and apply the ammonia evenly to the forage; (2) do not feed ammoniated hay alone—feed some grain with it; and (3) if toxicity signs develop when feeding ammoniated forage, discontinue feeding the forage immediately.

(Also see Chapter 5, Nutritional Disorders/Toxins, Table 5–3, Potential Poisons—Urea Toxicity [Ammonia Toxicity]; and Chapter 11, Protein Supplements, the several sections under "Nonprotein Nitrogen (NPN) Feedstuffs.")

Proteins

Chemically, proteins are complex organic compounds made up chiefly of amino acids. For each different protein there are specific amino acids and a specific number of amino acids which are joined in a specific order. Since amino acids always contain carbon, hydrogen, oxygen, and nitrogen, so do proteins. Moreover, the presence of nitrogen provides a tool for chemically estimating the amount of protein in a tissue, feed, or some other substance. Crude protein is routinely determined by the Kjeldahl process, which involves finding the nitrogen content and multiplying the result by 6.25, since the nitrogen content of all protein averages about 16% (100 ÷ 16 = 6.25). In addition, proteins usually contain sulfur and frequently phosphorus. Proteins are essential in all plant and animal life as components of the active protoplasm of each living cell.

In plants, the protein is largely concentrated in the actively growing portions, especially the leaves and seeds. Plants also have the ability to synthesize their own proteins from such relatively simple soil and air compounds and carbon dioxide, water, nitrates, and sulfates, using energy from the sun. Thus, plants, together with some bacteria which are able to synthesize these products, are the original sources of all proteins.

Proteins in animals are much more widely distributed than in plants. Thus, the proteins of the body are primary constituents of many structural and protective tissues—such as bone, ligaments, hair, hooves, skin, and the soft tissues which include the organs and muscles. The total protein content of animal bodies ranges from about 10% in very fat, mature animals to 20% in thin, young animals. By way of further contrast, it is also interesting to note that, except for the bacterial action in the rumen of ruminants, animals lack the ability of the plant to synthesize proteins from simple materials. They must depend upon plants or other animals as a source of dietary protein.

Animals of all ages and kinds require adequate amounts of protein of suitable quality. The protein requirements for growth, reproduction, and lactation are the greatest and most critical. Actually, the need for protein in the ration is really a need for amino acids.

Each protein has a distinctive function in the animal body, ranging from protection of the body surface (skin, hair) to defense against invading organisms. Structurally, proteins have important functions as components of muscle, cell membranes, skin, hair, and hooves. Metabolically important proteins are the blood serum proteins, enzymes, hormones, and immune bodies—all of which have important specialized functions in the body.

AMINO ACIDS

The basic structural components of protein are amino acids. Although more than 200 naturally occurring compounds have been classed as amino acids, most proteins contain about 20 of the amino acids shown in Fig. 4–10 (p. 52). Many of the amino acids can be synthesized within the body. These are called nonessential amino acids or dispensable amino acids. If the body cannot synthesize sufficient amounts of certain amino acids to carry out physiological functions, they must be provided in the ration; hence, they are referred to as essential or indispensable amino acids. Actually, it is not entirely correct to say that all indispensable amino acids need to be provided in the diet; the requirement is for the preformed carbon skeleton of the indispensable amino acids, except in the case of lysine and threonine.

The necessity of each amino acid in the diet of the experimental rat has been thoroughly tested, but less is known about the requirements of large animals or even the human. According to our present knowledge, based largely on work with the rat, the following division of amino acids as essential and nonessential seems proper:

Essential (indispensable)	Nonessential (dispensable)
Arginine	Alanine
Histidine	Asparagine
Isoleucine	Aspartic acid
Leucine	Cysteine
Lysine	Cystine
Methionine (may be replaced in part by cystine)	Glutamic acid
	Glutamine
	Glycine
Phenylalanine	Hydroxyproline
Threonine	Proline
Tryptophan	Serine
Valine	Tyrosine

Arginine is regarded as essential for animals, whereas it is not for humans; most young mammals cannot synthesize it in sufficient amounts to meet their needs for growth.

In practical animal nutrition, the amino acids most likely to be deficient are lysine, methionine, and tryptophan. This stems from the fact that the cereal grains, which are primary energy feeds, are quite low in these amino acids. So, it follows that rations based on a high percentage of these grains usually require supplementation with proteins which contain higher levels of these amino acids.

Fortunately, the amino acid content of proteins from various sources differs. Thus, the deficiencies of one protein may be improved by combining it with another, and the

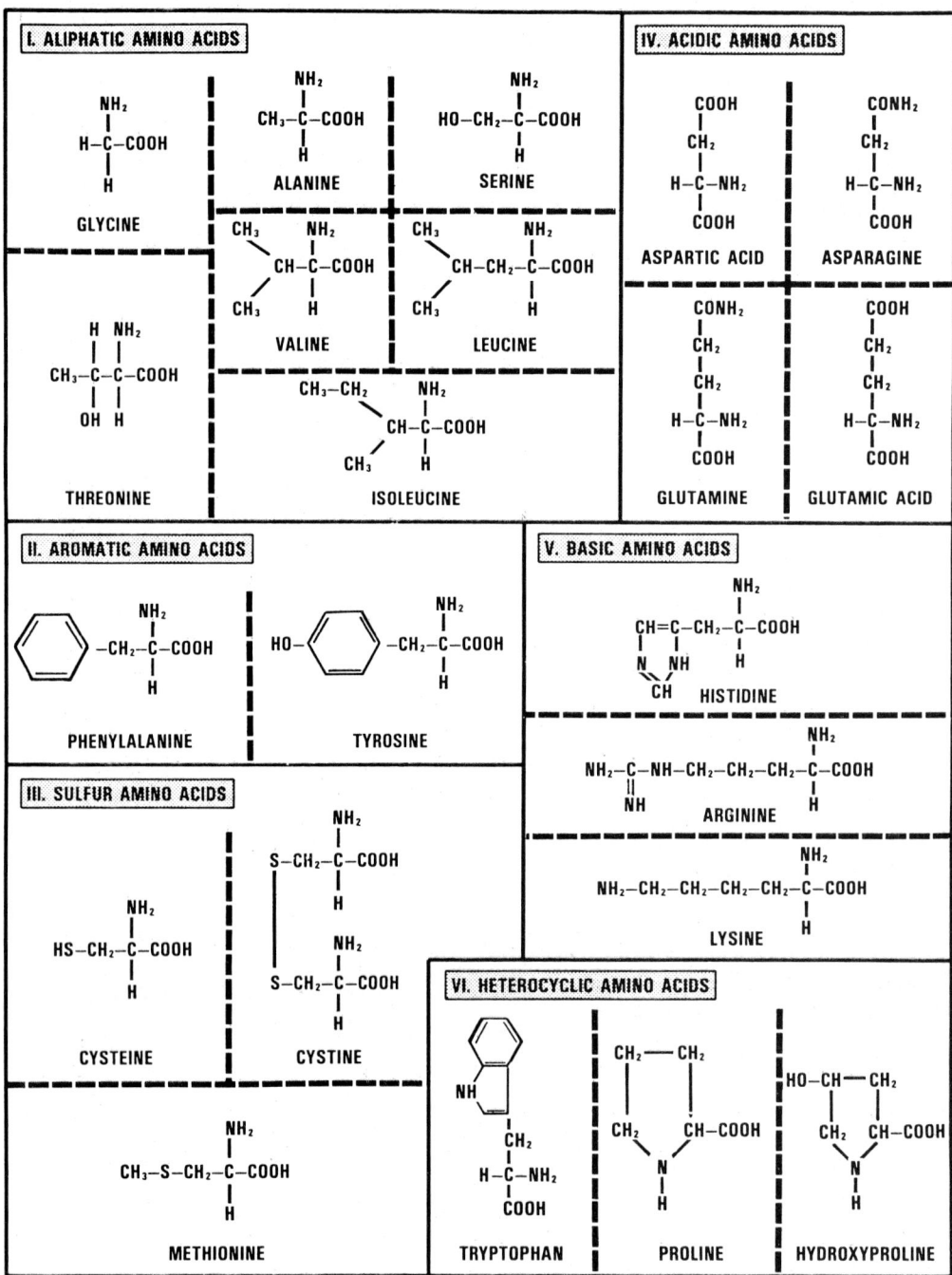

Fig. 4-10. Structures of the amino acids.

mixture of the two proteins often will have a higher feeding value than either one alone. It is for this reason that a considerable variety of feeds in the ration is usually recommended for monogastric animals.

The feed proteins are broken down into amino acids by digestion. They are then absorbed and distributed by the bloodstream to the body cells, which rebuild these amino acids into body proteins.

The various amino acids are then systematically joined together to form peptides and proteins. The amino nitrogen of one amino acid will combine with the carboxyl carbon atom of another amino acid to form a peptide linkage. When several of these junctions occur, the resulting molecule is called a protein. (See Fig. 4-11.)

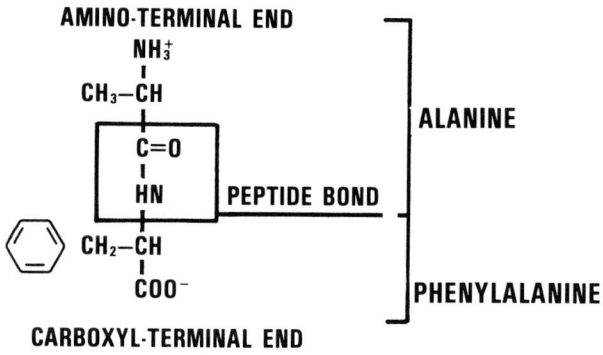

Fig. 4-11. A peptide bond between 2 amino acids.

SYNTHESIS OF AMINO ACIDS

Monogastric animals have the ability to synthesize a limited number of amino acids; hence, the term nonessential amino acids. Transamination (Fig. 4-12) is one metabolic process which makes this possible. Carbon skeletons are produced through various intermediates of carbohydrate metabolism; and a new amino acid can be produced when an amino group is transferred from an amino acid to the carbon skeleton. The deaminated molecule can then be used as an energy source.

TRANSAMINATION

```
   COOH                    COOH
    |                       |
   CH₂         COOH        CH₂          COOH
    |           |           |            |
   CH₂    +   C=O    →    CH₂     +   CHNH₂
    |           |           |            |
H—C—NH₂       CH₃          C=O          CH₃
    |                       |
   COOH                    COOH

GLUTAMIC ACID  PYRUVIC ACID  KETOGLUTARIC   ALANINE
                                ACID
```

Fig. 4-12. Process of transamination.

Ruminants (and some other herbivores) do not require dietary amino acids to the same extent as monogastrics. This is so because the vast majority of the dietary proteins are degraded and used as precursors for microbial protein. Thus, on a dietary level there is no differentiation between essential and nonessential amino acids in ruminants. However, on the cellular level, the ruminant needs the same amino acids as the nonruminant and possesses the same mechanisms for the interconversion of amino acids as the nonruminant animal.

There is some evidence, however, that high-producing animals, particularly dairy cows, may not receive optimal amounts of lysine or methionine and may benefit from dietary supplementation with them at peak production periods.

Protein synthesis is not a random process whereby a number of amino acids are joined together; rather, it is a detailed predetermined procedure. Within the cell, DNA serves as the information center concerning the sequences of the various proteins to be synthesized in the cell. When DNA is decoded, amino acids are linked to form a specific protein which has its own particular physiological function.

In order for a protein to be synthesized, all of its constituent amino acids must be available. If one amino acid is missing, the synthesis procedure is halted. When a particular amino acid is deficient, it is referred to as a limiting amino acid because it limits the synthesis of protein. This is the reason why protein quality is so important in the nutrition of monogastrics. High-quality proteins, upon digestion, provide balanced supplies of the various amino acids which can subsequently be absorbed as precursors for protein synthesis. Certain feeds are noted for being deficient in particular amino acids—for example, corn is known to be deficient in lysine. When these feeds are used, the livestock producer must make sure that other feeds are incorporated into the diet to compensate for the deficiencies of the particular amino acid(s).

Each sequence of amino acids is a different protein. Hence, different proteins are able to accomplish different functions in the body. With 22 amino acids, the different arrangements possible are endless, yielding a variety of proteins. For example, egg albumin, a small protein, contains approximately 288 amino acid units. Thus, if one assumes that there are about 20 different amino acids in the albumin molecule, then mathematical calculations show that the possible arrangements of this number of amino acids are in excess of 10^{300}. (For comparison, one million is equal to 10^6.)

Since the DNA of each cell carries the master plan for forming proteins, genetic mutations often are manifested by disorders in protein metabolism. Many metabolic defects—inborn errors of metabolism—are the result of a missing or modified enzyme or other protein. The misplacement or ommission of just one amino acid in a protein molecule can result in a nonfunctional protein. Therefore, when proteins are formed three requirements must be met: (1) the proper amino acids, (2) the proper number of amino acids, and (3) the proper order of the amino acid in the chain forming the protein. Meeting these requirements allows the formation of proteins which are very specific and which give tissues their unique form, function, and character.

BIOLOGICAL VALUE OF PROTEIN

Most chemical and microbiological tests for nutrient substances give information about the total amount of a nutrient present in a particular feedstuff or ration. However, they tell nothing about the digestibility and utilization of the feedstuff or ration in the digestive tract of the animal. Hence, biological tests directly involving animals are required to establish the true usefulness of feed in supplying the nutrient needs of animals. These biological tests are particularly important in evaluating protein. (They are also important in evaluating energy-yielding nutrients such as carbohydrates and fats.)

The biological value of a protein is the precentage of the digestible protein of a feed or feed mixture which is usable as a protein by the animal. It can be determined by a balance experiment in which a measured intake of protein is compared to the measured undigested protein in the feces of the animal. Thus, the biological value of a protein is a reflection of the kinds and amounts of amino acids available to the animal after digestion. If the amino acids available to the animal closely match those needed for body protein formation, the biological value of the protein is high. If, on the other hand, there are excesses of certain amino acids and deficiencies of other amino acids as a result of digestion, the biological value of the protein is low because of the increased number of amino acids which must be excreted via the kidney.

The biological values of animal proteins are generally much higher than those of plant proteins. For example, the biological value of whole milk is 85, compared to whole corn at 60. Because of this situation, the nutritional value of plant proteins (corn, beans, etc.) is substantially increased by feeding with them animal products rich in the amino acids in which plant proteins are deficient.

Other measures of protein quality are protein efficiency ratio (PER) and the net protein ratio (NPR). *The PER is, by definition, the number of grams of body weight gain of an animal per unit (lb or kg) of protein consumed. The NPR measures efficiency of growth by comparing body nitrogen resulting from feeding a test protein with that resulting from feeding a comparable group of animals a protein-free diet for the same period of time.*

Most feeds contain some protein, but the amount and quality of the protein varies considerably from one feed to another and the digestibility of the protein also varies. Legumes are particularly noted for high-quality protein—especially in ruminant rations. Young grasses may also be quite high in protein. When protein is deficient, it is necessary to consider other sources of protein. Both protein quality and quantity should be considered in the feeding of nonruminant, monogastric animals.

One of the most common supplemental sources of protein for farm animals is meal residue resulting from the extraction of oil from seeds of certain plants—primarily cotton, soybean, flax, safflower, and sunflower. Other excellent sources of high-quality proteins, although often more expensive than the plant sources, are the animal and fish meals—liver meal, fish meals (many different kinds of fish meals are available as single types of fish or blends of several), and offal (viscera and scraps of slaughtered animals). Meat meal and tankage are variable in nutritional quality, depending on the proportion of gelatin they contain, as well as the conditions used in cooking and drying them. Mill by-products, distillers' and brewers' by-products, and more recently, hydrolyzed feather meal, are high-protein by-products which are used to a lesser extent.

Processing of feeds may have an effect on the quality or availability of proteins. Heated forages show a decreased digestibility of proteins. Normally, such proteins are 60 to 70% digestible, compared to as low as 10% digestibility for hay or silage which has turned dark brown or black from heating. On the other hand, some protein feeds which are heated may be improved in digestibility for some monogastric animals. Heating cottonseed meal may not materially affect the utilization of protein, but it will destroy the gossypol which is toxic to some animals. Beans which contain the enzyme urease are heated when they are mixed with rations containing urea. If not, the urea will be converted to ammonia prior to reaching the rumen.

PROTEIN SOURCES

Rich feed sources of protein are shown in Fig. 4-13.

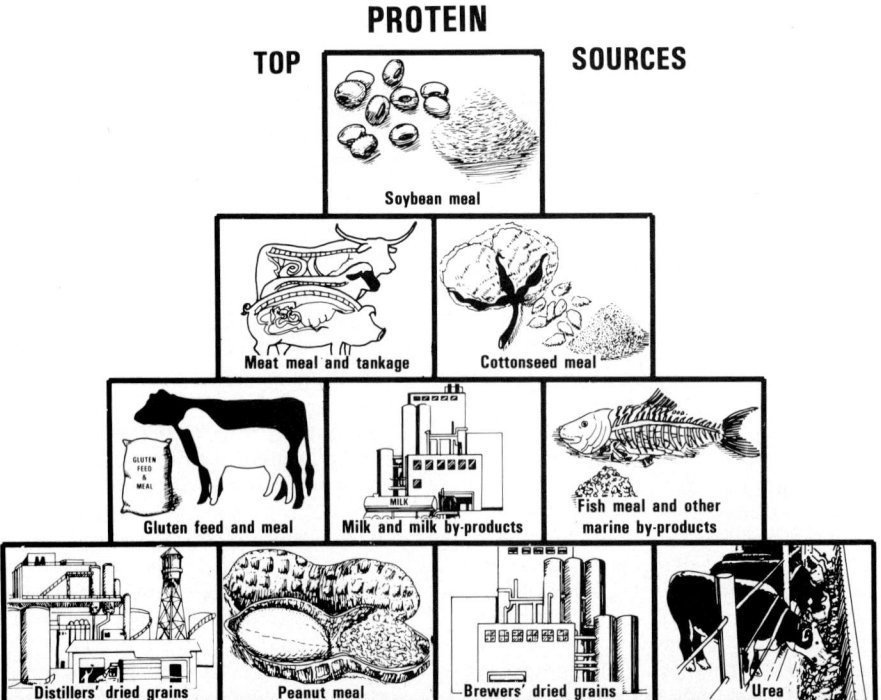

Fig. 4-13. Top sources of protein. No claim is made that all the best sources of protein are listed. Neither are the sources ranked. The evaluation of a protein depends on several factors, including amino acid composition, availability, and price.

MINERALS

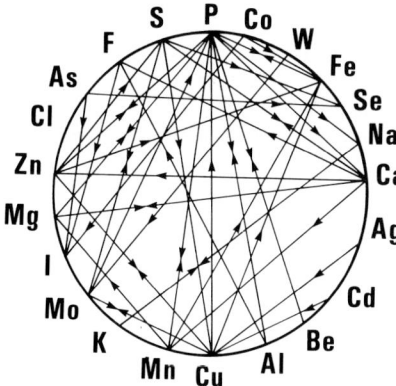

Fig. 4-14. Mineral interaction chart, showing the interrelationship of minerals. The importance of such relationships is evidenced by the following:
1. A great excess of dietary calcium and/or phosphorus interferes with the absorption of both minerals, and increases the excretion of the lesser mineral.
2. Excess magnesium upsets calcium metabolism.
3. Excess zinc interferes with copper metabolism.
4. Copper is required for the proper utilization of iron, but excess copper can markedly depress iron absorption.
And there are others! The maze of connecting lines in this figure shows the relation of each mineral to other minerals.

In nutrition, the term *mineral* denotes certain chemical elements which are found in the ash after a food or a body tissue is burned.

For a good understanding of minerals, it is very important that consideration be given to the following facts pertaining to them:

1. **The soil/plant/animal relationship of minerals.** There is a direct and most important relationship between the content and availability of mineral elements in the soil and the mineral composition of plants (forages and grains). Sometimes the concentration of an essential mineral in the soil is so low that plants growing on it will not contain enough of that mineral to meet the dietary requirements of animals. At other times, plants may contain such high concentrations of a certain mineral(s) that they are toxic to the animals that eat them. Such soil/plant/animal relationships are particularly important in trace elements.

2. **The interrelationship of minerals.** The interrelationship (Fig. 4-14) of minerals greatly complicates the task of determining minimum dietary requirements and tolerances for many elements; and the degree of interaction varies greatly. In addition to the interaction of mineral elements with other mineral elements, there are interactions (1) between mineral elements and the organic components of the diet, and (2) with factors other than nutrients.

3. **The role of minerals.** In addition to some general functions in which several minerals are involved, each essential mineral has at least one specific role. Although the metabolism of minerals does not produce energy, minerals are involved in many of the reactions of the body which comprise metabolism.

History/Discovery of Minerals

It is noteworthy that certain ancient peoples recognized and treated mineral deficiencies, although they did not understand the bases of their treatments. For example, a Chinese document dated about 3000 B.C. described goiter and recommended that afflicted people eat seaweed and burnt sponge, which are good sources of the trace element iodine. Another deficiency disease, anemia, was treated in ancient Greece (around the time of Hippocrates, or about the 4th century, B.C.) by giving the patient iron-containing water in which heated swords had been quenched. However, the effects of such treatments were often unpredictable, because there were no means of identifying or measuring the quantity of the active ingredients in the various medicinal substances.

The breakthroughs in our understanding of the functions of minerals in the body did not come for many centuries because laboratories for research were not developed until the Renaissance; although the medieval alchemists appear to have invented some of the techniques and tools of chemistry in their futile efforts to change base metals into gold.

The great French chemist Lavoisier, who is credited with being the founder of the science of nutrition, predicted in 1799 that such elements as sodium and potassium would soon be discovered because he believed that they were present in certain mineral compounds which were then known as *earths*. Sure enough, within a few years, the British chemist Davy discovered not only sodium and potassium, but also calcium, sulfur, magnesium, and chlorine. Davy's discoveries were so highly regarded by the French Academy that in 1806 they awarded him their new Volta medal, even though France and England were at war when the award was presented.

The pace of progress quickened as the noted Swedish chemist Berzelius added to our knowledge of minerals by (1) reporting his analysis of the calcium and phosphorus content of bone in 1801, and (2) concluding in 1838 that the iron in hemoglobin made it possible for the blood to absorb much oxygen. Similar contributions were made by the French chemist Boussingault, who (1) noted in 1822 that South American villagers who used salt containing iodine were protected from goiter which affected those who used plain salt; and (2) showed by means of animal feeding trials the necessity of providing dietary calcium and iron. Calcium studies were also conducted by Chossat, a physician in Switzerland, who won a prize in 1840 for his demonstration that the addition of calcium carbonate to a diet of wheat and water improved bone growth in pigeons.

Another half century elapsed before the value of iodine became widely accepted; thanks to (1) the discovery by the German biochemist Baumann in 1895 that the thyroid gland contained iodine, and (2) the demonstrations by the American medical scientist Marine and his coworkers between 1907 and 1918 that the administration of minute amounts of iodine prevented goiter in animals and in school children.

By the end of the 19th century only about one-third of the minerals now accepted as essential were known to be required in the diet. This state of affairs existed in spite of the many nutritional investigations which were conducted during that century because the need for vitamins had not yet been discovered. Hence, it was difficult to differentiate between the deficiency diseases due to lack of minerals and those due to lack of vitamins.

The first half of the 20th century was marked by finding that various minerals were required by animals and people, as indicated by the dates of the discoveries which follow: phosphorus, 1918; copper, 1925; magnesium, manganese, and molybdenum, 1931; zinc, 1934; and cobalt, 1935. However, there was often uncertainty as to the metabolic roles of various minerals, even though they were known to be essential. For example, it was not until 1948 that it was established that cobalt functions as a component of vitamin B-12. So, the two decades between the 1930s and the 1950s were marked by feverish activity aimed at learning how each of the newly discovered essential elements acted in the body.

The most recent chapter in this story began when Schwarz, a medical scientist who had emigrated to the United States from Germany, and his co-workers discovered the essentiality of selenium in 1957, and of chromium in 1959. They then developed isolation equipment for shielding their animals and their ultrapure diets from contamination by minute amounts of elements in the environment. Their painstaking work paid off; in 1972, they were able to show that fluorine and silicon were essential, also. It is ironic that only a short time ago all of these four trace elements—selenium, chromium, fluorine, and silicon—were considered to be only unwanted toxic contaminants of feeds, water, and air.

Much work remains to be done in the areas of (1) testing whether other elements might be essential, (2) defining the limits of safe and toxic doses for those already known, and (3) determining how the various elements interact with each other in the animal body.

Definitions/Classification/Functions

Minerals are inorganic elements, frequently found as salts with either inorganic elements or organic compounds. Their availability—and often their metabolic functions—are related to the form in which they are found. For example, in the presence of oxalates and citrates, calcium cannot be absorbed. When combined with phytin, phosphorus is available to some animals but not to others. Chelating agents have a selective attraction with various mineral elements, releasing one mineral element for another for which it has a greater attraction—sometimes creating deficiencies of an element which is present in otherwise adequate amounts.

In the chemical feed analysis, feed samples are burned in a furnace in order to destroy all organic matter, thus leaving only the minerals, or what is termed *ash*. At best, this analysis is a very crude measure of the mineral content in feed, because it (1) does not indicate what minerals are present; (2) does not include the volatile minerals such as iodine, chlorine, and selenium; (3) weighs the minerals as oxides or carbonates, with the weight of these elements included with the minerals; (4) does not indicate anything about the availability or form in which the minerals appear in the feed; (5) provides no information of the relative amounts of each mineral; and (6) ignores the importance of those minerals needed in such small quantities that they are not really contributing much to the overall ash content of feeds. In fact, the ash analysis is used quite often in forages to estimate the amount of dust and soil that has been harvested with the feed. It is also used by some feed inspectors to monitor the amount of *filler* in feedstuffs to which excessive amounts of limestone or other earthy materials are sometimes added.

Eighteen mineral elements are known to be required by at least some animal species. They can be divided into two groups based upon the quantity required in the ration.

- **Major or macrominerals**—These elements are required in amounts ranging from a few tenths of a gram to one or more grams per day.
- **Trace or microminerals**—These elements are required in minute quantities, ranging from a millionth of a gram (microgram) to a thousandth of a gram (milligram) per day.

The terms—major/macromineral and trace/micromineral—do not imply any lesser role for the latter group; rather, they represent quantity designations based on the amounts needed by animals.

The two groups, based on the amounts needed in the ration, follow:

Major/Macrominerals	Trace/Microminerals	
Salt (sodium & chlorine, NaCl)	Chromium (Cr)	Molybdenum (Mo)
Calcium (Ca)	Cobalt (Co)	Selenium (Se)
Phosphorus (P)	Copper (Cu)	Silicon (Si)
Magnesium (Mg)	Fluorine (F)	Zinc (Zn)
Potassium (K)	Iodine (I)	
Sulfur (S)	Iron (Fe)	
	Manganese (Mn)	

The percentages of the principal mineral constituents of the body are indicated by the following data, showing the average analyses of 18 steers of varying ages exclusive of the contents of the digestive tract.[3]

Element	Percent	Element	Percent
Calcium	1.33	Chlorine	0.11
Phosphorus	0.74	Magnesium	0.04
Potassium	0.19	Iron	0.01
Sodium	0.16		
Sulfur	0.15	Total	2.73

The dominant position of calcium and phosphorus in the above data becomes apparent when it is realized that calcium accounts for 49% of the total mineral; phosphorus, 27%; and all other minerals, 24%.

The general functions of minerals are as follows:
1. Give rigidity and strength to the skeletal structure.
2. Serve as constituents of the organic compounds, such as protein and lipid, which make up the muscles, organs, blood cells, and other soft tissues of the body.
3. Activate enzyme systems.
4. Control fluid balance—osmotic pressure and excretion.
5. Regulate acid-base balance.
6. Exert characteristic effects on the irritability of muscles and nerves.
7. Engage in mineral-vitamin relationships.

Information relative to minerals for each animal species is presented in the chapter devoted to each species, chapters 19 to 28.

---

[3]Hogan, A. G., and J. L. Nierman, *Studies of Animal Nutrition—VI, The Distribution of the Mineral Elements in the Animal Body as Influenced by Age and Condition*, Missouri Ag. Exp. Res. Bull. No. 107.

VITAMINS

Fig. 4-15. This shows an artist's dramatic impression of how British sailors, who ate little except salt meat and biscuits when on long voyages 200 years ago, suddenly collapsed and died of scurvy. (Reproduced with permission of *Nutrition Today*, Annapolis, Md.)

For proper physiological functions, the animal body requires some 40 to 50 dietary essentials, of which 16 are vitamins.

Throughout history, vitamin deficiencies have been a major cause of disease, morbidity, and death. Pellagra, scurvy, and beriberi decimated armies, ships' crews, and nations; they even reshaped the course of history. The importance of dietary factors in the genesis of diseases became recognized in the 18th century. But the significance of these observations was not fully understood until early in the 20th century, when scientists found it desirable in many types of investigations to use the biological approach—the use of laboratory animals (largely white, albino rats and mice; guinea pigs; and chicks) fed purified diets using pure protein such as casein or albumin, pure fat such as lard, and pure carbohydrate such as dextrin, plus minerals to supplement chemical analyses, in measuring the value of food. These diets were composed of relatively pure nutrients (proteins, carbohydrates, fats, and minerals) from which the unidentified factors were largely excluded. With these purified diets, all researchers shared a common experience—the animals not only failed to thrive, but they even failed to survive if the investigations were continued for any length of time. At first, many investigators explained such failures on the basis of unpalatability and monotony of diets. Finally, it was realized that these purified diets were lacking in certain factors, minute in amounts, the identities of which were unknown to science. These factors were essential for the efficient utilization of the main ingredients of the food and for the maintenance of health and life itself. The discovery, synthesis, and commercial production of vitamins followed. With these developments, the vitamin era of science was ushered in and the modern approach to nutrition was born.

History/Discovery of Vitamins

Until the early 1900s, if a diet contained proteins, fats, carbohydrates, minerals, and water, it was considered to be complete. True enough, the disease known as beriberi made its appearance in the rice-eating districts of the Orient when milling machinery was introduced from the West, having been known to the Chinese as early as 2600 B.C.; and scurvy was long known to occur among sailors fed salt meat and biscuits. However, for centuries these diseases were thought to be due to toxic substances in the digestive tract caused by pathogenic organisms, rather than food deficiencies; and more time elapsed before the discovery of vitamins.

Largely through the trial-and-error method, it was discovered that specific foods were helpful in the treatment of certain of these maladies. Hippocrates (460-377 B.C.), the Greek medical doctor, advocated liver as a cure for night blindness 400 years before the birth of Christ. At a very early date, the Chinese also used a concoction rich in vitamin A as a remedy for night blindness; and cod-liver oil was used in treating or preventing rickets long before anything was known about the cause of the disease. In 1747, James Lind, a British naval surgeon, in a study involving 12 sailors with scurvy on board the ship "Salisbury," showed that the juice of citrus fruits was a preventive and cure for the disease. Nicholas Lunin, as early as 1881, while a student of von Bunge at the University of Dorpat, had come to the conclusion that certain foods, such as milk, contain, beside the principal ingredients, small quantities of unknown substances essential to life. In 1882, Kanehiro Takaki, Director-General of the Japanese Navy, greatly reduced the number of beriberi cases among naval crews by adding meat and evaporated milk to their diet of rice. In 1897, Christiaan Eijkman, a Dutch medical officer, working in Java, had satisfied himself that the disease beriberi was due to the continued consumption of a diet of polished rice.

In 1912, Dr. Casimir Funk, a 28-year-old Polish biochemist working in London, coined the word *vitamine*. Funk postulated, as others had before him, that beriberi, scurvy, pellagra, and possibly rickets, were caused by a lack in the diet of "special substances which are of the nature of organic bases, which we will call vitamines." Presumably, the name vitamines alluded to the fact that they were vital to life, and that they were chemically of the nature of amines (nitrogen-containing). The name caught the popular fancy and persisted, despite the fact that the chemical assumption was not common to all such vital substances, with the result that the "e" was dropped in 1920; hence, the word *vitamin*. In 1922, Funk's book entitled *The Vitamins* was published.

The actual existence of vitamins has been known only since 1912, when Funk, the Polish biochemist working in London, coined the word. But it was much later before it was possible to see or touch any of them in pure form. Previously, they were merely mysterious invisible *little things* known only by their effects. In fact, most of the present fundamental knowledge relative to the vitamin content of both human foods and animal feeds was obtained through measuring their potency in promoting growth or in curing certain disease conditions in animals—a most difficult and tedious method. For the most part, small laboratory animals were used, especially rats, guinea pigs, pigeons, and chicks.

Today, there are 16 known vitamins. Additionally, there are at least nine other vitaminlike substances that have been proposed. But it is unlikely that all of them are distinct essentials. Yet, the probability that there are still undiscovered vitamins is recognized.

A chronological summary of the discovery/isolation and synthesis of the various vitamins is given in Table 4–1.

TABLE 4–1
CHRONOLOGY OF THE DISCOVERY/ISOLATION AND SYNTHESIS OF VITAMINS

Year	Discovery/Isolation	Synthesis
1849[1]	Choline	
1866–67		Choline
1913	Vitamin A[2]	
1926	Thiamin (B-1)	
1929	Vitamin K[3]	
1931	Vitamin A[2]	
1932	Vitamin C (Ascorbic acid)	
	Vitamin D_2	
1933	Riboflavin (B-2)	Vitamin C (Ascorbic acid)
1935		Riboflavin (B-2)
1936	Biotin	Thiamin (B-1)
	Vitamin E	
1937	Niacin (Nicotinic acid)	
1938	Vitamin B-6 (pyridoxine)	Vitamin E
1939	Pantothenic acid (B-3)	Vitamin B-6 (pyridoxine)
	Vitamin K[3]	Vitamin K[3]
1940		Pantothenic acid (B-3)
1943		Biotin
1945	Folacin (Folic acid)	Folacin (Folic acid)
1947		Vitamin A[2]
1948	Vitamin B-12	
1952		Vitamin D_3
1955		Vitamin B-12

[1]In 1844 and 1846, Gobley isolated a substance from egg yolk, which he called lecithin. In 1849, Strecker isolated a compound from hog bile, to which he subsequently (in 1862) applied the name *choline*.

[2]A more detailed chronology of vitamin A is: in 1913, it was discovered, independently, by McCollum and Davis of the University of Wisconsin, and Osborne and Mendel of the Connecticut Experiment Station; in 1931, its chemical formula was determined by P. Karrer, a Swiss researcher; and, in 1947, it was synthesized by Isler, working in Switzerland.

[3]Vitamin K was discovered in 1929, isolated in 1939, and synthesized in 1939.

Definitions/Classification/Functions

Vitamins are organic substances that are essential in small amounts for the health, growth, reproduction, and maintenance of one or more animal species, which must be included in the diet since they either (1) cannot be synthesized at all, or (2) cannot be synthesized in sufficient quantities in the body.

Each vitamin performs a specific function; hence, one cannot replace, or act for, another. In general, the body cannot synthesize them, at least in large enough amounts to meet its needs.

There is no universal agreement on the nomenclature of the vitamins. But the modern tendency is to use the chemical name, particularly in describing members of the B complex. In this book, the most common designations are used.

Today, vitamins are generally classed as (1) fat-soluble, (2) water-soluble, including the vitamin B complex, and (3) vitaminlike substances.

• **Fat-Soluble vs Water-Soluble Vitamins**—Many phenomena of vitamin nutrition are related to solubility—vitamins are soluble in either fat or water. Consequently, it is important that nutritionists (1) be well informed about solubility differences in vitamins and (2) make use of such differences in programs and practices. Based on solubility, vitamins may be grouped as follows:

The Fat-Soluble Vitamins
Vitamin A
Vitamin D
Vitamin E
Vitamin K

The Water-Soluble Vitamins
Biotin
Choline
Folacin (folic acid)
Inositol
Niacin (nicotinic acid, nicotinamide)
Pantothenic acid (vitamin B-3)
Para-aminobenzoic acid (PABA)
Riboflavin (vitamin B-2)
Thiamin (vitamin B-1)
Vitamin B-6 (pyridoxine, pyridoxal, pyridoxamine)
Vitamin B-12 (cobalamins)
Vitamin C (ascorbic acid, dehydroascorbic acid)

It is noteworthy that vitamin C is the only member of the water-soluble group that is not a member of the B family.

The two groups of vitamins exhibit the following several differences that distinguish them both chemically and biologically:

The fat-soluble vitamins contain only carbon, hydrogen, and oxygen, whereas the water-soluble B vitamins contain these three elements plus nitrogen and occasionally sulfur.

Vitamins originate primarily in plant tissues; with the exceptions of vitamins C and D, they are present in the animal tissues only if an animal consumes feed containing them or harbors microorganisms that synthesize them. Fat-soluble vitamins can occur in plant tissue in the form of a provitamin (or precursor of a vitamin), which can be converted into a vitamin in the animal body. Also, the B vitamins are universally distributed in all living tissues, whereas the fat-soluble vitamins are completely absent from some.

The fat-soluble vitamins are stored in appreciable quantities in the body, whereas the water-soluble vitamins are not. Any of the fat-soluble vitamins can be stored wherever fat is deposited; and the greater the intake, the greater the storage. By contrast, the water-soluble B vitamins are not stored in any appreciable amount. Moreover, the large amounts of water which pass through the body daily tend to carry out the water-soluble vitamins, thereby depleting the supply. Hence, they should be supplied in the diet on a daily basis. However, because all living cells contain all the B vitamins, and because the body conserves nutrients that are in short supply by using them only in vital reactions, deficiency symptoms do not appear immediately following their removal from the diet.

• **Vitamin B Complex**—With the exception of vitamin C, all of the water-soluble vitamins can be grouped together under the name vitamin B complex.

Under good practical conditions, the rations of farm animals usually contain adequate quantities of each of the several vitamins. However, deficiencies may occur during periods (1) of extended drought or in other conditions of restriction in diet, (2) when production is being forced, (3) when large quantities of highly refined feeds are being fed, or (4) when low-quality forages are utilized. Also, deficiencies may occur as a result of lack of availability of vitamins or because of the presence of antimetabolites. Both are important concepts. For example, analyses show corn to be adequate in niacin. Yet, due either to an antimetabolite or unavailability, there may be niacin deficiencies when corn is fed—deficiencies that can be remedied by niacin supplementation.

The absence of one or more vitamins in the ration may lead to a failure in growth or reproduction, or to characteristic disorders known as deficiency diseases. In severe cases, death may follow. Although the occasional deficiency symptoms are the most striking result of vitamin deficiencies, it must be emphasized that, in practice, mild deficiencies probably cause higher total economic losses than do severe deficiencies. It is relatively uncommon for a ration, or diet, to contain so little of a vitamin that obvious symptoms of a deficiency occur. When one such case does appear, it is reasonable to suppose that there must be several cases that are too mild to produce characteristic symptoms but which are sufficiently severe to lower the state of health and the efficiency of production. It is also to be emphasized that different species of animals vary in their needs for the vitamins. Further, not all animals suffer from the same deficiency diseases; thus, humans, monkeys, and guinea pigs react severely to the absence of vitamin C in the ration, whereas domestic pigs, fowl, and ruminants are unaffected.

Initially, vitamin nutrition was dependent upon incorporating into the ration feeds that were known to contain a high natural content of the needed vitamins. But this proved to be unsatisfactory, for the vitamin content of feeds varies considerably according to soil, climate conditions, and curing and storing. Through the years, however, methods for the laboratory synthesis of various vitamins were developed. So, today, most of the vitamins are available in pure crystalline form, and at prices that make them economical for supplementation of livestock.

The omission of a single vitamin from the diet of a species that requires it will produce specific deficiency symptoms. Many of the vitamins function as coenzymes (metabolic catalysts); others have no such role, but perform certain other essential functions.

Each of the vitamins is as much a distinct chemical compound as is cane sugar, for example. All of them contain carbon, hydrogen, and oxygen. In addition, all of the B vitamins except inositol contain nitrogen. Certain of the B vitamins also contain one or more of the mineral elements in their molecules. Even when added to the diet in very small amounts, vitamins are extraordinarily potent.

Information relative to vitamins for each animal species is presented in the chapter devoted to each species, chapters 19 to 28.

WATER

Chemically, water is the combination of 2 gases—hydrogen (H) and oxygen (O)—which are joined in the ratio of 2 hydrogen atoms to 1 oxygen (as H_2O). It is the most abundant chemical substance, and it performs endless functions in its three forms—liquid, solid, gas.

Water is vital to nutrition. It is the solvent wherein the metabolic reactions of the body take place. Also, as a solvent, water carries (1) the nutrients which are subjected to cellular metabolism, and (2) the waste products of metabolism. Also, it serves to disperse the heat generated by the metabolic reactions. In many of the metabolic reactions water is either added or subtracted. Subtracted water is termed metabolic water. The addition of water is termed hydrolysis.

Animals can survive for a longer period without feed than they can without water. Only oxygen is more important to animal life. Fortunately, under most conditions, water can be readily provided in abundance—and at little cost. In addition to what animals drink, water is found in all feeds, ranging from about 10% in air-dry feeds to over 80% in fresh green forage.

Water is one of the largest single constituents of the animal body, varying in amount from 40% in fat hogs to 80% in newborn pigs, 50% in a 1,000-lb steer to 70% in a newborn calf, and 50% in a fat lamb to 80% in a newborn lamb. In general, the percentage of water in the bodies of animals varies with species, condition, and age. The younger the animal, the more water it contains. Also, the fatter the animal, the lower the water content. Thus, as an animal matures, it requires proportionately less water on a weight basis because it consumes less feed per unit of weight and the water content of the body is being replaced by fat. This accounts for the fact that gains in older animals are more costly nutritionally than those in younger animals.

Water performs the following important functions in animals:

1. It is necessary to the life and shape of every cell and is a constituent of every body fluid.

2. It acts as a carrier for various substances, serving as a medium in which nourishment is carried to the cells and waste products are removed therefrom.

3. It assists with temperature regulation in the body, cooling the animal by evaporation from the skin as perspiration.

4. It is necessary for many important chemical reactions of digestion and metabolism.

5. It lubricates the joints, as a constituent of the synovial fluid; it acts as a water cushion for the nervous system, in the cerebrospinal fluid; it transports sound, in the perilymph in the ear; and it is involved with sight and provides a lubricant for the eye.

6. It acts as a solvent for a number of chemicals which can subsequently be detected by taste buds.

7. It aids in gas exchange in respiration by keeping the alveoli of the lungs moist.

The total body water involved in all of these functions is contained in two major compartments in the body: (1) the extracellular water outside the cells (about 20% of the body weight), and (2) the intracellular water inside each cell (about 45% of the body weight).

Deficits or excesses of more than a few percent of the total body water are incompatible with health, and large deficits, of about 20% of the body weight, lead to death. Under normal circumstances, thirst ensures that water intake meets or exceeds the requirement for water.

The specific water requirements of each class of animal receive further consideration in the section devoted to the respective species. In general, however, under practical conditions, the needs for water can best be taken care of by allowing the animals free access to plenty of clean, fresh water at all times.

Water Balance

In healthy animals, total body water remains reasonably constant. An increase or decrease in water intake brings about an appropriate increase or decrease in water output to maintain the balance. Water enters the body as a liquid, and as a component of feed—including metabolic water derived from the breakdown of feed. Water is lost from the body (1) by the skin as perspiration, (2) by the lungs as water vapor in expired air, (3) by the kidneys as urine, and (4) by the intestine in the feces. Therefore, under normal conditions the intake of water from the various sources is approximately equal to output of water by the various routes.

If water is withheld, the following compensatory mechanisms are initiated to provide enough water for maintenance:

1. Urine excretion is reduced as is the water content of the feces.
2. The animal oxidizes much of its tissue reserves to provide metabolic water. This results in a loss of body weight.
3. Animals become sedentary, seeking shade whenever possible to reduce the loss of water from surface evaporation and sweating.
4. There is a reduction in feed consumption except for feeds that are high in moisture.

If adequate water supplies are not made available to the animal, the blood eventually thickens, resulting in a decreased ability to transport nutrients and waste products. Animals can lose most of their fat, about half of their protein, and many other constituents of their body tissues, but a loss of one-tenth of the water in the body is lethal.

Water Sources

Water can be obtained from three sources: (1) drinking water, (2) water contained in feed, and (3) metabolic water derived from the breakdown of carbohydrates, fats, and proteins.

DRINKING WATER

Many livestock producers tend to underestimate the value of water. It is not enough to make water available. Quality, as dictated by the levels of pollutants, is as important as availability. With an increase in management of livestock in confinement, it becomes increasingly necessary to consider the use of water from approved, inspected sources, since the water is also used by the people working with the animals as well as for the washing of utensils used in the operation. It is advisable for the livestock manager to check local codes and regulations on water quality.

The pollution of streams is of increasing concern to our government and is monitored by the Environmental Protection Agency (EPA). Because of their regulations, it may become necessary to alter some practices in livestock operations, relating to accessibility of water to animals and subsequent waste removal procedures.

A particularly troublesome problem of drinking water found on ranges is that of salinity. Dissolved salts can affect the intake of water; but animals can eventually become conditioned to the salinity if it is not excessive. Production can also be affected by the use of salty water. Most livestock species can tolerate a total dissolved solid concentration of 15,000 to 17,000 ppm; but at these levels, production will, in all likelihood, be dramatically affected.

Nitrate pollution has received special emphasis recently as part of the environmental protection program. Quantities as small as 100 to 200 ppm can be dangerous while 3,000 ppm is potentially lethal.

The amount of water required by the various types of livestock varies considerably, as evidenced in Table 4-2. Several factors can affect the amount of water a particular animal will consume: (1) age, (2) body weight, (3) production, (4) weather (heat and humidity), and (5) type of ration.

TABLE 4-2
DAILY WATER CONSUMPTION BY LIVESTOCK

Species	Age	Body Weight	Condition	Water Consumption[1]
	(weeks)	(lb)		(gal)
Cattle	4	112	Growing	1.3- 1.5
	8	152	Growing	1.6- 2.0
	12	204	Growing	2.3- 2.5
	16	263	Growing	3.1- 3.5
	20	327	Growing	4.0- 4.5
	26	416	Growing	4.5- 6.0
	60	779	Growing	6.0- 8.0
	84	1,023	Pregnant	8.0-10.0
	1-2 yr	1,000-1,200	Fattening	8.0- 9.0
	2-8 yr	1,200-1,600	Lactating	10.0-25.0
	2-8 yr	1,200-1,600	Grazing	4.5- 9.0
Sheep		20	Growing	0.5
		50	Growing	0.4
		150-200	Grazing	0.5-1.5
		150-200	Grazing, salty feeds	2.1
		150-200	Hay and grain	0.1-0.8
		150-200	Good pasture	less than 0.5
Swine		30	Growing	0.3-1.0
		60- 80	Growing	0.7-1.2
		80-125	Growing	1.0-2.0
		200-400	Maintenance	1.5-3.5
		200-400	Pregnant	4.0-5.0
		200-400	Lactating	5.0-6.5
Chickens (per 100 birds)	1- 3		Growing	0.5-2.0
	3- 6		Growing	1.5-3.0
	6-10		Growing	3.0-4.0
	9-13		Growing	4.0-5.0
	Mature		Nonlaying hens	5.0
	Mature		Laying	5.0-7.5
	Mature (90°F)		Laying	9.0
Turkeys (per 100 birds)	1- 3			1.1- 2.6
	4- 7			4.0- 8.5
	9-13			9.0-14.5
	15-19			16.7-17.0
	21-26			13.5-15.0
	Mature			17.0
Horses	Mature			12.0

[1] A gallon of water weighs 8.3 lb.

The intensity of production dramatically affects the water requirement. A steer fed a maintenance ration will consume approximately 35 lb of water daily, whereas a steer fed a fattening ration will consume double this quantity. A cow that is not in lactation will have a daily water intake of about 90 lb. When she produces 20 to 50 lb of milk, this figure will increase to about 160 lb. When milk production reaches 80 lb per day, water intake will be near 200 lb.

WATER IN FEEDS

The water content of feeds is extremely important, especially for animals which do not have ready access to drinking water. The term succulence refers to the property of feeds whereby the composition of the feed is relatively high in water compared to the amount of dry matter. When water is scarce, succulent feeds become extremely important for production. In the case of the dairy cow in heavy production, however, the use of succulent feeds is curtailed because the intake of these feeds may severely limit the intake of adequate amounts of dry matter needed for milk production.

Water on the surface of the plant, such as dew, can serve as an important source for some animals. On arid ranges, cattle, goats, and sheep may rely heavily on this type of water; but this supply is rarely sufficient to meet the needs of livestock.

METABOLIC WATER

Metabolic water is produced from the catabolism of nutrients. When 100 g of carbohydrates are oxidized, 60 g of water are produced. The oxidation of proteins yields 42 g of water for every 100 g of protein. Fats can be said to be "wetter than water." For every 100 g of fat that are oxidized, nearly 110 g of water are produced. However, there are some losses of water in the oxidation of both proteins and fats. Water must be used to excrete nitrogen in the deamination process of protein—thus lowering the net availability of water. In fact, it requires more water to excrete nitrogen as urea than is formed in the deamination process. The oxidation of fats requires increased respiration. Water is lost from the lungs during this increased respiration, and the net yield of water produced from fat is less than that from the oxidation of carbohydrates.

Water Excretion

Water can be eliminated from the body through three routes: (1) urine, (2) feces, and (3) insensible perspiration.

Urine provides a means whereby water-soluble products of metabolism can be excreted. These substances can be end products of nutrient catabolism or products of the various detoxification and protective mechanisms of the body. Generally, when rations are high in protein or mineral content, urine flow is increased.

The amount of water lost in the feces is highly dependent on the species. Fecal pellets of sheep contain much less water than the feces of cattle. Likewise, when production demands are high, as in lactation, water can be a limiting nutrient, thereby necessitating conservation efforts within the animal. When cattle are first placed on succulent, fresh pasture or green chop, the feces become very watery—a reflection of the increased water intake.

Insensible perspiration is the loss of water through the lungs and skin. Animals that do not have sweat glands rely on the cooling effect of panting. During the exhalation process of respiration, water vapor is lost from the lungs. Water is also lost through the skin as sweat in order to act as a temperature regulator.

QUESTIONS FOR STUDY AND DISCUSSION

1. Define and discuss each of the following terms: *nutrients, metabolism, catabolism,* and *anabolism.*

2. Why are cells referred to as functional units of nutrition and metabolism?

3. Discuss the impact of the research work of Wilkins, Watson, and Crick, for which they shared the Nobel Prize for Physiology and Medicine, in 1962, on nucleic acids (DNA and RNA), and in turn, on nutrition and metabolism.

4. Discuss the synthesis of DNA and RNA. Why is the structure of DNA called a double helix?

5. What is gene-splicing?

6. What is the present and potential impact of genetic engineering (recombinant DNA) on nutrition and metabolism?

7. Define biotechnology.

8. Has biotechnology given rise to a major scientific revolution? Justify your answer by citing some examples.

9. Discuss the role of each of the following nutrients in meeting the energy requirements of the body: carbohydrates, fats, proteins, minerals, vitamins, and water.

10. Discuss each of the following pertaining to energy for the body: (a) what energy is required for, (b) the symptoms of energy deficiency, and (c) the role of carbohydrates, fats, and proteins in providing energy.

11. Discuss the following aspects of carbohydrates: (a) what they are, (b) their importance in feeds, and (c) how they are produced by photosynthesis.

12. What feed components are included (a) in nitrogen free extract, and (b) in crude fiber?

13. Discuss the availability and utilization of fiber as related to (a) ruminants, (b) stage of plant maturity, and (c) classes and ages of animals.

14. How may fats and oils be differentiated? When digested, why do fats and oils liberate more heat than carbohydrates?

15. Discuss the importance of each of the following chemical characteristics as related to fatty acids: (a) saturation, (b) iodine number, (c) rancidity, (d) hydrogenation, (e) carbon chain length, (f) saponification, and (g) emulsification.

16. Name the essential fatty acids, and describe their deficiency symptoms and primary functions.

17. Discuss the metabolism of fats.

18. What's the Total Digestible Nutrients (TDN) system? What are its main (a) advantages, and (b) disadvantages?

19. What's the Calorie System? Diagram the utilization of energy, showing the losses along the way.

20. Compare the advantages and disadvantages of the Total Digestible Nutrient and Net Energy Systems of expressing energy requirements. Which one do you favor? Support your answer.

21. Discuss the utilization of urea and ammoniated feeds by ruminants.

22. Chemically, (a) what are proteins, and (b) how may proteins be classified?

23. What is meant by (a) essential and (b) nonessential amino acids? List the amino acids in each of these categories.

(Continued)

24. Discuss the synthesis of amino acids.

25. Define and discuss the biological value of protein.

26. Discuss the history and discovery of minerals.

27. Define: (a) minerals, (b) major or macrominerals, and (c) trace or microminerals. List the 18 required minerals in either of two groups—major or trace.

28. Why is the ash fraction of the proximate analysis considered to be only a crude measurement of the mineral content of feeds?

29. List the general functions of minerals.

30. Summarize the history and discovery of vitamins.

31. Define (a) vitamins, (b) the fat-soluble vitamins, (c) the water-soluble vitamins, and (d) the vitamin B complex. List the vitamins in either of 2 groups—the fat-soluble vitamins, and the water-soluble vitamins.

32. Formerly, vitamin nutrition was dependent upon incorporating into the ration feeds that were known to contain a high natural content of the needed vitamins. Today, vitamin deficiencies may be averted by using vitamin supplements. What caused this transition?

33. Chemically, what is water? What is the comparative importance of feed and water?

34. List the functions of water in animals.

35. How does water enter the body; and how does it leave the body?

36. If water is withheld from an animal, what compensatory mechanisms are initiated to provide enough water for body maintenance?

37. List the 3 sources through which animals may obtain water.

38. Explain how fats are "wetter than water."

39. What are the 3 ways through which water is eliminated from the body?

Protein Quantity And Quality Made The Difference!

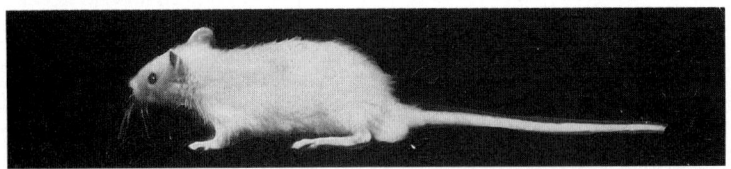

Fig. 4-16. The top rate ate too little of a good quality protein. The middle rat ate plenty of a poor quality protein. The bottom rate ate plenty of a good quality protein. (Adapted from USDA sources.)

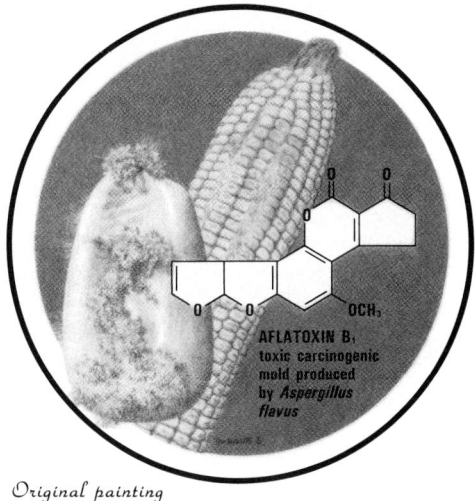

Original painting by Tom Phillips

NUTRITIONAL DISORDERS/TOXINS

Contents	Page
Nutritional Deficiencies and Imbalances	63
Energy	64
Protein	64
Minerals	65
Vitamins	65
Multiple Nutritional Deficiencies	65
Pica	65
Nutritional Diseases and Ailments	66
Nutrition and Disease/Parasite Interaction	90
Nutrition Related Disorders	91
Choking	91
Displaced Abomasum	91
Hardware Disease	92
Immunoglobulin Deficiency	92
Impaction In Horses	93
Poisonous Plants	93
Common Poisonous Plants	93
Treatment of Plant-Poisoned Animals	93
Preventing Losses From Poisonous Plants	94
Agricultural Chemicals and Drugs	94
Potential Poisons	94
Diagnosing and Treating Livestock Poisoning	109
National Animal Poison Control Center	109
Hair Analysis	110
Good Livestock Require Good Soils	110
Soil Testing	111
Organic Farming	111
Questions for Study and Discussion	112

Nutritional disorders may be caused by nutritional deficiencies and imbalances, by diseases and parasites, or by nutrition-related problems. Toxicity may be caused by poisonous plants, or by a host of agricultural chemicals and drugs.

NUTRITIONAL DEFICIENCIES AND IMBALANCES

Nutritional deficiency diseases may be brought about by (1) too little feed, (2) rations that are too low in one or more nutrients, (3) imbalance of nutrients, or (4) presence of antinutritional factors, such as antimetabolites or antivitamins.

Forced production (such as very high milk yields and marketing animals at early ages) and the feeding of forages and grains which may be produced on leached and depleted soils have created many problems in nutrition. This condition has been further aggravated through the increased confinement of stock, many animals being confined to corrals or buildings all or a large part of the year so that they have little or no opportunity to select their own diet. Under these unnatural conditions, nutritional diseases and ailments, along with abnormal behavior, have become increasingly common.

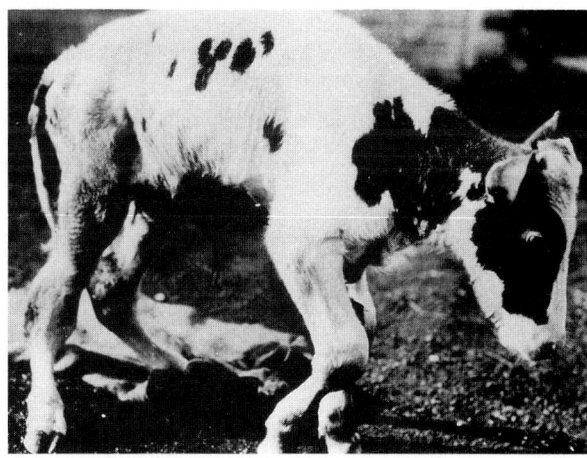

Fig. 5-1. Calf with severe rickets. Note the emaciation, humping of back, swelling of joints, knuckling of pasterns, and bowing of legs. Rickets may be caused by a lack of calcium, phosphorus, or vitamin D, or by an incorrect ratio of the two minerals. (Courtesy, Michigan State University, East Lansing)

Animals have rather narrow tolerances between what they need and what will cause problems. Whether nutrients are needed in small or large amounts, most of them have an upper tolerance level beyond which excesses will prevent the animals from performing normally, or may even cause death. The range between that which is necessary and that which can be tolerated varies among species and nutrients. Selenium, for example, which is needed in very minute quantities by most species is quite toxic when fed above a certain level. Water, on the other hand, which is needed by all animals, can cause problems if consumed in excess, but the range between what is needed and what will be troublesome is very great.

Although the cause, prevention, and treatment of most nutritional diseases and ailments are known, they continue to reduce profits in the livestock industry simply because the available knowledge is not put into practice. Moreover, those widespread nutritional deficiencies which are not of sufficient proportions to produce clear-cut deficiency symptoms cause even greater economic losses because they go unnoticed and unrectified. It is important, therefore, that livestock producers and those who counsel with them be able to recognize subacute nutritional deficiencies, as well as the more obvious symptoms of acute deficiencies or toxicities.

It is often difficult to isolate deficiencies or toxicities due to the deficiency of a single nutrient. It is one thing to control animals experimentally and to isolate a single factor, and quite another to distinguish it under field conditions where there may be many complicating factors. When an animal goes off feed for whatever reason, several nutrient deficiencies may enter into the symptoms that appear. What caused the animal to go off feed, rather than the fact that it is off feed, is the important thing in returning it to normal. This is why it is so important that caretakers detect the early symptoms of deficiencies or toxicities, thereby making it possible for them to rectify the causes.

Energy

Many animals throughout the world are underfed all or some part of the year. Thus, lack of sufficient energy is probably the most common deficiency suffered by animals, although it is frequently complicated by a concomitant shortage of protein and other nutrients.

Fig. 5-2. Lack of energy!

Restricted rations often occur during periods of drought, when pastures and ranges are overstocked, or when winter rations are skimpy. Fortunately, during such times of restricted energy intake, animals may have nutritive reserves —mainly body fat—upon which they may draw. Although they may survive for a considerable period of time under these conditions, there is an inevitable loss in body weight; and, varying with the degree of energy shortage, there may be slowing or cessation of growth, failure to conceive, and increased mortality. Low energy intake also commonly results in increased deaths from toxic plants and from lowered resistance to diseases and parasites.

During times of energy shortage when animals are withdrawing stored energy—mostly from body fat—the mobilized fat may not be completely metabolized, with the result that ketosis (incompletely metabolized fatty acids) develops (see Table 5-1, p. 79). Many high-producing animals are mildly ketotic, because it is nearly impossible for them to consume sufficient energy during maximum production. Mild cases of ketosis may not affect production or health. But animals with severe ketosis go off feed, which further aggravates the malady by lowering energy intake still more.

During cold weather, some energy—perhaps as much as 20% of the maintenance requirements—must be converted to heat to keep the body warm. Tables of nutrient requirements do not usually make provision for this added energy. The amount of added energy needed is proportional to the decrease in temperature below the comfort zone. It is further increased by humidity and wind. The combination of temperature, humidity, and wind is commonly referred to as the *chill factor*.

(Also see Chapter 4, section on "Energy," for more complete information relative to energy deficiencies and utilization.)

Protein

The protein allowance for animals, regardless of age or system of production, should be ample to replace the daily breakdown of the tissues of the body, including muscles, blood, hair, and hooves. Protein needs are increased by growth and gestation-lactation.

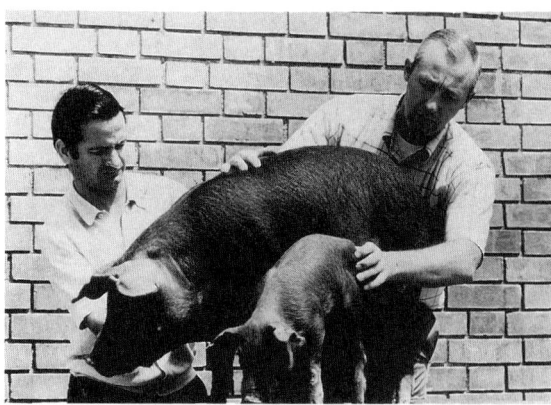

Fig. 5-3. Lysine made the difference! These littermate pigs were started on test at weaning, at a weight of 20 lb. The only difference in their ration was the kind of corn. The big pig (left) received high-lysine (opaque-2) corn; the little pig (right) got regular corn. During the 130-day trial, the pig fed opaque-2 gained a respectable 73.2 lb, whereas the pig eating ordinary corn gained only 6.6 lb. (Courtesy, The Rockefeller Foundation, New York, N.Y.)

Depressed appetite is the primary symptom of protein deficiency in livestock rations. Going off feed may, in turn, lead to an inadequate intake of energy; hence, protein deficiency and energy deficiency often occur together.

Other symptoms of protein deficiency are loss of weight, poor growth, irregular or delayed estrus, and reduced milk production.

Because of synthesis of essential amino acids by microorganisms, the quality of protein (or balance of essential amino acids) is of less importance in feeding ruminants than in feeding nonruminant animals. Thus, the latter must not only have sufficient amounts of available protein, but the protein must contain a balance of essential amino acids. When one or more of the essential amino acids is not available, the remaining amino acids cannot be used effectively, and may be converted to energy.

(Also see Chapter 4, Section on "Proteins," for more complete information relative to protein deficiencies and quality.)

Minerals

Animal bodies contain small amounts—only 2 to 5%—of inorganic elements, called minerals. But these constituents play a vital role in animal nutrition. They furnish structural materials for bones and teeth. Additionally, as constituents of the soft tissues, the blood, the fluids of the body, and certain of the secretions, they regulate many of the vital processes.

Fig. 5-4. Ewe showing goiter (big neck) due to a deficiency of one mineral—iodine. (Courtesy, Montana State University, Bozeman)

Fig. 5-5. Vitamin A made the difference! Left: Vitamin A-deficient chick, showing poor growth, watery eyes, unsteady gait, and ruffled appearance. Right: Chick that received the same basal ration, plus plenty of vitamin A. (Courtesy, University of Maryland, College Park)

Although acute mineral deficiency diseases and actual death losses are relatively rare, inadequate supplies of any one of the 18 essential mineral elements may result in lack of thrift, poor gains, inefficient feed utilization, lowered reproduction, and decreased production of meat, milk, eggs, wool, or work. Only when the mineral deficiency reaches such proportions that it results in excess emaciation, reproductive failure, or death is it likely to be detected.

With minerals, it is also very important that consideration be given to both nutrient relationships and availability.

(Also see Chapter 4, section on "Minerals," for more complete information relative to mineral deficiencies, relationships, and availabilities.)

Vitamins

Vitamin deficiencies in animals may occur as a result of lack of availability of vitamins, or because of the presence of antinutritional factors. Corn is an example of the latter concept. Analyses show corn to be adequate in niacin. Yet, due either to an antimetabolite or unavailability, there may be a niacin deficiency when corn is fed—a deficiency which can be rectified by niacin supplementation.

The fat-soluble vitamins (A, D, E, and K) are stored in appreciable quantities in the body, whereas the water-soluble vitamins are not. Thus, vitamin A and/or carotene may be stored by an animal in its liver and fatty tissue in sufficient quantities to meet its requirements for a period of 6 months or longer. By contrast, the large amounts of water which pass through most animals daily tend to carry out the water-soluble vitamins of the body, thereby depleting the supply. Thus, they must be supplied in the ration on a day-to-day basis for those animals having a simple stomach in which microbial synthesis is limited.

(Also see Chapter 4, section on "Vitamins," for more complete information relative to vitamin deficiencies and storage.)

Multiple Nutritional Deficiencies

Multiple nutritional deficiencies are altogether too common, making diagnosis difficult even for the trained observer. Among the nonspecific indicators of multiple nutritional deficiencies are the following: loss of appetite; depraved appetite; unthrifty appearance; failure to grow; rough coat; rough, scaly skin; lowered production of meat, milk, eggs, wool, or work; impaired reproduction (or egg hatchability); weakness; lack of coordination; depression; difficult breathing; increased pulse rate; and a host of others.

Pica

Pica, a perverted or depraved appetite—a craving for substances not ordinarily considered feed—may indicate a nutritional deficiency. Animal species differ in their preference for foreign material. Cattle have been known to ingest

Fig. 5-6. Bone chewing by cattle is a common sign of phosphorus deficiency. (From *Tex. Sta. Bull.* 344, courtesy of The Fertilizer Institute, Washington, D.C.)

cloth, leather, pieces of metal, wood, stone, and carcass material such as bone and hide. Horses may chew bones or eat dirt or sand. Sheep may eat dirt, wool, and bones.

Classically, pica is associated with a phosphorus deficiency. However, lambs may eat dirt to soothe abdominal pain caused by enterotoxemia (overeating disease). In all species, pica may occur as a result of a deficiency of protein, fiber, minerals, and/or vitamins. In addition, boredom, especially in stabled animals, may lead to excessive licking and chewing.

If pica is of nutritional origin, the deficiency or imbalance should be identified through the aid of blood analyses, and corrected.

NUTRITIONAL DISEASES AND AILMENTS

Nutritional diseases, which afflict both animals and humans, provide a dramatic and vivid way in which to relate the story of undernutrition and malnutrition.

It is noteworthy that nutritional deficiency areas throughout the world generally affect all species within the area, and that the manifestations of most deficiency diseases are similar regardless of species. For example, soils of an iodine-deficient area will produce iodine-deficient crops and result in many big-necked (goiterous) animals and people.

Table 5–1 lists the most common nutritional diseases and ailments of animals and summarizes pertinent information relative to each of them.

TABLE 5–1
NUTRITIONAL DISEASES AND AILMENTS

ACETONEMIA
(See KETOSIS.)

ACIDOSIS (LACTIC ACIDOSIS)
A metabolic disease of cattle and sheep.

Species Affected:

Cattle, especially feedlot cattle; and sheep, especially feedlot lambs.

Cause:

Acidosis is caused by an increase in lactic acid-producing bacteria and the rapid production of lactic acid (both the d- and l-forms). It commonly occurs when there is a sudden shift from a high-roughage to a high-concentrate ration. However, cattle maintained on high-energy rations may be in a marginal state of acidosis due to the formation of lactic acid by the rumen flora. Thus, ingredient changes, poor mixing of grain in the ration, or faulty feeding can produce acute acidosis.

Symptoms and Signs (or age group most affected):

Marginal acidosis is characterized by poor performance and inconsistent feed ingestion. If ingredient changes or erratic feeding persist, acute acidosis may result, creating laminitis—and eventually "ski shoe" cattle (founder). In severe cases, the rumen becomes immobilized, followed by increased pulse and respiration rate, variable rectal temperature, sunken eyes, loss of dermal elasticity (dehydration), staggering, coma, and death.

Distribution and Losses Caused By:

Acidosis occurs wherever beef or dairy cattle and lambs are fed, especially when consuming high-concentrate rations.
The annual loss from acidosis has been estimated at about 1% of the production.

Treatment:

Different treatments have been used with varying degrees of success; among them: (1) removing rumen contents and replacement by contents of an animal on a normal ration; (2) feeding a high level of an antibiotic to suppress lactic acid-producing bacteria; (3) drenching (or intravenous injection) with a solution of sodium bicarbonate to restore the acid-base balance; (4) administering intramuscularly antihistamines and cortical steroids daily for each of several days to help prevent intoxication and laminitis; and/or (5) backing the cattle down on both amount and kind of feed (lessening the total amount of the ration, and returning to a higher forage mix).

Control:

Acidosis is best controlled by (1) avoiding accidental access of cattle to large amounts of concentrates, (2) changing gradually and stepwise from a low to a high proportion of concentrate in the ration, and (3) adding buffer salts, such as sodium bicarbonate, to the ration.

Prevention:

Prevention consists of starting animals on a high-roughage ration and gradually reducing the roughage and increasing the grain; avoiding erratic feeding; and avoiding abrupt ration changes.

Remarks:

A feedlot history of deliberate or accidental starting of animals on high-energy feeds, or of sudden ration changes, helps establish the correct diagnosis.
A rapid field test can be used to diagnose and differentiate between rumen acidosis and urea poisoning. Samples can be collected by stomach tube or postmortem collection. In general, a rumen content pH of 5.0 or less is indicative of rumen acidosis; a rumen content pH greater than 7.5 is indicative of urea or NPN toxicosis.

(Continued)

Nutritional Disorders/Toxins

TABLE 5-1 *(Continued)*

ALKALI DISEASE
(See Table 5-3, POTENTIAL POISONS, SELENIUM TOXICITY.)

ANEMIA, NUTRITIONAL

Species Affected:

All warm-blooded animals, including humans.

Cause:

Commonly an iron deficiency, but it may be caused by a deficiency of copper, cobalt, and/or certain vitamins.
The baby pig is born with a total of about 40 mg of iron in the body. With an iron requirement of about 7 mg daily, it is apparent that without supplemental iron, body stores will not last very long.
Sow's milk is a good source of all nutrients the baby pig is known to require with the exception of iron.

Symptoms and Signs (or age group most affected):

Loss of appetite, progressive emaciation, and death.
Most prevalent in suckling young.
Pigs show listlessness, rough hair coat, wrinkled skins, drooping ears and tails, pale membranes around the mouth and eyes, labored breathing, and a swollen condition about the head and shoulders.

Distribution and Losses Caused By:

Worldwide.
Losses consist of slow and inefficient gains, and deaths.

Treatment:

Provide sources of the nutrient or nutrients, the deficiency of which is known to cause the condition. If iron deficiency is indicated, iron may be given by injection, in organic combination (iron dextran).

Fig. 5-7. Anemia, caused by an iron deficiency, characterized by listlessness, rough hair coat, and wrinkled skin. (Courtesy, University of Florida, Gainesville)

Control:

When nutritional anemia is encountered, it can usually be brought under control by supplying dietary sources of the deficient nutrient(s).

Prevention:

Supply dietary sources of iron, copper, cobalt, and certain vitamins (especially folacin, riboflavin, and vitamin B-6).
Keep confinement of suckling animals to a minimum and provide dry feeds at an early age.
Anemia in pigs can be prevented by providing supplemental iron in one of the following forms:
1. Inject intramuscularly 100 to 200 mg of iron from iron dextran into baby pigs at 2 to 3 days of age. If pigs remain in confinement and do not have access to creep feed at an early age, a second injection at 2 to 3 weeks of age is desirable. Injection is the method of choice, for it assures that every pig receives its requirement.
2. Orally administer iron dextran in a liquid or a solid preparation. To ensure daily intake by all pigs, it is important to have a preparation that is palatable and readily consumed. Also, placement of the oral preparation at the right location in the creep area is most important.
3. Give the pigs iron tablets or paste at 2 to 3 days of age. Repeat the treatment every 7 to 10 days until the pigs are eating the creep ration adequately. If pills are given, it is important to see that the pigs swallow them and not spit them out.
4. Place clean soil in the farrowing pen daily. Soil should not be contaminated with parasite eggs and other disease organisms. Iron sulfate can be sprinkled over the soil.
5. Swab sow's udder daily with a solution of 1 lb ferrous sulfate dissolved in 1 gal of warm water.
6. Provide pigs with access to a creep feed by the time they are 10 days old.

Remarks:

Anemia is a condition in which the blood is either deficient in quality or quantity. (A deficient quality refers to a deficiency in hemoglobin and/or red cells.)
Levels of iron in dry feed are generally believed to be ample, since feeds contain 40 to 400 mg/lb.

(Continued)

TABLE 5-1 (Continued)

AZOTURIA (HEMOGLOBINURIA, MONDAY MORNING DISEASE, BLACKWATER)

Species Affected:

Horses.

Cause:

Sudden exercise, following a day or two of rest during which time the horse has been on full feed, resulting in partial spasm or "tie-up." Azoturia is caused by an abnormal amount of glycogen being stored in the muscle. As the glycogen breaks down, lactic acid is formed, which builds up in the muscle causing severe muscle destruction and the release of myoglobin which manifests itself as partial spasm or "tie-up" and wine-colored urine.

Symptoms and Signs (or age group most affected):

Symptoms usually develop 15 to 60 minutes after the beginning of exercise. Azoturia is characterized by profuse sweating, elevated temperature and pulse, wine-colored urine (caused by the release of myoglobin—the red pigment in muscle tissue), tight (cramping) and sore loin hindquarter muscles—they're "tied up" due to semi-paralysis, stiff gait, reluctance to move due to pain, and knuckling over of the hind pasterns. Finally, the animal may assume a sitting position and, eventually, fall prostrate on its side. The breath and urine may have a peculiar odor.

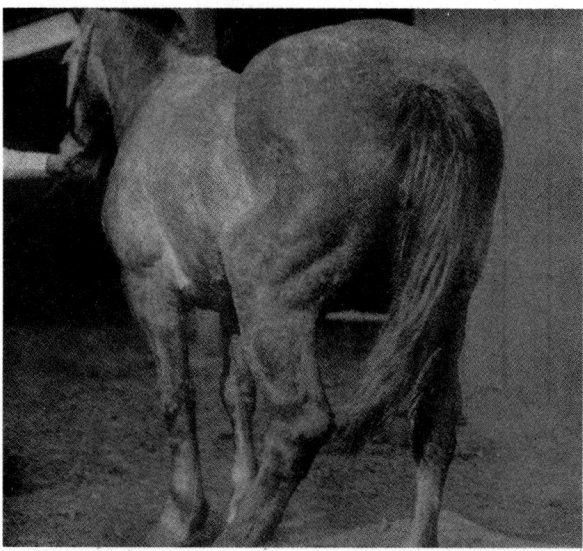

Fig. 5-8. Horse with Azoturia, evidencing sore loin and hindquarters—it is "tied up." (Courtesy, Pitman-Moore, Indianapolis, Ind.)

Distribution and Losses Caused By:

Worldwide, but the disease is seldom seen in horses at pasture and rarely in horses at constant work.

Treatment:

Absolute rest and quiet. While awaiting the veterinarian, apply heated cloths or blankets, or hot-water bottles to the swollen and hardened muscles, *but don't try to move the horse—don't take the horse back to the barn.* Keep it on its feet if possible, even if you have to use a sling.

The veterinarian should determine treatment. In mild cases, treatment may consist of the use of a tranquilizer or a sedative. In severe cases, the veterinarian may use (1) muscle relaxers or (2) sodium bicarbonate in solution to readjust the acid balance in the muscles.

Control:

When trouble is encountered, decrease the concentrate ration and increase the exercise on idle days.

Prevention:

Restrict the grain ration, increase good quality roughage, and provide daily exercise when the animal is idle. Give a wet bran mash the evening before an idle day or turn the idle horses to pasture.

Some believe that a diuretic (a drug which will increase the flow of urine) will prevent the tie-up syndrome. This is a common treatment of racehorses.

Others feel that increased B vitamins will prevent the lactic acid buildup.

Remarks:

The chances of recovery are good for horses that remain standing, are not forced to move after the signs are noticed, and whose pulse returns to normal within 24 hours. Azoturia and colic have some similar symptoms; hence, there is danger of misdiagnosis and the wrong treatment. Walking, a standard part of colic treatment, is the worst thing to do when a horse has azoturia.

BABY PIG SHAKES
(See HYPOGLYCEMIA.)

(Continued)

Nutritional Disorders/Toxins

TABLE 5-1 *(Continued)*

BLOAT—FEEDLOT

Species Affected:

All ruminants.

Cause:

Bloat is an excessive accumulation of gas in the rumen and reticulum of ruminants. High-concentrate rations, especially when finely ground, increase numbers of slime-producing bacteria in rumen. Slime traps fermentation gas and produces bloat. Both frothy and free gas bloat occur in feedlot bloat.
Genetic tendency or physiological abnormality.

Symptoms and Signs (or age group most affected):

Symptoms same as pasture bloat (see "Bloat—Pasture" which follows).
Occurs when cattle or sheep have been fed high-concentrate, low-roughage rations for approximately 60 days or longer.

Distribution and Losses Caused By:

A survey of Kansas feedlots showed the following losses from bloat: 0.1% died of bloat; 0.2% bloated severely; and 0.6% bloated mildly to moderately, with animal performance affected adversely.

Treatment:

Reduce intraruminal pressure as quickly as possible. This may be done by means of a large stomach tube, although this method is usually disappointing in foamy bloat.
Administer a defoaming agent immediately, such as (1) 1 pint of corn oil, peanut oil, or soybean oil; or (2) poloxalene administered according to the manufacturer's directions.
As a last resort, a trocar and cannula can be inserted on the left side of the animal at the center of a triangle formed by the backbone, the hipbone, and the last rib. The trocar is removed, but the cannula should stay in place until all gases have dissipated. If a trocar and cannula are not available, a knife may be used in emergencies.

Control:

If feasible, increase proportion of nonlegume roughage in ration. However, good-quality legume hay may increase incidence of feedlot bloat. In this case, poloxalene or oxytetracycline are effective preventives when used according to manufacturers' directions.

Prevention:

(1) Use poloxalene (Bloat Guard) or oxytetracycline (Terramycin and Neo-Terramycin) according to manufacturers' directions; and (2) proper management.

Remarks:

Feedlot bloat may occur during any month of the year; however, it is more common during hot, humid weather.
Two products are cleared by FDA for bloat control; namely, poloxalene (trade name, Bloat Guard) and oxytetracycline (trade names, Terramycin and Neo-Terramycin).

BLOAT—PASTURE (LEGUME BLOAT)

Species Affected:

All ruminants.

Cause:

Bloat is caused by the inability of the animal to get rid of ruminal gas. Lack of scabrous (rough) material in the rumen to stimulate eructation (belching), along with the formation of heavy foam bubbles, seems to be the main cause of pasture bloat. Pasture bloat is most common on immature, rapidly growing legumes and on wheat pasture. Pasture bloat is a frothy bloat caused by interaction of several factors—plant, animal, and microbial. Soluble plant proteins and the presence of saponins play a prominent role in permitting stable foam formation.
Heavy applications of urea fertilizer on pastures may also induce bloat.
Animals that will bloat on any feed are known as chronic bloaters. These animals, in which there may be a genetic tendency, are unable to eructate (belch) fermentation gases because of some physiological abnormality.

(Continued)

TABLE 5-1 *(Continued)*

BLOAT—PASTURE (LEGUME BLOAT) (Continued)

Symptoms and Signs (or age group most affected):

First observed as a distention of the paunch on the left side in front of the hipbone. This is followed by distention of the right side, protrusion of the anus, respiratory distress, cyanosis (bluish coloration) of the tongue, struggling, and death if not treated. The entire period of time from when a ruminant enters a pasture until death occurs can be as short as a half hour.

Distribution and Losses Caused By:

Widespread, although some areas appear to have more bloat than others.

It often results in death.

Bloat causes annual losses in beef and dairy cattle of more than $100 million from reduced weight gain and lower milk production.

Treatment:

Time permitting, severe cases of bloat should be treated by a veterinarian. The use of a stomach tube, carefully inserted, is usually very helpful in eliminating gases. Puncturing of the paunch with a trocar and cannula should be a last resort. A knife may be used in emergencies.

Mild cases may be home treated by (1) keeping the animal on its feet and moving; and (2) drenching cattle either with (a) 1 pint of corn oil, or soybean oil; or (b) 1-2 oz poloxalene.

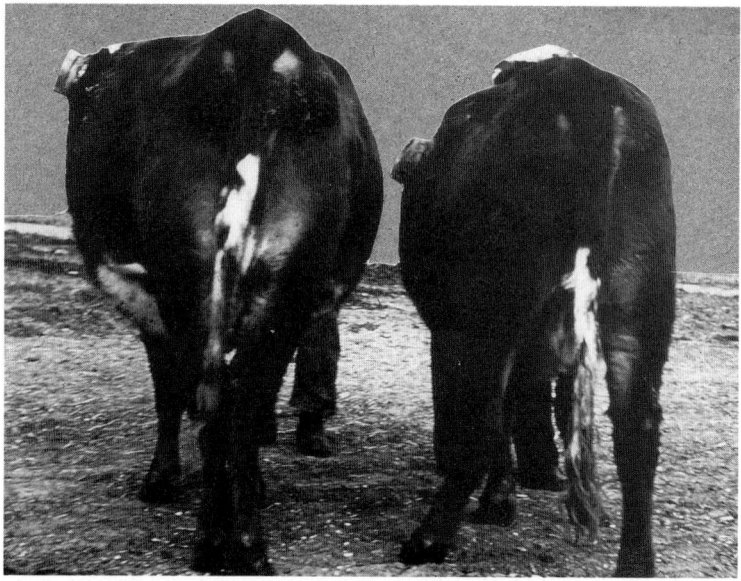

Fig. 5-9. Identical twins. *Left:* Bloated animal showing distention of the paunch on the left side in front of the hipbone. *Right:* Twin mate showing no bloating. (Courtesy, Kansas State University, Manhattan)

Control:

When there is a high incidence of bloat, it may be desirable to change the feed.

Where legume bloat is encountered, use poloxalene (Bloat Guard), oxytetracycline (antibiotic), or polyoxyethylene (23) lauryl ether (Laureth-23/Enproal Bloat Blox), according to the respective manufacturers' directions.

Prevention:

The incidence is lessened by (1) avoiding straight legume pastures and immature legumes, (2) feeding a coarse grass hay prior to turning onto lush pasture, (3) feeding dry forage along with pasture, (4) avoiding a rapid fill from an empty start, (5) keeping animals continuously on pasture after they are once turned out, (6) keeping salt and water conveniently accessible at all times, (7) avoiding frosted pastures, or (8) using poloxalene (Bloat Guard), oxytetracycline (Terramycin or Neo-Terramycin), or Laureth-23 (Enproal Bloat Blox) according to manufacturers' directions, including placing blocks containing these antifoaming agents in various parts of the pasture.

Oils and fats have been used successfully to control bloat in Australia and New Zealand.

Remarks:

Legume or cereal pastures, or alfalfa hay, appear to be associated with a higher incidence of bloat than any other feeds.

Legume pastures are particularly hazardous when immature, when moist, after a light rain or dew.

COBALT DEFICIENCY
(See SALT SICK.)

(Continued)

Nutritional Disorders/Toxins

TABLE 5-1 (Continued)

COLIC

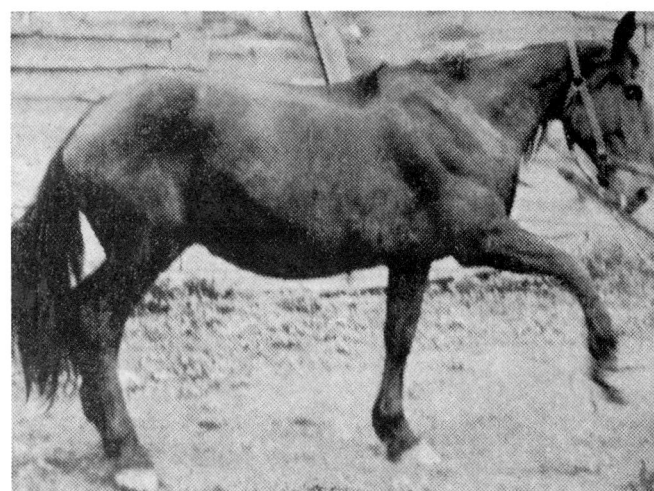

Fig. 5-10. Horse with colic, evidencing severe pain, but being taken for a slow walk. (Courtesy, Pitman-Moore, Indianapolis, Ind.)

Species Affected:

Horses.

Cause:

Internal parasites are the number one cause of colic; additional causes are improper feeding, working, or watering. There are more than 70 different things that can cause colic.

Symptoms and Signs (or age group most affected):

Severe pain, usually in the abdomen; and depending on the type of colic, other symptoms are: the horse looking at his belly, distended abdomen, increased intestinal rumbling, violent rolling and pawing, profuse sweating, constipation, and refusal of feed and water.

Distribution and Losses Caused By:

Worldwide.
Colic is the most common ailment among horses and is the leading cause of death. Livestock insurance companies report about 1/3 of all deaths of insured horses can be attributed to colic.

Treatment:

Call a veterinarian. To avoid danger of inflicting self-injury, (1) place the animal in a large, well-bedded stable, or (2) take the animal for a slow walk. Do not give the horse any type of drug, unless so advised by the veterinarian when telephoned. Painkillers may cover up symptoms which are vital for the veterinarian in making an accurate diagnosis.

Depending on the diagnosis, the veterinarian may use one or more of the following: sedatives; laxatives, such as mineral oil; drugs; or surgery. The surgeon may avoid recurrence of twists and displacements of the horse's colon by attaching it to other organs or the abdominal wall, thereby deliberately creating adhesions which prevent further twisting.

In the late 1980s, Colorado State University scientists developed a "scorecard" method of evaluating colic (1) to help differentiate between a colicking horse that needs surgery and one that should be treated medically, and (2) to predict how likely a horse in each category is to survive.

Control:

Follow a good management program, including parasite control.
Feed, work, and water horses properly.

Prevention:

Proper feeding (including adequate roughage), working, watering, and parasite control.

Remarks:

The word colic is not specific. There are many syndromes that can result in colic, and not all of them are gastrointestinal. For example, blood worms can cause colic due to damaging the walls of blood vessels, and mares with uterine torsions exhibit colic pain as do horses with urinary stones in the bladder. So, the first thing is to determine what is causing the animal to exhibit colic symptoms.

ENTEROTOXEMIA (OVEREATING DISEASE, PULPY-KIDNEY DISEASE)

Species Affected:

Sheep; less frequently goats, and rarely cattle.

Cause:

Clostridium perfringens type D. However, predisposing factors are essential; the most common of these are overconsumption of high-energy feeds, an abundant milk supply, and lush pastures. Under such conditions, the *Clostridium perfringens* bacteria grow rapidly and produce a powerful toxin.

Symptoms and Signs (or age group most affected):

Sudden death; frequently a lamb is found dead in the field or feedlot without having shown any previous signs of illness. Quite often it is the biggest lamb with the biggest appetite. The disease develops rapidly; and the animal becomes weaker and weaker and shows nervous disturbances such as circling, butting, or throwing the head from side to side or backwards. Finally, the animal collapses and may go into convulsions before dying. Enterotoxemia can be confirmed by laboratory tests if necropsy is performed shortly after death.

(Continued)

TABLE 5-1 (Continued)

ENTEROTOXEMIA (OVEREATING DISEASE, PULPY-KIDNEY DISEASE) (Continued)

Distribution and Losses Caused By:

Worldwide.

The death rate is a minimum of 1%, with an average of 3 to 4% in unvaccinated feedlot lambs. In explosive outbreaks, losses range from 10 to 40%.

Treatment:

None.

Control:

The method of control depends on the age of the lambs, the frequency with which the disease occurs, and the method of husbandry.

When an outbreak in feeder lambs occurs, for several days (1) increase the amount of roughage in the ration, and (2) add 200 g of chlortetracycline per ton of feed.

When an outbreak in nursing lambs occurs, injection of all susceptible lambs with enterotoxemia antiserum will provide protection for about 14 to 21 days, at which time the lambs can be vaccinated.

Prevention:

Along with proper feeding, vaccinate. Ewes should be vaccinated with type C and D toxoid. Lambs should be vaccinated with type D only.

Nursing lambs can be protected by vaccinating the ewes for enterotoxemia. For best results, previously unvaccinated ewes should be vaccinated with type C and D toxoid twice before lambing. The 2 doses should be spaced at least 1 month apart, with the second dose given 2 to 4 weeks before start of lambing. Ewes which have been vaccinated previously need only one booster shot before lambing. Vaccinating the ewes prior to lambing ensures that the lambs will receive colostral protection for 2 to 3 weeks, following which the lambs should be vaccinated at 4 to 6 weeks of age.

Feedlot lambs can be protected by giving them one dose of type D toxoid soon after their arrival in the feedlot. It takes about 10 days after vaccination for immunity to develop. Sometimes revaccination with the toxoid or bacterin (a booster shot) is required 2 to 4 weeks following the first vaccination.

Remarks:

The disease is caused by bacteria, but it is triggered by high-energy rations or excellent pastures.

FESCUE FOOT (FESCUE TOXICOSIS)

Species Affected:

Cattle, sheep, and horses.

Cause:

A fungus (endophyte), *Acremonium coenophialum*, which lives in the leaves, stems, and seeds of tall fescue, without adversely affecting the fescue plant.

Symptoms and Signs (or age group most affected):

The symptoms vary. Some animals show no apparent lameness, whereas others show varying degrees of sloughing (necrosis) on the ends of their tails. Mild fescue toxicosis is characterized by poor conception rates, low pasture gains, and depressed milk production.

The most common symptoms in horses, in the order of their occurrence, are a decrease or absence of milk production, prolonged gestation, abortion, and thickened placenta.

Fescue toxicity is more common in animals suffering from malnutrition and/or parasitism.

Distribution and Losses Caused By:

Fescue foot has occurred in the U.S., Australia, New Zealand, and Italy. In the U.S., it has been reported in California, Colorado, Florida, Kentucky, Missouri, and Tennessee.

Tall fescue is currently grown on about 35 million acres in the U.S.

The Mississippi Station reports that studies extending over several years show that each 10% fungus infection in a fescue pasture will lower the daily gains of cattle by about 10%.

University of Kentucky researchers report that dairy cows grazing 70% infected fescue produced an average of 11.2 lb less milk per day than cows grazing noninfected fescue.

Treatment:

There is no effective medication. Cattle usually recover if removed from fescue pasture or fescue hay.

Control:

Until, and unless, scientists find a way to remove the toxic factor(s) from fescue, the best control consists of good management, proper nutrition of animals, early detection of symptoms, and/or destroying toxic pastures and reseeding with endophyte-free seed.

(Continued)

Nutritional Disorders/Toxins

TABLE 5-1 *(Continued)*

FESCUE FOOT (FESCUE TOXICOSIS) *(Continued)*

Prevention:

The seeding of fungus-free fescue seed is the best way to prevent fescue foot. Also, interseeding alfalfa, clovers, or other grasses into fescue stands dilutes the amount of fescue eaten and helps to reduce the toxic effects.

Also, tall fescue selections low in, or free of, this endophytic fungus are evolving.

In areas where fescue toxicity is a problem, gestating mares should be removed from fescue pasture the last 2 or 3 months of pregnancy.

Remarks:

Most cases of fescue toxicity occur among cattle that graze pure stands of fescue during late fall and winter; and most toxic stands of fescue pasture are several years old.

FOUNDER (LAMINITIS)

Species Affected:

Horses, cattle, sheep, goats.

Cause:

A variety of causes have been recognized, including (1) overeating and too rapid increase in the ration (grain founder), (2) digestive disturbances (enterotoxemia), (3) retained afterbirth (foal founder), (4) lush pastures (grass founder), and (5) concussion (road founder).

Symptoms and Signs (or age group most affected):

Extreme pain, fever (103°–106°F), and reluctance to move—the animal appears to be "walking on eggs." If neglected, it causes an acute or chronic degeneration of the joining of the sensitive and insensitive laminae of the foot; and, if the degeneration is severe, the coffin bone may rotate and come through the bottom of the foot.

Distribution and Losses Caused By:

Worldwide.

Actual death losses from founder are not very frequent, but animal usefulness may be severely affected.

Treatment:

There is no widely accepted, standard method of treating founder. If known, the condition(s) that caused the problem should be alleviated. Treatment of acute horse founder usually involves one or more of the following procedures and medications:

1. *Mineral oil.* To aid passage of the excessive feed consumed and prevent further absorption of lactic acid and endotoxin into the bloodstream. One gallon should be given via stomach tube. **NOTE WELL**: A purgative should not be employed in cases involving pneumonia or parturient laminitis of mares.
2. *Analgesics (pain-killers).* To obtain pain relief.
3. *Injectable antihistamines.* To provide anti-inflammatory effects.
4. *Antibiotics.* To combat the subsequent formation of endotoxin which destroys tissue.
5. *Sodium bicarbonate (baking soda).* To neutralize the acidic toxicity of these inflammatory products.
6. *Temporarily deadening the nerve supply to the feet.* To alleviate pain and assist in restoring blood supply by allowing the horse to be walked.
7. *Water soaks.* To stimulate and massage the blood supply to the feet in an effort to open up previously constricted blood vessels, but there is disagreement as to whether the water should be warm or cold.
8. *Wraps applied to the affected feet.* To cushion the painful sole soreness which is evident in the toe region due to coffin bone rotation following breakdown of the laminae.

Other treatments for founder that are sometimes used, with varying degrees of reported success, are cortisones; and methionine, a sulfur containing amino acid.

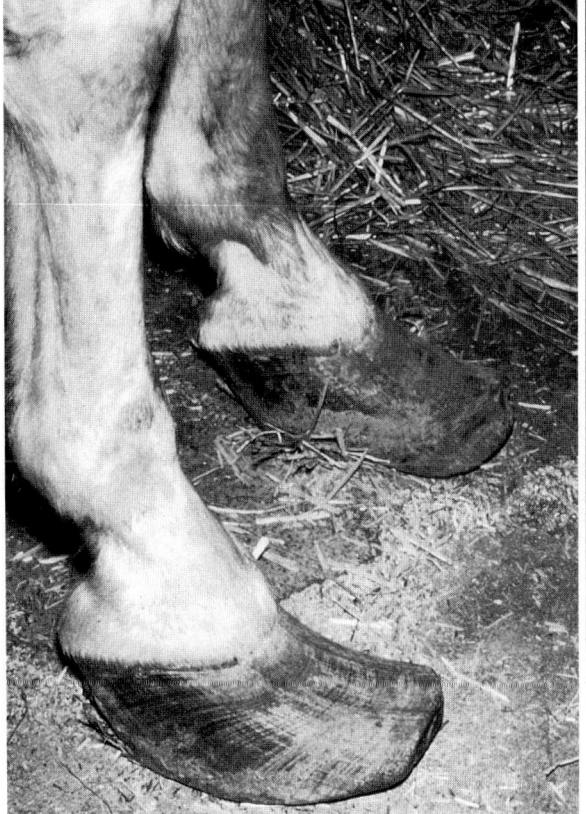

Fig. 5–11. "Snowshoe feet" of a horse due to chronic founder. (Courtesy, Colorado State University, Ft. Collins)

(Continued)

TABLE 5-1 *(Continued)*

FOUNDER (LAMINITIS) (Continued)

NOTE WELL: Due to the complexity of laminitis and its far-reaching effect on many of the horse's internal systems, the treatments described above may not be effective in controlling founder. If the disease is not diagnosed and treated early enough, the coffin bone rotation and damage to the associated hoof wall structure can become irreversible.

Treatment of chronic founder consists of attempting to restore the normal alignment of the rotated coffin bone by lowering the heels, removing excess toe, and protecting the dropped sole. This may be accomplished by a competent farrier through proper trimming and perhaps by using leather pads or a steel-plate shoe. Also, soft acrylic plastics are sometimes applied to the sole to replace the injured hoof area and help realign the rotated coffin bone. Because of the tendency to refounder, weight control is extremely important and overfeeding should be avoided.

Control:

Alleviate the causes.

Prevention:

Avoid (1) overeating, (2) overdrinking (especially when hot), and/or (3) inflammation of the uterus following parturition.
Veterinary attention should be given if mares retain the after-birth longer than 12 hours.
Careful management practices related to grain feeding will prevent many cases of founder in cattle, sheep, and goats.

Remarks:

Unless foundered animals are quite valuable, it is usually desirable to dispose of them following a case of severe founder.
Swine do not founder because they can unload their stomachs by vomiting.

GOITER
(See IODINE DEFICIENCY.)

GRASS TETANY (HYPOMAGNESMIC TETANY, GRASS STAGGERS, WINTER TETANY. In Europe, it's called FOG FEVER.)

Species Affected:

Cattle (beef and dairy; in the U.S., tetany is more common in beef herds than in dairy herds). Sometimes sheep and goats, in which the disease occurs under essentially the same conditions and has the same clinical signs as in cattle.

Cause:

Grass tetany is a nutritional disease caused by an inadequate level of magnesium (Mg) in the blood. It most commonly occurs among lactating animals grazing rapidly growing, lush spring pastures containing less than 0.2% magnesium and more than 3% potassium and 4% nitrogen (25% protein). Forage that is high in potassium and nitrogen should have a magnesium content of at least 0.25%. Such low magnesium pastures are most commonly encountered during the first two weeks of the pasture season, although somewhat later in the season outbreaks have been reported during rainy and foggy weather. Sometimes tetany is a problem when cattle are allowed to overgraze a field, then moved abruptly to a field of new lush growth. Small grain pastures (wheat/rye/oats/barley) are especially troublesome. Also, the disease may occur when animals are fed poor quality hay, straw, or corn stover—feeds that are low in magnesium. It is not common on legume pasture or in animals wintered on legume hay. (Legumes may contain twice the magnesium concentration of grasses grown on the same soil.)

Several factors adversely influence magnesium metabolism in cattle and may "trigger" grass tetany; among them, drastic fluctuations in spring temperatures, prolonged cloudy weather, organic acid content of plants, hormonal status of the animal, level of higher fatty acids in plants, energy intake of the animal, and additional stress—such as a dog chasing animals, parasites, or a cold rain.

Grass tetany is most likely to occur on pasture plants grown on soils that are low in available magnesium and high in available potassium. If calcium is low as well as magnesium, the hazard of tetany is even greater. Many state soil-testing laboratories provide information on the danger of tetany on pastures, and can recommend corrective fertilization or dolomitic liming (which contains magnesium). Also, the historical record of grass tetany in an area or on a specific pasture is important.

Symptoms and Signs (or age group most affected):

The initial signs of magnesium deficiency include nervousness, attentive ears, and decreased milk yield; signs which, to the experienced and observing caretaker, indicate the need for immediate preventive measures—before the animals become sick tomorrow. In more severe cases, affected animals may avoid the rest of the herd, walk with a stiff gait, lose their appetite, and urinate frequently. They are nervous, have staring eyes, and keep their head and ears in an erect position. Also, they stagger; have a twitching skin, especially on the face, ears, and flanks; and lie down and get up frequently. Animals may be irritable and behave aggressively; they may even charge or fight persons in the immediate area. After a time (as long as 2 to 3 days), extreme excitement and violent convulsions may develop. Animals lie flat on their sides, the fore legs pedal periodically, saliva flows freely, breathing is labored, and the heart pounds rapidly. If treatment is not given at this stage, animals usually die during or after a convulsion. The various symptoms of animals suffering from grass tetany indicate that the nervous system controlling both voluntary and involuntary muscles is affected. Quite often, clinical signs are not observed, and the only evidence is a dead animal.

Chronic grass tetany is generally slow to develop and muscular affection may be limited to twitching, a clumsy walk or exaggerated motions, but convulsions may occur if animals are driven or handled roughly.

(Continued)

Nutritional Disorders/Toxins

TABLE 5-1 *(Continued)*

GRASS TETANY (HYPOMAGNESMIC TETANY, GRASS STAGGERS, WINTER TETANY. In Europe, it's called *FOG FEVER*.) (Continued)

Older cows are more susceptible to grass tetany than those with their first or second calves, because of lowered magnesium stores and decreased absorption efficiency. Also, the disease is most likely to strike beef cows during early lactation, especially those with high levels of milk production. Dry cows and bulls are seldom affected.

Normal plasma magnesium levels range from 1.8 to 2.0 mg/100 ml; values below 1.0 to 1.2 mg/100 ml are indicative of magnesium deficiency. However, not all cattle with low plasma magnesium develop tetany. Also, plasma magnesium levels in affected animals may return to almost normal during the convulsive stage. So, diagnosis can not be based on blood tests alone. Since the kidneys apparently start conserving magnesium when the serum level reaches about 1.8 mg/100 ml, one of the better diagnostic aids to indicate grass tetany is low urinary magnesium.

Grass tetany should not be confused with nitrate toxicity or calcium deficiency. In nitrate toxicity, the blood is brown and there is a grayish to brownish discoloration of white areas on the skin and on nonpigmented mucous membranes of the mouth, nose, eyes, and vulva. In calcium deficiency, animals may be sluggish rather than nervous as they are when they have magnesium deficiency. Also, animals may have a calcium and magnesium deficiency at the same time, thus masking the signs of magnesium deficiency; this happens in wheat pasture poisoning, in which the animals may be deficient in both calcium and magnesium.

Distribution and Losses Caused By:

Grass tetany is a worldwide problem, with occurrence sporadic and unpredictable for any given area. It is generally considered to be the leading cause of cattle deaths in the U.S., killing an estimated 1 to 3% of the cattle in temperate regions.

Treatment:

Treatment of tetany cases can be successful if given early and without excessive handling of the affected animals. Chance of recovery is slight if treatment is delayed 8–12 hours; so, call the veterinarian immediately.

Under range conditions, 200 cubic centimeters (cc) of a sterile, saturated solution of magnesium sulfate (Epsom Salts) injected under the animal's skin (inject only 50 cc at any one place on the animal) places a high level of magnesium in the blood in 15 minutes.

Some veterinarians use intravenous injections of chloral hydrate or magnesium sulfate to calm excited animals, then follow with a calcium-magnesium gluconate solution. If the animal again goes into convulsions, a second dose of calcium-magnesium gluconate solution may be required. Intravenous injections should be administered slowly (allow about 15 minutes for a 500 cc bottle) by a trained person because there is a danger of heart failure if they are given too rapidly.

An enema of 60 g (*2 oz*) of magnesium chloride ($MgCl_2 \cdot 6H_2O$) in 10 oz of water is helpful. The enema may be given with an esophageal or oral calf feeder with the probe inserted 10 in. into the anus. Magnesium is absorbed through the walls of the large intestine and the lower bowel.

Oral administration of magnesium to sick animals, in place of intravenous injections or enemas, has not been effective because too much time is required for the magnesium to reach that part of the GI tract where it can be absorbed.

Herd treatment of the animals that are not down may involve adding magnesium sulfate (Epsom Salts) or magnesium acetate or chloride to the drinking water. Some diarrhea may occur, but this is not reason for concern. To be effective, the treated tanks should be the only source of drinking water. **NOTE WELL**: Production will be lowered by this treatment due to lowered consumption of water.

Follow-up treatment may involve removing all animals from the tetany-producing pasture and feeding alfalfa hay (plus concentrates if necessary). Additionally, each animal should consume 30 g of magnesium daily for 1 to 2 weeks, preferably through a highly palatable supplement; force-feeding should be resorted to if necessary.

Cattle that get tetany are likely to get it again later in the season or in later years; they are usually the high producers.

NOTE WELL: "Downer cows" should be turned daily—and more frequently if possible. (Also see Downer Cow Syndrome.)

• **Toxicity**—Magnesium toxicity does not occur in cattle fed normal rations. Supplemental levels of magnesium of 170 to 350 g have resulted in deleterious effects. Maximum tolerable levels have been established as 0.4% of the ration by the National Research Council. Feeding toxic levels has resulted in anorexia (loss of appetite), reduced performance, and occasional diarrhea. Also, cattle experiencing toxicity may exhibit lack of reflexes and respiration depression.

Control:

Commonly used feedstuffs vary widely in magnesium concentration and availability. The magnesium content of most cereal grains runs between 0.12 and 0.18%. Protein supplements of animal origin are low in magnesium, while those of plant origin usually contain 0.3 to 0.6%. Fat is not as beneficial as carbohydrate as a source of energy under tetany conditions, as fat tends to tie up calcium and magnesium in the digestive tract, rendering them less available to the animal. The magnesium content of forages varies greatly; normally, the legumes contain more than the grasses. Magnesium availability increases with plant maturity. Magnesium fertilization usually increases plant magnesium content. The inclusion of energy or protein supplements in high-magnesium mineral supplements will help to overcome palatability problems.

Prevention:

Prevention of grass tetany is always preferred to treatment. Prevention consists of providing magnesium daily throughout the high-risk period, because very little of it is stored in the body. **NOTE WELL**: Crash feeding programs begun after tetany appears in a herd are usually not adequate to stop the disease. A magnesium supplement should be started 30 days before grass tetany is usually observed in the area in order to get the animals accustomed to it. Since magnesium oxide or sulfate are not very palatable, cattle may not consume sufficient of them.

Meeting the magnesium requirements of beef cows calls for providing 10 g of magnesium daily for the dry cows, and 20 to 25 g daily for cows suckling calves. For dairy cows, 30 g of magnesium per day is recommended. For calves, 4 to 8 g per day is needed, depending on their ages.

Lactating ewes and does, just after parturition, which is the most tetany-susceptible period, should receive about 3 g of magnesium per day.

High levels of aluminum, potassium, phosphorus, or calcium decrease the efficiency of magnesium absorption and/or utilization; so, in areas where the levels of these elements are high, the magnesium allowance should be increased to overcome their antagonistic effect.

Normally, animals on pasture during the summer and early fall months receive an adequate supply of magnesium from the grasses on which they feed. However, during the late fall, winter, and spring months, many pastures are magnesium deficient. To prevent grass tetany during these months, cattle, sheep, and goats on pasture should receive a magnesium-rich feed in addition to pasture and/or have ready access to a magnesium mineral supplement.

(Continued)

TABLE 5-1 (Continued)

GRASS TETANY (HYPOMAGNESMIC TETANY, GRASS STAGGERS, WINTER TETANY. In Europe, it's called *FOG FEVER.*) (Continued)

One of the following high-magnesium feeds is commonly used:

1. Alfalfa, or other legume hay; 20 lb of average alfalfa hay will provide 30 g of magnesium. Additionally, the legume hay provides increased energy.
2. For self-feeding (with adequate available water and forage containing 10% protein), 65% ground grain (corn, barley, or grain sorghum), 20% magnesium oxide, and 15% iodized salt. Since early green grass is high in protein, a supplement containing cereal grain is preferred to a supplement containing an oilseed protein. When consumed at the rate of ½ lb/head/day, such a mix will provide 27 g of magnesium daily.
3. For self-feeding (with adequate available water and low-protein forage), 65% cottonseed meal or soybean meal and 35% magnesium oxide. When consumed at the rate of ⅓ lb/head/day, such a mix will provide 31 g of magnesium daily.
4. For hand-feeding of supplements, use (2) or (3) above, but omit the salt in (2). With the salt omitted, each ½ lb of mix No. 2 will provide 32 g of magnesium daily.
5. A liquid molasses supplement fortified with 4% magnesium sulfate[1] (80 lb of magnesium sulfate per ton of liquid molasses). When consumed at the rate of 2 lb/head/day, this will provide 7.2 g of magnesium.

Several good sources of inorganic magnesium may be used to supplement cattle; among them, those listed in the table that follows:

Magnesium Content of Various Magnesium Salts

Name	% Mg	Pounds of Mineral Required to Supply the Same Amount of Mg as 30 lb of MgO
Magnesium oxide (Magnesia)(available in light and heavy grades; heavy is more stable and easier to mix)	60.32	30.0
Magnesium hydroxide	41.69	43.4
Magnesium carbonate (Magnesite)	28.8	62.8
Magnesium carbonate hydroxide	27.0	67.0
Magnesium sulfate (Epsom Salts)	20.2	89.6
Potassium magnesium sulfate (Langbelinite)	11.6	156.0
Magnesium acetate (Cromosan)	11.34	159.6

Magnesium from dolomitic limestone is less readily available to cattle, sheep, and goats than some other salts. Any of the following mineral supplements are satisfactory:

1. A mineral mix made by mixing ⅓ each magnesium oxide, iodized salt, and either soybean oil meal or cottonseed meal. This mix should be made available as the only mineral. Each ⅓ lb of this mix will provide 30 g of magnesium.
2. A mineral mix made by mixing 30% magnesium oxide, 30% iodized salt, 30% bone meal, and 10% dried molasses. Each ⅓ lb of this mix will provide 27 g of magnesium.
3. A mineral mix made by mixing ⅔ (66⅔%) magnesium oxide and ⅓ (33⅓%) salt as the only source of salt is effective. Each ⅙ lb of this mix will provide 30 g of magnesium.
4. A commercial high-magnesium supplement, in blocks or mineral-salt mixtures which usually contain molasses, grain, and/or some other material to make them more palatable to animals. Generally, these are formulated to be fed at the rate of ½ to 1½ lb/head/day, and to provide 10 to 15 g of magnesium daily.

NOTE WELL: Blocks or mineral mixes usually give the best results if no additional salt is provided. Since the desire for salt varies among animals and with seasons, a high-salt mixture may not provide the required level of magnesium consumption. It should be emphasized that cattle must consume adequate magnesium on a regular basis. When using a supplement, one should pay particular attention to (1) the percentage of magnesium it contains and (2) the daily intake of the supplement. From these two factors the daily intake of magnesium can be determined and compared to the animal's requirement. The intake should be checked frequently as magnesium salts are generally unpalatable. The intake may be increased by adding grain or cottonseed meal/soybean meal.

On farms and ranches with a history of grass tetany, free-choice feeding of a mineral mix to insure a daily intake of 25 g of magnesium, with about half the daily intake coming from the natural magnesium in the pasture or other feed and half from the magnesium-containing mineral supplement, will usually provide protection from the development of grass tetany.

In high-risk situations—such as cows near calving grazing lush spring grass, highly fertilized with nitrogen or potassium, or both—the total magnesium requirement should be provided in the supplement. The reasons: (1) cows near calving are approaching lactation, when the magnesium demands are the highest; (2) nitrogen and potassium are antagonistic to magnesium; and (3) magnesium availability in early spring grass is low—besides, cattle cannot consume enough such grass to meet their energy requirements. A free-choice or hand-fed grain or protein-mineral supplement, providing 35 g of magnesium daily, will usually prevent the occurrence of grass tetany in such high-risk situations. If the magnesium supplement is self-fed, it should be located for easy access to the cattle, especially near water and shade where cattle tend to congregate and loiter. If the magnesium supplement is hand-fed, it is important that all animals have access to feeder space.

Where grass tetany is particularly troublesome, as an additional preventive measure, consideration should be given to (1) applying magnesium fertilizer and dolomitic limestone, or (2) dusting the pastures with magnesium oxide (MgO). Also, generally speaking, such pastures may be grazed without hazard by steers. But, before fertilizing or dusting, or shifting from a cow-calf program to a steer program, the counsel and advice of the local Farm Advisor/County Extension Agent should be sought.

Remarks:

Affected animals may be aggressive on getting up. So, watch out!

---

[1]Although magnesium sulfate contains only 20% Mg, it is commonly used as a magnesium additive to a molasses supplement because of its solubility. Magnesium oxide will not remain dispersed in liquid feed without the aid of suspending agents.

(Continued)

Nutritional Disorders/Toxins 77

TABLE 5-1 *(Continued)*

HEAVES

Species Affected:

Horses, mules.

Cause:

Exact cause unknown, but it is known that the condition is often associated with (1) the feeding of damaged, dusty, or moldy hay; and/or (2) the use of dusty bedding or paddocks.
It often follows severe respiratory infection such as strangles.
Probably an allergy.

Symptoms and Signs (or age group most affected):

Difficulty in forcing air out of the lungs resulting in a jerking of flanks (double flank action) and coughing. The nostrils are often slightly dilated, and there is a nasal discharge.
Heaves in horses is similar to emphysema in people.

Distribution and Losses Caused By:

Worldwide.
Losses are negligible.

Treatment:

Antihistamine granules can be administered in feed to control coughing due to lung congestion.

Control:

Affected animals are less bothered if turned to pasture, if used only at light work, if fed an all-pelleted ration, or if the hay is sprinkled lightly with water.

Prevention:

Avoid the use of damaged feeds.
Feed an all-pelleted ration, thereby alleviating dust.

Remarks:

Basically, heaves is a rupture of some of the alveoli in the lungs, the specific cause of which is unknown.

HYPOGLYCEMIA (BABY PIG SHAKES)

Species Affected:

Swine.

Cause:

Low blood-sugar level is characteristic of the trouble, but the cause of the low blood sugar is unknown.
Predisposition to piglet hypoglycemia occurs from any disease of the sow which decreases or inhibits milk production or let down.

Symptoms and Signs (or age group most affected):

Shivering, weakness, failure to nurse, with no evidence of scouring. If disturbed, the pigs emit a weak, crying squeal. Hair becomes erect and rough, and the heart action slow and feeble. Without treatment, death usually occurs in 24 to 36 hours after the first symptoms appear.
Confined to baby pigs only.

Fig. 5–12. Hypoglycemia in pig. Lack of milk during the first few days of life accompanies this condition; chilling speeds the process. Note the erect hair coat. (Courtesy, College of Veterinary Medicine, University of Illinois, Champaign-Urbana)

(Continued)

TABLE 5-1 (Continued)

HYPOGLYCEMIA (BABY PIG SHAKES) (Continued)

Distribution and Losses Caused By:

Worldwide.
Hypoglycemia accounts for 15 to 25% of total piglet mortality.

Treatment:

Provide heat lamps for pigs.
At earliest symptoms either (1) force feed at frequent intervals a mixture of 1 part corn syrup diluted with 2 parts of water, or (2) give intraperitoneal injections of 5% sterile glucose solution every 4-6 hours. Oxytocin may be administered to the sow to promote milk let down.
Consult the veterinarian.

Control:

Apparently not contagious.

Prevention:

Adequate rations and good care and management of the gestating sows may lessen the incidence of the disease.
Be sure there is adequate milk for baby pigs during first days of life.

Remarks:

One of the hazards of hypoglycemia is that the milk flow of the sow will not be stimulated or may even cease due to the inactivity of the affected pigs. In the latter case, the pigs may have to be either transferred to a foster mother or hand-fed.

IODINE DEFICIENCY (GOITER, BIG NECK)

Species Affected:

All farm animals and humans.

Cause:

A failure of the body to obtain sufficient iodine from which the thyroid gland can form thyroxin (an iodine-containing compound).

Symptoms and Signs (or age group most affected):

Goiter (big neck, which is a swelling under the chin) is the most characteristic symptom of iodine deficiency in calves, lambs, kids, and humans. Also, there may be reproductive failure and weak offspring that fail to survive. Pigs may be born hairless and show edema of the shoulders and neck. Foals may be born weak.

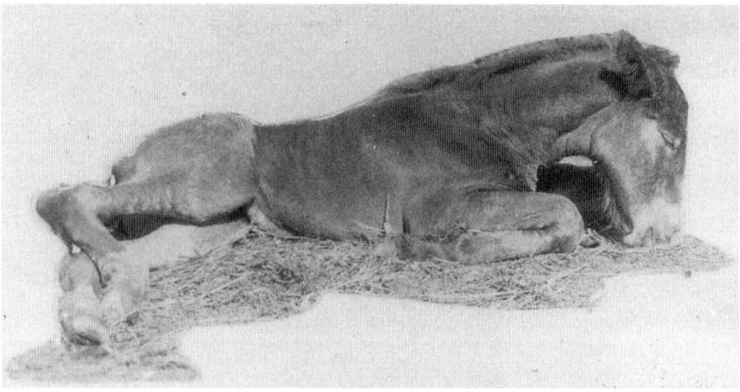

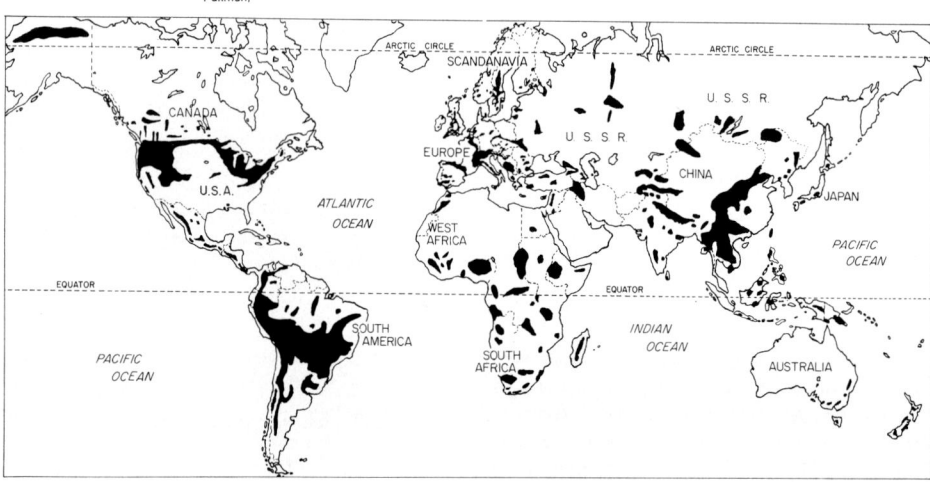

Fig. 5-13. Weak newborn foal due to iodine deficiency of the mare during pregnancy. (Courtesy, Washington State University, Pullman)

Distribution and Losses Caused By:

Iodine deficiencies occur worldwide; wherever feeds are grown on iodine-poor soil containing insufficient iodine to meet animal needs. The highest incidence has been observed in the Alps, the Pyrenees, the Himalayas, the Thames Valley of England, certain regions of New Zealand, a number of Central and South American countries, and the Great Lakes and Pacific Northwest regions of the U.S. Fig. 5-14 shows the goiter areas of the world.

Fig. 5-14. Goiter areas of the world. (Map prepared by the authors on the basis of information from the World Health Organization, Geneva, Switzerland)

(Continued)

Nutritional Disorders/Toxins

TABLE 5-1 *(Continued)*

IODINE DEFICIENCY (GOITER, BIG NECK) (Continued)

Treatment:

Occasionally borderline cases may survive; in these the moderate thyroid enlargement disappears in a few weeks.
Once the iodine deficiency symptoms appear, no treatment is very effective.

Control:

At the first signs of iodine deficiency, stabilized iodized salt should be fed to all farm animals.

Prevention:

In iodine-deficient areas, feed stabilized iodized salt containing 0.01% potassium iodide to all farm animals throughout the year.
Organic forms of iodide are also suitable sources of iodine, but they are usually more costly.

Remarks:

The enlarged thyroid gland (goiter) is nature's way of attempting to make sufficient thyroid hormone, thyroxin, under conditions when an iodine deficiency exists.
Mares fed excess iodine (48 mg or more) during late gestation will produce foals with hyperplastic goiter. Some mares will also develop goiter.

KETOSIS (ALSO KNOWN AS ACETONEMIA IN CATTLE AND PREGNANCY DISEASE IN SHEEP)

Species Affected:

Cattle, sheep, goats.

Cause:

A metabolic disorder of nutritional origin, characterized by hypoglycemia (low blood sugar). If the increased nutrient requirements are not met by more feed during the high-demand periods (in cows, 1–6 weeks after calving; in ewes, 2 weeks before lambing), the animal must draw on body fat reserves. If this is done too rapidly, and without adequate carbohydrates in the ration, ketosis follows.

Symptoms and Signs (or age group most affected):

In cows, ketosis or acetonemia is usually observed 2 to 6 weeks after calving. Affected animals show loss of appetite and condition, a marked decline in milk production, and the production of a peculiar sweetish chloroform-like odor of acetone that may be present in the milk and urine and pervade the barn. A positive diagnosis can be made by testing the milk or urine for the presence of ketones.

In ewes and goats, ketosis or pregnancy disease generally strikes during the last 2 weeks of pregnancy. Usually, affected ewes are carrying twins or triplets. Symptoms include going off feed suddenly, grinding of teeth, dullness, weakness, frequent urination, trembling when exercised, and blindness—with the final stage being complete collapse, followed by death in 90% of the cases. In dairy goats, lactation ketosis, which is similar to the ketosis of dairy cows, may be observed in high milk producers following kidding.

Fig. 5-15. Ewe with ketosis, or pregnancy disease. (Courtesy, College of Veterinary Medicine, University of Illinois, Champaign-Urbana)

Distribution and Losses Caused By:

Worldwide.
Ketosis or acetonemia affects cattle throughout the U.S.
Ketosis or pregnancy disease in sheep affects farm flocks more than range bands, the losses in the former sometimes being as high as 25%.

Treatment:

Cattle: ½–1 lb of either propylene glycol or sodium propionate daily, with the dose divided into 2 treatments for 5–10 days. Put treatment in grain if cow is eating; otherwise, give as drench.
Intravenous injection of glucose solution and glucocorticoids (to increase blood sugar levels temporarily) as well as the oral administration of propylene glycol. Numerous other treatments are sometimes used.
Sheep and goats before parturition: 3 to 4 oz of propylene glycol, given orally 3 times daily. Cesarean section early in the course of the disease usually leads to recovery and, if near term, the offspring may be saved.
Dairy goats after kidding: 6 to 8 oz of propylene glycol, given orally twice daily. Severe cases may be aided by intravenous injections of 50% dextrose solution. Cortocosteroid injection may be used in conjunction with either propylene glycol or dextrose solution in does that have kidded.

(Continued)

TABLE 5-1 *(Continued)*

KETOSIS (ALSO KNOWN AS ACETONEMIA IN CATTLE AND PREGNANCY DISEASE IN SHEEP) (Continued)

Control:

Cows: Maintain relatively high-energy intake before calving; increase energy intake substantially after calving.
Ewes: Avoid obesity in early pregnancy. Feed grains rather liberally the last 6 weeks of pregnancy.

Prevention:

Cows: The incidence of ketosis can be lessened by (1) avoiding excessively fat cows at calving; (2) increasing the level of concentrates gradually after calving; (3) feeding good-quality hay in preference to high-silage rations after calving, and avoiding abrupt changes in roughage; (4) feeding adequate proteins, minerals, and vitamins, and (5) providing comfort, exercise, and ventilation. In problem herds, feeding ¼ lb daily of propylene glycol or sodium propionate may be helpful.
Sheep and goats: Feed more hay and ½ to 1 lb of grain beginning a month before parturition. Good management is important, too, including exercise, freedom from parasites, and avoiding stress.

Remarks:

The clinical findings are similar in the case of affected cattle and sheep, but it usually strikes ewes just before lambing, whereas cows are usually affected within the first 2–6 weeks after calving.

LIVER ABSCESSES

Species Affected:

Cattle, especially feedlot cattle.

Cause:

Fusobacterium necrophorum, the same bacteria that causes foot rot, appears to be the major organism responsible for liver abscesses.
High concentrate (low roughage) finishing rations predispose cattle to a high incidence of liver abscesses.

Symptoms and Signs (or age group most affected):

Liver abscesses generally go undetected until cattle are slaughtered, at which time they appear as "walled-off" areas filled with pus. However, reduced feed intake and gain near the end of the feeding period may be indicative.

Distribution and Losses Caused By:

Liver abscesses occur in all countries where intensive beef production is practiced and are common in the Corn Belt and western U.S. and in western Canada.
The USDA records show that 12% of the beef livers condemned in the U.S. are due to abscesses.
The incidence of liver abscesses is highest in feedlot cattle, where an estimated 18 to 20% of the livers are affected. For a particular lot of feedlot cattle, liver abscesses may range from 1 to 90%.
Since the liver of a 1,000-lb steer weighs approximately 11 lb, liver condemnation represents a considerable monetary loss, but the loss from reduced feed efficiency and gains may be even greater.

Treatment:

Treatment of an acute liver abscess in feedlot cattle should be left to the veterinarian, who may administer (1) sulfapyridine, or (2) an antibiotic.

Control:

The low level feeding of certain antibiotics during the finishing period will markedly reduce the number of liver abscesses.

Prevention:

The incidence of liver abscesses in feedlot cattle can be reduced, but not entirely eliminated, by (1) changing from high-roughage to high-concentrate rations gradually, and (2) feeding an antibiotic (commonly chlortetracycline or tylosin) at the daily rate of 70 mg/animal.

Remarks:

Liver abscesses, as indicated by the name, are single or multiple abscesses of the liver observed at slaughter. Usually a liver abscess consists of a central mass of necrotic liver surrounded by pus and a wall of connective tissue. At slaughter, livers affected with abscesses are condemned for human food.

(Continued)

Nutritional Disorders/Toxins

TABLE 5-1 (Continued)

MILK FEVER (PARTURIENT PARESIS, HYPOCALCEMIA)

Species Affected:

Cattle, sheep, goats.

Milk fever, a metabolic disease, is similar in cows, ewes, and does. So, the discussion that follows pertaining to cows also applies to sheep and goats, with treatment and prevention adjusted for size of animal.

Cause:

Low blood calcium concentration. The name *milk fever* is a misnomer, because the animal does not have a fever.

Initiation of lactation places a severe strain on the calcium balance of the cow due to the amount of calcium secreted in the milk. All cows are slightly hypocalcemic at the time of calving, but some become so hypocalcemic that clinical signs of milk fever develop.

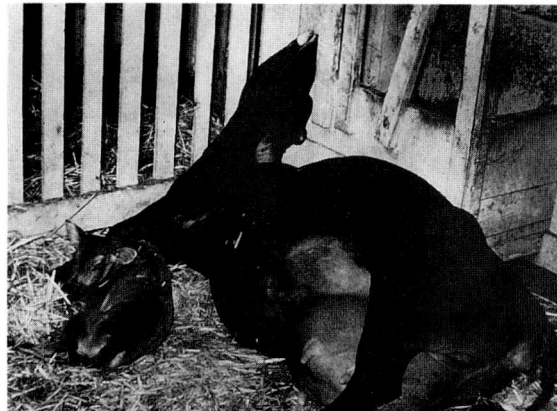

Fig. 5-16. Milk fever in Jersey cow, showing characteristic position of head—turned back over shoulder. (Courtesy, Washington State University, Pullman)

Symptoms and Signs (or age group most affected):

Commonly occurs within 3 days after calving and in high-producing cows. Rarely occurs at first calving. First symptoms are loss of appetite, constipation, and general depression. This is followed by nervousness and finally collapse and complete loss of consciousness. The head is usually turned back toward the flank.

The incidence of the disease increases with the age of the cow and is highest in Guernsey and Jersey breeds.

Distribution and Losses Caused By:

A common widespread disease of dairy cows. It is estimated that more than 8% of all dairy cows are stricken by milk fever; occasionally, with up to 80% affected in a single herd.

Losses are not great, although untreated animals are likely to die.

Causes estimated average annual losses in dairy cattle (including milk) of $100 million.

Treatment:

Milk fever should be regarded as an emergency and the affected animal treated as soon as possible.

The standard treatment is intravenous infusion of calcium borogluconate as soon as the first signs appear. *CAUTION:* Overdoses of calcium salts can result in acute heart damage; so, dose level should be carefully calculated, and intravenous administration should be performed slowly.

Cows that are already down will usually stand up within 1 to 2 hours following treatment.

Control:

(See Prevention.)

Prevention:

Each of the following measures will lessen the incidence of milk fever:

1. *Low calcium during the dry period.* Feeding low calcium (less than 100 g/day)—high phosphorus (more than 40 g/day) rations during the dry period is important. High calcium levels in dry cow rations aggravate the problem. So, feeding a low calcium ration (less than 0.1 lb/day) before calving has shown promise of preventing milk fever.

2. *Calcium shock treatment.* Feed low calcium, high phosphorus rations containing only 15 to 20 g of calcium per day for 2 weeks prior to the expected calving date, rather than a restricted, but somewhat higher, calcium intake during the entire dry period. This creates a mild calcium deficiency which stimulates production of the biologically active form of vitamin D in the animal's body. In turn, this form of vitamin D stimulates the bone and gut to supply more calcium and phosphorus. As a result, when the greater demand for calcium and phosphorus occurs at calving, the bone and gut are already activated and are able to meet the increased demands for calcium. Thus, milk fever is avoided.

3. *Calcium-phosphorus ratio and amounts.* Balancing cow rations to contain 0.5% calcium and 0.25% phosphorus on a dry matter basis will limit the incidence of milk fever.

4. *High vitamin D.* Feeding massive doses of 20 million I.U. of vitamin D/cow/day starting about 5 days before calving and continuing through the first day postpartum, with a maximum dosage period of 7 days, has been effective in controlling milk fever. However, difficulty in predicting calving dates accurately has reduced the effectivemess of this treatment under practical conditions.

5. *Avoid excessive fatness.* Excessive fatness, or any other conditions that reduce feed intake at calving, tends to cause more milk fever.

Remarks:

The name *milk fever* is a misnomer, because the disease is not accompanied by fever, the temperature really being below normal.

(Continued)

TABLE 5–1 *(Continued)*

OSTEOMALACIA
(An adult form of rickets)

Species Affected:

All species.

Cause:

Inadequate phosphorus; sometimes inadequate calcium; incorrect calcium:phosphorus ratio; or lack of vitamin D in confined animals.

Symptoms and Signs (or age group most affected):

Phosphorus deficiency symptoms are depraved appetite (gnawing on bones, wood, or other objects, or eating dirt); and lack of appetite, stiffness of joints, failure to breed regularly, decreased milk production, and an emaciated appearance.
Calcium deficiency symptoms are fragile bones, reproductive failures, and lowered lactations.
Mature animals most affected. Most acute cases occur during pregnancy and lactation.

Distribution and Losses Caused By:

Southwestern U.S. is classed as a phosphorus-deficient area, whereas calcium-deficient areas have been reported in parts of Fla., La., Nebr., Va., and W. Va.

Treatment:

Select natural feeds that contain sufficient quantities of calcium and phosphorus.
Feed a special mineral supplement or supplements.
If this disease is far advanced, treatment will not be successful.

Control:

(See Treatment.)

Prevention:

Feed balanced rations, and, if necessary, allow animals free access to a suitable phosphorus and calcium supplement.
Increase the calcium and phosphorus content of feed by fertilizing the soils.

Remarks:

Calcium deficiencies are much less frequent than phosphorus deficiencies in cattle, sheep, and horses.
Calcium deficiencies are fairly common in swine because grains, which are their chief feed, are low in this mineral.

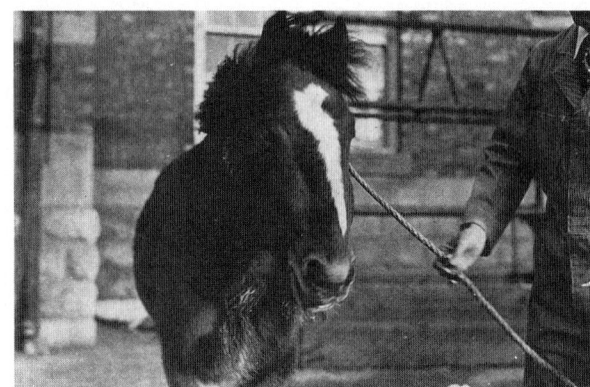

Fig. 5–17. Osteomalacia of the facial bones in a mature Hackney. (Courtesy, College of Veterinary Medicine, University of Illinois, Champaign-Urbana)

PHOTOSENSITIZATION

Species Affected:

Cattle, horses, sheep, swine.

Cause:

The sensitization of light-colored skin to sunlight. Some feeds, forages and certain medicines contain substances which may sensitize the skin (primary photosensitization). In other cases, products of metabolism, which normally would be removed from the body, accumulate because of faulty liver function (hepatogenous photosensitization). Primary photosensitization usually occurs in the spring when plants are lush, green, and growing rapidly. St. John's Wort (Klamath weed) and buckwheat are two of the most common sources of photosensitizing substances. Also, rape, kleingrass, kale, trefoil, alfalfa, alsike clover, Swedish clover, lamb's tongue, and plantain have been associated with photosensitization at one time or another.

(Continued)

Nutritional Disorders/Toxins

TABLE 5-1 *(Continued)*

PHOTOSENSITIZATION (Continued)

Symptoms and Signs (or age group most affected):

The signs of the disease are essentially those of severe sunburn. The lesions are confined to white, or lightly pigmented, exposed areas of skin. The muzzle, eyes, face, and light areas over the back are usually affected first. Areas of the belly and udder, which are exposed to the sun when the animal lies down, may also be affected. The earliest signs are redness and swelling of the skin. Later, tissue fluids ooze from the affected areas and crusting of the skin occurs, with resultant matting of the hair. In severe cases, the eyelids and nostrils may be swollen closed. In extreme cases, sloughing of the skin and gangrene result.

Distribution and Losses Caused By:

Photosensitization occurs worldwide. Death loss is higher in sheep than in cattle. But the monetary losses from the afflicted living are considerable, including loss in weight, damaged udders and teats, screwworms, secondary infections, eye damage, and stunted offspring.

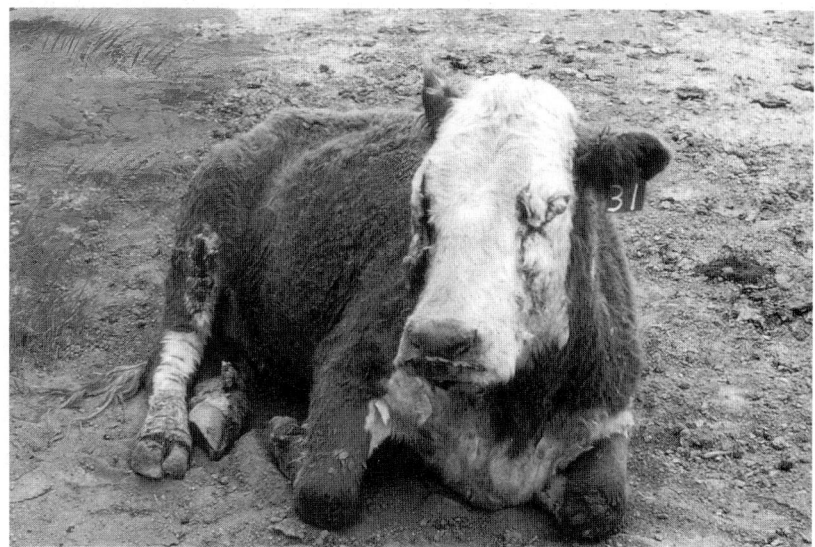

Fig. 5-18. Photosensitization in a cow, showing (1) swollen eyelids and nostrils, and (2) crusting of the skin and matting of the hair. (Courtesy, USDA-ARS Poisonous Plant Research Laboratory, Logan, Utah)

Treatment:

Discontinue the forage that's causing the trouble and keep the animal out of sunlight. In some cases, treatment of local lesions is warranted. In severe cases, supportive treatment, such as intravenous fluids, antibiotics, and other special medicines may be directed by the veterinarian.

Control:

Control of photosensitization can be achieved by preventing access to offending plants and keeping animals in the shade.

Prevention:

Good range and pasture management practices generally prevent the problem of photosensitization.

Remarks:

Photosensitization should be differentiated from sunburn in which the white or lightly pigmented skin of a normal animal becomes inflamed following overexposure to ultra violet rays.

POLIOENCEPHALOMALACIA (CEREBROCORTICAL NECROSIS, POLIO)

Species Affected:

Cattle, sheep, goats, deer.

Cause:

Although the cause is not known, it appears to be due to a deficiency. One hypothesis with considerable substantiation is that an acute thiamin deficiency is an important factor in the cause as evidenced by a favorable response to thiamin injection. Yet, the reasons why a thiamin deficiency should exist are not understood, because the thiamin intake appears to be more than adequate. One theory is that abnormally high concentrations of thiaminase enzyme from unusual plants or microflora destroy the vitamin before absorption takes place.

(Continued)

TABLE 5-1 (Continued)

POLIOENCEPHALOMALACIA (CEREBROCORTICAL NECROSIS, POLIO) (Continued)

Symptoms and Signs (or age group most affected):

Sudden deaths in animals. Sick animals are excitable, incoordinated, and have impaired vision. On driving, these animals go down into convulsions.

Affects feedlot and pasture cattle 3 months to 2 years of age.

In sheep, the incidence is highest in feedlot lambs 5 to 8 months of age. However, outbreaks of the disease may occur in farm flocks on pasture, especially in lambs following changing from overgrazed to lush pasture.

In goats, it may strike suckling young on pasture.

Distribution and Losses Caused By:

Most common in feedlot animals.

The disease occurs worldwide.

In sheep, the morbidity rate may range from a few cases up to 10% of the flock, and 50% of the affected animals may die.

Treatment:

Treatment consists of the IV or IM administration of thiamin at a dosage of 1 to 2 mg/lb. Twice daily treatment may be necessary for 2 days.

Rapidity of recovery relates directly to the speed of disease recognition and institution of thiamin treatment. Good nursing will help.

Usually, animals severely affected for more than 24 hours cannot be expected to respond well to treatment.

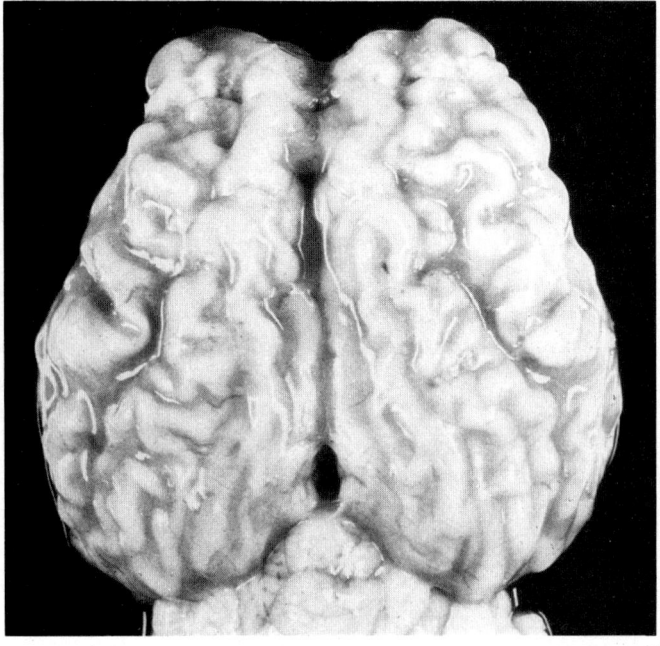

Fig. 5-19. Soft, swollen, yellowish cerebral gyri in brain of cow afflicted with polioencephalomalacia. (Courtesy, College of Veterinary Medicine, University of Florida, Gainesville)

Control:

Dietary cereal content should be decreased and additional good quality roughage supplied for a period of 5 days prior to a gradual return to higher energy rations.

Prevention:

Until the cause is discovered, little can be done to prevent the disease, except to provide a good ration.

Remarks:

Ruminants normally derive adequate thiamin from symbiotic ruminal activity; the inadequacy is thought to be a result of intraruminal thiamin destruction either by the enzymes of microbes or other dietary sources.

PREGNANCY DISEASE IN SHEEP
(See Ketosis)

PULMONARY EMPHYSEMA (BOVINE PULMONARY EMPHYSEMA, COW ASTHMA)

Species Affected:

Cattle.

Cause:

The condition can be traced to the quantities of the amino acid, tryptophan, consumed when cattle are pastured on lush, rapidly growing forage plants. The change from dry feed to lush pasture produces conditions favorable to an abnormal growth of clostridial organisms in the rumen. These organisms help convert tryptophan to 3-methylindole (3-Mi). When large quantities of 3-Mi are absorbed into the blood stream, pulmonary emphysema may result.

Also, the disease may be caused by pneumonia, allergic reactions to lungworm larvae or inhaled fungal organisms, or inhalation of irritating gases.

(Continued)

Nutritional Disorders/Toxins

TABLE 5-1 *(Continued)*

PULMONARY EMPHYSEMA (BOVINE PULMONARY EMPHYSEMA, COW ASTHMA) *(Continued)*

Symptoms and Signs (or age group most affected):

Difficult breathing. In severe cases, the common signs include panting, difficulty in exhaling air from the lungs, coughing, excessive salivation, reluctance to move, extreme weakness, and rapid loss in condition. Death may occur within a few hours after onset of the disease.

Distribution and Losses Caused By:

The disease occurs throughout the world. In affected herds, up to 20% of the cattle may develop emphysema and as many as 10% may die.

Treatment:

No specific medications are available. The recommended treatment consists of—
1. Removing the animals from lush pasture and placing them on hay.
2. Injecting antihistamines, steroids and other compounds to lessen the respiratory distress.
3. Using antibiotics and sulfonamides to prevent secondary bacterial infections.

Control:

The removal of cattle from lush pasture and placing them in a drylot and feeding hay will control the disease.

Prevention:

Measures that will prevent pulmonary emphysema include—
1. Removing cattle from summer range before feed becomes too dry.
2. Making a gradual transition from summer range to lush pasture.
3. Continuing to feed hay or straw while the cattle are on pasture.

Remarks:

Pulmonary emphysema is primarily a nutritional disease, although there are other causes.

Fig. 5–20. Lungs from a cow afflicted with pulmonary emphysema. The lungs are greatly enlarged, firm, and edematous. The lungs do not collapse normally, and they are pinkish gray in color. (Courtesy, College of Veterinary Medicine, University of Tennessee, Knoxville)

RICKETS

Species Affected:

All young farm animals and young humans.

Cause:

Lack of either calcium, phosphorus, or vitamin D; or an incorrect ratio of the 2 minerals.

In housed animals, vitamin D deficiency is not uncommon; grazing animals are more likely to be phosphorus-deficient.

Symptoms and Signs (or age group most affected):

Enlargement of the knee and hock joints, and the animal may exhibit great pain when moving. Irregular bulges (beaded ribs) at juncture of ribs with breastbone, and bowed legs.

Rickets is a disease of young animals—calves, foals, pigs, lambs, kids, pups, and chicks.

Poultry: Bones of growing birds become soft and rubbery.

Distribution and Losses Caused By:

Worldwide.

It is seldom fatal, but it can be severely debilitating and economically disastrous.

Fig. 5–21. Rickets (advanced case) caused by a deficiency of vitamin D. The pig was fed indoors, without exposure to sunlight. Because of leg abnormalities, it was unable to walk. (Courtesy, University of Saskatchewan, Saskatoon, Saskatchewan, Canada)

(Continued)

TABLE 5-1 *(Continued)*

RICKETS *(Continued)*

Treatment:

If the disease has not advanced too far, treatment may be successful by supplying adequate amounts of vitamin D, calcium, and phosphorus, and/or adjusting the ratio of calcium to phosphorus.

Control:

Control of rickets is usually achieved by providing a balanced ration, with special consideration given to calcium, phosphorus, and vitamin D.

Prevention:

Provide (1) sufficient calcium, phosphorus, and vitamin D, and (2) a correct ratio of the 2 minerals. Vitamin D_3, rather than D_2, is required by the chicken.

Remarks:

Rickets is characterized by a failure of growing bone to ossify, or harden, properly.
Hens fed rations deficient in vitamin D lay eggs with progressively thinner shells until production ceases.

SALT SICK (COBALT DEFICIENCY)

Fig. 5-22. Cobalt deficiency. *Left:* A heifer suffering from cobalt deficiency. Anemia, loss of appetite, and roughness of hair coat characterize the malady. *Right:* Illustrates the remarkable recovery in the same animal brought about by the administration of cobalt. (Courtesy, Michigan State University, East Lansing)

Species Affected:

Cattle, sheep, goats.

Cause:

Cobalt deficiency.
In Florida, cobalt deficiency is associated with copper deficiency.

Symptoms and Signs (or age group most affected):

Loss of appetite, emaciation, depraved appetite, scaliness of skin, rough hair coat, listlessness, and lack of thrift.

Distribution and Losses Caused By:

Cobalt deficiency is widespread. In different parts of the world, it is known as *Denmark disease, coast disease, enzootic marasmus, bush sickness, wasting disease, Nakuritis,* and *pining disease.* Fig. 5-23 shows cobalt-deficient areas of the U.S.

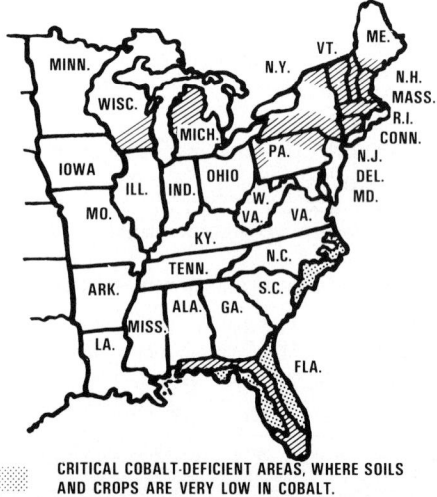

░░░ CRITICAL COBALT-DEFICIENT AREAS, WHERE SOILS AND CROPS ARE VERY LOW IN COBALT.

/// MARGINAL COBALT AREAS, WHERE SOILS AND CROPS ARE OFTEN LOW IN COBALT.

Fig. 5-23. Cobalt-deficient areas in eastern U.S., resulting from its deficiency in the soil and thus in the herbage produced thereon.

(Continued)

TABLE 5-1 (Continued)

SALT SICK (COBALT DEFICIENCY) (Continued)

Treatment:

Provide 0.2–0.5 oz cobalt salt/100 lb of salt—or feed a suitable trace mineral supplement. Injection of cobalt salts is not satisfactory, since ruminal action is needed to form vitamin B-12, the active form of cobalt.

Control:

Provide adequate cobalt in the ration—about 0.1 ppm. Deficiency symptoms appear when the level drops to the range of 0.04 to 0.07 ppm or lower.

Prevention:

Mix 0.2–0.5 oz of cobalt chloride, cobalt sulfate, or cobalt carbonate/100 lb of either (1) salt, or (2) whatever mineral mix is being used.

Remarks:

Cobalt is needed especially for rumen microbial synthesis of vitamin B-12. Nonruminants must be fed preformed vitamin B-12.

STIFF-LAMB DISEASE (WHITE MUSCLE DISEASE)
(See White Muscle Disease.)

SWEET CLOVER DISEASE

Species Affected:

Cattle; rarely affects sheep or horses.

Cause:

Usually produced only by moldy or spoiled sweet clover hay or silage.
In moldy or spoiled sweet clover hay, the harmless natural coumarins are converted to dicoumarol, which interferes with vitamin K in blood clotting.

Symptoms and Signs (or age group most affected):

Loss of clotting power of the blood. As a result, blood forms soft swellings beneath the skin on different parts of the body. Serious or fatal bleeding may occur at time of dehorning, castration, parturition, or following injury.
All ages affected. A newborn animal may also have the condition at birth.

Distribution and Losses Caused By:

Wherever sweet clover is grown and cured for hay.

Treatment:

Remove the offending materials and administer menadione (vitamin K_3).
The veterinarian usually gives the affected animal an injection of plasma or whole blood from a normal animal that was not fed on the same feed.

Fig. 5-24. Sweet clover disease in calf. Note the collection of blood at the point of the left shoulder. (Courtesy, College of Veterinary Medicine, University of Illinois, Champaign-Urbana)

Control:

When a case of sweet clover disease is observed in the herd, either (1) discontinue feeding the damaged product, or (2) alternate it with a better-quality hay—especially alfalfa.

Prevention:

Properly cure any sweet clover hay or ensilage.
Cultivars of sweet clover that are low in coumarin content, and hence safe to feed, have been developed.

Remarks:

The disease has also been produced from feeding moldy lespedeza hay and from sweet clover pasture.

(Continued)

TABLE 5-1 *(Continued)*

URINARY CALCULI (GRAVEL, STONES, WATER BELLY, UROLITHIASIS)

Species Affected:

Cattle, sheep, goats, horses, mink, and humans.

Cause:

The precipitation of various salts, usually inorganic, in the urine, frequently associated with rations high in cereal grains or grazing on the silica-rich soils of the northwest plains of Canada and the U.S. However, not all causative factors are known.

Experiments and experiences have shown a higher incidence of urinary calculi when there is (1) a high potassium intake, (2) a high phosphorus-low calcium ratio (from the standpoint of preventing urinary calculi, the Ca:P ratio should be about 2:1), (3) a high-silica content in the ration, or a high proportion of high-silica grains and forages, such as native grasses, wheat straw, sugar beet leaves or pulp, sorghums, and cottonseed meal. High dosages of diethylstilbestrol or a deficiency of vitamin A may be contributing factors.

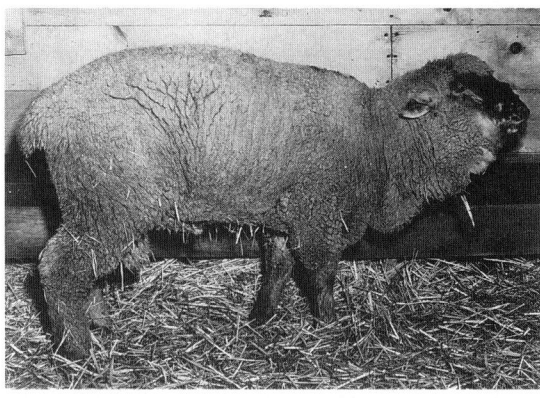

Fig. 5-25. Lamb suffering from urinary calculi. (Courtesy, Washington State University, Pullman)

Symptoms and Signs (or age group most affected):

Frequent attempts to urinate, dribbling or stoppage of the urine, pain and renal colic.

Usually only males affected; females are able to pass the concretions.

Bladder may rupture, with death following. Otherwise, uremic poisoning may set in.

Urinary calculi is one of the most important diseases in feedlot cattle and sheep, particularly in steers and wethers on full feed.

Distribution and Losses Caused By:

Worldwide. The economic loss may be considerable, since calculi formation frequently comes near the end of the feeding period.

Affected animals seldom recover completely.

Treatment:

(1) Add ammonium chloride at the rate of 1 oz (lambs) or 1¼–1½ oz (cattle) per head daily—or 50–60% more ammonium sulfate; (2) increase the phosphorus content of the ration (pasture) so that it equals the calcium content (by adding monosodium phosphate); (3) increase salt content of ration to 3 to 4% so as to increase water consumption (too much salt may lower feed intake); (4) incorporate 20% alfalfa in the ration; (5) administer muscle relaxants to help the passage of calculi from the bladder; or (6) surgically remove the calculi.

In cattle, surgical removal of the calculi is the most effective treatment, with the stone(s) removed at the point of blockage. In steers, the urethra may be bisected and brought to the outside of the body to bypass the constricted portion of the tract. After a short time to eliminate any tissue residue of urine, such animals are marketable.

In sheep, amputation of the urethral process is simple and allows the immediate passage of urine.

In horses, bladder calculi must be removed surgically.

Careful observation of susceptible animals by an experienced person several times daily will allow early detection and more successful treatment.

Control:

If severe outbreaks of urinary calculi occur in finishing steers or lambs, it is usually well to dispose of them if they are carrying acceptable finish.

Increase water consumption by including 3 to 4% sodium chloride (salt) in the ration.

The addition of a broad-spectrum antibiotic to the ration has been useful in controlling urinary calculi in some cases.

Prevention:

The basis of prevention is identification of the chemical composition of the calculi so that appropriate steps can be taken to reduce the concentration of the particular chemical in the urine by increasing urine volume, eliminating infection, changing urine pH, or altering the metabolism with drugs.

Good feed and management appear to lessen the incidence.

Delayed castration (castration of bull calves at 4–5 mo. of age) and high-salt rations for feedlot cattle (1–3% salt in the grain ration, using the upper limits in the winter months) in order to induce more water consumption, are effective preventive measures.

Avoid (1) a high potassium or phosphorus intake, (2) an incorrect Ca:P ratio, or (3) an excessive amount of beet pulp or grain sorghum in the ration.

Remarks:

Calculi are stonelike concretions in the urinary tract which almost always originate in the kidneys. These stones block the passage of urine, resulting in the condition commonly referred to as *water belly*.

The mineral deposits may be of variable sizes, shapes, and composition. In cattle, the phosphatic type predominates under feedlot conditions and the silicate type occurs most frequently in range cattle.

According to researchers at Canada's Lethbridge, Alberta, research station, the incidence of calculi in calves grazing native pasture is 10 times higher than in calves grazing Russian wild ryegrass.

(Continued)

TABLE 5-1 (Continued)

WHITE MUSCLE DISEASE (MUSCULAR DYSTROPHY; in sheep, STIFF-LAMB DISEASE)

Species Affected:
Calves, lambs, and foals. In lambs, it is commonly referred to as stiff-lamb disease.

Cause:
Selenium deficiency, due to the continuous consumption of a ration containing less than 0.02 ppm selenium.

Symptoms and Signs (or age group most affected):
In calves, white muscle disease is characterized by lameness or inability to stand, and heart failure. It most commonly affects calves 2 to 4 months of age.
In lambs, the symptoms and signs are: A stiff, stilted way of moving, chiefly in the hind legs, although the front legs and shoulders may be involved. The back is usually humped or "roached." Lambs that live are usually stunted. Young, rapidly growing lambs are especially susceptible.
It seems that more calves than lambs or foals develop heart damage, which may be fatal, especially if subjected to unusual exercise. Affected calves and lambs show similar pathological lesions—whitish areas or streaks in the heart and other muscles.

 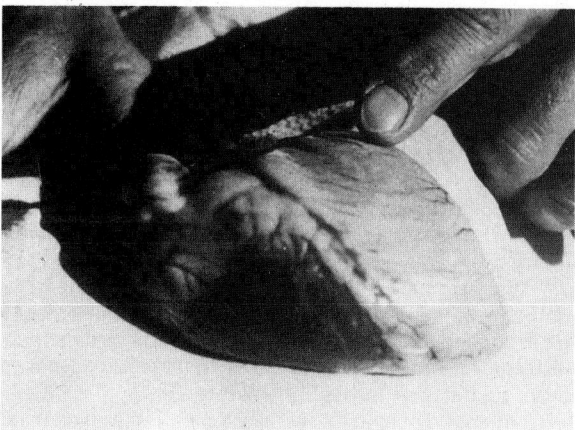

Fig. 5-26. White muscle disease in a calf. *Left:* Shows the generalized weakness of muscles, lameness, and difficulty in locomotion of an afflicted calf. Calf is about 3 months old. *Right:* Shows abnormal white areas in the heart muscles of a 6-week-old calf afflicted with white muscle disease. (Courtesy, Oregon State University, Corvallis)

Distribution and Losses Caused By:
Geographically, white muscle disease has been reported in Australia, Canada, Finland, Italy, Japan, New Zealand, Norway, Scotland, South Africa, Sweden, U.S., U.S.S.R., and Yugoslavia.
In the U.S., the disease is widely distributed, but the severity is greatest on the two coasts. Fig. 5-44 shows low, variable and adequate or high selenium areas in the U.S.
Economic losses result from the deaths of severely affected calves and lambs, the unthriftiness of survivors, and the cost of preventive programs.
Death losses range up to 50%, with an average of 15%; and the mortality of untreated animals may reach 80%.

Treatment:
Affected animals should receive early treatment—the intramuscular injection of sodium selenite/vitamin E in aqueous solution at the rate of 0.25 mg Se per pound of body weight. This may be repeated in 2 weeks, but should not exceed 4 doses. **NOTE WELL**: Federal law restricts injectable Se to the order of a licensed veterinarian. Do not use within 30 days of slaughter.

Control:
Control consists in meeting the selenium requirements by (1) supplementing the ration of cows or ewes during the last ⅓ of pregnancy and the first part of lactation with selenium in the form of sodium selenite at the rate of 0.3 ppm dry matter; or (2) injecting intramuscularly each cow or ewe 1 month before parturition with approved levels of selenium/vitamin E preparation.

Prevention:
Add selenium to the ration. In 1987, FDA approved the addition of 0.3 ppm selenium to the complete feed of cattle, sheep, swine, and poultry.
Also, mineral mixes containing selenium are available for free-choice feeding in areas of known selenium deficiency.

Remarks:
White muscle disease is most common (1) in rapidly growing calves and lambs, and (2) in calves and lambs on lush pastures on selenium-deficient soils.
Because of muscle failure, severely affected young animals die from starvation or heart failure.
White muscle disease cannot be produced in calves on vitamin E-free rations unless the rations are high in unsaturated fats.

NUTRITION AND DISEASE/PARASITE INTERACTION

In general, well-nourished animals are more resistant to bacterial, viral, and parasitic diseases than those poorly nourished. This is attributed to better body tissue integrity, more antibody production, more immunity to disease, greater detoxifying ability, increased blood regeneration, and other factors. Also, it is recognized that proper nutrition, which heads the list of what consitititues *good nursing,* is essential for fast recovery from diseases and the ravages of parasites.

It is estimated that, annually, diseases and parasites in the United States (1) decrease animal productivity by 15 to 20%, and (2) make for losses aggregating $10 billion; and that nutrition has some involvement in 85% of the veterinary cases. In the developing countries, diseases and parasites take an even greater toll—they decrease animal productivity by 30 to 40%.

A nutrition program can be fully effective only if the animals are healthy. The converse is equally true. Also, the higher the productivity level, the higher the nutritional and health requirements.

Some noteworthy interactions of nutrition and diseases/parasites follow:

1. Some nutrients (for example, vitamin A) are important in keeping the epithelial tissue (the skin and mucous membranes), the body's first line of defense, in a healthy condition.

2. Protein, certain B-complex vitamins, and some trace minerals and other nutrients are essential for the production of antibodies and phagocytes, which serve as defenders against infectious agents that enter the body through one of the openings or through the skin.

3. Adequate nutrition is essential for an animal to respond properly to a vaccination. Although many factors affect the effectiveness of a vaccination, studies are showing that the good nutritional health of the animal is one of the most important requisites for a vaccination to induce an immune response.

4. Certain diseases increase the need for various nutrients by animals. This may be due to reduced appetite resulting in inadequate nutrient intake, vomiting or diarrhea resulting in a loss of nutrients from the intestinal tract, fever, decreased absorption or utilization of nutrients, or other causes.

5. Certain minerals, vitamins, and proteins (amino acids) are required at higher than normal levels in order to produce maximum immune response and resistance to diseases. Thus, the decline of immunoglobulins in the milk, followed by early weaning of calves, pigs, and lambs necessitate superior nutrition in order to avoid outbreaks of diarrhea (scours).

6. Many extra nutrients are needed for the repair and restoration of tissues, red blood cells, vital organs, and other parts of the body destroyed by diseases and parasites.

7. Diseases and parasites that cause diarrhea (scours) or vomiting decrease intestinal absorption of nutrients and cause electrolyte loss and dehydration; hence, extra nutrients and electrolytes may be beneficial in treating these conditions.

8. Diseases and parasites that reduce appetite and decrease total feed intake increase the need for higher levels of the nutrients in the feed consumed as a means of ensuring that the total daily nutrient needs will be met.

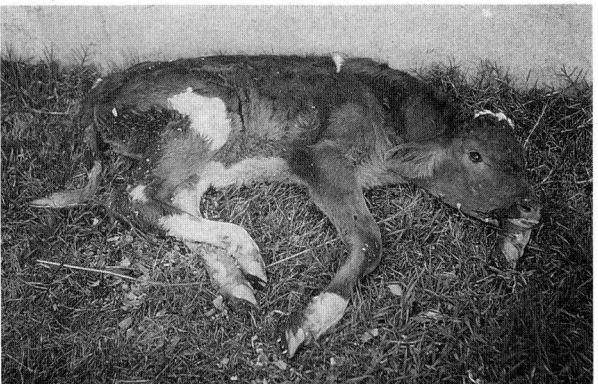

Fig. 5–27. Calf with severe scours. Scours may be (1) nutritional/management related (caused by overfeeding, irregular feeding, use of unclean utensils, too rapid changes in feed, or exposure to drafts and cold, damp floors), or (2) infections; and it is sometimes difficult to distinguish between the two causes. (Courtesy, North Carolina State University, Raleigh)

9. Parasites that cause severe damage to the digestive tract result in impairment of absorption of a number of essential nutrients. Thus, the protozoan parasites, particularly the coccidia, produce profound effects on the digestive physiology of animals.

10. Ketosis increases the need for niacin by dairy cows.

11. Certain nutrients are required at higher than normal levels during stress produced by such things as uncomfortable environmental conditions, vaccinations, crowding, loud noises, debeaking birds, and castrating and dehorning animals. Thus, it is noteworthy that shipping fever (bovine respiratory disease complex), which causes estimated losses of $500 million annually, is most frequently associated with animals in which resistance has been lowered due to change in weather and feed, overcrowding, hard driving, lack of rest, and improper shelter that accompany shipping.

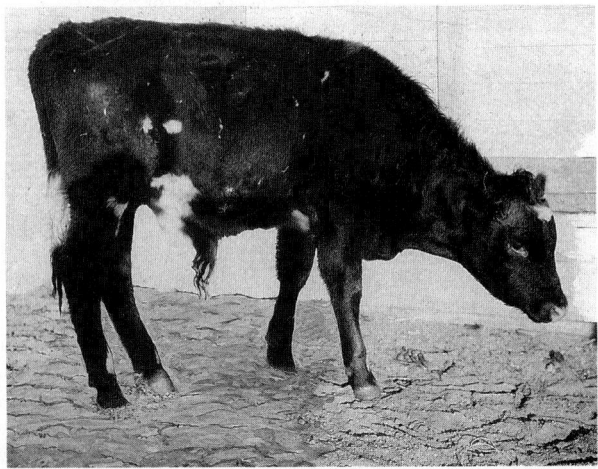

Fig. 5–28. Calf with shipping fever (bovine respiratory disease complex), most commonly associated with animals whose resistance has been lowered due to the stresses of travel. (Courtesy, USDA)

12. Well-fed animals have fewer parasites. For example, there is experimental evidence, substantiated by practical observation, that milk and its by-products are helpful in holding in check some of the internal parasites of swine. Milk-fed pigs make more rapid gains and have fewer parasites in the digestive tract than pigs not receiving milk.

13. Infectious diseases and parasites increase feed costs and lower feed efficiency.

The above list clearly shows that good nutrition is the first requisite of a disease and parasite prevention and control program. It also points up (1) that further intensive studies on these interrelationships hold great potential for discoveries of new ways to improve resistance to diseases and parasites, and (2) that there is a continuing need for cooperative effort between nutritionists and veterinarians.

NUTRITION RELATED DISORDERS

Not all noncontagious diseases are nutritionally responsive. Some of them are due to physical factors. Others are caused by faulty management. Regardless of the nature or the cause, however, all of them have an impact on the nutritional well-being of the affected animal; hence, they are nutrition related. Several of the most important of these disorders are discussed in the sections that follow.

Choking

Occasionally, feeds become lodged in the esophagus of cattle or horses, causing them to choke.

Cattle may choke on such feeds as beets, potatoes, apples, hay cubes, or ears of corn. Afflicted animals drool saliva from the mouth, make frequent attempts at swallowing, bloat rapidly due to closure of the outlet for gas from the stomach, and switch the tail. If the obstruction is in the region of the neck, it can be felt from the outside. Treatment consists in an attempt to work the object into the mouth, rather than force it into the stomach. A speculum should be applied to hold the mouth open, then the hand should be inserted into the animal's throat to grasp and remove the obstacle. (*CAUTION:* Beware of the hazard of a bitten hand or arm.) If this fails, a stomach tube or rubber hose lubricated with water or oil and pushed down the esophagus usually will free the object; but care must be taken not to damage the lining of the esophagus. Should there be marked bloating, the stomach may have to be punctured.

Horses choke most frequently from bolting (eating too rapidly) their grain, although they may choke on hay cubes, ears of corn, potatoes, and apples. Afflicted animals become excited, squeal, and thrust the head forward. Treatment consists in controlling the pain with sedatives, confining the animal, and allowing access to water but not feed. Passage of a stomach tube to the obstruction and repeated pumping and siphoning may relieve grain choke. As a last resort, the obstruction may be gently pushed into the stomach with a large stomach tube or rubber hose.

Minimizing choking in both cattle and horses consists in avoiding, to the extent practical, feeds that are most likely to cause choking. Such feeds as potatoes, apples, and roots are less apt to cause trouble if they are sliced or chopped. When feeding potatoes to cattle, choking can be materially lessened by forcing the animals to eat them from the ground level with their heads down. This can be accomplished by having a cable or pole arrangement at the top of their necks to keep their heads down. Choking can be lessened in gluttonous horses (horses that eat their feed too rapidly—that bolt their feed) by putting into the grain box several smooth stones about the size of a baseball, thereby slowing their eating.

Displaced Abomasum

This disorder is being diagnosed with increasing frequency in dairy cows.

Normally, the abomasum is located on the right side of the rumen of the cow and rests on the floor of the abdomen. It is attached at the front end to the omasum and at the back end to the small intestine. The mid-portion of the abomasum is not held rigidly in place by other structures; however, the weight of the material within it usually keeps it in place on the floor of the abdomen. Sometimes when the metabolism of the stomach is abnormal, the abomasum slides under the rumen, becomes partially filled with gas, and rises on the left side between the rumen and the body wall. When this happens, the afflicted cow goes off feed and exhibits many symptoms similar to ketosis. In fact, secondary ketosis may result from displaced abomasum because of decreased feed intake, thereby inducing a ketotic state. However, an experienced veterinarian can differentiate between displaced abomasum and primary ketosis.

REAR VIEW OF CROSS-SECTIONAL DIAGRAM

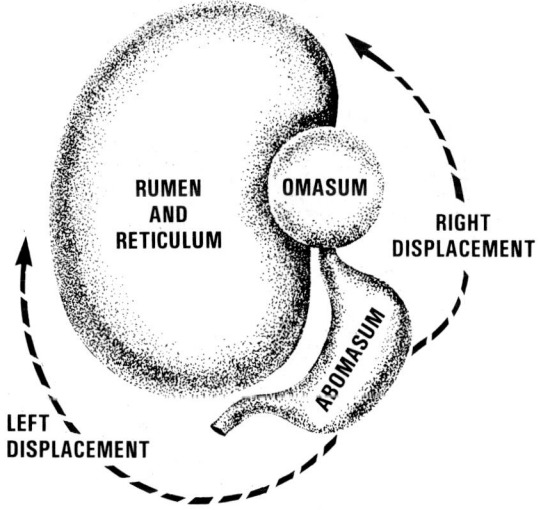

Fig. 5-29. Normally, the abomasum rests near the belly floor and to the right of the rumen. In most displaced abomasums, it shifts to the left.

Several treatments for displaced abomasum have been used with variable success; most commonly, the following two:

1. Rolling the cow onto her back and getting the abomasum back into normal position. Frequently, the problem recurs following this treatment.

2. Performing rather simple surgery, involving the veterinarian suturing the bottom of the rumen wall to the body

wall, thereby preventing the abomasum from slipping between the two. When properly done, this usually provides a permanent cure. The operation can be done using local anesthesia. Frequently, cows so treated do not miss any milkings.

The cause of displaced abomasum is not known; however, the following theories have been advanced:

1. High-concentrate rations and inadequate bulk, as a result of which the rumen pulls away from the body wall.

2. Reduced ruminal contents for various reasons, such as cows going off feed, resulting in a gap between the rumen and the body floor.

When displaced abomasum is a problem, the incorporation of more roughage in the ration is recommended. A rule of thumb is that at least half of the dry matter should be in the form of roughage-type feeds, such as hay, silage, or pasture.

Hardware Disease

The term *hardware disease* (traumatic gastritis) is used to describe the condition that results from swallowing foreign materials, usually metal (nails, wire, screws, and pins). Cattle are involved more than other classes of animals; however, cases have been reported in horses and goats. In most cases, the metal is found in the reticulum (second stomach). However, the material may puncture the reticulum lining and pass into the body cavity. Some objects may pierce the diaphragm and enter the thoracic cavity, thence work their way to the heart or lungs, where they may cause serious damage or death.

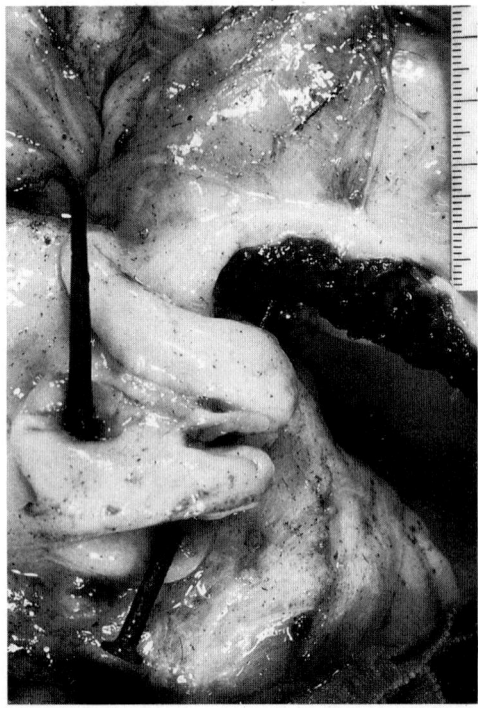

Fig. 5-30. Hardware disease. Note the nail puncturing the reticulum. (Courtesy, College of Veterinary Medicine, University of Tennessee, Knoxville)

Nearly 7,000 cattle are condemned each year by the Federal Meat Inspection Service as unfit for food because of hardware disease. Clinical reports indicate that the problem is increasing due to the use of more chopped feeds and more contamination. Sharp objects will injure the lining of the stomach and cause infection and inflammation, a condition known as traumatic gastritis.

Hardware disease is a problem in cattle because of their eating habits and stomach arrangement. The usual source of metals is the feed. The animals eat rapidly and are not able to sort foreign objects from their feed.

The most common symptoms of hardware disease are loss of appetite and digestive disturbance; slow and stiff movement and arched back; elbows that bow outward; decreased rumen movement and chewing; tendency to stand with the front feet elevated so as to lessen the pressure of the viscera on the inflamed area; rise in body temperature; and swellings under the jaw, at the brisket, and at the hock joints. Bulls may be reluctant to mate.

Prevention consists in avoiding foreign objects getting into the feed through good management. Also, it is recommended that strong magnets be installed, in keeping with the manufacturer's directions, (1) at the outlets of mechanical silo unloaders, and (2) in feed processing equipment.

Magnets may also be permanently placed in the cow's second stomach, for the purpose of holding objects that have not penetrated the stomach wall. However, the only sure cure for traumatic gastritis is veterinary surgery. Surgery will be successful only if performed before the condition has progressed to the point that damage has been done to the heart or other organs.

Immunoglobulin Deficiency

Immunoglobulin, found in colostrum, is the means by which most newborn farm mammals acquire passive immunity against pathogenic diseases and infections.

Deficiencies in immunoglobulins may result (1) when the young fail to nurse properly following birth, (2) when the mother is exhausted or sick after giving birth, (3) when the mother's colostrum contains low levels of specific antibodies, (4) when the caretaker fails to give colostrum to newborn, and/or (5) when other feeds are consumed prior to colostrum. Also, immunoglobulin deficiencies may exist in milk from first-calf heifers and in milk from cows that haven't dried up or have had less than a 30-day dry period. Older cows have been exposed to a wider range of diseases than first-calf heifers and therefore produce more immunoglobulins against them. So, if livability problems are encountered, colostrum from older cows should be fed to calves from heifers or from cows that haven't dried up or have had too short a dry period.

The highest absorption of intact immunoglobulins directly into the lymphatic system (thoracic duct) occurs immediately after birth (within 15 to 30 minutes) and during the first 6 hours after birth, but by 24 hours most of the absorption capacity is lost. Hence, for maximum protection against infection, newborn mammals should be fed colostrum very early in life.

Where the immunoglobulin phenomenon exists, as in the calf, the intestinal villi of the newborn are able to absorb the globulins by pinocytosis (engulfing). This enables those species which do not normally obtain adequate immune protection through placental transfer to acquire instant immunity by ingesting colostrum high in immunoglobulins. Aside from this unique stituation, protein must be digested. (Also see Colostrum.)

Impaction In Horses

Impaction is a form of colic caused by obstruction of the cecum or colon by fibrous feeds. It is usually caused by feeding horses large amounts of straw, cornstalks, or other coarse, high-fiber feeds, along with lack of water intake. Distention of the large colon with gas causes acute abdominal pain. The usual signs described by colic are seen.

Medical treatments, which should be administered by a veterinarian, meet with variable success. Mineral oil and magnesium sulfate are popular laxatives. Dioctyl calcium sulfosuccinate may be used to penetrate the impacted mass. Antiferments and oral antibiotics such as neomycin are helpful in preventing gas formation. Surgery should be the last resort.

POISONOUS PLANTS

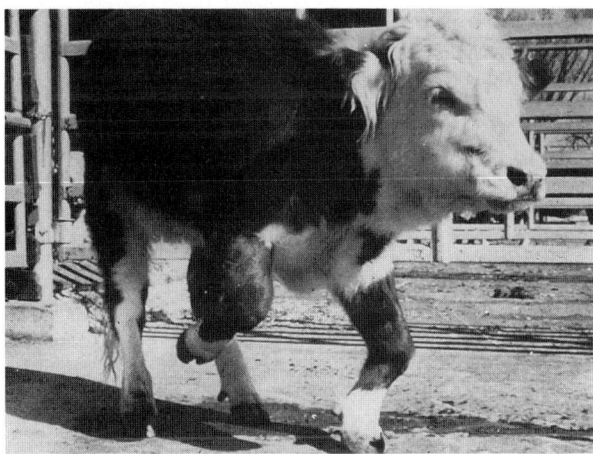

Fig. 5-31. Lupine toxicity. Calf showing abnormalities due to ingestion by dam of lupines during 40th to 70th days of gestation. (Courtesy, The American Institute of Nutrition, Bethesda, Md.)

Poisonous plants have been known since time immemorial. Biblical literature alludes to the poisonous properties of certain plants, and history records that hemlock (a poison made from the plant from which it takes its name) was administered by the Greeks to Socrates and other state prisoners.

No section of the United States is entirely free of poisonous plants, for there are hundreds of them. But the heaviest livestock losses from them occur on the western ranges because (1) there has been less cultivation and destruction of poisonous plants in range areas, and (2) the frequent overgrazing on some of the western ranges has resulted in the elimination of some of the more nutritious and desirable plants, and these have been replaced by increased numbers of the less desirable and poisonous species. It is estimated that poisonous plants account for 8 to 10% of all range animal losses each year, and even more in some areas. It is further estimated that poisonous plants cause average annual losses of beef cattle, sheep, goats, and horses of more than $200 million in the 17 western states. Additionally, toxic plants contribute to indirect losses, such as reduced weight gains; reduced calving, lambing, kidding, and foaling percentages; chronic illness; birth defects; fencing; abandoned ranges; and supplemental feeding.

Many plants contain substances which, if consumed in amounts above safe levels, may be toxic. These substances may be found in the leaves, stems, fruits, seeds, roots, and/or tubers. The type of substance, stages of highest concentration, destruction by heat or other processing, type and age of animals consuming the plant, and the amount of storage or excretion of the toxic substance, may influence the severity of toxicity.

Common Poisonous Plants

The list of poisonous plants is so extensive that no attempt is made herein to describe them in detail. Nevertheless, both producers and veterinarians should have a working knowledge of the principal poisonous species in the area in which they operate. The common poisonous plants of the intermountain ranges to which cattle and/or sheep are susceptible at certain times of the grazing season are listed in Table 5-2.

TABLE 5-2
TYPES OF RANGE ANIMALS SUSCEPTIBLE TO POISONOUS PLANTS AT DEFINITE SEASONS

Poisonous to Cattle	Time of Year	Poisonous to Sheep	Time of Year	Poisonous to Cattle and Sheep	Time of Year
Low larkspur	Spring	Death camas	Spring	Broomweed	Spring and summer
Oak	Spring and fall	Greasewood	Fall	Chokecherry	Spring
Tall larkspur	Early summer and early fall	Horsebrush	Spring	Copperweed	Summer
Timber milk vetch	Spring and summer	Rubberweed	Summer	Desert parsley	Spring
Water hemlock	Spring	Sneezeweed	Summer	Halogeton	All year
				Loco	All year
				Lupine	Summer and fall
				Milkweed	Summer
				Veratrum	Summer

Treatment of Plant-Poisoned Animals

Unfortunately, plant-poisoned animals are not generally discovered in sufficient time to prevent loss. Thus, prevention is decidedly superior to treatment.

When trouble is encountered, the owner or caretaker should *promptly* call a veterinarian. In the meantime, the animal should be (1) placed where adequate care and treatment can be given, (2) protected from excessive heat and cold, and (3) allowed to eat only feeds known to be safe.

The veterinarian may determine the kind of poisonous plant involved (1) by observing the symptoms, and/or (2) by finding out exactly what poisonous plant was eaten through looking over the pasture and/or hay and identifying leaves or other plant parts found in the animal's digestive tract at the time of autopsy.

It is to be emphasized, however, that many poisoned animals that would have recovered had they been left undisturbed, have been killed by attempts to administer home remedies by well-meaning but untrained persons.

Preventing Losses from Poisonous Plants

With poisonous plants, the emphasis should be on prevention of losses rather than on treatment, no matter how successful the latter. The following are effective preventive measures:

1. Follow good pasture or range management.
2. Know the poisonous plants common to the area.
3. Know the symptoms that generally indicate plant poisoning.
4. Avoid turning animals on pasture in very early spring.
5. Provide supplemental feed during droughts, after plants become mature, and after early frost.
6. Avoid turning out very hungry animals where there are poisonous plants.
7. Avoid driving animals too fast when trailing.
8. Remove promptly all animals from infested areas when plant poisoning strikes.
9. Treat promptly, preferably by a veterinarian.

AGRICULTURAL CHEMICALS AND DRUGS

In the everyday pursuit of modern agriculture, more and more chemicals and drugs are being used. Hand in hand with this development, there has been increased public concern over the use of these products, for fear of poisoning human food.

When properly used, agricultural chemicals are an important adjunct to providing feed for animals and food for people. However, improper use can result in toxicoses. Moreover, certain chemicals can accumulate in the body fat of animals, and be found in the meat or secreted in milk.

The vast majority of agricultural chemicals and drugs have been properly used. Of course, it shouldn't be too surprising that a few have been improperly used when it is realized that there are approximately 300,000 trade name products on the market.

When chemical poisoning or drug misuse happens, it can be both devastating and perplexing. Usually, the causative agent can be diagnosed after an investigation of the environment and the feed. However, few poisons can be diagnosed with certainty by clinical signs alone. When trouble is encountered, the producer should promptly call a veterinarian if animals are involved, or a medical doctor if people are involved.

A voluminous amount of information is available on the deleterious effects of poisonous chemicals and drugs. Because of space limitations, only a few of the more important ones will be covered in Table 5-3.

Potential Poisons

A poison is a substance which in sufficient quantities and/or over a period of time kills or harms living things. Toxic substances are chemical substances that may present an unreasonable risk of injury to health or to the environment. Many poisons are called toxins. The study of poisons is called *toxicology*. The discussion that follows and Table 5-3 pertain primarily to feed-related poisons that may be eaten by animals. For most of these, there is both a safe level and a poisonous level; and the severity of the effect depends upon (1) the amount taken, (2) the period of time over which the substance is taken (certain poisons are cumulative), and (3) the age and physical condition of the animal. This lends credence to the toxicological adage: "Only the dose makes the poison."

When poisoning happens, it can be both devastating and perplexing. No part of veterinary diagnostics is as difficult and complex as toxicology. First, what compound is being tested for out of the many thousands known? Second, detecting trace levels, such as parts per billion, of pesticides and other chemicals in feed and water by low level residue analysis can be as difficult as understanding them.

Table 5-3 lists the most common potentially toxic substances, both synthesized and naturally occurring, and presents pertinent facts pertaining to each.

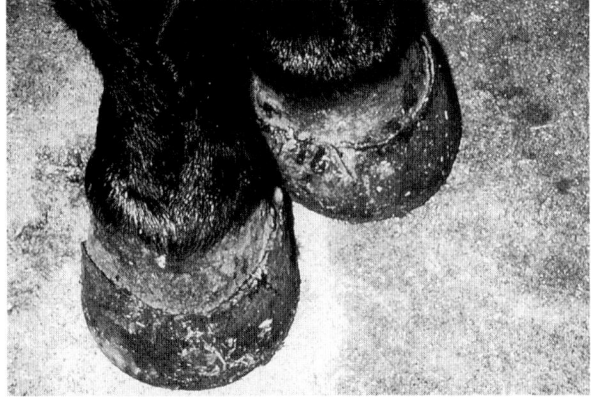 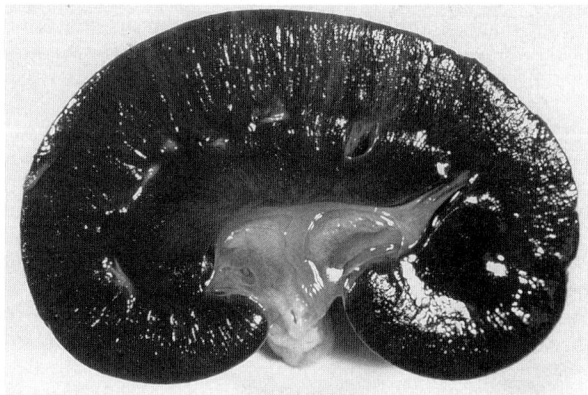

Fig. 5-32. For some substances, such as selenium and copper, a little is good but more may be poisonous. *Left:* Selenium toxicity resulting in severe hoof damage of a horse—note the horizontal cracks. (Courtesy, Colorado State University, Ft. Collins) *Right:* Copper toxicity evidenced by gunmetal-colored kidneys. (Courtesy, University of Tennessee, Knoxville)

TABLE 5-3
POTENTIAL POISONS

ACORN POISONING

It is caused by the tannin in oak buds, leaves, and acorns.

Source:

The oak buds, leaves, and acorns produced by the oak tree, *Quercus*. More than 60 species of oak have been identified in North America, and all should be considered potentially toxic. Cattle are poisoned in the spring by oak leaves and buds and in the fall by acorns. Young leaves and green acorns are more toxic than their mature counterparts. The tannin content diminishes as they mature. Most episodes of acorn poisoning occur soon after the acorns fall. Acorn shells are the major source of tannic acid.

Species Affected:

Cattle, although it occasionally occurs in horses, sheep, and swine.

Symptoms and Signs (or age group most affected):

Loss of weight, rough hair coat, loss of appetite, thin nasal discharge that may become red-tinged, excessive urination, and scouring. Calves that are severely affected will show signs of severe abdominal pain and appear bloated; often they will be found dead before any signs are noted.

Distribution and Losses Caused By:

Acorn poisoning may occur wherever oak trees grow, especially when there is a shortage of pasture.

Treatment:

Animals suspected of acorn poisoning should (1) be removed from pastures containing acorns, and (2) given a mild laxative, such as the magnesium hydroxide products, to purge the digestive tract and remove the digestive material as soon as possible. However, treatment is not very effective.

Prevention:

Prevention consists in not allowing cattle access to oak acorns or buds. Partial protection can be provided cows on oak-containing pastures by supplying supplemental feed, especially when there is a shortage of pasture.

Remarks:

Swine thrive and fatten on acorns, while cattle and calves often sicken and die from eating quantities of nuts.

ARSENIC (As) POISONING

Source:

Arsenic used to control insects and weeds, and to defoliate crops.
Overdosing of phenylarsonic compounds as feed additives to swine and poultry for growth and disease control.

Species Affected:

All farm animals.

Symptoms and Signs (or age group most affected):

The onset is sudden; characterized by groaning, restlessness, rapid breathing, muscular incoordination, blindness, and photosensitization. Death in 3–4 hours to a few weeks, depending on amount of arsenic consumed.

Necropsy reveals severe hemmorrhagic inflammation of the stomach and intestines, with perhaps areas of erosion on mucous membranes.

Distribution and Losses Caused By:

Arsenic has long been a leading cause of chemical poisoning.

Treatment:

Handled by the veterinarian. If caught in time, first remove the material from the animal. Sodium thiosulfate may be used, and supportive treatment may be indicated.

Fig. 5-33. Steer evidencing great pain from arsenic poisoning. This animal was accidentally poisoned by eating bran treated with arsenic for grasshopper bait. (Courtesy, College of Veterinary Medicine, University of Illinois, Champaign-Urbana)

(Continued)

TABLE 5-3 (Continued)

ARSENIC (As) POISONING (Continued)

British Anti-Lewisite (Dimercaprol) is a specific antidote for some forms of arsenic poisoning.

Prevention:

Keep animals away from arsenic.

Remarks:

Accumulation of arsenic in soils may sharply decrease crop growth and yields, but it is not hazardous to animals or humans that eat plants grown in these fields, provided they do not eat the foilage of the plants.

In spite of the recognized toxicity of many forms of arsenic, various arsenicals have been used in the practice of medicine. Also, animal feeds have been supplemented with growth-promoting organic arsenicals for many years. Another curious feature of arsenic biochemistry is the ability of the element partially to counteract the toxic effects of excess selenium.

BLACK WALNUT TOXICOSIS

It is thought to be due to the toxic compound *jugolene,* contained in black walnut trees.

Source:

Black walnut (*Juglans nigra*) shavings or sawdust used as bedding.

Species Affected:

Horses.

Symptoms and Signs (or age group most affected):

Founder (laminitis) occurs within 12 to 24 hours after horses have been exposed to black walnut shavings or bedding. Apparently, only skin contact is necessary; the material need not be eaten.

The temperature, pulse, and respiration are elevated; both front and hind legs may be affected, with edema from the knees and hocks downward; lameness may be severe or non-existent; some animals show sensitivity to hoof testers, especially over the toes; and the more severely affected horses may have to be destroyed because of severe founder. The horse's age may also be a factor; foals may be unaffected while their dams develop founder.

Distribution and Losses Caused By:

Black walnut toxicosis has been reported when black walnut shavings and sawdust have been used as bedding for horses. One-third or more of the horses bedded on black walnut materials may be affected. Usually, the condition is diagnosed promptly and the black walnut bedding removed, with the result that few animals must be destroyed because of founder.

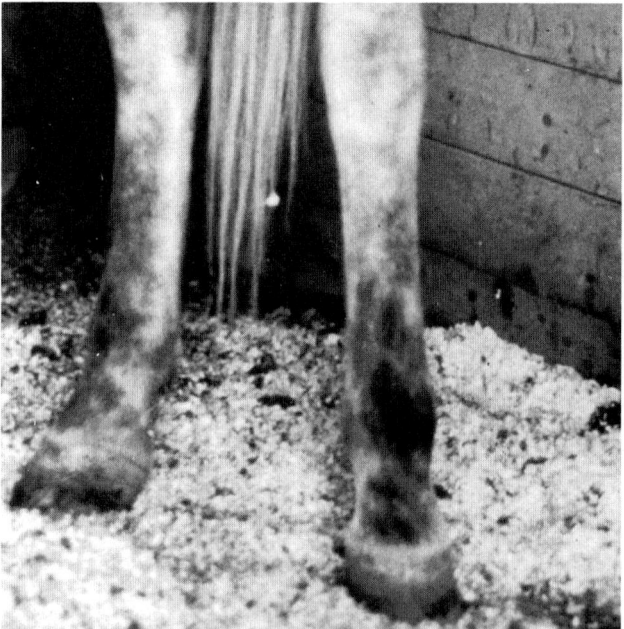

Fig. 5-34. Black walnut toxicosis, showing characteristic lower leg edema, caused by the use of black walnut bedding. (Courtesy, Colorado State University, Ft. Collins)

Treatment:

Remove *all* black walnut bedding. Call the veterinarian, who may administer medical treatment.

Prevention:

Do not use black walnut shavings or sawdust for bedding horses.

Remarks:

Exposure of black walnut shavings or sawdust to the air appears to diminish their toxic effect. Nevertheless, the use of this material for horse bedding is not recommended.

(Continued)

TABLE 5-3 (Continued)

BOTULISM

Botulism is a poisoning caused by ingestion of feed containing *Clostridium botulinum,* which organism proliferates in decomposing animal matter and sometimes in plant material. The toxins formed from these bacteria are the most potent poisons known; botulism Type A—the most lethal—is 10,000 times as deadly as cobra venom and millions of times more potent than strychnine or cyanide.

Source:

Ingestion of toxin in feed. The usual source of this toxin is decaying carcasses or vegetable materials such as decaying grass, hay, grain, or spoiled silage. In horses, the most common predisposing cause is moldy silage, haylage, or hay.

Species Affected:

All animals are subject to botulism. But it occurs most frequently in poultry, horses, and mink; it is relatively infrequent in cattle; and it is uncommon in other species.

Symptoms and Signs (or age group most affected):

The signs of botulism are associated with the paralysis of muscles, including motor paralysis, disturbed vision, difficulty in chewing and swallowing, and generalized progessive weakness. Death is usually due to respiratory or cardiac paralysis. Type B *Clostridium botulism* appears to cause "staggers," or the "shaker foal syndrome," in young equines.

Signs usually appear 3 to 7 days after animals gain access to toxic material, although rare, mild cases caused by the ingestion of small amounts of toxin may recover.

Distribution and Losses Caused By:

Botulism occurs worldwide.

The incidence in animals is not known with accuracy. Losses are particularly high in waterfowl; an estimated 10,000 to 50,000 birds are lost in most years, with losses reaching 1,000,000 during bad outbreaks in western U.S. Most botulism in cattle occurs in South Africa, where phosphorus-deficient cattle chew bones (and flesh) that they find on the range. Botulism in sheep has been encountered in Australia, where protein-deficient sheep eat the carcasses of rabbits and other small animals that they find on the range. Usually, botulism in mink is caused from eating meat or fish containing the toxin.

Treatment:

Botulism antitoxin has been used for treatment with varying degrees of success, depending upon the type of toxin involved and the animal species. Treatment of ducks and mink is often successful. In cattle, however, the treatment is rarely used.

Prevention:

Prevention consists of correcting dietary deficiencies and the proper disposal of all carcasses. Immunization of cattle with toxoid has been successful in South Africa and in Australia. Also, effective toxoids are available for immunizing mink and foals.

Remarks:

Botulism was first described as food poisoning of humans in Germany in 1817.

The toxin blocks transmission of the neuromuscular junctions.

ERGOT POISONING (ERGOTISM)
(caused by the parasitic fungus *Claviceps purpurea*)

Source:

The parasitic fungus replaces the seed in the heads of grasses and cereal grains, in which it appears as a purplish-black, hard banana-shaped dense mass from ¼ to ¾ in. long.

Most common in rye, wheat, wild rye, bromegrass, and dallisgrass.

Species Affected:

Cattle, sheep, swine, horses, and humans.

Symptoms and Signs (or age group most affected):

Acute ergot poisoning, caused by large quantities eaten at one time, may produce paralysis of the limbs and tongue, disturbance of the gastrointestinal tract, and abortion.

It is a cumulative poison; hence, poisoning may develop from lesser quantities eaten over a long period of time.

Chronic poisoning produces gangrene of the extremities, with subsequent sloughing off of hoofs, ears, and tail.

Delirium, spasms, and paralysis may occur before death.

Distribution and Losses Caused By:

Ergot is found throughout the world. However, seldom is sufficient of it ingested to cause poisoning.

(Continued)

TABLE 5-3 (Continued)

ERGOT POISONING (ERGOTISM) (Continued)

Treatment:

If noticed in time, stricken animals may recover if taken off the affected feed.

Tannin used as a drench is an antidote, and sedations such as chloral hydrate, may be given to nervous animals.

Prevention:

Never feed heavily ergot-infested hay or grain.

Remarks:

Poultry are more tolerant of ergot than other animals.

Grain containing 0.05% ergot will reduce gain and feed efficiency of finishing cattle.

Six different alkaloids are involved in ergot poisoning.

Grain containing even small amounts of ergot should not be fed to pregnant and lactating sows.

Fig. 5-35. A mixture of ergot sclerotia and wheat kernels as they appeared after harvest. (Courtesy, University of Idaho, Moscow)

FLUORINE (F) POISONING (FLUOROSIS)

Source:

Ingesting excessive quantities of fluorine through either the feed, water, or air; or some combination of these.

Toxic quantities of fluorides occur naturally in certain raw rock phosphates and in the superphosphates produced from them, and in partially defluorinated phosphates and in phosphatic limestones.

Species Affected:

All farm animals, poultry, fish, and humans.

Symptoms and Signs (or age group most affected):

Abnormal teeth (especially mottled enamel) and bones, roughened hair coat, stiffness of joints, loss of appetite, emaciation, reduction in milk flow, diarrhea, delayed maturity, and salt hunger (see tabulated symptoms below).

In most practical feeding situations, problems with fluorine have been associated with cattle. This is largely due to the fact that cattle have lower tolerances to fluorine than other classes of livestock.

SYMPTOMS OF FLUORINE TOXICITY IN CATTLE[1]

Symptom	Total Fluorine in Ration (ppm)			
	20–30	30–40	40–50	>50
Mottling of teeth[2]	Yes	Yes	Yes	Yes
Decreased growth of enamel[2]	No	No	Yes	Yes
Lameness	No	No	No	Yes
Decreased milk production	No	No	No	Yes

[1]Adapted by the authors from *Effects of Fluorides in Animals*, National Academy of Sciences, Washington, D.C.

[2]Occurs only when fluoride is present during the formative period of the tooth.

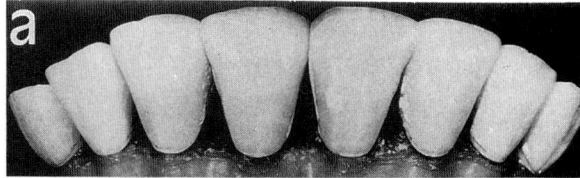
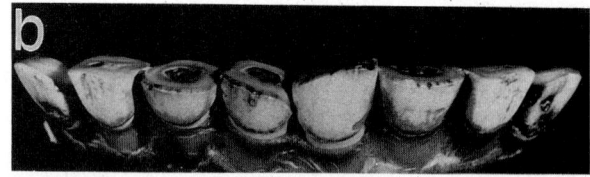
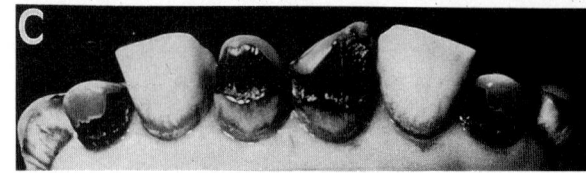

Fig. 5-36. Fluorosis evidenced in teeth of cow. A, normal teeth. B, fluorosis caused by continuous ingestion of high fluorine levels. C, fluorosis caused by intermittent ingestion of high fluoride levels. (Courtesy, Utah State University, Logan)

(Continued)

Nutritional Disorders/Toxins

TABLE 5-3 (Continued)

FLUORINE (F) POISONING (FLUOROSIS) (Continued)

Distribution and Losses Caused By:

The water in parts of Arkansas, California, South Carolina, and Texas has been reported to contain excessive fluorine. The largest high-fluorine area in the U.S. is the West Texas Panhandle. Occasionally, throughout the U.S. high-fluorine, raw-rock phosphates are used in mineral mixtures.

Areas near smeltering or metal-production industries which heat ores or burn high-fluoride coal may be a problem.

Treatment:

Any damage may be permanent, but animals which have not developed severe symptoms may be helped to some extent, if the source of excess flourine is eliminated. Some reduction in toxicity may be obtained by the addition of calcium, aluminum, or fat into the ration, as these elements seem to reduce the absorption of fluorine.

Prevention:

Avoid the use of feeds, water, or mineral supplements containing excessive fluorine (see Table, previous page).

The maximal safe level of fluorine in the total dry ration is as follows: Beef and dairy cattle, 50 to 100 ppm; sheep and swine, 100 to 200 ppm; and chickens, 300 to 400 ppm.

Remarks:

Fluorine is a cumulative poisoning.

Undefluorinated rock phosphate often contains 3.5 to 4.0% (35,000 to 40,000 ppm) of fluorine.

Phosphate clays (soft phosphates) are usually too high in fluorine to be used safely unless defluorinated.

LEAD (Pb) POISONING

Source:

Lead is discharged into the air from auto exhaust fumes and other sources.

Lead pollution of feed and food crops as a result of lead being deposited on the leaves and other edible portions of the plant by direct fallout.

Inhaling airborne lead.

Lead may get into feed or food and water from contact with lead pipes, lead utensils, discharged storage batteries, old lead base paint, or used motor oil.

Species Affected:

All animals, but cattle, sheep, and mink are especially susceptible.

Swine, goats, and chickens are relatively resistant.

There is no evidence that lead constitutes a health problem of fish in the U.S.

Symptoms and Signs (or age group most affected):

Symptoms develop rapidly in young animals, but slowly in mature animals.

Loss of appetite and evidence of gastroenteritis.

Feces may become very dark gray and be tinged with blood.

Salivation, champing of the jaws, frenzy, blindness, convulsions, coma, and death.

Mature animals usually have diarrhea and show incoordination, especially in the hind limbs, and prostration.

Distribution and Losses Caused By:

At one time, lead was a component of sprays used to control insects and plant diseases and of paints. But, leaded paint is now banned by law.

Lead poisoning is the most frequently diagnosed poisoning of domestic animals. However, its occurrence appears to be declining because lead is no longer used in paint and there is a growing tendency toward the use of nonleaded gasoline.

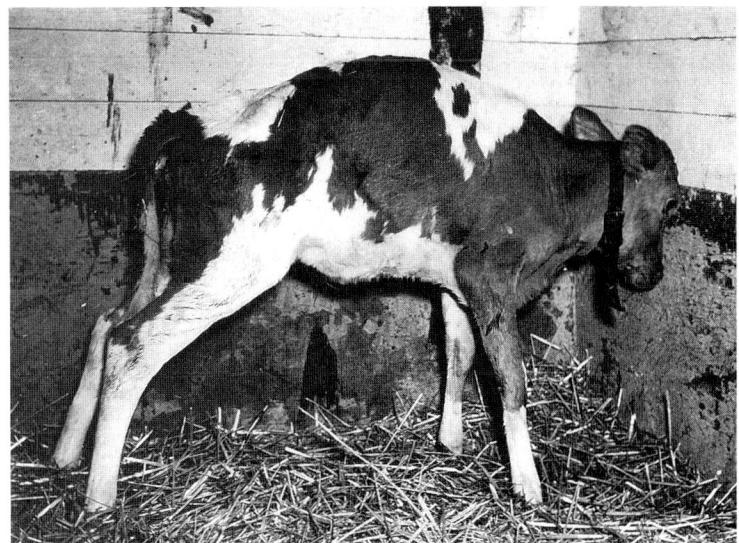

Fig. 5-37. Lead poisoning. Calf shows evidence of gastroenteritis and incoordination of the hind legs. (Courtesy, College of Veterinary Medicine, University of Illinois, Champaign-Urbana)

(Continued)

TABLE 5-3 (Continued)

LEAD (Pb) POISONING (Continued)

Treatment:

If damage to tissue has been extensive, treatment is of little value; in any event it should be handled by a veterinarian.
Magnesium sulfate (Epsom Salts) may be employed to remove any lead remaining in the digestive tract.
Calcium disodium EDTA should be given for several days to absorb lead from the tissues.

Prevention:

Avoid sources of lead.

Remarks:

Lead poisoning is cumulative.
When incorporated in the soil, nearly all the lead is converted into forms that are available to plants. Any lead taken up by plant roots tends to stay in the roots, rather than move up to the top of the plant.
Lead poisoning can be diagnosed by positively analyzing the blood tissue for lead content.

MOLYBDENUM TOXICITY (MOLYBDENOSIS)

Source:

Toxicity can show up at as little as 6 ppm, depending on the amount of copper available.

Species Affected:

Ruminants, especially calves and cows in milk.

Symptoms and Signs (or age group most affected):

Toxic levels of molybdenum interfere with copper metabolism, thus increasing the copper requirement and producing typical copper deficiency symptoms.
The toxicity signs in cattle are: diarrhea (scouring), loss of appetite, anemia, lack of coordination, bone malformation, and depigmentation of hair.
In sheep, there is depigmentation of the wool and loss of crimp.

Distribution and Losses Caused By:

Canada (Manitoba), England, and in California, Florida, Nevada, and other areas of the U.S.

Treatment:

For cattle 1 yr or older, 1 g of copper sulfate per head daily, added to the feed, will usually cure symptoms of molybdenum toxicity. For calves up to 1 year of age, add ½ g of copper sulfate to the feed daily.

Prevention:

Where the molybdenum content of the forage is below 5 ppm, add 1% copper sulfate to the salt.
Where the molybdenum content of the forage is above 5 ppm, add 2 to 5% copper sulfate to the salt, depending on the level of the molybdenum.

Remarks:

When feeds are high in sulfate, toxic symptoms will be produced on lower levels of molybdenum and, conversely, higher levels of molybdenum can be tolerated with low levels of sulfate.

MYCOTOXINS (TOXIN-PRODUCING FUNGI OR MOLDS)

Source:

Aflatoxin (most studied of the group) associated with peanuts, brazil nuts, silage, corn and most other cereals, hay, and grasses. The mold can produce toxic compounds on virtually any feed/food (even synthetic) that will support growth.
Aflatoxin is actually a group of toxins which are similar in chemical structure, the principal ones of which are aflatoxin B_1, aflatoxin B_2, aflatoxin G_1, and aflatoxin G_2. Aflatoxin B_1 is usually found in greatest abundance and is the most toxic form of aflatoxin.
While aflatoxin appears to cause most of the problem, it is not the only mycotoxin to be feared. Other mycotoxins are being studied, especially ochratoxin and T_2 (trichothecenes).

(Continued)

Nutritional Disorders/Toxins

TABLE 5-3 *(Continued)*

MYCOTOXINS (TOXIN-PRODUCING FUNGI OR MOLDS) (Continued)

Species Affected:

Cattle, chickens, ducklings, goats, horses, mink, pheasants, sheep, swine, trout, turkeys, and humans.

In all species, the young are far more susceptible than mature animals.

Generally, ruminants appear to tolerate higher levels of mycotoxins and longer periods of intake than simple-stomached animals.

Symptoms and Signs (or age group most affected):

Mold affects animals in a variety of ways, from decreased production to sudden death. Usually, the first sign is the loss of appetite and weight.

A few animals will abort, and an occasional animal will die.

Aflatoxin exposure has been associated with liver cancer in children, and with trout hepatoma.

With high intakes of mycotoxin, or with the several types of molds, any one or a combination of the following symptoms may develop: liver damage, hyperkeratosis, a typical interstitial pneumonia, bloody slimy scours, arched back, dry gangrene at the end of the tail or top of hoof, hemorrhagic hepatitis, renal damage, lameness and/or swollen legs. In swine, an estrogenic mycotoxin produced by *Fusarium graminearum* in corn produces swelling of the vagina and possibly mammary development in gilts and preputial enlargement in boars.

Generally, aflatoxin does not affect animals when the level in the ration is below 100 ppb (0.1 ppm). Typically, aflatoxin residues are short-lived in animals that don't develop liver lesions. In swine, levels of 500 ppb (0.5 ppm) have been eliminated from the tissues within 4 days.

There are many mycotic diseases. Only Aflatoxicosis (the response of animals to aflatoxin) is herewith summarized:

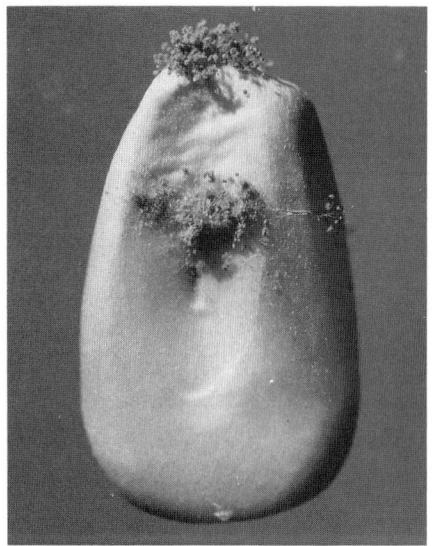

Fig. 5-38. Sporulation of *Aspergillus flavus* on a kernel of corn. This fungus produces aflatoxin. Aflatoxin (1) is associated with a high incidence of liver cancer, and (2) may be involved in some types of acute poisoning. (Courtesy, USDA, Agricultural Research, Peoria, Ill.)

Class of Animal	Level of Aflatoxin in Feed	Effects on Animals
Cattle:	10-20 ppm	Jaundice, hemorrhage, liver necrosis. Death in 1-2 weeks.
Calves	0.2 ppm	Fed 2-4 weeks will cause reduced weight gain and impaired blood coagulation.
Dairy	2-4 ppm	Off feed, reduced milk production.
Yearling Steers	0.7 ppm	Multiple doses (3-4) months may lead to liver damage, reduced rate of gain, and death.
Poultry:		
Broilers	5-10 ppm	Liver necrosis, hemorrhage, death.
	2-5 ppm	Impaired blood coagulation, reduced growth and feed efficiency.
	1.5-2.5 ppm	Decreased gain and feed efficiency.
Layers	2.0 ppm	Reduced egg production.
Ducks	0.3 ppm	Liver damage, death.
Turkeys	0.25 ppm	Decreased weight gain and impaired immunity to disease.
Swine	10-20 ppm	Single exposure, lethal. Hemorrhage, acute hepatitis.
	2-4 ppm	Lethal. Multiple doses required.
	0.8-2.0 ppm	Subacute, and may be lethal. Liver necrosis, jaundice, fibrosis (formation of fibrous tissue), hemorrhages.
	0.2-0.5 ppm	Reduced gain, impaired immunity to disease.

Distribution and Losses Caused By:

Widely distributed throughout the world.

In addition to the effect of mycotoxins on the animal's health, milk and eggs are contaminated by the residues or mycotoxins, or by their metabolic products.

Treatment:

Remove the source of the mold.

Animals suffering from molds frequently respond to vitamin B injections.

Iron therapy may be helpful, since hemorrhaging is a frequent problem.

Prevention:

The prime cause of aflatoxin is moisture, which favors mold growth; hence, proper harvesting, drying, and storage are important factors in lessening contamination and toxin production.

Propionic and acetic acids, and sodium propionate, will inhibit mold growth; hence, their use in preserving high-moisture grains is encouraged.

Increasingly, as a preventive measure, grains and processed feeds are being tested for mycotoxins and required to meet acceptable levels.

(Continued)

TABLE 5–3 (Continued)

MYCOTOXINS (TOXIN-PRODUCING FUNGI OR MOLDS) (Continued)

Remarks:

Certain molds produce toxins, or mycotoxins.
Aflatoxin has been clearly shown to be a carcinogen (tumor producing).
Ultraviolet irradiation and anhydrous ammonia under pressure will reduce the toxicity of aflatoxins and, if continued long enough, will deactivate them entirely.
Not all mold products are harmful. For example, zeranol is being commercially produced as a growth-promotant hormone for cattle.
Food and Drug Administration regulations do not permit feeding more than 100 ppb aflatoxins in the total ration of livestock and poultry.
(Also see Chapter 13, section on "Mold [Fungi] Inhibitors.")

NITRATE/NITRITE POISONING (OAT HAY POISONING, CORNSTALK POISONING)

Acute nitrate/nitrite poisoning is caused by the presence of nitrite in the blood at a level sufficient to cause anoxia (internal suffocation). Nitrate (NO_3) can be reduced to nitrite (NO_2) by microorganisms in the gastrointestinal tract, especially in the rumen, at a rate which overwhelms the body's defense system. Nitrite combines with the hemoglobin of the red blood cells to form methemoglobin, which cannot transport oxygen to the body tissues.

Source:

Nitrate is a naturally occurring form of nitrogen and a desirable part of our environment, found in most soils and in a number of fertilizers. Under normal conditions, plants use nitrates and other nitrogen compounds to form plant proteins. However, when nitrate concentrations are excessive and/or out of place, as may happen when normal growth is altered by the environment (such as drought) protein formation may be slowed and the nitrogen may remain in the plants as non-protein nitrogen—nitrates, nitrites, amides, free amino acids, and peptides. The nitrate is of special concern because of its potential toxicity when excessive amounts are ingested.

The three principle sources of nitrate for animals and humans—plants, water, and air—are interrelated. The sources, along with their relationships, are depicted in Fig. 5–39. As shown, nitrogen may be (1) fixed by microorganisms in ruminant animals and in legumes (natural nitrogen fixation), (2) formed when plant residues, animal manures, and human wastes decompose, or (3) added to soils as nitrogen fertilizer.

In order to maintain production, when nitrogen is removed from the cycle, it must either be (1) returned so that nature can reuse it, or (2) added as chemical fertilizer nitrogen. The cyclic nature of nitrogen in the environment is depicted in Fig. 5–40. The nitrogen cycle includes various changes from elemental atmospheric nitrogen to inorganic, to organic, and back to inorganic forms.

Crop production has been increased by making more nitrogen and other nutrients available to crops. But this practice also increases the chance of getting the nitrogen cycle out of balance. Even when proper fertility and crop selection decisions have been made, changes in the environment—the rainfall, temperature, sunlight, and shifting of seasons—can still alter the nitrate concentration of crops and the water supply. Nitrate poisoning of animals and people can result from ingesting plants or water high in nitrate content, commonly one or more of the following sources:

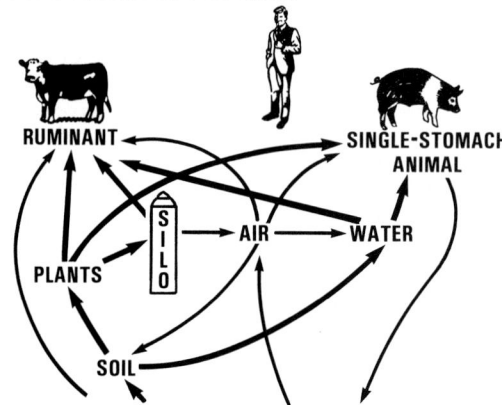
Fig. 5–39. Nitrates in relation to animals and humans.

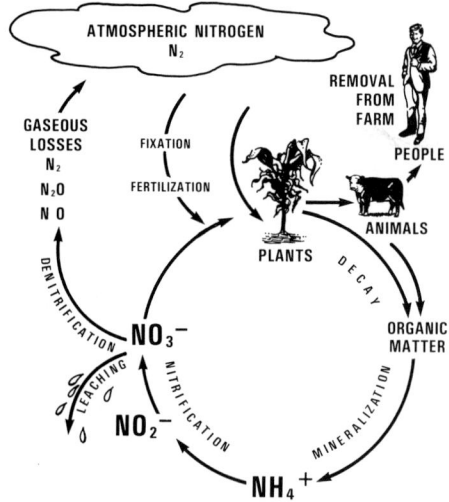
Fig. 5–40. The nitrogen cycle.

1. Forages (vegetative part) of most grain crops (oats, wheat, barley, rye, corn, sorghum), Sudangrass, and numerous weeds, especially (a) when under stress such as drought, insufficient sunlight, or after spraying with weed killer (herbicide); or (b) following heavy nitrate fertilization of soils (commercial, green manure crop, barnyard manure). Some nitrate may be formed after forage is stacked.
2. Inorganic nitrate or nitrite salts, or fertilizer left where animals have access to them, or where they may be mistaken for salt.
3. Pond or shallow well water into which surface runoff from barnyard or well-fertilized soil may drain.

Species Affected:

Primarily cattle. Sheep. Horses. Ruminants are most susceptible because of the conversion of nitrates to nitrites by the microorganisms in the rumen.

Symptoms and Signs (or age group most affected):

Accelerated respiration and pulse rate; diarrhea; frequent urination; loss of appetite; general weakness; trembling and staggering gait; frothing from mouth; lowered milk production; abortion; blue color of the mucous membrane, muzzle, and udder due to lack of oxygen in the blood; death within 4½–9 hours after consuming nitrates.

(Continued)

Nutritional Disorders/Toxins

TABLE 5-3 (Continued)

NITRATE/NITRITE POISONING (OAT HAY POISONING, CORNSTALK POISONING) (Continued)

A rapid and accurate diagnosis of nitrate poisoning may be made by examining blood. Normal blood is red and becomes brighter when exposed to air, whereas blood from cows toxic with nitrates is a brown color due to formation of the methemoglobin. Nitrates oxidize ferrous hemoglobin (oxyhemoglobin) to ferric hemoglobin (methemoglobin) which is not an efficient oxygen transporter. The animal essentially suffocates for lack of oxygen in tissues. When ¾ of the oxyhemoglobin is converted to methemoglobin, the animal will die.

Distribution and Losses Caused By:

Excessive nitrate content of feeds is an increasingly important cause of poisoning in farm animals, due primarily to more and more high nitrogen fertilization. But nitrate toxicity is not new, having been reported as early as 1850, and having occurred in semiarid regions of this and other countries for years.

Treatment:

Death usually occurs so suddenly that treatment is not possible, and few treated animals recover.

The most common treatment is a 4% solution of methylene blue (in a 5% glucose or a 1.8% sodium sulfate solution) administered by a veterinarian intravenously at the rate of 100 cc/1,000 lb liveweight.

Prevention:

More than 0.9% nitrate (dry basis) may be considered as potentially toxic. Feed should be analyzed when in question, by using a simple test to detect the presence of nitrates (qualitative); if present, follow with a quantitative test to determine how much is present.

Nitrate poisoning may be reduced by (1) feeding high levels of grains and other high-energy feeds (molasses) and vitamin A, (2) limiting the amount of high-nitrate feeds, (3) ensiling forages which are high in nitrates (fermentation reduces some nitrates to gas, but care must be taken to avoid nitric oxide and nitrogen dioxide released in early stages of fermentation), and avoid feeding until 3-4 weeks in storage.

Remarks:

The nitrate form of nitrogen does not appear to cause the actual toxicity. During digestion, the nitrate is reduced to nitrite, a far more toxic form (10-15 times more toxic than nitrates). In cows and sheep, this conversion takes place in the rumen (paunch); in horses, in the cecum.

Lethal dose varies with (1) nutritional state, size and type of animal; and (2) the consumption of feed other than nitrate-containing material.

Methods of reporting nitrates (dry basis) in rations in relation to death losses follow:

	Potentially Lethal Levels	
	(%)	(ppm)
Nitrate (NO_3)	over 0.9	9,000
Nitrate nitrogen (NO_3N)	over 0.21	2,100
Potassium nitrate (KNO_3)	over 1.5	15,000

OAT HAY POISONING

(See NITRATE POISONING.)

PESTICIDE POISONING

Poison:

Pesticides are chemicals used to destroy, prevent, or control pests, but they can also be toxic (poisonous) to animals, people, and plants.

Pests can be classified into 6 main groups: (1) insects (plus mites, ticks, and spiders); (2) snails and slugs; (3) vertebrates, including rats, mice, and certain birds (starling, linnets, English sparrows, crows, and blackbirds); (4) weeds; (5) plant diseases; and (6) nematodes.

Source:

Pesticides are chemicals. When properly used, they are beneficial; when improperly used, they may be hazards.

Pesticide poisoning may be caused by either (1) sudden exposure to lethal quantities, or (2) as a result of repeated exposure to nonlethal quantities (chronic poisoning) during a protracted period of time.

Species Affected:

Animals, people, and plants.

Many pesticides are so highly toxic that very small quantities can kill an animal and exposure to a sufficient amount of almost any pesticide can make an animal ill. Even fairly safe pesticides can irritate the skin, eyes, nose, or mouth.

Some pesticides produce unplanned and undesirable "side effects," particularly when they are not used properly. Among such effects are: reduction of beneficial species; drift; wildlife losses; honeybee and other pollenating insect losses; and pollution of air, soil, water, and vegetation.

(Continued)

TABLE 5-3 (Continued)

PESTICIDE POISONING (Continued)

Symptoms and Signs (or age group most affected):

Knowing something of the toxicity and symptoms or signs of poisoning of each type of pesticide may result in getting medical advice quickly and in saving a life. The overall symptoms of pesticide poisoning are:
1. Organophosphates—most toxic; they injure the nervous system.
2. Carbamates—safer than the organophosphates; they produce the same symptoms as the organophosphates, but they respond more easily to treatment.
3. Fumigants—cause poor coordination and confusion.
4. Plant-derived pesticides—some are very toxic; pyrethrum may cause allergic reaction, rotenone may irritate the respiratory tract, nicotine is a fast-acting nerve poison.

Distribution and Losses Caused By:

In every pursuit of modern agriculture, more and more pesticides are being used. They are the first line of defense against pests that affect human health and well-being and attack livestock, crops, and structures.

Pesticides are used to control many of the estimated 10,000 species of harmful insects; more than 160 bacteria, 250 viruses, and 8,000 fungi known to cause plant diseases; 2,000 species of weeds and brush; and 150 million rats.

Treatment:

Each label contains a "Statement of Practical Treatment." Read it before using a pesticide.

Prevention:

The first and most important precaution to observe when using any pesticide is to read and heed the directions on the label. In the event of an accident, the label becomes extremely important in remedial measures.

For public protection, all chemicals are rigidly controlled by federal laws. Each one is required to be registered by the Environmental Protection Agency before it can be sold in the U.S., and each one is issued a tolerance for residues that may result from its use on food or feed crops.

Remarks:

The Environmental Protection Agency (EPA) administers the following acts:
1. The Federal Insecticide, Fungicide, and Rodenticide Act (FIFRA) as amended in 1972. Every pesticide must be registered, and commercial applicators must be certfied—showing that they know the safe and correct way to use them.
2. The Toxic Substances Control Act (TOSCA) of 1976, which regulates chemicals that are not presently covered under other federal acts. It might more properly be described as the "chemical health and environmental regulation act."

U.S. consumers benefit greatly from the proper use of pesticides. They enjoy the world's most abundant food supply with quality and variety second to none; and they spend less than 20% of their income for food, compared with the 50% or more that people in many other countries spend for food.

PINE NEEDLE ABORTION

Source:

Pine needles or slash of *Pinus ponderosa* (ponderosa pine; western yellow pine).

Species Affected:

Cattle.

Symptoms and Signs (or age group most affected):

Pregnant cows, free of brucellosis, abort, especially during the last 3 months of pregnancy.

Pine needle abortion usually appears 1 to 3 days after pregnant cattle have eaten the needles or buds. Abortions will continue for up to 2 weeks though cattle are removed from the needles.

At calving, excessive hemorrhaging; retained placenta; septic metritis, often followed by peritonitis.

If cow is affected near parturition, calf may be born normal, but weak.

Fig. 5-41. Cow aborting, as a result of eating pine needles. (Courtesy, USDA-ARS Poisonous Plant Research Laboratory, Logan, Utah)

(Continued)

Nutritional Disorders/Toxins

TABLE 5-3 (Continued)

PINE NEEDLE ABORTION (Continued)

Distribution and Losses Caused By:

Wherever ponderosa pine trees are found; in the Northern Plains, Rocky Mountain, and Pacific Northwest regions of the U.S. and in western Canada.
Not all pregnant cows will abort after eating pine needles, but the disease has been known to affect as much as 50% of the herd.

Treatment:

None known.

Prevention:

Keep pregnant cows away from yellow pine, especially during the latter part of gestation.
Maintain the cow herd on the recommended level of nutrition and avoid stress conditions as much as possible.

Remarks:

Cows will eat pine needles even when well fed.
Generally, cattle eat pine needles or buds only under the following circumstances:
1. Sudden weather changes causing animals to seek shelter under ponderosa pine where they will eat the needles.
2. Severe storms which place large quantities of the needles on the ground.
3. Animals concentrated near ponderosa pine.
4. The cattle are hungry.
5. Hay is fed on the ground beneath the trees.
6. The cattle are changed to unfamiliar or poor quality feed.

PRUSSIC ACID (HCN) POISONING

The toxic material is hydrocyanic acid—HCN.

Source:

Most outbreaks of hydrocyanic acid poisoning are caused by the ingestion of plants which contain cyanogenetic glucosides. In this form, the acid is non-toxic but it may be liberated from the organic complex by the action of an enzyme which may also be present in the same plant or in another plant, or by the activity of rumen microorganisms. Among the common plants that may produce prussic acid poisoning are: Johnsongrass, Sudangrass, sorghum, wild black cherry, chokeberry, and arrow grass.
Some clovers, particularly white clover (*Trifolium repens*) and members of the *Brassica* genus (mustard) as well as plants of the flax family may also contain large amounts of the poison. A high soil nitrogen level increases the hydrocyanic acid content of the plants, especially when soil phosphorus levels are low. The prussic acid content is increased by heavy nitrate fertilization, excessive irrigation, wilting, trampling, and plant diseases. Very young, rapidly growing plants also contain greater quantities of the glycoside than more mature plants. Freezing does not ordinarily increase glycoside content in these plants but it does tend to increase the quantity of free hydrocyanic acid in the plants, thus resulting in a temporary increase in toxicity. Spraying cyanogenetic plants with plant herbicides apparently also increases the toxic hazard.

Species Affected:

All animals. However, ruminants are more susceptible to HCN poisoning from plants than nonruminants, because the rumen microflora and pH encourage greater glucoside breakdown than occurs in nonruminants.

Symptoms and Signs (or age group most affected):

If extremely large doses are consumed rapidly, generalized spasms develop and animals die within a few minutes. If small doses are consumed over a longer period, the more common clinical signs may be seen. The onset of this form is sudden and characterized by slobbering or frothing at the mouth and the gradual increase in the breathing rate. Within 5 to 15 minutes, this is developed to the point of open mouth breathing. Pulse becomes rapid and weak, muscle twitching occurs early, and progresses into generalized spasms just before death. Most animals stagger around and fall considerably before they go down. The mucous membranes may be bright red in color but become blue near death. Death from respiratory paralysis occurs during the convulsions. Bright red blood often passes from the nostrils and the mouth near the time of death. The course is rapid and does not usually exceed 30-45 minutes. A high percentage of animals that live for 2 hours after the onset of signs generally will recover. Diagnosis of prussic acid poisoning is made on the basis of history, cherry red color of the blood, signs and symptoms, post-mortem findings, and the demonstration of the presence of hydrocyanic acid in the stomach contents.

Distribution and Losses Caused By:

Prussic acid poisoning is worldwide. Approximately 1,000 plant species in 250 genera are known to contain HCN. Although reliable figures are not available, the losses are large.

(Continued)

TABLE 5-3 *(Continued)*

PRUSSIC ACID (HCN) POISONING *(Continued)*

Treatment:

Often when prussic acid poisoning is first noted, the animal is dead before there is time to administer an antidote. Animals which have not shown much evidence of toxicity may be injected intravenously with a mixture of sodium nitrite and sodium thiosulfate. The dose rates are 3 g of sodium nitrite and 15 g sodium thiosulfate in 200 ml water for cattle; for sheep, 1 g sodium nitrite and 2.5 g sodium thiosulfate in 50 ml water. Treatment may have to be repeated because of further liberation of hydrocyanic acid. Prussic acid poison is not cumulative; hence, on being removed from the feed, those animals not showing evidence of being poisoned will likely not be affected adversely.

Prevention:

Efforts to prevent prussic acid poisoning must be directed toward preventing the use of cyanogenetic plants for grazing. The risk of cyanide poisoning may also be decreased by the heavy feeding of ground grains before the animal is turned out to graze. Carbohydrates tend to inhibit the action of the enzyme that hydrolyzes the glucoside. Hungry cattle should not be allowed access to toxic plants, especially cultivated sorghum species when they are immature, wilted, frost bitten or growing rapidly after a stage of retarded growth. Plants of the sorghum family should be in flower before they are grazed or chopped to be fed green. If there is doubt as to the toxicity of a field, plants should be tested.

Observation of the following precautions will lessen prussic acid poisoning:
1. Allow 18 to 24 in. of growth before grazing Sudangrass. Sudangrass-sorghum crosses should be 25 to 30 in. high.
2. Do not graze frost-injured Sudangrass.
3. Do not allow hungry cattle to feed on pasture which is just recovering from a dry soil moisture condition and/or frost.
4. Watch out for Johnsongrass or sorghum which would be growing in Sudangrass fields. They contain higher concentrations of prussic acid (HCN) than Sudangrass.
5. Do not turn hungry animals into very short succulent growth.
6. Consider tissue as safe if it contains less than 500 ppm HCN, critical at 500 to 750 ppm, and dangerous above 750.

Remarks:

Care must be taken to distinguish between nitrate and HCN poisoning, because the treatment would cause death if the animal suffered from nitrate poisoning.
Sorghum, sorghum hybrids, and Sudangrasses should not be used for horse pasture; they are capable of producing a glycoside which converts to free cyanide in the horse and causes cystitis.
Piper Sudangrass is low in HCN.

SALMONELLOSIS

This is a toxic disease of all animals caused by many species of salmonellae.

Source:

Infected animals shed the bacteria in their feces, which are consumed by other animals in contaminated feedstuffs. Rodents and birds are also sources of infection.

Species Affected:

All animals, including humans.

Symptoms and Signs (or age group most affected):

Diarrhea, depression, dehydration, and fever. Often blood is seen in the feces. In pigs, a dark red-to-purple discoloration of the skin is common, especially of the ears and ventral abdomen. Nervous signs may appear in calves and pigs.

Distribution and Losses Caused By:

The disease is worldwide, and the incidence is increasing with the intensification of livestock production.
Usually 10 to 80% of the animals in a group are affected. Death losses vary with the severity of the infection and the treatment given.

Treatment:

Fluid therapy to correct acid-base balance and dehydration, and antibacterial drugs (broad-spectrum antibiotics, nitrofurans, or ampicillin); under the direction of a veterinarian.

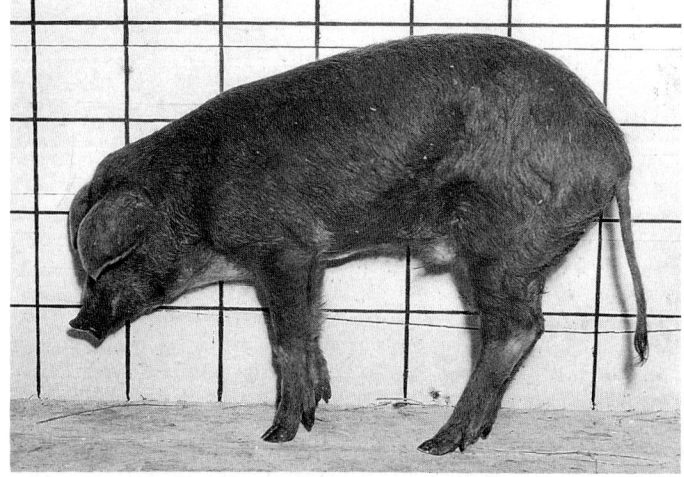

Fig. 5-42. Pig with salmonellosis. Note signs of marked depression and weakness. (Courtesy, Department of Veterinary Pathology and Hygiene, College of Veterinary Medicine, University of Illinois, Champaign-Urbana)

(Continued)

Nutritional Disorders/Toxins

TABLE 5-3 *(Continued)*

SALMONELLOSIS (Continued)

Prevention:

Quarantine new animals and avoid contaminated feed. Practice good sanitation. Clean and disinfect animal quarters, followed by drying. Control rodents and birds. Reduce stress, because stress triggers the disease. **NOTE WELL**: Producers could do away with salmonellosis by separating manure from animals. Also, salmonella in human foods can be destroyed by irradiation and proper cooking. **NOTE WELL**: Before fresh poultry can be irradiated, FDA must approve the treatment; and before it will be irradiated, consumers must be willing to buy the treated product.

Remarks:

Salmonellosis is named for the American bacteriologist and veterinarian, Daniel E. Salmon, who first isolated the organism in 1885.
Animals in close proximity to one another, such as veal calves or confined cattle, are of greater risk than isolated animals.
(Also see Chapter 13, section on Antibiotics, subsection headed "Safety And Future Of Antibiotics As Feed Additives.")

SELENIUM TOXICITY (ALKALI DISEASE, BLIND STAGGERS)

Source:

Chronic selenium poisoning, commonly called *alkali disease*, may result when animals consume forages and grains containing 5 to 40 ppm selenium.
Acute selenium poisoning, commonly called *blind staggers*, may result when animals consume high selenium content plants known as "selenium accumulators." Two of the accumulators in North America are: *Astragalas* and *Stanleya*. Also, cases of selenium toxicity have been caused by animals breaking into bags and consuming selenium supplements containing high levels of selenium.

Species Affected:

All farm animals and people. Young animals are especially susceptible. **NOTE WELL**: Confirmed cases of selenium toxicity in humans are rare because (1) the foods and beverages which are consumed by people are not likely to contain excesses of the element, and (2) well-nourished people are protected by metabolic processes that convert selenium into harmless substances which are excreted in the urine or in the breath. Nevertheless, a few cases of poisoning have occurred under unusual circumstances such as (1) very high levels of the element in the drinking water, or (2) the presence of malnutrition, parasitic infestation, or other factors which may make people highly susceptible to selenium toxicity.

Symptoms and Signs (or age group most affected):

In chronic selenium poisoning (alkali disease), there is a loss of hair from the mane and tail of horses, a loss of hair from the tail of cattle, and a general loss of hair in swine. The hoofs slough off, lameness occurs, feed consumption decreases, and death may occur from starvation. Chronic selenium poisoning in poultry and other birds is characterized by reduced egg production and hatchability, and by deformities in young, including lack of eyes and deformed wings and feet.
In acute selenium poisoning (blind staggers), vision is impaired. In cattle, blind staggers is manifested in 3 stages. In stage 1, there is a tendency to wander, the vision is poor, and there is a loss of appetite. In stage 2, the wandering increases, the vision becomes poorer, and the front legs become weak. In stage 3, the throat and tongue become paralyzed, the temperature is subnormal, and death follows from respiratory failure. In sheep, the 3 stages are not as clearly differentiated as in cattle.

Fig. 5-43. Selenium toxicity in cow grazing on forage produced on alkali soil containing excessive selenium. Note emaciated condition, curvature of back, and deformed hoofs. (Courtesy, Wyo. Ag. Exp. Sta., Laramie, Wyo.)

Distribution and Losses Caused By:

Most of the selenium poisoning in livestock occurs in Colorado, Nebraska, South Dakota, and Wyoming. Also, high selenium has been reported in Alberta, Saskatchewan, and Manitoba in Canada; in Mexico, Ireland, Israel, and China; in northern Queensland in Australia; and in parts of South America.
Losses result primarily from the failure to thrive on forages grown on seleniferous soils. But some deaths occur from the chronic and acute forms of the disease.
Fig. 5-44 shows the selenium level of U.S. soils and feeds. Note the deficient and high regions of the country.

Treatment:

The effect of chronic selenium toxicity may be reduced by feeding a high-protein ration, by the use of trace amounts of arsenic compounds, and/or by the oral administration of such compounds as naphthalene and bromobenzene; with such treatments under the direction of a veterinarian or nutritionist.
There is no known treatment for acute selenium poisoning.

(Continued)

TABLE 5-3 *(Continued)*

SELENIUM TOXICITY (ALKALI DISEASE, BLIND STAGGERS) *(Continued)*

Prevention:

Soils which contain more than 0.5 ppm selenium are potentially dangerous. Chronic toxicity is caused by rations containing as little as 8.5 ppm of selenium. In swine, levels as low as 10 ppm have been found to lower the conception rate and result in a higher percentage of small, weak, and dead pigs at birth. Acute cases of poisoning have been reported from levels of 500 to 1,000 ppm. So, prevention consists in the cautious use of soils and crops exceeding these levels. Some of these areas may be used by pasture rotation and the use of supplemental feeds.

The maximum and toxic levels of selenium for farm animals are:

	Maximum Total Recommended By FDA	Toxic Level	
	(mg/head/day)	(ppm in feed)	(mg/head/day)
Beef cattle	1.0	10-30	100-300
Dairy cattle	2.0	3-5	30-60
Sheep	0.23	3-20	7-50
Swine		5-10	8-16
Chickens		2	
All species	2.0 (or 2 ppm)[1]		

[1]Sugggested maximum tolerable level for all species (NRC, 1980).

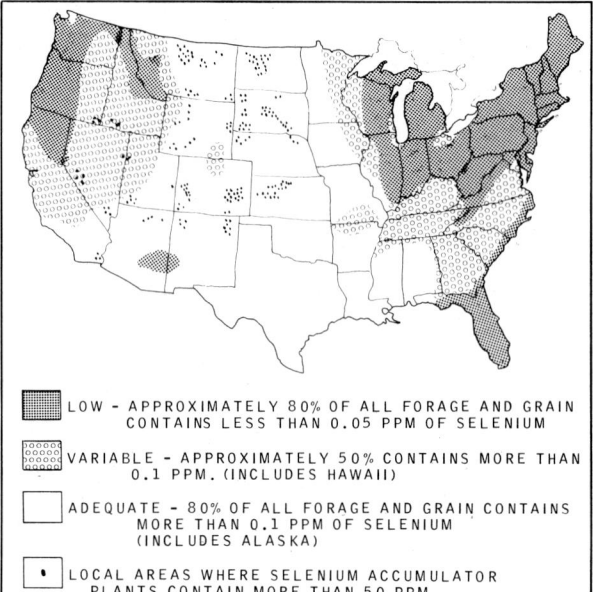

Fig. 5-44. A geographic distribution of low, variable and adequate or high selenium areas in the U.S. (Source: Kubota, J. and W. H. Allaway, "Geographic Distribution of Trace Element Problems," *Micronutrients in Agriculture*, Soil Science Society of America)

Remarks:

Although selenium is highly toxic in overdoses, it is essential in trace amounts to maintain life and to prevent such conditions as muscular dystrophies in many species and exudative diathesis in chicks.

The toxic effects of selenium were observed long before the existence of this element was known. In his travels in the mountains of western China at the end of the 13th century, Marco Polo recorded that consuming certain forages caused the hoofs of grazing animals to drop off.

UREA TOXICITY (AMMONIA TOXICITY/BOVINE BONKERS/CRAZY CATTLE DISEASE)

Ammonia is the actual toxic agent in urea poisoning.

Source:

Primarily urea, which dominates the market. But other nonprotein nitrogen (NPN) products, such as ammonium salts and ammoniated feeds, which are used in ruminant rations.

Urea is hydrolyzed by the urease activity of the rumen microorganisms with the production of ammonia as follows:

$$\text{(NH}_2\text{)}_2\text{C=O (urea)} + \text{H}_2\text{O (water)} \xrightarrow{\text{microbial urease}} 2\text{NH}_3 \text{ (ammonia)} + \text{CO}_2 \text{ (carbon dioxide)}$$

When urea is fed at excessive levels, large amounts of ammonia are liberated in the rumen. Eventually, the pH of the ruminal fluid increases, thus facilitating the passage of ammonia across the rumen wall. If the levels of ammonia absorbed are greater than the ability of the liver to convert ammonia to urea, ammonia accumulates in the blood. If blood ammonia levels reach toxic levels (80 mg per 100 ml), the animal shows signs of acute ammonia poisoning.

Species Affected:

Primarily ruminants, for little nonprotein nitrogen is used by nonruminants. There is hazard of ammonia toxicity (urea toxicity) when NPN is fed to young ruminants and young equines, due to their limited bacterial action.

(Continued)

TABLE 5-3 (Continued)

UREA TOXICITY (AMMONIA TOXICITY) (Continued)

Symptoms and Signs (or age group most affected):

The animal shows signs of nervousness, excessive salivation, muscular tremors, respiratory difficulty, and tetanic spasms. Death occurs within ½ to 2½ hours.

Distribution and Losses Caused By:

Urea or ammonia toxicity should never occur in practice, if feeds are thoroughly mixed and total intakes are moderate. Errors in formulation and improper mixing of urea with other ration ingredients are probably the major factors causing urea toxicity in the feeding of ruminants.

Treatment:

An effective treatment for urea toxicity of cattle, if applied before tetanic spasms occur, consists in administering, immediately, 5 to 10 gal of cold water orally. A gallon of either dilute acetic acid or vinegar given with cold water is more effective than cold water alone.

Prevention:

Prevention of urea toxicity consists in alleviating, or lessening, the following predisposing factors:
1. Poor mixing of feed.
2. Errors in ration formulation.
3. Inadequate period of adaptation.
4. Low intake of water.
5. Feeding urea in conjunction with poor quality roughages.
6. Low feed intake prior to feeding urea.
7. Treating hay with more than 1.5% anhydrous ammonia, and treating it unevenly.

Remarks:

Urea is less effective in young ruminants in which the rumen is not fully functional.

Urea and other NPN sources are not used to any appreciable extent for swine or poultry. The mature horse can utilize limited NPN (such as access to protein blocks containing urea); however, the efficiency of nitrogen utilization from NPN is considerably less than that of nitrogen from intake protein.

(Also see Chapter 11, Protein Supplements, sections on "Urea" and "Ammoniated Products.")

Diagnosing and Treating Livestock Poisoning

It is often difficult to make a definite diagnosis of an animal poisoning. Clinical signs are not usually specific, and all signs are not always seen in every poisoned animal. However, in a herd of poisoned animals, every sign or toxic effect will likely be seen in some animal. The recommended procedure for making a diagnosis of the cause of poisoning follows:

1. Check on the accessibility of a poisonous substance. A highly toxic substance may or may not be hazardous to livestock, depending on whether the animals could conceivably have come into contact with it.
2. Study the clinical signs. This may be difficult, especially with possible combinations of toxins or infectious agents.
3. Use a few test animals in a feeding trial.
4. Make a pathologic examination of the animal's internal organs and tissues.
5. Chemically analyze the feed, water, and animal tissues for the presence of suspected toxin. It is necessary to have enough information so that certain poisons or groups of poisons can be suspected, because the analytical methods are quite specific and certain tissues are required.
6. Use a specific antidote (where available) for the suspected poison. If it alleviates the clinical signs, it gives evidence of the cause.

The principles of treatment are directed toward accomplishing the following:

1. Preventing injury and controlling convulsions with a sedative, usually a barbiturate.
2. Relieving pain by use of chemical analgesics.
3. Removing or neutralizing the poison by—
   a. Washing off any surface poison.
   b. Using gastric lavage with activated charcoal for absorbing toxins in the stomach.
   c. Using cathartics to help fecal elimination of unabsorbed toxins.
   d. Using diuretics to help urinary elimination of absorbed toxins.
   e. Performing a rumenotomy for physical removal of unabsorbed toxins.
   f. Using a specific antidote, if available.
4. Maintaining the vital signs of respiratory, circulatory, and renal functions by physical or chemical resuscitation, fluid therapy, etc.
5. Observing the animal for further treatment needs, because the toxin may continue to be absorbed from the skin, gut, or respiratory system of the animal.

National Animal Poison Control Center

Established in 1978 and maintained at the University of Illinois, Urbana-Champaign, the National Animal Poison Control Center hotline number is: *217/333-3611*. Recognizing that accidents don't wait for business hours, the Center is open 24 hours a day, every day of the week. The toxicology group is staffed to answer questions about known or suspected cases of poisoning or chemical contaminations involving any species of animal. It is not intended to replace local veterinarians or state toxicology laboratories, but to complement them.

The toxicologists at the Center constantly update their files on chemicals, feed additives, human and veterinary drugs, pesticides, environmental contaminants, and plant and mold toxins. Their comprehensive file of information contains comparative species toxicity data, product ingredients, and recommended therapeutic and decontamination measures. The goal is a computer database containing 200,000 entries to facilitate quick and accurate responses to all types of poisoning/contamination incidents and inquiries.

Many times a proper treatment regimen can be recommended over the telephone. When telephone consultation is inadequate or the problem is of major proportions, a team of veterinary specialists can arrive at the scene of a toxic or contamination problem within a short time.

The cost of an investigation varies according to distance traveled, personnel time, and laboratory services required. Where consultation over the telephone is adequate, there is no charge to the veterinarian or producer.

HAIR ANALYSIS

Minerals are deposited in hair as it grows; and, in theory, the hair reflects the mineral status of an animal or person at the time of hair growth. Scientific analytical methods, such as atomic absorption spectometry, neutron activation analysis, and x-ray fluorescence spectometry, are sensitive enough to detect the levels of even trace minerals in hair samples. Because hair samples are easily and painlessly obtainable, and because hair samples are stable and store easily, there is considerable interest in the use of hair as a diagnostic tool for mineral deficiencies and/or toxicities. However, like many other diagnostic tests, hair analysis is only a tool to complement other tests and observations.

Clearly, some scientific findings demonstrate that if dietary intakes of certain minerals (chromium, copper, iron, manganese, and selenium) are extremely low, if the ration includes toxic minerals (arsenic, cadmium, lead, and mercury), hair analysis can detect these changes. Furthermore, some disorders such as anemia are reported to change the mineral levels of the hair.

In summary, while hair analysis has been used to detect certain types of heavy metal poisoning (e.g. lead, arsenic, mercury), its value in determining nutritional status remains to be established. There are a number of limitations to hair analysis, both in terms of analytical procedures and in interpretation of the results.

GOOD LIVESTOCK REQUIRE GOOD SOILS

Proper animal nutrition is obtained by feeding hays, silages, root crops, grains, and pastures grown on fertile soils. Good livestock require good soils.

Present research, together with practical observation, points to the fact that the mere evaluation of crop yields in terms of tons of forage or bushels of grain produced per acre is not enough. Neither does a standard feed analysis (a proximate analysis) tell the whole story. Rather, there is a direct and most important relationship between the fertility of the soil and the composition of the plant. Sometimes the concentration of an essential mineral in the soil is so low that plants growing on it will not contain enough of that mineral to meet the dietary requirements of animals. Under such circumstances, adding the deficient element to the soil may help the animals meet their requirements. At other times, feed crop plants may contain such high concentrations of certain minerals that they are toxic to the animals that eat them.

Fig. 5–45. Good farmers/ranchers, good livestock, good pastures and crops, and good soils go together. (Courtesy, *Holstein World*, Sandy Creek, N.Y.)

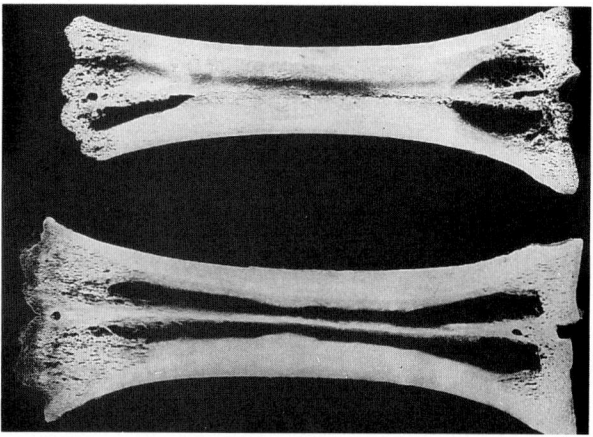

Fig. 5–46. Soil nutrients made the difference! Split bones from two calves of similar breeding and age. Small, fragile, pitted bone (top) obtained from calf pastured on belly-deep grass grown on highly weathered soil low in mineral content. Big, rugged, strong bone (bottom) from calf grown on moderately weathered, but highly mineralized soil. (Courtesy, University of Missouri, Columbia)

The land surface of the earth is covered with many kinds of soil. Some soils naturally contain an abundant supply of most of the elements needed by both plants and animals. Others may have an abundant supply of most required elements and yet be deficient in one or more essentials. For example, southwestern United States is known as a phosphorus-deficient area; northwestern United States and the Great Lakes area are iodine-deficient areas, and southeastern United States is a cobalt-deficient area.

Also, differences between plant species in their tendency to accumulate different elements are often important in determining the mineral status of animals that eat these

Nutritional Disorders/Toxins

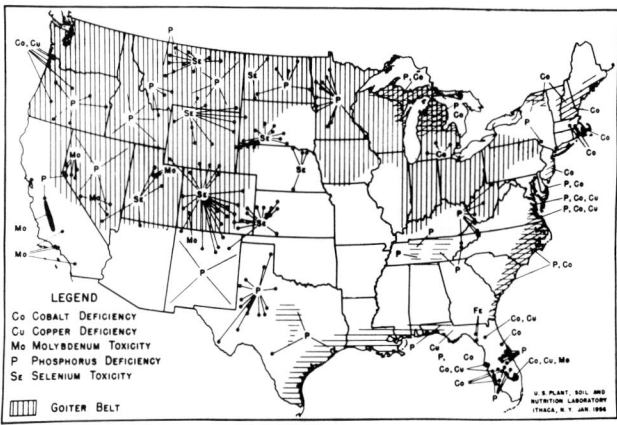

Fig. 5-47. Mineral deficiency areas of the U.S. and the excess selenium area of the northern and central Great Plains. (Courtesy, USDA)

plants. For example, in many places in the United States such forages as alfalfa and clover contain adequate levels of cobalt for cattle and sheep, whereas grass species in the same fields or pastures do not contain enough cobalt to meet the requirements of these animals. In places in the Great Lakes area, some native species of vetch accumulate toxic levels of selenium, yet farm crops and pasture grasses growing in the same field will contain substantially lower, and generally nontoxic, levels of this element.

At every step in the chain from soils to animals, the essential mineral elements interact with other elements, and these interactions may profoundly affect the availability of essential elements to plants or animals or the amount of the essential element required for normal growth or metabolic function. For example, a high level of soluble iron in the soil may depress the solubility of phosphorus and cause plants, and the animals that feed upon the plants, to suffer from phosphorus deficiency. The availability of zinc to animals may be depressed if the ration is high in calcium; and high levels of molybdenum may interfere with copper metabolism in animals. These and other interactions must be considered in assessing whether a given soil will supply plants with needed nutrients, and, in turn, whether plants will supply the animals that consume them with needed nutrients.

Thus, the transfer of essential nutrient elements from soils to plants, thence to animals, is a complicated process.

Soil Testing

Soil testing is the most important single guide to the profitable application of fertilizer and lime. When properly done, it provides a reliable basis for planning the fertility program for each field.

Traditionally, soil testing has been used to decide how much lime and fertilizer to apply. Today, it is important to determine where fertilizer should not be used for two reasons: (1) with high fertilizer prices, no more fertilizer should be applied than is needed; and (2) with increased concern about the environment, soil tests are a logical tool to determine areas where adequate or excessive fertilization has taken place.

If soils tests are to be of value, the samples must be taken properly and the results interpreted correctly. The local county agent (farm advisor) can provide instructions on how to take soil samples and recommend where to send them. Some land grant colleges run soil analyses at nominal cost.

ORGANIC FARMING

The U.S. Department of Agriculture's definition of organic farming follows:

> "Organic farming is a production system which avoids or largely excludes the use of synthetically compounded fertilizers, pesticides, growth regulators, and livestock feed additives. To the maximum extent feasible, organic farming systems rely upon crop rotations, crop residues, animal manures, legumes, green legumes, off-farm organic wastes, mechanical cultivation, mineral-bearing rocks, and aspects of biological pest control to maintain soil productivity and tilth, to supply plant nutrients, and to control insects, weeds, and other pests."[1]

Often, the meanings for *organic, natural,* and *health,* which people imply or conjure up in their minds, are misleading, harmful, and tend to polarize people. Frequently, it is proclaimed that such products are safer and more nutritious than conventionally grown and marketed foods. The growing interest of consumers in the safety and nutritional quality of the American diet is a welcome development. Regrettably, however, much of this interest has been colored by alarmists who state or imply that the American food supply is unsafe or somehow inadequate to meet our nutritional needs.

The FDA has taken no position on use of the terms *organic, natural,* and *health* in food labeling, because the terms are often used loosely and interchangeably. The Federal Trade Commission (FTC) in its proposed Food Advertising Rule would prohibit use of the words *organic* and *natural* in food advertising because of concern about the ability of consumers to understand the terms in the conflicting and confusing ways they are used.

- **Feed/food quality**—Undeniably, many benefits accrue from organic farming, not the least of which is the valuable exercise that the practicing advocates get from growing their own organically produced vegetables. However, no laboratory test or animal feeding trial can distinguish between crops fertilized with inorganic fertilizers and those fertilized with organic fertilizers. Moreover, experiments designed to compare the levels of different essential nutrients in crops produced with organic fertilizers against those produced with comparable amounts of nutrients supplied as inorganic materials have shown little difference, with the advantage in favor of the inorganic as often as the organic. In the few experiments in which the plants produced under the two systems have been fed to test animals, the small differences noted in animal growth have not consistently favored either the organic or inorganic sources of the nutrients.

The above results are as one would expect, based on the function of plants in the food chain—to convert inorganic compounds to organic compounds. If organic materials containing the nutrients are incorporated in the soil, the microorganisms in the soil must first break down the organic matter into inorganic forms. Inorganic ions of the essential nutrients are then taken up by the plant roots and manufactured into new organic materials. In the plants, and in the bodies of animals, these essential nutrient elements have the same effect, regardless of whether they were added to the soil in the form of organic or inorganic fertilizers.

---

[1]*Report and Recommendations on Organic Farming,* USDA, 1980.

- **Environmental control**—An important reason for adding organic materials to agricultural and garden soils is that this practice can be used to recycle the organic material without damaging the environment. On the other hand, too much organic matter can be added to the soil. Problems from excessive application of organic materials are generally confined to fields in close proximity to large cattle feedlots or poultry operations. Under such circumstances, nitrate toxicity and grass tetany have been serious problems where pastures have received excessive applications of manure.
- **Feed/food quantity**—The use of chemical fertilizers has been partly responsible for the abundance of food available. If all farmers were to adopt organic methods, there would be a decline in productivity as shown in Table 5-4.

TABLE 5-4
ESTIMATED NATIONAL AVERAGE CROP YIELDS UNDER CONVENTIONAL AND ORGANIC FARMING[1]

Crop	Bushels per Acre	
	Conventional	Organic
Corn	98	49
Wheat	43	20
Soybeans	40	20
Other grains	57	17

[1]*Organic and Conventional Farming Compared,* Council for Agricultural Science and Technology (CAST), Report No. 84, October 1980, p. 24, Table 6.

In a 1980 report issued by the Council for Agricultural Science and Technology (CAST), it was estimated that if organic farming were widely adopted the cost of food would increase because the total production from the land under cultivation would decrease. Those now practicing organic methods realize their yields are less, but they may receive higher prices for their goods because they sell to a specialty market.
- **Some inorganic fertilizers and mineral supplements are essential**—It is noteworthy that some nutritional problems can be corrected only by inorganic fertilizers or mineral supplements. For example, iodine deficiency in animals or people cannot be corrected by organic fertilizers, unless, of course, they contain kelp (seaweed) or marine by-products, good sources of iodine. The cobalt deficiency that plagued the cattle of the colonists in New Hampshire did not respond to organic fertilizers that contained little cobalt.
- **Summary**—Undoubtedly, world food production of the future will make use of a combination of both organic and inorganic fertilizers, with the nature and proportions of the combination for different farms and for different countries dependent on their access to fossil fuels, the availability and price of fertilizers, their soils, their food production requirements, their environmental control problems, and many other factors. Regardless of the combination of organic and inorganic fertilizers used, feed and food plants of adequate nutritional quality can be produced.

QUESTIONS FOR STUDY AND DISCUSSION

1. What are the primary differences between nutritional deficiencies brought about by (a) too little feed, and (b) rations too low in one or more nutrients? Which is the more serious?

2. What are the symptoms (a) of an energy deficiency, and (b) of a protein deficiency?

3. What vital roles in animal nutrition do minerals play?

4. Analyses show corn to be adequate in niacin. Yet, there may be niacin deficiencies when corn is fed—deficiencies which can be rectified by niacin supplementation. How do you explain this?

5. What is pica? Discuss the pica/behavior of cattle suffering from phosphorus deficiency and of lambs suffering from enterotoxemia.

6. List what you consider to be the 10 most important nutritional diseases and ailments. Summarize pertinent information relative to each of the 5 nutritional diseases and ailments which you feel cause the greatest economic loss, using the following headings: Species affected; Cause; Symptoms and Signs; Distribution and Losses caused by; Treatment; Control; and Prevention.

7. List and discuss 5 important interactions of nutrition and diseases/parasites.

8. Discuss the cause and importance of each of the following nutrition related disorders: choking, displaced abomasum, hardware disease, immunoglobulin deficiency, and impaction in horses.

9. Why do the heaviest livestock losses from poisonous plants occur on the western ranges?

10. Discuss each of the following points pertaining to poisonous plants:

  a. Economic importance
  b. Common poisonous plants
  c. Treatment
  d. Prevention

11. Have agricultural chemicals been good or bad? Justify your answer.

12. What is a poison? When it is suspected that an animal has been poisoned, what should the owner or caretaker do?

13. List what you consider to be the 10 most important potential animal poisons. Summarize pertinent information relative to each of the 5 potential poisons which you feel cause the greatest economic loss, using the following headings: Source; Species affected; Symptoms and Signs; Distribution and Losses caused by; Treatment; and Prevention.

14. Outline the recommended procedure for making a diagnosis of the cause of poisoning.

15. What is the purpose of the National Animal Poison Control Center? Where is it located?

16. What is your evaluation of hair analysis? Would you recommend that it be used for a determination of each of the following: heavy metal poisoning, mineral deficiencies, vitamin deficiencies?

17. Why is the evaluation of a feed crop in terms of yield and/or proximate analysis not complete enough?

18. Identify the major U.S. deficiency or excess areas of each of the following mineral elements: (a) phosphorus, (b) iodine, (c) cobalt, and (d) selenium.

19. Why should soils be tested? How would you go about getting a soil test?

20. Do you advocate organic farming, as opposed to chemical farming? Justify your answer.

Nutritional Disorders/Toxins

Fig. 5-48. Crooked calf, caused by either (1) lupine (a poison plant) toxicity, or (2) manganese deficiency. (Courtesy, Washington State University, Pullman)

Fig. 5-49. A pig showing gossypol toxicity, from eating too much cotton seed meal high in gossypol. (Courtesy, University of Arkansas, Fayetteville)

Fig. 5-50. Holstein cow in early stage of milk fever. (Courtesy, University of Ill., Champaign-Urbana)

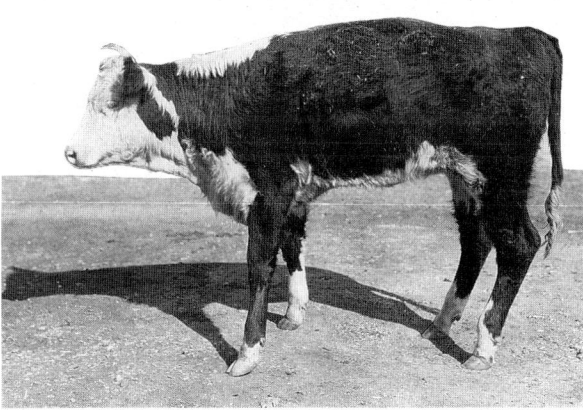

Fig. 5-51. Mercury poisoning. Heifer poisoned on seed corn that had been treated with a mercury compound. Note stiffness and weakness. (Courtesy, College of Veterinary Medicine, University of Illinois, Champaign-Urbana).

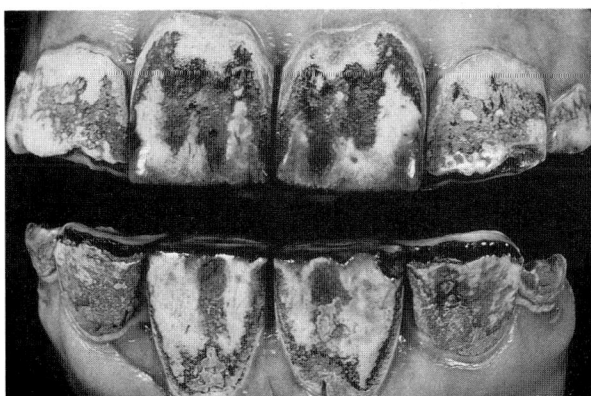

Fig. 5-52. Fluorosis (fluorine toxicity) evidenced in the teeth of a horse. Note mottling and irregular wear. (Courtesy, Utah State University, Logan)

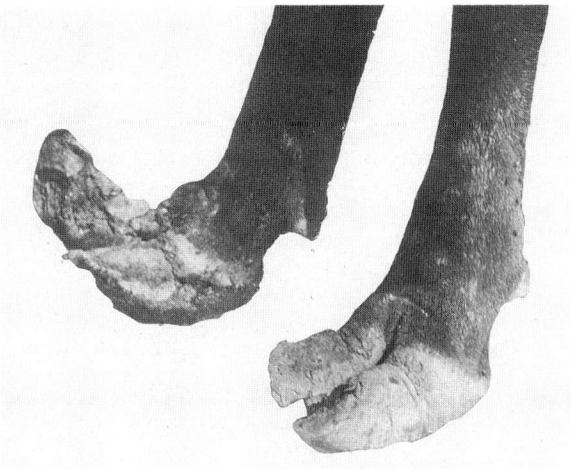

Fig. 5-53. Hind feet of a cow afflicted with selenium toxicity. Note the deformed hoofs. (Courtesy, The American Institute of Nutrition, Bethesda, Md.)

From an original painting by the noted artist, Tom Phillips (3333 17th Street, San Francisco, California 94110), prepared especially for this book. It portrays the artist's conception of what Chapters 6 to 16 are all about—the growing, harvesting, processing, and feeding of a great array of feeds. All flesh is grass!

SECTION II

FEEDS

Section II pertains to feeds in the broadest sense of the word. It includes natural and artificial products that are provided to animals for purposes of (1) sustaining them; (2) increasing production and/or efficiency, providing flavor, adding color, reducing stress; and/or (3) enhancing palatability, bulk, or the preservation of feeds.

Pasture and range forages, hays, silages, grains, and other high-energy feeds, protein supplements, and a host of by-product feeds and residues are the basis of successful livestock production. About two-thirds of the feeds consumed by animals are not suited to human comsumption. Many of them are produced on lands not adapted to the growth of bread grains or gardens.

Most feeds come directly or indirectly from plants that have their roots in the soil. Thus, we have the cycle as a whole—from the soil, through the plant, thence to the animal and back to the soil again.

The primary purpose of animals is to convert feeds into food, clothing, power, and recreation—hopefully, at a profit to the producer and with benefit to the consumer.

Chapter 6 presents a panoramic view of feeds and feedstuffs. It is followed by a chapter devoted to each of the major types of feedstuffs. Then Chapters 13 to 16—supplements, processing, analysis/evaluation, and buying feeds—are pertinent to all feeds.

6

TYPES AND ROLES OF FEEDSTUFFS

Original painting by Tom Phillips

Contents	Page
Classification of Feedstuffs	116
Economic Importance of Feeds for Livestock	116
Forages	116
Pasture and Range Forages	117
Green Chop (Soiling; Zero Grazing)	118
Hay	118
Silages and Haylages	118
Silage	119
Haylage	119
Concentrates	119
Grains and High-Energy Feeds (Carbonaceous Feeds)	120
Grains	120
High-Moisture Grain	120
Fats and Oils	120
Fruits, Roots, and Nuts	120
Mill By-product Feeds	120
Liquids and Semiliquids	121
Protein Supplements	121
Plant Proteins	121
Animal Proteins	121
Nonprotein Nitrogen (NPN)	121
Single-Cell Protein (SCP)	122
Complete Feeds	122
Premixes	122
By-product Feeds and Crop Residues	122
Plant By-products	122
Animal By-products	122
Industrial By-products	122
Fermentation By-products	122
By-products from the Wood and Paper Industry	123
Bakery By-products	123
Municipal Garbage	123
Specialty Feeds	123
Milk Replacers	123
Drought Area Feeds	123
Weeds	124
Supplements, Additives, and Implants	124
Mineral Supplements	124
Vitamin Supplements	124
Feed Additives and Implants	124
Wet Feeds (High-Moisture Feeds)	125
Questions for Study and Discussion	125

Chapter 6, Types and Roles of Feedstuffs, is an umbrella type of chapter, designed to present an overall view of feeds and feedstuffs. Detailed discussions of various feeds and feedstuffs are presented in subsequent chapters of this book.

Feeds are naturally occurring ingredients or materials fed to animals for the purpose of sustaining them. In many cases, the term feed connotes complete feeds, rations, or diets. The term feedstuff is generally synonymous with feed, except for one difference. *A feedstuff is any product, of natural or artificial origin, that has nutritional value in the ration when properly prepared.* Frequently, nonnutritive products are included in the ration for such purposes as increasing production and efficiency, providing flavor, adding color, reducing stress, or for reasons related to palatability, bulk, or preserving feeds. Hence, compounds such as butylated hydroxytoluene (commonly abbreviated as BHT, an antioxidant) and vitamin A acetate (a synthetic vitamin) can be considered feedstuffs but not feeds.

Different species of livestock utilize various feeds with different efficiency. That is, what may be a high-energy feed to a ruminant may not be so classified for a nonruminant due to the varying ability of the respective animals to digest, absorb, and utilize the nutrients within the feed. Thus, the livestock producer must be knowledgeable relative to the specific application of many types of feeds.

The relative importance of the principal U.S. livestock feeds is shown in Fig. 6-1.

RELATIVE IMPORTANCE OF PRINCIPAL LIVESTOCK FEEDS (% OF TOTAL TONNAGE FED)

PASTURE & GRAZING	40.0%
CORN	23.3%
HAY	12.2%
HIGH-PROTEIN FEEDS	8.9%
OTHER GRAINS	8.0%
SILAGE, STOVER, ETC.	5.2%
OTHER BY-PRODUCTS	2.4%

EACH SYMBOL = 5%

Fig. 6-1. These principal livestock feeds are converted into meat, eggs, milk, and wool. (Source of data: USDA, Economic Research Service)

CLASSIFICATION OF FEEDSTUFFS

In general, feedstuffs may be classified into one of the following categories: (1) forages, (2) concentrates, (3) complete feeds, (4) premixes, (5) by-product feeds and crop residues, (6) specialty feeds, and (7) supplements, additives, and implants. Regardless of the criteria used to classify them, there are always some feedstuffs that do not fit into any one category or which fit the criteria of more than one class. However, livestock producers need not be too concerned relative to the precise classification of each feedstuff. Rather, they should familiarize themselves with the characteristics, along with the conditions that affect the nutritive value of each available feed as it pertains to the species of animal they are feeding and their particular feeding program.

ECONOMIC IMPORTANCE OF FEEDS FOR LIVESTOCK

Feed costs generally account for 12 to 15% of the total cost of agricultural production. This represents the second greatest single item of farm production expenses, being exceeded only by replacement of capital equipment. Thus, feed is very important from an economic standpoint.

While a wide variety of feedstuffs is currently being used, relatively few of these products make up the great bulk of the U.S. supply, as shown in Table 6–1. From these figures, it is possible to ascertain the relative importance of the various feeds. For example, more alfalfa hay is used for feed than all other types of hays combined. Corn and soybean oil meal are the chief energy and protein sources, respectively.

Due to the efficiency of the American farmer, the United States is the largest exporter of agricultural commodities in the world—worth approximately $31.2 billion in 1985. Considerable amounts of feed grains are exported, with corn far in the lead (see Fig. 6–2). Of the $31.2 billion total farm product exports in 1985, feed grains accounted for $6.8 billion (with corn alone accounting for $5.8 billion of it), wheat and products for $4.5 billion, and oilseeds and prod-

TABLE 6–1
ANIMAL FEEDS CONSUMED IN THE UNITED STATES[1]

	Acreage Harvested	Use for Feed		Yield per Acre
	(1,000 acres)	(1,000 tons)		(tons)
Hay:	61.4	143,800		
Alfalfa	26.8	90,048		3.36
All other hay	34.6	53,752		1.75
Silage:				
Corn	7.5	104,000		13.9
Sorghum	0.6	6,000		10.6
Other	1.0	10,500		12.5
		(mil. tons)	(mil. bu)	(bu)
Grains:				
Corn	71.9	115,276	4,117	106.7
Sorghum	15.4	15,316	547	56.4
Barley	11.2	7,296	304	53.4
Oats	8.2	6,928	433	58.0
High-Protein:				
Oilseed meals:				
Soybean		19,501		
Cottonseed		1,782		
Sunflower		470		
Linseed		127		
Peanut		110		
Animal proteins:				
Tankage and meat meal		1,567		
Fish meal and solubles		251		
Feather meal		130		
Dried milk products		112		
Grain protein feeds:				
Gluten feed and meal		1,160		
Distillers' dried grains		551		
Brewers' dried grains		298		
Miscellaneous feeds:				
Wheat millfeeds		4,795		
By-products from hominy, oats, etc.		2,646		
Molasses		2,646		
Alfalfa meals		992		
Rice millfeeds		678		
Fats		644		
Urea[2]		265		

[1]USDA, Economic Research Service, estimates for the feed year 1984.
[2]Estimate for the feed year 1986–87 from *Feed Management*, June, 1987, p. 18.

ucts for $6.2 billion. These 3 commodities accounted for 53% of the total exports of farm products that year.

FORAGES

Forage is defined as vegetable material in a fresh, dried, or ensiled state (pasture, hay, or silage) which is fed to livestock. In the dry state, forages average more than 18% fiber. The term roughage is often used interchangeably with forage, although roughage usually implies a coarser, bulkier feed than forage. Forages are extremely important feeds for ruminants and nonruminant herbivores, not only for the nutrients they provide but also for the stimulatory effects of forages on the muscle tone and activity of the gastrointestinal tract. Because swine and poultry have very limited capacities to utilize forages, the role of forages in their nutritional program is, for the most part, minor.

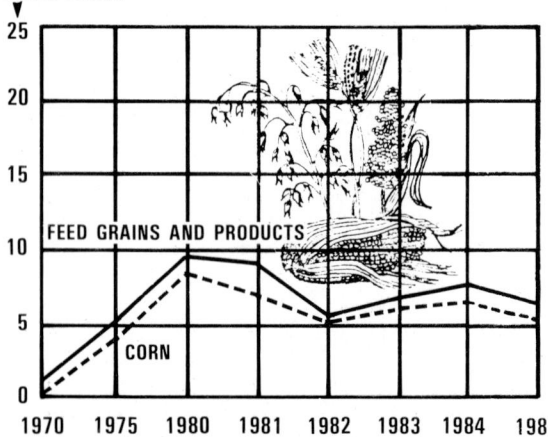

Fig. 6–2. Importance of feed grains as exports. Feed grains include corn, grain sorghum, barley, oats, and their products. Corn is also listed separately in order to show that it dominates feed grain exports. (From: *Statistical Abstracts of the United States 1987*, p. 796, No. 1408.)

From a feeding standpoint, the following general characteristics of forages are pertinent, although some well-known forages can be cited as an exception to each characteristic (for example, on a dry basis well-eared corn silage runs 18% crude fiber, but the TDN is high—about 70%):

1. **Bulk.** They are bulky feeds with a low weight per unit of volume.
2. **Fiber and energy.** They contain more than 18% crude fiber, and they are lower in energy than the concentrates.
3. **Digestibility.** They are generally lower in digestibility than concentrates, due to the lignin content.
4. **Minerals.** They are generally higher in calcium, potassium, and trace minerals than most concentrates; but phosphorus content is apt to be moderate to low.
5. **Vitamins.** They are higher in fat-soluble vitamins than most concentrates. Legumes are good sources of B vitamins.
6. **Protein.** They are variable in protein content. Legumes may run 20% or more crude protein, whereas other forages, such as straws, may have only 3 to 4% crude protein.

From an overall nutritional standpoint, forages may range from very good nutrient sources (such as lush young grass, legumes, and high-quality silage) to very poor feeds (such as straws, hulls, and some browse). Nevertheless, all of them can be used advantageously, provided (1) they are properly prepared and supplemented, and (2) the feeder uses judgment in selecting the species and class of animal to which the particular roughage is fed.

Forage is the natural feed of all herbivorous animals, including ruminants and horses. Swine can survive solely on forage, but their productivity thereon is generally too low to be economical.

Some form of forage is generally included in all balanced rations for ruminants—even in high-concentrate rations. Research and practical feedlot management have repeatedly shown that the addition of small amounts of forage (3 to 15%) to high-energy cattle and sheep finishing rations increases performance and reduces digestive disturbances. Forages appear to stimulate fermentation and promote good muscle tone and epithelial growth in the rumen. In many cases when a ration is 100% concentrate, parakeratosis of the rumen and ruminal atony develop.

Nonruminant herbivores can utilize forages reasonably well, although not as efficiently as ruminants. This is due to the fact that the enlarged cecum—the fermentation vat of the nonruminants—is located distally to the small intestine (the primary organ of absorption). Thus, the digested fiber in the nonruminant herbivore is not exposed to as great an absorptive surface as in ruminants. Rabbits practice coprophagy (eating of feces), resulting in increased forage utilization; but the efficiency of digestion and absorption is still lower than in ruminants. Swine have neither a rumen nor an enlarged cecum and are, therefore, relatively poor utilizers of forages. However, forages can provide an effective means of reducing the feed costs for swine fed maintenance or low-intensity production rations, such as brood sow rations.

Forages include, but are not limited to, pastures, hays, and silages. Table 6-2 shows that roughages account for 61.7% of all U.S. livestock feeds. Of course, the proportion of forages to concentrate consumption varies widely according to relative price and the class of animal. As shown in Table 6-2, sheep and goats head the list of forage consumers, with 93.8% of their total feed coming therefrom, including pasture. Beef cattle obtain 84.5% of their feed from forages. Poultry consume only a negligible amount of forage.

TABLE 6-2
PERCENTAGE OF FEED FOR DIFFERENT CLASSES OF U.S. LIVESTOCK DERIVED FROM (1) CONCENTRATES, AND (2) FORAGES, INCLUDING PASTURE[1]

Class of Animal	Concentrates	Roughages
	(%)	(%)
Beef cattle	15.5	84.5
Dairy cattle	41.3	58.7
Sheep and goats	6.2	93.8
Swine	95.7	4.3
Horses and mules	27.0	73.0
Poultry	100.0	0.0
All livestock	38.3	61.7

[1]USDA, Economic Research Service. Data for the feed year 1983-1984.

Pasture and Range Forages

A pasture is an area of land on which there is a growth of forage that animals may graze.

Broadly speaking, pastures are of two types: (1) cultivated pasture, and (2) native pasture. *Cultivated pastures are those which either receive in excess of 20 in. of rainfall annually or are irrigated.* They include the seeded (cultivated) pastures of the Corn Belt, the South, and the East, and the irrigated areas, and smaller scattered moderate- to high-rainfall areas throughout the West. Thus, cultivated pastures generally produce high yields and good-quality forages. *Native pastures include those which receive less than 20 in. of rainfall annually.* These pastures allow only a limited stocking rate, and much of the available forage is of low quality.

Pastures may also be classified as (1) permanent, (2) rotation, and (3) temporary and supplemental, depending on the topography of the land, the climate, and the cultivation program of the operation.

Numerous advantages accrue from using pasture and range forages for livestock—especially ruminants. But as with all types of feeds, certain disadvantages must also be recognized and weighed against the advantages. In rangelands and other relatively unproductive types of land, the stock grower generally has little choice as to the use of the land; hence, it becomes pasture.

The following are among the **advantages** of pasture and range forages:

1. They lessen feed cost. The amount of grain and protein supplements fed to livestock—both of which tend to be expensive—can be dramatically reduced when pasture and range forages are used wisely. Good pasture will produce 200 to 400 lb of beef per acre—superior pasture will produce even more. When a good pasture program is used in swine production, the cost of feeding brood sows can be reduced by 50%.

2. Pasture forages lessen the hazard of nutritional deficiencies. Well-managed pastures and ranges provide good sources of high-quality protein, certain vitamins and minerals, and unknown factors.

3. The threat of communicable diseases is reduced. When animals are reared in confinement, one sick animal poses

a potential threat to the entire group. Animals on pasture have less close contact with each other; hence, disease problems are generally minimal.

4. Pastures require less capital for buildings and equipment than confinement production. Generally, all one needs is fenced land, a source of water, simple shelter to protect the animals from the elements, and a good stand of forage.

5. The need for highly developed skills in management are generally not as critical in grazing programs as in confinement production.

6. Pastures make for improved soil conservation on rolling land.

7. Pastures and ranges provide a fairly uniform supply of feed throughout the entire season, provided they are properly cared for by avoiding overgrazing and keeping them relatively free from trash plants.

8. Animals get valuable exercise by grazing on pasture. This is particularly important for breeding animals.

9. A good pasture program permits maximum use of land that is not suited for crop production.

The **disadvantages** of pasture and range forages are:

1. Some land used for pasture may bring a higher return from other uses.

2. Some range areas require large amounts of land to support each animal. Hence, it is difficult to check the animals; and rounding them up usually takes considerable time and labor.

3. The nutritive value of the forage is directly related to the composition of the soils. Hence, a soil that is low in certain minerals produces forages that are low in those minerals. Conversely, toxicities can result in animals grazing crops grown on soils with high concentrations of certain trace minerals: for example, selenium.

(Also see Chapter 7, Pasture and Range Forages.)

GREEN CHOP (SOILING; ZERO GRAZING)

Green chop, or soilage or zero grazing, is fresh herbage cut and chopped in the field and fed to animals in confinement. This type of forage is most frequently fed to lactating dairy cattle. By using green chop, losses in moisture, color, and nutrients are minimized. However, this type of forage is labor intensive because the feed must be harvested daily.

Also, the crop must be harvested at the optimum stage of maturity for best results, thus presenting limitations for extended use.

(Also see Chapter 7, section on "Green Chop.")

Hay

Hay is forage harvested during the growing period and preserved by drying for later use in animal feeding. Hay provides the most permanent form of feed storage. While there is some loss of nutrients over a period of time, properly cured hay can be stored for years with little danger of spoiling. It is used as one of the primary feeds for ruminants and nonruminant herbivores. Also, good legume hay is frequently used as a vitamin and mineral supplement in swine and poultry rations.

The primary objective of haymaking is to lower the moisture content of the forage to a stage whereby enzymatic and

Fig. 6–3. Haystacks. Hay provides 12% of the feed consumed by U.S. livestock. (Courtesy, USDA)

microbial degradation in the plant is inhibited. This occurs when the moisture content is reduced to 15 to 20%. As forages mature, they contain more dry matter, but they also become less digestible. Thus, hay should be harvested when there is a balance between dry matter content and digestible nutrients.

Hay varies more in nutritive value than any other feed, primarily because of (1) differences in the crop from which it is made, (2) stage of maturity at which it is cut, (3) handling, and (4) possible weather damage during curing. Average-quality hay will run 25 to 35% crude fiber and 45 to 55% TDN.

The **advantages** of hay are:

1. It is the best form for long-term storage of forages. Silage can be stored for extended periods, but once the silo is opened, the storage life of the feed decreases.

2. It is an excellent source of certain vitamins and minerals. For example, poultry producers routinely use alfalfa meal as a source of vitamin A and for its pigment-producing value.

3. When animals are fed high-concentrate rations, hay can be added to facilitate digestion and prevent digestive disturbances.

4. Once hay has been harvested and properly packaged, it is easy to handle and feed.

But there are several **disadvantages** to hay; among them, the following:

1. Considerable labor and expensive equipment are needed to harvest it.

2. The physical process of harvesting hay creates considerable losses of material—for example, leaf shattering.

3. Throughout the curing process, dry matter and some nutrients are lost, with the magnitude of the losses determined largely by the weather (rain, air velocity, temperature).

4. If hay is not cured for a sufficiently long period, spontaneous combustion of the stored forage can occur.

5. Quite often, land that is used for the production of hay is also suited for more intensive crop production.

(Also see Chapter 8, Hay.)

Silages and Haylages

Silages and haylages are fermented forages stored under anaerobic (without air) conditions in a silo. Silos are designed so that anaerobic conditions prevail during the storage of high-moisture feedstuffs. These anaerobic conditions provide an environment whereby microorganisms can ferment the soluble carbohydrates of the feeds, thereby producing lactic acid, volatile fatty acids, and dicarboxylic acids. With the buildup of these end products of fermentation, the pH of the stored feed falls to a range of 3.8 to 5.0. The microorganisms in the feeds are sensitive to pH; and once these high hydrogen concentrations are achieved, microbial growth is inhibited. At this stage, lactic acid represents 8 to 12% of the feed.

So long as the ensiled feed is kept in anaerobic conditions, it can be stored for several years. However, once the silo is opened and silage is removed, it should be emptied within 12 months since exposure to air will speed up the deterioration of the feed.

(Also see Chapter 9, "Silage/Haylage/High Moisture Grain.")

SILAGE

Fig. 6-4. Silage stored in upright silos on a dairy farm. (Courtesy, USDA)

Silage is fermented forage plants. Usually, the forages made into silage are green crops, or dry crops to which moisture has been added, preferably stored at a level of 60 to 65% moisture. Corn is by far the most extensively used silage, followed by sorghum. A rule of thumb used by many producers to determine what crops can be ensiled is: Crops that are palatable and nutritious to animals as pasture, freshly harvested feed, or dry forage are suitable for ensiling.

Silages are very palatable, and they allow the farmer to get maximum yields per acre because the entire aerial part of the plant is used. The primary problems with silage are the high initial costs of constructing the silo and the increased managerial skills required of the producer—for example, knowing the procedure of harvesting, loading, and unloading the silage, and the problems inherent in feeding silage.

(Also see Chapter 9, Silage/Haylage/High-Moisture Grain.)

HAYLAGE

Haylage is low-moisture silage that is made from grasses and/or legumes that are wilted to 40 to 60% moisture content before ensiling. It is generally produced where the climate is too cool and the growing season too short for corn or sorghum silage. Haylage is higher in protein and carotene, but lower in TDN and vitamin D, than corn and sorghum silage. Animals will ingest more dry matter when offered haylage than when fed corn silage.

For haylage to be properly ensiled, it is very important that as much air as possible be excluded from the silo. Haylage is more difficult to ensile than conventional silage materials, because it is difficult to pack.

(Also see Chapter 9, Section on "Low-Moisture Silage (Haylage).

CONCENTRATES

Concentrates are feeds that are high in nitrogen-free extract and TDN and low in crude fiber (less than 18%). These feeds can be either high or low in protein content. Cereal grains, oil meals, and by-products of the milling industry—for example, corn gluten—are classified as concentrates. Concentrates can be broken down into (1) carbonaceous feeds, and (2) nitrogenous feeds.

Concentrates represented about 38.3% of the feeds consumed by all livestock in the United States in 1983–1984. While ruminant animals are fed considerable amounts of concentrate feeds—especially when in heavy production, monogastrics, such as poultry and swine, are fed almost entirely on concentrates.

The economic impact of concentrate feeds extends well beyond the farm level. Numerous jobs are involved in the processing and handling of these feeds. Additionally, feed grain exports give the United States a big assist in achieving a favorable balance of trade in the international market system.

Fig. 6-5 shows the tonnage of feed concentrates fed to U.S. livestock and poultry from 1975 to 1985. This figure points up (1) that corn is by far the most important livestock feed, (2) that by-products rank second, and (3) that wheat and rye are normally of minor importance as livestock feeds. In 1984, the following quantities of concentrates were fed (in million short tons): feed grains (corn, sorghum, oats, and barley), 145.3; wheat and rye, 8.2; by-products, 36.6; and total concentrates (including by-products), 190.1.

FEED CONCENTRATES FED TO LIVESTOCK AND POULTRY

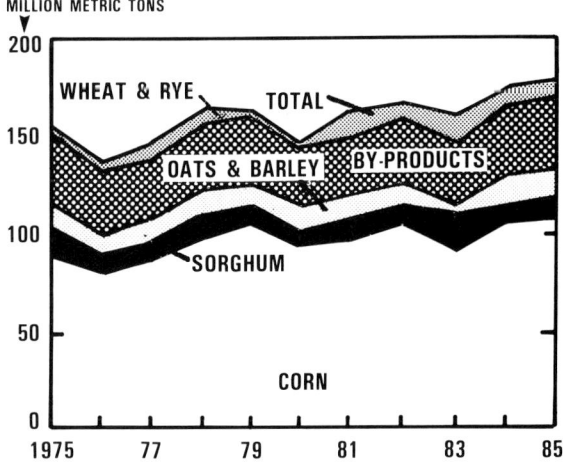

Fig. 6-5. Tonnages of concentrates fed to livestock and poultry. (From *1986 Handbook of Agricultural Charts*, USDA, p. 104, Chart 257)

Since concentrates are relatively low in fiber, most of them are highly digestible. Grains contain large amounts of starch that are easily digested and absorbed in the rumen of ruminants and in the small intestines of ruminants and monogastrics. The protein in most grains and protein concentrates is also highly digestible.

Grains and High-Energy Feeds (Carbonaceous Feeds)

High-energy feeds are feeds used primarily for their energy content. In most cases, high-energy feeds contain less than 20% protein and 18% crude fiber. However, many protein supplements can also be classified as high-energy feeds.

In addition to grains, numerous other feedstuffs are used as high-energy feeds. Price, availability, and nutrient composition determine which of these feeds are used and at what level.

(Also see Chapter 10, Grains/High-Energy Feeds.)

GRAINS

Fig. 6–6. A collection of the types of corn found in the western world. Corn is by far the most widely used feed grain in the U.S. Over 115.3 million short tons of corn were fed in the U.S. in 1984. (Courtesy, USDA)

Grains are the seeds from cereal plants. They constitute the bulk of the high-energy feeds. Some contain as much as 85% carbohydrate (starch) and 6% fat. Most harvested feed grains have relatively little moisture, about 10%, and are not as variable in composition as forages. Corn—representing more than 60% of the total tonnage of concentrate feed—is, by far, the most widely used high-energy feed in the United States. Normally, more than 8 times as much corn was fed to livestock as grain sorghums—the second most widely used grain.

(Also see Chapter 10, section on "Grains.")

HIGH-MOISTURE GRAIN

High-moisture grain is grain that contains 22 to 40% moisture. It includes considerable quantities of (1) ear corn, (2) shelled corn, and (3) small grains. Ensiling or acid treatment alleviates costs for drying and reduces the risks of spoilage due to molding or heating.

Corn is the most widely used high-moisture grain, but barley, milo, and other grains can also be used effectively. Although grain is ensiled whole, it is often ground or rolled before being fed, to increase digestibility.

By using high-moisture grain systems, the producer can harvest at an early date—often avoiding unfavorable weather thereby. Through this system, field and storage losses are minimized.

(Also see Chapter 9, section on "High Moisture Grain.")

FATS AND OILS

Fats and oils are the most potent feed energy source—having 2.25 times as much energy as carbohydrates. By incorporating fat into livestock rations, the producer can increase the caloric density of the ration, control dustiness, reduce wear on mixing equipment, provide a protective agent for some micronutrients, and increase palatability of the feed.

Animal and vegetable fats seem to be almost equally effective additions to rations; thus, selection should be determined solely by comparative price—based on energy content. Ordinarily, animal fats are much cheaper than such vegetable oils as soybean oil or cottonseed oil. Vegetable oils are generally priced out of the animal feed market, for use in margarine, and for use in paint and other industrial purposes.

Several different fat products are used as animal feed; among them, acidulated soap stock (foots), tallows, greases (white and yellow), blended feeding fat, house grease, brown grease, sewer grease, and modified yellow grease. Each of them should be stabilized with an antioxidant and should be bought by specifications and guarantees.

(Also see Chapter 10, section on "Fats and Oils.")

FRUITS, ROOTS, AND NUTS

If a livestock producer is located in an area where fruits, root crops, or nuts are grown, a limited amount of these feedstuffs or of their by-products may be incorporated in the ration. Since fruits and roots contain considerable quantities of water, their respective feeding values on an as-fed basis are low. However, on a moisture-free basis, these feeds are quite comparable in TDN to the conventional feed grains. Quite often, these feeds are used so as to provide variety, as well as for economy reasons. Nuts are seldom used in livestock rations unless they are off-quality.

(Also see Chapter 12, section on "Fruit, Vegetable, and Nut By-Product Feeds"; Chapter 10, section on "Roots and Tubers"; and Chapter 7, section on "Turnips For Fall/Winter Grazing.")

MILL BY-PRODUCT FEEDS

Since many of the cereal grains are processed routinely, mill by-products constitute an additional source of feedstuffs. By-products such as wheat mill run, corn gluten, and bran are used in all types of livestock feeds. Many of these feedstuffs are also excellent sources of protein, in addition to their high-energy content.

(Also see Chapter 10, section on "Grain Milling and Milling By-Products.")

LIQUIDS AND SEMILIQUIDS

Liquid and semiliquid products are often incorporated in livestock feeds. Most of these feedstuffs are by-products of some industry—for example, molasses, distillers' solubles, fish solubles, and corn fermentation solubles. Of these feedstuffs, molasses (including cane or blackstrap, beet, citrus, and wood molasses) is the most common; about 2.7 million tons are fed annually in the United States.

According to the American Feed Industry Association, nearly 25% of the nation's feedlot animals received a liquid feed supplement in 1986.

(Also see Chapter 10, Grains/High-Energy Feeds.)

Protein Supplements

Protein supplements are feedstuffs containing more than 20% protein or protein equivalent. Protein levels in livestock rations are extremely important, especially for young animals and animals in high production. Muscle growth, egg production, wool and hair growth, lactation, and gestation all require considerable quantities of protein because the products of these types of production are largely protein in composition.

The use of high-protein feeds is increasing. Fig. 6–7 shows the high-protein feeds fed to U.S. livestock and poultry in soybean meal equivalents from 1975 to 1985. The following points are pertinent to this figure: "other oilseed meals" includes cottonseed, linseed, peanut, and sunflower meals; "animal/marine proteins" includes tankage/meat meal, fish meal and solubles, dried milk products, and feather meal; and "grain proteins" include gluten feed and meal, distillers' grains, and brewers' grains. The figure illustrates that soybean meal is by far the most important high-protein feed. In 1984, in million short tons, the following quantities of different proteins were fed: soybean meal, 19.8; all oil meals, 21.8; animal protein, 2.7; grain protein 1.2; and total of all high-protein feeds, 25.8.

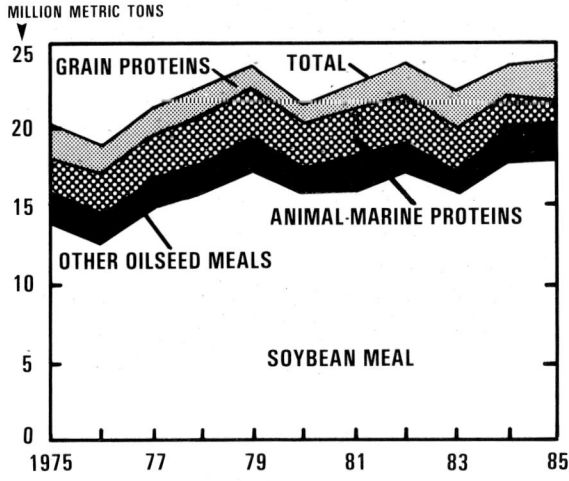

Fig. 6-7. Importance of high-protein feeds. (From *1986 Handbook of Agricultural Charts*, USDA, p. 104, Chart 258)

High-protein feeds are usually named and classified according to their origin and method of processing. On the basis of origin, they are usually grouped into the following categories: (1) plant, (2) animal (mammalian, avian, and marine), (3) nonprotein nitrogen (NPN), and (4) single-cell protein. (Also see Chapter 11, Protein Supplements.)

PLANT PROTEINS

This group includes the common oilseed by-products—soybean meal, cottonseed meal, linseed meal, peanut meal, safflower meal, sunflower meal, rapeseed meal, and coconut (or copra) meal. They vary in protein content and feeding value, depending on the seed from which they are produced, the amount of hull and/or seed coat included, and the method of oil extraction used.

Soybean meal is, by far, the most widely used protein supplement for livestock, with an estimated 19.8 million tons used in 1984. Approximately 2.0 million tons of other oilseed meals (including cottonseed, linseed, peanut, and sunflower meals) were fed to livestock in 1984.

(Also see Chapter 11, section on "Plant Proteins.")

ANIMAL PROTEINS

Animal protein supplements are generally high-quality protein feeds which are derived from inedible tissues from meatpacking or rendering plants, from surplus milk or milk products, and from marine sources. They include proteins from meat, fish, poultry, eggs, milk, and their processing by-products. Marine protein supplements—such as fish meal—are generally considered to be better quality feeds than the other animal protein feedstuffs because they contain an excellent balance of amino acids and are good sources of minerals and vitamins. In the past, hog and poultry rations almost always contained some type of animal protein source to provide an "unknown factor(s)." With the discovery and commerical availability of vitamin B–12, high-protein feeds of animal origin became less essential for swine and poultry.

Although most animal protein supplements are excellent feeds, the following concerns are inherent in the utilization of these feedstuffs:

1. **Susceptibility to autoxidation.** Some animal protein supplements contain large amounts of fat and are, therefore, vulnerable to autoxidation and rancidity.

2. **Sources of bacterial contamination.** Animal protein feedstuffs provide excellent media for the growth of bacteria. To prevent this contamination, many of the products must be processed and stored in such a way as to prevent bacterial growth.

3. **Cost.** Most animal sources are more costly than either plant protein or nonprotein nitrogen.

(Also see Chapter 11, section on "Animal Proteins.")

NONPROTEIN NITROGEN (NPN)

Protein quality is less important with ruminants and nonruminant herbivores than in omnivores or carnivores because of microbial synthesis in the gut. As a result, certain nonprotein nitrogen (NPN) sources may be substituted for all or much of the supplemental protein required in most ruminant rations, provided such rations are adequate in minerals and readily available carbohydrates. Among such products are urea, ammoniated molasses, ammoniated beet pulp, ammoniated cottonseed meal, ammoniated citrus pulp, and ammoniated rice hulls. In recent years, the use of liquid protein supplements containing NPN as the protein source has dramatically increased. Also, slow-release nonprotein nitrogen products have come on the market.

(Also see Chapter 11, section on "Nonprotein Nitrogen [NPN] Feedstuffs.")

SINGLE-CELL PROTEIN (SCP)

Some single-cell protein types, such as yeast, algae, and bacteria, can be useful sources of protein and vitamins for animal feeding.

(Also see Chapter 11, section on "Single-Cell [Microbial] Protein [SCP].")

COMPLETE FEEDS

Complete feeds are prepared products that provide all the nutrients, along with roughage, required to support the form of animal production for which they were designed.

Examples of complete feeds are a 17% layer mash or a 15% lactating dairy ration.

PREMIXES

Premixes are concentrated mixes that provide one or more micronutrients and/or specialty products to larger mixtures of feed ingredients.

Premixes are generally commercially prepared and consist of minerals, vitamins, and perhaps other additives, along with a carrier, formulated for blending with a larger mix. Rather than purchasing individual micronutrients (including minerals, vitamins, and other additives), then mixing the ration from the ground up, many commercial feed companies use premixes. Also, farmers may use premixes to blend with local or homegrown grains.

BY-PRODUCT FEEDS AND CROP RESIDUES

By-product feeds are roughages and concentrates other than the primary product from plant and animal processing and from industrial manufacturing. The term crop residue refers to that part of a crop which remains following harvesting. All processing industries produce some by-products which can be considered of secondary, or no value, compared to the primary product. Frequently, these by-products can provide cheap alternatives to other feeds. On many farms, crop residues, which were originally thought to be of little value, are now considered valuable feedstuffs.

(Also see Chapter 12, By-product Feeds/Crop Residues.)

Plant By-products

Plant by-products can be divided into two categories: (1) high-roughage by-products, and (2) high-energy by-products. The high-roughage by-products, which are generally low to moderate in energy and are fed to ruminants that can utilize the fiber, include cottonseed hulls, rice hulls, peanut hulls, soybean hulls, nut shells, sugarcane and bagasse, peelings,

Fig. 6–8. Cows grazing cornstalks (corn stover), the residue of corn, which is available in great abundance. (Courtesy, Ron Baker, C & B Livestock, Inc., Hermiston, Ore.)

vines, and corncobs. Included in the high-energy by-products are molasses, grain milling by-products, and dried beet pulp.

(Also see Chapter 12, section on "Plant By-Product Feeds.")

Animal By-products

Virtually everything that goes into a slaughterhouse can be processed for some purpose. Feathers, bones, connective tissues, organs, blood, meat scraps, and hoofs are used as either protein, vitamin, or mineral supplements.

In recent years, considerable attention has been devoted to the feeding of livestock manure and litter. In the past, feedlots and large poultry operations were often hard pressed to find ways of disposing of the mountainous volumes of manure and litter (manure with absorbant material). Today, these producers are turning this material—once thought to be of value only as fertilizer—into valuable feed.

(Also see Chapter 12 sections on "Animal By-Product Feeds" and "Manure as a Feed.")

Industrial By-products

Many feedstuffs are being derived from industrial by-products. Thus, many products which were once thought of as wastes are now considered valuable livestock feeds.

FERMENTATION BY-PRODUCTS

Both the brewing industry (beer and ale) and the distilling industry (liquor) produce by-products which are of high nutritive value to livestock. Brewers' dried yeast is 100% yeast solids and is rich in B complex vitamins, proteins, minerals, and unidentified factors. Brewers' dried grains contain about 65% TDN and 21% digestible protein. Distillers' solubles—a syrup produced in the distillation process—is high in protein, energy, linoleic acid, and unidentified factors. Likewise, distillers' dried grains are good sources of energy and protein.

(Also see Chapter 12, section on "Brewing and Distilling By-Product Feeds.")

BY-PRODUCTS FROM THE WOOD AND PAPER INDUSTRY

Many of the by-products from the wood and paper industry can be effectively used in limited amounts by ruminants and nonruminant herbivores—animals that can digest fiber effectively. Torula yeast and wood molasses are two commonly used by-products derived from the processing of wood.

(Also see Chapter 12, section on "Wood and Paper By-Product Feeds.")

BAKERY BY-PRODUCTS

Bakeries process many types of cereals, and the resultant by-products make excellent and highly digestible high-energy feeds.

(Also see Chapter 12, section on "Bakery Waste.")

MUNICIPAL GARBAGE

Vast quantities of foods unfit for human consumption, as well as garbage resulting from the wastage of food, pose serious disposal problems in many urban areas. One of the solutions to the problem is the cooking of garbage, so as to alleviate the hazard of trichinosis, and feeding it to hogs. All states now have laws requiring that commercial garbage be cooked.

(Also see Chapter 12, section on "Garbage.")

SPECIALTY FEEDS

The term *specialty feeds* as used herein refers to unusual or uncommon feeds, or feeds that are used for a specific purpose or under special circumstances. The following specialty feeds are discussed herein: milk replacers, drought area feeds, and weeds.

Milk Replacers

Milk replacers are formulated feeds designed to replace the mother's milk of young mammals during the critical, early suckling or milk-feeding stage of life. Milk replacers are available for calves, lambs, pigs, foals, and other animals.

Although scientists have not yet learned how to formulate a synthetic product that will alleviate the necessity of colostrum, in certain other respects they have been able to improve upon nature's product, milk. For example, it has long been known that milk is deficient in iron and copper, thus resulting in anemia in suckling young if proper precautions are not taken. In addition to correcting these deficiencies, milk replacers are fortified with minerals, vitamins, and antibiotics.

A good commercial milk replacer should contain (1) 22–25% protein, preferably derived from milk products (whey, casein, or nonfat dry milk, although about 25% of the milk protein can be replaced by a modified soybean protein); (2) 15 to 20% fat (preferably derived from animal fats, although homogenized soy lecithin is a very acceptable fat source), the higher fat level tends to reduce the severity of diarrhea and provide additional energy for growth; (3) carbohydrates from lactose (milk sugar) and dextrose, and *not* from starch and sucrose (table sugar); and (4) the essential minerals and vitamins. Acidified milk replacers for free-choice calf feeding systems, which were developed and tested in Europe, are gaining increased acceptance in large U.S. commercial operations.

Most milk replacers are formulated to be mixed and fed like normal whole milk. They should be diluted, mixed, and fed according to the directions of the manufacturer.

From the standpoint of livestock producers, synthetic milk is of interest in raising orphaned or early weaned animals of each class of livestock. For the dairy producers, it is generally more profitable to sell the whole milk and purchase a high-quality milk replacer for the young calves. Also, it is a valuable adjunct in certain disease control programs, especially those diseases that may be transmitted from dam to offspring; and, in some cases, it makes it practical to retain in production those valuable females which, due to injury or disease to the udder, cannot suckle their young.

The economics of using milk replacer instead of whole milk for calves can be determined by the following simple rule: 25 lb of milk replacer should not cost more than the selling price of the milk that it replaces. Thus, if 25 lb of milk replacer replaces 225 lb of whole milk selling at $12/100 lb, 25 lb of milk replacer has an equivalent value of $27 (2.25 × $12 = $27).

The raising of early-weaned or orphaned young of mammalian animals will be simplified if they have first received colostrum. During the first few days, it is generally best to feed the young mammal from a bottle with a rubber nipple. Later, it should be taught to drink from a suitable receptacle. It is important that all feeding utensils be kept absolutely clean and sanitary (cleaned and sterilized each time) and that feeding be at regular intervals. Also, young animals should be given grain (and high-quality, fine stemmed hay in the case of ruminants and foals) at the earliest possible time.

(Also see Chapter 20, section on "Colostrum."

Drought Area Feeds

In drought-stricken areas, finding enough feed is an age-old problem. Grain and by-product feeds can be shipped from surplus to deficit areas with relative ease. Silage cannot be shipped. Long hay can be shipped, but at great expense because of its bulk. Less desirable local feeds, such as weeds, prickly pear, and yucca, may be used.

The following alternative feeds have been used for ruminants and horses during droughts in different parts of the world:

• **Forage substitutes**—When hay is scarce and high in price, it may be replaced (preferably, partially only) by such forages as cereal straws, corn stalks, alfalfa straw, beet tops, bean straw, soybean hulls, cottonseed hulls, or poultry litter; adding protein, minerals, and vitamins as necessary to meet the nutrient requirements of the animals being fed.

• **Weeds and other unusual feeds**—During droughts, certain weeds and such unusual feeds as cactus (prickly pear) with the spines burned off, and yucca chopped or shredded, may be fed. (See section on "Weeds" in this chapter.) Before feeding any weeds, the veterinarian and/or county extension agent should be consulted.

• **Substituting grain for hay**—Half or more of the energy requirement of most rations may be furnished by substituting about 5 lb of grain for 8 lb of hay. (See Chapter 8, Hay, section headed "Stretching the Hay Supply.")

Weeds

Weeds may be defined as plants growing where they are not wanted and interfering with desired land use.

Despite the above degrading definition, it is noteworthy that many of today's most useful plants were once considered weeds, and that a number of weeds furnish forage for animals, especially cattle and sheep, when more palatable plants are scarce. This is frequently the case in late autumn or during droughts, when the more desirable vegetation is dried up but some of the deep-rooted perennial weeds may still be green.

When young and tender, some weeds, such as Russian thistles, pigweeds, sunflowers, ragweeds, lamb's-quarter, wild oats, and others, may be used for silage.

It is noteworthy that pigweed, which is the common name for *Amaranthus*, is used as a protein supplement by people in many protein-poor developing nations. Depending on the variety, pigweed plants may contain up to 28% protein and the seeds about 13% protein; and the protein is of high quality. Pigweeds are also a good source of calcium, iron, vitamin A, and vitamin C.

Precautions are necessary when weeds are harvested and used as feeds. For example, foxtail must be harvested and used before the bloom stage because the awns of the seeds are harmful to the mouth of the consuming animal. Also, some weeds are unpalatable, others are toxic.

NOTE WELL: Before feeding any weeds, the veterinarian and/or county extension agent should be consulted.

(Also see Chapter 12, section on "Aquatic Plants.")

SUPPLEMENTS, ADDITIVES, AND IMPLANTS

A feed supplement is a feedstuff that is mixed with a primary grain and/or roughage to provide all the nutrients required to support the form of production for which it is intended.

An additive is a substance of nonnutritive nature which when added to feed will improve feed efficiency and/or production of animals.

An implant is a substance that is implanted into the body for the purpose of growth promotion or controlling some physiological function. They are generally slow releasing in action and often contain a substance that would be destroyed before absorption if included in the feed.

(Also see Chapter 13, Feed Supplements/Additives/Implants.)

Mineral Supplements

Mineral supplements are rich sources of one or more of the inorganic elements needed to perform certain essential body functions.

Generally, the macrominerals of concern are NaCl (common salt), Ca, P, Mg, and sometimes S; and the trace elements that may be deficient are Cu, Fe, I, Mn, Zn, Co, and Se. Most feeds provide minerals in addition to basic organic nutrients, although fat and urea are marked exceptions. Nevertheless, most rations require more concentrated sources of one or more mineral elements.

Needed mineral supplements may be either home-mixed or provided by a commercial mineral product. Commercial mineral mixes are mixed by manufacturers who specialize in the mineral business, either handling minerals alone or in combination with a feed business. Because mineral mixes have become more complicated with the recognition of the importance of trace elements and interrelationships, and because most farmers and ranchers do not have the equipment with which to mix minerals properly, commercial minerals are finding a place of increasing importance in all livestock feeding.

Minerals may, in most cases, be either incorporated in the ration or self-fed.

(Also see Chapter 13, section on "Mineral Supplements.")

Vitamin Supplements

Vitamin supplements are rich synthetic or natural feed sources of one or more of the complex organic compounds, called vitamins, that are required in minute amounts by animals for normal growth, production, reproduction, and/or health.

Formerly, a wide variety of feed ingredients was added to livestock rations for their vitamin content. But it was found that the vitamin concentration of feedstuffs varied tremendously, being affected by plant species and part (leaf, stalk, or seed), harvesting, storing, and processing. Generally speaking, vitamins are easily destroyed by heat, sunlight, oxidation, and mold growth. So, today, nutritionists rely on vitamin supplements, which in many cases are chemically pure sources that need to be used only in very minute amounts. In modern feed formulation, premixes often represent the commonsense approach to providing vitamins.

For adult ruminants, vitamins A, D, and E are of concern, with A being the one most likely to be deficient. Under ordinary circumstances, ruminants synthesize adequate B vitamins, and vitamins C and K. Unless they are kept indoors, they usually receive sufficient exposure from direct sunlight to meet their needs for vitamin D.

Because of the greater prevalence of confinement feeding, along with limited gastrointestinal synthesis, swine are more apt to suffer from vitamin deficiencies than ruminants. Under practical conditions, special consideration should be given to the need for supplementing swine rations with vitamins A, D, E, riboflavin, niacin, pantothenic acid, B–12, and choline.

Vitamins A, D, (D_3 for poultry), B–12, and riboflavin are commonly low in poultry rations. Also, it is in the nature of good insurance to add the following vitamins to poultry rations, as they may be deficient: E and K, and the rest of the B vitamins (in addition to vitamin B–12 and riboflavin, already mentioned).

Because only limited amounts of water-soluble vitamins can be stored, they should be fed regularly in monogastric livestock rations in adequate amounts.

(Also see Chapter 13, section on "Vitamin Supplements.")

Feed Additives and Implants

These nonnutritive products are used to improve the rate and/or efficiency of production of animals, prevent certain diseases, or preserve feeds. But there is no evidence of a nutritional deficiency when they are omitted from a ration.

Most animal scientists and livestock producers agree that antibiotics and hormones stand out as the two nutritional discoveries since 1949 that have had the greatest impact on the livestock industry. In 1949, it was discovered that antibiotics were something new to be added to livestock feeds. Then, in 1954, stilbestrol was approved by the Food and Drug Administration for use in cattle finishing rations. In each case, a new era in livestock feeding was ushered in—comparable to the vitamin era which was born in 1912—and more feed additives and implants followed. Today, it is estimated that 75% of the nation's growing-finishing cattle, lambs, and pigs receive some form of feed additives or implants.

Some glowing reports to the contrary, there is no evidence to indicate that the use of these additives or implants can or will alleviate the need for vigilant sanitation, improved nutrition, and superior management. Instead, with the unfolding and applying of scientific information relative to these promotants, the producer will be able to achieve still greater efficiency of production. Also, practical producers should weigh the benefits of each one against its cost.

(Also see Chapter 13, section on "Additives, Implants and Injections.")

WET FEEDS (HIGH-MOISTURE FEEDS)

Wet feeds, or high-moisture feeds, can be defined as those containing more than 20% water, or less than 80% dry matter (DM). Some common categories of wet feeds are:

Feed	Dry Matter
	(%)
Fresh forages	15–30
Haylage	45–60
High-moisture grain	60–78
Milk, fresh	9–13
Molasses	60–80
Roots	9–30
Silages	25–40
Wet by-products	10–25

QUESTIONS FOR STUDY AND DISCUSSION

1. Differentiate between the two terms, feeds and feedstuffs.

2. How do you account for the fact that pasture or grazing and corn rank first and second, respectively, in total tonnage fed to livestock?

3. Classify feeds according to the seven usual categories.

4. Feed constitutes the second largest single expense in agricultural production, being exceeded only by replacement of capital equipment. As a percentage of total expenses, are feed costs showing an increasing or decreasing percentage? Do you think the trend will continue, if there is a trend? Justify your answer.

5. Since 1975, the trend has been to feed slightly more concentrates and slightly less harvested roughage and pasture. How do you account for this?

6. How do you account for the fact that more alfalfa hay is used for feed than all other types of hays combined and that corn and soybean oil meal are the chief animal energy and protein feed sources, respectively?

7. Why is the American farmer the largest exporter of agricultural commodities in the world?

8. Define the term roughage. What is the role of this type of feedstuff in monogastric rations? What are the characteristics of roughages?

9. How do you account for the fact that sheep and goats head the list for forage consumers whereas poultry consume the least forage?

10. Differentiate between cultivated and native pasture.

11. What are the advantages and the disadvantages of pasture and range forages?

12. What is green chop?

13. Define the term *hay*.

14. What are the advantages and disadvantages of growing and harvesting hay?

15. What are the basic principles of the ensiling of forages?

16. What is the difference between silage and haylage? Which would you prefer?

17. Define concentrate. How important is the feed grain industry? What are the recent trends of acreage harvested and yields?

18. What is meant by the terms (a) carbonaceous feeds, and (b) nitrogenous feeds?

19. What are grains? What are some of the criteria for evaluating the nutritive value of grains?

20. What is high-moisture grain? What advantages accrue from a high moisture grain system?

21. List the practical functions of added fat in livestock feed.

22. Discuss the role of protein quality in feeds for the feeding of ruminants and nonruminants.

23. List the various classes of high-protein feeds and give examples of each.

24. Why has there been an increased interest recently in using urea in rations for ruminants?

25. Why are slow-release NPN products advantageous?

26. What is single-cell protein?

27. What is a complete feed? What is a premix?

28. What is a crop residue?

29. For esthetic reasons, some consumers balk at purchasing products that came from animals that have been fed certain by-products and residues. How would you reassure them that their fears are unfounded?

30. What are milk replacers? How are they used?

31. Discuss the importance of using forage substitutes, weeds and other unusual feeds, and of substituting grain for hay during a severe drought.

32. Under what conditions are weeds sometimes fed?

33. Which minerals are most likely to be needed as supplements in livestock rations?

34. Why did the practice of selecting feed ingredients for their vitamin content give way to the use of vitamin supplements?

35. What vitamins are normally of most concern for (a) adult ruminants, (b) swine, and (c) poultry?

36. What two nutritional discoveries since 1949 have had the greatest impact on the livestock industry?

Fig. 6-9. Guernsey cows on pasture. Pasture and grazing provide 40% of U.S. livestock feed. (Courtesy, Union Pacific Railroad Company, Omaha, Nebr.)

Fig. 6-10. Hay in loaflike stack, 1-ton size, machine made. (Courtesy, Deere & Company, Moline, Ill.)

Fig. 6-11. Silage stored in oxygen-limiting silos, then fed out of the bottom of the silo by an unloader. (Courtesy, *Livestock Breeder Journal*, Macon, Ga.)

Fig. 6-12. Corn, the most important harvested feed in the U.S., ready for harvest. (Courtesy, USDA)

Fig. 6-13. Barley heads ready for harvest. Barley is used for livestock feed, human food, and beverages. (Courtesy, USDA)

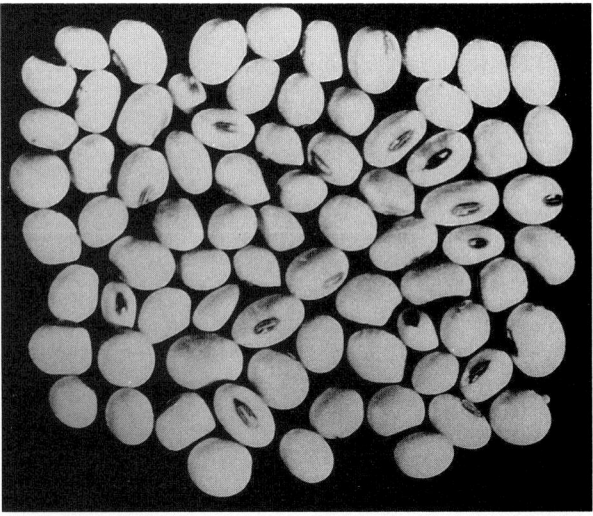

Fig. 6-14. Soybeans. The meal which remains after extracting the oil from soybeans is the most widely used protein supplement in U.S. rations. (Courtesy, J. I. Case Co., Racine, Wisc.)

7

PASTURE AND RANGE FORAGES[1]

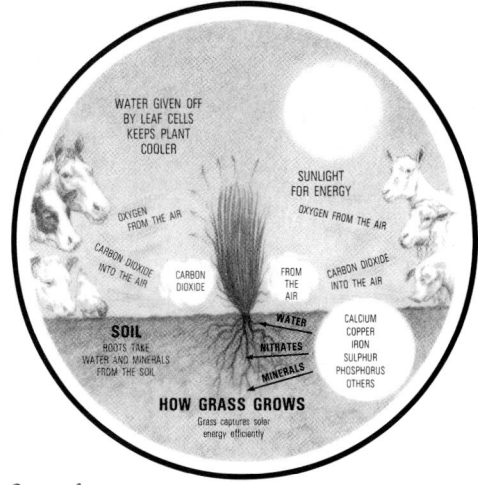

Original painting by Tom Phillips

Contents

Part I— **Pasture**   Page

Classes of Pasture	128
Pastures for Cattle, Sheep, and Horses	129
Adapted and/or Common Grasses and Legumes of U.S.	129
Pastures for Swine	131
Seeding and Management of Subhumid, Humid, and Irrigated Pastures	131
Establishing a New Pasture	131
Renovating An Old Pasture	131
Factors Affecting Value of Pasture	132
Pasture Management	133
Extending the Grazing Season	133
Turnips For Fall/Winter Grazing	134
Seeded Pasture Grazing Systems	134
Continuous Grazing	134
Rotation Grazing	134
Intensive Grazing	134
Creep Grazing	135
Strip Grazing	135
Green Chop	135
Irrigated Pastures	135

Part II— **The Western Range**

Range Area	137
Range Nutrient Deficiencies	137
Range Livestock Supplementation	138
Range Management Considerations	138
Stocking Rate	139
Season of Use	139
Kind of Livestock	139
Range Grazing Systems	140
Continuous Grazing	140
Rotation Grazing	140
Deferred Rotation Grazing Systems	140
Short Duration Grazing Systems	141
Range Improvement Methods	141
Conservative Stocking	141
Distribution of Animals on the Range	142
Range Reseeding	142
Brush and Weed Control	142
Grazing Publicly Owned Lands	143
Agencies Administering Public Lands	143

Part III— **Multiple Use/Conservation of Land**

Multiple Use Concept	144
Multispecies Grazing	144
Wildlife	145
Soil Erosion Control	145
Water	145
Questions for Study and Discussion	146

Fig. 7-1. "All flesh is grass"! (Isaiah 40:6) Hereford breeding herd on pasture. (Courtesy, American Hereford Assn., Kansas City, Mo.)

Grass is the largest and most remunerative crop in the United States, and the cornerstone of successful livestock production. Also, and most important, no method of harvesting has been devised which is as cheap as that which can be accomplished by animals.

As the ever-increasing human population of the world consumes a higher proportion of grains and seeds directly, there will be increased reliance on grass for meat, milk, and wool production. In this connection, it is noteworthy that petroleum is not needed to make wool, and that animals do not require fuel to graze the land and recover the energy that is stored in the grass. Noteworthy, too, is the fact that

---

[1]The authors gratefully acknowledge the helpful suggestions of the following authorities who reviewed this entire chapter: J. E. Baylor, Ph.D., Professor of Agronomy, the Pennsylvania State University, State College; R. A. Forsberg, Ph.D., et al., Department of Agronomy, University of Wisconsin-Madison; S. C. Fransen, Ph.D., Forage Agronomist, Western Washington Research and Extension Center, Washington State University, Puyallup; C. S. Hoveland, Ph.D., Professor of Agronomy, College of Agriculture, The University of Georgia, Athens; D. A. Miller, Ph.D., Professor of Plant Breeding and Genetics, College of Agriculture, University of Illinois at Urbana-Champaign; and W. E. McMurphy, Ph.D., Department of Agronomy, Oklahoma State University, Stillwater. Other authorities reviewed a section or sections, for which credit is given.

animals are completely recyclable; they produce a new crop each year and perpetuate themselves through their offspring. But it takes thousands of years to create coal, oil, and natural gas; and when they're gone, they're gone forever.

Grassland agriculture, better than any other type of agriculture, will continue in the face of economic and social changes to conserve the land and ensure a food supply of the desired quantity, variety, and quality. At its best, it calls for an interdisciplinary approach—for knowledge and application of soil, plant, and animal sciences. This joint focus characterizes the great livestock areas of the world.

From the standpoint of organization, this chapter is presented in three parts: Part I—Pasture; Part II—The Western Range; and Part III—Multiple Use/Conservation of Land.

Part I—Pasture

The term *pasture* is of Latin origin, from the word *pastus*, meaning *an area of land on which there is a growth of forage that animals may graze*.

The importance of pasture, rangeland, and forage in the United States is evidenced by the facts (1) that 29.1% of the total land area is devoted to pasture and rangeland (Fig. 7-2)—more than any other land use; and (2) that 61.7% of the feed for all livestock is derived from forage (Table 6-2).

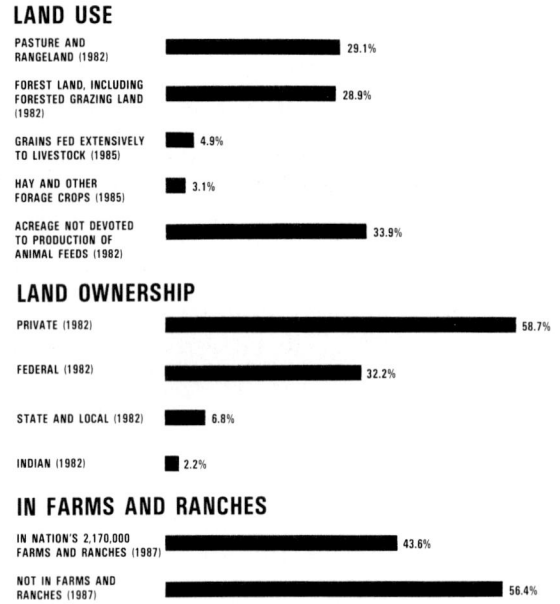

Fig. 7-2. Land use, ownership, and number of farms and ranches of the United States, including Alaska and Hawaii; with the year of the data given in parentheses. It is noteworthy (1) that 29.1% of the total U.S. land area is devoted to pasture and rangeland, exclusive of forested grazing land; (2) that 37.1% of the total U.S. land is devoted to the production of livestock feed (pasture and rangeland, hay and other forage crops, and grain); (3) that 58.7% of the U.S. land is privately owned vs 41.2% publicly owned; and (4) that there were 2,170,000 U.S. farms and ranches. (*Source:* Land use from *Agricultural Statistics 1986*, p. 373, Table 539; p. 388, Table 558; and p. 390, Table 559; and land ownership from *Statistical Abstracts of the United States 1987*, p. 182, No. 318.)

The value and use of pasture and other forages for each species is covered in the respective chapters devoted to each class of livestock, Chapters 19 through 25.

CLASSES OF PASTURE

Broadly speaking, all U.S. pastures may be classified as either (1) seeded pastures, or (2) native pastures (see Fig. 7-3). Although no sharp line of demarcation exists between

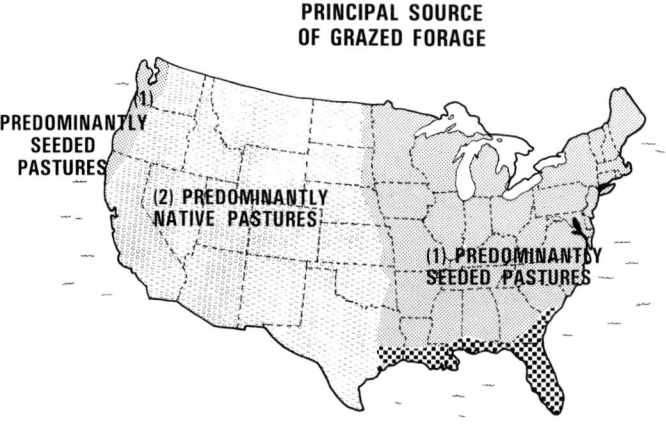

Fig. 7-3. The two major U.S. pasture areas—(1) seeded, and (2) native (range)—about equally divide the 48 contiguous states into east and west halves. (Courtesy, USDA)

the two groups, seeded pastures include those which either receive more than approximately 20 in. of rainfall annually or are irrigated. They are the seeded (cultivated) pastures of the Corn Belt, the South, the East, and the irrigated areas, and smaller and scattered moderate to high rainfall areas throughout the West. Seven grass species—Bahiagrass, Bermudagrass, orchardgrass, reed canarygrass, ryegrass, smooth bromegrass, and tall fescue—account for the major portion of seeded forage grass production in the United States.

The native pastures include those range pastures which receive less than 20 in. of rainfall annually. Their vegetative cover, known as native plants, consists of adapted plants developed by natural selection that have existed in the area for many years, and that were not intentionally introduced.

Pasture may be further classified as—

1. **Permanent pastures.** Those which, with proper care, last for many years. They are most commonly found on land that cannot be used profitably for cultivated crops, mainly because of topography, moisture, or fertility. The vast majority of the farms of the United States have one or more permanent pastures, and most range areas come under this classification.

2. **Semipermanent or rotation pastures.** Those that are used as a part of the established crop rotation. These are

seeded pastures that are generally used for 2 to 7 years before plowing.

3. **Temporary and supplemental pastures.** Those that are used for a short period; and they are usually annuals, such as Sudangrass, sorghum, millet, rye, barley, wheat, oats, ryegrass, arrowleaf clover, crimson clover, ball clover, or rape. They are generally seeded for the purpose of providing supplemental grazing during the season when the permanent or rotation pastures are relatively unproductive.

Pasture plants are classed as (1) grasses, (2) legumes, (3) browse, and (4) forbs. The definition of each of these terms follows:

1. **Grass.** *Botanically, any plant of the family, Gramineae.* In grassland agriculture, grass refers to the forage species of *Gramineae* when either grown alone or with a legume.

2. **Legume.** *Plants, such as alfalfa and the clovers, that obtain nitrogen through bacteria that live in their roots are known as legumes.* The nitrogen fixation aspect of legumes will be of increasing interest as energy sources become more scarce and costly.

3. **Browse.** *The edible parts of woody vegetation, such as leaves, stems, and twigs from bushes.*

4. **Forbs.** *Nongrasslike range herbs which animals eat (forbs are generally called weeds by western livestock producers).*

PASTURES FOR CATTLE, SHEEP, AND HORSES

The economic importance of pastures for beef cattle, dairy cattle, sheep, and horses has been demonstrated in many experiments and on thousands of livestock farms and ranches.

Facts pertinent to pastures for each of these classes of farm animals follow:

1. **Beef cattle.** It is estimated that 84.5% of the total feed supply of all U.S. beef cattle is derived from forage (see Table 6–2); in season, this means pasture. Good pasture alone will produce 200 to 400 lb of beef per acre annually (in weight of calves weaned, or in added weight of older cattle); superior pastures will do much better.

• **Custom cattle grazing (pasture leasing)**—*Custom cattle grazing is the grazing of cattle for a fee without taking ownership of the animals.* This practice has long existed in such noted tall-grass areas as the Flint Hills of Kansas, the Sand Hills of Nebraska, and the Osage Pastures of Oklahoma, where ranch owners grow out cattle, primarily yearling steers, owned by Corn Belt Feeders. Gains in weight depend upon the condition of the range, the length of the grazing season, the age and condition of the cattle, and the management. Daily gains of 2 lb on large-framed, thin yearling steers are not uncommon.

In some cases, custom grazers charge a flat fee per head per day for the season. In other contracts, the charge is based on per pound of gain. Also, some are using an incentive basis, in which the custom grazer is given a base rate for a modest gain (for example, 1 lb/day), then a bonus payment is made for increases above this figure.

(Also see Chapter 19, section on "Stocker and Grower Contracts.")

2. **Dairy cattle.** It is estimated that 59% of the nation's milk is produced from forages (see Table 6–2). Good pasture alone will provide cows with sufficient nutrients for body maintenance and for the production of about 20 lb of milk daily; superior pasture will provide for maintenance and more than 40 lb of milk daily.

As larger numbers of lactating cows are concentrated on smaller acreages and as milk production per cow increases, dairy producers depend less on pasture and more on other feeds. A major reason for this is the inability of high-producing cows to consume enough feed to supply their energy requirement when pasture is their main feed source. The physical form and volume of pasture fill the rumen to capacity before nutrient needs of high producers are fulfilled. However, pastures continue to be practical in smaller herds and for heifers and dry cows in both large and small herds.

3. **Sheep.** It is estimated that 94% of the total feed supply of all U.S. sheep is derived from forage (see Table 6–2); for the most part, this means pasture. No other class of farm animals is so well adapted to the utilization of maximum quantities of pasture as sheep. They are unique in that the vast majority of the young are marketed as milk-fed animals directly off grass.

4. **Horses.** When idle, horses do well on pasture and other forages as the only feed. Even when working, they can use some pasture, with the amount depending on the degree and severity of the work. Most pleasure horses are kept in pasture paddocks when not in use, and they receive supplemental grain and hay when at work.

Do not use Sudangrass for horses.

Adapted and/or Common Grasses and Legumes of U.S.

The specific grass or grass-legume mixture will vary from area to area, according to differences in soil, temperature, and rainfall. Fig. 7–4 shows the 10 generally recognized U.S. pasture areas; and Chart 7–1 (page 130) shows the best adapted and/or most common grasses and legumes for each of these areas.

In using Fig. 7–4 and Chart 7–1, bear in mind that many species of forages have wide geographic adaptation, but varieties often have rather specific adaptation.

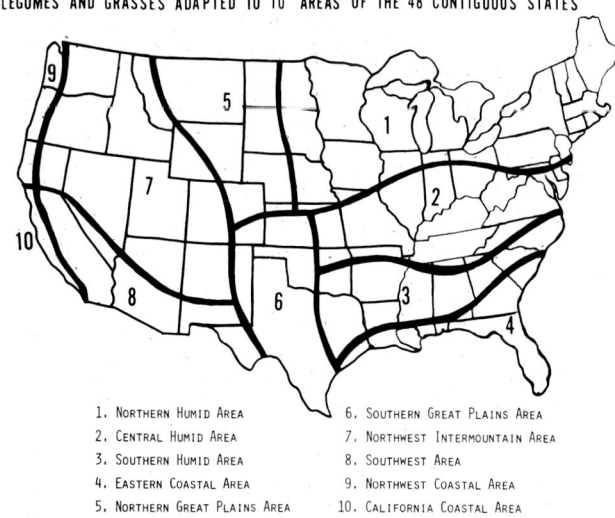

LEGUMES AND GRASSES ADAPTED TO 10 AREAS OF THE 48 CONTIGUOUS STATES

1. Northern Humid Area
2. Central Humid Area
3. Southern Humid Area
4. Eastern Coastal Area
5. Northern Great Plains Area
6. Southern Great Plains Area
7. Northwest Intermountain Area
8. Southwest Area
9. Northwest Coastal Area
10. California Coastal Area

Fig. 7–4. The 10 generally recognized U.S. pasture areas.

CHART 7-1
ADAPTED GRASSES AND LEGUMES (INCLUDING BROWSE AND FORBS) FOR CATTLE, SHEEP, AND HORSE PASTURES, BY 10 GEOGRAPHICAL AREAS OF THE UNITED STATES (SEE FIG. 7-4 FOR GEOGRAPHICAL AREAS)[1]

	Areas of the United States									
	1	2	3	4	5	6	7	8	9	10
Grasses, shrubs, forbs:										
Alfileria (filaree)								x	x	
Bahiagrass (a paspalum)			x	x						
Beardgrass (a bluestem)								x		
Bentgrass	x	x								
Bermudagrass		x	x	x		x		x		x
Bluegrass	x	x			x		x	x		
Bluestem	x	x			x	x		x		
Bristlegrass (a millet)								x		
Bromegrass	x	x			x		x	x	x	
Buckwheat (wild)								x		
Buffalograss						x	x			
Buffelgrass								x		
Chamiza (fourwing saltbush)								x		
Cottontop								x		
Curly mesquite (a Hilaria)								x	x	
Dallisgrass (a paspalum)			x	x						
Dropseed						x		x		
Fescue, tall	x	x	x				x	x	x	x
Foxtail							x		x	
Galleta (a Hilaria)					x			x		
Grama grass	x				x	x		x		
Hardinggrass									x	x
Indiangrass	x	x			x	x				
Indian ricegrass								x		
Indianwheat								x		
Johnsongrass (a sorghum)			x	x		x				
Junegrass					x	x	x			
Kleingrass						x				
Lovegrass	x	x				x		x		
Mesquite (vine; a panicum)						x		x		
Millet	x	x	x	x		x				
Mormon tea (ephedra, jointfir)								x		
Muhly								x		
Needlegrass (needle-and-thread)					x		x			
Oatgrass									x	
Oats	x	x	x	x	x	x			x	x
Orchardgrass	x	x	x	x	x		x		x	x
Pangola digitgrass			x	x						
Panicgrass (a panicum)						x		x		
Paragrass (malojillo)				x						
Pea bush								x		
Pearlmillet		x	x	x		x				
Ratany								x		
Redtop	x						x		x	
Reed canarygrass	x	x			x		x			
Rescuegrass			x	x				x		

	Areas of the United States									
	1	2	3	4	5	6	7	8	9	10
Rhodesgrass			x	x						
Rye	x	x	x	x	x	x		x	x	x
Ryegrass, annual			x	x	x		x			x
Ryegrass, perennial	x	x	x							x
Sacaton								x		
St. Augustine grass			x							
Sorghum-Sudan hybrids	x	x	x	x	x	x	x			
Stargrass			x							
Sudangrass	x	x	x	x	x	x	x	x	x	x
Switchgrass (a panicum)	x	x	x		x	x				
Three-awn (wiregrass)								x	x	
Timothy	x	x					x		x	
Tobosa (a Hilaria)								x		
Wheat	x	x	x		x	x	x	x	x	x
Wheatgrass						x	x	x	x	
Wild-rye							x	x	x	
Winterfat (white sage)								x		
Wintergrass, Texas								x		
Legumes:										
Alfalfa (lucerne)	x	x	x	x	x	x	x	x	x	x
Alyceclover			x	x						
Black medic (yellow trefoil)			x			x		x		
Bur-clover		x						x		x
Cicer milkvetch	x				x		x			
Clover, alsike	x	x			x		x	x	x	
Clover, arrowleaf			x							
Clover, crimson			x	x						
Clover, Hubam (white sweet clover)	x	x						x	x	
Clover, Kura	x	x			x		x		x	
Clover, Ladino	x	x	x	x			x	x	x	x
Clover, prairie						x		x		
Cover, red	x	x	x	x			x	x	x	
Clover, strawberry						x		x	x	x
Clover, subterranean			x	x				x	x	x
Clover, white	x	x	x	x			x	x	x	x
Cowpeas			x	x						
Crownvetch	x	x								
Field pea			x	x			x			
Hairy indigo				x						
Lespedeza (annual)		x	x	x						
Lespedeza (perennial, sericea)		x	x	x						
Peas (flat)									x	
Soybeans	x	x	x	x			x			
Sweet clover	x	x			x	x	x	x		
Trefoil, birdsfoot	x	x	x				x		x	x
Velvet beans			x	x						
Vetch		x	x	x	x			x	x	

[1]Authoritative recommendations for this chart were made by the following agronomists: J. E. Baylor, Ph.D., Professor Emeritus of Agronomy Extension, The Pennsylvania State University, State College; R. A. Forsberg, Ph.D., et al., Department of Agronomy, University of Wisconsin-Madison; J. R. Forwood, Ph.D., Research Agronomist, USDA-ARS, University of Missouri, Columbia; S. C. Fransen, Ph.D., Forage Agronomist, Western Washington Research and Extension Center, Washington State University, Puyallup; C. S. Hoveland, Ph.D., Professor of Agronomy, Department of Agronomy, The University of Georgia, Athens; W. E. McMurphy, Ph.D., Professor, Department of Agronomy, Oklahoma State University, Stillwater; D. A. Miller, Ph.D., Professor, Department of Agronomy, University of Illinois, Urbana-Champaign; R. R. Smith, Ph.D., Professor, Department of Agronomy, University of Wisconsin, Madison.

PASTURES FOR SWINE

Prior to 1950, pastures were considered essential for successful swine production. Subsequently, the importance of pastures for swine declined with increased knowledge of nutrition and escalated land and labor costs. These forces resulted in more and more confinement rearing, accompanied by increased labor efficiency and less use of pastures.

Today, only 4.3% of U.S. swine feed is derived from forage, including pasture (see Table 6–2). Yet, hogs, especially gestating sows, will often yield greater return from an acre of good pasture than any other class of farm animal.

Good pasture can reduce the concentrate requirements for the breeding herd by 75%, for sows and pigs by 20%, and for growing-finishing pigs by 15%.

The most common forages used in U.S. swine production are alfalfa and ladino pastures and dehydrated alfalfa meal. But many other pastures are used, including other clovers, lespedeza, birdsfoot trefoil, rape, winter rye, and certain grasses. In addition to suitability for swine, when choosing forage species and cultivars, consideration should be given to adaptation.

SEEDING AND MANAGEMENT OF SUBHUMID, HUMID, AND IRRIGATED PASTURES

This section, and the subsections under it, has reference to those pastures which either receive above approximately 20 in. of rainfall annually or are irrigated. This includes the pastures of the Corn Belt, the South, the East, and the irrigated valleys and smaller, scattered moderate-to-high-rainfall areas throughout the West.

Fig. 7-5. Hereford breeding herd on clover. (Courtesy, American Hereford Assn., Kansas City, Mo.)

Establishing a New Pasture

The following practices are usually adhered to in successfully establishing a new pasture in the subhumid, humid, and irrigated areas:
1. The species and cultivars are selected.
2. The soil is tested and limed/fertilized.
3. High-quality seed is purchased.
4. Scarified legume seed is used.
5. Legume seed is inoculated.
6. A good seedbed is prepared.
7. The seeding operation is timed and carried out properly.
8. A companion (nurse crop) or a preemergence herbicide may be used.

Renovating An Old Pasture

Fig. 7-6. No-till seeder in operation, overseeding a pasture in Caroline County, Maryland. (Courtesy, University of Maryland, College Park)

Renovation is the improvement of pasture by partial or complete destruction of a sod, plus liming, fertilizing, weed control, and seeding as required to establish desired species without an intervening crop. Extension of the grazing season by sod-seeding small grains or other winter annual species into permanent warm-season grass pastures was started in the early 1950s. About the same time, effective herbicides and sod-seeding machines evolved, making no-till pasture renovation a practical way to improve many grasslands. Today, both winter and summer annual species are used to increase yields and extend the grazing season.

Production from renovated permanent pasture may be increased 2- to 5-fold, depending on the soil characteristics and the condition of the sod.

The practice of sod-seeding works best with cool-season annuals seeded into warm-season perennial grasses. Winter annuals, such as small grains (rye, wheat, or oats), sod-seeded into Bermudagrass swards work especially well since there is minimum competition between the two species.

Successful seeding of summer annuals into perennial cool-season swards can also provide improved forage distribution.

Permanent pastures may be renovated by the following methods:

1. **Complete seedbed preparation with reseeding.** Complete seedbed preparation followed by reseeding has been superior to sod-seeding or overseeding in several areas of the North where perennial grass weeds in the sod are difficult to subdue. The establishment of a legume is markedly improved by plowing as compared to tilling with a field cultivator. Depending on soil conditions, the use of an intervening row crop for one year may or may not be desirable to reduce weed population and competition from the sod, and to help defray costs.

2. **Sod-seeding or overseeding (no-till seeding).** In humid regions, reseeding of legumes and/or grasses into thin permanent pasture sods by sod-seeding or overseeding is often preferred to complete seedbed preparations. The basic requirements for successful sod-seeding are: the partial or complete destruction of the existing plants (including invading weeds) by close grazing and the use of herbicides, thereby lessening top growth, surface resistance, and competition; liming when necessary; fertilizing; proper seed placement; and good pasture management following seeding.

In the North, sod-seedings are made in the spring at a time coinciding with the maximum growth period of grasses in the old sod.

In the South, sod-seeding is most successful when introduced legumes and grasses are seeded at a time that does not coincide with the maximum growth period of the existing sward. So, seedings are generally made in late summer or early fall. Winter annuals or perennial grasses and legumes can be incorporated into the sod to provide winter grazing. Overseeding with pelleted seed of legumes such as white clover and alfalfa from an airplane has also been successful in parts of the humid region.

3. **Fertilizing and liming.** Large areas of permanent pasture have adequate cover but are low in productivity because of lack of fertility. Fertilization can be an economical method of increasing the productivity of such pastures.

Legume-grass pastures with more than 50% legumes usually do not respond to N fertilization. However, such pastures usually respond to lime, P, and K; the lime and P-K fertilizer encourage legume growth and increase the N supply. Also, such fertilization of legume-grass enables earlier spring grazing and extends the pasture season.

Grass pastures and grass-legume pastures with less than 50% legumes respond to N fertilization. However, adequate pH, P, and K soil levels must be maintained to achieve top results from N.

Factors Affecting Value of Pasture

Many factors affect the value of pasture, including (1) soil and fertilizer, (2) plant species, (3) stage of maturity, (4) rate of growth and season of year, and (5) grazing.

• **Soil and fertilizer**—Soil and fertilizer affect the growth and composition of pasture crops. Many experiments have been conducted to determine the effect of soil fertility and fertilizer application on pasture. Generally, the following benefits accrue from pasture fertilization:
1. Increased yields.
2. Increased proportion of legumes.
3. Extended grazing season.
4. Increased protein and palatability.
5. Increased calcium and phosphorus.

• **Plant species**—Plant species affect the feeding value of pasture. Generally speaking, legumes contain a higher percentage of protein and calcium than nonlegumes. Also, there are marked differences between different kinds of pasture plants as growth advances. For example, bromegrass retains its palatability and nutritive value over a longer period than most grasses. By contrast, reed canarygrass is readily eaten when young, but becomes woody, high in alkaloids, and unpalatable with maturity. Most legumes retain their palatability and nutritive value as they mature better than most grasses.

• **Stage of maturity**—Many livestock producers are not aware of the great differences in nutritive value between young, immature pasture and the same plants when they are mature or even at the usual hay stage. These wide differences are shown in Section V—Composition of Feeds, and in Fig. 7-7.

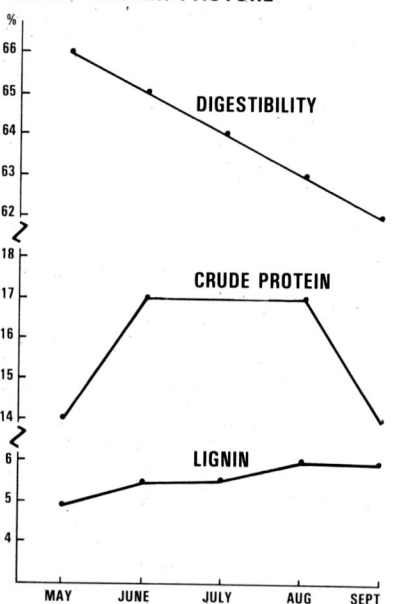

Fig. 7-7. Changes in digestibility, crude protein, and lignin content of orchardgrass *(Dactylis glomerata)*/ladino clover *(Trifolium repens)* pasture in Tennessee, during a 5-year period, from May to September. **Note well**: Digestibility decreased with maturity; it decreased steadily from 66% in May to 62% in September. Protein decreased with maturity; it dropped from 17% in June to 14% in September. Lignin increased with maturity; it increased from 4.8% in May to 5.8% in September. (Source of data: Lane, C. D., K. M. Barth, and J. B. McLaren, *Journal of Animal Science*, Vol. 34, No. 2, 1972, p. 351.)

Stage of maturity affects pasture composition as follows:
1. Protein decreases with maturity.
2. Fiber and lignin increase with maturity.
3. Calcium and phosphorus decrease with maturity.
4. Vitamin value decreases with maturity.

- **Rate of growth and season of year**—Rapidly growing grass is usually rich in protein and in other nutrients on a dry basis. It is important, therefore, that pasture plants be properly fertilized and managed so that they keep growing and be prevented from heading out.
- **Grazing**—When pastures are grazed closely throughout the season, the total yield of dry matter is usually 30 to 50% less than when they are allowed to grow to the normal hay stage. This is due to the smaller leaf surface and lowered photosynthesis. This explains why rotational, strip, and green chop grazing usually yield more than close continuous grazing.

The effect of frequent grazing will depend on the kind of plants. The yield of tall-growing plants—such as timothy, orchardgrass, alfalfa, and the erect clovers—is reduced much more than that of low-growing spreading plants, such as bluegrass, Bermudagrass, and white clover.

In contrast to the lowering of the yield of dry matter, frequent grazing usually results in greater total production of protein for the season than when the crop is cut for hay. Also, because immature plants are lower in fiber and more digestible than mature plants, the yield of total digestible nutrients is not reduced as much by frequent grazing as the dry matter yield—dry matter production is lowered by 30 to 50%, whereas digestibility is lowered only by 25 to 40%.

Also, it is noteworthy that plenty of available forage results in selective grazing—with the animals picking and choosing the leaves and finer parts of stems, which are more tender and more nutritious, and rejecting the coarser, stemmy parts. Thus, the portion consumed under such circumstances may differ appreciably from the chemical composition of the entire plant.

Pasture Management

Many good pastures have been established only to be lost through careless management. Good pasture management in the subhumid, humid, and irrigated areas involves the following practices:

1. **Controlled grazing.** Nothing contributes more to good pasture management than controlled grazing. At its best, it embraces the following:
   a. Protecting first-year seedings.
   b. Shifting the location of salt, shade, and water.
   c. Deferred spring grazing.
   d. Avoiding close late fall grazing.
   e. Avoiding overgrazing.
   f. Avoiding undergrazing.

2. **Clipping pastures and controlling weeds.** Pastures should be clipped at such intervals as necessary to control weeds (and brush) and to get rid of uneaten clumps and other unpalatable coarse growth left after incomplete grazing.

3. **Topdressing.** Like animals, for best results, grasses and legumes must be fed properly throughout a lifetime. It is not sufficient that they be fertilized (and limed if necessary) only at or prior to seeding time.

4. **Scattering droppings.** The droppings should be scattered at the end of each grazing season in order to prevent animals from leaving ungrazed clumps and to distribute the droppings over a larger area. This can best be done by the use of a brush harrow or chain harrow.

5. **Grazing by more than one kind of animal.** Grazing by two or more species of animals makes for more uniform pasture utilization and fewer weeds and parasites, provided the area is not overstocked. Different kinds of livestock have different habits of grazing; they show preference for different plants and they graze to different heights. For example, sheep consume shorter and finer forages and more forbs than cattle.

6. **Irrigating where practical and feasible.** Where irrigation is practical and feasible, it alleviates the necessity of depending on natural precipitation.

Extending the Grazing Season

In practically all U.S. pasture areas, the grazing season can be extended by grazing earlier in the spring and later in the fall/winter, thereby lessening the amount of stored feed needed for winter.

Fig. 7-8 illustrates in graphic form the growth period of each of the common pasture plants of area 2 (see Fig. 7-4 for areas), the northern part of the humid South. As shown, by selecting the proper combination of crops, pastures for each month of the year are assured. A similar chart for each area of the country can be developed.

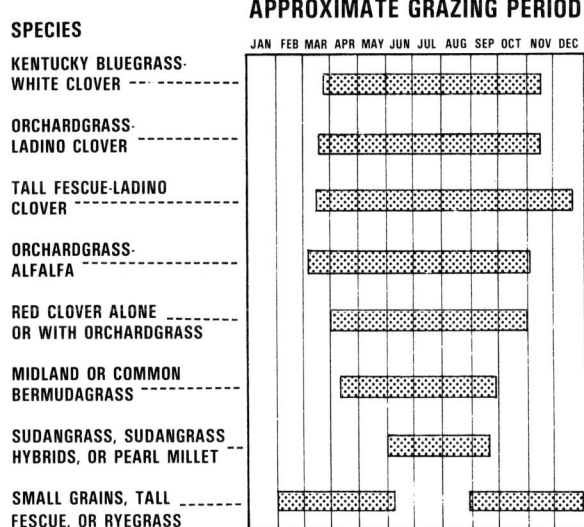

Fig. 7-8. Approximate grazing period of common pasture crops in area 2 (see Fig. 7-4 for areas), the northern part of the humid South. As shown, by selecting the proper crops year-round grazing can be achieved.

In addition to lengthening the grazing season through the selection of species, it may be extended as follows:

1. **By obtaining earlier spring pastures.** This can be accomplished by avoiding grazing too late in the fall and by applying nitrogen fertilizer in the fall or early spring. Nitrogen fertilizers will often stimulate the growth of grass so that it will be ready for grazing 10 days to 2 weeks earlier than unfertilized pastures.

2. **By saving fall growth for winter grazing.** For example, tall fescue that has been cut for winter feed can be used to stockpile regrowth in August through October for grazing

from November into the winter. Also, often the regrowth after hay harvest may be utilized.

3. **By using crop residues.** Following harvest, cornstalks will provide 2 months of winter grazing for dry cows and ewes. In addition to extending the grazing season, this utilizes feed which would otherwise be wasted or unmarketable.

TURNIPS FOR FALL/WINTER GRAZING

Turnips are a popular fall/winter forage for cattle and sheep in some northern areas, especially in the Pacific Northwest. They are commonly planted in July following harvest of early-maturing crops such as sweet corn, early potatoes, or small grains. Grazing usually begins in October or early November. Two varieties, Barive Cow Turnips and Purple Top White Globes, are used as forage.

Cattle generally remain healthy while grazing turnips, but occasionally they develop polioencephalomalacia, pulmonary emphysema, bloat, or hemolytic anemia; and sometimes they choke. Pregnant ewes grazing turnips without iodine supplementation may give birth to lambs with goiters and low survivability. These diseases may be lessened, or prevented, by the following dietary management:

1. Place animals on a high-quality diet 2–3 weeks prior to grazing turnips, thereby preparing the rumen microorganisms to digest turnips.
2. Immunize against enterotoxemia.
3. Avoid abrupt change to turnips.
4. Supplement the turnips with straw, hay, or other roughage.
5. Strip-graze turnips (small strips, moved weekly).
6. Feed an iodized salt-trace mineral mix at all times.

SEEDED PASTURE GRAZING SYSTEMS

Several systems of grazing management have been successfully applied to semipermanent and supplementary pastures. Generally speaking, the more intensive the system of management on such pastures, the higher the yield of forage and of livestock products.

It is noteworthy that pasture grazing systems have been changed/adapted by both researchers and farmers; and with such changes/adaptations, they have been given different names. Nevertheless, under whatever name, the basic types of rotation grazing, intensive grazing, creep grazing, strip grazing, and green chop are covered in the sections that follow.

(Also, see the section on Range Systems in this chapter. The principles involved in grazing seeded pastures and western ranges are similar. However, the application differs because western range pastures are generally much larger and have lower rainfall.)

Continuous Grazing

The name identifies the practice. *Continuous grazing is the uninterrupted grazing of a specific pasture by livestock throughout a year or grazing season.* It can be successful provided variable stocking is practiced, with some adjustment in animal numbers to reduce the severity of under- or overgrazing.

The **advantages** of continuous grazing as compared to rotational grazing are (1) lower costs for fencing and watering facilities, because there are fewer animals when they are confined to one pasture; and (2) fewer management decisions when animals are not moved from pasture to pasture.

The **limitations** of continuous grazing are (1) animal numbers are seldom flexible; and (2) pastures must be stocked lighter than desired when forage growth is maximal to avoid overgrazing during periods of minimal forage growth.

Rotation Grazing

Rotation grazing is that system in which two or more pastures are grazed and rested in a planned sequence. In this system, pastures are divided and fenced into two or more pastures.

The **advantages** of a rotation grazing system are:
1. It permits the farmer to match grazing more adequately to the growth habit of the forage species, condition of the pasture, and animal needs than does continuous grazing.
2. It improves stand persistence. Plants are given recovery periods during the growing season for more or less unhampered development of tillers, and leaves. This is essential to replenish root reserves. It is the only system of grazing which enables the tall-growing legumes and grasses to survive.
3. It increases carrying capacity. Greater amounts of feed nutrients can be removed in the form of herbage with reduced losses due to trampling, fouling, and herbage death and decay.
4. It encourages equalization of grazing. It prevents overgrazing and undergrazing, and results in maintaining a better balance of the legumes and grasses. Also, both the palatable and the inferior species will be grazed more nearly the same.
5. It provides more nutritious herbage since the herbage is at the most ideal pasture stage. It will be high in protein and low in crude fiber. The protein content will average 10 to 20% compared with 6 to 10% for more mature grasses.
6. It prevents the grasses from heading out, particularly if the mower is used each time the animals are shifted to new pasture. This allows new growth to come back uniformly and keeps it more palatable.
7. It helps control livestock parasites.
8. It makes it easier to harvest surplus forages as hay or silage.

The **limitations** of rotation grazing are:
1. It requires a higher input of capital and management than continuous grazing.
2. It results in a continuous day-to-day decline in the quality of the available forage. At first turn-on, animals have access to leafy, high quality forage, but the quality of the forage gets poorer and poorer during the grazing period.

Intensive Grazing

Several ingenious intensive grazing systems have evolved. All of them are designed to provide the maximum of high quality forage, to utilize the highest quality pastures for the highest producing animals, and/or to make more profit. They call for maximum spring and early summer grazing, and for using the best quality pastures for dairy cows, cows suckling calves, ewes with twin lambs, does with kids, or mares with foals.

- **First and second grazers**—This grazing system involves two herds: first grazers, and second grazers. Here is how it may work with dairy cows: High producing lactating cows, which have a high energy requirement, may be first grazers, they are allowed to graze the higher-quality (leafy) portion of pasture No. 1, following which they are moved to pasture No. 2—a fresh pasture. Dry cows, which have a low energy requirement, may be second grazers; they are turned onto pasture No. 1 immediately following the removal of the high producers. This progression is continued through all pastures, then the cycle is repeated. Also, it may involve three or more groups of animals. The groups may consist of any animal species or class. For example, the first grazers may consist of beef cows and suckling calves, and the second grazers of stocker steers; or the first grazers may be ewes with twin lambs, and the second grazers gestating ewes.

The chief **advantage** of the system of first and second grazers is the enhanced productivity of the first grazers. The main **limitations** are the necessity of maintaining (1) two groups of animals of different productivity levels, and (2) balanced stocking rates and pasture sizes.

- **Intensive early season stocking**—This grazing system calls for heavy stocking (perhaps twice the average year-round carrying capacity of the pasture) in the spring and early summer, when the pasture is of highest quality and most productive. Here is how it works with stocker cattle: Double the normal stocking rate in the spring and early summer. Then, around July 1, either sell half or more of the herd or place them in a feedlot.

The **advantages** of this system are (1) more pounds of beef per acre, (2) lower interest charges, because of owning the cattle for a shorter period of time, and (3) higher net returns. The main **limitation** is the lack of flexibility relative to removal of half the herd; they must either be sold or moved into the feedlot as scheduled—in mid-summer.

Creep Grazing

Creep grazing is a system of grazing nursing young on a high-quality pasture(s) (a grass-legume mixture, all legumes, or high-quality annuals) separated from their dams. This system may be used for cows and calves, ewes and lambs, goats and kids, and mares and foals. Creep grazing may be accomplished as follows:

1. Allowing young to forward graze ahead of their dams, then following with their dams later.
2. Keeping the dams and young on a base pasture, and providing an additional creep pasture for the young.

Strip Grazing

In this system, animals are allowed access to a strip which may be large enough for several days of grazing or small enough for one-half day to one-day of grazing. Heavy stocking rates of upwards to 50 animal units per acre are used by fencing each strip with a movable electric fence. The **advantages** claimed for this method are:

1. Increased utilization of herbage, with wastage reduced to 10 to 20%.
2. Increased meat and milk yields per acre up to 25%.
3. Improved stability of meat and milk yield because the nutritive value of the pasturage consumed is quite constant.

4. Improved utilization of the available forage. Less herbage is soiled by dung, urine, and treading. Under strip grazing, animals are quieter and settle down quickly for steady grazing rather than roaming about and tramping forage.
5. Increased animal units maintained on a given area, although animal productivity may not be increased.

Green Chop

Green chop is fresh herbage that is cut and chopped in the field, then transported and fed to animals in confinement.

Green chop, which is also called soilage or zero grazing, consists of growing a succession of forage crops, harvesting them with mechanized equipment, and hauling the green feed to the animals rather than allowing the animals to harvest their own forage.

Green chop minimizes the loss of moisture, color, nutrients, and wastage. Alfalfa, ladino clover, orchardgrass, bromegrass, grass-legume mixtures, Sudangrass, corn, sorghum, soybeans, and cereal grains are sometimes used in this manner. With tall-growing crops, more feed value may be realized from a given area than can be obtained by conventional pasturing. However, green chop requires special equipment and harvesting every day. Also, there are harvesting problems in wet weather.

Fig. 7-9. Harvesting forage as green chop. (Courtesy, Deere & Company, Moline, Ill.)

Most green chop is fed to lactating dairy cows, usually in combination with hay or silage because the total intake tends to be greater. Green chop has increased with herd size, with more intensive form of dairying, with drylotting of cows, and with high grain prices. Also, the use of green chop has been facilitated by the greater mechanization present on larger and more modern dairy farms.

IRRIGATED PASTURES

The irrigated land in conterminous United States fell from a high of 50.2 million acres in 1978 to 44.7 million in 1984, primarily due to the rise in energy costs and the fall in crop and livestock prices.[2]

---

[2]*1986 Agricultural Chartbook,* USDA, Agricultural Handbook No. 663, p. 27, Chart 68.

Irrigated pastures provide forage of high quality at a relatively low cost, often on land unsuited to other crops. Both perennial and annual irrigated pastures are important feed crops.

Successful pasture irrigation involves special decision making relative to (1) irrigation—the method, frequency, and amount of irrigating, and the removal of excess water; and (2) the kind and amount of fertilizer.

• **Method of water application**—Three basic methods of irrigating pastures are practiced: (1) border-flood, (2) furrow, and (3) sprinkler. Additionally, variations or combinations of the first two methods are sometimes used.

• **Frequency and amount of irrigation**—Many pasture plants, especially the clovers, are shallow-rooted and require more frequent and lighter irrigations than deep-rooted plants. The Washington Station reports that the highest yields per acre from an orchardgrass-ladino clover pasture can be obtained with a summer irrigation frequency of 7 to 11 days, rather than at less frequent intervals, and that more frequent irrigations also give the highest proportion of clover.

• **Excess water**—Excess water in irrigated pastures is caused by either (1) overirrigation and the inability of the excess water to drain from the soil, or (2) subsurface drainage from adjacent and higher land. Surface drains are necessary to remove excess irrigation. Drainage is particularly important wherever there is danger of salt accumulation.

• **Fertilization**—irrigated pastures require high soil fertility to be productive. The kind and amount of fertilizer should be determined by the level of the productivity desired, and the role of the legumes in the mixture. Production levels of irrigated pastures are increased more by N fertilizer than by other fertilizer elements, with responses also obtained from P and K where soils are deficient in these elements. Nitrogen stimulates grass growth, whereas P increases the legume component. Nitrogen can be supplied by either fertilizer or innoculated legumes. When legumes are a major component of pasture, economic returns from applied N, measured in increased animal production may not be obtained.

Fig. 7–10. Holstein cows on irrigated pasture. (Courtesy, Holstein Assn., Brattleboro, Vt.)

Part II—The Western Range[3]

The opening up of the western range accompanied the establishment of the Jesuit missions in what is now Arizona, New Mexico, and Texas in the period from 1670 to 1690. But livestock grazing did not spread throughout the West until the 19th century, with the annexation of Texas, the Southwest, and Oregon. Then, in 1848, came the magic news—*gold* had been discovered in California. The race was on! The trickle of adventurers turned into a flood. Trails and trading posts came alive; and there was need for cattle for meat.

The growth of the industrial East and the subsequent development and extension of the railroads provided the necessary stimulus for further expansion of the range cattle and sheep industries. The grass supply of the vast ranges seemed unlimited, and the region was regarded as a permanent paradise for cattle and sheep. About 1880, the lure of the grass bonanza fired the imagination of investors, big and little. Cowboys, sheepmen, lawyers, farmers, merchants, laborers, and bankers—many of them English and Scotch investors in great companies—rushed to seek their fortunes on the open ranges. Cattle barons and sheep kings reached a "high water mark." The number of cattle and sheep increased dramatically, joining the wild animals that were already there. Soon the range was overstocked. Regulations were few, and the guiding philosophy was that the grass belonged to the cattleman or sheepman who got there "firstest with the mostest." The cattle-sheep feuds waxed hot. Deadlines were set up and enforced by the side that could muster the greatest strength.

Suddenly, it became apparent that greed had taken its toll! The supply of tall grass was exhausted. Even more tragic, the winter of 1886–87 was ferocious. Few owners had made provisions for winter feed. Cattle and sheep perished by the thousands. In some herds, 85% of the animals starved to death. This marked the beginning of the end of the large livestock companies, the death knell of the open range, the gradual growth of smaller operations, and increased attention to range management, winter feed, and shelter.

The discipline of range management developed in the 1920s. By 1925, some 15 colleges were offering courses in range management.

The biggest event of the 1930s in range management was the passage of the Taylor Grazing Act of 1934, which placed the administration of remaining public lands under the Grazing Service, which later became the Bureau of Land Management. In 1933, the Soil Erosion Service was formed because of alarm over the drought in the Great Plains. In 1948, the Society for Range Management was organized.

During the 1960s, the multiple use concept of range management on federal lands developed. Wildlife, water, and recreation became recognized as important range products as well as red meat. Previously, range research and management had been geared towards producing forage for livestock.

Toward the end of the 1970s and into the present, concern over the world population explosion generated renewed interest in using public rangeland for livestock production. Simultaneously, energy costs spiraled, and it was recognized that lower energy inputs are required to produce red meat

---

[3]The authors are very grateful to the following specialists who reviewed this section: R. M. Williamson, Director of Range Management, Forest Service U.S. Department of Agriculture, Washington, D.C.; E. Moris, Assistantto the Director and Chief, Office of Public Affairs, Bureau of Land Management, U.S. Department of the Interior, Washington, D.C.

from rangeland than cropland. In addition, there was an awareness that range forage can only be converted by grazing animals into products usable by people. Scientific range management on private land accelerated during this period and replaced much of the frontier spirit and the romantic, adventurous life of the cowboy and sheep herder.

Fig. 7-11. On the range. (Courtesy, American Angus Assn., St. Joseph, Mo.)

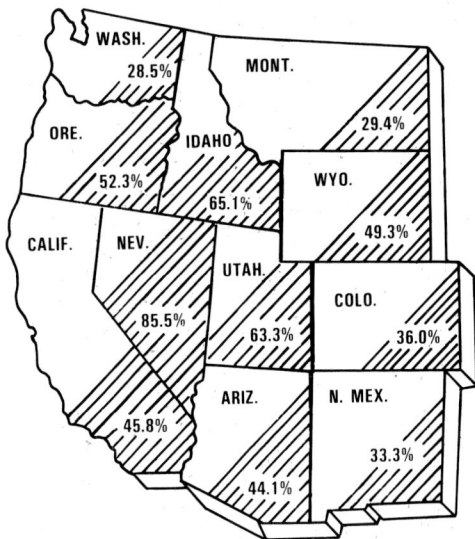

Fig. 7-12. A map showing the 11 western states and the proportion of land in each of these states that is owned by the U.S. Government. (Source: Public Land Statistics, 1984, Vol. 169, August 1985, p. 10.)

RANGE AREA

Range refers to large, naturally vegetated, mostly unfenced areas where animals harvest a rather sparse growth of grasses, legumes, and other edible parts of forbs and browse.

Various geographical divisions are assumed in referring to the western range area—the native pasture area. Sometimes reference is made to the 17 range states, embracing a land area of approximately 1.16 billion acres. At other times this larger division is broken down, chiefly on the basis of topography, into (1) the Great Plains area (the 6 states of Kansas, Nebraska, North Dakota, Oklahoma, South Dakota, and Texas); and (2) the 11 western states (Arizona, California, Colorado, Idaho, Montana, Nevada, New Mexico, Oregon, Utah, Washington, and Wyoming).

Almost half (47.6%) of the land area in the 11 western states is federally owned and administered. Domestic livestock graze on 89% of this area. Federal land is estimated to supply 17% of all grazing resources in the region.

Because of the magnitude of the range livestock industry and the fact that it is a highly specialized type of operation, considerable discussion will be devoted to the range area and the care and management of cattle and sheep in the range method.

The carrying capacity of much of the western range is low, and little of it provides yearlong grazing. Moreover, variation in vegetative types, climate, and topography in the range country is accompanied by great diversity in the seasonal use made of it. As a result, rangelands are usually grazed during different times of the year, and the herds and flocks migrate with the season, moving to the mountains and higher elevations in summer and returning to the lower ranges in winter.

From the standpoint of vegetation and utilization by livestock, ranges differ from cultivated pastures as follows:
1. They are less productive.
2. They are more likely to progress to less palatable plants.
3. They are more difficult to restore when depleted.
4. They often serve multiple uses.

RANGE NUTRIENT DEFICIENCIES

Nutrient deficiencies are rather common on the range. Many soils are deficient in certain nutrients, which affect the plants, and, in turn, the animals feeding on them. During droughts, and early or late in the season, forage is in short supply, limiting energy and other nutrients. Early spring pastures are washy. Later in the season they become leached or bleached—they increase in fiber and decrease in protein, phosphorus, and carotene.

• **Energy deficiencies**—Hunger, due to lack of feed, is the most common deficiency of range livestock. The most important requirement is sufficient feed for body maintenance. Over and above this, surplus energy is used for growth or fattening.

With bulky, low-quality roughages—such as range grass cured on the stalk, animals cannot consume sufficient quantities to meet their energy needs. The younger the animal, the more acute the problem. Under these circumstances, the low-energy intake is met by breaking down of body tissues. This results in loss of weight and condition and lack of growth.

In breeding animals, low-energy intake affects reproduction. Cows take longer to come in heat and require more services per conception, thus reducing the calf crop. Also, calves born from energy-deficient cows are lightweight at birth. Supplemental feeding is the practical way to eliminate energy deficiencies.

• **Protein deficiencies**—The protein intake of beef cattle must be adequate to develop muscle (meat) and replace worn body tissues. The protein need is most critical in young calves and in gestating-lactating cows.

Because protein supplements ordinarily cost more per ton than grains, the temptation is to feed too little of them. When grazing mature, weathered forage, cows should be fed a protein supplement.

Mature, weathered native range plants are almost always deficient in protein—as little as 3% (see Fig. 7-13). A deficiency of protein results in depressed appetite, poor growth, loss of weight, reduced milk production, irregular heat periods, and lowered calf crops.

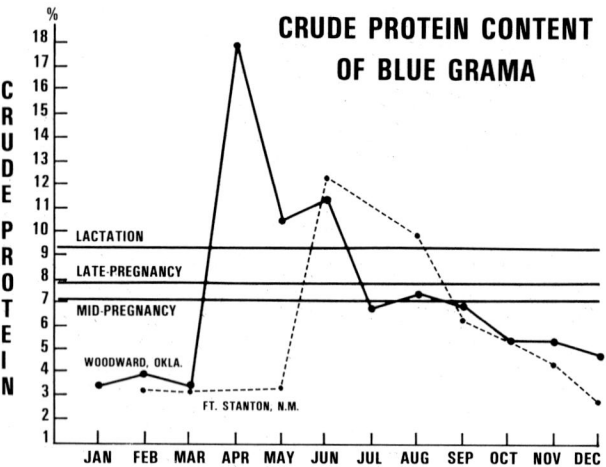

Fig. 7-13. Crude protein content (moisture-free basis) of blue grama (Bouteloua gracilis) grass. Woodward, Okla. data from: USDA Tech. Bull. 943, 1947, p. 53, Table 12, by Savage, D. A. and V. G. Heller. Ft. Stanton, N. Mex. data from: New Mexico Ag. Exp. Sta. Bull. 662, 1978, p. 22, Table 1, by Pieper, R. D., et al. NOTE: The peaks in the protein content of the vegetation in the two areas reflect the periods of greatest rainfall.

Superimposed on the chart are the crude protein requirements in a moisture-free ration of a 1,100-lb cow during mid-pregnancy, late-pregnancy, and lactation, respectively (taken from Table 19-2 of this book). Note that, in both areas, there is a protein deficiency except during periods of highest rainfall. (Chart provided by Ted McCollum, Ph.D., Animal Science Department, Oklahoma State University, Stillwater.)

• **Mineral deficiencies**—Phosphorus deficiencies are rather common on the range (see Fig. 7-14). A severe phosphorus deficiency results in depraved appetite, emaciation, retarded growth and development, failure to breed regularly, lowered calf crop, lowered milk production, and high death losses.

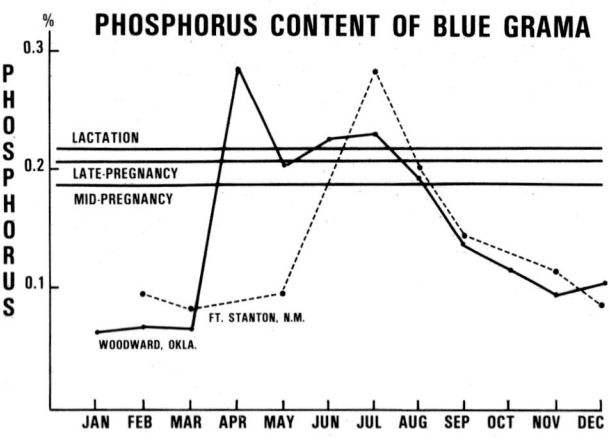

Fig. 7-14. Phosphorus content (moisture-free basis) of blue grama (Bouteloua gracilis). Woodward, Okla. data from: USDA Tech Bull. 943, 1947, p. 52, Table 11, by Savage, D. A. and V. G. Heller. Ft. Stanton, N. Mex. data from: New Mexico Ag. Exp. Sta. Bull. 662, 1978, p. 28, Table 7, by Pieper, R. D., et al. NOTE: The peaks in the phosphorus content of the vegetation in the two areas reflect the periods of greatest rainfall.

Superimposed on the chart are the phosphorus requirements in a moisture-free ration of a 1,100-lb cow during mid-pregnancy, late-pregnancy, and lactation, respectively (taken from Table 19-2 of this book). Note that, in both areas, there is a phosphorus deficiency except during periods of highest rainfall. (Chart provided by Ted McCollum, Ph.D., Animal Science Department, Oklahoma State University, Stillwater.)

• **Vitamin deficiencies**—Range plants are frequently low in carotene. A severe deficiency of carotene (vitamin A) may result in low conception rate, a small calf crop, many calves weak or stillborn, with some calves born blind, more cows with retained afterbirth, low gains, greater susceptibility to calf scours, and more respiratory troubles.

Severe vitamin A deficiency in bulls may result in decreased sexual activity and lowered semen quality.

When grazing dry range longer than 4 to 6 months, it is recommended that a supplement containing 20,000 to 30,000 IU of vitamin A per pound be fed to brood cows and bulls.

Range cattle usually receive sufficient vitamin D from exposure to direct sunlight or from sun-cured forages.

Range Livestock Supplementation

Energy, protein, phosphorus, and vitamin A are the major nutrients limiting the performance of range livestock.

Energy supplementation may be advantageous when range forage is in short supply, as during a drought or heavy snowfall. Also, energy supplementation may reduce nitrite toxicity problems and improve protein/energy ratios of livestock grazing lush, high protein pastures. Livestock perform best when energy (particularly grain) is provided on a daily basis.

Protein is the major supplement cost of most ranches. Economically, protein supplementation of livestock is most advantageous when the crude protein content of range forage drops below 6 to 7%. Protein supplements can be provided to livestock on an every-other-day or every-third-day basis without affecting their performance. Salt or fat may be used to limit the consumption of cottonseed meal or other protein supplements.

Phosphorus supplementation during periods of forage dormancy seems justified. Also, the routine inclusion of essential microminerals in salt blocks or mineral mixes is a good practice.

Vitamin A supplementation is recommended when livestock must be maintained for over 4 months without access to green grass or browse.

(Also see Chapter 11, Protein Supplements, Chapter 13, Feed Supplements/Additives/Implants; Chapter 19, Feeding Beef Cattle, Part II Feeding Breeding Beef Cattle, section on "Range and Pasture Feeding"; Chapter 21, Feeding Sheep, section on "Feeding Range Sheep"; and Chapter 22, Feeding Goats, "Part III Feeding Angora and Spanish [Meat] Goats.")

RANGE MANAGEMENT CONSIDERATIONS

Good range management may be achieved if an inventory or analysis is made of the forage resources and all contributing factors, followed by a sound plan of management based upon the analysis. Consideration should be given to such factors as proper stocking rate, and safe degree of use, season of use, kind of livestock, condition and trend of forage, soil stability, system of use, improvements needed, etc.

Stocking Rate

The key to successful long-term operation of rangeland lies in making (1) a reliable determination of the land that is suitable or adaptable to grazing use over a long period of time; (2) a realistic estimate of grazing capacity for this land; and (3) a flexible stocking rate, even within a single season, followed by (a) application of proper stocking intensity, and (b) frequent observations to determine the effect of the stocking rate upon changes in condition of the forage cover. Too light stocking wastes forage, while too heavy stocking results in a change of forage plant cover from an abundance of valuable forage plants to an abundance of less valuable, or worthless, plants.

The following rule of thumb, applied to the more heavily grazed key areas, may be used in arriving at the proper stocking rate: "Use half and save half, and the half you save will grow bigger and bigger." The rule refers to half the weight, which is concentrated at the bottom of the plant, and not to half the height. Thus, when the 50% rule of thumb is applied to bluebunch wheatgrass, a common range plant, it means that about 75% of the bunches have been grazed to an average stubble height of about 4 in., and the remaining 25% of the plants left relatively ungrazed.

Season of Use

A prime requisite of successful management for both cattle and sheep is that there shall be as nearly year-around grazing as possible and that both the animals and the range shall thrive. In some areas, especially in the southwestern Great Plains areas, these conditions are met without necessitating extensive migration of animals. The winter climate is mild, and the native forages cure well on the stalk, thus providing nutritious dry feed at times when green vegetation is not available. Generally speaking, however, most of the cattle and sheep from such areas are marketed via the feeder route rather than as grass-fat slaughter animals.

In general, the most desirable management, both from the standpoint of the animals and the vegetation, consists of the proper seasonal use of the range. Although there is wide variety in the customs and requirements for seasonal use of the range—because of the spread in climate, topography, and vegetative types included in the vast expanse of range country—seasonal-use ranges are usually placed in four major classes: (1) spring-fall, (2) winter, (3) spring-fall-winter, and (4) summer.

Because a range band of sheep can be moved and herded on unenclosed areas with greater ease than a herd of cattle and because investigations in range livestock management have been conducted more extensively with sheep, greater seasonal use of ranges is made with sheep. On the other hand, the more progressive cattle producers are finding ways and means of adopting many of the same methods.

Despite the value of yearlong grazing, it is recognized that the prevalence of severe winters in some parts of the West preclude winter grazing except to a limited degree, and stock must be fed during at least a part of the winter season. Where these conditions prevail, cattle and sheep are usually wintered in the irrigated valleys, close to the feed supply, especially a supply of alfalfa or meadow hay.

Kind of Livestock

Sheep and cattle share in the utilization of the western range. In fact, some ranges are simultaneously grazed by these two kinds of animals. This dual system of grazing is practical and beneficial provided the grazing capacity for each is properly adjusted so that the major forage plants are properly used, and that, at intervals, a careful determination is made of condition and trend of soil and forage. Some ranges, especially in Texas, are grazed by three classes of animals—cattle, sheep, and goats—with the goats controlling the brush without adversely affecting environment.

When sheep and/or goats are added to a cattle range (or vice versa), the increased numbers should not result in a total which exceeds the previous animal units of a single species by more than 10%; otherwise, overgrazing will likely result.

Actually, economic factors—often unrelated to range characteristics—probably have the greatest influence on the selection and popularity of kinds of livestock. The kind which the operator feels will return the greatest net profit is selected, and the choice changes with changing times. Nevertheless, range characteristics may be so specific as to favor one kind of livestock to the point that other kinds would be produced under handicap. Among such range characteristics which should be considered in choosing the kind of livestock are:

1. **Poisonous plants.** The presence of certain poisonous plant species may limit the use of the range to one kind or another of livestock. Thus, larkspur is a serious menace to cattle, but normally sheep are not affected by it. On the other hand, generally cattle may safely graze lupine-infested ranges, which are sometimes extremely dangerous to sheep. Many other examples of selective poisoning could be cited.

2. **Topography.** Cattle prefer level to gently rolling topography, whereas sheep and goats are better adapted than cattle to steep, rocky, or bushy ranges. The latter seem to have a natural instinct for climbing, and, through the efforts of the herder, they can be encouraged to graze the more difficult terrain. In addition, because of greater ease in herding, and moving about on unfenced public domain, sheep are trailed about more than cattle, thus more effectively utilizing seasonal ranges.

3. **Water.** Sheep and goats are much better adapted than cattle to more poorly watered ranges, because they can go for longer periods without water. Also, sheep utilize snow as a sole source of water more satisfactorily than cattle; therefore, sheep use range dependent on snow for water more efficiently than cattle.

4. **Vegetative cover.** In general, sheep do not utilize tall-growing grasses as effectively as cattle. Sheep and goats are weed eaters and browsers; and goats probably do better than sheep on a straight diet of browse. Horses are more selective than any other kind of livestock; they prefer grasses, although they will eat small amounts of other kinds of forage. Hogs do best on acorns, pods of certain leguminous shrubs, roots, and other concentrated feeds found on the range only during limited seasons and in certain areas, principally in the Southeast and Southwest.

5. **Predators, insects, and diseases.** Coyotes are serious predators of sheep, but bother cattle very little, comparatively speaking. Thus, heavy concentrations of coyotes, or other sheep-killing predators, may make sheep raising unprofitable, but present less serious problems to cattle production. The presence of certain insects and diseases may also become factors in the selection of the best suited kind of livestock.

6. **Big game population.** Deer compete more directly with sheep, and elk with cattle.

Theoretically, the most efficient use of most ranges can be made by two or more kinds of livestock grazing at the same time; by *common use* or *dual use*. The most popular combination is that of cattle and sheep. Destructive grazing often results therefrom, however, because common use requires much more critical grazing management than grazing by one kind of livestock only. This is so primarily because the most popular parts of the range—waterholes, creek bottom meadows, ridge tops—are the most preferred by all kinds of animals; and, in addition, many of the most valuable forage plants are preferred by both cattle and sheep. As a result, only very careful management can prevent the destruction of these most valuable range areas. In brief, a range unit cannot support its full quota of cattle in addition to its total capacity for sheep. Rather, a studied adjustment should be made to fit the particular unit, based on topography, water distribution, class of forage, and other considerations.

RANGE GRAZING SYSTEMS

Ranges may be grazed continuously throughout the entire grazing season without rest, or the area may be subdivided and the pasture grazed rotationally with interim rests varying from 3 to 4 months in the higher rainfall areas to 12 months in the lower rainfall areas.

Some of the basic range grazing systems, of which there are many variations and adaptations from ranch to ranch, follow:

Continuous Grazing
Rotation Grazing
  Deferred rotation grazing systems
    1. Two pastures—one herd system
    2. Three pastures—one herd system
    3. Four pastures—three herds
  Short duration grazing systems
    1. Conventional (rectangular) grazing system
    2. Savory (or cell) grazing system

Continuous Grazing

Continuous grazing is the simplest and most common grazing system on western ranges; and varying the number of animals allowed to graze on an acre of land is the most commonly used means for grazing management. In comparison with rotation systems, continuous grazing requires less fence, water, and pasture development, less labor in moving animals and fixing fences, and less knowledge of livestock and range management. Also, continuous grazing is more efficient on the less productive ranges, and more suitable and practical when used in conjunction with a seasonal range, than a complicated rotation system.

The major **disadvantages** of continuous grazing are: lower stocking rate; less animal gains per acre; poorer livestock distribution on the range caused by animals concentrating around water, bedding grounds, and feed grounds, and overgrazing such areas; and less opportunity to use such improvement practices as burning, brush control, fertilization, and livestock management.

Rotation Grazing

Rotation grazing is a system in which pastures are grazed and rested in a planned sequence. It gives the more desirable plants a chance to regrow, compete, and multiply, thus gradually increasing the number of high quality plants.

The objectives of any rotation grazing system are to favor the growth and survival of erect growing, easily grazed plants; to obtain greater use of the less palatable plants; and to improve range conditions by grazing some pastures while resting others. The improved range increases livestock production, improves the habitat of wildlife, reduces erosion, and conserves water.

The two main types of rotation grazing systems are deferred rotation grazing and short duration grazing.

DEFERRED ROTATION GRAZING SYSTEMS

In deferred rotation grazing systems, the range is usually divided into two to four units. There are different ways in which to apply deferred rotation grazing, with the following three basic systems, or some variations therefrom, most common in the Southwest:

1. **Two pastures—one herd system (switch-back system).** With this system one herd of livestock is rotated between two pastures. Each pasture is grazed or rested at a different time during the 2-year period required to complete the grazing cycle.

2. **Three pastures—one herd system.** This system is similar to the two pastures–one herd system, except the herd is moved through three pastures instead of two. In any one year, one pasture may be used during the growing season; the second pasture may be used at a later stage of vegetative maturity, such as seed-ripe; and the third pasture may be rested and not grazed by livestock. The length of each grazing period may be as short as 30 days or as long as 90 days. This sequence is rotated among years. By treating a unit in this manner each year, the entire area is rested, allowed to reseed itself, and grazed in rotation.

3. **Four pastures—three herds system (the Merrill system).** In Texas, and in much of the Southwest, ranges may be grazed throughout the year. Under these circumstances, a different type of rotation grazing system should be considered than where grazing is not year-round.

Where 4 pastures are available, or can be arranged, a 3-herds system is popular, with each pasture grazed 12 months and rested 4 months. This system is summarized in chart form (see chart 7-2).

The Texas station reports that, in comparison with conventional yearlong grazing on the same area, the "4 pastures—3 herds system" results in greater livestock gains and a 25% increase in carrying capacity.

Chart 7-2
FOUR PASTURES—THREE HERDS SYSTEM

Year and Season				Pastures			
				1	2	3	4
1979:							
Mar.	Apr.	May	June	Rest	Graze	Graze	Graze
July	Aug.	Sept.	Oct.	Graze	Rest	Graze	Graze
1980:							
Nov.	Dec.	Jan.	Feb.	Graze	Graze	Rest	Graze
Mar.	Apr.	May	June	Graze	Graze	Graze	Rest
July	Aug.	Sept.	Oct.	Rest	Graze	Graze	Graze
1981:							
Nov.	Dec.	Jan.	Feb.	Graze	Rest	Graze	Graze
Mar.	Apr.	May	June	Graze	Graze	Rest	Graze
July	Aug.	Sept.	Oct.	Graze	Graze	Graze	Rest
1982:							
Nov.	Dec.	Jan.	Feb.	Rest	Graze	Graze	Graze
Mar.	Apr.	May	June	Graze	Rest	Graze	Graze
July	Aug.	Sept.	Oct.	Graze	Graze	Rest	Graze
1983:							
Nov.	Dec.	Jan.	Feb.	Graze	Graze	Graze	Rest

SHORT DURATION GRAZING SYSTEMS

Short duration grazing employs frequent movement of animals, with the speed of the rotation adjusted according to the growth rate of the plants. During the peak of the growing season, animals are moved at shorter intervals, with longer intervals during the remainder of the year. In practice, animals are grazed for periods of about 1 to 30 days, and vegetation is rested about 120 to 150 days. In short duration grazing, it is important not to graze the regrowth, for the plants will be weakened by using up their reserve food supply. The short duration technique uses rest periods within a grazing season in order to restore plant vigor. Short duration grazing will usually give more rapid range improvement than deferred rotation grazing.

1. **Conventional (rectangular) grazing system.** This system involves the use of conventional, rectangular pastures, along with a 16 to 18 ft alley for cattle to get to water and minerals and move from one pasture to another. The principle and practices (grazing time, resting time, central watering, and easy movement of cattle between pastures) of the Savory (cell) system and the conventional (rectangular) system are similar. The only difference is the layout or design; the Savory (cell) system uses the wagon wheel design, whereas the conventional (rectangular) system uses the design identified by the name—rectangular pastures.

2. **Savory (cell) grazing system.** This system is named after Allan Savory, who originated and popularized it. Ideally, grazing is from 1 to 5 days, and resting is from 30 to 60 days. It usually involves 12 or more pastures, and generally, although not always, the pastures are arranged as a grazing cell, with pastures formed by fence lines fanning out from the hub like spokes on a wagon wheel, and with the water, minerals, and handling facilities at the hub. When the animals come to the center, or hub, for water and minerals, they can be moved between pastures by opening and closing gates. Producers using the Savory system generally use electric fences in order to reduce costs. The Savory system is designed to lessen movement stress. However, it can add to nutritional stress if (1) stock are held too long in pastures, (2) too low stock density is combined with too fast a move, or (3) accelerated grazing with rests become inadequate.

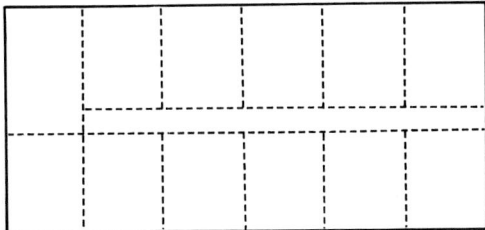

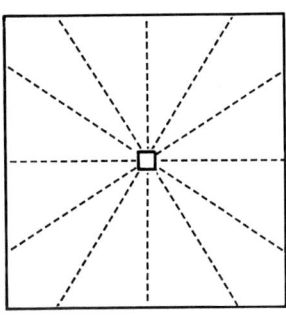

Fig. 7-15. Two basic layouts (designs) for short duration grazing: A, the conventional (rectangular) system; and B, the Savory (cell) system.

RANGE IMPROVEMENT METHODS

The warning signals of a range that is on the downgrade and that is in need of improvement are:

1. Desirable forage plants "going out" and being replaced by undesirable ones; the number of young, inferior plant species increasing.
2. Thinning of perennial grass cover, with the grass tufts breaking down and dying; and an increase in annual plants and perennial weeds. The poorer the condition of the range, the more rapidly this process takes place.
3. Weakened vitality of the important forage plants as shown by pale color and reduced height and yield in period favorable to good growth.
4. Increased soil erosion, by wind and/or water.
5. Excessive trampling damage.

The above warning signals have appeared in various intensities over part of the western range area of the United States.

There is no quick, easy, and inexpensive method by which poor ranges can be improved. However, one or more of the following methods should be employed.

Conservative Stocking

Usually controlled stocking and natural range reseeding are accomplished by employing one or more of the following practices: (1) rotation-deferred grazing; (2) rest-rotation grazing; (3) a lighter continuous grazing load; or (4) a shorter season of use, such as may be accomplished by using supplemental pastures.

Distribution of Animals on the Range

Next to the proper rate of stocking and proper seasonal use, distribution of the animals on the range is the most important feature in range management. Proper distribution of animals is reflected in more even utilization of the forage. This assignment is more difficult with cattle than with sheep, especially on rough mountainous land. Cattle have a strong tendency to utilize the flatter areas and to congregate around watering places. Also, sheep are usually herded.

Better distribution of animals on the range may be accomplished through (1) the grazing system (concentrating livestock into large herds and rotating the herds through 2 or more pastures tends to improve grazing distribution for each pasture); (2) fencing; (3) riding the range (or herding); (4) providing sufficient watering places (under ideal conditions, the distance between water in rough country should not exceed ½ mile for cattle or 1 to 2 miles for sheep and goats; in level country, 1 to 1½ miles for cattle or 2 to 3 miles for sheep and goats—water hauling on both cattle and sheep ranges is increasing); (5) systematically locating salt blocks or salt containers away from watering areas in underused areas; (6) building trails into inaccessible parts of the range; and (7) controlling livestock pests such as grubs and flies, which cause animals to congregate and seek protection.

Range Reseeding

Where improper management and overstocking have seriously reduced the quality of the forage and the grazing capacity of the range, some method of reseeding may be the only logical alternative.

Where considerable of the better forage plants remain, natural regeneration is preferred. The latter is accomplished by managing the grazing season so as to favor the propagation of the remaining desirable native forage and by controlling low-value brush and other competing vegetation. Often the recovery process can be speeded up by controlling most undesirable vegetation through the use of herbicides. In western United States, this process is especially successful on rangeland dominated by sagebrush and wyethia where a residual stand of native grasses is present as an understory. There are also other weed types which respond well to chemical treatment.

Where most of the desirable plants have been destroyed and the soils are suitable, artificial reseeding is advocated—even though it is expensive and subject to failure.

It is recognized (1) that only a relatively small proportion of the western range can be seeded, (2) that seeding is not a satisfactory substitute for good management practices needed to prevent further destruction of forage plants, (3) that if seeding is necessary, some practice(s) followed in the past has been faulty, and (4) that seeded areas require a high level of management to maintain them.

Brush and Weed Control

Brush and weeds compete with both native and introduced forage plant species for water, nutrients, and light. Their control is particularly difficult on rangelands, because they are not normally cultivated or rotated with other crops.

Brush is a primary problem on U.S. rangelands. An estimated 320 million acres of grazing land are predominantly brush. Mesquite, juniper, and sagebrush make for the major brush problems.

Brush can be controlled by the following methods:

1. **Chemical methods.** This involves applying select herbicides to the soil or the plant by ground equipment or by aircraft.

2. **Mechanical methods.** Where there is a dense stand of brush and few or no desirable grasses, mechanical methods may be most effective. Mowing, axing, root plowing, chaining and bulldozing have proven quite effective.

3. **Prescribed burning.** Fire is the oldest method of controlling unwanted woody plants, and it is still used. Prescribed burning refers to burning that is done at the end of the dormant season or at the time the desirable grasses are beginning spring growth; this suppresses certain undesirable plants, thereby giving an improved environment for the more desirable plants.

4. **Proper grazing management.** Proper grazing management avoids overgrazing and assures healthy, vigorous forage, in a healthy ecosystem, thereby reducing brush problems.

Fig. 7-16. Brush control would increase the carrying capacity of this range. (Courtesy, American Shorthorn Assn., Omaha, Nebr.)

To be effective, brush control must be followed by proper grazing management. During the first growing season following treatment, grazing should be deferred/limited in order to allow desirable grasses to become established. Reseeding of these areas may be necessary where a natural seed source of the desirable forage plants is not available.

• **Biological control of brush and weeds**—Interest in the biological control of brush and weeds is increasing. The classical example of the biological control of an undesirable species was the practical elimination of prickly pear cactus (*Opuntia* spp) on 30,000,000 acres of grazing land in Queensland, Australia, by using a moth introduced from Agrentina in 1925. In the United States, insects have been used to control St. Johnswort, lantana, puncturevine, and knapweed. The use of sheep and goats is also a form of biological control. Sheep will eat leafy spurge, tansy ragwort, tall larkspur, and spotted knapweed. Goats are effective for keeping some species of chaparral and other brushes under control.

GRAZING PUBLICLY OWNED LANDS

The ownership of U.S. land is summarized in Table 7-1.

TABLE 7-1
OWNERSHIP OF U.S. LAND (50 STATES)[1]

Ownership	Area		Percentage of Total
	(million acres)	(million ha)	(%)
Private ownership	1,329	538	58.7
Indian land	51	20.6	2.2
Public ownership	885	358.3	39.1
Federal	730	295.5	32.2
State and local governments	155	62.7	6.8

[1]*Statistical Abstract of the United States,* 1987, p. 182, Table 318.

About one-third of U.S. public lands are in Alaska. Because of its remoteness and northern location, land development has been slow in this state. As a result, the Federal Government still owns almost 67% of all the lands in Alaska.

The other two-thirds of the public lands are located in the 48 contiguous states, but are not evenly distributed across the country. About 93% of these federal lands outside Alaska are in the 11 western states.

Today, in the 11 western public land states, the Federal Government owns and administers approximately 320 million acres on which grazing is allowed. At one time or another during the year, domestic cattle and sheep graze on about half of these public lands. More of the public lands are used for this purpose than for any other economic activity. In 1983, lands in the 11 western states administered by the Bureau of Land Management and the U.S. Forest Service provided grazing all or part of the year for an estimated 4.5 million cattle and horses and 6.7 million sheep and goats, or a total of 11.2 million head of all classes, or a total of 17.6 million animal unit months.

Agencies Administering Public Lands

Because much of the grazing land that ranchers rely upon to maintain their cattle and sheep enterprises is built up into operating units by leasing or by obtaining use permits from several federal and state agencies, private corporations, and individuals, it is imperative that the owner have a working knowledge of the most important of these agencies. Some range operators are placed in the position of using range rented from as many as six landlords—either private, state, and/or federal.

The bulk of federal land is administered by the following six agencies: the Bureau of Land Management, the U.S. Forest Service, the Bureau of Indian Affairs, the Department of Defense, the National Park Service, and the Bureau of Reclamation. The largest land area from the standpoint of grazing permits and utilization of grazing areas by animals is administered by the first three of these agencies; hence, each of these three agencies is discussed at this point, followed by pertinent information relative to state and local government-owned lands, and railroad-owned lands.

1. **Bureau of Land Management.** The Bureau of Land Management of the U.S. Department of the Interior administers more than 40% of all federal lands. More than one-third of the land it manages is in Alaska. The remainder is almost entirely in the 11 western states.

From the standpoint of the livestock producers, the most important function of the Bureau of Land Management is its administration of the grazing district established under the Taylor Grazing Act of 1934 and of the unreserved public land situated outside of these districts which are subject to grazing lease under Section 15 of the Act. This federal act and its amendments authorize the withdrawal[4] of public domain from homestead entry and its organization into grazing districts administered by the Department of the Interior. Also, this legislation, as amended, allows the Bureau of Land Management to administer state and privately owned lands under a cooperative arrangement.

In 1984, the Bureau of Land Mangement had 52 grazing districts, operating in the 11 western states and totaling 157.3 million acres of public lands. In these districts, 12,000 operators were granted privileges to graze 3,973,000 head of livestock for an average of about 5 months each year. These operators paid the United States, as grazing fees for this range use, a total of $12,396,000. In addition to this livestock use, in 1984 the grazing districts supported, for approximately 5 months of the year, an estimated 1.8 million big game animals, of which approximately 1.1 million were deer.

In addition to, and outside of, the grazing districts, in 1984 the Bureau of Land Management supervised 17.7 million acres of public domain in the western states, most of which was leased to 7,300 livestock producers for 651,956 head of livestock for about 5 months. These operators paid rentals in the amount of $2,049,000 for the use of these lands.

Each district is administered by a District Manager, who is a technically trained employee of the Bureau of Land Management. The District Manager is responsible to the state bureau office for the proper use, management, and welfare of the public land resources of his district. In turn, the state office is responsible to the Director's office in Washington, D.C.

Grazing privileges are allocated to individual operators, associations, and corporations on the basis of (1) priority of use; (2) ownership or control of base property dependent on grazing district land for forage during certain seasons of the year, or control of permanent water needed to graze district land; (3) proximity of base property to public lands outside home ranch to the grazing district; and (4) adequate property to supply the feed needed along with grazing privileges, to maintain throughout the year the livestock permitted on public range. All of these lands are subject to classification and disposal under Sections 7 and 14 of the Taylor Grazing Act, for any higher use or other appropriate purpose. Grazing privileges may, therefore, be cancelled whenever such lands are determined to be more suitable for other purposes.

A fee is charged for grazing privileges. In 1989, the basic fee was equivalent to $2.29 per animal unit month (AUM). An AUM is the equivalent of the grazing of a mature cow, 5 sheep, or 1 horse, for 1 month.

2. **U.S Forest Service.** Almost one-fourth of the federal lands are administered by the Forest Service. Over 100 million acres of the national forests are used for grazing under a system of permits issued to local farmers and ranchers by the Forest Service of the U.S. Department of Agriculture. In 1985, about 1,482,000 mature sheep/goats and 1,469,000

---

[4]On May 28, 1954, a bill was signed by President Eisenhower lifting the 142 million acre limitation on public domain lands that can be included in Taylor Grazing Act districts.

mature cattle and horses (mostly cattle), owned by over 16,000 paid permit operators, were grazed on national forests for some part of the year. In addition, there were many calves and lambs for which no fee is charged and additional stock that were grazed under free permits to local settlers.

The forest service issues term grazing permits and annual permits. Among other things, the permit prescribes the boundaries of the range which they may use, the maximum number of animals allowed, the season in which grazing is permitted, and the expiration date of term permits.

Temporary permits may be waived back to the government when permittees sell livestock or base property. Then, the purchaser of the permitted livestock or base property may apply for and be issued a permit if qualified.

The requisites in order to qualify for a term permit are:
  a. U.S. citizenship.
  b. Ownership. The ownership of both the livestock and commensurate ranch property.

A term grazing permit is not a property right. Rather, it is approved for the exclusive use and benefit of the person to whom it is issued. Permits may be revoked in whole or in part for a clearly established violation of the terms of the permit, the regulations upon which it is based, or the instructions of forest officers issued thereunder.

A ranger administers the grazing use on each National Forest Ranger District. Several districts (usually 3 to 6 or more) comprise a national forest. A forest supervisor, with his staff, administers the national forest. Several national forests, under the direction of a regional forester and staff, comprise a forest service region. The Chief administers the Forest Service from Washington, D.C., under the supervision of the Secretary of Agriculture.

Local farmers and ranchers act in an advisory capacity in reviewing allotment management plans and the use of range betterment funds. About 800 such livestock associations are recognized and in operation.

Forest Service grazing fees are based on a formula which takes into account livestock prices over the past 10 years, the quality of forage on the allotment, and the cost of ranch operation. In 1989, average charges were $1.86 per animal unit month (AUM); or $1.86 for a mature cow or horse, or for 5 sheep, for a month. The use resulted in the payment of $9,039,100 in grazing fees in 1985.

3. **Bureau of Indian Affairs.** Most Indian lands, comprising 51 million acres, are really not public lands. Rather, these lands are held in trust for the benefit or use of the Indians and are merely administered by the Bureau of Indian Affairs of the Department of the Interior. Because over 80% of Indian lands are in the range area of the West, they are suited primarily to livestock. Thus, it is noteworthy that the sale of livestock and animal by-products regularly accounts for ⅔ of the total Indian agricultural income. Although the Indians themselves own most of the stock that graze these lands, animals owned by non-Indians utilize ¼ of the Indian lands devoted to grazing. Provision for such use is handled under lease agreement jointly approved by the Indian owners and the Bureau of Indian Affairs.

4. **State and local government-owned lands.** A total of 134 million acres are owned by state and local governments. For the most part, the management of these areas is diverse and confused, each state and local government having established different regulations relative to the lands under its ownership. In general, however, such lands are operated on a stipulated lease arrangement. On many such areas, range depletion has been severe.

5. **Railroad-owned lands.** Recognizing that the main deterrent to rapid settlement and development of the West was lack of adequate transportation facilities, the Federal Government very early encouraged the construction and westward extension of the railroads by means of large grants of land. It was intended that the railroads should sell or otherwise utilize these lands in financing their costs of construction. These initial grants, totaling 94,355,739 acres, consisted of alternate sections extending in a checkerboard fashion for a distance of from 10 to 40 miles on each side of the right-of-way. Today, less than 20 million acres of these lands are held by railroads. Many of these holdings are leased to livestock producers; but because of inconvenience, past abuses, or other reasons, some of these lands are considered worthless for grazing. In general, railroad lease agreements do not restrict the number of stock to be grazed or the season during which the land may be so used.

Part III—Multiple Use/Conservation of Land

The multiple use of publicly-owned lands evolved in response to public pressure. The multiple use of privately-owned lands followed, in response to economic pressure—the need to increase net returns. Soil and water conservation evolved on both public and private lands in recognition that they are national resources that should be preserved for posterity.

The sections that follow pertain to the multiple use/conservation aspects of both public and private lands—to both introduced (seeded) pastures and native (range) pastures.

MULTIPLE-USE CONCEPT

Multiple-use of land is the management of all the various resources of lands, both public and private, so that they are utilized in combination. With federal lands, multiple use is determined by the needs of people; and management decisions are made publicly. With private lands, multiple use is based on their most profitable use and management decisions are made privately by owners/managers. Important multiple uses include livestock grazing, mining, national heritage preservation, occupancy, recreation, water, wildlife, and wood/timber production.

The multiple use concept developed as a compromise relative to the use of public lands; it evolved as a result of attempting to placate individuals and groups who wish to have the land used for purposes which they consider desirable or to prevent others from using the land for purposes which they consider to be undesirable.

MULTISPECIES GRAZING

Grazing two or more species of livestock together or separately on the same land unit in a single growing season is known as multispecies grazing. Research indicates that multispecies grazing contributes to better and more uniform forage use and higher economic returns from livestock.

Multispecies grazing evolved in regions with diverse vegetation types and suitable climates. Grazing by a mix of domestic and wild animals can often result in more efficient use of forage and browse, more total animal gains, and a more vigorous plant community. While multispecies grazing is a common management practice on rangelands of the West, it is much less commonly practiced on the pasture lands of eastern United States.

Western rangelands are characterized by vast diversity in elevation, precipitation, temperature, and other climatic factors. These differences make for a multitude of range cites dispersed among several major vegetation or habitat types. It follows that great potential exists on these lands for multispecies grazing by livestock and wildlife to maintain forage production and species diversity.

Where multispecies grazing is practiced, cattle and sheep dominate. In the Southwest, goats are sometimes a component. Goats are without a peer in rough, unimproved areas and as browsers. Sheep prefer steeper terrain and eat more shrubs and forbs than cattle. Cattle stick to the more gentle slopes and prefer grasses. So, multispecies grazing can result in more complete and uniform utilization of multiplant species pastures and greater animal production. However, predators and labor problems have caused decreased sheep and goat numbers. In turn, this has resulted in lower income and the deterioration of many ranges, due to the invasion of undesirable plants such as bushy species.

In the past, wildlife has generally been incidental. Now, and in the future, economic pressures dictate that wildlife be an integral part of multiple land use.

WILDLIFE

Fig. 7-17. Deer at home on the range. Good range management is good for wildlife. (Courtesy, USDA-ARS, Cheyenne, Wyo.)

Wild animals and birds are becoming more valuable to today's landowner. Higher livestock production costs and demand for outdoor recreation have prompted practical landowners to seek means of increasing income by providing game for hunters. In some areas, wildlife income exceeds livestock income. This has caused landowners to include wildlife in farm and ranch planning.

There is a close association between kinds of plants and animals present in the habitat. Also, livestock numbers, domestic species, and grazing patterns can be manipulated to enhance wildlife habitat and maintain wildlife populations. Through proper habitat management, the farmer and rancher can maintain healthy, abundant wildlife populations. To accomplish this, land managers must place wildlife high on their priority list and consider wildlife in overall farm/ranch planning.

- **Kinds of wildlife**—Many kinds of wild animals and birds live on pastures and ranges. Identifying the kinds is necessary because management will vary for different species. Deer, for example, need browse, forbs, and grasses for feed, and timber and brushy areas for cover. Quail feed on weed seeds, nuts, and seeds of certain grasses and shrubs; and they prefer a mixture of wooded and open areas with small plots of low shrubs and vines for cover. Normally, management will involve meeting the needs of several different kinds of animals and birds.

- **Numbers of wild game**—Nationwide, in 1984 there were an estimated 10 to 11 million deer, 700 thousand pronghorn antelope, and 500 thousand elk.[5] Additionally, in 1984, in U.S. National Forests and Grasslands, there were an estimated 231,339 wild turkeys, 106,480 bear, 26,246 moose, 25,783 javelina (peccary), 25,152 mountain goats, 20,753 wild sheep, 11,091 mountain lion, 3,480 wild boar, 1,206 wolf, 323 caribou, and 206 bison.[6] Also, in 1984, hunters harvested 391,000 deer and 109,000 other big game in National Forests.[7]

In addition to the big game animals listed above, there are many species of small game animals. Also, there are numerous species of game birds, including quail, partridge, and pheasants.

- **Wildlife management**—Wildlife can exist in harmony with livestock operations provided (1) wildlife needs and species are inventoried and included in the management plan, and (2) good management aspects prevail.

SOIL EROSION CONTROL

Soil erosion control is any management plan to reduce soil and water losses.

Soil erosion is a natural occurrence. However, it may be increased by activities that disturb the natural balance of the pasture or range ecosystem. Poor grazing management is a major cause of erosion.

The primary method of controlling soil erosion is plant cover. Other practices to control erosion on grasslands include brush control, deferred grazing, reseeding, and mechanical land treatments. Among the latter are terracing, contour furrowing, pitting, small dams, and diversions.

WATER

Water is often a limiting factor in pasture productivity, affecting forage production and/or drinking water for grazing animals.

The water cycle is the never-ending movement of water from clouds to soil, through plants, and back to clouds again. The cycle begins when precipitation strikes the land and ends when the water leaves the land either through runoff

---

[5]Oldfield, J. E., Chairman, et al., *Forages*, Council for Agricultural Science and Technology, Ames, Iowa, 1986, p. 30.

[6]*Wildlife and Fish Habitat Management in the Forest Service*, USDA, 1985, pp. 52-53.

[7]*Agricultural Statistics 1986*, USDA, p. 492, Table 680.

or evaporation. During the intervening time, a livestock producer should store as much water as possible, in the soil and in reservoirs. The shortage of water over much of the West is particularly important. In addition to limiting livestock production, lack of water may limit stream flow for fish, cultivated crops, and industries.

There are various types of stock water developments. These include *natural* water supplies such as lakes, ponds, streams, springs, and seeps, and *made* developments such as wells, reservoirs, dugouts, sand tanks, and catchment basins. A combination of two or more types of water development is often more advantageous than one type only.

QUESTIONS FOR STUDY AND DISCUSSION

1. In a period of world food shortages, what characteristcs of an animal agriculture will favor its survival?

2. What is the difference between the terms pasture and forage?

3. Give facts and figures pointing up the magnitude and importance of pastures in the United States.

4. What are the primary differences between (a) seeded pastures, and (b) native pastures?

5. What are the primary differences between (a) permanent pastures, (b) semipermanent or rotation pastures, and (c) temporary and supplemental pastures?

6. Define each of the following terms: (a) grass, (b) legume, (c) browse, and (d) forbs.

7. Discuss the economic importance of pastures for each of the following classes of livestock: (a) beef cattle, (b) dairy cattle, (c) sheep, (d) swine, and (e) horses.

8. Define and describe custom cattle grazing.

9. Why do large dairies depend less and less on pastures and more and more on other feeds?

10. How may a livestock producer use Chart 7-1, Adapted Grasses and Legumes?

11. What local authorities, and what college could the farmers and ranchers of your area contact relative to grass/legume seeding recommendations?

12. Why are swine producers using less and less pasture? Are pastures outmoded in modern swine production?

13. List the practices that are usually adhered to in successfully establishing a new pasture in the subhumid, humid, and irrigated areas.

14. Define "pasture renovation." List and discuss three different methods by which permanent pastures may be renovated.

15. Discuss how each of the following factors affects the value of pasture: (a) soil and fertilizer, (b) plant species, (c) stage of maturity, (d) rate of growth and season of year, and (e) grazing.

16. List and discuss some of the important practices involved in good pasture management of subhumid, humid, and irrigated areas.

17. Why and how should a farmer/rancher attempt to extend the grazing season?

18. Discuss and compare each of the following grazing systems: (a) continuous grazing, (b) rotation grazing, (c) first and second grazers, (d) intensive early season stocking, and (e) creep grazing.

19. Define and compare strip grazing and green chop.

20. Is there a need and a place for more irrigated pastures in the United States? Discuss some of the special decisions that must be made if pasture irrigation is to be successful.

21. Discuss the history of the western range.

22. Why is so much of the range area of the West publicly owned and unenclosed? Is it good or bad to have so much public domain?

23. From the standpoint of vegetation and utilization by livestock, how do ranges differ from cultivated pastures?

24. List and discuss common range nutrient deficiencies. How are supplements of these nutrients usually provided?

25. Discuss the scientific basis for proper stocking rate on the western range.

26. Discuss the seasonal use of western ranges.

27. List and discuss the range characteristics which may affect the choice of the kind of livestock.

28. Discuss and compare each of the following range grazing systems: (a) continuous grazing, (b) rotation grazing, (c) deferred-rotation grazing, and (d) short duration grazing.

29. What are the warning signals of a range that is on the downgrade and in need of improvement?

30. List and discuss the methods that may be employed to improve a range.

31. Discuss each of the following types of ownership of U.S. land: (a) private ownership, (b) Indian land, (c) public ownership.

32. Discuss the role of each of the agencies administering public lands.

33. Some environmentalists are agitating for a ban of grazing rights of public lands. What are the pros and cons for such action, and what is your recommendation?

34. Define multiple-use of land. How and why did this concept evolve?

35. Define multispecies grazing. Where and how may it contribute to better and more uniform forage use and higher economic returns?

36. What forces have caused land owners to include wildlife in farm and ranch planning?

37. List and discuss the requisites for wildlife to exist in harmony with livestock operations.

38. Discuss the soil erosion control aspects of pasture and range management.

39. Discuss the water aspects of pasture and range management.

Original painting by Tom Phillips

HAY[1]

Contents	Page
History of Hay	148
Magnitude and Importance of Hay	148
Hay As an Energy Source	148
Comparative Value of Hay	149
Hay As a Grain Replacement	149
Kinds of Hay	150
Weeds and Other Potential Hay Crops	150
Hay Quality	150
Importance of Hay Quality	150
Visual Inspection	151
Chemical Analysis	151
NDF, ADF, and NIRS Analyses	152
Correct Sampling Necessary	152
What Tests to Make	153
Evaluating Test Results	153
Making Quality Hay	153
Growing Forage	153
Haying Equipment	153
Harvesting at Proper Stage	154
Retaining Adequate Plant Food Reserves	155
Cutting and Field Curing Hay	156
Cutting/Curing in the Swath and Windrow/ Raking or Cocking	156
Reducing Moisture Content/Shattering/Bleaching and Fermenting	156
Reducing Rain Damage	157
Considering Chemical Conditioning and Preserving Agents	157
Value of Different Cuttings of Alfalfa Hay	158
Nitrogenation of Low-Quality Hay	158
Haymaking Systems	158
Long, Loose Hay	159
Chopped Hay	159
Packaged Hay	159
Bales	159
Stacks	160
Cubes (Wafers)	160
Pellets	160
Artificial Drying	160
Mow Curing	161
Artificial Dehydrators	161
Wagon Dryers	161
Storing	161
Additives for Hay	162
Spontaneous Combustion	162
Buying and Selling Hay	162
Hay Sources	162
How Hay is Priced	162
Futuristic Hay Evaluating and Pricing	162
Freedom From Toxic Residues	164
Hay Shrinkage	164
Hidden Hay Costs	164
Hay Feeding Fundamentals	164

Contents	Page
Maximum Use of Homegrown Forages	164
Legume Vs Grass Hays	164
Hay Preparation	164
Hay Feeding Systems	165
Feeding Hay Packages	165
Baling Wire Danger	165
Hay Feeding Schedule	165
Proportion of Hay to Concentrate	165
Ruminants Need Hay	166
Different Qualities of Hay May Be Used	166
Hay Waste and Refusal	166
Supplementing the Hay Ration	166
Stretching the Hay Supply	166
Questions for Study and Discussion	167

Hay is forage harvested during the growing period and preserved by drying for subsequent use. Hays are made from legumes, grasses, and cereal crops. It is the most important harvested forage fed to livestock, and it ranks third among all livestock feeds, being exceeded only by pasture and corn. Hay is primarily a cattle, sheep, and horse feed, although alfalfa (especially ground, dehydrated alfalfa) may be included in swine and poultry rations. Average-quality hay runs 25 to 35% crude fiber and 45 to 55% TDN on an as-fed basis, whereas such concentrates as corn and wheat contain approximately 2 to 3% fiber and 80% TDN.

The object of haymaking is to (1) harvest the crop at the optimum stage of maturity which will provide the maximum yield of nutrients per acre without damage to the next crop, and (2) cure the crop properly by lowering the water content of the green herbage from 65–85% to 20% or less.

Drying, or making hay, is the most common method of preserving forage for storage, primarily because it is relatively easy to handle. It can be stored or transported long, chopped, pelleted, cubed, or packaged into various types and sizes of bales. Modern equipment and chemicals hasten drying time; and automated systems facilitate handling.

The great capacity and specialized functions of the rumen allow cattle and sheep to use hay, and other forages, in large amounts. Bacteria and protozoa in the rumen break down and make available to the host animal part of the nutrients in cellulose or fibrous material.

---

[1]The authors gratefully acknowledge the helpful suggestions of the following eminent authorities who reviewed this chapter: J. E. Baylor, Ph.D., Professor of Agronomy, The Pennsylvania State University, State College; R. A. Forsberg, Ph.D., et al., Department of Agronomy, University of Wisconsin-Madison; and S. C. Fransen, Ph.D., Forage Agronomist, Western Washington Research and Extension Center, Washington State University, Puyallup.

In addition to the nutrients that it contains, and to its value in providing feed throughout the year, hay has other values. Dry feed is essential for the proper functioning of the digestive tract; it acts as a stimulant in moving the feed through the intestines, and it maintains the proper conditions in the rumen for the microbial action which plays such a vital role in the digestion of the fibrous portions of feeds. Hay is often used as a supplement to "washy" pastures and succulent silages. Also, it speeds along the development of the rumen function of the young ruminant, lessens the incidence of displaced abomasum in cattle, and prevents a lowering of the fat content of the milk of lactating cows (unless it is finely ground). Also, and most important, good-quality hay is a hedge against high-concentrate prices, for when the price of such feeds increases disproportionately, increased amounts of hay may be fed and concentrates may be decreased, with a higher net return to the producer.

Despite its several advantages, hay has some shortcomings. It varies in nutrient content and palatability more than any other feed, because of differences in the (1) crops from which it is made, (2) stage of cutting, (3) handling, and (4) weather damage during curing. Not even ruminants can consume enough hay alone to meet the demands of high production; for example, when fed hay alone, dairy cows will produce only 50 to 70% as much milk as they would when fed a ration consisting of 50% concentrates. Also, fiber is poorly digested by monogastric animals, with the result that hay serves primarily as a source of minerals and vitamins for swine and poultry.

An estimated 80% of all hay is fed on the farms or ranches on which it is produced, rather than being purchased. It is important, therefore, that producers know how to produce good hay, as well as how to feed it, for most of them determine their own destiny from the standpoint of quality. For this reason, this chapter covers hay from production to feeding.

HISTORY OF HAY

Haymaking evolved with the domestication of animals, for from that remote day forward caretakers assumed responsibility for storing feed for them for use during times of scarcity. More than 2,000 years ago, the Roman agricultural writer, Columella, described haymaking as "throwing forage loosely together for a few days to heat and concoct itself and then cool before putting into the mow." But another 20 centuries were to pass before haymaking changed materially. As recently as 1850, it was cut with a scythe; and pitchfork haymaking persisted into the present century. Beginning about 1940, scientists and engineers pooled their efforts to transform roughages into high-quality hay. Haymaking went modern, with automated one-operator pick-up balers, field choppers, cubing machines, and other modern equipment, replacing the backbreaking, labor-intensive methods of old.

Automated haymaking and surplus grain were ushered in together. At the close of World War II, U.S. grain bins bulged. This spawned the era of high-energy, low-forage rations. Then, suddenly, in the early 1970s, there were world food shortages. The 20-year grain-feeding binge in the United States began reversing itself. Now, and in the future, more and more grain will be used for direct human consumption. Animals (especially cattle and sheep) will increasingly be "roughage burners."

MAGNITUDE AND IMPORTANCE OF HAY

The importance of the nation's hay crop is attested to by the fact that the total area devoted to hay in the United States exceeds 60 million acres, the total production averages about 150 million tons, and the annual crop is worth approximately $10 billion—it is worth more than any other crop except corn and soybeans. On an air-dry tonnage basis, about 3 times as much hay is produced as silage.

Despite the importance of hay, no other feed crop suffers a higher loss of nutrients from the time it is cut to the time it is fed. During the curing process, the quality and feeding value of hay decreases rapidly by rain, sun bleaching, raking, handling when too dry, and storing with too much moisture. Studies by the U.S. Department of Agriculture revealed that the following losses accrued in field-cured, second-cut alfalfa hay from the time of cutting to the time of feeding: leaves, 35%; dry matter, 20%; and proteins, 29% (see Fig. 8-1).

The longer hay remains in the field until it is dry enough to store, the greater the nutrient losses (Fig. 8-1). These losses have been estimated to have a feeding value of more than a billion dollars annually.

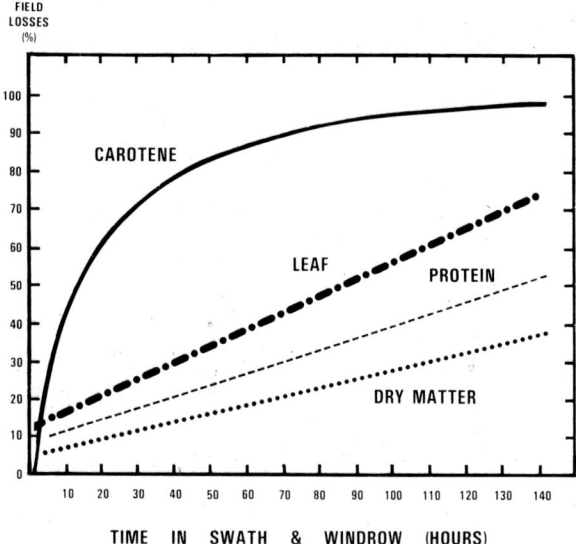

Fig. 8-1. Losses in sun-curing alfalfa hay as related to time in the field to reduce moisture to a safe level for storage. (Adapted by the authors from USDA data.)

Hay As an Energy Source

Hay is an important source of energy for cattle, sheep, and horses. Table 8-1 shows the percentage of total energy (TDN) intake provided to these species by hay and other kinds of feeds.

As shown in Table 8-1, hay is a more important source of energy for dairy cows than for any other class of farm animal. But it is also an important feed source for beef cattle and horses. It is noteworthy, too, that about one-half the total hay tonnage produced in North America is fed to dairy cattle, while beef cattle consume almost one-third of all hay produced. As increasing quantities of concentrates go to feed the world's hungry people, it is expected that livestock producers will depend even more on hay to meet a larger percentage of the total feed needs of ruminants.

TABLE 8-1
PERCENTAGE OF ENERGY SUPPLIED BY HAY AND OTHER KINDS OF FEEDS[1]

Animal	Concentrates	Hay	Other Harvested Forages	Pasture	All Forage	Total
	(%)	(%)	(%)	(%)	(%)	(%)
Lactating cows	37.9	23.1	19.4	19.6	62.1	100
Other dairy cows	19.4	29.0	5.9	45.7	80.6	100
Finishing beef cattle	69.8	16.3	8.7	5.2	30.2	100
Other beef cattle	8.7	15.5	4.1	71.7	91.3	100
Sheep and goats	10.4	4.7	3.1	81.8	89.6	100
Horses and mules	20.6	18.3	10.2	50.9	79.4	100

[1]Based on USDA data. From paper entitled, "Hay Production, Preservation and Quality," by J. E. Baylor, The Pennsylvania State University, *Beef Cattle Science Handbook*, Vol. 13, p. 199, published by Agriservices Foundation, edited by M. E. Ensminger.

Comparative Value of Hay

In ruminant rations, hay is primarily a source of energy, but the legumes also serve as a source of protein. For swine and poultry, ground hay (especially alfalfa) is fed primarily as a source of minerals and vitamins.

Table 8-2 shows the dry matter (DM), crude protein (CP), and total digestible nutrients (TDN) in 100 lb of dry matter from corn grain, corn silage, and three different types of alfalfa hay (mature, midbloom, and early bloom). Note, too, that this table compares these crops on a per acre basis.

TABLE 8-2
COMPARATIVE ECONOMICS OF THE NUTRIENTS IN CORN GRAIN, CORN SILAGE, AND ALFALFA HAY OF THREE QUALITIES[1]

Item	Corn Grain	Corn Silage	Alfalfa Hay		
			Mature	Mid-Bloom	Early Bloom
Analyses, DM basis, %					
Dry matter (DM)	88.0	33.0	90.0	90.0	90.0
Crude protein (CP)	10.1	8.1	12.9	17.0	18.0
Total Digestible Nutrients (TDN)	90.0	70.0	50.0	58.0	60.0
Value of 100 lb DM, $					
CP value[2]	1.72	1.38	2.19	2.89	3.06
TDN value[3]	5.04	3.92	2.80	3.25	3.36
Total value	6.76	5.30	4.99	6.13	6.42
Total value/acre,[2][3] $					
16 tons silage or 100 bu grain	333.00	560.00	—	—	—
21 tons silage or 150 bu grain	500.00	735.00	—	—	—
5 tons hay	—	—	449.00	552.00	578.00
8 tons hay	—	—	719.00	883.00	924.00

[1]Adapted by the authors from: *Haymaker's Handbook*, by J. E. Baylor, Professor of Agronomy, The Pennsylvania State University, and M. A. Balas, New Holland, Inc., published by Ford New Holland, Inc., New Holland, Pa., 1987, p. 140, Table 17.1.

[2]44% soybean meal used as a standard protein source, priced at $150 per ton or $.17 per pound of crude protein.

[3]Corn grain used as a standard TDN source, priced at $2.50 per bushel or $.056 per pound of TDN.

Of course, the economic comparisons in Table 8-2 are valid only at the stated prices of soybean meal and corn. However, in most practical feeding situations the comparisons are meaningful. It is noteworthy that, in terms of the economic value of the energy and protein provided by the different feeds listed in Table 8-2, early bloom alfalfa hay had a value nearly 95% (6.42 ÷ 6.76 × 100) that of corn grain. Note, too, that in terms of energy produced per acre, corn silage leads all other feeds.

Hay As a Grain Replacement

In the future, livestock producers will increasingly rely upon the ability of ruminants to convert coarse forage, grass, and by-product feeds, along with a minimum of concentrate, into food for human consumption, thereby competing less for humanly edible grains.

The U.S. Department of Agriculture conducted a study of all-forage rations for finishing cattle, the results of which are given in Table 8-3.

TABLE 8-3
FEEDLOT PERFORMANCE AND CARCASS EVALUATION OF STEERS FED ALL-FORAGE VS ALL-CONCENTRATE RATIONS[1]

Item	All-Forage Ration[2]	All-Concentrate Ration[2]
Average daily feed intake (lb)	23.3	16.0
(kg)	10.59	7.26
Average daily feed intake in % of body weight	3.23	2.15
Average daily gain (lb)	2.3	2.8
(kg)	1.05	1.27
Feed-gain ratio	10.06	5.71
Average carcass grade	Low Choice	Medium Choice
Dressing percentage (%)	55.4	59.9
Marbling score	Abundant	Abundant
Rib eye area (sq in.)	11.0	10.6
(sq cm)	71.1	68.5
Fat over rib eye (in.)	.37	.67
(mm)	9.4	17.0
Taste panel evaluation[3]	7.6	7.2

[1]Oltjen, R. R., T. S. Rumsey, and P. A. Putnam, "All-Forage Diets for Fattening Beef Cattle," *Journal of Animal Science*, Vol. 32, No. 2, 1971, pp. 327-333.

[2]Corn grain provided 90% of the all-concentrate ration; pelleted alfalfa provided 98% of the all-forage ration.

[3]Overall desirability rated on a scale of 1 to 9, with 9 being the most desirable.

As a result of the experiment summarized in Table 8-3, the U.S. Department of Agriculture researchers concluded that (1) beef cattle of an acceptable quality were produced on a pelleted, all-forage ration; (2) steers on an all-forage ration had to be fed a month longer than those on the all-concentrate ration; (3) the all-forage-fed steers consumed about 95% as much metabolizable energy and were about 86% as efficient converters of it to body weight gains as were the all-concentrate steers; (4) the forage-fed steers had only 55% as much fat over the rib eye as did the all-concentrate-fed steers; and (5) there was a 4.5% difference in dressing percentage in favor of the animals receiving all-concentrate ration. Based on this study, the following conclusion may be drawn: Since cattle fed high-roughage rations normally have lower dressing percentages, forages must be cheaper than grain in order for the feeder to obtain the same net return; this situation usually exists relative to pastures, but it doesn't always apply to dry forages.

KINDS OF HAY

Although there are favorite hays, a great variety of legumes, grasses, and cereal crops can be, and are, successfully used for hay. In terms of total tonnage produced annually, alfalfa (or lucerne), the "Queen of the Forages," accounts for approximately 57% of the U.S. hay crop. Many different kinds of hay make up the other 43% of the nation's hay supply; among them, other perennial legumes, cool season grasses, warm season grasses, cereal hays, summer annuals, and annual legumes.

The kind of forage grown should be determined by soil type, soil drainage, soil pH, topography, climatic conditions, preferred use, and the animals to which it will be fed. Also, more and more farmers are coming to appreciate the flexibility afforded by growing varieties of grasses and legumes that may be used three ways: for pasture (see Chapter 7, Pasture and Range Forages), for hay, or for silage (see Chapter 9, Silage/Haylage/High-Moisture Grain). With such an arrangement, surplus pasture may be converted into hay, or, if the weather is not favorable for haymaking, the crop can be ensiled.

Generally speaking, legumes should be used as hay crops wherever they are adapted, either alone or in combination with grass(es). There is one possible exception to this recommendation—where horses are involved, sometimes a good-quality grass hay may be preferable.

Whenever feasible, it is recommended that a legume be grown for hay, for the reasons that, in comparison with grasses, legumes are (1) higher in protein, vitamins, and minerals; (2) higher yielding; and (3) nitrogen-fixing when inoculated, because the bacteria (rhizobia) on their roots take free atmospheric nitrogen from the air. However, a mixture of grasses and legumes is often preferred for reasons of palatability, ease in curing, erosion control, and lessening bloat.

Weeds and Other Potential Hay Crops

In periods of serious drought, many plants are used for feed. Russian thistle was used to keep cattle alive during the drought of the 1930s. Even when harvested at a very early stage, it is a better feed when ensiled than when dried for hay. Its abrasive surface is less irritating to the mouths of livestock when softened as silage.

Pigweeds, sow thistles, and other weeds have been used as livestock feeds in emergencies. Even though their yield is low and their nutrient content is poor, they will sustain life.

In using weeds and other unusual crops for livestock feed in times of emergency, it is always advisable to obtain authoritative information relative to toxic substances in them and the stage at which it is best to harvest them for feeding.

HAY QUALITY

Hay quality is the degree of excellence, or the productive worth, that hay possesses. It refers to the nutritive value of hay. For hay to be of superior quality, it must be high in four factors: (1) nutrients, (2) palatability (intake), (3) digestibility, and (4) efficiency of utilization.

The most accurate method of determining hay quality involves live animal experiments on the farm or ranch where the forage is to be fed. However, this is often too costly, slow, and impractical. Therefore, forage value is predicted by visual inspection, chemical analysis, and/or new methods such as near infrared analysis.

Importance of Hay Quality

Hay is feed. Thus, as with any feed, it's the end results from feeding hay—the value as determined by animals—that count. Generally speaking, livestock producers recognize that the feeding value of hay varies according to quality. However, it is doubtful that they realize just how much returns in production—in meat, milk, wool, reproduction, and speed and endurance—are affected by quality.

The quality of hay greatly affects its consumption. High-quality forage is more digestible and passes through the digestive tract more rapidly than low-quality forage; hence, animals will consume more of it.

Feeding trials at the University of Wisconsin, Madison, showed, conclusively, the effect of hay quality on milk production. Alfalfa hay at four stages of maturity—prebloom, early bloom, midbloom, and full bloom—was fed to lactating cows. Table 8–4 summarizes the results. It gives for the four different hays the crude protein (CP), neutral detergent fiber (NDF), and acid detergent fiber (ADF); the digestible dry matter (DDM); the dry matter intake (DMI); and the 4% fat corrected milk (FCM) produced. Note that with maturity of the hay the crude protein, DDM, DMI, and milk production decreased, while the NDF and ADF increased. The increase in NDF and ADF with maturity is as expected, because NDF is inversely correlated with intake, whereas ADF is highly correlated with digestibility.

TABLE 8–4
EFFECT OF QUALITY OF ALFALFA HAY ON PERFORMANCE OF LACTATING COWS[1]

Stage of Harvest	Composition			DDM	DMI	4% FCM
	CP	NDF	ADF			
	(%)	(%)	(%)	(%)	(% BW[2])	(lb/day[3])
Prebloom	21.1	40.5	30.2	62.7	2.08	87.1
Early bloom	18.9	42.0	33.0	61.6	1.97	77.2
Midbloom	14.7	52.5	38.0	54.8	1.48	66.2
Full bloom	16.3	59.5	45.9	52.9	1.42	64.7

[1]Kawas, J. R., N. A. Jorgensen, A. R. Hardie, and J. L. Danelon, *Journal of Dairy Science*, Abstract, Supplement 1, 1983, Vol. 66, p. 181; and Kawas, J. R., N. A. Jorgensen, and D. A. Rohwede, *Proceedings Wisconsin Forage Council Eighth Forage and Use Symposium*, 1984, p. 21, Tables 3 and 4.
[2]80% hay and 20% concentrate (DM basis).
[3]46% hay and 54% concentrate (DM basis).

In feeding trials with lactating cows, Cornell workers compared alfalfa hay cut at two different stages of maturity—early bloom vs late bloom. They found that, in comparison with the late cut hay, the cows that were fed the early cut hay consumed 7 lb more of it per head per day and it was 16% more digestible and produced 12 lb more milk per day (see Fig. 8–2).

Thus, experiments and experiences show that, in addition to the low-nutrient content that characterizes poor-quality hay, a more serious loss may follow from feeding it. Studies show that part of the poor results obtained from feeding low-quality hay can be attributed to its failure to support maximum microflora in the rumen, with the result that the digestibility of the crude fiber suffers. Hand in hand with the decline in microflora activity, forage consumption goes down. Of course, if animals won't eat feed, it won't do them any good.

Hay

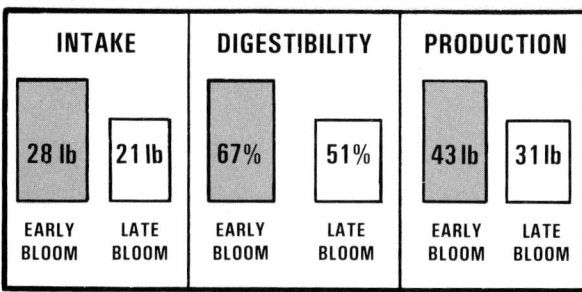

Fig. 8-2. Comparison of alfalfa hay cut at two different stages of maturity when fed to lactating cows (based on Cornell data).

Visual Inspection

Although not as reliable as chemical analysis, most hay is still bought and sold on the basis of visual appraisal.

Fortunately, hay quality and value can be estimated by certain characteristics. It is important, therefore, that those who grow hay, and those who buy and sell hay, be acquainted with the recognizable characteristics of hay which indicate high palatability and nutrient content. If in doubt, the animals will "tell" you, for they like and thrive on high-quality hay.

Guidelines for sensory evaluation of (1) legumes, (2) grasses, and (3) grass-legume mixtures are given in Table 8-5, Hay Scorecard.

TABLE 8-5
HAY SCORECARD[1]

	Hay Type		
Factor	Legumes	Grasses	Grass Legume Mixtures
	Point Score		
Leafiness. Legume hay should contain 40%, or more, leaves.	25		15
Color/aroma. Hay should be green, bright, and have a pleasant aroma.	25	30	25
Softness and pliability of stems. These qualities are indicative of harvest at early stage of maturity, and of high nutrients, palatability, and consumption.	15	30	20
Freedom from foreign material. Few weeds and little trash, no toxic substances, and minimum waste.	15	20	20
Condition. Hay cut, cured, and stored properly, with the result that it is free from dust and mold.	20	20	20
Totals	100	100	100

[1]Adapted by the authors from *Harvesting Quality Hay,* Circular R-624, North Dakota State University, Fargo.

The factors to look for in high-quality hay are:

1. **Species of plants.** Determine what plants are present and the proportion of each. Hay with a high percentage of legume is usually higher in feed value than pure grass hay.
2. **State of maturity when cut.** Plants should not be in full bloom, nor should they have formed seeds. Early cut hay assures the maximum content of protein, minerals, and vitamins, and the highest digestibility.
3. **Percentage of leaves present.** Leaves are the part of the plant of highest quality; hence, a high proportion of leaves relative to stems is indicative of high quality.
4. **Green color.** A bright green color indicates (a) minimum of bleaching and leaching losses of carotene and other nutrients, and (b) palatability.
5. **Aroma and fragrance.** High-quality hay has a pleasing, fragrant aroma. Moldy smells are undesirable.
6. **Stemminess.** Large stiff, woody stems make for low acceptability and quality. High-quality hay is fine stemmed and pliable.
7. **Foreign material.** High-quality hay is free from such foreign materials as weeds, stubble, sticks, dirt, etc.
8. **Condition.** Hay that has been cured and stored properly does not contain excess moisture, is not in layers or chunks due to excess moisture or heating, is not moldy, and is not dry and brittle.

Chemical Analysis

Visual estimates of hay quality are of value and should be used, but the most precise way to determine the nutrient value of hay is through chemical analysis. Analyses are not infallible, however; a Pennsylvania study revealed errors of as much as 5% in crude protein and 9% in TDN (energy) content of a forage, with evaluations made by trained individuals.

Visual inspection of hay is still needed. For example, hay cut at the right stage of maturity can become low-quality hay by poor hay-making practices and conditions, which only visual inspection can detect. Also, visual inspection is needed for (1) weed detection and color, (2) predicting palatability, and (3) detecting the effects of mold, rain damage, and brittleness. So, chemical analyses should supplement, but not replace, visual inspection. Also, it is recognized that any method of determining hay quality by means of chemical analyses is of value only if it is related to feeding value.

Most livestock and hay producers are aware that hay harvested at an early stage of maturity is high in protein and low in fiber (cellulose), although hay yields at immature stages are low. But the magnitude of the variation is usually greater than suspected. Thus, both chemical composition and yields must be considered in a practical management system. The highest yield of protein and of most of the other important chemical constituents is obtained at near the $\frac{1}{10}$ bloom stage of growth. Although the yield of hay may continue to increase between $\frac{1}{10}$ bloom and full bloom, it is due largely to an increase in the yield of cellulose (Fig. 8-3).

As shown in Fig. 8-3, the yield of protein is greatest when alfalfa is harvested at the first flower ($\frac{1}{10}$ bloom) stage of maturity.

Research has generally shown a good relationship between the chemical composition of hay and its feeding value. As a result, a growing number of states now have laboratories where, at a nominal charge, a quick determination can be made of the chemical composition of hay. As hay matures, protein decreases (pounds of protein per acre decreased from 935 lb in early cut hay to 605 lb in late cut hay, according to a Cornell study) and fiber increases. Likewise, weathering lowers the protein and raises the fiber content, since soluble nutrients are washed out by rain and leaves are lost during harvest. It is also noteworthy that palatability is nega-

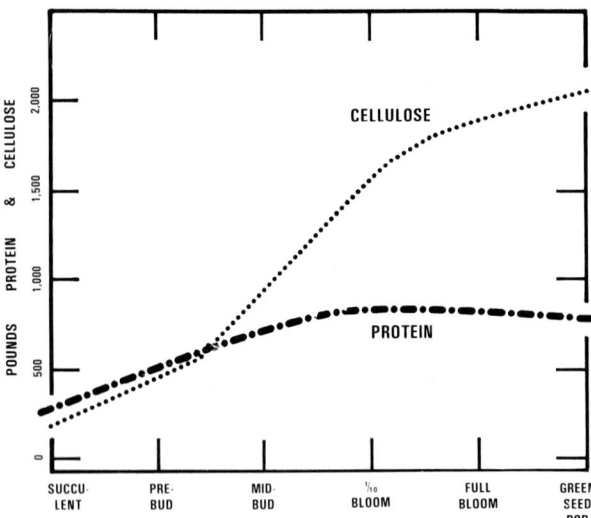

Fig. 8-3. Yield of crude protein and cellulose per acre from first cutting alfalfa. (*From:* Rohweder, D. A. and D. Smith, *Establishing and Managing Alfalfa*, Pub. A 1751, University of Wisconsin, 1982, p. 12.)

tively correlated with crude fiber levels—the higher the fiber content, the lower the palatability; this is important, for if the animals won't eat it, they can't produce. Cornell investigators found that cows ate 2 lb more of the early-cut hay per day than of late-cut hay.

NDF, ADF, and NIRS Analyses

Since 1865, fibrous materials traditionally have been analyzed by the *Weende proximate analysis* method. Although this method is still widely used, it is often supplemented with additional analyses. In a series of reports beginning in 1963,[2] Peter J. Van Soest proposed the *detergent analysis system*, better to evaluate the feeding value of fibrous materials; by using detergents, he separated the sample into two fibrous fractions: (1) a neutral detergent fibrous fraction (NDF), and (2) an acid detergent fibrous fraction (ADF). Further, he reported that, in comparison with traditional proximate analysis, NDF provided a better estimate of dry matter intake (consumption) by animals, and ADF provided a better estimate of the *in vivo* (inside the animal) dry matter digestibility. Van Soest partitioned forages into the following two fibrous fractions:

1. **Neutral detergent fiber (NDF),** which is the cell wall material or plant structure in feed, is comprised of hemicellulose, cellulose, lignin, lignified N, and insoluble ash. This constituent is insoluble in neutral detergent and is only partially available to animals. The lower the NDF percentage, the more the animal will eat; it is inversely related to voluntary intake (consumption). Thus, a low percentage of NDF is desirable.

The NDF content is also positively correlated with eating time and rumination, and this may be related to the rate of particle size reduction. Additionally, NDF is related to the proper function and health of the rumen; roughage value (the amount, source, and physical form of dietary NDF) is associated with chewing time, saliva flow (buffering and pH in the rumen), rumen fermentation patterns, milk fat test, and total energy output.

2. **Acid detergent fiber (ADF),** which is the highly indigestible plant material in a forage, is comprised of cellulose, lignin, and insoluble ash. This constituent is insoluble in acid detergent. ADF differs from crude fiber in that it contains silica. Silica and lignin in plants are associated with low *in vivo* (in animal) digestibility. The lower the ADF, the more feed an animal can digest. Thus, a low ADF percentage is desirable.

But chemical determinations are slow! The laboratory determination of crude protein, neutral detergent fiber, acid detergent fiber, *in vitro* (in a test tube or other artificial environment) dry matter digestibility, mineral, and vitamin analyses may take up to 2 weeks. This prompted the search for a more rapid, yet reliable, method of assessing the nutritive value of hay. In 1976, Norris, *et al.*, indicated (*Journal of Animal Science*, 47:747-759) that a relatively new procedure, known as *near infrared reflectance spectroscopy*, which had been applied successfully to grain quality evaluation, could be used for predicting forage quality, also. Additional research confirmed the initial findings of Norris and developed data bases and procedures for quickly analyzing hays.

The near infrared reflectance spectroscopy (NIRS) is a nonconsumptive instrumental method for fast, accurate, and precise evaluation of the chemical composition and associated feeding value attributes of forages and other feedstuffs. The instrument, known as a *near infrared analyzer*, produces infrared radiation over a given range of wavelengths and this radiation is focused onto the sample being tested. Because of the chemical structure of the sample material, certain combinations of infrared wavelengths are reflected and certain combinations are absorbed for each chemical characteristic tested, e.g., energy values, crude protein, digestibility, minerals, NDF, and ADF. By using a system of filters and detectors, the instrument senses these reflected wavelengths and passes this information on to the computer. The computer sorts out the appropriate wavelength combinations and their relative magnitudes for each chemical characteristic and transforms these data into percentages.

The near infrared reflectance measures hay quality by comparing the energy reflected back from a hay sample with computerized standards established by conventional laboratory analysis of a large number of reference samples.

The NIRS method of analysis has four main advantages: speed, simplicity of sample preparation, multiplicity of analysis with one operation, and nonconsumption of the sample (it can be analyzed again by the same or another procedure). With the NIRS method of analysis, it is possible to take a sample from a truckload of hay and provide, in less than 3 minutes, an analysis for crude protein, NDF, ADF, dry matter, lignin, and *in vitro* dry matter digestibility.

The chief disadvantages of the NIRS method are instrumentation requirements and costs, dependence on calibration procedures, complexity in the choice of data treatment, and lack of sensitivity for minor constituents.

Correct Sampling Necessary

No forage test is any better than the sample taken. Stated differently, a chemical analysis is valid only to the extent that the sample analyzed represents the lot of hay under

---
[2]Goldring, H. K. and P. J. Van Soest, *Forage Fiber Analyses*, Agriculture Handbook No. 379, ARS, USDA, 1970, p. 20 on which 12 papers are listed.

consideration. Thus, the most important single step in determining the chemical composition of hay is sampling.

With conventional, rectangular bales, at least 20 bales should be sampled at random, by probing every third bale, for example. The probe, or core sampler, should be at least ⅜ in. in diameter. The center of either end of a rectangular bale may be probed by inserting the probe at a right angle to the face of the bale and to a depth of 12 to 18 in. The hay testing laboratory should be consulted relative to sampling large round bales and stacks.

Hay samples should be placed in a closed plastic bag or freezer carton; otherwise, the moisture content will not be meaningful.

What Tests to Make

In modern haymaking, marketing, and feeding, proximate feed analysis no longer suffices. In order to predict the level of animal performance (milk, rate of gain) that may be obtained from hay, the following chemical components are commonly determined: crude protein (CP), neutral detergent fiber (NDF), acid detergent fiber (ADF), *in vitro* dry matter digestibility (IVDMD), crude fiber (CF), lignin, moisture, calcium, and phosphorus.

Moisture and crude protein are the most commonly analyzed components. NDF measures cell wall portion and is used to predict intake (consumption); and CF, ADF, and IVDMD give good estimates of the digestibility of the forage.

Other analyses which are often useful in evaluating hay are carotene and certain trace minerals. There are times when the amount of vitamins and amino acids in the feed might be useful, particularly when hay is used as a supplement in nonruminant rations. These analyses are costly to run, however, and it may not be economically feasible to run very many of them.

Evaluating Test Results

Test results can best be evaluated by comparing them with some standard. The testing laboratory may provide such information, possibly along with recommendations for applying the test results in balancing rations. For convenience, average crude protein and crude fiber values of some common hay crops are given in Table 8-6.

TABLE 8-6
APPROXIMATE CHEMICAL COMPOSITION (MOISTURE-FREE) OF VARIOUS SUN-CURED HAYS

Kind of Hay	Crude Protein		Crude Fiber	
	Average	Range	Average	Range
	(%)	(%)	(%)	(%)
Alfalfa	16.0	12.0–24.0	28.0	22.0–39.0
Bermudagrass	10.0	7.0–15.0	33.0	28.0–37.0
Bromegrass	10.5	6.0–15.0	28.0	24.0–31.0
Ladino clover	18.5	16.0–21.5	22.0	18.5–23.0
Red clover	12.0	10.5–18.5	27.0	18.0–34.0
Lespedeza	13.0	11.5–14.5	27.0	22.5–32.5
Oat hay	5.0	4.0–6.0	28.0	26.0–32.0
Orchardgrass	8.1	6.0–14.0	30.0	26.0–31.0
Soybean	14.5	9.0–16.5	28.0	20.5–41.0
Timothy	6.5	5.5–9.5	30.0	28.0–31.5
Sudangrass	8.8	6.5–11.0	28.0	26.0–30.5

If a chemically analyzed sample runs higher in protein and lower in fiber than the average figures given in Table 8-6, it means that the sample is better than average quality hay; conversely, if it is lower in protein and higher in fiber, the sample tested is below average quality.

Thus, a chemical test provides informed appraisal of hay values. Except for actual feeding trials, it is the best method of evaluating hay quality presently available.

MAKING QUALITY HAY

The object of haymaking is to (1) harvest the crop at the optimum stage of maturity which will provide the maximum yield of nutrients per acre without damage to the next crop; and (2) cure properly, which involves lowering the water content of the green herbage from 65 to 85% moisture to 20% or less.

Hay quality begins with the soil and ends with the manger, with many intermediate factors affecting it along the line. Once forage is cut, opportunities to increase nutrient content are over; from that point on, quality can only be preserved.

There is no one best haymaking method or kind of equipment. These must necessarily vary with the size of the operation, the kind of hay, the climate of the area, the individual farm or ranch conditions and buildings, and the available labor and machinery and their cost. Yet, the principles of good haymaking and the objectives sought are the same everywhere.

About 80% of all harvested forage is now baled. Only 10% is stored as loose hay and cubes, and the remaining 10% is stored as hay crop silage.

Growing Forage

Growing forage for hay has long been neglected. Average yields per acre are still well under one-half their potential. Little more than 1 acre of hay in 10 is fertilized on a regular basis; and the precious few acres that are fertilized get an average of only 12 to 15 lb per acre—a paltry amount compared to corn, which receives an average of about 200 lb of fertilizer per acre.

The steps in growing quality hay are:
1. Match crop to soil.
2. Choose quality seed, and proven varieties and mixtures.
3. Lime and fertilize.
4. Get good stand.
5. Irrigate where practical.
6. Control insects and diseases.

Haying Equipment

A brief rundown on today's haying tools follows:
1. **Mowers.** Simplicity, high speeds, and greater widths continue to make the conventional cutterbar the most common mechanism used in forage harvesting.
2. **Mower-conditioners.** These combine cutting and conditioning in one operation.
3. **Disc mowers/disc mower conditioners.** These mowers were developed for tough cutting conditions that often plugged sicklebar cutters. The addition of intermeshing rubber rolls to the disc mower permits conditioning the crop when cutting.

4. **Forage mat machine.** This newly developed machine mows alfalfa, shreds it, presses the shredded leaves and stems into a ¼-in.-thick mat, and spreads it on the stubble to dry—all in one operation. The forage mat machine squeezes out a lot of moisture followed by the forage drying quickly and uniformly.

5. **Windrowers.** Multipurpose, self-propelled windrowers provide up to 16 ft of cutting capacity. Most models feature hay conditioners as standard or optional equipment.

6. **Rakes.** Side-delivery rakes are still widely used to make hay crop windrows. Parallel bar rakes, which are most popular today, reduce impact between rake teeth and hay by moving hay from the outside of the swath to the windrow in less than 13 ft of travel.

7. **Tedders, rake-tedders.** In humid regions, making hay can be a problem because of the constant threat of showers and moisture from the morning dew. To help speed drying time, tedders have been developed to ted or fluff the hay crop either from a swath or windrow. Rake-tedders do a tedding operation in addition to raking.

8. **Balers.** Automatic pick-up balers package most hay today.

Balers that package small, rectangular bales, weighing 60 to 140 lb, continue to be common on farms and ranches that produce hay for their own use.

Balers that make large square bales are designed for custom operators or hay growers who have large volumes of hay or straw. Bale size is 3' × 4' × 8' long, and bale weight ranges from 1,000 to 1,500 lb. Large square bales are suitable packages for long distance transportation.

Balers that produce large round bales, weighing from 850 to 2,000 lb, are increasing in popularity, especially with cow-calf operators, and even with some dairy producers.

Fig. 8-4. Large round bales averaging 1,200 lb weight. (Courtesy, McArthur Farms, Inc., Okeechobee, Fla.)

9. **Bale handlers.** Mechanization took the backache out of bale handling. Today, the following types of sophisticated hay harvesting and handling equipment are available: bale throwers that throw bales into trailer wagons and eliminate lifting; bale conveyors for putting bales where desired; automatic bale wagons, which pick up bales from the field, load, transport, and stack them; and stack retrievers for transporting bales from storage to where they will be fed.

10. **Stack machines.** These are hydraulically operated field machines that compress long hay into stacks, weighing from 1 to 6 tons.

Harvesting at Proper Stage

Whether the crop is a grass or a legume, or a combination, the stage of maturity of the plants at the time of harvest affects digestibility, yield, and feeding value (see Fig. 8-5). Young, immature plants are high in protein and low in fiber or lignin. As hay crops mature, feeding value goes down and fiber content increases. Digestibility of the forage (TDN) declines about 0.5% each day cutting is delayed beyond the early bloom stage (Fig. 8-5); and the intake of forage decreases during this same period at more than 0.5% each day. Thus, in total, the feeding value of forage drops more than 1% for each day's delay after early bloom.

Forage dry weight yields increase until midbloom to late bloom stages (Fig. 8-5). Timothy and bromegrass fully headed, and red clover and alfalfa at full bloom, will give maximum yield of dry matter. However, maximum feeding value of first cutting forage is reached at least 10 days before the time of maximum dry weight yield (Fig. 8-5).

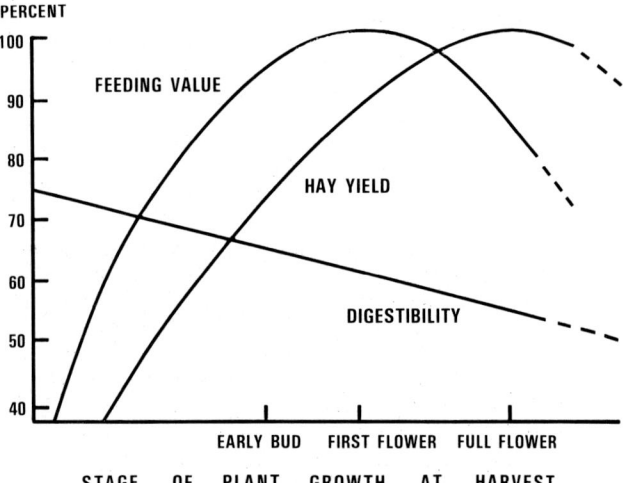

Fig. 8-5. Effect of advancing maturity on the feeding value, yield, and digestibility of alfalfa hay. Note that maximum feeding value per acre is reached 10 to 15 days before maximum yields.

Stage of maturity has a direct and dramatic effect on animal production, as reported in Table 8-7. When alfalfa hay that was harvested at four different stages of maturity was fed to yearling steers, the animals that were fed the very mature alfalfa (seed stage) made less than half the average daily gains and total gains of the animals that were fed early harvested alfalfa hay (bud stage). Moreover, they required more than twice as much hay per 100 lb gain.

TABLE 8-7
EFFECT OF STAGE OF MATURITY OF ALFALFA HAY AT HARVEST ON GAINS OF YEARLING STEERS[1]

Stage of Maturity	Average Daily Gains	Total Gains Per Steer	Feed/ 100 lb Gain
	(lb)	(lb)	(lb)
Bud	1.07	96	959
1/10 bloom	0.76	69	1,351
Full bloom	0.63	58	1,600
Seed	0.48	44	2,144

[1]Rohweder, D. A., *Maintaining Forage Stands for Efficient Production*, University of Wisconsin, Madison, A 2907, 1978, p. 9, Table 12; adapted from a study conducted by Kansas State University, with steers that had an initial weight of 440 lb.

In Wisconsin, first-growth alfalfa-bromegrass hay was harvested each year for 3 years at four stages of maturity. Dry matter digestibility, digestible energy, milk production, and animal gains all declined sharply with increasing maturity of the hay (Table 8-8).

TABLE 8-8
EFFECT OF STAGE OF MATURITY AT HARVEST OF ALFALFA-BROMEGRASS ON MILK AND MEAT PRODUCTION[1]

Stage of Maturity	Dry Matter Digestibility[2]	Digestible Energy[3]	4% Milk Production[4]	Lamb Gains[5]
	(%)	(kcal/g)	(lb/day)	(lb/day)
Vegetative	71.4	3.20	45	0.38
First flower	64.6	2.86	29	0.21
Full bloom	58.0	2.54	15	0.15
Green seed pod	55.2	2.43	4	0.05

[1]Adapted by the authors from University of Wisconsin data. The alfalfa-brome was first cutting, made at Arlington, Wisc.; and the data are an average of 3 years, except for lamb gains which are one year only.
[2]Animal digestion trial data, values similar to total digestible nutrients.
[3]Animal energy digestibility × forage gross energy.
[4]Estimated for a 1,200-lb cow fed hay alone.
[5]Lambs fed hay alone.

Stage of maturity also affects the vitamin content of hay. Carotene (precursor of vitamin A) and the B vitamins decrease as plants mature. Vitamin D content is the one exception—it increases as the forage is sun-cured.

Everything considered, there is a loss of about 1% in nutrient value for each day that hay harvest is delayed beyond the late vegetative stage of growth.

Table 8-9 gives guidelines relative to the proper forage-harvesting stage for maximum protein and minimum fiber.

TABLE 8-9
HAY CUTTING GUIDE

Kind of Hay	When to Cut
Alfalfa	Bud stage for first cutting; 1/10 bloom for second and later cuttings.
Alsike clover	Early bloom to 1/2 bloom stage.
Bermuda	When 16-18 in. tall, before lodging.
Birdsfoot trefoil	First flower to full bloom.
Bromegrass	Heads emerging.
Cowpeas	When pods are 1/2 to fully matured.
Crested wheatgrass	When the plants begin to head.
Crimson clover	From early bloom to 1/2 bloom.
Fescue	Boot to early head stage.
Grass-legume mixtures	When the legume is at the proper stage.
Johnsongrass, millet, Sudangrass, sorghum hybrids	40 in. height or early boot stage, whichever comes first.
Ladino clover	Few blooms to full bloom.
Lespedeza, annual	Early blossom.
Orchardgrass	Boot to early head stage.
Red clover	Late bud.
Sericea	When 12-15 in. high.
Small grains (oats, barley, wheat)	Boot stage to early dough stage.
Soybeans	Mid-to-full bloom and before bottom leaves begin to fall.
Sweet clover	Bud to very early flowering stage.
Timothy	Boot to early head stage.

RETAINING ADEQUATE PLANT FOOD RESERVES

Cutting or grazing plants when carbohydrate reserves are low may leave too little energy available to support new growth; e.g., continued cutting at immature stages of growth will eventually exhaust the plant and weaken it to the point of death. Plants weakened by too early, too close, or too frequent cutting or grazing are more susceptible to winter injury, drought, and disease. Usually, the closer to maturity that plants are cut or grazed, the higher the stored food reserves will be and the easier it is to maintain vigor for productivity. However, as emphasized in the preceding section, harvesting (by cutting or grazing) forage at a young growth stage makes for high feed value. So, a compromise must be reached between cutting or grazing forages for maximum quality of the forage and maximum vigor of the plants.

When alfalfa growth starts in the spring, food reserves in the roots and crowns are used to start new top growth from small crown buds or underground stems. Depletion of the food reserves continues until the plant is 6 to 8 in. tall. By this time, food is being synthesized in the leaves more rapidly than it is being used, and some food storage begins. Storage continues and reaches its highest level in the roots when alfalfa is about full bloom. The changes that occur in root energy reserves and dry matter yield in alfalfa during one growth period are illustrated in Fig. 8-6.

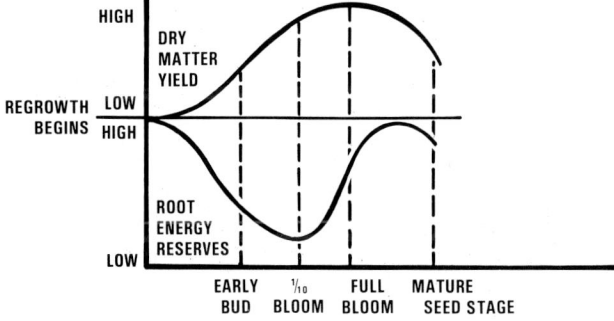

Fig. 8-6. Changes in root energy reserves and dry matter yield of alfalfa during one growth period of the crop. Note the changes by stages: early bud; 1/10 bloom; full bloom; mature seed stage. Note, too, that root reserves decrease with dry matter increases, with the use of root reserves preceding dry matter yields. (Adapted by the authors from Haymaker's Handbook, by J. E. Baylor, Professor of Agronomy, The Pennsylvania State University, and M. A. Balas, New Holland, Inc., published by Ford New Holland, Inc., New Holland, Pa., 1987, p. 121.)

So, the question: What is the best system to follow relative to fall cutting or grazing? *The answer:* To play it safe, either (1) harvest early enough to give the plants sufficient time to build up reserves before frost, or (2) delay harvest until near or after a killing frost so that there is little or no regrowth after cutting. *CAUTION:* If a late harvest is made in a cold area, leave at least 4 in. of stubble to help plants trap snow which acts as an insulator and shields plants from cold weather, and lessens heaving.

Cutting and Field Curing Hay

Proper cutting and field curing of hay embraces all the steps from cutting to ready-for-packaging or storing. In modern hay making, it includes (1) cutting, curing in the swath and windrow, and raking or cocking; (2) reducing moisture content, while minimizing shattering, bleaching and fermenting; (3) reducing rain damage; and (4) considering chemical and preserving agents.

CUTTING/CURING IN THE SWATH AND WINDROW/RAKING OR COCKING

The common steps, methods, and equipment used in cutting and field curing hay are as follows:

1. **Cutting and curing in the swath or windrow.** Cutting, followed by curing in the swath or windrow, is the first step in haymaking, regardless of the subsequent method or type of equipment employed.

Any one of several types of mowers may be used, for all of them are designed to get the hay down. The most important thing is that the hay be cut at the proper stage of maturity.

The following points are also pertinent to cutting hay and curing it in the swath or windrow:

   a. **Direction of mowing.** It is highly desirable to mow in the same direction as will be traveled in raking and in picking up the hay in subsequent operations.

   b. **Time of the day to mow.** Some opinions to the contrary, quality of hay is only slightly affected by either the time of day at which the forage is cut (the proportion of sugar does vary with the time of day), or the presence of dew.

   c. **Mechanical hay conditioning.** Mechanical conditioning reduces drying time and field losses each by about 50%. These machines are designed so that slow-drying stems are split, cracked, crushed, or broken as they pass through rollers or knives. As a result, the stems dry at about the same rate as the leaves; and there is more uniform curing and less leaf loss.

Although hay conditioners have considerable merit, farmers and ranchers with smaller hay acreages may not be able to justify the added cost of the equipment.

   d. **Length of swath curing.** Hay dries more rapidly in the swath than in the windrow, even if the windrow is small and fluffy. Therefore, it should be left to cure in the swath as long as is possible without damaging it; until the forage is wilted but before there is danger of the leaves shattering and/or of excessive bleaching and loss of carotene. At this stage, the moisture content will be about 40%. The point at which swath curing is sufficient should be carefully determined, because, without a conditioner, the leaves become dry and brittle long before the stems are cured, especially on legumes.

No definite period of time for swath curing can be assigned as it will vary according to the tonnage of hay per acre, temperature, sunshine, wind, and atmospheric humidity. Instead, the time should be determined entirely by the condition of the hay in the swath.

Sometimes curing in the swath is speeded up by using swath fluffers; machines which pick up the forage, lift it to a standing position, and then release it to fall loosely onto the stubble—thereby fluffing the hay.

Where it seldom rains, sometimes hay is not cured in the swath at all. Under such conditions, it may be cut and windrowed immediately. For the latter purpose, swathers—self-propelled mowers that cut and windrow hay in one simultaneous operation—may be used.

2. **Raking.** After the hay has wilted sufficiently in the swath, but while it is still tough and the leaves will not shatter, it should be windrowed.

If considerable shattering appears probable, it may be desirable to do the raking early in the morning when the dew makes the hay a bit tough.

Where windrowed hay is rained on, wait until the top half dries out, and then turn it upside down with the side-delivery rake (the use of the tedder for rewindrowing is not recommended because of excessive shattering).

3. **Cocking.** Formerly, well-made cocks, often adorned by hay caps, were considered a necessary part of good haymaking. However, this practice has greatly decreased, due primarily to higher labor costs and the advent of modern haymaking machinery. Today, the cocking of hay is confined almost entirely to use (a) in hot, arid regions where the leaves shatter if the hay is left in the swath or windrow for any appreciable length of time, and (b) as an emergency measure in order to protect hay when a storm is imminent.

REDUCING MOISTURE CONTENT/SHATTERING/ BLEACHING AND FERMENTING

Proper curing ensures that (1) the hay can be stored safely without heating excessively or becoming moldy; and (2) the maximum leafiness, green color, aroma, nutrient value, and palatability shall be retained. To the end that these desired objectives may be achieved, the following information is pertinent:

1. **Moisture content.** Freshly cut forage contains 75 to 80% moisture, whereas the maximum moisture content for safe hay storage is as follows:

   For loose hay—25% moisture.
   For baled hay—20 to 22% moisture (the lower figure for larger bales).
   For chopped hay—18 to 20% moisture.
   For cubes—16 to 17% moisture.

Hay of a higher moisture content than indicated should not be stored because (a) its value may be greatly lowered due to mold or to nutrient losses accompanying fermentations, and (b) of the ever-present danger of spontaneous combustion and a costly fire.

Two rule-of-thumb methods used by farmers in determining when hay is dry enough for storage are:

   a. **The twist method.** Twist a wisp of the hay in the hand. If the stems are slightly brittle and there is no evidence of moisture on the twisted stems, the hay can be stored safely.

   b. **The scrape method.** Scrape the outside of the stems with the thumbnail or a knife. If the epidermis can be peeled from the stem, the hay is not sufficiently cured. If the epidermis does not peel off, the hay is usually dry enough to stack or put in the mow.

2. **Shattering losses.** In field curing hay, losses from leaf shattering range from 2 to 5% for grass hay and 3 to 39% for legume hays, with as much as 15 to 20% for legume hays field cured under the most favorable conditions. Based on extensive experiments with field-cured alfalfa hay, the U.S. Department of Agriculture reported that leaf losses averaged 38.5% when none of the hay was wet; 47.3% when the hay

was wet by 2 showers; and 74.5% when the hay was wet by 3 showers— and milk production per acre was 19.7% less per acre when cows were fed rain damaged, field-cured alfalfa hay in comparison with field-cured hay without damage by rain.

3. **Bleaching and fermenting losses.** In general, the carotene or provitamin A content of freshly cured hay is proportional to the greenness. With severe bleaching, more than 90% of the vitamin A potency may be destroyed.

Even under the best of conditions, there is an unavoidable loss through fermentation, especially losses in sugars, starch, and carotene. With good weather and proper curing methods, however, these losses will not be excessive.

REDUCING RAIN DAMAGE

The leaching losses from rain are less severe soon after mowing, but increase in severity as curing progresses. Also, repeated showers are more damaging than one heavy rain. Experimental studies have revealed that damaging rains may lower the feeding value of hay by one-fourth to one-third, or even more with severe exposure.

The effects of rain on hay quality depend on the amount of rainfall and drying conditions. In a Wisconsin study, alfalfa and red clover were subjected to various rain treatments during field hay drying. The effects of rain damage on crude protein (CP) concentrations were small. The concentration of CP in both species decreased when harvest was delayed to full-late bloom compared with harvesting at an earlier state; based on these data, the reduction in CP due to a delay in harvest was larger than any change due to rain damage. The effects of rain on digestibility (*in vitro* dry matter disappearance—IVDMD) were greater than the effects on CP, and differed with drying conditions and amount of rain. The loss in IVDMD due to rain under poor drying conditions was from 73 to 57% for the alfalfa harvested at early maturity and from 62 to 39% for the alfalfa harvested at late maturity.

Losses from weather damage may be reduced (1) by using haymaking equipment that reduces the field drying time, (2) by understanding and using existing weather aides, and (3) by using proven chemical conditioning and preserving agents.

CONSIDERING CHEMICAL CONDITIONING AND PRESERVING AGENTS

Chemical hay drying agents and preservatives are giving haymakers a big assist in lessening hay making losses and improving hay quality. The big advantage of these products is that they speed up the haymaking process and reduce exposure to weather damage. In comparison with no treatment, the use of a desiccant, or drying agent, along with mechanical conditioning, can reduce the moisture content by an additional 2 to 10% during a 24-hour period. Adding a preservative to hay that is in the 25 to 35% moisture range will allow it to be baled and stored without undue heating.

• **Chemical conditioners**—Chemical drying agents, which are sprayed on the crop at mowing time, break down the waxy cutin layer on the wall of the stem and allow moisture to escape, thereby promoting faster drying time, with the drying rate of the stems approaching that of the leaves. Several chemicals can be, and are, used for conditioning, including potassium carbonate, sodium carbonate, and sodium silicate. Also, methyl esters of fats, vegetable oils, or animal fats have been mixed with potassium carbonate in an attempt to increase the effectiveness of chemical conditioning.

Chemical conditioners are effective on legumes such as alfalfa, birdsfoot trefoil, and red clover, but, generally, they are not effective on grasses. Although they will reduce drying time on all cuttings of legumes, they are most effective on second and third cuttings and least effective on first and late autumn cuttings. This situation is attributed to the fact that conditioners work best when drying conditions are best (in the summer), and that first cutting has heavier yields and heavier swaths than later cuttings—conditions that hamper drying, because the moisture movement inside the swath is inhibited.

Studies show that drying agents are more effective as an addition to, and not as a substitute for, mechanical conditioners. The chemical of choice is applied at the time of cutting, by either of two techniques: (1) a spray boom mounted ahead of the reel; or (2) spray nozzles are mounted behind the reel, but in front of the conditioning rollers so that the rollers help distribute the spray.

The recommended application rates range from 5.7 to 8.5 lb of the chemical powder per ton of hay, mixed and applied with 15 to 30 gal of water per acre in order to assure good coverage; at a cost of $5.00 to $10.00 per ton for the chemical in the late 1980s.

Additional costs are involved for labor and equipment. This prompts the question: Does chemical conditioning pay? The answer depends on the area and season. Michigan studies indicate that it pays for second and third cuttings, but not for first and last.

CAUTION: Since the crop is standing when desiccants are applied and won't be cut until a few milliseconds later, the EPA requires that some desiccants must be labeled as *pesticides*. So, before using a chemical conditioner, the haymaker should check with local regulatory authorities for their interpretation of this point.

• **Preserving agents**—Under normal conditions, for safe baling a moisture content of 20% or less is a must. Studies in many states have shown that, if properly treated with an adequate amount of the right preservative, alfalfa hay can be baled at 25 to 30% moisture, thereby speeding harvesting and lessening losses significantly. Preservatives act as fungicides and inhibit the growth and reproduction of the microorganisms that cause heating and molding in wet hay. A brief description of each of the common preservatives follows:

1. **Organic acids.** Propionic acid is the organic acid of choice. It is sometimes mixed with acetic acid, inorganic acids, formaldehyde, water, flavoring ingredients, and/or antioxidants. But, to be most effective, organic acid formulations should have at least 60% propionic acid.

To be effective, the organic acid(s) must be applied at the proper rate, depending on the moisture content of the hay; and it must be uniformly distributed throughout the hay mass. Table 8-10 gives the recommended rates of actual propionic acid to be applied at baling for different moisture levels.

To assure proper coverage, an applicator with a proper displacement pump and appropriately placed nozzles is needed. To date, preservatives have been more effective on conventional, rectangular bales than on large round

TABLE 8-10
RECOMMENDED RATE OF APPLICATION OF PROPIONIC ACID TO HAY[1]

High Moisture Level	Rate, Dry Weight Basis	Actual Acid/Ton Hay
(%)	(%)	(lb)
20–25	0.5	10
26–30	1.0	20
31–35	1.5	30

[1]Always follow the manufacturer's directions.

bales. The application of perservatives to large round bales is difficult because of the way hay is rolled in a round bale rather than compacted in a conventional bale chamber.

Organic acid preservatives are *acids;* hence, they can cause skin irritation and eye damage. Thus, rubber gloves and protective goggles must be used when handling these products. Some corrosion of equipment can be expected. Another disadvantage is that treated hay is usually somewhat bleached, thereby making it less attractive in the market place.

2. **Anhydrous ammonia.** When properly applied, anhydrous ammonia will stop bacteria and mold growth; and when applied to poor quality hay, it has the added advantages of increasing protein and digestibility. However, as a preservative it's not as effective as propionic acid. Also, unless large round bales are covered and/or contain less than 28% moisture, too much of the ammonia escapes. Excessive amounts of ammonia may cause animal disorders. (See Chapter 5, Nutritional Disorders/Toxins, section on "Nutrition Related Disorders.")

Where high-quality alfalfa hay is involved, which is already high in protein and digestibility, it's doubtful that the added expense of using ammonia can be justified.

3. **Bacterial inoculants.** Claims are made that most bacterial inoculants on the market will produce lactic acid, which acts as a fungicide and inhibits mold growth. More experimental work is needed, substantiating the effectiveness of bacterial inoculants as hay preservatives.

Value of Different Cuttings of Alfalfa Hay

In many areas, there is a decided preference among livestock producers in favor of a certain cutting of alfalfa hay.

Generally, first cutting alfalfa hay is coarser stemmed and less leafy than later cuttings, and therefore of somewhat lower feeding value when the different cuttings are equally well cured. Also, the weather is often less favorable for curing the first cutting. Yet, dairy producers participating in the Wisconsin Forage Analysis Superbowl Contest, sponsored by the Wisconsin Forage and Grassland Council, report (*Holstein World,* March 10, 1986, p. 90) that they obtain higher milk production from feeding higher fiber first cutting alfalfa than from lower fiber subsequent cuttings. For unknown reasons, this appears to contradict the expected performance based on average chemical analysis, from feeding different cuttings of alfalfa hay. As shown in Fig. 8-7, on the average, each successive cutting is lower in crude fiber and higher in crude protein.

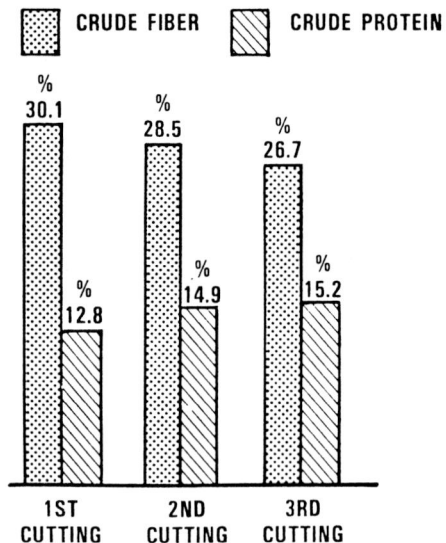

Fig. 8-7. Each successive cutting of alfalfa hay tends to be higher in nutritive value. Note that each successive cutting is lower in fiber and higher in protein. (Adapted by the authors from *Atlas of Nutritional Data on United States and Canadian Feeds,* NRC-National Academy of Sciences, 1972.)

Nitrogenation of Low-Quality Hay

Nitrogenation is the combining with nitrogen. Low-quality hay may be combined with ammonia or urea.

• **Ammoniation of low-quality hay**—In the late 1970s and the early 1980s, the ammoniation of low-quality forages increased greatly throughout North America, primarily as a means of providing an economical source of nitrogen (protein equivalent) for ruminants.

CAUTION: Anhydrous ammonia can be hazardous to people and animals if not handled properly. So, (1) use proper equipment, (2) take good care of equipment, and (3) follow safe practices.

• **Adding urea to low-quality hay**—Experimentally, University of Missouri scientists nitrogenated low-quality hay with urea; by injecting it into the bale, or by spraying it upon the hay at the time the hay was baled, at the rate of 2 to 4% of the dry weight of the hay. In some cases, the crude protein content was doubled, from 6 to 12%. In comparison with liquid ammonia, urea is easier to store and handle and may be applied at baling. Also, it is less hazardous to the applicator and to the animals consuming the treated hay.

Haymaking Systems

In haymaking, the term *system* refers to a team of processes and machines that does the work from field through feeding, saves crop nutrients, reduces manpower requirements, and eliminates drudgery. When each step is mechanized, it must be matched; otherwise, workers and machines end up waiting.

In recent years, automation has had great impact on haymaking. Some haymaking systems are completely mechanized from field to feeding.

There is no one best haymaking system for all conditions. Nevertheless, all good systems are fast, make handling easy, save labor and nutrients, and increase profits. Baling is the most popular hay-handling system in North America.

LONG, LOOSE HAY

The acreage harvested as long, loose hay has declined sharply in recent years, especially in the humid areas, because (1) of high labor cost, and (2) long hay is too bulky for mechanized feeding. Nevertheless, long, loose hay is still popular in many western areas where specialized handling equipment is used. Moreover, some of the newer systems of handling and self-feeding loose hay show promise.

The two common methods of handling long hay are:
1. Loading with hay-loader directly from windrows.
2. Hauling cured hay from windrows or cocks with buck rakes, sweep rakes, or sled.

CHOPPED HAY

Chopped dry hay fits into some feeding systems, particularly in the West.

For safe storage, the moisture of chopped hay should not exceed 18 to 20%.

Two common methods of chopping freshly cured hay follow:

1. **Field chopping cured hay directly from the windrow.** In this method, a field chopper gathers the cured hay from the windrow, chops it, and blows it into a truck or trailer. The chopped hay is then blown into the barn, stack, or other storage area.

Some pertinent facts relative to field chopping cured hay directly from the windrow are:

 a. **Equipment cost.** The equipment cost, on a per ton basis—for the chopper, the equipment for hauling, and the blower for unloading—is apt to be high unless a considerable acreage of hay is involved, or the operator can also use this equipment for corn or sorghum silage.

 b. **Moisture content.** The moisture content must be lower than is permissible for baling—from 18 to 20%.

 c. **Convenience in feeding.** Chopped hay is often more convenient to feed, since it can be handled mechanically.

 d. **Feeding value and wastage.** Chopping does not increase the feeding value of the hay, but it may make for less wastage, especially where low-quality hay is involved.

 e. **Length of cut.** A 1½- to 2-in. cut is recommended. Finer chopping requires more power, increases the tendency to heat in storage, and makes for dusty and unpalatable feed. Regardless of the length of cut, repeated blower action, first at the chopper and later at the barn, pulverizes the leaves and aggravates the dust nuisance.

 f. **Storage space.** Chopped hay requires only one-half to one-third as much storage space as long hay. This has the advantage of effecting a saving in space, but the disadvantage of requiring caution in order to prevent overloading mows.

2. **Chopping into the barn or other storage area.** In this method, the cured hay is generally hauled from the windrow or cock to the barn or other storage area where it is chopped by a hay chopper or silage cutter and blown directly into the storage area. This method is slower and requires more labor than where field-cured hay is chopped directly from the windrow, but less expensive equipment is necessary.

PACKAGED HAY

Great strides have been made in hay packaging in recent years, characterized by the advent of round bales, large rectangular bales, loaflike stacks, and cubes.

Although large round bales and compressed stacks are better adapted to outside storage than small round bales, unrestricted access at feeding will result in excess wastage. Large rectangular bales are suitable for commercial marketing and long distance hauling.

Compressing hay into cubes and pellets makes for many advantages; among them, (1) completely mechanized handling; (2) high density, with more economical transportation and storage; (3) easier self-feeding and higher intake by animals; and (4) lower feeding losses.

Bales

The following choices of bales are available:

1. **Rectangular bales.** Conventional, small, rectangular packages (often called *square bales*), weighing from 60 to 140 lb, were produced by the first baling machines; and they are still popular on farms and ranches that produce hay for their own use.

Fig. 8-8. Conventional rectangular baler in operation. (Courtesy, Deere & Company, Moline, Ill.)

Today, hay is also being packaged in large rectangular bales, weighing from 1,000 to 1,500 lb, designed for custom operators or hay growers with large volumes of hay or straw.

2. **Large round bales.** Many makes and models of large round balers are on the market. They range in weight from 850 to 2,000 lb, depending on the make of equipment.

If not handled properly, big round bales can be dangerous. Three types of bale handling hazards have surfaced: (a) a downhill roll; (b) tractor mounted front-end loaders carrying such a heavy, bulky load have a tendency to upset if not properly counterbalanced; and (c) a bale can roll out of the loader bucket back down the loader arm onto the operator when a loader handling a round bale is raised to full height.

Stacks

These are loaf-shaped (one system makes a circular stack), mechanically pressed haystacks. Long, loose hay is blown into a wagon and pressed down by a hydraulically operated canopy roof. Stacks range in size from 7 to 10 ft wide, 8 to 22 ft long, and 8 to 11 ft high, and weigh from 1 to 6 tons.

Fig. 8-9. Loaflike stacked hay; machine made, 1-ton size, sheds rain and snow. (Courtesy, USDA)

The stack system saves labor and permits nearly the same latitudes as loose hay, with the efficiencies of mechanization.

Limitations of the system are (1) Investment costs are high, (2) heavy rain or snow may seep down through the stacks if they are not formed properly, and (3) they are not efficiently transported over long distances. Nevertheless, the principles involved are good; hence, some of the problems will likely be overcome with more experience.

Cubes (Wafers)

Fig. 8-10. Hay cuber in operation, making cubes 1¼ in. square and 2 to 3 in. long. (Courtesy, Deere & Company, Moline, Ill.)

Field cubers are machines that move across hayfields, pick up windrows of forage, and produce dense, high-quality forage cubes or wafers. Stationary cubers are used to produce similar cubes from loose haystacks or bales.

The benefits derived from field cubing of hay are great. It (1) simplifies haymaking, (2) lessens transportation and storage space—cubed or wafered forages weigh between 45 and 55 lb per cubic foot (baled hay density is 8 to 15 lb), (3) reduces labor, (4) makes automatic hay feeding possible, (5) decreases nutrient losses, (6) eliminates dust, and (7) makes for increased feed comsumption, gains, and feed efficiency. Also, with cubing, the spread between high- and low-quality roughage is narrowed; that is, within reason the poorer the quality of the roughage, the greater the advantage from cubing or wafering. The latter is so because such preparation assures complete consumption of the roughage.

Hay cubes are of special interest to dairy producers because they have the advantages of pellets, without their disadvantages. Like finely ground forage that is pelleted, cube-feeding can be readily automated; and, in comparison with long hay, there is less transportation and storage cost. Besides, cubes will not lower the fat content of the milk as much as pellets when appreciable quantities of them are fed.

Pellets

Pelleted forages are finely ground, then condensed. The **advantages** of pellets are:

1. Pelleted feeds are less bulky than any other hay package (pelleted roughage requires one-fifth to one-third as much space as is required by the same roughage in loose or chopped form), and are easier to store and handle—thus lessening transportation, building, and labor costs.

2. Pelleting prevents animals from selectively refusing ingredients likely to be high in certain dietary essentials; each bite is a balanced feed.

3. Pelleting practically eliminates wastage. Since animals frequently waste up to 20% of long hay, less pelleted feed is required. Wastage of conventional feed is highest where low-quality hay is fed and/or feed containers are poorly designed.

4. Pelleting eliminates dustiness and lessens heaves.

The biggest deterrent to increased pelleting at the present time is the difficulty of processing chopped forage coarse enough so that it will not cause digestive disturbances. A minimum of a ¼-in. chop is recommended.

ARTIFICIAL DRYING

The use of forced air, either heated or unheated, for final drying is the most dependable way in which to preserve quality in hay. The application of heat provides faster drying, saves more leaves, and reduces losses of nutrients. Yet, artificial drying is on the decline in the United States because of the added cost in heating the air and the increased labor.

Hay-drying equipment permits handling while hay is green and tough enough to withstand mechanized processes without excessive leaf loss, and it minimizes weather damage. The most common methods of artificial drying are (1) mow curing, (2) artificial dehydrators, and (3) wagon driers.

Mow Curing

Mow curing, or drying, refers to the practice of curing partially dried hay—either long, chopped, or baled—in barn mows equipped with ventilation systems through which either unheated or heated air is forced. The pertinent facts relative to mow curing are:

1. **Equipment.** Although numerous variations in the design of mow-drying equipment exist, generally speaking the equipment consists of one of the following two types:

   a. A blower or fan (usually powered with an electric motor, but with provision for tractor operation in case of power failure) which forces air through a main duct or flue and out through a system of lateral ducts.

   b. A slotted floor built above the regular mow floor, with provision under this slotted floor for forced air circulation up through the hay.

2. **Moisture content of forage.** The following moisture conditions usually prevail in mow drying:

   a. Newly mown forage has a moisture content of 75 to 80%.

   b. When placed in the mow for further curing, the moisture content should not exceed 35 to 40%, which means that it is about half cured.

   c. For safe storage the moisture content of hay must be reduced to between 20 and 25%.

3. **Cost.** Although higher quality hay can generally be produced in mow curing than in field curing, the cost is also greater. The higher cost in mow curing is due to (a) the cost of the mow-drying equipment, (b) the cost of operating the equipment, (c) the added labor required in handling heavy forage of high-moisture content, and (d) the higher storage cost per ton, because in mow curing the hay cannot be piled too high over the ducts at any one time (it is recommended that not more than 4 to 8 ft of uncured hay be placed in the mow at any one time) without decreasing the all-important air circulation.

Artificial Dehydrators

Artificial dehydrating refers to a process in which forage is taken from the field as soon as it is cut (or in some instances after wilting), put through a hay chopper or silage cutter, and dried in large driers of various types, following which it is finely ground. For the most part, this method of curing is limited to large commercial operations which process early cut alfalfa (or its leaves) and other legume and/or grass crops chiefly as a supplement for swine and poultry. Occasionally, artificially dehydrated hay is produced for other classes of animals, especially in those areas which rarely have good haymaking weather.

Other pertinent facts relative to artificial dehydrating of hay are:

1. **Drier.** The most popular type of artificial dehydrator in use in this country is one that uses a high initial heat (1,200° to 1,400°F), and which is heated by gas or oil. In a good drier, the forage does not get sufficiently hot to be damaged, primarily because of the cooling effects produced when the water evaporates from the plant tissues, and because the forage is in contact with the hot air for only a few minutes.

2. **Nutritive value of artificially dehydrated hay.** Generally, artificially dehydrated forage is of high nutritive value, for few leaves are lost and the maximum content of protein, carotene, and riboflavin is retained. But the protein may be slightly less digestible due to the effect of the heat of dehydration, and the vitamin D content is low for the reason that in the curing process the hay is not exposed to the vitamin D-imparting ultraviolet rays of the sunlight.

3. **Cost.** Due primarily to high cost of equipment and fuel, artificial dehydration is usually considered practical only in commercial operations where large tonnages are processed.

The producer is well justified in paying a premium price for high-quality dehydrated forage for use in swine and poultry rations. Except in special circumstances or as a partial grain replacement, however, it is seldom economical to feed artificially dehydrated hay to cattle or sheep.

Wagon Dryers

Wagon dryers were developed to reduce the high labor requirements of batch or platform drying. Essentially, they are batch dryers on wheels.

Although good-quality hay can be produced by wagon drying, it requires more labor and handling than most other systems of haymaking.

Storing

Good hay should never be poorly stored. Naturally, the type of storage will vary from area to area. In the more arid sections where little rainfall comes during the fall and early winter, a good stack of loose or baled hay may provide entirely satisfactory storage. On the other hand, in high-rainfall areas, more expensive waterproof storage should be provided. At and between these two extremes, hay may be and is successfully stored in many different ways in different sections of the country.

In the West, a considerable amount of hay is chopped (either at the time of gathering from the windrow or adjacent to the stack) and stack stored. Most such stacks are round, and are built by sliding a snow fence toward the top as the stack is built. Generally these stacks are rounded off at the top and left uncovered. The advantages claimed for stack storage of chopped hay are (1) minimum labor in haymaking, (2) minimum stack storage space and spoilage, and (3) ease of feeding.

- **Round bale losses/storage**—Prior to 1970, most hay was packaged in rectangular bales, and in humid areas it was stored inside. With the advent of large round bales weighing 850 to 2,000 lb, most of them were left uncovered in the field or in a fence row.

The nutritional losses from bales stored outside and unprotected depend on the storage area (it should be high and dry), the amount of rainfall during the storage period, the hay type and condition when baled, the bale shape and density, and the method of feeding; and the monetary value of the losses depends on the price of hay and the nutritional loss. Legume hays do not form as tight a thatch as grass hays; consequently, they are not as weather-resistant.

Not all the deterioration that occurs is the result of rain falling on the bale. Moisture movement at the bottom of the bale, where it contacts the ground, can also cause considerable loss. Dry matter losses can be reduced by as much as 10% if the bales are stored on high ground on a well-drained site; set on racks or pallets, fence posts, railroad

Fig. 8-11. Round bales can be fed with a minimal of waste if proper equipment is used such as this round bale feeder for sheep. (Courtesy, Shalom Valley Sheep Equipment, Inc., Gary, S.D.)

ties, or a 3-in. base of rock; and spaced 12 to 28 in. between bales.

Large round bales stored outside and covered with plastic or canvas bonnets (or caps), sleeves, or bale bags sustain much less loss than unprotected bales.

Also, hay losses vary with the method of feeding. When animals have free access to hay, they trample and stand on it, with the result that losses may be as high as 40 to 50%. By using a barrier (a feeding rack, panels, feeding wagons, or gates), losses can be cut to 5 to 10%.

ADDITIVES FOR HAY

Farmers in many countries of the world have traditionally added about 20 lb of salt per ton of new hay at the time of stacking or putting it into the hay mow in the belief that the salt would prevent the hay from molding and heating. Carefully controlled experiments have failed to substantiate claims that salt will prevent excess heating or sweating; nor has it prevented spontaneous combustion of hay. However, when salt is used in moderate amounts, it may improve the color, aroma, and palatability of poor-quality hay. It is recognized, too, that much higher levels of salt—quantities sufficiently high to harm animals—may prevent mold.

Over the years a number of products, both liquids and powders, said to preserve hay have become commercially available. While claims have been made that there will be no heating or molding when these materials are used, the results have been highly variable.

Recent studies of several state experiment stations have shown that propionic acid or a combination of propionic and other organic acids can be used successfully to preserve baled or stacked hay stored at 35% or less moisture. Anhydrous ammonia and ammonium isobutyrate have also been found to be effective in preventing heating and preserving the quality of high-moisture hay. (See earlier section in this chapter on "Considering Chemical Conditioning and Preserving Agents.")

SPONTANEOUS COMBUSTION

Wet hay ferments and generates heat. Sometimes this results in spontaneous combustion and fire, usually about a month to 6 weeks after storing. Here are the facts:

1. **Symptoms of heating.** The warning signals are hay that feels hot to the hands, strong burning odor, and visible vapor.

2. **Temperature of hot hay.** Hot spots may be located by probing the hay with a steel rod. Then the temperature of the hot spots may be tested with a thermometer (a dairy thermometer or other type) attached to a wire and dropped down a pipe. If the hay is over 140°F, it should be checked periodically during the day. If the hay is 160°F, it should be checked hourly. If the hay is 180°F, there are apt to be fire pockets, and it should be removed from a barn.

3. **Cooling hay.** Hay that is heating may be cooled by discharging through pipes into the hot areas either dry ice or liquid carbon dioxide.

4. **Removing hot hay.** When a fire is imminent and hot hay must be removed, it is important to have plenty of help on hand including the fire department. Then the hay should be removed cautiously and without wetting unless necessary.

5. **Precautions.** Never walk on hay that is heating—place planks over it, and do not breathe hot and noxious fumes.

BUYING AND SELLING HAY

Historically, most hay has been fed on the farms where it was produced. But this practice is changing. Today, about 29 million tons, or about 25% of the U.S. production, with a cash value of over $2 billion, is sold off the farm.

Hay Sources

The five basic sources of hay in the United States are:
1. Hay dealers or brokers.
2. Neighbor to neighbor.
3. Associations of cooperatives.
4. Auctions.
5. Contract.

How Hay is Priced

Traditionally, most hay has been bought and sold by using the ancient art of bartering, based on visual evaluation.

Several states have long had programs for selling and buying hay on the basis of analysis, usually involving crude protein and moisture.

But there is need to eliminate the confusion caused by variation between states in the chemical components considered and the laboratory bases of determining them. The goal is a uniform, nationwide system that transcends state boundaries and is based on chemical analysis and visual appraisal that will project the feeding value of hay.

Futuristic Hay Evaluating and Pricing

Recognizing the problems of hay marketing, the American Forage and Grassland Council (AFGC) formed, in 1972, a Hay Marketing Task Force, made up of representatives of the AFGC, industry, and the National Hay Association, and charged them with the following primary responsibilities: (1) to develop improved measurements for reflecting the feeding value of hay, and (2) to develop improved standards (grades) for marketing hay.

The first challenge for the committee was to research and select the best practical methods of chemical analysis suitable for the widest range of hay species.

• **Proposed hay standards (grades)**—The committee developed new proposed hay standards, including five hay grades and one sample grade for all legumes and grasses.

But, for various reasons, these proposed standards were not accepted by the Federal Grain Inspection Service.

However, nine states, including Minnesota, Oregon, and Wisconsin, did implement the proposed AFGC standards in their hay marketing systems. (See Table 8-11 for the Wisconsin adaptation.) Note that Table 8-11 lists seven grades, and that the grades are based on maturity and species composition. Note, too, that legumes rank highest in the chemical components used in the grades, followed by legume-grass mixtures, grasses, and heavily weathered forage.

TABLE 8-11
MARKET HAY GRADES FOR LEGUMES, LEGUME-GRASS MIXTURES, AND GRASSES[1]

Grade	Description[2]	CP (%)	NDF (%)	ADF (%)	DDM (%)	DMI[3] gm/Wkg$^{0.75}$	DDMI gm/Wkg$^{0.75}$	RFV (%)
Prime	Leg.-Pre.Bl.	>19	<39	<30	>65.5	>143	>93.5	>143
1	Leg.-EBl., 20% grass-V.	17-19	40-46	31-35	62-65	134-143	82-93	126-143
2	Leg.-MBl., 30% grass-EH.	14-16	47-53	36-40	58-61	128-133	74-81	113-126
3	Leg.-FBl., 40% grass-Head	11-13	53-60	40-42	56-57	113-127	64-73[4]	97-113[4]
4	Leg.-FBl., 50% grass-Head	8-10	61-65	43-45	53-55	106-112	55-63	86-97
Fair	Grass-head, and/or rain damaged	<8	>65	>46	<53	<105	<55	<86
6	Sample	—	—	—	—	—	—	—

[1]Description and DDM adopted by National Alfalfa Hay Quality Committee.

[2]Prebloom, EBl. = early bloom, MBl. = midbloom, FBl. = mid- to full bloom, V. = vegetative, EH. = early head.

[3]DMI for sheep and goats => 82, 76-81, 72-75, 63-71, 52-62, and <56 for grades Prime through Fair, respectively.

[4]Reference hay mid- to full bloom alfalfa (Lema and Kawas and Jorgensen) DDM = 54.2, DMI = 120.2 gm/Wk$^{0.75}$, DDMI = 65.2 gm/Wk$^{0.75}$, and RFV = 100%.

In addition to leading to the development of the proposed hay grades listed in Table 8-11 the earlier effort of the Hay Marketing Task Force prompted the formation of the National Alfalfa Hay Quality Committee and the development of national alfalfa hay quality standards, with the intent of having uniform testing throughout the United States so that both buyer and seller can receive accurate and interpretable results on any given lot of hay. While these standards are based on a minimum test of alfalfa hay for dry matter (DM), crude protein (CP), acid detergent fiber (ADF), and estimated digestible dry matter (EDDM) calculated from the ADF, provision is also included for a description sheet of visual characteristics.

Procedures for the national alfalfa hay quality standards are carefully spelled out and include: (1) specific sampling procedures using an improved hay core sampler, (2) suggested visual factors to be used with chemical analysis to describe the sample, (3) approved testing procedures using either NIRS or wet chemistry, (4) development of an acceptance of acid detergent fiber (ADF) to predict estimated digestible dry matter, and (5) a voluntary laboratory certification program operated by the National Alfalfa Hay Test Association in conjunction with AFGC and the National Hay Association. See Table 8-12 for the estimated digestibility, intake, and relative feeding values of the grades given in Table 8-11.

TABLE 8-12
ESTIMATED DIGESTIBILITY, INTAKE AND RELATIVE FEED VALUES FOR THE MARKET HAY GRADES GIVEN IN TABLE 8-22

Grade	DDM[1] (%)	DMI[2] Sheep g/kgW$^{0.75}$	DMI[3] Cattle g/kgW$^{0.75}$	DDMI[4] Maintenance g/kgW$^{0.75}$	DDMI[5] 3 × Maint. g/kgW$^{0.75}$	Relative Feed Value[6] (%)
Prime	>65	>82	>143	>93	>89	>132
1	62-64	76-81	133-142	83-92	78-88	118-132
2	58-61	72-75	122-132	71-82	67-77	101-117
3	56-57	63-71	110-122	62-70	59-66	88-100
4	54-55	56-62	98-109	53-61	50-58	75-87
5	<53	<56	<98	<52	<49	<75
6	—	—	—	—	—	—

[1]DDM = digestible dry matter, *in vivo*, = 102 + .008 CP% - .382 ADF% - 4.63√ADF%.

[2]DMI = dry matter intake sheep = 96.4 - .0003 CP% - .0482 NDF% - .0085 NDF%^{2}.

[3]DMI = dry matter intake cattle = (DMI sheep × 1.75).

[4]DDMI = digestible dry matter intake cattle at maintenance level of intake = (DDM × DMI cattle) ÷ 100.

[5]DDMI for dairy cattle based on intake 3 × maintenance = (DDM × .95) • (DMI cattle) ÷ 100.

[6]Relative feed value = DDMI maintenance × 1.6.

Freedom From Toxic Residues

With the emphasis on residues in foods, it is important that hay be free from those residues which are prohibited. If meat or milk (or products derived from them) are found to have residues, the blame cannot be shifted to the hay grower by the livestock producer, unless there is a clear-cut case of fraudulent representation. The best assurance of freedom from such residue rests with the integrity of the hay growers, or those who represent them in selling their hay.

(Also see Chapter 5, Nutritional Disorders/Toxins, section on "Agricultural Chemicals and Drugs.")

Hay Shrinkage

Hay buyers should figure hay shrinkage closely. Here's why: If a ton of hay containing 90% dry matter is bought for $75, 1,800 lb of dry matter have been purchased at this price. However, if the $75 per ton hay contains 80% dry matter, only 1,600 lb of dry matter have been purchased for this same price. Purchase of the high-moisture, 80% dry matter hay has resulted in a loss of 200 lb of hay, or $\frac{1}{9}$ of the dry matter, worth $8.33 ($\frac{1}{9}$ × $75). Thus, if the 90% dry matter hay is worth $75 per ton, then the 80% dry matter hay is worth only $66.67 ($75 − $8.33) per ton. If 1,000 tons of hay are involved, that's a loss of $8,330 (1,000 × $8.33).

In addition to moisture losses, newly harvested hay may be expected to lose about 5% weight from going through the sweat.

Hidden Hay Costs

When buying and feeding hay, all costs and losses—handling, shrinkage and wastage, grinding costs and losses, insurance, interest, and storage—in getting it from the point of purchase to the feed bunk should be taken into consideration.

For example, producers should apply their own cost figures to baled hay and pellets in order to determine which is the better buy for them. The important thing is that all costs be accounted for—that in computing the cost of baled hay there be added such hidden costs and losses as handling, shrinkage and wastage, grinding costs and losses, insurance, interest, and storage. Additionally, allowance should be made for the added feeding value of the pellets. Also, the age and grade of cattle, other available feeds and prices, and starter vs finishing rations must be considered.

This same method of cost analysis can be applied to hay cubes, or wafers, or to any other alternate method of haymaking.

HAY FEEDING FUNDAMENTALS

Feeding is the end of the line for hay. No matter how carefully it has been grown, harvested, and stored, all that has gone before can be dissipated if it is improperly fed—unless hay feeding fundamentals are observed.

Monogastric animals, including swine and poultry, must eat a large percentage of grains and other concentrates and depend almost entirely on digestive enzymes to break down these compounds. But ruminants, with their four stomach compartments and the help of microorganisms, can subsist largely, or entirely, on bulky, high-fiber forages which, because of their low energy per unit weight of dry matter, must be consumed in large quantities to supply their nutrient needs. The horse, because of its greatly enlarged cecum and large intestine, can utilize quantities of hay intermediate between simple-stomached and ruminant animals.

The economics of the situation—the relative price of forage and grain—call for greater emphasis on forage accompanied by less grain feeding. With greater quantities of forage incorporated in rations, it is expected that performance—the production of meat and milk—will decrease. However, maximum net returns, rather than just maximum production, will be the primary objective.

Increasingly, forage testing will be used in two ways: (1) to purchase hay on a quality basis, and (2) to balance rations more precisely.

Maximum Use of Homegrown Forages

Although increased quantities of hay will be bought, especially by large dairies and feedlots, most forages will continue to be homegrown for the following reasons: (1) Hay is bulky and costly to transport, unless it is cubed or pelleted; (2) hay crops have an important place in most crop rotations; and (3) hay crops can be grown on wasteland not suited to other crop production.

Legume Vs Grass Hays

With the higher cost of both protein supplements and grains, legume hays have a particularly valuable place in rations. Legumes are also a rich source of many minerals and vitamins.

Hay Preparation

Hay is fed as long hay or in processed form. The common methods of processing are chopping, grinding, cubing, and pelleting.

Considerable hay is chopped in the West, for two reasons: (1) It facilitates handling, and (2) it lessens refusal and waste. Low-quality and coarse forages usually benefit more from chopping than high-quality forages.

Hay is usually finely ground when it is incorporated in mixed swine and poultry rations. Fine grinding is not desirable for ruminants; it results in reduced rumen acetate production and lower milk fat percentage.

Both cubing and pelleting (1) make automatic hay feeding feasible, (2) decrease nutrient losses, and (3) eliminate dust. Also, they narrow the spread between high- and low-quality forage; that is, the poorer the quality of the forage, the greater the advantage from cubing or pelleting. This is so because such preparation assures complete consumption. Also, cubing or pelleting, especially the latter, usually speeds up the passage of forage through the digestive system.

On the average, cattle fed high-roughage (above 80% roughage) or all-roughage rations will eat about one-third more pellets than long or chopped hay (due to increased density and more rapid passage through the digestive tract), make about ½ to ¾ lb faster daily gains, and require 200 to 250 lb less feed per 100 lb of gain. Also, it is recognized that the utilization of low-quality roughages is improved most by pelleting.

Cubes offer most of the advantages of pelleted forages, with few of the disadvantages. They alleviate fine grinding, and they facilitate automation in both haymaking and feeding, and they lower milk fat percentage only slightly, if at all.

Complete pelleted rations—in which the hay and grain are combined, then pelleted—are finding an increasing place for horses, and perhaps swine. Among the virtues ascribed to all-pelleted rations are (1) They prevent selective eating—if properly formulated, each mouthful is a balanced diet; (2) they alleviate waste; (3) they eliminate dust (thereby lessening heaves in horses); (4) they lessen labor and equipment; and (5) they lessen storage. (Also see Chapter 14, Feed Processing.)

Hay Feeding Systems

Hay may either be (1) self-fed, or (2) limited-fed.

Most hay is self-fed. With a manger full of hay in front of them, hay consumption is limited only by the capacity of animals—by the amount that they can hold. That's the reason that ruminants eat more cubes and pellets than long hay of equal quality.

Limited feeding of hay is accomplished either (1) by hand-feeding the hay and the concentrate allowances, or (2) by using a complete, mixed ration.

The vast majority of large cattle and sheep feedlots feed complete rations, in which the quantity of hay is limited. Also, an increasing number of large commercial dairies are switching to complete rations. Most experiments and experiences have not shown any difference between mixed rations and the feeding of roughage and concentrates separately insofar as rate and efficiency of gain are concerned. However, a mixed ration has the following **advantages**:

1. It makes for greater efficiency in feeding and lessens sorting at the feed bunk.
2. When the roughage is relatively unpalatable, a mixed ration forces consumption.
3. When concentrate consumption is to be limited, mixing with the roughage is desirable.

Feeding Hay Packages

Large round bales, large rectangular bales, and stacks are other alternatives to conventional, 60- to 140-lb, rectangular bales. There is little doubt that in many livestock operations large hay packages can greatly decrease labor and result in similar animal performance. However, special attention needs to be paid to methods of feeding these big hay packages; otherwise, waste can easily wipe out any saving in labor.

Under an in-field storage system, the bales are dropped where they are made, and remain there until needed for fall or winter grazing. Little or no labor is involved in making hay except for mowing, windrowing, and baling; and there is no manure to haul. Cattle graze the grass growth which occurs subsequent to baling and consume the bales in the field. There is little or no labor in feeding. The cattle go to the feed, rather than necessitating that the feed be taken to them.

The following methods are used in grazing round bales or small stacks, along with regrowth in the field:

1. **Continuous access to all bales.** In this system, the cattle are given continuous access to all the bales in a field.

2. **Strip grazing bales and stacks.** In this system, the cattle are given access only to those bales or stacks which are to be consumed within a given period of time. This is accomplished by using an electric fence and cross-stripping the field containing the bales or stacks. Such a strip-grazing program will increase the number of cattle days by at least 35%.

3. **Other feeding methods.** Other systems require more investment than in-field grazing of bales or small stacks, but the additional numbers of cattle carried per acre may justify the increased cost. Among such systems are the following:

    a. Hay packages (large round bales or small stacks) placed in rows and grazed with an electric fence.
    b. Portable feeding gates (fences).
    c. Feeding wagons.
    d. Three-sided or round feeders.

On a per cow basis, 18 to 24 in. of hay feeding space per cow should be allowed if all animals are to eat at the same time. Six to ten inches per cow will suffice when hay is available at all times.

BALING WIRE DANGER

Short pieces of baling wire are dangerous to cattle. If consumed, they are likely to pierce the stomach and damage the heart, probably killing the animal. Among cattle raisers, this condition is known as *hardware disease;* among veterinarians, it is more technically known as *traumatic pericarditis.* Because of this hazard, it is urged (1) that bales of hay and straw be broken by pulling the whole wire off rather than by cutting, and (2) that all used baling wire be carefully folded and placed in a barrel (and later disposed of) rather than left on the ground and allowed to get mixed with either the feed or the bedding.

Hay Feeding Schedule

In general, animals may be given as much nonlegume hay as they will consume, regardless of their previous ration. Thus, feedlot cattle and lambs are usually started with a full feeding of grass hay. Then, as grain is increased, the consumption of hay is decreased. With legume hay, however, it is necessary that they be gradually accustomed to it; otherwise, looseness and scouring will likely result.

Proportion of Hay to Concentrate

Cattle, sheep, and horses will eat 2 to 3 lb of hay per 100 lb of body weight if fed hay alone. Also, it is noteworthy that the higher the quality of the hay, the more of it they will eat, with the result that the grain requirement will be lessened.

The economics of the situation—the comparative price and quality of hay and concentrate—along with the management practices, will determine the proportion of hay to concentrate. Thus, during the period of low grain prices in relation to forage prices—from about 1950 to 1970—it was desirable to feed finishing cattle high-energy rations and to maximize gains. But the grain-fed cattle binge ended with the world grain shortages and high-priced grains of the early 1970s. In the years ahead, with grain becoming more scarce and higher in price than forages, comparatively speaking, more forage and less grain will be fed to finishing cattle, and net returns will be more important than high rate of gain. Cattle and sheep will increasingly be *roughage burners*. Livestock producers will rely upon the ability of the ruminant to convert coarse forage, grass, and by-product feeds, along with a minimum of concentrate, into palatable and nutritious food for human consumption, thereby competing less for humanly edible grains. Increasingly, the steer and the lamb of tomorrow will be produced on a maximum of milk and grass and a minimum of grain. More and more U.S. cereal grains will be used for human food, just as has been true, historically, in much of the rest of the world.

Ruminants can make the transition to more roughage with ease. For them, it is merely a return to nature, for they evolved as consumers of forage.

The best buy (hay vs grain) may be determined by calculating the cost per pound of TDN and of protein in the hay and grain being compared. Then, the proportion of hay to concentrate can be varied accordingly. If hay is the best nutrient buy, feed more hay and less grain. On the other hand, if grain is the best buy, feed more grain and less hay.

Ruminants Need Hay

Ruminants need some roughage. Hay fed early in life will develop the calf's rumen.

Many serious problems in high-producing dairy herds have been traced to lack of hay in the ration; among them, increased incidence of ketosis and displaced abomasums. In addition to these maladies attributable to no-hay rations, it is noteworthy that milk fat percent can be as much as 1% less on an all-silage and concentrate ration. The explanation: The amount and length of hay in the ration affects cud chewing time and percent milk fat. Cud chewing time is the biological response that indicates several desirable factors—rumen muscle tone and function, the supply of saliva going into the rumen to buffer rumen acids, and rumen fermentation with the production of acetic acid. It is estimated that about 15 minutes of cud chewing time is needed per pound of dry matter eaten.

Finishing cattle also require a minimum amount of roughage factor for normal rumen function. Feedlot cattle fed all-concentrate rations show a response to small amounts of roughage factor. As little as 1 lb per day of some low-quality roughage, such as alfalfa stems, rice hulls, or sorghum straw, has given improvement in efficiency ranging up to 15% of all-concentrate rations. To meet the *roughage factor* need, different roughage substitutes have been developed and used with varying degrees of success.

Different Qualities of Hay May Be Used

The type of ration which will be least costly and result in satisfactory performance will differ according to species, level of performance, reproductive status, age, etc. For example, the nutritive needs of a dry, pregnant beef cow are much lower than those of a high-producing dairy cow. Thus, a low-quality hay may be quite satisfactory for wintering a beef cow without calf at side, whereas high-producing,

lactating cows should always have high-quality hay. Also, high-quality hay is important for swine and poultry. Where forage is incorporated in monogastric rations, high-quality dehydrated alfalfa is most commonly used. High-quality hay is also essential for horses.

Hay Waste and Refusal

In a recent Texas study involving 10 different hay feeding racks, the cows at the best conventional feeder still wasted 14% of their hay. Dairy producers have commonly accepted 10% refusal as normal.

High-priced feeds and smaller margins are causing livestock producers to scrutinize hay losses, and to do something about them. Chopping hay and/or adding molasses will lessen wastage. But feeding high-quality hay is the best way in which to lessen waste and refusal.

Supplementing the Hay Ration

Hay is generally lower in energy and higher in fiber than most grains and concentrates. Legume hays have a high-calcium content, but they vary in available phosphorus. If sun-cured properly, they are high in carotene and vitamin D, along with many of the B vitamins. A supplement should supply the nutrients that are most likely to be lacking in the hay; thus, supplements for alfalfa hay should be high in energy and phosphorus and low in fiber. Also, carotene or vitamin A should be provided if the hay has been bleached or turned brown. Salt is lacking in hays and other natural feedstuffs and should be provided as a supplement, along with trace minerals that are deficient in the local area.

Stretching the Hay Supply

When hay is scarce and high in price, the supply of it for ruminants and horses may be stretched. As the amount of hay fed is reduced, it must be replaced with other feeds so that the total ration is still balanced and fulfills all the nutrient requirements.

Most grains contain 75 to 80% TDN, while most medium- to good-quality hays contain 45 to 50% TDN. Hence, as a general rule of thumb, about 5 lb of grain equal 8 lb of hay, provided they are of comparable quality. Thus, it follows that if corn can be bought for $100.00 per ton, hay should be bought at $62.50 per ton, or less.

If the price of hay is less than five-eighths the price of grain, relatively more of it should be fed; whereas, if the price of hay is higher than this, relatively more grain will make for cheaper production.

When hay is scarce and high in price, the hay supply for ruminants and horses can be stretched as follows:

1. Feed only ½ to ⅔ the normal ration of hay, but be on the alert for digestive disturbances. With lactating cows, too little hay or other forage, or hay that is chopped less than 0.25 in. in length, will result in low milk fat test and cow

health problems. A good rule of thumb for the dairy producer to follow is to feed at least 1.5% of the body weight of the cow daily as hay, or hay equivalent from other forages. For a 1,400-lb cow, this calls for a minimum of 21 lb of hay or hay equivalent. Under conditions of extreme shortages, hay consumption can probably be lowered to the 5 to 10 lb per cow range, but there is not general agreement on the minimum amount of hay necessary to maintain maximum feed intake and milk production.

2. Replace 1 lb of hay with 3 lb of silage. The basis for this substitution: hay is usually 90% dry matter, whereas silage runs about 30% dry matter (70% moisture); hence, 90% ÷ 30% = 3.

3. Replace 1 lb of hay with 4 lb of green chop. The basis for this substitution: green chop generally runs 22.5% dry matter (77.5% moisture); hence, 90% ÷ 22.5% = 4.

4. Replace each 2 lb of hay deleted with 1 lb of grain.

5. Make the maximum use of such feeds as cottonseed hulls, corncobs, straw, and grass aftermath in the ration for (a) all but 5% of the alfalfa (or other legume) hay of grower rations; and (b) all of the "hottest" finishing ration, adding such supplementary proteins, minerals, and vitamins as necessary to balance the ration.

6. Get finishing cattle and lambs, and animals being fitted for show or sale, on high-concentrate rations as expeditiously as possible. In cattle, eliminate the stocker feeding period—get weaned calves on full feed as quickly as possible.

7. Provide such supplementary proteins, minerals, and vitamins as necessary. This is especially important with gestating-lactating females or young, growing animals. This may be accomplished by (a) feeding some legume, either hay or silage, and/or (b) adding suitable protein, mineral, and vitamin supplements. For example, pregnant cows that are in medium to good condition can be wintered satisfactorily on 12 to 20 lb of straw or other low-quality roughage, plus 1 lb of oilseed cake or meal (or equivalent protein supplement), or on straw plus 4 to 5 lb of alfalfa or other legume hay. Unless cows have had good green pasture in the fall, and consequently have a store of vitamin A in their bodies, alfalfa pellets or a vitamin A supplement should be fed. In addition, straw-fed cattle should always have access to a mineral supplement high in calcium.

QUESTIONS FOR STUDY AND DISCUSSION

1. Why is making hay the most common method of preserving forage for storage?

2. Discuss the history of haymaking.

3. Tell of the magnitude and importance of the U.S. hay crop.

4. Discuss hay from each of the following standpoints:
   a. As an energy source.
   b. Camparable value to corn grain and corn silage.
   c. As a grain replacement.

5. Why is alfalfa known as "Queen of the Forages"?

6. What factors should be considered when determining the kind of forage to grow?

7. For hay to be of superior quality, it must be high in four factors. What are these factors?

8. Cite experimental evidence of the importance of hay quality.

9. What visual/sensory factors indicate high quality hay? Why should visual inspection supplement chemical analyses?

10. Discuss the per acre yield of crude protein and cellulose from first cutting alfalfa harvested at different stages of maturity.

11. Discuss the relationship between chemical composition of hay and its feeding value, palatability, and consumption as a result of (a) maturity and (b) weathering.

12. What is the difference between neutral detergent fiber (NDF) and acid detergent fiber (ADF)? In what ways are they superior to proximate analysis?

13. What is near infrared reflectance spectroscopy (NIRS)? How is it used in forage analyses? What are its advantages and disadvantages?

14. Discuss each of the following points with reference to determining hay quality by chemical composition: (a) proper sampling, (b) what tests to make, (c) evaluating the test results, and (d) using the results in buying hay and balancing rations.

15. List and discuss each of the steps in growing quality hay.

16. Give a brief rundown on today's haying equipment.

17. Describe the proper forage harvesting stage of alfalfa. How does harvesting stage of alfalfa affect chemical composition, gains of yearling steers, milk production, and lamb gains?

18. Why are plant food reserves important? Usually, the closer to maturity that plants are cut or grazed, the higher the stored food reserves. However, harvesting at a young growth stage makes for high feed value. So, when should alfalfa be harvested during the spring, the summer, and the fall?

19. Discuss modern hay cutting, curing in the swath and windrow, and raking or cocking.

20. Discuss reducing moisture content, shattering, bleaching and fermenting in modern haymaking.

21. Discuss the effect of rain damage on *in vitro* dry matter disappearance (IVDMD) and on crude protein (CP) of alfalfa and red clover.

22. What chemical hay drying agents and preservatives are being used? How and why are they being used?

23. Is there any basis for livestock producers favoring a certain cutting of alfalfa hay?

24. What products may be used in the nitrogenation of low-quality hays? What are the advantages and hazards of such treatments?

(Continued)

25. Describe and discuss the place of each of the following haymaking systems: long hay, chopped hay, packaged hay, and artificial drying.

26. What prompted the development of big round bales, large rectangular bales, and small mechanical stacks?

27. How do cubes and pellets compare?

28. How may the losses of large round bales due to weather and feeding be lessened?

29. What is meant by spontaneous combustion? What causes it in hay?

30. Discuss each of the following factors pertaining to buying and selling hay: (a) the amount, proportion, and value of the U.S. hay crop sold off the farm where it is produced, (b) hay sources, and (c) how hay was priced prior to the development of the near infrared reflectance spectroscopy.

31. The goal in hay pricing is a uniform, nationwide system based on chemical analysis and visual appraisal that will project the feeding value of hay. Is such a goal realistic and achievable?

32. What are the recommendations of the American Forage and Grassland Council "Hay Marketing Task Force" and the National Alfalfa Hay Test Association relative to (a) determining hay quality by infrared (NIRS) or wet chemistry, along with visual inspection; and (b) hay standards (grades)?

33. When buying and selling hay, of what importance are the following: (a) freedom from toxic residues, (b) hay shrinkage, and (c) hidden hay costs?

34. Discuss (a) hay preparation, (b) hay feeding systems, and (c) feeding hay packages.

35. What factors determine the proportion of hay to concentrate in a ration?

36. Cite proof of the following statement: Ruminants need hay. What desirable factors does cud chewing time indicate? How much chewing time is needed?

37. For what animals may low-quality hay be used? For what animals is high-quality hay essential?

38. Should livestock producers accept a 10% hay waste and refusal as normal? If not, what can they do to lessen it?

39. When hay is high in price and scarce, as may happen after a long, hard winter and much snow in the northern part of the United States or during a severe drought, how can producers *stretch* their hay supply?

Fig. 8–12. "Winter in Wyoming" from an original painting by artist Tom Phillips, 3333 17th St., San Francisco, Calif. 94110. Hay is still fed like this on some western snow-covered ranges.

SILAGE/HAYLAGE/HIGH-MOISTURE GRAIN[1]

Original painting by Tom Phillips

Contents	Page
The Ensiling Process	170
Economics/Advantages/Disadvantages of Silage	170
Silos	171
Conventional Upright (Tower) Silos	171
Gastight (Oxygen-Limiting) Silos	171
Pit Silos	172
Horizontal Silos	172
Trench Silos	172
Bunker or Self-Feeder Silos	172
Temporary Silos	172
Enclosed Stack Silos	172
Open Stack Silos	172
Modified Trench-Stack Silos	172
Plastic Silos	172
Enclosed Plastic Bag or Tube Silos	173
Round Bale Plastic Covered Silage	173
Advantages and Disadvantages of Round Bale Plastic Covered Silage	173
How to Determine the Size Silo to Build	173
Size of Tower Silo	174
Size of Trench Silo	175
Silage Storage Losses	176
Kinds of Silage	176
Corn and Sorghum Silage	176
Corn and Sorghum Residue Silage	176
Grass/Legume (Hay Crop) Silage	177
Direct-Cut Silage	177
Wilted Silage	178
Low-Moisture Silage (Haylage)	178
Corn or Sorghum Silage Vs Grass/Legume Silage	178
Other Silage Crops	178
Combining Crops for Silage	179
Rain-Damaged Hay Silage	179
Frosted Crop Silage	179
Drought Stricken Crop Silage	179
Silage Additives and Preservatives	179
Feed Additives	179
Acids	180
Fermentation Aids	180
Preservatives	180
Silage Additive and Preservative Considerations	180
Silage Additive and Preservative Recommendations	180
Harvesting Methods and Machinery	181
How to Make Good Silage	181
Harvest at Proper Stage of Maturity	181
Cut to Proper Length	181
Control the Moisture Content	182
How to Lower the Moisture Content	182
How to Increase the Moisture Content	182
How to Determine the Moisture Content	182
Fill Rapidly	182
Distribute Forage Uniformly in the Silo	182
Seal or Top-Off the Silo	183

Contents	Page
Feeding Value and Economy of Silage	183
Characteristics of Good Silage	183
Silage Pointers	183
Moldy Silage	183
Effect of Silage on Milk Odor and Flavor	183
Dangerous Silage Gases	183
High-Moisture Grain	184
Harvesting	184
Storage of High-Moisture Grain	184
Sealed (Airtight) Storage	184
Unsealed Storage	184
Preservation With an Organic Acid or Ammonia	185
Feeding Value of High-Moisture Grain	185
Reconstituted Grain	186
Buying and Selling High-Moisture Grain	186
Questions for Study and Discussion	186

Silage may be defined as fermented forage plants. It is a very old method of preserving feed. Columbus found that the American Indians used pits or trenches in which to store their grain, and, centuries earlier in the Old World, silos were used as a means of preserving both grain and green forage. The Frenchman, Auguste Goffart, preserved forage in a silo in 1865. The discovery was so highly regarded that the French government awarded him the Cross of the Legion of Honor. The first tower silo built in the United States by white man is said to have been erected by F. Morris in Maryland in 1876.

Silage making is one of the 3 common methods of utilizing forage crops, the other 2 methods being grazing and haying. Grazing is the least expensive of the 3 methods, but it is seasonal in nature. In the spring and early summer, forage plants generally grow faster than they can be utilized by normal grazing; and they become dormant in cold weather.

The surplus forage produced during the growing season may be preserved for feeding during the winter months and other periods of pasture scarcity by haymaking. But weather conditions are not always favorable to haymaking. Ensiling, on the other hand, can be done in inclement weather. Also,

---

[1] The authors gratefully acknowledge the helpful suggestions of the following authorities who reviewed this entire chapter: J. E. Baylor, Ph.D., Professor of Agronomy, Pennsylvania State University, State College; R. A. Forsberg, Ph.D., et al., Department of Agronomy, University of Wisconsin-Madison; S. C. Fransen, Ph.D., Forage Agronomist, Western Washington Research and Extension Center, Washington State University, Puyallup; D. A. Miller, Ph.D., Professor of Plant Breeding and Genetics, College of Agriculture, University of Illinois at Urbana-Champaign; and W. E. Murphy, Ph.D., Department of Agronomy, Oklahoma State University, Stillwater.

it has the added virtues of succulence and of preserving a higher proportion of the nutrients of the plant than can be accomplished in haymaking.

Silage is primarily a beef and dairy feed, where it is used as part or the only roughage in the ration. It is also a good sheep feed. Sometimes it is fed to brood sows. Very little silage is fed to horses.

The importance of silage in this country is attested to by the fact that more than 170 million tons are made annually, of which more than 100 million tons are corn silage. Further, there is ample evidence that silage making is on the increase. It is estimated that 2,000 to 3,000 silos are constructed in the United States each year,[2] with tower silos increasing more rapidly than other types. Most silage is fed on the farms or ranches on which it is produced, rather than being purchased. It is important, therefore, that livestock producers know how to produce good silage, as well as how to feed it, for most of them determine their own destiny from the standpoint of quality. For this reason, this chapter covers silage from production to feeding.

Fig. 9-1. Silage is a popular feed for lactating dairy cows. (Courtesy, Holstein-Friesian Assn. of America, Brattleboro, Vt.)

THE ENSILING PROCESS

The ensiling process refers to the changes which take place when forage or feed with sufficient moisture to cause fermentation is stored in a silo in the absence of air. Many different kinds of plants and plant products can be preserved in this way. Sauerkraut, for example, is the silage form of cabbage.

The ensiling process is governed by the interaction of three factors: (1) the chemical composition of the plant material placed in the silo, (2) the amount of air entrapped or allowed to enter the mass, and (3) the activity of the bacterial population.

The entire ensiling process requires 2 to 3 weeks, during which time the following phases of varying intensity occur:

1. **Aerobic phase (with air).** This is the respiration phase. In this phase, the living plant cells of the forage continue to respire, or take up oxygen, and the plant enzymes and aerobic bacteria use the readily available carbohydrates to produce heat, water, and carbon dioxide.

The oxygen supply is usually exhausted in 4 to 5 hours, but carbon dioxide continues to accumulate for about 48 hours; the temperature of the ensiled material increases over a period of about 15 days, but seldom exceeds 85 to 90°F, then decreases gradually. At this point, anaerobic conditions prevail.

2. **Anaerobic phase (without air).** This is the "pickling" phase. When the available oxygen of the entrapped air has been consumed, anaerobic bacteria—chiefly acid-forming and proteolytic—multiply at a prodigious rate. Simultaneously, the molds and the yeasts die, but continue in a minor way to provide enzyme systems which produce alcohol and other end products.

The combined anaerobic activity produces the following changes: (a) The complex carbohydrates and sugars (especially the sugars) are broken down into lactic acid (the acid in sour milk), some acetic acid (the acid in vinegar), and a small amount of other acids and alcohols; (b) small quantities of the proteins are broken down into ammonia, amino acids, amines, and amides; and (c) the acidity finally reaches a point when the bacteria themselves are killed, and the silage-making process is completed. At this stage, ideally the lactic acid concentration is equivalent to 4 to 10% of the dry matter.

3. **Stable phase.** When a pH of 4.2 or less is reached,[3] silage is stable and may be kept for years if air is excluded. Drier silages stabilize at a higher pH.

• **Browning reaction/Maillard reaction**—Lowering of the moisture content without excluding air may lead to undesirable effects on silage. It may cause high temperature in the silo and damage the protein and energy value of low-moisture silage as a result of the nonenzymatic browning reaction, also termed the *Maillard reaction,* evidenced by a tobacco-brown or black color and a caramelized or tobacco odor. In this reaction, heat causes carbohydrates to combine with protein to produce an insoluble product, decreasing the digestibility of both protein and energy sources. The loss of feeding value depends on the degree of the heat damage. Determination of heat damage to protein can be assessed by determining the residual insoluble N in acid detergent fiber.

ECONOMICS/ADVANTAGES/DISADVANTAGES OF SILAGE

Storing feed as silage often makes it possible to get more forage preserved from fewer acres. As a result, there is evidence that silage making in the United States is increasing. During the past 20 years, the total tonnage of corn silage, which comprises about ¾ of all the silage fed in the United States, has increased by approximately 25%. Corn, sorghum, and grass/legume silage production have benefited from improved technology, including higher yielding silage crops, improved harvesting and handling machinery, and lower

---

[2]Estimate made by International Silo Assn., Inc., 410 North Michigan Ave., Chicago, IL 60611, in a personal communication to the senior author. The 2,000 to 3,000 estimate includes moved and rebuilt structures and is exclusive of baled and bagged silage.

[3]The pH refers to the degree of acidity of the silage. Acidity or alkalinity is measured on a scale of 0 to 14, with 7 being neutral. Numbers that are below 7 indicate an acid condition, with the degree of acidity increasing as the numbers get smaller. Numbers that are above 7 indicate an alkaline condition, with the degree of alkalinity increasing as the numbers get larger. The preservation of silage involves only pH values that are on the acid side of the scale.

cost of storage. But silage has many other **advantages;** among them, the following:

1. It makes it possible to increase the livestock carrying capacity of a farm or ranch. Thus, corn, the chief U.S. silage crop, (a) yields more total digestible nutrients per acre than most other forage crops, and (b) has 30 to 50% higher feeding value as silage than when fed as grain and stover.

2. It retains a higher proportion of the nutrients of plants than can be accomplished by haymaking, even if the weather is satisfactory for the latter, chiefly because shattering and bleaching losses are held to a minimum. Thus, ensiling grass preserves 85% or more of the feed value of the crop, whereas haymaking under the best of conditions will preserve only 80%, and under poor conditions only 50 to 60%.

3. It is feasible to produce top quality hay crop silage during times of inclement weather when it would normally be impossible to cure the forage crop properly as hay.

4. It is the most economical form in which the whole stalk of corn or sorghum can be processed and stored.

5. It requires less storage space per pound of dry matter than dry hay, even when the latter is baled or chopped. A cubic foot of silage contains about 3 times more dry weight of feed than a cubic foot of long hay stored in the mow.

6. It practically eliminates the danger of loss by fire if stored within the recommended moisture range.

7. It is the most satisfactory and economical way in which to preserve a number of by-product feeds.

8. It makes it possible to remove forage crops from the land earlier than would otherwise be possible.

9. It is one of the best methods of controlling the European corn borer since the removal of cornstalks is required in making silage.

10. It helps to control weeds, which are often spread through hay or fodder.

11. It is the cheapest form in which a good succulent winter feed can be provided on most farms and ranches.

12. It is a better source of protein and of certain vitamins, especially carotene, and perhaps some of the unknown factors, than dried forage.

13. It is a very palatable feed and slightly laxative in nature.

14. It makes for less waste, the entire plant being consumed, which is an important consideration with coarse, stemmy forages.

15. It is without a peer from the standpoint of longtime storage, holding the feeding value of protein, carbohydrates, and carotene better than any other method of preservation, and providing a desirable backlog against drought or any other crop failure.

16. It may be completely mechanized as a feeding system, thereby eliminating much labor and time.

17. It offers many advantages over pasture, including (a) no fencing required, (b) approximately one-third more forage from the same acreage, (c) harvesting at optimum maturity, (d) more uniform quality, (e) little or no bloat, (f) closer observation of animals that are confined to a lot or corral, (g) reduced damage to the growing sward, and (h) lessened topsoil loss as a result of alleviating the hoof action of grazing animals.

Some of the **disadvantages** of silage are:

1. It requires a silo or storage structure and other special equipment, for best results. In comparison with the simpler methods of field curing and storing hay, this may mean higher costs for the small operator.

2. It contains considerably less vitamin D than sun-cured hay.

3. It necessitates that 2 to 3 times as much tonnage be handled as when the same forage is dried for hay, due to the high water content.

4. It incurs an added expenditure when preservatives are necessary.

5. It lessens the amount of organic material returned to the soil, which is needed in some soil types.

SILOS

Silos may be classified according to the five basic methods used for processing forages. Each method is associated with the shape and material of the structure, which also influences the efficiency of preserving the silage. Also, the different shaped structures are adapted to different methods of filling and unloading. Within each classification there are many variations of each type depending upon the manufacturer.

The kind of silo and the choice of construction material should be determined primarily by economics and the suitability to the particular needs of the farm or ranch.

Silos may be classified as follows:

I. Conventional Upright (Tower) Silos
   1. Concrete stave
   2. Galvanized steel
   3. Wood stave
   4. Monolithic concrete (poured in place)
   5. Tile block
   6. Brick
II. Gastight (Oxygen-Limiting) Silos
   1. Glass-lined structures
   2. Concrete stave
   3. Galvanized steel
   4. Monolithic concrete
III. Pit Silos
IV. Horizontal Silos
   1. Trench silos (belowground level)
   2. Bunker silos (aboveground level)
V. Temporary Silos
   1. Enclosed stack silos
   2. Open stack silos
   3. Modified trench-stack silos
   4. Plastic silos

Conventional Upright (Tower) Silos

The upright or tower silo, which is sometimes referred to as the "watch tower of prosperity," is a cylinder built aboveground. Its round shape withstands pressure well and is adapted to good packing.

Gastight (Oxygen-Limiting) Silos

These silos resemble conventional tower silos, but they are more expensive because of their construction.

Sealed silos are designed for storage of wilted or even overwilted forage with as little as 40 to 55% moisture content or for the storage of high-moisture grain containing 22 to 40% moisture. These structures may be partly filled on widely separated dates, provided they are sealed between

Fig. 9-2. Upright (tower) concrete stave silos. (Courtesy, USDA)

fillings. Packing and tramping of forage is not necessary or recommended although distribution is desirable.

Practically all outside air is kept out of the oxygen-limiting silo, and carbon dioxide formed during fermentation is kept in.

Pit Silos

The pit silo is shaped like the tower silo, but inverted into the ground. It resembles a well or cistern. The walls of a pit silo may or may not be lined. Where the water table is low enough that the silo will not fill with water, such as in semiarid areas, the pit silo is very satisfactory.

Horizontal Silos

Only two types of horizontal silos will be discussed herein; trench silos and bunker silos (or horizontal surface silos), both of which may be adapted to self-feeding.

TRENCH SILOS

The trench silo is a horizontal, trenchlike structure that can be built quickly and at low cost. It is most popular in areas where the weather is not too severe and where there is good drainage. The walls of a trench silo may or may not be lined, but for making good silage they should always be smooth; and there may or may not be a floor. A trench silo should be wider at the top than at the bottom, and the bottom should slope away from one end in order that excess juices will drain off if material with too high moisture content is ensiled.

Trench silos have the **advantages** of (1) low initial cost, (2) low cost of filling machinery, for a blower is not necessary, (3) relative freedom from freezing, and (4) ease of construction. The chief **disadvantages** of trench silos in comparison with tower silos are the (1) larger area to seal, (2) higher spoilage losses, and (3) inconvenience in feeding during inclement weather.

BUNKER OR SELF-FEEDER SILOS

As a laborsaving measure, some operators are now constructing horizontal silos aboveground (or slightly recessed) —usually with concrete floors, and side walls of wood, concrete, or other materials—and self-feeding silage to cattle by making use of either a feeding fence or an electrified pipe suspended 30 to 48 in. from the floor of the silo.

Temporary Silos

Several kinds of above ground temporary silos are used. Generally, this kind of storage is used to meet emergencies, to supplement permanent silos, or to ensile such by-product feeds as cannery refuse, pea vines, and beet tops or pulp. Above ground temporary silos are low in cost, can be erected at short notice, require no special foundation, and can be set up on almost any level site convenient for filling and feeding.

ENCLOSED STACK SILOS

These are built entirely aboveground, without trenches or holes. They are upright, are generally circular, and are enclosed by snow or picket fences, poles, wooden staves, heavy woven wire, or other materials. Most of them are lined with tar paper, plastic, or tough fiber-reinforced paper made especially for the purpose. Because of the relatively weak walls of these silos, their height should not be greater than twice their diameter unless poles are set at four to six points around their circumference and tied together at the top.

OPEN STACK SILOS

These are similar to enclosed stack silos, except that no supports or walls are used. As would be expected, greater spoilage is encountered in the open stack than in the enclosed stack, because of the greater evaporation and spoilage which accompanies the exposed sides. Less spoilage, percentagewise, occurs in stacks of considerable size—stacks that contain 500 to 1,000 tons or more silage—than in smaller stacks.

MODIFIED TRENCH-STACK SILOS

This silo, which is intermediate between a trench and stack silo, is adapted to areas where the ground-water level is high. It is constructed by excavating a shallow trench 12 to 18 in. deep, by piling the excavated earth on either side of the trench to support the silage and to keep out surface water, by packing silage thoroughly in and over the trench to a height of 10 to 15 ft and by covering the stack with any one of the materials recommended for covering the trench silo (see trench silo). The modified trench-stack silo is designed to give greater protection and less spoilage than can be accomplished by open or enclosed stacks. Also, this type of silo is easier to feed from than a trench silo.

PLASTIC SILOS

Plastic (polyethylene) is now available for use as temporary silos, and for use as covers for trench, bunker, and tower silos, and as silo liners. If not punctured, it is nearly airtight. Plastic thicknesses range from 4 to 9 mils. The thicker grades have better tear and puncture resistance, and low permeability by both air and moisture; however, they cost more and are difficult to tie tightly. Thinner grade plastics are less costly, more pliable, and easier to seal.

The two common types of plastic silos are: (1) enclosed plastic bag or tube silos, and (2) round bale plastic covered silage.

Enclosed Plastic Bag or Tube Silos

These temporary silos are made of heavy plastic in the form of a tube into which forage is forced by a special machine (much like stuffing sausage). The machine needed to pack the tube is generally rented or owned cooperatively. The filled structure is 8 ft in diameter and about 100 ft long. Preservation of silage is excellent provided the ends are kept sealed and the plastic is not torn or damaged by rodents or other animals. To remove or self-feed the silage, the plastic is cut and folded back at one end to expose as much silage as needed each day. The plastic cannot be re-used.

Fig. 9–4. Round bale silage properly fed in a steel rack, thereby alleviating the energy for chopping and lessening the labor for feeding. (Courtesy, Sperry New Holland, New Holland, Pa.)

Fig. 9–3. Plastic tube silos. (Courtesy, Montana State University, Bozeman)

Round Bale Plastic Covered Silage

The most common methods for using plastic material to produce round bale silage are:

1. **Individual bags.** Bags come in various lengths, diameters, and thicknesses. A tractor-mounted spear device is needed to lift the bale while applying the bag. Then the bale is placed in storage position before it is tied off. If possible, the bales should be stacked in cordwood fashion to reduce exposed surface area. Then, a plastic cover over the entire stack may reduce storage damage.

2. **Plastic tubes.** These consist of several round bales stuffed by a machine into a long plastic tube which is then sealed at both ends. The filled plastic tube resembles an "Enclosed Plastic Bag or Tube Silo" described earlier, except that it consists of a row of round bales covered with plastic rather than long, continuous sausage-type silage material. Plastic tubes can be effective and timesaving, but the multiple bales stored in one package tend to increase the loss if the bag is torn, punctured, or opened for feeding. However, the tube can be easily tied off into one-bale (or more) segments for feeding.

3. **Sheet plastic.** Several round bales can be stacked under two sheets of plastic, with the plastic ends on the ground covered with soil, sand, or other effective sealing procedure. The hazard with this type of storage is that there are more possibilities for air leaks to develop, which may result in a large number of bales being spoiled.

ADVANTAGES AND DISADVANTAGES OF ROUND BALE PLASTIC COVERED SILAGE

Round bale silage may serve as a supplement to, rather than a replacement for, other stored forages on most livestock farms. Some **advantages** are:

1. It doesn't require silo structures.
2. Hay-making equipment may be used to harvest it.
3. When silo capacity is lacking because of a surplus of forage, round bale silage can offer an effective method of storing excess forages.
4. The round bale silage system can be used to save a mowed field of hay when an anticipated rainstorm or extremely high humidity interfere with proper hay curing.
5. Round bale silage saves about one-third of the harvesting energy plus the fuel required for chopping silage.
6. Round baled silage can be self-fed if properly presented, thereby saving both labor and fuel by not requiring daily silage feeding.

But there are **disadvantages**; among them, the following:

1. Conditions associated with round bale silage are not optimum for fermentation.
2. Extreme care must be taken to eliminate air leaks.
3. The system requires prompt handling and storage of bales.
4. Machines for lifting and moving heavy, high-moisture bales must be available.
5. Plastic bags, storage tubes, or plastic sheets to cover group-stacked bales must be purchased.
6. Plastic is easily damaged, which can result in forage losses greater than in conventional silos.

How to Determine the Size Silo to Build

The size of silo to build should be determined by needs. With tower type and pit silos, this means (1) that the diameter should be determined by quantity of silage to be fed daily, and (2) that the height (depth in a pit silo) should be determined by the length of the silage feeding period. Similar consideration should be accorded trench silos.

SIZE OF TOWER SILO

If the diameter is too great, the silage will be exposed too long before it is fed; and, unless a quantity is discarded each day, spoiled silage will be fed.

The minimum recommended rate of removal of silage varies with the temperature. In most sections of the United States, it is desirable that a minimum of 1½ in. of silage be removed from tower silos daily during the winter feeding period, with the quantity increased to a minimum of 3 in. when summer feeding is practiced. Of course, the total daily silage consumption on any given farm or ranch will be determined by (1) the class and size of animals, (2) the number of animals, and (3) the rate of silage feeding. (Some suggestions on how much silage to feed are given in the chapters devoted to feeding each species of livestock.)

Silo height should be determined primarily by the length of the intended feeding period. In general, however, the height should not be less than twice, nor more than three and one-half times the diameter. The greater the depth, the greater the unit capacity. Extreme height is to be avoided because (1) of the excessive power required to elevate the cut silage material, and (2) of the heavier construction material required. Also, it is noteworthy that, with silos of large diameters, more labor is required in carrying the silage to the silo door if silage is manually removed.

Table 9-1 may be used as a guide in computing the proper diameter of tower silo for any given farm or ranch.

TABLE 9-1
MAXIMUM DIAMETER OF TOWER SILO TO BUILD IF SILAGE IS TO BE KEPT FRESH

Inches of Silage Removed Daily	Total Silage Removed Daily With An Inside Silo Diameter Of:											
	10 ft		12 ft		14 ft		16 ft		18 ft		20 ft	
	(lb)	(kg)	(lb)	(kg)	(lb)	(kg)	(lb)	(kg)	(lb)	(kg)	(lb)	(kg)
Summer: 3 in. (7.6 cm) daily will remove[1]	786	357	1,312	596	1,539	700	2,010	914	2,545	1,157	3,142	1,428
Winter: 1½ in. (3.8 cm) daily will remove[1]	393	179	656	298	770	350	1,005	457	1,272	578	1,571	714

[1]The lb(kg) listed in each of the columns are approximations based on an average constant weight of 40 lb of silage per cu ft (18.2 kg/.028 m³). Low-moisture silages to 40-55% moisture content will weigh somewhat less than 40 lb per cu ft (18.2 kg/.028 m³).

Fig. 9-5 shows capacities of tower silos of different heights and diameters. It is based on well-eared corn silage harvested in the early dent stage, cut in ¼-in. lengths, well-tramped when filled, and with the silo refilled once after settling for a day.

Fig. 9-5 can be adapted for corn silage of different stages of maturity and grain content, and for other kinds of silage, by applying the rules of thumb given in Table 9-2.

TABLE 9-2
EFFECT OF KIND OF SILAGE ON WEIGHT

Kind Of Silage	Changes To Be Made In The Number Of Tons Shown In Table 9-4
1. For corn silage ensiled when less mature than usual	Add 5-10%.
2. For corn ensiled when dry or overripe	Deduct 5-10%.
3. For corn very rich in grain	Add 5-10%.
4. For corn with very little grain	Deduct 5-10%.
5. For sorghum silage	Use the same weights as used for corn silage of comparable grain and maturity.
6. For sunflower silage	Add 5-10%.
7. For grass silage	Add 10-15%.[1]

[1]For this reason, a stronger structure is necessary where grass silage is stored.

The following example will serve to illustrate how to determine the size tower silo to build:

Over a period of years, a farmer plans to winter 34 head of 425-lb stocker calves on a ration of corn silage and protein supplement. There is a 240-day wintering period. No increase in the herd is planned. What size tower silo should be built?

The answer is obtained as follows:

1. First, here are the silage requirements:

 a. Stocker calves weighing 425 lb on a ration of corn silage and protein supplement should receive about 30 lb of silage per head daily.

 b. 34 × 30 = 1,020 lb of silage required daily for the 34 calves.

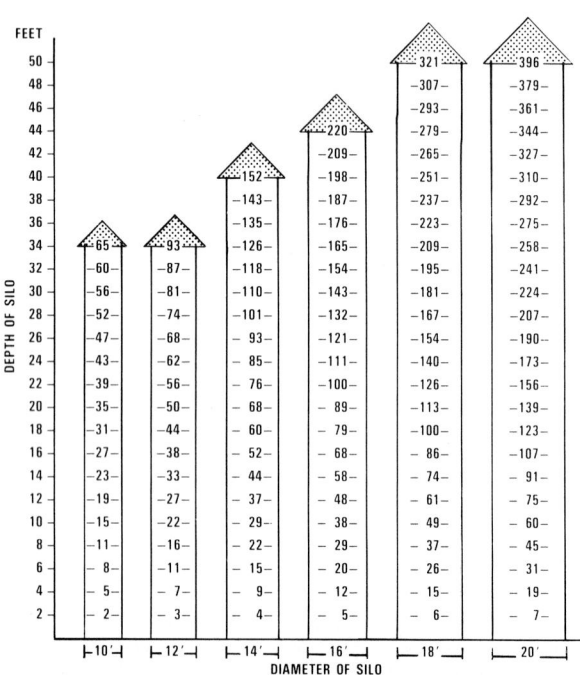

Fig. 9-5. Capacity in tons of settled corn silage in tower silos of varying sizes.

c. 1,020 × 240 = 244,800 lb, or 122.4 tons, of silage required for the 240-day wintering period for the 34 calves.

2. Next, here is the size silo to build:

a. Table 9-1 shows that in order to remove 1,005 lb of silage daily (which is only slightly less than the 1,020 lb needed daily), with 1½ in. removed from the top of the silo each day, the diameter of the silo should not be greater than 16 ft.

b. Fig. 9-5 can now be used as a guide in determining both the proper height (or depth) and diameter of the silo. Fig. 9-5 shows that a silo 16 ft in diameter and 27 ft high will hold 127 tons of silage, which would allow for 4.6 tons spoilage in excess of the required 122.4 tons. However, the height of a silo should not be less than twice the diameter. It appears best, therefore, to plan on a 14-ft diameter silo. As noted in Fig. 9-5, 34 ft of settled silage in a 14-ft diameter silo will provide 126 tons of silage, which would allow for 3.6 tons spoilage in excess of the required 122.4 tons. To allow for settling, an additional 4 to 6 ft should be added to the height, thus making a 38- to 40-ft height.

c. The size of silo to build to meet the needs outlined in this example, therefore, is one that is 14 ft in diameter and 38 to 40 ft high. Sufficient additional height, usually 4 to 6 ft, should be added to provide the necessary space required for the silage unloader.

SIZE OF TRENCH SILO

As in an upright silo, the cross-sectional area of a trench silo should be determined by the quantity of silage to be fed daily. The length is determined by the number of days of the silage feeding period. The only difference is that generally greater allowance for spoilage is made in the case of trench silos, though this factor varies rather widely.

Under most conditions, it is recommended that a minimum 4-in. slice be fed daily from the face (from the top to the bottom of the trench) of a trench silo during the winter months, with a somewhat thicker slice preferable during the summer months.

The dimensions, areas, and capacities given in Table 9-3 are based on the assumption that the silage weighs 35 lb per cubic foot,[4] which is an average figure for corn or sorghum silage. Thus, a trench silo 8 ft deep, 6 ft wide at the bottom, and 10 ft wide at the top has a cross-sectional area of 64 ft. This size silo will hold 747 lb of silage for each 4-in. slice, or 2,240 lb of silage for each 1-ft slice, or 112 tons in a trench 100 ft long.

For illustrative purposes, let us use the same example and silage requirements as were used in the section on "Size of Tower Silo," but this time determine the size trench silo to build. Briefly, the requirements are for 1,020 lb of silage daily for a 240-day wintering period. As noted in Table 9-3, one day's feed or 1,020 lb of silage (1,062 lb to be exact) can be obtained in each 4-in. slice of a trench silo 8 ft wide at the bottom, 14 ft 8 in. wide at the top, and 8 ft deep; or a 91 sq ft cross-sectional area. The cross-sectional area should not be larger than this if a 4-in. slice is to be removed daily in order to alleviate spoilage.

To obtain a 240-day feed supply, the filled trench must be 80 ft long (240 by ⅓—the ⅓ representing ⅓ ft or 4 in.).

The size trench silo to build to meet the specified needs, therefore, is one that is 8 ft wide at the bottom, 14 ft 8 in. wide at the top, 8 ft deep, and 80 ft long. In order to take care of spoilage and to provide a measure of safety, it is recommended that the actual length be from 85 to 90 ft.

About 8 ft is the most economical depth for a trench silo from the standpoint of cost and feeding. Of course, in filling it is desirable to pile silage 3 ft higher over the center of the trench and round it off, to provide for settling.

---

[4]Because the silage in trench silos is generally not as deep and well packed as that in tower silos, an average figure of 35 lb per cubic foot is used herein for trench silos and 40 lb for upright silos. With all types of silos—including above ground and belowground types—the weight of a cubic foot of silage varies with the kind and maturity of material, moisture content, length of cut, rate of filling, and depth of the silo. Corn silage harvested when about 74% of the grain has passed the milk stage and containing approximately 70% moisture is considered average silage. Volume for volume, sorghum silage weighs about the same as corn silage. Grass or grass-legume silage at 70% moisture is 10 to 15% heavier than corn silage, whereas grass haylage at about 50% moisture is 10 to 15% lighter than corn silage of equal volume.

TABLE 9-3
DIMENSIONS, CROSS-SECTION AREA OF TRENCH SILO, AND WEIGHT OF SILAGE IN 4-IN. (10.2-CM) SLICE AND PER LINEAL FOOT (0.3 M)[1]

Side Slope For Foot Of Depth		Depth		Bottom Width		Top Width		Cross-Sectional Area		Weight Of Silage			
										4-In. (10.2-cm) Slice		1-Ft (0.3-m) Slice	
(in.)	(cm)	(ft)	(m)	(ft)	(m)	(ft)	(m)	(ft²)	(m²)	(lb)	(kg)	(lb)	(kg)
3	7.6	4	1.2	5	1.5	7.0	2.1	24	2.2	280	127	840	382
4	10.2	4	1.2	6	1.8	8.7	2.6	29	2.6	338	154	1,015	461
5	12.7	4	1.2	7	2.1	10.3	3.1	33	3.0	385	175	1,155	525
3	7.6	6	1.8	6	1.8	9.0	2.7	45	4.0	525	239	1,575	716
4	10.2	6	1.8	7	2.1	11.0	3.3	54	4.9	630	286	1,890	859
5	12.7	6	1.8	8	2.4	13.0	3.9	63	5.7	735	334	2,205	1,002
3	7.6	8	2.4	6	1.8	10.0	3.0	64	5.8	747	340	2,240	1,018
4	10.2	8	2.4	7	2.1	12.3	3.7	77	6.9	898	408	2,695	1,225
5	12.7	8	2.4	8	2.4	14.7	4.4	91	8.2	1,062	483	3,185	1,447
3	7.6	10	3.0	6	1.8	11.0	3.3	85	7.7	992	451	2,975	1,352
4	10.2	10	3.0	8	2.4	14.7	4.4	113	10.2	1,318	599	3,955	1,798
5	12.7	10	3.0	10	3.0	18.3	5.5	142	12.8	1,657	753	4,970	2,259

[1]*Silos, Types and Construction*, Farmers' Bull. No. 1280, USDA, p. 55.

Silage Storage Losses

Tight structures, good distribution and packing, and the proper use of plastic covers minimize silage storage losses. Silage losses also vary widely between kinds of silos, as shown in Table 9-4. Losses in trench and open stack silos are also influenced by depth; less surface is exposed in deeper silos.

TABLE 9-4
ESTIMATED (1) AVERAGE, AND (2) RANGE OF SILAGE STORAGE LOSSES

Type Of Silo	Percent Of Loss	
	Average	Range
	(%)	(%)
Gastight upright	5	1-10
Conventional upright	6	2-12
Horizontal (trench)	15	8-25
Open stack	20	12-30

Losses in the silo are of four types: (1) surface or top spoilage, (2) seepage, (3) gaseous, and (4) heating (browning reaction and spontaneous combustion).

Surface or top spoilage losses of 20% or more may occur in stack silos and in any uncovered bunk, trench, or pit silo. These losses can be reduced by the use of suitable protection, such as a plastic cover.

Seepage losses can be high in high-moisture silage stored in upright silos. The higher the silo, the greater the pressure and the higher the losses through seepage. The seepage carries soluble feed nutrients with it. Horizontal silos have less seepage loss than upright (tower) silos because of lower vertical pressure. Seepage losses can be reduced by wilting forages to less than 65% moisture before ensiling.

Gaseous losses are unavoidable so long as the plant material respires and there is subsequent fermentation. However, these losses can be minimized by avoiding entry of air into the silo, by having the pH decline rapidly, and by encouraging favorable fermentations.

Lowering the moisture without excluding the air may lead to heat damage, known as the browning reaction or Maillard reaction. (See "Browning reaction/Maillard reaction" in the earlier section of this chapter headed "The Ensiling Process.")

Spontaneous ignitions sometimes occur in low-moisture silage (haylage). For such losses to occur, there must be a build-up of temperature to the combustion point in the silo mass, combined with a low transfer of heat. These fires are very difficult, and usually impossible, to extinguish. The addition of water may build up pressure and lead to an explosion. Most silo fires should be allowed to burn.

KINDS OF SILAGE

A great variety of crops can be and are made into silage. Generally speaking, crops that are palatable and nutritious to animals as pasture, as green chop, or as dry forage also make palatable and nutritious silage. Likewise, crops that are unpalatable and unnutritious as pasture, as green chop, or as dry forage make unpalatable and unnutritious silage. Most silage in the United States is made from either corn or sorghum, with corn silage far in the lead—over 16 times as much corn silage as sorghum silage is made. In 1985, 102.6 million tons of corn silage and 6.3 million tons of sorghum silage were produced in the United States. At the present time, it is estimated that 65% of the nation's silage is made from corn and sorghum and 35% from grasses, legumes, and other feeds. In addition to the kinds of silage already mentioned, silage is made from sunflowers, the small grains, sugar beet tops, crop residues, wastes from food processing (sweet corn, green beans, green peas), root crops, and various vegetable residues.

TABLE 9-5
COMPOSITION OF VARIOUS SILAGES

Type Of Silage	Analyses On A Dry Matter Basis			
	Crude Protein	TDN	Ca	P
	(%)	(%)	(%)	(%)
Corn	8.3	68.0	0.31	0.27
Grain sorghum	7.9	55.0	0.34	0.19
Forage sorghum	9.2	57.9	0.30	0.24
Oats	10.0	57.0	0.47	0.33
Alfalfa	17.4	59.0	1.75	0.27

Corn and Sorghum Silage

For the United States as a whole, corn ranks first in importance as a silage crop. Generally more total digestible nutrients can be obtained from an acre of corn as silage—which will yield from 5 to 25 tons of forage per acre, with an average of about 14 tons—than can be obtained from an acre of any other crop. Also, corn ensiles easily without the aid of a preservative, and keeps almost indefinitely in a good silo, is highly palatable, is well adapted to mechanized feeding, and may be fed with little waste.

There are four kinds of corn silage; namely,
1. The whole corn plant.
2. Ear corn silage.
3. Corn stover silage.
4. Shelled-corn silage.

The sorghums are more dependable and higher yielding than corn in certain areas, particularly in unirrigated, and relatively dry areas, of western and southwestern United States. Sorghum for silage is harvested with the same equipment as is used for corn silage. It should not be harvested for silage until the heads are soft to medium dough stage. Harvesting at this stage provides the highest yields of total feed material, enhances preservation, and makes silage that has good palatability.

On a dry-matter basis, corn silage contains an average of 8.3% crude protein, 68.0% total digestible nutrients, 0.31% Ca, and 0.27% P. Grain sorghum silage contains less protein and TDN than corn silage. Grass/legume silages contain more protein and less TDN than corn silage. The carotene content of corn silage is variable, but on the low side.

Corn and Sorghum Residue Silage

Corn and sorghum residues—the forages that remain after harvesting a grain crop of corn or sorghum—may be used as cattle feed three ways: (1) grazed, (2) harvested (stacked or baled) and fed dry, or (3) ensiled and fed as silage.

When ensiled, cornstalks (stover) produce a product known as corn stover silage or cornstalk silage. When stalks are processed as silage, the use of a forage harvester equipped with a screen or a recutter-blower at the silo is necessary in order to chop the material finely. Fine chopping will ensure good packing and improve consumption by avoiding selectivity.

Where corn stover silage is made, the residue should be harvested as soon as possible after the grain is taken off, before the residue loses any moisture. At that time, the grain moisture will generally be under 30% and the refuse moisture will be above 48%. In an airtight silo, 40 to 45% moisture is very satisfactory for ensiling. In an unsealed or bunker silo, the moisture content should be 48 to 55% for proper lactic acid formation. Water may be added at the silo if necessary. As a precaution, some authorities recommend the addition of 56 lb of corn meal (or other finely ground grain) per ton of corn stover silage, as a means of providing readily-fermentable carbohydrates from which acids will form and act as a preservative. With husklage, the latter precaution is not necessary since there is sufficient grain remaining in the husk and cob.

The biggest deterrent to harvesting stalklage, in either dry or ensiled form, is the cost—primarily for equipment. Rather than own such expensive equipment, which is used for a short period only, custom harvesting of stalklage is likely cheaper for most operators.

Husklage—the forage discharged from the rear of a corn combine, and consisting of the husks, cobs, and any grain carried through the combine—may also be ensiled. Ensiling husklage, along with recutting and adding water, results in increased cow consumption and less rejection of cobs.

Like corn, sorghum stover may either be grazed or harvested and stored either as dry feed or silage. Because the sorghum plant stays green late in the fall, good sorghum stover silage can be made without additional water.

Grass/Legume (Hay Crop) Silage

Grass/legume (hay crop) silage refers to silage made from any of the green crops which might otherwise be grazed or dried and made into hay. This includes grasses (such as timothy or fescues), legumes (such as alfalfa or clovers), grass-legume mixtures, and cereal grains (such as oats).

Grass/legume silage can be produced in areas where the climate is too cool and the growing season too short for corn or sorghum silage.

Although grass and legume crops have been ensiled in Europe for hundreds of years, the practice did not become widely used in the United States until the 1930s. At that time, interest in hay crops for silage increased as a result of farmers (1) becoming aware of the field losses that occur in haymaking, (2) being provided with the information necessary to make high-quality silages from grasses and legumes, and (3) having access to field choppers, which facilitated making silage from hay crops.

The following are the most important **advantages** of grass/legume silage:

1. It minimizes field, harvest, and storage losses of grass/legume forages. (See Fig. 9-6.)
2. It minimizes the dependence on favorable weather to harvest the crop.

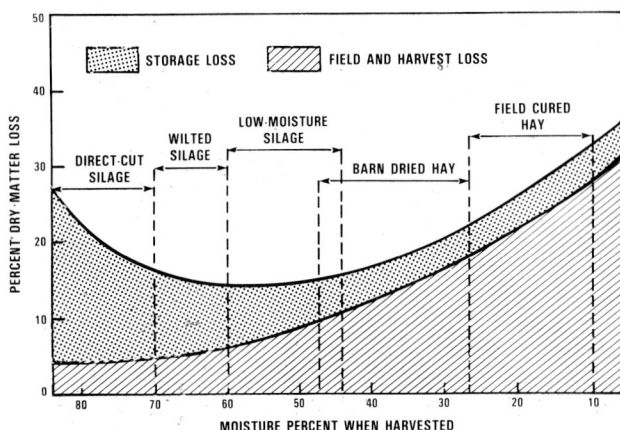

Fig. 9-6. Estimated total field, harvest, and storage loss when grass/legume forages are harvested at different moisture levels by alternative harvesting methods. (Courtesy, C. R. Hoglund, Professor Emeritus, Michigan State University, East Lansing)

3. It can be harvested with modern, efficient machinery and stored in large-volume structures.
4. It requires less supplemental feed than corn silage.
5. It can be handled and fed by mechanized methods, thereby reducing the labor requirements.
6. It kills weed seeds as a result of the fermentation.

Although grass/legume silage has important advantages, it also has the following **disadvantages**:

1. The initial investment costs for machinery, storage units, and feeding facilities are very high.
2. Silage-making machinery and storage and feeding facilities are highly specialized with the result that they have limited use for other purposes.
3. Inadequate fermentation may occur under certain conditions, resulting in poor-quality feed.
4. Storage and feeding losses may be high under poor management.
5. Like all silages, grass/legume silage is heavy, bulky, and costly to transport, thus its off-farm market value is limited.
6. Existing upright silos may not be in sufficiently good condition to store grass/legume silage.
7. Once grass/legume silage is removed from the silo, it must be fed within 12 to 24 hours to alleviate spoilage. Except for an oxygen-limiting silo, once feeding begins during warm weather, it is necessary to feed a minimum of 3 to 4 in. off the exposed surface daily to prevent spoilage in the silo.

Grass/legume silages are of three kinds based on moisture level:

1. Direct-cut silage, 70% moisture or above.
2. Wilted silage, 60 to 70% moisture.
3. Low-moisture silage (or haylage), 40 to 60% moisture.

DIRECT-CUT SILAGE

Direct-cut grass/legume silage is forage that is harvested and stored without field-drying, usually containing more than 70% moisture. Although direct-cut ensiling is the standard practice with mature corn and sorghum, it is not recommended for grass/legume crops because of (1) the difficulty in getting good preservation due to the high moisture content, and (2) the increased nutrient losses due to seepage.

WILTED SILAGE

Today, a high percentage of grass/legume forage is dried to some degree prior to ensiling. Wilting gets rid of some water in the field, so less weight is handled. Also, in comparison with direct-cut silage, odor and seepage problems are reduced and additives or preservatives are usually not needed.

Authorities generally agree on the following rules for making good wilted silage: (1) harvest at the proper stage of maturity; (2) allow the forage to wilt in the swath and/or windrow until the moisture reaches about 65%, which may take from 1 to 4 hours, or longer, depending on the weather, but which may be expedited by the use of a forage conditioner; (3) use a short cut (about ⅜ in.) on the forage harvester; (4) fill the silo rapidly and continuously; (5) distribute evenly and pack thoroughly; and (6) top off with 2 ft of unwilted material and cover with plastic or other suitable cover. No additive or preservative is needed with properly wilted silage, although it may be added if desired.

LOW-MOISTURE SILAGE (HAYLAGE)

This method of making silage is not new, contrary to common belief. It was developed many years ago at the Bacteriological Station at Crema, Italy.

Low-moisture grass/legume silage (or haylage), containing 40 to 60% moisture, is made with limited bacterial growth and fermentation. The term *oatlage* is sometimes used specifically to indicate low-moisture silage made from oats.

Fermentation is of minor concern in making low-moisture silage since little acid is produced and pH is not a useful criterion of quality. The most important factor is the establishment and maintenance of air-free conditions through fine chopping, rapid filling, and a good silo.

Properly made and stored low-moisture silage has a pleasant aroma and is a palatable, high-quality feed. Animals usually receive more dry matter and net feed value in low-moisture silage than in wilted silage made from the same cut. Low-moisture silage is increasing in popularity, especially as a dairy feed. It may be fed like wilted silage, with adjustment for difference in moisture content.

In addition to excluding air by providing an airtight silo and by thorough packing, the following directions should be observed to make the best quality low-moisture silage: (1) harvest at the proper stage of maturity; (2) wilt in the swath and/or windrow until the moisture reaches 40 to 60% (35 to 40% for a bunker silo), with the required time determined by the weather, but which may be expedited by the use of a forage conditioner; (3) chop short, a ¼ in. cut is best; (4) fill silo rapidly and continuously; (5) add an additive or preservative if desired; (6) distribute silage evenly in the silo; (7) apply a top seal of forage containing 65 to 70% moisture, level, and tramp to remove air; and (8) crown the center slightly and cover with a plastic silo cap.

Corn or Sorghum Silage Vs Grass/Legume Silage

Frequently livestock producers are confronted with choosing between corn or sorghum silage and grass/legume silage. Under these circumstances, the following facts are pertinent:

1. Where adapted, corn or sorghum will generally produce a greater tonnage of feed per acre than grass/legume silage.

2. Good-quality corn or sorghum silage can be made more consistently and with greater ease than good-quality grass/legume silage.

3. Corn or sorghum silage may be more palatable than grass/legume silage, even when the latter is carefully preserved.

4. Grass/legume silage is generally higher in protein and carotene but lower in total digestible nutrients and vitamin D (wilted grass/legume silage is higher in vitamin D than unwilted) than corn or sorghum silage (generally, grass/legume silage contains about 90% as much TDN as corn silage, but it will equal corn silage in TDN when 150 lb of grain per ton have been added). Thus, grass/legume silage generally requires the addition to the ration of less protein supplement but more total concentrates than corn or sorghum silage. This would indicate that corn or sorghum silage would be slightly preferable to grass/legume silage in high-forage finishing rations for beef cattle and sheep, whereas grass/legume silage would be preferable in high-forage rations for dairy animals and young beef cattle and sheep.

5. Grass/legume silage can be produced in areas where the climate is too cool and the growing season too short for corn or sorghum silage.

6. The production of grass/legume silage will result in less soil washing than the production of corn or sorghum silage on lands subject to erosion.

7. Grass/legume silage will freeze next to the silo wall more than corn or sorghum silage, especially if it is ensiled when too wet (unwilted).

8. The silo can be kept working full time by using both grass/legume and corn or sorghum silage; ensiling the first cutting of grass/legume silage for summer feeding, and ensiling corn or sorghum silage for winter feeding.

Other Silage Crops

In the Northwest and North Central states, where the weather is cool and the growing season is short, sunflowers are sometimes grown for silage. Although they yield and ensile well, sunflower silage is neither as palatable nor as nutritious as corn, sorghum, or grass silage. Pound for pound, sunflower silage is about 80 to 85% as valuable as corn silage.

Throughout the United States, a great array of by-product feeds are ensiled, especially in the less expensive and temporary types of silos. Among such by-products are grain chaff, pea and bean vines, beet tops and pulp, sunflower hulls and chaff, potatoes, cannery refuse, cull and surplus fruits and vegetables, pulp and trimming wastes from market vegetables and fruits, wet brewers' and distillers' grains, almond hulls, and poultry litter. Sometimes Russian-thistles and other weeds are ensiled.

When potatoes, which contain about 80% moisture, are ensiled for cattle, it is recommended either (1) that 20 to 25 lb of dry hay, straw, or chaff be run through the ensilage cutter with each 100 lb of potatoes, or (2) that 1 ton of corn or sorghum silage be chopped with each 500 lb of potatoes. Frozen and sprouted potatoes should not be ensiled. Potato silage intended for swine should be made from cooked or steamed potatoes ensiled alone in a shallow pit or silo. Potato processing wastes (cull potatoes, off-flavor french fries and chips, etc.) can be ensiled in the same manner as unprocessed potatoes.

Either of the methods recommended for ensiling potatoes for cattle is equally adapted for the preservation of other high-moisture crops, such as apples, beets, pears, tomatoes, cauliflower, broccoli, kale, and trimming wastes from market vegetables—provided the added forage is in proportion to their respective moisture contents.

Cabbage, rape, and turnips should not be ensiled, as they make unsatisfactory, watery, foul-smelling silage.

(Also see Chapter 12, By-product Feeds/Crop Residues.)

Combining Crops for Silage

Sometimes, in order to lower the moisture content, to alleviate the necessity of a preservative, and to assure better quality silage, forages of high sugar content are combined with forages of low sugar content. Thus, excellent silage can be made by mixing 1 ton of sorghum forage with each 3 tons of grass/legume silage material, or a ton of corn forage with each ton of grass/legume forage material (less sorghum forage is necessary than corn forage, because of the higher sugar content of the former).

At times such combination silage crops are even grown together; for example, corn and soybeans; millet or Sudangrass; and soybeans, oats, and peas.

A major difficulty in combining ensiling crops is that it is almost impossible to synchronize the stage of maturity of different crops so that they reach maximum yield and nutrient level at the same time.

Rain-Damaged Hay Silage

Partly cured hay that has been rained upon, but is not moldy, may be salvaged as silage (although it will not be of high quality), provided it is finely chopped, distributed evenly, and packed in the silo thoroughly enough to squeeze out the air. It is recommended that it be placed in the bottom of the silo, and, preferably, that alternate loads of a green crop be mixed with it. Otherwise, satisfactory packing can be obtained by putting a few loads of greener-than-ordinary material on top of it.

Frosted Crop Silage

Sometimes corn, sorghum, sunflowers, small grains, beans, and other crops, which may or may not have been intended for silage, are frosted before they reach the silage cutting stage. Corn that has been frosted before reaching maturity is commonly known as *soft corn*. Such frosted crops may be salvaged as silage. They should be cut at recommended moisture contents and ensiled according to directions. If they are too dry, water should be added.

Frosted crops, especially frosted sorghum, may be high in cyanide (HCN). (See *CAUTION* under the section on "Drought Stricken Crop Silage.")

Drought Stricken Crop Silage

Sometimes corn or sorghum, or other crops, are drought stricken to the extent that little or no grain will be produced. Such crops may be harvested for silage and used as an energy source for ruminants. They should be cut and ensiled like any other silage crop. If they are too dry, water should be added.

Drought stricken crop silage may be used in the same manner as any other low-energy source. It is well-suited for wintering breeding beef cattle and stockers, for backgrounding finishing cattle—to approximately 850-lb weights, and for dry dairy cows.

CAUTION: Danger of cyanide toxicity is much greater from sorghum than from corn. Drought stricken plants can accumulate cyanogenetic glycoside which hydrolizes to form free cyanide (HCN). The danger is increased when crops are grown on heavily nitrogen-fertilized soils or if any of the following have occurred: frosting, wilting, trampling, or hail. Any combination of these conditions can lead to a dangerous build-up or release of cyanide.

SILAGE ADDITIVES AND PRESERVATIVES

Silage additives are products that provide supplemental nutrients which enhance the feeding value of silage.

Silage preservatives are products that enhance the keeping qualities of silage.

Four types of additives or preservatives are used in silage making: (1) feed additives, (2) acids, (3) fermentation aids, and (4) preservatives.

Feed Additives

Feed additives may be used to provide a readily available source of carbohydrates for fermentation into lactic acid, to reduce the moisture content, to provide needed nutrients, and/or to enhance palatability.

Feedstuffs used as silage additives include:

1. Corn-and-cob meal, ground corn, barley, or oats; applied in amounts varying from 100 to 300 lb per ton, depending on the moisture content of the crop.

2. Beet pulp, citrus pulp, chopped corncobs, or chopped hay to reduce seepage losses from the silo if moisture is high.

3. Molasses, either liquid or dehydrated, at rates of 40 to 80 lb per ton of green forage.

4. Dried whey, a product of the dairy industry, applied at the rate of 30 to 300 lb per ton, as a source of fermentable carbohydrate, protein, and minerals.

5. Nonprotein nitrogen (NPN) products such as urea and anhydrous ammonia.

6. Ground limestone.

• **Urea, ammonia, and other NPN products**—Urea, ammonia, and other NPN products can be added to corn or sorghum silage at the time of ensiling as a source of nonprotein nitrogen.

Urea increases the crude protein content of the silage and the amount of lactic and acetic acids produced. The addition of 10 lb of urea per ton of ensiled corn material will make for the following approximate increases on a dry matter basis: the crude protein from 8.3 to 12.3%, the lactic acid from 4.2 to 5.4%, and the acetic acid from 0.9 to 1.2%. Since the amount of nonprotein nitrogen that can be converted to microbial protein by the organisms in the rumen is limited, no more than 10 lb of urea should be added to a ton of ensiled corn material. The urea can be added by spreading it over the top of each load of chopped corn or it can be added to the chopped corn through the blower by commercially manufactured metering equipment.

More recently, ammonia-containing materials have been added to corn silage as a source of nonprotein nitrogen, including ammonia-water solutions, ammonia-mineral solutions, ammonia-mineral-molasses solutions, anhydrous ammonia gas, and cold flow ammonia. For dairy cattle, 5 lb of actual nitrogen (about 6 lb of ammonia) may be added per ton of wet silage. Ammonia treated corn silage has been found to contain increased concentrations of true protein, lactic acid, and acetic acid. Also, it may have a higher pH and be more stable than untreated silage when exposed to air. Special equipment is required to add ammonia or ammonia-containing materials.

Acids

Both inorganic (mineral) and organic acids may be used as additives. Mineral acids lower the pH immediately, while organic acids have a limited effect on lowering pH. Both mineral and organic acids limit microbial growth and help to stabilize silage.

The use of inorganic acids, such as hydrochloric, sulfuric and phosphoric, for forage preservation was pioneered by A. I. Virtanen, Finnish biochemist, in the 1920s. He discovered the AIV method, named from his initials, for preserving silage by acidification, for which he was awarded the Nobel Prize in Chemistry in 1945. His work was highly regarded in that area of the world because hay-drying was difficult and dairying was, and still is, important.

• **Inorganic acids**—Inorganic acids (hydrochloric acid, sulfuric acid, phosphoric acid) have been used as silage preservatives, almost entirely in Europe, in connection with the ensiling of high-moisture material. These acids substitute for the acids produced by bacterial action. However, they are very corrosive, causing problems in their application, including problems with the silo walls and silage handling equipment. Of the three acids, phosphoric is preferred because (1) it is less corrosive than sulfuric acid or hydrochloric acid, (2) it may enhance the phosphorus content of the silage, and (3) it increases the residual manure value from the silage. But phosphoric acid may introduce a problem of proper calcium-phosphorus ratio. This can result in some abnormal conditions and unsatisfactory performance in the animals to which it is fed.

• **Organic acids**—Propionic, acetic, lactic, citric, and formic acids are included in this group. They are used in a manner similar to inorganic acids, but they are much less corrosive and not so difficult to handle, although precautions must be taken.

Organic acids will enhance the preservation of forage without the loss of palatability. Also, they serve as mold inhibitors. Even so, like all additives, they cost money; hence, the economics of using them in making silage must be considered. When an organic acid is used, the following guidelines should be observed:

1. Add 1% of the acid to the forage in the field at the time of harvest or at the chopper.
2. Limit the presence of oxygen by using a sound, well-built silo.
3. Prevent dilution of organic acid treated silage by rain or snow by covering it with plastic when it is stored outside or in a temporary silo.

It appears that organic acids will find their greatest use in the preservation of high-moisture grain. (See section on "High-Moisture Grain.")

Fermentation Aids

This group includes bacterial cultures, yeast cultures, and enzyme supplements. Controlled experiments support the claims made for some of these products, but not all of them. So, they should be purchased only from reputable sources that have valid research data to support the claims made for them.

Preservatives

This group includes antibiotics, salt, and sterilants. These products preserve silage by inhibiting microbial action or undesirable fermentations. All of them are of questionable value if air is properly excluded from the silage. If air is not excluded, they must be added at very high levels in order to be effective.

Silage Additive and Preservative Considerations

A variety of silage preservatives is presently on the market; and, no doubt, new ones will follow. Many of these products have been inadequately tested. Yet, farmers and ranchers are often in the position of having to decide whether an additive shall be used. In addition to understanding the silage-forming process and how different additives function, they should consider the following:

1. Additives and preservatives will not substitute for the proper exclusion of air.
2. Additives and preservatives do not produce new nutrients in silage, although they may aid in retention of those already present.
3. Additives and preservatives that add nutrients to the silage will be partially lost with any spoilage or seepage.
4. The cost of an additive or preservative is usually high in relation to the value of the silage.
5. Chemical analyses are of very limited use in evaluating silage additives and preservatives.

Silage Additive and Preservative Recommendations

When added to silage, the following materials will increase the amount of nutrients it contains:

1. Grain or grain by-products will increase total digestible nutrients and dry matter.
2. Molasses will increase the total digestible nutrients (TDN, or energy) and may improve fermentation in legumes and certain grasses.
3. Urea or other NPN products will increase the nitrogen (crude protein).
4. Limestone will increase the calcium content.

In order to assess the value of a silage additive or preservative, it is recommended that the following criteria be applied:

1. Does the product lower the ensiling temperature?
2. Does the product increase aerobic stability?
3. Does the product increase dry matter and nutrient recovery from the silo?
4. Does the product improve feed value and animal performance, particularly when silage is a major ingredient of the ration?

5. Does the product make for sufficient benefits to offset costs and give a return on investment?

HARVESTING METHODS AND MACHINERY

There is no one best silage-making method or kind of equipment. These must necessarily vary with the kind of forage, the kind of silo, the size of operation, and the labor and machinery cost.

Three principal kinds of machines are used for harvesting silage; namely, field forage harvesters, row-crop binders, and stationary silo fillers.

Field forage harvesters, which were first developed around 1936, are more widely used for harvesting silage than any other type of equipment. They tend to be concentrated in those states where the production of silage is most important. With different attachments, field forage harvesters can be used to harvest row crops for silage, grass silage as a standing crop or from the windrow, and hay from the windrow. Also, they can be used to harvest straw and other kinds of forage. Field choppers can even be adapted for grinding and blowing high-moisture cob corn for ensiling. With appropriate attachments, a modern field harvester can be used to harvest all major ensiled crops. Thus, with a minimum of complementary machinery, a forage harvester can be the major piece of equipment in providing a completely ensiled ration for beef cattle, dairy cattle, or sheep. But such equipment is expensive and may or may not be economical where a small operation is involved.

Field chopped forage is generally transported on wagons equipped with mechanical unloading devices or by means of dump trucks. Blowers and conveyors are used in filling both tower and horizontal silos. Frequently, trench silos are filled by dumping over the sides.

The use of row-crop binders reached a peak in 1942, following which they declined. Reduction in the numbers of these machines reflects the increased use of cornpickers, field forage harvesters, and grain combines. In addition to being used to harvest silage, row-crop binders are also used to harvest corn for grain, corn for fodder, and sorghum as bundle feed.

Beginning in 1951, the use of stationary silo fillers declined markedly. Today, few of them are used, primarily because of their high labor requirement.

HOW TO MAKE GOOD SILAGE

In addition to using a sound silo of proper size, those who make good silage generally harvest at the proper stage of maturity, cut to proper length, control the moisture content, add an additive or preservative when needed, fill rapidly, distribute forage uniformly in the silo, and seal or top-off the silo. Each of these factors will be discussed.

Harvest at Proper Stage of Maturity

Harvesting at the proper stage of maturity assures the maximum yield and nutrient content.

Fig. 9-7 shows the effect of stage of maturity of the corn plant on total dry matter accumulation.

The *black layer test* can be applied quickly and easily to determine when to harvest corn for maximum yield and nutrient quality (see Fig. 9-8).

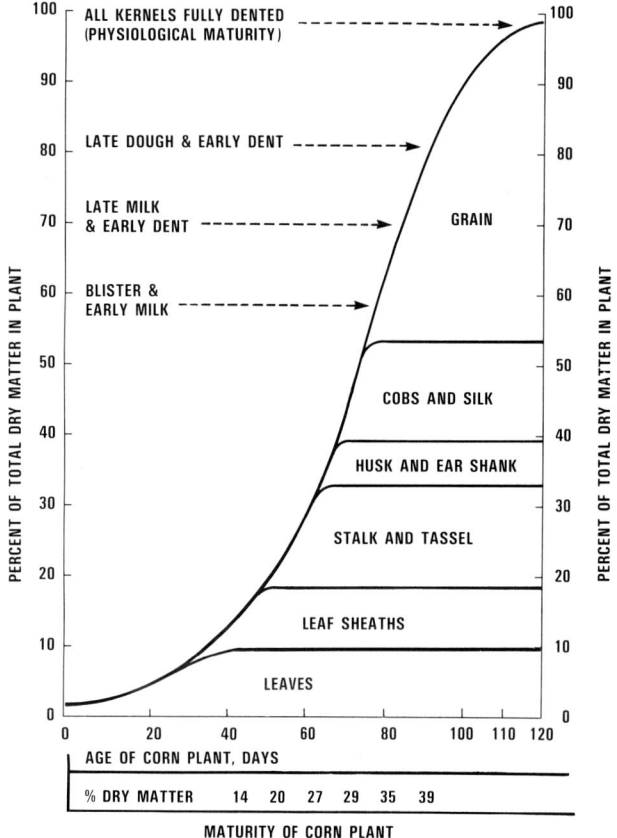

Fig. 9-7. Effect of maturity of corn plant on total dry matter accumulation.

Fig. 9-8. Black layer near the tip of the kernel indicates that the grain is physiologically mature and ready for the silo.

Sorghum should be cut for silage when the seeds are hard. Grass silage forages (grasses, legumes, and cereal crops) should be cut at the same stage at which they would make the best hay. (See Table 8-9 in Chapter 8, Hay.)

Cut to Proper Length

The length of the cut sections affects the packing and, hence, the quality of the silage. Also, the proper length of cut varies with the crop and the moisture content. Thus, for corn and sorghum crops, forage harvesters should be set to make a theoretical cut of ¼ to ⅜ in. If the knives are sharp and set up to the cutter bar, this will result in about 15% of the particles being 1½ in. and over, 25% of the particles being ¾ to 1½ in., and 60% being ⅛ to ¾ in. in length. Such a combination of particle size is necessary for high-quality feed. Grass silages should be more finely chopped than corn or sorghum silage. Also, wilted and dry forage and forage with hollow stems should be chopped more finely than forage of high-moisture content, thus permitting more thorough packing and eliminating most air pockets.

Control the Moisture Content

Moisture content is one of the most important factors in determining quality of silage. Experimental work and practical experience have indicated that 60 to 67% is the best moisture content for most crops to be ensiled. However, low-moisture silage of 40 to 60% moisture is now being preserved successfully in either oxygen-limiting silos, or tall conventional silos that are properly topped off with heavy, wet forage or sealed with a plastic cover.

HOW TO LOWER THE MOISTURE CONTENT

The moisture content of silage material may be lowered by any one or a combination of the following methods: by conditioning and/or wilting, by adding dry hay or straw, by combining with corn or sorghum silage, or by adding a dry additive/preservative of grain, dried molasses, or dried by-products of citrus or beets.

HOW TO INCREASE THE MOISTURE CONTENT

Drier material may be used for silage by cutting shorter and packing more thoroughly. If necessary, water should be added or the dry material should be mixed with very green, freshly cut material by alternating loads.

HOW TO DETERMINE THE MOISTURE CONTENT

Some methods of determining moisture content follow:
1. **The grab test (or squeeze method).** This test consists in taking a handful of the chopped forage and giving it a good hard squeeze for about 30 seconds. Then opening the hand slowly, noting the condition of the ball of forage in the hand, and referring to Fig. 9–9.
2. **The twist method.** Before chopping, the forage should be so well wilted that the stems may be twisted without breaking, but the limp leaves should show no signs of dryness. This test cannot be used for such coarse crops as sweet clover.
3. **The oven-drying method.** If in doubt or until more experience is obtained, the moisture content of a sample of any kind of forage may be obtained in about an hour's time by placing a sample on a tray, drying it in an oven with a vent or the door left open, and determining the loss in moisture by weighing the sample before and after drying.
4. **Other methods.** The heated-oil method and certain patented devices (as forced air dryers) may be used for moisture determination. Recently, several electronic testers which give instantaneous moisture readings of forages have come on the market. Also, microwave ovens can be used for moisture determination, provided the forage is spread in a thin layer to prevent burning. The latter two methods are expensive, but fast.

1.
Juice runs freely or shows between the fingers. The crop contains 75 to 85% moisture and is too wet to make high-quality silage without treatment. Silages made from crops in this condition will lose large quantities of juice. When possible, wilt these crops. If they must be ensiled without wilting, use an effective chemical preservative (not all of them are effective) or 200 lb of ground grain per ton of crop.

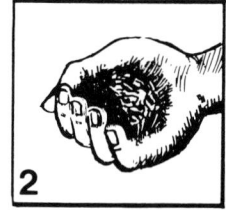

2.
The ball holds its shape—the hand is moist. The crop contains 68 to 75% moisture. Some juices will escape from tower silos. Additional wilting in the field is desirable. Where this is not done, use a chemical preservative or 150 lb of ground grain per ton of crop, or layer with wilted crops. Odors will be strong without some treatment.

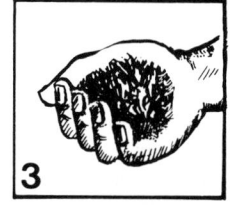

3.
The ball expands slowly—no dampness appears on the hand. The crop contains 60 to 67% moisture. This is the best condition for ensiling legumes without treatment.

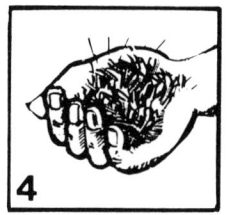

4.
The ball springs out in the opening hand. The crop contains less than 60% moisture. Only very young crops wilted in this condition can be safely ensiled. Others are likely to mold in the silo unless layered with wet crops or placed in gastight silos.

Fig. 9–9. The grab test.

Fill Rapidly

Once silo filling is started, it should be rapid, so as to avoid spoilage before the silo is filled and sealed. Generally speaking, a silo should be filled in 2 days or less.

Distribute Forage Uniformly in the Silo

In order to avoid the presence of air pockets and spoilage, it is essential that any kind of chopped forage be distributed uniformly in the silo and that it be packed well. Proper silage distribution is obtained by keeping the material nearly level or slightly higher at the center. Silage distributing equipment is available for keeping the material in an upright silo level. These devices are very helpful, especially in silos of 14-ft or larger diameters.

Where corn, sorghum, and sunflower forage is harvested at a green, immature stage and cut into short lengths, tramping in an upright silo will not be necessary; but uniform distribution is very important. The only filling precaution under these conditions is to see that the top is carefully leveled and well packed and covered whenever filling is completed.

Grass silage (especially when wilted), hollow-stemmed forages, and forages that have matured or dried beyond the best silage stage should always be trampled well, especially near the wall.

Packing in a trench silo should be obtained by use of a tractor.

Seal or Top-Off the Silo

Sealing or topping-off is necessary in order to avoid excess spoilage, especially with grass silage, which tends to dry on the surface and to shrink away from the silo walls. This may be accomplished by carrying out one or more of the following procedures:

1. Leveling off the top and thoroughly trampling the last few feet, especially near the walls.
2. Topping-off the silo with 2- to 3-loads of wetter material.
3. Covering the top with plastic cut to fit the silo diameter and turned up against the silo wall a distance of 5 to 8 in.

FEEDING VALUE AND ECONOMY OF SILAGE

A common rule of thumb is that 3 lb of 70% moisture grass silage or 2 lb of 40% haylage are equivalent to 1 lb of hay of similar kind and quality; a difference due primarily to the higher water content of silage or haylage. Suggested practical rations for different classes of livestock in which silage is incorporated, usually in combination with hay or some other dry forage, are given in this book in the chapters devoted to each species.

CHARACTERISTICS OF GOOD SILAGE

Fig. 9–10. Guernsey cows eating round bale alfalfa silage with relish, from an enclosed steel rack. (Courtesy, Sperry New Holland, New Holland, Pa.)

In order to make good-quality silage, producers need to know what constitutes silage quality. They need to be acquainted with those recognizable characteristics of silage which indicate high palatability and nutrient content. These are:

1. **Odor.** It has a "clean," rather pleasing acid odor, in contrast to the foul or objectionable odor of poor silage.

2. **Taste.** The taste is pleasing, not bitter or sharp.
3. **Absence of mold and rot.** There is no visible mold, and it is not musty or slimy.
4. **Moisture and color.** It is uniform in moisture and color. Very high-moisture silage is likely to be dark colored, slimy textured, and have a disagreeable odor. Generally, green or brownish silage is good; tobacco brown, dark brown, caramelized, or charred silage indicate excessive heat; and black silage is rotten and should not be fed.
5. **Animal acceptance.** Animals like and thrive on good silage.

SILAGE POINTERS

Some additional pointers which may be of value to the farmer or rancher who is making or feeding silage follow.

Moldy Silage

Moldy silage may be harmful. Any spoiled material that causes animals to go off feed, or that upsets the metabolic processes, should not be fed.

Some conditions cause certain molds to produce toxins. The toxins are called *mycotoxins* and the effects of the toxins on animals are called *mycotoxicoses*.

Mature ruminants appear to tolerate higher levels of mycotoxins than young ruminants, monogastric animals, or horses.

One way in which to determine the potential toxicity of moldy silage is to feed it to some less valuable animals for at least 2 weeks. Observe the animals daily for signs of toxicity—such as reduced gain and going off feed. If no toxic effects are noticed, it is probably safe to feed the suspect silage to other animals. If ill effects are noticed, switch them to other feed immediately and dispose of the suspect silage by spreading it on the land and plowing it under.

Effect of Silage on Milk Odor and Flavor

Silage sometimes affects the flavor and odor of milk, especially when ensiled too wet. This effect may be somewhat more pronounced with some silages than with others. The dairy producer will do well, therefore, to feed all silages after, rather than before, milking.

Dangerous Silage Gases

Two types of toxic gases may be formed when making silage: (1) carbon dioxide (CO_2), and/or (2) nitrogen dioxide (NO_2).

Carbon dioxide forms soon after filling begins and continues until fermentation stops. It is a colorless, suffocating gas, which is heavier than air and tends to collect in low places.

Under drought conditions, corn, sorghum, and other grass species may accumulate higher than normal levels of nitrates. When ensiled, nitrates are converted to nitrites, then nitrites are converted to nitrogen oxide by bacteria and plant cells. As the nitrogen oxide comes in contact with air, it is oxidized to form nitrogen dioxide, a reddish brown-colored gas, which is heavier than air. This gas is highly toxic to both humans and farm animals.

Precautions against hazards caused by silage gases include (1) operating the blower for a 15-minute period if it is still connected, (2) swinging a piece of canvas, a tree branch, or a burlap bag vigorously so as to agitate the air and dilute gases that may be present, or (3) taking proper life support equipment when entering an oxygen-limiting, or sealed, silo. Also, adequate provision for ventilation of the silo through the roof is essential.

A victim of silo gas should be moved into fresh air immediately, and artificial respiration should be applied. A physician should be called immediately.

HIGH-MOISTURE GRAIN

High-moisture grain refers to grain that is harvested at a moisture level of 22 to 40% and stored without drying.

Fig. 9-11. High-moisture corn being stored in one of the cement-lined trench storages at the Farr Feeders, Inc., cattle feedlot northeast of Greeley, Colorado. The high-moisture corn is stored in large quantities (60,000 tons) to provide a year's continuous supply of feed. Note truck unloading high-moisture corn at bottom of trench and tractor packing it at top of the silo. (Courtesy, William D. Farr, Farr Feeders, Inc., Greeley, Colo.)

The **advantages** of high-moisture grain in comparison with dried grain are:
1. It alleviates the high energy cost for drying grain.
2. It lessens field losses at harvest time.
3. It permits harvesting earlier, at higher moisture content, and usually during more desirable weather. As a result, it releases land for fall plowing in the North and for fall seeding of a second crop in the South.
4. It makes it practical to use later maturing, higher yielding varieties of corn and sorghum in the northern areas which often have early frost.
5. It usually requires less investment in processing equipment.
6. It improves the feeding value of grain for beef and dairy cattle.

The **disadvantages** of high-moisture grain in comparison with dried grain are:
1. It requires a large inventory of high-moisture grain, which may increase capital requirements.
2. It limits market flexibility for the grain, since it must be fed to livestock.
3. It may result in higher storage losses than for dry grain if proper ensiling or acid-treatment is not followed.
4. It may freeze in the bunk in the winter, and flies may be a problem in the summer.

Today, considerable quantities of high-moisture ear corn, shelled corn, sorghum (milo), and small grains (wheat, barley, and oats), containing about 30% moisture, are stored and fed. Some high-moisture grains are planned for and intended. Others are the result of happenstances, such as crops planted late, early frost damage, or harvesting when wet.

(Also see Chapter 14, Feed Processing, section on "High-Moisture Grain.")

Harvesting

A number of field equipment combinations may be used to harvest high-moisture grains. Harvest cost plus product quality should be considered in selecting the most desirable combination.

High-moisture grain should be harvested when it reaches physiological maturity and the moisture is 22 to 40%; with ear corn, sorghum (milo), and small grains higher in moisture than corn at harvest time. At this stage, grains yield the maximum available nutrients per acre and preservation conditions are best. If grain is harvested when too wet and immature, dry matter yield per acre will be less. If the grain is too dry, mold will be high and fermentation will be low; thus, grain containing less than 22% moisture should be reconstituted by adding water.

Storage of High-Moisture Grain

There are three basic storage methods for high-moisture grains: (1) ensiling in sealed (airtight) storage, (2) ensiling in nonsealed storage, and (3) preservation with an organic acid or ammonia.

SEALED (AIRTIGHT) STORAGE

High-moisture grain may be stored in an oxygen-limiting silo. It is not necessary to crack or grind grains before storing in this manner. Another advantage of sealed storage is that it can be unloaded from the bottom.

Sealed storage is the most popular method of storing high-moisture grains even though greater initial capital investment is required. This type of storage eliminates the 2 to 5% spoilage loss normally associated with unsealed storage.

UNSEALED STORAGE

Two types of unsealed high-moisture grain storage are used; (1) conventional upright silos made of concrete or steel, which are structurally adequate for storage of grass silage; or (2) horizontal silos. The grain should be ground into the storage unit, then firmly packed; otherwise, spoilage will result. With upright silos, about a 3 in. layer off the ground, high-moisture grain should be removed from the exposed surface each day during mild weather; with horizontal silos, 4 in. should be removed daily. Greater amounts should be removed from both types of storage units during warm weather to prevent spoilage.

PRESERVATION WITH AN ORGANIC ACID OR AMMONIA

Storage involving the treatment of high-moisture grain with an organic acid or ammonia to inhibit mold or spoilage is favored by many farmers because it alleviates artificial drying or the necessity to store in an airtight silo. Several different organic acids may be used—propionic, acetic, isobutyric, formic, benzoic, or a combination of these acids—but the most commonly used acids are propionic or propionic-acetic acid mixtures, marketed under various trade names. Anhydrous ammonia and other gaseous mixtures are also effective.

- **Amount of organic acid to apply**—The amount of acid required to treat high-moisture grain depends on moisture content of the grain, length of storage desired, and the temperature. The recommended rate of application of propionic acid to high-moisture grain to provide protection for 1 year is shown in Table 9-6. These rates are for corn, but they are suitable for other high-moisture grains.

**TABLE 9-6
AMOUNT OF 100% PROPIONIC ACID
REQUIRED FOR 1 YEAR OF STORAGE[1]**

% Of Moisture Content Of Grain	% By Weight	Lb (0.45 Kg) Per Wet Bushel[2] (35.2 l)	Lb (0.45 Kg) Per Ton (0.907 Metric Ton)	Gal (3.78 l) Per Ton (0.907 Metric Ton)
16-18	0.50	0.28	10	1.3
20	0.75	0.42	15	1.8
25	1.00	0.56	20	2.4
30	1.25	0.70	25	3.0
35	1.50	0.84	30	3.6

[1]The amounts of acid listed are for long term storage (1 year). For storage periods of 6 months or less, the amount of acid used could be reduced by ½.
[2]56 lb (25.5 kg) of high-moisture corn.

- **Application guidelines**—
  1. Check grain for moisture content so selected rate of acid application meets requirement for preservation.
  2. Treat the grain immediately after harvesting so as to eliminate heating and mold development.
  3. Make sure that all the grain is coated with the acid and that the application rate is correct.
  4. Treat outdoors if possible; otherwise, provide adequate ventilation.
  5. After treatment, flush equipment with water or untreated grain to prevent corrosion.
  6. Observe safety guidelines at all times.

- **Safe handling of grain preservatives**—When preservatives are applied, the following safety precautions should be observed by the user:
  1. Follow the manufacturer's instructions.
  2. Avoid storage of acids with fuels, lubricants, and pesticides. Store acids *only* in original container, tightly closed, with the bungs upright.

- **Advantages and disadvantages of acid-treated grain**— In comparison with ensiled high-moisture grain, acid-treatment has the following **advantages**: (1) Removal from the silo is eliminated; (2) it can be stored in a barn or other temporary storage facility; (3) treated grain can be transported over long distances; and (4) large batches of a ration which include acid-treated high-moisture grain can be mixed without risk of spoilage.

The major **disadvantages** to acid-treated grain are: (1) It must be fed to livestock—it cannot be marketed for any other purpose; (2) it will not germinate, so it canmnot be used for seed; (3) acids are corrosive to metal or concrete storage facilities; and (4) organic acids are costly, so for grain with over 30% moisture they may be uneconomical.

Feeding Value of High-Moisture Grain

The feeding value of high-moisture grain is equal or slightly superior to that of dry grain, with some variation according to class and productivity of livestock.

Beef cattle: Improvement is greater and more consistent with high-moisture ear corn than with high-moisture shelled corn. A summary of 14 experiments showed that high-moisture ear corn increased gains by 3% and feed efficiency by 10% over dried ear corn. Studies have also shown 3 to 5% improvement in the value of the dry matter in high-moisture shelled corn for cattle. Rate of gain has been similar for cattle fed dry or high-moisture shelled corn.

High-moisture storage improves the feeding value of sorghum (milo) more than it does corn for cattle. High-moisture harvested milo increases daily gains slightly (0 to 2%) and improves feed efficiency from 6 to 10% over dry sorghum grain for beef cattle.

High-moisture storage does not improve the feeding value of wheat for cattle.

Grain is stored whole in gastight silos, but it should be rolled or ground when removed for feeding. Grain should be ground or rolled when it is stored in horizontal or conventional upright silos in order for it to pack tightly and exclude air.

Dairy cows: Research studies have shown that the feeding value of properly ensiled or acid-treated high-moisture corn is equal to that of dry corn for lactating dairy cows; the milk yield and feed intake were similar. However, when high-moisture shelled corn supplies more than 50% of the total ration dry matter, depressed milk fat percentage may occur, with inadequate fiber given as the probable cause.

Some form of processing (e.g., rolling or grinding) of high-moisture corn improves utilization. Whole kernels appearing in feces indicate incomplete digestion. Processed corn is higher in digestible, metabolizable, and net energy for dairy cows than rations containing whole shelled corn.

Swine: When compared on an equal dry matter basis and in mixed rations, rate of gain and feed efficiency are essentially the same for hogs fed high-moisture or dry grains. Free-choice feeding of high-moisture grain may be used successfully for hogs weighing more than 60 lb provided proper intake of the protein supplement relative to grain intake is assured. But free-choice feeding is not recommended for pigs weighing under 60 lb.

There is no advantage to grinding or cracking high-moisture corn for growing-finishing pigs other than for mixing purposes. However, grinding or cracking increases the feeding value of high-moisture sorghum (milo), barley, and wheat.

Reconstituted Grain

Reconstituted grain is mature grain to which water has been added to raise the moisture content to 25 to 30% for storage.

Reconstituted sorghum (milo) stored whole in an airtight silo, then processed at feeding time and fed to finishing cattle, will improve rate of gain slightly and feed efficiency by 12 to 15%. However, if sorghum is ground before reconstituting, there is little improvement in feed value over dry milo for cattle. Reconstituted whole corn fed to finishing cattle will not improve rate of gain, but will improve feed efficiency by about 4.5%.

(Also see Chapter 14, Feed Processing, section on "Reconstituted Grain.")

Buying and Selling High-Moisture Grain

Most high-moisture grain is grown by livestock producers for livestock feed, or grown as a cash crop and sold at harvest to livestock producers. High-moisture grain is bought/sold at a lower figure than dry grain because of the water content. Also a greater quantity of it must be fed because of the water content. In most cases, the buying and selling transactions of high-moisture grains are based on an 87% dry matter and 13% moisture basis.

(Also see Chapter 16, Buying Feeds/Commercial Feeds/Feed Laws, section on "Moisture is Important.")

QUESTIONS FOR STUDY AND DISCUSSION

1. What is silage? How important is it as a U.S. livestock feed?

2. Since most silage is fed on the farms or ranches on which it is produced, how important is it that the farmer or rancher know how to produce it?

3. Define the *ensiling process*. Describe each of the 3 phases of the ensiling process. What is the browning reaction/Maillard reaction in silage?

4. Discuss the economics of silage production. List both the advantages and disadvantages of silage.

5. Classify and describe different kinds of silos. What are the advantages and disadvantages of each kind of silo?

6. List the advantages and the disadvantages of round bale plastic covered silage.

7. What factors determine the size silo to build of each (a) a tower type, and (b) a trench type?

8. List and discuss the importance of each of the 4 primary types of silage storage losses.

9. What accounts for corn ranking first in importance as a silage crop? Under what circumstances might it be desirable for a farmer to grow sorghum for silage rather than corn?

10. Compare the crude protein and TDN of the following five types of silages: corn, grain sorghum, forage sorghum, oats and alfalfa.

11. What is grass/legume (hay crop) silage? What are the advantages and the disadvantages of grass/legume (hay crop) silage?

12. What are the distinguishing characteristics of (a) direct-cut silage, (b) wilted silage, and (c) low-moisture silage (haylage)?

13. Give directions for making top-quality low-moisture silage.

14. Under what circumstances should a farmer or rancher make grass (hay crop) silage rather than corn or sorghum silage? Nutritionally, how do corn/sorghum silage and grass/legume silage compare?

15. Under what circumstances might it be desirable to make silage from each of the following types of products: (a) by-product feeds, (b) rain-damaged hay, (c) frosted crop, or (d) drought-stricken crop?

16. What are: (a) a silage additive, and (b) a silage preservative? List 4 types of additives or preservatives, and give examples and describe the role of each type.

17. Why, and under what circumstances, might nonprotein nitrogen (NPN) additives be mixed with silage?

18. What criteria for assessing the value of a silage additive or preservative would you recommend?

19. Why have silage harvesting methods and machinery changed so greatly in recent years, with field forage harvesters becoming more widely used than any other type of equipment, and with stationary silo fillers declining markedly?

20. Discuss the importance of each of the following practices from the standpoint of making good silage: (a) harvest at the proper stage of maturity, (b) cut to proper length, (c) control the moisture content, (d) fill rapidly, (e) distribute forage uniformly in the silo, and (f) seal or top-off the silo.

21. Discuss the feeding value and economy of silage.

22. What are the easily recognized characteristics of silage of high feeding value?

23. What types of toxic gases may be formed when making silage; what causes them; and what precautions may be taken against the hazards caused by them?

24. What is meant by high-moisture grain? How and why is it produced? Why has interest in high-moisture grain increased with higher energy cost?

25. What are the advantages and disadvantages of high-moisture grain in comparison with dried grain?

26. List and discuss the 3 basic storage methods for high-moisture grains.

27. Discuss the feeding value of high-moisture grain for beef cattle, dairy cows, and swine.

28. What is reconstituted grain? Under what circumstances might it be desirable to reconstitute grain?

29. What factors will determine whether a particular crop should be utilized as pasture, hay, or silage?

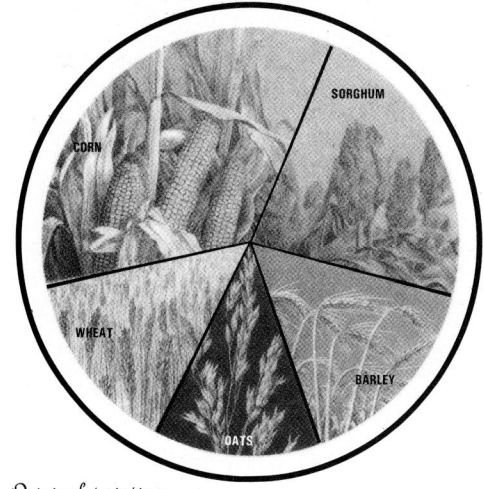

Original painting by Tom Phillips

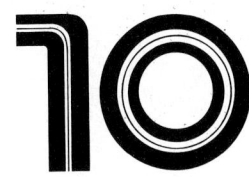

GRAINS/ HIGH-ENERGY FEEDS[1]

Contents	Page
Grains	188
Structure of Grain	188
Pericarp	190
Seed	190
Seed Coat (Testa) and Hyaline Layer (Nucellar Layer)	190
Endosperm	190
Aleurone	190
Starchy Endosperm	190
Embryo (Germ)	190
Grain Milling and Milling By-products	191
Grain Milling	191
Corn Milling	191
Wet Milling	191
Dry Milling	191
Sorghum Milling	192
Wheat Milling	192
Rye Milling	192
Rice Milling	192
Oat Milling	192
Barley Milling and Malting	192
Nutrient Composition of Grain	192
Carbohydrates	193
Lipids	193
Protein	193
High-Lysine Corn (Opaque-2)	194
Minerals	194
Vitamins	194
Effect of Stage of Maturity on the Nutritive Value of Grains	194
Grain Standards	194
Grain Storage	195
Moisture	195
Temperature	195
Biological Sources of Grain Damage During Storage	196
Insects and Mites	196
Rodents	196
Microbial Contamination	196
Other High-Energy Feeds	197
Other Seeds	197
Sprouted Grains	197
Cottonseed	197
Fats and Oils	198
Molasses	200
Cane Molasses	200
Beet Molasses	200
Citrus Molasses	200
Wood Molasses	200
Roots and Tubers	200
Processing Industry By-product Feeds as Energy Sources	202
Questions for Study and Discussion	202

All physiological processes require energy. Before any animal can reach maximum performance, energy must be expended to maintain routine body functions, such as blood pressure, body temperature regulation, respiration, transmission of nerve impulses, muscle tone, and metabolic processes. Only when these essential functions are provided can the animal synthesize tissues and secretions which we commonly call *production*. All production parameters—lactation, growth, pregnancy, egg laying—require substantial quantities of energy. To maximize production, the livestock producer must use a high-energy ration—that is, a ration with a high caloric density and digestibility. When low-energy rations are used—for example, high-roughage feeds—the animal cannot physically ingest sufficient feed to fill its energy needs for production. Hence, there is a suppression of production potential. A high-energy ration properly complemented with the other essential nutrients allows the animal to fulfill its energy requirements with the ingestion of considerably less feed.

Concentrates are feeds that are high in nitrogen-free extract and TDN and low in crude fiber (less than 18%). They can be broken down into two classes—(1) carbonaceous feeds, and (2) nitrogenous feeds. The feeds discussed in this chapter belong to the first category, as their main function is to supply energy. Sugars, various types of polysaccharides, fats, and oils—all consisting primarily of carbon, hydrogen, and oxygen—provide the necessary materials for energy production.

Of all the energy sources used in livestock and poultry rations, cereal grains fill the most important role, providing the bulk of the energy for animals as well as substantial portions of the protein.

Although grains are the most widely used high-energy feed, by-products of grain milling, fats, oils, fruits, nuts, roots, and specialized feeds such as molasses, often provide excel-

---

[1]The authors gratefully acknowledge the helpful suggestions of the following authorities who reviewed all or parts of, and furnished material for, this chapter: J. D. Axtel, Ph.D., Professor of Agronomy, Department of Agronomy, Purdue University, West Lafayette, Ind.; R. A. Forsberg, Ph.D., et al., Department of Agronomy, University of Wisconsin, Madison; R. G. Fulcher, Ph.D., Grain Quality Laboratory, Plant Research Centre, Agriculture Canada, Ottawa, Ontario; R. C. Hoseney, Ph.D., Department of Grain Science and Industry, Kansas State University, Manhattan; D. A. Miller, Ph. D., Professor, Department of Agronomy, University of Illinois, Urbana-Champaign; M. W. Phillips, Ph.D., Head, Department of Agronomy, Purdue University, West Lafayette, Ind.; Y. Pomeranz, Ph.D., Director, U.S. Grain Marketing Research Laboratory, Agricultural Research Service, USDA, Manhattan, Kan.; C. M. Wilson, Ph.D., Research Chemist and Professor of Plant Physiology, Agricultural Research Service, USDA, Urbana, Ill.; and V. L. Youngs, Ph.D., Research Food Technologist, Agricultural Research Service, USDA, Fargo, N.D.

lent alternatives. While nitrogenous concentrates are used primarily for the synthesis of protein in the body, a limited amount of the protein in these feeds can be used for energy. Nitrogenous concentrate feeds are discussed in Chapter 11, Protein Supplements.

Grains and high-energy feeds, being low in fiber, are generally considered to be highly digestible by all types of livestock. Since poultry, swine, and other nonruminants which do not have functional ceca cannot digest fiber efficiently, the bulk of their respective rations consists of concentrate feeds—even for maintenance and low production. Ruminants and herbivorous nonruminants, on the other hand, can obtain a large proportion of their energy needs through the consumption of forages. However, when production demands are increased in these animals, forages are generally partially replaced by concentrate feeds which have higher caloric densities and digestibilities. For example, many finishing rations for beef cattle contain more than 85% concentrate.

GRAINS

Grains are seeds from cereal plants—members of the grass family, Gramineae. In addition to the production of grains which contain large quantities of carbohydrates, the entire plant can be harvested prior to grain maturation and used for forage—pasture, hay, or silage. The primary use of cereal plants, however, is through the utilization of the highly digestible seeds.

All cereal crops are annuals. Depending on the type and variety of grain crop, they may be either (1) winter annuals—planted in the fall and harvested in the spring, or (2) summer annuals—seeded in the spring and harvested in the late summer or early fall.

Such by-products of harvested grains as chaff, stover, and straw can be utilized as low-quality forages for ruminant animals. Moreover, many of the grains are milled or processed in some manner, thereby creating additional by-products which can be fed to livestock with varying degrees of success. These by-products range from high-energy, high-protein wheat germ to relatively low-quality, coarse oat hulls.

Corn, oats, barley, and sorghums are the primary grains fed to livestock and poultry; hence, they are known as feed grains or coarse grains. The United States produces 41% of the world's corn, 30% of the grain sorghum, and more than 25% of all feed grains. Rice and wheat are consumed primarily by humans, but many of their milling by-products are fed to livestock. Additionally, wheat is sometimes fed to livestock when the price becomes competitive with the more widely used feed grains, or when it is contaminated or damaged by insects or weather. Barley and rye are fed to livestock in limited amounts, but a large proportion of these two grains is used in the brewing and distilling industries. Millet, emmer, spelt, and triticale are used occasionally, but the impact of these grains on the feed industry is very limited. With the exception of rye, grains are very palatable to livestock.

The utilization of grains by humans and animals varies from country to country. For example, most U.S. wheat is used for human food; in Europe, however, where there is a surplus of wheat, large quantities of it are fed to livestock. In Mexico, corn is considered as a food grain, where it is ground and made into tortillas or gruels, while sorghum is the primary feed grain. Although sorghum is almost exclusively a feed grain in the United States, it is the primary food grain for people in a number of African countries and in parts of India, Pakistan, and China.

Table 10-1 gives the yearly breakdown of the amounts of the main feed grains fed to livestock and poultry in the United States. Corn is, by far, the most widely used feed grain. Sorghum ranks second.

TABLE 10-1
FEED GRAINS FED TO LIVESTOCK AND POULTRY (MILLION TONS)[1]

Grain	Year				
	1965	1970	1975	1980	1985
Corn	94.1	100.6	99.6	115.9	114.7
Sorghum	15.9	19.0	14.3	8.6	18.6
Barley & Oats	10.8	19.0	13.2	11.1	14.0
Total	126.8	138.6	127.1	135.6	147.3

[1]*Agricultural Statistics,* USDA.

Grains provide the livestock producer an excellent source of highly digestible energy for animals that are either on a high level of production or unable to utilize forages effectively. However, several problems are inherent in the use of grains; among them, the following:

1. **In ruminant animals, high-concentrate rations may cause digestive disturbances, such as acidosis and parakeratosis of the rumen.** Ruminants need some *roughage factor* or *scratch factor* to stimulate the rumen papillae. In some cases, this is accomplished by rolling or coarse grinding the grain. Most cattle feedlot operators include 10 to 15% levels of some roughage source in finishing rations.

2. **Some grains must be processed before they can be fed.** The need for processing is primarily governed by the type of grain and the particular animal being fed. For example, grain should be ground when fed to very young or very old animals and animals which do not thoroughly masticate their feed.

3. **Grains are more costly than most fibrous feeds on a weight basis.** However, when comparing the costs of grain to roughage, the energy content, digestibility, and other nutrients must be considered on a per unit feed basis. Thus, a relatively expensive grain containing large amounts of highly digestible energy may in reality be a better buy than a low-cost, low-quality roughage.

4. **Grains are extremely deficient in calcium and certain vitamins.** Most grains contain less than 0.1% calcium. Adequate amounts of phosphorus are generally present in grain, but the calcium to phosphorus ratio is highly unbalanced. Additionally, grains are deficient in certain vitamins; for example, vitamin A is low in all grains except fresh yellow corn.

Structure of Grain

Grain develops from the ovary and its ovule after fertilization by pollen. The structures of the flowers of the various cereal plants differ, and these differences are reflected in the structure of the individual kernel.

In corn, the male and female structures are found in separate inflorescences (flower structures) on the plant. The male inflorescence, commonly referred to as the tassel, is

Grains/High-Energy Feeds

located at the top of the cornstalk. Pollen is shed from the tassels and subsequently comes into contact with the female inflorescence, thereby producing grain. The female inflorescence contains a central rachis, commonly referred to as the cob. The cob contains series of rows of sessile spikelets and is enclosed by overlapping bracts (husks). The silks which are found on the corncob structure are the stigmas—the pollen-receiving organs. As seen in Fig. 10-1 each spikelet contains two flowers—one fertile and one sterile.

In rye and wheat, the lemma and palea are loosely attached to the grain. During threshing, these particles are separated from the grain and constitute what is known as chaff.

Barley, oats, and rice retain their lemma and palea during threshing, thus giving rise to structures called husks or hulls. In barley, the lemma and palea fuse with the grain. In oats, the lemma and palea do not fuse with the kernel but enclose and adhere tightly to the entire grain. This hull structure can be removed during processing, resulting in dehulled oat grains called groats. Rice hulls are removed during processing.

Individual kernels of grain are called caryopses (see Fig. 10-3). Grains which contain husks (oats, barley, and rice) are called covered caryopses, whereas grains lacking husks (corn, wheat, rye, and sorghum) are referred to as naked caryopses. Each kernel (caryopsis), exclusive of the husk, is composed of two main parts—pericarp (fruit coat) and seed.

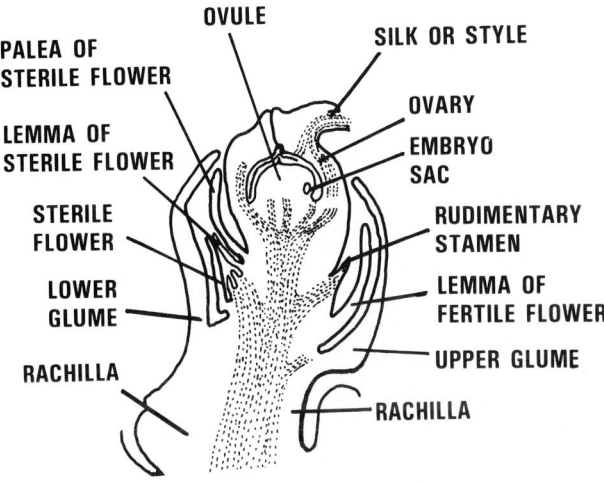

Fig. 10-1. Structure of a corn spikelet.

The grain from barley, oats, wheat, rice, and sorghum develops from flowers which contain the ovary, three stamens, and two scalelike lodicules. These structures are surrounded by a pair of bracts called the lemma (found on the dorsal side of the flower) and the palea (found on the ventral side of the flower) (see Fig. 10-2).

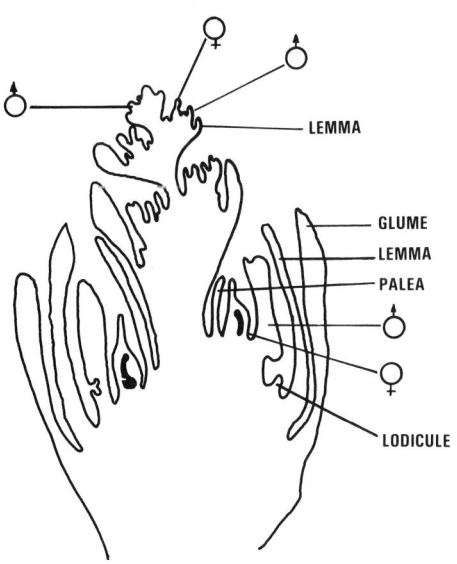

Fig. 10-2. Structure of the wheat flower.

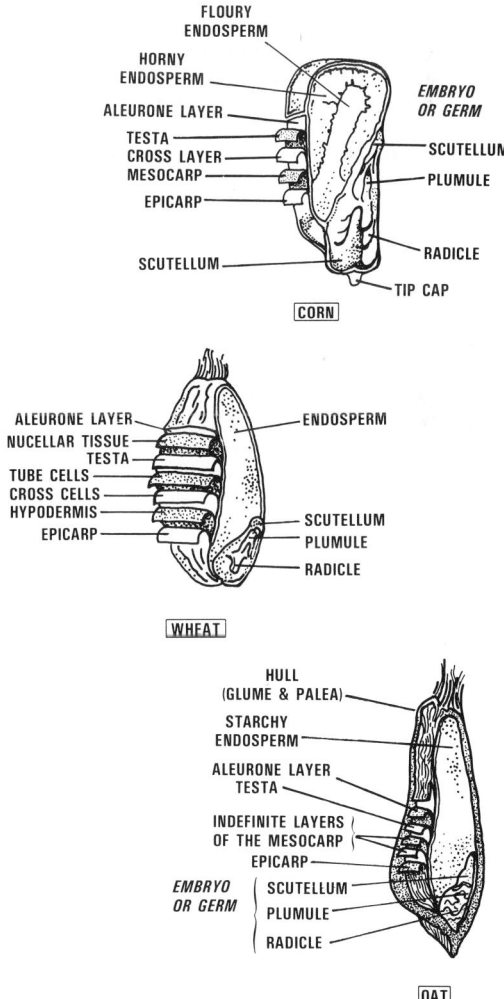

Fig. 10-3. Structure of corn, wheat, and oat caryopses.

PERICARP

The pericarp consists of two layers. The outer layer contains the epidermis and the hydroderm, collectively referred to as the beeswing. The cells of the epidermis are thin-walled and rectangular. The hydroderm is highly variable in thickness. These components of the outer layer are oriented lengthwise in the grain. The inner layer of the pericarp contains cross cells and tube cells. The cross cells are oriented transversely along the grain. Throughout the ripening process of grain, the innermost layer of the pericarp becomes distorted and torn, thus giving a tubelike appearance—hence, tube cells.

SEED

The seed portion of grain can be divided into three parts: (1) seed coat (testa) and hyaline layer (nucellar layer), (2) endosperm, (3) germ (embryo). When grain is processed in such a way that the germ and starch endosperm are removed, the composite of the remaining parts of the seed and the pericarp is called bran.

Seed Coat (Testa) and Hyaline Layer (Nucellar Layer)

The seed coat (testa) is generally either one or two layers thick. There is very little cellular structure in this region. Likewise, the hyaline layer lacks any cellular structure but, rather, acts as an embryo sac.

Endosperm

The endosperm of grain can be divided into two parts—the aleurone and the starchy endosperm.

ALEURONE

The aleurone surrounds the starchy endosperm of grain but does not encompass the scutellum of the embryo. The cells of the aleurone are thick-walled, cuboidal, and rich in oil, niacin, and mineral matter. Phytic acid is also produced in rather large quantities in this region.

The number of layers of cells in this region varies according to the type of grain. Wheat, rye, oats, and sorghum, generally have only one layer of cells in the aleurone. Depending on the particular variety, corn can contain anywhere from 1 to 6 layers; barley, 2 to 4; and rice, 2 to 6.

STARCHY ENDOSPERM

The starchy endosperm portion of grains contains thin-walled cells that are highly variable in shape, size, and contents. Pentosans are found in large amounts in the cell walls of this region, but starch and protein make up most of the cell contents. Starch is found primarily in the form of granules, with protein filling the intergranular spaces.

In wheat, the cells adjacent to the aleurone are relatively higher in protein and lower in starch than the rest of the starchy endosperm.

The concentration of starch in corn depends largely on the type of corn and the area of the kernel being analyzed. The endosperm of corn is divided into two regions—the crown (area opposite that of the embryo) and the horny region (area next to the embryo). The crown region contains loosely packed starch granules with little protein. In the horny region of the yellow varieties of corn are layers of proteinaceous material with starch granules interspersed. Hence, the protein content of the horny region is much higher than that of the crown region (about twice as high). Additionally, the oil content of the horny endosperm is high. In some varieties of corn, the crown region contracts during maturation, thereby creating an indentation in the kernel forming dent corn (see Fig. 10–4).

Fig. 10–4. Dent corn. At maturation, the crown region contracts, giving the kernel a characteristic indentation. (Courtesy, USDA)

Embryo (Germ)

The embryo (germ) of the seed consists essentially of an immature, undeveloped plant, surrounded by a quantity of stored food for its early nourishment and a protective seed coat.

Between the embryo proper and the endosperm is an organ called the scutellum. Upon germination, food reserves in the endosperm are mobilized and passed on to the embryo by the scutellum. The plumule of the embryo gives rise to the growing bud, and the radicle to the root system.

Grain Milling and Milling By-products

All grains can be fed to livestock with varying efficiencies in their intact state. However, routine processing, such as grinding, can often improve the feeding value of the grains. A detailed discussion on the processing of grains for livestock is given in Chapter 14, Feed Processing. Additionally, most of the grains are milled in some manner for the preparation of foods for human consumption. In these milling processes, a number of by-products are produced which are generally considered to be of little value to humans but which can be and are used extensively as livestock feeds.

GRAIN MILLING

The milling procedures of the grains discussed in the sections that follow illustrate the ways that by-product feeds for livestock are obtained. However, it is recognized that no two mills (no two corn mills, for example) are ever exactly alike—in the sequence, identity, or placement of machinery. Nevertheless, the accompanying figures show in a general way the basic steps involved in grain milling to obtain food for people and feed for farm animals.

Corn Milling

Between 5 and 6% of the corn grown in the United States is processed by wet milling for the production of starch, sugar, and corn oil. Slightly less corn, about 3.5% is dry milled in the production of flour, oil, and breakfast cereals. By-products from both of these processes are fed to livestock.

WET MILLING

As evident in Fig. 10-5, the wet milling of corn makes valuable use of each constituent of the corn kernel. In addition to starch and the various products derived from it, along with highly valued oil from the germ, 25 to 30% of the corn used by the wet-milling industry goes into the production of livestock feeds. Details relative to the wet-milling process follow.

Before storage, shelled corn is initially cleaned to remove extraneous material such as pieces of cob, foreign seeds, stray metal, fine dirt, or light, unwanted material. When the corn is taken from storage, it is cleaned again and steeped in a series of tanks where, at regular intervals, a solution of sulfurous acid water is recirculated. This procedure, which takes anywhere from 28 to 48 hours, disintegrates the protein that binds the starch within the kernel, thereby softening the kernel and facilitating grinding. The steep water which is removed from this process contains about 6% solids, of which about 35 to 45% consist of protein. The steep water then undergoes evaporation, whereupon the moisture level is reduced to 45 to 65%. This high-protein extract may be used as a nutrient for microorganisms in the production of enzymes, antibiotics, and other fermentation products. The major portion, however, is combined with fiber and gluten in the production of animal feed ingredients.

After steeping, the extracted corn is degermed by coarse grinding, yielding a pulpy material which contains the germ, hull, starch, and gluten. This material is then passed through a liquid cyclone which separates the germ from the rest of the material. Oil is then extracted from the germ through hydraulic, expeller, or solvent processes. The extracted germ is dried and sold as corn germ meal for animal feed.

Following the separation of the germ, the remaining materials containing starch, hulls, and gluten pass through a series of screens which separate the various fibrous fractions from starch. The fibrous fractions are dried and sold as mill by-products to the animal feed industry.

The water slurry of starch and gluten is separated by centrifugation. Typical operations yield a gluten stream containing over 60% protein, while the starch stream is over 99% starch. The gluten is dried and sold as gluten meal (60% protein) or it may be used as an ingredient in corn gluten feed (21% protein).

The white, nearly-pure starch slurry is further washed to remove small quantities of solubles, dewatered using filters or centrifuges, and dried to produce common starch (unmodified).

Various modified or derivatized starches may be produced by treating the slurry of washed starch with chemicals or enzymes. After treatment, the products are recovered by filtration or centrifugation and the starch is dried.

Thus, the wet milling of corn produces four major feed ingredients: corn gluten feed, corn gluten meal, corn germ meal, and condensed fermented corn extractives or corn steep liquor. Additionally, three further products used in feed manufacturing are obtained by wet millers: hydrol, a starch molasses; wet bran; and dried steep liquor concentrate.

DRY MILLING

The dry milling of corn is generally less involved than that of wet milling. Two basic processes—degerming and non-degerming—are used extensively.

In the degerming process, the hull, germ, and endosperm are separated before milling. This process is used for the production of grits, flakes, meal, flour, oil, and feeds.

The entire kernel is ground intact in the nondegerming process of dry milling. The resulting product is an oily flour

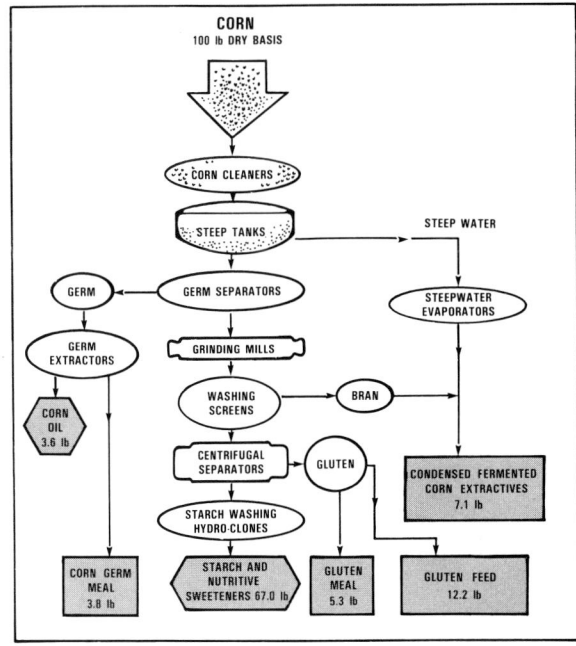

Fig. 10-5. Schematic outline of wet milling of corn.

which is subsequently used in baked products. Although some hulls and germ sift out in this process, the quantity of by-products is too insignificant to be considered as a useful source of animal feed.

Sorghum Milling

Sorghum is processed in much the same way as corn. It is used in the wet milling and dry milling industries as well as in the fermentation industries. The primary products of the wet milling of sorghum are starch, edible oil, and gluten feed. The primary products of dry milling are low-protein flour and feed by-product.

Wheat Milling

In the United States, wheat is cultivated primarily as a food grain for human consumption. As a result, most of the wheat fed to livestock is in the form of mill by-products. Worldwide, however, an average of about 19% of the global wheat use is for animal feed.

Prior to grinding, the cleaned wheat is tempered, to facilitate the removal of the bran from the endosperm. The tempered wheat then passes into an Entoleter machine which breaks and removes unsound wheat. The sound wheat then passes through a series of grinders, sifters, and purifiers to separate the various parts of the grain. The latter process is repeated over and over again—sifters, purifiers, reducing rolls—until the maximum amount of flour is separated. Generally, millers remove about 72% of the wheat kernel for wheat flour, and the other 28% goes into the production of livestock feeds, primarily wheat middlings, wheat bran, wheat shorts, wheat red dog, wheat screenings, wheat germ meal, and wheat germ oil.

Rye Milling

The same basic type of machinery is used in milling both rye and wheat; and the primary objectives are the same—to produce a granular or powdery product by pulverizing the seed. The final milled product is free of bran and germ.

Rice Milling

Rice, like wheat, has traditionally been one of the staple grains for human food. However, several rice milling by-products unsuitable for humans can be effectively fed to livestock.

The purpose of milling rice is to separate the outer portions from the inner endosperm with a minimum of breakage. In modern mills, the rough rice passes through several processes in the mill: cleaning, (sometimes parboiling), hulling, pearling, polishing, and grading.

Dried rice that still has its hull is referred to as *rough* or *paddy* rice. Following the removal of foreign material by means of a fan and screen separator and a magnetic separator, the rough rice passes into an awning machine where the awns are removed from the grain. The grain then passes through a series of shellers, and the hulled grain is separated from the unhulled in the paddy machine—a complex sifter.

At this stage, the hulled rice is referred to as *brown* rice. This brown rice undergoes scouring in a series of hullers to remove the outer bran and inner bran (polishings).

Throughout the entire milling process, a number of kernels are unavoidably broken. These broken kernels are sorted by size and used as by-product feeds. The largest particles, generally whole or ¾ kernels, are referred to as head rice. Broken rice consisting of ¾ to ⅓ size grains are termed second heads. Screenings—⅓ to ¼ size—are the next type of broken rice. The smallest fragments are used for brewing; hence, the term brewers' rice.

The following by-products of rice milling are used as feed ingredients: rice hulls, rice bran, and rice polishings. Additionally, the broken rice (milled rice or head rice, second heads, and screening) is generally used as livestock feed—as rice groats, or as rice mill by-product. Rice groats, which is rice with the hulls removed, may be used as either human food or livestock feed. Rice mill by-product is the total mixture obtained from the milling of rice (hulls, bran, polishings, and broken rice grains). Brewers' rice is occasionally used as animal feed, although it is used chiefly in the brewing industry.

Oat Milling

Cleaned oats are initially roasted to 212°F for 1 hour. This process reduces the moisture of the grain to about 6% and weakens the hulls to facilitate removal. After cooling, the oats are separated by size, and the hulls, which account for about 30% of the weight of the grain, are subsequently removed by passing the kernels through 2 large circular milling stones. The milled oats are then screened to separate the groats (hulless grain), hulls, and broken grains. The hulless grain can then be rolled or milled into a number of attractive products, including oatmeal.

Barley Milling and Malting

The most important uses of barley are for human food, beverages, and livestock feed. Normally, about 51% of the U.S. barley supply is used for food, alcohol, and seed, and 49% is used for feed. Additional barley by-product feeds are obtained from milling and from malting/brewing/distilling.

Barley can be milled through the usual milling procedures used for the other cereal grains. However, barley is also used extensively in the malting and brewing/distilling industries and can be processed in a manner unique to traditional grain processing. In the malting process, clean, graded barley is wetted and germinated under carefully controlled conditions to minimize the loss of weight due to respiration. Once germinated, the grain is dried to stop growth and to produce a storable product. Following drying, the malt sprouts are removed and subsequently used as a by-product feed. The resulting malt is then used in the brewing and distilling industries and in the specialty food industry.

Malting, brewing, and distilling result in the following valuable by-product feeds for livestock: malt cleanings, malt hulls, malt sprouts, malted barley, brewers' grains, distillers' grains, distillers' solubles, spent hops, and brewers' yeast.

Nutrient Composition of Grain

Grains are fed to livestock primarily for their high-energy content. Of the grains most widely used in livestock rations, corn contains the most energy (see Table 10–2). Although corn has the highest energy value, it is the grain with the lowest crude protein content (9 to 11%). The rest of the feed grains contain from about 12.8 to 14.7% crude protein with wheat having the highest amount of crude protein.

TABLE 10-2
COMPOSITION OF TYPICAL FEED GRAINS[1]

Grain	Digestible Energy (Mcal/lb moisture-free grain)		Crude Protein
	Ruminants	Swine	
			(%, dry basis)
Corn, No. 2	1.80	1.81	10.2
Barley	1.75	1.58	13.2
Wheat, all analyses	1.73	1.73	14.7
Rye	1.62	1.68	13.8
Oats	1.53	1.43	13.3
Sorghum	1.49	1.74	12.8

[1]*From:* This book, Table V-1A, Energy Feeds.

Whole grains contain large amounts of digestible energy, as illustrated in Table 10-3. Many of the by-products from these grains contain as much or more energy than the whole grain from which they are derived. Of particular note is the fact that whole oats contain about 90% of the digestible energy of groats—oats with the hulls removed.

TABLE 10-3
COMPARISON OF ENERGY VALUES OF GRAINS AND THEIR MILL PRODUCTS FED TO SWINE (AS-FED BASIS)[1]

Grain	Digestible Energy
	(kcal/lb)
Barley:	
Whole	1,396
Malt sprouts	1,219
Corn:	
Grain, No. 2	1,586
Hominy	1,620
Oats:	
Grain	1,278
Groats	1,410
Meal	1,641
Wheat:	
Grain, all analyses	1,544
Bran	1,119
Middlings	1,321
Shorts	1,413

[1]*From:* This book, Table V-1A, Energy Feeds.

CARBOHYDRATES

About 83% of the dry matter of wheat, corn, rye, barley, sorghum, millet, and rice consists of carbohydrates. In oats, this figure is roughly 79%. The carbohydrate fraction of grain is composed primarily of pentosans, starch, dextrins, sugars, cellulose, and hemicellulose.

Starch is the most abundant type of carbohydrate in grain. About 60% of the entire wheat grain (as-fed basis) and 70% of its endosperm are starch. Two forms of starch are found in grain—amylose and amylopectin. Amylose is a straight chained polymer of glucose, whereas amylopectin is a highly branched chain polymer of glucose. In wheat, about ¼ of the starch is in the form of amylose while the remainder is amylopectin. Starch units are stored in granules found throughout the endosperm. The size and shape of these granules tend to vary according to the type of cereal. Simple structured granules are found in corn, wheat, rye, barley, and sorghum. In rice, these granules are compound. For the most part, oats contain compound granules, although the simple type can be found, also.

Fiber is extremely low in grains—ranging from about 0.5% in polished rice to about 12% in whole oats. Once the hulls are removed from oats, the crude fiber content of the grain is lowered dramatically. Thus, grains serve as valuable, highly digestible feeds for nonruminants as well as ruminants.

LIPIDS

Wheat, rye, barley, and rice contain about 1 to 2% lipid (fatty) material. The lipid content of sorghum is approximately 3%; whereas corn, whole oats, and millet contain from 4 to 6%. When the hulls are removed from oats, the resulting grain (groats) contains in excess of 7% lipid.

Wheat and corn germ have a rather large lipid fraction, containing 10 to 35%, respectively. However, other cereal by-products tend to contain substantially lower amounts. Wheat germ, bran, and endosperm contain lipids in concentrations of 6 to 10%, 3 to 5%, and 0.8 to 1.5%, respectively. Corn bran contains less than 1% lipid.

For the most part, the fatty acids in cereals are unsaturated, abounding in both oleic and linoleic acid. However, relatively large amounts of the saturated fatty acid, palmitic acid, can be found in grains (ranging from 11.5% of total fatty acids in barley to about 25% of total fatty acids in millet). Linoleic acid composes more than 50% of the total fatty acids in wheat, barley, rye, corn, and millet, and more than 40% in oats and sorghum. Only about ⅓ of the fatty acids in rice is linoleic acid. Oleic acid ranges from about 11% of total fatty acids in wheat to about 50% in rice.

PROTEIN

Although whole grains are not used as protein feeds, the fact that they are incorporated in livestock rations in large amounts means that a sizable amount of protein is provided from them.

Protein is found in all parts of the grain, but the embryo, scutellum, and aleurone layer contain considerably higher protein concentrations than the endosperm, pericarp, and testa. Most of the protein in corn and wheat is located in the endosperm; but it must be remembered that the endosperm is the largest segment of the grain kernel, 79.6% in corn and 82.5% in wheat. In both types of grain, the embryo and the scutellum have high concentrations of protein. Of particular note is the extremely low-protein content of the pericarp in each of the two grains.

In nonruminant rations, the essential amino acid composition of the various feed ingredients warrants careful consideration. Since grains and grain products generally compose the bulk of these rations, their respective essential amino acid profiles can determine what type of protein supplement must be added to the ration. All the grains and their by-products, especially corn, are extremely low in tryptophan and lysine. Methionine—one of the sulfur amino acids—is also low in grains and their by-products. A large amount of data suggest that lysine is the first limiting amino acid in most nonruminant feeds. The grains are notably low in lysine. Additionally, many of the protein supplements of plant origin are low in lysine, thereby magnifying the problem. (See Chapter 11, Protein Supplements.)

High-Lysine Corn (Opaque-2)

It has been known for many years that corn, the world's third most important human food after rice and wheat, is nutritionally inadequate. In 1914, researchers at the Connecticut Agricultural Experiment Station induced starvation in laboratory rats by feeding them generous helpings of corn. Further, it was found that rats could be restored to health by supplementing the high-corn diet with two protein fractions—the amino acids lysine and tryptophan.

Although normal corn contains about 10% protein, half of the protein consists of zein, which is especially low in lysine and tryptophan, essential amino acids that the nonruminant cannot manufacture and must get from feed.

This deficiency of corn shows up in people wherever corn is a major source—if not the only source—of protein in the diet. Known by the name, *kwashiorkor*, this nutritional deficiency disease is the leading cause of mortality among infants and children in many parts of the world.

For years, plant scientists assayed the world's corn varieties one by one, looking for a strain with more nutritionally balanced protein. Finally, in 1963, a Purdue University team headed by biochemist Edwin T. Mertz analyzed an odd lot of corn characterized by soft, floury endosperm inside an opaque, chalk-white kernel. The Purdue scientists found that the opaque characteristic of corn, which had been noted for years without exciting much scientific interest, is associated with a recessive gene that replaces some of the kernel's nutritionally deficient zein with needed lysine and tryptophan. The mutant, labeled Opaque-2, was twice as high in lysine and tryptophan as ordinary corn. Later, another high-lysine strain was found, which was named Floury-2.

But the millenium that seemed so near with the discovery of Opaque-2 has remained frustratingly out of reach. Although the nutritional value of the high-lysine corn is recognized, two major hurdles between research discovery and application must yet be overcome: (1) The mutant gene is linked to Opaque-2's soft, floury kernel, which is both light in weight and vulnerable to pest attacks, producing lower yields for farmers; and (2) Opaque-2 has not been accepted by the majority of consumers, who are accustomed to the harder *flint* or *dent* kernels with a deeper, translucent color. But the need is great—human lives are at stake. So, plant breeders have set about crossing the Opaque-2 gene on corn varieties that better meet the demands of both farmers and consumers.

MINERALS

Grains generally contain more minerals than forages. However, all grains (especially corn) are extremely low in calcium but fair to good sources of phosphorus. Therefore, special attention must be given to mineral supplementation of rations that are predominantly grain.

Naked caryopses and grains with their coverings removed contain most of their minerals in the forms of phosphates and sulfates. The primary cations are potassium, magnesium, and calcium. Substantial portions of the phosphorus are bound in the form of phytic acid—a form of phosphorus that is of little value to certain livestock.

The hull portions of rice, barley, and oats are excellent sources of minerals, yielding ash contents of 22.6, 6.0 and 5.2%, respectively. However, the majority of this ash portion is silica.

VITAMINS

Vitamin A activity is low in all cereal grains except fresh yellow corn. Oats and wheat germ contain extremely high amounts of vitamin E.

Thiamin, riboflavin, pantothenic acid, and pyridoxine are found throughout the grain kernel. However, thiamin is found in relatively high concentrations in the embryo and scutellum. Niacin is found in relatively high concentrations in wheat, corn, and rice. However, in corn, most of the niacin is in the form of niocytin—a form of niacin that is biologically unavailable.

Effect of Stage of Maturity on the Nutritive Value of Grains

As grains mature, there is a steady decrease in moisture content paralleled by an increase in carbohydrate content. Workers at the Minnesota Agricultural Experiment Station demonstrated the effect of stage of maturity on the nutritive value of corn. They found that the dry matter content increased dramatically as the grain matured. Likewise, carbohydrate parameters (nitrogen-free extract and starch) showed increases. Fat content also increased from 3 to 4.9%. As corn matured, crude protein, crude fiber, ash, and cell wall constituents decreased.

On a dry matter basis, the Minnesota researchers also found that there was a decreasing trend in each amino acid concentration as the corn matured; but when concentration in protein was considered, there was relatively little change among the amino acids due to stage of maturity.

The Minnesota investigators also made interesting observations relative to the mineral composition of corn. When concentration in dry matter was used as a basis of comparison, both phosphorus and potassium showed dramatic declines as the corn matured. On the other hand, when the concentration in ash was used for comparison, potassium showed a dramatic decline, but phosphorus showed a dramatic increase. Magnesium content closely paralleled phosphorus. As the grain matured, trace minerals constituted larger proportions of the ash concentration.

Grain Standards

Grains, like all feeds, vary in quality. To assist grain producers, buyers/sellers, and users in recognizing this variability, the government has established a set of standards for the grading of grains. Grain grading is done by the Federal Grain Inspection Service (FGIS), on a per hour charge basis. Grades are based on the physical and biological factors present in a sample. In general, the criteria for grading are (1) test weights per bushel, (2) moisture content, (3) foreign material and other grains, (4) broken and damaged kernels, and (5) discoloration. Also, it is important that a visual inspection be made for mold and/or fungi. There are additional grading criteria for some grains. In all cases, the highest grade is

U.S. No. 1 and the lowest is U.S. Sample grade.

Grain prices are quoted and trading is done on the basis of a *trading grade* on each grain. The *trading grade* is established by trading custom rather than by law. Normally, it is based on the U.S. Standard for the No. 2 grade. The only important exceptions in the Midwest are: (1) the test weight on oats, which is usually 36 lb per bushel rather than 33 lb; and (2) the moisture content and foreign material on soybeans, which are usually the No. 1 specifications (13% and 1% respectively) rather than those for No. 2. The specifications for the U.S. No. 2 grade of grains are given in Table 10-4.

TABLE 10-4
U.S. STANDARDS FOR NO. 2 GRADE GRAINS[1]

Grain	Minimum Test Weight Per Bushel	Maximum Moisture	Minimum Sound Grain	Maximum Foreign Material	Maximum Heat-Damaged Kernels	Maximum Total Damaged Kernels	Maximum Discolored Grain
	(lb)	(%)	(%)	(%)	(%)	(%)	(%)
Barley[2]	45		94	2	0.3	4	1.0
Corn	54	15.5	—[3]	—[3]	0.2	5	
Oats[4]	33		94	3	0.3		
Rye[5]	54			6	0.2	4	
Sorghum[6]	55	14	—[7]	—[7]	0.5	5	
Soybeans[8]	54	14		2	0.5	3	2.0
Wheat, hard red or white club	57			1	0.2	4	

[1]Adapted by the authors from *The Official United States Standards for Grain*, USDA, January 1978.
[2]One additional grade requirement is used for barley—*thin* barley (%). The maximum limit of *thin* barley for U.S. Grade No. 2 is 15%.
[3]In the grading of corn, the criteria of broken kernels and foreign material are combined. In No. 2 corn, the maximum total of broken kernels and foreign material is 3%.
[4]Oats that are slightly weathered shall be graded no higher than U.S. No. 3. Oats that are badly stained or materially weathered shall be graded not higher than U.S. No. 4.
[5]One additional grade requirement is used for rye—foreign matter other than wheat (%); the maximum limit for rye grading U.S. No. 2 is 2%. The rye in grade U.S. No. 2 may contain not more than 15% *thin* rye, which *thin* rye shall consist of rye and other matter that wil pass readily through a sieve 0.032 in. thick with perforations 0.064 by 0.375 in.
[6]Under the U.S. Grades and Standards, there are four classes of sorghum: yellow, brown, mixed sorghum, and white. Sorghum which is distinctly discolored shall not be graded higher than U.S. No. 3.
[7]In the grading of sorghum, the criteria of broken kernels, foreign material, and other grains are combined. In No. 2 sorghum, the maximum total broken kernels, foreign material, and other grains is 8%.
[8]Soybeans which are purple mottled or stained shall be graded no higher than U.S. No. 3. Soybeans which are materially weathered shall be graded no higher than U.S. No. 4.
NOTE WELL: Effective May 1, 1988, the standards permit the presence of fewer insects before the grain is discounted or termed *infested*. Samples of corn, sorghum, oats, barley, or soybeans are discounted if they have 2 or more live weevils, or 1 live weevil and 5 or more other live insects injurious to stored grain, or 10 or more other live insects. Wheat, rye, or triticale samples will be penalized for 2 or more live weevils, or 1 live weevil and 1 or more live insects injurious to stored grain, or 2 or more live insects injurious to stored grain.

As is true of most standards, not everyone approves of the current grain standards; and they are subject to change. The chief criticisms of the present grain standards are:

1. Their enforcement is not rigid enough.
2. They need to define or describe quality more than by a grade number.
3. They do not provide bonuses (incentives) to encourage the production and sale of high-quality grain. Yet, others counter by saying that there is price incentive now—that quality is reflected in the prices of buying contracts.
4. They do not reflect consumer concerns relative to insect infestations; they do not force food processors to avoid grain purchases which have insect infestations or have been fumigated for insects.
5. They do not guarantee shipments of grain that meet the standards of nations to which U.S. grains are exported.
6. They are lower than the grain standards of two of our chief export competitors, notably, Australia and Canada.

Grain Storage

When grain is properly stored in well-designed facilities, losses due to spoilage and contamination can be held to a minimum. However, when little attention is given to the construction of the facilities or the condition of the grain to be stored, serious economic consequences can, and in all likelihood will, result.

When storing grain, special attention must be given to moisture, temperature, and to damage from molds, insects, and rodents.

MOISTURE

When grains are stored at moisture levels less than 12%, the growth of most microorganisms is kept in check. If the moisture level is reduced to 10%, development of most insects will be arrested. However, when moisture level is decreased, especially to levels below 12%, incidence of grain breakage is markedly increased.

Sprouting can occur in stored grains when moisture levels exceed 30%. In addition to the alteration of the physical form of the grain during sprouting, heat is generated in the stored grain, thus creating a potential fire hazard.

(Also see Chapter 14, section on "Drying [Dehydrating].")

TEMPERATURE

As the temperature of storage is decreased, the number of problems associated with storage likewise decreases. For this reason, small grain bins painted white or completely shaded are generally better than large bins that are unpainted or exposed to the sun.

When storage temperatures are decreased to about 40°F, mites do not develop. Temperatures below 60° and 32°F inhibit growth of insects and fungi, respectively. However, the effects of temperature are closely interrelated with moisture in the grain and air movements throughout the storage bins.

BIOLOGICAL SOURCES OF GRAIN DAMAGE DURING STORAGE

The actual physical destruction of stored grain by moisture and temperature themselves can be great, but the losses due to biological factors—such as mold growth and insect and rodent damage—can be devastating. Quite often, the careful control of moisture and temperature can reduce these losses. However, additional precautions involving the construction of the storage facilities and the treatment of stored grains with pesticides may be needed.

Because grain is an excellent source of nutrients for livestock, it follows that it is also relished by insects and rodents. In cases of excessive moisture and poor storage management, grain provides an excellent growth medium for toxin-producing microorganisms that can create morbidity or mortality in livestock which consume the contaminated feed. *CAUTION: For additonal assistance in choosing an insecticide or rodenticide, consult the local County Extension Agent; and always follow the manufacturer's label directions. Seed that is dyed bright pink or reddish purple has been chemically treated to control insects and other pests, or molds, and is for planting purposes only; so, do not use for animal feed or human food.*

(See Chapter 5, Table 5-3, Potential Poisons.)

Insects and Mites

Infestations by insects and mites is most prevalent when the stored grain is in poor condition—for example, when a large proportion of the kernels are broken. Grain is subject to exposure to these pests at all stages of handling—harvesting, storing, transporting, processing, and packaging. Hence, it is imperative that precautionary measures be employed throughout handling.

Two basic methods of controlling insects and mites are used: (1) fumigation, and (2) residual contact pesticides. Fumigants are particularly effective because they can diffuse throughout the intergranular spaces of the stored grain. Many of the fumigants that are used to kill pests are equally effective against humans; hence, it is imperative that only well-trained personnel undertake this procedure. Two factors should be considered when fumigants are used: (1) effective lethal concentration in the air, and (2) length of exposure time. Residual contact pesticides are used to kill pests present in the grain and to prevent further infestation. Little or no specialized equipment is necessary for the application of these pesticides, and the dangers to the applicator are not as great as with fumigants. Additionally, these pesticides retain their potency for several months. Five types of residual contact pesticides are used: (1) dusts, (2) wettable powders, (3) emulsions, (4) aerosols, and (5) smoke.

Both temperature and moisture affect the effectiveness of pesticides. High temperatures in stored grains can reduce the amount of adsorption and absorption of fumigants and increase the rate of metabolism of the insecticide by the grain. On the other hand, extremely low temperatures reduce the effectiveness of fumigants. As moisture in grain increases, the rate of penetration of the pesticide into the grain will be reduced.

When pesticides of any type are used, the applicator should follow the directions and specifications on the label. If too little pesticide is used, infestations will be neither eliminated nor prevented. If too much pesticide is used, serious consequences from residues can result.

Rodents

It has been estimated that rats consume 10% of their body weight in feed per day. In addition to the loss of feed, contamination from the feces and urine of rats constitutes further damage to grain. The United Nations Food and Agriculture Organization estimates that rats eat or contaminate 42.5 million tons of the world's grains each year—enough to feed 200 million people. For these reasons, rodenticides are commonly used where grain is stored.

The factors to be considered when selecting a rodenticide are (1) toxicity (for example, dosage levels), (2) acceptance, (3) reacceptance, (4) development of tolerances, (5) hazards to other animals, and (6) duration of potency. It is important to follow the label directions of the rodenticide carefully. Dosages recommended by the manufacturer are generally of sufficient strength to kill rodents with above-average tolerances. By using more than the recommended dosage level, the user may be decreasing the acceptability of the poison to the pest as well as increasing the risk of poisoning animals for which the poison is not intended.

Microbial Contamination

Both molds and bacteria can create serious problems in feeds. Molds will actively develop in grains stored at moisture contents in excess of 14.5%. Under certain temperature and moisture conditions, molds can produce mycotoxins which, if incorporated in feeds, can have an adverse effect on animals, *i.e.,* reduce weight gains, cause sickness, or even result in death.

Fig. 10-6. A mixture of ergot, a parasitic fungus (see black, banana-shaped masses), and wheat kernels as they appeared after harvest. (Courtesy, University of Idaho, Moscow)

The aflatoxin-producing fungi on feed grains in storage can be controlled, without affecting the feeding value for livestock and poultry, by treatment with an organic acid (propionic, acetic, or isobutyric; or a mixture of these). If applied properly to the grain as it is augered into the bins, organic acids will prevent growth of *A. flavus*. However, they will not remove any aflatoxins which formed within the grain before the fungus was killed.

Toxin-concentrated grain can be detoxified to levels below 20 ppb by using anhydrous ammonia. **NOTE WELL:** Ammonia is not registered by the Food and Drug Administration for use on grains to be shipped out of state, but it may be used on grains remaining in the state.

Animal feeds can be an important link in the *Salmonella Cycle*. Salmonella bacteria are responsible for feed-borne infections in animals and food-borne infections in humans.

Salmonella contamination can be prevented by good sanitation. This involves controlling rodents and birds and avoiding the contamination of grains by manure.

In all treatments, always follow manufacturer's directions.

(Also see Chapter 5, Table 5-3, under "Mycotoxins," "Pesticide Poisoning," and "Salmonellosis.")

OTHER HIGH-ENERGY FEEDS

Although feed grains and their milling by-products comprise the vast majority of the energy feeds, numerous other feeds are routinely used to supply energy to livestock. Seeds from plants other than *Graminae* can be used effectively (for example, beans). Fats provide an extremely concentrated source of energy. Molasses is a liquid energy feed that is highly palatable and digestible. When the price and availability are advantageous, roots, tubers, and certain other by-product feeds are fed to livestock.

Other Seeds

Seeds from plants other than cereal grains are used in livestock feeds when they are readily available and when the price is right. Legume seeds, or pulses, such as soybeans and peanuts, and whole cottonseed are used for their energy content in addition to protein. Many types of seeds are by-products of cash crop enterprises, representing culls of processing or marketing. On occasion, a surplus of a certain seed generally used for human consumption may reduce the cost to a level where it becomes economically feasible to incorporate it in livestock feeds. For this reason, livestock producers should become familiar with the feed substitution tables found in the feeding chapters devoted to each class of livestock.

Sprouted Grains

Sprouted grains are seeds that, following stimulation by water, air, and temperature, have germinated, or started to grow. Adverse moisture during harvest can cause grain to sprout in the head. Sorghum, barley, and wheat are especially affected, with the result that large quantities of these energy feeds become unsuited for milling purposes and become available for livestock feeding as sprouted grain.

U.S. Grade Standards discount grain on the basis of percent of sprouted kernels. For example, wheat showing more than 2% sprouted kernels is classified as sprouted wheat; and the grade is lowered with increased sprouting until, at 15%, the grain is classified as Sample grade. Likewise, in commercial trading channels, sorghum with 15% or more sprouted kernels is classed as Sample grade; and this, along with its light test weight per bushel, makes for a depressed price.

When sprouting drops the price of grain, the nutritive value of the sprouted grain should be evaluated to establish a fair price to both grain growers and livestock feeders. Fortunately, the nutritional value of sprouted grain is generally good, despite its dismal appearance. Also, it is noteworthy that the metabolic changes occurring in a grain kernel during germination (sprouting) are similar in many respects to what occurs in high-moisture grain or in reconstituted grain. It follows that experiments and experiences generally show that the sprouting of grain does not significantly affect its feeding value for cattle, sheep, swine, and poultry.

The Idaho Agricultural Experiment Station staff studied the value of sprouted wheat for growing-finishing hogs when the proportion of sprouted kernels represented 10, 20, and 30%, respectively, of the ration. Average daily gains were not affected, but more feed per pound of gain was required due to the lower energy value of the sprouted wheat. The Idaho researchers reported the following reduction in the energy value of the sprouted wheat for swine as compared to normal unsprouted wheat:[2]

Proportion of Sprouted Wheat Kernels	Energy Value of Sprouted Wheat Relative to Unsprouted Wheat For Swine
(%)	(%)
20	92.5
40	87.2
60	85.6

The Idaho researchers also conducted trials with cattle and chicks, feeding wheat that was classified as 60% sprouted. The rate and efficiency of gains of steers and chicks was not affected by the sprouted wheat that was fed.

North Dakota State University researchers reported that sprouted barley gave pig performance comparable to that of unsprouted barley. Kansas State University workers fed sprouted sorghum to growing-finishing pigs without significantly reducing daily gains, although slightly more feed was required per pound of gain.

It should be recognized that percentage of sprouted kernels alone is not an adequate measure of the nutritional value of sprouted grain for livestock and should not be the only criterion for discounting the price of sprouted grain. Substantial sprouting (involving most kernels and long sprouts) will reduce the energy available in the kernels, and require more feed per pound of gain. Also, when buying or selling sprouted grain the following additional factors should be considered: (1) the possible presence of molds and toxins, (2) the high moisture content of sprouted grains, and (3) storage problems.

Cottonseed

About 5 million tons of cottonseed are produced in the United States annually.

In recent years, the price of whole cottonseed has been favorable enough to include it in cattle feed, for both dairy and beef animals. Previously, the oil was extracted and used in human foods and industrial oils, leaving cottonseed meal and cottonseed hulls for animal feeds.

---

[2] Eide, W. D., et al., *Feeding Value of Sprouted Grain*, Circular AS-647, North Dakota State University, Fargo, 1978. (**Note:** This North Dakota publication reported the Idaho study. M.E.E.)

Whole cottonseed is very high in energy (95% TDN for ruminants, on a moisture-free [M-F] basis); high in protein (24% crude protein on M-F basis); high in phosphorus (0.76% M-F); and high in fiber (21.4% M-F); and it requires no feed processing.

Whole cottonseed is especially valuable as a hot weather feed for lactating dairy cows and finishing beef cattle because of its low heat increment, which means that less incidental heat is produced during its metabolism than from carbohydrates and proteins.

- **Whole cottonseed for lactating dairy cows**—Whole cottonseed is an important and rather unique feed for lactating cows, because it combines high energy and high-crude fiber with relatively high crude protein. In about 80% of the experimental feeding trials, whole cottonseed increased fat of milk about 0.3%. This increases milk values about 50¢ per cwt, which almost pays for the whole cottonseed. However, whole cottonseed tends to depress the protein fraction of solids-not-fat (SNF). So, where SNF is a milk price factor, lower protein in the milk may defeat the benefits of a high butterfat content.

Cottonseed can be fed to lactating cows "as-is," requiring no grinding or pelleting; and either linted or delinted seed may be used. It can be easily top-dressed over silage or green chop, mixed with other grains for bulk feeders, or processed with other grains for pelleted parlor mixes.

The amount of whole cottonseed to feed should be determined by comparative feed prices, cow weights, and the other ration components. For large animals, like Holsteins, most producers feed a minimum of 4 lb and a maximum of 7 lb per cow daily. There are a few reports favoring feeding up to 12 to 14 lb per cow daily. Also, some producers feed large amounts of whole cottonseed to the highest producers, a modest amount to average producers, and none at all to low producers.

- **Whole cottonseed for beef cattle**—Beef producers consider whole cottonseed a good buy when 100 lb of it costs less than the combined cost of 35 lb of cottonseed meal and 65 lb of sorghum. Pertinent points relative to the feeding of whole cottonseed to beef cattle follow:

1. Whole cottonseed may constitute up to 20% of the ration of finishing cattle. Because of the high oil content of cottonseed, weaned calves should not receive more than 4 lb per head daily, yearlings not more than 6 lb daily, and more mature animals not receive more than 7 lb daily.

2. As with any new feed, cattle may have to be enticed to eat cottonseed, which can be accomplished by top-dressing it with molasses or other palatable feed.

The following cautions should be observed when feeding whole cottonseed.

1. Cottonseed can combust spontaneously if stored too wet and stacked too high, so the moisture level should not exceed 14%, and it should be stored in a flat bulk container.

2. Aspergillus mold, which produces aflatoxins, grows in any moist feed, but cottonseed is more likely to grow aflatoxins than many other feeds. This emphasizes that cottonseed must be stored and kept dry.

3. Whole cottonseed containing more than 1.0% gossypol can be toxic when fed to young calves under 4 months of age or to monogastric animals.

NOTE WELL: (1) In mature ruminants, gossypol is detoxified by the action of the rumen contents; and (2) the new glandless varieties of cottonseed do not contain gossypol.

In 1988, the National Cottonseed Products Association developed quality standards/trading rules for whole cottonseed, with descriptions and specifications for prime feed grade cottonseed, delinted prime feed grade cottonseed, and feed grade cottonseed—off quality.

Fats and Oils

Feeding of fats was prompted in an effort to find a profitable outlet for surplus vegetable oils and packinghouse fats. For the most part, fats were formerly used for soapmaking, but they are not used extensively in manufacturing detergents. Thus, with the rise in the use of detergents in recent years, they became a competitive energy feed. Hand in hand with these circumstances, the use of fat in feeds was enhanced by (1) recognition of the effect of high-energy feeds on efficiency of feed utilization; (2) discovery of the role of amino acids in improving the energy utilization by animals; and (3) discovery of the role of antioxidants in maintaining the quality of feed grade fat and other renderers' products.

In 1984, 1,443,000,000 lb of fats and oils were used in animal feeds in the United States, 93% of which were tallow and grease.

Fats and oils are high-energy sources. They enable animals to meet a high-energy requirement with less feed, because they contain 2.25 times as much energy as carbohydrates. For ruminants, the energy value of fats in relation to carbohydrates is higher than the normal 2.25 to 1 for monogastric animals; for cattle feed, fat is worth about 3.2 times the energy value of corn.

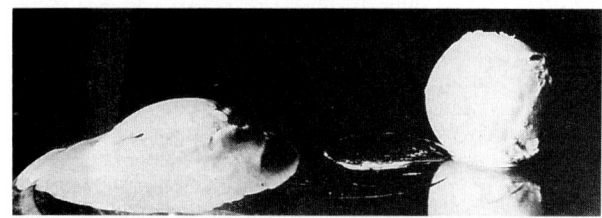

Fig. 10–7. Feed fats affect body fats. The soft lard sample (left) came from hogs fed a ration containing whole soybeans. Both samples of lard had been exposed to room temperature, 70°F, for 2 hours prior to photographing. (Courtesy, University of Illinois, Urbana)

Common belief to the contrary, animals can tolerate a rather high-fat content in the ration. As evidence of this, it is noteworthy that sucklings normally handle a relatively large amount of fat, for milk contains 25 to 40% of this nutrient on a dry matter basis. Also, except for the soft pork problem, no apparent difficulty is encountered in feeding hogs a rather high-fat ration, such as results when large quantities of peanuts or soybeans are fed.

A small amount of fat in the ration is desirable, as these fats are the carriers of the fat-soluble vitamins. Also, there is evidence that some species (humans, swine, rats, and dogs) require certain fatty acids. Although the fatty acid requirement of farm animals has not been determined, it is thought that ordinary farm rations contain ample quantities of these nutrients.

Because animal and vegetable fats seem to be equally effective additions to most livestock rations, with the exception of those for lactating cows, selection should be determined solely by comparative price. Ordinarily, animal fats are much cheaper than such vegetable fats as soybean or cottonseed oil.

There are many different feeding fats. Also, the terminology and standards differ. So, fats should be bought by specifications and guarantees. The quality of fat is based on free fatty acid content, moisture, insolubles, unsaponifiable matter, color, odor, and titer (temperature developed as a result of the heat of crystallization during cooling). Fats are designated as tallows or greases based on the titer value. The American Association of Feed Control Officials lists the following types of fats: animal fat; vegetable fat, or oil; hydrolyzed fat, or oil; fat product; corn endosperm oil; vegetable oil refinery lipid; corn syrup insolubles; beef fat; pork fat; poultry fat; and tall oil fatty acids (product obtained from making pulp from pine trees).

Generally, vegetable oils are used for human consumption; hence, their cost is usually prohibitive to the livestock producer. However, vegetable oils can be used for livestock when the following circumstances prevail:

1. When the size of the individual purchase is small, the costs of plant oils can sometimes be justified.

2. When animal fats are unavailable or scarce, crude vegetable oils are sometimes used.

3. When edible vegetable oils become off-color, their marketability for human consumption decreases; and the prices may be subsequently reduced to a level at which the livestock producer can afford to use them.

Fat serves the following functions when added to livestock rations: (1) increases the caloric density of the ration; (2) controls dust in feed processing; (3) lessens the wear and tear on feed mixing equipment; (4) facilitates pelleting of feeds; (5) increases palatability; (6) helps to homogenize and stabilize certain feed additives, especially those of a very fine particle size; and (7) enhances the digestibility of the other feed ingredients of the ration. Fats also provide needed caloric density in the ration for high-producing animals, such as high producing dairy cows in early lactation and horses in endurance trials. During hot and cold weather, fats provide added energy without overloading the digestive system. Results of cattle feedlot trials have also indicated that fat will improve the feeding value of wheat as a feed grain, making it more economically comparable to corn and sorghum.

Fats are used to some degree in all types of livestock operations. Of the livestock feeds, milk replacers, used in feeding suckling animals, contain the most fat—ranging from 15 to 30%.

The following general guidelines relative to the use of fats and oils in animal feeds are based on experiments and experiences:

1. Add supplemental fat when it is desired to increase energy density, lower heat increment during hot weather, and increase feed efficiency.

2. Make such adjustments as necessary in other dietary ingredients when feeding fats because other nutrients may be decreased as a result of the higher energy level and possibly decreased nutrient intake of the ration.

3. Exercise extreme care to prevent contamination of milk or meat from such items as pesticides, chlorinated hydrocarbons, and PCBs, which are fat-soluble and may be introduced with blended fats.

Additionally, the following species recommendations are pertinent when adding fats to rations:

- **Fat for beef cattle**—It is recommended that 2 to 5% fat be added to high concentrate cattle finishing rations in which milo, barley, and/or wheat are the chief grain sources. When corn is the major grain, the addition of fat is less effective than with small grains, since corn contains approximately 4% fat compared to 1 to 1½% fat in other feed grains.

- **Fat for dairy cows**—It is recommended that 1 to 1½ lb of blended fat be fed daily to high producing dairy cows in early lactation and/or during hot weather, to stimulate milk production and improve feed efficiency; and that the maximum content of the fat in the ration not exceed 6% of the dry matter content. But not all fats perform alike when fed to lactating cows! Most unsaturated types of fats such as corn oil, soybean oil, safflower oil, and cottonseed oil have been shown to depress fiber digestibility followed by lowering of the butterfat (BF) percentage and the protein fraction of the solid-not-fat (SNF). So, it is recommended that lactating cows be fed either (1) a blend of more-saturated animal fat and vegetable oils and soapstocks, or (2) whole cottonseed or raw or cooked soybeans, which break down and release fats slowly.

- **Fat for horses**—It is recommended that 5 to 10% fat be added to the ration of horses subjected to intense and prolonged exercise and stress, such as endurance trials, to increase energy intake and improve performance.

- **Fat for swine**—It is recommended that approximately 5% fat be added to brood sow rations during late gestation to increase pig survival at birth, milk yield of sows, and rebreeding conception.

It is recommended that 3 to 7% fat be added to pig starter rations for early-weaned pigs (pigs weaned at about 3 weeks of age) to improve daily gains and feed efficiency.

- **Fat for poultry**—It is recommended that 4 to 6% fat be added to broiler and turkey rations to increase growth rate, and to layer rations to increase egg production.

One of the more exciting nutritional developments is the so-called rumen protected fat system. In this system, the fat is emulsified with a protein. Then the protein is treated with formaldehyde to harden it. The product cannot be digested in the rumen due to the formaldehyde linkage on the protein, but the pH of the abomasum is such that this linkage is broken, with the result that the protein in fat will then be digested in the small intestine (similar to that of the monogastric animal). Studies have shown that the utilization of fat by the ruminant can be improved by the use of this system. Also, and most significant, by the use of the protein protected fat system, it is possible to alter the ratio of the saturated to the unsaturated fatty acids in beef depot fat. The same is true of butterfat. A beef fat which contains high levels of unsaturated fatty acids is especially attractive to people with certain types of heart conditions. Thus, in the future, we may see some beef cattle fed rations containing protein protected fat for the express purpose of producing Choice beef in which the fat contains a high level of unsaturated fatty acids.

(Also see Chapter 6, section on "Fats and Oils"; Chapter 11, section on "Protein Bypass (Protected Protein, Escaped Protein)"; Chapter 14, sections on "Fat Added," and "Slow-Release and Rumen Bypass Treatments"; and Chapter 25, section on "Fats.")

Molasses

Molasses (including cane or blackstrap, beet, citrus, wood, and starch molasses) is extremely palatable and an excellent source of energy. Approximately 638 million gal or 3,745,265 tons, of molasses of all kinds are used annually in the United States. About 73%, or 2,734,044 tons, is consumed in feeds and the balance is used for industrial purposes.

Cane and beet molasses are by-products of the manufacture of sugar from sugarcane and sugar beets, respectively. Citrus molasses is produced from the juice of citrus wastes. Wood molasses is a by-product of the manufacture of paper, fiberboard, and pure cellulose from wood; it is an extract from the more soluble carbohydrates of the wood material. Starch molasses is a by-product of the manufacture of dextrose from starch derived from corn or grain sorghums in which the starch is hydrolyzed by use of enzymes and/or acid. Cane or blackstrap is, by far, the most extensively used type of molasses. The different types of molasses are available in both liquid and dehydrated forms.

In addition to its use as an energy feed, molasses is used in the following ways: (1) as an appetizer, (2) to reduce the dustiness of a ration, (3) as a binder for pelleting, (4) to stimulate rumen microbial activity, (5) to supply unidentified nutrient factors, (6) in the case of cane molasses, to provide trace minerals, and (7) to provide a carrier for nonprotein nitrogen and vitamins in liquid supplements.

In ruminant rations, molasses is restricted to the level of 10 to 15% of the ration since it is most efficiently utilized when it does not exceed these levels. Excessive amounts of molasses (greater than 15%) will cause the feed to become messy and unmanageable as well as create digestive disturbances. Poultry are rather sensitive to molasses as excessive levels cause diarrhea. Hence, although molasses occasionally constitutes as much as 10% of poultry rations, levels are generally restricted to from 2 to 5%.

The quality of molasses is determined by its sugar content, which is expressed by the term, *Brix*. Brix is determined by measuring the specific gravity of molasses. After the specific gravity has been obtained, the value is applied to a conversion table from which the level of sucrose (or degrees Brix) can be determined. As sugar content increases, degrees Brix likewise increases. Since molasses also contains lipids, protein, inorganic salts, waxes, gums, and other material, the Brix classification can often be misleading, because each of these contaminants has an influence on the specific gravity of the solution. However, degrees Brix does give a relatively accurate indication as to the sugar content of molasses and is, therefore, a good means of determining quality. A summary of the minimum specifications for molasses is presented in Table 10-5.

TABLE 10-5
SPECIFICATIONS FOR MOLASSES

Type	Degrees Brix	Total Sugars as Invert	Weight per Gallon
	(minimum)	(minimum %)	(lb)
Beet	79.5	48	11.75
Cane	79.5	43	11.75
Citrus	71.0	45	11.29

(Also see Chapter 14, Feed Processing, section on "Molasses Added.")

CANE MOLASSES

As early as 1900, sugar planters in Louisiana fed cane molasses in long, open troughs to mules, cows, and hogs. The animals did their own mixing, alternating between a mouthful of hay, corn, oats, and molasses, and washing the molasses down with a drink of muddy water from a nearby bayou. Feeding was a sticky, messy business, and attracted swarms of flies. Even so, molasses was fed simply because it had to be disposed of, and the animals liked it. It couldn't be sold for the price of the barrel container, and there was still the matter of freight charges. Gradually, the feeding value of the product was appreciated, and, by 1914, the demand exceeded the domestic supply. Shipments were brought in from Cuba. Today, a large quantity of this carbohydrate feed is produced in the southern states, and an additional tonnage is imported.

Cane molasses weighs about 11.7 lb per gallon. One hundred pounds of cane molasses will yield about 50 to 55 lb of highly digestible sugars and 26 lb of water. Cane molasses contains an extremely low amount of protein (about 4.3% as-fed basis). In fact, large amounts of molasses in feed can depress protein utilization of other feeds. Molasses is deficient in thiamin, riboflavin, vitamin A, and vitamin D, but it is rich in niacin and pantothenic acid. Although cane molasses is low in phosphorus, it is an excellent source of other minerals. It may contain up to 10% ash.

Cane molasses can be used most effectively by adding small amounts to poor-quality roughages for ruminants.

BEET MOLASSES

The use of beet molasses is generally restricted to the western states which produce most of this country's sugar beets. When properly used, this type of molasses has the same feeding value as cane molasses. However, care should be taken when using beet molasses because it has highly laxative properties due to its high-mineral content. Total sugar content generally varies from about 48 to 53%. Protein values are higher than in cane molasses (6.6% vs 4.3%).

CITRUS MOLASSES

Citrus molasses contains about 41 to 43% total sugars—substantially less than either cane or beet molasses. Moisture content is also higher, ranging from 27 to 30%. Protein content is about 5.8% as-fed basis, which is intermediate between cane and beet molasses.

WOOD MOLASSES

This type of molasses is little used in the feed industry. The energy content of wood molasses is high since it contains at least 55% total carbohydrates. The protein content is very low, only 0.6% on an as-fed basis. Use of this feed has been limited to beef cattle rations, for the most part.

Roots and Tubers

A root crop consists of the fleshy subterranean parts of a harvested plant—for example, carrots and beets. Tubers are *short, thickened, fleshy stems, or terminal portions of stems or rhizomes, that are usually formed underground—for example, peanuts and potatoes.*

Root and tuber crops have traditionally been used more extensively as livestock feeds in Europe than in this country. They yield relatively large amounts of nutrients per acre; but the cost of the labor needed to harvest these crops has been prohibitive from a livestock feeding standpoint. Another limitation in their use is their high moisture content. However, in the production of roots and tubers for human consumption, considerable wastes become available which can be fed to livestock as by-product feeds.

Due to the large amounts of moisture in these feeds, only limited amounts are fed to high-producing animals which require considerable amounts of dry matter. Ingestion of highly succulent feeds decreases the amount of space in the stomach and small intestine, thus restricting the amount of dry matter that can be consumed.

When considered on a dry matter basis, roots and tubers are relatively good sources of energy (see Table 10-6). However, they generally have limited amounts of protein and, except for carrots and sweet potatoes, contain very little carotene. Calcium and vitamin D levels are also extremely low. Best results are secured in most animals when roots are limited to a replacement of not more than one-fourth of the grain ration.

TABLE 10-6
DIGESTIBLE ENERGY VALUES OF VARIOUS ROOTS AND TUBERS[1]

Feed	Dry Matter	Digestible Energy			
		Ruminants		Swine	
		As-Fed	Moisture-Free	As-Fed	Moisture-Free
	(%)	(Mcal/lb)	(Mcal/lb)	(kcal/lb)	(kcal/lb)
Carrots	11	0.19	1.69	208	1,805
Cassava	32	0.51	1.58	511	1,576
Chufa	27	0.39	1.47	480	1,812
Mangels	11	0.18	1.59	181	1,652
Potatoes	24	0.38	1.62	398	1,695
Sugar beets	20	0.33	1.67	355	1,786
Sweet potatoes	33	0.52	1.59	530	1,610
Turnips	9	0.16	1.70	146	1,594

[1]*From:* This book, Table V-1A, Energy Feeds.

Some of the most widely used root and tuber crops in livestock feeds are Irish potatoes, sweet potatoes, chufas, cassava, beets, mangels, carrots, and turnips.

• **Potatoes (Irish potatoes, *Solanum tuberosum*)**—In the United States, the culls of Irish potatoes are fed to livestock. In limited amounts and in properly balanced rations, they are a satisfactory energy feed for beef cattle, dairy cows, sheep, swine, horses, and poultry. They are rich in starch, fair in protein, and deficient in vitamins A and D—lacks which are made good when they are fed with well cured legume hay. About 400 to 450 lb of potatoes are equivalent in energy value to 100 lb of cereal grains. Stated in another way, potatoes are worth 22 to 25% as much as grain.

Irish potatoes should be cooked (steamed or boiled, preferably in salt water to increase palatability, with any cooking water discarded) for swine or poultry, but this is not necessary for other species.

Animals should be accustomed to potatoes gradually, as they are not very palatable. Also, raw potatoes should not be fed in too large amounts, as they may cause scours.

Sometimes cattle choke on potatoes, but this risk is minimized if the potatoes are fed from low troughs, with a cable or bar located so as to keep their heads down.

The sprouts of potatoes contain an alkaloid, solanin, which is toxic. Thus, long sprouts should be removed prior to feeding. Frozen potatoes should never be fed.

Potatoes may be ensiled (see Chapter 9, Silage/Haylage/High-Moisture Grain, section on "Other Silage Crops"). Potato silage is eaten readily by animals and is approximately equal to corn silage in value per ton.

Potatoes can be dehydrated, or dried, to make potato meal. This is a satisfactory substitute for 20 to 25% of the concentrate of beef cattle, dairy cattle, or sheep. If heated sufficiently in the drying process and cooked thoroughly, 10 to 20% potato meal can be incorporated in swine and poultry rations. Potato meal is equal to grain when substituted for part of it.

Dried potato pulp, or potato pomace, a by-product of starch production from potatoes, is slightly higher than potato meal in total digestible nutrients. It is palatable and nearly equal to hominy feed for dairy cows when forming 20% of the concentrate mixture.

Some by-products of processing potatoes for human food, including peelings, cull french fries, and discarded potato chips, are fed to animals kept in close proximity to the potato processing plants. These by-products may be fed as-is provided they are properly stored or ensiled.

• **Sweet potatoes (*Ipomea batatas*)**—Cull or unmarketable sweet potatoes are available for feeding. Also, in the South, some sweet potatoes are grown especially for livestock feed. Pigs are often turned into the field to harvest the crop.

Sweet potatoes are high in dry matter for a root crop—averaging 31.8%, rich in starch, and high in carotene. But they are low in protein, calcium, and phosphorus. For best results, they should not replace more than half of the grain in a ration; and the ration should be properly supplemented with protein, minerals, and vitamins. It requires about 4.3 lb of sweet potatoes to equal 1 lb of grain and other concentrates for pigs in drylot, and about 4.9 lb of potatoes to equal 1 lb of corn when the crop is hogged-down. Cooking increases the value of sweet potatoes for swine.

Sweet potatoes produce a hard pork, but pigs fed high levels of them tend to be paunchy and have a low dressing percentage.

Sweet potato meal, or dried sweet potato, is worth 90 to 100% as much as corn for beef cattle, dairy cattle, and sheep. But it is not palatable to horses, and it is least useful for swine and poultry. Also, the cost of dehydrating is rather high.

Satisfactory silage may be made from chopped sweet potatoes, without a preservative.

Green sweet potato vines are a nutritious forage. Sometimes animals are turned into a sweet potato field to graze on the vines after the crop is harvested. Wilted vines make a good silage, without a preservative. Sweet potato vine silage is equal to corn silage for finishing cattle or dairy cows.

• **Turnips**—Turnips are a popular fall/winter forage for cattle and sheep in some northern areas. Generally, they are harvested by grazing.

(See Chapter 7, section on "Turnips for Fall/Winter Grazing." Also see the feed substitution tables in the chapters devoted to each animal species.)

Processing Industry By-product Feeds as Energy Sources

Numerous by-products from various processing industries are used as energy feeds. Spent grains from the brewing and distilling industries are excellent sources of energy and other nutrients. The dairy processing industry produces numerous high-energy by-products, such as whey. Bakery residue provides a good source of highly digestible energy. Citrus pulp and beet pulp are relatively good energy feeds. These by-product feeds, along with several other types of by-products, are discussed in Chapter 12, By-product Feeds/Crop Residues. Additional information on the individual by-product feeds may be found in Chapter 30, Glossary of Feedstuffs.

QUESTIONS FOR STUDY AND DISCUSSION

1. What are concentrates? How are they classified?

2. Give your reasons for agreeing with or challenging the following statement: "Of all the energy sources used in livestock and poultry rations, cereal grains fill the most important role."

3. Define the term *grain*. In the U.S., what grains are grown primarily for food for humans; what grains are produced chiefly as feed for livestock?

4. How do you explain the following variations in the utilization of grains in other countries, in comparison with the U.S.: (a) In Europe, large quantities of wheat are fed to animals; (b) in Mexico, corn is favored as a food grain for humans; and (c) in Africa, sorghum is the primary human food grain?

5. Has the amount of grain fed to livestock in the U.S. increased or decreased in recent years? What are the reasons for this trend?

6. List the problems that are inherent in the use of grains.

7. Diagram and label the flower structures of corn or wheat. How are the grains formed?

8. What grains are called "covered caryopses"? What grains are referred to as "naked caryopses"?

9. The seed portion of grain can be divided into four parts. List them.

10. Why does a kernel of dent corn have an indentation?

11. Describe and/or diagram the wet milling of corn.

12. Name the valuable by-products commonly used as livestock and poultry feeds which result from the milling of each of the following grains: corn, wheat, rice, and barley.

13. Rank the commonly used feed grains in decreasing order of available energy content.

14. List some high energy mill by-products and name the cereal grains from which they are obtained.

15. Approximately how much of the dry matter of grains consists of carbohydrate? Of what is the carbohydrate fraction of grain composed?

16. Which grains contain the highest amount of lipid material? What part of the grain contains the most oil?

17. Are grain oils high in saturated fatty acids or in polyunsaturated fatty acids?

18. Which part of corn grain contains the highest concentration of protein?

19. What amino acids are found in very low concentrations in the cereal grains and their by-products? Are these essential amino acids?

20. Why is the nutritional deficiency disease kwashiorkor so prevalent where corn or sorghum is the leading cereal grain consumed by people?

21. What is Opaque-2 corn? What are its advantages? Why hasn't it been more widely accepted?

22. Discuss the calcium and phosphorus content, and the phosphorus availability, of cereal grains.

23. Niacin is found in relatively high concentrations in the aleurone of corn, yet corn is known to be a producer of pellagra (niacin deficiency disease). How can this be explained?

24. Briefly outline what changes in chemical composition occur as corn matures.

25. What basic criteria are used in the grading of grains?

26. What chief criticisms are being made of the current grain standards?

27. What level of insect infestation results in wheat and corn being termed *infested* and discounted in price? (See the *note well* footnote at the bottom of Table 10–4.)

28. Discuss the various factors that must be dealt with in the storage of grain.

29. Discuss the feeding value of sprouted grains.

30. Why has whole cottonseed evolved as a high-energy feed in recent years? Discuss the feeding value of whole cottonseed for beef cattle and for dairy cattle.

31. List the various types of fat products that are used as feeds.

32. What functions does the addition of fats into feeds serve?

33. Discuss the feeding of fats to each of the following species: beef cattle, dairy cattle, horses, swine, and poultry.

34. Explain the rumen protected fat system. How may this phenomenon be used?

35. Discuss molasses from the standpoints of kinds, advantages, level in the ration, and quality.

36. Differentiate between a root crop and a tuber crop. Discuss the feeding value, species to which fed and method of feeding potatoes, sweet potatoes, and turnips.

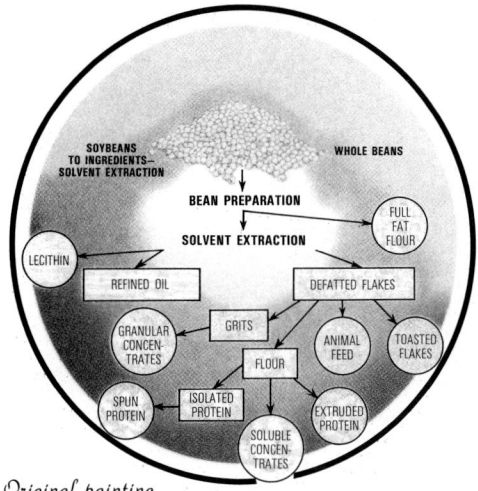

Original painting by Tom Phillips

11

PROTEIN SUPPLEMENTS

Contents	Page
Protein Requirements vs Amino Acid Requirements	204
Protein and Amino Acids for Nonruminants	204
Protein and Amino Acids for Ruminants	204
Protein Bypass (Protected Protein, Escaped Protein)	204
Importance of Protein Feeds	207
Plant Proteins	207
Oilseed Meals	208
Soybean Meal	208
Processing Soybean Meal	208
Solvent Extraction	208
Hydraulic Extraction	208
Expeller Extraction	208
Canola Meal (Rapeseed Meal)	209
Coconut Meal (Copra Meal)	209
Cottonseed Meal	209
Linseed Meal	209
Lupine	210
Peanut Meal, and Peanut Meal and Hulls	210
Safflower Meal	210
Sesame Meal	210
Sunflower Seed Meal	210
Feeding Value of Oilseed Meals	210
Corn Gluten Feed and Corn Gluten Meal	210
Pulse Proteins	211
Leaf Protein Concentrate (LPC)	211
Animal Proteins	211
Meat Packing By-products	212
Meat and Meat By-products	212
Tankage and Meat Meal	212
Meat and Bone Meal Tankage/Meat and Bone Meal	212
Blood Meal	212
Other Animal By-products	212
Poultry Wastes	212
Eggshell Meal	213
Feather Meal	213
Hydrolyzed Poultry By-products Aggregate	213
Poultry By-products	213
Poultry By-product Meal	213
Poultry Hatchery By-product	213
Dairy Products	213
Buttermilk	214
Cheese Rind (Cheese Meal)	214
Dairy Food By-products	214
Milk Protein Products	214
Milk, Whole Dried	214
Skimmed Milk	214
Whey	214
Marine By-products	214
Fish Processing	215
Marine Feeds	215
Fish Meal	215
Other Marine By-products	216

Contents	Page
Nonprotein Nitrogen (NPN) Feedstuffs	216
Urea	217
History of Urea as a Livestock Feedstuff	217
Conversion of Urea to Protein	217
Urea Fermentation Potential	218
Essentials for Optimum Use of Urea	218
Pertinent Facts About Urea	218
Methods of Feeding Urea	219
Other Nonprotein Nitrogen Feedstuffs	219
Ammoniated Products	219
Single-Cell (Microbial) Protein (SCP)	220
Types of Single-Cell Protein	220
Current Problems Associated with Single-Cell Protein	220
Types and Methods of Feeding Protein Supplements	221
Hand-feeding at Intervals	221
Liquid Protein Supplements	221
Protein Blocks	221
Range Cubes or Pellets	221
Self-feeding Salt-Feed Mixtures	222
High-protein By-product Feeds	223
Amino Acid Supplements	223
Protein Assessment Systems	223
Questions for Study and Discussion	224

The following definitions are pertinent to the discussion that follows in this chapter:

Proteins are complex organic compounds made up chiefly of combinations of amino acids present in characteristic proportions and arrangements for each specific protein; they always contain carbon, hydrogen, oxygen and nitrogen, and, in addition, usually sulfur and frequently phosphorus. Chemically, it has been determined that all proteins contain about 16% nitrogen; so, crude protein is determined by finding the nitrogen content, then multiplying the result by 6.25 (100 ÷ 16 = 6.25).

Amino acids are nitrogen-containing compounds that constitute the "building blocks," or units, from which more complex proteins are formed; they contain both amino (NH_2) and carboxyl (COOH) groups. At least 22 amino acids are found in proteins; and they occur in different combinations to form an almost limitless number of proteins.

Protein quality is a term used to describe the amino acid balance of protein. A protein is said to be of good quality when it contains all the essential amino acids in proper proportions and amounts needed by a specific animal; and it is said to be poor quality when it is deficient in either content or balance of essential amino acids.

Nonprotein nitrogen (NPN) is nitrogen which comes from other than protein sources, but which may be used by ruminants in building proteins. NPN feed sources include urea, anhydrous ammonia, biuret, ammonium sulfate, monoammonium phosphate, ammoniated molasses, ammoniated beet pulp, ammoniated cottonseed meal or hulls, and ammoniated rice hulls.

Protein equivalent is a term indicating the total nitrogenous contribution of a substance in comparison with the nitrogen content of protein; for example, the nonprotein nitrogen (NPN) compound urea contains approximately 45% nitrogen and has a protein equivalent of 281% (6.25 × 45%).

Protein supplements are feedstuffs that contain more than 20% protein or protein equivalent, and which are commonly used to improve the protein quantity and/or quality of the basal feed(s). Ideally, a protein supplement should provide the amino acids needed to match up the amino acids of the basal feed so that, combined, they meet the essential amino acid needs of the animal to which they are fed, without either a surplus or a deficiency of essential amino acids.

Proteins are found in most of the feeds commonly fed to animals. The amount, digestibility, and balance of essential amino acids, of the protein are important factors that must be considered in balancing rations. In general, animal proteins are superior to plant proteins for monogastric animals (including humans). For example, zein (a corn protein) is a low-quality or unbalanced protein, being deficient in the essential amino acids lysine and tryptophan. On the other hand, aminal proteins are excellent sources of lysine, and many (especially milk and eggs) are abundant in tryptophan. But not all animal proteins are superior and not all plant proteins are inferior; gelatins and soybeans are exceptions. Gelatin, an animal protein, is lacking or low in several amino acids. Soybean protein, a plant form, has good amounts of the essential amino acids.

Fortunately, the amino acid content of proteins varies from various plant and animal sources. Thus, the deficiencies of one protein source may be improved by combining it with another; and the mixture of the two proteins will often have a higher feeding value than either one alone. It is for this reason that a combination of protein feeds in the ration is usually recommended when the person formulating the ration does not have access to specific amino acid values of the feeds to be used or to supplements of individual amino acids.

In the past, it was common practice to use several sources of protein in nonruminant rations so that their respective amino acid profiles would complement each other. Today, the use of the computer in formulating rations and the increased availability of amino acid supplements, such as methionine hydroxy analog (MHA), allow nutritionists to formulate complete rations with a minimum number of protein feeds. The computer can rapidly determine what specific amino acids must be added and at what levels. Thus, the trend is toward fewer protein feed sources, properly supplemented with specific amino acids. A detailed discussion on the advantages and disadvantages to computer formulation is found in Chapter 18, Feeding Standards/Ration Formulation.

Also, it is noteworthy that researchers are bolstering plant protein sources by increasing, through genetic means, both the content and the nutritional quality of the protein in cereal grain. High-lysine corn is one such product of research.

(Also see Chapter 4, section on "Nitrogenous Feeds"; Chapter 6, section on "Protein Supplements"; Chapter 10, section on "High-Lysine Corn—Opaque 2, or O_2"; Chapter 13, sections on "Protein Supplements" and "Amino Acid Supplements.")

PROTEIN REQUIREMENTS VS AMINO ACID REQUIREMENTS

Many feeding standards list requirements for protein as determined by age, weight, sex, and production. However, the validity of listing total protein requirements has recently come under considerable scrutiny: Proteins are not always digestible or available to the animal. Certain processing methods, especially those involving heat, may reduce the digestibility or availability of proteins in feeds. In many cases, it is necessary to heat certain feeds to destroy a toxic substance, even when it is known that the protein value will be reduced. On the other hand, some proteins are made more readily available by processing. Another weakness of the total protein concept is the fact that it does not take into consideration protein quality—that is, the amino acid composition of feeds or their availability.

(Also see the section on "Protein Assessment Systems" at the end of this chapter in and Chapter 4, Nutrients/Metabolism.)

Protein and Amino Acids for Nonruminants

Amino acid availability is critical in the nutrition of nonruminants because they do not have the ability to synthesize certain amino acids. Therefore, in recent years nonruminant nutritionists have concentrated on establishing specific amino acid requirements for animals rather than establishing protein levels. A deficiency, and sometimes an excess, of a particular amino acid can severely affect production even though total protein levels appear to be adequate.

Protein and Amino Acids for Ruminants

Amino acids in the rations of ruminants is more important than formerly thought. It has been shown that the protein produced by ruminal synthesis does not supply all the amino acids in quality and quantity needed for maximum growth or peak milk production of ruminants. Moreover, it has been found that protein efficiency can be increased by protecting protein from the degradation of the microbes in the rumen and increasing the escape of the protein from the rumen to the intestine where it is digested and absorbed. Because protein is a costly ingredient, it follows that increasing its efficiency reduces cost since less of it is necessary. This technology of manipulating the quantity of dietary protein rumen fermentation, thereby increasing the supply of protein (amino acids) in the small intestine, is known as *protein bypass* (*protected* protein, *escaped* protein).

PROTEIN BYPASS (PROTECTED PROTEIN, ESCAPED PROTEIN)

Bypass protein (protected protein, escaped protein) is the feed protein which escapes digestion in the rumen and which passes into the lower digestive tract where it is digested and absorbed. Because all bypass protein resides in the rumen for a period of time, the term *bypass protein* is not factual; the terms *protected protein* and *escaped protein* are more accurate.

Bypass protein values of many feeds have not been determined. However, Table 11–1 shows the percentage bypass protein of some common feeds.

TABLE 11-1
BYPASS PROTEIN (PERCENT OF UNDEGRADED PROTEIN) IN COMMON FEEDS[1]

Feed[2,3]	Bypass Protein (%)	Feed[2,3]	Bypass Protein (%)	Feed[2,3]	Bypass Protein (%)	Feed[2,3]	Bypass Protein (%)
Alfalfa, fresh	20	Corn, high moisture[4]	80	Fish meal[4]	60	Soybean meal, dried 266°F (130°C)[4]	71
Alfalfa, artificially dehydrated, 17% protein[4]	59	Corn, steam flaked[4]	68	Grass silage[4]	29	Soybean meal, dried 284°F (140°C)[4]	82
Alfalfa hay, early bloom	18	Corn and cob meal	50	Linseed meal, solvent[4]	35	Soybean meal, HCHO (formaldehyde treatment)[4]	80
Alfalfa hay, midbloom	22	Corn cobs	50	Meat and bone meal[4]	49		
Alfalfa hay, full bloom	28	Corn gluten feed[4]	25	Meat meal	63		
Alfalfa hay, mature	35	Corn gluten meal[4]	55	Oats[4]	17		
Alfalfa cubes	35	Corn silage, milk stage	25	Oat groats	25	Sunflower meal, solvent[4]	26
Alfalfa stems	40	Corn silage, mature, well eared	40	Oat middlings	20		
Alfalfa silage[4]	23			Oat silage	25	Sunflower meal, with hulls	40
Alfalfa silage, wilted	22	Corn whole plant, pelleted	45	Orchardgrass, fresh, immature	25		
Bakery product, dried	20	Corn fodder	45	Orchardgrass hay	30	Timothy, fresh, pre-bloom	20
Barley[4]	27	Cottonseed meal, screw-pressed, 41% protein[4]	50	Peanut meal, solvent[4]	25	Timothy hay, early bloom	25
Barley, flaked[4]	67	Cottonseed meal, prepressed*	36	Peas[4]	22	Timothy hay, full bloom	35
Barley silage	25	Cottonseed meal, solvent, 41% protein[4]	41	Rapeseed meal, solvent (canola meal)[4]	28	Timothy silage	25
Barley silage, mature	35			Rye[4]	19	Triticale	25
Beet pulp, wet	30	Cottonseed hulls	40	Ryegrass, fresh[4]	48	Wheat[4]	22
Beet pulp, dried	35	Distillers' grain, barley	60	Sorghum grain (milo), ground	60	Wheat, hard	35
Beet pulp, wet with molasses	25	Distillers' grain, corn	57	Sorghum grain (milo), flaked	50	Wheat, soft	35
		Distillers' grain, with solubles	47	Sorghum silage	50	Wheat, sprouted	20
Beet pulp, dried with molasses	25	Distillers' stillage, corn	65	Soybeans, whole[4]	26	Wheat bran[4]	29
		Feathermeal, hydrolyzed[4]	71	Soybean meal, solvent, 44% protein	26	Wheat middlings[4]	21
Blood meal	82	Fescue, Kentucky 31, fresh	30			Wheat mill run	20
Brewers' grains, wet	57	Fescue, Kentucky 31, hay, early bloom	30	Soybean meal, solvent, 49% protein	23	Wheat shorts	20
Brewers' grains, dried[4]	49					Wheat straw	80
Coconut meal[4]	63	Fescue, Kentucky 31, hay, mature	35	Soybean meal, dried 248°F (120°C)[4]	59	Wheat straw, ammoniated	25
Corn, yellow dent[4]	52						

[1]In addition to the bypass protein values given in this table, more complete compositions of these and other feeds are given in Section V of this book.
[2]Feeds without an asterisk obtained from *Feedstuffs*, Oct. 12, 1987, pp. 18–26, by Dr. R. L. Preston, Animal Science Department, Texas Tech University, Lubbock; with the permission of Dr. Preston and *Feedstuffs*.
[3]Feeds followed by an asterisk obtained from *Ruminant Nitrogen Usage*, National Research Council, Washington, D.C., 1985, p. 33, Table 6.
[4]Adapted by the authors from *Nutrient Requirements of Dairy Cattle*, 6th rev. ed., NRC, National Academy Press, 1988, pp. 113–115, Table 7-3.

Ruminant bacteria can use various sources of nitrogen, energy (derived from fermentation), and minerals for growth. Lack of any of these factors can limit bacterial growth because requirements are interrelated. It is estimated that of the bacterial species normally present in the rumen, 80% can use ammonia as the sole source of nitrogen for growth, 26% require ammonia absolutely, and 55% can use either amino acids or ammonia. The supply of ammonia can be inadequate when either the intake of protein or the ruminal degradation of protein is low. Ammonia deficiency in the rumen reduces the efficiency of bacterial growth and may reduce the rate and extent of digestion of feed in the rumen, which, in turn, may reduce feed intake.

In addition to the microbial protein synthesized in the rumen, the animal has available another source of amino acids via the true protein in the feed which escaped breakdown in the rumen and arrived at the small intestine. On the average, approximately 40% of the true protein escapes degradation in the rumen. Under most production situations, the microbial protein synthesized plus the escape of dietary protein is adequate to meet the animal's protein requirement. However, there are practical production instances, such as rapidly growing, young ruminants and ruminants at peak lactation, in which microbial protein plus the normal dietary protein escaping ruminal breakdown may not be adequate to provide for optimal growth and feed efficiency. Present technology is being geared to manipulate the quantity of dietary protein rumen fermentation or breakdown, thereby increasing the supply of protein (amino acids) to the small intestine. In addition to increasing the amount of bypass protein that escapes degradation in the rumen, two other requisites are important: (1) The ruminant must be able to digest the bypass protein in the small intestine, and (2) the bypass protein must provide the amino acids that the animal needs.

The following definitions are pertinent to the discussion that follows:

1. *Degradability of protein refers to the rate and extent to which a feed protein may be broken down in the rumen.* This is important for two reasons: (1) Breakdown in the rumen supplies the microbes with peptides, amino acids, and ammonia; and (2) the nondegraded portion is swept out of the rumen to the abomasum and small intestine for possible breakdown there and absorption as peptides and amino acids. The degradability of protein is affected by a number of factors, including: (1) solubility of the various nitrogen fractions in the feed, (2) the particle size of the ingested material (small particles pass out of the rumen more quickly), and (3) the level of feeding (a high level of feeding pushes feed through the rumen more rapidly than a low level of feeding). Also, the ionophores (lasalocid [Bovatec] and monensin [Rumensin]), are very effective in decreasing the breakdown of natural protein by rumen bacteria (enhancing rumen bypass). (See Chapter 13, Section on "Ionophores.")

Urea (a source of non-protein N) is 100% degraded.

Casein, a very high-quality protein, is 90% degraded. However, fish meal, which is also a high-quality protein, is not degraded to any great extent in the rumen. Upon passing into the intestinal tract, where it is degraded rather completely, it supplies the animal with a great abundance of amino acids, such as lysine and methionine, which otherwise might be limiting.

2. *Solubility of protein is an estimate of the protein that breaks down rapidly in the rumen.* If a protein is extremely insoluble, it is less likely to be degraded. For example, zein, a protein found in corn, is extremely insoluble, with 40 to 60% escaping ruminal degradation. On the other hand, casein (milk protein) is highly soluble and is almost totally degraded in the rumen.

3. *Microbial protein is protein produced by the microbes in the rumen.* This involves a wide variety of microbe types including many species of anaerobic bacteria, protozoa, and even fungi, along with ammonia or, in some cases, peptides or amino acids. Ammonia may be provided in many ways, but it must be available for the microbes when needed and at the levels needed. The common recommendation in the use of NPN is that it can provide up to one-third of the total protein.

With ruminant (cud-chewing) animals, the amino acid dietary need is complicated. Swallowed food goes into the rumen (the largest compartment of the stomach) where it is partly digested by bacteria. It is then regurgitated to the mouth to be rechewed and reswallowed. During this process, the great bulk of dietary protein fed to ruminants is transformed into microbial protein in the rumen, with the result that the amino acids originally in the feed are not the ones actually used by the ruminant animal. But a portion of the dietary protein escapes (bypasses) rumen bacterial action, and, along with the microbes, passes to the lower gastrointestinal tract. So, the ruminant has two sources of protein: microbial protein and bypass protein, with the former predominating.

Until recent years, it was thought that ruminants were capable of producing the essential amino acids needed for microbial protein in sufficient amounts to meet all their needs, including those of rapid growth or peak milk production. This was likely true until improved genetics, nutrition, and management resulted in production increases demanding more protein (amino acids) than microbial synthesis could produce. Hand in hand with this new awareness, it was found that microbes were often wasteful. Much of the nitrogen liberated in the degradation of dietary protein is synthesized into nonprotein products. For example, it has been estimated that 20% of the nitrogen found in the crude protein fraction of ruminal microorganisms is actually nitrogen from nucleic acids. This fraction of the crude protein is essentially wasted as the nucleic acids are, for the most part, degraded to form urea in the postruminal section of the gastrointestinal tract. Additionally, the true protein fraction from ruminal microorganisms is not as digestible as the dietary protein in many cases.

Normally, the amount of protein that escapes ruminal degradation is dependent largely on the length of time the protein remains in the rumen. One can speed the rate of flow through the rumen by increasing dietary intake or frequency of feeding, or by reducing the feed particle size. Also, if large quantities of salt are fed, there is an increased water intake resulting in a faster rate of flow through the rumen. Thus, speeding feeds through the rumen ultimately increases the amount of dietary protein that reaches the lower gut.

• **Methods of protecting proteins from ruminant degradation**—Scientists have been developing ways of manipulating ruminant protein digestion to make more of the amino acids in the feed available to the animal after bacterial breakdown in the rumen. This technique is known as bypass proteins or protected proteins. Such proteins actually bypass the rumen, and their amino acids are not broken down by bacteria. Among the methods of protecting proteins from ruminant degradation are the following:

1. **Naturally protected proteins.** A few proteins are naturally protected by heat during normal processing and have good bypass characteristics. These include corn gluten meal, brewers' grains (wet or dry), distillers' grains (wet or dry), extruded or roasted soybeans, extruded soybean meal, meat meal, blood meal, fish meal, and hydrolyzed feather meal. Normally, solvent extracted soybean meal contains 28% bypass protein; by comparison, a modified expeller processed soybean meal contains 65% bypass protein.

2. **Heat and pressure treatment.** Heating protein feeds, usually under pressure, creates cross-linkages of free amino groups within and between protein molecules. In addition to altering the solubility properties of the protein, these cross-linkages reduce the surface of the protein that comes into contact with enzymes, thereby blocking the site of enzymatic attack.

3. **Treatment of protein with formaldehyde or other aldehydes, with or without heat.** This is the most widely

TABLE
COMMERCIAL FEEDS FED TO LIVESTOCK IN THE

Year Beginning October	Oilseed Cake and Meal						Animal Proteins			
	Soybean[3]	Cottonseed	Linseed	Peanut	Sunflower	Total	Tankage and Meat Meal	Fish Meal and Solubles	Dried Milk[4]	Total
1970	13,467	1,693	258	173	99	15,690	2,039	609	260	2,908
1975	15,612	1,266	94	313	119	17,404	2,001	508	147	2,656
1980	17,591	1,636	103	94	44	19,468	2,458	379	104	2,941
1984	19,480	1,732	120	122	338	21,792	2,889	228	98	2,987

[1]From USDA.
[2]Other mill products that are not listed include screenings, hominy, and oat feeds.
[3]Includes use in edible soy products and shipment to U.S. territories.
[4]Includes commercial dried milk products and noncommercial milk products.

used technique of protein protection. The aldehyde reacts with free amino groups and N-terminal groups to form Schiff bases and cross-linkages between protein chains. This process decreases the solubility of the protein and protects it from bacterial degradation. Once the treated protein enters into a highly acid environment, the reaction is reversed, and the protein is degraded. Combined aldehyde and heat treatment is also being used.

4. **Lipid (fat) treatment.** In this system, fat is emulsified with a protein, then the protein is treated with formaldehyde. The product cannot be digested in the rumen due to the formaldehyde linkage on the protein. However, the pH of the abomasum is such that the formaldehyde linkage of the protein is broken, with the result that the product is then digested in the same manner as in monogastric animals. Also, and most significant, by use of the protein protected fat system, it is possible to alter the ratio of the saturated to the unsaturated fatty acids in beef fat and in butterfat.

5. **Non-enzymatic browning.** Through a process known as non-enzymatic browning, University of Nebraska researchers have increased the by-pass content of soybean meal by as much as 2½ times. The process involves mixing the soybean meal, or other protein source, with xylose sugar, then heating the mixture to 200 to 250°F. A patent has been applied for on the process.

6. **Adding ionophores.** Bovatec (lasalocid) and Rumensin (monensin), two ionophores, are effective (a) in decreasing rumen degradation of natural protein and (b) in increasing rumen bypass—perhaps by as much as 25%. (See Chapter 13, section on "Ionophores.")

IMPORTANCE OF PROTEIN FEEDS

Since protein feeds are usually among the more costly components of animal rations, it is important to provide enough protein for the animal to perform its assigned function, but to avoid feeding more than is necessary. The primary functions of protein feeds are to supply (1) those amino acids not provided in adequate amounts by the cereal portion of nonruminant rations, or (2) nitrogen precursors of microbial protein in the case of ruminants. Some animals are equipped to utilize a large amount of protein for energy; for example, mink and fish may obtain more than half of their energy from protein. But usually there are more economical sources of energy than protein.

Protein supplements may be further categorized according to source of origin as (1) plant proteins, (2) animal proteins, (3) nonprotein nitrogen, and (4) single-cell (microbial) proteins.

Plants provide 79% of the protein feeds used in livestock rations (see Table 6-1). Most protein feeds of plant origin consist of processed oilseeds. Mill products generally make up the remainder of the plant protein feeds. Many protein feeds of animal origin are derived from sources that are considered unsuitable for human consumption. Also, animal protein feeds are generally more expensive than those of plant origin. It is noteworthy that in the 14-year period from 1970 to 1984 (see Table 11-2), the kinds and amounts of protein feeds used in livestock rations remained relatively constant.

PLANT PROTEINS

The bulk of the protein for ruminants (cattle, sheep, goats), whose requirements for specific amino acids are met by microbial fermentation of the material in the forestomach (rumen), comes from plant sources.

Even though they are not especially high in protein by comparison with other feedstuffs, the vegetative portions of many plants supply an extremely large portion of the protein in the total ration of livestock, simply because these feeds are consumed in large quantities. Needed protein not provided in these feeds is commonly obtained from one or more of the oilseed by-products—soybean meal, cottonseed meal, linseed meal, peanut meal, safflower meal, sunflower meal, rapeseed meal, or coconut (copra) meal. The protein content and feeding value of these products vary according to the seed from which they are produced, the geographical area in which they are grown, the amount of hull and/or seed coat included, and the method of oil extraction used. Sometimes, the unprocessed seed is used to provide both a source of protein and a concentrated source of energy. The unprocessed oil-bearing seeds are especially high in energy because of the oil that they contain.

Additional plant proteins are obtained as by-products from grain milling, brewing and distilling, and starch production. Most of these industries use the starch in grains and seeds, then dispose of the residue, which contains a large portion of the protein of the original plant seed.

11-2
UNITED STATES (1,000 SHORT TONS)[1]

	Mill Products[2]							Total Commercial Feeds
Wheat Millfeeds	Gluten Feed and Meal	Rice Millfeeds	Brewers' Dried Grains	Distillers' Dried Grains	Dried and Molasses Beet Pulp	Alfalfa Meal	Total	
4,499	1,237	436	361	382	1,509	1,584	10,008	28,606
4,933	1,477	547	321	400	1,861	1,569	11,108	31,168
5,114	839	778	320	499	1,310	1,096	9,956	32,365
5,556	2,065	607	156	967	1,207	891	11,449	36,228

Oilseed Meals

Several rich oil-bearing seeds are produced for vegetable oils for human food (oleomargarine, shortenings, and salad oil), and for paints and other industrial purposes. In processing these seeds, protein-rich products of great value as livestock feeds are obtained. Among such high-protein feeds are soybean meal, coconut meal, cottonseed meal, linseed (flax) meal, peanut meal, canola meal (rapeseed meal), safflower meal, sesame meal, and sunflower seed meal.

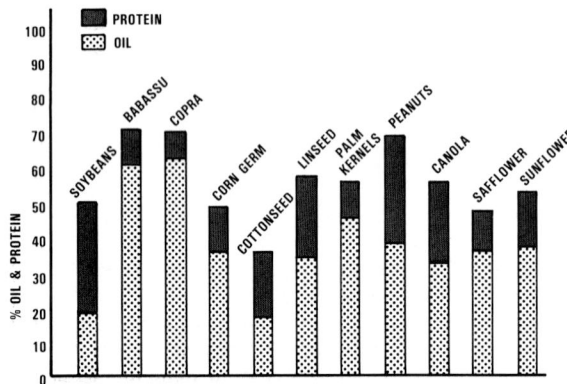

Fig. 11-1. The oil and protein content of commonly used protein supplements. (Adapted by the authors from USDA sources.)

Oil is extracted from these seeds by one of the following basic processes or modifications thereof: solvent extraction, hydraulic extraction, or expeller extraction. For purpose of illustration, all three processes are discussed relative to soybeans, since soybeans are the most widely used oilseed in the United States, and since the principles involved in processing all oilseeds are essentially the same. However, it should be noted that very little soybean meal is produced today via hydraulic or expeller extraction.

Although crude protein determination is the most frequently used chemical assay for comparing oilseed meals, the real quality measurement goes far beyond; there is more to oilseed meals than just protein. The availability and digestibility of amino acids, the concentration of minerals and vitamins, and the level of moisture, fiber, and urease can all affect oilseed meal performance in a livestock ration. Also, the feeding value of oilseed meals varies according to the method of extraction. Oilseeds vary in oil content, both between varieties and within a variety. Meals produced by mechanical extraction of the oil from the seed contain more fat and fiber and a lower percentage of protein than those produced by solvent extraction (hexane).

SOYBEAN MEAL

Soybean meal, processed from an oil-bearing seed that originated in the Orient, has the highest nutritive value of any plant protein source. It is now the most widely used protein supplement in the United States. Today, the United States produces about 55% of the world's soybeans.

Processing Soybean Meal

Raw whole soybeans are seldom fed to livestock. Rather, they are processed and utilized as either oil meal or whole, heat processed beans. Today, most soybeans are fed in oil meal form. But, as a result of increased processing costs, the use of heat processed whole soybeans may play an increasing role in livestock nutrition. Whole soybeans must be heated to deactivate the antinutritional factors of the seeds; but once properly heated, the whole seed provides a valuable source of energy and protein. Many of the problems associated with feeding this high-oil feed, such as soft pork, have been overcome by adjustments in feed formulations. (See Chapter 23, Feeding Swine, section headed "Full-Fat Soybeans.")

In the past, oil was extracted by the solvent, hydraulic, and expeller processes. But, today, almost all soybeans are solvent extracted. Soybean meal normally contains 41, 44, 48, or 50% protein, depending on the amount of hull removed. Because of its well balanced amino acid profile, the protein of soybean meal is of better quality than other protein-rich supplements of plant origin. However, it is low in calcium, phosphorus, carotene, and vitamin D.

SOLVENT EXTRACTION

In this method, the soybeans are first cracked, then heated to 140°F for about 10 minutes. After the cracked seeds are heated, they proceed through a series of grinding rollers where they are flaked. The flakes are allowed to cool to about 113°F and are then conveyed to the extraction equipment where the oil is removed by the solvent—hexane.

The extracted flakes then proceed to driers where the solvent is completely volatilized and removed. From the drier, the flakes are conveyed to a toaster, thence they are cooled and ground.

HYDRAULIC EXTRACTION

In the hydraulic extraction procedure, raw soybeans are cracked, ground, and flaked. The flakes (called meats) are then transported to cookers where they are exposed to both dry and steam heat. The cooking stage takes about 90 minutes.

After cooking, the meats are formed into cakes and wrapped in heavy cloth whereupon they are placed in hydraulic presses for the mechanical extraction of the oil. This procedure takes about 1 hour. Following extraction, the extracted cakes are ground. Hydraulic press cake may have 5 to 8% residual oil.

Since this form of extraction is labor intensive and inefficient in the removal of oil, very little meal is produced this way.

EXPELLER EXTRACTION

In this type of extraction, raw soybeans are initially cracked and dried to about 2% moisture. The dried soybeans are then transported hot to a steam jacketed tempering device which is directly above the expeller apparatus. The tempering apparatus stirs the cracked soybeans for about 10 to 15 minutes so that the seeds are heated uniformly.

From the tempering bin, the soybeans are fed into an expeller barrel (screw presses). A central revolving worm shaft creates pressure within the expeller barrel, thereupon extracting the oil from the ground soybeans. The extracted soybeans leave the expeller in the form of flakes, which are subsequently ground.

The expeller process tends to extract less oil than the solvent process; consequently, it is now used less frequently. Generally, expeller extracted soybean meal contains 4 to 5% oil while solvent extracted soybean meal contains less than 1%. However, expeller processed soybean meal is higher in by-pass protein than solvent extracted meal.

CANOLA MEAL (RAPESEED MEAL)

Canola was created from specially selected rapeseed by Canadian plant scientists in the 1970s. The old rapeseed was high in glucosinolate compounds, which, when fed to animals at high levels, made for palatability problems and lowered performance due to goitrogenic action. Canola changed this. The new canola is low in glucosinolates in the meal, and low in erucic acid (a long-chain fatty acid) in the oil. Canola is grown mainly in Canada, but it is increasing in the United States.

Canola meal averages about 36% crude protein and its amino acids compare favorably with soybean meal. When the price is favorable, canola meal may be used as a protein supplement for all classes of livestock and poultry. It may be used at maximum levels of 20% in most rations, but the amount in the total ration should be limited to about 10% for layers and breeding ducks, 15% for breeding turkeys, and 12% for young pigs and breeding swine.

COCONUT MEAL (COPRA MEAL)

This is a by-product from the production of oil from the dried meats of coconuts. The oil is generally extracted by either (1) the hydraulic process, or (2) the expeller process. Coconut meal averages about 21% protein content. The quality of the protein is not high; hence, its use should be restricted in nonruminant rations. If copra meal is fed to nonruminants, it should be supplemented with the amino acids lysine and methionine.

The lipid component of copra meal is very low in unsaturated fatty acids. Hence, the feeding of copra meal produces firm body fat in swine. Also, dairy producers use copra meal to produce a pleasant flavored, rather hard (highly saturated) butterfat. The maximum recommended level in dairy rations is 3.3 to 6.5 lb per day. Higher amounts tend to produce tallowy butter.

No copra meal is produced in the United States and imports have practically ceased since 1984.

COTTONSEED MEAL

Among the U.S. oilseed meals, cottonseed meal ranks second in tonnage to soybean meal. The processing steps in making cottonseed meal are as follows: (1) cleaning the seeds; (2) dehulling the seeds; (3) crushing the kernels; (4) extracting the oil by either (a) mechanical (or screw press), (b) solvent, or (c) partially mechanically extracted and then solvent extracted process; and (5) grinding the remaining residue or cake, thus forming cottonseed meal. Today, 80% of all processed cottonseed is solvent or combined mechanical-solvent extracted.

The protein content of cottonseed meal varies from about 22% in meal made from undecorticated seed to 50% in flour made from seed from which the hulls have been removed completely. Thus, by screening out the residual hulls, which are low in protein and high in fiber, the processor is able to make a cottonseed meal of the protein content desired—usually 36, 41, 44, and 48%. The protein content of cottonseed meal varies with the geographical location in which it was grown. Meals manufactured from cottonseed produced on the West Coast generally contain higher protein levels than those produced throughout the rest of the United States.

For monogastric animals, cottonseed meal is low in lysine and tryptophan and deficient in vitamin D, carotene (vitamin A value), and calcium. Also, it contains a toxic substance known as gossypol, varying in amounts with the seed and the processing. But, it is rich in phosphorus.

Today, glandless cottonseed, free of gossypol, is being grown and improved. Someday, meal made from it may replace conventional cottonseed meal in nonruminant rations and alleviate (1) many of the restrictions as to levels of meal, and (2) the need to add iron to tie up free gossypol. University of California researchers report (1) that glandless cottonseed meal caused no problems in egg production, hatchability, or egg yolk discoloration in laying hens; and (2) that glandless cottonseed meal supplemented with 0.3% or more lysine produced growth in broilers not statistically different from that obtained by feeding a practical starting ration.[1]

Today, considerable whole cottonseed, which averages about 21.8% crude protein, and 87% TDN for ruminants, is being fed to ruminants, as both a protein and energy feed source. (See Chapter 10, section on "Cottonseed.")

LINSEED MEAL

Linseed meal is a by-product of flax, a fiber plant which antedates recorded history. In this country, most of the flax is produced as a cash crop for oil from the seed and the resulting by-product, linseed meal. Practically none of the U.S. flax crop is grown for fiber, for it is more economical to import it from those countries where cheaper labor is available.

Most of the nation's flax is produced in North Dakota, South Dakota, and Minnesota. Normally, an additional quantity of seed is imported and processed in U.S. plants.

The oil is extracted from the seed by either of two processes: (1) the mechanical process, or (2) the solvent process. Horse caretakers prefer old process linseed meal, because it is palatable, and it imparts glossiness to the hair coat due to the higher oil content.

Linseed meal is the finely ground residue (known as cake, chips, or flakes) remaining after the oil extraction. It averages about 35% protein content (33 to 37%), about half as much as tankage. For swine, the proteins of linseed meal do not effectively supplement the deficiencies of the cereal grains; since linseed meal is also low in the amino acids lysine and tryptophan. Also, linseed meal is lacking in carotene and vitamin D, and is only fair in calcium and the B vitamins. It is laxative; hence, it may be used to regulate the bowels.

---

[1]Ryan, J. R., et al., "Glandless Cottonseed Meal for Laying and Breeding Hens and Broiler Chicks," *Poultry Science*, Vol. 65, No. 5, May 1986, p. 949.

LUPINE

Lupine is an ancient legume crop in which there is new interest. Sweet (or white) lupine produces less harmful alkaloids than the other approximately 100 varieties. Also, it grows well in marginal soils and tolerates both drought and cold temperature; conditions not suited to soybeans. The beans may be fed to animals as a protein supplement without heating.

PEANUT MEAL, AND PEANUT MEAL AND HULLS

Peanut meal, a by-product of the peanut industry, is ground peanut cake, the product which remains after the extraction of part of the oil of peanuts by pressure or solvents. It is a palatable, high-quality vegetable protein supplement used extensively in livestock and poultry feeds. Peanut meal ranges from 41 to 50% protein and from 4.5 to 8% fat. It is low in methionine, lysine, and tryptophan, and low in calcium, carotene, and vitamin D.

Peanut meal and hulls is ground peanut meal with added hulls, or the ground by-product remaining after extraction of part of the oil from whole or unshelled peanuts. Since about one-fourth of peanut meal and hulls consists of peanut hulls, it is high in fiber, averaging about 22.5%.

Since peanut meal tends to become rancid when held too long—especially in warm, moist climates—it should not be stored longer than 6 weeks in the summer or 2 to 3 months in the winter.

SAFFLOWER MEAL

India produces about half of the world safflower crop. In the United States, it is a minor crop—only about 100,000 tons of seed being produced annually.

A large proportion of the safflower seed is composed of hull—about 40%. Once the oil is removed from the seeds, the resulting product contains about 60% hulls and 18 to 22% protein. Various means have been tried to reduce this high-hull content. Most meals contain seeds with part of the hull removed, thereby yielding a product containing 35 to 50% protein and 10 to 15% fiber.

Dehulled (decorticated) safflower meal can be fed to swine and poultry, but it should be supplemented with soybean or meat meal to provide adequate levels of lysine and the sulfur-containing amino acids. High-fiber safflower meal (above 15%) should be fed only to ruminants.

SESAME MEAL

Worldwide, sesame ranks sixth among oilseed crops in the production of edible oilseeds and tenth in tonnage of vegetable oils. But, little sesame is grown in the United States despite the fact that it is one of the oldest cultivated oilseeds. The oil meal is produced from the entire seed. Solvent extraction yields higher protein (45%) but lower fat levels (1%) than either the screw press or hydraulic methods which produce meals containing about 38% protein and 5 to 11% oil.

SUNFLOWER SEED MEAL

The Russians pioneered in plant breeding research which resulted in sunflower seed which yielded more than 40% oil, and, in addition, increased the seed yield per acre. The improved Russian cultivars were brought to the United States in the 1960s.

Worldwide, sunflowers rank second as an oilseed crop, being exceeded only by soybeans. Twenty-eight percent of the world sunflower seed is produced in the U.S.S.R. In the United States, about 3 million acres of sunflowers are grown annually, with North Dakota the leading state, followed by South Dakota and Minnesota.

Sunflower meal—a relative newcomer to the commercial oilseed industry in the United States—is rapidly gaining acceptance as a high-quality source of plant protein. It competes with other oilseed meals. The oil meal varies considerably in nutrient content, depending on the extraction process and whether the seeds are dehulled. Meal from prepressed solvent extraction of dehulled seeds contains about 44% protein, as opposed to 28% for whole seeds. The hulls are not easily removed, with the result that sunflower meal is high in fiber, ranging from 15 to 24%. Meal produced from seed that has not been dehulled has a high fiber content; hence, it should be fed only to ruminants. When sunflower meal is used in nonruminant feeds, it should be combined with high-lysine supplements, such as meat scrap or fish meal.

FEEDING VALUE OF OILSEED MEALS

The feeding value of oilseed meals varies with the composition of the seed and the method of oil extraction. In general, meals produced by mechanical extraction of the oil from the seed contain more oil and fiber and a lower percentage of protein than those produced by solvent extraction using hexane.

Soybean meal is, by far, the most widely used oilseed meal in the United States. For this reason, it is generally used as the standard to which all the others are compared. Many of the oilseed meals compare favorably with soybean meal, but their use is generally restricted because of (1) limited supply, (2) antinutritional factors (such as gossypol in cottonseed meal), or (3) lack of palatability.

For information pertaining to the uses of oilseed meals in specific livestock rations, the reader is referred to the appropriate chapters of Section III—Feeding.

The quality of various oilseed proteins is compared in Table 11-3. Soybean protein is the best in all measurements of quality, with sunflower protein following closely.

TABLE 11-3
MEASUREMENTS OF OILSEED PROTEIN QUALITY[1]

Oilseed	Biological Value (BV)	Net Protein Utilization (NPU)	Protein Efficiency Ratio (PER)
Soybeans	73	61	2.32
Sunflower	70	58	2.10
Cottonseed	67	53	2.25
Peanuts	55	43	1.05

[1]Adapted by the authors from *Improvement of Protein Nutriture* by D. M. Hegsted, NRC-National Academy of Sciences, 1974.

Corn Gluten Feed and Corn Gluten Meal

Corn gluten is a mixture of bran (the high-fiber shell, or hull, of the kernel) and steepwater solubles (a molasseslike material derived from the water used to soak the corn during wet milling). More precisely, corn gluten feed and corn gluten meal may be defined as follows:

Corn gluten feed is that part of corn that remains after the extraction of the larger portion of the starch, gluten, and germ by the process employed in the wet milling manufacture of corn starch or corn syrup. Corn gluten feed may or may not contain fermented corn extractives and/or corn germ meal. It contains about 23% protein.

Corn gluten meal is the dried residue from corn after the removal of the larger part of the starch and germ, and the separation of the bran by the process employed in the wet milling manufacture of corn starch or syrup, or by enzymatic treatment of the endosperm. It may contain fermented corn extractives and/or corn germ meal. Corn gluten meal averages about 43% crude protein.

Corn gluten feed is an old product with a new look. The recent switch of the soft drink industry to corn sweeteners (high fructose) has made it abundant. But corn gluten feed has been around a very long time. A hundred years ago (in the 1880s), corn gluten feed was considered to be a worthless waste product of starch manufacture; a corn refining plant in Buffalo, New York dumped it into a nearby canal. In time, however, its value as a dairy feed became known.

Both corn gluten feed and corn gluten meal are by-products of the wet milling of corn (see Chapter 10, section on "Wet Milling"). In the dry form, corn gluten feed contains about 23% protein and has a bulky consistency due to its content of corn bran (hulls). Corn gluten feed is also available in wet form.

Corn gluten meal, which contains much less bran and is less bulky, averages about 43% crude protein. Both products are low in lysine and tryptophan. When made from yellow corn, as nearly all corn gluten feed and corn gluten meal are, they are quite rich in carotene and in the yellow pigment xanthophyll. The presence of xanthophyll makes corn gluten feed and corn gluten meal valued ingredients in poultry rations because it produces the yellow skin color, which is prized in dressed poultry.

Today, dairy cows and poultry consume most of the gluten feed and meal. Because of its bulk, gluten feed is better suited for dairy cows than for poultry. Beef cattle, sheep, and swine use lesser amounts of both feeds. Corn gluten feed and corn gluten meal are of special value as ruminant protein supplements because of their good protein bypass (escape) characteristics. Most of the U.S. annual production of more than 5 million tons of corn gluten feed and corn gluten meal is marketed in mixed commercial feeds as a substitute protein source for higher priced protein feeds.

Pulse Proteins

Pulses are the seeds of leguminous plants. They are used primarily for human consumption, but they can be fed to livestock effectively when the price is right. Although there are over 13,000 species within the family *Leguminosae,* only about 20 species are used for food and/or feed. Soybeans and peanuts are pulses, but they are used almost entirely as oilseed meals in livestock rations.

All of the pulses contain components which possess antinutritional properties. Fortunately, processing procedures, such as cooking, germination, and fermentation, can reduce the risks of feeding pulses to livestock. Among the chemical factors that can create problems in feeding pulses are protease inhibitors, goitrogens, cyanogens, antivitamins, metal-binding factors, lathyrogens, and phytohemagglutins.

Leaf Protein Concentrate (LPC)

In recent years, considerable attention has been focused on the extraction of protein from green, leafy plants. Green leaves are among the best sources of protein in feeds. To date, most of the research on leaf protein concentrate has centered around the processing of alfalfa; but there is a great potential for LPC development from other sources—for example, vegetable packinghouse by-products and field wastes.

Some leaf protein concentrate is commercially available, but due to the high cost of processing, little of it is produced; and most of that which is currently produced is used primarily as a supplement to provide high content protein (more than 40% crude protein), vitamins, xanthophylls, and unidentified factors. Until new developments lower the cost of processing, leaf protein concentrate will remain a protein supplement of minor importance.

One recent development showing promise for the future is the process whereby juice is obtained from the pressing of freshly cut alfalfa. About 25 to 50% of the protein in the leaves can be found in this juice. The high-protein liquid concentrate can then be fed to animals in its crude form or coagulated and used as a solid protein supplement. One such supplement that is currently available is marketed under the trade name of X-Pro. The extract can be coagulated and dried, forming a solid product that is 50% protein.

ANIMAL PROTEINS

Protein supplements of animal origin are derived from (1) meat packing and rendering operations, (2) poultry and poultry processing, (3) milk and milk processing, and (4) fish and fish processing. Before the discovery of vitamin B-12, it was generally considered necessary to include one or more of these protein supplements in the rations of hogs and chickens. With the discovery and increased availability of synthetic vitamin B-12, high-protein feeds of animal origin have become less essential, although they are still included to some extent in rations for most monogastric animals.

With improvements in the protein quality of plants, such as the development of high-lysine corn, the use of meat proteins may decline in the future. The cost of such proteins will need to be more competitive than they are at the present time if they are to be included in any major quantity in animal rations. Blending several proteins with complementary balances of amino acids and supplying more concentrated sources of individual amino acids may also be a factor affecting the future role of animal proteins in animal and poultry rations.

Many protein supplements of animal origin are difficult to process and store without some spoilage and nutrient loss. If they cannot be dried, they must usually be refrigerated. If not heated to destroy disease-producing (pathogenic) bacteria, they may be a source of infection. On the other hand, protein availability will be reduced and some nutrients lost if the feed is heated excessively.

Meat Packing By-products

Although the meat or flesh of animals is the primary object of slaughtering, modern meat packing plants process numerous and valuable by-products, including protein-rich livestock feeds.

MEAT AND MEAT BY-PRODUCTS

Meat is defined by the Association of American Feed Control Officials *as the clean, wholesome flesh derived from slaughtered animals and is limited to that part of striate muscle which is skeletal or which is found in the tongue, diaphragm, heart, or esophagus.*

Meat by-products are the nonrendered, clean, wholesome parts, other than meat. This feed classification includes blood, bones, brains, intestines, kidneys, lungs, spleens, and stomachs, but it does not include hair, hooves, or teeth.

TANKAGE AND MEAT MEAL

In 1984, 2.9 million tons of tankage and meat meal were fed to livestock. Tankage and meat meal are made from the trimmings that originate on the killing floor, inedible parts and organs, cleaned entrails, fetuses, residues from the production of fats, and certain condemned carcasses and parts of carcasses. In order to be used in animal feeds, carcasses condemned as unfit for human food because of antibiotic or biological residues, must meet the standards of the Food Safety and Inspection Service.

The end products, and the methods of processing each, are:

1. **Meat meal tankage (tankage)** is produced by the older wet-rendering method, in which all of the material is cooked by steam under pressure in large closed tanks; hence, the derivation of the name tankage. After cooking, the fat is skimmed off, the soupy liquid drained off, and the remaining residue pressed to remove as much of the fat and water as possible. The soupy liquid is then evaporated until it becomes gluey, at which state it is called *stick*. The stick is added to the pressed residue, following which the mixture is dried and ground.

The level of protein in tankage (generally 60%) is standardized during manufacture by the addition of enough blood to raise the total protein to the desired level. The protein content must be designated.

In addition to tanking under live steam (as previously described), tankage may also be made by the dry-rendering method (a description of which follows), or by mixing products containing both wet-rendered and dry-rendered materials.

2. **Meat meal** is produced by the newer and more efficient dry-rendering method, in which all of the material is cooked in its own grease by dry heat in open steam-jacketed drums until the moisture has evaporated. Then as much of the fat as possible is removed by draining off and the solid residue is passed through a screw press. Next the dry residue is granulated or ground into a meal. This product is lighter colored and does not have as strong an odor as wet-rendered tankage.

The level of protein of meat meal (generally 50 to 55%) was originally established because the normal proportions of raw materials available for rendering resulted in a product which, after being pressed and ground, contained approximately 50% protein. The protein content of meat meal is adjusted up or down by raising or lowering the quantity of bone and fat in the raw material. The protein content must be designated.

Tankage and meat meal are widely used as protein supplements. Generally, their proteins are of excellent quality, effectively correcting the deficiencies in the proteins of the cereal grains. They are also rich in minerals (especially calcium and phosphorus) and good sources of riboflavin, niacin, and vitamin B–12. However, they are lacking in vitamins A and D and pantothenic acid.

The protein contents of meat meal is somewhat lower than that of meat meal tankage; yet, the two products are about equal in feeding value, probably due to the greater digestibility and high nutritive value of the protein in the dry-rendered product, since it is subjected to less heat.

MEAT AND BONE MEAL TANKAGE/MEAT AND BONE MEAL

When, because of added bone, tankage or meat meal contains more than 4.4% phosphorus, the word *bone* must be inserted in the name; and they must be designated, according to the method of processing, as either (1) meat and bone meal tankage, or (2) meat and bone meal. Thus, when such high-phosphorus products are prepared by the older wet-rendering method, they are known as *meat and bone meal tankage*. Likewise, when products in excess of 4.4% phosphorus are prepared by the newer dry-rendering method, they are designated as *meat and bone meal*.

The protein content of meat and bone meal tankage and meat and bone meal must be designated. Generally, they contain less protein than tankage or meat meal, usually 45 to 50%. Except for noting this difference, the swine feeding recommendations given relative to tankage and meat meal are equally applicable to meat and bone meal tankage and meat and bone meal.

BLOOD MEAL

Today, dried blood meal may be, and is, prepared by any one of three processes: (1) spray drying, (2) cooker drying, or (3) flash drying.

Blood meal is low in the essential amino acid isoleucine and low in calcium and phosphorus.

Blood meal ranks very high as an escape (bypass) protein for ruminants (see Table 11–1). For this reason, it is finding an increasing place as a ruminant feedstuff.

OTHER ANIMAL BY-PRODUCTS

Meat meal tankage, meat meal, meat and bone meal tankage, and meat and bone meal are the leading meat packing by-products. Additional meat packing by-products are: animal by-product meal, animal digest, animal liver meal, blood protein, fleshings hydrolysate, glandular meal and extracted glandular meal, hydrolyzed hair, hydrolyzed leather meal, dried meat solubles, and unborn calf carcass (cattle fetus).

Poultry Wastes

By-product feedstuffs are derived from all segments of the poultry industry—from hatching all the way through processing for market; and they come from the broiler and turkey segments of the industry as well as from egg production. Centralization of these industries into large units, with enough volume of wastes to make it feasible to process the potential feeds, has opened new markets for what was previously a disposal problem. Certain precautions have had to be included to make the products most useful and safe, but considerable amounts of poultry products are currently being included in rations for various animals, both ruminants and monogastrics.

The three most extensively used high-protein by-products of the poultry industry are hatchery by-product, poultry by-products, and poultry feathers. Cull birds, unsalable eggs, eggshells, and slaughter wastes are also used in animal feeds.

In addition to the by-products of the poultry processing industry, poultry manure and litter are now being processed to produce a palatable, high-protein feed. A detailed discussion of this phase of by-product utilization is presented in Chapter 12, By-product Feeds/Crop Residues.

EGGSHELL MEAL

This is a mixture of eggshells, shell membranes, and egg content obtained by drying the residue of an egg-breaking plant in a dehydrator to an end product temperature of 180°F. It must be designated according to its protein and calcium content.

FEATHER MEAL

Feathers, a by-product that is nearly all protein, can be used in rations after they are hydrolyzed with heat and pressure to make the proteins available. *Hydrolyzed feather meal is defined as the product resulting from the treatment under pressure of clean, undecomposed feathers from slaughtered poultry, free of additives, and/or accelerators.* Not less than 75% of its crude protein content must be digestible by the pepsin digestibility method. Although hydrolyzed feather meal is high in protein (from 85 to 90%), it is rather low in nutritional value, being low in the amino acids histidine, lysine, methionine, and tryptophan. Those amino acids which are present are readily available.

Ruminants can utilize feather meal rather well; and benefit from its high bypass protein. Levels up to 10% have been used successfully in concentrated feeds for dairy cattle. Feather meal should be included in ruminant feeds gradually, since sudden addition of it may decrease feed intake.

Because of the deficiencies of several amino acids, care must be used when incorporating feather meal into nonruminant feed. The addition of fish meal, meat meal, or blood meal tends to complement feather meal and facilitates its use. In practice, feather meal rarely exceeds 5% of the ration for nonruminants.

HYDROLYZED POULTRY BY-PRODUCTS AGGREGATE

This is the product resulting from heat treatment under pressure of all by-products of slaughtered poultry, clean and undecomposed, including such parts as heads, feet, undeveloped eggs, intestines, feathers, and blood.

POULTRY BY-PRODUCTS

This consists of non-rendered clean parts of carcasses of slaughtered poultry such as heads, feet, and viscera, free from fecal content and foreign matter.

POULTRY BY-PRODUCT MEAL

This consists of the ground, dry-rendered or wet-rendered clean parts of carcasses of slaughtered poultry—such as heads, feet, undeveloped eggs, and viscera—free from feathers, fecal content and foreign matter except for such trace amounts as are unavoidable in good factory practice. It must contain not more than 16% ash and not more than 4% acid-insoluble ash.

Because of the heads and feet, poultry by-products are lower in nutritional value than the flesh of animals, including poultry. The biological value of the proteins is lower than the other animal proteins and the better plant proteins. They may be successfully used in animal rations, however, provided they are not the sole source of proteins.

POULTRY HATCHERY BY-PRODUCT

Of the various by-products from poultry, the most valuable is *hatchery by-product* consisting of eggshells, infertile and unhatched eggs, and culled chicks which have been cooked, dried, and ground, with or without removal of part of the fat. This product deteriorates quite rapidly if not cooled promptly—a factor which is true of most poultry, fish, and meat products.

Dairy Products

Skimmed milk and buttermilk have long been used on, or in close proximity to, the farms where they are produced. However, in their liquid form it is economically impossible to ship them long distances or to store them; and it is difficult to maintain sanitary feeding conditions in using them.

Along about 1910, processes were developed for drying buttermilk. Soon thereafter, special plants were built for dehydrating buttermilk, and the process was extended to skimmed milk and whey. Beginning about 1915, dried milk by-products were incorporated in commercial poultry feeds. In 1984, 98,000 tons (dry basis) of dairy products were fed to livestock.

The superior nutritive values of milk by-products are due to their high-quality proteins, vitamins, a good mineral balance, and the beneficial effect of the milk sugar, lactose. In addition, these products are palatable and highly digestible. They are an ideal feed for young animals and for balancing the deficiencies of the cereal grains. The chief limitation to their wider use is price, together with the perishability and bulkiness of the liquid products.

Although whole milk is an excellent feed—worth about twice as much as skimmed milk—it is usually too expensive to feed. In general, liquid milk products contain 10% dry matter; semisolid milk products, 30% dry matter; and dried milk products, 90% dry matter.

BUTTERMILK

Buttermilk and skimmed milk have approximately the same composition and feeding value, providing the buttermilk has not been diluted by the addition of churn washings. Accordingly, it may be considered that (1) 15 lb of buttermilk will replace 1 lb of tankage or meat meal, or (2) 6 lb of buttermilk will replace 1 lb of complete feed. Other buttermilk products are:

1. **Buttermilk, condensed.** This is made by evaporating buttermilk to about 1/3 of its original weight.

One pound of condensed buttermilk is worth about 3 lb of liquid buttermilk. But it requires about 3 lb of condensed buttermilk to equal 1 lb of dried milk. Another basis of evaluating condensed buttermilk is that, pound for pound, it is worth approximately 1/2 as much as tankage.

2. **Buttermilk, dried.** This is the residue obtained by drying buttermilk. It must contain less than 8% moisture, less than 13% ash, and more than 5% milk fat. One pound of dried buttermilk has about the same composition and feeding value as 10 lb of liquid buttermilk.

CHEESE RIND (CHEESE MEAL)

This is a by-product from the manufacture of processed cheese, consisting of the cheese trimmings from which most of the fat has been removed. It contains around 60% protein and 9% fat.

DAIRY FOOD BY-PRODUCTS

These products are derived from the collection of solids contained in the wash water from normal processing and packaging of various food manufacturing plants. Dairy products are the primary source, but non-dairy products may occasionally constitute a minor amount of the total volume. Dairy food by-products should be fed at levels less than 25% of the animal's total dry matter intake.

MILK PROTEIN PRODUCTS

The five milk protein products currently available are: dried milk albumin, casein, dried hydrolyzed casein, dried milk protein, and dried whey protein concentrate.

Although the quality and quantity of protein from milk protein products are excellent, these sources of protein are generally too expensive to be used routinely as major protein sources in livestock feeds.

MILK, WHOLE DRIED

This is the residue left following the drying of milk. It contains at least 26% milk fat but no more than 8% moisture. This product is generally too expensive to feed to livestock.

SKIMMED MILK

Because of the removal of the fat, skimmed milk has little vitamin A value. However, in comparison with whole milk, it is higher in content of protein, milk sugar, and minerals. Like all milk products, skimmed milk is low in vitamin D and iron.

Skimmed milk is the best single protein supplement for swine. It is especially valuable for young pigs prior to and immediately after weaning. The addition of either pasture or a choice legume hay will supplement skimmed milk with the needed vitamins A and D.

Skimmed milk should be fed consistently sweet or sour, because abrupt changes are apt to produce digestive disturbances. Where a choice is possible, fresh skimmed milk is recommended.

The amount of skimmed milk to feed will vary according to (1) the available supply, (2) the relative price of feeds, (3) the kind of grain ration fed, and (4) whether pasture is available.

The feeding value of skimmed milk varies with the age of the animal, the type of ration, the price of other feeds, and the amount of milk fed. Naturally, it has a higher value per pound when fed in limited amounts than when an excess is fed. On the other hand, when the supply of milk is abundant and cheap, a larger proportion may be fed profitably—especially when grain is scarce and high in price. Roughly, it can be figured that (1) 15 lb of skimmed milk will replace 1 lb of tankage or meat meal, or (2) 6 lb of skimmed milk will replace 1 lb of complete feed.

In addition to liquid skimmed milk, the following processed dairy products are valuable feeds:

1. **Skimmed milk, condensed.** This is the residue obtained by evaporating defatted milk. It contains 27% minimum total solids.

2. **Skimmed milk, condensed cultured.** This is the residue obtained by evaporating lactic acid bacteria cultured defatted milk. It contains 27% minimum total solids.

3. **Skimmed milk, dried.** As the name indicates, this product is dehydrated skimmed milk. It contains less than 8% moisture, and averages 32 to 35% protein. One pound of dried skimmed milk has about the same composition and feeding value as 10 lb of liquid skimmed milk. Although dried skimmed milk is an excellent swine feed, it is generally too high priced to be economical for this purpose except for limited use in pig starter rations.

Prior to World War II, dried skimmed milk was the most widely used dried milk product included in feeds. During and since the war, however, much of it has been marketed as a human food.

4. **Skimmed milk, dried cultured.** This is the residue obtained by drying lactic acid bacteria cultured defatted milk. It contains less than 8% moisture.

WHEY

This is the product obtained as a fluid by separating the cheese coagulum from milk, cream, or skimmed milk and from which a portion of the milk fat may have been removed. The following whey products, each involving different processing, are available: condensed whey, condensed hydrolyzed whey, condensed cultured whey, dried whey, dried hydrolyzed whey, fermented ammoniated condensed whey, condensed whey product, dried whey product, condensed whey solubles, condensed modified whey solubles, and dried whey solubles.

Marine By-products

Fish and marine by-products are most commonly used in swine, poultry, and mink rations. Increasingly, they are being used as a ruminant feedstuff to supply a dietary protein of proper amino acid balance which can survive ruminal degradation (see Table 11-1).

FISH PROCESSING

Processing of fish can be a difficult task. Many fish contain large amounts of fat which, if not properly processed, can create numerous problems with the processing machinery.

There are several methods of manufacturing fish products, each adaptable to a particular type of fish. These follow:

• **Simple direct drying**—Most fish containing little fat can be dried through this process. The raw material is cut extensively in a disintegrator and passed through a flame dryer. Once the fish material passes through the dryer, it is ground, cooled, and bagged.

• **Recirculation drying**—When whole fish are heated, large quantities of water mixed with lipids and other materials are liberated. Since this emulsion can create processing problems through clumping of the material subsequently clogging the machinery, material that has been partially dried is recirculated with the incoming material to dilute the concentration of the gluey water.

• **Traditional reduction processing**—Fig. 11-2 shows a flowsheet of this type of processing. Material is carried to a cooker via a screw conveyer. The cooking stage of this type of processing is the most critical because the fish must be properly coagulated if pressing is to be efficient. If the material is improperly cooked, the resulting presscake will not have a uniform texture due to the presence of oil which should have been previously extracted. Before the cooked material is pressed, it passes through a screen to remove much of the fine sludge which is produced during cooking. After screening, the material is passed through presses to remove moisture, oil, and dissolved solids. The solid material formed in this stage is called the presscake. It can contain as little as 50% moisture and 5% fat. The presscake is then pulverized to increase surface area for drying and subsequently dried, ground, and bagged. The liquid extracted during pressing is then separated into oil and gluewater.

When the glue water is condensed so that it contains about 50% total solids, it is marketed under the name "fish solubles," which is an excellent source of B vitamins and water-soluble proteins. Sometimes, this gluewater is added back to the meal to form what is known as *whole meal*.

• **Other processing methods**—Most fish used for feed are processed by the preceding methods. However, numerous other methods are being investigated; among them, (1) dry rendering, (2) solvent extraction, (3) ultrasonic extraction, and (4) solubilization processing.

The processing of fish meal and solubles varies widely, reflecting fish catch and competing feed prices. In 1970, 609,000 tons were processed; in 1984, the tonnage dropped to 228,000 tons.

MARINE FEEDS

Marine feeds consist of the by-product waste from processing fish, along with unmarketable species of fish and other marine animals.

On a worldwide basis, some 25 million metric tons of fish are processed into fish meal for animal feed each year, but less than ½ million metric tons of this amount is consumed in the United States. Nevertheless, fish by-products are generally considered by livestock producers to be excellent sources of nutrients. Proteins, vitamins, and minerals are all readily available in most fish products.

Fish Meal

Fish meal—a by-product of the fisheries industry—consists of dried, ground whole fish or fish cuttings—either or both—with or without the extraction of part of the oil. If it contains more than 3% salt, the salt content must be a part of the brand name. In no case shall the salt content exceed 7%.

The feeding value of fish meal varies somewhat, according to:

1. **The method of drying.** It may be either vacuum, steam, or flame dried. The older flame drying method exposes the product to a higher temperature, which makes the proteins less digestible and destroys some of the vitamins.

2. **The type of raw material used.** It may be made from the offal produced in fish packing or canning factories, or from the whole fish with or without extraction of part of the oil.

Fish meal made from offal containing a large proportion of heads is less desirable because of the lower quality and digestibility of the proteins. Although few feeding comparisons have been made between the different kinds of fish meals, it is apparent that all of them are satisfactory when properly processed raw materials of good quality and moderate fat content are used. A high-fat content may impart a fishy taste to eggs, meat, and milk.[2] Also, such meal is apt to become rancid in storage.

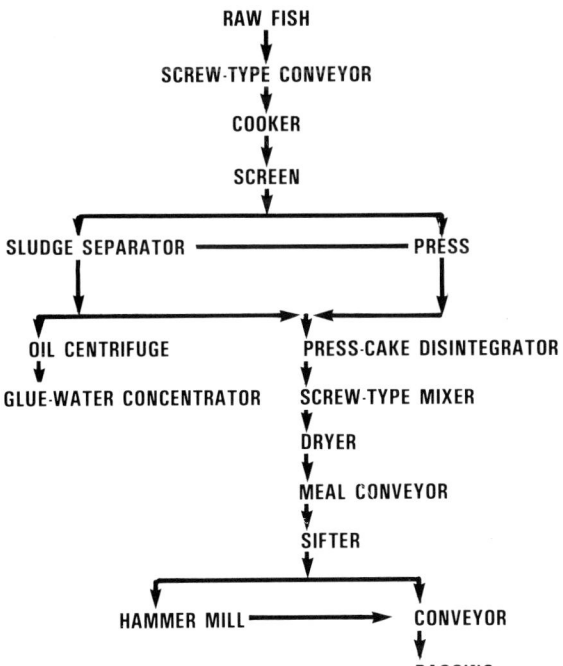

Fig. 11-2. Flow diagram of fish reduction processing.

---

[2]Research workers at the Indiana Station (Vestal, et al., Journal of Animal Science, Vol. 4, No. 1, 1945) found that the addition to the ration of 0.5 and 1.5% fish oil produced a fishy flavor in pork, which was more pronounced in the roasts and bacon than in the chops.

It is of interest to swine and poultry producers to know the sources of the commonly used fish meals. These are:
1. **Herring meal.** This is a high-grade product produced in the Pacific Northwest and Alaska.
2. **Menhaden fish meal.** Over 90% of the fish meal produced in the U.S. is made from menhaden (a very fat fish not suited for human food) caught in the Atlantic primarily for their body oil. The meal is the dried residue after most of the oil has been extracted.

Fig. 11-3. Menhaden are a fat fish unsuitable for human consumption. They are caught primarily for their body oil, but the dried residue after oil extraction provides an excellent high-quality protein feed for livestock. (Courtesy, National Fisheries Institute, Inc., Washington, D.C.)

3. **Salmon meal.** This is a by-product of the salmon canning industry in the Pacific Northwest and in Alaska.
4. **Sardine meal or pilchard meal.** This is made from sardine canning waste and from the whole fish, principally on the West Coast.
5. **White fish meal.** White fish meal, which ranks second to menhaden in U.S. production, is a by-product from fisheries making cod and haddock products for human food. Its proteins are generally of high quality.

Fish meal should be purchased from a reputable company on the basis of protein content. It varies in protein content from 57 to 77%, depending on the kind of fish from which it is made; and it must not contain over 10% moisture. When of comparable quality, fish meal is superior to tankage or meat meal as a protein supplement for nonruminant animals.

The protein of a good-quality fish meal is 92 to 95% digestible. If fish meal is poorly processed or improperly stored, the digestibility of protein decreases dramatically. Since fish meals are cooked, there is danger that certain amino acids—notably lysine, cystine, tryptophan, and histidine—will be denatured, but these losses are minimized when proper processing techniques are used.

Fish meals containing high levels of fat are considered to be low quality. If they are incorporated into feeds, they tend to impart a fishy flavor to the animal products. Also, problems of rancidity are greater in high-fat fish meals.

Fish meal is an excellent source of minerals. Calcium and phosphorus are especially abundant, being present in the amounts of 3 to 6% and 1.5 to 3.0%, respectively. Many of the trace minerals, especially iodine, required by livestock can be supplied in part by fish meal.

Fish meal is not a particularly good source of fat-soluble vitamins, which are lost during the extraction of oil, but a fair amount of the B vitamins remain. However, fish meal is one of the richest sources of vitamin B-12 and unidentified growth factors.

Historically, in the United States, fish meal was used largely in poultry feeds, where it supplied unidentified growth factors. But, in the 1980s, the use of fish meal changed significantly. The three most important new uses for fish meal are: aquaculture feeds, early weaned pig feeds, and ruminant bypass feeds.

Other Marine By-products

Other valuable marine by-products are: fish by-products, fish digest residue, fish liver and glandular meal, fish protein concentrate, condensed fish protein digest, dried fish protein digest, fish residue meal, condensed fish solubles, dried fish solubles, liquified seafood (fish) waste, shrimp meal and crab meal, and unprocessed fish and fish scraps.

NONPROTEIN NITROGEN (NPN) FEEDSTUFFS

Feedstuffs which contain nitrogen in a form other than proteins or peptides are termed nonprotein nitrogen (NPN). Since microorganisms in the rumen of ruminant animals degrade dietary protein to synthesize microbial protein, it follows that if one feeds readily available carbohydrate and nonprotein nitrogen sources, both precursors of amino acids, microbial protein can be successfully synthesized.

Although urea dominates the market, other nonprotein nitrogen (NPN) sources, such as ammonium salts, and ammoniated by-products, have been and are being used successfully in ruminant rations. Some of the nonprotein nitrogen sources are listed in Table 11-4.

Table 11-4
SOME NONPROTEIN NITROGEN SOURCES FOR RUMINANTS

	Formula	Nitrogen Content	Protein Equivalent[1]
		(%)	(%)
Ammonium acetate	$CH_3CO_2NH_4$	18	112
Ammonium bicarbonate	NH_4HCO_3	18	112
Ammonium carbamate	$NH_2CO_2NH_4$	36	225
Ammonium lactate	$CH_3CHOHCO_2NH_4$	13	81
Biuret	$NH_2CONHCONH_2$	35	219
Urea, pure	$(NH_2)_2CO$	46.7	292
Urea, feed grade[2]	—	42–45	262–281

[1]Nitrogen × 6.25.
[2]Feed grade urea is diluted with varying amounts of materials to prevent lumps forming.

(Also see Chapter 4, section on "Nonprotein Nitrogen.")

Urea

The structural formula of urea is given in Fig. 11-4.

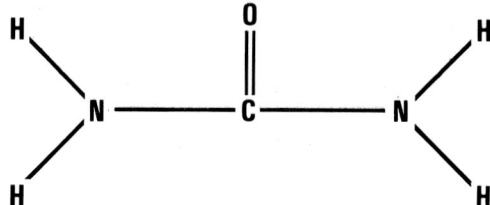

Fig. 11-4. Urea.

Crude protein is determined by multiplying the nitrogen content of a feedstuff by 6.25. Thus, feed grade urea (45% nitrogen) would have a crude protein equivalent of 281%. However, urea contains no precursors for the formation of a carbon skeleton for protein. Therefore, some readily available carbohydrate must be fed simultaneously to maximize protein production.

For specific recommendations as to the proper levels of urea or other NPN sources that can be used in livestock feeds, the reader is referred to the appropriate chapters of Section III in this book.

HISTORY OF UREA AS A LIVESTOCK FEEDSTUFF

The use of urea in ruminant rations was generally viewed with skepticism in the United States and Great Britain throughout the 1920s and 1930s. This resistance began to weaken in the late thirties. By then, a distinct difference in protein nutrition between ruminants and monogastric animals had been established. Unlike protein nutrition in most nonruminants, the emphasis of protein in ruminant rations had been placed on quantity, not quality. While this newborn interest in urea was being generated in the United States, Europeans were already using 10,000 tons of urea per year in sheep and cattle feeds by the mid-thirties.

In 1949, Loosli at Cornell, using a nearly protein-free purified diet with lambs, produced specific evidence (1) that microbial action in the rumen can synthesize from urea and associated carbohydrates all the 10 amino acids required for rat growth, and (2) that 3 to 10 times as much of each amino acid was excreted as was fed.[3]

The ability of ruminants to use rations containing large quantities of nonprotein nitrogen was dramatically illustrated in an experiment conducted by the Finnish Nobel laureate A. I. Virtanen.[4] Lactating cows averaged over 8,800 lb of milk per year despite the fact they were fed a purified ration containing no preformed protein. Urea and ammonium salts were the only nitrogen sources, and purified carbohydrates were the only energy sources. While these cows were managed under carefully controlled experimental conditions, and their production level was low in comparison with today's high producers, the study demonstrated conclusively the qualitative value of nonprotein nitrogen. Additionally, the Finland study showed that when 20% of the dietary nitrogen came from protein, milk production increased. Thus, depending entirely upon microbial protein is possible, but it is neither efficient nor practical.

Thus, the use of urea in ruminant rations was catapulted to the forefront of protein nutrition. Today, urea is routinely used in ruminant rations in Europe and the United States. It is estimated that about 265,000 tons of urea are fed annually in the United States (see Table 6-1).

CONVERSION OF UREA TO PROTEIN

When urea is fed to ruminants, it is first broken down in the rumen into ammonia (NH_3) and carbon dioxide by an enzyme—*urease*—produced by certain microorganisms. Paralleling the catabolism of urea in the rumen, carbohydrates (chiefly starch from cereal grains) are degraded by other microorganisms to produce volatile fatty acids (acetic, propionic, and butyric), which are used as an energy source.

In their multiplication and growth, rumen microorganisms use the ammonia released from the breakdown of protein and nonprotein nitrogen compounds (like urea), along with carbon skeletons and reduced sulfur, to form amino acids which are, in turn, manufactured into microbial protein.

For ruminants to use urea, the microorganisms in the rumen must be able to convert the ammonia released from the urea into microbial protein. The ammonia released from the urea can go via two pathways: (1) It can be made into microbial protein, or (2) it can be absorbed through the rumen wall into the blood stream, which carries it to the liver. The liver detoxifies ammonia by reconverting it to urea to be excreted in the urine. Some of the urea is recycled in the rumen, however, through saliva and by absorption from the blood through the rumen wall. If ammonia escapes the rumen too rapidly, the capacity of the liver is exceeded and ammonia spills into the main blood system. High levels of ammonia circulating in the blood can cause toxicity or even death. Thus, if urea is to be an effective source of true protein for ruminants, conditions in the rumen must be favorable for microorganisms to use the ammonia from urea before it escapes the rumen. So, whenever ammonia is overflowing the rumen and bacteria already have sufficient ammonia for metabolism, additional urea in the ration is of no value in meeting the protein needs of ruminants and may be harmful. This means that, for efficient microbial protein production, ammonia must be available when needed and at the levels needed.

The amount of urea that can be used by rumen microorganisms will depend upon the number of microbes and how rapidly they are growing and whether ammonia and other essential nutrients are available when needed. In addition to nitrogen, microbes need minerals, vitamins, and a readily available source of energy for fast growth. Therefore, more ammonia can be utilized when high energy feeds are fed. Cattle and sheep fed high grain rations can make greater use of urea than animals fed low energy roughage rations.

Throughout digestion, microorganisms are passed from the rumen to the more distal digestive organs. In the abomasum and small intestine, these microbes are hydrolyzed and digested to such a degree that the microbial protein is broken down to free amino acids which can then be absorbed by the host animal. So, once the protein, be it

---
[3] Loosli, J. K., et al., "Synthesis of Amino Acids in the Rumen," *Science*, Vol. 110, 1949, p. 144.

[4] Virtanen, A. I., "Milk Production of Cows on Protein-Free Feed," *Science*, Vol. 153, 1966, p. 1603.

microbial or undegraded true protein, passes into the small intestine, it is digested and absorbed by the ruminant much like that by the nonruminant. Thus, the host animal itself does not utilize urea directly.

UREA FERMENTATION POTENTIAL

Urea fermentation potential (UFP) is a term used to indicate the amount of urea that can be utilized in a ruminant ration.

The UFP system was developed by Iowa State University scientists for the purpose of evaluating the urea fermentation potential of feeds, and, in turn, estimating the amount of urea that can be useful in a ruminant ration. *A positive UFP value of a feed or ration can be defined as the grams of urea per kilogram of feed dry matter consumed (or pounds per 1,000 lb) that can be useful in fermentation by microorganisms in the rumen.* A positive UFP value implies that this quantity of urea feeding is a satisfactory level for achieving maximum or near maximum formation of urea nitrogen into microbial protein.

Calculation of a UFP value involves the amount of fermentable energy present in a feed and the amount of ammonia formed from breakdown of the protein in a feed by rumen fermentation. A feed with a positive UFP value is one that has more fermentable energy present than that needed for transforming the ammonia degraded from its own protein into rumen microbial protein.

A feed with a negative UFP value is one that has less fermentable energy than needed to transform into microbial protein all the ammonia arising from the breakdown of its protein during rumen fermentation. Urea addition to this type of feed would be of no value in satisfying the metabolizable protein requirements of the ruminant host; it would only add to the surplus nitrogen load in the rumen.

ESSENTIALS FOR OPTIMUM USE OF UREA

In this section, discussion centers on the proper use of urea. But the same principles apply to all NPN products.

Urea can be successfully and effectively used, or it can be abused. Observance of the following pointers will assure optimum results:
1. Feed urea only to ruminants.
2. Supply adequate energy.
3. Provide the necessary minerals.
4. Mix well.
5. Don't feed excessive levels of urea.
6. Do not use urea as the sole source of N for (a) young, rapidly growing ruminants or (b) high producing, lactating ruminants.
7. Give animals time to adjust to urea.
8. Give heavily stressed, new arrivals time to adjust.
9. Feed frequently.
10. Enhance urea utilization by proper fermentation, protein degradation, and length of time in the rumen.
11. Include alfalfa meal in urea rations.
12. Provide vitamin A.
13. Include adequate salt for palatability.
14. Use a free-flowing urea.
15. Never use urease-containing seeds in urea rations.

(Also see Chapters 19 and 20, relative to feeding urea to beef cattle and dairy cattle.)

PERTINENT FACTS ABOUT UREA

The following facts are pertinent to the informed use of urea:

1. **Feed grade urea.** Initially, the protein equivalent value of feed grade urea was 42% (nitrogen) times 6.25 (common protein factor), or 262%. Today, more concentrated 45% nitrogen (45 × 6.25 = 281) urea has replaced most of the 42% grade, at a lower unit cost.

2. **Feeding value of urea.** Attempts have been made to equate urea to oil meals by various thumb rules. One such thumb rule is that 1 lb of urea plus 6 lb of corn equals 7 lb of soybean (or cottonseed) meal. This combination of corn and urea supplies as much nitrogen as does soybean meal and, thus, could be considered equal to it in crude protein content. This is true if the rumen microorganisms can convert the urea nitrogen to protein.

3. **Urea is best utilized in well-balanced, high-energy rations.** Urea is not well utilized in supplements to low-quality roughages. The explanation is that the carbohydrates in grasses and hays appear to be so slowly available that the bacteria have difficulty in using the energy from roughages to make use of urea in preparing bacterial protein. It is generally held that some preformed protein should be present in the feed, also. Part of this will be provided by the grains, and frequently some oil meals are used in preparing the formula feeds.

Other components of a balanced feed include calcium, phosphorus, iron, copper, cobalt, manganese, iodine, and perhaps zinc, sulfur, and magnesium. The need for these minerals, as well as for vitamin A, will depend upon local conditions with respect to the types of roughages produced and the influence of weather upon the quality of such roughages.

4. **Toxicity.** When urea is fed at excessive levels, large amounts of ammonia are liberated in the rumen. Eventually, the pH of the ruminal fluid increases, thus facilitating the passage of ammonia across the rumen wall. If the levels of ammonia absorbed are greater than the ability of the liver to convert ammonia to urea, ammonia accumulates in the blood. If blood ammonia levels reach toxic concentrations (1mg/100 ml in cattle), the animal shows signs of acute ammonia poisoning.

(Also see Chapter 5, Table 5-3, Potential Poisons—Urea Toxicity.)

5. **Palatability.** Although various opinions exist relative to the palatability of urea and urea-containing feeds, most feeders feel that urea is not palatable and, therefore, that feed consumption may be lowered in comparison with rations in which oil meal protein supplements are used. For this reason, care should be exercised in selecting an appetizing urea-supplemented mixture.

Sometimes cattle will consume a urea-containing feed for a few days or weeks, then refuse it. This has occurred in drought areas where farmers have tried to extend their roughage supplies by feeding straw and other mineral poor, low quality roughages.

6. **How to compute how much urea is in a feed.** The level of urea in a feed may be determined in the following ways:

 a. **Percent of urea in the feed.** When the percent of urea is given, one can calculate the amount of protein furnished by urea by multiplying the percent urea by 281% (the protein equivalent of urea). For example, if a 40%

supplement contains 5% urea, then 14% protein is furnished by urea (281% × 5% = 14%). To determine the percent of the total protein furnished by urea, divide the percent of protein as urea by the percent of protein in the supplement (14% ÷ 40% = 35%). In this case, slightly more than ⅓ of the protein in the supplement is furnished by urea.

b. **Percent protein as urea.** When the urea in the supplement is expressed in percent protein as urea, one can determine the amount of urea by dividing this value by 281%. For example, if a 35% protein supplement has 12% protein as urea, it contains 4.3% urea (12 ÷ 281 = 4.3%). Slightly more than ⅓ of the protein in the supplement is furnished by urea (12% ÷ 35% = 34.3%).

METHODS OF FEEDING UREA

Urea can be provided by different methods and systems, with consideration given to the following factors: (1) protein needs of the animal as dictated by type of production; (2) availability of urea; (3) toxicity hazards; (4) cost of processing and mixing; and (5) availability of energy sources and amount of plant protein being used. The common methods of feeding urea follow:

1. **Urea mixed in concentrate.** Most of the urea fed to growing and lactating dairy cattle and to finishing beef cattle is incorporated into the concentrate portion of the ration. It can be supplied in a protein supplement or in the entire concentrate ration. Also, urea may be provided in a complete mixed feed.

Urea can be mixed in feed either as a powder or as an aqueous solution. Both methods are relatively simple and inexpensive. When urea is added as a powder, there is a chance that it will sift through the grain and be unevenly distributed, thus increasing the chance of toxicity. However, if careful mixing procedures are followed, this hazard can be kept to a safe minimum.

2. **Urea liquid supplements.** Liquid supplements combining molasses for energy and urea as a protein precursor are widely used. This type of supplement can also be used as a carrier for micronutrient and nonnutritive additives. It is fed primarily in conjunction with low-quality forages.

The **advantages** of liquid supplements are: (a) They provide a way to assure uniform distribution of urea throughout the supplement; (b) they are adapted to limited feeding—cattle consumption can be regulated by means of a lick-wheel feeder or the addition of a bitter principle in the liquid supplement; (c) they require less labor to feed; and (d) they eliminate dustiness of the ration and loss from blowing.

The **disadvantages** of liquid supplements are: (a) They require special mixing and feeding equipment; (b) they require greater transportation costs; (c) they are not well suited to keeping calcium and certain other nutrients in solution; (d) they may subject urea to degrading, once it is in solution, during prolonged storage; and (e) they tend to be highly corrosive.

3. **Urea salt blocks.** Another simple way of supplying protein precursors to livestock on pasture is through the use of urea in salt licks or blocks. This practice is used extensively where large and/or inaccessible ranges limit the amount of contact that ranchers have with their animals. Numerous combinations of salt and urea have been used.

4. **Urea mixed with silage.** Excellent utilization of urea can be obtained when it is ensiled with cereal grain silage, especially corn silage. If chopped, whole plant corn is being ensiled at 35 to 40% dry matter, urea can be added at a level of .5% of the wet material. This level should increase the crude protein level of the silage on a dry matter basis about 5 percentage points. Urea levels higher than .5% can create palatability problems as well as storage problems. When the silage contains little or no grain, the amount of urea to be added should be reduced.

Silage tends to be variable in moisture, and this variability can affect the benefits of added urea. Hence, one should have a reasonable estimate of the moisture content of the material that is to be ensiled. Likewise, water in silage will create some leaching of the urea. Also, ammonia is produced during the ensiling process, representing an additional loss of urea.

(Also see Chapter 9, section headed "Feed Additives.")

5. **Urea added to dry roughages.** A limited amount of research has been reported on the addition of a urea-molasses mixture to hay. Based on available information, the addition of urea to good-quality hay (over 10% crude protein) is generally not beneficial.

6. **Slow-release urea products.** Several products in which urea is bound in a slow-release complex have been developed in recent years; among them, urea combined with starch from grain, and urea combined with the sugars in molasses through heat and chemical treatment. These products are designed to decrease the solubility of urea in the rumen and thereby slow the release of ammonia. Slow ammonia release, or a more uniform ammonia level in the rumen throughout the day, is desirable, especially for urea used in low-energy rations. Additionally, there should be less danger of urea toxicity from overconsumption with slow-release products.

(Also see Chapter 14, Feed Processing, section headed "Slow-release and Rumen Bypass Treatment.")

Other Nonprotein Nitrogen Feedstuffs

Nonprotein nitrogen sources other than urea have been used to a limited extent. These products can be classified into two categories: (1) ammoniated products, and (2) biuret.

AMMONIATED PRODUCTS

Numerous ammonium salts have been used effectively as sources of nitrogen. Both organic and inorganic salts of ammonia can be utilized efficiently by ruminants; among them, ammonium acetate, ammonium bicarbonate, ammonium butyrate, ammonium carbonate, ammonium chloride, ammonium lactate, ammonium polyphosphate, ammonium propionate, ammonium sulfate, monoammonium and diammonium phosphate, and methionine hydroxy analog (MHA) (the latter is used as a source of methionine for poultry).

A wide range of feedstuffs has been treated with anhydrous ammonia for a variety of reasons. One of the first synthesized protein supplements for cattle and sheep was ammoniated molasses. The ammoniation of beet pulp and citrus pulp followed. Cottonseed meal has been ammoniated for two purposes: (1) to inactivate aflatoxin-contaminated feed, and (2) to provide a supplemental source of nitrogen for livestock. Aflatoxin-contaminated shelled corn has been detoxified and made safe for animals by treatment with anhydrous ammonia. Rice hulls and rice straw have also been ammoniated.

Experiments have demonstrated that the addition of anhydrous ammonia to corn silage can double the protein equivalent and increase the feeding value of the silage dramatically.

In the late 1970s and the early 1980s, the ammoniation of low-quality forages increased greatly throughout North America, with the anhydrous ammonia applied to forages in round bales, stacks, and barns. The reasons:

1. The ammonia provides an economical source of nitrogen (protein equivalent) which can be utilized by ruminants.

2. The ammonia treatment increases the digestibility of fiber, thereby increasing the amount of energy that animals can obtain from the forage.

3. The ammonia is inhibitory to many microbes. As a result, treated hay can be baled at a high moisture content (up to 30%) without mold and/or bacterial spoilage, and with reduced weather and harvesting losses.

All went well until 1983, when rare episodes of toxicity in animals consuming certain ammoniated hays was reported. The afflicted animals behaved abnormally, including running in circles, extreme nervousness, and convulsions. Many animals died, usually from injuries sustained while they were in an excited state. The problem was observed in cows, bulls, yearlings, newborn calves, and even in calves nursing cows fed ammoniated forage. The diagnosis: ammonia toxicity, also known as *bovine bonkers* or *crazy cattle disease*. After several cases were investigated and compared, it was determined that the syndrome resembled what had occurred 30 years earlier, in the 1950s, when ammoniated molasses was first fed to cattle and sheep. Subsequently, methods were developed to produce synthetic proteins that did not cause toxicity, and the problem was largely forgotten.

Ammoniated feedstuffs reported to have caused toxicity include alfalfa hay, barley hay, Bermudagrass hay, bromegrass hay, fescue hay, oat hay, orchardgrass hay, rice straw, sorghum forage, Sudangrass hay, wheat hay, and wheat straw. It is noteworthy that most of the toxic forages appeared to have been treated with high levels of ammonia (3% or more) and/or that the treatment was not evenly applied. It is noteworthy, too, that ammoniated beet pulp, ammoniated citrus pulp, ammoniated cottonseed meal, ammoniated shelled corn, and ammoniated corn silage have not been reported to have caused ammonia toxicity.

Although the specific toxin responsible for the malady has not been positively identified, the following preventive/control measures are recommended:

1. Do not ammoniate high-quality hay that is fed to lactating cows, sheep, or goats.

2. Do not treat hay with more than 1.0 to 1.5% ammonia by weight; and apply the ammonia evenly to the forage.

3. Do not ammoniate hay that contains more than 30% moisture.

4. Do not feed ammoniated hay alone—feed some grain with it.

5. If toxicity signs develop when feeding ammoniated forage, discontinue feeding the forage immediately. Affected animals may return to normal in 1 to 2 days; and the toxic hay will become less toxic with the passing of time.

- **Biuret**—In recent years, the cost of manufacturing biuret has been so high as to make it uneconomical as a feedstuff.

(Also see Chapter 5, Nutritional Disorders/Toxins, Table 5, Potential Poisons—Urea Toxicity [Ammonia Toxicity].)

SINGLE-CELL (MICROBIAL) PROTEIN (SCP)

Single-cell protein (SCP) is protein obtained from single-cell organisms, such as yeast, bacteria, fungi, and algae, that have been grown on specially prepared growth media. Production of this type of protein can be attained through the fermentation of petroleum derivatives or organic waste or through the culturing of photosynthetic organisms in special illuminated ponds.

Of course, yeast and bacteria have been used for centuries in the baking, brewing, and distilling industries, in making cheese and other fermented foods, and in the storage and preservation of foods. Dried brewers' yeast, a residue from the brewing industry, and torula yeast, resulting from the fermentation of wood residue and other cellulose sources, have been marketed as animal feeds for years.

A wide variety of materials can be used as substrates for the growth of these organisms. Current research deals with the use of products which otherwise would have little or no economic value; among them: straws; fodders, and other low quality cellulose wastes; wood and wood processing wastes; food, cannery, and food processing wastes; residues from alcohol production; and animal excreta.

The potential of single-cell protein as a high-protein source for both humans and livestock is enormous, but many obstacles must be overcome before it becomes widely used. It has been calculated that a single-cell protein fermenter covering .386 square mile could yield enough protein to supply 10% of the world's needs. To put this potential in a different perspective, a 1,000-lb steer will produce about 1 lb of protein per day. One thousand pounds of rapidly growing soybeans will produce 80 lb of protein per day. One thousand pounds of single-cell organisms might well produce up to 50 tons of protein per day. This is not to imply that we have solved the world's protein needs with single-cell protein because serious problems involving palatability, gastrointestinal disturbances, uric acid accumulation, and simple economics must be solved before wide-scale production of single-cell protein becomes a reality. There is a limited amount of single-cell protein on the market in the form of brewers' yeast and torula yeast, but these products are generally too expensive to use as a major protein supplement.

Types of Single-Cell Protein

Single-cell protein can be produced by nonphotosynthetic and photosynthetic organisms. Of the nonphotosynthetic organisms, yeasts are the most popular sources, but bacteria and fungi are also currently being investigated as potential sources. The photosynthetic organisms—algae—are grown in ponds that are illuminated and fortified with simple salts such as carbonates, nitrates, and phosphates.

Current Problems Associated with Single-Cell Protein

Although single-cell protein appears to be an excellent alternative source to the protein feeds currently used, several problems must be overcome before it becomes a widely used feedstuff; among them, palatability, digestibility, nucleic acid content, toxins, protein quality, and economics.

As the world's human population increases, there will be an increasing demand for cheap protein. This demand could dry up the sources of the tradional protein feeds for livestock, thereby opening the way for the intensive development of alternative sources of protein.

We are also entering an era of concern for the maintenance of the quality of our environment. This means that a greater emphasis will be placed on the transformation of industrial by-products into usable commodities. Single-cell protein is one way that this challenge can be met.

TYPES AND METHODS OF FEEDING PROTEIN SUPPLEMENTS

There is no one best type (or form) and method of feeding protein supplements for any and all conditions. When mixed feeds (either complete rations or concentrates) are fed, usually the protein ingredient(s) is incorporated in the mix. When the protein supplement is fed separately, such as on the pasture or range or in the corral, any one of the following types and methods of feeding may be used: (1) hand-fed at intervals, (2) liquid protein supplements, (3) protein blocks, (4) range cubes or pellets, or (5) self-fed in salt-feed mixtures.

Hand-feeding at Intervals

The authors recommend feeding nonurea range supplement to cattle twice weekly, allocating in each of the two feedings one-half as much supplement as would have been fed in a week on a daily feeding basis.

Protein cubes may be scattered on the ground—2 or 3 times a week. This offers a method of checking the animals because they are attracted by the sight or sound of the vehicle when they know that there is something to eat.

Twice weekly feeding has two distinct advantages over the use of salt-feed mixes: (1) It alleviates the cost of using excess salt, which has no nutritive value when so used; and (2) it forces inspection of the herd two times per week, which is as infrequent as is desirable.

Liquid Protein Supplements

Liquid supplements are the fastest growing type of protein supplement in the United States. They are fed primarily to beef cattle (both breeding and finishing cattle) and dairy cattle; they are fed in the corral and on the pasture and range; and they are incorporated in mixed feed, used as a top dressing, and self-fed.

When self-fed, consumption may be controlled by use of a *lick tank* and/or by incorporating in the formulation such ingredients as phosphoric acid, beet solubles, and citrus peel liquor.

The usual directions relative to the feeding of liquid protein supplements are:

1. Do not feed urea-containing supplement to animals that have been without feed for 36 hours until they have had a chance to fill the rumen with grass.

2. Restrict consumption to a desired level (1 to 2 lb/head/day), since some animals tend to overeat. This may be accomplished by a free-turning plastic or wood wheel (a lick wheel) dipped in a tank or other similar equipment, or by adding phosphoric acid to the mix at a level of 1% phosphorus. The phosphoric acid also serves as a source of phosphorus and keeps the molasses free-flowing in cold weather.

3. Allow several days for cattle grazing dry grass to become accustomed to a liquid protein supplement.

4. Never let the animals run out of the liquid supplement once feeding is started, until you discontinue feeding it entirely.

Protein Blocks

Protein blocks are just what the designation implies. They are compressed protein blocks, ranging in weight from 50 to 500 lb each. They are particularly adapted to supplementing cattle on rather inaccessable range when the grass dries.

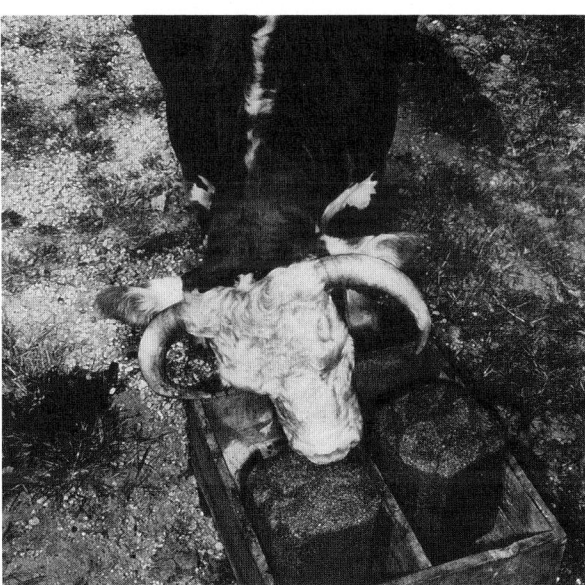

Fig. 11-5. Protein block in use on pasture—a means of lessening the labor attendant to the daily feeding of a protein supplement on pasture or range. (Courtesy, Moorman Mfg. Co., Quincy, Ill.)

Blocks may be placed in grazing areas where cattle have frequent access to them, with one 50-lb block provided to 15 cows. Intake will vary with the feed supply and the type of block. Generally, it is planned to limit feed consumption to about 2 lb/head/day by intake limiters such as hardness of block, salt, and/or fat content.

Range Cubes or Pellets

Traditionally, cattle have been supplemented either once or twice daily on pasture or range. Where this practice is followed, a urea-containing (preferably, a slow-release urea) range cube or pellet may be used. Cubes may be scattered on the ground.

Urea-containing supplements, particularly those containing high levels of urea, should not be fed at long intervals (such as twice weekly) on the range because (1) range forages are relatively low in energy, and (2) urea is extremely soluble and its nitrogen becomes available very quickly in the rumen. Also, it is important that NPN-containing range supplements provide readily available energy and needed minerals.

Fig. 11-6. Range cubes fed on pasture or range. Many producers prefer this method of supplementation, primarily for reasons of convenience and to reduce losses from wind blowing. (Courtesy, Ralston Purina Company, St. Louis, Mo.)

Self-feeding Salt-Feed Mixtures

The practice of using salt as a governor to limit feed consumption on pasture or range has been around a very long time. It was ushered in as a laborsaving device for cattle and sheep in inaccessible and rough areas.

Two suggested salt-meal supplements follow. Either 41% cottonseed or soybean meal may be used. Neither mix should be pelleted.

Ingredient:	Salt-Meal Mix No. 1	Salt-Meal Mix No. 2
	(lb)	(lb)
Salt	665	499
Meal (either 41% cottonseed or soybean meal)	1,331	1,497
Vitamin A (30,000 IU/g)	4	4
	2,000	2,000
Consumption level:	approx. 1½ lb daily	approx. 2 lb daily
Guarantee:		
Crude protein	min. 27%	min. 30%
Salt	max. 35%	max. 27%
Vitamin A	24,000 IU/lb	24,000 IU/lb

Based on experiments and experiences, the following points are pertinent to self-feeding salt-feed mixtures to range cattle and sheep, to cattle grazing stalk fields, or to cattle that are being grain-finished on pasture:

1. The practice need not be limited to any specific protein supplement or feed.

2. It is best that salt mixes be in meal form, rather than pelleted. If pellets are small and soft, they will work satisfactorily. However, there is always the hazard that they will be hard enough to permit animals to swallow them without the salt being fully effective as an inhibitor, with overeating resulting.

3. The proportion of salt and feed may vary anywhere from 5 to 40% (with 30 to 33⅓% salt content being most common), with the actual intake of feed supplement limited to 1 to 2½ lb daily. By varying the proportion of salt in the mixture, it is possible to hold the consumption of feed supplement to any level desired. In some range areas, a reduction of the salt level from 33⅓ to 24% will increase consumption by about 50%. When a liberal feeding of grain on pasture is desired, 5% salt may be sufficient.[5]

4. The quantity of salt and the proportion of salt to supplement required to govern supplement consumption varies according to (a) the daily rate of feed consumption desired, (b) the age and weight of animals (higher quantities of salt are required in the case of older animals), (c) the fineness of the salt grind (fine grinding lowers the salt requirement), (d) the salinity of the water, (e) the severity of the weather, (f) the quality and quantity of forage, and (g) the length of the feeding period (as animals become accustomed to the mixture, it may be necessary to increase the proportion of salt).

5. It is common practice to prepare the starting feed by mixing 1 lb of salt to 4 lb of feed supplement, and to increase the proportion of salt in the mixture as the animals become accustomed to the feed.

6. It lessens the difficulty in starting animals on a supplement, for sprinkling a little salt on the meal makes it more palatable.

7. It is recommended that animals be hand-fed a week or so before allowing free-choice to a salt-feed mixture; thus getting them on feed gradually.

8. It is necessary to regulate or limit (by hand-feeding for a few days) the supply of salt-feed mixture when it is desired to shift animals from a straight feed supplement (such as cottonseed meal alone) to a salt-feed mixture. Otherwise, hungry animals may consume too much.

9. It is estimated that the practice increases the total salt consumption to 8 to 10 times that required in conventional salt feeding, and doubles or triples the water consumption.

10. If the salt-feed mixture is placed in close proximity to the water supply, it will make for restricted grazing distribution on the range, because of the greater intake of water on a high-salt diet. On the other hand, if the salt-feed mixture is shifted about on the range, it will make for desirable distribution, because of the animals following the feed supply.

11. It reduces the labor required in feeding, promotes more uniform feed consumption (among the greedy and the timid), and permits animals to eat at their leisure with less disturbance during blizzards or cold weather.

12. It lessens the space required for feed equipment (bunks or feeders) to 20% of that required in conventional hand-feeding, but makes it desirable that the feeder be constructed so as to protect the mixture from the weather (especially wind and rain).

13. It is equally applicable to feeding during droughts, on dry summer range, and in the winter months.

14. It is commonly believed that under conditions of short feed supply (submaintenance) and relatively inaccessible water supply, animals may consume sufficient salt in this manner to produce toxic effects, especially during the winter months when low temperatures tend to lessen the water intake.

15. The practice of self-feeding salt-feed mixtures is well adapted to inaccessible and rougher areas, where daily feeding is difficult. In no case, however, should it be an

---

[5]At the Irrigated Agriculture Research and Extension Center, Prosser, Wash., it was found that 7½% salt limited grain consumption by yearling steers on pasture to 10 to 12 lb daily; 5% salt limited grain consumption to 12 to 14 lb.

excuse to neglect animals, for herds and flocks need to be checked often.

16. It reduces the consumption of minerals other than salt to practically nothing, with the result that correction of mineral deficiencies must be considered.

17. With adequate water, high salt intake has no effect on fertility, calf crop percentage, weaning weight, or bloom of animals.

18. *CAUTION:* Do not use trace-mineralized salt, because it may result in toxicity or mineral imbalances due to excessive intake of certain trace elements.

HIGH-PROTEIN BY-PRODUCT FEEDS

Many industrial by-products can be used as protein supplements. The distilling and brewing industries have done considerable research in the development of grain by-products as feeds for livestock. Likewise, many poultry operations are developing new ways of using poultry wastes as protein supplements. The meat and fish processing industries, discussed earlier in this chapter, have provided a variety of protein supplements. The grain milling industry now sells many of its by-products, such as corn gluten meal and wheat gluten meal, as protein feeds. (For additional information on by-product feeds which can be used as protein supplements, the reader is referred to Chapter 12, By-product Feeds/Crop Residues, and Chapter 30, Glossary of Feedstuffs.)

AMINO ACID SUPPLEMENTS

Amino acid supplements are supplements carrying large amounts of one or more pure amino acids which may be added to a ration to make up for an amino acid deficiency (or deficiencies).

The amino acids, of which there are 22, are the basic components, or building blocks, of proteins. But, according to our present knowledge, based largely on rat work, only the following 10 of these are essential in rations, because animals are able to synthesize sufficient quantities of others: arginine, histidine, isoleucine, leucine, lysine, methionine, phenylalanine, threonine, tryptophan, and valine. But all proteins are not created equal! Plant proteins often contain insufficient quantities of lysine, methionine, cystine, tryptophan, and /or threonine. Since the synthesis of body proteins (muscle, milk, eggs, wool, and hair/feathers/hooves/nails) is an "all or nothing" proposition, if any one of the essential amino acids needed to form the particular protein is deficient, that protein cannot be made.

Generally the first limiting amino acid (*i.e.*, the amino acid in lowest amount relative to need) in grain-based swine rations is lysine; and the second and third most limiting amino acids may be methionine and threonine, with their ranking depending on the grains used. So, normal swine rations may be deficient in lysine, methionine, and threonine.

Due to the large amount of vegetable proteins used, methionine is usually the first limiting amino acid in poultry rations. Also, lysine and cystine are often inadequate in normal poultry feedstuffs. So, most poultry rations call for supplementation with methionine, lysine, and cystine.

Ruminants have the same amino acid requirements as nonruminants at the tissue level. When ruminants are producing at average or low levels, they are capable of meeting any feed deficiencies of amino acids through microbial synthesis. But high-producing ruminants (young, rapidly growing ruminants, and ruminants in peak milk production) cannot achieve maximum production if certain amino acids are deficient. So, ruminants in high production have need for amino acids in excess of those furnished by microbial synthesis and the normal escape of approximately 40% of the dietary protein from the rumen. The needed limiting amino acids for ruminants may be provided by protected (escape) amino acids or by true proteins.

When formulating rations, therefore, it is necessary not only to provide certain levels of protein, but also to provide specific levels of the essential amino acids. Theoretically, a perfectly balanced protein is one in which the available amino acids meet the requirements of the animal, with neither a surplus nor a deficiency of any amino acid.

Synthetic amino acids are now available for feed supplementation. But, due to their high cost, supplementation to date has been largely limited to methionine in poultry rations and lysine in starter and prestarter pig rations. Beginning in the late 1980s, tryptophan and threonine were marketed in feed grade form and used to a limited extent. But, biotechnology has become more sophisticated, and amino acids have become less expensive to produce. Today, both American and Japanese firms are producing amino acids with the same technology as is used in manufacturing drugs. Microorganisms use corn or molasses as an energy source to produce amino acids through fermentation.

It is noteworthy that 97 lb of corn and 3 lb of lysine can replace 100 lb of soybean meal. As soybean prices rise and/or lysine prices fall, this alternative becomes more attractive.

PROTEIN ASSESSMENT SYSTEMS

The crude protein method of evaluating proteins by the century-old Kjeldahl method, or some minor variation of it, involves finding the nitrogen content of the sample, then multiplying it by 6.25, since the nitrogen content of all protein averages about 16% ($100 \div 16 = 6.25$). However, beginning about 1960, the poultry industry, and to a lesser extent the swine industry, developed rations on an amino acid balance basis rather than on a crude protein basis. Along the way, research evidence began to accumulate indicating that crude protein content was not an accurate evaluation of nitrogenous sources for ruminants. It was found that certain feed ingredients achieved better production than others in ruminants even though the rations were isonitrogenous. This led to the development of several proposed protein assessment systems; among them, (1) the metabolizable protein concept (the quantity of protein digested or amino acids absorbed in the postruminal portion of the digestive tract of ruminants) proposed by Burroughs, Trenkle, and Vetter of Iowa State, in 1972; and (2) the ammonia overflow system as related to feed ingredients developed by Satter and Roffler of Wisconsin, in 1978. Currently, there is no single procedure for determining protein efficiency or biological value for ruminants which is best for all purposes. However, the critical scientist can find methods that will serve the purpose at hand and which will yield reliable results provided they are properly interpreted.

The goal ahead is to develop a protein/carbohydrate evaluation system which predicts well for all types of feeding systems and all types of feed processing.

QUESTIONS FOR STUDY AND DISCUSSION

1. Define each of the following: proteins, amino acids, protein quality, nonprotein nitrogen (NPN), protein equivalent, and protein supplements.

2. Under what circumstances may protein produced by rumen synthesis not be sufficient?

3. Define and explain the significance of each of the following: bypass protein, degradability of protein, solubility of protein, and microbial protein.

4. List and discuss four methods for protecting proteins from ruminant degradation.

5. Discuss the relative importance of plant and animal protein sources. Which sources have become increasingly more important in recent years?

6. List and give pertinent information relative to each of the commonly used oilseed meals.

7. Describe the three processes of oil extraction of soybeans. What are the advantages and disadvantages of each?

8. What are corn gluten feed and corn gluten meal? Why have these feeds become so abundant in recent years?

9. Define pulses. List five types of pulses and give the relative crude protein content of each. What precautions should be taken when pulses are to be used as either feed or food?

10. Discuss the future of leaf protein concentrates as it pertains to livestock.

11. Thirty years ago, it was commonly believed that animal protein had to be incorporated into nonruminant feeds. Why? Why has this thinking changed?

12. Define, describe the processing, and tell of the importance and use of each of the following animal proteins: tankage, meat meal, meat and bone meal tankage, meat and bone meal, and blood meal.

13. Define, describe the processing, and tell of the importance and use of each of the following poultry wastes: feather meal, hydrolyzed poultry by-product aggregate, poultry by-products, poultry by-product meal, and poultry hatchery by-product.

14. What are the limitations to the use of dairy products? What is whey? Why have so many new whey products evolved in recent years?

15. Why are fish meals becoming of increasing importance as a ruminant feedstuff? What factors cause variation in the feeding value of fish meal? List the two main sources of fish meal.

16. The research on urea by Loosli of Cornell and by Virtanen of Finland had great impact on the subsequent use of urea. Tell about each of these studies, and explain why they were so significant.

17. Discuss the conversion of urea to protein and its subsequent utilization in the ruminant animal.

18. Define fermentation potential. What is the practicality of this concept?

19. List and discuss 10 pointers that should be observed in order to assure optimum results from urea.

20. If a 35% crude protein supplement contains 4% urea, what percent protein is furnished by urea?

21. What method would you use to feed urea to (a) dry dairy cows in a corral, (b) finishing steers in a dry lot, and (c) pregnant cows on an isolated range?

22. What are the advantages of slow-release NPN products?

23. List the precautions that should be observed if hay is ammoniated.

24. Briefly discuss the advantages and disadvantages inherent with single-cell protein.

25. What forms of protein supplements are fed to animals on range and pasture? Discuss the advantages and disadvantages of each pasture/range supplement.

26. Define amino acid supplements. What amino acid supplements are particularly important to each (a) swine, and (b) poultry? Why may they be deficient for these species?

27. Under what circumstances may there be amino acid deficiencies in ruminants? When amino acids are needed for ruminants, how should they be provided if they are to be effective?

28. Why has the century-old Kjeldahl crude protein method of evaluating protein come under considerable scrutiny in recent years? Is there a better and quicker system for protein evaluation?

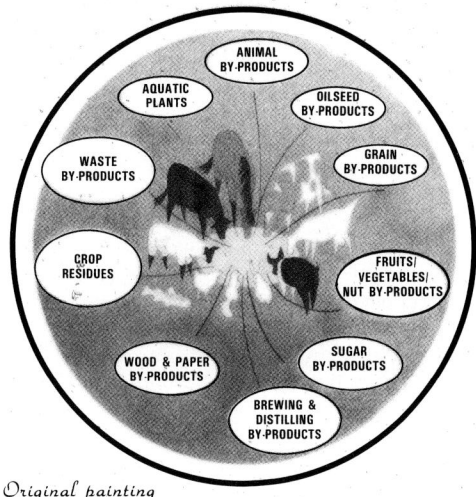

Original painting by Tom Phillips

12

BY-PRODUCT FEEDS/ CROP RESIDUES[1]

Contents	Page
Economy of By-product Feeds and Crop Residues	226
Animal By-product Feeds	226
Dairy By-product Feeds	226
Marine By-product Feeds	226
Meat Animal By-product Feeds	227
Poultry By-product Feeds	227
Plant By-product Feeds	228
Oilseed By-product Feeds	228
Grain By-product Feeds	229
Fruit, Vegetable, and Nut By-product Feeds	229
Sugar By-product Feeds	230
Sugarcane Processing	231
Sugar Beet Processing	231
Molasses and Other Sugar By-product Feeds	231
Brewing and Distilling By-product Feeds	231
Brewers' By-products	231
Distillers' By-products	232
Wood and Paper By-product Feeds	233
Crop Residues	234
Kind and Quantity of Residue Produced	234
Economy of Crop Residues	235
Harvesting	235
Nutrient Value of Crop Residues	235
Treating Crop Residues to Increase Digestibility	235
Transporting, Packaging, and Storing	236
Feeding Systems for Crop Residues	236
Some Crop Residues	236
Waste By-product Feeds	237
Animal Wastes (Manure and Litter)	237
Amount of Manure Produced	238
Daily Manure Production and Storage	238
Manure Gases	238
Yearly Manure Production	238
Manure Uses	239
Manure as a Feed	239
Chemical Composition of Animal Wastes	239
Nutrient Utilization	239
Performance of Animals Fed Animal Wastes	239
Cattle Waste	239
Swine Waste	240
Poultry Waste	240
Processing Animal Wastes for Feed	241
Dehydration	241
Ensiling	241
Other Processes	241
Animal and Human Health/Esthetic Aspects	241
Regulatory Aspects	242
Future Handling and Uses of Manure	242

Contents	Page
Municipal Wastes	242
Sewage Sludge	242
Garbage	242
Bakery Waste	243
Aquatic Plants	243
Quantity and Kinds of Aquatic Plants	243
Physical Characteristics of Aquatic Plants	243
Questions for Study and Discussion	244

Producing and processing animals and plants for food for people and feed for animals result in many by-products and crop residues which can be utilized as livestock feeds.

By-product feeds are concentrates and roughages other than the primary products from animal and plant processing and from industrial manufacturing.

Crop residues are the portions of crops that are normally left in the field following harvesting.

In addition to conventional by-product feeds and crop residues, many unusual and underutilized feeds are discussed in this chapter, including surplus and cull fruits and vegetables, certain weeds that are fed on occasion, wood and paper by-products, animal waste (excreta), and aquatic plants.

Animals provide a practical outlet for a host of products which are not suited for human consumption. Some of these products have been used for animal feeds so long, and so extensively, that they are commonly classed as feed ingredients, along with such things as the cereal grains, without reference to their by-product origin. Such ingredients include both animal and plant materials. More than 36 million tons of these by-product feeds are used as animal feeds each year in the United States. (See Chapter 6, Types and Roles of Feedstuffs, section on "Concentrates.")

In the United States, cereal grain residues are the principal crop residues, although some areas have significant amounts of cotton, soybean, sugar beet and sugarcane residues. Corn stover, wheat straw, and soybean residues account for more than 80% of the estimated 500 million tons of residues produced annually, with corn stover alone making up 50% of the total.

---

[1]The authors gratefully acknowledge the helpful suggestions of the following authority who reviewed this chapter: D. A. Miller, Ph.D., Department of Agronomy, College of Agriculture, University of Illinois, Urbana-Champaign.

The list of by-product feeds and crop residues is very long; and various classifications of them have evolved. In this chapter, by-product feeds and crop residues are classified in four major categories: animal by-product feeds, plant by-product feeds, waste by-product feeds, and aquatic plants. Regardless of the classification used, there are always some feedstuffs which do not fit well in any one category, or which may fit in more than one category. However, livestock producers need not be too concerned with the precise classification of each feedstuff. Rather, they should familiarize themselves with the characteristics and nutritive value of each feedstuff.

In addition to the by-product feeds and crop residues listed and discussed in this chapter, these and other by-product feeds and crop residues are listed in Part V, Composition of Feeds; and the replacement values for many of them are given in the feed substitution tables for each species, in Chapters 19 to 26. Additionally, many by-product feeds and crop residues are listed and discussed in Chapter 10, Grains/High-Energy Feeds; Chapter 11, Protein Supplements; and Chapter 30, Glossary of Feedstuffs. In order to avoid needless repetition in presenting such a great array of by-product feeds and crop residues, cross referencing and indexing are used.

ECONOMY OF BY-PRODUCT FEEDS AND CROP RESIDUES

Feed cost, the largest single expense in animal production, may often be lowered by including by-product feeds and crop residues in the ration. But more than cost per pound or per ton should be considered!

When determining the economy of by-product feeds and crop residues, consideration should be given to the cost of the nutrients supplied; transportation costs; storage costs and losses; possible variations in nutrient content because of different milling and processing procedures; palatability; possible toxicity or contamination with pesticides or heavy metals; effect upon the digestibility and utilization of the total ration; and labor costs in feeding. Reduced animal performance and less profit can result from improper feed substitution.

Increasingly, new uses for by-products will compete with their traditional use as livestock feeds. Fuel, plastics, paints, lubricants, road deicers, biodegradable plastic bags, and printing ink are among the products that can be made from crop by-products. Biotechnology will make these alternative uses more viable.

ANIMAL BY-PRODUCT FEEDS

Animal by-products are derived from animals; they include dairy by-products, marine by-products, meat animal by-products, and poultry by-products. As a whole, animal by-products are high quality protein feeds; they generally contain an excellent balance of amino acids and are good sources of minerals and vitamins. However, the following concerns are inherent in the utilization of these feedstuffs:

1. Some animal by-products contain large amounts of fat and are subject to oxidation and rancidity.
2. Animal by-products must be processed and stored properly to prevent bacterial contamination and growth.
3. Most animal by-products are more costly than plant by-products.

Dairy By-product Feeds

Dairy by-products, while excellent nutritionally, furnish a relatively small percentage of the total high-protein feeds consumed by livestock in the United States. In 1984, the several dairy by-products—in liquid, dry, and condensed forms—constituted less than 0.4% of the high-protein feeds fed to animals in the United States.

Milk is processed in numerous ways to provide a wide variety of products, for both human and animal consumption. Generally speaking, milk by-products are of high quality; they are good sources of protein and the essential amino acids, minerals, vitamins, and lactose (milk sugar), and they are palatable and highly digestible. The chief deterrent to their wider use in animal feeds is price.

A description, along with feeding recommendations of some common dairy by-product feeds follows:

- **Buttermilk**—This is the by-product of churning cream into butter. It is available in liquid, condensed, or dried forms. Its superior nutritive value is due to its content of high quality proteins, its vitamins, and its good mineral balance, along with its high palatability and digestibility.

Buttermilk is an excellent supplement for young pigs and chicks; it is ideal for balancing the deficiencies of the cereal grains.

- **Skimmed milk**—This is milk from which sufficient cream has been removed to reduce its milk fat content to less than 0.5% (usually less than 0.1%). It may be fed in liquid or dried forms. Dried skimmed milk contains less than 8% moisture and averages 32 to 35% protein. One pound of dried skimmed milk has about the same composition and feeding value as 10 lb of liquid skimmed milk.

Dried skimmed milk is generally too high priced to feed to livestock, except for limited use in starter pig, chick, and foal rations.

- **Whey**—Whey is the liquid by-product which results from cheese manufacture. It may be fed in the liquid, condensed, or dried form.

Fresh whey is high in moisture; hence, it is not economical to transport it a great distance. It is sometimes fed as a liquid to cattle and swine in close proximity to the plant where it is produced. Most dehydrated whey is incorporated in swine and poultry rations.

(Also see Chapter 11, section on "Dairy Products.")

Marine By-product Feeds

In the beginning, marine wastes were dumped into the sea. Eventually, some of them were dried at high temperatures—usually in an open flame dryer—and used as fertilizers. Then, about 1910, it was discovered that these waste materials from fish canning plants and fish species not used by humans were a desirable protein source for livestock feeding. Experiment stations led the way in determining the feeding value of these new products which are now incorporated extensively in swine and poultry rations.

Although fish meal and solubles are the leading marine by-products, in 1984 they accounted for only 1.0% of the high-protein feeds fed to animals. In addition to several fish by-products, marine products include crab meal and shrimp meal.

A description, along with feeding recommendations of some common marine by-product feeds follows:

• **Fish meal**—*Fish meal is the clean, dried, ground tissue of undecomposed fish or fish cuttings, either or both, with or without the extraction of part of the oil.*

Traditionally, fish meal has been extensively used in poultry feeds. But, in the 1980s, the use of fish meal changed significantly. The three most important new uses for fish meal are: acquaculture feeds, early weaned pig feeds, and rumen by-pass feeds. The bypass protein as a percent of crude protein of fish meal is 65%.

• **Shrimp meal**—*Shrimp meal is the undecomposed ground dried waste of shrimp, consisting of parts and/or whole shrimp.*

Shrimp meal is recommended for use as a protein and calcium supplement in swine, poultry, and fish feeds.

(Also see Chapter 11, section on "Marine By-Products.")

Meat Animal By-product Feeds

In the early days of the meat-packing industry, the only salvaged by-products were hides, wool, tallow, and tongue. The remainder of the offal was usually carted away and dumped into rivers, or burned or buried. In some instances, packers even paid for having the offal taken away.

Finally, in the early 1890s, the selected meat residues from lard rendering—known as cracklings—were sold locally for chicken feed. Then, about 1900, poultry producers in California started using dried tankage in chicken rations. Almost simultaneously, this new product was utilized as a hog feed. Thus was born a new era in the utilization of packinghouse by-products and a new era in livestock feeding.

Today, packinghouses produce numerous by-products; cattle slaughter alone produces approximately 80 by-products which have a great variety of uses. Some of these, such as fats and gelatin, are used as human food; others, like feathers and tankage, are only fed to animals.

Currently, tankage and meat meal provide 6.0% of the total supply of high-protein feeds available for livestock feeding. In 1986, 2,650,000 tons of tankage and meat meal were fed in the United States.

A description, along with feeding recommendations, of some common meat animal by-product feeds follows:

• **Animal fat**—Several different animal fats are used in feeds, including hydrolyzed animal fat, pork fat (lard), and beef or sheep fat (tallow). Quality is reflected by color, odor, rancidity, water content, and non-fat (foreign) material. Generally, the specifications of animal fats for feed state (1) a minimum of fatty acids (not less than 85% in hydrolyzed animal fat, and not less than 90% in lard or tallow), (2) a maximum of unsaponifiable matter (not more than 6% in hydrolyzed animal fat, not more than 2.5% in lard or tallow), and (3) a maximum of insoluble matter (not more than 1% in hydrolyzed animal fat, lard, or tallow). Fats should be stabilized with an antitoxidant.

Animal fats are used to increase the caloric density of feeds of any species.

• **Blood meal**—Blood meal is produced from clean, fresh animal blood. It may be prepared by any one of three processes: (1) spray drying, (2) cooker drying, or (3) flash drying. Blood meal is usually of dark black color and rather insoluble in water. It is low in the essential amino acid isoleucine and low in calcium and phosphorus.

Blood meal ranks very high as an escape (bypass) protein for ruminants. For this reason, it is finding an increasing place as a ruminant feedstuff. It is also used as an ingredient in milk replacers, especially for ruminants.

• **Meat meal**—Meat meal is the dry-rendered product from mammal tissues, exclusive of blood, hair, hoof, horn, hide trimmings, manure, stomach and rumen contents. It is a high protein feed, with protein of excellent quality. Also, it is rich in calcium and phosphorus and a good source of riboflavin, niacin, and vitamin B–12.

Meat meal is a widely used protein supplement, especially in swine and poultry feeds. Also, it ranks high as a rumen bypass protein.

• **Meat meal with bone**—Meat meal with bone is the dry-rendered product from mammal tissues, *including bone,* but exclusive of blood, hair, hoof, horn, hide trimmings, manure, stomach and rumen contents. It is similar to meat meal, but higher in calcium and phosphorus due to the added bone.

Meat and bone meal is widely used as a protein supplement, especially in swine and poultry feeds. It ranks high as a rumen by-pass protein.

• **Tankage**—Tankage, which is also known as meat meal tankage, is the wet-rendered product from mammal tissue exclusive of hair, horn, hide trimmings, manure, stomach and rumen contents. It may contain added blood or blood meal, however, to raise the protein to the desired level. The protein content of tankage is somewhat higher than meat meal. It is rich in riboflavin, niacin, and vitamin B–12.

Tankage is widely used as a protein supplement in swine and poultry feeds.

• **Tankage with bone**—Tankage with bone, which is also known as meat and bone meal tankage, is the wet-rendered product from mammal tissue, *including bone,* but exclusive of hair, hoof, horn, hide trimmings, manure, stomach and rumen contents. It may contain added blood or blood meal, however, to raise the protein to the desired level. It is a high protein/high calcium-phosphorus feed, and rich in riboflavin, niacin, and vitamin B–12.

Tankage with bone is widely used as a protein supplement in swine and poultry feeds.

(Also see Chapter 11, section on "Meat Packing By-Products.")

Poultry By-product Feeds

The by-products of the poultry industry consist of offal, dead birds deemed unfit for human consumption, feathers, blood, and eggs—including cracked eggs, those with blood spots, rots, and those that did not hatch. For many years, the chief problem in utilizing poultry by-products was that of assembling enough material in one place to make the operation profitable. But the growth and centralization of poultry production accomplished this.

Poultry by-products are widely used as protein supplements for four-footed animals and poultry, and in fish hatcheries.

A description, along with feeding recommendations, of some common poultry by-product feeds follows:

• **Feather meal**—Feather meal (hydrolyzed poultry feathers) consists of hydrolyzed, dried, ground feathers from slaughtered poultry. It is low in the amino acids histidine, lysine, methionine, and tryptophan.

Feather meal is used as a high protein feed for mature ruminants. It ranks high as an escape (bypass) feed for which levels up to 10% of the ration may be used.

• **Poultry by-product meal**—Poultry by-product meal consists of the ground, rendered, clean parts of the carcass of slaughtered poultry, such as necks, feet, undeveloped eggs, and intestines, exclusive of feathers.

Poultry by-product meal is widely used as a protein supplement in swine and poultry feeds.

• **Poultry fat**—Poultry fat (feed grade) is primarily obtained from the tissue of poultry in the commercial process of rendering or extracting. It must contain not less than 90% total fatty acids and not less than 3.0% of unsaponifiables and impurities. It shall have a minimum titer of 33 C. If an antioxidant is used, the common name or names must be indicated, followed by the word *preservative(s)*.

Poultry fat may be used to increase the caloric density of feeds for any species.

(Also see Chapter 11, section on "Poultry Wastes.")

PLANT BY-PRODUCT FEEDS

Almost every type of plant that is produced or processed for human food or animal feed yields one or more by-products which can be utilized as feed for animals. Additionally, some plants, such as trees, yield by-products which, when initially considered, seem inedible, but which upon proper processing yield valuable by-product feeds.

In the sections that follow, plant by-product feeds are classed and discussed in the following categories: oilseed by-product feeds; grain by-product feeds; fruit, vegetable, and nut by-product feeds; sugar by-product feeds; brewing and distilling by-product feeds; wood and paper by-product feeds; crop residues; waste by-product feeds; and aquatic plants.

The lexicon used in discussing plant by-product feeds and residues is often confusing. For this reason, some of the terms commonly applied to the various by-products from plant processing are defined as follows:

• **Bagasse**—Solid residue remaining after extraction of juice.

• **Cannery residue**—The waste remaining following canning vegetable products for human consumption.

• **Cull**—This denotes inferior quality. This should not be construed as implying spoiled products; rather, it refers to products which do not meet certain minimum quality standards for marketing.

• **In vitro**—Outside the living organism in an artificial environment.

• **In vivo**—Within the living organism.

• **Pit, shell, hull**—These terms are applied to the hard, woody coating which surrounds the seed.

• **Pulp, pomace, marc**—These terms refer to the solid residue, including such materials as skins and seeds, after the extraction of juices from the intact product. Pomace is often used to denote the dried product.

• **Stover**—Stalks and leaves of corn or sorghum after grain harvest.

• **Stubble**—Lower parts of plant stems that remain standing in the field after harvest.

Oilseed By-product Feeds

Several seeds high in oil are produced primarily as sources of vegetable oils for oleomargarine, shortening, and salad oil for human consumption, and for paints and other industrial purposes. In processing these seeds, residual protein-rich products of great value as livestock feeds are obtained.

A description, along with feeding recommendations, of some common oilseed by-product feeds follows:

• **Canola (rapeseed) meal**—Canola (rapeseed) meal is derived from canola, which was created from selected rapeseed by Canadian scientists in the 1970s. In comparison with the old rapeseed, canola is lower in glucosinolates and erucic acid (a long-chain fatty acid).

Canola meal may be used as a protein supplement for all classes of livestock and poultry. It may be used at maximum levels of 20% in most rations, but the amount in the total ration should be limited to about 10% for chicken layers and breeding ducks, 15% for breeding turkeys, and 12% for young pigs and breeding swine.

• **Cottonseed, whole**—As indicated by the name, this is whole cottonseed, with or without the lint. It is very high in energy, high in protein, high in phosphorus, and high in fiber. It does not require processing.

Whole cottonseed is used as a cattle feed, for both beef and dairy animals. Generally, whole cottonseed will increase the fat of milk by about 0.3%, making for an increase in the market value of the milk by about 50¢/cwt. Lactating Holstein cows are normally fed 4 to 7 lb of whole cottonseed/head/day.

• **Cottonseed meal**—Cottonseed meal is the finely ground flakes of the residue that remains after most of the oil from whole cottonseed has been extracted. It is low in lysine and tryptophan and deficient in vitamin D, carotene, and calcium. But it is rich in phosphorus.

Cottonseed meal is an excellent protein supplement for ruminants, since they do not require lysine and tryptophan and they can tolerate gossypol. However, it must be limited in the rations of monogastric animals, due to gossypol; unless derived from glandless seed.

• **Linseed (flaxseed) meal**—Linseed meal is the ground residue that remains after most of the oil from flaxseed has been extracted. It is low in the amino acids lysine and tryptophan. Also, it is lacking in carotene and vitamin D, and only fair in calcium and the B vitamins.

Linseed meal is the protein supplement of choice by horse caretakers. They especially prefer old process linseed meal, because it is palatable, and it imparts glossiness to the hair coat due to the high oil content. For swine and poultry, the amino acid deficiencies of linseed meal do not effectively compensate for the deficiencies of the cereal grains.

• **Peanut meal**—Peanut meal, a by-product of the peanut industry, is ground peanut cake, the product that remains after the extraction of part of the oil of peanuts. It is low in methionine, lysine, and tryptophan, and low in calcium, carotene, and vitamin D.

Peanut meal, which is palatable and contains high quality protein, is used extensively in livestock and poultry feeds. Since peanut meal tends to become rancid when held too long—especially in warm, moist climates—it should not be stored longer than 6 weeks in the summer or 2 to 3 months in the winter.

• **Safflower meal**—Safflower meal is the ground residue obtained after extracting oil from the whole safflower seed.

Dehulled (decorticated) safflower meal can be fed to swine and poultry, but it should be supplemented with soybean and meat meal to provide adequate levels of lysine and the sulfur-containing amino acids. High-fiber safflower meal (above 15%) should be fed only to ruminants.

- **Sesame meal**—Sesame meal is obtained by grinding the residue remaining after extraction of most of the oil from whole sesame seed.

Sesame meal is a low-fiber, excellent quality plant protein supplement suitable for all animal species.

- **Soybean meal**—Soybean meal is the product obtained by grinding the flakes which remain after extracting most of the oil from soybeans. It has the highest nutritive value of any plant protein source.

Soybean meal, which is the most widely used protein supplement in the U.S., is very palatable and suitable for all animal species.

- **Sunflower meal**—Sunflower meal is obtained by grinding the residue remaining after extraction of most of the oil from whole sunflower seed. It varies considerably in nutrient content, depending on the extraction process and whether the seeds are dehulled.

Sunflower meal is a suitable protein supplement for all animal species. Meal produced from seed that was not dehulled has a high fiber content; hence, it should be fed only to ruminants. When sunflower meal is used in monogastric feeds, it should be combined with high-lysine supplements, such as fish meal or meat scraps.

(Also see Chapter 10, section on "Cottonseed"; and Chapter 11, section on "Oilseed Meals.")

Grain By-product Feeds

Many of the cereal grains are processed to obtain a great array of foods for human consumption and by-product feedstuffs for animals. The traditional milling procedures of cereal grains to obtain food for people, and the by-products resulting therefrom, are illustrated and described in Chapter 10, in the section headed, "Grain Milling and Milling By-products."

A description, along with feeding recommendations, of some common grain by-product feeds follows:

- **Corn gluten feed**—Corn gluten feed is a major by-product feed that remains after the extraction of most of the starch and germ in the wet milling of corn to produce starch and syrup.

Corn gluten feed is generally not fed to monogastric animals because of its bulkiness, poor quality protein, and unpalatability. It is used extensively for dairy cows.

(Also see Chapter 11, section "Corn Gluten Feed and Corn Gluten Meal.")

- **Corn gluten meal**—Corn gluten meal is a major by-product feed obtained in the wet milling of corn to produce starch and syrup. It contains more than twice the crude protein content of corn gluten feed.

Corn gluten meal, like corn gluten feed, is used primarily as a feed for ruminants, especially for dairy cows. However, it is somewhat more valuable because of its higher protein content. Corn gluten meal is an excellent rumen by-pass feed.

(Also see Chapter 11, section on "Corn Gluten Feed and Corn Gluten Meal.")

- **Hominy feed**—Hominy feed is a mixture of corn bran, corn germ, and part of the starchy portion of either white or yellow corn kernels or mixture thereof, produced in the manufacture of pearl hominy, hominy grits, or table meal.

Hominy feed is used as a grain replacement in the rations of all animal species.

- **Wheat bran**—Wheat bran is the coarse outer covering of the wheat kernel which is separated from wheat in commercial milling. Bran is rich in niacin, vitamin B-1, phosphorus, and iron. A large part of the phosphorus is phytin phosphorus.

Wheat bran is widely used in horse rations because of its bulky nature and laxative effect. Also, it is a favored supplement for use in gestating cow, sheep, and swine rations.

- **Wheat middlings**—Wheat middlings, a by-product of the flour milling industry, consist of the fine particles of wheat bran, wheat shorts, wheat germ, wheat flour, and some of the offal from the "tail of the mill." Middlings are deficient in calcium, carotene, and vitamin D.

Wheat middlings are widely used as a potential grain replacement in rations for all animal species. When fed to monogastrics, middlings are most efficient when used along with an animal protein supplement.

- **Rice polishings**—Rice polishings are a by-product of rice milling, obtained in brushing the grain to polish the kernel. It is rich in many of the B-vitamins; it is especially high in thiamin.

Most rice polishings are incorporated in swine and poultry rations, as a rich source of the B-vitamins, especially thiamin. Because of their high fat content, (1) they should be limited in swine rations in order to avoid soft pork, and (2) they tend to become rancid when stored very long.

- **Screenings, cereal grain**—Grain screenings consist of 70% or more of light and broken grains, along with weed seeds, and other foreign material that is separated in the cleaning of grain with a screen. The quality varies according to the percentage of weed seeds and other foreign materials.

Grain screenings should be finely ground in order to destroy noxious weed seeds. Most grain screenings are incorporated in ruminant rations. Normally, screenings are limited to 15 to 20% of feedlot rations and 10% of concentrate mixes in lactating dairy rations.

Fruit, Vegetable, and Nut By-product Feeds

Fruit, vegetable, and nut by-product feeds may come from three sources: (1) cull, unmarketable, or damaged, but wholesome, commodities; (2) residues left in the field; or (3) canning, juicing, or processing wastes. These products can be used successfully in many feeding programs. But problems involving their continued availability, storage, and handling must be considered, because many of them are highly perishable. Most of these by-product feeds are restricted to areas where processing and canning operations are located.

A description, along with feeding recommendations, of some common fruit, vegetable, and nut by-product feeds follow.

- **Apple pomace**—Apple pomace, which contains pulp, peels, cores, and cull apples, is the by-product that remains (1) after expressing juice from apples or (2) after canning, drying, and freezing apples. It may be fed fresh (wet), as silage, or dry.

Apple pomace may be used as a feed for cattle and sheep, for which it is very palatable. Dried apple pomace may replace ¼ to ⅓ of the grain in the ration.

• **Carrots, roots, fresh**—Carrots are a succulent, thickened root, high in carotene—the precursor of vitamin A. They may be available as a result of surpluses, or culls, tips, crowns, and tops may be available as a by-product from dehydrating plants.

Carrots are an excellent cattle feed. They are usually fed in bunks or on the ground. A mature cow will eat about 35 lb of carrots daily, along with other feeds. Horses relish carrots as a treat.

• **Citrus molasses**—Citrus molasses is the partially dehydrated juices obtained from the manufacture of dried citrus pulp. It must contain not less than 45% total sugars expressed as invert; and its density determined by double dilution must not be less than 17.0° Brix.

Citrus molasses is used extensively as an energy feed for cattle, despite being bitter tasting. It is usually restricted to about 10% fo the ration.

(Also see Chapter 10, section on "Molasses.")

• **Citrus pulp, dehy**—Dried citrus pulp is the peel, seeds, pulp, and cull fruits, left from the canning of citrus juices and other citrus products, which have been dried. It is high in fiber and low in phosphorus.

Most of the dried citrus pulp is fed to ruminants. It is seldom used for monogastric animals because of its high fiber content.

• **Pear cannery residue, wet**—Pear cannery residue is the waste that remains from canning pears. The peels, cores, and screened solids are sometimes used as feed.

Based on feeding trials, pear cannery residue can replace up to 25% of the molasses and dried beet pulp in the concentrate ration of finishing steers, with the pear cannery residue having a value of 70 to 75% of the molasses and dried beet pulp. Because of the high water content and transportation costs, these wastes are fed fresh by farmers in close proximity to the cannery.

• **Peas, cull (split), dry**—Dry cull peas are the cull and split peas obtained from the production of yellow, green, and black peas grown for seed.

Dry cull peas may be used as a grain replacement and/or protein supplement for cattle, sheep, and hogs. Cull peas may replace up to 40% of the grain mixture in rations of finishing steers with good results. As a swine feed, 2 tons of cull peas are equivalent in feeding value to 1 ton of soybean meal plus 1 ton of grain.

• **Pineapple bran**—Pineapple bran is a by-product of the pineapple canning industry. It consists of the outer shells, and it may also include the cores of pineapples; sometimes, molasses is added. Its composition is similar to dried beet pulp. It is high in fiber and low in protein.

Pineapple bran may be used in the rations of ruminants and horses. Because of its low protein content, supplementation is needed. Levels up to 15 lb per cow daily have been fed in Hawaii with good results.

• **Potatoes, cull, fresh**—Potatoes, cull and surplus, are high in water (76%), low in fiber, high in carbohydrates, and comparable to whole grains in protein content on a dry basis, and deficient in vitamins A and D.

When properly processed, potatoes may be fed to all classes of livestock. Fresh potatoes may be fed to ruminants and horses. Choking of cattle can be minimized by feeding whole potatoes from low bunks, under a bar or cable, to keep the animals heads down, or by chopping the potatoes. Potatoes should be cooked for swine and poultry. About 400 to 450 lb of potatoes are equivalent in energy value to 100 lb of cereal grain.

(Also see Chapter 10, section on "Roots and Tubers.")

• **Potatoes, cull, cooked**—Cooked potatoes are potatoes that have been cooked (steamed or boiled, preferably in salt water to increase palatability, with the cooking water discarded).

Cooked potatoes may be fed to swine and poultry, but cooking is not necessary for other classes of livestock.

(Also see Chapter 10, section on "Roots and Tubers.")

• **Potatoes, dehy, meal**—Dehydrated potatoes—also called potato meal (if powdered) or potato flakes—are sun dried or artificially dried cull or surplus potatoes.

Dehydrated potatoes may replace 20 to 25% of the concentrate of beef cattle, dairy cattle, or sheep rations. If heated sufficiently in the drying process and cooked thoroughly, 10 to 20% can be incorporated in swine and poultry rations. Potato meal is equal to grain, pound for pound, when substituted for part of it.

(Also see Chapter 10, section on "Roots and Tubers.")

• **Sweet potato, dehy, meal**—Dehydrated sweet potatoes are sweet potatoes that have been dried, either in the sun or artificially. They are low in fiber, fat, and protein, but high in starch and vitamin A.

Dehydrated sweet potatoes are valuable as a partial substitute for grain in rations for cattle and sheep. It is worth 90 to 100% as much as corn for ruminants. But it is not palatable to horses, and it is least useful for swine and poultry.

• **Tomato pomace, wet**—Wet tomato pomace consists of skins, pulp, and crushed seeds that remain after the processing of tomatoes for juice, soup, or catsup. It is high in water (75%).

Wet tomato may constitute up to 15% of the ration of beef cattle, dairy cattle, sheep, and swine. At higher levels, its bitterness may make for unpalatability.

• **Turnip, roots, fresh**—Turnips are a root crop, popular as a winter forage in some northern areas, especially for sheep. On a moisture-free basis, they are high in energy and good in protein.

Turnips are a popular fall/winter forage grazing crop for cattle and sheep. Stocker cattle, weighing 600 to 700 lb, will make weight gains of 1.5 to 2.0 lb per head per day when grazing turnip fields. A roughage supplement of 2 to 3 lb of straw or hay per day is recommended.

Sugar By-product Feeds

Although the United States imports a considerable amount of sugar, the nation's sugar industry is large. Two crops—sugarcane and sugar beets—account for the bulk of refined sugar. Several by-product feeds are produced in various stages of processing these two crops for sugar.

SUGARCANE PROCESSING

In the initial steps of processing sugarcane, the cane is crushed by rollers to press out the sweet juices. The juice is then centrifuged and partially condensed. At this stage, the unrefined sucrose product is brown and sticky. Upon refinement of this product, the white, crystalline sucrose is separated from the molasses and brown sugars. This molasses product is referred to as blackstrap or final molasses.

After the juice has been extracted from the cane, the remaining by-product is known as bagasse. About 60% of the bagasse is used for fuel in the sugar mills. Being high in fiber, it has a low dry matter digestibility—only about 25%. Additionally, its TDN is extremely low, ranging from 20 to 25%. However, bagasse has been used effectively as a carrier of molasses, the combination of which yields a relatively high-fiber, high-energy feed. "Camola" is a term applied to a mixture of 4 parts bagasse pith to 10 parts cane molasses. "Molascuit" contains 1 part pith to 6.25 parts molasses.

SUGAR BEET PROCESSING

In the production of sugar from sugar beets, the beets are initially shredded into strips or slices (cossettes). The cossettes are then soaked in hot water. The sugar diffuses out of the cossettes producing a *raw juice* containing about 10 to 15% sugar. This juice is then processed in much the same manner as juice from sugarcane.

The resulting beet pulp can be fed wet if used within a short time, or it can be ensiled or dried when long-term storage is desired. Molasses is often added to beet pulp to increase the energy content; and, on occasion, beet pulp is ammoniated to provide a source of nonprotein nitrogen.

Beet tops and crowns are relished by livestock, but since they contain large quantities of oxalic acid, extra calcium must be added to the ration. Additionally, tops from beets grown on soils heavily fertilized with nitrogen are apt to create nitrate poisoning problems in cattle that do not have access to other feed.

MOLASSES AND OTHER SUGAR BY-PRODUCT FEEDS

Annually, about 638 million gallons, or 3,745,265 tons, of molasses of all kinds are used in the United States, of which about 73%, or 2,734,044 tons, is consumed in feeds or foods and the balance is used for industrial purposes. The United States produces about half of the molasses that it uses; the remainder is imported. Of the U.S. production, 48% is cane molasses and 52% is beet molasses.

A description, along with feeding recommendations, of some common sugar by-product feeds follows:

- **Beet molasses**—Beet molasses is a by-product of sugar beets, obtained in the process of manufacturing sugar. It must contain not less than 48% total sugars expressed as invert and its density determined by double dilution must be not less than 79.5° Brix. It is a good energy source, but it is low in protein. Beet molasses is very laxative when used at high levels.

Beet molasses may be fed to cattle, sheep, and swine. It may replace up to ¼ to ⅓ of the grain in finishing rations.

- **Beet pulp**—Beet pulp is the dried residue from sugar beets obtained in the manufacture of sugar. It is high in fiber, but the fiber is highly digestible. It is high in calcium and low in phosphorus. It may be fed wet or dried.

Wet beet pulp is high in moisture content; consequently, high transportation costs make it uneconomical to feed it very far from the processing plants. Dried beet pulp may replace up to 45% of the concentrate mix for dairy cows, finishing cattle, and finishing lambs. Beet pulp is an excellent ingredient for maintaining normal milk fat test when dairy cows are fed restricted roughage rations.

- **Beet tops**—Beet tops are the by-product obtained from the tops of sugar beets. The average production of beet tops is 5 to 6 tons of green weight per acre, of which 40% is in the crowns and 60% is in the leaves. Beet tops can be fed fresh, dried, or ensiled.

Beet top silage is a good, succulent winter feed for cattle and sheep. It is a fairly laxative feed. Mature cattle should not be fed more than 30 lb of beet top silage per day, and mature sheep not more than 3 lb per day.

- **Cane molasses**—Cane molasses is a by-product of the manufacture of sucrose from sugar cane. It must contain not less than 43% total sugars expressed as invert. If its moisture content exceeds 27%, its density determined by double dilution must not be less than 79.5° Brix. It is a good energy source but low in protein. Cane molasses is a very palatable feed—an excellent appetizer.

Cane molasses may be fed to any animal species. Normally, it is limited to 5 to 10% of rations.

(Also see Chapter 10, section on "Molasses.")

Brewing and Distilling By-product Feeds

Considerable quantities of grains are used in the brewing of beers and ales and in the distilling of liquors. After processing, the remaining by-products can be readily adapted to many feeding programs. In 1984, 298,000 tons of brewers' dried grains and 551,000 tons of distillers' dried grains were fed to livestock. In addition to these feeds, solubles and yeast products from these industries are used in livestock feeds.

BREWERS' BY-PRODUCTS

Barley is the primary grain used for the brewing of beers and ales. The initial step of brewing involves the malting of the barley. Special varieties of barley are steeped under carefully controlled conditions to germinate the grains. Upon germination, the enzymes which are activated convert the starch in the grain to dextrin and sugars. After the germination process is completed, the sprouted barley is dried by heating to halt enzymatic activity. The *malt sprouts* and *malt hulls* are then separated from the malted barley, forming two by-products which can be used as feed.

The clean malted barley is mixed with other grains (generally corn or rice) and a flavoring agent, hops, to form a mash. This mash is then cooked in water to enhance further enzymatic activity, and, following cooking, it is separated into liquid and solid fractions. The liquid—called *wort*—undergoes yeast fermentation to form the alcoholic beverage of either beer or ale.

An outline of the brewing process is shown in Fig. 12-1.

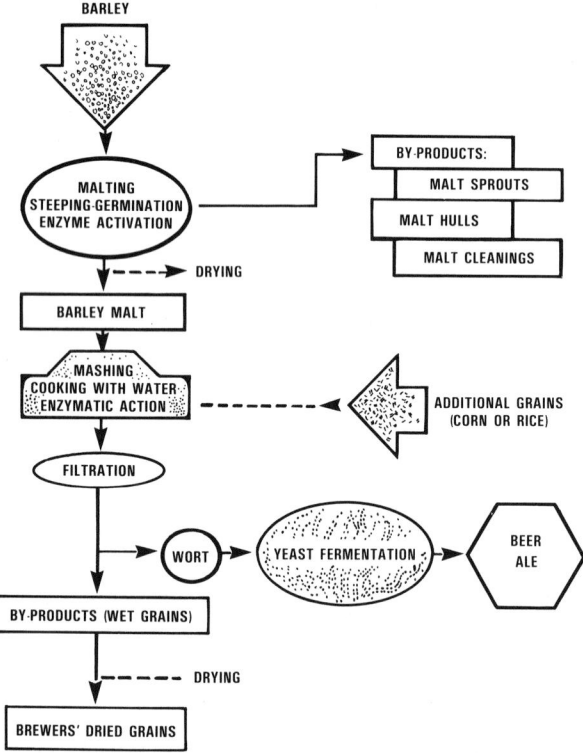

Fig. 12-1. An outline of the brewing of beer and ale, and the production of by-product feeds.

DISTILLERS' BY-PRODUCTS[2]

A large number of distilled spirits are produced throughout the world, each characterized by (1) the area of origin, (2) type of material used, (3) preparation of those materials, (4) proportions of materials, (5) fermentation conditions, (6) distillation processes, (7) maturation processes, and (8) mixture techniques. The art of distilling liquors has been largely a localized, traditional process; and many countries have developed their own characteristic spirit—for example, tequila is associated with Mexico; Scotch Whisky, with Scotland; bourbon, with the United States; and vodka, with the U.S.S.R.

It is necessary to review two basic distillation processes—British and American—because the stages from which by-products are produced vary (see Fig. 12-2).

---

[2]The section on Distillers' By-products was authoritatively reviewed by R. Hatch, Ph.D., Executive Director, Distillers Feed Research Council, Cincinnati, Ohio.

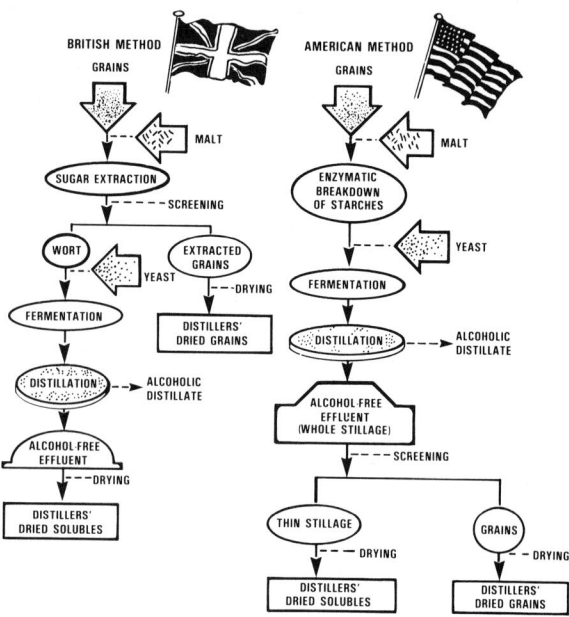

Fig. 12-2. An outline of the British and American processes of distillation, and the production of by-product feeds.

In the British method, grains are mixed with malt to facilitate the breakdown of starch to fermentable sugars. After the malting process, the resulting product is screened to separate the wort from the extracted grains. These grains can then be dried and used as feed. Yeast is added to the wort, and the mixture allowed to ferment. Following fermentation, the wort is distilled to remove the portion of the wort containing the alcohol. The product that remains is essentially alcohol free and can be dried to produce another by-product, *distillers' dried solubles.*

In the American process, there is no screening of the product formed during the enzymatic breakdown of the starches. Rather, the entire mixture is fermented and distilled. The alcohol-free effluent is termed whole stillage. The whole stillage is screened to separate the distillers' grain and the thin stillage. Drying procedures are applied to both products to produce distillers' dried grains from the grain portion and distillers' dried solubles from the thin stillage; or the grains and solubles may be combined prior to drying, then dried to produce distillers' dried grains with solubles.

A description, along with feeding recommendations, of some common brewing and distilling by-product feeds follows:

• **Brewers' dried grains**—Brewers' dried grains are the dried extracted residue of barley malt alone or in a mixture with other cereal grain or grain products resulting from the manufacture of wort or beer. They have about 80% the energy value of barley grain. Brewers' grains are not as palatable in the dried form as the original grain.

Brewers' grains are used chiefly as a cattle feed; they are used interchangeably with other feeds of similar bulk, fiber, and crude protein content. They are usually limited to 15 to 20% of cattle finishing rations and 25% of the dairy concentrate mix. Brewer's dried grains rank high as a rumen bypass protein.

- **Distillers' dried grains**—Distillers' dried grains are by-products of the production of distilled liquors from grains. Distillers' dried grains are high in fiber and high in protein. Corn is the most widely used grain in alcohol production. But rye, sorghum, and wheat are sometimes used for alcohol production. Distillers' dried grains are identified by the type of grain from which they are made.

  Distillers' dried grains are recommended for use by beef cattle, dairy cattle, and sheep. They may constitute up to 25% of the concentrate mix. Distillers' dried grains rank high as a rumen by-pass protein.

- **Distillers' dried grains with solubles**—Distillers' dried grains with solubles are by-products of the production of distilled liquors from grains, including solubles of fermentation which are added to the grain before drying. The predominant grain is declared as the first word in the name.

  Distillers' dried grains with solubles are used as a protein/B-vitamin supplement for all classes of animals, but especially for ruminants. For swine and poultry, they are usually limited to 5 to 10% of the ration.

- **Malt sprouts, barley, dehy**—Dried malt sprouts are obtained from malted barley by the removal of the rootlets and sprouts. It may include some of the hulls and other parts of the malt, but it must contain not less than 24% crude protein. Malt sprouts obtained from different grains are identified as barley malt sprouts, rye malt sprouts, or wheat malt sprouts.

  Because only about ½ of the protein is true protein, malt sprouts can be used most effectively in ruminant rations. They have a bitter flavor and can pose some palatability problems if fed at high levels. Dried malt sprouts are used chiefly in mixed feeds for dairy cattle, with an upper limit of about 3.0 lb per head daily.

- **Yeast, brewers', dried**—Dried brewers' yeast is the dried, nonfermentative, nonextracted yeast of the botanical classification *Saccharomyces* resulting as a by-product from the brewing of beer and ale. It must contain not less than 35% crude protein. It is an excellent source of highly digestible protein of good quality.

  Dried brewers' yeast can replace up to 80% of the animal protein portion of swine and poultry rations. However, in swine and poultry rations, it is used primarily as a source of B vitamins and unidentified growth factors.

Wood and Paper By-product Feeds[3]

The forest biomass is the world's largest storehouse of carbohydrates. In the past, it has been used primarily for traditional forest products and fuel; and its use for animal feed has been relatively minor, and for the most part limited to browse for goats, emergencies, and experimental work.

Wood residues, such as wood pulp and bark, have been used as energy sources for ruminants during periods of critical feed shortages, but they have never been considered as alternatives for conventional feedstuffs under normal economic conditions. Although more than 1.5 million tons of sulfate and sulfite pulps from spruce, pine, and fir were fed to cattle and horses in the Scandinavian countries during World War II when feed supplies were limited, the feeding of wood pulp to livestock ceased when conventional feedstuffs became available.

The primary structural components of wood are cellulose, hemicellulose, and lignin—all three of which are chemically bound to form a complex called lignocellulose. Cellulose in pure form is almost completely digestible by ruminants. Hemicellulose, composed primarily of 5-carbon sugars, varies in digestibility from about 45 to 90%. Lignin is completely indigestible. As a component of the lignocellulose complex, lignin can dramatically lower the digestibility of cellulose and hemicellulose by preventing cellulose-splitting enzymes from gaining access to the complex. So, whether the carbohydrates contained in wood lignocellulosic residues can be utilized by rumen microbes depends largely on how extensively the lignin-carbohydrate complex can be altered or opened up.

Of the woods tested, all of the coniferous species are essentially undigested by rumen microorganisms. Deciduous species, with a few exceptions, are only slightly digested. Aspen is the most digestible species tested, giving both an *in vitro* and *in vivo* digestibility of about 35%. Aspen bark is about 50% digestible.

Several chemical and physical treatments of woods are partially effective, but the effectiveness of each of them varies according to species. Hardwoods are generally more responsive to treatment than softwoods. Aspen is particularly responsive to treatment. Although several of the treatment methods can be readily adapted to commercial production, the cost of treatment has been the biggest deterrent to their use. Conventional feedstuffs must be relatively high priced before the treatment of wood to produce feed can be profitable. On occasion, such as during a war, the prices of conventional feedstuffs have been high enough to make processing woods into feeds attractive.

Wood residues are primarily an energy source, comparable to average or low-quality hay. But they are low in protein; so, they require protein supplementation. Nutritionally, treated wood residues are best suited for ruminants having low energy requirements, such as pregnant cows and ewes, and dry dairy cows.

Wood is effective as a roughage replacement. Concentrations of 5 to 15% screened sawdust in the ration of beef cattle appear practical. For lactating dairy cows, aspen sawdust may be used as a roughage extender or as a partial roughage substitute in high-grain rations; but some long hay appears to be necessary in the ration to stabilize feed intake.

Some of the pulp and papermaking residues which are already partially delignified, but which have little fiber value for paper manufacture, have excellent potential as feedstuffs. Care must be used in their selection, however, because some of them may contain toxic materials.

Both the research and the commercial production of wood and paper by-product feeds have lagged because of (1) an abundance of traditional feeds available at reasonable prices, (2) the lack of a steady market for these innovative by-product feeds, and (3) the rising cost of the energy and chemicals involved in treatment and processing.

A description, along with feeding recommendations, of some common wood and paper by-product feeds follows:

- **Ash leaves**—Ash leaves are the foliage from trees of the genus *Fraxinus*. Untreated ash leaves have a digestibility of 17%, in comparison with a digestibility of 31 to 37% of untreated aspen leaves. Forest foliage leaves are known as *muka*, a term that originated in the Soviet Union.

---

[3] In the preparation of this section on "Wood and Paper By-product Feeds" including most of the research cited, the authors drew heavily from the following source: Fontenot, J. P., Chairman of Subcommittee, National Research Council, "3. Forest Residues," *Underutilized Resources As Animal Feedstuffs*, NRC-National Academy of Sciences, 1983, pp. 69–120.

In the Soviet Union, where large quantities of muka are fed (about 100,000 metric tons), muka is fed to cattle, swine, and poultry, primarily as a source of carotene, trace elements, and vitamins.

• **Aspen wood, boiled, ground**—Aspen wood is obtained from trees of the genus *Populus*. Ground aspen wood is composed of the entire tree, including leaves, branches, trunk, and bark.

Boiled ground aspen wood is best suited for use in high-roughage, maintenance type rations, such as for wintering beef cows.

• **Paper waste**—Waste paper consists of discarded newspaper, office paper, wrapping paper, and cardboard, more than 20 million tons of which are produced annually. The dry-matter digestibility of waste paper by cattle is as follows: newspaper, 30%; wrapping paper and cardboard, 40 to 60%; and high-quality chemical pulp paper about 98%.

Waste paper may be used as a roughage for ruminants, especially cattle. When levels greater than 12% are incorporated in cattle feed, there may be a reduction in feed intake. One of the potential hazards of feeding waste paper is the lead in ink, but no toxicities have been reported to date.

• **Sawdust**—Sawdust is small fragments of wood made by a saw in cutting. Sawdust of the coniferous species is essentially undigested by rumen microorganisms. Deciduous species, with a few exceptions, are only slightly digested. Aspen is the most digestible; aspen sawdust has a digestibility of about 35%, and aspen bark has a digestibility of about 50%. Several chemical and physical treatments of sawdust are partially effective. But generally the cost of treatment exceeds the increased feeding value of the product.

Sawdust is effective as a roughage replacement. Concentrations of 5 to 15% screened sawdust in the ration of beef cattle appear practical. For lactating dairy cows, sawdust may be used as a roughage extender or as a partial roughage substitute in high grain rations; but some long hay in the ration appears to be necessary in order to stabilize feed intake.

• **Wood molasses**—Wood molasses is obtained from wood wastes of the timber and lumber industries. The wood is converted to sugar by boiling it under pressure with diluted acid, neutralizing it with alkali, and evaporating it to wood molasses. Up to 200 gal of molasses may be obtained from 1 ton of dry wood waste. Wood molasses is lacking in palatability.

Wood molasses can be substituted for cane molasses in ruminant rations, although it is not as palatable. For monogastric animals, it should be limited to upper levels of 2 to 2.5%; higher levels in nonruminant rations may cause digestive disorders.

• **Yeast, torula**—Torula yeast (*Torulopsis utilis*) is a hardy type of yeast that can be grown on a variety of substances, such as sulfite waste liquor and wood. While brewers' and distillers' yeast are truly by-product feeds, torula yeast is cultured specifically as a feed for livestock. It is usually dried. Dried torula yeast is an excellent source of high quality protein, minerals, and B vitamins. It is tasteless, rather than bitter like brewers' yeast.

Dried torula yeast can be included in the mixed feeds for all classes of livestock. However, its high price usually limits its use to providing a supplement to the amino acid and vitamin deficiencies of cereal grains. It is used primarily in the rations of monogastrics and young ruminants, at a level of 3 to 5% of the total ration.

CROP RESIDUES

Most of the by-product feeds result from some sort of industrial manufacturing; for example, the by-products from milling grains. Another excellent source of feed is crop residues—parts of plants that are normally left in the field following harvest of the primary crops.

As production costs increase, livestock producers become more interested in the enormous potential for animal production through feeding crop residues.

Kind and Quantity of Residue Produced

The quantity of crop residues produced may be estimated by multiplying the annual grain production by a grain weight: residue weight ratio. Normally, grain-producing plants produce as much (or more) weight of vegetative material as of grain. The crops, ratios used for conversion from grain to residue, and estimated annual production of residues are given in Table 12-1.

TABLE 12-1
ESTIMATED SUPPLY OF CROP RESIDUES[1]

Crop Source	U.S. Grain Production	World Grain Production	Canadian Grain Production	Ratio Residue/Grain[2]	U.S. Residue	Residue % of Total
	(mil. metric tons)	(mil. metric tons)	(mil. metric tons)		(mil. metric tons)	(%)
Barley	13.3	182.0	14.6	2.0	26.6	6.53
Corn	209.6	476.6	5.9	1.0	209.6	51.48
Cottonseed	4.79	30.63	—	3.0	14.37	3.53
Flax	0.21	2.36	0.90	3.0	0.63	0.15
Oats	5.6	47.5	3.3	1.0	5.6	1.37
Peanuts	1.87	19.99	—	1.5	2.8	0.69
Rice, rough	6.0	466.9	—	1.0	6.0	1.47
Rye	0.5	31.0	0.6	1.0	0.5	0.12
Sorghum	23.8	64.3	—	1.0	23.8	5.84
Soybeans	57.11	97.03	1.01	1.0	57.11	14.03
Sugarbeets[3]	22.9	285.7	0.94	0.14	3.21	0.79
Wheat	56.9	529.7	31.4	1.0	56.9	13.98
Total	402.58	2,233.71	58.65		407.12	

[1] U.S., World, and Canadian grain production from: *World Agricultural Production*, USDA, Foreign Agricultural Service, Circular Series WAP 5-88, May 1988, except for sugarbeets.
[2] Ratio residue/grain from: *Underutilized Resources as Animal Feedstuffs*, National Research Council, National Academy Press, 1983, p. 180, Table 48.
[3] Sugarbeets production from: *1986 FAO Production Yearbook*, Vol. 40, p. 163.

As shown in Table 12-1, corn, wheat, and soybeans account for nearly 80% of the total residues. It follows that the major U.S. corn-, wheat-, and soybean-producing states produce the most residues.

Corn is the most widely produced grain crop in the United States. It usually produces an amount of residue equal to the quantity of grain produced. So, as shown in Table 12-1, the production of 209 million tons of corn in 1986 resulted in 209 million tons of corn residue, which was over ½ the total available crop residue that year.

Fig. 12-3. The feeding of crop residues is becoming an important facet of livestock feeding. Here cornstalks are being harvested for feed. (Courtesy, Gehl Company, West Bend, Wisc.)

Wheat, which produces much less tonnage of grain per acre than corn, accounts for about 14% of the residue. About 57 million tons of wheat straw were produced in 1986 (see Table 12-1).

Soybean residue usually provides another 14% of the crop residue (57 million metric tons in 1986) and grain sorghum 6% (23.8 million metric tons in 1986). Other crops account for the remaining residue (see Table 12-1).

More than 400 million metric tons of straws, stalks, and stubble are available in the United States each year (see Table 12-1) and another 60 million tons in Canada. Worldwide, more than 2.2 billion tons of crop residues are produced annually.

In addition to being used as livestock feeds, crop residues may be, and are, used for bedding, soil improvement, and as a substitute for fossil fuels.

Economy of Crop Residues

Evaluating the nutrient content of crop residues, along with collecting, storing, treating, transporting, and feeding them, is much more difficult than determining the quantities available. Also, when evaluating crop residues for animal feeding, three major questions must be answered: (1) are they available in sufficient quantity to make their use as a feedstuff worthwhile, (2) do they have high enough nutrient content to justify feeding them to livestock, and (3) are they cost competitive? These and other important *field-to-feed* aspects of crop residues are discussed in the sections that follow.

HARVESTING

Because of differences in plant structure, grain harvesting methods, and moisture content, harvesting crop residues may not be easy. Straws from cereal grains are easily collected in dry state behind the combine. Corn and sorghum stovers often are too wet for dry storage, but they can be stored as silage. Soybean residue is difficult to harvest if allowed to drop on the ground behind the combine. Soil contamination during harvest may be a problem with all residues. Some of the residues, such as cottonseed hulls, rice milling by-products, and sugarcane bagasse, which are processed at central locations, have the advantage of being collected and available for treatment.

NUTRIENT VALUE OF CROP RESIDUES

Almost all crop residues are harvested after the plants reach physiological maturity; so, they are high in cell walls and lignin and low in protein and digestible dry matter. They are a feed energy source, but suitable only for ruminants—beef and dairy cattle, sheep, and goats.

Most of the energy in crop residues is in the lignocellulose complex (lignin, cellulose, and hemicellulose). The amount of the total lignocellulose and of its constituents varies widely with forage species and stage of maturity. Generally, the higher the lignin content the lower the digestibility of the cellulose material. The chemical and physical binding of cellulose and hemicellulose in the cell wall of plants is important; so, the amount of lignin *per se* is not always a good indicator of digestibility. Nutritionally, the lignocellulose complex consists of these fractions: (1) lignin, which is unavailable as an energy source, (2) a digestible energy source that can be utilized by rumen microorganisms, and (3) a fraction which is very resistant to bacterial action in the rumen, but which becomes an energy source after special treatment. The third fraction is of major interest because of the potential additional energy which can be made available in the cellulosic crop residues. In many of the crop residues, this third fraction is large enough that considerable research has been done on treatment to "unlock" its energy.

In addition to lack of available energy, crop residues may be deficient in protein, phosphorus and possibly other minerals, and vitamin A. Also, they are usually bulky and lacking in palatability. So, it is important to provide proper supplementation based on the performance expected of animals grazing or being fed crop residues.

TREATING CROP RESIDUES TO INCREASE DIGESTIBILITY

Crop residues are inefficiently utilized by animals because of the high content and poor digestibility of the fibrous fraction. This poor digestibility is related to the extent of lignification of the cell wall component of these low-quality forages. Although crop residues provide a satisfactory ration for dry gestating animals, they do not provide sufficient energy for either young or lactating ruminants—they simply cannot hold enough of these low-quality roughages to provide adequate energy. This prompts interest in increasing the digestibility of these crop residues.

Two approaches offer the best possibilities for increasing the digestibility of crop residues: (1) manipulation of plant genetics or harvest time to obtain higher quality residues, and (2) physical or chemical treatments to increase intake and digestibility. Among the physical and chemical treatments for delignifying and increasing digestibility are the following:
1. Grinding and pelleting.
2. Sodium hydroxide (NaOH).
3. Calcium hydroxide (CaOH).
4. Ammoniation.
5. High pressure steam.

The five treatments listed have been tried and found successful on corn and small grain residues. Soybean residues are very high in lignin and do not respond well to chemical treatment. Grain sorghum residue responds to sodium hydroxide and calcium hydroxide treatments, but probably not as well as corn residue or straws. Rice hulls are improved by ammoniation, and peanut hulls are improved by treatment with calcium hypochlorite. Chemical treatment of sugarcane bagasse with sodium hydroxide or calcium hydroxide dramatically increases its value as a feed for ruminants. Steam pressure treatment also increases *in vitro* digestibility of bagasse.

Summary: Because of the low nutritional value of crop residues, methods of increasing value by treatment are of interest. Sodium hydroxide has been widely studied. While it is effective, concerns about human safety and sodium residues may limit its ultimate usefulness. Calcium hydroxide and ammonia treatments offer the greatest long-term potential.

The potential of crop residue treatments becomes apparent when it is realized that straw, for example, is only 30 to 40% digestible before treatment. When pressure heated with water, it becomes 50 to 60% digestible; and digestibility increases to 70 to 80% when sodium hydroxide is added prior to cooking. By treating corn husklage and milo residue, workers at the Nebraska Station were able to increase the energy value of these residues to 90% that of corn silage.

Lowering the cost of treating crop residues to increase digestibility is the primary area which must be researched before these procedures can be applied to practical operations.

(Also see Chapter 14, Feed Processing, section headed "Treatment of High-Cellulose Feeds.")

TRANSPORTING, PACKAGING, AND STORING

The relatively low value of crop residues in comparison with grains necessitates low-cost harvesting, packaging, and storing. Because of their bulky nature and low market value per unit weight, it is not economical to move them very far. Also, the bulkiness of most crop residues makes packaging difficult.

In the past, the efficient management of the family farm generally included the use of most crop residues, such as straw and stalks, as bedding or feed, or both. However, the development of large volume, high density livestock and poultry units, which may not be located where the feed grains are produced, has created a situation in which there are large quantities of crop residues that are not being efficiently utilized. Moreover, many grain-producing farms no longer have fences or water for livestock; so, animals cannot graze the crop residues unless they are herded and water is hauled.

Most of the crop residues become available in the summer and fall, but the greatest need for them as livestock feeds is in the winter. So, some kind of packaging and storage is desirable, but may not be practical.

Because of the transportation, packaging, and storage costs involved, the increased use of crop residues in large volume, high density livestock operations is unlikely so long as conventional forages for use in high concentrate feeding systems are available at relatively low cost. However, crop residues have high potential as an alternative source of feed energy.

FEEDING SYSTEMS FOR CROP RESIDUES

Generally speaking, crop residues may be grazed, processed as dry feed, or made into silage. The important thing to remember is that their relatively low value, in comparison with grains, necessitates low-cost harvesting, storing, and feeding. Also, they must be fed to the right class of animals, and they must be properly supplemented.

The use of low quality crop residues is restricted primarily to ruminants, such as wintering beef cows. In many parts of the United States, cornstalks, sorghum stubble, and cereal grain stubble can be grazed, provided there are fences and an available supply of water. This is the most economical system of utilizing crop residues; there is essentially no harvesting machinery or energy cost, nothing is removed from the soil, and the manure is deposited on the land. However, weather problems are often encountered when winter grazing stalks and stubbles. In the higher rainfall areas of eastern United States, the residues deteriorate rapidly, and muddy fields may prevent continuous grazing. In the colder areas of the western Corn Belt and Plains States, snow cover may prevent grazing at times, with the result that an alternate reserve feed supply is needed.

Some Crop Residues

A description, along with feeding recommendations, of some crop residue feeds follows:
- **Almond hulls**—Almond hulls are a by-product of the almond nut industry. After the almonds are harvested, the nuts with attached hull are processed through a machine which removes the hull. Almond hulls are very variable in feeding value—from very poor to very good.

Almond hulls are very satisfactory for ruminants when fed in combination with such feeds as barley and alfalfa hay which compensate for the lack of protein in the hulls.
- **Bagasse, sugarcane**—Sugarcane bagasse, the fibrous residue of sugarcane stalks which remains after the juice is pressed out, is one of the principal by-products of the sugar-making process. Treatment of sugarcane bagasse with sodium hydroxide and/or steam pressure dramatically increases its feeding value.

Bagasse may be used in limited amounts as a low quality roughage for beef and dairy cattle.
- **Barley straw**—Barley straw is the product remaining after separating the seed from the mature plants. It is not equal to oat straw in feeding value, but usually it is superior to wheat straw. The feeding value of barley straw can be increased by treatment with calcium hydroxide, sodium hydroxide, or ammonium hydroxide. But the feeding value may not be increased sufficiently to pay for the treatment.

Barley straw may be fed to cattle on a maintenance ration, such as gestating beef cows, along with a legume forage.

- **Corn stover**—Corn stover is the mature corn plant from which the ears have been removed.

  Corn stover may be (1) grazed in the field, (2) stored as dry roughage, or (3) ensiled. Its highest and best use is for wintering beef cows. The traditional way of utilizing corn stover in the field is to allow the animals to do their own harvesting. However, grazing cornstalks results in considerable wastage. In an open field and winter, 2 acres of cornstalks will carry a pregnant cow for 100 to 120 days.

- **Corn husks**—Corn husks (husklage, shucklage) are the outer covering of ears of corn. It is the most digestible of the corn residues, which rank as follows in digestibility, in descending order: husks, leaf, cob, and stalks.

  Corn husks may be collected from the rear of the combine when harvesting corn, then stored. They are commonly fed to gestating cows, along with a protein supplement. As calving time approaches, a supplementary energy feed should be provided.

- **Corncobs**—Corncobs are the fibrous axis of the corn ear to which the kernels are attached, exclusive of the kernels and husk. They are a low quality roughage comparable to poor hay.

  Corncobs may be fed to cattle or sheep. Except for feedlot cattle, they should not replace more than half the roughage.

- **Cottonseed gin trash**—Cottonseed gin trash is composed of fragments of burs and stems, small amounts of immature cottonseed, lint, leaf fragments, and dirt. It is about 90% as valuable as cottonseed hulls in feeding value.

  Cottonseed gin trash is commonly fed to cattle, beef and dairy. It should be fed at low levels because of limited nutritional value.

- **Cottonseed hulls**—Cottonseed hulls are the outer covering of cottonseeds. The hulls are removed from the seed before the oil is extracted. They are a high-fiber, low-protein by-product.

  Cottonseed hulls may be used as a roughage source for beef cattle, dairy cattle, and sheep. As a roughage, they have about the same energy value as oat straw.

- **Oat straw**—Oat straw is the residue remaining after separation of the seeds by threshing of oats.

  Oat straw is the most nutritious and palatable of the cereal straws. It may constitute up to half the roughage of breeding and stocker beef cattle, provided the other half consists of a good legume hay.

- **Sorghum stover**—Sorghum stover is the stalks and leaves of sorghum after removing the mature heads.

  Sorghum stover may be (1) grazed in the field, (2) stored as dry roughage, or (3) ensiled. It is commonly fed to beef cattle. *Caution:* After harvest, sorghum will send up new shoots if the moisture is favorable. The prussic acid content of these shoots may be toxic to grazing animals. These shoots can be grazed safely 4 to 6 weeks after a hard killing frost.

- **Soybean hulls**—Soybean hulls are the outer covering of soybean seeds. They are high in fiber and fair in protein content. The hull constitutes 8% of the seed weight.

  Soybean hulls are incorporated in commercial feeds, primarily. Because of their high fiber, they are best suited for use by lactating cows, growing cattle and sheep. But they are not recommended for use in cattle finishing rations.

- **Sweet corn cannery refuse silage**—Sweet corn cannery refuse silage consists of the ensiled husks and cobs, plus some kernels not satisfactory for canning. It is low in carotene.

  Sweet corn cannery refuse silage can be fed liberally to gestating beef cows, finishing cattle, and finishing lambs. It is worth 85 to 95% as much as silage made from well-eared field corn.

- **Wheat chaff**—Wheat chaff includes the husks, hulls, joints, and small fragments of straw that are separated from the seeds in threshing of wheat. Wheat chaff is bulky, high in fiber, and low in protein.

  Wheat chaff may be used as a filler and to provide some nutrients for stocker and breeding cattle, and for dry dairy cows. It should be fed with a good legume hay at about a 1:1 ratio.

- **Wheat straw**—Wheat straw is the residue remaining after separation of the seeds by threshing of wheat.

  Wheat straw may constitute up to ½ the roughage of breeding and stocker beef cattle, provided the other half consists of a good quality legume hay.

(Also see Chapter 6, section on "Plant By-products"; Chapter 9, sections on "Corn and Sorghum Residue Silage" and "Other Silage Crops"; Chapter 11, section on "Ammoniated Products"; and Chapter 19, sections on "Crop Residues" and "Corn Residues.")

WASTE BY-PRODUCT FEEDS

Waste by-products represent a feed resource which, presently, is not being used to its nutritional and economical potential. Three major types of wastes are involved: (1) animal wastes (manure and litter), (2) municipal wastes (sewage sludge and garbage), and (3) bakery wastes.

Animal wastes refer to a mixture of animal excrements (consisting of undigested feeds plus certain body wastes), with or without bedding materials.

Municipal wastes refer to sewage sludge and garbage, both of which have been receiving increasing attention as feeds in recent years.

Bakery wastes refer to bread and other bakery products, along with dough, which make excellent energy feeds for cattle and swine.

Animal Wastes (Manure and Litter)

Historically, animal wastes have been used as fertilizer and feed, although their potential as a feed received little attention until recently.

The utilization of animal wastes as feedstuffs is not new, however. Coprophagy (eating of feces) is normal in rabbits and in many wild and domestic species. In 1914, Osborne and Mendel demonstrated that feeding 1% feces from normally fed rats to rats on a deficient purified diet prevented death.[4] In 1925, Evvard and Henness reported that on the average one pig following 1.9 steers recovered the equivalent of 212 lb of corn during a 120-day feeding period.[5] In

---

[4]Osborne, T. B. and L. B. Mendel, "The Contribution of bacteria to the feces after feeding diets free from indigestible components," *Journal of Biological Chemistry,* 1914, 18:177.

[5]Evvard, J. M. and K. K. Henness, "An experiment to study hogs following cattle," *Proceedings American Society of Animal Production,* 1925, p. 55.

1943, Bohstedt found that cow manure had nutritional value for pigs, in addition to the grain that it contained.[6] In 1956, Fuller reported that hydrolyzed poultry litter was as effective as fish meal in achieving growth from commercial type broiler diets.[7] Most animal wastes are fed to ruminants, especially to beef cattle.

AMOUNT OF MANURE PRODUCED

Figures relative to the amount of manure produced vary widely, primarily because of differences in (1) year-to-year numbers of animals, (2) estimated amount of bedding, (3) estimated recoverable manure and feces, and (4) estimated dry matter content of the wastes. Cattle (beef and dairy) account for almost 80% of the total manure production, but only about 40% is recoverable because of being voided on pastures and ranges, whereas virtually all of the poultry manure is collectable. Also, manure tonnage estimates may be on the basis of with or without bedding, and on the basis of moisture-free, 15% dry matter, or wet.

The authors' bases and estimated tonnages of animal wastes produced daily and annually follow.

Daily Manure Production and Storage

Table 12–2 shows the daily volume of manure produced by different kinds of animals. These figures can be used to determine storage needs. A rule of thumb for under the floor pits is to figure that the pit will fill at the rate of 1 ft per month. Of course, the amount of the excreta is influenced by the composition of the ration; for example, high-silage rations produce more manure than high-grain rations.

TABLE 12–2
APPROXIMATE DAILY MANURE PRODUCTION, WITHOUT BEDDING[1]

Animal	Cu Ft/Day Solids and Liquids[2]	Gallons/Day[3]
1,000-lb cow	1½	11
1,000-lb steer	1	7½
10 head of sheep	½	4
10 head of hogs:		
50 lb	⅔	5
100 lb	1⅓	10
150 lb	2¼	17
200 lb	2¾	20½
250 lb	3½	26
1,000-lb horse	¾	5½

[1]Adapted by the authors from *Michigan State University Circ. 231.*
[2]There are about 34 cu ft in a ton of manure.
[3]One cubic foot = 7½ gal.

Manure may be stored in a separate tank, in a nearby earthen dam lagoon, or it may be left to accumulate in a pit under slotted floors.

Storage capacity can be computed as follows:

Storage capacity = number of animals × daily manure production × desired storage time in days + extra water.

MANURE GASES

When stored inside a building, gases from liquid wastes create a hazard and undesirable odors. Most (95% or more) of the gas produced by manure decomposition is methane, ammonia, hydrogen sulfide, and carbon dioxide. Several have undesirable odors or may cause animal and human toxicity, and some promote corrosion of equipment. Table 12–3 gives some properties of the more abundant gases.

TABLE 12–3
PROPERTIES OF THE MORE ABUNDANT MANURE GASES[1]

Gas	Weight Air = 1	Physiologic Affect	Other Properties
CH_4 (Methane)	½	Anesthetic	Odorless, explosive.
NH_3 (Ammonia)	⅔	Irritant	Strong odor, corrosive.
H_2S (Hydrogen sulfide)	1 +	Poison	Rotten-egg odor, corrosive.
CO_2 (Carbon dioxide)	1⅓	Asphyxiant	Odorless, mildly corrosive.

[1]*Beef Housing and Equipment Handbook*, Midwest Plan Service, Iowa State University, Ames, 1968, p. 10.

Animals and people can be killed (asphyxiated) because methane and carbon dioxide displace oxygen.

Most gas problems occur when manure is agitated or when ventilation fans fail.

No one should enter a storage tank, unless (1) the space over the wastes is first ventilated with a fan, (2) another person is standing by to give assistance if needed, or (3) wearing self-contained breathing equipment—the kind used for fire fighting or scuba diving.

It is important that maximum building ventilation be provided when agitating or pumping wastes from a pit. Also, an alarm system (loud bell) to warn of power failures in tightly enclosed buildings is important, because there can be a rapid buildup of gases when forced ventilation ceases.

Yearly Manure Production

The quantity, composition, and value of manure produced vary according to species, weight, kind and amount of feed, and kind and amount of bedding. The computations herein are on a fresh manure basis (exclusive of bedding). Fig. 12–4 presents manure production data per year per 1,000 lb animal liveweight.

---

[6]Bohstedt, G., R. H. Grummer, and O. B. Ross, "Cattle manure and other carriers of B-complex vitamins in rations for pigs," *Journal of Animal Science*, 1943, 2:373.

[7]Fuller, H. L., "The value of poultry by-products as sources of protein and unidentified growth factors in broiler rations," *Poultry Science*, 1956, 35:1143.

Fig. 12-4. On the average, each class of confined animals produces per year per 1,000 lb weight the tonnages shown. (Drawing by R. F. Johnson)

Currently, we are producing manure (exclusive of bedding) at the rate of 1.35 billion tons annually. That is sufficient manure to add nearly ¾ ton each year to every acre of the total land area (1.9 billion acres) of continental United States.

MANURE USES

Historically, manure has been used as a fertilizer. But high feed prices and shortages of fossil fuels have made for new uses. Although it is expected that manure will continue to be used primarily as a fertilizer for many years to come, increasingly it will be recycled and used as a feed and converted into energy. Other manure-based products will continue to evolve, but it is expected that they will be of minor importance.

Manure as a Feed

Recycling manure as a livestock feed is the most promising of the nonfertilizer uses. Various processing methods are being employed; and some manure is being fed without processing. More and more feedlot manure will be either (1) incorporated in a grower ration, or (2) fed to breeding herds during periods when pasture supplementation is beneficial, with the residues distributed over grazing areas where they would have fertilizing value.

CHEMICAL COMPOSITION OF ANIMAL WASTES

Animal wastes contain several nutrients that are capable of being utilized when the material is recycled by feeding. Nitrogen, which is present in both protein and nonprotein forms, is a major constituent. Available energy is rather low. Fiber and ash are generally high. The high ash indicates that animal wastes are high in minerals; they are especially rich in phosphorus. Additionally, they contain certain vitamins synthesized in the digestive tract. The wastes possessing the highest nutritive value are broiler litter and layer waste.

One characteristic of all animal wastes is variability in composition due to diet regime, kind and amount of bedding, length of time before collecting, and processing method. The main difference in composition between raw and processed wastes is in moisture content; many of the processed wastes are low in moisture.

The high fiber and considerable nonprotein nitrogen of animal wastes indicate that they are best suited for feeding to ruminants, since they possess a digestive tract capable of efficiently utilizing high fiber and nonprotein nitrogen. Also, because of their low energy content, they are best adapted for use in maintenance and gestating rations, rather than in lactating and growing rations.

NUTRIENT UTILIZATION

Animal wastes processed by ensiling, dehydrating, and other methods can be fed successfully to a wide range of animals. But, for best results the rations in which they become a part should be well balanced following their incorporation. Several workers have shown that the inclusion of too high levels of waste in a ration results in an excessive level of fiber and/or minerals, followed by lowered animal performance. Because of this limitation, not more than 10 to 20% waste should be included in high-energy rations, such as in cattle finishing rations. However, much higher levels (up to 80%) can be incorporated in rations of gestating beef cows.

PERFORMANCE OF ANIMALS FED ANIMAL WASTES

When rations containing animal wastes are properly balanced and fed to the right species and class of animals, the performance of animals in growth, or in meat, milk, or egg production, is comparable to animals fed traditional feed ingredients of comparable chemical content.

Cattle Waste

Much of the early research on feeding cattle manure was conducted by Anthony at the Alabama Agricultural Experiment Station; in 1962, he established the feasibility of feeding steer manure to cattle. In a 1970 report, Anthony reappraised the feeding of manure-containing rations to finishing cattle and presented the results of a new experiment designed to determine if cooking the manure improved its feeding value.[8]

Anthony compared a typical corn-based finishing ration to rations containing untreated manure and cooked manure.

---

[8]Anthony, W. B., "Feeding Value of Cattle Manure for Cattle," *Journal of Animal Science*, Feb. 1970, Vol. 30, No. 2, p. 274.

The results of this experiment are summarized in Table 12-4. Steers fed the basal ration gained faster than the basal plus manure rations, but this difference was not statistically significant.

TABLE 12-4
RATE OF GAIN, FEED EFFICIENCY, AND FEED DIGESTIBILITY[1]

Treatment	Average Daily Gain		Feed Dry Matter Per Unit of Gain		Feed Digestibility (nylon bag)
	(lb)	(kg)	Concentrate	Manure	(%)
Basal	2.56	1.16	9.20	—	40.36
Untreated manure	2.20	1.00	8.52	1.66	46.61
Autoclaved manure	2.18	.99	8.75	1.70	45.56

[1]Adapted by the authors from "Feeding Value of Cattle Manure for Cattle," by W. B. Anthony, *Journal of Animal Science*, Vol. 30, 1970, p. 274.

When feed per unit gain was quantitated, less concentrate was used in the manure treatments. However, this was not statistically significant. Feed digestibility was significantly higher in the rations with manure than the basal ration. It was concluded that (1) manure could effectively be used in steer rations, and (2) cooking did not improve the feeding value of the manure.

Subsequently, several investigators have reported satisfactory performance of cattle when fed cattle waste that was ensiled, treated with sodium hydroxide, or fed fresh. But, in limited experiments, incorporating cattle waste in rations for swine and laying hens has lowered production and required more feed per unit of production, perhaps due to the cattle waste being too high in fiber for monogastric animals.

Swine Waste

The practice of fertilizing ponds with manure to promote the growth of fish has been followed extensively in China and Southeast Asia, where swine operations are often coordinated with fish production. The manure produced by the swine is flushed into nearby ponds where it promotes growth of vegetation and microbic animals which are subsequently eaten by the fish. Often the fish are fed solely on pig manure and grass. Hence, the Chinese have provided their own pollution control of pig manure and, at the same time, recycled and used the feed twice—first through hogs, and second through fish.

Limited U.S. studies with swine waste incorporated in cattle rations have not been encouraging; both the rate of gain and feed efficiency have been affected adversely.

Poultry Waste

Nearly 100 million tons of poultry wastes (from layers, broilers, and turkeys) are produced annually. Because poultry production is highly intensive, with many birds in a small area, waste disposal is a major problem. Most cage-layer operations produce manure free of litter as the primary form of waste. Broiler operations, generally, produce litter.

On a moisture-free basis, cage-layer manure generally contains 25 to 35% crude protein and minimal fiber, while broiler litter contains somewhat less protein—about 18 to 30% and substantially more fiber due to the presence of absorbent materials.

Poultry litter is the most collectable and the most nutritious of all animal wastes. It follows that many experiments have been conducted with it, involving feeding trials with different species. The results of numerous experiments are summarized in Table 12-5. The mean values for waste-fed animals reported therein were obtained by averaging all levels of feeding poultry wastes in the respective categories, though some of the levels were excessive. As shown in Table 12-5, the performance of animals fed wastes was generally slightly lower than that of the controls that were fed traditional feed ingredients. But, on a dry-matter basis, animal wastes generally make for least cost rations and highest net returns.

TABLE 12-5
PERFORMANCE OF ANIMALS FED RATIONS CONTAINING POULTRY WASTES[1]

Species of Experimental Animal Used	Kind of Poultry Waste Studied	Performance of Experimental Animals		
		Criteria	Control Group	Waste-Fed Group
Cattle	Dehydrated layer waste	Daily gain, lb	2.35 (1.07 kg)	2.31 (1.05 kg)
		Daily feed dry-matter intake, lb	15.82 (7.19 kg)	15.44 (7.02 kg)
		Feed/gain ratio	7.81	7.72
Lactating cows	Dehydrated layer waste	Milk yield, lb/day	41.8 (19.0 kg)	38.94 (17.7 kg)
		Milk fat, %	3.51	3.63
		Milk total solids, %	12.04	12.01
Sheep	Dehydrated layer waste	Daily gain, lb	0.42 (0.19 kg)	0.40 (0.18 kg)
		Feed/gain ratio	5.52	6.66
Swine	Dehydrated layer waste	Daily gain, lb	1.32 (0.60 kg)	1.14 (0.52 kg)
		Feed/gain ratio	4.12	4.82
Growing chicks	Dehydrated layer waste	Daily gain, grams	16.1	15.7
		Feed/gain ratio	2.36	2.60
Laying hens	Dehydrated layer waste	Egg production, % lay	71.9	72.8
		Feed/dozen eggs, lb	4.18 (1.90 kg)	4.18 (1.90 kg)
Cattle	Poultry litter[2]	Daily gain, lb	2.2 (1.0 kg)	1.91 (0.87 kg)
		Feed/gain ratio	10.18	11.58

[1]Adapted by the authors from *Unidentified Resources as Animal Feedstuffs*, NRC, National Academy Press, Wash., D.C., 1983, pp. 132-144, Tables 35-41.
[2]Also, dried poultry litter has been fed successfully to dry and lactating dairy cows, to growing and breeding sheep, to growing swine, and to broilers.

Also, dried poultry litter has been fed successfully to dry and lactating dairy cows, to growing and breeding sheep, to growing swine, and to broilers.

PROCESSING ANIMAL WASTES FOR FEED

Several methods have been used to process animal wastes for feed; among them, dehydration, ensiling, pelleting, preparation for liquid feeding, oxidation-ditch aerobic processing, commercial (patented) systems, and the use of wastes as substrates for single-cell protein production.

(Also see Chapter 14, section on "Animal Waste [Manure] Processing.")

Dehydration

When voided, layer waste contains about 75% water. Reduction of the moisture content from 75% to 15% in dehydrators (698 to 1,292°F) requires removal of 1,284 lb of water per ton of dry solids, at an energy cost of $25 to $50/ton of dehydrated material.

The main **advantages** to artificial dehydration are: (1) reducing pathogens to low levels or eliminating them entirely, (2) lessening or removing odors, and (3) facilitating storing and handling.

The chief **disadvantages** of artificial drying are: (1) the high energy cost, and (2) the considerable loss of nitrogen and certain other components due to heating.

It may be concluded that artificial dehydration of animal waste results in excellent products, but the process may not be economically feasible due to high energy costs.

Ensiling

Ensiling of animal waste is a controlled anaerobic fermentation process during which the carbohydrates in the mixture are converted to lactic and other acids. Once sufficient acids are produced, bacterial action ceases and the ensilage is stable. Heat is generated in the process, with an internal temperature of about 77°F achieved. Processing animal wastes by ensiling has the **advantages** of (1) being economical, (2) diminishing the hazards from certain potentially pathogenic organisms, (3) rendering the waste mixture more palatable, and (4) producing a product with a pleasant aroma.

Because of the considerable expense and energy required for drying animal wastes, the trend is toward ensiling. Except for the cost of the silo, ensiling of wastes involves few expenses. Poultry wastes can be ensiled in bunker-type silos as well as oxygen-limiting tower silos.

South Carolina Experiment Station researchers conducted extensive studies on ensiling poultry wastes. They found that manure mixed with forage or litter takes about 6 weeks to ensile adequately. In an experiment designed to test the proper moisture level for ensiling a manure-forage combination, they found that hay and manure ensiled at 44% moisture produced the most desirable combination from a pH standpoint (see Table 12-6). Based on their studies,

TABLE 12-6
PH OF MANURE-FORAGE COMBINATIONS AT VARYING MOISTURE LEVELS[1][2]

Moisture Level	Forage		
(%)	Hay	Peanut Hulls	Straw
33	5.91	6.09	6.31
36	5.84	5.79	6.16
39	5.70	5.68	5.92
44	5.52	5.80	5.98
49	5.72	6.05	6.44
54	5.77	6.25	6.77

[1]Adapted by the authors from "Feeding Ensiled Poultry Wastes to Cattle," by D. L. Cross, *Proceedings, 2nd Annual North Carolina Poultry Nutrition Conference*, 1975, p. 24.
[2]Ensiled in small-scale silos for 60 days.

the South Carolina Agricultural Experiment Station workers recommend the following practices for ensiling poultry litter in an upright silo:

1. The litter should be ensiled at about 37% moisture. Although maximum fermentation takes place at higher moisture levels, it is difficult to blow wet litter into a tall silo because it clogs the blower pipe. At 37% moisture, there is adequate moisture to promote good fermentation; and blower difficulties are minimized. Bunker or trench silos do not pose any moisture problem and can, therefore, be used to ensile litter of higher moisture content.

2. In order to remove metal objects which commonly get into the litter, a magnet should be included in the ensiling and the feeding systems.

3. The easiest place to add enough water to obtain the desired moisture is in the poultry house. A portable moisture tester can be used to check the moisture content. However, a preliminary check on the moisture content of litter may be obtained by squeezing it; litter first begins to stick together at 35% moisture.

4. A front-end loader can be used to clean out the poultry trucks. This clean-out process facilitates mixing of the litter, thereby evenly distributing the moisture.

Other Processes

Other processes that have been used follow:

• **Pelleting**—Pelleting animal wastes prevents ingredient-sorting by animals. However, the waste must be dried before pelleting; hence, pelleting is costly.
• **Commercial patented systems**—Several commercial (patented) systems have been developed for processing animal wastes for feeding, but details relative to these are proprietary to the companies involved.
• **Substrates for protein production**—The use of wastes as substrates for the production of protein supplements for livestock feeds is feasible, with systems using algae, yeasts, bacteria, and fungi all showing promise.

ANIMAL AND HUMAN HEALTH/ESTHETIC ASPECTS

There appears to be only minimal risk from feeding wastes that have been processed by ensiling or dehydration. However, spore-forming bacteria are not destroyed by either process.

Copper may present a mineral problem accompanying the recycling of animal wastes; and this occurs primarily in sheep. The problem of excess copper in sheep feeds is well known; and it is not connected solely with the feeding of animal waste.

Drug and chemical residues do not appear to present a major problem as a result of waste feeding. However, as a precautionary measure, it is recommended that waste feeding should be followed by a withdrawal period of at least 15 days when waste-fed animals are intended to provide meat, milk, or eggs for human consumption. Food quality is not affected by feeding waste.

Finally, there is the esthetic consideration. The vast majority of consumers are not knowledgeable about the physiological process of digestion and absorption—especially in the ruminant. To them, the subject of manure does not make for pleasant conversation. Worse yet, manure as a livestock feed is downright repugnant. Nevertheless, the esthetic angle does not appear to be unsurmountable. Many manure feeding trials have included meat quality evaluations. Livestock producers need to make a concerted effort to present facts and results; otherwise, consumer groups may seize the initiative and push for a ban on manure as a livestock feed.

REGULATORY ASPECTS

In December 1980, the U.S. Food and Drug Administration published a document revoking its earlier policy regarding the feeding of animal waste and leaving regulations governing the feeding of animal waste to the states. So, U.S. livestock producers using, or planning to use, animal wastes as feed ingredients should comply with the specific regulations of the state involved. In Canada, waste-feeding is governed by the Foods and Drugs Act.

Future Handling and Uses of Manure

Increasingly, what to do with manure and how to protect the environment will be a major problem for livestock producers. More and more laws and zoning will govern the handling of manure to prevent it from polluting water and air, and to minimize its potential to increase insects, odors, and dust. Responsible livestock producers need to be attuned to the worldwide awakening to the problem of pollution of the environment (air, water, and soil) and its effect on animal and human health, and science and technology need to evolve with new methods of handling manure, and new uses for it, so that animals and humans can continue to live happily together and enhance each other.

Municipal Wastes

As the world's population continues to expand, human waste will also increase—creating a major disposal problem. The U.S. Department of Energy estimated that, on a dry basis, 68 million tons of municipal solid waste were produced in 1975. The livestock industry offers one alternative to deal with this problem. The feeding value of two human waste products—sewage sludge and garbage—have been receiving increasing attention in recent years.

SEWAGE SLUDGE

One of the major problems with the feeding of sewage sludge is the danger of chemical contamination and subsequent residues in animals fed this by-product. Heavy metals, polychlorinated biphenyls (PCBs), pesticides, and other hazardous chemicals may be introduced into the sewage system, thereby creating a health risk to both livestock and people. As a result, the use of sewage sludge for animal feed is not considered viable at this time.

GARBAGE

From the remote day of domestication forward, swine have been considered scavengers—often fed table scraps and other wastes. Even today, in most of the developing countries, pigs, and, to a lesser extent other livestock, consume few products that are suitable for human food.

Municipal garbage has long been fed to hogs. But beginning about 1940, the practice declined because of a gradual lowering in the feeding value of garbage, along with other competition for it—notably, its manufacture into lawn, greenhouse, and garden fertilizer. By June, 1960, only 1.85% of the nation's hogs were being fed garbage.

Prior to 1940, the garbage feeder figured that a ton of city garbage would produce 60 to 100 lb of pork; whereas, at the present time, it is estimated that a similar quantity will not produce more than 30 lb of pork.[9] The change in feeding value may be largely attributed to improved refrigeration, the effective use of leftovers, and the change of the general eating habits of humans; for example, the increasing use of frozen and highly processed foods. Institutional, hotel, and restaurant garbage is superior to household garbage.

Garbage may be utilized either as a feed for a sow and pig enterprise or for finishing feeder pigs that are obtained from other sources. Usually, the venture seems most successful when a combination of grain and garbage feeding is practiced.

It is also observed that the most successful garbage feeders use concrete feeding floors, practice rigid sanitation, and take every precaution to prevent diseases and parasites. Unless considerable grain is fed to market hogs, especially after weights are over 100 lb, soft pork and paunchiness will result in garbage-fed hogs.

Swine are more likely to become infected with trichinosis, vesicular exanthema, and certain other diseases when fed raw garbage. For this reason, all states now have laws requiring that commercial garbage be cooked. Also, it is noteworthy that there is no danger of transmitting trichinosis from swine to humans provided pork and pork products are thoroughly cooked.

Because of high energy costs, very little garbage is artificially dehydrated.

Recent research has indicated that garbage can be fed successfully to ruminants. While the data are somewhat limited, it appears that garbage may be incorporated into maintenance or grower rations when availability and price

---

[9]Some Canadian authorities consider 4 lb of heavy garbage to be equivalent to 1 lb of concentrate (*Feeder's Guide and Formulae for Meal Mixtures*, 13th ed., published by the Quebec Provincial Feed Board for April 1959-61).

permit. As with activated sludge, the level of pesticides, PCBs, and heavy metals must be carefully monitored to avoid possible residues.

(Also see Chapter 6, section on "Municipal Garbage.")

Bakery Waste

Bakery waste consisting of bread, cookies, cake, crackers, flours, and doughs can be dried and sold as livestock feed under the name *dried bakery product*. Since it is high in digestible fat and carbohydrate, it is often used to replace grain when economically feasible. However, like grains, dried bakery product is low in vitamin A, protein, and minerals. It contains rather large amounts of salt; and if the salt content exceeds 3.5%, the Association of American Feed Control Officials stipulate that the maximum amount of salt must be so labeled in the name of the product. Because of these high levels of salt, poultry should not be fed dried bakery product in excess of 15% of the ration. On the other hand, it can replace all of the grain in swine rations and up to 30% in cattle rations without adversely affecting palatability.

A description, along with the feeding recommendations, of some common waste by-product feeds follows:

• **Bakery by-product**—Bakery by-product (bakery waste) consists of unsold bread, doughnuts, cakes, and other pastries. It is usually very low in fiber, high in fat, and fair in protein.

Bakery by-product is an excellent energy feed for cattle. It may constitute most of the maintenance of stocker and breeding cattle, provided it is properly supplemented with minerals and vitamin A. Up to 10% may be included in cattle finishing rations, and up to 15% may be included in lactating cow rations (due to its low fiber content, higher levels may depress milk fat).

• **Garbage, boiled, wet**—Boiled, wet garbage is animal and vegetable waste that has been boiled. It is usually collected by contractors sufficiently often to avoid decomposition, and injurious materials (crockery, glass, metal, string, and similar materials) are sorted out. Best quality garbage comes from restaurants. Boiled garbage must be cooked at a temperature high enough to destroy all organisms capable of producing animal diseases.

Boiled garbage may be utilized either as a feed for a sow and pig enterprise or for finishing feeder pigs. A combination of garbage and grain feeding is best.

• **Manure, poultry, with litter**—Poultry manure with litter is the type that is generally produced by broiler operations. Poultry litter is the most collectable and most nutritious of all animal wastes. It is fair in energy, but high in protein and minerals. Poultry manure with litter is usually used fresh, commercialy processed (patented), or ensiled.

Poultry manure with litter is best fed to beef cattle—breeding cattle, stockers, or finishing cattle.

• **Manure, poultry, without litter, dehy**—Dehydrated poultry manure without litter is the kind that is generally produced in caged layer operations. On a moisture-free basis, it is medium in fiber, fair in energy, and high in protein and minerals. When voided, it contains about 75% water. When dried, it contains about 15% moisture.

Dehydrated poultry manure without litter may be fed successfully to beef cattle, lactating cows, sheep, swine, and poultry.

(Also see Chapter 6, section on "Bakery By-products.")

AQUATIC PLANTS[10]

The word *aquatic* means of, or pertaining to, water. *Aquatic plants are plants that grow in water.*

Aquatic plants occur throughout the world; in the sea, saltwater marshes, rivers, lakes, and waste-treatment ponds.

To date, these plants have been regarded more as problems than resources; among their adverse effects are: blocking canals and pumps in irrigation projects, interfering with hydroelectric production, wasting water by evapotranspiration, hindering boat traffic, increasing waterborne diseases, impeding drainage, causing flooding, interfering with fish culture and fishing, and harboring mosquito breeding.

Presently, aquatic plants are being recognized as potentially valuable resources for animal feed, human food, and other uses.

There exist almost unlimited opportunities for increased sea farming. The water area of the world is many times greater than the land space; and phenomenal yields of aquatic plants are obtained—as much as 60 tons per acre. Not only that, aquatic farming does not suffer from drought or loss of crop through pests and disease; and aquatic plants require no planting, weeding, or fertilizing. Some scientists predict that soon after the turn of the century the world's sea crops will have to be farmed to ensure the survival of our teeming population.

With one of the world's densest populations and a long coastline, the Japanese already have great expertise in the art of sea farming. Teams of girls, all expert swimmers and skin divers, play a part in this form of cropping, which is also known as aqua-culture. These underwater laborers are specially trained and equipped to carry out the cutting of aquatic plants from cultivated beds off the seashores.

Quantity and Kinds of Aquatic Plants

In 1976, the Food and Agriculture Organization of the United Nations reported that, worldwide, 2,402,000 metric tons (wet weight) of aquatic plants were harvested, most of which was seaweed. Thus, the total tonnage harvested is very small, considering the huge production.

Data are not available on the quantity of aquatic plants other than seaweed produced, but the potential is very great. For example, growth rates of 713 lb dry matter/acre/day have been recorded for water hyacinth. If such growth could be sustained for 6 months a year, that would make for a yield of 64 tons of dry matter per acre. By comparison, on the average, alfalfa yields 4 to 7 tons of hay per acre on a 20% moisture basis.

The aquatic plants with greatest potential for feed production are: algae, seaweed, and water hyacinth, with certain other plants like duckweed (of the family *Lemnaceae*) having possibilities.

Physical Characteristics of Aquatic Plants

The physical characteristics of aquatic plants make for difficulties in harvesting and processing. Algae are small (5 to 15 microns, or 0.000195 to 0.000585 in.) in diameter.

---

[10]In the preparation of this section, the authors drew heavily from *Underutilized Resources as Feedstuffs*, NRC, National Academy Press, Washington, D.C., 1983.

Kelp varies in length from 20 ft to 200 ft. Water hyacinth plants are connected by stolons. Duckweed is a tiny free-floating vascular (tubular) plant. Harvested aquatic plants are slippery and tangled, with the result that it is difficult to handle them mechanically.

All aquatic plants contain much water—they are 85 to 95% water.

A description, along with feeding recommendations of some aquatic plants follows:

- **Algae, green *Scenedesmus quadricuada* whole, fan air dried**—Algae are primitive aquatic plants that contain chlorophyll and convert the sun's energy into single-cell protein by photosynthesis. Like most aquatic plants, algae are high in moisture. On a dry basis, however, algae are low in fiber, medium in energy, and high in crude protein and minerals.

  Algae may be fed to sheep, swine, and chicks. Algae meal should be limited to 10% of swine rations and 6% of poultry rations. Pellets in which 50% of the nitrogen was supplied by algae have been fed to sheep on summer range.

- **Seaweed (kelp)**—When botanists speak of seaweed, they usually mean one of the larger brown or red varieties. Seaweed is low in protein (5.7% on an as-fed basis), and the protein is of low biological value.

  Dehydrated seaweed may constitute up to 10% of cattle and sheep rations, and up to 6% of swine and poultry rations. Higher levels have been used experimentally in lactating dairy cow and sheep rations.

- **Water Hyacinth**—Water hyacinth is an unattached and free-floating waterplant with the leaves above the surface of the water and the roots in the water. Generally, it is considered a troublesome weed; it multiplies rapidly, clogs lakes, rivers, and ponds; and seriously obstructs traffic on waterways. Also, it is extremely difficult to eradicate. However, water hyacinth has considerable potential as a livestock feed.

  In China, water hyacinth is sometimes harvested, boiled, and fed to swine. Because it is a floating plant, it is easily harvested. The fresh plant contains prickly crystals which make it unpalatable; and the high water content of the plant imposes a limitation on the amount of dry matter an animal is capable of ingesting. For these reasons, normally not more than 25% boiled water hyacinth is included in swine rations in China.

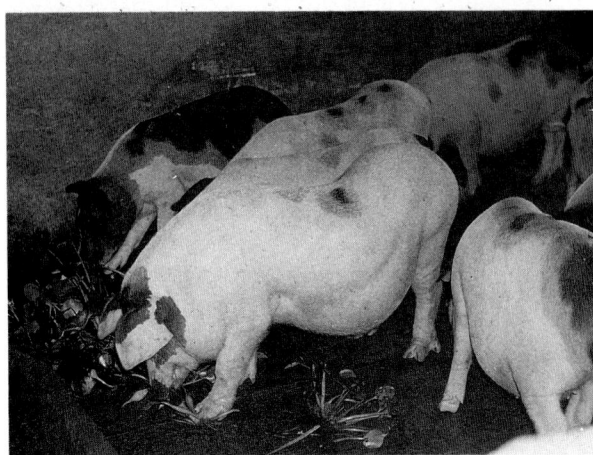

Fig. 12-5. Pigs eating water hyacinth in China. (Photo by A. H. Ensminger)

Boiled water hyacinth is used as a pig feed in Southeast Asia. The hyacinth is chopped and sometimes mixed with other vegetable wastes, such as banana stems, and boiled slowly for a few hours until the ingredients turn into a paste. To this paste, the following ingredients are commonly added: oil cake, rice bran, corn, and salt.

The following results of feeding trials with water hyacinth have been reported: Cattle and sheep readily consumed complete rations containing processed water hyacinth. The dry matter intake of pangolagrass silage by sheep was higher than that of water hyacinth silage.

QUESTIONS FOR STUDY AND DISCUSSION

1. Define (a) by-product feeds, and (b) crop residues. If by-product feeds and crop residues were not used for animals, would they be used for human food?

2. When determining the economy of by-product feeds and crop residues, what factors in addition to cost per pound or per ton should be considered?

3. List the four major categories of animal by-product feeds; and give the nutritional characteristics of animal by-product feeds as a whole.

4. Describe and give the feeding recommendations pertaining to each of the following dairy by-product feeds: (a) buttermilk, (b) skimmed milk, and (c) whey.

5. Describe and give the feeding recommendations pertaining to each of the following marine by-product feeds: (a) fish meal, and (b) shrimp meal.

6. Describe and give the feeding recommendations pertaining to each of the following meat animal by-product feeds: (a) animal fat, (b) blood meal, (c) meat meal, and (d) tankage.

7. Describe and give the feeding recommendations pertaining to each of the following poultry by-product feeds: (a) feather meal, (b) poultry by-product meal, and (c) poultry fat.

8. Define each of the following plant by-product terms: (a) bagasse, (b) cannery residue, (c) cull, (d) *in vitro* (e) *in vivo* (f) pit, (g) pulp, (h) stover, and (i) stubble.

9. Describe and give the feeding recommendations pertaining to each of the following oilseed by-product feeds: (a) canola (rapeseed) meal, (b) cottonseed, whole, (c) cottonseed meal, (d) linseed meal, and (e) soybean meal.

(Continued)

10. Describe and give the feeding recommendations pertaining to each of the following grain by-product feeds: (a) corn gluten feed and corn gluten meal, (b) hominy feed, (c) wheat bran, (d) rice polishings, and (e) screenings, cereal grain.

11. Describe and give the feeding recommendations pertaining to each of the following fruit, vegetable, and nut by-product feeds: (a) apple pomace, (b) carrot roots, fresh, (c) citrus pulp, dehydrated, (d) pineapple bran, (e) potatoes, cull, fresh or cooked, (f) tomato, pomace, wet, and (g) turnip, roots, fresh.

12. Describe and give the feeding recommendations pertaining to each of the following sugar by-product feeds: (a) beet molasses, (b) beet pulp, and (c) cane molasses.

13. Describe and give the feeding recommendations pertaining to each of the following brewing and distilling by-product feeds: (a) brewers' dried grains, (b) distillers' dried grains, (c) malt sprouts, barley, dehydrated, and (d) yeast, brewers' dried.

14. Although more than 1.5 million tons of sulfate and sulfite pulps from spruce, pine, and fir were fed to cattle and horses in the Scandinavian countries during World War II when feed supplies were limited, the feeding of wood pulp to livestock ceased when conventional feedstuffs became available. Why wasn't the feeding of wood and paper by-products continued in these countries after World War II?

15. Describe and give the feeding recommendations pertaining to each of the following wood and paper by-product feeds: (a) aspen wood, (b) paper waste, (c) sawdust, (d) wood molasses, and (e) torula yeast.

16. Discuss the following aspects of crop residues: (a) kind and quantity produced, (b) economy of crop residues, (c) harvesting, and (d) nutrient value.

17. Discuss the method and effectiveness of each of the following treatments for delignifying and increasing digestibility of crop residues by animals: (a) grinding and pelleting, (b) sodium hydroxide, (c) calcium hydroxide, (d) ammoniation, and (e) high pressure steam.

18. Discuss the advantages and the disadvantages of each of the following feeding systems for utilizing crop residues: (a) grazing, (b) processing as dry feed, or (c) ensiling.

19. Describe and give the feeding recommendations pertaining to each of the following crop residues: (a) almond hulls, (b) bagasse, sugarcane, (c) barley straw, (d) corn stover, (e) corncobs, (f) cottonseed hulls, (g) oat straw, (h) sorghum stover, (i) soybean hulls, (j) sweet corn cannery refuse silage, and (k) wheat straw.

20. Define each of the following terms: (a) animal wastes, (b) municipal wastes, and (c) bakery wastes.

21. On the average, how much manure (free of bedding) will be produced by each species per year per 1,000 lb weight?

22. Review the historic use of manure as an animal feed.

23. Discuss each of the following potential uses of manure: (a) fertilizer, (b) nonfeed energy source, or (c) feed. What is the highest and best use of manure?

24. Discuss the performance of animals in growth, or in meat, milk, or egg production, when fed (a) cattle manure, (b) swine manure, or (c) poultry manure, with and without litter.

25. What are the advantages and the disadvantages of each of the following methods for processing animal wastes: (a) dehydrating, and (b) ensiling.

26. When feeding manure, of what concern are the following aspects: (a) human health, (b) esthetic, and (c) regulatory.

27. What do you forsee relative to the future handling and uses of manure?

28. Why is the feeding of sewage sludge not considered to be viable at this time?

29. How is garbage normally prepared and fed to swine? How may humans alleviate the danger of trichinosis from eating pork derived from garbage-fed swine?

30. What is bakery waste? How may it be used as a feed?

31. What are aquatic plants? Currently, what are their adverse effects; and what are their potentials as a feed resource?

32. Discuss the physical characteristics, quantity, yield, harvesting, storing, and feeding of each (a) algae, (b) seaweed, and (c) water hyacinth.

Fig. 12-6. Cows grazing cornstalks. (Courtesy, Iowa State University)

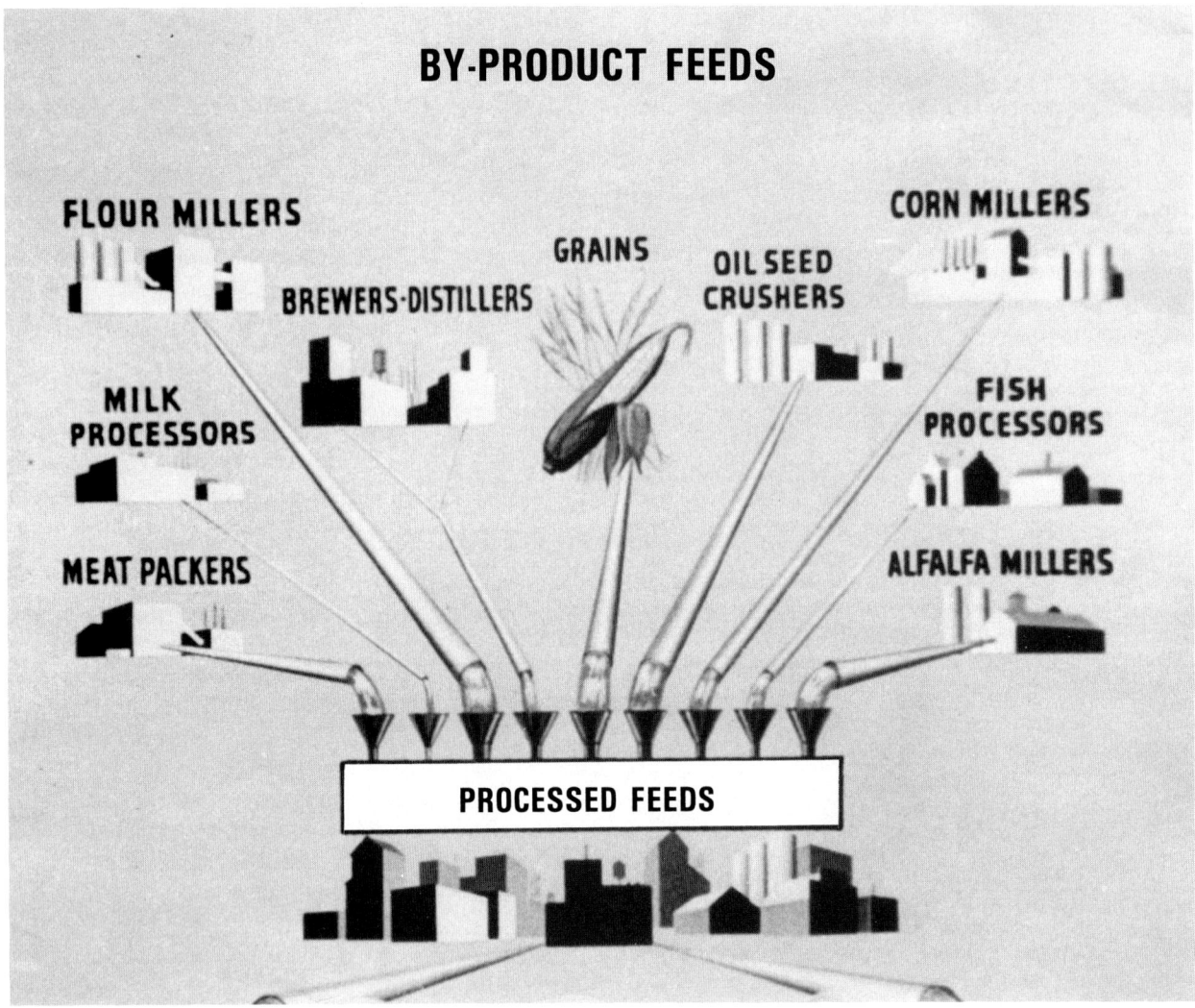

Fig. 12-7. A host of by-product feeds which are not suited for human consumption are used as animal feeds.

13

FEED SUPPLEMENTS/ ADDITIVES/IMPLANTS

Original painting by Tom Phillips

Contents	Page
Food and Drug Administration (FDA)	248
Delaney Clause	248
Use Feed Supplements, Additives, and Implants Safely	248
Feed Supplements	248
Protein Supplements	249
Amino Acid Supplements	249
Mineral Supplements	249
Classes of Mineral Supplements	250
Need for Mineral Supplementation	250
Guidelines for Mineral Supplementation	250
Supplementation of Mixed Feeds	250
Supplementation of Unmixed Feeds or Pasture	250
Salt (NaCl)	251
Calcium (Ca) and Phosphorus (P) Supplements	251
Chelates	251
Vitamin Supplements	251
Vitamin A and Carotene	251
Vitamin D	252
Vitamin E	252
Vitamin K	252
B-Complex Vitamins	252
Vitamin C (Ascorbic Acid, Dehydroascorbic Acid)	253
Vitamin Imbalances	253
Unidentified Factors	253
Additives, Implants, and Injections	253
Abortifacients	253
Additives That Enhance Market Value	253
Additives That Physically Aid Digestion	254
Grit	254
Roughage Substitutes	254
Anthelmintics (Wormers; Vermifuges)	254
Antibiotics	254
Mode of Action of Antibiotics	254
Importance and Use of Antibiotics in Animal Production	255
Safety and Future of Antibiotics as Feed Additives	255
Antioxidants	256
Arsenicals	256
Bloat Control Products	256
Buffers	257
Common Buffers and Their Uses	257
Chemotherapeutics	258
Copper (Cu)	258

Contents	Page
Electrolytes	259
Flavoring Agents	259
Hormone and Hormonelike Compounds	259
Somatotropin (Growth Hormone)	259
Bovine Somatotropin (BST) for Dairy Cattle	260
Porcine Somatotropin (PST) for Swine	261
Melengestrol Acetate (MGA)	261
Implants	261
Ionophores	261
Isoacids (IsoPlus)	262
Kelp	262
Medicated Feeds	262
Mold (Fungi) Inhibitors	262
Pellet Binders	263
Probiotics (Microbial Enhancers)	263
Steriods	263
Tranquilizers	264
Questions for Study and Discussion	264

From the remote day of animal domestication forward, people sought ways of improving their performance by using different feeds. Originally, it was not understood why two different feeds might elicit different responses, but the end result was recognized.

Today, the successful livestock producer uses supplements, additives, and implants to maximize performance; to improve animal health, to increase feed intake and hopefully feed efficiency, and/or to alter some physiological process in the animal that will stimulate production and/or improve the quality of the product.

The term, supplement, refers to feedstuffs that are used to improve the value of basal feeds. Thus, supplements are products that provide an additional nutrient or nutrients. They can be used in large quantities, such as protein supplements, or in extremely small quantities, such as trace minerals and vitamins.

An additive is a substance of non-nutritive nature which when added to feed will improve feed efficiency, production, and/or health of animals. In general, the term *feed additive* refers to a non-nutritive product that affects utilization of the feed or productive performance of the animal.

An implant is a substance that is implanted into the body for the purpose of growth promotion or controlling some physiological function. Compounds used in implants are generally materials (for example peptide hormones) that would be destroyed during digestion. Hence, implants enable the body to absorb the compounds intact. Additionally, implants provide a means of controlled continuous release of chemicals.

FOOD AND DRUG ADMINISTRATION (FDA)

The Food and Drug Administration (FDA), which is part of the U.S. Department of Health, Education, and Human Services, is charged with the responsibility of regulating the use and safety of additives and implants. Through the years, it has established rigorous testing policies requiring proof relative to product safety and efficacy. The requirements for new product approval by FDA have become more rigid, and enforcement has become more unrelenting.

Since 1960, manufacturers have been required to demonstrate both the efficacy and the safety of proposed new additives before they can be certified for use. The required demonstrations include tests at 3 different levels: (1) acute toxicity test to show the effects of a single lethal dose given to a variety of laboratory animals; (2) short-term (90-day) toxicity studies showing the effects of feeding different concentrations to 2 kinds of laboratory animals; and (3) long-term toxicity studies of 2 years or more to show the effects of lifetime consumption. Tolerance levels are based on the amount of chemical shown to be safe in long-term studies. The margin of safety is extremely wide. For example, if toxicity studies show that the use of 1,000 ppm of a particular chemical is safe, then the FDA will permit maximum use of 10 ppm (1% of the no-effect level). The one exception to this standard is that any substance demonstrated to be carcinogenic (cancer-producing) is not permitted in any amount, if such use leaves detectable residues in the product.

Wherever additives are involved, it is the responsibility of the producer to comply with the withdrawal periods prior to slaughter or milking, as established by the Food and Drug Administration.

Delaney Clause

In 1958, the food additives amendment, better known for its Delaney Clause (named after the congressman who sponsored it), was passed. This bill has proven to be one of the most controversial pieces of legislation ever to affect the American livestock industry. The Delaney Clause states:

> "*Provided,* That no additive shall be deemed safe if it is found to induce cancer when ingested by man or animal, or if it is found, after tests which are appropriate for the evaluation of the safety of food additives, to induce cancer in man or animal. . ."

This clause gave rise to the policy of *zero tolerance*—that is, no substance can be used as a feed additive, even in miniscule amounts, if it has been, in any way, implicated as an inducer of cancer in either human or beast. What, at the time, appeared to be a well-intentioned law aimed at protecting the consumer from potential health hazards proved to be a nightmare for the drug industry and livestock producers. The additive manufacturers must now prove a negative hypothesis which many feel is impossible. That is, an additive must be demonstrated to be 100% noncarcinogenic. Unfortunately, the lawmakers are in a tenuous position because, to repeal the Delaney Clause, they run the risk of being accused of supporting the addition of cancer-causing drugs to our food supply.

Only when those who have chosen sides become informed on the issues, realistic in their demands, and willing to listen to the opposition, will the issues be resolved. The Delaney Clause will, in all likelihood, be modified, but the debate of how safe is *safe* will, in all likelihood, continue without end.

With or without the Delaney Clause, *America's food supply is the safest in the world.*

Use Feed Supplements, Additives, and Implants Safely

The list of supplements, additives, and implants grows longer. So, also do the concerns of producers and consumers.

Producers use drugs to produce products faster, more abundantly, on less feed, and/or of improved quality. Consumers want to be assured of the wholesomeness of their meat, milk, and eggs.

Modern agriculture and a significant share of our present feed/food technology are dependent upon the use of chemicals. Yet our increasing awareness of food safety and environmental quality has resulted in increased scrutiny of many of the chemical compounds and their use patterns. Two primary concerns prevail: (1) to assure that the chemicals used in the production of food and fiber at every level are safe when used properly, and (2) to assure that the benefits of these chemicals in providing adequate food supplies and in protecting the health of consumers not be displaced by an abstract goal of zero risk.

The label holds the key to proper livestock drug use. But no matter how clear and accurate the label, it won't communicate anything if it is not read; and it won't prevent illegal residues if it is not followed. The alternatives to reading and heeding the label relative to the proper use of livestock drugs are: (1) the losses to non-complying producers through condemnation of their products, plus possible fines; and (2) the losses to the entire livestock industry as a result of some scare headlines in the news media, followed by the erosion of consumer confidence in the safety and wholesomeness of meat, milk, and eggs, the consequences of which no one in the livestock industry can afford.

FEED SUPPLEMENTS

In the formulation of rations, the energy and fiber feeds which are usually produced on the farm or ranch are the basal feeds. Such feeds are commonly deficient in protein and perhaps in one or more amino acids, and in minerals and vitamins.

A feed supplement is a feed that is used to improve the nutritional value of a basal feed. It follows that feed supplements are usually concentrated sources (1) of protein, or of one or more amino acids; (2) of one or more minerals; (3) of one or more vitamins; or (4) of a protein/mineral/

vitamin mix. Additionally, the following special supplements are sometimes used:

1. High fiber supplements, such as soybean hulls or corn bran, both of which are highly digestible, fed to dairy cows on low fiber rations to prevent lowered milk fat content.
2. High energy supplements, such as cereal grains, fat, and whole cottonseed, fed to animals to improve ration energy density and performance.
3. Protein/energy blocks; usually containing some molasses, natural protein and/or urea, minerals, vitamins, in some cases fat, and an intake limiter(s) such as salt; and weighing 50 to 500 lb; used primarily as winter range supplements.
4. Medicated health supplements.

Supplements may be fed (1) undiluted as an addition to other feeds; (2) further diluted and mixed to produce a complete feed; or (3) offered free-choice, with other parts of the ration fed separately. Also, supplements are provided in different forms—in meals, granules, pellets, cubes, blocks (ranging in weight from 50 to 500 lb), liquids (usually with molasses), or salt-limited.

The term *supplement* is descriptive within itself. Thus, in order properly to balance a basal high energy feed, additional protein, minerals, and vitamins are usually needed. Normally, these deficiencies are met by adding ingredients that are richer in the needed nutrient(s) than the basal feed. Consequently, such ingredients are commonly referred to as feed supplements.

Specific nutrient supplements are commercially available, thereby permitting the nutritionist to add small amounts of a specific nutrient or combination of nutrients to a ration without altering the general makeup of the initial formulation.

In general, little or no supplementation of specific amino acids is done for ruminants, although some research with methionine hydroxy analog (MHA) has indicated its possible use in the future. However, several amino acids, especially arginine, cystine, lysine, methionine, and tryptophan, warrant careful consideration in feeds for nonruminants; and sometimes individual amino acids or amino acid analogs are added to feed.

Mineral and vitamin supplementation is of paramount importance to *all* livestock rations. Imbalances, deficiencies, or excesses of minerals pose major problems. While toxicities of vitamins are rare, deficiencies are not; and in this era of highly refined scientific feeding, there can be no excuse for these occurrences.

Protein Supplements

Protein supplements are feedstuffs containing more than 20% protein or protein equivalent. They are obtained from animal, marine, plant, or microbial sources, as well as from nonprotein nitrogen sources such as urea, biuret, or ammoniated products. Additionally, a number of specially designed pasture and range protein supplements are available, many with added energy, minerals, and/or vitamins; and they are used in the following forms and systems that lessen the labor attendant to daily feeding: (1) cubes, hand-fed at intervals, (2) blocks, (3) salt-feed mixtures that are self-fed, and (4) liquid feed supplements that are self-fed.

Most high energy feeds (except for purified products like fat, starch, or sugar) supply some protein, but, except for adult animals during maintenance, they usually do not supply enough to meet total needs. Thus, supplementary sources are commonly needed in the formulation of rations for all species of animals. In particular, protein is a critical nutrient for young, rapidly growing animals and for high-producing dairy cows and layers. Animals cannot (1) develop their genetic potential, (2) produce the maximum of meat, milk, and eggs, or (3) produce the maximum power, animation, and speed in the case of horses unless their rations contain sufficient protein, along with the correct amino acid composition in the case of monogastrics.

Protein supplements are regularly in shorter supply and higher priced than the cereal grains and other high-energy feeds. As a result, the tendency is not to feed sufficient protein supplements to balance rations.

(Also see Chapter 6, Types and Roles of Feedstuffs, Section on "Protein Supplements"; Chapter 11, Protein Supplements; and Chapters 19 to 28, the chapters devoted to the respective classes of livestock.)

AMINO ACID SUPPLEMENTS

For monogastrics and very young ruminants (preruminants), the amino acids that make up proteins are really the essential nutrients, rather than the protein molecule itself. For ruminant species, the dietary need is (1) to nourish the rumen microorganisms, and (2) to have an adequate supply of digestible essential amino acids in the gut. High-producing ruminants (such as high-producing dairy cows) may have higher amino acid needs than can be satisfied by rumen synthesis; so, protein quality is more important under these circumstances than for animals producing at low levels and consuming much less feed.

Of the 22 amino acids, 5 are deemed critical, whereas the others are usually in sufficient supply from the combination of feedstuffs found in the rations of most monogastric species. The critical 5 are: arginine, cystine, lysine, methionine, and tryptophan. When a monogastric animal's ration is low in 1 or more of these 5 amino acids, protein supplements carrying large amounts of pure amino acids may be added to make up the deficiencies. Thus, swine feeds consisting chiefly of corn, wheat, and barley are deficient in the essential amino acid, lysine, and corn is also deficient in tryptophan. Due to the large amounts of vegetable proteins used, along with low levels of animal and fish proteins, poultry rations are most often lacking in methionine. Also, lysine and cystine are often inadequate in normal feedstuffs. So, poultry rations frequently call for the supplementation of DL-methionine; and lysine and cystine are often added, also.

(Also see Chapter 11, Protein Supplements.)

Mineral Supplements

The metabolic functions and interrelationships among the minerals are extremely varied and complex. An excessive amount of one mineral can create a deficiency of another. Additionally, several trace minerals have relatively narrow toxicity tolerances. For example, selenium can be legally added to feed to a maximum of 0.3 ppm (part per million), but toxicities can result when animals ingest 10 ppm. Therefore, supplementation with this and several other minerals must be carefully monitored.

Almost all feeds contain at least limited amounts of the various minerals, but these levels are highly variable and often reflect the profile of the soil on which they are grown and the genetic variations among its plant species. In addition to the variability of mineral levels in individual feeds, the mineral requirements of animals are highly variable, depending on such factors as age, size, sex, type of production, and stage of production.

CLASSES OF MINERAL SUPPLEMENTS

Mineral supplements can be divided into three basic categories: (1) packinghouse by-products, (2) naturally occurring mineral sources, and (3) synthetic mineral compounds.
- **Packinghouse Mineral By-products**—Packinghouse mineral by-products are exactly what the term implies. Bones and connective tissue resulting from the processing of various meat products represent an excellent source of quality calcium, phosphorus, and some trace minerals. Various processes—such as steaming, cooking, and precipitating—are applied to these products to (1) alter their relative mineral content, and (2) sterilize and stabilize the products for storage and use.
- **Naturally Occurring Mineral Sources**—These are mineral supplements which are obtained from our natural environment and processed in such manner as to render them safe for feeding. Such mineral sources as rock phosphate must be processed in some way to remove contaminants which can be toxic to livestock—in this particular case, fluorine.
- **Synthetic Mineral Compounds**—In recent years, chemists have developed processes for synthesizing mineral supplements that are cheap and yet of extremely high purity. Many of these compounds are sold as *Commercial Grade*. This term implies that there are trace amounts of impurities in the product which render it unusable for analytical purposes. This should not concern the buyer as the level of impurity is negligible. Rather, the buyer should be concerned about how much of each mineral is being bought.

NEED FOR MINERAL SUPPLEMENTATION

Only the specific minerals that are needed should be provided. Excesses and mineral imbalances should be avoided. Except for substances like fat and urea, most feeds provide some minerals. Nevertheless, many rations require more concentrated sources of one or more of the macro and/or microminerals.
- **Macrominerals**—Of the macrominerals demonstrated to be required by livestock, only salt (sodium chloride), calcium, and phosphorus are routinely added to all livestock rations. Two other macrominerals, magnesium and sulfur, are sometimes added to ruminant rations in specialized cases. Magnesium is sometimes provided in mineral mixes for cattle on pasture in areas where grass tetany is a problem. Sulfur is routinely added to rations containing urea, since urea replaces protein which is normally a source of sulfur. The levels of all the macrominerals should be carefully monitored for nonruminants.
- **Micro, or trace, minerals**—The following 7 trace minerals are common supplements: cobalt, copper, iodine, iron, manganese, selenium, and zinc. All 7 essential microminerals can be added to salt at a cost of 1¾¢ per pound, for the resulting trace mineralized salt. So, multiplying the yearly salt consumption figure in pounds for each class of animals by 1¾¢ gives the cost of protecting them for a whole year.

Even if the animal's ration is not deficient in all seven essential trace minerals, there is no harmful effect from their supplementation because there is a large safety factor between the level needed and the level that will cause a harmful effect. Also, a little extra of the trace minerals is in the nature of good insurance due to variations in the feed ingredients, level of animal productivity, stress, nutrient relationships, and other factors.

GUIDELINES FOR MINERAL SUPPLEMENTATION

No single plan can be proposed as being the best for mineral supplementation. Rather, livestock producers must tailor their supplement regimens to encompass the following considerations:
1. **Needs of the particular animal.** Age, sex, weight, and production parameters must all be considered.
2. **Types of feed.** An all- or high-concentrate ration will require a different mineral supplement than an all- or high-roughage ration.
3. **Region from which the feeds were obtained.** The mineral content of the feed will reflect the mineral composition of the soil and the genetic makeup of the specific plant.
4. **Facilities.** If the mineral mix is offered free-choice, containers protected from the elements—i.e., rain and wind—may have to be constructed.
5. **Free choice vs incorporating minerals in the ration.** The required minerals should be incorporated in the ration whenever possible. When feeding practices make this impractical, a complete mineral mix, including salt, should be fed free choice.

Supplementation of Mixed Feeds

When livestock are fed a mixed feed, totally or in part, the needed minerals are usually incorporated in the ration in keeping with the known requirements. In general, the following recommendations are applicable to the mineral supplementation of mixed feeds.
1. Salt is usually incorporated in the ration at levels of 0.25 to 0.50%. If less salt is added, it can also be made available *ad libitum*.
2. Calcium and phosphorus are added as needed to balance the ration. Numerous calcium and phosphorus supplements are available at reasonable cost; and the wide variety enables producers to select those which fit their particular needs.
3. If animals are housed in confinement where they receive little exposure to sunlight, careful attention must be given to providing adequate vitamin D, because vitamin D affects the assimilation and utilization of a number of minerals, especially calcium and phosphorus.
4. When the ration is suspected of being deficient in one or more minerals, a trace mineralized salt or specific minerals should be added to the ration.

Supplementation of Unmixed Feeds or Pasture

When animals are fed an unmixed ration or are on pasture, minerals may be provided as follows:
1. **When animals are on liberal grain feeding.** Provide free access to a 2–compartment mineral box, with (a) trace mineralized salt in one side; and (b) in the other side, a mix-

ture of ⅓ trace mineralized salt (salt included for purposes of palatability), ⅓ dicalcium or defluorinated phosphate or steamed bone meal, and ⅓ ground limestone or oystershell flour.

2. **When animals are primarily on roughage (pasture, hay, and/or silage).** Provide free access to a 2–compartment mineral box, with (a) trace mineralized salt in one side, and (b) in the other side, a mixture of ⅓ trace mineralized salt (salt included for purposes of palatability); and ⅔ dicalcium or defluorinated phosphate or steamed bone meal.

SALT (NaCl)

Salt, which serves both as a condiment and a nutrient, is needed by all classes of animals, but more especially by herbivora (grass-eating animals). Ratios of potassium to sodium may reach 17 to 1 in forage feeds; thus, salt is required to narrow this ratio to counteract the metabolic action of high levels of potassium. It may be provided in granulated, rock, or block form. In general, the form selected is determined by price and availability. It should be pointed out, however, that very hard block and rock salt are difficult for stock to eat, often resulting in sore tongues and inadequate consumption. Also, if there is much competition for the salt block, the more timid animals may not satisfy their requirements.

The amount of salt required by animals varies with their stage of growth, ration composition, level of production, and the temperature of their environment. Some animals sweat more than others and their salt requirements are reflected proportionately. Animals exposed to heat and those doing heavy work will need more salt than similar animals under unstressed conditions. Additionally, ruminants on pasture need salt to balance the high-potassium, low-calcium content of grass.

Carnivores usually require less supplemental salt than animals maintained largely on a plant-feed diet, because animal tissues and blood have higher salt concentrations than plants. Even with this lesser need, it is wise to make additional salt available to carnivorous animals.

Salt can be fed free-choice to cattle, sheep, swine, and horses provided they have not previously been salt starved.

CALCIUM (Ca) AND PHOSPHORUS (P) SUPPLEMENTS

When calcium alone is needed, ground limestone or ground oyster shells are commonly used. Other calcium supplements are: bone meal, calcium gluconate, calcium lactate, dicalcium phosphate, dolomite, and kelp.

The most common supplemental sources of phosphorus are: ammonium phosphate, bone meal, calcium phosphate, colloidal clay, dicalcium phosphate, monosodium phosphate, phosphoric acid, and defluorinated phosphate.

CHELATES

The word chelate is derived from the Greek *chelae*, meaning a claw or pincerlike organ. A chelate is a cyclic or complex ring structure in which a divalent or multivalent metal atom is held through two or more bonds in a coordination complex. Generally, chelates are chemically more stable than complexes in which the mineral element is held through only one chemical bond.

Those selling chelated minerals generally recommend a smaller quantity of them (but at a higher price per pound) and extol their "fenced-in" properties.

Many claims have been made for the benefit of chelated minerals: (1) greater physical stability, which reduces the tendency for trace mineral separation in feeds; (2) less oxidation of vitamins and their labile derivatives; and (3) better animal performance and/or higher bioavailability.

Perhaps the most important practical question concerning the use of specialty chelated mineral supplements should center around whether their use will make animal production more or less profitable than using alternative sources (inorganic) of mineral supplements.

Vitamin Supplements

As with mineral supplements, careful consideration must be given to the vitamin supplementation of livestock feeds. While the requirements of vitamins are extremely small in comparison with energy and protein, the omission of a single vitamin from the diet of a species that requires it will produce specific deficiency symptoms, thereby reducing production. Moreover, the cost for vitamin supplementation constitutes a very small fraction of the total feed bill.

Formerly, a wide variety of feed ingredients was added to livestock rations for their vitamin content. But it was found that the vitamin concentration of feedstuffs varied tremendously, being affected by plant species and part (leaf, stalk, or seed), harvesting, storing, and processing. Generally speaking, vitamins are easily destroyed by heat, sunlight, oxidation, and mold growth. So, today, nutritionists rely on vitamin supplements, which in many cases are chemically pure sources that need to be used only in very minute amounts. In modern feed formulation, premixes often represent the common sense approach to providing vitamins.

VITAMIN A AND CAROTENE

Vitamin A is required by all farm animals. No vitamin A is synthesized in plants; but carotene, a precursor of vitamin A, is found in varying quantities in plants. Since the animal body transforms carotene into vitamin A, it is often referred to as "provitamin A."

The degree of greenness in a roughage is a good index of its carotene content, provided it has not been stored too long. Early cut, leafy green hays are very high in carotene.

Aside from yellow corn, all cereal grains have little carotene or vitamin A value. Yellow corn has only about one-tenth as much carotene as well-cured hay, and even this small amount deteriorates over a period of time.

The most common supplemental sources of vitamin A are: cod and other fish liver oils and synthetic vitamin A. The latter is more stable and currently more widely used.

The vitamin A potency (whether due to the vitamin itself, to carotene, or to both) of feeds is usually reported in terms of IU or USP units. These two units of measurement are the same. They are based on the growth response of rats, in which several different levels of the test product are fed to different groups of rats, as a supplement to a vitamin A-free diet which has caused growth to cease. A USP or IU is the vitamin A value for rats of 0.30 microgram of pure vitamin A alcohol, or of 0.60 microgram of pure beta-carotene. The carotene or vitamin A content of feeds is commonly determined by colorimetric or spectroscopic methods.

VITAMIN D

As with vitamin A, vitamin D is required by all farm animals and humans.

For four-footed animals (cattle, sheep, swine, and horses), both D_2 (the plant form) and D_3 (the animal form) are equally effective, so there is no need to use some of each. With poultry, however, vitamin D_3 is more active than vitamin D_2, and should, therefore, be used.

The most potent supplemental products are obtained by irradiating plant or animal sterols that are subject to activation. Thus, ergosterol produced by plants is irradiated and sold for human use in a variety of forms. Irradiated animal sterol, activated 7–dehydrocholesterol, is most frequently used in poultry feeds in view of the superior value of the D_3 form of the vitamin for this species. Yeast is rich in ergosterol; thus, its irradiation results in a potent source that is used for farm animals other than poultry.

Young animals sometimes develop rickets because of insufficient vitamin D, calcium, or phosphorus. This condition can be prevented by exposing the animal to as much direct sunlight as possible, by allowing free access to a suitable mineral mixture, and/or providing good-quality sun-cured hay. In confinement operations, and in northern areas that do not have adequate sunshine, producers should provide young stock and lactating females with a vitamin D supplement.

The vitamin D requirement is less when a proper balance of calcium and phosphorus exists in the ration.

Other factors pertinent to vitamin D follow:

1. **Vitamin D, and cholesterol and ergosterol.** Most of the commonly used feeds contain little or no vitamin D; yet when animals are exposed to sunlight, there is no widespread need for special supplements containing this factor. Fortunately, the skins of animals and many feeds contain provitamins in certain forms of cholesterol and ergosterol, respectively, which, through the action of ultraviolet light (light of such short wavelength that it is invisible) from the sun, are converted into vitamin D. These certain forms of cholesterol and ergosterol themselves have no antirachitic effect.

2. **Vitamin D limited in feeds.** Of all the known vitamins, vitamin D has the most limited distribution in common feeds. Very little of this factor is contained in the cereal grains and their by-products, in roots and tubers, in feeds of animal origin, or in growing pasture grasses. The only important natural sources of vitamin D are sun-cured hay and other roughages. The chief vitamin D-rich concentrates include sun-cured hay, cod-liver and other fish oils, irradiated cholesterol and ergosterol, and irradiated yeast.

VITAMIN E

Vitamin E is required by a large number of animal species, but the deficiency signs may differ greatly among species and even within the same species. There is no experimental evidence that vitamin E deficiency will cause reproductive failure in cattle, sheep, and goats. Studies on the relation of vitamin E to reproduction in horses have been contradictory.

Vitamin E is widely distributed in plants; leafy forages (especially alfalfa) are a good source. Grains (particularly their germs) are fair sources of vitamin E. Germ oils, such as wheat germ oil, are rich supplemental sources of vitamin E.

Most practical rations contain liberal quantities of vitamin E, perhaps enough except under conditions of work, stress, or reproduction, or when there is interference with its utilization. Rather than buy and use costly vitamin E concentrates indiscriminately, the producer should only add them to the ration on the advice of a nutritionist or veterinarian.

The requirements for vitamin E are influenced by interrelationships with other essential nutrients—increased by the presence of interfering substances, and spared by the presence of other substances that may be protective or that may assume part of its functions, particularly the trace element selenium.

VITAMIN K

When vitamin K is deficient, the blood prothrombin level is decreased, and the coagulation time of the blood is increased. This is the main justification for adding this vitamin to the ration. However, vitamin K is widely distributed in normal farm feeds. Also, it appears that it is synthesized in adequate amounts by the intestinal microflora of ruminants and the horse.

When animals are on medication, additional vitamin K may be warranted as the medication may decrease the population of vitamin K-producing organisms in the gut. Additionally, animals consuming sweet clover may ingest sufficient amounts of the antimetabolite, dicoumarol, to create a vitamin K deficiency.

Menadione (vitamin K_3), the synthetic form of the vitamin, is the most widely used commercial source of vitamin K. Also, the following are rich natural sources: alfalfa meal, barley, corn, fish meal, hays (well-cured), milk, pasture (green), peas (green), sorghum grain, soybean meal, and wheat.

B-COMPLEX VITAMINS

Usually, deficiency symptoms of B vitamins are not observed in ruminants. In the case of niacin and choline, however, a response to dietary additions can be measured in terms of weight gain and feed efficiency. Despite the fact that rumen microbes synthesize thiamin and that whole grains contain it, thiamin deficiencies may develop in cattle and sheep and cause a deficiency disease known as polioencephalomalacia (PEM), a disorder of the central nervous system. PEM is thought to be caused by a severe deficiency of thiamin at the tissue level, created by the presence of the enzyme thiaminase, which destroys the thiamin that is in the feed or manufactured by the rumen organisms. (Also see Chapter 5, Nutritional Disorders/Toxins, Table 5–1, Nutritional Diseases and Ailments—Polioencephalomalacia.)

Unlike ruminants, pigs, mink, fish, and poultry have one stomach compartment and no large cecum. As a result, they do not synthesize enough of certain B vitamins. Consequently, these factors must be provided regularly in the ration in adequate amounts if deficiencies are to be averted. This means that the nutritionist and the caretaker should provide a dietary source of water-soluble B vitamins on a daily basis for simple-stomached animals—pigs, mink, fish, and poultry.

Commercially produced, synthetic sources of each of the B vitamins are available. Also, the following natural feeds are excellent sources of most of the B vitamins: green pastures, green hays, yeast, distillers' solubles, and animal and marine products.

VITAMIN C (ASCORBIC ACID, DEHYDROASCORBIC ACID)

A dietary need for vitamin C is generally limited to humans, guinea pigs, monkeys, and certain fruit bats. Hence, there is no need to add this vitamin to most animal rations. However, since it has been demonstrated that some types of fish, such as the catfish, require dietary vitamin C when placed in stressed conditions, it may be advisable to supplement vitamin C for fish in intensive production.

VITAMIN IMBALANCES

Experiments have shown that the amounts needed of certain vitamins may be affected by the supply of another vitamin or of some other nutritive essential. Also, it is known that excess fortification of the animal's diet with certain vitamins may prove more detrimental than helpful. Thus, harmful imbalances should be avoided; and vitamins should be provided on the basis of recommended allowances. Also, when fortifying with vitamins, consideration should be given to the vitamins provided by the ingredients of the normal ration, for it is the total composition of the feed that counts.

UNIDENTIFIED FACTORS

In addition to the vitamins as such, certain unidentified or unknown factors are important in animal nutrition. They are referred to as *unidentified* or *unknown* because they have not yet been isolated or synthesized in the laboratory. Nevertheless, rich sources of these factors and their effects have been well established. A ration that supplies the specific levels of all the known nutrients but which does not supply the unidentified factors may be inadequate for best performance. There is evidence that these production factors exist in dried whey, marine and packinghouse by-products, distillers' solubles, antibiotic fermentation residues, yeasts, alfalfa meal, and certain green forages. There is also evidence that at least one unknown hatchability factor is in fish solubles and green forage. Most of the unidentified factor sources are added to the ration at a level of 1 to 3%.

ADDITIVES, IMPLANTS, AND INJECTIONS

More than 1,000 drug products are approved by the Food and Drug Administration (FDA) for use by livestock and poultry producers. This includes additives, implants, and injectables, along with other drugs that are used to fight diseases and protect animals from infections. Two other statistics which point up the important role of drugs in animal production are: (1) 8 out of every 10 animals raised for food in the United States receive some drugs during their lifetime; and (2) chemicals that regulate growth, modify the rumen's activity, and/or improve feed efficiency increase U.S. meat, milk, and egg production approximately 15% each year. Used properly, these drugs enable livestock producers to provide safe and wholesome meat, eggs, and milk to consumers at lower costs than would otherwise be possible. Used improperly, however, these drugs can be hazardous to consumers.

Consumers are very much aware of what is in their food. While they enjoy the price and supply benefits of modern food production technology, they want to be assured of the wholesomeness of the food they eat.

Thus, livestock producers have the unenviable task of choosing the right drug(s) to maximize rate and efficiency of production, while, at the same time, observing FDA regulations and protecting the consumer. Under such circumstances, they should carefully analyze all the information presented by each company in support of its product. Also, they should study the results of unbiased experimental work, as reported in both scientific literature and popular articles; and they should sound out reliable users of the product. Finally, in the United States, they must comply with FDA regulations.

Feed additives and implants constitute a diverse group—so diverse that it is difficult to classify all of them as to mode of action or function. For this reason, some are grouped whereas others are merely listed alphabetically in the sections that follow. **NOTE WELL**: Recommended additives or implants for each animal species, along with pertinent details relative to their use, are presented in the separate chapters devoted to each class of animal, Chapters 19 to 28 of this book.

Abortifacients

An abortifacient is a drug or other agent that induces abortion. In the livestock industry, the primary use of abortifacients is to abort feedlot heifers. Prostaglandins and prostaglandin analogues are the abortifacients of choice during the first 150 days of pregnancy; beyond 150 days pregnancy, additional products, such as dexamethasone or estradiol, may be used.

The following management options may be considered when feeding heifers of unknown pregnancy status:

1. Feed heifers the same as steers, and meet the calving problems (difficult births, and caring for newborn calves) as they occur.
2. Buy only open or spayed heifers, the supply of which is limited.
3. Pregnancy examine all heifers and use an abortive agent on the pregnant ones, according to directions.

By pregnancy testing and the use of an abortifacient, the termination of early pregnancy can be brought about. However, the cost and setback in performance may not justify such action.

(Also see Chapter 19, Feeding Beef Cattle, section on "Spayed Heifers.")

Additives That Enchance Market Value

Among the types of product-enhancing additives currently being used or tested experimentally are the following:

• **Xanthophylls and carotenoids in poultry**—Many consumers believe that a deep yellow color of broiler skin/shanks and egg yolks is indicative of top quality. Consequently, the poultry producer may receive a premium price for such cosmetically esthetic products. Xanthophylls or carotenoid additives are commonly used for this purpose. (See Chapter 24, Feeding Poultry, section on "Additives.")

• **More lean/less fat**—Several products for the purpose of producing more muscle/protein and less fat in beef cattle, lambs, hogs, and broilers are being tested.

• **Lower cholesterol**—Consumers are cholesterol conscious; so, scientists are developing and testing products that will lower the cholesterol in meat, milk, and eggs.

Additives That Physically Aid Digestion

In some types of livestock—notably poultry and ruminants—the physical characteristics of the feed can markedly alter its digestibility. Two products, which have received considerable attention as physical aids to digestion are grit for poultry and roughage substitutes for ruminants.

GRIT

Since poultry do not have teeth to facilitate grinding of feed, most grinding takes place in the thick-muscled gizzard. The more thoroughly feed is ground, the more surface area is provided for digestion and subsequent absorption. Hence, when hard, coarse, or fibrous feeds are fed to poultry, grit is sometimes added to supply additional surface for grinding within the gizzard. Additionally, grit serves to break down ingested feathers and litter which can sometimes lead to gizzard impaction. When mash or finely ground feeds are used, the value of grit is greatly diminished.

Oyster, clam, coquina shells, and limestone are sometimes used for grit. Being relatively soft and calcareous, they provide a source of calcium as they, too, are ground in the process. Gravel and pebbles have been used successfully as long-lasting sources of grit. Several granite products are available commercially.

ROUGHAGE SUBSTITUTES

Roughages are bulky, coarse feeds which are high in fiber (cellulose and related compounds) and thus low in digestibility; such as hay, straw, silage, wheat bran, corncobs, and cottonseed hulls. In human nutrition, roughage has been a long-standing means of alleviating constipation. In ruminants, roughage promotes rumination and rumen function and the production of milk of normal fat content. The capacity of feed sources of roughage to elicit these effects varies with the composition and physical character of the fiber source. Thus, there must be a balance between lignification and digestible cellulose; for example, one cannot feed plastic hay to a lactating cow and maintain milk fat because acetate-producing and cellulose-digesting bacteria must be maintained. Also, fine grinding and pelleting decrease the value of roughage for ruminants because of the loss of the scratch factor and reduced rumination.

The above statement, which has been well documented by research, clearly shows that ruminants require some form of roughage or scratch factor in their feed in order to maintain a healthy, functioning rumen and produce milk with normal fat content. However, as a source of energy, roughages are sometimes too expensive in comparison with concentrate feeds. Under such circumstances, ruminants may be fed high-concentrate rations.

The most common practice in feeding high-energy feeds is to incorporate in the ration some source of natural roughage, such as hay, cottonseed hulls, almond hulls, or ground corncobs.

(Also see Chapter 4, section on "Availability of Fiber"; and the section on "Buffers" in this chapter.)

Anthelmintics (Wormers; Vermifuges)

Anthelmintics are drugs used to control worms.

The prevention and control of parasitic infection is one of the quickest, cheapest, and most dependable methods of increasing production with no extra animals, no additional feed, and little more labor. This is important, for, after all, the farmer or rancher bears the brunt of this reduced production, wasted feed, and damaged products.

Knowing what kind of worm(s) is present in an animal is the first requisite to the choice of the proper anthelmintic. Since no one drug is appropriate or economical for all conditions, the next requisite is to select the right one; the one which, when used according to directions, will be most effective and produce a minimum of side effects on the animal treated. Coupled with knowledge of the kind of worm(s) present, an individual assessment of each animal is necessary. Among the factors to consider are age, pregnancy, illnesses and medications, and the method by which the drug is to be administered. Some drugs characteristically put animals off performance for several days after treatment, whereas others have less tendency to do so. Some drugs are unnecessarily harsh or expensive for the problem at hand, whereas a safe inexpensive alternative would be more suitable.

Each livestock establishment should, in cooperation with the local veterinarian, evolve with a parasite control program and schedule. It is recommended that several different wormers be used, and that they be rotated. Also, a schedule of treatments should be prepared, based on knowledge of the life cycles of the various parasites.

From time to time, new vermifuges, or wormers, are approved and old ones banned or dropped. When parasitism is encountered, therefore, it is suggested that the producer obtain from local authorities the current recommendation relative to the choice and concentration of the vermifuge to use. This information can be obtained from a county extension agent, entomologist, veterinarian, or agricultural consultant.

Antibiotics

Antibiotics are substances which are produced by living organisms (molds, bacteria, fungi, or green plants) and which have bacteriostatic or bactericidal properties. They are the most widely used of the microbial drugs.

In addition to their use as growth stimulators, antibiotics are used as nutritional stimulants to promote better feed efficiency in ruminants and swine, and to increase egg production, hatchability, and shell quality in poultry. They are also added to feed in substantially higher quantities to remedy pathological problems.

MODE OF ACTION OF ANTIBIOTICS

Numerous theories, each with convincing support, have been hypothesized with respect to the mode of action of growth-stimulating antibiotics.

One fact is well substantiated. Antibiotics are effective in controlling certain environmental stresses, as evidenced by the fact that animals raised under germ-free conditions exhibit no improved responses to them. Additionally, it has been observed that pigs raised in clean pens which had not previously housed other pigs, did not exhibit any improved responses to antibiotic supplementation. These facts indicate that exposure to everyday stresses may reduce performance. However, antibiotics are not a means whereby one can ignore good sanitation and management practices.

The following six mechanisms by which antibiotics increase rate of gain and feed efficiency have been suggested:

1. **Disease control.** There is evidence that antibiotics exert a "disease defense effect" by suppressing microorganisms which might otherwise produce subclinical diseases in the animals. Although the effect of the microorganisms might be so mild that there would be no clinical symptoms of disease, it could still slow the animal's growth.

2. **Nutrient-sparing effect.** When antibiotics are fed, the populations of microorganisms in the animal's digestive tract change. It follows that if the number of microorganisms which manufacture nutrients which the animal can use increases, or the number of microorganisms which compete with the animal for nutrients decreases, the animal will be able to grow more on the same amount of feed. Studies indicate that antibiotics have a sparing effect on some vitamins and amino acids.

3. **Metabolic effect.** According to this theory, there is a "metabolic effect," by which the antibiotic affects the body functions. Since some of the antibiotics used as growth promotants are not absorbed from the digestive tract into the animal's body, they could not act in this way. However, other antibiotics could exert a metabolic effect.

4. **Feed and water intake.** Antibiotics usually increase feed and/or water intake.

5. **Toxic waste products or toxins.** Antibiotics may inhibit the growth of organisms which produce toxic waste products or toxins.

6. **Digestion and absorption.** Antibiotics may improve the digestion and subsequent absorption of certain nutrients. Thus, it has been noted that the intestinal wall of animals fed antibiotics is thinner than the intestinal wall of those not fed antibiotics, which may promote better utilization of nutrients.

More than likely, all six modes of action apply to antibiotics in general, but not to any specific antibiotic—for antibiotics differ. Also, there are dose-related responses. Since antibiotics are most effective in stressful conditions where animals are more likely to be exposed to disease and more susceptible to it, the disease-control effect is probably the most important of the six modes of action.

IMPORTANCE AND USE OF ANTIBIOTICS IN ANIMAL PRODUCTION

Antibiotics are used in the following two ways in livestock and poultry production:

1. **Low levels in feeds.** Low, subtherapeutic doses of antibiotics are included in livestock and poultry feeds to increase growth rate, and/or feed efficiency, and help prevent bacterial diseases. The effects of antibiotics on rate of gain or feed efficiency of each animal species is presented in the respective chapters devoted to each class of livestock.

2. **High (therapeutic) levels in feeds.** High levels of antibiotics are used to treat diseases, just as they are in human medicine. Used for short periods, high levels have been quite effective in treating anaplasmosis and controlling shipping fever in cattle, bacterial enteritis in swine, and respiratory diseases, diarrheas, fowl cholera, typhoid, and breast blisters in poultry. Also, high levels of antibiotics have been useful for preventing and treating stresses associated with transporting animals and their adjustment to a new environment.

Economically, antibiotics are of benefit to both producers and consumers of animal foods. They reduce the cost of animal production; therefore, they reduce the retail price of meat and poultry (and to a lesser extent, milk and eggs). The Council for Agricultural Science and Technology (CAST) is authority for the statement that banning the use of penicillin and tetracyclines in animal feeds would make for added food costs to U.S. consumers of more than $3.5 billion per year, in the short term, if no substitutes were available.[1]

SAFETY AND FUTURE OF ANTIBIOTICS AS FEED ADDITIVES

Although the practice of low level (subtherapeutic) feeding of antibiotics has been routine for decades, it has recently come under fire. The concern: Are animals serving as the *training ground* for bacteria that, in effect, learn how to disarm the antibiotics that we use against them? Are antibiotic-laced feeds encouraging the growth of antibiotic-resistant bacteria, which can be transmitted to people and cause sickness?

Those favoring low-level feeding of antibiotics (at a dose much lower than is required to treat an illness caused by bacteria) claim (1) that antibioitics are essential to raising healthy animals under today's crowded, confined, stressful conditions; and (2) that banning antibiotic feeding would cost American consumers hundreds of millions of dollars in higher food prices yearly. Opponents counter (1) that human health is being sabotaged by the feeding of antibiotics, and (2) that antibiotics are being substituted for poor livestock sanitation and management.

Scientists generally agree that the use of antibiotics (whether in livestock or people) results in the proliferation of bacteria that are resistant to one or more antibiotics. Currently, for example, bacteria causing everything from diarrhea to sore throat to meningitis in people, can be killed only when antibiotics are given in much higher doses than were originally required, and in some cases they are completely unaffected by the antibiotics that once destroyed them. Since some bacteria are common to both animals and humans, the concern is that disease organisms resistant to treatment by antibiotics will develop in animals and spread to humans. But scientists disagree on the key issue involved in the low-level feeding of antibiotics: Are antibiotic-resistant organisms causing diseases in people as a direct consequence of feeding antibiotics to animals?

---

[1] Hays, Virgil W., et al., *Antibiotics In Animal Feeds*, Report No. 88, Council for Agricultural Science and Technology (CAST), Ames, Iowa, 1981.

Although the cause is elusive, the fact remains that more and more drug-resistant Salmonella bacteria are infecting animals and moving up through the food chain to humans, as is illustrated in Fig. 13-1.

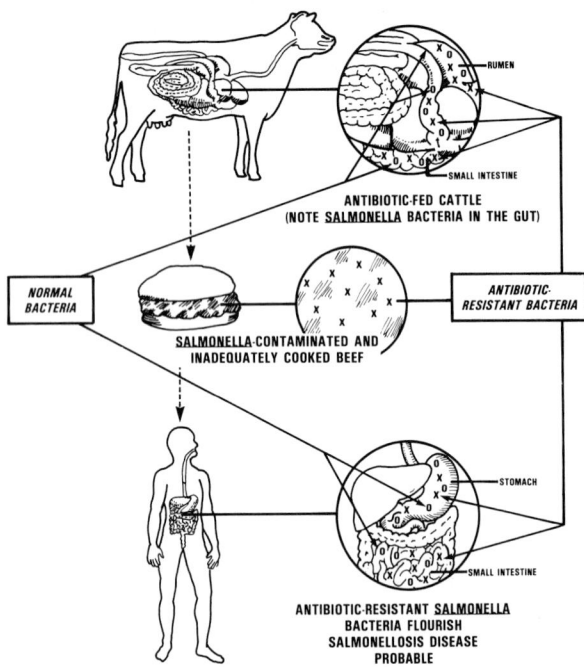

Fig. 13-1. Theory of antibiotic resistance—how it could happen. This shows how antibiotic-resistant Salmonella could cause Salmonellosis disease in humans. The steps: (1) antibiotic-fed cattle produce antibiotic-resistant Salmonella bacteria; (2) at slaughter, Salmonella-contaminated beef may be produced; and (3) when Salmonella-contaminated beef is inadequately cooked and consumed by people, antibiotic-resistant Salmonella may flourish and produce Salmonellosis disease.

(**NOTE WELL**: The U.S. meat supply is the world's safest. The above figure is for the purpose of illustrating a theory. M.E.E.)

Since there is risk to human health from the use of antibiotics in feed, the careful monitoring and periodic review of the human health impact of the subtherapeutic use of antibiotics in animal feeds is prudent. But decisions should be based on scientific facts rather than political expediency.
• **Some facts pertinent to the antibiotic issue**—The following facts are pertinent to the issue of antibiotics in feed:
1. *Old* **antibiotics still potent growth promotants.** Poultry scientists at the University of Wisconsin have published papers documenting the continued effectiveness in chicken and turkey feeds of penicillin and tetracycline, two *old* antibiotics that have been around and used since the early 1950s. The effects of the *old* antibiotics are comparable or superior to the *new* antibiotics, although it is known that bacteria resistant to penicillin and the tetracyclines have been present in the environment of chickens and turkeys for more than 30 years.
2. **Antibiotics specific for animals are increasing.** A number of new antibiotics specifically for animals have been developed, and more will come. These products improve livestock performance in rate of gain and/or feed efficiency, but have no medical applications, and, therefore, pose no health hazard with regard to bacteria becoming resistant to them. Among such antibiotics designed specifically for animals, but which will not cause resistance to antibiotics in humans are virginiamycin, bambermycin, bacitracin, and the ionophores, monensin and lasalocid.

Antioxidants

Antioxidants are compounds that prevent oxidative rancidity of polyunsaturated fats. It is important that rancidity of feeds be prevented because it may cause destruction of vitamins A, D, and E, and several of the B-complex vitamins. Also, the breakdown products of rancidity may react with the epsilon amino group of lysine and thereby decrease the protein and energy values of the ration. These effects can be prevented by inclusion in the ration of an effective antioxidant such as ethoxyquin (6-ethoxy-1, 2-dihydro-2, 4-trimethylquinoline), BHT (butylated hydroxytoluene), or BHA (butylated hydroxyanisole); with these products used singly or in combination with each other.

Vitamin E can serve as an antioxidant both in the feed and in the cell of the animal ingesting the feed. Such antioxidants as ethoxyquin, BHT, or BHA are unable to prevent peroxidation within the cell; consequently, they cannot reduce the dietary requirements for vitamin E.

Oxidation reactions are accelerated by high temperature, light (ultraviolet and blue), ionizing radiation, peroxides (including oxidized fats), lipoxidase enzymes, organic ion catalysts (hemoglobin), and trace metals (copper).

Oxidation reactions are inhibited by refrigeration, exclusion of light, exclusion of oxygen, destruction of enzymes, metal deactivators, and antioxidants.

Arsenicals

Arsenilic acid and sodium arsanilate are the most widely used growth-promoting arsenicals. They are FDA-approved for use alone or in certain drug combinations for chickens, turkeys, and swine. When used alone according to directions, they increase rate of gain and improve feed efficiency of chickens, turkeys, and swine; improve pigmentation in growing chickens and turkeys; increase egg production in layers; and prevent coccidiosis in chickens and dysentery in swine. Arsenic compounds are toxic; hence, they should be used with care and according to manufacturer's directions.

Bloat Control Products

Three products are approved by FDA for bloat control; namely, poloxalene (trade name, Bloat Guard), oxytetracycline (trade names, Terramycin or Neo-Terramycin), or laureth-23 (Enproal Bloat Blox). (See Chapter 5, Table 5-1, Nutritional Diseases and Ailments—Bloat [Feedlot] and Bloat [Pasture]).

Also, two ionophores (lasalocid [trade name, Bovatec] and monensin [trade name, Rumensin]) inhibit gas formation and decrease methane production in the rumen, thereby reducing feedlot bloat. (See section on "Ionophores" in this chapter.)

These products are effective preventives when used according to manufacturers' directions, but they should be accompanied by proper management.

Buffers

Buffers are substances which lessen the change in hydrogen ion concentration produced by adding acids or alkalis to ruminant rations. The counterpart substances for humans are called *antacids*.

When used in beef, dairy, and sheep rations, a buffer chemically maintains a balanced pH in the animal's digestive system. The pH, which is a measure of the level of acid and alkali in a solution, is expressed in numbers from 0 to 14. Acids are on the low end of the scale—below 7; alkalis (bases) are on the upper end—7 or above. A rumen pH between 6.2 and 6.8 is generally considered best for optimum digestion and rumen function in ruminants, although this will vary with feeding regimen and time after feeding.

When the rumen pH drops from a normal of about 6.5 to a range of 5.0 to 6.0, milk fat decreases; and when it goes on down to 4.0, there is acidosis.

Ruminants were developed as forage utilizers. The unique arrangement of four-compartment stomachs enables them to consume fibrous feedstuffs which are digested and used to meet the energy demands of the animal. This is accomplished by a large microbial population that inhabits the rumen. However, when today's ruminants are in high production, especially feedlot cattle and lambs and high-producing dairy cows, they are routinely given high-concentrate and low-forage rations to maximize production. Characteristically, such rations lower the pH of the rumen. The lower ruminal pH of high-concentrate rations can be attributed to decreased saliva secretion, increased energy intake, and increased rate of carbohydrate degradation. The saliva contains bicarbonates and phosphates that buffer the ruminal contents and help maintain the pH between 6.5 and 7.0. The increased feed intake of high-concentrate rations along with more rapid carbohydrate degradation make for more rapid feed conversion to volatile fatty acids in the rumen.

Fig. 13-2 summarizes the important changes encountered in rumen conditions. Obviously, this chart cannot be applied to any specific case; rather, it portrays the general findings of research. When high-roughage rations are fed, the pH of rumen fluid tends to remain relatively high, with acetic acid as the predominant fatty acid. As more concentrates are fed, propionic acid and butyric acid tend to increase relative to acetic acid. Finally, as the concentrates are increased still further, lactic acid is formed. In some instances, increased propionic acid levels accompany the increased lactic acid concentrations. Presumably, there is an optimum ratio of these acids based upon the ultimate production goal, i.e., milk or meat. Presumably, too, the effect of buffer additions depends on (1) the status of the rumen prior to addition of the buffer and (2) the amount and kind of buffer added.

Reducing the particle size of the feeds by chopping, grinding, and/or pelleting also decreases ruminal pH, primarily because of the reduction in saliva secretion and a more rapid breakdown of feed to volatile fatty acids. Likewise, fermented feeds (silage or haylage) increase the *acid load*.

Also, a sharp drop in pH is noted when animals are switched from high-roughage feeds to high-concentrate feeds without allowing a period of gradual adjustment.

So, high-concentrate rations, smaller feed particle size, fermented feeds, and rapid shifts from high-roughage to high-concentrate rations all lower the pH of the rumen. Since many of the microorganisms in the rumen cannot tolerate low pH concentrations, the normally heterogeneous, balanced population of the microbes becomes skewed, favoring the acidophilic (acid-loving) bacteria. This condition often leads to upsets. However, the addition of feed buffers can prevent dramatic changes of pH in the rumen, thereby stabilizing the microbial population.

COMMON BUFFERS AND THEIR USES

Many buffer products are available. However, the following six products make up the most common ingredients: sodium bicarbonate, magnesium oxide, sodium bentonite, sodium sesquicarbonate, limestone, and whey. The recommended feeding level for the four most used of these products is given in Table 13-1.

TABLE 13-1
RECOMMENDED FEEDING LEVELS FOR COMMON BUFFERS

Buffer	Percent of total ration[1]	Percent of Grain Mix	Pounds/ cow/day
	(%)	(%)	(lb)
Sodium bicarbonate	0.6-1.0	1.2-2.0	0.3-0.5
Magnesium oxide	0.2-0.35	0.4-0.7	0.1-0.2
Sodium bentonite	2.0-3.0	4.0-6.0	1.0-1.5
Sodium sesquicarbonate	0.6-1.0	1.2-2.0	0.3-0.5

[1]Assumes about 50% forage and 50% grain on a dry matter basis in a total mixed ration. Note that recommended buffer levels are given for either total ration (column 2) or grain mix (column 3).

The common uses of buffers follow:
- **Buffers for feedlot cattle and feedlot lambs**—High-grain rations fed to feedlot cattle and lambs make for rapid and efficient weight gains, but they may also result in rumen acidity. Too much acidity in the rumen can depress appetite, reduce feed efficiency, and lower gains.

High-energy grain rations or acidic, ensiled crops (silage or haylage) can cause cattle and sheep to produce more acid than their own supply of sodium bicarbonate can neutralize naturally. A buffer supplement may be the answer. It can help reduce excess acid and maintain the normal balance of acidity and alkalinity.

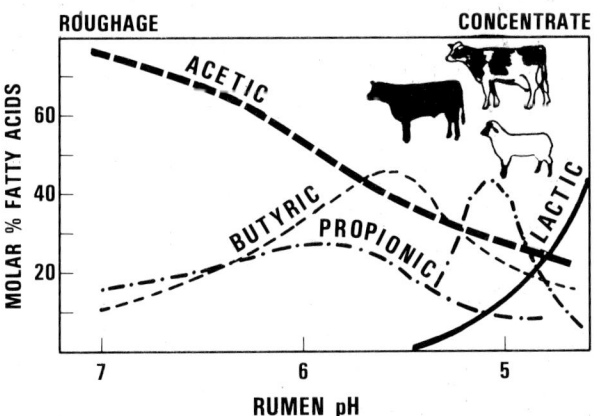

Fig. 13-2. Interrelationships among volatile fatty acids (acetic, butyric, and propionic), lactic acid, and pH during transition from high-roughage to high-concentrate rations. (Source: Adapted by the authors from *Buffers in Ruminant Physiology and Metabolism*, Edited by M. S. Weinberg and A. L. Sheffner, Published by Church & Dwight, Inc., New York, N.Y., 1976, p. 92.)

• **Buffers for dairy cattle**—Studies indicate that buffers can replace a portion of the forage or fiber in the ration of lactating dairy cows even though forage is an essential component thereof. The conditions under which buffers will be of greatest benefit to the dairy cow are: (1) when large amounts of concentrates are fed, as in early lactation; (2) when a fermented forage (such as silage or haylage) is the major or only forage in the ration; (3) when the particle size of the ration has been reduced by chopping, grinding, or pelleting, with the result that it increases the rate of ruminal fermentation and depresses saliva secretion and buffering capacity; (4) when cows are suddenly switched from high-forage to high-concentrate rations; (5) when the milk fat test is low, or (6) when cows are off-feed due to feeding rapidly fermentable rations.

None of the additives will increase butterfat above normal levels. The bicarbonates or magnesium oxide lower palatability and feed consumption, resulting in lower milk production; however, total fat may be greater.

Despite the role and importance of buffers, feeding lactating cows rations with adequate fiber (a minimum of 17% crude fiber or 21% acid-detergent fiber in the ration dry matter) is the best preventive of depressed fat known at the present time.

• **Buffers for poultry**—There is some indication that sodium bicarbonate will reduce the incidence of rough shells and improve eggshell quality during hot weather, but the evidence is not conclusive.

Sodium bentonite is often added to pelleted feed for poultry in order to improve the hardness of the pellets. Additionally, it absorbs water from processed pellets, tends to reduce wet droppings, and improves the growth of young chicks.

The bottom line: Any economic analysis of buffer feeding should take into consideration the net returns based on improved production from feeding buffers and the price of the buffer.

Chemotherapeutics

Chemotherapeutics are organic compounds with bacteriostatic or bactericidal properties similar to those of antibiotics. But, unlike antibiotics, these compounds are produced chemically rather than microbiologically.

The chemotherapeutics are used primarily for disease control. However, arsenicals, carbadox, furaxolidone, and roxarsone are also used as growth promotants and to improve feed conversion efficiency.

The arsenicals and the sulfas, two of the most widely used chemotherapeutics, have been around for a very long time. Pertinent information about each of them follows:

• **Arsenicals**—Various compounds containing arsenic, when used at carefully-controlled levels, have been found to increase rate of gain and improve feed efficiency in chickens, turkeys, and swine; and to prevent coccidiosis in chickens and turkeys and prevent dysentery (bloody scours) in swine.

• **Sulfonamides**—The therapeutic use of the sulfonamides precedes that of antibiotics. The coming of the antibiotics, many of which were more specific in action and easier to administer than the sulfas, markedly reduced the use of the sulfas. In time, however, bacteria emerged that were resistant to many of the antibiotics, which led to a resurgence in the use of the sulfonamides, in feed or water.

All went well until the 1960s and the early 1970s, when the U.S. Department of Agriculture became aware of sulfa residues in pork carcasses. Then, in the late 1980s, a preliminary FDA study showed that massive doses of sulfamethazine caused thyroid tumors in laboratory mice. Then, FDA followed with a warning that sulfamethazine, a drug used subtherapeutically to prevent respiratory problems and as a growth promotant in swine, may be carcinogenic.

Animal products are deemed to be in violation of sulfa residue regulations if levels of 0.1 ppm sulfa are found in the muscle, liver, kidney, eggs, or milk. The 0.1 ppm level is based on toxicological studies with rats and dogs and provides a 2,000-fold safety factor for humans.

Research at the University of Kentucky indicates that as little as 1 g of sulfa per ton of complete feed can cause 100% volatile liver residues and 63% kidney condemnations. This means that as little as ¼ teaspoon of sulfa per ton of complete feed can result in pork carcasses that are in violation.

The sulfonamides are also capable of premise contamination. If excreted in the feces or urine, they may contaminate the premises. Thus, if the pens are not thoroughly cleaned between groups of pigs, untreated animals reared in these pens may show sulfonamide residues.

In order to alleviate sulfa residues in excess of tolerance levels, it is essential (1) that producers observe pre-slaughter withdrawal periods and avoid contaminated premises and animals, (2) that producers use one of the new tests to detect the presence of sulfa in the serum or urine of hogs, and (3) that feed companies avoid cross contamination of sulfa-containing feeds. For the latter reason, feed companies should not use the powder form of sulfa, because it can attach itself more easily to the sides of mixers, bins, augers, and feeders than granulated sulfa products.

Although many of the antibiotics and chemotherapeutics have similar properties, the expertise of the nutritionist should be sought when increased growth rate and improved feed efficiency is the primary objective, and the veterinarian should be consulted where disease control and/or treatment is involved. Correctly (1) determining the objective and/or problem and (2) matching the additive to the objective and/or problem are the first and most important decisions when choosing a feed additive.

NOTE WELL: Chemotherapeutics for each class of livestock for which approved, along with pertinent information relative to their use, and the results to expect, are presented in the separate chapters devoted to each species in this book, Chapters 19 to 28.

Copper (Cu)

Copper is widely used as a swine feed growth additive in Europe, where numerous experiments have shown growth responses and increased feed efficiency comparable to those obtained with antibiotics.

A supplemental level of 175 to 200 ppm copper is recommended. No toxicity problems are expected when copper is added at a level up to 250 ppm in a well-mixed, balanced ration adequate in zinc and iron. Levels above 250 ppm may be toxic. As with antibiotics, copper is more effective with younger pigs than older hogs. European reports indicate that copper will increase rate of gain by about 8.0% and decrease feed per unit of gain by about 5.0%.

Copper carbonate, copper chloride, copper oxide, copper sulfate, copper glycinate, and copper methionine are effective sources of copper, with copper sulfate the preferred form. They cost less than antibiotics.

Electrolytes

An electrolyte is a substance which when dissolved in water enables the resulting solution to conduct an electric current. The most common electrolytes in the animal body are salts of such minerals as sodium, potassium, magnesium, calcium, phosphorus, sulfur, and chlorine.

The most commonly used electrolyte solution is 0.9% sodium chloride, which is also called *physiological saline* because it has the same solute strength (tonicity) as the body fluids. Some solutions also contain potassium and magnesium salts because these minerals are also highly essential in the maintenance of a variety of vital functions.

Solutions of electrolytes are administered when it is necessary to replace the mineral salts and water that have been lost under circumstances such as dehydration, diarrhea, hemorrhage, excess urination, and vomiting.

The volume of the electrolyte administered must be adequate, with the veterinarian determining the amount. Up to 7 to 10% of the body weight may be administered over a 24-hour period.

The oral route of administering the electrolyte should be chosen whenever the condition of the animal so permits. The IV route of administration is indicated in life-threatening situations (severe vomiting, diarrhea, impending circulatory failure). Subcutaneous or intraperitoneal administration may be used effectively if circulation is adequate to ensure absorption.

Flavoring Agents

If the flavor is not enticing to the animal, the feed will not be eaten; hence, no matter how important the feed may be nutritionally, it won't contribute to the ration unless consumed.

Flavoring agents are feed additives that are designed to increase palatability and feed intake.

There are four primary taste sensations—sweet, bitter, salty, and sour. Additional feed flavors are the result of the sense of smell. In humans (and perhaps in farm animals, too), smells arouse emotions; they evoke fear, sadness, disgust, longing, love, and/or passion.

Taste and smell play an important role in the sensory apparatus of animals. Most animals will eat only what they can first smell. Then, if it tastes good, they will consume it with relish.

Today, chemists can make chemicals in the laboratory which, alone or in various combinations, can imitate many of the natural feed flavors. In many cases, the synthetic flavors are superior to natural flavors in terms of (1) withstanding processing, (2) cost, (3) availability, and (4) consistent quality.

Hand in hand with improved animal nutrition and care have come feed palatability problems. Antibiotics, wormers, growth stimulants, and waste products often taste bad. Besides, today's animals are in forced production and stressed. There is a place, therefore, for feed flavors and aromas that will overcome palatability problems, attract animals to feed and keep them on feed, and increase feed consumption and performance. To meet this need, a wide array of feed flavor products is on the market; and they are available in both dry and liquid forms, usually under alluring brand names. However, additional research work is needed in order to establish the true value of flavoring agents in feeds.

Hormone and Hormonelike Compounds

Hormones are chemicals released by a specific area of the body that are transported to another region within the animal where they bring about a physiological response. The word *hormone* comes from the Greek word *hormon,* which means "to spur on, to set in motion, to excite to action." All these phrases are very descriptive of the hormones.

Scientists have identified many of the hormones of the body and have successfully synthesized several hormone or hormonelike compounds which produce the same physiological responses as those from naturally produced hormones. In some cases, this has made it possible to administer these products to obtain a specific response; for example, increased growth, milk production, or meat production.

Each hormone or hormonelike compound elicits a different response. In separate sections that follow, the nutrition-related responses of each of the following products are presented:

1. Somatotropin (growth hormones), including the bovine somatotropin (BST) for dairy cattle and porcine somatotropin (PST) for swine.

2. Melengestrol alcetate (MGA).

(Also see Chapter 19, section on "Growth Stimulants and Implants.")

SOMATOTROPIN (GROWTH HORMONE)

Initially, this product was known as *growth hormone.* Today, the term *somatotropin* is preferred, because the general public associates the word *hormone* with estrogens, progestins, and testosterone, all of which are steroids, whereas growth hormone or somatotropin is a peptide (protein) hormone.

Somatotropin is secreted naturally by the anterior pituitary gland, located in the skull at the base of the brain of all vertebrates. It is a peptide, and it is species-specific. Thus, bovine somatotropin (BST) and porcine somatotropin (PST) are distinctly different.

Pure somatotropins, obtained from the pituitaries of many species, have been around for many years, but they were far too costly to use routinely. For example, it required the pituitaries of 200 cows to obtain enough bovine growth hormone to treat one cow for one day. However, the "new biology," using recombinant DNA technology (genetic engineering), has made it possible for scientists to identify the gene responsible for producing the bovine growth hormone, isolate it from experimental animals, splice it onto bacteria, and use the altered bacteria to synthesize somatotropins cheaply and in quantity. In turn, research to find new and exciting uses for somatotropins has been accelerated.

Many questions must still be answered before growth hormones become commercially available and widely used, including their long-term effect. Also, a practical method of administering somatotropin to animals must be devised. Since it is a protein, as is insulin, it cannot be taken orally in the feed because it would be digested in the gut; and, unlike commercially-available beef cattle implants (Compudose, Ralgro, Steer-oid and Heifer-oid, and Synovex), a somatotropin implant would break down too quickly. Therefore, a specially protected implant or long-lasting, slow-release injection is required. Finally, as with all such products, FDA approval must precede commercial use of growth hormones.

Currently, scientists are working feverishly away perfecting and testing growth hormones. For dairy cattle, somatotropin is being tested as a stimulator of milk production, and as a growth promotant for replacement heifers. For beef cattle, sheep, swine, and poultry, somatotropin is being tested as a promotant to improve gains and feed efficiency, and, hopefully, produce leaner meat.

More recently, researchers at Michigan State University reported preliminary studies with a growth hormone-releasing factor (GRF), which increased milk production by 23%. GRF is a natural hormone, identical to the hormone produced by the cow. Unlike bovine growth hormone (BST), which adds growth hormone directly to the cow, GRF stimulates the cow's pituitary gland to release more growth hormone, which, in turn, increases milk production.

The full impact of these products on the livestock industry won't be felt until the 1990s—and beyond.

It appears, that growth hormones will be one of the early significant commercial successes of DNA technology; and that they will usher in a new era in livestock production.

Bovine Somatotropin (BST) for Dairy Cattle

In 1985, Bauman and Eppard of Cornell University, and DeGeeter and Lanza of the Monsanto Company in St. Louis, reported that responses of high-producing dairy cows to long-term treatment with pituitary somatotropin and recombinant somatotropin.[2] Daily injection of graded doses of somatotropin were started 84 days after calving and continued for 188 days. Cows received either 13.5, 27, or 40.5 mg per day of genetically-engineered bovine somatotropin, or 27 mg per day of pituitary-derived bovine somatotropin, while a control group was injected with a placebo. Cows injected with either type of somatotropin produced from 16 to 41% more 3.5% fat-corrected milk (FCM) than the control group. Cows injected with somatotropin consumed more feed on a body weight basis than the control group, but their relative feed energy efficiency was higher. The increased feed efficiency could be accounted for by the smaller proportion of total intake required to meet the cows' maintenance energy requirements.

This report showing the dramatic effect of the hormone on milk production sparked much controversy. Some dairy groups questioned the need or desirability to develop new technology for increasing milk production at a time when there are tremendous milk surpluses, nationally and worldwide; lawsuits were filed to prevent further research with bovine somatotropin and institutions conducting such research were picketed by activist groups. Also, fear was expressed that the commercial application of growth hormone technology would result in the demise of the family dairy farm because of fewer and fewer cows being needed to produce the nation's milk supply. Even some U.S. Congressmen got into the act; an informational House Agriculture Subcommittee hearing on genetically engineered bovine somatotropin was held on Capitol Hill.

Before the commercial use of bovine somatotropin can become a reality, more research is needed; a practical method of administering the product must be developed; and FDA approval must be secured. But, given time, there is every reason to expect that all these requisites will be forthcoming.

Some additional pertinent information relative to bovine somatotropin follows:

• **Historical**—Somatotropin is not new. Cows have been producing it ever since there were cows, and dairy researchers have been studying it since the late 1930s.

In the early days of BST research, which began in the U.S.S.R. about 1937, scientists showed that the growth hormone (then known as pituitary extract) would boost mild production of dairy cows. But the only source of the hormone was the pituitary gland; so, supplies of the product were extremely short and incredibly expensive. But genetic engineering has changed this! Scientists reprogrammed bacteria to produce large quantities of BST.

• **Human safety**—BST is a natural protein produced by all cows, which is always present in milk. It is not a steroid or sex hormone; and it is not related to the steroid hormones which are used as growth promotants. There is no evidence that the BST normally found in milk is harmful when consumed by humans. Moreover, milk levels do not increase significantly following treatment. As a peptide, when the BST molecule is consumed by humans, it is broken down and inactivated in the digestive tract in the same manner as any other protein. Both the U.S. Food and Drug Administration and England's Milk Marketing Board have approved the sale of milk from BST-treated cows. Thus, there does not appear to be any negative effect on human health.

• **How to administer**—Because BST is a peptide and would be broken down in the digestive tract, it cannot be fed. But daily injections don't seem practical. (In the experimental stage, it was injected, daily, in the hip area.) So, a long-lasting time-released injection or implant appears to be the answer.

• **Increased feed efficiency and milk production**—In a commercial dairy and over a full lactation, bovine growth hormone may be expected to increase feed efficiency from 10 to 20%, and milk production by 10 to 25%. But scientists still don't know exactly how it works.

• **Growth rate of heifers**—Bovine growth hormone will increase the growth rate of heifers by 8 to 10% and stimulate the development of secretory tissue in the mammary gland.

• **No adverse effect on cows**—It appears that the growth hormone is a safe product to use on dairy animals; treated cows have about the same somatic cell count, disease incidence, post-calving behavior, and rebreeding performance as untreated cows. However, because of their higher production, treated cows will require greater feed intake and more intensive management and observation.

• **Sire evaluation**—Future sires may require two proofs, one developed with and the other developed without hormonally-treated daughters; or correction equations may be developed to equalize bulls regardless of how the proof was developed.

• **Progress can be slowed, but not stopped**—In the future, higher milk yields per cow and fewer dairy cows in the U.S. are inevitable, with or without bovine growth hormone. This will come about as a result of improved genetics, feeding, and management. However, use of the bovine growth hor-

---

[2]Bauman, Dale E., et al., Responses of High-Producing Dairy Cows to Long-Term Treatment with Pituitary Somatotropin and Recombinant Somatotropin, Journal of Dairy Science, Vol. 68, No. 6, 1985, p. 1352.

mone will speed the process. In this connection, it is noteworthy that the following developments were products of research: artificial insemination, balancing rations, somatic cell testing, sire proving formulas, new milking systems, and various disease control measures.

• **Progressive dairymen will survive**—The survivors of these changes are likely to be dairymen who adopt technology which increases production efficiency. Milking fewer cows with higher milk yields and higher efficiency appears to be in the best interest of both producers and consumers in the future.

• **Price supports are more responsible for dairy surpluses than new technology**—Government policies which maintain price supports above market-clearing levels and prevent milk prices from fully adjusting to increased supplies are more responsible for dairy surpluses than new dairy technology.

Porcine Somatotropin (PST) for Swine

Porcine somatotropin (PST) is the scientific name for the growth hormone in swine, which is the counterpart of bovine somatotropin (BST) in dairy cattle; and it appears to be equally as effective. Both products are normally produced by the anterior pituitary at the base of the brain. But each is species specific; that is, PST works only on swine, and BST works only on cattle. PST is present in all pigs. It stimulates protein synthesis and growth in most tissues of the body, but it causes breakdown of fat deposits in adipose tissue.

Scientists have known for sometime that providing extra somatotropin to a pig will cause it to grow more rapidly and produce leaner pork. However, it was not possible to take advantage of this knowledge practically, because the only way to get porcine somatotropin was to isolate it from the pituitary glands of slaughtered hogs, which was extremely expensive because of the very low yield of hormone from each hog. But new techniques in molecular genetics (gene splicing, by inserting the pig gene that causes production of somatotropin into the genetic material of the bacteria) make it possible for bacteria to make large quantities of porcine somatotropin. As a result, it is now possible to utilize commercially the positive effects of extra somatotropin. When injected, lactating sows produce more milk and wean heavier pigs; and growing-finishing hogs grow faster and more efficiently, and produce leaner pork. Under field conditions, it is estimated that the use of PST will produce 15 to 20% (1) more rapid gain, (2) greater feed efficiency, and (3) increase in lean muscle.

A major deterrent to the commercial use of PST is lack of an appropriate method of administering the product to swine. Daily injections, as used in early experimental studies, are not practical in commercial hog production. A suitable method of administering PST will be developed; the only question is when and by whom. Also, FDA approval will be necessary.

MELENGESTROL ACETATE (MGA)

Melengestrol acetate (MGA) a synthetic progestogen hormone, is approved by FDA for use as a feed additive for nonpregnant heifers.

Melengestrol acetate is similar in structure and activity to progesterone, the naturally-occurring hormone of pregnancy. It suppresses estrus (heat) and ovulation and promotes growth.

MGA is fed at a very low level to feedlot heifers—0.25 to 0.40 mg/head/day. On the average, it will increase daily rate of gain by 10% and feed efficiency by 6%. A 48-hour withdrawal period prior to slaughter is required.

MGA also appears promising for use in synchronizing estrus as well as inducing estrus in non-cycling females.

Implants

An implant is a small pellet that is deposited underneath the skin behind the ear of an animal for the purpose of promoting growth.

The idea of implants evolved from watching the response of animals to alfalfa. For years, livestock producers have known that alfalfa produces faster gains and higher production than other feeds with similar nutrient content. This phenomenon prompted researchers to attempt to isolate the factor in alfalfa responsible for stimulating growth and production. It was found that alfalfa contained substances that had estrogenic activity. Subsequent research led to the development of implants that utilized sex hormones as growth promotants.

The presently available and FDA-approved growth implants are: Compudose, Finaplex, Ralgro, Steer-oid, Heifer-oid, and Synovex-S, Synovex-H, and Synovex-C.

Except for the implant, Finaplex (trenbolene acetate [TBA]), all implants contain the estrogen, estradiol, or an estrogenlike compound. Although their mode of action is not entirely clear, estrogens seem to affect protein synthesis indirectly by altering the animals' endocrine system. Acting primarily on the anterior pituitary gland, estrogen implants increase growth hormone production, which stimulates skeletal muscle growth. Some research indicates that estrogens may also alter the production of other hormones, such as insulin, which might also contribute to additional protein deposition.

Finaplex (TBA) is a synthetic analog of testosterone. It appears to increase growth and protein deposition by acting directly on skeletal muscle and other tissues.

NOTE WELL: Implants for beef cattle, along with pertinent information relative to their use and the results to expect, are presented in Chapter 19, Feeding Beef Cattle, in the section on "Growth Stimulants and Implants."

Ionophores

Ionophores are feed additives that change the metabolism within the rumen by altering the rumen microflora to favor propionic acid production.

Currently, two ionophores—Bovatec (lasalocid) and Rumensin (monensin)—are approved for cattle. But both products are antibiotics that have been around for a long time. However, they were initially approved as anticoccidial drugs for poultry; monensin was marketed under the trade name Coban, and lasalocid as Avatec. There are at least 76 known polyether ionophores; so, more than likely additional ones will be approved as feed additives.

The following facts are pertinent to the use of ionophores:
• **Bovatec (lasalocid)**—In 1982, Bovatec (lasalocid) was approved by the Food and Drug Administration for use in feedlot cattle, both to improve feed efficiency and daily gain. Lasalocid is a polyether antibiotic produced by *Streptomyces lasaliensis*. Research has shown that lasalocid is an effective

coccidiostat in both cattle and sheep. Lasalocid is toxic to horses, but higher doses are required than of monensin to cause death.

- **Rumensin (monensin)**—Rumensin, an antibiotic fermentation product produced by a strain of *Streptomyces cinnamonensis,* alters the metabolism within the rumen. Extensive testing has shown Rumensin to be a highly effective improver of feed efficiency in feedlot cattle and in stocker and feeder cattle and beef and dairy replacement heifers on pasture; Rumensin-fed cattle eat less feed per day but gain about the same amount of weight per day as cattle not fed Rumensin. Hence, higher feed efficiency makes for higher profits.

NOTE WELL: Ionophores for each class of livestock for which approved, along with pertinent information relative to their use and the results to expect, are presented in the separate chapters devoted to each species in this book, Chapters 19 to 28.

Isoacids (IsoPlus)

Isoacids provide three branched-chain fatty acids (isobutyric acid, isovaleric acid, and 2-methyl butyric acid) and valeric acid—the same fatty acids that are made by ruminant bacteria and are present naturally in the rumen of cattle. Isoacids are essential for the growth of some rumen organisms that digest fiber. Adding isoacids to the cow's ration results in higher milk production.

The U.S. Food and Drug Administration has approved a blend of isoacids as a feed additive for lactating cows, sold as the calcium salts, under the trade name IsoPlus, manufactured by Eastman Chemical Co., a division of Eastman Kodak. IsoPlus is marketed as (1) a dry, flowable powder, and (2) a carrier containing protein and minerals.

Pertinent information relative to isoacids as a feed additive to dairy cows follows:

- **Results**—Use of isoacids may boost milk production 8 to 10%, or 4 to 6 lb per day, with little or no increase in feed consumption.
- **Cost**—The feed additive is relatively expensive; it costs 25 to 30¢ per cow per day. A response of 2 to 3 lb more milk per day is needed to break even.
- **A time lag in activity**—A 30- to 60-day time lag occurs from the time isoacids are first fed until an economic response occurs.
- **Not all herds respond**—Generally, response has been positive, with approximately 85% of the herds showing increases in milk. However, about 15% do not improve.
- **Isoacids deliver the greatest response early in the lactation period**—The manufacturer recommends that it be discontinued 220 to 250 days after calving, because the economic response is marginal at this time.

Kelp

Kelp, or seaweed, grows in the sea. Botanically, it is a member of the algae family.

For centuries, kelp has been promoted as a natural feed for animals and food for humans, prized for its minerals and vitamins.

The nutritive value of seaweed is quite variable; it is affected by species, geographic area, season of year, and temperature of water. The Norwegian Seaweed Institute reports an assortment of 60 different mineral elements in seaweed. Additionally, it contains carotene, vitamin D, vitamin K, and most of the water-soluble vitamins, including vitamin B-12.

Kelp is always a rich source of iodine; dried kelp contains more than eight times as much iodine as iodized salt. So, dried kelp may be harmful if fed to animals in large amounts and over a prolonged period.

Kelp has long been promoted for its therapeutic properties for both animals and people, but with few of these claims substantiated by properly conducted and controlled experiments. When used at proper levels, it is an excellent source of certain minerals and vitamins.

Medicated Feeds

Medication is the administration of remedies for the prevention or healing of disease. It follows that medicated feeds are feeds that contain remedies to prevent or heal disease.

Various routes of administering medications to animals are used. Incorporating coccidiostats, histomonostats, and anthelmintics, and other medicants in feed or water is often the most convenient method. Medications and/or feed additives are also being incorporated in molasses/salt blocks and given to animals free-choice.

Coccidiostats are drugs that are used to prevent coccidiosis. Histomonostats are used to prevent blackhead disease in turkeys. A host of anthelmintic drugs are used to treat cattle, sheep, goats, swine, and horses suffering from a large variety of worm parasites. In each category—coccidiostats, histomonostats, and anthelmintics—a wide variety of drugs sold under many trade names are available. But medicated feeds are not limited to these three groups. Other common feed additives are: ethylenediamine dihydriodide (EDDI) used to treat bovine foot rot, soft tissue, lumpy jaw, and wooden tongue; bloat control products; and ammonium chloride used in the prevention of urinary calculi.

CAUTION: Medicated feeds can be a major source of residues, especially of antibiotics and sulfas because of their wide use; 75% of all animals receive one or both of these products during their lives. So, medicated feeds should be used in keeping with the instructions on the label, especially as they pertain to (1) the drug withdrawal period from animals in advance of marketing products, and (2) proper mixing, handling, and storage of medicated feeds.

Mold (Fungi) Inhibitors

Molds are fungi distinguished by the formation of mycelium (a network of filaments or threads), or by spore masses.

Mold inhibitors are substances that prevent the growth of molds.

All feeds are suitable mediums for growth of a wide variety of molds, provided temperature and moisture conditions are favorable.

More than 100 different molds which grow on standing crops or in feeds are known to produce toxins (mycotoxins), and about 20 of these mycotoxins have been associated with diseases in animals or humans.

In recent years, nutritionists have been giving increasing attention to the effects of fungal infestations of feeds. It has been speculated that perhaps many nutritional problems of the past (for example, suspected nutrient deficiencies) were, in fact, caused by feeds contaminated with fungi.

Fungi can affect feed intake and subsequent production through contamination at one or more of four stages in the feeding chain: (1) in the field (preharvest), (2) during storage, (3) at mixing, and (4) in the animal itself. Fungal contamination can pose problems through the production of toxins, alterations of the chemical composition of the feed, or alterations of the metabolic functioning of the animal ingesting or harboring the fungus.

Certain fungi, most notably the organism *Aspergillus flavus*, produce toxins; the toxin produced by *Aspergillus flavus* is known as aflatoxin. Aflatoxin, which has clearly been shown to be a carcinogen (cancer-producing), causes much trouble in livestock. But it is not the only mycotoxin to be feared. Mycotoxins affect all species, especially the young. Generally, ruminants appear to tolerate higher levels of mycotoxins over longer periods of intake than simple-stomached animals. Growing chickens are markedly less susceptible to aflatoxins than ducklings, goslings, pheasants, or turkey poults. Fish are probably one of the most susceptible species of animals to aflatoxin poisoning.

Of all the mold inhibitors currently available at an economical price, propionic acid is the most efficacious. It is Generally Recognized As Safe (GRAS), by FDA, available in liquid or dry form, of low toxicity to animals, and economical for addition to feeds at an effective level. Sorbic acid is also very effective in preventing mold growth. Mold inhibitors should be applied in keeping with the directions of the manufacturer.

The toxicity of aflatoxin-contaminated feed can be reduced when irradiated by ultraviolet light or exposed to anhydrous ammonia under pressure.

(Also see Chapter 5, Table 5-3, Potential Poisons—Mycotoxins [toxin-producing fungi or molds]; Chapter 9, section on "Acid Preservation of High-Moisture Grain"; and Chapter 14, section on "Organic Acids.")

Pellet Binders

Pellet binders are products that enhance the firmness of pellets. Several feed additives are known to produce a marked increase in the firmness of pellets; among them, (1) sodium bentonite (clay), (2) cellulose products from the wood pulp industry, (3) lignin derivatives, and (4) grain industry by-products. Although bentonite has no nutritive value, several reports indicate that at the level of common usage (2 to 2.5% of the ration) it may even improve the growth and/or feed utilization of animals. Hemicellulose preparations at levels up to 2.5% may serve as good energy sources for ruminants, but lignin has practically no nutritive value.

Molasses or fat are sometimes added to feed as an aid in pelleting, as well as being a concentrated source of energy.

Probiotics (Microbial Enhancers)

Probiotics are substances that contain desirable gastrointestinal microbial cultures and/or ingredients that enhance the growth of desirable gastrointestinal microbes. They establish a desirable balance of gastrointestinal organisms and/or the substances which contribute toward the balance.

The concept of probiotics is not unlike the use of bacterial cultures as a silage preservative or the inoculation of legume seeds.

As knowledge of the types and functions of microorganisms in the digestive tract unfolded, there evolved with it the concept of inoculating animals with beneficial microorganisms and/or giving them substances to encourage the growth of beneficial microorganisms.

Without doubt, young animals and animals under stress—adverse weather, changes in ration, weaning, transporting, co-mingling, or other stressful situations—will be more likely to respond to probiotics. It follows that the greater the need, the greater the response from probiotics.

Microbial products are available in powder, granular, or liquid form; for use as feed additives, water additives, or drenches. They should be used in accordance with the label of the manufacturer.

Steroids

Steroids are a group of fat-related organic compounds. They include cholesterol, 7-dehydrocholesterol and ergosterol, bile acids, and steroid hormones.

• **Cholesterol**—Cholesterol occurs free or in combination with fatty acids in all animal cells, in blood, and in wool grease (lanolin). Also, it is an important constituent of the brain, where it may form up to 17% of the dry matter. It is synthesized from acetate in the liver; hence, it is not a dietary essential. In recent years, cholesterol has been implicated in atherosclerosis of humans, which involves a thickening of the cell walls due to deposits containing cholesterol.

• **7-dehydrocholesterol**—This is the animal-derived precursor of vitamin D_3, which is produced when the sterol is exposed to ultra-violet light.

• **Ergosterol**—This is the principal plant sterol of importance in animal nutrition. It is important as the precursor of ergocalciferol or vitamin D_2, into which it is converted by ultra-violet irradiation.

• **Bile acids**—These are polar derivatives of cholesterol which are synthesized in the liver. They are either stored in the gall bladder from which they are released after eating, or released continuously from animals having no gall bladder (e.g., horses and rats). They aid in the emulsification of fats.

• **Steroid hormones**—These hormones are complex alcohols synthesized from cholesterol, primarily at the following sites: androgens from the testicles, estrogens from the ovaries, and glucocorticoids and mineral corticoids from the adrenal cortex.

Varying types of steroids have been used to assist human and equine athletes increase muscle mass and eliminate pain from creaky joints. Additionally, steroids are sometimes used on horses to assist in the healing of broken bones, to fight against parasitic anemias, to correct low blood counts, to combat respiratory infections, to increase appetites, and to improve performance of race horses.

Steroids used to increase muscle mass are based on testosterone, the male hormone secreted by the testicles in males, and are known as *anabolic steroids*. Steroids used to treat inflamed joints, along with a variety of other ailments, are based on the hormone secreted by the adrenal gland, and are known as *corticosteroids*. Steroids heavily involved in pregnancy are produced by the ovaries; they include progesterone and estrogen. Estrogens are also used to stimulate muscle growth in meat animals (see earlier section in this chapter on "Implants").

But the use of anabolic steroids may reduce fertility in both stallions and mares. The use of corticosteroids carries the risk of the horse using the injured part more than it should and making the injury worse. Also, on most U.S. race tracks the use of steroids is forbidden within a certain period prior to a race (usually within 48 hours of a race). Because horses may be used for human consumption in a number of European countries, notably in France and Belgium, the use of steroid drugs is banned in the European Economic Community (EEC).

While anabolic steroids and corticosteroids are controversial because of some of their side effects, no such argument surrounds the proper use of progesterone and estrogen.

Tranquilizers

Several drugs such as reserpine, aspirin, ethylene glycol, and other tranquilizers, have been fed to animals for the purpose of quieting and curbing activity.

Tranquilizers are sometimes added to poultry feeds to quiet birds being moved from place to place, to reduce the incidence of cannibalism, or to calm flocks affected with hysteria. But the use of these drugs as feed additives for other animal species is essentially nonexistent.

QUESTIONS FOR STUDY AND DISCUSSION

1. Define each of the following: (a) supplement, (b) additive, and (c) implant.

2. What is the responsibility of the FDA relative to additives and implants?

3. What is the Delaney Clause? How does it affect the feed industry?

4. Why do livestock producers use feed additives and implants?

5. Why is compliance with the feed additive/implant label of importance to both producers and consumers?

6. What is a feed supplement? List the common types of feed supplements, and discuss the importance of each type.

7. Under what circumstances are amino acid supplements needed?

8. Why must considerable caution be exercised when choosing a mineral premix?

9. List the three classes of mineral supplements and give examples of each.

10. What macrominerals and what microminerals are commonly needed as mineral supplements?

11. List the pertinent guidelines for mineral supplementation.

12. Outline the guidelines for a mineral supplementation program for mixed feeds. For unmixed feeds or pasture.

13. What are chelates? Would you recommend the use of chelated mineral supplements instead of alternative sources of inorganic mineral supplements?

14. Why do nutritionists rely on vitamin supplements to meet vitamin needs rather than select feed ingredients for vitamin content?

15. Discuss the need for, and the common supplemental sources of, each of the following vitamins: vitamin A, vitamin D, vitamin E, vitamin K, B-complex vitamins, and vitamin C.

16. Give three examples of products believed to contain unidentified factors.

17. Cite statistics that show the important role of additives, implants, and injections in modern animal production.

18. What management options may be considered when feeding heifers of unkown origin?

19. Why is so much research time and money being devoted to the development of additives that will produce meat (a) with more lean and less fat, and (b) with lower cholesterol?

20. What is the role of roughage or scratch factor for ruminants?

21. What are anthelmintics? Why is a parasite control program and schedule important?

22. What is an antibiotic? Through what mechanisms do they appear to increase rate of gain and feed efficiency?

23. Economically, how do antibiotics benefit both producers and consumers?

24. Discuss the theory of antibiotic resistance and explain how it could happen. Why and how are more and more drug-resistant *Salmonella* bacteria infesting animals and moving up the feed chain to humans?

25. Should the use of penicillin and tetracyclines be discontinued as livestock feeds; are suitable alternatives available? Justify your answer.

26. Discuss the role and importance of each of the following additives: (a) antioxidants, (b) arsenicals, and (c) bloat control products.

27. What are buffers? List and discuss the factors that commonly lower the pH of the rumen.

28. Explain the mode of action and importance of buffers in dairy cattle.

29. What are chemotherapeutics? How do they differ from antibiotics? When chemotherapeutics and antibiotics perform a similar role, what bases should be used in making a choice between them?

30. Discuss the role and importance of each of the following additives: (a) copper, (b) electrolytes, and (c) flavoring agents.

31. What are hormones? Discuss the mode of action and the importance of each of the following hormone and hormonelike compounds: (a) growth hormones, including the bovine growth hormone (BGH) for dairy cattle and porcine somatotropin (PST) for swine; and (b) melengestrol acetate (MGA).

32. What are implants? List the presently available and FDA-approved implants.

33. What are ionophores? List the presently available ionophores, explain the mode of action of each of them, and briefly summarize how to use them and the results to expect.

34. Discuss the role and importance of each of the following types of additives: (a) isoacids (IsoPlus), (b) kelp, (c) medicated feeds, (d) mold (fungi) inhibitors, (e) pellet binders, (f) probiotics, (g) steroids, and (h) tranquilizers.

14

FEED PROCESSING

Original painting by Tom Phillips

Contents	Page
Purposes of Processing	266
Selecting the Processing Method	268
Nutritional Considerations	268
Nonnutritional Considerations	268
Concentrate Processing Methods	268
Mechanical Alterations	269
Dehulling	269
Extruding (Gelatinization)	269
Grinding	269
Rolling	270
Dry Rolling (Cracking, Crushing)	270
Steam Rolling (Crimping, Steam Crimping)	270
Heat Treatments	270
Dry Heat Processing	270
Micronizing	270
Popping (Jet-sploding)	270
Roasting	270
Moist Heat Processing	271
Cooking	271
Exploding	271
Flaking	271
Steam Flaking	271
Pressure Flaking	271
Pelleting	271
Crumbling	272
Moisture Alterations	272
Bran Mash	272
Drying (Dehydrating)	272
High-Moisture Grain	272
Reconstituted Grain	272
Watered Feeds	273
Blocks	273
Liquid Supplements	273
Fermenting	273
Hydroponics	274
Unprocessed (Whole) Corn	274
Some Nonnutritive Feed Additives	275
Effect of Storage on Feedstuffs	275
Forage Processing Methods	275
Chopping, Grinding, or Shredding	276
Cubing (Wafering)	276
Drying	276
Ensiling	276
Pelleting	277
Miscellaneous Processing Methods	277
Ammoniation	277
Animal Waste (Manure) Processing	277
Fat Added	277
Freezing	277
Irradiation	277
Molasses Added	277
Organic Acids	278
Preservatives	278
Self-feeding Governors	278
Slow-Release and Rumen Bypass Treatments	278
Treatment of High-Cellulose Feeds	279
Complete (all-in-one) Rations	279
Choice of Processing Method	280
Farm Processed Feed	280
Questions for Study and Discussion	281

Feed processing refers to the operations necessary to achieve the maximum potential nutritional value of a feedstuff. Practically speaking, it involves changing ingredients in such manner as to maximize their natural value and the net returns from their use.

Feed processing may be physical and/or chemical. Physical changes result from such things as moisture addition or removal, heat, pressure, agglomeration (a clustering together), and particle reduction. Chemical changes may include structural changes in the starch and disrupting the protein matrix, resulting in changes in digestibility and metabolic end products. In some cases, both physical and chemical alterations, known as physiochemical changes, occur simultaneously.

Accelerated rate of ingesta passage and altered site of digestion within the G.I. tract are both likely end results of physiochemical changes in processed grains.

Feed processing has been practiced for a very long time. The "swill barrel" and slop bucket evolved with the domestication of swine. Grinding, crushing, rolling, and soaking of cereal grains for livestock have been practiced by livestock producers for many years. Cattle exhibitors have long cooked grain, and horse fanciers have long used wet bran mashes.

New technology in feed preparation for feedlot cattle was pioneered by W. H. Hale, of the University of Arizona, beginning about 1963.[1] But much of it is applicable to all ruminants. Feed preparation for swine and poultry has remained relatively simple as compared to the variety of methods available and in use for ruminant feeds. The major

---
[1]Hale, W. H., et al., "The Effect of Steam Processing Milo and Barley and the Use of Various Levels of Fat with Steam Processed Milo and Barley in High-Concentrate Steer Fattening Rations," *Arizona Cattle Feeders' Day*, May 6, 1965, p. 16.

change in horse feed preparation has been the increased use of all-pelleted rations (hay and grain combined). Complete, dry pellets for mink that can be dispensed from a self-feeder are only now being realized.

Feed is a major cost in animal production. It accounts for 70% of the cost of finishing cattle, 55% of the cost of producing milk, 50% of the cost of finishing lambs, and 65 to 75% of the cost of producing pork. Hence, it is economically important that feed be processed in such a manner as to make for maximum efficiency.

Improvements of even 5 to 10% in feed efficiency (feed/unit of production) make for large increases in profit. Assuming a total feed requirement of 3,000 lb per head in feedlot cattle during the finishing period and a feed cost of $4 per 100 lb, 5 and 10% improvements in feed efficiency would result in feed cost savings of $6 and $12 per head, respectively. These would translate to 50 and 100% greater profits per head (assuming an average feeding return of $12 per head).

Feed processing influences the nutritional value of feedstuffs —enhancing some, lowering others. Many of the methods that have been developed for the commercial processing of grains have resulted in decreases in the nutritional value of some of the ingredients. Thus, it is well known that the outer bran portion of a cereal grain is higher in most vitamins than the inner floury portion. Heat treatment of grain results in gelatinization of the starch and denaturing (changing structurally) the protein. Some processes to which feedstuffs are subjected result in destruction of certain vitamins, others enhance the stability of some vitamins, and still others bring about an improvement in availability of vitamins from certain feedstuffs. For example, fine grinding and pelleting of forages tend to increase rate of passage through the gut, which lowers fiber digestibility. However, overall animal response to pelleted forages is usually increased over the same forage fed in long or chopped form, because the slightly lower digestibility is more than offset by increased feed consumption.

Feed processing may also affect handling. For example, hay cubes facilitate automation. Since labor is the second greatest cost item of animal production, it is important that feed processing be evaluated from the standpoint of mechanization.

But feed processing costs money! Thus, recent increases in energy costs have forced producers to take a hard look at the economics of grain processing. Since it is net returns that count, improvements accruing from feed preparation in production and feed efficiency and in laborsaving must be weighed against the cost of processing. Is the processing sufficiently beneficial to cover all costs and return a profit? The economics of processing are well illustrated by the following example: A cattle feeder is considering installation of steam flaking equipment vs feeding whole shelled corn. At best, it can be expected that the steam flaker will improve the feeding value of corn 4%. If shelled corn costs $2.75 per bushel or $4.91 per 100 lb, a 4% improvement would be worth 20¢ per 100 lb or $3.93 per ton. Further investigation into the cost of installing, maintaining, and operating a steam flaker would indicate to the feeder that it would cost about $4.00 per ton, to get $3.93 worth of improvement, or there would be a net loss to the feeding operation of 7¢ per ton as a result of installing the steam flaking equipment. On the other hand, a cattle feeder using ground or dry rolled milo could install steam flaking equipment which would make for an 8% improvement in $3.50-per-100 lb milo. This would represent a 28¢ per 100 lb or a $5.60 per ton improvement for a processing cost of $4.00 per ton, or a net return of $1.60 per ton by switching from grinding or dry rolling to steam flaking. Thus, under the above circumstances, the feeder could not afford to flake corn. However, flaking milo, rather than grinding or dry rolling, would be profitable, because it enhances its value much more than similar processing of corn.

Generally speaking, the higher the level of feeding and the greater the production desired, the more important proper feed preparation becomes. This is so because (1) the higher the level of feeding, the more selective animals become in their eating habits; and (2) in ruminants, digestibility decreases as level of feeding increases, primarily because the feed does not remain in the digestive tract long enough for maximal effect of the various digestive processes.

Also, farm and feedlot differences make for differences in feed preparation. The sheer size of an operation is a major factor in amortizing equipment costs. Thus, greater animal numbers, along with using equipment to capacity, make it easier to amortize expensive equipment. Smaller operations may be able to afford and justify sophisticated feed processing methods through a feed processing cooperative or by having the work done on a custom basis on the farm or at the mill.

Then, too, alternative uses of capital must be considered. The owner must weigh the return from a large investment in processing equipment against a similar investment in more animals or in a confinement slotted-floor facility. Small feeders who have limited capital may opt to omit processing equipment if the estimated return therefrom will likely be less than that which could be obtained if the same money were invested in animals, or in a confinement building, or in other alternative investments.

Finally, energy requirements and costs for processing must be considered. It is inevitable that energy costs will continue to rise. Thus, those processing methods that have the lowest energy requirements will be favored in the years ahead.

PURPOSES OF PROCESSING

The primary reasons for processing feeds are:

1. **To make more profit.** Most people are in business to make money—and livestock producers are people.

Feed efficiency can be routinely improved by as much as 10%, and occasionally by as much as 15 to 20% by changing the method of grain processing. It is more difficult to increase production (such as increased rate of gain or milk production) by grain processing. Some processing methods may even improve feed efficiency, but reduce gain. It follows that the relative importance of feed efficiency and rate of gain should be known.

2. **To alter particle size.** Some feeds need to be reduced in size so that they can be consumed, or so that they may be more digestible. In some instances, particle size is increased (agglomerated) by pelleting or cubing, such as (a) when feeding cattle or sheep on the range; (b) when fines need to be alleviated to control waste; (c) when there is selectivity, with the animals consuming the more palatable ingredients and leaving those that are least palatable; or (d) when improved handling efficiency is desired.

3. **To change moisture content.** The moisture content of a feedstuff may need to be changed to make it safer to store, more palatable, more digestible, or to prepare it for

other processes. Also, with very high moisture content, such as root crops, the total dry matter intake may be reduced; which may be undesirable when maximum production is desired, but desirable for control of obesity.

Basically, there are two ways to remove moisture from grain: (a) with heat (heated air drying), or (b) without heat (aeration)—or some combination of the two.

Water additions to feeds stored as high-moisture grain, silage, or haylage may be desirable. For best results, it is recommended that high-moisture grain be stored at about 30% moisture content, although the moisture may be as high as 40% or as low as 20%. But water should be added when it falls below 26%.

It may be desirable to add water to a finely ground meal mixture at the time of feeding in order to lessen dustiness and increase palatability; and added water may be necessary in preparing feedstuffs for other processes, such as adding moisture in the form of steam to ground feeds that are to be pelleted.

4. **To change density of feed.** The weight per unit volume, or the bulk, of the ration affects total intake; hence, bulky rations which make for "fill" reduce dry matter consumption. For this reason, very bulky feeds are sometimes pelleted or cubed in order to increase energy density and feed consumption. Also, pelleted hay or cubed hay lessen transportation costs and storage space.

On the other hand, horse caretakers favor less dense rations. They prefer that grains be flaked, rather than ground or pelleted, because it makes for a lighter ration and fewer digestive disturbances.

Bulky rations are sometimes prepared for the purpose of limiting energy intake.

5. **To change acceptability (palatability).** In most instances, feeds are processed in such manner as to increase acceptability (palatability) and feed intake. To this end, molasses, flavors, and fats may be added. However, processing may be used for the purpose of decreasing palatability and limiting feed consumption. Salt-feed mixtures are an example of the latter.

6. **To change nutrient content.** When used alone and in their natural state, few feedstuffs meet the nutrient requirements of the animals to which they are fed. Even milk, when fed to young animals over an extended period of time, may be improved by enrichment with certain minerals and vitamins. Also the vitamin D content of ergosterol-containing feeds can be changed through irradiation, with the provitamin ergosterol converted into vitamin D_2.

7. **To increase nutrient availability and digestibility.** Milo that has been dry ground or rolled, and not processed in some more sophisticated manner, has a relative TDN value of only 90% that of corn, and a net energy value of only 80 to 89% that of corn. Yet, chemical composition data indicate that there should be less difference in the feeding value of milo and corn than suggested above. The starch, which represents 70 to 80% of the total dry matter, and perhaps other fractions such as protein, appear to be less available in milo than in other grains. However, some of the newer processing techniques have produced dramatic improvements in the feeding value of milo, with it responding more than other grains. This is attributed to a gelatinization of the starch granules (hydration or rupturing of the complex starch molecules), rendering them more digestible, with the level of gelatinization appearing to be influenced by such factors as steaming time, temperature, grain moisture, roller size and tolerance, processing rate, and variety of milo.

Although feed processing is not known to influence calcium and magnesium utilization, pelleting of feeds increases the utilization of phosphorus by chickens and pigs.

The processing of grains for ruminants enhances availability and digestibility by (1) increasing surface area for greater microbial activity, and (2) giving rumen microorganisms and digestive enzymes easier access to starches and readily utilizable nutrients.

8. **To detoxify or remove undesirable ingredients.** Some feeds may contain toxic substances, the excesses of which will prevent them from performing normally, or even cause death. Fortunately, some of these toxic principles can be removed by processing. Considerable control of gossypol, the yellow pigment of cottonseeds which is toxic to simple-stomached animals, is possible by heating; and the addition of iron salts ($FeSO_4 \cdot 7H_2O$) will protect against egg discoloration. Heating soybeans destroys the factors which inhibit the digestive enzymes, trypsin and chymotrypsin. The toxicity of linseed meal can be removed by adding 2 or 3 parts of water to the meal and allowing it to stand for 12 to 18 hours at a temperature between 72 and 99°F. Through heating at high temperatures under certain conditions, the excess fluorine of rock phosphate can be removed.

9. **To improve keeping qualities.** Because feeds are seasonally produced, some of them must be stored for use in the nongrowing season. Usually, this involves the application of some type of preservation.

Feeds carrying more than 14% moisture cannot be stored in bulk. They are likely to mold, and spontaneous combustion may also take place. High-moisture grains may be processed by either drying or chemical treatment (adding an organic acid), or they may be stored in oxygen-limiting silos.

Forages may either be dried to safe storage levels, ranging from 25% moisture in loose hay to 16 to 17% for cubes, or preserved by ensiling (preferably, at 60 to 67% moisture content).

10. **To reduce storage and transportation space and cost.** Sometimes feeds are processed in a certain way in order to reduce storage and transportation space. This is particularly true of forages. Thus, alfalfa hay averages about the following pounds per cubic foot in different forms: loose hay, 4.2; chopped hay, 6.2; and baled hay, 7.5. Cubes and pellets require even less storage space; hence, it is economically feasible to ship cubed or pelleted hay from mainland United States to Hawaii.

11. **To improve mechanization.** Because it is difficult to mechanize the feeding of long hay, pitchfork haymaking persisted in the present century. But, beginning about 1940, scientists and engineers pooled their efforts to automate haymaking and feeding. Hay went modern with one person pick-up balers, field choppers, cubing machines, and other equipment.

Likewise, silage feeding has been automated. It is possible to achieve complete push button controlled feeding in an upright silo. With horizontal silos, silage is handled with a front-end loader, or with a mechanical unloading wagon or truck. Self-feeding of silage is also being achieved.

12. **To lessen molds, salmonella, and other harmful substances in feeds.** Sometimes feeds are subjected to a certain process in order to ensure safety and avoid contamination, especially from molds and salmonella.

Molds on feeds and foods have long been a problem. Aflatoxins, part of a larger group called mycotoxins and classified as carcinogens, are toxic substances produced as a result of mold (fungus) growing on grains and other feedstuffs. Fortunately, through processing, much can be done to prevent the formation of molds in feeds, and to reduce the level of aflatoxin or deactivate it in contaminated feeds. Proper harvesting, drying, and storage are important factors in lessening aflatoxin contamination and toxin production. Propionic and acetic acids will inhibit mold growth; hence, they are finding increasing use in preserving high-moisture grains. Treatment with ammonia or ammonium hydroxide will detoxify feeds.

Salmonella, rod-shaped bacteria, the incidence of which is high in meat meal, may be destroyed by pelleting.

13. **To enhance rumen bypass.** Processing, especially heat and pressure treatment, may be used for the purpose of increasing rumen bypass (protected or escape protein).

SELECTING THE PROCESSING METHOD

Many grain processing methods have been developed. Obviously, no single method will be suited to all feeds, all species, and all functions (maintenance, growth, finishing, reproduction, eggs, lactation, work, and wool). Both nutritional and nonnutritional factors should be considered in arriving at a choice.

Nutritional Considerations

The following important nutritional factors should be considered in choosing grain processing methods:

1. **Type of grain.** Grains differ in their response to processing. For example, milo is more responsive to processing than any other grain.

2. **Uniformity and quality of finished product.** Will processing influence the amount of fines, the separation of the components, the palatability, the surface area, and the density or weight per bushel?

3. **Moisture content.** After processing, can the grain be stored or self-fed if desired? Addition of moisture during processing requires that the processed feed be fed soon after treatment; otherwise, it must be dried prior to storage.

Steam heating may add 6 to 8% moisture to milo, while dry heat or pressure may decrease moisture content by a like amount. This point is important to both feed suppliers and custom feeders.

Handling cereals as high-moisture harvested grain will produce nearly the same results as more costly processes; e.g., steam flaking.

Reconstituted (moisture added) grain produces nearly the same results as high-moisture harvested grain.

4. **Percentage of concentrate in the ration.** Grain processing for ruminants gives greater returns when feed intake of grains is high. Thus, cattle fed maintenance rations are not normally fed much grain; hence, the increase in feed efficiency may not return the added processing cost.

5. **Change in structure of the starch.** Milo responds to some of the newer processing techniques (heat provided either by steam, dry heat, microwaves, or pressure) which gelatinize the starch granules and increase digestibility and production of volatile fatty acids.

6. **Feed intake, rate of production, and feed efficiency.** What effect will the processing have upon feed intake, rate of production, and feed efficiency? Will increased performance more than offset any increase in cost?

7. **Effect on health.** Do the animals go on feed and stay on feed satisfactorily? Will the processed feed increase acidosis or other metabolic disorders?

8. **Influence on end product.** Will processing affect the end product—the carcass grade and yield, the milk, wool, eggs, or power?

Nonnutritional Considerations

The following important nonnutritional factors should be considered in choosing the grain processing method:

1. **Time of grain purchase.** Will grain be purchased at harvest time or throughout the year?

2. **Size of operation.** A small operation may not be able to justify a large investment in sophisticated processing equipment.

3. **Effect on hauling costs.** Are transportation costs increased or decreased according to density? Does the process in question increase or decrease the number of tons or loads hauled per day from the mill and/or storage area to the bunks?

4. **Type of ration and kind of operation.** The type of processing should vary according to the type of ration and the kind of operation, with these factors determining the relative merits of such choices as (a) hay cubes vs big round bales, (b) grass silage vs hay, and (c) all-concentrate vs low-concentrate rations.

5. **Capacity of mill.** Some grain processing methods are slower than others. Thus, hourly volume may be a limiting factor when the mill is already operating to capacity if the process is more time-consuming.

6. **Initial investment in equipment.** There is a wide range in the initial cost of processing equipment. Also, the greater the quantity of feed processed, the lower the per ton processing cost. It follows that the larger the operation and the greater the tonnage of feed processed, the greater the initial investment in equipment that can be justified.

7. **Maintenance, repair, and operating cost.** Before buying and installing processing equipment, consideration should be given to the normal maintenance, repair, availability of replacement parts, and operating cost. Some types of processing equipment are more expensive to use than others.

8. **Labor requirements.** Feed processing equipment varies from the standpoints of both (a) hours of labor per ton of feed processed, and (b) the caliber of labor required, with the more complicated types of equipment requiring an engineer or expert mechanic for operation.

9. **Energy requirements.** Increasing energy shortages and rising energy costs necessitate that careful consideration be given to the energy requirements when choosing the feed processing method.

CONCENTRATE PROCESSING METHODS

Most concentrate processing methods have as their primary objective, improvement in starch availability in cereal grains, which, in turn, enhances digestion and feed efficiency. Since 70 to 80% of the dry matter in cereal grains is composed of starch, this is a logical approach. However, the method of accomplishing this is complicated because (1) the type of starch varies among grains, with the starch in some grains more digestible than in others; and (2) availability of starch even varies from one grain variety to another, particularly in milo.

Feed Processing

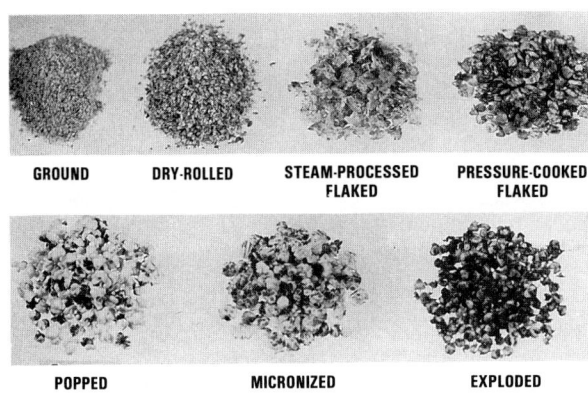

Fig. 14-1. Milo processed by several different methods. (Courtesy, Department of Animal Science, University of Arizona, Tucson)

Grain processing gives greater returns when feed intake of grains is high. Animals fed maintenance rations are not normally fed much grain; hence, the increase in feed efficiency may not return the added processing cost.

Several grain processing methods have evolved. A survey of the literature reveals that no process prior to the development and use of steam flaking for feedlot cattle improved performance so dramatically. Thus, most modern processing techniques have been developed in an attempt to obtain similar, and hopefully better, performance than steam flaking—with possibly a reduction in cost. Some are physical, others are chemical; some are dry processing, others are wet processing. Despite some overlapping, the authors have evolved with the following classification of grain processing methods:

Mechanical alterations
    Dehulling
    Extruding (Gelatinization)
    Grinding
    Rolling
        Dry rolling (cracking, crushing)
        Steam rolling (crimping, steam crimping)

Heat treatments
    Dry heat processing
        Micronizing
        Popping —Jet-sploding
        Roasting
    Moist heat processing
        Cooking
        Exploding
    Flaking
        Steam flaking
        Pressure flaking
    Pelleting
        Crumbling

Moisture alterations
    Bran mash
    Drying (dehydration)
    High-moisture grain (early harvested)
    Reconstituted grain
    Watered feeds

Blocks
Liquid supplements
Fermenting
Hydroponics (sprouted grain)
Unprocessed (whole) corn

Mechanical Alterations

The oldest and most widely used methods for processing grains are those which merely cause physical disruption of the cells by mechanical means. The fact that the more nutritious portions of the grain are surrounded by an outside coating or hull makes it easy to understand how the exposure of these nutrients to the action of digestion processes would increase the utilization of the nutrients. The mechanical methods by which the grain kernel is broken vary, but generally speaking, they involve either shearing, cutting, or mashing. In the milling of grains, there is also the abrasive action to scrub off the outer coats in processes referred to as burring, pearing, polishing, dehulling, and other similar terms.

When any of the dry processing methods are used, it is important that the kernel be broken, but that there be coarseness and relative freedom from fines.

DEHULLING

Dehulling is the process of removing the outer coat of grain, nuts, and some fruits. The hulls are high in fiber and low in digestibility by swine, poultry, and other monogastric animals.

EXTRUDING (GELATINIZATION)

Extruding is a process by which feed is pressed, pushed, or protruded through constrictions under pressure.

Extruding usually involves grinding the grain, followed by heating with steam in order to soften it, then forcing the material through a steel tube by an auger. The softened material is then extruded through cone-shaped holes which are smaller where the feed enters and gradually enlarge where the feed is expelled. The expansion causes disruption, or granulation, of the starch granules.

GRINDING

Grinding is that process by which a feedstuff is reduced in particle size by impact, shearing, or attrition. It may change the digestibility of cellulose and protein.

Grinding is the most common, cheapest, and simplest method of feed preparation. It is usually accomplished by means of a hammer mill, which, by impact, reduces the particle size of the grain until it passes through a screen of a certain size.

A major advantage of grinding compared with more sophisticated processing methods is the economic feasibility of having a hammer mill on the farm, or of having a custom grinder come to the farm or ranch periodically to process grain.

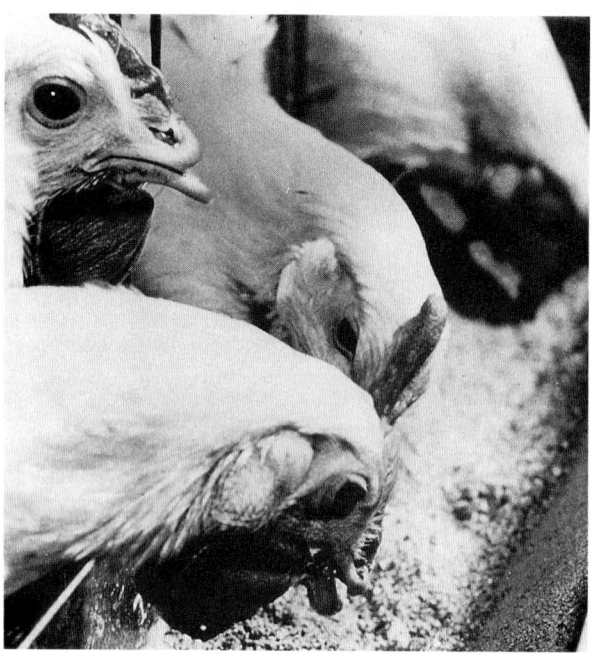

Fig. 14-2. Hens eating mash. (Courtesy, Ralston Purina Company, St. Louis, Mo.)

ROLLING

Rolling refers to the process by which grain is compressed into flat particles by passing it between rollers. The rolling may be accomplished without the addition of water (dry rolling) or after subjecting the grain to steam (steam rolling).

Dry Rolling (Cracking, Crushing)

Dry rolling, which is also called cracking or crushing, refers to passing grain, without steam, between a closely fitted set of steel rollers which are usually grooved on the surface. It breaks the hull and/or seed coat and results in an end product resembling coarsely ground grain.

Steam Rolling (Crimping, Steam Crimping)

Steam rolling, which is also called crimping or steam crimping, refers to exposing grain to steam for a short period of time, usually 1 to 8 minutes, followed by rolling. The steam softens the kernel, producing a more intact, crimped-appearing product than that produced by dry rolling. The moisture content of steam rolled grain is increased by an average of 6 to 8%.

Heat Treatments

Excess heating damages some nutrients, such as the amino acids, and vitamins, whereas proper heating of protein sources (such as soybeans) and of carbohydrate sources (such as cereal grains, potatoes, and beans) results in better availability of nutrients. Heating soybeans destroys the trypsin inhibitor or a possible active protein fraction in raw soybeans, increases the amino acid availability, results in better availability of the fat, and increases metabolizable energy.

Proper heating of cereal grains, such as corn, barley, and milo, will make for partial gelatinization and improve rate and efficiency of gains of cattle.

Mink may suffer from thiamin deficiency when fed certain types of raw fish that contain the enzyme thiaminase; but since thiaminase is heat labile, cooking the fish or dehydrating the fish meal averts the problem. Biotin deficiency can be produced experimentally in different animals by feeding raw egg white; but, avidin, the causative factor in egg white, is destroyed by thorough cooking.

Prolonged heat treatment in processing destroys several of the vitamins. The fat-soluble vitamins and biotin, folacin, pantothenic acid, and thiamin are particularly susceptible to destruction by heat. This is especially true if the material contains high levels of polyunsaturated fat.

Losses of most of the water-soluble vitamins occur during cooking when the cooking water is drained off and not dried with the feed material. Thus, the B vitamin content of fish stick water is much higher than that of fish meal, and the final fish meal dried without the fish-soluble fraction is very low in water-soluble vitamin content.

In general, heat treatments do not improve the nutritional value of most feedstuffs for monogastrics. However, they are the most successful of the newer feed processing techniques for ruminants.

DRY HEAT PROCESSING

Dry heat processing consists of surrounding the feed with dry air. The common methods of processing by dry heat are micronizing, popping, and roasting.

Micronizing

Micronizing is a coined word used to describe a dry heat treatment of grain by microwaves emitted from infrared burners. In micronizing, grain is heated to 300°F by gas-fired infrared generators as it passes along an oscillating steel plate or skillet, thence is dropped into knorling rolls. Micronized grain is not popped. It is reduced to about 7% moisture, then rolled to produce a uniform, stable, dry, free-flowing product. The product has an intact, flakelike appearance, resembling some steam flaked grains.

Popping (Jet-sploding)

Popping is the exploding, or puffing, of grain resulting from the rapid application of dry heat. Popping grain for livestock involves the same principle as processing popcorn for people, and the end results are similar.

Roasting

Roasting is a simple process of heating feed to the desired temperature in some form of oven for a period of time. It is another method of heat treatment.

Corn and soybeans are the principle feeds that are processed by roasting.

• **Roasting corn**—In roasting corn, the grain is heated to about 300°F. The roasted grain has a pleasant, "nutty" aroma and a puffed, carmelized appearance. Purdue University reports that for fattening cattle roasting improves feed efficiency by 10% and increases weight gains by 14% over ground corn.

• **Whole cooked (or roasted) soybeans**—Raw soybeans are poorly utilized by pigs and poultry. Numerous experiments have shown that the nutritive value of soybeans and of soybean oil meal for monogastric animals is much improved by proper cooking. The explanations are that cooking soybeans does the following things: (1) It destroys a substance known as the *trypsin inhibitor,* which depresses the growth of nonruminants and prevents the action of the protein digestive enzymes, trypsin and erepsin; (2) it greatly increases the availability and value of protein for these animals, through making the sulfur-containing amino acids more readily available; and (3) it improves the palatability of soybeans for monograstics.

It is noteworthy that hogs fed cooked whole soybeans have a softer carcass than those fed a soybean meal ration. Whether this condition (soft pork) will influence the price packers are willing to pay for live hogs is yet to be determined.

MOIST HEAT PROCESSING

Moist heat processing consists in surrounding the feed by water or steam and (1) cooking either in a conventional vessel or under pressure, or (2) compressing.

The common moist heat processing methods are cooking, exploding, flaking (steam flaking, pressure flaking), pelleting, and crumbling.

Cooking

Cooking is processing by applying heat.

Professional beef cattle fitters have long cooked feed (especially barley) for show cattle. However, the practice declined with the deemphasis of livestock shows, better milking dams of young show cattle, and fewer hard-working caretakers willing to cook grain.

Farmers have long known that potatoes, beans, and soybeans should be cooked for pigs. In general, cooking feedstuffs does not improve their nutritional value for monogastrics—swine and poultry. However, soybeans are an exception. Their nutritional value is greatly enhanced by heating.

Garbage is cooked to control *Trichinella,* which causes trichinosis in humans.

Exploding

Exploding is the swelling of grain, produced by steaming under pressure followed by releasing to the air. Under the high pressure, moisture is forced into the kernels, which, when released into the air, swell to several times the original size. The product resembles puffed breakfast cereals.

Flaking

Flaking is a modification of steam rolling in which the grain is subjected to steam either for a longer period of time or under pressure. The end product has a distinct and pleasant aroma, resembling cooked cereal. Proper flaking of grains renders the starch fraction more readily available to rumen microorganisms and enzyme degradation than conventional methods of steam or dry rolling.

The flaking process varies according to the grain. The grain that responds the most to flaking is milo. In comparison with dry rolling or grinding, cattle fed flaked milo will gain from 0.25 to 0.5 lb more, or about 10% more, per head per day and require 5 to 10% less feed. Steam processing and flaking of barley and wheat appear to improve gain but not utilization of the grain. This is probably due to improved palatability and intake of the flaked product as compared to dry rolled or ground product. Flaking is the preferred method of processing grains for horses; it produces light, fluffy particles which result in fewer digestive disturbances than any other method of feed preparation.

Flaking is rolling into flat pieces following either (1) steaming at atmospheric pressure, or (2) steaming under pressure.

STEAM FLAKING

This was the first modern technique which markedly increased feed efficiency and rate of gain in the case of milo. This process differs from steam rolling or crimping in that the grain is subjected to steam under atmospheric conditions for a longer period of time, usually 15 to 30 minutes, prior to rolling. Large, heavy roller mills set at near zero tolerance produce a very thin, flat flake which usually weighs from 22 to 28 lb per bushel and contains 16 to 20% moisture. The flaking process causes gelatinization of the starch granules (hydration or rupturing of the complex starch molecule), rendering them more digestible. The degree of flaking and level of gelatinization are influenced by such factors as steaming time, temperature, grain moisture, roller size and tolerance, processing rate, and type and variety of grain.

PRESSURE FLAKING

In pressure flaking, the grain is subjected to steam under pressure for a short time, such as 50 psi for 1 to 2 minutes. A continuous flow cooker is operated by air lock valves to inject and eject grain. In comparison with steam flaking, flakes produced by pressure are less brittle and less subject to fragmenting during the mixing and feeding operation.

Pelleting

Fig. 14-3. Broilers eating bite-sized pellets. (Courtesy, Ralston Purina Company, St. Louis, Mo.)

Pelleting is the agglomerating of feed by compacting and forcing it through die openings by a mechanical process. Pellets can be made into small chunks or cylinders of different diameters, lengths, and degrees of hardness. Large pellets—especially those large enough to be fed on pasture or range—are commonly called range cubes.

Fig. 14-4. Range cubes fed to replacement heifers wintered on low-quality pasture. (Courtesy, Ralston Purina Company, St. Louis, Mo.)

Grains and other concentrates are pelleted for the purposes of (1) facilitating mechanization in handling; (2) eliminating fines and dust, and increasing palatability; (3) alleviating separation of ingredients and sorting; (4) increasing feed density—thereby lessening transportation and labor costs; (5) reducing storage space; (6) making it possible to feed on the ground or in windy areas with little loss; and (7) improving the nutritional value of certain feedstuffs through the instantaneous heat and pressure.

On rations containing a low level of crude fiber, there is no advantage in pelleting feed for beef cattle or swine. However, with more fibrous feeds—especially barley—there is a decided advantage in pelleting feed for swine. Pelleted feeds are popular with horse owners and caretakers; and this includes pelleted concentrates, pelleted hay, and pelleted complete feeds (concentrates and hay combined).

Pelleting broiler and turkey feeds improves performance. But there is no advantage from pelleting layer feeds, with the result that layers are usually fed mash feeds. Pelleting results in marked improvement in the nutritional value for chicks and poults of certain feedstuffs, such as wheat bran, wheat germ meal, dehydrated alfalfa meal, rye, rapeseed oil meal, and field peas. Much of the improvement is still apparent when the pellets are reground and fed as mash; hence, most of the improvement is due either to enhanced nutrient availability or destruction of heat-labile toxins. But pelleting brings about much less enhancement in the value of certain other feedstuffs. High-temperature steam pelleting may even be detrimental in rations that are borderline in protein or in certain critical amino acids.

Pelleting feeds may destroy vitamins A, E, and K, especially if the ration does not contain sufficient antioxidants to prevent the accelerated oxidation of these vitamins under conditions of moisture and high temperature.

CRUMBLING

Crumbles are crushed pellets. They are made by crushing pellets into a coarse, granular form. In comparison with pellets, crumbles are preferred by many poultry producers and are better adapted to mechanical feeders. Crumbles retain the heating and density advantages of pellets, but alleviate the sometimes disadvantages of pellets being difficult to chew, swallow, and digest. In comparison with ground feeds, crumbles have the advantage of being dust-free, irregular, and granular.

Moisture Alterations

Water is important in feed preparation and processing. Sometimes the water content of a feed must be altered for proper feed storage, and sometimes it must be changed for feeding purposes.

BRAN MASH

A bran mash is steamed wheat bran. It is the traditional feed for use in regulating the bowels of horses on idle days and at such other times as required. The wet mash is prepared by filling a 2- to 2½-gal bucket with wheat bran, pouring enough boiling water over it to make it the consistency of breakfast oatmeal, covering the bucket with a blanket and allowing it to steam until cooked, then feeding it to the horse.

DRYING (DEHYDRATING)

Drying is the removal of moisture by artificial or natural means. To avoid spoilage in storage, grains must be dry enough to prevent the growth of bacteria and molds.

Generally speaking, shelled or threshed grains stored in unventilated bins should not have more than about 14% moisture; preferably, it should not exceed 10 to 12%. Grain may be dried (1) by the use of fuel—artificially; (2) by natural air drying; or (3) by a combination of the two methods.

Energy shortages and costs favor delaying harvest until grain is lower in moisture, along with maximum natural air drying. Also, the following alternatives to drying should be considered: (1) the immediate feeding of high-moisture grain, (2) Storing it as high-moisture grain in an oxygen-limiting silo, or (3) treating it with an organic acid(s).

HIGH-MOISTURE GRAIN

High-moisture grain refers to grain that is harvested at a moisture level of 22 to 40% and stored without drying. Optimum conditions for ensiling high-moisture grain appear to be 25 to 32% moisture content. Correctly speaking, high-moisture grain does not involve moisture alteration.

High-moisture grain may be successfully stored in either of three ways:

1. It may be ensiled (fermented) in an oxygen-limiting silo.
2. It may be ensiled in unsealed storage (in conventional upright silos, or in horizontal silos).
3. It may be preserved by the addition of an organic acid, most commonly propionic acid (or a mixture of propionic acid with either acetic acid or formic acid), or ammonia.

(Also see Chapter 9, Silage/Haylage/High-Moisture Grain, section on "High-Moisture Grain.")

RECONSTITUTED GRAIN

Reconstituted grain is mature grain that is harvested at the normal moisture level (10 to 14% moisture), following which water is added to bring the moisture level to 25 to 30% and the wet product is stored in a suitable structure for 15 to 21 days prior to feeding. Thus, reconstituted grain involves processing that resembles soaking, and which results in an end product similar to high-moisture grain.

When stored in upright silos, the grain is stored whole, then rolled or ground at the time of removal. Reconstituted grain cannot be satisfactorily stored in horizontal silos as compaction cannot be obtained.

(Also see Chapter 9, Silage/Haylage/High-Moisture Grain, section on "Reconstituted Grain.")

WATERED FEEDS

Water is frequently added to feed, with the amount varying from just enough for dust control to making a slop.

Ground and dry rolled grains, and finely ground alfalfa, tend to be dusty. The palatability of such feeds may be improved by adding a small amount of water at the time of feeding.

- **Soaking**—Sometimes hard grains that are not mechanically processed are soaked for 12 to 24 hours. The soaking softens and swells the grain. Also, dried beet pulp and soybean flakes may be fed in wet form.
- **Liquid and paste feeding**—Liquid feeding usually involves mixing predetermined amounts of feed and water prior to, or at the time of, feeding. When properly used, this method can practically eliminate feed dust in the feeding area and minimize wastage. Ratios of feed and water can be varied to produce a free-flowing liquid or a thick paste.

Some swine producers feed a slop (slurry, gruel, or swill), especially to early weaned pigs and pigs being fitted for show or sale, feeling that they get greater feed consumption and gains thereby.

Blocks

Blocks are compressed packages, generally weighing from 30 to 50 lb each, although high-energy blocks (high in fat content) weighing up to 500 lb are now available. Mineral blocks have been used for a very long time. These were followed by the development of protein blocks, primarily for supplementing cattle on the range and horses on pastures or in corrals. More recently, high-energy blocks evolved.

Blocks may be placed in grazing areas where cattle have frequent access to them, with one block provided to 15 mature cattle. Intake will vary with the feed supply and the type of block. Generally, it is planned to limit feed consumption to about 2 lb per head per day by hardness of block and salt and/or fat content.

Range cattle producers use blocks as a means of (1) lessening the labor attendant to the daily feeding of a range supplement, (2) alleviating the loss that accompanies feeding a meal, and (3) distributing cattle on the range.

Liquid Supplements

Liquid supplements are supplements in liquid form. Many of them contain water, molasses, and urea, usually with added trace minerals and vitamins. This is a convenient way of feeding supplements to cattle on pasture or in a corral. Also, liquid supplements are sometimes added to complete ration mixes, either as part of the mix or as a top dressing.

The amount of molasses in most liquid supplements varies from 50 to 70% of the total weight. Most liquid supplements contain ½ to 2% phosphorus, often phosphoric acid. Other compounds that may be present in liquid feed supplements are fat, either animal or vegetable, to increase the amount of energy; alcohols—both ethyl alcohol and propylene glycol are used; and/or a product(s) to govern consumption.

Liquid supplements in a "lick" tank can be offered free-choice. This is a convenient and satisfactory way in which to supply protein, energy, and other nutrients, so long as the cattle do not consume more than they need.

Fermenting

Two fermentation processes are of practical importance in livestock feeding: (1) ensiling; and (2) improvement of the nutritional value of feeds, either by fermenting the feedstuff itself, or by fermenting other materials that may be used as feed additives to supplement the original feed.

- **Ensiling**—The earliest use of fermentation in animal feeding, and still the most extensive one, involves the ensiling process which takes place when certain feeds with sufficient moisture are stored in a silo in the absence of air. The entire ensiling process requires 2 to 3 weeks, during which time a small amount of oxygen is deleted with aerobic respiration, and anaerobic fermentation occurs.

(Also, see Chapter 9, Silage/Haylage/High-Moisture grain.)

- **Improvement of feed nutrient content by fermentation**—The age-old practice of slop-feeding pigs was a continuous-batch fermentation process. Today, more sophisticated and better controlled fermentation techniques are being used; among them, the following:

1. **Proteins and amino acids.** Yeasts, which are a fermentation product, probably have the most favorable characteristics of all microorganisms as major protein sources. Brewers' dried yeast, the most common kind of yeast used in stock feeding, is made from the yeast filtered from beer or ale after the fermentation is completed. Torula-type yeast is made by the fermentation of pentoses, such as the waste sulfite liquor from making paper pulp. In addition to its protein value, yeast carries with it some nutritionally important bonuses, especially the B complex vitamins.

Fig. 14-5. Block in use on pasture—a means of lessening the labor attendant to the daily feeding of a supplement on pasture or range. (Courtesy, Moorman Manufacturing Co., Quincy, Ill.)

It is frequently advantageous to supplement livestock rations with specific amino acids, rather than intact proteins. Thus, it is noteworthy that fermentations are already operated on a commercial scale to produce lysine and glutamic acid.

2. **Vitamins.** Certain vitamin supplements may be provided by fermentation. Feasible processes have been developed for synthesis of carotene and vitamin A, riboflavin, and vitamin B-12. Also, the B complex vitamins are provided by a multitude of microorganisms.

3. **Antibiotics.** While conducting nutrition studies with poultry in 1949, Jukes and Stokstad of Lederle Laboratory and McGinnis of Washington State University, demonstrated that much of the growth-promoting activity in certain fermentation solubles was due to their antibiotic content. This ushered in the era of feeding antibiotics to livestock.

4. **Enzymes.** Microbiological processes for producing various enzymes are available, and the thought of improving the digestive efficiency of animals by adding the appropriate enzyme(s) to the ration is intriguing. However, the enzymatic output of the digestive system of animals is adequate for maximum digestion of starches, fats, and proteins.

The above discussion identifies some of the contributions that have been made by fermentation processes to the feeding of animals. Beyond these accomplishments there exists an area of significant impact of fermentative processes in converting by-products and waste materials into livestock feeds, and, at the same time, lessening the pollutants in the environment of humans. For example, considerable research is in progress to convert manure from poultry and other animals through bacterial fermentation into animal protein feed. This recycling process could produce much more protein and help solve a pollution problem.

Hydroponics

Hydroponics is the growing of plants with their roots immersed in an aqueous solution containing the essential mineral nutrient salts, instead of soil. This means that the plants are produced with water and chemicals, but without soil.

The Wisconsin Alumni Research Foundation chemically analyzed and compared the composition of oat grain and 5-day oat grass on a dry matter basis (see Table 14-1).

TABLE 14-1
COMPOSITION OF OAT GRAIN AND 5-DAY OAT GRASS, MOISTURE-FREE BASIS[1]

Constituent		Oat Grain	Oat Grass
Dry Matter	(%)	100.00	100.00
Protein	(%)	15.00	21.00
Ether extract (fat)	(%)	4.21	5.20
Nitrogen-free extract	(%)	65.86	42.79
Fiber	(%)	11.71	26.11
Ash	(%)	3.22	3.90
Calcium	(%)	0.063	0.238
Phosphorus	(%)	0.360	0.509
		(mg/kg)	(mg/kg)
Carotene[2]		0	39.067
Vitamin E		17.95	48.87
Niacin		7.18	103.96
Riboflavin		1.96	22.29
Thiamin		3.14	12.86
Vitamin C		0	218.3

[1]Analyses by Wisconsin Alumni Research Foundation.
[2]Each mg of beta carotene was considered to be equivalent to 1,556 IU of vitamin A.

As shown in Table 14-1, the 5-day oat grass is a better source than oat grain of calcium, phosphorus, carotene, vitamin E, the B vitamins (riboflavin, thiamin, and niacin), and vitamin C. In addition to these comparative figures, however, the following facts are pertinent:

1. Supplemental quantities of calcium and phosphorus can be provided in many forms at a relatively low cost.

2. Sprouting greatly increases the carotene content; hence, if carotene, or vitamin A, is deficient, sprouted grains are a good supplemental source. However, supplemental vitamin A can be provided in a dry, stabilized form at a low cost.

3. Most rations are adequate in vitamin E. The B vitamins (riboflavin, thiamin, and niacin) are produced by the microorganisms in cattle, sheep, and horses; hence, supplemental quantities of them are not normally needed. Vitamin C is not required in the ration of farm animals.

The Michigan Station researchers made a study of hydroponically-grown oat forage as a feed for dairy cows. They reported that the cost of the hydroponically-produced oat forage was over four times that of the original oats or similar grains; and that there was a loss in nutrients during sprouting, a decrease in digestibility of sprouted oats, and no observed increase in milk production when sprouted oats were added to an adequate ration.[2]

Based on studies conducted by the different universities, sprouting results in an average loss of 83% of the dry matter of the oat grain. One study showed a reduction in TDN from 75.7% in the oat grain to 70.2% for the sprouted oats. Also, the digestibility of dry matter, energy, protein, ether extract, and nitrogen-free extract was lower for the sprouted oats than for the oat grain. The composition of hydroponically-grown forage will vary according to the growth stage of the plants, the temperature, the nutrients in the aqueous solution, and several other variables

In arriving at a decision whether to produce feeds hydroponically, consideration should be given to (1) the needs of different classes of livestock for each nutrient, and (2) the cost of supplying these nutrients hydroponically.

Unprocessed (Whole) Corn

Unprocessed (whole) corn refers to shelled corn, the kernels of which have not been broken.

It is generally recognized that young cattle (both beef and dairy animals under 6 months of age) masticate their feed well. Thus, although the digestibility of corn may be increased when it is processed for young bovines, the increased feeding value may not be sufficient to offset the added cost of processing. With the exception of young cattle, it has been assumed that corn should be ground, or otherwise processed, for cattle. Recent experiments at a number of experiment stations have indicated that there are exceptions—that the proportion of concentrate to forage is a factor in determining whether corn should be processed for cattle. Cattle fed dry, whole shelled corn gain an average of 5% faster and require 7% less feed per pound of gain than cattle fed ground, rolled, or crimped corn *when high-concentrate rations are fed*. However, processing appears to have some value for dry shelled corn in rations with 20% or more roughage content or when corn is very dry—less than 12% moisture.

Eliminating processing costs is the main advantage of feeding whole corn.

[2]Report from *Quarterly Bulletin*, Vol. 44, No. 4, Michigan State University, East Lansing, May 1962, pp. 654–665.

SOME NONNUTRITIVE FEED ADDITIVES

In processing practical livestock feeds, nutritionists may include those nonnutritive feed additives that will improve the ingestion, digestion, protection, absorption, and/or transport of the nutrients to an extent that will increase the nutritive value of the feed and decrease the feed cost for production. No blanket recommendation is possible as to which additives are most useful. Each nutritionist must decide which feed additive(s) is needed under each specific set of circumstances. Among such nonnutritive additives are the following:

1. **Antifungals (mold inhibitors).** Used to prevent harmful molds in feeds and/or in the digestive tract.
2. **Antioxidants.** Used to protect the polyunsaturated fatty acids and the fat-soluble vitamins from destruction by peroxidation.
3. **Enzymes.** Used to improve the digestibility of certain feedstuffs.
4. **Flavoring agents.** Used in an effort to improve the palatability of feed.
5. **Pellet binders.** Used to improve firmness of pellets.

Of course, many other additives than those listed above are incorporated in modern livestock rations; among them, antibiotics, arsenicals, and hormones. These are covered in Chapter 13, Feed Supplements/Additives/Implants, and in the respective chapter devoted to each of the several species; hence, the reader is referred thereto.

EFFECT OF STORAGE ON FEEDSTUFFS

The extent of the effects of storage on feedstuffs is dependent upon a number of factors, including moisture content, temperature, degree of maturity when harvested, the manner of handling until it is placed in storage, the type of construction of storage bins or containers, and the length of time stored. Also, whole grains generally withstand storage better than the same grains after processing or milling.

It is generally recognized that several of the vitamins are unstable in storage—that they encounter many antagonists in storage, with oxygen heading the list. All the obvious factors contributing to the destruction of vitamins—time, moisture, heat, trace minerals, and low pH—hasten the oxidation of vitamins. This instability of vitamins, along with their recognized very great nutritional importance, prompted many commercial vitamin manufacturers to improve their storage qualities in two ways: (1) by enveloping the vitamin or vitamins in a stable fat or gelatin, forming small beads, thereby preventing most of the vitamins from coming in contact with oxygen until consumed by the animal, and (2) by the use of effective antioxidants that will delay oxidation.

Based on limited work, it may be tentatively concluded that storage affects the nutritive value of feeds as follows:

1. **Loss of carotene.** Carotene content decreases significantly in storage, with the consequent loss of vitamin A activity. Experiments have shown that the loss of carotene in alfalfa meal during storage is only 10% in 6 months at a temperature of −9° to −15°F, whereas the loss is 60 to 73% over the same period of storage at room temperature. Also, studies show that the carotene of alfalfa is completely preserved by storage *in vacuo* or under nitrogen.
2. **Trace minerals destroy vitamins.** Of the many ingredients present in feeds or vitamin premixes, trace mineral elements probably have the greatest effect on vitamin losses. When heat and moisture are encountered, vitamin destruction is accelerated; and in the presence of some trace minerals, the rate of loss is compounded.
3. **Destruction by light.** Riboflavin, pyridoxine, and ascorbic acid are readily destroyed by light. It is important, therefore, that premixes of feeds containing these vitamins be stored in a dark place.
4. **Vitamins A, D, and E lowered.** Vitamins A, D, and E are unstable in storage.
5. **Thiamin (B-1) little affected.** Thiamin (B-1) content is little changed by longtime storage under favorable conditions.
6. **Fats deteriorate.** Fats deteriorate, with the formation of free fatty acids, which may affect palatability. These effects can be prevented by proper use of an effective antioxidant, like ethoxyquin or butylated hydroxytoluene (BHT).
7. **Proteins may deteriorate.** Proteins may deteriorate, especially if grains are placed in storage bins immediately following harvesting.
8. **Insects may destroy.** Unless proper steps are taken to control insects through sanitation and the use of insecticides, insects may destroy feedstuffs.

In view of the above, adequate overage of the various vitamins should be provided in all feeds or premixes subjected to processing and/or time and temperature storage conditions conducive to rapid loss of vitamins. Also, trace minerals should be used with discretion, light should be subdued, antioxidants should be used, and insects should be eliminated.

FORAGE PROCESSING METHODS

In recent years, researchers and feeders have been much interested in improved processing of grains. But little study has been made of forage preparation, except from the standpoint of mechanizing and ease in mixing. With the increased competition of grain for human consumption around the world, it is expected that roughage preparation will assume greater importance.

Before discussing each of the common methods of forage preparation, the following generalizations are pertinent to all of them:

1. Most forages are roughages; and ruminants need roughages.
2. In preparing forages, avoid processing those (a) with high moisture, which may heat and produce spontaneous combustion, and (b) in which there are foreign objects (wire and other hardware) which animals may not be able to avoid, and which may generate sparks and ignite a fire during processing (grinding-chopping, conveying, or mixing).
3. Processing forages does increase cost from $2 to $10 a ton, depending on the method of processing.
4. Processed forages result in the forced feeding of the entire plant, including stems which may be of low nutritional value. With high-producing animals, this may be a disadvantage.
5. The preparation of forages does not increase the value of the initial product.

The common methods of forage preparation are chopping, grinding, shredding, cubing (wafering), drying, ensiling, and pelleting.

Chopping, Grinding, or Shredding

Chopping, grinding, or shredding result in forages divided into smaller particles; but they differ in how they section it, and in the size of the particles. In comparison with a similar forage fed in long form, a forage subjected to any one of these three processes (1) is easier to handle and mechanize, (2) can be stored in a smaller area at less cost, (3) is fed with less feed refusal and waste, and (4) may make for slightly greater production.

Low-quality, coarse forages are usually improved more from chopping than high-quality, fine forages. This should not be construed as license to make poor-quality forage, then improve it by processing. Rather, processing makes for less waste, and perhaps some improvement in digestibility, but it does nothing to improve the nutrient content.

- **Chopping**—This refers to cutting forage not less than 2 in. in length. (The 2 in. refers to the set of the choppers. Some of the material will be cut longer than this, and some shorter).

Chopping has the disadvantage of being dusty. Also, there may be considerable leaf loss, or shattering, in field chopping because the hay must be drier than when it is baled or put up as long hay.

(Also see Chapter 8, Hay, section on "Chopped Hay.")

- **Grinding**—This refers to processing forage less than 1 in. in length. Usually grinding is accomplished by means of a hammer mill, in which the forage is beaten by revolving metal hammers until it is small enough in size to pass through the screen placed in the grinder. Generally, screens with holes ¼ in. or larger are used so as to avoid pulverizing the hay. Chopping to a length of less than 1 in. is also referred to as grinding, even without the hammer mill treatment.

Fine grinding is more costly than coarse chopping; hence, it is less appealing from a practical standpoint. Yet, fine grinding is sometimes desirable when the material (either sun-cured or dehydrated) is to be incorporated in the rations of swine or poultry. Ground forages are less digestible for ruminants because they pass through the paunch more rapidly, with only limited bacterial action. When finely ground hay is fed to lactating cows, the fermentation in the rumen produces less acetic acid and more propionic acid than when coarse forage is fed, and, in turn, this results in the fat content of the milk being substantially reduced.

When it is advantageous to use ground hay in a ration, the addition of molasses, fat, or water will lessen the dustiness and reduce the air pollution by nutrients. Some commercial mills spray a small amount of liquid fat on bales of hay just before they enter the grinder. Fat is easier to work with in a grinder or mixer than molasses, for the latter has a tendency to be sticky and "gum up" the equipment.

- **Shredding**—This process is similar to chopping, except shredding tends to separate the stems longitudinally rather than cut them crosswise. Coarse forages, such as fodder and stover, are better suited to shredding than to chopping and grinding. In some ways, it may be superior to chopping, because of exposing more of the inner part of the stem to fermenting bacteria in the rumen, thereby increasing the likelihood of better digestion. Shredding necessitates that hay be as dry as when it is chopped (10% or less); hence, it may result in as much leaf shattering as chopping. Shredding appears to have a more desirable image than chopping, with the result that hay chopping is sometimes referred to as shredding, when it is really nothing more than conventional chopping.

Cubing[3] (Wafering)

When applied to forages, the term cubing (wafering) refers to the practice of compressing long or coarsely cut hay into cubes about 1¼ in. square and 2 in. long, with a bulk density of 30 to 32 lb per cubic foot. They do not necessitate fine grinding, and they facilitate automation in both haymaking and feeding. Cubing costs about $5 per ton more than baling.

This method of haymaking is increasing, because it offers most of the advantages of pelleted forages, with few of the disadvantages. Because cubed forage is relatively coarse, it lowers milk fat percentage only slightly, if at all.

It is noteworthy, however, that horses occasionally choke when fed cubes.

(Also see Chapter 8, Hay, section on "Cubes [Wafers].")

Drying

For safe storage, the moisture of hay must be lowered to the following levels: loose hay, 25%; baled hay, 20 to 22% (the lower figure for larger bales); field chopped hay, 18 to 20%; and cubes (wafers), 16 to 17%. These figures must be modified according to temperature; higher temperatures necessitate lower moisture.

Generally, hay moisture is lowered by field curing. However, artificial drying—artificial dehydrators, mow curing, and wagon dryers—may be used during times of inclement weather or when very high-quality forage is desired.

Artificial dehydrating refers to that process in which forage is taken from the field as soon as it is cut (or in some instances after wilting), put through a hay chopper or silage cutter, and dried in large rotating drum driers of different types. For the most part, this method of drying is limited to processing forage for swine and poultry feeds. The most popular type of artificial dehydrator in use in this country is one that uses a high initial heat (1,200° to 1,400°F), and which is usually heated by natural gas or fuel oil. Due to high equipment and fuel cost, along with the added cost of moving heavy high-moisture forage from field to dryer and in operating the dehydrator, artificially dried forages must command a premium price over field-cured forage.

(Also see Chapter 8, Hay, section on "Reducing Moisture Content/Shattering/Bleaching and Fermenting.")

Ensiling

Ensiling refers to the changes which take place when forage or feed with sufficient moisture to allow fermentation is stored in a silo in the absence of air. The entire ensiling process requires 2 to 3 weeks, during which time a small amount of oxygen is deleted with aerobic respiration, and anaerobic fermentation occurs.

---

[3]There is overlapping in the use of the word *cube*. The compressed long or coarsely cut hay packages about 1¼ in. square and 2 in. long are known as cubes. Also, pellets that are large enough to be fed on pasture or range are commonly called cubes.

Pelleting[4]

When applied to forages, the term pelleting refers to the process of forcing ground forage (usually with some added moisture) through a thick steel die and compressing it into a circular or rectangular mass which is cut at predetermined lengths. They can be formed into shapes of varying thickness, length, and hardness. The larger shapes, commonly fed to cattle and sheep on the range, are referred to as cubes.

(Also see Chapter 8, Hay, section on "Pellets.")

MISCELLANEOUS PROCESSING METHODS

There is hardly any limit to the number of processing methods—some old, others new. Some preserve quality, others increase consumption and lessen labor, and still others change the chemical composition and feeding value. In addition to the processing methods already covered, several miscellaneous, but important, methods are discussed in the sections that follow.

Ammoniation

Ammonium salts and anhydrous ammonia (gas or liquid) have been used for ammoniating feeds that contain high levels of carbohydrates and low levels of nitrogen. Among such ammoniated feeds are: citrus pulp, beet pulp, molasses, sugarcane bagasse, and rice hulls. Also, low quality roughages may be ammoniated.

(Also see Chapter 11, Protein Supplements, section on "Ammoniated Products.")

Animal Waste (Manure) Processing

Animal waste (manure) has nutritive value for ruminants because these animals are capable of utilizing nonprotein nitrogen and fiber. So, proper processing is important.

Broiler and layer litter has been successfully used as an ingredient of cattle feed for many years. However, wastes from all species may be, and are, used. Among the methods employed to process animal wastes prior to feeding are: deep-stacking, ensiling (fermentation), dehydration, and pelleting. The two most common and practical methods of processing are:

1. **Deep-stacking.** In this method, the litter is deep-stacked for several weeks, during which it generates temperatures of 160°F or higher, which render it free of any potentially pathogenic microorganisms that might be present. (Pathogenic bacteria do not grow at temperatures over 80°F, and they are killed at 145°F in a matter of minutes.) It follows that there have been no documented animal health problems associated with feeding broiler or layer litter processed in this manner.

2. **Ensiling (fermentation).** Ensiling is a controlled fermentation process during which carbohydrates in the mixture are converted to lactic, acetic, and other acids. Once sufficient acids are produced, bacterial action ceases and the ensilage is stable. During the fermentation, heat is generated, thereby diminishing the hazard from certain pathogenic organisms that might be present.

Fat Added

Typical livestock rations contain relatively small quantities of fat. Although most animals require minimal amounts of certain fatty acids in their diets, these minimals are low and easily met by normal rations. Nevertheless, fat serves the following practical functions when added to livestock rations:

1. It increases the caloric density of the ration.
2. It improves palatability.
3. It facilitates absorption of vitamins A and D and provides fatty acids.
4. It delays the sensation of hunger.
5. It controls dust.
6. It lubricates feed processing equipment.
7. It improves handling qualities.

However, the following problems are inherent in the incorporation of fats in feeds:

1. Animal fats tend to solidify in cold weather.
2. Fats can coat and clog mixing and distribution equipment.
3. High levels of fat in pelleted feeds can cause soft pellets.
4. Fats can become rancid.

(Also see Chapter 10, Section on "Fats and Oils"; and Chapter 13, section on "Mold (Fungi) Inhibitors.")

Freezing

A considerable amount of mink food, consisting of meat and fish, is frozen. Freezing inhibits bacterial growth and slows the enzymatic processes which can destroy the product. Proper freezing is the key to good mink nutrition.

Irradiation

Upon irradiation, ergosterol, a plant sterol, yields ergocalciferol, commonly known as vitamin D_2.

The ultraviolet radiation in sunlight serves as a source of radiant energy necessary to convert 7-dehydrocholesterol (an animal sterol stored beneath the skin surface) into biologically active vitamin D_3.

Vitamin D_2, the plant form of the vitamin, and vitamin D_3, the animal form, have the same antirachitic value for the rat, dog, pig, ruminant, and human, but vitamin D_3 is more active for poultry.

Molasses Added

Molasses (including cane or blackstrap, beet, citrus, wood and starch molasses) is extensively used as a livestock feed. When used at levels of 5 to 15% of the ration, it has about ¾ the energy value of corn. However, molasses has added values as an appetizer, to reduce dustiness of a ration, as a binder for pelleting, to stimulate rumen microbial activity, and as a source of unidentified factors. Also, cane molasses is a good source of certain trace minerals.

(Also see Chapter 10, Grains—High-Energy Feeds, section on "Molasses.")

---

[4]Pellets may refer to (a) the entire concentrate in pellet form, (b) the fines of the concentrate in pellet form, which are usually added back to the grain for feeding, (c) the forage in pellet form, (d) the protein supplement, or (e) the range supplements in pellet form.

Organic Acids

The proper use of organic acids provides another way in which to preserve high-moisture grains. The organic acid treatment involves the application of 1 to 1½% acid (*i.e.,* propionic, acetic, formic, ammonium isobutyric, etc.) at time of harvest, followed by storage in a pile. Commercial applicators have been developed whereby the grain flows into a hopper and is sprayed with acid during spiral action of exposed auger flights. The acid flow rate is adjustable and is dependent on auger throughput and moisture content of the grain. The acid is absorbed into the kernel surface, giving it a shiny or glazed appearance, and becomes absorbed into the kernel. Properly treated high-moisture corn has the same appearance after 18 months as it has immediately after treatment.

Acid treatment inhibits the growth of molds and bacteria. Research has shown that propionic acid alone or a mixture of 75% acetic and 25% propionic acid are quite effective. Acetic acid should not be used alone. Limited research indicates that sodium propionate, formalin, ammonium isobutyrate and citric acids have been successful, as well as combinations of propionic acid and formic acid or formalin.

Experimental studies indicate that acid-treated grain has approximately the same feeding value as high-moisture grain stored in an oxygen-limiting silo. Also, it alleviates the cost of drying. Thus, organic acid treatment of grain may be a practical way in which to preserve high-moisture grains.

(Also see Chapter 9, Silage/Haylage/High-Moisture Grain.)

Preservatives

A preservative is a material added at the time of mixing or storing to enhance the keeping qualities of a feed.

Various methods of preserving feeds have been used, including drying, oxygen-limiting silos (and smaller containers), freezing, and organic acids. Several of these preservatives are covered earlier in this chapter. A brief description of hay and silage preservatives is in order, however.

- **Hay preservatives**—Preservatives are available commercially which can be applied to hay. Usually the directions (1) recommend the addition of 1 to 3 lb of these products for each ton of damp hay, and (2) claim that there will be no heating or molding.

More experimental work is needed relative to chemical hay preservatives. But, available data indicate that propionic acid is the hay preservative of choice.

Anhydrous ammonia is one of the most recent materials being studied as a hay preservative. In Indiana trials, applying this material at the rate of 1.0% to hay baled at 30% moisture successfully prevented molding, heating, and quality deterioration.

(Also see Chapter 8, Hay, section on "Additives for Hay.")

- **Silage preservatives**—Two types of additives have generally been used in silage making: (1) feed additives, and (2) chemical additives.

Feed additives supply a readily available source of carbohydrates for bacterial fermentaion of the silage. Some feed additives, such as corn-and-cob meal, when mixed with high-moisture forages, also absorb water and help to reduce runoff. When used as preservatives, approximately 75 to 85% of the feed nutrients added may be recovered as feed.

A large number of chemical additives have been used in silage making, with variable results. These are fully discussed in Chapter 9, Silage/Haylage/High-Moisture Grain, section on "Silages, Additives, and Preservatives"; hence, the reader is referred thereto.

Self-feeding Governors

The commonly used self-feeding governors are (1) bulky, fibrous feeds; (2) salt-feed or fat-feed mixtures; (3) fat content of block; and (4) liquid supplements.

- **Bulky, fibrous feeds**—Bulk can be used as a self-feeding governor. This consists in adding to the bulkiness of the ration, such as can be achieved by increasing the amount of chopped hay and lessening the concentrate. Actually, this is a way in which to lower the energy content of the ration. Since an animal can hold only so much, it is an effective control of feed intake.

- **Salt-feed mixtures**—The practice of using salt as a governor to limit feed consumption on pasture or range has been used for a very long time. It was ushered in as a labor-saving device for cattle and sheep in inaccessible and rough areas. Today, salt-feed mixtures are used in either meal or block form.

- **Fat content of block**—Since animals tend to eat until a certain caloric intake is reached, they consume less total weight when fed high-fat rations. Thus, pounds of feed consumed can be governed by the amount of fat in a block. It is noteworthy, too, that fat serves as a needed feed nutrient, whereas consuming more salt than required (as happens when a salt-feed mixture is used) makes for a waste of salt.

- **Liquid supplements**—When self-fed, the consumption of liquid supplements is generally controlled by (1) the use of a lick tank, and/or (2) incorporating in the formulation phosphoric acid, beet solubles, and/or citrus peel liquor.

(Also see Chapter 11, section on "Types and Methods of Feeding Protein Supplements.")

Slow-Release and Rumen Bypass Treatments

Two feed processing techniques—slow-release nonprotein nitrogen, and rumen bypass protein—are designed to delay digestion.

- **Slow-release nonprotein nitrogen**—Among the slow-release nonprotein nitrogen products that liberate nitrogen slowly are a combination of urea and gelatinized starch, and urea combined with gelatinized corn.

- **Rumen bypass**—This refers to bypass protein (also known as protected or escaped protein) in feed that escapes digestion in the rumen and passes into the lower digestive tract where it is digested and absorbed. Feed processors have developed treatments through which the bypass proteins in certain feeds can be increased; among them, heat and pressure treatment, treatment with tannins, treatment with formaldehyde or other aldehydes, lipid (fat) treatment, complexing with bentonite clay, use of amino acid analogs, increasing microbial metabolism in the rumen, and adding ionophores.

(Also see Chapter 11, section on "Protein Bypass (Protected Protein, Escaped Protein.")

Treatment of High-Cellulose Feeds

High feed prices and more stringent burning regulations have spurred research to find a practical method of improving the feeding value of several high-cellulose products, such as rice, wheat, barley and oat straws; bagasse; tree bark; corncobs; gin trash; newspaper; and seed hulls.

In their natural state, these products make poor feedstuffs because lignin or silica, or a combination of the two, (1) encrust the energy-rich carbohydrates, cellulose, and hemicellulose; and (2) keep the microbes in the ruminant's stomach from breaking them down to release energy.

The answer to this problem lies in some treatment that opens up the fibers enough to permit increased digestion in the rumen. Several methods of chemical and/or physical treatment are being investigated; among them, alkali treatment, ammoniation, hydrogen peroxide treatment, and high pressure steam.

• **Alkali treatment**—Sodium hydroxide is the common alkali treatment of high-cellulose products (crop residues), although calcium hydroxide and potassium hydroxide are sometimes used. The effectiveness of alkali treatment depends on the residue or waste being treated and the technique employed. Treatment level ranges from about 2 to 10% of the chemical based on the total dry matter content of the material being treated; and treatment time is about 24 hours. There is indication that mild heat can be used to reduce chemical levels.

• **Ammoniation treatment**—This method involves placing the high-cellulose products (crop residues) in an air-tight enclosure (such as black plastic) and adding either anhydrous gas or liquid ammonia. It is important to add the correct amount of ammonia; too much makes for unnecessary expense, and too little makes for poor quality feed. Optimum treatment level appears to be 3.0 to 3.5% anhydrous ammonia based on the total dry matter content of the material being treated; and optimum treatment time is about 20 days vs 24 hours for the sodium hydroxide treatment. The two major advantages of using ammonia in comparison with the alkali treatment are: (1) it adds nonprotein nitrogen to the product, and (2) no mineral residue remains that might be detrimental to the animal or to the soil to which the manure is added. Ammoniation produces the following benefits:

1. It increases the crude protein equivalent by 3 to 10%.
2. It increases the digestible energy (TDN) by 3 to 23%.
3. It increases animal intake by 20 to 27%.
4. It prevents molding of crop residues and high-moisture forages.

Although the ammoniation treatment of low-quality forages will, at low cost, increase both protein equivalent and fiber digestibility, the toxicity problem cannot be ignored.

Ammonia toxicity (characterized by hyperexcitability, circling, convulsions, and some deaths) occurs among some cattle and sheep receiving ammoniated feeds and among some young suckling lactating mothers fed ammoniated feeds. Ohio researchers reported that 3 out of 9 sheep (33%) fed ammoniated hay (4% ammonia on a dry basis) developed signs of toxicity, and 2 of them (22%) died; and that 4 out of 5 calves (80%) that received milk from cows fed ammoniated hay (5% ammonia on a dry basis) developed signs of toxicity, and 1 calf (20%) died.[5]

CAUTIONS relative to ammoniation: Anhydrous ammonia in liquid form is very toxic to the skin and eyes; so, when a person's skin/eyes come in contact with anhydrous ammonia, it should be flushed away with water immediately; otherwise, serious injury may result. Also, ammonia can be flammable and even explosive; so, never smoke or light a flame near it.

(Also see Chapter 4, section headed "Urea and Ammoniated Feeds"; Chapter 11, section headed "Ammoniated Products"; and Chapter 5, Table 5-3, Potential Poisons—Urea [Ammonia Toxicity].)

• **Hydrogen peroxide**—In this treatment, the crop residue (corn stalks, wheat straw, or other crop residue) is shredded and mixed with a 1% solution of hydrogen peroxide. The pH level of the mixture is brought up to 11.5 with an alkali material. The crop residue literally disintegrates into a mushy substance, which is rinsed off; following which it can be dried or handled wet, in the same way as corn gluten feed. When dried, the material is light and flaky. It can be pelleted to increase its density and ease of handling. Limited experiments with cattle and sheep show that corn stalks and wheat straw treated with hydrogen peroxide has feed value higher than corn silage.

• **High pressure steaming**—High pressure steaming, with or without added chemicals, has been used to a limited extent in treating crop residues and wood. Aspen, which is the most digestible of the woods, has been shown to reach a digestibility of 56.6% after steaming for 2 hours at 165°C, with the treated product readily accepted by sheep at up to 60% of the ration and producing normal weight gains and carcass yields. The treating of wood products is limited because: (1) the cost of treatment is high, (2) conventional feedstuffs need to be relatively high priced before treated wood residues can compete in the marketplace, and (3) the lack of a steady market for the treated products.

Decisions as to type and amount of chemicals and treatment system must be based on evaluation of processing costs in relation to the value of the finished feedstuff.

(Also see Chapter 12, By-product Feeds/Crop Residues, section headed "Treating Crop Residues to Increase Digestibility.")

COMPLETE (ALL-IN-ONE) RATIONS

Most experiments and experiences have not shown any difference between mixed rations and the feeding of roughage and concentrates separately insofar as efficiency and production are concerned. However, a mixed ration has the following advantages:

1. It makes for greater efficiency in feeding and lessens the sorting at the feed bunk.
2. When the roughage is relatively unpalatable, a mixed ration forces consumption.

---

[5]Weiss, W. P., et al., Etiology of Ammoniated Hay Toxicosis, *Journal of Animal Science*, Vol. 63, No. 2, 1986, p. 525.

3. When it is desired to limit concentrate consumption, mixing with the roughage is desirable.

4. A mixed ration makes it easier to get animals on full feed.

Thus, each feeder must decide on the matter of mixed feed vs feeding roughage and concentrate separately, with relative costs and other factors considered. Most large cattle and sheep feedlots use completely mixed rations. Also, the trend is toward complete feeds for both dairy cows and swine, primarily because such complete feeds (1) lend themselves better to automation, and (2) provide better control of nutrient intake.

- **All-pelleted rations (grain and forage combined)**—Increasingly, complete pelleted rations are being used for horses, swine, and fish. Among the virtues ascribed to all-pelleted rations are (1) They prevent selective eating—if properly formulated, each mouthful is a balanced diet; (2) they alleviate waste; (3) they eliminate dust (thereby lessening heaves in horses); (4) they lessen labor and equipment; (5) they lessen storage; and (6) they facilitate automation.

CHOICE OF PROCESSING METHOD

The choice of a processing method is highly dependent on the feedstuff to be fed. It is clear that a given processing technique may be very desirable for one grain, but quite detrimental to another. Corn may be fed without any processing, but not milo. Pressure treating appears to be desirable for milo, but harmful to wheat.

Comparison of grain processing techniques is difficult because there are a number of interactions between processing technique and roughage level or type of ration fed. For example, data from Ohio State University have shown that whole shelled corn was superior to crimped corn in very low-roughage rations, whereas crimped corn was clearly superior in high-roughage rations.

Interactions cause results on new grain processes to be biased by the kind of control ration fed. Consequently, feeders should consider only those comparative tests which involve rations and conditions very similar to those they intend to use in their own feeding program.

FARM PROCESSED FEED

There are two alternative sources of most feeds and rations—home mixed vs commercially mixed; and the able manager will choose wisely between the two.

The value of farm-grown grains—plus the cost of ingredients which need to be purchased to balance the ration, and the cost of processing and mixing—as compared to the cost of commercial ready-mixed feeds laid down on the farm, should determine whether it is best to mix feeds at home or depend on ready-mixed feeds.

Although there is nothing about the mixing of feeds which is beyond the capacity of the intelligent farmer or rancher, under many conditions a commercially mixed feed supplied by a reputable dealer may be the most economical and the least irksome, especially when many ingredients and additives are involved, such as in young stock and poultry rations. The commercial dealer has the distinct advantages of (1) purchase of feeds in quantity lots, making possible price advantages, (2) economical and controlled mixing, and (3) the hiring of scientifically trained personnel for use in determining the rations.

Modern feed mixing is much more complicated than in the era when hog rations consisted primarily of corn and tankage. Today, micronutrients are added, and this necessitates both sophisticated equipment and skillful mixing. For example, a recommended swine ration may call for the addition of 1 mg of riboflavin per pound of mixed ration. When one considers that a grain of wheat weighs 60 mg, some perspective of the mixing problem comes into focus. Baby chick rations are even more exacting. With a consumption of only 2 to 3 oz of feed per day, it is important that the chick get all the trace amounts of the various required nutrients in that quantity. This necessitates that the feed be blended so well that a tablespoon from one sack will be almost identical in chemical composition to a tablespoon taken from the same batch 20 bags later.

Because of these several advantages, commercial feeds are finding a place of increasing importance in American agriculture. Nevertheless, practical considerations favor the use of much homegrown feed. Many feed manufacturers formulate supplements for this particular purpose. For example, a supplement may be prepared for a corn-soybean ration for hogs.

Fig. 14-6. Turkey eating pellets. (Courtesy, Ralston Purina Company, St. Louis, Mo.)

QUESTIONS FOR STUDY AND DISCUSSION

1. Explain the difference between physical and chemical feed processing.

2. Cite examples of feed processing lowering the nutritional value of feedstuffs.

3. Why does the importance of proper feed preparation become increasingly important with higher levels of feeding and greater production?

4. How does the size of feeding operation affect the choice of feed preparation?

5. List and discuss what you consider to be the six most important reasons for processing feeds.

6. List and discuss what you consider to be the four most important nutritional factors in choosing the grain processing method.

7. List and discuss what you consider to be the four most important nonnutritional factors in choosing the grain processing method.

8. Briefly describe each of the following processing methods:

 Extruding   Steam flaking
 Popping   Pelleting
 Roasting   Crumbling
 Cooking

9. What is meant by *extruding* feed? How is this process accomplished?

10. What does roasting do for soybeans? Why is there concern lest the feeding of whole soybeans produce soft pork?

11. Are beef cattle fitters justified in cooking barley for animals being fitted for show?

12. Why do poultry producers tend to favor crumbles?

13. How is a bran mash prepared and used for horses? Why is a bran mash used?

14. How will energy shortage and high energy cost affect the choice between the following preparation and processing methods of corn harvested at 30% moisture?
  a. Storage in a silo
  b. Artificially dried
  c. The use of an organic acid

15. Can the use of hydroponics be justified for race and show horses?

16. Why are antioxidants incorporated in feeds?

17. List and describe how storage may affect the nutritive value of feeds.

18. Explain why forage should not be finely ground when fed to lactating dairy cows. Why do dairy operators favor cubed (wafered) forage over pellets for lactating cows?

19. Identify and describe the two most common and practical methods of processing broiler litter.

20. Of what practical value is irradiation in the production of vitamin D?

21. Under what circumstances would you use a preservative for each (a) hay, and (b) silage? Which preservative(s) would you use?

22. Under what circumstances would you use self-feeding governors, such as salt or fat, on the western range?

23. What makes rumen bypass such an exciting area for research?

24. What's the practicality and future of improving the feeding value of high-cellulose feeds through alkali treatment, ammoniation treatment, hydrogen peroxide treatment, and high pressure steaming?

25. Discuss the difference between complete mixed rations vs feeding roughage and concentrate separately.

26. Under what circumstances would you recommend commercially prepared feed? Under what conditions would you recommend home mixing?

Fig. 14-7. Chopping cured hay directly from the windrow, using a forage harvester. (Courtesy, Gehl Company, West Bend, Wisc.)

Fig. 14-8. Forage may be processed by fermenting (ensiling) in a silo. (Courtesy, The American Guernsey Cattle Club)

Fig. 14-9. Alfalfa hay cubes, 1¼ in. long. (Courtesy, Deere & Company, Moline, Ill.)

Fig. 14-10. Alfalfa pellets. (Courtesy, American Dehydration Assoc., Mission, Kan.)

Fig. 14-11. A complete (all-in-one) ration being fed to finishing cattle. The ingredients: corn silage, high moisture corn, steam flaked corn, sugar beet pulp pellets, alfalfa pellets, brewers' malt pellets, and molasses. (Courtesy, Wm. R. Farr, Farr Feeders, Inc., Greeley, Colo.)

Fig. 14-12. A complete (all-in-one) ration self-fed to turkeys in feeders. (Courtesy, *Turkey World*, Mount Morris, Ill.)

Fig. 14-13. This advanced electronic feed mill, which houses the computerized central control facility and utilizes a continuous flow mixing process, is capable of producing more than a ton of mixed feed per minute for an 80,000 head per year cattle feedlot near Greeley, Colorado. The continuous flow system, incorporating the use of augers and weigh belts, ensures ration accuracy within 1% on each ingredient. (Courtesy, Farr Feeders, Inc., Greeley, Colo.)

Original painting by Tom Phillips

15

FEED ANALYSIS/ FEED EVALUATION

Contents	Page
Importance of Monitoring Feed Quality	283
How to Take Feed Samples	284
Mixed Feeds	284
Grains	284
Silage	284
Hay	284
Pasture	284
Feed Analysis	284
Physical Evaluation of Feedstuffs	284
Microscopic Examination	285
Proximate Analysis	285
Moisture	286
Ash	286
Crude Protein (CP)	286
Ether Extract (Fat)	287
Crude Fiber (CF)	287
Nitrogen Free Extract (NFE)	287
Neutral Detergent Fiber (NDF) and Acid Detergent Fiber (ADF)	288
Bomb Calorimetry	288
Chromatography	288
Colorimetry and Spectrophotometry	289
Protein and Amino Acid Analyses	289
Biological Analysis	289
Microbiological Assay	289
Nutrient-Deficient Animals	289
Near Infrared Reflectance Spectroscopy (NIRS)	289
Molds or Fungi Assays	290
Techniques Utilized in Feed Evaluation	290
Cannulation and Fistulation (Nylon Bag Technique)	290
Cannulation	290
Fistulation	290
Indicators	290
Types of Feed Evaluation Trials	291
Digestion and Metabolism Trials	291
Nutrient Balance Trials	291
Nitrogen Balance Trials	292
Energy Balance Trials	292
Calorimetric Systems	293
Direct Calorimetry	293
Indirect Calorimetry	293
Comparative Slaughter Method of Determining Net Energy	294
In Vitro Digestion Determination	295
Feeding Trials	295
Conducting Applied Feedlot Tests	295
Feed Grades	296
Questions for Study and Discussion	296

At first glance, a feed may appear to be nutritious and of high quality; but unless it is systematically analyzed—through physical, chemical, and/or biological means—there is no way that one can be sure of its true value to livestock. Two hays may look alike, yet one may contain 12% protein and the other 18%. Thus, chemical analysis of the two hays can be justified. But the chemical composition of feed is not enough! Experiments must be designed to determine the availability of the nutrients in the feed to the animal. This involves the digestibility of the nutrients within an ingredient. A feed that is high in nutrients, but low in digestibility, is merely a filler and of little value to livestock.

The most accurate method of determining feed value would involve live animal trials on the farm or ranch where the feed is to be fed. However, this is too costly, slow, and impractical. Therefore, laboratory methods have been developed to predict feed value, and new ways are evolving.

IMPORTANCE OF MONITORING FEED QUALITY

Profit is the ultimate criterion of success in any livestock operation, and the cost of nutrients is an important factor in determining success. The feed composition values presented in this and other books merely represent an averaging of an accumulation of data concerning the nutritive value of feeds. Considerable variation is inherent in the nutrient content of different samples of feeds. Thus, the successful producer must recognize the value of a well-planned feed analysis program.

Nutrient variations of 10 to 15% are normal; due to differences from farm-to-farm and year-to-year; due to differences in samplings, varieties, soil fertility, weather, maturity of harvest, and storage; and due to differences in laboratory testing techniques, and differences in reporting results—some report the results on an as-fed basis whereas others report on a moisture-free basis. Because of such spread in feed compositions, the actual analysis of a given lot of feed should be obtained and used wherever practical. In most cases, however, there is insufficient time to sample and analyze a given lot of feed, and/or the lot of feed may be too small to justify the expense. So, feed composition tables based on large numbers of samples tested, serve well as the bases on which (1) to conduct buying and selling transactions, and (2) to formulate rations to meet the requirement of animals.

HOW TO TAKE FEED SAMPLES

An analysis of a feed is only as good as the sample it represents. If the sample is not representative of the entire batch, the usefulness of the evaluation is limited, no matter how extensively the sample is analyzed. Feed tends to be highly variable in composition. Thus, one bale of hay may be very different in feeding value from another bale immediately next to it. For this reason, several samples are taken from various representative areas, then the samples are thoroughly mixed together to form a representative sample of the entire feed; and an aliquot (part of a whole) is taken therefrom so that the final analysis will represent an overall average of the feed being sampled.

All feed samples should be put in plastic containers or insulated bags and immediately sent for analysis. If, for some reason, there is a delay in submitting samples that have high moisture content, such as silages, they should be frozen until analyzed. The date, place of sampling, and an identification number should be supplied with each sample.

Quite often, handlers may want to send split samples for analysis. In this procedure, the final sample is carefully split in half and sent as two separate samples. This gives evidence of the accuracy of the laboratory analysis. While this procedure increases cost, it does test the reliability of the laboratory. If the reported results of the split samples differ materially, the producer can deduce that one or both of the feed analyses were faulty.

Keeping feed samples can be very helpful should any feed-related problems arise. So, a 1- to 2-lb sample of feed from each lot should be placed in a plastic bag, properly labeled, and stored in a cool, dry place for 30 to 60 days. If no problems arise during that period and the feed has not become moldy, it can be fed.

Mixed Feeds

Feeds that are mixed in either a horizontal or a vertical mixer can be sampled with relative ease. When it is certain that the feed is thoroughly mixed, samples can be taken periodically as the feed comes down the chute. Samples should be taken at random intervals.

Since the various ingredients in mixed feeds tend to separate out according to size and hygroscopic properties, accurate sampling of mixed feeds can sometimes be difficult. In many cases, an analysis profile of the individual ingredients used in the mixed feed is more accurate, assuming care is taken in the weighing and mixing of the feed.

Grains

Grain samples are generally obtained with a grain probe. A minimum of five cores should be taken from various well separated places in the truck or bin. Grain obtained from these cores should be mixed thoroughly, thence a sample (about 1 lb) obtained from the mixture should be bagged and labeled.

Silage

Silage is difficult to sample for the following reasons:
1. It is highly variable in moisture content.
2. It is usually harvested at different times; hence, the silage within a silo will vary in stage of maturity.
3. It is difficult to obtain a sample from various locations throughout the silo, due to the physical structure of the silo.

To obtain a representative sample of silage from a tower silo that is equipped with a mechanical unloader, the person collecting the sample should catch at least 12 handfuls of silage. It is best to let the unloader run until fresh, clean silage is available. When a pit silo is to be sampled, the sampler should collect about 20 handfuls of silage from various areas of the freshly cut face. These samples should then be thoroughly mixed together; and a single sample should be taken from the mixture. This procedure should be repeated every third or fourth face cut in order to take into account variation within the silo. All samples of silage must be placed in an airtight plastic bag and either sent immediately for analysis or frozen.

Hay

With standard hay bales, core samples should be taken from the ends of bales. With large round bales, 2 to 3 core samples should be taken from each side. It is advisable to take random samples of about 5% of the bales.

If loose or chopped hay is to be analyzed, core samples must be taken from numerous locations of the stack, thence the samples should be mixed and sealed in a plastic bag.

Pasture

When sampling a pasture, it is advisable to move through the pasture in a "Z" or "X" pattern. At predetermined intervals, for example every 50 paces, a 12- by 12-in. sample is cut at mowing height. If there is very little forage at this predetermined sampling point, cut what is available and move on. Do not move off the path to cut a more densely populated area. Upon completion of the walking pattern, mix the cut forage thoroughly and bag a sample for analysis.

FEED ANALYSIS

Feeds are analyzed by physical, chemical, or biological procedures. Although physical evaluation may be the least accurate, it provides a quick and easy means of obtaining considerable information about the overall quality of a feed. The chemical procedure is more accurate than a physical evaluation, but it takes time. The biological method necessitates considerable time and expense, and the results are often variable, but it helps to assess the availability of the feed nutrients to animals.

Physical Evaluation of Feedstuffs

In order to produce or buy superior feeds, producers need to know what consititutes feed quality, and how to recognize it. They need to be familiar with those recognizable characteristics of feeds which indicate high palatability and nutrient content. If in doubt, observation of the animals consuming the feed will tell them, for livestock prefer and thrive on high-quality feed.

The physical evaluation of feedstuffs, especially forages, is based largely on visual and smell appeal. Does it look good and smell good?

• **Characteristics of good hay**—The easily recognizable characteristics of hay of high feeding value are:

1. It is made from plants cut at an early stage of maturity, thus assuring the maximum content of protein, minerals and vitamins, and the highest digestibility.
2. It is leafy, thus giving assurance of high content of protein and other nutrients.
3. It is bright green in color, indicating proper curing, a high carotene or provitamin A content, and palatability.
4. It is free from foreign material, such as weeds, dirt, and stubble.
5. It is free from mold and dust.
6. It is fine stemmed and pliable—not coarse, stiff, and woody. This is particularly true when comparing first cut hay with subsequent cuttings. Later cuttings tend to be less fibrous.
7. It has a pleasing, fragrant aroma; it smells good enough to eat.

(Also see Chapter 8, Hay.)

• **Characteristics of good silage**—The easily recognized characteristics of silage of high feeding value are:

1. It has a *clean*, rather pleasing lactic acid odor, in contrast to the foul or objectionable butyric acid odor of poor silage.
2. It has a pleasing taste—it is not bitter or sour.
3. It is not moldy, musty, or slimy.
4. It is uniform in moisture and color. Generally, green or brownish silage is good; tobacco brown or dark brown silage indicates excessive heat; and black silage is spoiled and should not be fed.

(Also see Chapter 9, Silage/Haylage/High-Moisture Grain.)

• **Characteristics of good grains and other concentrates**—The easily recognizable physical characteristics of good grains and other concentrates are:

1. Seeds are not split or cracked.
2. Seeds are of low moisture content—generally containing about 88% dry matter.
3. Seeds have a good color, characteristic of the species.
4. Concentrates and seeds are free from mold.
5. Concentrates and seeds are free from rodent and insect damage.
6. Concentrates and seeds are free from foreign material, such as iron filings.
7. Concentrates and seeds are free from rancid odor.

(Also see Chapter 10, Grains/High-Energy Feeds.)

MICROSCOPIC EXAMINATION

Many laboratories use microscopes to aid in feed evaluation. Through the use of this instrument, a trained technician can identify what ingredients are in the feed mix, along with the quantities of each. Also, the microscopist can detect adulteration and variation in quality more quickly and economically than with any other known technique. Weed seeds, foreign objects—such as iron filings and rodent excreta—mold, and damaged feed (for example, split grains), can all be observed under the microscope.

Fig. 15-1. Microscopic examination of feed. (Courtesy, Michigan State University, East Lansing)

Proximate Analysis

For more than 100 years, feeds have been analyzed by a method developed by 2 scientists, Henneberg and Stohmann, at the Weende Experiment Station in Germany. This method is called the proximate analysis, or the Weende System, of feed analysis. Feeds are evaluated in terms of 6 components: (1) moisture, (2) ash, (3) crude protein, (4) ether extract, (5) crude fiber, and (6) nitrogen-free extract (see Table 15-1).

TABLE 15-1
THE FRACTIONS OF PROXIMATE ANALYSIS

Fraction	Procedure[1]	Major Components
1. Moisture (dry matter).	Heat sample to constant weight at temperature just above boiling point of water. Loss in weight equals water.	**W**ater and any volatile compounds (100% − H_2O = DM%).
2. Ash (mineral matter).	Burn at 500° to 600°C for 2 hours.	**M**ineral elements.
3. Crude protein (protein averages 16% N; hence, N × 6.25 = crude protein).	Determine nitrogen by Kjeldahl process.	**P**roteins, amino acids, nonprotein nitrogen.
4. Ether extract (fat).	Extraction with diethyl ether.	**F**ats, oils, waxes, resins, pigments.
5. Crude fiber (CF).[2]	Residue after boiling in weak acid and weak alkali.	**C**ellulose, hemicellulose, lignin.
6. Nitrogen-free extract (NFE).[2]	Remainder; *i.e.*, 100 minus sum of the other fractions.	**S**tarch, sugars, some cellulose, hemicellulose, and lignin.

[1]Each procedure can be applied to a separate sample, of standard weight, of the feedstuff to be analyzed; or a single sample can be used to determine dry matter, crude fat, and crude fiber. In the latter case, separate samples would be run for ash and crude protein.

[2]Carbohydrates (CHO = CF + NFE).

A chemical analysis gives a solid foundation on which to start in the evaluation of feeds. Thus, feed composition tables serve as a basis for ration formulation and for feed purchasing and merchandising. Commercially prepared feeds are required by state law to be labeled with a list of ingredients and a guaranteed analysis. Although state laws vary slightly, most of them require that the feed label (tag) show in percent the minimum crude protein and fat; and maximum crude fiber and ash. Some feed labels also include maximum salt, minimum TDN, and/or minimum calcium and phosphorus. These figures are the buyer's assurance that the feed contains the minimal amounts of the higher cost items—protein and fat; and not more than the stipulated amounts of the lower cost, and less valuable, items—the crude fiber and ash.

Fig. 15-2. Weighing samples of feed for analyses. (Courtesy, International Multifoods, Minneapolis, Minn.)

MOISTURE

In order to compare the nutritional value of feeds, the first step is to determine how much water is contained in them. Many high-moisture products, such as beets, compare very favorably with corn and other more traditional feedstuffs on a dry matter basis; but the high amounts of water contained in them tend to restrict the dry matter intake of livestock. Due to this variation in moisture of many of the feeds used in livestock production, feeding standards are now being converted to a moisture-free (dry matter) basis. **NOTE WELL:** The feed composition tables in Part V of this book give compositions on both as-fed (A-F) and moisture-free (M-F) bases. Thus, fresh alfalfa pasture, all analyses, is listed as having 26% dry matter. So, the moisture is determined by subtracting the dry matter from 100 (100 − 26 = 74% moisture content).

The determination of moisture of various compounds or feeds requires the use of different techniques. The appropriate method is selected by the analyst based on such characteristics as: (1) presence of volatile matter, (2) possibility of browning of some ingredients, (3) need for low temperatures and vacuum, and (4) the presence of some compounds which might be chemically altered during drying—for example, sugars.

Moisture is now being determined in five ways:

1. **Oven dried.** According to the official methods of the Association of Official Analytical Chemists (AOAC), samples are heated at 275°F to a constant weight. The loss in weight represents the amount of water contained in the feed. This method is not exactly correct for the determination of water content, since many short chain fatty acids and organic acids become volatilized and are lost in addition to the evaporation of water.

2. **Vacuum dried.** Samples can be dried *in vacuo* at lower temperatures. The boiling point of water is lowered when the samples are placed in a vacuum. Therefore, vacuum ovens are sometimes used to minimize the loss of compounds other than water.

3. **Distillation with toluene.** The dry matter of samples having large quantities of volatile acids and bases can be determined through distillation with toluene.

4. **Freeze drying.** This method of dry matter determination is receiving increasing attention. The freeze drier basically consists of heated shelves surrounded by a refrigerated condenser. Samples are frozen prior to freeze drying. After the frozen sample is placed in the freeze drier, the apparatus is evacuated to an extremely low pressure. Under these conditions, sublimation takes place. That is, the frozen water crystals within the sample pass directly into a vapor phase without first becoming a liquid. This technique prevents the loss of many of the volatile organic compounds in the sample.

5. **Rapid moisture testing equipment.** Quite often, farmers or feeders need to know the moisture content of a feed immediately. For example, if they are buying high-moisture grain, they may not be able to wait for some chemical laboratory to run a moisture analysis. For this purpose, several types of rapid moisture testing equipment have been developed and are available at reasonable prices.

ASH

The ash fraction of the proximate analysis represents the inorganic constituents of the feed. Samples in porcelain crucibles are placed in a muffle furnace and ignited at temperatures in excess of 500 to 600°C.

The residue that is left after burning is termed *ash*. In plants, ash composition is highly variable since soil conditions determine the makeup of this fraction.

CRUDE PROTEIN (CP)

Protein, on the average, contains about 16% nitrogen. Thus, theoretically, if we know the nitrogen content of feed, we can estimate the amount of protein that it contains by multiplying the nitrogen content by 6.25 (100% ÷ 16%).

The commonly used procedure for determining the nitrogen content of feeds is called the *Kjeldahl process,* after the discoverer, the Danish chemist, Johan Kjeldahl. The organic matter of the sample is destroyed by digestion with sulfuric acid. The nitrogen is then converted to an ammonia compound which is quantitatively released as ammonia during alkaline distillation. The precise amount of ammonia is titrated against a standard solution; and the figure therefrom derived is converted to nitrogen content, and finally to protein.

It should be remembered that the figure derived from the analysis is rather crude. This procedure involves the following two basic assumptions which are generally applicable to feeds:

1. **Protein contains approximately 16% nitrogen.** In reality, this merely represents an average. Some feeds contain protein that averages more than 16% nitrogen, whereas others contain protein with lesser amounts of nitrogen.

2. **All nitrogen is in the form of protein.** Practically speaking, this may be true. But there are a number of other compounds that may be present in feeds that contain nitrogen; for example, yeasts contain a large quantity of nitrogen as nucleic acids.

• **Heat damaged protein (unavailable protein)**—Unless carefully controlled, the heating that occurs in many commercial feed processes can injure protein quality. Special care must be exercised to avoid heat damage when drying fish meal and milk products, when removing oil from oil-bearing seeds by the expeller process, and when baling, stacking, or barn-storing hay high in moisture. Severely heat-damaged protein is indigestible. The amount of indigestible protein in a feed can be estimated by measuring the nitrogen content of the acid detergent fiber, then multiplying it by 6.25 (ADF − N × 6.25).

• **Available crude protein**—Crude protein minus heat-damaged protein equals available crude protein. This is the protein figure that should be used when balancing rations, if there is heat-damaged protein in the feed.

ETHER EXTRACT (FAT)

Many people refer to this fraction as the fat portion of the sample. This tends to be an oversimplification as the ether extract also contains organic acids, oils, pigments, alcohols, and the fat-soluble vitamins. Many of the complex lipids, such as phospholipids, are not completely extracted in this procedure.

The procedure is exactly what the name implies. The sample is continuously extracted with ether, using a specially designed apparatus. After the extraction is completed, the ether solvent is evaporated and the residue that remains constitutes the ether extract.

CRUDE FIBER (CF)

The crude fiber fraction is an indicator of the relative indigestibility and bulkiness of the sample. The feed sample is first boiled in dilute acid and then in dilute alkali to simulate the digestive action of gastric secretions. The residue of the sample that remains undigested after these boiling procedures is then weighed and ashed. The difference between the initial residue weight and the ashed weight indicates the amount of fiber present in the sample.

After being used more than 100 years (its development dating back to 1865), crude fiber (CF) is declining in popularity as a negative measure of feed quality; negative because, typically, digestibility, and thus energy value, of a

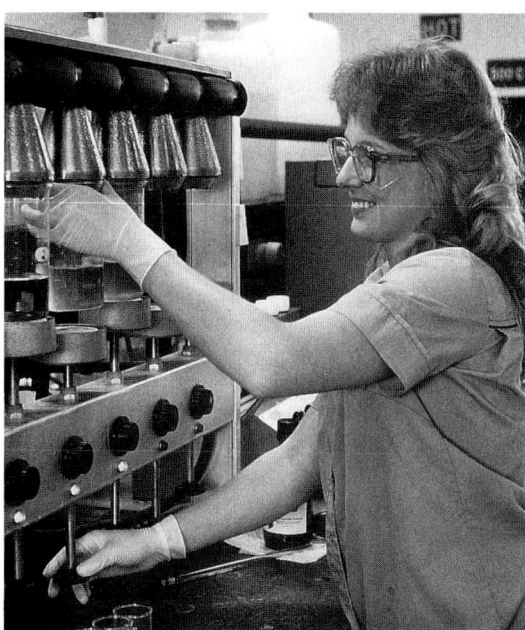

Fig. 15-3. Crude fiber analysis. (Courtesy, International Multifoods, Minneapolis, Minn.)

feed decreases as crude fiber percentage increases. But this isn't always true, especially in ruminants; they are able to digest much of the cellulose and hemicellulose in feeds and are only limited in their ability to utilize crude fiber by lignification and the capacities of their digestive tracts. So, crude fiber is only a rough approximation of differences in the digestibility of feeds. For example, the fiber of immature plants is mostly cellulose, which is highly digestible by ruminants. By contrast, the fiber of mature plants contains large amounts of lignin, which is indigestible. Another major problem is that variable amounts of hemicellulose and lignin end up in other chemical fractions (i.e., in the ether extract and ash). This is important because lignin is essentially indigestible whereas hemicellulose is partially digestible. Hence, CF does not accurately measure all of the lignin. Nevertheless, CF has remained in the scheme of feedstuff analysis because of its requirement for the determination of TDN.

In recent years, a system developed by Van Soest, of Cornell, for evaluating the digestible fraction of fiber has increased in usage. (See later section in this chapter on "Neutral Detergent Fiber [NDF] and Acid Detergent Fiber [ADF].")

NITROGEN-FREE EXTRACT (NFE)

The nitrogen-free extract (NFE) fraction in the proximate analysis program is determined by the following formula:

NFE = 100 − (% moisture + % crude fiber + % ash + % ether extract + % crude protein)

This fraction, represents a catchall for the organic material for which there is no specific analysis. The vast majority of components in this fraction are carbohydrates (starch, fructans, pectins, cellulose, hemicellulose, and lignin), but other substances, such as pigments, organic acids, and water-soluble vitamins, are also present.

Neutral Detergent Fiber (NDF) and Acid Detergent Fiber (ADF)

The inadequacies of proximate analysis gave rise to the detergent system of feed analysis for estimating energy content of forages, developed by Peter J. Van Soest at the USDA Beltsville National Research facility, in the 1960s. By using detergents, the Van Soest system separates fibrous feeds into two fractions: a neutral detergent fibrous fraction; and an acid detergent fibrous fraction (see Fig. 15-4).

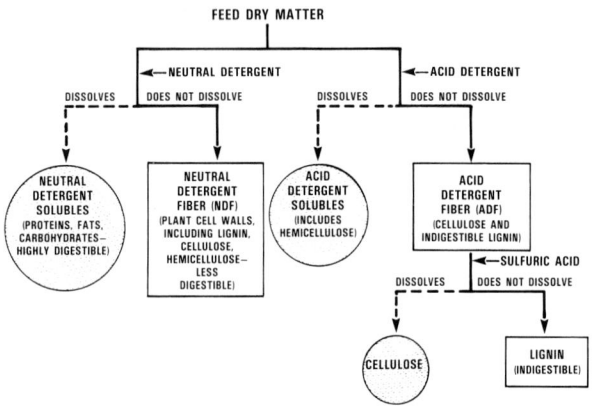

Fig. 15-4. Van Soest method of feed analysis.

• **Neutral detergent fiber (NDF)**—Neutral detergents are used to separate the feed into two fractions: (1) neutral detergent solubles, representing the highly digestible portion of the feed and consisting of proteins, fats, and carbohydrates (along with nonprotein nitrogen, pectin, and soluble materials); and (2) neutral detergent fiber (NDF), representing the less digestible portion of the feed, consisting of plant cell walls, including lignin, cellulose, and hemicellulose.

NDF is closely related to feed intake because it contains all the fiber components that occupy space in the rumen and are slowly digested. Thus, the lower the NDF percentage, the more the animal will eat; it is inversely related to voluntary feed consumption. Hence, a low percentage of NDF is desirable. It is noteworthy that milk production of lactating cows is more highly correlated with the NDF portion of the ration than with the ADF in the ration.

• **Acid detergent fiber (ADF)**—Acid detergent solutions are used to separate the feed into two fractions: (1) acid detergent solubles, containing the more readily digestible hemicellulose; and (2) acid detergent fiber (ADF), representing the less digestible portion of the feed and consisting of lignin (indigestible) and cellulose (digestible).

ADF is an indicator of forage digestibility because it contains a high proportion of lignin which is the indigestible fiber fraction. NDF will always be a higher number than ADF because ADF does not contain hemicellulose. The lower the ADF, the more feed an animal can digest. Hence, a low ADF percentage is desirable.

• **Acid detergent lignin (ADL)**—Sulfuric acid may be used further to separate the ADF into (1) cellulose, which is digestible; and (2) lignin, which is indigestible.

(Also see Chapter 8, section on "NDF, ADF, and NIRS Analyses.")

Bomb Calorimetry

When compounds are burned completely in the presence of oxygen, the resulting heat is referred to as gross energy or heat of combustion. The bomb calorimeter is an instrument used to determine the gross energy of feed, waste products from feed (for example, feces and urine), and tissues.

It should be reiterated that the calorie is defined as the amount of heat required to raise the temperature of 1 g of water 1°C (precisely from 14.5 to 15.5°C).

Briefly stated, the procedure is as follows: An electric wire is brought in contact with the material being tested, so that it can be ignited by remote control; 2,000 g of water are poured around the bomb; 25 to 30 atmospheres of oxygen are forced into the bomb; the material is ignited; the heat given off from the burned material warms the water; and a thermometer registers the change in temperature of the water.

It is noteworthy that the determination of the heat of combustion with a bomb calorimeter is not as difficult or time-consuming as the chemical analyses used in arriving at TDN values.

Chromatography

In 1903, Tswett, a Russian botanist, first described his attempts to separate colored substances; hence, the origin of the term chromatography. Today, many of the compounds that are separated and identified by chromatographic techniques are colorless; but new refinements in these techniques enable the feed evaluator to quantitate extremely minute amounts of many compounds.

Numerous materials from feeds, such as proteins, amino acids, sugars, fatty acids, minerals, and many other components, are routinely identified and quantitated through the

Fig. 15-5. Liquid chromatograph being used to test the amino acid and vitamin content of feeds. (Courtesy, International Multifoods, Minneapolis, Minn.)

chromatographic technique. In addition to nutrient analysis, chromatography can be adapted to the detection of drug residues, hormones, pesticides, and other feed contaminants.

Colorimetry and Spectrophotometry

Colorimetry and spectrophotometry are chemical analyses whereupon light is passed through solutions to yield information about the concentration of certain compounds. A particular wavelength of light is passed through the samples, and the amount of light absorbed by the sample gives an indication of the concentration of the compound being tested. Colorimetry differs from spectrophotometry in that colorimetry is useful for measuring wavelengths in the visible region of the light spectrum whereas spectrophotometry utilizes wavelengths in the ultraviolet, visible, and infrared regions of the spectrum.

The analytical procedures for many nutrients and drugs involve either colorimetry or spectrophotometry. Vitamin A is a good example of a colorimetric procedure. The standard assay for vitamin A determination involves the treatment of the sample with antimony trichloride. A deep blue colored solution is produced, the intensity of which is dependent on the amount of vitamin A in the sample. The solution of unknown concentration is measured in the colorimeter and compared to a series of standards of known concentrations. Spectrophotometric assays are essentially the same as colorimetric assays except the researcher has a more versatile machine with which to work.

The atomic absorption spectrophotometer is one of the most widely used instruments for mineral analysis, having the ability to detect many minerals at concentrations less than 1 part per billion (1 mcg/kg of sample). In addition to its high sensitivity, this machine is readily adaptable to automation, thus presenting the chemist a rapid, accurate method of feed analysis. The atomic absorption spectrophotometer works on a slightly different principle from the regular spectrophotometer. The main principle behind this machine is that when certain compounds (for example, minerals) are volatilized, they emit light of a characteristic wavelength. The machine is calibrated to detect this light.

Protein and Amino Acid Analyses

In the past, the Kjeldahl procedure was considered to be the most efficient way to quantitate protein content in feed. In recent years, several new approaches to protein determination have been developed.

In rations for monogastric animals, the amino acid composition of the feed must be considered. While most procedures for determining the amino acid profiles of feeds are automated, considerable time and labor is still involved. In order to determine the amino acid profile of a feed, the protein within the feed must be completely hydrolyzed into its constituent amino acids.

Regression equations are available for estimating the amino acid levels of selected feed ingredients. Although these equations are not a replacement for other procedures for determining amino acid profiles of feeds, they do provide a better estimate of amino acid content for formulation purposes than can be calculated from simple ratios of protein alone.

(Also see Chapter 4, section on "Biological Value of Protein.")

Biological Analysis

Quite often, biological assays are used in the analysis of micronutrients in feeds. There are two basic types of biological assays—(1) microbiological assays, and (2) the use of nutrient-deficient animals.

(Also see Chapter 4, section on "Biological Value of Protein.")

MICROBIOLOGICAL ASSAY

In microbiological assay, a microorganism is selected that is known to require the nutrient in question. Therefore, if the nutrient is unavailable, the selected microorganism will not grow. The growth medium is prepared so that it is nutritionally complete except for the nutrient to be tested. Graded levels of the nutrient are then added to the media and a growth response curve is prepared. The sample to be assayed can then be tested and compared to the growth response curve to determine the concentration of the nutrient. Many of the micronutrients, such as the B complex vitamins, are assayed in this manner.

NUTRIENT-DEFICIENT ANIMALS

In this type of assay, experimental animals, such as the rat and the chick, are fed diets deficient in a specific nutrient. Growth response curves are developed by the feeding of known amounts of the nutrient to some of the deficient animals. Other deficient animals are given the product to be assayed, and their responses are compared to the growth curves. In addition, the evaluator can observe changes in specific tissues as various levels of the specific nutrients are supplied.

Near Infrared Reflectance Spectroscopy (NIRS)

Chemical determinations are slow! This prompted the search for a more rapid, yet reliable, method of evaluating feeds and led to the use of the near infrared reflectance spectroscopy (NIRS).

The infrared technique, discovered in 1954, introduced for the analysis of grains and seeds in 1965, perfected in the 70s, and used in analyzing forages and other feedstuffs today, is based on the fact that near infrared light is absorbed variously by different molecular bonds. NIRS is a method of quantitative analysis, which comes under the category of applied analytical chemistry. When the instrument, known as an *infrared analyzer,* directs different wavelengths of infrared light at a sample of material, information can be gathered about the chemicals in the sample. The instrument passes this information on to a computer. In turn, the computer can automatically determine the percentage of nutrients in the sample. In a hay sample, for example, NIRS can determine the percentage of moisture, crude protein, calcium, phosphorus, sulfur, magnesium, potassium, neutral detergent fiber (NDF), and acid detergent fiber (ADF).

Today, modern NIRS technology is widely used in the feed, grain, and food industries. Instruments capable of measuring starch, oil, sugar, fiber, moisture, and protein are commercially available; and research has shown that many other applications are possible, such as the determination of amino acids and mineral constituents. Also, NIRS is being used to detect impurities. Furthermore, it is quicker and easier than other tests, and it doesn't use caustic chemicals like sulfuric acid or sodium hydroxide.

(Also see Chapter 8, section on "NDF, ADF, and NIRS Analyses.")

Molds or Fungi Assays

Certain types of molds or fungi have caused death losses among various animal species for many years. A very damaging mold is *Aspergillus flavus*, which produces aflatoxin.

The three basic tests for molds are mold counts, culture and identification of the organisms, and assay for the mycotoxins that may have been liberated in the growth of the mold. Of the three tests, assay for mycotoxins is preferred since the toxic effects are known for some of them. Biological assays for the toxin are available in which ducklings or poults are used. Chemical assays are also available, using fluorescent or chromatographic techniques. A practical test consists in feeding some of the moldy feed to 2 or 3 less valuable animals and observing their health and production.

(Also see Chapter 5, Table 5-3, "Mycotoxins.")

TECHNIQUES UTILIZED IN FEED EVALUATION

Two commonly used techniques that have been developed to aid in feed evaluation are the surgical techniques of cannulation and fistulation of the digestive systems of animals and the use of indicators.

Cannulation and Fistulation (Nylon Bag Technique)

Quite often in the evaluation of feeds and their utilization in the body, it is necessary to obtain or introduce samples into various parts of the digestive tract. Through surgical procedures, permanent openings can be made in the body which give the researcher ready access to the desired segment of the gastrointestinal tract.

Cannulation and fistulation techniques enable the researcher to (1) determine flow rates, (2) collect digestive fluids, (3) infuse materials in various parts of the digestive tract, and (4) observe movement of the organs.

CANNULATION

Cannulation is a procedure whereby a cannula (tube) is inserted in the cavity of an organ and subsequently brought to the exterior of the body.

Quite often, a re-entry cannula is used in the investigation of digestion in the intestine. After the intestine is cut, a cannula is inserted and brought to the exterior of the body. It is then secured to the skin and reinserted into the body cavity, thus forming a loop on the outside of the body. Finally, the other piece of the cut intestine is attached to the cannula. This type of cannulation does not impede the flow of ingesta but does give the researcher ready access to intestinal fluids with a minimum of trauma to the animal.

- **Mobile bag technique**—Scientists at the University of Alberta, Edmonton, Canada, have adapted the nylon bag technique to swine, in a procedure which they refer to as the *mobile bag technique*. It consists of the following procedure: Approximately a 1 g sample of the feedstuff is weighed and placed in a small nylon bag. Then, the bag is inserted in the pig's duodenum via a duodenal cannula. The bag travels through the intestinal tract and is voided with the feces 36 to 48 hours following insertion in the duodenum.

FISTULATION

Fistulation is a procedure in which an opening is made in a hollow organ with the edges of this opening exteriorized by being sewn to the body wall. A plug is then attached to the opening to prevent leakage of the contents from within the organ and to maintain an anaerobic environment. When a sample is to be taken from the organ, the plug is removed, thereby giving the investigator ready access to the interior of the organ. A fistula differs from a cannula in that a fistula is a permanent opening in the body wall and the organ while a cannula is a tube which leads to the interior of the body.

Rumen fistulas are commonly employed in the study of ruminant digestion. By fistulating the rumen, researchers can clearly observe the movements of the rumen and collect ingesta for analysis or for use in *in vitro* digestion trials. One such adaptation of fistulation is a procedure developed by Heinemann and VanKeuren at the Washington Station. In this procedure, the bags containing samples are then attached to a ¾- × 8-in. plastic stick with a lead weight attached to one end; and the entire apparatus is inserted in the rumen of a fistulated steer and imbedded in the ingesta. Here the forage samples are exposed to the ruminal fluids and microorganisms for digestion. At predetermined intervals, the sticks are removed; and forage samples are placed in nylon bags measuring 2 × 4½ in. The nylon bags plus digested samples are dried, weighed, and analyzed to determine how much of the sample has disappeared. This technique offers advantages to the traditional techniques of *in vitro* digestion trials where the feed samples are digested in test tubes under conditions which simulate the environment of the rumen. By using the nylon bag technique, the researchers can determine digestibility parameters in a natural environment, thereby eliminating many factors which can inherently affect the results of an *in vitro* trial involving an artificial rumen.

Indicators

Traditionally, indicators, or tracers, have been used to study rate of passage, forage intake, and digestibility of feed. Numerous compounds have been used for this purpose; among them, chromic oxide, ferric oxide, various dyes, silica, lignin, and chromogen (a pigment found in plants). Polyethylene glycol has been used to trace fluid changes in the digestive tract. Indicators can be broken down into two classes: (1) internal indicators, those inert portions of the plant themselves (for example, lignin, silica, and plant chromogen); and (2) external indicators, inert chemicals which are fed to animals in addition to the feed (for example, chromic oxide and certain dyes).

Digestion trials tend to be laborious and cumbersome since a total collection of the feces is necessary to determine

Feed Analysis/Feed Evaluation

Fig. 15-6. A steer with a harness and bag for the collection of feces. (Courtesy, University of California)

the digestibility of the feed. Indicators provide a quick, indirect way of determining the digestibility of feed.

Indicators are used to estimate forage intake and digestibility in grazing animals. Since the animals are not in confinement, it is often difficult to determine how much feed is consumed. By placing a feces collection bag on the animal (see Fig. 15-6), the researcher can determine feed intake through the use of internal indicators. The total collection of fecal material is weighed and the dry matter content determined. Samples of the forages being grazed are analyzed to determine the content of the internal indicator (i.e., lignin). The fecal material is then analyzed for the content of the internal indicator, and from this information the amount of dry matter consumed can be calculated.

Through the use of external indicators, it is possible to estimate the total amount of fecal dry matter without having to worry about total collections. All the researcher has to do is obtain grab samples—that is, small samples of fecal material taken directly from the rectum. From the information of how much external indicator was fed and how much was present in the feces, the fecal dry matter output can be calculated from the following equation:

$$\text{Fecal dry matter output} = \frac{\text{weight of external indicator fed}}{\text{\% external indicator in fecal dry matter}}$$

TYPES OF FEED EVALUATION TRIALS

Once the proper techniques have been developed for feed analysis, it is necessary to set up trials to evaluate the feed. This can be done *in vivo* (through the use of animals) or *in vitro* (through the use of artificial means).

Digestion and Metabolism Trials

Animals are not able to extract all the nutrients present in feeds. The actual value of ingested nutrients is dependent upon the extent to which they are metabolized. The first consideration here is digestibility, since undigested nutrients are not incorporated into the body proper.

A digestion trial is made by determining the percentage of each nutrient in the feed through chemical analysis; giving the feed to the test animal for a preliminary period (usually 7 to 10 days), so that all residues of former feeds will pass out of the digestive tract; giving weighed amounts of the feed during the test period (7 to 10 days); collecting, weighing, and analyzing the feces; determining the difference between the amount of the nutrient fed and the amount found in the feces; and computing the percentage of each nutrient digested. The latter figure is known as the *digestion coefficient* for that nutrient in the feed.

Various techniques and equipment may be used to make the fecal collections; among them, a specially designed digestion stall (see Fig. 15-7); collection harness and bag; markers (such as carmine, ferric oxide, chromic oxide, or soot), fed with the ration at the beginning and the end of the collection period; and indicators of an inert reference subject.

It is important that the urine and feces be collected separately, since only the feces are analyzed in a digestion trial. Anatomical problems prevent the use of female animals in most digestion trials, since it is difficult to prevent the contamination of their fecal samples with urine.

Nutrient Balance Trials

Nutrient balance trials are experiments which attempt to account for all losses of a particular nutrient. If an animal is losing more of the nutrient being tested (for example, energy) than it is ingesting, the animal is said to be in a negative balance. Conversely, if the animal is ingesting more of the nutrient than it is losing, it is said to be in a positive balance. A negative balance indicates that catabolism of stored nutrients in tissues is occurring to maintain the animal. Positive balances indicate that the excess amounts of the particular nutrient are being incorporated into body tissues and fluids.

Balance trials can be run for any nutrient, but the two most widely used types of balance trials are for the measurement of protein and energy.

Fig. 15-7. Steers in a nutrient balance trial. Note that the stalls are designed so that fecal material and urine are collected separately. (Courtesy W. W. Heinemann, Washington State University, Prosser)

Nitrogen Balance Trials

Nitrogen balance trials are used to determine the availability of protein in feeds. Fig. 15-8 illustrates the organization and breakdown of the various components in a nitrogen balance trial.

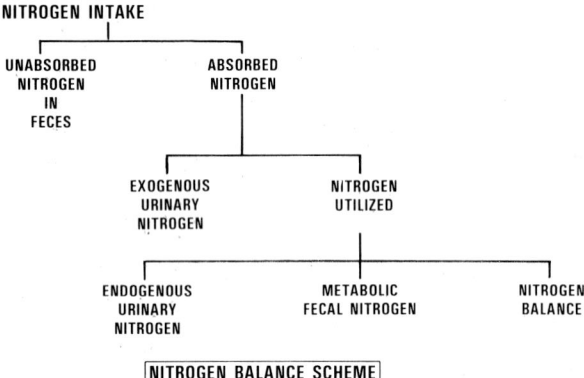

Fig. 15-8. Format of the nitrogen balance scheme.

Both feces and urine are collected in this type of experiment. Since metabolic processes involve considerable amounts of nitrogen, much of the nitrogen found in the feces and urine comes from sources other than ingested feed. To correct for this fact, feed is withheld from animals so that their bodies enter into a catabolic condition. The nitrogen excreted in the urine and feces during this period represents the nitrogen lost through metabolic processes. This urine production represents endogenous urinary nitrogen. Likewise, the fecal nitrogen excreted during the fasting period is called metabolic fecal nitrogen. These values are then subtracted from the nitrogen values obtained in the collection of the urine and feces during the feeding stage of the balance trial.

Through the use of the nitrogen balance scheme, the *biological value* (BV) of a feedstuff can be determined. *Biological value can be defined as the percent of absorbed nitrogen retained.*

Another method of evaluating protein sources is the calculation of the protein efficiency ratio (PER). Animals are weighed prior to commencement of the feeding period and upon completion of the feeding period. The gain in body weight is then computed as a ratio with the amount of protein consumed, as follows:

$$\text{PER} = \frac{\text{gain in body weight}}{\text{protein intake}}$$

PER is dependent on the level of protein fed. That is, at a dietary level of 10% protein, the experimenter may get a completely different PER than if a 20% protein diet was fed. Therefore, PER values are usually related to a standard; or the value is obtained by taking the maximum obtained through the use of several levels.

A third method of protein evaluation is the determination of net protein utilization (NPU). Through the use of NPU, it is possible to determine biological value in the following manner:

$$\text{Biological value} = \frac{\text{NPU}}{\text{digestibility}}$$

Energy Balance Trials

The gross energy of feed is the total energy content of the feed as determined by bomb calorimetry. When feed is consumed, the energy contained therein can be partitioned into fecal energy and apparent digestible energy. The energy content of feces is quantitated and subtracted from the gross energy to give apparent digestible energy. Since some of the energy in the feces represents energy derived from metabolic end products, the digestible energy term must be labeled apparent digestible energy. True digestible energy is the gross energy of a feed minus fecal energy from feed alone. The fecal energy from feed alone can be quantitated by subtracting the fecal energy of a fasting animal from the fecal energy of the same animal when it is properly fed. Most researchers, however, feel that apparent digestible energy is a relatively accurate and simple means of evaluating the energy content of feeds.

Apparent digestible energy is then partitioned into metabolizable energy, urinary energy, and gas loss (for example, methane). Urine and gases are collected, and the energy values derived from these two sources are substracted from the apparent digestible energy value to yield metabolizable energy.

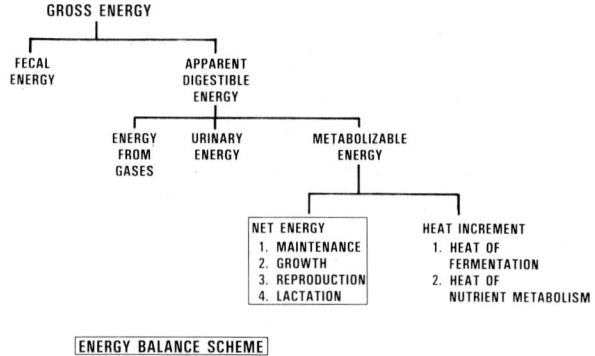

Fig. 15-9. Format of the energy balance scheme.

Metabolizable energy can then be broken down into heat increment and net energy. Heat increment is the term used to describe the heat produced in the processes of nutrient metabolism and fermentation. Net energy is the energy remaining for maintenance, growth, and production. When the values for heat increment and net energy for maintenance are combined, we can obtain an indication of the heat production in the body.

• **True metabolizable energy**—Some of the leading research centers of Europe and Canada are now using the term True Metabolizable Energy (TME) in the evaluation of feed for poultry. TME is metabolizable energy that has been corrected for metabolic and endogenous losses. It takes cognizance of the fact that the feces contain both (1) undigested feed, and (2) metabolic and endogenous products.

Calorimetric Systems

Measurement of animal heat began 200 years ago, when Lavoisier and LaPlace, the great French scientists, enclosed a guinea pig in a chamber surrounded by ice. The amount of ice melted by the heat of the animal multiplied by the latent heat of the ice indicated the heat given off by the guinea pig, thus giving rise to direct calorimetry. They also noted that the melting of the ice was directly related to the amount of carbon dioxide given off by the animal. This was the basis of indirect calorimetry. Since that time, more sophisticated types of equipment for measuring body heat have been developed, but the basic principles remain the same.

DIRECT CALORIMETRY

Fig. 15-10. Dr. R. W. Swift, former head of the Department of Animal Nutrition at Pennsylvania State University, is shown operating the respiration calorimeter built by H. P. Armsby. (Courtesy, Pennsylvania State University, University Park)

In direct calorimetry (see Fig. 15-10), the animal is confined to a well-insulated chamber and the heat losses (by radiation, convection, and conduction from the body surface; by evaporation of water from the skin and lungs; and by excretion of urine and feces) are measured either by (1) the increase in temperature of a known volume of water, or (2) electrical current generated as heat passes across thermocouples (gradient layer calorimetry). Brody[1] estimated that about one-fourth of the heat lost by the body resulted from moisture vaporization. The remaining three-fourths of the body heat is lost by radiation, conduction, and convection.

---

[1]Brody, S. B., *Bioenergetics and Growth*, Reinhold Publishing Company, New York, N.Y., 1945.

Direct calorimetry is the most accurate method of measuring the heat production of animals, but it is costly and arduous. The machine itself is very expensive to build and operate; and considerable labor is involved in controlling the animal, running the machine, and analyzing the results. For these reasons, indirect calorimetry is usually the method of choice for the evaluation of energy in feeds.

INDIRECT CALORIMETRY

In indirect calorimetry, heat of production is calculated from measurement of the respiratory exchange—the O_2 consumption, and usually the CO_2 production—of the animal. This method is based on the fact that O_2 consumption and CO_2 production are closely related to heat production. The ratio of carbon dioxide produced to oxygen consumed, which is known as the respiratory quotient (RQ), is distinctive for each compound; hence, it served to indicate the type of nutrient being metabolized. Thus, carbohydrates have an RQ of 1.0; fats have an RQ of 0.70; and proteins have an intermediate value between carbohydrates and fats—about 0.81.

It should be noted that in stress conditions, the RQ value can be greater than one. This can occur when there is hyperventilation where large quantities of carbon dioxide are exhaled while an oxygen debt exists in the body. Metabolic acidosis creates an excess exhalation of carbon dioxide as a compensatory mechanism.

The total heat production of the animal may be computed by either (1) measuring RQ along with oxygen, then making readings from tables; or (2) using a single equation relating heat production to the respiratory exchange.

Comparative measurements of direct and indirect calorimetry reveal that the two methods give results that are in close agreement.

The actual measurement of the respiratory exchange of animals may be accomplished by several different types of apparatus; among them, (1) the open circuit gravimetric system (see Fig. 15-11), (2) the open circuit chamber system involving gas analysis (see Fig. 15-12), (3) the open circuit mask system, and (4) the closed circuit mask system (spirometer).

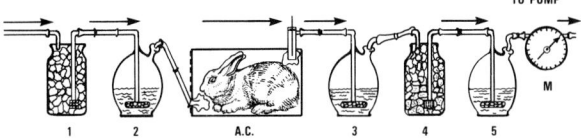

Fig. 15-11. Haldane respiration apparatus for the open circuit gravimetric system of indirect calorimetry: A.C., animal chamber. M, meter for measuring rate of ventilation. Bottles 1 and 4 contain soda lime (or caustic alkali) for absorption of carbon dioxide. Bottles 2, 3, and 5 contain sulfuric acid for absorption of water. Air entering the animal chamber is freed of carbon dioxide and moisture by passage through bottles 1 and 2. The animal gives off carbon dioxide and moisture, and these are collected in the bottles of the outgoing chain. Bottle 5 is necessary because soda lime gives off moisture. The gain in weight of bottles 4 and 5 represents the carbon dioxide production. The gain in weight of bottles 3, 4, and 5 minus the loss in weight of the animal and chamber represents the oxygen consumption.

Fig. 15-12. A cow in an open circuit respiration chamber. The gas meter to the left of the chamber is used to measure the respiratory exchange of the cow. These data, plus the gas composition, provide the information needed to calculate the heat production (HP) of the cow. The HP of an animal consuming feed in a thermoneutral environment is composed of the heat increment (heat of fermentation plus heat of nutrient metabolism) plus heat used for maintenance (basal metabolism plus voluntary activity). (Courtesy, USDA)

The comparative slaughter method requires relatively large numbers of animals. A random sample, or check group, is selected and slaughtered at the beginning of an experiment to determine the initial body composition. Then, at the close of the experiment, the remaining animals are slaughtered and analyzed. The difference in the calorie content of the two groups represents the energy storage or gain, which is a far more accurate measure of the true value of feed than liveweight gain.

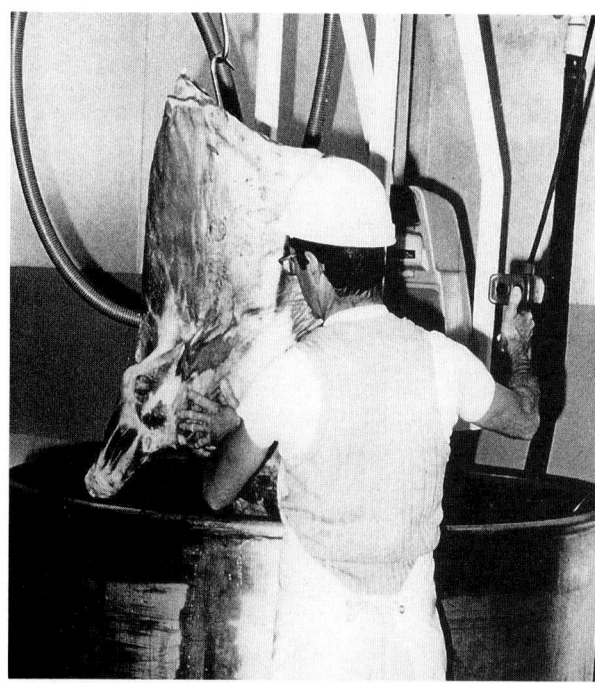

Fig. 15-13. This shows the dipping procedure being used to obtain carcass density from which specific gravity is computed, which, in turn, is used to estimate carcass fat content. By use of an initial and a final slaughter group of animals in each feeding trial, the energy gain can be measured, giving a more accurate measure of the true feed value than does just liveweight gain. Further, the method provides a measure of energy content without grinding and chemical analysis of the carcass. (Courtesy, University of California, Davis)

Comparative Slaughter Method of Determining Net Energy

The comparative slaughter method of determining net energy (energy storage and heat production) is an old technique with a new look. It was first employed by Mitchell and co-workers, of the Illinois Agricultural Experiment Station, in 1926. But, for the most part, it was discontinued, because chemically analyzing the body (carcass) is slow, tedious, and expensive. Then, in 1959, Lofgreen and Garrett, of the California Agricultural Experiment Station, reported an ingenious modification of the comparative slaughter technique. By making use of established relationships between carcass density and the composition of the animal, they were able to estimate the energy content of the carcass without analyzing the body chemically. The density of the carcass is determined by weighing in water—using a dipping procedure (see Fig. 15-13), which can be done quickly and without affecting the sale value of the carcass.

The comparative slaughter method is especially well suited to studies involving growing and fattening animals—cattle, lambs, hogs, and broilers, in which the amount of energy stored in the carcass can be measured. However, it is not adapted to use with dairy animals.

The comparative slaughter method of feed evaluation is unique in the following respects:

1. It provides a relatively inexpensive way in which to determine net energy values.
2. The animals can be kept under more natural conditions, similar to those found in a commercial enterprise.
3. Feeds can be assigned NE_m (net energy for maintenance) and NE_g (net energy for gain) values in keeping with the efficiency of metabolizable energy utilization for these different physiological processes.

The modified comparative slaughter technique has had a major impact on feed evaluation and ration formulation throughout the United States.

In Vitro Digestion Determination

Many laboratories are using *in vitro* digestion techniques to estimate the digestibility of various feeds. Some investigators refer to this as the artificial rumen evaluation. Ruminal fluid is obtained by either a stomach pump or collection through a rumen fistula and strained through cheesecloth. Test tubes containing buffers and the feed samples are inoculated with the fluid that contains, in theory, a representative sample of the microflora of the rumen. The samples are then maintained at the body temperature of the animal in a carbon dioxide environment (anaerobic). In this environment, microbes digest the feed samples.

Feeding Trials

Each method of evaluating feedstuffs, discussed earlier in this chapter, has a place and is valuable. But none of them takes into consideration all the factors which determine the true value of any feed for a particular class of livestock. The "court of last judgment" for determining the true value of a feedstuff is the animal. How well do animals eat the feed? How does it affect their health and well-being? How are they producing? Answers to these questions call for feeding the ingredient or ration under controlled conditions to the particular class of livestock.

CONDUCTING APPLIED FEEDLOT TESTS

When carefully conducted and properly interpreted and used, feedlot trials can be a valuable adjunct in the operation of a large feedlot. Among their virtues, the feedlot operator can study area and feed differences. Among their limitations, usually most applied feedlot trials can afford fewer controls than most university-conducted experiments thereby resulting in less accuracy. For the latter reason, most of them should be looked upon as applied tests or demonstrations *per se*, rather than carefully controlled, basic experiments; terminology which does not detract from their value, but which does place them in proper perspective.

The number of pens which a feedlot should devote to test work will vary according to the size of the operation and the number of treatments planned at one time.

There should always be a minimum of 2 lots for controls, plus 2 lots for each treatment evaluated. Generally, the 2 control lots should be fed the standard feedlot ration, and each treatment to be evaluated should be given to 2 lots.

The local county extension agent should be invited to participate in the test; and will usually welcome the opportunity.

The following procedure is recommended in conducting tests in commercial cattle feedlots:

1. **Cattle.** The animals should be of uniform breeding, background, age and weight, and of the same sex. Use cattle owned by the operator, rather than custom-fed animals.

2. **Number per lot.** Ten head if individually weighed; 20 to 40 head, or more, if group weighed.

3. **Randomization.** Gate or chute cut; one per treatment, or not more than five at a time.

4. **Identity.** Preferably (a) apply a different brand to each lot, and (b) individually identify each animal with duplicate numbers—one in each ear. For the latter, use plastic ear tags, the numbers on which can be easily read at a distance.

5. **Variables.** Have as few variables in each treatment group as possible. Let us suppose, for example, that in a certain feedlot, cattle are given a standard ration without implants. But there is a choice of two implant products that claim to promote feed efficiency and rate of gain. So, the owner would like to determine (a) if either product does, in fact, promote feed efficiency and rate of gain; and (b) if so, which product is better. The design would be as follows:

	Lot	Treatment
Control	1	Standard ration, no implants
	2	Standard ration, no implants
Treatment 1, Implant A	3	Standard ration + Implant A
	4	Standard ration + Implant A
Treatment 2, Implant B	5	Standard ration + Implant B
	6	Standard ration + Implant B

6. **Adjustment period.** After sorting cattle into test lots, allow a minimum adjustment period of 7 days, during which the animals should be individually tagged and handled as necessary, and gradually accustomed to their new rations. In case of sickness, a longer adjustment period may be necessary—sometimes as much as 2 to 4 weeks.

7. **Weighing conditions.** Keep off feed and water overnight, then weigh the next morning. Weigh (preferably using a self-recording beam, so as to eliminate the human error) pens in the same order and at the same time each morning when (a) initiating the experiment, (b) at 28-day intervals, and (c) at the close of the test.

Also, weigh and record the amount of feed given to each lot of cattle. In some experiments, it is best to limit all lots (both controls and treatments) to the level of the lot consuming the least, although this will vary according to the treatment being evaluated.

8. **Carcass data.** Sell, or have animals custom slaughtered, with the stipulation that the slaughter plant provide individual (according to individual ear tags) (a) carcass weight and yield, and (b) Federal grade. If slaughter data cannot be obtained on all cattle, get it for as many as possible and of the same number from each lot.

9. **Summarize results.** At the end of the trial, summarize the results, using as criteria (a) rate of gain, (b) feed efficiency, and (c) carcass results.

10. **Determine the application.** If both lots of a given treatment are considerably better than the controls, decide (a) whether to repeat the test, or (b) adopt and use the new treatment throughout the feedlot. The first course of action should be taken under the following circumstances: (1) when there is wide variation in response within the treatments; (2) when the responses are considerably less than expected; (3) when unusual circumstances prevail—for example, disease problems; and (4) when the experimenter has any doubts as to the validity of the results. If the new treatment becomes the standard, continue with it until a new and superior treatment evolves, based on new trials.

FEED GRADES

Grades have been established for many of the commonly marketed grains and other feeds. These grades allow for a more uniform marketing system whereby purchasers of feeds can get a reasonable idea of the quality of the feed—even though they may not have seen it.

Federal grades for grains are based on weight per bushel, moisture content, percentage of damaged grains, and amount of foreign material.

On a voluntary basis, near infrared reflectance spectroscopy (NIRS) is being used currently in the NIRS Forage Research Project Network; and either NIRS or wet chemistry tests may be used in the laboratory certification program operated by the National Alfalfa Hay Test Association in conjunction with the American Forage Grassland Council and the National Hay Association.

(Also see Chapter 8, sections on "Hay Quality" and "Buying and Selling Hay"; and Chapter 10, section on "Grain Standards.")

QUESTIONS FOR STUDY AND DISCUSSION

1. What is the most accurate method of determining feed value? Why isn't it used more?

2. What causes such a wide variation in the composition of the same kind of feed?

3. Why is not an actual analysis of a given lot of feed always obtained?

4. Why must care be taken in the sampling of feeds?

5. Why should feed samples be kept for 30 to 60 days?

6. How would you go about obtaining a representative sample of each: silage, baled hay, and pasture grass?

7. What are the qualities to look for in the physical evaluation of each of the following feedstuffs: hay, silage, and grains and other concentrates?

8. What evaluations of feed can a trained microscopist make?

9. The Proximate Analysis, or Weende, system of feed analysis divides feeds into several fractions. What are they?

10. For routine moisture determination of grain, which procedure would you choose? Why?

11. What is heat damaged protein? Why should it be considered when balancing rations?

12. What is ether extract? How is it determined?

13. Why is the determination of crude fiber by the proximate analysis declining in usefulness?

14. Outline the Van Soest procedure of fiber analysis. Discuss the two fractions: (a) neutral detergent fiber (NDF) and (b) acid detergent fiber (ADF).

15. Briefly describe energy determination by use of the bomb calorimeter.

16. Briefly discuss the principles of chromatography.

17. Why is spectrophotometry a good technique in the determination of vitamins in tissues and feed?

18. What are the two basic types of biological assays? Discuss their relative merits.

19. Explain how near infrared reflectance spectroscopy (NIRS) works. How is it being applied in feed analysis?

20. Why should livestock producers be concerned about molds or fungi in feed? Identify the most common damaging mold.

21. Define cannulation and fistulation. How are these techniques useful in the evaluation of feeds?

22. Why are indicators, or tracers, used in digestion trials?

23. Explain the difference between *in vivo* and *in vitro* feed evaluation trials.

24. Explain how a digestion trial is usually made.

25. How does one determine endogenous fecal nitrogen and metabolic fecal nitrogen?

26. Differentiate between the following terms used in protein evaluation: biological value, protein efficiency ratio, and net protein utilization.

27. Outline the energy balance scheme. Which energy value do you think gives the best indication as to the energy needs of the animal? Why?

28. Why is indirect calorimetry used preferentially over direct calorimetry?

29. Describe the comparative slaughter method of determining net energy.

30. List three respects in which the comparative slaughter method of feed evaluation is unique.

31. Briefly describe *in vitro* fermentation trials.

32. Outline the recommended procedure for conducting feeding tests in commercial cattle feedlots.

BUYING FEEDS/ COMMERCIAL FEEDS/ FEED LAWS

Original painting by Tom Phillips

Contents	Page
Homegrown vs Purchased Feeds	298
Buying Feeds	298
What the Livestock Producer/Feed Buyer Should Know	298
Other Feed Requisites	299
Futures Trading in Feed	299
Feed Substitutions	299
Best Buy in Feeds	300
Cost per Unit of Nutrients	300
Moisture is Important	300
Home-Mixed vs Commercial Feeds	300
Commercial Feeds	301
Importance and Nature of Commercial Feeds	301
Types of Commercial Feed Formulations	302
How to Select Commercial Feeds	302
Cooperatives	303
Feed Quality	303
Feed Laws	303
Questions for Study and Discussion	304

TABLE 16-1
KINDS AND QUANTITIES OF FEED CONSUMED BY LIVESTOCK AND POULTRY, FEEDING YEARS 1965-66 and 1984-85[1]

Feed Materials	1965-66 Feeding Year	Percent Of Total	1984-85 Feeding Year	Percent Of Total
	(million tons)	(%)	(million tons)	(%)
Grains:				
Corn	81.5	16	115.2	22.6
Wheat and rye	3.0	1	13.7	2.7
Other feed grains	32.1	7	40.0	7.9
Protein feeds	31.7	6	28.7	5.6
By-product feeds	11.5	3	13.3	2.6
Total concentrates	160.4	33	203.8	40.0
Hay	49.4	10	59.6	11.7
Other harvested roughages	26.3	5	20.0	3.9
Pasture	249.1	52	226.1	44.4
Total roughage	324.8	67	305.7	60.0
Total, all feeds	485.3	100	509.5	100.0

[1]Measured in feed units (corn equivalents). From *1987 Fact Book of U.S. Agriculture*, p. 19, Table 8.

Providing feeds for, and feeding, livestock and poulty are important parts of today's agriculture, involving farmers and ranchers, feed and food processors, and commercial feed industries. Feed is the major item of expense in producing animals; it generally accounts for 60% or more of the cost of production. About 28% of the grains fed to animals are used on the farms where grown; the other 72% moves through commercial channels—it is bought and sold.

In the crop year 1984-85, the livestock and poultry industries consumed 509.5 million tons of feed, 5% more than the 485.3 million tons fed in 1965-66 (see Table 16-1). The quantity of concentrates fed increased 27%, while roughage consumption declined 6%, reflecting higher concentrate rations.

In 1984-85, pasture, hay, and other roughage provided 60% of the total tonnage of feed used, while concentrates furnished 40%.

Some significant shifts (not shown in Table 16-1) occurred in the feed consuming animal species from 1965 to 1985. Poultry accounted for 22% of the grain consuming animal units in 1965-66, compared with 26% of the total in 1984-85, reflecting the expanding poultry industry.

Increasing numbers of livestock and poultry are produced on big and intensive livestock operations, along with increased confinement; resulting in more feeds being purchased, rather than homegrown, and more feeds being milled at or near the feeding location.

The economic impact of purchased feed is clearly reflected in the following figures: In 1975, $12.9 billion worth of feed was purchased by the nation's farms, ranches, and feedlots; in 1985, the feed purchase bill totaled $18 billion (see Fig. 16-1), representing 13.5% of the farm expenditures that year (see Fig. 16-2).

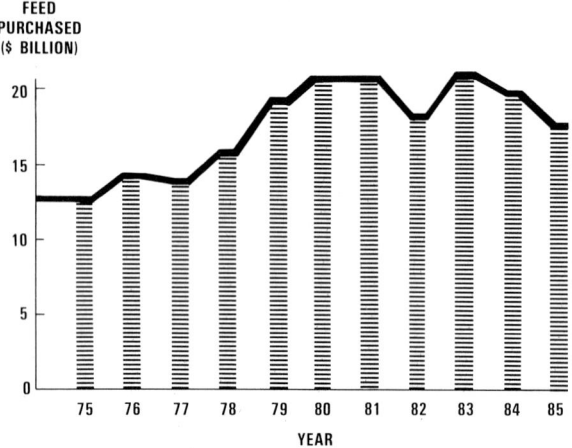

Fig. 16-1. Feed expenditures for 1975-1985. (*Agricultural Statistics 1987*, USDA, p. 414, Table 585)

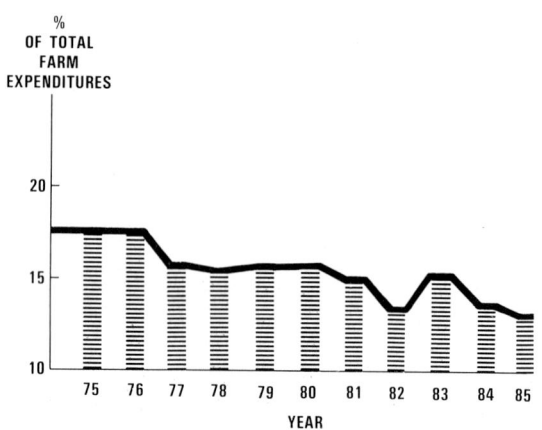

Fig. 16-2. Feed costs as a percent of total farm expenditures, 1975-1985. (*Agricultural Statistics 1987*, USDA, p. 414, Table 585)

HOMEGROWN VS PURCHASED FEEDS

A major factor in determining whether feeds shall be homegrown or purchased is the system of farming, three broad systems of which are practiced in the United States: (1) crop farming, (2) livestock farming, and (3) crop and livestock farming combined.

Generally speaking, part of or all animal feeds are purchased under the following circumstances:

1. When land, water, and/or labor are limited.
2. Where the soil and climate of the farm are not suited to the production of a desired feed(s).
3. When feeds of comparable quality can be bought on a delivered basis more cheaply than they can be produced.
4. Where storage facilities are limited, with the result that feeds produced on the farm cannot be held for year-long use.
5. When season-long financing for crop production cannot be secured, but when it is possible to "pay as you go," such as a dairy producer using his monthly paycheck to buy feeds.
6. Where homegrown feeds need to be supplemented by certain nutrients, such as protein, minerals, and/or vitamins. This need exists on almost all livestock operations, for few of them produce all the nutrients needed for a balanced ration.
7. When special feeds are desired, such as molasses and fats, which are not farm-produced.
8. Where a purchased feed can be substituted for a homegrown feed on an equal feeding value basis, but with the purchased feed acquired at a lower cost than the selling price of the homegrown product.
9. Where it is more profitable to use available capital, management, and labor for maximum animal production. This frequently applies to large operations among cattle and sheep feedlots, dairies, confinement swine units, and poultry operations, where profits may be maximized by increasing livestock output rather than spending time and capital on raising crops.
10. Where the operator is not knowledgeable relative to crop production.

Generally speaking, forages are more apt to be homegrown than grains, primarily due to their greater bulkiness and higher cost of transportation. However, increased pelleting and cubing have made it economically feasible to transport forages greater distances.

BUYING FEEDS

Buying feeds is an integral part of modern livestock production. Moreover, the trend to purchase feeds, rather than to grow them on location, will continue. In a broad sense, modern sophisticated buying involves knowledgeable buyers, futures trading, consideration of feed substitutions, volume buying, storage, capital outlay, and how to determine the best buy in feeds.

What the Livestock Producer/Feed Buyer Should Know

The vast majority of feeds used on the farms and ranches, and in the feedlots, of America, is purchased by producers—by practical operators who subsequently feed it. So, no one has a greater incentive to purchase wisely and well than they do.

Successful feed buying necessitates knowledge of all the factors that affect net returns, from the time a deal is made to buy the feed until the end product is marketed. Today, sophisticated livestock producers-feed buyers need to know the following:

1. The nutritive requirements of animals.
2. The language of the feed trade; feed buyers should know feed terms and speak the language.
3. The production and economic trends; shrewd buyers follow price trends.

4. The business aspects, including sources of credit, interest rates, contracts, futures, and possible tax savings to accrue from purchasing feeds before the end of the year.
5. The different feed grades and quality classifications.
6. The restrictive use of certain feedstuffs; for example, buyers should know the upper limit of molasses and fat in feeds, and that more than 10% peanuts in the ration of growing-finishing hogs may cause soft pork.
7. The associative, or additive, effects of certain feedstuffs.
8. The origin of the feed ingredients; for example, if the ingredients were grown in a low phosphorus or in a high selenium area.
9. The local potential to grow certain feeds.
10. The longtime availability of certain feeds.
11. The moisture content of the feed ingredients.
12. The transportation cost of feeds.
13. The storage capabilities of feeds.
14. The characteristic shrinkage of feeds.
15. The risks; for example, storage of wet hay makes for a fire hazard.
16. The processing that will be involved.
17. The truth about certain feedstuffs affecting the product produced.
18. The facts about toxic residues.
19. The government regulations; for example, the regulations pertaining to incorporation and withdrawal of feed additives.
20. The impact of foreign feed purchases.

OTHER FEED REQUISITES

In addition to the considerations already noted, it is important that all feeds—both bought and homegrown—meet the following requisites:
1. Palatability; if animals don't eat it, they won't produce; and if they don't eat enough, feed efficiency will be poor.
2. Variety, which makes for increased palatability and balance of nutrients.
3. Minimize digestive disturbances; bloat, colic, scours, and constipation can be minimized by the choice of feeds.
4. Bulk; ruminants can consume bulkier feeds than monogastric animals.
5. Be free from poisonous plants and poisonous feeds, such as feeds containing prussic acid, ergot, lead, and mercury.

Futures Trading in Feed[1]

Futures trading is not new. It is a well-accepted, century-old procedure used in many commodities, for reducing risks, protecting profits, stabilizing prices, and smoothing out the flow of merchandise.

Futures trading allows producers, processors, and merchandizers to provide a continuous supply of agricultural goods to consumers at smaller overall profit margins than would otherwise be possible for commodities with such volatile prices.

---

[1]The entire section on "Futures Trading in Feed" was reviewed by the following authorities: P. J. Catania, Vice President, Chicago Board of Trade, La Salle at Jackson, Chicago, Ill.; and W. J. Brodsky, President, Chicago Mercantile Exchange, 30 South Wacker Drive, Chicago, Ill.)

A commodity exchange is a place where buyers and sellers meet on an organized market and transact business on paper, without the physical presence of the commodity. The exchange neither buys nor sells; rather, it provides the facilities, establishes rules, serves as a clearinghouse, holds the margin money deposited by both buyers and sellers, and guarantees delivery on all contracts. Buyers and sellers either trade on their own account or are represented by brokerage firms.

A futures contract is a standardized, transferable agreement to buy (take delivery of) or to sell (make delivery of) a specific amount and type of commodity at a future date up to 30 months ahead, at a price established at the time of trading.

Since feed represents such a large proportion of the cost of certain types of livestock production, it may be wise to set the price months in advance whenever possible.

Usually, feed can be bought most advantageously at harvest time. Thus, cattle feedlot owners who have adequate storage and finances generally buy their main feed ingredients at that time. By so doing, they can project with reasonable accuracy what it will cost them to feed cattle. Futures permit the cattle feeder to accomplish the same thing without actually taking delivery on the feed and incurring storage costs and risks of physical deterioration. Cattle feeders can use such futures to protect against increases in feed prices—increases in prices of corn, oats, soybeans, soybean meal, and wheat.

• **Futures trading terms**—*Hedgers are farmers, processors, or merchandizers who sell a futures contract when a commodity is produced or bought, and who buy an equivalent contract at the market price when a commodity is sold; thus, hedging is a form of insurance.*

Speculators seek to assume risks for the sake of profit. They provide necessary liquidity to offset the trading of hedgers.

Basis is the difference between the cash price at a particular location and the futures price established at the commodity exchange. Some of the factors affecting basis include:

1. The overall supply and demand of the commodity.
2. The overall supply and demand of substitute commodities and comparable prices.
3. Geographic disparities in supply and demand.
4. Transportation pricing structures.
5. Storage space available.
6. Carrying costs.

A strong basis is a narrow difference between cash and future prices. A weak basis represents a wide difference in prices and generally indicates an over-supply situation. Ideally, prices of cash grain and futures should move simultaneously in parallel patterns, but this is rarely the case. However, when basis trends are plotted over time, the pattern becomes more predictable.

Feed Substitutions

In arriving at feed substitutions, two primary factors besides cost, chemical composition, and feeding value should be considered—namely, palatability and quality of product produced.

Special feed substitution tables have been prepared for each animal species. In order to facilitate their use, they are presented in the feeding chapters devoted to each class of livestock, Chapters 19 through 26.

Best Buy in Feeds

Feed prices vary widely. For profitable production, therefore, feeds with similar nutritive properties should be interchanged as price relationships warrant.

Purchase of feed nutrients at least cost, coupled with knowledge of the nutritive needs of animals and production results, provides a sound basis for arriving at the best buy in feeds.

Two different methods of arriving at the best nutrient buy in feeds are: (1) the computer method, and (2) the cost per unit of nutrients. The computer method is presented in Chapter 18, in the section on "Computer Method." The cost per unit of nutrients is presented in the section that follows.

COST PER UNIT OF NUTRIENTS

One method of arriving at the best buy in feeds is to compute and compare the cost per unit of nutrients, based on feed composition. Where a chemical analysis of a specific feed is not available, feed composition tables, such as those in Section V of this book, may be used as good indicators. Thus, feed composition tables may serve as a basis of feed purchasing and merchandising, as well as for ration formulation.

The use of the cost per unit of nutrients method can best be illustrated by the examples that follow:

- **Cost per pound of protein and TDN**— If 44% protein (crude) soybean meal is selling at $9.88 per 100 lb whereas 35% protein (crude) linseed meal sells for $6.25 per 100 lb, which is the better buy? Divide $9.88 by 44 to get 22.4¢ per pound of crude protein for the soybean meal. Then divide $6.25 by 35 and get 17.8¢ per pound of crude protein for the linseed meal. Thus, at these prices linseed meal is the better buy—by 4.6¢ (22.4 − 17.8 = 4.6) per pound of crude protein.

When buying energy feed, one can compare the cost per pound of total digestible nutrients (TDN). For example, if corn is priced at $3.63 per 100 lb and has a TDN of 91%, divide $3.63 by 91 and the result is 3.99¢ per pound of TDN. If milo with 86% TDN sells for $3.25 per 100 lb, divide $3.25 by 86, and the price is 3.78¢ per pound of TDN. Thus milo would be the better buy by 0.21¢ (3.99 − 3.78 = 0.21) per pound of TDN.

- **Factors other than price affect feeding value of feeds**— It is recognized that many factors affect the actual feeding value of each feed, such as (1) palatability, (2) grade of feed, (3) preparation of feed, (4) ingredients with which each feed is combined, and (5) quantities of each feed fed. It follows that, from the standpoint of the producer, the most important measurement of a feed's usefulness is in terms of *net returns*, rather than cost per bag or cost per ton.

Moisture is Important

When buying grains, feeders should never lose sight of how much water they may be purchasing. Table 16–2 illustrates the relative value (dry matter purchased) when paying for corn on a 15.5% moisture basis while actually receiving corn of another moisture content. Thus, if feeders were receiving 19% moisture corn and paying for 15.5% moisture, they would receive only 95.86% of the dry matter for which they paid. On the other hand, if corn is delivered with 7% moisture, while paying on a 15.5% moisture basis, feeders would receive 110.06% of that for which they paid.

TABLE 16–2
RELATIVE VALUE OF U.S. NO. 2 CORN (15.5% MOISTURE) AS AFFECTED BY CHANGES IN MOISTURE[1]

Moisture (%)	DM Basis Multiplier	Moisture (%)	DM Basis Multiplier
0	1.1834	19	.9586
1	1.1716	20	.9467
2	1.1598	21	.9349
3	1.1479	22	.9231
4	1.1361	23	.9112
5	1.1243	24	.8994
6	1.1124	25	.8876
7	1.1006	26	.8757
8	1.0888	27	.8639
9	1.0769	28	.8521
10	1.0651	29	.8402
11	1.0533	30	.8284
12	1.0414	31	.8166
13	1.0296	32	.8047
14	1.0178	33	.7929
15	1.0059	34	.7811
16	.9941	35	.7691
17	.9822	36	.7574
18	.9704		

[1]If 15.5% moisture corn is the purchase basis, it will require 1.1834 units of purchase base corn to make 1 unit of 100% dry matter base corn.

So, when high-moisture grain is bought, it should be purchased at a lower figure than dry grain. Likewise, a greater quantity of it must be fed to compensate for the higher moisture content.

(Also see Chapter 6, section on "High-Moisture Grain.")

HOME-MIXED VS COMMERCIAL FEEDS

The value of farm-grown grains—plus the cost of ingredients which need to be purchased in order to balance the ration, and the cost of grinding and mixing—as compared to the cost of commercial ready-mixed feeds laid down on the farm, should determine whether it is best to mix feeds at home or depend on commercial feeds. Of course, the ultimate criterion for choosing between home-mixed and commercial feeds is which program will make for maximum returns to producers for their labor, management, and capital. Generally speaking, the use of commercial feeds makes it possible for the producer to have more animals and concentrate on production, whereas home mixing restricts animal numbers and necessitates that part of the time and capital be devoted to feed formulating and manufacturing.

Fig. 16-3. Home processing feed center for swine unit, consisting of three bins: *large bin,* corn storage; *middle-sized bin,* soybean meal; and *small bin,* mineral-viamin premix. (Courtesy, Iowa State University, Ames)

Many farmers and ranchers are faced with the choice of home-mixing vs the purchase of commercially prepared feed. Even though the economics may favor the use of homegrown feeds, the following searching questions should be asked before launching a home processing feed operation:

1. Do I have the necessary equipment to process my feed efficiently, effectively, and without segregation?

2. Do I have a reliable, cost-competitive, quality source of the ingredients which I must buy?

3. Do I need to get FDA approval to add certain medications; must my feed processing operation meet FDA inspection?

4. Do I have the necessary facilities in which to store bulk ingredients and/or finished feed?

5. Do I have the expertise to go it alone without the help of my commercial feed representative?

COMMERCIAL FEEDS

Commercial feeds are feeds that are mixed by commercial feed manufacturers who specialize in the business, instead of being home-mixed.

In the United States, commercial feeds had their beginning as horse *tonics, conditioners, potions,* and *cure-alls.* Claims were made for increased growth and development, improved breeding, more speed, and increased stamina. The feeding directions called for a cup or for 3 to 4 Tbsp per horse daily.

In 1860, the first of these horse tonics was shipped to the United States from London. It contained beans, barley, flax seed, Peruvian bark, and quinine tonic. The price: $14.00/100 lb, at a time when oats were selling at $1.40/100 lb.

Pelleted feeds were introduced in Europe in 1870, during the Franco-Prussian War, when a compressed product was needed to save space. The German armies used pelleted feeds extensively during World Wars I and II, because they required only one-fifth as much transportation and storage space as loose hay and bulk grain.

One of the most famous products in the United States in the early 1900s was the International Stock Food Tonic. The developer and promoter was W. W. Savage, owner of the great Standardbred horse, Dan Patch, holder of the mile pacing record from 1906 to 1933 and the idol of his day. The advertisements for the tonic claimed that Dan Patch was fed the tonic every day and included a picture of the great horse suitable for framing.

But tonics were not for horses only! The labels on these secret and magic formulas were expanded to include cattle, sheep, swine, and poultry.

Not all commercial horse feeds of the Dan Patch era were tonics. Numerous high-quality feeds were also available. Typical commercial horse feeds during the period 1915 to 1930 consisted of corn, oats, barley, alfalfa, wheat bran, and molasses. During the 1930s, calcium and vitamins were added to commercial feeds.

Today, the commercial feed business is a complex and highly respected industry.

Importance and Nature of Commercial Feeds

The importance and the nature of commercial feeds in the United States is attested to by the following statistics for the year 1984, based on a survey made by the USDA:[2] A total of 109.6 million tons of commercial feeds were produced, which represented 21.5% of the 509.5 million tons of feed consumed that year. Of the commercial production, 87% was primary feed (which the researchers defined as a mixed feed, but to which a premix is sometimes added at the rate of less than 100 lb/ton) and 13% consisted of secondary feed (which the researchers defined as feed to which a formula feed supplement is added at a rate of 100 lb per ton or more). Primary feed production had the following breakdown by type of livestock feed: poultry, 35.5%; dairy, 22%; beef and sheep, 21.2%; hogs, 15%; and others, 6.5%. Also, the survey revealed that 51.7% of the primary feed was pelleted. Corporations owned 58.5% of all U.S. feed mills and produced 71.1% of the commercial feed; farmer cooperatives owned 28.5% of the feed mills and produced 21.8% of the feed tonnage; and the remaining 13% of the mills that produced 7.1% of the feed were privately owned.

Unfortunately, the terms identifying the types or classes of commercial feeds are not standardized; that is, there are no universally accepted definitions of kinds of commercial feeds. So, the authors have evolved with the definitions that follow.

---

[2]The statistics presented, and the terms used, herein were taken from a nationwide survey made by the USDA's Economic Research Service (ERS) in 1984, to which there was more than a 40% response. The report was written by ERS economists Mark Ash, William Lin, and Mae Dean Johnson and reported in *Feedstuffs,* March 14 and March 21 issues, 1988.

- **Complete feed**—*A complete feed is a ready-to-use feed which is nutritionally balanced and requires no additional ingredients, except it may or may not contain processed hay and/or other roughage.*

Fig. 16-4. Complete feed automated to hog finishing pens. In recent years, there has been an increasing trend toward the use of complete mixed rations for growing-finishing hogs and baby pigs. (Courtesy, University of Illinois, Urbana)

- **Supplement**—*A supplement is a formula feed which contains a substantial portion of the protein, minerals, and vitamins required in the final ration.*

- **Base mix**—*A base mix differs from a supplement in that it provides less of the animal's protein requirements.*

- **Premix**—*A combination of one or more trace minerals, vitamins, and/or performance-enhancing compounds with a carrier.*

- **Specialty feeds**—*Specialty feeds refer to a number of products that are produced by feed manufacturers for special purposes, including such products as milk replacers and range cubes.*

The commercial feed manufacturer has the distinct advantages of (1) purchasing feed in quantity lots, making possible price advantages; (2) using computers for purchasing and least-cost formulating; (3) having the knowledge to manufacture medicated feeds; (4) having the knowledge and the facilities to manufacture specialty feeds, such as milk replacers; (5) processing and mixing economy and control; (6) hiring scientifically trained personnel for use in determining the rations; and (7) controlling quality. Most producers have neither the know-how nor the quantity of business to provide these services on their own.

In summary, it may be said that there exist two good alternative sources of most feeds and rations—home-mixed or commercial—and the able manager will choose wisely between them, or use some of each.

Types of Commercial Feed Formulations

Commercial feed can be classified under three types of feed formulations—closed formulas, open formulas, and custom formulas.

- **Closed formulas**—Closed formula feeds are commercial feed preparations that do not list the proportions of the respective ingredients. Instead, they either (1) list the ingredients on the feed tag in decreasing order of incorporation, or (2) use group terms such as animal proteins, *etc.*

- **Open formulas**—Open formula feeds are those that have a statement on the tag that tells how many pounds of each ingredient are incorporated in the mix.

- **Custom formulas**—Custom formula feeds are those that are mixed according to the specifications of the purchaser.

How to Select Commercial Feeds

There is a difference in commercial feeds! That is, there is a difference from the standpoint of what livestock producers can purchase with their feed dollars. Smart operators will know how to determine what constitutes the best in commercial feeds for their specific needs. They will not rely solely on the appearance or aroma of the feed, nor on the salesperson. They will consider (1) the reputation of the manufacturer, (2) the specific needs of the animal, (3) labeling of feeds, and (4) flexible formulas.

- **Reputation of the manufacturer**—The reputation of the manufacturer can be determined by (1) conferring with other producers who have used the particular products, and (2) checking on whether the commercial feed under consideration has a good record for meeting its guarantees. The latter can be determined by reading the bulletins and reports prepared by the state department in charge of monitoring feed quality and enforcing feed laws.

- **Specific needs of the animals**—Feed requirements vary according to (1) the class, age, and productivity of the animals; and (2) whether the animals are fed primarily for maintenance, growth, finishing (or show-ring fitting), reproduction, lactation, or work (running). The wise producer will buy different formula feeds for different needs.

- **Labeling of feeds**—Most states require that feeds carry labels guaranteeing the ingredients and the chemical makeup of the feed, along with directions for its use. The feed tag should contain the following information:

1. Net weight of the feed.
2. Brand name and product name.
3. Guaranteed analysis.
4. Listing of ingredients.
5. Directions for use.
6. Name and mailing address of the manufacturer.
7. Warnings.

Fig. 16-5 shows a typical feed tag.

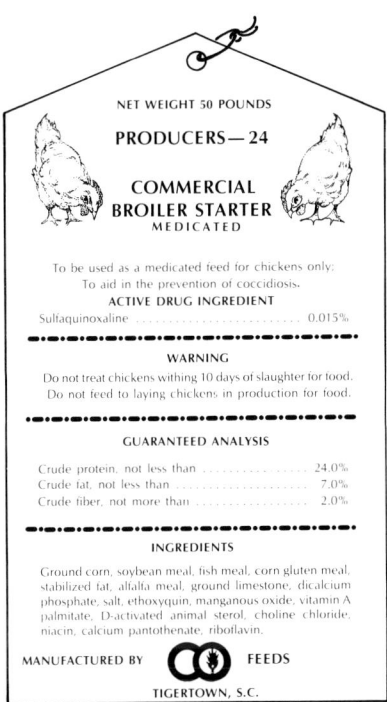

Fig. 16-5. Example of a feed tag.

- **Flexible formulas**—Feeds with flexible formulas are generally the best buy. This is because the price of feed ingredients in different source feeds varies considerably from time to time. Thus, good manufacturers, having access to least-cost computer programs, will shift their formulas as prices change, in order to give the producers the most for their money. This is as it should be, for (1) there is no one best ration; and (2) if substitutions are made wisely, the price of the feed can be kept down and the feeder will continue to get equally good results.

Cooperatives

Cooperatives are business organizations that are formed, operated, and financed by patron members. The members themselves are responsible for the decisions of the organization and subsequently share the risks and reap the benefits. For example, a group of livestock feeders may decide to form a cooperative for the procurement of feeds.

Cooperatives enjoy substantial tax advantages. Among their additional advantages are pride of ownership, collective bargaining, influence in decision making, and sharing in any profits.

To form a cooperative, members must invest considerable capital. This means that they take certain risks. Although they share in profits, they must also assume losses.

FEED QUALITY

Feed quality is the degree of excellence of a feed.
Quality control is a system of assuring the maintenance of standards by checking.

Feed quality and quality control are important in both home-mixed and commercial feeds. High-quality feed must be palatable—the animals must like and eat it. Also, finished (mixed) feeds must meet the nutritional needs of the animals to which they are fed, without deficiencies.

Feed quality begins with the selection of feed ingredients. High quality mixed feed cannot be produced from poor quality feed ingredients. To assure high quality feed ingredients, they should be subjected to the following tests and meet high standards in each test:

1. They should be physically evaluated for quality; including color, smell, taste, texture, and microscopic examination.

2. They should be tested by proximate analysis and/or other methods for components; and roughages should be tested for NDF and ADF.

3. They should be examined for rodent contamination, salmonella bacteria, molds and mycotoxins, and undesirable chemicals such as PCBs and pesticides.

Quality control involves—

1. Buying (or producing) feed ingredients that meet high standards.

2. Inspecting and sampling ingredients at receiving.

3. Monitoring processing, including keeping inventory of drugs, flushing out the feed system after preparing certain medicated feeds, isolating some ingredients, and keeping records of ingredients and finished feeds.

4. Providing finished feed storage and transportation, with rigid sanitation and pest control.

FEED LAWS

The U.S. Food and Drug Act was passed in 1906, giving the Federal Government authority to regulate and inspect feeds shipped in interstate commerce. Additional controls were authorized in the Food, Drug, and Cosmetic Act of 1938. In addition to the Federal laws, nearly all states have laws regulating the sale of commercial feeds. These benefit both producers and reputable feed manufacturers. In most states the laws require that every brand of commercial feed sold in the state be licensed, and that the chemical composition be guaranteed.

(Also see Chapter 13, section on "Food and Drug Administration [FDA].")

QUESTIONS FOR STUDY AND DISCUSSION

1. Why do feed costs often determine the success or failure of a livestock operation?

2. Why did the use of concentrates increase while the use of roughages declined during the 20-year period 1965 to 85?

3. Poultry accounted for 22% of the grain consuming animal units in 1965-66, compared with 26% of the total in 1984-85. What accounted for this increase in poultry feeding?

4. Fig. 16-1 shows that total feed expenditures for feed trended upward from 1975-1985, while Fig. 16-2 shows the feed costs as a percent of total farm expenditures trended downward from the mid-1970s to the mid-1980s. How and why did total feed expenditures trend upward at a time when feed costs as a percent of total farm expenditures trended downward?

5. On a livestock farm or on a crop and livestock farm combined, what circumstances will determine whether feeds will be homegrown or purchased?

6. List and discuss 10 factors about which producers-feed buyers need to be knowledgeable in order to be successful.

7. When buying feed, how important are the following factors: palatability, variety, digestive disturbances, bulk, residues, and poisonous plants?

8. Define commodity exchange futures contracts, hedgers, speculators, and basis.

9. List some of the factors that affect basis.

10. What precautions should be taken when using a feed substitution table?

11. A producer has the opportunity to select from among the following feeds, to be used for supplying energy:

    a. Milo—80% TDN at $3.40 cwt
    b. Corn—89% TDN at $3.75 cwt
    c. Barley—83% TDN at $3.50 cwt

Which is the best buy as determined by cost per unit of TDN?

12. Formerly, feeds were evaluated largely on the basis of energy and/or protein. Computer technology makes it possible to evaluate feeds on the basis of a number of nutrients simultaneously. Does this fact justify the purchase of a computer on a medium-sized livestock farm?

13. List five factors other than price that affect the feeding value of each feed.

14. Why is moisture important when buying feeds and balancing rations?

15. What factors should determine whether it is best to mix feeds at home or depend on commercial feeds?

16. What are commercial feeds? Discuss the early history of commercial feeds in the United States.

17. Discuss the magnitude of U.S. commercial feed industry. Rank primary feed production according to type of livestock to which it is fed.

18. Define the following terms: complete feed, supplement, base mix, premix, and specialty feed.

19. List the main advantages of commercial feed manufacturers in comparison with home-mixing.

20. Differentiate between closed, open, and custom formula feeds.

21. How important are the following when selecting commercial feeds: (a) the reputation of the manufacturer, and (b) the specific needs of the animals.

22. What information can be found on a feed tag?

23. Discuss the advantages and disadvantages of cooperatives vs private feed companies.

24. Define (a) feed quality, and (b) quality control.

25. How may a commercial feed company assure high quality feed ingredients?

26. What four steps are involved in feed quality control?

27. Who monitors the commercial feed industry; and who benefits therefrom?

Buying Feeds/Commercial Feeds/Feed Laws

OLD MACDONALD'S FARM, above, is from an original painting by the noted artist, Tom Phillips (3333 17th Street, San Francisco, California 94110), prepared especially for this book. It portrays the artist's conception of OLD MACDONALD'S FARM, immortalized in the Old English Folk Song. And on that FARM he had a menagerie of chatty animals: some cows that *MOO*ed; some sheep that *BAA*ed; some pigs that *OINK*ed; some horses that *NAY*ed; some chickens that *CLUCK*ed; some turkeys that *GOBBLE*d; and some ducks that *QUACK*ed.

Those of us who grew up on such a farm learned barnyard talk. The animals kept up a running conversation with us as we gave a handful of hay to a still-hungry cow, treated a favorite horse to an apple, took a last look at a litter of newborn pigs, closed the gate on the sheep corral, and shut the poultry house door. It was a sign language, but it spoke louder than words; it told us how the animals felt and what they wanted. Every movement and every sound conveyed a message of well-being, distress, or disease. Lack of interest, dull eyes, sluggishness, rough coat, poor appetite, and/or abnormal feces or urine spelled trouble. Alertness, stretching on rising, yawning, vocalizing, eating with relish, and frisking were good omens and told us that all was well in the barnyard.

Artist Tom Phillips' *OLD MACDONALD'S FARM* captures the marvels and mysteries of animal behavior and brings on a bad case of nostalgia among those of us who remember *the good old days*.

SECTION III
FEEDING

Animals inherit certain genetic possibilities, but how well these potentialities develop depends upon the environment to which they are subjected; and the most important influence in the environment is the feed.

Also, feeding is important from an economic standpoint; it is the major item of expense in producing livestock. Normally, it accounts for approximately the following proportions of the cost of livestock production: finishing cattle, 70%; feedlot lambs, 50%; pork, 65 to 75%; milk production, 55%; and poultry, 55 to 75%, with the production of eggs toward the lower side of this range and the production of broilers and turkeys toward the upper side. It is important, therefore, that feeding practices be as efficient and economical as possible. To this end, livestock producers should endeavor to provide rations that are both satisfactory and inexpensive—rations that make for maximum production of a quality product per unit of feed consumed and for maximum net returns.

Section III is devoted to the art and the science of livestock feeding. In recognition of the effect of artificial environments on animal nutrition, the innovative Chapter 17, Animal Behavior/Environment, is presented. Chapter 18, Feeding Standards/Ration Formulation, tells how feeding programs may be tailored to specific operations. Then, Chapters 19 to 28 embrace an in-depth discussion of the feeding of each species.

17

ANIMAL BEHAVIOR/ ENVIRONMENT[1]

Original painting by Tom Phillips

Contents	Page
Animal Behavior	308
How Animals Behave	308
Behavioral Systems	308
Social Relationships	312
Social Order (Dominance)	313
Leader—Follower	314
Interspecies Relationships	315
People—Animal Relationships	315
Anamalous (Abnormal) Animal Behavior	315
Abnormal Cattle Behavior	315
Abnormal Sheep Behavior	315
Abnormal Swine Behavior	315
Abnormal Horse Behavior (Vices)	316
Abnormal Chicken Behavior	317
Abnormal Turkey Behavior	318
Animal Environment	318
Effect of Environmental Factors on Animals	319
Feed/Environmental Interactions	319
Water/Environmental Interactions	322
Weather/Environmental Interactions	324
Thermoneutral Zone (Comfort Zone)	324
Adaptation, Acclimation, Acclimatization, and Habituation of Species/Breeds to the Environment	325
Facilities/Environmental Interactions	326
Recommended Environmental Controls	326
Health/Environmental Interactions	327
Antibody Production	327
Disease Defense	328
Colostral Defense	328
Immune Suppression (Altered Immune Response)	328
Pollution	328
Muddy Lots	328
Stray Voltage (Tingle Voltage, Neutral-to-Earth Voltage)	329
Porcine Stress Syndrome (PSS)	329
Ulcers	330
Stress/Environmental Interaction	330
Bleeders—A Stress/Environmental Interaction of Equines	330
Pollution Laws and Regulations	331
Sustainable Agriculture	332
Animal Welfare/Animal Rights	333
Questions for Study and Discussion	333

Fig. 17-1. A serious case! Knowledge of behavioral norms is necessary in order to detect and treat abnormal situations—especially illness. (Courtesy, The Bettmann Archive)

"Wherefore, come one, O young husbandman
Learn the culture proper to each kind."

Human life, animallike and stagnant for a million years, came alive with the domestication of animals a mere 11,000 years ago. Ever since, admonitions like the above quote

---

[1]The authors gratefully acknowledge the helpful suggestions of the following eminent scientists who reviewed this chapter: J. V. Craig, Ph.D., Department of Animal Sciences and Industry, Kansas State University, Manhattan; R. Kilgour, D.Phil., Ruakura Animal Research Centre, Ministry of Agriculture and Fisheries, Private Bag, Hamilton, New Zealand; J. J. McGlone, Ph.D., Department of Animal Science, Texas Tech University, Lubbock; and W. R. Stricklin, Ph.D., Department of Animal Sciences, The University of Maryland, College Park. Additionally, the authors are grateful to S. E. Curtis, Ph.D., Professor of Animal Sciences, University of Illinois, Urbana, for arranging for one of his former students J. J. McGlone, to serve as a reviewer. Also, the senior author obtained many helpful ideas from the authoritative books by Drs. Craig, Curtis, and Kilgour.

by Virgil, the great Roman poet, have characterized the relationship of caretakers with their herds and flocks. But as the art of husbandry evolved into science, caretakers' fluency with the language of their beasts declined, and human-made environments became increasingly artificial, with emphasis on economic and biological conversion mechanisms. Today, ethologists (those who study animal behavior) are assisting practical caretakers, scientifically.

Presently, there is great interest in animal behavior and environment. Those who grew up around animals and dealt with them in practical ways already have accumulated substantial workaday knowledge about their reaction to certain stimuli or their environment. But those who are less familiar with them may need to acquaint themselves with animal behavior in a people-shaped environment, better to feed and care for them, and in order to recognize the signs when all is not well.

Neolithic (New Stone Age) man showed some understanding of animal behavior from the remote day of their domestication, beginning about 11,000 years ago. It required knowledge of basic behavior patterns to capture, confine, and herd animals, as did the subsequent breeding, feeding, watering, and sheltering them. Without this understanding, domestication would have failed and animals might not have survived.

Caretakers were little concerned with environment so long as animals roamed over pastures and ranges. But concentration of animals in smaller spaces and concentration of people in urban areas changed all this. It produced profound changes in the environment, many of which are unfavorable to the quality of both human and animal life. Husbandry that reduces labor, land, and housing costs often results in physical and social conditions that increase behavioral problems. Some stressors acting on animals kept as sources of meat, milk, eggs, fiber, and power are unavoidable. Nevertheless, means of reducing behavioral stress may be needed so that decreased labor, land, and housing costs are not offset by losses in productivity and profits. Also, for humanitarian reasons the stress of animals serving people should be minimal. So, today, the keepers of herds and flocks are giving attention to environmental control, involving space requirements, light, air temperature, humidity, air velocity, wet bedding, ammonia buildup, dust, odors, and manure disposal.

The relationship between animal behavior and environment is evidenced by tail biting in swine, which is sometimes referred to as an anticomfort syndrome, because any part of the pig's environment that makes it uncomfortable may lead to tail biting. Among such environmental factors causing this abnormality are malnutrition; lack of space in confinement; abrasive floors and lack of bedding; improper temperature, high humidity, and poor ventilation; inadequate feed and water space; lack of uniformity within pens; infestation by parasites; or just plain boredom.

ANIMAL BEHAVIOR

Animal behavior is the reaction of animals to certain stimuli, or the manner in which they react to their environment. Stated simply, behavior is the movements animals make. It embraces much more than locomotion; it includes the movements animals make when feeding, when breathing, and when mating. Also, it involves such things as pricking up the ears or making a sound.

Most scientists agree that natural selection favors animal behaviors that increase chances of survival and reproduction, and that information gained from wild and feral relatives provides clues about the behaviors of domestic animals.

The term *ethology* is now used to describe *the study of the behavior of animals.* More recently, animal behavior studies have come to be termed *applied ethology*, which covers such things as feeding, care and management, stress, transporting, and welfare. In this chapter, emphasis will be placed on feeds and feed-related aspects, although it is recognized that there is interaction between all factors affecting animals.

Through the years, behavior has received less attention than the quantity and quality of the meat, milk, eggs, fiber, and power produced by animals. But modern breeding, feeding, and management have brought renewed interest in behavior, especially as a factor in obtaining maximum production and efficiency.

How Animals Behave

Animals behave in a great variety of ways, differing by species and environments.

Our ancestors had practical knowledge, gained firsthand or learned from others with experience, on how animals behaved. In primitive societies, knowledge of how animals behave was necessary when hunting or when capturing animals for domestication. Later, domesticated animals were maintained in an open pastoral environment, with freedom of movement, where they behaved under relatively natural conditions.

Then came the industrial age! Cheap labor and cheap land were no longer available, and marginally productive animals could no longer be afforded. Intensive production, confinement, and artificial environments evolved. Today, mother-care of young is being removed at earlier and earlier ages; and human care is being replaced by mechanical devices. Animals are segregated by age, size, and sex. This calls for a special fit between the behavior of animals and their artificial environments. Ethology, the scientific study of animal behavior, coupled with physiological evaluations of stress, will make for innovative management and housing methods in the decades to come.

BEHAVIORAL SYSTEMS

Some behavioral systems or patterns are better developed in certain species than in others. Ingestive and sexual behavior systems have been most extensively studied because of their importance commercially. Nevertheless, most animals exhibit the following nine general functions or behavioral systems:[2]

1. **Ingestive (eating and drinking) behavior.** This is characteristic of animals of all species and all ages. Without feed and water, animals cannot survive.

2. **Allelomimetic behavior.** This is mutual mimicking behavior. Thus, when one member of a group does something, another tends to do the same thing; and because others are doing it, the original individual continues.

---
[2]Adapted by the authors from: Scott, J. P., *Animal Behavior*, 2nd ed., The University of Chicago Press, Chicago, Ill., 1972.

Animal Behavior/Environment

Fig. 17-2. Allelomimetic (mimicking) behavior exhibited by chickens. Feed consumption is facilitated by the presence of feeding companions. (Courtesy, Ralston Purina Co., St. Louis, Mo.)

3. **Eliminative behavior (defecation and urination).** Nature ordained that if animals eat, they must eliminate wastes—they produce manure (feces and urine). Properly handled, manure is an asset; improperly handled, it may be a pollution problem.

4. **Gregarious behavior—this refers to the flocking, or herding, of certain species.** It is closely related to allelomimetic behavior. If animals imitate each other, they must stay together. If they stay together as a mobile unit, they use either allelomimetic behavior or leader-follower relationship to do so.

5. **Sexual behavior.** This involves courtship and mating. It is largely controlled by hormones. Each animal species has a special pattern of sexual behavior.

6. **Care-giving and care-seeking (mother-young) behavior.** The care-giving behavior among domestic animals is largely confined to females, where it is usually described as *maternal*. The care-seeking behavior, which extends until the young are weaned, is normal for young animals. Expressions of care-giving and care-seeking vary widely among different species of farm animals.

Fig. 17-3. "Madonna." Care-giving and care-seeking (mother-young) behavior. (An original painting by artist Tom Phillips, 3333 17th St., San Francisco, CA 94110)

7. **Agonistic behavior (combat).** This type of behavior includes fighting, flight, submission, and other related reactions associated with conflict. Among all species of farm mammals, males are more likely to fight than females. Nevertheless, females may exhibit fighting behavior under certain conditions. Castrated males are usually quite passive, which indicates that hormones (especially testosterone), are involved in this type of behavior.

8. **Investigative behavior.** All animals are curious and have a tendency to explore their environment. Investigation takes place through seeing, hearing, smelling, tasting, and touching. Whenever an animal is introduced into a new area, its first reaction is to explore it.

Fig. 17-4. Investigative behavior evidenced by a curious pig. (Courtesy, USDA).

9. **Shelter-seeking behavior.** This is the type of behavior which causes all species of animals to seek shelter—protection from the sun, wind, rain and snow, insects, and predators.

Fig. 17-5. Shelter-seeking behavior, showing cattle in a ravine, seeking protection from the storm. Note that they are grouped together and facing away from the storm. (Courtesy, American Hereford Assn., Kansas City, Mo.)

Of the nine behavioral systems, the following four involve feeding, either directly or indirectly; hence, they are crucial to animal production and are detailed by species in Table 17-1. (See pages 310–311.)

1. Ingestive behavior (eating and drinking).
2. Eliminative behavior (defecation and urination).
3. Allelomimetic behavior (mimicking).
4. Gregarious behavior (flocking).

TABLE
HOW ANIMALS

Behavior	Cattle (Beef and Dairy)	Sheep
Ingestive behavior (eating and drinking): This type of behavior includes eating and drinking; hence, it is characteristic of animals of all species and all ages. It is very important because animals cannot live without feed and water. *Rumination is the act of chewing the cud, characteristic of herbivorous animals with split hoofs—cattle, sheep, and goats.* It involves regurgitation of ingesta from the reticulo-rumen, swallowing of regurgitated liquids, remastication of the solids accompanied by reinsalivation, and reswallowing of the bolus. The first ingestive behavior trait, common to all young mammals, is suckling. Each species has its own particular method of ingesting feed.	The natural feeding (grazing) position of cattle is heads down. In this position, they produce more saliva; and saliva aids digestion. When grazing, cattle wrap their tongues around grass, then jerk their heads forward so that the vegetation is cut off by the lower incisor teeth. (There are no upper incisor teeth; only the thick hard dental pad.) When grazing, cattle also move their heads from side to side. This movement gives them a continuous view of their entire surroundings, an essential for wild cattle in an environment containing dangerous predators. It is important that artificial feeding devices and arrangements not depart too far from this natural pattern. Rumination occupies about 8 hours of the cow's time each day. In addition, the harvesting or grazing time may take another 8 hours. This means that a cow may be active for a 16-hour day. The Iowa Station reported that steers in lots on self-feeders spent 12 hours per day lying down, and that this time was unaffected by shelter or season. When the cow regurgitates, a soft mass of coarse feed particles, called a bolus, passes from the rumen through the esophagus in a fraction of a second. She chews each bolus for about 1 minute, then swallows the entire mass again. Originally, it was thought that the regrinding which occurred during rechewing helped the digestion by exposing a greater surface area to fiber-digesting microflora. But recent studies indicate that rechewing does not improve digestibility. Instead, rumination has an important effect on the amount of feed the animal can utilize. Feed particle size must be reduced to allow passage of the material from the rumen. It follows that high-quality forages require much less rechewing and pass out of the rumen at a faster rate; hence, they allow a cow to eat more. This concept is very important to the production of beef and milk because a cow will eat only as much coarse material as she can grind up by ruminating.	Sheep graze very much like cattle, but their cleft upper lip allows them to graze vegetation closer to the ground. As in cattle, the incisors are in the lower jaw only. The total grazing time of sheep ranges from 10 to 12 hours per day, which is normally divided into two periods and highly correlated with sunrise and sunset. Sheep are very selective grazers; they will consistently select plants higher in protein and lower in crude fiber than can be obtained by harvesting the forage. In general, the selectivity is proportional to the amount of herbage available; thus, when the feed supply is short, sheep become less discriminating. Sheep also form preferences for plant species based on previous experience. Sheep are one of the most drought-resistant of domestic animals. In some cases, they graze without access to water, relying on dew, snow, and/or moisture in plants. Goats graze in the same manner as cattle and sheep, but they are very fond of browse—the young shoots of shrubs and trees.
Eliminative behavior (defecation and urination): In recent years, elimination has become a most important phenomenon, and pollution has become a dirty word. Nevertheless, nature ordained that if animals eat, they must eliminate.	Cattle deposit their feces in a random fashion. Although cows can defecate while walking, with the result that their feces are scattered, generally they deposit their "chips" in neat piles. Most cows elevate their tails and hump up to urinate, whereas bulls are inclined to stand squarely on all "fours."	The eliminative behavior in sheep is very similar to that of cattle. However, they differ from cattle in that ewes usually assume a squat position when they urinate, and their feces are relatively dry and pelleted.
	A full understanding of the eliminative behavior will make for improved animal building design and give a big assist in handling manure. Right off, it should be recognized that the eliminative behavior in farm animals tends to follow the general pattern of their wild ancestors; but it can be influenced by the method of management.	
Allelomimetic behavior (mimicking): Allelomimetic behavior is mutual mimicking behavior. Thus, when one animal in a group commences to eat, this triggers a response for others to do likewise. In the wild state, this trait was advantageous in detecting the enemy, and in providing protection therefrom. In wolves and coyotes, this behavior is important in attacks on prey, since a pack working together is much more likely to be successful than when working alone.	Cows moving across a pasture toward a milking barn often display allelomimetic behavior. One cow starts toward the barn, and the others follow. Because the rest of the herd is following, the first cow proceeds on. Because of stimulating and competing with each other, there is usually higher per steer feed consumption among a group of steers than by one steer alone. Thus, one steer penned alone may consume "X" pounds of feed per day. However, when he is placed with other steers, his intake may be "X + Y" pounds. But, of course, the feed consumption advantage can be nullified when animals are placed together too closely, with the result that the agonistic behavior comes into play.	Sheep walk, run, graze, and bed down together. Sheep graze when they observe others in adjacent paddocks doing so.
Gregarious behavior (flocking): Gregarious behavior refers to the flocking, or herding, of certain species. It is closely related to allelomimetic behavior. If animals imitate each other, they must stay together. If they stay together as a mobile group, they must use allelomimetic behavior to do so. All such behavior arises out of the process of social attachment. Gregarious behavior differs among species.	Cattle tend to roam in groups of various sizes when a large herd is placed on a pasture or range. However, there is usually considerable space between the members of the herd. Moreover, on close observation it is evident that there are several small groups within a herd, each ranging from 3-5 head.	The gregarious, or flocking, behavior is particularly strong in sheep. Moreover, it is more evident in some breeds than in others. The Merino, and animals carrying Merino breeding, are noted for their flocking behavior. This makes it possible to herd them on the range.
	It is noteworthy that the gregarious instinct of sheep diminishes to some extent when they are placed within fenced holdings, instead of herded. As a result, those who handle western range bands do not try to switch back and forth from fenced range to herding, for the reason that the band becomes unmanageable from the standpoint of herding once they have been in a fenced holding for an extended period of time. Packers use the gregarious behavior of sheep by having an old goat, appropriately called a *Judas*, lead sheep to slaughter. A well-trained Judas will lead group after group to slaughter.	

Animal Behavior/Environment

17-1
BEHAVE

Swine	Horses	Poultry
Swine possess incisor teeth in both the upper and lower jaws; hence, they bite off grass or take a mouthful of grain, then chew and swallow it. Pigs have a single stomach, whereas ruminants have a 4-compartment stomach. By nature, pigs love to root. If given the opportunity, they will stick their noses into the ground and lift forward and upward, moving earth out of the way and exposing earthworms, grubs, and roots.	The mobile upper lip of the horse is used in gathering grass and other feed, in the same way that a cow uses her tongue or an elephant uses his trunk. The upper lip is very sensitive. The long, strong, and roughened teeth of the horse are well suited to grinding common feeds. A foal may have difficulty in getting its head to the ground for grazing, because, in the evolutionary process, the proportionate elongation of the legs and head was not always perfect. As a result, a long-legged foal with a short head and neck may have to stand with the front legs spread apart in order to reach the ground. When snow covers the pasture or range, horses will paw with a front foot (either right or left) through the snow and clear an area so that they can reach the grass. The horse has a blind spot in front of its nose. Hence, it cannot see the feed as it is eaten. Horses rarely browse; they will not eat the leaves of trees or shrubs provided grass is available. Horses prefer grazing in an open area, where they can watch for their enemies. Also, they prefer young, tender grass to coarse-stemmed plants, a preference which often causes them to overgraze certain areas. The horse evolved to graze small amounts almost continuously, rather than large amounts infrequently. This explains why horses do better if fed small amounts—often. In areas where water is scarce, wild horses usually graze 6–8 miles from water and come to water every other night. A thirsty horse may lower its head in the water deep enough to cover the nostrils.	Chickens and turkeys ingest their feed by pecking; ducks scoop their feed with their broad, soft bills. Except for geese (which graze on grass), poultry do not eat much forage. Chicks do not peck much until their second day after hatching, presumably due to ingestion of the yolk sac during hatching. Normal pecking experience requires some light. Initially, chicks peck and ingest both nutritive and non-nutritive substances. When foraging for food, a mother hen will cluck to her chicks and call them each time a choice morsel is found. All the chicks wil come running to participate in the find.
If given an opportunity, pigs are of very clean habits. They like to keep their nesting area clean and dry. Hence, they usually deposit their feces and urine away from the feeding and resting areas. By proper building design, it is possible to facilitate feeding, resting, and defecating in separate areas.	Horses tend to deposit their feces in certain locations, such as along well-traveled paths, like those leading to waterholes. Hence, if given the opportunity, they often return to these locations for defecation. Stallions are much more prone to deposit droppings on the same old mound than mares or geldings. Mares and geldings are inclined to use the border of their defecating area, with the result that they enlarge it each time. There is disagreement whether a feral or range stallion marks the outside boundary of his home range with his feces, thereby staking it out for himself and his harem of mares, with most horse owners believing and most ethologists disbelieving.	Except when on the roosts at night, chickens and turkeys deposit their excreta at random.
Swine exhibit allelomimetic behavior in their eating habits. Thus, when one pig eats, there is a tendency for the rest to join it. As a result, pigs in a group usually average higher feed consumption than one pig alone.	A timid horse will follow behind the pack, in order not to be left behind. When kept alone, high-strung racehorses may become nervous and fail to eat properly. To alleviate this situation, a companion—known as a mascot, is often provided. All sorts of mascots are used—a goat, a sheep, a chicken, a duck, or a pony.	Pecking and feeding are facilitated by social stimulation; e.g., the presence of a feeding companion. Social facilitation of feeding also occurs in adult chickens. For example, when a solitary hen eats from a large portion of feed until satisfied, she will immediately resume eating if a second hungry hen is introduced and starts consuming some of the feed.
In the wild state, swine roved through the forest in herds. Usually these wild groups consisted of 1 to 4 females, along with their young of the year and their yearlings. Adult males join these groups during the mating season, but range separately during the rest of the year. Under domestication, swine retain their gregarious nature. However, it has been altered a great deal. Today, hogs are usually confined to a very limited area. Also, under domestication, they have lost most of their ferocity and are usually gentle and easily handled.	In the wild state, horses ran in bands; thus, they were gregarious by nature. These bands seldom consisted of more than 40 animals, and always there was a stallion in each group. Under domestication, horses show definite preferences for their herdmates; they will even avoid certain horses in the herd. In the draft horse era, animals that were worked together usually stayed together when they were turned to pasture.	Chickens, turkeys, ducks, and geese tend to flock together. Of course, under domestication, humans have interfered with normal flocking of both chickens and turkeys. Even under domestication, however, ducks and geese exhibit their gregarious or flocking nature as they walk Indian file, one behind the other.

SOCIAL RELATIONSHIPS

Social behavior may be defined as any behavior caused by or affecting another animal, usually of the same species, but also, in some cases, of another species.

Social organization may be defined as an aggregation of individuals into a fairly well integrated and self-consistent group in which the unity is based upon the interdependence of the separate organisms and upon their respsonses to one another.

The social structure and infrastructure in herds and flocks are of great practical importance. Livestock producers should be knowledgeable relative to the social relationships of each species with which they work. Then, if this social relationship is disturbed and/or modified under intensive, confined conditions, they will be better able to feed, care, and manage the animals with maximum consideration accorded to both economy of production and animal welfare.

• **Social organization among cattle**—Breed affects social stratification in cattle. For example, on pasture or range, Angus tend to dominate Shorthorns, while Herefords tend to submit to both breeds. Older cows generally dominate the younger ones, and the heavier animals (usually the older animals) tend to dominate the lighter; for this reason, 2-year-old heifers should be segregated from older cows. However, among cows of similar age and breed, the smaller and more aggressive ones are most dominant. Also, cows with more seniority in the group and cows with horns tend to be of higher social rank. Aggression in cows appears to be ritualized, with most encounters taking place in the following sequence: approach, threat, and physical contact (or fighting).

Limited studies indicate a high relationship between social status of cattle and spacing, or social distance. The higher the social rank, the more likely cows are to be found near other members of the herd. Also, dominant cows tend to allow close approach by other cows more often than subordinates.

On the range, the following rank orders are evident in large heterosexual herds of cattle: (1) adult males, (2) adult females, and (3) juveniles. Adult males dominate adult females, which, in turn, dominate juveniles. However, at about 1½ years of age, young males begin to fight with adult females, and by 2½ years of age, they dominate all the females, and join the adult male rank order.

When grouped together on the range, bulls are loners, and do not organize socially.

When moving from the paddock to the milking parlor, dairy cows travel in a consistent leadership order. Mid-dominant cows tend to be in front of the group. However, the same individuals are seldom consistent leaders; instead, there is a pool of animals which tends to be in or near the lead. More consistency is found in the cows bringing up the rear of the moving herd. So, *rearship* is a more distinctive feature than *leadership*. The animals at the rear are usually the younger subordinate heifers. Also, in most herds there is a definite order in which cows enter the milking parlor; the mid-age, mid-dominant cows tend to be milked first, followed by the older cows. Social dominance orders (called *bunt order* in polled cows and *hook order* in horned cows) become more complex as herd sizes increase. Generally, not more than 100 dairy cows should be group-fed.

Within a corral of feedlot (finishing) cattle, a linear-tending (a linear-tending hierarchy is a type of social hierarchy in which dominance-subordinance ranking includes a triangle or some more complex hierarchy loop) "peck order," or dominance order, can be determined by observing agonistic encounters between pairs of animals within the group. The degree of linearity is greater with increases in heterogeneity of such factors as age, sex, weight, breed, and background. Linearity and stability of the dominance order tend to increase the longer the group is together, and linearity is greater among smaller groups.

• **Social organization among sheep**—When grazing, large flocks (bands) of sheep generally split into subgroups and occupy separate areas. Different breeds vary in their tendency to move or flock together, with the gregarious trait being strongest in the Merino and Rambouillet breeds.

The social structure of sheep is dependent upon visual contact and a flocking tendency. Also, group size is important. It has been shown that Merino sheep kept in pairs (just 2 sheep) gained less liveweight and produced less wool than their counterparts in groups ranging from 4 to 30 animals. The pairs spent less time grazing and more time walking along the fence line trying to keep contact with the flock in the next pasture; thus, pairing produced a level of stress which affected production. The need for visual contact during grazing may well account for this stress. It should be noted, however, that preferred group size and overall flock dispersion are dependent on breed, age, stability of flock membership, and vegetation.

Studies have not shown any correlation between the rank in which sheep reach feed troughs and their competitive ability at the troughs. However, certain individuals are constantly among the first few sheep to reach the troughs. Thus, it appears that going to feed is initiated by a few sheep, then others follow; and that dominance is not involved. Aggression is rare during normal grazing and social hierarchies; *i.e.,* dominance of one sheep over another is far from absolute.

• **Social organization among goats**—A herd of goats forms smaller groups than sheep, usually built up from the extended family group. Dominance is a relatively mild phenomenon in goat societies; when competing for feed, they may rear up and head clash with a downward stroke of the head.

The principles and practices of good goat herding are very similar to those with sheep, with one exception: Rarely do sheep herders work ahead of a range band. However, it is common practice for goat herders to work in front, turning the lead goats back to avoid unnecessary travel.

• **Social organization among swine**—Wild and feral pigs prefer to live in herds in scrub brush and light forest areas, rather than on the open range. The basis of the social structure is the matriarchal herd, consisting of one or several females with their offspring. Juvenile males are tolerated in the family group. Mature boars join matriarchal herds when sows become sexually receptive. But, apart from the breeding season, boars move about together in bachelor groups, although older ones often range as solitary animals. Under domestication, the pig has been transformed from a pugnacious, free-ranging animal to a docile animal which is readily handled in large groups under confinement. Shortly after birth, pigs show preference for the front teats of the udder. The teat order is effective until weaning, after which it is superseded by adult dominance.

• **Social organization among horses**—Przewalsky's Horse, a wild species, discovered in 1879 in the northwestern corner of Mongolia, has been preserved and propagated in captivity in Europe and America. Also, feral horses thrive in semiarid environments in western United States and southwestern Canada. Studies of feral horses, along with historical records of wild horses, give clear evidence of the social organization of horses in their natural habitat: It centers on a dominant stallion and his harem of 1 to 3 mares and their immature offspring. Each group has an alpha or leading female, and the other members respond as followers. However, the stallion maintains his partriarchal position until displaced by another adult male; sometimes he is at the front of his group, at other times he assumes a defensive position between intruders and his band, and at still other times he herds or drives his harem during the breeding season. Immature males are very submissive to the dominant stallion. The stallion and his harem are a closed society; animals not belonging to a group are rejected by either the dominant stallion or the mares.

Excess stallions live apart from the family/harem groups, either singly or in bachelor groups of up to eight males. The groups are organized with the dominant individual herding the other stallions in the same manner as the harem stallion herds mares.

Home range behavior exists, with each home range including at least one watering hole and a large grazing area. But there is considerable overlapping of the home ranges, with more than one group using the same area. The spacing between groups is controlled primarily by dominating stallions, which approach each other with threats, which occasionally result in pushing and kicking matches, during which time the animals within groups move closer together. Following such encounters, the stallions return to their respective harems and move them apart. When approaching a watering hole, the group whinneys; if the watering site is already occupied by a band, they await their turn until the watering group moves away.

Wild and feral mares commonly foal in the spring and summer and usually mate a few days after foaling. In an amazingly short time after birth, foals run almost as fast as their mothers—and on legs almost as long. Foals stay near the sides of their mothers; and the mares are very protective of their young. Harem stallions also look after the foals and herd them back to the group when they become separated. Barren mares are sometimes protective of other mares' foals.

• **Social organization among poultry, including chickens, turkeys, ducks, and geese—**

Chickens. Among both wild and domestic chickens, a dominant male will organize a harem, prevent fighting, and mate with several females. Subordinate cocks will remain at a distance.

Chicks develop aggressive tendencies a couple of days after hatching, but the brood's dominance order is not established until sometime between six and ten weeks after hatching.

Chickens recognize each other and stratify themselves socially mainly by the configuration of the head and neck. Consequently, hens with combs and wattles trimmed have significantly more fights than do undubbed hens, because individual recognition is more difficult. Also, dubbed hens tend to be subordinate to those with combs and wattles.

Debeaking reduces injuries, but debeaked hens are able to maintain normal social relations with intact flockmates.

Dominance orders of laying hens are established more quickly when the hens are in cages than when kept on the floor. Competition for space, feed, and water is reduced in cages, primarily because the social group is smaller.

Turkeys. During the first 3 months after hatching, turkey poults spar; during the fourth and fifth months, this is replaced by fighting.

Like chickens, turkeys recognize each other mainly by the sight of the head and neck structure. Dominance orders are maintained by specific displays of higher ranking turkeys when they meet lower ranking flockmates. Also, dominant turkey hens peck the backs of necks of subordinates; and the larger the denuded area, the lower the social rank.

As with chickens, a gobbler will mate with several hens. Adult male and female turkeys spend much of the nonbreeding season in separate flocks, males in bachelor groups and hens in groups with their young.

Ducks. All breeds of domestic ducks, except the Muscovy, descend from the wild Mallard. Domestic ducks neither fly nor migrate. They form pair-bonds during the mating season, but the drakes are noted for being unfaithful; most pair-bonds break while the female is incubating.

Geese. All breeds of geese, except the Chinese goose, descend from the Graylag goose. Unlike ducks, geese are known for their fidelity; mates usually form pair-bonds lasting for life. The gander remains with the goose during incubation; parents and young live together until the next breeding season, at which time the gander drives the young away. No social hierarchy is apparent within a brood, but family units may interact; thus, the members of a dominant family may chase a subordinate family.

Social Order (Dominance)

Within most groups of farm animals of the same species, there is a well-organized social rank. When we restrict or confine them and force them into spaces that bring them within the natural, individual distance that has been established (the distance between each other when moving as a herd or flock), we immediately create stress throughout the herd or flock. Thereupon, the dominants have to pay more attention to maintaining their dominance. They have to be more aggressive in their reactions. The subservients become far more nervous, and their nervousness spreads throughout the herd.

Fig. 17-6. Dominance. This shows a dominant cow (right) attacking the neck of a subordinate. The latter submits and avoids a fight.

In chickens, in which it was first observed, the social rank order is called the *peck order*. It is established as follows: When a number of strange birds are placed together in a pen, fights ensue by twos until each bird has engaged all the others. The winner of each initial contest thereafter has the right to peck the loser, the latter usually avoiding the former. Some individuals give way without a fight and others may challenge the winner again before dominance relations are settled. At subsequent meetings, one member of each pair pecks or threatens the other, definite dominance-subordination patterns become habitual, and thus the peck order is estalished. Unaccustomed chickens establish dominance-subordinate relations soon after they meet.

A social rank order similar to the peck order in chickens exists in most species of farm animals, especially in older females and larger groups. Thus, the alpha animal in the herd or flock will be dominant over all other individuals, and the omega animal will be subordinate to all. In between, some animals will be subordinate in some relationships and dominant in others. Moreover, once these relationships are established, they seldom change. The social rank order is usually important only in females, because mature male animals are seldom run together in groups.

Once the social rank order is established, it results in a peaceful coexistence of the herd or flock. Thereafter, when the dominant one merely threatens, the subordinate animal submits and avoids conflict. Of course, there are some pairs that fight every time they chance to meet. Also, if strange animals are introduced into such a group, social disorganization results in the outbreak of new fighting, as a new social rank order is established.

Social rank among farm animals exists, but does not affect production adversely, so long as they are on pasture or range, and if there is plenty of feed and water. But it becomes of very great importance when animals are placed in confinement. When cows are moved into winter quarters, social dominance decrees that replacement heifers be sorted out and fed separately, that young bulls be cared for in separate quarters, and that old cows with poor teeth be fed separately; otherwise, these animals will not get enough feed.

Social rank becomes of importance when a group of animals is fed in confinement; and it becomes doubly important if limited feeding is practiced. Under such circumstances, the dominant individuals crowd the subordinate ones away from the feed bunk, with the result that the subordinates may go hungry. This happens both in feedlot cattle and in breeding cattle being wintered.

Several factors influence social rank; among them, (1) age—both young animals and those that are senile rank toward the bottom; (2) early experience—once a subordinate in a particular relationship, usually always a subordinate; (3) weight and size; and (4) aggressiveness or timidity. Also, social rank is influenced by hormones; for example, a capon (castrated male chicken) automatically goes to the bottom of the social rank, whereas the injection of roosters and hens with the male sex hormone, testosterone, increases their social rank.

In feedlots and other confinement operations, social facilitation is of great practical importance. Since dominance tends to conflict or interfere with social facilitation, dominant animals should be sorted out and, if possible, grouped together. Of course, they will fight it out until a new social order is established. In the meantime, both feed efficiency and gains will suffer. But, as a result of removing the dominants, the feed intake of the rest of the animals will be improved, followed by greater feed conversion efficiency and profit. Among the more settled animals, social facilitation will become more evident. After the dominants have been removed, the rest of the animals will settle down into a new hierarchy, but within the limits of their dominance.

Their interaction or social facilitation will be far more likely to have a calming effect on this group, to both the economic and practical advantages of the operator.

Dominance and subordination are not inherited as such; they are developed by experience. Rather, the capacity to fight (agonistic behavior) is inherited, and, in turn, this determines dominance and subordination. Hence, when combat has been bred into a herd, such herds never have the same settled appearance and docility that is desired of high-producing animals under intensive management.

Leader-Follower

The leader is the animal that is frequently at the head of a moving column and often initiates a new activity; the other animals in the group are followers. "Followership" appears to be a fairly strong phenomenon in many species, as most animals resist being left behind.

Fig. 17-7. Leader-follower social relationship exhibited by mares on the range, with the gray mare the leader. (Courtesy, Denise Marquiss, Beaver Creek Ranch, Box 668, Gillette, WY 82716)

If the lead animal can be controlled, generally the remainder of the group can be moved easily.

Leader-follower relationships are particularly strong in sheep, where lambs follow their mothers from birth. In a naturally formed flock of sheep, the oldest ewes lead, followed immediately by their young lambs. Each is followed less closely by her descendants, with the females followed by their own lambs. Thus, the leader in the flock is usually the oldest ewe with the largest number of descendants. This type of leadership is broken up in flocks where unrelated animals are brought together.

In wild horse bands, the stallion was the defender-protector of his harem of mares from rival stallions and predators. He herded and kept the mares together, fought off other studs and chased other horse groups from his territory, and brought up the rear. But, generally, a dominant mare, with special qualities of leadership, served as the leader of the band. She possessed an intimate knowledge of the area in which the group lived; she knew the location of the best grass and water; she knew where to seek protection from storms and predators; and she knew the trails and the best escape routes.

Domesticated groups of horses also exhibit the leader-follower relationship. But the leadership may be shared between the stallion and the dominant mare, although some activities may be initiated by other animals in the herd.

When the latter happens, usually the activator soon pauses and the leader proceeds to lead.

Leadership develops in ducks and geese, which show an early following reaction.

Interspecies Relationships

Social relationships are normally formed between members of the same species. However, they can be developed between two different species. In domestication this tendency is important (1) because it permits several species to be kept together in the same pasture or corral, and (2) because of the close relationship between caretakers and animals. Such interspecies relationships can be produced artificially, generally by taking advantage of the maternal instinct of females and using them as foster mothers.

All sorts of bizarre interspecies relationships have been arranged—including cows raising pigs; bitches (dogs) raising pigs, rabbits, and cats; and cats raising mice.

Interspecies relationship is being used to protect sheep from predators. By raising puppies, young llamas, or young donkeys with sheep, at maturity they become their protectors (guards).

Fig. 17–8. Llama serving as guard animal for sheep. (Courtesy, F. C. Hinds, Department of Animal Science, the University of Wyoming, Laramie)

People—Animal Relationships

Social relationships can also be transferred to human beings. Thus, animal caretakers usually form care-dependency relationships with the animals under their care. This is particularly true with pets—horses, dogs, and cats—and with single housed animals such as calves.

People need pets and pets need people! Both groups desire to love and be loved. This relationship is especially valuable for children, shut-in, handicapped, and elderly people. The Delta Society, a movement ably spearheaded by Leo K. Bustad, DVM, Ph.D, a former student of the senior author of whom he is very proud, is contributing richly to the happiness and well-being of people through furthering the human-animal bond and animal-facilitated therapy.

Anamalous (Abnormal) Animal Behavior

Some behaviors that are useful in natural environments may be maladaptive to the individual animal or to a group of animals in artificial environments. Also, we have learned from studies of captured wild animals that when the amount and quality, including variability, of the surroundings of an animal are reduced, there is increased probability that abnormal behaviors will develop.

It is recognized that confinement of animals makes for limited space, and that this often leads to unfavorable changes in habitat and social interactions for which the species have become adapted and best suited over thousands of years of evolution. This is due to a genetic time lag; the keepers of domestic herds and flocks have altered the environment faster than the genetic makeup of animals. As a result, abnormal behaviors have evolved in domestic animals in confinement to plague those who raise them.

Homosexual behavior in all species is common where adult animals of one sex are confined together. Intersucking sometimes occurs in young calves, lambs, and pigs that are reared artificially. Pica, the eating of unnatural material(s), occurs in several species, but it is more prevalent in cattle and horses. Other abnormal behaviors in different species that frequently develop among domestic animals are listed and described in the sections that follow.

ABNORMAL CATTLE BEHAVIOR

Abnormal behavior in cattle may take many forms, including those that follow.

• **Kicker**—Milk cows may kick because they are in pain or frightened, or because they have been mistreated.

• **"Mean bull" complex**—There are inherited differences in the temperaments of cattle. Nevertheless, constant stress can change the temperament of an animal. Thus, when a bull is kept for hand-mating in a corral by which the cow herd passes each day, cows in heat stimulate his sexual behavior. Since he cannot respond naturally by mating, he may become a mean bull.

• **Pica**—The eating of unnatural material(s) is called *pica*. Cattle may develop pica (consumption of dirt, hair, bones, and/or feces) due to boredom, nutritional deficiencies, or physiological stress. Even after these conditions have been rectified, it may be disconcerting to find that pica persists among certain animals—perhaps due to habit.

ABNORMAL SHEEP BEHAVIOR

Abnormal behavior of sheep may take many forms, including wood or metal bar chewing, head banging, and wool-pulling. Only the latter is of economic importance.

• **Wool-pulling**—This abnormal trait may occur among sheep, especially animals kept in confinement and fed a finely-processed or pelleted ration and limited roughage. It decreases the value of the fleece; and it may have a permanent, adverse effect because coarse, black fibers replace the wool fibers that have been pulled.

ABNORMAL SWINE BEHAVIOR

Abnormal behavior in swine may take many forms, including those that follow.

• **Fighting (agonistic behavior)**—Fighting is a vicious act of aggression, which invites immediate retaliation from the victim. It occurs among pigs as a result of (1) establishing *teat order* (the front teats give more milk—they are worth fighting for); (2) aggression around the feeder; (3) tail biting;

(4) mixing pigs together, as a means of establishing dominance hierarchy; and (5) bringing together sexually mature animals, especially boars. The following factors appear to increase the incidence of fighting: physical discomfort, frustration, and high light intensity. Limited experimental work indicates that spraying *androstenone,* a chemical extracted from pork fat which has a scent like a boar, will stop fighting among pigs, but the product has not been approved by FDA for preventing swine fighting. Also, researchers in Sweden recently presented impressive experimental evidence showing the effectiveness of *amperozide* in reducing agonistic behavior of newly grouped pigs.

• **Intersucking**—Newly weaned pigs may nuzzle their penmates' abdomens, mimicking early stages of nursing. Persistent nuzzling may cause ulcers and necrosis (destruction of the tissue).

• **Savage sow syndrome**—Occasionally, a sow becomes savage during or soon after parturition, at which time she may injure, kill, or even eat some or all of her pigs. The cause of this behavior is unknown, but it seems to be more prevalent among gilts farrowing their first litter or where sows are maladjusted to their farrowing quarters. Where this syndrome is encountered, it is recommended (1) that sows be moved to their farrowing quarters several days ahead of parturition, thereby giving them an opportunity to adjust to the new environment before the pigs come; and (2) that the attendant remove each pig just as soon as it is born, keep the pigs away until the sow has settled down (usually soon after parturition), then return the entire litter to the sow. Some producers have found tranquilizers helpful.

• **Tail biting**—Tail biting usually accompanies close confinement. Bad environment, poor nutrition, inadequate management, and a host of other factors appear to increase tail biting. Docking most of the tail soon after birth is the most effective means of preventing tail biting in swine. Odor-masking sprays sometimes appear to be effective in reducing fighting and tail biting, especially when pigs are regrouped or moved to a different pen. Some producers have provided different objects for pigs to bite and play with, such as chains or old automobile tires suspended from the ceiling, and bowling balls. Initially, these are of some benefit, but the pigs seem to become habituated to their novelty quickly.

It is recommended that tail docking be a part of the regular management program, with the tails docked at the same time that the needle teeth are cut, when the pigs are about 3 days old. The side cutting-type pliers will work for both jobs, but tails will bleed less when they're cut with a dull blade. Emasculators and poultry debeakers also work well.

To dock the tail, clean it first, then cut it ½ to ¾ in. from its base, lifting it gently so as not to stretch the skin. The skin won't heal over the end bone as rapidly if you pull the tail away from the body. Don't cut the tail shorter than ½ in. because it will cause excessive bleeding and slow healing.

ABNORMAL HORSE BEHAVIOR (VICES)

Domestication and confinement of horses have spawned many abnormal behaviors, which are commonly referred to as *vices* (bad habits). Horses have more abnormal behaviors than any other species; not because they are naturally bad, but because there are a lot of spoiled horses—horses that have received too much tender, loving care (including lumps of sugar and carrots) and too little discipline.

• **Barn sour**—This refers to a horse that refuses to leave the barn—the horse refuses to leave home, friends, security, and feed. There are no easy cures for barn-sour horses; and each individual is different. But the most used, and the most successful, treatment for barn-sour horses consists of giving a bit of feed along the trail, with the feed given further and further from the barn.

• **Biting**—This vice is acquired as a result of incompetent handling. Generally, it is started in either of two ways: (1) The horse has been accustomed to treats (such as a lump of sugar), and nips as an expression of disappointment when there is no treat (treats should always be placed in the feed manger); or (2) as a result of rubbing the horse's nose while petting it (never rub the nose; it teaches a horse to bite).

• **Bolting feed**—Bolting feed is the name given to the habit of eating too fast (gulping the feed down without chewing). This condition can be controlled by spreading the concentrate thinly over the bottom of a large grain box, so that the horse cannot get a large mouthful; by adding chopped hay to the grain ration; or by placing some large, round stones, as big or bigger than baseballs, in the feed box.

• **Charging (attacking)**—This refers to a deliberate attack on a person, with the horse's mouth wide open. In the beginning stage, this vice can usually be corrected. The technique is to discipline without inflicting pain. The horse must be taught to obey, but it must have confidence in the handler. If mature stallions have had this vice for a long time, it is difficult, and perhaps impossible, to break.

• **Cribbing (wind-sucker, stump-sucker)**—A horse that has the vice of biting or setting the teeth against some object, such as the manger or a post, while sucking air is known as a cribber. This causes a bloated appearance and hard keeping; and such horses are more subject to colic. The common remedy for a cribber is a *cribbing strap* buckled around the neck in such a way that it will compress the larynx when the head is flexed, but not cause any discomfort when the horse is not indulging in the vice. A surgical operation to relieve cribbing has been developed and used with some success.

Fig. 17-9. A cribber (wind-sucker, or stump-sucker) in action. This is the vice of biting or setting the teeth against some object, such as a post or manger, while sucking in air. (Courtesy, Dr. George H. Waring, Department of Zoology, Southern Illinois University, Carbondale)

- **Eating bedding**—Sometimes gluttonous horses eat their bedding. This is undesirable because (1) most bedding materials are low in nutritional value, and (2) feces soiled bedding will likely add to the parasite problem. This habit can be prevented by muzzling the horse.
- **Halter pulling**—A confirmed halter puller breaks halters and lead ropes (straps) as it pulls back, then escapes. Either of two methods of tying will likely break the habit: (1) Tie a strong rope that the horse cannot break around the throatlatch, using a bowline knot so that the rope cannot slip and choke the horse; or (2) run a strong rope around the chest just back of the withers (some prefer to run it around the back in the area of the rear flanks), using a bowline knot. After a few struggles and attempts to break away, the horse will give up.
- **Handler aversion**—As an aversion to handlers, some horses display aggressive vices such as bucking, or objection to catching, harnessing, saddling, and grooming. Flight responses occur in the form of backing, balking, bolting, or running away. Most of these vices originate with incompetent handling. Nevertheless, they may be difficult to cope with or to correct, especially in older animals.
- **Kicking**—Two types of kickers are encountered: (1) the kicker that kicks the stall wall or door, and (2) the kicker that kicks people. The stall kicker kicks for no other excuse or satisfaction than to strike something and make a noise (kicking hollow tile makes a loud noise). Padding the stall will stop some stall kickers. Also, a chain or stick strapped to the back of the leg will usually break the habit; when the horse kicks, the chain or stick strikes the leg.

A horse that kicks people is dangerous. In the formative stage, usually the vice can be eliminated by prompt attention. But it is difficult to correct a confirmed and seasoned kicker.
- **Pawing**—This refers to a horse that digs at the stall floor with its front feet. Heavy rubber mats on the stall floor and under the bedding will discourage stall digging.
- **Pica**—This refers to a depraved appetite, or the eating of unnatural materials. This type of behavior is most common among horses that are kept in stalls and fed highly concentrated rations.
- **Rearing**—Rearing is a very dangerous vice. When the horse rears up, the flailing forelegs can inflict injury on the handler. Such horses can usually be corrected by proper use of a lead shank or whip.

A horse that rears while being ridden should be handled by an experienced trainer, who can usually find the cause and correct the vice.
- **Shying**—Shying at unfamiliar objects makes a horse dangerous to ride. The only solution is, patiently and gently, to take the horse over new trails and in new surroundings, again and again until there is no more shyness.
- **Stall walking**—This is a stereotypic movement about the stall. A mascot, e.g. a goat, may calm a stall walker.
- **Striking**—Striking with the front feet is a dangerous vice, because the handler is always vulnerable. The handler should always stay at the side of such a horse, never in front of it. Each time that a horse attempts to strike, it should be punished with a war bridle or whip.
- **Tail rubbing**—This refers to persistent rubbing of the tail against the side of the stall or other objects, resulting in the loss of hair and an unsightly tail. Tail rubbing is a common vice of Saddlebred horses that wear tail sets. Also, the presence of parasites may cause animals to acquire this vice. Installation of a tail board (a 2" × 12" shelf that runs around the stall at a height just above the point of the horse's buttock), or an electric wire similarly placed, may be necessary to break an animal of this habit.
- **Weaving**—This is a rhythmical swaying back and forth while standing at the stall door. The prevention and cure are exercise, ample room, and freedom from stress. A mascot, e.g. a goat, may calm a weaver.
- **Wood chewing**—This is the chewing of wood, usually a wood manger or a board fence. It is generally caused by (1) boredom, (2) nutritional inadequacies, or (3) psychological stress and habit. There is only one foolproof way in which to prevent wood chewing; to have no wood on which horses can chew—to use metal, or other similar materials, for barns and fences. But wood chewing can be lessened, although it cannot be entirely prevented, through one or more of the following practices:

1. Stepping up the exercise.
2. Feeding three times daily, rather than the normal two times; without increasing the total daily allowance.
3. Spreading out the feed in a large feed container and placing a few large, smooth stones about the size of a baseball in the feed container, thereby making the horse work longer and harder to obtain its feed.
4. Providing 2 to 4 lb of straw or coarse grass hay per animal per day, thereby giving the horse something to nibble on.

ABNORMAL CHICKEN BEHAVIOR

With the restriction, or confinement, of flocks, many abnormal behaviors evolved, including those that follow.
- **Cannibalism**—This is the most common abnormal behavior observed in chickens in confinement. It may be encountered among birds of all ages. Many types of cannibalism occur, with the following being most common:

1. **Toe picking.** This type of cannibalism is most commonly seen in baby chicks. It may be brought on by hunger.
2. **Vent picking.** Picking of the vent, or of the area below the vent, is the most severe form of cannibalism. This type is generally seen in pullet flocks in high production. Predisposing factors are prolapse or tearing of the tissues caused by the passage of a very large egg.
3. **Head picking.** This type of cannibalism usually follows injuries to the comb or wattles caused by freezing or by fighting.

Cannibalism appears to be brought on by boredom and too much light, accentuated by deficiencies in management and nutrition. The best way to control cannibalism is by debeaking and dubbing.

Debeaking is the removal of a portion of the upper (and often a lesser portion of the lower) beak of the fowl. Many broiler producers, and some egg producers, have their chicks debeaked at the hatchery. Additionally, layers can be debeaked at the time of placement in the laying house.

Dubbing is the removal of the comb, and in some cases the wattles of chickens. It may be done to either males or females. Generally, dubbing is done when chicks are a day old, using curved manicure scissors. At this age, very little bleeding or discomfort is noted.
- **Egg eating**—Egg eating is a costly vice. It is usually predisposed by factors favoring egg breakage, including insufficient nests, insufficient nesting material, not collecting eggs frequently enough, and soft-shelled or thin-shelled eggs.

Prevention consists in alleviating these conditions. Once the egg eating habit has started, it is very difficult to stop and it usually spreads quickly throughout the flock. If the birds have not been debeaked, this should be done immediately. Also, nests should be darkened and eggs should be collected frequently.

• **Feather pulling**—The term *feather pulling* is used to describe feather loss and hemorrhaging of skin in chickens.

• **Hysteria**—Sometimes, excessive fright occurs among growing pullets or layers. With floor-housed birds, it may take the form of flight and piling up in a corner(s) of the house, with the result that many birds may be suffocated. Caged layers may attempt to fly, resulting in injuries to wings and legs or broken necks. Ordinarily, an episode of hysteria only lasts about a minute, but the losses can be devastating. The cause of hysteria is unknown. However, it appears to be triggered by such, things as loud noise, quick movements, and sharp changes in light intensity.

• **Polydipsia**—This refers to excess water drinking. It may occur in caged birds which, due to boredom, play excessively with drinking nipples. Polydipsia leads to regurgitation of food and water.

ABNORMAL TURKEY BEHAVIOR

Abnormal behavior in turkeys may take many forms, including those that follow.

• **Blueback**—Blueback is a permanent, dark discoloration of the skin on the back and sometimes the sides and breast of turkeys with dark plumage, but not turkeys with white plumage. It is caused by feather picking, followed by exposure to sunlight. Blueback may result from overcrowding in the brooder, keeping the poults on the sun porch too long, or lack of sufficient fiber in the ration.

• **Feather pulling (cannibalism)**—Feather pulling is a mild form of cannibalism to which turkeys may become addicted, especially during the growing period. It results in unsightly appearance, more trouble from pinfeathers when the birds are marketed, and blueback in varieties with dark plumage. It can develop into flesh picking and become serious enough to retard growth. Feather pulling usually reaches serious proportions only in turkeys raised in confinement.

Feather pulling and cannibalism of turkeys raised in confinement can be prevented almost completely by debeaking poults, preferably by the hatchery when poults are 1 day old.

Additional management practices that tend to prevent feather pulling include: (1) avoiding overcrowding in confinement rearing quarters, (2) feeding an adequate diet, (3) feeding pelleted feed rather than loose mash, and (4) avoiding confinement of turkeys to roosts or other closely restricted quarters, particularly in the early morning.

ANIMAL ENVIRONMENT[3]

An animal is the result of two forces—heredity and environment. Heredity has already made its contribution at the time of fertilization, but environment works ceaselessly away until death. Since most animal traits are only 30 to 50% heritable, the expresssion of the rest (more than 50%) depends on the quality of all of the components of the environment. Thus, it is very important that the keepers of herds and flocks have enlightened knowledge of, and apply expert management to, animal environment.

Environment may be defined as all the conditions, circumstances, and influences surrounding and affecting the growth, development, and production of animals. The most important influences in the environment are the feed and quarters (space and shelter).

The branch of science concerned with the relation of living things to their environment and to each other is known as ecology.

Through the years, the domesticated animals best suited to a particular environment survived, and those that were poorly adapted either moved to a more favorable environment or perished. During the past two centuries, livestock producers have made great strides in the selection and propagation of animals suited to a particular environment, and during the past 50 years they have made progress in modifying the environment for the benefit of their animals and themselves.

The keepers of herds and flocks were little concerned with the effect of environment on animals so long as they grazed on pastures or ranges. But rising feed, land, and labor costs, along with the concentration of animals into smaller spaces, changed all this. Today, most layers and broilers are maintained in confinement throughout life; many layers are kept in cages, and essentially all broilers are on litter floors. Turkeys are shifting rapidly from range to confinement. Water is important for ducks, but even with ducks the trend is toward higher population densities and more confinement. Many swine are raised partially or totally in confinement; and confinement production is increasing with beef cattle, dairy cattle, and sheep.

Among animals, environmental control involves space requirements, light, air temperature, relative humidity, air velocity, wet bedding, ammonia buildup, dust, odors, and manure disposal, along with proper feed and water. Control or modification of these factors offers possibilities for improving animal performance. Although there is still much to be learned about environmental control, the gap between awareness and application is becoming smaller. Research on animal environment has lagged, primarily because it requires a melding of several disciplines—nutrition, physiology, genetics, engineering, and climatology. Those engaged in such studies are known as ecologists.

Samuel Brody, brilliant pioneer in the field of environmental physiology, growth, and energetics, at the Missouri Agricultural Experiment Station from 1921 to 1956, frequently reminded his students that scientists often prove what farmers have long known. This idea is well illustrated by the practical observation that when exposed to cold weather farm animals tend to increase their heat production by increasing the amount of feed consumed. Thus, cattle increase forage consumption during cold weather, thereby taking advantage of the warming effect of the heat increment of the feed. They also tend to increase their tolerance to cold by increasing their deposition of an insulating subdermal fat layer. Also, farm animals may increase heat production by shivering and conserve heat by huddling together.

---

[3]In the preparation of this section, and the sub-sections under it, the authors adapted considerable material, including experimental results, from the following source: Ames, D. R., Chairman, et al., *Effect of Environment on Nutrient Requirements of Domestic Animals*, NRC, National Academy Press, Washington, D.C., 1981.

Effect of Environmental Factors on Animals

The effect of environment on dairy cattle was clearly demonstrated in an experiment in New Zealand. It involved the selection of 20 calves from low-producing herds and 20 calves from high-producing herds. All of them were sired artificially by outstanding bulls. The 40 head were assembled at the Ruakura Experiment Station, raised and milked together for the first lactation. Under these conditions, no significant difference between the production of the 2 groups was observed. Then, they were sent back to the respective herds from whence they came, whereupon their production was comparable to that of the cows with which they were being milked. Then, for a second time, they were returned to the Ruakura Experiment Station, where again there was no significant difference in their production. The Ruakura Station then went one step further; they confirmed these results by using identical twins, with both twins milked at the Ruakura Station, and then later divided between high- and low-producing herds for subsequent lactation.

The New Zealand experiment underscores the importance of environment. No matter how good the genetics, a good environment is essential to obtain high production.

Also, selection of genetically superior individuals to be parents of the next generation is hampered by environmental factors that tend to mask the actual breeding values of individuals being selected. Thus, the contribution of these environmental factors to the total phenotypic variation should be minimized before estimating the genetic parameters.

The environmental factors affecting animals vary. The factors known to predominate in determining sheep productivity under arid conditions are year of birth, age of dam, type of birth and rearing, sex, and breed.[4]

In collaboration with the U.S. Department of Agriculture, Cornell University scientists studied the effect of summer weather on performance of Holstein cows in 3 stages of lactation. They reported that, for all stages of lactation, 9% of the variation in milk yield, 13% in milk fat, 5% in feed intake, and 65% in rectal temperature were attributable to weather conditions.[5]

We now know that controlled environment must embrace far more than an air-conditioned chamber, along with ample feed and water. The producer needs to be concerned more with the natural habitat of animals. Nature ordained that they do more than eat, sleep, and reproduce. For example, studies on the behavior of swine show that they spend much of their day in active investigative behavior, primarily rooting and manipulating their environment. When free ranging, pigs may spend 40% of their day resting, 35% investigating novel surroundings, 15% eating, and 10% in other activities. What happens when pigs are confined in a building on slotted floors? How is the nervous energy dissipated that would normally be used to satisfy the drives for investigating and rooting? Evidently, environmental deficiencies are manifested by tail biting, gastric ulcers, poor maternal care and loss of young, or other physiological functions resulting in a sudden death syndrome or tissue degeneration.

Preventing disorders by merely cutting off the tails of pigs to alleviate tail biting, debeaking poultry to prevent cannibalism, and using choke collars on horses to inhibit cribbing, is not unlike trying to control malaria fever in humans by the use of drugs without getting rid of mosquitoes. Rather, we need to recognize these disorders for what they are—warning signals that conditions are not right. Correcting the cause of the disorder is the best solution. Unfortunately, this is not usually the easiest. Correcting the cause may involve trying to emulate the natural conditions of the species, such as altering space per animal and group size, providing training and experience at opportune times, promoting exercise, and gradually changing rations. Over the long pull, selection provides a major answer to correcting confinement and other behavioral problems; we need to breed animals adapted to people-made environments.

The following factors are of special importance in any discussion of animal environment:

1. Feed
2. Water
3. Weather
4. Facilities
5. Health
6. Stress

FEED/ENVIRONMENTAL INTERACTIONS

Animals may be affected by either (1) too little or too much feed, (2) rations that are too low in one or more nutrients, (3) an imbalance between certain nutrients, or (4) objection to the physical form of the ration—for example, it may be ground too finely.

Forced production (such as growth, milk production, and racing 2-year-old horses) and the feeding of forages and grains which are often produced on leached and depleted soils have created many problems in nutrition. These conditions have been further aggravated through the increased confinement of animals, many animals being confined to stalls or lots all or a large part of the year. Under these unnatural conditions, nutritional diseases and ailments have become increasingly common.

Also, nutritional reproductive failures plague livestock operations. Generally speaking, energy supply tends to be more limiting than protien in reproduction. The level and kind of feed before and after parturition will determine how many females will show heat and conceive. After giving birth, feed requirements increase tremendously because of milk production; hence, a female suckling young needs approximately 50% greater feed allowance than during the pregnancy period. Otherwise, she will suffer a serious loss in weight, and she may fail to come in heat and conceive. This basic fact, along with other pertinent findings, was confirmed by researchers at the Montana Agricultural Experiment Station. Based on 12 years research at the Havre and Miles City Stations, they concluded that beef cattle size and milk production should be tailored to fit the environment. Big size and more milk are not better unless the range forage supply is better. The best sized cow is one that fits the range conditions. Small cows do best on poor range because they can usually get 100% of their daily feed requirement for maintenance and milk production, whereas big cows on a poor range are borderline hungry all the time. Also, cows that give a lot of milk must have a good range; otherwise,

---

[4]Eltawil, E. A., et al., "Evaluation of Environmental Factors Affecting Birth, Weaning and Yearling Traits in Navajo Sheep," *Journal of Animal Science*, Vol. 31, No. 5, Nov. 1970, p. 823.

[5]Maust, L. E., R. E. McDowell, and N. W. Hoover, "Effect of Summer Weather on Performance of Holstein Cows in Three Stages of Lactation," *Journal of Dairy Science*, Vol. 55, No. 8, Aug. 1972, p. 1133.

they are stressed by lack of feed; and their fertility rate and calf crops drop. So, cow size and milk production should match their environment.

The next question is whether a breeding program can make maximum progress under conditions of suboptimal nutrition (such as is often found under some farm and range conditions). One school of thought is that selection for such factors as body form and growth rate in animals can be most effective only under nutritive conditions promoting the near maximum development of those characters of which the animal is capable. The other school of thought is that genetic differences affecting usefulness under suboptimal conditions will be expressed under such suboptimal conditions, and that differences observed under forced conditions may not be correlated with real utility under less favorable conditions. Those favoring the latter thinking argue, therefore, that the production and selection of breeding animals for the range should be under typical range conditions and that the animals should not be highly fitted in a box stall.

In general, the results of a 10-year experiment conducted by the senior author and his colleagues at Washington State University, designed to study the effect of plane of nutrition on meat animal improvement, support the contention that selection of breeding animals should be carried on under the same environmental conditions as those under which commercial animals are produced.[6]

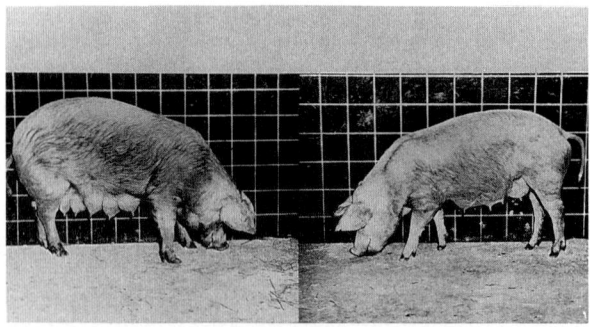

Fig. 17-10. Feed made the difference! The 2 sows are of the same age and breeding, but the sow shown at left received all she could eat from birth; whereas the gaunt sow shown at right was limited to 70% of the ration consumed by the better-fed animal. (Courtesy, Washington State University, Pullman)

The following additional feed/environmental factors are pertinent:

• **Appetite/intake**—Animals control their feed intake through a combination of the following mechanisms:

1. **The hypothalmus.** The two primary theories pertaining to hypothalmic control are: (a) the chemostatic hypothesis, which reasons that when blood nutrient levels, such as sugar and lipid, become too low, the hypothalmus sends signals to begin feeding; and (b) the thermostatic hypothesis, which theorizes that a decrease in hypothalmic temperature will induce feeding.

(Also see Chapter 3, section on "Hunger and Appetite.")

---

[6]Fowler, S. H., and M. E. Ensminger, *Relationship of Plane of Nutrition to the Improvement of Swine for Meat Production Through Selection*, Tech. Bull. 34, Washington State University, Pullman, 1961.

2. **The volatile fatty acids.** The appetite in ruminants has been shown to be sensitive to the volatile fatty acid levels in the rumen. Thus, increased acetate and propionate result in satiety and reduced feed consumption.

(Also see Chapter 3, section on "Gastric Influences on Appetite Regulation.")

3. **The physical size of the digestive tract.** Generally, animals eat to satisfy their energy requirements. However, if the caloric density of the ration is low, as in the case of certain roughages, the bulk may limit the animal's ability to hold sufficient of the feed to meet its energy needs.

4. **The thermostatic control.** According to the thermostatic theory, in hot environments animals reduce metabolic rate by reducing feed intake; and in cold environments animals increase metabolic rate by increasing feed intake.

5. **The fiber content of the ration.** Low-fiber high-energy rations stimulate animal growth and production in hot environments, whereas high-fiber rations increase the heat increment and keep the body warm in cold weather.

6. **Disease.** Feed intake is reduced quickly during most metabolic diseases such as pregnancy toxemia, acetonemia, D-lactic acidosis, ketosis, or bloat. Also, most gastrointestinal disorders of either infectious or parasitic origin as well as many systemic diseases result in decreased feed intake.

• **Calving time affected by feeding time**—Limited research indicates that partial calving control can be achieved by night feeding (feeding at 5:00 p.m. to 10:00 p.m. starting about 2 to 4 weeks before calving), with about 15–35% more daytime calving than when pregnant cows are fed in the morning, between 8:00 and 9:00 a.m. The reason that night feeding causes daytime calving has not been established.

(Also see Chapter 19, section on "Calving Control [Daytime Calving].")

• **Compensatory growth**—Compensatory growth is a cattle feeding term, which refers to increased growth rate in one time period as a result of growth restriction imposed during an earlier time period. This phenomenon is evidenced by stocker cattle that are roughed through the winter with limited daily weight gains, followed by feeding more liberally on a higher energy ration in the spring, when they are turned to lush pasture or put in a feedlot. Cattle so managed gain faster and more efficiently than similar cattle that are fed more liberally during the winter stocker stage.

(Also see Chapter 19, section on "Compensatory Growth.")

• **Competition**—When one animal eats (or grazes) or drinks, others may be stimulated to do likewise, even when they are not hungry or thirsty. This behavior is called social facilitation (competition). It follows that group-fed animals generally consume more feed per animal than individually-fed animals, due to competition. However, if there is fighting (agonistic behavior), the feed consumption of subordinates is likely to be reduced.

• **Familiarity of feed**—When given a choice, animals usually continue eating those feeds with which they are familiar. However, by gradually adding new and novel feeds, they learn to accept them (animals adapt). Also, early exposure of young animals to a feed may result in ready consumption of the feed later in life. When grazing, cattle and sheep develop preferences for certain pasture plants, with the result that they may not readily accept supplementary concentrates on pasture, and they usually have even greater difficulty in changing from grazing to eating dry forages and concentrates in confinement.

• **Feed as a reward**—Feed rewards are particularly common with cattle and horses.

Feed rewards may be used in gentling cattle as follows: Place up to 12 cattle in a relatively small pen where you can walk among and stay fairly close to them. Then, (1) carry a small bucket half-full of pellets, and shake it to attract the attention of the animals; (2) hold one pellet between your thumb and forefinger, and feed each animal that will take a pellet from you, one by one; and (3) repeat this procedure daily, or more often, until the animals get sufficiently gentle that they can be hand-fed and back-rubbed; and (4) *never, never* touch the animals on the forehead or horns, as this will teach them to butt. This procedure substitutes the application of animal behavior and human patience for rough handling.

Satisfying a horse's liking for a lump of sugar or a carrot can be most effective. To be effective, rewards must not be given promiscuously, only when deserved; for example, after carrying out a training command.

• **Frequency of feeding**—Except for swine, all species of animals fed for maximum production seem to respond to frequent feeding or to self-feeding (most poultry are fed on a free-choice basis), along with regularity of feeding. Thus, when fed three or more times daily (or self-fed), rather than twice daily, beef cattle, feeder lambs, and broilers make more rapid gains, lactating dairy cows produce more milk, and layers produce more eggs; and their increased production requires less feed per unit of product (meat, milk, or eggs) produced. However, unless self-fed, more frequent feeding generally makes for increased labor; so, the practicality of increased frequency of feeding should be determined by net returns—the value of the increased product, less the cost of producing it.

There is little evidence that frequency of feeding affects the rate of gain and feed efficiency of swine. Also, animals fed limited rations, such as dry beef cows, dry dairy cows, dry ewes, and gestating sows, may be fed less frequently than the traditional twice daily. Dry cows and ewes on the range may be fed as infrequently as three times a week without appreciably affecting performance.

• **Flushing**—*Flushing is the practice of having females gain in condition 2 to 3 weeks before breeding.* This may be accomplished by grain feeding, or cows and ewes may be turned to more lush pasture or range. Under most circumstances, the following amounts per head daily of a suitable concentrate are added to the ration that the females were receiving prior to flushing: cows, 2 to 5 lb; ewes, 1 to 2 lb; and sows, about 2 lb. Flushing fat animals is not effective. Immediately after breeding, females should be returned to normal rations.

Although it is not likely that all the benefits ascribed to flushing will be fully realized under all conditions, the general feeling persists that the practice will result in (1) more eggs being shed, (2) the females coming in heat more promptly, (3) more certain and prompt conception—with the young arriving more nearly at the same time, and (4) a 15 to 30% increase in lamb and pig crops.

• **Hair growth**—Hair growth on cattle and horses is greatly affected by three environmental factors—feed, temperature, and health. Cattle and horses fed maintenance and submaintenance rations do not shed as early in the spring as animals on a higher nutritional plane. In cold areas and during the winter months, long, shaggy coats of hair are nature's way of protecting animals against excessive energy loss. Sick animals tend to retain their winter hair coats longer than normal.

Fig. 17-11. Hair growth of Hereford cattle is nature's way of protecting them against excessive energy loss in the winter months. (Courtesy, American Hereford Assn., Kansas City, Mo.)

• **Milk composition affected by feed**—Some feeds reduce the fat content of milk. Among such feeds are cod-liver oil and other fish oils, early spring (lush) pasture, and pearl millet. Also, too small an amount of roughage, fine grinding of forage, or heated starch will lower the butterfat content of milk. On the other hand, such feeds as whole cottonseed, soybeans, and coconut oil increase the fat content of milk.

The amount of fat-soluble vitamins A, D, and E in milk are influenced by the amount of these particular vitamins in the ration; and in the case of vitamin D, exposure to sunlight is a factor, also.

• **Nutrient deficiencies**—The nutrient requirement tables (feeding standards) presented for each species in this book list values for animals under conditions presumed to be relatively free from environmental stress and for animals expected to perform near their genetic potential. In practice, environmental factors are not always ideal. Stresses are produced by weather (temperature, humidity, air movement, and solar radiation); diseases and parasites; surgery, dehorning, and castrating; altitude; sound; density and confinement; and pollution. As a result, animal performance often falls short of genetic potential. Animal shelters and housing are intended to eliminate or moderate the impact of some of the environment, but if they are poorly designed, they may create a new array of stresses with which the animal must contend.

So, although nutrient requirement tables are excellent and needed guides, nutrient deficiencies may result even when they are followed because the environment in which the animals are produced can modify the requirements. For this reason, along with making provision for variations in feed composition, *nutrient allowances*, which provide margins of safety, are presented in this book for each species, in addition to the requirements.

• **Overfeeding**—Too much feed is wasteful. Besides, it creates a health hazard; there is usually lowered reproduction in breeding animals, and a higher incidence of digestive disturbances (acidosis, bloat, founder, and scours)—and even death. Animals that suffer from mild digestive disturbances are commonly referred to as *off feed.*

• **Palatability**—*Palatability refers to the combination of factors that result in a feed being well liked and eaten with relish.* If animals don't eat their feed, they won't produce; and if they don't eat enough feed, production efficiency will be poor. Only an animal can assess the palatability of a feed. Palatability is the result of the following factors: taste, appearance, odor, texture, temperature, and in some cases auditory properties of the feed (like the sound of pigs eating corn).

• **Preconditioning**—*Preconditioning, which is a beef cattle term, is a way of preparing calves to withstand the stress and rigors of leaving their mothers, learning to eat new kinds of feeds, and shipping from the farm or ranch where they were raised to markets, feedlots, or other farms or ranches.* To the cow-calf producer (the seller), it is a program of management (including castration, dehorning, and weaning), nutrition, and immunization. To the feedlot operator (the buyer), preconditioning is a way in which to prepare calves to fit into the program and to minimize costly and unnecessary procedures. Economically, complete preconditioning is difficult to justify for both the cow-calf producer and the cattle feeder. But limited creep feeding plus presale vaccination may be economically feasible.

(Also see Chapter 19, section on "Preconditioning.")

• **Regularity of feeding**—Animals are creatures of habit; hence, they should be fed at regular times each day.

• **Selective grazing**—Different species of animals have different habits of grazing; they show preference for different plants and graze to different heights. Cattle are less selective in their grazing habits than other animal species; they will eat many kinds of vegetation, and at all stages of maturity. Sheep consume shorter and finer forages and more forbs than cattle; and their cleft upper lip allows them to graze vegetation close to the ground. Goats are fond of browse—the young shoots of shrubs and trees. Swine prefer legume pastures and love to root in the soil. Horses prefer young, tender grass to coarse-stemmed plants, a preference which often causes them to overgraze cetain areas. Except for geese, which graze grass, poultry do not eat much forage. Because of selective and preference grazing, grazing by two or more classes of animals makes for more uniform pasture utilization and fewer weeds and parasites, provided the area is not overstocked.

• **Stress**—Stress should be considered when formulating rations. All environmental stressors, whether physiological, immunological, or behavioral in nature, require energy expenditure on the part of the animal. Stress especially increases the animal's need for metabolizable energy, which, in turn, affects the optimal ratio of metabolizable energy to protein and other nutrients. So, when under extreme stress, nutrients will be diverted to maintenance, which is high priority, with the result that production, reproduction, and disease resistance will be reduced. So, it is important that rations be formulated to meet stressful conditions.

• **Underfeeding**—Too little feed results in slow and stunted growth of young stock; in loss of weight, poor condition, and excessive fatigue of mature animals; and in poor reproduction, failure of some females to show heat, more services per conception, lowered young crop, and light birth weights.

WATER/ENVIRONMENTAL INTERACTIONS

Animals can survive for a longer period without feed than without water. Water is one of the largest constituents in the animal body, ranging from 40% in very fat, mature animals to 80% in newborn animals. Deficits or excesses of more than a few percent of the total body water are incompatable with health, and large deficits of about 20% of the body weight lead to death.

The total water requirement of animals varies primarily with the weather (temperature and humidity); feed (kind and amount); the species, age, and weight of the animal; and the physiological state. The need for water increases with increased intakes of protein and salt, and with increased milk production of lactating animals. Water quality is also important, especially with respect to the content of salts and toxic compounds.

It is generally recognized that animals consume more water in summer than in winter. Based on 5 summer and 4 winter trials, the Iowa Agricultural Experiment Station reported that yearling cattle consumed an average of 8.5 gal per day in summer vs 5 gal per day in winter.

Pertinent details relative to important water/environmental interactions follow.

(Also see Chapter 4, section on "Water"; and see Chapter 4, Table 4–2, for daily water consumption.)

• **Air temperature**[7]—Numerous experiments have shown a very positive correlation between water intake and ambient temperature. Some species differences are noted:

Cattle. Under controlled temperature conditions, it has been demonstrated that cattle tend to increase water intake as temperature rises above 81°F. Below this point, water consumption is largely a function of dry matter intake.

British investigators reported that at daily temperatures around 46°F, water intake of lactating cows was significantly correlated with daily milk yield and dry matter content of the forage, but was not significantly related either to air temperature or relative humidity.

Sheep. The relation of drinking water intake to ambient temperature for sheep appears to parallel that for cattle. The water intake of growing-finishing lambs at 32 to 50°F was 2.0 lb/lb dry matter (DM) consumed; at 59 to 68°F, it was 2.5 lb/lb DM; and above 68°F, it was 3.0 lb/lb DM.

Swine. Experiments under controlled temperature conditions have shown an inconsistent relationship between ambient temperature and water intake of swine.

Poultry. As ambient temperatures rise, chickens consume increasing amounts of water; when compared to water consumption at 70°F, intake at 90°F is 2.0-fold, and intake at 99°F is 2.5-fold. The hen drinks water at a ratio of 2 to 3 g of water per gram of feed. Water availability for poultry is important for survival under heat stress.

• **Feed**—The water content of feeds ranges from about 10% in air-dry feeds to more than 80% in fresh, green forage.

---

[7]Adapted by the authors from: Ames, D. R., Chairman Subcommittee on Environmental Stress, et al., *Effect of Environment on Nutrient Requirements of Domestic Animals,* NRC, National Academy Press, Washington, D.C., 1981, pp. 44–50.

Feeds containing more than 20% water are known as *wet feeds*.

The water content of feeds is especially important for animals which do not have ready access to drinking water. Also, the water on the surface of plants, such as dew, may serve as an important source for cattle, sheep, and goats on arid ranges, but this supply is rarely sufficient to meet their needs.

• **Frequency of watering**—Home range behavior exists among most wild and feral animals, with each home range including at least one watering hole—generally a stream, spring, lake, or pond. The frequency of watering wild and feral animals, as well as of domestic animals on extensive pastures or ranges, is determined primarily by temperature and humidity—the higher the temperature and humidity, the more frequent the watering. However, under average conditions, the frequency of watering of different species is about as follows: Cattle, 1 to 4 times per day. Sheep, 1 time per day, although when grazing desert ranges in early spring, they may go for weeks without drinking water. Goats, 1 time per day, although goats approach camels in their water requirements. When hand-fed, pigs generally eat all their feed, then drink; when self-fed, they alternate between eating and drinking. Horses, usually drink once each day, but during extreme heat they may return a second time. Poultry drink frequently, alternating between eating and drinking. The frequency of watering of cattle and sheep decreases as distance to water increases.

Under practical conditions, the frequency of watering is best determined by the animals, by allowing them access to clean, fresh water at all times.

• **Physiological state**—The physiological state of animals dramatically affects the water requirements. A steer fed a maintenance ration will consume approximately 35 lb of water daily, whereas a steer fed a fattening ration will double this quantity. A dry cow will drink about 90 lb of water daily; during the last 4 months of pregnancy, she will consume 30% more water than when dry and open; when she produces 20 to 50 lb of milk, the daily water consumption will increase to about 160 lb; and when she produces 80 lb of milk per day, water intake will be near 200 lb. Young calves generally drink 1¼ to 1½ times more water per pound of dry matter consumed than older cattle.

The pounds of water/pound dry matter consumed by pregnant ewes increase from about 2.0 in the first month of pregnancy to 4.3 in the 5th month. Ewes carrying twins will consume over twice the amount of water of nonpregnant ewes; and those carrying single lambs will consume 138% more water than nonpregnant ewes.

The daily water intake of sows is estimated to be about as follows: nonpregnant sows, 11 lb; pregnant sows, 11 to 18 lb; and lactating sows, 33 to 44 lb. Weanling pigs will consume approximately 20 lb of water daily per 100 lb of body weight, whereas hogs near market weight will consume only 7 lb per 100 lb of body weight.

• **Quality of water**—Quality of water is determined by the level of dissolved substances, or pollutants. Water quality is almost as important as availability.

A particularly troublesome problem of water found on ranges is salinity. Most species can tolerate a total dissolved solid concentration of 15,000 to 17,000 ppm, but at these levels production will be affected.

Nitrate pollution may also be a problem. Quantities as small as 100 to 200 ppm can be dangerous, while 3,000 ppm is potentially lethal.

• **Species differences**—The amount of water consumed varies by species.

The water consumption per pound of dry matter consumed may be as much as 40% more for cattle than for sheep. Over the temperature range of 1 to 81°F, the estimated requirements for cattle are 3.5 to 5.5 lb water/pound dry matter consumed, whereas in about the same temperature range sheep need only 2.0 to 3.0 lb water/pound dry matter consumed. The estimated water needs for swine are near those for sheep, 2.0 to 2.5 lb/pound dry matter.

(Also see Chapter 4, Table 4-2, showing the daily water consumption of different species, along with the effect of age, body weight, and condition.)

• **Water excretion**—Water is excreted from the body through three routes: (1) urine, (2) feces, and (3) evaporation from the body surface and respiratory tract.

Urine provides a means whereby water-soluble products of metabolism can be excreted. Generally, when rations are high in protein or mineral content, urine flow is increased.

The amount of water lost in the feces is highly dependent on the animal species and the ration. Sheep, goat, and horse feces contain only 60 to 65% water and are relatively dry, whereas cattle feces contain 75 to 85% water. When cattle are first placed on early spring pasture, the feces become very watery—a reflection of the increased water intake.

Water loss from the respiratory tract is very variable, depending on humidity and respiration rate. Expired air is over 90% saturated; hence, when the humidity is low, respiratory losses are high. Conversely, respiratory losses are low when inhaled air is near saturation. When respiration rate is increased in response to high temperature or other behavioral stimulus, the rate of respiratory water loss is increased.

There are large differences among species in the importance of sweating, with animals ranked in descending order as follows: horses, donkeys, cattle, buffalos, goats, sheep, and swine.

Swine and poultry depend more on the respiratory than the skin route for water loss. In addition to panting, if given the opportunity swine wallow in water and mud to cool themselves.

• **Water sources**—The water needs of animals are filled from three major sources: (1) drinking water, (2) water contained in feed, and (3) metabolic water produced by oxidation of carbohydrates, fats, and proteins.

The water contained in or on feed is very variable; it may range from a low of 5% in dry grains to more than 80% in lush, young grasses. In addition, the amount of dew or precipitation on the grass at the time of grazing is subject to wide fluctuations. In the case of swine and poultry, the water in feed accounts for about 10% of the total feed intake.

• **Water temperature**—Findings on the effect of temperature on water intake are variable. However, cows in a cold environment generally drink more water when it is heated, but cooling water in a warm environment has no effect. Unlike cattle, heating water for sheep in a cold environment does not influence consumption.

WEATHER/ENVIRONMENTAL INTERACTIONS

Webster defines weather as a state of the atmosphere with respect to heat or cold, wetness or dryness, calm or storm, clearness and cloudiness.

Extreme weather can cause wide fluctuations in animal performance. The difference in weather impact from one year to the next, and between areas of the country, causes difficulty in making a realistic analysis of buildings and management techniques used to reduce weather stress.

The research data clearly show that winter shelters and summer shades improve production and feed efficiency. The issue is clouded only because the additional costs incurred by shelters have frequently exceeded the benefits gained by the improved performance, particularly in those areas with less severe weather and climate.

The animal kingdom may be divided into two kinds of animals: (1) *poikilotherms,* or cold-blooded animals, in which the body temperature fluctuates according to the environmental temperature, and (2) the *homeotherms,* or warm-blooded animals, which maintain almost constant body temperature despite wide fluctuations of environmental conditions. The giant dinosaurs which once dominated the earth were poikilotherms, as are the lizzards of today that resemble them. Since farm animals are homeotherms, they require a delicate balance between the heat produced within the animal, the heat or cold gained from the environment, and the heat lost by the animal to the environment. Thus, hot and cold weather are of great importance to them. This becomes evident when it is realized that the annual atmospheric temperature range in Minnesota may be more than 150°F (−40 to 110°F), yet the body temperatures of cattle, sheep, and horses wintering outdoors are constant within a few degrees.

The maintenance requirement of animals increases as temperature, humidity, and air movement depart from the comfort zone. Likewise, the heat loss from animals is affected by these three factors. Animals adapt to weather as follows.

In cold weather, the heating mechanisms are employed, including (1) increased insulation from growth of hair and more subcutaneous fat; (2) increase in thyroid activity; (3) seeking protective shelter and warming solar radiations (the animals sun themselves); (4) huddling together; (5) consumption of more feed, which increases the heat increment and warms animals; and (6) increasing activity. The most important animal body heating mechanisms are amount of feed consumed and body activity, which are also evidenced in people. For example, after skiing in bitter cold weather, a skier feels comfortable after eating a beefsteak; and during a marathon race, a runner may feel quite warm when the temperature is near freezing (30°F).

In hot weather, the cooling mechanisms are employed, including (1) moisture vaporization (from the skin and lungs), (2) avoidance of the heating solar radiation (the animals seek shade), (3) depression of thyroid activity, and (4) loafing (including lessening the production of meat, milk, and eggs, since they increase heat production).

Thermoneutral Zone (Comfort Zone)

Fig. 17-12 and the definitions that follow it are pertinent to an understanding of thermal zones.

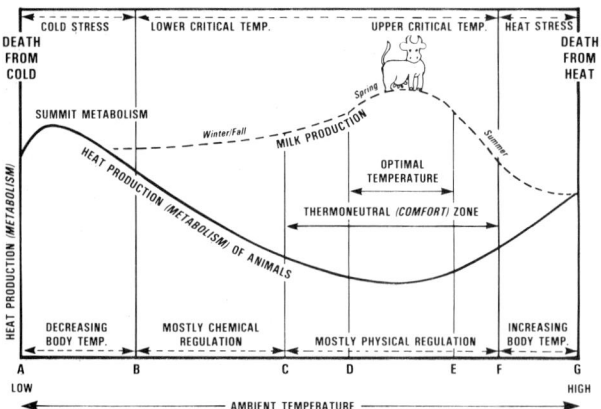

Fig. 17-12. Diagram showing (1) the influence of thermal zones and temperature on homeotherms (warm-blooded animals), and (2) the peak of milk yields in the spring, followed by the summer slump due to high (hot) summer temperature and lignification of forage.

In Fig. 17-12, *heat production (metabolism)* is plotted against *ambient temperature* to depict the relationship between chemical and physical heat regulation. Note, too, the broad range of accommodation to low (cool) temperatures in contrast to the restricted range of accommodation to high (warm) temperatures. Definitions of terms pertaining to Fig. 17-12 follow.

• **Thermoneutral (comfort) zone** *(C to F) is the range in temperature within which the animal may perform with little discomfort, and in which physical temperature regulation is employed.*
• **Optimum temperature** *(D to E) is the temperature at which the animal responds most favorably, as determined by maximum production (gains, milk, wool, work, eggs) and feed efficiency.*
• **Lower critical temperature** *(B) is the low point of the cold temperature beyond which the animal cannot maintain normal body temperature.* The chemical temperature regulation is employed in the zone below C. When the environmental temperature reaches point B, the chemical-regulating mechanism is no longer able to cope with the cold, and the body temperature drops, followed by death. The French physiologist, Giaja, used the term *summit metabolism* (maximum sustained heat production) to indicate the point beyond which a decrease in ambient temperature causes the homeothermic mechanisms to break down, resulting in a decline in both heat production and body temperature and eventually in death of the animal.
• **Upper critical temperature** *(F) is the high point on the range of the comfort zone, beyond which animals are heat stressed and physical regulation comes into play to cool them.*

The cow produces the maximum yield of milk during the spring when the temperature is optimum (D to E), and the minimum yield in the summer when it is hot (F to G).

Two facts about Fig. 17-12 should be noted: (1) The animal's cool range is wider than its hot range, which is as nature ordained it—the animal has more body regulators which it can call upon in a cool environment than in a hot environment; and (2) the comfort zone may not be best for the highest productivity, because productive processes involve a *heat increment* which is not easily dissipated in a warm environment, so a cooler environment is more stimulating to high productivity and activity than a warmer environment.

At temperatures immediately below optimum, but still within the comfort zone, there is a *cool zone* (see Fig. 17-12) in which animals invoke mechanisms to conserve body heat, including postural adjustments, changes in hair or feathers, and vasoconstriction of peripheral blood vessels. When the evnironmental temperature goes below the *lower critical temperature* (see Fig. 17-12), the animal is *cold stressed* and chemical regulation comes into play. As the environmental temperature continues to drop, the homeothermic mechanisms fail, resulting in a decline in both heat production and body temperatue, followed by the tissues freezing and eventually death of the animal.

At temperatures immediately above optimum, but still within the comfort zone, there is a *warm* zone (see Fig. 17-12) in which animals invoke limited thermoregulatory reactions, including decreasing tissue insulation by vasodilation and increasing effective surface area by changing posture (for example, hogs sleep stretched out full length so as to expose the maximum body surface to the air) to facilitate rate of heat loss. When the environmental temperature goes above the *upper critical temperature* (see Fig. 17-12), the animal is *heat stressed* and employs physical regulation, especially evaporative heat loss mechanisms such as sweating and panting; when these become insufficient to cool the body, death follows. Also, the immediate response of animals to heat stress is reduced feed intake, to attempt to bring heat production in line with heat dissipation capabilities. Higher producing animals with greater metabolic heat from product synthesis tend to be more susceptible to heat stress than lower producing animals. Evaporation of moisture from the skin surface or respiratory tract is the primary mechanism used by animals to lose excess body heat in a hot environment; this mechanism is limited by air vapor pressure but enhanced by air movement.

The comfort zone, optimum temperature, and both upper and lower critical temperatures vary with different species, breeds, ages, body sizes, physiological and productive status, acclimatization, feed consumed (kind and amount), the activity of the animal, and the opportunity for evaporative cooling.

The temperature varies according to age, too. For example, the comfort zone of newborn lambs is 75 to 80°F, whereas the comfort zone of mature sheep is 45 to 75°F.

Animals that consume large quantities of roughage or high-protein feeds produce more heat during digestion; hence, they have a different critical temperature than the same animals fed a high-concentrate, moderate-protein ration. Because of this, experienced cattle feeders decrease the roughage and increase the concentrate of finishing cattle during the hot summer months.

Stresses at both high and low temperatures are increased with high humidity. The cooling effect of evaporating sweat is minimized and the respired air has less of a cooling effect. As humidity of the air increases, discomfort at any temperature, and nutrient utilization, decrease proportionately.

Air movement (wind) results in body heat being removed at a more rapid rate than when there is no wind. In warm weather, air movement may make the animal more comfortable, but in cold weather it adds to the stress temperature. At low temperatures, the nutrients required to maintain body temperature are increased as the wind velocity increases. In addition to the wind, a drafty condition where the wind passes through small openings directly onto some portion or all of the animal body will usually be more detrimental to comfort and nutrient utilization than the wind itself.

Adaptation, Acclimation, Acclimatization, and Habituation of Species/Breeds to the Environment

Every discipline has developed its own vocabulary. The study of adaptation/environment is no exception. So, the following definitions are pertinent to a discussion of this subject:

Adaptation refers to the adjustment of animals to changes in their environment.

Acclimation refers to the short-term (over days or weeks) response of animals to their immediate environment.

Acclimatization refers to evolutionary changes of a species to a changed environment which may be passed on to succeeding generations.

Habituation is the act or process of making animals familiar with, or accustomed to, a new environment through use or experience.

Species differences in response to environmental factors result primarily from the kind of thermoregulatory mechanism provided by nature, such as type of coat (hair, wool, feathers), and sweat glands. Thus, hogs, which have a light coat of hair, are very sensitive to extremes of heat and cold. On the other hand, nature gave cattle an assist through growing more hair for winter and shedding hair for summer, with the result that they can withstand higher and lower temperatures than hogs. The long-haired, shaggy yak of Tibet and the wooly Scotch Highland cattle of Scotland are as cold tolerant as the arctic-dwelling caribou and reindeer.

From time to time, American buffalo *(Bison bison)* and domestic beef cattle *(Bos taurus)* have been crossed to obtain a more hardy beast than cattle. The most publicized early work of this type was the development of the Cattalo (bison x domestic cattle), the initial cross for which was made at the Dominion Experiment station, at Scott, Saskatchewan, in Canada, in 1915.

Fig. 17-13. Cattalo (¼ buffalo, ¾ domestic cattle) cow. The initial Cattalo breeding experiment was started by the Dominion Experimental Station, Scott, Saskatchewan, Canada, in 1915. The foundation herd consisted of 16 female and 4 male hybrids. (Courtesy, Research Station, Canada Department of Agriculture, Lethbridge, Alberta, Canada)

Male fertility and female reproductive rate have remained a problem in Cattalo. Although unquestionably hardy, bison x cattle crosses can be outperformed in nearly all environments by the currently available cattle breeds or crosses; and management procedures.

Also, there are breed differences, which make it possible to select animals well adapted to specific environments. Thus, the breeds of cattle that originated in the British Isles and Northern Europe are cold tolerant, whereas the Indian-evolved Zebu, or Brahman, cattle are heat tolerant. The long-fibered Black-faced Highland sheep are cold tolerant; the haired sheep are suited to hot, desert areas; and the fat-tailed sheep are adapted to arid conditions. The Shetland Pony, native to the Shetland Isles, not more than 400 miles from the Arctic Circle, evolved in the rigors of the northland climate and on sparse vegetation, which imparted that hardiness for which the stocky breed is famed. The long-legged donkey is adapted to hot, desert areas.

In recent years, attempts have been made to combine the heat tolerance characteristics of tropical breeds with the high productive capacity of European stock. The best known of these planned beef breeds is the Santa Gertrudis, developed on the famed King Ranch of Texas, in the early 1900s, which carry approximately five-eighths Shorthorn and three-eighths Brahman breeding.

Animals have the ability to become acclimated to high temperature. Thus, two trials of 46-day-old broilers were used in an experiment at Mississippi State University. One group was acclimated to a diurnal cycle of 75–95–75°F for a 4-day period, while the controls were kept at a constant 70°F. During the fourth day, both groups of broilers were exposed to 106°F temperature for 210 minutes. Acclimated broilers had significantly lower heat stress mortality. At the end of Trial 1, total mortality was 30% for the controls and 10% for the acclimated broilers. In Trial 2, mortality was 60% and 10% for the control and acclimated broilers, respectively.[8]

The Missouri Station workers conducted classical studies designed to show breed differences between Shorthorn, Santa Gertrudis (five-eighths Shorthorn and three-eighths Brahman), and Brahman cattle.[9] The animals were housed in *climatic chambers,* in which the temperature, humidity, and air movements were regulated as desired. The ability of representatives of the different breeds to withstand different temperatures was then determined by studying the respiration rate and body temperature, the feed consumption, and the productivity in growth, milk, beef, etc. Dr. Samuel Brody, who directed the study, reported the following pertinent points:

1. The most comfortable temperature for the Shorthorns was in the range of 30 to 60°F, while for the Brahmans it was 50 to 80°F, and for the Santa Gertrudis it was intermediate between the ideal temperatures for the two parent breeds.

2. The Brahman cattle could tolerate more heat—they could withstand higher temperature better than Shorthorns, whereas the Santa Gertrudis approached the Brahman in heat tolerance.

3. The Shorthorn cattle could tolerate more cold—they could withstand a lower temperature better than the other two breeds, while the Santa Gertrudis were more cold-tolerant than Brahman cattle.

Dr. Brody attributed the higher heat tolerance of Brahman cattle to their lower heat production, greater surface area (their loose skin) per unit weight, shorter hair, and "other body temperature-regulating mechanisms not visually apparent."

FACILITIES/ENVIRONMENTAL INTERACTIONS[10]

Optimum facility environments can only provide the means for animals to express their full genetic potential of production, but they do not compensate for poor management, health problems, or improper rations.

Research has shown that animals are more productive and feed-efficient when raised in an ideal environment. The primary reason for having facilities, therefore, is to modify the environment. Proper barns and other shelters, shades, sprinklers, insulation, ventilation, heating, air conditioning, and lighting can be used to approach the desired environment. Also, increasing attention needs to be given to other stress sources such as space requirements, and the grouping of animals as affected by class, age, size, and sex.

The principal scientific and practical criteria for decision making relative to the facilities for animals in modern, intensive operations is the productivity and cost of production of animals, which can be achieved only by healthy animals under minimal stress. So, the investment in environmental control facilities is usually balanced against the expected increased returns.

Recommended Environmental Controls

Temperature, humidity, and ventilation recommendations for different classes of livestock are given in Table 17–2, Recommended Environmental Control. This table will be helpful in obtaining a satisfactory environment in confinement livestock buildings, which require careful planning and design.

---

[8]May, J. D., J. W. Deaton, and S. L. Branton, "Body Temperature of Acclimated Broilers During Exposure to High Temperature," *Poultry Science,* Vol. 66 (2), 1987, pp. 378–380.

[9]Brody, Samuel, *Climate Physiology of Cattle,* Jour. Series No. 1607, Mo. Ag. Exp. Sta.

[10]In the preparation of this section, the authors adapted considerable material, including experimental results from: *Scientific Aspects Of The Welfare Of Food Animals,* F. H. Baker, task force chairman, CAST, Ames, Iowa, 1981.

TABLE 17-2
RECOMMENDED ENVIRONMENTAL CONTROL

Class of Animal	Temperature				Acceptable Humidity	Commonly Used Ventilation Rates[1]					Drinking Water			
	Comfort Zone		Optimum			Basis		Winter[2]		Summer[2]		Winter		Summer
	(°F)	(°C)	(°F)	(°C)	(%)	(lb)	(kg)	(cfm)	(m³/minute)	(cfm)	(m³/minute)	(°F)	(°C)	(°F)	(°C)
Beef cow	40–70	5–21	50–60	10–15	50–75	1,000	454	100	2.8	200	5.7	35–37	1.7–2.8	60–75	15–24
Steer, enclosed building on slotted floor	40–70	5–21	50–60	10–15	50–75	1,000	454	100	2.1–2.3	200	14.2	35–37	1.7–2.8	60–75	15–24
Dairy cow	40–70	5–21	50–60	10–15	50–75	1,000	454	100	2.8	200	5.7	35–37	1.7–2.8	60–75	15–24
Dairy calves	50–75	10–24	65	17	—	per 100 lb (45 kg)		10	—	25	—	—	—	—	—
Sheep:															
Ewe	45–75	7–24	55	13	50–75	—	—	20–25	0.6–0.7	40–50	1.1–1.4	35–37	1.7–2.8	60–75	15–24
Feeder lamb	40–70	5–21	50–60	10–15	50–75	—	—	15	0.3	30	0.65	35–37	1.7–2.8	60–75	15–24
Newborn lamb	75–80	24–27	—	—	—	—	—	—	—	—	—	—	—	—	—
Swine:															
Sow, farrowing house	60–70	15–20	65	17	60–85	Sow & litter		80	1.4	210	2.8	35–37	1.7–2.8	60–75	15–24
Newborn pigs[3] (pig level)	79–85	27–32	85	29	60–85	—	—	—	—	—	—	—	—	—	—
Growing-finishing hogs	60–65	15–17	60	15	60–85	125	57	15	0.7	75	2.1	35–37	1.7–2.8	60–75	15–24
Horse	45–75	7–24	55	13	50–75	1,000	454	60	1.7	160	4.5	35–37	1.7–2.8	60–75	15–24
Newborn foal	75–80	24–27	—	—	—	—	—	—	—	—	—	—	—	—	—
Poultry:															
Layers	50–75	10–24	62–68	17–20	50–75	per bird		2	—	5	—	50	10	60–75	15–24
Broilers	85–95 (baby chicks)	21–27	70	24	50–75	per lb body weight		½	—	1	—	50	10	60–75	15–24
Turkeys	95–100 (beginning poults)	35–38	70	24	—	per lb body weight		½	—	1	—	50	10	60–75	15–24

[1]Generally two different ventilating systems are provided; one for winter, and an additional one for summer. Hence, as shown in this table, the winter ventilating system in a beef cow barn should be designed to provide 100 cfm (cubic feet/minute) *(2.8 m³/minute [cubic meters/minute])* for each 1,000-lb *(454-kg)* cow. Then, the summer system should be designed to provide an added 100 cfm *(2.8 m³/minute)*, thereby providing a total of 200 cfm *(5.7 m³/minute)* for summer ventilation.
In practice, in many buildings, added summer ventilation is provided by opening (1) barn doors, and (2) high-up hinged walls.

[2]The ventilation rates for winter and summer are designed to meet the needs in the column headed *Basis*. For example, the commonly used winter ventilation rate of a beef cow weighing 1,000 lb *(454 kg)* is 100 cubic feet per minute (cfm) *(2.8 cubic meters per minute [m³/min.])*. Approximately ¼ of the winter rate should be provided continuously for moisture removal.

[3]The nursery temperature at pig level should be 85°F *(20°C)* for the first week following birth, then decrease to 79°F *(26°C)* at the fifth week. See Chapter 23, Feeding Swine, Table 23-29, for additional baby pig guidelines.

HEALTH/ENVIRONMENTAL INTERACTIONS

Health is the state of complete well-being, and not merely the absence of disease.

Environment embraces the forces and conditions, both physical and biological, that (1) surround animals, and (2) interact with heredity to determine behavior, growth, and development.

Disease is defined as any departure from the state of health.

Parasites are organisms living in, on, or at the expense of another living organism.

Feed, air quality, lighting, noise, other animals, and weather are among the many factors that constitute an animal's environment. Extremes or alterations in the environment may subject an animal to stress; and stress may affect health and lead to more diseases and parasites.

The importance of good animal health is underscored by the following statistics: It is estimated that animal diseases and parasites in the United States (1) decrease animal productivity by 15 to 20%, and (2) make for annual losses of $10 billion. Further, there is evidence that nutrition has some involvement in 85% of the veterinary cases. In the developing countries, diseases and parasites take an even greater toll—they decrease animal productivity by 30 to 40%.

Some important health/environmental interactions not covered elsewhere in this book are discussed in the sections that follow.

(Also see Chapter 5, Nutritional Disorders/Toxins.)

Antibody Production

An antibody is a protein substance (a modified type of blood-serum globulin) developed or synthesized by lymphoid tissue of the body in response to an antigenic stimulus.

Antigens may be (1) components of certain drugs or feed; (2) infectious microorganisms or parasites; (3) substances from the environment, such as chemicals, dusts, pollen grains, etc.; or (4) tissues of the body itself.

Each antigen elicits production of a specific antibody. In disease defense, the animal must have an encounter with the pathogen (antigen) before a specific antibody is developed in its blood.

Normally, antibodies react with antigens and render them harmless. However, certain antibodies may attack body tissue. This abnormal condition is called autoimmunity. Generally, repeated exposure to a specific type of antigen increases the rate at which antibodies against the substance are produced.

Antibody production may be impaired under such conditions as (1) malnutrition, (2) oversecretion of stress hormones, (3) advanced aging, or (4) inherited inability to produce certain antibodies.

Disease Defense

Pathogens affecting animals are ever present in the environment. But, in order to produce disease, they must overcome the body's first line of defense—they must first gain entrance to the animal by one of the body openings or through the skin. Then, they usually multiply and attack the tissues. To accomplish this, they must be sufficiently powerful (virulent) to overcome the defenses of the animal body. The defenses of the animal body vary and may be weak or entirely lacking, especially under conditions of a low nutritional plane and poor management and sanitation practices.

COLOSTRAL DEFENSE

Colostrum is the first milk secreted by mammalian females following parturition.

Newborn mammals are unable to produce antibodies within their own bodies for sometime after birth; they acquire these antibodies from their mothers either while in the uterus before birth or through colostrum after birth. Newborn calves, lambs, kids, pigs, and foals do not acquire passive immunity while in utero; so, the transfer of immunoglobulins via colostrum is of special importance to them. However, they should receive colostrum during the first 12 to 24 hours after birth if they are to acquire passive immunity. After about 24 hours following birth, gut closure occurs, following which the newborn animal digests these proteins, which then lose their immunization properties. Apparently, this results from the newborn not being able to absorb the large protein molecule.

Orphan calves may receive colostrum from any fresh cow or from frozen colostrum that has been stored. Also, calves can absorb antibodies in ewe or mare colostrum, but certain diseases are species specific; hence, colostrum from another species may not afford the desired protection.

(Also see Chapter 5, section on "Immunoglobulin Deficiency.")

IMMUNE SUPPRESSION (ALTERED IMMUNE RESPONSE)

People blush when they are embarrassed, and their hearts race when they are frightened—reactions that show the linkage of the brain and body. Recent studies in both humans and animals link the brain to the body's immune system—the complex array of organs, glands, and cells that comprise the body's principal mechanism for repelling invaders.

Studies show that people's emotions have a great impact on their health—that people stressed by bereavement (such as the loss of a loved one) or by loneliness (social isolation) suffer high incidences of disease and mortality. Also, research has shown that chronic stress causes the adrenal gland to pump increased amounts of corticosteroids into the bloodstream, and that these chemical messengers inhibit immune action.

Although animals do not blush, they react to stress. Thus, for many years, livestock producers have tried to minimize stress among their animals by providing a clean sanitary environment, and by keeping them warm in the winter and cool in the summer. Yet, animals still experience unavoidable and harmful stresses, especially during weaning and shipping. Following such stressful times, outbreaks of animal diseases like shipping fever, coccidiosis, scours, and transmissible gastroenteritis are more common.

Many stressors increase corticosteroid production and alter the animal's immune system by depressing antibody production and leucocyte levels in the blood, both of which are important in fighting diseases. The most common stressors that can alter animals' immune systems are: heat, cold, crowding, mixing, weaning, fatigue, limit-feeding, noise, and restraint.

Pollution

Anything that defiles, desecrates, or makes impure or unclean the surroundings pollutes the environment and can have a detrimental effect on animal health and performance. Thus, gases, odorous vapors, and dust particles from animal wastes (feces and urine) in buildings directly affect the quality of the environment. Muddy lots and stray electrical voltage may also pollute the environment. For healthy and productive animals, each of these pollutants must be maintained at an acceptable level.

Fig. 17-14. Pollution control. The runoff from the cattle feedlots (left) on the Carl Embry farm, in Arkansas, flows into a basin (right). Solids remain in the basin, from which they are removed from time to time, and liquids are pumped on surrounding fields. (Courtesy, USDA, Soil Conservation Service)

MUDDY LOTS

Muddy lots often plague beef cow-calf producers, dairy producers, and cattle feedlot operations, especially during the winter months. Mud increases calf scours and other diseases in newborn calves and reduces gains and feed efficiency in feedlot cattle.

California Agricultural Experiment Station studies show that mud can reduce finishing cattle gains by as much as 10 to 35%, and increase the feed required per pound of gain by a like amount. Thus, it is important that the problem be minimized, especially in high rainfall areas. Good drainage is the first essential. This should be assured at the time the lot is located and constructed.

Mounds 6 to 12 ft high, preferably perpendicular to the feed bunk, will provide finishing cattle a dry place on which to lie. Concrete aprons 10 X 12 ft wide and sloping 1 in.

per foot along the bunk will provide them with solid footing on which to stand and feed. Also, lessening of cattle density during the winter months—fewer animals per lot—is an effective method of controlling the mud problem. Thus, many feedlots plan to feed fewer cattle during the muddy season.

To cope with mud and alleviate calf scours, cow-calf operators should move the cows to a clean pasture during the calving season. If no pasture is available, dirt mounds and a coarse straw bedding are recommended.

STRAY VOLTAGE (TINGLE VOLTAGE, NEUTRAL-TO-EARTH VOLTAGE)

Stray electrical voltage has caused serious problems on many dairy farms—affecting animal behavior and lowering milk production, although it may affect other animal species also. Contrary to popular belief, stray voltage is not new; it is as old as electricity itself. However, it has become a problem on many farms recently for two reasons: (1) There is more electrical load on today's farms; and (2) in the last 20 years we have used more equipment grounding for safety purposes.

Stray voltage is excessive voltage between two animal contact points. The conditions that cause stray voltage are, electrically, quite simple: If sufficient voltage is present, it may force a current through any available conductor, including a cow's body. Cows are good conductors because of their body design (the length from mouth to front and rear legs); cows bridge the gaps between electrically grounded objects and "true earth." The cow doesn't feel the voltage as such; she feels the tingling current running through her body.

People seldom feel the current for several reasons. Usually, caretakers wear rubber-soled shoes when in the barn, whereas the bare-footed cow stands on concrete that is often wet. Also, humans have only two legs instead of four like the cow, and human's legs touch the floor near the same vicinity.

• **Signs of stray voltage**—One or more of the following signs may indicate that stray voltage exists in a dairy:

1. **Cows reluctant to enter the parlor.** When cows are subjected to stray voltages in the parlor stalls, they soon become reluctant to enter the parlor.

2. **Cows nervous in parlor.** Cows often dance or step around almost constantly while in the milking parlor.

3. **Uneven milk let-down and milk-out.** When milk let-down and milk-out are uneven, more machine stripping is required and longer milking time becomes apparent.

4. **Increased mastitis.** When milk-out is incomplete, more mastitis is likely to occur; all that is required is the presence of infectious bacteria. In turn, this will result in increased somatic count.

5. **Reduced feed intake in the parlor.** If cows encounter stray voltage while eating from the grain feeders, a reluctance to eat and reduced feed intake usually follow.

6. **Reluctance to drink water.** If stray voltage reaches the cows in stall barns through the water supply or metal drinking cups, the animals soon become reluctant to drink.

7. **Lowered milk production.** Each of the symptoms listed above is associated with stress and reduced feed intake, followed by a drop in daily milk production.

But detection of stray voltage is not easy! Other factors such as mistreatment of animals, milking machine problems, disease, sanitation, and nutritional disorders can create problems which manifest themselves in the seven symptoms listed.

• **Use voltmeter to monitor voltage**—The only sure method to determine if significant stray voltage is present is to have a qualified person perform a stray voltage survey, using approved equipment and monitoring the voltage through one, and preferably two, milkings. Point to point measurements between cow contact points will determine if the voltage is actually getting to the cow. Generally, stray voltage is not constant throughout the day; so, readings should be taken over a long period.

Most milking machine company representatives, many power supplier employees, some milking equipment dealers, and some veterinarians and county extension agents have equipped themselves with suitable voltmeters and are prepared to lend assistance. *Someone familiar with electrical systems, wiring, and equipment should be present when measurements are made.*

• **Install neutral isolators**—Neutral isolators may be installed to eliminate off-farm sources of stray voltage. They must be installed by a licensed electrician and must meet the code requirements.

As a result of a study of 395 Minnesota dairy farms on which neutral isolators had been installed, Minnesota Agricultural Experiment Station workers reported that the isolators increased milk production by 700 lb per cow over a 12-month period. Today, neutral isolators are commonly used to eliminate off-farm sources of stray voltage.

• **Establish equipotential planes**—This method of control can be applied to existing facilities or new facilities. But it must be expertly engineered and installed.

Porcine Stress Syndrome (PSS)

Swine with porcine stress syndrome (PSS) cannot cope with stressful situations. When present, the disorder is usually associated with heavily muscled animals and results in sudden unexplained death losses. Animals afflicted with PSS often show signs of nervousness and may have muscle tremors indicated by a rapid tremor of the tail. When exposed to a stressful situation such as a change in surroundings, a sudden change in the weather, vaccination, castration, estrus or mating, they often become overly excited and develop blotches on their skin, along with muscle rigidity, followed by rapid labored breathing. Their body temperature also rises and they show signs of heat stress even in cold weather. Death losses from PSS usually occur when hogs are being sorted and delivered for slaughter, with the highest losses occurring during the summer months. Further details follow:

1. **Cause.** Although the true cause of PSS is not known, it appears to be genetic—inherited as a recessive. It afflicts all breeds; it is common in animals with superior muscling; and its symptoms are triggered by stress. When exposed to stress, PSS pigs undergo a very rapid depletion of their muscle energy stores (glycogen) and a corresponding increase in lactic acid in both the muscle and blood. Normal pigs can remove the lactic acid from the muscle and blood fast enough to prevent build-up, but PSS pigs cannot. The excess lactic acid results in acidosis, which, in turn, leads to the production of pale, soft, exudative (PSE) pork at slaughter in affected animals.

2. **Tests.** Two different tests may be made for the presence of PSS for the purpose of evaluating animals for the breeding herd:

    a. **The creatine phosphokinase (CPK) test.** This involves obtaining a sample of blood and analyzing it for CPK, a serum enzyme that is abnormally high in PSS swine.

    b. **The halothane test.** This involves anesthetizing the animals with halothane, to which PSS swine respond by showing extreme muscle rigidity within 3 minutes.

3. **Relation of PSS to pork quality.** Most pale, soft, and exudative (PSE) pork is the end result of PSS. The two problems are closely related. But not all PSS pigs produce PSE pork; if there is little stress, the pork may be normal.

4. **Preslaughter handling and prevention of PSE pork.** Some environmental conditions may be comfortable to a normal animal, but stressful to a PSS pig. Moreover, it is impossible to handle pigs under practical conditions without imposing some stress. Nevertheless, the amount of PSE pork may be lessened by following good management practices when preparing and shipping hogs to market.

Ulcers

Gastric ulcers are of special importance among swine and foals, and abomasal ulcers appear to be increasing in mature cattle and young calves.

• **Swine ulcers**—Stomach ulcers in growing-finishing pigs are widespread throughout the world, with the incidence higher in England and Sweden than in the United States. Finely ground feeds and stresses (caused by such things as long duration of shipment, crowding, and fasting) increase the incidence. The economic loss to the industry is restricted mainly to deaths among affected pigs, as growth rate and feed efficiency are not adversely affected.

• **Foal ulcers**—In a 5-year study conducted at the University of Florida, gastric and/or duodenal ulcers were found in 25% (129) of the 511 foals that were presented for necropsy. A single cause of foal ulcer was not found, although stress and diet seemed to be especially important. At the University of Kentucky's Department of Veterinary Science, where gastric ulcers in foals have been studied for 15 years, the 3 main predisposing factors causing the gastric ulcer syndrome in foals, in order of incidence, were: (1) excessive milk from the mare, (2) stress on the foal, and (3) drug-induced ulcers. The mares that produced the most milk tended to have the most foals with ulcers—the foals overindulged in milk. Stress-related ulcers were caused by sudden weather changes, especially the onset of hot weather, by mares coming in heat and being bred, and by weaning. Drug-induced ulcers tended to follow the drug treatments of foals for diseases, especially diarrhea and pneumonia.

STRESS/ENVIRONMENTAL INTERACTION

Stress, is defined by Hans Selye, M.D., the world's leading authority on stress, as *the nonspecific response of the body to any demand*.

The authors use the term stress to indicate an environmental condition that is adverse to an animal's well-being, either external (nutritional, weather, social) or internal (disease, parasites).

Stresses of many kinds affect animals; among them, cold stress, heat stress, drafts, poor ventilation, excitement, presence of strangers, fatigue, mixing animals, number of animals together, space, changing corral and corral mates, weaning, previous nutrition, hunger, thirst, poor sanitation, disease, parasites, surgical operations, injury, and management.

Race and show horses are always under stress; and the greater the speed and the more tired they become, the greater the stress. Also, the greater the stress, the more exacting the nutritive requirements. Thus, the ration of race and show horses should be scientifically formulated.

Animals can be prepared, or adapted to the environment, in such a manner as to reduce stress. For example, if calves are properly *preconditioned* (started on feed, vaccinated, treated for parasites, etc.) prior to weaning, the stress of subsequent weaning and movement to a feedlot will be minimized.

In the life of an animal, some stresses are normal, and they may even be beneficial—they can stimulate favorable action on the part of an individual. Thus, we need to differentiate between stress and distress. Distress—not being able to adapt—is responsible for the harmful effects. The trick is to manage stress so that it doesn't become distress and cause damage and to recognize the warning signals of distress. For example, Texas Agricultural Experiment Station workers recently reported that added vitamin C, in either the feed or water, may reduce many of the health hazards associated with various kinds of stress to chickens such as hot weather, interaction with other birds in crowded conditions, and exposure to diseases.

The principal criteria used to evaluate, or measure, the well-being or stress of people are: increased blood pressure, increased muscle tension, body temperature, rapid heart rate, rapid breathing, and altered endocrine gland function. In the whole scheme, the nervous system and the endocrine system are intimately involved in the response to stress and the effects of stress.

The principal criteria used to evaluate, or measure, the well-being or stress of animals are: growth rate or production, efficiency of feed use, efficiency of reproduction, body temperature, pulse rate, breathing rate, mortality, and morbidity. Other signs of animal well-being, any departure from which constitutes a warning signal, are: contentment, alertness, eating with relish (and cudding by ruminants), sleek coat and pliable and elastic skin, bright eyes and pink eye membranes, and normal feces and urine.

Stress is unavoidable. Wild animals were often subjected to great stress; there were no caretakers to modify their weather, often their range was overgrazed, and sometimes malnutrition, predators, diseases, and parasites took a tremendous toll.

Domestic animals are subjected to different stresses than their wild ancestors, especially to more restricted areas and greater animal density. However, in order to be profitable, their stresses must be minimal.

Bleeders—A Stress/Environmental Interaction of Equines

Stress triggers different syndromes in different species and in different individuals. When subjected to heavy exercise, such as in racing, horses are prone to a condition known as *bleeders.*

It has been estimated that 70 to 80% of Thoroughbred racehorses are bleeders. In a recent major U.S. race, it was reliably reported that 5 horses in the field of 11 ran with the aid of Lasix, a diuretic used to combat bleeding. So, bleeders are a serious problem!

Bleeders are horses afflicted by blood flowing from their nostrils or bronchial tubes, but originating in the lungs, following strenuous exercise. But the problem is not confined to Thoroughbreds, nor is it limited to racehorses. It also afflicts Quarter Horses and Standardbreds, and it afflicts horses when they are subjected to maximum stress by strenuous physical activity of whatever kind; for example, in endurance rides. But it does not affect racing dogs or humans who race competitively. The species dissimilarity is attributed to the difference in the horse's anatomy. The horse has a sloping diaphragm and is primarily an abdominal breather, inhaling by movement of the diaphragm. This type of breathing appears to create stress in the equine lung.

Current indications are that the blood originates in the lung tissue, and not within the nasal cavity. Recent findings indicate that the bleeding is caused by ruptured capillaries (pulmonary hemorrhaging) rather than nosebleed; and that, although a large number of horses bleed from the lungs following racing, only a small percentage actually show external evidence of blood at the nostrils—it is swallowed into the gastrointestinal tract.

Bleeding appears to affect the performance of racehorses. But the cause and cure remain elusive. The condition is exercise-induced; and there appears to be a higher frequency of bleeders among older horses. When bleeding is excessive, the most common treatment is to discontinue training and racing temporarily or permanently.

Furosemide, popularly known by the trade name Lasix, which reduces (but does not prevent) pulmonary edema almost instantly, was approved by the FDA as an equine medication in 1967, and is a regulated raceday medication permitted in a majority of states. According to researchers at Washington State University, Lasix promotes the clotting of blood.

Research studies indicate that bleeders are a stress/environmental interaction of equines induced by heavy exercise, but the findings do not signal any restriction in racing horses; because galloping is a very natural behavior in horses, inherited from their wild ancestors who escaped from the attacks of their predators by flight.

Pollution Laws and Regulations[11]

Invoking an old law (the Refuse Act of 1899, which gave the Corps of Engineers control over runoff or seepage into any stream which flows into navigable waters), the U.S. Environmental Protection Agency (EPA) launched a program to control water pollution by requiring that all cattle feedlots which had 1,000 head or more the previous year must apply for a permit by July 1, 1971. The states followed suit; although differing in their regulations, all of them increased legal pressures for clean water and air. Then followed the Federal Water Pollution Control Act Amendments, enacted by Congress in 1971, charging the EPA with developing a broad national program to eliminate water pollution.

The current federal requirements, which are rather broad, follow:

1. **Who must apply.** Owners/operators of animal feeding facilities with more than 1,000 animal units. Animal units are computed as follows: multiply number of slaughter and feeder cattle by 1.0; multiply number of mature dairy cattle by 1.4; multiply number of swine weighing over 55 lb by 0.4; multiply the number of sheep by 0.1; and multiply the number of horses by 2.0. (See Table 17–3, footnote 1, for what constitutes 1,000 animal units.)

2. **Definition of *animal feeding operation.*** *An animal feeding operation is defined as a lot or facility where animals have been, or will be, stabled or confined and fed for a total of 45 days or more in any 12-month period and where crops, vegetation, forage growth, or post-harvest residue are not sustained over any portion of the lot or facility during the normal growing season.*

3. **Who does not need a permit.** Permits are required only if there is a discharge to a waterway. Totally enclosed units without pollutant discharge do not require a permit regardless of size. No permit is required if all pollutants are recycled or spread on land. Runoff control ponds or filter strips can be employed to prevent discharges.

4. **How to apply.** Fill out an Application Form For Permit To Discharge Wastewater, and submit as directed. Forms may be obtained from the offices of EPA and state environmental agencies, the county agent, or the Soil Conservation Service district offices. Then, either the federal EPA or the state agency will make an on-the-site inspection. They will draft a proposed permit, put it on public notice, and give the applicant and the public 30 days to comment on it. Then, if there are no protests, the federal discharge permit will be issued.

TABLE 17–3
SUMMARY OF REGULATIONS

Feedlots with 1,000 or More Animal Units[1]	Feedlots with Less than 1,000 but with 300 or More Animal Units[2]	Feedlots with Less than 300 Animal Units
Permit required for all feedlots with discharges[3] of pollutants.	Permit required if feedlot— 1. Discharges[3] pollutants through an unnatural conveyance, or 2. Discharges[3] pollutants into waters passing through or coming into direct contact with animals in the confined area. Feedlots subject to case-by-case designation requiring an individual permit only after on-site inspection and notice to the owner or operator.	No permit required unless— 1. Feedlot discharges pollutants through an unnatural conveyance, or 2. Feedlot discharges pollutants into waters passing through or coming into direct contact with the animals in the confined area, and 3. After on-site inspection, written notice is transmitted to the owner or operator.

[1]More than 1,000 feeder or slaughter cattle, 700 mature dairy cows (milked or dry), 2,500 swine weighing over 55 lb *(24.9 kg)*, 500 horses, 10,000 sheep or lambs, 55,000 turkeys, 100,000 laying hens or broilers with continuous overflow watering, 30,000 laying hens or broilers with liquid manure handling, 5,000 ducks; or any combination of these animals adding up to 1,000 animal units.

[2]More than 300 slaughter or feeder cattle, 200 mature dairy cows (milked or dry), 750 swine weighing over 55 lb *(24.9 kg)*, 150 horses, 3,000 sheep, 16,500 turkeys, 30,000 laying hens or broilers with continuous overflow watering, 9,000 laying hens or broilers with liquid manure handling, 1,500 ducks; or any combination of these animals adding up to 300 animal units.

[3]Feedlot not subject to requirement to obtain permit if discharge occurs only in the event of a 25-year, 24-hour storm event.

---

[11]This section was authoritatively reviewed and updated by Lawrence J. Jensen, U.S. Environmental Protection Agency, Washington, D.C.

Before constructing an *animal feeding operation*, the owner should become familiar with both state and federal regulations. The state regulations can be obtained from the state water board. They differ from state to state, but most states require a catch basin (detention pond) sufficient to contain the runoff from a storm of the magnitude of the largest rainfall during a 48-hour period of the most recent 10 years. A feedlot may minimize runoff by locating near the top of the slope and, if necessary, by using diversion embankments to divert runoff from other areas.

Cattle feedlots located near centers of population are having an increasing number of complaints lodged against them because of manure, dust, and odor. Lawsuits, based on the nuisance law, have been filed against them.

Sustainable Agriculture

Endangered species—and more! Today, it is endangered planet, endangered people and animals, and endangered agriculture. Among the deluge of warnings of environmental catastrophes are:

• Pollution-caused warming of the atmosphere, known as the *greenhouse effect*, threatening weather changes that could render large areas of the planet unproductive and uninhabitable.
• Toxic and radioactive wastes and dumped garbage that could poison drinking water and despoil the land.
• Chemical pollution that is depleting the atmosphere's protective ozone layer.
• Slashing and burning of tropical rain forests, driving thousands of species to extinction, increasing the amount of carbon dioxide in the atmosphere, and contributing to the greenhouse effect that warms the earth.

Is ¼ lb hamburger worth ½ ton of Brazil's rain forest? Is 67 sq ft of rain forest (an area about the size of one small kitchen) too much to pay for 1 hamburger? Should we form cattle pastures to produce hamburgers in the Amazon, or should we retain the rain forest and the natural environment? These and other similar questions are being asked too little and too late to preserve much of the great tropical rain forest of the Amazon and its environment. It took nature thousands of years to form the rain forest, but it took a mere 25 years for people to destroy much of it. And when a rain forest is gone, it is gone forever![12]

Although less dramatic, the Amazon rain forest story has been, or is being, repeated all over the world in the form of the greenhouse effect, toxicities, polluted streams, and/or other harbingers of threats to our environment. Too long we have managed our nonrenewable resources like there is no tomorrow! Now, the situation is being righted. Worldwide, environmental quality and economic efficiency are in vogue. In the United States, this movement is called *Sustainable Agriculture*.

Sustainable agriculture is often described as farming that is ecologically sound and economically viable. It may be high or low input, large scale or small scale, a single crop or diversified farm, and use either organic or conventional inputs and practices. Obviously, the actual practices will differ from farm to farm. A definition follows.

A sustainable agriculture is farming with reduced off-farm purchased inputs of pesticides, herbicides, and fertilizers, along with reduced negative impact on natural resources and improved environmental quality and economic efficiency, while producing and distributing abundant, nutritious, affordable, high-quality foods and fibers for American and world markets.

The development of improved crops, cropping systems, irrigation, farm management, and marketing will be needed to make farms more profitable and sustainable. Typically, such farms will rely more on biological resources and management than on nonrenewable inputs of energy and chemicals. The foundation of a sustainable farm system is a comprehensive understanding of the land, the farm resources and operations, and potential short- and long-term markets.

In the Soviet Union, the counterpart to United States' *Sustainable Agriculture* is called *Perestroika;* and the movement is being spearheaded by a newly established "State

---
[12]Uhl, C. and G. Parker, "Is a One-Quarter Pound Hamburger Worth A Half-Ton of Rain Forest?," *Interciencir*, 1986, Sept.-Oct., Vol. II, No. 5, p. 213.

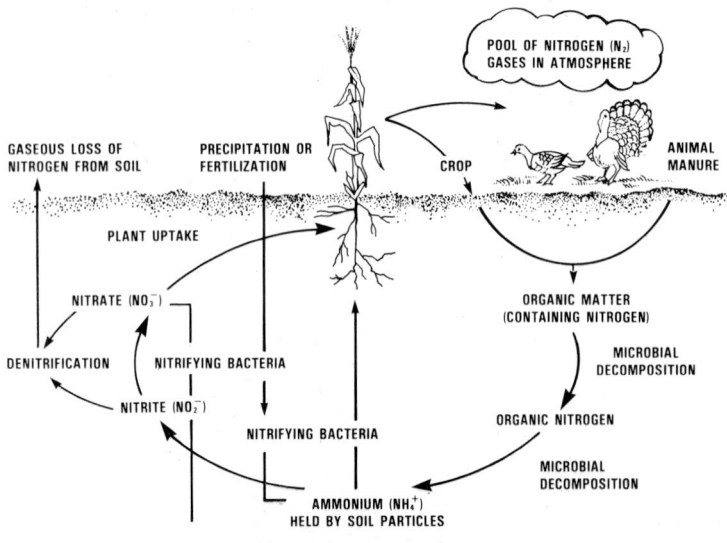

Fig. 17-15. At its best, a sustainable agriculture enhances the *cycle of nature*: Manure is applied to soils and decomposed by microbes; complex protein in manure is broken down to release nitrogen as ammonia (NH_4^+); aerobic microbes convert ammonium nitrogen to nitrite (NO_2^-), thence to nitrate (NO_3^-) nitrogen; nitrate is either (1) taken up by plants and built back into protein compounds; (2) leached downward when the soil is saturated—contaminating surface and groundwater if excessive nitrogen has been applied to the soil; and/or (3) released into the atmosphere when soils are wet for extended periods of time and the absence of air causes anaerobic microbes to convert the nitrate nitrogen to gaseous form. **NOTE:** Nitrates may be derived from a breakdown of fertilizers, animal wastes, human wastes, crop residues, and sewage sludges.

Committee for Nature Conservation," charged with the responsibility of protecting their natural resources—land, waters, forests, and other national assets. The Committee Chairman, Mr. Fyodor Morgun, is quoted in *Soviet Life* as saying:

> Consideration is being given to (imposing) fines for the discharge of pollutants. .... There can be no profit in the destruction of nature. .... We intend to introduce the experience of all humankind in the development of ecologically clean, low-waste or waste-free technologies, nonhazardous types of energy, wholesome foodstuffs, and so on. .... We have been entrusted with full responsibility for the protection of nature and for the organization of rational use and reproduction of natural resources in the U.S.S.R. .... Perestroika in nature conservation is a must.[13]

Many of the practices advocated under a sustainable agriculture are not new; they involve such timeless agricultural practices as soil erosion control, the protection of groundwater, the use of legumes as a source of nitrogen, biological insect and weed control, and the use of pastures as a primary feed source.

ANIMAL WELFARE/ANIMAL RIGHTS

In recent years, the behavior and environment of animals in confinement have come under increased scrutiny of animal welfare/animal rights groups all over the world. For example, in 1987 Sweden passed legislation designed (1) to phase out layer cages as soon as a viable alternative can be found; (2) to discontinue the use of sow stalls and farrowing crates; (3) to provide more space and straw bedding for slaughter hogs; and (4) to forbid the use of genetic engineering, growth hormones, and other drugs on farm animals except for veterinary therapy. Also, the law provides for fining and imprisoning violators.

Animal welfarists see many modern practices as unnatural, and not conducive to the welfare of animals. In general, they construe animal welfare as the well-being, health, and happiness of animals; and they believe that certain intensive production systems are cruel and should be outlawed. The animal rightists go further; they maintain that humans are animals, too, and that all animals should be accorded the same moral protection. They contend that animals have essential physical and behavioral requirements, which, if denied, lead to privation, stress, and suffering; and they conclude that all animals have the right to live.

---

[13]"Perestroika In Nature Conservation Is Essential," an interview with Fydor Morgun, reported in *Soviet Life*, Nov., 1988, No. 11, p. 16.

Livestock producers know that abuse of animals in intensive/confinement systems leads to lowered production and income—a case in which decency and profits are on the same side of the ledger. They recognize that husbandry that reduces labor and housing costs often results in physical and social conditions that increase animal problems. Nevertheless, means of reducing behavioral and environmental stress are needed so that decreased labor and housing costs are not offset by losses in productivity. The welfarists/rightists counter with the claim that the evaluation of animal welfare must be based on more than productivity; they believe that there should be behavioral, physiological, and environmental evidence of well being, too. And so the arguments go!

But wild animals were often more severely stressed than domesticated animals. They didn't have caretakers to store feed for winter or to irrigate during droughts; to provide protection against storms, extreme temperatures, and predators; and to control diseases and parasites. Often survival was grim business. In America, the entire horse population died out during the Pleistocene Epoch. Fossil remains prove that members of the horse family roamed the plains of America (especially the area that is now known as the Great Plains of the United States) during most of tertiary period, beginning about 58 million years ago. Yet no horses were present on this continent when Columbus discovered America in 1492. Why they perished, only a few thousand years before, is still one of the unexplained mysteries. As the disappearance was so complete and so sudden, many scientists believe that it must have been caused by some contagious disease or some fatal parasite. Others feel that perhaps it was due to multiple causes, including (1) climatic changes, (2) competition, and/or failure to adapt. Regardless of why horses disappeared, it is known that conditions were favorable to them at the time of their reestablishment by the Spanish conquistadores about 500 years ago.

To all animal caretakers, the principles and application of animal behavior and environment depend on understanding; and on recognizing that they should provide as comfortable an environment as feasible for their animals, for both humanitarian and economic reasons. This requires that attention be paid to environmental factors that influence the behavioral welfare of their animals as well as their physical comfort, with emphasis on the two most important influences of all in animal behavior and environment—feed and confinement.

Animal welfare issues tend to increase with urbanization. Moreover, fewer and fewer urbanites have farm backgrounds. As a result, the animal welfare gap between town and country widens. Also, both the news media and the legislators are increasingly informed from urban centers. It follows that the urban views that are propounded will have greater and greater impact in the years ahead.

QUESTIONS FOR STUDY AND DISCUSSION

1. Why did caretakers' fluency with the language of animals decline, and why did environments become increasingly artificial, as the art of husbandry evolved into science?

2. Why was it necessary for Neolithic (New Stone Age) man to have knowledge of animal behavior in order to capture and domesticate wild animals?

3. How did concentration of animals in smaller spaces and concentration of people in urban areas prompt interest in the environment?

4. Cite evidence that animal behavior and environment are closely related.

5. Define the term *animal behavior*.

(Continued)

6. How may information gained from wild and feral relatives provide clues about the behavior of domestic animals?

7. Define the term *applied ethology*.

8. List nine general functions of behavioral systems that most animals exhibit.

9. Discuss the following behavioral systems as they pertain to each major farm animal species:
    a. Ingestive behavior
    b. Eliminative behavior
    c. Allelomimetic behavior
    d. Gregarious behavior

10. Define the following terms: *social behavior* and *social organization*.

11. Describe the social organization in your choice of any two of the following species: cattle (beef and dairy), sheep, goats, swine, horses, poultry.

12. How is the "peck order" of chickens established?

13. Explain the difference between dominance and leader-follower.

14. Cite examples of interspecies relationships being used to protect sheep from predators.

15. Discuss the importance of people-animal relationships.

16. Describe each of the following anamalous (abnormal) animal behaviors, and tell what you would do to rectify each condition:
    a. Kicker in milk cow
    b. Pica in beef cattle
    c. Wool-pulling in sheep
    d. Fighting (agonistic behavior) in pigs
    e. Tail-biting in pigs
    f. Bolting feed in horses
    g. Wood chewing in horses
    h. Cannibalism in chickens
    i. Feather pulling in chickens and turkeys

17. An animal is the result of two forces—heredity and environment. Which is the more important?

18. Define the following terms: *environment* and *ecology*.

19. Why were the keepers of herds and flocks little concerned about the effect of environment on animals so long as they were kept on pastures and ranges most of the time?

20. Describe the classic New Zealand experiment showing the effect of environment on dairy cattle. How did it underscore the importance of environment?

21. Discuss how each of the following feed/environmental interactions affects animals:
    a. Size and milk production of beef cattle and the quality of the range
    b. Compensatory growth of beef cattle
    c. Flushing
    d. Milk composition
    e. Palatability
    f. Preconditioning of beef cattle
    g. Stress

22. Discuss how each of the following water/environmental interactions affects animals:
    a. Air temperature
    b. Physiological state
    c. Quality of water

23. Define each of the following terms:
    a. Thermoneutral zone (comfort zone)
    b. Optimum temperature
    c. Lower critical temperature
    d. Upper critical temperature

24. Discuss how each of the following weather/environmental interactions affects animals:
    a. Cold weather animal heating mechanisms
    b. Hot weather animal cooling mechanisms
    c. Adaptation, acclimation, acclimatization, and habituation of species/breeds to the environment, including pertinent facts about the bison x domestic cattle cross made in Canada and the classic beef cattle study of breed differences (Shorthorn, Santa Gertrudis, Brahman) conducted by Dr. Samuel Brody at the University of Missouri

25. Discuss the similarities and the differences in the recommended temperature, humidity, and ventilation for different classes of livestock.

26. Define each of the following terms: *health, disease, parasite*.

27. Cite statistics that underscore the importance of good animal health.

28. Discuss how each of the following health/environmental interactions affects animals:
    a. Antibody production
    b. Disease defense
    c. Colostral defense
    d. Immune suppression (altered immune response)
    e. Pollution, including muddy lots and stray voltage
    f. Porcine stress syndrome (PSS)
    g. Ulcers

29. Define the term *stress*; and give examples of six common kinds of animal stresses.

30. What criteria are commonly used to evaluate, or measure, the well being or stress of people?

31. What criteria are commonly used to evaluate, or measure, the well being or stress of animals?

32. Define the term *bleeders* as applied to horses; and discuss the magnitude and concerns of this malady.

33. Summarize the current requirements of the pollution laws and regulations.

34. Define the term *sustainable agriculture*. Are the following harbingers really dangerous threats to our environment; if so, what can and should be done about each of them?
    a. The conversion of the Amazon rain forest to cattle pastures
    b. Toxicities
    c. Polluted streams

35. Define the terms *animal welfare* and *animal rights*.

36. How do the stresses of wild and feral animals compare with those of domestic animals?

37. What more can livestock producers do to reduce behavioral and environmental stress of animals without increasing costs and/or lowering productivity to the point at which consumers will no longer benefit from an abundance of animal products?

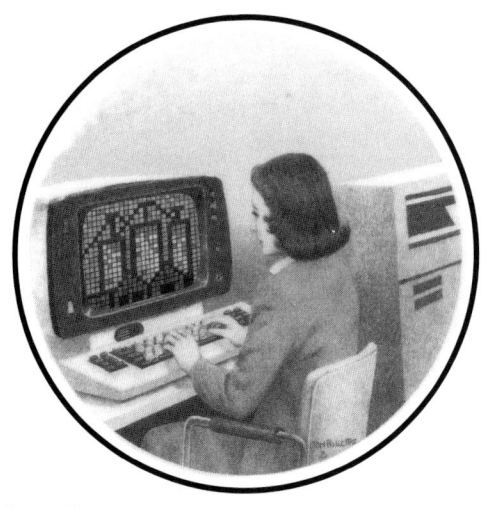

Original painting by Tom Phillips

18

FEEDING STANDARDS/ RATION FORMULATION[1]

Contents	Page
Feeding Standards	335
National Research Council Feeding Standards	336
Limitations of Feeding Standards	336
Ration Formulation	337
Consideration of Ration Ingredients	336
Health Considerations in Ration Formulation	336
How to Balance Rations	337
Steps in Ration Formulation	337
Adjusting Moisture Content	337
Methods of Formulating Rations	338
Square (or Pearson Square) Method	338
Trial-and-Error Method	339
Computer Methods	340
Trial-and-Error Formulation With the Computer	340
Linear Programming (LP)	340
Procedure for Use of Linear Programming (LP)	341
Selection of Computer Software and Hardware for Ration Formulation	343
Use of the Flexible Formula	343
Formulation Worksheet	343
How to Apply the Net Energy Method	344
Questions for Study and Discussion	345

During the last half of the 20th century, livestock production has been transformed from a secondary source of farm income to a highly sophisticated technological industry. Pigs are no longer scavengers, and chickens are no longer tenderly cared for by the farmer's wife and fed on table scraps and unaccounted for grain from the crib. With the expanding knowledge of the nutrient requirements of animals, new and refined feeding standards have been developed, thereby allowing producers to tailor their feeding program to their particular operations.

Hand in hand with the revision of feeding standards, came new analytical techniques whereby feedstuffs can be rapidy and accurately analyzed to determine their nutritive values. Thus, with an accurate estimate of what the nutritive requirements of their animals are and the nutrient value of their feeds, producers are now able to formulate balanced rations which result in efficient utilization of feed.

In the past, producers balanced rations by hand calculations, often using long and tedious trial-and-error methods. But in the past two decades, American industries have adapted computers to almost every conceivable task, and the livestock industry is no exception. Today, computers are used to formulate rations utilizing a wide variety of feeds; and the days of pencil-pushing formulations by big operations and commercial feed companies are, for the most part, only a memory. However, sophisticated livestock producers realize that this new computerized technology is a tool, which must be used wisely, and that they must clearly understand the principles on which computer calculations are based.

FEEDING STANDARDS

Feeding standards are tables listing the amounts of one or more nutrients required by different species of animals for specific productive functions, such as maintenance, growth, finishing, lactation, work, wool, or mohair. They are necessary guides in balancing rations. Most feeding standards are expressed in either (1) quantities of nutrients required per day, and/or (2) concentration in the ration; the first type is used where animals are provided a given amount of a feed during a 24 hour period, and the second is used where animals are provided a ration without limitation on the time in which it is consumed.

Today, the most widely used feeding standards in the United States are those published by the National Research Council (NRC) of the National Academy of Sciences. In England, similar standards are issued by the Agricultural Research Council (ARC). Other countries have similar bodies which make recommendations relative to the nutritive requirements of animals.

In the United States, the older TDN system is gradually giving way to newer energy evaluation systems, particularly net energy. England uses metabolizable energy (ME), adjusted according to the efficiency with which a feedstuff or ration is used for a particular purpose. Other European standards are based on starch equivalents, Scandinavian feed units, and other methods.

---

[1]The authors gratefully acknowledge the authoritative review accorded this chapter by L. M. Larsen, Ph.D., Consultant, Nutri-Systems, 426 E. Shields, Fresno, CA 93704.

National Research Council Feeding Standards

Today, the NRC recommended nutrient requirements for each species of farm animals are the most authoritative feeding standards in the United States. Periodically, a specific committee, composed of outstanding researchers who have worked extensively with the class of animal whose requirements are being reviewed, revises the nutrient requirements of each species for different functions. Thus, the nutritive needs of each type of livestock are dealt with separately and in depth.

The National Research Council (NRC) has established feeding standards for beef cattle, dairy cattle, sheep, goats, swine, poultry, horses, rabbits, mink and foxes, coldwater fishes, warmwater fishes and shellfishes, dogs, and cats.

The NRC standards pertaining to the classes of animals covered in this book are reproduced in the respective chapters pertaining to each species, in Chapters 19 to 28. In order to make for greater convenience and enhance the usefulness of the standards, the authors adapted and standardized the formats and included both U.S. customary and metric values.

Limitations of Feeding Standards

Although feeding standards are almost indispensable guides for meeting the nutritive needs of animals, numerous factors are not considered in setting the standards, especially their economy. For example, dairy producers are interested in obtaining that level of milk production which will make for the largest net returns in light of current feed costs and the market price of milk. Moreover, feeding standards tell nothing about the palatability, physical nature, or possible digestive disturbances associated with a ration. Neither do they give consideration to individual animal differences, management differences, and the effects of such stresses as weather, disease, parasitism, and surgery (e.g., dehorning and castrating). Thus, there are many variables that alter the nutrient needs and utilization of animals—variables that are difficult to include quantitatively in feeding standards, even when feed quality is well known. The experiences of the ration formulator and the feeder are invaluable in adjusting for such variables.

RATION FORMULATION

To supply all the needs for maintenance, growth, finishing, reproduction, lactation, work (or running), egg production and/or wool production, the different classes of animals must receive sufficient feed to furnish the necessary quantity of energy, proteins, minerals, vitamins, and water. Perhaps under certain conditions nonnutritive feed additives may be desirable, although they are not essential. A ration that meets all these needs is said to be balanced. More specifically, by definition, *a balanced ration is one which provides an animal the proper proportions and amounts of all the required nutrients for a period of 24 hours.*[2]

Consideration of Ration Ingredients

When rations are formulated, feeds are initially divided into three categories: (1) concentrates, (2) roughages, and (3) supplements. Additionally, when formulating rations for lactating cows, consideration should be given to the amount and kind of fiber.

- **Concentrates**—Most of the protein and energy of the ration is supplied by the concentrate feeds. For ruminants in heavy production (*i.e.*, finishing, or lactation) and almost all nonruminants, the concentrate portion of the ration constitutes most of the dry matter intake.

- **Roughages**—The amount of roughage incorporated in a ration depends largely upon intensity of production of the particular animal being fed. Nonruminants without functional cecums utilize very little roughage. On the other hand, ruminants and animals with functional cecums are generally given at least a small amount of roughage to maintain healthy, functional gastrointestinal activity.

- **Supplements**—When rations are formulated, supplements are generally considered after the macronutrients (for example, protein and energy) have been balanced. The ration is then checked for any deficiencies or imbalances of micronutrients; and supplements are added to correct the deficiencies. The supplements can be in the form of either a premix (combination of many micronutrients) or individual micronutrients (for example, lysine). The producer may also want to incorporate some feed additives in the ration.

- **Amount and Kind of Forage to Feed Lactating Cows (ADF and NDF)**—The amount and kind of fiber should be considered when formulating rations for lactating cows. The forage should constitute a minimum of 40% of the total dry matter of the ration and account for an intake of approximately 1.5% of the body weight daily. The acid detergent fiber (ADF) should constitute 19% of the ration dry matter, increased to 21% during the first 3 weeks of lactation. The neutral detergent fiber (NDF) should constitute 25% of the ration dry matter, increased to 28% during the first 3 weeks of lactation. However, these general guidelines may need to be modified due to either the source of roughage or its physical form. The NDF and ADF feeding recommendations are given in Chapter 20, Feeding Dairy Cattle, Table 20-5.

(Also see Chapter 8, section headed "NDF, ADF, and NIRS Analyses"; and Chapter 15, sections headed "Neutral Detergent Fiber [NDF] and Acid Detergent Fiber [ADF]," and "Near Infrared Reflectance Spectroscopy [NIRS].")

---

[2]Although Webster defines the noun *ration* as *the amount of food (feed) supplied to an animal for a definite period, usually for a day,* to most livestock producers the word implies the feeds fed to an animal or animals, without limitation to the time in which they are consumed. In this and other sections of *Feeds and Nutrition,* the authors accede to the common usage of the word, rather than to dictionary correctness.

Health Considerations in Ration Formulation

All animals are susceptible to health-related problems when radical changes are made in the composition of their feed. Therefore, it is recommended that any great changes in feed composition be done gradually over a period of time. Chapter 5, Table 5-1, includes a number of animal disorders, or diseases, attributable to the feeding regimen.

Also, a number of toxic substances may be found in feeds. Chapter 5, Table 5-3, includes a number of these potential poisons.

Whether caused by faulty management or undetected poisons, feed-related disorders are costly and should be guarded against when formulating rations.

How to Balance Rations

When in confinement, animals have access only to the feed provided by the caretaker. Therefore, it is important to provide balanced rations.

Suggested rations for different classes of livestock are given in Chapters 19 to 28. Generally these rations will suffice, but it is recognized that rations should vary with ingredient availability and cost, and that many times they should be formulated to meet the conditions of a specific farm or ranch or the practices common to an area.

Good livestock producers should know how to balance rations. They should be able to select and buy feeds with informed appraisal; to check on how well their manufacturers, dealers, or consultants are meeting their needs; and to evaluate the results.

Ration formulation consists of combining feeds that will be eaten in the amount needed to supply the daily nutrient requirements of the animal. This may be accomplished by the methods presented later in this chapter, but first the following pointers are necessary:

1. In computing rations, more than simple arithmetic should be considered, for no set of figures can substitute for experience and livestock intuition. Formulating rations is both an art and a science—the art comes from animal know-how, experience, and keen observation; the science is largely founded on mathematics, chemistry, physiology, bacteriology, and nutrition. Both are essential for success.

2. Before attempting to balance a ration, the following major points should be considered:

   a. Availability and cost of the different feed ingredients.
   b. Moisture content of ingredients.
   c. Composition of the feeds under consideration.
   d. Quality of feed.
   e. Degree of processing of the feed.
   f. Soil analysis where ingredients are grown.
   g. Nutrient requirements and allowances.

3. In addition to providing a proper quantity of feed and to meeting the nutritive requirements, a well-balanced and satisfactory ration should be:

   a. Palatable and digestible.
   b. Economical.
   c. Suited to the unique needs of the species involved.
   d. One that will enhance, rather than impair, the quality of the product (meat, milk, eggs, or wool) produced.

4. In addition to considering changes in availability of feeds and feed prices, ration formulation should be altered at stages to correspond to changes in the animal life cycle, weight, and productivity.

STEPS IN RATION FORMULATION

The ideal ration is one that will maximize production at the lowest cost. A costly ration may produce phenomenal gains in livestock, but the cost per unit of production may make the ration economically infeasible. Likewise, the cheapest ration is not always the best since it may not allow for a satisfactory level of production.

Therefore, the cost per unit of production is the ultimate determinant of what constitutes the best ration at any given time. Awareness of this fact separates successful producers from marginal or unsuccessful ones.

The following four steps should be taken in an orderly fashion in order to formulate an economical ration:

1. Find and list the nutrient requirements and/or allowances for the specific animal to be fed.

2. Determine what feeds are available and list their respective nutrient compositions.

3. Determine the cost of the feed ingredients under consideration.

4. Consider the limitations of the various feed ingredients and formulate the most economical ration.

ADJUSTING MOISTURE CONTENT

A careful feeder must constantly monitor the moisture content when purchasing feeds, and consider the effect of moisture on nutritional quality control. Most good feeders will readjust feeding formulas whenever moisture in a leading ingredient changes more than 2 or 3%.

Moisture changes may cause imbalances, as pointed up in the following example:

> Let's assume that a cattle feeder is using a ration which has as one of its main ingredients corn silage with 68% moisture content, and that this ration requires 1.9% protein supplement on an *as-fed* basis. Now, assume that the moisture of the silage suddenly decreased to 55%, and with it the necessary supplement to balance the ration increased to 2.62%. Obviously, if the feeder did not adjust the feeding formula, a serious shortage of protein could result. In this case, the cattle would receive only 72.5% as much supplement as they should have since the mixing formula was not recalculated.

The simplest way to avoid errors in ration formulation is to formulate on a 100% dry matter basis.

The multipliers in Table 18-1 may be used to convert feeds of various moisture contents to a 100% dry matter basis.

TABLE 18-1
CORRECTION FACTORS TO USE WHEN CONVERTING FEEDS OF VARIOUS MOISTURE CONTENTS TO A 100% DRY MATTER BASIS (0% MOISTURE)

Percent Moisture	100% DM Basis Multiplier	Percent Moisture	100% DM Basis Multiplier	Percent Moisture	100% DM Basis Multiplier
0	1.0000	29	1.4084	58	2.3809
1	1.0101	30	1.4285	59	2.4390
2	1.0204	31	1.4492	60	2.5000
3	1.0309	32	1.4705	61	2.5641
4	1.0416	33	1.4925	62	2.6315
5	1.0526	34	1.5151	63	2.7020
6	1.0638	35	1.5384	64	2.7777
7	1.0752	36	1.5625	65	2.8571
8	1.0869	37	1.5873	66	2.9411
9	1.0989	38	1.6129	67	3.0303
10	1.1111	39	1.6393	68	3.1250
11	1.1235	40	1.6666	69	3.2258
12	1.1363	41	1.6949	70	3.3333
13	1.1494	42	1.7241	71	3.4482
14	1.1627	43	1.7543	72	3.5714
15	1.1765	44	1.7857	73	3.7037
16	1.1904	45	1.8181	74	3.8461
17	1.2048	46	1.8518	75	4.0000
18	1.2195	47	1.8867	76	4.1666
19	1.2345	48	1.9231	77	4.3478
20	1.2500	49	1.9607	78	4.5454
21	1.2658	50	2.0000	79	4.7619
22	1.2820	51	2.0408	80	5.0000
23	1.2987	52	2.0833	81	5.2631
24	1.3157	53	2.1276	82	5.5555
25	1.3333	54	2.1739	83	5.8824
26	1.3513	55	2.2222	84	6.2500
27	1.3698	56	2.2727	85	6.6666
28	1.3889	57	2.3255		

The majority of feed composition tables are listed on an "as-fed" basis, while most of the National Research Council nutrient requirement tables are on either an "approximate 90% dry matter" or "moisture-free basis." Since feeds contain varying amounts of dry matter, it would be much simpler, and more accurate, if both feed composition and nutrient requirement tables were on a dry basis. In order to facilitate ration formulation, the authors list both the "as-fed" and "moisture-free" contents of feeds in Section V—Composition of Feeds.

The significance of water content of feeds becomes obvious in the examples given in Table 18-2. When using total digestible nutrients (TDN) as a measure of energy, the two high-moisture feeds, carrots and milk, have a higher energy value than oats on a moisture-free basis. The same principle applies to other nutrients, also.

TABLE 18-2
COMPARATIVE ENERGY VALUE OF THREE FEEDS ON (1) AS-FED, AND (2) MOISTURE-FREE BASIS

Feed	Water	Dry Matter	Energy Value (TDN)	
			As-Fed	Moisture-Free Basis
	(%)	(%)	(%)	(%)
Oats, grain	11	89	69	77
Carrots, roots	84	16	12	78
Milk	88	12	16	128

(Also see Chapter 16, Section headed "Moisture is Important.")

METHODS OF FORMULATING RATIONS

In the sections that follow, three different methods of ration formulation are presented: (1) the square method, (2) the trial-and-error method, and (3) the computer method.

An exercise in ration formulation follows for purposes of illustrating the application of each of these three methods:

1. **Square method,** applied to a swine ration.
2. **Trial-and-error method,** applied to a lactating cow ration.
3. **Computer method,** applied to a lactating cow ration.

Square (or Pearson Square) Method

The square method is a simple, direct, and easy way in which to figure proportions between two ingredients. It permits quick substitution of feed ingredients in keeping with market fluctuations, without disturbing the protein content.

In balancing rations by the square method, it is recognized that one specific nutrient alone receives major consideration. Correctly speaking, therefore, it is a method of balancing one nutrient requirement, with no consideration given to the other nutritive requirements.

To compute rations by the square method, or by any other method, it is first necessary to have available both feeding standards (see the nutrient requirement tables in the respective chapters devoted to each class of livestock, Chapters 19 to 28) and feed composition tables (Section V—Composition of Feeds).

The following example shows how to use the square method in formulating a swine ration:

Example. *A swine producer has 40-lb pigs to which it is desired to feed a 16% protein ration until they reach 120 lb weight. Corn containing 8.9% protein is on hand. A 36% protein supplement, which is reinforced with minerals and vitamins, can be bought. What percent of the ration should consist of corn and of the 36% protein supplement?*

Step by step, the procedure in balancing this ration is as follows:

1. Draw a square, and place the number 16 (desired protein level) in the center.
2. At the upper left-hand corner of the square, write *protein supplement* and its protein content (36); at the lower left-hand corner, write *corn* and its protein content (8.9).

Feeding Standards/Ration Formulation

3. Subtract diagonally across the square (the smaller number from the larger number), and record the difference at the corners on the right-hand side (36 − 16 = 20; 16 − 8.9 = 7.1). The number at the upper right-hand corner gives the parts of concentrate by weight, and the number at the lower right-hand corner gives the parts of corn by weight to make a ration with 16% protein.

```
Protein         36.0  \    /  7.1  Parts protein
supplement            \  /        supplement
                      16
                      /  \
         Corn 8.9  /      \ 20.0  Parts corn
                              27.1  Total parts
```

4. To determine what percent of the ration would be corn, divide the parts of corn by the total parts and multiply by 100: 20.0 ÷ 27.1 × 100 = 73.8% corn. The remainder, 26.2%, would be supplement.

Trial-and-Error Method

In the example that follows, the trial-and-error method is used, with consideration given to energy and protein. Also, crude protein rather than digestible protein is used because (1) this is what feed manufacturers want to know as they plan feed formulas, and (2) this is what livestock producers see on the tag when they purchase feed. In most mixed feeds, approximately 80% of the total protein is digestible.

Example. *Let's assume that a dairy producer has a 1,433-lb cow producing 65 lb of milk testing 4.0% fat. The producer is feeding 14 lb of alfalfa hay and 40 lb of corn silage per day. Corn, oats, and soybean meal are available. What concentrate mix should the producer use to meet the needs of this lactating cow, from the standpoint of energy and protein?*

The available feeds have approximately the following composition (as-fed basis):

	TDN (%)	Crude Protein (%)
Alfalfa hay, all analyses	51.0	16.0
Corn silage, all analyses	18.0	2.2
Corn, all analyses	80.0	9.9
Oats, all analyses	69.0	11.9
Soybean meal, solv extd, 44%	76.0	44.4

Here are the steps in balancing this ration:

Step 1. The daily TDN and crude protein requirements of this cow (1,433 lb body weight, 65 lb of milk testing 4% fat) are:[3]

Requirements of cow for—

	TDN (lb)	Crude Protein (lb)	(g)
Maintenance	9.94	0.94	428
Milk production	20.9	5.87	2,665
Total	30.84	6.81	3,093

[3]From Chapter 20, Table 20–3, "Daily Nutrient Requirements of Lactating and Pregnant Dairy Cattle."

Step 2. The forage (14 lb alfalfa hay, 40 lb corn silage) is supplying:

	TDN (lb)	Crude Protein (lb)
Alfalfa hay, 14 lb	7.14	2.24
Corn silage, 40 lb	7.20	0.88
Total from forage	14.34	3.12

Step 3. Remainder, to be supplied by concentrate:

TDN (lb)	Crude Protein (lb)
16.5	3.69

Step 4. Let's try out (that's why it is called the *trial-and-error method*) a grain mix of 700 lb corn, 280 lb oats, 10 lb monosodium phosphate, and 10 lb salt, and determine the amounts of TDN and crude protein in 1,000 lb of the grain mix:

	TDN (lb)	Crude Protein (lb)
Corn, 700 lb	560.0	69.30
Oats, 280 lb	193.2	33.3
Monosodium phosphate, 10 lb	—	—
Salt, 10 lb	—	—
Total	753.2	102.60
or in percent	75.3%	10.3%

Step 5. Divide the TDN needed from concentrate (16.5 lb) by the percent TDN in the mixture (75.3%). Thus, feeding 21.9 lb of the concentrate will meet the energy needs.

Step 6. Will this level of grain mix (21.9 lb) also meet the crude protein needs? By multiplying the pounds of concentrate mixture by the percent crude protein (21.9 × 10.3%), we find that the proposed concentrate would supply 2.26 lb of crude protein, whereas 3.69 lb are needed. Therefore, a high-protein supplement must be substituted for some of the homegrown grain.

Step 7. Let's substitute 175 lb of soybean meal for 175 lb of corn. Hence, the concentrate mix as now proposed will consist of:

	TDN (lb)	Crude Protein (lb)
Corn, 525 lb	420.0	52.0
Oats, 280 lb	193.2	33.3
Soybean meal, 175 lb	133.0	77.7
Monosodium phosphate, 10 lb	—	—
Salt, 10 lb	—	—
Total	746.2	163.0
or in percent	74.6%	16.3%

Step 8. By referring back to Step 3, we can divide the pounds of TDN and crude protein needed from the concentrate, by the percentage of TDN and crude protein found in the grain mix in Step 7. We find that 16.5 ÷ .746 = 22.1 lb needed to supply 16.5 lb TDN; and 3.69 ÷ .163 = 22.63 lb needed to supply 3.69 lb crude protein. Thus, we find that the following ration will supply the needed TDN (with a slight overage) and crude protein for a 1,433-lb lactating cow producing 65 lb of milk testing 4% fat:

	TDN	Crude Protein
	(lb)	(lb)
Alfalfa hay, 14 lb	7.1	2.2
Corn silage, 40 lb	7.2	0.9
Concentrate mix (Steps 7 & 8), 22.63 lb .	16.9	3.7
Total	31.2	6.8

In many sections of the country, especially in grain-deficient areas and on highly specialized dairies where little or no grain is grown, the dairy producer may find it most economical to purchase a commercial dairy feed to augment the roughage that is being fed.

Computer Methods[4]

Most large livestock establishments and feed companies now use computers in ration formulation. Also, many of the state universities, through their Federal-State Extension Services, are offering ration balancing computer services to farmers within their respective states on a charge basis. Consulting nutritionists are available throughout the United States and provide computer services, as well as other services. With the recent advent of the low cost personal computer, this powerful technology is available to almost everyone.

Despite their sophistication, there is nothing magical or mysterious about the use of computers in ration balancing. Their primary advantages are accuracy and speed of computation. In addition, computer programs (software) used in ration balancing provide a means of organizing needed information in a logical and systematic manner. The computer should be viewed as an extension of the knowledge and skills of the formulator.

At this time, there is no "push-button" system of feed formulation available. The degree of success realized is very dependent on the management of data put into the computer, and on the evaluation of the resulting formulations that the computer generates. In the hands of experienced users, the computer enables the producer and nutritionist to be more precise in carrying out ration formulation.

Two basic approaches to ration formulation are practiced with computers:
1. Trial-and-error formulation.
2. Linear programming (LP).

---

[4]The section on "Computer Methods" was prepared especially for this book by L. M. Larsen, Ph.D., Consultant, Nutri-Systems, 426 E. Shields, Fresno, CA 93704.

TRIAL-AND-ERROR FORMULATION WITH THE COMPUTER

For a discussion of the trial-and-error method of ration balancing, see the earlier section in this chapter headed "Trial-and-Error Method." Many ration balancing software programs written for the computer allow for trial-and-error ration balancing. Feed mill nutritionists frequently use this technique to enter into the computer rations that are given to them by other nutritionists or by a producer. The objective in this case is to confirm the nutrient values for the ration based on the specific ingredients used by the feed manufacturer. In many cases, these rations are not to be altered without permission. In other cases, the number of ingredients for a specific ration may be limited so that the trial-and-error technique is just as fast as using linear programming to arrive at the desired nutrient levels in the ration.

NOTE: It does not take specialized computer software to use the trial-and-error method. Spreadsheet (or Financial Spreadsheet) programs, for instance, organize data into rows and columns. Information, such as nutrient values for a feedstuff, may be entered into data cells (see Fig. 18-1). Simple and complex arithmetic operations can be controlled by the user to the extent that rather large trial-and-error method rations can be programmed and run.

Spreadsheets have been developed with specific microcomputers in mind; and there are a great number of them on the market.

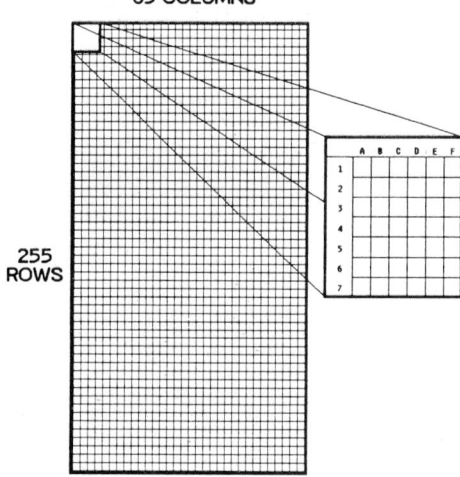

Fig. 18-1. Graphic representation of a spreadsheet (From Lane, R. J. and T. L. Cross, *Spreadsheet Applications for Animal Nutrition and Feeding,* Reston Publishing Co., Inc., Reston, Va., 1985).

LINEAR PROGRAMMING (LP)

The most common technique for computer formulation of rations is the linear programming (LP) technique. At times, this is referred to as *least cost* ration formulation. This designation results from the fact that most LP techniques for ration formulation have as their objective *minimization of cost.* A few LP programs are in use that solve for *maximization of income over feed costs.* Regardless, the livestock producer and nutritionist should always keep in mind that maximizing net profit is the only true objective of most ration formulations. A skilled user of the LP system will control ration quality by writing specifications that lead to rations that will maximize profit.

Briefly described, the LP program is a mathematical technique in which a large number of simultaneous equations are solved in such a way as to meet the minimum and maximum levels of nutrients and levels of feedstuffs specified by the user at the lowest possible cost. It is not necessary to understand the inner workings of the computer program to use LP, though it does take experience to use it to good advantage and to avoid certain pitfalls. The most common pitfalls are incorrectly entered or missing data and the specification of minimums and maximums that cannot be met with the feedstuffs available. The latter is called an infeasible solution. When an infeasible solution is encountered, the user must determine (1) if this is due to incorrect or missing data, or (2) if the specifications must be relaxed.

PROCEDURE FOR USE OF LINEAR PROGRAMMING (LP)

Before using the LP approach to ration formulation, the user should become familiar with the specific software package to be used. (See later section on "Selection of Computer Software and Hardware for Ration Formulation.") It is also desirable to study the LP technique as applied to feed formulation. After users are familiar with LP and their computer software, they are ready to begin using the computer for ration formulation by LP. It must first be understood that all data entered into the computer is directed to files. In most cases, these files are located on disks, or perhaps on tapes. Currently, most computers use keyboards and CRT (cathode ray tube) displays for entry of data. The necessary data files are generally created in steps as follows:

1. **Enter names of available feed ingredients, and the cost of each.** It is necessary that all of the available feeds be listed along with the unit cost. It matters little if the formulator uses cost per ton, cost per hundred weight (cwt), or cost per pound, but the same method of cost input must be used for all feeds. The computer software may call for a specific form for entering costs.

2. **Enter nutrient values for feeds.** Tables of feed composition using average or typical values, like those in Section V of this book, may be used, but, because of the wide variation in the composition of feeds, a chemical analysis of a representative sample of each lot of feed is more precise and should be used if available. This is especially true of forages, in which composition may be affected in a major way by cultural conditions and stage of maturity.

3. **Enter ration specifications.** Ration specifications are generally broken into two parts: (1) Nutrient limits, and (2) Ingredient limits. In each case, the formulator specifies either a lower limit and/or an upper limit for each item. If no specification for the particular item is desired, it may be specified as zero (0) or left blank, depending on the circumstances. It is also appropriate to list feedstuffs available, but not currently on hand (with an upper limit of zero). Most LP solutions will then tell the user the highest cost at which such feeds would enter the solution if allowed to do so.

Ratios between nutrients (such as a calcium/phosphorus ratio) or feedstuffs (corn/barley ratio) may also be specified in most LP software packages. The experienced formulator usually deals with palatability or feedstuff quality considerations by setting an upper limit on the amounts of problem feeds or a lower limit on feeds that contribute a positive quality to the ration. Nonnutritive attributes, such as bulk density, may also be programmed into the LP system. The LP technique is a very flexible and powerful ration balancing tool.

> **NOTE:** Important additional items to consider when creating ration specifications are upper limits on the use of nonprotein nitrogen (or urea) and limits on the usage of feed additives, such as drugs, feed flavors, and the like.

Fig. 18-2 illustrates, by means of a worksheet, a logical method of organizing the restrictions for a ration.

LEAST-COST FORMULATION WORKSHEET

Specifications	Ingredient A	Ingredient B	Ingredient C	Restrictions
Cost				Minimize
Total weight				1,000 lb
Crude protein				133 lb
Digestible protein				100 lb
Ether extract				25 to 80 lb
Net energy lactation				900 Mcal
Calcium				5 to 10 lb
Phosphorus				7 lb
Vitamin A equivalent				35,000
Vitamin D				60,000
Limits on ingredients				
Minimum				
Maximum				

Fig. 18-2. Sample worksheet for a least-cost formulation. The first column lists the specifications. The various feedstuffs to be considered are listed in the succeeding columns with their respective costs and nutritive values. The last column lists the restrictions desired on the final formulation.

4. **Submit all of the above information to the matrix building and solving portion of the LP software package.** Matrix building and solving are generally accomplished automatically by the computer software once the specifications have been entered into the computer. Mathematically, the procedure involves the solution of a complex algebraic problem, with an answer being derived in seconds or minutes. Using the LP program, the computer produces a mix that will meet the desired specifications at the lowest possible cost.

5. **Examine the solution provided by the computer software.** The end result should be feasible, both from a mathematical standpoint and from a nutritional standpoint. The feedstuff mixture should be acceptable to the animals for which it is intended. In most cases the first solution provided to the user is not acceptable. Repeat runs may be necessary to obtain the best solution.

342   Feeds & Nutrition Digest

Fig. 18-3 is a computer printout of an LP solution. The various columns of the report have been numbered for indentification. Similar columns have been given the same number. The three sections of the report are each identified with a Roman numeral.

```
16-JAN-89     ANYCO GRAIN AND MILLING                        ( 7) ID. NO. 29704
              P. O. BOX 1234                    COSTS.. ...$/CWT ...$/TON
              ANYTOWN, USA 90909   (123) 456-7890 BASE LP  8.069    161.38
* FEASIBLE*  RATION: L507 LAYER MASH, 17 PCT.      BATCH   8.030    161.60
  (1)    (2)         (3)     (4)     (5)    (6)     (7)    (8)      (9)
                                                  STABLE          COST
   #  INGREDIENT   AMOUNT   BATCH   COST  LIMITS  COSTS  RANGE  PER UNIT
  I
   4  GROUND CORN    33.333  665.00  6.40           6.15  64.50   0.0024I
                                         10.000L    6.47  25.05   0.0007D
   6  GROUND MILO    33.848  675.00  6.00           5.94  43.26   0.0006I
                                                    6.22   0.00   0.0022D
  10  WHEAT MILLRUN   2.085   40.00  6.00  15.000U  4.84  28.64   0.0116I
                                                    6.23   0.00   0.0029D
  19  SOYBEAN MEAL,47.5 16.117 320.00 14.95        11.04  18.80   0.0391I
                                                   26.37  14.80   0.1142D
  20  MEAT SCRAP,50   6.000  120.00 15.03   6.000U  7.54         -0.0510I
                                                   20.13   3.94   0.0510D
  21  DICAL PHOS(22CA/18P) 0.396 10.00 17.00        0.41   2.03   0.1659I
                                                  139.27   0.27   1.2227D
  23  LIMESTONE       7.564  150.00  1.45           0.00   7.84   0.0309I
                                                   12.89   7.53   0.1144D
  24  SALT, PLAIN     0.250    5.00  2.35  0.250U   0.00   0.93   0.0393I
                                           0.250L          0.00  -0.0393D
  27  POULTRY PREMIX  0.250    5.00 56.37  0.250U   0.00   0.93   0.5795I
                                           0.250L          0.00  -0.5795D
  33  DL-METHIONINE,99 0.086   1.73 157.00         76.28   0.09   0.8072I
                                                  540.27   0.07   3.8327D
  37  SELENIUM, 90.8  0.050    1.00 17.50  0.050U   0.00   0.73   0.1903I
                                           0.050L          0.00  -0.1903D
       TOTALS        100.000 1992.73  REQUESTED BATCH WEIGHT IS 2000.00

  (1)   (2)        (3)              (6)         (9)              (8)
                                  LIMITS     COST PER UNIT   EFFECTIVE RANGE
   #  NUTRIENT    AMOUNT     LOWER    UPPER  DECREASE INCREASE DECREASE INCREASE
  II
   1  WEIGHT        100.000  100.000  100.000   0.016  -0.016  99.3204  104.2085
   8  CRUDE PROTEIN  17.000   17.000          -0.243   0.243  15.8640   18.1362
  12  CRUDE FAT       2.930    0.000          -0.054   0.053   2.8216    5.7481
  13  CRUDE FIBER     2.308             4.000  0.137   0.097   2.2498    2.3682
  14  ASH            11.963    0.000           0.657   0.033  11.9068   12.2025
  15  CALCIUM         3.600    3.600    3.700 -0.080   0.080   2.0008    3.8582
  16  PHOSPHORUS      0.671    0.650           0.557   0.938   0.6609    0.9609
  17  AVAIL. PHOSPHORUS 0.470  0.470          -0.910   0.910   0.4480    0.7686
  31  M. E. (POULTRY)/LB 1270.000 1270.000    -0.003   0.003   1204.    1281.
  47  LYSINE          0.798    0.700          20.036   1.694   0.6856    0.8010
  48  METHIONINE      0.350    0.350          -1.359   1.359   0.3309    1.7376
  49  METHIONINE + CYSTINE 0.619 0.600         3.532   1.380   0.6172    2.0342
  61  LINOLEIC ACID   1.134    1.000           0.061   0.107   1.0381    2.5310
  62  XANTHOPHYLL /LB 3.000    3.000          -0.026   0.026   1.6701    5.8047

   (1)    (2)            (5)        (10)    (6)       (9)        (8)
      ** NOT USED **             RELATIVE  UPPER  COST/UNIT   INCREASE
    #   INGREDIENT       COST     WORTH    LIMIT  INCREASE    RANGE
   III
    8  GROUND BARLEY    6.500     5.782            0.0072     6.531
   12  ALFALFA, DEHY.,17 8.000    7.374    2.500   0.0063     0.932
   31  VEGETABLE FAT   17.200    11.914            0.0529     2.835
   32  CANE MOLASSES    4.450     2.384    0.000   0.0207     1.738
```

Fig. 18-3. Example Leghorn layer ration processed by computer linear programming. See text for explanation of marked ( ) columns. (Courtesy, Nutri-Systems, Fresno, Calif.)

An explanation of the information contained in each column of Fig. 18-3 follows:

Column (1)—Ingredient and nutrient numbers.
Column (2)—Ingredient and nutrient names.
Column (3)—Solution *amounts* given in percentage for feed ingredients (Section I) and nutrients (Section II).
Column (4)—The percentage solution for ingredients has been converted to a "ton" *batch* using prespecified rounding factors for each ingredient. (The batch totals 1,992.73 lb, rather than an exact 2,000 lb because of the rounding requirements.)
Column (5)—Ingredient *costs* in dollars per hundred weight (cwt)
Column (6)—*Lower(L)* and *upper(U) limits* specified for each ingredient (Section I) and each nutrient (Section II).

Column (7)—A pair of values, one under the other, gives the *stable costs*. The first value is the feed cost below which the present optimal solution would no longer be valid. Similarly, the second value is the feed cost above which the present solution would no longer be valid. The *stable cost* figures let the LP user know when it is desirable to reprocess the ration.

Column (8)—The *range* values are related to the *stable costs* and the *cost per unit* columns. The values delineate the limits over which the *stable cost* and *unit decrease/ increase* columns are applicable. (An example: If the cost of ground corn decreases to $6.163/cwt [Column 7], then the usage of corn would increase to 64.50% [Column 8]. Of course, there would be changes in the usage of other ingredients as corn increases in amount.)

Feeding Standards/Ration Formulation

Column (9)—Another pair of values gives the *cost per unit decrease(D)* and *increase(I)*. These values indicate how much the cost of the ration would be changed if either an ingredient or nutrient is increased or decreased by one unit in the percent solution. A positive value means that cost would be increased and a negative value means that cost would be decreased.

Column (10)—Section III contains information about the *ingredients not used* in the solution. The *relative worth* column indicates the cost at which each of these ingredients would enter the solution.

6. **Reformulate with LP at periodic intervals.** Changes in ingredient costs, in ingredient availability, and in the needs of animals dictate the need for reprocessing the ration. The good formulator monitors all these items on a regular basis. It is also critical to evaluate the feeding results to confirm that production goals and cost objectives are being met with the ration. Computers don't feed animals—people do!

SELECTION OF COMPUTER SOFTWARE AND HARDWARE FOR RATION FORMULATION

Numerous companies market computer software for ration formulation. The software varies from the very simple and straight-forward to very complex packages intended for large feed manufacturers. The latter packages include applications for formula costing, inventory control, control of usage of ingredients in limited availability, production of feed tags, etc. Most software is intended for use on a single computer model or at least a certain family of computers (IBM-PC tm, for example). It is therefore most desirable to select the software desired before purchasing the computer hardware. Computer type and size of memory and disk drive storage capacity must meet the criteria of the software developer or the software may not be usable.

Directories which list software by application are a good place to start looking. Other sources of information are feed and livestock trade publication advertisements, university personnel, and nutritionists who use feed blending software. Nutritionists are a good source of information as to how well a certain software package performs.

Ration formulation software may be generalized so that it can be made applicable to all species of animals or it may be designed with the unique requirements of specific species such as poultry, dairy cattle, etc. When the software has been designed for a certain species, it may incorporate tables of nutrient requirements and tables of typical feedstuffs and their nutrient values. This can save the user time, but it does not mean that the software will run itself without the judgment of the user. No one has yet developed software that will anticipate all the conditions under which livestock will be fed. Computers are not able to assess all aspects of ingredient quality, environment, and animal management. The judgment of the producer and formulator must be imposed on the computer software. Look for the freedom to make changes as needed. When in doubt, seek advice from those with experience.

USE OF THE FLEXIBLE FORMULA

The flexible formula is a ration formulation that allows for the substitution of various feeds on the basis of price and availability. The overall formula does not change. The only changes that take place are substitutions of feeds—for example oats for corn—within the formula to supply the same nutrient levels.

Flexible formulas are extremely useful in ration formulation. No one ration is best and, if substitutions are made wisely, the prices of feed can be minimized and the feeder will continually obtain good results. Common sense is an invaluable tool when using flexible formulas because the feed manufacturer must always keep in mind the factors which cannot always be quantitated, such as palatability and bulkiness of feed.

FORMULATION WORKSHEET

When formulating rations, it is advisable to record the ration on a worksheet similar to that in Fig. 18-4. This worksheet is merely an example of the format that should be used. A similar sheet can be developed for micronutrient

MACRONUTRIENT WORKSHEET

Ration Number: Date:

Ingredient	✓ if mixed	Amount	Proximate Analysis				Energy	Minerals and Vitamins		
			Crude Fiber	Ether Extract (Fat)	N-Free Extract	Crude Protein		Calcium (Ca)	Phosphorus (P)	Vitamin A
		(lb)	(lb)	(lb)	(lb)	(lb)	TDN = lb ME = Mcal NE = Mcal	(lb)	(lb)	(IU)
TOTAL										
NUTRIENT REQUIREMENTS										
NUTRIENT BALANCE (Total−Nutrient Requirements)										

Fig. 18-4. Formulation worksheet for macronutrients.

composition of premixes for minerals and vitamins as well as for amino acids. In modern feed formulation practice, computer printouts provide convenient worksheets. The worksheet serves three purposes:

1. It provides a means of reviewing and double checking the calculations used to formulate the ration. If there is a gross error, it will become obvious when listed on the worksheet.

2. It can be used to organize mixing procedures. It is vital that the person mixing feed be able to refer to a worksheet on which can be recorded what has been mixed and what mixing order should be followed.

3. The worksheet can be filed for future reference. If any questions should arise when the feed is fed, the worksheet provides an orderly record of the content of the feed and its mixing.

HOW TO APPLY THE NET ENERGY METHOD

In order to apply the net energy method to the feeding of livestock, the following net energy values must be available:

1. A table showing the net energy requirements of the particular class of animal. Chapter 19, Table 19-3, Daily Nutrient Requirements of Growing and Finishing Cattle, shows the net energy requirements for growing-finishing beef cattle (in megacalories [Mcal] per animal per day), with a breakdown into steers, bulls, and heifers.

2. A table showing the nutrient composition of feeds, with the net energy of each feed partitioned into energy used for body maintenance and for gain; thus, the net energy values in megacalories (Mcal) per unit (lb or kg) are needed for each feed for maintenance (NE_m) and for gain (NE_g)(see Section V—Composition of Feeds).

The two examples that follow will show how to apply the net energy method. In the first example, net energy values of feeds are used to calculate the number of pounds of a given ration that a steer would need to consume to make a specified daily gain. In the second example, net energy is used to predict average daily gain based on consuming a certain number of pounds of a specified ration. Bear in mind that the ration in both cases (in these examples, the ration in Table 18-3) must be balanced for protein, minerals, and vitamins, in order for these net energy values to have validity for calculating daily consumption and predicting average daily gain.

Example 1. *Using net energy values to calculate the number of pounds of the ration that must be consumed to produce a specific gain*—How many calories would a 770-lb medium-frame steer calf need to consume to gain 2.6 lb daily?

Step 1. Calculate the net energy for maintenance (NE_m) and gain (NE_g) values for a pound of the ration shown in Table 18-3.

By referring to Section V—Composition of Feeds, it is determined that l lb of the Table 18-3 ration supplies 0.7620 megacalories of net energy for maintenance (Mcal NE_m) and 0.5015 megacalories of net energy for gain (Mcal NE_g).

Step 2. From Chapter 19, Table 19-3 of this book, we find that the requirement for a 770-lb medium-frame steer calf to gain 2.6 lb daily is as follows:

	Mcal/day
NE_m	6.24
NE_g	5.50

Step 3. Pounds of feed to meet the daily maintenance requirement:

6.24 Mcal ÷ .7620 Mcal = 8.19 lb

Step 4. Pounds of feed to meet the requirement for 2.6 lb daily gain:

5.50 Mcal ÷ .5015 Mcal = 10.97 lb

Step 5. Total pounds of feed that the steer calf must eat daily to gain 2.6 lb:

8.19 lb + 10.97 lb = 19.16 lb

Example 2. *Using net energy to predict the average daily gain of a 770-lb medium-frame steer calf that is consuming a certain number of pounds of a specified ration*—Let's assume that we have a 770-lb steer that is consuming 18 lb of the ration shown in Table 18-3. What daily gain should be expected?

Step 1. Pounds of feed to meet the daily maintenance requirement = 8.19 lb (see prior example).

Step 2. Pounds of feed left for gain:

18 lb − 8.19 lb = 9.81 lb

Step 3. Mcal of NE_g supplied by remaining feed:

9.81 lb × .5015 Mcal = 4.92 Mcal

TABLE 18-3
RATION FOR FINISHING CATTLE

Ration Ingredient	(lb)	(kg)	Composition of Ingredients (as-fed basis) NE_m[1]		Ration Supplies NE_m[1]	Composition of Ingredients (as-fed basis) NE_g[2]		Ration Supplies NE_g[2]
			(Mcal/lb)[3]	(Mcal/kg)[3]	(Mcal)[3]	(Mcal/lb)[3]	(Mcal/kg)[3]	(Mcal)[3]
Shelled corn, all analyses	68.60	31.14	0.86	1.90	59.00[4]	0.59	1.89	40.47[5]
Soybean meal (solvent), 44%	4.00	1.82	0.79	1.74	3.16	0.53	1.17	2.12
Alfalfa hay (mid-bloom)	27.00	12.26	0.52	1.15	14.04	0.28	0.62	7.56
Salt	0.40	0.18	—	—	—	—	—	—
Total	100.00	45.4	—	—	76.20	—	—	50.15

[1]NE_m = net energy for maintenance. [2]NE_g = net energy for gain. [3]Mcal stands for megacalorie. [4]68.60 lb × 0.86 = 59.00 [5]68.60 lb × 0.59 Mcal = 40.47

Step 4. Daily gain expected from 4.92 Mcal of NE$_g$ (Table 19-3):

4.51 Mcal produces 2.2 lb gain

Therefore, 4.92 Mcal will produce 2.40 lb daily gain

$$\left[\frac{4.92\,(2.2)}{4.51}\right]$$

QUESTIONS FOR STUDY AND DISCUSSION

1. What are feeding standards? How are they expressed?

2. What are the National Research Council feeding standards? Of what value are they? How do the NRC feeding standards compare with the feeding standards of other countries?

3. What are the limitations of feeding standards?

4. Define "balanced ration."

5. When rations are formulated, why are feeds divided into the following three categories: (a) concentrates, (b) roughages, and (c) supplements?

6. Why should special consideration be given to the amount and kind of fiber, including ADF and NDF, for lactating cows?

7. What major points should be considered before attempting to balance a ration?

8. Outline the four steps involved in formulating an economical ration.

9. In order to feed livestock efficiently and economically, one must understand thoroughly the nutrients furnished by the available feeds, the extent to which livestock can utilize each feed, and the actual feeding value of these feeds. This can be accomplished only through careful and thorough study of the different feeds.

The following exercises are designed to acquaint the student with the feeds commonly fed to ruminants.

a. **A study of available forages.** It is well known that a relatively large portion of the feed consumed by ruminants is used in meeting the energy needs. Thus, a convenient and reasonably accurate way of determining which forages are most economical under the conditions existing in a particular area at any given time is to compute the cost at which each of the available forages furnishes 100 lb of total digestible nutrients. This is a measure of the economy with which the various feeds furnish fuel or energy. Refer to Section V—Composition of Feeds in this book for analyses, and obtain prices of available forages locally. Then fill out the following table of "Available Roughages":

AVAILABLE ROUGHAGES (AS-FED)

Feed	Farm Price Per Ton	TDN Per 100 Lb	Cost Per 100 Lb TDN	Total Protein Per 100 Lb	Calcium Content (%)	Phosphorus Content (%)	Carotene Mg Per Lb

b. **A study of available grains and by-products.** At least one of the cereal grains is grown in every section of the country, and all of them are used quite widely as feeds. As a group, the cereals and their by-products are high in energy. However, they possess certain nutritive deficiencies which may prove to be quite limiting if they are not properly used. Refer to Section V—Composition of Feeds in this book for analyses, and obtain prices of available grain and by-product feeds locally. Then fill out the following table of "Available Grain and By-product Feeds":

AVAILABLE GRAIN AND BY-PRODUCT FEEDS (AS-FED)

Feed	Retail Price Per Cwt	TDN Per 100 Lb	Cost Per 100 Lb TDN	Crude Protein Per 100 Lb	Cost Per 100 Lb Crude Protein	Calcium Content (%)	Phosphorus Content (%)	Carotene Mg Per Lb

Also, in studying the by-product feeds, be sure to understand the source of the feed and just what part of the original grain or seed goes into the by-product.

10. Why should consideration be given to the moisture content of ingredients when buying feeds and formulating rations? What are the advantages in formulating a ration on a moisture-free basis?

11. List three methods of balancing rations.

12. Using the square method, and corn (8.9% protein) and a protein supplement (44% protein), balance out a ration for 6-week-old broilers from the standpoint of protein content (refer to Chapter 24, Feeding Poultry, for requirements).

13. A farmer is feeding some hay and silage to cattle but calculates that a corn and soybean meal supplement must also be supplied. The TDN that must be supplied by the supplement is 5.8 lb/cow/day. The protein that must be supplied is 1.7 lb/cow/day. The TDN content of the No. 2 corn (as-fed) is 80% and the protein content is 8.9%. The TDN content of the soybean meal (as-fed) has been estimated at 76% and the protein content 44.4%. How much corn and soybean meal should be given to each cow? To answer the question, use one of the 3 methods of balancing rations presented in this chapter.

(Continued)

14. Select a cow of a certain body weight and milk production, then balance a ration for her from the standpoint of TDN and protein, using available feeds and the trial-and-error method. See Chapter 20, Feeding Dairy Cattle, Table 20–5 for requirements; and see Section V, Composition of Feeds, for the content of the feeds selected.

15. Select a specific class of sheep and prepare a balanced ration, using those feeds that are available at the lowest cost. See Chapter 21, Feeding Sheep, for requirements; and see Section V, Composition of Feeds, for the content of the feeds selected.

16. Select a specific class of swine and formulate a balanced ration using those feeds that are available at the lowest cost. Present the ration in terms of both as-fed and moisture-free bases. See Chapter 23, Feeding Swine, for requirements; and see Section V, Composition of Feeds, for the content of the feeds selected.

17. Will the *least-cost* ration always make for the greatest net returns?

18. Why are the nutritionist and the producer necessary when rations are formulated by computers?

19. What is a *spreadsheet*?

20. List the various steps involved in formulating rations by the computer.

21. What is "computer software and hardware"? How would you go about selecting computer software and hardware?

22. What is a flexible formula? How does it work?

23. Design a worksheet for formulating a mineral and vitamin premix formula. Why is such a worksheet necessary?

24. Using the net energy method, formulate a ration for a 500–lb medium-frame steer calf, which is gaining 2 lb/day.

Fig. 18–5. A modern feed mill in which computer technology is used to formulate rations. (Courtesy, American Feed Industry Association, Arlington, Va.)

Fig. 18–6. Bulk feed being covered from trucks to self-feeder. (Photo courtesy Simonsen Manufacturing Co., Quimby, Iowa; and American Feed Industry Association, Arlington, Va.)

Original painting by Tom Phillips

FEEDING BEEF CATTLE[1]

Contents	Page
Part I—Nutritive Needs of, and Feeds for, Beef Cattle	
Economic Importance of Feed for Beef Cattle	348
Nutritive Needs of Beef Cattle	348
National Research Council (NRC) Requirements	349
Energy	372
Protein	373
Minerals	374
Beef Cattle Mineral Chart	375
Vitamins	380
Beef Cattle Vitamin Chart	381
Water	384
Growth Stimulants and Implants	384
Feeds for Beef Cattle	386
Pastures	386
Hays and Other Dry Roughages	386
Silages and Root Crops	386
Concentrates	387
Urea	387
By-product Feeds	387
Mineral Supplements	387
Vitamin Supplements	388
Feed Substitution Table	388
Feed Preparation	393
Feed Allowance and Some Suggested Rations	393
Fitting for Show and Sale	394
Fitting Rations	395
Nutritional Diseases and Ailments	396
Part II—Feeding Breeding Beef Cattle	
Feeding Brood Cows	396
Nutritional Requirements of Brood Cows	396
Nutritional Reproductive Failure in Cows	397
Winter Feeding	397
Rations for Dry Pregnant Cows	397
Calving Control (Daytime Calving)	398
Rations for Cows Nursing Calves	398
Crop Residues and Winter Pastures	398
Crop Residues	398
Corn Residues	398
Winter Pasture	399
Range and Pasture Feeding	399
Range Nutrient Deficiencies	400
Pasture and Range Supplementation	400
Sorting Pasture and Range Cattle	400
Choosing a Pasture or Range Supplement	400
Types and Systems of Pasture and Range Supplementation	400
Confinement (Drylot) Beef Cows	401
Rations for Drylot Cows	401
Semiconfinement (or Partial Confinement) Cow Herds	402

Contents	Page
Feeding Bulls	402
Feeding Sale Bulls	402
Feeding Young Bulls	402
Feeding Mature Bulls	403
Feeding Calves	403
Feeding at Birth	403
Feeding Orphan and Multiple Birth Calves	403
Creep Feeding	403
Feeding Early Weaned Calves	405
Weaning	405
Preconditioning	405
Feeding Replacement Heifers	405
Calving Two-Year-Olds	407
Part III—Feeding Stocker (Feeder) Cattle	
Types of Stocker Programs	408
Rations for Stockers	409
Level of Wintering	410
Compensatory Growth	410
Facts Pertinent to Stocker Programs	411
Stocker and Grower Contracts	411
Part IV—Feeding Finishing (Fattening) Cattle	
Kinds of Cattle to Feed	412
Age and Weight of Cattle	412
Sex of Cattle	413
Spayed Heifers	413
Bulls	413
Grade of Cattle	413
Breeding and Type of Cattle	414
Crossbreds	414
Dairy Beef	414
Boxed Beef	414
Feedlot Finishing Facilities	415
Open Pen Feedlot	415
Confinement Feeding; Slotted (or Slatted) Floors	415
Cold Confinement	415
Warm Confinement	416
Feeds and Rations for Finishing Cattle	416
Feed Preparation	416
Mixed Rations vs Feeding Roughages and Concentrates Separately	416

(Continued)

---

[1] Key sections, tables, and charts in this chapter were authoritatively reviewed by the following beef cattle specialists: D. R. Gill, Ph.D., Department of Animal Science, Oklahoma State University, Stillwater; W. H. Hale, Ph.D., Department of Animal Sciences, The University of Arizona, Tucson; H. W. Newland, Ph.D., Department of Animal Science, The Ohio State University, Columbus; T. W. Perry, Ph.D., Department of Animal Sciences, Purdue University, West Lafayette, Ind.; and L. M. Schake, Ph.D., Department of Animal Science, The University of Connecticut, Storrs.

Contents	Page
How to Use the Net Energy Method	416
Managing Feedlot Cattle	416
Handling Newly Arrived Cattle	417
Amount to Feed; Rate of Gain	417
Cull Out; Top Out	418
Overfinishing	418
Feed Bunk Management	418
Weather Affects Eating and Drinking Habits	418
Mud Problem	418
Records	419
Make 28-Day Test Weights	419
Custom (Contract) Feeding	419
Types of Custom Feeding Contracts	419
Conducting Applied Feedlot Tests	420
Pasture Finishing Cattle	420
Systems of Pasture Finishing	420
Basic Considerations in Utilizing Pastures for Finishing Cattle	420
Nutritional Feedlot Diseases	421
Questions for Study and Discussion	421

Pastures and range forages, along with other roughages, are the very foundation of successful beef cattle production. In fact, it may be said that the principal function of beef cattle is to harvest vast acreages of forages, and, with or without supplementation, to convert these feeds into more nutritious and palatable products for human consumption. It is estimated (1) that 85.7% of the total feed of beef cattle is derived from roughages, and (2) that 31% of the land area of continental United States is used for grassland pasture and range, with much of this area utilized by beef cattle. If produced on well-fertilized soils, green pasture and well-cured, green, leafy hay can supply all of the nutrient requirements of beef cattle, except the need for common salt and whatever energy-rich feeds may be necessary for additional conditioning or drylot finishing.

In order to cover *Feeding Beef Cattle* with a minimum of repetition and a maximum of cohesiveness, and to enhance readership, the authors opted to present this important subject in one large chapter (rather than four separate chapters), with four parts as follows:

Part I—Nutritive Needs of, and Feeds for, Beef Cattle

Part II—Feeding Breeding Beef Cattle

Part III—Feeding Stocker (Feeder) Cattle

Part IV—Feeding Finishing (Fattening) Cattle

PART I—Nutritive Needs of, and Feeds for, Beef Cattle

ECONOMIC IMPORTANCE OF FEED FOR BEEF CATTLE

The feeding of beef cattle constitutes the greatest single cost item of their production. It is important, therefore, that feeding practices be as satisfactory and economical as possible.

Feed affects total profit and cow productivity. It accounts for 65 to 75% of the total cost of keeping cows, and it exerts a powerful influence on cow fertility and calf weaning weight—the two biggest success factors in the cattle business. Without doubt, faulty feeding is a major factor in the nutritional reproductive failure of cows—resulting (1) in only an 88% calf crop (calves born alive or dead), and (2) an average calf death loss of 6% from birth to weaning.

Also, feed is a major item of expense in finishing cattle. It accounts for 70 to 80% of the cost of feedlot finishing, exclusive of the purchase price of the animals.

NUTRITIVE NEEDS OF BEEF CATTLE

The nutritive requirements of beef cattle have become more critical with the shift in beef production practices. Steers were formerly permitted to make their growth primarily on roughages—pastures in the summertime and hay and other forages in the winter. After making moderate and unforced growth for 2 to 4 years, usually the animals were either turned into the feedlot or placed on more lush pastures for a reasonable degree of finishing. With this system, the growth and finishing requirements of cattle came largely at two separate periods in the life of the animal.

Under the old system of moderate growth rate, reasonably good pastures and good-quality hay fully met the nutritive needs. Such older cattle finished quickly when it was desired to ready them for market. Because of their mature size, there was little need for supplemental protein or minerals; and sufficient vitamin A was stored in the liver for the finishing period.

But fashions and sizes of beef cattle shifted radically during the present century—and with them the nutritive requirements changed. Cattle producers moved from the unhurried production and marketing of 2- to 4-year-old steers to the crowding and marketing of calves and yearlings.

In recent years, the introduction of crossbreeding and of the exotic breeds has resulted in heavier milking cows and faster gaining calves. Also, more and more heifers are being bred to calve as 2-year-olds. Hand in hand with these developments, scarce and high-priced grains in the early 1970s provided a preview of world food shortages in the years ahead, punctuated by some years of plenty and other years of scarcity; but, increasingly and relentlessly, with the scarce years dominating the world situation. During periods when grains are in short supply and high in price, feeder cattle will be carried to heavier weights on milk and grass before going into feedlots, then grain fed for a shorter period of time. Conversely, when grains are abundant and a better buy than forages, cattle will be fed more grain and less forage. Also, in the future, (1) the beef industry will breed and feed cattle for the most desirable combination of muscling, marbling, and external finish, and (2) biotechnology will increase production per animal dramatically. In this chapter, provision has been made for the nutritive needs created by these changes.

Feeding Beef Cattle

As feeds represent by far the greatest cost item in beef production, it is important that there be a basic understanding of the nutritive requirements. For convenience, these needs will be discussed under the following categories: (1) National Research Council Requirements, (2) energy, (3) protein, (4) minerals, (5) vitamins, and (6) water.

National Research Council (NRC) Requirements

Efficient beef production cannot be achieved unless nutrient requirements are met, and these are influenced by a number of factors: size (body weight) and reproduction/lactation are especially important in breeding cattle; and body weight and frame size are especially important in growing and finishing cattle.

Although the basic biology of all beef cattle remains the same, differences in rate of maturity and in mature size have a marked influence on the application of the basic nutrition principles to the wide range of environmental and management conditions to which beef cattle are subjected. In recent years, types of beef cattle have changed in response to economic pressures and consumer demand for leaner cuts—larger and faster gaining cattle have evolved. For this reason, frame size has been considered in calculating the nutritive requirements presented in the tables in this section. The medium-frame steer is projected to have 990 to 1,144 lb liveweight at usual market finish, and the medium-frame heifer 800 to 1,045 lb. The finished weight is projected to be over 1,144 and 1,045 lb for large-frame steers and heifers, respectively.

Tables 19-1 to 19-6 were adapted by the authors from *Nutrient Requirements of Beef Cattle,* sixth revised edition, 1984. In using these tables, note the following:

1. **Protein requirements.** These do not provide safety margins.
2. **Nutrient requirements of breeding beef cattle.** Table 19-1 presents the daily nutrient requirements; and Table 19-2 presents the nutrient requirements as a percentage of the ration.
3. **Nutrient requirements of growing and finishing cattle.** These are presented in both daily requirements (Table 19-3) and nutrient requirements in the ration (Table 19-4).
4. **Mineral requirements and maximum tolerable levels for beef cattle.** These are presented in Table 19-5.
5. **Maximum tolerable levels of certain toxic elements.** These are presented in Table 19-6.

TABLE 19-1
DAILY NUTRIENT REQUIREMENTS OF BREEDING CATTLE [1] (See footnotes at end of table.)

Weight[2]		Daily Gain[3]		Daily Consumption[4]				Energy					Total Protein		Calcium	Phosphorus	Vitamin A[6]
				As-Fed[5]		Moisture-Free Dry Matter		TDN		ME	NE_m	NE_g					
(lb)	(kg)	(lb)	(kg)	(lb)	(kg)	(lb)	(kg)	(lb)	(kg)	(Mcal)	(Mcal)	(Mcal)	(lb)	(kg)	(g)	(g)	(1,000 IU)
Pregnant yearling heifers—Last third of pregnancy																	
700	318	0.9	0.4	17.0	7.7	15.3	7.0	8.5	3.9	13.9	7.95	NA[7]	1.3	0.6	19	14	19
700	318	1.4	0.6	17.6	8.0	15.8	7.2	9.6	4.4	15.7	7.95	0.87	1.4	0.6	24	15	20
700	318	1.9	0.9	17.6	8.0	15.8	7.2	10.6	4.8	17.4	7.95	1.89	1.5	0.7	27	16	20
750	341	0.9	0.4	17.9	8.1	16.1	7.3	8.9	4.0	14.6	8.25	NA	1.3	0.6	20	14	20
750	341	1.4	0.6	18.4	8.4	16.6	7.5	10.0	4.5	16.4	8.25	0.92	1.5	0.7	24	16	21
750	341	1.9	0.9	18.4	8.4	16.6	7.5	11.1	5.0	18.2	8.25	1.99	1.6	0.7	28	17	21
800	364	0.9	0.4	18.7	8.5	16.8	7.6	9.2	4.2	15.2	8.56	NA	1.4	0.6	21	15	21
800	364	1.4	0.6	19.3	8.8	17.4	7.9	10.4	4.7	17.1	8.56	0.96	1.5	0.7	25	16	22
800	364	1.9	0.9	19.4	8.8	17.5	8.0	11.6	5.3	19.0	8.56	2.09	1.6	0.7	28	17	22
850	386	0.9	0.4	19.6	8.9	17.6	8.0	9.6	4.4	15.7	8.85	NA	1.4	0.6	21	16	22
850	386	1.4	0.6	20.2	9.2	18.2	8.3	10.8	4.9	17.8	8.85	1.01	1.6	0.7	25	17	23
850	386	1.9	0.9	20.3	9.2	18.3	8.3	12.1	5.5	19.8	8.85	2.19	1.7	0.8	28	18	23
900	409	0.9	0.4	20.3	9.2	18.3	8.3	9.9	4.5	16.3	9.15	NA	1.5	0.7	22	17	23
900	409	1.4	0.6	21.1	9.6	19.0	8.6	11.3	5.1	18.5	9.15	1.05	1.6	0.7	26	18	24
900	409	1.9	0.9	21.3	9.7	19.2	8.7	12.5	5.7	20.6	9.15	2.28	1.7	0.8	28	19	24
950	432	0.9	0.4	21.1	9.6	19.0	8.6	10.3	4.7	16.9	9.44	NA	1.5	0.7	23	17	24
950	432	1.4	0.6	22.0	10.0	19.8	9.0	11.7	5.3	19.1	9.44	1.09	1.7	0.8	26	19	25
950	432	1.9	0.9	22.2	10.1	20.0	9.0	13.0	5.9	21.3	9.44	2.38	1.8	0.8	29	19	25
Dry pregnant mature cows—Middle third of pregnancy																	
800	364	0.0	0.0	17.0	7.7	15.3	7.0	7.5	3.4	12.3	6.41	NA	1.1	0.5	12	12	19
900	409	0.0	0.0	18.6	8.5	16.7	7.6	8.2	3.7	13.4	7.00	NA	1.2	0.5	14	14	21
1,000	454	0.0	0.0	20.1	9.1	18.1	8.2	8.8	4.0	14.5	7.57	NA	1.3	0.6	15	15	23
1,100	500	0.0	0.0	21.7	9.9	19.5	8.9	9.5	4.3	15.6	8.13	NA	1.4	0.6	17	17	25
1,200	545	0.0	0.0	23.1	10.5	20.8	9.5	10.1	4.6	16.6	8.68	NA	1.4	0.6	18	18	26
1,300	591	0.0	0.0	24.4	11.1	22.0	10.0	10.8	4.9	17.7	9.22	NA	1.5	0.7	20	20	28
1,400	636	0.0	0.0	25.9	11.8	23.3	10.6	11.4	5.2	18.7	9.75	NA	1.6	0.7	21	21	30
Dry pregnant mature cows—Last third of pregnancy																	
800	364	0.9	0.4	18.7	8.5	16.8	7.6	9.2	4.2	15.0	8.56	NA	1.4	0.6	20	15	21
900	409	0.9	0.4	20.2	9.2	18.2	8.3	9.8	4.5	16.2	9.15	NA	1.5	0.7	22	17	23
1,000	454	0.9	0.4	21.8	9.9	19.6	8.9	10.5	4.8	17.3	9.72	NA	1.6	0.7	23	18	25
1,100	500	0.9	0.4	23.3	10.6	21.0	9.5	11.2	5.1	18.3	10.28	NA	1.6	0.7	25	20	26
1,200	545	0.9	0.4	24.8	11.3	22.3	10.1	11.8	5.4	19.4	10.83	NA	1.7	0.8	26	21	28
1,300	591	0.9	0.4	26.2	11.9	23.6	10.7	12.5	5.7	20.4	11.37	NA	1.8	0.8	28	23	30
1,400	636	0.9	0.4	27.7	12.6	24.9	11.3	13.1	6.0	21.5	11.90	NA	1.9	0.9	29	24	32

(Continued)

TABLE 19-1 (Continued)

Weight[2]		Daily Gain[3]		Daily Consumption[4]				Energy					Total Protein		Cal-cium	Phos-phorus	Vitamin A[6]
				As-Fed[5]		Moisture-Free Dry Matter		TDN		ME	NE_m	NE_g					
(lb)	(kg)	(lb)	(kg)	(lb)	(kg)	(lb)	(kg)	(lb)	(kg)	(Mcal)	(Mcal)	(Mcal)	(lb)	(kg)	(g)	(g)	(1,000 IU)
Two-year-old heifers nursing calves—First 3-4 months postpartum—10 lb (4.5 kg) milk/day																	
700	318	0.5	0.2	17.7	8.0	15.9	7.2	10.3	4.7	17.0	9.20[8]	0.87	1.8[9]	0.8	26	17	28
750	341	0.5	0.2	18.6	8.5	16.7	7.6	10.8	4.9	17.7	9.51	0.92	1.8	0.8	26	18	30
800	364	0.5	0.2	19.6	8.9	17.6	8.0	11.2	5.1	18.4	9.81	0.96	1.9	0.9	27	19	31
850	386	0.5	0.2	20.4	9.3	18.4	8.4	11.6	5.3	19.1	10.11	1.01	1.9	0.9	27	19	33
900	409	0.5	0.2	21.3	9.7	19.2	8.7	12.0	5.5	19.8	10.40	1.05	2.0	0.9	28	20	34
950	432	0.5	0.2	22.2	10.1	20.0	9.0	12.5	5.7	20.5	10.69	1.09	2.0	0.9	28	21	35
1,000	454	0.5	0.2	23.1	10.5	20.8	9.5	12.9	5.9	21.1	10.98	1.14	2.1	1.0	29	22	37
Cows nursing calves—Average milking ability—First 3-4 months postpartum—10 lb (4.5 kg) milk/day																	
800	364	0.0	0.0	19.2	8.7	17.3	7.9	10.1	4.6	16.6	9.81	NA	1.8	0.8	23	17	31
900	409	0.0	0.0	20.1	9.1	18.8	8.5	10.8	4.9	17.7	10.40	NA	1.9	0.9	24	19	33
1,000	454	0.0	0.0	22.4	10.2	20.2	9.2	11.5	5.2	18.8	10.98	NA	2.0	0.9	25	20	36
1,100	500	0.0	0.0	24.0	10.9	21.6	9.8	12.1	5.5	19.9	11.54	NA	2.0	0.9	27	22	38
1,200	545	0.0	0.0	25.6	11.6	23.0	10.5	12.8	5.8	21.0	12.09	NA	2.1	1.0	28	23	41
1,300	591	0.0	0.0	27.0	12.3	24.3	11.0	13.4	6.1	22.0	12.63	NA	2.2	1.0	30	25	43
1,400	636	0.0	0.0	28.4	12.9	25.6	11.6	14.0	6.4	23.0	13.15	NA	2.3	1.0	31	26	46
Cows nursing calves—Superior milking ability—First 3-4 months postpartum—20 lb (9.1 kg) milk/day																	
800	364	0.0	0.0	17.4	7.9	15.7	7.1	12.1	5.5	19.9	13.22	NA	2.2	1.0	34	22	28
900	409	0.0	0.0	20.8	9.5	18.7	8.5	13.1	6.0	21.5	13.81	NA	2.4	1.1	35	24	33
1,000	454	0.0	0.0	22.9	10.4	20.6	9.4	13.8	6.3	22.7	14.38	NA	2.5	1.1	36	25	37
1,100	500	0.0	0.0	24.8	11.3	22.3	10.1	14.5	6.6	23.8	14.94	NA	2.6	1.2	38	27	40
1,200	545	0.0	0.0	26.4	12.0	23.8	10.8	15.2	6.9	24.9	15.49	NA	2.7	1.2	39	28	42
1,300	591	0.0	0.0	28.1	12.8	25.3	11.5	15.9	7.2	26.0	16.03	NA	2.8	1.3	41	30	45
1,400	636	0.0	0.0	29.7	13.5	26.7	12.1	16.5	7.5	27.1	16.56	NA	2.9	1.3	42	31	47
Bulls, maintenance and slow rate of growth (regain body condition)																	
<1,300	For growth and development use requirements for bulls in Table 19-3																
1,300	591	1.0	0.5	28.2	12.8	25.4	11.5	14.2	6.5	23.3	9.22	2.20	1.9	0.9	25	22	45
1,300	591	1.5	0.7	29.0	13.2	26.1	11.9	15.6	7.1	25.5	9.22	3.43	2.0	0.9	28	23	46
1,300	591	2.0	0.9	29.1	13.2	26.2	11.9	16.8	7.6	27.6	9.22	4.71	2.2	1.0	31	24	46
1,400	636	1.0	0.5	29.8	13.5	26.8	12.2	15.0	6.8	24.6	9.75	2.33	2.0	0.9	26	23	48
1,400	636	1.5	0.7	30.7	14.0	27.6	12.5	16.5	7.5	27.0	9.75	3.63	2.1	1.0	29	24	49
1,400	636	2.0	0.9	30.8	14.0	27.7	12.6	17.8	8.1	29.1	9.75	4.98	2.2	1.0	31	25	49
1,500	682	0.0	0.0	28.0	12.7	25.2	11.5	12.2	5.5	20.0	10.26	NA	1.7	0.8	23	23	45
1,500	682	1.0	0.5	31.4	14.3	28.3	12.9	15.8	7.2	25.9	10.26	2.45	2.1	1.0	27	24	50
1,500	682	1.5	0.7	32.2	14.6	29.0	13.2	17.3	7.9	28.4	10.26	3.82	2.2	1.0	29	25	51
1,600	727	0.0	0.0	29.4	13.4	26.5	12.0	12.8	5.8	21.0	10.77	NA	1.8	0.8	23	24	47
1,600	727	1.0	0.5	33.0	15.0	29.7	13.5	16.6	7.5	27.2	10.77	2.57	2.2	1.0	29	26	53
1,600	727	1.5	0.7	33.8	15.4	30.4	13.8	18.2	8.3	29.8	10.77	4.01	2.3	1.0	31	27	54
1,700	773	0.0	0.0	30.8	14.0	27.7	12.6	13.4	6.1	22.0	11.28	NA	1.9	0.9	26	26	49
1,700	773	0.5	0.2	32.9	15.0	29.6	13.5	15.4	7.0	25.3	11.28	1.26	2.1	1.0	27	26	52
1,800	818	0.0	0.0	32.1	14.6	28.9	13.1	14.0	6.4	23.0	11.77	NA	2.0	0.9	27	27	51
1,800	818	0.5	0.2	34.3	15.6	30.9	14.0	16.1	7.3	26.4	11.77	1.31	2.2	1.0	28	28	55
1,900	864	0.0	0.0	33.4	15.2	30.1	13.7	14.6	6.6	23.9	12.26	NA	2.0	0.9	29	29	53
1,900	864	0.5	0.2	35.8	16.3	32.2	14.6	16.8	7.6	27.5	12.26	1.37	2.2	1.0	29	29	57
2,000	909	0.0	0.0	34.8	15.8	31.3	14.2	15.2	6.9	24.9	12.74	NA	2.1	1.0	30	30	55
2,100	955	0.0	0.0	36.1	16.4	32.5	14.8	15.7	7.1	25.8	13.21	NA	2.2	1.0	32	32	58
2,200	1,000	0.0	0.0	37.3	17.0	33.6	15.3	16.3	7.4	26.7	13.68	NA	2.3	1.0	33	33	60

[1]Adapted from *Nutrient Requirements of Beef Cattle*, 6th revised edition, National Research Council—National Academy of Sciences, 1984, p. 85, Table 11.
[2]Average weight for a feeding period.
[3]Approximately 0.9 ± 0.2 lb of weight gain/day over the last third of pregnancy is accounted for by the products of conception. Daily 2.15 Mcal of NE_m and 0.1 lb of protein are provided for this requirement for a calf with a birth weight of 80 lb.
[4]Consumption should vary depending on the energy concentration of the ration and environmental conditions. These intakes are based on the energy concentration shown in the table and assuming a thermoneutral environment without snow or mud conditions. If the energy concentrations of the ration to be fed exceed the tabular value, limit feeding may be required.
[5]As-Fed was calculated using an average figure of 90% dry matter. When using silages, roots, and other wet feeds, these feeds should be converted to a moisture-free basis and the ration calculated using the moisture-free data.
[6]Vitamin A requirements per pound of ration are 1,273 IU for pregnant heifers and cows, and 1,773 IU for lactating cows and breeding bulls.
[7]Not applicable.
[8]Includes 0.34 Mcal NE_m/lb of milk produced for all heifers and cows nursing calves.
[9]Includes 0.03 lb protein/lb of milk produced for all heifers and cows nursing calves.

TABLE 19-2
NUTRIENT REQUIREMENTS IN RATION FOR BREEDING CATTLE [1] (See footnotes at end of table.)

Weight[2]		Daily Gain[3]		Daily Consumption[4]				Moisture Basis[5] A-F (as-fed) M-F (moisture-free)	Energy							Total Protein	Calcium	Phosphorus	Vitamin A (IU per)	
				As-Fed		Moisture-Free Dry Matter			TDN	ME (Mcal per)		NE_m (Mcal per)		NE_g (Mcal per)						
(lb)	(kg)	(lb)	(kg)	(lb)	(kg)	(lb)	(kg)		(%)	(lb)	(kg)	(lb)	(kg)	(lb)	(kg)	(%)	(%)	(%)	(lb)	(kg)
	Pregnant yearling heifers—Last third of pregnancy
700	318	0.9	0.4	17.0	7.7	15.3	7.0	A-F	49.9	0.82	1.80	0.47	1.03	NA[a]	NA	7.6	0.24	0.18	1,146	2,521
								M-F	55.4	0.91	2.00	0.52	1.14	NA	NA	8.4	0.27	0.20	1,273	2,801
700	318	1.4	0.6	17.6	8.0	15.8	7.2	A-F	54.3	0.89	1.96	0.54	1.19	0.31	0.63	8.1	0.30	0.19	1,146	2,521
								M-F	60.3	0.99	2.18	0.60	1.32	0.34	0.75	9.0	0.33	0.21	1,273	2,801
700	318	1.9	0.9	17.6	8.0	15.8	7.2	A-F	60.3	0.99	2.18	0.63	1.39	0.39	0.86	8.8	0.30	0.19	1,146	2,521
								M-F	67.0	1.10	2.42	0.70	1.54	0.43	0.95	9.8	0.33	0.21	1,273	2,801
750	341	0.9	0.4	17.9	8.1	16.1	7.3	A-F	49.6	0.84	1.78	0.47	1.03	NA	NA	7.5	0.24	0.17	1,146	2,521
								M-F	55.1	0.90	1.98	0.52	1.14	NA	NA	8.3	0.27	0.19	1,273	2,801
750	341	1.4	0.6	18.4	8.4	16.6	7.5	A-F	53.9	0.88	1.94	0.54	1.19	0.30	0.66	8.0	0.29	0.19	1,146	2,521
								M-F	59.9	0.98	2.16	0.60	1.32	0.33	0.73	8.9	0.32	0.21	1,273	2,801
750	341	1.9	0.9	18.4	8.4	16.6	7.5	A-F	59.9	0.98	2.16	0.62	1.37	0.38	0.83	8.6	0.33	0.21	1,146	2,521
								M-F	66.5	1.09	2.40	0.69	1.52	0.42	0.92	9.5	0.37	0.23	1,273	2,801
800	364	0.9	0.4	18.7	8.5	16.8	7.6	A-F	49.3	0.81	1.78	0.46	1.01	NA	NA	7.4	0.25	0.18	1,146	2,521
								M-F	54.8	0.90	1.98	0.51	1.12	NA	NA	8.2	0.28	0.20	1,273	2,801
800	364	1.4	0.6	19.3	8.8	17.4	7.9	A-F	53.6	0.88	1.94	0.53	1.17	0.30	0.66	7.9	0.30	0.19	1,146	2,521
								M-F	59.6	0.98	2.16	0.59	1.30	0.33	0.73	8.8	0.33	0.21	1,273	2,801
800	364	1.9	0.9	19.4	8.8	17.5	8.0	A-F	59.5	0.97	2.14	0.62	1.37	0.38	0.83	8.4	0.32	0.19	1,146	2,521
								M-F	66.1	1.08	2.38	0.69	1.52	0.42	0.92	9.3	0.35	0.21	1,273	2,801
850	386	0.9	0.4	19.6	8.9	17.6	8.0	A-F	49.1	0.80	1.76	0.46	1.01	NA	NA	7.4	0.23	0.18	1,146	2,521
								M-F	54.5	0.89	1.96	0.51	1.12	NA	NA	8.2	0.26	0.20	1,273	2,801
850	386	1.4	0.6	20.2	9.2	18.2	8.3	A-F	53.4	0.87	1.92	0.53	1.17	0.29	0.63	7.7	0.27	0.19	1,146	2,521
								M-F	59.3	0.97	2.13	0.59	1.30	0.32	0.70	8.6	0.30	0.21	1,273	2,801
850	386	1.9	0.9	20.3	9.3	18.3	8.3	A-F	59.1	0.97	2.14	0.61	1.35	0.37	0.81	8.2	0.31	0.20	1,146	2,521
								M-F	65.7	1.08	2.38	0.68	1.50	0.41	0.90	9.1	0.34	0.22	1,273	2,801
900	409	0.9	0.4	20.3	9.3	18.3	8.3	A-F	48.9	0.80	1.76	0.46	1.01	NA	NA	7.3	0.23	0.18	1,146	2,521
								M-F	54.3	0.89	1.96	0.51	1.12	NA	NA	8.1	0.26	0.20	1,273	2,801
900	409	1.4	0.6	21.1	9.6	19.0	8.6	A-F	53.2	0.87	1.92	0.52	1.15	0.29	0.63	7.7	0.27	0.19	1,146	2,521
								M-F	59.1	0.97	2.13	0.58	1.28	0.32	0.70	8.5	0.30	0.21	1,273	2,801
900	409	1.9	0.9	21.3	9.7	19.2	8.7	A-F	58.9	0.96	2.12	0.61	1.35	0.37	0.81	8.1	0.29	0.19	1,146	2,521
								M-F	65.4	1.07	2.35	0.68	1.50	0.41	0.90	9.0	0.32	0.21	1,273	3,801
950	432	0.9	0.4	21.1	9.6	19.0	8.6	A-F	48.7	0.80	1.76	0.45	0.99	NA	NA	7.2	0.24	0.18	1,146	2,521
								M-F	54.1	0.89	1.96	0.50	1.10	NA	NA	8.0	0.27	0.20	1,273	2,801
950	432	1.4	0.6	22.0	10.0	19.8	9.0	A-F	53.0	0.87	1.92	0.52	1.15	0.29	0.63	7.6	0.26	0.19	1,146	2,521
								M-F	58.9	0.97	2.13	0.58	1.28	0.32	0.70	8.4	0.29	0.21	1,273	2,801
950	432	1.9	0.9	22.2	10.1	20.0	9.0	A-F	58.6	0.96	2.12	0.60	1.32	0.36	0.79	7.9	0.29	0.19	1,146	2,521
								M-F	65.1	1.07	2.35	0.67	1.47	0.40	0.88	8.8	0.32	0.21	1,273	2,801
	Dry pregnant mature cows—Middle third of pregnancy
800	364	0.0	0.0	17.0	7.7	15.3	7.0	A-F	43.9	0.72	1.58	0.38	0.83	NA	NA	6.4	0.15	0.15	1,146	2,521
								M-F	48.8	0.80	1.76	0.42	0.92	NA	NA	7.1	0.17	0.17	1,273	2,801
900	409	0.0	0.0	18.6	8.5	16.7	7.6	A-F	43.9	0.72	1.58	0.38	0.83	NA	NA	6.3	0.16	0.16	1,146	2,251
								M-F	48.8	0.80	1.76	0.42	0.92	NA	NA	7.0	0.18	0.18	1,273	2,801
1,000	454	0.0	0.0	20.1	9.1	18.1	8.2	A-F	43.9	0.72	1.58	0.38	0.83	NA	NA	6.3	0.16	0.16	1,146	2,521
								M-F	43.8	0.80	1.76	0.42	0.92	NA	NA	7.0	0.18	0.18	1,273	2,801
1,100	500	0.0	0.0	21.7	9.9	19.5	8.9	A-F	43.9	0.72	1.58	0.38	0.83	NA	NA	6.3	0.17	0.17	1,146	2,521
								M-F	48.8	0.80	1.76	0.42	0.92	NA	NA	7.0	0.19	0.19	1,273	2,801
1,200	545	0.0	0.0	23.1	10.5	20.8	9.5	A-F	43.9	0.72	1.58	0.38	0.83	NA	NA	6.2	0.17	0.17	1,146	2,521
								M-F	48.8	0.80	1.76	0.42	0.92	NA	NA	6.9	0.19	0.19	1,273	2,801
1,300	591	0.0	0.0	24.4	11.1	22.0	10.0	A-F	43.9	0.72	1.58	0.38	0.83	NA	NA	6.2	0.18	0.18	1,146	2,521
								M-F	48.8	0.80	1.76	0.42	0.92	NA	NA	6.9	0.20	0.20	1,273	2,801
1,400	636	0.0	0.0	25.9	11.8	23.3	10.6	A-F	43.9	0.72	1.58	0.38	0.83	NA	NA	6.2	0.18	0.18	1,146	2,521
								M-F	48.8	0.80	1.76	0.42	0.92	NA	NA	6.9	0.20	0.20	1,273	2,801

(Continued)

TABLE 19-2 (Continued)

Weight[2]		Daily Gain[3]		Daily Consumption[4]				Moisture Basis[5] A-F (as-fed) M-F (moisture-free)	Energy							Total Protein	Calcium	Phosphorus	Vitamin A (IU per)	
				As-Fed		Moisture-Free Dry Matter			TDN	ME (Mcal per)		NE$_m$ (Mcal per)		NE$_g$ (Mcal per)						
(lb)	(kg)	(lb)	(kg)	(lb)	(kg)	(lb)	(kg)		(%)	(lb)	(kg)	(lb)	(kg)	(lb)	(kg)	(%)	(%)	(%)	(lb)	(kg)
Dry pregnant mature cows—Last third of pregnancy																				
800	364	0.9	0.4	18.7	8.5	16.8	7.6	A-F	49.1	0.80	1.76	0.46	1.01	NA	NA	7.4	0.23	0.18	1,146	2,521
								M-F	54.5	0.89	1.96	0.51	1.12	NA	NA	8.2	0.26	0.20	1,273	2,801
900	409	0.9	0.4	20.2	9.2	18.2	8.3	A-F	48.6	0.80	1.76	0.45	0.99	NA	NA	7.2	0.24	0.19	1,146	2,521
								M-F	54.0	0.89	1.96	0.50	1.10	NA	NA	8.0	0.27	0.21	1,273	2,801
1,000	454	0.9	0.4	21.8	9.9	19.6	8.9	A-F	48.2	0.79	1.75	0.45	0.99	NA	NA	7.1	0.23	0.18	1,146	2,521
								M-F	53.6	0.88	1.94	0.50	1.10	NA	NA	7.9	0.26	0.20	1,273	2,801
1,100	500	0.9	0.4	23.3	10.6	21.0	9.5	A-F	47.9	0.78	1.72	0.44	0.97	NA	NA	7.0	0.23	0.19	1,146	2,521
								M-F	53.2	0.87	1.91	0.49	1.08	NA	NA	7.8	0.26	0.21	1,273	2,801
1,200	545	0.9	0.4	24.8	11.3	22.3	10.1	A-F	47.6	0.78	1.72	0.44	0.97	NA	NA	7.0	0.23	0.19	1,146	2,521
								M-F	52.9	0.87	1.91	0.49	1.08	NA	NA	7.8	0.26	0.21	1,273	2,801
1,300	591	0.9	0.4	26.2	11.9	23.6	10.7	A-F	47.4	0.78	1.72	0.43	0.95	NA	NA	6.9	0.23	0.19	1,146	2,521
								M-F	52.7	0.87	1.91	0.48	1.06	NA	NA	7.7	0.26	0.21	1,273	2,801
1,400	636	0.9	0.4	27.7	12.6	24.9	11.3	A-F	47.3	0.77	1.70	0.43	0.95	NA	NA	6.8	0.23	0.19	1,146	2,521
								M-F	52.5	0.86	1.89	0.48	1.06	NA	NA	7.6	0.26	0.21	1,273	2,801
Two-year-old heifers nursing calves—First 3-4 months postpartum—10 lb (4.5 kg) milk/day																				
700	318	0.5	0.2	17.7	8.0	15.9	7.2	A-F	58.6	0.96	2.12	0.60	1.32	0.36	0.79	10.2	0.32	0.22	1,596	2,521
								M-F	65.1	1.07	2.35	0.67	1.47	0.40	0.88	11.3	0.36	0.24	1,773	2,801
750	341	0.5	0.2	18.6	8.5	16.7	7.6	A-F	58.0	0.95	2.10	0.59	1.31	0.36	0.79	9.9	0.31	0.22	1,596	2,521
								M-F	64.4	1.06	2.33	0.66	1.45	0.40	0.88	11.0	0.34	0.24	1,773	2,801
800	364	0.5	0.2	19.6	8.9	17.6	8.0	A-F	57.4	0.95	2.08	0.59	1.31	0.35	0.77	9.7	0.31	0.22	1,596	2,521
								M-F	63.8	10.5	2.31	0.66	1.45	0.39	0.86	10.8	0.34	0.24	1,773	2,801
850	386	0.5	0.2	20.4	9.3	18.4	8.4	A-F	56.9	0.94	2.06	0.59	1.29	0.34	0.76	9.5	0.30	0.21	1,596	2,521
								M-F	63.2	1.04	2.29	0.65	1.43	0.38	0.84	10.6	0.33	0.23	1,773	2,801
900	409	0.5	0.2	21.3	9.7	19.2	8.7	A-F	56.4	0.93	2.04	0.58	1.27	0.33	0.73	9.4	0.29	0.21	1,596	2,521
								M-F	62.7	1.03	2.27	0.64	1.41	0.37	0.81	10.4	0.32	0.23	1,773	2,801
950	432	0.5	0.2	22.2	10.1	20.0	9.0	A-F	56.1	0.92	2.02	0.57	1.25	0.33	0.73	9.2	0.28	0.21	1,596	2,521
								M-F	62.3	1.02	2.24	0.63	1.39	0.37	0.81	10.2	0.31	0.23	1,773	2,801
1,000	454	0.5	0.2	23.1	10.5	20.8	9.5	A-F	55.7	0.92	2.02	0.56	1.22	0.32	0.71	9.0	0.28	0.21	1,596	2,521
								M-F	61.9	1.02	2.24	0.62	1.36	0.36	0.79	10.0	0.31	0.23	1,773	2,801
Cows nursing calves—Average milking ability—First 3-4 months postpartum—10 lb (4.5 kg) milk/day																				
800	364	0.0	0.0	19.2	8.7	17.3	7.9	A-F	52.4	0.86	1.90	0.51	1.13	NA	NA	9.2	0.27	0.20	1,596	3,511
								M-F	58.2	0.96	2.11	0.57	1.25	NA	NA	10.2	0.30	0.22	1,773	3,901
900	409	0.0	0.0	20.1	9.1	18.8	8.5	A-F	51.6	0.85	1.86	0.50	1.09	NA	NA	8.9	0.25	0.20	1,596	3,511
								M-F	57.3	0.94	2.07	0.55	1.21	NA	NA	9.9	0.28	0.22	1,773	3,901
1,000	454	0.0	0.0	22.4	10.2	20.2	9.2	A-F	50.9	0.84	1.85	0.50	1.09	NA	NA	8.6	0.25	0.20	1,596	3,511
								M-F	56.6	0.93	2.05	0.55	1.21	NA	NA	9.6	0.28	0.22	1,773	3,901
1,100	500	0.0	0.0	24.0	10.9	21.6	9.8	A-F	50.4	0.83	1.82	0.49	1.07	NA	NA	8.5	0.24	0.20	1,596	3,511
								M-F	56.0	0.92	2.02	0.54	1.19	NA	NA	9.4	0.27	0.22	1,773	3,901
1,200	545	0.0	0.0	25.6	11.6	23.0	10.5	A-F	50.0	0.82	1.80	0.48	1.05	NA	NA	8.4	0.24	0.20	1,596	3,511
								M-F	55.5	0.91	2.00	0.53	1.17	NA	NA	9.3	0.27	0.22	1,773	3,901
1,300	591	0.0	0.0	27.0	12.3	24.3	11.0	A-F	49.6	0.81	1.78	0.47	1.03	NA	NA	8.2	0.24	0.20	1,596	3,511
								M-F	55.1	0.90	1.98	0.52	1.14	NA	NA	9.1	0.27	0.22	1,773	3,901
1,400	636	0.0	0.0	28.4	12.9	25.6	11.6	A-F	49.2	0.81	1.78	0.46	1.01	NA	NA	8.1	0.24	0.20	1,596	3,511
								M-F	54.7	0.90	1.98	0.51	1.12	NA	NA	9.0	0.27	0.22	1,773	3,901
Cows nursing calves—Superior milking ability—First 3-4 months postpartum—20 lb (9.1 kg) milk/day																				
800	364	0.0	0.0	17.4	7.9	15.7	7.1	A-F	70.0	1.14	2.51	0.77	1.68	NA	NA	12.8	0.43	0.28	1,596	3,511
								M-F	77.3	1.27	2.79	0.85	1.87	NA	NA	14.2	0.48	0.31	1,773	3,901
900	409	0.0	0.0	20.8	9.5	18.7	8.5	A-F	62.8	1.04	2.28	0.67	1.47	NA	NA	11.6	0.37	0.25	1,596	3,511
								M-F	69.8	1.15	2.53	0.74	1.63	NA	NA	12.9	0.41	0.28	1,773	3,901
1,000	454	0.0	0.0	22.9	10.4	20.6	9.4	A-F	60.3	0.99	2.18	0.63	1.39	NA	NA	11.1	0.35	0.24	1,596	3,511
								M-F	67.0	1.10	2.42	0.70	1.54	NA	NA	12.3	0.39	0.27	1,773	3,901
1,100	500	0.0	0.0	24.8	11.3	22.3	10.1	A-F	58.7	0.96	2.12	0.60	1.32	NA	NA	10.7	0.34	0.24	1,596	3,511
								M-F	65.2	1.07	2.35	0.67	1.47	NA	NA	11.9	0.38	0.27	1,773	3,901
1,200	545	0.0	0.0	26.4	12.0	23.8	10.8	A-F	57.3	0.95	2.08	0.59	1.29	NA	NA	10.4	0.32	0.23	1,596	3,511
								M-F	63.7	1.05	2.31	0.65	1.43	NA	NA	11.5	0.36	0.26	1,773	3,901

(Continued)

Feeding Beef Cattle

TABLE 19-2 *(Continued)*

Weight[2]		Daily Gain[3]		Daily Consumption[4]				Moisture Basis[5] A-F (as-fed) M-F (moisture-free)	Energy							Total Protein	Cal- cium	Phos- phorus	Vitamin A (IU per)	
				As-Fed		Moisture-Free Dry Matter			TDN	ME (Mcal per)		NE$_m$ (Mcal per)		NE$_g$ (Mcal per)						
(lb)	(kg)	(lb)	(kg)	(lb)	(kg)	(lb)	(kg)		(%)	(lb)	(kg)	(lb)	(kg)	(lb)	(kg)	(%)	(%)	(%)	(lb)	(kg)
Cows nursing calves—Superior milking ability—First 3-4 months postpartum—20 lb *(9.1 kg)* milk/day (Continued)																				
1,300	*591*	0.0	*0.0*	28.1	*12.8*	25.3	*11.5*	A-F	56.3	0.93	*2.04*	0.58	*1.27*	NA	*NA*	10.1	0.32	0.23	1,596	*3,511*
								M-F	62.6	1.03	*2.27*	0.64	*1.41*	NA	*NA*	11.2	0.36	0.26	1,773	*3,901*
1,400	*636*	0.0	*0.0*	29.7	*13.5*	26.7	*12.1*	A-F	55.5	0.91	*2.00*	0.56	*1.22*	NA	*NA*	9.9	0.32	0.23	1,596	*3,511*
								M-F	61.7	1.01	*2.22*	0.62	*1.36*	NA	*NA*	11.0	0.35	0.25	1,773	*3,901*
Bulls, maintenance and slow rate of growth (regain body condition)																				
<1,300		For growth and development, use requirements for bulls in Tables 19-3 and 19-4																		
1,300	*591*	1.0	*0.5*	28.2	*12.8*	25.4	*11.5*	A-F	50.2	0.83	*1.82*	0.48	*1.05*	0.25	*0.56*	6.8	0.20	0.17	1,596	*3,511*
								M-F	55.8	0.92	*2.02*	0.53	*1.17*	0.28	*0.62*	7.6	0.22	0.19	1,773	*3,901*
1,300	*591*	1.5	*0.7*	29.0	*13.2*	26.1	*11.9*	A-F	53.7	0.88	*1.94*	0.53	*1.17*	0.30	*0.66*	7.1	0.22	0.17	1,596	*3,511*
								M-F	59.7	0.98	*2.16*	0.59	*1.30*	0.33	*0.73*	7.9	0.24	0.19	1,773	*3,901*
1,300	*591*	2.0	*0.9*	29.1	*13.2*	26.2	*11.9*	A-F	57.6	0.95	*2.08*	0.59	*1.29*	0.35	*0.77*	7.4	0.23	0.18	1,596	*3,511*
								M-F	64.0	1.05	*2.31*	0.65	*1.43*	0.39	*0.86*	8.2	0.26	0.20	1,773	*3,901*
1,400	*636*	1.0	*0.5*	29.8	*13.5*	26.8	*12.2*	A-F	50.2	0.83	*1.82*	0.48	*1.05*	0.25	*0.56*	6.8	0.19	0.17	1,596	*3,511*
								M-F	55.8	0.92	*2.02*	0.53	*1.17*	0.28	*0.62*	7.5	0.21	0.19	1,773	*3,901*
1,400	*636*	1.5	*0.7*	30.7	*14.0*	27.6	*12.5*	A-F	53.7	0.88	*1.94*	0.53	*1.17*	0.30	*0.66*	6.9	0.21	0.17	1,596	*3,511*
								M-F	59.7	0.98	*2.16*	0.59	*1.30*	0.33	*0.73*	7.7	0.23	0.19	1,773	*3,901*
1,400	*636*	2.0	*0.9*	30.8	*14.0*	27.7	*12.6*	A-F	57.6	0.95	*2.08*	0.59	*1.29*	0.35	*0.77*	7.2	0.23	0.18	1,596	*3,511*
								M-F	64.0	1.05	*2.31*	0.65	*1.43*	0.39	*0.86*	8.0	0.25	0.20	1,773	*3,901*
1,500	*682*	0.0	*0.0*	28.0	*12.7*	25.2	*11.5*	A-F	43.6	0.71	*1.57*	0.37	*0.81*	NA	*NA*	6.2	0.18	0.18	1,596	*3,511*
								M-F	48.4	0.79	*1.74*	0.41	*0.90*	NA	*NA*	6.9	0.20	0.20	1,773	*3,901*
1,500	*682*	1.0	*0.5*	31.4	*14.3*	28.3	*12.9*	A-F	50.2	0.83	*1.82*	0.48	*1.05*	0.25	*0.56*	6.7	0.19	0.17	1,596	*3,511*
								M-F	55.8	0.92	*2.02*	0.53	*1.17*	0.28	*0.62*	7.4	0.21	0.19	1,773	*3,901*
1,500	*682*	1.5	*0.7*	32.2	*14.6*	29.0	*13.2*	A-F	53.7	0.88	*1.94*	0.53	*1.17*	0.30	*0.66*	6.8	0.20	0.17	1,596	*3,511*
								M-F	59.7	0.98	*2.16*	0.59	*1.30*	0.33	*0.73*	7.6	0.22	0.19	1,773	*3,901*
1,600	*727*	0.0	*0.0*	29.4	*13.4*	26.5	*12.0*	A-F	43.6	0.71	*1.57*	0.37	*0.81*	NA	*NA*	6.2	0.17	0.18	1,596	*3,511*
								M-F	48.4	0.79	*1.74*	0.41	*0.90*	NA	*NA*	6.9	0.19	0.20	1,773	*3,901*
1,600	*727*	1.0	*0.5*	33.0	*15.0*	29.7	*13.5*	A-F	50.2	0.83	*1.82*	0.48	*1.05*	0.25	*0.56*	6.6	0.20	0.17	1,596	*3,511*
								M-F	55.8	0.92	*2.02*	0.53	*1.17*	0.28	*0.62*	7.3	0.22	0.19	1,773	*3,901*
1,600	*727*	1.5	*0.7*	33.8	*15.4*	30.4	*13.8*	A-F	53.7	0.88	*1.94*	0.53	*1.17*	0.30	*0.66*	6.7	0.20	0.18	1,596	*3,511*
								M-F	59.7	0.98	*2.16*	0.59	*1.30*	0.33	*0.73*	7.4	0.22	0.20	1,773	*3,901*
1,700	*773*	0.0	*0.0*	30.8	*14.0*	27.7	*12.6*	A-F	43.6	0.71	*1.57*	0.37	*0.81*	NA	*NA*	6.1	0.19	0.19	1,596	*3,511*
								M-F	48.4	0.79	*1.74*	0.41	*0.90*	NA	*NA*	6.8	0.21	0.21	1,773	*3,901*
1,700	*773*	0.5	*0.2*	32.9	*15.0*	29.6	*13.5*	A-F	46.8	0.77	*1.63*	0.42	*0.93*	0.20	*0.43*	6.3	0.18	0.17	1,596	*3,511*
								M-F	52.0	0.85	*1.87*	0.47	*1.03*	0.22	*0.48*	7.0	0.20	0.19	1,773	*3,901*
1,800	*818*	0.0	*0.0*	32.1	*14.6*	28.9	*13.1*	A-F	43.6	0.71	*1.57*	0.37	*0.81*	NA	*NA*	6.1	0.19	0.19	1,596	*3,511*
								M-F	48.4	0.79	*1.74*	0.41	*0.90*	NA	*NA*	6.8	0.21	0.21	1,773	*3,901*
1,800	*818*	0.5	*0.2*	34.3	*15.6*	30.9	*14.0*	A-F	46.8	0.77	*1.63*	0.42	*0.93*	0.20	*0.43*	6.3	0.18	0.18	1,596	*3,511*
								M-F	52.0	0.85	*1.87*	0.47	*1.03*	0.22	*0.48*	7.0	0.20	0.20	1,773	*3,901*
1,900	*864*	0.0	*0.0*	33.4	*15.2*	30.1	*13.7*	A-F	43.6	0.71	*1.57*	0.37	*0.81*	NA	*NA*	6.1	0.19	0.19	1,596	*3,511*
								M-F	48.4	0.79	*1.74*	0.41	*0.90*	NA	*NA*	6.8	0.21	0.21	1,773	*3,901*
1,900	*864*	0.5	*0.2*	35.8	*16.3*	32.2	*14.6*	A-F	46.8	0.77	*1.63*	0.42	*0.93*	0.20	*0.43*	6.2	0.18	0.18	1,596	*3,511*
								M-F	52.0	0.85	*1.87*	0.47	*1.03*	0.22	*0.48*	6.9	0.20	0.20	1,773	*3,901*
2,000	*909*	0.0	*0.0*	34.8	*15.8*	31.3	*14.2*	A-F	43.6	0.71	*1.57*	0.37	*0.81*	NA	*NA*	6.1	0.19	0.19	1,596	*3,511*
								M-F	48.4	0.79	*1.74*	0.41	*0.90*	NA	*NA*	6.8	0.21	0.21	1,773	*3,901*
2,100	*955*	0.0	*0.0*	36.1	*16.4*	32.5	*14.8*	A-F	43.6	0.71	*1.57*	0.37	*0.81*	NA	*NA*	6.1	0.20	0.20	1,596	*3,511*
								M-F	48.4	0.79	*1.74*	0.41	*0.90*	NA	*NA*	6.8	0.22	0.22	1,773	*3,901*
2,200	*1,000*	0.0	*0.0*	37.3	*17.0*	33.6	*15.3*	A-F	43.6	0.71	*1.57*	0.37	*0.81*	NA	*NA*	6.1	0.20	0.20	1,596	*3,511*
								M-F	48.4	0.79	*1.74*	0.41	*0.90*	NA	*NA*	6.8	0.22	0.22	1,773	*3,901*

[1]Adapted from *Nutrient Requirements of Beef Cattle*, 6th revised edition, National Research Council—National Academy of Sciences, 1984, p. 85, Table 11.

[2]Average weight for a feeding period.

[3]Approximately 0.9 ± 0.2 lb of weight gain/day over the last third of pregnancy is accounted for by the products of conception. Daily 2.15 Mcal of NE$_m$ and 0.1 lb of protein are provided for this requirement for a calf with a birth weight of 80 lb.

[4]Consumption should vary depending on the energy concentration of the ration and environmental conditions. These intakes are based on the energy concentration shown in the table and assuming a thermoneutral environment without snow or mud conditions. If the energy concentrations of the ration to be fed exceed the tabular value, limit feeding may be required.

[5]As-fed was calculated using an average figure of 90% dry matter. When using silages, roots, and other wet feeds, these feeds should be converted to a moisture-free basis and the ration calculated using the moisture-free data.

[6]Not applicable.

TABLE 19-3
DAILY NUTRIENT REQUIREMENTS OF GROWING AND FINISHING CATTLE [1] [2] (See footnotes at end of table.)

Weight		Daily Gain		Energy		Total Protein	Calcium	Phosphorus
				NE_m	NE_g			
(lb)	(kg)	(lb)	(kg)	(Mcal/day)	(Mcal/day)	(g/day)	(g/day)	(g/day)
Medium-frame steer calves								
330	150	0.4	0.2	3.30	0.41	343	11	7
		0.9	0.4	3.30	0.87	428	16	9
		1.3	0.6	3.30	1.36	503	21	11
		1.8	0.8	3.30	1.87	575	27	12
		2.2	1.0	3.30	2.39	642	32	14
		2.6	1.2	3.30	2.91	702	37	16
440	200	0.4	0.2	4.10	0.50	399	12	9
		0.9	0.4	4.10	1.08	482	17	10
		1.3	0.6	4.10	1.69	554	21	12
		1.8	0.8	4.10	2.32	621	26	13
		2.2	1.0	4.10	2.96	682	31	15
		2.6	1.2	4.10	3.62	735	35	16
550	250	0.4	0.2	4.84	0.60	450	13	10
		0.9	0.4	4.84	1.28	532	17	12
		1.3	0.6	4.84	2.00	601	21	13
		1.8	0.8	4.84	2.74	664	25	14
		2.2	1.0	4.84	3.50	720	29	16
		2.6	1.2	4.84	4.28	766	33	17
660	300	0.4	0.2	5.55	0.69	499	14	12
		0.9	0.4	5.55	1.47	580	18	13
		1.3	0.6	5.55	2.29	646	22	14
		1.8	0.8	5.55	3.14	704	25	15
		2.2	1.0	5.55	4.02	755	29	16
		2.6	1.2	5.55	4.90	794	32	17
770	350	0.4	0.2	6.24	0.77	545	15	13
		0.9	0.4	6.24	1.65	625	19	14
		1.3	0.6	6.24	2.57	688	22	15
		1.8	0.8	6.24	3.53	743	25	16
		2.2	1.0	6.24	4.51	789	28	17
		2.6	1.2	6.24	5.50	822	31	18
880	400	0.4	0.2	6.89	0.85	590	16	15
		0.9	0.4	6.89	1.82	668	19	16
		1.3	0.6	6.89	2.84	728	22	17
		1.8	0.8	6.89	3.90	780	25	17
		2.2	1.0	6.89	4.98	821	27	18
		2.6	1.2	6.89	6.69	848	29	19
990	450	0.4	0.2	7.52	0.93	633	17	16
		0.9	0.4	7.52	1.99	710	20	17
		1.3	0.6	7.52	3.11	767	22	18
		1.8	0.8	7.52	4.26	815	24	19
		2.2	1.0	7.52	5.44	852	26	19
		2.6	1.2	7.52	6.65	873	28	20
1,100	500	0.4	0.2	8.14	1.01	675	19	18
		0.9	0.4	8.14	2.16	751	21	18
		1.3	0.6	8.14	3.36	805	23	19
		1.8	0.8	8.14	4.61	849	24	20
		2.2	1.0	8.14	5.89	882	26	20
		2.6	1.2	8.14	7.19	897	27	21
1,210	550	0.4	0.2	8.75	1.08	715	20	19
		0.9	0.4	8.75	2.32	790	22	20
		1.3	0.6	8.75	3.61	842	23	20
		1.8	0.8	8.75	4.95	883	24	21
		2.2	1.0	8.75	6.23	911	25	21
		2.6	1.2	8.75	7.73	921	26	21

(Continued)

TABLE 19-3 (Continued)

Weight		Daily Gain		Energy		Total Protein	Calcium	Phosphorus
				NE_m	NE_g			
(lb)	(kg)	(lb)	(kg)	(Mcal/day)	(Mcal/day)	(g/day)	(g/day)	(g/day)
Large-frame steer calves, compensating medium-frame yearling steers, and medium-frame bulls[3]								
330	150	0.4	0.2	3.30	0.36	361	11	7
		0.9	0.4	3.30	0.77	441	17	9
		1.3	0.6	3.30	1.21	522	22	11
		1.8	0.8	3.30	1.65	598	28	13
		2.2	1.0	3.30	2.11	671	33	14
		2.6	1.2	3.30	2.58	740	38	16
		3.1	1.4	3.30	3.06	806	44	18
		3.5	1.6	3.30	3.53	863	49	20
440	200	0.4	0.2	4.10	0.45	421	12	9
		0.9	0.4	4.10	0.96	499	17	10
		1.3	0.6	4.10	1.50	576	22	12
		1.8	0.8	4.10	2.06	650	27	14
		2.2	1.0	4.10	2.62	718	32	15
		2.6	1.2	4.10	3.20	782	37	17
		3.1	1.4	4.10	3.79	842	42	18
		3.5	1.6	4.10	4.39	892	47	20
550	250	0.4	0.2	4.84	0.53	476	13	10
		0.9	0.4	4.84	1.13	552	18	12
		1.3	0.6	4.84	1.77	628	23	13
		1.8	0.8	4.84	2.43	698	27	15
		2.2	1.0	4.84	3.10	762	31	16
		2.6	1.2	4.84	3.78	822	36	18
		3.1	1.4	4.84	4.48	877	40	19
		3.5	1.6	4.84	5.19	919	44	20
660	300	0.4	0.2	5.55	0.61	529	14	12
		0.9	0.4	5.55	1.30	603	19	13
		1.3	0.6	5.55	2.03	676	23	15
		1.8	0.8	5.55	2.78	743	27	16
		2.2	1.0	5.55	3.55	804	31	17
		2.6	1.2	5.55	4.34	859	35	18
		3.1	1.4	5.55	5.14	908	38	20
		3.5	1.6	5.55	5.95	943	42	21
770	350	0.4	0.2	6.24	0.68	579	16	13
		0.9	0.4	6.24	1.46	651	19	15
		1.3	0.6	6.24	2.28	722	23	16
		1.8	0.8	6.24	3.12	786	27	17
		2.2	1.0	6.24	3.99	843	30	18
		2.6	1.2	6.24	4.87	895	34	19
		3.1	1.4	6.24	5.77	938	37	20
		3.5	1.6	6.24	6.68	967	40	21
880	400	0.4	0.2	6.89	0.75	627	17	15
		0.9	0.4	6.89	1.61	697	20	16
		1.3	0.6	6.89	2.52	766	24	17
		1.0	0.8	6.89	3.45	828	27	18
		2.2	1.0	6.89	4.41	881	30	19
		2.6	1.2	6.89	5.38	929	33	20
		3.1	1.4	6.89	6.38	967	36	21
		3.5	1.6	6.89	7.38	989	38	22
990	450	0.4	0.2	7.52	0.82	673	18	16
		0.9	0.4	7.52	1.76	742	21	17
		1.3	0.6	7.52	2.75	809	24	18
		1.8	0.8	7.52	3.77	867	27	19
		2.2	1.0	7.52	4.81	918	29	20
		2.6	1.2	7.52	5.88	961	32	21
		3.1	1.4	7.52	6.97	995	34	22
		3.5	1.6	7.52	8.07	1,011	37	22

(Continued)

TABLE 19-3 (Continued)

Weight		Daily Gain		Energy		Total Protein	Calcium	Phosphorus
				NE_m	NE_g			
(lb)	(kg)	(lb)	(kg)	(Mcal/day)	(Mcal/day)	(g/day)	(g/day)	(g/day)
Large-frame steer calves, compensating medium-frame yearling steers, and medium-frame bulls[3] (Continued)									
1,100	500	0.4	0.2	8.14	0.89	719	19	18
		0.9	0.4	8.14	1.01	785	22	19
		1.3	0.6	8.14	2.98	850	24	20
		1.8	0.8	8.14	4.08	906	27	20
		2.2	1.0	8.14	5.21	953	29	21
		2.6	1.2	8.14	6.37	993	31	22
		3.1	1.4	8.14	7.54	1,022	33	22
		3.5	1.6	8.14	8.73	1,031	35	23
1,210	550	0.4	0.2	8.75	0.96	762	20	20
		0.9	0.4	8.75	2.05	827	23	20
		1.3	0.6	8.75	3.20	890	25	21
		1.8	0.8	8.75	4.38	944	27	22
		2.2	1.0	8.75	5.60	988	29	22
		2.6	1.2	8.75	6.84	1,023	30	23
		3.1	1.4	8.75	8.10	1,048	32	23
		3.5	1.6	8.75	9.38	1,052	34	24
1,320	600	0.4	0.2	9.33	1.02	805	22	21
		0.9	0.4	9.33	2.19	867	24	22
		1.3	0.6	9.33	3.41	930	25	22
		1.8	0.8	9.33	4.68	980	27	23
		2.2	1.0	9.33	5.98	1,021	28	23
		2.6	1.2	9.33	7.30	1,053	30	24
		3.1	1.4	9.33	8.64	1,073	31	24
		3.5	1.6	9.33	10.01	1,071	32	24
Large-frame bull calves and compensating large-frame yearling steers									
330	150	0.4	0.2	3.30	0.32	355	11	7
		0.9	0.4	3.30	0.69	438	17	9
		1.3	0.6	3.30	1.07	519	23	11
		1.8	0.8	3.30	1.47	597	28	13
		2.2	1.0	3.30	1.87	673	34	15
		2.6	1.2	3.30	2.29	745	40	17
		3.1	1.4	3.30	2.71	815	45	18
		3.5	1.6	3.30	3.14	880	51	20
		4.0	1.8	3.30	3.56	922	56	22
440	200	0.4	0.2	4.10	0.40	414	12	9
		0.9	0.4	4.10	0.85	494	18	11
		1.3	0.6	4.10	1.33	574	23	12
		1.8	0.8	4.10	1.82	649	28	14
		2.2	1.0	4.10	2.32	721	34	16
		2.6	1.2	4.10	2.84	789	39	17
		3.1	1.4	4.10	3.36	854	44	19
		3.5	1.6	4.10	3.89	912	49	21
		4.0	1.8	4.10	4.43	942	54	22
550	250	0.4	0.2	4.84	0.47	468	13	10
		0.9	0.4	4.84	1.01	547	19	12
		1.3	0.6	4.84	1.57	624	23	14
		1.8	0.8	4.84	2.15	697	28	15
		2.2	1.0	4.84	2.75	765	33	17
		2.6	1.2	4.84	3.36	830	38	18
		3.1	1.4	4.84	3.97	890	42	20
		3.5	1.6	4.84	4.60	943	47	21
		4.0	1.8	4.84	5.23	962	51	22
660	300	0.4	0.2	5.55	0.54	519	15	12
		0.9	0.4	5.55	1.15	597	19	13
		1.3	0.6	5.55	1.80	672	24	15
		1.8	0.8	5.55	2.47	741	28	16
		2.2	1.0	5.55	3.15	807	33	18
		2.6	1.2	5.55	3.85	868	37	19
		3.1	1.4	5.55	4.56	924	41	20
		3.5	1.6	5.55	5.28	971	45	22
		4.0	1.8	5.55	6.00	980	49	23

(Continued)

TABLE 19-3 (Continued)

Weight		Daily Gain		Energy		Total Protein	Calcium	Phosphorus
				NE_m	NE_g			
(lb)	(kg)	(lb)	(kg)	(Mcal/day)	(Mcal/day)	(g/day)	(g/day)	(g/day)
Large-frame bull calves and compensating large-frame yearling steers (Continued)								
770	350	0.4	0.2	6.24	0.60	568	16	13
		0.9	0.4	6.24	1.29	644	20	15
		1.3	0.6	6.24	2.02	718	24	16
		1.8	0.8	6.24	2.77	795	28	18
		2.2	1.0	6.24	3.54	847	32	19
		2.6	1.2	6.24	4.32	904	36	20
		3.1	1.4	6.24	5.11	956	40	21
		3.5	1.6	6.24	5.92	998	44	23
		4.0	1.8	6.24	6.74	997	47	23
880	400	0.4	0.2	6.89	0.67	615	17	15
		0.9	0.4	6.89	1.43	689	21	16
		1.3	0.6	6.89	2.23	761	25	18
		1.8	0.8	6.89	3.06	826	29	19
		2.2	1.0	6.89	3.91	885	32	20
		2.6	1.2	6.89	4.77	939	36	21
		3.1	1.4	6.89	5.65	986	39	22
		3.5	1.6	6.89	6.55	1,024	42	23
		4.0	1.8	6.89	7.45	1,013	45	24
990	450	0.4	0.2	7.52	0.73	661	18	17
		0.9	0.4	7.52	1.56	733	22	18
		1.3	0.6	7.52	2.44	803	25	19
		1.8	0.8	7.52	3.34	866	29	20
		2.2	1.0	7.52	4.27	922	32	21
		2.6	1.2	7.52	5.21	973	35	22
		3.1	1.4	7.52	6.18	1,016	38	23
		3.5	1.6	7.52	7.15	1,048	41	24
		4.0	1.8	7.52	8.13	1,028	44	25
1,100	500	0.4	0.2	8.14	0.79	705	20	18
		0.9	0.4	8.14	1.69	776	23	19
		1.3	0.6	8.14	2.64	844	26	20
		1.8	0.8	8.14	3.62	905	29	21
		2.2	1.0	8.14	4.62	958	32	22
		2.6	1.2	8.14	5.64	1,005	35	23
		3.1	1.4	8.14	6.68	1,045	37	24
		3.5	1.6	8.14	7.74	1,072	40	25
		4.0	1.8	8.14	8.80	1,043	42	25
1,210	550	0.4	0.2	8.75	0.85	747	21	20
		0.9	0.4	8.75	1.82	817	24	21
		1.3	0.6	8.75	2.83	884	27	22
		1.8	0.8	8.75	3.88	942	29	22
		2.2	1.0	8.75	4.96	994	32	23
		2.6	1.2	8.75	6.06	1,037	34	24
		3.1	1.4	8.75	7.18	1,072	36	25
		3.5	1.6	8.75	8.31	1,005	30	25
		4.0	1.8	8.75	9.46	1,057	41	26
1,320	600	0.4	0.2	9.33	0.91	789	22	21
		0.9	0.4	9.33	1.94	857	25	22
		1.3	0.6	9.33	3.02	923	27	23
		1.8	0.8	9.33	4.15	979	30	24
		2.2	1.0	9.33	5.30	1,027	32	24
		2.6	1.2	9.33	6.47	1,067	34	25
		3.1	1.4	9.33	7.66	1,099	36	26
		3.5	1.6	9.33	8.87	1,117	38	26
		4.0	1.8	9.33	10.10	1,071	39	26
Medium-frame heifer calves								
330	150	0.4	0.2	3.30	0.49	323	10	7
		0.9	0.4	3.30	1.05	409	15	9
		1.3	0.6	3.30	1.66	477	20	10
		1.8	0.8	3.30	2.29	537	25	12
		2.2	1.0	3.30	2.94	562	29	13

(Continued)

TABLE 19-3 (Continued)

Weight		Daily Gain		Energy		Total Protein	Calcium	Phosphorus
				NE$_m$	NE$_g$			
(lb)	(kg)	(lb)	(kg)	(Mcal/day)	(Mcal/day)	(g/day)	(g/day)	(g/day)
Medium-frame heifer calves (Continued)								
440	200	0.4	0.2	4.10	0.60	374	11	9
		0.9	0.4	4.10	1.31	459	16	10
		1.3	0.6	4.10	2.06	522	20	11
		1.8	0.8	4.10	2.84	574	23	12
		2.2	1.0	4.10	3.65	583	27	14
550	250	0.4	0.2	4.84	0.71	421	12	10
		0.9	0.4	4.84	1.55	505	16	11
		1.3	0.6	4.84	2.44	563	19	12
		1.8	0.8	4.84	3.36	608	23	13
		2.2	1.0	4.84	4.31	603	26	14
660	300	0.4	0.2	5.55	0.82	465	13	11
		0.9	0.4	5.55	1.77	549	16	12
		1.3	0.6	5.55	2.79	602	19	13
		1.8	0.8	5.55	3.85	640	22	14
		2.2	1.0	5.55	4.94	621	24	15
770	350	0.4	0.2	6.24	0.92	508	14	13
		0.9	0.4	6.24	1.99	591	17	14
		1.3	0.6	6.24	3.13	638	19	14
		1.8	0.8	6.24	4.32	670	21	15
		2.2	1.0	6.24	5.55	638	23	16
880	400	0.4	0.2	6.89	1.01	549	16	14
		0.9	0.4	6.89	2.20	630	17	15
		1.3	0.6	6.89	3.46	674	19	16
		1.8	0.8	6.89	4.78	700	20	16
		2.2	1.0	6.89	6.14	654	22	16
990	450	0.4	0.2	7.52	1.11	588	17	16
		0.9	0.4	7.52	2.40	669	18	16
		1.3	0.6	7.52	3.78	708	19	17
		1.8	0.8	7.52	5.22	728	20	17
		2.2	1.0	7.52	6.70	670	20	17
1,100	500	0.4	0.2	8.14	1.20	626	18	17
		0.9	0.4	8.14	2.60	706	19	18
		1.3	0.6	8.14	4.10	741	19	18
		1.8	0.8	8.14	5.65	755	19	18
		2.2	1.0	8.14	7.25	685	19	18
1,210	550	0.4	0.2	8.75	1.29	662	19	19
		0.9	0.4	8.75	2.79	742	19	19
		1.3	0.6	8.75	4.40	773	19	19
		1.8	0.8	8.75	6.07	781	19	19
		2.2	1.0	8.75	7.79	700	19	19
Large-frame heifer calves and compensating medium-frame yearling heifers								
330	150	0.4	0.2	3.30	0.43	342	11	7
		0.9	0.4	3.30	0.93	426	16	9
		1.3	0.6	3.30	1.47	500	21	10
		1.8	0.8	3.30	2.03	568	26	12
		2.2	1.0	3.30	2.61	630	31	14
		2.6	1.2	3.30	3.19	680	35	15
440	200	0.4	0.2	4.10	0.53	397	12	9
		0.9	0.4	4.10	1.16	480	16	10
		1.3	0.6	4.10	1.83	549	21	12
		1.8	0.8	4.10	2.62	613	25	13
		2.2	1.0	4.10	3.23	668	29	14
		2.6	1.2	4.10	3.97	708	33	16

(Continued)

TABLE 19-3 (Continued)

Weight		Daily Gain		Energy		Total Protein	Calcium	Phosphorus
				NE_m	NE_g			
(lb)	(kg)	(lb)	(kg)	(Mcal/day)	(Mcal/day)	(g/day)	(g/day)	(g/day)
Large-frame heifer calves and compensating medium-frame yearling heifers (Continued)								
550	250	0.4	0.2	4.84	0.63	449	13	10
		0.9	0.4	4.84	1.37	530	17	11
		1.3	0.6	4.84	2.16	596	21	13
		1.8	0.8	4.84	2.98	654	24	14
		2.2	1.0	4.84	3.82	703	28	15
		2.6	1.2	4.84	4.69	734	31	16
660	300	0.4	0.2	5.55	0.72	497	14	12
		0.9	0.4	5.55	1.57	577	17	13
		1.3	0.6	5.55	2.47	639	21	14
		1.8	0.8	5.55	3.41	693	24	15
		2.2	1.0	5.55	4.38	735	27	16
		2.6	1.2	5.55	5.37	758	30	17
770	350	0.4	0.2	6.24	0.81	543	15	13
		0.9	0.4	6.24	1.76	622	18	14
		1.3	0.6	6.24	2.78	681	21	15
		1.8	0.8	6.24	3.83	730	23	16
		2.2	1.0	6.24	4.92	767	26	17
		2.6	1.2	6.24	5.03	781	28	17
880	400	0.4	0.2	6.89	0.90	588	16	15
		0.9	0.4	6.89	1.95	665	19	15
		1.3	0.6	6.89	3.07	721	21	16
		1.8	0.8	6.89	4.24	765	23	17
		2.2	1.0	6.89	5.44	797	25	18
		2.6	1.2	6.89	6.67	803	27	18
990	450	0.4	0.2	7.52	0.98	631	17	16
		0.9	0.4	7.52	2.13	707	19	17
		1.3	0.6	7.52	3.35	759	21	17
		1.8	0.8	7.52	4.63	799	23	18
		2.2	1.0	7.52	5.94	826	24	18
		2.6	1.2	7.52	7.28	824	25	19
1,100	500	0.4	0.2	8.14	1.06	672	18	18
		0.9	0.4	8.14	2.31	747	20	18
		1.3	0.6	8.14	3.63	796	21	19
		1.8	0.8	8.14	5.01	833	22	19
		2.2	1.0	8.14	6.43	854	23	19
		2.6	1.2	8.14	7.88	844	24	20
1,210	550	0.4	0.2	8.75	1.14	712	20	19
		0.9	0.4	8.75	2.47	787	21	20
		1.3	0.6	8.75	3.90	832	22	20
		1.8	0.8	8.75	5.38	865	22	20
		2.2	1.0	8.75	6.91	881	23	20
		2.6	1.2	8.75	8.47	864	23	20
1,320	600	0.4	0.2	9.33	1.21	751	21	21
		0.9	0.4	9.33	2.64	825	22	21
		1.3	0.6	9.33	4.16	867	22	21
		1.8	0.8	9.33	5.74	896	22	21
		2.2	1.0	9.33	7.37	907	22	21
		2.6	1.2	9.33	9.03	883	22	21

[1]Adapted from *Nutrient Requirements of Beef Cattle*, 6th revised edition, National Research Council—National Academy of Sciences, 1984, pp. 40–43, Tables 1, 2, and 3.

[2]Shrunk liveweight basis.

[3]In *Nutrient Requirements of Beef Cattle*, 1984, the energy requirements for large-frame steers, compensating medium-frame yearling steers, and medium-frame bulls are listed together. However, the protein requirements for medium-frame bulls are listed separately from the protein requirements for large-frame steers and compensating medium-frame yearling steers, because the protein requirements for medium-frame bulls are somewhat lower than for large-frame steer calves and compensating medium-frame yearling steers. In Table 19-3, however, the protein requirements of all three groups are listed together in order to facilitate use and save space.

TABLE 19-4
NUTRIENT REQUIREMENTS IN RATION FOR GROWING AND FINISHING CATTLE[1,2,3,4] (See footnotes at end of table.)

| Weight | | Daily Gain | | Daily Consumption | | | | Moisture Basis[5] A-F (as-fed) M-F (moisture-free) | Energy | | | | | | | | Total Protein | Calcium | Phosphorus |
| | | | | As-Fed | | Moisture-Free Dry Matter | | | TDN | | ME (Mcal per) | | NE_m (Mcal per) | | NE_g (Mcal per) | | | | |
(lb)	(kg)	(lb)	(kg)	(lb)	(kg)	(lb)	(kg)		(%)	(lb)	(kg)	(lb)	(kg)	(lb)	(kg)		(%)	(%)	(%)
								Medium-frame steer calves											
300	136	0.5	0.2	8.7	4.0	7.8	3.5	A-F	48.6	0.80	1.76	0.45	1.00	0.23	0.50		8.6	0.28	0.18
								M-F	54.0	0.89	1.96	0.50	1.10	0.25	0.55		9.6	0.31	0.20
		1.0	0.5	9.3	4.2	8.4	3.8	A-F	52.7	0.90	1.90	0.51	1.13	0.28	0.61		10.3	0.41	0.22
								M-F	58.5	0.96	2.11	0.57	1.25	0.31	0.68		11.4	0.45	0.24
		1.5	0.7	9.7	4.4	8.7	4.0	A-F	56.7	0.94	2.06	0.58	1.26	0.34	0.76		11.9	0.52	0.25
								M-F	63.0	1.04	2.29	0.64	1.40	0.38	0.84		13.2	0.58	0.28
		2.0	0.9	9.9	4.5	8.9	4.0	A-F	60.8	1.00	2.20	0.63	1.39	0.40	0.87		13.3	0.65	0.29
								M-F	67.5	1.11	2.44	0.70	1.54	0.44	0.97		14.8	0.72	0.32
		2.5	1.1	9.9	4.5	8.9	4.0	A-F	66.2	1.10	2.42	0.71	1.57	0.46	1.01		15.0	0.78	0.33
								M-F	73.5	1.21	2.66	0.79	1.74	0.51	1.12		16.7	0.87	0.37
		3.0	1.4	8.8	4.0	8.0	3.6	A-F	76.5	1.30	2.86	0.86	1.88	0.58	1.27		18.0	1.02	0.42
								M-F	85.0	1.39	3.06	0.95	2.09	0.64	1.41		19.9	1.13	0.47
400	182	0.5	0.2	10.8	4.9	9.7	4.4	A-F	48.6	0.80	1.76	0.45	1.00	0.23	0.50		8.0	0.24	0.16
								M-F	54.0	0.89	1.96	0.50	1.10	0.25	0.55		8.9	0.27	0.18
		1.0	0.5	11.6	5.3	10.4	4.7	A-F	52.7	0.90	1.90	0.51	1.13	0.28	0.61		9.3	0.34	0.19
								M-F	58.5	0.96	2.11	0.57	1.25	0.31	0.68		10.3	0.38	0.21
		1.5	0.7	9.7	4.4	10.8	4.9	A-F	56.7	0.94	2.06	0.58	1.26	0.34	0.76		10.4	0.42	0.22
								M-F	63.0	1.04	2.29	0.64	1.40	0.38	0.84		11.5	0.47	0.25
		2.0	0.9	12.2	5.5	11.0	5.0	A-F	60.8	1.00	2.20	0.63	1.39	0.40	0.87		11.4	0.50	0.23
								M-F	67.5	1.11	2.44	0.70	1.54	0.44	0.97		12.7	0.56	0.26
		2.5	1.1	12.2	5.5	11.0	5.0	A-F	66.2	1.10	2.42	0.71	1.57	0.46	1.01		12.8	0.61	0.27
								M-F	73.5	1.21	2.66	0.79	1.74	0.51	1.12		14.2	0.68	0.30
		3.0	1.4	11.1	5.0	10.0	4.5	A-F	76.5	1.30	2.86	0.86	1.88	0.58	1.27		14.9	0.77	0.33
								M-F	85.0	1.39	3.06	0.95	2.09	0.64	1.41		16.6	0.86	0.37
500	227	0.5	0.2	12.8	5.8	11.5	5.2	A-F	48.6	0.80	1.76	0.45	1.00	0.23	0.50		7.7	0.23	0.15
								M-F	54.0	0.89	1.96	0.50	1.10	0.25	0.55		8.5	0.25	0.17
		1.0	0.5	13.7	6.2	12.3	6.0	A-F	52.7	0.90	1.90	0.51	1.13	0.28	0.61		8.6	0.29	0.18
								M-F	58.5	0.96	2.11	0.57	1.25	0.31	0.68		9.5	0.32	0.20
		1.5	0.7	14.2	6.5	12.8	5.8	A-F	56.7	0.94	2.06	0.58	1.26	0.34	0.76		9.5	0.36	0.20
								M-F	63.0	1.04	2.29	0.64	1.40	0.38	0.84		10.5	0.40	0.22
		2.0	0.9	14.6	6.6	13.1	6.0	A-F	60.8	1.00	2.20	0.63	1.39	0.40	0.87		10.3	0.42	0.22
								M-F	67.5	1.11	2.44	0.70	1.54	0.44	0.97		11.4	0.47	0.24
		2.5	1.1	14.4	6.5	13.0	6.0	A-F	66.2	1.10	2.42	0.71	1.57	0.46	1.01		11.3	0.50	0.24
								M-F	73.5	1.21	2.66	0.79	1.74	0.51	1.12		12.5	0.56	0.27
		3.0	1.4	13.1	6.0	11.8	5.4	A-F	76.5	1.30	2.86	0.86	1.88	0.58	1.27		13.0	0.62	0.29
								M-F	85.0	1.39	3.06	0.95	2.09	0.64	1.41		14.4	0.69	0.32
600	273	0.5	0.2	14.7	6.7	13.2	6.0	A-F	48.6	0.80	1.76	0.45	1.00	0.23	0.50		7.4	0.20	0.16
								M-F	54.0	0.89	1.96	0.50	1.10	0.25	0.55		8.2	0.23	0.18
		1.0	0.5	15.7	7.1	14.1	6.4	A-F	52.7	0.90	1.90	0.51	1.13	0.28	0.61		8.1	0.25	0.17
								M-F	58.5	0.96	2.11	0.57	1.25	0.31	0.68		9.0	0.28	0.19
		1.5	0.7	16.3	7.4	14.7	6.7	A-F	56.7	0.94	2.06	0.58	1.26	0.34	0.76		8.8	0.31	0.19
								M-F	63.0	1.04	2.29	0.64	1.40	0.38	0.84		9.8	0.35	0.21
		2.0	0.9	16.7	7.6	15.0	6.8	A-F	60.8	1.00	2.20	0.63	1.39	0.40	0.87		9.5	0.36	0.20
								M-F	67.5	1.11	2.44	0.70	1.54	0.44	0.97		10.5	0.40	0.22
		2.5	1.1	16.6	7.5	14.9	6.8	A-F	66.2	1.10	2.42	0.71	1.57	0.46	1.01		10.3	0.41	0.21
								M-F	73.5	1.21	2.66	0.79	1.74	0.51	1.12		11.4	0.46	0.24
		3.0	1.4	15.0	6.8	13.5	6.1	A-F	76.5	1.30	2.86	0.86	1.88	0.58	1.27		11.6	0.51	0.26
								M-F	85.0	1.39	3.06	0.95	2.09	0.64	1.41		12.9	0.57	0.29
700	318	0.5	0.2	16.4	7.5	14.8	6.7	A-F	48.6	0.80	1.76	0.45	1.00	0.23	0.50		7.1	0.20	0.16
								M-F	54.0	0.89	1.96	0.50	1.10	0.25	0.55		7.9	0.22	0.18
		1.0	0.5	17.6	8.0	15.8	7.2	A-F	52.7	0.90	1.90	0.51	1.13	0.28	0.61		7.7	0.24	0.16
								M-F	58.5	0.96	2.11	0.57	1.25	0.31	0.68		8.6	0.27	0.18

(Continued)

TABLE 19-4 (Continued)

Weight		Daily Gain		Daily Consumption				Moisture Basis[5] A-F (as-fed) M-F (moisture-free)	Energy								Total Protein	Calcium	Phosphorus
				As-Fed		Moisture-Free Dry Matter			TDN	ME (Mcal per)		NE$_m$ (Mcal per)		NE$_g$ (Mcal per)					
(lb)	(kg)	(lb)	(kg)	(lb)	(kg)	(lb)	(kg)		(%)	(lb)	(kg)	(lb)	(kg)	(lb)	(kg)		(%)	(%)	(%)
Medium-frame steer calves (Continued)

700 (Continued)	318	1.5	0.7	18.3	8.3	16.5	7.5	A-F	56.7	0.94	2.06	0.58	1.26	0.34	0.76	8.3	0.28	0.18
								M-F	63.0	1.04	2.29	0.64	1.40	0.38	0.84	9.2	0.31	0.20
		2.0	0.9	18.7	8.5	16.8	7.6	A-F	60.8	1.00	2.20	0.63	1.39	0.40	0.87	8.8	0.31	0.19
								M-F	67.5	1.11	2.44	0.70	1.54	0.44	0.97	9.8	0.34	0.21
		2.5	1.1	18.6	8.5	16.7	7.6	A-F	66.2	1.10	2.42	0.71	1.57	0.46	1.01	9.5	0.36	0.20
								M-F	73.5	1.21	2.66	0.79	1.74	0.51	1.12	10.5	0.40	0.22
		3.0	1.4	16.9	7.7	15.2	6.9	A-F	76.5	1.30	2.86	0.86	1.88	0.58	1.27	10.5	0.44	0.23
								M-F	85.0	1.39	3.06	0.95	2.09	0.64	1.41	11.7	0.49	0.26
800	364	0.5	0.2	18.2	8.3	16.4	7.5	A-F	48.6	0.80	1.76	0.45	1.00	0.23	0.50	6.9	0.20	0.15
								M-F	54.0	0.89	1.96	0.50	1.10	0.25	0.55	7.7	0.22	0.17
		1.0	0.5	19.4	8.8	17.5	8.0	A-F	52.7	0.86	1.90	0.51	1.13	0.28	0.61	7.5	0.22	0.17
								M-F	58.5	0.96	2.11	0.57	1.25	0.31	0.68	8.3	0.24	0.19
		1.5	0.7	20.2	9.2	18.2	8.3	A-F	56.7	0.94	2.06	0.58	1.27	0.34	0.76	7.9	0.25	0.17
								M-F	63.0	1.04	2.29	0.64	1.41	0.38	0.84	8.8	0.28	0.19
		2.0	0.9	20.7	9.4	18.6	8.5	A-F	60.8	1.10	2.20	0.63	1.39	0.40	0.87	8.3	0.28	0.18
								M-F	67.5	1.11	2.44	0.70	1.54	0.44	0.97	9.2	0.31	0.20
		2.5	1.1	20.6	9.4	18.5	8.4	A-F	66.2	1.09	2.40	0.71	1.57	0.45	1.01	8.8	0.32	0.19
								M-F	73.5	1.21	2.66	0.79	1.74	0.51	1.12	9.8	0.35	0.21
		3.0	1.4	18.7	8.5	16.8	7.6	A-F	76.5	1.25	2.75	0.86	1.88	0.58	1.26	9.7	0.38	0.23
								M-F	85.0	1.39	3.06	0.95	2.09	0.64	1.40	10.8	0.42	0.25
900	409	0.5	0.2	19.9	9.0	17.9	8.1	A-F	48.6	0.80	1.76	0.45	0.90	0.23	0.50	6.8	0.19	0.16
								M-F	54.0	0.89	1.96	0.50	1.10	0.25	0.55	7.6	0.21	0.18
		1.0	0.5	21.2	9.6	19.1	8.7	A-F	52.7	0.86	1.90	0.51	1.13	0.28	0.61	7.2	0.21	0.16
								M-F	58.5	0.96	2.11	0.57	1.25	0.31	0.68	8.0	0.23	0.18
		1.5	0.7	22.1	10.0	19.9	9.0	A-F	56.7	0.94	2.06	0.58	1.26	0.34	0.76	7.6	0.23	0.17
								M-F	63.0	1.04	2.29	0.64	1.40	0.38	0.84	8.4	0.25	0.19
		2.0	0.9	22.6	10.3	20.3	9.2	A-F	60.8	1.10	2.20	0.63	1.39	0.40	0.87	7.9	0.25	0.18
								M-F	67.5	1.11	2.44	0.70	1.54	0.44	0.97	8.8	0.28	0.20
		2.5	1.1	22.4	10.2	20.2	9.2	A-F	66.2	1.09	2.40	0.71	1.57	0.46	1.01	8.4	0.28	0.18
								M-F	73.5	1.21	2.66	0.79	1.74	0.51	1.12	9.3	0.31	0.20
		3.0	1.4	20.3	9.2	18.3	8.3	A-F	76.5	1.25	2.75	0.86	1.88	0.58	1.26	9.0	0.33	0.21
								M-F	85.0	1.39	3.06	0.95	2.09	0.64	1.40	10.1	0.37	0.23
1,000	454	0.5	0.2	21.4	9.7	19.3	8.8	A-F	48.6	0.80	1.76	0.45	0.90	0.23	0.50	6.8	0.19	0.16
								M-F	54.0	0.89	1.96	0.50	1.10	0.25	0.55	7.5	0.21	0.18
		1.0	0.5	23.0	10.5	20.7	9.4	A-F	32.7	0.86	1.90	0.51	1.13	0.28	0.61	7.0	0.19	0.16
								M-F	58.5	0.96	2.11	0.57	1.25	0.31	0.68	7.8	0.21	0.18
		1.5	0.7	23.9	10.9	21.5	9.8	A-F	56.7	0.94	2.06	0.58	1.27	0.34	0.76	7.3	0.22	0.16
								M-F	63.0	1.04	2.29	0.64	1.41	0.38	0.84	8.1	0.24	0.18
		2.0	0.9	24.4	11.1	22.0	10.0	A-F	60.8	1.10	2.20	0.63	1.39	0.40	0.87	7.6	0.23	0.17
								M-F	67.5	1.11	2.44	0.70	1.54	0.44	0.97	8.4	0.25	0.19
		2.5	1.1	24.3	11.0	21.9	10.0	A-F	66.2	1.09	2.40	0.71	1.57	0.46	1.01	7.9	0.24	0.17
								M-F	73.5	1.21	2.66	0.79	1.74	0.51	1.12	8.8	0.27	0.19
		3.0	1.4	22.0	10.0	19.8	9.0	A-F	76.5	1.25	2.75	0.86	1.88	0.58	1.26	8.6	0.29	0.20
								M-F	85.0	1.39	3.06	0.95	2.09	0.64	1.40	9.5	0.32	0.22
Large-frame steer calves and compensating medium-frame yearling steers
300	136	0.5	0.2	9.1	4.1	8.2	3.7	A-F	47.3	0.77	1.70	0.43	0.95	0.21	0.50	8.6	0.27	0.17
								M-F	52.5	0.86	1.89	0.48	1.06	0.23	0.51	9.5	0.30	0.19
		1.0	0.5	9.7	4.4	8.7	3.9	A-F	50.4	0.83	1.82	0.49	1.07	0.25	0.56	10.2	0.41	0.21
								M-F	56.0	0.92	2.02	0.54	1.19	0.28	0.62	11.3	0.46	0.23
		1.5	0.7	10.1	4.6	9.1	4.1	A-F	53.6	0.88	1.94	0.53	1.17	0.30	0.66	11.6	0.52	0.24
								M-F	59.5	0.98	2.16	0.59	1.30	0.33	0.73	12.9	0.58	0.27
		2.0	0.9	10.4	4.7	9.4	4.3	A-F	57.2	0.94	2.06	0.58	1.26	0.34	0.76	13.1	0.63	0.27
								M-F	63.5	1.04	2.29	0.64	1.40	0.38	0.84	14.6	0.70	0.30

(Continued)

TABLE 19-4 (Continued)

Weight		Daily Gain		Daily Consumption				Moisture Basis[5] A-F (as-fed) M-F (moisture-free)	Energy								Total Protein	Calcium	Phosphorus
				As-Fed		Moisture-Free Dry Matter			TDN	ME (Mcal per)		NE_m (Mcal per)		NE_g (Mcal per)					
(lb)	(kg)	(lb)	(kg)	(lb)	(kg)	(lb)	(kg)		(%)	(lb)	(kg)	(lb)	(kg)	(lb)	(kg)		(%)	(%)	(%)

Large-frame steer calves and compensating medium-frame yearling steers (Continued)

Weight (lb/kg)	Daily Gain (lb/kg)	As-Fed (lb/kg)	DM (lb/kg)	Basis	TDN %	ME (lb/kg)	NE_m (lb/kg)	NE_g (lb/kg)	Protein %	Ca %	P %
300 / 136 (Continued)	2.5 / 1.1	10.7 / 4.9	9.6 / 4.4	A-F	60.8	1.10 / 2.20	0.63 / 1.39	0.40 / 0.87	14.7	0.77	0.31
				M-F	67.5	1.11 / 2.44	0.70 / 1.54	0.44 / 0.97	16.3	0.85	0.34
	3.0 / 1.4	10.7 / 4.9	9.6 / 4.4	A-F	64.8	1.06 / 2.34	0.69 / 1.52	0.44 / 0.97	16.2	0.89	0.35
				M-F	72.0	1.18 / 2.60	0.77 / 1.69	0.49 / 1.08	18.0	0.99	0.39
	3.5 / 1.6	10.3 / 4.7	9.3 / 4.2	A-F	70.7	1.16 / 2.56	0.77 / 1.70	0.51 / 1.13	18.3	1.04	0.41
				M-F	78.5	1.29 / 2.84	0.86 / 1.89	0.57 / 1.25	20.3	1.16	0.45
400 / 182	0.5 / 0.2	11.2 / 5.1	10.1 / 4.6	A-F	47.3	0.77 / 1.70	0.43 / 0.95	0.21 / 0.46	8.0	0.23	0.15
				M-F	52.5	0.86 / 1.89	0.48 / 1.06	0.23 / 0.51	8.9	0.26	0.17
	1.0 / 0.5	9.7 / 4.4	10.8 / 4.9	A-F	50.4	0.83 / 1.82	0.49 / 1.07	0.25 / 0.56	9.2	0.33	0.18
				M-F	56.0	0.92 / 2.02	0.54 / 1.19	0.28 / 0.62	10.2	0.37	0.20
	1.5 / 0.7	12.6 / 5.7	11.3 / 5.1	A-F	53.6	0.88 / 1.94	0.53 / 1.17	0.30 / 0.66	10.3	0.42	0.21
				M-F	59.5	0.98 / 2.16	0.59 / 1.30	0.33 / 0.73	11.4	0.47	0.23
	2.0 / 0.9	13.0 / 5.9	11.7 / 5.3	A-F	57.2	0.94 / 2.06	0.58 / 1.26	0.34 / 0.76	11.4	0.51	0.23
				M-F	63.5	1.04 / 2.29	0.64 / 1.40	0.38 / 0.84	12.7	0.57	0.26
	2.5 / 1.1	13.2 / 6.0	11.9 / 5.4	A-F	60.8	1.10 / 2.20	0.63 / 1.39	0.40 / 0.87	12.5	0.59	0.27
				M-F	67.5	1.11 / 2.44	0.70 / 1.54	0.44 / 0.97	13.9	0.65	0.30
	3.0 / 1.4	13.2 / 6.0	11.9 / 5.4	A-F	64.8	1.06 / 2.34	0.59 / 1.52	0.44 / 0.97	13.7	0.68	0.30
				M-F	72.0	1.18 / 2.60	0.77 / 1.69	0.49 / 1.08	15.2	0.76	0.33
	3.5 / 1.6	12.8 / 5.8	11.5 / 5.2	A-F	70.7	1.16 / 2.56	0.77 / 1.70	0.51 / 1.13	15.2	0.81	0.32
				M-F	78.5	1.29 / 2.84	0.86 / 1.89	0.57 / 1.25	16.9	0.90	0.36
500 / 227	0.5 / 0.2	13.3 / 6.0	12.0 / 5.5	A-F	47.3	0.77 / 1.70	0.43 / 0.95	0.21 / 0.46	7.7	0.22	0.15
				M-F	52.5	0.86 / 1.89	0.48 / 1.06	0.23 / 0.51	8.5	0.24	0.17
	1.0 / 0.5	14.2 / 6.5	12.8 / 5.8	A-F	50.4	0.83 / 1.82	0.49 / 1.07	0.25 / 0.56	8.6	0.30	0.17
				M-F	56.0	0.92 / 2.02	0.54 / 1.19	0.28 / 0.62	9.5	0.33	0.19
	1.5 / 0.7	14.9 / 6.8	13.4 / 6.1	A-F	53.6	0.88 / 1.94	0.53 / 1.17	0.30 / 0.66	9.4	0.35	0.19
				M-F	59.5	0.98 / 2.16	0.59 / 1.30	0.33 / 0.73	10.4	0.39	0.21
	2.0 / 0.9	15.3 / 7.0	13.8 / 6.3	A-F	57.2	0.94 / 2.06	0.58 / 1.27	0.34 / 0.76	10.3	0.41	0.22
				M-F	63.5	1.04 / 2.29	0.64 / 1.41	0.38 / 0.84	11.4	0.46	0.24
	2.5 / 1.1	15.6 / 7.1	14.0 / 6.4	A-F	60.8	1.00 / 2.20	0.63 / 1.39	0.40 / 0.87	11.2	0.50	0.23
				M-F	67.5	1.11 / 2.44	0.70 / 1.54	0.44 / 0.97	12.4	0.55	0.25
	3.0 / 1.4	15.6 / 7.1	14.0 / 6.4	A-F	64.8	1.06 / 2.34	0.69 / 1.52	0.44 / 0.97	12.1	0.57	0.25
				M-F	72.0	1.18 / 2.60	0.77 / 1.69	0.49 / 1.08	13.4	0.63	0.28
	3.5 / 1.6	15.1 / 6.9	13.6 / 6.2	A-F	70.7	1.16 / 2.56	0.77 / 1.70	0.51 / 1.13	13.2	0.66	0.29
				M-F	78.5	1.29 / 2.84	0.86 / 1.89	0.57 / 1.25	14.7	0.73	0.32
600 / 273	0.5 / 0.2	15.3 / 7.0	13.8 / 6.3	A-F	47.3	0.77 / 1.70	0.43 / 0.95	0.21 / 0.46	7.4	0.20	0.16
				M-F	52.5	0.86 / 1.89	0.48 / 1.06	0.23 / 0.51	8.2	0.22	0.18
	1.0 / 0.5	16.2 / 7.4	14.6 / 6.6	A-F	50.4	0.83 / 1.82	0.49 / 1.07	0.25 / 0.56	8.1	0.26	0.16
				M-F	56.0	0.92 / 2.02	0.54 / 1.19	0.28 / 0.62	9.0	0.29	0.18
	1.5 / 0.7	17.0 / 7.7	15.3 / 7.0	A-F	53.6	0.88 / 1.94	0.53 / 1.17	0.30 / 0.66	8.7	0.32	0.18
				M-F	59.5	0.98 / 2.16	0.59 / 1.30	0.33 / 0.73	9.7	0.35	0.20
	2.0 / 0.9	17.6 / 8.0	15.8 / 7.2	A-F	57.2	0.94 / 2.06	0.58 / 1.27	0.34 / 0.76	9.5	0.36	0.20
				M-F	63.5	1.04 / 2.29	0.64 / 1.41	0.38 / 0.84	10.5	0.40	0.22
	2.5 / 1.1	17.9 / 8.1	16.1 / 7.3	A-F	60.8	1.00 / 2.20	0.63 / 1.39	0.40 / 0.87	10.2	0.42	0.21
				M-F	67.5	1.11 / 2.44	0.70 / 1.54	0.44 / 0.97	11.3	0.47	0.23
	3.0 / 1.4	17.9 / 8.1	16.1 / 7.3	A-F	64.8	1.06 / 2.34	0.69 / 1.52	0.44 / 0.97	10.9	0.47	0.23
				M-F	72.0	1.18 / 2.60	0.77 / 1.69	0.49 / 1.08	12.1	0.52	0.26
	3.5 / 1.6	17.3 / 7.9	15.6 / 7.1	A-F	70.7	1.16 / 2.56	0.77 / 1.70	0.51 / 1.13	11.9	0.55	0.25
				M-F	78.5	1.29 / 2.84	0.86 / 1.89	0.57 / 1.25	13.2	0.61	0.28
700 / 318	0.5 / 0.2	17.1 / 7.8	15.4 / 7.0	A-F	47.3	0.77 / 1.70	0.43 / 0.95	0.21 / 0.46	7.1	0.19	0.15
				M-F	52.5	0.86 / 1.89	0.48 / 1.06	0.23 / 0.51	7.9	0.21	0.17
	1.0 / 0.5	18.2 / 8.3	16.4 / 7.5	A-F	50.4	0.83 / 1.82	0.49 / 1.07	0.25 / 0.56	7.7	0.24	0.17
				M-F	56.0	0.92 / 2.02	0.54 / 1.19	0.28 / 0.62	8.6	0.27	0.19

(Continued)

TABLE 19-4 (Continued)

| Weight | | Daily Gain | | Daily Consumption | | | | Moisture Basis[5] A-F (as-fed) M-F (moisture-free) | Energy | | | | | | | Total Protein | Calcium | Phosphorus |
| | | | | As-Fed | | Moisture-Free Dry Matter | | | TDN | ME (Mcal per) | | NE$_m$ (Mcal per) | | NE$_g$ (Mcal per) | | | | |
(lb)	(kg)	(lb)	(kg)	(lb)	(kg)	(lb)	(kg)		(%)	(lb)	(kg)	(lb)	(kg)	(lb)	(kg)	(%)	(%)	(%)
\multicolumn{19}{c}{Large-frame steer calves and compensating medium-frame yearling steers (Continued)}																		
700 (Continued)	318	1.5	0.7	19.1	8.7	17.2	7.8	A-F	53.6	0.88	1.94	0.53	1.17	0.30	0.66	8.3	0.28	0.17
								M-F	59.5	0.98	2.16	0.59	1.30	0.33	0.73	9.2	0.31	0.19
		2.0	0.9	19.8	9.0	17.8	8.1	A-F	57.2	0.94	2.06	0.58	1.27	0.34	0.76	8.8	0.32	0.19
								M-F	63.5	1.04	2.29	0.64	1.41	0.38	0.84	9.8	0.36	0.21
		2.5	1.1	20.0	9.1	18.0	8.2	A-F	60.8	1.00	2.20	0.63	1.39	0.40	0.87	9.5	0.36	0.20
								M-F	67.5	1.11	2.44	0.70	1.54	0.44	0.97	10.5	0.40	0.22
		3.0	1.4	20.0	9.1	18.0	8.2	A-F	64.8	1.06	2.34	0.69	1.52	0.44	0.97	10.0	0.41	0.21
								M-F	72.0	1.18	2.60	0.77	1.69	0.49	1.08	11.1	0.45	0.23
		3.5	1.6	19.4	8.8	17.5	8.0	A-F	70.7	1.16	2.56	0.77	1.70	0.51	1.13	10.8	0.47	0.23
								M-F	78.5	1.29	2.84	0.86	1.89	0.57	1.25	12.0	0.52	0.26
800	364	0.5	0.2	19.0	8.6	17.1	7.8	A-F	47.3	0.77	1.70	0.43	0.95	0.21	0.46	6.9	0.19	0.16
								M-F	52.5	0.86	1.89	0.48	1.06	0.23	0.51	7.7	0.21	0.18
		1.0	0.5	20.2	9.2	18.2	8.3	A-F	50.4	0.83	1.82	0.49	1.07	0.25	0.56	7.5	0.22	0.16
								M-F	56.0	0.92	2.02	0.54	1.19	0.28	0.62	8.3	0.24	0.18
		1.5	0.7	21.1	9.6	19.0	8.6	A-F	53.6	0.88	1.94	0.53	1.17	0.30	0.66	7.9	0.25	0.17
								M-F	59.5	0.98	2.16	0.59	1.30	0.33	0.73	8.8	0.28	0.19
		2.0	0.9	21.8	9.9	19.6	8.9	A-F	57.2	0.94	2.06	0.58	1.27	0.34	0.76	8.4	0.29	0.18
								M-F	63.5	1.04	2.29	0.64	1.41	0.38	0.84	9.3	0.32	0.20
		2.5	1.1	22.1	10.0	19.9	9.0	A-F	60.8	1.00	2.20	0.63	1.39	0.40	0.87	8.8	0.32	0.19
								M-F	67.5	1.11	2.44	0.70	1.54	0.44	0.97	9.8	0.35	0.21
		3.0	1.4	22.1	10.0	19.9	9.0	A-F	64.8	1.06	2.34	0.69	1.52	0.44	0.97	9.4	0.36	0.20
								M-F	72.0	1.18	2.60	0.77	1.69	0.49	1.08	10.4	0.40	0.22
		3.5	1.6	21.4	9.7	19.3	8.8	A-F	70.7	1.16	2.56	0.77	1.70	0.51	1.13	10.0	0.41	0.22
								M-F	78.5	1.29	2.84	0.86	1.89	0.57	1.25	11.1	0.45	0.24
900	409	0.5	0.2	20.7	9.4	18.6	8.5	A-F	47.3	0.77	1.70	0.43	0.86	0.21	0.46	6.8	0.18	0.16
								M-F	52.5	0.86	1.89	0.48	0.96	0.23	0.51	7.6	0.20	0.18
		1.0	0.5	22.0	10.0	19.8	9.0	A-F	50.4	0.83	1.82	0.49	1.07	0.25	0.56	7.2	0.20	0.16
								M-F	56.0	0.92	2.02	0.54	1.19	0.28	0.62	8.0	0.23	0.18
		1.5	0.7	23.1	10.5	20.8	9.5	A-F	53.6	0.88	1.94	0.53	1.17	0.30	0.66	7.7	0.24	0.16
								M-F	59.5	0.98	2.16	0.59	1.30	0.33	0.73	8.5	0.27	0.18
		2.0	0.9	23.8	10.8	21.4	9.7	A-F	57.2	0.94	2.06	0.58	1.26	0.34	0.76	8.0	0.26	0.18
								M-F	63.5	1.04	2.29	0.64	1.40	0.38	0.84	8.9	0.29	0.20
		2.5	1.1	24.2	11.0	21.8	9.9	A-F	60.8	1.00	2.20	0.63	1.39	0.40	0.87	8.4	0.28	0.18
								M-F	67.5	1.11	2.44	0.70	1.54	0.44	0.97	9.3	0.31	0.20
		3.0	1.4	24.1	11.0	21.7	9.9	A-F	64.8	1.06	2.34	0.69	1.52	0.44	0.97	8.8	0.32	0.19
								M-F	72.0	1.18	2.60	0.77	1.69	0.49	1.08	9.8	0.36	0.21
		3.5	1.6	23.4	10.6	21.1	9.6	A-F	70.7	1.16	2.56	0.77	1.70	0.51	1.13	9.4	0.30	0.21
								M-F	78.5	1.29	2.84	0.86	1.89	0.57	1.25	10.4	0.40	0.23
1,000	454	0.5	0.2	22.4	10.2	20.2	9.2	A-F	47.3	0.77	1.70	0.43	0.86	0.21	0.46	6.8	0.18	0.15
								M-F	52.5	0.86	1.89	0.48	0.96	0.23	0.51	7.5	0.20	0.17
		1.0	0.5	23.9	10.9	21.5	9.8	A-F	50.4	0.83	1.82	0.49	1.07	0.25	0.56	7.0	0.20	0.15
								M-F	56.0	0.92	2.02	0.54	1.19	0.28	0.62	7.8	0.23	0.17
		1.5	0.7	25.0	11.4	22.5	10.2	A-F	53.6	0.88	1.94	0.53	1.17	0.30	0.66	7.4	0.23	0.16
								M-F	59.5	0.98	2.16	0.59	1.30	0.33	0.73	8.2	0.25	0.18
		2.0	0.9	25.8	11.7	23.2	10.5	A-F	57.2	0.94	2.06	0.58	1.26	0.34	0.76	7.7	0.24	0.16
								M-F	63.5	1.04	2.29	0.64	1.40	0.38	0.84	8.6	0.27	0.18
		2.5	1.1	26.2	11.9	23.6	10.7	A-F	60.8	1.00	2.20	0.63	1.39	0.40	0.87	8.0	0.26	0.17
								M-F	67.5	1.11	2.44	0.70	1.54	0.44	0.97	8.9	0.29	0.19
		3.0	1.4	26.2	11.9	23.6	10.7	A-F	64.8	1.06	2.34	0.69	1.52	0.44	0.97	8.4	0.29	0.18
								M-F	72.0	1.18	2.60	0.77	1.69	0.49	1.08	9.3	0.32	0.20
		3.5	1.6	25.3	11.5	22.8	10.4	A-F	70.7	1.16	2.56	0.77	1.70	0.51	1.13	8.8	0.32	0.19
								M-F	78.5	1.29	2.84	0.86	1.89	0.57	1.25	9.8	0.35	0.21

(Continued)

TABLE 19-4 (Continued)

Weight		Daily Gain		Daily Consumption				Moisture Basis[5] A-F (as-fed) M-F (moisture-free)	Energy								Total Protein	Calcium	Phosphorus
				As-Fed		Moisture-Free Dry Matter			TDN	ME (Mcal per)		NE$_m$ (Mcal per)		NE$_g$ (Mcal per)					
(lb)	(kg)	(lb)	(kg)	(lb)	(kg)	(lb)	(kg)		(%)	(lb)	(kg)	(lb)	(kg)	(lb)	(kg)	(%)	(%)	(%)
Large-frame steer calves and compensating medium-frame yearling steers (Continued)
1,100	500	0.5	0.2	24.1	11.0	21.7	9.9	A-F	49.3	0.77	1.70	0.43	0.86	0.21	0.46	6.7	0.17	0.16
								M-F	52.5	0.86	1.89	0.48	0.96	0.23	0.51	7.4	0.19	0.18
		1.0	0.5	25.7	11.7	23.1	10.5	A-F	50.4	0.83	1.82	0.49	1.07	0.25	0.56	6.9	0.19	0.16
								M-F	56.0	0.92	2.02	0.54	1.19	0.28	0.62	7.7	0.21	0.18
		1.5	0.7	26.8	12.2	24.1	11.0	A-F	53.6	0.88	1.94	0.53	1.17	0.30	0.66	7.2	0.21	0.16
								M-F	59.5	0.98	2.16	0.59	1.30	0.33	0.73	8.0	0.23	0.18
		2.0	0.9	27.7	12.6	24.9	11.3	A-F	57.2	0.94	2.06	0.58	1.26	0.34	0.76	7.5	0.23	0.16
								M-F	63.5	1.04	2.29	0.64	1.40	0.38	0.84	8.3	0.25	0.18
		2.5	1.1	28.1	12.8	25.3	11.5	A-F	60.8	1.00	2.20	0.63	1.39	0.40	0.87	7.7	0.23	0.16
								M-F	67.5	1.11	2.44	0.70	1.54	0.44	0.97	8.5	0.26	0.18
		3.0	1.4	28.1	12.8	25.3	11.5	A-F	64.8	1.06	2.34	0.69	1.52	0.44	0.97	8.0	0.26	0.17
								M-F	72.0	1.18	2.60	0.77	1.69	0.49	1.08	8.9	0.29	0.19
		3.5	1.6	27.2	12.4	24.5	11.1	A-F	70.7	1.16	2.56	0.77	1.70	0.51	1.13	8.4	0.29	0.19
								M-F	78.5	1.29	2.84	0.86	1.89	0.57	1.25	9.3	0.32	0.21
Medium-frame bulls
300	136	0.5	0.2	8.7	4.0	7.8	3.5	A-F	48.2	0.79	1.75	0.44	0.97	0.22	0.48	8.7	0.28	0.18
								M-F	53.5	0.88	1.94	0.49	1.08	0.24	0.53	9.7	0.31	0.20
		1.0	0.5	9.2	4.2	8.3	3.8	A-F	51.8	0.85	1.86	0.50	1.11	0.27	0.59	10.4	0.43	0.22
								M-F	57.5	0.94	2.07	0.56	1.23	0.30	0.66	11.6	0.48	0.24
		1.5	0.7	9.6	4.4	8.6	3.9	A-F	55.4	0.91	2.00	0.56	1.22	0.32	0.69	12.1	0.56	0.25
								M-F	61.5	1.01	2.22	0.62	1.36	0.35	0.77	13.4	0.62	0.28
		2.0	0.9	9.8	4.5	8.8	4.0	A-F	59.0	0.97	2.14	0.61	1.35	0.40	0.81	13.7	0.68	0.30
								M-F	65.5	1.08	2.38	0.68	1.50	0.41	0.90	15.2	0.75	0.33
		2.5	1.1	9.9	4.5	8.9	4.0	A-F	63.0	1.04	2.28	0.67	1.47	0.42	0.93	15.3	0.83	0.33
								M-F	70.0	1.15	2.53	0.74	1.63	0.47	1.03	17.0	0.92	0.37
		3.0	1.4	9.7	4.4	8.7	4.0	A-F	68.9	1.13	2.49	0.76	1.67	0.49	1.07	17.4	0.98	0.39
								M-F	76.5	1.26	2.77	0.84	1.85	0.54	1.19	19.3	1.09	0.43
400	182	0.5	0.2	10.7	4.9	9.6	4.4	A-F	48.2	0.79	1.75	0.44	0.97	0.22	0.48	8.1	0.25	0.16
								M-F	53.5	0.88	1.94	0.49	1.08	0.24	0.53	9.0	0.28	0.18
		1.0	0.5	11.4	5.2	10.3	4.7	A-F	51.8	0.85	1.86	0.50	1.11	0.27	0.59	9.4	0.35	0.19
								M-F	57.5	0.94	2.07	0.56	1.23	0.30	0.66	10.4	0.39	0.21
		1.5	0.7	11.9	5.4	10.7	4.9	A-F	55.4	0.91	2.00	0.56	1.22	0.32	0.69	10.6	0.44	0.23
								M-F	61.5	1.01	2.22	0.62	1.36	0.35	0.77	11.8	0.49	0.25
		2.0	0.9	12.2	5.5	11.0	5.0	A-F	59.0	0.97	2.14	0.61	1.35	0.37	0.81	11.8	0.54	0.25
								M-F	65.5	1.08	2.38	0.68	1.50	0.41	0.90	13.1	0.60	0.28
		2.5	1.1	12.3	5.6	11.1	5.0	A-F	63.0	1.04	2.28	0.67	1.47	0.42	0.93	13.0	0.63	0.29
								M-F	70.0	1.15	2.53	0.74	1.63	0.47	1.03	14.4	0.70	0.32
		3.0	1.4	9.7	4.4	10.8	4.9	A-F	68.9	1.13	2.49	0.76	1.67	0.49	1.07	14.5	0.76	0.33
								M-F	76.5	1.26	2.77	0.84	1.85	0.54	1.19	16.1	0.84	0.37
500	227	0.5	0.2	12.7	5.8	11.4	5.2	A-F	48.2	0.79	1.75	0.44	0.97	0.22	0.48	7.7	0.23	0.15
								M-F	53.5	0.88	1.94	0.49	1.08	0.24	0.53	8.6	0.25	0.17
		1.0	0.5	13.4	6.1	12.1	5.5	A-F	51.8	0.85	1.86	0.50	1.11	0.27	0.59	8.7	0.32	0.18
								M-F	57.5	0.94	2.07	0.56	1.23	0.30	0.66	9.7	0.35	0.20
		1.5	0.7	14.1	6.4	12.7	5.8	A-F	55.4	0.91	2.00	0.56	1.22	0.32	0.69	9.6	0.38	0.21
								M-F	61.5	1.01	2.22	0.62	1.36	0.35	0.77	10.7	0.42	0.23
		2.0	0.9	14.4	6.5	13.0	5.9	A-F	59.0	0.97	2.14	0.61	1.35	0.37	0.81	10.5	0.44	0.23
								M-F	65.5	1.08	2.38	0.68	1.50	0.41	0.90	11.7	0.49	0.25
		2.5	1.1	14.5	6.6	13.1	6.0	A-F	63.0	1.04	2.28	0.67	1.47	0.42	0.93	11.5	0.53	0.24
								M-F	70.0	1.15	2.53	0.74	1.63	0.47	1.03	12.8	0.59	0.27
		3.0	1.4	14.2	6.5	12.8	5.8	A-F	68.9	1.13	2.49	0.76	1.67	0.49	1.07	12.7	0.62	0.28
								M-F	76.5	1.26	2.77	0.84	1.85	0.54	1.19	14.1	0.69	0.31

(Continued)

TABLE 19-4 (Continued)

Weight		Daily Gain		Daily Consumption				Moisture Basis⁵ A-F (as-fed) M-F (moisture-free)	Energy							Total Protein	Calcium	Phosphorus
				As-Fed		Moisture-Free Dry Matter			TDN	ME (Mcal per)		NE_m (Mcal per)		NE_g (Mcal per)				
(lb)	(kg)	(lb)	(kg)	(lb)	(kg)	(lb)	(kg)		(%)	(lb)	(kg)	(lb)	(kg)	(lb)	(kg)	(%)	(%)	(%)
Medium-frame bulls (Continued)																		
600	273	0.5	0.2	14.6	6.6	13.1	6.0	A-F	48.2	0.79	1.75	0.44	0.97	0.22	0.48	7.5	0.22	0.17
								M-F	53.5	0.88	1.94	0.49	1.08	0.24	0.53	8.3	0.24	0.19
		1.0	0.5	15.4	7.0	13.9	6.3	A-F	51.8	0.85	1.86	0.50	1.11	0.27	0.59	8.3	0.27	0.17
								M-F	57.5	0.94	2.07	0.56	1.23	0.30	0.66	9.2	0.30	0.19
		1.5	0.7	16.1	7.3	14.5	6.6	A-F	55.4	0.91	2.00	0.56	1.22	0.32	0.69	9.0	0.32	0.19
								M-F	61.5	1.01	2.22	0.62	1.36	0.35	0.77	10.0	0.36	0.21
		2.0	0.9	16.6	7.5	14.9	6.8	A-F	59.0	0.97	2.14	0.61	1.35	0.37	0.81	9.7	0.39	0.22
								M-F	65.5	1.08	2.38	0.68	1.50	0.41	0.90	10.8	0.43	0.24
		2.5	1.1	16.7	7.6	15.0	6.8	A-F	63.0	1.04	2.28	0.67	1.47	0.42	0.93	10.4	0.45	0.23
								M-F	70.0	1.15	2.53	0.74	1.63	0.47	1.03	11.6	0.50	0.25
		3.0	1.4	16.3	7.4	14.7	6.7	A-F	68.9	1.13	2.49	0.76	1.67	0.49	1.07	11.4	0.51	0.26
								M-F	76.5	1.26	2.77	0.84	1.85	0.54	1.19	12.7	0.57	0.29
700	318	0.5	0.2	16.3	7.4	14.7	6.7	A-F	48.2	0.79	1.75	0.44	0.97	0.22	0.48	7.2	0.21	0.16
								M-F	53.5	0.88	1.94	0.49	1.08	0.24	0.53	8.0	0.23	0.18
		1.0	0.5	17.3	7.9	15.6	7.0	A-F	51.8	0.85	1.86	0.50	1.11	0.27	0.59	7.9	0.25	0.18
								M-F	57.5	0.94	2.07	0.56	1.23	0.30	0.66	8.8	0.28	0.20
		1.5	0.7	18.1	8.2	16.3	7.4	A-F	55.4	0.91	2.00	0.56	1.22	0.32	0.69	8.5	0.29	0.18
								M-F	61.5	1.01	2.22	0.62	1.36	0.35	0.77	9.4	0.32	0.20
		2.0	0.9	18.6	8.5	16.7	7.6	A-F	59.0	0.97	2.14	0.61	1.35	0.37	0.81	10.0	0.34	0.20
								M-F	65.5	1.08	2.38	0.68	1.50	0.41	0.90	10.1	0.38	0.22
		2.5	1.1	18.7	8.5	16.8	7.6	A-F	63.0	1.04	2.28	0.67	1.47	0.42	0.93	9.7	0.39	0.22
								M-F	70.0	1.15	2.53	0.74	1.63	0.47	1.03	10.8	0.43	0.24
		3.0	1.4	18.3	8.3	16.5	7.5	A-F	68.9	1.13	2.49	0.76	1.67	0.49	1.07	10.5	0.44	0.23
								M-F	76.5	1.26	2.77	0.84	1.85	0.54	1.19	11.7	0.49	0.25
800	364	0.5	0.2	18.0	8.2	16.2	7.4	A-F	48.2	0.79	1.75	0.44	0.97	0.22	0.48	7.0	0.20	0.17
								M-F	53.5	0.88	1.94	0.49	1.08	0.24	0.53	7.8	0.22	0.19
		1.0	0.5	19.2	8.7	17.3	7.9	A-F	51.8	0.85	1.86	0.50	1.11	0.27	0.59	7.6	0.23	0.17
								M-F	57.5	0.94	2.07	0.56	1.23	0.30	0.66	8.4	0.25	0.19
		1.5	0.7	20.0	9.1	18.0	8.2	A-F	55.4	0.91	2.00	0.56	1.22	0.32	0.69	8.1	0.26	0.18
								M-F	61.5	1.01	2.22	0.62	1.36	0.35	0.77	9.0	0.29	0.20
		2.0	0.9	20.6	9.4	18.5	8.4	A-F	59.0	0.97	2.14	0.61	1.35	0.37	0.81	8.6	0.30	0.19
								M-F	65.5	1.08	2.38	0.68	1.50	0.41	0.90	9.5	0.33	0.21
		2.5	1.1	20.7	9.4	18.6	8.5	A-F	63.0	1.04	2.28	0.67	1.47	0.42	0.93	10.0	0.34	0.21
								M-F	70.0	1.15	2.53	0.74	1.63	0.47	1.03	10.1	0.38	0.23
		3.0	1.4	20.2	9.2	18.2	8.3	A-F	68.9	1.13	2.49	0.76	1.67	0.49	1.07	9.7	0.40	0.22
								M-F	76.5	1.26	2.77	0.84	1.85	0.54	1.19	10.8	0.44	0.24
900	409	0.5	0.2	19.7	9.0	17.7	9.0	A-F	48.2	0.79	1.75	0.44	0.97	0.22	0.40	6.9	0.19	0.17
								M-F	53.5	0.88	1.94	0.49	1.08	0.24	0.53	7.7	0.21	0.19
		1.0	0.5	21.0	9.5	18.9	8.6	A-F	51.8	0.85	1.86	0.50	1.11	0.27	0.59	7.4	0.23	0.17
								M-F	57.5	0.94	2.07	0.56	1.23	0.30	0.66	8.2	0.25	0.19
		1.5	0.7	21.9	10.0	19.7	9.0	A-F	55.4	0.91	2.00	0.56	1.23	0.32	0.69	7.7	0.25	0.17
								M-F	61.5	1.01	2.22	0.62	1.36	0.35	0.77	8.6	0.28	0.19
		2.0	0.9	22.4	10.2	20.2	9.2	A-F	59.0	0.97	2.14	0.61	1.35	0.37	0.81	8.2	0.28	0.19
								M-F	65.5	1.08	2.38	0.68	1.50	0.41	0.90	9.1	0.31	0.21
		2.5	1.1	22.6	10.3	20.3	9.2	A-F	63.0	1.04	2.28	0.67	1.47	0.42	0.93	8.6	0.31	0.20
								M-F	70.0	1.15	2.53	0.74	1.63	0.47	1.03	9.6	0.34	0.22
		3.0	1.4	22.1	10.0	19.9	9.0	A-F	68.9	1.13	2.49	0.76	1.67	0.49	1.07	9.2	0.35	0.21
								M-F	76.5	1.26	2.77	0.84	1.85	0.54	1.19	10.2	0.39	0.23
1,000	454	0.5	0.2	21.3	9.7	19.2	8.7	A-F	48.2	0.79	1.75	0.44	0.97	0.22	0.48	6.8	0.19	0.16
								M-F	53.5	0.88	1.94	0.49	1.08	0.24	0.53	7.5	0.21	0.18
		1.0	0.5	22.7	1.03	2.04	9.3	A-F	51.8	0.85	1.86	0.50	1.11	0.27	0.59	7.2	0.22	0.16
								M-F	57.5	0.94	2.07	0.56	1.23	0.30	0.66	8.0	0.24	0.18

(Continued)

TABLE 19-4 *(Continued)*

| Weight | | Daily Gain | | Daily Consumption | | | | Moisture Basis[5] A-F (as-fed) M-F (moisture-free) | Energy | | | | | | | Total Protein | Calcium | Phosphorus |
| | | | | As-Fed | | Moisture-Free Dry Matter | | | TDN | ME (Mcal per) | | NE$_m$ (Mcal per) | | NE$_g$ (Mcal per) | | | | |
(lb)	(kg)	(lb)	(kg)	(lb)	(kg)	(lb)	(kg)		(%)	(lb)	(kg)	(lb)	(kg)	(lb)	(kg)	(%)	(%)	(%)
Medium-frame bulls *(Continued)*																		
1,000 (Continued)	454	1.5	0.7	23.7	10.8	21.3	9.7	A-F	55.4	0.91	2.00	0.56	1.23	0.32	0.69	7.6	0.23	0.17
								M-F	61.5	1.01	2.22	0.62	1.36	0.35	0.77	8.4	0.26	0.19
		2.0	0.9	24.2	11.0	21.8	9.9	A-F	59.0	0.97	2.14	0.61	1.35	0.37	0.81	7.8	0.25	0.17
								M-F	65.5	1.08	2.38	0.68	1.50	0.41	0.90	8.7	0.28	0.19
		2.5	1.1	24.4	11.1	22.0	10.0	A-F	63.0	1.04	2.28	0.67	1.47	0.42	0.93	8.2	0.28	0.18
								M-F	70.0	1.15	2.53	0.74	1.63	0.47	1.03	9.1	0.31	0.20
		3.0	1.4	23.9	10.9	21.5	9.8	A-F	68.9	1.13	2.49	0.76	1.67	0.49	1.07	8.6	0.32	0.20
								M-F	76.5	1.26	2.77	0.84	1.85	0.54	1.19	9.6	0.35	0.22
1,100	500	0.5	0.2	22.9	10.4	20.6	9.4	A-F	48.2	0.79	1.75	0.44	0.97	0.22	0.48	6.7	0.18	0.17
								M-F	53.5	0.88	1.94	0.49	1.08	0.24	0.53	7.4	0.20	0.19
		1.0	0.5	24.3	11.0	21.9	10.0	A-F	51.8	0.85	1.86	0.50	1.11	0.27	0.59	7.0	0.20	0.17
								M-F	57.5	0.94	2.07	0.56	1.23	0.30	0.66	7.8	0.22	0.19
		1.5	0.7	25.4	11.5	22.9	10.4	A-F	55.4	0.91	2.00	0.56	1.23	0.32	0.69	7.3	0.22	0.17
								M-F	61.5	1.01	2.22	0.62	1.36	0.35	0.77	8.1	0.24	0.19
		2.0	0.9	26.0	11.8	23.4	10.6	A-F	59.0	0.97	2.14	0.61	1.35	0.37	0.81	7.6	0.23	0.17
								M-F	65.5	1.08	2.38	0.68	1.50	0.41	0.90	8.4	0.26	0.19
		2.5	1.1	26.2	11.9	23.6	10.7	A-F	63.0	1.04	2.28	0.67	1.47	0.42	0.93	7.8	0.25	0.18
								M-F	70.0	1.15	2.53	0.74	1.63	0.47	1.03	8.7	0.28	0.20
		3.0	1.4	25.7	11.7	23.1	10.5	A-F	68.9	1.13	2.49	0.76	1.67	0.49	1.07	8.3	0.29	0.19
								M-F	76.5	1.26	2.77	0.84	1.85	0.54	1.19	9.2	0.32	0.21
Large-frame bull calves and compensating large-frame yearling steers																		
300	136	0.5	0.2	8.8	4.0	7.9	3.6	A-F	48.2	0.77	1.70	0.43	0.95	0.21	0.46	8.7	0.28	0.18
								M-F	52.5	0.86	1.89	0.48	1.06	0.23	0.51	9.7	0.31	0.20
		1.0	0.5	9.3	4.2	8.4	3.8	A-F	50.4	0.83	1.82	0.49	1.07	0.25	0.56	10.5	0.42	0.22
								M-F	56.0	0.92	2.02	0.54	1.19	0.28	0.62	11.7	0.47	0.24
		1.5	0.7	9.8	4.5	8.8	4.0	A-F	53.6	0.88	1.95	0.53	1.17	0.30	0.66	12.2	0.57	0.25
								M-F	59.5	0.98	2.16	0.59	1.30	0.33	0.73	13.5	0.63	0.28
		2.0	0.9	10.0	4.5	9.0	4.1	A-F	56.3	0.93	2.04	0.57	1.25	0.33	0.73	13.6	0.68	0.29
								M-F	62.5	1.03	2.27	0.63	1.39	0.37	0.81	15.1	0.76	0.32
		2.5	1.1	10.2	4.6	9.2	4.2	A-F	59.9	0.98	2.16	0.62	1.37	0.38	0.83	15.3	0.82	0.32
								M-F	66.5	1.09	2.40	0.69	1.52	0.42	0.92	17.0	0.91	0.36
		3.0	1.4	10.2	4.6	9.2	4.2	A-F	63.5	1.04	2.30	0.68	1.49	0.42	0.93	16.9	0.97	0.39
								M-F	70.5	1.16	2.55	0.75	1.65	0.47	1.03	18.8	1.08	0.43
		3.5	1.6	10.1	4.6	9.1	4.1	A-F	68.0	1.12	2.46	0.74	1.62	0.48	1.05	18.8	1.12	0.43
								M-F	75.5	1.24	2.73	0.82	1.80	0.53	1.17	20.9	1.24	0.48
		4.0	1.8	9.1	4.1	8.2	3.7	A-F	77.4	1.27	2.79	0.86	1.90	0.59	1.31	22.2	1.38	0.53
								M-F	86.0	1.41	3.10	0.96	2.11	0.66	1.45	24.7	1.53	0.59
400	182	0.5	0.2	10.9	5.0	9.8	4.5	A-F	47.3	0.77	1.70	0.43	0.95	0.21	0.46	8.1	0.24	0.16
								M-F	52.5	0.86	1.89	0.48	1.06	0.23	0.51	9.0	0.27	0.18
		1.0	0.5	11.6	5.3	10.4	4.7	A-F	50.4	0.83	1.82	0.49	1.07	0.25	0.56	9.5	0.36	0.19
								M-F	56.0	0.92	2.02	0.54	1.19	0.28	0.62	10.5	0.40	0.21
		1.5	0.7	12.1	5.5	10.9	5.0	A-F	53.6	0.88	1.94	0.53	1.17	0.30	0.66	10.7	0.46	0.22
								M-F	59.5	0.98	2.16	0.59	1.30	0.33	0.73	11.9	0.51	0.24
		2.0	0.9	12.4	5.6	11.2	5.1	A-F	56.3	0.93	2.04	0.57	1.25	0.33	0.73	11.8	0.55	0.25
								M-F	62.5	1.03	2.27	0.63	1.39	0.37	0.81	13.1	0.61	0.28
		2.5	1.1	12.7	5.8	11.4	5.2	A-F	59.9	0.98	2.16	0.62	1.37	0.38	0.83	13.1	0.65	0.28
								M-F	66.5	1.09	2.40	0.69	1.52	0.42	0.92	14.5	0.72	0.31
		3.0	1.4	12.8	5.8	11.5	5.2	A-F	63.5	1.04	2.30	0.68	1.49	0.42	0.93	14.3	0.74	0.32
								M-F	70.5	1.16	2.55	0.75	1.65	0.47	1.03	15.9	0.82	0.35
		3.5	1.6	12.6	5.7	11.3	5.1	A-F	68.0	1.12	2.46	0.74	1.62	0.48	1.05	15.8	0.86	0.35
								M-F	75.5	1.24	2.73	0.82	1.80	0.53	1.17	17.5	0.96	0.39
		4.0	1.8	11.3	5.1	10.2	4.6	A-F	77.4	1.27	2.79	0.86	1.90	0.59	1.31	18.3	1.07	0.43
								M-F	86.0	1.41	3.10	0.96	2.11	0.66	1.45	20.3	1.19	0.48

(Continued)

Feeding Beef Cattle

TABLE 19-4 *(Continued)*

Weight		Daily Gain		Daily Consumption				Moisture Basis[5] A-F (as-fed) M-F (moisture-free)	Energy							Total Protein	Calcium	Phosphorus
				As-Fed		Moisture-Free Dry Matter			TDN	ME (Mcal per)		NE_m (Mcal per)		NE_g (Mcal per)				
(lb)	(kg)	(lb)	(kg)	(lb)	(kg)	(lb)	(kg)		(%)	(lb)	(kg)	(lb)	(kg)	(lb)	(kg)	(%)	(%)	(%)
Large-frame bull calves and compensating large-frame yearling steers *(Continued)*																		
500	227	0.5	0.2	12.9	5.9	11.6	5.3	A-F	47.3	0.77	1.70	0.43	0.95	0.21	0.46	7.7	0.23	0.17
								M-F	52.5	0.86	1.89	0.48	1.06	0.23	0.51	8.6	0.25	0.19
		1.0	0.5	13.7	6.2	12.3	5.6	A-F	50.4	0.83	1.82	0.49	1.07	0.25	0.56	8.8	0.32	0.19
								M-F	56.0	0.92	2.02	0.54	1.19	0.28	0.62	9.8	0.36	0.21
		1.5	0.7	14.3	6.5	12.9	5.9	A-F	53.6	0.88	1.94	0.53	1.17	0.30	0.66	9.8	0.39	0.20
								M-F	59.5	0.98	2.16	0.59	1.30	0.33	0.73	10.9	0.43	0.22
		2.0	0.9	14.7	6.7	13.2	6.0	A-F	56.3	0.93	2.04	0.57	1.25	0.33	0.73	10.6	0.47	0.23
								M-F	62.5	1.03	2.27	0.63	1.39	0.37	0.81	11.8	0.52	0.25
		2.5	1.1	15.0	6.8	13.5	6.1	A-F	59.9	0.98	2.16	0.62	1.37	0.38	0.83	11.6	0.53	0.25
								M-F	66.5	1.09	2.40	0.69	1.52	0.42	0.92	12.9	0.59	0.28
		3.0	1.4	15.1	6.9	13.6	6.2	A-F	63.5	1.04	2.30	0.68	1.49	0.42	0.93	12.6	0.61	0.28
								M-F	70.5	1.16	2.55	0.75	1.65	0.47	1.03	14.0	0.68	0.31
		3.5	1.6	14.9	6.8	13.4	6.1	A-F	68.0	1.12	2.46	0.74	1.62	0.48	1.05	13.8	0.69	0.32
								M-F	75.5	1.24	2.73	0.82	1.80	0.53	1.17	15.3	0.77	0.35
		4.0	1.8	13.3	6.0	12.0	5.5	A-F	77.4	1.27	2.79	0.86	1.90	0.59	1.31	15.8	0.87	0.36
								M-F	86.0	1.41	3.10	0.96	2.11	0.66	1.45	17.5	0.97	0.40
600	273	0.5	0.2	14.8	6.7	13.3	6.0	A-F	47.3	0.77	1.70	0.43	0.95	0.21	0.46	7.5	0.21	0.16
								M-F	52.5	0.86	1.89	0.48	1.06	0.23	0.51	8.3	0.23	0.18
		1.0	0.5	15.7	7.1	14.1	6.4	A-F	50.4	0.83	1.82	0.49	1.07	0.25	0.56	8.3	0.28	0.18
								M-F	56.0	0.92	2.02	0.54	1.19	0.28	0.62	9.2	0.31	0.20
		1.5	0.7	16.4	7.5	14.8	6.7	A-F	53.6	0.88	1.94	0.53	1.17	0.30	0.66	9.1	0.33	0.19
								M-F	59.5	0.98	2.16	0.59	1.30	0.33	0.73	10.1	0.37	0.21
		2.0	0.9	16.9	7.7	15.2	6.9	A-F	56.3	0.93	2.04	0.57	1.25	0.33	0.73	9.8	0.40	0.21
								M-F	62.5	1.03	2.27	0.63	1.39	0.37	0.81	10.9	0.44	0.23
		2.5	1.1	17.2	7.8	15.5	7.0	A-F	59.9	0.98	2.16	0.62	1.37	0.38	0.83	10.6	0.46	0.23
								M-F	66.5	1.09	2.40	0.69	1.52	0.42	0.92	11.8	0.51	0.26
		3.0	1.4	17.2	7.8	15.5	7.0	A-F	63.5	1.05	2.30	0.68	1.49	0.42	0.93	11.4	0.52	0.24
								M-F	70.5	1.16	2.55	0.75	1.65	0.47	1.03	12.7	0.58	0.27
		3.5	1.6	17.0	7.7	15.3	7.0	A-F	68.0	1.12	2.46	0.74	1.62	0.48	1.05	12.3	0.59	0.27
								M-F	75.5	1.24	2.73	0.82	1.80	0.53	1.17	13.7	0.66	0.30
		4.0	1.8	15.3	7.0	13.8	6.3	A-F	77.4	1.27	2.79	0.86	1.90	0.59	1.31	14.0	0.73	0.33
								M-F	86.0	1.41	3.10	0.96	2.11	0.66	1.45	15.6	0.81	0.37
700	318	0.5	0.2	16.6	7.5	14.9	6.8	A-F	47.3	0.77	1.70	0.43	0.95	0.21	0.46	7.2	0.20	0.16
								M-F	52.5	0.86	1.89	0.48	1.06	0.23	0.51	8.0	0.22	0.18
		1.0	0.5	17.7	8.0	15.9	7.2	A-F	50.4	0.83	1.82	0.49	1.07	0.25	0.56	7.9	0.26	0.17
								M-F	56.0	0.92	2.02	0.54	1.19	0.28	0.62	8.8	0.29	0.19
		1.5	0.7	10.4	0.4	16.6	7.5	A-F	53.6	0.88	1.94	0.53	1.17	0.30	0.66	8.6	0.32	0.19
								M-F	59.5	0.98	2.16	0.59	1.30	0.33	0.73	9.6	0.35	0.21
		2.0	0.9	18.9	8.6	17.0	7.7	A-F	56.3	0.93	2.04	0.57	1.25	0.33	0.73	9.2	0.35	0.20
								M-F	62.5	1.03	2.27	0.63	1.39	0.37	0.81	10.2	0.39	0.22
		2.5	1.1	19.3	8.8	17.4	7.9	A-F	59.9	0.98	2.16	0.62	1.37	0.38	0.83	10.0	0.40	0.22
								M-F	66.5	1.09	2.40	0.69	1.52	0.42	0.92	11.0	0.44	0.24
		3.0	1.4	19.4	8.8	17.5	8.0	A-F	63.5	1.05	2.30	0.68	1.49	0.42	0.93	10.5	0.45	0.23
								M-F	70.5	1.16	2.55	0.75	1.65	0.47	1.03	11.7	0.50	0.25
		3.5	1.6	19.1	8.7	17.2	7.8	A-F	68.0	1.12	2.46	0.74	1.62	0.48	1.05	11.3	0.50	0.25
								M-F	75.5	1.24	2.73	0.82	1.80	0.53	1.17	12.5	0.56	0.28
		4.0	1.8	17.2	7.8	15.5	7.0	A-F	77.4	1.27	2.79	0.86	1.90	0.59	1.31	12.7	0.63	0.30
								M-F	86.0	1.41	3.10	0.96	2.11	0.66	1.45	14.1	0.70	0.33
800	364	0.5	0.2	18.3	8.3	16.5	7.5	A-F	47.3	0.77	1.70	0.43	0.95	0.21	0.46	7.1	0.19	0.17
								M-F	52.5	0.86	1.89	0.48	1.06	0.23	0.51	7.9	0.21	0.19
		1.0	0.5	19.4	8.8	17.5	8.0	A-F	50.4	0.83	1.82	0.49	1.07	0.25	0.56	7.7	0.23	0.17
								M-F	56.0	0.92	2.02	0.54	1.19	0.28	0.62	8.5	0.26	0.19
		1.5	0.7	20.3	9.2	18.3	8.3	A-F	53.6	0.88	1.94	0.53	1.17	0.30	0.66	8.2	0.28	0.18
								M-F	59.5	0.98	2.16	0.59	1.30	0.33	0.73	9.1	0.31	0.20

(Continued)

TABLE 19-4 (Continued)

| Weight | | Daily Gain | | Daily Consumption | | | | Moisture Basis[5] A-F (as-fed) M-F (moisture-free) | Energy | | | | | | | Total Protein | Calcium | Phosphorus |
| | | | | As-Fed | | Moisture-Free Dry Matter | | | TDN | ME (Mcal per) | | NE$_m$ (Mcal per) | | NE$_g$ (Mcal per) | | | | |
(lb)	(kg)	(lb)	(kg)	(lb)	(kg)	(lb)	(kg)		(%)	(lb)	(kg)	(lb)	(kg)	(lb)	(kg)	(%)	(%)	(%)
								Large-frame bull calves and compensating large-frame yearling steers (Continued)										
800	364	2.0	0.9	20.9	9.5	18.8	8.5	A-F	56.3	0.93	2.04	0.57	1.25	0.33	0.73	8.7	0.32	0.19
Continued								M-F	62.5	1.03	2.27	0.63	1.39	0.37	0.81	9.7	0.35	0.21
		2.5	1.1	21.3	9.7	19.2	8.7	A-F	59.9	0.98	2.16	0.62	1.37	0.38	0.83	9.3	0.36	0.21
								M-F	66.5	1.09	2.40	0.69	1.52	0.42	0.92	10.3	0.40	0.23
		3.0	1.4	21.4	9.7	19.3	8.8	A-F	63.5	1.04	2.30	0.68	1.49	0.42	0.93	9.8	0.41	0.22
								M-F	70.5	1.16	2.55	0.75	1.65	0.47	1.03	10.9	0.45	0.24
		3.5	1.6	21.1	9.6	19.0	8.6	A-F	68.0	1.12	2.46	0.74	1.62	0.48	1.05	10.4	0.45	0.23
								M-F	75.5	1.24	2.73	0.82	1.80	0.53	1.17	11.6	0.50	0.26
		4.0	1.8	19.0	8.6	17.1	7.8	A-F	77.4	1.27	2.79	0.86	1.90	0.59	1.31	11.7	0.55	0.28
								M-F	86.0	1.41	3.10	0.96	2.11	0.66	1.45	13.0	0.61	0.31
900	409	0.5	0.2	20.0	9.1	18.0	8.2	A-F	47.3	0.77	1.70	0.43	0.95	0.21	0.46	6.9	0.20	0.16
								M-F	52.5	0.86	1.89	0.48	1.06	0.23	0.51	7.7	0.22	0.18
		1.0	0.5	21.3	9.7	19.2	8.7	A-F	50.4	0.83	1.82	0.49	1.07	0.25	0.56	7.5	0.23	0.16
								M-F	56.0	0.92	2.02	0.54	1.19	0.28	0.62	8.3	0.25	0.18
		1.5	0.7	22.2	10.1	20.0	9.1	A-F	53.6	0.88	1.94	0.53	1.17	0.30	0.66	7.9	0.26	0.18
								M-F	59.5	0.98	2.16	0.59	1.30	0.33	0.73	8.8	0.29	0.20
		2.0	0.9	22.9	10.4	20.6	9.4	A-F	56.3	0.93	2.04	0.57	1.25	0.33	0.73	8.9	0.30	0.18
								M-F	62.5	1.03	2.27	0.63	1.39	0.37	0.81	9.2	0.32	0.20
		2.5	1.1	23.3	10.6	21.0	9.5	A-F	59.9	0.98	2.16	0.62	1.37	0.38	0.83	8.8	0.32	0.19
								M-F	66.5	1.09	2.40	0.69	1.52	0.42	0.92	9.8	0.36	0.21
		3.0	1.4	23.4	10.6	21.1	9.6	A-F	63.5	1.04	2.30	0.68	1.49	0.42	0.93	9.3	0.36	0.21
								M-F	70.5	1.16	2.55	0.75	1.65	0.47	1.03	10.3	0.40	0.23
		3.5	1.6	23.1	10.5	20.8	9.5	A-F	68.0	1.12	2.46	0.74	1.62	0.48	1.05	9.8	0.41	0.22
								M-F	75.5	1.24	2.73	0.82	1.80	0.53	1.17	10.9	0.45	0.24
		4.0	1.8	20.8	9.5	18.7	8.5	A-F	77.4	1.27	2.79	0.86	1.90	0.59	1.31	10.9	0.48	0.25
								M-F	86.0	1.41	3.10	0.96	2.11	0.66	1.45	12.1	0.53	0.28
1,000	454	0.5	0.2	21.7	9.9	19.5	8.9	A-F	47.3	0.77	1.70	0.43	0.95	0.21	0.46	6.8	0.19	0.16
								M-F	52.5	0.86	1.89	0.48	1.06	0.23	0.51	7.6	0.21	0.18
		1.0	0.5	23.0	10.5	20.7	9.4	A-F	40.4	0.83	1.82	0.49	1.07	0.25	0.56	7.3	0.23	0.17
								M-F	56.0	0.92	2.02	0.54	1.19	0.28	0.62	8.1	0.25	0.19
		1.5	0.7	24.1	11.0	21.7	9.9	A-F	53.6	0.88	1.94	0.53	1.17	0.30	0.66	7.7	0.24	0.17
								M-F	59.5	0.98	2.16	0.59	1.30	0.33	0.73	8.5	0.27	0.19
		2.0	0.9	24.8	11.3	22.3	10.1	A-F	56.3	0.93	2.04	0.57	1.25	0.33	0.73	7.9	0.27	0.18
								M-F	62.5	1.03	2.27	0.63	1.39	0.37	0.81	8.9	0.30	0.20
		2.5	1.1	25.2	11.5	22.7	10.3	A-F	59.9	0.98	2.16	0.62	1.37	0.38	0.83	8.4	0.30	0.18
								M-F	66.5	1.09	2.40	0.69	1.52	0.42	0.92	9.3	0.33	0.20
		3.0	1.4	25.3	11.5	22.8	10.4	A-F	63.5	1.04	2.30	0.68	1.49	0.42	0.93	8.7	0.32	0.19
								M-F	70.5	1.16	2.55	0.75	1.65	0.47	1.03	9.7	0.36	0.21
		3.5	1.6	25.0	11.4	22.5	10.2	A-F	68.0	1.12	2.46	0.74	1.62	0.48	1.05	9.3	0.36	0.22
								M-F	75.5	1.24	2.73	0.82	1.80	0.53	1.17	10.3	0.40	0.24
		4.0	1.8	22.4	10.2	20.2	9.2	A-F	77.4	1.27	2.79	0.86	1.90	0.59	0.31	10.2	0.43	0.24
								M-F	86.0	1.41	3.10	0.96	2.11	0.66	1.45	11.3	0.48	0.27
1,100	500	0.5	0.2	23.2	10.5	20.9	9.5	A-F	47.3	0.77	1.70	0.43	0.95	0.21	0.46	6.8	0.19	0.17
								M-F	52.5	0.86	1.89	0.48	1.06	0.23	0.51	7.4	0.21	0.19
		1.0	0.5	24.8	11.3	22.3	10.1	A-F	50.4	0.83	1.82	0.49	1.07	0.25	0.56	7.1	0.21	0.17
								M-F	56.0	0.92	2.02	0.54	1.19	0.28	0.62	7.9	0.23	0.19
		1.5	0.7	25.9	11.8	23.3	10.6	A-F	53.6	0.88	1.94	0.53	1.17	0.30	0.66	7.5	0.23	0.17
								M-F	59.5	0.98	2.16	0.59	1.30	0.33	0.73	8.3	0.26	0.19
		2.0	0.9	26.6	12.1	23.9	10.9	A-F	56.3	0.93	2.04	0.57	1.25	0.33	0.73	7.7	0.25	0.17
								M-F	62.5	1.03	2.27	0.63	1.39	0.37	0.81	8.6	0.28	0.19
		2.5	1.1	26.9	12.2	24.2	11.0	A-F	59.9	0.98	2.16	0.62	1.37	0.38	0.83	8.1	0.27	0.18
								M-F	66.5	1.09	2.40	0.69	1.52	0.42	0.92	9.0	0.30	0.20

(Continued)

TABLE 19-4 (Continued)

Weight		Daily Gain		Daily Consumption				Moisture Basis[5] A-F (as-fed) M-F (moisture-free)	Energy								Total Protein	Calcium	Phosphorus
				As-Fed		Moisture-Free Dry Matter			TDN	ME (Mcal per)		NE$_m$ (Mcal per)		NE$_g$ (Mcal per)					
(lb)	(kg)	(lb)	(kg)	(lb)	(kg)	(lb)	(kg)		(%)	(lb)	(kg)	(lb)	(kg)	(lb)	(kg)		(%)	(%)	(%)
Large-frame bull calves and compensating large-frame yearling steers (Continued)
1,100 (Continued)	500	3.0	1.4	27.2	12.4	24.5	11.1	A-F	63.5	1.04	2.30	0.68	1.49	0.42	0.93		8.4	0.29	0.19
								M-F	70.5	1.16	2.55	0.75	1.65	0.47	1.03		9.3	0.32	0.21
		3.5	1.6	26.8	12.2	24.1	11.0	A-F	68.0	1.12	2.46	0.74	1.62	0.48	1.05		8.8	0.32	0.20
								M-F	75.5	1.24	2.73	0.82	1.80	0.53	1.17		9.8	0.36	0.22
		4.0	1.8	24.1	11.0	21.7	9.9	A-F	77.4	1.27	2.79	0.86	1.90	0.59	1.31		9.6	0.39	0.23
								M-F	86.0	1.41	3.10	0.96	2.11	0.66	1.45		10.7	0.43	0.25
Medium-frame heifer calves
300	136	0.5	0.2	8.3	3.8	7.5	3.4	A-F	50.4	0.83	1.82	0.49	1.07	0.25	0.56		8.6	0.26	0.19
								M-F	56.0	0.92	2.02	0.54	1.19	0.28	0.62		9.6	0.29	0.21
		1.0	0.5	8.9	4.0	8.0	3.6	A-F	55.8	0.92	2.03	0.57	1.25	0.32	0.71		10.3	0.40	0.20
								M-F	62.0	1.02	2.25	0.63	1.39	0.36	0.79		11.4	0.44	0.22
		1.5	0.7	9.1	4.1	8.2	3.7	A-F	61.7	1.02	2.24	0.65	1.42	0.40	0.87		11.8	0.53	0.24
								M-F	68.5	1.13	2.49	0.72	1.58	0.44	0.97		13.1	0.59	0.27
		2.0	0.9	8.9	4.0	8.0	3.6	A-F	69.3	1.13	2.50	0.76	1.67	0.50	1.09		13.6	0.67	0.30
								M-F	77.0	1.26	2.77	0.84	1.85	0.55	1.21		15.1	0.74	0.33
400	182	0.5	0.2	10.3	4.7	9.3	4.2	A-F	50.4	0.83	1.82	0.49	1.07	0.25	0.56		8.0	0.23	0.17
								M-F	56.0	0.92	2.02	0.54	1.19	0.28	0.62		8.9	0.26	0.19
		1.0	0.5	11.0	5.0	9.9	4.5	A-F	55.8	0.92	2.03	0.57	1.25	0.32	0.71		9.2	0.32	0.18
								M-F	62.0	1.02	2.25	0.63	1.39	0.36	0.79		10.2	0.36	0.20
		1.5	0.7	11.3	5.1	20.2	4.6	A-F	61.7	1.02	2.24	0.65	1.42	0.40	0.87		10.3	0.41	0.22
								M-F	68.5	1.13	2.49	0.72	1.58	0.44	0.97		11.4	0.45	0.24
		2.0	0.9	11.1	5.0	10.0	4.5	A-F	69.3	1.13	2.50	0.76	1.67	0.50	1.09		11.6	0.51	0.26
								M-F	77.0	1.26	2.77	0.84	1.85	0.55	1.21		12.9	0.57	0.29
500	227	0.5	0.2	12.2	5.5	11.0	5.0	A-F	50.4	0.83	1.82	0.49	1.07	0.25	0.56		7.7	0.22	0.16
								M-F	56.0	0.92	2.02	0.54	1.19	0.28	0.62		8.5	0.24	0.18
		1.0	0.5	13.1	6.0	11.8	5.4	A-F	55.8	0.92	2.03	0.57	1.25	0.32	0.71		8.5	0.27	0.19
								M-F	62.0	1.02	2.25	0.63	1.39	0.36	0.79		9.4	0.30	0.21
		1.5	0.7	13.4	6.1	12.1	5.5	A-F	61.7	1.02	2.24	0.65	1.42	0.40	0.87		9.3	0.34	0.20
								M-F	68.5	1.13	2.49	0.72	1.58	0.44	0.97		10.3	0.38	0.22
		2.0	0.9	13.1	6.0	11.8	5.4	A-F	69.3	1.13	2.50	0.76	1.67	0.50	1.09		10.3	0.41	0.22
								M-F	77.0	1.26	2.77	0.84	1.85	0.55	1.21		11.4	0.45	0.24
600	273	0.5	0.2	14.0	6.4	12.6	5.7	A-F	50.4	0.83	1.82	0.49	1.07	0.25	0.56		7.3	0.21	0.16
								M-F	56.0	0.92	2.02	0.54	1.19	0.28	0.62		8.1	0.23	0.18
		1.0	0.5	15.0	6.8	13.5	6.1	A-F	55.8	0.92	2.03	0.57	1.25	0.32	0.71		7.9	0.25	0.18
								M-F	62.0	1.02	2.25	0.63	1.39	0.36	0.79		8.8	0.28	0.20
		1.5	0.7	15.3	7.0	13.8	6.3	A-F	61.7	1.26	2.24	0.65	1.42	0.40	0.87		8.6	0.29	0.19
								M-F	68.5	1.13	2.49	0.72	1.58	0.44	0.97		9.5	0.32	0.21
		2.0	0.9	15.0	6.8	13.5	6.1	A-F	69.3	1.13	2.50	0.76	1.67	0.50	1.09		9.4	0.34	0.21
								M-F	77.0	1.26	2.77	0.84	1.85	0.55	1.21		10.4	0.38	0.23
700	318	0.5	0.2	15.7	7.1	14.1	6.4	A-F	50.4	0.83	1.82	0.49	1.07	0.25	0.56		7.1	0.20	0.17
								M-F	56.0	0.92	2.02	0.54	1.19	0.28	0.62		7.9	0.22	0.19
		1.0	0.5	16.8	7.6	15.1	6.9	A-F	55.8	0.92	2.03	0.57	1.25	0.32	0.71		7.6	0.23	0.17
								M-F	62.0	1.02	2.25	0.63	1.39	0.36	0.79		8.4	0.25	0.19
		1.5	0.7	16.9	7.7	15.5	7.0	A-F	61.7	1.26	2.24	0.65	1.42	0.40	0.87		8.1	0.25	0.18
								M-F	68.5	1.13	2.49	0.72	1.58	0.44	0.97		9.0	0.28	0.20
		2.0	0.9	16.9	7.7	15.2	7.0	A-F	69.3	1.13	2.50	0.76	1.67	0.50	1.09		8.6	0.29	0.20
								M-F	77.0	1.26	2.77	0.84	1.85	0.55	1.21		9.6	0.32	0.22
800	364	0.5	0.2	17.3	7.9	15.6	7.0	A-F	50.4	0.83	1.82	0.49	1.07	0.25	0.56		6.9	0.19	0.16
								M-F	56.0	0.92	2.02	0.54	1.19	0.28	0.62		7.7	0.21	0.18
		1.0	0.5	18.6	8.5	16.7	7.6	A-F	55.8	0.92	2.03	0.57	1.25	0.32	0.71		7.3	0.20	0.16
								M-F	62.0	1.02	2.25	0.63	1.39	0.36	0.79		8.1	0.22	0.18

(Continued)

TABLE 19-4 (Continued)

Weight		Daily Gain		Daily Consumption				Moisture Basis[5] A-F (as-fed) M-F (moisture-free)	Energy							Total Protein	Calcium	Phosphorus
				As-Fed		Moisture-Free Dry Matter			TDN	ME (Mcal per)		NE$_m$ (Mcal per)		NE$_g$ (Mcal per)				
(lb)	(kg)	(lb)	(kg)	(lb)	(kg)	(lb)	(kg)		(%)	(lb)	(kg)	(lb)	(kg)	(lb)	(kg)	(%)	(%)	(%)
Medium-frame heifer calves (Continued)
800	364	1.5	0.7	19.1	8.7	17.2	7.8	A-F	61.7	1.26	2.24	0.65	1.42	0.40	0.87	7.7	0.22	0.17
(Continued)								M-F	68.5	1.13	2.49	0.72	1.58	0.44	0.97	8.5	0.24	0.19
		2.0	0.9	18.7	8.5	16.8	7.6	A-F	69.3	1.13	2.50	0.76	1.67	0.50	1.09	8.1	0.25	0.18
								M-F	77.0	1.26	2.77	0.84	1.85	0.55	1.21	9.0	0.28	0.20
900	409	0.5	0.2	19.0	8.6	17.1	7.8	A-F	50.4	0.83	1.82	0.49	1.07	0.25	0.56	6.8	0.19	0.16
								M-F	56.0	0.92	2.02	0.54	1.19	0.28	0.62	7.5	0.21	0.18
		1.0	0.5	20.3	9.2	18.3	8.3	A-F	55.8	0.92	2.03	0.57	1.25	0.32	0.71	7.0	0.20	0.16
								M-F	62.0	1.02	2.25	0.63	1.39	0.36	0.79	7.8	0.22	0.18
		1.5	0.7	20.9	9.5	18.8	8.5	A-F	61.7	1.26	2.24	0.65	1.42	0.40	0.87	7.3	0.20	0.17
								M-F	68.5	1.13	2.49	0.72	1.58	0.44	0.97	8.1	0.22	0.19
		2.0	0.9	20.3	9.2	18.3	8.3	A-F	69.3	1.13	2.50	0.76	1.67	0.50	1.09	7.7	0.23	0.17
								M-F	77.0	1.26	2.77	0.84	1.85	0.55	1.21	8.5	0.25	0.19
1,000	454	0.5	0.2	20.6	9.4	18.4	8.4	A-F	50.4	0.83	1.82	0.49	1.07	0.25	0.56	6.7	0.18	0.17
								M-F	56.0	0.92	2.02	0.54	1.19	0.28	0.62	7.4	0.20	0.19
		1.0	0.5	22.0	10.0	19.8	9.0	A-F	55.8	0.92	2.03	0.57	1.25	0.32	0.71	6.8	0.18	0.16
								M-F	62.0	1.02	2.25	0.63	1.39	0.36	0.79	7.6	0.20	0.18
		1.5	0.7	22.6	10.3	20.3	9.2	A-F	61.7	1.26	2.24	0.65	1.42	0.40	0.87	7.0	0.19	0.16
								M-F	68.5	1.13	2.49	0.72	1.58	0.44	0.97	7.8	0.21	0.18
		2.0	0.9	22.0	10.0	19.8	9.0	A-F	69.3	1.13	2.50	0.76	1.67	0.50	1.09	7.3	0.20	0.17
								M-F	77.0	1.26	2.77	0.84	1.85	0.55	1.21	8.1	0.22	0.19
Large-frame heifer calves and compensating medium-frame yearling heifers
300	136	0.5	0.2	8.7	4.0	7.8	3.5	A-F	48.6	0.80	1.76	0.45	1.00	0.23	0.50	8.6	0.28	0.18
								M-F	54.0	0.89	1.96	0.50	1.10	0.25	0.55	9.5	0.31	0.20
		1.0	0.5	9.3	4.2	8.4	3.8	A-F	53.1	0.88	1.94	0.52	1.15	0.29	0.63	10.2	0.41	0.22
								M-F	59.0	0.98	2.16	0.58	1.28	0.32	0.70	11.3	0.45	0.24
		1.5	0.7	9.8	4.5	8.8	4.0	A-F	57.6	0.95	2.08	0.59	1.29	0.35	0.77	11.7	0.52	0.23
								M-F	64.0	1.05	2.31	0.65	1.43	0.39	0.86	13.0	0.58	0.25
		2.0	0.9	9.9	4.5	8.9	4.0	A-F	62.6	1.03	2.26	0.67	1.47	0.41	1.00	13.1	0.62	0.27
								M-F	69.5	1.14	2.51	0.74	1.63	0.46	1.01	14.6	0.69	0.30
		2.5	1.1	9.7	4.4	8.7	4.0	A-F	69.3	1.13	2.49	0.76	1.67	0.50	1.09	15.0	0.77	0.32
								M-F	77.0	1.26	2.77	0.84	1.85	0.55	1.21	16.7	0.86	0.35
400	182	0.5	0.2	10.8	4.9	9.7	4.4	A-F	48.6	0.80	1.76	0.45	1.00	0.23	0.50	8.0	0.24	0.16
								M-F	54.0	0.89	1.96	0.50	1.10	0.25	0.55	8.9	0.27	0.18
		1.0	0.5	11.7	5.3	10.5	4.8	A-F	53.1	0.88	1.94	0.52	1.15	0.29	0.63	10.0	0.32	0.19
								M-F	59.0	0.98	2.16	0.58	1.28	0.32	0.70	10.1	0.36	0.21
		1.5	0.7	12.1	5.5	10.9	5.0	A-F	57.6	0.95	2.08	0.59	1.29	0.35	0.77	10.2	0.41	0.20
								M-F	64.0	1.05	2.31	0.65	1.43	0.39	0.86	11.3	0.45	0.22
		2.0	0.9	12.3	5.6	11.1	5.0	A-F	62.6	1.03	2.26	0.67	1.47	0.41	0.91	11.3	0.49	0.23
								M-F	69.5	1.14	2.51	0.74	1.63	0.46	1.01	12.6	0.54	0.26
		2.5	1.1	9.7	4.4	10.8	4.9	A-F	69.3	1.13	2.49	0.76	1.67	0.50	1.09	12.7	0.59	0.28
								M-F	77.0	1.26	2.77	0.84	1.85	0.55	1.21	14.1	0.65	0.31
500	227	0.5	0.2	12.8	5.8	11.5	5.2	A-F	48.6	0.80	1.76	0.45	1.00	0.23	0.50	7.6	0.21	0.15
								M-F	54.0	0.89	1.96	0.50	1.10	0.25	0.55	8.4	0.23	0.17
		1.0	0.5	13.8	6.3	12.4	5.6	A-F	53.1	0.88	1.94	0.52	1.15	0.29	0.63	8.5	0.27	0.18
								M-F	59.0	0.98	2.16	0.58	1.28	0.32	0.70	9.4	0.30	0.20
		1.5	0.7	14.3	6.5	12.9	5.9	A-F	57.6	0.95	2.08	0.59	1.29	0.35	0.77	9.3	0.34	0.18
								M-F	64.0	1.05	2.31	0.65	1.43	0.39	0.86	10.3	0.38	0.20
		2.0	0.9	14.6	6.6	13.1	6.0	A-F	62.6	1.03	2.26	0.67	1.47	0.41	0.91	10.1	0.40	0.22
								M-F	69.5	1.14	2.51	0.74	1.63	0.46	1.01	11.2	0.44	0.24
		2.5	1.1	14.2	6.5	12.8	5.8	A-F	69.3	1.13	2.49	0.76	1.67	0.50	1.09	11.2	0.48	0.23
								M-F	77.0	1.26	2.77	0.84	1.85	0.55	1.21	12.4	0.53	0.26

(Continued)

TABLE 19-4 (Continued)

Weight		Daily Gain		Daily Consumption				Moisture Basis[5] A-F (as-fed) M-F (moisture-free)	Energy								Total Protein (%)	Calcium (%)	Phosphorus (%)
				As-Fed		Moisture-Free Dry Matter			TDN (%)	ME (Mcal per)		NE$_m$ (Mcal per)		NE$_g$ (Mcal per)					
(lb)	(kg)	(lb)	(kg)	(lb)	(kg)	(lb)	(kg)			(lb)	(kg)	(lb)	(kg)	(lb)	(kg)				
								Large-frame heifer calves and compensating medium-frame yearling heifers (Continued)											
600	273	0.5	0.2	14.7	6.7	13.2	6.0	A-F	48.6	0.80	1.76	0.45	1.00	0.23	0.50	7.3	0.20	0.16
								M-F	54.0	0.89	1.96	0.50	1.10	0.25	0.55	8.1	0.22	0.18
		1.0	0.5	15.7	7.1	14.1	6.4	A-F	53.1	0.88	1.94	0.52	1.15	0.29	0.63	8.0	0.25	0.17
								M-F	59.0	0.98	2.16	0.58	1.28	0.32	0.70	8.9	0.28	0.19
		1.5	0.7	16.4	7.5	14.8	6.7	A-F	57.6	0.95	2.08	0.59	1.29	0.35	0.77	8.6	0.30	0.17
								M-F	64.0	1.05	2.31	0.65	1.43	0.39	0.86	9.6	0.33	0.19
		2.0	0.9	16.7	7.6	15.0	6.8	A-F	62.6	1.03	2.26	0.67	1.47	0.41	0.91	9.3	0.34	0.20
								M-F	69.5	1.14	2.51	0.74	1.63	0.46	1.01	10.3	0.38	0.22
		2.5	1.1	16.2	7.4	14.6	6.6	A-F	69.3	1.13	2.49	0.76	1.67	0.50	1.09	10.1	0.40	0.22
								M-F	77.0	1.26	2.77	0.84	1.85	0.55	1.21	11.2	0.44	0.24
700	318	0.5	0.2	16.4	7.5	14.8	6.7	A-F	48.6	0.80	1.76	0.45	1.00	0.23	0.50	7.1	0.19	0.16
								M-F	54.0	0.89	1.96	0.50	1.10	0.25	0.55	7.9	0.21	0.18
		1.0	0.5	17.7	8.0	15.9	7.2	A-F	53.1	0.88	1.94	0.52	1.15	0.29	0.63	7.7	0.23	0.16
								M-F	59.0	0.98	2.16	0.58	1.28	0.32	0.70	8.5	0.25	0.18
		1.5	0.7	18.4	8.4	16.6	7.5	A-F	57.6	0.95	2.08	0.59	1.29	0.35	0.77	8.1	0.26	0.17
								M-F	64.0	1.05	2.31	0.65	1.43	0.39	0.86	9.0	0.29	0.19
		2.0	0.9	18.7	8.5	16.8	7.6	A-F	62.6	1.03	2.26	0.67	1.47	0.41	0.91	8.6	0.30	0.18
								M-F	69.5	1.14	2.51	0.74	1.63	0.46	1.01	9.6	0.33	0.20
		2.5	1.1	18.2	8.3	16.4	7.5	A-F	69.3	1.13	2.49	0.76	1.67	0.50	1.09	9.3	0.34	0.20
								M-F	77.0	1.26	2.77	0.84	1.85	0.55	1.21	10.3	0.38	0.22
800	364	0.5	0.2	18.2	8.3	16.4	7.5	A-F	48.6	0.80	1.76	0.45	1.00	0.23	0.50	6.9	0.18	0.15
								M-F	54.0	0.89	1.96	0.50	1.10	0.25	0.55	7.7	0.20	0.17
		1.0	0.5	20.0	9.1	17.6	8.0	A-F	53.1	0.88	1.94	0.52	1.15	0.29	0.63	7.4	0.22	0.16
								M-F	59.0	0.98	2.16	0.58	1.28	0.32	0.70	8.2	0.24	0.18
		1.5	0.7	20.3	9.2	18.3	8.3	A-F	57.6	0.95	2.08	0.59	1.29	0.35	0.77	7.7	0.23	0.16
								M-F	64.0	1.05	2.31	0.65	1.43	0.39	0.86	8.6	0.25	0.18
		2.0	0.9	20.7	9.4	18.6	8.5	A-F	62.6	1.03	2.26	0.67	1.47	0.41	0.91	8.1	0.25	0.17
								M-F	69.5	1.14	2.51	0.74	1.63	0.46	1.01	9.0	0.28	0.19
		2.5	1.1	20.1	9.1	18.1	8.2	A-F	69.3	1.13	2.49	0.76	1.67	0.50	1.09	8.6	0.30	0.19
								M-F	77.0	1.26	2.77	0.84	1.85	0.55	1.21	9.6	0.33	0.21
900	409	0.5	0.2	19.8	9.0	17.8	8.1	A-F	48.6	0.80	1.76	0.45	1.00	0.23	0.50	6.8	0.18	0.16
								M-F	54.0	0.89	1.96	0.50	1.10	0.25	0.55	7.5	0.20	0.18
		1.0	0.5	21.3	9.7	19.2	8.7	A-F	53.1	0.88	1.94	0.52	1.15	0.29	0.63	7.1	0.20	0.16
								M-F	59.0	0.98	2.16	0.58	1.28	0.32	0.70	7.9	0.22	0.18
		1.5	0.7	22.2	10.1	20.0	9.1	A-F	57.6	0.95	2.08	0.59	1.29	0.35	0.77	7.4	0.21	0.16
								M-F	64.0	1.05	2.31	0.65	1.43	0.39	0.86	8.2	0.23	0.18
		2.0	0.9	22.6	10.3	20.3	9.2	A-F	62.6	1.03	2.26	0.67	1.47	0.41	0.91	7.7	0.23	0.16
								M-F	69.5	1.14	2.51	0.74	1.63	0.46	1.01	8.6	0.26	0.18
		2.5	1.1	22.0	10.0	19.8	9.0	A-F	69.3	1.13	2.49	0.76	1.67	0.50	1.09	8.1	0.26	0.18
								M-F	77.0	1.26	2.77	0.84	1.85	0.55	1.21	9.0	0.29	0.20
1,000	454	0.5	0.2	21.4	9.7	19.3	8.8	A-F	48.6	0.80	1.76	0.45	1.00	0.23	0.50	6.7	0.17	0.16
								M-F	54.0	0.89	1.96	0.50	1.10	0.25	0.55	7.4	0.19	0.18
		1.0	0.5	23.1	10.5	20.8	9.5	A-F	53.1	0.88	1.94	0.52	1.15	0.29	0.63	6.9	0.19	0.16
								M-F	59.0	0.98	2.16	0.58	1.28	0.32	0.70	7.7	0.21	0.18
		1.5	0.7	24.1	11.0	21.7	9.9	A-F	57.7	0.95	2.08	0.59	1.29	0.35	0.77	7.2	0.19	0.16
								M-F	64.0	1.05	2.31	0.65	1.43	0.39	0.86	8.0	0.21	0.18
		2.0	0.9	24.4	11.1	22.0	10.0	A-F	62.6	1.03	2.26	0.67	1.47	0.41	0.91	7.4	0.21	0.16
								M-F	69.5	1.14	2.51	0.74	1.63	0.46	1.01	8.2	0.23	0.18
		2.5	1.1	23.9	10.9	21.5	9.8	A-F	69.3	1.13	2.49	0.76	1.67	0.50	1.09	7.7	0.23	0.16
								M-F	77.0	1.26	2.77	0.84	1.85	0.55	1.21	8.6	0.25	0.18

(Continued)

TABLE 19-4 (Continued)

Weight		Daily Gain		Daily Consumption				Moisture Basis[5] A-F (as-fed) M-F (moisture-free)	Energy								Total Protein	Calcium	Phosphorus
				As-Fed		Moisture-Free Dry Matter			TDN	ME (Mcal per)		NE$_m$ (Mcal per)		NE$_g$ (Mcal per)					
(lb)	*(kg)*	(lb)	*(kg)*	(lb)	*(kg)*	(lb)	*(kg)*		(%)	(lb)	*(kg)*	(lb)	*(kg)*	(lb)	*(kg)*		(%)	(%)	(%)
Large-frame heifer calves and compensating medium-frame yearling heifers (Continued)																			
1,100	*500*	0.5	*0.2*	23.1	*10.5*	20.8	*9.5*	A-F	48.6	0.80	*1.76*	0.45	*1.00*	0.23	*0.50*		6.6	0.17	0.16
								M-F	54.0	0.89	*1.96*	0.50	*1.10*	0.25	*0.55*		7.3	0.19	0.18
		1.0	*0.5*	24.8	*11.3*	22.3	*10.1*	A-F	53.1	0.88	*1.94*	0.52	*1.15*	0.29	*0.63*		6.8	0.18	0.16
								M-F	59.0	0.98	*2.16*	0.58	*1.28*	0.32	*0.70*		7.5	0.20	0.18
		1.5	*0.7*	25.9	*11.8*	23.3	*10.6*	A-F	57.6	0.95	*2.08*	0.59	*1.29*	0.35	*0.77*		6.9	0.18	0.16
								M-F	64.0	1.05	*2.31*	0.65	*1.43*	0.39	*0.86*		7.7	0.20	0.18
		2.0	*0.9*	26.2	*11.9*	23.6	*10.7*	A-F	62.6	1.03	*2.26*	0.67	*1.47*	0.41	*0.91*		7.1	0.19	0.16
								M-F	69.5	1.14	*2.51*	0.74	*1.63*	0.46	*1.01*		7.9	0.21	0.18
		2.5	*1.1*	25.7	*11.7*	23.1	*10.5*	A-F	69.3	1.13	*2.49*	0.76	*1.67*	0.50	*1.09*		7.4	0.20	0.16
								M-F	77.0	1.26	*2.77*	0.84	*1.85*	0.55	*1.21*		8.2	0.22	0.18

[1]Adapted from *Nutrient Requirements of Beef Cattle*, 6th revised edition, National Research Council—National Academy of Sciences, 1984, pp. 77–83, Table 10.

[2]Shrunk liveweight basis. This refers to weight after an overnight shrink without feed and water (generally equivalent to 96% of unshrunk weights taken in early morning).

[3]Vitamin A requirements are 1,000 IU per lb (*2,200 IU per kg*) of ration.

[4]This table gives reasonable examples of nutrient concentrations that should be suitable to formulate rations for specific management goals. It does not imply that rations with other nutrient concentrations when consumed in sufficient amounts would be inadequate to meet nutrient requirements.

[5]As-fed was calculated using an average figure of 90% dry matter. When feeding silages, roots, and other wet feeds, these feeds should be converted to a moisture-free basis and the ration calculated using the moisture-free data.

TABLE 19-5
MINERAL REQUIREMENTS AND MAXIMUM TOLERABLE LEVELS FOR BEEF CATTLE[1]

Mineral		Requirements		Maximum Tolerable Level[3]	Mineral		Requirements		Maximum Tolerable Level[3]
		Suggested Value	Range[2]				Suggested Value	Range[2]	
Calcium	(%)	—	See Tables 19-1, 19-2, 19-3, 19-4	2	Phosphorus	(%)	—	See Tables 19-1, 19-2, 19-3, 19-4	1
Cobalt	(ppm)	0.10	0.07–0.11	5	Potassium	(%)	0.65	0.5–0.7	3
Copper	(ppm)	8	4–10	115	Selenium	(ppm)	0.20	0.05–0.30	2
Iodine	(ppm)	0.5	0.20–2.0	50	Sodium	(%)	0.08	0.06–0.10	10[4]
Iron	(ppm)	50	50–100	1,000	Chlorine	(%)	—	—	—
Magnesium	(%)	0.10	0.05–0.25	0.40	Sulfur	(%)	0.10	0.08–0.15	0.40
Manganese	(ppm)	40	20–50	1,000	Zinc	(ppm)	30	20–40	500
Molybdenum	(ppm)	—	6						

[1]Adapted from *Nutrient Requirements of Beef Cattle*, 6th revised edition, National Research Council—National Academy of Sciences, 1984, p. 43.

[2]The listing of a range in which requirements are likely to be met recognizes that requirements for most minerals are affected by a variety of dietary and animal factors (body weight, sex, rate of gain). Thus, it may be better to evaluate rations based on a range of mineral requirements and for content of interfering substances than to meet a specific dietary value.

[3]From National Research Council (1980). Maximum tolerable levels are given on the basis of the ration dry matter.

[4]10% sodium chloride.

TABLE 19-6
MAXIMUM TOLERABLE LEVELS OF CERTAIN TOXIC ELEMENTS[1,2]

Element	Maximum Tolerable Level, ppm
Aluminum	1,000
Arsenic	50 (100 for organic forms)
Bromine	200
Cadmium	0.5
Fluorine	20–100
Lead	30
Mercury	2
Strontium	2,000

[1]Adapted form *Nutrient Requirements of Beef Cattle*, 6th revised edition, National Research Council—National Academy of Sciences, 1984, p. 43.

[2]National Research Council (1980), Table 4, Mineral Requirements and Maximum Tolerable Levels for Beef Cattle.

Energy

Carbohydrates, which constitute about 75% of all the dry matter of plants, are the chief sources of energy in cattle feed. Next to carbohydrates, fats are important as energy sources. In addition to supplying nitrogen, natural plant protein compounds also supply a certain amount of energy.

A relatively large portion of the feeds consumed by beef cattle is used in meeting the energy needs, regardless of whether the animals are merely being maintained (as in wintering) or fed for growth, finishing, or reproduction.

The first and most important function of feeds is that of meeting the maintenance needs. If there is not sufficient feed, as is frequently true during periods of drought or when winter rations are skimpy, the energy needs of the body are met by the breakdown of body tissue. This results in loss of condition and body weight.

After the energy needs for body maintenance have been met, any surplus energy may be used for reproduction, lactation, growth, or finishing. When cattle are finished at early ages, growth and finishing are concurrent in most instances and, therefore, not easily separated.

In the finishing process, the percentage of protein, ash, and water steadily decreases as the animal matures and fattens, whereas the percentage of fat increases. Thus, the body of a calf at birth may contain about 70% water and 4% fat; whereas the body of a fat 2-year-old steer may contain only 45 to 50% water but from 30 to 35% fat. This storage of fat requires a liberal allowance of energy feeds.

Through bacterial action in the rumen, cattle are able to utilize a considerable portion of roughages as sources of energy. Yet it must be realized that with extremely bulky rations, the animal cannot consume sufficient quantities to produce the desired amount of gain. For this reason, finishing rations generally contain a considerable proportion of concentrated feeds, mostly cereal grains. On the other hand, when the energy requirements are primarily for maintenance, roughages are usually the most economical sources of energy for beef cattle.

At times, fats may be sufficiently economical to merit consideration as partial substitutes for standard energy feeds. Also, very small amounts of fatty acids are essential for beef cattle, as is true in certain other species, but no exact requirements have thus far been established. Normal cattle rations probably meet such fatty acid requirements.

- **Symptoms of energy deficiency (underfeeding)**—Many cattle throughout the world are underfed all or part of the year. In fact, lack of sufficient total feed is probably the most common deficiency suffered by beef cattle, although it is recognized that underfeeding is frequently complicated by a concomitant shortage of protein and other nutrients. Restricted rations often occur during periods of drought, when pastures or ranges are overstocked, or when winter rations are skimpy. Also, many range producers regularly plan that cows in good flesh should lose some condition during the winter months; they feel that it is uneconomical to feed enough to retain the fleshy condition. Fortunately, during such times of restricted feed intake, animals have nutritive reserves upon which they can draw. Although they may survive for a considerable period of time under these conditions, there is an inevitable loss in body weight and condition; and, varying with the degree of underfeeding, there may be a slowing or cessation of growth (including skeletal growth), failure to conceive, and increased mortality. Low feed intake also commonly results in increased deaths from toxic plants and from lowered resistance to parasites and diseases.

Protein

The protein allowance for beef cattle, regardless of age or system of production, should be ample to replace the daily breakdown of the tissues of the body and to provide for the growth of hair, horns, and hooves. In general, the protein needs are greatest for the growth of the young calf and for the gestating-lactating cow.

The protein requirements listed in Tables 19-1 to 19-4 are estimated needs for optimal production. They can be exceeded without toxicity or reduced animal performance. As noted, the requirements are expressed on the basis of both total and digestible protein. Nitrogen values were converted to digestible protein values by multiplying by the factor 6.25 and using an average biological value of 77.5. Cattle fed these levels of protein have gained and reproduced at optimum rates. Methods of feeding, feed preparation, and various feed additives do not appear to alter protein requirements. Feed consumption is reduced when all-concentrate rations are fed. As consumption declines, the percentage of protein in such rations should be increased proportionally.

As protein supplements ordinarily cost more per ton than grains, normally beef cattle should not be fed larger quantities of these supplements than actually are needed to balance the ration.

With stocker cattle, or in the maintenance of the beef breeding herd, it usually does not pay to add a protein supplement when a legume hay is fed. With feedlot cattle fed high-concentrate rations, or when the breeding herd is being wintered on a nonlegume roughage, sufficient protein supplement—usually 1 to 2 lb daily—should be added to the ration.

Because of rumen synthesis of essential amino acids by microorganisms, the quality of protein (or balance of essential amino acids) is of less importance in the feeding of beef cattle than in feeding some other classes of stock. Protein from plant sources, therefore, is quite satisfactory. Also, these microorganisms are able to use inorganic compounds such as ammonia, just as plants utilize chemical fertilizers—build body protein of high quality in their cells from sources of inorganic nitrogen that nonruminants cannot use. Since the life-span of these bacteria is short, further on in the digestive tract, the ruminant digests the bacteria and obtains good protein therefrom. In ruminant nutrition, therefore, even such nonprotein sources of combined nitrogen as urea and ammonia have a protein replacement value. An exception is the very young ruminant in which the rumen and its ability to synthesize are not yet well developed. For such an animal, high-quality protein in the ration is requisite to normal development. In the suckling calf, milk provides such protein.

- **Symptoms of protein deficiencies and toxicities**—Depressed appetite is the primary symptom of protein deficiency in beef cattle rations. Depressed appetite may, in turn, lead to an inadequate intake of energy; hence, protein deficiency and energy deficiency often occur together.

Other symptoms of protein deficiency are loss of weight, poor growth, irregular or delayed estrus, and reduced milk production.

Rations containing up to 40% protein have been fed to steers. Feed intake was reduced for several days when protein was added, but no signs of ammonia toxicity were evident.[2] However, excesses of nonprotein nitrogen or soluble protein may precipitate ammonia toxicity.

---

[2]Fenderson, C. L. and W. G. Bergen, "Effect of Excess Dietary Protein on Feed Intake and Nitrogen Metabolism in Steers," *Journal of Animal Science*, Vol. 42, 1976, p. 1323.

Minerals

Beef cattle are susceptible to the usual inefficiencies and ailments when exposed to (1) prolonged and severe mineral deficiencies, or (2) excesses of fluorine, selenium, or molybdenum (see Chapter 5, Nutritional Disorders/Toxins).

Needed minerals may be incorporated in beef cattle rations or in the water. In addition, it is recommended that all classes and ages of cattle be allowed free access to a two-compartment mineral box, with (1) salt (iodized salt in iodine-deficient areas) in one side, and (2) a suitable mineral mixture in the other side. Free-choice feeding is in the nature of cheap insurance, with the animals consuming the minerals if they are needed.

TABLE
BEEF CATTLE MINERAL

Mineral; Absorption; Excretion	Conditions Usually Prevailing Where Deficiencies Are Reported	Functions of Mineral	Deficiency Symptoms; Toxicity*
Major or Macrominerals: **SALT (NaCl, sodium and chloride)**—The requirements for sodium and chlorine are commonly expressed as salt requirements because salt is an effective, economical way of supplementing rations with these elements. **Absorption**—Sodium and chlorine are mainly absorbed from the proximal portion of the small intestine, but they may also be absorbed from the distal section of the small intestine and from the large intestine. Also, some absorption of sodium and chloride may occur from the rumen. **Excretion**—Excess salt is excreted in the urine.	Negligence; for salt is inexpensive. Deficiencies of sodium and chlorine may occur because plants have low sodium contents, because sodium losses caused by perspiration may occur in animals maintained in warm environments or used for hard work, and because sodium needs increase during lactation and during periods of rapid growth.	Sodium (Na) functions in maintaining osmotic pressure, acid-base balance, and body-fluid balance; is involved in nerve transmission and active transport of amino acids; is required for cellular uptake of glucose through activation of the glucose carrier protein; and is a major cation of extracellular fluid and provides a majority of the alkaline reserve in plasma. Chlorine (Cl) is necessary for the activation of amylase; is essential for the formation of gastric hydrochloric acid; and is involved in respiration and regulation of blood pH, through the chloride shift.	**Deficiency symptoms**—Intensive craving of salt, manifested by the animals chewing and licking various objects, and by muscle cramps. Prolonged deficiency results in lack of appetite, unthrifty appearance, and decreased production. High-producing milk cows may collapse and die when salt deficiency has been of long duration. It is noteworthy that when salt is omitted, sodium expresses its deficiency first. **Toxicity**—*The NRC gives the maximum tolerable level of salt (NaCl) as 10% of ration dry matter. As much as 3 lb (1.4 kg) can be consumed per cow daily without harm provided animals have access to plenty of water.
CALCIUM (Ca)—Calcium is the most abundant mineral in the body. Most of the calcium in the body is found in the bones and teeth. It constitutes 2% of the body weight. In blood, calcium is found mostly in the plasma, with a controlled concentration of 10 mg/100 ml. **Absorption**—Calcium is absorbed actively from both the duodenum and the jejunum; but most calcium is absorbed in the proximal portion of the duodenum. **Excretion**—Calcium is excreted mainly in feces with only small quantities appearing in urine.	A calcium deficiency may occur when finishing cattle are fed heavily on concentrates and limited quantities of nonlegume roughage, especially young cattle on a long feed. Adding calcium to such a ration increases the rate of gain, improves feed utilization, results in heavier, stronger bones, and enhances market grades. Also, a calcium deficiency may occur when the ration consists chiefly of dried mature grasses or cereal straws, and when cows are in heavy lactation. Osteomalacia may occur when there are high metabolic demands on calcium and phosphorus stores, such as occur during pregnancy and lactation.	Essential for bone formation, development of teeth, production of milk, transmission of nerve impulses, maintenance of normal muscle excitability (along with sodium and potassium), regulation of heart beat, movement of muscles, blood clotting (conversion of prothrombin to thrombin), and activation and stabilization of enzymes (i.e., pancreatic amylase).	**Deficiency symptoms**—A deficiency of calcium results in rickets in young animals and osteomalacia in older animals. Rickets may be caused by a deficiency of calcium, phosphorus, or vitamin D. It is characterized by improper calcification of the organic matrix of bones of young, growing animals. Thus, the bones are weak, soft, and lack density. Signs include swollen tender joints; enlargement of the ends of bones; and arched back; stiffness of the legs; and development of beads on the ribs. If the cause is not corrected, calves develop bowed and deformed legs. Also, rachitic bones are highly susceptible to fracture. Osteomalacia is the result of demineralization of the bones of adult animals. This condition is characterized by weak, brittle bones that may break when stressed. **Toxicity**—Ruminants tolerate high levels of caclium. *The NRC gives the maximum calcium level as 2% of ration dry matter. However, when high levels of calcium are fed, there may be reduced feed consumption and daily gains; reduced protein and energy digestibilities; reduced absorption of tetracyclines, manganese, and zinc; and stimulation of the production of calcitonin of the thyroids. Calcitonin inhibits bone resorption, with the result that the bones may thicken (osteopetrosis) because of continued deposition but limited resorption.
PHOSPHORUS (P)—Phosphorus has varied, but extremely important, biochemical and physiological roles **Absorption**—Phosphorus absorption is dependent on source, intestinal pH, age of animal, and ration levels of sodium, calcium, iron, aluminum, manganese, potassium, magnesium, and fat. **Excretion**—Excess phosphorus is excreted primarily in the feces.	Semiarid regions are commonly associated with soils deficient in phosphorus. The phosphorus content of plants generally decreases markedly with maturity, with the result that deficiencies often occur in cattle subsisting for long periods on mature dried forage. High iron levels result in the formation of insoluble iron phosphate. Also, aluminum forms insoluble, unavailable phosphates.	Phosphorus is deposited in bones. It is also found in high concentrations in brain, muscle, liver, spleen, and kidneys. Phosphorus, as a component of phospholipids, influences cell permeability and is a component of myelin sheathing of nerves. Also, many energy transfers in cells involve the high-energy phosphate bonds in ATP. Phosphorus plays an important role in blood buffer systems. Activation of several B-vitamins (thiamin, niacin, pyridoxine, riboflavin, biotin, and pantothenic acid) to form coenzymes requires their initial phosphorylation. Phosphorus is also a part of the genetic materials DNA and RNA.	**Deficiency symptoms**—Phosphorus deficiencies in cattle are widespread. A deficiency of phosphorus results in decreased growth rates, in inefficient feed utilization, and in a depraved appetite (chewing of wood, soil, and bones—called pica); anestrus, low conception rate, and reduced milk production; low plasma phosphorus levels, and weak, fragile bones and stiffness of joints. **Toxicity**—*The NRC gives the maximum phosphorus level as 1% of ration dry matter. High phosphorus intakes may cause bone resorption, elevated plasma phosphorus levels, and urinary calculi.

Feeding Beef Cattle

The calcium and phosphorus requirements of cattle are presented in Tables 19-1, 19-2, 19-3, and 19-4. The requirements and maximum tolerable levels of other minerals are presented in Table 19-5. Maximum tolerable levels of several elements that are known to be toxic to cattle are presented in Table 19-6. (Also see Chapter 5, Nutritional Disorders/Toxins, Table 5-1, for additional information relative to mineral toxicities.)

BEEF CATTLE MINERAL CHART

Table 19-7, Beef Cattle Mineral Chart, presents in summary form pertinent information pertaining to the mineral needs of beef cattle.

19-7 CHART (See footnotes at end of table.)

Nutrient Requirements[1]		Recommended Allowances[1]	Practical Sources of the Mineral	Comments
Daily Nutrients/ Animal	Percentage of Ration			
For young, growing animals: 2-3 g of sodium, and less than 5 g of chlorine. For lactating cows: 11 g of sodium, and 15 g of chlorine.	*Sodium concentrations of 0.06-0.10% of ration dry matter for nonlactating yearlings and calves, and no more than 0.1% dry matter for lactating beef cows. (See Table 19-5.)	Cows on pasture or fed high-roughage winter rations will consume from 1-3 lb (0.45-1.36 kg) salt per head per month; finishing steers fed heavy grain rations in drylot will consume 1-3.5 lb (0.45-1.59 kg) per head per month; a wide range due to differences in age, rations, form of salt (rock, coarse ground, or block). Most ranchers compute the yearly salt requirements on the basis of 25 lb (10 kg) per cow. The careful location of the salt supply is an important adjunct in range management.	Salt should be available at all times. It should be both (1) self-fed, free-choice, and (2) mixed with other ration ingredients. Free access to salt in the form of loose rock, coarse ground, or block salt. Cattle prefer loose salt to block salt, because it can be eaten more rapidly and with less effort. However, experiments with lactating cows have shown fully as good results with block salt as with loose salt even though smaller quantities were consumed. This means that the additional intake of loose salt over block salt does not appear to benefit cattle. Commercial mineral mixes (in block, or loose form) may contain ½ or more salt.	The salt requirements of cattle differ (1) between individuals, (2) according to whether milk is produced (being higher for lactating cows than for dry cows, because of the salt in the milk), (3) from season to season, (4) between block and loose salt (animals often consuming twice as much easy-to-get loose salt as block salt), and (5) according to the salt content of the soil, feed, and water (being higher when vegetable proteins are fed than when animal proteins are fed, higher on predominantly forage rations than on predominantly concentrate rations, and higher on lush early pasture than on more mature grasses). These are some of the reasons why free-choice feeding of salt is advocated.
*Variable, according to age weight, and type and level of production of cattle. (See Tables 19-1 and 19-3.) Because true digestibilities of calcium in feedstuffs vary, the dietary calcium requirements shown in the tables may in some instances need to be adjusted.	*Variable, according to age, weight, and type and level of production of cattle (See Tables 19-2 and 19-4.) Because true digestibilities of calcium in feedstuffs vary, the dietary calcium requirements shown in the tables may in some instances need to be adjusted.	Free access to a calcium supplement, or a calcium supplement incorporated in the ration.	Legumes are high in calcium. Also, several of the oilseed meals are good sources of calcium. Sources of supplemental calcium include calcium carbonate, ground limestone, bone meal, dicalcium phosphate, defluorinated phosphate, monocalcium phosphate, and calcium sulfate. Where both calcium and phosphorus need to be supplemented, they should be provided in a readily available and palatable form such as dicalcium phosphate, defluorinated phosphate, or bone meal.	In addition to an adequate supply of calcium, proper utilization is dependent upon (1) a highly available source of the mineral, (2) a suitable ratio between calcium and phosphorus (somewhere between 1 and 2 parts of calcium to 1 part of phosphorus). Calcium-phosphorus ratios of 2:1 have been shown to be beneficial in reducing urinary calculi. When calculi problems are encountered, even higher levels of calcium may be advisable. Ratios between calcium and phosphorus of 7:1 have been reported to be satisfactory for cattle. Generally, when cattle receive at least ⅓ of a legume forage, ample calcium will be provided. But even nonlegume forages contain more calcium than cereal grains. Plants grown on calcium-rich soils are high in calcium. Calcium availability of 70% is generally assumed for all feedstuffs.
*Variable, according to age, weight, and type and level of production. (See Tables 19-1 and 19-3.)	*Variable according to age, weight, and type and level of production. (See Tables 19-2 and 19-4.)	Free access to a phosphorus supplement, or a phosphorus supplement added to the daily ration in keeping with the nutrient requirements. Where phosphorus is added to water, either of the following methods may be employed: 1. Added by hand at rate of ¼ oz (7 g) of monosodium phosphate/8 gal (30 liter) water, or ¼ oz/head/day. 2. Added by dispenser, using stock solution of 2½ lb (1.13 kg) of monosodium phosphate/gal (3.8 liter) water (or 100 lb/ 40 gal [45 kg/151 liter] water).	Common sources of phosphorus are: dicalcium phosphate, defluorinated phosphate, bone meal, soft phosphate, sodium phosphate, ammonium polyphosphate, orthophosphates, metaphosphates, pyrophosphates, and tripolyphosphate. Oilseed meals and animal and fish products contain large amounts of phosphorus. Phytate phosphorus is not well utilized by nonruminants, but ruminants appear to use considerable quantities of this form of phosphorus.	Grains, grain by-products and high-protein supplements are fairly high in phosphorus; hence, rations high in such ingredients require little or no phosphorus supplementation. Calcium-phosphorus ratios of 2:1 are beneficial in reducing urinary calculi; and even higher levels of calcium may be necessary when urinary calculi is encountered. Ratios between calcium and phosphorus of 7:1 have been reported to be satisfactory for cattle.

(Continued)

TABLE 19-7

Mineral; Absorption; Excretion	Conditions Usually Prevailing Where Deficiencies Are Reported	Functions of Mineral	Deficiency Symptoms; Toxicity*
Major or Macrominerals: (Continued) **MAGNESIUM (Mg)**—Magnesium is the fourth most abundant cation in the body. **Absorption**—Absorption of magnesium occurs prior to the intestines, from the small intestine, and some from the large intestine. **Excretion**—Excretion of endogenous magnesium is primarily via feces. However, excess magnesium is disposed of primarily via urine.	When milk feeding of calves is prolonged without grain or hay. (Milk is rather low in magnesium.) When there is grass tetany, which is most likely to occur when beef cows in early lactation graze early spring pastures containing less than 0.2% magnesium.	Magnesium is required for skeletal development as a constituent of bone; plays an important role in neuromuscular transmission and activity; is required to activate many enzyme systems, including those involving ATP; and is required as a cofactor in decarboxylation and an activator of many peptidases. Approximately 65% of total body magnesium is contained in bone; the other 35% is distributed among various tissues and organs.	**Deficiency symptoms**—Grass tetany or grass staggers, characterized by anorexia, hyperemia, hyperirritability, convulsions, and death. Magnesium-deficient cattle exhibit loss of appetite and reduced dry matter digestibilities. Deficiencies in young cattle may result in defective bones and teeth. **Toxicity**—Normal rations will not cause toxicity. The maximum tolerable level of magnesium is considered to be 0.4% of the ration. Toxicity is characterized by loss of appetite,

reduced performance, and occasional diarrhea. Also, cattle experiencing toxicity may exhibit lack of reflexes and respiration depression.

POTASSIUM (K)—Potassium is the third most abundant mineral element in the body. **Absorption**—Potassium is primarily absorbed in the small intestine. **Excretion**—Excretion is mainly via the kidneys.	When drylot finishing cattle receive high- or all-concentrate rations.	Essential for proper enzyme, muscle, and nerve function, rumen microorganism activity, and appetite.	**Deficiency symptoms**—Poor appetite and feed conversion, slow growth, stiffness, and emaciation. **Toxicity**—*The NRC gives the maximum tolerable level of potassium as 3% of ration dry matter. Toxicity from excessive intake is unlikely except (1) when water intake is restricted or water is saline, or (2) when the kidneys are not functioning properly.
SULFUR (S)—Sulfur is a component of protein, some vitamins, and several important hormones.	Cattle fed high-grain rations supplemented with nonprotein nitrogen.	Body functions that involve sulfur include protein synthesis and metabolism, fat and carbohydrate metabolism, blood clotting, endocrine function, and intra- and extracellular fluid acid-base balance. Sulfur has both structural and metabolic functions; it is found in virtually every tissue and organ of the body. Muscle has a fairly constant nitrogen to sulfur ratio of 15.3:1. The total body content of sulfur is approximately 0.15%.	**Deficiency symptoms**—Depressed appetite, loss of weight, weakness, excessive salivation, watery eyes, dullness, emaciation, and death. A lack of sulfur also results in a microbial population that does not utilize lactate. **Toxicity**—*The NRC gives the toxic level of sulfur as 0.40% of the ration dry matter. Sulfur toxicity is characterized by restlessness, diarrhea, muscular twitching, dyspnea, and in prolonged cases of inactivity followed by death.
Trace or Microminerals: **COBALT (Co)**—The cobalt requirement of cattle is actually a cobalt requirement of rumen microorganisms. The microbes incorporate cobalt into vitamin B-12, which is utilized by both microorganisms and animal tissues. **Absorption**—Vitamin B-12 is absorbed in the lower part of the small intestine. **Excretion**—Cobalt and vitamin B-12 are mainly excreted in the feces, although variable amounts are excreted in urine.	In cobalt-deficient soils where this element is not provided. Cobalt-deficient soils occur in many parts of the world, with large deficient areas in Australia, New Zealand, and along the southeast Atlantic Coast of the U.S.	The main function of cobalt is to serve as an integral part of vitamin B-12 (cobalamin). Vitamin B-12 is of importance in the metabolism of propionic acid, needed for the activity of the enzyme methylmalonyl-CoA isomerase. Vitamin B-12 is also a part of the enzyme that catalyzes the recycling of methionine from homocysteine after the loss of its labile methyl group. Vitamin B-12 is also needed for normal liver folate metabolism.	**Deficiency symptoms**—Loss of appetite and body weight, muscular wasting, severe anemia, followed by death. In severe deficiency, the mucous membranes become blanched, the skin turns pale, a fatty liver develops, and the body becomes almost totally devoid of fat. **Toxicity**—Cobalt toxicity is rare because toxic levels are about 300 times requirement levels. *The NRC gives the maximum tolerable level of cobalt as 5% of the ration dry matter.
COPPER (Cu) **Absorption**—Copper is absorbed from the upper portion of the duodenum. Zinc and silver are antagonistic to copper absorption. **Excretion**—Copper is released into bile, thence into feces. Trace amounts of copper are excreted in urine, perspiration, and milk.	In copper-deficient areas (soils), as in Florida and the Coastal Plain region. On peat and muck soils, or where soil molybdenum levels are high. Deficiencies have occurred in calves kept on an exclusive milk diet for long periods.	Copper is necessary in hemoglobin formation, iron absorption from the small intestine, iron mobilization from tissue stores, and for the oxidation of iron, permitting it to bind with the iron transport—transferrin. Copper is essential in enzyme systems, hair development and pigmentation, bone development, reproduction, and lactation.	**Deficiency symptoms**—Emaciation, depigmentation (cattle turn yellowish) and loss of hair, stunted growth, anemia, and brittle and malformed bones. Also, heat periods are suppressed, and there may be depraved appetite and diarrhea. Young calves may have straight pasterns and stand forward on their toes. Low copper intake reduces the synthesis and activity of the copper-containing enzyme, tyro-

sinase, which is required for pigmentation of hair, wool, and feathers.
Toxicity—*Maximum tolerable levels for cattle are 115 ppm of dry matter. Acute toxicity may cause nausea, vomiting, salivation, abdominal pain, convulsions, paralysis, collapse, and death. Also, high copper levels may predispose animals to anemia, muscular dystrophy, decreased growth, and impaired reproduction.

Feeding Beef Cattle

(Continued)

Nutrient Requirements[1]		Recommended Allowances[1]	Practical Sources of the Mineral	Comments
Daily Nutrients/ Animal	**Percentage of Ration**			
*Young calves and growing-finishing cattle, 12–30 mg/kg body weight. *Beef cows, 7–9 g/day during gestation and 21, 22, and 18 g/day during early, mid, and late lactation, respectively. Magnesium requirements are increased by feeding high levels of aluminum, potassium, phosphorus, or calcium; by younger cattle and magnesium-deficient cattle; and by high levels of milk production.	*0.10% of dry matter, with a range of 0.05–0.25.		Commonly used feedstuffs vary widely in magnesium content and availability. Magnesium carbonate, oxide, and sulfate are good sources of supplemental magnesium.	Supplemental feeding of magnesium (20 g/day) reduces the incidence of grass tetany in many outbreaks.
	*0.65% of the total ration dry matter, with a range of 0.5–0.7%. The needs for potassium vary with amounts of protein, phosphorus, calcium, and sodium consumed.	*0.7–1.0% of the total ration dry matter.	Roughages usually contain ample potassium. Potassium chloride is the supplement of choice.	Grains often contain less than 0.5% potassium. Excessive levels of potassium have been found to interfere with magnesium absorption. Also, excessive levels of potassium, along with high levels of phosphorus, increase the incidence of phosphatic urinary calculi.
	*0.10% of ration dry matter, with a range of 0.08–0.15%.	*The NRC suggested maximum level of sulfur in the ration is 0.4%.	Feeds high in protein are usually high in sulfur. The microbial population of the rumen has the ability to convert inorganic sulfur into organic sulfur compounds that can be used by the animal. So, either organic or inorganic sulfur can be utilized by cattle. Most feedstuffs provided to beef cattle contain sufficient sulfur to meet their needs.	Copper requirements are increased by both sulfur and molybdenum. Selenium can replace sulfur in some organic compounds.
	*0.10 ppm of ration dry matter, with a range of 0.07–0.11 ppm of dry matter.	Free access to a cobaltized mineral mixture in cobalt-deficient areas; or administering a cobalt pellet.	A cobaltized mineral mixture may be prepared by adding cobalt at the rate of 0.2 oz/100 lb (1.25 mg/kg) of salt as cobalt chloride or cobalt sulfate, cobalt carbonate, cobalt oxide, or a good commercial mineral mixture or salt product may be used. Also, cobalt sulfate and cobalt oxide are effective as a drench; and a cobalt pellet (composed of cobalt oxide and finely divided iron) that lodges in the reticulum is an effective preventive.	Several good commercial cobalt-containing minerals are on the market. A vitamin B–12 injection will relieve a cobalt deficiency.
	*8 ppm of dry matter, with a range of 4–10 ppm of dry matter. For presence of high levels of molybdenum and inorganic sulfate, increase the copper requirements.	*Copper deficiency can be prevented by adding 0.25–0.5% copper sulfate to salt fed free-choice. *Copper (Cu) added to total feed (dry basis) 4 ppm. Copper may also be injected as glycinate to meet the nutritional needs for the mineral.	*Salt containing 0.25–0.5% copper sulfate.	Copper deficient cattle can be returned to normal by feeding 3 g of copper sulfate or blue vitriol every 10 days. An interesting interrelation exists between copper and molybdenum. An excess of molybdenum (in the presence of sulfate) causes a condition which can be cured only by administering copper. Excess copper is toxic; it accumulates in the liver, and death may result.

(Continued)

TABLE 19-7

Mineral; Absorption; Excretion	Conditions Usually Prevailing Where Deficiencies Are Reported	Functions of Mineral	Deficiency Symptoms; Toxicity*
Trace or Microminerals (Continued) **IODINE (I)** **Absorption**—In ruminants, the rumen is the primary absorption site. **Excretion**—Two-thirds of ingested inorganic iodine is excreted by the kidneys.	In iodine-deficient areas (soils) where iodized salt is not fed (in northwestern U.S. and in the Great Lakes Region). Where feeds come from iodine-deficient areas Substances that interfere with iodine metabolism. Rapeseed meal, soybean meal, and cottonseed meal have goitrogenic effects.	Inorganic iodine is taken up by the thyroid gland for the synthesis of thyroid hormones. Thyroid hormones have an active role in thermoregulation, intermediary metabolism, reproduction, growth and development, circulation, and muscle function.	**Deficiency symptoms**—Goiter, hairlessness in the young; retarded growth and maturity, lowered metabolic rate, and increased water retention. Occasional borderline cases may survive; in these, the moderate thyroid enlargement disappears in a few weeks. **Toxicity**—*50 ppm is the maximum tolerable level for calves. Symptoms of iodine toxicity include loss of appetite, coma, and death.
IRON (Fe) **Absorption**—Iron may be absorbed from all sections of the small intestine, but the principal site of absorption is the duodenum. Ferrous iron is absorbed to a much greater extent than ferric iron. **Excretion**—Excretion of iron occurs in urine, feces, sweat, dermis, and blood.	Calves on an exclusive milk ration (milk contains less than 10 ppm iron). Animals with excessive blood loss.	Iron has important biochemical functions in animals since it is a component of hemoglobin, myoglobin, cytochrome, and the enzymes catylase and peroxidase. Iron in these materials exists in porphyrin rings. Iron is involved in the transport of oxygen to cells and in cellular respiration.	**Deficiency symptoms**—Signs of lack of iron include anemia, reduced saturation of transferrin, listlessness, pale mucous membrane, reduced appetite and weight gain, and atrophy of the papillae of the tongue. **Toxicity**—*An iron level of 1,000 ppm is considered as the maximum tolerable level for cattle. Iron toxicity is characterized by reduced feed intake, reduced daily gain, diarrhea, hypothermia, and metabolic acidosis.
MANGANESE (Mn) **Absorption**—Ruminants regulate manganese levels in blood and tissue via intestinal absorption. **Excretion**—Manganese is excreted via feces, with little in the urine.	In northwestern U.S. All-concentrate rations based on corn supplemented with nonprotein nitrogen.	Manganese is essential for normal reproduction in both males and females, for bone formation, and for the functioning of the central nervous system. Also, manganese is a preferred metal cofactor for many enzymes involved in carbohydrate metabolism and in mucopolysaccharide synthesis.	**Deficiency symptoms**— In males: impaired spermatogenesis, testicular and epididymal degeneration, sex hormone inadequacy, and sterility. In females: irregular and absent estrus, delayed conception, abortion, and deformed young at birth—crooked calves. **Toxicity**—For ruminants, manganese is among the least toxic of required minerals. *With balanced rations, about 1,000 ppm is the maximum tolerable level on a short-term basis for cattle.
MOLYBDENUM (Mo)—Molybdenum is found in nearly all body cells and fluids. **Absorption**—Molybdenum is well absorbed by cattle, chiefly from the small intestine. **Excretion**—Excretion of molybdenum is primarily via urine, with small amounts excreted in bile and milk.	Molybdenum toxicity occurs only occasionally in cattle and appears to be an area problem.	Molybdenum is a constituent of the enzymes xathine oxidase, aldehyde oxidase, and sulfide oxidase; enzymes involved in the oxidation of purines and reduction of cytochrome C.	**Deficiency symptoms**—Molybdenum deficiencies have not been demonstrated in cattle. **Toxicity**—*The NRC gives the maximum tolerable level as 6 ppm. Clinical signs of molybdenum toxicity in cattle are diarrhea, loss of appetite, anemia, ataxia, and bone malformation.
SELENIUM (Se)—Initially, interest in selenium was confined to the problem of toxicity in animals. **Absorption**—Most selenium is absorbed in the duodenum. **Excretion**—Selenium is excreted in feces and urine; fecal excretion is greater than urinary excretion in ruminants.	Low selenium forage and low vitamin E. It is an area problem, but it occurs in many parts of the U.S.	Selenium functions (1) as a component of glutathione peroxidase, an enzyme that destroys peroxides in tissues, and (2) intertwined with vitamin E in a mutual sparing effect.	**Deficiency symptoms**—White muscle disease; characterized by white muscle, heart failure, and paralysis evidenced by lameness or inability to stand. Depression of glutathione peroxidase in tissues of selenium-deficient animals may account for many of the manifestations of selenium deficiency.
	Toxicity—The NRC suggests that 2 mg/kg (*2 ppm*) dry weight ration is the maximum tolerable levels for all species. Signs of toxicity include loss of appetite, loss of tail hair, sloughing of hoofs, and eventual death. Two types of selenium poisoning have been observed: (1) acute, blind staggers; and (2) chronic, alkali disease. Selenium toxicity can be counteracted by feeding some forms of sulfur. Toxic levels reported in South Dakota, North Dakota, Montana, Wyoming, Utah, Nebraska, Kansas, and Colorado.		
ZINC (Zn) **Absorption**—Absorption of zinc occurs primarily from the abomasum and lower small intestine. **Excretion**—The primary route of excretion is the feces.	Zinc deficiencies have been reported in ruminants grazing forages low in zinc or high in compounds interfering with zinc utilization.	Zinc functions as both an activator and a constituent of several dehydrogenases, peptidases, and phosphates that are involved in nucleic acid metabolism, protein synthesis, and carbohydrate metabolism.	**Deficiency symptoms**—Deficiencies are characterized by decreased performance and listlessness, followed by development of swollen feet and a dermatitis that is most severe on the neck, head, and legs. Deficiencies may also result in vision impairment, excessive salivation, decreased rumen volatile fatty acid production, failure of wounds to heal normally, and impaired reproductive performance in both bulls and cows. **Toxicity**—The maximum zinc tolerance level is dependent on the ration, particularly concentrations of minerals that affect zinc absorption and utilization. *The NRC lists the maximum tolerable level of zinc as 500 ppm, but the NRC also reports that steers have been fed rations containing 1,000 ppm zinc for 13–18 months without marked reduction of performance.

[1]As used herein, the distinction between *nutrient requirements* and *recommended allowances* is as follows: In nutrient requirements, no margins of safety are included intentionally; whereas in recommended allowances, margins of safety are provided to compensate for variations in feed composition, environment, and possible losses during storage or processing.

*Where preceded by an asterisk, the toxicity levels, nutrient requirements, and recommended allowances listed herein were taken from *Nutrient Requirements of Beef Cattle*, sixth revised edition, National Research Council–National Academy of Sciences, Washington, D.C., 1984.

(Continued)

Nutrient Requirements[1]		Recommended Allowances[1]	Practical Sources of the Mineral	Comments
Daily Nutrients/ Animal	**Percentage of Ration**			
*1 mg/day for a 1,100-lb (*500-kg*) cow.	*Iodized salt at rate of 0.10% of dry ration. *0.5 ppm iodine in dry matter, with a range of 0.20–2.0 ppm iodine in dry matter.	Free access to stabilized iodized salt containing 0.01% potassium iodide (0.0076% iodine).	Stabilized iodized salt containing 0.01% potassium iodide. Feed additives that supply iodine are: ethylenediamine dihydroiodide (EDDI), calcium iodate, cuprous iodide, potassium iodate, sodium iodate, potassium iodide, sodium iodide, and pentacalcium periodate.	The enlargement of the thyroid gland (goiter) is nature's way of trying to make enough thyroxin, when there is insufficient iodine in the feed. Eighty percent of hormonal iodine stored in the thyroid is thyroxin. The amount of iodine in milk is influenced by iodine intake, season, level of milk production, and use of iodine disinfectants.
	*100 ppm for calves; 50 ppm for older cattle.		Levels of iron in common feed believed to be ample. Sources of supplemental iron in decreasing order of availability are: ferrous sulfate, ferrous carbonate, ferric chloride, and ferric oxide.	After calves are past 20 weeks of age, iron does not seem to be beneficial. About 30% of all calves are affected by prenatal iron deficiency. In cattle, a majority of body iron is in the form of hemoglobin, with lesser amounts existing as protein-bound stored iron, myoglobin, and cytochrome.
	*40 ppm for mature cows and bulls and 20 ppm for growing-finishing cattle, with a range of 20–50 ppm. **Note well**: Requirements for manganese are increased by elevated dietary levels of calcium and phosphorus.	An intake of 40 ppm for mature breeding cattle and 20 ppm for growing-finishing cattle.	Most forages contain high levels of manganese. Manganous oxide, sulfate, and carbonate are good sources of supplemental manganese.	The manganese levels in pastures, grains, and forages are variable because of variations in plant species, soil types, soil pH, and fertilization practices. A deficiency of manganese exists in northwestern U.S., where it has been shown to cause *crooked calves*.
Requirements for molybdenum are not established. Because copper and sulfate alter molybdenum metabolism, arriving at the molybdenum requirement is impossible.	Requirements for molybdenum are not established. Because copper and sulfate alter molybdenum metabolism, arriving at the molybdenum requirement is impossible.	As a feed additive, molybdenum is not cleared by the Food and Drug Administration.	Many feeds contain 6.8–13.6 mg/lb of ration dry matter.	Excess molybdenum may cause a copper deficiency. Sulfur, in the absence of molybdenum, also may cause a copper deficiency. Increasing copper level in ration to 1 g/head daily is effective in overcoming molybdenum toxicity in beef cattle.
	The selenium requirement of beef cattle depends on the amount of vitamin E in the ration, but ranges are suggested as follows by the NRC: *Growing-finishing steers and heifers, 0.10 mg/kg (*0.1 ppm*) dry weight ration. *Breeding bulls and pregnant and lactating cows, 0.05–0.10 mg/kg (*0.05–0.10 ppm*) dry weight ration.	*0.05–0.30 mg/kg (*0.1–0.3 ppm*) dry weight of ration.	In 1979, the FDA approved the addition of selenium as either sodium selenite or sodium selenate at the rate of 0.1 ppm complete feed for beef cattle, dairy cattle, and sheep. In 1987, FDA increased the allowance of selenium in complete feeds for cattle (beef and dairy), sheep, swine, chickens, turkeys, and ducks from 0.1 ppm to 0.3 ppm.	Selenium toxicity may occur when cattle consume feeds containing 10–30 ppm of selenium on a dry matter basis for an extended period. In Israel, in a series of experiments extending over 3 years, low doses of selenium injected intramuscularly reduced the incidence of retained placenta to half that of the controls. (Eger, S., et al., "Effect of Selenium and Vitamin E on the Incidence of Retained Placenta," *Journal of Dairy Science*, Vol. 68, No. 8, Aug. 1985, p. 219.)
	*30 ppm of ration dry matter, with a range of 20–40 ppm of dry matter. Beef cows with high levels of milk production have higher requirements, because milk contains 300–500 mg (*300–500 ppm*) of zinc per liter. Requirements vary according to age and growth rate, since zinc absorption decreases with age and as growth rate decreases. Requirements may be altered by dietary levels of cadmium, calcium, iron, magnesium, manganese, molybdenum, and selenium, since these minerals affect zinc absorption and/or utilization.	*20–40 ppm zinc in the total feed (air-dry basis).	Feedstuffs vary widely in zinc concentrations, with legumes usually having higher concentrations than grasses, and with protein supplements of animal origin being higher than other protein supplements.	Mild zinc deficiency in feedlot cattle results in lowered weight gains without the development of a specific syndrome.

Mineral recommendations for all classes and ages of cattle: Provide free access to a two-compartment mineral box, with (1) salt (iodized salt in iodine-deficient areas) in one side and (2) dicalcium phosphate, defluorinated phosphate, or a mixture of ⅓ salt (salt added for purposes of palatability) and ⅔ steamed bone meal in the other side. Also, the mineral requirements may be met by using a good commercial mineral, in either block or loose form. If desired, the mineral supplement may be incorporated in the ration in keeping with the recommended allowances given in this table.

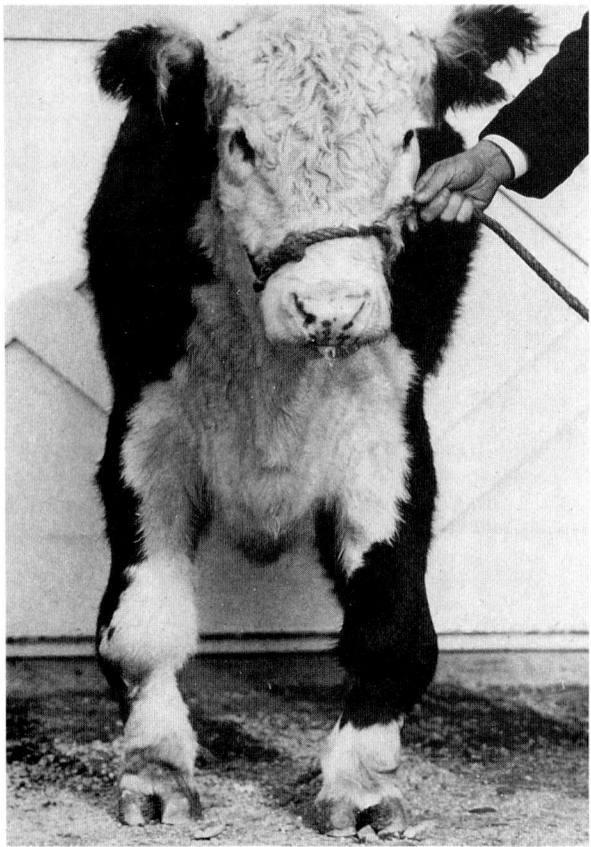

Fig. 19-1. This heifer developed rickets early in life due to a deficiency of calcium and phosphorus. Note the bowed front legs and enlarged joints. (Courtesy, USDA)

Fig. 19-3. Low phosphorus semi-arid range in southwestern U.S. (Courtesy, *Livestock Weekly*, San Angelo, Texas)

Fig. 19-4. Mineral/protein blocks in use on pasture. (Courtesy, Moorman Mgf. Co., Quincy, Ill.)

Fig. 19-2. Copper deficiency in calf. Note rough coat and bleaching of hair. (Courtesy, University of Florida, Gainesville)

Vitamins

The absence of one or more vitamins in the ration may lead to a failure in growth or reproduction, or to characteristic disorders known as vitamin deficiency diseases. In severe cases, death itself may follow. Although the occasional deficiency symptoms are the most striking result of vitamin deficiencies, it must be emphasized that in practice, mild deficiencies probably cause higher total economic losses than do severe deficiencies. It is relatively uncommon for a ration, or diet, to contain so little of a vitamin that obvious symptoms of a deficiency occur. When one such case does appear, it is reasonable to suppose that there must be several cases that are too mild to produce characteristic symptoms but which are sufficiently severe to lower the state of health and the efficiency of production.

Cattle have physiological requirements for most vitamins needed by other mammals. Synthesis by microorganisms in the rumen, supplies in natural feedstuffs, and synthesis in tissues meet most of the usual requirements. Although colostrum is rich in vitamins, providing immediate protection to the newborn calf, calves have minimal stores of vitamins at birth. The ability of the calf to synthesize B vitamins and vitamin K in the rumen develops rapidly when solid feed is introduced into the ration. Vitamin D is synthesized by animals exposed to direct sunlight and is found in large amounts in sun-cured forages. High-quality forages contain large amounts of vitamin A precursors and vitamin E.

Table 19-8 lists the vitamin requirements of beef cattle.

TABLE 19-8
VITAMIN REQUIREMENTS OF BEEF CATTLE
(in Percentage or Amount per Kilogram of Dry Ration)[1]

Nutrient	Growing and Finishing Steers and Heifers	Dry Pregnant Cows	Breeding Bulls and Lactating Cows
	◄------ % per kg dry ration ------►		
Vitamin A activity .. (IU)[2]	2,200	2,800	3,900
Vitamin D .... (IU)	275	275	275
Vitamin E .... (IU)	15-60	—	15-60

[1]From *Nutrient Requirements of Beef Cattle*, 6th rev. ed., National Research Council—National Academy of Sciences, Washington, D.C., 1984.
[2]May be vitamin A or provitamin A equivalent.

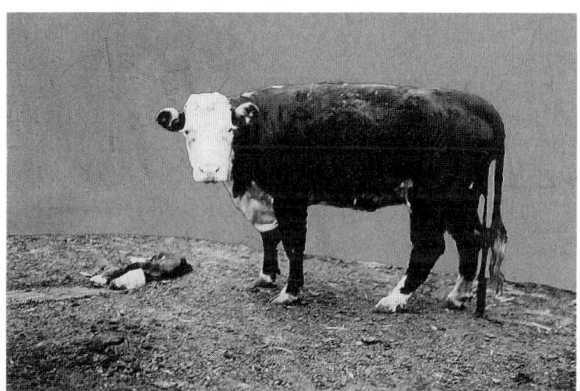

Fig. 19-5. Cows in drought area, deficient in energy and Vitamin A. (Courtesy, FAO, United Nations)

Fig. 19-7. Effect of vitamin A deficiency on reproduction. The heifer in the upper picture received a ration deficient in vitamin A, but otherwise complete. She became night blind and aborted during the last month of pregnancy; also, note the retained placenta. The heifer in the lower picture received the same ration, but during the latter part of the gestation period, a daily supplement of 1 lb of dehydrated alfalfa meal containing 50 mg of carotene was added. She produced a normal vigorous calf. (Courtesy, Calif. Ag. Exp. Sta., Davis, Calif.)

BEEF CATTLE VITAMIN CHART

Table 19-9, Beef Cattle Vitamin Chart (see next page), presents in summary form pertinent information pertaining to the vitamin needs of beef cattle.

(See Appendix Tables V-1A, V-1B, V-1C, V-1D, and V-1E for the vitamin content of feeds commonly used in beef cattle rations.)

Fig. 19-6. Tall grass cured on the stalk, deficient in Vitamin A. (Courtesy, *Livestock Weekly*, San Angelo, Texas)

TABLE
BEEF CATTLE

Vitamins Which May Be Deficient Under Normal Conditions	Conditions Usually Prevailing Where Deficiencies Are Reported	Functions of Vitamin	Deficiency Symptoms
Fat-Soluble Vitamins: **A**—Vitamin A is found only in animals; plants contain the precursor—carotene. **V**itamin A is the vitamin most likely to be of practical importance in feeding cattle.	**V**itamin A deficiency is most likely to occur when cattle are fed (1) high-concentrate rations; (2) bleached pasture or hay grown under drought conditions; (3) feeds that have had excess exposure to sunlight, air, and high temperature; (4) feeds that have been heavily processed or mixed with oxidizing materials such as minerals; and (5) feeds that have been stored for long periods of time. **C**attle particularly susceptible to vitamin A deficiency are: newborn calves deprived of colostrum; cattle that have been prevented from establishing or maintaining good liver stores through exposure to drought; cattle wintered without high-quality forage, and cattle exposed to stresses such as high temperatures or elevated nitrate intake.	**V**itamin A functions as a component of the visual purple required for dim-light vision, and is essential for normal growth, reproduction, and maintenance of healthy epithelial tissue.	**S**igns of vitamin A deficiency include reduced feed intake, rough hair coat, edema of the joints and brisket, lacrimation, xerophthalmia, night blindness, slow growth, diarrhea, convulsive seizures, improper bone growth, blindness, low conception rates, abortion, stillbirths, blind calves, abnormal semen, reduced libido, and susceptibility to respiratory and other infections. Of these symptoms, only night blindness has proved unique to vitamin A deficiency. Clinical verification may include ophthalmoscopic examination, liver biopsy and assay, blood assay, testing spinal fluid pressure, conjunctival smears, and response to vitamin A therapy.
D	**Y**oung calves kept indoors, especially in the wintertime. **F**inishing cattle in northern U.S. on high silage and grain rations and a minimum of sun-cured hay.	**V**itamin D is required for calcium and phosphorus absorption, normal mineralization of bone, and mobilization of calcium from bone.	**R**ickets in young calves, the symptoms of which are: decreased appetite, lowered growth rate, digestive disturbances, stiffness in gait, labored breathing, irritability, weakness, and occasionally, tetany and convulsions. Later, enlargement of the joints, slight arching of the back, bowing of the legs, and the erosion of the joint surfaces cause difficulty in locomotion. Posterior paralysis may follow fracture of vertebrae. **I**n older animals with vitamin D deficiency, bones become weak and easily fractured, and posterior paralysis may accompany vertebral fractures. **V**itamin D deficiency in the pregnant animal may result in dead, weak, or deformed calves at birth.
E	**W**here soils are very low in selenium. **W**hen unsaturated fats are fed.	**V**itamin E is an antioxidant. It has been widely used to protect and to facilitate the uptake and storage of vitamin A. In metabolism, it is linked closely with selenium. Some deficiency signs, particularly in white muscle disease, may respond to either selenium or vitamin E, or may require both.	**M**uscular dystrophy (commonly called white muscle disease) in calves 2 to 12 weeks of age; characterized by heart failure and paralysis varying in severity from slight lameness to inability to stand. Also, a dystrophic tongue is often seen in affected animals. **A** deficiency of vitamin E may be precipitated or accentuated by feeding unsaturated fats.
K	**W**hen moldy sweet clover hay high in dicoumarol content is fed. Vitamin K deficiency results from the antagonistic action of dicoumarol that is formed in moldy sweet clover hay.		**S**weet clover disease, characterized by prolonged blood clotting. Mild cases can be treated effectively with vitamin K.
Water-Soluble Vitamins: B Vitamins	**W**hen an antagonist is present. **W**hen ruminal synthesis is limited by lack of precursors or other problems. **V**itamin B-12 is of special interest because of its role in propionate metabolism, and the practical incidence of vitamin B-12 deficiency as a secondary result of cobalt deficiency. **N**iacin has been reported to enhance protein synthesis by ruminal microorganisms.	**M**ost of the established metabolic functions of B vitamins are important to cattle, as well as to other animals. Consequently, a physiological need for most B vitamins can be assumed for cattle of all ages.	**D**eficiency signs in young calves have been clearly demonstrated for thiamin, riboflavin, pyridoxine, pantothenic acid, biotin, nicotinic acid, vitamin B-12, and choline. **P**olioencephalomalacia in grain-fed cattle has been linked to thiaminase activity or production of a thiamin antimetabolite in the rumen. Affected animals have responded to intravenous administration of thiamin (2.2 mg/kg body weight).

[1]As used herein, the distinction between *nutrient requirements* and *recommended allowances* is as follows: in nutrient requirements, no margins of safety are included intentionally; whereas in recommended allowances, margins of safety are provided in order to compensate for variations in feed composition, environment, and possible losses during storage or processing.

19-9
VITAMIN CHART

Nutrient Requirements[1]		Recommended Allowances[1]	Practical Sources of the Vitamin	Comments
Daily Nutrients/ Animal (or Injection)	Amount/Lb (or/kg) of Feed			
*Variable according to class, age, and weight of cattle. (See Table 19-1.) Injection of 1 million IU of vitamin A intramuscularly will prevent deficiency symptoms for 2-4 months in growing or breeding cattle.	*Variable according to class, age, and weight of cattle. (See Tables 19-2 and 19-8.) On a dry ration basis, the vitamin A requirements are about as follows: *1. Growing-finishing steers and heifers, 1,000 IU/lb (*2,200 IU/kg*). *2. Pregnant heifers and cows, 1,270 IU/lb (*2,800 IU/kg*). *3. Lactating cows and breeding bulls, 1,770 IU/lb (*3,900 IU/kg*).	Inject newborn calves (at birth) with 250,000–1,000,000 IU of vitamin A (use the higher level under confinement production or when scours may be a problem). Legume hays (including alfalfa), av. quality.......... 9–14 (lb) / 20–31 (kg) Nonlegume hays, av. quality.......... 4–8 / 9–18 Dehydrated alfalfa meal, av. quality.......... 50–70 / 110–154 Yellow corn.......... 0.8–1.0 / 1.8–2.2 Silages, corn, or sorghum.......... 2–10 / 4–22	**S**tablized vitamin A. **G**reen pasture. Grass or legume silages. **Y**ellow corn. **G**reen hay not over 1 year old. The average carotene content of some common feeds is as follows: mg Carotene per (lb) (kg)	**C**arotene is rapidly destroyed by exposure to sunlight and air, especially at high temperatures. **H**ay over 1 year old, regardless of green color, is usually not an adequate source of carotene or vitamin A activity. **E**nsiling effectively preserves carotene, but the availability of carotene from corn silage may be low. **T**he younger the animal, the quicker vitamin A deficiencies will occur. Mature animals may store sufficient vitamin A to last 6 months. **W**hen deficiency symptoms appear, they can be corrected (1) by increasing carotene intake through the introduction of high-quality forage, or (2) by supplying vitamin A in the feed or by injection.
	*125 IU/lb (*275 IU/kg*) of dry ration.	**N**ormally, beef cattle receive sufficient vitamin D from exposure to direct sunlight or from sun-cured hay.	**E**xposure to direct sunlight. **S**un-cured hay. **I**rradiated yeast.	**S**un-cured alfalfa hay contains 300–1,000 IU/lb (*661–2,204 IU/kg*).
	*dl-alpha-tocopherol acetate added to dry ration at level of 6.8 to 27.3 IU/lb (*15 to 60 IU/kg*).	**G**enerally natural feeds supply adequate quantities of alpha-tocopherol for mature cattle, although muscular dystrophy in calves occurs in certain areas.	**A**lpha-tocopherol, added to the ration or injected intramuscularly. **C**ommercial vitamin E supplements. **G**rains contain 6–15 mg vitamin E/lb (*13–33 mg/kg*).	**T**he incidence of white muscle disease appears to be lower when the cows receive 2–3 lb (*.91–1.36 kg*) of grain during last 60 days of pregnancy. **W**here supplemental vitamin E is needed, it may be added to the ration or injected intramuscularly.
			Vitamin K_1 is abundant in pasture and green roughage. **V**itamin K_2 is synthesized in large amounts in the rumen. **E**ither K_1 or K_2 effectively fulfill the vitamin K role in blood clotting mechanism.	**E**xcept when the dicoumarol content of hay is excessively high (as in moldy sweet clover hay), sufficient vitamin K is synthesized in the rumen of cattle.
		Usually, no dietary B vitamins need be supplied to cattle.	**B**-vitamins are abundant in milk and many other feeds, and synthesis of B vitamins by ruminal microorganisms is extensive. Calves begin microbial synthesis of B vitamins very soon after the introduction of dry feed in the ration.	

*Where preceded by an asterisk, the nutrient requirements listed herein were taken from *Nutrient Requirements of Beef Cattle,* sixth revised edition, National Research Council—National Academy of Sciences, Washington, D.C., 1984.

Water

Water is needed for all the essential processes of the body, such as the digestion and absorption of food nutrients, the removal of waste, and in regulating body temperature. Animals can survive for a longer period without feed than they can without water. Yet, under ordinary conditions, it can be readily provided in abundance and at little cost. Beef cattle should have an abundant supply of water before them at all times.

Water need is influenced by environmental temperature. Intake of water is fairly consistent up to about 40°F. Above that temperature, water intake varies as water becomes increasingly used for cooling. Where sub-zero winter temperatures occur, it may be necessary to provide heaters to make water available—i.e. to keep it from freezing. No further warming of the water is necessary, however.

Table 19-10 may be used as a guide to the water requirements of beef cattle.

TABLE 19-10
DAILY WATER INTAKE OF BEEF CATTLE[1]

Weight		Temperature in °F (°C)[2]											
		40 (4.4)		50 (10.0)		60 (14.4)		70 (21.1)		80 (26.6)		90 (32.2)	
(lb)	(kg)	(gal)	(liter)	(gal)	(liter)	(gal)	(liter)	(gal)	(liter)	(gal)	(liter)	(gal)	(liter)
Growing heifers, steers, and bulls													
400	182	4.0	15.1	4.3	16.3	5.0	18.9	5.8	22.0	6.7	25.4	9.5	36.0
600	273	5.3	20.1	5.8	22.0	6.6	25.0	7.8	29.5	8.9	33.7	12.7	48.1
800	364	6.3	23.8	6.8	25.7	7.9	29.9	9.2	34.8	10.6	40.1	15.0	56.8
Finishing cattle													
600	273	6.0	22.7	6.5	24.6	7.4	28.0	8.7	32.9	10.0	37.9	14.3	54.1
800	364	7.3	27.6	7.9	29.9	9.1	34.4	10.7	40.5	12.3	46.6	17.4	65.9
1,000	454	8.7	32.9	9.4	35.6	10.8	40.9	12.6	47.7	14.5	54.9	20.6	78.0
Wintering pregnant cows[3]													
900	409	6.7	25.4	7.2	27.3	8.3	31.4	9.7	36.7	—	—	—	—
1,100	500	6.0	22.7	6.5	24.6	7.4	28.0	8.7	32.9	—	—	—	—
Lactating cows													
900+	409+	11.4	43.1	12.6	47.7	14.5	54.9	16.9	64.0	17.9	67.8	16.2	61.3
Mature bulls													
1,400	636	8.0	30.3	8.6	32.6	9.9	37.5	11.7	44.3	13.4	50.7	19.0	71.9
1,600+	727+	8.7	32.9	9.4	35.6	10.8	40.9	12.6	47.7	14.5	54.9	20.6	78.0

[1]From *Nutrient Requirements of Beef Cattle*, sixth revised edition, National Research Council-National Academy of Sciences, 1984.
[2]Water intake of a given class of cattle in a specific management regime is a function of dry matter intake and ambient temperature. Water intake is quite constant up to 40°F (4.4°C).
[3]Dry matter intake has a major influence on water intake. Heavier cows are assumed to be higher in body condition and to require less dry matter and, thus, less water intake.

Fig. 19-8. Windmill and water tank on a New Mexico Ranch. (Courtesy, National Cottonseed Products, Assoc.)

Growth Stimulants and Implants

Table 19-11 summarizes the growth stimulants that are presently available and can be used. All these products have been shown to improve gain and feed efficiency of feedlot cattle significantly.

In considering the additives listed in Table 19-11, it should be noted that there is no evidence to indicate that the use of these products can or will alleviate the need for vigilant sanitation, improved nutrition, and superior management. Also, the benefits of each one must be weighed against its cost.

(Also see Chapter 13, sections on "Hormone and Hormonelike Compounds" and "Implants.")

TABLE 19-11
GROWTH STIMULANTS AND IMPLANTS FOR CATTLE[1]

Additive	Method of Administering	Dosage	Increase in Daily Rate of Gain	Increase in Feed Efficiency	Effect on Carcass Quality	Other Comments	Withdrawal Period Prior to Slaughter
Finishing Steers							
Antibiotic	Oral.	10 mg/100 lb body wt. daily; or 70 to 75 mg/head daily.	6%	4%	Improves carcass quality slightly; more fat deposition and marbling. Decreases liver and rumen condemnations.	Antibiotics will also reduce the disease level. More effective with high-roughage rations than with high-concentrate rations.	No withdrawal required.
Bovatec (lasalocid)	Oral.	150 to 360 mg/day. 250 to 360 mg/day.	8%	8% 5%	No effect.	Alters rumen fermentation similar to monensin.	No withdrawal required.
Rumensin (monensin)	Oral.	50–360 mg/head/day, drylot. 50–200 mg/head/day, pasture.		10%	No effect.	Not a hormone. It results in more propionic acid and less butyric and acetic acids; hence, more energy.	No withdrawal required.
Compudose	Implant.	24 mg estradiol.	10–15%	5–10%	No effect.	Only one implant. Effective for 200 days.	No withdrawal required.
Finaplex	Implant.	140 mg.	7%	7%		The active ingredient is trenbolene acetate (TBA). It is a synthetic analog of testosterone. It gives an additional response when used in combination with an estrogen.	
Ralgro (Zeranol)	Implant.	36 mg resorcyclic acid lactone.	10%	5–10%	No effect.	Nonestrogenic.	65 days.
Steer-oid	Implant.	200 mg progesterone, 20 mg estradiol.	10–15%	5–10%	No effect.	Effective period of 140 days.	No withdrawal.
Synovex-S (for steers)	Implant.	200 mg progesterone, 20 mg estradiol benzoate.	10–15%	5–10%	No effect.	Effective period of 90 to 120 days.	No withdrawal required.
Finishing Heifers							
Antibiotic	Oral.	10 mg/100 lb body wt. daily; or 70 to 75 mg/head daily.	6%	4%	Improves carcass quality slightly; more fat deposition and marbling.	Antibiotics will also reduce the disease level. More effective with high-roughage than with high-concentrate rations.	No withdrawal required.
Bovatec (lasalocid)	Oral.	150 to 360 mg/day.	5%	8%	No effect.	Alters rumen fermentation similar to monensin.	No withdrawal.
MGA	Oral.	0.25 to 0.50 mg daily melengestrol acetate.	10%	6%	MGA will lower the incidence of estrus in heifers and increase rate and efficiency of gain. It is not effective with pregnant heifers.	MGA is effective for heifers, but not for steers.	48 hours.
Rumensin	Oral.	50–360 mg/head/day, drylot. 50–200 mg/head/day, pasture.		15%	No effect.	Not a hormone. It results in more propionic acid and less butyric and acetic acids; hence, more energy.	No withdrawal required.
Heifer-oid	Implant.	Follow label directions.	10%	5–10%	No effect.	Effective period of 140 days. For heifers over 400 lb.	No withdrawal.
Ralgro (zeranol)	Implant.	36 mg resorcyclic acid lactone.	10%	5–10%	No effect.	Nonestrogenic.	65 days.
Synovex-H (for heifers)	Implant.	200 mg testosterone propionate. 20 mg estradiol benzoate.	10%	5–10%	No effect.	Recommended for use in heifers during last 60 to 150 days of the finishing period.	No withdrawal required.
Suckling Calves							
Antibiotic	Oral (in creep feed).	15 to 20 mg/100 lb body wt. daily.	6%	4%		Antibiotics will also reduce the disease level. Administer in creep feed.	No withdrawal required. **Note Well**: If fed at level of 350 mg or over/day, 48-hour withdrawal required.
Ralgro (zeranol)	Implant.	36 mg resorcyclic acid lactone.	10%	5–10%	No effect.	Nonestrogenic.	65 days.
Synovex-C	Implant.	110 mg.	5–10%		No effect.	For calves of either sex.	No withdrawal required.

[1]*CAUTIONS*: FDA regulations are subject to change. Always follow the manufacturer's directions on the use of these products.

FEEDS FOR BEEF CATTLE

Beef cattle feeding practices vary according to the relative availability of grasses, dry roughages, and grains. Where roughages are abundant and grain is limited, as in the western range states, cattle are primarily grown out on roughages. On the other hand, where grain is relatively more abundant, as in the Corn Belt and in the High Plains area of Texas and Oklahoma, finishing with more concentrates is common.

Pastures

Pastures include all crops that are harvested directly by animals.

Good pasture is the cornerstone of successful beef cattle production. It has been said that good cattle producers can be recognized by the character of their pastures, and that good cattle graze good pastures. Thus, the three go hand in hand—good producers, good pastures, and good cattle. The historic relationship and importance of cattle and pastures was extolled in an old Flemish proverb: "No grass, no cattle; no cattle, no manure; no manure, no crops."

Fig. 19-9. Good grass and plenty of milk make for heavy calves. (Courtesy, International Braford Assn., Ft. Pierce, Fla.)

The type of pasture, as well as its carrying capacity and seasonable use, varies according to topography, soil, and climate. Because of the hundreds of species of grasses and legumes that are used as beef cattle pastures, each with its own best adaptation, no attempt is made to discuss the respective virtues of each variety. Instead, it is recommended that the farmer or rancher seek the advice of the local county agricultural agent, or write to the crop science department of the state university.

During the winter months, and in periods of drought, the pasture utilized by beef cattle may consist of dried grass cured on the stalk. On a dry basis, the crude protein content of some mature, weathered grasses may be 3% or less. To supplement such feed, cattle producers commonly feed cake or cubes. The use of cake or cubes instead of meal reduces losses from wind, an especially important factor on the range. Also, dried grass may be supplemented by means of "lick tanks" or molasses, along with urea, minerals, and vitamin A.

Hays and Other Dry Roughages

Hay is the most important harvested roughage fed to beef cattle, although many other dry roughages can be and are utilized.

Roughages, as with concentrates, may be classified as carbonaceous or nitrogenous, depending on their protein content. The principal dry carbonaceous roughages used by cattle include hay from the grasses, the straws and hays from cereal grains, corncobs, the stalks and leaves of corn and the grain sorghums, and cottonseed hulls. Cured nitrogenous roughages include the various legume hays such as alfalfa, the clovers, peanut, soybean, cowpea, and velvet bean.

Although leguminous roughages are preferable, weather conditions and soils often make it more practical to produce the nonlegumes. Also, in many areas, such feeds as dry grass cured on the stalk, cereal straws, corncobs, and cottonseed hulls are abundantly available and cheap. Under such circumstances, these feeds should be used as part of the ration for wintering beef cows, for wintering stockers that are more than 1 year of age, or for supplying the limited roughage needed for finishing beef cattle.

In comparison with good-quality legume hays, the carbonaceous roughages are lower in protein content and in quality of protein, lower in calcium, and generally deficient in carotene (provitamin A). Thus, where nonlegume roughages are used for extended periods, these nutritive deficiencies should be corrected; this is especially true with the gestating-lactating cow or the young, growing calf.

Silages and Root Crops

Silage is a valuable adjunct to pastures in beef cattle production, since it is possible to use a combination of the two forages to furnish green, succulent feeds on a year-round basis.

Corn was the first and still remains the principal crop used in the making of silage, but many other crops are ensiled in various sections of the country. The sorghums are the

Fig. 19-10. Upright (tower) silos used for storing feed for beef cattle. (Courtesy, A. O. Smith Harvestore Products, Inc., Arlington Heights, Ill.)

leading ensilage crops in the Southwest, and grasses and legumes are the leading ensilage crops in the Northeast. Also, in different sections of the country to which they are adapted, the following feeds are ensiled: cereal grains, field peas, cowpeas, soybeans, potatoes, and numerous fruit and vegetable refuse products.

When silage is fed to cattle, it must be remembered that, because of its high-moisture content, about 3 lb of silage are generally considered equivalent to 1 lb of dry roughage of comparable quality. A ration of 55 to 60 lb of corn silage plus ½ to ¾ lb of a protein concentrate daily will carry a dry cow through the winter. The ration may be improved, however, by replacing ⅓ to ½ of the silage with an equivalent amount of dry roughage, adding 1 lb of dry roughage for each 3 lb of silage replaced.

Silage may be successfully used for finishing steers. Long-yearling steers will eat 50 to 55 lb of a 37% dry matter corn silage plus 2 lb of a 32% protein supplement per day at the beginning of the feeding period. The amount of silage is gradually decreased as the concentrates are increased. At the end of the feeding period, the cattle should be getting around 10 to 12 lb of silage. Because of their more limited digestive capacity, the allowance of silage fed to calves should be correspondingly less.

Usually, silage provides a much cheaper succulent feed for beef cattle than roots. For this reason, the use of roots for beef cattle is very limited, and confined almost entirely to the northern areas.

Concentrates

The concentrates include those feeds which are low in fiber and high in energy. For purposes of convenience, concentrates are often further classified as (1) carbonaceous feeds, and (2) nitrogenous feeds.

In general, the use of concentrates for beef cattle is limited to (1) the finishing of cattle, (2) the development of young stock, and (3) use as limited supplements in the winter ration. Over most of the United States, the cereal grains are the chief concentrates fed to beef cattle—these grains being combined, if necessary, with protein supplements to balance the ration.

The chief carbonaceous concentrates used for beef cattle are the grains, chiefly corn, and such processed feeds as hominy feed, beet pulp, and molasses. The choice of the particular feeds is usually determined primarily by price and availability.

The nitrogenous feeds are usually subdivided into (1) natural proteins, including the oilseed meals and animal proteins, and (2) nonprotein nitrogen, consisting primarily of urea and ammonia.

UREA

Urea may constitute up to one-third of the total protein of the ration of cattle, provided additional fermentable energy is added in the form of molasses or grain to compensate for the lack of energy in the urea, in order to feed the rumen bacteria properly. Producers can tell the nonprotein nitrogen content (NPN) of commercial feed by looking at the feed tag, which must list the contents.

Common guidelines relative to the use of urea for beef cattle are given in Table 19-12.

TABLE 19-12
COMMON GUIDELINES TO THE USE OF UREA FOR CATTLE

	For Finishing Cattle	For Grower (stocker) Cattle	For Wintering Pregnant and Lactating Cows
Percent of total protein in ration from urea .... (%)	33⅓	25.0	25.0
Maximum urea/animal/day .... (lb)	0.22 *(100 g)*	0.15 *(68 g)*	—
Percent of urea, by weight of total air-dry feed consumed ...... (%)	1.0	1.0	1.0
Percent of urea, by weight, of concentrate mix (grain plus protein supplement)[1] .... (%)	2.0-3.0	3.0	3.0
Percent of urea, by weight, of the protein supplement .... (%)	20-30[2]	10.0[3]	10.0
Percent of supplemental nitrogen in high-protein supplement from urea[4] ... (%)	60-90[5]	30.0	30.0
Pounds of urea added/ton of corn silage at ensiling time[6] .... (lb)	10.0 *(4.5 kg)*	10.0 *(4.5 kg)*	10.0 *(4.5 kg)*

[1]Feed intake may be depressed if over 1% is used. Yet, many beef producers are successfully using 2%.

[2]This means that as much as 60-90% of the protein value of the supplement may come from nonprotein sources. However, because such a supplement will constitute only 2-5% of the total ration fed, the first rule of thumb given in Table 19-12 still applies; namely, only ¼-⅓ of the total protein in the ration will be supplied from a nonprotein source.

[3]A protein supplement containing 10% urea provides 28.1% of the protein equivalent (281% × .10) from nonprotein nitrogen.

[4]High-urea supplements are best fed in complete mixed rations, which are *thoroughly* mixed. Supplements containing 20-30% urea require extreme caution when being hand-fed.

[5]In a feedlot ration, this may be equivalent to 25-40% of the total nitrogen from all sources.

[6]On a dry matter basis, corn silage ensiled at the well-dented stage contains about 8% protein. The addition of 10 lb of urea per ton (or *5 kg/1,000 kg*) of silage increases the protein content from 8-13%. However, there is loss of flexibility in feeding such a ration, and the rate of gain will be less than can be secured from higher energy, more dense rations. Also, it is extremely important that the urea be well mixed in the silage; otherwise, there is hazard of toxicity.

(Also see Chapter 11, Protein Supplements, section on "Urea.")

By-product Feeds

Innumerable by-products—both roughages and concentrates—from plant and animal processing, and from industrial manufacturing, are available and used as cattle feeds in different areas. Also, small grain stubble fields and cornstalks are utilized, and cotton fields may be grazed following harvest. Then, there are such additional by-product feeds as cull potatoes, cottonseed hulls, corncobs, cull citrus, cannery refuse, beet tops, and a host of other similar products.

(Also see Chapter 12, By-product Feeds/Crop Residues.)

Mineral Supplements

The mineral recommendations for all classes and ages of cattle, especially those fed unmixed rations or on pasture, are:

1. **When animals are on liberal grain feeding.** Provide free access to a 2-compartment mineral box, with (a) trace mineralized salt in one side, and (b) in the other side, a

mixture of ⅓ trace mineralized salt (salt included to improve palatability), ⅓ defluorinated phosphate, dicalcium phosphate, or steamed bone meal, and ⅓ ground limestone or oystershell flour.

2. **When animals are primarily on roughage (pasture, hay, and/or silage).** Provide free access to a 2-compartment mineral box, with (a) trace mineralized salt in one side, and (b) in the other side a mixture of ⅓ trace mineralized salt (salt included for purposes of palatability), ⅓ defluorinated phosphate, dicalcium phosphate, or steamed bone meal.

Salt should always be available on a free-choice basis in addition to whatever mineral mix is provided.

Vitamin Supplements

Table 19-13 gives the estimated carotene content of feeds in relation to appearance and methods of conservation.

Beef cattle usually receive sufficient vitamin D from exposure to direct sunlight or from sun-cured roughages.

When supplemental vitamin E is needed, it may be added to the ration or injected intramuscularly.

Vitamin K is synthesized in the rumen of cattle in adequate amounts under most feeding conditions.

The dietary requirements of the young calf for the B vitamins (thiamin, biotin, niacin, pyridoxine, pantothenic acid, riboflavin, and vitamin B-12) during the first 8 weeks of life, prior to the development of the functioning rumen, are usually met by milk supplied by the cow during early lactation. Later, B vitamins are synthesized by rumen bacteria in sufficient quantities in most feeding regimens.

TABLE 19-13
ESTIMATED CAROTENE CONTENT OF FEEDS IN RELATION TO APPEARANCE AND METHODS OF CONSERVATION[1]

Feedstuff	Carotene	
	(mg/lb)	(mg/kg)
Dehydrated alfalfa meal, fresh, dehydrated without field curing, very bright green color[2]	110-135	242-298
Alfalfa leaf meal, bright green color	60-80	132-176
Dehydrated alfalfa meal after considerable time in storage, bright green color	50-70	110-154
Legume hays, including alfalfa, very quickly cured, with minimum sun exposure, bright green color, leafy	35-40	77-88
Legume hays, including alfalfa, good green color, leafy	18-27	40-60
Fresh green legumes and grasses, immature	15-40	33-88
Legume hays, including alfalfa, partly bleached, moderate amount of green color	9-14	20-31
Nonlegume hays, including timothy, cereal, and prairie hays, well cured, good green color	9-14	20-31
Legume silage	5-20	11-44
Legume hays, including alfalfa, badly bleached or discolored, traces of green color	4-8	9-18
Nonlegume hays, average quality, bleached some green color	4-8	9-18
Corn and sorghum silages, medium to good green color	2-10	4.4-22
Grains, mill feeds, protein concentrates, and by-product concentrates, except yellow corn and its by-products	.01-0.2	0.2-0.4

[1]From *Nutrient Requirements of Beef Cattle*, NRC-National Academy of Sciences, No. 579, with metric system added by the authors.

[2]Green color is not uniformly indicative of high-carotene content.

FEED SUBSTITUTION TABLE

Table 19-14, Feed Substitution Table for Beef and Dairy Cattle, is a summary of the comparative values of the most common U.S. feeds. In arriving at these values, two primary factors besides chemical composition and feeding value have been considered—namely, palatability and product quality.

The comparative values of feeds shown in the feed substitution table are not absolute. Rather, they are reasonably accurate approximations based on average-quality feeds, together with experiences and experiments.

TABLE 19-14
FEED SUBSTITUTION TABLE FOR BEEF AND DAIRY CATTLE (AS-FED BASIS) (See footnotes at end of table.)

Feedstuff	Relative Feeding Value (lb for lb) In Comparison With The Designated (underlined) Base Feed Which = 100	Maximum Percentage of Base Feed (or comparable feed or feeds) Which It Can Replace For Best Results	Remarks
GRAINS, BY-PRODUCT FEEDS, ROOTS AND TUBERS:[1] (Low and Medium Protein Feeds)			
Corn, No. 2	*100*	*100*	The most important concentrate for cattle in the U.S. Grind coarsely or flake.
Almond hulls, dried, no shells	70-75	15-30	
Almond hulls and shell meal	35	15-20	
Apple pomace, air-dry	78	33⅓	Values given are for apple pomace with paper or rice hulls as press aids.
Bakery products, dried	110	15-30	

(Continued)

TABLE 19-14 (Continued)

Feedstuff	Relative Feeding Value (lb for lb) In Comparison With The Designated (underlined) Base Feed Which = 100	Maximum Percentage of Base Feed (or comparable feed or feeds) Which It Can Replace For Best Results	Remarks
GRAINS, BY-PRODUCT FEEDS, ROOTS AND TUBERS:[1] (Continued)			
Bakery waste, not dried (30% water)	75	15-30	
Barley	90	25-100	The heavier the barley and the smaller the proportion of hulls, the higher the feeding value. Grind coarsely or roll for cattle. In Canada, where considerable barley is fed, it is often used as the only basal feed in the ration once animals are accustomed to it.
Beans (cull)	80	10	Best when cooked, but can also be fed raw. Beans should be ground. When cooked, 3-4 lb (1.4-1.8 kg)/head daily; when raw, 1-2 lb (0.45-0.91 kg). Scouring may occur if they constitute more than 15% of total ration.
Beet pulp, dried	90	50	
Beet pulp, molasses, dried	90-95	50	
Beet pulp, wet	25	40	50% the value of corn silage. May compose 40% of ration on dry matter basis.
Brewers' dried grains	80	33⅓	Not very palatable. Fed chiefly to dairy cattle.
Brewers' grains (wet)	13-15	33⅓	Too bulky and usually too costly to be used in finishing rations. Grains usually come from barley. Best to haul and feed directly. Can be stored in silo if salt is added at rate of 25 lb (11.4 kg) per ton of grains.
Buckwheat	55-75	33⅓	Should be ground and mixed with other grains.
Carrots (cull)	10-15	20-25	Store 3-4 weeks before using; fresh carrots cause scouring. Feed whole or sliced.
Citrus pulp, dried	80-88	25-50	
Corn-and-cob meal	85-90	100	
Corn gluten feed (gluten feed)	85-90	50	
Distillers' dried grains	73-90	33⅓	Rye distillers' dried grains are of lower value than similar products made from corn or wheat. Distillers' dried grains are used chiefly for dairy cattle.
Distillers' dried solubles	73-90	33⅓	The chief difference between distillers' dried grains and distillers' dried solubles is the higher B vitamin content of the latter. Normally this is not important for cattle.
Fat (animal or vegetable)	225	5	Fat has 203 megacalories energy/100 lb (45.4 kg) for maintenance and 127 megacalories for weight gain, as compared to 92 and 60, respectively, for corn.
Hominy feed	100	50	
Manure, cattle, without bedding	75	50	Approximately 80% of the total nutrients of feeds is excreted as animal manure. However, the feeding value of manure will vary according to (1) the nutritive value of the feeds initially fed, (2) the class, age, and individuality of the animal to which the feeds were initially fed, and (3) the handling and processing of the manure.
Manure, poultry (see poultry house litter)			
Molasses, beet	75	10-40	Value is highest when used as an appetizer. May be laxative if fed at levels above 6 lb (2.7 kg) daily.
Molasses, cane	75	10-40	Value is highest when used as an appetizer.
Molasses, citrus	65-75	10-40	
Molasses, wood	26-30	10-20	Unpalatable.
Oats	70-90	10-100	Valuable for young stock, for breeding stock and for getting animals on feed. Oats have lowest value for finishing cattle and should be limited to ⅓ of such rations. Also, the feeding value of oats varies according to the test weight per bushel. Grind or roll for cattle.
Paunch, dried (also see "paunch-blood" under Protein Supplements of this table)	90	5-10	Dried paunch is not palatable, with the result that it depresses appetite. Rate of gain is not affected, but feed efficiency is slightly lowered.
Peas (cull), dried	88	40	Because of lack of palatability, peas will lower feed intake if they constitute more than 20% of the total ration. Also, there is bloat hazard if they exceed 40% of the ration.
Pear waste, air-dry	75	40	
Potatoes (Irish), wet	20-25	85	When fed with alfalfa hay, they are worth about 80% as much per ton as corn silage. Do not feed frozen. Sunburned, decomposed, or sprouted potatoes should not make up more than 10% of potatoes fed. Keep steers' heads down while eating to prevent choking.

(Continued)

TABLE 19-14 (Continued)

Feedstuff	Relative Feeding Value (lb for lb) In Comparison With The Designated (underlined) Base Feed Which = 100	Maximum Percentage of Base Feed (or comparable feed or feeds) Which It Can Replace For Best Results	Remarks
GRAINS, BY-PRODUCT FEEDS, ROOTS AND TUBERS:[1] (Continued)			
Potatoes (Irish), dehydrated	88	50	Excellent source of energy, but deficient in protein, minerals, and vitamins.
Potatoes (sweet)	25	85	
Potatoes (sweet), dehydrated	95–100	50	Dehydrated sweet potatoes are more palatable than dehydrated Irish potatoes.
Poultry house litter	10–40	15–25	Poultry house litter may also be used as a protein source (see Protein Supplements, this table).
Prunes	62	15	Because of the laxative quality of prunes, they should be limited to 7% of the total ration.
Raisins (cull)	70	33⅓	
Raisin pulp	53	25	
Rice (rough rice)	80	100	
Rice bran	66⅔–75	33⅓	
Rice polishings	88	25	
Rye	96	33⅓	Not palatable when fed in large amounts.
Screenings, refuse	62–70	25–35	Should be finely ground in order to kill noxious weed seeds. Quality varies; good-quality screenings are equal to oats whereas poor-quality screenings resemble straw.
Sorghum (milo, kafir), grain	90–95	100	Varieties vary in protein content. Grind or roll for cattle.
Spelt and emmer	70–90	30–100	Similar to oats.
Wheat	100–105	50	Grind coarsely, or roll.
Wheat bran	70–90	25–33⅓	Because of its bulk and fiber, bran is not desirable for finishing rations. Bran is valuable for young animals, for breeding animals, and for starting animals on feed.
Wheat-mixed feed (mill run)	95	33⅓	Sometimes fed to the breeding herd, to young calves, and to finishing cattle being started on feed.
Wheat screenings	85	50	
Wood (cooked)	75–80	70	Wood products, which are largely cellulose and lignin, must be cooked before animals can digest them.
PROTEIN SUPPLEMENTS:			
Soybean meal (41%)	*100*	*100*	Slightly laxative effect.
Alfalfa or clover screenings	70–75	50	Grind finely to destroy weed seeds.
Brewers' dried grains	55–65	50	Not very palatable. Fed chiefly to dairy cattle.
Copra meal (coconut oil meal), 21%	90–100	50	
Corn gluten feed (gluten feed)	65–75	50–100	
Corn gluten meal (gluten meal)	90–100	50	Somewhat unpalatable.
Cottonseed meal (41%)	100	100	
Distillers' dried grains	65–70	100	Rye distillers' grains are about 10% lower in protein than similar products made from corn or wheat.
Distillers' dried solubles	70	100	Low in palatability.
Feather meal (hydrolyzed; 84% protein)	175	50	Feather meal is unpalatable; hence, cattle must be accustomed to it gradually and it must be limited in quantity. It is best used for wintering brood cows and stocker cattle.
Legume screenings	75	75	Satisfactory, but less palatable than soybean or cottonseed meal.
Linseed meal (35%)	95	100	Linseed meal has laxative effect. Some cattle will not tolerate more than 5–8% linseed meal in the ration.
Paunch-blood feed (also see "paunch, dried" under Grains section of this table)	100	100	At slaughter, each bovine yields about 20 lb (9.1 kg) of paunch and 20 lb (9.1 kg) of blood. Dried paunch runs around 10% protein, dried blood around 80%, and a 50-50 mixture of the 2 products, around 45%.
Peanut meal (45%)	100	100	Peanut meal may become rancid if stored too long, especially in warm, moist climates.
Peas (cull), dried	65–75	50	
Poultry house litter	50–55	25	Poultry house litter may also be used as an energy source (see Grains section of this table).
Rapeseed meal (Canola meal) (37%)	88	75	Rapeseed meal should be limited to not more than 2 lb (0.91 kg) per cow.

(Continued)

TABLE 19-14 (Continued)

Feedstuff	Relative Feeding Value (lb for lb) In Comparison With The Designated (underlined) Base Feed Which = 100	Maximum Percentage of Base Feed (or comparable feed or feeds) Which It Can Replace For Best Results	Remarks
PROTEIN SUPPLEMENTS: (Continued)			
Safflower meal, well hulled (42%)	92	100	
Safflower meal, with hulls (20%)	40–45	100	**S**afflower meal with hulls is unpalatable. Thus, it should be mixed with more palatable feeds.
Sesame meal	90–95	25	
Soybeans, whole	95–100	95	**N**ot satisfactory for finishing calves. **S**oybean allowance should be limited to amount necessary to balance the ration. Larger amounts may be unduly laxative and cause cattle to go off feed.
Sunflower meal (39%)	95–100	100	**I**f poorly hulled and lower protein content than 39%, feeding value will be lowered accordingly. It is well liked by cattle and keeps well in storage.
DRY FORAGES AND SILAGES:[2]			**A**ll the dry nonlegume forages listed herein are satsifactory when needed minerals and either a limited amount of legume hay or a protein supplement are supplied to balance the ration.
Alfalfa hay, all analyses	*100*	*100*	**D**oes away with or lessens protein supplement requirements.
Alfalfa silage	33⅓–50	50–85	**W**hen alfalfa silage replaces corn silage, more energy feed must be provided but less protein.
Alfalfa straw	37	50	**F**eed with good hay.
Apple pomace silage	17–25	50–85	**U**sually fed as a substitute for corn or grass silage. **5**0% the value of corn silage. **S**ometimes fed out of a stack or trench silo.
Apples	17–25	50–85	**D**o not feed more than 25 lb (11.4 kg)/mature bovine. **N**ot recommended for finishing cattle. **D**anger of choking when fed whole. **R**elatively high handling cost.
Bagasse, dried; sugarcane or sorghum	10–20	5–10	
Barley hay	70	100	**A**void bearded varieties.
Barley silage	25–40	50–80	**I**n silage, there is no problem with bearded varieties, which usually outyield beardless.
Barley straw	40	70	**O**f the cereal straws, barley ranks next to oat straw in feeding value. Feed to dry pregnant cows. Supplement daily with 5–6 lb (2.3–2.7 kg) alfalfa hay or 1–2 lb (0.45–0.91 kg) of 30–40% protein supplement.
Bean straw	34	50	**F**eed with good hay.
Beet tops, fresh	20	33⅓–50	**I**n the West, large acreages of fresh beet tops are grazed by cattle and sheep. **B**loat may be a problem when tops are frozen. **T**ops are laxative. **A**dd 2½ lb (1.1 kg) of ground limestone/ton of feed.
Beet top silage, sugar	17–25	33⅓–50	**F**eed 2 oz (56.7 g) of finely ground limestone or chalk with each 100 lb (45.4 kg) of tops, as calcium changes the oxalic acid to insoluble calcium oxalate.
Clover hay, crimson	90–100	100	**C**rimson clover hay has a considerably lower value if not cut at an early stage.
Clover hay, red	90–100	100	**I**f the rest of the ration is adequate in protein, clover hay will be equal to alfalfa in feeding value; otherwise, it will be lower.
Clover straw	37	50	**F**eed with good hay.
Clover-timothy hay	80–90	100	**V**alue of clover-timothy mixed hay depends on the proportion of clover present and the stage of maturity at which it is cut.
Corncobs, ground	70	90	**G**round corncobs can be used as the only roughage for beef cattle if properly supplemented with proteins, minerals, and vitamins.
Corn fodder	75	80–90	
Corn husklage (shucklage)	50	80–90	**H**ighest and best use is for dry pregnant cows. It is slightly higher in energy and more palatable than corn stover.
Corn silage	33⅓–50	50–85	
Corn (sweet) silage, cannery waste	26–40	50–85	
Corn stover	45	70–90	**C**orn stover will meet the energy needs of dry pregnant cows, but is deficient in protein and low in phosphorus and vitamin A. Two acres of cornstalks will carry a cow 100–120 days.
Corn (sweet) stover	50	80–90	
Cottonseed hulls	66⅔	75	**U**se for dry pregnant cows. **S**upplement daily with 4–6 (1.8–2.7 kg) lb of good legume hay or 1–2 lb (0.45–0.91 kg) of a 30–40% protein supplement.
Cowpea hay	90–100	100	

(Continued)

TABLE 19–14 *(Continued)*

Feedstuff	Relative Feeding Value (lb for lb) In Comparison With The Designated (underlined) Base Feed Which = 100	Maximum Percentage of Base Feed (or comparable feed or feeds) Which It Can Replace For Best Results	Remarks
DRY FORAGES AND SILAGES:[2] *(Continued)*			
Gin trash, cotton	75	75	
Grape pomace or meal	5–15	10–15	Pomace including stems is of little value as a feed.
Grass-legume mixed hay	80–90	100	Value depends on the proportion of legume present and the stage of maturity at which it is cut.
Grass-legume silage	32–47	50–85	Unless grain is added as a preservative, grass silage requires more energy feed, but less protein supplement than corn silage when fed to finishing cattle.
Grass silage	30–45	50–85	For finishing cattle, grass silage must be supplemented with additional energy feeds, such as cereal grain or molasses, to be of the same value as corn silage.
Hop vine silage	20	50–75	It should be chopped when placed in the silo.
Hops, spent, dehydrated	80	50–65	Devoid of carotene; feed with legume hay.
Johnsongrass hay	70	100	
Lespedeza hay	80–100	100	Feeding value of lespedeza hay varies considerably with stage of maturity at which it is cut.
Mint hay	70–80	75	Cattle tire of mint hay when it is fed as the only roughage for extended periods.
Oat hay	75	100	
Oat silage	32–47	50–85	Must be chopped finely to exclude air from silo.
Oat straw	50	75	Oat straw is the best of the cereal straws. Use for dry pregnant cows. Supplement daily with 4–6 lb *(1.8–2.7 kg)* of good legume hay or 1–2 lb *(0.45–0.91 kg)* of 30–40% protein supplement.
Paper (newspaper; waste paper)	66⅔	50	Paper varies in feeding value in proportion to the cellulose (most paper is 60–90% cellulose) and lignin content. Magazine and bookstock papers are higher in cellulose and lower in lignin than newspapers; hence, of higher feeding value. Pelleting or cubing may increase the value of paper. *Caution:* Some newspapers contain heavy metals (boron, lead, barium, and antimony), sometimes used as a dye carrier in printer's ink, which may be toxic to animals. This is especially true of "funny" papers because of the quantity of heavy metals carried on the colored ink of the comics.
Pea straw	45–75	60–75	
Pea-vine hay	100–110	75–90	Can constitute the only roughage for finishing cattle.
Pea-vine silage	33⅓–50	50–85	Unless grain is added as a preservative, pea-vine silage requires more energy feed, but less protein supplement than corn silage when fed to finishing cattle.
Potato silage	25–30	50–75	About 75% the value of corn silage.
Prairie hay	65–70	100	
Reed canarygrass hay	70	100	
Rice straw	47	70	High levels of rice straw can be used for wintering cattle if the straw is properly fortified.
Sawdust	75–80	70	Feeding value varies among species of trees. Digestibility is increased by cooking and other treatments.
Sorghum fodder	70	100	
Sorghum silage (grain varieties)	32–47	50–85	For finishing cattle, 85–90% as valuable as corn silage and must be supplemented in the same manner as corn silage.
Sorghum silage (sweet varieties)	25–30	50–85	Nearly equal to grain varieties in value per acre because of greater yield.
Sorghum (milo) stover	35	70–90	Can be grazed or harvested and stored either as dry feed or silage. About 2% higher in protein, but less palatable, than corn stover.
Soybean hay	85–90	50–75	Lower value than alfalfa hay, largely due to greater wastage in feeding. It may cause scouring when fed alone.
Sudangrass hay	70	100	
Sunflower silage	25–35	50–85	65–75% value of corn silage. Somewhat unpalatable and may cause constipation. Harvest for silage when ½–⅔ of heads are in bloom.
Sweet clover hay	100	100	Value of sweet clover hay varies widely. Moldy or spoiled sweet clover hay may cause sweet clover disease.
Timothy hay	70	100	
Vetch-oat hay	80–90	100	The higher the proportion of vetch, the higher the value.
Wheat hay	70	100	
Wheat straw	35	65	Of the cereal straws, wheat ranks third in nutritive value, behind oat straw and barley straw. Highest and best use is for dry pregnant cows. Supplement daily with 6 lb *(2.7 kg)* of alfalfa or 2 lb *(0.91 kg)* of a 30–40% protein supplement.

[1] Roots and tubers are of lower value than the grain and by-product feeds due to their higher moisture content.
[2] Silages are of lower value than dry forages due to their higher moisture content.

FEED PREPARATION

The physical preparation of cereal grains for cattle by soaking and cooking has been practiced by cattle exhibitors for a very long time. In recent years, many sophisticated techniques for the processing of grains have been developed, especially for feedlot cattle. Basically, however, grain is either soaked, cooked, ground, or rolled (wet or dry), and hay is either cut, shredded, ground, pelleted, or cubed.

The subject of feed preparation for all classes of livestock is fully covered in Chapter 14, Feed Processing.

Fig. 19-13. Protein blocks used as range supplement.

Fig. 19-11. Forage harvester field chopping cured hay directly from the windrow. (Courtesy, Sperry New Holland, New Holland, Pa.)

Fig. 19-14. Chopped forage stored in tower silos. (Courtesy, Dave Brown and Associates, Oak Brook, Ill.)

Fig. 19-12. Feed truck with automatic auger system continuously mixes feed rations at Farr Feeders, Inc. (Courtesy, Tom R. Farr, Greeley, Colo.)

FEED ALLOWANCE AND SOME SUGGESTED RATIONS [3]

Some general rules of feeding may be given, but it must be remembered that *"the eye of the master fattens his cattle."*

Table 19-15 (see page 394), Daily Rations For Beef Cattle, contains suggested rations for different classes and ages of cattle. These are merely intended as general guides. Variations can and should be made in the rations used. The feeder should give consideration to (1) the supply of homegrown feeds, (2) the availability and price of purchased feeds, (3) the class and age of cattle, (4) the health and condition of the animals, and (5) the length of the grazing season.

---

[3] Insofar as possible, these rations were computed from the requirements reported by the National Research Council and applied by the authors.

TABLE
DAILY RATIONS
(As-Fed

Suggested Rations With all rations and for all classes and ages of cattle, provide free access in separate containers to (1) salt (iodized salt in iodine-deficient areas), and (2) a suitable mineral mixture.	Wintering Mature Pregnant Beef Breeding Cows (av. wt. 1,100 lb or *499 kg*)		Wintering Mature Lactating Beef Breeding Cows (av. wt. 1,100 lb or *499 kg*)		Wintering Replacement Heifers (weighing 400–500 lb or *181–227 kg* start of wintering)	
	Per Day		Per Day		Per Day	
	(lb)	*(kg)*	(lb)	*(kg)*	(lb)	*(kg)*
1. Legume hay or grass-legume mixed hay, good quality	18–20	*8.2–9.1*	30	*13.6*	13–15[3]	*5.9–6.8*[3]
Grain	—	—	—	—	2–3	*0.91–1.36*
Protein supplement	—	—	—	—	—	—
2. Grass hay or other nonlegume dry roughage	18–20	*8.2–9.1*	24–26	*10.9–11.8*	12–18[3]	*5.4–8.2*[3]
Grain	—	—	2	*0.91*	2½–4½	*1.13–2.04*
Protein Supplement	½–1	*0.23–0.45*	3	*1.36*	1¼–1½	*0.57–0.68*
3. Legume hay or grass-legume mixed hay, good quality	7–11	*3.2–5.0*	26–28	*11.8–12.7*	8–12[3]	*3.6–5.4*[3]
Grass hay or other nonlegume dry roughage	9–11	*4.1–5.0*	—	—	4–6	*1.8–2.7*
Grain	—	—	1	*0.45*	2½–4	*1.13–1.81*
Protein supplement	—	—	1	*0.45*	½–1	*0.23–0.54*
4. Corn or sorghum silage	50–55	*22.7–25*	55	*25*	25–40	*11.3–18.2*
Grain	—	—	2	*0.91*	—	—
Protein supplement	0–½	*0–0.23*	3	*1.36*	1½–1¾	*0.68–0.79*
5. Grass silage, half or more legume	50	*22.7*	50	*22.7*	25–40	*11.3–18.2*
Grain	—	—	4	*1.81*	3–4	*1.36–1.81*
Protein supplement	—	—	—	—	½	*0.23*
6. Silage (corn or sorghum silage fed with legume hay or legume silage fed with grass hay)	35	*15.9*	40	*18.1*	15–30	*6.8–13.6*
Hay	5–6	*2.3–2.7*	10	*4.5*	3–4	*1.4–1.8*
Grain	—	—	—	—	1–2	*0.45–0.91*
Protein supplement	0–½	*0–0.23*	—	—	½–1	*0.22–0.45*

[1]If stocker calves are late or the roughage is fair to poor quality, it may be desirable to add 2–4 lb *(0.91–1.81 kg)* of grain per head daily. If farm scales are available, monthly weights may be used as the criterion for grain feeding. Keep in mind that calves should gain ¾–1 lb *(0.34–0.45 kg)* daily.

[2]In general, the experienced feeder plans that cattle on full feed shall consume (1) feeds in amounts (daily: air-dry basis) equal to about 2.5–3.0% of their liveweight, (2) 70–90% concentrates, and (3) a minimum of 2–4 lb *(0.9–1.8 kg)* roughage for each 100 lb *(45 kg)* liveweight. In areas where roughage is more abundant and comparatively cheaper than grain, the proportions of roughage to grain should be somewhat higher than indicated. In computing roughage consumption, 3 lb *(1.36 kg)* of silage are considered equivalent to 1 lb *(0.45 kg)* of hay.

FITTING FOR SHOW AND SALE

All animals intended for show purposes, including both breeding animals and steers, should be placed in the proper state of condition—they should be neither too fat nor too thin. Requirements differ slightly among breeds and lines of cattle. For example, the larger, leaner continental European breeds do not mature as quickly nor fatten as readily as the British breeds. The essentials in feeding cattle for show might be described as similar to those in feeding steers for market, except that more attention must be given to the smallest details. A suitable ration must be selected and the animal or animals must be fed with care over a sufficiently long period.

Fig. 19-15. Spot, grand champion steer at the National Western Stock Show, Denver, 1986. The 1,261-lb Shorthorn steer was shown by Brandon Horn, Lookeba, Okla. (Courtesy, National Western Stock Show, Denver, Colo.)

19-15
FOR BEEF CATTLE
(Basis)

Wintering Stocker Calves Roughed Through Winter and Grazed the Following Summer. Fed for winter gain of ¾–1 lb (0.34–0.45 kg) per head daily (weighing 400–500 lb or 181–227 kg start of wintering)[1]		Finishing Calves in Drylot, Generally in Winter (weighing 400–500 lb or 181–227 kg start of feeding and 750–850 lb or 340–386 kg at marketing)[2]		Wintering Yearlings; Roughed Through the Winter, and Generally Pasture Finished the Following Summer. Fed for winter gains of 1–1¼ lb or 0.45–0.57 kg per head daily (weighing about 600 lb or 227 kg start of wintering)		Finishing Yearlings in Drylot, Generally in Winter (weighing about 600 lb or 272 kg start of feeding, and 900–1,050 lb or 409–477 kg at marketing)[2]		Finishing Long-yearling Steers in Drylot Generally in Winter (weighing about 850 lb or 386 kg start of feeding and 1,000–1,100 lb or 454–499 kg at marketing)[2]	
Per Day		Per Day		Per Day		Per Day		Per Day	
(lb)	(kg)	(lb)	(kg)	(lb)	(kg)	(lb)	(kg)	(lb)	(kg)
12–18[3]	5.4–8.2	4–6	1.8–2.7	16–24	7.2–10.9	4–8	1.8–3.6	6–12	2.7–5.4
—	—	12–15	5.4–6.8	—	—	15–19½	6.80–8.8	16–22	7.2–10.0
—	—	1–1½	0.45–0.68	—	—	1–1½	0.45–0.68	—	—
12–18[3]	5.4–8.2	4–5	1.8–2.3	16–24	7.2–10.9	4–8	1.8–3.6	6–12	2.7–5.4
—	—	12–15	5.4–6.8	—	—	15–20	6.8–9.1	16½–22¾	7.5–10.3
¼–1½	0.57–0.68	1¾–2	0.79–0.91	1½–1¾	0.68–0.79	1½–2½	0.68–1.1	1½–1¾	0.68–0.79
8–12[3]	5.4–8.2	2–3	0.91–1.36	6–8	2.7–3.6	2–4	0.91–1.81	3–6	1.4–2.7
4–6	1.8–2.7	2–3	0.91–1.36	10–16	4.5–7.2	2–4	0.91–1.81	3–6	1.4–2.7
—	—	12–15	5.4–6.8	—	—	15–19¾	6.8–9.0	16–22	7.2–10.0
¼–1	0.11–0.45	1½–1¾	0.68–0.79	1–1½	0.45–0.68	1¼–1¾	0.57–0.79	½–¾	0.23–0.34
25–40	11.3–18.1	6–16	2.7–7.3	40–55	18.2–24.9	6–25	2.7–11.3	6–35	2.7–5.9
—	—	8–12	3.6–5.4	—	—	11–16	5.0–7.3	15–21	6.8–9.5
1–1¼	0.45–0.57	2	0.91	1¼–1½	0.57–0.68	2	0.91	1¼–1½	0.57–0.68
25–40	11.3–18.1	6–16	2.7–7.3	40–55	18.1–24.9	6–25	2.7–11.3	6–35	2.7–15.9
2–3	0.91–1.36	8–12	3.6–5.4	4–5	1.8–2.3	11–16	5.0–7.3	15–21	6.8–9.5
½	0.23	1–2	0.45–0.91	½	0.23	1–1½	0.45–0.68	1	0.45
15–30	6.8–13.6	3–8	1.4–3.6	20–35	9.1–15.9	3–15	1.4–6.8	3–15	1.4–6.8
3–4	1.4–1.8	1–3	0.45–1.4	7	3.2	1–4	0.45–1.8	1–7	0.45–3.2
1–2	0.45–0.91	8–12	2.6–5.4	—	—	11–16	5.0–7.2	15–21	6.8–9.5
½	0.23	1–2	0.45–0.91	½–¾	0.23–0.34	1–1¾	0.45–0.79	1–1¼	0.45–0.57

[3]With calves (both replacement heifers and stockers) an extra 2 lb (0.91 kg) of hay daily, over and above requirements, are herewith indicated to allow for wastage. Practical operators generally feed stemmy or other hay left over by calves to the cow herd.

Fitting Rations

Variations can and should be made in fitting rations, depending upon the individual animal, the relative prices of feeds, and the supply of homegrown feeds. To attain the correct state of condition, a suitable ration must be selected and the animal or animals must be fed with care over a sufficiently long period. The rations listed in Table 19–16 (see page 396) have been used by successful fitters. They are higher in protein content than rations normally used in commercial-finishing operations, but most experienced caretakers feel that by such means they get more bloom. In general, when show animals are being force-fed on any one of these concentrate mixtures, experienced caretakers prefer to feed a grass hay or a grass-legume mixed hay to a straight legume, because of the laxative effect and possible bloat hazard of the latter.

Ration 11 is the one which the senior author has used in fitting show steers. The cooked barley is prepared by (1) adding water in the proportion of 2 to 2½ gal to each gallon of dry barley, and (2) cooking until the kernels are thoroughly swollen and can be crushed easily between the thumb and forefinger. Each young steer also receives 4 lb daily of a supplement high in milk by-products. As the animal approaches show finish, the ration is changed by decreasing the rolled barley by 7 lb and increasing the rolled oats by 5 lb and the wheat bran by 2 lb.

TABLE 19-16
FITTING RATIONS FOR SHOW AND SALE CATTLE
(As-Fed Basis)

Rations 1 to 5 are bulky. They are recommended for use (1) by the inexperienced feeder, and (2) in starting prospective show animals on feed.

Rations 6 to 11 are less bulky and higher in energy. They are recommended for use (1) by the experienced feeder, and (2) during the latter part of the fitting period.

Ration No. 1	(lb)	(kg)	Ration No. 4	(lb)	(kg)	Ration No. 7	(lb)	(kg)	Ration No. 10	(lb)	(kg)
Rolled barley	50	22.7	Crushed oats	30	13.6	Flaked corn	55	25.0	Rolled barley	35	15.9
Crushed oats	20	9.1	Rolled barley	30	13.6	Crushed oats	20	9.1	Crushed oats	20	9.1
Wheat bran	20	9.1	Wheat bran	20	9.1	Dried beet pulp	10	4.5	Rolled wheat	20	9.1
Protein supplement[1]	10	4.5	Flaked corn	10	4.5	Protein supplement[1]	15	6.8	Dry beet pulp	15	6.8
			Protein supplement[1]	10	4.5	**Ration No. 8**			Protein supplement[1]	10	4.5
Ration No. 2						Flaked corn	40	18.1	**Ration No. 11**		
Rolled barley	30	13.6	**Ration No. 5**			Rolled barley	20	9.1			
Flaked corn	20	9.1				Crushed oats	10	4.5	Rolled barley	20	9.1
Crushed oats	20	9.1	Flaked corn	55	25.0	Dried beet pulp	10	4.5	Flaked corn	20	9.1
Wheat bran	20	9.1	Crushed oats	30	13.6	Wheat bran	10	4.5	Crushed oats	20	9.1
Protein supplement[1]	10	4.5	Protein supplement[1]	15	6.8	Protein supplement[1]	10	4.5	Whole barley (dry wt. basis but cooked before feeding)	13	5.9
						Ration No. 9			Commercial supplement	8	3.6
Ration No. 3			**Ration No. 6**			Crushed oats	25	11.3	Linseed meal	8	3.6
						Rolled barley	20	9.1	Wheat bran	6	2.7
Flaked corn	40	18.1	Flaked corn or sorghum	50	22.7	Rolled wheat	20	9.1	Beet pulp, dried molasses	4	1.8
Crushed oats	30	13.6	Rolled barley	40	18.1	Flaked corn	20	9.1			
Wheat bran	20	9.1	Protein supplement[1]	10	4.5	Wheat bran	10	4.5			
Protein supplement[1]	10	4.5				Protein supplement[1]	5	2.3	Salt	1	.5

[1]The protein supplement may consist of linseed, soybean, cottonseed, or peanut meal. With most caretakers, linseed meal is the preferred protein supplement. It gives the animal a sleek hair coat and a pliable hide. Because it is a laxative feed, however, caution should be used in feeding it. Although it is true that an animal getting good clover or alfalfa hay needs less protein supplement than does one eating nonleguminous roughage, it is not possible to supply all the needed protein with hay and still get enough grain into young animals to finish them quickly.

NUTRITIONAL DISEASES AND AILMENTS

Nutritional deficiencies may be brought about by either (1) too little feed, or (2) rations that are too low in one or more nutrients.

Chapter 5, Nutritional Disorders/Toxins, of this book contains a summary of the important nutritional diseases and ailments affecting cattle and other animals; hence, the reader is referred thereto.

PART II—Feeding Breeding Beef Cattle

The beef breeding herd must be properly fed if a good calf crop is to be produced. The size of the calf crop, the vigor and size the calves attain by market time, and the feeding efficiency of the herd largely determine the profit realized.

FEEDING BROOD COWS

Feed affects total profit and cow productivity. It accounts for 65 to 75% of the total cost of keeping cows, and it exerts a powerful influence on cow fertility and calf weaning weight —the two biggest success factors in the cattle business.

Nutritional Requirements of Brood Cows

The nutrient allowances should be adequate to provide for maintenance, growth (if animals are immature), and reproduction and lactation. Fortunately, these needs can be met largely through feeding roughages—pasture in season, and dry forages and silages during the winter months.

The nutritional requirements of beef cows are influenced by weight and size of the female, milk production, age, and climate. Size of the cow has more influence on feed needs than any other item. Also, bigger cows produce bigger calves. Research has shown that 5 to 15 lb extra weaning weight is obtained with each 100 lb increase in cow weight. U.S. Department of Agriculture research at Clay Center, Nebraska showed that large frame brood cows produced 9% more calf weight than medium frame cows. But the large frame cows required 28% more feed than the medium frame cows. So, large cows eat more feed and wean heavier calves. The two-pronged question that each producer must determine is: Will added calf weight be sufficient to offset the added cost of maintaining a larger cow; and will the calf be so large that it causes difficulty at calving time?

Experiments and practical observations reveal that the period during which calf crop percentage is affected most by nutrition extends from 30 days before calving until 70 days after calving—until after rebreeding; a period of approximately 100 days. This, then, is the most critical period in the cow-calf business. It's when life begins—that period within which one calf is born and another is conceived. The needs for the cow during this more critical production period are approximately equal to her needs for the remainder of the year.

The average daily energy requirement is about 14.5 lb of TDN. However, the requirements are above the average for nearly 6½ months of the year. This means that, for reasons of economy, the calving season should be timed so that much of the feed can be supplied by pasture and other economical sources of homegrown energy and protein.

A second important requisite of a sound beef cattle nutrition program is to feed animals according to their requirements. It is impossible to feed the herd properly when calving occurs year-around, or when dry pregnant cows, replacement heifers, and cows nursing calves are run together.

Weight makes a difference, as shown in Table 19-17, which gives the daily nutrient requirements at various weights of (1) dry pregnant cows, and (2) cows nursing calves.

TABLE 19-17
DAILY NUTRIENT REQUIREMENTS OF BEEF COWS[1]

Body Weight		TDN		Total Protein		Calcium	Phosphorus
(lb)	(kg)	(lb)	(kg)	(lb)	(kg)	(g)	(g)
Dry pregnant mature cows (middle third of pregnancy)							
770	350	9.02	4.1	1.34	.61	20	15
880	400	9.68	4.4	1.45	.66	22	16
990	450	10.56	4.8	1.55	.70	23	18
1,100	500	11.22	5.1	1.64	.74	25	20
1,210	550	11.88	5.4	1.74	.79	26	21
1,320	600	12.54	5.7	1.83	.83	28	23
1,430	650	13.20	6.0	1.92	.87	30	25
Cows nursing calves, first 3 to 4 months after calving (superior milking ability)							
770	350	11.2	5.1	2.22	1.01	36	24
880	400	13.0	5.9	2.42	1.10	37	25
990	450	14.1	6.4	2.61	1.19	39	26
1,100	500	15.0	6.8	2.74	1.24	40	28
1,210	550	15.6	7.1	2.86	1.30	42	30
1,320	600	16.5	7.5	2.97	1.35	43	31
1,430	650	17.2	7.8	3.07	1.39	45	33

[1]Adapted by the authors from *Nutrient Requirements of Beef Cattle*, sixth revised edition, National Research Council-National Academy of Sciences, Washington, D.C., 1984, pp. 45 and 46; with U.S. Customary added by the authors.

Of course, these are minimum requirements. Hence, it would be well to add 1 percentage unit to the crude protein requirement and 3.0 percentage units to the TDN requirement, and to self-feed the minerals. This would take care of variations in feedstuffs and differences in requirements among individual cows within a herd.

Nutritional Reproductive Failure in Cows

A review of the literature clearly points to 3 important reproductive difficulties: (1) the small number of cows in heat and bred the first 21 days of the breeding season, (2) the low conception rate at first service, and (3) the excessive calf losses at birth or within the first 2 weeks of age. Also, it is noteworthy that each of the causes is more marked in young cows (first-calf heifers) than in mature cows.

Research throughout the country gives ample evidence that the real cause of most beef cow reproductive failure is a deficiency of one or more essential nutrients just before and immediately following calving—nutritive deficiencies during that critical 100-day period when life begins—a deficiency of energy, protein, minerals, and/or vitamins.

Winter Feeding

In a country as large and diverse as the United States, wide variations exist in both the length of the winter season and the available feeds. But the same principles are applicable to all areas and enterprises, and the chief objective remains the same—economically to produce high percentage calf crops with heavy birth and weaning weights.

Winter feeding is the most expensive time in cow-calf operations because the feed must be processed and brought to the animals. From an economic standpoint, therefore, it is important that wintering practices be both knowledgeable and wise. Cheaper home-grown roughages should constitute the bulk of the winter ration for dry pregnant cows. Most of the grain and the higher class roughages may be used for other classes of livestock. A practical ration may consist of silage and/or dry roughages (legume or grass hays) combined with a small quantity of protein-rich concentrates (such as soybean meal or cottonseed meal). With the use of a legume roughage, the protein-rich concentrate may be omitted. Dusty or moldy feed and frozen silage should be avoided in feeding all cattle—especially in the case of the pregnant cow, for such feed may produce complications and possible abortion.

The best calf crop is produced by cows that are kept in vigorous breeding condition—that are neither overfat nor thin and run-down. Generally speaking, this calls for winter feeding, with the maximum use of roughage. The kind and amount of concentrate needed will depend upon (1) the amount and kind of roughage given, (2) the age and condition of the cattle, and (3) whether the cows are dry or suckling calves. In total, it is important that the ration provides the kinds and amounts of nutrients needed, along with sufficient bulk to satisfy the appetite reasonably well. On a dry-feed basis, the daily requirement of dry pregnant cows is about as follows: thin cows, 2¼% of their liveweight; cows in average flesh, 2% of liveweight, and cows in good condition, 1¾% of liveweight. Cows suckling calves should receive approximately 50% more feed than dry cows of comparable weight and condition. From this it should be concluded that, unless the herd is so small as to make it impractical, dry cows should be wintered separately from those that are suckling calves. This makes it possible to limit the feed of dry cows and to effect certain other economies in their handling.

RATIONS FOR DRY PREGNANT COWS

Dry pregnant cows in average condition should gain in weight sufficient to account for the growth of the fetus (60 to 90 lb) plus sufficient increase in weight and condition to carry them through the suckling period. In total, they should gain 100 to 150 lb during the pregnancy period, or at the rate of approximately ½ lb daily. Of course, the size and condition of the cow is the best gauge as to the feed allowance and desired gain.

When winter grazing is not possible, the rations in Table 19-18 may be used to meet the daily needs for energy and protein of a 1,100-lb dry pregnant cow. A combination of legume roughage with lower quality roughage (such as stalklage, straw, corncobs, or cottonseed hulls) will meet both the energy and protein requirements without the use of a supplement.

TABLE 19-18
WINTERING RATIONS FOR A 1,100-LB (500-KG) DRY PREGNANT COW

	Rations				
	1	2	3	4	5
	(lb /kg/)day)				
Legume-grass hay	18 (8)				10 (4.5)
Legume-grass haylage[1]		30 (12)			
Corn or grain sorghum silage			35 (15)		
Stalklage or husklage				45 (20)	
Straw, cobs, or cottonseed hulls					10 (4.5)
Supplement[2]			.5 (0.2)	1 (0.45)	

[1]Haylage figured at 55% dry matter, corn or grain sorghum silage at 35% dry matter, stalklage or husklage at 45% dry matter.

[2]Supplement figured at 48% crude protein. Quantity to be adjusted in keeping with the protein content of the supplement. For example, if a 24% crude protein supplement is fed, the quantity of supplement should be doubled.

CALVING CONTROL (DAYTIME CALVING)

The following benefits would accrue if the majority of cows would calve during the daylight hours:
1. Improved calf survival due to birth during warmer daylight hours and readily available assistance.
2. Reduced nighttime labor from fewer cows calving.

Limited research supports the theory of altering calving time by late feeding (feeding 5 p.m. to 9 p.m., starting about 2 weeks before calving), with the bulk of calves born during daylight. Scientists at the Fort Keogh Livestock and Range Research Laboratory, Miles City, Montana, report that heifers fed between 8 and 9 p.m. had 17% more daytime births than heifers fed between 8 and 9 a.m. Other researchers report a 10 to 15% increase in daytime calving from nighttime feeding. The reason late feeding causes daytime calving has not been established.

RATIONS FOR COWS NURSING CALVES

Cows with calves at side should be fed for the production of milk, which requirements are more rigorous than those during pregnancy.

The energy requirement of a cow nursing a calf is about 50% higher than that of a dry pregnant cow; and the protein, calcium, and phosphorus requirements are nearly double.

The vast majority of the nation's cows with calves at side are on pasture most, if not all, of the lactation period. So, it is important that producers recognize that the quantity and quality of pastures used for lactating cows and their calves affect milk production, daily gains and weaning weights of calves, and rebreeding of cows.

The rations in Table 19-19 may be used for drylot feeding of beef cows nursing calves.

TABLE 19-19
WINTERING RATIONS FOR A 1,100-LB (500-KG) COW NURSING A CALF

	Rations				
	1	2	3	4	5
	(lb /kg/)day)				
Legume-grass hay	30 (13)			20 (9)	10 (4.5)
Legume-grass haylage[1]		50 (22)			
Corn or grain sorghum silage			60 (27)		40 (18)
Grain				5 (2)	
Supplement[2]				1.5 (0.6)	

[1]Haylage figured at 55% dry matter; corn or grain sorghum silage figured at 35% dry matter.

[2]Supplement figured at 48% crude protein. Quantity to be adjusted in keeping with the content of the supplement. For example, if a 24% crude protein supplement is fed, the quantity of the supplement should be doubled.

CROP RESIDUES AND WINTER PASTURES

Two requisites are important in wintering the cow herd: (1) bringing them through the winter in proper condition for calving, and (2) keeping feed costs to the minimum consistent with nutritional demands. Meeting these requirements has prompted increased use of crop residues and winter pastures for brood cows. As the ever-increasing human population of the world consumes a higher proportion of grains and seeds, and their by-products, directly, cattle will utilize increasing amounts of crop residues and pastures and a minimum of products suitable for human consumption. Thus, more and more farmers with crops will include a beef herd in their operations and realize a fair return from feeds which would otherwise be wasted.

Crop Residues

Crop residues are the parts of forages that remain after harvesting a grain or seed crop. Among such crop residues are cornstalks and husklage, sorghum stalks, soybean refuse, small grain straws and chaff, and legume and grass seed straws.

CORN RESIDUES

Fig. 19-16. Cows grazing cornstalks. (Courtesy, Iowa State University, Ames)

Of all crop residues, that from corn is produced in greatest abundance and offers the greatest potential for expansion in cow numbers. Of course, knowing the feeding value and proper supplementation of corn residues are pertinent to their profitable use.

Table 19-20 lists the daily nutritive requirements of a dry pregnant cow, middle third of pregnancy, weighing 1,100 lb. Table 19-21 gives the nutritive composition of air-dry corn stover and husklage.

TABLE 19-20
NUTRITIVE REQUIREMENTS OF A DRY PREGNANT COW (MIDDLE THIRD OF PREGNANCY) WEIGHING 1,100 LB (500 KG)[1]

Dry matter, daily	20.9 lb (9.5 kg)
TDN, daily	11.2 lb (5.1 kg)
Total protein	7.8%
Calcium	0.26%
Phosphorus	0.21%
Vitamin A	27,000 IU

[1]*Nutrient Requirements of Beef Cattle*, sixth revised edition, National Research Council-National Academy of Sciences, Washington, D.C., 1984.

TABLE 19-21
ANALYSIS OF AIR-DRY CORN STOVER AND HUSKLAGE[1]

	Corn Stover	Husklage
	(%)	(%)
TDN	48	57
Crude Protein	4.5	3.4
Calcium	0.4	0.02
Phosphorus	0.07	0.05
Vitamin A	—	—

[1]*Cow-Calf Information Roundup*, University of Illinois, 1971, p. 10, Table 2.

Studies show that a 1,100-lb cow will eat approximately 22–24 lb per day of palatable, air-dry stover, or about 2 lb or more of air-dry stover per hundred pounds body weight per day. She will eat slightly larger amounts of husklage. This consumption, along with the information presented in Tables 19-20 and 19-21, suggests that stover and/or husklage rations will meet the daily energy (TDN) needs of dry pregnant cows, but such rations will be slightly deficient in protein, and low in phosphorus and vitamin A. Nevertheless, the highest and best use for corn residue is for dry pregnant cows for the period following conception to about 30 days before calving.

For nursing cows, the protein deficiency of stover and/or husklage may be corrected by supplementation, on a per head per day basis, with 2 lb of a 40% protein supplement, or 6 lb of a good legume hay. If desired, the protein supplement may be provided in the form of protein blocks, with one block provided for each 15 cows. Where hay is fed, it should be taken to the field, rather than fed in a feedlot, as this will encourage the cows to stay in the field and graze the cornstalks.

Phosphorus should be provided to all cattle fed corn residue. Calcium may be deficient, especially for lactating cows. Also, some of the trace elements may be deficient. Hence, it is recommended that all cattle on high corn refuse have free access to a complete mineral supplement. A mineral mixture with a Ca:P ratio of 1:2 is recommended for gestating cows, and a 1:1 ratio for lactating cows.

Corn residue is deficient in vitamin A, which should be supplemented. The precalving and postcalving (heavy milking) needs of approximately 27,000 and 39,000 IU per head per day, respectively (NRC-1984), may be met by feeding vitamin A supplement, by intramuscular injection of vitamin A solution, or by feeding adequate levels of green, leafy hay.

It is important that corn residue be tailored to match the cow's nutritional needs. This is relatively simple with dry pregnant cows, where supplementation with a high-phosphorus mineral and vitamin A will usually suffice. Beginning 4 to 6 weeks before calving and continuing through the lactation period, much heavier supplementation is necessary; in addition to phosphorus and vitamin A, protein must be added, and preferably some energy and calcium for nursing cows.

(Also see Chapter 12, By-product Feeds/Crop Residues.)

Winter Pasture

Where feasible, winter pasture offers cattle producers a means of reducing costs. By accumulating the feed in the field, rather than harvesting, storing, and handling the forage, the cost and labor of winter feeding can be substantially reduced. Also, costs of bedding and manure hauling can be eliminated.

Tall fescue is used as a winter pasture in the area to which it is adapted—Missouri, Illinois, Indiana, and Ohio. Usually, the new regrowth is baled in late June into round bales and left in the field. The round bales shed rain and snow and, together with the regrowth, make excellent late fall and winter grazing. Experience shows that field-stored forage has adequate quality to maintain beef cows in good condition.

Fig. 19-17. Cows on winter fescue pasture, supplemented with round bales of fescue harvested the previous June and left in the field. (Courtesy, University of Illinois, Urbana)

Range and Pasture Feeding

Western pastures that receive less than 20 in. of rainfall annually are classified as the western range. The carrying capacity of much of the range is low, and little of it provides yearlong grazing. Moreover, variation in vegetative types, climate, and topography in the range country is accompanied by great diversity in the seasonal use made of it. As a result, rangelands are usually grazed during different times of the year, and the herds migrate with the season, moving to the mountains and higher elevations in summer and returning to the lower ranges in winter.

From the standpoint of vegetation and utilization by livestock, ranges differ from cultivated pastures as follows:

1. They are less productive.
2. They are more likely to progress to less palatable plants.
3. They are more difficult to restore when depleted.
4. They often serve multiple use—for wildlife production, recreation, timber production, and mineral production, as well as use by cattle and other domestic animals.

Improved ranges and pastures should be the first goal of cattle raisers, without using supplemental feeding as a substitute for good grass or as a crutch for poor range. Instead, the two—good range and proper supplemental feeding—go hand in hand.

RANGE NUTRIENT DEFICIENCIES

Growing grasses provide adequate nutrients for beef cattle in unforced production when (1) produced on fertile soils, (2) available in sufficient quantities, (3) not washy, and (4) not weathered, leached, or bleached. However, the simultaneous fulfillment of all these conditions is the exception, rather than the rule. Every cattle producer worthy of the name, forces young stock for an early market; most soils are deficient in certain nutrients, which, in turn, affect the plants and the animals feeding thereon; during droughts and early and late in the season, feed may be in short supply (thereby limiting energy and other nutrients); early spring pastures are washy and lacking in energy; and during droughts and late in the season, grasses become mature, leached, and bleached—they increase in fiber and decrease in protein, phosphorus, and carotene. To meet these conditions, a supplemental source of energy, protein, phosphorus, and vitamin A is necessary.

Fig. 19-18. Drought on the range, causing a short forage supply along with limited energy and other nutrients. (Courtesy, Santa Gertrudis Breeders International, Kingsville, Tex.)

PASTURE AND RANGE SUPPLEMENTATION

Where dried grass cured on the stalk is grazed, or where insufficient pasture is available—perhaps due to drought or overstocking—supplemental feeding is necessary. Also, supplemental feeding is a way in which to extend the grazing season, both early and late.

Sorting Pasture and Range Cattle

When supplemental feeding is planned, it is strongly recommended that cattle first be sorted by age and condition groups.

Heifers should not be supplemented at the same levels as older cattle. Because they are growing, they must be fed more liberally. Also, heifers have need for more protein, and they must be fed for a longer period. But heifers should not be overfed to the point that reproduction is adversely affected.

Thin cows should be placed where they can be given extra feed and special care. Most of the ration of pregnant cows should consist of pasture plus such supplements as required—with emphasis on proteins, minerals, and vitamin A; and the kind and level of supplementation should be varied according to the quality and quantity of grass available.

Immediately before and after calving, cows that are not on pasture or range should be fed lightly and with laxative feeds. At this time, the amount of supplementation should be governed by the milk flow, the condition of the udder, the demands of the calf, and the appetite and condition of the cow. More energy, proteins, and minerals, are required for a cow suckling a calf than for a pregnant or dry cow. The nutritional requirement of cows nursing calves is approximately 50% higher than for pregnant cows.

Until weaning time, the growth of the calf is determined chiefly by the amount of milk available from its dam, plus whatever assist is given through grass or creep feeding.

Choosing a Pasture or Range Supplement

Every cattle producer faces the question of what supplement to use, when to feed, and how much to feed under existing conditions.

In supplying a supplement to range cattle, the following requisites should be observed:

1. It should balance the ration of the animal(s) to which it is fed, which means that it should supply all the nutrients needed by the animal(s) which are missing in the forage.
2. It should be fed in such a way that each animal gets its proper portion.
3. It should be fed in a form that is convenient and practical from the standpoint of the feeder, and that will least disturb the animal.

Types and Systems of Pasture and Range Supplementation

Many different feeds may be, and are, used; among them, (1) ranch- or locally-produced hay, (2) alfalfa pellets or cubes, with or without fortification, and (3) supplements of various kinds. Likewise, many different systems of range supplementation are used, including (1) range cubes or pellets; (2) hand-feeding at intervals, rather than daily; and (3) protein blocks, liquid protein supplements, and self-feeding salt-feed mixtures. Where these feeding systems do not result in the neglect of the herd, there is no adverse effect upon the health and weight of the cows, percent calf crop, or weaning weight of calves.

Today, on an increasing commercial basis, beef cows are being confined to small quarters—to drylots, all or part of the year. This is a viable alternative management system for marketing forage, feeds, and crop residues through the cow herd. Low feed prices and by-product feeds, available breeding cattle, labor, facilities, and equipment may lead beef producers to consider a drylot or partial drylot beef cow/calf enterprise. Granted, the drylot herd will not replace the grazing cow, but, under some circumstances, such as high land values, it can complement the more traditional beef cow operation.

Fig. 19-19. Protein block in use on pasture—a means of lessening the labor attendant to the daily feeding of a protein supplement on pasture or range. (Courtesy, Moorman Mfg. Co. Inc., Quincy, Ill.)

(Also see Chapter 11, Protein Supplements, section headed, "Types and Methods of Feeding Protein Supplements.")

Fig. 19-20. Part of a herd of 80 brood cows in confinement in a deep, open shed on Circle S Ranch, Rockwood, Ontario, Canada. (Courtesy, *Country Guide*, Winnipeg, Manitoba, Canada)

Confinement (Drylot) Beef Cows

Confining (drylotting) beef cows refers to the practice of confining beef cows to small quarters—to drylots, all or part of the year.

From a feeding standpoint, the following points are pertinent in drylot beef cow operations:

1. All feed must be mechanically harvested and moved to the feedlot, rather than being harvested directly by the cows.
2. An assured, adequate, and economic feed supply must be available. The capital tied up in harvesting equipment and stored feeds may be quite large.
3. More knowledge of beef cow nutrition and ration formulation is needed.

RATIONS FOR DRYLOT COWS

Rations for drylot cows generally consist of low-cost roughages—such as crop refuse, straw, cottonseed hulls, and gin trash—supplemented with protein, grain, vitamins, and minerals as required. Where available, higher quality roughages—such as silages, hays, and haylages—may be used, especially (1) during the critical 100 days, beginning 30 days before calving and extending 70 days after calving, and (2) for heifers calving as 2-year-olds. Also, during the summer and fall, green chop is frequently fed. Cows in partial confinement may, in season, graze such forages as cornstalks, grain stubble, or irrigated or native pastures.

Phase feeding according to stage of production and age of animals is recommended.

It is relatively easy to meet the nutritive requirements of a dry pregnant cow. Generallly speaking, low-quality roughages, properly supplemented, or a combination of low-quality roughages and high-quality roughages, will suffice.

The lactation requirements are much more rigorous than the dry pregnancy requirements; and the higher the milk production, the higher the nutritive requirements. Thus, requirements for two levels of milk production for lactating cows—(1) average milk production, and (2) superior milk production—are presented in Tables 19-1 and 19-2.

Under a drylot system, heifers are commonly calved as 2-year-olds, when they are still growing. Thus, during lactation they have a nutritional requirement for both growth and lactation. The nutritional requirements for these animals will be large, particularly if they are crossbreds and bred for rapid growth, considerable size, and high milk production.

The mineral needs of confinement cows may be met either by incorporating the needed minerals in the supplement which is fed, or by feeding the required minerals free-choice.

Vitamin A supplementation is extemely important for drylot cows. The carotene content of the dry forage should be disregarded and the total vitamin A requirement met by supplementation. This can be done by feeding a supplement, such as 2 lb of mill waste, containing 1 million IU of vitamin A per animal—feeding this vitamin A supplement once a month to heifers, and every other month to older cows. With older cows receiving high levels of dry forages containing normal amounts of carotene, it is probable that the vitamin A requirements are being met. However, it has been demonstrated under range conditions that (1) percent calf crop is markedly increased by supplementing with vitamin A during drought years, and (2) calves respond to vitamin A treatments given their dams 90 days prior to calving.

SEMICONFINEMENT (OR PARTIAL CONFINEMENT) COW HERDS

A semiconfinement (or partial confinement) operation takes advantage of grazing during part of the year, such as winter grazing of corn or sorghum stalks or seasonal grazing of pastures. In addition to providing low-cost feed and allowing the animals to do their own harvesting, breeding may be timed so that the calves will be dropped on clean pasture as a means of (1) preventing calf scours, and (2) stimulating milk flow.

FEEDING BULLS[4]

Frequently, little thought is given to the management and feeding of bulls except during the breeding season. Instead, the feeding program for herd bulls should be such as to keep them in a thrifty, vigorous condition at all times. They should neither be overfitted nor in thin, run-down condition. Also, exercise is necessary for the normal well-being of the bull.

The feeding and management of bulls differ according to age and condition. For this reason, sale bulls, young bulls, and mature bulls are treated separately in the sections that follow.

---

[4]The nutritive requirements of bulls for growth and maintenance at different weights, are given in Tables 19-1, 19-2, 19-3, and 19-4.

Feeding Sale Bulls

Most bull sales are held in late winter and early spring, at which time mostly yearling and 2-year-old bulls are sold. In order to attract buyers, they have usually been grain fed since calfhood. Most bull buyers—especially commercial cattle producers in rougher range areas—would rather have their new bulls in less than fitted sale condition. They find that such bulls are more fertile and more apt to range with the cows when turned to pasture during the breeding season.

Sale and show bulls should be acquired 2 to 3 months ahead of the breeding season, so that they may be conditioned, or let down. Also, bear in mind that it takes about 40 days from the time a sperm cell is formed until it is ready to be ejaculated. Since the stress of handling and hauling a bull can reduce his fertility for about 40 days (it may be even longer where there is a change in elevation and climate), the rest period lets his body overcome these problems.

Feeding Young Bulls

Lack of fertility in a bull may often be traced back to his early care and feeding. From weaning to three years of age, bulls should be kept separate by age groups. Young bulls should be fed more liberally than mature bulls because their growth requirements must be met before any improvement in condition can take place.

Following weaning, bulls should be fed and developed sufficiently to show their inherited characteristics, but without excessive finishing. Simultaneously, they should be given plenty of exercise. Overfeeding and lack of exercise are apt to result in infertility, low-quality sperm, and unsound feet and legs.

To achieve proper development, young bulls should gain at least 2½ lb daily from weaning to 12 to 15 months of age. This will necessitate a daily feed allowance equal to about 2½% of their body weight, with a ration comprised of 50% or more concentrate. From 15 months to 3 years old, they should make a daily gain of 2 to 2¼ lb and receive a daily feed allowance equal to 2 to 2¼% of their body weight, with the proportion of roughage increased after the first year.

Fig. 19-21. Yearling Hereford bulls on pasture. (Courtesy, American Hereford Assn., Kansas City, Mo.)

Bulls handled as recommended above will generally attain half their mature weight by the time they are 14 to 15 months of age and may be used in limited service.

During the breeding season, young bulls should be fed a grain ration consistent with pasture quality and number of cows to be bred in order to promote proper growth and development. Drought, overgrazing, and poor-quality pastures are situations in which grain supplementation is particularly needed. Heavy service and poor pasture with no supplemental feeding may shorten the breeding career of a young bull.

After the breeding season, yearling bulls generally need 5 to 6 lb of grain along with good roughage.

Feeding Mature Bulls

Winter is the proper time to condition bulls for the next breeding season. Bulls that have been running on pasture with the cows are likely to be thin; thus, they require sufficient concentrate to put them in proper flesh. Mature bulls will consume daily amounts of feeds equal to 1½ to 3% of their liveweight, depending upon condition and individuality.

The importance of having bulls in proper condition at the opening of the breeding season cannot be overemphasized. Nothing is quite so disheartening or costly as a small calf crop, with many of the calves coming late. Lack of fertility in the bull often may be traced back to his care and feeding.

Feed mature bulls all the legume hay they will eat plus 3 to 5 lb of ground or rolled grain and 1 lb of a 32% protein supplement (or equivalent) per head per day. Also, provide free access to a suitable mineral mixture. About 60 days before the bulls are turned with the cows, increase the concentrate allowance by 25 to 50%, with the amount of the increase determined by the condition of the bulls.

The mature herd bull needs no additional feed when running with the cow herd on good summer pasture.

FEEDING CALVES

Beef producers, as a whole, have lagged in applying much of what we know about feeding and managing calves. They're inclined to let mother cows and mother nature fend for the calves. More good proven practices, based on both successful experience and research, need to be put to use in feeding and handling calves.

Feeding at Birth

Losing a calf means losing the profit on the cow for a whole year. Proper feeding and managing at birth can make the difference.

Recommended calving-time feeding practices follow:

1. See that the newborn calf nurses within 2 hours after birth. It is essential that it receive colostrum. The caretaker may have to assist a calf to nurse a dam that has very large teats or an udder that hangs very low. Also, weak calves should be helped to nurse.

2. Keep a close watch for signs of mastitis or injury to udders. It may be necessary to milk out a few cows for the first 2 or 3 days after calving.

3. Be sure cows have access to plenty of clean, fresh water.

Feeding Orphan and Multiple Birth Calves

Occasionally a cow dies during or immediately after parturition, leaving an orphan calf to be raised. Also, there are times when cows fail to give a sufficient quantity of milk for the newborn calf. Sometimes, there are multiple births.

If there are only a few orphans, usually they can be grafted onto other cows (or adopted)—either cows that have lost their calves or that give sufficient milk to raise two calves. When such calves cannot be grafted, they must be raised by artificial methods—without a cow.

Regardless of whether orphans are grafted or raised artificially, the problem will be simplified if the calf receives colostrum, the first milk produced by a cow after giving birth to a calf, during the first 24 hours, and preferably for the first 3 days, of its life—from its mother, from another fresh cow, or from frozen-stored colostrum. Colostrum is higher than normal milk in dry matter, protein, vitamins, and minerals. Also, it contains antibodies (a modified type of serum globulin) that give newborn calves a passive immunity against common calfhood diseases.

Because colostrum is so important for the newborn calf, producers should store a surplus of it from time to time. It can be frozen and stored for a period of 1 year or longer, then, as needed, thawed and warmed to 100° to 105°F, and fed. Also, colostrum may be fermented and stored.

Orphan calves can now be raised successfully on a milk replacer and calf starter ration, using them as directed. The milk replacer may be fed by using a bottle or pail equipped with a rubber nipple, or the calf may be taught to drink from a pail. It is important that all receptacles be kept absolutely clean and sanitary (clean and scald each time) and that feeding be at regular intervals. Dry feed should be started at the earliest possible time; not later than 1 week of age. With proper management, healthy calves may be switched entirely to a suitable dry feed at 4 to 5 weeks of age.

Basically, calves are fed according to one of 3 systems: (1) the whole milk system, (2) the combination whole milk-milk replacer system, or (3) the combination whole milk-calf starter system. Also, various combinations of these 3 systems are used. Further information on the subject of raising calves is presented in Chapter 20, Feeding Dairy Cattle.

Creep Feeding

Creep feeding is the supplementation of calves while they are nursing their dams. It increases weaning weight. The basis for this response is related to the lactation curve of beef cows, the increasing nutrient requirements of the calf during the nursing period, and the decline in feed quality and quantity typical of most pastures or ranges which support the cows and calves during lactation. Studies reveal that milk production of dairy cows increases up to the fourth or sixth month following freshening, then declines gradually. By contrast, maximum milk production of beef cows occurs during the first 2 months after calving, then declines.

Fig. 19-22 shows why creep feeding is important. From birth to weaning, the protein and energy requirements of a growing calf increase well beyond the ability of most beef cows to meet those needs. For example, to meet the protein and energy requirements for growth, a 100-lb calf needs 10 lb of milk, whereas a 500-lb calf needs 50 lb of milk. Since the average beef cow gives only 13 lb of milk per day throughout a 7-month suckling period, a 500-lb calf lacks 37 lb of getting enough milk from its dam at this stage of lactation to meet its needs—that's the *hungry calf gap*.

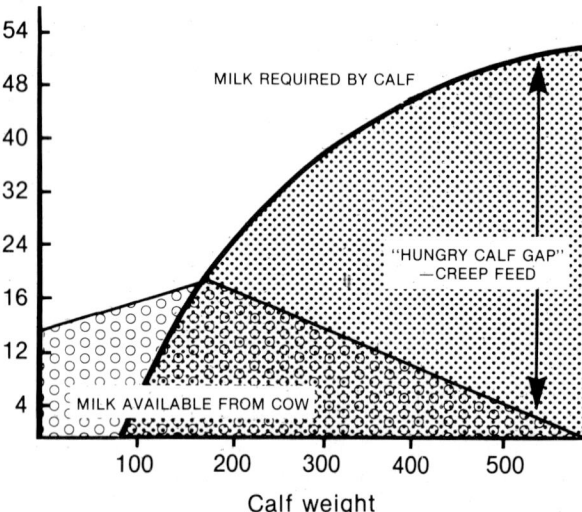

Fig. 19-22. Milk yield of a typical beef cow vs nutrient requirements of a nursing calf. This points up the need for creep feeding.

To fill the *hungry calf gap*—the nutrient requirements over and above those provided by 13 lb of milk—would require the consumption of 50 lb of green grass daily. Of course, that's a physical impossibility, because a 500-lb calf simply cannot hold that much bulk. So, the best way to fill the hungry calf gap is to creep feed with concentrate mixes.

• **Creep**—*A creep is an enclosure or feeder for feeding purposes which is accessible to the calves but through which the cows cannot pass.* It allows for the feeding of the calves but not their dams.

Fig. 19-23. Movable calf creep, with openings that will permit the calves to enter and keep the cows out. (Courtesy, National Cottonseed Products Assn., Inc., Memphis, Tenn.)

• **Creep rations**—Tables 19-22 and 19-23 show two creep rations, formulated by the authors, that have been widely and successfully used. A simple, yet very satisfactory, creep ration may be made by grinding and pelleting 75% alfalfa and 25% cereal grain.

TABLE 19-22
CALF CREEP RATION # 1[1] (AS-FED BASIS)

Ingredient	Precent	Per Ton	
	(%)	(lb)	(kg)
Oats	39.60	800.0	363.2
Corn # 2	14.80	300.0	136.2
Barley	8.90	177.5	80.7
Wheat bran	9.90	200.0	90.8
Dried molasses beet pulp	9.90	200.0	90.8
Soybean meal, 44%	9.90	200.0	90.8
Molasses	4.90	100.0	45.4
Salt	.50	10.0	4.5
Dicalcium phosphate	.50	10.0	4.5
Trace minerals[2]	.04	1.0	0.45
Vitamin A (30,000 IU/g)	.06	1.5	0.68
Total	100.00	2,000.0	907.2
Proximate analysis:	(%)		
Crude protein	14.30		
Fat	3.20		
Fiber	8.30		
Calcium	.32		
Phosphorus	.50		
TDN	69.60		

[1]*Feed preparation:* Preferably ⅛- or 3/16-in. pellets. Otherwise, steam roll and flake grains, or grind grains coarsely.

[2]See Table 19-7 for recommended trace mineral levels. Follow manufacturer's directions.

TABLE 19-23
CALF CREEP RATION # 2 (AS-FED BASIS)

Ingredient	Percent	Per Ton	
	(%)	(lb)	(kg)
Corn # 2	24.25	485	220.2
Alfalfa meal, 15%	22.50	450	204.3
Oats	20.00	400	181.6
Alfalfa hay (all analyses)	10.00	200	90.8
Soybean meal, 44%	6.20	124	56.3
Bran	5.00	100	45.4
Linseed meal, 35%	5.00	100	45.4
Molasses	5.00	100	45.4
Dicalcium phosphate	2.00	40	18.2
Trace minerals[1]	.05	1	0.45
Vitamin A (325,000 IU/g)[2]	—	63 g	
Total	100.00	2,000	908.0
Proximate analysis:	(%)		
Crude protein	15.10		
Fat	3.00		
Fiber	12.70		
Calcium	1.04		
Phosphorus	.73		
TDN	64.90		

[1]See Table 19-7 for recommended trace mineral levels. Follow manufacturer's directions.

[2]When 4 lb/head/day of the calf creep ration is consumed, 40,950 IU of vitamin A will be obtained in the feed.

• **Limited creep feeding**—Instead of allowing creep-fed calves to consume all they will eat, limited creep feeding, is gaining in popularity. Generally, limiting creep feed to about 3 lb/head/day is recommended. This will supply enough supplemental energy and protein to the dam's milk and forage to meet the requirements for normal growth of young calves. Feeding more than 3 lb/head/day may result in excessive fat deposition instead of skeletal and muscle growth.

Limited creep feeding can be accomplished by hand-feeding. But, the disadvantages of this practice are: (1) the high labor cost of feeding the calves each day; and (2) the larger calves tend to overeat and the smaller ones not to get enough, unless adequate bunk space is provided.

A common salt-limited creep feed consists of a mixture of 5 to 10% salt and 90–95% cottonseed meal. Also, either of the creep rations listed in Tables 19-22 and 19-23 may be limit-fed by adding approximately 10% salt.

• **Creep grazing**—This refers to the practice of grazing nursing calves on separate pastures from their dams. They may either graze before the cows do, getting first choice of the more succulent, highly nutritious pastures; or they may have access to special pastures. The calves enter the special pastures through gates with openings large enough for calves, but too small for the cows to get through. In an alternative method, electric fences are positioned high enough (36 to 42 in.) for calves to pass under, but low enough to keep cows out. Limited studies indicate that as much as ½ lb extra weight gain per day may be obtained by creep grazing.

Feeding Early Weaned Calves

Early weaning refers to the practice of weaning calves earlier than the usual weaning age of about 7 months, usually at about 35 days of age. Although it is not common practice among U.S. beef producers, dairy producers have been weaning 3-day-old calves for years. Also, early weaning has long been an integral part of many of the beef programs of Europe.

Currently, there is much interest in early weaning because (1) it fits into a drylot cow-calf management system; and (2) it can give a big assist in getting females, especially 2-year-old heifers, to rebreed in a short period of time.

Where early weaning is successful, the only responsibility of the beef cow is to produce a calf and give it a good start in life for a brief period, then go on a maintenance ration the rest of the year.

As with many good things in life, early weaning does have some disadvantages. To be successful, superior nutrition and management are essential; and the earlier the weaning age, the more exacting these requirements.

• **Rations for early weaned calves**—From 35 days of age on, early weaned calves can be fed any good starter rations, most of which contain dry skim milk. Two such rations are given in Chapter 20, Feeding Dairy Cattle. Most commercial feed companies manufacture a starter ration. The starter ration should be made available to the calves well ahead of weaning in order that they will be accustomed to it, thereby avoiding any setback.

Weaning

Normally, calves should be weaned when they are 7 to 8 months old. Weaning at this age fits in well with the weight record-keeping requirements of most performance testing programs. Also, calves will be about the right age and weight for fall feeder calf sales.

The best way to wean is to remove the calves from their dams and keep them out of sight of each other. Cows and calves should never be turned together once the separation has been made. Such a practice will only prolong the weaning process, and it may also cause digestive disorders in the calf. Provide calves with plenty of water, free-choice hay, and 3 to 4 lb of grain per head per day. If calves were creep fed, continue their rations during the weaning period.

Preconditioning

Preconditioning is a way of preparing the calf to withstand the stress and rigors of leaving its mother, learning to eat new kinds of feed, and shipping from the farm or ranch to the feedlot. To the cow-calf producer, it is a program of management, nutrition, and immunization. It, along with improved breeding based on production testing, is the trademark of the producer of feeder calves. To the feedlot operator, preconditioning is a way in which to prepare calves to fit into the program and to minimize costly and unnecessary procedures.

Changed environment; excitement of sorting, loading, and shipping; long periods without feed and/or water; movement through one or more assembly points; change of feed; and exposure to disease—all add up to *fatigue, stress, shrink,* and *lowered disease resistance*.

The term *preconditioning* generally consists of the following practices being conducted on the farm or ranch of origin and certified to by a licensed veterinarian: weaned; bunk broke and water tank or fountain trained; castrated and dehorned; vaccinated for IBR, PI-3, BVD, and *Haemophilus somnus;* and, depending on local conditions, additional vaccinations may be required for blackleg and malignant edema, and for brucellosis of heifers.

It is important that the program be written down, adhered to rigidly, then certified to by both the owner and the veterinarian. The producer should take the lead in developing such a program, but the counsel of the veterinarian and potential buyers should be sought.

Studies show that preconditioning (1) increases the weight of calves when they leave the farm or ranch, and (2) improves the performance of calves during the first month in the feedlot. But the premium price needed to pay for preconditioning may be larger than the buyer is willing to pay. Nevertheless, the surgical component of preconditioning—castrating and/or dehorning—should be performed prior to weaning and shipment.

FEEDING REPLACEMENT HEIFERS

The feed and management program of replacement heifers will have a lifelong effect on their productivity. It will determine how young they may be bred, whether they calve early or late, whether they are good milkers or poor milkers, the weaning weight of their calves, and how long they remain in the herd. Also, feed accounts for 40 to 70% of the cost of raising replacement heifers; hence, it is important to know whether it is possible to effect savings on feed during the growing period without affecting reproduction adversely. It is even more important to know whether their performance as adult animals can be enhanced by proper nutrition and management.

Fig. 19-24 shows optimum growth rates and weights at different ages, based on a summary of research. Heifers should not be bred before reaching ⅔ of their mature weight, 9/10 of their mature height, and ⅘ of their mature heart girth and width at hips. Naturally, these measurements vary with different frame sizes and breeds. But once the needed growth rate is established, rations can be formulated that will allow these growth rates to be achieved.

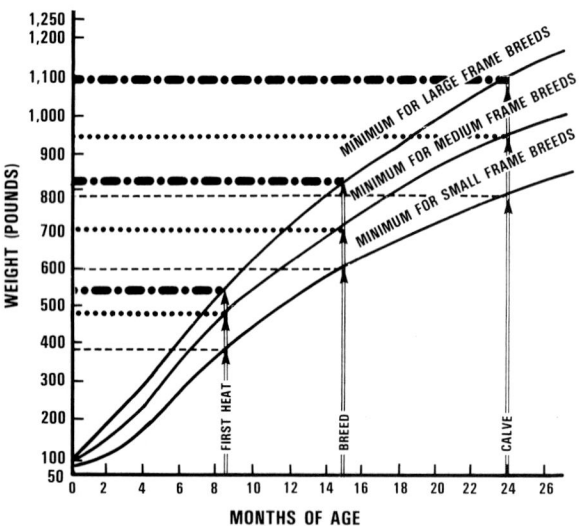

Fig. 19-24. Minimum weights in the reproductive cycle of replacement heifers of different frame sizes.

- **Nutrient requirements of replacement heifers**—Meeting the nutrient requirements of heifers from weaning to first calving is of great importance. The requirements of heifers of different body weights and growth rates are given in Table 19-24.

TABLE 19-24
DAILY NUTRIENT REQUIREMENTS
OF MEDIUM-FRAME GROWING HEIFERS[1]

Body Weight		Daily Gain		TDN[2]		Protein[2]	Calcium	Phosphorus
(lb)	(kg)	(lb)	(kg)	(lb)	(kg)	(lb)	(g)	(g)
400	182	2.00	0.9	7.7		1.29	26	13
500	227	2.00	0.9	9.1		1.34	24	13
600	273	2.00	0.9	10.4		1.40	23	14
700	318	2.00	0.9	11.7		1.45	22	15
800	364	1.40	0.6	10.4		1.60	25	16
900	409	1.40	0.6	11.3		1.60	26	18

[1]The above requirements for 800- and 900-lb (364- and 409-kg) weights are for pregnant yearling heifers last third of pregnancy. Adapted by the authors from *Nutrient Requirements of Beef Cattle*, sixth revised edition, National Research Council-National Academy of Sciences, Washington, D.C., 1984.

[2]Pounds protein and TDN can be converted to kg by dividing by 2.2.

- **Rations for replacement heifers**—In season, good pasture plus mineral supplements fed free-choice will meet the nutrient requirements for proper growth and development of heifers.

On winter range, when dry forage is of low quality, and sometimes not too abundant, 1 to 2 lb of a protein supplement should be provided in the form of cubes, blocks, meal-salt, or liquid. When consumed at the intended level, the supplement should contain sufficient vitamin A to meet the requirements. Mineral supplements should also be provided, preferably free-choice.

Where winter grazing is not available, heifers must be drylotted and fed a complete ration. Sufficient nutrients should be provided to meet the requirements and to keep heifers in a thrifty condition, neither too fat nor too thin.

Fig. 19-25. Yearling Polled Hereford replacement heifers bunk fed a winter ration in a corral. (Courtesy, *Polled Hereford World*, Kansas City, Mo.)

The wintering rations in Table 19-25 for 500-lb heifer calves should result in a rate of gain of 1 to 1.5 lb per day.

The wintering rations in Table 19-26 for 800- to 900-lb bred yearling heifers should allow a gain of 0.75 to 1 lb per day during the wintering period prior to calving.

TABLE 19-25
DAILY RATIONS FOR HEIFER CALVES (500 LB [227 KG])
(AS-FED BASIS)

	Rations								
	1		2		3		4		5	
	(lb)	(kg)	(lb)	(kg)	(lb)	(kg)	(lb)	(kg)	(lb)	(kg)
Legume-grass haylage	25	11								
Legume-grass hay			10	4.5	10	4.5			5	2.3
Corn or sorghum silage							30	13.6	20	9.1
Ground ear corn			4	1.8						
Corn, grain sorghum, or barley					3	1.4				
Supplement[1]							1	.45		

[1]Supplement contains 48% crude protein. Quantity to be adjusted in keeping with the protein content of the supplement.

TABLE 19-26
RATIONS FOR BRED YEARLING HEIFERS
(800-900 LB [364-409 KG]) (AS-FED BASIS)

	Rations								
	1		2		3		4		5	
	(lb)	(kg)	(lb)	(kg)	(lb)	(kg)	(lb)	(kg)	(lb)	(kg)
Corn or sorghum silage	45	20.5	25	11					15	2.3
Legume-grass hay			10	4.5	20	9.1				
Legume-grass haylage							35	15.9		
Corn, grain sorghum, or barley									3	1.4
Supplement[1]	1.5	0.7								

[1]Supplement contains 48% crude protein. Quantity to be adjusted in keeping with the protein content of the supplement.

- **Separate heifers by ages**—The nutritive requirements of heifers differ according to body weight and expected daily gain (Table 19-24). Consequently, the recommended ration for a 500-lb heifer calf (Table 19-25) differs from that of an 800- to 900-lb bred heifer (Table 19-26). It is important, therefore, that replacement heifers be separated by ages for wintering, with coming yearlings in one group and coming twos in another.

Calving Two-Year-Olds

From the above, it may be concluded that more yearling heifers can and should be bred to calve as 2-year-olds. But, in doing so, the following practices should be observed in order to lessen calving difficulties:

1. Keep heifers separate from older cows.

2. Give consideration to the increased nutritional requirements of calves with increased growth rates.

3. Feed replacement heifers for gains of approximately 1 lb/head/day from weaning to first breeding. Following the breeding season, heifers should be managed to assure continued growth and achieve 80 to 85% of expected mature weight at the time of first calving. From breeding until calving, 1¼ lb gain/day is about right.

4. Breed only well-developed heifers, weighing 700 to 750 lb (depending on breed) at 13 to 14 months of age. Size at breeding is more important than age. Also, some breeds come in heat and mature a little earlier than others.

5. "Flush" feed heifers to gain approximately 2.0 lb per head daily beginning 20 days before the start of and continuing through the breeding season.

6. Breed heifers to a bull known to sire small calves at birth.

7. Feed a well-balanced ration, and feed for continuous gain of 1.25 lb during the pregnancy period; but don't get them too fat.

8. Feed heifers to weigh at least 800 lb by 120 days before calving.

9. Feed heifers to gain 100 to 120 lb from 120 days prior to calving. Heifers should weigh at least 875 lb just before calving and approximately 775 lb shortly after calving.

10. Give heifers special care at calving time.

11. Provide superior nutrition—well balanced, and rather liberal—during the lactation period, because a heifer's nutritional requirements double after calving. This requires a good ration—one containing adequate energy and proteins, and fortified with the necessary vitamins and minerals.

12. If practical, wean early; at 2 to 6 months of age, rather than the normal 7 months. Otherwise, creep feed the calves.

13. Run heifers that calved as 2-year-olds in a separate herd until after they have had their second calf.

14. Try it (calving 2-year-olds) out on half of your replacement heifers to start with; make sure that you know what is involved before going all out.

Some breeders may wish to take another year and stick to calving out 3-year-olds. But more and more progressive, commercial cattle raisers will calve out 2-year-olds from the standpoint of cutting production costs and increasing profits.

PART III—Feeding Stocker (Feeder) Cattle

Fig. 19-26. Thin, yearling Santa Gertrudis stocker cattle shipped from Texas and turned to lush pastures in Pennsylvania, followed by a short grain feed. (Courtesy, Pennsylvania Millers and Feed Dealers Assn., Ephrata, Pa.)

Currently used stocker and feeder terms, along with their definitions, follow:

- **Stockers** are calves and yearlings, both steers and heifers, that are intended for eventual finishing and slaughtering, and which are being fed and cared for in such manner that growth rather than finishing will be realized. They are generally younger and thinner than feeder cattle.

- **Feeders** are calves and yearlings, both steers and heifers, carrying more weight and/or finish than stockers, which are ready to be placed on high-energy rations for finishing and slaughtering.

- **Replacement heifers** are the top end of the heifer calves selected to replace the older cows that are culled from the herd.

- **Preconditioning** refers to preparing the calf to withstand the stress and rigors of leaving its mother, learning to eat new kinds of feeds, and shipping from the farm or ranch to the feedlot or stocker grower.

- **Backgrounding** is an old practice with a new emphasis and a new name. Actually, backgrounding and the stocker stage are one and the same. Both refer to that period in the life of a calf from weaning to around an 800-lb weight, when it is ready to go on a high-energy finishing ration. However, the term *backgrounding*, which was ushered in with the development of large feedlots, indicates a shift in emphasis. The term *stocker stage* connotes emphasis on marketing roughages through thin cattle, whereas *backgrounding* connotes emphasis on growing out feeder calves ready to go on a high-energy finishing ration. Backgrounding may be done on pasture or in the drylot, or some combination of both. At its best, the animals should be in good health, bunk broke, and ready to go on full feed.

From the above, it may be concluded that in the variable period of a calf's life between weaning and finishing, it is usually classed as either a stocker, a feeder, or a replacement heifer. Prior to weaning, calves may or may not be preconditioned.

The dividing line between stockers and feeders is not always as clear-cut as the above definitions would indicate. That is, not all thin cattle are suitable for stockers. For example, very large yearlings and most heifers are usually sold as feeders, to be placed on high-energy feeds. Also, "Okie-" type cattle are usually backgrounded for 50 to 60 days, then placed on a finishing ration.

TYPES OF STOCKER PROGRAMS

Sometimes the stocker operation is the only cattle enterprise on a farm or ranch, but more frequently it is conducted in conjunction with a cow-calf operation or it precedes the finishing program.

When the stocker enterprise is the only cattle enterprise on a farm or ranch, it is usually conducted using one of the following plans:

1. Calves or light yearlings are bought in the fall to be wintered on high-roughage rations in dry lot and sold in the spring to buyers either (a) to go on grass for the summer, or (b) to go on a drylot finishing program.

2. Lightweight calves are bought in the fall to be wintered on roughage rations, then, under the same ownership, grazed throughout the following pasture season and sold in the fall. Under this plan, usually lighter weight calves are acquired and they are wintered at a lower rate of gain than in plan 1.

3. In Kansas, Oklahoma, and Texas, calves or light yearlings are bought in the fall and grazed on small winter grains, chiefly wheat. Good wheat pastures will produce very acceptable stocker gains. The main disadvantage to the program is that, due to weather conditions, winter wheat pasture cannot always be counted upon. When it fails, the stockers must either be sold or fed a higher cost roughage.

4. In southeastern United States, which is primarily a cow-calf area, winter oats and fescue are used extensively in stocker programs. This area is turning to stocker programs in order to utilize winter pastures profitably, and to satisfy the demand for 600- to 800-lb feeder steers as a result of the expansion of feedlots.

There is a trend for more and more calves (not yearlings)

TABLE DAILY RATION FOR STOCKER (AS-FED

	Rations (fed for gains of							
	1		2		3		4	
	(lb)	(kg)	(lb)	(kg)	(lb)	(kg)	(lb)	(kg)
Legume hay or grass-legume mixed hay............	12–18	5.4–8.2	—	—	8–12	3.6–5.4	—	—
Grass hay	—	—	12–18	5.4–8.2	4–6	1.8–2.7	—	—
Straw, corncobs, cornstalks, stalklage, cottonseed hulls	—	—	—	—	—	—	—	—
Corn or sorghum silage	—	—	—	—	—	—	25–40	11.4–18.1
Legume-grass silage, or oat silage	—	—	—	—	—	—	—	—
Legume-grass haylage, or oat haylage	—	—	—	—	—	—	—	—
Grain (corn, sorghum, barley, or oats)	—	—	—	—	—	—	—	—
Protein supplement (41% or equivalent)	—	—	1¼–1½	0.6–0.7	¼–1	0.1–0.5	1–1¼	0.5–0.6

[1]With all rations, provide suitable minerals (see Tables 19-5 and 19-7).

TABLE DAILY RATION FOR STOCKER YEAR- (AS-FED

	Rations (fed for gains of							
	1		2		3		4	
	(lb)	(kg)	(lb)	(kg)	(lb)	(kg)	(lb)	(kg)
Legume hay or grass-legume mixed hay............	16–24	7.3–10.9	—	—	6–8	2.7–3.6	—	—
Grass hay	—	—	16–24	7.3–10.9	10–16	4.5–7.3	—	—
Straw, corncobs, cornstalks, stalklage, cottonseed hulls	—	—	—	—	—	—	—	—
Corn or sorghum silage	—	—	—	—	—	—	45–55	20.4–25.0
Legume-grass silage, or oat silage	—	—	—	—	—	—	—	—
Legume-grass haylage, or oat haylage	—	—	—	—	—	—	—	—
Grain (corn, sorghum, barley, or oats)	—	—	—	—	—	—	—	—
Protein supplement (41% or equivalent)	—	—	1½–1¾	0.7–0.8	1–1½	0.5–0.7	1¼–1½	0.6–0.7

[1]With all rations, provide suitable minerals (see Tables 19-5 and 19-7).

Feeding Beef Cattle

to be handled according to plan 1—that is, bought in the fall, wintered on roughage, and sold directly into a finishing program. This trend will be accelerated because of heavier calves being weaned in the fall, and because it is more profitable either to use presently available pasture areas (1) for brood cows to produce more calves, and (2) for crop production.

The most common type of operation is a combination stocker-feeder program, typical of the Corn Belt and the irrigated sections of the West, where high-yielding corn and sorghum crops are produced for silage. In these areas, cattle feeders usually purchase steer calves or light yearlings in the fall or late winter; fall-graze stalk fields and small grain stubble where available; move into the drylot for the winter and feed corn or sorghum silage, supplemented with a legume hay or protein supplement; then finish on a high-energy ration either in the drylot or on pasture and sell for slaughter in the summer or fall.

An increasing number of feeders are grown under contract for, and delivered to, a feedlot for finishing. This trend has been prompted by the competition between feedlots. It is their way of assuring a continuous supply of feeders of the desired weights and quality. As a further inducement, many of the feedlots finance the grower (backgrounding) operation.

RATIONS FOR STOCKERS

For a stocker operation to be profitable, the grower must be ever aware of the following reasons back of it and feed stockers accordingly: (1) to provide a supply of the kind of cattle desired by finishing lots at the time needed; (2) to utilize roughages and other low-cost feeds, and (3) to "cheapen" the cattle.

Tables 19-27 and 19-28 contain some recommended rations for stocker cattle. Variations can and should be made in the rations used. The grower should give consideration to (1) the supply of homegrown feeds, (2) the availability and price of purchased feeds, (3) the class and age of cattle, (4) the health and condition of animals, and (5) the kind of feeder cattle in demand by feedlots.

In using Tables 19-27 and 19-28 as guides, it should be recognized that feeds of similar nutritive properties may be interchanged as price relationships warrant. Thus, (1) the cereal grains may consist of corn, barley, wheat, oats, and/or sorghum; (2) the protein supplement may consist of soybean, cottonseed, peanut, linseed, safflower, and/or sunflower meal; (3) the roughage may include many varieties of hays and silages; and (4) a vast array of by-product feeds may be utilized.

19-27
CALVES (400–500 LB [181–227 KG])[1] BASIS)

1.25 lb [0.6 kg]/head/day

5		6		7		8		9		10	
(lb)	(kg)	(lb)	(kg)	(lb)	(kg)	(lb)	(kg)	(lb)	(kg)	(lb)	(kg)
2–4	0.9–1.8	—	—	8–10	3.6–4.5	—	—	—	—	—	—
—	—	2–4	0.9–1.8	—	—	—	—	10–12	4.5–5.4	—	—
—	—	—	—	2–4	0.9–1.8	2–3	0.9–1.4	—	—	2	0.9
20–30	9.1–13.6	20–30	9.1–13.6	—	—	—	—	—	—	—	—
—	—	—	—	—	—	20–25	9.1–11.4	—	—	—	—
—	—	—	—	—	—	—	—	—	—	20–25	9.1–11.4
—	—	—	—	4–5	1.8–2.3	—	—	4–5	1.8–2.3	4–5	1.8–2.3
¾–1	0.3–0.5	1¼–1½	0.6–0.7	—	—	1–1½	0.5–0.7	1–1½	0.5–0.7	—	—

19-28
LINGS (600–700 LB [273–318 KG])[1] BASIS)

0.9 lb [0.4 kg]/head/day

5		6		7		8		9		10	
(lb)	(kg)	(lb)	(kg)	(lb)	(kg)	(lb)	(kg)	(lb)	(kg)	(lb)	(kg)
2–4	0.9–1.8	—	—	6–8	2.7–3.6	—	—	—	—	—	—
—	—	2–4	0.9–1.8	—	—	—	—	16–20	7.3–9.1	—	—
—	—	—	—	12–15	5.4–6.8	10–12	4.5–5.4	—	—	2	0.9
40–50	18.2–22.7	40–50	18.2–22.7	—	—	—	—	—	—	—	—
—	—	—	—	—	—	20	9.1	—	—	—	—
—	—	—	—	—	—	—	—	—	—	35–40	15.9–18.2
—	—	—	—	5–6	2.3–2.7	—	—	5–6	2.3–2.7	5–6	2.3–2.7
¾–1	0.3–0.5	1¼–1½	0.6–0.7	—	—	1	0.5	1–1½	0.5–0.7	—	—

The following points are pertinent to the success of a stocker operation and should be kept in mind:

- **Recommended nutrient allowances**—When grower rations are formulated on the basis of percentage of nutrients in the ration, the following allowances are recommended:

Protein:
For up to 1.5 lb daily gain 10.5%
For 1.5 lb daily gain or more 11.0%

Calcium and Phosphorus:
For up to 500-lb liveweight 0.3–0.5%
For over 500-lb liveweight 0.25%

Vitamin A:
Air-dry feed (10% moisture) ...... 1,200 to 1,500 IU per lb
.... 15,000 IU daily per head

Implant:
Gains of more than 1.5 lb
per head daily .... Include growth stimulant

- **Urea**—Urea is not well utilized as a protein supplement when fed to cattle on high-roughage rations. Because of this, usually it is best to use a plant protein supplement or a slow-release urea product in growing rations.
- **Stubble and stalk fields** furnish much feed for stocker cattle, especially yearlings, in the late fall and early winter. Unless there is access to a good winter pasture, cattle grazing stalk fields should be fed 4 to 6 lb of legume hay or 1½ lb of protein concentrate daily. In addition, minerals should be provided.
- **Grain**—Calves are unable to consume enough dry roughage to gain more than a pound a day. Thus, grain should be added in the quantity necessary to achieve the desired gains. Bear in mind that with calves of the British breeds and crossbreds it takes a gain of about 1.25 lb daily to maintain condition; yearlings of the British breeds or crossbreds will maintain condition on a gain of about 0.9 lb daily.

Some grain should be included in the ration of stocker cattle when (1) they are to be finished immediately after the wintering period, (2) they weigh less than 350 lb when started on winter feeding, and (3) heifers are to be bred when they are 13 to 15 months old.

Calves that are full-fed corn or sorghum silage high in grain content, plus 1 lb of protein concentrate or 4 to 5 lb of legume hay, need not be fed grain.

- **Pasture supplement**—Following wintering, many stocker cattle graze throughout the pasture season. Supplementing grass with a high protein feed at the rate of I lb/head/day when summer pastures drop off in quantity and protein content (generally beginning about mid-July) will usually boost average daily gain by 0.4 lb. For convenience, pasture supplements are usually fed in cube form. Usually they run 38 to 41% protein and consist primarily of cottonseed or soybean meal.
- **Growth stimulants or implants**—An approved growth stimulant or implant may increase growth in stocker cattle, on either winter rations or summer pastures. In cattle that are gaining 0.75 to 2.00 lb/head/day, a growth stimulant or implant will boost average daily gains by an additional 0.2 lb. (See Table 19-11 for FDA approved stimulants and implants.)

LEVEL OF WINTERING

The level of wintering stockers affects the gains in the next stage. Thus, calves gaining the most during the winter make the least gains on pasture the following summer.

Calves wintered to gain 1.0 lb daily make satisfactory summer pasture gains. This level is recommended for calves that are to graze season-long the following summer, provided the same ownership is retained all the way through. Somewhere in the range of 1.0 to 2.0 lb daily gain during the winter is usually desirable if calves (1) are to be sold in the spring, (2) will be on full feed 2 to 3 months after going to grass, (3) will be receiving a limited feed of grain on grass, or (4) are replacement heifers that are to be bred at 13 to 15 months of age.

Since yearlings are not growing as rapidly as calves, they may be fed for smaller gains than calves, and yet show comparable condition. Thus, for maximum growth without fattening (for just holding their condition) calves should gain approximately 1.25 lb daily, whereas yearlings need to gain only about 0.9 lb daily.

Compensatory Growth

Compensatory growth is increased growth rate in one time period as a result of growth restriction imposed during an earlier time period.

It is common practice for stocker cattle to be *roughed through* the winter as cheaply as possible, with limited daily gains. Then, in the spring, the animals are turned to lush pasture or put in a feedlot and fed a high-energy ration. Animals so managed exhibit the phenomenon of *compensatory growth;* that is, on the high-energy ration they gain faster and more efficiently than similar cattle which were fed more liberally during the wintering period. Feedlot operators were quick to sense this situation, and to take advantage of it. This is the chief reason for the popularity of Okie-type cattle. Usually, they are animals whose growth has been held back to less than their genetic potential. When fed more liberally, they exhibit a surge in growth rate and feed efficiency. Large compensatory growth usually indicates that someone (the stocker operator) has lost money while someone else (the feeder) has made money. It is noteworthy that Holsteins and the larger exotics should never be handled so as to exhibit compensatory gains. If they're held back in the winter, they're too heavy when they finish.

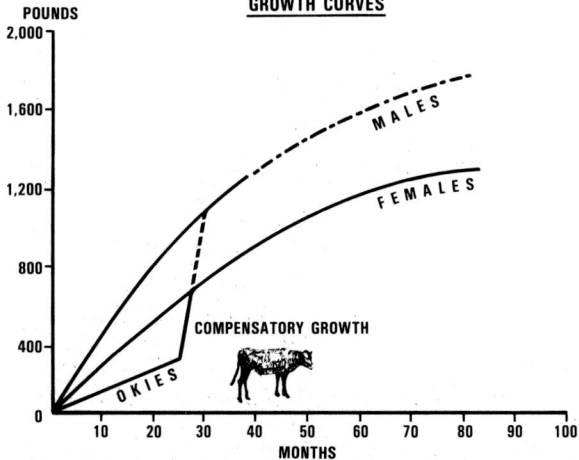

Fig. 19-27. The two curved lines show the normal growth of beef cattle under proper environment. Note that females grow most rapidly from birth to 10 months and males from birth to 15 months. Okie-type cattle grow more slowly during the first part of their lives when their environment is poor and they are under stress. When put on feedlot rations, their growth curve is steep. This rapid, economical gain is called *compensatory growth.*

FACTS PERTINENT TO STOCKER PROGRAMS

The following points are pertinent to stocker programs:

• **Stocker programs are used by large cattle feedlots to ensure a continuous supply of feeder replacements**—Many large feedlots, which feed on a year-round basis, are effectively using stocker programs to ensure a continuous supply of feeder replacements. They usually accomplish this by buying both calves and yearlings when they are available at favorable prices, then putting them on different stocker programs, often on a contract basis, designed to stagger their readiness to move to the feedlot at the weight and time desired.

• **Lightweight calves gain more efficiently**—The feed requirement for calf gain is directly related to the animal's weight. Because of low maintenance requirements, lightweight calves gain more efficiently than heavyweight calves. This is illustrated in Fig. 19-28, which shows the amount of TDN required for each 100-lb gain when calves are gaining at the rate of 1½ lb per day. All things being equal (health, genetics, management, environment), calves will gain 100 lb on less feed at the lighter weights, which is an advantage for which most stocker buyers are looking.

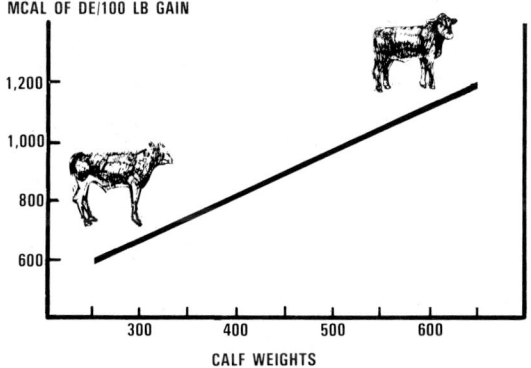

Fig. 19-28. Feed required per 100-lb gain for calves at different weights gaining 1½ lb per day. (Source: *Keys to Profitable Stocker Calf Operations*, MP-964, Texas A&M Univ. Agr. Ext. Service)

• **Avoid excess feeder condition**—If cattle get too fleshy as feeders, (1) they may reach market finish before they attain desirable market weight, and (2) they will tend to gain more slowly than desirable during the feedlot finishing period. The planned rate of gain should be determined by feeder finish, growth potential, sex, and the beginning weight of the feeder calves. It will also depend upon how much condition is acceptable to the buyer. British breed or crossbred steer calves gaining at the rate of 1.25 lb daily will about maintain their condition; at 1.5 lb daily they will add some condition; at 2.0 lb daily they may be too fleshy by the time they reach 700- to 750-lb feeder weight. British breed or crossbred yearlings making daily gains of 0.9 lb will maintain growth without fattening. Larger cattle of some of the exotic breeds may make larger gains without fattening.

• **Buy and sell carefully**—Because the original weight purchased in a stocker program is a high percentage of the weight sold by the grower, any mistake made in buying or selling has a greater impact on profits or losses than it would in a finishing program. If you pay too much, buy with too much fill, pay for quality that the feeder is not willing to pay for, or sell too low, profits will be eroded away rapidly.

• **Steers are heavier**—At weaning, steer calves normally weigh about 5% more than heifer calves.

STOCKER AND GROWER CONTRACTS

Hand in hand with the development of big feedlots and year-round feeding came the need for an assured supply of feeder cattle of the desired kind on a continuous basis. To meet this need, more and more feedlots have turned to contractual arrangements with stocker growers, with numerous kinds of contracts. Usually, the cattle are owned by the feedlot, most of which are large and in a stronger financial position than the majority of stocker growers. The two most common kinds of contracts are based on either (1) a fixed cost for the gain, or (2) an agreed feed cost plus an extra charge for labor and lot rental. Usually, there is provision of adjusting for death loss. Such contracts should always be in writing, with all provisions, including weighing conditions, spelled out.

Although the use of stocker and grower contracts has increased in recent years, the concept is not new. Many Kansas bluestem pasture owners have long grown out yearlings owned by Iowa and other Corn Belt feeders.

Today, many corn farmers in the fertile irrigated area around Greeley, Colorado, make corn silage and feed stocker cattle on a contract basis, with the stockers owned by one of several large feedlots in the vicinity. Stocker cattle are also being grown under contract on the wheat pastures of Kansas, Oklahoma, and Texas; on hay and other roughages in the irrigated valleys of the West; and on sorghum silage and stalk fields throughout the Southwest.

(Also see Chapter 7, section on "Pastures for Cattle, Sheep, and Horses," • Custom cattle grazing [pasture leasing].)

PART IV—Feeding Finishing (Fattening) Cattle

The finishing of cattle is what the name implies, the laying on of fat. Additionally, there is an increase in the total muscle (red meat) mass. The ultimate aim of the finishing process is to produce beef that will best answer the requirements and desires of the consumer. This is accomplished through an improvement in the flavor, tenderness, and quality of the lean beef which results from marbling (intramuscular fat).

Cattle feeders are commonly classed as either commercial feeders, or farmer-feeders, based largely on numbers. From the standpoint of statistical reporting, the U.S. Department of Agriculture commonly draws the line at 1,000 head. A commercial cattle-feeding operation is defined as one having a capacity of 1,000 head or more, at any one time.

Traditional farmer-feeders evolved with Corn Belt farming, in the north central region of the United States. Generally speaking, they market their crop, usually corn, through cattle (or hogs, or lambs), and spread the manure on the land. The purchase of feeder cattle for these enterprises is generally in the fall, with the actual feeding done during the winter

Fig. 19-29. Open pen cattle feedlot, with shades—the cheapest and most common type of feedlot. This is the Alta Verde Industries feedlot near Eagle Pass, Texas. (Courtesy, *West Texas Livestock Weekly*, San Angelo, Tex.)

months when labor is available due to limited field work. This type of operation has persisted to the present time, although it has been modernized through the years.

In addition to being larger, commercial cattle feeders generally differ from farmer-feeders in the following respects: (1) They usually feed cattle on a year-round basis, rather than during the winter months only; (2) they may grow little, or none, of their feed; (3) they are highly mechanized; (4) they are knowledgeable of costs and returns, skillful buyers and sellers, and aware of market trends; and (5) they usually do some custom feeding. Today, commercial feedlots with more than 1,000-head capacity dominate cattle feeding.

There are 2 methods of finishing cattle for market: (1) cattle feedlots, including confinement (sheltered) finishing; and (2) pasture finishing.

The principles of beef cattle nutrition are covered in Part I of this chapter; hence, they will not be repeated at this point. Instead, the application of nutrition to cattle finishing (fattening) will be discussed.

The major nutritional requirements of finishing cattle are energy, protein, minerals, vitamins, and water. The greatest need is for energy. Of course, net profit depends on how much of that energy can be converted to pounds of gain—and how efficiently.

About 75% of the cost of finishing cattle, exclusive of the purchase price of the feeders, is feedstuffs—grain, hay, silage, and miscellaneous wastes and by-products.

KINDS OF CATTLE TO FEED

All kinds of cattle may be, and are, fed. But, for maximum success, it is imperative that the right kind of cattle be selected for a particular feedlot. The cattle should match the operator's available feed, labor, shelter, and credit. Also, it is imperative that there be a suitable market outlet following finishing; for example, it would be unwise to feed lightweight heifers in an area where the strongest slaughter market is for heavy steers; nor should one finish heavy Holstein steers where the primary interest of packers is for Choice grade beef. But, assuming that a satisfactory slaughter outlet exists for different kinds of cattle, the general guides that follow will be helpful in determining what kind of cattle to feed in a given lot.

Age and Weight of Cattle

Today, cattle are referred to by ages as calves, yearlings, and 2-year-olds. This shift to younger cattle has been brought about primarily by consumer demand for smaller and lighter cuts of meat and improved feeding and management practices.

The age of cattle to feed is one of the most important questions to be decided upon by every practical cattle producer. The following factors should be considered in reaching an intelligent decision on this point.

• **Rate of gain**—When cattle are fed liberally from the time they are calves, the daily gains will reach their maximum the first year and decline with each succeeding year thereafter. On the other hand, when in comparable condition, thin but healthy 2-year-old steers will make more rapid gains in the feedlot than yearlings; likewise, yearlings will make more rapid gains than calves.

• **Economy of gain**—Calves require less feed to produce 100 lb of beef than do older cattle. This may be explained as follows:

1. The increase in body weight of older cattle is largely due to the deposition of high-energy fat, whereas the increase in body weight of young animals is due mostly to the growth of muscles, bones, and organs. Thus, the body of a calf at birth usually consists of more than 70% water, whereas the body of a fat 2-year-old steer will contain only 45% water. In the latter case, a considerable part of the water has been replaced by fat.

2. Calves consume a larger proportion of feed in proportion to their body weight than do older cattle.

3. Calves masticate and digest their feed more thoroughly than older cattle. Despite the fact that calves require less feed per 100 lb gain—because of the high-energy value of fat—older cattle store as much energy in their bodies for each 100 lb of total digestible nutrients consumed as do younger animals.

From the above, it is apparent that age of cattle affects the pounds of feed required to produce 100 lb of gain—that the younger the cattle, the greater the feed efficiency.

• **Flexibility in marketing**—Unless they have been crowded early in life, calves will continue to make satisfactory gains at the end of the ordinary feeding period, whereas the efficiency of feed utilization decreases very sharply when mature steers are held past the time that they are finished. Therefore, under unfavorable market conditions, calves can be successfully held for a reasonable length of time, whereas prolonging the finishing period of older cattle is usually unprofitable.

• **Length of feeding period**—Calves require a somewhat longer feeding period than older cattle to reach comparable finish. To reach Choice condition, steer calves are usually full fed about 7 to 8 months; yearlings, 4 to 5 months; and 2-year-olds only about 3 to 4 months.

• **Total gain required to finish**—Calves must put on more total gain in the feedlot than older animals to attain the same degree of finish. In terms of initial weight, calves practically double their weight in the feedlot. On the average, yearlings increase in weight about 400 lb, and 2-year-olds increase their initial feedlot weight about 320 lb.

- **Total feed consumed**—Because of their smaller size, the daily feed consumption of calves is considerably less than for older cattle. However, as calves must be fed for a longer feeding period, the total feed requirement for the entire finishing period is approximately the same for cattle of different ages.
- **Kind and quality of feed**—Because calves are growing, it is necessary that they have more protein in the ration. Since protein supplements are higher in price than carbonaceous feeds, the younger the cattle, the more expensive the ration. Also, because of smaller digestive capacity, calves cannot utilize as much coarse roughage, pasture, or cheap by-product feeds as older cattle.

Calves also are more likely to develop peculiar eating habits than older cattle. They may reject coarse, stemmy roughages or moldy or damaged feeds that would be eaten readily by older cattle. Calves also require more elaborate preparation of the ration and attention to other small details designed to increase their appetite.
- **Comparative costs**—Calves generally cost more per 100 lb as feeders than do older cattle.
- **Dressing percentage and quality of beef**—Older cattle have a slightly higher dressing percentage than calves or baby beef. Moreover, many consumers have a decided preference for the greater flavor of beef obtained from older animals.

From the above discussion, it should be obvious that there is no best age of cattle to feed under any and all conditions. Rather, each situation requires individual study and all factors must be weighed and balanced.

Sex of Cattle

More steers than heifers are fed, because more of them are available. A portion of the heifers is held back for replacement purposes. In the future, more young bulls may be fed, because they make more rapid and efficient gains than steers or heifers. Thus, the feedlot operator must give consideration to the sex of cattle fed. First, and foremost, consideration must be given to market outlets.

Table 19-29 shows the effect of sex on rate of gain and feed efficiency. It is noteworthy that bulls gain more rapidly on less feed than steers, and that steers gain more rapidly on less feed than heifers.

TABLE 19-29
EFFECT OF SEX ON GAIN OF MEDIUM-FRAME CATTLE[1]

Sex	Average for U.S. Feedlots		Top 5% of U.S. Feedlots	
	Daily Gain	Feed/Lb Gain[2]	Daily Gain	Feed/Lb Gain[2]
	(lb)	(lb)	(lb)	(lb)
Heifers	2.4	7.5	2.7	6.9
Steers	2.8	6.9	3.2	6.6
Bulls	3.0	6.7	3.4	6.4

[1]To convert lb to kg, divide by 2.2.
[2]Air dry basis (approximately 90% dry matter).

On the market, cattle are divided into five sex classes: steers, heifers, cows, bullocks, and bulls. The sex of feeder cattle is important to the producer from the standpoint of cost and selling price (or margin), the contemplated length of feeding period, quality of feeds available, and ease of handling. The consumer is conscious of sex differences in cattle and is of the impression that it affects the quality, finish, and conformation of the carcass.

Steers are by far the most important sex class on the market, both from the standpoint of numbers and their availability throughout the year, whereas heifers are second.

SPAYED HEIFERS

In females, the operation corresponding to castration is known as spaying. Under most conditions, desexing heifers by the traditional surgical method is not recommended because (1) the operation is complicated and difficult, requiring a very experienced technician; (2) it is attended with more danger than castration; (3) it lowers both rate and efficiency of gains; (4) it eliminates the heifers for possible replacement purposes or sale as breeding stock; and (5) experiments and practical operations with spayed heifers have generally shown that the selling price obtained is not sufficiently higher to compensate for the lower and less efficient gains plus the attendant risk of the operation.

However, spaying by one of the new methods may be justified because of the following advantages of spayed heifers: (1) They may move freely across state lines without being tested for brucellosis when destined for a feedlot, (2) they are easier to manage in the presence of bulls or steers, (3) it avoids the losses generally associated with pregnancy, and (4) they generally bring a premium price.

(Also see Chapter 13, Feed Supplements/Additives/Implants, section on "Abortifacients.")

BULLS

The feeding of bulls (uncastrated males) instead of steers has been standard practice throughout Europe for many years. For example, since about 1954 Germany has fed and slaughtered bulls as yearlings, instead of steers, because they obtain 10 to 15% greater rate of gain and feed efficiency thereby. The practice will increase gradually in the United States, now that carcasses from young bulls are federally graded as *bullock beef* rather than *bull beef*, thereby removing the connotation that the meat is inferior to or different from steer or heifer beef.

The carcasses from older bulls are still labeled *bull beef,* to differentiate them from the carcasses of younger bulls. Bullock beef from young bulls is graded according to the same quality standards as beef from steers and heifers.

Also, the economics of the situation favors the feeding of bulls instead of steers. The male hormones secreted by the testicles are excellent growth stimulants and will improve gain and feed efficiency by 10 to 15%. Also, bulls will produce more healthful lean meat than steers, and research has shown that bull meat is equal in value, quality, and palatability to steer meat.

Grade of Cattle

The most profitable grade of cattle to feed will generally be that kind of cattle in which there is the greatest spread or margin between their purchase price as feeders and their selling price as finished cattle. This cannot be determined by merely comparing the existing price between the various grades at the time of purchase. Rather, it is necessary to project the differences that will probably exist, based on past records, when the animals are finished and ready for market.

The length of the feeding period and the type of feed available should also receive consideration in determining the grade of cattle to feed. Thus, for a long feed and when a liberal allowance of grain is to be fed, only the better grades of feeders should be purchased. On the other hand, when a maximum quantity of coarse roughage is to be used and a short feed is planned, cattle of the medium or lower grades are most suitable. Thus, successful cattle feeders match the quality of the cattle selected with the quality of the available feed; the better the feed, the higher the grade of cattle.

Certainly producers who raise their own feeders should always strive to breed high-quality cattle, regardless of whether they finish them or sell them as feeders. On the other hand, the purchasers of feeder steers can well afford to appraise the situation fully prior to purchasing any particular grade.

Breeding and Type of Cattle

Although supporting data are limited, it is realized that there is considerable difference between individual animals insofar as rate and economy of gain is concerned. Also, it is recognized that these differences are greater within breeds than between breeds.

CROSSBREDS

Good crossbreds will likely show 2 to 4% improvement over the average of the parent breeds for rate and efficiency of feedlot gains. Additionally, even larger advantages accrue to the cow-calf producer. Thus, it is inevitable that an increasing number of crossbred feedlot cattle will be seen.

DAIRY BEEF

Dairy beef accounts for about 25% of the beef consumed in this country, with these animals marketed as veal calves, cull dairy cows and bulls, and finished dairy heifers and steers. Improvements in the science and technology of feeding and processing favor growing and finishing dairy beef, and minimum slaughter of veal calves.

Fig. 19-30. Holstein steers on feed in Wisconsin. (Courtesy, Ralston Purina Co., St. Louis, Mo.)

Dairy beef is just what the term implies—*beef derived from cattle of dairy breeding, or from dairy X beef crossbreds.*

Some pertinent points relative to dairy beef follow:

- **High-growth thrust essential**—For a dairy beef program to be most successful, scientists and producers in both Britain and the United States agree that the animals should have a high growth potential, as evidenced by heavy birth weight and heavy weight at maturity. Since Holsteins are heavy at birth and mature at around 1,400 lb, in comparison with mature weights of 1,000 to 1,200 lb of the European beef breeds, it is obvious that Holsteins are ideal when it comes to producing dairy beef.
- **High-energy rations; light market weights**—If dairy steers are to be slaughtered at young ages and light weights, high-energy (low-roughage) rations are imperative. Under this system, usually young calves of either dairy or dairy X beef breeding are fed in confinement—in barns; and are fed milk replacers from 1 to 4 days of age to 200 to 300 lb, and full fed a high-concentrate ration from about 300 lb to market weight of 750 to 950 lb.

Crowding for market at an early age takes advantage of the fact that growth is generally most economical when most rapid, and that young gains are cheap gains. Also, experience shows that when Holstein calves are started on super-energy rations at around 350 to 450 lb weight and marketed under 1,100 lb, (1) there's excellent marbling with very little outside fat, and (2) many of these animals will grade Choice.

- **High-roughage rations; heavy market weights**—If roughages are relatively more abundant and cheaper than concentrates, then it may be more profitable to feed dairy beef more roughage and market at heavier weights—and to expect slower and less efficient gains. In any event, it's net returns that count, rather than rate of gain and pounds of feed required per pound of gain.

Under the high-roughage system, steers of dairy breeding are grown on maximum roughage to 600 to 750 lb weight, following which the ratio of concentrate to roughage is increased. Most dairy steers fed according to this system are marketed at weights of 1,150 to 1,400 lb, grading Select or Commercial. (Most of them are too old to grade Standard, and too lacking in marbling to grade Choice.)

Before producing beef from dairy steers, a market should be established. This calls for a steady flow of cattle of a predictable quality.

BOXED BEEF

With the advent of boxed beef, the steer must fit the box. This calls for steers produced to specification in size and quality.

More and more packers are fabricating and boxing beef in their plants, thereby freeing the back rooms of 200,000 supermarkets. After chilling, the carcass is subjected to a disassembly process, in which it is fabricated or broken into counter-ready cuts; vacuum-sealed; moved into storage by an automated system; loaded into refrigerated trailers; and shipped to retailers across the nation.

Beef that is not suitable for sale over the block is (1) boned out and disposed of as boneless cuts, (2) canned, (3) made into sausage, or (4) cured by drying and smoking. It is estimated that about ⅕ of all slaughter cattle are disposed of as processed meats.

FEEDLOT FINISHING FACILITIES

Feedlot finishing refers to feeding cattle in a restricted area, with the feed conveyed to the animals; and it may involve either an open pen feedlot, or confinement (sheltered) feeding.

Cattle feeding facilities and equipment are a manufacturing plant, wherein animate objects (cattle) convert feed into beef. Hence, they merit the same level of competence in planning and design as any other sophisticated manufacturing plant.

Open Pen Feedlot

An open pen feedlot is, as indicated by the name, a lot in which the cattle are in the open—usually it is without shelter (except for such natural protection as may be afforded by trees, hills, or wind fences, or perhaps a roof over the feed bunks or shades).

An open lot without shelter is the cheapest type of feedlot construction. In the Southern Plains area, where the weather is mild and shelters are unnecessary, investment costs range from $100 to $125 per head of capacity.

Confinement Feeding; Slotted (or Slatted) Floors

Currently, there is much interest in cattle confinement feeding and slotted floors. The main deterrent is cost; construction costs vary with type of structure and may range up to $300 per steer space.

Fig. 19-31. Confinement feeding on a slotted floor. (Courtesy, Pioneer Hi-Bred Corn Co., Des Moines, Iowa)

Confinement cattle feeding refers to feeding in limited quarters, generally 20 to 25 sq ft per yearling animal, which is about ⅛ the space normally allotted to a yearling in an unsurfaced lot and ⅓ that of a paved lot. The confinement is usually under roof on slotted floors.

Slotted floors are floors with slots (space) through which the feces and urine pass to a storage area immediately below or nearby.

Interest in confinement feeding and slotted floors was ushered in for the purpose of (1) automating and saving labor, (2) cutting down on bedding and facilitating manure handling, (3) lessening mud, dust, odor, and fly problems, (4) increasing gains and saving feed, (5) lessening land requirements, (6) lessening pollution, and (7) lessening complaints from area residents.

Research has shown conclusively that cattle fed during the winter months in cold areas gain faster and more efficiently if they are sheltered. However, as pointed out earlier in this chapter under the section headed "Open Pen Feedlot," the per head cost is much higher for confined or sheltered cattle. Thus, the decision on whether cattle confinement can be justified, even in the northern part of the United States, should be determined by economics. Will the cattle in confinement quarters gain sufficiently more rapidly and efficiently to justify the added cost? Of course, manure disposal and pollution control should also be considered.

COLD CONFINEMENT [5]

Cold confinement refers to a more or less open shed for confining cattle; hence, winter temperatures therein are within a few degrees of outdoor temperatures. Open sheds should be faced away from the direction of the prevailing winds. Additionally, doors or other openings in the closed walls should be provided for summer ventilation.

Fig. 19-32. Finishing cattle feed in an open shed—cold confinement. (Courtesy, *Feedlot Management,* Minneapolis, Minn.)

---

[5] The terms *cold confinement* and *warm confinement* refer to winter conditions. Without mechanical cooling, both systems are *warm* during the summer months.

WARM CONFINEMENT[6]

Warm confinement refers to a confinement building for cattle which is sufficiently insulated and ventilated to maintain inside winter conditions above 35°F in severe weather, and in the range of 50° to 60°F most of the time. Warm confinement makes for the maximum in feed efficiency.

FEEDS AND RATIONS FOR FINISHING CATTLE

The commonly used feeds for finishing cattle consist of (1) concentrates, (2) by-product feeds, (3) roughages, (4) protein supplements, (5) minerals, (6) vitamins, and (7) water. Also, feed additives and implants are part of modern cattle feeding. These feeds and additives/implants are discussed/detailed in this book in the following chapters and sections: In this chapter, in the section on "Feeds for Beef Cattle"; in Chapter 7, Pasture and Range Forages; in Chapter 8, Hay; in Chapter 9, Silage/Haylage/High-Moisture Grain; in Chapter 10, Grain/High-Energy Feeds; in Chapter 11, Protein Supplements; in Chapter 12, By-Product Feeds/Crop Residues; and in Chapter 13, Feed Supplements/Additives/Implants.

The NRC nutritional requirements for growing and finishing cattle are given in this chapter in Tables 19-3 and 19-4. Suggested rations for finishing cattle are given in this chapter in Table 19-15. Also, methods for balancing rations are given in Chapter 18, Feeding Standards/Ration Formulation. Growth stimulants and implants are presented in this chapter, in Table 19-11.

Because feedlot cattle have access only to the rations provided by the caretaker, it is important that their rations be balanced, and that they make for maximum net returns. Also, in addition to considering availability and prices of feeds, ration formulations should be altered at stages to correspond to weight increases in the cattle.

Fig. 19-33. Drylot steers eating field-cured dry beet tops. (Courtesy, The Great Western Sugar Co., Denver, Colo.)

FEED PREPARATION

Prior to 1960, very little attention was given to feed processing for commercial cattle production, other than grinding or crushing grain and chopping forage. But in recent years great progress has been made and many new techniques have been developed.

Feed preparation is fully covered in Chapter 14, Feed Processing.

Mixed Rations Vs Feeding Roughages and Concentrates Separately

Most experiments and experiences have not shown any difference between mixed rations and the feeding of roughage and concentrates separately insofar as rate and efficiency of gain are concerned. However, a mixed ration has the following **advantages**:

1. It makes for greater efficiency in feeding and lessens the sorting at the feed bunk.
2. When the roughage is relatively unpalatable, a mixed ration forces consumption.
3. When it is desired to limit concentrate consumption, mixing with the roughage is desirable.
4. After cattle have become adjusted to the feedlot, a mixed ration makes it easier to get them on full feed.

Thus, each feeder must make a decision on the matter of mixed vs feeding roughage and concentrate separately, with relative costs and other factors considered. Most large feedlots use completely mixed rations.

HOW TO USE THE NET ENERGY METHOD

Use of the net energy method necessitates that the following net energy values be available:

1. The net energy requirements for growing and finishing cattle (in Mcal per animal per day), with a breakdown for steers and heifers (see Table 19-3).
2. The nutrient composition of feeds, with the net energy of each feed partitioned into energy used for body maintenance and for gain; thus, the net energy values in Mcal per unit (lb or kg) are needed for each feed for maintenance (NE_m) and for gain (NE_g) (see Part V, Composition of Feeds).

With such information available, the net energy system may be used two ways: (1) to calculate the quantity of feed necessary to meet the animal's energy needs and to compound a ration to supply the needed concentration of energy per unit of dry matter, and (2) to predict weight gains and to determine whether cattle have gained in accordance with expectations. Instructions on how to use the net energy method for these purposes, along with examples, are given in Chapter 18, Feeding Standards/Ration Formulation.

MANAGING FEEDLOT CATTLE

Although it is not possible to arrive at any overall, certain formula for success in operating a cattle feedlot, those operators who have made money have paid close attention to the details of management.

There are many facets of cattle management. Only those that are unique to cattle feedlots, and that have not been covered elsewhere in this book, will be discussed in the sections that follow.

---

[6]*Ibid.*

Handling Newly Arrived Cattle

The most critical period for feeder cattle is the first 21 to 28 days in the feedlot. The following recommendations pertaining to incoming cattle will minimize death losses and maximize performance:

• **Provide clean, dry, comfortable quarters**—Whether it be an open lot or a building, incoming cattle should be provided with clean, dry, comfortable quarters. A dry and comfortable bed for resting is very essential because cattle are tired and have a low resistance to respiratory diseases.

• **Delay processing for 28 days**—Recent studies by researchers at both the University of California and Colorado State University showed that delaying processing for 28 days causes less stress to animals than processing upon arrival or 14 days later—that delaying processing 14 days is not long enough.

• **Provide clean, fresh water**—Give the cattle easy access to clean, fresh water because they are usually dehydrated and thirsty upon arrival and will drink water before they eat feed. Open water tanks are preferable to automatic water bowls because most farm and ranch cattle are accustomed to drinking from tanks or ponds.

• **Provide a palatable ration**—Feeding a palatable ration—one that cattle will start eating soon after they are unloaded in the feedlot—will reduce the incidence of shipping fever and make the cattle recover their weight loss more rapidly.

1. **Roughage.** The best roughage for newly arrived feedlot cattle is *long grass* hay, because it is very similar in composition and taste to the grass to which most range cattle have been accustomed. Thus, cattle will usually eat long grass hay more quickly than any other roughage. In areas where grass hays are not available, or are too expensive to feed, any other nonlegume roughage can be fed, such as corn silage, sorghum silage, cottonseed hulls, corncobs, or grass-legume hay that contains more grass than legumes. Above all, do not feed high-quality alfalfa hay because it is too laxative and it will cause scouring which will trigger shipping fever. The same may be said relative to alfalfa haylage or alfalfa silage.

Corn silage of approximately 65% moisture content is an excellent feed for new cattle. If cattle do not eat the corn silage too well at the outset, the feeder should sprinkle a little grass hay on the top of it to encourage them to start eating. Also, buffering corn silage to a pH of 7.0 will increase feed intake.

2. **Concentrate.** Incoming cattle may be fed approximately 4 lb of concentrate per head daily, with a breakdown between protein supplement and grains as follows:

   a. Two pounds of a high bypass natural protein supplement, such as blood meal, corn gluten meal, or protected soybean meal, or a mixture of high bypass ingredients, preferably with a little cane molasses added to improve palatability. The protein supplement should be fortified so as to provide 50,000 IU of vitamin A daily. For heavily stressed cattle, the protein supplement should also contain a high level of antibiotic, or a combination of antibiotic and a bactericidal agent such as sulfamethazine. The following level of antibiotic-sulfamethazine is recommended:

   Feed 350 mg of antibiotic plus 350 mg of sulfamethazine per head daily to newly arrived cattle for a period of 28 days. With the antibiotic-sulfamethazine treatment, shipping fever is practically alleviated.

   Do not feed urea for the first 28 days after the cattle arrive. Starvation destroys the ability of the rumen to utilize urea or other nonprotein nitrogen and makes cattle more sensitive to urea toxicity. Therefore, it is not wise to put extra stress on cattle by feeding urea during this adjustment period.

   b. Two pounds of cereal grain or beet pulp per head daily, with the grain processed in the usual manner. The grain level can be raised at the rate of 1.0 lb/head/day if it seems desirable.

It has been, and still is, common practice to start cattle on a high-roughage ration, then gradually change them to a high-concentrate ration as they progress through the feeding program. However, based on California and New Mexico studies, it appears that for calves (not older cattle), a starting ration consisting of 75% concentrate and 25% hay (roughage) is best.

• **Satisfy mineral hunger**—Incoming cattle are usually hungry for minerals, especially if they have been on dry range forage. Thus, they should have access either to a mineral mixture consisting of two parts of dicalcium phosphate and one part of salt, or to a good commercial mineral.

• **Consider vitamin needs**—Unless incoming cattle come off green feed, vitamin A (50,000 IU per head per day) and vitamin E (100 IU per head per day) may improve gains and feed efficiency.

• **Observe, isolate, and treat sick animals**—Newly arrived cattle should be observed at least twice daily. Sick animals should be removed and treated. Treating sick animals promptly, rather than waiting until tomorrow, may mean the difference between life and death. Animals that show clinical signs of shipping fever—sunken eyes, runny nose, drooling at the mouth, labored breathing, and/or weaving (unsteady gait)—should be isolated in a separate *sick pen* or *hospital*.

Rest, fresh water, good feed, proper medication, and TLC (tender, loving care) are the cardinal essentials for preventing shipping fever and death losses.

Amount to Feed; Rate of Gain

There are two schools of thought relative to the amount to feed cattle after they are placed in the finishing lot. The traditional method calls for getting them on full feed as quickly as possible, then keeping them on full feed until marketing. The newer concept calls for limit (controlled) feeding in the early part of the finishing period, followed by full feeding the latter part. The wise manager will choose between the two programs, recognizing that neither is adapted to every operation.

• **Full feeding (ad lib)**—Those favoring full feeding in the finishing lot from beginning to end point out that, once a sufficient amount of ration is consumed to meet the maintenance needs of a finishing animal, the remainder is converted to gain with remarkable efficiency. Conversely, low feed intake results in too high a percentage of the total nutrients being expended for maintenance.

Although full feeding does result in animals gaining rapidly early in the finishing period, the rate of gain tapers off as the finishing period progresses. It is noteworthy, too, that high ownership costs (labor, interest on investment, etc.) favor pushing cattle for maximum gains and shortening the finishing period.

• **Limit (controlled) feeding**—Instead of giving cattle all the feed that they will consume, limit feeding calls for restricted feed intake during the early part of the finishing period—

for feeding only enough to produce the desired gains, followed by full feeding. This concept evolved in the early 1980s as the result of grain prices being comparatively more favorable than forage prices. During periods of favorable grain prices, limit feeding makes it possible to feed high energy grain rations in the early part of the finishing period without getting the cattle too fleshy. Generally, limit feeding results in fewer digestive upsets, lessens feed storage and labor, minimizes feed wastage, and improves feed efficiency for the entire finishing period by about 5% because of compensatory gain.

In general, cattle will consume daily an amount (on an air-dry basis) equal to 2.5 to 3.0% of their liveweight. Feed intake will vary according to the condition of the cattle, the palatability of the feeds, the energy of the ration (in general, animals eat to meet their energy needs), the weather conditions, and the management practices. For example, older and more fleshy cattle consume less feed per 100 lb than do younger animals carrying less condition; thus, mature, overfinished steers will consume feeds in amounts equal to about 1.5% of their liveweight, whereas thin steers under 2 years of age will consume fully twice as much feed per unit liveweight.

Overfeeding is undesirable, being wasteful of feeds and creating a health hazard. When overfeeding exists, there is usually considerable leftover feed and wastage, and there is a high incidence of bloat, founder, scours, and even death. Animals that suffer from mild digestive disturbances are commonly referred to as *off feed*.

Cull Out; Top Out

Obvious poor doers should be taken out early and marketed at Standard grade. Where individual weighing can be made, consideration should be given to the practicality of individually tagging (with duplicate tags, one in each ear) and weighing incoming calves; weighing them again at the end of the grower-ration period and prior to going on finishing rations; then culling out the bottom 10%.

Also, cattle should be sold when they make their grade, thereby avoiding loss in efficiency, excess finish, and too heavy weights. Usually, it is unwise to challenge a sagging market by holding and feeding for a higher market. There is no need to put feed and labor into heavy cattle at a cash discount when younger cattle will use these resources more efficiently.

Overfinishing

Experienced cattle feeders are fully aware of the fact that to carry finishing cattle to an unnecessarily high finish is usually prohibitive from a profit standpoint. This is true because the gains in weight then consist chiefly of fat but little lean tissue and water. In addition, a very fat animal eats less heartily, with the result that a small proportion of the nutrients, over and above the maintenance requirement, is available for making body tissue.

Fig. 19-34 shows that the heavier the cattle, the more expensive the gains. Also, this points up (1) the importance of topping out finished cattle, rather than waiting until the entire lot is ready; and (2) the reason it is generally wise to sell cattle when they are ready to go, rather than to hold for a higher market.

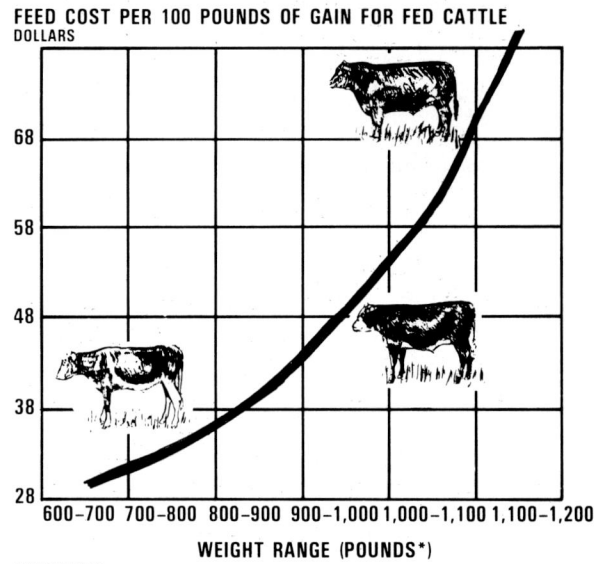

Fig. 19-34. This graph illustrates changes in feed conversion efficiency for cattle from normal feeder weights to slaughter weights. Note that feed costs per 100 lb gain more than double from 600-700 lb to 1,000-1,100 lb, and that the conversion efficiency ratio changes even more sharply when cattle pass 1,100 lb.

Feed Bunk Management

Feed bunk management is a combination of management factors involved in obtaining maximum performance, minimum digestive disorders, and keeping cattle on feed. Feed bunk management and quality control are directly involved with obtaining maximum and economical performance from cattle. It should be every feeder's goal to obtain maximum feed intake of a consistently high-quality ration, since both rate and efficiency of gain are directly related to nutrient intake.

WEATHER AFFECTS EATING AND DRINKING HABITS

During hot weather, feedlot cattle *peak* their eating during early morning and again during the evening hours—when it is cool. With heat, night drinking increases. In cool weather, they eat more during midday. Feeders should sense these changes in cattle eating habits and program their feeding accordingly.

Cattle eat more following a bad storm or a hot spell. Thus, at such times the bunks may be *slick* for 2 to 3 hours and the cattle may line up waiting to be fed. When this happens, the ration should be increased. By going to a higher roughage ration at these times, the problems from acidosis and laminitis can be minimized.

MUD PROBLEM

University of California studies show that mud can reduce cattle gains by as much as 25 to 35%. Thus, it is important that the problem be minimized, especially in high rainfall areas. Good drainage is the first essential. This should be assured at the time the feedlot is located and constructed.

Mounds, preferably perpendicular to the feed bunk, will provide cattle a dry place on which to lie down. Concrete aprons along the bunk will provide them with solid footing on which to stand and feed. Also, lessening of cattle density during the winter months—fewer animals per lot—is an effective method of controlling the mud problem. Thus, many feedlots plan to feed fewer cattle during the muddy season.

RECORDS

Complete and well-kept records are a must in the operation of a cattle feedyard, even though they require a lot of time and expense. Deficient records and deficient managers generally go hand in hand. Modern computer programs make record processing easier and more efficient.

There is no limit to the number of different kinds of record forms that can be, and are, kept in a given feedlot. Also, there is little similarity in record forms between lots, due to differences between people, primarily managers and bookkeepers. The important things are that (1) record forms be so designed as to facilitate record keeping, with as much ease, efficiency, and accuracy as possible; and (2) records be kept. Two basic record forms are needed: (1) a daily record, and (2) a monthly, cumulative and final feed summary.

Among other necessary records are the following:
1. Feed costs, with this record kept by individual pens.
2. Grain inventory.
3. Roughage inventory.
4. Feed projections ahead.
5. Cattle receiving and movement records.
6. Sick pen and movement records.
7. Sick pen costs.
8. Mortality slips, proofs of deaths, and post-mortem reports.
9. Maintenance and repair costs.
10. Routine office bookkeeping.
11. Customer billing for feed.
12. Closeout records.

Make 28-Day Test Weights

Taking 28-day test weights will not adversely affect the performance of feedlot cattle, provided the cattle are handled properly. Check weights should include a representative cross section of the cattle in the yard, including age, weight, type, background, and sex. Where it is not convenient, or it is not desired, to weigh an entire lot of cattle, "markers"—cattle of certain odd colors, animals with tail switches clipped, etc.—may be weighed. Also, it is important that weighing be done at the same time of day, and that the lots be weighed in the same order, due to the effect of rumen fill.

CUSTOM (CONTRACT) FEEDING

Custom cattle feeding is the feeding of cattle for a fee, without taking ownership of the animals.

Capital requirements, periods of severe economic conditions (like scarce money and high interest), times of depressed feeder cattle prices, and adverse pasture conditions caused custom feeding to grow following World War II. These same forces, along with the need for high occupancy (full feedlots) and increased integration, have resulted in further expansion of custom feeding.

Most custom feeders have developed large, highly mechanized, and very efficient plants. Usually, they have on their staffs highly trained nutritionists and veterinarians who are charged with the responsibility of formulating rations and of obtaining maximum gains and feed efficiency at the lowest possible cost. Through custom feeding, they sell the use of their facilities, services, and know-how to cattle owners, usually with profit to each party.

The proportion of custom-fed cattle to cattle owned by the feedlot varies (1) in period of time—it increases in times of financial stress (when cattle feeding is not profitable, money is scarce, and interest is high); (2) according to area—for example, there is more custom feeding in Texas than in any other state; and (3) according to size of feedlot—generally speaking, the larger the feedlot, the greater the percentage of custom feeding. Some feedlots do not do any custom feeding whatsoever; others are almost wholly on a custom basis; but most lots have part of each. Feedlots that do both—those in the dual role of custom feeding and owning cattle—vary in the proportion of cattle in each category, but most of them seem to prefer about ⅔ custom-fed cattle and ⅓ ownership. It's a good bread-and-butter division; in times when fed cattle lose money, such a feedlot has sufficient assured income to pay its bills.

The ownership of custom-fed cattle is diverse. It includes (1) cow-calf producers (farmers and ranchers) who wish to retain ownership of the cattle that they produce through the feedlot phase, (2) stocker operators, (3) packers, and (4) investors, including limited partnerships, corporations, cattle buyers, cattle dealers, and others.

Custom feeding contracts should always be detailed and in writing, for a good understanding is the best way to avoid a misunderstanding. Also, contracts should be fair to both parties—to both the feedlot owner and the cattle owner.

Types of Custom Feeding Contracts

The services rendered vary from feedlot to feedlot and according to the type of contract. In some instances, the services may be so complete that the customer never sees the cattle. The feedlot operator may buy the feeder cattle, feed them, market them, and send the customer (the client) a check for the balance, after deducting input costs, interest charges, and custom feeding charges. Less complete services are usually available to suit the customer.

Both the feedlot owner and the cattle owner should analyze different types of contracts and determine which best fits their respective circumstances. Some feedlots offer several types of contracts, thereby according the cattle owner a choice.

Competition may dictate the type of contract and the changes made. But by knowing the variables and managing them correctly, the feedlot owner can write and carry out a contract that will be fair to both parties.

Generally speaking, contracts with fixed charges are the most satisfactory and the most common, primarily because there is less room for misunderstanding.

Although there are many types of custom cattle-feeding contracts, and many variations of each kind exist, most of

them can be classified under one of the following types:
1. Feed cost plus daily yardage fee per head.
2. Feed cost plus markup.
3. Feed cost plus (a) daily yardage per head, and (b) markup per ton of feed.
4. Agreement to purchase contract.
5. Payment for weight gained.
6. The incentive basis contract—the higher the gain, the higher the charges.

CONDUCTING APPLIED FEEDLOT TESTS

When carefully conducted, and properly interpreted and used, feedlot trials can be a valuable adjunct in the operation of a large feedlot. Among their virtues, the feedlot operator can study area and feed differences. Among their limitations, usually fewer controls can be afforded than in most university-conducted experiments, thereby resulting in less accuracy. For the latter reason, most of them should be looked upon as applied tests or demonstrations *per se,* rather than carefully controlled, basic experiments; terminology which doesn't detract from their value, but which does place them in proper perspective.

The procedure for conducting tests in a commercial cattle feedlot is outlined in Chapter 15, Feed Analysis/Feed Evaluation, under the heading, "Conducting Applied Feedlot Tests"; hence, the reader is referred thereto.

PASTURE FINISHING CATTLE

Fig. 19-35. Hereford steers on bromegrass pasture near Lincoln, Nebr. (Courtesy, C. B. & Q. Railroad Co., Chicago, Ill.)

When grains are scarce and high in price, more cattle are grass finished. But, because young cattle grow and do not reach market finish under usual pasture conditions, it is impossible to finish them at early ages and light weights without supplemental feeding on pasture and/or lot finishing at the end of the grazing season.

Systems of Pasture Finishing

When cattle are finished on pasture, any one of the following systems may be employed:
1. Finishing on pastures alone—no concentrates being fed.
2. Limited grain allowance during the entire pasture period, usually by adding 15% salt or 10% fat to a concentrated mixture fed in self-feeders.
3. Full feeding during the entire pasture period.
4. Full or limited grain feeding on pasture following the period of peak pasture growth.
5. Short feeding (60 to 120 days) in the feedlot at the end of the pasture period.

The system of pasture finishing that will be decided upon will depend upon the age of the cattle, the quality of the pasture, the price of concentrates, the rapidity of gains desired, and the market conditions.

Basic Considerations in Utilizing Pastures for Finishing Cattle

The following points are basic in utilizing pastures for finishing cattle:
- **Moderate winter feeding makes for most effective pasture utilization**—The more liberally beef cattle are fed during the winter, the less will be their effective utilization of pasture the following summer—the less the compensatory gains. Generally speaking, for maximum utilization of pasture, stocker calves should be fed for winter gains not in excess of 1.25 lb/head/day, and yearlings not in excess of 0.9 lb.
- **Time of starting grain feeding on pastures is determined by condition of cattle and quality of pastures**—Cattle that have been fed grain rather liberally through the winter and are in good condition should usually be fed grain from the beginning of the grazing period. On the other hand, if they have been roughed through the winter, it may be just as well to feed the grain only during the last 80 to 120 days of the grazing season, after the season of peak pasture growth. The latter recommendation is made because it is sometimes difficult to get animals to consume grain when an abundance of palatable forage is available. At peak pasture growth, the animals should be started on feed and brought to full feed as rapidly as possible.
- **Grain supplements on pastures usually make for larger daily gains and earlier marketing**—Young cattle (calves and yearlings) on summer pasture usually do not grow at their maximum potential due to energy and protein deficiencies in the feed at various times of the season. Thus, the addition of a grain supplement for cattle on pasture makes for larger daily gains and earlier marketing—either directly off grass or with a shorter drylot finishing period.
- **Whole corn preferred to rolled corn**—When self-feeding steers on pasture, whole corn is preferred to rolled corn for the following reasons: (1) Slightly less feed is required per 100 lb gain; (2) it alleviates processing cost; and (3) it results in less incidence of founder and rumen parakeratosis because whole corn supplies some *roughness factor* in the ration to stimulate the rumen.
- **Protein supplement not needed on good pastures**—So long as pasture is green and growing, no supplemental protein is required. During drought periods and late fall when the grass matures, extra protein is needed. At such times, it is good business to add 1 lb of protein supplement to each 8 to 12 lb of grain. Usually this will increase the rate and efficiency of gain.
- **Carrying capacity of pastures will vary**—The carrying capacity of pastures will vary with the amount of grain supplement, the quality of pasture, and the age and condi-

tion of the cattle. Because of these factors, the acreage per steer will vary from ⅓ to 10.

- **Age is a factor**—Young cattle (yearlings) tend to grow as well as to fatten. Thus, older cattle will reach a high degree of finish on pastures alone. As good as the pastures are, it must be remembered that grass is still a roughage.
- **Minerals for cattle on pasture**—Salt is especially necessary when grass is being utilized. Finishing steers consume from ¾ to 1½ oz of salt per head daily. Also, cattle on pasture should have free access to a 2-compartment mineral box, with (1) trace mineralized salt in one side (salt included for purposes of palatability), and (2) in the other side, a mixture of ⅓ trace mineralized salt and ⅔ defluorinated phosphate or steamed bone meal.
- **Species of grasses or legumes will vary**—The best species of grasses or legumes or grass-legume mixtures to be seeded will vary according to the area, especially according to the soil and climatic conditions. Pasture yields vary greatly from area to area and season to season (see Chapter 7, Pasture and Range Forages, for recommended grass and/or legume species).

Temporary or supplemental pastures, such as Sudangrass or millet, are used for short periods and are usually more productive and palatable than permanent pastures. They are seeded for the purpose of providing supplemental grazing during the season when the regular permanent or rotation pastures are relatively unproductive.

- **Self-feeding vs hand-feeding on pasture**—Self-feeding grain on pasture has generally proven superior to hand-feeding, as the animals consume more feed, make more rapid gains, and return more profit. Adding salt (usually about 15%) or fat (usually about 10%) to a concentrate mixture fed in self-feeders has proven to be a satisfactory method of controlling grain consumption at the desired level. Fresh water should be readily accessible.
- **Economy of grain feeding on pasture**—Whether it will be profitable to feed grain on pasture will depend primarily upon the price of grain, the premium paid for cattle of higher finish and grade, the season in which it is desired to market, and the area and quality of pasture.
- **Pasture bloat can be controlled**—The following preventive measures will reduce pasture bloat:

1. Avoid straight legume pasture (except for trefoil) and immature legumes.
2. Feed a coarse grass hay prior to turning onto lush pasture.
3. Feed dry forage along with pasture.
4. Avoid a rapid fill from an empty start.
5. Keep animals continuously on pasture after they are once turned out.
6. Keep salt and water conveniently accessible at all times.
7. Avoid frosted pasture.
8. Use poloxalene (Bloat Guard), oxytetracycline (Terramycin or Neo-Terramycin), or Laureth-23 (Enproal Bloat Blox) according to manufacturers' directions.

(Also see Chapter 5, Table 5-1, Nutritional Diseases and Ailments—Bloat, pasture.)

NUTRITIONAL FEEDLOT DISEASES

Several metabolic disorders, or diseases, in feedlot cattle are attributable wholly or in part to the feeding regimen. Among the more prevalent ones are acidosis, bloat, liver abscesses, and urinary calculi (water belly; urolithiasis). These are fully covered in Chapter 5, Nutritional Disorders/Toxins.

Fig. 19-36. Yearling Hereford steers on Sudangrass pasture near Weeping Water, Nebr. (Courtesy, Soil Conservation Service, USDA)

QUESTIONS FOR STUDY AND DISCUSSION

1. Do you agree or disagree with the following statement: "Pastures and other roughages are the very foundation of successful beef cattle production." Justify your answer.

2. Discuss the economic importance of feed for beef cattle.

3. In recent years, crossbreeding and the exotic breeds have produced larger and heavier milking cows, faster-gaining calves, and larger-framed growing/finishing cattle. How has this changed nutritive needs? Do the National Research Council (NRC) Requirements reflect these changes and needs?

4. Describe the symptoms of (a) energy deficiency and (b) protein deficiency in cattle.

5. For beef cattle, discuss the functions and the deficiency symptoms of each of the following minerals: salt, calcium, phosphorus, cobalt, iodine, and selenium.

6. For beef cattle, list the vitamins most apt to be deficient; then give (a) some of the deficiency symptoms, and (b) practical sources of each vitamin for use on a farm or ranch.

7. How does water facilitate the essential processes of the body?

8. It has been said that feed additives and implants lowered the retail price of beef by 5 to 6¢ per pound, because of the increased rate and efficiency of gains that producers obtained on the treated cattle. Do you agree with this statement? Justify your stand.

9. Will the proportion of concentrates and roughages used in beef cattle rations change in the years ahead? If so, why?

10. Why is hay the most important harvested roughage fed to beef cattle?

11. Discuss the impact of the expanding world human population on the use of urea and other nonprotein sources for cattle.

12. Give free-choice mineral feeding recommendations (a) for cattle that are on liberal grain feeding, and (b) for cattle that are primarily on roughage (pasture, hay, and/or silage).

13. How may the vitamin A, D, and E requirements of beef cattle be met?

14. Table 19-14, Feed Substitution Table for Beef and Dairy Cattle, is a summary of the comparative values of common feeds, based upon chemical composition, palatability, and product quality. Why include the last two bases?

15. Some suggested rations and general rules of feeding are given. Yet, the following statement is added: "The eye of the master fattens his cattle." Why this statement?

16. Wherein does feeding cattle for show differ from feeding cattle in a commercial cattle feedlot?

17. How may feed costs, which account for 65 to 75% of the total cost of keeping cows, be lowered?

18. In comparison with medium-frame cows, large-frame cows produce 9% more calf weaning weight, but they require 28% more feed. How can a producer determine which size cow is more profitable?

19. How may a practical cattle producer meet the added energy requirements of a brood cow during the critical 100 days, extending from 30 days before calving to 70 days after calving?

20. Compare and discuss the nutritive requirements of (a) dry pregnant cows, (b) bred heifers, and (c) lactating cows. From a practical standpoint does this mean that a producer should separate different classes and ages of cattle, then feed them according to needs?

21. What is meant by the terms "calving control"/"daytime calving"? How can this phenomenon be brought about by time of feeding?

22. Discuss the interrelationship between pasture quality and lactating cow/nursing calf requirements and performance.

23. Discuss the analysis of corn stover and husklage in relation to the nutritive requirements of dry pregnant cows. Why is not more corn residue fed to cattle?

24. List and discuss the feeding value of crop residues other than corn residue that can be used for feeding cows.

25. Discuss nutrient deficiencies of cattle that are frequently encountered on U.S. ranges. How would you rectify each one?

26. When pasture and range supplemental feeding is planned, what sorting and grouping of cattle should be done at the outset?

27. What type and system of pasture and range supplementation would you recommend? Justify your answer.

28. What factors favor increased confinement production?

29. Discuss phase-feeding confinement cows according to stage of production and age of animals.

30. Discuss the advantages and disadvantages of (a) a semiconfinement vs (b) a year-round (total) confinement cow-calf operation.

31. How would you handle a heavily fitted sale bull from auction time to breeding season, which we shall assume to be a period of 60 days?

32. Wherein does the feed and management of a young bull 13 to 14 months old differ from the feed and management of a mature bull?

33. Discuss the feeding of a newborn calf.

34. How would you raise orphan calves or multiple birth calves?

35. Define: (a) creep feeding, (b) "hungry calf gap," and (c) creep. Under what conditions would you recommend creep feeding; under what conditions would you recommend against creep feeding?

36. Why and how may creep feeding be limited?

37. What are the advantages and disadvantages of early weaning? Outline a program for early weaning.

38. How would you go about weaning calves, from the standpoint of both the cows and the calves?

39. Who benefits the most from a preconditioning program, the cow-calf operator or the cattle feeder?

40. Detail a preconditioning program.

41. Discuss the weight and age of small-frame, of medium-frame, and of large-frame heifers at first heat, breeding, and calving.

42. Why should heifers be separated according to age, body weight, and expected daily gain? List and discuss the practices that should be observed when calving two-year-old heifers.

43. Define the following terms: (a) stockers; (b) feeders; (c) replacement heifers; (d) preconditioning; and (e) backgrounding.

44. What are the common types of stocker programs, and what are the characteristics of each?

45. Discuss pasture supplement for stocker cattle.

46. Discuss growth stimulants or implants for stocker cattle.

47. Tell how the level of wintering stockers affects the next stage of feeding.

48. For the farmer/rancher who produces stocker calves, and who finishes them on a custom basis, is compensatory growth good or bad? Justify your answer.

49. Why do lightweight calves gain more efficiently than heavyweight calves?

50. What provisions should be incorporated in a stocker and grower contract?

51. What factors should determine the following alternative choices of a cattle feeder: (a) cattle feedlot (including confinement feeding) vs pasture finishing; and (b) farmer-feeder vs commercial cattle feeder?

52. What percentage of the cost of finishing cattle, exclusive of the purchase price of the feeders, is feedstuffs? How may this cost be lowered?

53. Discuss how each of the following enter into the choice of the kind of cattle to feed in a given lot: (a) age and weight of cattle; (b) sex of cattle; (c) grade of cattle; and (d) breeding and type of cattle.

54. Discuss the impact of each of the following on cattle feeding: (a) dairy beef, and (b) boxed beef.

55. When feeding dairy beef, what will determine the choice between (a) high-energy rations and light market weight cattle; vs (b) high-roughage rations and heavy market weight?

56. With the advent of boxed beef, the steer must fit the box. How does this affect the cattle feeding industry?

57. Select a certain area for feeding cattle. Then, give for that particular area the pros and cons for each (a) an open feedlot, (b) cold confinement, and (c) warm confinement. Finally, give your recommendation.

58. What is a slotted floor? Why and how is it sometimes used in cattle feeding?

59. How do the feeds and rations for finishing cattle compare to the feeds and rations for (a) breeding beef cattle, and (b) stocker and feeder cattle?

60. Outline, step by step, a recommended program for handling newly arrived feedlot cattle that will minimize losses and maximize performance.

61. Outline, step by step, a program for getting (a) yearling cattle, and (b) calves on full feed.

62. Why is overfinishing undesirable?

63. How may a feedlot mud problem be alleviated?

64. Discuss each of the types of custom feeding contracts.

65. Under what conditions would you recommend pasture finishing of cattle rather than feedlot finishing?

66. Discuss the alternate systems of pasture finishing.

67. As grain becomes scarcer and higher in price, is it likely that more cattle will be grass finished, perhaps by supplemental grain feeding?

68. Discuss each of the following basic points as they apply to pasture finishing: (a) moderate winter gains; (b) when to start grain feeding on pasture; (c) effect of supplementing young cattle on pasture on subsequent feedlot performance; (d) self-feeding whole corn vs rolled corn on pasture; (e) the use of a protein supplement; (f) age of cattle; (g) species of grasses or legumes; (h) self-feeding vs hand-feeding; and (i) bloat control.

69. Using current feed prices, compute the value of grass on a per steer basis if it effects a saving of 100 lb of dry feed per 100 lb of gain. What additional advantages accrue from grain feeding cattle on pasture, compared to drylot finishing?

Fig. 19–37. A snowmobile gives Howard Gestrin, Cascade, Idaho, new motility to check cattle in deep winter snow. (Courtesy, USDA. Photo by Nicholas deVore.)

Fig. 19-38. How it used to be done! This shows The White Heifer That Traveled, a noted Shorthorn heifer, at 7 years of age and weighing 2,300 lb, with the caretaker hand-preparing a feed of mangels, a root crop, for her. (Painting by Thomas Weaver, 1811. Courtesy, Harding and Harding, Geneva, Ill.)

Fig. 19-39. How it is done today! Hereford steers eating a complete, mechanically-mixed feed from a concrete fenceline feed bunk. (Courtesy, Polled Hereford World, Kansas City, Mo.)

Original painting by Tom Phillips

20

FEEDING DAIRY CATTLE[1]

Contents	Page
Economic Importance of Feed	426
Nutritive Needs of Dairy Cattle	426
National Research Council (NRC) Requirements	426
Dry Matter Intake (DMI)	430
Energy	430
Protein	439
Degraded Intake Protein (DIP) and Undegraded Intake Protein (UIP)	439
Urea and Other NPN Products	440
Minerals	440
Major or Macrominerals	440
Trace or Microminerals	441
Vitamins	443
Fat-Soluble Vitamins	443
Water-Soluble Vitamins	444
Water	445
Feeds for Dairy Cattle	445
Forages	445
Root Crops	446
Concentrates	446
By-products	446
Special Feeds and Feed Additives for Dairy Cattle	446
Feed Additives	447
Commercial Feeds	447
Feed Considerations	448
Feed Substitutions	448
Feed Preparation	448
Rations	448
Feeding Lactating Cows	448
Thumb Rules for Feeding Lactating Cows	448
Amount and Kind of Forage to Feed	449
Amount and Kind of Concentrate (Grain) to Feed	449
Rations for Lactating Cows	450
How to Balance a Dairy Ration	450
Computer-Formulated Rations/Diskette	451
Feeding Systems	454
Phase Feeding	454
Challenge Feeding (Lead Feeding)	455

Contents	Page
Group Feeding	455
Feeding on Pasture	458
Feeding Dry Cows	458
Feeding Calves	458
Dairy Calf Production	458
Colostrum	458
Calf Starters	459
Hay or Silage for Calves	459
Water	459
Amount and Method of Feeding, Frequency of Feeding, and Age of Weaning	459
Veal Calf Production	460
Preventing Calf Scours	460
Feeding Replacement Heifers/Normal Growth for Dairy Heifers	460
Feeding Dairy Bulls	461
Feeding Dairy Beef	461
Feeding Show and Sale Animals	461
Feed and Management Aspects of Dairy Production	461
Time and Frequency of Feeding and Watering	461
Frequency of Milking	462
Milk and Fat Records; Testing Programs	462
Health Disorders of Dairy Cows	462
Questions for Study and Discussion	463

Feeding lactating dairy cows differs from feeding other classes of farm animals. There is limited time between calvings in which to get whatever milk the cow will produce. Milk production reaches a peak 2 to 6 weeks after a cow freshens, then declines during the remaining portion of the lactation period. If feeding is limited, the rate of decline is more rapid than normal and total production is lowered. Thus, reducing the ration and concomitantly lowering production is almost certain to be economically unsound.

As labor became scarcer, and more expensive, dairy producers automated—they replaced part of the labor force by machines. Eventually, it became advantageous to divide

---

[1]The authors gratefully acknowledge the helpful suggestions of the following dairy specialists who reviewed this chapter: D. L. Bath, Ph.D., Extension Dairy Nutritionist, University of California, Davis; L. R. Brown, Ph.D., Extension Dairyman, The University of Connecticut, Storrs; W. H. Brown, Ph.D., Professor, Department of Animal Sciences, The University of Arizona, Tucson; R. R. Grummer, Ph.D., Department of Dairy Science, University of Wisconsin, Madison; L. M. Larsen, Ph.D., Consultant, Nutri-Systems, 426 E. Shields Ave., Fresno, CA 93704; J. W. Thomas, Ph.D., Consultant and Professor Emeritus, Department of Animal Science, Michigan State University, East Lansing; and W. H. Van Horn, Ph.D., Professor, Institute of Food and Agricultural Sciences, University of Florida, Gainesville.

Also, the following dairy specialists responded liberally to the call of the authors for counsel and advice, literature, and/or pictures: L. E. Chase, Ph.D., Associate Professor, Dairy Cattle Nutrition, Department of Animal Science, Cornell University, Ithaca, N.Y.; D. A. Hartman, Ph.D., Department of Dairy Science, Virginia Tech, Blacksburg; A. J. Heinrichs, Ph.D., Dairy Specialist, Pennsylvania State University, University Park; T. W. Howard, Ph.D., Extension Dairy Specialist, University of Wisconsin, Madison; M. F. Hutjens, Ph.D., Extension Dairyman, University of Illinois, Urbana-Champaign; J. G. Linn, Ph.D., Extension Animal Scientist, Dairy Nutrition, University of Minnesota, St. Paul; M. E. McCullough, Ph.D., Consulting Nutritionist/ Professor Emeritus, University of Georgia, Experiment; O. E. Otterby, Professor, Department of Animal Science, University of Minnesota, St. Paul; and J. Tappan, Arizona Dairy Company, Higley, Ariz.

the herd into groups of cows producing at about the same level, with each group of 50 to 100 cows managed as a unit. The group concept led to changes in feeding practices, including group feeding based on the needs of a corral of cows, rather than the needs of an individual cow, and complete rations (grain and forage combined), with a different combination of feeds for each group. Next came confinement and environmental control. Other innovations will follow.

ECONOMIC IMPORTANCE OF FEED

Feed, more than any other one factor, determines the productivity and profitability of dairy cows. Within a herd, approximately 25% of the difference in milk production between cows is due to heredity; the remaining 75% is determined by environmental factors, with feed making up the largest portion. Feed accounts for about 55% (with a range from 45% to 65%) of the cost of milk production. Therefore, a good feeding program is necessary for profitable milk production.

Although it costs more to feed high producers than low producers, milk income is much higher and the net return above the cost of feed is also increased. Fig. 20-1 shows how income over feed cost improves as production per cow increases.

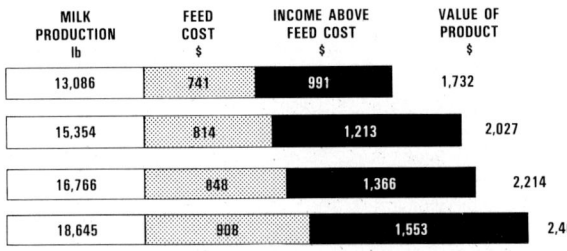

Fig. 20-1. It costs more to feed high-producing cows, but the practice pays handsome dividends. The reason: feed and overhead costs for maintenance are practically the same, regardless of the level of production. (Based on data from New York Dairy Records Laboratory; courtesy, L. R. Brown, Extension Dairyman, The University of Connecticut, Storrs)

NUTRITIVE NEEDS OF DAIRY CATTLE

The first consideration in any dairy feeding program is to determine the nutritive needs for body maintenance, growth, pregnancy or reproduction, and milk production.

When cows are underfed, they cannot produce at their most efficient and profitable level. It is more profitable to feed on the generous side than to feed so little that production drops to an uneconomical level. Of course, it is also unprofitable to waste nutrients. So, cows that put on flesh and fail to return it later in increased milk production should be fed somewhat different amounts and types of feeds than cows which have a higher priority for milk production.

National Research Council (NRC) Requirements

The current recommended nutritive requirements for dairy cattle are contained in *Nutrient Requirements of Dairy Cattle,* Sixth Revised Edition, Update 1989, prepared by the Subcommittee on Dairy Nutrition, National Research Council, and published by the NRC, Washington, D.C. These are listed in Tables 20-1 to 20-5. It should be noted that the requirements listed in these tables do not allow for any margin of safety; that is, they do not provide for animal differences, feed differences, losses of certain nutrients in storage, and stresses. Accordingly, in the formulation of rations, margins of safety should be provided.

Table 20-6, shows NRC's maximum tolerable dietary levels of certain elements.

Table 20-7, Composition of Feeds Commonly Used in Dairy Cattle Rations, consists of selected feeds from *Nutrient Requirements of Dairy Cattle,* Sixth Revised Edition, Update 1989, Table 7-1. In their feed compositions, the NRC committee assumed an average decrease of 4% per unit of dry matter intake above maintenance in calculating NE_{lc} values for feed ingredients, or an average discount of 8% based on their assumption that lactating cows are fed at 3 X maintenance. For the convenience of those dairy producers and dairy nutritionists who wish to use these values, the authors selected from *Nutrient Requirements of Dairy Cattle,* Sixth Revised Edition, Update 1989, Table 7-1, the feeds most commonly used in dairy cattle rations, and herein reproduced them in Table 20-7. **NOTE WELL:** Table 20-7 is a 3-page spread; see pages 436-438.

TABLE 20-1
DRY MATTER INTAKE REQUIREMENTS TO FULFILL NUTRIENT ALLOWANCES FOR MAINTENANCE, MILK PRODUCTION, AND NORMAL LIVE WEIGHT GAIN DURING MID- AND LATE LACTATION[1]

Live Wt.: (lb) / (kg)	882 / 400	1,103 / 500	1,323 / 600	1,544 / 700	1,764 / 800
FCM (4%)[2]	\multicolumn{5}{c}{Percent Live Weight[3][4]}				
(lb) / (kg)	(%)	(%)	(%)	(%)	(%)
22 / 10	2.7	2.4	2.2	2.0	1.9
33 / 15	3.2	2.8	2.6	2.3	2.2
44 / 20	3.6	3.2	2.9	2.6	2.4
55 / 25	4.0	3.5	3.2	2.9	2.7
66 / 30	4.4	3.9	3.5	3.2	2.9
77 / 35	5.0	4.2	3.7	3.4	3.1
88 / 40	5.5	4.6	4.0	3.6	3.3
99 / 45	—	5.0	4.3	3.8	3.5
110 / 50	—	5.4	4.7	4.1	3.7
121 / 55	—	—	5.0	4.4	4.0
132 / 60	—	—	5.4	4.8	4.3

[1]Adapted by the authors from *Nutrient Requirements of Dairy Cattle,* 6th rev. ed., update 1989, NRC, National Academy Press, p. 78, Table 6-1.

[2]4% fat-corrected milk (kg) = (0.4)(kg of milk) + (15)(kg of milk fat).

[3]The probable DMI may be up to 18% less in early lactation.

[4]DMI as a percentage of live weight may be 0.02% less per 1% increase in ration moisture content above 50% if fermented feeds constitute a major portion of the ration.

NOTE: The following assumptions were made in calculating the DMI requirements shown in this table:

1. The basic or reference cow used for the calculations weighed 1,323 lb *(600 kg)* and produced milk with 4% milk fat. Other live weights in the table and corresponding fat percentages were 882 lb *(400 kg)* and 5% fat; 1,103 lb *(500 kg)* and 4.5% fat; and 1,544 and 1,764 lb *(700 and 800 kg)* and 3.5% fat.

2. The concentration of energy in the ration for the reference cow was 1.42 Mcal of NE_{lc}/kg of DM for milk yields equal to or less than 22 lb *(10 kg)*/day. It increased linearly to 1.72 Mcal of NE_{lc}/kg for milk yields equal to or greater than 88 lb *(40 kg)*/day.

3. The energy concentrations of the rations for all other cows were assumed to change linearly as their energy requirements for milk production, relative to maintenance, changed in a manner identical to that of the 1,323-lb *(600-kg)* cow as she increased in milk yield from 22 to 88 lb *(10 to 40 kg)*/day.

4. Enough DM to provide sufficient energy for cows to gain 0.055% of their body weight daily was also included in the total. If cows do not consume as much DM as they require, as calculated from this table, their energy intake will be less than their requirements. The result will be a loss of body weight, reduced milk yields, or both. If cows consume more DM than what is projected as required from this table, the energy concentration of their ration should be reduced or they may become overly fat.

TABLE 20-2
DAILY NUTRIENT REQUIREMENTS OF GROWING DAIRY CATTLE AND MATURE BULLS[1] (See footnotes at end of table.)

Live Weight		Gain		Dry Matter Intake[2]		Energy					Protein			Minerals		Vitamins	
						TDN		DE	ME	NE$_m$	NE$_g$	CP	DIP[3]	UIP[4]	Ca	P	A	D
(lb)	(kg)	(lb)	(kg)	(lb)	(kg)	(lb)	(kg)	(Mcal)	(Mcal)	(Mcal)	(Mcal)	(g)	(g)	(g)	(g)	(g)	(1,000 IU)	(1,000 IU)
Growing Large-Breed Calves Fed Only Milk or Milk Replacer																		
88	40	0.4	0.2	1.06	0.48	1.37	0.62	2.73	2.54	1.37	0.41	105	—	—	7	4	1.70	0.26
99	45	0.7	0.3	1.19	0.54	1.54	0.70	3.07	2.86	1.49	0.56	120	—	—	8	5	1.94	0.30
Growing Large-Breed Calves Fed Milk Plus Starter Mix																		
110	50	1.1	0.5	2.87	1.30	3.22	1.46	6.42	5.90	1.62	0.72	290	—	—	9	6	2.10	0.33
165	75	1.8	0.8	4.37	1.98	4.90	2.22	9.78	8.98	2.19	1.30	435	—	—	16	8	3.20	0.50
Growing Small-Breed Calves Fed Only Milk or Milk Replacer																		
55	25	0.4	0.2	0.84	0.38	1.08	0.49	2.16	2.01	0.96	0.37	84	—	—	6	4	1.10	0.16
66	30	0.7	0.3	1.12	0.51	1.46	0.66	2.90	2.70	1.10	0.52	112	—	—	7	4	1.30	0.20
Growing Small-Breed Calves Fed Milk Plus Starter Mix																		
110	50	1.1	0.5	3.15	1.43	3.53	1.60	7.06	6.49	1.62	0.72	315	—	—	10	6	2.10	0.33
165	75	1.3	0.6	3.88	1.76	4.34	1.97	8.69	7.98	2.19	0.96	387	—	—	14	8	3.20	0.50
Growing Veal Calves Fed Only Milk or Milk Replacer																		
88	40	0.4	0.2	0.99	0.45	1.04	0.47	2.07	1.89	1.37	0.55	100	—	—	7	4	1.70	0.26
110	50	0.9	0.4	1.26	0.57	1.30	0.59	2.63	2.39	1.62	0.57	125	—	—	9	5	2.10	0.33
132	60	1.2	0.5	1.76	0.80	1.57	0.71	3.17	2.84	1.85	0.81	176	—	—	13	8	2.60	0.40
165	75	2.0	0.9	3.00	1.36	2.67	1.21	5.39	4.82	2.19	1.47	300	—	—	16	9	3.20	0.50
221	100	2.8	1.3	4.41	2.00	3.48	1.58	7.06	6.22	2.72	2.26	440	—	—	20	11	4.20	0.66
276	125	2.8	1.3	5.25	2.38	4.15	1.88	8.40	7.40	3.21	2.44	524	—	—	22	13	5.30	0.82
331	150	2.4	1.1	6.00	2.72	4.74	2.15	9.60	8.46	3.69	2.29	598	—	—	24	15	6.40	0.99
Large-Breed Growing Females																		
221	100	1.3	0.6	5.80	2.63	4.02	1.84	8.13	7.03	2.72	1.22	421	57	317	17	9	4.24	0.66
221	100	1.5	0.7	6.22	2.82	4.37	1.98	8.72	7.54	2.72	1.44	452	75	346	18	9	4.24	0.66
221	100	1.8	0.8	6.66	3.02	4.65	2.11	9.32	8.06	2.72	1.66	483	92	374	18	10	4.24	0.66
331	150	1.3	0.6	7.74	3.51	5.31	2.41	10.61	9.14	3.69	1.45	562	150	283	19	11	6.36	0.99
331	150	1.5	0.7	8.27	3.75	5.67	2.57	11.33	9.76	3.69	1.71	600	173	307	19	12	6.36	0.99
331	150	1.8	0.8	8.80	3.99	6.04	2.74	12.07	10.39	3.69	1.97	639	196	331	20	12	6.36	0.99
441	200	1.3	0.6	9.68	4.39	6.50	2.95	12.99	11.14	4.57	1.65	699	239	254	20	14	8.48	1.32
441	200	1.5	0.7	10.32	4.68	6.92	3.14	13.84	11.87	4.57	1.95	749	267	274	21	14	8.48	1.32
441	200	1.8	0.8	10.96	4.97	7.36	3.34	14.71	12.62	4.57	2.25	796	295	294	22	15	8.48	1.32
551	250	1.3	0.6	11.71	5.31	7.67	3.48	15.33	13.10	5.41	1.84	718	326	229	22	16	10.60	1.65
551	250	1.5	0.7	12.46	5.65	8.16	3.70	16.32	13.94	5.41	2.18	787	359	246	23	17	10.60	1.65
551	250	1.8	0.8	13.21	5.99	8.67	3.93	17.32	14.79	5.41	2.51	857	393	263	24	17	10.60	1.65
662	300	1.3	0.6	13.80	6.26	8.84	4.01	17.69	15.05	6.20	2.02	752	413	209	23	17	12.72	1.98
662	300	1.5	0.7	14.69	6.66	9.42	4.27	18.81	16.00	6.20	2.39	814	452	223	24	18	12.72	1.98
662	300	1.8	0.8	15.57	7.06	9.97	4.52	19.95	16.97	6.20	2.77	884	490	236	25	19	12.72	1.98
772	350	1.3	0.6	16.07	7.29	10.05	4.56	20.09	17.01	6.96	2.20	874	501	193	24	18	14.84	2.31
772	350	1.5	0.7	17.09	7.75	10.67	4.84	21.36	18.09	6.96	2.60	930	545	204	25	19	14.84	2.31
772	350	1.8	0.8	18.10	8.21	11.33	5.14	22.64	19.18	6.96	3.01	985	590	214	26	20	14.84	2.31
882	400	1.3	0.6	18.50	8.39	11.29	5.12	22.58	19.03	7.69	2.37	1,007	592	182	25	19	16.96	2.64
882	400	1.5	0.7	19.67	8.92	12.00	5.44	24.00	20.23	7.69	2.80	1,070	641	190	26	20	16.96	2.64
882	400	1.8	0.8	20.86	9.46	12.72	5.77	25.44	21.44	7.69	3.24	1,135	692	198	26	21	16.96	2.64
992	450	1.3	0.6	21.15	9.59	12.59	5.71	25.18	21.12	8.40	2.53	1,151	686	176	28	19	19.08	2.97
992	450	1.5	0.7	22.49	10.20	13.38	6.07	26.78	22.46	8.40	2.99	1,224	742	182	28	20	19.08	2.97
992	450	1.8	0.8	23.86	10.82	14.20	6.44	28.40	23.81	8.40	3.46	1,298	799	187	29	21	19.08	2.97
1,103	500	1.3	0.6	24.10	10.93	13.98	6.34	27.96	23.32	9.09	2.69	1,311	785	175	28	20	21.20	3.30
1,103	500	1.5	0.7	25.64	11.63	14.88	6.75	29.74	24.81	9.09	3.18	1,395	848	179	28	20	21.20	3.30
1,103	500	1.8	0.8	27.18	12.33	15.79	7.16	31.55	26.32	9.09	3.68	1,480	913	182	29	21	21.20	3.30
1,213	550	1.3	0.6	27.39	12.42	15.48	7.02	30.95	25.67	9.77	2.84	1,490	891	180	28	20	23.32	3.63
1,213	550	1.5	0.7	29.15	13.22	16.47	7.47	32.95	27.33	9.77	3.37	1,587	963	183	28	20	23.32	3.63
1,213	550	1.8	0.8	30.96	14.04	17.51	7.94	34.99	29.02	9.77	3.90	1,685	1,035	185	29	21	23.32	3.63
1,323	600	1.3	0.6	31.11	14.11	17.13	7.77	34.24	28.23	10.43	3.00	1,694	1,007	193	28	20	25.44	3.96
1,323	600	1.5	0.7	33.19	15.05	18.26	8.28	36.50	30.09	10.43	3.55	1,805	1,088	194	28	21	25.44	3.96
1,323	600	1.8	0.8	35.26	15.99	19.40	8.80	38.79	31.98	10.43	4.11	1,919	1,170	195	29	21	25.44	3.96
Small-Breed Growing Females																		
221	100	0.9	0.4	5.31	2.41	3.68	1.67	7.35	6.34	2.72	0.91	386	38	249	15	8	4.24	0.66
221	100	1.1	0.5	5.82	2.64	4.01	1.82	8.03	6.92	2.72	1.16	422	59	275	16	8	4.24	0.66
221	100	1.3	0.6	6.31	2.86	4.37	1.98	8.71	7.51	2.72	1.40	458	80	300	17	9	4.24	0.66

(Continued)

TABLE 20-2 (Continued)

Live Weight		Gain		Dry Matter Intake[2]		Energy						Protein			Minerals		Vitamins	
						TDN		DE	ME	NE$_m$	NE$_g$	CP	DIP[3]	UIP[4]	Ca	P	A	D
(lb)	(kg)	(lb)	(kg)	(lb)	(kg)	(lb)	(kg)	(Mcal)	(Mcal)	(Mcal)	(Mcal)	(g)	(g)	(g)	(g)	(g)	(1,000 IU)	(1,000 IU)
Small-Breed Growing Females (Continued)																		
331	150	0.9	0.4	7.30	3.31	4.90	2.22	9.78	8.39	3.69	1.09	529	129	222	17	10	6.36	0.99
331	150	1.1	0.5	7.94	3.60	5.31	2.41	10.63	9.12	3.69	1.39	575	156	243	18	11	6.36	0.99
331	150	1.3	0.6	8.58	3.89	5.76	2.61	11.50	9.86	3.69	1.69	622	185	263	19	11	6.36	0.99
441	200	0.9	0.4	9.35	4.24	6.09	2.76	12.16	10.38	4.57	1.26	578	217	201	19	13	8.48	1.32
441	200	1.1	0.5	10.14	4.60	6.59	2.99	13.19	11.25	4.57	1.60	648	251	217	20	13	8.48	1.32
441	200	1.3	0.6	10.94	4.96	7.12	3.23	14.23	12.14	4.57	1.95	718	286	232	20	14	8.48	1.32
551	250	0.9	0.4	11.55	5.24	7.28	3.30	14.57	12.36	5.41	1.41	629	305	185	21	15	10.60	1.65
551	250	1.1	0.5	12.52	5.68	7.89	3.58	15.78	13.38	5.41	1.80	682	346	197	21	16	10.60	1.65
551	250	1.3	0.6	13.49	6.12	8.51	3.86	17.01	14.43	5.41	2.20	753	389	209	22	16	10.60	1.65
662	300	0.9	0.4	13.98	6.34	8.53	3.87	17.06	14.38	6.20	1.56	761	395	176	22	16	12.72	1.98
662	300	1.1	0.5	15.15	6.87	9.24	4.19	18.48	15.57	6.20	1.99	824	445	184	23	17	12.72	1.98
662	300	1.3	0.6	16.32	7.40	9.97	4.52	19.92	16.79	6.20	2.43	888	495	192	23	17	12.72	1.98
772	350	0.9	0.4	16.69	7.57	9.86	4.47	19.71	16.50	6.96	1.71	909	490	173	23	17	14.84	2.31
772	350	1.1	0.5	18.08	8.20	10.67	4.84	21.35	17.87	6.96	2.18	985	548	178	23	18	14.84	2.31
772	350	1.3	0.6	19.51	8.85	11.51	5.22	23.03	19.28	6.96	2.66	1,062	608	183	24	18	14.84	2.31
882	400	0.9	0.4	19.80	8.98	11.29	5.12	22.58	18.77	7.69	1.84	1,078	592	177	24	18	16.96	2.64
882	400	1.1	0.5	21.48	9.74	12.26	5.56	24.50	20.36	7.69	2.35	1,169	661	181	24	19	16.96	2.64
882	400	1.3	0.6	23.20	10.52	13.23	6.00	26.45	21.98	7.69	2.87	1,263	730	183	25	19	16.96	2.64
992	450	0.9	0.4	23.46	10.64	12.90	5.85	25.80	21.27	8.40	1.98	1,276	706	191	27	18	19.08	2.97
992	450	1.1	0.5	25.49	11.56	14.02	6.36	28.04	23.12	8.40	2.52	1,387	786	193	28	19	19.08	2.97
992	450	1.3	0.6	27.56	12.50	15.17	6.88	30.33	25.01	8.40	3.08	1,500	867	194	28	19	19.08	2.97
Large-Breed Growing Males																		
221	100	1.8	0.8	6.17	2.80	4.32	1.96	8.66	7.48	2.72	1.42	448	65	401	18	10	4.24	0.66
221	100	2.0	0.9	6.55	2.97	4.59	2.08	9.16	7.92	2.72	1.60	475	79	433	19	10	4.24	0.66
221	100	2.2	1.0	6.90	3.13	4.83	2.19	9.67	8.36	2.72	1.79	501	93	465	20	11	4.24	0.66
331	150	1.8	0.8	7.94	3.60	5.51	2.50	11.03	9.52	3.69	1.64	576	155	364	20	12	6.36	0.99
331	150	2.0	0.9	8.38	3.80	5.82	2.64	11.63	10.03	3.69	1.85	607	172	393	21	13	6.36	0.99
331	150	2.2	1.0	8.80	3.99	6.11	2.77	12.22	10.55	3.69	2.07	639	190	422	22	13	6.36	0.99
441	200	1.8	0.8	9.77	4.43	6.68	3.03	13.34	11.48	4.57	1.84	709	241	333	22	15	8.48	1.32
441	200	2.0	0.9	10.28	4.66	7.01	3.18	14.02	12.06	4.57	2.08	745	262	359	23	15	8.48	1.32
441	200	2.2	1.0	10.78	4.89	7.36	3.34	14.71	12.66	4.57	2.33	782	284	385	24	16	8.48	1.32
551	250	1.8	0.8	11.62	5.27	7.78	3.53	15.58	13.37	5.41	2.03	843	325	305	24	17	10.60	1.65
551	250	2.0	0.9	12.19	5.53	8.18	3.71	16.35	14.03	5.41	2.30	885	350	329	25	18	10.60	1.65
551	250	2.2	1.0	12.79	5.80	8.58	3.89	17.13	14.70	5.41	2.57	927	375	352	26	18	10.60	1.65
662	300	1.8	0.8	13.52	6.13	8.91	4.04	17.80	15.22	6.20	2.21	863	408	281	25	19	12.72	1.98
662	300	2.0	0.9	14.18	6.43	9.33	4.23	18.66	15.96	6.20	2.51	934	436	302	25	19	12.72	1.98
662	300	2.2	1.0	14.84	6.73	9.77	4.43	19.53	16.70	6.20	2.80	1,004	464	323	26	20	12.72	1.98
772	350	1.8	0.8	15.48	7.02	10.01	4.54	20.02	17.06	6.96	2.38	885	490	261	26	20	14.84	2.31
772	350	2.0	0.9	16.23	7.36	10.50	4.76	20.98	17.88	6.96	2.70	956	522	280	26	20	14.84	2.31
772	350	2.2	1.0	16.98	7.70	10.98	4.98	21.94	18.70	6.96	3.02	1,027	554	298	27	21	14.84	2.31
882	400	1.8	0.8	17.55	7.96	11.14	5.05	22.27	18.91	7.69	2.55	955	572	244	26	21	16.96	2.64
882	400	2.0	0.9	18.39	8.34	11.66	5.29	23.32	19.80	7.69	2.89	1,001	608	260	27	21	16.96	2.64
882	400	2.2	1.0	19.23	8.72	12.19	5.53	24.39	20.71	7.69	3.24	1,056	644	277	28	22	16.96	2.64
992	450	1.8	0.8	19.73	8.95	12.28	5.57	24.56	20.78	8.40	2.71	1,074	656	230	29	21	19.08	2.97
992	450	2.0	0.9	20.66	9.37	12.86	5.83	25.72	21.76	8.40	3.08	1,125	696	245	29	22	19.08	2.97
992	450	2.2	1.0	21.61	9.80	13.45	6.10	26.89	22.75	8.40	3.44	1,176	736	259	29	23	19.08	2.97
1,103	500	1.8	0.8	22.05	10.00	13.47	6.11	26.92	22.69	9.09	2.87	1,201	742	220	29	21	21.20	3.30
1,103	500	2.0	0.9	23.11	10.48	14.09	6.39	28.19	23.76	9.09	3.25	1,257	786	233	29	22	21.20	3.30
1,103	500	2.2	1.0	24.14	10.95	14.73	6.68	29.47	24.84	9.09	3.64	1,314	830	346	29	23	21.20	3.30
1,213	550	1.8	0.8	24.56	11.14	14.69	6.66	29.38	24.66	9.77	3.02	1,336	831	213	29	21	23.32	3.63
1,213	550	2.0	0.9	25.71	11.66	15.39	6.98	30.76	25.82	9.77	3.43	1,399	879	225	29	22	23.32	3.63
1,213	550	2.2	1.0	26.88	12.19	16.07	7.29	32.16	27.00	9.77	3.84	1,463	927	236	30	23	23.32	3.63
1,323	600	1.8	0.8	27.25	12.36	15.99	7.25	31.95	26.71	10.43	3.17	1,483	923	211	29	21	25.44	3.96
1,323	600	2.0	0.9	28.55	12.95	16.74	7.59	33.47	27.97	10.43	3.60	1,554	976	221	29	22	25.44	3.96
1,323	600	2.2	1.0	29.86	13.54	17.51	7.94	34.99	29.25	10.43	4.03	1,624	1,029	231	30	23	25.44	3.96
1,433	650	1.8	0.8	30.19	13.69	17.33	7.86	34.67	28.86	11.07	3.32	1,643	1,020	212	29	21	27.56	4.29
1,433	650	2.0	0.9	31.64	14.35	18.17	8.24	36.33	30.24	11.07	3.77	1,722	1,078	222	29	22	27.56	4.29
1,433	650	2.2	1.0	33.10	15.01	19.01	8.62	38.00	31.63	11.07	4.22	1,801	1,137	230	30	23	27.56	4.29

(Continued)

Feeding Dairy Cattle

TABLE 20-2 *(Continued)*

Live Weight		Gain		Dry Matter Intake[2]		Energy						Protein			Minerals		Vitamins	
						TDN		DE	ME	NE_m	NE_g	CP	DIP[3]	UIP[4]	Ca	P	A	D
(lb)	(kg)	(lb)	(kg)	(lb)	(kg)	(lb)	(kg)	(Mcal)	(Mcal)	(Mcal)	(Mcal)	(g)	(g)	(g)	(g)	(g)	(1,000 IU)	(1,000 IU)
Large-Breed Growing Males (Continued)																	
1,544	700	1.8	0.8	33.43	15.16	18.79	8.52	37.59	31.14	11.70	3.46	1,820	1,124	219	29	22	29.68	4.62
1,544	700	2.0	0.9	35.06	15.90	19.71	8.94	39.40	32.64	11.70	3.93	1,907	1,187	227	29	22	29.68	4.62
1,544	700	2.2	1.0	36.67	16.63	20.62	9.35	41.23	34.16	11.70	4.40	1,996	1,252	235	30	23	29.68	4.62
1,654	750	1.8	0.8	37.02	16.79	20.37	9.24	40.73	33.59	12.33	3.60	2,015	1,235	232	29	22	31.80	4.95
1,654	750	2.0	0.9	38.85	17.62	21.37	9.69	42.73	35.23	12.33	4.09	2,114	1,305	239	29	23	31.80	4.95
1,654	750	2.2	1.0	40.68	18.45	22.38	10.15	44.74	36.89	12.33	4.58	2,213	1,376	246	30	23	31.80	4.95
1,764	800	1.8	0.8	38.72	17.56	21.30	9.66	42.59	35.12	12.94	3.74	2,107	1,303	216	29	22	33.92	5.28
1,764	800	2.0	0.9	40.59	18.41	22.34	10.13	44.67	36.83	12.94	4.25	2,210	1,377	221	29	23	33.92	5.28
1,764	800	2.2	1.0	42.51	19.28	23.40	10.61	46.76	38.55	12.94	4.76	2,313	1,451	227	30	23	33.92	5.28
Small-Breed Growing Males																	
221	100	1.1	0.5	5.40	2.45	3.79	1.72	7.56	6.54	2.72	1.02	392	41	287	16	8	4.24	0.66
221	100	1.3	0.6	5.82	2.64	4.08	1.85	8.15	7.04	2.72	1.23	422	58	316	17	9	4.24	0.66
221	100	1.5	0.7	6.24	2.83	4.37	1.98	8.74	7.55	2.72	1.45	453	75	345	18	9	4.24	0.66
331	150	1.1	0.5	7.23	3.28	4.96	2.25	9.92	8.55	3.69	1.20	525	129	257	18	11	6.36	0.99
331	150	1.3	0.6	7.76	3.52	5.31	2.41	10.64	9.16	3.69	1.46	563	151	282	19	11	6.36	0.99
331	150	1.5	0.7	8.29	3.76	5.69	2.58	11.36	9.78	3.69	1.71	601	174	306	19	12	6.36	0.99
441	200	1.1	0.5	9.11	4.12	6.09	2.76	12.18	10.45	4.57	1.37	630	213	232	20	13	8.48	1.32
441	200	1.3	0.6	9.70	4.40	6.50	2.95	13.02	11.17	4.57	1.66	699	241	252	20	14	8.48	1.32
441	200	1.5	0.7	10.34	4.69	6.95	3.15	13.87	11.90	4.57	1.96	751	268	273	21	14	8.48	1.32
551	250	1.1	0.5	11.00	4.99	7.21	3.27	14.41	12.31	5.41	1.53	648	296	210	21	16	10.60	1.65
551	250	1.3	0.6	11.73	5.32	7.70	3.49	15.38	13.14	5.41	1.86	718	328	228	22	16	10.60	1.65
551	250	1.5	0.7	12.48	5.66	8.18	3.71	16.35	13.97	5.41	2.19	787	361	245	23	17	10.60	1.65
662	300	1.1	0.5	12.99	5.89	8.31	3.77	16.64	14.15	6.20	1.68	707	378	193	23	17	12.72	1.98
662	300	1.3	0.6	13.85	6.28	8.86	4.02	17.74	15.09	6.20	2.04	754	415	207	23	17	12.72	1.98
662	300	1.5	0.7	14.73	6.68	9.44	4.28	18.85	16.04	6.20	2.41	814	453	221	24	18	12.72	1.98
772	350	1.1	0.5	15.13	6.86	9.46	4.29	18.91	16.01	6.96	1.82	823	461	180	23	18	14.84	2.31
772	350	1.3	0.6	16.12	7.31	10.08	4.57	20.15	17.06	6.96	2.22	877	503	191	24	18	14.84	2.31
772	350	1.5	0.7	17.11	7.76	10.72	4.86	21.41	18.13	6.96	2.62	932	547	203	25	19	14.84	2.31
882	400	1.1	0.5	17.42	7.90	10.63	4.82	21.25	17.91	7.69	1.96	947	545	171	24	19	16.96	2.64
882	400	1.3	0.6	18.54	8.41	11.33	5.14	22.64	19.08	7.69	2.39	1,010	594	180	25	19	16.96	2.64
882	400	1.5	0.7	19.71	8.94	12.04	5.46	24.06	20.27	7.69	2.82	1,073	644	189	26	20	16.96	2.64
992	450	1.1	0.5	19.91	9.03	11.84	5.37	23.70	19.87	8.40	2.10	1,083	634	166	28	19	19.08	2.97
992	450	1.3	0.6	21.21	9.62	12.63	5.73	25.26	21.18	8.40	2.55	1,155	689	174	28	19	19.08	2.97
992	450	1.5	0.7	22.56	10.23	13.43	6.09	26.84	22.51	8.40	3.01	1,227	744	180	28	20	19.08	2.97
1,103	500	1.1	0.5	22.67	10.28	13.14	5.96	26.29	21.93	9.09	2.23	1,233	726	167	28	19	21.20	3.30
1,103	500	1.3	0.6	24.17	10.96	14.02	6.36	28.04	23.39	9.09	2.71	1,315	788	173	28	20	21.20	3.30
1,103	500	1.5	0.7	25.69	11.65	14.91	6.76	29.81	24.87	9.09	3.20	1,398	851	177	28	20	21.20	3.30
1,213	550	1.1	0.5	25.73	11.67	14.55	6.60	29.08	24.12	9.77	2.36	1,400	825	174	28	19	23.32	3.63
1,213	550	1.3	0.6	27.47	12.46	15.52	7.04	31.05	25.75	9.77	2.87	1,495	895	178	28	20	23.32	3.63
1,213	550	1.5	0.7	29.24	13.26	16.52	7.49	33.03	27.40	9.77	3.39	1,591	966	181	28	20	23.32	3.63
1,323	600	1.1	0.5	29.22	13.25	16.07	7.29	32.14	26.50	10.43	2.48	1,590	933	187	28	19	25.44	3.96
1,323	600	1.3	0.6	31.22	14.16	17.18	7.79	34.35	20.32	10.43	3.02	1,600	1,012	190	28	20	25.44	3.96
1,323	600	1.5	0.7	33.25	15.08	18.30	8.30	36.59	30.17	10.43	3.57	1,810	1,091	192	28	21	25.44	3.96
Maintenance of Mature Breeding Bulls																	
1,103	500	—	—	17.40	7.89	9.57	4.34	19.15	15.79	9.09	—	789	472	161	20	12	21.20	3.30
1,323	600	—	—	19.96	9.05	10.98	4.98	21.95	18.10	10.43	—	905	573	155	24	15	25.44	3.96
1,544	700	—	—	22.40	10.16	12.33	5.59	24.64	20.32	11.70	—	1,016	670	148	28	18	29.68	4.62
1,764	800	—	—	24.76	11.23	13.63	6.18	27.24	22.46	12.94	—	1,123	764	142	32	20	33.92	5.28
1,985	900	—	—	27.06	12.27	14.88	6.75	29.76	24.53	14.13	—	1,227	854	135	36	22	38.16	5.94
2,205	1,000	—	—	29.28	13.28	16.10	7.30	32.20	26.55	15.29	—	1,328	943	129	41	25	42.40	6.60
2,426	1,100	—	—	31.44	14.26	17.31	7.85	34.59	28.52	16.43	—	1,426	1,029	122	45	28	46.64	7.26
2,646	1,200	—	—	33.56	15.22	18.46	8.37	36.92	30.44	17.53	—	1,522	1,113	115	49	30	50.88	7.92
2,867	1,300	—	—	35.63	16.16	19.60	8.89	39.21	32.32	18.62	—	1,616	1,196	108	53	32	55.12	8.58
3,087	1,400	—	—	37.68	17.09	20.73	9.40	41.45	34.17	19.68	—	1,709	1,277	102	57	35	59.36	9.24

[1] Adapted by the authors from *Nutrient Requirements of Dairy Cattle*, 6th rev. ed., update 1989, NRC, National Academy Press, pp. 81–84, Table 6-2.

[2] The data for DMI are not requirements *per se*, unlike the requirements for net energy maintenance, net energy gain, and absorbed protein. They are not intended to be estimates of voluntary intake but are consistent with the specified dietary energy concentrations. The use of rations with decreased energy concentrations will increase dry matter intake needs; metabolizable energy, digestible energy, and total digestible nutrient needs; and crude protein needs. The use of rations with increased energy concentrations will have opposite effects on these needs.

[3] DIP = degraded intake protein.

[4] UIP = undegraded intake protein.

Dry Matter Intake (DMI)

Dry matter intake is of the utmost importance, especially for high-yielding lactating cows. High-producing cows must consume very large amounts of highly nutritious, digestible dry matter if they are to maximize their genetic potential for production; otherwise, they will lose weight and, subsequently, produce less milk. Conversely, middle- and late-lactation cows should be fed rations with reduced energy; otherwise, they may become too fat—a condition which can adversely affect reproduction and health. So, it is important that the caretaker monitor dry matter and energy intake closely and adjust amounts fed to fit the stage and intensity of production.

The many variables that affect feed intake make it difficult to predict maximum DMI with accuracy. So, NRC's Subcommittee on Dairy Cattle Nutrition prepared Table 20-1, which gives the DMI requirements for cows weighing between 880 and 1,760 lb and yielding from 22 to 132 lb of 4% fat corrected milk (FCM) daily. If cows do not consume as much DM as they require (as projected in Table 20-1), and energy concentration is not increased, energy intake will be less than required; as a result, there will be loss of liveweight or reduced milk yield, or both. If cows consume more DM than they need (as projected in Table 20-1), they may become too fat unless the energy concentration of the ration is reduced; but this is likely to happen only at lower milk yields.

The amount of DM calculated from Table 20-1 should be reduced about 18% during the first 3 weeks of lactation to reflect the fresh cow's low appetite in this period. Within reason, a cow will meet her additional energy needs from body reserves. Also, if fermented feeds, like silage, constitute a major portion of the ration, the amount of DM should be reduced by 0.02 lb/100 lb of liveweight for each 1% increase in ration moisture content above 50%. In both cases, reduced DMI is likely to result in loss of body weight or reduced milk yield, or both, unless the cow's energy requirement can be fulfilled by increasing the energy concentration of the ration.

Energy

Cows use energy for a variety of functions. A certain amount is used for body maintenance; heifers need energy for growth; pregnant cows need additional energy for development of the fetus; and lactating cows require energy to produce milk. For optimal milk production, minimum health disorders, and optimum reproductive efficiency, cows must not be too fat or too thin.

Lack of energy is the most common deficiency of dairy rations. In young animals, an insufficient supply of energy results in retarded growth and a delay in the onset of puberty. In lactating cows, it results in a decline in milk yields and a loss in liveweight; and severe and prolonged energy deficiency also depresses reproductive performance.

Most of the energy required is supplied by carbohydrates, although fats and protein are also used as energy. All cows, except low-producing ones—those producing less than 25 to 35 lb of milk per day, need some grain if they are to produce at top levels.

The NRC energy requirements for dairy cattle, which are presented in this chapter as Tables 20-2, 20-3, and 20-4, are expressed as digestible energy (DE), metabolizable energy (ME), net energy for maintenance (NE_m), net energy for body gain (NE_g), net energy for lactation (NE_{lc}), and total digestible nutrients (TDN). Separate energy values for each maintenance (NE_m) and gain (NE_g) are given because animals use energy for maintenance more efficiently than for growth. However, the efficiency of energy use by lactating cows for maintenance, pregnancy, and milk production is similar; so, only one energy value, net energy for lactation (NE_{lc}), is used for these functions.

In both the NRC requirement tables which are reproduced in this chapter (as Tables 20-2, 20-3, and 20-4), and in the feed composition tables in Section V of this book, energy is also expressed as total digestible nutrients (TDN).

The energy requirement for maintaining a lactating cow is affected by a number of factors, especially the following: (1) *body size*—the larger the animal, the higher the maintenance energy requirement; (2) *activity*—to support grazing activity, the maintenance allowance may be increased by 10% on good pasture and up to 20% on poor pasture; and (3) *cold temperature*—under severe winter conditions without access to dry shelter, the maintenance feed allowance may be increased up to 8%. Also, during the first lactation, when a heifer is still growing, her energy needs are about 20% greater than a mature cow; and during the second lactation, her energy needs are 10% greater than a mature cow. The energy requirement for gestation is 30% of that required for maintenance alone, with most of the increase during the last 8 weeks of pregnancy.

Table 20-3 includes allowances for liveweight changes during lactation. These values will aid the user in identifying the extent of dietary energy insufficiency during weight loss in early lactation and in estimating feed required to regain body weight in later lactation. The desired rate of liveweight gain will depend on the animal's body condition and stage of pregnancy.

The NRC maintenance requirements for growing replacement heifers are 12% higher than for beef cattle. Also, NRC (1) recommends that milk or milk replacer be fed to replacement calves for at least the first month of life, (2) states that longer periods (up to 2 months) of liquid feeding may be beneficial under some conditions because of decreased disease and death losses, and (3) recommends that veal calves be fed maximum *ad libitum* amounts of milk (or milk replacer).

The most common high energy dairy feeds are: barley, beet or sugar cane molasses, beet pulp, citrus pulp, corn, corn silage, fats, high-moisture corn, high quality legume forage (hay or silage), lush pasture, oats, sorghum grain, wheat, whole cottonseeds, and whole soybeans.

• **NRC energy values of feeds vs *Feeds & Nutrition* energy value of feeds**—In *Nutrient Requirements of Dairy Cattle*, Sixth Revised Edition, energy values of feeds for TDN, digestible energy (DE), and metabolizable energy (ME) have been determined at the maintenance level of intake, whereas values for net energy for lactation are adjusted to three times (3X) the maintenance level. The NRC dairy cattle requirements for TDN, DE, and ME for animals fed above maintenance (lactating animals) have been increased to allow for the depression of digestibility of energy at these higher levels of intake. As long as formulators use both NRC energy values for feeds and NRC nutrient requirements, satisfactory rations will result. (This section is continued on p. 439)

TABLE 20-3
DAILY NUTRIENT REQUIREMENTS OF LACTATING AND PREGNANT COWS[1]

Live Weight		Energy					Total Crude Protein	Minerals		Vitamins	
		TDN		DE	ME	NE$_{lc}$		Ca	P	A	D
(lb)	(kg)	(lb)	(kg)	(Mcal)	(Mcal)	(Mcal)	(g)	(g)	(g)	(1,000 IU)	(1,000 IU)
Maintenance of Mature Lactating Cows[2]											
882	400	6.90	3.13	13.80	12.01	7.16	318	16	11	30	12
992	450	7.54	3.42	15.08	13.12	7.82	341	18	13	34	14
1,103	500	8.16	3.70	16.32	14.20	8.46	364	20	14	38	15
1,213	550	8.75	3.97	17.53	15.25	9.09	386	22	16	42	17
1,323	600	9.35	4.24	18.71	16.28	9.70	406	24	17	46	18
1,433	650	9.94	4.51	19.86	17.29	10.30	428	26	19	49	20
1,544	700	10.50	4.76	21.00	18.28	10.89	449	28	20	53	21
1,654	750	11.07	5.02	22.12	19.25	11.47	468	30	21	57	23
1,764	800	11.60	5.26	23.21	20.20	12.03	486	32	23	61	24
Maintenance Plus Last 2 Months of Gestation of Mature Dry Cows[3]											
882	400	9.15	4.15	18.23	15.26	9.30	875	26	16	30	12
992	450	9.99	4.53	19.91	16.66	10.16	928	30	18	34	14
1,103	500	10.80	4.90	21.55	18.04	11.00	978	33	20	38	15
1,213	550	11.62	5.27	23.14	19.37	11.81	1,027	36	22	42	17
1,323	600	12.39	5.62	24.71	20.68	12.61	1,074	39	24	46	18
1,433	650	13.16	5.97	26.23	21.96	13.39	1,120	43	26	49	20
1,544	700	13.91	6.31	27.73	23.21	14.15	1,165	46	28	53	21
1,654	750	14.66	6.65	29.21	24.44	14.90	1,209	49	30	57	23
1,764	800	15.39	6.98	30.65	25.66	15.64	1,254	53	32	61	24

Fat	Energy							Total Crude Protein		Minerals				Vitamins	
	TDN		DE		ME		NE$_{lc}$			Ca		P		A	D
(%)	(lb)	(kg)	(Mcal/lb)	(Mcal/kg)	(Mcal/lb)	(Mcal/kg)	(Mcal/lb)	(Mcal/kg)	(g/lb)	(g/kg)	(g/lb)	(g/kg)	(g/lb)	(g/kg)		
Milk Production—Nutrients/2.2 lb or/kg of Milk of Different Fat Percentages															
3.0	0.616	0.280	0.56	1.23	0.49	1.07	0.29	0.64	35	78	1.24	2.73	0.76	1.68	—	—
3.5	0.662	0.301	0.60	1.33	0.52	1.15	0.31	0.69	38	84	1.35	2.97	0.83	1.83	—	—
4.0	0.708	0.322	0.64	1.42	0.56	1.24	0.34	0.74	41	90	1.46	3.21	0.90	1.98	—	—
4.5	0.755	0.343	0.69	1.51	0.60	1.32	0.35	0.78	44	96	1.57	3.45	0.98	2.13	—	—
5.0	0.801	0.364	0.73	1.61	0.64	1.40	0.38	0.83	46	101	1.68	3.69	1.04	2.28	—	—
5.5	0.847	0.385	0.77	1.70	0.69	1.48	0.40	0.88	49	107	1.78	3.93	1.10	2.43	—	—
Live Weight Change During Lactation—Nutrients/kg of Weight Change[4]															
Weight loss	-0.99	-2.17	-4.34	-9.55	-3.75	-8.25	-2.25	-4.92	-145	-320	—	—	—	—	—	—
Weight gain	1.03	2.26	4.52	9.96	3.88	8.55	2.32	5.12	145	320	—	—	—	—	—	—

[1] Adapted by the authors from *Nutrient Requirements of Dairy Cattle*, 6th rev. ed., update 1989, NRC, National Academy Press, p. 84, Table 6-3.
[2] To allow for growth of young lactating cows, increase the maintenance allowances for all nutrients except vitamins A and D by 20% during the first lactation and 10% during the second lactation.
[3] Values for calcium assume that the cow is in calcium balance at the beginning of the last 2 months of gestation. If the cow is not in balance, then the calcium requirement can be increased from 25 to 33%.
[4] No allowance is made for mobilized calcium and phosphorus associated with live weight loss or with live weight gain. The maximum daily nitrogen available from weight loss is assumed to be 30 g or 234 g of crude protein.

TABLE 20-4
DAILY NUTRIENT REQUIREMENTS OF LACTATING COWS USING ABSORBABLE PROTEIN[1] (See footnotes at end of table.)

Live Weight		Fat	Milk		Live Weight Change		Dry Matter Intake		Energy				Protein		Minerals	
									TDN		NE$_{lc}$	NE$_{lcdm}$[2]	DIP[3]	UIP[4]	Ca	P
(lb)	(kg)	(%)	(lb)	(kg)	(lb)	(kg)	(lb)	(kg)	(lb)	(kg)	(Mcal)	(Mcal/lb)	(Mcal/kg)	(g)	(g)	(g)	(g)
Intake at 100% of the Requirement for Maintenance, Lactation, and Weight Gain																
882	400	4.5	17.6	8.0	0.485	0.220	22.36	10.14	14.20	6.44	14.55	0.65	1.43	753	511	44	28
882	400	4.5	30.9	14.0	0.485	0.220	27.92	12.66	18.70	8.48	19.26	0.69	1.52	1,052	710	65	41
882	400	4.5	44.1	20.0	0.485	0.220	32.88	14.91	23.17	10.51	23.96	0.73	1.61	1,355	880	85	54
882	400	4.5	57.3	26.0	0.485	0.220	37.35	16.94	27.65	12.54	28.67	0.77	1.69	1,662	1,026	106	67
882	400	4.5	70.6	32.0	0.485	0.220	42.80	19.41	32.15	14.58	33.37	0.78	1.72	1,962	1,220	127	80
882	400	5.0	17.6	8.0	0.485	0.220	22.84	10.36	14.55	6.60	14.94	0.65	1.44	778	525	46	30
882	400	5.0	30.9	14.0	0.485	0.220	28.67	13.00	19.34	8.77	19.93	0.69	1.53	1,096	730	68	43
882	400	5.0	44.1	20.0	0.485	0.220	33.85	15.35	24.10	10.93	24.93	0.74	1.62	1,419	902	90	57
882	400	5.0	57.3	26.0	0.485	0.220	38.46	17.44	28.82	13.07	29.92	0.78	1.72	1,745	1,048	112	71
882	400	5.0	70.6	32.0	0.485	0.220	44.76	20.30	33.63	15.25	34.91	0.78	1.72	2,061	1,277	134	84
882	400	5.5	17.6	8.0	0.485	0.220	23.31	10.57	14.93	6.77	15.32	0.66	1.45	803	538	48	31
882	400	5.5	30.9	14.0	0.485	0.220	29.39	13.33	20.00	9.07	20.61	0.70	1.55	1,140	748	71	45

(Continued)

TABLE 20-4 (Continued)

Live Weight		Fat	Milk		Live Weight Change		Dry Matter Intake		Energy			Protein		Minerals	
									TDN		NE_{lc}	NE_{lcdm}[2]	DIP[3]	UIP[4]	Ca	P
(lb)	(kg)	(%)	(lb)	(kg)	(lb)	(kg)	(lb)	(kg)	(lb)	(kg)	(Mcal)	(Mcal/lb)	(Mcal/kg)	(g)	(g)	(g)	(g)

Intake at 100% of the Requirement for Maintenance, Lactation, and Weight Gain (Continued)

Live Weight (lb)	(kg)	Fat (%)	Milk (lb)	(kg)	LWC (lb)	(kg)	DMI (lb)	(kg)	TDN (lb)	(kg)	NE_{lc} (Mcal)	(Mcal/lb)	(Mcal/kg)	DIP (g)	UIP (g)	Ca (g)	P (g)
882	400	5.5	44.1	20.0	0.485	0.220	34.77	15.77	25.00	11.34	25.89	0.74	1.64	1,483	923	95	60
882	400	5.5	57.3	26.0	0.485	0.220	39.98	18.13	30.03	13.62	31.17	0.78	1.72	1,826	1,091	118	75
882	400	5.5	70.6	32.0	0.485	0.220	46.75	21.20	35.10	15.92	36.45	0.78	1.72	2,160	1,334	142	89
1,103	500	4.0	19.8	9.0	0.606	0.275	25.56	11.59	16.10	7.30	16.49	0.64	1.42	883	540	49	32
1,103	500	4.0	37.5	17.0	0.606	0.275	32.59	14.78	21.74	9.86	22.38	0.69	1.51	1,257	797	75	48
1,103	500	4.0	55.1	25.0	0.606	0.275	38.85	17.62	27.34	12.40	28.27	0.73	1.61	1,635	1,015	101	64
1,103	500	4.0	72.8	33.0	0.606	0.275	44.41	20.14	32.92	14.93	34.15	0.77	1.70	2,018	1,201	126	80
1,103	500	4.0	90.4	41.0	0.606	0.275	51.35	23.29	38.57	17.49	40.04	0.78	1.72	2,392	1,453	152	95
1,103	500	4.5	19.8	9.0	0.606	0.275	26.11	11.84	16.52	7.49	16.92	0.65	1.43	911	556	51	33
1,103	500	4.5	37.5	17.0	0.606	0.275	33.52	15.20	22.51	10.21	23.20	0.69	1.53	1,310	821	79	50
1,103	500	4.5	55.1	25.0	0.606	0.275	40.04	18.16	28.49	12.92	29.47	0.74	1.62	1,715	1,043	107	68
1,103	500	4.5	72.8	33.0	0.606	0.275	45.84	20.79	34.42	15.61	35.74	0.78	1.72	2,124	1,230	134	85
1,103	500	4.5	90.4	41.0	0.606	0.275	53.89	24.44	40.46	18.35	42.02	0.78	1.72	2,519	1,526	162	102
1,103	500	5.0	19.8	9.0	0.606	0.275	26.64	12.08	16.93	7.68	17.36	0.65	1.44	939	571	53	35
1,103	500	5.0	37.5	17.0	0.606	0.275	34.40	15.60	23.31	10.57	24.01	0.70	1.54	1,364	844	83	53
1,103	500	5.0	55.1	25.0	0.606	0.275	41.19	18.68	29.64	13.44	30.67	0.74	1.64	1,795	1,069	113	71
1,103	500	5.0	72.8	33.0	0.606	0.275	47.87	21.71	35.96	16.31	37.33	0.78	1.72	2,226	1,289	142	89
1,103	500	5.0	90.4	41.0	0.606	0.275	56.40	25.58	42.36	19.21	43.99	0.78	1.72	2,646	1,599	172	108
1,323	600	3.0	22.1	10.0	0.728	0.330	27.61	12.52	17.35	7.87	17.79	0.64	1.42	974	533	52	34
1,323	600	3.0	44.1	20.0	0.728	0.330	35.72	16.20	23.53	10.67	24.18	0.68	1.49	1,375	845	79	51
1,323	600	3.0	66.2	30.0	0.728	0.330	42.71	19.37	29.61	13.43	30.58	0.72	1.58	1,784	1,102	106	68
1,323	600	3.0	88.2	40.0	0.728	0.330	48.97	22.21	35.70	16.19	36.98	0.76	1.67	2,198	1,323	133	84
1,323	600	3.0	110.3	50.0	0.728	0.330	55.63	25.23	41.78	18.95	43.38	0.78	1.72	2,608	1,565	161	101
1,323	600	3.5	22.1	10.0	0.728	0.330	28.36	12.86	17.82	8.08	18.27	0.64	1.42	1,004	557	54	35
1,323	600	3.5	44.1	20.0	0.728	0.330	36.82	16.70	24.43	11.08	25.15	0.69	1.51	1,438	874	84	54
1,323	600	3.5	66.2	30.0	0.728	0.330	44.19	20.04	31.00	14.06	32.03	0.73	1.60	1,879	1,137	113	72
1,323	600	3.5	88.2	40.0	0.728	0.330	50.72	23.00	37.51	17.01	38.90	0.77	1.69	2,326	1,360	143	90
1,323	600	3.5	110.3	50.0	0.728	0.330	58.72	26.63	44.10	20.00	45.78	0.78	1.72	2,763	1,654	173	109
1,323	600	4.0	22.1	10.0	0.728	0.330	29.11	13.20	18.30	8.30	18.75	0.64	1.42	1,034	581	56	37
1,323	600	4.0	44.1	20.0	0.728	0.330	37.90	17.19	25.36	11.50	26.11	0.69	1.52	1,501	902	89	57
1,323	600	4.0	66.2	30.0	0.728	0.330	45.62	20.69	32.37	14.68	33.47	0.74	1.62	1,975	1,170	121	77
1,323	600	4.0	88.2	40.0	0.728	0.330	52.43	23.78	39.34	17.84	40.83	0.78	1.72	2,454	1,395	153	96
1,323	600	4.0	110.3	50.0	0.728	0.330	61.81	28.03	46.42	21.05	48.19	0.78	1.72	2,918	1,744	185	116
1,544	700	3.0	26.5	12.0	0.849	0.385	31.88	14.46	20.04	9.09	20.54	0.64	1.42	1,154	607	61	40
1,544	700	3.0	52.9	24.0	0.849	0.385	41.34	18.75	27.43	12.44	28.21	0.68	1.50	1,638	968	94	60
1,544	700	3.0	79.4	36.0	0.849	0.385	49.57	22.48	34.75	15.76	35.89	0.73	1.60	2,129	1,269	127	81
1,544	700	3.0	105.8	48.0	0.849	0.385	56.89	25.80	42.01	19.05	43.57	0.77	1.69	2,627	1,525	159	101
1,544	700	3.0	132.3	60.0	0.849	0.385	65.73	29.81	49.37	22.39	51.25	0.78	1.72	3,114	1,857	192	121
1,544	700	3.5	26.5	12.0	0.849	0.385	32.77	14.86	20.59	9.34	21.11	0.64	1.42	1,190	636	64	42
1,544	700	3.5	52.9	24.0	0.849	0.385	42.64	19.34	28.53	12.94	29.37	0.69	1.52	1,713	1,002	100	64
1,544	700	3.5	79.4	36.0	0.849	0.385	51.29	23.26	36.38	16.50	37.62	0.74	1.62	2,244	1,309	135	86
1,544	700	3.5	105.8	48.0	0.849	0.385	58.92	26.72	44.19	20.04	45.88	0.78	1.72	2,781	1,567	171	108
1,544	700	3.5	132.3	60.0	0.849	0.385	69.41	31.48	52.15	23.65	54.13	0.78	1.72	3,300	1,964	207	130
1,544	700	4.0	26.5	12.0	0.849	0.385	33.52	15.20	21.17	9.60	21.69	0.65	1.43	1,227	658	67	44
1,544	700	4.0	52.9	24.0	0.849	0.385	43.92	19.92	29.64	13.44	30.52	0.69	1.53	1,789	1,035	105	68
1,544	700	4.0	79.4	36.0	0.849	0.385	52.96	24.02	38.04	17.25	39.35	0.74	1.64	2,359	1,347	144	91
1,544	700	4.0	105.8	48.0	0.849	0.385	61.81	28.03	46.42	21.05	48.19	0.78	1.72	2,930	1,648	182	115
1,544	700	4.0	132.3	60.0	0.849	0.385	73.12	33.16	54.93	24.91	57.02	0.78	1.72	3,485	2,071	221	139
1,764	800	3.0	30.9	14.0	0.970	0.440	36.07	16.36	22.69	10.29	23.24	0.64	1.42	1,331	682	71	46
1,764	800	3.0	59.5	27.0	0.970	0.440	46.15	20.93	30.67	13.91	31.56	0.69	1.51	1,857	1,064	106	68
1,764	800	3.0	88.2	40.0	0.970	0.440	55.01	24.95	38.59	17.50	39.88	0.73	1.60	2,390	1,388	142	90
1,764	800	3.0	116.9	53.0	0.970	0.440	62.93	28.54	46.48	21.08	48.20	0.77	1.69	2,928	1,665	177	112
1,764	800	3.0	145.5	66.0	0.970	0.440	72.48	32.87	54.44	24.69	56.51	0.78	1.72	3,457	2,022	213	134
1,764	800	3.5	30.9	14.0	0.970	0.440	37.00	16.78	23.33	10.58	23.92	0.64	1.42	1,374	710	74	49
1,764	800	3.5	59.5	27.0	0.970	0.440	47.61	21.59	31.91	14.47	32.86	0.69	1.52	1,942	1,102	113	72
1,764	800	3.5	88.2	40.0	0.970	0.440	56.93	25.82	40.42	18.33	41.80	0.74	1.62	2,517	1,432	151	96
1,764	800	3.5	116.9	53.0	0.970	0.440	65.20	29.57	48.88	22.17	50.75	0.78	1.72	3,099	1,711	190	120
1,764	800	3.5	145.5	66.0	0.970	0.440	76.56	34.72	57.48	26.07	59.69	0.78	1.72	3,661	2,140	228	144
1,764	800	4.0	30.9	14.0	0.970	0.440	37.86	17.17	23.99	10.88	24.59	0.65	1.43	1,418	734	77	51

(Continued)

TABLE 20-4 (Continued)

Live Weight		Fat	Milk		Live Weight Change		Dry Matter Intake		Energy				Protein		Minerals	
									TDN		NE$_{lc}$	NE$_{lcdm}$[2]	DIP[3]	UIP[4]	Ca	P
(lb)	(kg)	(%)	(lb)	(kg)	(lb)	(kg)	(lb)	(kg)	(lb)	(kg)	(Mcal)	(Mcal/lb)	(Mcal/kg)	(g)	(g)	(g)	(g)
					Intake at 100% of the Requirement for Maintenance, Lactation, and Weight Gain (Continued)												
1,764	800	4.0	59.5	27.0	0.970	0.440	49.04	22.24	33.14	15.03	34.16	0.70	1.54	2,027	1,139	119	76
1,764	800	4.0	88.2	40.0	0.970	0.440	58.78	26.66	42.25	19.16	43.73	0.74	1.64	2,644	1,474	161	102
1,764	800	4.0	116.9	53.0	0.970	0.440	68.36	31.00	51.33	23.28	53.29	0.78	1.72	3,263	1,800	203	128
1,764	800	4.0	145.5	66.0	0.970	0.440	80.61	36.56	60.55	27.46	62.86	0.78	1.72	3,865	2,259	244	154
					Intake at 85% of the Requirement for Maintenance and Lactation												
882	400	4.5	44.1	20.0	-1.535	-0.696	25.62	11.62	18.72	8.49	19.41	0.76	1.67	1,066	687	85	54
882	400	4.5	57.3	26.0	-1.852	-0.840	30.91	14.02	22.58	10.24	23.41	0.76	1.67	1,310	931	106	67
882	400	4.5	70.6	32.0	-2.168	-0.983	36.18	16.41	26.44	11.99	27.41	0.76	1.67	1,554	1,187	127	80
882	400	5.0	44.1	20.0	-1.601	-0.726	26.70	12.11	19.51	8.85	20.23	0.76	1.67	1,118	720	90	57
882	400	5.0	57.3	26.0	-1.936	-0.878	32.30	14.65	23.62	10.71	24.47	0.76	1.67	1,377	987	112	71
882	400	5.0	70.6	32.0	-2.271	-1.030	37.93	17.20	27.69	12.56	28.72	0.76	1.67	1,635	1,255	134	84
882	400	5.5	44.1	20.0	-1.665	-0.755	27.78	12.60	20.31	9.21	21.05	0.76	1.67	1,169	761	95	60
882	400	5.5	57.3	26.0	-2.020	-0.916	33.71	15.29	24.63	11.17	25.54	0.76	1.67	1,443	1,042	118	75
882	400	5.5	70.6	32.0	-2.375	-1.077	39.65	17.98	28.97	13.14	30.03	0.76	1.67	1,717	1,323	142	89
1,103	500	4.0	55.1	25.0	-1.806	-0.819	30.14	13.67	22.03	9.99	22.83	0.76	1.67	1,286	810	101	64
1,103	500	4.0	72.8	33.0	-2.201	-0.998	36.76	16.67	26.86	12.18	27.83	0.76	1.67	1,590	1,134	126	80
1,103	500	4.0	90.4	41.0	-2.597	-1.178	43.35	19.66	31.69	14.37	32.84	0.76	1.67	1,894	1,458	152	95
1,103	500	4.5	55.1	25.0	-1.887	-0.856	31.49	14.28	23.02	10.44	23.85	0.76	1.67	1,350	864	107	68
1,103	500	4.5	72.8	33.0	-2.309	-1.047	38.54	17.48	28.16	12.77	29.18	0.76	1.67	1,674	1,205	134	85
1,103	500	4.5	90.4	41.0	-2.730	-1.238	45.58	20.67	33.30	15.10	34.52	0.76	1.67	1,998	1,546	162	102
1,103	500	5.0	55.1	25.0	-1.967	-0.892	32.83	14.89	23.99	10.88	24.87	0.76	1.67	1,414	917	113	71
1,103	500	5.0	72.8	33.0	-2.414	-1.095	40.31	18.28	29.46	13.36	30.53	0.76	1.67	1,758	1,275	142	89
1,103	500	5.0	90.4	41.0	-2.862	-1.298	47.78	21.67	34.91	15.83	36.19	0.76	1.67	2,103	1,633	172	108
1,323	600	3.0	66.2	30.0	-1.943	-0.881	32.44	14.71	23.68	10.74	24.56	0.76	1.67	1,399	860	106	68
1,323	600	3.0	88.2	40.0	-2.373	-1.076	39.60	17.96	28.93	13.12	30.00	0.76	1.67	1,728	1,223	133	84
1,323	600	3.0	110.3	50.0	-2.803	-1.271	46.79	21.22	34.18	15.50	35.44	0.76	1.67	2,057	1,585	161	101
1,323	600	3.5	66.2	30.0	-2.396	-0.925	34.05	15.44	24.87	11.28	25.79	0.76	1.67	1,476	924	113	72
1,323	600	3.5	88.2	40.0	-2.503	-1.135	41.76	18.94	30.52	13.84	31.63	0.76	1.67	1,830	1,308	143	90
1,323	600	3.5	110.3	50.0	-2.964	-1.344	49.48	22.44	36.16	16.40	37.48	0.76	1.67	2,184	1,692	173	109
1,323	600	4.0	66.2	30.0	-2.137	-0.969	35.65	16.17	26.06	11.82	27.01	0.76	1.67	1,552	988	121	77
1,323	600	4.0	88.2	40.0	-2.631	-1.193	43.92	19.92	32.08	14.55	33.27	0.76	1.67	1,932	1,393	153	96
1,323	600	4.0	110.3	50.0	-3.127	-1.418	52.19	23.67	38.12	17.29	39.52	0.76	1.67	2,311	1,798	185	116
1,544	700	3.0	79.4	36.0	-2.280	-1.034	38.06	17.26	27.81	12.61	28.83	0.76	1.67	1,669	1,054	127	81
1,544	700	3.0	105.8	48.0	-2.796	-1.268	46.68	21.17	34.11	15.47	35.36	0.76	1.67	2,064	1,489	159	101
1,544	700	3.0	132.3	60.0	-3.312	-1.502	55.30	25.08	40.40	18.32	41.88	0.76	1.67	2,458	1,924	192	121
1,544	700	3.5	79.4	36.0	-2.397	-1.087	40.02	18.15	29.24	13.26	30.30	0.76	1.67	1,761	1,131	135	86
1,544	700	3.5	105.8	48.0	-2.952	-1.339	49.28	22.35	36.01	16.33	37.32	0.76	1.67	2,186	1,591	171	108
1,544	700	3.5	132.3	60.0	-3.506	-1.590	58.54	26.55	42.78	19.40	44.34	0.76	1.67	2,611	2,052	207	130
1,544	700	4.0	79.4	36.0	-2.434	-1.140	41.96	19.03	30.65	13.90	31.78	0.76	1.67	1,853	1,208	144	91
1,544	700	4.0	105.8	48.0	-3.107	-1.409	51.86	23.52	37.90	17.19	39.28	0.76	1.67	2,308	1,694	182	115
1,544	700	4.0	132.3	60.0	-3.700	-1.678	61.78	28.02	45.14	20.47	46.79	0.76	1.67	2,764	2,180	221	139
1,764	800	3.0	88.2	40.0	-2.529	-1.147	42.23	19.15	30.05	13.99	31.08	0.76	1.67	1,871	1,176	142	90
1,764	800	3.0	110.3	50.0	-2.959	-1.342	49.41	22.41	36.10	16.37	37.42	0.76	1.67	2,200	1,538	169	107
1,764	800	3.0	132.3	60.0	-3.389	-1.537	56.58	25.66	41.34	18.75	42.86	0.76	1.67	2,529	1,900	196	124
1,764	800	3.5	88.2	40.0	-2.659	-1.206	44.39	20.13	32.44	14.71	33.62	0.76	1.67	1,973	1,261	151	96
1,764	800	3.5	110.3	50.0	-3.122	-1.416	52.10	23.63	38.08	17.27	39.46	0.76	1.67	2,327	1,645	181	114
1,764	800	3.5	132.3	60.0	-3.583	-1.625	59.82	27.13	43.70	19.82	45.31	0.76	1.67	2,682	2,028	211	133
1,764	800	4.0	88.2	40.0	-2.787	-1.264	46.55	21.11	34.00	15.42	35.25	0.76	1.67	2,075	1,346	161	102
1,764	800	4.0	110.3	50.0	-3.283	-1.489	54.82	24.86	40.04	18.16	41.51	0.76	1.67	2,455	1,751	193	122
1,764	800	4.0	132.3	60.0	-3.777	-1.713	63.06	28.60	46.08	20.90	47.76	0.76	1.67	2,835	2,156	225	142

[1]Adapted by the authors from *Nutrient Requirements of Dairy Cattle*, 6th rev. ed., update 1989, NRC, National Academy Press, pp. 85-86, Table 6-4.

[2]NE$_{lcdm}$ = net energy for lactation/kg of dry matter.

[3]DIP = degraded intake protein.

[4]UIP = undegraded intake protein.

TABLE 20-5
RECOMMENDED NUTRIENT CONTENT OF RATIONS FOR DAIRY CATTLE[1] (See footnotes at end of table.)

Cow Weight		Fat	Weight Gain			Lactating Cow Rations										Early Lactation (Weeks 0-3)	Dry, Pregnant Cows	Calf Milk Replacer	Calf Starter Mix	Growing Heifers and Bulls[2]			Mature Bulls	Maximum Tolerable Levels[3,4]
						Milk Yield		Milk Yield		Milk Yield		Milk Yield		Milk Yield						3-6 Months	6-12 Months	Over 12 Mos		
(lb)	(kg)	(%)	(lb/day)	(kg/day)		(lb/day)	(kg/day)	(lb/day)	(kg/day)	(lb/day)	(kg/day)	(lb/day)	(kg/day)	(lb/day)	(kg/day)									
882	400	5.0	0.485	0.220		15.4	7	28.7	13	44.1	20	57.3	26	72.8	33									
1,103	500	4.5	0.606	0.275		17.6	8	37.5	17	55.1	25	72.8	33	90.4	41									
1,323	600	4.0	0.728	0.330		22.1	10	44.1	20	66.2	30	88.2	40	110.3	50									
1,544	700	3.5	0.849	0.385		26.5	12	52.9	24	79.4	36	105.8	48	132.3	60									
1,764	800	3.5	0.970	0.440		28.7	13	59.5	27	88.2	40	116.9	53	147.7	67									

		Lactating Cow Rations									Early Lactation	Dry, Pregnant Cows	Calf Milk Replacer	Calf Starter Mix	3-6 Months	6-12 Months	Over 12 Mos	Mature Bulls	Max Tolerable
Energy:																			
NE_lc	(Mcal/lb)	0.64		0.69		0.74		0.78		0.78	0.76	0.57	—	—	—	—	—	—	—
NE_lc	(Mcal/kg)	1.42		1.52		1.62		1.72		1.72	1.67	1.25	—	—	—	—	—	—	—
NE_m	(Mcal/lb)	—		—		—		—		—	—	—	1.09	0.86	0.77	0.72	0.64	0.52	—
NE_m	(Mcal/kg)	—		—		—		—		—	—	—	2.40	1.90	1.70	1.58	1.40	1.15	—
NE_g	(Mcal/lb)	—		—		—		—		—	—	—	0.70	0.54	0.49	0.44	0.37	—	—
NE_g	(Mcal/kg)	—		—		—		—		—	—	—	1.55	1.20	1.08	0.98	0.82	—	—
ME	(Mcal/lb)	1.07		1.15		1.23		1.31		1.31	1.27	0.93	1.72	1.41	1.18	1.12	1.03	0.91	—
ME	(Mcal/kg)	2.35		2.53		2.71		2.89		2.89	2.80	2.04	3.78	3.11	2.60	2.47	2.27	2.00	—
DE	(Mcal/lb)	1.26		1.34		1.42		1.50		1.50	1.46	1.12	1.90	1.60	1.37	1.31	1.22	1.10	—
DE	(Mcal/kg)	2.77		2.95		3.13		3.31		3.31	3.22	2.47	4.19	3.53	3.02	2.89	2.69	2.43	—
TDN	(% of DM)	63		67		71		75		75	73	56	95	80	69	66	61	55	—
Protein equivalent:																			
Crude protein	(%)	12		15		16		17		18	19	12	22	18	16	12	12	10	—
DIP[5]	(%)	7.8		8.7		9.6		10.3		10.4	9.7	—	—	—	4.6	6.4	7.2	—	—
UIP[6]	(%)	4.4		5.2		5.7		5.9		6.2	7.0	—	—	—	8.2	4.4	2.1	—	—
Fiber content (minimum):[7]																			
Crude fiber	(%)	17		17		17		15		15	17	22	—	—	13	15	15	15	—
Neutral detergent fiber (NDF)	(%)	28		28		28		25		25	28	35	—	—	23	25	25	25	—
Acid detergent fiber (ADF)	(%)	21		21		21		19		19	21	27	—	—	16	19	19	19	—
Ether extract (minimum)	(%)	3		3		3		3		3	3	3	10	3	3	3	3	3	—
Major or Macrominerals:																			
Calcium (Ca)	(%)	0.43		0.51		0.58		0.64		0.66	0.77	0.39[8]	0.70	0.60	0.52	0.41	0.29	0.30	2.00
Chlorine (Cl)	(%)	0.25		0.25		0.25		0.25		0.25	0.25	0.20	0.20	0.20	0.20	0.20	0.20	0.20	—
Magnesium (Mg)[9]	(%)	0.20		0.20		0.20		0.25		0.25	0.25	0.16	0.07	0.10	0.16	0.16	0.16	0.16	0.50
Phosphorus (P)	(%)	0.28		0.33		0.37		0.41		0.41	0.48	0.24	0.60	0.40	0.31	0.30	0.23	0.19	1.00
Potassium (K)[10]	(%)	0.90		0.90		0.90		1.00		1.00	1.00	0.65	0.65	0.65	0.65	0.65	0.65	0.65	3.00
Sodium (Na)	(%)	0.18		0.18		0.18		0.18		0.18	0.18	0.10	0.10	0.10	0.10	0.10	0.10	0.10	—
Sulfur (S)	(%)	0.20		0.20		0.20		0.20		0.20	0.25	0.16	0.29	0.20	0.16	0.16	0.16	0.16	0.40
Trace or Microminerals:																			
Cobalt (Co)	(ppm)	0.10		0.10		0.10		0.10		0.10	0.10	0.10	0.10	0.10	0.10	0.10	0.10	0.10	10.00
Copper (Cu)[11]	(ppm)	10		10		10		10		10	10	10	10	10	10	10	10	10	100
Iodine (I)[12]	(ppm)	0.60		0.60		0.60		0.60		0.60	0.60	0.25	0.25	0.25	0.25	0.25	0.25	0.25	50.00[13]
Iron (Fe)	(ppm)	50		50		50		50		50	50	50	100	50	50	50	50	50	1,000
Manganese (Mn)	(ppm)	40		40		40		40		40	40	40	40	40	40	40	40	40	1,000
Selenium (Se)	(ppm)	0.30		0.30		0.30		0.30		0.30	0.30	0.30	0.30	0.30	0.30	0.30	0.30	0.30	2.00
Zinc (Zn)	(ppm)	40		40		40		40		40	40	40	40	40	40	40	40	40	500

(Continued)

Feeding Dairy Cattle

TABLE 20-5 (Continued)

Cow Weight		Fat	Weight Gain		Lactating Cow Rations									
					Milk Yield		Milk Yield		Milk Yield		Milk Yield		Milk Yield	
(lb)	(kg)	(%)	(lb/day)	(kg/day)	(lb/day)	(kg/day)	(lb/day)	(kg/day)	(lb/day)	(kg/day)	(lb/day)	(kg/day)	(lb/day)	(kg/day)
882	400	5.0	0.485	0.220	15.4	7	28.7	13	44.1	20	57.3	26	72.8	33
1,103	500	4.5	0.606	0.275	17.6	8	37.5	17	55.1	25	72.8	33	90.4	41
1,323	600	4.0	0.728	0.330	22.1	10	44.1	20	66.2	30	88.2	40	110.3	50
1,544	700	3.5	0.849	0.385	26.5	12	52.9	24	79.4	36	105.8	48	132.3	60
1,764	800	3.5	0.970	0.440	28.7	13	59.5	27	88.2	40	116.9	53	147.7	67

Vitamins:[14]					Early Lactation (Weeks 0-3)	Dry, Pregnant Cows	Calf Milk Replacer	Calf Starter Mix	Growing Heifers and Bulls[2]			Mature Bulls	Maximum Tolerable Levels[3,4]
									3-6 Months	6-12 Months	Over 12 Mos		
Vitamin A				(IU/lb)	1,816	1,816	1,725	999	999	999	999	1,453	29,964
Vitamin A				(IU/kg)	4,000	4,000	3,800	2,200	2,200	2,200	2,200	3,200	66,000
Vitamin D				(IU/lb)	454	545	272	136	136	136	136	136	4,540
Vitamin D				(IU/kg)	1,000	1,200	600	300	300	300	300	300	10,000
Vitamin E				(IU/lb)	7	7	18	11	11	11	11	7	908
Vitamin E				(IU/kg)	15	15	40	25	25	25	25	15	2,000

[1]Adapted by the authors from *Nutrient Requirements of Dairy Cattle*, 6th rev. ed., update 1989, NRC, National Academy Press, p. 87, Table 6-5.

[2]The approximate weight for growing heifers and bulls at 3-6 months is 331 lb (150 kg); at 6-12 months, it is 551 lb (250 kg); and at more than 12 months, it is 882 lb (400 kg). The approximate average daily gain is 25 oz/day (700 g/day).

[3]The maximum safe levels for many of the mineral elements are not well defined and may be substantially affected by specific feeding conditions. Additional information is available in Table 20-6 and in *Mineral Tolerance of Domestic Animals* (NRC, 1980).

[4]Vitamin tolerances are discussed in detail in *Vitamin Tolerance of Animals* (NRC, 1987b).

[5]DIP = degraded intake protein.

[6]UIP = undegraded intake protein.

[7]It is recommended that 75% of the NDF in lactating cow rations be provided as forage. If this recommendation is not followed, a depression in milk fat may occur.

[8]The value for calcium assumes that the cow is in calcium balance at the beginning of the dry period. If the cow is not in balance, then the dietary calcium requirement should be increased by 25 to 32%.

[9]Under conditions conducive to grass tetany, magnesium should be increased to 0.25 or 0.30%.

[10]Under conditions of heat stress, potassium should be increased to 1.2%.

[11]The cow's copper requirement is influenced by molybdenum and sulfur in the ration.

[12]If the ration contains as much as 25% strongly goitrogenic feed on a dry basis the iodine provided should be increased 2 times or more.

[13]Although cattle can tolerate this level of iodine, lower levels may be desirable to reduce the iodine content of milk.

[14]The following minimum quantities of B-complex vitamins are suggested per unit of milk replacer: Niacin (Nicotinic Acid, Nicotinamide), 2.6 ppm; Pantothenic Acid (Vitamin B-1), 13 ppm; Riboflavin (Vitamin B-2), 6.5 ppm; Vitamin B-6 (Pyridoxine, Pyridoxal, Pyridoxamine), 6.5 ppm; Folacin (Folic Acid), 0.5 ppm; Biotin, 0.1 ppm; Vitamin B-12 (Cobalamins), 0.07 ppm; and Choline, 0.26%. It appears that adequate amounts of these vitamins are furnished w/en calves have functional rumens (usually at 6 weeks of age) by a combination of rumen synthesis and natural feedstuffs.

TABLE 20-6
MAXIMUM TOLERABLE DIETARY LEVELS OF CERTAIN ELEMENTS[1]

Element	Maximum Tolerable Level
	(ppm)
Aluminum	1,000[2]
Arsenic:	
Inorganic	50
Organic	100
Bromine	200
Cadmium	0.5[3]
Fluorine	40[4]
Lead	30[3]
Mercury	2[3]
Molybdenum	10[5]
Nickel	50
Vanadium	50

[1]Adapted by the authors from *Nutrient Requirements of Dairy Cattle*, 6th rev. ed., update 1989, NRC, National Academy Press, p. 88, Table 6-6.

[2]As soluble salts of high bioavailability. Higher levels of less soluble forms found in natural substances can be tolerated.

[3]Levels are based on human food residue considerations.

[4]As sodium fluoride or fluorides of similar toxicity. The maximum safe level of fluorine for growing heifers and bulls is lower than for other dairy cattle. Somewhat higher levels are tolerated when fluorine is from less available sources such as phosphates. Morphological lesions in cattle teeth may be seen when dietary fluoride for the young exceeds 20 ppm, but a relationship between the lesions caused by fluoride levels below the maximum tolerable levels and animal performance has not been established.

[5]Toxicity related to the dietary level of copper.

TABLE
COMPOSITION OF FEEDS COMMONLY USED IN DAIRY CATTLE RATIONS

Entry No.	Feed Name Description	International Feed Number [2]	Dry Matter (%)	TDN (%)	DE (Mcal/kg)	ME (Mcal/kg)	Production Growing Dairy Cattle NEM (Mcal/kg)	NEG (Mcal/kg)	Lactating Cows NEL (Mcal/kg)	Production Growing Dairy Cattle NEM (Mcal/lb)	NEG (Mcal/lb)	Lactating Cows NEL (Mcal/lb)	Crude Protein (%)
	ALFALFA *Medicago sativa*												
1	—hay, sun-cured, early vegetative	1-00-050	90	66	2.91	2.49	1.51	0.92	1.50	0.69	0.42	0.68	23.0
2	—hay, sun-cured, late vegetative	1-00-054	90	63	2.78	2.36	1.41	0.83	1.42	0.64	0.38	0.65	20.0
3	—hay, sun-cured, early bloom	1-00-059	90	60	2.65	2.22	1.31	0.74	1.35	0.60	0.34	0.61	18.0
4	—hay, sun-cured, midbloom	1-00-063	90	58	2.56	2.13	1.24	0.68	1.30	0.57	0.31	0.59	17.0
5	—hay, sun-cured, full bloom	1-00-068	90	55	2.43	2.00	1.14	0.58	1.23	0.52	0.26	0.56	15.0
6	—silage, wilted, 25-45% dry matter (see similar maturity descriptions of hays)	—	—	—	—	—	—	—	—	—	—	—	—
	ALMOND *Prunus amygdalus*												
7	—hulls	4-00-	90	59	2.60	2.18	1.27	0.70	1.33	0.58	0.32	0.60	2.7
	BARLEY *Hordeum vulgare*												
8	—grain	4-00-549	88	84	3.70	3.29	2.06	1.40	1.94	0.94	0.64	0.88	13.5
9	—grain, Pacific Coast	4-07-939	89	86	3.79	3.38	2.12	1.45	1.99	0.96	0.66	0.90	10.8
	BEET, SUGAR *Beta vulgaris altissima*												
10	—pulp, dehydrated	4-00-669	91	78	3.44	3.02	1.88	1.24	1.79	0.86	0.57	0.81	9.7
11	—pulp w/molasses, dehydrated	4-00-672	92	78	3.44	3.02	1.88	1.24	1.79	0.86	0.57	0.81	10.1
	BLOOD												
12	—meal	5-00-380	92	66	2.91	2.49	1.51	0.92	1.50	0.69	0.42	0.68	87.2
	BREWERS' GRAINS												
13	—dehydrated	5-02-141	92	66	2.91	2.49	1.51	0.91	1.50	0.69	0.41	0.68	25.4
	BROME *Bromus* spp												
14	—fresh, early vegetative	2-00-892	34	74	3.26	2.85	1.75	1.13	1.69	0.80	0.51	0.77	18.0
	CASEIN												
15	—dehydrated (cattle)	5-01-162	91	89	3.92	3.51	2.20	1.52	2.06	1.00	0.69	0.94	92.7
	CITRUS *Citrus* spp												
16	—pulp w/o fines, dehydrated (dried citrus pulp)	4-01-237	91	77	3.40	2.98	1.86	1.22	1.77	0.85	0.55	0.80	6.7
	CORN, DENT YELLOW *Zea mays indentata*												
17	—distillers' grains, dehydrated	5-28-235	94	86	3.79	3.38	2.12	1.45	1.99	0.96	0.66	0.90	23.0
18	—ears, ground (corn and cob meal)	4-28-238	87	83	3.66	3.25	2.03	1.37	1.91	0.92	0.62	0.87	9.0
19	—gluten, meal	5-28-241	91	86	3.79	3.38	2.12	1.45	1.99	0.96	0.66	0.90	46.8
20	—gluten, meal, 60% protein	5-28-242	90	89	3.92	3.51	2.20	1.52	2.06	1.00	0.69	0.94	67.2
21	—gluten w/bran (corn gluten feed)	5-28-243	90	83	3.66	3.25	2.03	1.37	1.91	0.92	0.62	0.87	25.6
22	—grain, flaked	4-28-244	89	88	3.88	3.47	2.18	1.50	2.04	0.99	0.68	0.93	10.0
23	—grain, high-moisture	4-20-770	77	88	3.88	3.47	2.18	1.50	2.04	0.99	0.68	0.93	10.0
24	—grits by-product (hominy feed)	4-03-011	90	87	3.84	3.42	2.16	1.48	2.01	0.98	0.67	0.91	11.5
25	—silage, few ears	3-28-245	29	62	2.73	2.31	1.38	0.80	1.40	0.63	0.36	0.64	8.4
26	—silage, well-eared	3-28-250	33	70	3.09	2.67	1.63	1.03	1.60	0.74	0.47	0.73	8.1
	COTTON *Gossypium* spp												
27	—hulls	1-01-599	91	45	1.98	1.55	0.78	0.25	0.98	0.35	0.11	0.45	4.1
28	—seeds, w/lint	5-01-614	92	96	4.23	3.83	2.41	1.69	2.23	1.10	0.77	1.01	23.0
29	—seeds, w/o lint	5-01-	90	96	4.23	3.82	2.41	1.69	2.23	1.10	0.77	1.01	25.0
30	—seeds, meal prepressed, solv-extd, 41% protein	5-07-872	91	76	3.35	2.93	1.82	1.19	1.74	0.83	0.55	0.79	45.6
31	—seeds, meal prepressed, solv-extd, 44% protein	5-07-873	91	75	3.31	2.89	1.79	1.16	1.72	0.81	0.53	0.78	48.9
	FATS AND OILS (not exceeding 3% of diet)												
32	—fat, animal, hydrolyzed	4-00-376	99	177	7.30	7.30	5.84	5.84	5.84	2.65	2.65	2.65	—
	FESCUE, KENTUCKY 31 *Festuca arundinacea*												
33	—fresh, vegetative	2-01-902	29	67	2.91	2.49	1.51	0.92	1.50	0.69	0.42	0.68	14.5
	FISH, MENHADEN *Brevoortia tyrannus*												
34	—meal mech-extd	5-02-009	92	73	3.22	2.80	1.73	1.11	1.67	0.79	0.50	0.76	66.7
	FLAX *Linum usitatissimum*												
35	—seeds, meal solv-extd (linseed meal)	5-02-048	90	78	3.44	3.02	1.88	1.24	1.79	0.85	0.56	0.81	38.3
	MILK												
36	—skimmed dehydrated (cattle)	5-01-175	94	85	3.75	3.34	2.10	1.43	1.96	0.95	0.65	0.89	35.8
	MOLASSES AND SYRUP												
37	—beet, sugar, molasses, more than 48% invert sugar, more than 79.5 degrees brix	4-00-668	78	75	3.31	2.89	1.79	1.16	1.72	0.81	0.53	0.78	8.5
38	—citrus, syrup (citrus molasses)	4-01-241	68	75	3.31	2.89	1.79	1.16	1.72	0.81	0.53	0.78	8.2
39	—sugarcane, molasses, dehydrated	4-04-695	94	70	3.09	2.67	1.63	1.03	1.60	0.74	0.47	0.73	10.3
40	—sugarcane, molasses, more than 46% invert sugar, more than 79.5 degrees brix (Black strap)	4-04-696	75	72	3.17	2.76	1.69	1.08	1.64	0.77	0.49	0.75	5.8
	OATS *Avena sativa*												
41	—grain	4-03-309	89	77	3.40	2.98	1.86	1.22	1.77	0.85	0.55	0.80	13.3
	PEANUT *Arachis hypogaea*												
42	—kernels, meal solv-extd (peanut meal)	5-03-650	92	77	3.40	2.98	1.86	1.22	1.77	0.85	0.55	0.80	52.3
	PINEAPPLE *Ananas comosus*												
43	—process residue, dehydrated (pineapple bran)	4-03-722	87	68	3.00	2.58	1.57	0.97	1.55	0.71	0.44	0.70	4.6
	RAPE *Brassica* spp												
44	—seeds, meal solv-extd	5-03-871	91	69	3.04	2.62	1.60	1.00	1.57	0.73	0.45	0.71	40.6
	SORGHUM *Sorghum bicolor*												
45	—grain, 8-10% protein	4-20-893	87	80	3.53	3.12	1.94	1.30	1.84	0.88	0.59	0.84	9.7
46	—silage	3-04-323	30	60	2.65	2.22	1.31	0.74	1.35	0.60	0.34	0.61	7.5
47	—silage, dough stage	3-04-321	28	55	2.43	2.00	1.14	0.58	1.23	0.52	0.26	0.56	6.0
	SOYBEAN *Glycine max*												
48	—hulls	1-04-560	91	77	3.40	2.98	1.86	1.22	1.77	0.85	0.55	0.80	12.1
49	—seeds, heat-processed	5-04-597	90	94	4.14	3.74	2.35	1.64	2.18	1.07	0.75	0.99	42.2
50	—seeds, meal solv-extd, 44% protein	5-20-637	89	84	3.70	3.29	2.06	1.40	1.94	0.94	0.64	0.88	49.9
	SUNFLOWER, COMMON *Helianthus annuus*												
51	—seeds w/o hulls, meal solv-extd	5-04-739	93	65	2.87	2.45	1.47	0.88	1.47	0.67	0.40	0.67	49.8
	TIMOTHY *Phleum pratense*												
52	—hay, sun-cured, early bloom	1-04-882	90	61	2.69	2.27	1.35	0.77	1.38	0.61	0.35	0.63	15.0
	TRITICALE *Triticale hexaploide*												
53	—grain	4-20-362	90	84	3.70	3.29	2.06	1.40	1.94	0.94	0.64	0.88	17.6
	UREA												
54	—45% nitrogen, 281% protein equivalent	5-05-070	99	0	0.0	0.0	0.0	0.0	0.0	0.0	0.0	0.0	281.0
	WHEAT *Triticum aestivum*												
55	—bran	4-05-190	89	70	3.09	2.67	1.63	1.03	1.60	0.74	0.47	0.73	17.1
56	—flour by-product, less than 7% fiber (wheat shorts)	4-05-201	88	73	3.22	2.80	1.73	1.11	1.67	0.79	0.50	0.76	18.6
57	—flour by-product, less than 9.5% fiber (wheat middlings)	4-05-205	89	69	3.04	2.62	1.60	1.00	1.57	0.73	0.45	0.71	18.4
58	—grain	4-05-211	89	88	3.88	3.47	2.18	1.50	2.04	0.99	0.68	0.93	16.0
	WHEY												
59	—dehydrated (cattle)	4-01-182	93	81	3.57	3.16	1.97	1.32	1.87	0.90	0.60	0.85	14.2

[1] Selected feeds from *Nutrient Requirements of Dairy Cattle*, 6th rev. ed., update 1989, NRC, National Academy Press, p.90, Table 7-1.
[2] Some specific numbers have not been assigned by the USDA Feed Composition Data Bank.

20-7
(ON A 100% DRY MATTER BASIS)[1] NOTE WELL: This is a 3-page spread. So, see pp. 436, 437, and 438.

Entry No.	Ether Extract (%)	Total Ash (%)	Crude Fiber (%)	Neutral Detergent Fiber (%)	Acid Detergent Fiber (%)	Cellulose (%)	Lignin (%)	Macrominerals (%)						
								Calcium	Chlorine	Magnesium	Phosphorus	Potassium	Sodium	Sulfur
1	4.0	10.2	20.5	38	28	22	5	1.80	0.34	0.26	0.35	2.21	0.22	0.33
2	3.8	9.2	22.0	40	29	23	7	1.54	0.34	0.24	0.29	2.56	0.15	0.31
3	3.0	9.6	23.0	42	31	24	8	1.41	0.38	0.33	0.22	2.52	0.14	0.28
4	2.6	9.1	26.0	46	35	26	9	1.41	0.38	0.31	0.24	1.71	0.12	0.28
5	2.0	8.9	29.0	50	37	28	10	1.25	0.35	0.31	0.22	1.53	0.11	0.27
6	—	—	—	—	—	—	—	—	—	—	—	—	—	—
7	3.6	7.6	11.0	25	20	14	6	0.23	—	0.13	0.11	0.53	0.02	0.11
8	2.1	2.6	5.7	19	7	5	2	0.05	0.18	0.15	0.38	0.47	0.03	0.17
9	2.0	3.1	7.1	21	9	—	—	0.06	0.17	0.14	0.39	0.58	0.02	0.16
10	0.6	4.4	19.8	54	33	31	2	0.69	0.04	0.27	0.10	0.20	0.21	0.22
11	0.6	6.1	16.5	44	25	22	3	0.61	—	0.16	0.10	1.78	0.53	0.42
12	1.4	5.8	1.1	—	—	—	—	0.32	0.30	0.24	0.26	0.10	0.35	0.37
13	6.5	4.8	14.9	46	24	18	6	0.33	0.17	0.16	0.55	0.09	0.23	0.32
14	3.7	10.7	24.0	56	31	27	3	0.50	—	0.18	0.30	2.30	0.02	0.20
15	0.7	2.4	0.2	0	0	0	0	0.67	—	0.01	0.90	0.01	0.01	—
16	3.7	6.6	12.7	23	22	18	3	1.84	—	0.17	0.12	0.79	0.09	0.08
17	9.8	2.4	12.1	43	17	12	5	0.11	0.08	0.07	0.43	0.18	0.10	0.46
18	3.7	1.9	9.4	28	11	9	2	0.07	0.05	0.14	0.27	0.53	0.02	0.16
19	2.4	3.4	4.8	37	9	8	1	0.16	0.07	0.06	0.50	0.03	0.10	0.39
20	2.4	1.8	2.2	14	5	4	1	0.08	0.10	0.09	0.54	0.21	0.06	0.72
21	2.4	7.5	9.7	45	12	—	—	0.36	0.25	0.36	0.82	0.64	1.05	0.23
22	4.3	1.6	2.6	9	3	2	1	0.03	0.05	0.14	0.29	0.37	0.03	0.12
23	4.3	1.6	2.6	9	3	2	1	0.02	0.05	0.14	0.32	0.35	0.01	0.14
24	7.7	3.1	6.7	55	13	10	2	0.05	0.06	0.26	0.57	0.65	0.09	0.03
25	3.0	7.2	32.3	53	30	23	5	0.34	—	0.23	0.19	1.41	—	0.08
26	3.1	4.5	23.7	51	28	24	4	0.23	—	0.19	0.22	0.96	0.01	0.15
27	1.7	2.8	47.8	90	73	59	24	0.15	0.02	0.14	0.09	0.87	0.02	0.09
28	20.0	4.8	24.0	44	34	24	10	0.21	—	0.46	0.64	1.00	0.01	0.26
29	23.8	4.5	17.2	37	26	12	14	0.12	—	0.41	0.54	1.18	0.01	—
30	1.3	7.0	14.1	26	19	12	6	0.22	0.04	0.55	1.21	1.39	0.04	0.34
31	1.7	6.7	12.1	28	21	13	7	0.17	0.04	0.55	1.00	1.39	0.04	0.34
32	99.5	—	—	—	—	—	—	—	—	—	—	—	—	—
33	5.5	9.9	24.6	—	—	—	—	0.51	—	—	0.37	—	—	—
34	10.5	20.8	1.0	—	—	—	—	5.65	0.60	0.16	3.16	0.76	0.43	0.49
35	1.5	6.5	10.1	25	19	13	6	0.43	0.04	0.66	0.89	1.53	0.15	0.43
36	0.9	8.4	0.2	—	—	—	—	1.36	0.96	0.13	1.09	1.70	0.49	0.34
37	0.2	11.3	—	—	—	—	—	0.17	1.64	0.29	0.03	6.07	1.48	0.60
38	0.3	7.9	—	—	—	—	—	1.72	0.11	0.21	0.13	0.14	0.41	0.23
39	0.9	13.3	6.7	—	—	—	—	1.10	—	0.47	0.15	3.60	0.20	0.46
40	0.1	13.1	—	—	—	—	—	1.00	3.10	0.43	0.11	3.84	0.22	0.47
41	6.4	3.4	12.1	32	16	11	3	0.07	0.11	0.14	0.38	0.44	0.08	0.23
42	1.4	6.3	10.8	—	—	—	—	0.29	0.03	0.17	0.68	1.23	0.08	0.33
43	1.5	3.5	20.9	73	37	—	7	0.23	—	—	0.13	—	—	—
44	1.8	7.5	13.2	—	—	—	—	0.67	0.11	0.60	1.04	1.36	0.10	1.25
45	3.4	2.1	2.0	18	9	8	1	0.04	0.10	0.18	0.34	0.40	0.01	0.09
46	3.0	8.7	27.9	—	38	—	6	0.35	0.13	0.29	0.21	1.37	0.02	0.11
47	3.3	9.3	28.5	—	—	—	—	0.29	0.11	0.27	0.26	1.02	0.03	0.14
48	2.1	5.1	40.1	67	50	46	2	0.49	—	—	0.21	1.27	0.01	0.09
49	20.0	5.1	5.6	—	11	—	—	0.28	—	0.23	0.66	1.89	0.03	0.24
50	1.5	7.3	7.0	—	10	—	—	0.30	0.08	0.30	0.68	1.98	0.03	0.37
51	3.1	8.1	12.2	—	—	—	—	0.44	0.11	0.77	0.98	1.14	0.24	—
52	2.9	5.7	28.0	61	32	31	4	0.53	—	0.14	0.25	1.62	0.18	—
53	1.7	2.0	4.4	—	8	—	—	0.06	—	—	0.33	0.40	—	0.17
54	0.0	—	0.0	0	0	0	0	—	—	—	—	—	—	—
55	4.4	6.9	11.3	51	15	11	3	0.13	0.05	0.60	1.38	1.56	0.04	0.25
56	5.2	4.9	7.7	—	—	—	—	0.10	0.08	0.28	0.91	1.06	0.03	0.22
57	4.9	5.2	8.2	37	10	—	—	0.13	0.04	0.40	0.99	1.13	0.19	0.20
58	2.0	1.9	2.9	—	8	8	—	0.04	0.08	0.16	0.42	0.42	0.05	0.18
59	0.7	9.8	0.2	0	0	0	0	0.92	0.08	0.14	0.82	1.23	0.70	1.12

(Continued)

TABLE 20-7 (Continued)

Entry No.	Feed Name Description	International Feed Number[2]	Microminerals (mg/kg)							Vitamins		
			Cobalt	Copper	Iodine	Iron	Manganese	Selenium	Zinc	A Activity (1,000 IU/kg)	D (1,000 IU/kg)	E (IU/kg)
	ALFALFA Medicago sativa											
1	—hay, sun-cured, early vegetative	1-00-050	0.10	11	0.19	253	45	0.37	24	80	1.9	—
2	—hay, sun-cured, late vegetative	1-00-054	0.09	9	0.18	227	34	0.35	27	81	—	—
3	—hay, sun-cured, early bloom	1-00-059	0.16	11	0.17	192	31	0.34	25	56	2.0	26
4	—hay, sun-cured, midbloom	1-00-063	0.36	14	0.16	134	28	0.32	23	46	2.0	11
5	—hay, sun-cured, full bloom	1-00-068	0.33	14	0.13	150	37	0.29	25	26	2.0	11
6	—silage, wilted, 25-45% dry matter (see similar maturity descriptions of hays)		—	—	—	—	—	—	—	—	—	—
	ALMOND Prunus amygdalus											
7	—hulls	4-00-	0.30	11	—	301	21	—	24	—	—	—
	BARLEY Hordeum vulgare											
8	—grain	4-00-549	0.10	9	0.05	85	18	0.22	19	1	—	25
9	—grain, Pacific Coast	4-07-939	0.10	9	—	97	18	0.11	17	—	—	30
	BEET, SUGAR Beta vulgaris altissima											
10	—pulp, dehydrated	4-00-669	0.08	14	—	329	38	—	10	—	0.6	—
11	—pulp w/molasses, dehydrated	4-00-672	0.23	16	—	207	27	—	10	—	—	—
	BLOOD											
12	—meal	5-00-380	0.10	11	—	4,064	6	0.80	5	—	—	—
	BREWERS' GRAINS											
13	—dehydrated	5-02-141	0.08	23	0.07	266	40	0.76	30	0	—	29
	BROME Bromus spp											
14	—fresh, early vegetative	2-00-892	0.08	11	—	200	142	—	27	184	—	—
	CASEIN											
15	—dehydrated (cattle)	5-01-162	—	4	—	15	5	—	30	—	—	—
	CITRUS Citrus spp											
16	—pulp w/o fines, dehydrated (dried citrus pulp)	4-01-237	0.16	6	—	378	7	—	15	—	—	—
	CORN, DENT YELLOW Zea mays indentata											
17	—distillers' grains, dehydrated	5-28-235	0.09	48	0.05	223	23	0.48	35	1	—	—
18	—ears, ground (corn and cob meal)	4-28-238	0.31	8	0.03	91	14	0.09	14	2	—	20
19	—gluten, meal	5-28-241	0.08	30	—	423	8	1.11	29	7	—	34
20	—gluten, meal, 60% protein	5-28-242	0.05	29	0.02	313	7	0.92	35	14	—	26
21	—gluten w/bran (corn gluten feed)	5-28-243	0.10	52	0.07	471	26	0.30	72	3	—	14
22	—grain, flaked	4-28-244	0.05	4	—	30	5	0.08	14	1	—	25
23	—grain, high-moisture	4-20-770	0.05	4	—	30	6	0.08	18	1	—	25
24	—grits by-product (hominy feed)	4-03-011	0.06	15	—	75	16	0.11	3	—	—	—
25	—silage, few ears	3-28-245	—	—	—	—	—	—	—	5	—	—
26	—silage, well-eared	3-28-250	0.06	10	—	260	30	—	21	18	0.1	—
	COTTON Gossypium spp											
27	—hulls	1-01-599	0.02	13	—	131	119	—	22	—	—	—
28	—seeds, w/lint	5-01-614	—	9	—	151	19	—	33	—	—	—
29	—seeds, w/o lint	5-01-	—	11	—	108	14	—	36	—	—	—
30	—seeds, meal prepressed, solv-extd, 41% protein	5-07-872	0.82	20	—	223	23	—	69	—	—	—
31	—seeds, meal prepressed, solv-extd, 44% protein	5-07-873	0.82	20	—	223	23	10	69	—	—	—
	FATS AND OILS (not exceeding 3% of diet)											
32	—fat, animal, hydrolyzed	4-00-376	—	—	—	—	—	—	—	—	—	—
	FESCUE, KENTUCKY 31 Festuca arundinacea											
33	—fresh, vegetative	2-01-902	—	—	—	—	—	—	—	—	—	—
	FISH, MENHADEN Brevoortia tyrannus											
34	—meal mech-extd	5-02-009	0.17	12	1.19	524	37	2.40	1.62	—	—	13
	FLAX Linum usitatissimum											
35	—seeds, meal solv-extd (linseed meal)	5-02-048	0.21	29	—	354	42	0.91	—	—	—	15
	MILK											
36	—skimmed dehydrated (cattle)	5-01-175	0.12	1	—	10	2	0.13	41	—	0.4	—
	MOLASSES AND SYRUP											
37	—beet, sugar, molasses, more than 48% invert sugar, more than 79.5 degrees brix	4-00-668	0.46	22	—	87	6	—	18	—	—	5
38	—citrus, syrup (citrus molasses)	4-01-241	0.16	108	—	508	38	—	137	—	—	—
39	—sugarcane, molasses, dehydrated	4-04-695	1.21	79	2.10	250	57	—	33	—	—	—
40	—sugarcane, molasses, more than 46% invert sugar, more than 79.5 degrees brix (Black strap)	4-04-696	1.21	79	2.10	250	56	—	30	—	—	7
	OATS Avena sativa											
41	—grain	4-03-309	0.06	7	0.11	85	42	0.26	41	—	—	15
	PEANUT Arachis hypogaea											
42	—kernels, meal solv-extd (peanut meal)	5-03-650	0.12	17	0.07	154	29	—	22	—	—	—
	PINEAPPLE Ananas comosus											
43	—process residue, dehydrated (pineapple bran)	4-03-722	—	—	—	561	—	—	—	22	—	—
	RAPE Brassica spp											
44	—seeds, meal solv-extd	5-03-871	—	—	—	—	—	1.07	—	—	—	—
	SORGHUM Sorghum bicolor											
45	—grain, 8-10% protein	4-20-893	0.29	11	—	50	17	—	16	—	—	12
46	—silage	3-04-323	0.30	35	—	285	73	0.22	32	6	0.7	—
47	—silage, dough stage	3-04-321	0.29	27	—	187	49	0.19	27	5	0.7	—
	SOYBEAN Glycine max											
48	—hulls	1-04-560	0.12	18	—	324	11	—	24	—	—	—
49	—seeds, heat-processed	5-04-597	—	18	—	89	33	0.12	60	—	—	—
50	—seeds, meal solv-extd, 44% protein	5-20-637	0.20	24	—	175	35	0.11	66	—	—	—
	SUNFLOWER, COMMON Helianthus annuus											
51	—seeds w/o hulls, meal solv-extd	5-04-739	—	4	—	33	20	—	—	—	—	12
	TIMOTHY Phleum pratense											
52	—hay, sun-cured, early bloom	1-04-882	—	11	—	200	103	—	62	21	—	13
	TRITICALE Triticale hexaploide											
53	—grain	4-20-362	—	7	—	44	45	—	25	—	—	—
	UREA											
54	—45% nitrogen, 281% protein equivalent	5-05-070	—	—	—	—	—	—	—	—	—	—
	WHEAT Triticum aestivum											
55	—bran	4-05-190	0.11	14	0.07	128	125	0.43	128	1	—	21
56	—flour by-product, less than 7% fiber (wheat shorts)	4-05-201	0.12	13	—	82	132	0.49	124	—	—	61
57	—flour by-product, less than 9.5% fiber (wheat middlings)	4-05-205	0.10	22	0.12	93	126	0.83	116	—	—	—
58	—grain	4-05-211	0.14	7	0.10	61	42	0.30	50	—	—	17
	WHEY											
59	—dehydrated (cattle)	4-01-182	0.12	50	—	181	6	—	3	—	—	—

In *Feeds & Nutrition,* Section V, the energy values of feeds are not identical to the energy values in *Nutrient Requirements of Dairy Cattle,* Sixth Revised Edition, due to varying conclusions from interpreting analytical information. However, the differences are minor and should not materially affect ration formulation. If, however, formulators are primarily interested in rations for dairy cattle, and if TDN is used as the energy measure, they may wish to refer to the NRC publication for the appropriate TDN values. As a convenience, Table 20-7, which contains TDN and NE_{lc} values from *Nutrient Requirements of Dairy Cattle,* Sixth Revised Edition, for the most common feeds used in dairy rations, is presented in this chapter. (Table 20-7 is a 3-page spread; so, see pages 436, 437, and 438.

> **NOTE WELL:** Net energy for lactation values (NE_{lc}) that conform to those used in *Nutrient Requirements of Dairy Cattle* may be calculated from TDN values of *Feeds & Nutrition,* Section V, by using the following equation:
>
> NE_{lc} (Mcal/kg of DM) = 0.0245 × TDN (% of DM) − 0.12
>
> This equation is based on an average 4% reduction in digestibility for each multiple increase in intake over maintenance intake and assumes the intake to be 3X that of maintenance.

(For a more complete and in-depth discussion of energy, also see Chapter 4, section headed "Energy [Carbohydrates and Lipids]" and the subsections under it.)

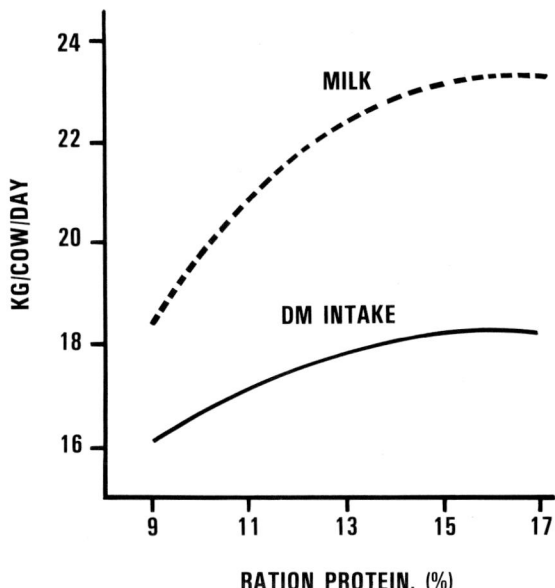

Fig. 20-2. This depicts the increase in milk yield as the protein present in the total ration DM increases. (Courtesy, J. W. Thomas, Ph.D., Professor Emeritus and Consultant, Department of Animal Science, Michigan State University, East Lansing)

Protein

Protein is essential for dairy cattle for maintenance, growth, milk production, and the development of the fetus. Also, it is required for the formulation of enzymes and certain hormones that control or regulate chemcial reactions in the body. The protein requirement is really a requirement for amino acids.

The protein composition of feeds, and the protein requirements of dairy cattle, may be expressed as crude protein, digestible protein, degraded intake protein, undegraded intake protein, and/or nonprotein nitrogen (NPN).

• **Amount of protein needed**—The amount of protein needed in the total ration of lactating cows is determined primarily by the amount of milk produced. Milk is a rich source of high-quality protein; so, as milk production increases, a substantial amount of dietary protein is necessary. Thus, a high-producing 1,320-lb cow yielding 88 lb of 3.5% protein milk daily secretes 3.08 lb of milk protein. A deficiency of protein results in lowered milk production and may depress the protein content of milk. Excess protein usually results in high cost rations.

Fig. 20-2 depicts the increase in milk yield as the protein percent of the total ration DM increases. At some point, the value of increased milk will not exceed the cost of the additional protein, with this point determined by the relative milk price and the cost of the additional protein.

The amount of protein needed in the concentrate mix depends on the kind and quality of forage fed. As the amount of legume increases, the percentage of protein in the concentrate can be lowered. For most lactating cows, the total ration (forage plus grains and protein and energy supplements) should have 19% crude protein during the first ⅓ of lactation, lowered to 14% in midlactation and 12% during the dry period. The interaction of the protein and energy supply in the rumen of the high-producing cow is very important; so, an adequate supply of degradable intake protein (DIP) is essential to maximize both feed intake and ruminal digestibility. Moe, of the USDA, Agricultural Research Service, found that high-producing cows receiving 17% protein digested feed better than those fed 14% protein.[2]

DEGRADED INTAKE PROTEIN (DIP) AND UNDEGRADED INTAKE PROTEIN (UIP)

Approximately 60% of the crude protein in the typical dairy cow ration is broken down (degraded) by microbial digestion to ammonia. The rumen microbes must convert the ammonia to microbial protein in their own cells if the dairy animal is to receive any benefit. Fermentable energy must be available for the microorganisms to grow and synthesize the necessary amino acids. If rumen ammonia levels are excessively high, the ammonia is absorbed into the blood and either recycled or excreted in the urine as urea.

All feed protein sources are not degraded in the rumen to the same extent. The optimal ration will meet both the nitgrogen requirement of rumen microorganisms for maximum synthesis of microorganism protein and allow for

---
[2]*Research News,* ARS-USDA, p. 20, May 1977.

maximum escape or *bypass* of high quality feed protein for digestion in the small intestine. Protein synthesis by rumen microbes depends on feed intake, organic matter digestibility, feed type, protein level, and feeding system. Since 3.5 lb of microbial protein synthesis per day is near the maximum, the remainder of the protein must be derived from nondegraded (escape) protein sources. Young, fast-growing heifers and high-producing cows generally require additional nondegraded protein sources beyond their normal ration to meet total protein requirements; and the more rapid the growth and the higher the milk production, the greater the quantities of undegradable protein needed. Brewers' grain, distillers' grain, corn gluten meal, fish meal, meat meal, and heat-treated soybeans are examples of feeds with reduced rumen degradability that may be substituted in rations in which excess rumen ammonia exists and less than optimal amounts of quality protein (undegraded) pass into the small intestine.

(For an in-depth discussion of protein, also see Chapter 4, section headed "Proteins"; and for an in-depth discussion of degraded intake protein [DIP] and undegraded intake protein [UIP], also see Chapter 11, section headed "Protein Bypass [Protected Protein, Escaped Protein]," including Table 11-1, showing the percent of undegraded protein in common feeds.)

UREA AND OTHER NPN PRODUCTS

Using urea in the ration is similar to using degradable intake protein. It and other nonprotein nitrogen (NPN) compounds, such as ammonium salts, can be used to replace part of the protein required in dairy cattle rations after rumen function has become established.

The following guidelines should be observed for the successful use of urea in dairy rations:

1. All rations should be assessed for protein content before either supplemental NPN or natural protein is added to the ration. Protein may not be needed.
2. Feeds most successfully supplemented with NPN are high in energy, low in protein, and low in natural NPN (such as grains and corn silage).
3. Maximum amounts of urea to feed are:

   a. 1% urea in the grain mix; 0.5% urea in the total ration.

   b. 0.5% urea in corn silage (10 lb/ton). If 0.5% is added to corn silage, the amount in the grain should be no more than 0.5%. The addition of 10 lb of urea per ton of corn silage will increase the protein content from 8 to 12% on a dry matter basis (depending on losses incurred).

   c. 0.4 lb urea per head per day, with cows in early lactation limited to 0.2 lb of urea per head per day.

Since NPN products do not provide any energy, minerals, or vitamins, these nutrients must be provided through other sources.

(For an in-depth discussion of urea and NPN, also see Chapter 4, section headed "Nonprotein Nitrogen"; and Chapter 11, section headed "Nonprotein Nitrogen [NPN] Feedstuffs.")

Minerals

Tables 20-2, 20-3, and 20-4 show the daily calcium and phosphorus requirements for different classes of dairy animals; Table 20-5 shows the recommended content of major and trace minerals of rations for different classes of dairy animals; and Table 20-6 shows the maximum tolerable levels of certain elements. Chapter 19, Table 19-7, presents in summary form the mineral requirements of all cattle; the mineral requirements of dairy cattle and beef cattle are similar except for the higher milk production of lactating dairy cows. Section V, Composition of Feeds, presents the macro- and micromineral content of a great array of feeds.

MAJOR OR MACROMINERALS

The major or macrominerals of importance in dairy cattle nutrition are: salt (sodium chloride), calcium, phosphorus, magnesium, potassium, and sulfur.

Salt (sodium chloride [NaCl]). The current NRC recommendation is for a minimum of 0.43% sodium chloride in the total dairy ration, including that contributed by the feeds (Table 20-5). Excessive levels of chlorine without sodium or potassium can contribute to an acidosis condition in dairy cattle.

Calcium (Ca). Whole milk contains 0.12% calcium. The NRC subcommittee based the dietary calcium requirements for dry pregnant and lactating cows (Table 20-3) on 38% availability. Minimum calcium percentage for the complete

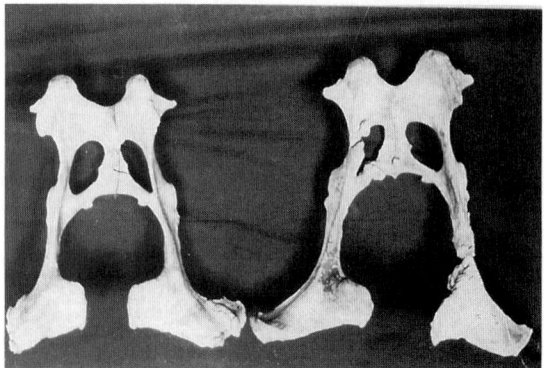

Fig. 20-3. Calcium deficiency. Lactating cows need calcium. Both hips of the cow shown above have been broken (knocked down) as a result of feeding a low-calcium ration. At lower left, the pelvis of a cow which had three breaks while the cow received a low-calcium ration. At lower right, the pelvis of the cow pictured above, showing the breaks involving both hipbones. (From *Fla. Ag. Exp. Sta. Tech. Bull. 262*, through the courtesy of R. B. Becker)

ration (dry matter basis) recommended by the NRC for lactating cows varies from 0.43 to 0.66%, depending on level of milk production (Table 20–5). A deficiency of calcium may cause rickets, slow growth and poor bone development, easily fractured bones, reduced milk yield and increased incidence of milk fever. Feeding calcium at more than 0.95 to 1.00% (DM basis) in mixed rations may reduce dry matter intake and lower performance. The effects of variations in the calcium-to-phosphorus ratios have been overemphasized, as evidenced by studies showing that dietary calcium-to-phosphorus ratios of between 1:1 and 7:1 result in nearly equal performance, provided the animal's phosphorus intake meets its requirement.

Phosphorus (P). Whole milk contains 0.09% phosphorus. The NRC subcommittee assumed a phosphorus availability from mixed rations fed to lactating cows of 45 to 50%. A deficiency of phosphorus may result in fragile bones, stiff joints, poor growth, low blood P (less than 4–6 mg/100 ml), depraved appetite (chewing wood, hair, and bones), and poor reproductive performance. Excessive phosphorus intakes may cause bone resorption, elevated plasma phosphorus levels, and urinary calculi.

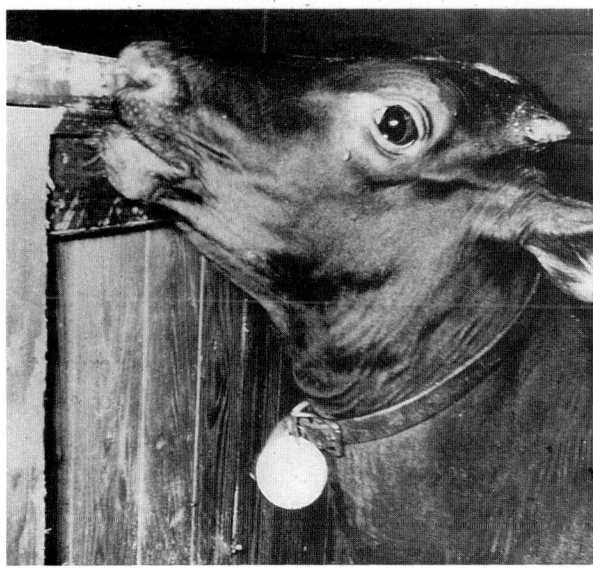

Fig. 20–4. Phosphorus-deficient calf chewing wood, a manifestation of depraved appetite. (Courtesy, Dr. S. E. Smith, Department of Animal Science, Cornell University, Ithaca, N.Y.)

Magnesium (Mg). Milk contains a substantial amount of magnesium (about 0.015%). Thus, when expressed as a percentage of the ration, the magnesium requirement increases with the cow's level of milk production. Under practical conditions, magnesium deficiencies may occur (1) when calves are fed an all-milk diet for extended periods, during which their body reserves of magnesium are depleted; or (2) when dairy cattle, especially older and lactating cows, are grazing lush, rapidly growing pastures that have been highly fertilized with nitrogen or potassium, or both, during cool seasons. Under conditions conducive to grass tetany and for high-producing cows in early lactation, the suggested requirement is 0.25 to 0.30% dietary magnesium, with the supplemental magnesium provided in a readily available form such as magnesium oxide. Magnesium toxicity is not known to be a practical problem in dairy cattle.

Potassium (K). Milk contains about 0.15% potassium. The NRC minimum dietary potassium requirement for lactating cows is 0.90%, increased to 1.00% for high-yielding and early-lactation cows; for dry cows and young stock, it is 0.65%. Stress, especially heat stress, appears to increase the need for potassium, perhaps due to greater loss of potassium through sweat. The signs of relatively severe potassium deficiencies in lactating cows include a marked decrease in feed intake, loss in weight, decreased milk yield, pica, loss of hair glossiness, decreased pliability of hide, lower plasma and milk potassium, and higher hematocrit readings. Generally, forages contain considerably more potassium than is required by dairy cattle.

High levels of potassium (3% or above) in very lush forages grown on high potassium soils in cool weather appear to interfere with magnesium metabolism and utilization and are considered to be a factor in causing grass tetany of lactating cows.

Sulfur (S). Milk contains 0.03% sulfur, much of which is in the form of the amino acids methionine and cystine. Sulfur is needed for microbial protein synthesis, especially when nonprotein nitrogen is fed. The NRC estimated minimum sulfur requirement for lactating cows is 0.20% of the ration; with the sulfur needs of other dairy cattle calculated from the minimum protein requirement for these animals based on a nitrogen-to-sulfur ratio of 12:1.

TRACE OR MICROMINERALS

The trace or microminerals of importance in dairy cattle nutrition are: cobalt, copper, iodine, iron, manganese, molybdenum, selenium, and zinc.

Cobalt (Co). Normal cow's milk averages 0.38 to 1.04 mcg of cobalt/qt. Colostrum contains 4 to 10 times more cobalt than milk. Since cobalt is a component of vitamin B–12, ruminal microorganisms are able to synthesize this vitamin only when adequate cobalt is in the ration of the cow. NRC recommends a minimum of 0.1 ppm of cobalt in the total ration. Supplements of 30 to 45 g of cobalt sulfate or 20 to 25 g of cobalt carbonate with 100 lb of salt have prevented any cobalt deficiency problems. Also, a heavy pellet containing cobalt oxide and finely divided iron, which is administered orally and remains in the reticulo-rumen, will prevent cobalt deficiency for extended periods in cattle that graze cobalt-deficient pastures.

Copper (Cu). Colostrum contains more copper than milk. The amount of copper in milk decreases with the length of lactation. Copper is needed for hemoglobin formation, although it is not actually contained in it. A deficiency of copper will result in anemia and bleaching of the hair.

Black hair turns gray and red hair becomes yellow. NRC recommends a minimum of 10 ppm copper in the ration, with the caution that higher levels may be required for cattle grazing pastures or consuming feedstuffs that contain high levels of molybdenum or other interfering substances.

Fig. 20-5. Copper deficiency. *Top:* Registered Jersey heifer showing copper deficiency. *Bottom:* Same heifer after receiving copper. (Courtesy, R. B. Becker, University of Florida, Gainesville)

Iodine (I). About 10% of the iodine intake of lactating cows is normally excreted in milk.

Goiter (an enlargement of the thyroid gland) occurs in newborn calves if their mothers were fed iodine-deficient rations; necks of the calves are swollen and they are weak at birth or born dead. Much of the small amount of iodine in the body is contained in the thyroid gland as thyroxin and diiodotyrosine, both of which are contained in the protein thyroglobulin, a part of the thyroid hormone. The principal function of the thyroid gland is to regulate the metabolic rate. Many protein supplements (including soybean meal and cottonseed) are mildly goitrogenic because they reduce the availability of dietary iodine, and *Brassica* forages (cabbage, kale, rape) are highly goitrogenic. The NRC recommends that cows in lactation receive a dietary iodine concentration of 0.6 ppm, and that cows in the last 2 months of gestation be fed 0.6 ppm of iodine. When stabilized iodine is used, a level of 0.0076% in salt is adequate. The Northwest and Great Lakes regions are the most iodine deficient areas of the United States. Lactating cows should not receive excessive dietary iodine because the resulting high iodine milk content is considered undesirable for humans. The use of iodine disinfectants as teat dips or udder washes can increase the iodine content of milk, but the main cause of high iodine levels in milk is dietary iodine.

Iron (Fe). Iron is essential because it is a constituent in hemoglobin, the oxygen carrier in the blood. Cow's milk is low in iron—about 10 ppm. The iron requirements of a young calf are higher than those of a mature cow and are thought to be about 100 ppm until 3 months of age, and 50 ppm thereafter; as recommended by the NRC.

Manganese (Mn). Manganese deficiency in dairy cattle is seldom a problem. In general, forages contain higher levels of manganese than grains. The manganese requirement for cattle is higher for reproduction than for growth. Little experimental work has been done on the manganese requirements of dairy cattle, but rations containing 40 ppm are recommended by the NRC.

Molybdenum (Mo). Molybdenum is an indispensable component of the enzyme xanthine oxidase, which is found in milk and distributed widely in animal tissue. Yet a deficiency of molybdenum has never been developed or observed in cattle. Molybdenum is known largely for its toxic characteristics; molybdenum toxicosis is a practical problem in grazing cattle in several areas of the world.

The NRC has set the maximum tolerable level of molybdenum for cattle for relatively short feeding periods at 10 ppm. *CAUTION:* As a feed additive, molybdenum is not approved by the Food and Drug Administration.

Selenium (Se). Selenium, like molybdenum, was known for its toxic characteristics long before it was discovered to be an essential nutrient. However, research has firmly established the essentiality of selenium for ruminants; it is needed in trace amounts to prevent retarded growth, reproductive problems, retained placenta, white muscle disease—a condition that occurs in calves and lambs in selenium-deficient areas, and some mastitis problems. Also, it is closely associated with vitamin E; both selenium and vitamin E protect cells from the detrimental effects of peroxidation, but each takes a different approach.

The current NRC recommended requirements for all cows and heifers are 0.3 ppm, which is the maximum level permitted by FDA. Deficient or toxic selenium areas are widely scattered throughout the United States and the world.

(Also see Chapter 5, Table 5-1—White Muscle Disease.)

Zinc (Zn). Milk generally contains about 4 ppm of zinc, but this level has been doubled by increasing the intake of zinc in the ration. Zinc is involved in several enzyme systems and is affected adversely when excess quantities of calcium are present. The NRC recommendation of 40 ppm of zinc for all classes and ages of dairy cattle is based upon limited data.

Feeding Dairy Cattle

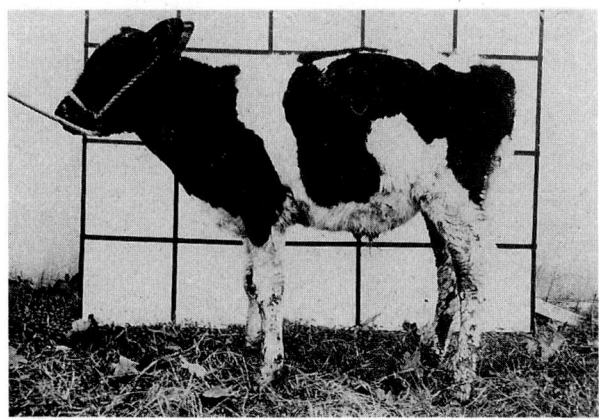

Fig. 20-6. Zinc deficiency. *Top:* Calf showing loss of hair on legs and severe scaliness, cracking, and thickening of the skin as a result of zinc deficiency. *Bottom:* The same calf after receiving supplemental zinc. (Courtesy, W. J. Miller, The University of Georgia, Athens)

Vitamins

Dairy cattle, like other animals, require vitamins for optimum performance and health.

Tables 20-2 and 20-3 present the daily vitamin A and vitamin D requirements for different classes of dairy animals; and Table 20-5 shows the recommended vitamin content of rations for different classes of dairy animals. Chapter 19, Table 19-9, presents in summary form the vitamin requirements of all cattle; the vitamin requirements of dairy cattle and beef cattle are similar except for the higher milk production of lactating cows. Section V, Composition of Feeds, shows the fat-soluble vitamin and water-soluble vitamin content of feeds.

FAT-SOLUBLE VITAMINS

Dairy cattle require fat-soluble vitamins A, D, E, and K. Generally, all classes of dairy cattle require a dietary source of vitamins A and E. Vitamin D must either be synthesized in the skin by the action of ultraviolet radiation or be included in the ration. Rumen microbes synthesize adequate amounts of vitamin K to meet the needs of most dairy cattle with the exception of young calves, whose rumen has not begun all of its functions.

Fortunately, under normal conditions, natural feeds furnish most fat-soluble vitamins or their precursors in adequate amounts. High-quality forages contain large amounts of vitamin A precursors, and vitamin E is abundant in most feeds. Vitamin D is found in large quantities in sun-cured forages. Additionally, cattle can store adequate reserves of the fat-soluble vitamins to meet their needs for several months. Yet, when dairy producers feed limited or low-quality forage, use high levels of ensiled forage, expose cattle to little sunlight, or use milk replacers for young calves, additional vitamins will probably be needed for optimum health and high performance.

Vitamin A. Vitamin A supplementation may be desirable when (1) poor-quality or limited amounts of forage are fed, (2) forage that has been stored for a long period loses its carotene through oxidation, or (3) high levels of corn silage and low-carotene concentrates are fed.

The daily vitamin A requirements for different classes of dairy cattle are shown in Tables 20-2 and 20-3; and Table 20-5 shows the recommended vitamin A content of rations for different classes of dairy animals. There are no allowances of vitamin A for milk production as intakes above those required for normal reproduction do not increase milk yield. **NOTE WELL:** The vitamin A requirements given in Tables 20-2, 20-3, and 20-5 are adequate under most practical conditions, but may be increased when animals are under certain stressful conditions such as low environmental temperature or exposure to infective bacteria.

Vitamin D. Cows fed sun-cured forage or exposed to sunlight do not need supplemental vitamin D.

Fig. 20-7. Calf with severe rickets. Note the bowed legs and swollen joints. Rickets may be caused by a lack of vitamin D, calcium, or phosphorus; or by an incorrect ratio of the two minerals. (Courtesy, Michigan State University, East Lansing)

A vitamin D deficiency leads to a failure of bones to calcify normally, resulting in rickets in calves and osteomalacia in adults. Vitamin D deficiencies in calves kept indoors do occur, but deficiencies in mature cattle under normal conditions are extremely unlikely because exposure to sunlight provides adequate vitamin D.

Vitamin E. Vitamin E is an antioxidant associated with selenium. It stimulates the immune system and reduces the incidence of oxidized flavor when fed at high levels (400 to 1,000 mg/cow/day); and it may aid in protection against white muscle disease, caused by deficiency of selenium.

For young calves, the vitamin E requirements are estimated to range from 11 to 18 IU of vitamin E/pound of total feed (Table 20-5).

(Also see Chapter 5, Table 5-1—White Muscle Disease.)

Vitamin K. Normally, vitamin K_2 is synthesized in large amounts in the rumen; so, dietary supplementation is not recommended.

When cows consume moldy sweetclover hay, which is high in dicoumarol, blood coagulation may be impaired, followed by generalized hemorrhaging. This syndrome, commonly called *sweetclover disease* or *sweetclover poisoning*, responds to treatment with vitamin K.

WATER-SOLUBLE VITAMINS

The water-soluble vitamins include biotin, choline, folacin (folic acid), inositol, niacin (nicotinic acid, nicotinamide), pantothenic acid (vitamin B-3), para-aminobenzoic acid (PABA), riboflavin (vitamin B-2), thiamin (vitamin B-1), vitamin B-6 (pyridoxine, pyridoxal, pyridoxamine), vitamin B-12 (cobalamins), and vitamin C (ascorbic acid, dehydroascorbic acid). However, a physiological need for cattle of all of these vitamins has not been demonstrated.

Until recently, it was assumed that dairy cattle with a functional rumen did not require supplemental B vitamins. The rumen microflora were believed to synthesize adequate amounts of these nutrients for the host's requirements. Besides, the B vitamins are relatively abundant in dairy feeds. But recent evidence suggests a need for supplemental niacin under certain conditions and possibly supplemental choline and thiamin in the case of mature cattle, for which microbial synthesis and quantities in feeds may be inadequate, especially during diseased conditions or periods of stress. It is assumed that dairy cattle of all ages have a physiological need for most of the B vitamins, especially biotin, choline, niacin, pantothenic acid, riboflavin, thiamin, vitamin B-6, and vitamin B-12. In young calves, deficiency signs have been demonstrated when there is inadequate intake of these vitamins, but, even without a functioning rumen, their needs for these B vitamins appear to be met when they are fed whole milk. When young calves are fed milk replacers, however, it is advisable to ascertain the adequacy of vitamin intakes until their rumens are functional. Table 20-5, footnote 14, shows the minimum quantities of B vitamins that should be provided in milk replacers.

Biotin. A biotin deficiency in calves, characterized by paralysis of the hindquarters, has been produced. Signs of deficiency did not develop when synthetic milk was supplemented with 4.5 mcg of biotin/pound of feed and fed at 10% of liveweight.

Choline. Researchers have produced choline deficiency in calves by using a synthetic ration containing 15% casein. Within 6-8 days, the calves developed extreme weakness and labored breathing and were unable to stand. Supplementation of the ration with 236 mg of choline/quart of synthetic milk prevented the development of these signs. Adding choline to the ration may increase the percentage of milk fat in lactating cows.

Niacin (Nicotinic Acid, Nicotinamide). Niacin is required by the young preruminant calf. In order to prevent a niacin deficiency, it is recommended that niacin be added to milk replacers at a level of 2.6 ppm.

Although research results are not consistent, there appears to be significant benefits from feeding niacin to dairy cows with above average incidence of ketosis and that are overconditioned. Research studies indicate that supplementation of 6 to 12 g of niacin/day should begin 2 weeks before calving, then continue for 8 to 12 weeks after calving.

Pantothenic Acid (Vitamin B-3). Pantothenic acid deficiency in the calf is characterized by a scaly determatitis around the eyes and muzzle, loss of appetite, diarrhea, weakness (unable to stand), and convulsions.

Pantothenic acid deficiency in animals with functioning rumens is unlikely due to microbial production of pantothenic acid.

Riboflavin (Vitamin V-2). Riboflavin deficiency in the calf is characterized by hyperemia (presence of blood) of the mucosa of the mouth, lesions in the corners of the mouth and along the edges of the lips, loss of hair—especially on the belly, and excess salivation.

A riboflavin deficiency in lactating cattle is unlikely because of the amounts of riboflavin that are present in feedstuffs and synthesized in the rumen.

Thiamin (Vitamin B-1). Thiamin deficiency in the calf may cause polioencephalomalacia, characterized by listlessness, muscular incoordination, progressive blindness, convulsions, and sudden death; which may be accompanied by diarrhea and dehydration. The condition is found primarily in cattle, sheep, and goats fed high-concentrate rations; and it has been linked to increased microbial thiaminase activity and the production of thiamin analogs in the rumen. Treatment consists of the IV or IM administration of thiamin at a rate of 1 mg/pound liveweight.

(Also see Chapter 5, Table 5-1—Polioencephalomalacia.)

Vitamin B-6 (Pyridoxine, Pyridoxal, Pyridoxamine). Vitamin B-6 deficiency has been produced in calves fed a synthetic diet. It is characterized by loss of appetite, cessation of growth; and after about 3 months, epileptic fits in some, but not all, calves. Calves respond to vitamin B-6 therapy if it is initiated in the early stages of the disease.

Vitamin B-12 (Cobalamins). Vitamin B-12 deficiency has been produced in calves under 6 weeks of age by feeding them a diet containing no animal protein. Deficiency signs include poor appetite and growth, muscular weakness, and poor general condition. It has been suggested that the vitamin B-12 requirement for dairy cattle is between 0.15 and 0.30 g/pound liveweight.

Vitamin B-12 is of special interest in the mature ruminant because of its role in propionate metabolism and because of the incidence of B-12 deficiency as a secondary result of cobalt deficiency. Certain soils have insufficient cobalt to produce levels of the element in plants that are adequate to support optimum vitamin B-12 synthesis in the rumen.

Water

Cows drink an average of 100 to 200 lb of water per day, with heavy producers drinking up to 300 lb per day (1 gal water = 8.33 lb). Cows need 4 to 5 lb of water for each pound of milk produced. The amount of water a cow will drink depends on her size and milk yield, the quantity of dry matter consumed, the temperature and relative humidity of the air, the temperature of the water, the quality of the water, and the amount of moisture in her feed.
(Also see Chapter 4, section on "Water.")

FEEDS FOR DAIRY CATTLE

For convenience, in the sections that follow the commonly used dairy cattle feeds are grouped in eight general categories: (1) forages, (2) root crops, (3) concentrates, (4) by-products, (5) special feeds and additives for dairy cattle, (6) commercial feeds, (7) feed considerations, and (8) feed substitutions. Additional information pertaining to feed ingredients and feed preparation is contained in Section II of this book.

Forages

Legumes and grasses are the major sources of forages for dairy animals. They may be harvested and fed as pasture, green chop, hay, silage, or haylage. When properly grown, harvested, and stored, they are excellent sources of nutrients. High-quality forage can make up to ⅔ of the ration dry matter with cows consuming 2½ to 3% of their body weight in forage dry matter. High-quality forages fed in balanced rations will supply much of the energy, protein, minerals, and vitamins needed for milk production.

Average cows can produce up to 70% of their potential milk yield when fed excellent-quality forage without grain, whereas poor cows may produce their total potential without supplemental feed. The higher the level of production, the higher the percentage of the total ration which should come from grains and other concentrate feeds.

In using any kind of forage, three important points should be kept in mind: (1) To obtain the most nutrients from forage, it must be of high quality; (2) the better the forage, the smaller the requirement for grains; and (3) the cow is, by nature, a good consumer of forage.
(For methods of evaluating hay, see Chapter 8, Hay; and for methods of evaluating silage, see Chapter 9, Silage/Haylage/High-Moisture Grain.)

• **Pasture**—Of all the feedstuffs for dairy cows, pasture is the oldest, and today the most controversial. Since the 1960s, pastures have played a decreasing role in the total feed program of U.S. dairy farms. Nevertheless, pastures are effectively and efficiently used on some dairy farms. Grazing may supply 10 to 50% of the total dry matter, and at times an even higher percentage of the protein, of the ration of lactating cows. Also, good pastures are excellent for replacement heifers.

Large herds are not easily handled in a pasture system. Where land prices are high, it may not be economically feasible to maintain pastures in close proximity to the corrals.

When a herd reaches 100 cows or more, land may be unavailable within a radius of ½ mile of the corrals, which is considered the maximum practical distance that cows can be trailed to and from pasture. With cured feed, transportation is much more practical than with green feed, either grazed or green chopped. For these reasons, most commercial dairy producers are shifting from pasture to stored feed, usually produced on land located some distance from the corrals.
(Also see Chapter 7, Pasture and Range Forages.)

• **Green Chop (Soilage or Zero Grazing)**—Many producers harvest and feed green chop daily. It reduces wastage from trampling and manure fouling which are inherent in grazing systems. With tall-growing crops, it can be used in place of pastures. However, cutting green chop every day can be a major problem during wet weather or during peak work periods.
(Also see Chapter 7, Pasture and Range Forages, Section headed "Green Chop.")

• **Hays**—Legume hays are best for dairy cattle. Alfalfa is by far the most popular hay crop. Some of the grass-legume combinations (such as clover and timothy) are more popular than straight grass hay. To make the best hay, alfalfa and grasses should be cut at the proper stage of maturity. With advancing maturity, plants decrease in energy, protein, calcium, phosphorus, and digestible dry matter, and increase in fiber. As crude fiber, neutral detergent fiber (NDF), and acid detergent fiber (ADF) increase, the lignin content of the plant also increases. Lignin is indigestible and makes other nutrients less digestible.

Quality of hay and level of production go hand in hand. Since, in recent years, good-quality hay has been sold at a premium price, it is mandatory that producers utilize available hay as efficiently as possible. The best quality hay should be fed to the best cows only. To feed a low-producing cow or a dry cow top quality alfalfa in times of hay shortages is wasteful and inefficient. Therefore, the feeder needs to separate the hay and feed according to quality if at all possible.

Recently the relationship of fiber to milk and butter fat production has received considerable attention. Excessive fiber levels limit intake and energy concentration, while shortage of fiber reduces rumen digestibility and milk fat test. This is due primarily to the fact that these feeds alter the microflora of the rumen in such manner that propionic acid levels are increased, thereby lowering the amount of butterfat produced.

Because fiber is important, and because all fibers are not the same, the new NRC requirements presented in Table 20-5 give fiber minimums for crude fiber, acid detergent fiber, and neutral detergent fiber.
(Also see Chapter 8, Hay.)

• **Silages**—It is often possible to produce more nutrients per acre—and thereby more milk per acre—from silage than from hay crops. This is especially true for corn, where yields are usually high. Merely putting the crop into silage does not ensure this, however. It takes special care and a proper storage facility to make this possible.

The feeding value of silage is no better than the forage that goes into the silo. Good silage is easy to make if the crop is harvested at the proper stage of maturity, cut fine, and ensiled as quickly as possible at 55 to 70% moisture. In the silo, it must be evenly distributed, firmly packed, and protected from the air. If the crop does not contain sufficient readily fermentable carbohydrate, it is important that there should be added either (1) a carbohydrate, or (2) an acid, to preserve it without fermentation.

Generally speaking, a higher percentage of good-quality silage is made from corn and sorghum than is made from hay-crop, but this need not be so. Consistently good hay-crop silage can be made, provided these procedures are followed: (1) it is cut early, (2) it is wilted to 55 to 70% moisture content, and (3) a suitable preservative is added. The chief reason these special methods are necessary in ensiling hay-crop silages is that they contain a much smaller percentage of sugars than corn or sorghum silage.

(Also see Chapter 9, Silage/Haylage/High-Moisture Grain.)

- **Haylage (Low-Moisture Silage)**—Haylage—grass and/or legumes that are wilted to 40 to 50% moisture before ensiling—is popular. Its feeding value depends on the stage of maturity when the crop is harvested and the percentage of dry matter in the haylage. Unlike wilted or direct-cut silage, there need be no limitation on how much haylage is fed. It may be fed free-choice if it is good quality.

(Also see Chapter 9, Silage/Haylage/High-Moisture Grain.)

Root Crops

The per acre yield of nutrients from roots is higher than for most other feed crops, but labor costs in the United States make it impractical to grow, process, and feed root crops to dairy cattle. However, in Europe mangels (fodder beets) and stock carrots have been among the more popular root crops used in dairy rations. These crops should be sliced, or in some other way reduced in size to pieces which can be consumed safely by dairy cattle. They are very high in moisture; hence, they may need to be limited for cows producing at very high levels so that there will be sufficient room in the digestive tract for nutrients other than water.

Concentrates

Concentrate feeds are those which are high in energy and low in fiber. Many different kinds of concentrate feeds are used in dairy cattle feeding. They are usually classed according to total (crude) protein content as (1) low-protein, (2) medium-protein, or (3) high-protein feeds. The chemical analysis of various feeds is shown in Section V of this book.

Three factors besides chemical composition are important in evaluating concentrates for milk cows—palatability, quality of milk produced, and cost. The most infallible way in which to appraise the first two factors is through actual feeding trials. Consideration of the third factor—cost—necessitates that dairy producers be keen students of values. They must change the formulations of their ration(s) in keeping with comparative feed prices, and do so without causing the animals to go off feed.

Corn and barley are the two chief grains used in dairy rations, although oats, sorghum grains, and wheat are also used when there is a price advantage.

(Also see Chapter 10, Grains/High-Energy Feeds.)

By-products

By-product feeds are important in dairy rations. The milling, sugar, vegetable oil, and fermentation industries provide by-products of special significance to most commercial dairy rations, and for many home-mixed feeds. However, if producers plan to incorporate by-product feeds in their rations, they should first determine the moisture content of the product, its relative feeding value and price, and the appropriate amount to feed. By-products are extremely variable in feeding value. For example, almond hulls—a widely used by-product feed in California—can range from excellent to poor as a feedstuff. Additionally, some by-product feeds may contain pesticide residues that can be excreted in the milk. For the latter reason, the producers should make sure that their feeds are free from environmental contaminants.

(Also see Chapter 12, By-product Feeds/Crop Residues.)

Special Feeds and Feed Additives for Dairy Cattle

Certain feeds and additives are especially adapted to dairy cattle, primarily to increase milk production and/or to affect milk composition; among such products are: (1) fats and oils, (2) fiber, and (3) the following additives: antibiotics, bovine somatotropin (BST), buffers, ionophores, and isoacid (branched-chain fatty acids and valeric acid).

- **Fats and Oils**— Most forages and grains are low in lipids—they contain less than 2 to 3% fat. In general, dairy cows should be able to utilize 1 to 1½ lb of fat per day in addition to the fat present in natural feedstuffs. This means that about 3% more fat can be added to the total ration (forage plus concentrate), or that 5 to 6% fat can be added to the grain ration.

Added fat is especially effective in early lactation. Because of the increased caloric concentration provided by dietary fat and because high-producing cows are usually in negative energy balance during early lactation, fat is frequently added to the ration to increase the cow's energy intake and provide fatty acids to the udder. It may also be beneficial to provide supplemental fat when the capacity of the gastrointestinal tract limits energy intake.

The type of fat (saturated or unsaturated) added to the ration greatly influences the animal's nutrient utilization, milk production, feeding behavior, ration acceptability, the amount of fat that can be fed, and milk composition. Unsaturated fats are less desirable for dairy cows because of their inhibitory effects on rumen fermentation and digestion. Animal fats (which are more saturated) and blended animal-vegetable fats have generally given the most positive responses in animal performance.

Added ration fat, including feeding whole cottonseed and soybeans, decreases the protein content of milk by about 0.1%, primarily because of the lower casein content. Whole cottonseed also increases the proportion of long-chain fatty acids in milk.

Until the rumen becomes functional, young dairy calves require some fat in the diet. A level of 10% fat in milk replacers appears to be sufficient to supply essential fatty acids, carry fat-soluble vitamins, but insufficient to supply adequate energy for normal gains under optimum environmental temperatures. For veal production, a higher fat milk replacer (15 to 20% or more), will increase fat deposition in the carcass and is desirable. Also, 15 to 20% fat in milk replacers is needed for normal gains when calves are exposed to cold environmental temperatures.

(Also see Chapter 10, sections on "Cottonseed" and "Fats and Oils.")

• **Fiber**—Fiber is important in dairy rations. Excessive fiber levels limit intake and energy concentration, while a shortage of fiber reduces rumen digestibility and milk fat test. Suggested minimums for crude fiber, acid detergent fiber, and neutral detergent fiber are given in Table 20-5.

Feed factors, such as small particle size of forages, that reduce the pH of ruminal fluid, decrease the number and activity of fiber-degrading bacteria and cause a depression in fiber degradation. Feeding an insufficient amount of fiber or feeding forages that have a poor buffering capacity in the rumen may have undesirable effects on rumen fermentation, fiber degradation, and milk fat percentage that are similar to those caused by reducing the particle size of the forage.

So, the general recommendation is that lactating dairy cows should receive at least one-third of the total ration dry matter as long hay or as its DM equivalent in medium-to-coarse chopped silage or other forage. A minimum of 5 lb of forage dry matter measuring 1 to 2 in. in length will meet the fiber need of most lactating cows.

(Also see Chapter 8, section on "NDF, ADF, and NIRS Analyses"; and Chapter 15, sections on "Neutral Detergent Fiber (NDF) and Acid Detergent Fiber (ADF)," and "Near Infrared Reflectance Spectroscopy [NIRS].")

Feed Additives

Many additives are used by dairy producers to increase milk production, affect milk composition, and/or improve feed efficiency; and new products are constantly evolving. Among such additives are those which follow:

• **Antibiotics**—Antibiotics, which are widely used in the diet of young dairy calves, are especially beneficial for calves exposed to adverse conditions of housing, sanitation, and disease.

Those using antibiotics should always read and follow the label directions on any antibiotic container before slaughtering animals or selling milk from cows treated with antibiotics.

(Also, for an in-depth discussion, see Chapter 13, section on "Antibiotics.")

• **Bovine Somatotropin (BST)**—Experimentally, milk, lactose, milk fat, and protein yields have been increased significantly when exogenous bovine somatotropin (BST) has been injected into lactating cows. To meet this higher production, cows consume more total feed when BST is administered. However, the efficiency of milk production by cows is improved because a smaller proportion of the ingested nutrients in feed is needed to take care of maintenance requirements.

(Also, for an in-depth discussion, see Chapter 13, section on "Bovine Somatotropin [BST] For Dairy Cattle.")

• **Buffers (Mineral Salts)**—Buffers are used primarily to improve the feed intake, rumen function, milk production, milk composition, and health of lactating cows. When used for young calves, buffers have given inconsistent results; they will likely be most beneficial when the calf diet being fed results in higher than normal acidity.

The common buffers are: sodium bicarbonate ($NaHCO_3$), magnesium oxide (MgO), sodium bentonite, sodium sesquicarbonate, and calcium carbonate or limestone ($CaCO_3$). Buffers function to maintain the hydrogen ion concentration in the rumen, intestines, tissues, and body fluids, or to increase the rate of passage of liquids from the rumen, or both.

(Also, for an in-depth discussion of buffers, see Chapter 13, section on "Buffers.")

• **Ionophores**—Ionophores are feed additives that change the metabolism within the rumen by altering the rumen microflora to favor propionic acid production. Currently, two ionophores—Bovatec (lasalocid) and Rumensin (monensin)—are FDA-approved for replacement heifers. Both are antibiotics. Feeding Bovatec or Rumensin to replacement heifers improves liveweight gains and the efficiency of feed utilization.

(Also, for an in-depth discussion of ionophores, including approved use levels for replacement heifers, see Chapter 13, section on "Ionophores.")

• **Isoacids (Branched-Chain Fatty Acids and Valeric Acid)**—The isoacids provide three branched-chain fatty acids (isobutyric acid, isovaleric acid, and 2-methyl butyric acid)—the same fatty acids that are made by ruminant bacteria and are present naturally in the rumen of cattle. Isoacids are essential for the growth of some rumen organisms that digest fiber. The use of isoacids may boost milk production by 8 to 10%, or 4 to 6 lb per day, with little or no increase in feed consumption. The mode of action of isoacids is not entirely clear, but it appears to be due to enhancing fiber digestion and acetate production without stimulating insulin secretion. Not all dairy cattle benefit from the use of isoacids. Moreover, the benefits of a profitable response are delayed 30 to 60 days following the initiation of isoacids.

(Also, for an in-depth discussion of isoacids, see Chapter 13, section on "Isoacids [IsoPlus].")

Commercial Feeds

Several different types of commercial feeds are available for dairy cattle; among them, (1) complete dairy concentrates, (2) dry cow rations, (3) fitting rations, (4) growing or young stock rations, (5) calf starters, (6) milk replacer feeds, (7) protein supplements, and (8) mineral and vitamin premixes.

Dairy producers are major users of U.S. commercial feeds; 22% of the primary feed (mixed feed, to which a premix is sometimes added) is fed to dairy cattle.

One of the important roles of commercial feed manufacturers is that of assuring a uniform mix of all ingredients, including those added in very minute amounts. Usually, commercial feed mills offer supplements which can be combined with farm grains, thereby coupling the advantages of both commercial and home mixing.

(Also see Chapter 16, section on "Commercial Feeds.")

Feed Considerations

In addition to being nutritionally complete, the following factors should receive consideration in dairy rations:

1. Palatability.
2. Variety.
3. Bulk.
4. Laxativeness.
5. Cost.
6. Effect on milk flavor.

The following control measures are recommended to alleviate feed flavors:

   a. Avoid sudden change to fresh, lush pasture.
   b. Control and avoid undesirable weeds.
   c. Feed silage after milking.

Feed Substitutions

Feed substitutions for dairy cattle and beef cattle are similar. These are presented in Chapter 19, Table 19-14, Feed Substitution Table for Cattle.

FEED PREPARATION

Most grains for dairy cattle should be processed before feeding, although calves under 6 months of age can be fed whole corn or oats. Coarse grinding or flaking are the preferred methods of processing grains for dairy cattle.

Cows produce as well on chopped or ground hay as on long hay. However, finely ground, pelleted hay affects the amount and proportion of volatile fatty acids in the rumen, with the result that the percentage fat content of the milk is lowered. Wafering or cubing, on the other hand, has less of a depressing effect on the fat content of the milk, and will increase intake and maintain or increase the milk production slightly.

(Also see Chapter 14, Feed Processing.)

RATIONS

Dairy producers must put together the available feeds so as to achieve the most profitable production. At its best, developing a dairy ration involves combining the art and the science of feeding. For small herds, individual animal response may be satisfactory. With large commercial herds, the formulating of rations must be more precise, because small costs per cow become large costs when multiplied by many cows. Yet, the most sophisticated computer must be augmented by the good judgment of the manager if the rations are to be successful in meeting the nutrient needs of individual cows and of the herd as a whole. Producers must always keep in mind that the best formula on paper is not always the best feed. A feed is of no value if it is not actually consumed.

Also, there should be a specific ration for every need—for lactating cows, dry cows, calves, replacement heifers, dairy beef, and show and sale animals.

FEEDING LACTATING COWS

Few animal stresses are as great as those involved in the production of a large volume of milk. For each gallon of milk produced, 400 to 500 gal of blood must pass through the udder. Thus, if a cow is producing 10 gal (86 lb) of milk daily, 15 to 20 tons of blood course the udder each 24 hours. This 10 gal of milk contains more than 3 lb of fat, more than 3 lb of protein, more than 4 lb of lactose (milk sugar), and more than ½ lb of minerals. All these must be supplied in the ration over and above the nutrients needed for the body processes, wastes, and energy to sustain all of the operation.

Also, producers realize greatest profits from feeding when cows convert the maximum proportion of their feed into milk. The nutrient requirements for production depend primarily on the amount and composition of the milk. These needs for cows of all sizes and levels of production are shown in Tables 20-1 to 20-5. Rations that fulfill these requirements, plus a margin of safety, can be formulated based on composition of feeds listed in Section V. The primary concern in feeding lactating cows is to provide a ration adequate in energy, protein, fiber and roughage factor, salt, calcium, phosphorus, and vitamin A (or carotene). When allowances for these nutrients are met, other minerals and vitamins usually are present in sufficient amounts, also.

Additional considerations in feeding lactating cows include palatability of the ration; physical form; protein and mineral content of concentrates; proportion of concentrate to roughage; relative prices of ingredients; voluntary feed intake; and frequency and regularity of feeding. Thus, the proper feeding of lactating cows necessitates that producers have sufficient knowledge relative to basic nutrient requirements and principles to plan an efficient feeding program, and the experience and management ability to apply it.

Dry matter consumption is very important in feeding dairy cows. The best ration formulation on paper will not make for profitable production if the cows either fail to eat it or are given insufficient amounts of it. Also, high-producing cows must consume very large amounts of a balanced ration if they are to produce to their maximum. For this reason, producers should adjust the feeding recommendations given in Table 20-1 to fit the needs of their cows, with consideration given to body weight and condition of cows, milk production, stage of lactation, weather and other environmental factors, and type and quality of feed.

Thumb Rules for Feeding Lactating Cows

The feed requirements of lactating cows are significantly influenced by the volume and composition of the milk that they produce. Although knowledge of the nutrient requirements of the animals and of the composition of feeds is essential in order to feed properly, the ability of the cows to consume sufficient volume of feed complicates adequate feeding. Table 20-8 may be used as a guide for dry matter intake; and the two sections that follow give some thumb rules relative to the amount and kind of forage and the amount and kind of concentrate to feed.

TABLE 20-8
DAILY DRY MATTER INTAKE GUIDELINES[1]

Live Wt.: (lb)	900	1,100	1,200	1,300	1,500
(kg)	*409*	*499*	*545*	*590*	*681*
Milk[2]	Percent of Body Weight[3]				
(lb/day) *(kg/day)*	(%)	(%)	(%)	(%)	(%)
20 *9.1*	2.6	2.3	2.2	2.1	2.0
30 *13.6*	3.0	2.7	2.6	2.5	2.3
40 *18.2*	3.4	3.1	2.9	2.8	2.5
50 *22.7*	3.8	3.4	3.2	3.1	2.8
60 *27.2*	4.1	3.7	3.5	3.4	3.1
70 *31.8*	4.6	4.0	3.8	3.6	3.3
80 *36.3*	5.1	4.3	4.1	3.8	3.5
90 *40.9*		4.7	4.4	4.1	3.7
100 *45.4*		5.0	4.7	4.4	3.9

[1]Adapted by the authors from: Linn, J. G., M. F. Hutjens, W. T. Howard, L. H. Kilmer, and D. E. Otterby, *Feeding the Dairy Herd,* Cooperative Extension Services, Universities of Illinois, Iowa State, Minnesota, and Wisconsin, 1988, p.27, Table 19.

[2]Fat-corrected milk = (milk lb × .4) + (fat lb × 15).

[3]Intakes may be up to 18% less for cows in early lactation.

TABLE 20-9
AMOUNT OF CONCENTRATE (GRAIN) TO FEED BY PERIODS (1,400 LB *[636 KG]* COW, 4% MILK)[1]

	Milk Production Ability of the Cow[2]			
Average Daily 1st Period .. (lb)	50	60	80	90–100
Average Daily 1st Period .. (kg)	*23*	*27*	*36*	*41–45*
Lactation Total .... (lb)	10,000	12,000	15,000	18,000
Lactation Total .... (kg)	*4,540*	*5,448*	*6,810*	*8,172*
Phase of Lactation	Grain to Milk Ratio			
1 (1st 10 weeks)	1:4	1:3	1:3	1:2.5
2 (2nd 10 weeks)	1:4	1:3	1:3	1:3
3 (last 24 weeks)	1:4	1:4	1:2.5	1:2.5
	Daily	Daily	Daily	Daily
4 (dry, 6–8 weeks) .... (lb)	0–4	0–4	0–4	0–6
(dry, 6–8 weeks) .... (kg)	*0–1.8*	*0–1.8*	*0–1.8*	*0–2.7*
Total grain (approximate) .... (lb)	3,000	4,000	5,000	6,000
Total grain (approximate) .... (kg)	*1,362*	*1,816*	*2,270*	*2,724*

[1]Adapted by the authors from: Linn, J. G., M. F. Hutjens, W. T. Howard, L. H. Kilmer, and D. E. Otterby, *Feeding the Dairy Herd,* Cooperative Extension Services, Universities of Illinois, Iowa State, Minnesota, and Wisconsin, 1988, p.26, Table 18.

[2]Ratios based on 100% dry matter basis, grain containing 80 Mcal, and forage 60 Mcal of NE_{lc} per 100 lb *(45 kg)*.

AMOUNT AND KIND OF FORAGE TO FEED

The common thumb rules for forage feeding of lactating cows follow.

1. **Forage dry matter and intake.** The forage should constitute a minimum of 40% of the total dry matter of the ration and account for an intake of approximately 1.5% of the body weight daily.

2. **Hay consumption.** If good quality hay only is fed, a cow will eat about 3 lb per 100 lb of body weight.

3. **Silage.** Depending on the moisture content, 2.5 to 4.5 lb of silage are equal to (and may replace) 1 lb of hay; the lower feeding value of silage is due to its high moisture content—hay runs 10 to 15% moisture, whereas silage runs 65 to 75% moisture.

4. **Hay/grain equivalent.** It takes about 3 lb of good hay to supply the same amount of usable energy as 2 lb of grain.

5. **Pasture (grass) consumption.** Cows will consume 100 to 200 lb of pasture per day; since pasture normally contains 70 to 85% moisture, that's 15 to 60 lb of dry matter per day.

6. **Forage:concentrate ratio.** If forage is very high quality, cows will eat more of it, with the result that the grain requirement will be lessened. However, over and above meeting the minimum forage requirement, the proportion of forage to concentrate should be determined primarily by the economics of the situation—that is, it should be decided on the basis of the relative price of available forage and concentrate, the milk production, and the net returns.

AMOUNT AND KIND OF CONCENTRATE (GRAIN) TO FEED

The common thumb rules for concentrate feeding of dairy cows follow.

1. **Amount of concentrate (grain).** The concentrate (grain) should constitute a maximum of 60% of the total dry matter of the ration and account for an intake of not to exceed 2.3% of the body weight daily. Table 20-9 can be used as a guide for feeding concentrate (grain) according to milk production.

2. **Amount and kind of protein.** Feed protein according to requirements (19% in early lactation, decreased thereafter according to milk production—see Table 20-5). A low rumen degradable protein source is recommended for high-producing cows in early lactation. Limit urea to 0.4 lb per day, and preferably to 0.2 lb per day, in phases 1 and 2.

3. **Added fat.** In addition to the fat present in natural feedstuffs, lactating cows may be fed 1 to 1½ lb of *added fat* per day; which translates into about 6% added fat to the concentrate (grain) ration, or 3% added fat to the total mixed ration (grain and forage combined). Fats in oilseeds (soybeans or whole cottonseed) should be considered as added fat. When feeding added fat, increase the calcium to 0.9 to 1.0%, the magnesium to 0.3%, and the acid detergent fiber to 20%.

4. **Salt.** Include 1% salt in the concentrate (grain) mix, or 0.5% salt in the total ration (concentrate and forage combined); which will provide for a salt intake of 2 to 3 oz per cow per day.

5. **Calcium/phosphorus and trace minerals.** A calcium/phosphorus mineral source should constitute 1 to 2% of the grain mix, or be fed at a rate of 1 oz per 10 lb of milk. Trace minerals should be incorporated in the ration or self-fed in trace mineralized salt to meet the requirements.

6. **Vitamins.** Vitamins A, D, and E should be added to the ration to meet the requirements.

Any thumb rules, and even calculated values (calculated by hand or computer), are estimates to be used with some judgment by the feeder. The rule of the successful feeder is to increase concentrates so long as cows respond with extra milk at a profit. The commonly used profit indicator is the "milk-feed price ratio," which is the pounds of 16% protein dairy concentrate equal in value to 1 lb of milk.

Generally the cost of 16% protein dairy concentrate per pound is about 70% of the price received for milk; of course there are yearly, seasonal, and area variations. For the United States as a whole, the milk-feed price ratio was $1.31 in 1975 and $1.79 in 1986.[3] Thus, the cost per pound of 16% protein dairy concentrate was 76% the price received per pound of milk in 1975, and 56% in 1986.

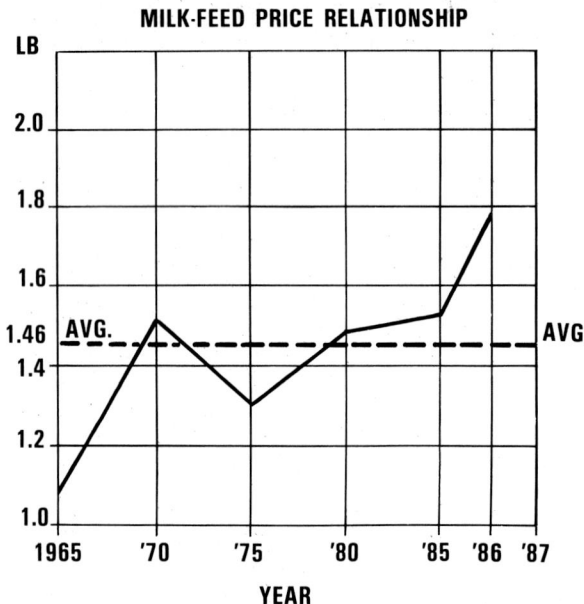

Fig. 20-8. Pounds of 16% protein dairy concentrate equal in value to 1 lb of milk. (Adapted by the authors from USDA sources)

The amount of grain that it will pay to feed milking cows depends upon several factors: (1) quality of forage, (2) price of forage, (3) price and quality of concentrates, (4) price of milk, and (5) inherent producing ability of cows. Of these, the most important is the inherent milk-producing ability of the cow. To determine the inherent ability to produce milk, dairy producers must rely on (1) the use of milk and fat production records of each cow, and (2) weighing and feeding the proper proportions of concentrates, with consideration given to the stage of lactation of the cows.

Rations for Lactating Cows

Since proteins are always the most expensive part of a ration, for practical reasons the dairy producer should not feed more of them than necessary. Nevertheless, the protein level should be in keeping with the production—the higher the yield, the higher the protein requirement. Generally speaking, this is best accomplished by varying the protein content of the grain mix according to the protein content of the roughage being fed.

The concentrate ration needed to supplement the available roughage on the dairy farm may be either home mixed or commercially manufactured. Home-mixing involves the mixing of homegrown grains and such purchased feeds as necessary to balance the ration. Commercially manufactured feeds are generally nutritionally complete, concentrate feeds. On small dairies where grain is home-grown or abundantly available locally, home mixing of concentrates is a widely used practice. However, there has been an increase in the use of commercial feeds on small dairies. Large specialized dairies purchase individual feed ingredients in bulk, then mix and feed total mixed rations.

Not all dairy producers, in the United States or in other countries, who home-mix feeds balance rations (1) on the basis of the chemical analyses of their feed ingredients, (2) with the use of computers, or (3) by combining the concentrates and forage into a complete ration. For these producers, Table 20-10, Feeding Guide For Lactating Cows, may serve as a useful guide. It shows how ingredients partitioned into four approximate protein levels (columns 2, 3, 4, and 5) may be combined to make concentrates suitable for feeding with three different qualities of roughages—excellent, medium, and poor.

Variations can and should be made in the rations listed in Table 20-10. Producers should give consideration to the supply of homegrown feeds, and to the availability and price of ingredients. Feeds of similar nutritive properties can and should be interchanged as price relationships warrant. Thus, the cereal grains may consist of corn, barley, wheat, oats, and/or sorghum; the protein supplements may consist of soybean, cottonseed, peanut and/or linseed meal; and a vast array of by-product feeds may be utilized.

Here is how to use Table 20-10: Let's assume that a producer has (1) medium-quality forage, and (2) both low- and medium-high protein (columns 2 and 4) ingredients from which to choose. How many pounds each of the low- and medium-high protein ingredients will be required in a 1,000-lb concentrate mix? Step by step, here is the answer:

1. Look under "Medium roughage—medium protein forage" 15-17% (column to the left).
2. Mix No. 6, containing 650 lb of low-protein ingredients (under column 2, under 12% ingredients) and 350 lb of medium-high protein ingredients (column 4, 18 to 28% protein), will meet the needs. The concentrates may be chosen from among those listed at the top of the respective columns of Table 20-10—the low-protein concentrates from column 2 (under 12%) and the medium-high protein concentrates from column 4 (18 to 28%).

HOW TO BALANCE A DAIRY RATION

Good dairy producers have nutrient analyses made of all major ration ingredients and use computers to balance their rations for at least 20 nutrients.

It is recognized that rations should vary with conditions, and that many times they should be formulated to meet the conditions of a specific dairy farm. Also, good producers should know how to balance rations. Complete instructions on how to balance rations (including examples) are given in Chapter 18, Feeding Standards/Ration Formulation. By (1) following these instructions, and (2) using the nutrient requirement tables, Tables 20-1 to 20-5, it is possible to balance rations for specific weights of animals and levels of production.

---

[3]USDA sources.

TABLE 20-10
FEEDING GUIDE FOR LACTATING COWS (AS-FED BASIS)[1]

Note: This shows how ingredients of 4 protein levels may be combined to make different concentrate mixes of approximate protein content to match 3 different qualities of roughages.

(1) Suggested Grain Mix, Based on Kind of Roughage Available	(2) Low-Protein (Under 12%) Ingredients		(3) Low-Medium Protein (12–18%) Ingredients		(4) Medium-High Protein (18–28%) Ingredients		(5) High-Protein (Over 32%) Ingredients	
Feeds	(% protein)		(% protein)		(% protein)		(% protein)	
	Barley, all analyses ...... 11.7 Beet pulp w/molasses, dried . 9.3 Corn-and-cob meal .... 7.8 Corn #2 .... 8.9 Dairy feed, 12% .... 12.0 Hominy feed .... 10.3 Molasses, cane* .... 4.3 Oats, all analyses .... 11.9 Rye, all analyses* .... 12.0 Sorghum (milo) .... 10.1 Wheat, all analyses 13.1		Dairy feed, 16% .... 16.0 Wheat bran .... 15.5 Wheat middlings 16.4		Brewers' dried grains* .... 27.3 Copra (coconut) meal .... 21.3 Corn gluten feed .... 23.0 Dairy feed, 18–24% .... 18–24 Distillers' dried grains .... 27.3 Malt sprouts .... 22.9 Peas, field* 23.2		Dairy feed, 32–34% .. 32–24 Corn gluten meal ..... 60.8 Cottonseed meal* ..... 41.2 Linseed meal .... 35.7 Peanut meal .... 49.0 Soybean meal 44.4	
	(lb)	(kg)	(lb)	(kg)	(lb)	(kg)	(lb)	(kg)
Excellent roughage—High-protein forage, 18%: (1) legume, or (2) legume and nonlegume mixed forages of *high quality*; consisting of dry forages and/or silage.								
Mix No. 1	1,000	454						
Mix No. 2	900	409					100	45
Mix No. 3	800	363			200	91		
Mix No. 4	850	386	100	45			50	23
Medium roughage—Medium-protein forage, 15–17%: (1) legume, or (2) legume and nonlegume mixed forages of *medium quality*; consisting of dry forages and/or silage.								
Mix No. 5	800	363					200	91
Mix No. 6	650	295			350	159		
Mix No. 7	700	318	100	45	100	45	100	45
Mix No. 8	Straight 16% dairy feed, or ½ Mix No. 9 and ½ 16% dairy feed							
Poor roughage—Low-protein forage, under 14%: nonlegume forage; consisting of dry forages and/or silage.								
Mix No. 9	700	318	300	136				
Mix No. 10	600	272			200	91	200	91
Mix No. 11	600	272	100	45	100	45	200	91
Mix No. 12	500	227	and 500 lb (227 kg) 32% dairy feed					

[1]The protein compositions in columns 2 to 5 were obtained from Section V, Composition of Feeds.

Comments:

Add—To all rations (1) 1% iodized or trace-mineralized salt; (2) 1% steamed bone meal, dicalcium phosphate, or the equivalent (use monosodium phosphate or a high-phosphorus commercial mineral where alfalfa is fed liberally); (3) 1,000 IU of vitamin A/lb *(2,205 IU of vitamin A/kg)* of concentrate and, unless cows are in sunlight, add 150 IU of vitamin D/lb *(331 IU of vitamin D/kg)* of concentrate.

*Limitations—Wheat, not more than 50% of the ration; dried molasses beet pulp, 20%; molasses, 15%; peas and brewers' dried grains, 30%; rye, 10%; and cottonseed meal, 20% of the mix for calves, but as needed for mature cows.

COMPUTER-FORMULATED RATIONS/DISKETTE [4]

Until the late 1970s, only those dairy producers with access to a large main-frame computer could formulate a ration using the computer. Usually this was limited to those associated with a university (Extension Service) or subscribing to a time-sharing system. Many rations were formulated by using a pencil, an eraser, paper, and a calculator. Balancing rations was time-consuming, and options were limited. Then came the microcomputer! By the mid–1980s, most dairy producers owned or had access to a microcomputer. Rations could be, and were, adjusted with changes in availability, price, composition, and moisture content of ingredients.

[4]This section on "Computer-Formulated Rations/Diskette" was prepared specially for this book by L. M. Larsen, Ph.D., Consultant, Nutri-Systems, 426 E. Shields, Fresno, CA 93704.

In the past, computer programs were written by software companies and universities to convert the tabular NRC requirements into equations that computed the animal requirements. With the publication of *Nutrient Requirements of Dairy Cattle,* sixth revised edition, a new step was taken by the NRC committee. In addition to providing nutrient requirements for dairy cattle, as reproduced in this chapter in Tables 20-1 to 20-5, the committee published prediction equations. These equations may easily be translated into computer programs by programmers. Nutrient requirements for growing, lactating, or dry cows are generated by these equations.

Also, the NRC committee commissioned a software company, Microsoft® corporation, to develop a FORTRAN program that is supplied with the dairy cattle publication in the form of a 5¼ in. diskette. This program will calculate nutrient requirements from the prediction equations and from specific information supplied by the producer or nutritionist. The results of the calculations may either (1) be displayed on the computer screen, or (2) be printed. The printout is in two parts: The first part lists the requirements for the major nutrients and for vitamins A and D and is useful to the producer or nutritionist formulating rations. The second part is a very involved breakdown predicting utilization of dietary protein, information of value primarily to those who conduct research. Fig. 20-9 is a sample printout of the first page generated by this program.

```
NRC DAIRY (1988) REQUIREMENTS CALCULATED ON  3- 8-1989 AT 12:58
PREGNANT OR LACTATING CATTLE
ENERGY CONCENTRATION FED/NRC ASSUMED IS      1.000
LIVE WEIGHT IN LB IS                         1400.
MILK PRODUCTION IN KG IS                       75.
MILK FAT TEST % IS                              3.65
NUMBER OF DAYS PREGNANT IS                      0.
LACTATION NUMBER IS                             3
PROPORTIONAL FEED NEL/REQUIRED NEL IS           1.00
WEIGHT CHANGE IN LACTATION IS                    .750 LB
DRY MATTER INTAKE IS                           48.39  LB   OR     3.46  % LW
NEL NEEDED IS                                  35.76  MCAL OR      .74  MCAL/LB
ME NEEDED IS                                   59.89  MCAL OR     1.24  MCAL/LB
DE NEEDED IS                                   69.11  MCAL OR     1.43  MCAL/LB
BASELINE TDN NEEDED IS                         34.56  LB   OR    71.41  % DM
CRUDE PROTEIN INTAKE NEEDED IS                  7.902 LB   OR    16.33  % DM
UNDEGRADED INTAKE PROTEIN NEEDED IS             2.761 LB   OR     5.70  % DM
DEGRADED INTAKE PROTEIN NEEDED IS               4.678 LB   OR     9.67  % DM
INTAKE PROTEIN (IP) NEEDED IS                   7.439 LB   OR    15.37  % DM
CALCIUM NEEDED IS                                .285 LB   OR      .589 % DM
PHOSPHORUS NEEDED IS                             .181 LB   OR      .373 % DM
VITAMIN A NEEDED IS                           48263.   IU  OR   997.    IU/LB
VITAMIN D NEEDED IS                           19051.   IU  OR   394.    IU/LB
UNDEGRADED INTAKE PROTEIN IN IP IS                                37.11 % IP
```

Fig. 20-9. NRC dairy requirements calculated by using the data diskette supplied with *Nutrient Requirements of Dairy Cattle,* sixth revised edition, 1988.

The program runs on IBM-PC™ or PC-compatible computers. Various computer software companies have written the equations into their dairy ration formulation software, making the procedure of ration formulation more rapid and convenient. Fig. 20-10 is a computer printout of a ration for lactating cows formulated with the use of the NRC prediction equations.

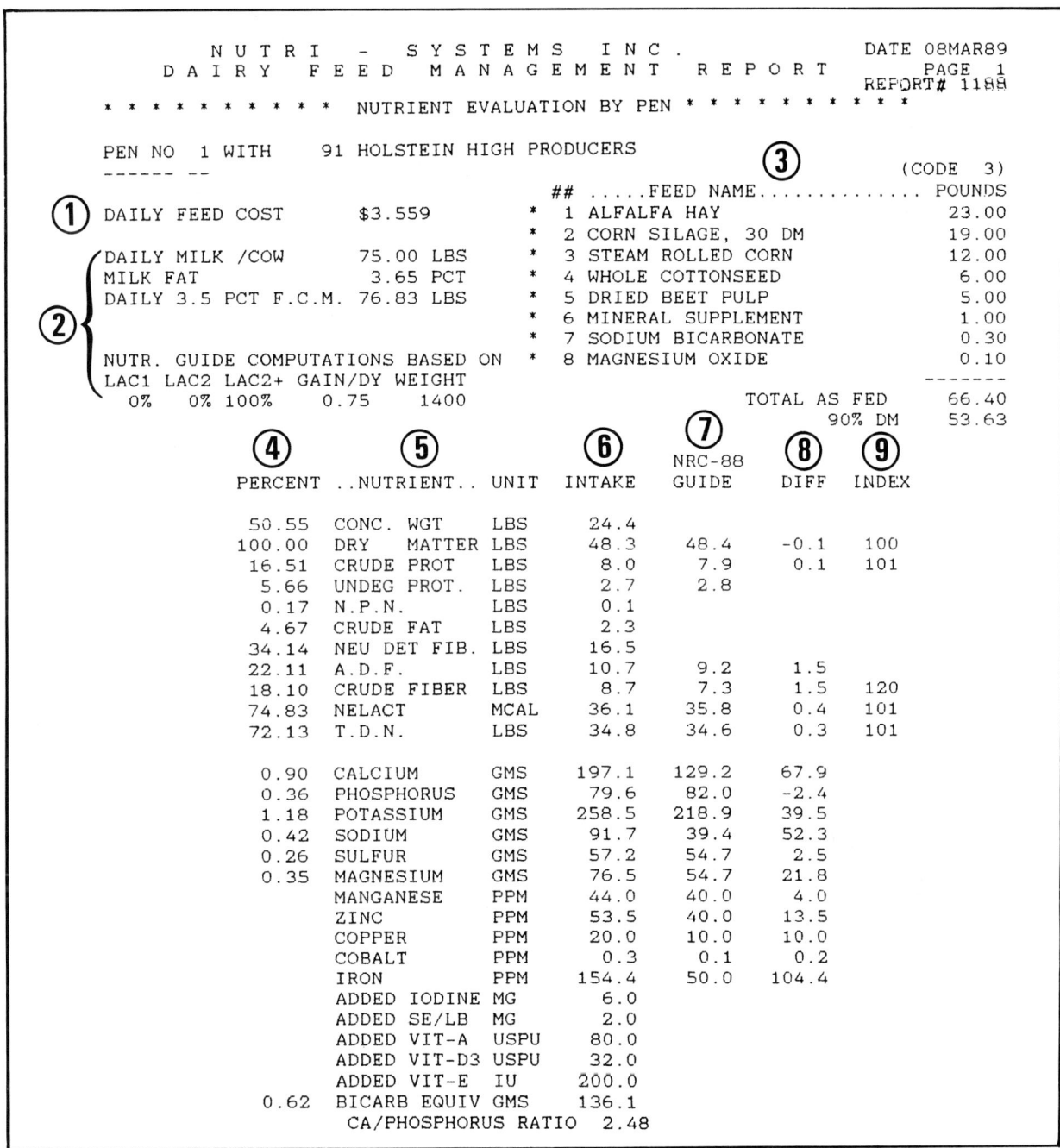

Fig. 20-10. Lactating dairy ration comparing nutrient intake (column 6) with the NRC requirement (column 7).

An explanation of the numbered sections (columns) of Fig. 20-10 follows:

1. Feed cost per cow per day.
2. Input data used for computing the requirements.
3. Feedstuffs amounts given in pounds per cow per day.
4. Nutrient composition of the ration on a 100% dry matter basis.
5. Nutrient names and units.
6. Daily intake of nutrients per cow per day.

7. The NRC requirements computed from the prediction equations.

8. Deficiencies (negative values) or excesses (positive values). These values are computed by subtracting the requirements from the daily intake figures.

9. A computed index for dry matter, crude protein, NE_{lc}, and TDN. This value is computed as the percentage that nutrient intake is of the nutrient requirement.

(For a more detailed discussion of computer formulation of rations, see Chapter 18, section on "Computer Methods.")

Feeding Systems

Traditional individual feeding of lactating cows in stanchioned barns or milking parlors is giving way to new feeding systems. Although the newer methods are not as effective as feeding cows individually, they are much more economical than feeding all cows in the herd the same amount of grain, regardless of production. Additionally, they make for considerable saving in labor and facilities.

PHASE FEEDING

Phase feeding is a feeding program that is divided into periods based on milk production, milk fat percentage, feed intake, and body weight. Fig. 20-11 illustrates the shape and relationship of curves for milk production, fat percentage, dry matter intake, and body weight. Based on these curves, four distinct feeding phases of lactating cows can be identified.

Producers should formulate rations to match each of these phases in order to optimize milk yield, minimize metabolic disorders, increase longevity, and increase profits. The four phases illustrated in Fig. 20-11 are:

1. Phase 1, early lactation, 0 to 70 days postpartum.
2. Phase 2, peak dry matter intake, second 10 weeks postpartum.
3. Phase 3, mid- to late-lactation, 140 to 305 days postpartum.
4. Phase 4, dry period, 45 to 60 days before parturition.

The example rations presented in Table 20-11 are suitable for the three lactating phases.

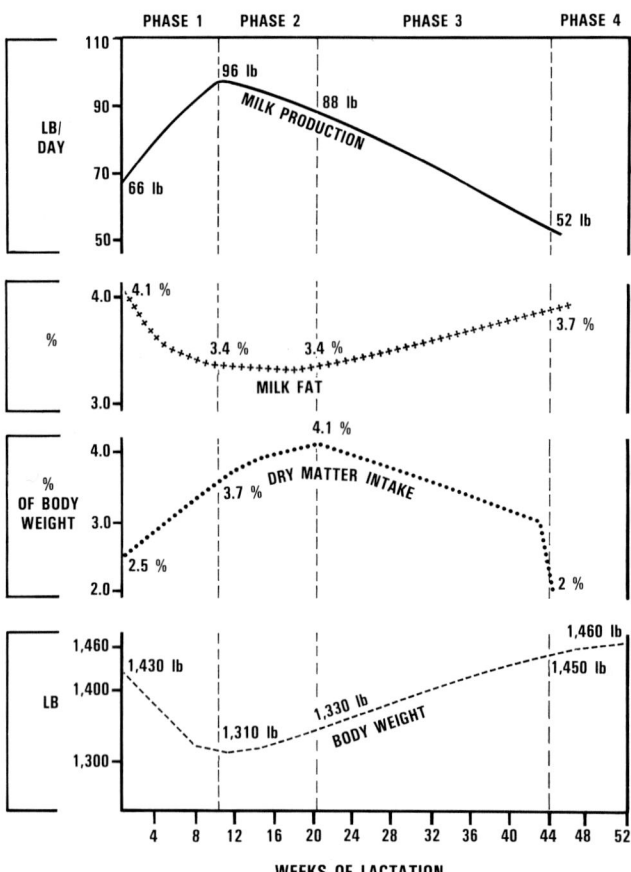

Fig. 20-11. Lactation cycle phases with corresponding changes in milk production, milk fat percentage, dry matter intake, and body weight. (*Source:* Linn, J. G., M. F. Hutjens, W. T. Howard, L. H. Kilmer, and D. E. Otterby, *Feeding the Dairy Herd,* Cooperative Ext. Services, Ill., Iowa State, Minn., and Wisc., 1988, p. 15, Fig. 6)

**TABLE 20-11
EXAMPLE RATIONS
FOR VARIOUS MILK PRODUCTION PHASES,
1,350 LB *(613 KG)* COW, 3.8% FAT TEST[1]**

Item		Phase 1	Phase 2	Phase 3
Milk (lb/day)		90	80	50
Milk *(kg/day)*		*40.9*	*36.3*	*22.7*
DM intake[2] (lb/day)		49	51	38
DM intake[2] *(kg/day)*		*22.2*	*23.2*	*17.3*
		As-Fed		
		(lb/day) *(kg/day)*	(lb/day) *(kg/day)*	(lb/day) *(kg/day)*
Ration 1				
Alfalfa hay (88% DM), 140 RFV, 20% crude protein		28 *12.71*	34 *15.44*	27 *12.26*
Corn-oats[3]		21 *9.53*	24 *10.90*	16 *7.26*
Soybean meal, 44%		5.0 *2.27*	— —	— —
Dical, 18% phosphorus		0.5 *0.23*	0.45 *0.20*	0.30 *0.14*
Salt, vitamins, trace mineralized		0.30 *0.14*	0.25 *0.11*	0.25 *0.11*
Weight change		-1.5 *-0.68*	— —	+.5 *+.23*
Ration 2 (corn silage limit fed)				
Alfalfa hay, 140 RFV, 20% CP		19 *8.63*	34 *15.44*	23 *10.44*
Corn silage (35% DM)		25 *11.35*	25 *11.35*	25 *11.35*
Corn-oats		18 *8.17*	12 *5.45*	10 *4.54*
Soybean meal, 44%		7.5 *3.41*	0.3 *0.14*	— —
Dical, 18% phosphorus		0.45 *0.20*	0.50 *0.23*	0.3 *0.14*
Salt, vitamins, trace mineralized		0.30 *0.14*	0.25 *0.11*	0.25 *0.11*
Weight change		-1.2 *-0.54*	— —	+.5 *+.23*
Ration 3 (hay limit fed)[4]				
Alf-grass hay, 113 RFV, 16% CP		10 *4.54*	10 *4.54*	10 *4.54*
Corn silage		41 *18.61*	70 *31.78*	57 *25.88*
Corn-oats		16 *7.26*	11 *4.99*	6 *2.72*
Soybean meal, 44%		11.5 *5.22*	8.2 *3.72*	4.5 *2.04*
Dical, 18% phosphorus		0.40 *0.18*	0.30 *0.14*	0.25 *0.11*
Limestone		0.40 *0.18*	0.30 *0.14*	0.15 *0.07*
Salt, vitamins, trace mineralized		0.30 *0.14*	0.25 *0.11*	0.25 *0.11*
Weight change		-1.4 *-0.64*	+.7 *+.32*	+.5 *+.23*
Ration 4				
Alf-grass hay, 113 RFV, 16% CP		23 *10.44*	32 *14.53*	24 *10.90*
Corn-oats		22 *9.99*	22 *9.99*	19 *8.63*
Soybean meal, 44%		8.5 *3.86*	3.5 *1.59*	1.1 *0.50*
Dical, 18% phosphorus		0.45 *0.20*	0.40 *0.18*	0.25 *0.11*
Limestone		0.20 *0.09*	— —	— —
Salt, vitamins, trace mineralized		0.30 *0.14*	0.25 *0.11*	0.25 *0.11*
Weight change		-1.9 *-0.86*	— —	+.5 *+.23*

[1]*Source:* Linn, J. G., M. F. Hutjens, W. T. Howard, L. H. Kilmer, and D. E. Otterby, *Feeding the Dairy Herd,* Cooperative Ext. Services, Ill., Iowa State, Minn., and Wisc., 1988, p. 16, Table 6.
[2]Estimated average intake during the phase.
[3]85% corn–15% oats mix.
[4]Feed amounts may have to be limited during phase 2 and 3 to avoid over-conditioning.

Table 20-12 lists examples of dry cow rations.

TABLE 20-12
EXAMPLE DRY COW RATIONS, 1,400 LB (636 KG) DRY COW[1]

Forage	As-Fed	
	(lb/day)	(kg/day)
Grass forage		
Orchard grass hay, 12% crude protein	25.0	11.35
Corn	3.0	1.36
Soybean meal	0.5	0.23
Limestone	0.15	0.07
Trace mineralized salt and vitamins	0.1	0.05
Limited legume forage[2]		
Alfalfa hay, RFV 140, 20% crude protein	12.0	5.45
Corn silage	43.0	19.52
Monosodium phospshate	0.1	0.05
Trace mineralized salt and vitamins	0.1	0.05
Limited corn silage		
Alfalfa-grass hay, RFV 113, 16% crude protein	21	9.53
Corn silage	20	9.08
Dicalcium phosphate	0.1	0.05
Trace mineralized salt and vitamins	0.1	0.05

[1]*Source:* Linn, J. G., M. F. Hutjens, W. T. Howard, L. H. Kilmer, and D. E. Otterby, *Feeding the Dairy Herd,* Cooperative Ext. Services, Ill., Iowa State, Minn., and Wisc., 1988, p. 17, Table 7.
[2]Ration contains excess energy as formulated and may over-condition cows in some situations.

CHALLENGE FEEDING (LEAD FEEDING)

Challenge feeding, or lead feeding, refers to feeding the lactating cow so that she is challenged to reach her peak (summit) production level early in lactation.

Because of the strong relationship between peak (summit) milk yield and the total milk production for the entire lactation period, emphasis should be placed on attaining maximum yield between weeks 3 and 8. According to figures provided by Iowa State University extension dairy specialist Ron Orth, for every 5 lb increase in the summit peak, rolling herd average will increase 1,000 lb.

Challenge feeding helps a cow reach peak production earlier than she otherwise might, thus taking advantage of the fact that her system, at the time, is physiologically adapted to heavy production.

After peak production is reached, the amount of concentrate fed should be determined by a concentrate feeding guide based on body weight, milk production, and fat test (see Table 20-9).

GROUP FEEDING

Individual feeding of lactating cows has largely given way to mechanized group feeding. The latter was developed for convenience and saving of labor, rather than for improved animal well-being or feed efficiency. Today, lactating herds with several hundred cows are common; and some herds number several thousand. In order to design a nutritional program for such large numbers that can be adapted to the specific needs of the cows, they are separated into groups according to production (and, therefore, nutritional needs).

When producers decide to go to group feeding, they must decide on the number of groups into which to divide the herd. To answer this question, consideration should be given to the following: (1) herd size; (2) types and costs of available feeds; (3) current type of housing, feeding, and milking system; and (4) overall economic integration of the operation—for example, labor, machinery, etc.

In large herds (more than 250 milking cows), a commonly used system is one in which a minimum of 5 groups are established: (1) high-production cows (about 90 lb of milk/head/day), (2) medium-production cows (about 65 lb of milk/head/day), (3) low-production cows (about 45 lb of milk/head/day), (4) dry cows, and (5) first calf heifers. More groups are desirable in very large herds if corrals and facilities are available. Because of feeding and social considerations, a maximum of 100 cows per group is advisable. With this program, there can be a maximum of two moves during the lactation cycle. In many cases, only one move is necessary; and in a few cases, no moves are required. This system allows each group to be fed according to need. The high-producing groups should be fed the highest quality ingredients at maximum levels. The middle-producing cows should be fed in such a way as to reduce feed cost, increase butterfat test, improve rumen function, and promote lactation persistency. The same holds true for the low-producing cows as for the medium producers except that considerable care must be exercised to avoid excessive fattening.

When group feeding programs are followed, grain is seldom fed in the milking parlor. This is commonly referred to as corral or bunk feeding since feeding generally takes place in bunks along the fenceline of the corrals or pens. Studies have demonstrated that cows fed their grain as a group in a common manger do as well as those fed individually in the milking parlor, but some cows may not always come into the parlor as easily when there is no grain to attract them.

• **Example rations for group feeding**—Example rations for group feeding mature lactating cows at each of 3 milk production levels (90, 65, and 45 lb) are presented in Tables 20-13, 20-14, and 20-15; and example rations suitable for feeding dry cows are presented in Table 20-16.

When group feeding, first-calf heifers should be handled in a separate group and fed for both milk production and growth. Their nutrient requirements for milk production are similar to the requirements of their older counterparts producing milk at the same level, but, because of their growth, they should receive about 20% more nutrients than are required for maintenance.

Although variations can and should be made in Tables 20-13, 20-14, 20-15, and 20-16 rations (see pages 456 and 457), they are excellent guides. When milk yields and/or body weights differ from those used in these tables, a suitable computer program or hand calculations should be employed to obtain amounts to be fed, and to give consideration to costs of alternative feeds.

TABLE 20-13
COMPLETE RATIONS FOR 1,300 TO 1,400 LB *(591 TO 636 KG)* COWS, IN EARLY LACTATION—HIGH-PRODUCTION GROUP (90-LB MILK, 3.6% FAT AVERAGE)[1]

Feeds	All Alfalfa		¾ Alfalfa ¼ Corn Silage		⅔ Alfalfa ⅓ Corn Silage		½ Alfalfa ½ Corn Silage		⅓ Alfalfa ⅔ Corn Silage		¼ Alfalfa ¾ Corn Silage	
	All Amounts on Dry Matter Basis											
	(lb)	(kg)	(lb)	(kg)	(lb)	(kg)	(lb)	(kg)	(lb)	(kg)	(lb)	(kg)
Alfalfa, 17% crude protein, 59% TDN	24.2	10.99	19.2	8.72	17.28	7.85	13.59	6.17	9.24	4.19	7.03	3.19
Corn silage, 8% crude protein, 68.7% TDN	—	—	6.4	2.91	8.63	3.92	13.59	6.17	18.51	8.40	21.10	9.58
Corn, 10% crude protein, 88% TDN	18.9	8.58	16.1	7.31	15.32	6.96	13.11	5.95	11.22	5.09	10.16	4.61
44% crude protein supplement, 49% crude protein, 86% TDN	5.5	2.50	6.6	3.00	7.00	3.18	7.83	3.55	8.82	4.00	9.34	4.24
Fat, 182% TDN	1.2	0.54	1.2	0.54	1.19	0.54	1.19	0.54	1.18	0.54	1.19	0.54
Dicalcium phosphate, 22% calcium, 19% phosphorus	0.42	0.19	0.52	0.24	0.51	0.23	0.49	0.22	0.46	0.21	0.44	0.20
Limestone, 34% calcium	—	—	0.07	0.03	0.14	0.06	0.25	0.11	0.41	0.19	0.48	0.22
Trace mineralized salt	0.25	0.11	0.25	0.11	0.25	0.11	0.25	0.11	0.25	0.11	0.25	0.11
Mineral-vitamin mix	0.08	0.04	0.08	0.04	0.08	0.04	0.09	0.04	0.10	0.05	0.11	0.05
	Ration Nutrient Information											
Dry matter	50.5	22.93	50.4	22.88	50.4	22.88	50.4	22.88	50.2	22.79	50.1	22.75
Crude protein	8.72	3.96	8.71	3.95	8.72	3.96	8.72	3.96	8.72	3.96	8.72	3.96
Crude protein (%)	17.26		17.29		17.29		17.30		17.36		17.40	
TDN	37.52	17.03	37.39	16.98	37.43	16.99	37.39	16.98	37.33	16.95	37.30	16.93
TDN (%)	74.29		74.19		74.26		74.18		74.36		74.446	
NE$_{lc}$ (Mcal)	38.72		38.66		38.72		38.72		38.72		38.72	
NE$_{lc}$ (Mcal/lb or kg)	0.767	1.69	0.767	1.69	0.768	1.69	0.768	1.69	0.771	1.70	0.773	1.70
Calcium (g)	208		206		206		206		205		205	
Calcium (%)	0.91		0.90		0.90		0.90		0.90		0.90	
Phosphorus (g)	103		114		114		114		114		114	
Phosphorus (%)	0.45		0.50		0.50		0.50		0.50		0.50	
Acid detergent fiber (%)	18.04		18.12		18.07		18.24		18.09		18.05	
Ether extract (%)	5.66		5.54		5.50		5.41		5.30		5.28	
Forage:grain ratio	48:52		51:49		51:49		54:46		55:45		56:44	

[1] Ration calculations based on 1988 NRC recommendations, and on the use of either alfalfa or alfalfa and corn silage (CS) forage. Table 20-13 was prepared specially for this book by D. E. Otterby, Professor, and J. G. Linn, Extension Animal Scientist, Department of Animal Science, University of Minnesota, St. Paul; with metric added by the authors.

TABLE 20-14
COMPLETE RATIONS FOR 1,300 TO 1,400 LB *(591 TO 636 KG)* COWS, IN MID-LACTATION—MEDIUM-PRODUCTION GROUP (65-LB MILK, 3.6% FAT AVERAGE)[1]

Feeds	All Alfalfa		¾ Alfalfa ¼ Corn Silage		⅔ Alfalfa ⅓ Corn Silage		½ Alfalfa ½ Corn Silage		⅓ Alfalfa ⅔ Corn Silage		¼ Alfalfa ¾ Corn Silage		All Corn Silage	
	All Amounts on Dry Matter Basis													
	(lb)	(kg)	(lb)	(kg)	(lb)	(kg)	(lb)	(kg)	(lb)	(kg)	(lb)	(kg)	(lb)	(kg)
Alfalfa, 17% crude protein, 59% TDN	21.93	9.96	17.93	8.14	16.43	7.46	11.98	5.44	8.07	3.66	6.15	2.79	—	—
Corn silage, 8% crude protein, 68.7% TDN	—	—	5.98	2.71	8.20	3.72	—	—	—	—	—	—	—	—
Urea-corn silage, 12% crude protein, 70% TDN	—	—	—	—	—	—	11.98	5.44	16.16	7.34	18.46	8.38	26.39	11.98
Corn, 10% crude protein, 91% TDN	17.12	7.77	14.29	6.49	13.23	6.01	13.25	6.02	12.48	5.67	11.80	5.36	6.28	2.85
44% CP supplement, 49% CP, 86% TDN	2.33	1.06	3.21	1.46	3.54	1.61	3.62	1.64	4.10	1.86	4.33	1.97	7.34	3.33
Limestone	0.34	0.15	0.32	0.15	0.31	0.14	—	—	0.01	0.01	0.07	0.03	0.28	0.13
Trace mineralized salt	0.21	0.10	0.21	0.10	0.21	0.10	0.21	0.10	0.21	0.10	0.21	0.10	0.20	0.09
Mineral-vitamin mix	0.08	0.04	0.07	0.03	0.08	0.04	0.08	0.04	0.07	0.03	0.10	0.05	0.10	0.05
	Ration Nutrient Information													
Dry matter	42.0	19.07	42.0	19.07	42.0	19.07	41.5	18.84	41.5	18.84	41.5	18.84	40.9	18.57
Crude protein	6.60	3.00	6.60	3.00	6.60	3.00	6.60	3.00	6.60	3.00	6.60	3.00	6.60	3.00
Crude protein (%)	15.72		15.72		15.72		15.91		15.91		15.91		16.14	
TDN	29.90	13.57	29.85	13.55	29.84	13.55	29.56	13.42	29.73	13.50	29.70	13.48	29.60	13.44
TDN (%)	71.18		71.08		71.05		71.23		71.63		71.58		72.37	
NE$_{lc}$ (Mcal)	30.73		30.73		30.73		30.73		30.73		30.73		30.73	
NE$_{lc}$ (Mcal/lb or kg)	0.73	1.61	0.73	1.61	0.73	1.61	0.74	1.63	0.74	1.63	0.74	1.63	0.75	1.65
Calcium (g)	146		129		122		139		122		122		122	
Calcium (%)	0.77		0.68		0.64		0.74		0.65		0.65		0.66	
Phosphorus (g)	86		86		86		85		85		85		80	
Phosphorus (%)	0.45		0.45		0.45		0.45		0.45		0.45		0.43	
Acid detergent fiber (%)	19.00		19.57		19.78		19.73		19.59		19.69		19.03	
Ether extract (%)	3.51		3.36		3.34		3.00		2.84		2.74		2.95	
Forage:grain ratio	52:48		57:43		59:41		58:42		58:42		59:41		65:35	

[1] Ration calculations based on 1988 NRC recommendations, and on the use of either alfalfa or alfalfa and corn silage (CS) forage. Table 20-14 was prepared specially for this book by D. E. Otterby, Professor, and J. G. Linn, Extension Animal Scientist, Department of Animal Science, University of Minnesota, St. Paul; with metric added by the authors.

Feeding Dairy Cattle

TABLE 20-15
COMPLETE RATIONS FOR 1,300 TO 1,400 LB *(591 TO 636 KG)* COWS, IN LATE LACTATION—LOW-PRODUCTION GROUP (45-LB MILK, 3.8% FAT AVERAGE)[1]

Feeds	All Alfalfa		¾ Alfalfa ¼ Corn Silage		⅔ Alfalfa ⅓ Corn Silage		½ Alfalfa ½ Corn Silage		⅓ Alfalfa ⅔ Corn Silage		¼ Alfalfa ¾ Corn Silage		All Corn Silage	
	\multicolumn{14}{c}{All Amounts on Dry Matter Basis}													
	(lb)	(kg)	(lb)	(kg)	(lb)	(kg)	(lb)	(kg)	(lb)	(kg)	(lb)	(kg)	(lb)	(kg)
Alfalfa, 17% crude protein, 59% TDN	26.14	11.87	21.53	9.77	19.74	8.96	14.99	6.81	10.46	4.75	7.98	3.62	—	—
Corn silage, 8% crude protein, 68.7% TDN	—	—	7.18	3.26	9.86	4.48	—	—	—	—	—	—	—	—
Urea-corn silage, 12% crude protein, 66% TDN	—	—	—	—	—	—	14.99	6.81	20.95	9.51	23.94	10.87	31.10	14.12
Corn, 10% crude protein, 91% TDN	10.76	4.89	8.21	3.73	6.97	3.16	6.74	3.06	4.79	2.17	3.97	1.80	2.60	1.18
44% CP supplement, 49% CP, 81% TDN	—	—	—	—	0.34	0.15	0.13	0.06	0.63	0.29	0.94	0.43	2.20	1.00
Dicalcium phosphate, 22% Ca, 19% P	0.27	0.12	0.26	0.12	0.25	0.11	0.30	0.14	0.31	0.14	0.31	0.14	0.30	0.14
Limestone, 34% calcium	—	—	—	—	—	—	—	—	—	—	—	—	0.12	0.05
Trace mineralized salt	0.19	0.09	0.19	0.09	0.19	0.09	0.19	0.09	0.19	0.09	0.19	0.09	0.18	0.08
Mineral-vitamin mix	0.06	0.03	0.06	0.03	0.06	0.03	0.07	0.03	0.08	0.04	0.09	0.04	0.20	0.09
	\multicolumn{14}{c}{Ration Nutrient Information}													
Dry matter	37.42	16.99	37.42	16.99	37.42	16.99	37.42	16.99	37.42	16.99	37.42	16.99	36.7	16.66
Crude protein	5.52	2.51	5.11	2.32	5.09	2.31	5.09	2.31	5.09	2.31	5.09	2.31	5.09	2.31
Crude protein (%)	14.75		13.65		13.60		13.60		13.60		13.60		13.87	
TDN	24.91	11.31	24.85	11.28	24.83	11.27	24.78	11.25	24.73	11.23	24.70	11.21	24.58	11.16
TDN (%)	66.52		66.40		66.35		66.22		66.10		66.02		66.98	
NE$_{lc}$ (Mcal)	25.37		25.37		25.37		25.37		25.37		25.37		25.37	
NE$_{lc}$ (Mcal/lb or kg)	0.677	1.49	0.678	1.50	0.678	1.50	0.678	1.50	0.678	1.50	0.678	1.50	0.691	1.52
Calcium (g)	195		173		165		148		129		117		96	
Calcium (%)	1.15		1.02		0.97		0.87		0.76		0.69		0.58	
Phosphorus (g)	64		64		64		64		64		64		64	
Phosphorus (%)	0.38		0.38		0.38		0.38		0.38		0.38		0.38	
Acid detergent fiber (%)	23.92		24.63		24.91		25.42		26.02		26.16		25.4	
Ether extract (%)	3.47		3.38		3.32		2.86		2.59		2.45		2.09	
Forage:grain ratio	70:30		77:23		79:21		80:20		84:16		85:15		85:15	

[1]Ration calculations based on 1988 NRC recommendations, and on the use of either alfalfa or alfalfa and corn silage (CS) forage. Table 20-15 was prepared specially for this book by D. E. Otterby, Professor, and J. G. Linn, Extension Animal Scientist, Department of Animal Science, University of Minnesota, St. Paul; with metric added by the authors.

TABLE 20-16
COMPLETE RATIONS FOR 1,300 TO 1,400 LB *(591 TO 636 KG)* DRY COWS[1]

Feeds	Alfalfa Corn Silage		Alfalfa-Grass Hay, Corn Stover		Alfalfa-Grass Hay		Oatlage		Urea-Corn Silage, Grass Hay		Grass Hay	
	\multicolumn{12}{c}{All Amounts on Dry Matter Basis}										
	(lb)	(kg)	(lb)	(kg)	(lb)	(kg)	(lb)	(kg)	(lb)	(kg)	(lb)	(kg)
Alfalfa, 17% CP, 59% TDN	11.09	5.03	—	—	—	—	—	—	—	—	—	—
Alfalfa-grass hay, 16.5% CP, 58% TDN	—	—	14.59	6.62	22.52	10.25	—	—	13.50	6.13	17.93	8.14
Grass hay, 12% CP, 60% TDN	—	—	—	—	—	—	—	—	—	—	—	—
Oatlage, 12.8% CP, 59% TDN	—	—	—	—	—	—	25.74	11.69	—	—	—	—
Corn silage, 8.7% CP, 68.7% TDN	12.66	5.75	—	—	—	—	—	—	—	—	—	—
Urea-corn silage, 12% CP, 66% TDN	—	—	—	—	—	—	—	—	12.28	5.58	—	—
Corn stover, 6.7% CP, 66% TDN	—	—	11.59	5.26	—	—	—	—	—	—	—	—
Corn, 10% CP, 88% TDN	—	—	—	—	3.47	1.58	1.09	0.49	—	—	5.46	2.48
44% CP supp., 49.0% CP, 81% TDN	—	—	—	—	—	—	0.01	0.01	—	—	0.53	0.24
Dicalcium phosphate, 22% Ca, 19% P	—	—	—	—	—	—	0.05	0.02	0.05	0.02	0.09	0.04
Limestone, 34% Ca	—	—	—	—	—	—	—	—	—	—	—	—
Monosodium phosphate, 22.5% P	0.05	0.02	0.10	0.05	0.03	0.01	—	—	—	—	—	—
Trace mineralized salt	0.07	0.03	0.07	0.03	0.07	0.03	0.07	0.03	0.07	0.03	0.07	0.03
Mineral-vitamin mix	0.03	0.01	0.02	0.01	0.02	0.01	0.06	0.03	0.06	0.03	0.06	0.03
	\multicolumn{12}{c}{Ration Nutrient Information}										
Dry matter	25.07	11.38	26.37	11.97	26.11	11.85	27.00	12.26	25.96	11.79	24.14	10.96
Crude protein	3.05	1.38	3.24	1.47	4.07	1.85	3.40	1.54	3.09	1.40	2.97	1.35
Crude protein (%)	12.18		12.29		15.58		12.60		11.92		12.31	
TDN	16.17	7.34	16.10	7.31	16.09	7.30	16.14	7.33	16.08	7.30	15.99	7.26
TDN (%)	64.49		61.08		61.64		59.78		61.98		66.27	
Calcium (g)	88		91		87		48		48		48	
Calcium (%)	0.78		0.77		0.73		0.39		0.41		0.44	
Phosphorus (g)	30		30		30		30		30		37	
Phosphorus (%)	0.27		0.25		0.25		0.25		0.27		0.34	
Acid detergent fiber (%)	30.16		37.31		28.96		33.48		35.05		31.32	
Forage:grain ratio	99:1		99:1		87:13		95:5		99:1		74:26	

[1]Ration calculations based on 1988 NRC recommendations. Table 20-16 was prepared specially for this book by D. E. Otterby, Professor, and J. G. Linn, Extension Animal Scientist, Department of Animal Science, University of Minnesota, St. Paul; with metric added by the authors.

FEEDING ON PASTURE

On the average, a lactating cow on pasture spends 8 hours daily in harvesting, or grazing. In addition, she spends another 8 hours each day in rumination—the act of chewing the cud. This means that a cow spends two-thirds of her time eating. If she is a high producer, meeting her nutritive needs within this period of time calls for good pastures plus supplemental feeding.

Problems of milk production are at a minimum during the early pasture season, when plant growth is lush. However, when the weather gets hot—the period known as the "summer slump"—it is a different story. High temperatures actually affect pasture growth more than the well-being of the cows. Many dairy producers have discontinued pasture grazing for two reasons: (1) It is difficult to keep milk production uniform when cows are on pasture, because of changing temperatures and pasture growth; and (2) with larger herds, it is not possible to have sufficient pastures in close proximity to headquarters.

Pasture of high quality has a value that is intermediate between concentrate and hay. Thus, on good pasture, it is possible to sustain a high level of production with less grain than is needed for conventional winter forage supplementation.

FEEDING DRY COWS

Dry cows have three important needs: (1) regression of old milk producing cells in the udder and regeneration of new milk producing cells, (2) developing the unborn calf, and (3) storing body reserves for the next milking period. This necessitates that they be properly fed in late lactation and during the dry period.

Proper dry cow feeding is one of the most critical practices for successful dairy operations. It involves management practices (1) to have the cow dry 50 to 65 days, and (2) to have the cow in proper condition when beginning and finishing the dry period. Proper dry cow feeding begins during the last half to one-third of the lactation period when body reserves should be replaced. At drying off time, the cow's body condition should be classed as 2 or 3 on an arbitrary scale of 1 as too thin and 5 as too fat. At calving, the body condition should be in the 3.5 to 4.0 class. During the 60–day dry period, cows should be fed to gain 120–200 lb.

FEEDING CALVES

One of the most important phases of dairy production is that of feeding and managing dairy calves. Statistics reveal that more than 20% of the dairy calves die of sickness or disease before reaching maturity. With good management, these losses may be reduced to 3 to 5%. Many calf deaths are caused by faulty nutrition or poor housing and management.

A carefully planned and executed feeding program is necessary to produce growthy, vigorous, and healthy calves.

The following feeding program is recommended:

Day 1 . . . . Dam's colostrum
Day 2 . . . . Dam's colostrum
Day 3 . . . . Dam's colostrum
Day 4 . . . . Liquid feed of choice, introduce starter and water
Day 5 to weaning . . . Continue feeding program
Weaning to 12 weeks . . Starter (up to 5 lb daily), introduce forage

Dairy Calf Production

Physiologically, the newborn calf is not a functioning ruminant. The abomasum, which represents the largest portion of the stomach of the newborn calf, is the primary functional unit of the gastric region. The sucking reflex allows colostrum or milk to bypass the rudimentary rumen and reticulum via the esophageal groove directly into the abomasum where digestion is initiated. As the calf grows older, it consumes solid feedstuffs which serve as a mechanical stimulator on the other sections of the gastric region, thereby hastening their development. Additionally, the rumen becomes inoculated with microorganisms from the immediate environment.

COLOSTRUM

Colostrum is the milk which is high in antibodies, and which is secreted by cows, and other mammalian females, for the first few days following parturition.

Colostrum (either dam's colostrum or mixed colostrum from first milking of older cows) should be fed to calves as soon after birth as possible (ideally within 15 minutes and certainly within 4 hours) to protect against disease.

Early feeding of colostrum is necessary because:

1. Newborn calves have no antibodies to provide natural protection against disease until colostrum is received.

2. The calves' ability to absorb immunoglobulins (the disease protecting component) is substantially reduced after 24 to 36 hours.

3. Calves may become infected with highly pathogenic bacteria immediately after birth.

Calves should receive a total of 10 to 12% of their birth weight as first milking colostrum, with half of this amount received 4 to 6 hours after birth.

Surplus colostrum can be frozen and stored for a period of 1 year or longer without losing its antibody value. It may then be thawed, warmed to about 100°F, and fed as needed.

CALF STARTERS

A high quality, palatable calf starter should be offered when the calf is 4 days old, and not later than 10 to 12 days of age.

Calf starters should be fed until calves are about 12 weeks of age, with intake limited to 5 to 7 lb per calf daily.

Many good commercial starters are on the market. Also, calf starters may be home-mixed. Table 20-17 presents examples of some good grain calf starters.

TABLE 20-17
GRAIN STARTER RATIONS FOR CALVES[1]

	Grain Starters[2]		
	1	2	3
Ingredients (air dry basis)			
Corn (cracked or coarse ground) ... (%)	50	30	
Ear corn (coarse ground) .... (%)			50
Oats (rolled or crushed) .... (%)	22	18	
Barley (rolled or coarse ground) ... (%)		20	21
Wheat bran .... (%)		8	
Soybean meal .... (%)	20	16	21
Molasses .... (%)	5	5	5
Dicalcium phosphate .... (%)	0.5	0.5	0.5
Limestone .... (%)	1.5	1.5	1.5
Trace mineralized salt and vitamins ... (%)	1	1	1
Composition (dry matter basis)			
ADF .... (%)	7.0	6.9	9.1
Crude protein .... (%)	18.1	18.0	18.4
TDN .... (%)	80.0	78.8	78.0
Calcium (Ca) .... (%)	0.80	0.80	0.82
Phosphorus (P) .... (%)	0.48	0.56	0.47
Vitamin A .... (IU/lb)	1,000	1,000	1,000
Vitamin A .... (IU/kg)	2,205	2,205	2,205
Vitamin D .... (IU/lb)	150	150	150
Vitamin D .... (IU/kg)	331	331	331
Vitamin E .... (IU/lb)	11	11	11
Vitamin E .... (IU/kg)	24	24	24

[1]*Source:* Linn, J. G., M. F. Hutjens, W. T. Howard, L. H. Kilmer, and D. E. Otterby, Feeding the Dairy Herd, Cooperative Ext. Services, Ill., Iowa State, Minn., and Wisc., 1988, p. 20, Table 11.

[2]Hay may be offered free choice with grain starters.

HAY OR SILAGE FOR CALVES

While calves may begin nibbling on good quality hay as early as 5 to 10 days of age, it is not necessary to feed forage before 8 to 10 weeks of age. If the housing and management system makes it inconvenient to provide forage, it may be desirable to incorporate a forage factor (more fiber) in the starter ration. Table 20-18 presents examples of suitable rations for calves not receiving hay or silage. Corn silage or pasture should not be fed before 3 months of age because of their high moisture content which can limit intake and growth. Low moisture haylage is acceptable if it is kept fresh.

TABLE 20-18
COMPLETE STARTER RATIONS FOR CALVES[1]

	Complete Starters		
	1	2	3
Ingredients (air dry basis)			
Corn (cracked or coarse ground) ... (%)	40	25	30
Oats (rolled or crushed) .... (%)	14.5	8	18
Beet pulp .... (%)		25	25
Alfalfa hay (ground) .... (%)		10	
Corn cobs (ground) .... (%)	15		
Soybean meal .... (%)	23	18	20
Molasses .... (%)	5	5	5
Dried whey .... (%)		7	
Dicalcium phosphate .... (%)	0.5	0.5	0.5
Limestone .... (%)	1	0.5	0.5
Trace mineralized salt and vitamins ... (%)	1	1	1
Composition (dry matter basis)			
ADF .... (%)	13.3	15.8	14.2
Crude protein .... (%)	18.3	18.0	18.2
TDN .... (%)	75.5	78.0	79.4
Calcium (Ca) .... (%)	0.63	0.72	0.58
Phosphorus (P) .... (%)	0.45	0.44	0.43
Vitamin A .... (IU/lb)	1,000	1,000	1,000
Vitamin A .... (IU/kg)	2,205	2,205	2,205
Vitamin D .... (IU/lb)	150	150	150
Vitamin D .... (IU/kg)	331	331	331
Vitamin E .... (IU/lb)	11	11	11
Vitamin E .... (IU/kg)	24	24	24

[1]*Source:* Linn, J. G., M. F. Hutjens, W. T. Howard, L. H. Kilmer, and D. E. Otterby, Feeding the Dairy Herd, Cooperative Ext. Services, Ill., Iowa State, Minn., and Wisc., 1988, p. 20, Table 11.

WATER

Clean, fresh water in clean pails may be offered free choice starting on day four. Calves fed limited liquid (such as when fed once-a-day) should receive supplemental water, especially during warm weather. Calves offered water during the liquid feeding period (birth to 4 weeks) tend to consume more starter and perform better than calves fed liquid only.

AMOUNT AND METHOD OF FEEDING, FREQUENCY OF FEEDING, AND AGE OF WEANING

Calves may be separated from their dams at birth, or within 12 to 24 hours after birth. In any case, they should receive their dam's colostrum for the first 3 days of life, following which they may be shifted to a liquid feed of the feeder's choice.

In order to obtain proper growth, calves must be provided adequate dry matter. For an 80- to 100-lb calf, this calls for 1 lb of dry matter (solids) daily from milk, surplus colostrum, or milk replacer, from birth to weaning at 4 weeks.

Milk or milk replacer may be fed by open pail, by nipple feeding from a pail or bottle, or by automated feeding equipment. Each method of feeding is satisfactory, provided it is accompanied by cleanliness and sanitation.

Most calf raisers feed twice daily. Weak or unthrifty calves may benefit from more frequent feedings.

When calves are housed in hutches and the weather is extremely cold, they should be fed a 20% fat replacer 3 times daily and the daily feed allowance should be increased by 1¼ to 1½ times in order to meet their increased energy requirements. Young calves that are doing poorly should be moved to warmer quarters.

Most producers wean calves between 4 and 8 weeks of age. A good practice is to wean according to starter intake—wean when the starter intake is 1 to 1½ lb/day.

Veal Calf Production

A veal calf is defined as a young bovine animal, usually not over 4 months of age, that has subsisted largely on milk or milk replacers, thus making the color of the lean meat light, grayish pink. The majority of veal calves are of dairy breeding, consisting of bull calves and heifer calves not retained as replacements. Producers receive a premium if the lean meat of veal is light grayish pink in color (due to reduced muscle myoglobin), characteristic of feeding milk, which is naturally low in iron. Some producers attempt to enhance the desired grayish pink color by restricting iron intake and exercise. Research has shown that a dietary iron concentration of 11.4 to 13.6 mg per pound of dry matter in milk replacers is sufficient for the well being of veal calves and produces desirable grayish pink carcasses, with and without exercise.

A conversion rate of 10 lb of whole milk for 1 lb of body weight gain is normal. If milk replacer is used, the conversion rate is generally about 1.3 to 1.5 lb of dry replacer per pound of gain.

Profitable veal production depends on: (1) a low mortality rate, (2) economical housing, (3) plenty of inexpensive labor, and (4) an established market.

Preventing Calf Scours

To prevent calf scours, the producer must prevent primary infection of the newborn. This rests on strict sanitary measures and isolation, along with other preventive measures. Observance of the following practices will lessen the incidence of calf scours:

1. Make certain that the calf gets first-milk colostrum.
2. Augment natural resistance with vitamins.
3. Avoid overfeeding and irregularity of feeding.
4. Keep feeding utensils clean and sanitary.
5. Don't overcrowd.
6. Provide adequate ventilation.
7. Avoid damp, wet calves.

• **Use of electrolytes**—Feeding an oral electrolyte solution usually is beneficial when a calf has a mild case of scours (not off feed, not depressed, and no fever).

FEEDING REPLACEMENT HEIFERS/ NORMAL GROWTH FOR DAIRY HEIFERS

Between weaning and calving (12 weeks to 2-year-olds), the nutrition of heifer replacements is often neglected. At its best, the feeding and management program during this period involves 3 distinct phases: (1) weaning (about 12 weeks of age) to 1 year; (2) 1 year to 2 months before calving at 2 years; and (3) 2 months before calving to calving.

• **Replacement heifers, weaning (about 12 weeks of age) to 1 year**—During this period, replacement heifers may be fed forage free-choice and limited grain. The amount, and the protein content, of the grain mix needed will be determined by the quality of the forage being fed. Pasture can be used successfully in the feeding program of replacement heifers, provided it is supplemented with a grain mix and some dried forage, along with suitable minerals (incorporated in the grain mix or offered free-choice). Also, there should be access to clean, fresh water.

During the yearling stage, replacement heifers should not be overfed and become too fat. Overconditioning has an inhibitory effect on mammary secretory tissue development during the critical period of its maximum development between 3 and 9 months of age and results in lower milk production later in life. Overconditioning of heifers after 15 months of age does not affect mammary secretory tissue.

• **Replacement heifers, 1 year to 2 months before calving at 2 years**—If good quality forage is available, it may be the only feed required for heifers over 1 year of age. A suitable mineral mix should be provided on a free-choice basis. Heifers should gain 1.6 to 1.8 lb per day. If growth is not satisfactory, some grain should be provided.

First estrus in heifers is dependent on size and weight, primarily weight. A general guide is that heifers will show their first estrus at 40% of their mature weight, which should be before 12 months of age. Heifers fed high planes of nutrition will show estrus at an earlier age than heifers grown at recommended rates, but underfeeding of heifers will delay estrus. Underfed or very slow growing heifers may ovulate, but estrus signs are often suppressed. Heifers in good condition and gaining weight at breeding time generally show more definite signs of estrus and have improved conception rates over heifers in poor condition and/or losing weight. Overconditioned heifers require more services per conception than heifers of normal size and weight. Table 20–19 shows desirable weights for first breeding at 15 months of age along with weights for other age categories.

TABLE 20-19
DESIRABLE WEIGHTS FOR DAIRY HEIFERS TO BE BRED AT 15 MONTHS OF AGE[1]

Age in Months	Brown Swiss or Holstein		Ayshire, Shorthorn or Guernsey		Jersey	
	(lb)	(kg)	(lb)	(kg)	(lb)	(kg)
Birth	90–100	41–45	65–75	30–34	55–60	25–27
1	120	54	90–100	41–45	70–80	32–36
2	170	77	135–145	61–66	110–120	50–54
4	270	123	225–235	102–107	190–200	86–91
6	370	168	315–325	143–148	270–280	123–127
12	670–700	304–318	585–600	266–272	510–520	232–236
15[2]	800–875	363–397	720–750	327–341	630–650	286–295
18	970–1,000	440–454	850–875	363–397	750–775	341–352
22	1,150–1,200	527–545	1,025–1,075	465–488	900–950	409–431

[1]*Source:* Linn, J. G., M. F. Hutjens, W. T. Howard, L. H. Kilmer, and D. E. Otterby, *Feeding the Dairy Herd*, Cooperative Ext. Services, Ill., Iowa State, Minn., and Wisc., 1988, p. 21, Table 13.
[2]Breed heifers in this weight range. Heifers should weigh about 60% of their mature weight when bred. With proper feeding, heifers should reach these weights and have good skeletal growth at 14 to 16 months of age.

- **Two months before calving to calving**—The feeding of heifers during this period can affect milk production during the first lactation. During the last 2 months of gestation, heifers should make daily gains of about 2.0 lb per day, in comparison with 1.7 lb during early pregnancy. Heifers that are growing rapidly at calving time, and continuing to grow during the first lactation, are more persistent milkers than full-sized heifers at calving.

The amount of grain to feed before calving will depend on forage quality, size, and condition of the heifer. A good thumb rule is to feed grain at 1% of body weight starting about 6 weeks before calving. The ration should have adequate protein, minerals, and vitamins. Excess salt intake can contribute to udder edema and should be avoided the last 2 weeks before calving.

Well-grown heifers will have a minimum of problems at calving time. But plane of nutrition can affect ease of calving in two ways: (1) calf size, and (2) fatness of the dam. Fat heifers have higher incidents of dystocia because of small pelvic openings and usually a larger than normal sized calf at birth. Underfed or poorly grown heifers will require more assistance at calving and have a higher death rate at calving than normal sized heifers.

FEEDING DAIRY BULLS

Bull calves raised for breeding purposes should be fed and handled much the same as heifers. But, since they grow slightly faster than heifers, they should receive somewhat more feed than heifers of the same age.

Older bulls should be kept in thrifty, vigorous condition, but they should not be permitted to become too fat. Mature bulls can be fed the same grain ration as the lactating cows. Depending on the quality of the roughage, usually about ½ lb of grain per 100 lb of body weight will suffice for the mature bull. Also, individual differences must be considered, for some bulls are easier keepers than others.

FEEDING DAIRY BEEF

Dairy beef is beef derived from cattle of dairy breeding. Dairy heifers and steers finished at market weights comparable to the finished market weights of heifers and steers of beef breeding are commonly referred to as dairy beef. Generally speaking, there are two different finishing programs and market weights for dairy beef:

1. **High-energy rations; light market weights.** These animals are full fed a high-concentrate ration from about 300 lb to market weights of 750 to 950 lb.
2. **High-roughage; heavy market weights.** These animals are grown on a maximum of roughage to 600- to 750-lb weight, following which the proportion of concentrate is increased; and they are marketed at weights of 1,150 to 1,400 lb.

(Also see Chapter 19, section on "Dairy Beef.")

FEEDING SHOW AND SALE ANIMALS

Dairy animals intended for show or sale should be fed so as to achieve a certain amount of finish or bloom, but they should not be too fat. Linseed meal, beet pulp, oats, barley, molasses, and wheat bran are popular feeds in a fitting and showing ration. Linseed meal, in particular, is used to impart a healthy bloom or shine to the hair. Likewise, good quality forages are always very important.

In fitting show and sale animals, it is important that the grain mixture be palatable and light, and that they not go off feed. Feeds are sometimes soaked before feeding to make them more palatable.

FEED AND MANAGEMENT ASPECTS OF DAIRY PRODUCTION

Higher levels of milk production per cow, larger dairy herds, and increased facility and labor costs have focused attention on the need for more efficient management.

There are innumerable management aspects of great importance to dairy production. When disregarded, many of them will materially lessen production and make the enterprise unprofitable, no matter how good the genetic capability of the animals or the feed being used. Still others are important tools from the standpoint of enhancing good management. Among the management aspects of importance in dairy production are time and frequency of feeding and watering, and use of milk and butterfat records.

Time and Frequency of Feeding and Watering

Where complete rations are fed, cows are generally fed twice or three times daily. Where the forage and grain are fed separately, most dairy farms provide hay free-choice at all times, feed silage once or twice daily, and feed concentrates twice daily.

Silage is usually fed once daily, because of the added cost of more frequent feeding. However, some of the automated systems do not have the added labor cost factor, with the result that they may be used to feed twice daily, or more frequently. Silage should be fed soon after milking, so that any residual feed flavors will disappear before the next milking. Cows should not have access to silage, or other feeds that cause off-flavored milk, for at least 2 to 3 hours prior to milking.

Grain should be fed twice, or more frequently, daily. When high-producing cows are fed in the milking parlor, they are not in the milking parlor long enough to eat all the grain that they need; hence, part of the concentrate of high producers should be fed in the manger along with the hay and/or silage. Group or corral feeding of grain has evolved, as a means of lowering labor costs. Under this system, lactating cows are grouped according to level of production. Grain is fed twice daily in the manger, right after milking, either (1) as a top dressing on the silage and/or hay, or (2) mixed with the silage and/or hay, with the grain allocation for each group or corral determined by the average level of production of each cow in the group. The advent of electronic/computerized grain feeders gives the producer the opportunity to feed concentrates frequently and to control the amount of feed that cows consume.

Calves should be fed milk or milk replacer once or twice daily, at regular intervals. It is more harmful to overfeed than to underfeed a young calf.

Water should be clean, fresh, and available at all times. If there is too little water or if the cows must stand in line to get it, milk production will decrease. In cold weather, it may be necessary to protect water from freezing, as cows cannot get sufficient water by licking ice.

Frequency of Milking

Most cows are milked 2X per day. Increasing milking frequency to 3X per day increases milk production by 10 to 25% and milking 4X per day will stimulate milk yield another 5 to 15%. Whether these increases in milk production are worth the extra expense in labor, feed, utilities, and milking supplies depends upon economic conditions on each particular dairy farm. Also, with increasing genetic capability of cows, more frequent milking (3X and 4X) appears to improve udder health and lower the incidence of mastitis.

Milk and Fat Records; Testing Programs

Individual cow production and reproduction records are a must in any progressive dairy operation. They necessitate that each cow be identified. In addition to milk weight and fat test, many programs test for protein and somatic cell count. Also, most states and/or processing centers can store and provide information via computer relative to individual cow and herd records, reproduction problems, breeding dates, calving dates, genetics, heifer management, feeding, culling, health, and sales. With good and complete records, producers can identify the strengths and weaknesses of their operations, and the areas upon which improvements will reap the greatest financial rewards.

The three testing programs sponsored by Federal and state research extension services are Dairy Herd Improvement Association (DHIA), Owner-Sampler Records (OS), and Weigh-A-Day-A-Month (WADAM). Also, there is a wide range of sampling, such as sampling only one milking, alternate A.M. and P.M. weighing and sampling, etc. Each system has advantages and disadvantages. More important than the type of system used is the producer's total understanding and use of the system to obtain maximum benefits and willingness to keep records.

HEALTH DISORDERS OF DAIRY COWS

Milk production is a very intensive and demanding form of production, accompanied by much stress. Therefore, certain health problems are associated with it; among them, acidosis, bloat, displaced abomasum, fat cow syndrome, grass tetany, hardware disease, ketosis, milk fever, retained placenta, and udder edema.

- **Acidosis (Lactic Acidosis)**—*Acidosis is a metabolic disease of cattle and sheep.* Afflicted animals show signs of weakness, diarrhea, and abdominal discomfort, and in many cases die.

Prevention consists in starting animals on a high-roughage ration and gradually reducing the roughage and increasing the grain; avoiding erratic feeding; and avoiding abrupt ration changes. Also, the addition of buffers to a high-grain ration aids in the prevention of acidosis.

Various treatments have been used with different degrees of success. Perhaps the most successful treatment consists in decreasing both the amount and kinds of feeds fed (lessening the total amount of the ration), then returning to a higher forage mix.

(Also see Chapter 5, Table 5-1—Acidosis [Lactic Acidosis].)
- **Bloat**—Bloat affects all ruminants, dairy cattle included. When dairy animals are first introduced to concentrates or lush, legume pastures, bloat problems arise. Normally, the eructation process allows for the expulsion of gases that are produced in the rumen. In cases of bloat, frothing (the trapping of gas in the ingesta) prevents this process and intraruminal pressure builds, thereupon distending the left side of the abdomen until the animal is thrown off feed and goes down due to pain and the buildup of toxic metabolites.

Treatment, control, and prevention of bloat are detailed in the two bloat sections referred to in the parentheses that follow.

(Also see Chapter 5, Table 5-1—Bloat, Feedlot; and Bloat, Pasture.)
- **Displaced Abomasum**—In recent years, dairy producers have encountered increasing numbers of cows with displaced abomasums. The practice of feeding dry cows liberal amounts of grain has been pointed to as one contributing factor. It has been suggested that the lack of bulk and rumen fill promotes flabby muscle tone of the rumen, thereby permitting the abomasum to become displaced. The feeding of some effective roughage is recommended.

(Also see Chapter 5, section on "Displaced Abomasum.")
- **Fat Cow Syndrome (Fatty Liver Syndrome)**—Cows afflicted with fat cow syndrome have lowered level of liver function due to enlarged liver infiltrated with fat. General herd signs include very fat cows in the dry cow group, decreased resistance to infection, increased incidence of metabolic diseases such as ketosis, reduced feed intake, and reduced milk production and body weight.

The incidence of the fat cow syndrome may be reduced by avoiding overconditioning of cows during late lactation and the dry period and by formulating rations that maximize feed intake after calving.
- **Grass Tetany (Hypomagnesemia)**—Grass tetany is a metabolic condition that affects cows (especially lactating cows) grazing lush pasture high in nitrogen, resulting in low absorption of magnesium, and is most common in the spring. Afflicted animals develop tetany, walk with a stiff gait, go into convulsions, and may die. During the danger period, cows grazing lush pasture should be supplemented with magnesium oxide.

(Also see Chapter 5, Table 5-1—Grass Tetany.)
- **Hardware Disease**—The condition in which the collection of foreign objects irritates or punctures the reticulum is called reticulitis, or *hardware disease*. Ruminants are grazers by nature and are sometimes indiscriminant in their selection of feed. Often they will consume nails, pieces of wire, and other foreign objects. Due to the motility patterns of the gastric region, these objects tend to accumulate in the reticulum; and the presence of these sharp objects can pose serious problems, especially if the reticulum should be punctured.

(Also see Chapter 5, section on "Hardware Disease.")
- **Ketosis**—Ketosis is a metabolic disease characterized by a drop in milk production, hypoglycemia, ketonuria, and a rapid loss of weight. In general, the disorder develops within the first 30 days of lactation. While no preventative measures have proven to be 100% effective, the feeding of propylene glycol or sodium propionate has been successful in some cases; and the inclusion of niacin in the ration at a level of 6 g per head per day in the last 2 weeks of the dry period and in the fresh cow ration may aid in reducing the incidence of ketosis. Additionally, starting limited grain feeding during the latter part of the dry period is helpful in preventing ketosis.

If a cow comes down with ketosis, a glucose solution can be administered intavenously to promote rapid recovery. (Also see Chapter 5, Table 5-1—Ketosis.)

• **Mastitis**—Mastitis is an infection of the mammary gland caused by any one of several bacterial organisms, most frequently *staphylococcus* or *streptococcus*. Symptoms vary with degree of inflammation. Acute cases show a swollen and painful udder, and frequently cause the cow to go off feed. Chronic cases result in slightly swollen udders and small flakes in the milk.

No feed is known to cause or cure mastitis. However, the sudden addition of nutrients may result in a marked increase in milk production and cause more stress; in turn, this may cause subclinical cases. Also, feeding recommended levels of selenium and vitamin E may be helpful in preventing mastitis.

• **Milk Fever**—At or soon after calving (generally within 48 to 72 hours), a sharp decrease in blood calcium (hypocalcemia) occurs in some cows, resulting in loss of appetite, subnormal temperature, and an unsteady gait. This is followed by nervousness, and, finally, collapse or complete loss of consciousness. The head is usually turned back. The name *milk fever* is a misnomer, because the body temperature is below normal.

The triggering mechanism for this drop in blood calcium is the onset of lactation—an intensive mobilization of calcium.

Feeding practices involving dry cows can markedly reduce the incidence of milk fever. When certain cows are known to have a history of milk fever, excessive calcium intake during the dry period should be avoided. If the problem persists, some nutritionists recommend the limited feeding of a high-energy, low-calcium (less than 15 g of calcium per day) ration. After calving, the calcium levels should be raised rapidly to meet the high requirements of lactation.

Recent studies have indicated that the addition of certain anions (negatively charged ions) to the ration may reduce the incidence of milk fever by aiding calcium absorption and mobilization. But more experimental work is needed relative to this method.

(Also see Chapter 5, Table 5-1—Milk Fever.)

• **Retained Placenta**—Normally, the placenta is expelled within 3 to 6 hours after parturition. If it is retained as long as 12 hours after calving, competent assistance should be obtained.

Retained placenta occurs in about 10% of dairy cattle. It is more common following abnormally short or abnormally long pregnancies, among older cows, and following twinning. Experimentally, it has been found that a high incidence of retained afterbirth occurs when premature calving is induced by the administration of glucocorticoid drugs.

While infections such as brucellosis, vibriosis, and others have been associated with abortion and retained afterbirth, these are by no means the only causes. Its incidence increases with parturient hypocalcemia and appears to be related to the fat cow syndrome. Nutritionally, deficiencies of vitamin A, selenium, copper, and iodine have been incriminated. The prepartum injection of selenium at low doses has been shown to reduce the incidence of retained placenta. Also, it appears that fewer cases of retained placenta occur (1) when calves stay with their dams and nurse for 12 to 24 hours, and (2) when cows are kept on pasture year-round. Among cows which have previously retained the placenta, 20% are likely to do so again.

• **Udder Edema**—Udder edema, characterized by excessive accumulation of fluid in the intercellular spaces of the udder and forward of it, is sometimes of serious magnitude before calving. The cause is not well understood, but a reduction of blood proteins at calving time and increased blood flow without compensatory lymph removal have been suggested. It appears that high intakes of sodium chloride or potassium chloride increase the severity of udder edema, and that restriction of the salt intake will reduce the severity.

Severe edema may reduce milk production and may be one of the causes of pendulous udder.

• **Drugs and Pesticides**—Many drugs used in the treatment of cattle diseases, along with many pesticides, are excreted in milk. Such milk should be discarded to prevent the drugs from entering the human food supply. The presence of antibiotics, sulfas, and pesticides in milk is illegal. Dairy producers should follow a residue avoidance program.

(Also see Chapter 13, section on "Food and Drug Administration [FDA].")

QUESTIONS FOR STUDY AND DISCUSSION

1. How and why does the feeding of lactating cows differ from the feeding of other classes of farm animals?

2. What motivating forces were back of group feeding? How is group feeding accomplished?

3. Discuss the economic importance of dairy cattle feeding.

4. Because it costs more to feed high producers than low producers, how can high producers return the most net profit?

5. Of what value are the National Research Council (NRC) Requirements? Why should one provide added margins of safety when using them?

6. Why is dry matter intake of such practical importance for high-producing cows?

7. Discuss dairy ration energy functions and deficiency symptoms.

8. What is meant by the *net energy of lactation* (NE_{lc})? Why is this measure of feed energy so important in lactating cow rations?

9. List and discuss the factors that affect the energy requirements for maintaining a lactating cow.

10. The NRC net energy value of feeds for lactation are adjusted to three times (3X) the maintenance level. What is the explanation for this?

11. Discuss proteins for lactating cows from the standpoints of amount needed, degraded and undegraded, and replacement by urea and other NPN products.

(Continued)

12. For dairy cattle, discuss the importance of each of the following minerals: salt, calcium, phosphorus, iodine, selenium, and zinc.

13. For dairy cattle, discuss the importance of each of the following vitamins: vitamin A, vitamin D, vitamin E, niacin, riboflavin, thiamin, and vitamin B-12.

14. How much water will a lactating dairy cow drink in a day? Why is the water requirement for a lactating cow so much higher than it is for other classes of farm animals?

15. Discuss the importance of each of the following forages for dairy cattle: (a) pasture, (b) green chop, (c) hays, (d) silages, and (e) haylage.

16. Discuss the importance and role of concentrate and by-product feeds for dairy cattle.

17. Discuss the functions, amount, and type of fat to add to lactating cow rations.

18. Why is fiber so important in the rations of lactating cows? What proportion of the total ration of lactating dairy cows should consist of long hay?

19. Why and how are each of the following feed additives used, practically or experimentally, in dairy rations: antibiotics, bovine somatotropin (BST), buffers, ionophores, and isoacid?

20. What types of commercial dairy feeds are available? Why are dairy producers major users of U.S. commercial feeds?

21. Discuss the importance of each of the following feed factors in dairy nutrition: palatability, variety, bulk, laxativeness, cost, and effect on milk flavor.

22. How should grain be prepared for dairy cattle?

23. Under what sort of stress is the lactating cow placed? Discuss the importance of monitoring dry matter intake in dairy cows.

24. What thumb rules should be followed when feeding forages to lactating cows?

25. What thumb rules should be followed when feeding concentrates to lactating cows?

26. Why should a dairy producer be well informed relative to the milk-feed price relationship?

27. What factors determine the amount of grain that it will pay to feed milking cows?

28. Is there need for, or a better alternative to, Table 20-10 (or a similar table) for use in balancing lactating cow rations by producers who (a) home-mix their dairy rations, (b) are without chemical analyses of feed ingredients, (c) are without access to a computer, and (d) do not have processing and mixing equipment with which to combine their roughage(s) with their concentrates to make a complete ration?

29. Discuss the history and the present status of computers in dairy ration formulation.

30. Define phase feeding; and discuss the four lactation cycle phases with corresponding changes in milk production, milk fat percentage, dry matter intake, and body weight.

31. Define and describe challenge feeding. Discuss its advantages and disadvantages.

32. Study the group feeding program presented in this chapter under the section headed "Group Feeding," and analyze it from the following standpoints: (a) the number and types of groups; (b) milking parlor versus corral feeding, or a combination of both; and (c) the example rations presented.

33. When on pasture, how does a lactating cow divide her time between harvesting and ruminating?

34. Outline the routine for feeding and managing dry cows.

35. Outline and discuss a feeding program for young dairy calves, including colostrum, a starter ration, hay or silage, method of feeding, frequency of feeding, and age of weaning.

36. Why is it important that young calves get a good feeding of colostrum as soon after birth as possible?

37. What are veal calves? How are they fed? Why do consumers prefer grayish pink veal?

38. List recommended practices to lessen the incidence of calf scours.

39. Discuss the feeding of replacement heifers. What is the relationship of growth of dairy heifers to time of breeding?

40. Discuss the feeding of dairy bulls, dairy beef, and show and sale animals.

41. Discuss the time and frequency of feeding and watering lactating cows.

42. What factors determine whether 3X or 4X per day milking will be profitable?

43. Why, and how, should dairy producers keep individual milk, butter fat, and other records?

44. Discuss the cause, prevention, and treatment of each of the following health disorders of dairy cows: acidosis, bloat, displaced abomasum, fat cow syndrome, grass tetany, hardware disease, ketosis, mastitis, milk fever, retained placenta, and udder edema.

45. Why should dairy producers follow a drug and pesticide residue avoidance program?

Original painting by Tom Phillips

21

FEEDING SHEEP[1, 2]

Contents	Page
Economic Importance of Feed for Sheep	466
Nutritive Needs of Sheep	466
National Research Council (NRC) Requirements	466
Energy	470
Protein	471
Minerals	473
Sheep Mineral Chart	473
Vitamins	478
Sheep Vitamin Chart	478
Water	480
Additives	480
Feeds For Sheep	481
Feed Substitution Table For Sheep	481
Rations For Sheep	484
Feeding Rams	484
Feeding Breeding Ewes	484
Feeding Range Sheep	485
Feeding Growing-Finishing Lambs	486
Feeding Finishing Lambs	487
Feeding Show Sheep	489
Nutritional Disorders and Toxins	489
Questions for Study and Discussion	490

The world sheep population is 1.1 billion head. Australia leads in numbers, with 150 million head; and the U.S.S.R. ranks second, with 143 million head.

Sheep produce 4.2% of the world's meat and 1.7% of the world's milk. Additionally, they produce more than 6.6 billion lb of grease wool (4 billion lb scoured wool), annually. The annual world per capita consumption is about 2.8 lb of lamb and mutton, 3.9 lb of sheep milk, and 0.8 lb of scoured wool.

In the mid-1980s, there were about 10.4 million sheep in the United States, with a total value of $638 million; and these sheep produced 380 million lb of meat and 93 million lb of grease wool. About 78% of the gross income from sheep is from meat and 22% from wool.

Although sheep are produced in every state, the 17 western states account for 82% of the nation's total production. Texas is the leading state in sheep production, with more than 18% of the nation's breeding sheep. Following Texas, the other top ranking states in numbers, in order, are: California, Wyoming, South Dakota, Utah, New Mexico, Montana, and Colorado.

United States sheep production and management systems vary greatly. Western range sheep bands number from 1,000 to 5,000 head and are handled in extensive operations in terms of land area. The intensive sheep operations of the rest of the U.S. vary from flocks of 50 ewes to 3,000 ewes. Lamb feeding operations vary from 1,000 to 50,000 head on feed at the same time.

Sheep consume a higher proportion of forages than any other class of livestock, it being estimated that 94% of the total feed supply of the U.S. sheep production is derived from forages. They are naturally adapted to grazing pastures and ranges which supply a variety of forage plants, and they thrive best on forage that is short and fine rather than high and course. Although sheep will eat considerable quantities of weeds and brush, they prefer choice grasses and legumes.

Except at lambing season, breeding sheep seldom need much grain, although when grain is less costly than forage, on the basis of energy and/or protein, it may be fed to advantage. In the northern latitudes, farm-flock ewes are frequently given ½ to 1 lb daily of a grain ration, in addition to the forage allowance, beginning about 6 weeks before lambing; and the grain allowance is usually doubled for the first 8 weeks after lambing.

Also, ewes suckling twins or on an accelerated lambing schedule are generally fed relatively high grain levels. However, many of the farm flocks of the South and the range bands of the Southwest are kept in good thrifty condition, and the lambs are raised to the marketing stage, without the feeding of any grain. In still other areas, the ewes are fed only during periods of deep snows or extended droughts. The range bands in the colder regions of the West are normally fed alfalfa hay and grain during the period of about 3 to 4 weeks that they are confined to the lambing camp.

---

[1]The authors gratefully acknowledge the helpful suggestions of the following authorities who reviewed this chapter: R. R. Jordan, Ph.D., Extension Animal Scientist, Department of Animal Science, University of Minnesota, St. Paul; J. B. Outhouse, Ph.D., Professor Emeritus, Department of Animal Science, Purdue University, West Lafayette, Ind.; and C. E. Terrill, Ph.D., Collaborator, USDA, Beltsville, Md.

[2]The statistics presented in the introductory section were for the mid-1980s and were obtained from: *FAO Production Yearbook*, published by FAO, Vol. 39, 1985; and *Agricultural Statistics*, USDA, 1985.

In general, for practical reasons, the ration of ewes should consist of pastures as nearly year-round as possible, with well-cured hay and other forages available the balance of the year, plus a limited grain allowance under certain conditions. Good-quality sun-cured hay and lush pastures will not only provide sufficient protein, but they are excellent sources of most of the minerals and vitamins, also.

ECONOMIC IMPORTANCE OF FEED FOR SHEEP

Feed is economically important in sheep production; it accounts for 50 to 65% of the total cost of producing market lambs, and it accounts for about two-thirds of the cost of finishing feedlot lambs. This leads to the deduction that there are two ways to make money in the sheep business: (1) increase the value of the products (lamb and wool), and/or (2) reduce the cost of production. Of the two, the latter is more feasible for the vast majority of producers.

Feed is the major item of expense whether consumed as range vegetation, as permanent or improved pasture, or as rations in confinement; and whether the sheep producer be a western rancher, a purebred breeder, or a farm flock operator. It is noteworthy, too, that 5 to 10% of all ewes bred fail to lamb, and that 15 to 20% of all lambs die between birth and weaning. Although there are many causes of reproductive failure, it is recognized that faulty nutrition is a major contributing factor.

Nutritional deficiencies and diseases in sheep are of special concern because they have such widely-differing uncontrolled diets as a consequence of the great variety of conditions under which they are produced.

NUTRITIVE NEEDS OF SHEEP

As with other classes of livestock, the nutritive requirements of sheep may be classified as (1) energy, (2) protein, (3) minerals, (4) vitamins, and (5) water.

The nutritive requirements are the values considered necessary for maintenance, optimum production, and prevention of all signs of nutritional deficiency.

National Research Council (NRC) Requirements

The National Research Council (NRC) nutritive requirements of sheep are given in the following tables: Table 21-1, 21-2, 21-3, 21-4, 21-5, 21-6, 21-7, and 21-10.

The NRC requirements are adequate for average, or below average, animals. In practical rations, margins of safety should be added to provide for below-average feeds, deterioration of feeds during transportation and storage, conditions of stress (bad weather, shipment, disease, or parasitism), and above-average animals in size, stage of production, and level of production.

TABLE 21-1
DAILY NUTRIENT REQUIREMENTS OF SHEEP (PER ANIMAL)[1] (See footnotes at end of table.)

Body Weight		Weight Gain/Loss Per Day		Daily Consumption				% Body Weight	Nutrients Per Animal									
				As-Fed[2]		Moisture-Free Dry Matter[3]			Energy[4]			Crude Protein		Ca	P	Vitamin A Activity	Vitamin E Activity
									TDN		DE	ME						
(lb)	(kg)	(lb)	(g)	(lb)	(kg)	(lb)	(kg)	(%)	(lb)	(kg)	(Mcal)	(Mcal)	(lb)	(g)	(g)	(g)	(IU)	(IU)
								EWES[5]										
								Maintenance										
110	50	0.02	10	2.4	1.1	2.2	1.0	2.0	1.2	0.55	2.4	2.0	0.21	95	2.0	1.8	2,350	15
132	60	0.02	10	2.7	1.2	2.4	1.1	1.8	1.3	0.61	2.7	2.2	0.23	104	2.3	2.1	2,820	16
154	70	0.02	10	2.9	1.3	2.6	1.2	1.7	1.5	0.66	2.9	2.4	0.25	113	2.5	2.4	3,290	18
176	80	0.02	10	3.2	1.4	2.9	1.3	1.6	1.6	0.72	3.2	2.6	0.27	122	2.7	2.8	3,760	20
198	90	0.02	10	3.4	1.6	3.1	1.4	1.5	1.7	0.78	3.4	2.8	0.29	131	2.9	3.1	4,230	21
								Flushing—2 Weeks prebreeding and first 3 weeks of breeding										
110	50	0.22	100	3.9	1.8	3.5	1.6	3.2	2.1	0.94	4.1	3.4	0.33	150	5.3	2.6	2,350	24
132	60	0.22	100	4.1	1.9	3.7	1.7	2.8	2.2	1.00	4.4	3.6	0.34	157	5.5	2.9	2,820	26
154	70	0.22	100	4.4	2.0	4.0	1.8	2.6	2.3	1.06	4.7	3.8	0.36	164	5.7	3.2	3,290	27
176	80	0.22	100	4.7	2.1	4.2	1.9	2.4	2.5	1.12	4.9	4.0	0.38	171	5.9	3.6	3,760	28
198	90	0.22	100	4.9	2.2	4.4	2.0	2.2	2.6	1.18	5.1	4.2	0.39	177	6.1	3.9	4,230	30
								Nonlactating—First 15 weeks gestation										
110	50	0.07	30	2.9	1.3	2.6	1.2	2.4	1.5	0.67	3.0	2.4	0.25	112	2.9	2.1	2,350	18
132	60	0.07	30	3.2	1.4	2.9	1.3	2.2	1.6	0.72	3.2	2.6	0.27	121	3.2	2.5	2,820	20
154	70	0.07	30	3.4	1.6	3.1	1.4	2.0	1.7	0.77	3.4	2.8	0.29	130	3.5	2.9	3,290	21
176	80	0.07	30	3.7	1.7	3.3	1.5	1.9	1.8	0.82	3.6	3.0	0.31	139	3.8	3.3	3,760	22
198	90	0.07	30	3.9	1.8	3.5	1.6	1.8	1.9	0.87	3.8	3.2	0.33	148	4.1	3.6	4,230	24
								Last 4 weeks gestation (130–150% lambing rate expected) or last 4–6 weeks lactation suckling singles[6]										
110	50	0.40 (0.10)	180 (45)	3.9	1.8	3.5	1.6	3.2	2.1	0.94	4.1	3.4	0.38	175	5.9	4.8	4,250	24
132	60	0.40 (0.10)	180 (45)	4.1	1.9	3.7	1.7	2.8	2.2	1.00	4.4	3.6	0.40	184	6.0	5.2	5,100	26
154	70	0.40 (0.10)	180 (45)	4.4	2.0	4.0	1.8	2.6	2.3	1.06	4.7	3.8	0.42	193	6.2	5.6	5,950	27
176	80	0.40 (0.10)	180 (45)	4.7	2.1	4.2	1.9	2.4	2.4	1.12	4.9	4.0	0.44	202	6.3	6.1	6,800	28
198	90	0.40 (0.10)	180 (45)	4.9	2.2	4.4	2.0	2.2	2.5	1.18	5.1	4.2	0.47	212	6.4	6.5	7,650	30

(Continued)

Feeding Sheep

TABLE 21-1 (Continued)

Body Weight		Weight Gain/Loss Per Day		Daily Consumption				% Body Weight	Nutrients Per Animal					Crude Protein		Ca	P	Vitamin A Activity	Vitamin E Activity
				As-Fed[2]		Moisture-Free Dry Matter[3]			Energy[4]										
									TDN		DE	ME							
(lb)	(kg)	(lb)	(g)	(lb)	(kg)	(lb)	(kg)	(%)	(lb)	(kg)	(Mcal)	(Mcal)		(lb)	(g)	(g)	(g)	(IU)	(IU)
								EWES Continued											
							Last 4 weeks gestation (180–225% lambing rate expected)												
110	50	0.50	225	4.1	1.9	3.7	1.7	3.4	2.4	1.10	4.8	4.0		0.43	196	6.2	3.4	4,250	26
132	60	0.50	225	4.4	2.0	4.0	1.8	3.0	2.6	1.17	5.1	4.2		0.45	205	6.9	4.0	5,100	27
154	70	0.50	225	4.7	2.1	4.2	1.9	2.7	2.8	1.24	5.4	4.4		0.47	214	7.6	4.5	5,950	28
176	80	0.50	225	4.9	2.2	4.4	2.0	2.5	2.9	1.30	5.7	4.7		0.49	223	8.3	5.1	6,800	30
198	90	0.50	225	5.1	2.3	4.6	2.1	2.3	3.0	1.37	6.0	5.0		0.51	232	8.9	5.7	7,650	32
							First 6–8 weeks lactation suckling singles or last 4–6 weeks lactation suckling twins[6]												
110	50	-0.06 (0.20)	-25 (90)	5.1	2.3	4.6	2.1	4.2	3.0	1.36	6.0	4.9		0.67	304	8.9	6.1	4,250	32
132	60	-0.06 (0.20)	-25 (90)	5.7	2.6	5.1	2.3	3.8	3.3	1.50	6.6	5.4		0.70	319	9.1	6.6	5,100	34
154	70	-0.06 (0.20)	-25 (90)	6.1	2.8	5.5	2.5	3.6	3.6	1.63	7.2	5.9		0.73	334	9.3	7.0	5,950	38
176	80	-0.06 (0.20)	-25 (90)	6.3	2.9	5.7	2.6	3.2	3.7	1.69	7.4	6.1		0.76	344	9.5	7.4	6,800	39
198	90	-0.06 (0.20)	-25 (90)	6.6	3.0	5.9	2.7	3.0	3.8	1.75	7.6	6.3		0.78	353	9.6	7.8	7,650	40
							First 6–8 weeks lactation suckling twins												
110	50	-0.13	-60	5.9	2.7	5.3	2.4	4.8	3.4	1.56	6.9	5.6		0.86	389	10.5	7.3	5,000	36
132	60	-0.13	-60	6.3	2.9	5.7	2.6	4.3	3.7	1.69	7.4	6.1		0.89	405	10.7	7.7	6,000	39
154	70	-0.13	-60	6.9	3.1	6.2	2.8	4.0	4.0	1.82	8.0	6.6		0.92	420	11.0	8.1	7,000	42
176	80	-0.13	-60	7.3	3.3	6.6	3.0	3.8	4.3	1.95	8.6	7.0		0.96	435	11.2	8.6	8,000	45
198	90	-0.13	-60	7.8	3.6	7.0	3.2	3.6	4.6	2.08	9.2	7.5		0.99	450	11.4	9.0	9,000	48
								EWE LAMBS											
							Nonlactating—First 15 weeks gestation												
88	40	0.35	160	3.4	1.6	3.1	1.4	3.5	1.8	0.83	3.6	3.0		0.34	156	5.5	3.0	1,880	21
110	50	0.30	135	3.7	1.7	3.3	1.5	3.0	1.9	0.88	3.9	3.2		0.35	159	5.2	3.1	2,350	22
132	60	0.30	135	3.9	1.8	3.5	1.6	2.7	2.0	0.94	4.1	3.4		0.35	161	5.5	3.4	2,820	24
154	70	0.28	125	4.1	1.9	3.7	1.7	2.4	2.2	1.00	4.4	3.6		0.36	164	5.5	3.7	3,290	26
							Last 4 weeks gestation (100–120% lambing rate expected)												
88	40	0.40	180	3.7	1.7	3.3	1.5	3.8	2.1	0.94	4.1	3.4		0.41	187	6.4	3.1	3,400	22
110	50	0.35	160	3.9	1.8	3.5	1.6	3.2	2.2	1.00	4.4	3.6		0.42	189	6.3	3.4	4,250	24
132	60	0.35	160	4.1	1.9	3.7	1.7	2.8	2.4	1.07	4.7	3.9		0.42	192	6.6	3.8	5,100	26
154	70	0.33	150	4.4	2.0	4.0	1.8	2.6	2.5	1.14	5.0	4.1		0.43	194	6.8	4.2	5,950	27
							Last 4 weeks gestation (130–175% lambing rate expected)												
88	40	0.50	225	3.7	1.7	3.3	1.5	3.8	2.2	0.99	4.4	3.6		0.44	202	7.4	3.5	3,400	22
110	50	0.50	225	3.9	1.8	3.5	1.6	3.2	2.3	1.06	4.7	3.8		0.45	204	7.8	3.9	4,250	24
132	60	0.50	225	4.1	1.9	3.7	1.7	2.8	2.5	1.12	4.9	4.0		0.46	207	8.1	4.3	5,100	26
154	70	0.47	215	4.4	2.0	4.0	1.8	2.6	2.5	1.14	5.0	4.1		0.46	210	8.2	4.7	5,950	27
							First 6–8 weeks suckling singles (wean by 8 weeks)												
88	40	-0.22	-100	5.1	2.3	4.6	2.1	5.2	3.2	1.45	6.4	5.2		0.67	306	8.4	5.6	4,000	32
110	50	-0.22	-100	5.7	2.6	5.1	2.3	4.6	3.5	1.59	7.0	5.7		0.71	321	8.7	6.0	5,000	34
132	60	-0.22	-100	6.1	2.8	5.5	2.5	4.2	3.8	1.72	7.6	6.2		0.74	336	9.0	6.4	6,000	38
154	70	-0.22	-100	6.7	3.0	6.0	2.7	3.9	4.1	1.85	8.1	6.6		0.77	351	9.3	6.9	7,000	40
								REPLACEMENT EWE LAMBS[7]											
66	30	0.50	227	2.9	1.3	2.6	1.2	4.0	1.7	0.78	3.4	2.8		0.41	185	6.4	2.6	1,410	18
88	40	0.40	182	3.4	1.6	3.1	1.4	3.5	2.0	0.91	4.0	3.3		0.39	176	5.9	2.6	1,880	21
110	50	0.26	120	3.7	1.7	3.3	1.5	3.0	1.9	0.88	3.9	3.2		0.30	136	4.8	2.4	2,350	22
132	60	0.22	100	3.7	1.7	3.3	1.5	2.5	1.9	0.88	3.9	3.2		0.30	134	4.5	2.5	2,820	22
154	70	0.22	100	3.7	1.7	3.3	1.5	2.1	1.9	0.88	3.9	3.2		0.29	132	4.6	2.8	3,290	22
								REPLACEMENT RAM LAMBS[7]											
88	40	0.73	330	4.4	2.0	4.0	1.8	4.5	2.5	1.1	5.0	4.1		0.54	243	7.8	3.7	1,880	24
132	60	0.70	320	5.9	2.7	5.3	2.4	4.0	3.4	1.5	6.7	5.5		0.58	263	8.4	4.2	2,820	26
176	80	0.64	290	6.9	3.1	6.2	2.8	3.5	3.9	1.8	7.8	6.4		0.59	268	8.5	4.6	3,760	28
220	100	0.55	250	7.3	3.3	6.6	3.0	3.0	4.2	1.9	8.4	6.9		0.58	264	8.2	4.8	4,700	30

(Continued)

TABLE 21-1 (Continued)

Body Weight		Weight Gain/Loss Per Day		Daily Consumption				% Body Weight	Nutrients Per Animal										
				As-Fed[2]		Moisture-Free Dry Matter[3]			Energy[4]				Crude Protein		Ca	P	Vitamin A Activity	Vitamin E Activity	
									TDN		DE	ME							
(lb)	(kg)	(lb)	(g)	(lb)	(kg)	(lb)	(kg)	(%)	(lb)	(kg)	(Mcal)	(Mcal)	(lb)	(g)	(g)	(g)	(IU)	(IU)	
LAMBS FINISHING—4 TO 7 MONTHS OLD[8]																			
66	30	0.65	295	3.2	1.4	2.9	1.3	4.3	2.1	0.94	4.1	3.4	0.42	191	6.6	3.2	1,410	20	
88	40	0.60	275	3.9	1.7	3.5	1.6	4.0	2.7	1.22	5.4	4.4	0.41	185	6.6	3.3	1,880	24	
110	50	0.45	205	3.9	1.7	3.5	1.6	3.2	2.7	1.23	5.4	4.4	0.35	160	5.6	3.0	2,350	24	
EARLY WEANED LAMBS—MODERATE GROWTH POTENTIAL[8]																			
22	10	0.44	200	1.2	0.6	1.1	0.5	5.0	0.9	0.40	1.8	1.4	0.38	127	4.0	1.9	470	10	
44	20	0.55	250	2.4	1.1	2.2	1.0	5.0	1.8	0.80	3.5	2.9	0.37	167	5.4	2.5	940	20	
66	30	0.66	300	3.2	1.4	2.9	1.3	4.3	2.2	1.00	4.4	3.6	0.42	191	6.7	3.2	1,410	20	
88	40	0.76	345	3.7	1.7	3.3	1.5	3.8	2.6	1.16	5.1	4.2	0.44	202	7.7	3.9	1,880	22	
110	50	0.66	300	3.7	1.7	3.3	1.5	3.0	2.6	1.16	5.1	4.2	0.40	181	7.0	3.8	2,350	22	
EARLY WEANED LAMBS—RAPID GROWTH POTENTIAL[8]																			
22	10	0.55	250	1.4	0.7	1.3	0.6	6.0	1.1	0.48	2.1	1.7	0.35	157	4.9	2.2	470	12	
44	20	0.66	300	2.9	1.3	2.6	1.2	6.0	2.0	0.92	4.0	3.3	0.45	205	6.5	2.9	940	24	
66	30	0.72	325	3.4	1.6	3.1	1.4	4.7	2.4	1.10	4.8	4.0	0.48	216	7.2	3.4	1,410	21	
88	40	0.88	400	3.7	1.7	3.3	1.5	3.8	2.5	1.14	5.0	4.1	0.51	234	8.6	4.3	1,880	22	
110	50	0.94	425	4.1	1.9	3.7	1.7	3.4	2.8	1.29	5.7	4.7	0.53	240	9.4	4.8	2,350	25	
132	60	0.77	350	4.1	1.9	3.7	1.7	2.8	2.8	1.29	5.7	4.7	0.53	240	8.2	4.5	2,820	25	

[1]Adapted by the authors from *Nutrient Requirements of Sheep*, sixth revised edition, NRC-National Academy of Sciences, 1985, pp. 45–47.

[2]As-fed was calculated using an average figure of 90% dry matter. When using silages, roots, and other wet feeds, these feeds should be converted to a moisture-free basis and the ration calculated using the moisture-free data.

[3]To convert dry matter to an as-fed basis, divide dry matter values by the percentage of dry matter in the particular feed.

[4]One kilogram TDN (total digestible nutrients) = 4.4 Mcal DE (digestible energy); ME (metabolizable energy) = 82% of DE. Because of rounding numbers, values in Table 1 and Table 2 may differ.

[5]Values are applicable for ewes in moderate condition. Fat ewes should be fed according to the next lower category and thin ewes at the next higher weight category.

[6]Values in parentheses are for ewes suckling lambs the last 4–6 weeks of lactation.

[7]Lambs intended for breeding; thus, maximum weight gains and finish are of secondary importance.

[8]Maximum weight gains expected.

TABLE 21-2
NUTRIENT CONCENTRATION IN RATIONS FOR SHEEP[1] [2](See footnotes at end of table.)

Body Weight		Weight Gain/Loss Per Day		Moisture Basis[3] A-F (as-fed) M-F (moisture-free)	Energy[4]					Example Diet Proportions		Crude Protein	Calcium	Phosphorus	Vitamin A Activity (IU/)		Vitamin E Activity (IU/)	
					TDN[5]	DE (Mcal/)		ME (Mcal/)		Concentrate	Forage							
(lb)	(kg)	(lb)	(g)		(%)	(lb)	(kg)	(lb)	(kg)	(%)	(%)	(%)	(%)	(%)	(lb)	(kg)	(lb)	(kg)
EWES[6]																		
Maintenance																		
154	70	0.02	10	A-F	50	4.8	2.2	4.0	1.8	0	100	8.5	0.18	0.18	5,442	2,468	31	14
				M-F	55	5.3	2.4	4.4	2.0	0	100	9.4	0.20	0.20	6,046	2,742	33	15
Flushing—2 weeks prebreeding and first 3 weeks of breeding																		
154	70	0.22	100	A-F	53	5.1	2.3	4.1	1.9	15	85	8.2	0.29	0.16	3,627	1,645	31	14
				M-F	59	5.7	2.6	4.6	2.1	15	85	9.1	0.32	0.18	4,031	1,828	33	15
Nonlactating—First 15 weeks gestation																		
154	70	0.07	30	A-F	50	4.8	2.2	4.0	1.8	0	100	8.4	0.23	0.18	4,664	2,115	31	14
				M-F	55	5.3	2.4	4.4	2.0	0	100	9.3	0.25	0.20	5,182	2,350	33	15
Last 4 weeks gestation (130–150% lambing rate expected) or last 4–6 weeks lactation suckling singles[7]																		
154	70	0.40 (0.10)	180 (0.45)	A-F	53	5.1	2.3	4.1	1.9	15	85	9.6	0.32	0.21	6,561	2,975	31	14
				M-F	59	5.7	2.6	4.6	2.1	15	85	10.7	0.35	0.23	7,290	3,306	33	15
Last 4 weeks gestation (180–225% lambing rate expected)																		
154	70	0.50	225	A-F	59	5.8	2.6	4.6	2.1	35	65	10.2	0.36	0.22	6,215	2,819	31	14
				M-F	65	6.4	2.9	5.1	2.3	35	65	11.3	0.40	0.24	6,906	3,132	33	15
First 6–8 weeks lactation suckling singles or last 4–6 weeks lactation suckling twins[7]																		
154	70	-0.06 (0.20)	-25 (90)	A-F	59	5.8	2.6	4.8	2.2	35	65	12.1	0.29	0.23	4,723	2,142	31	14
				M-F	65	6.4	2.9	5.3	2.4	35	65	13.4	0.32	0.26	5,248	2,380	33	15
First 6–8 weeks lactation suckling twins																		
154	70	-0.13	-60	A-F	59	5.8	2.6	4.8	2.2	35	65	13.5	0.35	0.26	4,962	2,250	31	14
				M-F	65	6.4	2.9	5.3	2.4	35	65	15.0	0.39	0.29	5,513	2,500	33	15

(Continued)

TABLE 21-2 (Continued)

Body Weight		Weight Gain/Loss Per Day		Moisture Basis[3] A-F (as-fed) M-F (moisture-free)	Energy[4]					Example Diet Proportions		Crude Protein	Cal-cium	Phos-phorus	Vitamin A Activity (IU/)		Vitamin E Activity (IU/)	
					TDN[5]	DE (Mcal/)		ME (Mcal/)		Con-centrate	Forage							
(lb)	(kg)	(lb)	(g)		(%)	(lb)	(kg)	(lb)	(kg)	(%)	(%)	(%)	(%)	(%)	(lb)	(kg)	(lb)	(kg)
							EWE LAMBS											
							Nonlactating—First 15 weeks gestation											
121	55	0.30	135	A-F	53	5.1	2.3	4.1	1.9	15	85	9.5	0.32	0.20	3,310	1,501	31	14
				M-F	59	5.7	2.6	4.6	2.1	15	85	10.6	0.35	0.22	3,678	1,668	33	15
							Last 4 weeks gestation (100–120% lambing rate expected)											
121	55	0.35	160	A-F	57	5.6	2.5	4.6	2.1	30	70	10.6	0.35	0.20	2,550	5,622	31	14
				M-F	63	6.2	2.8	5.1	2.3	30	70	11.8	0.39	0.22	6,247	2,833	33	15
							Last 4 weeks gestation (130–175% lambing rate expected)											
121	55	0.50	225	A-F	59	5.8	2.6	4.8	2.2	40	60	11.5	0.43	0.23	5,622	2,550	31	14
				M-F	66	6.4	2.9	5.3	2.4	40	60	12.8	0.48	0.25	6,247	2,833	33	15
							First 6–8 weeks lactation suckling singles (wean by 8 weeks)											
121	55	0.22	−50	A-F	59	5.8	2.6	4.8	2.2	40	60	11.8	0.27	0.20	4,217	1,913	31	14
				M-F	66	6.4	2.9	5.3	2.4	40	60	13.1	0.30	0.22	4,686	2,125	33	15
							First 6–8 weeks lactation suckling twins (wean by 8 weeks)											
121	55	−0.22	−100	A-F	62	5.9	2.7	5.0	2.3	50	50	12.3	0.33	0.23	4,549	2,063	31	14
				M-F	69	6.6	3.0	5.5	2.5	50	50	13.7	0.37	0.26	5,054	2,292	33	15
							REPLACEMENT EWE LAMBS[8]											
66	30	0.50	227	A-F	59	5.8	2.6	4.8	2.2	35	65	11.5	0.48	0.20	2,332	1,058	31	14
				M-F	65	6.4	2.9	5.3	2.4	35	65	12.8	0.53	0.22	2,591	1,175	33	15
88	40	0.40	182	A-F	59	5.8	2.6	4.8	2.2	35	65	9.2	0.38	0.16	2,665	1,209	31	14
				M-F	65	6.4	2.9	5.3	2.4	35	65	10.2	0.42	0.18	2,961	1,343	33	15
110–	50–	0.25	115	A-F	53	5.1	2.3	4.1	1.9	15	85	8.2	0.28	0.15	3,110	1,410	31	14
154	70			M-F	59	5.7	2.6	4.6	2.1	15	85	9.1	0.31	0.17	3,455	1,567	33	15
							REPLACEMENT RAM LAMBS[8]											
88	40	0.73	330	A-F	57	5.6	2.5	4.6	2.1	30	70	12.2	0.39	0.19	2,332	1,058	31	14
				M-F	63	6.2	2.8	5.1	2.3	30	70	13.5	0.43	0.21	2,591	1,175	33	15
132	60	0.70	320	A-F	57	5.6	2.5	4.6	2.1	30	70	9.9	0.32	0.16	3,292	1,493	31	14
				M-F	63	6.2	2.8	5.1	2.3	30	70	11.0	0.35	0.18	3,658	1,659	33	15
176–	80–	0.60	270	A-F	57	5.6	2.5	4.6	2.1	30	70	8.6	0.27	0.14	3,928	1,781	31	14
220	100			M-F	63	6.2	2.8	5.1	2.3	30	70	9.6	0.30	0.16	4,364	1,979	33	15
							LAMBS FINISHING—4 TO 7 MONTHS OLD[9]											
66	30	0.65	295	A-F	65	6.4	2.9	5.0	2.3	60	40	13.2	0.46	0.22	2,153	977	31	14
				M-F	72	7.1	3.2	5.5	2.5	60	40	14.7	0.51	0.24	2,392	1,085	33	15
88	40	0.60	275	A-F	68	6.6	3.0	5.4	2.4	75	25	10.4	0.38	0.19	2,332	1,058	31	14
				M-F	76	7.3	3.3	6.0	2.7	75	25	11.6	0.42	0.21	2,591	1,175	33	15
110	50	0.45	205	A-F	69	6.8	3.1	5.6	2.5	80	20	9.0	0.32	0.17	2,915	1,322	31	14
				M-F	77	7.5	3.4	6.2	2.8	80	20	10.0	0.35	0.19	3,239	1,469	33	15
							EARLY WEANED LAMBS—MODERATE AND RAPID GROWTH POTENTIAL[9]											
22	10	0.55	250	A-F	72	6.9	3.2	5.8	2.6	90	10	23.4	0.74	0.34	1,866	846	40	18
				M-F	80	7.7	3.5	6.4	2.9	90	10	26.2	0.82	0.38	2,073	940	44	20
44	20	0.66	300	A-F	70	6.8	3.1	5.6	2.5	85	15	15.2	0.49	0.22	1,866	846	40	18
				M-F	78	7.5	3.4	6.2	2.8	85	15	16.9	0.54	0.24	2,073	940	44	20
66	30	0.72	325	A-F	70	6.6	3.0	5.4	2.4	85	15	13.6	0.46	0.22	2,153	977	31	14
				M-F	78	7.3	3.3	6.0	2.7	85	15	15.1	0.51	0.24	2,392	1,085	33	15
88–	40–	0.88	400	A-F	70	6.6	3.0	5.4	2.4	85	15	13.1	0.50	0.25	2,487	1,128	31	14
132	60			M-F	78	7.3	3.3	6.0	2.7	85	15	14.5	0.55	0.28	2,763	1,253	33	15

[1]Adapted by the authors from *Nutrient Requirements of Sheep*, sixth revised edition, NRC-National Academy of Sciences, 1985, p. 48.
[2]Values in Table 2 are calculated from daily requirements in Table 1 divided by DM intake. The exception, vitamin E daily requirements/head, are calculated from vitamin E/kg diet × DM intake.
[3]As-fed was calculated using an average figure of 90% dry matter. When using silages, roots, and other wet feeds, these feeds should be converted to a moisture-free basis and the ration calculated using the moisture-free data.
[4]One kilogram TDN = 4.4 Mcal DE (digestible energy); ME (metabolizable energy) = 82% of DE. Because of rounding numbers, values in Table 1 and Table 2 may differ.
[5]TDN calculated on following basis: hay DM, 55% TDN and on as-fed basis 50% TDN; grain DM, 83% TDN and on as-fed basis 75% TDN.
[6]Values are for ewes in moderate condition. Fat ewes should be fed according to the next lower weight category and thin ewes at the next higher weight category. Once desired or moderate weight condition is attained, use that weight category through all production stages.
[7]Values in parentheses are for ewes suckling lambs the last 4–6 weeks of lactation.
[8]Lambs intended for breeding; thus, maximum weight gains and finish are of secondary importance.
[9]Maximum weight gains expected.

Energy

Lack of energy—hunger—is probably the most common nutritional deficiency of sheep. It may result from lack of feed or from the consumption of poor-quality feed.

Inadequate amounts of feed may result from (1) overgrazing, (2) drought, (3) snow covering the feed, (4) a low dry matter content of lush, washy feeds, or (5) a low level of range feed, with the result that sheep have to walk too far to obtain adequate intake. Also, poorly digested low-quality forage leads to reduced feed intake.

The energy needs of sheep are largely met through the consumption and digestion of forages—pasture, hay, and silage. Grains, such as corn, barley, milo, wheat, and oats, are used to raise the energy level of the ration during periods when supplementation is necessary. In general, sheep subsist on an even higher proportion of forages to concentrates than do beef cattle, and this applies to finishing lambs. The bacterial action in the rumen of sheep efficiently converts forages into suitable sources of energy.

Energy intake can be controlled by limiting the amount of feed offered, by adding fiber or bulk to the ration, by feeding every other day, or by limiting the time of eating.

In addition to size, age, pregnancy, lactation, and growth, covered in the nutrient requirement tables—Tables 21-1 and 21-2, and their relationship to such nutrients as protein, which must be supplied in adequate amounts, the following factors can affect energy requirements and diet concentration:

1. **Mature size of the breed (large mature genotype).** Lambs of the larger breeds (of the larger mature genotypes) grow more rapidly, have a higher energy (feed) requirement, and utilize energy (feed) more efficiently than lambs of the smaller breeds. Table 21-3 points up the comparative efficiency in the net energy for gain (NE$_g$) requirements of ram lambs of small, medium, and large genotypes with projected mature weights of 209, 254, and 298 lb, respectively. Note

TABLE 21-3
NET ENERGY REQUIREMENTS FOR LAMBS OF SMALL, MEDIUM, AND LARGE MATURE WEIGHT GENOTYPES[1,2]

Body Weight[3]	Lb	22	44	55	66	77	88	99	110
	Kg	10	20	25	30	35	40	45	50
Daily Gain[3]		kcal/d	kcal/d	kcal/d	kcal/d	kcal/d	kcal/d	kcal/d	kcal/d
(lb)	(g)								
NE$_m$ REQUIREMENTS[4]
		315	530	626	718	806	891	973	1,053
NE$_g$ REQUIREMENTS — Small mature weight lambs[5]
.22	100	178	300	354	406	456	504	551	596
.33	150	267	450	532	610	684	756	826	894
.44	200	357	600	708	812	912	1,008	1,102	1,192
.55	250	446	750	886	1,016	1,140	1,261	1,377	1,490
.66	300	535	900	1,064	1,219	1.368	1,513	1,652	1,788
Medium mature weight lambs[6]
.22	100	155	261	309	354	397	439	480	519
.33	150	233	392	463	531	596	658	719	778
.44	200	310	522	618	708	794	878	960	1,038
.55	250	388	653	771	884	993	1,097	1,199	1,297
.66	300	466	784	926	1,062	1,191	1,316	1,438	1,557
.77	350	543	914	1,080	1,238	1,390	1,536	1,678	1,816
.88	400	621	1,044	1,234	1,415	1,589	1,756	1,918	2,076
Large mature weight lambs[7]
.22	100	132	221	262	300	337	372	407	439
.33	150	197	332	392	450	505	558	610	660
.44	200	263	442	524	600	674	744	813	880
.55	250	329	553	654	750	842	930	1,016	1,099
.66	300	394	663	785	900	1,010	1,016	1,220	1,320
.77	350	461	775	916	1,050	1,179	1,303	1,423	1,540
.88	400	526	885	1,046	1,200	1,347	1,489	1,626	1,760
.99	450	592	996	1,177	1,350	1,515	1,675	1,830	1,980

[1]Adapted by the authors from *Nutrient Requirements of Sheep*, sixth revised edition, NRC-National Academy of Sciences, 1985, p. 49.
[2]Approximate mature ram weights of 209 lb (*95 kg*), 254 lb (*115 kg*), and 298 lb (*135 kg*), respectively.
[3]Weights and gains include fill.
[4]NE$_m$ = 56 kcal • W$^{0.75}$ • d^{-1}.
[5]NE$_g$ = 317 kcal • W$^{0.75}$ • LWG, kg • d^{-1}.
[6]NE$_g$ = 276 kcal • W$^{0.75}$ • LWG, kg • d^{-1}.
[7]NE$_g$ = 234 kcal • W$^{0.75}$ • LWG, kg • d^{-1}.

that when weighing 110 lb and making daily gains of 0.66 lb, the NE_g requirements of small, medium, and large mature weight rams are 1,788, 1,557, and 1,320 kcal/day, respectively.

2. **Last 6 weeks of gestation.** Ewes need more energy during the last 6 weeks of gestation to meet increased requirements for fetal growth and the development of the potential for high milk production.

3. **Multiple births.** At various stages of gestation, ewes carrying twins require more net energy for pregnancy (NE_{preg}) than ewes carrying singles, and ewes carrying triplets require still more NE_{preg} than ewes carrying twins. Moreover, the spread in the energy required between ewes carrying singles, twins, and triplets increases with advancing gestation. This is pointed up in Table 21–4.

TABLE 21–4
NE_{PREG} (NE_Y) REQUIREMENTS OF EWES CARRYING DIFFERENT NUMBERS OF FETUSES AT VARIOUS STAGES OF GESTATION[1]

Number of Fetuses Being Carried	Stage of Gestation (days)[2]					
	100	%[3]	120	%[3]	140	%[3]
	NE_{preg} Required (kcal/day)					
1	70	100	145	100	260	100
2	125	178	265	183	440	169
3	170	243	345	238	570	219

[1]Adapted by the authors from *Nutrient Requirements of Sheep*, sixth revised edition, NRC-National Academy of Sciences, 1985, p. 49. The (NE_Y) refers to reproductive process.
[2]For gravid uterus (plus contents) and mammary gland development only.
[3]As a percentage of a single fetus's requirement.

4. **Lactation.** The lactation requirements are higher than the maintenance or gestation requirements. At the peak of lactation, it is estimated that the net energy requirements of ewes suckling twins are 1.7 to 1.9 times the maintenance requirement. Also, ewes nursing twin lambs produce 20 to 40% more milk than ewes nursing singles. However, milk production during the 3rd and 4th months of lactation is only about ½ of the production during the first 2 months; hence, following peak milk production, the feed allowance should be reduced in order to maximize profits.

• **Symptoms of energy deficiency**—An energy deficiency is characterized by slowing and cessation of growth, loss of weight, reduced fertility or reproductive failure, lowered milk production and shortened lactation period, reduced quantity and quality of wool (including breaks in the fiber), lowered resistance to infection with internal parasites, and increased mortality.

Protein

Green pastures and legume hays (alfalfa, clover, soybeans, lespedeza, and others) are excellent practical sources of proteins for sheep in most areas. When the ranges are bleached and dry for an extended period, or legume hays cannot be produced for winter feeding, however, it may be desirable to provide sheep with such protein-rich supplements as soybean meal, cottonseed meal, linseed meal, peanut meal, canola (rapeseed) meal, sunflower meal, or a commercial protein supplement, at the rate of about ¼ to ⅓ lb per ewe per day.

The protein requirements of sheep are affected by age, growth, pregnancy, lactation, mature size, weight for age, body condition, rate of gain, and protein-energy ratio. Though correspondingly less because of their smaller body size and lower milk production, the protein requirements of ewes nursing lambs are similar to those of lactating cows.

The lamb is born with a nonfunctional rumen; hence, dietary protein must be provided through milk or a milk replacer until the rumen becomes functional. The rumen develops some degree of functionality by 2 weeks of age, but during early rumen development creep feed should be provided to supplement milk or milk replacer. By 6 to 8 weeks of age, the functioning rumen develops into a culture system for anaerobic bacteria, protozoa, and fungi.

Additional factors that can affect the protein requirements and utilization by sheep follow:

1. **Protein bypass (protected protein, escaped protein).** Quality and degradability of protein fed to sheep is more important than formerly thought. This is detailed in Chapter 11, Protein Supplements, under the heading "Protein Bypass (Protected Protein, Escaped Protein)."

2. **Nonprotein nitrogen (NPN).** Urea or other nonprotein nitrogen, in either liquid or dry supplements, can be used to provide all the supplemental nitrogen that may be needed in high-energy, grain-based rations, provided the diets are properly formulated and fed continuously.

3. **Mature size of genotype (breed).** Table 21–5 (see next page) gives the crude protein requirements of ram lambs of small, medium, and large mature weight genotypes, with projected mature weights of 209, 254, and 298 lb, respectively. It shows that the daily crude protein requirements of lambs increase with the size of the mature genotype (from small to medium, to large), and with the daily weight gains; but that the daily crude protein requirement per pound of body weight decreases as lambs become heavier because of the decrease in the protein content and the concomitant increase in the fat content of the body tissues.

• **Symptoms of protein deficiency**—A protein deficiency is characterized by reduced appetite, lowered feed intake, and poor feed efficiency. In turn, this makes for poor growth, poor muscular development, loss of weight, reduced reproductive efficiency, and reduced wool production. Under extreme conditions, there are severe digestive disturbances, nutritional anemia, and edema.

TABLE 21-5
CRUDE PROTEIN REQUIREMENTS FOR LAMBS OF SMALL, MEDIUM, AND LARGE MATURE WEIGHT GENOTYPES[1,2]

Body Weight[3]	Lb	22	44	55	66	77	88	99	110
	Kg	*10*	*20*	*25*	*30*	*35*	*40*	*45*	*50*
Daily Gain[3]		g/d	g/d	g/d	g/d	g/d	g/d	g/d	g/d
(lb)	*(g)*								
				Small mature weight lambs					
.22	*100*	84	112	122	127	131	136	135	134
.33	*150*	103	121	137	140	144	147	145	143
.44	*200*	123	145	152	154	156	158	154	151
.55	*250*	142	162	167	168	168	169	164	159
.66	*300*	162	178	182	181	180	180	174	168
				Medium mature weight lambs					
.22	*100*	85	114	125	130	135	140	139	139
.33	*150*	106	132	141	145	149	153	151	149
.44	*200*	127	150	158	160	163	166	163	160
.55	*250*	147	167	174	175	177	179	175	171
.66	*300*	168	185	191	191	191	191	186	181
.77	*350*	188	203	207	206	205	204	198	192
.88	*400*	209	221	224	221	219	217	210	202
				Large mature weight lambs					
.22	*100*	94	128	134	139	145	144	150	156
.33	*150*	115	147	152	156	160	159	164	169
.44	*200*	136	166	170	173	176	174	178	182
.55	*250*	157	186	188	190	192	189	192	195
.66	*300*	179	205	206	207	208	204	206	208
.77	*350*	200	224	224	224	224	219	220	221
.88	*400*	221	243	242	241	240	234	234	234
.99	*450*	242	262	260	256	256	249	248	248

[1]Adapted by the authors from *Nutrient Requirements of Sheep,* sixth revised edition, NRC-National Academy of Sciences, 1985, p. 50.
[2]Approximate mature ram weights of 209 lb *(95 kg),* 254 lb *(115 kg),* and 298 lb *(135 kg),* respectively.
[3]Weights and gains include fill.

TABLE
SHEEP MINERAL

Minerals Which May Be Deficient Under Normal Conditions	Conditions Usually Prevailing Where Deficiencies Are Reported	Function of Mineral	Deficiency Symptoms; Toxicity
Major or Macrominerals:			
Salt (sodium and chlorine—NaCl)	Negligence; for salt is inexpensive.	Sodium and chlorine are known to have regulatory functions in the body. They maintain osmotic pressure in cells, regulate the acid-base balance, and control water metabolism in tissues.	**Deficiency symptoms**—A deficiency of salt may result in an abnormal appetite, with the sheep trying to satisfy their craving by licking dirt, or eating toxic amounts of poisonous plants; decreased feed consumption; and decreased efficiency in the utilization of nutrients. ***Toxicity**—The maximum tolerable level of salt is 9.0% of the ration.
Calcium (Ca)	Lack of vitamin D. When finishing lambs are fed heavily on concentrates and limited quantities of legume roughage. When the feed consists largely of dried mature grasses or corn silage. Calcium-deficient areas (where pasture and range forages are deficient in Ca) are Fla., La., Neb., Va. Chronic internal parasite infections. Where there is magnesium deficiency.	Essential for development and maintenance of normal bones and teeth. Important in blood coagulation and lactation. Enables heart, nerves, and muscles to function. Regulates permeability of tissue cells. Affects availability of phosphorus and zinc.	**Deficiency symptoms**—Subnormal development of bone; rickets in young animals, and osteomalacia in adults. A high incidence of urinary calculi when there is a low calcium:high phosphorus ratio. To lessen the incidence of urinary calculi, the Ca:P ratio should be about 2:1. ***Toxicity**—If there is adequate phosphorus, sheep can tolerate a calcium-to-phosphorus ratio of 7:1 and as much as 2% calcium in the ration.
Phosphorus (P)	Lack of vitamin D. When sheep subsist for long periods on mature forages (such as dry range or grass or cereal hays). When the ration consists of a high proportion of beet by-products. When sheep subsist on pastures in phosphorus-deficient areas. Chronic internal parasite infections.	Essential for sound bones and teeth, and for the assimilation of carbohydrates and fats. A vital ingredient of the proteins in all body cells. Necessary for enzyme activation. Acts as a buffer in blood and tissue. Occupies a key position in biologic oxidation and reactions requiring energy.	**Deficiency symptoms**—Slow growth, depraved appetite, unthrifty appearance, listlessness, low level of phosphorus in the blood (less than 4 mg/100 ml of plasma), and development of knock-knees. *Caution:* A high level of phosphorus in the blood is not always an indication of adequacy in the diet; it may result from loss of weight. ***Toxicity**—Phosphorus at levels of 2 to 3 times the requirement can cause increased bone resorption in mature sheep.

Minerals

Although the body contains approximately 40 mineral elements, only 16 have been demonstrated to be essential for sheep—7 major mineral constituents, and 9 trace elements. Four of the 16 essential minerals are toxic when consumed in excessive amounts, so they must be fed carefully. Tables 21-6 and 21-7 list the essential minerals, present the requirements, and give the toxic levels if known.

TABLE 21-6
MACROMINERAL REQUIREMENTS OF SHEEP
(PERCENTAGE OF RATION)[1]

Nutrient	Requirement	
	As-fed[2]	Moisture-free
	(%)	(%)
Sodium	0.08-0.16	0.09-0.18
Chlorine	—	—
Calcium	0.18-0.74	0.20-0.82
Phosphorus	0.14-0.34	0.16-0.38
Magnesium	0.11-0.16	0.12-0.18
Potassium	0.45-0.72	0.50-0.80
Sulfur	0.13-0.23	0.14-0.26

[1]Adapted by the authors from *Nutrient Requirements of Sheep*, sixth revised edition, NRC-National Academy of Sciences, 1985, p. 48.

[2]As-fed was calculated using 90% dry matter (moisture-free).

TABLE 21-7
MICROMINERAL REQUIREMENTS OF SHEEP
AND MAXIMUM TOLERABLE LEVELS (PPM OR MG/KG OF RATION)[1]

Nutrient	Requirement		Maximum Tolerable Level	
	As-fed[2]	Moisture-free	As-fed	Moisture-free
	(ppm or mg/kg)	(ppm or mg/kg)	(ppm or mg/kg)	(ppm or mg/kg)
Cobalt	0.09-0.18	0.1-0.2	9	10
Copper	6-10	7-11[3]	23	25[4]
Fluorine	—	—	54-135	60-150
Iodine	0.09-0.72	0.10-0.80[5]	45	50
Iron	27-45	30-50	450	500
Manganese	18-36	20-40	900	1,000
Molybdenum	0.45	0.5	9	10[4]
Selenium	0.09-0.18	0.1-0.2	1.8	2
Zinc	18-30	20-33	675	750

[1]Adapted by the authors from *Nutrient Requirements of Sheep*, sixth revised edition, NRC-National Academy of Sciences, 1985, p. 50.

[2]As-fed was calculated using 90% dry matter (moisture-free).

[3]Requirement when dietary Mo concentrations are <1 mg/kg DM.

[4]Lower levels may be toxic under some circumstances.

[5]High level for pregnancy and lactation in rations not containing goitrogens; should be increased if rations contain goitrogens.

Table 21-8, Sheep Mineral Chart, gives, in summary form, the following pertinent information relative to each mineral listed: (1) conditions usually prevailing where deficiencies are reported, (2) function, (3) deficiency symptoms/toxicity, (4) nutrient requirements, (5) recommended allowances, and (6) practical sources.

21-8
CHART (See footnotes at end of table.)

Mineral Requirements[1]		Recommended Allowances[1]	Practical Sources of the Mineral	Comments
Minerals/ Animal/Day	Mineral Content of Ration, in % or ppm			
	As-fed[2] M-F			
*Lambs in drylot consume about 9 g of salt daily. Mature sheep in drylot may consume more.	*Salt for growing lambs, %: 0.38 0.42 *Na requirement of sheep, %: 0.08-0.16 0.09-0.18 (See Table 21-6.)	*Salt for mature sheep: 0.5% of the complete feed, or 1.0% to the concentrate portion. *Range operators commonly provide ½-¾ lb (¼-⅓ kg) salt/ewe/month. Mature sheep in drylot may consume more.	Free access to salt. Loose salt, rather than block salt, should be provided, for the reason that sheep bite at salt blocks, rather than lick, with the result that their teeth may be broken. In iodine-deficient areas, stabilized iodized salt should always be provided.	Sheep consume about 5 times more salt/ 100 lb body weight than cattle, which is attributed to their high forage consumption. Sheep can consume high quantities of salt without apparent harm provided water is freely available. In alkaline areas, the water may contain enough salt to meet the requirements, and supplemental salt may not be needed.
Variable, according to class, age, and weight of sheep (see Table 21-1).	*0.18-0.74% *0.20-0.82% (See Tables 21-2 and 21-6.)	Self-feed suitable mineral, or add calcium to the ration as required to bring level of total ration slightly above requirements.	Ground limestone, or oystershell flour. Where both calcium and phosphorus are needed, use bone meal, dicalcium phosphate, or defluorinated phosphate.	Most pasture and range forage contains adequate amounts of calcium. Forage containing from 0.24-0.32% calcium is considered adequate. Calcium requirements are usually met when sheep receive at least ⅓ of a legume forage. *Blood calcium levels below 9 mg/100 ml of plasma suggest chronic low calcium intake.
Variable, according to class, age, and weight of sheep (see Table 21-1).	*0.14-0.34% *0.16-0.38% (See Tables 21-2 and 21-6.)	Self-feed suitable mineral, or add phosphorus to the ration as required to bring level of total ration slightly above requirements.	Monosodium phosphate or diammonium phosphate. Where both calcium and phosphorus are needed, use bone meal, dicalcium phosphate, or defluorinated phosphate.	The proper calcium-phosphorus ratio should be maintained. Forage containing below 0.16% phosphorus is usually considered deficient for ewes during gestation, and 0.20% borderline during lactation. *A phosphorus deficiency may be manifested when the blood phosphorus level falls below 4 mg/100 ml of plasma.

(Continued)

TABLE 21-8

Minerals Which May Be Deficient Under Normal Conditions	Conditions Usually Prevailing Where Deficiencies Are Reported	Function of Mineral	Deficiency Symptoms; Toxicity
Major or Macrominerals (Continued):			
Magnesium (Mg)	Tetany most frequently occurs in nursing ewes shortly after they are turned to pasture in the spring (grass tetany), when the magnesium requirements for lactation are high and grass is low in magnesium.	It is a constituent of bone. Also, it is necessary for many enzyme systems and for proper functioning of the nervous system. Closely associated with the metabolism of calcium and phosphorus.	**Deficiency symptoms**—Hypomagnesemic tetany, a hyperirritability of the neuromuscular system. Sometimes this condition is accompanied by hypocalcemia. Acute tetany may occur as a result of insufficient dietary magnesium or inability to mobilize skeletal magnesium. *Toxicity—Oral administration of 0.8% magnesium in the ration will produce toxicosis.
Potassium (K)	When finishing lambs are fed high-concentrate and urea rations and limited amounts of dry roughage. When sheep are grazing mature range forage during winter or drought periods. The potassium level of such forage may decrease to less than 0.2%.	It affects osmotic pressure and acid-base balance within the cell. It also aids in activating several enzyme systems involved in energy transfer and utilization, protein synthesis, and carbohydrate metabolism.	**Deficiency symptoms**—Poor appetite and feed conversion, progressive stiffness from front to rear, and dry wool. *Toxicity—The maximum tolerable level of potassium for sheep is about 3% of the ration DM.
Sulfur (S)	When finishing lambs are fed high-concentrate and urea rations and limited amounts of roughage.	Functions in synthesis of sulfur-containing amino acids (methionine and cystine) in the rumen and various compounds of the body. Wool is high in sulfur; hence, sulfur is closely related to wool production.	**Deficiency symptoms**—Loss of appetite, reduced weight gains and feed efficiency, and reduced wool growth. Also, excessive salivation, lacrimation, and shedding of wool. *Toxicity—It appears that 0.4% is the maximum tolerable level of dietary sulfur as sodium sulfate.
Trace or Microminerals:			
Cobalt (Co)	Cobalt-deficient areas or soils in the U.S. and Canada. The most severely deficient U.S. areas include portions of New England and the lower Atlantic Coastal Plain. Moderately deficient areas include the rest of New England, northern N.Y., northern Mich., and parts of the Central Plains.	Promote synthesis of vitamin B-12 in the rumen.	**Deficiency symptoms**—Cobalt deficiency signs are actually signs of vitamin B-12 deficiency. They are: Lack of appetite, lack of thrift, severe emaciation, weakness, anemia, decreased fertility, and decreased milk and wool production. *Toxicity—Approximately 204.5 mg/100 lb live weight.
Copper (Cu)	In copper-deficient areas (soils), as in Fla. and in the coastal plains region of the Southeast. Also, in several of the western states, there are areas where an excess of molybdenum induces copper deficiency.	Anemia is associated with copper deficiency. Animals suffering from inadequate copper intake appear unable to absorb iron at a normal rate, and a deficiency in hemoglobin synthesis results. Steely wool and depigmentation of black sheep. Sheep suffering from a copper deficiency may produce "steely" or "stringy" wool, which is lacking in crimp, tensile strength, affinity for dyes, and elasticity. Depigmentation of the wool of black sheep has been noted as a sign of severe deficiency.	**Deficiency symptoms**—Signs in suckling lambs include "swayback," muscular incoordination, partial paralysis of the hindquarters, and degeneration of the myelin sheath of the nerve fibers. Lambs may be born weak and may die because of their inability to nurse. *Toxicity—23 ppm As-fed or 25 ppm M-F (see Table 21-7), but Mo level of the ration is a factor.
Fluorine (F)	*Conditions which may result in fluorine toxicity:* High fluorine in the water supply. Use of rock phosphate that contains 3-4% fluorine.		Fluorine deficiency not reported. Rather, the hazard is fluorine toxicity. *Toxicity—Acute toxicity can occur at 200 ppm.
Iodine (I)	Iodine-deficient areas or soils (in northwestern U.S. and in the Great Lakes and Rocky Mountain regions) where iodized salt is not fed. Feeds from iodine-deficient areas.	Formation of thyroxin, a hormone of the thyroid gland. In mature sheep an iodine deficiency may result in reduced wool yield and reduced rate of conception.	**Deficiency symptoms**—Lambs born with goiter; usually stillborn or die soon after birth. Usually, such lambs have very little wool. *Toxicity—Maximum tolerable level for sheep is 45 ppm As-fed basis or 50 ppm M-F (see Table 21-7). However, much higher tolerable levels have been reported.
Iron (Fe)	Iron-deficiency anemia sometimes occurs in lambs raised on slotted floors. Loss of blood from parasite infestation can produce a secondary iron-deficiency anemia.	Hemoglobin formation.	**Deficiency symptoms**—Anemia, poor growth, lethargy, increased respiration rate, decreased resistance to infection, and in severe cases high mortality. *Toxicity—Signs of chronic toxicity are reduction in feed intake, growth rate, and feed efficiency. In acute toxicosis, animals exhibit loss of appetite, scanty urination, diarrhea, below normal temperature, shock, acidosis, and death.

Feeding Sheep

(Continued)

Mineral Requirements[1]				
Minerals/ Animal/Day	Mineral Content of Ration, in % or ppm	Recommended Allowances[1]	Practical Sources of the Mineral	Comments
	As-Fed **M-F** *0.11, 0.14, and 0.16% for growing lambs, ewes in late pregnancy, and ewes in early lactation, respectively. *0.12, 0.15, and 0.18% for growing lambs, ewes in late pregnancy, and ewes in early lactation, respectively. *Where ewes in early lactation are grazing forage with high nitrogen and potassium content, the minimum level of magnesium in the ration is 0.2%.		Plant protein supplements are excellent sources of Mg. Likewise, by-product feedstuffs derived from plants tend to be good sources. The common magnesium supplements are magnesium carbonate, magnesium oxide, and magnesium sulfate.	*Blood serum normally contains about 2.5 mg/100 ml.
	*0.45% for growth of lambs. 0.63–0.72 for lactation and stress. *0.5% for growth of lambs. 0.7–0.8 for lactation and stress.	0.7 to 1.0% of total air-dry ration.	Roughages usually contain adequate potassium, with the possible exception of nonlegume silage. Potassium chloride and potassium sulfate are the supplements of choice.	The feeding of potassium chloride appears to reduce the incidence of urinary calculi in feedlot lambs. This is especially true with high-milo rations.
	*Mature ewes: 0.13–0.16% 0.14–0.18% *Young lambs: 0.16–0.23% 0.18–0.26%	*It is recommended that a dietary nitrogen-sulfur ratio of 10:1 be maintained.	Sulfate sulfur, elemental sulfur, or sulfur-containing proteins or amino acids. Inorganic compounds are generally more convenient and economical for supplemental feeding.	*Practically all common feedstuffs contain more than 0.1% sulfur. However, mature grass and grass hays are sometimes low in sulfur. Where forages are low in sulfur or high in urea, increased weight gains and wool growth can be obtained by feeding sulfur.
	*0.09–0.18 ppm *0.1–0.2 ppm However, young, rapidly growing lambs may have a slightly higher requirement.	*Feed cobalt at the rate of 1.4 g/100 lb (2.5 g/100 kg) of salt as cobalt chloride or cobalt sulfate.	A cobalt mineral mixture. Other effective methods of providing cobalt are (1) to add cobalt to the soil, or (2) to place cobalt pellets into the rumen.	Several good commercial cobalt-containing minerals are on the market in either block or loose form. Cobalt is much more effective when given by mouth than when given intravenously.
	As-fed[2] **M-F** *6.3–20.7 ppm *7–23 ppm The Cu requirement varies with (1) the Mo content of the feed, and (2) the growth, pregnancy, lactation, and breed involved. (Also, see Table 21-7.)	*Add copper sulfate to the salt at rate of 0.5%.	Salt containing 0.5% of copper sulfate.	Copper deficiencies may exist alone or along with deficiencies of cobalt and iron. An interesting interrelation exists between copper, molybdenum and sulfur. An excess of molybdenum causes a pathological condition which can be cured only by administering copper. Stores of copper in the liver, kidney, heart, lungs, pancreas, and spleen serve as a reserve for as long as 4–6 months when animals are grazing copper-deficient forage. Sheep are much more susceptible to copper toxicity than cattle. As much as 25 mg of copper in the daily ration of sheep is considered toxic; and about 9 mg/day is considered the safe tolerance level. Copper toxicity may result from feeding poultry wastes or mineral supplements designed for other species.
	*Breeding sheep should not be fed a ration containing more than 55 ppm (As-fed) or 60 ppm of fluorine on a moisture-free basis. *Finishing lambs can tolerate up to 135 ppm (As-fed) or 150 ppm of fluorine in the ration on a moisture-free basis.			Symptoms of fluorine toxicity are loss of appetite; the normal ivory color of bones changes to chalky white; bones thicken, and the teeth, especially the incisors, may become pitted and eroded to such an extent that the nerves are exposed.
	*0.09–0.72 ppm *0.1–0.8 ppm The higher levels are indicated for pregnancy and lactation. When goitrogens are present increase the iodine.	*Free access to stabilized iodized salt containing 0.0078% iodine.	Stabilized iodized salt containing 0.0078% iodine. Calcium iodate.	Do not use iodized salt in a mixture with a concentrate to limit feed intake, as the animals may consume an excessive amount of iodine.
	*27–45 ppm *30–50 ppm	*Intramuscular injections of iron-dextran; 2 injections, 150 mg of iron in each, given 2 to 3 weeks apart.	Ferrous gluconate, ferrous succinate, or ferrous sulfate given orally. Iron dextran injection.	A primary iron deficiency in grazing sheep is very unlikely.

(Continued)

TABLE 21-8

Minerals Which May Be Deficient Under Normal Conditions	Conditions Usually Prevailing Where Deficiencies Are Reported	Function of Mineral	Deficiency Symptoms; Toxicity
Trace or Microminerals (Continued)			
Manganese (Mn)	Lambs on a purified diet containing less than 1 ppm of manganese over a 5-month period. High calcium and iron may increase manganese requirements.	Skeletal development and reproduction.	**Deficiency symptoms**—Bone abnormalities, lack of coordination in newborn lambs, impaired growth, and depressed or disturbed reproduction. *****Toxicity**—It appears that 1,000 ppm of dietary Mn is the maximum tolerable level.
Molybdenum (Mo)	The major concern about molybdenum is that in excess it may induce a copper deficiency. Excess molybdenum in the soil such as is found in areas of Calif., Nev., and England.	It is believed that the molybdenum binds and inactivates the copper in the intestine.	**Deficiency symptoms**—A low intake of molybdenum causes excess copper to accumulate in tissues, especially the liver, even when the copper
	intake is moderate, thus producing fatal jaundice (easily detected in the eyes). This disease can be prevented by increasing the molybdenum intake. *****Toxicity**—High levels of molybdenum (10 to 20 ppm in forage plants) will induce copper deficiency characterized by stringy wool, lack of pigmentation in black wool, anemia, bone disorders, and infertility. Also, sheep start to scour after being turned to high molybdenum pasture (5 to 20 ppm M-F basis). The scouring can be controlled by increasing the copper level in the diet to 5 ppm.		
Selenium (Se)	Areas where selenium content of crops is below 0.1 ppm, such as northwestern, northeastern, and southeastern U.S. Parts of S.D., Wyo., and Utah produce forage containing excess selenium which causes toxicity in farm animals.	Component of the enzyme glutathione peroxidase, the metabolic role of which is to protect against oxidation of polyunsaturated fatty acids and resultant tissue damage. Interrelation with vitamin E—they spare each other.	**Deficiency symptoms**—The most commonly noticed lesion from a deficiency of selenium is white muscle disease, which affects lambs 0-8 weeks of age, along with reduced growth of lambs. Additional signs of inadequate selenium are unthriftiness, infertility, early embryonic death, and periodontal disease. *****Toxicity**—Chronic selenium toxicity occurs when
	sheep consume feeds containing more than 3 ppm of selenium on a dry basis over a prolonged period. Toxicity signs include loss of wool, soreness and sloughing of the hooves, and marked reduction in reproductive performance.		
Zinc (Zn)	Diets high in calcium adversely affecting zinc utilization.		**Deficiency symptoms**—Loss of appetite, reduction in growth rate, excessive salivation, parakeratosis, wool loss, and delayed wound healing. Ram
	lambs show reduced testicular development and defective spermatogenesis. In females, all phases of the reproductive process from estrus to parturition and lactation may be adversely affected. *****Toxicity**—There is a wide margin of safety between zinc requirements and zinc toxicity. However, 0.1% zinc in the diet reduced feed consumption and gain in lambs; and 0.075% induced severe copper deficiency in pregnant ewes and caused a high incidence of abortions and still births.		

[1]As used herein, the distinction between "mineral requirements" and "recommended allowances" is as follows: In mineral requirements, no margins of safety are included intentionally; whereas in recommended allowances, margins of safety are provided to compensate for variations in feed composition, environment, and possible losses during storage or processing.
Where preceded by an asterisk, the requirements, recommended allowances, and other facts presented herein were taken from *Nutrient Requirements of Sheep*, 6th rev. ed., NRC-National Academy of Sciences, 1985.
[2]Estimated 90% dry matter.

Fig. 21-1. Lamb fed a ration deficient in phosphorus. Note the knock-kneed conformation. (Courtesy, University of Idaho, Moscow)

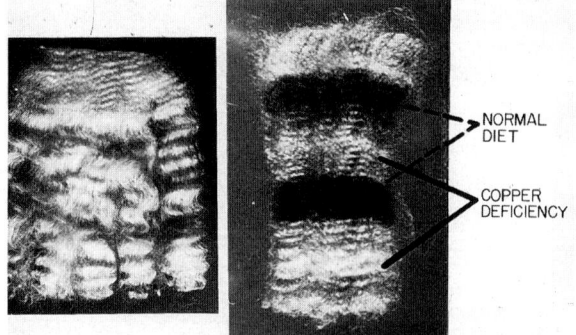

Fig. 21-2. Two samples of Australian wool, both of which show what may happen when sheep are on a copper-deficient ration. *Left:* The outer (bottom) ⅔ of this sample was produced by a sheep on a copper-deficient ration, resulting in hairlike or "steely" wool. Then copper was added to the sheep's ration, and normal, well-crimped wool was produced. *Right:* Wool sample from a normally black sheep. The white bands appeared at intervals when copper was deficient in the ration, because copper is essential for melanin or pigment production. Where such deficiencies occur under natural conditions, it is recognized that copper deficiencies result in the production of wool of lowered elasticity, tensile strength, and affinity for dyes.

(Continued)

Mineral Requirements[1]		Recommended Allowances[1]	Practical Sources of the Mineral	Comments
Minerals/ Animal/Day	Mineral Content of Ration, in % or ppm			
	As-Fed **M-F** *18–36 ppm *20–40 ppm	18 ppm in As-fed ration, or 20 ppm in M-F ration.	Manganese gluconate.	
	*The minimum dietary requirement of molybdenum is not known. *The Food and Drug Administration does not recognize molybdenum as safe; hence, the law prohibits adding it to feed for sheep.	The two contrasting situations—(1) high molybdenum and copper deficiency, or (2) low molybdenum and excess copper accumulation—make it very difficult to define nutrient requirements of molybdenum and copper.	A high-molybdenum intake induces a copper deficiency even when the copper content of pasture is quite high; the scouring effect can be prevented by providing an increased copper intake. Sheep are less affected than cattle by high-molybdenum intakes. *In treating copper toxicity, both molybdenum and sulfate should be administered. Drench each lamb with 100 mg of ammonium molybdate and 1 g of sodium sulfate in 20 ml of water.	
	*0.09–0.18 ppm 0.1–0.2 ppm	Selenium as either sodium selenite or sodium selenate at the rate of 0.3 ppm of complete feed. Selenium added to salt-mineral mixture fed free-choice at rate of 90 ppm. Selenium in the limit feeding (feed supplements and salt-mineral mixtures) consumption rate for sheep of 0.7 mg per head per day.		Plants grown on the same seleniferous soils vary greatly in their uptake of selenium, with a range of 1,000 ppm to only 10–25 ppm. The most practical way to prevent livestock losses from selenium poisoning is to manage the grazing so that animals alternate between high-selenium and low-selenium areas. Selenium is a cumulative poison, but mild chronic signs can be overcome readily by feeding low-selenium forage.
	*Growth: 18 ppm 20 ppm *Reproduction: 30 ppm 33 ppm			

NOTE: Mineral recommendations for all classes and ages of sheep, especially those fed unmixed rations or on pasture, are—

1. *When sheep are on liberal grain feeding*—Provide free access to a 2-compartment mineral box, with (a) trace mineralized salt in one side, and (b) in the other side, a mixture of ⅓ trace mineralized salt (salt included for purposes of palatability), ⅓ defluorinated phosphate or steamed bone meal, and ⅓ ground limestone or oystershell flour.

2. *When sheep are primarily on roughage (pasture, hay, and/or silage)*—Provide free access to a 2-compartment mineral box, with (a) trace mineralized salt in one side, and (b) in the other side, a mixture of ⅓ trace mineralized salt and ⅔ defluorinated phosphate or steamed bone meal.

Additionally, in those areas where cobalt and/or copper deficiencies exist in the soil (and plants), add cobalt and/or copper sulfate to either the salt or salt-phosphorus mixture in the proportions indicated. If desired, the mineral supplement may be incorporated in the ration in keeping with the recommended allowances given in this table.

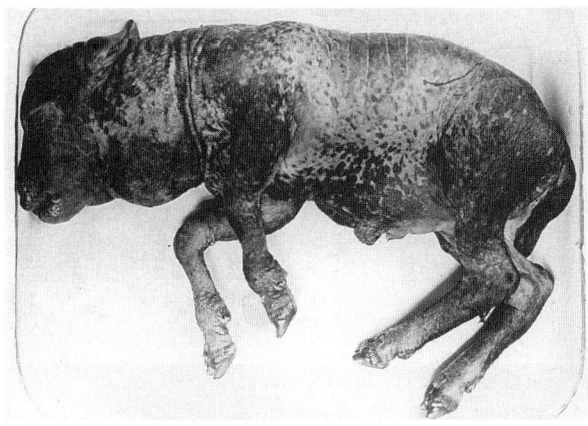

Fig. 21-3. Goiter. Woolless, goitered (big-necked) lamb stillborn due to iodine deficiency. (Courtesy, Montana State University, Bozeman)

Fig. 21-4. Lamb afflicted with white muscle disease (stiff lamb disease). Note the stiff hind legs and the humped or "roached" back. Such lambs have a stilted way of moving and are usually stunted. (Courtesy, James E. Oldfield, Oregon State University, Corvallis)

Vitamins

Mature sheep require the fat-soluble vitamins A, D, E, and K, but they do not need added sources of the B vitamins, since the latter are normally synthesized in adequate amounts

**TABLE
SHEEP VITA**

Vitamin Which May Be Deficient Under Normal Conditions	Conditions Usually Prevailing Where Deficiencies Are Reported	Function of Vitamin	Deficiency Symptoms/Toxicity
Fat-Soluble Vitamins:			
A	Vitamin A deficiencies may occur when—(1) extended drought results in dry, bleached pastures; (2) winter feeding on bleached hays (especially over-ripe cereal hays or straws) with little or no green hay or silage; (3) drylot finishing on rations with little or no green forage or yellow corn, especially for feeding periods longer than 2–3 months; and (4) there is high nitrate intake, in either water or feed.	Necessary for maintaining normal epithelial tissue.	**Deficiency symptoms**—Keratinization of the respiratory, alimentary, reproductive, urinary, and ocular epithelia; lowered resistance to infection; abnormal development of bone; birth of lambs that are weak, malformed, or dead; and night blindness. **Toxicity**—Changes in bone composition.
D	When ration consists predominantly of dehydrated hays, green feeds and seeds and their by-products. Prolonged cloudy weather or when kept inside, especially in fast-growing young lambs. When the vitamin D in feed is lost by oxidation.	Prevention of rickets in young lambs and osteomalacia in older sheep.	**Deficiency symptoms**—Rickets in young lambs and osteomalacia in older sheep. Congenital malformations in newborn lambs from extreme deficiencies. **Toxicity**—Abnormal deposition of calcium in soft tissues and brittle bones subject to deformation and fractures.
E	When lambs are making rapid growth, although this isn't always the case. When old hay is fed, as oxidation destroys vitamin E.	Prevention of white muscle disease (stiff lamb disease). As an antioxidant. Closely associated with selenium in metabolism.	**Deficiency symptoms**—Stiff lamb disease, or white muscle disease; characterized by a stiff, stilted way of moving and a "roached" back. Sometimes a paralysis of hind legs.
K	Vitamin K deficiency may occur when the dicumarol content of hay is excessively high, as when moldy sweet clover hay is fed.	Vitamin K_1 or K_2 is necessary in the blood clotting mechanism.	
Water-Soluble Vitamins:	B vitamin deficiencies may be evident in poorly fed and unhealthy animals.		

[1]As used herein, the distinction between "vitamin requirements" and "recommended allowances" is as follows: In vitamin requirements, no margins of safety are included intentionally; whereas in recommended allowances, margins of safety are provided in order to compensate for variations in feed composition, environment, and possible losses during storage or processing.

Feeding Sheep

by rumen microorganisms.

Table 21-9, Sheep Vitamin Chart, presents in summary form pertinent information relative to the vitamin needs of sheep.

21-9 MIN CHART

Vitamin Requirements[1]		Recommended Allowances[1]	Practical Sources of the Vitamin	Comments
Vitamins/Animal/Day	Vitamin Content of Ration			
Variable, according to class, age, and weight of sheep, as indicated by the following:	Variable, according to class, age, and weight of sheep (see Table 21-2).	Sheep that are deficient in vitamin A and weigh 70 lb (32 kg) or more should receive 100,000 IU of vitamin A by injection, and their rations should be adjusted to provide recommended levels of vitamin A or carotene. Ewes deficient in vitamin A should be given vitamin A either orally or by injection prior to breeding.	Stabilized vitamin A. Green pasture. Grass or legume silages. Yellow corn. Green hay not over 1 year old. *The vitamin A value of carotene from two common feeds is as follows: IU/mg Dehydrated alfalfa meal ... 254–520 Silage, corn .... 436 Where sheep are grazing forage low in carotene for extended periods, vitamin A deficiency can be prevented by (1) intramuscular injection of vitamin A, or (2) the addition of vitamin A to the ration as a pasture supplement or part of a salt mixture.	*Sheep do not convert carotene to vitamin A as efficiently as rats. For sheep, 1 mg of feed carotene is equivalent to 400–500 IU of vitamin A. *It requires 200 days to deplete entirely liver storage of ewe lambs previously grazing on green feed. Because of this storage, animals that normally graze on green forage during the growing season are able to do reasonably well on a low-carotene ration of dry feed for periods of 4–6 months.
	Mcg of B-carotene/day			
Categories	/lb body wt.	/kg body wt.			
All categories	31	69			
Late gestation and in lactation ..	57	125			
First 6–8 weeks of lactation of ewes suckling twins	67	147			
	IU of Vitamin A/day			
Categories	/lb body wt.	/kg body wt.			
All categories	21	47			
Late gestation and in lactation ..	39	85			
First 6–8 weeks of lactation of ewes suckling twins	45	100			
(See Table 21-1.)				
*For all sheep except early-weaned lambs: 555 IU per 220 lb (100 kg) body weight. *For early-weaned lambs: 666 IU per 220 lb (100 kg) body weight.	Variable, according to class, age, and weight of sheep.	Breeding sheep, 500 to 800 IU/head/day. Feeder lambs, 500 IU/head/day.	Exposure to sunlight, through irradiation. Sun-cured hays. Irradiated yeast. Vitamin D_2 or vitamin D_3, which sheep use equally well.	Newborn lambs are provided with enough vitamin D from their dams to prevent rickets for 4 to 6 weeks if the ewes have adequate storage. Sheep with white skin and short wool receive more vitamin D activity from irradiation by sunlight than do animals with dark skin or long wool.
	*IU Vitamin E As-Fed basis		Meet the requirements given in the column headed "Vitamin Content of Ration."	Wheat germ meal, dehydrated alfalfa, some green feeds, and vegetable fats are good sources of vitamin E. Alpha-tocopherol (either in dl or d forms).	Experiments have failed to relate vitamin E deficiency with reproductive failure in sheep. The need for vitamin E in the ration of young nursing lambs is related to the selenium level in the ration. Selenium has a sparing effect on the vitamin E requirement; the higher the selenium level in the diet, the lower the vitamin E requirement, and vice versa.
Categories	/lb	/kg			
Lambs under 44 lb (20 kg) live weight	8	18			
Lambs over 44 lb (20 kg) live weight and pregnant ewes	6	13.5			
	*IU Vitamin E M-F basis				
Categories	/lb	/kg			
Lambs under 44 lb (20 kg) live weight	9	20			
Lambs over 44 lb (20 kg) live weight and pregnant ewes	7	15			
		Vitamin K_2 is normally synthesized in large amounts in the rumen; no need for dietary supplementation has been established.	Green leafy materials of any kind, fresh or dry, are good sources of K_1.	
		*The B vitamins are not required in the diet of sheep with functioning rumens, because the microorganisms synthesize these vitamins in adequate amounts.		Addition of B vitamins has not been shown to be beneficial to mature sheep. However, young lambs (to about 2 months of age) with undeveloped rumens have been shown to have a dietary need for vitamin B-12, thiamin, pyridoxine, riboflavin, niacin, folic acid, and possibly some of the other B vitamins, since they will not be receiving these in the milk from their dams. Cobalt is necessary for the synthesis of vitamin B-12 in the rumen. *No supplementary dietary need for vitamin C has been shown.

[1] Where preceded by an asterisk, the vitamin requirements, recommended allowances, and other facts presented herein were taken from *Nutrient Requirements of Sheep*, 6th rev. ed., NRC-National Academy of Sciences, 1985.

Fig. 21-5. Bilateral bent leg in yearling Rambouillet ram due to a deficiency of vitamin D. (Courtesy, Utah Ag. Exp. Sta.)

• **Vitamin E**—The need for vitamin E in the diet of young nursing lambs is related to the selenium level in the diet. Selenium has a sparing effect on the vitamin E requirement; the higher the selenium level in the diet, the lower the vitamin E requirement, and vice versa.

White muscle disease in lambs is prevented by adding alpha-tocopherol and selenium to the diet. The suggested dietary levels of vitamin E are as follows: Lambs under 44 lb in weight should recieve 8 IU/lb of as-fed ration; lambs over 44 lb in weight and pregnant ewes should receive 6 IU/lb of as-fed ration. (The IU is defined as 1 mg of dl-alpha-tocopherol acetate; 1 mg dl-alpha-tocopherol has the biological potency of 1.5 IU of vitamine E activity.) The above recommendations assume that dietary selenium levels are 0.05 ppm.

Values for the vitamin E requirements of sheep are presented in Tables 21-1, 21-2, and 21-10. The values presented in Table 21-1 were calculated from values per kilogram of dry feed consumed given in Table 21-2.

Table 21-10 presents daily vitamin E requirements for lambs and the suggested amounts of alpha-tocopherol acetate to add to rations to provide 100% of the requirements.

TABLE 21-10
VITAMIN E REQUIREMENTS OF GROWING-FINISHING LAMBS AND SUGGESTED LEVELS OF FEED FORTIFICATION TO PROVIDE 100% OF REQUIREMENTS[1]

Body Weight		Alpha-Tocopherol Acetate			Feed Intake Per Lamb		Amount of Vitamin E Added to Concentrate			Amount of Vitamin E Added to Protein Supplement[2]		
(lb)	(kg)	(mg/lamb/day)[3]	(mg/lb ration)	(mg/kg ration)	(lb)	(kg)	(mg/lb)	(mg/kg)	(mg/ton)	(mg/lb)	(mg/kg)	(mg/ton)
22	10	5.0	44	20	0.50	0.23	9.1	20	18,200	133	60	120,000
44	20	10.0	44	20	1.00	0.45	9.1	20	18,200	60	133	120,000
66	30	15.0	33	15	2.10	0.96	6.8	15	13,600	45	100	90,000
88	40	20.0	33	15	2.86	1.30	6.8	15	13,600	45	100	90,000
110	50	25.0	33	15	3.50	1.60	6.8	15	13,600	45	100	90,000

[1]Adapted by the authors from *Nutrient Requirements of Sheep*, sixth revised edition, NRC-National Academy of Sciences, 1985, p. 51.
[2]Assumes the concentrate diet contains 15% protein supplement.
[3]Rounded values based on approximate diet intake containing recommended vitamin E levels.

Water

Sheep get water by drinking, and from snow, dew, and feed. The amount of water that sheep voluntarily consume is affected by temperature, rainfall, snow and dew covering, age, breed, stage of production, number of lambs carried, wool covering, respiratory rate, frequency of watering, kind and amount of feed, and exercise. On the average, mature animals consume approximately a gallon of water per day, whereas feeder lambs require about half this amount. However, sheep may go for weeks without drinking water when foraging on grasses and other feeds of high-moisture content. This condition often prevails on desert ranges in the early spring and on many of the mountain ranges during the summer months.

Additives

Table 21-11 summarizes the growth stimulants that are presently available and can be used. All of these products have been shown to improve gain and feed efficiency of sheep.

TABLE 21-11
SHEEP FEED ADDITIVES AND IMPLANTS

Type of Additive	Method of Administering	Dosage	Effect On			Comments
			Daily Rate of Gain	Feed Efficiency	Carcass Quality	
			(% increase)	(% increase)		
Antibiotics (chlortetracycline and oxytetracycline)	Feeding (oral)	Aureomycin (chlortetracycline) 10 to 25 mg/lb of feed. Terramycin (oxytetracycline) 5 to 10 mg/lb of feed.	Range: 0–31 Average: 11	Range: 4–27 Average: 10	No effect to slight improvement.	Antibiotics (especially chlortetracycline and oxytetracycline) may improve performance when added to creep and lamb finishing rations. Response to antibiotics varies markedly according to differences in management and degree of stress to which lambs are subjected. There is some evidence that antibiotics reduce the incidence of enterotoxemia.
Bovatec (lasalocid)	Feeding (oral)	10–15 mg/lb complete feed, fed at rate of 15–70 mg lasalocid/day.	Range: 0–20 Average: 6–8	Range: 5–15 Average: 8–10	No effect.	Bovatec is an ionophore. In addition to increasing rate of gain and feed efficiency, Bovatec reduces rumen protein degradation and increases the amount of bypass protein. Greatest response is obtained where coccidiosis is a problem, for which purpose Bovatec was initially approved by FDA.
Ralgro (zeranol)	Implant	12 mg/head	Range: 0–25 Average: 10	Average: 6		Do not implant animals within 40 days of slaughter. Do not implant breeding animals.

FEEDS FOR SHEEP

Sheep are adapted to the consumption of a great variety of feeds, most of which are of plant origin and bulky in nature. The feeding of concentrates is usually limited to the finishing of lambs and for use by the breeding flock at such special periods as the lambing season or just before and during the breeding season. Forages constitute 100% of the ration of the vast majority of sheep during most seasons of the year.

• **Pasture**—No other class of farm animals is so well adapted to the utilization of maximum quantities of pasture as sheep. Although cattle compete with sheep for many of the same grazing areas and are also ruminants, sheep are unique in that the vast majority of the young are marketed as milk-fat animals directly off pastures. Also, in their grazing habits, sheep differ from cattle in that (1) they show a decided preference for short, fine forages, and (2) they have the gregarious or flocking instinct.

• **Hays and Other Dry Forages**—Hays are the standard winter feed for sheep when they cannot be on the pasture or range or when the condition of the pastures is such as to require supplemental feeding. The choicest hay for sheep is a legume which has been produced on fertile soil, cut at the proper stage, and well cured. Such hay is palatable and rich in protein (and the quality of protein is good), calcium, and vitamins A and D. If legume hay cannot be secured, a high-quality grass-legume mixed hay will be entirely satisfactory and much superior to a straight grass hay. Sheep may do very well for a considerable period of time when fed only a good-quality legume hay, salt, and water.

• **Silages and Root Crops**— Silage for sheep may be made from a great variety of plants, including corn, sorghums, cereal grains, legumes, grasses, cannery refuse, pea vines, potatoes, beets, beet tops, sunflowers, and other materials. When properly preserved and fed, silages made from any of these materials are quite satisfactory.

Roots include all plants whose roots, tubers, bulbs, or other underground vegetative parts are used for feed. The important root crops for sheep are mangels (stock beets), rutabagas (swedes), turnips, and carrots.

• **Quality of Forage**—The quality of forage—pasture, hay or other dry roughage, or silage—greatly affects its consumption. High-quality forage is more digestible and passes through the digestive tract more rapidly than low-quality forage; hence, sheep will consume more of it.

• **Concentrates**—Except when they are abundant and cheap, few concentrates are fed to sheep, except immediately before and after lambing, in conditioning ewes and rams for breeding, or when finishing lambs. During these periods, the most frequently used concentrates consist of the common farm grains —oats, corn, barley, wheat, rye, and the grain sorghums. Numerous by-product feeds are also utilized for sheep, including those from the flour- and corn-milling industries, beet by-products, and oil meals or cake made from soybeans, cottonseed, and flaxseed.

• **Feed Preparation**—Sheep masticate grain more thoroughly than cattle, with the result that feed preparation is of less value for them than for cattle. Processing of grains for sheep is not necessary unless seeds are hard (such as sorghum or millet) or the sheep have "broken mouths." Pellets are increasingly being used by lamb feeders, especially when lower quality feed ingredients are used.

FEED SUBSTITUTION TABLE FOR SHEEP

Successful sheep raisers are keen students of feed values. They recognize that feeds of similar nutritive properties can and should be interchanged in the ration as price relationships warrant, thus making it possible at all times to obtain a balanced ration at the lowest cost.

Table 21-12, Feed Substitution Table for Sheep and Goats, is a summary of the comparative values of the most common U.S. feeds. In arriving at these values, two primary factors besides chemical composition and feeding value have been considered; namely, palatability and carcass quality.

TABLE 21-12
FEED SUBSTITUTION TABLE FOR SHEEP AND GOATS (AS-FED BASIS) (See footnotes at end of table.)

Feedstuff	Relative Feeding Value (lb for lb) in Comparison with the Designated Base Feed (in bold italic) Which = 100	Maximum Percentage of Base Feed (or comparable feed or feeds) Which it Can Replace for Best Results	Remarks
GRAINS, BY–PRODUCT FEEDS, ROOTS AND TUBERS:[1] **(Low and Medium Protein Feeds)**			**NOTE WELL:** Although sheep terms are used in this column, the replacement feeding value (column 2), maximum percentage replacement (column 3), and remarks (column 4) of each feed listed pertain to goats, also.
Corn, No. 2	*100*	*100*	Grinding not necessary unless (1) for old ewes with poor teeth, (2) for lambs under 5-6 weeks, (3) for incorporation in a mixed ration.
Apple pomace, dehydrated	82–86	33⅓	
Barley	90	100	It does not pay to grind barley for sheep.
Beans (cull)	80	15	
Beet pulp, dried	90	33⅓–50	Value of about 80% when used as the only concentrate for finishing lambs.
Beet pulp, molasses, dried	95	33⅓–50	Value of about 80% when used as the only concentrate for finishing lambs.
Beet pulp, wet	25	33⅓–50	
Brewers' dried grains	80–95	33⅓	Not very palatable. Fed chiefly to dairy cattle.
Citrus pulp, dried	95	25–50	
Corn gluten feed (gluten feed)	85–90	50	
Distillers' dried grains	95–100	33⅓–50	
Distillers' dried solubles	95–100	33⅓–50	
Fat	225	5	
Hominy feed	100	100	
Molasses, beet	75	10	Actual value may be higher as an appetizer.
Molasses, cane	75	10	Actual value may be higher as an appetizer.
Molasses, citrus	75	10	
Oats	80	10–100	Lower value when used as the only grain for finishing lambs. Highest value for young lambs, for breeding animals, and for starting lambs on feed. Need not be ground for sheep. Should not constitute more than ⅓ of finishing rations. Feeding value varies according to the test weight per bushel.
Peas, dried	100	40	
Rice (rough rice)	55–75	100	
Rice bran	66⅔–75	33⅓	
Rice polishings	85–90	25	
Potatoes (Irish)	25	85	Contrary to popular belief, potatoes can be fed successfully through the pregnancy and lactation periods.
Roots (chiefly mangles or stock beets, rutabagas or swedes, turnips, and carrots)	15	50	Some sheep producers believe that the feeding of high levels of roots over a long period will produce urinary calculi. Therefore, caution should be exercised in feeding them to rams and wethers (females not affected). Keep the Ca level higher than the P level. Many shepherds add roots to the ration of show sheep, for conditioning purposes.
Rye	83–87	50–100	Apparently rye is more palatable to sheep than to other classes of animals. Rye may be fed whole to sheep.
Sorghum, milo	85–100	100	All varieties have about the same feeding value. There is no advantage in grinding sorghum for sheep.
Wheat	100–110	50	May be fed as the only grain, but it is improved by mixing with another grain. Wheat may be fed whole. Wheat-fed sheep appear to be especially susecptible to founder.
Wheat bran	90	10–33⅓	Because of its bulk and fiber, wheat bran should not constitute more than 10–15% of a finishing ration. Bran is valuable for young animals, for breeding animals, and for starting animals on feed.
Wheat mixed feed (mill-run)	90–95	10–33⅓	Can be used in about the same way and in the same quantities as wheat bran for sheep.
PROTEIN SUPPLEMENTS:			
Soybean meal (41%)	*100*	*100*	
Alfalfa or clover screenings	70–75	50	Grind finely to destroy weed seeds.
Brewers' dried grains	75	100	
Copra meal (coconut meal)(21%)	90–100	50	
Corn gluten feed (gluten feed)	60–65	50–100	
Corn gluten meal (gluten meal)	100	50	
Cottonseed meal (41%)	100	100	Unlike the situation with finishing cattle, cottonseed meal is about equal to linseed meal for finishing lambs.
Distillers' dried grains	90	100	Rye distillers' dried grains are about 10% lower in protein than similar products made from corn or wheat.

(Continued)

TABLE 21-12 (Continued)

Feedstuff	Relative Feeding Value (lb for lb) in Comparison with the Designated Base Feed *(in bold italic)* Which = 100	Maximum Percentage of Base Feed (or comparable feed or feeds) Which it Can Replace for Best Results	Remarks
Distillers' dried solubles	90	100	
Linseed meal (35%)	90	100	
Peanut meal (45%)	100	100	
Peas, dried	65–75	50	
Safflower meal, with hulls (42%)	40–45	100	
Soybeans	95–100	100	It does not pay to grind soybeans for sheep.
Sunflower meal (35%)	85	100	
DRY FORAGES AND SILAGES:[2]			
Alfalfa hay, all analyses	*100*	*100*	
Alfalfa silage	33⅓–50	50–85	When alfalfa silage replaces corn silage, more energy feed must be provided but less protein, unless grain is used as a preservative.
Barley hay	70	50	The beards may be harmful, especially to woolly faced sheep.
Beet tops, fresh	16–25	33⅓–50	In the West, large acreages of fresh beet tops were formerly grazed by sheep and cattle.
Beet tops, dry	70	50	
Beet top silage, sugar	17–25	33⅓–50	Either provide some dry forage or feed 2 oz *(56.7 g)* of finely ground limestone to each 100 lb *(45.4 kg)* of silage.
Clover, alsike	90–100	100	
Clover hay, crimson	90–100	100	Crimson clover hay has a considerably lower value if not cut at an early stage.
Clover hay, red	90–100	100	If the rest of the ration is adequate in protein, clover hay will be equal to alfalfa in feeding value; otherwise, it will be lower.
Clover-timothy hay	80–90	100	
Corn fodder	75	100	Should be chopped.
Corn silage	33⅓–50	50–85	Although a ration in which corn silage is the only forage is sometimes fed to sheep, most feeders prefer to limit the silage and use some hay. Avoid silage that is contaminated with listeriosis.
Corn stover	35	50	Unsatisfactory for finishing lambs, but cut or shredded stover may be used as a part of the roughage for breeding ewes if fed along with a good legume.
Cowpea hay	95–100	100	
Grass hay (bluegrass, bromegrass, canarygrass, orchardgrass, timothy, and a variety of native grasses)	75	50–75	In comparison with legume hays, grass hays contain almost as much energy, have 50 to 75% less protein (they run 6–10% crude protein), and have only about 25% as much calcium (about 0.2% calcium).
Grass-legume mixed hay	80–90	100	Value depends on the proportion of legumes present and the stage of maturity at which it is cut.
Grass-legume silage	32–45	50–85	Although a ration in which grass silage is the only forage is sometimes fed to sheep, most feeders prefer to limit the silage and use some hay.
Grass silage	30–45	50–85	
Haylage (alfalfa-grass)	50	50–85	
Johnsongrass hay	70	50	
Lespedeza hay	80–100	100	Feeding value varies considerably with stage of maturity at which it is cut.
Mint hay	80–95	75	
Oat hay	75	50	Oat hay is equal to alfalfa hay in energy, but substantially lower in protein.
Pea-vine silage	33⅓–50	50–85	Unless grain is added as a preservative, pea-vine silage requires more energy feed, but less protein supplement than corn silage when fed to finishing lambs.
Pea-vine hay	100–110	75	
Prairie hay	65–70	100	
Reed canarygrass	70	100	
Sorghum fodder	70	100	
Sorghum silage (grain varieties)	32–47	50–85	Although a ration in which sorghum silage is the only forage is sometimes fed to sheep, most feeders prefer to limit the silage and use some hay. Nearly equal to grain varieties in value per acre because of greater yields.
Sorghum silage (sweet varieties)	25–30	50–85	Unsatisfactory for finishing lambs, but cut or shredded stover may be used as part of the roughage for breeding ewes if fed along with a good legume.
Sorghum stover	35	50	The lower value is for finishing lambs. For other classes of sheep, it is equal to alfalfa hay.
Soybean hay	85–100	100	
Soybean straw	68	50–75	
Sudangrass hay	50–60	50	
Sunflower hulls	40	50	
Sweet clover hay	100	100	Value of sweet clover hay varies widely. Second year sweet clover hay is less desirable than first year clover hay and is more apt to cause sweet clover disease.
Vetch-oat hay	80–90	100	The higher the proportion of vetch, the higher the value.
Wheat hay	70	50	

[1] Roots and tubers are of lower value than the grain and by-product feeds due to their higher moisture content.
[2] Silages are of lower value than dry forages due to their higher moisture content.

RATIONS FOR SHEEP

Sheep rations are closely associated with the age and stage of production of the animals. Except at lambing time or when emergencies occur as a result of drought or inclement weather, western bands receive little supplemental feed. Even with farm flocks, a minimum of grain is fed to breeding animals. Grain feeding usually is limited to the latter part of gestation and to the lactation period prior to turning to pasture. However, when grain is less costly than forage, on the basis of energy and/or protein, it may be fed to advantage to animals of any age or stage of production.

FEEDING RAMS

In general, rams are fed the same kind of feeds as ewes, but in slightly larger quantities. (See Table 21-13.)

FEEDING BREEDING EWES

Table 21-13 contains suggested rations for ewes during various stages of production—maintenance, early gestation, late gestation, and lactation (continued on page 485).

TABLE 21-13
DAILY RATIONS FOR BREEDING EWES AT VARIOUS STAGES OF PRODUCTION

Ration No.	Moisture Basis[1] A-F (as-fed) M-F (moisture-free)	Hay[2] (lb)	(kg)	Corn Silage (lb)	(kg)	Haylage (lb)	(kg)	Corn Straw (lb)	(kg)	Stover (Stalks) (lb)	(kg)	Grain[3] (lb)	(kg)	Protein Supplement[4] (lb)	(kg)
Maintenance															
1	A-F	3.0	1.4												
	M-F	3.3	1.5												
2	A-F			6.0	2.7									0.20	0.09
	M-F			6.7	3.0									0.22	0.10
3	A-F					6.0	2.7								
	M-F					6.7	3.0								
4	A-F							3.0	1.4					0.40	0.18
	M-F							3.3	1.5					0.44	0.20
Gestation, early (first 15 weeks)															
1	A-F	3.5	1.6												
	M-F	3.9	1.8												
2	A-F	2.0	0.9									1.0	0.45		
	M-F	2.2	1.1									1.1	0.49		
3	A-F	1.8	0.8									0.6	0.27	0.20	0.09
	M-F	2.0	0.9									0.7	0.31	0.22	0.10
4	A-F			8.0	3.6									0.20	0.09
	M-F			8.9	4.0									0.22	0.10
5	A-F					7.0	3.2					0.20	0.09		
	M-F					7.8	3.5					0.22	0.10		
6	A-F	2.0	0.9							2.0	0.9	0.5	0.23		
	M-F	2.2	1.0							2.2	1.0	0.6	0.27		
7	A-F	1.0	0.45							2.0	0.9	0.5	0.23	0.30	0.14
	M-F	1.1	0.49							2.2	1.0	0.6	0.27	0.33	0.15
Gestation, late (last 4 weeks): Add 0.5-1.0 lb (0.23-0.45 kg) grain per ewe daily to any of the above rations.															
Lactation															
1	A-F	4.0	1.8									2.0-3.0	0.9-1.4	—	—
	M-F	4.4	2.0									2.2-3.3	1.1-1.5	—	—
2	A-F			10.0	4.5							1.5	0.7	0.25	0.11
	M-F			11.1	5.0							1.7	0.8	0.28	0.13
3	A-F	1.0	0.45	8.0	3.6							1.5	0.7	0.20	0.09
	M-F	1.1	0.49	8.9	4.0							1.7	0.8	0.22	0.10
4	A-F					8.9	3.6					2.0-3.0	0.9-1.4		
	M-F					8.9	4.0					2.2-3.3	1.1-1.4		

[1]As-fed was calculated using an average figure of 90% dry matter. When using silages, roots, and other wet feeds, these feeds should be converted to a moisture-free basis and the ration calculated using the moisture-free data.

[2]Alfalfa hay, midbloom, preferred.

[3]Grain may consist of corn, barley, wheat, oats, and/or grain sorghum.

[4]Protein supplement may consist of soybean, cottonseed, linseed, sunflower, safflower, or rapeseed meal.

Feeding Directions:

1. These rations are formulated to meet the requirements of a 154 lb (70 kg) ewe in average condition, and are designed for hand-feeding. The daily feed allowance can be increased or decreased, depending on the actual size of the ewe and the body condition.

2. Some of these rations are deficient in calcium and/or phosphorus; therefore, a supplement containing 50% trace mineral salt (for sheep) and 50% dicalcium phosphate should be fed free choice. The consumption of 0.05 lb (0.02 kg) per sheep per day of this mixture will provide the amounts of calcium and phosphorus needed for maintenance and the first 15 weeks of gestation; and 0.10 lb (0.05 kg)/day will provide the needed Ca and P for late gestation and lactation. Vitamins A and E should be added to the salt-mineral mix when sheep are fed the wheat straw and corn stover rations.

3. Ewes should gain 15 to 25 lb (6.8 to 11.4 kg) during gestation. During early gestation (first 15 weeks), they should gain 0.05 lb (0.02 kg)/day. During late gestation (last 4 weeks), they should gain 0.5 lb (0.23 kg)/day. If, the last 4 weeks of pregnancy, ewes are fed 0.5 to 1.0 lb (0.23 to 0.45 kg) grain per head daily and gain 8 to 15 lb (3.6 to 6.8 kg), ketosis (lambing paralysis) can be prevented almost entirely.

4. During maintenance and early gestation, each ewe should have 14 in. (36 cm) of bunk feed space. In late gestation and during lactation, bunk space should be increased to 15 to 18 in. (38 to 46 cm).

- **Drylot (Confinement) Feeding**—The vast majority of the nation's sheep utilize pasture in season. However, some ewe-lamb producers are drylotting all or part of the year. So, now there are two alternatives, and the producer may choose between the two.

Without doubt, many sheep producers can advantageously combine pasture and confinement production—using pastures for the breeding flock and confinement production for young lambs. So, sheep confinement will not eliminate good pasture systems; rather, confinement production, all or in part, will increase and will replace more and more of the conventional practice of grazing all sheep throughout the pasture season.

- **Flushing Ewes**—Flushing is the practice of conditioning or having thin ewes gain in weight just prior to breeding. Its purpose is to increase the ovulation rate and, consequently, the lambing rate.

Although it is not likely that all the benefits ascribed to flushing will be fully realized under all conditions, the general feeling persists that the practice will result in a 15 to 20% increase in the lamb crop, and that the ewes will breed both earlier and more nearly at the same time. Hence, it follows that the lamb crop will be earlier and more uniform in age and size.

- **Feeding Pregnant Ewes**—If a strong, healthy crop of lambs is to be expected, the ewes must be properly fed and cared for throughout the period of pregnancy. In general, this means the feeding of a suitable and well-balanced ration, together with the necessary minerals and vitamins as required for maintenance (and growth, if the ewe is not fully mature), growth of the fleece, and development of the fetus. In addition, plenty of exercise must be provided.

During the last 4 to 5 weeks of pregnancy, the fetus develops very rapidly and the demands on the ewe are rather heavy. Also, ewes carrying twins or triplets, especially if they are a bit fat, are very prone to ketosis (lambing paralysis), which can be prevented by feeding a high-energy ration. So, during the last 4 to 5 weeks of pregnancy, ewes should be fed 0.5 to 1.0 lb of grain per head daily and gain 8 to 15 lb. Besides, ewes so fed milk better. The concentrate given to the farm flock usually consists of homegrown grains, whereas range bands are often given pelleted or cubed protein supplements.

- **Feeding at Lambing Time**—At lambing, or immediately after lambing, each ewe should be placed in an individual holding or lambing pen. At this time, the grain allowance should be materially reduced, but dry roughage may be fed free-choice, when it is certain that it is of good quality and palatable. Usually, about 5 to 7 days should elapse before ewes are placed on full feed following parturition. In general, feeds of a bulky and laxative nature should be provided during the first few days. A mixture of equal parts of oats and wheat bran is excellent. Soon after lambing, the ewe should be given water with the chill removed but should not be allowed to gorge.

- **Feeding Lactating Ewes**—In general, it is considered good practice to feed lactating ewes rather liberally for the first month or two after lambing, for lambs make the most economical gains when suckling. It is a good plan to separate the ewes with twins from those with singles, giving the former more liberal rations or the benefit of the better pastures or ranges. In fact, some large sheep operators find this practice so advisable that they regularly separate out the twin bands.

Weaning age varies greatly in the sheep industry. Lambs may be weaned as early as 3 to 4 weeks of age and as late as 5 to 6 months. When pasture is plentiful and lamb growth is satisfactory, there is little reason to wean lambs before they reach market weight and condition at 5 to 6 months of age.

FEEDING RANGE SHEEP

Because of the magnitude of the range sheep industry and the fact that it is a highly specialized type of operation, in the sections that follow special discussion is devoted to the feeding and management of sheep on the range.

- **Nutrient Deficiencies of Range Forage**—Hunger, due to lack of feed, is the most common deficiency on the western range. In particular, there may be a shortage of energy during droughts, late in the season, or early in the spring when grass is washy. Under such energy-deficient conditions, sheep lose weight and condition and lambs fail to grow. Also, reproduction is adversely affected.

Mature, weathered native range grass is almost always deficient in protein—being as low as 3%, or less. Protein-leaching losses due to fall and winter rains may range from 37 to 73%.

Phosphorus deficiencies are rather common among range sheep, but calcium deficiency is seldom encountered.

Of the vitamins, vitamin A is most likely to be deficient in range forage, because dry, bleached range grass is very low in carotene (the precursor of vitamin A).

- **Range Supplements**—Four suggested range supplements, ranging from high to low protein, are given in Table 21-14 (see page 486).

(Also see Chapter 11, Protein Supplements, section headed, "Self-feeding Salt-Feed Mixtures.")

- **How to Choose a Range Supplement**—In choosing a supplement for range sheep, the following requisites should be observed:

1. It should balance the ration of the animals to which it is fed. This means that it should supply all the nutrients needed by the animal that are missing in the forage.

2. It should be fed in such a way that each animal gets its proper portion.

3. It should be fed in a form that is convenient and practical from the standpoint of the feeder, and that will least disturb the animals.

- **Rate of Supplemental Feeding**—The normal range of supplementation for sheep is ¼ to ½ lb per head per day. Rates above ½ lb approach a level that will result in reduced intake of range forage. Where range vegetation is so short as to require supplementation in excess of ½ lb per head per day, consideration should be given either to moving the sheep into a drylot or to moving them to a better grazing area.

Some managers divide their sheep according to age, condition, and twins vs single lambs. Of course, this is facilitated where there are several bands. By so doing, it is possible (1) to give the animals that require the highest level of nutrition the best pasture or range, and/or (2) to supplement according to need.

TABLE 21-14
FORMULAS FOR RANGE SHEEP SUPPLEMENTS[1]

Feed[2]	Recommended Level of Protein			
	High	Medium-High	Medium-Low	Low
	◄──────────────────── (%) ────────────────────►			
Barley, grain or corn, dent yellow, grain, grade 2 US, minimum 54 lb (24.5 kg)/bu	5	40	75	65
Beet, sugar, molasses, or sugar cane molasses, 48% invert sugar, minimum 79.5° Brix	5	5	5	5
Cottonseed with some hulls, solvent extracted, ground, minimum 41% protein, maximum 14% fiber, minimum 0.5% fat (cottonseed meal)	66	36	—	16
Soybean, seeds, solvent extracted, ground, maximum 7% fiber, 44% protein (soybean meal)	10	10	10	10
Urea, technical, 282% protein equivalent	—	—	5	—
Alfalfa, aerial parts, dehydrated, ground, minimum 17% protein or alfalfa, hay, sun-cured, early bloom	10	5	—	—
Vitamin A .... (IU/lb)	—	1,818	3,636	3,636
Vitamin A .... (IU/kg)	—	4,000	8,000	8,000
Calcium phosphate, monobasic, commercial	1	1	2	1
Sodium phosphate, monobasic, technical	2	2	2	2
Salt or trace mineralized salt	1	1	1	1
Total	100	100	100	100

Composition[3]	As-Fed[4]	M-F	As-Fed[4]	M-F	As-Fed[4]	M-F	As-Fed[4]	M-F
Digestible energy .... (Mcal/lb)	1.4	*1.5*	1.4	*1.5*	1.4	*1.5*	1.3	*1.4*
Digestible energy .... (Mcal/kg)	3.0	*3.3*	3.0	*3.3*	3.0	*3.3*	2.8	*3.1*
Protein (N × 6.25) .... (%)	30.4	*33.8*	21.9	*24.3*	23.6	*26.2*	15.9	*17.7*
Phosphorus .... (%)	1.8	*2.0*	1.4	*1.5*	0.8	*0.9*	1.1	*1.2*
Carotene .... (mg/lb)	9.0	*10.0*	4.1	*4.5*	—	—	—	—
Carotene .... (mg/kg)	19.8	*22.0*	9.0	*10.0*	—	—	—	—
Vitamin A .... (IU/lb)	—	—	1,636.0	*1,818.0*	3,273.0	*3,636.0*	3,273.0	*3,636.0*
Vitamin A .... (IU/kg)	—	—	3,600.0	*4,000.0*	7,200.0	*8,000.0*	7,200.0	*8,000.0*
Rate of feeding .... (lb/day)	0.20–0.40	*0.22–0.44*	0.20–0.40	*0.22–0.44*	0.20–0.40[5]	*0.22–0.44[5]*	0.20–0.40[5]	*0.22–0.44[5]*
Rate of feeding .... (kg/day)	0.09–0.18	*0.1–0.2*	0.09–0.18	*0.1–0.2*	0.09–0.18[5]	*0.1–0.2[5]*	0.09–0.18[5]	*0.1–0.2[5]*

[1]Adapted by the authors from *Nutrient Requirements of Sheep*, sixth revised edition, NRC-National Academy of Sciences, 1985, p. 52, Table 11.
[2]Feeds mixed and fed in meal or pellet form.
[3]Molasses and alfalfa hay, sun-cured, early bloom not included.
[4]Estimated 90% dry matter.
[5]In emergency situations, up to 1.1 lb (0.5 kg) may be fed.

FEEDING GROWING-FINISHING LAMBS

The growing-finishing stage of lambs refers to that period extending from birth to weaning at 4 to 6 months of age. At no other period in the life of the sheep is the promotion of growth and the prevention of disease so important.

Where succulent pastures are available, most practical sheep producers, including producers with both farm flocks and range bands, consider that a combination of such green forage plus the ewe's milk is ample for growing-finishing lambs. In fact, lambs are unique among farm animals, inasmuch as they may be marketed at top prices off grass.

• **Creep Feeding**—*The practice of supplemental feeding of nursing lambs in a separate enclosure away from their dams is known as creep feeding.* Lambs will usually consume some creep feed at 7 to 10 days of age.

Creep rations can either be hand-fed or self-fed. Many sheep producers hand-feed until the lambs begin to eat regularly, then self-feed from this point on.

Suggested creep rations are given in Table 21-15.

TABLE 21-15
SOME EXCELLENT CREEP RATIONS (AS-FED BASIS)[1]

	Unpelleted		Pelleted	
	First 2 Months	2 Months To Market	First 2 Months	2 Months To Market
	(%)	(%)	(%)	(%)
Ground corn	80	60	40	50
Ground oats	—	20	15	—
Soybean meal	20	10	20	10
Alfalfa hay	—	—	10	35
Bran	—	10	10	10
Molasses	—	—	5	5
Trace mineral salt	.5	.5	.5	.5
Limestone	1.0	1.0	1.0	1.0
Antibiotic (mg/lb)[2]	50	20	50	15
Vitamin A (IU/lb)	1,000	1,000	1,000	1,000
Vitamin D (IU/lb)	200	200	200	200
Vitamin E (mg/lb)	20	20	20	20

[1]The addition of 0.25 to 0.50% ammonium chloride will minimize urinary calculi.
[2]Chlortetracycline (Aureomycin) or oxytetracycline (Terramycin).

Feeding Directions:

1. Lambs should be started on creep feed about 10 days after birth. Although they will not consume significant amounts of feed until 3–4 weeks of age, the small amounts consumed at earlier ages are critical for establishing both rumen function and the habit of eating.

2. Feed high quality legume hay in a separate rack. Feed hay and creep ration twice daily to keep them fresh.

3. The amount of creep feed consumed by lambs 2 to 6 weeks of age is affected by the palatability of the ration (ration composition and ration form) and the location and environment of the creep area. A well-bedded, well-lighted area located close to where the ewes congregate is preferred.

- **Feeding Orphan ("Bummer") Lambs (Artificial Rearing)**—Observance of the following principles and practices will increase the chances of raising orphan lambs artificially:

  1. **Select the strong rather than the weak.** Where a multiple birth is involved and 1 or 2 lambs may be left with the mother, the authors recommend choosing the strongest and most aggressive lamb for artificial rearing for the reason that weak orphan lambs will not do well in a self-feeding system, although they can be raised satisfactorily if they are bottle fed by hand.

  2. **Give colostrum.** Colostrum makes for a good start in life. A newborn lamb needs 3.2 oz of colostrum per pound body weight during the first 18 hours after birth. Colostrum may be stored for this purpose.

  3. **Inject orphans.** When orphan lambs are placed in the nursery, inject them with (1) vitamins A, D, E, (2) iron-dextran, and (3) selenium in selenium-deficient areas.

  4. **Use milk replacer.** A number of commercially prepared milk replacers are on the market. Best results will be obtained by using a replacer containing 25 to 30% fat, 20 to 25% protein provided by spray-dried milk products, and not to exceed 30 to 35% lactose; with the milk replacer diluted, mixed, and fed according to the manufacturer's directions.

  5. **Provide a good dry starter feed and water from day one.** A good starter ration follows:

Lamb Starter Ration:

Ingredients	%
Soybean meal (49% CP)	40.0
Ground corn	27.0
Alfalfa meal	15.0
Dextrose (corn sugar)	10.0
Fat (e.g. vegetable oil)	5.0
Limestone	2.0
Trace mineral salt	0.7
Vitamin premix	0.3
Total	100.0

Once the lambs have fully adjusted to the starter ration, they can be slowly switched onto the regular creep or grower ration. (See Table 21-15.)

  6. **Maximize sanitation, observation, and TLC.** The successful artificial rearing of orphan lambs necessitates that the caretaker maximize sanitation, observation, and TLC—tender loving care.

- **Feeding Early Weaned Lambs**—*Early weaning refers to the practice of weaning lambs earlier than usual*—at 6 to 8 weeks of age. Among practical sheep producers, the rule of thumb relative to early weaning is that lambs may be weaned at 45 to 60 days of age or 45 lb, whichever comes first. Currently, some flock owners are weaning lambs at 45 to 60 days of age with good results.

For successful early weaning, superior nutrition and management are essential; and the earlier the weaning age, the more exacting these requirements.

FEEDING FINISHING LAMBS

The primary objective of the sheep producer is that of producing milk-fat lambs suitable for slaughter at weaning time. Only when the pasture is inadequate are lambs sold via the feeder route. Almost all feeder lambs come from the range area. Some range areas produce only a small percentage of lambs which are classed as feeders, whereas in other areas almost all the lambs must be sold as feeders because the vegetation is not sufficient to promote rapid growth and finishing. It is estimated that, for the range area as a whole, an average of at least 50% of all lambs produced in any year receive additional feed after they are removed from the range and prior to slaughter. Because such lambs are fed and marketed out of season, they are sometimes referred to as *old crop lambs*.

- **Areas and Types of Lamb Feeding**—Numerous feeding practices and a great variety of feeds are used in lamb-finishing operations. In general, however, all methods may be classified as either (1) field finishing, or (2) drylot finishing.

Fig. 21-6. Lambs finishing on winter wheat in western Kansas. On a dry-matter basis, wheat pasture contains 60 to 80% TDN and 20 to 30% protein. Also, wheat pastures are generally parasite-free. (Courtesy, Kansas State University, Manhattan)

- **Field Finishing**—Field finishing refers to the partial or complete finishing of lambs on pastures or on such field crops as sorghum stubble and sorghum crops, corn fields, alfalfa pasture, winter cereal grain pasture (wheat, oats, barley, rye), sugar beet tops, and vegetable crops. In good years and on good feed, many of these lambs are finished enough to go to the packer. However, most of them are marketed as heavy feeders weighing 90 to 100 lb.

Fig. 21-7. "Sheeping down" corn on an Iowa farm. First, the crabgrass, weeds, and lower corn leaves are eaten. (Photo by A. M. Wettach, Mt. Pleasant, Iowa)

- **Drylot Finishing**—Drylot feeding is, as the name indicates, feeding under restricted conditions. This may either be (1) shelter or barn feeding, or (2) open-yard feeding.

Because of inclement weather in the fall and early winter, most of the lamb-feeding operations in the central and eastern states are in drylots which afford shelter. In some instances, the lambs are kept under cover without an exercising lot. These barns may consist of anything from open sheds to more costly and elaborate structures, including slotted floors.

Open-yard feeding is the common method of finishing lambs in the irrigated areas of the West, though a few eastern lamb feeding operations are in open yards. In this system, equipment costs are kept to a minimum—the facilities merely consisting of an enclosed and well-drained yard which may or may not have a natural or constructed windbreak, and the necessary feed bunks. Open-yard feeding is often used by large operators who feed thousands of lambs.

- **Finishing Lamb Rations**—The rations given in Table 21–16 are typical of those that are widely used to finish lambs. The corn/alfalfa hay/soybean meal rations are used in areas of the United States where these ingredients are raised. The milo/cottonseed hulls/cottonseed meal rations are typical of those used in the Southwest. These rations increase in energy value (grain content) as they go from 1 to 5 in each series. Generally, lambs should be started on rations 1 or 2, then switched to ration 3, thence 4, and thence 5; allowing a 4–7 day adjustment period on each ration before stepping up to the next higher energy level.

TABLE 21-16
GROWING–FINISHING RATIONS FOR LAMBS[1]

Ingredient	Moisture Basis[2] A-F (as-fed) M-F (moisture-free)	Rations Using Corn/Alfalfa Hay/Soybean Meal					Rations Using Milo/Cottonseed Hulls/Cottonseed Meal				
		1	2	3	4	5	1	2	3	4	5
		(%)	(%)	(%)	(%)	(%)	(%)	(%)	(%)	(%)	(%)
Corn grain (dent yellow)		31.0	41.5	51.7	63.0	73.3	—	—	—	—	—
Sorghum grain (milo)		—	—	—	—	—	19.5	32.7	46.2	60.7	73.7
Alfalfa hay (mature)		55.0	45.0	35.0	25.0	15.0	15.0	15.0	15.0	15.0	15.0
Cottonseed hulls		—	—	—	—	—	40.0	30.0	20.0	10.0	—
Soybean meal (solvent 44% CP)		7.0	6.5	6.0	5.5	5.0	—	—	—	—	—
Cottonseed meal (solvent 41% CP)		—	—	—	—	—	17.5	14.0	10.5	7.0	4.0
Molasses (cane)		6.0	6.0	6.0	5.0	5.0	6.0	6.0	6.0	5.0	5.0
Calcium carbonate		—	—	.3	.5	.7	1.0	1.3	1.3	1.3	1.3
Trace mineral salt (sheep)		.5	.5	.5	.5	.5	.5	.5	.5	.5	.5
Ammonium chloride		.5	.5	.5	.5	.5	.5	.5	.5	.5	.5
Nutritional content											
Dry matter (%)	A-F	87.5	87.5	87.5	87.6	87.6	89.0	88.9	88.7	88.7	88.5
	M-F	100	100	100	100	100	100	100	100	100	100
TDN (%)	A-F	75.3	80.1	84.6	89.2	93.1	67.9	72.3	77.2	82.0	86.8
	M-F	65.9	70.1	74.0	78.1	82.0	60.4	64.3	68.5	72.7	76.8
Net energy for maintenance (Mcal/lb)	A-F	.80	.86	.91	.98	1.04	.71	.76	.82	.88	.94
	M-F	.70	.75	.80	.86	.91	.63	.68	.73	.78	.83
Net energy for maintenance (Mcal/kg)	A-F	.36	.39	.41	.44	.47	.32	.35	.37	.40	.43
	M-F	.32	.34	.36	.39	.41	.29	.32	.33	.35	.38
Net energy for gain (Mcal/lb)	A-F	.40	.47	.53	.59	.66	.33	.38	.45	.52	.59
	M-F	.35	.41	.46	.52	.58	.29	.34	.40	.46	.52
Net energy for gain (Mcal/kg)	A-F	.18	.21	.24	.27	.30	.15	.17	.20	.24	.27
	M-F	.16	.19	.21	.24	.26	.13	.15	.18	.21	.24
Crude protein (%)	A-F	17.1	16.6	16.0	15.4	14.8	17.0	16.3	15.8	15.2	14.8
	M-F	15.0	14.5	14.0	13.5	13.0	15.1	14.5	14.0	13.5	13.1
Protein bypass (%)	A-F	42.3	44.1	45.8	47.9	49.7	44.6	47.6	51.0	54.7	57.9
	M-F	37.0	38.6	40.1	42.0	43.5	39.7	42.3	45.2	48.5	51.2
Calcium (%)	A-F	.86	.71	.70	.67	.66	.85	.96	.94	.91	.89
	M-F	.75	.62	.61	.59	.58	.76	.85	.83	.81	.79
Phosphorus (%)	A-F	.29	.30	.31	.31	.32	.37	.37	.36	.36	.36
	M-F	.25	.26	.27	.27	.28	.33	.33	.32	.32	.32

[1]Adapted by the authors from *The Sheepman's Production Handbook*, published by the Sheep Industry Development Program, Inc., Denver, Colo., 1986, p. 44, Table 13.

[2]As-fed was calculated using an average figure of 90% dry matter. When using silages, roots, and other wet feeds, these feeds should be converted to a moisture-free basis and the ration calculated using the moisture-free data.

Feeding Directions:

1. These rations can be fed once daily in troughs or bunks if there is capacity for a day's feed. They can also be self-fed if the feeders are designed to handle such feed without bridging.
2. Offering lambs a good quality hay for 1-3 days along with rations 1 or 2 (provided free choice) can be used to start lambs on feed.
3. About 3 in. (7.6 cm) of self-feeder or trough space must be provided per lamb for self-feeding and about 12 in. (30.5 cm) if hand fed.
4. Gradually adapt the lambs to the higher energy rations by allowing 4-7 days on a ration before switching to the ration with the next higher energy level.
5. Complete mixing to prepare a uniform ration is important.
6. Lambs must not be allowed to be without feed even for a short period of time.

- **Basic Considerations in Finishing Lambs**—Although no rules of success are applicable to any and all conditions, the following basic considerations in finishing lambs are worth noting:

  1. **Lamb feeding is seasonal in nature.** It usually extends from August to about the following May.

  2. **Range feeder lambs preferred.** Range feeder lambs are more plentiful than native feeders, thus allowing for greater selection; and usually they are more uniform and have fewer parasites.

  3. **Shearing feeder lambs may or may not be profitable.** Among the factors determining whether feeder lambs should be shorn are: (a) season of the year—heat or cold stress, (b) price at slaughter for shorn versus unshorn lambs, and (c) wool incentive price.

  4. **Purchase price of lambs constitutes main cost.** In lamb feeding operations, the purchase price of the lambs represents 60 to 70% of all costs.

  5. **Expect death loss of about 1 to 2%.** Experienced feeders normally expect to lose about 1 to 2% of lambs on feed. This is higher than occurs in commercial cattle feeding operations.

  6. **Highly stressed lambs should not be processed.** Incoming lambs should not be drenched, vaccinated, shorn, and/or implanted until they have rested and are eating and drinking.

  7. **Exercise care when starting on feed.** Special care is necessary in starting heavily stressed lambs on feed. Immediately upon arrival, let them have a good rest, in a comfortable place. Also, provide them free access to a good-quality grass or grass-legume hay (preferably, not over 50% legume), clean water (preferably running water), and minerals.

  After they have rested for several days, start them on concentrate by feeding about ¼ lb per head daily. Gradually, this allowance should be increased until they are getting a full feed 2 to 4 weeks later. Commercial feeders place lambs on feed as rapidly as possible, without throwing them off feed.

  8. **Feed efficiency is lower and feed cost higher than for younger lambs.** Most incoming feeder lambs range from 5 to 8 months of age and from 65 to 90 lb weight; and they are marketed 60 to 90 days after they are placed on feed, at a weight of 105 to 110 lb. Feed efficiency is lower and cost of gains is higher than with younger lambs, as indicated by the following comparisons:

Age	Lb feed/Lb gain
Preweaned lambs	2.0 to 2.5
Early weaned lambs	2.5 to 4.0
Late weaned lambs	6.0 to 8.0

  9. **Feed costs account for ⅔ of finishing cost.** Feed accounts for about 66% of the cost of finishing feedlot lambs, exclusive of the intital purchase price of the feeder lambs.

  10. **Most grains need not be ground.** Unless such extremely hard seeds as millet are included in the ration, it does not pay to grind feeds for finishing lambs.

  11. **Wether lambs make more rapid gains than ewe lambs.** Wether lambs appear to make slightly more rapid gains than ewe lambs, but they do not finish quite so early as ewe lambs.

FEEDING SHOW SHEEP

The most important pointers in feeding and handling show sheep are as follows:

1. **Keep show sheep healthy and free from parasites.** It is impossible to obtain proper growth, finish, and bloom in sheep that are unthrifty or infested with parasites.

2. **Provide a suitable ration.** In addition to being reasonably economical and well balanced, the ration for show sheep must be palatable.

Some suggested grain-fitting rations are listed in Tables 21-17 and 21-18. To each of these grain rations should be added (1) good-quality roughage, and (2) salt and other minerals on a free-choice basis.

The Table 21-17 concentrate mixture is recommended for show lambs, with judicious ingredient substitutions made as determined by availability and price.

TABLE 21-17
FITTING CONCENTRATE MIX FOR SHOW LAMBS

Ingredient	%
Cracked corn	50
Whole or rolled oats	35
Soybean meal	10
Molasses	4
Mineral (limestone/sheep salt-mineral mix)	1
Total	100

The rations in Table 21-18 are for fitting yearlings and mature sheep. Show yearlings and mature sheep will eat 3 to 5 lb of grain per head daily. Additionally, they should receive a good legume hay. It is important that yearlings and mature sheep not be "fed off their feet," and not become soft and flabby as a result of lack of exercise.

TABLE 21-18
RATIONS FOR FITTING YEARLING AND MATURE SHEEP

Ingredient	Ration Number							
	1		2		3		4	
	(lb)	(kg)	(lb)	(kg)	(lb)	(kg)	(lb)	(kg)
Barley, rolled	—	—	40	18.2	—	—	10	4.5
Corn, cracked	—	—	—	—	40	18.2	—	—
Oats, rolled	50	22.7	40	18.2	40	18.2	60	27.2
Peas (split)	40	18.2	—	—	—	—	10	4.5
Protein supplement[1]	—	—	10	4.5	10	4.5	10	4.5
Wheat bran	10	4.5	10	4.5	10	4.5	10	4.5
Total	100	45.4	100	45.4	100	45.4	100	45.4

[1]Cottonseed, linseed, rapeseed (canola), soybean, and/or sunflower meal.

NUTRITIONAL DISORDERS AND TOXINS

There are numerous other feed and nutritional disorders of great importance in sheep production; among them, enterotoxemia, polioencephalomalacia, pregnancy disease (ketosis), urinary calculi, and poisonous plants. These and other similar problems affecting sheep are fully discussed in Chapter 5, Nutritional Disorders/Toxins.

QUESTIONS FOR STUDY AND DISCUSSION

1. In what ways may a sheep producer take practical advantage of the fact that sheep consume a higher proportion of forages than any other class of livestock?

2. What is the economic importance of feed for sheep?

3. Why should margins of safety be provided over and above the NRC requirements?

4. List and discuss each of the factors that can affect the energy requirements of sheep.

5. Discuss each of the following as they relate to the energy needs of sheep: (a) chief sources, (b) how intake may be controlled, (c) factors affecting the energy requirements, and (d) symptoms of deficiency.

6. List and discuss each of the factors that can affect the protein requirements of sheep.

7. Discuss each of the following as they relate to the protein needs of sheep: (a) chief sources, (b) the nonfunctional rumen of newborn lambs, (c) factors affecting the protein requirements, and (d) symptoms of deficiency.

8. Is quality of protein important in sheep nutrition? Justify your answer.

9. For sheep, for each of the following minerals (a) describe the deficiency symptoms and tell how you could distinguish between them, and (b) give practical sources: salt, calcium, phosphorus, magnesium, potassium, sulfur, cobalt, copper, iodine, selenium, and zinc.

10. What are the manifestations of toxicity in sheep of each of the following: fluorine, molybdenum, and selenium?

11. For sheep, for each of the following vitamins (a) describe the deficiency symptoms and tell how you could distinguish between them, and (b) give practical sources: vitamin A, vitamin D, and vitamin E.

12. List, and discuss the effects of each of the presently FDA approved sheep feed additives and implants.

13. Most lambs are marketed as milk-fat lambs directly off pastures or ranges without having had any grain. What are the advantages of this, from an economic standpoint? Can this practice be applied to cattle or swine?

14. How do sheep differ from cattle in their grazing habits?

15. What is the physiological explanation of why sheep can consume more pounds of a high-quality forage than of a low-quality forage?

16. How does the preparation of feed for sheep and cattle differ?

17. Of what value is a feed substitution table, such as Table 21–12?

18. Will drylot operations obsolete the use of pastures for sheep?

19. Discuss each of the following as they relate to feeding breeding ewes: (a) flushing, (b) feeding ewes carrying multiple fetuses during late pregnancy, and (c) feeding lactating ewes.

20. What are the most common nutrient deficiencies of range forage for sheep?

21. Step by step, how would you go about determining the kind of range supplement needed for sheep?

22. Range sheep operators generally agree that where range vegetation is so short as to require supplementation in excess of ½ lb per head per day, consideration should be given either to moving the sheep into a drylot or to moving them to a better grazing area. What prompts sheep producers to set an upper limit of ½ lb of supplement per head per day?

23. Discuss each of the following as they pertain to growing-finishing lambs: (a) creep feeding, (b) feeding orphan lambs, and (c) early weaning.

24. Discuss each of the following as they pertain to feeding finishing lambs: (a) field finishing, (b) drylot finishing, and (c) finishing lamb rations.

25. Discuss each of the following basic considerations in finishing lambs: (1) the seasonal nature, (2) range feeder lambs preferred, (3) normal death losses, (4) processing upon arrival at the feedlot, and (5) starting on feed.

26. It requires 6 to 8 lb of feed to produce 1 lb of on-foot feeder lamb. How does this efficiency of feed utilization compare with finishing cattle and growing-finishing hogs?

27. Under what conditions might it be preferable that a feedlot operator feed lambs instead of cattle?

28. List and discuss the most important pointers in feeding and handling show sheep.

22

Original painting by Tom Phillips

FEEDING GOATS[1,2]

Contents	Page
PART I—NUTRITIVE NEEDS AND FEEDS	
Economic Importance of Feed for Goats	492
Nutritive Needs of Goats	492
National Research Council (NRC) Requirements	492
Energy	494
Protein	495
Minerals	495
Goat Mineral Chart	495
Vitamins	498
Goat Vitamin Chart	498
Water	500
Nonnutritive Factors	500
Feeds for Goats	500
Browse, Forbs, and Grasses/Legumes	500
Hays and Other Dry Roughages	501
Silages, Haylages, and Root Crops	501
Concentrates	501
Milk and Milk Replacers	501
Commercial Feeds	501
Feed Preparation	501
Feed Substitution Table	502
Nutrition Related Disorders	502
Abortion	502
Caprine Arthritis-Encephalitis (CAE)	502
Enterotoxemia (Overeating Disease)	502
PART II—FEEDING DAIRY GOATS	
Feeding Lactating and Dry Dairy Does	503
Feeding Dairy Kids	504
Feeding For Show	504
PART III—FEEDING ANGORA AND SPANISH (MEAT) GOATS	
Feeding Kids and Yearlings	505
Questions for Study and Discussion	506

Fig. 22-1. Mother and kid. (Courtesy, Jodi Frediani, Santa Cruz, Calif.)

There are more than 450 million goats in the world, of which about ⅓ are in Africa. They contribute 1.4% of the world meat supply and 1.5% of the world milk supply.

Goats provide nearly ⅓ of the total meat produced in India and from 7 to 16% of the total meat produced in Turkey, Morocco, Indonesia, Nigeria, and Cyprus. In a number of countries, goat meat is preferred to other meats.

The dairy goat has long been a popular milk animal in the Old World, where it is often referred to as the cow of the poor. In some countries, goat milk accounts for up to 50% of the total milk production. Southeast Asia, Africa, and the Near East lead in the production of goat milk.

The goats of the world also produce 36 million lb of mohair

---

[1]The authors gratefully acknowledge the helpful suggestions of the following eminent authorities who reviewed this chapter: E. K. Cassel, Ph.D., Extension Dairy Specialist, the University of Maryland, College Park; L. D. Guthrie, Ph.D., Head, Extension Dairy Science Department, the University of Georgia, Athens; G. F. W. Haemlein, Ph.D., Professor, University of Delaware, Newark; J. E. Huston, Ph.D., Texas A&M University, San Angelo; C. N. Lee, Ph.D., University of Hawaii at Manoa, Honolulu; C. D. Lu, Ph.D., American Institute of Goat Research, Langston University, Langston, Okla.; S. J. Lyford, Jr., Ph.D., Professor, University of Massachusetts at Amherst; B. R. Moss, Ph.D., Auburn Univeristy, Auburn, Ala.; J. C. Porter, Ph.D., Extension Dairy Specialist, University of New Hampshire, Penacook; M. Shelton, Ph.D., the Texas Agricultural Experiment Station, the Texas A&M System, San Angelo; T. H. Teh, Ph.D., Prairie View A&M University, the Texas A&M System, Prairie View; and C. S. F. Williams, B.V.Sc., M.R.C.V.S., Professor, College of Veterinary Medicine, Michigan State University, East Lansing.

[2]The statistics presented in the introductory section were for the mid-1980s and were obtained from: *FAO Production Yearbook*, Published by FAO, Vol. 39, 1985; and *Agricultural Statistics*, USDA, 1985.

and cashmere and 33 million skins, annually. The three leading mohair-producing countries of the world, by rank, are South Africa, Turkey, and the United States.

There are approximately 2,950,000 goats in the United States, consisting of about 1,600,000 Angoras, 850,000 dairy goats, and 500,000 Spanish goats. Ninety-five percent of the Angoras are located in Texas. About 85% of the gross income from Angoras is from mohair and 15% from meat. Most of the Spanish goats are also in Texas. California is the most important dairy goat state.

So, the goat industry of America can be divided into three distinct types of production: (1) dairy, (2) mohair, and (3) meat. Also, mention should be made of the pygmy goat.

The dairy goat, which is kept primarily for milk production, is gaining in popularity in the United States. Presently, it consists of the following important breeds: Alpine, American La Mancha, Nubian, Oberhasli (Swiss Alpine), Saanen, and Toggenburg. According to the American Dairy Goat Association, 41,153 purebred dairy goats were registered in 1986. In most areas, goat's milk and cheese command premium prices as specialty foods.

The most numerous goat breed in America is the mohair-bearing Angora, the heavy-coated creatures kept for fiber production and brush control. Yet, few people outside the Angora district, characterized by rugged grazing lands, know what they look like. Angora goats are used to produce (1) a beautiful, long, lustrous fiber known as mohair and (2) meat, and to augment other brush-control methods.

The Spanish goat is kept for meat production and brush control. It is of uncertain origin, but in all probability its ancestors were brought from Spain by early explorers. Subsequently, dairy goat breeds have been infused. Colors vary from solid black, brown, or white to striped, to spotted. In Mexico, the meat from young milk-fed Spanish goat kids, known as *cabrito* (Spanish for little goat), is considered a delicacy. In the United States, Spanish goats are usually slaughtered when a little older at which time the meat is known as *chevon*. Spanish goats are usually left to survive on range forages with little or no feed supplementation.

Also, small numbers of pygmy goats are found in the United States. The American pygmies are descended mostly from West African dwarf goats found in Nigeria, Ghana, and the Cameroons. They are small, adaptable animals, used for meat, milk, and research. The pygmy is smaller than the other recognized types and breeds in the United States; full grown bucks stand about 20 in. high at the withers, and does are even smaller. Pygmies may be fed the same rations as their larger counterparts, but in smaller quantities.

Because the goat industry is so diverse, the feeding methods vary accordingly. For this reason this chapter, devoted to feeding goats, is presented in three parts: Part I—Nutritive Needs and Feeds, covering the principles and practices of feeding that are applicable to all goats, regardless of type. Part II—Feeding Dairy Goats. Part III—Feeding Angora and Spanish Goats, with Angora and Spanish goats discussed together because both are produced under similar conditions.

PART I—NUTRITIVE NEEDS AND FEEDS

ECONOMIC IMPORTANCE OF FEED FOR GOATS

Angora and Spanish goats utilize rough, brushy range areas that are not suited to other species—many of these ranges would not otherwise be utilized and would revert to brush and wilderness. On such ranges, Angora goats are supplemental fed to a limited degree only, whereas most Spanish goats live entirely off the land and are rarely supplemented. This does not mean, however, that Angora and Spanish goats would not benefit from, and increase production, with supplemental feeding, especially during the critical periods—just before breeding (for flushing), just before and after kidding, and when feed is short.

Modern dairy goat producers generally feed well-balanced rations that are high in energy and protein and contain adequate minerals and vitamins. Many of them use commercially prepared feeds during lactation and for the young kids.

NUTRITIVE NEEDS OF GOATS

In the past, efforts to set nutritional requirements for goats have relied heavily on the extrapolation of values derived from cattle and sheep studies. Despite their similarities as ruminants, goats exhibit significant differences from cattle and sheep in grazing habits, feed selection, water requirements, physical activities, milk composition, carcass composition, metabolic disorders, and parasites. So, the nutrient requirements of goats should be treated separately from those of other ruminants.

The hearty appetite of goats makes for a significant species difference. Lactating and growing goats will consume from 3.5 to 5.0% of their body weight (moisture-free basis) in one day, while cattle and sheep normally eat only 2.5 to 3.0%. It follows that their large feed capacity in relation to body weight makes it possible for them to consume large quantities of low quality materials. This characteristic, along with their ability to select the high-quality parts of plants, makes it possible to maintain goats successfully on poor pastures.

Since the nutritional requirements of the goat are distinctly different for milk, mohair, and meat production, specific requirements and allowances are discussed in separate feeding sections. Despite these distinctly different quantitative needs, the basic nutritional physiology of all goats is similar; hence, certain fundamentals relative to their nutritive needs—energy, protein, minerals, vitamins, and water—apply to all goats regardless of the purpose for which they are kept.

National Research Council (NRC) Requirements

The most up-to-date feeding standards for goats in the United States are those published by the National Research Council (NRC) of the National Academy of Sciences. Through the use of these standards, rations can be formulated for the different classes and categories of goats by proper use of available feedstuffs.

The nutritive requirements of goats for maintenance, various levels of activity, late pregnancy, and growth, are given in Table 22-1. Additional nutrient requirements for milk production at different fat percentages are given in Table 22-2; and additional nutrient requirements for mohair production are given in Table 22-3.

TABLE 22-1
DAILY NUTRIENT REQUIREMENTS OF GOATS[1]

Body Weight (BW)		Dry Matter per Animal[2]					Feed Energy					Protein		Minerals		Vita-min A	Vita-min D
		0.9 Mcal ME/lb (*2 Mcal ME/kg*)		1.09 Mcal ME/lb (*2.4 Mcal ME/kg*)													
		Total	% of BW	Total		% of BW	TDN		DE	ME	NE	TP	DP	Ca	P		
(lb)	(*kg*)	(lb) (*kg*)	(lb)	(lb) (*kg*)		(kg)	(lb)	(*kg*)	(Mcal)	(Mcal)	(Mcal)	(g)	(g)	(g)	(g)	(1,000 IU)	(IU)
Maintenance only (includes goats under stable-fed conditions, minimal activity, and early pregnancy)																	
22	*10*	0.6 *0.28*	2.8	0.5 *0.24*		2.4	0.4	*0.16*	0.70	0.57	0.32	22	15	1	0.7	0.4	84
44	*20*	1.1 *0.48*	2.4	0.9 *0.40*		2.0	0.6	*0.27*	1.18	0.96	0.54	38	26	1	0.7	0.7	144
66	*30*	1.4 *0.65*	2.2	1.2 *0.54*		1.8	0.8	*0.36*	1.59	1.30	0.73	51	35	2	1.4	0.9	195
88	*40*	1.8 *0.81*	2.0	1.5 *0.67*		1.7	1.0	*0.45*	1.98	1.61	0.91	63	43	2	1.4	1.2	243
110	*50*	2.1 *0.95*	1.9	1.7 *0.79*		1.6	1.2	*0.53*	2.34	1.91	1.08	75	51	3	2.1	1.4	285
132	*60*	2.4 *1.09*	1.8	2.0 *0.91*		1.5	1.3	*0.61*	2.68	2.19	1.23	86	59	3	2.1	1.6	327
154	*70*	2.7 *1.23*	1.8	2.2 *1.02*		1.5	1.5	*0.68*	3.01	2.45	1.38	96	66	4	2.8	1.8	369
176	*80*	3.0 *1.36*	1.7	2.5 *1.13*		1.4	1.7	*0.75*	3.32	2.71	1.53	106	73	4	2.8	2.0	408
198	*90*	3.3 *1.48*	1.6	2.7 *1.23*		1.4	1.8	*0.82*	3.63	2.96	1.67	116	80	4	2.8	2.2	444
220	*100*	3.5 *1.60*	1.6	3.0 *1.34*		1.3	2.0	*0.89*	3.93	3.21	1.81	126	86	5	3.5	2.4	480
Maintenance plus low activity (basic plus 25% increment, includes goats under intensive management, tropical range, and early pregnancy)																	
22	*10*	0.8 *0.36*	3.6	0.7 *0.30*		3.0	0.4	*0.20*	0.87	0.71	0.40	27	19	1	0.7	0.5	108
44	*20*	1.3 *0.60*	3.0	1.1 *0.50*		2.5	0.7	*0.33*	1.47	1.20	0.68	46	32	2	1.4	0.9	180
66	*30*	1.8 *0.81*	2.7	1.5 *0.67*		2.2	1.0	*0.45*	1.99	1.62	0.92	62	43	2	1.4	1.2	243
88	*40*	2.2 *1.01*	2.5	1.9 *0.84*		2.1	1.2	*0.56*	2.47	2.02	1.14	77	54	3	2.1	1.5	303
110	*50*	2.6 *1.19*	2.4	2.2 *0.99*		2.0	1.5	*0.66*	2.92	2.38	1.34	91	63	4	2.8	1.8	357
132	*60*	3.0 *1.36*	2.3	2.5 *1.14*		1.9	1.7	*0.76*	3.35	2.73	1.54	105	73	4	2.8	2.0	408
154	*70*	3.4 *1.54*	2.2	2.8 *1.28*		1.8	1.9	*0.85*	3.76	3.07	1.73	118	82	5	3.5	2.3	462
176	*80*	3.7 *1.70*	2.1	3.1 *1.41*		1.8	2.1	*0.94*	4.16	3.39	1.91	130	90	5	3.5	2.6	510
198	*90*	4.1 *1.85*	2.1	3.4 *1.54*		1.7	2.3	*1.03*	4.54	3.70	2.09	142	99	6	4.2	2.8	555
220	*100*	4.4 *2.00*	2.0	3.7 *1.67*		1.7	2.4	*1.11*	4.91	4.01	2.26	153	107	6	4.2	3.0	600
Maintenance plus medium activity (basic plus 50% increment, includes goats on semiarid rangeland, slightly hilly pastures, and early pregnancy)																	
22	*10*	0.9 *0.43*	4.3	0.8 *0.36*		3.6	0.5	*0.24*	1.05	0.86	0.48	33	23	1	0.7	0.6	129
44	*20*	1.6 *0.72*	3.6	1.3 *0.60*		3.0	0.9	*0.40*	1.77	1.44	0.81	55	38	2	1.4	1.1	216
66	*30*	2.2 *0.98*	3.3	1.8 *0.81*		2.7	1.2	*0.54*	2.38	1.95	1.10	74	52	3	2.1	1.5	294
88	*40*	2.7 *1.21*	3.0	2.2 *1.01*		2.5	1.5	*0.67*	2.97	2.42	1.36	93	64	4	2.8	1.8	363
110	*50*	3.2 *1.43*	2.9	2.6 *1.19*		2.4	1.8	*0.80*	3.51	2.86	1.62	110	76	4	2.8	2.1	429
132	*60*	3.6 *1.64*	2.7	3.0 *1.37*		2.3	2.0	*0.91*	4.02	3.28	1.84	126	87	5	3.5	2.5	492
154	*70*	4.1 *1.84*	2.6	3.4 *1.53*		2.2	2.2	*1.02*	4.52	3.68	2.07	141	98	6	4.2	2.8	552
176	*80*	4.5 *2.03*	2.5	3.7 *1.69*		2.1	2.5	*1.13*	4.98	4.06	2.30	156	108	6	4.2	3.0	609
198	*90*	4.9 *2.22*	2.5	4.1 *1.85*		2.0	2.7	*1.24*	5.44	4.44	2.50	170	118	7	4.9	3.3	666
220	*100*	5.3 *2.41*	2.4	4.4 *2.01*		2.0	3.0	*1.34*	5.90	4.82	2.72	184	128	7	4.9	3.6	723
Maintenance plus high activity (basic plus 75% increment, includes goats on arid rangeland, sparse vegetation, mountainous pastures, and early pregnancy)																	
22	*10*	1.1 *0.50*	5.0	0.9 *0.42*		4.2	0.6	*0.28*	1.22	1.00	0.56	38	26	2	1.4	0.8	150
44	*20*	1.9 *0.84*	4.2	1.5 *0.70*		3.5	1.0	*0.47*	2.06	1.68	0.94	64	45	2	1.4	1.3	252
66	*30*	2.5 *1.14*	3.8	2.1 *0.95*		3.2	1.4	*0.63*	2.78	2.28	1.28	87	60	3	2.1	1.7	342
88	*40*	3.1 *1.41*	3.5	2.0 *1.10*		3.0	1.7	*0.78*	3.46	2.82	1.59	108	75	4	2.8	2.1	423
110	*50*	3.7 *1.67*	3.3	3.1 *1.39*		2.7	2.1	*0.93*	4.10	3.34	1.89	128	89	5	3.5	2.5	501
132	*60*	4.2 *1.92*	3.2	3.5 *1.60*		2.7	2.3	*1.06*	4.69	3.83	2.15	146	102	6	4.2	2.9	576
154	*70*	4.7 *2.14*	3.0	3.9 *1.79*		2.6	2.6	*1.19*	5.27	4.29	2.42	165	114	6	4.2	3.2	642
176	*80*	5.2 *2.37*	3.0	4.4 *1.98*		2.5	2.9	*1.32*	5.81	4.74	2.68	182	126	7	4.9	3.6	711
198	*90*	5.7 *2.59*	2.9	4.8 *2.16*		2.4	3.2	*1.44*	6.35	5.18	2.92	198	138	8	5.6	3.9	777
220	*100*	6.2 *2.81*	2.8	5.2 *2.34*		2.3	3.4	*1.56*	6.88	5.62	3.17	215	150	8	5.6	4.2	843
Additional requirements for late pregnancy (for all goat sizes)																	
		1.6 *0.71*		1.3 *0.59*			0.9	*0.40*	1.74	1.42	0.80	82	57	2	1.4	1.1	213
Additional requirements for growth—weight gain at 1.75 oz (*50 g*) per day (for all goat sizes)																	
		0.4 *0.18*		0.3 *0.15*			0.2	*0.10*	0.44	0.36	0.20	14	10	1	0.7	0.3	54
Additional requirements for growth—weight gain at 3.5 oz (*100 g*) per day (for all goat sizes)																	
		0.8 *0.36*		0.7 *0.30*			0.4	*0.20*	0.88	0.72	0.40	28	20	1	0.7	0.5	108
Additional requirements for growth—weight gain at 5.3 oz (*150 g*) per day (for all goat sizes)																	
		1.2 *0.54*		1.0 *0.45*			0.7	*0.30*	1.32	1.08	0.60	42	30	2	1.4	0.8	162

[1] Adapted by the authors from *Nutrient Requirements of Goats*, No. 15, NRC-National Academy of Sciences, 1981, pp. 10–11.

[2] Good-quality roughages furnish about 0.9 Mcal ME/lb (*2 Mcal ME/kg*) of dry matter. Roughage-concentrate mixed rations are sometimes necessary to increase the energy content of the diet to 1.09 Mcal/lb (*2.5 or 3.0 Mcal ME/kg*) of dry matter when early weaned kids or high-producing dairy goats are being fed.

TABLE 22-2
ADDITIONAL NUTRIENT REQUIREMENTS FOR MILK PRODUCTION PER POUND AT DIFFERENT FAT PERCENTAGES[1]

Fat	Feed Energy					Protein		Minerals		Vitamins	
	TDN		DE	ME	NE	TP	DP	Ca	P	A	D
(%)	(lb)	(kg)	(Mcal)	(Mcal)	(Mcal)	(g)	(g)	(g)	(g)	(1,000 IU)	(IU)
2.5	0.33	0.151	0.67	0.54	0.31	26.7	19.1	0.9	0.6	1.7	345
3.0	0.34	0.153	0.68	0.55	0.31	29.0	20.4	0.9	0.6	1.7	345
3.5	0.34	0.155	0.68	0.56	0.31	30.8	21.8	0.9	0.6	1.7	345
4.0	0.35	0.157	0.69	0.57	0.32	32.7	23.1	1.4	1.0	1.7	345
4.5	0.35	0.159	0.70	0.57	0.32	34.9	24.5	1.4	1.0	1.7	345
5.0	0.36	0.161	0.71	0.58	0.33	37.2	25.9	1.4	1.0	1.7	345
5.5	0.36	0.163	0.72	0.59	0.33	39.0	27.2	1.4	1.0	1.7	345
6.0	0.37	0.166	0.73	0.59	0.34	40.8	28.6	1.4	1.0	1.7	345

[1]Adapted by the authors from *Nutrient Requirements of Goats*, No. 15, NRC-National Academy of Sciences, 1981, p. 11. These requirements are in addition to those listed in Table 22-1. They include requirements for nursing single, twin, or triplet kids at the respective milk production level. To convert to requirements for milk production per kg, multiply by 2.205.

TABLE 22-3
ADDITIONAL NUTRIENT REQUIREMENTS FOR MOHAIR PRODUCTION BY ANGORA GOATS AT DIFFERENT FLEECE PRODUCTION LEVELS[1]

Annual Fleece Yield		Feed Energy					Protein	
		TDN		DE	ME	NE	TP	DP
(lb)	(kg)	(lb)	(kg)	(Mcal)	(Mcal)	(Mcal)	(g)	(g)
4.4	2	0.035	0.016	0.07	0.06	0.03	9	6
8.8	4	0.075	0.034	0.15	0.12	0.07	17	12
13.2	6	0.110	0.050	0.22	0.18	0.10	26	18
17.6	8	0.146	0.066	0.29	0.24	0.14	34	24

[1]Adapted by the authors from *Nutrient Requirements of Goats*, No. 15, NRC-National Academy of Sciences, 1981, p. 12. These requirements are in addition to those listed in Table 22-1.

Energy

Efficient utilization of nutrients depends on an adequate supply of energy, which is of paramount importance in determining the productivity of goats. Energy deficiency retards kid growth, delays puberty, reduces fertility, and depresses milk production. With continued deficiency the animals show a concurrent reduction in resistance to infectious diseases and parasites. The problem may be further complicated by deficiencies of proteins, minerals, and vitamins.

In Table 22-1, in addition to the energy requirement for maintenance only, energy values are given for three different levels of activity: (1) light, (2) medium, and (3) high. Also, additional energy allowances are made for: (1) late pregnancy (the last 60 days); and (2) three different levels of growth—50, 100, 150 g per day.

TABLE
GOAT MINERAL

Mineral Which May Be Deficient Under Normal Conditions	Conditions Usually Prevalent Where Deficiencies Are Reported	Function of Mineral	Deficiency Symptoms/Toxicity
Major or Macrominerals:			
Salt (NaCl)	Negligence, for salt is inexpensive. Lactating does may require additional salt as milk contains high amounts of sodium.	Sodium chloride helps maintain osmotic pressure in body cells, upon which depends the transfer of nutrients to the cells, the removal of waste materials, and the maintenance of water balance among the tissues. Also, sodium is important in making bile, which aids in the digestion of fats and carbohydrates; and chlorine is required for the formation of hydrochloric acid in the gastric juice so vital to protein digestion. It is noteworthy that when salt is omitted, sodium expresses its deficiency first.	**Deficiency symptoms**—Loss of appetite, depraved appetite and consumption of soil and debris, emaciation, decline in milk production, a general rough appearance with poor coat and lusterless eyes. **Acute deficiency symptoms** include shivering, weakness, cardiac disturbances, and ultimately death. **Toxicity**—The maximum tolerable level of salt for sheep is 9.0%. For goats, a similar level of salt will likely be toxic.
Calcium (Ca)	Goats in heavy lactation. Lack of vitamin D. Calcium-deficient areas (where pasture and range forages are deficient in calcium) are Fla., La., Neb., Va., and W. Va. Feeds that contain primarily cereal grains.	Essential for the development and maintenance of good strong bones and teeth; maintains the contractability, rhythm, and tonicity of the heart muscles; antagonizes the action of the sodium and potassium on the heart; is required for normal coagulation of the blood; is necessary for proper nerve irritability; and appears to be essential for selective cellular permeability.	**Deficiency symptoms**—In young kids, retarded growth and abnormal bone development. Also, a deficiency of calcium may cause rickets in young animals and osteomalacia in adults. In lactating does, depressed milk yields and fragile bones. Milk fever can occur when calcium levels in the blood drop. **Toxicity**—If there is adequate phosphorus, sheep can tolerate a calcium:phosphorus ratio of 7:1 and as much as 2% calcium in the diet. It is postulated that goats can tolerate a similar level of calcium.

In Table 22-2, energy requirements in addition to those listed in Table 22-1 are given for milk production per pound at different fat percentages.

In Table 22-3, energy requirements in addition to those listed in Table 22-1 are given for mohair production of Angora goats of different fleece weights.

Protein

Proteins are the principal constituents of the animal body and are continuously needed in the feed for growth and cell repair. The transformation of feed protein into body protein is an important process of nutrition and metabolism. Proteins consist of amino acids, which are the building blocks of all body cells. Secretions such as enzymes, hormones, mucin, and milk make for additional amino acid requirements. Proteins are, therefore, vital for animal maintenance, growth, reproduction, and milk production. However, in goats as in other ruminants, nonprotein nitrogen (NPN) can substitute for parts of the required proteins for these functions.

In Table 22-1, protein requirements are given for maintenance, activity, late pregnancy, and growth, along with the energy requirements.

In Table 22-2, protein requirements in addition to those listed in Table 22-1 are given for milk production per pound at different fat percentages.

In Table 22-3, protein requirements in addition to those listed in Table 22-1 are given for mohair production of different fleece weights of Angora goats.

Total protein (TP) is considered to be the most accurate guide for converting proteins from feed composition tables to the quantities required, but digestible protein (DP) values are also used.

Protein deficiencies in the ration deplete stores in the blood, liver, and muscles, and predispose animals to a variety of serious and even fatal ailments. Below a minimum level of 6% crude protein (CP) in the ration, feed intake will be reduced, which leads to a combined deficiency of energy and protein. This deficiency further reduces rumen function and lowers the efficiency of feed utilization. Long-term protein deficiencies retard fetal development, lead to low birth weights, affect kid growth, and depress milk production.

Minerals

If goats are fed a good concentrate, along with a good-quality hay produced on land that has been properly fertilized, few problems arising from mineral deficiencies occur.

The mineral requirements of goats are given in Tables 22-1, 22-2, and 22-4.

GOAT MINERAL CHART

Table 22-4, Goat Mineral Chart, gives a summary of the different factors involved with mineral nutrition in the goat. Further elucidation of certain minerals is contained in the accompanying narrative. Fluorine is discussed because of its toxicity.

22-4
CHART [1](See footnotes at end of table.)

Nutrient Requirements[2]	Recommended Allowances[2]	Practical Sources	Comments
	Salt should be provided free-choice or as a component of the ration. In a complete feed, 0.5 to 1.0% salt is recommended, with proportionately higher levels in supplements.	Iodized salt in iodine-deficient areas. Can be offered free-choice or incorporated into the ration. In alkaline areas, water may contain enough salt to meet the requirements.	In range areas, salt may be added to feed to limit feed intake. If self-feeders are located near water, the level of salt in the ration should be high (25–40%). If self-feeders are some distance from water, the level of salt in the ration should be reduced. In arid regions, the salt content of some water sources can reduce intake of water and feed.
Variable according to age, sex, and class (see Tables 22-1, and 22-2).	Because milk is high in calcium, lactating does need rations with high calcium levels. In % of ration: 0.78 M-F 0.70 A-F	Ground limestone, steamed bone meal, dicalcium phosphate, and oyster shell.	The recommended ratio of calcium to phosphorus ranges from 2:1 to 4:1. If the ratio falls below 2:1, urinary calculi may develop in males. Under grazing conditions, calcium is seldom a problem with either Angora or meat-type goats.

(Continued)

TABLE 22-4

Mineral Which May Be Deficient Under Normal Conditions	Conditions Usually Prevalent Where Deficiencies Are Reported	Function of Mineral	Deficiency Symptoms/Toxicity
Major or Macrominerals (Continued):			
Phosphorus (P)	When goats subsist on pastures in phosphorus-deficient areas. When goats subsist for long periods on mature, dry forages. Lack of vitamin D.	Essential for sound bones and teeth, and for the assimilation of carbohydrate and fats. A vital ingredient of the proteins in all body cells. Necessary for enzyme activation. Acts as a buffer in blood and tissue. Occupies a key position in biologic oxidation and reactions requiring energy.	**Deficiency symptoms**—Slowed growth, depraved appetite (chewing bones, wood, hair), unthrifty appearance, rickets in young animals, osteomalacia in mature animals, and depressed milk yields in lactating does. **Toxicity**—There is no known phosphorus toxicity in goats. However, excess phosphorus consumption may decrease the absorption of calcium. Also, when phosphorus is high in relation to calcium, urinary calculi may be formed.
Magnesium (Mg)	Animals grazing lush green grass or winter cereal pastures fertilized with nitrogen and potassium.	Required for many enzyme systems and for proper functioning of the nervous system. Also, closely associated with the metabolism of calcium and phosphorus.	**Deficiency symptoms**—Loss of appetite, excitability, and calcification of soft tissues. The most noted problem associated with low magnesium is grass tetany. **Toxicity**—Magnesium toxicity of goats has not been reported under practical conditions.
Potassium (K)	When goats are grazing mature range forage during winter or drought periods. High concentrate rations.	It (1) affects osmotic pressure and acid-base balance within the cells, and (2) aids in activating several enzyme systems involved in energy transfer and utilization, protein synthesis, and carbohydrate metabolism.	**Deficiency symptoms**—Marginal deficiencies result in reduced feed intake, retarded growth, and reduced milk production. Severe deficiencies cause emaciation and poor muscular tone. **Toxicity**—The maximum tolerable level of potassium for sheep is about 3% of the ration DM. It is postulated that the toxicity level of goats is similar.
Sulfur (S)	Possibly with liberal intake of tannic acid-containing plants. This is of concern with range goats, which liberally graze and browse such plants.	Essential for synthesis of the sulfur amino acids (cystine and methionine). Sulfur is particularly high in goat hair.	**Deficiency symptoms**—Depressed appetite, loss of weight, poor growth, excessive salivation, tearing, loss of mohair, depressed milk yields. **Toxicity**—Elemental sulfur is practically devoid of toxicity.
Trace or Microminerals:			
Cobalt (Co)	In cobalt deficient areas when the cobalt level in the feed drops to 0.04 to 0.07 ppm or lower.	The only function of cobalt is that of being an integral part of vitamin B-12.	**Deficiency symptoms**—The deficiency symptoms are actually vitamin B-12 deficiencies. They are: loss of appetite, emaciation, weakness, anemia, and decreased production. **Toxicity**—In sheep, about 204.5 mg cobalt/100 lb live weight is toxic. Likely, the same applies to goats.
Copper (Cu) and Molybdenum (Mo)	Copper and molybdenum are interrelated in animal metabolism; hence, they should be considered together. The most common problem occurs when a normal or low level of copper is accompanied by a high level of molybdenum, resulting in copper being excreted and producing a copper deficiency. This condition can be corrected by adding copper.	Copper and iron are mutually involved in the formation of hemoglobin—the red pigment which carries oxygen.	Few studies on copper and molybdenum have been conducted with goats. It appears that sheep are sensitive to copper toxicity and resistant to molybdenosis, but it is not known whether this is also the case with goats.
Fluorine (F)		Necessary for sound bones and teeth.	**Deficiency symptoms**—Fluorine deficiency appears to be rare. Rather, the hazard is fluorine toxicity. **Toxicity**—With sheep, fluorine toxicity occurs at levels above 200 ppm. So, it is postulated that the toxicity level for goats is similar.
Iodine (I)	Iodine-deficient areas or soils (in northwestern U.S., and in the Great Lakes and Rocky Mountain Regions), unless iodized salt is fed.	Formation of thyroxin, a hormone of the thyroid gland.	**Deficiency symptoms**—Enlarged thyroid gland, a condition called goiter. Kids born weak or dead. **Toxicity**—The maximum tolerable level for sheep is 45 ppm A-F or 50 ppm M-F. It is postulated that the toxicity level for goats is similar.
Iron (Fe)	Iron deficiency may occur in young goat kids because of their minimal body stores at birth and the low iron content of milk.	As a component of blood hemoglobin required for oxygen transport. Iron is also required for some enzyme systems.	**Deficiency symptoms**—Anemia, poor growth, lethargy, increased respiration rate, decreased resistance to infection, and in severe cases high mortality. **Toxicity**—Free iron ions are very toxic, causing loss of appetite, diarrhea, below normal temperature, shock, acidosis, and death.

(Continued)

Nutrient Requirements[2]	Recommended Allowances[2]	Practical Sources	Comments
Variable according to age, sex, and class (see Table 22–1, and 22–2).	Can be offered free-choice or incorporated into the ration. In % of ration: 0.45 M-F 0.40 A-F	Cereal grains. Defluorinated phosphate, dicalcium phosphate, steamed bone meal, monosodium phosphate.	Phosphorus is the mineral most likely to be deficient in range forages. It is, therefore, recommended that it be supplied in range supplements. *The calcium-to-phosphorus ratio should not drop below 1.2:1.
	In % of ration: 0.25 M-F 0.22 A-F	Plant protein supplements and plant by-product feeds are excellent sources of magnesium. The common magnesium supplements are magnesium carbonate, magnesium oxide, and magnesium sulfate.	*Goats have a marginal ability to compensate for low dietary magnesium by reducing the rate of excretion.
*In growing sheep, the potassium requirement is 0.5% of the ration. In lactating dairy cattle, the requirement is 0.8% of the complete ration. These levels are also postulated as the requirements of growing and lactating goats, respectively.	In % of ration: 1.0 M-F 0.9 A-F	Roughage-based rations. Common potassium supplements are potassium chloride, potassium bicarbonate, and potassium sulfate.	
	In % of ration: 0.20 M-F 0.18 A-F A sulfur-to-nitrogen ratio of 1:10 is recommended.	Sulfates, such as sodium sulfate and ammonium sulfate, are the most available forms of sulfur for ration formulation.	Because of mohair production, Angora goats may have an elevated sulfur requirement.
*A level of 0.1 ppm in the M-F ration.	In % of ration: 0.1 to 0.2 ppm M-F 0.09 to 0.18 ppm A-F	*Cobalt sulfate or cobalt chloride added at the rate of 5.45 g per 100 lb *(12 g per 100 kg)* of salt.	
	Add copper sulfate to the salt at the rate of 0.5% Copper in total ration: 5.0 ppm M-F 4.5 ppm A-F	Salt containing 0.5% copper sulfate.	
Iodine in the ration: A-F, 0.09–0.72 ppm; M-F, 0.1–0.8 ppm. The higher levels are indicated for pregnancy and lactation.	Free access to stabilized iodized salt containing 0.0078% iodine. In total ration: 0.5 ppm M-F 0.45 ppm A-F	Iodized salt.	Iodized salt should not be used as a feed-limiter because it could lead to excessive intakes of iodine.
*0.03% ferrous iron in the ration.	*Iron-dextran (150 mg) may be injected in kids at 2 to 3 week intervals if iron deficiencies are observed. In total ration: 50 ppm M-F 45 ppm A-F	Iron-dextran is recommended as an injection; and ferrous sulfate and ferric citrate are recommended for incorporating in rations.	Iron deficiency seldom occurs in mature grazing goats.

(Continued)

TABLE 22-4

Mineral Which May Be Deficient Under Normal Conditions	Conditions Usually Prevalent Where Deficiencies Are Reported	Function of Mineral	Deficiency Symptoms/Toxicity
Trace or Microminerals (Continued):			
Manganese (Mn)	High calcium and iron may increase manganese requirements.	Skeletal development and reproduction.	**Deficiency symptoms**—Reluctance to walk, deformity of the forelegs, delayed estrus, more inseminations per conception, more abortions, and 20% reduction in birth weights. **Toxicity**—1,000 ppm appears to be the maximum tolerance level for sheep; so, it is postulated that the toxicity level for goats is similar.
Selenium (Se)			**Deficiency symptoms**—White muscle disease in young kids from birth to a few months of age, which may take one of two forms: (1) sudden unexplained death, or (2) muscular paralysis, particularly of the hind limbs, or stiffness and inability to rise. **Toxicity**—All livestock species, including goats, are susceptible to selenium toxicity. Selenium toxicity in sheep occurs from prolonged consumption of plants containing over 3 ppm Se. It is postulated that the toxicity level for goats is about the same as for sheep.
Zinc (Zn)	Rations excessively high in calcium adversely affect zinc utilization.	Needed for normal skin, bones, and hair. A component of several enzyme systems involved in digestion and respiration.	**Deficiency symptoms**—Reduced feed intake, weight loss, parakeratosis, stiffness of joints, excessive salivation, swelling of the feet and horny overgrowth, small testicles, and low libido. **Toxicity**—Levels of 1,000 ppm may be toxic.

[1]Where preceded by an asterisk, the requirements, recommended allowances, and other facts presented herein were adapted from *Nutrient Requirements of Goats*, No. 15, NRC-National Academy of Sciences, 1981.

[2]As used herein, the distinction between "nutrient requirements" and "recommended allowances" is as follows: In nutrient requirements, no margins of safety are included intentionally, whereas in recommended allowances, margins of safety are provided in order to compensate for variations in feed composition, environment, and possible losses during storage or processing.

Vitamins

Typical range or pasture diets of goats usually contain adequate levels of vitamins or vitamin precursors to maintain the normal health of goats. However, young kids, goats kept in confinement, and high-producing dairy goats may need supplemental vitamins.

TABLE
GOAT VITA

Vitamin Which May Be Deficient Under Normal Conditions	Conditions Usually Prevalent Where Deficiencies Are Reported	Function of Vitamin	Deficiency Symptoms
Fat-Soluble Vitamins:			
A	During extended dry periods when the supply of green forage is limited.	Required for normal vision. Aids in reproduction and lactation. Needed for maintaining normal epithelial tissue. Aids in resistance to infection.	Keratinization of the epithelia of the respiratory, alimentary, reproductive, and urinary tracts, and of the eye. Multiple infections, poor bone development, birth of abnormal offspring, and vision impairment, including night blindness.
D	Young goats kept in confinement where they have little or no access to sunlight.	Absorption of calcium and phosphorus.	Bone abnormalities, including rickets. Depressed growth.
E	Abnormally high levels of nitrates may produce vitamin E deficiencies. Where soils are very low in selenium.	Serves as a physiological antioxidant. In dairy goats, the vitamin E transferred to the milk is important because of the antioxidant properties that aid in milk storage.	Evidence of spontaneous vitamin E deficiency signs in goats is lacking. The probability of lowered productivity in goats as a result of a vitamin E deficiency is remote.
K	Vitamin K deficiency may occur when the dicoumarol content of hay is excessively high, as when moldy sweet clover hay is fed.	Vitamin K or K$_2$ is necessary in the blood clotting mechanism.	
Water-Soluble Vitamins: B vitamins Vitamin C	B vitamin deficiencies may be evident in poorly fed and unhealthy animals. B-12 may be deficient if cobalt is absent or at extremely low levels, as cobalt is required for the synthesis of vitamin B-12.	B-1 participates as a coenzyme in the utilization of carbohydrates.	

[1]As used herein, the distinction between "nutrient requirements" and "recommended allowances" is as follows: In nutrient requirements, no margins of safety are included intentionally; whereas in recommended allowances, margins of safety are provided to compensate for variations in feed composition, environment, and possible losses during storage or processing.

Nutrient Requirements[2]	Recommended Allowances[2]	Practical Sources	Comments
	In total ration: 40 ppm M-F 36 ppm A-F	Manganese gluconate.	
	In total ration: 0.15 ppm M-F 0.13 ppm A-F		
*Direct and indirect evidence indicates minimum requirements of 10 ppm in the ration.	In total ration: 50 ppm M-F 45 ppm A-F	Zinc carbonate. Zinc sulfate.	

GOAT VITAMIN CHART

Table 22-5, Goat Vitamin Chart, gives, in summary form, the following pertinent information relative to each vitamin listed: (1) conditions usually prevailing where deficiencies are reported, (2) function, (3) deficiency symptoms, (4) nutrient requirements, (5) recommended allowances, and (6) practical sources.

22-5 MIN CHART

Nutrient Requirements[1]	Recommended Allowances	Practical Sources	Comments
Variable according to size, sex, age, and class (see Table 22-1 and 22-2).	The recommended allowances should provide margins of safety over and above the requirements. So, add 10 to 20% to the requirements given in Tables 22-1 and 22-2.	Synthetic vitamin A. Injectable vitamin A. Yellow corn. Green forages.	Young animals, which have not built up vitamin A reserves, are more susceptible to a vitamin A deficiency than are mature animals. Goats that have had access to green feed can store sufficient vitamin A in the liver and fat to last for 3 months on a low carotene ration without showing signs of vitamin A deficiency.
Variable according to size, sex, age, and class (see Table 22-1 and 22-2).	Add 10 to 20% to the requirements given in Tables 22-1 and 22-2 to provide a margin of safety.	Sunlight action on ergosterol, a plant sterol, and on 7-dehydrocholesterol, a sterol of animal origin. Sun-cured hays. Irradiated yeast. Vitamin D_2 or vitamin D_3, which goats use equally well.	Vitamin D should be of little concern when goats are maintained on pasture or range.
		Alpha-tocopherol, added to the diet or injected intramuscularly. Grains are generally high in vitamin E.	Most goat rations contain adequate amounts of vitamin E. Hence, there is little need for vitamin E supplementation.
		Green leafy materials of any kind, fresh or dry are good sources of K_1. Vitamin K_2 is normally synthesized in large amounts in the rumen; no need for dietary supplementation has been established.	
		Only vitamin B-12 (cobalamin) is likely to be deficient in goats with functioning rumens, because the microorganisms synthesize these vitamins in adequate amounts. Adequate vitamin C is synthesized in body tissues to satisfy requirements.	The B vitamins should be included in the diets of very young kids, animals with poorly functioning rumens, sick animals, and those with radically changed diets.

Water

The water requirement may be met by water consumption (drinking), but other important sources include water contained in the feed ingested and metabolic water resulting from oxidation of feed energy sources. The major avenues of water losses are those from urine, lactation, evaporation, and perspiration.

Factors affecting the water intake of goats are lactation level, environmental temperature, water content of the forage consumed, amount of exercise, and the salt and other minerals in the ration.

Normally, goats consume 1.4 to 1.7 lb of water per 1 lb of dry matter, whereas cattle consume 2.1 lb of water per 1 lb of dry matter.

Nonnutritive Factors

Nonnutritive factors are substances that cannot be classified as metabolic nutrients but can aid in the utilization of the nutrients in the feed, such as bulk and feed additives.

- **Bulk**—Forages and browse-type feeds, as well as coarse textured concentrates, contain considerable bulk. In many cases, animals can utilize bulky feed more efficiently than if the feed were finely ground because finely ground feeds pass through the digestive tract more rapidly. Finely ground feeds are not exposed to microbial fermentation and enzymatic degradation for periods sufficient to maximize utilization. Hence, they tend to be digested somewhat inefficiently.

- **Antibiotics and Other Feed Additives**—As in the case with most other farm animals, antibiotics and other compounds are often added to the rations of goats in order to improve production performance and health.

The withdrawal and milk-discard periods of the various drugs are constantly being reviewed and revised; hence, the producer must always read and heed the label of the drug that is being used. **The label is the ultimate guide to the producer as to the proper use of the drug.** Unless it is followed, costly condemnations or seizures can result.

(Also see Chapter 13, Feed Supplements/Additives/Implants, section on "Antibiotics.")

FEEDS FOR GOATS

Goats can effectively use the same kinds of feeds as are consumed by other ruminants—grasses and legumes; hays and other dry roughages; silages, haylages, and root crops; concentrates; milk and milk replacers; and commercial feeds. Additionally, goats have a unique preference for, and succeed in feeding on, a wide assortment of browse and forbs on which other species fail.

Browse, Forbs, and Grasses/Legumes

Browse refers to the edible parts of woody vegetation, such as leaves, stems, and twigs from bushes.

Forbs refers to nongrasslike range herbs which animals eat (forbs are commonly called weeds by western ranchers).

For using browse and forbs, goats are without a peer. Mohair and meat-type goats are used extensively to graze unimproved pastures and range areas where vegetation is generally of low quality. Since goats are good browsers, they can be used effectively to control brush and undergrowth. As a result, they have been exploited *mobile pruning weapons* against encroaching browse and forbs in range areas.

While goats can utilize a number of types of browse and forbs that other livestock refuse, poisonous plants must be avoided. (See Chapter 5 for additional information relative to poisonous plants.) Also, goat producers should be aware that many palatable browse species are limited in value because of one or more inhibitors that bind or otherwise prevent utilization of the nutrients contained in plants. Among such inhibitors are high levels of (1) lignin in woody twigs, which is practically indigestible; (2) essential oils (terpene-based organic compounds), which inhibit growth of rumen bacteria; and (3) tannins, which depress digestion by binding and/or inhibiting enzyme activity.

Fig. 22-2. Goats are superb browsers. (Courtesy, West Virginia University, Morgantown)

Improved pasture is a necessity for lactating dairy does. In order to prevent overgrazing, grass should be allowed to get 3 to 4 in. high before animals are allowed to graze. A good management practice with goats is to divide the pasture into lots and rotate the animals from the various lots every 10 to 14 days. An electric fence provides an easy way of setting up pasture lots. Not only does this practice prevent overgrazing, it helps to break up the life cycle of internal parasites which can create health problems as well as reduce production. Some of the grasses and legumes that can be effectively used in pasture management for goats are alfalfa, alfalfa-brome mix, clover, clover and grass, timothy, and bluegrass. Since goats are ruminants, care should be exercised when fresh, lush legume pastures are first used, as bloat problems can result.

When on pasture, high-producing dairy goats should be properly supplemented with concentrates and minerals.

Hays and Other Dry Roughages

With the exception of pasture and range feeds, hay and dry roughages are usually the most economical feeds for goats. A good-quality legume hay or a mixed legume and grass hay provide an excellent source of highly digestible nutrients. Mixed hays should be at least 50% legume, especially if hay is to be the primary source of feed. Grass hays require supplementation with concentrate. Except for dairy goats in lactation, it is not necessary to provide large amounts of concentrates to goats, especially if they are on maintenance or low production levels.

Silages, Haylages, and Root Crops

Silages and haylages have never been used extensively for feeding goats because of the following practical reasons: (1) their high water content makes it impractical to feed them at great distances, thereby alleviating their use on most Angora and Spanish (meat) goat operations; and (2) the small number of animals in most dairy goat operations makes it impossible to feed the top 3–4 in. of silage or haylage that must be removed from the silo daily to prevent spoilage.

Goats are quite fond of root crops and garden products; and these types of feeds can be effectively incorporated in the ration for a change of routine. Carrots, beets, turnips, and cabbage are especially relished by goats. These types of feeds are high in moisture and should be fed in the same manner as silage. Roots should be chopped in order to lessen choking.

In order to prevent off-flavors in milk, it is recommended that silage, haylage, and root crops such as turnips be fed either after milking or in amounts that will be consumed 3 to 4 hours prior to milking. (For more details on silage and haylage, see Chapter 9.)

Concentrates

The concentrates used in goat rations can be classified as either energy feeds or protein supplements.

• **Energy Feeds**—The common energy ingredients are corn, oats, barley, milo, and wheat, along with their by-products. The amount of energy feed to include in the ration should be determined by the production demands. A dry doe requires little or no supplementation, while a doe at the peak of lactation requires substantial amounts of energy.

Molasses, an excellent energy source, is commonly used to reduce the dustiness of feed and to increase palatability. If too much molasses is included in the ration, the feed becomes sticky and lumpy; so, it is usually limited to 5 to 8% of the mixture.

• **Protein Feeds**—A wide variety of protein supplements can be used in rations for goats. As is the case with most other species of livestock, the oil meals are used extensively. Cottonseed meal and soybean meal are probably the most widely used sources of protein for goats, but other meals can be and are used, depending on their respective prices and availability. Among the alternative sources are copra meal, peanut meal, sunflower meal, safflower meal, rapeseed (canola) meal, corn gluten feed, brewers' dried grains, and distillers' dried grains.

Urea and other nonprotein nitrogen (NPN) sources are often used in rations for goats, but several precautions should be taken when they are incorporated in the ration. Urea can constitute up to 1% by weight of the total concentrate mix, or supply ⅓ of the protein equivalent in the total ration. Rations containing NPN should be introduced very gradually in the feeding scheme of goats, as a period of adaptation is required by the microorganisms of the rumen. In addition to urea, other NPN sources are ammoniated cottonseed meal, ammoniated rice hulls, ammoniated citrus pulp, and ammoniated beet pulp.

Milk and Milk Replacers

Unless extenuating circumstances prevail, newborn kids are seldom taken from Angora or Spanish goat mothers. However, with dairy goats, normally kids are allowed to nurse for 2 to 4 days only, or not at all. It is usually easier to train kids to nipple- or pan-feeding if they have never nursed their mothers.

When newborn kids are to be raised separately from their dams, they may be fed cow's milk or a milk replacer. The most desirable milk replacers for kids contain a minimum of 20% fat and 24% crude protein; are skimmed milk-based, rather than grain-based; and are fortified with minerals and vitamins. Because of varying formulations, care should be taken to follow the manufacturer's directions.

(Also see Chapter 6, Types and Roles of Feedstuffs, section on "Milk Replacers.")

Commercial Feeds

Commercial feeds, containing a variety of ingredients, including minerals and vitamins, are used by many goat producers. Special ingredients such as molasses and/or fat may be added to reduce dustiness and increase palatability. Cottonseed hulls may be added to improve texture and provide fiber. In some cases, the commercial feeds are pelleted.

Commercial feeds may be available as (1) complete feeds (roughage and concentrate combined), (2) concentrates, or (3) protein supplements.

Feed Preparation

Grain can best be utilized by goats when it is processed to a limited extent. Cracking, rolling, crimping, flaking, or coarse grinding all aid in making grains more digestible. Grinding feed to a fine powder form is not desirable, especially in low-roughage rations, because it generally results in lowered palatability and digestibility. Large grains, such as corn, can be fed whole on the ground in range areas and effectively utilized with little wastage. When grains are to be mixed with hay and other ingredients or to be used on the range, pelleting is advisable.

In rangeland areas, feed supplements are sometimes offered free-choice, using salt as a governor to limit feed consumption. The proportion of salt to feed may vary from 5 to 40%. By varying the proportion of salt in the mixture, it is possible to hold the consumption of feed supplement to any level desired.

Pelleting and cubing are used when roughages and grains are mixed together. Unless these feeds are pelleted or cubed, the various ingredients can separate out, resulting in waste and inefficient utilization of feed. Hay is sometimes chopped to reduce waste. It should be emphasized that any processing of feed creates additional feed costs; hence, feed should not be processed if the additional treatment does not increase feed efficiency or reduce waste sufficiently to offset the added cost.

Feed Substitution Table

Sometimes, goat producers have an opportunity to obtain a feed ingredient at a favorable price, but they may not know the relative feeding value of the product with respect to the feed currently in use. In order to assist producers in making these managerial decisions, a feed substitution table giving the comparative value of feeds is needed.

Feed substitutions for goats and sheep are similar. Hence, the reader is referred to Chapter 21, Feeding Sheep, Table 21-12, Feed Substitution Table for Sheep and Goats.

NUTRITION RELATED DISORDERS

Goats are subject to several nutrition-related disorders, most of which occur in sheep, and many of which occur in other species. Many such disorders are named and discussed in Chapters 4 and 5 of this book. However, abortion, caprine arthritis-encephalitis (CAE), and enterotoxemia of goats are discussed in the sections that follow.

Abortion

Infectious diseases such as brucellosis are capable of causing abortions in goats. However, herein reference is made to a particular type of abortion caused by a metabolic disturbance of the functional corpus luteum, to which Angora goats are predisposed. Under normal production conditions, this malady commonly causes a low level of abortion, but catastrophic losses sometimes occur. Most abortions occur in response to stress between 90 and 110 days of gestation. Undernutrition during the critical stage of rapid fetal development and competition for nutrients between fetal and maternal organisms appear to be one explanation. It is noteworthy that the incidence of abortion is reduced in herds in which replacement does are fed for proper size and development prior to the first breeding season and during gestation.

Caprine Arthritis-Encephalitis (CAE)

It is estimated that more than 80% of the dairy goats of the United States are infected by the retrovirus that causes CAE. In kids, it causes paralysis; in adults, it causes arthritis, which, in the late stages, is similar to rheumatoid arthritis in humans.

In most herds, the expression of the disease ranges from 0 to 25%. In those animals that do show clinical symptoms, the rate of progression and the severity of the disease varies markedly. In kids, the disease may vary from a barely noticeable unsteadiness of gait to a rapid fatal paralysis. In mature goats, the joints (front knees, hocks, and stifle joints) become swollen and disfigured, accompanied by a loss in body weight and a drop in production. The severity of the arthritis varies from years of intermittent lameness or stiffness to complete debilitation.

CAE is caused by a retrovirus, a slow acting virus that is latent, with the result that many animals may not show signs of the disease until after a long incubation period. The virus is transmitted from the doe to the kid(s) through the colostrum and milk. Does not showing symptoms of CAE may carry the virus and transmit it.

The rate of the infection in newborn goats can be reduced by more than 90% by (1) removing kids from infected does at the time they pass from the birth canal; (2) providing them colostrum that is from does identified as negative with the agar-gel immuno-diffusion (AGID) test or that has been heated to 132°F for 1 hour; (3) raising them on pasteurized goat's milk or a milk replacer; and (4) keeping them in isolation from infected goats. The AGID test can be used to monitor infection.

Enterotoxemia (Overeating Disease)

Enterotoxemia, also known as *overeating disease* or *toxic indigestion,* may be the most insidious, most often undiagnosed, and, in many herds, the most important of all goat diseases. It is a toxic reaction to *Clostridium perfringens,* Types C and D, characterized by diarrhea, depression, lack of coordination, digestive upsets, coma, and death.

In baby kids, excess feeding or sudden access to palatable feed, changes in feed, or feeding following an unusual period of starvation, may cause acute enterotoxemia and death. Causative factors for enterotoxemia in mature goats include sudden changes in concentrate feed, excessive feeding of concentrates to animals not accustomed to such feeds, sudden access to highly palatable forage, sudden change to lush pasture, sudden access to feed, and overeating by hungry goats.

Prevention consists in proper feeding along with a vaccination program, using *Clostridium perfringens* toxoid, Types C and D. Initially, all goats should be given two separate doses of the toxoid, at 2-4 week intervals. Then, all does should be given an annual booster toxoid about 1 month before kidding; all kids should be vaccinated at 3-4 weeks of age, followed by a second dose of toxoid 2 weeks later. All goats in the herd, including bucks, should receive at least two doses of toxoid annually; one when does are in late pregnancy, and another when the kids are 4 to 5 months old.

PART II—FEEDING DAIRY GOATS

A good dairy doe will average 5 lb of milk, or more, per day over a lactation period of 10 months, whereas superior animals will average 10 lb or more. Based on 200,000 official lactation records since 1968, average milk production per goat per lactation is 1,643 lb with 3.8% fat. The highest official milk production on record in the United States was made by a Saanen doe that produced 6,850 lb of milk in 305 days, in 1984. The butterfat record is held by a Nubian doe that produced 384 lb in 305 days.

FEEDING LACTATING AND DRY DAIRY DOES

Fig. 22-4. Alpine does at feed at The Coach Dairy Goat Farm, Pine Plains, N.Y. (Courtesy, The Coach Dairy Goat Farm, Pine Plains, N.Y.)

The nutritional demands upon lactating does are tremendous. It is essentially impossible for the doe to consume enough to meet the demands for body maintenance and milk production during the first few months of lactation; so, she must draw upon her body reserves to augment the nutrients consumed.

Fig. 22-3. All-time, all-breed world record milk production holder. This Saanen doe produced 6,850 lb milk and 296 lb fat, in 305 days, in 1984. Bred by Gary and Sharon Swanson, Renton, Wash.; owned by Gary Lee Cox, Eagle Point, Ore. (Courtesy, T. H. Teh, Ph.D., Prairie View A&M University Research Center, Prairie View, Tex.)

Some suggested rations for lactating dairy goats are given in Table 22-6. Note that when a legume hay (alfalfa or clover) is fed, a 12–14% crude protein grain mix will suffice. But, when a grass hay is fed, a 16–18% crude protein grain mix should be fed.

TABLE 22-6
RATIONS FOR LACTATING DAIRY GOATS (160-LB BODY WEIGHT)[1]

Feedstuffs	Daily Milk Production (lb)							
	2.5		5.0		10.0		15.0	
	Daily Feed							
	(lb)	(kg)	(lb)	(kg)	(lb)	(kg)	(lb)	(kg)
Ration # 1:								
Alfalfa clover hay (16% CP)	2.0	0.9	3.0	1.4	3.5	1.6	4.5	2.0
Grain mix to be selected from 4 mixes listed below (14–16% CP, 70% TDN)	3.0	1.4	4.0	1.8	6.0	2.7	8.0	3.6
Ration # 2:								
Grass hay (7% CP)	2.5	1.1	2.5	1.1	3.0	1.4	4.0	1.8
Grain mix to be selected from 4 mixes listed below (16–18% CP, 65% TDN)	3.4	1.5	4.6	2.1	7.0	3.2	9.0	4.1
Grain mixes:								
Ration	# 1		# 2		# 3		# 4	
Level of crude protein (CP) in grain mix	14%		16%		18%		20%	
Lb or kg per ton	(lb)	(kg)	(lb)	(kg)	(lb)	(kg)	(lb)	(kg)
Ingredients								
Rolled corn	900	409.0	800	363.6	720	327.3	656	298.2
Crimped oats	421	191.0	300	136.0	240	109.1	200	91.0
Beet-citrus pulp	200	91.0	200	91.0	200	91.0	200	91.0
Dried brewers' grain	—	—	150	68.0	200	91.0	200	91.0
40% protein supplement	300	136.0	—	—	516	234.5	—	—
Soybean meal	—	—	356	161.8	—	—	600	272.7
Molasses	150	68.0	150	68.0	100	45.5	100	45.5
Trace mineralized salt	10	4.5	20	9.1	10	4.5	20	9.1
Dicalcium phosphate	—	—	10	4.5	10	4.5	20	9.1
Monosodium phosphate	15	7.0	10	4.5	—	—	—	—
Magnesium oxide	4	1.8	4	1.8	4	1.8	4	1.8
Vitamins[2]								

[1]Adapted by the authors from *Extension Goat Handbook*, "Feeding," by R. S. Adams, Pennsylvania State University; B. Harris, University of Florida; M. F. Hutjens, University of Illinois; and E. T. Oleskie and F. A. Wright, Rutgers University.

[2]During winter, all mixtures should be supplemented with 6 million IU of vitamin A and 3 million IU of vitamin D per ton of grain mix.

It is important to dry off dairy goats about 6 to 8 weeks prior to kidding. This gives the doe a brief rest from the heavy demands of lactation; enables her to meet the nutrient needs of the rapidly growing fetus, and allows her to build body reserves with which to meet the rigorous requirements of lactation which follow.

Depending on the kind and quality of the forage and the size and condition of the doe, 1 to 2 lb of 12 to 16% protein concentrate ration should be fed during the dry period. Trace-mineralized salt and water should also be available.

In the last 6 weeks of gestation, the fetus gains 70% of its birth weight, so the nutrition of the doe during this period is critical.

FEEDING DAIRY KIDS

It is important to get 2 to 4 oz of colostrum in a kid as quickly as possible after birth.[3] Colostrum contains higher levels of total protein, milk solids, globulins, fat, and vitamin A than normal milk. It is also a laxative. Most important, it contains antibodies against disease to which the doe has immunity. Young kids are able to absorb this antibody protection effectively at birth, but by the time they are 3 days old, this ability almost disappears.

During the first 2 days of life, kids should receive at least three colostrum feedings per day. A kid will consume 1½ to 2 pt daily.

If the kid is to be raised without its mother's milk, it should be allowed to nurse for 2 or 3 days only or not at all. Most commercial dairy goat producers favor putting the newborn kid directly on bottle- or pan-feeding without allowing it to nurse its mother; after a kid nurses its mother a couple of times, it is very difficult to get it to accept a bottle or a pan. Initially, the nipple bottle is generally preferred to pan feeding as it tends to be more natural, thereby resulting in less ingested air. However, as soon as kids are strong enough and can drink milk easily, experienced dairy goat producers prefer to train them to bucket or pan feeding, which is faster and allows for easier cleaning and maintenance of utensils.

After the kid is removed from colostrum feeding, it can be fed cow's milk or a milk replacer, along with a starter ration, with the change made gradually.

Kids can be weaned from milk as early as 5 to 6 weeks of age, although most goat breeders delay it until 3 to 4 months of age. As young kids approach weaning age, gradually add warm water to their milk diet. This will provide them with the necessary fluids for rumen development and ease the stress of weaning them. After the kids are weaned from the milk, feed them all the bright green forage they will eat, plus ¾ to 1 lb of any good grower ration.

---

[3]Because of the hazard of caprine arthritis-encephalitis (CAE) virus, scientists recommend that kids be fed (a) colostrum only from does negative to the CAE test, (b) heat-treated colostrum (132°F for 1 hour), or (c) no colostrum at all. (See section headed Caprine Arthritis-Encephalitis [CAE].)

Suggested starter and grower rations for kids are given in Table 22-7.

TABLE 22-7
STARTER AND GROWING RATIONS FOR DAIRY KIDS[1]

	Kid Starter[2]	Growing Ration[3]
	(%)	(%)
Corn	27.6	12.9
Crimped oats	37.9	10.0
Soybean meal (44%)	10.0	8.6
Alfalfa leaf meal	18.0	10.0
Cane molasses	5.0	5.0
Cottonseed hulls	—	51.9
Trace mineralized salt	1.0	1.0
Limestone	0.3	0.4
Vitamins A, D, and E (premix)	0.2	0.2

[1]From: *Raising Goat Kids*, by T. H. Teh et al., Texas A&M, International Dairy Goat Research Center, Prairie View A&M Research Center, Prairie View, Tex., Vol. A1, No. 1, Bull. March, 1985.

[2]On a dry matter basis, the kid starter should contain a minimum of 80% TDN, 16% protein, 0.6% calcium, and 0.4% phosphorus.

[3]The grower ration may be fed free-choice after 4 months.

FEEDING FOR SHOW

For success in the show-ring, an animal should be fed and managed to attain maximum development in body conformation. In general, the feeding of dairy show goats differs from normal operations only in that greater effort and expense and more liberal feed allowance may be justified in order to produce a winner. In addition to being well balanced,

Fig. 22-5. Dairy goat properly fitted and shown by a 4-H Club member. (Courtesy, University of New Hampshire, Penacook)

the ration for show goats must be palatable. No definite set of rules relative to feed allowances can be followed satisfactorily. Rather, the judgment of the skillful feeder must prevail.

PART III—FEEDING ANGORA AND SPANISH (MEAT) GOATS

Fig. 22-6. Angoras on typical range on a west Texas ranch. (Courtesy, *Sheep & Goat Raiser*)

The basic nutrient requirements established by the National Research Council (NRC) are suited for both Angora and Spanish goats (see Table 22-1). These requirements are based upon the weight of the animals being fed, the level of activity, the stage of pregnancy, and the growth rate. The type of range generally governs the activity of goats. For example, goats that range on sparsely vegetated grassland and on seasonal mountainous ranges must travel long distances daily for grazing and watering; whereas, slightly less activity is required by goats that graze on semiarid rangeland or on slightly hilly land. For mohair production, nutrient requirements in excess of those in Table 22-1 are needed. These are provided in Table 22-3 according to the level of mohair production. Furthermore, the nutrient requirements of lactating does as given in Table 22-2 must also be considered for goats nursing kid(s). Therefore, based on Table 22-1 and 22-3, a 44-lb Angora goat kid gaining 0.22 lb per day, having medium activity, and producing 4.4 lb of mohair per year has the following requirements for energy and protein:

	Energy	Total Protein
	(Mcal DE/day)	(g/day)
Maintenance (medium activity)	1.77	55
Growth	0.88	28
Mohair	0.07	9
Total	2.72	92

The feed recommendations listed herein should be considered to be minimum levels. Thus, goat producers should adapt the recommendations to fit the particular operation; and, in many cases, it is advisable to use higher levels than are listed.

Available browse and forages will satisfy many of the nutritive needs of goats that are raised on ranges, but, for maximum performance, it is advisable to provide supplemental feed when range conditions become adverse. Twenty percent protein range cubes are a popular supplement. Also, shelled corn can be fed on the ground to goats, with very little waste. Usually, about ¼ to 1 lb of supplement per head per day is adequate in the winter or during dry periods when green feed is scarce. Some examples of concentrate supplements for range goats are found in Table 22-8.

TABLE 22-8
CONCENTRATE SUPPLEMENTS FOR RANGE GOATS[1]

Ingredients	Supplement A 20% Protein	Supplement B 30% Protein	Supplement C 40% Protein
	(%)	(%)	(%)
Corn or sorghum	82	58	25
Cottonseed meal or soybean meal	14	37	70
Urea	2	3	3
Dicalcium phosphate	2	2	2
Vitamin A supplement	—[2]	—[2]	—[2]
Approximate nutrient composition			
Energy:			
TDN .... (%)	75	72	70
DE .... (Mcal/lb)	1.50	1.45	1.40
Protein:			
Crude .... (%)	20	30	40
Digestible .... (%)	16	24	32
Phosphorus .... (%)	0.55	0.65	0.77

[1]Adapted by the authors from *Nutrient Requirements of the Angora Goat* by J. E. Huston, M. Shelton, and W. C. Ellis, Tex. Ag. Exp. Sta. Bull. B-1105.

[2]Sufficient supplement to provide 2,500 IU of vitamin A per lb of feed (5 million IU/ton of total concentrate supplement mix).

The same rations and feeding practices used for Angora goats are suited for Spanish goats, also. However, supplemental feeding of meat goats is not the norm. Yet, Spanish goats do respond well to supplemental feeding during critical periods—just before breeding, just before and following kidding, and when range feed is short.

FEEDING KIDS AND YEARLINGS

As long as kids are receiving adequate amounts of milk from their mothers, they do very well on good range. Additional supplementation, however, makes for more rapid growth and better prepares them for market or breeding. One pound of supplement for each 2½ to 3 kids should be provided. Older and larger kids may have their supplement reduced to 1 lb daily for each 5 kids if the range conditions are good. When range is poor, the grain and supplement should be increased to provide 1 to 1⅓ lb of grain daily per kid. In addition, kids should have access to good-quality hay.

A suitable ration for kids and yearlings is given in Table 22-9.

TABLE 22-9
RATION FOR ANGORA AND SPANISH KIDS AND YEARLINGS

Age	Ingredients	Amount
		(%)
Kids/yearlings	Alfalfa hay	32
	Cottonseed hulls	28
	Sorghum grain	18
	Barley grain	8
	Molasses	6
	Cottonseed oilmeal	6
	Salt/mineral mix	2
	Total	100

QUESTIONS FOR STUDY AND DISCUSSION

1. Can experimental work and instruction relative to goats be justified on the basis of their contributions in terms of animal numbers, meat, milk, and fiber; (a) worldwide, and (b) in the United States?

2. Name and describe the three major types of goats, and tell how they are similar and how they are different from the standpoint of nutrition/feeding.

3. Discuss the economic importance of feed for goats.

4. Angora goats are supplementally fed on a limited basis whereas Spanish goats are seldom supplemented at all. Does this mean that it would not pay to supplement Spanish goats during critical periods—just before breeding, before and after kidding, and during feed scarcity?

5. Since cattle, sheep, and goats are all ruminants, why go to the trouble and expense of evolving with separate nutritional requirements for goats?

6. Why did the National Research Council (NRC) evolve with different nutrient requirements of goats for maintenance, various levels of activity, late pregnancy, growth, milk production, and mohair?

7. What are the signs of energy deficiency?

8. Why did the NRC evolve with different protein requirements for maintenance, milk production, and mohair production?

9. Discuss what happens to goats when they are fed protein-deficient rations.

10. Give the functions, deficiency symptoms, and practical sources of each of the following minerals for goats: salt, calcium, phosphorus, cobalt, iodine, and selenium.

11. Give the functions, deficiency symptoms, and practical sources of each of the following vitamins for goats: vitamin A, vitamin D, and vitamin E.

12. What are the important ways of meeting the water requirements of goats, and what are the major avenues of water losses from the body?

13. Define (a) browse, and (b) forbs. If goats can effectively use grasses and legumes, why provide them with browse and forbs?

14. What are the advantages to dividing goat pastures into lots and rotating their use every 10 to 14 days?

15. Discuss the need for, and the importance of, hay and other dry roughages for goats.

16. Why haven't silages and haylages been used extensively for goats?

17. What precautions should be taken when feeding silage and root crops to dairy goats?

18. How much molasses can safely be incorporated into a goat ration? Why is it not advisable to incorporate more?

19. What is the maximum recommended amount of non-protein nitrogen that can be added to goat feeds?

20. What's a milk replacer? How and why may a milk replacer be used for dairy kids?

21. What determines the concentration of salt when it is used as a governor in rations that are to be offered free-choice to goats?

22. Discuss the cause and control/prevention of each of the following nutrition related disorders: abortion, caprine arthritis-encephalitis, and enterotoxemia.

23. How do the top milk and butterfat records of dairy goats and dairy cows compare?

24. Discuss the nutritional demands upon lactating dairy does, and tell how they may be met.

25. What is the reason for feeding dry dairy does liberally during the last six weeks of gestation?

26. Outline a feeding program for dairy kids.

27. How does the feeding of dairy show goats differ from the feeding of other goats?

28. Discuss the feeding of Angora and Spanish does.

23

FEEDING SWINE[1]

Original painting by Tom Phillips

Contents	Page
Economic Importance of Feed for Swine	507
Nutritive Needs of Swine	508
National Research Council (NRC) Requirements	508
Recommended Nutrient Allowances	511
Energy	512
Protein	513
Minerals	514
Swine Mineral Chart	515
Feeds as a Source of Minerals	518
Method of Feeding Mineral Supplements	519
Vitamins	519
Swine Vitamin Chart	519
Unidentified Factors	519
Water	522
Additives	522
Feeds for Swine	525
Concentrates	525
Energy Feeds	525
Corn	525
Fats and Oils	526
Other Energy Feeds for Swine	526
Protein Feeds	526
Soybean Meal	526
Full-Fat Soybeans	527
Other Protein Feeds for Swine	527
Pastures	527
Dry Forages	527
Silages	527
Hogging Down Crops	528
Garbage	528
Feed Preparation	528
Swine Feeding Program	528
Feed Allowances and Suggested Rations	529
Pointers in Formulating Rations	538
Feed Substitution Table	539
Feeding Baby Pigs	541
Feeding Orphan Pigs	541
Feeding Growing-Finishing Pigs	541
Feeding Prospective Breeding Gilts	541
Feeding Boars	541
Feeding Brood Sows	541
Fitting Rations for Show and Sale Swine	542
Feeding Systems	542
Feeder Pig Production	543
Other Feed/Management Related Aspects	543
Docking Tails	543
Nutritional Anemia	543
Scouring (Diarrhea, Enteritis)	544
Weaning Pigs	544
Feed Required to Produce a Pound of Market Hog	544

Contents	Page
Effect of Sex on Performance of Growing-Finishing Hogs	544
Soft Pork	544
Corn-Hog Ratio	545
Contract Hog Production (Custom Feeding/Leasing)	545
Poisons and Toxins	546
Environmental Effects of Swine	546
Questions for Study and Discussion	546

In the natural state, the wild boar and his kind and kin roved through the forests, gleaning the feeds provided by nature. On a modern farm, the range is restricted and frequently entirely devoid of vegetation. More than half of the hogs marketed in this country are raised from farrow to finish in some type of confinement system—ranging all the way from simple shelters to environmentally controlled pig palaces. As a result of this confinement, domestic swine have less choice in their selection of feed than any other class of four-footed animals. For the most part they are able to consume only what the caretaker provides. This consists largely of concentrate feeds with only a small proportion of roughage. These conditions are made more critical because hogs grow much faster in proportion to their body weight than the larger farm animals, and they produce young at an earlier age. Thus, a knowledge of the nutritional needs of swine is especially important.

ECONOMIC IMPORTANCE OF FEED FOR SWINE

Knowledge of feeding swine is important from an economic standpoint, because feed accounts for approximately 65 to 75% of the total cost of producing pork. For this reason, every swine producer should endeavor to provide rations that are both satisfactory and inexpensive—rations that make for the maximum production of quality pork per unit of feed consumed—and at least cost.

---

[1]The authors gratefully acknowledge the assistance of the following authorities who contributed helpful information and suggestions for this chapter: R. A. Easter, Ph.D., Swine Nutritionist, Department of Animal Sciences, College of Agriculture, University of Illinois, Urbana; V. C. Speer, Ph.D., Swine Nutritionist, Department of Animal Science, Iowa State University of Science and Technology, Ames; and R. F. Wilson, Ph.D., Swine Nutritionist, Department of Animal Science, The Ohio State University, Columbus.

Also, extensive surveys indicate that 15% of all sows bred fail to give birth to young, and that 25 to 30% of all pigs farrowed fail to live to weaning age. Although reproductive failure is heavy, baby pig losses are due to many and variable factors, certainly nutritional deficiencies play a major role.

NUTRITIVE NEEDS OF SWINE

The nutrient needs of swine are influenced by age, function, disease level, nutrient interaction, and environment. It has been established that the pig has a requirement for over 40 different nutrients. Fortunately, not all of them are of practical concern.

National Research Council (NRC) Requirements

The NRC nutritional requirements, or standards, are presented in Tables 23–1 to 23–7. It is not intended that these standards impart the impression that such figures are absolute, final, and unchangeable. Rather, they should be used as guides based on research. Also, these figures are, for the most part, requirements (rather than allowances); hence, they do not provide for margins of safety to compensate for variations in feed compositions, environment, and possible losses of nutrients during storage or processing.

TABLE 23–1
DAILY NUTRIENT INTAKES AND REQUIREMENTS OF SWINE ALLOWED FEED *AD LIBITUM*[1]

		Swine Liveweight								
Intake and Performance Levels	Lb Kg	2.2–11 1–5		11–22 5–10		22–44 10–20		44–110 20–50		110–242 50–110	
		(lb)	(g)	(lb)	(g)	(lb)	(g)	(lb)	(g)	(lb)	(g)
Expected weight gain per day		0.4	200	0.6	250	1.0	450	1.5	700	1.8	820
Expected feed intake per day		0.6	250	1.0	460	2.1	950	4.2	1,900	6.9	3,110
Expected efficiency (gain/feed)		0.800		0.543		0.474		0.368		0.264	
Expected efficiency (feed/gain)		1.25		1.84		2.11		2.71		3.79	
Digestible energy intake (kcal per day)		850		1,560		3,230		6,460		10,570	
Metabolizable energy intake (kcal per day)		805		1,490		3,090		6,200		10,185	
Energy concentration (kcal ME per lb ration)		1,461		1,470		1,474		1,479		1,486	
Energy concentration (kcal ME per kg ration)		3,220		3,240		3,250		3,260		3,275	
Protein per day		0.1	60	0.2	92	0.4	171	0.6	285	0.9	404
Nutrient		**Requirement (Amount Per Day)**									
Indispensable amino acids:											
Arginine (g)		1.5		2.3		3.8		4.8		3.1	
Histidine (g)		0.9		1.4		2.4		4.2		5.6	
Isoleucine (g)		1.9		3.0		5.0		8.7		11.8	
Leucine (g)		2.5		3.9		6.6		11.4		15.6	
Lysine (g)		3.5		5.3		9.0		14.3		18.7	
Methionine + cystine (g)		1.7		2.7		4.6		7.8		10.6	
Phenylalanine + tyrosine (g)		2.8		4.3		7.3		12.5		17.1	
Threonine (g)		2.0		3.1		5.3		9.1		12.4	
Tryptophan (g)		0.5		0.8		1.3		2.3		3.1	
Valine (g)		2.0		3.1		5.3		9.1		12.4	
Linoleic acid (g)		0.3		0.5		1.0		1.9		3.1	
Major or macrominerals:											
Calcium (g)		2.2		3.7		6.6		11.4		15.6	
Chlorine (g)		0.2		0.4		0.8		1.5		2.5	
Magnesium (g)		0.1		0.2		0.4		0.8		1.2	
Phosphorus, total (g)		1.8		3.0		5.7		9.5		12.4	
Phosphorus, available (g)		1.4		1.8		3.0		4.4		4.7	
Potassium (g)		0.8		1.3		2.5		4.4		5.3	
Sodium (g)		0.2		0.5		1.0		1.9		3.1	
Trace or microminerals:											
Copper (mg)		1.50		2.76		4.75		7.60		9.33	
Iodine (mg)		0.04		0.06		0.13		0.27		0.44	
Iron (mg)		25		46		76		114		124	
Manganese (mg)		1.00		1.84		2.85		3.80		6.22	
Selenium (mg)		0.08		0.14		0.24		0.28		0.31	
Zinc (mg)		25		46		76		114		155	
Fat-soluble vitamins:											
Vitamin A (IU)		550		1,012		1,662		2,470		4,043	
Vitamin D (IU)		55		101		190		285		466	
Vitamin E (IU)		4		7		10		21		34	
Vitamin K (menadione) (mg)		0.02		0.02		0.05		0.10		0.16	
Water-soluble vitamins:											
Biotin (mg)		0.02		0.02		0.05		0.10		0.16	
Choline (g)		0.15		0.23		0.38		0.57		0.93	
Folacin (Folic acid) (mg)		0.08		0.14		0.28		0.57		0.93	
Niacin (Nicotinic acid, Nicotinamide), available (mg)		5.00		6.90		11.88		19.00		21.77	
Pantothenic acid (Vitamin B-3) (mg)		3.00		4.60		8.55		15.20		21.77	
Riboflavin (Vitamin B-2) (mg)		1.00		1.61		2.85		4.75		6.22	
Thiamin (Vitamin B-1) (mg)		0.38		0.46		0.95		1.90		3.11	
Vitamin B-6 (Pyridoxine, Pyridoxal, Pyridoxamine) (mg)		0.50		0.69		1.42		1.90		3.11	
Vitamin B-12 (Cobalamins) (mcg)		5.00		8.05		14.25		19.00		15.55	

[1]Adapted by the authors from *Nutrient Requirements of Swine*, 9th rev. ed., National Research Council, National Academy Press, 1988, p. 51, Table 5–2.

TABLE 23-2
DAILY NUTRIENT INTAKES AND REQUIREMENTS OF INTERMEDIATE-WEIGHT BREEDING ANIMALS[1]

Intake and Performance Levels		Mean Gestation or Farrowing Weight of:	
		Bred Gilts, Sows, and Adult Boars	Lactating Gilts and Sows
	Lb	358.3	363.8
	Kg	162.5	165.0
		(lb) (kg)	(lb) (kg)
Daily feed intake		4.2 1.9	11.7 5.3
Digestible energy (Mcal per day)		6.3	17.7
Metabolizable energy (Mcal per day)		6.1	17.0
		(g)	(g)
Crude Protein per day		0.5 228	1.5 689

Nutrient	Requirement (Amount Per Day)	
Indispensable amino acids:		
Arginine (g)	0.0	21.2
Histidine (g)	2.8	13.2
Isoleucine (g)	5.7	20.7
Leucine (g)	5.7	25.4
Lysine (g)	8.2	31.8
Methionine + cystine (g)	4.4	19.1
Phenylalanine + tyrosine (g)	8.6	37.1
Threonine (g)	5.7	22.8
Tryptophan (g)	1.7	6.4
Valine (g)	6.1	31.8
Linoleic acid (g)	1.9	5.3
Major or macrominerals:		
Calcium (g)	14.2	39.8
Chlorine (g)	2.3	8.5
Magnesium (g)	0.8	2.1
Phosphorus, total (g)	11.4	31.8
Phosphorus, available (g)	6.6	18.6
Potassium (g)	3.8	10.6
Sodium (g)	2.8	10.6
Trace or microminerals:		
Copper (mg)	9.5	26.5
Iodine (mg)	0.3	0.7
Iron (mg)	152	424
Manganese (mg)	19	53
Selenium (mg)	0.3	0.8
Zinc (mg)	95	265
Fat-soluble vitamins:		
Vitamin A (IU)	7,600	10,600
Vitamin D (IU)	380	1,060
Vitamin E (IU)	42	117
Vitamin K (menadione) (mg)	1.0	2.6
Water-soluble vitamins:		
Biotin (mg)	0.4	1.1
Choline (g)	2.4	5.3
Folacin (Folic acid) (mg)	0.6	1.6
Niacin (Nicotinic acid, Nicotinamide), available (mg)	19.0	53.0
Pantothenic acid (Vitamin B-3) (mg)	22.8	63.6
Riboflavin (Vitamin B-2) (mg)	7.1	19.9
Thiamin (Vitamin B-1) (mg)	1.9	5.3
Vitamin B-6 (Pyridoxine, Pyridoxal, Pyridoxamine) (mg)	1.9	5.3
Vitamin B-12 (Cobalamins) (mcg)	28.5	79.5

[1]Adapted by the authors from *Nutrient Requirements of Swine*, 9th rev. ed., National Research Council, National Academy Press, 1988, p. 52, Table 5-4.

TABLE 23-3
DAILY ENERGY AND FEED REQUIREMENTS OF PREGNANT GILTS AND SOWS[1]

Intake and Performance Levels		Weight of Bred Gilts and Sows at Mating[2]					
	Lb	265		309		353	
	Kg	120		140		160	
		(lb)	(kg)	(lb)	(kg)	(lb)	(kg)
Mean gestation weight[3]		314.2	142.5	358.3	162.5	402.3	182.5
Energy required:							
Maintenance[4] (Mcal DE per day)		4.53		5.00		5.47	
Gestation weight gain[5] (Mcal DE per day)		1.29		1.29		1.29	
Total (Mcal DE per day)		5.82		6.29		6.76	
Feed required per day[6]		4.0	1.8	4.2	1.9	4.4	2.0

[1]Adapted by the authors from *Nutrient Requirements of Swine*, 9th rev. ed., National Research Council, National Academy Press, 1988, p. 53, Table 5-6.
[2]Requirements are based on 55-lb (25-kg) maternal weight gain plus 44-lb (20-kg) increase in weight due to the products of conception; the total weight gain is 99 lb (45 kg).
[3]Mean gestation weight is weight at mating + (total weight gain/2).
[4]The animal's daily maintenance requirement is 110 kcal of DE/kg$^{0.75}$.
[5]The gestation weight gain is 1.10 Mcal of DE/day for maternal weight gain plus 0.19 Mcal of DE/day for conceptus gain.
[6]The feed required/day is based on a corn-soybean meal ration containing 3.34 Mcal of DE/kg.

TABLE 23-4
DAILY ENERGY AND FEED REQUIREMENTS OF LACTATING GILTS AND SOWS[1]

Intake and Performance Levels		Weight of Lactating Gilts and Sows at Postfarrowing					
	Lb	320		364		408	
	Kg	145		165		185	
		(lb)	(kg)	(lb)	(kg)	(lb)	(kg)
Milk yield		11.0	5.0	13.78	6.25	16.5	7.5
Energy required:							
Maintenance[2] (Mcal DE per day)		4.5		5.0		5.5	
Milk production[3] (Mcal DE per day)		10.0		12.5		15.0	
Total (Mcal DE per day)		14.5		17.5		20.5	
Feed required per day[4]		9.7	4.4	11.7	5.3	13.5	6.1

[1]Adapted by the authors from *Nutrient Requirements of Swine*, 9th rev. ed., National Research Council, National Academy Press, 1988, p. 53, Table 5-7.
[2]The animal's daily maintenance requirement is 110 kcal of DE/kg$^{0.75}$.
[3]Milk production requires 2.0 Mcal of DE/kg of milk.
[4]The feed required/day is based on a corn-soybean meal ration containing 3.34 Mcal of DE/kg.

Fig. 23-1. Lactating sows require a liberal allowance of a ration rich in energy, protein, minerals, and vitamins. (Courtesy, Hampshire Swine Registry, Peoria, Ill.)

TABLE 23-5
NUTRIENT REQUIREMENTS IN THE RATION OF SWINE ALLOWED FEED *AD LIBITUM* (90% DRY MATTER)[1]

Intake and Performance Levels	Lb / Kg	Swine Liveweight 2.2-11 / 1-5		11-22 / 5-10		22-44 / 10-20		44-110 / 20-50		110-242 / 50-110	
		(lb)	(g)	(lb)	(g)	(lb)	(g)	(lb)	(g)	(lb)	(g)
Expected weight gain per day		0.4	200	0.6	250	1.0	450	1.5	700	1.8	820
Expected feed intake per day		0.6	250	1.0	460	2.1	950	4.2	1,900	6.9	3,110
Expected efficiency	(gain/feed)	0.800		0.543		0.474		0.368		0.264	
Expected efficiency	(feed/gain)	1.25		1.84		2.11		2.71		3.79	
Digestible energy intake	(kcal per day)	850		1,560		3,230		6,460		10,570	
Metabolizable energy intake	(kcal per day)	805		1,490		3,090		6,200		10,185	
Energy concentration	(kcal ME per lb ration)	1,461		1,470		1,474		1,479		1,486	
Energy concentration	(kcal ME per kg ration)	3,220		3,240		3,250		3,260		3,275	
Protein	(%)	24		20		18		15		13	

Nutrient		Requirement (Percent or Amount/Lb *[Kg]* Ration)[2]									
		(lb)	(kg)	(lb)	(kg)	(lb)	(kg)	(lb)	(kg)	(lb)	(kg)
Indispensable amino acids:											
Arginine	(%)	0.60		0.50		0.40		0.25		0.10	
Histidine	(%)	0.36		0.31		0.25		0.22		0.18	
Isoleucine	(%)	0.76		0.65		0.53		0.46		0.38	
Leucine	(%)	1.00		0.85		0.70		0.60		0.50	
Lysine	(%)	1.40		1.15		0.95		0.75		0.60	
Methionine + cystine	(%)	0.68		0.58		0.48		0.41		0.34	
Phenylalanine + tyrosine	(%)	1.10		0.94		0.77		0.66		0.55	
Threonine	(%)	0.80		0.68		0.56		0.48		0.40	
Tryptophan	(%)	0.20		0.17		0.14		0.12		0.10	
Valine	(%)	0.80		0.68		0.56		0.48		0.40	
Linoleic acid	(%)	0.1		0.1		0.1		0.1		0.1	
Major or macrominerals:											
Calcium	(%)	0.90		0.80		0.70		0.60		0.50	
Chlorine	(%)	0.08		0.08		0.08		0.08		0.08	
Magnesium	(%)	0.04		0.04		0.04		0.04		0.04	
Phosphorus, total	(%)	0.70		0.65		0.60		0.50		0.40	
Phosphorus, available	(%)	0.55		0.40		0.32		0.23		0.15	
Potassium	(%)	0.30		0.28		0.26		0.23		0.17	
Sodium	(%)	0.10		0.10		0.10		0.10		0.10	
Trace or microminerals:											
Copper	(mg)	2.7	6.0	2.7	6.0	2.3	5.0	1.8	4.0	1.4	3.0
Iodine	(mg)	0.06	0.14	0.06	0.14	0.06	0.14	0.06	0.14	0.06	0.14
Iron	(mg)	45	100	45	100	36	80	27	60	18	40
Manganese	(mg)	1.8	4.0	1.8	4.0	1.4	3.0	0.9	2.0	0.9	2.0
Selenium	(mg)	0.14	0.30	0.14	0.30	0.11	0.25	0.07	0.15	0.05	0.10
Zinc	(mg)	45	100	45	100	36	80	27	60	23	50
Fat-soluble vitamins:											
Vitamin A	(IU)	2,200		2,200		1,750		1,300		1,300	
Vitamin D	(IU)	220		220		200		150		150	
Vitamin E	(IU)	16		16		11		11		11	
Vitamin K (menadione)	(mg)	0.2	0.5	0.2	0.5	0.2	0.5	0.2	0.5	0.2	0.5
Water-soluble vitamins:											
Biotin	(mg)	0.04	0.08	0.02	0.05	0.02	0.05	0.02	0.05	0.02	0.05
Choline	(g)	0.27	0.6	0.23	0.5	0.18	0.4	0.14	0.3	0.14	0.3
Folacin (Folic acid)	(mg)	0.1	0.3	0.1	0.3	0.1	0.3	0.1	0.3	0.1	0.3
Niacin (Nicotinic acid, Nicotinamide), available	(mg)	9.1	20.0	6.8	15.0	5.7	12.5	4.5	10.0	3.2	7.0
Pantothenic acid (Vitamin B-3)	(mg)	5.4	12.0	4.5	10.0	4.1	9.0	3.6	8.0	3.2	7.0
Riboflavin (Vitamin B-2)	(mg)	1.8	4.0	1.6	3.5	1.4	3.0	1.1	2.5	0.9	2.0
Thiamin (Vitamin B-1)	(mg)	0.7	1.5	0.5	1.0	0.5	1.0	0.5	1.0	0.5	1.0
Vitamin B-6 (Pyridoxine, Pyridoxal, Pyridoxamine)	(mg)	0.9	2.0	0.7	1.5	0.7	1.5	0.5	1.0	0.5	1.0
Vitamin B-12 (Cobalamins)	(mcg)	9.1	20.0	7.9	17.5	6.8	15.0	4.5	10.0	2.3	5.0

[1] Adapted by the authors from *Nutrient Requirements of Swine*, 9th rev. ed., National Research Council, National Academy Press, 1988, p. 50, Table 5-1.

[2] These requirements are based upon the following types of pigs and rations: 2.2- to 11-lb *(1- to 5-kg)* pigs, a ration that includes 25 to 75% milk products; 11- to 22-lb *(5- to 10-kg)* pigs, a corn-soybean meal ration that includes 5 to 25% milk products; 22- to 242-lb *(10- to 110-kg)* pigs, a corn-soybean meal ration. In the corn-soybean meal rations, the corn contains 8.5% protein; the soybean meal contains 44%.

acid, only the latter need be a dietary constituent.

A deficiency of essential fatty acids in the young, growing pig results in a dull, dry hair coat and a scaly, dandrufflike dermatitis. In later stages, a brownish, gummy exudate and necrotic areas appear about the ears and under the flanks. Also, retarded sexual maturity, an undeveloped digestive system, and an abnormally small gallbladder have been reported.

Fig. 23-2. Fat deficiency. Littermate pigs. *Top:* Pig received fat in the ration (5.0% ether extract). *Bottom:* Pig received little fat in the ration (0.06% ether extract). Note loss of hair and scaly dandrufflike dermatitis. (Courtesy, Purdue University, West Lafayette, Ind.)

(Also see Chapter 4, section on "Energy [Carbohydrates and Lipids].")

Protein

While carbohydrates and fats are the principal sources of energy, proteins supply the building materials from which body tissue and many body regulators, such as enzymes and hormones, are made. Each protein is made up of several nitrogen compounds called amino acids.

There are 22, or more, amino acids. Of these, 10 assume unusual importance for the growing pig and are called essential amino acids. *An essential amino acid is one which the body cannot manufacture (synthesize) in sufficient quantity to permit maximum growth and performance.* Therefore, it must be supplied in the ration. The essential and amino acids for swine are: arginine, histidine, isoleucine, leucine, lysine, methionine, phenylalanine, threonine, tryptophan, and valine.

Quality of protein is a term used to describe the amino acid balance of protein. A protein is said to be of good quality when it contains all the essential amino acids in proper proportions and amounts, and to be poor quality when it is deficient in either content or balance of essential amino acids. From this it is evident that the usefulness of a protein source depends upon its amino acid composition, because the real need of the pig is for amino acids and not for protein as such.

Fig. 23-3. Lysine deficiency. *Top:* Pig gained 25 lb in 28 days after lysine was added to the basal ration (2.0% DL-lysine). *Bottom:* A lysine-deficient pig that received the basal ration only. This pig lost 2.0 lb in 28 days. (Courtesy, Purdue University, West Lafayette, Ind.)

From a practical standpoint, the problem of building a balanced ration for swine is centered around correcting the deficiencies in the cereal grains. Athough corn, wheat, and barley may contain from 8 to 12% protein, their protein is seriously deficient in the essential amino acid, lysine; corn is also deficient in tryptophan. Meat and bone meal is deficient in tryptophan. Moreover, the digestive tract of the pig is not adapted to extensive synthesis of proteins by microorganisms like the paunch of ruminants. Also, since protein supplements are more expensive than grain, the tendency is to feed too little of them.

Symptoms of protein (amino acid) deficiency are reduced feed intake accompanied by increased feed wastage, stunted growth, poor hair and skin condition, and lowered reproduction.

Minerals

Of all common farm animals, the pig is most likely to suffer from mineral deficiencies. This is due to the following peculiarities of swine husbandry:

1. Hogs are fed principally upon cereal grains and their by-products, all of which are relatively low in mineral matter, particularly in calcium.
2. The skeleton of the pig supports greater weight in proportion to its size than that of any other farm animal.
3. Hogs are fed to grow at a maximum rate for an early market, before they are mature.
4. Hogs reproduce at a younger age than other classes of livestock.
5. The trend toward increased confinement rearing, without access to soil or forage, which would tend to balance the mineral deficiencies of the grains.

The functions of minerals are extremely diverse. They range from structural functions in some tissues to a wide variety of regulatory functions in other tissues.

Swine require at least 13 known inorganic elements, including calcium, chlorine, copper, iodine, iron, magnesium,

TABLE
SWINE MINERAL

Minerals Which May Be Deficient Under Normal Conditions	Conditions Usually Prevailing Where Deficiencies Are Reported	Functions of Mineral	Deficiency Symptoms/Toxicity
Major or Macrominerals: Salt (NaCl)	Salt deficiencies may exist when the protein supplement is all or chiefly of plant origin, although herbivorous animals require more salt than swine.	Sodium and chlorine are the principal extracellular cation and anion, respectively, in the body. Chlorine is the chief anion in gastric juice. Improves appetite, promotes growth, helps regulate body pH, and is essential for hydrochloric acid formation in the stomach.	**Deficiency symptoms**—Poor and depraved appetite, unthrifty condition, and failure to grow. **Toxicity**—Nervousness, weakness, staggering, epileptic seizures, paralysis, and death.
Calcium (Ca)	When the protein supplements are chiefly of plant origin and little forage is used. When swine are raised in confinement without vitamin D added to the ration. When feed intake is restricted during gestation. When there is a poor calcium-phosphorus ratio. Retention of calcium is affected by source of dietary protein (or phytic acid content) and the level of magnesium.	Bone and teeth formation; nerve function; muscle contraction; blood coagulation; cell permeability. Essential for milk production.	**Deficiency symptoms**—Loss of appetite and poor growth, lack of thrift, lameness and stiffness, weakened bone structure, and impaired reproduction. Severe cases may show reduced serum calcium and tetany. Rickets may develop in young pigs, or osteomalacia in older animals. Paralysis of the hind legs. **Toxicity**—An excess level of calcium tends to reduce the performance of pigs and increase the pig's zinc requirement.
Phosphorus (P)	Rations containing only plant ingredients; late gestation; lactation; high-calcium rations; swine in confinement without vitamin D added to the ration; poor calcium to phosphorus ratio. Retention of phosphorus is affected by source of dietary protein (or phytic acid content) and the level of magnesium.	Bone and teeth formation; a component of phospholipids which are important in lipid transport and metabolism and cell-membrane structure. In energy metabolism. A component of RNA and DNA, the vital cellular constituents required for protein synthesis. A constituent of several enzyme systems.	**Deficiency symptoms**—Loss of appetite and poor growth, lameness and stiffness, weakened bone structure, reduced inorganic blood phosphorus, depraved appetite, breeding difficulties, and rickets in young pigs, or osteomalacia in older animals. Paralysis of the hind legs, which is called posterior paralysis. **Toxicity**—An excess of phosphorus tends to reduce the performance of pigs, but is not toxic as such.
Magnesium (Mg)	Some research suggests that the magnesium in natural ingredients is only 50–60% available to the pig.	Essential for normal skeletal development, as a constituent of bone, cofactor in many enzyme systems, primarily in the glycolytic system.	**Deficiency symptoms**—Hyperirritability, muscular twitching, reluctance to stand, weak pasterns, loss of equilibrium, and tetany, followed by death. **Toxicity**—The toxicity level of magnesium is not known.
Potassium (K)		Major cation of intracellular fluid where it is involved in osmotic pressure and acid-base balance. Electrolyte balance and neuromuscular function. Required in enzyme reaction involving phosphorylation of creatine. Influences carbohydrate metabolism.	**Deficiency symptoms**—Loss of appetite, slow growth, poor hair and skin condition, decreased feed efficiency, inactivity, lack of coordination, and cardiac impairment. **Toxicity**—The toxic level of potassium is not well established. Pigs can tolerate up to 10 times the requirement if plenty of drinking water is provided.
Sulfur (S)		For synthesis of sulfur-containing compounds, such as glutathione, taurocholic acid, and chondroitin sulfate.	

manganese, phosphorus, potassium, selenium, sodium, sulfur, and zinc. Also, cobalt is required in the synthesis of vitamin B-12. Pigs may also require other trace elements, such as arsenic, boron, bromine, cadmium chromium, fluorine, lead, lithium, molybdenum, nickel, silicon, tin, and vanadium, which have been shown to have a physiological role in one or more species. These elements are required at such low levels, however, that their dietary essentiality for the pig has not been proven.

Several elements can be toxic to swine; among them, antimony, arsenic, cadmium, fluorine, lead, mercury, and selenium.

SWINE MINERAL CHART

Table 23-9, Swine Mineral Chart, gives in summary form the following pertinent information relative to each mineral listed: (1) conditions usually prevailing where deficiencies are reported, (2) functions, (3) deficiency symptoms/toxicity, (4) mineral requirements, (5) recommended allowances, and (6) practical sources.

23-9
CHART (See footnote at end of table.)

Mineral Requirements[1]		Recommended Allowances[1]	Practical Sources of the Mineral	Comments
Minerals/Animal/Day	Mineral Content of Ration			
*Na, variable according to class, age, and weight of swine (see Tables 23-1 and 23-2).	*Na, variable according to class, age, and weight of swine (see Table 23-5 and 23-6).	Salt variable according to class, age, and weight of swine (see Table 23-8).	Salt in loose form.	In iodine-deficient areas, stabilized iodized salt should be used. When pigs are salt starved, precaution should be taken to prevent overconsumption of salt.
*Variable according to class, age, and weight of swine (see Tables 23-1 and 23-2).	*Variable according to class, age, and weight of swine (see Table 23-5, 23-6, and 23-7).	Variable according to class, age, and weight of swine (see Table 23-8).	Ground limestone, gypsum, or oyster-shell flour. Where both Ca and P are needed, use monocalcium phosphate, dicalcium phosphate, tricalcium phosphate, defluorinated phosphate, or bone meal.	Because cereal grains (which largely form the ration of swine) are low in Ca, swine are more apt to suffer from Ca deficiencies than from any of the other minerals except salt. *Most favorable Ca:P ratio is between 1:1 and 1.5:1. Sow's milk contains a Ca:P ratio of 1.3:1.
*Variable according to class, age, and weight of swine (see Tables 23-1 and 23-2).	*Variable according to class, age, and weight of swine (see Tables 23-5, 23-6, and 23-7).	Variable according to class, age, and weight of swine (see Table 23-8).	Where both Ca and P are needed, use monocalcium phosphate, dicalcium phosphate, tricalcium phosphate, defluorinated phosphate, or bone meal.	*About 60-75% of the P in cereal grains and their by-products, and in oilseed meals, is organically bound in the form of phytate and poorly available to the pig. *Most favorable Ca:P ratio is between 1:1 and 1.5:1. Sow's milk contains a Ca:P ratio of 1.3:1. Excess levels of P reduce performance of pigs.
*Variable according to class, age, and weight of swine (see Tables 23-1 and 23-2).	*0.04% of as-fed ration.	Practical rations are adequate in magnesium.	Magnesium oxide, magnesium sulfate, or magnesium carbonate. Dolomitic limestone.	Milk contains adequate magnesium for suckling pigs.
*Variable according to class, age, and weight of swine (see Tables 23-1 and 23-2).	*Variable for young pigs (see Tables 23-5 and 23-6). *No estimates available for finishing and breeding swine.	Practical rations are adequate in potassium.	Corn contains 0.33% potassium, and other cereals contain 0.42-0.49% potassium.	Potassium is the third most abundant mineral in the body of the pig, exceeded only by calcium and phosphorus.
		The addition of inorganic sulfate to low-protein rations has not been beneficial.		

(Continued)

TABLE 23-9

Minerals Which May Be Deficient Under Normal Conditions	Conditions Usually Prevailing Where Deficiencies Are Reported	Functions of Mineral	Deficiency Symptoms/Toxicity
Trace or microminerals:			
Cobalt (Co)	If vitamin B-12 is limited.	An essential component of vitamin B-12.	**Deficiency symptoms**—No deficiency symptoms reported in swine. However, supplemental cobalt prevents lesions associated with zinc deficiency. **Toxicity**—A level of 400 ppm of cobalt is toxic to the young pig and may cause loss of appetite, stiff-leggedness, humped back, incoordination, muscle tremors, and anemia.
Copper (Cu)	Suckling pigs kept off soil.	Essential element in a number of enzyme systems and necessary for synthesizing hemoglobin and preventing nutritional anemia. Hemoglobin serves as a carrier of oxygen throughout the body. When fed at 100 to 250 ppm, copper stimulates growth in pigs.	**Deficiency symptoms**—Slow growth, poor hair and skin condition, lameness and stiffness, weakened bone structure, weak and crooked legs, anemia, and cardiac and vascular disorders. **Toxicity**—Depressed hemoglobin levels and jaundice.
Iodine (I)	Iodine-deficient areas/soils (in northwestern U.S. and in the Great Lakes region) when iodized salt is not fed. Where feeds come from iodine-deficient areas.	Needed by the thyroid gland for making thyroxin, an iodine-containing hormone which controls the rate of body metabolism or heat production. **Toxicity**—An 800-ppm iodine level in the ration depresses growth, hemoglobin level, and liver iron concentration in growing pigs. During the last 30 days of gestation and during lactation, 1,500 to 2,500 ppm of iodine was found harmful to sows.	**Deficiency symptoms**—Loss of appetite, slow growth, poor hair and skin condition, impaired breeding or gestation, offspring dead or weak at birth, pigs hairless at birth, and/or enlarged thyroid.
Iron (Fe)	Suckling pigs kept off soil.	*Iron is required as a component of hemoglobin in red blood cells. Iron also is found in muscle as myoglobin, in serum as transferrin, in the placenta as interoferrin, in milk as lactoferrin, and in the liver as ferritin and hemosiderin. Iron also plays an important role in the body as a constituent of a number of metabolic enzymes. indicates borderline anemia; a level of 7 g or less/100 ml indicates anemia. **Toxicity**—In 3- to 10-day-old pigs, the toxic oral dose of iron from ferrous sulfate is approximately 273 mcg/lb *(601 mcg/kg)* body weight.	**Deficiency symptoms**—Loss of appetite, slow growth, poor hair and skin condition, paleness of mucous membranes, high mortality in young pigs, susceptibility to disease, thumps (characterized by labored breathing), and anemia. The number of grams of hemoglobin per 100 ml of blood is a rapid, reliable indicator of the iron status of the pig. A hemoglobin level of 8 g/100 ml
Manganese (Mn)		Functions as a component of several enzymes involved in carbohydrate, lipid, and protein metabolism. A component in the organic matrix of bone.	**Deficiency symptoms**—Abnormal skeletal growth, increased fat deposition, irregular or absent estrus cycles, resorbed fetuses, small and weak pigs at birth, and reduced milk production. **Toxicity**—The toxic level of manganese is not clearly defined. But high levels result in depressed feed intake, reduced growth, and limb stiffness.
Selenium (Se)	When rations consist almost exclusively of ingredients grown on selenium-deficient soils.	Functions as a part of glutathione peroxidase, an enzyme which enables the tripeptide glutathione to perform its role as a biological antioxidant in the body. The mutual sparing effect of selenium and vitamin E stems from their shared antiperoxidant roles. But high levels of vitamin E do not completely eliminate the level for selenium.	**Deficiency symptoms**—Sudden death, impaired reproduction, reduced milk production, and impaired immune response. **Toxicity**—Loss of appetite, loss of hair, fatty infiltration of the liver, degenerative changes in the liver and kidney, edema, and occasional separation of the hoof and skin at the coronary band. Dietary arsenicals help to alleviate selenium toxicity.
Zinc (Zn)	High levels of calcium in relation to zinc levels impair zinc utilization and increase the requirements.	Zinc is a component of many metalloenzymes and the hormone insulin. So, it plays an important role in protein, carbohydrate, and lipid metabolism. Zinc deficiency results in gilts producing fewer and smaller pigs; in boars with retarded testicular development; and in young pigs with retarded thymic development. **Toxicity**—Growth depression, arthritis, hemorrhage in axillary spaces, gastritis, and enteritis. High dietary calcium reduces the severity of zinc toxicity.	**Deficiency symptoms**—Parakeratosis or swine dermatitis, pigs have a mangy appearance, reduced appetite, unthriftiness, poor growth rate, and diarrhea, and there may be vomiting. It affects swine of all ages.

[1]As used herein, the distinction between "mineral requirements" and "recommended allowances" is as follows: In mineral requirements, no margins of safety are included intentionally; whereas in recommended allowances, margins of safety are provided in order to compensate for variations in feed compositions, environment, and possible losses during storage or processing.

Where preceded by an asterisk, the mineral requirements, recommended allowances, and other facts presented herein were taken from *Nutrient Requirements of Swine*, 9th rev. ed., NRC-National Academy of Sciences, 1988.

(Continued)

Mineral Requirements[1]				
Minerals/ Animal/Day	**Mineral Content of Ration**	**Recommended Allowances**[1]	**Practical Sources of the Mineral**	**Comments**
No requirements for cobalt have been established.		Practical rations are adequate in cobalt.	Cobalt chloride, cobalt sulfate, cobalt oxide, or cobalt carbonate. Also, several good commercial minerals containing cobalt are on the market.	
*Variable according to class, age, and weight of swine (see Tables 23-1 and 23-2).	*Variable according to class, age, and weight of swine (see Tables 23-5 and 23-6).	Variable according to class, age, and weight of swine (see Table 23-8).	Copper sulfate, copper carbonate, and copper chloride are about equally effective. The copper in copper sulfide and copper oxide is poorly available to the pig.	Beyond the suckling period, natural feedstuffs usually contain enough copper. *When fed at a level of 100 to 250 ppm, copper will increase rate and efficiency of gains of pigs to breeding age.
*Variable according to class, age, and weight of swine (see Tables 23-1 and 23-2).	*0.06 mg/lb *(0.14 mg/kg)* as-fed ration.	Variable according to class, age, and weight of swine (see Table 23-8).	Stabilized iodized salt containing 0.007% iodine. Calcium iodate, potassium iodate, and pentacalcium orthoperiodate.	The majority of the iodine in the bodies of swine is present in the thyroid.
*Variable according to class, age, and weight of swine (see Tables 23-1 and 23-2). *Newborn pigs require 7 to 16 mg of absorbed iron daily for normal growth.	*Variable according to class, age, and weight of swine (see Tables 23-5 and 23-6). *The iron requirement of young pigs fed milk or purified liquid diets is 23 to 68 mg/lb *(50 to 150 mg/kg)* of milk solids. The iron requirement of pigs fed a dry, casein-based diet is about 50% higher/unit of dry matter than for those fed a similar diet in liquid form.	Variable according to class, age, and weight of swine (see Table 23-8).	A single intramuscular injection of 100 to 200 mg of iron, in the form of iron dextran, iron dextrin, or gleptoferron given in the first 3 days of life; or Oral administration of iron from iron chelates within the first few hours of life. Ferrous sulfate, ferric chloride, ferric citrate, ferric choline citrate, and ferric ammonium citrate are effective in preventing iron deficiency anemia. The iron in ferric oxide is largely unavailable.	*Pigs are born with about 50 mg of iron. *Iron has a detoxifying effect when added to gossypol-containing diets. Add iron from soluble source to free gossypol at a weight ratio of 1:1. *Milk is deficient in iron (sow's milk contains an average of 1 mg of iron/liter). Pigs should be encouraged to eat grain ration as soon as old enough. *Natural feed ingredients usually supply enough iron to meet postweaning requirements.
*Variable according to class, age, and weight of swine (see Tables 23-1 and 23-2).	*Variable according to class, age, and weight of swine (see Tables 23-5 and 23-6).	Variable according to class, age, and weight of swine (see Table 23-8).	Manganous oxide.	Manganese is usually present in adequate amounts in most swine rations, but it may not be adequate for the optimum reproductive performance of sows.
*Variable according to class, age, and weight of swine (see Tables 23-1 and 23-2).	*The dietary requirement for selenium is between 0.1 and 0.3 ppm (see Tables 23-5 and 23-6). *In 1987, the FDA approved up to 0.3 ppm selenium in the ration of all pigs.	Variable according to class, age, and weight of swine (see Table 23-8).	Sodium selenite or sodium selenate.	Environmental stress may increase the incidence and degree of selenium deficiency. *Caution:* Toxic level of selenium is in range of 2.27–3.63 mg/lb *(5–8 mg/kg)* selenium in the feed.
*Variable according to class, age, and weight of swine (see Tables 23-1 and 23-2).	*Variable according to class, age, and weight of swine (see Tables 23-5 and 23-6). The zinc requirement is increased when excessive levels of calcium are fed.	Variable according to class, age, and weight of swine (see Table 23-8).	Zinc carbonate or zinc sulfate.	It has been shown that parakeratosis is caused by zinc and calcium forming an unavailable complex.

NOTE: *Mineral recommendations for all classes and ages of swine, especially those fed unmixed rations or on pastures are—*

1. *Where animals are on liberal grain feeding*—Provide free access to a 2-compartment mineral box, with (a) trace mineralized salt in one side, and (b) in the other side, a mixture of ⅓ trace mineralized salt (salt included for purposes of palatability), ⅓ defluorinated phosphate or steamed bone meal, and ⅓ ground limestone or oystershell flour.

2. *Where animals are primarily on roughage (pasture, hay, and/or silage)*—Provide free access to a 2-compartment mineral box, with (a) trace mineralized salt in one side (salt included for purposes of palatability), and (b) in the other side, a mixture of ⅓ trace mineralized salt and ⅔ defluorinated phosphate or steamed bone meal.

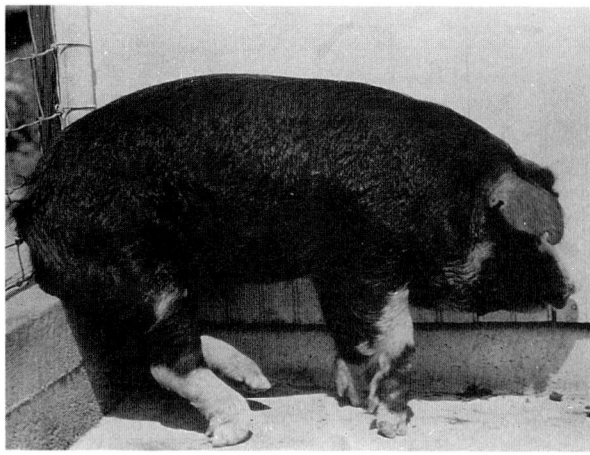

Fig. 23-4. Calcium deficiency. Note abnormal bone development and rachitic condition in advanced stage of deficiency. Lack of calcium retards normal skeletal development, but it does not usually depress total gain. (Courtesy, USDA)

Fig. 23-5. Phosphorus deficiency. Left: Typical phosphorus-deficient pig in advanced stage of deficiency. Leg bones are weak and crooked. Right: This pig received the same ration as the one on the left, except that the ration was adequate in available phosphorus. (Courtesy, Purdue University, West Lafayette, Ind.)

Fig. 23-6. Hairlessness in pigs caused by a deficiency of iodine. In iodine-deficient areas, farm animals should receive iodized salt throughout the year. (Courtesy, Department of Veterinary Pathology and Hygiene, College of Veterinary Medicine, University of Illinois, Urbana)

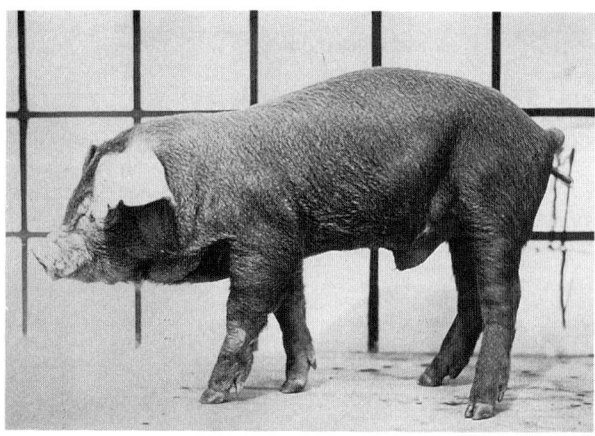

Fig. 23-7. Suckling pig with nutritional anemia, caused by a lack of iron, characterized by swollen condition about the head and paleness of the mucous membranes. (Courtesy, College of Veterinary Medicine, University of Illinois, Urbana)

Fig. 23-8. Zinc deficiency. Left: Pig received 17 ppm of zinc and gained only 3 lb in 74 days. Note severe dermatosis ("mangy look"), or parakeratosis. Right: Pig received the same ration as the pig on the left, except that the ration contained 67 ppm of zinc. This pig gained 111 lb in 74 days. (Courtesy, Purdue University, West Lafayette, Ind.)

FEEDS AS A SOURCE OF MINERALS

The most satisfactory source of minerals for hogs is in the feed consumed. Thus, it is important to know whether the minerals in the ration are of the right kind and sufficient in amount. Certain general characteristics of feeds in regard to calcium and phosphorus (the two predominating mineral elements of the body) are worth noting:

1. The cereal grains and their by-products and protein supplements of plant origin are low in calcium but fairly good in phosphorus. However, as mentioned earlier, the phosphorus in plants is not fully utilizable by swine.

Feeding Swine

2. The protein supplements of animal origin (skim milk, buttermilk, tankage, meat scraps, fish meal), legume forage (pasturage and hay), and rape, are all rich in calcium.

3. Most protein-rich supplements are high in phosphorus.

METHOD OF FEEDING MINERAL SUPPLEMENTS

Generally, supplementary minerals are incorporated in rations when needed (see Table 23-9). Additionally, hogs may be given free access to a suitable mineral mix. This is cheap insurance against possible needs beyond what's provided in the ration. It's a way of hedging against (1) individual pig differences in mineral requirements, and (2) feeds varying in both mineral content and availability.

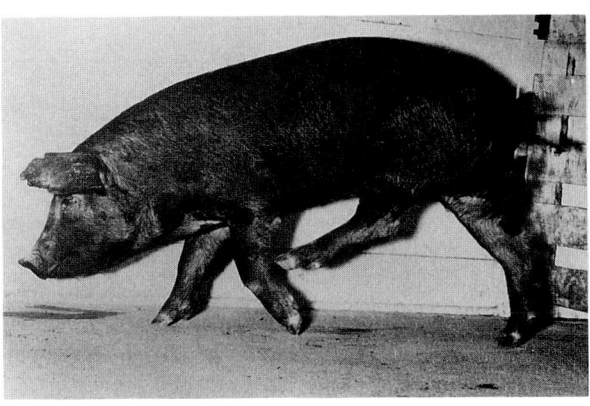

Fig. 23-10. Pig showing pantothenic acid-deficiency symptoms. Note high, goose-stepping gait. (Courtesy, University of California, Davis)

Vitamins

Vitamins are complex organic compounds needed in minute amounts, which are essential for health and normal body functions. Like amino acids, each vitamin has a specific function to perform. Vitamins are classified into two groups—fat-soluble and water-soluble. The body can store reserves of the fat-soluble vitamins for a considerable period of time. But stores of the water-soluble vitamins are depleted rapidly.

Because of the greater prevalence of confinement feeding, swine are more likely to suffer from vitamin deficiencies than any other class of four-footed animals.

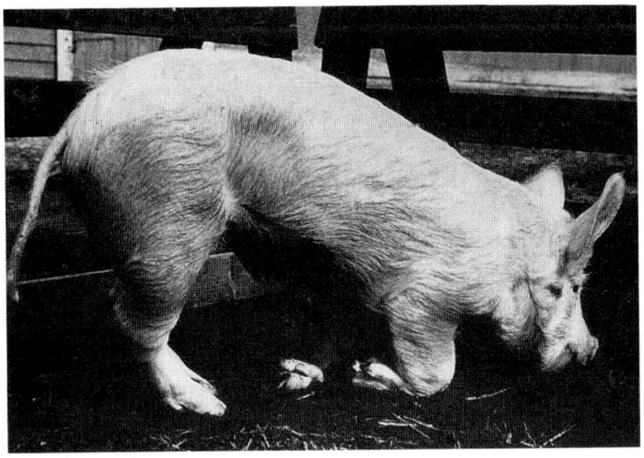

Fig. 23-11. Thiamin deficiency in littermate pigs. *Right:* Pig received no thiamin. *Left:* Pig received the equivalent of 2 mg thiamin/100 lb liveweight. Otherwise, their diets were the same. (Courtesy, USDA)

SWINE VITAMIN CHART

Table 23-10, Swine Vitamin Chart (pages 520 to 523), gives in summary form the following pertinent information relative to each vitamin listed: (1) conditions usually prevailing where deficiencies are reported, (2) functions, (3) deficiency symptoms/toxicity, (4) vitamin requirements, (5) recommended allowances, and (6) practical sources.

UNIDENTIFIED FACTORS

Some unidentified factor or factors may, under certain circumstances, be involved in securing optimum results during the critical periods (early growth and gestation-lactation). Sources of the unknown factor or factors are distillers' dried solubles, fish solubles, dried whey, grass juice concentrate, soil, alfalfa meal, brewers' dried yeast, liver, and pasture.

Fig. 23-9. Rickets (advanced case) caused by a deficiency of vitamin D. The pig was fed indoors, without exposure to sunlight and without adequate vitamin D. Because of leg abnormalities, it was unable to walk. Later, the pig responded to vitamin D. (Courtesy, University of Saskatchewan, Saskatoon, Canada)

TABLE
SWINE VITAMIN

Vitamins Which May Be Deficient Under Normal Conditions	Conditions Usually Prevailing Where Deficiencies Are Reported	Functions of Vitamin	Deficiency Symptoms/Toxicity
Fat-soluble vitamins:			
A	Absence of green forages, either pasture or green hay—especially under drylot conditions. Where the ration consists chiefly of white corn, milo, barley, wheat, oats, or rye; or by-products of these grains; or yellow corn that has been stored more than one year.	Essential for normal maintenance and functioning of the epithelial tissues, particularly of the eye and the respiratory, digestive, reproductive, nerve, and urinary systems.	**Deficiency symptoms**—Night and day blindness, very irritable, poor appetite and slow growth, lameness, incoordination of movement, loss of control of the hind legs, and weakness of the back. Low resistance to respiratory infections. Sows may fail to come in heat, may resorb their fetuses, and may have young born dead with various deformities and defects. **Toxicity**—A roughened hair coat, scaly skin, hyperirritability and sensitivity to touch, bleeding from the cracks which appear in the skin above the hooves, blood in the urine and feces, loss of control of the legs accompanied by inability to rise, periodic tremors, and death.
D	Limited sunlight and/or limited quantities of sun-cured hay in drylot rations.	Aids in assimilation and utilization of calcium and phosphorus, and necessary in the normal bone development of animals—including the bones of the fetus.	**Deficiency symptoms**—Rickets in young pigs, or osteomalacia in mature hogs. Both conditions result in large joints and weak bones. In severe vitamin D deficiency, pigs may exhibit signs of calcium and magnesium deficiency, including tetany. **Toxicity**—Reduced feed intake and growth rate; and death. Vitamin D_3 is more toxic than D_2 in swine.
E	Rations containing excessive amounts of highly unsaturated fatty acids or oxidized fats. Swine feeds low in selenium, especially where swine are raised in confinement without access to forages.	Antioxidant. Muscle structure. Reproduction. High levels of vitamin E in the diet may increase the immune response.	**Deficiency symptoms**—Loss of appetite and slow growth. Increased embryonic mortality and muscular incoordination in suckling pigs from sows fed vitamin E-deficient rations during gestation and lactation. A wide variety of pathological conditions. **Toxicity**—Vitamin E toxicity in swine has not been demonstrated.
K	Moldy feed. High antibiotic levels, which may make for inadequate intestinal synthesis of vitamin K.	Essential for prothrombin formation and blood clotting.	**Deficiency symptoms**—Bleeding condition in young pigs, which responds to injection or oral administration of vitamin K. Slow growth and hyperirritability. **Toxicity**—Concentrations of 50 mg of menadione pyrimidinol bisulfite (MPB)/lb *(110 mg of MPB/kg)* of ration were not toxic to weanling pigs.
Water-soluble vitamins:			
Biotin	When pigs are fed dried, raw egg white or given sulfa drugs. Marginal deficiency may exist when hogs are fed cereal grain rations, housed in individual stalls or on slotted floors (which lessens coprophagy), and/or have no access to green forage.	Biotin is important metabolically as a cofactor for several enzymes.	**Deficiency symptoms**—Excessive hair loss, skin ulcerations and dermatitis, exudate around the eyes, cracking of the hooves, and cracking and bleeding of the footpads.
Choline	Baby pigs fed a synthetic milk ration containing not more than 0.8% methionine.	Involved in nerve impulses; a component of phospholipids; donor of methyl groups; and involved in the mobilization and oxidation of fatty acids in the liver.	**Deficiency symptoms**—Unthriftiness, lack of coordination, spraddled hind legs at birth, fatty infiltration of the liver, poor reproduction, poor lactation, and decreased survival of the young. **Toxicity**—No signs of choline toxicity have been reported in swine.
Folacin (Folic Acid)		Metabolic reactions involving incorporation of single carbon units into larger molecules. Folacin is involved in the conversion of serine to glycine and homocystine to methionine.	**Deficiency symptoms**—Poor growth, fading hair color, and anemia.
Niacin (Nicotinic Acid, Nicotinamide)		Niacin is a component of the coenzymes which are essential for the metabolism of carbohydrates, proteins, and lipids.	**Deficiency symptoms**—Loss of appetite and decreased gain, followed by diarrhea, occasional vomiting, dermatitis, and loss of hair.

Feeding Swine

23-10
CHART (See footnote at end of table.)

Vitamin Requirements[1]		Recommended Allowances[1]	Practical Sources of the Vitamin	Comments
Vitamins/ Animal/Day	Vitamin Content of Ration			
*Variable according to class, age, and weight of swine (see Tables 23-1 and 23-2).	*Variable according to class, age, and weight of swine (see Tables 23-5 and 23-6).	Variable according to class, age, and weight of swine (see Table 23-8).	Either vitamin A or various provitamins.	**B**ased upon liver storage, the biopotency of 1 mg of carotene in corn fed to weanling pigs is 261 IU of vitamin A. **M**eals from artificially dehydrated forages are much higher in carotene than sun-cured products. **T**aken together, liver storage, levels of plasma vitamin A, and pressure of cerebrospinal fluid give reliable estimates of the vitamin A status of the pig.
*Variable according to class, age, and weight of swine (see Tables 23-1 and 23-2).	*Variable according to class, age, and weight of swine (see Tables 23-5 and 23-6).	Variable according to class, age, and weight of swine (see Table 23-8). **I**rradiated yeast. **E**xposure to sunlight. **S**un-cured hay (10% alfalfa in the total ration will normally supply sufficient vitamin D).	Vitamin D_2 (ergocalciferol) and vitamin D_3 (cholecalciferol) are similar in biological activity for swine. The action of ultraviolet light on the ergosterol that is present in plants forms ergocalciferol; and the photochemical conversion of 7-dehydrocholesterol in the skin of animals forms cholecalciferol.	**G**rains, grain by-products, and high-protein feedstuffs are practically devoid of vitamin D; therefore, unless swine are exposed daily to the ultraviolet rays of the sun, the ration should be fortified with vitamin D. **T**he vitamin D requirement is less when a proper balance of calcium and phosphorus exists in the ration. **O**ne IU vitamin D is defined as the biological activity of 0.025 mcg of cholecalciferol.
*Variable according to class, age, and weight of swine (see Tables 23-1 and 23-2). **M**any dietary factors affect the vitamin E requirement, including the selenium level, unsaturated fatty acids, sulfur amino acids, retinol, copper, iron, and synthetic antioxidants.	*Variable according to class, age, and weight of swine (see Tables 23-5 and 23-6).	Variable according to class, age, and weight of swine (see Table 23-8). **P**redicting the amount of vitamin E activity in feed is difficult.	Alpha tocopherol.	**T**he 8 naturally occurring tocopherols differ in their biological activities, with *d*-alpha-tocopherol being the most active. One IU of vitamin E is the equivalent in biopotency of 1 mg *dl*-alpha-tocopherol acetate.
*Variable according to class, age, and weight of swine (see Tables 23-1 and 23-2).	*Supplement the as-fed ration with menadione at level of 0.2 mg/lb (0.5 mg/kg).	**U**nder practical conditions, the vitamin K requirement is met by vitamin K in feedstuffs and by intestinal synthesis.	The following water-soluble forms of menadione are commonly used to supplement swine diets: menadione sodium bisulfite complex (MSB), menadione sodium bisulfate complex (MSBC), and menadione pyrimidinol bisulfite (MPB).	**V**itamin K exists in 3 forms: phylloquinone (K_1), menaquinone (K_2), and menadione (K_3).
*Variable according to class, age, and weight of swine (see Tables 23-1 and 23-2).	*Variable according to class, age, and weight of swine (see Tables 23-5 and 23-6).	Variable according to class, age, and weight of swine (see Table 23-8).	Biotin.	**T**he protein *avidin* in raw egg white makes biotin unavailable to pigs. Heat treatment inactivates avidin and makes egg white safe for feeding to pigs.
*Variable according to class, age, and weight of swine (see Tables 23-1 and 23-2).	*Variable according to class, age, and weight of swine (see Tables 23-5 and 23-6).	Practical rations are adequate in choline.	Choline chlorides or choline dihydrogen.	**C**holine does not qualify as a true vitamin, because it is required at far greater levels than true vitamins and is not known to participate in any enzyme system. **S**tudies have shown that more live pigs are born and weaned when sows receive supplemental choline throughout gestation.
*Variable according to class, age, and weight of swine (see Tables 23-1 and 23-2).	*Variable according to class, age, and weight of swine (see Tables 23-5 and 23-6).	Practical rations plus intestinal synthesis is adequate.	Synthetic folacin.	**F**olacin includes a group of compounds with folic acid activity.
*Variable according to class, age, and weight of swine (see Tables 23-1 and 23-2).	*Variable according to class, age, and weight of swine (see Tables 23-5 and 23-6).	Variable according to class, age, and weight of swine (see Table 23-8).	Nicotinamide. Nicotinic acid.	**N**iacin occurs in corn, wheat, and milo in bound form; hence, it may be unavailable to the pig. Also, the dietary tryptophan level affects the niacin requirement because of the conversion of tryptophan to niacin.

(Continued)

TABLE 23-10

Vitamins Which May Be Deficient Under Normal Conditions	Conditions Usually Prevailing Where Deficiencies Are Reported	Functions of Vitamin	Deficiency Symptoms/Toxicity
Water-soluble vitamins (Continued):			
Pantothenic Acid (Vitamin B-3)	Long periods of inadequate pantothenic acid intake.	As a component of coenzyme A, pantothenic acid is important in the catabolism and synthesis of 2–carbon units evolved during carbohydrate and fat metabolism.	**Deficiency symptoms**—A goose-stepping gait, loss of appetite, poor growth, diarrhea, loss of hair, reduced fertility, and breeding failure.
Riboflavin (Vitamin B-2)		A component of 2 coenzymes. It is important in the metabolism of proteins, fats, and carbohydrates.	**Deficiency symptoms**—Loss of appetite, poor growth, rough hair coat, diarrhea, cataracts, vomiting, reproductive failure in the sow, pigs dead or weak at birth, and crooked legs and incoordination.
Thiamin (Vitamin B-1)	Thiamin is heat labile. So, excess heat can reduce the thiamin content of the ration ingredients.	As a coenzyme in energy and protein metabolism. Promotes appetite and growth, required for normal carbohydrate metabolism, and aids reproduction.	**Deficiency symptoms**—Loss of appetite and poor growth, diarrhea, dead or weak offspring, slow pulse, low body temperature, and flabby heart.
Vitamin B-6 (Pyridoxine, Pyridoxal, Pyridoxamine)		An important cofactor for many amino acid enzyme systems. Vitamin B-6 also plays a crucial role in central nervous system function.	**Deficiency symptoms**—Loss of appetite and poor growth, unsteady gait, anemia, exudate around the eyes, and epilepticlike fits (convulsions).
Vitamin B-12 (Cobalamins)	Pigs fed ingredients of plant origin and housed on slotted floors.	Vitamin B-12 contains the trace element cobalt in its molecule. As a coenzyme, B-12 is involved in the synthesis of methyl groups derived from formate, glycine, or serene, and their transfer to homocystine to reform methionine. It is also important in the methylation of uracil to form thymine, which is converted to thymidine and used for the synthesis of DNA.	**Deficiency symptoms**—Loss of appetite, reduced weight gain, rough skin and hair coat, irritability, hypersensitivity, hind leg incoordination, and anemia.
Vitamin C (Ascorbic Acid, Dehydroascorbic Acid)		Vitamin C is a water-soluble antioxidant that is involved in the formation and maintenance of collagen, absorption and movement of iron, metabolism of fats and lipids, cholesterol control, sound teeth and bones, strong capillary walls and healthy blood vessels, metabolism of folic acid, and as a general antioxidant.	**Deficiency symptoms**—No specific symptoms noted when there is a deficiency.

[1]As used herein, the distinction between *vitamin requirements* and *recommended allowances* is as follows: In vitamin requirements, no margins of safety are included intentionally; whereas in recommended allowances, margins of safety are provided in order to compensate for variations in feed composition, environment, and possible losses during storage or processing.

Water

Water is so common that it is seldom thought of as a nutrient. However, it is the largest single part of nearly all living things. The body of a baby pig is about three-fourths water.

In general, swine will consume ¼ to ⅓ gal of water for every pound of dry feed. The higher the temperature, the greater the water consumption. It is preferable that swine have access to automatic waterers, with cool, clean water available at all times. Otherwise, they should be hand watered at least twice daily. During winter, the drinking water should not be permitted to fall below 40°F. The estimated water needs of various classes of swine are:

Class of Swine	Water Consumption (gal/head/day)
Gestating sows	2–3
Lactating sows	4–5
Weaned pigs (15–40 lb)	0.5–1
Growing pigs (40–110 lb)	1
Finishing pigs (110–240 lb)	1.5–2

Fig. 23-12. Pigs drinking from nipple waterers. (Courtesy, National Pork Producers Council, Des Moines, Iowa)

Additives

Certain feed additives have become standard ingredients of swine rations, especially for pigs from birth to market weight. They are not nutrients as such; hence, they should not be considered as dietary essentials. Although many

(Continued)

Vitamin Requirements[1]		Recommended Allowances[1]	Practical Sources of the Vitamin	Comments
Vitamins/ Animal/Day	Vitamin Content of Ration			
*Variable according to class, age, and weight of swine (see Tables 23–1 and 23–2).	*Variable according to class, age, and weight of swine (see Tables 23–5 and 23–6).	Variable according to class, age, and weight of swine (see Table 23–8).	**C**alcium pantothenate (only the D isomer has vitamin activity). **D**ried milk products, condensed fish solubles, and alfalfa meal.	**W**idely distributed and occurs in practically all feedstuffs. However, the quantity present may not always be sufficient to meet the needs of the pig.
*Variable according to class, age, and weight of swine (see Tables 23–1 and 23–2).	*Variable according to class, age, and weight of swine (see Tables 23–5 and 23–6).	Variable according to class, age, and weight of swine (see Table 23–8).	**S**ynthetic riboflavin. **Y**east. **M**ilk and milk products. **M**eat scraps and fish meal.	**R**iboflavin is apt to be lacking in swine rations that do not contain animal product sources.
*Variable according to class, age, and weight of swine (see Tables 23–1 and 23–2).	*Variable according to class, age, and weight of swine (see Tables 23–5 and 23–6).	Practical rations are adequate in thiamin.	**T**hiamin hydrochloride. **T**hiamin mononitrate. **R**ice polish. **W**heat germ meal. **Y**east. **T**he oilseed meals. **D**istillers' solubles.	
*Variable according to class, age, and weight of swine (see Tables 23–1 and 23–2).	*Variable according to class, age, and weight of swine (see Tables 23–5 and 23–6).	Practical rations are adequate in vitamin B-6.	**P**yridoxine hydrochloride. **R**ich supplemental sources include rice polish, wheat germ, and yeast.	The vitamin B-6 content of normal feed is usually sufficient.
*Variable according to class, age, and weight of swine (see Table 23–1 and 23–2).	*Variable according to class, age, and weight of swine (see Tables 23–5 and 23–6).	Variable according to class, age, and weight of swine (see Table 23–8).	**S**ynthetic B-12, which is produced commercially by microbial fermentation. **P**rotein supplements of animal origin. **F**ermentation products.	**V**itamin B-12 is apt to be lacking in swine rations. **S**ynthesis of vitamin B-12 by intestinal flora may supplement dietary sources.
		hence, the synthesis of B-12 in the intestines is dependent on the presence of cobalt in the feed. This may be the major, if not the only, function of cobalt as an essential nutrient.		**B**-12 contains the trace element cobalt;
		Practical rations plus intestinal synthesis is adequate.	**V**itamin C (ascorbic acid).	**N**ormally, pigs are able to synthesize vitamin C in amounts sufficient to meet their requirements. However, there is limited evidence that dietary ascorbic acid is beneficial under some conditions.

Where preceded by an asterisk, the vitamin requirements, recommended allowances, and other facts presented herein were taken from *Nutrient Requirements of Swine*, 9th rev. ed., NRC-National Academy of Sciences, 1988.

different additives are used in swine rations, antibiotics and sulfas are most common.

• **Antibiotics**—Antibiotics are widely used as feed additives to stimulate growth, improve feed efficiency, secure uniformity of performance, and control infections. The response secured from their use depends on (1) the age of the pig, (2) the sanitary conditions, (3) the level fed, (4) the health and environment of the animal, (5) the type of ration, and (6) the season of the year. When fed to young, unthrifty pigs, antibiotics have increased growth rate by over 200%. For growing-finishing hogs under good sanitary conditions, antibiotics generally result in about 10% faster gains on 5% less feed. Pigs up to 100 lb weight give the greatest response to antibiotic feeding. Experimental results have been inconsistent relative to the value of antibiotics in brood sow rations, but it appears that breeding herds with a high disease level may show a favorable response.

• **Other antimicrobial compounds**—In addition to antibiotics, certain other antimicrobial compounds can be used as feed additives, either (1) at low levels for promotion of growth and improvement in feed efficiency, or (2) for treatment and prevention of disease. Among these are nitrofurans, sulfonamides, copper, and arsenicals. Such compounds, alone or in combination with other antimicrobial compounds, should be used only at approved levels and for the specific purpose for which they are authorized.

• **Sulfonamides (sulfas)**—*Sulfonamides are organic compounds with bactericidal and growth promotant properties similar to those of the antibiotics.* But, unlike the antibiotics, they are produced chemically rather than microbiologically. Also, the therapeutic use of sulfonamides preceded that of the antibiotics; they have been widely used in human and veterinary medicine since the mid-1930s.

As a result of finding sulfa residues in pork, followed by an experiment showing that sulfamethazine produced thyroid tumors in mice, in 1987 the USDA and FDA launched a campaign to alleviate sulfa residues in pork.

Animal products are deemed to be in violation of sulfa residue regulations if levels of 0.1 ppm, or more, sulfa are found in the muscle, liver, kidney, eggs, or milk. Research at the University of Kentucky showed that as little as 1 g of sulfa per ton of complete feed can cause 100% violative liver residues and 63% kidney contaminations. This means that as little as ¼ tsp of sulfa per ton of complete feed can result in pork carcasses that are in violation. The sulfonamides are also capable of premise contamination.

Pertinent facts about sulfas, their uses and resulting concerns, follow:

1. **Prevalence of use.** It is estimated that 60% of all the feeds that pigs receive in starter and grower-finishing rations contain a sulfa, and that 80% of all pigs receive a sulfa sometime during their lives.

2. **Approved sulfas for swine.** Of the more than 5,000 sulfa compounds that have been synthesized, only two—sulfamethazine and sulfathiazole—are approved for use in swine feeds. Two additional sulfa drugs, sulfamerazine and sulfapyridine, are approved, along with sulfamethazine and sulfathiazole, for use as water medicants for swine.

When used properly and withdrawn at the specified time before marketing, sulfa drugs have been shown to be safe.

- **Porcine somatotropin (PST) for swine**—Porcine somatotropin (PST) is the scientific name for the growth hormone in swine. It is normally produced by the anterior pituitary at the base of the brain. It stimulates protein synthesis and growth in most tissues of the body, and it causes breakdown of fat deposits in adipose tissue.

The major remaining barrier to the use of PST is a method of administering. Presently, daily injections are required to capture the benefits of growth and carcass changes. PST is a naturally occurring protein that is broken down immediately by digestion; so, an orally active feed ingredient won't work.

Experiments have consistently shown that PST can make the following great leaps forward in the swine industry: 15–20% higher average daily gain, 20–30% improvement in feed conversion efficiency, 10–15% improvement in muscle mass, and 25–30% less backfat.

Before PST can be used, it must have FDA approval. Then, its acceptance and use will be determined by a practical method of administering the product, and producer and consumer acceptance.

(Also see Chapter 13, section on "Porcine Somatotropin [PST] for Swine.")

Table 23-11 is a partial list of approved additives for use in swine rations.

TABLE 23-11
PARTIAL LIST OF FEED ADDITIVES IN SWINE RATIONS[1,2]

Additive Chemical Name	Trade Name	Withdraw Before Slaughter	A	B	C	D	E	F	G	H	I	J	K	L	M	N
Apramycin	Aparlan	28 days						x								
Arsanilic acid or sodium arsanilate	Pro-Gen	5 days	x			x				x	x					
Arsanilic acid/bacitracin		5 days	x						x							
Arsanilic acid/chlortetracycline		5 days	x			x			x							
Arsanilic acid/furazolidone/oxytetracycline	Furox-O-A 390	5 days	x		x				x			x				
Arsanilic acid/oxytetracycline	Furox/OXTC	5 days	x			x			x							
Arsanilic acid/penicillin		5 days	x			x	x		x	x						
Arsanilic acid/streptomycin/penicillin		5 days	x			x			x							
Bacitracin[4]	BMD; Baciferm; Albac	None	x		x				x							
Bambermycin	Flavomycin	None	x													
Carbadox	Mecadox	10 weeks	x					x	x							
Chlortetracycline	Aureomycin; CLTC; PhiChlor	None	x		x	x	x						x		x	x
Chlortetracycline/sulfamethazine/penicillin	Aureo SP 250; PhiChlor 250	15 days	x		x	x							x			x
Chlortetracycline/sulfathiazole/penicillin	CSP 250	7 days	x			x							x			x
Furazolidone	Furox; nf-180	5 days	x	x		x	x		x	x						
Furazolidone/oxytetracycline		5 days	x													
Lincomycin	Lincomix	6 days								x	x	x				
Neomycin base[5]	Neomix; Neomycin sulfate	Varies[5]				x					x					
Nitrofurazone	nfz; Amifur	5 days							x							
Oxytetracycline[6]	Terramycin; OXTC	Varies[6]	x			x	x		x		x		x		x	
Oxytetracycline/neomycin base	Neo-terramycin; Neo/OXTC	10 days		x	x	x			x		x					
Penicillin	Penicillin P-100	None	x													
Penicillin/streptomycin		None	x		x	x										
Roxarsone	3-Nitro	5 days	x									x				
Roxarsone/bacitracin		5 days	x													
Roxarsone/chlortetracycline		5 days	x													
Tiamulin	Denagard	2 days	x							x						
Tylosin	Tylan	None	x							x	x	x	x			
Tylosin/sulfamethazine	Tylan/sulfa	15 days								x			x	x	x	
Virginiamycin	Stafac	None	x							x	x					

[1]Adapted by the authors from *Life Cycle Swine Nutrition*, Iowa State University, Ames, PM-489, June, 1988.

[2]This is only a partial list of the feed additives and combinations available. This list does not contain the use level approved for each additive. Users are advised to read the product label and adhere to use recommendations of the additive manufacturer. The regulations governing the use of these additives are subject to change.

[3]The manufacturer claims, use levels and limitations are those for which FDA clearance was obtained. It is essential that the rules and regulations governing the use of feed additives be followed. When used properly, feed additives can greatly benefit the swine producer. The following claims are indicated for the appropriate additive:
  a. Promote growth and improve feed efficiency.
  b. Prevention of bacterial scours of baby pigs when fed in sow's ration.
  c. Treatment of bacterial scours of baby pigs when fed in sow's ration.
  d. Prevention of bacterial enteritis (scours).
  e. Treatment of bacterial enteritis (scours).
  f. Control of bacterial enteritis (scours).
  g. Prevention of swine dysentery (bloody scours).
  h. Control of swine dysentery (bloody scours).
  i. Treatment of swine dysentery (bloody scours).
  j. Control of swine pneumonias caused by bacterial pathogens (*P. multocida* and/or *C. pyogenes*).
  k. Maintenance of weight gain in presence of atrophic rhinitis.
  l. Lower incidence and severity of Bordetella bronchiseptica rhinitis.
  m. Aid in treatment and reducing spreading of Leptospirosis.
  n. Reduce cervical abcesses.

[4]Bacitracin is available in several forms including zinc and methylene disalicylate (MD) derivatives.

[5]Neomycin levels are expressed as neomycin base, which is equivalent to 70% of the neomycin sulfate level (*i.e.* 4.9 oz *(140 g)* neomycin base is equal to 7.1 oz *(200 g)* of neomycin sulfate). Withdraw from feed 20 days before slaughter when neomycin base level is 4.9 oz *(140 g)*/ton and 5 days before slaughter when neomycin base is below 4.9 oz *(140 g)*/ton.

[6]At 17.6 oz *(500 g)*/ton level, withdraw 5 days before slaughter.

FEEDS FOR SWINE

Throughout the world, swine are raised on a great variety of feeds, including numerous by-products. Except when on pasture or when ground dry forages are incorporated in the ration, they eat relatively little roughage; only 4.3% of the total feed consumed by swine in the United States is derived from roughages.

Although corn is the chief concentrate fed to swine, the agriculture of the 50 states is very diverse, and the diet of the pig is readily adapted to feeds produced locally. A similar adaptation in feeding practices is found in other countries. Thus, in most sections of the world, swine are fed predominantly on homegrown feeds. Ireland depends largely upon potatoes and dairy by-products in combination with barley; the swine industry of Denmark has been built up to augment the dairy industry, and milk and whey supplementing homegrown and imported cereals (mostly barley); in Germany, the pig is fed on such crops as potatoes, sugar beets, and green forage; and in China, the pig is primarily a scavenger, competing very little for grains suitable for human consumption.

Concentrates

Because of their simple monogastric stomach, swine consume more concentrates and less roughages than any other class of large farm animals. This characteristic gives pigs limited opportunity to consume large quantities of calcium and of vitamin-rich and better quality protein roughages, with the result that they suffer from more nutritional deficiencies than any other species except poultry.

Although most concentrate feeds are not suitable as the sole ration for hogs, it must be realized that swine can utilize a larger variety of feeds to greater advantage than other farm animals. In general, the grain crops—corn, barley, wheat, oats, rye, and the sorghums—constitute the major component of the swine ration. However, sweet potatoes and peanuts are successfully and extensively used in the South, soybeans in the central states, and peas in the Northwest. In those districts where they are grown, potatoes (cull) also are utilized in considerable quantities in feeding hogs. In addition, in almost every section of the country one or more by-product feeds are fed to hogs—including the by-products of the fishing, meat-packing, milling, and dairy industries. Human food wastes, such as garbage, are also fed extensively.

The protein and vitamin requirements of the monogastric pig differ very greatly from those of the ruminant, for the latter improves the quality of proteins and creates certain vitamins through bacterial synthesis.

ENERGY FEEDS

Carbohydrates and fats may be classed together as energy feeds for swine. Three essential fatty acids—linoleic, linolenic, and arachidonic acid—are required by swine, but cereal grains contain adequate quantities of fat to meet the fatty acid requirements. The ingredients commonly used as sources of energy are corn, barley, sorghum, wheat, and fats and oils. Full replacement of corn with barley, oats, sorghum, or wheat will decrease dressing percentage by approximately 1% and decrease backfat by about 0.1 in.

(Also see Table 23–28, Feed Substitution Table for Swine, in this chapter; Chapter 10, Grains/High-Energy Feeds; and Chapter 12, By-product Feeds/Crop Residues.)

Corn

In this country, corn and swine production have always gone hand in hand. Normally, about one-fourth of the U.S. corn crop is fed to hogs. Because of the dominant position of corn as a swine feed, it is herein singled out for further elucidation.

Corn is an excellent energy feed for all classes of swine. It is an ideal finishing feed because it is high in digestible carbohydrate (starch), low in fiber, and very palatable. Also, it can be fed in a variety of ways. It may be fed shelled, ground, mixed, or free-choice, or even as ear corn. It may be dry or high-moisture.

In spite of its virtues, corn alone will not keep pigs alive. It contains 7 to 9% protein, but the protein is deficient in some of the essential amino acids required by the weanling pig, especially lysine and tryptophan (see Fig. 23–13). It is also so deficient in calcium and other minerals, and so inadequate in vitamin content, that pigs will die if they are limited to a ration containing only corn. So, corn must be supplemented with a protein that makes up its amino acid deficiencies. Equally important are the needed minerals and vitamins. When properly supplemented, corn is an excellent energy feed for all classes of swine.

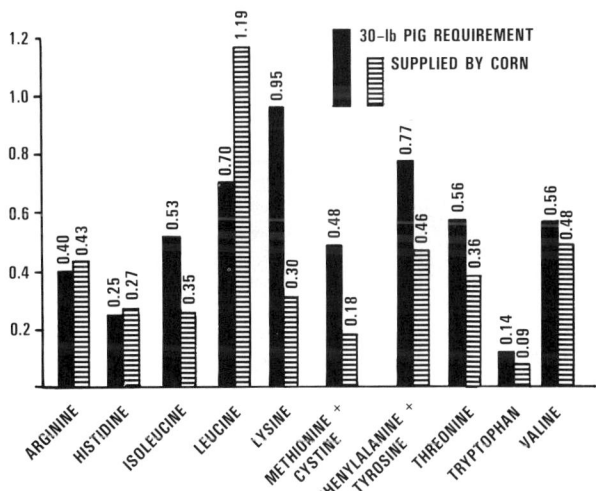

Fig. 23–13. Hogs cannot live by corn alone! This shows the amino acid content of corn (in %) in comparison with the ration requirements (in %) of a 30-lb pig. Note that corn is especially deficient in lysine and tryptophan. (Sources of data: Amino acid ration requirements [in % as-fed] of a 30-lb pig from *Nutrient Requirements of Swine*, 9th rev. ed., National Research Council, National Academy Press, 1988, p. 50, Table 5-1 [see Table 23-5 in this chapter]. Amino acid content of corn from Table V-2, Amino Acid, Composition of Feeds, in this book.)

It should be noted that corn is higher in fat than barley or wheat (4% vs less than 2%). Fat not only contributes to the high-energy content of corn, but it also improves its palatability and feeding properties in general.

- **High-moisture corn**—High-moisture corn may be substituted for dry corn on a dry matter basis with little effect on overall performance, in either feed conversion or rate of gain.
- **High-lysine corn (Opaque–2)**—Corn is now being bred that is much higher than normal corn in lysine and tryptophan; hence, it has a better balance of the amino acids for swine. Also, the high-lysine corn (called Opaque–2) is higher in total protein than normal corn. However, Opaque–2 is lower in leucine than regular corn.

Fig. 23-14. Amino acid deficiency. Littermate pigs. *Big pig (B):* Pig fed an adequate diet containing Opaque–2 corn (high-lysine corn). *Little pig (C):* Pig fed inadequate amounts of lysine and tryptophan. (Courtesy, Cornell University, Ithaca, N.Y.)

Fats and Oils

The following types of fats and oils are available and may be added to swine rations as high-energy feeds: animal fat, poultry fat, tallow, lard, corn oil, soybean oil, and other plant seed oils. When of comparable quality, there is little difference in swine performance resulting from the type of fat or oil used.

When fat is included in growing-finishing rations, it is usually added at a level of 4 to 5%. As a result of added fat, daily feed intake usually decreases, daily gain increases slightly, and feed efficiency improves. Each 1% added fat produces about 2% improvement in feed efficiency. Response to fat is greater in the summer than in the winter, because less heat is produced by pigs when they digest fat than when they digest starch in grain. High levels of fat (more than 5%) will increase backfat thickness slightly and reduce overall carcass lean. As little as 2½% added fat (50 lb/ton) reduces dust in mixing and feeding. More than 6 to 8% added fat may cause flowability problems.

When fat is added to rations, the caloric density of the ration increases. So, dietary nutrient density should increase in proportion to the increase in caloric density to maximize swine performance.

When fat is added to brood sow rations at a level of 7½% or more 10 to 14 days before farrowing, it increases the fat content of colostrum and milk and improves baby pig survival by 2 to 3% in herds where survival rates are less than 80%. However, added fat has little effect on litter size, birth weight, or weaning weight.

When oils are fed to pigs, they tend to make the body fat softer. To alleviate this problem, the level of oil must be limited and the oil must be deleted from the ration for a "hardening off" period prior to slaughter. Feeding fish oils gives a disagreeable fishy taste to the pork.

Other Energy Feeds for Swine

Yellow corn is usually the cheapest source of energy over much of the United States. But price fluctuations frequently justify consideration of other feeds; among them, the following: bakery waste, barley, beans (cull), molasses (cane or beet), oats, potatoes (cull, Irish), rye, sorghum (milo), spelt, triticale, and wheat.

Although barley, oats, wheat, milo, and rye are higher in protein than corn, it is noteworthy that their protein is generally of the same poor quality as corn. It is noteworthy, too, that all of the cereal grains have about the same vitamin and mineral deficiencies.

PROTEIN FEEDS

Protein is made up of nitrogenous compounds called amino acids. Protein feeds vary in the kind and amount of amino acids they contain. During the digestion process, the protein in feed is broken down into the various amino acids and the pig recombines them into the kind of protein needed for muscle development and repair of worn-out tissue. Thus, the real need of the pig is for amino acids, not protein as such.

The pig can synthesize some of the amino acids, with the result that they are not required in the ration. However, 10 of the amino acids are termed *essential*, because the body cannot manufacture them in sufficient quantity to permit maximum growth and performance. It is important that ingredients rich in the essential amino acids be used in formulating the ration.

Although it is a common practice to refer to *percent protein* in a ration, this term has little meaning in swine feeds unless there is knowledge concerning the amino acids present. A protein feed is considered to be of good quality when it contains all the essential amino acids in the proportions and amounts needed by the pig.

(Also see Table 23-28, Feed Substitution Table for Swine, in this chapter; Chapter 11, Protein Supplements; and Chapter 12, By-product Feeds/Crop Residues.)

Soybean Meal

Soybean meal is by far the leading high-protein supplement of the United States. In 1984, 75% of the total high-protein animal feed (including both oilseed meals and animal proteins) consisted of soybean meal.

Although soybean meal is marginal in methionine, it is otherwise very well balanced in amino acids. It must be supplemented with minerals and vitamins. Usually, it is not fed free-choice because of its high palatability, which results in pigs eating more than is needed to meet their protein needs.

Producers generally have a choice of buying soybean meal of different protein content, usualy 44 or 49%. The higher protein content meal is the more desirable to use in prestarter and starter rations because much of the hull has been removed. Thus, it is lower in fiber (not more than 3%), higher in energy, and more palatable. Growing-finishing pigs can utilize both meals about equally well; hence, the choice should be on the basis of which is the best buy as determined by the price per unit of protein.

Full-Fat Soybeans

The protein of raw soybeans is poorly utilized by young, growing pigs due to the presence of antitrypsin, a powerful growth inhibitor that affects young swine.

Cooking or roasting at the proper temperature (250°F for 2½ to 3½ minutes in a roaster) destroys this factor and makes soybeans a satisfactory feed for young pigs. However, cooking whole soybeans for brood sows is not necessary; comparable performance results when raw soybeans replace soybean meal on an equal protein basis in rations fed to gestating and/or lactating sows.

Research has shown that whole cooked soybeans can be used to replace soybean meal or other forms of protein supplement in growing-finishing rations. They will increase daily gain up to 5%, and improve feed efficiency by 5 to 10% due to the higher fat content of the whole soybeans (17 to 18%), which makes for a higher energy ration. However, this improvement in feed efficiency may be offset by the lower protein content of whole soybeans; whole cooked beans average about 37% crude protein, whereas soybean meal usually runs 44%. Also, due to the higher energy of the beans, the protein content of the ration must be 1 to 2% higher than in a soybean meal ration in order to maintain the same protein-to-energy ratio.

The feeding of full-fat, cooked soybeans to growing-finishing pigs has little effect on grade and yield if the proper protein-to-energy ratio is maintained, but softer pork may be produced.

It is noteworthy that hogs fed cooked whole soybeans have a softer carcass than those fed a soybean meal ration. Whether this condition will influence the price packers are willing to pay for live hogs should be determined.

Other Protein Feeds for Swine

Although soybean meal is the most widely used protein supplement, many other protein feeds are suitable, and are used; among them, cottonseed meal, dried skim milk, fish meal, linseed meal, meat scraps, meat and bone scraps, peanut meal, rapeseed meal, and tankage.

Pastures

Good pasture is an excellent feed for swine, but it is not indispensable. In comparison with confinement raising, it can lower sow feed costs, help maintain high reproductive capacity for boars, and increase litter size in many cases.

Pasture was formerly thought to be an absolute essential for a successful swine operation. However, in recent years producing hogs in confinement has become a reality because of vastly improved rations, along with greater disease and parasite control. But it is still possible to utilize large amounts of forage effectively for the breeding herd.

Research reports on feed savings by swine on pasture vary considerably, depending on type of pasture, class and age of hogs, and management system. On the average, however, good pasture for growing-finishing hogs will effect a saving of 3 to 10% of the grain and of as much as 33% of the supplement, in comparison with confinement production.

Bred sows and gilts on legume pasture require about half as much grain and much less supplemental protein than those in drylots. Thus, the decision on whether to raise swine on pasture or in drylot should be based primarily on (1) net returns, and (2) whether the land can be put to a more profitable alternative use.

Fig. 23-15. Pigs in alfalfa pasture in Nebraska. (Courtesy, Agricultural Agent, Burlington Northern, Inc., St. Paul, Minn.)

(Also see Chapter 7, Pasture and Range Forages, section on "Pastures for Swine.")

Dry Forages

High-priced grains caused alfalfa to return in favor as a feedstuff, particularly in rations for gestating sows. If the price of alfalfa is right—if it is a cheaper source of protein and energy than corn and other grains—it may, to advantage, be incorporated in swine rations at the following levels: up to 5% for grower-finishing hogs, up to 50% for gestating sows, and up to 10% for lactating sows. These levels serve as a safety factor to ensure the presence of certain minerals, vitamins, and possibly unidentified factors.

Alfalfa is especially well suited for winter feeding of swine. Because of its low energy and high fiber, it generates considerable heat when digested. The extra heat can be utilized to help maintain body temperature during the winter. Thus, alfalfa is more cost effective as a feed for swine during winter than during summer.

Silages

Good-quality corn or alfalfa-grass silage is an excellent feed for brood sows; hence, it may be used to advantage where it is available on dairy and beef farms. Unless the sow herd is very large, however, it will not likely pay to construct a silo especially for hogs.

Silage should be fed fresh daily in amounts the sows will clean up in 2 to 3 hours. Sows will usually eat 10 to 15 lb, and gilts 8 to 12 lb. Some wastage can be expected. It is important that the silage be supplemented with 1.0 to 1.5 lb of a good protein supplement per head daily. Additional grain should be fed if necessary to keep sows in proper condition.

Hogging Down Crops

Sometimes pigs are permitted to do their own harvesting. Corn is the principal crop so used, the animals being turned into the field when the grain is in the dent stage. Also, small grain crops that have been badly lodged or otherwise damaged, or that cannot be harvested because of weather, may be harvested by hogs. Soybeans and field peas may be hogged off. In the South, such crops as peanuts, sweet potatoes, chufas, and other root and tuber crops, are sometimes harvested by hogs.

Space will not permit a full discussion of this method of utilizing the various crops. As corn is the main feed hogged down, comments will be limited to this crop; but the same general principles apply to other crops, when and if they are so utilized.

Fig. 23-16. Hogging down corn. The animals are usually turned into the field when the grain is in the dent stage. (Courtesy, USDA)

Garbage

Municipal garbage has long been fed to finishing hogs, but, following World War II, the practice declined because of (1) a gradual lowering in the feeding value of garbage, and (2) other competition for garbage—notably its manufacture into lawn, greenhouse, and garden fertilizer. The recent development of garbage recycling processes, along with high-priced grains, has created renewed interest in garbage as a hog feed.

As a rule of thumb, about 4 lb of heavy garbage may be considered as equivalent to 1 lb of concentrate.

All states now have laws requiring that commercial garbage be cooked to prevent trichinosis in humans.

(Also see Chapter 12, section on "Garbage.")

FEED PREPARATION

Grinding grains improves feed efficiency; hence, it is recommended. Fine grinding will cause some bridging in self-feeders, however.

The recommended fineness for grinding various grains follows:

Grain	Fineness of Grind
Corn	Medium
Wheat	Medium to coarse
Barley	Fine
Grain sorghum	Fine to medium
Oats	Fine

As noted above, corn should be medium grind; gastric ulcers in pigs have been associated with finely ground corn. Neither should wheat be finely ground; fine grinding makes it pasty and less palatable. All other cereal grains should be finely ground.

Pelleting corn-soybean meal rations will generally improve feed utilization and increase rate of gain by at least 4 to 5%. Part of the improvement in feed efficiency is probably due to less feed wastage. Also, rations containing considerable amounts of fiber are improved by pelleting because of increased consumption, improved carbohydrate digestibility, and reduced sorting and wastage compared to meal rations.

Other methods of feed preparation, such as cooking, soaking, or fermenting, have not been shown to be of value when swine are full fed, although there are exceptions—for example, the cooking of soybeans and potatoes does improve efficiency.

The results from liquid or paste feeding have been very inconsistent. Some research has indicated increased feed consumption and rate of gain, whereas other studies have shown a deleterious effect from liquid feeding. Therefore, it is generally recommended that liquid feeding be considered only as a mechanical means of dispensing the feed.

Although high-moisture corn does improve efficiency with some classes of livestock, there is no improvement when fed to swine. Therefore, the relative value of high-moisture corn as compared to regular corn should be computed on a dry matter basis.

SWINE FEEDING PROGRAM

In order to compute balanced rations, it is necessary to have available feeding standards, or allowances, and feed composition tables. The NRC requirements are given in Tables 23-1 to 23-7, and recommended allowances with margins of safety are given in Table 23-8. Feed composition tables are given in Section V.

For purposes of convenience and facilitating ration formulation, Table 23-12, Swine Feeding Program, is presented. This gives, in summary form, the recommended types of rations and feeding programs for all classes of swine—a ration for every need.

Feeding Swine

TABLE 23-12
SWINE FEEDING PROGRAM[1]

Sex/Stage	Length of Feeding Program	Season	Complete Ration[2] Protein (%)	Complete Ration[2] Amount (lb/day)	Complete Ration[2] Amount (kg/day)	Corn or Grain (lb/day)	Corn or Grain (kg/day)	Supplement[3] (lb/day)	Supplement[3] (kg/day)
Boars	From time purchased at 5–6 months of age.	Summer	12–16	4–6	1.8–2.7	3–5	1.3–2.3	0.8–1.0	0.4–0.5
		Winter	12–16	5–7	2.3–3.2	4–6	1.8–2.7	0.8–1.0	0.4–0.5
	Increase intake 1–2 lb (0.5–0.9 kg) during the heavy breeding season.								
Gilts Pregestation	From time selected at 5–6 months until breeding at 7–9 months of age.	Summer	12–16	4–6	1.8–2.7	3–5	1.3–2.3	0.8–1.0	0.4–0.5
		Winter	12–16	5–7	2.3–3.2	4–6	1.8–2.7	0.8–1.0	0.4–0.5
Flushing and breeding	For 3 weeks before breeding (do not continue after breeding).		12–14	6–9	2.7–4.1	5–8	2.3–3.6	0.8–1.0	0.4–0.5
Gestation		Summer	12–14	4–5	1.8–2.3	3–4	1.3–1.8	0.8–1.0	0.4–0.5
		Winter	11–14	5–6	2.3–2.7	4–5	1.8–2.3	0.8–1.0	0.4–0.5
	Increase intake 1–2 lb (0.5–0.9 kg) last 3 to 5 weeks of gestation if gilts seem to be too thin.								
Lactation	Wean at 3–5 weeks after farrowing.		13–16	10–14	4.5–6.4	Full feed complete ration.			
Sows Breeding and gestation	Breed back at first heat period after weaning (flushing is not beneficial with sows).	Summer	12–14	3–4	1.3–1.8	2–3	0.9–1.3	0.8–1.0	0.4–0.5
		Winter	11–14	4–6	1.8–2.7	3–4	1.3–1.8	0.8–1.0	0.4–0.5
	Increase intake 1–2 lb (0.5–0.9 kg) last 3 to 5 weeks of gestation if sows seem to be too thin.								
Lactation	Wean at 3–5 weeks after farrowing.		13–16	11–15	5.0–6.8	Full feed complete ration.			
Pigs Prestarter	Use only if weaned before 3 weeks and feed until pigs weigh 12 lb (5.5 kg).		20–24			Full feed complete ration.			
Starter	Use as a creep feed and continue after weaning until pigs are 8 weeks of age or 40 lb (18.2 kg).		18–20			Full feed complete ration.			
Grower-finisher	From 8 weeks of age or 40 lb (18.2 kg) until market weight.		13–17	Full feed (may limit feed after 125 lb [56.8 kg]).		With free-choice, corn consumption varies depending on weight of pig. Daily supplement intake should be approximately 0.75 lb (0.34 kg) regardless of the weight of the pig.			

[1]Adapted by the authors from *Life Cycle Swine Nutrition*, Iowa State University, Ames, PM-489, June, 1988.
[2]Assumes corn based rations.
[3]See supplements in Table 23-25.

FEED ALLOWANCES AND SUGGESTED RATIONS

In most instances, the nutrient allowances should be higher than the minimum requirements established by the National Research Council (see Tables 23-1 to 23-7). This is desirable to obtain maximum performance and reduce the risk of nutrient deficiencies that might occur because of the differences in ingredient quality, environment, health, genetics, and performance of individual animals.

Following the selection of feed ingredients, they may be fed to swine in either combined form or cafeteria style, with balanced rations achieved by any of the following mixing/feeding procedures.

• **Complete feed vs feeding grain and supplement separately**—The swine producer has two primary feeding options from which to choose: feeding complete feeds vs feeding the grain and supplement separately.

• **Commercial vs home mixed complete feeds**—Complete feeds may be commercially or home mixed. Commercial feed companies specializing in swine feeds generally manufacture and sell complete rations suitable for each class of swine. Because of the preciseness needed in formulating and mixing prestarter and starter rations, most swine producers purchase commercially prepared feeds for these needs. However, the home mixing of complete feeds for other classes of swine, and even for baby pigs, may be accomplished either (1) by "building from the ground up" (by adding each ingredient, one by one); or (2) by mixing a complete supplement, base mix, or premix with the primary ingredients.

• **Recommended nutrient allowances of the swine specialists at Iowa State University**—In this chapter, the authors present the nutrient allowances, along with premixes and rations, of the swine specialists of Iowa State University, the Land Grant University in the state of Iowa, the leading swine state in the nation.

The recommended allowances and rations formulated by the swine specialists at Iowa State University consist of three premixes (Tables 23-13, 23-14, and 23-15. See pp. 530-531.) and nine finishing rations (Tables 23-16 to 23-24. See pp. 532-536.). This makes for a popular and convenient method of building swine rations by blending the proper premix with the main ingredients of the ration.

TABLE 23-13
COMPOSITION AND ANALYSIS OF TRACE MINERAL PREMIX[1]

Element	Source[2]	Amount (lb)	Amount (kg)	Percent in Premix (%)	Parts Per Million When Added to a Complete Ration At The Following Pounds Per Ton:			
					2 (ppm)	3 (ppm)	4 (ppm)	5 (ppm)
Copper (Cu)	Copper sulfate	1.500	0.681	0.38	4	6	8	10
Iodine (I)	Potassium iodide[3]	0.010	0.005	0.008	0.08	0.11	0.15	0.19
Iron (Fe)	Ferrous sulfate	25.000	11.350	5.03	50	75	101	126
Manganese (Mn)	Manganese sulfate	2.500	1.135	0.57	6	9	11	14
Selenium (Se)	Sodium selenite[3]	0.025	0.011	0.011	0.11	0.17	0.23	0.29
Zinc (Zn)	Zinc sulfate	25.000	11.350	5.68	57	85	114	142
	Carrier	45.965	20.868					
	Total	100.000	45.400					

[1]Adapted by the authors from *Life Cycle Swine Nutrition*, Iowa State University, Ames, PM-489, June, 1988.
[2]Other sources of trace minerals may be substituted. Iodine may be omitted if iodized salt is used.
[3]Iodine and selenium probably will be added in a separate premix form.

TABLE 23-14
COMPOSITION OF VITAMIN PREMIX[1][2]

Vitamins	Amount	Unit	Units Per Pound of Complete Ration When Added At The Following Pounds Per Ton:			
			3	5	8	10
Essential[3]						
Vitamin A .... (million IU)	5.0	(IU)	750.00	1,250.00	2,000.00	2,500.00
Vitamin D .... (million IU)	0.6	(IU)	90.00	150.00	240.00	300.00
Vitamin E .... (thousand IU)	26.0	(IU)	3.90	6.50	10.40	13.00
Niacin (nicotinic acid, nicotinamide) .... (g)	25.0	(mg)	3.75	6.25	10.00	12.50
d-Pantothenic acid (vitamin B-3) .... (g)	20.0	(mg)	3.00	5.00	8.00	10.00
Riboflavin (vitamin B-2) .... (g)	6.0	(mg)	0.90	1.50	2.40	3.00
Vitamin B-12 (cobalamins) .... (mg)	25.0	(mcg)	3.75	6.25	10.00	12.50
Optional[4]						
Biotin .... (g)	0.3	(mg)	0.05	0.08	0.12	0.15
Menadione (source of vitamin K) .... (g)	4.0	(mg)	0.60	1.00	1.60	2.00
Carrier	?					
Total[5] .... (lb)	10.0					

[1]Adapted by the authors from *Life Cycle Swine Nutrition*, Iowa State University, Ames, PM-489, June, 1988.
[2]A feed additive may be included in the vitamin premix.
[3]Most natural feedstuffs contain very little vitamin D or B-12. The amount of provitamin A (beta-carotene) in feedstuffs will depend on processing and storage, while niacin in most grains is relatively unavailable for swine. Riboflavin and pantothenic acid in natural feedstuffs can meet part of the requirement.
[4]Supplemental biotin is not necessary with corn-soybean meal based rations. It should be included in sow rations based on other grains. The vitamin K requirement is normally met by the level present in natural feedstuffs and by intestinal synthesis. A hemorrhagic or bleeding syndrome has been diagnosed which is probably due to a vitamin K antimetabolite. The antimetabolite is thought to be produced by mold occuring in one or more of the ration ingredients. When this has occurred, adding menadione has been helpful in preventing or overcoming the problem.
[5]If this premix is used in Table 23-15, dilute to 2 lb *(0.9 kg)* only.

Feeding Swine

TABLE 23-15
COMPLETE MINERAL-VITAMIN PREMIXES
FOR CORN-SOYBEAN MEAL RATIONS[1,2,3,4]

Ingredients	1		2	
	(lb)	(kg)	(lb)	(kg)
Calcium carbonate	540	245	300	136
Dicalcium phosphate	1,030	468		
Defluorinated phosphate			1,060	481
Salt	250	114	250	114
Trace mineral premix (Table 23-13)	80	36	80	36
Vitamin premix (Table 23-14)[5]	40	18	40	18
Carrier (corn, middlings, or grain by-products)	60	27	270	123
Total	2,000	908	2,000	908

Calculated Analyses:

	(%)	(lb)	(kg)
Salt (NaCl)	12.5		
Calcium (Ca)	22.8		
Phosphorus (P)	9.5		
Copper (Cu)	0.015		
Iodine (I)	0.00031		
Iron (Fe)	0.201		
Manganese (Mn)	0.023		
Selenium (Se)	0.00046		
Zinc (Zn)	0.227		
Vitamin A (IU)		50,000	110,000
Vitamin D (IU)		6,000	13,200
Vitamin E (IU)		260	572
Vitamin K (optional) (mg)		40	88
Biotin (optional) (mg)		3	7
Niacin (Nicotinic Acid, Nicotinamide) (mg)		250	550
Pantothenic Acid (Vitamin B-3) (mg)		200	440
Riboflavin (Vitamin B-2) (mg)		60	132
Vitamin B-12 (Cobalamins) (mcg)		250	550

[1]Adapted by the authors from *Life Cycle Swine Nutrition*, Iowa State University, Ames, PM-489, June, 1988.

[2]Due to the instability of vitamins in the presence of trace minerals this premix should be used within 30 days of preparation.

[3]Mixing directions:

Stage	Premix		Soybean Meal, 44%		Corn	
	(lb)	(kg)	(lb)	(kg)	(lb)	(kg)
Gestation:						
3 lb (1.4 kg)/day	100	45	270	123	1,630	740
4 lb (1.8 kg)/day	80	36	160	73	1,760	799
5 lb (2.3 kg)/day	65	30	100	45	1,835	833
Lactation	80	36	270	123	1,650	749
Starter rations	65	30	690	313	1,245	565
Grower-finisher rations	50	23	300–450	136–204	1,700–1,500	772–681

[4]These premixes can be fed free-choice to sows on pasture or in other instances where free choice minerals and vitamins are needed. Do not add or feed free-choice additional minerals or vitamins with any of the ration formulas in Tables 23-16 to 23-24, because they contain sufficient minerals and vitamins.

[5]Table 23-14 premix diluted to 2 lb (0.9 kg) instead of 10 lb (4.5 kg). See footnote 5 under Table 23-14.

Fig. 23-17. Growing-finishing pigs fed added minerals and vitamins in a self-fed complete ration. (Courtesy, *National Hog Farmer*, St. Paul, Minn.)

Suggested rations are shown for each of the following classes of swine:

Table 23-16, Pregestation, Breeding, and Gestation Rations—for boars, sows, or gilts fed 3 lb per day.

Table 23-17, Pregestation, Breeding, and Gestation Rations—for boars, sows, or gilts fed 4 lb per day.

Table 23-18, Pregestation, Breeding, and Gestation Rations—for boars, sows, or gilts fed 5 lb per day.

Table 23-19, Lactation Rations.

Table 23-20, Prestarter Rations—for baby pigs before 3 weeks of age.

Table 23-21, Starter Rations.

Table 23-22, Recommended Rations for Performance Testing Boars.

Table 23-23, Swine Conditioner Rations—for newly received feeder pigs, for stress periods and convalescence.

Table 23-24, Grower-Finisher Rations—for 40 to 240 lb.

As previously indicated, the rations shown in Tables 23-16 to 23-24 call for the use of the mineral and vitamin premixes shown in Tables 23-13 to 23-15.

TABLE 23-16
PREGESTATION, BREEDING, AND GESTATION RATIONS (FOR BOARS, SOWS AND GILTS FED 3 LB (1.4 KG) PER DAY)[1,2,3]

Ingredient	Ration # 1 (lb)	(kg)	Ration # 2 (lb)	(kg)	Ration # 3 (lb)	(kg)	Ration # 4 (lb)	(kg)
Corn, yellow (8.4% protein)[4]	1,636	743	1,559	708	1,640	744	1,562	709
Soybean meal, solvent extracted (44.0% protein)[5]	270	123	250	113	194	88	175	80
Alfalfa meal, dehydrated (17.0% protein)			100	45			100	45
Meat and bone meal (50.0% protein)					100	45	100	45
Dicalcium phosphate	50	22	51	23	31	14	31	14
Limestone	19	8	15	7	10	5	7	3
Iodized salt	10	5	10	5	10	5	10	5
Trace mineral premix (Table 23-13)	5	2	5	2	5	2	5	2
Vitamin premix (Table 23-14)	10	5	10	5	10	5	10	5
Feed additives[6]								
Total	2,000	908	2,000	908	2,000	908	2,000	908
Calculated analyses:								
Metabolizable energy (kcal/lb)	1,470		1,434		1,469		1,433	
Metabolizable energy (kcal/kg)	3,234		3,155		3,232		3,153	
Protein (%)	12.81		12.90		13.66		13.76	
Lysine (%)	0.60		0.60		0.60		0.60	
Threonine (%)	0.49		0.49		0.50		0.51	
Tryptophan (%)	0.14		0.15		0.13		0.14	
Calcium (Ca) (%)	1.01		1.01		1.00		1.01	
Phosphorus (P) (%)	0.75		0.75		0.75		0.75	

[1]Adapted by the authors from *Life Cycle Swine Nutrition*, Iowa State University, Ames, PM-489, June, 1988.

[2]See Table 23-12 for recommended feeding levels in drylot or confinement. These rations can be used for sows on pasture since they require only supplemental minerals or at the most 2 to 3 lb (0.9 to 1.4 kg) of complete feed. If less than 3 lb (1.4 kg) is fed per day, free-choice minerals should be available.

[3]These rations can also be used as a silage balancer. Gestating sows will eat 5 to 7 lb (2.3 to 3.2 kg) of corn silage daily which should be supplemented with 2 to 3 lb (0.9 to 1.4 kg) of one of these rations.

[4]Ground oats can replace corn up to 20% of the total ration. Ground milo, wheat, or barley can replace the corn.

[5]To replace 44% soybean meal with 47% soybean meal or whole soybeans, use the following ratios:
Each 100 lb (45 kg) SBM (44%) = 93 lb (42 kg) SBM (47%) + 7 lb (3 kg) corn.
Each 100 lb (45 kg) SBM (44%) + 35 lb (16 kg) corn = 135 lb (61 kg) whole soybeans.

[6]Feed additives are not generally recommended during gestation or for gilts during the developer period after selection unless specific disease problems exist. High levels (100–300 g/ton) may be beneficial when fed 2 weeks before and after breeding and 2 weeks before farrowing.

TABLE 23-17
PREGESTATION, BREEDING, AND GESTATION RATIONS (FOR BOARS, SOWS, OR GILTS FED 4 LB (1.8 KG) PER DAY)[1,2]

Ingredient	Ration # 1 (lb)	(kg)	Ration # 2 (lb)	(kg)	Ration # 3 (lb)	(kg)	Ration # 4 (lb)	(kg)
Corn, yellow (8.4% protein)[3]	1,769	803	1,692	768	1,767	802	1,692	768
Soybean meal, solvent extracted (44.0% protein)[4]	160	72	140	64	90	41	68	31
Alfalfa meal, dehydrated (17.0% protein)			100	45			100	45
Meat and bone meal (50.0% protein)					100	45	100	45
Dicalcium phosphate	36	16	36	16	16	7	16	7
Limestone	15	7	12	5	7	3	4	2
Iodized salt	8	4	8	4	8	4	8	4
Trace mineral premix (Table 23-13)	4	2	4	2	4	2	4	2
Vitamin premix (Table 23-14)	8	4	8	4	8	4	8	4
Feed additives[5]								
Total	2,000	908	2,000	908	2,000	908	2,000	908
Calculated analyses:								
Metabolizable energy (kcal/lb)	1,493		1,457		1,492		1,456	
Metabolizable energy (kcal/kg)	3,285		3,205		3,282		3,203	
Protein (%)	10.95		11.04		11.90		11.95	
Lysine (%)	0.45		0.45		0.46		0.46	
Threonine (%)	0.41		0.41		0.43		0.43	
Tryptophan (%)	0.11		0.12		0.10		0.11	
Calcium (Ca) (%)	0.75		0.76		0.75		0.76	
Phosphorus (P) (%)	0.60		0.60		0.60		0.60	

[1]Adapted by the authors from *Life Cycle Swine Nutrition*, Iowa State University, Ames, PM-489, June, 1988.

[2]See Table 23-12 for recommended feeding levels when hand-fed in drylot or confinement. These rations can be used for gilts on pasture during gestation since they require 3 to 4 lb (1.4 to 1.8 kg) of feed daily. These rations can also be used for interval-fed sows or gilts if the average daily intake is approximately 4 lb (1.8 kg).

[3]Ground oats can replace corn up to 20% of the total ration. Ground milo, wheat, or barley can replace the corn.

[4]To replace 44% soybean meal with 47% soybean meal or whole soybeans, use the following ratios:
Each 100 lb (45 kg) SBM (44%) = 93 lb (42 kg) SBM (47%) + 7 lb (3 kg) corn.
Each 100 lb (45 kg) SBM (44%) + 35 lb (16 kg) corn = 135 lb (61 kg) whole soybeans.

[5]Feed additives are not generally recommended during gestation or for gilts during the developer period after selection unless specific disease problems exist. High levels (100–300 g/ton) may be beneficial when fed 2 weeks before and after breeding and 2 weeks before farrowing.

Feeding Swine

TABLE 23-18
PREGESTATION, BREEDING, AND GESTATION RATIONS (FOR BOARS, SOWS, OR GILTS FED 5 LB (2.3 KG) PER DAY)[1]

Ingredient	Ration # 1		Ration # 2		Ration # 3		Ration # 4	
	(lb)	(kg)	(lb)	(kg)	(lb)	(kg)	(lb)	(kg)
Corn, yellow (8.4% protein)[2]	1,845	838	1,764	801	1,853	841	1,824	828
Soybean meal, solvent extracted (44.0% protein)[3]	100	45	85	38	20	9		
Alfalfa meal, dehydrated (17.0% protein)			100	45			50	23
Meat and bone meal (50.0% protein)					100	45	100	45
Dicalcium phosphate	26	12	26	12	6	3	7	3
Limestone	14	6	10	5	6	3	4	2
Iodized salt	6	3	6	3	6	3	6	3
Trace mineral premix (Table 23-13)	3	1	3	1	3	1	3	1
Vitamin premix (Table 23-14)	6	3	6	3	6	3	6	3
Feed additives[4]								
Total	2,000	908	2,000	908	2,000	908	2,000	908
Calculated analyses:								
Metabolizable energy (kcal/lb)	1,508		1,473		1,507		1,489	
Metabolizable energy (kcal/kg)	3,318		3,241		3,315		3,276	
Protein (%)	9.95		10.13		10.72		10.59	
Lysine (%)	0.38		0.38		0.37		0.36	
Threonine (%)	0.37		0.38		0.38		0.37	
Tryptophan (%)	0.10		0.10		0.08		0.08	
Calcium (Ca) (%)	0.61		0.60		0.61		0.61	
Phosphorus (P) (%)	0.50		0.50		0.50		0.50	

[1]Adapted by the authors from *Life Cycle Swine Nutrition*, Iowa State University, Ames, PM-489, June, 1988.

[2]Ground oats can replace corn up to 20% of the total ration. Ground milo, wheat, or barley can replace the corn.

[3]To replace 44% soybean meal with 47% soybean meal or whole soybeans, use the following ratios:
   Each 100 lb (45 kg) SBM (44%) = 93 lb (42 kg) SBM (47%) + 7 lb (3 kg) corn.
   Each 100 lb (45 kg) SBM (44%) + 35 lb (16 kg) corn = 135 lb (61 kg) whole soybeans.

[4]Feed additives are not generally recommended during gestation or for gilts during the developer period after selection unless specific disease problems exist. High levels (100-300 g/ton) may be beneficial when fed 2 weeks before and after breeding and 2 weeks before farrowing.

TABLE 23-19
LACTATION RATIONS[1] [2]

Ingredient	Ration # 1		Ration # 2		Ration # 3		Ration # 4	
	(lb)	(kg)	(lb)	(kg)	(lb)	(kg)	(lb)	(kg)
Corn, yellow (8.4% protein)	1,658	752	1,545	701	1,580	717	1,469	667
Soybean meal, solvent extracted (44.0% protein)[3]	270	122	283	128	177	80	260	118
Fat or oil source			100	45				
Meat and bone meal (50.0% protein)					100	45		
Oats (11.5% protein)							200	90
Beet pulp (8.0% protein)					100	45		
Dicalcium phosphate	35	16	35	16	15	7	33	15
Limestone	15	7	15	7	6	3	16	7
Iodized salt	10	5	10	5	10	5	10	5
Trace mineral premix (Table 23-13)	4	2	4	2	4	2	4	2
Vitamin premix (Table 23-14)	8	4	8	4	8	4	8	4
Feed additives[4]								
Total	2,000	908	2,000	908	2,000	908	2,000	908
Calculated analyses:								
Metabolizable energy (kcal/lb)	1,487		1,409		1,468		1,457	
Metabolizable energy (kcal/kg)	3,271		3,100		3,230		3,205	
Protein (%)	12.90		12.72		13.43		13.04	
Lysine (%)	0.60		0.60		0.60		0.60	
Threonine (%)	0.49		0.49		0.48		0.49	
Tryptophan (%)	0.14		0.14		0.13		0.15	
Calcium (Ca) (%)	0.75		0.76		0.76		0.76	
Phosphorus (P) (%)	0.61		0.60		0.60		0.60	

[1]Adapted by the authors from *Life Cycle Swine Nutrition*, Iowa State University, Ames, PM-489, June, 1988.

[2]These rations may be limit-fed from a few days before farrowing and full-fed during lactation.

[3]To replace 44% soybean meal with 47% soybean meal or whole soybeans, use the following ratios:
   Each 100 lb (45 kg) SBM (44%) = 93 lb (42 kg) SBM (47%) + 7 lb (3 kg) corn.
   Each 100 lb (45 kg) SBM (44%) + 35 lb (16 kg) corn = 135 lb (61 kg) whole soybeans.

[4]High levels of feed additives (100-300 g/ton) may be beneficial when fed 2 weeks before and after breeding and 2 weeks before farrowing.

TABLE 23-20
PRESTARTER RATIONS (FOR BABY PIGS BEFORE 3 WEEKS OF AGE)[1,2]

Ingredient	Ration # 1		Ration # 2		Ration # 3		Ration # 4		Ration # 5	
	(lb)	(kg)	(lb)	(kg)	(lb)	(kg)	(lb)	(kg)	(lb)	(kg)
Corn, yellow (8.4% protein)	917	416	878	399	866	393	621	282	774	351
Soybean meal, solvent extracted (44.0% protein)					635	288				
Soybean meal, solvent extracted, dehulled (47.0% protein)	680	309	580	263			580	263	725	329
Oat groats (dehulled oats) (16.0% protein)							200	91		
Skim milk, dried (33.0% protein)	100	45	100	45	200	91	200	91		
Whey, dried (12.0% protein)	200	91	300	136	200	91	200	91	400	182
Fish meal, menhaden (62.0% protein)			50	23						
Sugar							100	45		
Fat or oil source	40	18	40	18	40	18	40	18	40	18
Dicalcium phosphate	26	12	19	9	25	12	25	12	25	12
Limestone	19	9	15	7	16	7	16	7	18	8
Iodized salt	5	2	5	2	5	2	5	2	5	2
Trace mineral premix (Table 23-13)	5	2	5	2	5	2	5	2	5	2
Vitamin premix (Table 23-14)	8	4	8	4	8	4	8	4	8	4
Feed additives[3]	◄——————————————— 100 to 300 g per ton ———————————————►									
Total	2,000	908	2,000	908	2,000	908	2,000	908	2,000	908
Calculated analyses:										
Metabolizable energy (kcal/lb)	1,458		1,459		1,441		1,387		1,441	
Metabolizable energy (kcal/kg)		3,208		3,210		3,170		3,051		3,170
Protein (%)	22.68		22.32		22.11		22.34		22.64	
Lysine (%)	1.40		1.40		1.40		1.40		1.40	
Threonine (%)	0.94		0.95		0.94		0.93		0.96	
Tryptophan (%)	0.29		0.28		0.29		0.29		0.30	
Calcium (Ca) (%)	0.91		0.91		0.91		0.90		0.91	
Phosphorus (P) (%)	0.70		0.70		0.70		0.71		0.70	

[1]Adapted by the authors from *Life Cycle Swine Nutrition,* Iowa State University, Ames, PM-489, June, 1988.

[2]The prestarter ration is normally fed in only limited amounts. It should be fed to pigs weaned before 3 weeks of age until they reach approximately 12 lb *(5.4 kg).* They can then be switched to a starter ration. These are good rations for orphan pigs, when extreme disease outbreaks (TGE) occur, or when the sow fails to produce sufficient milk.

[3]The feed additive may be part of the vitamin premix, or if a separate premix, it should replace an equal amount of corn.

TABLE 23-21
PIG STARTER RATIONS[1,2]

Ingredient	Ration # 1		Ration # 2		Ration # 3		Ration # 4		Ration # 5	
	(lb)	(kg)	(lb)	(kg)	(lb)	(kg)	(lb)	(kg)	(lb)	(kg)
Corn, yellow (8.4% protein)	1,072	487	939	426	1,032	468	1,273	578	865	393
Soybean meal, solvent extracted (44.0% protein)					610	277	689	313		
Soybean meal, solvent extracted, dehulled (47.0% protein)	570	259	580	263						
Soybeans, full-fat, cooked (37.0% protein)									800	363
Oat groats, dehulled (16.0% protein)			200	91						
Whey, dried (12.0% protein)	300	136	200	91	300	136			300	136
Fat or oil source			20	9						
Dicalcium phosphate	24	11	25	12	25	12	3	1	2	1
Limestone	16	7	18	8	15	7	17	8	15	7
Iodized salt	5	2	5	2	5	2	5	2	5	2
Trace mineral premix (Table 23-13)	5	2	5	2	5	2	5	2	5	2
Vitamin premix (Table 23-14)	8	4	8	4	8	4	8	4	8	4
Feed additives[3]	◄——————————————— 100 to 300 g per ton ———————————————►									
Total	2,000	908	2,000	908	2,000	908	2,000	908	2,000	908
Calculated analyses:										
Metabolizable energy (kcal/lb)	1,483		1,472		1,460		1,473		1,525	
Metabolizable energy (kcal/kg)		3,263		3,238		3,212		3,241		3,355
Protein (%)	19.70		20.37		19.55		20.39		20.13	
Lysine (%)	1.15		1.16		1.15		1.15		1.20	
Threonine (%)	0.82		0.82		0.82		0.81		0.85	
Tryptophan (%)	0.25		0.25		0.26		0.26		0.28	
Calcium (Ca) (%)	0.80		0.81		0.81		0.80		0.84	
Phosphorus (P) (%)	0.65		0.65		0.65		0.65		0.68	

[1]Adapted by the authors from *Life Cycle Swine Nutrition,* Iowa State University, Ames, PM-489, June, 1988.

[2]The pig starter ration can be used as a creep ration before weaning and fed after weaning until the pigs reach approximately 40 lb *(18 kg).* They can then be switched to a grower-finisher ration.

[3]The feed additive may be part of the vitamin premix, or if a separate premix, it should replace an equal amount of corn.

Fig. 23-18. Good feeding and management of the gestating sow give newborn pigs a good start in life. (Courtesy, Pennsylvania State University, State College, Pa.)

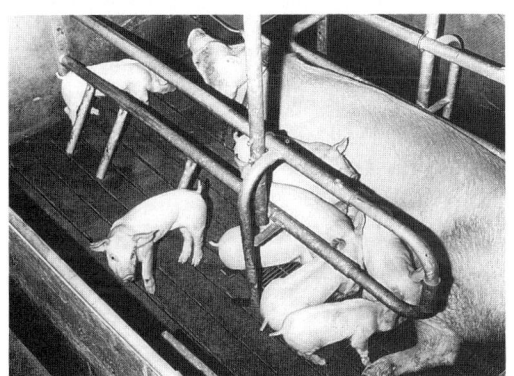

Fig. 23-19. Pigs 3 weeks of age and ready for early weaning. (Courtesy, Watt Publishing Co., Mount Morris, Ill.)

TABLE 23-22
RECOMMENDED RATIONS FOR PERFORMANCE TESTING OF BOARS[1,2]

Ingredient	Conditioner Ration		Test Ration	
	(lb)	(kg)	(lb)	(kg)
Corn, yellow (8.4% protein)[3]	1,115	506	1,359	617
Soybean meal, solvent extracted (44.0% protein)[4]	500	227	550	250
Wheat middlings (15.5% protein)	200	91		
Whey, dried (12.0% protein)	100	45		
Dicalcium phosphate	50	23	60	27
Limestone	15	7	11	5
Iodized salt	5	2	5	2
Trace mineral premix (Table 23-13)	5	2	5	2
Vitamin premix (Table 23-14)	10	5	10	5
Feed additives (g/ton)	(100–300)		(0–100)	
Total	2,000	908	2,000	908
Calculated analyses:				
Metabolizable energy (kcal/lb)	1,438		1,459	
Metabolizable energy (kcal/kg)	3,164		3,210	
Protein (%)	17.83		17.81	
Lysine (%)	0.98		0.97	
Threonine (%)	0.71		0.70	
Tryptophan (%)	0.23		0.22	
Calcium (Ca) (%)	1.01		1.01	
Phosphorus (P) (%)	0.87		0.89	

[1]Adapted by the authors from *Life Cycle Swine Nutrition*, Iowa State University, Ames, PM-489, June, 1988.

[2]These rations normally would be used for growing boars from 40 to 250 lb (18 to 114 kg) body weight.

[3]If the ration is to be pelleted, 25 to 50 lb (11 to 23 kg) of molasses or binder can replace an equal amount of corn.

[4]To replace 44% soybean meal with 47% soybean meal or whole soybeans, use the following ratios:
Each 100 lb (45 kg) SBM (44%) = 93 lb (42 kg) SBM (47%) + 7 lb (3 kg) corn.
Each 100 lb (45 kg) SBM (44%) + 35 lb (16 kg) corn = 135 lb (61 kg) whole soybeans.

TABLE 23-23
SWINE CONDITIONER RATIONS[1,2]

Ingredient	Ration # 1		Ration # 2		Ration # 3	
	(lb)	(kg)	(lb)	(kg)	(lb)	(kg)
Corn, yellow (8.4% protein)	882	401	840	382	979	444
Oats (11.5% protein)	300	136	600	272	600	272
Wheat middlings (15.5% protein)	300	136				
Soybean meal, solvent extracted (44.0% protein)[3]	170	77	300	136	350	159
Whey, dried (12.0% protein)	200	91	200	91		
Alfalfa meal, dehydrated (17.0% protein)	50	23				
Fish meal, menhaden (62.0% protein)	50	23				
Dicalcium phosphate	14	6	26	12	30	14
Limestone	14	6	14	6	10	7
Iodized salt	5	2	5	2	10	5
Trace mineral premix (Table 23-13)	5	2	5	2	5	2
Vitamin premix (Table 23-14)	10	5	10	5	10	5
Feed additives[4]	← 100 to 300 g/ton →					
Total	2,000	908	2,000	908	2,000	908
Calculated analyses:						
Metabolizable energy (kcal/lb)	1,398		1,387		1,391	
Metabolizable energy (kcal/kg)	3,076		3,051		3,060	
Protein (%)	14.67		14.78		15.26	
Lysine (%)	0.75		0.75		0.75	
Threonine (%)	0.59		0.60		0.58	
Tryptophan (%)	0.18		0.18		0.19	
Calcium (Ca) (%)	0.73		0.71		0.71	
Phosphorus (P) (%)	0.61		0.60		0.60	

[1]Adapted by the authors from *Life Cycle Swine Nutrition*, Iowa State University, Ames, PM-489, June, 1988.

[2]These rations are recommended as the first ration fed to newly received feeder pigs, for stress periods and convalescence.

[3]To replace 44% soybean meal with 47% soybean meal use the following ratio:
Each 100 lb (45 kg) SBM (44%) = 93 lb (42 kg) SBM (47%) + 7 lb (3 kg) corn.

[4]Be certain that only approved feed additives and levels are used for therapy. The feed additive may be a part of the vitamin premix, or if a separate premix, it should replace an equal amount of corn.

TABLE 23-24
GROWER-FINISHER RATIONS (FOR PIGS FROM 40 TO 240 LB *(18 TO 109 KG)*)[1,2,3,4]

Ingredient	For Pigs 40-120 Lb *(18-54 Kg)*						For Pigs 121-240 Lb *(55-109 Kg)*[5]					
	1		2		3		4		5		6	
	(lb)	(lb)	(lb)	(lb)	(lb)	(lb)	(lb)	(lb)	(lb)	(lb)	(lb)	(lb)
Corn, yellow (8.4% protein)[6,7]	1,571	1,497	1,422	1,343	1,561	1,492	1,692	1,613	1,575	1,481	1,694	1,620
Soybean meal, solvent extracted (44.0% protein)[8]	380	455			310	380	265	345			190	264
Soybeans, full-fat, cooked (37.0% protein)[9]			530	610					380	475		
Meat and bone meal (50.0% protein)					110	110					100	100
Dicalcium phosphate	21	19	19	18			18	17	20	20		
Limestone	15	16	16	16	6	5	15	15	15	14	6	6
Iodized salt	5	5	5	5	5	5	5	5	5	5	5	5
Trace mineral premix (Table 23-13)	3	3	3	3	3	3	2	2	2	2	2	2
Vitamin premix (Table 23-14)	5	5	5	5	5	5	3	3	3	3	3	3
Feed additives[10]	◀────── 0 to 100 g per ton ──────▶						◀────── 0 to 50 g per ton ──────▶					
Total	2,000	2,000	2,000	2,000	2,000	2,000	2,000	2,000	2,000	2,000	2,000	2,000
Calculated analyses:												
Metabolizable energy (kcal/lb)	1,500	1,497	1,543	1,547	1,498	1,495	1,510	1,507	1,538	1,543	1,508	1,505
Metabolizable energy *(kcal/kg)*	*3,300*	*3,293*	*3,395*	*3,403*	*3,296*	*3,289*	*3,322*	*3,315*	*3,384*	*3,395*	*3,318*	*3,311*
Protein (%)	14.96	16.30	15.78	16.93	16.13	17.38	12.94	14.36	13.65	15.01	13.79	15.11
Lysine (%)	0.75	0.85	0.81	0.90	0.77	0.86	0.60	0.70	0.65	0.76	0.60	0.70
Threonine (%)	0.58	0.63	0.61	0.66	0.60	0.65	0.49	0.55	0.52	0.58	0.51	0.56
Tryptophan (%)	0.17	0.20	0.20	0.21	0.17	0.19	0.14	0.17	0.16	0.18	0.13	0.15
Calcium (Ca) (%)	0.60	0.60	0.61	0.61	0.61	0.60	0.55	0.55	0.59	0.58	0.55	0.56
Phosphorus (P) (%)	0.50	0.50	0.51	0.51	0.51	0.53	0.46	0.46	0.49	0.51	0.47	0.49

[1] Adapted by the authors from *Life Cycle Swine Nutrition,* Iowa State University, Ames, PM-489, June, 1988.

[2] Feed the ration with the higher level of soybean meal (lower level of corn) to lighter pigs in each group and decrease the soybean meal (increase the corn) until you reach the lower level as pig weights increase. If preferred, one level of protein can be fed from 40 to 240 lb *(18 to 109 kg)* with similar results as with varying the levels. To accomplish this, use the lower protein formulations from rations 1, 2, or 3 (for example, in ration No. 1 use 1571 lb *(713 kg)* of corn and 380 lb *(173 kg)* of soybean meal).

[3] If barrows and gilts are separated, use the higher range for soybean meal for the gilts and the lower range for the barrows.

[4] To convert lb to kg, divide by 2.2. To convert g/ton (short) to g/ton (metric), divide by 0.907.

[5] For potential replacement gilts, the level of dicalcium phosphate should be increased by 10 lb *(4.5 kg)* per ton. This will provide a minimum dietary level of 0.67% calcium and 0.55% phosphorus.

[6] Ground milo, wheat, or barley can replace the ground corn. Ground oats can replace corn up to 20% of the total ration.

[7] If the ration is to be pelleted, 25 to 50 lb *(11 to 23 kg)* of molasses or binder can replace an equal amount of corn.

[8] To replace 44% soybean meal with 47% soybean meal or synthetic lysine, use the following ratios:
  Each 100 lb *(45 kg)* SBM (44%) = 93 lb *(42 kg)* SBM (47%) + 7 lb *(3 kg)* corn.
  Each 100 lb *(45 kg)* SBM (44%) = 3 lb *(1.4 kg)* 98% lysine hydrochloride + 1 lb *(0.45 kg)* dicalcium phosphate + 96 lb *(43.6 kg)* corn.

[9] The fat content of whole soybeans increases the energy content of the ration. For maximum utilization of the ration, the protein content has been increased to maintain a similar energy-to-protein ratio.

[10] The feed additive may be part of the vitamin premix, or if it is a separate premix, it should replace an equal amount of corn.

Fig. 23-20. Growing-finishing hogs self-fed a complete ration on slotted floor. (Courtesy, American Landrace Association, Lebanon, Ind.)

• **Formulation and use of complete protein supplements**—In addition to the formulations given in Tables 23-13 to 23-24, Table 23-25 gives the formulas for complete protein supplements, ranging from 34.49% to 43.64% protein content, which may be used to make growing-finishing, gestation, and lactation rations; and Table 23-26 gives the mixing directions for incorporating the Table 23-25 protein supplements in complete rations.

TABLE 23-25
COMPLETE PROTEIN SUPPLEMENTS[1,2,3,4]

Ingredient	1	2	3	4	5	6	7	8
	(lb)	(lb)	(lb)	(lb)	(lb)	(lb)	(lb)	(lb)
Wheat middlings (15.5% protein)[5]	192	102	59	202				
Soybean meal, solvent extracted (44.0% protein)	1,500	1,495	1,150	1,200				
Soybean meal, solvent extracted, dehulled (47.0% protein)					1,636	1,415	1,420	1,193
Alfalfa meal, dehydrated (17.0% protein)		100	200			100		
Meat and bone meal (50.0% protein)[6]			400	400			300	500
Fish meal, menhaden (62.0% protein)						150		100
Dicalcium phosphate	145	145	70	70	169	153	111	59
Limestone	75	70	33	40	90	77	64	43
Iodized salt	40	40	40	40	45	45	45	45
Trace mineral premix (Table 23-13)	16	16	16	16	20	20	20	20
Vitamin premix (Table 23-14)	32	32	32	32	40	40	40	40
Feed additives[7] (g/ton)								
Total	2,000	2,000	2,000	2,000	2,000	2,000	2,000	2,000
Calculated analyses:								
Metabolizable energy (kcal/lb)	1,228	1,203	1,167	1,222	1,260	1,243	1,249	1,254
Metabolizable energy (kcal/kg)	2,702	2,647	2,567	2,688	2,772	2,735	2,748	2,759
Protein (%)	34.49	34.53	37.46	37.97	38.45	38.75	40.87	43.64
Lysine (%)	2.24	2.24	2.22	2.26	2.54	2.59	2.54	2.65
Threonine (%)	1.40	1.41	1.44	1.46	1.55	1.57	1.59	1.66
Tryptophan (%)	0.49	0.50	0.45	0.45	0.52	0.52	0.49	0.48
Salt, added (%)	2.00	2.00	2.00	2.00	2.25	2.25	2.25	2.25
Calcium (%)	3.39	3.36	3.39	3.39	3.94	3.94	3.94	3.95
Phosphorus (%)	1.87	1.84	1.86	1.91	2.09	2.10	2.10	2.10

[1]Adapted by the authors from *Life Cycle Swine Nutrition,* Iowa State University, Ames, PM-489, June, 1988.

[2]These supplements can be used to make growing-finishing, gestation, or lactation rations. See Table 23-26, including the table footnote, for mixing directions.

[3]Supplements with meat and bone meal may be self-fed free-choice with shelled corn for growing-finishing pigs.

[4]To convert lb to kg, divide by 2.2. To convert g/ton (short) to g/ton (metric), divide by 0.907.

[5]The wheat middlings may be replaced with corn, corn distillers' grains with solubles, or other grain by-products.

[6]The meat and bone meal was assumed to contain 8.10% calcium and 4.10% phosphorus. If meat and bone meal with a higher concentration of calcium and phosphorus is used, the amount of dicalcium phosphate should be reduced accordingly.

[7]The concentration of feed additives will depend on the type of ration in which the supplement will be used. The concentration should be 3 to 5 times higher in supplements 1 to 4 and 4 to 6 times higher in supplements 5 to 8 than desired in the complete ration.

TABLE 23-26
COMPLETE RATIONS (USING SUPPLEMENTS IN TABLE 23-25)[1,2,3]

Ingredient	Protein Level in Complete Rations									
	13%		14%		15%		16%		17%	
	(lb)	(lb)	(lb)	(lb)	(lb)	(lb)	(lb)	(lb)	(lb)	(lb)
Corn, yellow (8.4% protein)	1,635	1,685	1,555	1,625	1,480	1,550	1,400	1,490	1,320	1,425
Supplements 1-4 (See Table 23-25.)	365		445		520		600		680	
Supplements 5-8 (See Table 23-25.)		315		375		450		510		575
Total	2,000	2,000	2,000	2,000	2,000	2,000	2,000	2,000	2,000	2,000
Calculated analyses:[4]										
Metabolizable energy (kcal/lb)	1,485	1,506	1,470	1,497	1,456	1,485	1,440	1,475	1,425	1,465
Metabolizable energy (kcal/kg)	3,267	3,313	3,234	3,293	3,203	3,267	3,168	3,245	3,135	3,223
Protein (%)	13.04	13.13	14.05	14.04	15.01	15.16	16.02	16.06	17.04	17.04
Lysine (%)	0.61	0.61	0.68	0.68	0.76	0.76	0.83	0.83	0.91	0.91
Threonine (%)	0.50	0.50	0.54	0.54	0.58	0.58	0.62	0.62	0.67	0.66
Tryptophan (%)	0.14	0.13	0.15	0.15	0.17	0.16	0.18	0.17	0.19	0.19
Salt, added (%)	0.37	0.38	0.45	0.42	0.52	0.51	0.60	0.57	0.68	0.65
Calcium (%)	0.62	0.63	0.76	0.75	0.88	0.89	1.02	1.01	1.15	1.14
Phosphorus (%)	0.54	0.54	0.60	0.60	0.66	0.66	0.72	0.72	0.78	0.78

[1]Adapted by the authors from *Life Cycle Swine Nutrition,* Iowa State University, Ames, PM-489, June, 1988.

[2]Suggested stages of production for using the above rations:

Grower	(— Protein and lysine low —)				+	+	(— Calcium high —)			
Finisher	+	+	+	+	+	+	(— Calcium high —)			
Gestation:										
3 lb (1.4 kg) per day	−	−	−	−	−	−	(Phosphorus marginal)		+	+
4 lb (1.8 kg) per day	−	−	(Phosphorus marginal)		+	+	+	+	+	+
5 lb (2.3 kg) per day	+	+	+	+	+	+	+	+	+	+
Lactation	−	−	(Phosphorus marginal)		+	+	+	+	+	+

[3]To convert lb to kg, divide by 2.2.

[4]Expected analysis using minimum analyses for each nutrient in the supplements from Table 23-25.

Also, commercial supplements (usually a combined protein-mineral-vitamin-additive supplement) may be bought and mixed with the locally available grain. Table 23-27 shows the proportion of grain (of 8.4% protein) to mix with supplements ranging from 30 to 45% protein to obtain finished rations ranging from 10 to 18% protein content.

Where commercial supplements are bought to use with farm-grown grains, they may be utilized in the following ways:

1. Mixed with ground, farm-grown grain in the approximate amounts shown in Table 23-27 to make a complete ration (see the Table 23-27 footnote example for mixing directions).

TABLE 23-27
GRAIN AND SUPPLEMENT COMBINATIONS (POUNDS) NEEDED TO FORMULATE RATIONS OF DIFFERENT PROTEIN LEVELS (GRAIN VALUED AT 8.4% PROTEIN)[1,2,3]

Protein in Supplement (%)		Percent Protein in Total Ration								
		10	11	12	13	14	15	16	17	18
		(lb)	(lb)	(lb)	(lb)	(lb)	(lb)	(lb)	(lb)	(lb)
30	Grain	1,852	1,759	1,667	1,574	1,481	1,389	1,296	1,204	1,111
	Supplement	148	241	333	426	519	611	704	796	889
31	Grain	1,858	1,770	1,681	1,593	1,504	1,416	1,327	1,239	1,150
	Supplement	142	230	319	407	496	584	673	761	850
32	Grain	1,864	1,780	1,695	1,610	1,525	1,441	1,356	1,271	1,186
	Supplement	136	220	305	390	475	559	644	729	814
33	Grain	1,870	1,789	1,707	1,626	1,545	1,463	1,382	1,301	1,220
	Supplement	130	211	293	374	455	537	618	699	780
34	Grain	1,875	1,797	1,719	1,641	1,563	1,484	1,406	1,328	1,250
	Supplement	125	203	281	359	438	516	594	672	750
35	Grain	1,880	1,805	1,729	1,654	1,579	1,504	1,429	1,353	1,278
	Supplement	120	195	271	346	421	496	571	647	722
36	Grain	1,884	1,812	1,739	1,667	1,594	1,522	1,449	1,377	1,304
	Supplement	116	188	261	333	406	478	551	623	696
37	Grain	1,888	1,818	1,748	1,678	1,608	1,538	1,469	1,399	1,329
	Supplement	112	182	252	322	392	462	531	601	671
38	Grain	1,892	1,824	1,757	1,689	1,622	1,554	1,486	1,419	1,351
	Supplement	108	176	243	311	378	446	514	581	649
39	Grain	1,895	1,830	1,765	1,699	1,634	1,569	1,503	1,438	1,373
	Supplement	105	170	235	301	366	431	497	562	627
40	Grain	1,899	1,835	1,772	1,709	1,646	1,582	1,519	1,456	1,392
	Supplement	101	165	228	291	354	418	481	544	608
41	Grain	1,902	1,840	1,779	1,718	1,656	1,595	1,534	1,472	1,411
	Supplement	98	160	221	282	344	405	466	528	589
42	Grain	1,905	1,845	1,786	1,726	1,667	1,607	1,548	1,488	1,429
	Supplement	95	155	214	274	333	393	452	512	571
43	Grain	1,908	1,850	1,792	1,734	1,676	1,618	1,561	1,503	1,445
	Supplement	92	150	208	266	324	382	439	497	555
44	Grain	1,910	1,854	1,798	1,742	1,685	1,629	1,573	1,517	1,461
	Supplement	90	146	202	258	315	371	427	483	539
45	Grain	1,913	1,858	1,803	1,749	1,694	1,639	1,585	1,530	1,475
	Supplement	87	142	197	251	306	361	415	470	525

[1] Adapted by the authors from *Life Cycle Swine Nutrition*, Iowa State University, Ames, PM-489, June, 1988. To convert lb to kg, divide by 2.2.

[2] The grain common to the area may be substituted in Table 23-27, with the 8.4% protein content changed in keeping with the protein content of the grain used, and the proportions of grain and supplement adjusted to obtain the desired percent protein in the total ration.

[3] **Example**: In order to obtain a total ration with 15% protein, each 2,000 lb *(908 kg)* of feed should contain 1,389 lb *(631 kg)* of the 8.4% protein grain and 611 lb *(277 kg)* of the 30% supplement.

2. Self-fed in separate feeders, with the ground or whole grain also being self-fed in separate self-feeders.

3. Hand-fed; with the supplement and the grain each being hand-fed in the proportions recommended in Table 23-27.

Pointers in Formulating Rations

In formulating rations and in feeding swine, the following points are noteworthy:

1. Feeds of similar nutritive properties can be interchanged in the ration as price relationships warrant.

2. If wheat, barley, oats, or grain sorghum is used instead of corn as the grain in a ration, the protein supplement may be slightly reduced.

3. When proteins of animal origin predominate, adequate mineral protection can be obtained by allowing hogs free access to a 2-compartment box or self-feeder with (a) salt (trace mineralized) in one side, and (b) a mixture of ⅓ salt (salt added for purposes of palatability) and ⅔ monosodium phosphate or other phosphorus supplement, in the other side. When supplements of plant origin constitute most of the source of proteins, add a third compartment to the mineral box and place in it a mixture of ⅓ salt (trace mineralized) and ⅔ ground limestone or oystershell flour.

Feeding Swine

4. When hogs are not exposed to sunlight or when dehydrated alfalfa meal is fed, vitamin D should be added in keeping with the recommended allowances (see Table 23-8).

5. Where the ration consists chiefly of white corn, barley, wheat, oats, rye, kafir, or by-products of these grains, there may be a deficiency of vitamin A (see Table 23-8 for recommended allowances).

6. Except for gestating sows and boars of breeding age, hogs are generally self-fed.

7. An exception should be made to the cafeteria-style feeding when the grain ration consists of barley, oats, rye, or kafir. These feeds are higher in protein content than corn, and for this reason are generally fed as a mixed ration. Otherwise, the pigs will often eat more protein supplement than is necessary to balance the ration. Likewise, when corn is fed as the grain, sometimes such protein supplements as (a) roasted soybeans, (b) soybean meal, and (c) peanut meal are too palatable to be fed separately from the corn, especially if the corn is not of good quality.

8. Full-fed finishing hogs will consume 4 to 5 lb of feed daily per 100 lb liveweight until they weigh 100 lb. They will eat 3 to 4 lb daily per 100 lb weight from this stage until marketing.

FEED SUBSTITUTION TABLE

Table 23-28, Feed Substitution Table for Swine, is a summary of the comparative values of the most common U.S. and Canadian feeds. In arriving at these values, two primary factors besides chemical composition and feeding value have been considered; namely, palatability and carcass quality.

TABLE 23-28
FEED SUBSTITUTION TABLE FOR SWINE (AS-FED BASIS) (See footnote at end of table.)

Feedstuff	Relative Feeding Value (lb for lb) In Comparison With The Designated (underlined) Base Feed Which = 100	Maximum Percentage of Base Feed (or comparable feed or feeds) Which It Can Replace for Best Results	Remarks
ENERGY FEEDS: GRAINS, BY-PRODUCT FEEDS, ROOTS AND TUBERS:[1]			
Corn, No. 2	*100*	*100*	Corn is the leading U.S. swine feed, about 25% of the total production being fed to hogs. Corn is high in energy, but low in lysine and tryptophan. It does not pay to grind corn for growing-finishing pigs, but it should be ground for older hogs.
Alfalfa hay, early bloom, or Alfalfa meal, dehy	> 75-85	10-50	Low energy, good source of carotene and B vitamins, unpalatable to baby pigs. None in starter, 0-5% grower-finishing, 0-50% gestation, 0-10% lactation.
Bakery waste	95-110	20-40	Bakery wastes average about 10% protein and 13% fat. Variable salt content. May constitute up to 20% of starter rations and up to 40% of grower-finisher, gestation, and lactation rations.
Barley	90-95	100	Of variable feeding value due to wide spread in test weight per bushel. Should be ground or rolled. Low lysine.
Beans (cull)	90	33-66	Cook thoroughly; on a dry weight basis, limit to ½ the grain ration for pigs under 100 lb (45.4 kg) and ⅔ of the grain ration for pigs above 100 lb (45.4 kg); supplement with animal protein.
Beet pulp	70-80	10	Bulky, high fiber, laxative. May constitute up to 10% of gestation and lactation rations.
Carrots (or beets, mangels, or turnips)	12-20	25	
Cassava, dried meal	85	33⅓	Low in methionine. Available as a swine feed in the tropics.
Corn and cob meal	80-90	0-70	Bulky, low energy. May constitute up to 70% of gestation rations.
Corn gluten feed (23% protein)	100	5-40	Corn gluten feed is bulky, low in lysine, and not too palatable. It may constitute up to 5% of starter rations, 10% of grower-finisher and lactation rations, and 40% of gestation rations.
Corn, high lysine	100-105	100	Superior to corn in lysine.
Corn meal	100	20	
Corn silage (25-30% D.M.)	20-30	0-90	Bulky, low energy, feed to sows only.
Emmer	80-90	80	Emmer may be used about like barley.
Fat (stabilized)	185-210	5	High energy, reduces dust.
Hominy feed	95	50	Hominy feed will produce soft pork if it constitutes more than ½ the grain ration. Hominy feed is subject to rancidity.
Millet (Proso)	85-90	50	Low lysine.
Molasses, beet	70	5	Used in pelleting. Laxative at high levels.
Molasses, cane (74% D.M.)	70	5	Used in pelleting. High levels (above 15% of the ration) cause soft, watery feces.
Molasses, citrus	70	5	It takes pigs 5-7 days to get used to the bitter taste of citrus molasses.
Oats	70-80	15-70	Grind for swine. Feeding value varies according to test weight per bushel. Oats may constitute up to 15-20% of grower-finishing rations and up to 70% of gestation rations.
Oats, groats	110-115	20	Palatable, but expensive. Primarily used in starter rations in which it may consititute up to 20%.
Peanuts	120-125	100	Peanuts are usually fed by hogging off.
Peas, dried	90-100	100	Normally peas should be fed to swine as a protein supplement. Two tons of peas equal 1 ton of grain plus 1 ton of soybean meal. Peas are low in methionine.
Potatoes, Irish (24% D.M.)	25-28	25-50	Not palatable in the raw state; must be cooked. When cooked and fed in a ratio of 3 lb (1.4 kg) of potatoes to 1 lb (0.45 kg) of grain, they are worth 25-28% as much as corn.
Potatoes (Irish), dehy	100	33⅓	May constitute 25% of grower-finisher rations and 50% of gestation rations.
Potatoes (sweet)	20-25	33⅓-50	Cooking also improves the feeding value of sweet potatoes.
Potatoes (sweet), dehy	90	33⅓	

(Continued)

TABLE 23-28 (Continued)

Feedstuff	Relative Feeding Value (lb for lb) In Comparison With The Designated (underlined) Base Feed Which = 100	Maximum Percentage of Base Feed (or comparable feed or feeds) Which It Can Replace for Best Results	Remarks
ENERGY FEEDS: GRAINS, BY-PRODUCT FEEDS, ROOTS AND TUBERS:[1] (Continued)			
Rice (rough rice)	80–85	50	Low energy, low lysine. Rice should be ground.
Rice bran	100	33⅓	If more than ⅓ of the grain consists of rice bran, soft pork will result.
Rice polishings	100–120	33⅓	Limited because feed becomes rancid in storage and soft pork will be produced.
Rice screenings	95	50	
Rye	90	20–30	Should be limited because it is unpalatable. Grind for swine. Watch for ergot.
Sorghum, grain	95	100	Check protein content and add supplement as necessary. Both very dry grain and bird-resistant varieties should be ground. Grain sorghum is low in lysine.
Spelt	65–80	25	Low energy, low lysine. Value varies according to the amount of hulls.
Sugar	70–80	0–5	High palatability, no protein. Normally, used only in starter rations.
Sunflower seed	100	50	High energy, high fiber.
Triticale	90–95	50	Higher levels not palatable. Watch for ergot.
Wheat	100–105	100	Feed whole if self-fed. Otherwise, grind coarsely; fine grinding makes it pasty and unpalatable. Low in lysine. Wheat-corn mixtures are more efficient than wheat alone.
Wheat bran	65–75	15–25	Bulky, high fiber, laxative. Bran is particularly valuable at farrowing time.
Wheat middlings	103	20	May be used as a partial substitute for grain.
Wheat standard middlings	85–100	10–30	Use as a partial grain substitute.
Wheat red dog and wheat white shorts	115–120	25	
Whey, dry	50	5–20	High lactose, very palatable. Use for baby pigs.
Whey, liquid	15	5–20	High lactose, very palatable. Use for baby pigs.
PROTEIN FEEDS:			
Soybean meal (41–50%)	**_100_**	**_100_**	Well balanced in amino acids. Best quality of all plant protein supplements. Very palatable.
Blood meal (80%)	120–130	20	High in protein (above 80%), high in lysine, but low in isoleucine. Not very palatable.
Buttermilk, dry	90–105	100	Good amino acid balance.
Buttermilk, liquid	15	100	Pound for pound, worth ⅒ as much as dried buttermilk.
Buttermilk, semisolid	33⅓–50	100	Pound for pound, worth ⅓ as much as dried buttermilk.
Canola meal (32–44%)	90	75	Low in goitrogenic compounds.
Copra meal (coconut meal)(21%)	50	25	
Corn, distillers' dried grains w/solubles	65–75	5	Used primarily as B-vitamin and unidentified sources, usually at about 5% of the ration.
Corn gluten meal (60% protein)	90	50	Bulky, low in lysine, and not too palatable.
Cottonseed meal (36–48%)	85	33⅓	Except when new glandless cottonseed meal is used, high levels may produce gossypol poisoning; hence, the level of cottonseed meal in swine rations should not exceed 8–9% of the total ration. Cottonseed meal is low in lysine.
Fish meal (60%)	115	100	Excellent balance of amino acids, and good source of calcium and phosphorus.
Linseed meal (35%)	80	25–50	Low in lysine; slightly laxative.
Malt sprouts	100	10	Malt sprouts contain a growth factor(s). They result in increased feed intake and gain.
Meat and bone meal (50%)	100	100	Low in tryptophan; good source of calcium and phosphorus.
Meat scraps (50–55%)	100	100	
Peanut meal (45%)	95	50	Becomes rancid when stored too long. Low in lysine; very palatable.
Peanuts	60–70	50	Peanuts are usually fed by hogging off. High levels will produce soft pork.
Peas, dried	50	50	
Rapeseed meal (32–44%)	85–90	33⅓	Rather unpalatable. Contains goitrogenic compounds that can be hazardous. But is usually detoxified.
Shrimp meal	90–100	50	
Skim milk, dried	90–120	100	Excellent-quality protein; very palatable; expensive. Especially good in prestarter and starter rations, of which it may constitute up to 10%.
Skim milk, liquid			Pound for pound, worth ⅒ as much as dried skim milk.
Soybeans, full fat, cooked	90–100	25–40	High energy. At high levels, will produce soft pork.
Sunflower meal (36–45%)	90–95	50	For swine, it should be combined with high-lysine supplements such as meat scraps or fish meal.
Tankage (60%)	110	100	Good source of calcium and phosphorus. Low in tryptophan. Not palatable.
PASTURES AND DRY LEGUMES:			
Pasture, good		5–20% of grain, and 20–50% of protein supplement.	Pasture and dry legumes are sources of good-quality proteins, of minerals, and of vitamins.
Alfalfa meal		It can replace all of pasture, in drylot rations.	Low energy, good source of carotene and B vitamins, unpalatable to baby pigs. For drylot rations, include 5–10% alfalfa in ration of grower-finishing pigs, up to 50% in ration of gestating sows, and up to 10% for lactating sows.

[1] Roots and tubers are of lower value than the grain and by-product feeds due to their higher moisture content.

FEEDING BABY PIGS

Baby pigs should have access to a creep feed beginning at 7 to 10 days of age. Commercial prestarters and starters are readily available, or farm-mixed rations can be used (see Table 23-20 for suggested prestarter rations, and see Table 23-21 for suggested starter rations). Pigs should receive a prestarter ration until they are about 3 weeks of age, after which they should be switched to a starter ration until they weigh about 40 lb.

For successful baby pig feeding, the following pointers are pertinent:

1. Begin by giving the baby pigs a mere handful of creep feed, replenish daily. The creep feed should not be allowed to become stale or contaminated. Place feed in flat pans.

2. Once the pigs have started to eat readily, place the ration in a creep feeder so that they have access to the ration at all time. One linear foot of feeder space should be provided for each five pigs. The edge of the feeder trough should not be more than 4 in. above the floor. For maximum consumption of the creep ration, the feeder should be located near the waterer for the baby pigs.

3. Make clean, fresh water available to the young pigs in a separate waterer. It is not sufficient to rely on the sow's waterer to furnish water to the baby pigs.

4. The creep area should be light, warm, dry, and draft-free. It should be located in an area where the pigs are the least disturbed. Excitement, noise, and a change in feeding routine affect eating habits and subsequent feed consumption. Having the creep area near the sleeping area encourages more frequent eating. Arrange the creep area in such a way that it can be easily cleaned, and so that feed and water can be supplied conveniently without the producer getting into the area.

5. Individual litter creep areas are preferable. Where several litters have access to one creep area, it is advisable to limit the number of pigs to about 40 per creep.

6. If postweaning scours are encountered, the substitution of 200 to 400 lb of ground oats for a like amount of corn in rations Table 23-21 (Nos. 1, 3, 4, and 5) may be helpful.

7. By the time pigs reach a weight of 40 lb, they will have consumed about 54 lb of feed (4 lb of prestarter and 50 lb of starter).

FEEDING ORPHAN PIGS

There is no replacement for the sow's colostrum. If the newborn pig does not receive colostrum within 4 to 6 hours of birth, it has a lesser chance for survival. An orphan pig can obtain colostrum by being placed with another sow (a foster sow) that has just farrowed. If no such sow is available, the orphan can be started on a good commercial milk replacer, prepared and fed according to directions.

From 5 to 7 days of age until about 3 weeks of age, the orphan pig can be fed a dry 22 to 23% crude protein prestarter (see Table 23-20). At this time, it can be switched to a 20 to 21% crude protein pig starter (see Table 23-21).

FEEDING GROWING-FINISHING PIGS

In the practical swine enterprise, growing-finishing generally refers to that period from weaning to market weight of about 240 lb. Because hogs are finished at an early age, the process really consists of both growing and finishing. In a general way, there are 2 methods of finishing hogs for market: (1) full feeding all the time until the animals attain a market weight, and (2) limited feeding early in the period, with full feeding the last 60 to 75 days of the period before marketing.

Neither system, full feeding nor limited feeding, can be recommended as being best for all conditions. The plan to follow should be determined by (1) market conditions, (2) type and breeding of the pigs, (3) price of feeds, (4) feeds available on the farm, (5) kind and extent of pastures available, (6) available labor, and (7) capital invested in facilities. Self-feeders are well adapted to a system of full feeding, but hand-feeding or interval feeding are necessary in any plan for limiting the ration.

Suggested growing-finishing rations are given in Table 23-24. Note that provision is made for meeting the nutritional needs at two different stages of growth—40 to 120 lb, and 121 to 240 lb—with three different rations suggested for each stage.

FEEDING PROSPECTIVE BREEDING GILTS

Prospective breeding gilts should be kept from getting too fat. Meat-type animals can usually be left on a high-energy ration until they reach 150 to 200 lb without becoming too fat. It is neither necessary nor desirable that females intended for breeding purposes carry the same degree of finish as market animals. After selecting replacement gilts, they should be fed as follows:

1. Give about 5 lb per head per day through their second heat period.

2. Flush—full feed—after the second heat period until breeding on the third heat period.

3. After breeding, limit the feed intake to 3 to 5 lb per day. Overfeeding during gestation can cause embryonic death and thus decrease litter size.

FEEDING BOARS

The feed requirements of the herd boar are about the same as those of a female of equal weight. He should always be kept in thrifty, vigorous condition and virile. In no case should boars be overfat, nor should they be in a thin rundown condition. Normally, the following feed allowances will suffice: for boars weighing 120 to 150 lb, 6 to 9 lb of feed daily; for mature boars, 5 to 7 lb of feed daily. A more liberal ration must be provided in the wintertime and when the sire is in heavy service. The feed allowance should be varied with the age, development, temperament, breeding demands, and roughage consumed.

Boars and pregestating/gestating sows may be fed the same rations. Rations, formulated for feeding 3, 4, or 5 lb per head daily, are presented in Tables 23-16, 23-17, and 23-18 (see pp. 532-533).

FEEDING BROOD SOWS

The nutrition of brood sows is critical, for it may materially affect conception, reproduction, and lactation. Proper feeding of sows should begin with replacement gilts and continue through each stage of the breeding cycle—flushing, gestation, farrowing, and lactation.

• **Flushing Sows**—*The practice of conditioning or having the sows gain in weight just prior to breeding is known as* *flushing.* The purpose of flushing is to increase the number of ova shed during estrus. About 10 to 14 days prior to expected breeding, the female should be fed a ration that will produce gains of 1 to 1¼ lb per day. Generally 6 to 8 lb per head per day of a high-energy feed that is well balanced in minerals and vitamins, is adequate. Immediately after breeding, the females should be put back on limited feeding. Continuation of a high level of feeding after breeding will result in a higher embryo mortality.

• **Gestation Period**—The nutrients fed the pregnant gilt or sow must first take care of the usual maintenance needs. If the gilt is not fully mature, nutrients are required for both maternal growth and growth of the fetus. Quality and quantity of proteins, minerals, and vitamins become particularly important in the ration of young pregnant gilts, for their requirements are much greater and more exacting than those of the mature sow.

Approximately two-thirds of the growth of the fetus is made during the last month of the gestation period. It may be said, therefore, that the demands resulting from pregnancy are particularly accelerated during the latter third of the gestation period. Again, the increased needs are primarily for proteins, minerals, and vitamins.

It is important that the condition of dry sows be regulated so that they are neither too fat nor too thin at farrowing time. Overly fat sows may have difficulty in farrowing and give birth to weak or dead pigs. Sows that are too thin at farrowing tend to become suckled down during lactation. Thus, one way or another, limited feeding is a must for gestating gilts and sows. This may be accomplished by any one of the following feeding systems (these feeding systems are further detailed later in this chapter in the section headed "Feeding Systems"):

1. By adding sufficient bulk.
2. By interval feeding.
3. By group hand-feeding.
4. By individual feeding.

In addition to the above limited feeding systems, the use of pasture should be considered. Where available, a leguminous pasture is the ideal way in which to limit-feed gestating gilts and sows. Dry sows on good legume pasture are usually fed ½ lb less supplement and 2 lb less corn per day. In addition to limiting the feed intake, the pasture system provides valuable quality protein, minerals, vitamins, and exercise.

Suggested gestation rations, formulated for feeding 3, 4, or 5 lb per head daily, are given in Tables 23-16, 23-17, and 23-18.

• **Farrowing Time**—It is considered good practice to feed lightly and with bulky laxative feeds from 4 to 5 days before and after farrowing. Wheat bran or oats may constitute half of the limited ration, and a small amount of linseed meal may be added.

The sow may be watered at frequent intervals before or after farrowing, but in no event should she be allowed to overeat. It is also a good plan to take the chill off the drinking water in the wintertime.

• **Lactation Period**—The nutritive requirements of a lactating sow are more rigorous than those during gestation. They are very similar to those of a milk cow, except they are more exacting relative to quality proteins and the B vitamins because of the absence of rumen synthesis in the pig. A good lactating sow will produce an average of about 1 gal of milk daily during the suckling period. A sow's milk is also richer than cow's milk in all nutrients, especially in fat. Thus, sows suckling litters need a liberal allowance of concentrates rich in protein, calcium, phosphorus, and vitamins.

The lactating sow should be provided with a liberal feed allowance—ranging from 2½ to 4½ lb daily for each 100 lb weight. Generous feeding during lactation, with a small shrinkage in weight, is more economical than a stingy allowance of feed, for the nutrients in milk must come either from the feed or from the sow's body. Lactating sows are commonly self-fed, because even when hand-fed they are practically on full feed.

Suggested lactation rations are given in Table 23-19.

FITTING RATIONS FOR SHOW AND SALE SWINE

Any of the rations listed in Tables 23-16, 23-17, 23-18, 23-22, or 23-24 for the respective classes and ages of swine are suitable for use in fitting show animals of similar classification.

In fitting show barrows, it may be necessary to decrease or discontinue slop feeding 2 to 4 weeks before the show to avoid paunchiness and lowering of the dressing percentage.

When oatmeal (oat groats, rolled hulled oats) is not too high priced, many successful hog caretakers replace up to 50% of the grain (corn, wheat, barley, oats, and/or sorghum) in the ration with oatmeal. They do this especially when fitting hogs—both breeding animals and barrows—in the younger age groups. Oatmeal is highly palatable, lighter, and less fattening than corn.

Suitable minerals and vitamins should always be provided.

FEEDING SYSTEMS

The choice of the feeding system(s) and the choice of the ration(s) must go hand in hand. For example, if the grain and the protein supplements are to be self-fed in separate feeders or compartments, it is important that they be of equal palatability; otherwise, pigs will consume too much of one and too little of the other. A listing, along with a discussion, of each of the common feeding systems follows.

• **Complete self-fed rations**—The trend is toward the use of complete self-fed rations for baby pigs and growing-finishing hogs, because, in comparison with free-choice feeding, they (1) lend themselves better to automation, (2) provide better control of nutrient intake, and (3) result in faster gains.

• **Floor or drop feeding**—Floor or drop feeding is particularly suited to the controlled feeding of growing-finishing swine or the breeding herd. Feeding in the sleeping area encourages cleanliness, since pigs are less inclined to defecate where they eat. Feed wastage is reduced to a minimum when the animals do not have more feed available than they will consume at one eating. Even though automated, restricted feeding requires close attention, because the daily feed intake of pigs is affected by weather.

• **Free-choice**—Grain and protein supplements may be fed separately and free-choice.

• **Liquid feeding**—Liquid feeding usually involves mixing predetermined amounts of feed and water prior to, or at the time of, feeding.

- **Limit feeding**—With gestating sows, limit feeding to 4 to 6 lb per head daily is a must in order to keep them from getting too fat. A discussion of limit feeding of (1) gilts and sows, and (2) growing-finishing pigs follows.

  1. **Gilts and sows.** Replacement gilts should be started on a limited feeding program at 180 to 200 lb; and all gestating sows and gilts should be limit fed. Limit feeding may be accomplished by any one of the following methods:

     a. **By feeding bulky, fibrous feeds,** such as silage, haylage, or alfalfa, with such feed constituting at least one-third of the ration.

     b. **By interval feeding,** in which gilts or sows are turned to self-feeders for 2 to 8 hours every second or third day.

     c. **By group hand-feeding** a limited ration to several sows. This is apt to result in the "bossy" sows getting too much and the "timid" sows getting too little. This problem can be partially alleviated by feeding over a large area.

     d. **By individual feeding** in either individual stalls or in tie stalls, tethered by a neck collar or belt.

Fig. 23-21. Landrace sows tethered by neck collar and individually fed. (Photo by A. H. Ensminger)

  2. **Growing-finishing pigs.** Sometimes growing-finishing pigs are limit-fed in order to produce leaner carcasses. Usually, it is started when pigs weigh around 100 lb and feed is limited to about 85 to 95% of what pigs of comparable age consume when self-fed. Limit feeding of market hogs results in slower gains, increased labor, and more mechanization. Thus, unless sufficient premium is paid for the modestly leaner carcasses, it cannot be justified.

FEEDER PIG PRODUCTION

Feeder pig production refers to the production and sale of immature pigs weighing 30 to 60 lb, usually throughout the year, for growing and finishing on other farms. It makes for a two-phase system in swine production, with some swine operators specializing in feeder pig production and others in growing-finishing market hogs. Several important scientific and technological developments which occurred in the swine industry in the 1950s and early 1960s ushered in considerable two-phase production of hogs. Among such developments were: (1) specific pathogen-free (SPF) herds and other improved disease control measures; (2) confined and continuous production—which increased specialization in breeding, in farrowing, and in finishing; and (3) increased mechanization.

OTHER FEED/MANAGEMENT RELATED ASPECTS

In addition to the subject matter covered earlier in this chapter, there are other feed/management related aspects of importance in swine production. Some of them will be discussed in the sections that follow.

Docking Tails

Tail biting accompanies close confinement. It results when pigs are prevented from rooting, nibbling, and chewing—from disturbing the pig's normal behavior pattern. Tail docking has become a common management practice to prevent subsequent tail biting of pigs in confinement. It should be done on all market hogs. Tails should be cut ¼ to ½ in. from the body with side-cutting pliers or another blunt instrument; the crushing action stops bleeding. The tail stump should be disinfected with a good antiseptic, and the instrument should be disinfected after docking each pig. (See Chapter 17, Animal Behavior/Environment, for additional information on tail biting and docking.)

Nutritional Anemia

Anemia is a blood condition in which there is a deficiency of hemoglobin which transports oxygen to various parts of the body. It is caused by a deficiency of iron and/or copper in the diet, and it is most likely to occur in nursing pigs that do not have access to soil.

Anemic pigs show listlessness, rough hair coat, wrinkled skins, drooping ears and tails, pale membranes around the mouth and eyes, and labored breathing.

Any of the following anemia prevention measures may be used:

1. Inject intramuscularly in the neck or ham muscle 100 to 200 mg of iron, in the form of iron dextran, iron dextrin, or gleptoferron into baby pigs at 2 to 3 days of age. If pigs remain in confinement and do not have access to creep feed at an early age, a second injection at 2 to 3 weeks of age is desirable. Injection is the method of choice, for it assures that every pig receives its requirement.

2. Orally administer iron from iron chelates within the first few hours of life; early administration before gut closure to large molecules is crucial. To ensure daily intake by all pigs, it is important to have a preparation that is palatable and readily consumed. Also, placement of the oral preparation at the right location in the creep area is most important.

3. Give the pigs iron tablets or paste at 2 to 3 days of age. Repeat the treatment every 7 to 10 days until the pigs are eating their creep ration adequately. If pills are given, it is important to see that the pigs swallow them and not spit them out.

4. Place clean soil in the farrowing pen daily. Soil should not be contaminated with parasite eggs and other disease organisms. Iron sulfate can be sprinkled over the soil.

5. Swab sow's udder daily with a solution of 1 lb ferrous sulfate dissolved in 1 gal of warm water.

Scouring (Diarrhea, Enteritis)

Scouring is one of the major problems facing swine producers. It is estimated that about 40% of U.S. swine herds are affected with scouring, and that about 20% of pig losses between farrowing and weaning are caused by scouring. Many different etiological agents cause scouring, including nutrition, management, environment, stress, bacteria, viruses, and parasites. Although the nutritional aspects of scouring are frequently discussed, they are poorly understood. However, it is generally recognized that good nutrition, along with good management and minimal stress, will lessen the incidence of scouring and be effective treatment when an outbreak of the disease occurs.

Weaning Pigs

The optimum age to wean pigs varies considerably, depending on nutritional programs, facilities, environment, health, and management. The average age at weaning for pigs in the United States is about 4 weeks, with a range of 2 to 7 weeks. Most early weaned pigs are weaned at 3 to 4 weeks of age. The earlier pigs are weaned, the better the required feeding and management practices.

For best results, the guidelines given in Table 23-29 should be observed when planning for early weaning.

TABLE 23-29
GUIDELINES TO SUCCESSFUL EARLY WEANING[1]

Guideline	Age in Weeks				
	1	2	3	4	5
Minimum pig weight .... (lb)	5	9	12	15	21
Nursery temperature at pig level .... (°F)	85	85	83	81	79
Minimum floor space per pig .... (sq ft)[2]	3	3	3	3	3
Maximum number of pigs per linear foot of feeding space	5	5	5	5	5
Maximum number of pigs per nipple waterer[3]	8	8	8	8	8
Maximum number of pigs per group	10	10	10	15	25

[1]See Appendix for conversion of U.S. customary to metric.
[2]The figures given herein are for solid floors. On slotted floors, this may be lowered 2 sq ft per pig from 1 to 5 weeks of age.
[3]Where bowls are used instead of nipples, there should be one bowl for each 12 pigs.

The age of pigs at weaning may vary from herd to herd, according to the facilities available, intensity of operation, and managerial skills of the producer. Generally, pigs can be weaned over a wide age range; however, the younger the pigs, the more demanding the management required to do it successfully. Observance of the following guides will reduce the stress at weaning:

1. Wean only pigs weighing more than 12 lb.

2. Wean over a 2- to 3-day period, weaning the larger pigs in the litter first.

3. For 3-week-old pigs provide an environmental temperature of 80–85°F.

4. Group pigs according to size.

5. Limit numbers in a pen to 30.

6. Limit feed intake for 48 hours if post-weaning scours are a problem.

7. Provide 1 feeder hole for 4 to 5 pigs and 1 waterer for each 20 to 25 pigs.

8. Medicate drinking water if scours develop.

Feed Required to Produce a Pound of Market Hog

Nationally, it has been estimated that it requires 4.0 lb of feed to produce 1 lb of on-foot hog (live) from birth to market weight (see Chapter 1, Table 1-2), exclusive of the feed required by sows and boars to produce pigs. This is high. But remember that 25 to 30% of all pigs farrowed die before weaning. Remember, too, that many swine producers are inefficient.

Table 23-30 shows realistic goals for well-managed swine operations.

TABLE 23-30
ESTIMATED FEED REQUIRED TO PRODUCE 240-LB MARKET PIG[1]

Stage of Production	Feed Required per 240-Lb Market Pig
	(lb)
Sow gestation ration (includes pregestation and breeding)	110
Boar ration	8
Lactation ration	45
Starter ration (creep to 40 lb)	54
Grower-finisher ration (40 to 240 lb)	700
Total, lb	917

Per 100 lb of pork produced $\frac{917}{240} \times 100 = 382$ lb

[1]See Appendix for conversion of U.S. customary to metric.

Effect of Sex on Performance of Growing-Finishing Hogs

When full fed, boars consume 10 to 15% less feed daily than barrows or gilts and are 10 to 15% more efficient in feed conversion. Also, boars gain faster than barrows and gilts. Barrows gain approximately $\frac{1}{10}$ lb faster per day than gilts, which reduced their age at slaughter by 10 days. Feed per pound of gain is similar for barrows and gilts. Gilts yield carcasses having .11 in. less backfat, .52 sq in. larger loin eye area, and 1.8% more lean cuts than barrows. Dressing percentage usually favors barrows, which is consistent with their greater depth of backfat.

Soft Pork

Feed fats are laid down in the body of pigs without undergoing much change. Thus, when finishing hogs are liberally fed high fat content feeds in which the fat is liquid at ordinary temperatures, soft pork results. This condition prevails when hogs are liberally fed such feeds as soybeans, peanuts, mast, or garbage. The oil of the cereal grains is also liquid at ordinary temperatures, but fortunately the fat content in these feeds is relatively low. When such feeds are liberally fed to swine, most of the pork fat is actually formed from the more abundant carbohydrates.

Soft pork is undesirable from the standpoint of both the processor and the consumer. It remains flabby and oily even under refrigeration. In soft pork, there is a higher shrinkage in processing; the cuts do not stand up and are unattractive in the showcase; it is difficult to slice the bacon; and the cooking losses are higher through loss of fat. For these reasons, hogs that are liberally fed those feeds known to produce soft pork are heavily discounted on the market.

Experimental evidence and practical observation have shown, however, that when a ration producing hard fat is given following a period of feeds rich in unsaturated fats, the body fat gradually becomes harder. It has also been found that this process takes place more rapidly if the animals are first fasted for a period before the change in ration is made. This practice is called *hardening off*. Thus, many hogs that are, for practical reasons, finished primarily on such feeds as soybeans, peanuts, or garbage, are hardened off with a ration of corn or some other suitable grain.

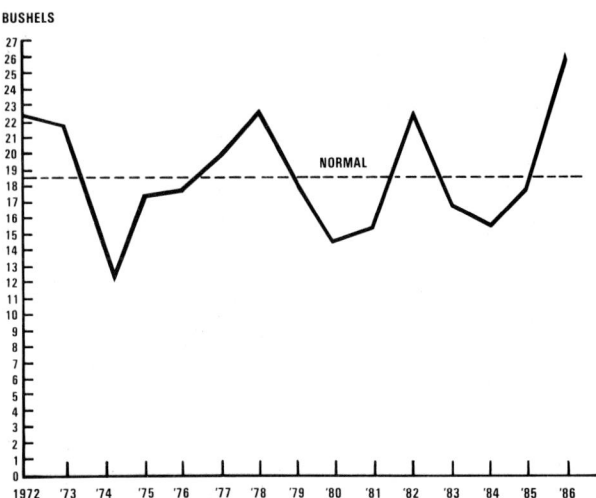

Fig. 23-23. The corn-hog ratio, 1972 to 1986. As shown, it averaged 18.6. (Based on data from *Agricultural Statistics*, 1987, USDA, p. 278, Table 412.)

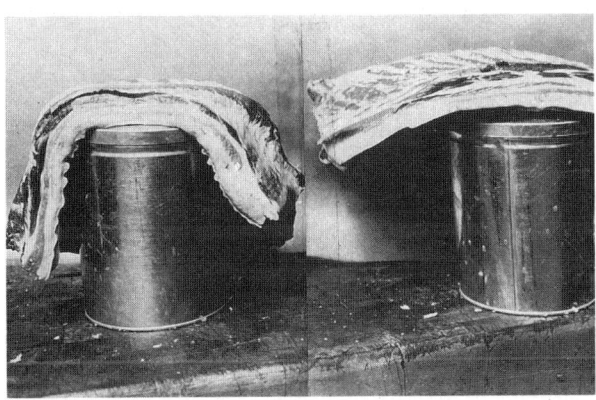

Fig. 23-22. Soft pork. Feed fats do affect body fats. The bacon belly on the left came from a hog liberally fed soybeans. (Courtesy, University of Illinois, Urbana)

Corn-Hog Ratio

The corn-hog ratio refers to the number of bushels of corn required to be equivalent in value to 100 lb of live hogs at local markets, based on average prices received by farmers for corn and hogs. During the 15-year period, 1972 to 1986, the hog-corn ratio averaged 18.6. This means that the price relationship was such that 18.6 bushels of corn equalled in value 100 lb of hogs.

A high corn-hog ratio—one above 18.6 in recent years—indicates cheap corn and high-priced hogs and likely profit to the producer—conditions that stimulate more breeding and more feeding to heavier weights. On the other hand, a low ratio, one which is below 18.6 means high-priced corn and low-priced hogs—conditions that result in less breeding and feeding of swine.

Contract Hog Production (Custom Feeding/Leasing)

A contract is an agreement between two or more persons to do or refrain from doing certain things. In recent years, swine producers have shown increasing interest in contract hog production due to (1) the high cost of capital, (2) the difficulty of many producers in obtaining adequate financing, and (3) the desire to forego the possibility of large profits for the assurance of more reliable returns. In the late 1980s, an estimated 8 to 10% of U.S. hogs were under some kind of production contract, with a much smaller number under a marketing contract.

An overview of the most common types of contract used in the swine industry follows.

• **Production Contracts**—Investors, feed dealers, farmers and others are often interested in producing hogs, but are unwilling or unable to provide the necessary labor, facilities, and equipment. Some producers have also found contract production to be an effective method of expanding their operations. So, these entrepreneurs find producers who are willing to furnish the labor and equipment in exchange for a fixed wage or share of the profits. The resulting contracts vary considerably in form and responsibility of each party involved.

The more popular contracts provide for fixed payment, direct feeding, or profit sharing. These contracts are most commonly used for feeder pig production and hog finishing.

• **Farrow-to-Finish Contracts**—Most farrow-to-finish programs are set up on a percentage basis to reflect the relative amount of input supplied by each person or firm.

1. **Option 1.** Based on input costs, the following participants may share in a percentage of gross sales: The breeding stock supplier; the feed and medications supplier; the management consultant and computerized record services; and the producer-supplier of facilities, labor, veterinary care, utilities, and insurance.

2. **Option 2.** The current hog inventory is purchased by a limited partnership, which supplies sow replacements. Each of the following contract participants receives a percentage of the proceeds when hogs are marketed: The feed and medications supplier; the management agency-supplier of production and marketing guidance; and the producer-supplier of facilities, labor, utilities, veterinary costs, repairs, and manure disposal. The remaining percentage is split between the limited partnership and the general partner who manages the partnership.

• **Breeding Stock Leasing**—Under a breeding stock lease, the contractor furnishes the producer with breeding age gilts and/or boars. The rent paid by the producer for the breeding stock may be either a specified number of pigs or an equivalent amount of money at designated times. The popularity and use of breeding stock leases has declined in recent years.

• **Marketing Contracts**—Marketing contracts are of two kinds: (1) market hog contracts, and (2) feeder pig contracts.

• **Market hog contract**—A market hog contract is a forward sale contract between a buyer (normally a meat packer or a marketing agent) and a seller (normally a producer), in which the producer agrees to sell, at a specified date, a specified number of hogs to a buyer at a certain price. Normally, the following terms are detailed in a forward marketing contract: (1) the quantity, with the minimum ranging from 5,000 lb to 30,000 lb; (2) the date and location of delivery; (3) acceptable weights and grades, including premiums and discounts; (4) a description of the pricing mechanism (some contracts now price hogs on a grade and yield basis); (5) provisions for non-deliverable hogs and unacceptable carcasses; (6) provisions outlining the credit requirements of the seller and inspection of the hogs by the buyer; and (7) provisions for breach of contract.

Under a forward sale contract, the producer retains all risks of production, other than selling price. The producer uses the forward sale contract to reduce the risk of price fluctuations and to lock in an acceptable profit. But a forward sale contract may also cause the producer to miss out on greater profits if prices rise. Sometimes, a minimum price (a floor price) is used for hogs, in which the buyer guarantees the seller a minimum price (a floor price), with the seller receiving whichever price is higher at market time—the floor price, or the market price.

• **Feeder pig contracts**—Typically, feeder pig marketing contracts are between a marketing agency, often a cooperative, and a pig producer, in which the marketing agency agrees to market the pigs of a producer for a fee. A feeder pig marketing contract may contain the following provisions: (1) the producer agrees to market exclusively through the agency; and (2) the marketing agency specifies management practices and weight of feeder pigs at marketing. Essentially, producers are hiring market expertise through feeder pig contracts.

Poisons and Toxins

Swine are susceptible to a number of poisons, any one of which may be disastrous in a herd. Among them are the following: moldy feed, including three species of mycotoxins—aflatoxins, ergot poisoning, and estrogenic syndrome; pitch poisoning; lead poisoning; mercury poisoning; pesticides; plant poisoning, involving several plants that are toxic to swine; and blue-green algae.

(Also see Chapter 5, Nutritional Disorders/Toxins.)

Environmental Effects of Swine

Pollution potential, affecting the environment of both people and swine, increased as the U.S. swine industry moved toward specialization, mechanization, high animal density, and confinement.

To operate compatably within the community, to provide maximum self-protection, and to avoid neighbor complaints and legal actions seeking either monetary damages or court injunctions, swine producers must be aware of some basic information and strategy concerning pollution and apply pollution control measures appropriate to the location.

(Also see Chapter 12, section on "Animal Wastes [Manure and Litter].")

QUESTIONS FOR STUDY AND DISCUSSION

1. How has the shift to confinement rearing affected the nutritional well-being of pigs?

2. What is the economic importance of feed for swine?

3. Why aren't nutritional requirements, or standards, such as NRC Tables 23-1 to 23-7, absolute, final, and unchangeable?

4. What is the difference between nutrient requirements and nutrient allowances? Why are the allowances generally higher than the requirements?

5. Discuss carbohydrates and fats/oils as energy supplying nutrients. Describe the symptoms in swine of (a) energy deficiency, and (b) essential fatty acid deficiency.

6. Define (a) essential amino acid, and (b) quality of protein. List the essential amino acid deficiencies of small grains (wheat, barley, oats), corn, and meat and bone meal.

7. What are the symptoms of protein (amino acid) deficiency?

8. What peculiarities of swine husbandry are conducive to swine suffering from mineral deficiencies?

9. Give the (a) function, (b) deficiency symptoms, (c) practical sources, and (d) prevention of deficiency of each of the following minerals in the pig: salt, calcium, phosphorus, iodine, and iron.

10. Discuss (a) feeds as a source of minerals, and (b) methods of feeding mineral supplements to swine.

11. Give the (a) function, (b) deficiency symptoms, (c) practical sources, and (d) prevention of deficiency of each of the following vitamins in the pig: A, D, E, choline, pantothenic acid, riboflavin, and thiamin.

12. List feed sources of unidentified factors for swine.

Feeding Swine

13. Discuss the relationship of water consumption of swine to (a) feed consumption, and (b) temperature.

14. Why are antibiotics used as feed additives? What factors affect the response of swine from feeding antibiotics?

15. Discuss the history, prevalence of use, benefits, and concerns stemming from the use of sulfonamides (sulfas) for swine.

16. What is porcine somatotropin (PST)? What great leaps forward do experiments show from the use of PST?

17. Why is so much corn fed to hogs in the United States, despite the fact that it is deficient in both quantity and quality of protein?

18. In what amino acids is corn particularly deficient? In what ways would Opaque-2 corn contribute to improved nutrition of swine?

19. Discuss the effect of feeding fats/oils (a) to hogs in the summer vs the winter, and (b) to gestating sows beginning 10 to 14 days before farrowing.

20. How do you account for the fact that soybean meal is the leading high-protein supplement in the United States?

21. What are "full-fat soybeans"? How, and under what circumstances, may they be used in swine rations?

22. Is the use of pastures for swine outmoded in the United States? If not, how, and under what circumstances, may they be used in a practical way?

23. Will world food shortages cause a trend back to the use of more alfalfa meal for hogs? Why is alfalfa especially well suited for winter feeding of swine?

24. On your home farm (or a farm with which you are familiar), how, and under what circumstances, would you (a) utilize silage for swine, and (b) "hog off" certain crops?

25. Why has garbage declined in feeding value and importance as a swine feed in recent years? Why must commercial garbage be cooked?

26. Why should there be a swine ration for every need? What classes of swine need rations suited to their particular needs?

27. Following the selection of feed ingredients, they may be fed to swine in either combined form or cafeteria style. How may balanced rations be achieved by each method through mixing/feeding procedures?

28. Under what circumstances would you (a) buy a commercial hog feed, or (b) home mix a hog feed?

29. Under what circumstances should feed substitutions be made?

30. How would you recommend that a 4-H club or FFA member feed (a) baby pigs, and (b) orphan pigs?

31. Discuss what is unique and different about feeding each of the following classes of swine: (a) growing-finishing pigs, (b) prospective breeding gilts, (c) boars, (d) gestating sows, and (e) lactating sows.

32. List and discuss each of the common hog feeding systems. Which system would you recommend for (a) brood sows, and (b) growing-finishing hogs?

33. Why did tail biting become a problem with the advent of confinement swine production?

34. Discuss the cause, symptoms, and prevention of nutritional anemia in pigs.

35. What constitutes early weaning of pigs? Give specific guidelines for early weaning.

36. Estimate the total feed required to produce a 240 lb market pig, then break it down into the amount required for each of the following stages of production: (a) sow gestation ration (including pregestation and breeding), (b) boar ration, (c) lactation ration, (d) starter ration (creep to 40 lb), and (e) grower-finisher ration (40 to 240 lb).

37. Since boars gain faster and are more efficient in feed conversion than barrows or gilts, why don't we feed more boars for market slaughter?

38. Smithfield Hams are derived from peanut-fed hogs; they are soft pork. Yet, they are highly advertised and sold at a premium. So, why be concerned about soft pork?

39. Of what significance is the corn-hog ratio?

40. What is contract hog production? Why has it increased in recent years? What are the common types of contracts and the characteristics of each?

41. Discuss poisons and toxins of swine.

42. Discuss the increased pollution potential, affecting both people and swine, which accompanied the swine industry as it moved toward specialization, mechanization, high animal density, and confinement.

Fig. 23-24. How it used to be done! Hogs ere hand-fed ear-corn scooped from the crib, and self-fed tankage in a self-feeder. (Courtesy, J. C. Allen and Son, West Lafayette, Ind.)

Fig. 23-25. Today! Hogs in a modern, environmentally controlled hog house, with automatic self-feeders. (Courtesy, J. C. Allen and Son, West Lafayette, Ind.)

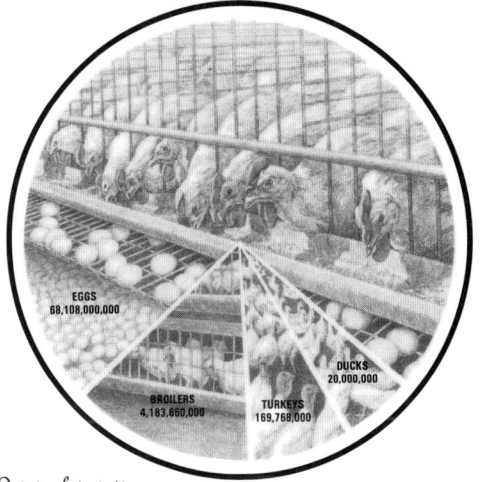

Original painting by Tom Phillips

24

FEEDING POULTRY[1]

Contents	Page
Economic Importance of Feed for Poultry	550
Changes in Feed Efficiency	550
Nutrient Needs of Poultry	550
National Research Council (NRC) Requirements	550
Energy	559
Protein	560
Minerals	561
Poultry Mineral Chart	561
Vitamins	564
Poultry Vitamin Chart	564
Unidentified Growth Factors (UGF)	569
Water	569
Additives	569
Feeds for Poultry	570
Presence of Substances Affecting Product Quality	571
Feed Preparation	571
Poultry Rations	571
Factors Involved in Formulating Poultry Rations	572
Formulating Rations	572
Feed Substitution Table	573
Special Feeding Programs; Feed Intake and Suggested Rations	574
Feeding Chickens	574
Feeding Layers	574
Phase Feeding	576
Molting	576
Feeding Breeders	577
Feeding Replacement Pullets	578
Feeding Broilers	580
Feeding Turkeys	582
Feeding Ducks	584
Feeding Geese	586
Feeding Pigeons	587
Feeding Game Birds: Pheasants, Bobwhite Quail, Japanese Quail, Chukars, Partridges, Grouse, and Doves	587
Feeding Guinea Fowl	588
Feeding Ostriches	588
Feeding Peafowl	589
Salmonella/Toxic Inorganic Elements	589
Questions for Study and Discussion	590

Poultry feeding has changed more than the feeding of any other species—it has paced the entire livestock field. Today, the vast majority of commercial poultry is produced in large units wherein the maximum of science and technology exist. Confinement production is rather commonplace, and well-balanced rations containing adequate sources of all known nutrient materials are fed for maximum production. The current trend in poultry production is toward controlled environment, which usually results in lowered feed consumption. Under such conditions, the daily feed consumption is often taken into consideration and the nutrient content of the feed (energy, amino acids, minerals, and vitamins) varied so as to compensate for the reduced feed intake and meet the requirements.

The changes in the poultry industry have resulted in greater efficiency of production, along with favorable product prices and increased human consumption. As shown in Table

Fig. 24–1. Eggs and feed! (Courtesy, H & N, Inc., Redmond, Wash.)

[1]The authors gratefully acknowledge the helpful suggestions of the following authorities who reviewed this chapter: J. D. Garlich, Ph.D., Professor, Department of Poultry Science, North Carolina State University, Raleigh; D. M. Hooge, Ph.D., Hooge Consulting Service, 1117 Sycamore Street, Turlock, CA 95380; L. S. Jensen, Ph.D., Professor, Department of Poultry Science, The University of Georgia, Athens; L. M. Larsen, Ph.D., Consultant, Nutri-Systems, 426 E. Shields, Fresno, CA 93704; P. L. Potts, Sr., Ph.D., Poultry Science Department, California Polytechnic State University, San Luis Obispo; and F. L. Stephenson, Ph.D., Professor, Department of Animal Sciences, University of Arkansas, Fayetteville.

24-1, during the 30-year period 1956-1986, the consumption of all red meats (beef, veal, lamb and mutton, and pork) decreased, while the consumption of chicken and turkey increased dramatically—by 141 and 156%, respectively.

TABLE 24-1
CHANGES IN PER CAPITA CONSUMPTION OF SELECTED MEAT PRODUCTS[1]

Food	1956 Consumption		1986 Consumption		Change
	(lb)	(kg)	(lb)	(kg)	(%)
Meats:					
Beef	85.4	38.8	78.4	35.6	− 8.0
Veal	9.5	4.3	1.9	0.9	− 80.0
Lamb and mutton	4.5	2.0	1.4	0.6	− 68.0
Pork	67.3	30.6	58.6	26.6	− 13.0
Poultry products:					
Chicken	24.4	11.1	58.8	26.7	+141.0
Turkey	5.2	2.4	13.3	6.0	+156.0

[1]U.S. Department of Agriculture sources.

Nutrition of poultry is more critical than that of other farm animals with regard to a number of factors. This is so because birds are quite different from four-footed animals; their digestion is more rapid, their respiration and circulation are faster, their body temperature is 8 to 10°F higher (about 106°F), they are more active, they are more sensitive to environmental influences, they grow at a more rapid rate, and they mature at an earlier age. Also, egg production is an all-or-none phenomenon—that is, birds must have enough nutrients to produce an egg; otherwise, no egg is produced.

ECONOMIC IMPORTANCE OF FEED FOR POULTRY

The economic importance of poultry feeding becomes apparent when it is realized that 55 to 75% of the total production cost of poultry is from feed, with the production of eggs toward the lower side of this range and the production of broilers and turkeys toward the upper side. For this reason, the efficient use of feed is extremely important to poultry producers.

CHANGES IN FEED EFFICIENCY

Table 24-2 shows the marked lowering of feed required

TABLE 24-2
FEED UNITS REQUIRED PER UNIT OF EGGS AND POULTRY PRODUCED, SELECTED YEARS, 1940-90

Year Ending October	Per Dozen Eggs[1][2]	Per Pound Liveweight	
		Turkey[3]	Broiler[1][2]
	(feed units)	(feed units)	(feed units)
1940	7.4	4.50	4.7
1950	7.2	3.56	3.7
1960	6.4	3.37	3.0
1970	4.6	3.21	2.6
1980	4.1[4]	2.90[4]	2.1[4]
1990	3.75[4]	2.70[4]	1.9[4]

[1]Feed units used per dozen eggs or per pound of liveweight broiler produced. A feed unit is the economic equivalent of 1 lb of corn. 1940-1960 from *Handbook of Agricultural Charts 1965*, Ag. Hdbk. No. 300, USDA, Oct. 1965, p. 58.

[2]1970 from *Agricultural Statistics 1974*, USDA, p.358, Table 518.

[3]The turkey data are based on estimates presented in *Efficiency in Animal Feeding with Particular Reference to Nonnutritive Feed Additives*, Council for Agricultural Science and Technology, Report No. 22, Jan. 18, 1974.

[4]Estimates by the authors.

to produce a unit of eggs, turkeys, and broilers since 1940. In 1940, it required 4.7 lb of feed to produce 1 lb of weight gain in broilers; in 1990, it took only 1.9 lb.

NUTRIENT NEEDS OF POULTRY

The nutrient composition of chickens and eggs shown in Table 24-3 is indicative of the relative importance of these nutrients as body and egg constituents.

TABLE 24-3
NUTRIENT COMPOSITION OF BROILERS AND EGGS

Nutrient	Broilers[1]	Egg
	(%)	(%)
Water	65.7	66
Protein	18.4	13
Fat	12.2	10[2]
Minerals	3.7	11[3]

[1]From Table 2-3 of this book. [2]Chiefly in the yolk. [3]Nearly all is calcium in shell.

The nutritive requirements of poultry presented in this chapter do not provide for margins of safety. Rather, the values reported represent adequacy, using as criteria growth, health, reproduction, feed efficiency, and quality of products produced.

National Research Council (NRC) Requirements

The National Research Council (NRC) nutritive requirements are given in the following tables:

For Chickens:

Table 24-4, Nutrient Requirements of Leghorn-Type Chickens as Percentages or as Milligrams or Units per Kilogram of Diet.

Table 24-5, Body Weight and Feed Requirements of Leghorn-Type Pullets and Hens.

Table 24-6, Nutrient Requirements of Broilers as Percentages or as Milligrams or Units per Kilogram of Diet.

Table 24-7, Body Weights and Feed Requirements of Broilers.

Table 24-8, Nutrient Requirements of Meat-Type Hens for Breeding Purposes.

Table 24-9, Typical body Weights and Feed Allowances for Male and Female Meat-Type Chickens (Replacement Stock).

Table 24-10, Metabolizable Energy Required Daily by Chickens in Relation to Body Weight and Egg Production.

For Turkeys:

Table 24-11, Nutrient Requirements of Turkeys as Percentages or as Milligrams or Units per Kilogram of Feed.

Table 24-12, Growth Rate, Feed and Energy Consumption of Large-Type Turkeys.

Table 24-13, Body Weights and Feed Consumption of Large-Type Turkeys During Holding and Breeding Periods.

For Ducks:

Table 24-14, Nutrient Requirements of Pekin Ducks as Percentages or as Milligrams or Units per Kilogram of Diet.

Table 24-15, Typical Body Weights and Feed Consumption of Pekin Ducks to 8 Weeks of Age.

For Geese:
Table 24-16, Nutrient Requirements of Geese as Percentages or as Milligrams or Units per Kilogram of Diet.

For Pheasants and Bobwhite Quail:
Table 24-17, Nutrient Requirements of Pheasants and Bobwhite Quail as Percentages or as Milligrams or Units per Kilogram of Diet.

For Japanese Quail:
Table 24-18, Nutrient Requirements of Japanese Quail (Coturnix) as Percentages or as Milligrams or Units Per Kilogram of Diet.

In establishing the NRC requirements for poultry, it was further assumed that the environmental temperature in which poultry of various species and ages are grown is ideal or as near optimum as possible for efficient growth and reproduction. Therefore, the energy level of the diet was first established for each species and age of poultry, then the other nutrients were determined based upon the established level of energy. If a higher level of energy than the NRC requirement is used in the diet, feed consumption will decrease; hence, the minimum level of the other nutrients should be increased in proportion to the energy content. Similarly, if a lower dietary energy level is used, then proportionately lower levels of other nutrients should be used in the diet.

TABLE 24-4
NUTRIENT REQUIREMENTS OF LEGHORN-TYPE CHICKENS AS PERCENTAGES OR AS MILLIGRAMS OR UNITS PER KILOGRAM OF DIET[1]

	Growing			Laying		Breeding
Energy Base:	0-6 Weeks	6-14 Weeks	14-20 Weeks		Daily Intake Per Hen (mg)[3]	
kcal ME/lb Diet[2]	1,315	1,315	1,315	1,315		1,315
kcal ME/kg Diet[2]	2,900	2,900	2,900	2,900		2,900
Protein (%)	18	15	12	14.5	16,000	14.5
Amino acids:						
Arginine (%)	1.00	0.83	0.67	0.68	750	0.68
Glycine and serine (%)	0.70	0.58	0.47	0.50	550	0.50
Histidine (%)	0.26	0.22	0.17	0.16	180	0.16
Isoleucine (%)	0.60	0.50	0.40	0.50	550	0.50
Leucine (%)	1.00	0.83	0.67	0.73	800	0.73
Lysine (%)	0.85	0.60	0.45	0.64	700	0.64
Methionine + cystine (%)	0.60	0.50	0.40	0.55	600	0.55
Methionine (%)	0.30	0.25	0.20	0.32	350	0.32
Phenylalanine + tyrosine (%)	1.00	0.83	0.67	0.80	880	0.80
Phenylalanine (%)	0.54	0.45	0.36	0.40	440	0.40
Threonine (%)	0.68	0.57	0.37	0.45	500	0.45
Tryptophan (%)	0.17	0.14	0.11	0.14	150	0.14
Valine (%)	0.62	0.52	0.41	0.55	600	0.55
Linoleic acid (%)	1.00	1.00	1.00	1.00	1,100	1.00
Major or macrominerals:						
Calcium (Ca) (%)	0.80	0.70	0.60	3.40	3,750	3.40
Chlorine (Cl) (%)	0.15	0.12	0.12	0.15	165	0.15
Magnesium (Mg) (mg)	600	500	400	500	55	500
Phosphorus (P), available (%)	0.40	0.35	0.30	0.32	350	0.32
Potassium (K) (%)	0.40	0.30	0.25	0.15	165	0.15
Sodium (Na) (%)	0.15	0.15	0.15	0.15	165	0.15
Trace or microminerals:						
Copper (Cu) (mg)	8	6	6	6	0.88	8
Iodine (I) (mg)	0.35	0.35	0.35	0.30	0.03	0.30
Iron (Fe) (mg)	80	60	60	50	5.50	60
Manganese (Mn) (mg)	60	30	30	30	3.30	60
Selenium (Se) (mg)	0.15	0.10	0.10	0.10	0.01	0.10
Zinc (Zn) (mg)	40	35	35	50	5.50	65
Fat-soluble vitamins:						
Vitamin A (IU)	1,500	1,500	1,500	4,000	440	4,000
Vitamin D (ICU)	200	200	200	500	55	500
Vitamin E (IU)	10	5	5	5	0.55	10
Vitamin K (mg)	0.50	0.50	0.50	0.50	0.055	0.50
Water-soluble vitamins:						
Biotin (mg)	0.15	0.10	0.10	0.10	0.011	0.15
Choline (mg)	1,300	900	500	?	?	?
Folacin (Folic Acid) (mg)	0.55	0.25	0.25	0.25	0.0275	0.35
Niacin (Nicotinic Acid, Nicotinamide) (mg)	27.0	11.0	11.0	10.0	1.10	10.0
Pantothenic Acid (Vitamin B-3) (mg)	10.0	10.0	10.0	2.20	0.242	10.0
Riboflavin (Vitamin B-2) (mg)	3.60	1.80	1.80	2.20	0.242	3.80
Thiamin (Vitamin B-1) (mg)	1.8	1.3	1.3	0.80	0.088	0.80
Vitamin B-6 (Pyridoxine, Pyridoxal, Pyridoxamine) (mg)	3.0	3.0	3.0	3.0	0.33	4.50
Vitamin B-12 (Cobalamins) (mg)	0.009	0.003	0.003	0.004	0.00044	0.004

[1]Adapted by the authors from *Nutrient Requirements of Poultry*, 8th rev. ed., NRC, National Academy Press, Washington, D.C., 1984, p. 12, Table 4.
[2]These are typical dietary energy concentrations.
[3]Assumes an average daily intake of 110 g of feed/hen daily.

TABLE 24-5
BODY WEIGHTS AND FEED REQUIREMENTS OF LEGHORN-TYPE PULLETS AND HENS[1]

Age	Body Weight[2]		Feed Consumption[3]		Typical Egg Production (Hen/Day)
(weeks)	(lb)	(kg)	(lb/week)	(kg/week)	(%)
0	0.07	0.03	0.11	0.05	—
2	0.31	0.14	0.20	0.09	—
4	0.60	0.27	0.40	0.18	—
6	0.99	0.45	0.57	0.26	—
8	1.37	0.62	0.73	0.33	—
10	1.74	0.79	0.86	0.39	—
12	2.09	0.95	0.95	0.43	—
14	2.34	1.06	1.01	0.46	—
16	2.56	1.16	1.01	0.46	—
18	2.78	1.26	1.01	0.46	—
20	3.00	1.36	1.01	0.46	—
22	3.13	1.42	1.17	0.53	10
24	3.31	1.50	1.30	0.59	38
26	3.48	1.58	1.48	0.67	64
30	3.81	1.73	1.70	0.77	88
40	4.01	1.82	1.70	0.77	80
50	4.12	1.87	1.70	0.77	74
60	4.19	1.90	1.65	0.75	68
70	4.19	1.90	1.63	0.74	62

[1]Adapted by the authors from *Nutrient Requirements of Poultry*, 8th rev. ed., NRC, National Academy Press, Washington, D.C., 1984, p. 13, Table 5.

[2]Pullets and hens of Leghorn-type strains are generally fed *ad libitum* but are occasionally control-fed to limit body weights. Values shown are typical but will vary with strain differences, season, and lighting. Specific breeder guidelines should be consulted for desired schedules of weights and feed consumption.

[3]Based on rations containing 1,315 ME kcal/lb *(2,900 ME kcal/kg)*. Consumption will vary depending upon the caloric density of the ration, environmental temperature, and rate of production (see Table 24-9).

TABLE 24-6
NUTRIENT REQUIREMENTS OF BROILERS AS PERCENTAGES OR AS MILLIGRAMS OR UNITS PER KILOGRAM OF DIET[1]

Energy Base:	Weeks 0-3	Weeks 3-6	Weeks 6-8
kcal ME/lb Diet[2]	1,452	1,452	1,452
kcal ME/kg Diet[2]	*3,200*	*3,200*	*3,200*
Protein (%)	23.0	20.0	18.0
Amino acids:			
Arginine (%)	1.44	1.20	1.00
Glycine and serine (%)	1.50	1.00	0.70
Histidine (%)	0.35	0.30	0.26
Isoleucine (%)	0.80	0.70	0.60
Leucine (%)	1.35	1.18	1.00
Lysine (%)	1.20	1.00	0.85
Methionine + cystine (%)	0.93	0.72	0.60
Methionine (%)	0.50	0.38	0.32
Phenylalanine + tyrosine (%)	1.34	1.17	1.00
Phenylalanine (%)	0.72	0.63	0.54
Threonine (%)	0.80	0.74	0.68
Tryptophan (%)	0.23	0.18	0.17
Valine (%)	0.82	0.72	0.62
Linoleic acid (%)	1.00	1.00	1.00
Major or macrominerals:			
Calcium (Ca) (%)	1.00	0.90	0.80
Chlorine (Cl) (%)	0.15	0.15	0.15
Magnesium (Mg) (mg)	600	600	600
Phosphorus (P), available (%)	0.45	0.40	0.35
Potassium (K) (%)	0.40	0.35	0.30
Sodium (Na) (%)	0.15	0.15	0.15
Trace or microminerals:			
Copper (Cu) (mg)	8.0	8.0	8.0
Iodine (I) (mg)	0.35	0.35	0.35
Iron (Fe) (mg)	80.0	80.0	80.0
Manganese (Mn) (mg)	60.0	60.0	60.0
Selenium (Se) (mg)	0.15	0.15	0.15
Zinc (Zn) (mg)	40.0	40.0	40.0
Fat-soluble vitamins:			
Vitamin A (IU)	1,500	1,500	1,500
Vitamin D (ICU)	200	200	200
Vitamin E (IU)	10	10	10
Vitamin K (mg)	0.50	0.50	0.50
Water-soluble vitamins:			
Biotin (mg)	0.15	0.15	0.10
Choline (mg)	1,300	850	500
Folacin (Folic Acid) (mg)	0.55	0.55	0.25
Niacin (Nicotinic Acid, Nicotinamide) (mg)	27.0	27.0	11.0
Pantothenic Acid (Vitamin B-3) (mg)	10.0	10.0	10.0
Riboflavin (Vitamin B-2) (mg)	3.60	3.60	3.60
Thiamin (Vitamin B-1) (mg)	1.80	1.80	1.80
Vitamin B-6 (Pyridoxine, Pyridoxal, Pyridoxamine) (mg)	3.0	3.0	2.5
Vitamin B-12 (Cobalamins) (mg)	0.009	0.009	0.003

[1]Adapted by the authors from *Nutrient Requirements of Poultry*, 8th rev. ed., NRC, National Academy Press, Washington, D.C., 1984, p. 13, Table 6.

[2]These are typical dietary energy concentrations.

Fig. 24-2. Leghorn-type layers. (Courtesy, University of Georgia, Athens)

TABLE 24-7
BODY WEIGHTS AND FEED REQUIREMENTS OF BROILERS[1,2]

Age (weeks)	Body Weights				Weekly Feed Consumption				Cumulative Feed Consumption				Weekly Energy Consumption		Cumulative Energy Consumption	
	Male		Female		Male		Female		Male		Female		Male	Female	Male	Female
	(lb)	(g)	(lb)	(g)	(lb)	(g)	(lb)	(g)	(lb)	(g)	(lb)	(g)	(ME kcal/bird)	(ME kcal/bird)	(ME kcal/bird)	(ME kcal/bird)
1	0.29	130	0.26	120	0.26	120	0.24	110	0.26	120	0.24	110	385	350	385	350
2	0.71	320	0.66	300	0.57	260	0.53	240	0.84	380	0.77	350	830	770	1,215	1,120
3	1.23	560	1.14	515	0.86	390	0.78	355	1.70	770	1.55	705	1,250	1,135	2,465	2,255
4	1.90	860	1.74	790	1.18	535	1.10	500	2.88	1,305	2.66	1,205	1,710	1,600	4,175	3,855
5	2.76	1,250	2.45	1,110	1.63	740	1.42	645	4.51	2,045	4.08	1,850	2,370	2,065	6,545	5,920
6	3.73	1,690	3.15	1,430	2.16	980	1.76	800	6.67	3,025	5.84	2,650	3,135	2,560	9,680	8,480
7	4.63	2,100	3.85	1,745	2.41	1,095	2.01	910	9.08	4,120	7.85	3,560	3,505	2,910	13,185	11,390
8	5.56	2,520	4.54	2,060	2.67	1,210	2.14	970	11.75	5,330	9.99	4,530	3,870	3,105	17,055	14,495
9	6.45	2,925	5.18	2,350	2.91	1,320	2.23	1,010	14.66	6,650	12.21	5,540	4,225	3,230	21,280	17,725

[1]Adapted by the authors from *Nutrient Requirements of Poultry*, 8th rev. ed., NRC, National Academy Press, Washington, D.C., 1984, p. 14, Table 7.
[2]Typical for broilers fed well-balanced rations containing 1,452 ME kcal/lb *(3,200 ME kcal/kg)*.

Fig. 24-3. Broilers with automatic feed and water facilities. (Photo by J. C. Allen & Son, Inc., West Lafayette, Ind.)

TABLE 24-8
NUTRIENT REQUIREMENTS OF MEAT-TYPE HENS FOR BREEDING PURPOSES[1,2]

Energy Base: kcal ME/lb Diet *kcal ME/kg Diet*		1,293[3] *2,850[3]*	Daily Intake Per Hen (mg)
Protein	(%)	14.5	22,000
Amino acids:			
Arginine	(%)	0.74	1,110
Glycine + serine	(%)	0.62	932
Histidine	(%)	0.14	205
Isoleucine	(%)	0.57	850
Leucine	(%)	0.83	1,250
Lysine	(%)	0.51	765
Methionine + cystine	(%)	0.55	820
Methionine	(%)	0.35	520
Phenylalanine + tyrosine	(%)	0.75	1,112
Phenylalanine	(%)	0.41	610
Threonine	(%)	0.48	720
Tryptophan	(%)	0.13	190
Valine	(%)	0.63	950
Major or macrominerals:			
Calcium (Ca)	(%)	2.75	4,125
Phosphorus (P), available	(%)	0.25	375
Sodium (Na)	(%)	0.10	150

[1]Adapted by the authors from *Nutrient Requirements of Poultry*, 8th rev. ed., NRC, National Academy Press, Washington, D.C., 1984, p. 14, Table 8.

[2]Rations are generally fed on a limited intake basis to control body weight gains. Adjust quantity of feed offered based on desired body weights and egg production levels for specific breed or strain.

[3]Rations for laying hens generally are fed to provide daily energy intakes of 375 to 450 ME kcal/day based on body weight, environmental temperature, and rate of egg production. Percentage of nutrients shown is typical of hens given 425 ME kcal/day.

TABLE 24-9
TYPICAL BODY WEIGHTS AND FEED ALLOWANCES FOR MALE AND FEMALE MEAT-TYPE CHICKENS (REPLACEMENT STOCK)[1,2]

Age	Male Body Weight[3]		Male Feed Consumption[4]		Female Body Weight[3]		Female Feed Consumption[4]		Typical Egg Production
(weeks)	(lb)	(g)	(lb/week)	(g/week)	(lb)	(g)	(lb/week)	(g/week)	(hen/day %)
0	0.09	40	0.22	100	0.09	40	0.17	75	—
2	0.55	250	0.55	250	0.50	225	0.56	225	—
4	1.20	545	0.77–0.85	350–385	1.00	455	0.69–0.73	315–330	—
6	1.75	795	0.86–0.94	390–425	1.46	660	0.73–0.77	330–350	—
8	2.25	1,020	0.89–1.03	405–475	1.85	840	0.77–0.88	350–400	—
10	2.98	1,250	1.03–1.21	475–550	2.20	1,000	0.85–0.98	385–445	—
12	3.26	1,480	1.19–1.38	540–625	2.60	1,180	0.94–1.06	425–480	—
14	3.75	1,700	1.27–1.54	575–700	3.00	1,360	1.01–1.21	460–550	—
16	4.25	1,930	1.38–1.69	625–765	3.42	1,550	1.09–1.32	495–600	—
18	4.74	2,150	1.47–1.82	665–825	3.81	1,730	1.16–1.48	525–670	—
20	5.29	2,400	—[5]	—[5]	4.25	1,930	1.26–1.61	570–730	—
22	5.82	2,640	—	—	4.65	2,110	1.40–1.75	635–795	10
24	7.05	3,200	—	—	5.40	2,450	1.76–2.04	800–925	15
26	7.80	3,540	—	—	6.02	2,730	2.09–2.31	950–1,050	30
28	8.27	3,750	—	—	6.35	2,880	2.38–2.52	1,078–1,141	56
30	8.60	3,900	—	—	6.61	3,000	2.38–2.52	1,078–1,141	75
32	9.02	4,090	—	—	6.81	3,090	2.38–2.52	1,078–1,141	80
34	9.30	4,220	—	—	6.90	3,130	2.38–2.52	1,078–1,141	78
36	9.57	4,340	—	—	6.97	3,160	2.38–2.52	1,078–1,141	76
38	9.81	4,450	—	—	7.01	3,180	2.36–2.50	1,071–1,134	73
40	10.01	4,540	—	—	7.01	3,180	2.35–2.48	1,064–1,127	72

[1] Adapted by the authors from *Nutrient Requirements of Poultry*, 8th rev. ed., NRC, National Academy Press, Washington, D.C., 1984, p. 15, Table 9.
[2] Broiler-breeder strains must be grown on a controlled feeding program to limit weight. Values shown are typical but will vary according to strain. Specific breeder guidelines should be consulted for desired schedule of weights and feed allotments.
[3] Values are typical for fall-hatched chicks. Spring-hatched chicks will have decreasing natural daylight during the time of sexual maturity and usually need to be heavier to attain sexual maturity at the desired age.
[4] Adjust as required to maintain desired body weight.
[5] Males and females intermingled.

Fig. 24-4. Meat-type hens being pen-mated to one male. (Courtesy, J. C. Allen & Son, Inc., West Lafayette, Ind.)

TABLE 24-10
METABOLIZABLE ENERGY REQUIRED DAILY BY CHICKENS IN RELATION TO BODY WEIGHT AND EGG PRODUCTION[1,2]

Body Weight		Rate of Egg Production (%)					
		0	50	60	70	80	90
(lb)	(kg)	◄──── Metabolizable Energy/Hen Daily (kcal)[3] ────►					
2.2	1.0	130	192	205	217	229	242
3.3	1.5	177	239	251	264	276	289
4.4	2.0	218	280	292	305	317	330
5.5	2.5	259	321	333	346	358	371
6.6	3.0	296	358	370	383	395	408
7.7	3.5	333	395	408	420	432	445

[1] Adapted by the authors from *Nutrient Requirements of Poultry*, 8th rev. ed., NRC, National Academy Press, Washington, D.C., 1984, p. 15, Table 10.
[2] A number of formulas have been suggested for prediction of the daily energy requirements of chickens. The formula used here was derived from that in *Effect of Environment on Nutrient Requirements of Domestic Animals* (NRC, 1981).

$$\text{ME/hen daily} = W^{0.75}(173 - 1.95T) + 5.5 \Delta W + 2.07 EE$$

where: W = body weight (kg),
T = ambient temperature (°C),
ΔW = change in body weight in g/day, and
EE = daily egg mass (g).

[3] Temperature of 22°C, egg weight of 60 g, and no change in body weight were used in calculations.

TABLE 24-11
NUTRIENT REQUIREMENTS OF TURKEYS AS PERCENTAGES OR AS MILLIGRAMS OR UNITS PER KILOGRAM OF FEED[1]

		Age (Weeks)							
Male		0-4	4-8	8-12	12-16	16-20	20-24	Holding	Breeding Hens
Female		0-4	4-8	8-11	11-14	14-17	17-20		
Energy Base:									
kcal ME/lb Diet[2]		1,270	1,315	1,361	1,406	1,452	1,497	1,315	1,315
kcal ME/kg Diet[2]		*2,800*	*2,900*	*3,000*	*3,100*	*3,200*	*3,300*	*2,900*	*2,900*
Protein	(%)	28	26	22	19	16.5	14	12	14
Amino acids:									
Arginine	(%)	1.60	1.5	1.25	1.1	0.95	0.8	0.6	0.6
Glycine and serine	(%)	1.0	0.9	0.8	0.7	0.6	0.5	0.4	0.5
Histidine	(%)	0.58	0.54	0.46	0.39	0.35	0.29	0.25	0.3
Isoleucine	(%)	1.1	1.0	0.85	0.75	0.65	0.55	0.45	0.5
Leucine	(%)	1.9	1.75	1.5	1.3	1.1	0.95	0.5	0.5
Lysine	(%)	1.6	1.5	1.3	1.0	0.8	0.65	0.5	0.6
Methionine + cystine	(%)	1.05	0.9	0.75	0.65	0.55	0.45	0.4	0.4
Methionine	(%)	0.53	0.45	0.38	0.33	0.28	0.23	0.2	0.2
Phenylalanine + tyrosine	(%)	1.8	1.65	1.4	1.2	1.05	0.9	0.8	1.0
Phenylalanine	(%)	1.0	0.9	0.8	0.7	0.6	0.5	0.4	0.55
Threonine	(%)	1.0	0.93	0.79	0.68	0.59	0.5	0.4	0.45
Tryptophan	(%)	0.26	0.24	0.2	0.18	0.15	0.13	0.1	0.13
Valine	(%)	1.2	1.1	0.94	0.8	0.7	0.6	0.5	0.58
Linoleic acid	(%)	1.0	1.0	0.8	0.8	0.8	0.8	0.8	1.0
Major or macrominerals:									
Calcium (Ca)	(%)	1.2	1.0	0.85	0.75	0.65	0.55	0.5	2.25
Chlorine (Cl)	(%)	0.15	0.14	0.14	0.12	0.12	0.12	0.12	0.12
Magnesium (Mg)	(mg)	600	600	600	600	600	600	600	600
Phosphorus (P), available	(%)	0.6	0.5	0.42	0.38	0.32	0.28	0.25	0.35
Potassium (K)	(%)	0.7	0.6	0.5	0.5	0.4	0.4	0.4	0.6
Sodium (Na)	(%)	0.17	0.15	0.12	0.12	0.12	0.12	0.12	0.15
Trace or microminerals:									
Copper (Cu)	(mg)	8	8	6	6	6	6	6	8
Iodine (I)	(mg)	0.4	0.4	0.4	0.4	0.4	0.4	0.4	0.4
Iron (Fe)	(mg)	80	60	60	60	50	50	50	60
Manganese (Mn)	(mg)	60	60	60	60	60	60	60	60
Selenium (Se)	(mg)	0.2	0.2	0.2	0.2	0.2	0.2	0.2	0.2
Zinc (Zn)	(mg)	75	65	50	40	40	40	40	65
Fat-soluble vitamins:									
Vitamin A	(IU)	4,000	4,000	4,000	4,000	4,000	4,000	4,000	4,000
Vitamin D[3]	(ICU)	900	900	900	900	900	900	900	900
Vitamin E	(IU)	12	12	10	10	10	10	10	25
Vitamin K	(mg)	1.0	1.0	0.8	0.8	0.8	0.8	0.8	1.0
Water-soluble vitamins:									
Biotin	(mg)	0.2	0.2	0.15	0.125	0.100	0.100	0.100	0.15
Choline	(mg)	1,900	1,600	1,300	1,100	950	800	800	1,000
Folacin (Folic Acid)	(mg)	1.0	1.0	0.8	0.8	0.7	0.7	0.7	1.0
Niacin (Nicotinic Acid, Nicotinamide)	(mg)	70.0	70.0	50.0	50.0	40.0	40.0	40.0	30.0
Pantothenic Acid (Vitamin B-3)	(mg)	11.0	11.0	9.0	9.0	9.0	9.0	9.0	16.0
Riboflavin (Vitamin B-2)	(mg)	3.6	3.6	3.0	3.0	2.5	2.5	2.5	4.0
Thiamin (Vitamin B-1)	(mg)	2.0	2.0	2.0	2.0	2.0	2.0	2.0	2.0
Vitamin B-6 (Pyridoxine, Pyridoxal, Pyridoxamine)	(mg)	4.5	4.5	3.5	3.5	3.0	3.0	3.0	4.0
Vitamin B-12 (Cobalamins)	(mg)	0.003	0.003	0.003	0.003	0.003	0.003	0.003	0.003

[1]Adapted by the authors from *Nutrient Requirements of Poultry*, 8th rev. ed., NRC, National Academy Press, Washington, D.C., 1984, p. 17, Table 11.
[2]These are typical ME concentrations for corn-soya rations. Different ME values may be appropriate if other ingredients predominate.
[3]These concentrations of vitamin D are satisfactory when the dietary concentrations of calcium and available phosphorus conform with those in this table.

TABLE 24-12
GROWTH RATE, FEED AND ENERGY CONSUMPTION OF LARGE-TYPE TURKEYS[1]

Age	Body Weight				Feed Consumption				Cumulative Feed Consumption				ME Consumption	
	Male		Female		Male		Female		Male		Female		Male	Female
(wks)	(lb)	(kg)	(lb)	(kg)	(lb/week)	(kg/week)	(lb/week)	(kg/week)	(lb)	(kg)	(lb)	(kg)	(Mcal/wk)	(Mcal/wk)
1	0.24	0.11	0.24	0.11	0.22	0.10	0.22	0.10	0.22	0.10	0.22	0.10	0.30	0.30
2	0.60	0.27	0.53	0.24	0.44	0.20	0.37	0.17	0.66	0.30	0.60	0.27	0.60	0.50
3	1.28	0.58	1.04	0.47	0.99	0.45	0.86	0.39	1.65	0.75	1.46	0.66	1.1	0.80
4	2.21	1.0	1.54	0.70	1.35	0.61	1.01	0.46	3.00	1.36	2.47	1.12	1.7	1.2
5	3.31	1.5	2.43	1.1	1.54	0.70	1.32	0.60	4.54	2.06	3.79	1.72	2.3	1.6
6	4.41	2.0	3.53	1.6	1.90	0.86	1.68	0.76	6.44	2.92	5.47	2.48	2.9	2.1
7	5.73	2.6	4.63	2.1	2.38	1.08	1.96	0.89	8.82	4.00	7.43	3.37	3.5	2.6
8	7.28	3.3	5.73	2.6	2.87	1.30	2.29	1.04	11.69	5.30	9.72	4.41	4.1	3.1
9	8.82	4.0	6.84	3.1	3.33	1.51	2.60	1.18	15.02	6.81	12.33	5.59	4.8	3.6
10	10.36	4.7	8.16	3.7	3.92	1.78	2.95	1.34	18.94	8.59	15.28	6.93	5.2	4.1
11	12.13	5.5	9.48	4.3	4.39	1.99	3.24	1.47	23.33	10.58	18.52	8.40	5.7	4.6
12	13.89	6.3	10.58	4.8	4.96	2.25	3.51	1.59	28.29	12.83	22.03	9.99	6.3	5.1
13	15.66	7.1	11.69	5.3	5.53	2.51	3.75	1.70	33.82	15.34	25.78	11.69	7.1	5.5
14	17.64	8.0	12.79	5.8	5.87	2.66	3.86	1.75	39.69	18.00	29.64	13.44	7.8	5.8
15	19.40	8.8	13.89	6.3	6.37	2.89	4.01	1.82	46.06	20.89	33.65	15.26	8.4	6.1
16	21.39	9.7	14.77	6.7	6.73	3.05	4.23	1.92	52.79	23.94	37.88	17.18	8.8	6.4
17	23.15	10.5	15.66	7.1	6.90	3.13	4.48	2.03	59.60	27.03	42.36	19.21	9.6	6.7
18	24.92	11.3	16.54	7.5	7.21	3.27	4.56	2.07	66.90	30.34	46.92	21.28	10.2	6.9
19	26.68	12.1	17.20	7.8	7.56	3.43	4.74	2.15	74.46	33.77	51.66	23.43	10.9	7.1
20	28.22	12.8	17.86	8.1	7.94	3.60	4.92	2.23	82.40	37.37	56.58	25.66	11.6	7.3
21	29.77	13.5	—	—	8.18	3.71	—	—	90.58	41.08	—	—	12.5	—
22	31.31	14.2	—	—	8.42	3.82	—	—	99.00	44.90	—	—	12.9	—
23	32.63	14.8	—	—	8.69	3.94	—	—	107.69	48.84	—	—	13.2	—
24	33.96	15.4	—	—	8.93	4.05	—	—	116.62	52.89	—	—	13.5	—

[1]Adapted by the authors from *Nutrient Requirements of Poultry*, 8th rev. ed., NRC, National Academy Press, Washington, D.C., 1984, p. 18, Table 12.

TABLE 24-13
BODY WEIGHTS AND FEED CONSUMPTION OF LARGE-TYPE TURKEYS DURING HOLDING AND BREEDING PERIODS[1,2]

Age	Hens					Toms			
	Weight		Egg Production	Feed		Weight		Feed	
(weeks)	(lb)	(kg)	(%)	(lb/day)	(g/day)	(lb)	(kg)	(lb/day)	(g/day)
20	15.4	7.0	—	0.44	200	26.5	12.0	0.88	400
25	17.6	8.0	—	0.47	215	29.8	13.5	0.93	420
30	19.8	9.0	Start light Stimulation	0.51	230	35.3	16.0	0.97	440
35	20.9	9.5	66	0.57	260	37.5	17.0	0.99	450
40	20.5	9.3	63	0.56	255	39.7	18.0	1.01	460
45	20.1	9.1	60	0.55	250	40.1	18.2	1.06	480
50	19.8	9.0	50	0.53	240	40.8	18.5	1.10	500
55	19.8	9.0	40	0.51	230	41.5	18.8	1.12	510
60	19.8	9.0	35	0.49	220	41.9	19.0	1.15	520

[1]Adapted by the authors from *Nutrient Requirements of Poultry*, 8th rev. ed., NRC, National Academy Press, Washington, D.C., 1984, p. 18, Table 13.

[2]These values are based on experimental data involving "in season" egg production (i.e., November through July) of commercial stock. It is estimated that summer breeders would produce 70-90% as many eggs and consume 60-80% as much feed, respectively, as "in season" breeders.

TABLE 24-14
NUTRIENT REQUIREMENTS OF PEKIN DUCKS AS PERCENTAGES OR AS MILLIGRAMS OR UNITS PER KILOGRAM OF DIET[1,2]

	Starting (0-2 Weeks)	Growing (2-7 Weeks)	Breeding
Energy Base:			
kcal ME/lb Diet[3]	1,315	1,315	1,315
kcal ME/kg Diet[3]	2,900	2,900	2,900
Protein (%)	22.0	16.0	15.0
Amino acids:			
Arginine (%)	1.1	1.0	—
Lysine (%)	1.1	0.9	0.7
Methionine + cystine (%)	0.8	0.6	0.55
Major or macrominerals:			
Calcium (Ca) (%)	0.65	0.6	2.75
Chlorine (Cl) (%)	0.12	0.12	0.12
Magnesium (Mg) (mg)	500	500	500
Phosphorus (P), available (%)	0.40	0.35	0.35
Sodium (Na) (%)	0.15	0.15	0.15
Trace or microminerals:			
Manganese (Mn) (mg)	40.0	40.0	25.0
Selenium (Se) (mg)	0.14	0.14	0.14
Zinc (Zn) (mg)	60.0	60.0	60.0
Fat-soluble vitamins:			
Vitamin A (IU)	4,000	4,000	4,000
Vitamin D (ICU)	220	220	500
Vitamin K (mg)	0.4	0.4	0.4
Water-soluble vitamins:			
Niacin (Nicotinic Acid, Nicotinamide) (mg)	55.0	55.0	40.0
Pantothenic Acid (Vitamin B-3) (mg)	11.0	11.0	10.0
Riboflavin (Vitamin B-2) (mg)	4.0	4.0	4.0
Vitamin B-6 (Pyridoxine, Pyridoxal, Pyridoxamine) (mg)	2.6	2.6	3.0

[1]Adapted by the authors from *Nutrient Requirements of Poultry,* 8th rev. ed., NRC, National Academy Press, Washington, D.C., 1984, p. 20, Table 15.
[2]For nutrients not listed, see requirements for chickens as a guide.
[3]These are typical dietary energy concentrations.

Fig. 24-5. White Pekin drake. (Courtesy, USDA)

TABLE 24-15
TYPICAL BODY WEIGHTS AND FEED CONSUMPTION OF PEKIN DUCKS TO 8 WEEKS OF AGE[1]

Age (weeks)	Body Weight Male (lb)	(kg)	Body Weight Female (lb)	(kg)	Feed Consumption By 1-Week Periods Male (lb)	(kg)	Female (lg)	(kg)	Cumulative Feed Consumption Male (lb)	(kg)	Female (lb)	(kg)
0	0.11	0.05	0.11	0.05	—	—	—	—	—	—	—	—
1	0.60	0.27	0.60	0.27	0.49	0.22	0.49	0.22	0.49	0.22	0.49	0.22
2	1.72	0.78	1.63	0.74	1.70	0.77	1.61	0.73	2.18	0.99	2.09	0.95
3	3.04	1.38	2.82	1.28	2.47	1.12	2.45	1.11	4.65	2.11	4.52	2.05
4	4.32	1.96	4.01	1.82	2.82	1.28	2.82	1.28	7.50	3.40	7.34	3.33
5	5.49	2.49	5.07	2.30	3.26	1.48	3.15	1.43	10.74	4.87	10.50	4.76
6	6.53	2.96	6.02	2.73	3.59	1.63	3.51	1.59	14.33	6.50	14.00	6.35
7	7.36	3.34	6.75	3.06	3.70	1.68	3.59	1.63	18.04	8.18	17.60	7.98
8	7.96	3.61	7.25	3.29	3.70	1.68	3.59	1.63	21.74	9.86	21.19	9.61

[1]Adapted by the authors from *Nutrient Requirements of Poultry,* 8th rev. ed., NRC, National Academy Press, Washington, D.C., 1984, p. 20, Table 16.

TABLE 24-16
NUTRIENT REQUIREMENTS OF GEESE AS PERCENTAGES OR AS MILLIGRAMS OR UNITS PER KILOGRAM OF DIET[1,2]

	Starting (0-6 Weeks)	Growing (After 6 Weeks)	Breeding
Energy Base:			
kcal ME/lb Diet[3]	1,315	1,315	1,315
kcal ME/kg Diet[3]	*2,900*	*2,900*	*2,900*
Protein (%)	22.0	15.0	15.0
Amino acids:			
Lysine (%)	0.9	0.6	0.6
Methionine + cystine (%)	0.75	—	—
Major or macrominerals:			
Calcium (Ca) (%)	0.8	0.6	2.25
Phosphorus (P), available (%)	0.4	0.3	0.3
Fat-soluble vitamins:			
Vitamin A (IU)	1,500	1,500	4,000
Vitamin D (ICU)	200	200	200
Water-soluble vitamins:			
Niacin (Nicotinic Acid, Nicotinamide) (mg)	55.0	35.0	20.0
Pantothenic Acid (Vitamin B-3) (mg)	15.0	—	—
Riboflavin (Vitamin B-2) (mg)	4.0	2.5	4.0

[1]Adapted by the authors from *Nutrient Requirements of Poultry*, 8th rev. ed., NRC, National Academy Press, Washington, D.C., 1984, p. 19, Table 14.

[2]For nutrients not listed, see requirements for chickens as a guide.

[3]These are typical dietary energy concentrations.

Fig. 24-6. Toulouse gander. (Photo by J. C. Allen & Son, Inc., West Lafayette, Ind.)

TABLE 24-17
NUTRIENT REQUIREMENTS OF PHEASANTS AND BOBWHITE QUAIL AS PERCENTAGES OR AS MILLIGRAMS OR UNITS PER KILOGRAM OF DIET[1,2,3]

	Pheasant			Bobwhite Quail		
Energy Base:	Starting	Growing	Breeding	Starting	Growing	Breeding
kcal ME/lb Diet[4]	1,270	1,225	1,270	1,270	1,270	1,270
kcal ME/kg Diet[4]	*2,800*	*2,700*	*2,800*	*2,800*	*2,800*	*2,800*
Protein (%)	30.0	16.0	18.0	28.0	20.0	24.0
Amino acids:						
Glycine + serine (%)	1.8	1.0	—	—	—	—
Lysine (%)	1.5	0.8	—	—	—	—
Methionine + cystine (%)	1.1	0.6	0.6	—	—	—
Linoleic acid (%)	1.0	1.0	1.0	1.0	1.0	1.0
Major or macrominerals:						
Calcium (Ca) (%)	1.0	0.7	2.5	0.65	0.65	2.3
Chlorine (Cl) (%)	0.11	0.11	0.11	0.11	0.11	0.11
Phosphorus (P), available (%)	0.55	0.45	0.40	0.55	0.45	0.50
Sodium (Na) (%)	0.15	0.15	0.15	0.15	0.15	0.15
Trace or microminerals:						
Iodine (I) (mg)	0.30	0.30	0.30	0.30	0.30	0.30
Water-soluble vitamins:						
Choline (mg)	1,500.0	1,000.0	—	1,500.0	—	1,000.0
Niacin (Nicotinic Acid, Nicotinamide) (mg)	60.0	40.0	—	30.0	—	20.0
Pantothenic Acid (Vitamin B-3) (mg)	10.0	10.0	—	13.0	—	15.0
Riboflavin (Vitamin B-2) (mg)	3.5	3.0	—	3.8	—	4.0

[1]Adapted by the authors from *Nutrient Requirements of Poultry*, 8th rev. ed., NRC, National Academy Press, Washington, D.C., 1984, p. 21, Table 17.

[2]For Pheasant values not listed see requirements for turkeys as a guide.

[3]For Bobwhite Quail values not listed see requirements for Leghorn-type chickens as a guide.

[4]These are typical dietary energy concentrations.

TABLE 24-18
**NUTRIENT REQUIREMENTS
OF JAPANESE QUAIL (COTURNIX) AS PERCENTAGES
OR AS MILLIGRAMS OR UNITS PER KILOGRAM OF DIET**[1]

		Starting and Growing	Breeding
Energy Base:			
kcal ME/lb Diet[2]		1,361	1,361
kcal ME/kg Diet[2]		*3,000*	*3,000*
Protein	(%)	24.0	20.0
Amino acids:			
Arginine	(%)	1.25	1.26
Glycine and serine	(%)	1.20	1.17
Histidine	(%)	0.36	0.42
Isoleucine	(%)	0.98	0.90
Leucine	(%)	1.69	1.42
Lysine	(%)	1.30	1.15
Methionine + cystine	(%)	0.75	0.76
Methionine	(%)	0.50	0.45
Phenylalanine + tyrosine	(%)	1.80	1.40
Phenylalanine	(%)	0.96	0.78
Threonine	(%)	1.02	0.74
Tryptophan	(%)	0.22	0.19
Valine	(%)	0.95	0.92
Linoleic acid	(%)	1.0	1.0
Major or macrominerals:			
Calcium (Ca)	(%)	0.8	2.5
Chlorine (Cl)	(%)	0.20	0.15
Magnesium (Mg)	(mg)	300	500
Phosphorus (P), available	(%)	0.45	0.55
Potassium (K)	(%)	0.4	0.4
Sodium (Na)	(%)	0.15	0.15
Trace or microminerals:			
Copper (Cu)	(mg)	6	6
Iodine (I)	(mg)	0.3	0.3
Iron (Fe)	(mg)	100	60
Manganese (Mn)	(mg)	90	70
Selenium (Se)	(mg)	0.2	0.2
Zinc (Zn)	(mg)	25	50
Fat-soluble vitamins:			
Vitamin A	(IU)	5,000	5,000
Vitamin D	(ICU)	1,200	1,200
Vitamin E	(IU)	12	25
Vitamin K	(mg)	1	1
Water-soluble vitamins:			
Biotin	(mg)	0.3	0.15
Choline	(mg)	2,000	1,500
Folacin (Folic Acid)	(mg)	1	1
Niacin (Nicotinic Acid, Nicotinamide)	(mg)	40	20
Pantothenic Acid (Vitamin B-3)	(mg)	10	15
Riboflavin (Vitamin B-2)	(mg)	4	4
Thiamin (Vitamin B-1)	(mg)	2	2
Vitamin B-6 (Pyridoxine, Pyridoxal, Pyridoxamine)	(mg)	3	3
Vitamin B-12 (Cobalamins)	(mg)	0.003	0.003

[1]Adapted by the authors from *Nutrient Requirements of Poultry,* 8th rev. ed., NRC, National Academy Press, Washington, D.C., 1984, p. 22, Table 18.

[2]These are typical dietary energy concentrations.

Energy

The energy requirement may be defined as the amount of available energy that will provide for growth or egg production at a high enough level to permit maximal economic return for the production unit.

Although each primary energy source—carbohydrates, fats, proteins—has specific functions, all of them can be used to provide energy for maintenance and production of poultry. From the standpoint of providing the normal energy needs, however, the carbohydrates are by far the most important, whereas the fats rank next as an energy source.

Of the various systems of expressing energy values of feeds and nutrient requirements—gross energy (GE), digestible energy (DE), metabolizable energy (ME), and net energy (NE)—the poultry industry has found metabolizable energy to be the most reliable expression of energy needs. In general, metabolizable energy represents 25 to 90% of gross energy. Metabolizable energy, as a portion of gross energy, is usually lowest for fibrous feedstuffs and highest for fats and oils, with grain and protein supplements intermediate.

Carbohydrates, which constitute about 75% of the dry weight of plants and grain, make up a large part of poultry rations. They are composed of carbon, hydrogen, and oxygen. The group includes sugars, starch, cellulose, gums, and related substances. Carbohydrates serve as a source of heat and energy in the bird's body. A surplus taken into the body may be transformed into fat and stored as a reserve supply of heat and energy.

Although fats are used primarily to supply energy in poultry diets, they also improve the physical consistency of rations and the dispersion of microingredients in feed mixtures. The fats used for feeding poultry are derived from three sources: animal or poultry fats obtained from the rendering industry, restaurant greases, acidulated soapstocks from the vegetable oil industry, and/or mixtures thereof. The nutritional value of fats for poultry feed is determined by moisture, impurities, unsaponifiables, free fatty acids, total fatty acids, and fatty acid composition. The polyunsaturated linoleic and arachidonic acids are considered to be *essential fatty acids.* They have specific functions in the body that are not related to energy production. Birds exhibit poor growth, fatty livers, reduced egg size, and poor hatchability without these essential fatty acids. Fats for poultry feed should be stabilized against oxidation.

Because the primary function of both carbohydrates and fats is to serve as a source of energy for the body, an insufficient supply of these nutrients results in reduced growth rate or egg production in poultry.

Feed intake is governed by the energy concentration of the feed. While the bulkiness of feed can alter feed intake, the bird, for the most part, will eat to satisfy its energy needs. Because of this, special attention must be given to nutrient ratios, especially the ratio of energy to various nutrients such as amino acids and minerals. So, the ME values given in the NRC tables that are reproduced in this chapter are not intended as *requirements*. Rather, they are provided to give perspective to the other nutrient requirement levels. Using the energy level for reference will enable the nutritionist to form a ratio of the amount of the nutrient per unit of energy, thereby keeping the nutrients in balance with available energy. In eating to satisfy its energy need, a bird will eat less of a high-energy diet and more of a low-energy diet. Having the nutrients in relation to dietary energy will ensure proper intake on a daily basis.

- **Apparent metabolizable energy (AME)**—The term *apparent metabolizable energy* (AME) refers to the traditionally determined energy values which include metabolic and endogenous fractions.

- **True metabolizable energy (TME)**—*The TME for poultry is the gross energy of the feed minus the gross energy of the excreta of food origin.* A correction for nitrogen retention may be applied to give a TME$_n$ value.

In simple formulas AME and TME are: **AME = feed energy − (fecal + urinary + gaseous energy),** whereas **TME = AME + (metabolic + endogenous energy).** In poultry, the gaseous losses are negligible and usually ignored.

TME values are easier and much less expensive to determine than the traditional ME values.

(Also see Chapter 3, section on "Hunger and Appetite"; and Chapter 4, section on "Energy Systems.")

Protein

Typical broiler starter rations contain from 21 to 24% protein, and typical laying rations from about 15 to 17% protein. Grain and millfeeds supply approximately ½ of the protein needs for most poultry rations. Additional protein is supplied from the high-protein concentrates of either animal or vegetable origin.

From the standpoint of poultry nutrition, the amino acids that make up proteins are really the essential nutrients, rather than the protein molecule itself. Hence, protein content as a measure of the nutritional value of a feed is becoming less important, and each amino acid is being considered individually. The essential amino acid requirements of each class of poultry are given in the NRC tables.

The energy content of the diet must be considered in formulating to meet the desired intake of all essential nutrients other than energy itself, including the intake of the essential amino acids. For example, if the producer uses a high-energy feed, the protein content of the feed must be high if the bird is to ingest adequate amounts of protein. Conversely, if the energy content of the feed is low, the protein content should be low, also; otherwise, the bird will consume excessive amounts of expensive protein.

It has been determined that the chick requires dietary sources of protein to furnish 13 different amino acids. These amino acids are referred to as *essential,* since the chicken cannot produce them in sufficient amounts for maximum growth or egg production, and because a dietary deficiency of any one of them interferes with body protein formation and affects growth or egg production. The primary object of protein feeding, therefore, is to furnish the bird with protein which, upon digestion, will yield sufficient quantities of the 13 essential amino acids needed for top performance.

When formulating poultry rations, they must be so designed as to supply all the essential amino acids in ample amounts. Special attention needs to be given to supplying the amino acids lysine, methionine and cystine, and tryptophan, which are sometimes referred to as the "critical amino acids" in poultry nutrition. Additionally, there must be sufficient total nitrogen for the chicken to synthesize the other amino acids needed.

The consequences of a protein or amino acid deficiency vary with the degree of the deficiency. A *borderline deficiency* is characterized by poor growth, deformed primary wing feathers, reduced egg size, poor egg production (but hatchability is not affected), tendency toward greater deposition of carcass and liver fat, poor feed conversion into eggs or meat, and lack of melanin pigment in black- or reddish-colored feathers with low lysine. A *severe protein deficiency* is marked by stopping of feed intake, stopping of egg production, loss of body weight, resorption of ova, a tongue deformity with leucine, isoleucine, and phenylalanine deficiency, stasis of the digestive tract, and death.

TABLE POULTRY MIN-

Minerals Which May Be Deficient Under Normal Conditions	Conditions Usually Prevailing Where Deficiencies Are Reported	Function of Mineral	Some Deficiency Symptoms	Types of Poultry Rations Usually Requiring Supplementation			
				Starting	Growing	Laying	Breeding
Major or macrominerals:							
Salt (sodium and chlorine—NaCl)	Omitted at the mill.	Improves appetite, promotes growth, helps regulate body pH, and is essential for hydrochloric acid formation in the stomach.	Chloride-deficient chicks show poor growth, high mortality, nervous symptoms, and reduced blood chloride level. Sodium deficiency in layers results in decreased egg production, poor growth, and cannibalism.	Yes	Yes	Yes	Yes
Calcium (Ca)	Imbalance of Ca:P ratio. Presence of interfering elements.	Bone formation; eggshell formation; blood clotting; and neuromuscular function.	Anorexia, thin eggshells, rickets, or osteoporosis, tetany, abnormal walk.	Yes	Yes	Yes	Yes

- **True digestibility of amino acids**[2]—Recently, there has been much interest in determining amino acid availability, using some of the same techniques that were originally developed for determination of true metabolizable energy. This interest stemmed from the fact that, frequently, amino acids cannot be completely utilized due to inherent characteristics of the feedstuff. Details follow.

The available amino acids are the amino acids in the diet that are not combined with compounds interfering with their digestion, absorption, or utilization by the bird; they are actually supplied to the sites of protein synthesis.

The digestible amino acids are those that are absorbed from the gut lumen; they are calculated as the difference between the amount of amino acids in the feed and the excreta.

Differentiation must be made between apparent and true digestibility. In addition to nonabsorbed feed, the excreta also contain materials originating in the tissue of the bird, e.g., cells sloughed from the gut wall, mucus, bile, unabsorbed digestive juices, etc. Apparent digestibility makes no correction for these endogenous components. In simple formulas, *apparent digestibility coefficient* and *true digestibility coefficient* follow:

$$\text{Apparent Digestibility Coefficient} = \frac{\text{amino acid consumed} - \text{amino acid excreted}}{\text{amino acid consumed}} \times 100$$

---

[2]In the preparation of this section, the authors adapted selected material from the publication: *True Faecal Digestibility of Essential Amino Acids in Feedstuffs for Poultry,* published by Eurolysine, 16 Rue Ballu 75009 Paris, France, with the permission of Eurolysine's "sister company," Heartland Lysine, Inc., 8430 West Bryn Mawr, Suite 650, Chicago, IL 60631.

$$\text{True Digestibility Coefficient} = \frac{\text{amino acid consumed} - \text{amino acid excreted} + \text{endogenous amino acid}}{\text{amino acid consumed}} \times 100$$

The availability of amino acids in feedstuffs is estimated by three main methods: (1) *in vitro* tests, (2) growth tests, and (3) digestibility tests. Most of the recent research has focused on digestibility determinations.

- **Table V-4, True Digestibility of Essential Amino Acids for Poultry, and True Digestible Amino Acid Recommendations for Poultry Feed Formulation**—Along with the total content of amino acids in feedstuffs, knowledge of their digestibility is essential to enable a formulator to satisfy the requirements for the projected performance. Section V, Table V-4, is presented for this purpose.

Minerals

Minerals are required for the formation of the skeleton, as components of various compounds with particular functions within the body, as activators of enzymes, and for the proper maintenance of necessary osmotic relationships within the body of the bird.

The minerals which have been shown to be essential for chickens and turkeys are calcium, chlorine, copper, iodine, iron, magnesium, manganese, molybdenum, phosphorus, potassium, selenium, sodium, sulfur, and zinc. Of these, calcium, chlorine, iodine, manganese, phosphorus, selenium, sodium, and zinc are considered to be of most practical importance since outside sources of them must be added to practical feed formulation for chickens and turkeys. Most of the pertinent facts relative to poultry minerals are summarized in Table 24-19.

24-19

ERAL CHART (See footnote at end of table.)

Mineral Requirements[1]				
Mineral Content (%) of Ration (Variable According to Class, Age, and Weight of Poultry)				
For	See Table	**Recommended Allowances**[1]	**Practical Sources of the Mineral**	**Comments**
Layers	24-4	0.2-0.5% of the diet.	Common table salt.	Sodium level is sometimes reduced to minimal to control the moisture level of the feces.
Broilers	24-6			
Breeding hens	24-8			
Turkeys	24-11			
Ducks	24-14			
Geese	24-16			
Pheasants/Bob White	24-17			
Japanese Quail	24-18			
Layers	24-4	The calcium allowance should vary with level of production and temperature. A minimum of 3.4% is believed to be optimum for layers in moderate climates. Growing rations should contain 0.8-0.9% of Ca and 0.4-0.7% of P.	Dicalcium phosphate. Limestone. Oystershell.	For young poultry, the Ca:P ratio should be about 1.2:1. However, ratios of 1:1 to 1.5:1 are well tolerated. An excess of calcium interferes with the utilization of magnesium, manganese, and zinc.
Broilers	24-6			
Breeding hens	24-8			
Turkeys	24-11			
Ducks	24-14			
Geese	24-16			
Pheasants/Bob White	24-17			
Japanese Quail	24-18			

(Continued)

TABLE 24-19

Minerals Which May Be Deficient Under Normal Conditions	Conditions Usually Prevailing Where Deficiencies Are Reported	Function of Mineral	Some Deficiency Symptoms	Types of Poultry Rations Usually Requiring Supplementation			
				Starting	Growing	Laying	Breeding
Major or macrominerals:							
Phosphorus (P)		Bone formation; metabolism of carbohydrates and fats; a component of all living cells; maintenance of the acid-base balance of the body; and calcium transport in egg formation.	Anorexia, weakness, rickets. Cage-layer fatigue, characterized by birds being paralyzed and unable to rise from a recumbent position. But there is evidence that this condition is not due to this factor alone.	Yes	Yes	Yes	Yes
Magnesium (Mg)	Diets containing high levels of Ca and P.	Essential for normal skeletal development, as a constituent of bone, enzyme activator, primarily in glycolytic system.	Decreased egg production, depressed growth and lethargy, convulsions.	No	No	No	No
Potassium (K)	Low plant protein level/high animal protein level.	Major cation of intracellular fluid where it is involved in osmotic pressure and acid-base balance. Muscle activity. Required in enzyme reaction involving phosphorylation of creatine. Influences carbohydrate metabolism.	*Chicks:* Retarded growth and high mortality.	No	No	No	No
Trace or microminerals:							
Copper (Cu)		Essential element in a number of enzyme systems and necessary for synthesizing hemoglobin and preventing nutritional anemia.	Anemia, depigmentation of feathers, digestive disorders. *Poults:* Marked cardiac hypertrophy.	Yes	Yes	Yes	Yes
Iodine (I)	Feeds produced on iodine-deficient soils.	Needed by the thyroid gland for making thyroxin, an iodine-containing hormone which controls the rate of body metabolism or heat production.	Enlarged thyroid. Eggs produced from deficient breeder hens have a lowered hatchability, prolonged hatching time, and a subsequent retardation of yolk-sac absorption.	Yes	Yes	Yes	Yes
Iron (Fe)		Necessary for formation of hemoglobin, an iron-containing compound which enables the blood to carry oxygen. Iron is also important to certain enzyme systems.	*Chicks and poults:* Microcytic, hypochromic anemia. In red-feathered chickens, complete depigmentation of the feathers occurs.	Yes	Yes	Yes	Yes
Manganese (Mn)		Necessary for growth, bone structure, and reproduction.	*Chicks and poults:* Perosis, shortened leg bones, skull deformation, parrot beak. Poor egg production, shell quality, and hatchability.	Yes	Yes	Yes	Yes
Selenium (Se)	Feeds that are grown in selenium-deficient areas.	Involved in the destruction of peroxides within the cell as a constituent of glutathione peroxidase. Useful in preventing exudative diathesis.	Exudative diathesis. Pancreatic fibrosis. Steatitis. Muscular dystrophy. With a severe selenium deficiency, growth rate is reduced and mortality is increased. *Turkeys:* Myopathies of the gizzard and heart.	Yes	Yes	Yes	Yes
Zinc (Zn)		Zinc is a component of several enzyme systems, including peptidases and carbonic anhydrase. Also, zinc is required for normal protein synthesis and metabolism and is a component of insulin.	Bone problems, poor feathering (feather fraying occurs near the ends of the feathers), retarded growth, and loss of appetite. A zinc deficiency in breeder diets reduces egg production and hatchability.	Yes	Yes	Yes	Yes

[1]As used herein, the distinction between *mineral requirements* and *recommended allowances* is as follows: In *mineral requirements,* no margins of safety are included intentionally; whereas in *recommended allowances,* margins of safety are provided in order to compensate for variations in feed composition, environment, and possible losses during storage or processing.

Feeding Poultry

(Continued)

Mineral Requirements[1] — Mineral Content (%) of Ration (Variable According to Class, Age, and Weight of Poultry) For .... See Table	Recommended Allowances[1]	Practical Sources of the Mineral	Comments
Layers 24-4 Broilers 24-6 Breeding hens 24-8 Turkeys 24-11 Ducks 24-14 Geese 24-16 Pheasants/Bob White 24-17 Japanese Quail 24-18	Dependent on production. Laying hens require diets containing at least 3% P. Growing rations should contain 0.8–0.9% of Ca and 0.4–0.7% of P.	Defluorinated phosphate. Dicalcium phosphate. Monosodium phosphate. Phosphoric acid. Steamed bone meal.	Organic phosphorus (present in plants) is poorly utilized by growing birds, but is satisfactory for adult birds. Only about 30 to 40% of the phosphorus in plant products is available to the young chick, poult, or duckling.
Layers 24-4 Broilers 24-6 Turkeys 24-11 Ducks 24-14 Japanese Quail 24-18	600 ppm in the ration of broilers should suffice.	Magnesium oxide or magnesium sulfate.	Requirements are affected by Ca and P levels in the diet. Not normally deficient in poultry rations.
Layers 24-4 Broilers 24-6 Turkeys 24-11 Japanese Quail 24-18		Corn contains 0.33% potassium (as-fed), and other cereals contain 0.42–0.49% potassium.	Potassium is not deficient in normal rations, due to large amounts of plant products in poultry feeds.
Layers 24-4 Broilers 24-6 Turkeys 24-11 Ducks 24-14 Geese 24-16 Japanese Quail 24-18	Rations containing 6 to 8 ppm of copper should be adequate.	Copper sulfate and copper carbonate are about equally effective.	
Layers 24-4 Broilers 24-6 Turkeys 24-11 Pheasants/Bob White 24-17 Japanese Quail 24-18	Laying hens require feed containing 300 ppb. Growing chicks require feed containing 350 ppb.	Stabilized iodized salt containing 0.007% iodine. Trace mineral mixes.	
Layers 24-4 Broilers 24-6 Turkeys 24-11 Japanese Quail 24-18	Rations should contain about 80 ppm of iron.	Alfalfa meal. Meat, liver, and fish meals.	Iron salts are used as a means of detoxifying gossypol from cottonseed meal.
Layers 24-4 Broilers 24-6 Turkeys 24-11 Ducks 24-14 Japanese Quail 24-18	Rations should contain at lease 30 ppm for layers, and 60 ppm for broilers.	Alfalfa meal, distillers' solubles, or grain by-products. Manganese sulfate, manganous chloride, manganous carbonate, and manganous dioxido.	
Layers 24-4 Broilers 24-6 Turkeys 24-11 Ducks 24-14 Japanese Quail 24-18	0.2 to 0.3 ppm added selenium.	Fish meal and brewers' dried yeast. Sodium selenite or sodium selenate.	Selenium can pose toxicity problems. Hence, care should be taken when adding it to poultry rations. In 1987, FDA provided for an increase in the maximum allowance of selenium in complete feeds for chickens, turkeys, and ducks to 0.3 ppm. Added selenium is often put into the vitamin premix to avoid separation.
Layers 24-4 Broilers 24-6 Turkeys 24-11 Ducks 24-14 Japanese Quail 24-18		Zinc carbonate or zinc sulfate.	

Fig. 24-7. Perosis, or slipped tendon, due to manganese deficiency. (Courtesy, Department of Poultry Science, Cornell University, Ithaca, N.Y.)

Vitamins

The vitamins required by poultry, along with their deficiency symptoms and dietary sources, are shown in Table 24-20.

The column of Table 24-20 headed, "Types of Poultry Rations Usually Requiring Supplementation" indicates the types of poultry rations in which special attention must be paid to the inclusion of dietary sources of the vitamins. As shown, vitamin A, vitamin D, riboflavin, and vitamin B-12 are commonly low in most poultry rations. It is also to be emphasized that vitamin D_3, the animal form of vitamin D (made by the irradiation of 7-dehydrocholesterol), is more active for poultry, and should, therefore, be used instead of vitamin D_2, the plant form of the vitamin.

TABLE
POULTRY VITA-

Vitamins Which May Be Deficient Under Normal Conditions	Conditions Usually Prevailing Where Deficiencies Are Reported	Function of Vitamin	Some Deficiency Symptoms	Types of Poultry Rations Usually Requiring Supplementation			
				Starting	Growing	Laying	Breeding
Fat-soluble vitamins:							
Vitamin A	Old vitamin premix.	Essential for normal maintenance and functioning of the epithelial tissues, particularly of the eye and the respiratory, digestive, reproductive, nerve, and urinary systems.	*Chicks:* Depressed growth, weakness; loss of coordination; xerophthalmia; anorexia, lowered resistance to infection; alterations in mucous membranes. *Adults:* Depressed production; low hatchability; discharge from nose and eyes; lowered resistance to infection; alterations in mucous membranes.	Yes	Yes	Yes	Yes
Vitamin D_3	Birds that are in confinement.	Aids in assimilation and utilization of calcium and phosphorus, and necessary in normal bone development.	*Chicks:* Rickets, poor feathering, reduced growth. *Adults:* Weak bones, poor eggshell formation, reduced production and hatchability.	Yes	Yes	Yes	Yes
Vitamin E	Destruction by oxidation of the diet.	Antioxidant. Muscle structure. Reproduction.	*Chicks:* Encephalomalacia, exudative diathesis, muscular dystrophy. *Adults:* Poor reproductive performance; prolonged vitamin E deficiency results in permanent sterility in the male and reproductive failure in the female. *Poults:* Myopathy of the gizzard.	Yes	Yes	Yes	Yes
Vitamin K	Coccidiosis. When high levels of antibodies or sulfa drugs are fed. Newly hatched chicks from deficient females.	Essential for prothrombin formation and blood clotting.	Hemorrhaging. Increased clotting time.	Yes	Yes	Yes	Yes
Water-soluble vitamins:							
Biotin	Broilers fed a milo, wheat, or wheat-barley based diet. Feeding avidin, a protein in uncooked egg white, which binds biotin and renders it unavailable nutritionally.	Involved in carbohydrate, lipid, and protein metabolism.	*Chicks:* Cracking and degeneration of skin on feet, around beak, and perosis (slipped tendon). *Adults:* Reduced hatchability. *Poults:* Broken flight feathers, bending of the metatarsus, and dermatitis of the footpads and toes, base of beak, eye ring and vent.	Yes	No	No	Yes

Feeding Poultry

The fat-soluble vitamins (A, D, E, and K) can be stored and accumulated in the liver and other parts of the body, while only very limited amounts of the water-soluble vitamins (biotin, choline, folacin, niacin, pantothenic acid, riboflavin, thiamin, vitamin B-6, and vitamin B-12) are stored. For this reason, it is important that the water-soluble vitamins be fed regularly in the ration in adequate amounts.

Vitamin C is synthesized by poultry; hence, it is not considered as a required dietary nutrient. There is some evidence, nevertheless, of a favorable response to vitamin C by birds under stress.

Requirements for some of the vitamins may be met by the amounts occurring in natural feedstuffs. However, formulators of poultry feeds should be alert to the need for dietary supplementation with vitamins usually assumed to be supplied by the feedstuffs.

Fig. 24-8. A chick deficient in vitamin D, showing ungainly manner of balancing body. The beak is also soft and rubbery. (Courtesy, Department of Poultry Science, Cornell University, Ithaca, N.Y.)

24-20
MIN CHART (See footnote at end of table.)

Vitamin Requirements[1] Vitamin Content (%) of Ration (Variable According to Class, Age, and Weight of Poultry) For .... See Table	Recommended Allowances[1]	Practical Sources of the Vitamin	Comments
Layers .... 24-4 Broilers .... 24-6 Turkeys .... 24-11 Ducks .... 24-14 Geese .... 24-16 Japanese Quail 24-18	Variable according to class, age, and weight (see the suggested rations under each respective class).	Green forage, alfalfa meal, corn gluten meal, yellow corn, fish oils, synthetic vitamin A.	Toxicities can occur. The toxic level is on the order of 500 times the requirement. Symptoms of a vitamin A toxicity are weight loss, depressed feed intake, inflammation of epithelial tissue, and bone abnormalities. Requirements for vitamin A are expressed in either International Units (IU) or U.S. Pharmacopeia units (USP) per kilogram of diet.
Layers .... 24-4 Broilers .... 24-6 Turkeys .... 24-11 Ducks .... 24-14 Geese .... 24-16 Japanese Quail 24-18	Variable according to class, age, and weight (see the suggested rations under each respective class).	Irradiated animal sterols, fish liver oils, vitamin A and D feeding oils, synthetic vitamin D_3.	Vitamin D_3 is more than 30 times as efficient for preventing rickets in chickens as vitamin D_2. Hypervitaminosis can occur.
Layers .... 24-4 Broilers .... 24-6 Turkeys .... 24-11 Japanese Quail 24-18	Variable according to class, age, and weight (see the suggested rations under each respective class).	Alfalfa meal, vegetable oils, wheat germ, and pure vitamin concentrates such as alpha-tocopherol.	Vitamin E and selenium have a close interrelationship. In many cases, selenium can reduce the dietary requirement of vitamin E.
Layers .... 24-4 Broilers .... 24-6 Turkeys .... 24-11 Ducks .... 24-14 Japanese Quail 24-18	Variable according to class, age, and weight (see the suggested rations under each respective class).	Green pasture, alfalfa meal, synthetic vitamin K (menadione sodium bisulfite).	
Layers .... 24-4 Broilers .... 24-6 Turkeys .... 24-11 Japanese Quail 24-18	Variable according to class, age, and weight (see the suggested rations under each respective class).	Grains, soybean meal, alfalfa meal, dried yeast, milk products, green pasture.	Availability in wheat and barley is extremely low.

(Continued)

TABLE 24-20

Vitamins Which May Be Deficient Under Normal Conditions	Conditions Usually Prevailing Where Deficiencies Are Reported	Function of Vitamin	Some Deficiency Symptoms	Types of Poultry Rations Usually Requiring Supplementation			
				Starting	Growing	Laying	Breeding
Water-soluble vitamins (Continued): Choline		Involved in nerve pulses. A component of phospholipids. Donor of methyl groups.	*Chicks, poults, ducklings:* Retarded growth and perosis (slipped tendon). *Adults:* Increased mortality, lowered egg production, and increased abortion of egg yolks from ovaries.	Yes	Yes	No	No
Folacin (Folic Acid)		Related to B-12 metabolism. Metabolic reactions involving incorporation of single carbon units into larger molecules.	*Chicks:* Poor growth, poor feathering, perosis, and anemia. *Adults:* Reduced hatchability and egg production. *Turkey poults:* Nervousness, droopy wings, and a stiff extended neck. *Turkey breeder hens:* Normal egg production, but reduced hatchability.	Yes	Yes	Yes	Yes
Niacin (Nicotinic Acid, Nicotinamide)	A predominately corn-soybean ration.	Required by all living cells, and an essential component of important metabolic enzyme systems involved in glycolysis and tissue respiration.	*Chicks:* Enlargement of hock joints and perosis, retarded growth, and inflammation of mouth and tongue ("black tongue"). *Adults:* No symptoms observed in hen except on protein-deficient diet. *Turkey poults:* A hock disorder similar to perosis.	Yes	Yes	Yes	Yes
Pantothenic Acid (Vitamin B-3)	Use of artificially dried corn (heating destroys the pantothenic acid) and the omission of milk by-products from the diet.	Part of coenzyme A, a necessary factor for intermediary metabolism.	*Chicks:* Poor growth, ragged feather development, degeneration of skin around beak, eyes, and vent, and liver damage. *Adults:* Reduced hatchability. Mortality is high in newly hatched chicks from pantothenic acid-deficient hens.	Yes	Yes	Yes	Yes
Riboflavin (Vitamin B-2)		A component of enzyme systems essential to normal metabolic processes.	*Chicks:* Curled toe paralysis, reduced growth, and diarrhea. *Adults:* Poor hatchability with many dying during 2nd week of incubation.	Yes	Yes	Yes	Yes
Thiamin (Vitamin B-1)		As a coenzyme in energy metabolism. Promotes appetite and growth, and required for normal carbohydrate metabolism.	*Chicks:* Anorexia, loss of coordination, poor feathering, polyneuritis. *Adults:* Blue comb, paralysis.	No	No	No	No
Vitamin B-6 (Pyridoxine, Pyridoxal, Pyridoxamine)		As coenzyme in protein and nitrogen metabolism. Involved in red blood cell formation. Important in endocrine systems.	*Chicks:* Poor growth, lack of coordination, and convulsions. *Adults:* Reduced body weight, egg production, and hatchability.	No	No	No	No
Vitamin B-12 (Cobalamins)		Numerous metabolic functions, and essential for normal growth and reproduction in poultry.	*Chicks:* Poor growth, perosis, mortality. *Adults:* Reduced hatchability, fatty heart, liver, and kidneys.	Yes	Yes	Yes	Yes

[1]As used herein, the distinction between *vitamin requirements* and *recommended allowances* is as follows: In *vitamin requirements*, no margins of safety are included intentionally; whereas in *recommended allowances*, margins of safety are provided in order to compensate for variations in feed compositions, environment, and possible losses during storage or processing.

(Continued)

Vitamin Requirements[1] Vitamin Content (%) of Ration (Variable According to Class, Age, and Weight of Poultry) For .... See Table	Recommended Allowances[1]	Practical Sources of the Vitamin	Comments
Layers .... 24–4 Broilers .... 24–6 Turkeys .... 24–11 Pheasants/Bob White .... 24–17 Japanese Quail 24–18	Variable according to class, age, and weight (see the suggested rations under each respective class).	Fish products and pure vitamin.	**R**ecent evidence indicates that choline is synthesized by mature chickens in quantities adequate for egg production. **C**holine may be a factor in egg size of quail. Dietary requirements of growing quail appear to be higher than those for chicks or poults.
Layers .... 24–4 Broilers .... 24–6 Turkeys .... 24–11 Japanese Quail 24–18	Variable according to class, age, and weight (see the suggested rations under each respective class).	Alfalfa meal, wheat, soybean meal, and liver preparations.	
Layers .... 24–4 Broilers .... 24–6 Turkeys .... 24–11 Ducks .... 24–14 Geese .... 24–16 Pheasants/Bob White .... 24–17 Japanese Quail 24–18	Variable according to class, age, and weight (see the suggested rations under each respective class).	Chemically synthesized nicotinic acid, liver, yeast, and fermentation products, and most grasses.	**S**ome niacin can be synthesized in the body through the conversion of tryptophan. **T**he niacin in cereal grains and by-products is virtually unavailable and should not be included in the available niacin calculation.
Layers .... 24–4 Broilers .... 24–6 Turkeys .... 24–11 Ducks .... 24–14 Geese .... 24–16 Pheasants/Bob White .... 24–17 Japanese Quail 24–18	Variable according to class, age, and weight (see the suggested rations under each respective class).	Pure calcium pantothenate, alfalfa meal, dried milk products, and fermentation residues.	
Layers .... 24–4 Broilers .... 24–6 Turkeys .... 24–11 Ducks .... 24–14 Geese .... 24–16 Pheasants/Bob White .... 24–17 Japanese Quail 24–18	Variable according to class, age, and weight (see the suggested rations under each respective class).	Alfalfa meal, milk products, distillers' solubles, fermentation products, and pure vitamin.	
Layers .... 24–4 Broilers .... 24–6 Turkeys .... 24–11 Japanese Quail 24–18	Variable according to class, age, and weight (see the suggested rations under each respective class).	Cereal grains and their by-products. Synthetic thiamin.	
Layers .... 24–4 Broilers .... 24–6 Turkeys .... 24–11 Ducks .... 24–14 Japanese Quail 24–18	Variable according to class, age, and weight (see the suggested rations under each respective class).	Milk products, meat and fish by-products, soybean meal.	
Layers .... 24–4 Broilers .... 24–6 Turkeys .... 24–11 Japanese Quail 24–18	Variable according to class, age, and weight (see the suggested rations under each respective class).	Fish meal, fish solubles, meat meal, liver preparations, fermentation products, and commercial vitamin B–12 concentrate.	**B**ody reserves are rapidly depleted in hens that are fed high-protein diets.

Fig. 24-9. A chick with nutritional encephalomalacia, due to a lack of vitamin E. Note head retraction and loss of control of legs. (Courtesy, Department of Poultry Science, Cornell University, Ithaca, N.Y.)

Fig. 24-10. Biotin deficiency. Note the severe lesions on the bottom of the feet. (Courtesy, Department of Poultry Science, University of Wisconsin, Madison)

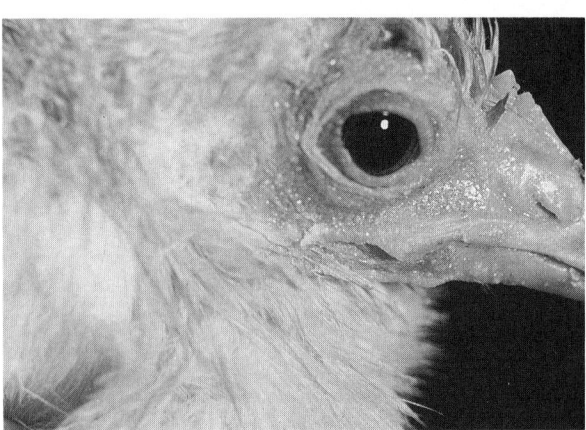

Fig. 24-11. "Spectacled eye" in a niacin-deficient chick. Also, note the loss of feathers around the eye. (Courtesy, University of Wisconsin, Madison)

Fig. 24-12. Thiamin (B-1) deficiency, resulting in acute stage of polyneuritis. Note characteristic head retraction. (Courtesy, Department of Poultry Science, University of Wisconsin, Madison)

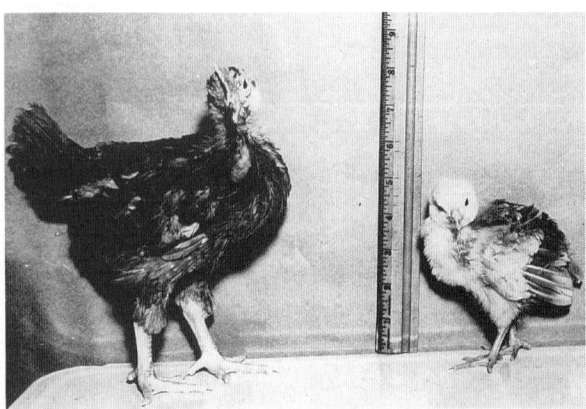

Fig. 24-13. Pyridoxine (B-6) deficiency. *Left:* Normal, control chick. *Right:* Chick shows retarded growth and abnormal feathering due to vitamin B-6 deficiency. (Courtesy, University of Georgia, Athens)

Fig. 24-14. Vitamin B-12 deficiency. The perky chick at right and his smaller companion are both 3½ weeks old. *Left:* The small chick, fed a ration deficient in vitamin B-12, weighed 157 g. *Right:* The larger chick, fed the same ration plus vitamin B-12, weighed 280 g. (Courtesy, Merck Chemical Division, Rahway, N.J.)

Unidentified Growth Factors (UGF)

In addition to the vitamins listed in Table 24-20, certain unidentified growth factors (UGF) are important in poultry nutrition. They are referred to as *unidentified* or *unknown* because they have not yet been isolated or synthesized in the laboratory. Nevertheless, numerous reports attest to the favorable responses to the dietary inclusion of these products in poultry rations, including stimulation of growth, increased egg production and hatchability, reduced liver fat, improved product quality, and reduced toxicity of minerals. A diet that supplies the specific levels of all the known nutrients but which does not supply the unidentified factors is inadequate for best performance. Rich sources of unidentified factors are egg yolk, whey, yeast, fish and meat by-products, soybeans, green forages, and fermentation by-products. Most nutritionists recognize the possible benefits of adding some UGF supplementation to the diets of broilers and breeding hens. Commonly, the UGF source is added to the diet at a level of 1 to 3%, although antibiotic fermentation residue products may be used at levels ranging from 2 to 8 lb per ton. This practice may have a twofold advantage: (1) providing possible unidentified growth factor responses, and (2) supplying additional amounts of some of the known vitamins.

(Also see Chapter 13, section on "Unidentified Factors.")

Water

Poultry should have free access to clean, fresh water at all times. It is needed as a solvent, a lubricant, and a temperature control device.

A general rule is that chickens drink approximately twice as much water by weight as the feed they consume.

The daily water consumption of chickens and turkeys is given in Table 24-21.

TABLE 24-21
DAILY WATER CONSUMPTION BY CHICKENS AND TURKEYS OF DIFFERENT AGES[1,2]

	Per 1,000 Birds					
Age	Leghorn-Type Pullets		Chicken Broilers[3]		Turkeys[3]	
(week)	(U.S. gal)	(liter)	(U.S. gal)	(liter)	(U.S. gal)	(liter)
1	5	19	5	19	10	38
2	10	38	13	50	20	76
3	12	45	24	90	30	114
4	17	64	37	140	40	151
5	22	83	53	200	50	189
6	25	95	69	260	60	227
7	28	106	85	320	75	284
8	30	114	100	380	95	360
9	35	132			115	435
10	38	144			125	473
12	40	151			150	568
15	42	158			160	606
20	45	170			200	757
			Laying or Breeding			
35	50	189			M 240	908
					F 130	492

[1]Adapted by the authors from *Nutrient Requirements of Poultry*, 8th rev. ed., NRC, National Academy Press, Washington, D.C., 1984, p. 8, Table 2.
[2]Will vary considerably depending on temperature and ration composition.
[3]Mixed sexes.

Of course, water consumption varies according to the kind of feed, temperature, humidity, activity of the bird, and rate of egg production. During hot weather, chickens will consume about twice as much water as they do under conditions of average temperature.

In addition to being readily available, water quality is important. Water should be tested to determine that salts, pesticides, and microorganisms are at acceptable levels and that the water is palatable to poultry. Water that adversely affects growth, reproduction, or productivity should not be used.

(Also see Chapter 4, section on "Water.")

Additives

Modern poultry feeds commonly contain one or more nonnutritive additives. These additives are used for a variety of reasons. They are not nutrients, but some of them improve production under certain circumstances. Others prevent rancidity in the feed. There is no evidence of a nutritional deficiency when they are omitted from a ration. Among such additives are the following:

1. **Antibiotics.** The primary reasons for using antibiotics in poultry feeds are for their growth-stimulating and improved efficiency of feed conversion effects, for which purpose they are generally used in both broiler and market turkey rations. Also, egg production is frequently improved by dietary supplementation with antibiotics. The reasons for the beneficial effects of antibiotics still remain obscure, but the best explanation for their growth-stimulating activity is the disease level theory, based on the fact that antibiotics have failed to show any measurable effect on birds maintained under germ-free conditions.

Antibiotics are generally fed to poultry at levels of 0.5 to 25 mg/pound of diet, depending upon the particular antibiotic used. Higher levels of antibiotics (100 to 400 g per ton of feed) are used for disease control purposes.

2. **Antifungal agents.** Feeds provide an excellent environment for the growth of fungi (molds), such as *Aspergillus flavus, Fusarium,* and *Candida albicans,* which are detrimental to the health of poultry. *Aspergillus flavus* produces a potent toxin which is referred to as aflatoxin. *Candida albicans* is the causative agent of a condition in poultry called thrust or moniliasis.

Several compounds have been introduced as feed additives to prevent the growth of molds in feeds. The products of choice are: propionic acid, acetic acid, and sodium propionate.

For an in-depth discussion of fungi (molds), see Chapter 5, Table 5-3—Mycotoxins [Toxin producing Fungi or Molds]; and Chapter 13, section on "Mold [Fungi] Inhibitors.")

3. **Antioxidants.** Antioxidants are compounds that prevent oxidative rancidity in polyunsaturated fats. They are used to prevent rancidity in poultry feeds. The antioxidants which are presently accepted for addition to fat in poultry feeds are butylated hydroxyanisole (BHA), butylated hydroxytoluene (BHT), and ethoxyquin. They are used at a level of ¼ lb per ton.

Antioxidants are added to feed fats to stabilize them against rancidity. BHT and BHA are commonly used to stabilize fat.

(Also see Chapter 13, section on "Antioxidants.")

4. **Arsenicals.** These products exert much the same effects as the antibiotics; hence, they are often added to poultry feeds to improve performance. It would appear that the action of arsenicals and antibiotics is very similar, since the effects of the two are not considered to be additive. Arsanilic acid and sodium arsanilate are the most widely used growth-promoting arsenicals. They are FDA-approved for use alone or in certain drug combinations for chickens and turkeys. When used according to directions, they increase rate of gain and improve feed efficiency of chickens and turkeys; improve pigmentation in growing chickens and turkeys; and increase egg production in layers.

(Also see Chapter 13, section on "Arsenicals.")

5. **Drugs for disease prevention and control.** Poultry rations frequently contain drugs designed to prevent specific diseases. For example, a wide variety of chemical substances, sold under various trade names, is available for use in the prevention of coccidiosis. These drugs are known as coccidiostats.

Turkey rations are frequently formulated with drugs for the prevention of blackhead. This class of drugs, known as histomonostats, also contains a wide variety of chemical substances sold under various trade names.

(Also see section in this chapter on "Salmonella.")

6. **Flavor additives.** *Flavoring agents are feed additives that are designed to increase palatability and feed intake.*

Chickens have the ability to differentiate between sucrose solutions, for which they show a preference, and saccharin solutions which they avoid. Other studies indicate that chickens possess a sense of taste, but little or no ability to smell.

(Also see Chapter 13, section on "Flavoring Agents.")

7. **Grit.** The use of grit is controversial. Some research indicates that hens fed an all-mash ration do not need grit, but there is evidence that as a component of or supplement to all-mash it may improve feed utilization and increase production under certain conditions. The primary function of grit is to help the gizzard grind food materials that pass through it. It is definitely needed when birds consume whole grains or coarse, fibrous feedstuffs. Crushed granite or other hard, insoluble material can be used for grit.

(Also see Chapter 13, section on "Grit.")

8. **Xanthophylls.** Feeds that contain large amounts of xanthophylls produce a deep yellow color in the beak, skin, shanks, feet, fat, and egg yolks of poultry. The consumer associates this pigmentation with quality and in many cases is willing to pay a premium price for a bird of this type. Also, processors of egg yolks are frequently interested in producing dark-colored yolks to maximize coloration of egg noodles and other food products. The latter can be accomplished by adding about 60 mg of xanthophyll per kilogram of diet. In recognition of these consumer preferences, many producers add ingredients that contain xanthophylls to poultry rations.

It is not necessary to incorporate high levels of xanthophyll in starter and grower rations. Low levels can be maintained through these periods of feeding, but finishing rations should contain high levels of these pigment-producing compounds. Alfalfa meal, yellow corn, and corn gluten meal are sources that are commonly used. Common dried algae and marigold petal meal are even richer sources of xanthophyll. The synthetic carotenoid canthaxanthin (red) is approved by the Food and Drug Administration and is now widely used by producers in the United States.

Table 24-22 lists xanthophyll-rich feedstuffs, along with the quantities that they contain.

TABLE 24-22
XANTHOPHYLL-RICH FEEDSTUFFS[1]

Feedstuffs	Xanthophylls	
	(mg/lb)	(mg/kg)
Alfalfa meal, 17% protein	118	*260*
Alfalfa meal, 20% protein	127	*280*
Alfalfa meal, 22% protein	150	*330*
Alfalfa juice protein, 40% protein	363	*800*
Algae, common, dried	908	*2,000*
Corn, yellow	8	*17*
Corn gluten meal, 41% protein	79	*175*
Corn gluten meal, 60% protein	132	*290*
Marigold petal meal	3,178	*7,000*

[1]Adapted by the authors from *Nutrient Requirements of Poultry,* 8th rev. ed., NRC, National Academy Press, Washington, D.C., 1984, p. 9, Table 3.

(Also see Chapter 13, section on "Additives That Enhance Market Value.")

FEEDS FOR POULTRY

A wide variety of feedstuffs can be, and is, used in poultry rations. Broadly speaking, these may be classed as energy feedstuffs, protein supplements, mineral supplements, and vitamin supplements.

1. **Energy feedstuffs.** The major energy sources of poultry feeds are the cereal grains and their by-products and fats. Corn is the most important grain used by poultry, supplying about one-third of the total feed which they consume.

Oats, barley, and the sorghum grains are also used extensively in poultry rations. Oats are lower in energy than corn and are generally too expensive for broiler and layer rations. But oats can be used very effectively in feeds for replacement birds. Barley is less palatable than corn and is lower in vitamin A and energy. The sorghum grains can be readily substituted in the place of corn as an energy feed; hence, in the southern states, they are being used extensively. Wheat is used when the price is right. Although millet is seldom used in poultry rations, it can be freely substituted for corn.

Molasses can be used effectively in poultry rations provided that its level of usage is closely monitored. Excessive amounts cause wet droppings.

Animal and vegetable fats are now used extensively in poultry feed. In addition to their high energy value, fats reduce the dustiness of feed mixtures, increase their palatability, and improve the texture and appearance of the feed. However, the use of fats in poultry feeds requires good mixing equipment. Also, it is necessary that the fat be properly stabilized in order to prevent rancidity.

Many other energy feedstuffs, including a great array of milling by-products (for example, the corn gluten and bran feeds and the wheat processing by-products), are used in poultry feeds.

2. **Protein supplements.** The usefulness of a protein feedstuff depends upon its ability to furnish the essential amino acids required by the bird, the digestibility of the protein, and the presence or absence of toxic substances. As a general rule, several different sources of protein produce better results than single protein sources. Both animal and plant protein supplements are used for poultry. Most of the protein supplements of animal origin contribute minerals

and vitamins which significantly affect their value in poultry rations, but they are generally more variable in composition than the plant protein supplements.

Among the animal protein supplements commonly used in poultry rations are meat and fish by-products, milk by-products, and such miscellaneous animal by-products as blood meal, hydrolyzed poultry feathers, and poultry by-product meal.

The common plant protein supplements used in poultry feeding include the oilseed meals (soybean meal, cottonseed meal, peanut meal, rapeseed meal, and limited amounts of linseed meal), corn gluten meal, and alfalfa meal and other legume meals. Soybean meal is, by far, the most widely used protein supplement in poultry rations.

3. **Mineral supplements.** Mineral supplements are required by poultry for skeletal development in growing birds, for eggshell formation in laying hens, and for certain other regulatory processes in the body.

The common calcium supplements used in poultry feeding are ground limestone, crushed oystershells or oystershell flour, bone meal, calcite, chalk, and marble.

Most of the phosphorus in plant products is in organic form and not well utilized by young chicks or turkey poults. Hence for poultry, emphasis is placed upon inorganic phosphorus sources in feed formulation. Bone meal, dicalcium phosphate, defluorinated phosphate, and colloidal phosphate are used where phosphorus is needed in the ration.

Salt is added to most poultry rations at a 0.2 to 0.5% level. Too much salt will result in increased water consumption and wet droppings.

4. **Vitamin supplements.** A great many vitamins are important in present-day poultry feed formulation. Formerly, a wide variety of crude feedstuffs were added to poultry formulas primarily for their vitamin content. Today, many of these have been replaced by special vitamin supplements, which in many cases are chemically pure sources that need to be used only in very minute amounts. In modern poultry feed formulation and production, premixes often represent the common sense approach to providing both mineral and vitamin needs for poultry.

Presence of Substances Affecting Product Quality

The composition of the feed can affect the product. The color of the skin or shanks of a broiler or of the yolk of an egg is primarily due to the carotenoid pigments consumed in the feed. Corn, alfalfa meal, and corn gluten meal are the main feeds used to contribute these pigments.

Screw process cottonseed meal, which is high in gossypol, when fed to laying hens may cause egg yolk discoloration in stored eggs. Some fish products may impart off-flavors to poultry meat or eggs. Thus, certain feedstuffs may be undesirable simply because of the effect they produce on the end product.

• **Cholesterol in eggs**—Without doubt, the per capita consumption of eggs has decreased because of relating the fat and cholesterol content of eggs to heart disease, although changes in breakfast eating habits have also contributed to the decline. But scientists and producers are coming to the rescue!

Recently, low-cholesterol eggs have been produced (1) by manipulating the feed ingredients in layer rations, in some cases with added lighting changes; (2) by additives to the ration; and (3) by solvent extraction of the yolk. At this stage, most of these techniques are secretive; some of them are in the process of being patented. More experimental work is needed; and the profitability (the price that consumers are willing to pay for low-cholesterol eggs, less the cost of production) must be determined.

(Also see Chapter 13, section on "Additives That Enhance Market Value.")

FEED PREPARATION

The usual end product resulting from mixing poultry feedstuffs is a ground feed known as a mash. While this mash is usually in the form of a ground mixture, it can be processed to produce pellets or crumbles.

Pellets or crumbles cost slightly more than the same ration in mash form. Yet, they are used for broilers and turkeys because of improved feed efficiency (fewer pounds of feed to produce a pound of gain).

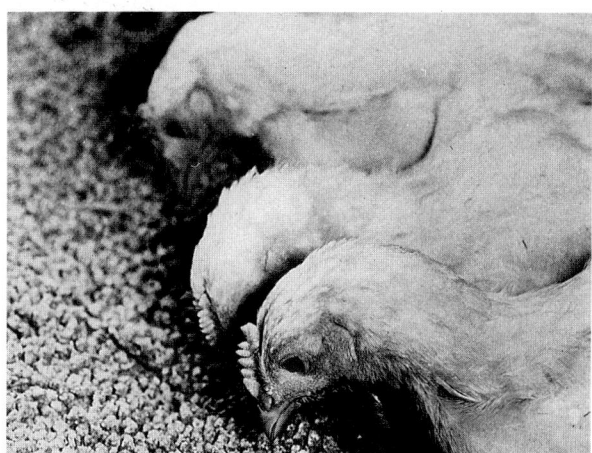

Fig. 24–15. Six-week-old White Rock broilers eating pellets. (Courtesy, Ralston Purina Company, St. Louis, Mo.)

(Also see Chapter 14, Feed Processing.)

POULTRY RATIONS

In 1984, 109.6 million tons of commercial feeds were produced in the United States, of which 87% was primary feed (mixed feed) and 13% consisted of secondary feed (feed to which a formula feed supplement is added at the rate of 100 lb per ton or more). A total of 35.5%—more than one-third—of the primary feed was fed to poultry.

The poultry producer has the following alternatives for purchasing and preparing rations:

1. Purchase of a commercially prepared complete feed.
2. Purchase of a commercially prepared protein supplement, reinforced with vitamins and minerals, which may be blended with local grain.
3. Purchase of a commercially prepared vitamin-mineral premix which may be mixed with an oil meal, and then blended with local grain.
4. Purchase of individual ingredients (including vitamins and minerals) and mixing the feed from the "ground" up.

Factors Involved in Formulating Poultry Rations

Before anyone can intelligently formulate a poultry ration, it is necessary to know (1) the nutrient requirements of the particular birds to be fed; (2) the availability, nutrient content, and cost of feedstuffs; (3) the acceptability and physical condition of feedstuffs; (4) the average daily consumption of the birds to be fed; and (5) the presence of substances harmful to product quality.

Formulating Rations

The increasing complexity of poultry rations, along with larger and larger enterprises, makes it imperative that producers who choose to mix feed be absolutely sure that they will have a nutritionally balanced and adequate ration.

When fed free-choice, birds tend to eat to satisfy their energy requirements. Consequently, it is possible, within limits, to regulate the intake of all nutrients, except water, by including them in the diet in specific ratios to available energy. Thus, the energy content of the diet must be considered in formulating to meet a desired intake of all essential nutrients other than energy itself.

The larger commercial feed companies, and the larger poultry producers who do their own mixing or formulating, generally rely on the services of a nutritionist and the use of a computer in formulating their rations. **(NOTE:** See Chapter 18, section on "Computer Methods" for computer ration formulation.) Even though they are more time-consuming, and fewer factors can be considered simultaneously, a good job can be done in formulating rations by the hand method.

(Also see Chapter 18, Feeding Standards/Ration Formulation.)

Fig. 24-16. A number of forces may affect the nutritive requirements of poultry, among them, (1) temperature and humidity, and (2) genetic differences. Also, the nutrient composition of feedstuffs is variable. (Courtesy, Jamesway Division, Butler Manufacturing Company, Kansas City, Mo.)

FEED SUBSTITUTION TABLE

Successful poultry producers are well educated in economics. They know that feed conversion rate alone does not determine success. The cost of producing each pound of broiler, turkey, or dozen eggs is the determinant of success. For this reason, they want to obtain a balanced ration that will make for the highest net returns. To help poultry producers decide what feeds might be interchangeable with the ones they are using and to what extent certain feeds can be replaced, the authors prepared Table 24-23, Feed Substitution Table for Poultry.

TABLE 24-23
FEED SUBSTITUTION TABLE FOR POULTRY (AS-FED)

Feedstuff	Relative Feeding Value (lb for lb) in Comparison With the Designated (underlined) Base Feed Which = 100	Maximum Percentage of Base Feed (or comparable feed or feeds) Which It Can Replace for Best Results	Remarks
GRAINS, BY-PRODUCT FEEDS:			
Corn, No. 2	*100*	*100*	Corn is the most widely used grain in poultry feed.
Bakery wastes	75	50	Bakery wastes are very similar to cereal grains in composition.
Barley	80–85	50	Barley is very low in vitamin A; less palatable than corn.
Beans (cull)	90	5	
Cassava	85	20	Extremely low in protein.
Hominy	95	50	Not generally used in poultry rations; low energy value, but is good source of linoleic acid.
Millet	95–100	65	Best when used as a 50:50 mix with barley, corn, or oats; can be used as a scratch feed.
Molasses	70	5	High levels of molasses will produce wet droppings.
Oats	70–80	50	Usually too expensive for broiler and layer rations, but is used extensively in replacement rations.
Peas, dried	90–100	5–10	
Rice, rough	80–85	20–50	Rice may be deficient in vitamin A.
Rice, bran	50	5–10	Rice bran is high in fat and susceptible to oxidation.
Rice polishings	85–90	5–10	
Rye	90	25–30	Rye is not a very good poultry feed as it can depress growth and cause sticky droppings.
Sorghum	100	100	
Triticale	80–90	30	Triticale is somewhat lower in feeding value than either wheat or corn.
Wheat	90–95	100	Wheat is the most variable of the cereal grains in protein; very low in xanthophylls; should be processed to increase palatability.
Wheat bran	75	10–15	
PROTEIN SUPPLEMENTS:			
Soybean meal (48%)	*100*	*100*	Well balanced in amino acids. Best quality of all plant protein supplements. Very palatable.
Babassu oil meal	50	20	Similar to copra meal.
Blood meal, flash- or ring-dried	120	5–20	Excellent source of lysine.
Copra meal (coconut meal)	50	25	Copra meal should be supplemented with some animal protein.
Corn gluten meal	50–75	25	Animal protein must be used along with corn gluten meal.
Cottonseed meal	85	80[1]	Gossypol, a compound found in cottonseed meal, can cause discoloration of egg yolks. The addition of iron in the ration may minimize the dangers of gossypol poisoning. Cyclopropenoid fatty acids in cottonseed meal can cause discoloration of egg white. Glandless cottonseed meal is recommended.
Feather meal	50	5	Feather protein is deficient in methionine, lysine, histidine, and tryptophan.
Fish meal, anchovy and menhaden	115	50–65	Expensive. Excellent balance of amino acids and good source of calcium and phosphorus. Most poultry rations incorporate some fish meal at levels of about 2–5% of the ration. Fish meal can impart fishy flavors if fed at too high levels.
Linseed meal	80	10	Linseed meal depresses growth and can cause diarrhea. Levels should be restricted to a maximum of 5% of the ration.
Meat and bone meal	100	20–50	Tends to be deficient in tryptophan.
Meat meal (50–55% protein)	100	50–65	Meat meal is high in phosphorus and low in methionine and cystine. It is recommended that the maximum level of usage not exceed 10% of the ration.
Peanut meal (41% protein)	95	75–100	If peanut meal is to be substituted for soybean meal, lysine must be added.
Poultry by-product meal	100	20–50	
Rapeseed meal	85–90	33⅓	It may contain goitrogenic compounds that may be hazardous. But the newer varieties of rape (known as canola) are much lower in goitrogenic compounds.
Safflower seed meal (decorticated)	95	50–100	Safflower seed meal can be incorporated at levels up to 15% of the ration. Deficient in lysine and methionine.
Sesame meal	95–100	100	Extremely deficient in lysine.
Sunflower seed meal	95–100	100	
Yeast, torula	100	60	Yeast is a good source of the B-complex vitamins.

[1]When cottonseed meal is substituted for soybean meal at this level, it must be degossypolized.

SPECIAL FEEDING PROGRAMS; FEED INTAKE AND SUGGESTED RATIONS

The nutritive requirements of poultry vary according to species (between chickens, turkeys, ducks, and geese), according to age, and according to the type of production—whether the birds are kept for meat production, layers, or breeders. For this reason, many different rations are required.

To be successful, rations must meet the nutritive requirements of the birds to which they are fed. The National Research Council nutritive requirements are given in Tables 24-4 through 24-18. In using these tables to formulate practical rations, it must be remembered that they are minimum requirements, which means that they do not provide for any margins of safety. Further, the protein-energy relationships shown therein should be retained.

Birds eat primarily to satisfy their energy needs. Also, the temperature of the environment has an important influence on feed intake—the warmer the environment, the less the feed intake. Therefore, the requirement of all nutrients, expressed as a percent of the diet, is dependent upon the environmental temperature. Other factors affecting feed intake are health, genetics, form of feed, nutritional balance, stress, body size, and rate of egg production or growth.

It is believed that feed intake is, in part, controlled by the amount of glucose in the blood. It has been observed that the addition of fat to the diet results in overconsumption on the part of the bird. As a result, some variation in the protein:energy ratio may be tolerated. In general, when free-choice feeding dietary protein levels that are low in relation to energy, fat deposition is markedly increased; with higher levels of protein, less fat is deposited. Increasing the protein level above that required for maximum growth rate reduces fat deposition still further.

FEEDING CHICKENS

Today, most commercial chickens are provided complete rations with all the needed nutrients available in the quantities necessary, and with the feed and water available on an *ad libitum* or free-choice basis. Formulations are varied according to the type of production—whether the birds are bred and kept for egg production (layers), hatchery production (breeders), or meat production (broilers). Also, consideration is given to age; sex; stage and level of production; temperature, disease level, and other stresses; management; and other factors.

Feeding Layers

Chickens kept for the production of eggs for human consumption (Leghorn-type) have small body size and are prolific layers. They are generally fed *ad libitum* during the laying period. Occasionally, layers will consume excess feed during the latter phases of egg production (following peak production) with resultant obesity, reduced feed efficiency, and higher incidence of fatty liver syndrome. When this situation is detected, limiting feed intake to 85 to 95% of full feed consumption is desirable. Data on feed consumption in a particular flock, together with information on body weight, ambient temperature, and rate of egg production, may be used to determine the degree of feed restriction.

Fig. 24-17. A modern cage facility in a commercial laying house. Note the 3 tiers of cages, with a feed trough and an automatic egg collecting belt in front of each row of cages. (Courtesy, P. L. Potts, Sr., Ph.D., Poultry Department, California Polytechnic University, San Luis Obispo)

These additional pointers are pertinent to feeding layers:

1. The largest item of cost in the production of eggs is feed. It normally constitutes 50 to 70% of the total cost; though, in exceptional cases, it can run as low as 45% or as high as 75%.

2. In the final analysis, the objective of feeding laying hens is to produce a dozen eggs of good quality at the lowest possible feed cost. Thus, the actual cost of the feed that a layer eats in producing a dozen eggs—not the price per pound of feed—determines the economy of the ration.

3. Feed consumption per bird varies primarily with egg production and body size. It is also influenced by the health of the birds and the environment, especially the temperature.

4. Normally, a mature Leghorn, or other lightweight bird, eats about 82.5 lb of feed per year and produces about 22 doz eggs in that same period of time. Hence, it requires about 3.75 lb of feed to produce 1 doz eggs. A bird of the heavier breeds eats 95 to 115 lb of feed per year; hence, they are not as efficient egg producers. With lightweight layers, the producer should aim for a feed efficiency of 3.5 to 4.0 lb of feed per dozen eggs.

5. Feed may affect egg quality. Deficiencies of calcium, phosphorus, manganese, and vitamin D_3 lead to poor shell quality. Yolk color is almost entirely dependent on the bird's diet. Low vitamin A levels may increase the incidence of blood spots.

Example rations for *layers* producing eggs for human consumption are given in Table 24-24.

TABLE 24-24a
EXAMPLE LAYER RATIONS[A,B,1]

Ingredient	Protein Level of Rations[C]											
	15%		16%		17%		18%		19%		20%	
	(lb)	(kg)	(lb)	(kg)	(lb)	(kg)	(lb)	(kg)	(lb)	(kg)	(lb)	(kg)
Ground yellow corn[2,3]	1,457	662.3	1,403	637.7	1,339	608.6	1,242	564.5	1,177	535	1,120	509.1
Alfalfa meal, 17%	25	11.3	25	11.3	25	11.3	25	11.3	25	11.3	25	11.3
Soybean meal, dehulled	292.2	132.8	340.6	155	393.6	179	451.6	205.3	504.6	229.4	554	251.8
Meat and bone meal, 47%[5]	50	23.0	50	23.0	50	23.0	50	23.0	50	23.0	50	23.0
DL-Methionine or equivalent	1.0	0.5	1.0	0.5	1.0	0.5	1.0	0.5	1.0	0.5	1.0	0.5
Dicalcium phosphate[6]	9	4.1	8	3.6	8	3.6	7	3.1	7	3.1	7	3.1
Ground limestone[7]	159	72.3	159	72.3	159	72.3	174	79.1	174	79.1	174	79.1
Iodized salt[4]	7	3.1	7	3.1	7	3.1	7	3.1	7	3.1	7	3.1
Stabilized yellow grease, or equivalent	—	—	7	3.1	18	8.2	43	19.5	55	25	62	28.2
Antioxidant	g	g	g	g	g	g	g	g	g	g	g	g
Mineral and vitamin supplements:[12]												
Calcium pantothenate (mg)	5,000		4,500		4,500		4,500		4,000		4,000	
Manganese[11] (Mn) (g)	52		52		52		52		52		52	
Selenium[25] (Se) (mg)	90.8		90.8		90.8		90.8		90.8		90.8	
Zinc[17] (Zn) (g)	16		16		16		16		16		16	
Vitamin A (IU)	6,000,000		6,000,000		6,000,000		6,000,000		6,000,000		6,000,000	
Vitamin D_3 (IU)	2,000,000		2,000,000		2,000,000		2,000,000		2,000,000		2,000,000	
Vitamin K[20]	—		—		—		—		—		—	
Choline (mg)	274,000		231,000		184,000		140,000		94,000		50,000	
Niacin (Nicotinic Acid, Nicotinamide) (mg)	12,000		12,000		12,000		12,000		12,000		12,000	
Riboflavin (Vitamin B-2) (mg)	2,000		2,000		2,000		2,000		2,000		2,000	
Vitamin B-12 (Cobalamins) (mg)	6		6		6		6		6		6	
Totals[21]	2,000	909.4	2,000	909.3	2,000	909.3	2,000	909.4	2,000	909.4	2,000	909.1
Calculated analysis:[27]												
Metabolizable energy (kcal)	1,306.2	2,873.6	1,303.9	2,868.6	1,303.4	2,867.5	1,304.1	2,869	1,304.5	2,870	1,301	2,862.2
Protein (%)	15.08		16.02		17.03		18.01		19.03		20.00	
Lysine (%)	0.68		0.75		0.83		0.91		0.98		1.06	
Methionine (%)	0.31		0.32		0.33		0.34		0.35		0.36	
Methionine + cystine (%)	0.54		0.57		0.59		0.62		0.64		0.67	
Fat (%)	3.29		3.54		3.98		5.05		5.54		5.76	
Fiber (%)	2.20		2.20		2.21		2.18		2.18		2.19	
Calcium (Ca) (%)	3.25		3.24		3.24		3.50		3.50		3.51	
Total phosphorus (P) (%)	0.52		0.52		0.53		0.52		0.53		0.54	
Available phosphorus[13] (P) (%)	0.45		0.45		0.45		0.45		0.45		0.45	
Vitamins (units or mg/lb or kg):												
Vitamin A activity (IU)	5,904	12,988.8	5,842	12,852.4	5,770	12,694	5,660	12,452	5,586	12,239.2	5,522	12,148.4
Vitamin D_3, added (IU)	1,000	2,200	1,000	2,200	1,000	2,200	1,000	2,200	1,000	2,200	1,000	2,200
Choline (mg)	500.13	1,100.3	500.34	1,100.7	500.05	1,100.1	500.39	1,100.9	500.48	1,101.1	500.66	1,101.5
Niacin (Nicotinic Acid, Nicotinamide) (mg)	15.40	33.9	15.48	34.1	15.55	34.2	15.50	34.1	15.56	34.2	15.64	34.4
Pantothenic Acid (Vitamin B-3) (mg)	5.01	11	4.88	10.7	4.99	10.9	5.07	11.2	4.95	10.8	5.01	11
Riboflavin (Vitamin B-2) (mg)	1.84	4.0	1.85	4.1	1.86	4.1	1.86	4.1	1.87	4.1	1.88	4.1

[A]Adapted by the authors from *NECC Chicken and Turkey Rations*, prepared by the New England College Poultry Conference Board, by poultry specialists from, and distribution by, the New England Land-Grant Universities: University of Connecticut, Storrs; University of Maine, Orono; University of Massachusetts, Amherst; University of New Hampshire, Durham; University of Rhode Island, Kingston; and University of Vermont, Burlington.

[B]See footnotes following Table 24-30, p. 584.

[C]Six rations varying in protein levels are presented. The ration which best meets the needs of a particular flock may be determined from the factors presented below. See "Factors to Consider in Determining Which Layer Ration to Feed," which follows:

FACTORS TO CONSIDER IN DETERMINING WHICH LAYER RATION TO FEED

A. Feed consumption of the hens must be known to select the appropriate layer ration. Feed consumption may vary depending on type of bird, bird weight, pen temperature, rate of production, energy content of ration, disease problems and many other factors.

B. **Table 24-24b. SUGGESTED MINIMUM DAILY PROTEIN INTAKE.**

Production Status	Daily Protein Required*
	(g)
Coming into production	19–20
90% hen day egg production	18–19
80% hen day egg production	17–18
70% hen day egg production	16–17
60% and under egg production	15–16

*Under conditions of higher production, disease, or stress, add 2 g protein to the above.

D. When feed consumption is not known, we suggest using not less than 17% protein in the diet.

C. **Table 24-24c. A GUIDE FOR ESTIMATING DIETARY PROTEIN LEVEL BASED ON FEED INTAKE.**

Lbs Feed Consumed Per/100 Birds/Day	% Protein In Laying Ration					
	15	16	17	18	19	20
	Protein Intake					
	(g)	(g)	(g)	(g)	(g)	(g)
18–19	12	13	14	15	16	17
20–21	14	15	15	16	17	18
22–23	15	16	17	18	19	20
24–25	16	17	18	20	21	22
26–27	18	19	20	21	22	24
28–29	19	20	22	23	24	26
30–31	20	22	23	25	26	27

Example—Your birds are consuming 24 lb per 100 birds per day and laying at a rate of 80% hen day egg production. Their need for protein is 17–18 g (Table 24-24b). Therefore to supply 18 g of protein you should use the 17% protein ration (Table 24-24c).

WEIGHT CONVERSION TABLE

1 pound	= 453.57 grams or	0.4536 kilogram
1 ounce	= 28.349 grams	1 kilogram = 2.2046 pounds
1 kilogram (kg)	= 1,000 grams	1 gram (g) = 1,000 milligrams
1 milligram (mg)	= 100 micrograms (gammas)	
1 part per million (ppm)	= 1 milligram/kilogram or 1 microgram/gram	
1 part per million (ppm)	= 0.454 mg/lb or 0.907 g/ton	

PHASE FEEDING

The trend is to phase feeding laying hens. Phase feeding refers to changes in the laying hen's diet (1) to adjust to age and state of production of the hen, (2) to adjust for season of the year and for temperature and climatic changes, (3) to account for differences in body weight and nutrient requirements of different strains of birds, and (4) to adjust one or more nutrients as other nutrients are changed for economic or availability reasons. Research has shown, for example, that a hen laying at the rate of 60% has different nutritional requirements than one laying at the rate of 80%; hens have different requirements in summer and in winter; a 24-week-old layer has different needs than one 54 weeks old. The main objective, therefore, of phase feeding is to reduce the waste of nutrients caused by feeding more than a bird actually needs under different sets of conditions. In this way, feed efficiency can be improved and the cost of producing a dozen eggs reduced.

MOLTING

Molting is the process of shedding and renewing of feathers. It is a normal process of chickens and other feathered species; and it occurs in both sexes. In the wild state, birds usually shed and renew old, worn plumage before the beginning of the cold weather and their migratory flights. Since undomesticated birds lay only a few eggs, molting and reproduction are not usually associated.

Chickens kept for commercial egg production have a different molting pattern. They have been bred for high performance, and their environment, with respect to temperature and light, is usually modified to remove major seasonal influences. A natural molt does not normally occur until after a period of 8 to 12 months of egg production. If nothing is done to alter the normal molting cycle, it requires about 4 months for a hen to drop her feathers and grow a new set. It is possible, however, to speed the process through a program of forced molting, thereby recycling the hens for another period of egg production and improved egg quality.

Molting is controlled by the gonads and the thyroid gland and is associated with a drop in estrogen levels and a decreasing rate of egg production. Egg production is not greatly affected by the process of natural molting, but molting is prolonged when the birds are kept in production. High producers tend to molt late; but once production ceases, molting is rapid.

Several factors affect the onset and length of molting; among them, (1) weight and physical condition of the bird, (2) length of light exposure, (3) nutrition of the bird, and (4) environmental influences, such as temperature and humidity. Thus, if one drastically reduces the amount of light or starves the birds in such a way as to knock them out of production, molting can be induced or speeded up.

In a natural molt, a chicken loses feathers from various sections of its body in the following order: head; neck; feather tracts of the breast, thighs, and back; and wing and tail feathers. Some birds molt earlier than others; and some molt more slowly than others. A high producing flock generally molts late and rapidly.

Decreasing day length is the normal trigger for molting. Therefore, lighting programs for layers should provide either constant or increasing day length. Minor stresses, such as temporary feed or water shortage, disease, cold temperature, or sudden changes in the lighting program, can also initiate a partial or premature molt, evidenced by a considerable number of feathers on the floor of the poultry house. In these cases, chickens lose some head and neck feathers. If the molt continues beyond this point, a more severe drop in performance can be expected.

- **Force (induced) molting**—Under force molting, a layer flock is induced to shed and replace its feathers at a time selected by the flock manager. This may come near the end of a normal laying cycle, or a flock may be force molted earlier as part of a multiple molting program.

An induced molt causes all of the hens in a flock to go out of production for a period of time. During this period, regression and rejuvenation of the reproductive tract occur, accompanied by the loss and replacement of feathers. After a molt, the hen's reproductive rate usually peaks slightly below the previous peak rate, and egg quality is improved.

Forced molting has been practiced in a limited way since the turn of the century, but always as economic conditions dictated. Today, it is estimated that approximately 60% of the nation's laying flocks are recycled, and that 90% or more of the California flocks are recycled. The first and most important reason for induced molting is that it usually improves profits and, therefore, is part of a planned replacement policy. The question of whether it pays to recycle depends primarily upon the relative performance of all-pullet versus recycled flocks.

- **Methods of molting**—Many satisfactory molting programs are in use, but most of them are simply modifications of two basic concepts—feed withdrawal, followed by a low nutrient feed intake period. Modifications involve the number of days without feed, the composition of the feed, and the duration of the low-nutrient intake period. Other minor factors include the choice of lighting programs during the molt and whether water should be restricted.

1. **Conventional force molting program.** This procedure, which is sometimes referred to as an on-again/off-again program, is outlined in Table 24-25.

TABLE 24-25
CONVENTIONAL FORCE MOLTING PROGRAM
(ON-AGAIN/OFF-AGAIN PROGRAM)

Day	Feed	Water	Light
1	None	None	8 hours
2	None	None	8 hours
3	Egg-type layers: 10 lb/100 hens	Water	8 hours
4	None	None	8 hours
5	Egg-type layers: 10 lb/100 hens	Water	8 hours
6	None	None	8 hours
7	Egg-type layers: 10 lb/100 hens	Water	8 hours
8	None	None	8 hours
9	Egg-type layers: 10 lb/100 hens	Water	8 hours
10 through 55-60	Restricted feeding— about 75% of full feed intake	Water	8 hours
61	Full-feed layer ration	Water	14-16 hours

2. **California force molting program.** California has a higher percentage of force molting than any other state. For this reason, the California method is presented. It is characterized by simplicity, low cost and high subsequent performance, and with all birds getting equal treatment and having uniform recovery. It is presented in Table 24-26.

TABLE 24-26
CALIFORNIA FORCE MOLTING PROGRAM

Day	Feed	Water	Light
1 through 10 to 14	None	Water	Discontinue artificial light or limit to 8 hours
11 to 15 through 35	Full feed cracked grain or low-protein, low-calcium molt mash	Water	Discontinue artificial light or limit to 8 hours
36 through 68	Full feed laying mash	Water	14-16 hours

Additional pertinent information about the California molting program follows:

a. The 10 to 14 days without feed will usually result in 25 to 30% loss in body weight, with less than 1% mortality.

b. The feed withholding period (from day 1 through up to day 10 to 14) should end when the accumulated mortality reaches 1.25%, when body weight reduction for a 3.6-lb hen reaches 30%, or when it has gone 14 days—whichever comes first.

3. **"Fast-molting" and "fast-fast molting."** The *fast-molting* involves elimination of the resting period or low nutrient period and a return to 50% rate of lay in 4 to 6 weeks instead of the typical 8 weeks of the conventional system. The *fast-fast molt* goes one step further; it involves reducing the initial feed withholding period to 4 to 6 days (instead of 10 to 14) and elimination of the resting period. The fast-fast molt is extremely fast; flocks can be back to 50% production by the third week, instead of the typical 8 weeks of the conventional system. Preliminary studies indicate that the early production achieved by the fast-molt and by the fast-fast molt is at the expense of later production and egg quality. But more experimental work is needed.

• **Will force molting pay?**—There is no simple answer to this question, since it depends on a variety of economic factors. However, the following general guidelines should be taken into consideration:

1. Recycling becomes less profitable as egg prices increase and as price differential between egg sizes decreases.

2. As the cost of replacement pullets increases, it becomes more advantageous to recycle.

3. The profitability of recycling increases as the price paid for *spent* hens decreases.

4. Using replacement pullets instead of recycling ties up additional capital.

Feeding Breeders

Fig. 24-18. Single sire pedigree pens. (Courtesy, Indian River Company, Nacogdoches, Tex.)

The following pointers are pertinent to feeding breeders:

1. The nutritive requirements for breeding flocks are more rigorous than those for commercial laying flocks. Breeders require greater amounts of vitamins A, D, E, and B-1, and of riboflavin, pantothenic acid, niacin, and manganese than do laying flocks. Rations with these added ingredients in the right proportions give high hatchability and good development of chicks. Such rations cost more than normal layer rations.

2. Broiler breeder replacement pullets should receive low energy diets, in the range of 1090 to 1135 kcal/pound and/or the feed intake should be restricted, to avoid excess fat accumulation at the time they reach sexual maturity.

3. Broiler breeder hens, which are heavy and have a high energy requirement for maintenance, require approximately 400 to 450 kcal ME per hen per day, for maximum egg production. Since these hens tend to become over-fat when fed high energy diets, it appears best to limit the energy content of their diets to approximately 1200 to 1250 kcal/pound and/or to restrict their feed intake in some way so that they do not obtain much more than about 420 kcal ME per hen per day.

4. Male breeders require slightly less energy than females during growth. The lower fat deposition in the male compared with the female is offset by the energy needs for more rapid growth. Being larger, the adult male cock requires considerably more energy than the hen for maintenance, but this is largely offset by the hen's need for energy for egg production.

Example rations for chicken breeders producing eggs for hatching are given in Table 24-27 (see next page).

TABLE 24-27
EXAMPLE CHICKEN BREEDER RATIONS[A,B,1,14]

Ingredient		Body Weight			
		Egg-Type Breeders 3½–5 Lb		Meat-Type Breeders 5–8 Lb	
		(lb)	(kg)	(lb)	(kg)
Ground yellow corn[2,3]		1,305	593.2	1,379	627
Wheat middlings		—	—	70	32
Alfalfa meal, 17%		50	23	50	23
Soybean meal, dehulled		324	147.3	248	113
Fish meal, herring, 65%[4,5]		50	23	50	23
Meat and bone meal, 47%[5]		50	23	50	23
Dicalcium phosphate[6]		6	2.7	4	1.8
Ground limestone[7]		157	71.4	142	64.5
DL-Methionine, or equivalent		0.5	0.2	0.4	0.2
Stabilized yellow grease, or equivalent		51	23	[16]	[16]
Iodized salt[4]		7	3.1	7	3.1
Antioxidant		[9]	[9]	[9]	[9]
Mineral and vitamin supplements:[12]					
Calcium pantothenate	(mg)	6,000		6,000	
Manganese[11] (Mn)	(g)	52		52	
Selenium[26] (Se)	(mg)	90.8		90.8	
Zinc[17] (Zn)	(g)	16		16	
Vitamin A	(IU)	4,000,000		4,000,000	
Vitamin D$_3$	(IU)	2,000,000		2,000,000	
Vitamin E	(IU)	2,000		2,000	
Vitamin K[20]		—		—	
Choline	(mg)	172,000		208,000	
Niacin (Nicotinic Acid, Nicotinamide)	(mg)	10,000		10,000	
Riboflavin (Vitamin B-2)	(mg)	3,000		3,000	
Vitamin B-12 (Cobalamins)	(mg)	6		6	
Totals[21]		2,000.5	909.9	2,000.4	910.6
Calculated analysis:[27]					
Metabolizable energy	(kcal)	1,337	2,941	1,295	2,849
Protein	(%)	17.01		16.03	
Lysine	(%)	0.87		0.78	
Methionine	(%)	0.33		0.31	
Methionine + cystine	(%)	0.59		0.56	
Fat	(%)	5.81		3.57	
Fiber	(%)	2.42		2.67	
Calcium	(%)	3.27		2.98	
Total phosphorus (P)	(%)	0.54		0.54	
Available phosphorus[13] (P)	(%)	0.46		0.46	
Vitamins (units or mg/lb or kg):					
Vitamin A activity	(IU)	5,983	13,163	6,067	13,347
Vitamin D$_3$, added	(IU)	1,000	2,200	1,000	2,200
Choline	(mg)	500.40	1,101	500.57	1,101
Niacin (Nicotinic Acid, Nicotinamide)	(mg)	15.21	33.5	17.00	37.4
Pantothenic Acid (Vitamin B-3)	(mg)	5.67	12.5	5.74	12.6
Riboflavin (Vitamin B-2)	(mg)	2.49	5.5	2.51	5.5

[A]Adapted by the authors from *NECC Chicken and Turkey Rations*, prepared by the New England College Poultry Conference Board, by poultry specialists from, and distribution by, the New England Land-Grant Universities: University of Connecticut, Storrs; University of Maine, Orono; University of Massachusetts, Amherst; University of New Hampshire, Durham; University of Rhode Island, Kingston; and University of Vermont, Burlington.

[B]See footnotes following Table 24-30, p. 584.

Feeding Replacement Pullets

Pullets generally perform well during their laying year when their nutrient requirements have been met during the growing period.

Leghorn-type pullets are seldom restricted-fed during the growing period because varying the lighting during growth from 6 to 20 weeks of age can be used to control feed consumption and sexual development.

However, pullets of heavy breeds tend to accumulate excessive amounts of body fat; so, it is common practice to restrict the feed intake of these birds to produce pullets with leaner bodies at the time of sexual maturity. This is beneficial because: (1) it produces healthier pullets, and (2) it reduces feed costs during the pullet rearing period.

Also, some producers restrict the feed intake of light breeds, based on research which shows that, even among the light breeds, restricted feed intake results (1) in slightly higher mortality during the rearing period, but (2) in lower mortality and higher egg production during the laying year.

Most feed restriction programs are started at 9 to 12 weeks of age. The methods commonly employed follow:

1. Skip-a-day method.
2. Daily restriction of feed.

Feeding Poultry

3. Bulky, low-energy or low-protein and/or amino acid imbalanced rations.

These further pointers are pertinent to feeding replacement chicks (pullets):

1. Feed accounts for approximately 60% of the cost of raising replacement pullets.

2. Replacement chicks are usually fed a diet lower in energy than broiler chicks. Also, feed and daily light periods may be restricted, so as to permit the pullets to reach larger body size before they start to lay than would be the case were they full fed, and fully lighted.

3. Always use complete starter feeds for chicks, and give chicks starter feeds without grain supplement until they are 5 weeks old.

4. When chicks are 5 weeks old, change to the growing ration.

The NRC requirements for replacement pullets are presented in tables in the earlier section of this chapter headed, "National Research Council (NRC) Requirements."

Example *starter* and *grower* rations (1) for egg-type and meat-type strains, from days 1 to 35, and (2) for egg- and meat-type strains, from day 36 until egg production begins are given in Table 24-28.

TABLE 24-28
EXAMPLE REPLACEMENT CHICKEN RATIONS[A][B][C][1][14][26]

Ingredient	Starter (For egg- and meat-type strains, from days 1 to 35)				Grower (For egg- and meat-type strains, from day 36 until egg production begins)					
	20% Protein		18% Protein		15% Protein		14% Protein		12% Protein	
	(lb)	(kg)	(lb)	(kg)	(lb)	(kg)	(lb)	(kg)	(lb)	(kg)
Ground yellow corn[2][3]	1,267	576	1,310	595.4	1,412	641.8	1,438	653.6	1,481	673.2
Wheat middlings	130	59.1	200	90.9	254	115.5	254	115.5	323	146.8
Alfalfa meal, 17%	25	11.3	25	11.3	25	11.3	25	11.3	25	11.3
Soybean meal, dehulled	422	192	309	140.4	217	98.6	217	98.6	104.8	47.6
Fish meal, herring, 65%[4][5]	50	23	50	23	—	—	—	—	—	—
Meat and bone meal, 47%[5]	50	23	50	23	50	23	—	—	—	—
Lysine	—	—	—	—	—	—	1	0.5	1.2	0.5
Dicalcium phosphate[6]	10	4.5	9	4.1	14	6.4	30	13.6	29	13.2
Ground limestone[7]	19	8.6	20	9.1	21	9.5	28	12.7	29	13.2
Stabilized yellow grease, or equivalent	20	9.1	20	9.1	16	16	16	16	16	16
Iodized salt[4]	7	3.1	7	3.1	7	3.1	7	3.1	7	3.1
Antibiotic supplement	8	8	—	—	—	—	—	—	—	—
Antioxidant	9	9	9	9	9	9	9	9	9	9
Coccidiostat	10	10	10	10	10	10	10	10	10	10
Mineral and vitamin supplements:[12]										
Calcium pantothenate (mg)	4,000		4,000		3,000		3,000		3,000	
Manganese[11] (Mn) (g)	52		52		52		52		52	
Selenium[25] (Se) (mg)	90.8		90.8		90.8		90.8		90.8	
Vitamin A (IU)	3,000,000		3,000,000		3,000,000		3,000,000		3,000,000	
Vitamin D_3 (IU)	1,000,000		1,000,000		1,000,000		1,000,000		1,000,000	
Vitamin K[20]	—		—		—		—		—	
Choline (mg)	213,000		298,000		84,000		125,000		209,000	
Niacin (Nicotinic Acid, Nicotinamide) (mg)	10,000		10,000		10,000		10,000		10,000	
Riboflavin (Vitamin B-2) (mg)	1,500		1,500		1,500		1,500		1,500	
Vitamin B-12 (Cobalamins) (mg)	6		6		6		6		6	
Totals[21]	2,000	909.7	2,000	909.4	2,000	909.2	2,000	908.9	2,000	908.9
Calculated analysis:[27]										
Metabolizable energy (kcal)	1,361	2,994	1,362	2,996	1,343	2,955	1,341	2,950	1,342	2,952
Protein (%)	20.03		18.01		15.03		14.01		12.01	
Lysine (%)	1.04		0.89		0.63		0.63		0.49	
Methionine (%)	0.34		0.32		0.25		0.24		0.21	
Methionine + cystine (%)	0.64		0.59		0.48		0.46		0.41	
Fat (%)	4.48		4.70		3.76		3.54		3.74	
Fiber (%)	2.67		2.83		3.01		3.00		3.15	
Calcium (%)	0.90		0.90		0.89		0.90		0.90	
Total phosphorus (%)	0.66		0.66		0.66		0.66		0.65	
Available phosphorus[13] (%)	0.41		0.41		0.40		0.40		0.40	
Vitamins (units or mg/lb or kg):										
Vitamin A activity (IU)	4,188	9,214	4,237	9,321	4,354	9,579	4,381	9,638	4,430	9,746
Vitamin D_3 added (IU)	500	1,100	500	1,100	500	1,100	500	1,100	500	1,100
Choline (mg)	600.20	1,320	600.11	1,320	420.48	925	420.20	924	419.76	923
Niacin (Nicotinic Acid, Nicotinamide) (mg)	19.08	42	20.49	45	20.87	45.9	20.53	45.2	21.84	48
Pantothenic Acid (Vitamin B-3) (mg)	5.34	11.7	5.27	11.6	4.69	10.3	4.72	10.4	4.64	10.2
Riboflavin (Vitamin B-2) (mg)	1.78	3.9	1.76	3.9	1.68	3.7	1.64	3.6	1.62	3.6

[A]Adapted by the authors from *NECC Chicken and Turkey Rations*, prepared by the New England College Poultry Conference Board, by poultry specialists from, and distribution by, the New England Land-Grant Universities: University of Connecticut, Storrs; University of Maine, Orono; University of Massachusetts, Amherst; University of New Hampshire, Durham; University of Rhode Island, Kingston; and University of Vermont, Burlington.

[B]See footnotes following Table 24-30, p. 584.

[C]Equivalent to 14% protein + 1 lb lysine.

Feeding Broilers

The following pointers pertaining to feeding broilers, roasters, and capons are pertinent:

1. Feed is the largest cost item in broiler production, representing 60 to 75% of the total cost.

2. Producers aim for broilers with an average weight of over 4.0 lb at 6 to 7 weeks of age, feed conversion of less than 2.0, and mortality under 3.0%. Many good producers are achieving feed conversion of about 1.9.

3. Some operations use a 2-stage ration program (starter and finisher) for broilers, but most are using at least 3 stages in their feeding programs (starter, grower, and finisher) to reduce costs and make more efficient use of the nutrients. In a 3-stage program the starter feed should be used for 3 to 4 weeks, the grower for about 2 weeks, and the finisher for the remainder of the feeding period.

Fig. 24-19. Six-week-old broilers. Note 2 continuous pan feeders and 4 rows of waterers. (Courtesy, P. L. Potts, Sr., Ph.D., Poultry Department, California Polytechnic University, San Luis Obispo)

Fig. 24-20. Nutrition research has played a major role in the development of the modern broiler industry. Battery brooders such as these are extensively used in research to evaluate feeds and additives. (Courtesy, University of Georgia, Athens)

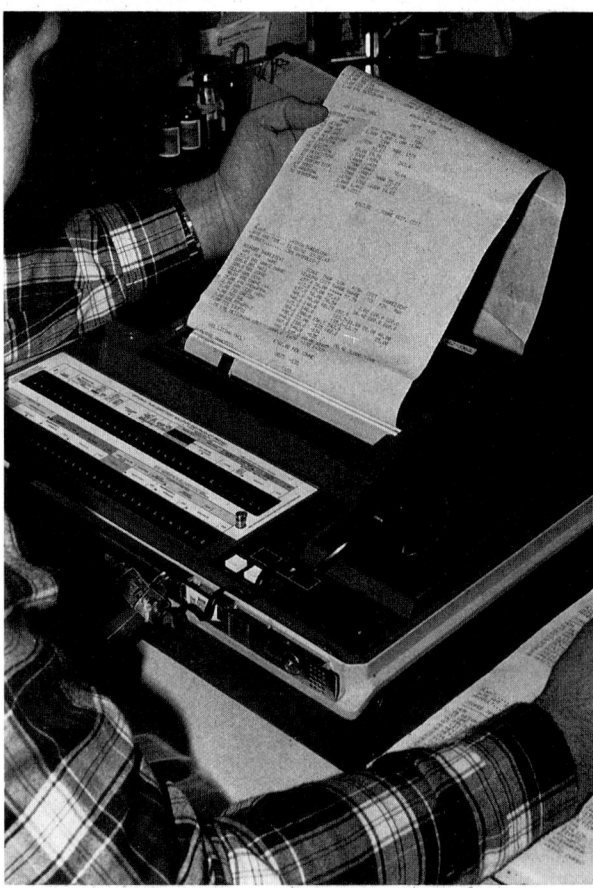

Fig. 24-21. Broiler rations are routinely computed by least cost formulation. The computer determines the combination of feedstuffs that will meet the nutrient specifications set by the nutritionist at the lowest cost per unit of feed. (Courtesy, University of Georgia, Athens)

The NRC requirements for broilers are presented in the tables in the earlier section of this chapter headed, "National Research Council (NRC) Requirements."

Example broiler *starter* and *finisher* rations are given in Table 24-29.

TABLE 24-29
EXAMPLE BROILER RATIONS[A,B,1]

Ingredient	Starter[18] (For Broiler Chicks Until 24 Days of Age)		Finisher[C] (For Broilers From Day 25 to Market)	
	(lb)	(kg)	(lb)	(kg)
Ground yellow corn[3]	1,106	503	1,235	561
Alfalfa meal, 17%	–	–	25	11
Soybean meal, dehulled	605	275	420	191
Corn gluten meal, 60%	50	23	75	34
Fish meal, herring, 65%[4,5]	50	23	50	23
Meat and bone meal, 47%[5]	50	23	50	23
Dicalcium phosphate[6]	10	4	9	4
Ground limestone[7]	16	7	14	6.3
DL-methionine, or equivalent	0.8	0.3	–	–
Stabilized yellow grease, or equivalent	106	48	115	52.2
Iodized salt[4]	7	3	7	3
Antibiotic supplement	8	8	8	8
Antioxidant	9	9	9	9
Coccidiostat	10	10	10	10
Mineral and vitamin supplement:[12]				
Calcium pantothenate (mg)	5,000		5,000	
Manganese[15] (Mn) (g)	75		75	
Organic arsenical supplement[19]	0.1		0.1	
Selenium[25] (Se) (mg)	90.8		90.8	
Zinc[24] (Zn) (g)	30		30	
Vitamin A (IU)	4,000,000		4,000,000	
Vitamin D₃ (IU)	1,000,000		1,000,000	
Vitamin E (IU)	2,000		2,000	
Vitamin K[20] (mg)	2,000		1,000	
Choline (mg)	503,000		672,000	
Niacin (Nicotinic Acid, Nicotinamide) (mg)	20,000		20,000	
Riboflavin (Vitamin B-2) (mg)	3,000		3,000	
Vitamin B-12 (Cobalamins) (mg)	12		12	
Totals[21]	2,000.9	909.3	2,000.1	908.5
Calculated Analysis:[27]				
Metabolizable energy (kcal)	1,399	3,078	1,494	3,287
Protein (%)		24.08		21.00
Lysine (%)		1.30		1.05
Methionine (%)		0.45		0.38
Methionine + cystine (%)		0.81		0.71
Fat (%)		8.20		8.92
Fiber (%)		1.97		2.22
Calcium (%)		0.84		0.80
Total phosphorus (%)		0.64		0.60
Available phosphorus[13] (%)		0.40		0.38
Vitamins (units or mg/lb or kg):				
Vitamin A activity (IU)	3,769	8,292	5,424	11,933
Vitamin D₃ (added) (IU)	500	1,100	500	1,100
Choline (mg)	800.03	1,760	800.48	1,761
Niacin (Nicotinic Acid, Nicotinamide) (mg)	21.36	47	21.33	46.9
Pantothenic Acid (Vitamin B-3) (mg)	5.69	12.5	5.51	12.1
Riboflavin (Vitamin B-2) (mg)	2.44	5.4	2.49	5.5
Xanthophyll[28] (mg)	9.50	20.9	14.05	30.9

[A]Adapted by the authors from *NECC Chicken and Turkey Rations*, prepared by the New England College Poultry Conference Board, by poultry specialists from, and distribution by, the New England Land-Grant Universities: University of Connecticut, Storrs; University of Maine, Orono; University of Massachusetts, Amherst; University of New Hampshire, Durham; University of Rhode Island, Kingston; and University of Vermont, Burlington.

[B]See footnotes following Table 24-30, p. 584.

[C]For preslaughter drug withdrawal times see section headed "Regulations Regarding Feed Additives," following "Footnotes."

FEEDING TURKEYS

Fig. 24-22. Turkeys grown in confinement. (Courtesy, P. L. Potts, Sr., Ph.D., Poultry Department, California Polytechnic University, San Luis Obispo)

Feeding turkeys involves two distinct areas of emphasis: feeding market turkeys, and feeding turkey breeders.

• **Feeding market turkeys**—Most market turkeys are of the large type. Males (toms) are usually marketed at 19 to 25 weeks of age and at liveweights of 23 to 35 lb. Younger toms are often sold as oven-ready dressed birds; older toms are generally further processed or used in the restaurant trade. Females (hens) are commonly marketed at 16 to 17 weeks of age and at about 14 lb weight. Medium- and small-type turkeys (roasters/fryers) are often sold at younger ages and lighter liveweights—normally, about 10 lb.

The formulations of the rations fed to market turkeys should be changed as the birds grow. Thus, the nutrient requirements shown in Table 24-11 provide for such changes at 3- to 4-week intervals; in practice, however, the changes may occur more or less frequently than indicated in Table 24-11. Nutritional adjustments are often made for expected ambient temperature variations in order to assure that the birds consume the necessary amount of protein, minerals, and vitamins, regardless of changes in feed consumption.

• **Feeding turkey breeders**—The feeding programs for breeder stock are usually divided into prebreeding (holding) and breeding periods. The prebreeding, or holding, rations may be fed from the time the breeders are selected, at about 16 weeks of age. Holding rations are usually formulated at medium energy levels in order to stabilize development and weight gains after market age. Hens are fed the holding ration until the time of light stimulation, at about 30 weeks of age; thereafter, breeder rations are fed. Toms may be fed a nutritionally balanced holding ration from the time of breeder selection throughout the breeding season. In some programs, the body weight of toms is controlled by limited feeding. The light stimulation of toms is normally initiated at about 26 weeks of age.

It is not necessary to feed low-energy rations or to restrict feed intake of turkey hens in the prebreeding, or holding, period. Corn-soybean meal type rations may be fed *ad libitum*. Growth restriction does not result in any consistent improvement in reproductive performance. Nevertheless, the use of holding rations for turkey breeders is common practice. These rations usually contain medium energy concentrations so as to stabilize development and weight gains after mature body weight is attained. Care should be taken to provide turkey breeder hens adequate intake of minerals and vitamins during the holding period so that they are not depleted of these nutrients prior to the onset of lay.

These further pointers are pertinent to feeding turkeys:

1. Prevent poult "starve out." Upon arrival, poults should be encouraged to consume feed and water as soon as possible. It may be necessary to dip the beaks of some of them in feed and water to start them eating and drinking.

2. Turkeys grow faster than chickens; hence, they have relatively higher feed and protein requirements.

3. Young turkeys use feed efficiently. Small White turkeys raised to a liveweight of 6 to 8 lb at 14 to 16 weeks of age require 3 lb of feed per pound of live turkey produced. Large White turkeys require about 3.5 to 3.75 lb of feed to produce 1 lb of liveweight, when grown to a market weight of about 30 lb and 24 weeks of age.

4. A high-fiber, low-energy holding ration retards sexual maturity and may result in some desirable effects upon later reproductive performance. The holding ration limits energy intake, but should not limit protein, vitamins, and minerals. When a holding ration is used, the birds should be switched to the breeder ration 2 weeks prior to egg production.

5. Good range provides green feed and tends to reduce feed costs. However, it may make for higher losses from blackhead and other diseases, and predators; and range turkey operations may make the neighbors unhappy because of dust, odors, and noise.

6. As they approach maturity, turkeys fed for market purposes should be fed rations that are quite different from those that are fed to turkey breeders.

The NRC requirements for turkeys are presented in Tables 24-11, 24-12, and 24-13 in the earlier section of this chapter headed, "National Research Council (NRC) Requirements."

Example rations for market turkeys and breeders are given in Table 24-30, Example Turkey Rations.

Feeding Poultry

TABLE 24-30
EXAMPLE TURKEY RATIONS[A][B][C][D][1]

Ingredient	Starter (0-8 Weeks)		Grower (8-16 Weeks)		Finisher (16-Market)[C][22]		Breeder	
	(lb)	(kg)	(lb)	(kg)	(lb)	(kg)	(lb)	(kg)
Ground yellow corn[2][3]	929	422.3	1,199	545	1,490	677.2	1,218	554
Wheat middlings	100	45.4	50	23	—	—	250	114
Alfalfa meal, 17%	25	11.3	25	11.3	25	11.3	60	27.3
Soybean meal, dehulled	675	307	570	259	335	152.3	190	86.4
Fish meal, herring, 65%[4][5]	100	45.4	—	—	—	—	100	45.4
Meat and bone meal, 47%[5]	100	45.4	50	23	50	23	50	23
Dicalcium phosphate[6]	13	6	32	14.5	23	10.5	10	4.5
Ground limestone[7]	10	4.5	14	6	17	8	92	42
DL-Methionine, or equivalent	0.6	0.3	—	—	—	—	—	—
Stabilized yellow grease, or equivalent	40	18.2	50	23	50	23	20	9
Iodized salt[4]	8	3.6	10	4.5	10	4.5	10	4.5
Antibiotic supplement	8	8	—	—	—	—	—	—
Coccidiostat or antihistomonal	10	10	10	10	10	10	—	—
Antioxidant	9	9	9	9	9	9	9	9
Mineral and vitamin supplements:[D][12]								
Calcium pantothenate (mg)	4,500		4,500		6,000		10,000	
Manganese[23] (Mn) (g)	30		30		30		30	
Selenium[25] (Se) (mg)	181.6		181.6		181.6		181.6	
Zinc[24] (Zn) (g)	30		30		30		30	
Vitamin A (IU)	7,500,000		7,500,000		7,500,000		4,000,000	
Vitamin D_3 (IU)	1,700,000		1,700,000		1,700,000		1,700,000	
Vitamin E (IU)	10,000		5,000		—		30,000	
Biotin (mg)	100		100		100		100	
Choline (mg)	674,000		388,000		417,000		427,000	
Niacin (Nicotinic Acid, Nicotinamide) (mg)	42,000		46,000		48,000		50,000	
Riboflavin (Vitamin B-2) (mg)	4,000		5,000		5,000		5,000	
Vitamin B-12 (Cobalamins) (mg)	6		6		6		6	
Totals[21]	2,000.6	909.4	2,000	909.3	2,000	909.8	2,000	910.1
Calculated analysis:[27]								
Metabolizable energy (kcal)	1,322	2,908	1,371	3,016	1,440	3,168	1,291	2,840
Protein (%)	27.31		21.09		16.22		17.01	
Lysine (%)	1.60		1.10		0.75		0.87	
Methionine (%)	0.48		0.33		0.27		0.32	
Methionine + cystine (%)	0.88		0.65		0.53		0.57	
Fat (%)	5.31		5.43		5.84		4.88	
Fiber (%)	2.60		2.47		2.28		3.25	
Calcium (%)	1.17		1.00		0.93		2.26	
Total phosphorus (P) (%)	0.91		0.80		0.69		0.70	
Available phosphorus[13] (P) (%)	0.65		0.55		0.46		0.61	
Vitamins (units or mg/lb or kg):								
Vitamin A activity (IU)	6,054	13,319	6,361	13,994	6,691	14,720	6,382	14,040
Vitamin D_3, added (IU)	850	1,870	850	1,870	850	1,870	850	1,870
Choline (mg)	1,000.07	2,200	700.08	1,540	600.00	1,320	650.24	1,430
Niacin (Nicotinic Acid, Nicotinamide) (mg)	35.94	79.1	34.51	75.9	33.84	74.4	42.07	92.6
Pantothenic Acid (Vitamin B-3) (mg)	6.07	13.4	5.59	12.3	5.63	12.4	8.11	17.8
Riboflavin (Vitamin B-2) (mg)	3.19	7	3.44	7.6	3.38	7.4	3.65	8

[A]Adapted by the authors from *NECC Chicken and Turkey Rations*, prepared by the New England College Poultry Conference Board, by poultry specialists from, and distribution by, the New England Land-Grant Universities: University of Connecticut, Storrs; University of Maine, Orono; University of Massachusetts, Amherst; University of New Hampshire, Durham; University of Rhode Island, Kingston; and University of Vermont, Burlington.

[B]See footnotes following this table.

[C]For preslaughter drug withdrawal times, see section following footnotes entitled "Regulations Regarding Feed Additives."

[D]Folacin may be required under certain conditions. It may be added at the rate of 1,000 mg per ton of feed.

(Continued)

FOOTNOTES FOR TABLES 24-27, 24-28, 24-29, AND 24-30

[1] Wherever substitutions are made in the rations, the total nutrient content should be adjusted to meet established requirements.

[2] Two to four hundred pounds of coarsely ground wheat or yellow hominy may be used to replace an equal amount of corn. If wheat is used, add 200,000 IU of vitamin A for each 100 lb of corn removed.

[3] There is usually some loss of provitamin A activity in corn and alfalfa meal during storage. If stored ingredients are used, it may be advisable to increase the added vitamin A level of the ration by 1,000 or 2,000 IU/lb. This can be accomplished by increasing the recommended supplement by 2,000,000 or 4,000,000 IU/ton of feed.

[4] The added salt level should be reduced by the amount supplied by the fish meal and other by-product ingredients.

[5] Poultry by-product meal may be substituted for all of the meat and bone scrap and up to 50% of the fish meal. Correct for calcium and phosphorus loss due to substitutions of poultry by-product meal.

[6] Based on an 18.5% phosphorus product, steamed bone meal or defluorinated rock phosphate may replace the dicalcium phosphate on a phosphorus basis.

[7] Based on 35% calcium, low magnesium limestone.

[8] An antibiotic may be used in these rations at the level recommended by the manufacturer (see section following footnotes entitled "Regulations Regarding Feed Additives").

[9] 1,2-dihydro-6-ethoxy-2,2,4-trimethylquinoline (ethoxyquin) is recommended in the chick starter, broiler and breeder rations at the 0.0125% level to help prevent the appearance of encephalomalacia (crazy chick disease). If desired, it, or an equivalent antioxidant, may be added to help prevent the oxidation of dietary components. Total ethoxyquin from all sources must not exceed 0.25 lb per ton.

[10] A coccidiostat or antihistomonal drug may be used in these rations, as required, at levels recommended by the manufacturer (see section following footnotes entitled "Regulations Regarding Feed Additives").

[11] This amount of manganese will be furnished by 0.5 lb of manganese sulfate or 0.21 lb manganous oxide (70% feeding grades). An equivalent amount of manganese may be added from other acceptable sources.

[12] Caution should be used when high potency vitamin mixes are involved. It is recommended that 10 lb be the minimum amount of any item added to a ton of feed to insure proper mixing. Thus, high potency vitamin, mineral, or drug mixes should be premixed with a carrier (such as corn meal) to such a dilution that 10 lb of the final mix will be added for each ton of feed mixed. Minerals and vitamins should not be premixed together.

[13] Available phosphorus has been taken as 30% of total phosphorus from plant sources for chicks, and 75% of total phosphorus from plant sources for adult birds. Phosphorus from other than plant sources is considered to be 100% utilized.

[14] For those persons wanting a specific restricted feeding program, specific programs are available from individual breeders or Extension specialists.

[15] This amount of manganese will be furnished by 0.7 lb manganese sulfate or 0.3 lb manganous oxide (70% feeding grades). An equivalent amount of manganese may be added from other acceptable sources.

[16] Stabilized fats may replace an equal amount of cereal grains to provide a higher energy level, control dust, and aid pelleting. Where maintaining body weight in layers is a problem, increase fat by 1 or 2% during the winter by replacing an equal amount of cereal grains.

[17] Approximately this amount of zinc will be furnished by 29 g of zinc carbonate or 20 g of zinc oxide. An equivalent amount of zinc may be used from other acceptable sources.

[18] Feed starting ration until birds are 35 days old.

[19] Based on 3-nitro-4 hydroxyphenylarsonic acid at a level of 45 g (0.1 lb) per ton. Other compounds that may be used at a level recommended by the manufacturer are sodium arsanilate or arsanilic acid (see section following footnotes entitled "Regulations Regarding Feed Additives").

[20] In the absence of alfalfa or if the birds are raised on wire, 2 g of vitamin K activity should be added. Values in the broiler rations are based on menadione. Other compounds supplying equivalent levels of vitamin K may be used.

[21] If an even 2,000 lb is desired, adjust by removing or adding ground yellow corn.

[22] May be fed with grain after 20 weeks.

[23] This amount of manganese will be furnished by approximately 0.3 lb of manganese sulfate or 0.13 lb of manganous oxide (70% feeding grades). An equivalent amount of manganese may be added from other acceptable sources.

[24] This amount of zinc will be furnished by approximately 53 g of zinc carbonate or 37 g of zinc oxide. An equivalent amount of zinc may be added from other acceptable sources.

[25] Federal law, which strictly regulates the addition of selenium to poultry rations, should be consulted. Selenium, as sodium selenite or sodium selenate, may be added to complete feed for chickens at a level not to exceed 0.1 ppm, and to complete feed for turkeys at a level not to exceed 0.2 ppm. It shall be incorporated into each ton of complete feed for chickens by a premix containing not more than 90.8 mg of added selenium and weighing not less than 1 lb. It shall be incorporated into each ton of complete feed for turkeys by a premix containing no more than 181.6 mg of added selenium and weighing not less than 2 lb.

[26] For heavy caged layer pullets we suggest feeding 18% protein 0-6 weeks, 14% protein 7-12 weeks, and 12% protein 13-20 weeks of age.

[27] Any discrepancies in calculated analysis that occur in the decimal part of the figures are due to rounding errors.

[28] These are not highly pigmented diets. The xanthophyll of natural ingredients is variable, so if more pigment is desired use a high potency source of xanthophyll.

REGULATIONS REGARDING FEED ADDITIVES

The Food and Drug Administration of the U.S. Department of Health and Human Services has published a series of regulations concerning the use of additives, such as arsenicals, antibiotics, coccidiostats and other drugs, in animal feeds. For information concerning the use of any additive, consult the feed tag or label. If you still have questions about proper use of the additive, especially in conjunction with other additives, see your veterinarian, feed dealer or drug supplier.

FEEDING DUCKS

Until 1975, the annual production of ducks in the United States was approximately 10 million. Then, between 1975 and 1985, annual production doubled. In 1985, 21.6 million ducks were marketed in the United States. In 1989, U.S. per capita consumption of duck was estimated to be 0.66 lb.

Ducks are grown successfully in two different types of environments: (1) in an open rearing system in which the growing house opens onto an exercise yard with water for wading or swimming; and (2) in a confinement system in which they are raised in environmentally controlled houses, with litter or with a combination of litter and wire floors.

Typically, ducks are provided with two or three feeds during the growing period: when only two feeds are provided during the growing period, a 22% protein starter ration is usually fed the first 2 weeks, followed by a grower-finisher ration. When 3 feeds are provided during the growing period, they consist of a 22% protein starter ration, an 18% protein grower ration, and a 16% protein finisher ration.

The following pointers are pertinent to feeding ducks:

1. Ducks should be fed pellets rather than mash. Use ⅛ in. pellets for starter rations, and 3⁄16 in. pellets for older ducks. Pellets will make for a saving of 15 to 20% in the feed required to produce a market duck.

2. Ducks are very susceptible to aflatoxicosis; so, monitoring feeds, for aflatoxin is important.

3. Ducks are nearly as good foragers as geese.

4. Ducks should be ready for market between 7½ and 8 weeks of age.

5. When used, holding rations are designed to maintain breeding ducks from about 8 weeks of age until the breeding

Feeding Poultry

season commences, without their getting too fat. It is recommended that birds fed holding rations be limited to about ½ lb per bird per day.

6. When a holding ration is used, breeder diet should be substituted for it about 4 weeks before eggs are desired for hatching purposes.

7. When feeding ducks, pellet quality, proper feather development, and limiting carcass fat disposition are concerns, in addition to proper growth, and satisfactory feed conversion.

8. Commercial ducks grow as rapidly and efficiently as commercial broilers.

The NRC requirements for ducks are presented in Tables 24-14 and 24-15, in the earlier section of this chapter headed, "National Research Council (NRC) Requirements."

Examples of three-phase grower rations, and of a breeder ration, for ducks are given in Table 24-31.

TABLE 24-31a
EXAMPLE MARKET DUCK AND BREEDER DUCK RATIONS[1,2]

Ingredients and Analysis	Starter (0-2 weeks)		Grower (2-4 weeks)		Finisher (4-8 weeks)		Breeder	
	(lb)	(kg)	(lb)	(kg)	(lb)	(kg)	(lb)	(kg)
Ingredients:								
Yellow corn	1,209	548.9	1,420	644.8	1,489	676.0	1,309.5	594.5
Soybean meal, 48.5%	510	231.5	320	145.3	260	118.1	318	144.4
Meat and bone meal, 50%	80	36.3	80	36.3	80	36.3	76	34.5
Fish meal, 60%	56	25.4	65	29.5	50	22.7	60	27.3
Dried whey, delactosed	—	—	—	—	—	—	45	20.4
Animal-vegetable fat	50	22.7	30	13.6	40	18.2	—	—
Dicalcium phosphate	—	—	—	—	3	1.4	—	—
Limestone	13	5.9	10	4.5	8	3.6	112	50.8
Salt	7	3.2	6	2.7	6	2.7	6	2.7
Trace mineral mix (Table 24-32a)	2	0.9	2	0.9	2	0.9	2	0.9
Vitamin mix (Table 24-32b)	20	9.1	15	6.8	10	4.5	20	9.1
Methionine, hydroxy analogue	3	1.4	2	0.9	2	0.9	1.5	0.7
Pellet binder	50	22.7	50	22.7	50	22.7	50	22.7
Total	2,000.0	908.0	2,000.0	908.0	2,000.0	908.0	2,000.0	908.0
Calculated analysis:								
Metabolizable energy (cal./lb)	1,400		1,425		1,450		1,300	
Metabolizable energy (cal./kg)	3,087		3,142		3,197		2,867	
Protein (%)	21.4		18.0		16.4		17.6	
Fat (%)	5.6		5.0		5.5		3.3	
Fiber (%)	2.6		2.6		2.6		2.4	
Calcium (%)	0.85		0.80		0.75		2.75	
Available phosphorus (%)	0.4		0.4		0.4		0.4	
Total phosphorus (%)	0.60		0.58		0.57		0.57	
Ash (%)	5.3		4.7		4.4		10.1	

Footnotes Table 24-31c.

TABLE 24-31b
THE TRACE MINERAL MIX[1]

Mineral	Percent per Kg of Mix
	(%)
Trace or microminerals:	
Copper (Cu)	0.30
Iodine (I)	0.06
Iron (Fe)	3.00
Manganese (Mn)	6.50
Zinc (Zn)	6.50

TABLE 24-31c
THE VITAMIN MIX[1]

Vitamin	Amount per Lb or Kg of Mix	
	(lb)	(kg)
Fat-soluble vitamins:		
Vitamin A (IU)	400,000	880,000
Vitamin D$_3$ (ICU)	60,000	132,000
Vitamin E (IU)	500	1,100
Vitamin K (msb)[3] (mg)	200	440
Water-soluble vitamins:		
Choline (mg)	13,018	28,639.6
Niacin (Nicotinic Acid, Nicotinamide) (mg)	2,500	5,500
Pantothenic Acid (Vitamin B-3) (mg)	300	660
Riboflavin (Vitamin B-2) (mg)	300	660
Vitamin B-12 (Cobalamins) (mg)	0.4	0.88

Footnotes for Tables 24-31a, 24-31b, and 24-31c:

[1]*From: Complete Duck Grower and Breeder Rations,* Purdue University, West Lafayette, Ind.

[2]For best results, all rations should be pelleted.

[3]Menadione sodium bisulfite.

Fig. 24-23. White Pekin ducks swimming in a creek. (Photo by J. C. Allen & Son, West Lafayette, Ind.)

FEEDING GEESE

Fig. 24-24. Goose. (Courtesy, USDA)

Geese are very hardy, highly disease resistant, are the closest grazers known, and can live almost entirely on good pasture. Yet, the production of geese for meat purposes has never enjoyed the popularity in the United States that it has in some European countries.

Geese are raised under the following variety of feeding programs:

1. The production of *farm geese,* with the goslings given a starter feed for about 2 weeks, followed by foraging the farm for a variety of pasture and grain feedstuffs; then, marketed at about 18 weeks of age after liberal grain feeding for the last 2 or 3 weeks.

2. The goslings are limit-fed prepared feed throughout the growing period, but are allowed considerable foraging in addition; then, marketed at about 14 weeks of age following liberal feeding of a high-energy finishing ration.

3. The goslings are full-fed in confinement and marketed as *junior* or *green geese* at about 10 weeks of age.

4. The raising and use of geese for weeding purposes. Weeder geese are used with great success to control and eradicate troublesome grass and certain weeds in a great variety of crops and plantings, including cotton, hops, mint, onions, garlic, strawberries, nurseries, corn, orchards, groves, and vineyards. The geese eat grass and young weeds as quickly as they appear, but they do not touch certain cultivated plants. They will work continuously from daylight to dark, 7 days a week (even on bright moonlight nights) nipping off the grass and weeds as promptly as new growth appears.

At the end of the weeding season, geese are generally brought from the field and placed in pens for fattening for 3 or 4 weeks, until they weigh 10 to 12 lb or more. Markets are highest during the 4 to 6 weeks prior to Thanksgiving and Christmas.

The carrying of geese over from one season to the next for weeding purposes is not recommended, because older geese are less active in hot weather than young birds.

5. The production in some European countries of goose livers for *pate de foie gras.* In this program, the geese are grown to about 12 weeks of age, following which they are force-fed a high-grain ration for the production of livers of high-fat content.

For breeding purposes, geese are fed a prebreeding (holding) ration beginning 6 to 8 weeks before the breeding season, followed by a breeding ration formulated for the intensive production of fertile eggs.

The following additional pointers are pertinent to feeding geese:

1. Rations for geese should be pelleted, with $\frac{3}{32}$- or $\frac{3}{16}$-in. pellets preferred. Mash and crumbles cause too much feed wastage and should not be used.

2. Although all rations may be home-mixed, a commercially prepared ration is recommended for young goslings and breeders during the laying season.

3. Succulent green feed should provide the bulk of the ration for young growing geese.

The NRC requirements for geese are presented in Table 24-16, in the earlier section of this chapter headed, "National Research Council (NRC) Requirements."

Example rations for geese are given in Table 24-32.

TABLE 24-32
EXAMPLE GEESE RATIONS[1]

Ingredient	Starter 0-3 Weeks		Grower 3 Weeks to Market		Breeder Layer	
	(lb)	(kg)	(lb)	(kg)	(lb)	(kg)
Ground yellow corn	600	272.3	600	272.3	300	136.2
Ground oats	400	181.5	500	227.0	700	317.8
Ground barley	375	170.2	375	170.2	210	95.3
Dehydrated alfalfa	72	32.6	72	32.6	250	113.5
Soybean oil meal, 44% protein	500	227.0	400	181.5	450	204.3
Limestone	20	9.1	20	9.1	50	22.7
Dicalcium phosphate	20	9.1	20	9.1	25	11.4
Iodized salt	10	4.5	10	4.5	10	4.5
Vitamin premix	3	1.7	3	1.7	5	2.3
Total	2,000	908.0	2,000	908.0	2,000	908.0

[1]From: *Goose Production in North Dakota,* North Dakota State University, Fargo.

FEEDING PIGEONS

Pigeons are raised primarily for sport and hobby—being widely used in racing, shows, and training to perform tricks. But many people consider pigeon (squab) to be an epicurean delight, and a limited demand for pigeon meat exists. Squabs are marketable as early as 28 days of age, at which time their dressed weight is about 1 lb. At this age, squabs are tender and self-basting due to the fat under the skin.

Pigeons grow more rapidly than most birds during the first 20 days of life. They receive their first nourishment from "pigeon milk" regurgitated from the parent pigeon's crop. Pigeon milk is a thick, creamy, semi-digested substance high in protein and fat, but low in carbohydrate. When 20 to 40 days of age, squabs may be fed a pigeon feed. Unlike other forms of poultry, pigeons will not eat mash, so pigeon feed either consists of whole or cracked grains or commercially prepared pellets.

Fig. 24-25. Pigeons at feed. (Courtesy, Ralston Purina Company, St. Louis, Mo.)

Commercial pigeon feeds are available. But the fancier may prepare a suitable ration of grains, plus a free-choice mineral mixture, similar to the ration shown in Table 24-33.

TABLE 24-33
EXAMPLE RATION FOR PIGEONS[1]

Grain Mix (Whole Grains)	Amount	
	(lb)	(kg)
Yellow corn	35	15.8
Grain sorghum	20	9.1
Cowpeas or field peas	20	9.1
Wheat	15	6.8
Oat groats	5	2.3
Hempseed	5	2.3
Total	100	45.4
Mineral mix (fed in separate hopper free-choice):		
Medium-sized ground oyster shells	50	22.7
Grit (appropriate size)	25	11.3
Bone meal or dicalcium phosphate	20	9.1
Salt	5	2.3
Total	100	45.4

[1]Adopted by the authors from *Managing Game Birds*, p. 6, Table 2, published by Michigan State University, East Lansing.

FEEDING GAME BIRDS: PHEASANTS, BOBWHITE QUAIL, JAPANESE QUAIL, CHUKARS, PARTRIDGES, GROUSE, AND DOVES

Game birds may be propagated for many different purposes. Fanciers may keep game birds as pets. Some growers produce dressed birds, especially pheasants and Bobwhite quail, for specialty markets in stores and restaurants. Other growers produce pheasants, Bobwhite quail, chukars and various other types of partridges and/or grouse for game release farms (shooting preserves). Still others produce birds for release to the wild.

Japanese quail are fed for egg production. They mature at 5 to 6 weeks of age and lay up to 300 eggs per year. As is true of other game birds, Japanese quail must be fed higher protein rations than chickens so as to achieve fast early growth.

Fig. 24-26. Pheasant.

Fig. 24-27. Bobwhite quail.

Fig. 24-28. Japanese quail.

The NRC requirements for pheasants and Bobwhite quail are presented in Table 24-17, and the NRC requirements for Japanese quail are given in Table 24-18; in an earlier section of this chapter headed, "National Research Council (NRC) Requirements."

Commercial game bird feeds are available in most areas. Generally, these are of the following types: *starter ration* containing 28% protein, for the first 6 weeks; *grower ration* containing 20% protein, for 7 to 14 weeks of age; and, depending on whether they are being marketed for game-release for dressed game birds, *finisher ration* or *flight conditioner ration* containing 15% protein, from 15 weeks until market. Commercial game bird feeds may be in the form of mash, crumbles, or pellets. **NOTE:** Game birds require a higher level of protein in early life than chickens.

Example rations for home-mixing of game bird rations are given in Table 24-34. Grit should be available in a separate feeder, with the size grit determined by the size of the birds.

TABLE 24-34
EXAMPLE RATIONS FOR GAME BIRDS

Item	Starter	Grower	Breeder
	(%)	(%)	(%)
Alfalfa meal, sun-cured	7.5	5.5	5.0
Corn, yellow, ground	14.0	44.0	56.7
Meat and bone meal	7.0	0.0	0.0
Fish meal	7.0	0.0	0.0
Sorghum, ground	8.0	0.0	0.0
Soybean meal, 44% protein	41.0	35.0	14.7
Wheat, ground	12.0	0.0	0.0
Wheat middlings	2.0	0.0	0.0
Wheat bran	0.0	12.0	16.8
Limestone, ground	0.0	1.0	4.1
$CaHPO_4 \cdot 2H_2O$	0.0	1.5	1.5
Salt, iodized	0.7	0.4	0.5
DL-methionine	0.3	0.1	0.2
Premix[1]	0.5	0.5	0.5
Calculated Analysis:			
Protein (%)	28.1	20.8	15.1
Metabolizable energy (kcal/kg)(or kcal/2.2 lb)	2,720	2,660	2,570
Calcium (%)	1.0	0.94	2.15
Total phosphorus (%)	0.76	0.76	0.74
Available phosphorus (%)	0.52	0.45	0.44

[1]Premix should contain: in mg per kg (or per 2.2 lb) diet—$MnSO_4 \cdot H_2O$, 40; ZnO, 60; vitamin B-12, (cobalamins) 0.005; menadione sodium bisulfite, 2; riboflavin (vitamin B-2), 6; niacin (nicotinic acid, nicotinamide), 40; calcium pantothenate, 20; folacin (folic acid), 0.5; antioxidant, 100; antibiotic, 10; in IU—vitamin A, 5,000; vitamin D_3, 1,500; vitamin E, 20. An equivalent commercial premix can be used, but follow the directions of the supplier.

Specific information relative to feeding pheasants, Bobwhite quail, and Japanese quail follows:

• **Feeding pheasants**—following the starter phase, a grain supplement consisting of equal portions of (1) corn or grain sorghum, and (2) oats, barley, or wheat can be effectively used in limited amounts; but the grain supplement should not consititute more than one-fourth of the entire ration. Pheasants can be fed and managed to produce fertile eggs at any time of the year. Egg production requires proper feeds and light stimulation.

• **Feeding Bobwhite quail**—Wild quail survive on native grass seeds and insects. Confinement-reared quail require nutritionally balanced rations to promote growth and health. After 6 weeks of age, Bobwhites should be fed according to whether they will be utilized as breeders, flight conditioned for shooting preserves, or processed for meat purposes.

• **Feeding Japanese quail**—Laying rations for Japanese quail should contain about 25% protein; additionally, a free-choice supply of calcium (limestone or oystershell) should be available. Some growers may wish to supplement commercial feeds with small seeds or cracked grain. When Japanese quail are fed whole seeds, fine grit should be provided.

FEEDING GUINEA FOWL

In recent years, there has been a growing interest, worldwide, in the production of guinea fowl. In France and Italy, the commercial production of guinea fowl is a highly profitable industry.

Fig. 24-29. Guinea fowl. (Courtesy, USDA)

Hens lay for 35 weeks and produce 175 eggs. Baby guineas are called *guinea poults* or *keets*. At 86 days of age, the keets weigh about 3.3 lb, made with a feed conversion of 2.7 to 3.2.

A 3-phase feeding program is normally followed, consisting of a 24% protein starter, a 19–21% grower, and a 17–18% finisher.

FEEDING OSTRICHES

Feeding ostriches is different! The ostrich is the largest bird in the world. At maturity, it may stand 10 ft tall and weigh more than 330 lb. Young ostriches grow very rapidly; they reach full size in about 6 months, but they do not attain sexual maturity until 3 to 4 years of age. They may live to 70 years of age. The ostrich is the only bird that eliminates its urine and feces separately.

Ostriches are valued primarily for their skins, which are made into fine quality leather. Spasmodically, the plumes are popular for decorations and accessories. The eggs may be used for human food, but the meat is seldom consumed because it is tough and has an unpleasant taste.

The nutritive requirements of ostriches and turkeys appear to be very similar. So, the rations presented in Table 24-31, Example Turkey Rations, may be used for ostriches. Because of the very rapid growth of ostrich chicks, however, it is suggested that they be continued on the high-protein turkey *starter ration* given in 24-31 longer than the 8 weeks suggested for turkey poults.

be substituted for the starter, and small amounts of cracked corn, wild bird seed, or chopped green grass (lawn clippings) may be added to the ration.

When roaming free, adult peafowl eat a variety of seeds, insects, and plants. Additionally, they should be provided some turkey or game-bird feed, bird seed, or grain, with the allowance increased in cold weather.

Fig. 24-30. Ostrich.

FEEDING PEAFOWL

The peafowl belongs to the same family as pheasants and chickens, differing primarily in plumage. Although they are prized as ornamental birds (blue, white, black, or green—with blue most common), peafowl are edible and are regarded as a delicacy on special occasions, perhaps more for rareness than for taste.

The care and management of peafowl is similar to that of turkeys. Peachicks may be fed a high-protein (28 to 30%) turkey or game-bird starter feed, preferably crumbles. At about 6 to 8 weeks of age, a game bird grower diet may

SALMONELLA/TOXIC INORGANIC ELEMENTS

Salmonella and/or toxic inorganic elements may be problems in poultry.

• **Salmonella**—*Salmonella* is a family of bacteria that consists of more than 2,000 different strains, which may be found in feeds—especially animal by-products (meat meal, poultry by-product meal, and fish meal), and in human foods—including broilers, turkeys, and eggs. The disease, known as salmonellosis, may occur if foods contaminated with *Salmonella* are eaten raw, not properly cooked, or mishandled after cooking. The symptoms of the illness are diarrhea, nausea, vomiting, and sometimes fever. The illness may occur within 6 to 72 hours after eating the contaminated food and may last 2 to 6 days.

Salmonella in human foods can be destroyed by proper cooking at a temperature of 160°F, or more, at the centermost part of the thickest item being cooked. The bacteria may also be inactivated by treating dressed poultry with 1% lactic acid, but it will color the meat slightly; and the bacteria may be completely destroyed by irradiation, but such treatment is not approved by FDA, presently. So, consumers should be urged (1) to refrigerate all animal products, (2) to fully cook all foods of animal origin, and (3) to avoid the recontamination of all foods after cooking. Additionally, producers should use all animal drugs and medications in compliance with FDA regulations and in accord with the directions on the label; and all producers and processors should continue their vigilance in reducing *Salmonella* in all feeds and facilities by rigid sanitation and heat treatment, and by preventing humanly edible animal products from becoming contaminated by *Salmonella*.

Despite all the scare stories, however, American consumers have the blessed assurance that they enjoy the safest and most abundant high quality animal products in the world.

(For an in-depth discussion of *Salmonella* and salmonellosis, see Chapter 5, Table 5-3—Salmonellosis; Chapter 13, sections on "Antibiotics" and "Chemotherapeutics"; and Chapter 23, section on "Additives.")

• **Toxic inorganic elements**—Poultry are susceptible to a number of toxins, any one of which may prove disastrous in a flock. Among them are a number of inorganic elements such as arsenic, lead, and selenium. It is important, therefore, that the poultry producer guard against toxic levels of inorganic elements.

For an in-depth discussion of toxic inorganic elements, see Chapter 5, section on "Potential Poisons," including Table 5-3, Potential Poisons.

Fig. 24-31. Peafowl.

QUESTIONS FOR STUDY AND DISCUSSION

1. How do you account for the fact that poultry feeding has changed more than the feeding of any other species?

2. During the last half of the 20th century, per capita red meat consumption declined, while chicken and turkey consumption increased. What forces caused this difference?

3. What is the economic importance of feed for poultry?

4. In 1940, it required 4.7 lb of feed to produce 1 lb of weight gain of broilers vs 1.9 lb of feed in 1990; whereas, during this same period of time, feed required to produce 1 lb of beef has been lowered by about a pound only. Why have broiler producers made more marked progress in feed efficiency than beef cattle producers?

5. Explain how temperature/humidity and genetics may affect the nutritive requirements of poultry.

6. Why do nutritionists add a margin of safety to *requirements*, known as *allowances*, in ration formulation?

7. What are the NRC requirements? How are they used?

8. Define *energy requirement*. What are the primary energy sources? Discuss the relative importance of each energy source.

9. How does the nutritionist use *metabolizable energy (ME)* in ration formulation?

10. What are (a) apparent metabolizable energy (AME), and (b) true metabolizable energy (TME)? How do they differ?

11. From the standpoint of poultry nutrition, the amino acids that make up proteins are really the essential nutrients, whereas in cattle nutrition the protein molecule suffices. Why the difference?

12. What is the relationship of the energy content of the diet and the protein/amino acid content of the diet?

13. What is meant by *true digestibility of amino acids*? Explain the difference between *apparent digestibility coefficient* and *true digestibility coefficient*.

14. What minerals are of most practical importance for chickens and turkeys because outside sources of them must be added to most feed formulations?

15. Discuss the function, deficiency symptoms, and practical sources of each of the following major minerals in the nutrition of poultry: salt/(NaCl), calcium, and phosphorus.

16. Discuss the function, deficiency symptoms, and practical sources of each of the following trace minerals in the nutrition of poultry: iodine, manganese, selenium, and zinc.

17. What vitamins are commonly low in poultry rations? Discuss the relative effectiveness of vitamin D_2 (the plant form) and vitamin D_3 (the animal form) in meeting the vitamin D needs of poultry.

18. Discuss the function, deficiency symptoms, and practical sources of each of the following fat-soluble vitamins in the nutrition of poultry: vitamins A, D, and E.

19. Discuss the function, deficiency symptoms, and practical sources of each of the following water-soluble vitamins in the nutrition of poultry: biotin, riboflavin, thiamin, vitamin B–6, and vitamin B–12.

20. What components of water should poultry producers be concerned about when they are in high concentration?

21. What are the primary reasons for using antibiotics in poultry feeds? What is the best explanation for the growth-stimulating activity of antibiotics?

22. Why do poultry producers use each of the following additives: antifungal agents, antioxidants, arsenicals, grit, and marigold petal meal?

23. What major energy feedstuffs are fed to poultry? What major protein feedstuffs are fed to poultry?

24. Why are many consumers concerned about the cholesterol content of eggs? Should egg producers evolve with low-cholesterol eggs?

25. Why do poultry producers use more feed in the form of crumbles than producers of other classes of livestock?

26. In 1984, more than one-third of the primary commercial feed used in the United States was fed to poultry. Why is so much U.S. commercially-prepared feed fed to poultry?

27. Is it possible to formulate good poultry rations by the hand method, without a computer?

28. How may a poultry producer use a *feed substitution table* advantageously?

29. What are the main controls of feed intake of poultry?

30. Describe *phase feeding* and *molting* of layers. Why do so many poultry producers apply these practices?

31. Why may it be necessary to restrict the feed intake of pullets of heavy breeds?

32. List three methods of restricted (limited) feeding commonly employed in feeding pullets of the heavy breeds.

33. In modern broiler production, what is considered average (a) market weight, (b) market age, (c) feed conversion, and (d) mortality?

34. What are the primary differences between feeding market turkeys and feeding turkey breeders?

35. Why do the protein requirements of turkeys from 1 day of age to marketing differ so widely?

36. What forces caused the production of ducks in the United States to double from 1975 to 1985?

37. Discuss the two major types of environments in which ducks are grown.

38. List and discuss the variety of programs under which geese are raised in the United States.

39. What is pigeon milk?

40. Outline a feeding program for a game bird farm raising pheasants, Bobwhite quail, Japanese quail, and other game birds.

41. Discuss the feeding of guinea fowl, ostriches, and peafowl.

42. Discuss the importance of *toxic inorganic elements* and *Salmonella,* from the standpoints of (a) poultry and (b) people consuming poultry products.

43. Is the farm poultry flock a thing of the past? If so, why?

Original painting by Tom Phillips

25

FEEDING HORSES[1]

Contents	Page
Economic Importance of Feed for Horses	592
Nutritive Needs of Horses	592
National Research Council (NRC) Requirements	593
Nutrient Requirements Vs Allowances	596
Recommended Nutrient Allowances	596
Energy	597
Carbohydrates	598
Fats	598
Protein	598
Protein Poisoning	599
Minerals	599
Horse Mineral Chart	600
Mineral Imbalances	605
Feeding Minerals	605
Vitamins	605
Horse Vitamin Chart	606
Vitamin Imbalances	610
Unidentified Factors	610
Water	610
Antibiotics	610
Feeds For Horses	611
Forages	611
Concentrates	611
Protein Supplements	612
Special Feeds	612
Treats	612
Palatability	612
Feed Preparation	613
Feed Allowances and Suggested Rations	613
Feed Allowances	613
Suggested Rations	614
Overfeeding	615
Home-Mixed Feeds	615
Commercial Horse Feeds	615
Sweet Feed	615
Feed Substitution Table	615
Feeding Pleasure Horses	617
Feeding Horses in Training (Equine Athletes)	617
Feeding Racehorses	617
Feeding Broodmares	617
Feeding Stallions	618
Feeding Foals	618
Feeding Weanlings	619
Feeding Yearlings	619
Feeding Two- and Three-Year-Olds	619
Fitting for Show or Sale	619
Feeding Systems	619
Art of Feeding	620
Starting Horses on Feed	620
General Feeding Rules	620
Signs of a Well Fed Healthy Horse	621
Other Feed and Management Aspects	621
Nutritional Diseases	621
Questions for Study and Discussion	621

Unless horses are fed properly, their maximum potential in reproduction, growth, body form, speed, endurance, style, and attractiveness cannot be achieved. Additionally, the following conditions make it imperative that the nutrition of horses be the best that science and technology can devise:

1. **Confinement.** Many horses are kept in stables or corrals.

2. **Fitting yearlings.** When forcing young equines, it is important to their development and soundness that the ration be nutritionally balanced.

3. **Racing 2-year-olds.** In the United States, we race more 2-year-olds than any other nation in the world; our richest races are for them. If the nutrient content of the ration is not adequate, there is bound to be more breakdown on the track than with older horses. This is costly.

Fig. 25-1. Some horses fared better than others in their development. The diminutive Shetland Pony (left) evolved on the scanty vegetation of the Shetland Isles, whereas the giant Shire (right) was the product of fertile soils and abundant vegetation of Flanders.

---

[1]The authors gratefully acknowledge the helpful suggestions of the following eminent authorities who reviewed this chapter: H. F. Hintz, Ph.D., Professor of Animal Nutrition, Cornell University, Ithaca, New York; R. D. Scoggins, D.V.M., Equine Extension Veterinarian, University of Illinois, Urbana-Champaign; W. J. Tyznik, Ph.D., Animal Science Department, The Ohio State University, Columbus; David McGlothlin, Manager, Horse Division, Harris Farms, Coalinga, Calif.; and Charles Pollard, Pollard Ranch, Clovis, Calif.

4. **Stress.** Stress may be caused by excitement, temperament, fatigue, number of horses together, previous nutrition, breed, age, and management. Race and show horses are always under stress; and the more tired they become and the greater the speed, the greater the stress. Thus, the ration for race and show horses should be scientifically formulated, rather than based on fads, foibles, and trade secrets. The greater the stress, the more exacting the nutritive requirements.

5. **Horses are unique.** They differ from other farm animals in that they have greater individual value; are kept for recreation, sport, and work; are fed for a longer life of usefulness; have a smaller digestive tract; should not carry surplus weight; and are fed for nerve, mettle, animation, character of muscle, and athletic performance.

ECONOMIC IMPORTANCE OF FEED FOR HORSES

Although feed constitutes the greatest single cost item in the horse business, its relative economic importance varies more widely than with any other class of livestock. This is so because many horses are kept for recreation, sport, or hobby purposes, whereas the vast majority of other species are kept for strictly business reasons—for making money. It follows that the cost of land, buildings and equipment, and labor for horses is usually much higher in relation to the cost of feed than for other animals.

Where no pasture is available, a 1,000-lb horse will, on the average, consume about 25 lb of feed (hay and grain) daily, or 4.6 tons per year. The cost of horse feed will vary in keeping with prevailing feed prices, the quality of the ration, and the proportion of concentrate and roughage fed. Also, caretakers are prone to feed a good many additives, all of which cost money. In total, U.S. horse owners spend $9 billion each year on feed, medication, and services for their horses.

In addition to the cost of horse feed as such, faulty nutrition makes for hidden costs of great economic importance. Horse owners are losing millions of dollars through inefficiency. On the average, they are producing a 50% foal crop —which means that they are keeping two mares a whole year to produce one foal. They are retiring an appalling number of horses from tracks, shows, and other uses due to unsoundnesses. Many of the bone ailments that plague breeders and trainers—the sprains, spavins, splints, and ringbones—are the tragic result of improper skeletal development during the fetal and early growth stages.

NUTRITIVE NEEDS OF HORSES

Meeting the nutrient needs of horses is a major factor in determining their efficiency and years of service. As with other animals, the horse needs nutrients for maintenance, growth, fattening, reproduction, and production. With horses, the production need is for work—mostly for recreation and sport, and as cow ponies. Unlike most animals, however, the work is usually irregular and often very strenuous— characteristics which create a particular stress on the animal and make the job of feeding according to the nutritive needs very difficult.

- **The digestive tract of horses makes for differences**—The digestive tract of equines makes for difficulties in meeting their nutritive needs. It differs anatomically and physiologically from that of the cow (ruminant) as follows:

1. It is smaller, with the result that the horse cannot eat as much roughage as cattle.

2. Without feed, the horse's stomach will empty completely in 24 hours, whereas it takes about 72 hours (3 times as long) for the cow's stomach to empty.

3. The cow has four compartments (rumen, reticulum, omasum, and the abomasum or true stomach), whereas the horse has one.

4. There is comparatively little microbial action in the stomach of the horse, but much such action in the first compartment (rumen) of the cow.

5. The primary seats of microbial activity in ruminants and horses occupy different locations in the digestive system in relation to the small intestine. In cattle (and sheep), the rumen precedes the small intestine; in horses, the cecum follows it. As a result, the efficiency of absorption of nutrients synthesized by the microorganisms is likely to be lower in a horse than in a ruminant.

The limited protein synthesis in the horse (limited when compared with ruminants), and the lack of efficiency of absorption due to the cecum being on the lower end of the gut (thereby not giving the small intestine a chance at the ingesta after it leaves the cecum), clearly indicate that horse rations should contain high-quality proteins, adequate in amino acids.

In comparison to a cow, therefore, a horse should be fed less roughage, more and higher quality protein, and perhaps added B vitamins. Actually, the nutrient requirements of a horse more nearly parallel those of a pig than a cow.

- **Horses have special nutritive needs for bone formation and maintenance**—In addition to meeting the nutritive needs common to all species—the needs for maintenance, growth, fattening, reproduction, and production—the formation and maintenance of sound bones in horses is extremely important and very complicated. The 205 bones of the horse's skeleton contain 25% ash or minerals, 20% protein, 10% fat, and 45% water.

But bone formation and maintenance require certain vitamins, too. Vitamin A is essential for sound bones and teeth; and vitamin D, along with calcium and phosphorus, is vital. A deficiency of vitamin D in the horse's ration results in reduced bone calcification, swollen joints, stiffness of gait, softness of bones, bone deformities, and frequent cases of fracture.

Vitamin C is involved in the formation and maintenance of intracellular material, including collagen and related substances in the bones and soft tissues. Also, vitamin C supplementation may be beneficial for horses under stress such as during rapid growth and high-level performance, in hot weather, or when something interferes with vitamin synthesis.

Other minerals and vitamins may also be beneficial for the proper formation and maintenance of the bones of the horse, as well as for a healthy horse. So, many owners with high-level performance animals supplement with additional minerals and vitamins to ensure adequate nutrition.

- **Horses under stress have special nutritive needs**—Some scientists and horse owners feel that higher nutritive levels are helpful under high stress conditions, such as when horses are subjected to training, transporting, racing, performing, crowds, and various environmental conditions.

Feeding Horses

• **Horses should be fed as individuals**—Most high-performing horses are *prima donnas*; hence, they require individual attention. They must eat their feed; otherwise, no matter how nutritionally complete it may be, it won't do them any good.

Meeting the nutritive needs of horses is further complicated because these needs do not necessarily remain the same from day to day or from period to period. The age and size of the animal; the stage of gestation or lactation of a mare; the kind and degree of activity; climatic conditions; the kind, quality, and amount of feed; the system of management; and the health, condition, and temperament of the animal are all continually exerting a powerful influence in determining its nutritive needs. How well the caretaker understands, anticipates, interprets, and meets these requirements usually determines the success or failure of the ration. For these reasons, no set of instructions, calculator, or book of knowledge can substitute for experience and born horse intuition. Skill and good judgment are essential.

National Research Council (NRC) Requirements

The nutrient requirements of horses are given in Tables 25-1, 25-2, 25-3, 25-4, and 25-5, which follow.

In addition to the NRC tables presented in this section, *Nutrient Requirements of Horses*, 5th rev. ed., 1989, also presents tables giving the daily nutrition requirements of horses with mature weights of 1540 lb *(700 kg)*, 1760 lb *(800 kg)*, and 1980 lb *(900 kg)*. Also, the new NRC computer program calculates energy requirements based on the foal's body weight, age, and average daily gain; then, it calculates the requirements for most of the other nutrients based on this energy requirement.

TABLE 25-1
DAILY NUTRIENT REQUIREMENTS PER PONY, 440 LB *(200 KG)* MATURE WEIGHT[1] (See footnotes at end of table.)

Animal	Body Weight		Daily Gain	Energy		Protein		Calcium	Phosphorus	Magnesium	Potassium	Vitamin A Activity
				TDN	Digestible Energy	Crude Protein	Lysine					
	(lb)	*(kg)*	(lb)	(lb)	(Mcal)	(lb)	(g)	(g)	(g)	(g)	(g)	(1,000 IU)
Mature horses												
Maintenance	440	*200*		3.7	7.4	0.65	10	8	6	3.0	10.0	6
Stallions (breeding season)	440	*200*		4.7	9.3	0.81	13	11	8	4.3	14.1	9
Pregnant mares[2]												
9 months	440	*200*		4.1	8.2	0.79	13	16	12	3.9	13.1	12
10 months	440	*200*		4.2	8.4	0.81	13	16	12	4.0	13.4	12
11 months	440	*200*		4.5	8.9	0.86	14	17	13	4.3	14.2	12
Lactating mares												
Foaling to 3 months	440	*200*		6.9	13.7	1.51	24	27	18	4.8	21.2	12
3 months to weaning	440	*200*		6.1	12.2	1.16	18	18	11	3.7	14.8	12
Working horses												
Light work[3]	440	*200*		4.6	9.3	0.81	13	11	8	4.3	14.1	9
Moderate work[4]	440	*200*		5.5	11.1	0.97	16	14	10	5.1	16.9	9
Intense work[5]	440	*200*		7.4	14.8	1.30	21	18	13	6.8	22.5	9
Growing horses												
Weanling, 4 months	165	*75*	0.88	3.7	7.3	0.80	15	16	9	1.6	5.0	3
Weanling, 6 months												
Moderate growth	209	*95*	0.66	3.8	7.6	0.83	16	13	7	1.8	5.7	4
Rapid growth	209	*95*	0.88	4.4	8.7	0.95	18	17	9	1.9	6.0	4
Yearling, 12 months												
Moderate growth	309	*140*	0.44	4.4	8.7	0.86	17	12	7	2.4	7.6	6
Rapid growth	309	*140*	0.66	5.2	10.3	1.01	19	15	8	2.5	7.9	6
Long yearling, 18 months												
Not in training	375	*170*	0.22	4.2	8.3	0.82	16	10	6	2.7	8.8	8
In training	375	*170*	0.22	5.8	11.6	1.14	22	14	8	3.7	12.2	8
2-year-old, 24 months												
Not in training	408	*185*	0.11	4.0	7.9	0.74	13	9	5	2.8	9.4	8
In training	408	*185*	0.11	5.7	11.4	1.06	19	13	7	4.1	13.5	8

[1]Adapted by the authors from *Nutrient Requirements of Horses*, 5th rev. ed., NRC-National Academy of Sciences, 1989, p. 42. To convert lb to kg, divide by 2.2.
[2]Mares should gain weight during late gestation to compensate for tissue deposition. However, nutrient requirements are based on maintenance body weight.
[3]Examples are horses used in Western and English pleasure, bridle path hack, equitation, etc.
[4]Examples are horses used in ranch work, roping, cutting, barrel racing, jumping, etc.
[5]Examples are horses in race training, polo, etc.

TABLE 25-2
DAILY NUTRIENT REQUIREMENTS PER HORSE, 880 LB *(400 KG)* MATURE WEIGHT[1]

Animal	Body Weight (lb)	Body Weight (kg)	Daily Gain (lb)	Energy TDN (lb)	Energy Digestible Energy (Mcal)	Protein Crude Protein (lb)	Protein Lysine (g)	Calcium (g)	Phosphorus (g)	Magnesium (g)	Potassium (g)	Vitamin A Activity (1,000 IU)
Mature horses												
Maintenance	880	*400*		6.7	13.4	1.17	19	16	11	6.0	20.0	12
Stallions (breeding season)	880	*400*		8.4	16.8	1.47	23	20	15	7.7	25.5	18
Pregnant mares[2]												
9 months	880	*400*		7.5	14.9	1.43	23	28	21	7.1	23.8	24
10 months	880	*400*		7.6	15.1	1.46	23	29	22	7.3	24.2	24
11 months	880	*400*		8.1	16.1	1.55	25	31	23	7.7	25.7	24
Lactating mares												
Foaling to 3 months	880	*400*		11.5	22.9	2.50	40	45	29	8.7	36.8	24
3 months to weaning	880	*400*		9.9	19.7	1.84	29	29	18	6.9	26.4	24
Working horses												
Light work[3]	880	*400*		8.4	16.8	1.47	23	20	15	7.7	25.5	18
Moderate work[4]	880	*400*		10.5	20.1	1.76	28	25	17	9.2	30.6	18
Intense work[5]	880	*400*		13.4	26.8	2.35	38	33	23	12.3	40.7	18
Growing horses												
Weanling, 4 months	320	*145*	1.87	6.8	13.5	1.48	28	33	18	3.2	9.8	7
Weanling, 6 months												
Moderate growth	397	*180*	1.21	6.5	12.9	1.41	27	25	14	3.4	10.7	8
Rapid growth	397	*180*	1.54	7.3	14.5	1.53	30	30	16	3.6	11.1	8
Yearling, 12 months												
Moderate growth	584	*265*	1.29	7.8	15.6	1.57	30	23	13	4.5	14.5	12
Rapid growth	584	*265*	1.29	8.6	17.1	2.12	33	27	15	4.6	14.8	12
Long yearling, 18 months												
Not in training	728	*330*	1.60	7.9	15.9	1.57	30	21	12	5.3	17.3	15
In training	728	*330*	1.60	10.3	21.6	2.12	41	29	16	7.1	23.4	15
2-year-old, 24 months												
Not in training	805	*365*	1.77	7.7	15.3	1.42	26	19	11	5.7	18.7	16
In training	805	*365*	1.77	10.8	21.5	2.00	37	27	15	7.9	26.2	16

[1]Adapted by the authors from *Nutrient Requirements of Horses,* 5th rev. ed., NRC-National Academy of Sciences, 1989, p. 43. To convert lb to kg, divide by 2.2.
[2]Mares should gain weight during late gestation to compensate for tissue deposition. However, nutrient requirements are based on maintenance body weight.
[3]Examples are horses used in Western and English pleasure, bridle path hack, equitation, etc.
[4]Examples are horses used in ranch work, roping, cutting, barrel racing, jumping, etc.
[5]Examples are horses in race training, polo, etc.

TABLE 25-3
DAILY NUTRIENT REQUIREMENTS PER HORSE, 1,100 LB *(500 KG)* MATURE WEIGHT[1]

Animal	Body Weight (lb)	Body Weight (kg)	Daily Gain (lb)	Energy TDN (lb)	Energy Digestible Energy (Mcal)	Protein Crude Protein (lb)	Protein Lysine (g)	Calcium (g)	Phosphorus (g)	Magnesium (g)	Potassium (g)	Vitamin A Activity (1,000 IU)
Mature horses												
Maintenance	1,100	*500*		8.2	16.4	1.44	23	20	14	7.5	25.0	15
Stallions (breeding season)	1,100	*500*		10.3	20.5	1.79	29	25	18	9.4	31.2	22
Pregnant mares[2]												
9 months	1,100	*500*		9.1	18.2	1.75	28	35	26	8.7	29.1	30
10 months	1,100	*500*		9.3	18.5	1.78	29	35	27	8.9	29.7	30
11 months	1,100	*500*		9.9	19.7	1.89	30	37	28	9.4	31.5	30
Lactating mares												
Foaling to 3 months	1,100	*500*		14.2	28.3	3.12	50	56	36	10.9	46.0	30
3 months to weaning	1,100	*500*		14.2	24.3	2.29	37	36	22	8.6	33.0	30
Working horses												
Light work[3]	1,100	*500*		10.3	20.5	1.79	29	25	18	9.4	31.2	22
Moderate work[4]	1,100	*500*		12.3	24.6	2.15	34	30	21	11.3	37.4	22
Intense work[5]	1,100	*500*		16.4	32.8	2.87	46	40	29	15.1	49.9	22

(Continued)

TABLE 25-3 (Continued)

Animal	Body Weight (lb)	(kg)	Daily Gain (lb)	Energy TDN (lb)	Energy Digestible Energy (Mcal)	Protein Crude Protein (lb)	Protein Lysine (g)	Calcium (g)	Phosphorus (g)	Magnesium (g)	Potassium (g)	Vitamin A Activity (1,000 IU)
Growing horses												
Weanling, 4 months	385	175	1.87	7.2	14.4	1.58	30	34	19	3.7	11.3	8
Weanling, 6 months												
Moderate growth	473	215	1.43	7.5	15.0	1.64	32	29	16	4.0	12.7	10
Rapid growth	473	215	1.87	8.6	17.2	1.88	36	36	20	4.3	13.3	10
Yearling, 12 months												
Moderate growth	715	325	1.10	9.5	18.9	1.86	36	29	16	5.5	17.8	15
Rapid growth	715	325	1.43	10.7	21.3	2.09	40	34	19	5.7	18.2	15
Long yearling, 18 months												
Not in training	880	400	0.77	9.9	19.8	1.95	38	27	15	6.4	21.1	18
In training	880	400	0.77	13.6	26.5	2.61	50	36	20	8.6	28.2	18
2-year-old, 24 months												
Not in training	990	450	0.44	9.4	18.8	1.75	32	24	13	7.0	23.1	20
In training	990	450	0.44	13.2	26.3	2.44	45	34	19	9.8	32.2	20

[1] Adapted by the authors from *Nutrient Requirements of Horses*, 5th rev. ed., NRC-National Academy of Sciences, 1989, p. 43. To convert lb to kg, divide by 2.2.
[2] Mares should gain weight during late gestation to compensate for tissue deposition. However, nutrient requirements are based on maintenance body weight.
[3] Examples are horses used in Western and English pleasure, bridle path hack, equitation, etc.
[4] Examples are horses used in ranch work, roping, cutting, barrel racing, jumping, etc.
[5] Examples are horses in race training, polo, etc.

TABLE 25-4
DAILY NUTRIENT REQUIREMENTS PER HORSE, 1,320 LB (600 KG) MATURE WEIGHT [1]

Animal	Body Weight (lb)	(kg)	Daily Gain (lb)	Energy TDN (lb)	Energy Digestible Energy (Mcal)	Protein Crude Protein (lb)	Protein Lysine (g)	Calcium (g)	Phosphorus (g)	Magnesium (g)	Potassium (g)	Vitamin A Activity (1,000 IU)
Mature horses												
Maintenance	1,320	600		9.7	19.4	1.70	27	24	17	9.0	30.0	18
Stallions (breeding season)	1,320	600		12.5	24.3	2.12	34	30	21	11.2	36.9	27
Pregnant mares [2]												
9 months	1,320	600		10.8	21.5	2.07	33	41	31	10.3	34.5	36
10 months	1,320	600		10.9	21.9	2.11	34	42	32	10.5	35.1	36
11 months	1,320	600		11.7	23.3	2.24	36	44	34	11.2	37.2	36
Lactating mares												
Foaling to 3 months	1,320	600		16.9	33.7	3.74	60	67	43	13.1	55.2	36
3 months to weaning	1,320	600		14.5	28.9	2.75	44	43	27	10.4	39.6	36
Working horses												
Light work [3]	1,320	600		12.6	24.3	2.12	34	30	21	11.2	36.9	27
Moderate work [4]	1,320	600		14.6	29.1	2.55	41	36	25	13.4	44.2	27
Intense work [5]	1,320	600		19.4	38.8	3.40	54	47	34	17.8	59.0	27
Growing horses												
Weanling, 4 months	440	200	2.20	8.3	16.5	1.80	35	40	22	4.3	13.0	9
Weanling, 6 months												
Moderate growth	539	245	1.65	8.5	17.0	1.86	36	34	19	4.6	14.5	11
Rapid growth	539	245	2.09	9.6	19.2	2.10	40	40	22	4.9	15.1	11
Yearling, 12 months												
Moderate growth	825	375	1.43	11.4	22.7	2.24	43	36	20	6.4	20.7	17
Rapid growth	825	375	1.76	12.6	25.1	2.47	48	41	22	6.6	21.2	17
Long yearling, 18 months												
Not in training	1,045	475	0.99	12.0	23.9	2.36	45	33	18	7.7	25.1	21
In training	1,045	475	0.99	16.0	32.0	3.13	60	44	24	10.2	33.3	21
2-year-old, 24 months												
Not in training	1,188	540	0.66	11.8	23.5	2.18	40	31	17	8.5	27.9	24
In training	1,188	540	0.66	16.7	32.3	3.00	55	43	24	11.6	38.4	24

[1] Adapted by the authors from *Nutrient Requirements of Horses*, 5th rev. ed., NRC-National Academy of Sciences, 1989, p. 44. To convert lb to kg, divide by 2.2.
[2] Mares should gain weight during late gestation to compensate for tissue deposition. However, nutrient requirements are based on maintenance body weight.
[3] Examples are horses used in Western and English pleasure, bridle path hack, equitation, etc.
[4] Examples are horses used in ranch work, roping, cutting, barrel racing, jumping, etc.
[5] Examples are horses in race training, polo, etc.

TABLE 25-5
NUTRIENT CONCENTRATIONS IN RATIONS FOR HORSES AND PONIES[1]

Animal	Ration Proportions[2] Conc. (%)	Ration Proportions[2] Hay (%)	Moisture Basis A-F (As-fed) M-F (Moisture-free)	TDN (lb)	Digestible Energy (Mcal/lb)	Digestible Energy (Mcal/kg)	Crude Protein (%)	Lysine (%)	Calcium (%)	Phosphorus (%)	Magnesium (%)	Potassium (%)	Vitamin A Activity (IU/lb)	Vitamin A Activity (IU/kg)
Mature horses														
Maintenance	0	100	A-F	0.40	0.80	1.80	7.2	0.25	0.21	0.15	0.08	0.27	750	1,650
			M-F	0.45	0.90	2.00	8.0	0.28	0.24	0.17	0.09	0.30	830	1,830
Stallions	30	70	A-F	0.50	1.00	2.15	8.6	0.30	0.26	0.19	0.10	0.33	1,080	2,370
			M-F	0.55	1.10	2.40	9.6	0.34	0.29	0.21	0.11	0.36	1,200	2,640
Pregnant mares														
9 months	20	80	A-F	0.45	0.90	2.00	8.9	0.31	0.39	0.29	0.10	0.32	1,510	3,330
			M-F	0.50	1.00	2.25	10.0	0.35	0.43	0.32	0.10	0.35	1,680	3,710
10 months	20	80	A-F	0.45	0.90	2.00	9.0	0.32	0.39	0.30	0.10	0.33	1,490	3,280
			M-F	0.50	1.00	2.25	10.0	0.35	0.43	0.32	0.10	0.36	1,660	3,650
11 months	30	70	A-F	0.50	1.00	2.15	9.5	0.33	0.41	0.31	0.10	0.35	1,490	3,280
			M-F	0.55	1.10	2.40	10.6	0.37	0.45	0.34	0.11	0.38	1,660	3,650
Lactating mares														
Foaling to 3 months	50	50	A-F	0.55	1.10	2.35	12.0	0.41	0.47	0.30	0.09	0.38	1,130	2,480
			M-F	0.60	1.20	2.60	13.2	0.46	0.52	0.34	0.10	0.42	1,250	2,750
3 months to weaning	35	65	A-F	0.53	1.05	2.20	10.0	0.34	0.33	0.20	0.08	0.30	1,240	2,720
			M-F	0.58	1.15	2.45	11.0	0.37	0.36	0.22	0.09	0.33	1,370	3,020
Working horses														
Light work[3]	35	65	A-F	0.53	1.05	2.20	8.8	0.32	0.27	0.19	0.10	0.34	1,100	2,420
			M-F	0.58	1.15	2.45	9.8	0.35	0.30	0.22	0.11	0.37	1,220	2,690
Moderate work[4]	50	50	A-F	0.55	1.10	2.40	9.4	0.35	0.28	0.22	0.11	0.36	970	2,140
			M-F	0.60	1.20	2.65	10.4	0.37	0.31	0.23	0.12	0.39	1,100	2,420
Intense work[5]	65	35	A-F	0.60	1.20	2.55	10.3	0.36	0.31	0.23	0.12	0.39	800	1,760
			M-F	0.65	1.30	2.85	11.4	0.40	0.35	0.25	0.13	0.43	890	1,950
Growing horses														
Weanling, 4 months	70	30	A-F	0.63	1.25	2.60	13.1	0.54	0.62	0.34	0.07	0.27	650	1,420
			M-F	0.70	1.40	2.90	14.5	0.60	0.68	0.38	0.08	0.30	720	1,580
Weanling, 6 months														
Moderate growth	70	30	A-F	0.63	1.25	2.60	13.0	0.55	0.50	0.28	0.07	0.27	760	1,680
			M-F	0.70	1.40	2.90	14.5	0.61	0.56	0.31	0.08	0.30	850	1,870
Rapid growth	70	30	A-F	0.63	1.25	2.60	13.1	0.55	0.55	0.30	0.07	0.27	670	1,470
			M-F	0.70	1.40	2.90	14.5	0.61	0.61	0.34	0.08	0.30	740	1,630
Yearling, 12 months														
Moderate growth	60	40	A-F	0.58	1.15	2.50	11.3	0.48	0.39	0.21	0.07	0.27	890	1,950
			M-F	0.65	1.30	2.80	12.6	0.53	0.43	0.24	0.08	0.30	980	2,160
Rapid growth	60	40	A-F	0.58	1.15	2.50	11.3	0.48	0.40	0.22	0.07	0.27	790	1,730
			M-F	0.65	1.30	2.80	12.6	0.53	0.45	0.25	0.08	0.30	870	1,920
Long yearling, 18 months														
Not in training	45	55	A-F	0.53	1.05	2.30	10.1	0.43	0.31	0.17	0.07	0.27	930	2,050
			M-F	0.58	1.15	2.50	11.3	0.48	0.34	0.19	0.08	0.30	1,030	2,270
In training	50	50	A-F	0.55	1.10	2.40	10.8	0.45	0.32	0.18	0.08	0.27	740	1,620
			M-F	0.60	1.20	2.65	12.0	0.50	0.36	0.20	0.09	0.30	820	1,800
2-year-old, 24 months														
Not in training	35	65	A-F	0.50	1.00	2.20	9.4	0.38	0.28	0.15	0.08	0.27	1,080	2,380
			M-F	0.58	1.15	2.45	10.4	0.42	0.31	0.17	0.09	0.30	1,200	2,640
In training	50	50	A-F	0.55	1.10	2.40	10.1	0.41	0.31	0.17	0.09	0.29	840	1,840
			M-F	0.60	1.20	2.65	11.3	0.45	0.34	0.20	0.10	0.32	930	2,040

[1]Adapted by the authors from *Nutrient Requirements of Horses*, 5th rev. ed., NRC-National Academy of Sciences, 1989, pp. 46 and 47.
[2]Values assume a concentrate feed containing 1.5 Mcal/lb *(3.3 Mcal/kg)* and hay containing 0.91 Mcal/lb *(2.00 Mcal/kg)* of dry matter.
[3]Examples are horses used in Western and English pleasure, bridle path hack, equitation, etc.
[4]Examples are horses used in ranch work, roping, cutting, barrel racing, jumping, etc.
[5]Examples are horses in race training, polo, etc.

Nutrient Requirements Vs Allowances

In ration formulation, two words are commonly used—*requirements* and *allowances*. Requirements do not provide for margins of safety; allowances do.

Scientists normally list the mineral and vitamin requirements of horses on the basis of body weight or units per day. To facilitate application by horse owners, the authors of this book give recommended nutrient allowances in Tables 25-8 and 25-9 three ways: (1) per horse daily, (2) nutrient concentration in ration, and (3) nutrient concentration per ton of feed (as-fed basis).

Recommended Nutrient Allowances

Presently available information indicates that the recommended nutrient allowances given in Table 25-6 will meet the minimum requirements for horses and provide reasonable margins of safety. Additional recommended allowance figures for minerals are given in Table 25-8, Horse Mineral Chart, under the heading "Nutrient Allowances"; and additional recommended allowance figures for vitamins are given in Table 25-9, Horse Vitamin Chart, under the heading "Nutrient Allowances."

Feeding Horses

TABLE 25-6
RECOMMENDED NUTRIENT ALLOWANCES FOR HORSES (TOTAL RATION/AS-FED BASIS)[1] (See footnotes at end of table.)

	Mature Horses (Consuming 25 lb feed/horse/day. Idle horses require less feed and/or consume more roughage than heavily worked horses or lactating mares.)					Young Horses, Based on Mature Weight 1,000 lb				
	Idle Horses/ Light Work/ Moderate Work (1,000 lb Wt.)	Heavy Training/ Heavy Work (1,000 lb Wt.)	Stallions in Breeding Season (1,000 lb Wt.)	Mares, Last 90 Days Gestation (1,000 lb Wt.)	Mares, Peak of Lactation (1,000 lb Wt.)	Creep Feed (250 lb Body Wt/11 lb Feed Daily)	Weanlings (450 lb Body Wt/12 lb Feed Daily)	Yearlings (650 lb Body Wt/13 lb Feed Daily)	2-Yr-Olds & 3-Yr-Olds (800 lb Body Wt/14 lb Feed Daily)	2-Yr-Olds in Light Training (800 lb Body Wt/15 lb Feed Daily)
Digestible Energy:										
TDN[2] (%)	55	62.60	75	62.50	75	75	75	70	60	65
Mcal per (lb)	0.8	1.2	1.0	0.90	1.10	1.25	1.25	1.15	1.00	1.10
Mcal per (kg)[3]	1.80	2.55	2.15	2.0	2.35	2.60	2.60	2.50	2.20	2.40
Crude Protein (%)	9.0	11.0	14.0	13.0	14.0	18.0	16.0	14.0	13.0	13.0
Lysine (%)	0.25	0.36	0.30	0.32	0.41	0.54	0.55	0.48	0.38	0.41
Major or Macrominerals:										
Salt (%)	0.75	0.75	0.75	0.75	0.75	0.75	0.75	0.75	0.75	0.75
Calcium (%)	0.21	0.31	0.26	0.29	0.47	0.62	0.55	0.40	0.28	0.31
Phosphorus (%)	0.15	0.23	0.19	0.30	0.30	0.34	0.30	0.22	0.15	0.17
Magnesium (%)	0.08	0.12	0.10	0.10	0.09	0.07	0.07	0.07	0.08	0.09
Potassium (%)	0.27	0.39	0.33	0.33	0.38	0.27	0.27	0.27	0.27	0.29
Sulfur (%)	0.15	0.15	0.15	0.15	0.15	0.15	0.15	0.15	0.15	0.15
Trace or Microminerals:										
Cobalt (ppm)[4]	0.11	0.11	0.11	0.11	0.11	0.11	0.11	0.11	0.11	0.11
Copper (ppm)	25	25	25	25	30	40	40	30	25	25
Iodine (ppm)	0.11	0.11	0.11	0.11	0.11	0.11	0.11	0.11	0.11	0.11
Iron (ppm)	40	60	90	90	90	90	80	60	60	60
Manganese (ppm)	46	46	46	46	46	46	46	46	46	46
Selenium (ppm)	0.11	0.11	0.11	0.11	0.11	0.11	0.11	0.11	0.11	0.11
Zinc (ppm)	80	90	90	100	100	100	100	100	90	90
	(/lb)	(/lb)	(/lb)	(/lb)	(/lb)	(/lb)	(/lb)	(/lb)	(/lb)	(/lb)
Fat-soluble Vitamins in Feed:										
Vitamin A (IU)	1,045	1,045	1,045	1,569	1,569	1,045	1,045	1,045	1,045	1,045
Vitamin D (IU)	156	156	156	314	314	419	419	419	419	419
Vitamin E (IU)	26	41	41	41	41	41	41	41	41	41
Vitamin K (mg)	0.32	0.32	0.32	0.32	0.32	0.30	0.30	0.30	0.30	0.30
Water-soluble Vitamins in Feed:										
Biotin (mg)	0.1	0.1	0.1	0.1	0.1	0.1	0.1	0.1	0.1	0.1
Choline (mg)	20	30	30	30	30	62.5	62.5	62.5	62.5	62.5
Folacin (mg)	0.8	1.2	1.2	1.2	1.2	3.0	3.0	3.0	3.0	3.0
Niacin (mg)	10	20.8	10	10	10	10	10	10	10	10
Pantothenic acid (mg)	10	20.8	10	10	10	10	10	10	10	10
Riboflavin (mg)	1.6	1.6	1.6	1.6	1.6	1.6	1.6	1.6	1.6	1.6
Thiamin (B-1) (mg)	1.57	2.61	1.57	1.57	1.57	1.57	1.57	1.57	1.57	1.57
Vitamin B-6 (mg)	1.0	1.0	1.0	1.0	1.0	0.5	0.5	0.5	0.5	0.5
Vitamin B-12 (mg)	0.005	0.006	0.006	0.006	0.006	0.007	0.007	0.007	0.007	0.007
Vitamin C (Ascorbic acid) (mg)	2.4	4.0	4.0	4.0	4.0	3.75	3.75	3.75	3.75	3.75

[1] Where hay is fed separately, double the amounts shown in this table should be added to the concentrate.
[2] 1 lb TDN = 2 Mcal or 2,000 Kcal.
[3] 1 kg = 2.2 lb or 1,000 g.
[4] 1 ppm (parts per million) = 1 mg/kg.

In the discussion that follows, the nutrient requirements and recommended allowances of the horse are discussed under these headings: (1) energy (carbohydrates and fats), (2) protein, (3) minerals, (4) vitamins, and (5) water.

Energy

The energy requirements of horses for various activities are hard to develop because it is difficult to express quantitatively the type of exercise, the intensity and duration of work, the condition and training of the animals, the ability and weight of the rider and driver, the degree of fatigue, and the environmental temperature—all of which influence energy requirements. Based on Cornell studies by Pagan, Hintz reported the energy requirements given in Table 25-7 (see next page). The Cornell researchers found that the amount of energy expended was proportional to the body weight of the riderless horse or the combined weight of the horse plus the rider, and that the amount of energy expended was exponentially related to speed. Additional studies are needed to determine the energy expenditures at speeds faster than the 13 miles per hour reported in Table 25-7.

TABLE 25-7
DIGESTIBLE ENERGY REQUIREMENTS FOR VARIOUS ACTIVITIES OF LIGHT HORSES[1]

Gait	Speed (Miles/Hour)[2]	DE/Hour (Kcal/kg of Wt.)[3]
Slow walk	2.2	1.7
Fast walk	3.6	2.5
Slow trot	7.5	6.5
Medium trot	9.3	9.5
Fast trot/slow canter	11.2	13.7
Medium canter	13.0	19.5

[1]Hintz, H. F., Energy Requirements of Horses, *Feed Management*, Vol. 37, No. 2, Feb. 1986, p. 15.
[2]To convert to metric, see Appendix, Weights and Measures.
[3]Body weight of horse plus weight of rider and tack.

Fig. 25-2. When racing, horses may require up to 100 times more energy when running than at rest. (Courtesy, *The Backstretch*, Detroit, Mich.)

A lack of energy intake may cause slow and stunted growth in foals and loss of weight, poor condition, and excessive fatigue in mature horses. Excess energy may result in obese horses, which are more susceptible to stress and founder and have lowered reproductive efficiency and decreased longevity.

The caretaker may base individual horse energy requirements on observation. If the horse is too thin, increase the energy intake; if too fat, decrease energy intake.

CARBOHYDRATES

Increased energy for horses is generally met by increasing the grain and decreasing the roughage. But, since the horse naturally evolved as a grazing animal, its digestive system often has difficulty in handling large quantities of cereal grains (starchy material). Evidence of the horses' difficulty with high grain rations is manifested clinically as founder, colic, and/or loss of appetite. So, to promote normal physiological activity of the gastrointestinal tract, some coarse roughage is necessary; finely ground roughage (or pelleted forage) will not suffice. But the precise amount of fiber needed has not been determined. It is generally recommended that horses be fed 1.0 to 2.0 lb of roughage per 100 lb of body weight, daily. Young horses and working (or running) horses must have rations in which a large part of the carbohydrate content is low in fiber, and in the form of nitrogen-free extract.

FATS

Some fat in the ration is desirable, because fat is the carrier of the fat-soluble vitamins (vitamins A, D, E, and K), and because the horse needs some linoleic acid (an essential fatty acid) in the ration. A ration devoid of fats results in reduced growth, scaly skin, and a rough, thin hair coat. Although the fatty acid requirements of horses have not been determined, it is thought that most ordinary farm rations contain ample quantities of these nutrients.

In the past, most horse owners and scientists were of the opinion that horses could not tolerate high-fat rations. But, studies indicate that they will readily consume 10 to 20% added fat in the ration, without difficulty—and even with benefit.[2] In an endurance trial conducted at the Colorado Station, horses fed fat supplemented rations (9% added fat) outperformed their counterparts that were fed either (1) starch supplemented rations, or (2) protein supplemented rations.[3] Subsequent experiments and experiences indicate that the amount of the energy and the energy density needed in the ration increase dramatically as work intensity increases.

(Also see Chapter 10, section on "Fats and Oils.")

Protein

Except for the proteins built by bacterial action in the cecum, horses must have amino acids or more complex protein compounds in the ration.

Horses of all ages and kinds require adequate amounts of protein of suitable quality for maintenance, growth, finishing, reproduction, and work. Of course, the protein requirements for growth, reproduction, and lactation are the greatest and most critical. (See Table 25-6.) The protein requirements for work are minimal and are not increased by work load; they're the same for maintenance, medium exercise, and intense exercise, according to researchers at Washington State University.[4] Therefore, the protein requirements for work are essentially the same as the maintenance requirements. Of course, when total feed intake increases to meet the added energy requirements of working horses, total protein intake also increases, even though the percent protein in the ration remains the same.

A deficiency of protein in the ration of the horse may result in the following deficiency symptoms: depressed appetite, poor growth, loss of weight, reduced milk production, irregular estrus, lowered foal crops, loss of condition, and lack of stamina.

The limited protein synthesis in the horse (as compared with ruminants) and the lack of efficiency of absorption due to the cecum being on the lower end of the gut (thereby not giving the small intestine a chance at the digesta after

---

[2]Tyznik, W. J., "Energy for Horses," paper presented at the 1975 California Livestock Symposium.
[3]Slade, L. M., and P. L. Hambleton, "Feeding the Horse for Endurance," *Stud Managers Handbook*, Vol. 12, 1976, p. 140, published by Agriservices Foundation, edited by M. E. Ensminger.
[4]Patterson, P. H., C. N. Coon, and I. M. Hughes, "Protein Requirements for Mature Working Horses," *Journal of Animal Science*, Vol. 61, No. 1, July 1985, p. 187.

it leaves the cecum), clearly indicate that horse rations should contain high-quality proteins, adequate in essential amino acids. Most cereal grains, such as oats, corn, or barley, are deficient in lysine, tryptophan, and methionine for optimum growth. Some protein supplements, such as linseed and cottonseed meal, do not contain adequate lysine.

In recognition that lysine is the first limiting amino acid of horses and is thus an indicator of the quality of protein which horses require, the lysine requirements of horses are given in Tables 25-1 through 25-6.

The extent to which the horse's ration is supplemented with proteins depends on the age of the horse and on the quality of the forage fed. Growing or lactating animals require somewhat more protein than horses that are idle, gestating, or working. Also, grass hays are generally low in quality and quantity of proteins and require more supplementation than legumes.

PROTEIN POISONING

Some opinions to the contrary, protein poisoning as such has never been documented. Some horses do appear to be allergic to certain proteins or to excesses of specific amino acids, as a result of which they may develop *protein bumps*.

Minerals

In an amazingly short time after birth, a healthy foal can run almost as fast as its mother—and on legs almost as long. In fact, the cannon bones (the lower leg bones extending from the knees and hocks to the fetlocks) are as long at the time of birth as they will ever be. This indicates that important development of the skeleton takes place in the fetus, before the foal is born. It is evident, therefore, that adequate minerals must be provided the broodmare if the bones of her offspring are to be sound.

The mineral requirements of mares in lactation are even more rigorous than those during gestation. Mares weighing about 1,000 lb will produce an average of 2 gal, or more, of milk per day throughout the 7-month suckling period. That's a total of 3,612 lb of milk. Since fresh mare's milk contains 0.7% ash, this amount of mare's milk contains 25.28 lb of mineral (3,612 × 0.7% = 25.28). Here's how this phenomenon works: When properly fed before breeding, in early pregnancy, and when barren, mineral deposits are made in the mare's skeleton. Then at those times when the mineral demands are greater than can be obtained from the feed—the last of pregnancy, and during lactation—the mare draws from the stored reserves in her skeleton. Of course, if there hasn't been proper storage in the mare's skeleton, something must "give"—and that something is the mother. Nature has ordained that growth of the fetus, and the lactation that follows, shall take priority over the maternal requirements. Hence, when there is a mineral deficiency, the mare's body will be deprived, or even stunted if she is young, before the developing fetus or milk production will be materially affected.

The proper development of the bone is particularly important in the horse, as evidenced by the stress and strain on the skeletal structure of the racehorse, especially when racing the 2-year-old. Since the greatest development of the skeleton takes place in the young, growing animal, it is evident that adequate minerals must be provided at an early age if the bone is to remain sound.

- **Metabolic bone disease (MBD)**—In recent years, there has been a great increase in metabolic bone disease in growing horses, especially epiphysitis, contracted tendons, and osteochondritis dissecans (OCD). A brief description of each of these conditions follows:

1. **Epiphysitis.** This is an inflammation of the growth plate of the long bones, primarily found at the lower end of the radius above the knee, but it may be noticeable at the distal tibial and the distal metacarpal and metatarsal bones. Epiphysitis results in a firm and painful swelling.

2. **Contracted tendons.** This involves a shortening of the flexor tendons, causing the heels to be raised and the pasterns to be straight or, in severe cases, to knuckle forward with the horse walking on its toe. Contracted tendons may be present at birth, or they may be acquired during growth.

3. **Osteochondritis dissecans (OCD).** This is a condition in which the cartilage in a growing foal does not properly convert into bone. It may appear in either of two forms: (a) The form in which it is localized in one or a few joints (most commonly the stifle and hock joints, although any joint may be involved), usually without any clinical signs; and (b) the second and less common form, which most commonly affects the more distal limb joints such as the pastern and fetlock, although it may affect any joint, including those of the back.

At this time, the cause of the increase in the incidence of the above bone diseases is not entirely clear. However, it appears that the major factors are: (1) rapid growth and excess weight, (2) injury to the epiphysis, (3) nutritional imbalances, (4) genetic predisposition, (5) limited forced exercise, (6) exercise on hard ground, and (7) faulty conformation.

Based on field observations and a study conducted by Ohio State University, involving 384 yearlings raised on 19 breeding farms in Ohio and Kentucky, and including the Thoroughbred, Standardbred, Arabian, and Quarter Horse breeds, there is strong evidence that calcium, phosphorus, copper, and zinc deficiencies/imbalances and/or masking are involved. They reported that the average calcium content of the rations on farms with the fewest skeletal problems was 1.16% ± 0.09 and the phosphorus content was 0.72% ± 0.08.[5]

Based on experiments (including unpublished work at both Ohio State and Cornell) and experiences, the authors of *Feeds & Nutrition* recommend (1) that breeders continue to feed alfalfa hay to pregnant mares and growing horses, and (2) that the levels of calcium, phosphorus, copper, iron, manganese, and zinc be in keeping with the recommendations given in Table 25-6 of this chapter.

Table 25-8, Horse Mineral Chart, pp. 600–603, lists the minerals required by horses and gives pertinent information pertaining to each. Minerals may be incorporated in the ration in keeping with the recommended allowances given in this table and in Table 25-6. Additionally, horses should have free access to salt.

---

[5]Knight, Debra A., et al, *Correlation of Dietary Minerals to Incidence and Severity of Metabolic Bone Disease in Ohio and Kentucky,* College of Veterinary Medicine, The Ohio State University; paper presented at the 1985 American Association of Equine Practitioners meeting in Toronto, Canada.

TABLE
HORSE MINERAL

Minerals Which May Be Deficient Under Normal Conditions	Conditions Usually Prevailing Where Deficiencies Are Reported	Function of Mineral	Some Deficiency Symptoms	Practical Sources of the Mineral
Major or macrominerals:				
Salt (NaCl)	Negligence, for salt is cheap. The salt requirement is greatly increased under conditions which cause heavy sweating, thereby resulting in large losses of this mineral from the body. Unless it is replaced, fatigue will result. For this reason, when engaged in hard work and perspiring profusely, horses should receive liberal allowances of salt.	Salt serves as both a condiment and a nutrient. Sodium and chlorine help maintain osmotic pressure in body cells, upon which depends the transfer of nutrients to the cells and the removal of waste materials. Sodium is associated with muscle contraction and is important in making bile, which aids in the digestion of fats and carbohydrates. Chlorine is required for the formation of hydrochloric acid in the gastric juice so vital to protein digestion.	In warm or hot weather, workhorses show heat stress. Long-term symptoms of sodium deficiency are depraved appetite, rough hair coat, reduced growth of young animals, and decreased milk production.	Salt provided free-choice, preferably in loose form, or 0.5–1.0% salt added to the ration. It is very difficult for horses to eat very hard block or rock salt. This often results in inadequate consumption. Also, if there is much competition for a salt block, the more timid animals may not get their requirements. Iodized salt should be used in iodine-deficient areas.
Calcium (Ca)	The typical horse ration of grass hay and farm grains—usually deficient in calcium.	Builds strong bones and sound teeth. Very important during lactation. Affects availability of phosphorus. Calcium and phosphorus comprise ¾ of the ash of the skeleton and from ⅓–½ of the minerals of milk.	A deficiency of calcium in young animals is generally characterized by poorly formed, soft bone, which may bend or bow; and a severe deficiency may cause rickets. A deficiency of calcium in older animals results in porous, fragile bones. Because deficiency conditions may not be completely reversible, prevention is imperative.	Ground limestone or oystershell flour. When both calcium and phosphorus are needed, use steamed bone meal or dicalcium phosphate. Horses absorb 55 to 75% of the calcium in a typical ration.
Phosphorus (P)	Horses grazed on phosphorus-deficient areas or fed for a long period on mature, weathered forage.	Important in the development of bones and teeth. Essential to metabolism of carbohydrates and fats, and enzyme activation.	Rickets in young horses; osteomalacia in mature horses.	Monosodium phosphate, disodium phosphate, or sodium tripolyphosphate. Where both calcium and phosphorus are needed, use steamed bone meal or dicalcium phosphate. Horses absorb 35–55% of the phosphorus in a typical ration.
Magnesium (Mg)	Horses fed high grain-low forage ration, which characterizes most horses at hard work (as in racing and showing). Lactating mares grazing on lush spring pastures low in magnesium or in which Mg is unavailable.	Reduces stress and irritability. Magnesium is important in enzyme systems, bone formation, and calcium and phosphorus metabolism.	Horses under stress are keyed up, high-strung, and jumpy. Foals fed a purified ration deficient in magnesium develop nervousness, muscular tremors, convulsive paddling of the legs and, in some cases, die. Grass tetany.	Magnesium sulfate. Magnesium oxide.
Potassium (K)	When stabled horses are fed high-concentrate rations. Excessive sweating.	Major cation of intracellular fluid where it is involved in osmotic pressure and acid-base balance. Muscle activity. Required in enzyme reaction involving phosphorylation of creatine. Influences carbohydrate metabolism.	Reduced appetite, growth retardation, unsteady gait, general muscle weakness, pica, diarrhea, distended abdomen, emaciation, followed by death. Fatigue. Abnormal electrocardiograms.	Potassium chloride. Roughages usually contain ample potassium.
Sulfur (S)		Sulfur is an integral part of the amino acids methionine and cystine.		

Feeding Horses

25-8
CHART (See footnotes at end of table.)

Classes/Function	Nutrient Requirements[1][2]				Nutrient Allowances[1][2]				Comments
	Per Horse Daily	In Ration A-F	Per Ton Ration A-F		Per Horse Daily	In Ration A-F	Per Ton Ration A-F		
	(g)	(%)	(lb)	(kg)	(g)	(%)	(lb)	(kg)	
Maintenance: 1,000-lb (454-kg) horse	85	0.5-1.0	10-20	4.5-9.1	85	0.75	15	6.8	Horses require both sodium and chlorine, but the requirement for chlorine is approximately half that of sodium. Generally, the chlorine requirements will be met if the sodium needs are adequate.
Gestation/Lactation: 1,000-lb (454-kg) mare	85	0.5-1.0	10-20	4.5-9.1	85	0.75	15	6.8	Sodium and chlorine are low in feeds of plant origin.
Growth: 450-lb (204.5-kg) weanling	41	0.5-1.0	10-20	4.5-9.1	41	0.75	15	6.8	There is little danger of overfeeding salt unless a salt-starved animal is suddenly exposed to too much salt, or if liberal amounts of water are not available.
Working: 1,000-lb (454-kg) horse	85	0.5-1.0	10-20	4.5-9.1	85	0.75	15	6.8	Excessive salt intake may result in high water intake, excessive urine excretion, digestive disturbances, or death from salt cramps.
Maintenance: 1,000-lb (454-kg) horse	20	0.175	3.5	1.6	23	0.21	4.1	1.8	The calcium-phosphorus ratio should be maintained close to 1.1:1 although 2:1 is acceptable. Narrower ratios may cause osteomalacia in mature horses. When there is a shortage of calcium in the ration, it is withdrawn from the bones.
Gestation/Lactation: 1,000-lb (454-kg) mare	56	0.495	9.9	4.5	64	0.57	11.3	5.1	
Growth: 450-lb (204.5-kg) weanling	36	0.66	13.2	6.0	41	0.76	15.1	6.8	Feeding excess calcium interferes with the utilization of magnesium, manganese, and iron—and perhaps in the utilization of zinc.
Working: 1,000-lb (454-kg) horse	40	0.35	7.0	3.2	46	0.41	8.1	3.7	
Maintenance: 1,000-lb (454-kg) horse	14	0.125	2.5	1.1	16.1	0.14	2.8	1.3	For the growing horse, the calcium-phosphorus ratio should be maintained close to 1.1:1, although 2:1 is acceptable.
Gestation/Lactation: 1,000-lb (454-kg) mare	36	0.315	6.3	2.9	41.4	0.37	7.3	3.3	The mature horse can tolerate a Ca:P ratio as wide as 4:1 or 5:1 provided adequate levels of phosphorus are available.
Growth: 450-lb (204.5-kg) weanling	19	0.35	7.0	3.2	21.9	0.4	8.0	3.7	Excess phosphorus can cause bighead.
Working: 1,000-lb (454-kg) horse	29	0.255	5.1	2.3	33.4	0.3	5.9	2.7	If plenty of vitamin D is present, the ratio of calcium to phosphorus becomes less important.
Maintenance: 1,000-lb (454-kg) horse	7.5	0.065	1.3	0.6	8.6	0.08	1.5	0.7	Excess of magnesium upsets calcium and phosphorus metabolism.
Gestation/Lactation: 1,000-lb (454-kg) mare	10.9	0.096	1.92	0.9	12.5	0.11	2.2	1.0	Rations containing 50% forage will likely contain sufficient magnesium for unstressed horses.
Growth: 450-lb (204.5-kg) weanling	5.7	0.105	2.1	1.0	6.6	0.12	2.4	1.1	
Working: 1,000-lb (454-kg) horse	15.1	0.135	2.7	1.2	17.4	0.16	3.1	1.4	
Maintenance: 1,000-lb (454-kg) horse	25.0	0.22	4.4	2.0	28.8	0.26	5.1	2.3	A ration that contains at least 50% forage can be expected to meet potassium requirements.
Gestation/Lactation: 1,000-lb (454-kg) mare	46.0	0.405	8.1	3.7	52.9	0.47	9.3	4.2	
Growth: 450-lb (204.5-kg) weanling	18.2	0.335	6.7	3.0	20.9	0.39	7.7	3.5	
Working: 1,000-lb (454-kg) horse	49.9	0.44	8.8	4.0	57.4	0.51	10.1	4.6	
Maintenance: 1,000-lb (454-kg) horse					17.0	0.15	3.0	1.36	The precise sulfur requirement is not known, but an allowance of 0.15% of the total ration appears to be adequate.
Gestation/Lactation: 1,000-lb (454-kg) mare					17.0	0.15	3.0	1.36	If the protein requirement of the ration is met, the sulfur intake will usually be at least 0.15%, which appears to be adequate.
Growth: 450-lb (204.5-kg) weanling					8.2	0.15	3.0	1.36	
Working: 1,000-lb (454-kg) horse					17.0	0.15	3.0	1.36	

(Continued)

TABLE 25-8

Minerals Which May Be Deficient Under Normal Conditions	Conditions Usually Prevailing Where Deficiencies Are Reported	Function of Mineral	Some Deficiency Symptoms	Practical Sources of the Mineral
Trace or microminerals: Cobalt (Co)	Animals grazed in cobalt-deficient areas, such as Australia, Western Canada, and the following states of U.S.: Fla., Mich., Wisc., N.H., Penn., and N.Y.	Cobalt is required for the synthesis of vitamin B-12 in the intestinal tract of the horse.	Anemia. Severe weight loss.	Cobaltized mineral mix made by adding cobalt at the rate of 0.2 oz/100 lb (*5.7 g/45.4 kg*) of salt as cobalt chloride, cobalt sulfate, cobalt oxide, or cobalt carbonate. Also, several good commercial cobalt-containing minerals are on the market.
Copper (Cu)	Suckling foals. Mare's milk, along with milk from other species, is low in copper. Deficiency occurs in regions where soils contain too little copper or where horses are getting an excess of molybdenum, sulfur, or zinc.	Copper, along with iron and vitamin B-12 is necessary for hemoglobin formation, although it forms no part of the hemoglobin molecule (or red blood cells). Closely associated with normal bone development in young growing animals.	Anemia, characterized by fewer than normal red cells and less than normal amount of hemoglobin. Abnormal bone development in young equines, including an increased incidence of epiphysitis, contracted tendons, and osteochondritis dissecans (OCD).	Trace mineralized salt containing copper sulfate or copper carbonate.
Iodine (I)	Iodine-deficient areas or soils (in Northwestern U.S. and in the Great Lakes region) when iodized salt is not fed. Use of feeds that come from iodine-deficient areas.	Iodine is needed by the thyroid gland in making thyroxin, an iodine-containing compound which controls the rate of body metabolism or heat production.	Foals born dead, or very weak with enlarged thyroid glands (goiter) and unable to stand or nurse. Higher than normal incidence of navel ill.	Stabilized iodized salt containing 0.01% potassium iodide (0.0076% iodine). Calcium iodate.
Iron (Fe)	Suckling foals kept away from soil and feed other than milk. Horses subjected to pressure from racing, showing, or other heavy use. Such animals require added iron in their daily ration. Excessive blood loss from a wound or heavy parasite infestation.	Necessary for formation of hemoglobin, an iron-containing compound which enables the blood to carry oxygen. Also, important to certain enzyme systems.	Iron-deficiency anemia, characterized by fewer than normal red cells and less than normal amount of hemoglobin. Anemic horses tire easily. **NOTE WELL**: Iron deficiency anemia may also result from heavy parasitization.	Ferrous sulfate administered orally. Trace mineralized salt. Cane molasses. Iron oxide should not be used as a source of iron for horses because it is poorly absorbed.
Manganese (Mn)	Excess calcium and phosphorus which decreases absorption of manganese.	Essential for normal bone formation (as a component of the organic matrix). Thought to be an activator of enzyme systems. Growth and reproduction.	Poor growth. Lameness, shortening and bowing of legs, and enlarged joints. Impaired reproduction (testicular degeneration of males; defective ovulation of females).	Trace mineralized salt containing 0.25% manganese (or more).
Selenium (Se)	Muscle disorders and lowered serum selenium.		Infertility. Myositis (muscular discomfort or pain).	Forages or grains grown on soils known to have adequate selenium. Sodium selenate. Sodium selenite.
Zinc (Zn)	Feeds low in zinc. Excess calcium may reduce the absorption and utilization of zinc.	Important in many enzyme systems. Required for normal protein synthesis and metabolism. Imparts gloss or *bloom* to the hair coat.	Rough, dull hair coat. Loss of appetite.	Zinc carbonate. Zinc sulfate.

[1] All "nutrient requirements" given in this table were adapted by the authors from *Nutrient Requirements of Horses*, 5th rev. ed., NRC—National Academy of Sciences, 1989. The "nutrient allowances" given in this table represent the authors' best judgment based on current research; it is intended that they meet the nutrient requirements, and provide adequate margins of safety in addition.

[2] Feed consumption of a mature 1,000-lb (*454-kg*) horse estimated at 25 lb (*11.36 kg*) per day. Feed consumption of a 450-lb (*204.5-kg*) weanling estimated at 12 lb (*5.45 kg*) per day.

(Continued)

Classes/Function	Nutrient Requirements[1][2]			Nutrient Allowances[1][2]			Comments
	Per Horse Daily	In Ration A-F	Per Ton Ration A-F	Per Horse Daily	In Ration A-F	Per Ton Ration A-F	
	(mg)	(ppm, or mg/kg)	(g/ton)	(mg)	(ppm, or mg/kg)	(g/ton)	
Maintenance: 1,000-lb *(454-kg)* horse	1.13	0.1	0.091	1.3	0.11	0.104	The disease called *salt sick* in Florida is due to a cobalt deficiency associated with a copper deficiency.
Gestation/Lactation: 1,000-lb *(454-kg)* mare	1.13	0.1	0.091	1.3	0.11	0.104	The cobalt requirement for horses is very low, for horses have remained in good health while grazing pastures so low in cobalt that ruminants confined to them have died.
Growth: 450-lb *(204.5-kg)* weanling	0.54	0.1	0.091	0.6	0.11	0.100	
Working: 1,000-lb *(454-kg)* horse	1.13	0.1	0.091	1.3	0.11	0.104	
Maintenance: 1,000-lb *(454-kg)* horse	113.4	10.0	9.070	283.4	25	22.675	A copper deficiency in horses has been reported in Australia.
Gestation/Lactation: 1,000-lb *(454-kg)* mare	113.4	10.0	9.070	340.1	30	27.210	In high-molybdenum areas, more copper may be added to horse rations; but excesses and toxicity should be avoided.
Growth: 450-lb *(204.5-kg)* weanling	54.4	10.0	9.070	217.7	40	36.280	
Working: 1,000-lb *(454-kg)* horse	113.4	10.0	9.070	283.4	25	22.675	
Maintenance: 1,000-lb *(454-kg)* horse	1.13	0.1	0.091	1.3	0.11	0.104	Enlargement of the thyroid gland (goiter) is nature's way of trying to make enough thyroxin (an iodine-containing hormone) when there is insufficient iodine in the feed. Feeding excess iodine continuously will also produce goiter in foals.
Gestation/Lactation: 1,000-lb *(454-kg)* mare	1.13	0.1	0.091	1.3	0.11	0.104	
Growth: 450-lb *(204.5-kg)* weanling	0.54	0.1	0.091	0.6	0.11	0.100	Iodine deficiency seldom occurs in coastal areas because of the abundance of iodine from spray drift from ocean or sea water.
Working: 1,000-lb *(454-kg)* horse	1.13	0.1	0.091	1.3	0.11	0.104	
Maintenance: 1,000-lb *(454-kg)* horse	453.5	40	36.280	453.5	40	36.280	The horse's body contains about 0.004% iron. Milk is deficient in iron, and the iron content of the mother cannot be increased through feeding iron. Thus, foals should be individually or creep fed as soon as they are old enough. A variable store of both iron and copper is located in the liver and spleen, and some iron is found in the kidneys. Too much iron may be harmful.
Gestation/Lactation: 1,000-lb *(454-kg)* mare	566.9	50	45.350	1,020.4	90	81.630	
Growth: 450-lb *(204.5-kg)* weanling	272.1	50	45.350	489.8	90	81.630	
Working: 1,000-lb *(454-kg)* horse	453.5	40	36.280	680.3	60	54.420	
Maintenance: 1,000-lb *(454-kg)* horse	453.5	40	36.280	521.5	46	41.720	Most natural feedstuffs are rich in manganese.
Gestation/Lactation: 1,000-lb *(454-kg)* mare	453.5	40	36.280	521.5	46	41.720	
Growth: 450-lb *(204.5-kg)* weanling	217.7	40	36.280	250.4	46	41.734	
Working: 1,000-lb *(454-kg)* horse	453.5	40	36.280	521.5	46	41.720	
Maintenance: 1,000-lb *(454-kg)* horse	1.13	0.1	0.091	1.3	0.11	0.104	Excess selenium results in selenium poisoning, or alkali disease (see Chapter 5, Nutritional Disorders/Toxins).
Gestation/Lactation: 1,000-lb *(454-kg)* mare	1.13	0.1	0.091	1.3	0.11	0.104	
Growth: 450-lb *(204.5-kg)* weanling	0.54	0.1	0.091	0.6	0.11	0.100	
Working: 1,000-lb *(454-kg)* horse	1.13	0.1	0.091	1.3	0.11	0.104	
Maintenance: 1,000-lb *(454-kg)* horse	453.5	40	36.280	907.0	80	72.560	If zinc in the feed is on the low side, the addition of zinc should improve the hair coat.
Gestation/Lactation: 1,000-lb *(454-kg)* mare	453.5	40	36.280	1,133.8	100	90.700	Excess zinc prevents calcium utilization and produces signs of calcium deficiency.
Growth: 450-lb *(204.5-kg)* weanling	217.7	40	36.280	544.2	100	90.700	The toxicity level exceeds 1,000 ppm.
Working: 1,000-lb *(454-kg)* horse	453.5	40	36.280	1,020.4	90	81.630	

The daily mineral requirements and recommended allowances vary with the mature weight, age, and type and level of productivity of the horse. Likewise, the mineral requirements and recommended allowances of the total ration vary with the percent dry matter in the ration, and with the age and the type and level of productivity of the horse.

Generally speaking, legume forages, such as alfalfa hay or pasture, are rich in calcium; cereal grains and their by-products—oats, corn, barley, and wheat bran—are fair to good sources of phosphorus; and the protein supplements—linseed meal, soybean meal, and dried skim milk—are good sources of both calcium and phosphorus. So, by selecting and combining the common horse feeds properly, the maintenance needs of most horses can be met. (See Fig. 25-3.)

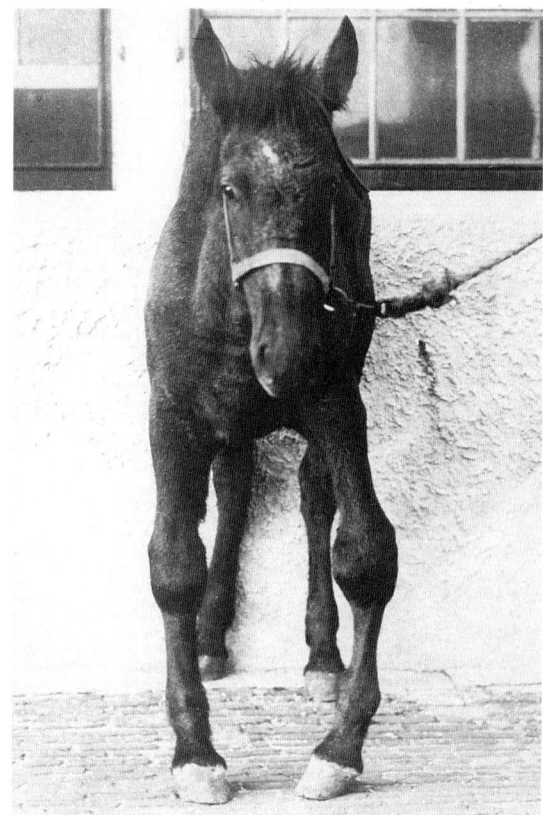

Fig. 25-4. Foal with severe rickets. Note the enlarged joints and crooked legs. (Courtesy, Department of Veterinary Pathology and Hygiene, College of Veterinary Medicine, University of Illinois, Urbana)

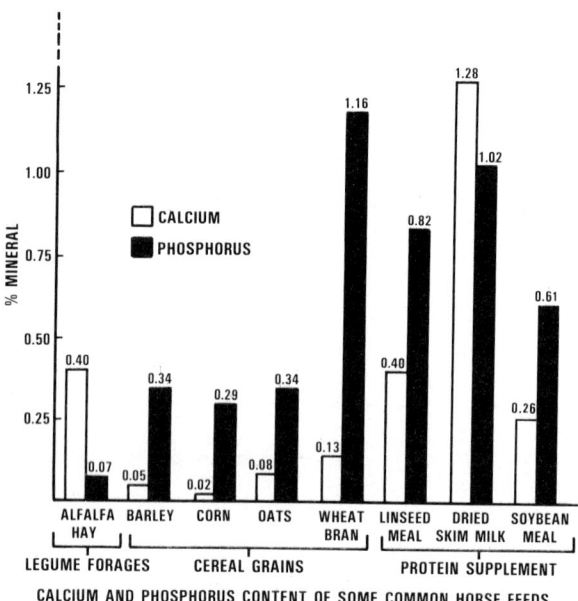

Fig. 25-3. Calcium and phosphorus content of some common horse feeds (as-fed basis).

Fig. 25-5. Calcium-phosphorus imbalance. Horse with *big head disease* (nutritional secondary hyperparathyroidism) resulting from feeding a ration low in calcium and high in phosphorus, commonly associated with a ration excessively high in wheat bran. Note that the upper jaw is enlarged because calcium is replaced by fibrous connective tissue. (Courtesy, National Academy of Sciences, and College of Veterinary Medicine, Texas A&M University, College Station)

When both calcium and phosphorus supplementation are needed, the authors favor the use of high quality steamed bone meal for horses, because bone meal contains many ingredients in addition to calcium and phosphorus. It is a good source of iron, manganese, and zinc, and it contains such trace minerals as copper and cobalt. However, it is increasingly difficult to get good bone meal. Some of the imported products are high in fat, rancid, and/or odorous and unpalatable. Where good bone meal is not available, dicalcium phosphate is generally recommended.

MINERAL IMBALANCES

Having the right balance and forms of minerals can be very important. The more calcium you feed, the more phosphorus you need. The more copper you feed, the more manganese you need.

Also, minerals can be fed in several different forms. For example, iron can be fed as an oxide, sulfite, sulfate, or as a proteinate. Oxides may be absorbed at about 2 to 5%, while sulfites may be absorbed at up to 10%, and sulfates at 25%.

Thus, the requirements of any mineral may be modified (1) by another mineral which enhances or interferes with its utilization, or (2) by the form of the mineral.

From the above, it is apparent that excess fortification of the horse's ration with one or more mineral elements may prove more detrimental than helpful. Thus, caretakers who know and care will avoid harmful imbalances; they will provide minerals on the basis of *recommended allowances* (see Table 25–6). Also, when fortifying rations with minerals, consideration should be given to the minerals provided by the ingredients of the normal ration, for it is the total composition of the feed that counts.

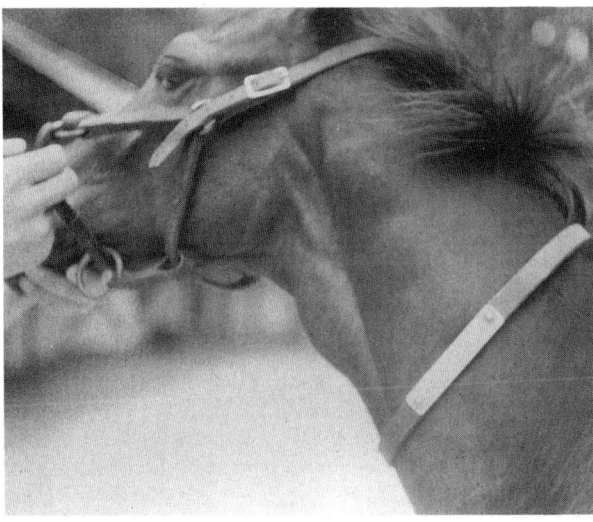

Fig. 25–6. Goiter caused by feeding excess iodine. Goiter is usually caused by an iodine deficiency, but it may result from feeding too much iodine over a long period of time. (Courtesy, Dr. D. E. Cooperrider, Chief, Diagnostic Laboratories, Division of Animal Industry, Kissimmee, Fla.)

FEEDING MINERALS

With the exception of sodium, the self-feeding of the major minerals cannot be relied upon to meet the needs of horses. This is so because horses consume such supplements on the basis of palatability, rather than because of dietary need. As a result, the free-choice intake of minerals among individual horses will vary from too little to too much. Sometimes minerals are incorporated in a salt mix, but salt consumption is erratic and variable according to the sodium content of the feedstuffs being fed. So, the only way to ensure that each horse receives the needed major minerals is to incorporate the proper amounts in the animal's feed and/or water.

Trace minerals may be added to the ration and/or incorporated in the salt. In either case, the amounts and proportions of trace minerals should be selected with care because the improper use of trace minerals can lead to induced deficiencies. Theoretically, the total ration (grain plus forage) should be balanced in trace mineral content, with the trace mineral mix providing only the minerals needed and with each one in the right amount. Of course, this isn't practical. Therefore, a trace mineral mix must contain an array of minerals in adequate levels to meet a wide variety of conditions. Fortunately, the horse is tolerant to most trace mineral excesses.

When horses are on pasture and no grain or protein supplement is being fed, minerals may be self-fed, usually as either a commercially manufactured mineral block or as a mineral mixture. A suitable home-mixed mineral for self-feeding on pasture may be prepared as follows:

1. **Where the pasture is primarily grass.** Prepare a mixture containing two parts of calcium to one part of phosphorus.

2. **Where the pasture is primarily a legume.** Prepare a mixture containing one part of calcium to one part of phosphorus.

To each of the above mixes, add one-third trace-mineralized salt to provide the microminerals and improve the palatability.

Vitamins

Unfortunately, there are no warning signals to tell a caretaker when a horse is not getting enough of certain vitamins. But a continuing inadequate supply of any one of several vitamins can produce illness which is very difficult to diagnose until it becomes severe, at which time it is difficult and expensive—if not too late—to treat. The important thing, therefore, is to ensure against such deficiencies occurring. But caretakers should not shower a horse with mistaken kindness through using shotgun-type vitamin preparations. Instead, the quantity of each vitamin should be based on available scientific knowledge.

It has long been known that the vitamin content of feeds varies considerably according to soil, climatic conditions, and curing and storing.

Deficiencies may occur during periods (1) of extended drought or in other conditions of restriction in diet, (2) when production is being forced, or during stress, (3) when large quantities of highly refined feeds are being fed, or (4) when low-quality forages are utilized.

Table 25–6 lists the vitamins most commonly involved in horse nutrition; and Table 25–9, Horse Vitamin Chart (pp 606 to 609), gives pertinent information pertaining to each. Although there is no evidence of deficiencies of certain vitamins, it is possible that more of them may be destroyed or used by horses during stress or strain than can be obtained through normal feeds or synthesized by the intestinal microflora of the horse; hence, adding them to the ration may assure maximum performance.

TABLE
HORSE VITAMIN

Vitamins Which May Be Deficient Under Normal Conditions	Conditions Usually Prevailing Where Deficiencies Are Reported	Function of Vitamin	Some Deficiency Symptoms	Practical Sources of the Vitamin
Fat-soluble vitamins:				
A	Extended drought, bleached hays. Stall feeding where there is little or no green forage or yellow corn. Following great stress, as when race or show horses are put in training. The younger the animal, the quicker vitamin A deficiencies will show up. Mature animals may store sufficient vitamin A to last 6 months.	Promotes growth and stimulates appetite. Assists in reproduction and lactation. Keeps the mucous membranes of respiratory and other tracts in healthy condition. Makes for normal vision. Prevents night blindness.	Loss of appetite, poor growth, reproductive problems, nerve degeneration, night blindness, lachrymation (tears), keratinization of the cornea and skin, uneven and poor hoof development, a predisposition to respiratory infection, incoordination, progressive weakness, convulsive seizures, certain bone disorders, and finicky appetite.	Stabilized vitamin A. Green grass. Green grass or legume hay not over 1 year old. Carrots, yellow corn.
D	Limited sunlight and/or limited sun-cured hay, especially when horse is kept inside most of the time.	Assimilation and utilization of calcium and phosphorus, necessary in normal bone development—including the bones of the fetus.	Rickets in foals, osteomalacia in mature horses. Both conditions result in large joints and weak bones. Rickets is characterized by reduced bone calcification, stiff and swollen joints, stiffness of gait, irritability, and reduction in serum calcium and phosphorus. Osteomalacia results in bones which soften, become distorted, and fracture easily.	Either vitamin D_2 (the plant form) or D_3 (the animal form) is equally effective for the horse. Exposure to sunlight. Sun-cured hays.
E	More vitamin E may be destroyed or used by horses during times of stress or strain than can be obtained through normal feeds.	As an antioxidant. As an occasional replacement for selenium. Improves reproduction. Prevents anhidrosis.	Lowered breeding preformance in both mares and stallions. Anhidrosis—a dry, dull hair coat; elevated temperature; and high blood pressure. Anhidrosis has been successfully treated by the oral administration of 1,000 to 3,000 IU of vitamin E daily.	Alpha-tocopherol acetate, a stable form of vitamin E. Wheat germ meal and wheat germ oil. Green plants. Green hays.
K	Following intestinal disorders.	Concerned with blood coagulation. It converts precursor proteins to the active blood clotting factors.	Increased clotting time of the blood and lowered level of prothrombin.	Green pasture. Well-cured hays. Cereal grains. Milk. Menadione (vitamin K_3).
Water-soluble vitamins:				
Biotin	Sulfa drugs kill intestinal organisms; hence, when they are used an extended period, there may be a deficiency of biotin.	Biotin plays an important role in the metabolism of carbohydrates, fats, and proteins.	In all animals, a deficiency of biotin will depress growth and cause a loss of hair and/or a dermatitis.	Alfalfa hay, blackstrap molasses, cottonseed meal, soybean meal, peanut meal, milk, wheat bran, synthetic biotin, and yeast (brewers', torula).
Choline	Ration low in methionine, an amino acid.	Prevention of fatty livers, the transmitting of nerve impulses, and the metabolism of fat.	Slow growth and fatty livers are the deficiency symptoms.	Feed sources, such as alfalfa hay, blackstrap molasses, and cereal grains. Body manufacture of choline from excess of the amino acid methionine. Choline chloride. Choline dihydrogen.
Folacin (Folic Acid)		In all vertebrates, folacin is essential for normal growth and reproduction, for the prevention of blood disorders, and for important biochemical mechanisms in each cell.	Poor growth. Anemia.	Alfalfa hay, the oil meals (soybean, cottonseed, and linseed), skimmed milk, and wheat and wheat by-products. Synthetic folacin, wheat germ, and yeast (brewers', torula).

Feeding Horses

25-9
CHART (See footnotes at end of table.)

Classes/Function	Nutrient Requirements[1][2]			Nutrient Allowances[1][2]			Comments
	Per Horse Daily	In Ration A-F	Per Ton Ration A-F	Per Horse Daily	In Ration A-F	Per Ton Ration A-F	
	(IU)	(IU/lb)	(IU/ton)	(IU)	(IU/lb)	(IU/ton)	
Maintenance: 1,000-lb *(454-kg)* horse	22,725	909	1,818,000	26,134	1,045	2,090,700	**V**itamin A is not synthesized in the cecum. **H**ay over 1 year old, regardless of green color, is usually not an adequate source of carotene or vitamin A activity. **W**hen deficiency symptoms appear, add stabilized vitamin A to the ration. **I**t is wasteful to feed more vitamin A than is needed. Also exceedingly high levels over an extended period of time may cause bone fragility, hyperostosis, and exfoliated epithelium.
Gestation/Lactation: 1,000-lb *(454-kg)* mare	34,100	1,364	2,728,000	39,215	1,569	3,137,200	
Growth: 450-lb *(204-kg)* weanling	10,908	909	1,818,000	12,544	1,045	2,090,700	
Working: 1,000-lb *(454-kg)* horse	22,725	909	1,818,000	26,134	1,045	2,090,700	
Maintenance: 1,000-lb *(454-kg)* horse	3,400	136	272,000	3,910	156	312,800	**T**he vitamin D requirement is less when a proper balance of calcium and phosphorus exists in the ration. **W**hen animals are exposed to direct sunlight, the ultraviolet light produces vitamin D from traces of cholesterol in the skin. Stabled horses, exercised in the early morning, will not get sufficient vitamin D in this manner. **T**oo much vitamin D may harm a horse. Vitamin D toxicity is characterized by calcification of the blood vessels, heart, and other soft tissues, and by bone abnormalities. Toxic level of vitamin D has not been established in the horse, but a level 50 times the requirement may be harmful.
Gestation/Lactation: 1,000-lb *(454-kg)* mare	6,825	273	546,000	7,849	314	627,900	
Growth: 450-lb *(204.5-kg)* weanling	4,368	364	728,000	5,023	419	837,200	
Working: 1,000-lb *(454-kg)* horse	3,400	136	272,000	3,910	156	312,800	
Maintenance: 1,000-lb *(454-kg)* horse	575	23	46,000	661	26	52,900	**U**tilization of vitamin E is dependent on adequate selenium.
Gestation/Lactation: 1,000-lb *(454-kg)* mare	900	36	72,000	1,035	41	82,800	
Growth: 450-lb *(204.5-kg)* weanling	432	36	72,000	497	41	82,800	
Working: 1,000-lb *(454-kg)* horse	900	36	72,000	1,035	41	82,800	
	(mg)	(mg/lb)	(mg/ton)	(mg)	(mg/lb)	(mg/ton)	
Maintenance: 1,000-lb *(454-kg)* horse				8.0	0.32	640	**H**igh levels of vitamin K will overcome bleeding due to dicoumarol. **V**itamin K is generally (1) widely distributed in normal feeds, and/or (2) synthesized in adequate amounts by the intestinal microflora of the horse.
Gestation/Lactation: 1,000-lb *(454-kg)* mare				8.0	0.32	640	
Growth: 450-lb *(204.5-kg)* weanling				3.6	0.30	600	
Working: 1,000-lb *(454-kg)* horse				8.0	0.32	640	
Maintenance: 1,000-lb *(454-kg)* horse				2.5	0.1	200	**B**iotin is closely related metabolically to folacin, pantothenic acid, and vitamin B-12.
Gestation/Lactation: 1,000-lb *(454 kg)* mare				2.5	0.1	200	
Growth: 450-lb *(204.5-kg)* weanling				1.2	0.1	200	
Working: 1,000-lb *(454-kg)* horse				2.5	0.1	200	
Maintenance: 1,000-lb *(454-kg)* horse				500	20.0	40,000	**C**holine content of normal feeds is usually sufficient.
Gestation/Lactation: 1,000-lb *(454-kg)* mare				750	30.0	60,000	
Growth: 450-lb *(204.5-kg)* weanling				750	62.5	125,000	
Working: 1,000-lb *(454-kg)* horse				750	30.0	60,000	
Maintenance: 1,000-lb *(454-kg)* horse				20	0.8	1,600	**F**olacin is widely distributed in horse feeds. **A**lso, folacin is synthesized in the lower gut.
Gestation/Lactation: 1,000-lb *(454-kg)* mare				30	1.2	2,400	
Growth: 450-lb *(204.5-kg)* weanling				36	3.0	6,000	
Working: 1,000-lb *(454-kg)* horse				30	1.2	2,400	

(Continued)

TABLE 25-9

Vitamins Which May Be Deficient Under Normal Conditions	Conditions Usually Prevailing Where Deficiencies Are Reported	Function of Vitamin	Some Deficiency Symptoms	Practical Sources of the Vitamin
Niacin (Nicotinic Acid, Nicotinamide)		Constituent of two important coenzymes. They are involved in the release of energy from carbohydrates, fats, and proteins, and in the synthesis of fatty acids, protein, and DNA.	Reduced growth and appetite. Skin rashes, diarrhea, nerve disorders.	Green alfalfa. Niacin is widely distributed in feeds; fermentation solubles and certain oil meals are especially good sources. Synthetic niacin.
Pantothenic Acid (Vitamin B-3)		Part of coenzyme A, which plays a key role in body metabolism.	Poor growth, skin rashes, poor appetite, nervous disorders.	Safflower meal, blackstrap molasses, wheat bran, and milk. Calcium pantothenate.
Riboflavin (Vitamin B-2)	When green feeds (pasture, hay, or silage) are not available.	Riboflavin has an essential role in the oxidative mechanisms of the cells.	Periodic ophthalmia (or moon blindness), characterized by catarrhal conjunctivitis in one or both eyes, accompanied by photophobia, and lachrymation. Decreased rate of growth and feed efficiency. Porous and weak bones; ligaments and joints impaired.	Green pasture. Green hay. Milk and milk products. Synthetic riboflavin. Yeast.
Thiamin (Vitamin B-1)	Poor-quality hay and grain. When sulfa drugs or antibiotics are given to the horse, the synthesis of B vitamins is impaired. Consumption of bracken fern (*Pteris aquilina*) and horsetail (*Equisetum* spp) will cause thiamin deficiency due to the antithiamin compounds that they contain.	In energy metabolism. Without thiamin, there would be no energy. In the working of the peripheral nerves. Promotes appetite and growth.	A thiamin deficiency has been produced experimentally. Decreased feed consumption (loss of weight), anemia, incoordination (especially in the hindquarters), lowered blood thiamin, elevated blood pyruvic acid, enlarged heart, and nervous symptoms.	Wheat and wheat by-products. Oilseed meals. Oat grain and groats. Thiamin hydrochloride. Yeast (brewers', torula).
Vitamin B-6 (Pyridoxine, Pyridoxal, Pyridoxamine)		In its coenzyme forms, it is involved in a large number of physiologic functions, particularly protein, carbohydrate, and fat metabolism.	No deficiency symptoms of vitamin B-6 have been reported in the horse. So, it is thought to be synthesized in the cecum.	Green pasture, alfalfa hay, wheat bran, wheat germ, and yeast (brewers', torula).
Vitamin B-12 (Cobalamins)	When few, or no feeds of animal origin are fed. Where cobalt is not present in the feed, thereby precluding the synthesis of vitamin B-12 in the gastrointestinal tract.	Coenzyme in several enzyme systems. Closely linked with choline, folacin, and pantothenic acid.	Loss of appetite and poor growth.	Protein supplements of animal origin. Fermentation products. Cobalamins, yeast.
Vitamin C (Ascorbic Acid, Dehydroascorbic Acid)	The vitamin C requirements of fish and humans have been observed to increase in periods of stress. So, it is conjectured that heavily stressed horses may require more vitamin C than they can synthesize.	Formation and maintenance of collagen. More rapid healing of wounds. Sound bones.	No deficiency symptoms in horses noted. In humans and monkeys, scurvy is the main deficiency symptom. Also, in humans sudden death from severe internal hemorrhage and heart failure are always a danger.	Ordinary rations and body synthesis provide adequate vitamin C for horses. Well-cured hays and green pastures are good sources of vitamin C.
Unidentified factors	Since the U.S. foal crop is only around 50%, it is obvious that there is room for improvement somewhere along the line; and perhaps unidentified factors are involved. Also, optimal results with horses during the critical periods (growth, gestation-lactation, and when under stress as in racing or showing) appear to be dependent upon providing unidentified factors through such ingredients as distillers' dried solubles, dehydrated alfalfa meal, condensed fish solubles, brewers' dried yeast, antibiotic fermentation residues, dried whey, and corn fermentation solubles.			

[1]As used herein, the distinction between "nutrient requirements" and "nutrient allowances" is as follows: In nutrient requirements, no margins of safety are included intentionally; whereas in nutrient allowances, margins of safety are provided in order to compensate for variations in feed composition, environment, and possible losses during storage or processing. The "nutrient requirements" in Table 25-9 were adapted by the authors from *Nutrient Requirements of Horses,* 5th rev.ed., NRC-National Academy of Sciences, 1989. The "nutrient allowances" were developed by the authors, based on experiments and experiences; it is intended that they meet the nutrient requirements, and provide adequate margins of safety in addition.

(Continued)

Classes/Function	Nutrient Requirements[1,2]			Nutrient Allowances[1,2]			Comments
	Per Horse Daily	In Ration A-F	Per Ton Ration A-F	Per Horse Daily	In Ration A-F	Per Ton Ration A-F	
	(mg)	(mg/lb)	(mg/ton)	(mg)	(mg/lb)	(mg/ton)	
Maintenance: 1,000-lb *(454-kg)* horse				250	10.0	20,000	There is some evidence that niacin is synthesized by the horse.
Gestation/Lactation: 1,000-lb *(454-kg)* mare				250	10.0	20,000	The horse can convert the essential amino acid tryptophan into niacin. Hence, it is important to make certain that the ration is adequate in niacin; otherwise, the horse will use tryptophan to supply niacin needs.
Growth: 450-lb *(204.5-kg)* weanling				250	20.8	41,600	
Working: 1,000-lb *(454-kg)* horse				250	10.0	20,000	
Maintenance: 1,000-lb *(454-kg)* horse				250	10.0	20,000	Grain is very deficient in pantothenic acid. Of all the B vitamins, pantothenic acid is most likely to be deficient under stable (confinement) conditions.
Gestation/Lactation: 1,000-lb *(454-kg)* mare				250	10.0	20,000	
Growth: 450-lb *(204.5-kg)* weanling				250	20.8	41,600	
Working: 1,000-lb *(454-kg)* horse				250	10.0	20,000	
Maintenance: 1,000-lb *(454-kg)* horse	22.8	0.91	1,820	40.0	1.6	3,200	Lack of vitamin B-2 is not the only cause of moon blindness. Sometimes, moon blindness follows leptospirosis, and it may be caused by an allergic reaction.
Gestation/Lactation: 1,000-lb *(454-kg)* mare	22.8	0.91	1,820	40.0	1.6	3,200	
Growth: 450-lb *(204.5-kg)* weanling	10.9	0.91	1,820	19.2	1.6	3,200	
Working: 1,000-lb *(454-kg)* horse	22.8	0.91	1,820	40.0	1.6	3,200	
Maintenance: 1,000-lb *(454-kg)* horse	34.1	1.36	2,720	39.2	1.57	3,140	Thiamin is synthesized in the lower gut of the horse by bacterial action, but there is some doubt as to its sufficiency. When neither green pasture nor high-quality roughage is available, thiamin hydrochloride should be added to the ration. Since carbohydrate metabolism is increased during physical exertion, it is important that B-1 be available in quantity at such times.
Gestation/Lactation: 1,000-lb *(454-kg)* mare	34.1	1.36	2,720	39.2	1.57	3,140	
Growth: 450-lb *(204.5-kg)* weanling	16.4	1.36	2,720	18.9	1.57	3,140	
Working: 1,000-lb *(454-kg)* horse	56.8	2.27	4,540	65.3	2.61	5,220	
Maintenance: 1,000-lb *(454-kg)* horse				25.0	1.0	2,000	Normally, horse rations contain adequate vitamin B-6. Also, it appears to be synthesized in the cecum. Yet, these sources may not be adequate for the maximum performance of the horse.
Gestation/Lactation: 1,000-lb *(454-kg)* mare				25.0	1.0	2,000	
Growth: 450-lb *(204.5-kg)* weanling				6.0	0.5	1,000	
Working: 1,000-lb *(454-kg)* horse				25.0	1.0	2,000	
Maintenance: 1,000-lb *(454-kg)* horse				0.125	0.005	10	It is reported that horses in poor nutritional condition showing anemia respond to the administration of vitamin B-12.
Gestation/Lactation: 1,000-lb *(454-kg)* mare				0.150	0.006	12	
Growth: 450-lb *(204.5-kg)* weanling				0.084	0.007	14	
Working: 1,000-lb *(454-kg)* horse				0.150	0.006	12	
Maintenance: 1,000-lb *(454-kg)* horse				60	2.4	4,800	Dietary need is clearly evident for humans, monkeys, guinea pigs, fruit-eating bats, and bulbul birds. However, vitamin C is probably required by other species, but synthesized in the body; the only question is whether the horse can synthesize enough vitamin C when under stress.
Gestation/Lactation: 1,000-lb *(454-kg)* mare				100	4.0	8,000	
Growth: 450-lb *(204.5-kg)* weanling				45	3.75	7,500	
Working: 1,000-lb *(454-kg)* horse				100	4.0	8,000	

[2] Feed consumption of mature 1,000-lb *(454-kg)* horse estimated at 25 lb *(11.36 kg)* per day. Feed consumption of 450-lb *(204.5-kg)* weanling estimated at 12 lb *(5.45 kg)* per day.

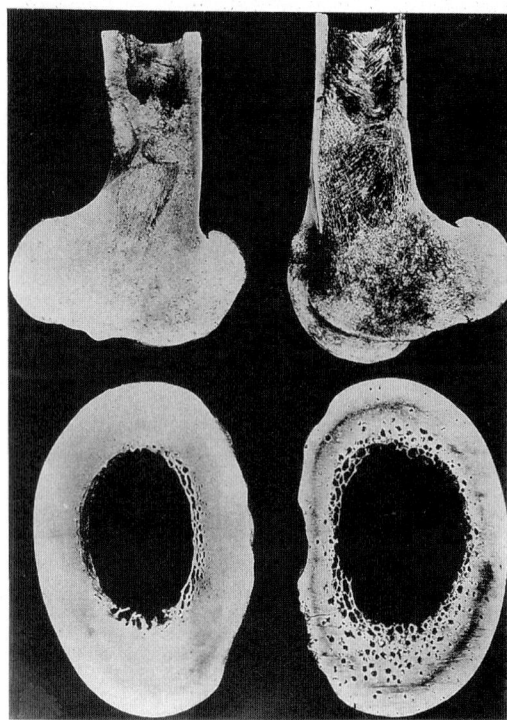

Fig. 25-7. (Left) Vitamin A made the difference! *Upper:* On the right is shown the sagittal section of the distal end of the femur of a vitamin-deficient horse compared to normal bone (left). *Lower:* On the right is shown the cross section of the cannon bone from a vitamin-deficient horse compared to normal bone (left). (Courtesy, Calif. Ag. Exp. Sta.)

VITAMIN IMBALANCES

Experiments have shown that the amounts needed of certain vitamins may be affected by the supply of another vitamin or of some other nutritive essential. Also, it is known that excess fortification of the horse's ration with certain vitamins may prove more detrimental than helpful. Thus, caretakers should avoid harmful imbalances; they should provide vitamins on the basis of recommended allowances. Also, when fortifying with vitamins, consideration should be given to the vitamins provided by the ingredients of the normal ration, for it is the total composition of the feed that counts.

UNIDENTIFIED FACTORS

Unidentified factors include those vitamins which the chemist has not yet isolated and identified. For this reason, they are sometimes referred to as the vitamins of the future. There is mounting evidence of the importance of unidentified factors for animals, including humans. Among other things, they lower the incidence of ulcers in humans and swine. For horses, they appear to increase growth and improve feed efficiency and breeding performance when added to rations thought to be complete with regard to known nutrients. The anatomical and physiological mechanism of the digestive system of the horse, plus the stresses and strains to which modern horses are subjected, would indicate the wisdom of adding unidentified factor sources to the ration of the horse. Unidentified factors appear to be of special importance during breeding, gestation, lactation, and growth.

Three highly regarded unidentified factor sources are dried whey product, corn fermentation solubles, and dehydrated alfalfa meal.

Water

Horses can survive for a longer period without feed than they can without water. The loss of 10% body water will result in disorders; the loss of 20% body water will cause death.

Water is essential for the various physiological processes of the horse, such as the production of saliva and sweating.

Normally, horses will drink about 1 gal of water per 100-lb body weight per day; so, on an average day, a 1,000-lb horse will drink about 10 gal of water. But, in addition to size, the amount of water consumed depends on the following:

1. **Age.** The younger the horse, the more water it will drink per unit of body weight. This is because the bodies of younger animals contain more water than older animals.

2. **Temperature and humidity.** A rise in the environmental temperature from 55° to 75°F will increase water requirements by 15 to 20%. Also, humidity is a factor.

3. **Work.** Depending on the kind and severity, work may increase water consumption by 20 to 300%.

4. **Ration.** Water consumption increases with the amount of feed consumed daily, the dry matter of the ration, and the kind of feed—it's higher on a high-fiber hay ration than on a high-grain ration. Impaction colic may occur with limited water intake and a dry, high-fiber ration.

5. **Lactation.** Lactation may increase water consumption by 50 to 100% above maintenance.

Fig. 25-8. Automatic waterer of good design positioned in a stall. (Courtesy, *Equus Magazine,* Gaithersburg, Md.)

Antibiotics

Certain antibiotics, at stipulated levels, are approved by the FDA for growth promotion and for the improvement of feed efficiency of young equines up to 1 year of age. Unless there is a disease, however, there is no evidence to warrant the continuous feeding of antibiotics to mature horses. Such practice may even be harmful. Hence, when antibiotics are needed for therapeutic purposes, it is best to seek the advice of a veterinarian.

It appears that antibiotics may be especially helpful for young foals which suffer setbacks from infections, digestive disturbances, inclement weather, and other stress factors. Also, horses may benefit from antibiotics (1) when being transported from one location to another—for example, when being moved to a new show or track; (2) when there is a low disease level in the herd; or (3) when mares are foaling.

The poorer the sanitation and management, the greater the response from antibiotics. It follows, therefore, that there is a temptation to use antibiotics as a *crutch,* rather than improve the regimen.

When added to feed, the level of antibiotics should be in keeping with the directions of the manufacturer and with the Food and Drug Administration regulations.

(Also see Chapter 13, section on "Antibiotics.")

FEEDS FOR HORSES

Individual feeds vary widely in feeding value. Oats and barley, for example, differ in feeding value according to the hull content and weight per bushel, and forages vary according to the stage of maturity at which they are cut and how well they are cured and stored. Also, the feeding value of certain feeds is materially affected by preparation.

Regardless of the feeds selected, they should be of sound quality, and not moldy, spoiled, or dusty. This applies to both hay and grain. The careful selection of feeds is more important for horses than for any other class of livestock.

More than one kind of hay makes for appetite appeal. In season, any good pasture can replace part or all of the hay unless work or training conditions make substitution impractical.

Good-quality oats and timothy hay always have been considered standard feeds for light horses. However, feeds of similar nutritive properties can be interchanged in the ration as price relationships warrant; among them, the grains —oats, corn, barley, wheat, and sorghum; the protein supplements—linseed meal, soybean meal, cottonseed meal, and canola meal; and hays of many varieties. Feed substitution makes it possible to obtain a balanced ration at lowest cost.

During the winter months, it is well to add a few sliced carrots to the ration, an occasional bran mash, or a small amount of linseed meal. Also, a bran mash or linseed meal may be used to regulate the bowels.

The proportion of concentrates must be increased and the roughages decreased as energy needs rise with the greater amount, severity, or speed of work. A horse that works at a trot needs considerably more feed than one that works at a walk. For this reason, riding horses in medium to light use require somewhat less grain and more hay in proportion to body weight than light horses that are racing. Also, from an esthetic standpoint, large, paunchy stomachs are objectionable on horses that are used for recreation and sport.

In addition to making for a nutritionally complete ration, the following factors should be considered when choosing horse feeds: cost, palatability, preparation, variety, bulk, and laxativeness.

Forages

Forages may be classified as (1) pasture, (2) hay, or (3) silage.

• **Pastures**—In season, there is no finer forage for horses than superior pastures—pastures that are much more than gymnasiums. This is especially true of idle horses, broodmares, and young stock. In fact, pastures have a very definite place for all horses, with the possible exception of animals at heavy work or in training. Even with the latter groups, pastures may be used with discretion. Horses in use may be turned to pasture at nights or over the weekend. Certainly, the total benefits derived from pasture are to the good, although pasturing may have some laxative effects and produce a greater tendency to sweat.

• **Hays**—Under most conditions, the roughage requirement of horses ranges from 0.5 to 1.0% of body weight, or from 5 to 10 lb of roughage daily for a 1,000-lb horse.

Usually, young horses and idle horses can be provided with an unlimited allowance of hay. In fact, much good will result from feeding young and idle horses more roughage and less grain. But one should gradually increase the grain and decrease the hay as work or training begins.

Racehorses should receive a minimum of roughage, since they need a maximum of energy. When limiting the allowance of roughage, it is sometimes necessary to muzzle greedy horses (gluttons) to prevent them from eating the bedding.

Hay native to the locality is usually fed. However, horseowners everywhere prefer good-quality timothy. With young stock and breeding animals especially, it is desirable that a sweet grass-legume mixture of alfalfa or clover hay be fed. The legume provides a source of high-quality proteins and certain minerals and vitamins.

(Also see Chapter 8, Hay.)

• **Silages**—Well-preserved silage of good quality, free from mold and not frozen, affords a highly nutritious succulent forage for horses during the winter months. *Because horses are more susceptible than cattle or sheep to botulism or other digestive disturbances resulting from the feeding of poor quality silage, nothing but choice, fresh silage, free from mold or spoilage, should ever be fed.*

Various types of silages may be fed successfully to horses, but corn silage and grass-legume silage are most common. If the silage contains much grain, the concentrate allowance should be reduced accordingly.

Concentrates

Horses cannot handle as large quantities of roughages as ruminants. When used for heavy work, for pleasure, or for racing, they must be even more restricted in their roughage allowance and should receive a higher proportion of concentrates.

Because of less bulk and lower shipping and handling costs, the concentrates used for horse feeding are less likely to be locally grown than the roughages. Even so, the vast majority of grains fed to horses are homegrown, thus varying from area to area, according to the grain crops best adapted.

Of all the concentrates, heavy oats most nearly meet the needs of horses; and, because of the uniformly good results obtained from their use, they have always been recognized as the preferred grain for horses. Corn is also widely used as a horse feed, particularly in the central states. Despite occasional prejudice to the contrary, barley is a good horse feed. As proof of the latter assertion, it is noteworthy that the Arab—who was a good horseman—fed barley almost exclusively. Also, wheat, wheat bran, molasses, and commercial-mixed feeds are extensively used. It is to be emphasized, therefore, that careful attention should be given to the prevailing price of feeds available locally, for many feeds are well suited to horses. Often substitutions can be made that will result in a marked saving without affecting the nutritive value of the ration. So, the primary consideration in selecting the cereal grain(s) is the cost per unit of energy.

(Also see Chapter 10, Grains/High-Energy Feeds.)

• **Bran Mash**— Feeding a bran mash is the traditional way of regulating the bowels of horses on idle days and at such other times as required.

The mash is prepared by filling a 2- to 2½-gal bucket with wheat bran, pouring enough boiling hot water over it to make it the consistency of breakfast oatmeal, covering the bucket with a blanket and allowing it to steam until cool, then feeding it to the horse.

Protein Supplements

The extent to which the horse's ration is supplemented with proteins depends primarily on the age of the horse and on the quality of the forage fed. Growing or lactating animals require somewhat more protein than horses that are idle, gestating, or working. Also, grass hays and farm grains are generally low in quality and quantity of proteins and require more supplementation than legumes.

In practical horse feeding, foals should be provided with some protein feeds of animal origin in order to supplement the proteins found in grains and forages. In feeding mature horses, a safe plan to follow is to provide plant protein from several sources.

The following oil meals are most commonly used as protein supplements for horses: linseed meal, soybean meal, cottonseed meal, sunflower meal, and rapeseed meal (canola meal).

Because of its laxative nature, linseed meal in limited quantities is a valuable addition to the ration of horses. Also, if prepared by the old process, it imparts a desirable *bloom* to the hair of show and sale animals.

(Also see Chapter 11, Protein Supplements.)

Special Feeds

Special feeds may be needed from time to time for imparting bloom or gloss to the hair, or for promoting growth of young stock.

• **Bloom-Imparting Feeds**—Bloom or gloss is important in horses. But sometimes they lack this desired quality—their hair is dull and dry. Feeding a well-balanced ration will usually rectify this situation. Also, feeding either of the following products will make for an attractive, shiny coat:

1. **Corn oil.** Feed at the rate of 2 oz (4 Tbsp) per horse per day.

2. **Whole flaxseed soaked.** Put a handful of whole flaxseed in a teacup, cover it with water, let it stand overnight, then pour it over the morning feed. Repeat twice each week.

Unless the horse is afflicted with lice, mange, or some other ailment, either of the above treatments will impart bloom or gloss to the coat.

• **Milk By-Products**—The superior nutritive values of milk by-products are due to their high-quality proteins, vitamins, a good mineral balance, and the beneficial effect of the milk sugar, lactose. In addition, these products are palatable and highly digestible. They are an ideal feed for young equines and for balancing the deficiencies of the cereal grains. Most foal rations contain one or more milk by-products, primarily dried skim milk, with some dried whey and dried buttermilk included at times. The chief limitation to their wider use is price.

• **Milk Replacer**—As indicated by the name, *a milk replacer is a replacement for milk.*

Foals suckling their dams generally develop very satisfactorily up to weaning time. But the most critical period in the entire life of a horse is that interval from weaning time (about 6 months of age, or earlier) until 1 year of age. This is especially so in the case of young horses being fitted for shows or sales, where condition is so important. Thus, where valuable weanlings or yearlings are to be shown or sold, the use of a milk replacer may be practical.

(Also see Chapter 6, Types and Roles of Feedstuffs, section on "Milk Replacers.")

Treats

Horses are fed a great variety of treats. On a government horse breeding establishment in Brazil, the senior author saw a large, well-manicured vegetable garden growing everything from carrots to melons, just for horses. Also, trainers recognize that most racehorses, which are the *prima donnas* of the equine world, don't "eat like a horse"; they eat like people—and sometimes they're just as finicky. Their menus may include a choice of carrots or other roots, fruit, pumpkins, squashes, or melons, sugar or honey, and innumerable other goodies.

Ask caretakers why they feed treats to their horses and you'll get a variety of answers. However, high on the list of reasons will be (1) appetizers; (2) a source of nutrients and as conditioners; (3) rewards; (4) alleviating obesity (dieting the horse); or (5) folklore.

Palatability

Palatability is important, for horses must eat their feed if it's to do them any good. But many horses are finicky simply because they're spoiled. For the latter, stepping up the exercise and halving the ration will usually effect a miraculous cure.

Also, it seems possible that well-liked feeds are digested somewhat better than those which are equally nutritious, but less palatable.

Palatability is particularly important when feeding horses that are being used hard, as in racing or showing. Unless the ration is consumed, such horses will not obtain sufficient nutrients to permit maximum performance. For this reason, lower quality feeds, such as straw or stemmy hay, should be fed to mature, idle horses.

Familiarity and habit are important factors concerned with the palatability of horse feeds. For example, horses have to learn to eat pellets, and very frequently they will back away from feeds with new and unfamiliar odors. For this reason, any change in feeds should be made gradually.

Occasionally, the failure of horses to eat a normal amount of feed is due to a serious nutritive deficiency. For example, if horses are fed a ration made up of palatable feeds, but deficient in one or more required vitamins or minerals, they may eat normal amounts for a time. Then when the body reserves of the lacking nutrient(s) are exhausted, they will usually consume much less feed, due to an impairment of their health and a consequent lack of appetite. If the deficiency is not continued until the horses are injured permanently, they will usually recover their appetites if some feed is added which supplies the nutritive lack and makes the ration complete.

FEED PREPARATION

The physical preparation of cereal grains for horses has been practiced by caretakers for a very long time. Basically, grain is either soaked, cooked, ground, or rolled (wet or dry), and hay is either fed long, pelleted, or cubed.

For horses, flaking or extruding are the preferred methods of grain preparation; they make for light rations and fewer digestive disturbances. For animals with good teeth, the value of oats is increased only 5% by processing.

Hay for horses is usually fed long or incorporated in an all-pelleted ration (with the grain and hay combined).

Further elucidation of feed preparation for horses is contained in Chapter 14, Feed Processing.

FEED ALLOWANCES AND SUGGESTED RATIONS

Proper feeding of horses calls for giving them (1) the right allowance, and (2) balanced rations. The ration may be either a home-mixed feed or a commercial feed.

In formulating rations, the following points are pertinent:

1. The effects of the rations on horses should be observed. Calculations alone will not ensure success.
2. Rations should be palatable, economical, and practical. They need not be complicated mixtures.
3. The roughage should be selected first. Then, the concentrates (grains) may be selected which, when combined with the roughage, will provide a balanced ration.

Feed Allowances

When given all the feed that they will consume, mature horses will generally eat an amount equivalent to about 2.5% of their body weight, daily. Growing foals and lactating mares eat more heartily—they will consume up to 3% of their body weight.

Because the horse has a limited digestive capacity, the amount of concentrates must be increased and the roughages decreased when the energy needs rise with more work. The following general guides may be used for the daily ration of horses under usual conditions.

• **Horses at light work (1 to 3 hours per day of riding or driving)**—Allow ⅖ to ½ lb of grain and 1¼ to 1½ lb of hay per day per 100 lb of body weight.

• **Horses at medium work (3 to 5 hours per day of riding or driving)**—Allow about ¾ lb of grain and 1 to 1¼ lb of hay per 100 lb of body weight.

• **Horses at hard work (5 to 8 hours per day of riding or driving)**—Allow about 1¼ to 1⅓ lb of grain and 1 to 1¼ lb of hay per 100 lb of body weight.

As will be noted from these recommendations, the total allowance of both concentrates and hay should be about 2 to 2½ lb daily per 100 lb of body weight.

The recommended feed allowances on the basis of animal weight are equally applicable to equines of all sizes, including ponies and donkeys; simply vary as necessary according to the work performed and the individuality of the animal.

About 6 to 12 lb of grain daily is an average grain ration for a light horse at medium or light work. Racehorses in training usually consume 10 to 16 lb of grain per day; the exact amount varies with the individual requirements and the amount of work. The hay allowance averages about 1 to 1¼ lb daily per 100 lb of body weight, but it is restricted as the grain allowance is increased. Light feeders should not be overworked.

The quantities of feeds recommended are intended as guides only. The allowance, especially of the concentrates, should be increased when the horse is too thin and decreased when the horse is too fat.

Fig. 25–9. Racing is "hard work." This shows Albatross, speedy harness racehorse, with Stanley Dancer in the sulky. (Courtesy, *The Harness Horse*, Harrisburg, Penn.)

Suggested Rations

A ration is the amount of feed given to a horse in a day, or a 24-hour period. To most caretakers, however, the word implies the feeds fed to an animal without limitation of the time in which they are consumed.

Several suggested rations are given in Table 25-10.

TABLE 25-10
RATIONS FOR HORSES[1] (AS-FED BASIS)

Age, Sex, and Use	Daily Allowance	Kind of Hay (More than one kind of hay makes for variety and appetite appeal. In season, any good pasture can replace part or all of the hay except for horses at work or in training.)	Suggested Grain Rations (With all rations, and for all classes and ages of horses, provide suitable minerals; see Table 25-8, Horse Mineral Chart, for recommendations.)		
			Ration No. 1	Ration No. 2	Ration No. 3
Mature idle horses; Stallions, mares, and geldings (weighing 900–1,400 lb; or *409–636 kg*)	1½–1¾ lb *(0.7–0.8 kg)* of hay per 100 lb *(45 kg)* body weight.	Pasture in season; or grass-legume mixed hay.	(With grass hay, add ¾ lb *(0.34 kg)* daily of a high-protein supplement.)		
Light horses at work, in riding, driving, and racing (weighing 900–1,400 lb; or *409–636 kg*)	**Light use**—⅖–½ lb *(0.18–0.2 kg)* grain and 1¼–1½ lb *(0.6–0.7 kg)* hay per 100 lb *(45 kg)* body weight. **Medium use**—¾–1 lb *(0.3–0.5 kg)* grain and 1–1¼ lb *(0.5–0.6 kg)* hay per 100 lb *(45 kg)* body weight. **Hard use**—1¼–1⅓ lb *(0.5–0.6 kg)* grain and 1–1¼ lb *(0.5–0.57 kg)* hay per 100 lb *(45 kg)* body weight.	Grass hay.	(lb) (kg) Oats 100 45	(lb) (kg) Oats .... 70 32 Corn 30 14	(lb) (kg) Oats .... 70 32 Barley 30 14
Stallions in breeding season (weighing 900–1,400 lb; or *409–636 kg*)	¾–1½ lb *(0.3–0.7 kg)* grain per 100 lb *(45 kg)* body weight, together with a quantity of hay within same range.	Grass-legume mixed (or ⅓–½ legume hay, with balance grass hay).	Oats .... 55 25 Wheat .... 20 9 Wheat bran .. 20 9 Linseed meal . 5 2	Corn .... 35 16 Oats .... 35 16 Wheat .... 15 7 Wheat bran .. 15 7	Oats 100 45
Pregnant/lactating mares (weighing 900–1,400 lb; or *409–636 kg*)	¾–1½ lb *(0.3–0.7 kg)* grain per 100 lb *(45 kg)* body weight, together with a quantity of hay within same range.	Grass-legume mixed; or ⅓–½ legume hay, with balance grass hay (straight grass hay may be used first half of pregnancy).	Oats .... 80 36 Wheat bran .. 20 9	Barley ...... 45 20 Oats .... 45 20 Wheat bran .. 10 5	Oats .... 95 43 Linseed meal . 5 2
Foals before weaning (weighing 100–350 lb, or *45–159 kg*; with projected mature weights of 900–1,400 lb; or *409–636 kg*)	½–¾ lb *(0.2–0.3 kg)* grain per 100 lb *(45 kg)* body weight, together with a quantity of hay within same range.	Legume hay.	Oats .... 50 23 Wheat bran .. 40 18 Linseed meal . 10 5	Oats .... 30 14 Barley ...... 30 14 Wheat bran .. 30 14 Linseed meal . 10 5	Oats .... 80 36 Wheat bran .. 20 9
		(Rations balanced basis of following assumptions: Mares of mature weights of 600, 800, 1,000, and 1,200 lb *[or 273, 364, 455, and 545 kg]* may produce 36, 42, 44, and 49 lb *[or 16, 19, 20, and 22 kg]* of milk daily.)			
Weanlings (weighing 350–450 lb; or *159–204 kg*)	1–1½ lb *(0.5–0.7 kg)* grain and 1½–2 lb *(0.7–0.9 kg)* hay per 100 lb *(45 kg)* body weight.	Grass-legume mixed (or ½ legume hay, with balance grass hay).	Oats .... 30 14 Barley ...... 30 14 Wheat bran .. 30 14 Linseed meal . 10 5	Oats .... 70 32 Wheat bran .. 15 7 Linseed meal . 15 7	Oats .... 80 36 Linseed meal . 20 9
Yearlings: 2nd summer (weighing 450–700 lb; or *204–317 kg*)	Good luxuriant pastures. (If in training or for other reasons without access to pastures, the ration should be intermediate between the adjacent upper and lower groups.)				
Yearling or rising 2-year-old; 2nd winter (weighing 700–1,000 lb; or *317–454 kg*)	½–1 lb *(0.2–0.5 kg)* of grain and 1–1½ lb *(0.5–0.7 kg)* hay per 100 lb *(45 kg)* body weight.	Grass-legume mixed; or ⅓ to ½ legume hay, with remainder grass hay.	Oats .... 80 36 Wheat bran ... 20 9	Barley ...... 35 16 Oats .... 35 16 Wheat bran .. 15 7 Linseed meal . 15 7	Oats 100 45

[1]Mineral recommendations for all classes and ages of horses, especially those fed unmixed rations or on pasture:

  a. *When animals are on liberal grain feeding*—Provide free access to a 2-compartment mineral box, with (1) trace mineralized salt in one side; and (2) in the other side, a mixture of ⅓ trace mineralized salt (salt included for purposes of palatability), ⅓ defluorinated phosphate or steamed bone meal, and ⅓ ground limestone or oystershell flour.

  b. *When animals are primarily on roughage (pasture, hay, and/or silage)*—Provide free access to a 2-compartment mineral box, with (1) trace mineralized salt in one side (salt included for purposes of palatability); and (2) in the other side, a mixture of ⅓ trace mineralized salt and ⅔ defluorinated phosphate or steamed bone meal.

Overfeeding

Overfeeding may result in the following bad consequences: If done suddenly, it may cause founder (laminitis), colic, or enterotoxemia; if prolonged, it will likely result in obesity (too fat).

The main qualities desired in horses are trimness, action, spirit, and endurance. These qualities cannot be obtained in horses that are overfed and fat. The latter is especially true with horses used for racing, where the carrying of any surplus body weight must be avoided.

Home-Mixed Feeds

When selecting rations, one should compare the cost of home-mixed vs commercial feeds. If only small quantities are required or little storage space is available, it may be more satisfactory to buy ready-mixed feeds.

In general, horses may be given as much nonlegume roughage as they will eat. But they must be accustomed gradually to legumes because legumes may be laxative.

Commercial Horse Feeds

Commercial horse feeds are feeds mixed by manufacturers who specialize in the feed business.

Commercial feed manufacturers are able to purchase feed in quantity lots, making possible price advantages and the scientific control of quality. Many caretakers have found that because of the small quantities of feed usually involved, and the complexities of horse rations, they have more reason to rely on good commercial feeds than do owners of other classes of farm animals.

SWEET FEED

A sweet feed is a feed to which has been added one or more ingredients that are sweet. Most commonly, it is molasses (approximately 10%); although brown sugar (about 5%) is sometimes used, and occasionally honey.

The horse has a *sweet tooth*; hence, it's not easy to switch an animal from a sweet feed to what may be a more nutritious ration.

In addition to enhancing palatability, sweet feed controls dust and keeps the minerals and vitamins blended.

FEED SUBSTITUTION TABLE

Successful horse owners are keen students of values. They recognize that feeds of similar nutritive properties can and should be interchanged in the ration as price relationships warrant, thereby making it possible at all times to obtain a balanced ration at the lowest cost.

Table 25–11, Feed Substitution Table for Horses, is a summary of the comparative values of the most common U.S. horse feeds. In arriving at these values, chemical composi-

TABLE 25–11
FEED SUBSTITUTION TABLE FOR HORSES (AS-FED BASIS) (See footnotes at end of table.)

Feedstuffs	Relative Feeding Value (lb for lb) in Comparison with the Designated Base Feed (in bold italic) Which = 100	Maximum Percentage of Base Feed (or comparable feed or feeds) Which it Can Replace for Best Results	Remarks
GRAINS, BY-PRODUCT FEEDS, ROOTS AND TUBERS:[1] (Low and Medium Protein Feeds)			
Oats	*100*	*100*	The leading grain for horses. The feeding value of oats varies according to the hull content and test weight per bushel. Need not be ground.
Barley	110	100	Most caretakers feel that it is preferable to feed barley along with more bulky feeds; for example, 25% oats or 15% wheat bran. Crush for horses.
Beet pulp, dried	100	33⅓	Not palatable to horses.
Beet pulp, molasses, dried	100	33⅓	Not palatable to horses.
Brewers' dried grains	100	50	
Carrots	15–25	10	Horses are very fond of carrots.
Corn, No. 2	115	100	Ranks second to oats as a light horse feed.
Corn gluten feed (gluten feed)	100	50	It has a lower value than indicated when forage is of low-protein content.
Distillers' dried grains	90–100	25	
Distillers' dried solubles	90–100	25	
Hominy feed	115	100	
Molasses, beet	80–95	10	In hot, humid areas, molasses should be limited to 5%; otherwise, mold may develop unless an inhibitor is used. Cane molasses is slightly preferred to beet molasses.
Molasses, cane	80–95	10	In hot, humid areas, molasses should be limited to 5%; otherwise, mold may develop unless an inhibitor is used.
Peas, dried	100	40	
Rice (rough rice)	115	50	Grind for horses.
Rye	115	33⅓	Higher levels or abrupt changes to rye may cause digestive disturbances. Not palatable.
Sorghum, grain	110–115	85	All varieties have about the same feeding value. Crush for horses.
Wheat	115	50	Wheat should be mixed with a more bulky feed in order to prevent colic.
Wheat bran	100	20	Valuable for horses because of its bulky nature and laxative properties.
Wheat-mixed feed (mill run)	105	20	Excessive quantities will cause colic or other digestive upsets.

(Continued)

TABLE 25-11 (Continued)

Feedstuffs	Relative Feeding Value (lb for lb) in Comparison with the Designated Base Feed (in bold italic) Which = 100	Maximum Percentage of Base Feed (or comparable feed or feeds) Which it Can Replace for Best Results	Remarks
PROTEIN SUPPLEMENTS:			
Linseed meal (35%)	*100*	*100*	Linseed meal (old process) is the preferred protein supplement for horses. It is valued because of its laxative properties and because of the sleek hair coat which it imparts.
Brewers' dried grains	65–70	50	
Buttermilk, dried	100	100	May be used in place of dried skimmed milk for foals.
Copra meal (coconut meal); (21%)	90–100	50	
Corn gluten feed (gluten feed)	70	100	
Corn gluten meal (gluten meal)	100	50	Somewhat unpalatable to horses.
Cottonseed meal (41%)	100	100	Satisfactory if limited to amount necessary to balance ordinary rations.
Peanut meal (45%)	100	100	
Peas, dried	75	50	
Skimmed milk, dried	100	100	Especially valuable for young equines; for creep feeding until past weaning.
Soybean meal (41%)	100	100	
Soybeans	100	100	Soybeans should be limited to ⅓ of the concentrate ration.
Sunflower meal (41%)	100	33⅓	Sunflower meal should not constitute more than ⅓ of the protein supplement for palatability reasons.
Whey, dried	50	50	Whey may be laxative.
DRY FORAGES AND SILAGES:[2]			
Timothy hay	*100*	*100*	The preferred hay of caretakers.
Alfalfa hay, all analyses	133⅓	100	Good quality alfalfa is excellent for horses. Alfalfa may be ground and pelleted. It provides high-quality proteins, and certain minerals and vitamins. It is somewhat laxative. Contrary to some "old wives' tales," it will not damage the kidneys.
Barley hay	100	100	Lower value if not cut at the early dough stage.
Bromegrass hay	100	100	
Clover hay, crimson	125	100	Crimson clover hay has considerably lower value if not cut at an early stage.
Clover hay, red	125	100	Clover hay should be well cured and free from dust and mold.
Clover-timothy hay	110–115	100	Value of clover-timothy mixed hay depends on the proportion of clover present and the stage of maturity at which it is cut.
Corn fodder	100	50	Preferably fed along with a good legume hay. It is best to shred the fodder.
Corn silage	45–55	33⅓–50	
Corn stover	60	50	Preferably fed along with a good legume hay. It is best to shred the stover.
Cowpea hay	110	100	
Grass-legume mixed hay	110–115	100	
Grass-legume silage	45–50	33⅓–50	
Grass silage	40–45	33⅓–50	
Johnsongrass hay	90–95	100	
Lespedeza hay	115	100	
Oat hay	100	100	Lower value if not cut at the early dough stage.
Orchardgrass	100	100	Should be cut before maturity.
Prairie hay	100	100	
Reed canarygrass	90–95	100	
Sorghum fodder	100	50	Preferably fed along with a good legume hay. It is best to shred the fodder.
Sorghum silage	40–45	33⅓–50	
Sorghum stover	60	50	Preferably fed along with a good legume hay. It is best to shred the stover.
Soybean hay	110	100	
Sudangrass hay	90–95	100	
Vetch-oat hay	110–115	100	The higher the proportion of vetch, the higher the value.
Wheat hay	100	100	

[1] Roots and tubers are of lower value than the grain and by-product feeds due to their higher moisture content.

[2] Well preserved silage of good quality, free from mold and not frozen, affords a highly nutritious succulent forage for horses during the winter months—especially for idle horses, broodmares, and growing colts. Silages are of lower value than dry forages due to their higher moisture content.

tion, feeding value, and palatability have been considered. The comparative values shown are not absolute. Rather, they are reasonably accurate approximations based on average quality feeds.

FEEDING PLEASURE HORSES

It is difficult to feed pleasure horses because they are used irregularly. Also, most pleasure horses are worked lightly, perhaps 1 to 3 hr of riding per day. Others are worked medium hard, as when ridden 3 to 5 hr per day. Still others are worked very hard, as when raced or when ridden 5 to 8 hr per day. The recommended daily feed allowance per 100 lb body weight of pleasure horses in light, medium, and hard use follows:

	Lb Daily/100 Lb Weight of Horse		
	Light Use	Medium Use	Hard Use
Hay	1¼–1½	1–1¼	1–1¼
Grain	⅖–½	¾–1	1¼–1⅓

As shown, the roughage content of the ration decreases and the concentrate content increases as the amount of work increases. This is because the digestibility and the efficiency of conversion are greater for high-energy concentrates than for roughages.

In season, pasture may replace hay, all or in part, according to the quality of the pasture. But the concentrate allowance of the working horse should remain about the same on pasture as in the stable or dry corral. There is a tendency of the pastured working horse to sweat and tire more easily (be *soft*), probably due to the high water content of green forage.

In addition to forage and grain, pleasure horses should be provided with suitable minerals and vitamins.

FEEDING HORSES IN TRAINING (EQUINE ATHLETES)

Fig. 25-10. Cutting horse in action. Horses in heavy training have a higher nutritional requirement than most pleasure horses. (Courtesy, Vim Mathews, Orland, Calif.)

Strenuous exercise of equine athletes in heavy training—such as training for racing, endurance trials, cutting, roping, jumping, or hunting—does not appear to alter significantly the requirement for any specific nutrient except calories. Exercise creates little or no increase in demand by the horse for protein, calcium, phosphorus, most trace minerals, and most fat soluble vitamins. So, if the ration contains adequate nutrients for maintenance, any increase in the requirements for heavy training may be met by the increased feed intake required to meet the increased energy requirements for exercise. Of course, if the ration is deficient for maintenance, exercise could worsen the situation. Likewise, if the ration is marginal for young animals in training, heavy exercise could make for unsoundnesses.

Horses in training will eat about 1½ to 1¾ lb of grain and ¾ to 1 lb of hay per 100 lb liveweight.

FEEDING RACEHORSES

High-strung and highly stressed, racehorses need special rations just as human athletes do—and for the same reasons; and, the younger the age, the more acute the need. This calls for rations adequate in protein, rich in readily available energy, fortified with vitamins, minerals, and unidentified factors—and with all nutrients in proper balance.

A racehorse is asked to develop a large amount of horsepower in a period of 1 to 3 minutes. The oxidations that occur in a racehorse's body are at a higher pitch than in an idle or little-worked horse, and, therefore, more vitamins are required.

During the racing season, the hay fed to a racehorse should be limited to 7 or 8 lb, whereas the concentrate allowance may range up to 16 lb. Heavy roughage eaters may have to be muzzled, to keep them from eating their bedding. A bran mash is commonly fed once a week.

FEEDING BROODMARES

Broodmares require good quality, balanced rations. The nutrients of primary concern are energy, protein, minerals, and vitamins. The nutrient requirements of a broodmare change considerably as she advances from being open (not pregnant), through pregnancy, and into lactation. In general, during the first 7 months of pregnancy the requirements are very similar to those for maintenance at the time of breeding. However, from the eighth month on, the requirements of the pregnant mare increase by 20 to 50%. During lactation, the mare requires even more nutrients, up to twice the amounts that she was receiving at breeding time.

Broodmares need a ration that will meet their body needs plus (1) the needs of the fetus, or (2) furnishing the nutrients required for milk production. If work is also being performed, additional energy feeds must be provided. Moreover, for the young, growing mare additional proteins, minerals, and vitamins, above the ordinary requirements, must be provided; otherwise, the fetus will not develop normally, or milk will be produced at the expense of the tissues of the dam. Also, protein deficiency may affect undesirably the fertility of the mare.

In comparison with geldings or unbred mares, the following differences in feeding gestating-lactating broodmares should be observed:

1. A greater quantity of feed is necessary—from 20 to 100% more—the highest requirement being during lactation.

2. Dusty or moldy feed and frozen silage should be avoided in feeding all horses, but especially in feeding the broodmare, for such feed may produce complications and possible abortion.

3. More proteins are necessary for the broodmare.

4. More attention must be given to supplying the necessary minerals and vitamins.

5. The bowels should be carefully regulated through providing regular exercise and feeding such laxative feeds as bran, linseed meal, and alfalfa hay.

6. A few days before and after foaling, the ration should be (a) decreased, and (b) lightened by using wheat bran.

7. Regular and ample exercise is a necessary adjunct to proper feeding of the broodmare.

FEEDING STALLIONS

The program throughout the entire year should be such as to keep the stallion in a vigorous, thrifty condition at all times. Immediately before the breeding season, the feed might very well be increased in quantity so that the stallion will gain in weight. The quantity of grain fed will vary with the individual temperament and feeding ability of the stallion, the work and exercise provided, services allowed, available pastures, and quality of roughage. Usually this will be between ¾ and 1½ lb daily of the grain mixture per 100 lb body weight, together with a quantity of hay within the same range.

During the breeding season, the stallion's ration should contain more protein and additional minerals and vitamins than are given in rations fed work horses or stallions not in service. During the balance of the year (when not in service), the stallion may be provided a ration like that of other horses similarly handled. In season, pastures are an excellent source of both nutrients and exercise.

FEEDING FOALS

At birth, foals have little, if any, natural immunity to disease and infection. Because of this, it is important that they nurse soon after birth so that they obtain colostrum (the first milk secreted for 12 to 24 hours after birth) from their dams. Consumption of colostrum is critical because it is a rich source of antibodies which will provide the foal with temporary immunity. Additionally, colostrum is very nutritious. It is much higher in energy, protein, mineral, and vitamin content than later milk.

When the foal is between 10 days and 3 weeks of age, it will begin to nibble on grain and hay. In order to promote thrift and early development and to avoid setback at weaning time, it is important to encourage the foal to eat supplementary feed as early as possible. For this purpose, a low-built grain box should be provided especially for the foal. If on pasture, the supplemental feeding of foals can be accomplished in either of the following two ways: (1) by bringing the mares and foals to a central area where the mares may be tied while the foals eat, or (2) by feeding the foals in a creep. The choice between individual feeding and creep feeding may be left to the caretaker; the important thing is that foals receive supplemental feed.

A creep is an enclosure for feeding purposes, made accessible to the foal(s), but through which the dam cannot pass. For best results, the creep should be built at a spot where the mares are inclined to loiter. The ideal location is on high ground, well drained, in the shade, and near the place of watering. Keeping the salt supply nearby will be helpful in holding mares near the creep.

Table 25-12 gives the formulation of an excellent foal ration, which may be either individually fed or creep fed.

TABLE 25-12
FOAL RATION

Ingredients	Percent	Amount in 500-lb Mix	
	(%)	(lb)	(kg)
Corn (flaked)	37.4	187.0	87.9
Soybean meal (41%)	33.0	165.0	74.9
Oats (rolled)	23.0	115.0	52.2
Brewers' yeast	0.5	2.5	1.1
Molasses	3.0	15.0	6.8
Steamed bone meal or dicalcium phosphate	2.0	10.0	4.5
Salt (trace mineralized)	1.0	5.0	2.3
Vitamins A and D premix	0.1	0.5	0.2
Total	100.0	500.0	227.0

Because of the difficulty in formulating and home mixing a foal ration, the purchase of a good commercial feed usually represents a wise investment.

In addition to its grain ration, the foal should be given good-quality hay (preferably a legume), unless it is on good pasture.

Giving creep-fed foals all they want to eat is not recommended. Although they will not founder or overeat when fed in this manner, a creep feed may increase body weight to the point that it causes bone disorders. The heavy body weight may lead to problems with epiphysitis and other bone abnormalities. So, creep feeding should be limited as follows: Feed 1 lb of concentrate per month of age per day, with a maximum allowance of 6 lb per day.

• **Orphaned foal**—Occasionally a mare dies during or immediately after parturition, leaving an orphan foal to be raised. Also, there are times when a mare rejects her foal or fails to give sufficient milk for the newborn foal. Sometimes there are twins. In such cases, it is necessary to resort to other milk supplies. The problem will be simplified if the foal has at least received the colostrum from the dam, for it does play a very important part in the well-being of the newborn young.

If at all possible, the foal should be shifted to another mare. Some breeding establishments regularly follow the plan of breeding a mare that is a good milk producer but whose foal is expected to be of little value. Her own foal is either destroyed or raised on a bottle, and the mare is used as a foster mother or nurse mare.

Some nurseries keep a supply of colostrum on hand. They remove colostrum from mares (1) whose foals were stillborn, or (2) that produce excess milk, then store it in a freezer for future use for foals that do not receive colostrum from their dams. When needed, it can be removed from the freezer, heated, and fed. This is an excellent practice.

If no colostrum is available, the foal should be placed on either (1) cow's milk made as nearly as possible to the same composition as mare's milk; or (2) a synthetic milk replacer.

Orphan foals may also be raised on a synthetic milk replacer made especially for foals, fed according to the directions of the manufacturer. Here again the situation is simplified if the foal has first received colostrum.

For the first few days, the milk (either cow's milk or milk replacer) may be fed by using a bottle and rubber nipple. Later, the foal should be taught to drink from a pail. It is important that all receptacles be kept absolutely clean and sanitary (cleaned and scalded each time), and that feeding be at regular intervals. Grain feeding should be started at the earliest possible time with the orphan foal.

• **Shaker foal (equine botulism)**—This is the name given to the disease in foals in which *Cl botulism* is present in the tissues, where it produces toxins. Most commonly, horses contract botulism by ingestion of a preformed toxin, which can be present in contaminated feed. One well known clinical sign of the disease is quivering or shaking knees, from which the name is derived. Other symptoms include depression, loss of appetite, and muscle paralysis. The disease is fatal in 80% to 90% of the cases; and afflicted foals may die within 24 hours. Equine botulism strikes adult horses as well as foals. The disease may be effectively prevented by vaccinating broodmares with 3 doses of a toxoid, given one month apart, during the eighth, ninth, and tenth months of pregnancy, followed by annual boosters. An annual single-dose booster 2 to 4 weeks before parturition is all that is required to provide immunity to the mare and subsequent foals. Adult horses should also receive an initial 3-dose vaccination and a yearly booster.

(Also see Chapter 5, Table 5-3—Botulism.)

FEEDING WEANLINGS

No great setback or disturbances will be encountered at weaning time provided the foals have developed a certain independence as a result of proper grain feeding during the suckling period. Generally, weanlings should receive 1 to 1½ lb of grain and 1½ to 2 lb of hay daily per each 100 lb of liveweight. The amount of feed will vary somewhat with the individuality of the animal, the quality of roughage, available pastures, the price of feeds, and whether the weanling is being developed for show, race, or sale. Naturally, animals being developed for early use or sale should be fed more liberally, although it is equally important to retain clean, sound joints, legs, and feet—a condition which cannot be obtained so easily in heavily fitted animals.

FEEDING YEARLINGS

When on pasture, yearlings that are being grown for show or sale should receive grain in addition to grass.

The winter feeding program for the rising 2-year-olds should be such as to produce plenty of bone and muscle rather than fat. From ½ to 1 lb of grain and 1 to 1½ lb of hay should be fed for each 100 lb of liveweight. The quantity will vary with the quality of the roughage, the individuality of the animal, and the use for which the animal is produced. In producing for sale, more liberal feeding may be economical. Minerals should be incorporated in the ration; and an abundance of fresh, pure water should be available.

FEEDING TWO- AND THREE-YEAR-OLDS

Except for the fact that the 2- and 3-year-olds will be larger, and, therefore, will require more feed, a description of their proper care and management would be merely a repetition of the principles that have already been discussed for the yearling.

With the 2-year-old that is to be raced, however, the care and feeding at this time become matters of extreme importance. Once the young horse is placed in training, the ration should be adequate enough to allow for continued development and to provide necessary maintenance and additional energy for work. This means that special attention must be given to providing adequate proteins, minerals, and vitamins in the ration. Overexertion must be avoided, the animal must be well groomed, and the feet must be cared for properly. In brief, every precaution must be taken if the animal is to remain sound—a most difficult task when animals are raced at an early age, even though the right genetic makeup and the proper environment are present.

FITTING FOR SHOW OR SALE

Each year, many horses are fitted for shows or sales. In both cases, a fattening process is involved, but exercise is doubly essential.

For horses that are being fitted for shows, the conditioning process is also a matter of hardening, and the horses are used daily in harness or under saddle. Regardless of whether a sale or a show is the major objective, fleshing should be obtained without sacrificing action or soundness or without causing filling of the legs and hocks.

In fattening horses, the animals should be brought to full feed rather gradually, until the ration reaches a maximum of about 2 lb of grain daily for each 100 lb of liveweight. When on full feed, horses make surprising gains. Daily weight gains of 4 to 5 lb are not uncommon. Such animals soon become fat, sleek, and attractive. This is probably the basis for the statement that "fat will cover up a multitude of sins in a horse."

Breeders who fit and sell yearlings or younger animals may feed a palatable milk replacer or commercial feed to advantage.

FEEDING SYSTEMS

Most horses are hand-fed. The grain ration usually is divided into 3 equal feeds given morning, noon, and night. Because a digestive tract distended with hay is a hindrance

Fig. 25-11. A round feed container suspended by eye hooks and located in one corner of the stall, at shoulder level or lower, will keep the horse from accidentally cutting or puncturing itself while it eats. (Courtesy, *Equus Magazine*, Gaithersburg, Md.)

in hard work, most of the hay should be fed at night. The common practice is to feed ¼ of the daily hay allowance at each of the morning and noon feedings and the remaining ½ at night when the animals have plenty of time to eat leisurely.

Usually the grain ration is fed first and then the roughage. This way, the animals usually eat bulky forages more slowly.

A few caretakers do self-feed high-energy rations, but, sooner or later, those who do usually founder a valuable horse. Except for the use of reasonably hard salt-protein blocks, salt-feed mixes in meal form (never in pellet form), or high-roughage rations, the self-feeding of horses is not recommended.

Fig. 25-12. A good hayrack. Note the rack's nearly vertical bars and curved form. These features, along with positioning the rack in the corner of the stall, are designed to alleviate injury to the horse. (Courtesy, *Equus Magazine*, Gaithersburg, Md.)

ART OF FEEDING

Feeding horses is both an art and a science. The art is knowing how to feed and how to take care of each horse's individual requirements. The science is meeting the nutritive requirements with the right combination of ingredients.

Starting Horses on Feed

Horses must be accustomed to changes in feed gradually. In general, they may be given as much nonlegume roughage as they will consume. But they must be accustomed gradually to high-quality legumes, which may be very laxative. This can be done by slowly replacing the nonlegume roughage with greater quantities of legumes. Also, as the grain ration is increased, the roughage is decreased.

Starting horses on grain requires care and good judgment. Usually it is advisable first to accustom them to a bulky type of ration; a starting ration with considerable rolled oats is excellent for this purpose.

The keenness of the appetite and the consistency of the feces are an excellent index of a horse's capacity to take more feed. In all instances, scouring should be avoided.

General Feeding Rules

Observance of the following rules will help avoid some of the common difficulties that result from poor feeding practices:

1. Know the approximate weight and age of each animal.
2. Feed by weight of feed, not by volume (volume as determined by a coffee can or marked bucket). Horses do not require a certain volume of feed; rather, they require a certain weight of nutrients based on their body weight.
3. Avoid sudden changes in the ration.
4. Never feed moldy, musty, dusty, or frozen feed.
5. Feed regularly. Horses anticipate their feed.
6. Look for problems at feeding time; don't just dump the feed and run. Look for injuries and abnormalities.
7. Check the feces. Any change in quantity, odor, color, or composition may presage trouble.
8. Inspect the feedbox frequently to see if the horse goes off feed. Feed refusal means (1) the horse was over-fed, (2) something is wrong with the feed, or (3) the horse is sick.
9. Keep the feed and water containers clean. Scrub them periodically to insure proper sanitation.
10. Do not overfeed. Some horses suffer from obesity, while others suffer from deficiency. Fat horses not receiving adequate exercise are predisposed to colic and founder. An old Arab proverb cautions: "The two greatest enemies of horses are fat and rest."
11. Force aggressive eaters to slow down. Some horses may bolt their feed when fed in deep narrow feed boxes. Their eating may be slowed by scattering the feed in a larger box, or by placing large round stones, bricks, or salt blocks in the feed container.
12. Accord timid eaters solitude to eat. Feed them where it is quiet and they will not be disturbed.
13. Do not feed from the hand; this can lead to *nibbling*.
14. Exercise stalled horses daily. It improves their appetite, digestion, and overall well being. This may be accomplished by riding, longeing, walking, ponying, swimming, treadmilling.
15. Avoid excessive exercise (to the point of fatigue and stress), rough treatment, noise, and excitement.
16. Do not feed concentrates 1 hour before or within 1 hour after hard work.
17. Feed horses as individuals; consider their likes and temperaments. Learn the peculiarities and desires of each animal because each one is different.
18. Gradually decrease the condition of horses that have been fitted for show or sale. Many caretakers accomplish this difficult task, and yet retain strong vigorous animals, by cutting down gradually on the feed and increasing the exercise.
19. Prevent wood chewing. This habit usually results from boredom, lack of exercise, lack of adequate roughage, or lack of phosphorus; so, alleviate the causes.
20. Make certain that the horse's teeth are sound.
21. Know the signs of a well-fed, healthy horse, any departure from which constitutes a warning signal.

Signs of a Well Fed, Healthy Horse

Fig. 25-13. Signs of a well-fed and healthy horse—it is bright-eyed and "bushy-tailed." (Courtesy, Ruth White, White Horse Ranch, Naper, Nebr.)

The signs of a well-fed, healthy horse are given in Chapter 17, Animal Behavior/Environment.

OTHER FEED AND MANAGEMENT ASPECTS

Several other feed and management aspects are pertinent to feeding horses; among them, (1) bolting feed, (2) eating bedding, (3) pica—wood chewing, and (4) soil analyses. For further information on bolting feed, eating bedding, and pica—wood chewing, see Chapter 17, Animal Behavior/Environment; and for further information on soil analysis, see Chapter 5, Nutritional Disorders/Toxins.

NUTRITIONAL DISEASES

Chapter 5, Nutritional Disorders/Toxins, contains a summary of the important nutritional diseases and ailments and nutrition related disorders affecting horses, as well as other classes of farm animals.

QUESTIONS FOR STUDY AND DISCUSSION

1. What conditions pertaining to the care and use of horses make it imperative that their nutrition be the best that science and technology can devise?

2. Discuss the economic importance of feed for horses.

3. Explain why the nutritive needs of horses differ from other species primarily because of their (a) digestive tract, and (b) special needs for bone formation and maintenance.

4. What is the difference between "nutritive requirements" and "nutritive allowances"?

5. Note that Table 25-6 recommends much higher allowances of several nutrients for lactating mares and young horses than for mature horses that are idle or performing light or moderate work. Why is this so?

6. How do you explain (a) that the energy requirements of a horse increase with weight (with or without rider) and speed, and (b) that the energy requirements in racing may be 100 times the energy requirements at rest?

7. Recent experimental work indicates that horses will readily consume 10 to 20% added fat to the ration, with benefit. When, and for what reasons, should fat be added to a horse ration; and what type of fat should be used?

8. Are the protein requirements of the horse increased by work load? Explain.

9. What is meant by "protein poisoning"? Is there such a thing?

10. How can a lactating mare produce milk that contains more minerals than the feed that she is consuming at the time?

11. Describe the following bone diseases and the factors thought to be responsible for each of them: epiphysitis, contracted tendons, and osteochondritis dissecans (OCD). What preventive measures would you recommend?

12. Give (a) the deficiency symptoms, and (b) practical sources of each of the following minerals for horses: salt, calcium, phosphorus, and iodine.

13. Upon what three factors does the proper utilization of calcium and phosphorus depend?

14. Why is incorporating needed minerals in the feed or water preferable to free-choice feeding?

(Continued)

15. List the vitamins that are most apt to be deficient in horse rations; then, for each of them give (a) the deficiency symptoms, and (b) practical sources.

16. If there is no evidence of deficiencies of certain vitamins, can adding them to horse rations be justified?

17. What is meant by "unidentified factors"? List three rich sources of unidentified factors.

18. What factors influence the amount of water consumed by a horse?

19. What roles may antibiotics play as an additive?

20. Is there a need and a place for pastures in modern horse production?

21. Why do caretakers prefer timothy hay and oats to all other feeds?

22. Why do caretakers use each of the following: old process linseed meal, bran mash, corn oil, treats, and sweet feed?

23. Why is feed palatability so important in horse rations?

24. What are the preferred methods of preparing feeds for horses?

25. Give recommended allowances for grain and hay for horses at (a) light work, (b) medium work, and (c) heavy work.

26. Under what circumstances would you recommend that a caretaker use (a) a home-mixed feed, or (b) a commercial feed?

27. Of what value is a feed substitution table?

28. Discuss the proper feeding of each of the following classes of horses: (a) pleasure horses, (b) horses in training, (c) racehorses, (d) broodmares, (e) stallions, (f) foals, (g) weanlings, (h) yearlings, (i) 2- and 3-year-olds, and (j) horses being fitted for show or sale.

29. Discuss (a) free-choice feeding of foals, and (b) feeding the orphaned foal.

30. Discuss the advantages and disadvantages of (a) hand-feeding, and (b) self-feeding of horses.

31. List what you consider to be the six most important horse feeding rules.

32. Write down the ration that is fed to a certain horse (your horse or a friend's horse). Then, evaluate it from the standpoints of (a) rate of feeding; (b) protein content; (c) content of salt, iodine, calcium, and phosphorus; and (d) content of vitamins A, D, and riboflavin.

33. List all the ways in which the nutritive requirements and the feeding of horses are different from those for cattle, sheep, swine, and poultry.

Fig. 25-14. A magnificent eight-horse hitch of Clydesdale horses. (Courtesy, Anheuser-Busch, Inc., St. Louis, Mo.)

26

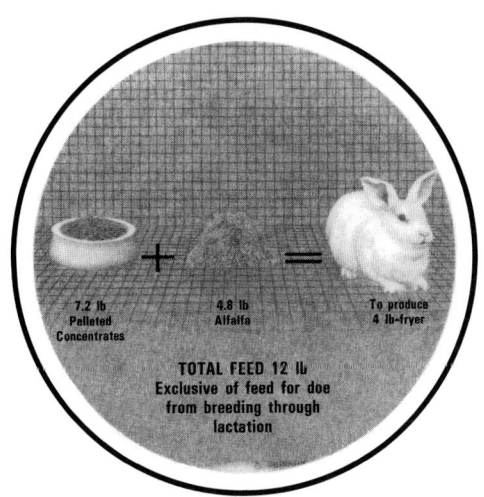

Original painting by Tom Phillips

FEEDING RABBITS[1]

Contents	Page
Economic Importance of Feed for Rabbits	624
Nutritive Needs of Rabbits	625
National Research Council (NRC) Requirements	625
Energy	625
Protein	626
Minerals	627
Rabbit Mineral Chart	627
Major or Macrominerals	627
Trace or Microminerals	627
Vitamins	630
Rabbit Vitamin Chart	631
Fat-Soluble Vitamins	632
Water-Soluble Vitamins	632
Water	632
Nonnutritive Factors	632
Fiber	632
Feed Additives	632
Anticoccidial Drugs	632
Antibiotics	632
Antioxidants	633
Feeds for Rabbits	633
Energy Feeds	633
Protein Supplements	633
Dry Forages	633
Grasses (Green)	633
Miscellaneous Feeds	633
Feed Preparation	633
Feed Allowances and Suggested Rations	634
Feeding Guide and Rations	634
Feed Substitution Table	635
Feeding Rabbits	637
Feeding Methods	637
Other Feed and Management Aspects	638
Questions for Study and Discussion	638

Rabbit meat is a staple food in European diets. Estimated annual per capita consumption is: Hungary, 8.8 lb; France, 7.9 lb; Spain, 7.9 lb; Italy, 6.2 lb; and Portugal, 4.4 lb. By comparison, the estimated annual per capita rabbit consumption in the United States is a mere 2 oz.

While rabbit production in the United States is rather limited in comparison with other animal industries, it warrants serious consideration because about 200,000 people are either directly or indirectly engaged in it, and because of its potential.

Fig. 26–1. Angora rabbits in China, used for wool production. The better types of Angoras will produce from 2 to 2½ in. of wool in approximately 11 weeks, and an annual growth of 8 to 10 in. weighing 12 to 16 oz. When fed properly balanced rations, 100 lb of feed will, on the average, produce 1 lb of wool. (Photo made in China by Audrey H. Ensminger)

Commercial operations produce rabbits primarily for meat, with much of the demand centered around the ethnic eating habits of immigrants in urban areas. Prices for fur are currently so low that it should be considered only as a supplemental income. The vast majority of rabbits are raised by part-time backyard operators as a sideline or hobby.

---

[1]The authors gratefully acknowledge the helpful suggestions of the following authoritative reviewers of this chapter: P. R. Cheeke, Ph.D., Professor of Comparative Nutrition, Department of Animal Science, Oregon State University, Corvallis; L. B. Daniels, Ph.D., Department of Animal Sciences, University of Arkansas Division of Agriculture, Fayetteville; D. J. Harris, Ph.D., General Manager, Farms Division, Pel-Freez Rabbit Meat, Inc., Rogers, Ark.; L. J. Heppler, D.D.S., Heppler Farms, Oregon City, Ore.; S. D. Lukefahr, Ph.D., Assistant Professor, Department of Food Science and Animal Industries, Alabama Agricultural and Mechanical University, Normal; J. I. McNitt, Ph.D., Rabbit Production Specialist, Southern University, Baton Rouge, La.; and T. E. Reed, D.V.M., President, The American Rabbit Breeders Assn., Inc., Markle, Ind.

Commercial rabbit raising is an extremely intensive form of animal production. Californian and New Zealand Whites are the most widely used breeds for meat production because of their large litters, good mothering ability, large size, and the preference of many processors for white rabbits. New Zealand Whites and Florida Whites are the most widely used rabbits for research. The American Rabbit Breeders Association recognizes more than 50 different breeds.

Most full-time commercial operations involving a family with a minimum of hired help require 500 to 1,000 does in order to be an economic unit. In the past, rabbit producers routinely produced 4 litters per doe annually. Today, with the high costs of labor, feed, and facilities, most commercial operations must produce 8 litters per doe annually to make a profit. To attain the 8 litters per doe level of production, young rabbits are weaned at 4 to 5 weeks of age and raised separately from their mother until they reach market weight, generally at 8 to 9 weeks of age. These intensive breeding programs require that does be rebred anywhere from 7 to 14 days after kindle (after giving birth). Since 1 doe can produce 8 litters per year, she has the ability to produce from 130 to 250 lb of live weight per year through her reproductive ability. In large operations, 1 buck is kept for as many as 20 to 30 does and produces more than 1,000 offspring annually. From this, it may be concluded that 4 does can produce 700 lb live weight, or 400 lb of dressed carcass weight, annually—as much as an average beef cow can produce in 1½ years—and produce it on less feed.

Also, it is noteworthy that rabbits have the potential for a constant state of reproduction. Does can be rebred on the day they kindle, and the young can be weaned at 28 days. Since the gestation period is 31 days, a doe can have a litter 3 days after weaning. Thus, it is theoretically possible to have a doe with three litters simultaneously: one weaned litter (4 weeks), one litter in the nest box, and one litter in utero. **NOTE WELL**: Such intensive breeding is still experimental and is not recommended until further studies are completed.

The meat of the domestic rabbit is white, high in protein (25%), low in fat (4%), and low in cholesterol and sodium—facts that should please the discriminating American consumer. When the public is made aware of the high quality of rabbit meat, consumption of rabbit should increase; but this will happen only after vigorous educational promotion by the industry, extolling the virtues of rabbit meat and telling how to prepare and serve it.

Rabbits are also used extensively in biomedical teaching and research. Over 600,000 rabbits are used annually for these purposes, helping researchers investigate a host of maladies, including cardiac diseases, hypertension, antibody production, endocrinology, venereal disease, and virology. Instructors find rabbits invaluable for laboratory demonstrations. Since large numbers of adult rabbits are needed for these purposes, a market exists. But producing for such an outlet is risky as the demand is unpredictable. The producer must have on hand enough rabbits to meet the sporadic demands, yet, at the same time, keep the number of adult breeder rabbits at as low a level as possible in order to minimize feed costs.

Despite their prolificacy, rapid growth rate, good feed conversion, use of humanly inedible foods, and nutritious meat, there are many failures in the rabbit business; primarily, because of the intensive labor requirements, disease problems, and unstable markets.

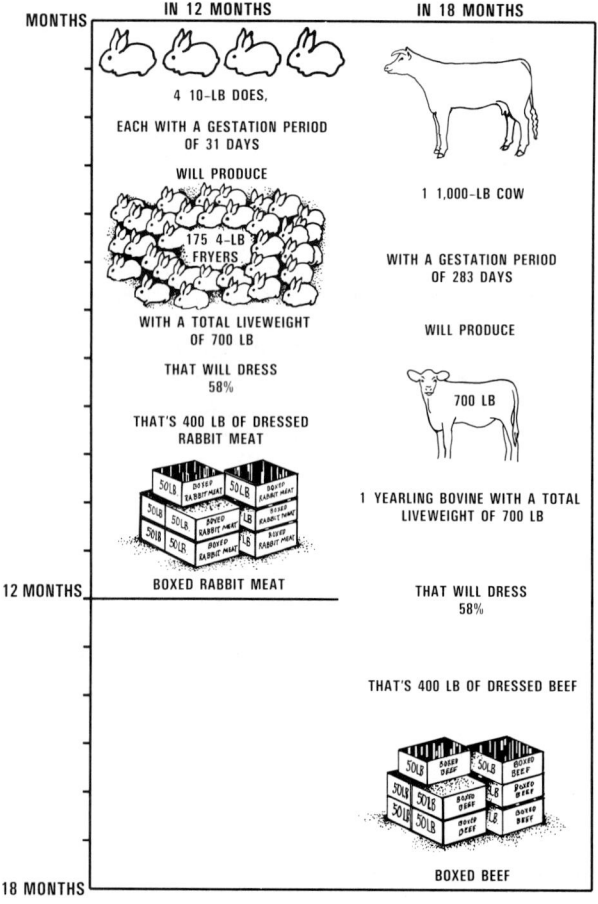

Fig. 26-2. Four 10-lb does will produce 400 lb of dressed meat in 12 months, whereas it takes one 1,000-lb cow 18 months to produce the same amount of dressed (carcass) meat. And that's not all! Rabbits produce the same amount of meat on less feed; and the ready-to-eat yield of rabbit meat after deboning and cooking is 80% of the initial dressed (carcass) weight vs 49% for the beef animal.

ECONOMIC IMPORTANCE OF FEED FOR RABBITS

Rabbit feed is very important nutritionally and economically.

Confined rabbit production precludes the choice of feeds by rabbits. It follows that big differences in their performance can be expected as a result of different diet compositions and feeding methods.

Annually, an estimated 250,000 tons of commercial feeds are fed to the more than 10 million rabbits produced in the United States. Most rabbitries use commercially pelleted rabbit feeds.

As a meat-type animal, the rabbit compares very favorably in feed conversion with the more traditional animals. With a balanced ration, feed conversions of 3:1 can be obtained in fryer rabbits. While this conversion rate is not as favorable as the 2.1:1 achieved in broilers, it does compare very favorably with the steer (9:1). Protein efficiency in rabbits is approximately 6:1, compared to 1.9:1 in broilers and 10.6:1 in steers. (See Chapter 1, Food and Animals—a global perspective, Table 1-2, for the feed to food efficiency rating of rabbits in comparison with other animal species.)

Since the rabbit possesses an enlarged cecum, similar to that of the horse, approximately 40% of commercial rabbit rations consist of alfalfa or other roughage material. So, it follows that rabbits are not in direct competition with humans for foodstuffs. In the future, if competition for human food becomes more acute, the use of roughages will assume added importance in the livestock industry; and the rabbit will out-compete many other animals in the ability to utilize roughages.

NUTRITIVE NEEDS OF RABBITS

Rabbits are nonruminant herbivores. The digestive anatomy and physiology of the rabbit closely resemble those of the horse, and in many ways the nutritive requirements are similar. Several differences, such as the habit of coprophagy (the ingestion of fecal material) and decreased fiber utilization in rabbits, alter the requirements somewhat. Nevertheless, the types of feeds used for rabbits and horses are very similar.

Rabbits excrete two types of feces—hard and soft. The hard feces (or day feces, containing about 40% water), which are produced in the large intestine, are the fecal pellets which are most commonly seen. The soft feces (or night feces, containing about 70% water), are produced in the cecum, excreted in grape-like clusters, and consumed directly from the anus in a peculiar type of coprophagic behavior displayed by rabbits as early as 3 weeks of age. **NOTE WELL**: Night feces is really a misnomer, because soft feces are often produced during the day.

Fig. 26-3. Two types of feces. *Left:* regular hard pellets. *Right:* soft feces in a grapelike cluster. (Courtesy, Pel-Freez Rabbit Meat, Inc., Rogers, Ark.)

Coprophagy by rabbits plays an important role in the modification of their nutritive requirements. It has been speculated that the act of coprophagy may aid in the absorption of some of the essential amino acids and certain vitamins—including vitamin K and the B complex vitamins. By recycling the digesta, certain feeds are digested and absorbed that were not utilized the first time. The small intestine is the main site of absorption for these nutrients, but a good deal of this synthesis takes place in the cecum. Since the cecum lies behind the small intestine, much of that which is digested in the cecum is not absorbed but is passed in the feces. Coprophagy enables the rabbits to recycle these nutrients which may not have been absorbed in the cecum or the large intestine. In addition to the microbial synthesis of many of these nutrients, some fiber digestion takes place in the cecum.

National Research Council (NRC) Requirements

Although rabbits have been used extensively in biological and medical research, very limited research data have been published concerning their nutritive needs. However, the National Research Council (NRC) has established requirements for energy, crude protein, crude fiber, essential amino acids for growth, and some minerals and vitamins for rabbits. These are given in Table 26-1 (see page 626). The NRC requirements should be used only as guides; they are subject to further refinement as the results of additional research become available.

The nutritional requirements of rabbits differ according to type and level of production; hence, Table 26-1 makes provisions for such differences. Additionally, the nutritional requirements are affected by (1) breeds, largely as a matter of size and feed capacity, (2) environmental temperature, and (3) stress.

Energy

Rabbit production is intensive in nature, and the energy demands on the doe are high if she is to produce up to eight litters a year. Likewise, growth of young juniors creates a high energy requirement.

Carbohydrates (starch and cellulose) and fats are the primary sources of energy for rabbits. Starch is found in cereal grains and tubers; cellulose is the structural component (fiber) of plants.

Good-quality legume hays and supplemental concentrates or commercial pellets are routinely used to supply a high level of energy during peak production periods. The energy requirements for maintaining dry does or young bucks not in service are low, with the result that good-quality hay should be sufficient.

While no requirement for dietary fat has been established, rations routinely contain 2 to 5% fat. It has been suggested that higher fat levels might be feasible, but caution should be exercised in order to prevent digestive disturbances, such as scours. Also, high levels of fat may reduce feed pellet quality, resulting in pellets that break apart easily.

TABLE 26-1
NUTRIENT REQUIREMENTS OF RABBITS FED AD LIBITUM (PERCENTAGE OR AMOUNT PER LB OR KG OF DIET)[1]

Nutrients[2]	Unit	Growth		Maintenance		Gestation		Lactation	
		(lb)	(kg)	(lb)	(kg)	(lb)	(kg)	(lb)	(kg)
Energy and protein:									
TDN	(%)	65	65	55	55	58	58	70	70
Digestible energy	(kcal)	1,134	2,500	952	2,100	1,134	2,500	1,134	2,500
Fat[3]	(%)	2	2	2	2	2	2	2	2
Crude protein	(%)	16	16	12	12	15	15	17	17
Crude fiber[3]	(%)	10-12	10-12	14	14	10-12	10-12	10-12	10-12
Minerals:									
Calcium	(%)	0.4	0.4	—	—[4]	0.45	0.45[3]	0.75	0.75[3]
Phosphorus	(%)	0.22	0.22	—	—[4]	0.37	0.37[3]	0.5	0.5
Magnesium	(mg)	136-181	300-400	136-181	300-400	136-181	300-400	136-181	300-400
Potassium	(%)	0.6	0.6	0.6	0.6	0.6	0.6	0.6	0.6
Sodium[3,5]	(%)	0.2	0.2	0.2	0.2	0.2	0.2	0.2	0.2
Chlorine[3,5]	(%)	0.3	0.3	0.3	0.3	0.3	0.3	0.3	0.3
Copper	(mg)	1.4	3.0	1.4	3.0	1.4	3.0	1.4	3.0
Iodine[3]	(mg)	0.1	0.2	0.1	0.2	0.1	0.2	0.1	0.2
Iron[4]		—	—	—	—	—	—	—	—
Manganese	(mg)	3.9	8.5	1.1	2.5	1.1	2.5	1.1	2.5
Zinc[4]		—	—	—	—	—	—	—	—
Vitamins:									
Vitamin A	(IU)	263	580	—	—[4]	4,545[6]	10,000[6]	—	—[4]
Vitamin A as carotene	(mg)	0.38	0.83[3,7]	—	—[8]	0.38	0.83[3,7]	—	—[8]
Vitamin D[8]		—	—	—	—	—	—	—	—
Vitamin E	(mg)	18	40[10]	—	—[4]	18	40[10]	18	40[10]
Vitamin K	(mg)	—	—[11]	—	—[11]	0.1	0.2[3]	—	—[11]
Choline	(g)	0.5	1.2[3]	—	—[12]	—	—[12]	—	—[12]
Niacin	(mg)	82	180	—	—[12]	—	—[12]	—	—[12]
Vitamin B-6	(mg)	18	39	—	—[12]	—	—[12]	—	—[12]
Amino acids:									
Arginine	(%)	0.6	0.6	—	—[9]	—	—[9]	—	—[9]
Glycine	(%)	—	—[4]	—	—[9]	—	—[9]	—	—[9]
Histidine	(%)	0.3	0.3	—	—[9]	—	—[9]	—	—[9]
Isoleucine	(%)	0.6	0.6	—	—[9]	—	—[9]	—	—[9]
Leucine	(%)	1.1	1.1	—	—[9]	—	—[9]	—	—[9]
Lysine	(%)	0.65	0.65	—	—[9]	—	—[9]	—	—[9]
Methionine + cystine	(%)	0.6	0.6	—	—[9]	—	—[9]	—	—[9]
Phenylalanine + tyrosine	(%)	1.1	1.1	—	—[9]	—	—[9]	—	—[9]
Threonine	(%)	0.6	0.6	—	—[9]	—	—[9]	—	—[9]
Tryptophan	(%)	0.2	0.2	—	—[9]	—	—[9]	—	—[9]
Valine	(%)	0.7	0.7	—	—[9]	—	—[9]	—	—[9]

[1] Adapted by the authors from *Nutrient Requirements of Rabbits,* 2nd rev. ed., NRC-National Academy of Sciences, 1977, p. 14, Table 1.
[2] Nutrients not listed indicate dietary need unknown or not demonstrated.
[3] May not be minimum but known to be adequate.
[4] Quantitative requirement not determined, but dietary need demonstrated.
[5] May be met with 0.5% NaCl.
[6] Based on studies at the Oregon State University Rabbit Research Center.
[7] Converted from amount per rabbit per day using an air-dry feed intake of 60 g per day for a 1-kg rabbit.
[8] Quantitative requirement not determined.
[9] Probably required, amount unknown.
[10] Estimated.
[11] Intestinal synthesis probably adequate.
[12] Dietary need unknown.

Protein

All production parameters—fur, growth, reproduction, and lactation—require high levels of good-quality protein.

Rabbits require certain amino acids in the diet. The essential amino acid profile is very similar to that of the chick and the pig. Table 26-1 shows the essential amino acid requirements for the growing rabbit. Of the essential amino acids, lysine is most likely to be deficient in rabbit feeds.

Studies have shown that nonprotein nitrogen (NPN), such as urea, is of little value in rabbit rations. This is attributed to the fact that NPN sources are degraded and absorbed in the small intestine and subsequently eliminated as waste products before the NPN ever reaches the cecum where it might be transformed into bacterial protein.

Coprophagy has been shown to increase the biological value of certain low-quality proteins. There are indications that some increase in value is observed with the coprophagy of high-quality protein, but the magnirute of the increase is not large.

Legumes are excellent sources of protein, and alfalfa is

used extensively in rabbit rations. Oilseed meals are widely used as protein supplements when high-protein levels are required. Animal and fish products are seldom included in rabbit diets because of their high costs.

Minerals

Little research has been reported concerning the mineral requirements of rabbits. It is generally thought that the requirements involve the same elements as are required by other animals; but for many elements, quantitative requirements have not yet been established.

RABBIT MINERAL CHART

A summary of the various minerals, along with their respective requirements and nutritional considerations, is presented in Table 26-2 (see pages 628 and 629).

MAJOR OR MACROMINERALS

In addition to the summary presented in Table 26-2, pertinent information relative to the role of certain major minerals in rabbit nutrition follows.

- **Calcium and Phosphorus**—Calcium absorption in the rabbit is extremely efficient; and an excess of calcium in the diet may affect the requirements of other minerals—notably magnesium and phosphorus. The suggested dietary calcium levels have been set at 0.4% for growth, 0.45% for gestation, and 0.75% for lactation. The recommended levels of phosphorus are 0.22% for growth, 0.37% for gestation, and 0.5% for lactation. When formulating rations, the producer should make sure that the calcium to phosphorus ratio falls in the range of 1:1 to 1.5:1. When the ratio falls below 1:1 or rises above 1.5:1, imbalances can occur.

It is noteworthy that from one-half to two-thirds of the phosphorus in seeds and grains, and their by-products, is present as phytate, the salt of myoinositol hexaphosphate, which is unavailable to monogastric animals. But phytate can be hydrolyzed by ruminants because of the presence of microbial phytase in the rumen. Arkansas researchers reported that phytate phosphorus is hydrolyzed in the digestive tract of, and made available to, rabbits (pseudo-ruminants) much like it is in ruminants; thus, formulas for rabbit rations can be based on total phosphorus, rather than on available phosphorus.[2]

- **Magnesium**—Rabbits suffering from the lack of dietary magnesium exhibit poor growth, fur-chewing habits, and hyperirritability. A number of minerals are interrelated; so what might seem to be an adequate level of magnesium may actually be a deficiency due to an excessive amount of another mineral with which it is competing for absorption and utilization—for example, calcium. Under normal conditions, .03 to .04% magnesium in the ration should be adequate.

- **Potassium**—A form of muscular dystrophy which resembles a deficiency of vitamin E occurs in rabbits suffering from a deficiency of potassium. It has been suggested that the ration should contain about .6% potassium. Generally, a diet consisting of 50% roughage is adequate in fulfilling the potassium requirement of rabbits.

TRACE OR MICROMINERALS

In addition to the summary presented in Table 26-2, pertinent information relative to the role of cobalt in rabbit nutrition follows.

- **Cobalt**—Vitamin B-12 contains cobalt as the central element of a complex porphyrin structure. Since the microorganisms in the cecum of the rabbit have the ability to synthesize vitamin B-12, all of the precursors—including cobalt—must be supplied.

In cobalt-deficient areas, most commercial rabbit pellets contain approximately 1.0 ppm cobalt.

Fig. 26-4. Rabbit with bowed legs and enlarged joints resulting from eating alfalfa produced on low-phosphorus soils. (Courtesy, Dr. Wilton W. Heinemann, Washington State University, Prosser)

---

[2]Nelson, T. S., L. B. Daniels, L. A. Schriver, and L. K. Kirby, "Hydrolysis of Phytate Phosphorus by Young Rabbits," *Arkansas Farm Research*, July-August, 1985, p.8.

TABLE
RABBIT MIN

Minerals Which May Be Deficient Under Normal Conditions	Conditions Usually Prevailing Where Deficiencies Are Reported	Functions of Mineral	Some Deficiency Symptoms
Major or macrominerals:			
Salt (NaCl)	Negligence, for salt is cheap.	Sodium and chlorine help maintain osmotic pressure in body cells, upon which depends the transfer of nutrients to the cells and the removal of waste materials. Also, sodium is important in making bile, which aids in the digestion of fats and carbohydrates; and chlorine is required for the formation of hydrochloric acid in the gastric juice so vital to protein digestion.	Depressed growth.
Calcium (Ca)	Rations of grass hay and farm grains.	Essential for development and maintenance of normal bones and teeth. Important in blood coagulation and lactation. Enables heart, nerves, and muscles to function. Regulates permeability of tissue cells. Affects availability of phosphorus and zinc.	Rickets; tetany; brittle bones.
Phosphorus (P)	High legume forage ration without supplemental phosphorous.	Essential for sound bones and teeth, and for the assimilation of carbohydrates and fats. A vital ingredient of the proteins in all body cells. Necessary for enzyme activation. Acts as a buffer in blood and tissue. Occupies a key position in biologic oxidation, and reactions requiring energy.	Rickets; tetany; brittle bones.
Magnesium (Mg)		Necessary for many enzyme systems and for proper functioning of the nervous system. Closely associated with the metabolism of calcium and phosphorus.	Poor fur growth; fur chewing; hyperirritability.
Potassium (K)		Essential for proper enzyme, muscle and nerve function, cecal mircroorganism activity, and appetite.	Muscular dystrophy.
Trace or microminerals:			
Cobalt (Co)	Feeds from cobalt-deficient areas.	Constituent of vitamin B-12.	Anemia.
Copper (Cu)		Copper, along with iron, is necessary for hemoglobin formation, although it forms no part of the hemoglobin molecule of red blood cells.	Anemia; graying of hair.
Iodine (I)	Feeds grown in iodine-deficient areas.	Needed for the production of thyroxin, an iodine-containing hormone that regulates metabolic rate.	
Iron (Fe)		Necessary for formation of hemoglobin, an iron-containing compound which enables the blood to carry oxygen. Also, important to certain enzyme systems.	Microcytic and hypochromic anemia.
Manganese (Mn)	Excess calcium and phosphorus decreases absorption of manganese.	Considered essential in utilization of calcium and phosphorus, for proper functioning of mammary glands and normal reproduction.	Abnormalities of the skeletal system.
Zinc (Zn)		Component of several enzyme systems and also required for normal protein synthesis.	Weight loss, alopecia, graying of hair, dermatitis, low hematocrit, reproductive problems.

[1]As used herein, the distinction between "mineral requirements" and "recommended allowances" is as follows: In mineral requirements, no margins of safety are included intentionally, whereas in recommended allowances, margins of safety are provided in order to compensate for variations in feed composition, environment, and possible losses during storage or processing. Where preceded by an asterisk, the mineral requirements, allowances, and other facts presented herein were taken from *Nutrient Requirements of Rabbits,* 2nd rev. ed., NRC-National Academy of Sciences, 1977.

26-2
ERAL CHART

Mineral Requirements[1] Mineral Content of Ration	Recommended Allowances[1]		Practical Sources of the Mineral	Comments
	Daily Nutrients/ Animal	Percent of Total Ration		
*0.5%.		0.5-1.0	Salt spools. Can be added to feed.	Sodium and chlorine are low in feeds of plant origin. There is little danger of overfeeding salt unless a salt-starved animal is suddenly given access to too much salt or if liberal amounts of water are not available.
Variable according to age and production (see Table 26-1).		0.4-1.0	Ground limestone or oystershell flour. When both Ca and P are needed, use bone meal, dicalcium phosphate, or defluorinated phosphate.	The Ca:P ratio should be maintained close to 1:1, although 2:1 is acceptable when the higher calcium content is due to the presence of legume. Where there is a shortage of calcium in the ration, it is withdrawn from the bones.
Variable according to age and production (see Table 26-1).		0.22-0.50	Monosodium phosphate. Monoammonium phosphate. When both Ca and P are needed, use bone meal, dicalcium phosphate, or defluorinated phosphate.	(Same as stated for Ca under "Comments" above.) If plenty of vitamin D is present, the ratio of Ca:P becomes less important. Apparently phosphorus cannot be withdrawn from the bone.
*300-400 ppm.		0.03-0.04	Magnesium sulfate. Magnesium oxide.	In high Ca diets, Mg may become deficient due to interference from the Ca in absorption.
*0.6%.		0.6	Potassium chloride. Roughages.	
			Cobalt chloride, cobalt sulfate, cobalt oxide, or cobalt carbonate.	Cobalt-deficient areas are: Fla., Mass., Mich., N.H., N.Y., N.C., Pa., S.C., and Wisc. Also, Australia and western Canada.
*3 ppm.		0.00003	Copper carbonate. Copper sulfate.	Trace mineralized salt containing copper is satisfactory.
*0.2 ppm.		0.2 ppm	Iodized salt.	
		0.0050	Trace mineralized salt.	
Variable according to age and production (see Table 26-1).	1 mg for growing animals; 0.3 mg for mature animals.	0.000025-0.000085	Trace mineralized salt.	Manganese is needed for growth and reproduction of most animals.
			Zinc carbonate. Zinc sulfate.	

Vitamins

As with mineral nutrition in rabbits, the amount of information concerning the vitamin requirements is very limited.

TABLE
RABBIT VITA

Vitamins Which May Be Deficient Under Normal Conditions	Conditions Usually Prevailing Where Deficiencies Are Reported	Functions of Vitamins	Some Deficiency Symptoms
Fat-soluble vitamins:			
A	When on high-forage rations that have been stored a long time and have lost their carotene value.	Promotes growth and stimulates appetite. Assists in reproduction and lactation. Keeps the mucous membranes of respiratory and other tracts in healthy condition. Makes for normal vision. Prevents night blindness.	Retarded growth, incoordination and paralysis, blindness, drooping ears, and hydrocephalus (enlarged head) of fetuses from vitamin A-deficient does.
D	In confinement rearing where the rabbits do not have access to sunlight or sun-cured hay.	Assimilation and utilization of calcium and phosphorus, necessary in normal bone development—including the bones of the fetus.	Rickets.
E		Serves as insurance against destruction of vitamin A. Makes for improved reproduction. Protection of cellular lipids from oxidation.	Nutritional muscular dystrophy of skeletal and cardiac muscle, paralysis, and fatty livers.
K	Intestinal disorders or when antibiotics are used.	Concerned with blood coagulation.	Prolonged bleeding following a minor injury and abortion and placental hemorrhage in does.
Water-soluble vitamins:			
Biotin	The presence in feeds of avidin, a biotin antagonist.	Important in the metabolism of carbohydrates, fats, proteins.	Loss of hair and dermatitis.
Choline	Rations low in methionine, an amino acid.	Prevention of fatty livers, transmitting nerve impulses, and the metabolism of fat.	Depressed growth, fatty and cirrhotic liver, and necrotic kidneys.
Folacin (Folic Acid)		Essential for normal growth and reproduction, for the prevention of blood disorders, and for important biochemical mechanisms within each cell.	
Niacin (Nicotinic Acid, Nicotinamide)		Constituent of coenzymes which produce energy within the cells.	Loss of appetite, diarrhea, and emaciation.
Pantothenic Acid (Vitamin B-3)	No deficiency ever produced in rabbits.	Constituent of coenzyme A (CoA). It plays a key role in energy metabolism.	
Riboflavin (Vitamin B-2)	Destruction of riboflavin by light or by heat in an alkaline solution.	In the oxidative mechanisms in the cells.	Retarded growth and lowered feed efficiency.
Thiamin (Vitamin B-1)	Prolonged feeding of a thiamin-deficient diet.	A cofactor of certain enzymes involved in carbohydrate and fat metabolism.	Loss of appetite, muscle paralysis, and accumulation of pyruvic acid in the blood.
Vitamin B-6 (Pyridoxine, Pyridoxal, Pyridoxamine)		Key constituent of cofactors involving amino acid and energy metabolism.	Depressed growth, acrodynia (dermatitis), convulsions, and paralysis.
Vitamin B-12 (Cobalamins)	Lack of cobalt.	Two coenzyme forms—coenzyme B-12, and methyl B-12.	Retarded growth.

[1]As used herein, the distinction between "vitamin requirements" and "recommended allowances" is as follows: In vitamin requirements, no margins of safety are included intentionally, whereas in recommended allowances, margins of safety are provided in order to compensate for variations in feed composition, environment, and possible losses during storage or processing. Where preceded by an asterisk, the vitamin requirements, allowances, and other facts presented herein were taken from *Nutrient Requirements of Rabbits,* 2nd rev. ed., NRC-National Academy of Sciences, 1977.

Feeding Rabbits

RABBIT VITAMIN CHART

Table 26-3, Rabbit Vitamin Chart, lists the vitamins that have been studied in rabbits, and gives pertinent information pertaining to each.

26-3
MIN CHART

Vitamin Requirements[1] Vitamin Content of Ration	Recommended Allowances[1] Daily Nutrients Per Rabbit	Per Lb of Ration	Practical Sources of the Vitamin	Comments
Variable according to age and production (see Table 26-1).	23 mcg of carotene per pound of ration has been shown to prevent symptoms of vitamin A deficiency.	For does in production, 4,545 IU/lb feed.	Stabilized vitamin A. Green grass or legume.	Vitamin A is not synthesized in the cecum. A vitamin A level of 86,355 IU/lb of diet, which is only 19 times the vitamin A requirement, is toxic to rabbits.
	No allowances have been recommended for vitamin D supplementation.		Sun-cured hays. Exposure to sunlight. Commercially available supplements of either D_2 or D_3.	The vitamin D requirement is directly related to the Ca:P ratio and their respective levels. Vitamin D toxicity is of greater concern than deficiency in rabbits. Toxicity symptoms are loss of appetite, impaired movement, and calcification of soft tissues such as the kidneys and the arteries.
*18 mg/lb for rabbits in production.	0.5 mg/lb of body weight.	25 IU	Stabilized vitamin E. Germ or gum oils of plants. Green plants and hays. Cereals.	Selenium does not exert a vitamin E-sparing effect in rabbits; instead, rabbits depend entirely on vitamin E for protection against peroxides.
*0.1 mg/lb for gestating does.	No allowances have been recommended for vitamin K supplementation.		Synthetic vitamin K. Green grass. Well-cured hays.	Studies indicate that dietary vitamin K is required for reproduction but not growth.
			Widely distributed in nature.	Deficiencies are not associated with normal rations.
*0.5 g/lb for growing rabbit.		0.12%	Alfalfa hay and meal. Choline chloride. Rice polishings.	
			Alfalfa hay/meal and oil seed meals.	
*82 mg/lb for growing rabbit.	5 mg/lb of body weight.		Nicotinamide (synthetic). Nicotinic acid (synthetic). Rice polishings. Yeast (brewers', torula).	Limited amounts of niacin can be synthesized from the amino acid tryptophan. Supplemental niacin in some cases can increase growth rate.
			Calcium pantothenate (synthetic), rice polishings, yeast (brewers', torula), alfalfa hay and meal.	
Not known.			Synthetic riboflavin, alfalfa hay, whey, yeast.	
Not known.			Thiamin hydrochloride, rice polishings.	
*18 mg/lb for growing rabbit.		0.5 mg	Pyridoxine hydrochloride. Rice polishings. Alfalfa hay/meal.	Normally, rabbit rations contain adequate vitamin B-6.
			Meat animal by-products. Marine by-products.	

FAT-SOLUBLE VITAMINS

Vitamin A and Vitamin E are probably the only two vitamins for which there are serious needs for dietary supplementation.

- **Vitamin A**—The vitamin A requirement of does in production is 4,545 IU/lb of feed. Generally, either a deficiency or a toxicity of vitamin A will affect the reproductive performance of the female before other symptoms can be recognized. Additional symptoms of vitamin A deficiency are: retarded growth, incoordination and paralysis, blindness, drooping ears, and hydrocephalus (enlarged head) of fetuses born to vitamin A-deficient does.

Research indicates that 23 mcg of carotene per pound of body weight per day are adequate to prevent the symptoms of vitamin A deficiency.

A vitamin A level of 86,355 IU/lb of diet, which is only 19 times the vitamin A requirement, is toxic to rabbits. Toxicity symptoms in does are: reproductive problems, including abortion; fetal resorption; small, weak litters with a high mortality the first week; and kits with hydrocephalus.

- **Vitamin E**—Unlike other animal species, selenium does not exert a vitamin E-sparing effect in rabbits; instead, rabbits depend entirely on vitamin E for protection against peroxides.

Symptoms of a vitamin E deficiency in rabbits are muscular dystrophy, reproductive failure, and fatty livers.

WATER-SOLUBLE VITAMINS

Through bacterial synthesis in the cecum, along with coprophagy, the rabbit is able to fulfill many of the vitamin B complex requirements. The requirements for pantothenic acid, riboflavin, biotin, folic acid, and B-12 are met through this route. Research has indicated that some supplementation of niacin (11 mg/kg body weight), pyridoxine (39 ppm of the diet), and choline (.12% of the diet) will aid in production. Cobalt must be supplied in the diet as it is a precursor of vitamin B-12. Vitamin B-12 is synthesized in the gut of the rabbit if cobalt is present.

Water

The water requirement of rabbits is influenced by a number of factors; among them, the following:

1. **Temperature and humidity.** When the temperature and humidity increase to levels above the zone of thermoneutrality, water within the body becomes an important means of heat dissipation.

2. **Stage of production.** A doe with a litter of seven can drink up to 1 gal of water per day.

3. **Composition of feed.** Feeds that are high in protein and fiber increase the need for water because of the increased need to excrete end products produced in the digestion and metabolism of these feed components. The inclusion of succulent feeds into the ration provides an additional source of water. But care should be taken to restrict the use of this type of feed as the high water content may restrict the level of nutrient intake.

Nonnutritive Factors

Two nonnutritive factors—fiber and feed additives—are discussed with regards to their respective influences on rabbit nutrition.

FIBER

Roughages make up a large percentage of rabbit rations, and the digestibility of fiber plays an important role in the utilization of feed by rabbits. Experiments have shown that rabbits do not digest fiber efficiently. Rabbits digest the same amount of crude protein on a percentage basis as the horse and the pony; but fiber digestibility in the rabbit is less than one-half that of either the horse or the pony.

Although fiber is not a useful energy source for rabbits, it is a very important component of rabbit feeds. Numerous studies have shown that low fiber diets cause increased enteritis. Also, when fiber is lacking in the ration, rabbits will sometimes resort to eating their own fur. Fur then accumulates in the digestive tract, resulting in obstruction. The inclusion of some roughage in the ration will remedy the problem in many cases.

Nonproducing animals can be fed rations that contain up to 16 to 22% fiber while animals in production should be limited to 12 to 14% fiber because of the greater need for digestible nutrients.

The form of fiber can affect its value to the rabbit. There is some evidence indicating that finely ground fiber can cause diarrhea. Thus, for proper utilization, fiber should be fed in a coarse form.

FEED ADDITIVES

Feed additives in rabbit rations can be divided into three groups: coccidiostats, antibiotics, and antioxidants.

Anticoccidial Drugs

Coccidiosis is the most prevalent parasitic disease in rabbits. Treatment for either the intestinal or liver forms of coccidiosis in rabbits is as follows: sulfaquinoxaline (1) administered in the drinking water at 0.04% for 14 to 30 days; or (2) given in the feed at 0.025% for 20 days, or for 2 days out of every 8. Rabbits should not be treated for coccidiosis within 10 days of slaughter if they are to be used for food. Since most rabbit producers feed only one feed or at the most two, the addition of anticoccidial drugs to the feed becomes impractical. The best method to administer these drugs is through the drinking water. Withdrawing these drugs can then be facilitated by merely changing the water.

Antibiotics

In some cases, the addition of low levels of antibiotics in the rations of various livestock has been shown to exert growth-promoting effects. Oxytetracycline is approved by the Food and Drug Administration as a feed additive for rabbits at a use level of 10 g/ton, as an aid in stimulating growth and improving feed efficiency. (See Chapter 13, Feed Supplements & Additives/Implants.)

Antioxidants

Several antioxidants can be added to rabbit rations in order to prevent spoilage due to the autoxidation of fats within the feed. Also, vitamin E is an effective, natural antioxidant. Compared to the composition of rations for other livestock, rabbit rations are relatively low in fat; but the addition of antioxidants to the ration can ensure against feed spoilage and increase the storage life of the feed. (See Chapter 13, Feed Supplements/Additives/Implants.)

FEEDS FOR RABBITS

Generally speaking, a good-quality legume hay is sufficient to meet most of the maintenance needs of the rabbit; but as production pressures intensify, there arises a need for additional feed sources that are higher in energy. A variety of feedstuffs has been used successfully in feeding rabbits.

Energy Feeds

Production demands necessitate high-energy levels in rations. Traditionally, cereal grains and their by-products have filled this need. Whole grains from barley, buckwheat, corn, oats, rye, wheat, and the grain sorghums have been widely used in rabbit rations.

The digestibility of certain grains can be increased with minimal processing. For example, corn is more easily digested when it is cracked.

Cereal by-products, such as red dog flour, wheat bran, mill run, and middlings, offer excellent alternatives to the feeding of whole grains; but since these by-products tend to have laxative properties, the level of usage in the ration should be carefully monitored, and sufficient fiber should be present. Economic considerations should be the determining factor as to which feed should be used.

Protein Supplements

Fish and animal by-products are seldom used as protein supplements in rabbit rations. But various oilseed meals are routinely added. Peanut meal, the protein supplement most readily eaten by rabbits, provides an excellent source of both protein and energy. Soybean meal, brewers' dried grains, sunflower meal, rapeseed meal (canola meal), safflower meal, sesame meal, linseed meal, and cottonseed meal can all be used interchangeably to different extents.

Amino acid balance in dietary protein is important in the growing rabbit.

Dry Forages

Rabbits have the ability to utilize dry forages as a large portion of their diet. A good-quality hay will alleviate much of the expense of feeding more costly grains. When using hays, the producer must make certain that they are clean and free from dirt and mold, since it does not take much nutritionally related stress to cause digestive disturbances in rabbits. Once a young rabbit develops scours, precious weight is lost, lowering the efficiency of feed utilization, and the animal becomes highly susceptible to disease and other stresses.

Typical commercial rabbit feeds contain about 40% alfalfa meal, or occasionally some other legume. However, when the price is right, and when incorporated in nutritionally balanced rations, high quality grass hays can be used successfully in rabbit rations, often at a saving in feed costs.

Grasses (Green)

A wide variety of grasses can be fed to rabbits—from high-quality alfalfa to lawn clippings. Since fresh grasses are high in water and bulk, it is wise to use a supplemental feed to ensure against any nutritional deficiencies.

Miscellaneous Feeds

A wide variety of feeds can be used in rabbit rations. Often many feedstuffs can come from the garden or table scraps—especially in backyard production. Fats, meats, and spoiled foods should not be used, but garden trimmings and vegetables are low cost and may be efficient.

- **Miscellaneous Roughages**—Scraps and by-products from various friuts can be fed to rabbits. Potato peelings are readily eaten. The pulp and peelings of citrus fruits as well as apples are often good sources of cheap, supplemental feed. Caution should be exercised when using fruit pulp, lest the rabbits consume too much fiber.

- **Roots and Tubers**—The addition of roots and tubers to the rations of rabbits can be beneficial, especially in the winter when fresh greens are not available. They are highly palatable and are good sources of vitamins and minerals. However, the water content of roots and tubers tends to be extremely high (about 90%), and the protein level is quite low (1 to 4%). Thus, the producer should not incorporate too high a level of them in the ration. A deficiency in some of the nutrients may result when feeding roots and tubers, because rabbits preferentially eat this type of feed first, subsequently neglecting the higher quality feeds. For this reason, the daily allowance of roots and tubers in a maintenance ration should be limited to 1.5% of body weight. Rabbits in production should not be fed any roots or tubers.

- **Shrubs and Trees**—Occasionally, twigs from woody plants can be given to rabbits. While the nutritional value of such feeds may be doubtful, they provide the rabbit with something on which to chew, and some additional fiber.

FEED PREPARATION

The size, type, and intensity of operation will, in most cases, determine the type of feed preparation. Since most backyard operations are small and do not have access to feed mixing facilities, they generally use commercially available complete pelleted rations, along with some additional hay.

Fig. 26-5. *Left:* pelleted feed. *Right:* unpelleted feed. (Courtesy, Pel-Freez Rabbit Meat, Inc., Rogers, Ark.)

If rabbit production is a sideline to a farm operation, homegrown grains and hays are fed with very little processing involved. Commercial operations must obtain feed in large quantities; hence, the processing and preparation of feed for them warrants considerable attention.

Nutritionally complete pelleted rations are available commercially and are being used extensively and successfully. They generally come in 2 types: (1) production rations, which are high in protein and energy; and (2) maintenance rations, which are somewhat lower in protein and energy. The complete pelleted rations generally contain 50 to 60% concentrate and 40 to 50% roughage, with micronutrients supplied at nutritionally adequate levels. The size of the pellet is important. It should be $\frac{1}{8}$ to $\frac{5}{32}$ in. in diameter and $\frac{1}{8}$ to $\frac{1}{4}$ in. long. If the pellets are larger than these specifications, the rabbits will bite off pieces of them, with the result that there will be fines and considerable waste. Also, smaller pellets result in kits getting on dry feed more quickly.

Grains can be fed whole to rabbits, but oats and barley should be rolled and corn cracked in order to maximize digestibility. Rabbits are extremely sensitive to moldy and dusty feeds, so care should be taken to provide only clean, quality grains.

Protein supplements should be in cake, pelleted, or crumble form. If the supplement is fed in mash form, there is the possibility that it will settle out from the rest of the feed and be wasted. Since protein supplements are probably the most costly components of the ration, it is imperative that they be used as efficiently as possible. Cottonseed meal must be degossypolized before feeding it to rabbits, and soybean seeds must be heat-treated in order to make them palatable.

Forages should be clean, leafy, and free from mold. In order to maximize the utilization of forages, they should be cut in lengths of about 3 in. and fed in a hay rack. If forage is fed in a form longer than 3 in., rabbits will pull the feed from the hay rack, drag it out on the floor of the hutch, chew a piece from the end, then leave the remainder.

FEED ALLOWANCES AND SUGGESTED RATIONS

Generally, the term nutrient requirements implies a rigid, inflexible level of a certain nutrient for a particular nutritional demand. In reality, the actual nutrient requirement for a particular animal in a fixed set of environmental influences is seldom as precise as has been stated by a group of scientists agreeing on a hypothetical figure. Rather, nutrient allowances are given in order that the producer might have a general idea as to the range in which the real nutrient requirement should fall.

A listing of the nutrient allowances for rabbits is given in Table 26-4

TABLE 26-4
NUTRIENT ALLOWANCES FOR RABBITS[1]

	Crude Protein[2]	Fat	Fiber	Nitrogen-Free Extract	Ash
	(%)	(%)	(%)	(%)	(%)
Pregnant does	15-17 (16)	3-6	12-16	44-52	5-6.5
Does with litters	24-26 (25)	3-6	12-16	44-52	5-6.5
Dry does	12-15 (13)	2-4	16-22	42-50	5-6.5
Herd bucks	12-15 (13)	2-4	16-22	42-50	5-6.5
Developing young (after weaning)	16-18 (17)	3-6	12-16	44-52	5-6.5

[1]Allowances are on the basis of air-dried feed (as-fed basis).
[2]Values in parentheses represent what the authors feel is optimal for most rations.

Feeding Guide and Rations

A rabbit feeding guide, along with suggested rations, is presented in Table 26-5.

TABLE 26-5
RABBIT FEEDING GUIDE AND SUGGESTED RATIONS[1]

Age, Sex, Production	Total Daily Feed Per Day As-fed		Total Daily Feed As % Live Weight As-fed	Suggested Rations[2]		
					Total Diet	
				Ingredient	A	B
	(oz)	(g)	(%)		(%)	(%)
Normal growth, does or bucks, average 6.5 lb (2.95 kg)	5.1	145	5.8	Alfalfa hay Corn, grain Barley, grain Soybean meal Salt	60 21.5 15 3 0.5	
Normal growth and fattening, does or bucks,[3] at body weight of:				Alfalfa hay Corn Wheat bran Barley, grain Oats, grain Soybean meal Salt	40 — 5 31.5 18 5 0.5	50 23.5 5 11 — 10 0.5
4 lb (1.8 kg)	4.0	113	6.2			
5 lb (2.3 kg)	4.8	136	6.0			
6 lb (2.7 kg)	5.4	154	5.7			
7 lb (3.2 kg)	6.1	172	5.4			
Maintenance, does or bucks, at body weight of:				Alfalfa hay Clover Oats, grain Wheat, grain Salt	70 — 19.5 10 0.5	— 70 29.5 — 0.5
5 lb (2.3 kg)	3.2	91	4.0			
10 lb (4.5 kg)	5.3	150	3.3			
15 lb (6.8 kg)	7.2	204	3.0			
Pregnant does, at body weight of:				Alfalfa Clover hay Oats, grain Soybean meal Salt	— 50 43.5 6 0.5	50 — 45.5 4 0.5
5 lb (2.3 kg)	4.0	113	5.0			
10 lb (4.5 kg)	6.6	186	4.1			
15 lb (6.8 kg)	9.0	254	3.7			
Lactating doe, at body weight of 10 lb (4.5 kg) with a litter of 7	18.3	520	3.4	Alfalfa hay Wheat, grain Sorghum, grain Soybean meal Salt	40 25 24.5 10 0.5	40 25 22.5 12 0.5

[1]Adapted by the authors from *Nutrient Requirements of Rabbits*, No. 9, 1st and 2nd rev. ed., NRC-National Academy of Sciences, 1966 and 1977.
[2]In iodine-deficient areas, use iodized salt.
[3]Ration can be used for pregnant and lactating does, also.

FEED SUBSTITUTION TABLE

In order to utilize effectively feeds which might become available to the producer as alternatives to traditional feeds, a feed substitution table for rabbits has been developed. (See Table 26-6, page 636.) **NOTE WELL:** Listings for *protein supplements* and *dry forages* are not included in Table 26-6 because, for the most part, only the traditional feedstuffs under these classifications are fed to rabbits.

TABLE 26-6
RABBIT FEED SUBSTITUTION TABLE (See footnote at end of table.)

Feedstuff	Relative Feeding Value (lb for lb) in Comparison with the Designated Base Feed (in bold italic) Which = 100	Maximum Percentage of Base Feed (or comparable feed or feeds) Which it Can Replace for Best Results	Remarks
GRAINS, BY-PRODUCT FEEDS, ROOTS AND TUBERS:[1]			
(Low and Medium Protein Feeds)			
Oats	*100*	*100*	Most preferred concentrate by rabbits; should be rolled.
Barley	110	100	Should be rolled.
Beets:			Do not feed to rabbits in production or young less than 3 months old.
Garden	10	10	Do not feed to other rabbits in excess of 1½% of body weight.
Mangel	10	10	
Sugar	10	25	
Buckwheat	95	100	
Cabbage, aerial	10	10	Same precautions as with beets; can cause goiter if fed in high amounts.
Carrots, roots	10	10	Same precautions as with beets.
Chicory	65	10	Same precautions as with beets.
Corn	125	100	Should be cracked in order to maximize digestibility.
Kale	15	10	In excessive amounts, kale produces a strong odor in the urine.
Kohlrabi	15	10	
Potato, roots	25	10	Cut out green buds before use.
Potato, peelings	25	10	
Rutabagas, roots	10	10	Same precautions as with beets.
Rye	100	35	May have palatability problems.
Sorghum	125	100	
Sunflower seeds	115	100	Excellent feed, but it is commonly used for other purposes than rabbit feed.
Sweet potato, roots	25	10	Same precautions as with beets.
Turnips, roots	10	10	Same precautions as with beets.
Wheat, grain	120	100	
Wheat, bran	120	100	Has laxative properties; make sure the ration has adequate fiber.
Wheat, middlings	130	100	Has laxative properties; make sure the ration has adequate fiber.
Wheat, mill run	115	100	Has laxative properties; make sure the ration has adequate fiber.
Wheat, red dog	125	100	Has laxative properties; make sure the ration has adequate fiber.
GRASSES:			
Alfalfa	*100*	*100*	
Bermudagrass	90	33⅓	
Canadian bluegrass	100	50	
Carpetgrass	90	33⅓	
Clover, red	90	100	
Colonial bentgrass	115	100	
Crotolaria	65	33⅓	
Dallisgrass	75	33⅓	
Dandelion	70	33⅓	
Foxtail millet	80	33⅓	
Kentucky bluegrass	95	50	
Kudzu	95	100	
Lespedeza	95	100	
Meadow fescue	90	50	
Napiergrass	90	50	Low crude protein but high TDN values.
Oatgrass	70	33⅓	Low crude protein content.
Orchardgrass	90	50	
Pigeon pea	100	100	
Red fescue	100	50	
Redtop	90	50	
Rhodesgrass	65	33⅓	Low crude protein.
Sudangrass	50	33⅓	

[1] Roots and tubers are of lower value than grain and by-product feeds due to their higher moisture content.

FEEDING RABBITS

When good nutrition and proper management procedures are followed, growing rabbits should obtain feed conversion rates of 2.8 to 4.0. Most producers estimate that it takes about 100 lb of feed to grow one litter to market weight of fryers. This figure includes the feed consumed by the doe from the time of mating to the weaning of the litter.

When including working does, their young, bucks, and replacement does, the average feed efficiency for rabbits ranges from 4.5:1 to 5:1.

Rabbits should be fed according to their level of production. Thus, lactating does have a much greater feed requirement than dry or pregnant does.

Table 26–5, Rabbit Feeding Guide and Suggested Rations, presents pertinent feeding information for the following age, sex, and production classes: (1) growth, (2) growth and fattening, (3) dry does and herd bucks, (4) pregnant does, and (5) lactating does.

Fig. 26–6. Breeding age (20 weeks) New Zealand White Does. (Courtesy, Pel-Freez Rabbit Meat, Inc., Rogers, Ark.)

Fig. 26–7. New Zealand White fryers at 8 weeks of age. (Courtesy, Pel-Freez Rabbit Meat, Inc., Rogers, Ark.)

• **Feeding Orphan Litters**—When a doe dies at or after kindling, the producer must make provisions for feeding the orphan litter. If a recently kindled doe with a small litter is available, some of the young may be transferred (fostered) to her. This method is far more practical than feeding each kit by hand. However, since this practice is not always possible, a procedure like the following may be used:

1. For the first two weeks of life, feed cow's or goat's milk, or a commercial milk replacer. Heat the milk to body temperature, then feed it by using an eyedropper or a doll's nursing bottle. The eyes of the young rabbits will open at about day 10.

2. After the initial nursing period, solid food, such as fresh grass and rolled oats, can be offered in addition to milk. This will stimulate development of the gut.

3. When the young are about 17 days of age, they can be taught to drink from a pan and offered small quantities of a good growing ration.

4. Gradually, the quantity of solid feed can be increased.

FEEDING METHODS

A standard rule is that rabbits should never be exposed to radical changes in their diets. If feed changes are to be made, they should be made gradually, taking as long as 5 to 10 days. If, for example, the ration is to be lowered from 100% alfalfa to 40% alfalfa, a gradual daily reduction in the alfalfa content of the feed is recommended until the desired feeding level has been attained. Occasionally, when making a slow transition in the ration, some rabbits will waste a large amount of feed by digging and searching for the old, familiar feed. When this happens, an abrupt switch to the new ration may be made.

• **Hand-feeding**—While this method of feeding involves considerable time and labor, it enables producers to keep a close watch relative to the general condition and feeding habits of their animals.

Rabbits can be fed once, twice, or three times daily. However, in most operations, a once-a-day feeding practice is the best. Less labor is involved; and the rabbits are more likely to maintain an active appetite if they are allowed to clean up their feed before the next feeding. Since rabbits eat about 2½ times as much at night as during the day, a once-a-day feeding should be offered in the evening. If more than one daily feeding is used, the last feeding of the day should provide the largest quantity of feed. The number of feedings is not critical, but it is important that the time of feeding be regular from day to day. Rabbits are creatures of habit, and any break of routine can cause digestive problems.

If rabbits are hand fed only the amount of feed that they will clean up in a day, it is easy to detect an animal that is off feed; hence, the daily allowance may serve as an important health check.

• **Self-feeding**—Growth tends to be more rapid and efficient in self-fed rabbits than those that are fed by hand. This improved growth rate is due to the fact that self-fed rabbits have access to feed at all times; consequently, they eat more frequently and chew their food more thoroughly.

Animals that are on maintenance rations may consume more feed than necessary if feed is provided *ad libitum*. Thus, only lactating does and market rabbits 4 to 8 weeks of age should be fed free-choice.

Filling self-feeders with enough pelleted feed to last several days may cause problems in areas with high humidity, resulting in pellets absorbing moisture, expanding, lodging in the feeders, and molding.

Fig. 26-8. Two common rabbit feeders. *Left:* A solid-bottom feeder. *Right:* A self-cleaning feeder with a screened bottom which allows the fines to filter out without clogging the feeder. (Courtesy, Pel-Freez Rabbit Meat, Inc., Rogers, Ark.)

OTHER FEED AND MANAGEMENT ASPECTS

In any discussion of rabbit nutrition, certain managerial and nutritionally related health aspects should be considered.

• **Fur Eating**—Quite often, if there is a deficiency of fiber, protein, or some minerals in the feed, rabbits will resort to eating their fur. If too much fur is eaten, there is a danger of gastrointestinal obstruction which, if unnoticed and unrectified, can result in death. When fur eating is a problem, the ration should be reviewed and corrective measures should be taken.

Rabbits suffering from a blockage in the gastrointestinal tract should be given 10 cc of fresh pineapple juice for three consecutive days, along with free-choice hay or straw. The pineapple juice can be administered with an eyedropper, spoon, or stomach tube—preferably, the latter.

• **Enteric Diseases**—Enteric diseases are a major cause of death in young rabbits worldwide. Previously, most diarrheal diseases were called enteritis complex or mucoid enteritis. Currently, three specific enteric diseases have emerged—enterotoxemia, mucoid enteropathy, and Tyzzer's disease.

• **Enterotoxemia**—This is an explosive diarrheal disease, primarily of rabbits 4 to 8 weeks of age, although it may affect rabbits of any age. The clinical signs are profuse diarrhea, loss of appetite, rough hair coat, and death within 48 hours. Several bacteria have been implicated as causing this disease—especially *Clostridium spiroforme, C. perfringens,* and *Escherichia coli.* Because of the rapidity of death, treatment is seldom attempted. The diet is a big factor in the development of the disease; much less enterotoxemia is seen when high fiber diets are fed. The feeding of hay, straw, and whole oats appears to lessen the incidence of the disease.

• **Mucoid enteropathy**—This is a diarrheal disease of rabbits of any age. While the cause of the disease is unknown, it has been shown to be the result of constipation—an impaction in the digestive tract. The clinical signs of the disease are gelatinous or mucous-covered feces, loss of appetite, loss of weight, drinking of large quantities of water, and often a bloated abdomen due to excess water in the stomach. There is no effective treatment. Changing to a new batch of feed will usually check the disease, for unknown reason.

• **Tyzzer's disease**—Tyzzer's disease, named after the man who discovered it in Japanese Waltzing mice in 1917, is caused by *Bacillus piliformis* and is associated with poor sanitation and stress. It results in severe diarrhea and death in young rabbits 6–12 weeks of age. The disease also affects rodents, cats, and foals. It is characterized by profuse diarrhea, loss of appetite, dehydration, and death within 1 to 3 days. Post-mortem signs are very similar to those of enterotoxemia, except for white spots of salt-grain size on the liver. No treatment is effective, but control is accomplished by not allowing dogs access to the area where feed and nesting materials are stored.

Fig. 26-9. Dutch breed, which originated in Holland. Rabbits are nonruminant herbivores. Their digestive anatomy and physiology closely resemble those of the horse. (Courtesy, Pel-Freez Rabbit Meat, Inc., Rogers, Ark.)

QUESTIONS FOR STUDY AND DISCUSSION

1. Why is per capita rabbit consumption so much higher in Europe than in the United States?

2. Discuss the efficiency of rabbits versus beef cattle in the production of meat.

3. What characteristics of rabbit meat should appeal to health conscious consumers?

4. Discuss the economic importance of feed for rabbits.

5. Of what importance is the cecum of the rabbit from the standpoint of the kind of feed eaten?

6. What is coprophagy and what role does it play in rabbit nutrition?

7. Of what value are the NRC requirements for rabbits?

8. What factors affect the nutritional requirements of rabbits?

9. Do rabbits have an essential amino acid requirement? If so, what amino acids are most likely to be deficient?

10. Why doesn't the rabbit utilize nonprotein nitrogen effectively

11. Can coprophagy increase the biological value of protein for rabbits?

12. What considerations should be reviewed when supplying calcium and phosphorus to rabbits?

13. Why can rabbit ration formulas be based on total phosphorus rather than on available phosphorus?

14. Analyze the following situation: A rabbit producer is certain that the ration being fed contains adequate magnesium. Yet, the animals are showing all the symptoms of a magnesium deficiency. What is likely the problem, and how can it be remedied?

15. Cobalt is not required in most nonruminant rations. Why is it of concern in the nutrition of rabbits?

16. Describe vitamin A deficiencies in rabbits. Should there be concern about vitamin A toxicity?

17. If a rabbit shows symptoms of muscular dystrophy, what nutrient(s) might be deficient in the ration?

18. What factors influence the water requirements of rabbits?

19. How does the rabbit compare with the horse in the digestibility of fiber? Why is fiber a very important component of rabbit feed?

20. What is coccidiosis, and how is it treated in rabbits?

21. List and discuss the rabbit feeds commonly used as sources of (a) energy, (b) protein, and (c) dry forages.

22. Discuss the preparation of feeds for rabbits.

23. What's the difference between nutrient requirements and nutrient allowances?

24. Suggest a ration, including the daily amount, for a 10 lb pregnant doe.

25. List and discuss some alternatives to traditional feeds that a rabbit producer may consider when selecting—(1) grains, by-products, and roots/tubers; and (2) grasses.

26. A producer has an orphan litter. How may they be fed?

27. What are the advantages and disadvantages of hand-feeding in rabbit operations?

28. If fur eating is encountered, what would you do about it?

29. Discuss the feed-related aspects of enteric diseases.

Fig. 26-10. New Zealand White doe and her 8-weeks-old litter. This is the leading meat breed, worldwide. (Courtesy, Pel-Freez Rabbit Meat, Inc., Rogers, Ark.)

Fig. 26-11. California x New Zealand White hybrid market fryers at 8 weeks of age. In recent years, commercial producers have more widely adopted crossbreeding practices. (Courtesy, Department of Animal Science, Alabama Agricultural and Mechanical University, Normal, Ala.)

Fig. 26-12. The raising of rabbits for show, and as 4-H and FFA projects, is very popular. (Courtesy, USDA)

Fig. 26-13. Rabbits eating pellets. (Courtesy, Ralston Purina Co., St. Louis, Mo.)

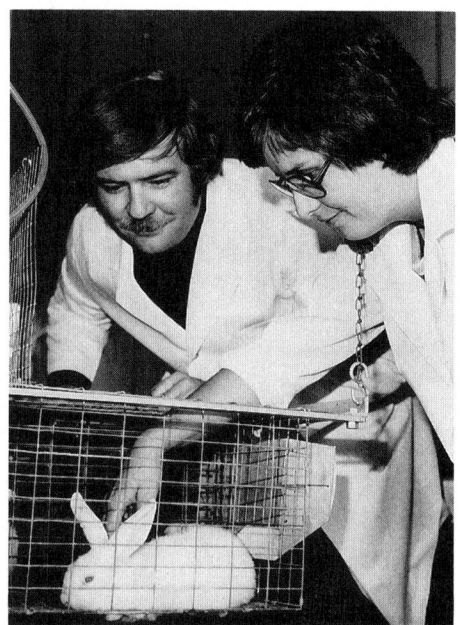

Fig. 26-14. Rabbits used in biomedical research. About 600,000 rabbits are used annually in research, ranging from parasitology to hypertension studies. (Courtesy, Clemson, University, Clemson, S.C.)

Fig. 26-15. Doe with eight bunnies in a wire cage. (Courtesy, Ralston Purina Co., St. Louis, Mo.)

27

FEEDING MINK[1]

Original painting by Tom Phillips

Contents	Page
Economic Importance of Feed for Mink	641
Nutritive Needs of Mink	642
National Research Council (NRC) Requirements	642
Energy	643
Carbohydrates	644
Fats	644
Protein	645
Minerals	645
Mink Mineral Chart	645
Vitamins	647
Mink Vitamin Chart	647
Water	650
Nonnutritive Additives	650
Nutritional Disorders and Toxins	650
Disorders Related to Nutrition	650
Chastek Paralysis (Thiamin Deficiency)	650
Cotton Fur (Iron Deficiency)	651
Gray Underfur (Biotin Deficiency)	651
Nursing Sickness	651
Urinary Calculi (Gravel, Stones, Water Belly, Urolithiasis)	651
Wet-Belly Disease	652
Yellow Fat Disease (Steatitis)	652
Disorders Caused by Toxic Substances	652
Botulism	652
Histamine	652
Mycotoxins	652
Salmonella	652
Thyroid Glands	652
Feeds for Mink	652
Feed Preparation	653
Feed Allowances and Suggested Rations	654
Feeding Breeding Stock	654
Feeding Kits	655
Methods of Feeding and Watering	655
Feed Contaminants	656
Questions for Study and Discussion	656

---

[1]The authors gratefully acknowledge the helpful suggestions of the following eminent authorities who reviewed this chapter: J. Adair, Senior Instructor—Mink, Department of Animal Science, Oregon State University, Corvallis, Ore.; E. Alden, Funbo-Lovsta Research Station, Department of Animal Nutrition and Management, Swedish University of Agricultural Sciences, Uppsala, Sweden; Dr. W. L. Leoschke, Ph.D., Fur Animal Nutrition Research, Chemistry Department, Valparaiso University, Valparaiso, Ind.; B. Smith, Editor, Fur Rancher, Eden Prairie, Minn.; and Dr. W. Wustenberg, DVM, Bay City, Ore.

Foxes were raised commercially for fur earlier than mink (*Mustela vison*), but, in the 1940s U.S. fashions for women changed to slimmer, less bulky styles, for which fox fur was not suited but mink fur was ideal. Thus, the decline of the fox industry and the rise of the mink industry in the 1940s was primarily related to fashions. Hand in hand with this development, the sciences of genetics and nutrition teamed up to make for tremendous growth in commercial mink production. More than 20 different fur colors were created, ranging from white (a Scandinavian specialty) to demibuff (a rich mahogany) to *standard* (formerly brown, now black); and new diets were formulated and fed to maximize production, fur color, and quality.

Today, U.S. fur sales are booming with retail sales of approximately $1.5 billion annually and mink the most popular fur by far. About 2,000 U.S. mink ranches produce more than 4 million pelts each year, with half of the production centered in three states—by rank: Wisconsin, Minnesota, and Utah.

A large part of the world mink output comes from Scandinavia. In 1984, Denmark, Norway, Finland, and Sweden together produced 13.8 million mink pelts, nearly half of the world's production of 28 million.

ECONOMIC IMPORTANCE OF FEED FOR MINK

Feed costs represent the largest single cost in mink production.

Mink are carnivorous animals. While limited amounts of grains and cereal by-products are used in mink diets, meat and fish and their by-products compose the greatest portion (80–85%) of the traditional standard mink feed. However, the supplies of fresh meat (including horsemeat) and fish are becoming both scarce and expensive; additionally, storage, handling, and feeding of these high-moisture feeds makes for additional costs. The high cost of feed has prompted pursuit of two alternatives: (1) greater dependence on meat and fish wastes and by-products (scraps, trimmings, viscera), and (2) formulation of diets totally from dry ingredients because of the ease with which they can be stored and fed.

In the mid-1980s, most ranchers with a four kit average produced a pelt from 110–120 lb of wet feed (cereal + fresh frozen ingredients, containing 65–67% water); and this included the feed consumed by the parents of the kits. This wet feed cost 10–15¢/lb delivered to the ranch. So, the feed cost of each pelt was $11.50 to $17.25. During this same period, in the mid-1980s, pelleted mink feed cost 30–40¢/lb (i.e., 33–44¢/lb on a dehydrated, moisture-free basis), but it required little more than ⅓ as much poundage of the pellets as of the wet feed to produce each pelt; which, along with the saving in storage, handling, and feeding costs, made the dry feed competitive.

In the mid-1980s, more than 10 to 15% of the North American mink were raised on 100% pellets and another 10% received some pellets along with the wet ranch mix. Some of the latter were on a "5 + 2 program," i.e., 7 days of pellets in the hoppers with concurrent redi-mix, or ranch mix, on the wire, and with no feeding of the mink on weekends *except* pellets in the hoppers.

As we learn more about the nutritional requirements and feeding behavior of mink, more and more pellets will be used.

NUTRITIVE NEEDS OF MINK

Mink require the same nutrients as all other animals. The main difference lies in the amounts of each nutrient and the types of feeds which best supply them.

Mink are flesh eaters and have a simple digestive system much like that found in the pig, dog, fish, monkey, and humans. They have the shortest digestive tract of any animal other than fish, along with limited digestive capacity. Also, feed passage through the digestive tract is rapid, averaging 2⅓ hours vs 72 hours for the ruminant cow. So, mink need diets composed of highly digestible ingredients.

National Research Council (NRC) Requirements

Since the mink industry is of comparatively recent origin, less is known about the nutrient requirements of mink than of other domesticated animals. Many of the requirements for growth have been established, but little information has been reported on the requirements for maintenance, gestation, and lactation. Tables 27–1, 27–2, and 27–3 list the nutrient requirements for mink as established by the National Research Council.

Fig. 27-1. An adult male mink will consume 200 lb of wet feed per year, and an adult female, 130 lb. (Courtesy, Ralston Purina Company, St. Louis, Mo.)

TABLE 27–1
NUTRIENT AND ENERGY REQUIREMENTS OF MINK: PERCENTAGE OR AMOUNT PER KILOGRAM OF DRY MATTER[1][2]

Constituent	Unit	Growth		Maintenance (Mature)	Gestation	Lactation
		Weaning to 13 Weeks	13 Weeks to Maturity			
Energy:						
Males	kcal ME[3]	4,080	4,080	3,600	—	—
Females	kcal ME	3,930	3,930	3,600	3,930	4,500
Crude protein	%	38[4]	32.6–38.0[4]	21.8–26.0[4]	38[4]	45.7[4]
Fat-soluble vitamins:						
Vitamin A	IU	5,930	[5]	[5]	[5]	[5]
Vitamin E	mg	27	[5]	[5]	[5]	[5]
Water-soluble vitamins:						
Thiamin	mg	1.3	[5]	[5]	[5]	[5]
Riboflavin	mg	1.6	[5]	[5]	[5]	[5]
Pantothenic acid	mg	8.0	[5]	[5]	[5]	[5]
Vitamin B-6	mg	1.6	[5]	[5]	[5]	[5]
Niacin	mg	20.0	[5]	[5]	[5]	[5]
Folic acid	mg	0.5[6]	[5]	[5]	[5]	[5]
Biotin	mg	0.12	[5]	[5]	[5]	[5]
Vitamin B-12	mcg	32.6	[5]	[5]	[5]	[5]
Minerals:						
Calcium	%	0.4	0.4	0.3	0.4	0.6
Phosphorus	%	0.4	0.4	0.3	0.4	0.6
Ca:P ratio		1:1 to 2:1	1:1 to 2:1	1:1 to 2:1	1:1 to 2:1	1:1 to 2:1
Salt	%	0.5	0.5	0.5	0.5	0.5

[1]Adapted by the authors from *Nutrient Requirements of Mink and Foxes*, No. 7, 2nd rev. ed., NRC-National Academy of Sciences, 1982, p. 33.
[2]Nutrient requirements are based on an energy level of 5,300 kcal E, or 4,080 kcal ME. Nutrient requirements will increase with higher ME levels and decrease with lower ME levels.
[3]For method of calculating ME, see *Nutrient Requirements of Mink and Foxes*, No. 7, NRC, 1982, p. 38, Table 9.
[4]Based on average quality protein with calculated digestibility of 83%. Higher-quality protein and higher digestibility will decrease the requirement, and lower-quality protein and lower digestibility will increase the requirement.
[5]Quantitative requirement not determined, but dietary need demonstrated.
[6]May not be minimum, but known to be adequate.

Feeding Mink

TABLE 27-2
NUTRIENT AND ENERGY REQUIREMENTS OF MINK: DAILY AMOUNTS[1]

Constituent	Unit	Growth		Maintenance (Mature)	Gestation	Lactation
		Weaning to 13 Weeks	13 Weeks to Maturity			
Energy:						
Males	kcal ME/kg BW	275–330	250–140	140	—	—
Females	kcal ME/kg BW	280–355	280–150	140	200	200–500
Amount per 100 kcal metabolizable energy[2]:						
Digestible protein	kcal ME[3]	35	30–35	20–24	35	42
Fat-soluble vitamins:						
Vitamin A	IU	145	[4]	[4]	[4]	[4]
Vitamin E	mg	0.66	[4]	[4]	[4]	[4]
Water-soluble vitamins:						
Thiamin	mg	0.033	[4]	[4]	[4]	[4]
Riboflavin	mg	0.04	[4]	[4]	[4]	[4]
Pantothenic acid	mg	0.2	[4]	[4]	[4]	[4]
Vitamin B-6	mg	0.04	[4]	[4]	[4]	[4]
Niacin	mg	0.5	[4]	[4]	[4]	[4]
Folic acid	mcg	13.0[5]	[4]	[4]	[4]	[4]
Biotin	mcg	3.0	[4]	[4]	[4]	[4]
Vitamin B-12	mcg	0.8	[4]	[4]	[4]	[4]

[1]Adapted by the authors from *Nutrient Requirements of Mink and Foxes*, No. 7, 2nd rev. ed., NRC-National Academy of Sciences, 1982, p. 34.
[2]When original data were not presented on the basis of ME, requirements were calculated according to the following formula: ME = 0.77 E.
[3]For method of calculating ME, see *Nutrient Requirements of Mink and Foxes*, No. 7, 1982, p. 38, Table 9.
[4]Quantitative requirement not determined, but dietary need demonstrated.
[5]May not be minimum, but known to be adequate.

TABLE 27-3
AVERAGE DAILY REQUIREMENTS FOR METABOLIZABLE ENERGY (ME) AND DRY FEED FOR GROWTH OF MINK[1]

	Unit	Age (Weeks)												
		7	9	11	13	15	17	19	21	23	25	27	29	31
Male:														
Body weight[2]	g	630	930	1,240	1,520	1,730	1,900	2,040	2,160	2,260	2,330	2,350	2,380	2,380
ME daily[3]	kcal	173	307	394	445	435	439	441	436	387	336	323	284	278
Daily dry feed containing:														
3,500 kcal ME/kg	g	49	88	113	127	124	125	126	125	111	96	92	81	79
4,000 kcal ME/kg	g	43	77	99	111	109	110	110	109	97	84	81	71	70
4,500 kcal ME/kg	g	38	68	88	99	97	98	98	97	86	75	72	63	62
Female:														
Body weight[2]	g	450	650	810	930	1,030	1,110	1,180	1,240	1,280	1,320	1,325	1,320	1,300
ME daily[3]	kcal	126	231	284	323	289	273	260	266	260	231	210	197	196
Daily dry feed containing:														
3,500 kcal ME/kg	g	36	66	81	92	83	78	74	76	74	66	60	56	56
4,000 kcal ME/kg	g	32	58	71	81	72	68	65	67	65	58	53	49	49
4,500 kcal ME/kg	g	28	51	63	72	64	61	58	59	58	51	47	44	44

[1]Adapted by the authors from *Nutrient Requirements of Mink and Foxes*, No. 7, 2nd rev. ed., NRC-National Academy of Sciences, 1982, p. 36.
[2]Average body weights of animals reared on Oregon State University (OSU) stock diet, reported by Oldfield, *et al.* (1971).
[3]Calculated from average daily intakes of OSU diet, reported by Oldfield, *et al.* (1978). The OSU diet had a calculated E content of 5,790 kcal/kg dry matter and a calculated ME content of 4,410 kcal/kg dry matter. (The ME calculation employed: (a) the specific digestibility coefficients reported by Glem-Hansen and Joergensen [1978] for the proximate constituents of similar diet ingredients, and (b) the values of 4.0, 9.5, and 4.5 kcal ME/g of digested carbohydrate, fat and protein, respectively [Glem-Hansen, 1978]. The calculated ME in this instance is 76.2% of the calculated E.)

Energy

Metabolizable energy (ME) is the common energy base used in feeding standards and ration formulation of mink. It is determined by subtracting energy losses in the urine and combustible gases from the digestible energy (DE) consumed.

The quantity of energy consumed by mink is directly related to the need to satisfy their energy requirement. So, energy density is of practical importance. *Energy density refers to the concentration of the energy in the diet.* A diet of high-energy density (a high-energy diet) provides more kilocalories per gram than does a diet of low-energy density (a low-energy diet).

Like most animals, mink tend to eat to satisfy their energy requirements; so, they generally consume smaller amounts of high-energy feeds than of low-energy feeds. It follows that if the protein and micronutrient levels are not relatively high in a high-energy diet, the amounts of the respective nutrients that are actually consumed may be less than recommended for normal growth or production. Thus, it is mandatory that the producer be certain that the other essential nutrients are being supplied in adequate amounts in high-energy diets. Conversely, if a low-energy diet is fed to mink, protein and micronutrient levels should be reduced because more feed will be consumed. A high-protein, low-energy diet could be expensive due to the excessive amounts of costly protein consumed in the process of meeting the energy requirement. Also, it should be recognized that the capacity and the digestive capability of the gastrointestinal tract of mink are limited, with the result that it may be physically impossible for the animal to consume sufficient amounts of a low-energy diet to satisfy energy demands, particularly in periods of high-energy requirements such as lactation and early kit growth.

The energy requirements of mink are affected by many different factors such as age, gestation, lactation, environmental temperature, and nutritional deficiencies.

A deficiency of energy in mink causes retardation or cessation of growth accompanied by varying stages of emaciation, dull fur—lacking in sheen, and depressed milk yields in lactating females.

CARBOHYDRATES

Mink, in common with other species, do not have a carbohydrate requirement as such. The function of carbohydrates in mink diets is to supply energy.

Cereal grains and their by-products are the chief sources of carbohydrates, primarily starch. Carbohydrates may supply 15–30% of the ME for growth and for fur development, and 10–20% of the ME for pregnancy and lactation.

The digestibility of cereal grains can be significantly increased either by (1) heat treatment (cooking, popping, steam rolling, toasting), or (2) fine grinding (almost to flour consistency).

FATS

Fat (lipid), which is the most concentrated source of energy, is most commonly supplied in mink diets by meat and fish and their by-products. For high-energy diets, rendered fats and oils (tallow, lard, fish oils, and/or vegetable oils) are usually added. The cereals and their by-products are generally low in fat.

In addition to furnishing energy, fat provides essential fatty acids; notably, linoleic, linolenic, and arachidonic acids, small quantities of which need to be supplied in the feed. Linoleic and linolenic acids occur mainly in vegetable oil, while arachidonic acid is found in animal fat. Danish researchers report that, for optimal growth from birth to weaning, kits have a linoleic acid requirement of 5% of the ME. Since mink diets traditionally include high levels of fat, there have been no reported cases of deficiencies of essential fatty acids in practical rations.

The digestibility of fat depends largely upon the composition of the fatty acids. But it generally ranges from 80–90%, with an average of about 85%.

Dr. Wm. L. Leoschke, Valparaiso University, Valparaiso, Indiana, noted mink authority, recommends that fat supply

TABLE
MINK MIN-

Minerals Which May Be Deficient Under Normal Conditions	Conditions Usually Prevailing Where Deficiencies Are Reported	Function of Mineral	Deficiency/Excess/Symptoms
Major or macrominerals: Salt (NaCl)	Lactation, when high amounts of sodium are secreted in milk.	Sodium chloride helps maintain osmotic pressure in body cells, upon which depends the transfer of nutrients to the cells, the removal of waste materials, and the maintenance of water balance among the tissues. Also, sodium is important in making bile, which aids in the digestion of fats and carbohydrates; and chlorine is required for the formation of hydrochloric acid in the gastric juice so vital to protein digestion. It is noteworthy that when salt is omitted, sodium expresses its deficiency first.	**Deficiency**: Nursing sickness in lactating females, characterized by loss of appetite, emaciation, weakness, incoordination, and coma. **Excess**: 1.5% salt (dry basis) in diet of kits results in reduced reproduction during the following breeding period.
Calcium (Ca)	Diets that have a serious imbalance of calcium and phosphorus or are deficient in vitamin D. When sunlight is restricted.	Essential for the development and maintenance of good strong bones and teeth; maintains the contractability, rhythm, and tonicity of the heart muscles; antagonizes the action of the sodium and potassium on the heart; is required for normal coagulation of the blood; is necessary for proper nerve irritability; and appears to be essential for selective cellular permeability.	**Deficiency**: Abnormal bone growth. **Excess**: On a high Ca and low vitamin D diet, mink kits show the following signs, progressively: difficulty in walking, they crawl, unable to stand, knobs on the ribs, concave spinal column (lordosis), and leg bones bend and enlarge at the ends.
Phosphorus (P)	Diets that have a serious imbalance of calcium and phosphorus.	Essential for sound bones and teeth, and for the assimilation of carbohydrate and fats. A vital ingredient of the proteins in all body cells. Necessary for enzyme activation. Acts as a buffer in blood and tissue. Occupies a key position in biologic oxidation and reactions requiring energy.	**Deficiency**: Abnormal bone growth.
Magnesium (Mg)			**Deficiency**: Magnesium deficiency is not a problem. **Excess**: High magnesium may decrease the absorption of Ca and P.
Potassium (K)			

the following percentages of the total ME of the diet: for growth, 44–53%; for fur development (including late growth), 42–47%; for pregnancy, 34–37%; and for lactation, 47–50%.[2] In the Scandinavian countries, an average of 35–45% of the metabolizable energy in mink diets is of fat origin (Eva Alden, Funbo-Lovsta Research Station, Department of Animal Nutrition and Management, Swedish University of Agricultural Sciences, Uppsala, Sweden, personal communication to the senior author).

Protein

The protein requirement of mink is actually the requirement for essential amino acids. It follows that the protein requirement is related to the protein quality in a given feedstuff.

The protein of mink feed should be of high biological value, especially during the critical periods of gestation, lactation, and immediately following weaning. That is, the protein should contain relatively balanced and desirable levels of the essential amino acids which are available to the animal. Muscle tissue contains large amounts of lysine and methionine, while fur contains considerable amounts of methionine, arginine, and cystine. Requirements for essential amino acids in mink have not been established, but it can safely be assumed that mink have an essential amino acid requirement somewhat similar to those of other nonruminant, mammalian livestock—for example, swine.

Meat is a high quality protein feedstuff for mink, as it (1) contains an amino acid pattern similar to the actual amino acid requirements of the animal, and (2) is highly digestible. Likewise, properly processed fish meal is an excellent source of high-quality protein for mink. However, excessive heating of fish products during dehydration can result in impairment of protein quality by the well known Maillard or browning reaction, resulting in the destruction of the amino acids lysine, cystine/methionine, and tryptophan, and the bonding of the amino acid arginine in an indigestible form.

Feed consumption of mink is determined primarily by the taste appeal and caloric density of the diet. Because of the critical role of dietary energy density in the determination of mink feed intake, it is logical to relate the protein requirements of the mink to the energy content of the diet rather than merely to list them as a percentage of protein in the diet.

Minerals[3]

Most of the research on the mineral needs of mink has been limited to salt, calcium, phosphorus, and iron, which is indicative of the main mineral problem areas. The proper combination of common mink feed ingredients will reduce or eliminate the need for supplementation with other minerals.

A summary of pertinent information concerning mineral nutrition in mink is presented in Table 27-4, Mink Mineral Chart; and additional information is presented in the narrative that follows.

---

[2]*Nutrient Requirements of Mink and Foxes*, No. 7, 2nd rev. ed., NRC-National Academy of Sciences, 1982, p. 9.

[3]All mineral guidelines in this section are on a dry matter basis unless otherwise stated.

27-4
ERAL CHART

	Mineral Requirements[1]			
Minerals/ Animal/day	Mineral Content of Dry (M-F) Diet	Recommended Allowances in Dry (M-F) Diet[1]	Practical Sources of the Mineral	Comments
	*1.3 to 1.5% in the dry diet (0.5% salt in wet feed) will prevent nursing sickness.	*0.5% salt is routinely added to wet feeds, especially for mink in lactation.	Salt (sodium chloride).	Fresh meat contains very little salt. Higher levels of salt will increase water intake. Salt is inexpensive and is routinely added to mink diets.
	*Requirements are variable according to age, sex, and production (see Table 27-1). *For growing mink, 0.4 to 1.0% Ca.[2]	0.5 to 1.0%	Bone meal Defluorinated rock phosphate. Dicalcium phosphate. Ground limestone. Oystershell. Fish meals.	For growing mink, the Ca:P ratio should be 1:1 to 1.2:1. Meat and meat by-products are extremely low in calcium. Vitamin D is necessary for proper calcium absorption.
	*Requirements are variable according to age, sex, and production (see Table 27-1). *For growing mink, 0.4 to 0.8 P.[2]	0.4 to 0.9%	Monosodium phosphate. Dicalcium phosphate. Steamed bone meal.	Meat products are generally good sources of phosphorus. On the average, only 30% availability of P should be assumed.
	*440 mg/kg of diet.			Excess of calcium or phosphorus in the diet may decrease the absorption of magnesium, and vice versa.
		*Add 0.3% potassium to breeder and grower diets.		Potassium is adequate in diets containing 10 to 30% cereal, because cereals are rich in potassium.

(Continued)

TABLE 27-4

Minerals Which May Be Deficient Under Normal Conditions	Conditions Usually Prevailing Where Deficiencies Are Reported	Function of Mineral	Deficiency/Excess/Symptoms
Trace or microminerals: Copper (Cu)			
Iodine (I)			
Iron (Fe)	Some ocean fish (Pacific hake, Atlantic whiting, and coalfish) can induce an iron deficiency.	Iron is a part of the hemoglobin molecule, essential for oxygen transport.	**Deficiency**: Cotton fur—an almost complete lack of pigmentation in the under fur; depressed growth; emaciation; anemia.
Manganese (Mn)		ciency, where it results in "screw neck" or head tilting—a birth defect. The condition can be prevented by supplementing pregnant mink with 1,000 ppm of manganese.	**Deficiency**: Pastel mink are prone to manganese defi-
Selenium (Se)			
Zinc (Zn)			**Deficiency**: There is no evidence of zinc deficiency.

[1]As used herein, the distinction between "mineral requirements" and "recommended allowances" is as follows: In mineral requirements, no margins of safety are included intentionally, whereas in recommended allowances, margins of safety are provided in order to compensate for variations in feed composition, environment, and possible losses during storage or processing. Where preceded by an asterisk, the mineral requirements, recommended allowances, and other facts presented herein were taken from *Nutrient Requirements of Mink and Foxes*, No. 7, 2nd rev. ed., NRC-National Academy of Sciences, 1982.

- **Salt**—During lactation, females sometime deplete their salt reserves because milk is relatively high in sodium. When this occurs, the female becomes thin and emaciated, and her fur has an unthrifty appearance. This condition is commonly called *nursing sickness*. Although most mink diets are fortified with salt, a lactating female that becomes extremely thin should be immediately separated from her kits if they are old enough to eat solid food.
- **Calcium and Phosphorus**—The ratio of calcium to phosphorus should be 1:1 to 1.2:1, but mink can get along satisfactorily on levels up to 2:1. Generally, meats and their by-products are extremely low in calcium but high in phosphorus. For example, on an as-fed basis, beef kidney contains 0.01% calcium and 0.22% phosphorus—a ratio of 1 part of calcium to 22 parts of phosphorus. On the other hand, fish meals and whole fish or animal products containing bone provide a good balance of calcium and phosphorus.

It is generally recommended that calcium constitute 0.5 to 1.0% of the dry diet, and that phosphorus make up 0.4 to 0.9% on a dry basis, depending on the age and type of production of the animal.

On a rachitogenic diet high in calcium and low in phosphorus and vitamin D, mink kits develop rickets, characterized by difficulty in walking, knobs on the ribs, lordosis (the spinal column in the thoracic region becomes concave), and the leg bones bending and enlarging at the ends.

TABLE
MINK VITAMIN

Vitamin Which May Be Deficient Under Normal Conditions	Conditions Usually Prevailing Where Deficiencies Are Reported	Function of Vitamin	Deficiency/Excess/Symptoms
Fat-soluble vitamins: Vitamin A		Essential for (1) normal maintenance and functioning of the epithelial tissues, particularly of the eye and the respiratory, digestive, reproductive, nerve, and urinary systems; (2) the production of visual pigments in the eye, which are necessary for vision in dim light; and (3) growth of bony structures.	**Deficiency**: Night blindness, lack of coordination, xerophthalmia, metaplasia of epithelial tissues, depressed growth, fatty livers, and abnormal development of the skull. **Excess**: Loss of appetite; bone change with bony outgrowth (spurs) from bone, decalcification, and fractures; loss of fur; protruding eyes; very sensitive skin; and lowered reproduction.
Vitamin D	Confinement, where animals do not have access to sunlight.	Vitamin D is associated with calcium absorption, transportation, and deposition. Vitamin D, phosphorus, and calcium all play a role in the prevention of rickets, and their effectiveness depends upon proper amounts of each.	**Deficiency**: Rickets and abnormal bone development. **Excess**: Loss of appetite, nausea, loss of weight, and digestive disorders.
Vitamin E (alpha-tocopherol)	Feeds containing large amounts of fish scrap, horsemeat, or poultry offal.	Necessary for reproduction. Protection of vitamin A and carotene from oxidation. Hence, it *stretches* the supply of vitamin A. As an antioxidant in feed.	**Deficiency**: Yellow fat disease: abnormal eating behavior, blood in urine, and anemia. Cotton fur may accompany yellow fat disease if rancid fat is fed. On a synthetic diet (without consumption of rancid fat), the signs are: sudden death due to minor stress, and erythrocyte (red blood corpuscle) fragility.

Feeding Mink 647

(Continued)

Mineral Requirements[1]		Recommended Allowances in Dry (M-F) Diet[1]	Practical Sources of the Mineral	Comments
Minerals/ Animal/day	Mineral Content of Dry (M-F) Diet			
	*4.5–6.0 ppm.			The copper requirement is adequately met by diets containing fish.
	*0.2 ppm for breeder and growth diets.			Fish containing diets provide adequate iodine.
	*The minimum requirement is not known.	*44 and 40 ppm for breeder and grower diets, respectively.		
	*The minimum requirement is not known.	*0.1 ppm.		
	*66 and 59 ppm on a dry matter basis for breeder and grower diets, respectively.			There is no evidence of zinc deficiency in mink.

[2]These requirements apply provided (1) the vitamin D concentration is 820 IU/kg dry feed, and (2) the calcium-to-phosphorus ratio is between 0.75:1.0 and 1.7:1.0.

- **Iron**—When certain raw ocean fish (Pacific hake, Atlantic whiting, or coalfish) are fed to mink, severe anemia and *cotton fur* may result. In cotton fur, there is a depigmentation of the underfur which severely reduces the value of the pelt.

The malfunction of iron is caused by very high levels of trimethylamine oxide (TMAO) present in the species of fish named, which is broken down by enzymes present in the fish digestive track to yield several products, including formaldehyde (FA). Both TMAO and FA interfere with iron absorption and may cause anemia and cotton fur. Freezing raw fish accentuates the problem while cooking the fish at 200°F destroys or inactivates the causative factor.

Cotton fur may also be caused by feeding rancid fat, which interferes with iron absorption. The problem can be corrected by administering iron parenterally. However, feeding iron supplements has met with mixed success.

Vitamins

The experimental basis for specifying the vitamin requirements and allowances of mink is rather limited. However, based on experiments and experiences, some guidelines have been established. Vitamins A and E have received the most attention in mink nutrition. The vitamin requirements are presented in Table 27-5.

27-5 CHART (See footnote at end of table.)

Vitamin Requirements[1]		Recommended Allowances of Dry (M-F) Diet[1]	Practical Sources of the Vitamin	Comments
Vitamins/Animal/Day	Vitamin Content of Dry (M-F) Diet			
*For growing mink, 91 IU/lb (200 IU/kg) of body weight, or 145 IU/100 kcal ME.	*Young mink require 2,695 IU/lb (5,930 IU/kg) dry matter.	For young mink, 2,965 IU/lb (6,523 IU/kg) dry matter.	Synthetic vitamin A. Liver. Milk fat.	Mink are poor converters of carotene; hence, the carotene content of the diet should be disregarded in supplying the vitamin A requirements of mink.
		Where mink are not exposed to sunlight, add 227 IU of vitamin D/lb feed (500 IU/kg feed).	Synthetic vitamin D. Sunlight. Fish-liver oils.	A deficiency of vitamin D is unlikely when mink are allowed exposure to the sun. Supplementation of vitamin D will not overcome the effects of an acute calcium-phosphorus imbalance.
*Young mink require 0.66 mg/100 kcal ME.	*Young mink require 12.3 mg/lb (27 mg/kg) dry matter.	For young mink, 13.5 mg/lb (29.7 mg/kg) dry matter.	Alpha-tocopherol. Wheat germ oil. Liver. Cereal grains.	Selenium can spare vitamin E to a limited degree. Iron and copper destroy vitamin E. Synthetic antioxidants added to the diet can aid in the prevention of yellow fat disease.

(Continued)

TABLE 27-5

Vitamin Which May Be Deficient Under Normal Conditions	Conditions Usually Prevailing Where Deficiencies Are Reported	Function of Vitamin	Deficiency/Excess/Symptoms
Fat-soluble vitamins: (Continued)			
Vitamin K	Little experimental work, but a deficiency of vitamin K in practical diets is unlikely.		
Water-soluble vitamins:			
Biotin	Diets containing raw egg or turkey offal, due to the presence of avidin. Avidin is a protein found in egg white and oviduct tissue, which binds biotin and prevents its absorption. Including as little as 5% spray-dried eggs in the diet without supplementing with biotin.	As an essential component of a coenzyme concerned in carbon dioxide fixation. Biotin enters into several reactions of intermediary metabolism.	**Deficiency**: Graying or banding of underfur in dark mink and sometimes loss of hair. On biotin-free purified diets, deficient animals show "spectacle eyes," crusty feet, yellow or bloody exudate, and a dermatitis of the foot pads, in addition to the gray underfur.
Folacin (Folic Acid)		Enters into certain enzyme systems that are concerned with nucleic acid metabolism. Essential for the formation of blood cells.	**Deficiency**: Loss of appetite and weight, anemia, and bloody diarrhea.
Niacin (Nicotinic Acid, Nicotinamide)		As a constituent of coenzymes I and II. These coenzymes are essential in biologic oxidation, especially in the oxidation of carbohydrates.	**Deficiency**: Loss of appetite and weight and diarrhea.
Pantothenic Acid (Vitamin B-3)		Part of coenzyme A, a necessary factor for intermediary metabolism. Maintenance of hair and skin.	**Deficiency**: Loss of appetite, diarrhea, weakness, emaciation, and dehydration.
Riboflavin (Vitamin B-2)		Primarily in protein metabolism. Essential for normal growth, maintaining a healthy condition of the skin, and reproduction.	**Deficiency**: Loss of appetite, loss of weight, and extreme weakness. Subsequently, riboflavin-deficient kits reproduce poorly even when fed adequate riboflavin thereafter.
Thiamin (Vitamin B-1)	Diet containing raw fish that have thiaminase.	Carbohydrate metabolism of all living cells. Promotes growth, appetite, and digestion.	**Deficiency**: Chastek paralysis. Loss of appetite and weight, diarrhea, and paralysis.
Vitamin B-6 (Pyridoxine, Pyridoxal, Pyridoxamine)		Functions in the metabolism of protein. Also, necessary for nerves, proper heart function, blood regeneration, and prevention of anemia.	**Deficiency**: In growing kits, there is reduced feed intake, loss of weight, diarrhea, brown exudate around the nose, excessive tears, swelling and puffiness around the nose and face region, listlessness, muscular incoordination, convulsions, and finally death. In breeding females, there is lowered conception and number of kits per litter. In males, there is degeneration of the testes.
Vitamin B-12 (Cobalamins)		Necessary for growth. Red blood formation. It is essential to prevent and cure anemia, and to facilitate the development of lots of erythrocytes capable of carrying ample oxygen from the lungs to the muscle. Increased food utilization.	**Deficiency**: Loss of appetite, loss of weight and fatty degeneration of liver.
Vitamin C (Ascorbic Acid, Dehydroascorbic Acid)	No requirement for vitamin C for mink has been demonstrated.		

[1]As used herein, the distinction between "vitamin requirements" and "recommended allowances" is as follows: In vitamin requirements, no margins of safety are included intentionally; whereas in recommended allowances, margins of safety are provided in order to compensate for variations in feed composition, environment, and possible losses during storage or processing. In this table, most allowances are 10% higher than the requirements.

Feeding Mink

(Continued)

Vitamins/Animal/Day	Vitamin Requirements[1] Vitamin Content of Dry (M-F) Diet	Recommended Allowances of Dry (M-F) Diet[1]	Practical Sources of the Vitamin	Comments
*Young mink require 3.0 mcg/ 100 kcal ME.	*Young mink require 0.05 mg/lb *(0.12 mg/kg)* dry matter.	For young mink, 0.05 mg/lb *(0.12 mg/kg)* dry matter.	Synthetic biotin. Yeast. Liver. Milk. Molasses.	Cooking raw eggs and turkey offal at 196°F *(91°C)* for 5 minutes destroys the antivitamin avidin. Deficiencies of biotin are not normally encountered on conventional mink diets.
*Young mink require 13.0 mcg /100 kcal ME.	*For young mink, 0.23 mg/lb *(0.5 mg/kg)* of dry feed, or 0.135 mg/100 kcal ME, may not be minimum, but is adequate.	For young mink, 0.23 mg/lb *(0.51 mg/kg)* dry feed.	Synthetic folic acid. Yeast. Liver. Meat.	Typical ranch feeds contain adequate folic acid.
*Young mink require 0.5 mg/ 100 kcal ME.	*Young mink require 9.1 mg/lb *(20 mg/kg)* dry matter.	For young mink, 10.01 mg/lb *(22.02 mg/kg)* dry matter.	Synthetic niacin. Meat. Yeast. Cereals. Fish. Liver.	Not likely to be deficient because fish, meat, liver, and mink milk are good sources. Mink are unable to convert sufficient tryptophan into niacin.
*Young mink require 0.2 mg/ 100 kcal ME.	*Young mink require 3.64 mg/lb *(8.0 mg/kg)* dry matter.	For young mink, 4.0 mg/lb *(8.8 mg/kg)* dry matter.	Calcium pantothenate. Organ meats. Eggs. Cereals. Fish solubles. Certain vegetables.	
*Young mink require 0.04 mg/ 100 kcal ME.	*Young mink require 0.73 mg/lb *(1.6 mg/kg)* dry matter.	For young mink, 0.80 mg/lb *(1.76 mg/kg)* dry matter.	Synthetic riboflavin. Liver. Milk products. Kidney. Heart. Eggs. Soybean.	
*Young mink require 0.033 mg/100 kcal ME.	*Young mink require 0.59 mg/lb *(1.3 mg/kg)* dry matter.	For young mink, 0.65 mg/lb *(1.43 mg/kg)* dry matter.	Synthetic thiamin. Pork. Cereal grains. Oilseed meals. Milk products. Brewers' yeast.	Fish known to contain thiaminase should be cooked before feeding.
*Young mink require 0.04 mg/ 100 kcal ME.	*Young mink require 0.73 mg/lb *(1.6 mg/kg)* dry matter.	For young mink, 0.8 mg/lb *(1.76 mg/kg)* dry matter.	Vitamin B-6. Yeast. Liver. Cereal grains. Meat. Egg yolk. Milk.	
*Young mink require 0.8 mcg/ 100 kcal ME.	*Young mink require 14.8 mcg/lb *(32.6 mcg/kg)* dry matter.	For young mink, 16.3 mcg/lb *(35.86 mcg/kg)* dry matter.	Vitamin B-12. Fish meal. Meat scraps. Liver. Dried skim milk.	Generally not needed as a supplement since animal protein is high in B-12.

Where preceded by an asterisk, the vitamin requirements, recommended allowances, and other facts presented herein were taken from *Nutrient Requirements of Mink and Foxes*, No. 7, 2nd rev. ed., NRC-National Academy of Sciences, 1982.

Fig. 27-2. Biotin deficiency in dark mink. *Left to right:* normal; gray-banded; and gray underfur. (Courtesy, John Adair, Department of Animal Science, Oregon State University, Corvallis)

Water

Clean, fresh water should be available free-choice to mink at all times.

The water requirements of mink depend on various factors, including the water content of the feed, the stage of the life cycle of the mink, and the weather.

On dry feed containing about 10% water, mink drink more water than when on wet feed containing 65-75% water. Experiments show that mink need 2.8 oz of water per 1 oz of dry feedstuff, and that it does not matter whether the water is mixed with the feed or given separately. This points up the critical need for water when mink are on dry feed.

In the last part of the growth period, a male mink weighing 4.5 lb and a female mink weighing 2.25 lb have a daily water requirement of approximately 0.6 lb and 0.4 lb, respectively. The most critical stages are during lactation and right after suckling kits start eating solid feed.

In the winter, it may be necessary to heat the water to prevent freezing. If the watering system is not automatic, the waterers should be checked often during the cold weather to make sure that fresh, unfrozen water is always available to mink. In the summer, heat can place heavy stresses on mink, which can be partially alleviated by clean, cool water. The water systems should be cleaned routinely to prevent the growth of pathogenic organisms.

Nonnutritive Additives

Most mink feeds contain one or more nonnutritive factors, which are not nutrients as such. Nevertheless, some of them may improve production and prevent or control diseases under certain circumstances, whereas others prevent rancidity in feed. Among such factors are antibiotics and antioxidants.

1. **Antibiotics.** Antibiotics may be fed to mink (1) at low levels to modify bacterial populations, and (2) at high levels to suppress disease-producing organisms.

The response of mink to dietary antibiotics has been variable and may depend upon (1) the quality of the diet, (2) the health of the animal, and (3) the resistance of the bacteria harbored by the mink.

Although the results have not been consistent, most experiments and experiences attribute the following to the feeding of antibiotics: (1) increased growth rate of weaned kits, (2) improved pelt quality, and (3) increased weaning weights and reduced kit mortality when fed to female breeders.

When adding antibiotics (or any drug) to a feed, the producer should read the drug label and follow the directions carefully. Never exceed the recommended level.

(Also see Chapter 13, Feed Supplements/Additives/Implants, section on "Antibiotics.")

2. **Antioxidants.** Antioxidants are commonly used in mink feeds to prevent rancidity due to the autoxidation of fats. When improperly stored, or when stored for prolonged periods, horsemeat and fish are especially prone to oxidation; and when fed to mink, may cause yellow fat disease (steatitis). The compounds which are currently being used for addition to fat in mink rations are Santoquin (ethoxyquin), BHT (butylated hydroxytoluene), and BHA (butylated hydroxyanisole). When levels of 55 mg per pound of wet feed of these compounds are used, yellow fat disease (steatitis) can be prevented. These compounds are capable of inhibiting the detrimental effects of oxygen on unsaturated fats, fat-soluble vitamins, and other related components of the diet, but are not capable of reducing the physiological vitamin E requirement.

(Also see Chapter 13, Feed Supplements/Additives/Implants, section on "Antioxidants.")

NUTRITIONAL DISORDERS AND TOXINS

All animals are subject to certain nutritional disorders and toxins, a subject that is fully covered in this book in Chapter 5, Nutritional Disorders/Toxins. The nutritional disorders and toxins covered in the sections that follow are peculiar to mink, although they are not always limited to mink.

Disorders Related to Nutrition

Since mink are by nature carnivorous, their feeding has involved different types of ingredients and different storage and handling practices and problems than those of most other species of animals. Mink are prone to the nutrition-related disorders described in the sections that follow.

CHASTEK PARALYSIS (THIAMIN DEFICIENCY)

Freshwater fish and some species of saltwater fish contain the enzyme thiaminase, which inactivates thiamin (vitamin B-1). Among such fish are: whitefish, freshwater smelt, carp, goldfish, creek chub, fathead minnow, buckeye shiner, sucker, channel catfish, bullhead and minnow, white bass, sugar pike, burbot, and saltwater herring.

If mink are fed raw fish containing thiaminase, thiamin is destroyed and thiamin deficiency known as Chastek paralysis, may occur.

Thiamin deficiency is ushered in by loss of appetite and extreme weakness, which is followed by a lurching, rolling gait; convulsions; and paralysis (Chastek paralysis), either in the hindquarters or in the whole animal. In adult mink, diarrhea is usually present in the last stage of the disease. In young kits, death occurs very soon after the first symptoms are observed.

Sick animals should be given an IP injection of thiamin (1 mg/lb), and the thiaminase-containing fish should either be cooked (at 181°F for at least 5 minutes) or deleted from the ration. If mink are still eating, supplementing feed with thiamin will restore the animals to good health.

(Also see Table 27-5, section on "Thiamin.")

COTTON FUR (IRON DEFICIENCY)

Cotton fur, which is caused by a deficiency of iron, is characterized by an almost complete lack of pigmentation of the underfur. This condition generally indicates anemia in mink. It is usually caused by feeding certain raw fish (Pacific hake, coalfish, whiting) containing high levels of trimethylamine oxide (TMAO) which interferes with iron absorption. The condition can be prevented by cooking the offending fish, or by feeding it less frequently.

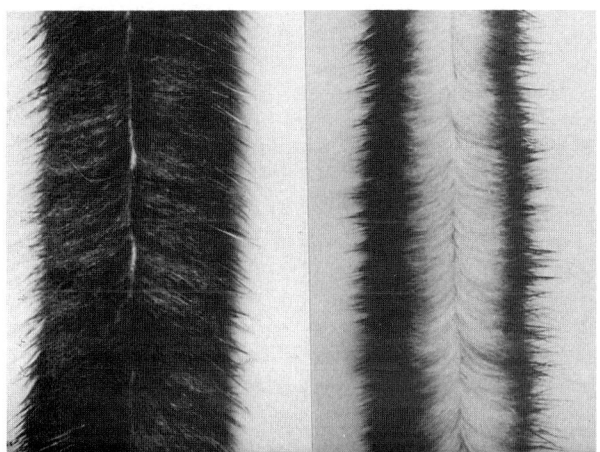

Fig. 27-3. Cotton fur in mink. Pelts parted to show underfur. *Left:* normal fur. *Right:* cotton fur. (Courtesy, John Adair, Department of Animal Science, Oregon State University, Corvallis)

(Also see Table 27-4 section on "Iron.")

GRAY UNDERFUR (BIOTIN DEFICIENCY)

Gray underfur generally indicates a biotin deficiency. It is usually caused by feeding high levels of uncooked eggs or turkey offal to young mink. These feed ingredients contain avidin, which inactivates biotin, a vitamin required for pigmentation. Affected mink may be injected with 1 mg of biotin twice weekly for 4 weeks. Prevention consists of cooking eggs and turkey offal and/or adding biotin to the ration.

(Also see Table 27-5, section on "Biotin.")

NURSING SICKNESS

Nursing sickness is a disease of lactating female mink. It is most frequently seen in females with large litters—more than seven kits. Females nursing large litters produce about 2.5 qt of milk a day throughout the lactation period. If a female loses a lot of weight during lactation, it indicates (1) that she has not been able to consume enough feed to prevent loss of weight, and (2) that she has had to convert her own body fat and muscle tissue to make up for the deficit not provided by the feed.

Nursing sickness occurs near the end of, or just after, the lactation period. It is characterized by loss of appetite—partially or totally, loss of weight, fatigue, dehydration, and kidney failure. Death occurs within a few days after the onset of clinical signs. Nursing kits of affected females should be weaned or fostered as soon as possible. To minimize the psychological stress due to loss of her kits, a male kit should be left with the ailing mother. Also, affected females should be tempted to eat by providing them with fresh liver, freshly killed sparrows, or similar feed. Treatment by the veterinarian may include injecting an electrolyte or saline solution.

The following preventive measures will lessen the incidence of nursing sickness: (1) encouraging kits to start eating solid food as early as possible by placing trays containing soft food in the pen; (2) including 0.5% salt (NaCl) in wet feed, or 1.3 to 1.5% in dry feed; (3) feeding lactating females a high-energy ration; and (4) providing plenty of fresh water and feed to nursing females at all times.

URINARY CALCULI (GRAVEL, STONES, WATER BELLY, UROLITHIASIS)

Urinary calculi is sometimes a problem, with losses exceeding 10%.

The disorder is most frequently seen in male kits during the summer, and in pregnant and lactating females during the spring. Males are more prone to urinary calculi than females, due to the greater difficulty in passing stones caused by the inelasticity of the penis bone.

Mink urinary calculi consist primarily of magnesium ammonium phosphate hexahydrate ($MgNH_4PO_4 \cdot 6H_2O$, struvite stones). The cause is unknown. However, the formation of the bladder stones appears to be favored by the presence of alkaline urine—above a pH of 6.0.

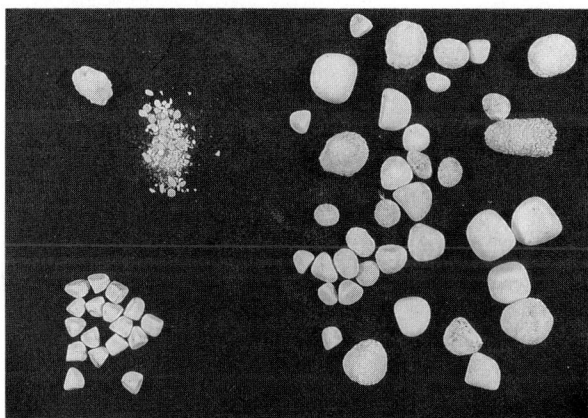

Fig. 27-4. Mink bladder stones (urinary calculi). (*Photo:* University of Wisconsin. *Print:* Courtesy, W. L. Leoschke, Fur Animal Nutrition Research, Valparaiso University, Valparaiso, Ind.)

The disease is characterized by dribbling of the urine and staining of the pelt around the urinary orifice. However, there may be sudden death, without prior symptoms, caused by uremia (urea in the blood).

A urine pH level below 6.0 appears to prevent urinary calculi. So, the following preventive measures may be used: (1) the addition to the diet of 1 g of ammonium chloride per mink per day; (2) the addition to the diet of 2.0% phosphoric acid (75% acid concentration) on a dry matter basis, or the addition of 0.8% phosphoric acid (75% feed grade) to diets containing 15 to 20% fortified cereal and 80 to 85% fresh/frozen feedstuffs; or (3) the acid preservation (with sulfuric, acetic, or formic acid) of fish silage, with the level of the feeding of the silage limited to (a) 10% during reproduction and early growth and (b) 30% during the late growth and furring periods.

(Also see Chapter 5, Nutritional Disorders/Toxins, Table 5-1, Nutritional Diseases and Ailments—Urinary Calculi.)

WET-BELLY DISEASE

This is a condition, primarily affecting male mink in the late summer and autumn, characterized by dribbling of urine and staining of the pelt around the urinary orifice. In some cases, the pelt is only damp and the urine can be washed out; in other cases, affected areas of the pelt must be discarded. Hence, the condition is of economic importance.

Although the cause of this condition is not known, some researchers have reported little or no problem with certain strains of animals, indicating a genetic difference. Other researchers have implicated the following causes: (1) high calorie density diets; (2) high calcium diets with wide calcium-to-phosphorus ratios; and (3) high levels of raw poultry waste, whole fish, and tripe, due to the presence of *Proteus* organisms in these feedstuffs and triggered by the stress of cold weather.

The incidence of wet-belly may be reduced by restricting feed intake from mid-October to pelting, and by cooking poultry waste, whole fish and tripe. Males with wet-belly should not be used for breeding.

YELLOW FAT DISEASE (STEATITIS)

This nutritional disease is common in young, rapidly growing kits. Affected mink may exhibit slight locomotor disturbances followed by death, or they may be found dead without prior symptoms. Postmortem examination reveals yellow internal or subcutaneous fat.

Yellow fat disease is caused by feeding rancid, unsaturated fatty acids, in which all the vitamin E is used as an antioxidant.

Treatment consists in discontinuing the feeding of rancid fat, adding stabilized vitamin E to the feed, and injecting affected kits with 10 to 20 mg of vitamin E for several days. Because of adding antioxidants to raw products, the disease has become relatively rare.

Disorders Caused by Toxic Substances

Toxicities may result from a host of substances in mink feeds. Some of the more common nonnutritional toxic substances are listed and discussed in the sections that follow.

BOTULISM

Botulism is a food poisoning caused by ingestion of food containing *Clostridium botulinum*. The toxins produced by these bacteria are among the most potent poisons known. There are several types of botulism toxin; mink are especially susceptible to Type C.

Occasionally, botulism causes heavy losses in unvaccinated mink. When an animal consumes feed that contains the toxin, the onset of the disease is rapid; usually many mink are found dead within 24 hours. The toxin affects the nerve centers, causing muscular incoordination and stiffness, followed by paralysis and death.

Treatment is usually ineffective. The only successful control is annual vaccination of all kits with botulism toxoid (generally Type C vaccine) soon after weaning.

(Also see Chapter 5, Nutritional Disorders/Toxins, Table 5-3, Potential Poisons—Botulism.)

HISTAMINE

Histamine is formed by decarboxylation (removal of the COOH groups) of histidine, one of the essential amino acids, by the enzymatic activity of certain bacteria (*Clostridium, Proteus, Salmonella,* and *Escherichia coli*). Histamine is poisonous to mink. The symptoms: decreased feed consumption, diarrhea, reduced gains, vomiting, and bloated stomachs. Diets that contain high levels of acid-preserved feedstuffs are especially prone to histamine formation.

MYCOTOXINS

Mycotoxins are poisons produced by fungi or molds that grow on grains, forages, and other feedstuffs. Although there are many mycotoxin-producing fungi, *Aspergillus flavus* which produces aflatoxin, is the most important from the standpoint of mink production. Kits are more susceptible than mature mink. Aflatoxin is fatal in doses of more than 227 mcg/lb.

Aflatoxin can cause serious economic loss and mortality when consumed by mink. It is carcinogenic and has a toxic effect on the liver. The symptoms: loss of appetite, loss of weight, liver damage, followed by death.

(Also see Chapter 5, Nutritional Disorders/Toxins, Table 5-3, Potential Poisons—Mycotoxins.)

SALMONELLA

Several species of salmonella may cause the toxic disease of animals known as salmonellosis. In mink, the infection is commonly caused by *Salmonella dublin* which may be found in raw meat, or *Salmonella typhimurium,* which may be found in raw eggs and poultry.

Salmonella infection is particularly dangerous to mink during gestation because it causes abortion. At other times, salmonella do not produce any specific symptoms in mink, although they may contribute to unthrifty animals and deaths.

Salmonella infection can be treated with antibiotics, given in the feed. In addition, it is important to avoid raw slaughterhouse and poultry offal during gestation. Heat treatment kills salmonella bacteria.

(Also see Chapter 5, Nutritional Disorders/Toxins, Table 5-3, Potential Poisons—Salmonellosis.)

THYROID GLANDS

Slaughterhouse products that contain thyroid glands, and thus the hormone thyroxin, may cause reproductive problems when fed to mink. Diets containing 15% gullet trimmings (including the thyroid/parathyroid tissue) cause a marked decrease in the number of females that whelp, the number of kits whelped, and the kit birth weight and survival.

FEEDS FOR MINK

In the early days of the mink industry, there was always a good supply of horsemeat available to provide a relatively cheap feed for mink. As the industry grew, and the role of

Feeding Mink

Fig. 27-5. At home on land or water; in his native habitat, the mink's varied diet includes fish and mice.

Fig. 27-6. Mink raised in captivity have a far better life than their counterparts in the wild. Feeds are formulated to meet all of their nutritional requirements, and the environment is altered so as to minimize stress. (Courtesy, Michigan State University, East Lansing)

the horse in American agriculture gradually declined, horsemeat became more expensive and more difficult to obtain. Mink producers then turned to fish and meat animal by-products as sources of feed. More recently, with the great expansion of the U.S. poultry industry, poultry offal has become abundant. Many of these products are economical and of good feeding value, but the problems inherent with wet, bulky feeds remain. Additionally, the nutritional value of the feed varies from batch to batch.

To avoid problems inherent in wet feeds, the mink industry developed new dry feeds which are acceptable to fussy eaters. Today, mink can be raised on 100% dry ingredients, which generally consist of dehydrated combinations of fish meal, poultry meal, liver, cereals, molasses, and other ingredients, fed in pellet form. In addition to the convenience and ease of storage, dry feed makes it possible to use special mixes for different life cycles; for example, breeding animals and kits can be fed mixes designed for their special needs, and energy-deficient feed mixes can be designed for use in periods when the breeder wants mink to lose weight. Because of these several advantages of dry feed mixes, the shift from predominantly wet feed to partly or all dry feed will likely continue.

FEED PREPARATION

Because of the relatively small digestive and absorptive area in mink, coarse fibrous feeds cannot be utilized effectively. Fiber should be minimal, and feed should be ground and well mixed.

Unground feeds do not maintain a uniform consistency. Besides, fine grinding of cereal grains increases their digestibility. So, cereals and other nonmeat and nonfish ingredients should be finely ground.

Cooking is essential in the preparation of many mink feeds. Fish that contain either thiaminase, which destroys thiamin, or trimethylamine oxide (TMAO), which interferes with iron metabolism, causing cotton fur, should be cooked for 20 minutes at low pressure (less than 15 lb per square inch). If pressure cooking is not used, fish should be simmered for at least 2 hours. If cooking is impractical, alternate feeding of thiaminase- and nonthiaminase-containing fish, and of TMAO-containing and non-TMAO-containing fish, may be practiced. Also, eggs and turkey hen offal should be cooked and/or supplemented with biotin in order to destroy avidin, which inactivates biotin.

In order to improve digestibility, cereals should be cooked if they are not finely ground.

Proper freezing of wet feeds is the key to good mink nutrition. Freezing inhibits bacterial growth in the feed and slows the enzymatic processes which can destroy the product. Fresh meat and fish should be chilled immediately, then quick frozen at −10°F, or colder, then stored at 0°F.

Fish and fish offal may be ensiled, usually with the addition of an acid preservative (sulfuric, acetic, or formic acid) or of an enzyme-impeding or enzyme-destroying substance. When good quality fish products are processed rapidly, properly preserved with an acid or an enzyme inhibitor, and stored in a good tank or silo, they will keep for several months.

Dry diets are now being fed successfully to mink. Because of their several advantages, they will be used more widely in ranch practices in the future. Dry feeds are easier to transport, store, and feed than fresh meat and fish products, and they can be standardized more effectively with regard to nutrient composition.

FEED ALLOWANCES AND SUGGESTED RATIONS

The various stages of mink production can be broken down on a calendar basis (see Table 27-6).

TABLE 27-6
TYPES OF RATIONS FED TO MINK THROUGHOUT THE YEAR[1]

Stage of Production	Type of Ration		
	Maintenance	Breeder-Grower	Grower-Pelter
Breeder females	Dec. 1–March 1	March 1–July 15	July 15–Dec. 1
Breeder males	Dec. 1–March 1	March 1–April 1	April 1–Dec. 1
Kits		June 1–Sept. 15	Sept. 15–Dec 1
Nonpregnant females			May 1–Dec. 1

[1]Adapted by the authors from *Fundamentals of Mink Ranching*, H. F. Travis and P. J. Schaible, Circ. Bull. 229, Michigan State University, p. 73.

The National Research Council (NRC) has developed some basic diet formulations, which are presented in Table 27-7. These formulas do not give specific recommendations with regard to the various levels of production.

Sometimes producers need to convert rations or ingredients from wet to dry basis, or from dry to as-fed. Formulations for accomplishing these conversions are presented in Chapter 18, Feeding Standards/Ration Formulation.

TABLE 27-7
SUGGESTED RANGES OF COMPOSITION OF PRACTICAL DIETS FOR MINK ON A "DRY MATTER" (MOISTURE-FREE) BASIS[1]

Ingredients	Percent
Fortified cereal[2]	15–30
Liver	0–10[3]
Quality protein feedstuffs (cooked eggs, whole poultry, whole fish, horsemeat, rabbits, nutria, etc.)	0–30[4]
Beef by-products (tripe, lungs, lips, udders, spleen, etc.)	10–30
Poultry by-products (heads, entrails, feet)	10–70
Fish scrap	10–50
Fat supplementation (rendered animal fat or vegetable oils)	0–6[5]
Proximate analysis[6] of diet	**Percent**
Protein	25–40
Fat	18–30
Carbohydrate	20–50
Ash	6–12

[1]Adapted by the authors from *Nutrient Requirements of Mink and Foxes*, No. 7, 2nd rev. ed., NRC-National Academy of Sciences, 1982, p. 38.

[2]May consist of single-cooked grains such as oat groats or wheat in combination with vitamin and trace mineral supplementation or commercially prepared fortified cereal mixtures.

[3]Reproduction-lactation diets (March-May) often contain 5–10% beef liver, although necessity for this has not been accepted universally.

[4]Level of quality protein feedstuffs is often increased during the critical fur development and reproduction-lactation phases—a practice consistent with the higher protein requirements of the mink during these critical periods.

[5]That level of fat supplementation that provides proper protein/energy balance for each phase of the life cycle.

[6]That proximate analysis consistent with the optimum nutritional balance for each phase of the life cycle.

FEEDING BREEDING STOCK

Fig. 27-7. As the breeding season approaches, the animals should be kept a little hungry to stir activity. They should be alert and inquisitive, as shown. (Courtesy, Michigan State University, East Lansing)

Fig. 27-8. The mother of a fast-growing litter should be fed to meet the rigorous nutritional demands of lactation. (Courtesy, Maple Leaf Mills, Ltd., Ontario, Canada)

All successful operations require that animals should be fed according to their nutritional demands. If production is to be maximized, each animal should be fed according to needs—no more, no less. For breeding stock, this calls for rations designed to meet the special needs for maintenance, breeding, gestation, and lactation (see Tables 27-1, 27-2, and 27-3 for nutrient requirements).

FEEDING KITS

The early growth of kits is very rapid (see Fig. 27-9). It follows that this is one of the most critical periods in the life of a mink and requires the best possible nutrition.

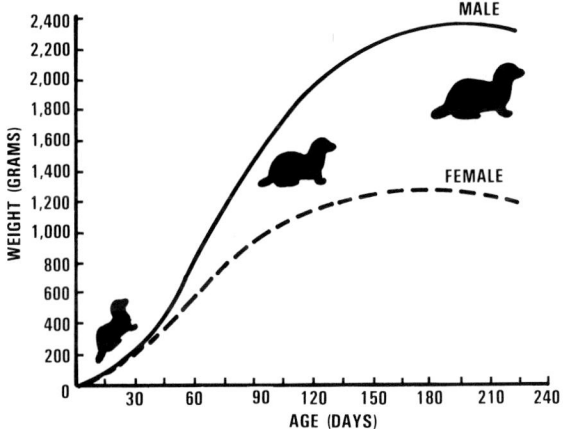

Fig. 27-9. Standard dark mink growth curve. In this figure, the various color types of mink are not differentiated; rather, it is assumed that they will grow similarly if given adequate nutrition in satisfactory environments. (Source: Adapted by the authors from Nutrient Requirements of Mink and Foxes, No. 7, 2nd rev. ed., NRC-National Academy of Sciences, 1982, p. 4. Original source: Wehr, N. B., J. E. Oldfield, and J. Adair, Oregon State University, Corvallis)

Kits will begin to eat solid feed at 4 weeks of age. They should be weaned at 6 to 8 weeks of age. At weaning, they are generally separated from the mother, but are housed as a group for an additional couple of weeks. When kits are

Fig. 27-10. Kits are normally weaned from their mother at 6 to 8 weeks of age but are usually housed together for an additional couple of weeks. (Courtesy, Ralston Purina Company, St. Louis, Mo.)

grouped together, they compete for feed and eat extremely fast, sometimes resulting in digestive upsets. It is important that their feed be thoroughly thawed before feeding, as partially frozen feed can create digestive disorders in young kits. Demands for high-quality feeds are high in kits up to 16 weeks of age. The protein content of the feed can be reduced after 16 weeks of age, but the kits should remain on a full feed regimen.

In order to correct for overfeeding and to sharpen their appetites, kits can be allowed to go unfed one day a week following separation. This should only be done after they are housed singly; otherwise, fighting, sometimes resulting in death, may occur.

METHODS OF FEEDING AND WATERING

Since mink have a short digestive tract and cannot accommodate large quantities of feed at one time, it is necessary to feed them more frequently than many other species. It is recommended that mink be fed once or twice daily during breeding and gestation. Lactating females should be fed two or three times a day, while adult mink on maintenance rations can be fed once daily if the weather isn't too cold. Young, growing kits need to be fed four times a day.

Fig. 27-11. Violet variety of mink begging for food. (Courtesy, USDA)

In addition to feeding conventional wet feed, some mink ranchers are now providing pellets (dry feed) free-choice in hoppers. This allows the feeder to be less precise in the amount of wet feed given at each feeding. Also, under this system, the wet feed may be omitted on weekends without throwing the mink off feed during the 2 days that they are provided free-choice pellets only. Other ranchers are self-feeding pellets only, without any wet feed whatsoever.

Many mink ranchers feed by hand. While this method involves more time and labor, the producer is able to keep close watch over the animals. Overfeeding and underfeeding can be carefully monitored by hand-feeding. In addition, any unhealthy animals can be spotted and treated quickly before the disease has a chance to spread throughout the operation. Automatic feeding is much more time-saving, but considerable wastage of feed can occur, and the personal touch in management is diminished.

Several types of watering systems can be adapted to mink ranching. One of the most important factors to be considered is the prevention of freezing in the cold, winter months. This can be accomplished by frequent draining of the water vessels and pipes or by the use of electric heating cables. The simplest, but the most laborious, method of watering is the use of the sprinkling can. In large operations, the time involved with this system is too great to be economical. Manually operated mechanical systems, whereby pipes are laid out so that there is a hole over the water containers, can be used effectively. Periodically a water valve is opened, and water flows to the drinking cups. Fully automatic watering systems—for example, the nipple valve—are excellent in warm weather but tend to freeze up in cold weather.

FEED CONTAMINANTS

Many animal and fish by-products contain compounds that can be detrimental to the production of mink. For this reason, it is important that the producer obtain feed from reliable sources that routinely check the quality of their ingredients.

Currently, the Scandinavian countries are using chemical and microbiological methods for evaluating the quality of feedstuffs and mixtures. This concept is also being used in some of the larger U.S. units.

(Also see earlier section on "Disorders Caused by Toxic Substances.")

QUESTIONS FOR STUDY AND DISCUSSION

1. What caused the rise of the U.S. mink industry in the 1940s?

2. It takes an average of 70 skins to make one full length mink coat; it requires an average of 115 lb of wet feed to produce one mink ready for pelting; and wet feed costs an average of about 12½¢/lb. What is the feed cost alone to produce one mink coat?

3. Why are mink feeds generally more expensive than the feeds normally fed to other animals?

4. Define (a) metabolizable energy, and (b) energy density.

5. The quantity of feed consumed by mink is directly related to the need to satisfy their energy requirement. What is the practical significance of this fact?

6. In what ways may the digestibility of cereal grains be increased for mink?

7. Discuss the nutritive importance of fat in mink diets.

8. Why should the protein of mink feed be of high biological value?

9. Present pertinent information relative to the needs of mink for each of the following minerals: salt, calcium, phosphorus, and iron.

10. Present pertinent information relative to the function, deficiency symptoms, and practical sources of each of the following vitamins for mink: vitamin A, vitamin E, biotin, and thiamin.

11. For what purposes are antibiotics and antioxidants used in mink production?

12. Discuss the cause, prevention, and treatment of each of the following disorders related to nutrition: Chastek paralysis, cotton fur, gray underfur, nursing sickness, urinary calculi, wet-belly disease, and yellow fat disease.

13. Discuss the symptoms, prevention, and treatment of each of the disorders caused by the following toxic substances: botulism, histamine, mycotoxins, salmonella, and thyroid glands.

14. In order and in period of time, there has been a trend shift in the wet feed of mink from horsemeat, to fish, to poultry offal. What forces were back of these shifts?

15. Discuss the advantages of dry feeds. Discuss the proportion of dry feed that may be fed to mink, and the frequency and method of feeding dry feed.

16. Why is cooking essential in the preparation of many mink feeds?

17. What are the advantages of each of the following methods of feeding mink: hand-feeding, self-feeding, and a combination of hand-feeding and self-feeding?

28

FEEDING FISH[1]

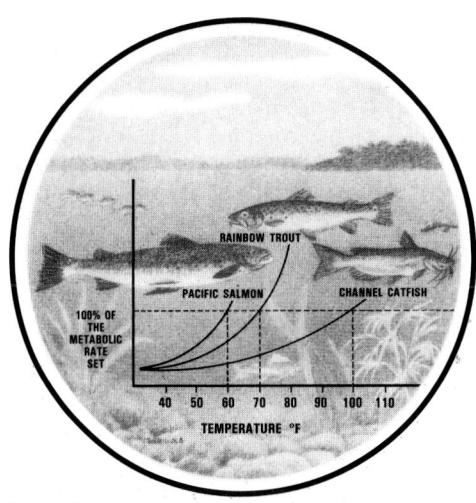

Original painting by Tom Phillips

Contents	Page
Economic Importance of Fish	658
Nutritive Needs of Fish	658
Energy	659
Carbohydrates	660
Fats (Lipids)	660
Protein as an Energy Source	660
Protein	660
Minerals	661
Fish Mineral Chart	662
Vitamins	662
Fish Vitamin Chart	663
Fat-Soluble Vitamins	663
Water-Soluble Vitamins	663
Water	664
Nonnutritive Factors	664
Feeds for Fish	665
Natural Foods and Feeds	665
Artificial Wet Feeds	665
Formulated Feeds	665
Feed Handling and Storage	666
Rations for Fish	666
Trout and Salmon Diets	666
Catfish Diets	668
Feeding Coldwater Fish	669
Feeding Warmwater Fish	670
Feeding Catfish	670
Feeding Carp	671
Feeding Systems	671
Other Feed and Management Aspects	672
Feed Ingredients Affecting Fish Quality and Flavor	672
Specialized Aquatic Production	672
Nutritional Toxicants of Fish	673
Questions for Study and Discussion	673

Although recent research has shown the advantage of scientific fish feeding, aquaculture is an old term with a new look. It is derived from the latin word *aqua*, which means water, and from the work culture, which means to till, to cultivate, and to grow. So, *aquaculture is the controlled cultivation and harvest of aquatic animals and plants.* It involves the production of a marketable crop of fish or other commodity in water.

Aquaculture dates back more than 4,000 years to China, Japan, and Egypt. The practice of fish farming can also be found in the societies of India and Java 3,000 years ago, and in Europe 2,500 years ago. Until recent years, most fish feeds were the result of trial and error. Today, the nutritional requirements of fish are based on research, although some gaps in our knowledge still exist.

Aquaculture can be divided into three types of production—freshwater, brackish water, and marine. Within these three types of production, three systems of management can be listed as follows:

1. **Hatchery propagation.** Young fish are hatched and raised to a size where they can be released into natural populations to grow and reproduce.

2. **Capture of young fish.** In this system, wild fish are captured, transferred to managed water, and grown to market weight by supplemental feeding or on natural foods.

3. **Management of entire life cycle.** Catfish and trout production in the United States are examples of this system. Young fish are hatched and grown either to be marketed or kept as replacement breeder stock.

Within the classification of vertebrates, there are more species of fish than any other animal. Estimates of the number of fish species range from 15,000 to 17,000, as compared to 8,600 avian species and 4,500 mammalian species. Among this vast number of species of fish, representatives of almost every type of feeding behavior can be found.

---

[1]The authors gratefully acknowledge the helpful suggestions of the following authorities who reviewed this chapter: H. K. Dupree, Ph.D., Scientific Director, Fish Farming Experiment Station, U.S. Department of the Interior, Fish and Wildlife Service, Stuttgart, Ark.; L. G. Fowler, Ph.D., Assistant Director, Abernathy Salmon Cultural Development Center, U.S. Department of the Interior, Fish and Wildlife Service, Longview, Wash.; D. M. Gatlin III, Ph.D., Assistant Professor, Department of Agriculture, University of Arkansas at Pine Bluff; B. P. Grant, Ph.D., Director, International Aquaculture Research Center, Hagerman, Idaho; J. E. Halver, Ph.D., School of Fisheries, University of Washington, Seattle; J. D. Hendricks, Ph.D., Department of Food Science and Technology, Oregon State University, Corvallis; and D. Wenger, Wenger Manufacturing, Sabetta, Kans.

Presently, fish contribute less than 1% of the world food supplies in terms of dietary energy, less than 5% of total protein, and less than 14% of animal protein. But as health concerns and dieting increase throughout the world, more fish will be included in human diets.

ECONOMIC IMPORTANCE OF FISH

Today, U.S. fish farming is a big and growing business. In 1984, U.S. Aquaculture production had a value of $503,030,000, with catfish leading, baitfish ranking second, and trout ranking third. Fig. 28-1 shows the comparative value of all U.S. Aquaculture production in 1980 and in 1984.

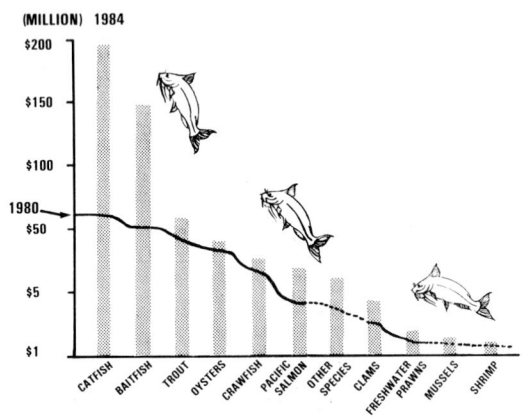

Fig. 28-1. In 1980, U.S. Aquaculture production had a total value of $203,178,000, with a breakdown by species as shown by the curve. In 1984, the total value was $503,030,000, representing a 2½ fold increase, with a breakdown by species as shown by the bars. **NOTE:** "Other species" include abalone, buffalo, carp, mullet, sturgeon, tilapia, and lesser species. (*From:* Annual series of *Current Fisheries Statistics, Fisheries of the United States,* National Marine Fisheries Service, U.S. Department of Commerce)

Fish are very efficient feed converters. It takes only 1.5 to 1.7 lb of feed to produce 1 lb of fish; this compares with 2.1 lb of feed to produce 1 lb of broiler. Also, protein efficiency (pounds of protein in feed required to produce 1 lb of fish protein) is very excellent, 2.1:1, although it is exceeded by broilers (1.9:1). Besides, fish consume proteins which are not generally used for human consumption; for example, trash fish and animal by-products—they recycle protein. For these reasons, plus the fact that fish do not compete with crop production, it is expected that both freshwater and brackish-water fisheries, including fish production in small ponds and paddy fields, will receive increasing attention in the future.

Fish are poikilotherms (cold-blooded animals); that is, their body temperature changes to that of the environment. From a production standpoint, this offers both advantages and disadvantages when comparing fish farming with the production of homothermal (warm-blooded animal) species. Since the body temperature of fish is the same as the temperature of their environment, little or no energy is required to maintain their body temperature. By contrast, considerable energy is required to maintain the body temperature of warm-blooded animals. However, fish are very susceptible to stresses which result from environmental changes, especially rapid fluctuations in water temperature. It follows that the production of freshwater fish is more extensive than that of marine fish because it is easier to monitor and control the environmental stresses which affect their growth rate.

Freshwater fish production can be divided into two classifications—coldwater culture (40 to 60°F); and warmwater culture (70 to 100°F). In the United States, channel catfish is the predominant warmwater species, whereas trout and Pacific salmon represent the bulk of coldwater fish production. In other parts of the world, the carps are the major warmwater fish.

NUTRITIVE NEEDS OF FISH

Fish in the wild seldom show signs of nutrient deficiency: natural foods are relatively well-balanced nutritionally and growth rate is proportional to quantity, not quality, of food available. In confinement, where natural food is limited or absent, nutritional requirements become critical. Although fish require essentially the same nutrients for the same metabolic functions as land animals, there are the following differences:

1. The energy requirements of fish are lower, resulting in superior feed efficiency.
2. Vitamin C (ascorbic acid) is synthesized poorly, or not at all, by most fish.
3. Most fish require dietary omega-3 fatty acids, whereas warm-blooded animals do not.
4. Fish can absorb some soluble minerals from the water, thereby reducing the need for those minerals in the diet.

The nutritional requirements of fish do not vary greatly among species. There are exceptions, however, usually associated with coldwater or warmwater fish, finfish or crustaceans, and freshwater or marine species. These differences generally relate to the requirement of essential fatty acids, and to carbohydrate utilization.

Fish can be divided into three types of eaters—carnivores, herbivores, and omnivores.

1. **Carnivores.** Consume primarily animal material. Foods consumed by this type of fish may be as small as a microscopic crustacean or insect or as large as an amphibian or a small mammal.
2. **Herbivores.** Consume primarily available vegetation and decayed organic materials in the environment.
3. **Omnivores.** Consume both animal organisms and plant materials.

Some anatomic adaptations in the mouth and digestive system, relating to their foods and feeding habits, are observed in some fishes.

Based on well-defined differences in the organization and structure of the mouth, it is possible to classify fish according to their feeding habits into the following categories:

1. **Predators.** These fish, of which trout and salmon are examples, feed on organisms that are generally large enough to be seen with the naked eye. Teeth are well developed and act as a means of grasping and holding the prey. Some predators rely primarily on sight to hunt while others rely on the senses of taste and touch or on lateral-line sense organs.
2. **Grazers.** These fish, of which the mullet is an example, graze in the same sense as grazing animals. Generally, they

feed continuously at the bottom of the water habitat on either plants or small animal organisms. Ingested food is taken by well-defined bites.

3. **Strainers.** These fish, of which menhaden are an example, select food primarily by size, rather than by type. An adult menhaden can strain in excess of 6 gal of water per minute through its gill rakers. Through this process of rapid straining, the menhaden is able to concentrate a relatively large mass of plankton and other organisms.

4. **Suckers.** This group of fish, of which buffalo fish are an example, feeds primarily on the bottom of their habitat—sucking in mud, filtering, and extracting disgestible material.

5. **Parasites.** Some fish, notably the sea lamprey, attach themselves to other animals and exist on the body fluids of the host.

In addition to the anatomical adaptations to eating, fish have developed behavioral feeding patterns which are sensitive to environmental stimuli. The fish farmer should recognize the influences of environment on feeding behavior if the efficiency of production is to be maximized. By knowing the behavioral patterns of the particular species of fish, the producer can adapt a system of feeding which will best utilize labor and feed. Some environmental influences on feeding behavior follow:

1. **Time of day.** Some fish depend largely on their sense of sight in locating food, while others rely primarily on their senses of taste, touch, and smell. One would expect the fish using sight for feed to be active feeders during the daylight hours. Night feeders rely on the other senses to locate food.

2. **Season of the year.** Some fish, such as largemouth bass, cease feeding activity during their spawning season. Most fish increase feed intake in the spring in the temperate climatic regions when the water temperature starts to rise. As a result, the peak growth period for most fish occurs in the spring and early summer.

3. **Rapid changes of light intensity.** Some fish, such as the yellow perch, show peak feed activity at dawn and dusk.

4. **Physical contact with the food.** Quite often the texture of a potential food source is felt before the fish will consume it.

5. **Water temperature.** Apart from feed and environmental qualities (dissolved oxygen, etc.), water temperature is probably the most important factor affecting appetite and amount of feed consumed. Sharply reduced feed consumption in trout occurs when the temperature decreases to 38°F. The lower limit for feed consumption by catfish is about 50°F.

Size and age of fish are also important factors in determining the nutrient requirements of fish.

One very important factor in establishing the nutrient requirements of fish is the effect of water temperature. To deal with this problem, the National Research Council (NRC) reports Standard Environmental Temperatures (SET) for various species of fish in order that fish producers will have an idea of the applicability of the requirements to their particular systems. The Standard Environmental Temperatures (SET's) used by the National Research Council are: 59°F for chinook salmon, 50°F for rainbow trout, and 86°F for channel catfish. As water temperature deviates from these standards, nutrient requirements increase. If the temperature of the water is lower or higher than the SET for a particular species of fish, feed intake is usually reduced.

A discussion of the nutrient requirements of fish follows, with separate sections devoted to energy, protein, minerals, vitamins, water, and nonnutritional factors.

Energy

The energy requirements of fish are lower than the energy requirements of warm-blooded animals due to the following differences in energy metabolism:

1. Fish have a low basal energy need, because, being cold-blooded, they do not expend energy to maintain body temperature. By not having to regulate body temperature, more energy is available for growth, fattening, and reproduction. It has been estimated that only 70% of the dietary calories are used for maintenance, thus leaving a sizeable number of calories for growth, fattening, and reproduction.

2. Fish have a low energy need for locomotion and voluntary activity; they have no need for large antigravitation muscles because (1) of the buoyancy of their water environment, and (2) a streamlined fish moving through water represents one of nature's most efficient modes of locomotion.

3. Fish use little energy requirement for protein catabolism and waste nitrogen excretion, since ammonia is the principal end product of their protein catabolism as contrasted with the more complex urea and uric acid formed by terrestrial homeotherms. Fishes excrete ammonia principally through the gills and expend little energy in the excretion process.

4. Fish use proteins and fats for energy, preferentially, and carbohydrates sparingly; however, carbohydrates are used more efficiently by warmwater fish than by coldwater fish. By contrast, warm-blooded farm animals and poultry use carbohydrates for energy, preferentially.

The amount of energy required by fish is affected by species, diet, size, age, sex, reproductive stage, water temperature, water quality, light exposure (darkness and the rest period that accompanies it, decreases the energy requirement in some species), and activity.

• **Energy value of feedstuffs for fish**—The standard methods of measuring and expressing energy value of feedstuffs for farm animals and poultry is fully covered (1) in Chapter 4, Nutrients/Metabolism, and (2) in the preliminary section of Section 5, Composition of Feeds. Additionally, four standard energy values for warm-blooded animals—total digestible nutrients (TDN), digestible energy (DE), metabolizable energy (ME), and net energy (NE)—for a great array of feeds are reported in the Composition of Feeds tables in this book. So, only the differences between fish and warm-blooded animals relative to the energy value of feedstuffs will be discussed in this section.

Digestible energy (DE) and metabolizable energy (ME) are used to express feed values and energy requirements of fishes.

• **Energy values of fish feeds**—Suitable feeds for fish are included in this book in Section V—Composition of Feeds. Since only limited energy values for fish foods are available, it is suggested that the metabolizable energy values for poultry be used. As a result of a comparison of efficiencies of energy and protein retention between rainbow trout and broiler chickens, Cowley et al. concluded that the retention of dietary energy and protein by the rainbow trout is only moderately superior to that of broiler chickens.[2]

---

[2]Cowley, C. B., A. M. Mackie, and J. G. Bell, *Nutrition and Feeding in Fish*, Academic Press, 1985, p. 113.

Table 28-1 lists the limited, presently available feed ingredient ME values for trout and the DE values for channel catfish and tilapia.

TABLE 28-1
METABOLIZABLE ENERGY VALUES FOR TROUT AND DIGESTIBLE ENERGY VALUES FOR CHANNEL CATFISH AND TILAPIA OF SELECTED FEED INGREDIENTS[1,2]

Feed Name	Metabolizable Energy–Trout	Digestible Energy	
		Channel Catfish	Tilapia
	(kcal/lb)	(kcal/lb)	(kcal/lb)
Alfalfa meal, 17% protein	577	304	459
Blood meal, spray dehy	1,182		
Brewers' grains, dehy	1,086		
Casein	1,777		
Corn, dent yellow grain	682		
Nonheated		500	1,118
Heated		1,150	1,372
Corn, distillers' solubles	954		
Corn gluten meal	1,364		
Cottonseed meal, solv extd 41%	945	1,159	
Cottonseed meal, w/o hulls 50%	1,123		
Crab meal, process residue	1,345		
Fishmeal, anchovy	1,682		
Fishmeal, herring	1,727		
Fishmeal, menhaden		1,773	1,836
Fishmeal, white fish	1,245		
Fish solubles, dehy	1,523		
Gelatin	1,841		
Meat meal w/bone	1,209	1,577	1,336
Poultry by-product meal	1,164		
Poultry, feathers, hydrolyzed	1,309	1,550	
Rapeseed meal (Canola meal)	1,123		
Soybean meal, solv extd		1,173	1,518
Soybean meal, w/o hulls	1,314		
Soybean seeds, heat processed, 175°C	1,654		
Wheat, flour by-product	559		
Wheat, hard red winter, grain		1,159	1,314
Wheat, middlings	559		
Whey, dehy, low lactose	964		
Yeast, brewers', dehy	1,232		
Yeast, torula	1,404		
Oil, menhaden		4,014	
Oil, soybean		4,059	

[1]Adapted by the authors from *Nutrient Requirements of Warmwater Fishes and Shellfishes*, NRC-National Academy Press, 1983, pp. 41–42, Table 12.

[2]To convert kcal/lb to kcal/kg, multiply by 2.2.

As with traditional livestock, fish derive their energy from three sources—carbohydrates, fats, and proteins.

CARBOHYDRATES

Carbohydrates are a major source of energy for humans and domestic animals, but not for fish. The primary sources of carbohydrates in fish feeds are plant feedstuffs, including cereal grains, wheat by-products, soybean meal, and cottonseed meal.

• **Carbohydrates in salmonid feeds**—Nutritionists have placed the maximum carbohydrate level of salmonid diets at 12 to 20%.

• **Carbohydrates in feeds for warmwater fish**—Studies have shown that channel catfish and carp can utilize higher levels of carbohydrates than trout—they can use up to 25% carbohydrates as effectively as fats as an energy source. Starches are more readily utilized than sugars.

FATS (LIPIDS)

In nature, fat is the major source of energy for fish. In addition to providing energy, fats serve several other functions for fish such as reserve energy storage, insulation of the body, cushion for vital organs, lubrication, transport of fat-soluble vitamins, and maintenance of neutral buoyancy.

• **Fat requirements for coldwater fishes**—When there is little or no fat in the feed, trout form their own fat from carbohydrates and proteins. Feed manufacturers use fish oil and vegetable oil in fish feeds as the primary energy source.

Linolenic fatty acids (omega-3 type) are essential for trout and salmon and should be incorporated at a level of at least 1% of the ration for maximum growth response.

The level of dietary fat required for trout and salmon depends on the size of the fish, the protein level in the feed, and the kind of supplemental fat. The recommended percent of fat and protein in the ration dry matter for different ages of trout and salmon follows:

Feed/Size of Fish	Protein	Fat
	(%)	(%)
Starter feeds (fry)	50	19
Grower feeds (fingerlings)	40	15
Production feeds (older fish)	35	12

Some fatty acid deficiency signs reported for trout include poor growth, necrosis of the caudal fin, pale and fatty livers, dermal pigmentation, increased muscle water content, and reduced hemoglobin.

• **Fat requirements for warmwater fishes**—The available data suggest that warmwater fish feeds should contain 10 to 15% fat, and that more than 15% fat in the feed will not improve growth or increase protein deposition.

The level of the omega-3 type fatty acids required by warmwater fishes is unknown, but it is believed to be between 0.3 and 0.5% of the dry diet.

Feed fats affect the taste and storage quality of catfish products. Rancid or off-flavor fish oil has an adverse effect on the flavor of fresh and frozen fish. Fish reared on soybean oil, safflower oil, or corn oil have a better flavor than those fed beef tallow or fish oil.

The principal EFA deficiency signs of warmwater fishes are reduced growth rate, reduced feed efficiency, and in some cases increased mortality.

PROTEIN AS AN ENERGY SOURCE

In fish feeds, fats and carbohydrates are the primary sources of energy, but excess protein, or protein that is deficient in one or more essential amino acids, is also utilized for energy. In order to metabolize protein for energy, it must be deaminated and the carbon skeleton altered in such a manner as to permit it to enter the energy metabolic pathway. However, fish are relatively efficient in using protein, deriving 3.9 of the 4.65 gross kilocalories per gram from protein, for an 84% efficiency.

Protein

The primary objective of fish husbandry is to produce fish flesh, which contains more than 50% protein on a dry weight basis.

Weight gain of fish is essentially proportional to the protein content of the diet at a range of 20 to 40%, but above 50% does not improve growth.

Fish can synthesize some amino acids, but usually not in sufficient quantity to satisfy their total requirement. So, certain amino acids must be supplied in the feed due to the inability of fish to synthesize them. Fish require the same ten essential amino acids as higher mammals, and they require them in about the same amounts as do broilers. When fish are fed feeds lacking in one or more of the essential amino acids, they become inactive, and lose both appetite and weight. The National Research Council amino acid requirements for fish are given in Table 28-2.

TABLE 28-2
COMPARATIVE AMINO ACID REQUIREMENTS OF GROWING ANIMALS, FOR SELECTED SPECIES[1]

Amino Acid	Japanese Eel	Common Carp	Channel Catfish	Chinook Salmon	Chick	Swine	Rat
Arginine	4.5 (1.7/37.7)	4.2 (1.6/38.5)	4.3 (1.03/24)	6.0 (2.4/40)	5.6 (1.00/18)	1.2 (0.16/13)	5.0 (0.60/12)
Histidine	2.1 (0.8/37.7)	2.1 (0.8/38.5)	1.5 (0.37/24)	1.8 (0.7/40)	1.4 (0.26/18)	1.2 (0.15/13)	2.5 (0.30/12)
Isoleucine	4.0 (1.5/37.7)	2.3 (0.9/38.5)	2.6 (0.62/24)	2.2 (0.9/41)	3.3 (0.60/18)	3.4 (0.41/13)	4.2 (0.50/12)
Leucine	5.3 (2.0/37.7)	3.4 (1.3/38.5)	3.5 (0.84/24)	3.9 (1.6/41)	5.6 (1.00/18)	3.7 (0.48/13)	6.3 (0.75/12)
Lysine	5.3 (2.0/37.7)	5.7 (2.2/38.5)	5.0 (1.5/30)	5.0 (2.0/40)	4.7 (0.85/18)	4.4 (0.57/13)	5.8 (0.70/12)
Methionine[2]	5.0 (1.9/37.7)[3]	3.1 (1.2/38.5)	2.3 (0.56/24)	4.0 (1.6/40)[3]	3.3 (0.60/18)[3]	2.3 (0.30/13)	5.0 (0.60/12)
Phenylalanine[4]	5.8 (2.2/37.7)	6.5 (2.5/38.5)	5.0 (1.2/24)	5.1 (2.1/41)	5.6 (1.00/18)	4.4 (0.57/13)	6.7 (0.80/12)
Threonine	4.0 (1.5/37.7)	3.9 (1.5/38.5)	2.0 (0.53/24)	2.2 (0.9/40)	3.1 (0.56/18)	2.8 (0.37/13)	4.2 (0.50/12)
Tryptophan	1.1 (0.4/37.7)	0.8 (0.3/38.5)	0.5 (0.12/24)	0.5 (0.2/40)	0.9 (0.17/18)	0.8 (0.10/13)	1.3 (0.15/12)
Valine	4.0 (1.5/37.7)	3.6 (1.4/38.5)	3.0 (0.71/24)	3.2 (1.3/40)	3.4 (0.62/18)	3.2 (0.41/13)	5.0 (0.60/12)

[1]From: *Nutrient Requirements of Warmwater Fishes and Shellfishes*, NRC-National Academy Press, 1983, p. 4, Table 3. **NOTE:** Requirements are expressed as percent of protein. In parentheses the numerators are requirements as percent of diet dry matter and the denominators are percent total protein in the diet DM. Data for chinook salmon, chick, swine, and rat are from NRC (1981), NRC (1977a), NRC (1979), and NRC (1978), respectively; data for eel and carp are from information of Nose and Arai (Cowey and Sargent, 1979); data for catfish are from Wilson and coworkers (Harding et al., 1977; Wilson et al., 1977, 1978, 1980; Robinson et al., 1980a,b and 1981).

[2]In the absence of cystine.
[3]Methionine plus cystine.
[4]In the absence of tyrosine.

Several factors determine the requirement for protein in fish feeds; among them, temperature, age, species, energy content of the feed, feeding rate, and size of fish.

Table 28-3 gives the recommended protein levels for different species and sizes of fish.

TABLE 28-3
PROTEIN REQUIREMENTS (PERCENT OF TOTAL RATION, BY WEIGHT) FOR RAPID GROWTH IN DIFFERENT KINDS OF FISH[1]

Species	Fry To Fingerlings (%)	Fingerlings To Subadults (%)	Adults and Brood Fish (%)
Trout and salmon	50	35-40	30-32
Channel catfish	35-40	25-36	28-32
Common carp	43-47	37-42	28-32
Largemouth bass	40	40	35
Striped bass	40	36	35
Eel	50-60	45-50	

[1]Adapted by the authors from: Dupree, H. K. and J. V. Huner, Editors, *Third Report to the Fish Farmers*, U.S. Department of the Interior, Fish and Wildlife Service, 1984, p. 142, Table 11.1.

The quality of protein, reflecting the amino acid content, is most important in optimizing the utilization of dietary proteins. If a ration is deficient in any of the ten essential amino acids, poor growth and decreased efficiency of feed conversion will result, despite a high total protein level in the feed.

Fish meal appears to be a highly desirable feed ingredient in fish formulas. Substitutions can be made for most of the ingredients in standard fish formulations, but whenever fish meal is left out poorer growth and feed conversion usually result. Starter diets usually contain at least 15% fish meal (starter trout and salmon diets generally contain at least 50% fish meal), and production and brood stock diets normally contain more than 5% fish meal. Other commonly used protein ingredients for fish diets are: soybean meal, cottonseed meal, corn gluten meal, meat meal, poultry by-product meal, hydrolyzed feather meal, dried blood meal, and dried skimmilk.

• **Protein in salmonid feeds**—The protein requirements for salmon and trout are similar (see Table 28-3).

The level of protein required in feed varies with the quality and proportion of natural proteins that make up the feed. Between 0.5 and 0.7 lb of dietary protein, in a balanced hatchery feed, is required to produce a pound of trout. The requirement for protein is also temperature-dependent. The optimal protein level in the feed for chinook salmon is 40% at 47°F and 55% at 58°F.

• **Protein in catfish feeds**—Protein utilization of catfish is affected by the protein source and water temperature.

Channel catfish convert the best protein source, fish meal, better than they do the best plant source, soybean meal. In catfish feeds, generally 20% of the dietary protein requirement consists of animal protein.

Better efficiency is obtained for all proteins when they are fed to catfish at temperatures between 75°F and 88°F than at 65°F or below.

Minerals

Fish, like domesticated animals, require minerals. However, across the gills and skin, fish have the ability to absorb from, and excrete into, the water a number of minerals, thereby reducing the mineral requirement in the diet. For this reason, research concerning the dietary mineral requirements of fish has been difficult to conduct, and the results have been inconclusive. Most researchers agree that fish require all of the macro- and micro-elements required by other animals for enzymes and cofactors. Calcium and cobalt are readily absorbed and excreted through the gills. While phosphate, chloride, and sulfate ions may be absorbed from the water, they are more readily absorbed in the digestive tract.

FISH MINERAL CHART

More research is necessary on the requirements, functions, and interactions of minerals in the diet and water of fish. In the meantime, based on presently available information, Table 28-4 summarizes the mineral deficiency symptoms and gives (1) the *requirements* for coldwater fish, and (2) the recommended *allowances* for warmwater fish. **NOTE WELL:** The figures for coldwater fish are *requirements*, whereas the figures for warmwater fish are *allowances*.

TABLE 28-4
FISH MINERAL CHART

	Deficiency Symptoms	Coldwater Fish Requirement in Feed[1]	Warmwater Fish Recommended Allowance in Feed	
		(%)	(%)	(mg/kg)[2]
Major or Macrominerals:				
Calcium	*Trout:* Poor growth, appetite, and feed efficiency. *Catfish:* Reduced growth and lower carcass ash, calcium, and phosphorus.	0.2-1.0	0.35-0.45[3]	3,500-4,500[3]
Phosphorus	*Coldwater and warmwater fish:* Reduced growth, appetite, feed conversion, and bone ash. Additionally, salmonids show skeletal abnormalities and bone deformities.	0.7-0.8 (inorganic P)	0.45[3] (available P)	4,500[3] (available P)
Magnesium	*Coldwater and warmwater fish:* Poor growth, loss of appetite, sluggishness, muscle flacidity, high mortality, and depressed magnesium levels in the body. Additionally, trout show renal calculi, vertebral curvature, degeneration of the muscles, and high mortality.	>.006	0.04[3]	400[3]
Trace or Microminerals:				
Cobalt	*Coldwater and warmwater fish:* Required for the synthesis of vitamin B-12 by gut bacteria.		0.000005[4]	0.05[4]
Copper	*Warmwater fish:* Depressed growth.		0.0005[4]	5.0[4]
Iodine	*Salmonids:* Thyroid hyperplasia (goiter).	0.00006-0.00011	0.0005[4]	5.0[4]
Iron	*Coldwater and warmwater fish:* Hypochromic, microcytic anemia.		0.003[4]	30.0[4]
Manganese	*Coldwater and warmwater fish:* Depressed growth and loss of appetite. Additionally, trout show abnormal tail growth and shortening of the body.		0.001[4]	10.0[4]
Selenium	*Salmonids:* Muscular dystrophy and exudative diathesis.	0.1-0.35	0.1[4]	1,000[4]
Zinc	*Salmonids:* Cataracts, caudal fin erosion, and depressed growth. *Channel catfish:* Poor growth and appetite.	0.0015-0.003	0.015[4]	150[4]

[1]From: Hilton, J. W. and S. J. Slinger, *Nutrition and Feeding of Rainbow Trout*, Department of Nutrition, College of Biological Science, University of Guelph, Guelph, Ontario; published by Department of Fisheries and Oceans, Ottawa, 1981, p. 6, Table 3.

[2]1 mg is the same as 1 ppm.

[3]From: Lovell, R. T., *Nutrition and Feeding of Channel Catfish* (Revised), Southern Cooperative Series Bul. No. 296, 1984, p. 28. These are dietary mineral requirements for catfish.

[4]From: Gatlin, D. M. III, University of Arkansas at Pine Bluff, Pine Bluff, Ark., in a personal communication to the senior author, 1986. These are recommendations for channel catfish.

Fig. 28-2. Thyroid hyperplasia (goiter) induced by iodine deficiency. When thyroid tumors in fish were first described in 1891, it was believed that the condition represented a form of throat cancer. After considerable debate and research, it was found that the tumors resulted from a dietary iodine deficiency—goiter. Today, iodine is routinely added to fish diets, and thyroid tumors are rarely seen in commercial fish production. (From Gaylord & Marsh; print provided by J. E. Halver, School of Fisheries, University of Washington, Seattle)

Because of the difficulties in determining the mineral requirements of fish, along with the lack of sufficient fish mineral research, some nutritionists suggest that a trace mineral mixture suitable for poultry be included in fish feeds.

Vitamins

As the digestive system of the fish is monogastric in structure and function, there is definite need for vitamins in fish diets. In general, the vitamin requirements of fish resemble those of monogastric animals, with a few exceptions. Fish represent one of the few types of higher animals that have been shown to have a requirement for vitamin C. There is not enough bacterial activity in the gut of the fish to satisfy either the B complex or the vitamin K requirements.

FISH VITAMIN CHART

The presently recommended vitamin requirements and allowances for fish are given in Table 28-5. Vitamin losses in fish feeds occur due to oxidation and reaction with other feed components. For this reason, amounts in excess of the requirements should be added—the latter are known as recommended allowances.

TABLE 28-5
FISH VITAMIN CHART

	Coldwater Fishes[1]				Warmwater Fishes[2]			
	Requirements[3]		Recommended Allowances[4]		Supplemental Allowances[5]		Complete Diet Allowance[6]	
	(per lb body wt./day)	(per kg body wt./day)	(per lb dry feed)	(per kg dry feed)	(per lb dry feed)	(per kg dry feed)	(per lb dry feed)	(per kg dry feed)
Fat-Soluble Vitamins:								
A .... (IU)	34	75	1,136	2,500	1,000	2,200	2,500	5,500
D .... (IU)	33	72	1,091	2,400	100	220	454	1,000
E .... (IU)	.45	1	13.6	30	5	11	23	50
K .... (mg)	.05	0.1	4.5	10	2.3	5	4.5	10
Water-Soluble Vitamins:								
Biotin .... (mg)	0.02	0.05	0.45	1	0	0	0.05	0.1
Choline .... (mg)	14-23	30-50	1,364	3,000	200	440	250	550
Folacin (Folic Acid) .... (mg)	0.09	0.20	2.3	5	0	0	2.3	5
Inositol .... (mg)	8-9	18-20	182	400	0	0	45	100
Niacin (Nicotinic Acid, Nicotinamide) .... (mg)	1.4-2.7	3-6	68	150	7.7-12.7[6]	17-28[6]	45	100
Pantothenic Acid (Vitamin B-3) .... (mg)	0.45	1	18	40	3.2-5[6]	7-11[6]	23	50
Riboflavin (Vitamin B-2) .... (mg)	0.2-0.5	0.5-1.0	9	20	0.9-3.2[6]	2-7[6]	9.1	20
Thiamin (Vitamin B-1) .... (mg)	0.07-0.09	0.15-0.20	4.5	10	0	0	9.1	20
Vitamin B-6 (Pyridoxine, Pyridoxal, Pyridoxamine) ... (mg)	0.09-0.18	0.2-0.4	4.5	10	5	11	9.1	20
Vitamin B-12 (Cobalamins) .... (mg)	0.0003	0.0006	0.009	0.02	0.9-4.5	2-10	9.1	20
Vitamin C (Ascorbic Acid, Dehydroascorbic Acid) ..... (mg)	1.4-2.7	3-6	45	100	0-45[6]	0-100[6]	14-45[6]	30-100[6]

[1]From: *Nutrient Requirements of Coldwater Fishes*, NRC-National Academy of Sciences, 1981, p. 41, Table 7.
[2]From: *Nutrient Requirements of Warmwater Fishes*, NRC-National Academy of Sciences, 1977, p. 18; and *Third Report to the Fish Farmers*, edited by Dupree, H. K. and J. V. Huner, published by the U.S. Department of the Interior, Fish and Wildlife Service, 1984, p. 147, Table 11.5.
[3]Based on young fish.
[4]Total vitamin contribution from all sources. Other amounts may be more appropriate to offset losses resulting from the effects of formulation and storage, or when feeding other than small fish at SET.
[5]These amounts do not allow for processing or storage losses. Other amounts may be more appropriate for various species and under various environmental conditions.
[6]Highest amounts probably appropriate when "standing crop" of fish exceeds 500 kg/hectare of water surface.

FAT-SOLUBLE VITAMINS

The fat-soluble vitamins, A, D, E, and K, are all of practical importance in fish feeds.

• **Vitamin A**—Vitamin A is essential for the normal structure and functions of eye and gill and for general maintenance of differentiated epithelia in various physiological systems.

Fish are susceptible both to deficiencies and excesses of vitamin A. At certain temperatures and levels some species of fish are capable of using pro-vitamin, beta-carotene, to fulfill their requirements, while other species must have vitamin A *per se* added to the diet.

Vitamin A can be supplied by rich feed sources, synthetic vitamin A, or fish oils.

• **Vitamin D**—Dietary vitamin D is needed by fish for facilitating mobilization, transport, absorption, and utilization of calcium and phosphorus. Vitamin D_3 has greater activity for fish than vitamin D_2.

Channel catfish on a deficient diet show reduced weight gain, along with lower body ash, calcium, and phosphorus.

• **Vitamin E**—Vitamin E has two functions for fish: (1) as an intracellular antioxidant and (2) metabolic functions unrelated to cellular antioxidants.

The requirement for vitamin E in fish diets is dependent on several factors—the amount and type of dietary polyunsaturated fatty acids, storage facilities, vitamer used, and diet preparations. Vitamin E is generally provided in the form of synthetic dl-alpha-tocopherol.

• **Vitamin K**—Vitamin K has been shown to be a dietary essential, involved in electron transport and oxidative phosphorylation reaction of cellular metabolism. The normal supplemental source is menadione (vitamin K_3).

WATER-SOLUBLE VITAMINS

All of the B vitamins and vitamin C are soluble in water. Water-soluble vitamins must be replenished almost daily, for there is little tissue storage. Fish require biotin, choline, folacin, inositol, niacin, pantothenic acid, riboflavin, thiamin, vitamin B-6, vitamin B-12, and vitamin C.

- **Biotin**—Biotin is synthesized in some fish by intestinal flora, but supplemental biotin in the diet may be required for maximum growth. Most diets containing fish meal will probably contain sufficient biotin for normal growth of young fish.
- **Choline**—Choline, which is a key part of lecithin, is vital for the prevention of fatty livers, the transmission of nerve impulses, and the metabolism of fat. Choline hydrochloride is routinely added as a supplement to fish diets.
- **Folacin (Folic Acid)**—Folacin is essential for the synthesis of nucleic acids, DNA, and RNA, and thus necessary for normal erythrocyte formation.
- **Inositol**—The classification of inositol as a vitamin is disputed; more properly perhaps, it should be classified as an essential nutrient.
- **Niacin (Nicotinic Acid, Nicotinamide)**—Supplementary niacin is necessary in the diet to maximize fish growth. Nicotinic acid and nicotinamide provide about the same biological activity for fish, and both forms are stable in multivitamin premixes. However, losses in nicotinic acid activity due to processing of extruded feeds may be 20% or more.
- **Pantothenic Acid (Vitamin B-3)**—Pantothenic acid is a dietary essential for fish. Nutritional gill disease (caused by pantothenic acid deficiency) is one of the more common nutritional disorders found in trout hatcheries. The free form of pantothenic acid is heat and pH sensitive, and processing and storage can cause substantial losses of the vitamin. Pantothenic acid as salts of Na+ or Ca++ should be added to the diet as these forms are relatively stable.
- **Riboflavin (Vitamin B-2)**—Riboflavin is a component of two flavoprotein coenzymes, needed for the breakdown of pyruvate, fatty acids, and amino acids.

In most fish diets, the producer does not have to be concerned with the loss of riboflavin in storage if the feed is protected from light. Storage of fish feed in light-proof bags is recommended.
- **Thiamin (Vitamin B-1)**—Thiamin is essential for the metabolism of carbohydrates and lipids, and for the direct oxidative cellular metabolism of glucose.

Several dietary factors can be instrumental in determining the thiamin requirement of fish. Many fish contain relatively high levels of thiaminase—an enzyme that breaks thiamin into two inactive molecules. Since many fish diets contain fish products, the producer should be aware of the potential risks of feeding raw fish. Cooking destroys the enzyme. The dietary level of fat can also affect the thiamin requirement. A high-fat diet can have a thiamin-sparing effect, but it will not totally alleviate the requirement for thiamin.

Thiamin hydrochloride and thiamin mononitrate are the usual synthetic sources of thiamin used in diet supplementation.
- **Vitamin B-6 (Pyridoxine, Pyridoxal, Pyridoxamine)**—Vitamin B-6 is a key coenzyme in protein metabolism. It follows that fish consuming high-protein diets—fish that are in the growing stage—have high requirements for pyridoxine. In carnivores, the pyridoxine requirement is especially high, and the body stores are rapidly depleted. In general, when needed, it is recommended that the vitamin be added to feed in the form of pyridoxine hydrochloride.
- **Vitamin B-12 (Cobalamins)**—Vitamin B-12, also known as cyanocobalamin, is a large molecule that contains cobalt. Animal products have traditionally been considered excellent sources of vitamin B-12 but care should be taken in processing and storage since B-12 is easily destroyed. Cool and dry storage, as well as rapid turnover of stored feeds, is recommended in order to ensure adequate dietary levels of vitamin B-12.
- **Vitamin C (Ascorbic Acid, Dehydroascorbic Acid)**—The amount of vitamin C required by fish is dependent on several factors—degree of stress, size of fish, and dietary nutrients. In the past, citrus fruits, cabbage, and some animal products were used for vitamin C supplementation, but synthetic ascorbic acid is now widely used. Vitamin C is very unstable to heat, moisture, and oxidation.

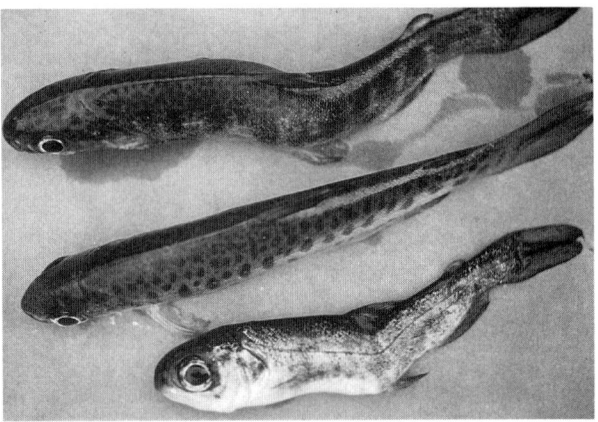

Fig. 28-3. Vitamin C deficiency symptoms in coho salmon (top and bottom). *Top:* SCOLIOSIS—a lateral curvature of the spine. *Bottom:* LORDOSIS—a forward curvature of the spine. *Center:* Normal (control) fish that received adequate vitamin C. (Courtesy, Dr. J. E. Halver, School of Fisheries, University of Washington, Seattle)

Water

In addition to its role as a nutrient carrier, water serves additional functions in fish. In freshwater fish, the concentration of ions in the blood is stronger than in the water of the environment. Since water always diffuses from the area of weakest ionic concentration to the strongest, fresh water readily diffuses through the gills and digestive tract into the fish. In saltwater fish, the blood ion concentration is weaker than that of marine water, consequently forcing the fish to absorb nutrients from the environment.

Water also acts as a medium for carrying oxygen. The amount of dissolved oxygen in water depends upon the movement of air over the water's surface, the movement of the water itself, the population of aquatic plants, the amount of sunshine, the population and activity of aquatic animals, and water temperature. Water temperature is critical because less oxygen is dissolved as the temperature increases and, equally important, as the temperature increases, the metabolic rate of aquatic animals accelerates, increasing the oxygen demand. On a physiological basis, if the water temperature is increased and there is less oxygen in the water, the fish has to increase its respiratory rate; and the demand for energy in the fish to sustain this activity is increased.

Nonnutritive Factors

While nonnutritive factors do not directly contribute to the maintenance, growth, or reproduction of fish, they should be considered in the formulation of rations as they can affect feed efficiency or the quality of the final marketable product. Three nonnutritive factors—antioxidants, fiber, and pigment-producing factors—warrant discussion concerning fish nutrition.

- **Antioxidants**—Due to the high-fat content of fish diets, along with the highly unsaturated nature of the fats, the dangers of oxidation and subsequent spoilage of feeds can present major problems. One means of controlling these problems is the addition of antioxidants to the feed, such as ethoxyquin, butylated hydroxyanisole (BHA), butylated hydroxytoluene (BHT), or vitamin E.
- **Fiber**—It is recommended that crude fiber not exceed 10% of fish diets—preferably not more than 5 or 6%.
- **Pigment-Producing Factors**—Quite often, producers wish to enhance the skin and tissue color in order to make the product more attractive to the consumer. This can be achieved through the addition of certain ingredients to the diet. Paprika fed at a level of 2% of the diet will improve the coloration of brook trout. Xanthophylls from alfalfa meal, corn gluten meal, and dried egg products will increase yellow pigmentation in brown trout skin. The use of shrimp or prawn wastes, which contain carotinoids, will produce a healthy "rosy" color when fed to trout. Also, commercial sources of coloring agents are now available for addition to fish feeds. Species differences with regard to the amount of color change have been observed, so it is possible that what works with one type of fish will not be effective with another.

FEEDS FOR FISH

Feed for commercial fish production accounts for 30 to 50% of the total expense. So, if fish farmers wish to maximize production at the lowest cost, they must know the nutritional requirements of the fish species which they produce. Also, they must recognize and, to the extent practical, maintain the environment to which fish are exposed. Producers have the good fortune of being able to select a type of fish which can adapt to their environment.

Good quality feeds, along with proper processing and feeding techniques, are essential in the intensive production of fish. Feeds that are not eaten do not produce fish; worse yet, they often reduce production and fish quality by contributing to oxygen demand and water quality deterioration. Several types of fish feeds are available, with the choice determined by the formulation and nutrient content, the cost of the feed, the behavioral characteristics of the species, and the size of fish. Also, the feed preferences of different species often change during their life span, and are often influenced by the type of culture system being used and the level of production. Suitable feed ingredients are included in this book in Section V—Composition of Feeds.

The processing methods used in preparing fish feeds are similar to those used in preparing feeds for domestic animals (see Chapter 14, Feed Processing).

Natural Foods and Feeds

As the term implies, natural foods are obtained from the immediate environment. Small fish feed upon algae and zooplankton. As fish grow, they devour progressively larger natural foods—insects, worms, mollusks, crustaceans, small fish, tadpoles, frogs, and plants.

Pondfish culturalists take advantage of the natural foods present. The insects, worms, and forage fish which pond fish consume are high in water—containing 75 to 80%. The remaining components are: protein, 12–15%; fat, 3–7%; ash, 1–4%; and carbohydrate, less than 1%. During warm weather when insects hatch and bottom organisms are abundant, a pond can provide a considerable amount of food for fish. The food production can be increased by pond fertilization with chemical fertilizers, organic materials, and animal manures. Because the environment tends to be highly variable in its production of biomass, this method of providing food is inefficient unless the producer is utilizing large bodies of water. However, natural food organisms are relied upon to provide nutrients lacking in supplemental feeds used in pond culture.

Artificial Wet Feeds

Artificial wet feeds contain various organs, meats, and by-products from animals, poultry, and fish. The most common ingredients in wet feeds are liver, spleen, ovaries, intestines, blood, testicles, condemned meat, trash fish, kidneys, fish scraps, mollusks, brain, meat trimmings, heart, poultry offal and by-products, and milk by-products. When feeding artificial wet diets, it is important to ensure that sufficient amounts of the omega-3 fatty acid (linolenic acid family) is present. Also, fish products contain thiaminase, necessitating heat treatment of these fish ingredients or supplemental thiamin.

Formulated Feeds

Most commercial fish producers are now using dry formulated feeds, that contain a combination of both plant and animal ingredients. The most commonly used grains are wheat and corn. The most commonly used grain by-products are brewers' grains, corn gluten meal, cottonseed meal, peanut meal, soybean meal, rice bran, and various wheat by-products. The most commonly used animal by-products are blood meal, feather meal (hydrolyzed), fish meal, poultry by-product meal, shrimp meal, and whey.

Formulated feeds may be either supplemental or complete rations.

- **Supplemental fish feeds**—These feeds are formulated to provide adequate energy and protein, but they may be deficient in minerals and vitamins which the fish are expected to obtain from natural foods. Such feeds are fed to fish reared in low densities in ponds.
- **Complete fish feeds**—These feeds are formulated to provide all the essential nutrients required by fish for optimal growth. If high densities of fish are being reared, a complete feed must be provided, as natural feeds will be limited or absent.

Formulated feeds are manufactured in compressed (sinking) pellets, expanded (floating) pellets, moist or semi-moist pellets, crumbles (granules), meals, or flakes.

- **Compressed or sinking pellets**—These pellets are made by adding steam to the feed as it is pelleted. The steam increases the moisture content by 5 to 6% and raises the temperature to 150–180°F during processing. The feed mixture is forced through a die (dies are available in different sizes) to extrude a compressed, dense pellet. The pellets are cooled and air dried to no more than 10% moisture immediately after pelleting.

- **Expanded or floating pellets**—These pellets require higher temperatures and pressures than compressed or sinking pellets. Under these conditions, raw starch is gelatinized; and bonds are formed within the gelatinized starch to give a durable, water-stable pellet. The sudden release of pressure following extrusion allows water vapor to expand and the ensuing entrapment of gas creates a buoyant, floating food particle.

Recent studies with catfish have shown that feeding 15% of the ration as floating feed and 85% as sinking feed gives better feed utilization and is more economical than feeding either alone.

- **Moist and semi-moist pellets**—Moist pellets, which contain 30–50% moisture are made from variable amounts of either fresh or frozen pasteurized fish, together with some dry ingredients. No heat is required for pelleting moist feeds. Refrigeration must be used to protect moist feeds against spoilage. After extrusion, moist pellets should be quick-frozen and stored at −14°F. Moist pelleted feed spoils rapidly when thawed. Also, moist feeds cost more to manufacture, ship, and store than dry pelleted feeds because they must be kept frozen. Salmon producers are the major users of moist feeds.

Semi-moist pellets contain 20–25% moisture, which is intermediate in moisture between moist and dry pellets. They do not require refrigeration, but they must be protected by adequate mold inhibitors and preservatives.

- **Crumbles (granules)**—These are made by crushing pellets, followed by screening out the granules to the desired sizes. The finished feed should be sized and contain not more than 15% oversize or undersize granules.

- **Meals and flakes**—Meals are often coated with vegetable oils or animal fats to increase energy level and improve flotation. They are usually scattered over the surface of the water. Meals are fed to bait minnows, goldfish, and fry of striped bass, grass carp, and sunfish.

Flake feed is prepared for aquarium fishes. It is usually sprinkled on the water surface.

FEED HANDLING AND STORAGE

Two rules should be followed in the handling and storage of pelleted dry feed for fish. They are:

1. **Handle with care.** Pellets are fragile and easily broken. When breakage occurs, fines from the pellet represent lost feed and can cause water pollution.

2. **Store feed in a cool, dry place and use it within 90 days.** Due to the high protein and fat levels in fish diets, the potential of ingredient spoilage is extremely great. The storage area should be kept clean and adequately ventilated; and the feed should be used within 90 days.

(Also see Chapter 14, Feed Processing.)

RATIONS FOR FISH

Because the cost of feed is the largest single item of expense in fish production, the fish farmer must utilize feed as efficiently as possible. An excess of a certain ingredient could well be as economically detrimental to overall production expenses as a deficiency. If producers are to obtain efficient feed conversions, they must adapt the feed to the particular needs of the fish. Water temperature, size and species of fish, and stocking density are probably the most critical factors to consider when formulating a ration.

Trout and Salmon Diets

In trout and salmon diets, protein levels should not be less than 45% in starter diets, not less than 40% in production diets, and not less than 35% in brood stock diets. Fat levels should be 15 to 20% in starter diets, 10 to 15% in production diets, and 10 to 15% in brood stock diets. Crude fiber should not exceed 4% in starter diets, and not exceed 5% in production and brood stock diets. The starter diet should not be fed longer than necessary to get the fish off to a good start; prolonged feeding of the starter can cause gill irritation, so it is important to switch to larger size particles as soon as possible. A list of commonly used feed ingredients for trout, along with their recommended feeding levels, is found in Table 28-6.

TABLE 28-6
SUGGESTED INGREDIENTS FOR TROUT FEEDS[1]

Ingredient	Quality	Percent of Diet; Recommended Level in Parentheses
Soybean oil meal	Solvent extracted; dehulled; minimum 47.5% protein	0–25 (20)
Cottonseed meal	Prime quality; solvent extracted; dehulled; minimum 48% protein; less than 0.4% free gossypol	0–20 (10)
Corn gluten meal	Maximum 3% fiber; minimum 60% protein	0–20 (10)
Meat meal	Maximum 7% fat; maximum 2.5% fiber; minimum 50% protein	0–15 (10)
Blood meal	Maximum 3.5% fiber; minimum 0.5% fat; minimum 80% protein	0–10 (5)
Hydrolyzed feather meal	Maximum 3% fiber; minimum 1.0% fat; minimum 85% protein	0–15 (10)
Poultry by-product meal	Maximum 2% fiber; minimum 12.5% fat; minimum 60% protein	0–15 (10)
Crab meal	Maximum 11% fiber; minimum 31% protein	0–10 (10)
Shrimp meal	Minimum 40% protein	0–10 (5)
Fish solubles (condensed)	Maximum 0.5% fiber; maximum 7% fat; minimum 30% protein	0–10 (5)
Whey, dried	Maximum 6% water; maximum 10% ash; maximum 3% salt; minimum 12% protein	0–10 (10)
Yeast, brewers' dried or torula	Maximum 3% fiber; minimum 35% protein	0–15 (10)
Fish meal:	Maximum 12% fat; maximum 10% moisture; maximum 5% salt	10–50 (25–40) —diet should contain minimum of 7% fish meal protein
Herring	Minimum 67.5% protein	
Anchovy	Minimum 65% protein	
Menhaden	Minimum 60% protein	

[1]Prepared especially for this book by L. Orme, Director, U.S. Department of the Interior, Fish and Wildlife Service, Diet Development Center, Spearfish, S.D.

Feeding Fish

Formulation specifications for trout are presented in Table 28-7; and diets for Pacific salmon are given in Table 28-8.

TABLE 28-7
FORMULATION SPECIFICATIONS FOR TROUT DIETS[1]

	Percent of Diet		
	Starter	Production (Grower)	Brood Stock
	(%)	(%)	(%)
Fish meal (herring, anchovy, mackerel, capelin):			
—minimum crude protein 65%			
—stabilized with ethoxyquin			
—maximum fat level 12%	45–50	25–35	30–35
—maximum moisture 10%			
—maximum salt level not to exceed 3%			
—maximum ash 15%			
Wheat middlings:			
—minimum crude protein 16%	0–15	10–30	15–35
—maximum crude fiber 9.5%			
Wheat gluten meal:			
—minimum crude protein 80%	0–3	0–2	0–1
Soybean meal[2]:			
—minimum crude protein 48%	5–10	5–15	5–20
Corn gluten meal:			
—minimum crude protein 60%	0–10	0–10	0–10
Dehydrated alfalfa meal:			
—minimum crude protein 17%	—	0–3	0–5
—maximum crude fiber 27%			
Dried whey:			
—minimum crude protein 13%			
—partially delactosed	0–5	0–3	0–5
Yeast, dried brewers':			
—45% crude protein	0–5	0–5	0–5
Corn distillers' dried solubles:			
—minimum crude protein 27%	0–10	0–10	0–10
Animal by-product meals:			
—hydrolyzed feather meal, minimum crude protein 85%	0–5	0–7	0–7
—Poultry by-product meal, minimum crude protein 60%	0–5	0–7	0–7
—blood meal, ring dried or spray dried, minimum crude protein 80%	0–5	0–7	0–7
—meat meal, minimum crude protein 50%	0–5	0–7	0–7
Fat supplement:			
—marine oil (salmon, capelin, herring, mackerel, etc.)	5–15	5–15	5–15
—vegetable oil (soybean oil, canola oil)	0–5	0–8	0–5
—animal fat and grease, all oils and fats to be stabilized	0–5	0–5	0–5
Vitamin premix	2–4	2–4	2–4
Mineral premix	2–4	2–4	2–4

[1]Adapted by the authors from *Nutrition and Feeding of Rainbow Trout*, by Hilton, J. W., and S. J. Slinger, Department of Nutrition, College of Biological Science, University of Guelph, Guelph, Ontario; published by Department of Fisheries and Oceans, Ottawa, 1981, p. 8.

[2]Soybean meal is not commonly used in Pacific salmon diets due to poor growth, probably because of unpalatability.

TABLE 28-8
DIETS FOR PACIFIC SALMON— OREGON MOIST AND ABERNATHY DRY FORMULATIONS[1]

Ingredients	Starter		Small Pellets or Crumbles[2]		Large Pellets	
	Moist	Dry	Moist	Dry	Moist	Dry
	(%)	(%)	(%)	(%)	(%)	(%)
Herring meal (70% crude protein)[3]	49.9	58.0	47.5	55.0	28.0	50.0
Wheat germ meal (25% crude protein)	10.0	—	Remainder	5.0	Remainder	5.0
Dried whey (12% crude protein)	8.0	5.0	4.0	5.0	5.0	5.0
Trace mineral premix	0.1	0.1	0.1	0.1	0.1	0.1
Vitamin premix	1.5	1.5	1.5	1.5	1.5	1.5
Wet fish (pasteurized)	20.0	—	30.0	—	30.0	—
Fish oil (stabilized)	10.0	12.0	7.0	9.0	7.75	9.0
Vitamin C	0.15	0.1	0.15	0.1	0.15	0.1
Choline chloride (70% product)	0.5	0.5	0.5	0.5	0.5	0.5
Binder	—	2.0	3.0	2.0	—	2.0
Cottonseed meal (47% crude protein)	—	—	—	—	15.0	—
Corn distillers' solubles (25% crude protein)	—	—	—	—	4.0	—
Dried blood meal (spray or flash, 80% crude protein)	—	10.0	—	10.0	—	10.0
Wheat middlings (15% crude protein)	—	Remainder	—	Remainder	—	Remainder

[1]Formulations provided for this book by L. G. Fowler, U.S. Fish and Wildlife Service, Longview, WA 98632

[2]Small pellets are moist feeds, and crumbles are dry feeds.

[3]Anchovy meal (65% crude protein) allowed in dry crumbles and large pellets.

Fig. 28-4. Feeding trout. (Courtesy, USDA)

Catfish Diets

Catfish feeds are generally lower in protein than those of coldwater fish. Fingerlings require 35 to 40% protein, production (grower) fish 25-36%, and brood stock 28-32% protein. Examples of various catfish feeds are found in Tables 28-9 and 28-10.

TABLE 28-9
FEED INGREDIENTS AND PERCENT COMPOSITION OF SOME PRACTICAL FEED FORMULATIONS FOR CHANNEL CATFISH[1]

Ingredient	Percent Protein	Fish Farming Experimental Station[2]	Texas A&M University 1	Texas A&M University 2	Auburn University Fingerlings (36% protein)	Auburn University Production (32% protein)
Alfalfa meal	17.5	3.5				
Blood meal	75.3	5.0				
Bone meal	11.2					
Brewers' grains	25.0					
Corn, distillers' grains	27.1					
Corn, distillers' solubles	27.3	8.0			5.0	
Corn grain, yellow	9.6		30.0	30.5	23.5	29.1
Corn grain, flint	9.9					
Cottonseed meal	40.8	10.0				
Cottonseed meal (without hulls)	50.0					
Dicalcium phosphate			0.25	1.5	1.5	1.0
Fats, animal				2.0	2.5	1.5
Fish meal, catfish	55.3					
Fish meal, menhaden	61.1	12.0		9.0	10.0	8.0
Fish meal, tuna	59.4					
Fish meal, white	61.9					
Grains, distillers'	27.4					
Liver meal, animal	66.5					
Meat meal, whole animal	54.3					
Meat meal (with bone)	50.5		15.0			
Oats, cereal by-product	14.6					
Peanut meal	44.0				18.0	
Poultry by-product meal	57.8					
Poultry feathers, hydrolyzed	85.4	5.0				
Rice bran	12.7	25.0				10.0[3]
Rice polishings (dust)	12.1	10.0				
Soybean meal	44.0		47.5			
Soybean meal (without hulls)	48.8	20.0		54.4	37.0	48.3
Wheat bran	15.1					
Wheat middlings	16.7		0.85			
Wheat grain	14.9					
Wheat shorts	16.4					
Whey (low lactose)	16.5		2.5			
Yeast (brewers')	45.1					
Vitamin premix			[4]	[4]	[5]	[5]
Mineral premix			0.50	0.50	[6]	[6]
Limestone			0.90			

[1]Adapted by the authors from *Third Report to the Fish Farmers*, by Dupree, H. K., and J. V. Huner, U.S. Department of the Interior, Fish and Wildlife Service, 1984, p. 151, Table 11.6.
[2]Herring fish meal (70% protein) may be substituted at 10% of the formulation. Wheat shorts, wheat middlings, or cereal grains may be substituted for the rice bran.
[3]Wheat shorts may be substituted for the rice bran.
[4]Vitamin premixes as published by the National Research Council (footnote 1, above).
[5]Quantities of vitamins per ton (grams, unless otherwise indicated): vitamin A, 4 million IU; vitamin D, 2 million IU; vitamin B-12, 8 mg; vitamin E, 50; menadione, 10; choline chloride (70%), 500; niacin, 80; riboflavin, 12; pyridoxine, 10; thiamin, 10; pantothenic acid, 32; folic acid, 2; and ethoxyquin, 125. Ascorbic acid (335 g) is added to the feed during pelleting, or top-dressed on extruded feed.
[6]Quantities of minerals per ton (milligrams): manganese, 110; zinc, 105; iron, 36; copper, 4.5; iodide, 2.3; and cobalt, 0.5.

TABLE 28-10
EXTRUDED (FLOATING) AND HARD (SINKING) PELLET FORMULA FOR CATFISH IN PONDS, RACEWAYS, AND CAGES[1]
(ALSO MINNOWS, GOLDFISH, CARP, AND BUFFALO)

Ingredient	Percent
Fish meal, menhaden, minimum 61% protein	10.0
Soybean meal, solv extd, w/o hulls, minimum 49% protein	35.0
Cottonseed meal, solv extd, minimum 41% protein	12.0
Wheat, whole grain (ground)[2]	31.7
Rice bran, with germs, solv extd	3.5
Fat, animal or plant[3]	5.0
Dicalcium phosphate	1.0
Vitamin premix[4]	0.8
Mineral premix[5]	1.0
Analysis:	
Crude protein, more than	32.0
Crude fiber, less than	3.5
Crude fat, more than	7.0

[1]Feed formula prepared for this book by H. Dupree, Fish Farming Experimental Station, U.S. Fish and Wildlife Service, Stuttgart, Ark.

[2]Wheat may be replaced up to 25% with corn.

[3]Sprayed on after manufacture.

[4]A. If the feed is to be used as a supplemental pond feed (i.e., low intensity pond culture), the premix should be as follows (per ton of feed basis): vitamin A, 900,000–1,800,000 IU; vitamin D_3, 450,000–900,000 IU; vitamin E, 27 g; thiamin, 0.9 g; riboflavin, 8.1 g; pyridoxine, 2.7 g; pantothenic acid, 9–18 g; nicotinic acid, 12 g; and ascorbic acid, 55 g.
B. If the feed is to be used as a complete feed (i.e., raceways, cages, and high intensity pond culture), the premix should be as follows (per ton of feed basis): vitamin A, 4,000,000 IU; vitamin D_3, 2,000,000 IU; vitamin E, 50 g; vitamin K, 10 g; thiamin, 10 g; riboflavin, 12 g; pyridoxine, 10 g; pantothenic acid, 32 g; nicotinic acid, 80 g; folic acid, 2.0 g; choline chloride, 500 g; ascorbic acid, 350 g; and vitamin B-12, 8 mg.

[5]Mineral premix should provide the following (per ton of feed basis): manganese, 100 g; iodide, 2.5 g; copper, 3.9 g; zinc, 80 g; iron, 40 g; and cobalt, 45 mg.

FEEDING COLDWATER FISH

In the United States, trout and salmon are the two most common species of coldwater fish grown commercially.

The art and science of feeding coldwater fish has progressed dramatically over the last 100 years. Originally, producers relied on natural feeds to grow trout, followed years later by the addition of wet diet supplements to natural feeds, and finally to the recent development of complete diets.

Variations in feeding habits occur among the different types of coldwater fish. For example, rainbow trout are surface feeders whereas brown trout are bottom feeders. Therefore, the type of feed pellet to be used must be given careful consideration. Since trout consume their feed in about 5 to 10 minutes, the producer does not have to be concerned with the pellets falling apart unless too much feed is being used and remains uneaten.

The producer should feed according to the stocking rate, the size of the fish, the type of pond or culture facility, the water temperature, and the energy content of the feed. Feed consumption is markedly affected by water temperature, decreasing in cold weather. Also, feed intake is affected by the energy content of the feed, since fish, like terrestrial animals, eat to satisfy their needs. Feed consumption is reduced in polluted water.

A feeding guide for salmonid fish is given in Table 28-11.

TABLE 28-11
FISH FEED GUIDE FOR SALMONIDS[1][2]

Number Fish		Granule/Pellet	Water Temperature °F/°C									
			43/6	45/7	46/8	48/9	50/10	52/11	54/12	55/13	57/14	59/15
(per lb)	(per kg)	(Granule No.)	◄――――――――――――――― % body weight per day ―――――――――――――――►									
1,182	2,600	1	2.9	3.4	3.7	3.9	4.6	4.8	5.2	5.8	6.0	6.4
591	1,300	1	2.8	3.3	3.6	3.8	4.4	4.7	4.9	5.6	5.9	6.1
318	700	2	2.7	3.0	3.3	3.6	4.1	4.5	4.8	5.1	5.6	5.8
182	400	2	2.6	2.8	3.0	3.2	3.9	4.0	4.6	4.9	5.0	5.1
91	200	3	2.3	2.6	2.8	3.0	3.6	3.8	4.3	4.5	4.6	4.7
59	130	3–4	2.1	2.3	2.5	2.8	3.3	3.6	3.7	3.9	4.0	4.1
41	90	4	1.9	2.0	2.1	2.4	2.7	2.9	3.0	3.2	3.6	3.8
		(Pellet No.)										
18	40	3/32	1.6	1.7	1.8	1.9	2.0	2.1	2.4	2.6	3.0	3.2
14	30	3/32	1.5	1.6	1.7	1.8	1.8	1.9	2.0	2.2	2.8	2.9
9	20	1/8	1.3	1.4	1.5	1.6	1.7	1.8	1.9	2.1	2.4	2.5
7	15	1/8	1.2	1.3	1.4	1.5	1.6	1.7	1.8	2.0	2.3	2.4
4	10	3/16	1.1	1.2	1.3	1.4	1.5	1.6	1.7	1.8	1.9	2.0
2	5	3/16	1.0	1.1	1.2	1.3	1.4	1.5	1.6	1.7	1.8	1.9
1	2	1/4	0.8	0.9	1.0	1.0	1.1	1.1	1.2	1.3	1.5	1.6

[1]Adapted by the authors from *Nutrition and Feeding of Rainbow Trout*, by Hilton, J. W., and S. J. Slinger, Department of Nutrition, College of Biological Science, University of Guelph, Guelph, Ontario; published by the Department of Fisheries and Oceans, Ottawa, 1981.

[2]Feeding rates based on a single strain of rainbow trout fed dry diets containing about 3,000 kcal digestible energy/kg.

In addition to feeding rates, the following feeding practices are important:

1. Frequency of feeding, with swim-up fry being fed small amounts of feed 20–24 times per day, gradually reduced to one to three times per day.

2. Feed particle size, hardness and texture, palatability, and placement of feed in relation to fish size are important. Very small fish will not travel far for feed.

3. Changes in feed intake and feed size should be made gradually, extending over a few days.

4. Most salmonids should be selected for brood stock at 2 to 3 years of age. At this time, they should be switched from the production or grower diet to a brood stock diet and fed only once a day. Depending on the temperature of the water, breeders should be taken off feed 3 to 6 weeks before spawning, then gradually brought back to full feed over a period of 2 to 4 weeks after spawning. Overfeeding before spawning reduces reproductive performance.

FEEDING WARMWATER FISH

Warmwater fish production involves numerous species of fish, but the major industries are catfish and carp farming. There have been attempts to produce predacious fish commercially, but due to the food requirements and behavioral characteristics of the fish, success has been very limited.

Feeding Catfish

Fig. 28-5. A basketful of farm-raised channel catfish ready for market, weighing 2 to 2½ lb. (Courtesy, Mississippi State University, Mississippi State)

Catfish raised for food and for fee-fishing in the Southwest are fed to liveweights of 2 to 2½ lb. Most catfish marketed in the Southeast—the area of greatest production—are processed at liveweights of ¾ to 2 lb. If good management practices are followed, fingerlings will have feed conversion rates of 0.9 to 1.0, and marketed fish will have conversion rates of 1.5 to 1.7.

Catfish, being omnivores, have well-defined stomachs and can effectively digest meat; but in commercial production the use of fresh meat diets is extremely limited. Most feeds are in dry pellet form, which make for economical storage and handling. Catfish diets generally contain the oil meals, distillers' solubles, and fish meal, in addition to vitamin and mineral premixes.

Since a sizable portion of the diet of catfish can come from the natural feed chain, ponds containing fingerlings should be fertilized in order to establish an optimum level of plankton growth. But fingerling producers must avoid over fertilizing.

It is generally recommended that 50 lb of 16-20-4 or 16-20-0 inorganic fertilizer per acre be applied every 10 days until an adequate plankton bloom is present. The producer can tell if the pond needs more fertilizer by sticking an arm into the water up to the elbow. If the hand can be seen, more fertilizer is needed.

In general, catfish are bottom feeders, but they can be taught to feed at the surface. While floating pellets are more expensive, some producers justify the added cost because they can routinely observe the condition of the fish at feeding.

The water temperature, fish weight, and water quality are the primary factors affecting the feed consumption of warmwater fish.

Table 28-12 presents a feeding guide for channel catfish in ponds. Table 28-13 gives a feeding guide for catfish in raceways. The pond feed is a supplemental, 36% protein diet; the raceway feed is a complete formulation.

TABLE 28-12
TYPICAL SPRING-SUMMER-FALL FEEDING SCHEDULE FOR CHANNEL CATFISH IN PONDS STOCKED AT 2,000-3,000 FINGERLINGS PER ACRE AS 5-IN. FISH AND HARVESTED AS 1.1-LB FOOD FISH, IN SOUTHEASTERN UNITED STATES[1,2]

Date	Water Temperature		Fish Weight		Feed Allowance Per Day
	(°F)	(°C)	(lb)	(kg)	(% of fish wt.)
April 15	68	20.0	0.04	0.02	2.0
April 30	72	22.2	0.06	0.03	2.5
May 15	78	25.5	0.11	0.05	2.8
May 30	80	26.6	0.16	0.07	3.0
June 15	83	28.3	0.21	0.10	3.0
June 30	84	28.8	0.28	0.13	3.0
July 15	85	29.4	0.35	0.16	2.8
July 30	85	29.4	0.42	0.19	2.5
August 15	86	30.0	0.60	0.27	2.2
August 30	86	30.0	0.75	0.34	1.8
September 15	83	28.3	0.89	0.40	1.6
September 30	79	26.1	1.01	0.46	1.4
October 15	73	22.7	1.10	0.50	1.1

[1]Adapted by the authors from *Third Report to the Fish Farmers*, by Dupree, H. K., and J. V. Huner, U.S. Department of the Interior, Fish and Wildlife Service, 1984, p. 156, Table 11.8.

[2]Feed allowances are based on data obtained with rations containing 36% protein and about 2.88 kcal of digestible energy per gram of protein. If feeds of lower protein and energy concentrations are used, daily allowances should be increased proportionally. Data adapted from R. T. Lovell, 1977. Feeding practices. Pages 50-55 in Nutrition and feeding of channel catfish. *Southern Cooperative Series Bul. 218*. Alabama Ag. Exp. Sta., Auburn University, Auburn, Ala.

TABLE 28-13
FEEDING RATES (PERCENT BODY WEIGHT PER DAY) FOR CHANNEL CATFISH FED A COMPLETE FEED (25% FLOATING, 75% SINKING FEED) IN RACEWAYS[1]

Water Temperature	Size and Weight		
	1-2 in. 0.001-0.004 lb	2-5 in. 0.004-0.04 lb	over 5 in. over 0.04 lb
(°F)	(% body weight)	(% body weight)	(% body weight)
Below 55°	1	1	1
At 55°	3	2	1.5
Above 55°	5	3	2

[1]Adapted by the authors from *Fish Hatchery Management*, by Piper, R. G., I. B. McElwain, L. E. Orme, J. P. McCraren, L. G. Fowler, and J. R. Leonard, U.S. Department of the Interior, Fish and Wildlife Service, Washington, D.C., 1982, p. 252, Table 29.

Multiple daily feeding can increase growth rate. In such a production system, feed is offered in the maximum amounts that can be metabolized in the pond. Feeding frequency varies with water temperature (see Table 28-14). A rule of thumb is that 90% of the feed should be eaten in 15 minutes or less.

TABLE 28-14
SUGGESTED MAXIMUM FEEDING RATES AND FREQUENCIES FOR CHANNEL CATFISH[1]

Water Temperature		Feeding Frequency	Feeding Rates
(°F)	(°C)	(times per day)	(% of total fish wt.)
>90	>32.2	1	1
80-86	26.6-30.0	2	3
68-80	20.0-26.6	1	2½
58-68	14.4-20.0	1	1½
50-58	10.0-14.4	0.5[2]	¾-1
<50	<10.0	0.3[3]	½-1

[1]Adapted by the authors from *Third Report to the Fish Farmers*, by Dupree, H. K., and J. V. Huner, U.S. Department of the Interior, Fish and Wildlife Service, 1984, p. 156, Table 11.9.

[2]Feed once on alternate days. [3]Feed once every 3 to 4 days.

• **Rules of thumb for feeding catfish**—The following rules of thumb for feeding catfish are generally followed:

1. Be aware of water temperature; catfish make their most efficient gains in water temperatures of around 84°F.
2. Do not full-feed on rainy days or when it has been overcast for extended periods (more than 4 days).
3. Until catfish reach a weight of 1 lb, the maximum feeding rate should be 3% of body weight per day.
4. The feeding rate of catfish weighing more than 1 lb should not exceed 2% of body weight per day.

• **Feeding catfish fry and fingerlings**—Catfish feeding begins with the fry and fingerling stages. Newly hatched catfish, which are called fry, live on nutrients from the yolk sac for 3 to 10 days (depending upon the water temperature), following which they accept food from a variety of sources.

Feeding Carp

Carp are the most extensively cultured fish in the world. They grow well under a variety of cultural conditions, use natural foods efficiently, and respond well to supplemental feeding.

The feeding principles of carp production can be generally considered to be applicable to the less commonly produced warmwater food and bait fish—for example, minnows, buffalo fish, and barbs.

Carp grow fastest in water which is 77° to 86°F. When the temperature drops to 60°F, growth is inhibited; and if the temperature falls below 55°F, feeding activity is greatly decreased. However, carp still require supplemental feed in cold water to avoid excessive weight loss and debilitation. For every 18°F above 55°F, there is a two- to threefold increase in feed consumption by carp.

Some researchers recommend that at least 50% of the feed for carp consist of natural feeds. Carp eat plankton primarily, along with small animals close to the shore and the bottom. But they will also eat some of the natural vegetation. With emphasis on the utilization of natural feeds, pond fertilization becomes an important production technique.

Artificial feeding—supplementary feeding is probably a more appropriate term—is commonly practiced. Soybean, corn, and wheat are the most widely used feedstuffs; but barley, oats, rye, beans, potatoes, millet, rice bran, vetch, and grass seeds may be used.

Carp are slow eaters. It generally takes them 30 minutes to an hour to finish eating a dry feed as compared to 5 minutes for trout. The pellets should be water stable in order to prevent leaching of nutrients, wastage of feed, and possible reduction in water quality.

Feeds for fry carp and related species are generally in a fine, flourlike consistency. It is best to feed on all sides of the pond in order to ensure that all the young fish have adequate access to feed.

In the warm regions of the world, where rice is commonly grown, carp are sometimes raised in rice fields. The warm water provides an excellent environment for rapid growth of both the rice and carp. The waste products from the carp fertilize the rice, and the carp consume some of the aquatic organisms that can interfere with the growth of the rice. Carp grow most efficiently during the summer, which is the time that rice will have its greatest growth.

In most countries, fertilization of ponds with manure is a common practice. Manure and organic wastes can be used to fertilize ponds, but the producer should be aware of possible nitrate-nitrite toxicity and oxygen depletion in the ponds. By using manure as a pond fertilizer, two purposes are served. First, it provides a convenient method of dealing with the potential pollution problem resulting from the feeding of livestock. Second, nutrients from the manure can be utilized to support the natural food system of the pond.

The Chinese Academy of Agricultural and Forestry Sciences recommends the following practices when feeding manure to fish: (1) mixing the manure with plant materials, then composting it before feeding (although manure may be, and is, fed to fish without composting, pollution may be lessened by first composting); (2) fertilizing and fermenting the composted manure; and (3) feeding manure to the little fingerlings, rather than to the big fish.

Fig. 28-6. A hog-fish combination on Kwang Li People's Commune, Kaw Yao County, Kwang Tung Province, where, in 1971, they produced 49,700 pigs and had a fish catch of 863,000 lb. (Photo by A. Ensminger)

FEEDING SYSTEMS

In all feeding systems, the producer must compare the costs of labor to the costs of automation. In general, larger operations can justify more automation. The producer must also take into consideration the frequency of feeding to be followed. Small fish must be fed frequently—8 to 20 times daily for trout—in order to prevent cannibalism and irregular growth. As the fish become large, the frequency of feeding can eventually be reduced to once or twice daily.

There are three methods of feeding fish—hand-feeding, semiautomatic feeding, and automatic feeding.

• **Hand-feeding**—This is the oldest form of feeding. It in-

Fig. 28-7. While hand-feeding requires considerable labor, it enables the farmer to check the health and general condition of the fish. (Courtesy, Mississippi State Univeristy, Mississippi State)

volves the fish culturist walking along the water banks and spreading the feed from a ladle or by hand. It is best to feed along all sides of the pond. Overfeeding will reduce profits as well as cause pollution.

• **Semiautomatic Feeding**—This involves feeding fish (1) from boats, (2) by blowing feed from mechanical equipment traveling along the edge of the water, or (3) by releasing feed from airplanes flying close to the water.

• **Automatic Feeding (Mechanical Feeders)**—Automatic feeding involves two types of feeder systems: (1) the demand type, which is activated by the fish; and (2) the automatic type, which is activated by a time clock.

Fig. 28-8. An automatic, demand type feeder. With this type of feeder, fish obtain the food by tripping a feed release trigger in the water. (Courtesy, Ralston Purina Co., St. Louis, Mo.)

OTHER FEED AND MANAGEMENT ASPECTS

When discussing the various nutritional aspects of fish culture, two additional areas should be considered—feeds affecting the quality and flavor of fish, and types of specialized aquatic production.

Feed Ingredients Affecting Fish Quality and Flavor

In many cases, fish farmers can alter the texture and flavor of their product merely by altering the composition of the ration. With this in mind, they should be aware of the demands of the consumer and attempt to produce a uniform product.

Trout texture may be affected by the moisture content of the diet; and trout flavor may be affected by feeding sea fish, spoiled feed, or rancid oils, or by algae in the water.

Carp fed bread and potatoes develop a wet, soft meat consistency which is considered undesirable.

Catfish flavor may be affected by excessive plankton bloom, feeds high in fish oil, presence of muskgrass, overfeeding, chemicals, or organic debris.

Specialized Aquatic Production

Variety in fish foods may be attained by such specialized productions as aquarium fish, crayfish, eel, lobster, prawn, predacious fish, shrimp, sturgeon, and tilapia.

• **Aquarium Fish (Hobby Fish/Tropical Fish) Culture**—This group includes representatives of several families, and over 100 species of small, colorful, and unique fishes. Although aquarium fish are not cultured for food, they represent a major segment of the pet industry.

There are many commercial feeds for aquarium fish, in a variety of shapes and formulations, most of which are nutritionally adequate.

• **Crayfish (Crawfish) Culture**—Feeds for crayfish include natural food and agricultural by-products. The feeding of specific formulations manufactured for crayfish is neither necessary nor economically sound.

• **Eel (Elvers) Culture**—In the past, the natural feeds were supplemented only by fresh fish diets. As a result, feed conversion rates were extremely poor (5 to 15:1). Today, modern Japanese diets make for conversion rates of 1.2 to 1.6:1 using a paste feed, and of 1.0 to 1.6:1 using a dry pellet. The diets generally range from 20 to 30% protein.

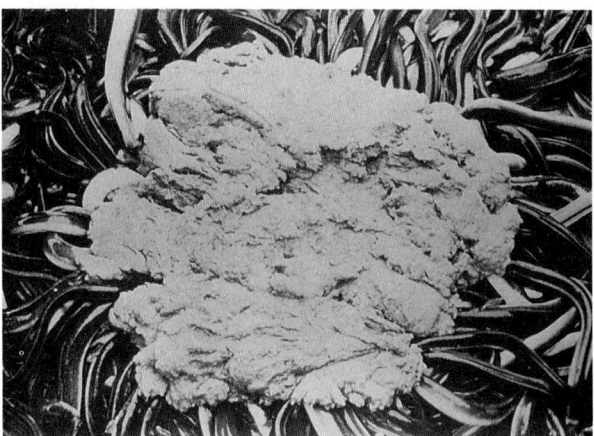

Fig. 28-9. Young Japanese eel feeding on paste food. (Courtesy, Dr. P. Ghittino, Torino, Italy)

• **Lobster (Shellfish) Culture**—Lobsters are predators. They eat crabs, snails, small fish, and even other lobsters. At night, lobsters walk along the ocean bottom seeking food.

Attempts to substitute cereal-based or purified diets for these natural foods have usually resulted in poor growth and low survival.

• **Prawn (Large Shrimp) Culture**—Feeding need not begin until the small prawns reach a length of 1 to 1½ in.—usually in 30 to 40 days. Special shrimp rations or the cheaper sinking catfish feeds, offered at about 3% of the body weight of the prawns, are suitable.

- **Predacious Fish Culture**—Feed conversion rates are highly variable (2:1 to 10:1), and high rates of cannibalism have been reported. In order to reduce cannibalism effectively, stocking densities must be extremely low, and the ponds must have heavy vegetation accompanied by a large population of forage fish.
- **Shrimp (Shellfish) Culture**—*Shrimp is the common name applied to about 2,000 species of crustaceans, whereas the term prawn may refer to any large shrimp.*

Because shrimp are slow feeders, feeds that remain stable in water for several hours must be used. The pellets should be small (0.08 to 0.16 in.). Shrimp farmers who culture shrimp in large tanks or small earthen ponds generally feed a mixture containing 40% dry pellets (formulations similar to broiler rations), 40% shrimp waste, and 20% clam waste.
- **Sturgeon Culture**—Both wild and cultured fish are used in caviar production. Dry diets using fish and animal meals, distillers' solubles, yeast, and vitamins have been developed and are being used extensively.
- **Tilapia Culture**—Research has indicated that tilapia utilize many plant ingredients to about the same degree as trout utilize animal products. Thus, a program combining pond fertilization with the feeding of a dry diet and chopped leaves and grasses should provide a suitable ration for culturing tilapia.

NUTRITIONAL TOXICANTS OF FISH

In addition to being susceptible to excesses and deficiencies of certain nutrients, fish are highly susceptible to a number of heavy metal toxicities and organic compound toxicities that often are found in water—either naturally or as a result of soil pollution due to past agricultural or industrial practices. In the past, many of these compounds have been routinely dumped in waterways.
- **Heavy Metal Toxicities**—Several metals have been shown to be toxic to fish. Mercury poisoning in fish has received the most attention, but cases of toxicities from arsenic, chromium, copper, and selenium have been documented.
- **Organic Compound Toxicities**—Numerous naturally occurring and synthetic organic compounds elicit toxic responses in fish; among them, aflatoxin, nitrosamines, cyclopropenoid fatty acids, tannic acid, and gossypol.

QUESTIONS FOR STUDY AND DISCUSSION

1. Rank in order the five leading species groups in monetary value of aquaculture production; and explain why they rank as they do. What accounts for the tremendous increases since 1980?

2. Why is freshwater fish production more prevalent than marine fish production?

3. List the primary differences in the nutrient requirements of fish and land animals.

4. List and discuss five environmental influences on the feeding behavior of fish.

5. What is meant by Standard Environmental Temperature (SET)? What are the SET values for salmon, trout, and catfish?

6. How do the energy requirements of fish differ from the energy requirements of warm-blooded animals?

7. What factors affect the energy requirements of fish as such?

8. Since only limited energy values of fish feeds are available, what metabolizable energy values for another species may a fish producer use?

9. Discuss the relative importance of carbohydrates, fats, and proteins as energy sources for fish.

10. Discuss the comparative maximum levels of carbohydrates that may be used as energy for each salmonids and warmwater fish.

11. Discuss the comparative fat requirement of coldwater fishes and of warmwater fishes.

12. Compare the amino acid requirements of fish and broilers.

13. List the factors that determine the requirement for protein in fish feeds.

14. Explain how fish and warm-blooded animals differ in their ability to absorb and excrete a number of minerals.

15. What causes thyroid hyperplasia (goiter) in fish? How can it be prevented?

16. What are the major differences between the vitamin requirements of fish and higher animals? Describe the symptoms of vitamin C deficiency in fish.

17. Physiologically, how do freshwater and saltwater fishes differ in their use of water?

18. Discuss the role and importance of each of the following nonnutritive factors in fish rations: antioxidants, fiber, and pigment-producers.

19. Define each of the following terms as they apply to fish: natural foods and feeds, artificial wet feeds, formulated feeds, supplemental feeds, complete feeds, compressed or sinking pellets, expanded or floating pellets, moist pellets, crumbles, and meals/flakes.

20. Why and how are farm ponds fertilized?

21. Give four rules of thumb for feeding catfish.

22. What are the primary differences between the rations of salmonids, catfish, and carp?

23. What three characteristics account for carp being the most extensively cultured fish in the world?

24. How and why may manure be used to fertilize ponds populated by carp?

25. List three systems of feeding fish, and give the advantages and the disadvantages of each system.

26. What's unique about each of the following specialized types of aquatic production: aquarium fish, crayfish, eel, lobster, prawn, predatory fish, shrimp, sturgeon, and tilapia?

27. List and discuss three heavy metals and three organic compounds that may be toxic to fish.

ABOUT THE WORLD OF ANIMALS. This is from an original painting by the noted artist, Tom Phillips (3333 17th Street, San Francisco, California 94110), prepared especially for this book. It portrays the artist's conception of what is in Chapter 29, Glossary of Nutrition Terms, and Chapter 30, Glossary of Feedstuffs.

SECTION IV

GLOSSARY

It is important that nutritionists and keepers of herds and flocks use the correct feed and nutrition terms and know what they mean. They need to know old terms, many of which have been redefined by the relentless wheels of progress; and they need to know a host of new terms. They need to know about new grains and forages; and about new by-products, recycled feeds, and crop residues. They need to know about new technology in feed processing; and about new supplements, additives, and implants.

To meet these needs, two glossary chapters are presented: Chapter 29, Glossary of Nutrition Terms; and Chapter 30, Glossary of Feedstuffs. These glossaries cover a wide range of terms. Each term has been defined as simply as possible, yet every effort has been made to retain technical correctness. Because of space limitations, many terms that are covered elsewhere in the text are not repeated in Chapters 29 and 30. Instead, the reader may refer to them in the index. It is the authors' fond hope that this compilation of terms will provide a common understanding in communication in feeds and nutrition.

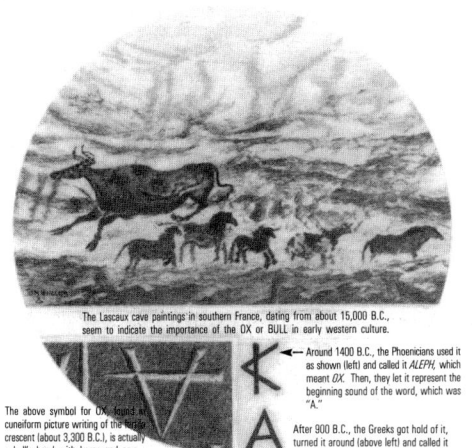

The Lascaux cave paintings in southern France, dating from about 15,000 B.C., seem to indicate the importance of the OX or BULL in early western culture.

Around 1400 B.C., the Phoenicians used it as shown (left) and called it *ALEPH*, which meant *OX*. Then, they let it represent the beginning sound of the word, which was "A."

The above symbol for OX, found in cuneiform picture writing of the *crescent* (about 3,300 B.C.), is actually a bull's head with horns and ears.

After 900 B.C., the Greeks got hold of it, turned it around (above left) and called it *ALPHA*. So, the *OX* stands at the head of the alphabet.

29

GLOSSARY OF NUTRITION TERMS

Original painting by Tom Phillips

A glossary of terms frequently used in discussing matters related to feeds and nutrition is presented in this chapter. (See Chapter 30 for a glossary of feedstuffs.)

Because of space limitations, most terms that are defined or explained elsewhere in this book are not repeated in this chapter. Thus, if a particular term is not listed herein, the reader should look in the index or in the particular chapter where it is discussed.

A

ABLACTATION. The act of weaning.

ACETONE. A by-product of the breakdown of fats for energy. It builds up when the body's glycogen stores are depleted, which happens when carbohydrate is not available for fuel.

ACETYL-CoA. The chief precursor of lipids, and important intermediary in the Kreb's cycle; formed by an acetyl group attaching itself to coenzyme A during the oxidation of amino acids, fatty acids, or pyruvate.

ACHROMOTRICHIA. Loss of pigment in hair.

ACTIN. Muscle fiber fraction that complexes with myosin to bring about muscle contraction.

ADDITIVE. An ingredient or combination of ingredients added to a basic feed mix to help fulfill a specific need.

ADIPOSE TISSUE. Fatty tissue.

AD LIBITUM. Free-choice access to feed.

ADRENAL GLAND. One of the endocrine (ductless) glands of the body, located near the kidney, which secretes hormones needed to utilize nutrients.

AERIAL PART. The aboveground part of a plant.

AEROBE. The term usually applied to microorganisms that require oxygen to live and reproduce.

AFTERMATH. The regrowth of range or seeded pasture forage after grazing or mowing.

AGAR. An extract of seaweed that forms a firm gel at temperatures of 104°F and lower. It is used as a gelling and stabilizing agent in foods and as a microbiological plating medium.

AGGLOMERATED FEED. A mixture of feeds in compacted or extruded form.

AGROFORESTRY. A production system in which trees for timber are grown in forage crops.

AIR DRY (approximately 90% dry matter). This refers to feed that is dried by means of natural air movement, usually in the open. It may be either an actual or an assumed dry matter content; the latter is approximately 90%. Most feeds are fed in the air dry state.

AIV PRESERVATIVE. A combination of hydrochloric and sulfuric acids used to increase the acidity of silage and to improve its keeping quality and nutritive value, developed by Dr. A. I. Virtanin of Finland.

ALGAE. Single-celled plants which contain about 50% protein on a dry basis and offer an attractive possibility as a protein source. They synthesize proteins by the use of solar energy. Cultivated freshwater algae will produce about 10 times as much protein per unit of land area as soybeans.

ALLERGY. A severe reaction, or sensitivity, which occurs in some individuals following the introduction of certain antigens into their bodies.

ALOPECIA. Loss of hair.

ALPHA-TOCOPHEROL EQUIVALENT. A standard unit of measurement (in milligrams) for designating vitamin E requirements, since potencies of the other members of the vitamin E group vary.

AMBIENT TEMPERATURE. The prevailing or surrounding temperature.

AMERICAN FEED INDUSTRY ASSOCIATION (AFIA). The address is 1701 N. Ft. Myer Drive, Arlington, VA 22209. The American Feed Industry Association is a nationwide organization of feed manufacturers banded together (1) to improve the quality and promote the use of commercial feeds,

(2) to encourage high standards on the part of its members, and (3) to protect the best interests of the feed manufacturer and the livestock producer in legislative programs.

AMMONIATED. Combined or impregnated with ammonia or an ammonium compound.

AMPHOTERIC. Having properties of both an acid and a base and therefore able to function as either. Amino acids have this dual chemical nature because of their structure—they contain both an acid (carboxyl, COOH) and a base (amino, NH_2) group.

AMYLOPECTIN. A polysaccharide, the insoluble part of starch, which forms a paste in hot water and thickens during cooking.

ANABOLISM. The conversion of simple substances into more complex substances by living cells (constructive metabolism).

ANAEROBE. A microorganism that does not require air or free oxygen to live and reproduce.

ANALOGUE (ANALOG). Anything that is analogous or similar to something else.

ANAPHYLAXIS. An acute allergic reaction.

ANASARCA (DROPSY). Generalized edema.

ANEURISM. The dilation of an artery wall.

ANIMAL NUTRITIONIST. A professional who possesses an earned bachelor's, or higher, degree with a major in animal science, chemistry, biochemistry, or related sciences from an accredited college or university, and who is able to apply the knowledge of feeds and their nutrients to the growth, maintenance, production, and health of animals.

ANIMAL PROTEIN. Protein derived from meat-packing or rendering plants, surplus milk or milk products, and marine sources. It includes proteins from meat, milk, poultry, eggs, fish, and their products.

ANION. An ion carrying a negative (−) electrical charge.

ANOREXIA. A lack or loss of appetite for food.

ANOXIA. Lack of oxygen in the blood or tissues. This condition may result from various types of anemia, reduction in the flow of blood to tissues, or lack of oxygen in the air at high altitudes.

ANTACID. Any substance which counteracts or neutralizes acidity.

ANTAGONIST. A substance that counteracts the action of another substance. The antagonist prevents the normal action because its molecular structure is so like that of the first substance that it almost fits into its position in a metabolic process; it gets in the way and prevents the reaction from taking place.

ANTHELMINTIC (VERMIFUGE). A product which expels or destroys internal parasites.

ANTIBODY. A protein substance (modified type of blood-serum globulin) developed or synthesized by lymphoid tissue of the body in response to an antigenic stimulus. Each antigen elicits production of a specific antibody. In disease defense, the animal must have an encounter with the pathogen (antigen) before a specific antibody is developed in its blood.

ANTIGEN. A high molecular weight substance (usually protein) which, when foreign to the bloodstream of an animal, stimulates formation of a specific antibody and reacts specifically *in vivo* or *in vitro* with its homologous antibody.

ANTIHISTAMINES. Any of various compounds used to treat certain allergic reactions in the body.

ANTIMETABOLITE. A substance bearing a close structural resemblance to one required for normal physiological functioning, which exerts its effect by replacing or interfering with the utilization of the essential metabolite.

ANTIOXIDANT. A compound that prevents oxidative rancidity of polyunsaturated fats. Antioxidants are used to prevent rancidity in feeds and foods.

ANTIVITAMIN. Any substance which inhibits the normal function of a vitamin.

APPETIZER. A substance that stimulates the appetite.

AS-FED. This refers to feed as normally fed to animals. It may range from 0 to nearly 100% dry matter. Moisture-free feed consists of 100% dry matter.

ASH. The mineral matter of a feed. The residue that remains after complete incineration of the organic matter.

ASPIRATING. The process of removing chaff, dust, or other light materials by use of air.

ASSAY. Determination of (1) the purity or potency of a substance, or (2) the amount of any particular constituent of a mixture.

ASSIMILATION. A physiological term referring to a group of processes by which the nutrients in feed are made available to and used by the body; includes digestion, absorption, distribution, and metabolism.

ATAXIA. Lack of coordination.

ATOM. A particle of matter indivisible by chemical means. It is the fundamental building block of the chemical elements. The elements, such as iron and sulfur, differ from each other because they contain different kinds of atoms.

ATROPHY. A wasting away of a part of the body, usually muscular, induced by injury or disease.

AVAILABLE NUTRIENT. A nutrient which can be digested, absorbed, and/or used in the body.

AVERAGE DAILY GAIN (ADG). The average daily liveweight increase of an animal.

AVIDIN. A protein in egg white (albumen) which combines with biotin and makes it unavailable; an antivitamin.

B

BACTERIA. Microscopic, single-cell plants, found in most environments, often referred to as microbes; some are beneficial, others are capable of causing disease.

BACTERICIDE. A product that destroys bacteria.

BASAL DIET. A diet common to all groups of experimental animals to which the experimental substance(s) is added.

BASAL METABOLIC RATE (BMR). The heat produced by an animal during complete rest (but not sleeping) following fasting, when using just enough energy to maintain vital cellular activity, respiration, and circulation, the measured value of which is called the basal metabolic rate (BMR). Basal conditions include thermoneutral environment, resting, postabsorptive state (digestive processes are quiescent), consciousness quiescence, and sexual repose. It is determined in humans 14 to 18 hours after eating and when at absolute rest. It is measured by means of a calorimeter and is expressed in calories per square meter of body surface.

BASE. A chemical substance that is capable of neutralizing acid by accepting hydrogen ions from the acid. *Alkali* is a synonymous term.

BATTERY. A series of pens or cages used to house animals in concentrated confinement rearing systems.

BIOASSAY. Determination of the relative effective strength of a substance (as a vitamin, hormone, or drug) by comparing its effect on a living test organism with that of a standard preparation.

BIOCHEMISTRY. The chemistry of living things — plants and animals.

BIOLOGIC ACTIVITY. Degree of effect in an organism of a specific vitamin; means of measuring required amount of vitamin to prevent a deficiency.

BIOLOGICAL. Of or pertaining to life and life processes, including the study of structure, functioning, growth, origin, evolution, and distribution of living organisms.

BIOLOGICAL VALUE OF A PROTEIN. The percentage of the protein of a feed or feed mixture which is usable as a protein by the animal. Thus, the biological value of a protein is a reflection of the kinds and amounts of amino acids available to the animal after digestion. A protein which has a high biological value is said to be of *good quality*.

BIOSYNTHESIS. The production of a new material in living cells or tissues.

BLENDED. Combined or mixed so as to render the constituent parts indistinguishable from one another, such as when two or more feed ingredients are mixed.

BLOCK. Feed or feeds compressed into a solid mass and cohesive enough to hold its form, usually weighing from 15 to 500 lb.

BLOOM. Said of an animal that has beauty and freshness. An animal in bloom has a glossy hair coat and presents an attractive appearance.

BOLTING FEED. Animals that ingest or eat too rapidly or greedily are said to be bolting their feed.

BOLUS.
• Regurgitated feed that has been chewed and is ready to be swallowed, known as the cud.
• A large pill for dosing animals.

BOMB CALORIMETER. An instrument used to measure the gross energy content of any material, in which the feed (or other substance) tested is placed in an enclosed chamber and burned in the presence of oxygen.

BRAND NAME. Any word, name, symbol, or device, or any combination of these, often registered as a trademark or name, which identifies a product and distinguishes it from others.

BRITISH THERMAL UNIT (BTU). The amount of energy required to raise 1 lb of water 1°F; equivalent to 252 calories.

BRIX. A term commonly used to indicate the sugar (sucrose) content of molasses. It is expressed in degrees and was originally used to indicate the percentage by weight of sugar in sucrose solutions, with each degree Brix being equal to 1% sucrose.

BROILER. A young chicken.

BUFFER. A substance in a solution that makes the degree of acidity (hydrogen-ion concentration) resistant to change when an acid or base is added.

BY-PRODUCT FEEDS. The innumerable roughages and concentrates obtained as secondary products from plant and animal processing, and from industrial manufacturing.

C

CACHEXIA. A general lack of nutrition and wasting of tissues due to malnutrition, a chronic disease, or a terminal illness.

CAKE (PRESSCAKE). The mass resulting from the pressing of seeds, meat, or fish in order to remove oils, fats, or other liquids.

CALCIFICATION. The process by which organic tissue becomes hardened by a deposit of calcium salts.

CALCITONIN. One of two hormones responsible for fine regulation of blood calcium level; the other is parathyroid hormone. Calcitonin, secreted by the thyroid gland, inhibits bone resorption and release of calcium to the blood, thereby lowering blood calcium.

CALCULI. Mineral deposits that occur in the kidney or urinary tract.

CALORIC. Pertaining to heat or energy.

CALORIMETER. An instrument for measuring the amount of energy in a substance.

CANNIBALISM.
• The habit of one animal pecking at or eating on another animal, such as a fowl pecking at or eating on another fowl or one pig biting the tail of another.
• The eating of young, such as a sow may do after farrowing or a doe rabbit may do if disturbed soon after kindling.

CANNULA (CANULA). A tubular device used to connect an internal structure with the outside of the animal. A metal-, rubber-, or glass-tube cannula may be inserted into the body cavity to allow the escape of fluids or gas. Liquids and other materials may be introduced into the body through a cannula.

CARCASS WEIGHT. Weight of the carcass of an animal following slaughter, as it hangs on the rail, expressed either as warm (hot) or chilled (cold) carcass weight.

CARCINOGEN. Any cancer-producing substance or agent.

CARCINOGENIC. Cancer-producing.

CARNIVORE. A flesh-eating animal.

CARRIER. An edible material to which ingredients are added to facilitate their uniform incorporation into feeds. The active particles are absorbed, impregnated, or coated into or onto the edible material in such a way as to carry the active ingredient physically.

CARRYING CAPACITY. The number of animal units (one cow, plus suckling calf—if there is a calf, or one heifer two years old or over; or equivalent) a property or area will carry on a year-round basis. This includes the land grazed plus the land necessary to produce the winter feed.

CATABOLISM. The conversion or breaking down of complex substances into more simple compounds by living cells (destructive metabolism).

CATALYST. Any substance which speeds up the rate of a chemical reaction without being destroyed or inactivated in the process. Enzymes are organic catalysts.

CATION. Ion that carries a positive (+) electric charge.

CHAFF. Glumes, husks, or other seed covering, together with other plant parts, separated from seed in threshing or processing.

CHELATE. The word *chelate* is derived from the Greek word meaning *claw*. It refers to a cyclic compound which is formed between an organic molecule and a metalic ion, the latter being held within the organic molecule as if by a claw. Examples of naturally occurring chelates are the chlorophylls, cytochromes, hemoglobin, and vitamin B–12.

CHEMOSTATIC CONTROL OF APPETITE. Certain cells located at the base of the brain in the lateral region of the hypothalamus regulate feed intake by sensing mechanisms. Factors shown to influence these cells and thereby alter feed intake include the volatile fatty acids. (Also see THERMOSTATIC CONTROL OF APPETITE.)

CHLOROPHYLL. The green coloring material in plants which is essential for one phase of the conversion of solar energy into organic material, the process known as photosynthesis.

CHOLESTEROL. A white, fat-soluble substance found in animal fats and oils, bile, blood, brain tissue, nervous tissue, the liver, kidneys, and adrenal glands. It is important in metabolism and is a precursor of certain hormones. Some have implicated cholesterol as a factor in arteriosclerosis.

CHYME. Semifluid feed mass in the gastrointestinal tract following gastric digestion.

CIRRHOSIS OF THE LIVER. Progressive destruction of the liver cells and an abnormal increase of connective tissue.

COAGULATED. Curdled, clotted, or congealed, usually brought about by the action of a coagulant.

COEFFICIENT OF DIGESTIBILITY. The percentage value of a food nutrient that is absorbed. For example, if a food contains 10 g of nitrogen and it is found that 9.5 g are absorbed, the digestibility is 95%.

COENZYME. A substance, usually containing a vitamin, which works with an enzyme (protein mainly) to perform a certain function.

COLLAGEN. A white, papery transparent type of connective tissue which is of protein composition. It forms gelatin when heated with water.

COLLOID. Glutinous, gluelike; a dispersion of matter throughout a medium.

COMBUSTION. The combination of substances with oxygen accompanied by the liberation of heat.

COMFORT ZONE. The temperature range within which no demand is made on temperature regulating mechanism.

COMMERCIAL FEEDS. Feeds mixed by manufacturers who specialize in the feed business.

COMPACTION. Closely packed feed in the stomach and intestines of an animal causing constipation and/or digestive disturbances.

COMPENSATORY GROWTH. Accelerated growth following a period of limited feed intake.

COMPLETE RATION. All feedstuffs (forages, grains, and processed feeds) combined in one feed. A complete ration fits well into mechanized feeding and the use of computers to formulate least-cost rations.

COMPOSITE SAMPLE. A combination of individual samples taken at selected intervals to minimize the effect of the variability of an individual sample.

CONCENTRATE. A broad classification of feedstuffs which are high in energy and low in crude fiber (under 18%). For convenience, concentrates are often broken down into (1) carbonaceous feeds, and (2) nitrogenous feeds.

CONDITION.
- The state of health, as evidenced by the hair-coat, and general appearance.
- The amount of flesh or finish (fat covering).

CONDITIONING. Achieving predetermined moisture characteristics and/or temperature of ingredients or a mixture of ingredients prior to further processsing.

CONGENITAL. Malformation existing at birth; acquired during development in the uterus and not through heredity.

COPROPHAGY. The ingestion of fecal material.

CREATININE. A nitrogenous compound arising from protein metabolism and excreted in the urine.

CRUDE FAT. Material that is extracted from moisture-free feeds by ether. It consists largely of fats and oils with small amounts of waxes, resins, and pigments. In calculating the energy value of a feed, the fat is considered to have 2.25 times as much energy as either nitrogen-free extract or protein.

CRUDE FIBER (CF). Crude fiber contains cellulose, hemicellulose, and lignin. It is determined by treating a sample of feed with ether to remove fats, then the residue is boiled alternately in a weak acid and a weak base. The remaining residue contains crude fiber and ash. Then, the ash is subtracted out and the crude fiber is what remains.

CRUDE PROTEIN (CP). This is a mixture of true protein and nonprotein nitrogen. It indicates the capacity of a feed to meet an animal's protein needs, although it is of little value in predicting energy availability.

CUBING. Refers to the practice of compressing long or coarsely cut hay into cubes about 1¼ in. square and 2 in. long, with a bulk density of 30 to 32 lb per cu ft.

CUD. A bolus of previously eaten feed which is regurgitated by a ruminant animal for further chewing.

CURD. The coagulated or thickened part of milk. Curd from whole mild consists of casein, fat, and whey, whereas curd from skim milk contains casein and whey, but only traces of fat.

CYANOCOBALAMIN. Same as Vitamin B-12.

CYSTITIS. Inflammation of the urinary bladder.

D

DATABASE. An organized body of information placed in computers for systematic storage and retrieval; usually deals with a specific topic or project.

DEBEAKING. The removal of part of the beak of chickens and poults to prevent cannibalism. Many broiler growers and some egg producers have chicks debeaked at the hatchery, with about ⅔ of the beak removed. Turkeys are usually debeaked at 2 to 3 weeks of age, with at least ½ of the upper beak removed.

DECORTIFICATION.
- Removal of the bark, hull, husk, or shell from a plant, seed, or root.
- Removal of portions of the cortical substance of a structure or organ, as in the brain, kidney, and lung.

DEFECATION. The evacuation of fecal material from the rectum.

DEFICIENCY DISEASE. A disease caused by a lack of one or more basic nutrients, such as a vitamin, a mineral, or an amino acid.

DEFLUORINATED. Processed in such manner that the fluorine content is reduced to a level which is nontoxic under normal use.

DEHYDRATE. To remove most or all moisture from a substance for the purpose of preservation, primarily through artificial drying.

DEPRAVED APPETITE (PICA). A craving for and eating of unnatural substances, such as dirt, hair, dung, wood, etc.

DESICCATE. To dry completely.

DIALYSIS. Separating substances in solution by taking advantage of the different rates at which they pass through a semipermeable membrane.

DICOUMAROL. A chemical compound found in spoiled sweet clover hay or made synthetically. It is an anticoagulant and can cause internal hemorrhages when eaten by animals. Medically, it is used as an anticoagulant. The trade name is Dicumarol.

DIGESTIBLE DRY MATTER (DDM). This refers to digestiblity as predicted by ADF or measured by feeding trials with animals.

DIGESTIBLE DRY MATTER INTAKE (DDMI). This is an estimate of how much DDM an animal will consume.

DIGESTIBLE NUTRIENT. The part of each feed nutrient that is digested or absorbed by the animal.

DIGESTIBLE PROTEIN. That protein of the ingested food protein which is absorbed.

DIGESTION COEFFICIENT (coefficient of digestibility). The difference between the nutrients consumed and the nutrients excreted expressed as a percentage.

DILUENT. An edible substance used to mix with and reduce the concentration of nutrients and/or additives to make them more acceptable to animals, safer to use, and more capable of being mixed uniformly in a feed. (It may also be a carrier.)

DISACCHARIDE. Compound sugars composed of two molecules of a monosaccharide. The three common members are: sucrose (table sugar), lactose (milk sugar), and maltose (grain sugar).

DIURESIS. An increased excretion of urine.

DIURETIC. An agent that increases the production of urine.

DORMANT. Plants cured on the stem; applied to most nongrowing range plants after the seeds have formed.

DOUGH STAGE. Stage at which grain seeds are soft and immature.

DRESSED. Made uniform in texture by breading or screening of lumps from feed and/or the application of liquid(s).

DRUGS. Substances of mineral, vegetable, or animal origin used in the relief of pain, for the cure of disease, or for the enhancement of production.

DRYLOT. A relatively small enclosure without vegetation, either (1) with shelter, or (2) an open yard, in which animals may be confined.

DRY MATTER (DM). That part of a feed which is not water. It is computed by determining the percentage of water and subtracting the water content from 100%.

DRY MATTER BASIS. A method of expressing the level of a nutrient contained in a feed on the basis that the material contains no moisture.

DRY MATTER INTAKE (DMI). This is an estimate of the amount of forage an animal will eat with only forage fed.

DRY-RENDERED. Residues of animal tissues cooked in open steam-jacketed vessels until the water has evaporated. Fat is removed by draining and pressing the solid residue.

DUNG. The feces (manure) or excrement of animals and birds.

DYSTROPHY. Degeneration.

E

EARLY LEAF. Stage at which the plant reaches one-third of its growth before blooming.

EARLY WEANING. The practice of weaning young animals earlier than usual; weaning beef calves at 35 days to 5 months of age, weaning lambs at 3 to 4 weeks of age, weaning pigs at 3 to 4 weeks of age, and weaning foals under 5 months of age.

EASY KEEPER. An animal that grows or fattens rapidly on limited feed.

EATING WITH RELISH. In healthy animals, the appetite is good and the feed is attacked with relish (as indicated by eagerness to get to the trough, wagging the tail, etc.).

ECZEMA. A condition involving the inflammation of the skin with lesions of either a dry or weeping nature. It is most commonly caused by allergies or exposure to chemicals or other irritants, but it may be due to deficiencies of such nutrients as essential fatty acids and vitamins.

EDEMA. Swelling of a part or all of the body due to the accumulation of excess water.

EFFICIENCY OF FEED CONVERSION. This is expressed as units of feed per unit of product—meat, milk, or eggs.

ELECTROLYTE. A chemical compound which in solution dissociates by releasing ions. (An ion is an atomic particle that carries a positive (+) or a negative (-) charge.

ELEMENT. One of the 103 known chemical substances that cannot be divided into simpler substances by chemical means.

EMACIATED. An excessively thin condition of the body.

EMBDEN-MEYERHOF PATHWAY. The final pathway within cells by which glucose is metabolized to pyruvic acid. Pyruvic acid is metabolized by the rumen microorganisms to form one of the volatile fatty acids. Animal cells oxidize pyruvic acid to carbon dioxide, energy, water, and heat.

EMULSIFIER. An agent that breaks down large fat globules to smaller, uniformly-distributed particles. This action is accomplished in the intestine chiefly by the bile acids, which lower surface tension of the fat particles. Emulsification greatly increases the surface area of fat, facilitating contact with fat-digesting enzymes.

ENDOCRINE. Pertaining to glands and their secretions that pass directly into the blood or lymph instead of into a duct (secreting internally). Hormones are secreted by endocrine glands.

ENDOGENOUS. Originating within the body; *e.g.*, hormones and enzymes.

ENDOSPERM. The carbohydrate portion of seed.

ENEMA. The introduction of a liquid into the intestines by way of the anus.

ENSILED. Materials that have been subjected to anaerobic fermentation to form silage.

ENTERITIS. Inflammation of the intestines.

ENZYMATIC. Related to an enzyme.

ERGOSTEROL. A plant sterol which, when activated by ultraviolet rays, becomes vitamin D_2. It is also called provitamin D_2 and ergosterin.

ERGOT.
- A fungus disease of plants.
- The horny growth at the back of the fetlock joint; the spurs of a horse's hoofs.

ERUCTATION. The act of belching gas from the stomach.

ESTER. A compound produced by the reaction between an acid and an alcohol with elimination of a molecule of water. This process is called esterification. For example, a triglyceride is a glycerol ester of fatty acids.

ESTROGEN. A general term for the principal female sex hormones.

ETHER EXTRACT (EE). Fatty substances of feeds and foods that are soluble in ether.

EXCRETA. The products of excretion—primarily feces and urine.

EXOGENOUS. Provided from outside of the organism.

EXPELLER PROCESS. A process for the mechanical extraction of oil from seeds, involving the use of a screw press.

EXPERIMENT. The word *experiment* is derived from the Latin *experimentum,* meaning proof from experience. It is a procedure used to discover or to demonstrate a fact or general truth.

EXTRACELLULAR FLUID. Fluid outside the cell. It comprises about one-third of the total body fluid and includes tissue fluid, blood plasma, cerebrospinal fluid, fluid in the eye, and fluid of the gastrointestinal tract.

EXTRINSIC FACTOR. A dietary substance which was formerly thought to interact with the intrinsic factor of the gastric secretion to produce the antianemic factor, now known to be vitamin B–12. (Also see INTRINSIC FACTOR.)

EXTRUDED. A type of feed preparation in which the feed is forced through a die under pressure.

EXUDATE. Material that escapes from blood vessels and is deposited in tissues or tissue surfaces; characterized by a high content of protein, cells, or other cellular solid matter.

F

FACULTATIVE ANAEROBES. Microorganisms capable of living either in the presence or absence of oxygen.

FAT SOLUBLE. Soluble in fats and fat solvents but generally not soluble in water.

FDA. Food and Drug Administration, a federal regulatory agency.

FEBRILE. Condition characterized by fever.

FECES. The excreta discharged from the digestive tract through the anus.

FEED (or FEEDSTUFF). Any naturally occurring ingredient, or material, fed to animals for the purpose of sustaining them.

FEED EFFICIENCY. The ratio expressing the number of units of feed required for one unit of production (meat, milk, eggs) by an animal. Also known as feed conversion.

FEED GRADE. Feedstuffs suitable for animals, but not for human consumption.

FEED GRAIN. Any of several grains most commonly used for livestock or poultry feed, such as corn, sorghum, oats, and barley.

Glossary of Nutrition Terms

FEEDER'S MARGIN. The difference between the cost per hundredweight of feeder animals and the selling price per hundredweight of the same animals when finished.

FEEDSTUFF. Any product, of natural or artificial origin, that has nutritional value in the ration when properly prepared.

FERMENTATION. Chemical changes brought about by enzymes produced by various microorganisms.

FERMENTED. Acted upon by yeasts, molds, or bacteria in a controlled aerobic or anaerobic process in the manufacture of such products as alcohols, acids, vitamins of the B complex group, or antibiotics.

FERRITIN. Protein-iron compound in which iron is stored in tissues; the storage form of iron in the body.

FIBER CONTENT OF A FEED. The amount of hard-to-digest carbohydrates. Most fiber is made up of cellulose and lignin.

FIBROUS. High in cellulose and/or lignin content.

FILL.
- A term designating the fullness of the digestive tract of an animal.
- With market animals, the fill refers to the amount of feed and water consumed upon their arrival at the market and prior to selling.

FINES. Any material which will pass through a screen whose openings are smaller than the specified minimum size.

FINISH. To fatten a slaughter animal. Also, the degree of fatness of such an animal.

FINISHING ANIMALS. The laying on of fat.

FISTULA. An abnormal tubelike passage from some part of the body to another part or to the exterior—sometimes surgically inserted. Rumen fistulas consist of a capped opening to the rumen which permits concentrates to be removed. (Also see CANNULA.)

FITTING. The conditioning of an animal for show or sale, which usually involves a combination of special feeding plus exercise and grooming.

FLATULENCE. Excessive formation of gas in the stomach or intestines.

FLAVORING AGENTS. Feed additives that are supposed to increase palatability and feed intake.

FLORA. Plant life. In nutrition it generally refers to the bacteria present in the digestive tract (microflora).

FLUSHING. The practice of feeding females more generously 2 to 3 weeks before breeding. The beneficial effects attributed to this practice are (1) more eggs (ova) are shed, and this results in more offspring; (2) the females come in heat more promptly; and (3) conception is more certain.

FODDER. Coarse feeds, such as corn or sorghum stalks.

FOOD. When used in reference to animals, it is synonymous with feed.

FOOD AND DRUG ADMINISTRATION (FDA). The federal agency in the Department of Health, Education, and Welfare that is charged with the responsibility of safeguarding American consumers against injury, unsanitary food, and fraud. It protects industry against unscrupulous competition, and it inspects and analyzes samples and conducts independent research on such things as toxicity (using laboratory animals), disappearance curves for pesticides, and long-range effects of drugs.

FORAGE. The vegetative portion of plants in a fresh, dried, or ensiled state, which is fed to livestock (as pasture, hay, or silage). Although most forages are roughages, a finely ground forage is not. Conversely, many roughages) such as corncobs, straws, and plastic tabs) are not forages.

FORMULA FEED. A feed consisting of two or more ingredients mixed in specified proportions.

FORTIFY. Nutritionally, to add one or more nutrients or feedstuffs.

FOWLER'S SOLUTION. A tonic containing arsenic which is sometimes used in conditioning show animals, particularly those breeds that have a very heavy hair coat. Poisoning can result from its excessive or continued use.

FREE-CHOICE. Free to eat two or more feeds at will.

FREEZE DRYING. See LYOPHILIZATION.

FRESH.
- A cow that has recently given birth to a calf and is lactating.
- Newly produced or gathered feed material; not stored, cured, or preserved.

FULL-FEED. The term indicating that animals are being provided as much feed as they will consume safely without going off feed.

FUNGI. Plants that contain no chlorophyll, flowers, or leaves, such as molds, mushrooms, toadstools, and yeasts. They may get their nourishment from either dead or living organic matter.

FUNGICIDE. Any substance used to kill fungi.

FUTURES TRADING. The futures market is a way in which to provide (1) an insurance medium in the marketing field, and (2) the facilities and machinery for underwriting price risks.

G

GASTRIC. Pertaining to the stomach.

GASTRIC JUICE. A clear liquid secreted by the wall of the stomach. It contains hydrochloric acid and the enzymes rennin, pepsin, and gastric lipase.

GASTRITIS. An inflammation of the stomach, especially of the lining, or mucous membrane.

GASTROENTERITIS. Inflammation of the membranes of the stomach and the intestine.

GASTROINTESTINAL. Pertaining to the stomach and intestines.

GAVAGE. Introduction of material (as nutrients) into the stomach by means of a stomach tube.

GELATINIZED. Having had the starch granules completely ruptured by a combination of moisture, heat, and pressure, and in some instances, by mechanical shear.

GERM. Embryo of a seed.

GESTATION. The period of embryonic and fetal development from fertilization to birth; pregnancy.

GLOSSITIS. A swollen, reddened tongue; a symptom of riboflavin deficiency.

GLUCONEOGENESIS. Formation of glucose from protein or fat.

GLUTEN. The tough, viscid, nitrogenous substance remaining when the flour of wheat or other grain is washed to remove the starch.

GLYCERIDE. An ester of glycerol and fatty acids.

GLYCEROL. An alcohol containing three carbons and three hydroxy groups. It is most commonly found in chemical combinations with fats in compounds called triglycerides.

GLYCOGENESIS. Conversion of glucose to glycogen.

GLYCOGENOLYSIS. Conversion of glycogen to glucose.

GLYCOLYSIS. Conversion of carbohydrate to lactic acid in animals or pyruvic acid in enzymatic reactions.

GOITROGENIC. Producing or tending to produce goiter.

GOSSYPOL. A toxic yellow pigment found in cottonseed, which is toxic to swine and certain other nonruminants, and which may cause discoloration of egg yolks during cold storage.

GRAS (GENERALLY RECOGNIZED AS SAFE). A designation of food additives that have been judged as safe for human consumption by a panel of expert pharmocologists and toxicologists who consider available data, including experience of common use in food. The common use factor is often the major criterion on which judgment is based.

GRAZE. To consume standing vegetation, as by livestock or wild animals.

GRAZING FEE. A charge, usually on a monthly basis, for grazing of livestock.

GREEN REVOLUTION. Refers to the adoption of high-yielding varieties of wheat and rice. It all started with a short-strawed wheat developed by Dr. Norman Borlaug, a Rockefeller Foundation scientist stationed in Mexico, who evolved new spring wheats that helped Mexico and Southeast Asia close the food gap, a development that brought him fame and a Nobel Prize. The first of these improved wheat varieties was released in 1948.

GRITS. Coarsely ground grain from which the bran and germ have been removed, usually screened to uniform particle size.

GROATS. Grain from which the hulls have been removed.

GROW OUT. To feed animals so that they attain a certain desired amount of growth with little or no fattening.

GROWTHY. Describes an animal that is large and well developed for its age.

GRUEL. A feed prepared by mixing ground ingredients with hot or cold water.

H

HARD KEEPER. An animal that is unthrifty and grows or fattens slowly regardless of the quantity or quality of feed.

HAY BELLY. Said of a horse with a distended barrel due to excessive feeding of bulky rations, such as hay, straw, or grass. Also called *grass belly*.

HAY QUALITY. Refers to the physical and chemical characteristics of hay commonly associated with palatability and abundance of feed nutrients.

HEALTH. A state of complete physical, mental, and social well being; not merely the absence of disease and infirmity.

HEAT INCREMENT (HI). The increase in heat production following consumption of feed when the animal is in a thermoneutral environment.

HEAT LABILE. Unstable to heat.

HEAT OF FERMENTATION (HF). The heat produced in the digestive tract as a result of microbial action.

HEAT-PROCESSED. Subjected to a method of preparation involving the use of elevated temperatures, with or without pressure.

HEAT-RENDERED. Melted, extracted, or clarified through use of heat. Usually water and fat are removed.

HEME. Iron-containing, nonprotein portion of the hemoglobin.

HEMICELLULOSE. A carbohydrate classified in the crude fiber fraction of feedstuffs that is similar to cellulose, except that it contains pentoses (5–carbon sugars) and uronic acid in addition to hexoses.

HEMOGLOBIN. The oxygen-carrying, red-pigmented protein of the red blood cells.

HEMORRHAGE. Escape of blood from the vessels; bleeding.

HERB. A flowering plant with one or more stems that die back to the ground each year without persistent stems aboveground; ferns, grasses, and forbs as distinct from shrubs and trees.

HERBIVORE. A plant-eating animal.

HIGH-NUTRIENT DENSITY. A high level of a specific nutrient in a given amount (weight) of food.

HIVES. Small swellings under or within the skin. They appear suddenly over large portions of the body and can be caused by feed allergies.

HOME RANGE. The area around an animal's established home which is traversed in its normal activities.

HOMOGENIZED. Particles broken down into evenly distributed globules small enough to remain emulsified for long periods of time.

HORMONE. A body-regulating chemical secreted by an endocrine gland into the bloodstream, thence transported to another region within the animal where it elicits a physiological response.

HORSEPOWER. A unit of power equal to 746 watts.

HYDRAULIC PROCESS. A process for the mechanical extraction of oil from seeds, involving the use of a hydraulic press. Sometimes referred to as *old process*.

HYDROGENATION. The chemical addition of hydrogen to any unsaturated compound.

HYDROLYSIS. The splitting of a substance into smaller units by chemically adding water to the material.

HYGROSCOPIC. The degree of tendency to absorb and retain moisture. For example, nonfat dry milk is hygroscopic in that it tends to absorb moisture.

HYPERTHYROIDISM. Overactivity of the thyroid gland.

HYPERTROPHIED. Having increased in size beyond the normal growth.

HYPERVITAMINOSIS. An abnormal condition resulting from the intake of an excess of one or more vitamins.

HYPOCALCEMIA. Below normal concentration of ionic calcium in blood resulting in convulsions, as in tetany or parturient paresis (milk fever).

HYPOGLYCEMIA. A reduction in concentration of blood glucose below normal.

HYPOKALEMIA. Low potassium levels in the blood.

HYPOMAGNESEMIA. An abnormally low level of magnesium in the blood.

HYPONATREMIA. Decreased level of sodium in the blood.

I

ICU (INTERNATIONAL CHICK UNIT). The vitamin D requirements of poultry are generally expressed as ICU, based on using the chick as the assay animal because of the unequal activity of vitamin D_2 and vitamin D_3 for this species.

IMMUNOGLOBULINS. A family of proteins found in body fluids which have the property of combining with antigens, and when the antigen is pathogenic, sometimes inactivating it and producing a state of immunity. Also called antibodies.

IMPACTION. See COMPACTION.

IMPLANT. A substance that is implanted into the body for the purpose of growth promotion or controlling some physiological function.

INERT. Relatively inactive.

INGEST. To eat or take in through the mouth.

INGESTA. Food or drink taken into the stomach.

INGESTION. The taking in of food and drink.

INGREDIENT. A constituent feed material.

INOCULATION. The addition of the proper bacteria to legume seed to enable the plant to use nitrogen from the air.

INSULIN. A hormone secreted by the pancreas into the blood, which regulates sugar metabolism.

INTESTINAL JUICE. A clear liquid secreted by glands in the wall of the small intestine. It contains the enzymes lactase, maltase, and sucrase, and several peptidases.

INTOLERANCE. Sensitivity or allergy to certain foods.

INTRACELLULAR. Inside the cells.

INTRADERMAL. Into, or between, the layers of the skin.

INTRAMUSCULAR. Within the muscle.

INTRAPERITONEAL. Within the peritoneal cavity.

INTRAVENOUS. Within the vein or veins.

INTRINSIC FACTOR. A chemical substance secreted by the stomach which is necessary for the absorption of vitamin B-12. The exact chemical nature of intrinsic factor is not known, but it is thought to be a mucoprotein or mucopolysaccharide. A deficiency of this factor may lead to a deficiency of vitamin B-12, and, ultimately, to pernicious anemia.

IN VITRO. Occuring in an artificial environment, as in a test tube.

IN VIVO. Occuring in the living body.

IODINE NUMBER. A number which denotes the degree of unsaturation of a fat or fatty acid. It is the amount of iodine in grams which can be taken up by 100 g of fat.

ION. Molecular constituent of one or more atoms that is a free-wandering particle in solution. An ion carries a positive or negative electric charge. Ions carrying negative charges are called anions; those carrying positive charges are called cations.

IRRADIATED YEAST. Yeast that has been treated with ultraviolet light. Yeast contains considerable ergosterol, which, when exposed to ultraviolet light, produces vitamin D_2.

IRRADIATION. Exposing to ultraviolet light.

ISOLATED SOYBEAN PROTEIN. Protein obtained from soybeans that can be spun into fibers, flavored, colored, and fabricated into meatlike products, including beef steaks, chicken, pork chops, ham, bacon, lamb chops, or sausage—all difficult to distinguish from the real products.

ISOMER. The possession of two or more distinct compounds of the same molecular formula, each molecule possessing an identical number of atoms of each element but in different arrangement.

ISOTOPE. Element that has the same number of protons (atomic number) as another element but a different number of neutrons (atomic mass).

IU (INTERNATIONAL UNIT). A standard unit of potency of a biologic agent (e.g., a vitamin, hormone, antibiotic, antitoxin) as defined by the International Conference for Unification of Formulae. Potency is based on the bioassay that produces a particular effect agreed on internationally. Also called a *USP unit*.

J

JOULE. A proposed international unit (4,184 j = 1 calorie) for expressing mechanical, chemical, or electrical energy, as well as the concept of heat. In the future, energy requirements and feed values will likely be expressed by this unit.

K

KERATIN. A sulfur-containing protein which is the primary component of epidermis, hair, wool, hoof, horn, and the organic matrix of the teeth.

KERNEL. The whole grain of a cereal. The meats of nuts and drupes (single-stoned fruits).

KETOGENIC. Conducive to the production of ketones, products of fatty acid oxidation that eventually are broken down to carbon dioxide and water.

KETOSIS. A condition characterized by an abnormally high concentration of keytone (acetone) bodies in fluids and tissues—acetonemia.

KIBBLED. Cracked or crushed baked dough, or extruded feed that has been cooked prior to or during the extrusion process.

KJELDAHL METHOD. A method of determining the amount of nitrogen in an organic compound. The quantity of nitrogen measured is then multiplied by 6.25 to calculate the protein content of the feed or compound analyzed. The method was developed by the Danish chemist, J. G. C. Kjeldahl, in 1883.

KREBS CITRIC ACID CYCLE. Also called TCA (tricarboxylic acid) cycle and the Krebs cycle. The pathway by which animal cells use oxygen to oxidize carbohydrates to carbon dioxide, energy, water, and heat.

KWASHIORKOR. A syndrome produced by a severe protein deficiency, with characteristic changes in pigmentation of the skin and hair, a bulging belly, edema, skin lesions, anemia, and apathy.

L

LABILE. Unstable. Easily destroyed.

LACHRYMATION. Tearing; secreting and conveying tears.

LACTIC ACID. A compound formed in the body in the metabolism of carbohydrates.

LACTOSE (MILK SUGAR). A disaccharide found in milk having the formula $C_{12}H_{22}O_{11}$. It hydrolyzes to glucose and galactose. Commonly known as milk sugar.

LAMINITIS (or FOUNDER). An inflamation of the sensitive laminae under the horny wall of the hoof, characterized by ridges running around the hoof, often associated with overfeeding. All feet may be affected, but the front feet are most susceptible.

LAYER. An adult chicken which produces eggs.

LECITHIN. A phospholipid found in many animal tissues.

LIGNIFICATION. The process of impregnating cell walls of a plant with lignin.

LIGNIN. A practically indigestible compound which along with cellulose is a major component of the cell wall of certain plant materials, such as wood, hulls, straws, and overripe hays.

LIMITED FEEDING. Feeding animals less than they would like to eat. Giving sufficient feed to maintain weight and growth, but not enough for their potential production or finishing.

LIMITING AMINO ACID. The essential amino acid of a protein which shows the greatest percentage deficit in comparison with the amino acids contained in the same quantity of another protein selected as a standard.

LIPASE. A fat-splitting enzyme, present in gastric juice and pancreatic juice. It acts on fats to produce fatty acids and glycerol.

LIPOLYSIS. The hydrolysis of fats by enzymes, acids, alkalis, or other means to yield glycerol and fatty acids.

LIQUID PROTEIN SUPPLEMENTS. Protein products which usually contain molasses and urea, with added vitamins and trace minerals.

LIVER ABSCESSES. Single or multiple abscesses on the liver, observed at slaughter. Usually the abscess consists of a central mass of necrotic liver surrounded by pus and a wall of connective tissue. At slaughter, those livers affected with abscesses are condemned for human food.

LIVEWEIGHT. Weight of an animal on foot.

LYMPH. The slightly yellow, transparent fluid occupying the lymphatic channels of the body.

LYOPHILIZATION. The evaporation of water from a frozen product with the aid of high vacuum. Also called freeze drying.

M

MAINTENANCE REQUIREMENT. A ration which is adequate to prevent any loss or gain of tissue in the body when there is no production.

MAIZE. A cereal grain, commonly called corn in the United States.

MALIGNANT. A cancerous growth, as distinguished from a benign growth.

MALNUTRITION. Any disorder of nutrition. Commonly used to indicate a state of inadequate nutrition.

MANURE. A mixture of animal excrements (consisting of undigested feeds plus certain body wastes) and bedding.

MARASMUS. From the Greek *marasmos*, meaning a *dying away*. A progressive wasting and emaciation. *Enzootic marasmus* is a condition of malnutrition in domestic animals due to a deficiency of one or more trace elements, especially cobalt and copper. In human infants, the condition results primarily from lack of calories (energy foods).

MASH. A mixture of ingredients in meal form.

MASTICATION. The chewing of feed.

MEAL.
- A feed ingredient having a particle size somewhat larger than flour.
- Mixtures of concentrate feeds, usually in which all of the ingredients are ground.

MEAT ANALOGS. Food material usually prepared from vegetable protein to resemble specific meats in texture, color, and flavor.

MECHANICALLY EXTRACTED. A method of extracting the fat content from oilseeds by the application of heat and mechanical pressure. The hydraulic and expeller processes are both methods of mechanical extraction.

MECONIUM. Excrement accumulated in the bowels during fetal development.

MEDICATED FEED. Any feed which contains drug ingredients intended or represented for the cure, mitigation, treatment, or prevention of diseases of animals (other than humans).

METABOLISM. Refers to all the changes that take place in the nutrients after they are absorbed from the digestive tract, including (1) the building-up processes in which the absorbed nutrients are used in the formation or repair of body tissues, and (2) the breaking-down processes in which nutrients are oxidized for the production of heat and work.

METABOLITE. Any substance produced by metabolism or by a metabolic process.

MICROBE. Same as microorganism.

MICROFLORA. Microbial life characteristic of a region, such as the bacteria and protozoa populating the rumen.

MICROINGREDIENT. Any ration component, such as minerals, vitamins, antibiotics, and drugs, normally measured in milligrams or micrograms per kilogram or in parts per million.

MICROORGANISM. Any organism of microscopic size, applied especially to bacteria and protozoa.

MILK EJECTION OR *LET-DOWN*. The process, controlled by oxytocin, in which milk is forced from the alveoli, where it is stored, into the larger ducts and cisterns, where it is available to sucklings or the milker.

MILK STAGE. The plant period after bloom when the seeds begin to form.

MILL BY-PRODUCT. A secondary product obtained in addition to the principal product in milling practice.

MILL RUN. The state in which a by-product material comes from the mill, ungraded and usually uninspected.

MINERAL SUPPLEMENT. A rich source of one or more of the inorganic elements needed to perform certain essential body functions.

MIXING. To combine by agitation two or more materials to a specific degree of dispersion.

MOISTURE. A term used to indicate the water contained in feeds—expressed as a percentage.

MOISTURE-FREE (M-F, OVEN-DRY, 100% DRY MATTER). This refers to any substance that has been dried in an oven at 221°F until all the moisture has been removed.

MOLDS (FUNGI). Fungi which are distinguised by the formation of mycelium (a network of filaments, or threads), or by spore masses.

MORBIDITY. A state of sickness.

MUCUS. Viscid fluid secreted by mucous membranes and glands, consisting mainly of mucin (a glycoprotein), inorganic salts, and water. Mucus serves to lubricate and protect the gastrointestinal mucosa and helps to move the feed mass along the digestive tract.

MYCOTOXINS. Toxic metabolites produced by molds during growth. Sometimes present in feed materials.

MYOSIN. Muscle fiber fraction that complexes with actin to bring about muscle contraction.

N

NATIONAL RESEARCH COUNCIL (NRC). A division of the National Academy of Sciences established in 1916 to promote the effective utilization of scientific and technical resources. Periodically, this private, nonprofit organization of scientists publishes bulletins giving nutrient requirements and allowances of domestic animals, copies of which are available on a charge basis through the National Academy of Sciences, National Research Council, 2101 Constitution Avenue, N.W., Washington, DC 20418.

NECROSIS. Cell death.

NEPHRITIS. Inflammation of the nephrons of the kidneys.

NEUTRAL DETERGENT FIBER (NDF). This indicates the amount of cell wall material or plant structural fiber in a feed. NDF contains ADF plus hemicellulose. The lower the NDF the more forage the animal will eat; hence, low NDF is desired.

NITROGEN. A chemical element essential to life. Animals get it from protein feeds; plants get it from the soil; and some bacteria get it directly from the air.

NITROGEN BALANCE. The nitrogen in the feed intake minus the nitrogen in the feces, minus the nitrogen in the urine.

NITROGEN FIXATION. Conversion of free nitrogen of the atmosphere to organic nitrogen compounds by symbiotic or nonsymbiotic microbial activity.

NITROGEN-FREE EXTRACT (NFE). It consists principally of sugars, starches, pentoses, and nonnitrogenous organic acids. The percentage is determined by subtracting the sum of the percentages of moisture, crude protein, crude fat, crude fiber, and ash from 100.

NONPROTEIN NITROGEN (NPN). Nitrogen which comes from other than a protein source but may be used by a ruminant in the building of protein. NPN sources include compounds like urea and anhydrous ammonia, which are used in feed formulations for ruminants only.

NOURISH. To provide food or other substances necessary for life and growth.

NOXIOUS. Harmful, not wholesome.

NUTRIENT ALLOWANCES. Nutrient recommendations that allow for variations in feed composition; possible losses during storage and processing; day-to-day and period-to-period differences in needs of animals; age and size of animal; stage of gestation and lactation; the kind and degree of activity; the amount of stress; the system of management; the health, condition, and temperament of the animal; and the kind, quality, and amount of feed—all of which exert a powerful influence in determining nutritive needs.

NUTRIENT REQUIREMENTS. This refers to meeting the animal's minimum needs, without margins of safety, for maintenance, growth, fitting, reproduction, lactation, and work. To meet these nutritive requirements, the different classes of animals must receive sufficient feed to furnish the necessary quantity of energy (carbohydrates and fats), protein, minerals, and vitamins.

NUTRIENTS. The chemical substances found in feed materials that can be used, and are necessary, for the growth, maintenance, production, and health of animals. The chief classes of nutrients are carbohydrates, fats, proteins, minerals, vitamins, and water.

NUTRITION. This is the science of the use of feed (food) as it relates to health. It includes all the processes by which the living organism ingests, digests, absorbs, and uses the nutrients in feeds (foods) for maintenance, growth, work, and reproduction.

NUTRITIVE RATIO (NR). The ratio of digestible protein to other digestible nutrients in a feedstuff or ration. (The NR of shelled corn is about 1:10.)

NUTRITURE. Nourishment.

O

OBESE. Overweight due to a surplus of body fat.

OFFAL. It usually refers to that part of the animal carcass not sold for meat, consisting principally of the digestive tract.

OFF FEED. Not eating with a normal, healthy appetite.

OIL CROPS. Crops grown primarily for oil, including soybeans, cottonseed, rape (canola), peanuts, flaxseed, sunflower, safflower, and castor bean.

OLD PROCESS. Pertains to the extraction of oil from seeds. Same as hydraulic process.

OLFACTORY. Pertaining to the sense of smell.

ORAL. Pertaining to the mouth.

ORTS. Leftover feed which an animal refuses to eat.

OSMOTIC PRESSURE. The pressure that causes water or another solvent to move from a solution with a low concentration of solid (solute) to one having a high concentration of solute.

OSSIFICATION. The process of bone formation; the calcification of bone with advancing maturity.

OSTEITIS. Inflammation of a bone.

OSTEOMALACIA. A bone disease of adult animals caused by lack of vitamin D, inadequate intake of calcium or phosphorus, or an incorrect dietary ratio of calcium and phosphorus.

OSTEOPOROSIS. Abnormal porosity and fragility of bone as the result of (1) a calcium, phosphorus, and/or vitamin D deficiency, or (2) an incorrect ratio between the two minerals.

OVERFEEDING. Fed to excess.

OVERFINISHING. Excess finishing or fatness—a wasteful practice.

OVERRIPE. Stage after the plant is mature and the seeds are ripe (applies mostly to range plants).

OXIDATION. The combination with oxygen, or the loss of a hydrogen, or the loss of an electron, all of which render an ion more electropositive. The animal combines carbon from feedstuffs with inhaled oxygen to produce carbon dioxide, energy (as ATP), water, and heat. (Also see KREBS CITRIC ACID CYCLE.)

OXYTOCIN. The hormone that controls milk let-down.

OYSTER SHELL. Consists mainly of calcium salts; it is a useful and attractive source of calcium for layers to aid egg shell strength.

P

PALATABILITY. The result of the following factors sensed by the animal in locating and consuming feed: appearance, odor, taste, texture, temperature, and, in some cases, auditory properties of the feed (like the sound of pigs eating corn). These factors are affected by the physical and chemical nature of the feed.

PAPILLAE. Small nipple-shaped projections located on the interior of the rumen wall.

PARATHYROID GLANDS. Four small endocrine glands situated beside the thyroid gland, concerned chiefly with maintaining normal calcium in the body.

PARATHYROID HORMONE. The parathyroid hormone, secreted by the parathyroid glands, stimulates the release of calcium from bone and the elevation of the blood calcium level.

PARTS PER BILLION (PPB). It equals micrograms per kilogram or microliters per liter.

PARTS PER MILLION (PPM). It equals milligrams per kilogram or milliliters per liter.

PEARLED. Dehulled grains reduced into smaller smooth particles by machine brushing, or abrasion.

PECTIN. A nondigestible polysaccharide found in the soft tissue of fruits and vegetables. One of the richest sources is lemon or orange rind. Pectin possesses the ability to gel, and it is often used as a base for fruit jellies.

PELLET BINDERS. Products that enhance the firmness of pellets.

PEPTIDE LINKAGE. The characteristic joining of amino acids to form proteins. Such a chain of amino acids is called a peptide.

PER ORAL. Administration through the mouth.

PER OS. Oral administration (by the mouth).

pH. A measure of the acidity or alkalinity of a solution. Values range from 0 (most acid) to 14 (most alkaline), with neutrality at pH 7.

PHASE FEEDING. Refers to changes in the animal's diet (1) to adjust for age and stage of production, (2) to adjust for season of the year and for temperature and climatic changes, (3) to account for differences in body weight and nutrient requirements of different strains of animals, or (4) to adjust one or more nutrients as other nutrients are changed for economic or availability reasons.

PHOSPHOLIPIDS. Fatlike substances containing phosphorus and nitrogen, along with fatty acids and cholesterol.

PHOTOSYNTHESIS. The process whereby green plants utilize the energy of the sun to build complex organic molecules containing energy.

PHYSIOLOGICAL FUEL VALUES. Units, expressed in calories, used in the United States to measure food energy in human nutrition. It is similar to metabolizable energy.

PHYSIOLOGICAL SALINE. A salt solution (0.9% NaCl) having the same osmotic pressure as the blood plasma.

PHYTATE. A salt of phytic acid (a phosphorus compound occurring in the outer layers of cereal grains. It binds phosphorus and other minerals [e.g. zinc] making them unavailable to monogastric animals. The calcium of the insoluble calcium phytate is unabsorbable from the intestine).

PICA. Depraved appetite characterized by eating dirt, sand, hair, feces, etc. This condition is usually caused by nutritional deficiencies and is particularly prevalent among animals in confinement.

PLANT PROTEINS. This group includes the common oilseed by-products—soybean meal, cottonseed meal, linseed meal, peanut meal, safflower meal, sunflower meal, rapeseed meal, and coconut (or copra) meal. They vary in protein content and nutrient value, depending on the seed from which they are produced, the amount of hull and/or seed coat included, and the method of oil extraction used.

PLASMA. The colorless fluid portion of the blood in which the corpuscles are suspended. It is often used as a basis for measurement of bloodborne nutrients and their metabolites.

POLAR. Miscible in water.

POLY. Many.

POLYDIPSIA. Excessive or abnormal thirst.

POLYNEURITIS. Neuritis of several peripheral nerves at the same time, caused by metallic and other poisons, infectious disease, or vitamin deficiency. In humans, alcoholism is also a major cause of polyneuritis.

POLYUNSATURATED FATTY ACIDS. Fatty acids having more than one double bond. Linoleic and linolenic acids, which contain 2 and 3 double bonds, respectively, are essential in the diet of humans.

POT-BELLIED. Designating any individual that has developed an abnormally large abdomen.

POULT. A young turkey of either sex from a day old to a few weeks.

PRECONDITIONING. A way of preparing an animal (usually referring to calves) to withstand the stress and rigors of leaving its mother, learning to eat new kinds of feed, and shipping from the farm or ranch to the feedlot.

PRECURSOR. A compound that can be used by the body to form another compound; for example, carotene is a precursor of vitamin A.

PREHENSION. The seizing (grasping) and conveying of feed to the mouth.

PREMIX. A uniform mixture of one or more microingredients and a carrier, used in the introduction of microingredients into a larger mixture.

PRESERVATIVES. A number of materials are available to incorporate into silage and hay, with claims made that they will improve the preservation of nutrients, nutritive value, and/or palatability of the feed.

PRESSED. Compacted or molded by pressure; having fat, oil, or juice extracted under pressure.

PRESSURE COOKER. An airtight container for the cooking of feed at high temperature under steam pressure.

PRODUCT. A substance produced from one or more other substances as a result of chemical or physical change.

PROLACTIN. The lactating-stimulating hormone of the anterior lobe of the pituitary gland.

PROSTAGLANDIN. A group of naturally occurring long-chain fatty acid derivatives having local hormonelike actions of widely diverse forms.

PROTEIN-BOUND IODINE (PBI) TEST. Test used to measure thyroid activity by determining the amount of iodine bound to thyroxin and in transit in the plasma.

PROTEIN BUMPS. Some horses appear to be allergic to certain proteins or to excesses of specific amino acids, as a result of which they may develop *protein bumps*.

PROTEIN EQUIVALENT. A term indicating the total nitrogenous contribution of a substance in comparison with the nitrogen content of protein (usually plant protein). For example, the nonprotein nitrogen (NPN) compound urea contains approximately 45% nitrogen and has a protein equivalent of 281% (6.25 × 45%).

PROTEIN-SPARING. An effect in which less protein is used by the animal to meet the animal's glucose needs in times of glucose shortage. Propionic acid is protein-sparing in that it can be converted to glucose. Acetic and butyric acids cannot be converted to glucose. Likewise, fat cannot be converted. The glucogenic amino acids may be converted to glucose.

PROTEIN SUPPLEMENTS. Products that contain more than 20% protein or protein equivalent.

PROTHROMBIN. Plasma fraction that is converted to thrombin during the clotting of blood.

PROTOPLASM. The essential protein sustance of living cells.

PROVITAMIN. The material from which an animal may produce vitamins; e.g., carotene (provitamin A) in plants is converted to vitamin A in animals.

PROVITAMIN A. Carotene.

PULLET. A female chicken from a day old up to 12 months of lay when it becomes a hen.

PULP. The solid residue remaining after extraction of juices from fruits, roots, or stems.

PURGATIVE. A laxative.

PURIFIED DIET. A mixture of the known essential dietary nutrients in a pure form that is fed to experimental (test) animals in nutrition studies.

PUTREFACTION. The decomposition of proteins by microorganisms under anaerobic conditions.

Q

QUALITY. A term used to denote the desirability and/or acceptance of an animal or feed product.

QUALITY OF PROTEIN. A term used to describe the amino acid balance of protein. A protein is said to be of good quality when it contains all of the essential amino acids in proper proportions and amounts needed by a specific animal; and it is said to be poor quality when it is deficient in either content or balance of essential aminc acids.

R

RACHITIC. Adjective used to denote the condition of rickets.

RADIANT HEAT. Heat transmitted by radiation (as from the sun) in contrast to that transmitted by conduction or convection.

RADIATION. Electromagnetic phenomena that have properties combining both wave and particle functions; spans the entire spectrum from low frequency radio waves through white light to high frequency gamma rays.

RADIOACTIVE. Giving off atomic energy in the form of alpha, beta, or gamma rays.

RADIO ACTIVE ^{131}I UPTAKE TESTS. Tests of thyroid function using a radioactive isotope of iodine, ^{131}I.

RADIOACTIVE ISOTOPE. A chemical element that changes into another with the emission of rays (alpha, beta, or gamma). Among the naturally radioactive elements are uranium and radium; some other elements can be made radioactive by bombardment with neutrons or deutrons, or by other means.

RADIOACTIVE TRACER. A small quantity of radioactive isotope used to follow biological, chemical, or other processes by detection, determination, or localization of the radioactivity.

RANCID. A term used to describe fats that have undergone hydrolytic or oxidative decomposition.

RANGE CUBES. Large pellets produced for feeding on the ground on pasture or range.

RANGE FORAGES. Native forages suitable for grazing by livestock that are produced in the western range area.

RANGELAND. Land that, for the most part, produces native forage suitable for grazing by livestock. It embraces rather extensive areas of land suitable for grazing, but not suitable for cultivation—especially in arid, semiarid, or forested regions.

RATE OF PASSAGE. The time taken by undigested residues from a given meal to reach the feces. (A stained indigestible material is commonly used to estimate rate of passage.)

RATION(S). The amount of feed supplied to an animal for a definite period, usually for a 24-hour period. However, by practical usage, the word ration implies the feed fed to an animal without limitation to the time in which it is consumed.

RATIOS. *Weight ratio, gain ratio,* and *conformation score ratio* are used to indicate the performance of an individual in relation to the average of all animals of the same group. It is calculated as follows:

$$\frac{\text{Individual record}}{\text{Average of animals in group}} \times 100$$

It is a record or index of individual deviation from the group average expressed in terms of percentage. Thus, if the average bull test station gain was 3.00 lb per day, the gain ratio of a bull gaining 3.30 lb per day would be 110. A ratio of 100 is average for a particular group. Thus, ratios above 100 indicate animals above average, whereas ratios below 100 indicate animals below average.

RED MEAT. Meat that is red when raw, due to the red coloration of myoglobin, the pigment of muscle. Red meats include beef, veal, pork, mutton, and lamb muscle tissue with attendant fat and bone.

REGURGITATION. The casting up (backward flow) of undigested food from the stomach to the mouth, as by ruminants.

RELATIVE FEED VALUE (RFV). This is a measure of forage intake and energy value that takes into account CP, ADF, and NDF. RFV is expressed as a percent compared to full bloom alfalfa at 100% RFV. High RFV reflects high quality, greater intake, higher digestibility, and improved performance.

RENNIN. Milk-curdling enzyme of the gastric juice found in young animals such as calves.

RESIDUE. That which remains of any particular substance.

RESPIRATORY QUOTIENT (RQ). A ratio indicating the relation of the volume of carbon dioxide given off in respiration to that of the oxygen consumed. The RQ is used to indicate the *type* of feed being metabolized. This is possible because carbohydrates, fats, and proteins differ in the relative amounts of oxygen and carbon contained in their molecules. Thus, the RQ is near 1 when the body is burning chiefly carbohydrates; near 0.7 when chiefly fats; near 0.8 when chiefly proteins; and sometimes exceeding 1 when carbohydrates are being changed to fats for storage.

RESTRICTED FEEDING. Reduction of the *ad libitum* intake to avoid over-fatness and/or to improve productive efficiency.

RETARDED GROWTH. Slower than normal rate of growth.

RETINOL EQUIVALENT (RE). Measure of vitamin A activity currently adopted by FAO/WHO and U.S. National Research Council's recommendations of vitamin A, replacing the term IU (international unit). The measure accounts for dietary variances in preformed vitamin A (retinol) and its precursor, carotene. One RE (retinol equivalent) equals 3.33 IU or 1 mcg retinol.

RETORT-CHARRED. Material partially burned in a closed retort, as is done in the manufacture of bone black.

ROOTS. Subterranean parts of plants.

ROUGHAGE. A coarse, bulky feed that is high in fiber content and coarse textured. If a high-fiber feed is finely ground, it is no longer a roughage even though the fiber concentration is unchanged.

RUMEN CULTURE. A liquid or dried preparation consisting of microbes from the rumen.

RUMEN FILL. The *fill* in the rumen places a ceiling on feed intake. Since low-quality roughages pass through the rumen at a slower rate than high-quality roughages, they can limit the amount of total feed that the animal can consume in a 24-hour period.

RUMEN FLORA. The microorganisms of the rumen.

RUMINATION. The act of regurgitating previously eaten feed and chewing a soft mass of coarse feed particles, called a bolus or cud. Each bolus is chewed for about a minute, then swallowed again. Ruminants may spend 8 hours or more per day in rumination, the amount of time varying according to the nature of the ration. Coarse, fibrous rations result in more time ruminating. Rechewing does not improve digestibility. Rather, rumination has an important bearing on the amount of feed the animal can eat and utilize. Feed particle size must be reduced to allow passage of the material from the rumen. It follows that high-quality forages require much less rechewing and pass out of the rumen at a faster rate; hence, they allow the ruminant to eat more.

S

SACCHARIDES. Referring to sugars. The prefixes mono-, di-, tri-, and poly- denote the number of sugars contained in the saccharide.

SALINE. Consisting of or containing salt.

SALMONELLA. A pathogenic, diarrhea-producing organism, of which there are over 100 known strains, sometimes present in contaminated feeds.

SALMONELLOSIS. Infection caused by *Salmonella* (a genus of microbes), characterized by violent diarrhea, abdominal cramps, painful straining or defecation, and fever.

SAPONIFICATION. The formation of soap and glycerol from the reaction of fat with alkali.

SATIETY. Full satisfaction of desire; may refer to satisfaction of appetite.

SATURATED FAT. A completely hydrogenated fat—each carbon atom is associated with the maximum number of hydrogens; there are no double bonds.

SCALPING. The removal of larger material by screening.

SCRATCH. A poultry feed term referring to whole, cracked, or coarsely cut grain.

SCREENED. A feedstuff that has been separated into various sized particles by passing over or through screens.

SCURVY. A hemorrhagic disease caused by a lack of vitamin C.

SELF-FED. Provided with a part or all of the ration on a continuous basis, thereby permitting the animal to eat at will.

SELF-FEEDER. A feed container by means of which animals can eat at will. See *AD LIBITUM*.

SEMIPERMEABLE MEMBRANE. A membrane that allows the passage of only certain solids but is freely permeable to water.

SERUM. The colorless fluid portion of blood remaining after clotting and removal of corpuscles. It differs from plasma in that the fibrinogen has been removed.

SHORT FEED. To feed for a short period, usually less than 120 days for cattle.

SHREDDED. Cut into long, narrow pieces.

SHRINKAGE.
- A term indicating the amount of loss in body weight when animals are exposed to adverse conditions, such as being transported, severe weather, or shortage of feed.
- The loss in carcass weight during the aging process.

SIFTED. Materials that have been passed through wire sieves to separate particles of different sizes. The common connotation is the separation of finer material than would be done by screening.

SLEEK COAT AND PLIABLE AND ELASTIC SKIN. A sleek, oily coat and a pliable and elastic skin characterize healthy animals. When the hair coat loses its luster and the skin becomes dry, scurfy, and hidebound, there is usually trouble.

SLOTTED FLOORS. Floors with slots through which the feces and urine pass to a storage area below or nearby.

SMALL GRAIN REFUSE. This refers to (1) straw, and (2) tailings—the chaff and grain behind the combine.

SOAP. A compound formed along with glycerol from the reaction of fat with alkali.

SODIUM BENTONITE (CLAY). Used as a pellet binder. Also shows promise for improving the nitrogen utilization of ruminants.

SODIUM BICARBONATE (NaHCO₃). A chemical compound which is known to function as a buffer and pH agent, maintaining sufficient alkaline reserves (buffering capacity) in the animal to ensure normal physiological and metabolic functions.

SOFT PORK. Feed fats are laid down in the body without undergoing much change. Thus, when finishing hogs are liberally fed on high fat content feeds in which the fat is liquid at ordinary temperatures, soft pork results. This condition prevails when hogs are liberally fed such feeds as soybeans, peanuts, mast, or garbage.

SOLUBLES. Liquid containing dissolved substances obtained from processing animal or plant materials. It may contain some fine suspended solids.

SOLUTION. A uniform liquid mixture of two or more substances molecularly dispersed within one another.

SOLVENT-EXTRACTED. Fat or oil removed from materials (such as oilseeds) by organic solvents. Also called *new process*.

SORGHUM. A cereal grass used mainly for feed grain or silage. Often grown in the drier areas of the United States.

SPECIFIC DYNAMIC ACTION (SDA). The increased production of heat by the body as a result of a stimulus to metabolic activity caused by ingesting food.

SPECIFIC GRAVITY. The ratio of the weight of a body to the weight of an equal volume of water.

$$\text{Specific gravity} = \frac{\text{Weight of body in air}}{\text{Wt. of body in air minus wt. in H}_2\text{O}}$$

SPECIFIC HEAT.
- The heat-absorbing capacity of a substance in relation to that of water.
- The heat expressed in calories required to raise the temperature of 1 g of a substance 1°C.

SPF. Specific pathogen-free.

SPRAY-DEHYDRATED. Material which has been dried by spraying onto the surface of a heated drum. It is recovered by scraping it from the drum.

STABILIZED. Made more resistant to chemical change by the addition of a particular substance.

STALKS. The main stem of a herbaceous plant; often with its dependent parts, as leaves, twigs, and fruit.

STEAMED. Treatment of ingredients with steam to alter physical and/or chemical properties.

STEARIN. The fat formed from the reaction of stearic acid with glycerol.

STEROL. An alcohol of high molecular weight, such as cholesterol and ergosterol.

STOCKING RATE. This is a range management term pertaining to animal numbers in relation to carrying capacity of a unit. Stocking too lightly wastes forage, while stocking too heavily results in reduced plant vigor and less forage produced per plant, as well as a change of forage plant cover from an abundance of valuable forage plants to an abundance of worthless plants.

STOMATITIS. Inflammation of the oral mucosa.

STOOL (FECES). Fecal material; evacuation from the digestive tract.

STOVER. Fodder; mature cured stalks from which seeds have been removed, such as stalks of corn without ears or stalks of sorghum without heads.

STRAW. The plant residue remaining after separation of the seeds in threshing.

STRESS. Any physical or emotional factor to which an animal fails to make a satisfactory adaptation. Stress may be caused by excitement, temperament, fatigue, shipping, disease, heat or cold, nervous strain, number of animals together, previous nutrition, breed, age, or management. The greater the stress, the more exacting the nutritive requirements.

STUBBLE. The basal portion of herbaceous plants remaining after the top portion has been harvested either mechanically or by grazing animals.

SUBCUTANEOUS. Situated or occurring beneath the skin.

SUCCULENCE. A condition of plants characterized by juiciness, freshness, and tenderness, making them appetizing to animals. Succulence is provided in such feeds as silage, root crops, and grasses.

SUCKLE. To nurse at the breast or mammary glands.

SUCRASE. An enzyme present in intestinal juice which acts on sucrose to produce glucose and fructose.

SUCROSE. A disaccharide having the formula $C_{12}H_{22}O_{11}$. It is hydrolyzed to glucose and fructose. Commonly known as cane, beet, or table sugar.

SUN-CURED. Material dried by exposure in open air to the direct rays of the sun.

SUPPLEMENT. A feed or feed mixture used to improve the nutritional value of a basal feed (e.g., protein supplement—soybean meal). Supplements are usually rich in protein, minerals, vitamins, antibiotics, or a combination of part or all of these; and they are usually combined with basal feeds to produce a complete feed.

SWEET FEED. Refers to a commercial horse feed which is characterized by its sweetness due to the addition of molasses.

SYMBIOSIS. The living together in intimate association of two dissimilar organisms, with a resulting mutual benefit. The ruminant animal and its microorganisms in the paunch are a well-known example of symbiosis.

SYNDROME. A combination of symptoms resulting from a single cause.

SYNTHESIS. The bringing together of two or more substances to form a new material.

SYNTHETICS. Artificially produced products that may be similar to natural products.

T

TAIL BITING. An abnormal behavior, characterized by one pig biting the tail of another.

TDN. See TOTAL DIGESTIBLE NUTRIENT.

TDN TO MCAL. One pound of TDN = 2.0 Mcal or 2,000 kcal. It is recognized, however, that the roughage component in a ration affects its energy value. Thus, when converting all-roughage rations from TDN to calories, some scientists figure that 1 lb of TDN = 1,500 kcal, instead of 2,000.

TEART. Molybdenosis of farm animals caused by feeding on vegetation grown on soil that contains high levels of molybdenum.

TEMPERING. See CONDITIONING.

TETANY. A condition in an animal in which there are localized, spasmodic muscular contractions.

TETHER. To tie an animal with a rope or chain to allow grazing but prevent straying.

THERM. The amount of heat required to raise the temperature of 1,000 kg of water 1°C, or 1,000 lb of water 4°F. One therm is 1,000 large Calories, or 1 megacalorie (Mcal).

THERMAL. Refers to heat.

THERMOSTATIC CONTROL OF APPETITE. The theory that food intake is altered by special receptors in the hypothalamus that sense heat and cold.

THIAMINASE. An enzyme that splits the thiamin molecule into two biologically inactive molecules, found especially in raw freshwater fish and in bracken fern.

THIAMIN-STIMULATING HORMONE (TSH). Hormone secreted by the anterior pituitary gland that regulates uptake of iodine and synthesis of thyroxin by the thyroid gland.

THRIFTY. Healthy and vigorous in appearance.

THROMBOSIS. The obstruction of a blood vessel by the formation of a blood clot.

THYROXIN. The iodine-containing hormone produced by the thyroid gland.

TOASTED. Browned, dried, or parched by exposure to a fire, or to gas or electric heat.

TOCOPHEROL. Any of various forms of an alcohol having the properties of vitamin E.

TONIC. A drug, medicine, or feed designed to stimulate the appetite.

TOTAL DIGESTIBLE NUTRIENTS (TDN). A term which indicates the energy value of a feedstuff. The TDN is computed by use of the following formula:

% TDN = % DCP + % DCF + % DNFE + (% DEE × 2.25)
where DCP = digestible crude protein; DCF = digestible crude fiber; DNFE = digestible nitrogen-free extract; and DEE = digestible ether extract.

TOTAL MILK SOLIDS. Primarily milk fat, proteins, lactose, and minerals.

TOXEMIA. A condition produced by the presence of poisons (toxins) in the blood.

TOXIC. Of a poisonous nature.

TRACE ELEMENT. A chemical element used in minute amounts by organisms and held essential to their physiology. The essential trace elements are cobalt, copper, iodine, iron, manganese, selenium, and zinc.

TRACE MINERAL. A mineral nutrient required by animals in micro amounts only (measurable in milligrams per pound or smaller units).

TRACER ISOTOPE. An isotope of an element, a small amount of which may be incorporated into a sample of material (the carrier) to follow (trace) the sites of its deposition in the tissues and its paths of excretion. The tracer may be radioactive, in which case observations are made by means of a geiger counter. If the tracer is stable, mass spectrometers, density measurements, or neutron activation analysis may be employed to determine isotopic composition. Tracers are also called labels, or tags, and materials are said to be labeled, or tagged, when radioactive tracers are incorporated in them.

TRANSFERRIN. An iron-bearing protein complex, a serum beta-globulin; the transport form of iron in the body.

TREATS. This refers to such well-liked things as carrots or other roots, apples and other fruits, pumpkins, squashes, melons, molasses, sugar, honey, or innumerable other goodies.

TRIGLYCERIDE. A compound of 3 fatty acids esterified to glycerol. A neutral fat, synthesized from carbohydrate, stored in adipose tissue. It releases free fatty acids into the blood on being hydrolyzed by enzymes.

TRUE PROTEIN. A nitrogenous compound which will hydrolyze completely to amino acids.

TRYPSIN. A protein-splitting (proteolytic) enzyme secreted by the pancreas that acts in the small intestine to reduce proteins to shorter chain polypeptides and dipeptides.

TUBER. A short, thickened, fleshy stem or terminal portion of a stem or rhizome that is usually formed underground, bears minute scale leaves each with a bud capable under suitable conditions of developing into a new plant, and constitutes the resting stage of various plants such as the potato or the Jerusalem artichoke.

TWENTY-EIGHT-HOUR LAW IN RAIL SHIPMENTS. This law prohibits transporting livestock by rail for a longer period than 28 consecutive hours without unloading, feeding, watering, and resting 5 consecutive hours before resuming transportation. On request of the owner, the period can be extended to 36 hours.

U

ULTRA-VIOLET. Rays in sunlight which enable vitamin D to be synthesized under the skin.

UNDERFEEDING. Usually, this refers to providing insufficient energy. The degree of lowered production therefrom is related to the extent of underfeeding and the length of time it exists.

UNSATURATED FAT. A fat having one or more double bonds; not completely hydrogenated.

UNSATURATED FATTY ACID. Any one of several fatty acids containing one or more double bonds, such as oleic, linoleic, linolenic, and arachidonic acids.

UNTHRIFTINESS. Lack of vigor, poor growth or development; the quality or state of being unthrifty in animals.

UREASE. An enzyme which acts on urea to produce carbon dioxide and ammonia. It is found in the jackbean and the soybean, and is produced by certain microorganisms in the rumen.

UREMIA. A toxic accumulation of urinary constituents in the blood.

URINE. Liquid or semisolid matter produced in the kidneys and discharged through the ureters to the urinary bladder, thence voided via the urethra. Normally, it is a clear, transparent, amber-colored, slightly acid fluid.

USP (UNITED STATES PHARMACOPOEIA). A unit of measurement or potency of biologicals that usually coincides with an international unit. (Also see IU.)

V

VACUUM-DEHYDRATED. Freed of moisture after removal of surrounding air while in an airtight enclosure.

VALENCE. Power of an element or a radical to combine with or to replace other elements or radicals. Atoms of various elements combine in definite proportions. The valence number of an element is the number of atoms of hydrogen with which one atom of the element can combine.

VEALER. Calves fed for early slaughter, usually under 3 months of age.

VEGETABLE PROTEIN. The form of protein in all feed sources of vegetable (plant) origin. These are not as well balanced in essential amino acids as animal protein sources; hence, supplementation is required.

VFA. See VOLATILE FATTY ACIDS.

VILLI. Small threadlike projections attached to the interior side of the wall of the small intestine, which increase its absorptive capacity.

VITAMIN SUPPLEMENTS. Rich synthetic or natural feed sources of one or more of the complex organic compounds, called vitamins, that are required in minute amounts by animals for normal growth, production, reproduction, and/or health.

VOID. To evacuate feces and/or urine.

VOLATILE FATTY ACIDS (VFA). Commonly used in reference to acetic, propionic, and butyric acids found especially in rumen contents and/or silage.

VOMITING. The forcible expulsion of the contents of the stomach through the mouth.

W

WAFERS. A form of compressed feed based on fibrous ingredients in which the finished form usually has a diameter or cross-section measurement greater than its length.

WARM CONFINEMENT. Refers to a confinement building which is sufficiently insulated and ventilated to maintain inside winter conditions above 35°F in severe weather, and in the range of 50° to 60°F most of the time.

WEANING. The stopping of young animals from suckling their mothers.

WEIGHT RATIO. See RATIOS.

WET MILLED. Subjected to a milling process while containing moisture.

WET RENDERED. Cooked with steam under pressure in closed tanks.

WILTED. To become limp because of water loss.

WOOD CHEWING. Abnormal behavior in horses characterized by chewing on wood.

X

XANTHOPHYLL. Pigment in feed that imparts color to chicken egg yolks, and to the beak, skin, and shanks of yellow-skinned breeds of chickens. Synthetic pigments are also used.

XEROSIS. Abnormal dryness.

Y

YEAST. A source of protein.

Z

ZERO GRAZING. Feeding of green forage as green chop in a lot or stall.

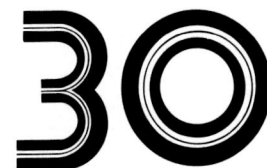

Original painting by Tom Phillips

GLOSSARY OF FEEDSTUFFS

Table 30 (1) lists feedstuffs by the common names most generally used, followed by the genus and species—the latin, or scientific, names; and (2) gives, in summary form, pertinent information relative to each of them, including a brief description, place of origin, adaptation, cultural characteristics, importance, and use. Because of space limitations, many of the feeds described elsewhere in this book are not repeated in this chapter. For example, pertinent information about grasses and legumes is presented in Chapter 7; pertinent information about the cereal grains is presented in Chapter 10; and pertinent information about the pulse proteins is presented in Chapter 11. Thus, if a particular feedstuff is not listed in Chapter 30, the reader should look in the index or in the particular chapter and section where it is discussed. But no claim is made to listing all feedstuffs, or to listing the most important feedstuffs of a given area.

The authors recognize that a number of range plants which are generally considered to be poisonous may, in some cases, be consumed in limited amounts, especially by some species and at certain seasons. However, those poisonous plants that are responsible for the most extensive U.S. livestock losses are not listed in this chapter in order that the reader will not be misled relative to their feeding value.

The chemical composition of many of the feeds listed in this chapter can be found in Section V—Composition of Feeds.

TABLE 30-1
GLOSSARY OF FEEDSTUFFS

Feedstuff	Description	Place of Origin Geographical Adaptation Cultural Characteristics	Importance/Use	Comments
A				
ACACIA *Acacia* spp	Leguminous trees or shrubs. Orange-yellow flowers are found in dense heads, spikes, or auxiliary racemes.	Most of the 400 or more species are native to Africa and Australia. Widely distributed throughout the tropics and subtropics, occasionally in temperate regions.	Many species with low tannin content are considered to be useful browse. Leaves and young shoots are readily eaten by livestock. Almost all of the species in the U.S. provide palatable browse.	Pods are too tough and/or too bitter to be used for browse. CATCLAW is the most important species in the U.S.
ACORNS *Quercus* spp	Nuts of the oak species.	Acorns of the white oak group are the most widely used feed.	Very nutritious; high in fats and oils. Can be an important feed for domestic and wild animals.	Acorns are high in tannin; hence, excess levels may produce toxicities which result in the following abnormalities: cows drying up; deformed calves, known as *acorn calves;* and hens producing eggs with olive-colored yolks and lowered hatchability. Acorns have been used throughout history as human food.
AFRICAN MARIGOLD (AZTEC MARIGOLD) *Tagetes erecta*	Stout, branching annual herb. Yellow to orange flower heads are 2 to 4 in. (5 to 10 cm) across.	Introduced to the U.S. from Mexico.	Used to produce zempa meal—a meal composed of dried flower petals.	Zempa meal enhances skin pigmentation and egg yolk color.
ALFILERIA (FILAREE, HERONSBILL, PINCLOVER, PINGRASS) *Erodium cicutarium*	Annual or biennial herb of the geranium family. Hairy weed of the rosette type, but may sometimes form clumps. Low, spreading red stems.	Native to Europe. Introduced in the U.S. from the Mediterranean region. Common in all of the western range states. Thrives in desert and foothill areas. Grows well on sandy, barren soils. Spreads extremely rapidly.	Considered to be one of the best winter forages in Arizona. Provides excellent spring forage for all classes of range livestock.	Troublesome weed in lawns, gardens, and cultivated fields.
ALKALI SACATON (FINE-TOP SALT GRASS) *Sporobolus airoides*	Hardy bunchgrass. Grows to 3 ft (91 cm) in height. Deep, coarse roots.	Found on alkali flats, rocky soils, bottom lands, and desert soils.	Highly palatable in early stages of growth. Tough and unpalatable upon maturing.	Mineral content is extremely high. Good forage grass in alkali regions.

(Continued)

TABLE 30-1 (Continued)

Feedstuff	Description	Place of Origin Geographical Adaptation Cultural Characteristics	Importance/Use	Comments
ALMOND *Prunus dulcis*	Nut-bearing tree. Nuts are surrounded by a fleshy husk that dries and splits at maturity.	The two main regions where almonds are cultivated are California and the Mediterranean.	Nuts are used primarily for human consumption. Hulls are commonly dried and fed to ruminants with such feeds as alfalfa and barley. They may also be ensiled.	Almond hulls are extremely variable in feeding value.
ALYCE CLOVER *Alysicarpus vaginalis*	Summer annual legume. Not a true clover. Low-spreading, coarse-stemmed plant.	Native to Asia. Widely used in the Gulf Coast states. Adapted to moderately fertile soil. Intolerant of wet soils.	Used to a limited extent for pasture, hay, and soil conservation in southern U.S.	Very susceptible to root-knot nematode.
ANISE *Pimpinella anisum*	Annual herb related to the carrot. Grows up to 2 ft (*61 cm*) in height. Leaves resemble parsley. Flowers and seeds are produced in large, loose clusters.	Native to the Mediterranean region.	Leaves are used for flavoring and garnishing in human foods. Seeds are used to flavor curry powder. Oil extracted from the seeds is used in beverages, drugs, and cosmetics. By-products can be fed to livestock.	Anise oil is sometimes used to increase the palatability of feeds.
ARROWROOT (YUQUILLA) *Maranta arundinacea*	Perennial herb. Multibranched stems may reach 6 ft (*1.8 m*) in height. Roots are rich in starch.	Native to South America and the West Indies.	Starch is extracted from the roots to form tapioca. The residue, called bittie, is sometimes fed to cattle and swine.	Fine bittie is fed to swine, coarse bittie to cattle.
ARTICHOKE *Cynara scolymus*	Tall perennial herb. Produces tuberous roots and edible flower buds.	Primarily cultivated in the coastal areas of California.	Tubers can be fed to livestock either fresh or ensiled.	Flower buds are used primarily for human consumption.
ASPEN *Populus grandidentata*	Deciduous tree.	Temperate and cool subtropical regions.	Leaves and treated wood scraps can be fed to livestock.	Used primarily for pulpwood.
B				
BAMBOO *Arundinaria* spp, *Chusquea* spp, *Bambusa* spp	Diverse and multispecied type of plant. Ranges in height from 2 ft (*61 cm*) to over 100 ft (*30.5 m*).	Found throughout the world, especially in the tropics and subtropics.	Sprouts used as human food. Whole plants and shoots used as forage.	Bamboo is the most useful and versatile plant in the Orient.
BEECHNUTS *Fagus* spp	Seeds (covered by a spiny husk) from the beech tree.	Found throughout the eastern U.S.; usually in association with sugar maple.	Of little importance as a feed. Used primarily as forage for hogs and wild animals.	
BEET *Beta vulgaris*	Biennial plant. Food is stored in a fleshy taproot one season to provide nutrients for flowering and fruiting in the second.	Descended from the wild beet of southern Europe. Adapted to cool climates.	Three varieties of beets are commonly cultivated, all of which can be, and are, fed to livestock in some form; (1) sugar beet, (2) mangel, and (3) common beet. Beet tops can be fed fresh or ensiled. The fleshy root is a succulent energy feed.	For more information on sugar beet and its by-products, the reader is referred to Chapters 10 and 12. Mangels are cultivated primarily as a livestock feed, while the common beet is primarily used as a human food.
BENTGRASS (REDTOP) *Agrostis* spp	Moderately tall fine-stemmed, creeping annual or perennial grass.	Grown over much of the northern half of the U.S. Thrives on soils too low in lime or fertility for bluegrass.	Most species cultivated for lawn cover and golf putting greens. Redtop is grown to a limited extent for pasture and hay. It is among the lowest in palatability of the northern grasses.	To all but one species, the common name *bent* is applied, the one exception being REDTOP.
BERMUDAGRASS (BAHAMA GRASS, DEVIL'S GRASS, STAR GRASS, WIREGRASS, JOINT GRASS) *Cynodon dactylon*	Long lived, warm season, perennial grass; forms a dense turf.	Probably originated in Africa, but widespread throughout tropical, subtropical, and mild temperate areas of the world. Prefers heavy soils, but when well fertilized, will grow on deep sands. Hardy grass capable of withstanding drought or waterlogging.	Used primarily for pasture and lawn cover. Also used for hay. Produces excellent yields of palatable, nutritious forage when managed properly. Rated as an excellent forage for cow-calf operations, good for growing, but poor for finishing.	COASTAL BERMUDAGRASS, a sterile hybrid, is now widely grown in the southern U.S. COASTCROSS-1, another hybrid, is less winter-hardy, but more digestible and produces higher gains than Coastal Bermudagrass. Hybrids are established vegetatively by planting sprigs of rhizomes and stolons or clippings from mature forage.
BLUEGRASS *Poa* spp	Grass distinguished by small awnless spikelets. Lemmas with a heavy midnerve.	Approximately 200 species are distributed throughout the world, primarily in cool temperate regions. Sixty-nine species are found in North America.	Among the most palatable of the grasses. Widely used for pasture, hay, and lawns.	KENTUCKY BLUEGRASS and CANADA BLUEGRASS are the most important species of bluegrass. Other well-known species are ANNUAL BLUEGRASS, BIG BLUEGRASS, MUTTON BLUEGRASS, ROUGHSTALK BLUEGRASS, SANDBERG BLUEGRASS, and TEXAS BLUEGRASS.

(Continued)

Glossary of Feedstuffs

TABLE 30-1 *(Continued)*

Feedstuff	Description	Place of Origin Geographical Adaptation Cultural Characteristics	Importance/Use	Comments
BLUESTEM (BEARDGRASS, TURKEY-FOOT, BROOMSEDGE) *Andropogon* spp	Large genus of grasses. Stems are either solid or pithy. Two spikelets are produced at each node of the rachis—one is fertile, one sterile.	Well-represented throughout the warmer regions of the world.	Several species are considered to be good forage grasses. They are palatable when young and tender, but become coarse and tough with maturity. Bluestem cures well and makes a good dry forage for cattle and horses.	BIG BLUESTEM and LITTLE BLUESTEM are probably the most prevalent constituent of wild hay in the prairies of the U.S. Although rather unpalatable, BROOMSEDGE produces vegetative cover on soils of very low fertility.
BROMEGRASS *Bromus* spp	Annual and perennial grasses; flat leaf blades; spreading panicles; open seed heads.	Large genus; about 36 species native to the U.S. Most species are found in north temperate zone. Vary greatly in adaptability.	Some species are important forages; others are considered troublesome weeds.	The species of major agricultural importance are CALIFORNIA BROMEGRASS, RESCUEGRASS, SMOOTH BROME, and MOUNTAIN BROME.
BUCKWHEAT *Fagopyrum* spp	Summer annual. Coarse, branched stems with large, broadly arrow-shaped leaves. Seeds are broad at the base and triangular to nearly round in cross section. Taproot has many short laterals.	Believed to have originated in China. Grown primarily in the northeastern U.S. Best adapted to uplands and mountainous sections. Thrives in a cool, moist climate but is sensitive to freezing temperatures.	Most buckwheat is processed for flour. Whole seeds can be used for poultry scratch feed. Hulls (roughage) and middlings (protein feed) are mill by-products that can be fed to livestock. The whole plant can be used for forage.	Relatively minor crop in the U.S. Not classified as a cereal.
BUFFALOGRASS *Buchloe dactyloides*	Low-growing perennial grass. Spreads by stolons. Seeds are enclosed in a hard bur. Grayish green foliage turns to a light straw color at maturity. Plants are unisexual.	Grows primarily in the central and southern Great Plains of the U.S. Usually found on hard clay soils. Tolerates alkaline soils but not sandy soils. Very drought-resistant.	Widely regarded as a range pasture grass. It is highly palatable and nutritious in the green summer stages. Also provides valuable winter grazing in its dry cured stage.	Tolerant of heavy grazing.
BUFFELGRASS (ANJAN, BLUE BUFFALOGRASS, AFRICAN FOXTAIL, RHODESIAN FOXTAIL) *Cenchrus ciliaris*	Tufted perennial grass; up to 2 ft *(61 cm)* tall. Large, strong root system. Occurs in bunch or spreading types.	Native to South Africa. In the U.S., it is grown primarily in South Texas for pasture. Thrives on light, sandy soil. Drought-resistant, but sensitive to cold and waterlogging.	Good pasture grass in dry areas. High protein content and highly digestible when young, but quality decreases with maturity.	Can withstand close grazing and fire once established.
BUR-CLOVER *Medicago* spp	Annual legume resembling clover except that the pods are spiny and coiled.	Adapted to warm temperate regions. Prefers moist, well-drained soils. Very tolerant of alkali. Does poorly on soils of low fertility.	Used mostly for pasture. Animals must acquire a taste for bur-clover. Sheep are fond of ripe pods. In the southern U.S., it is commonly combined with Bermudagrass.	Burs are eaten when softened by rain. Burs may get tangled in the wool of sheep and reduce its value.
C				
CACTUS, PRICKLY PEAR *Opuntia engelmanni*	Semierect, spiny, branched perennial. Grows in dense patches. Flat stems. Showy flowers are bright yellow.	Found from western Texas to California and throughout Mexico. Drought- and heat-resistant.	Of greatest value as an emergency feed in drought when the spines are burned off and the plant fed to livestock. Valuable source of water in arid regions. High mineral content. Will not maintain livestock as a sole source of feed.	Fruit is sweet and sometimes used for human consumption.
CANARYGRASS *Phalaris* spp	Annuals or perennial grasses. In the southern U.S., most species are summer annuals. Some species in the northern U.S. are winter annuals.	Widely distributed throughout the world.	Most species are used for forage. The name, canarygrass, was derived from *P. canariensis*, which is commonly used for bird seed.	REED CANARYGRASS and HARDINGGRASS are the only species cultivated for forage in the U.S. Both are perennials and must be harvested or grazed in the early stages of growth to obtain palatable, high quality forage.
CANTALOUPE (MUSK MELON) *Cucumis melo*	Creeping annual. Produces round to oval fleshy fruit about 5 in. *(13 cm)* in diameter. Large leaves.	Believed to have originated in Africa, Persia, or India. Prefers hot weather.	Leaves and fruit are used as feed.	High water content limits its value as a feed.
CARIB GRASS (ALEMAN GRASS, MALOJILLA) *Eriochloa polystachya*	Branching, grass with trailing stems that root at the lower nodes.	Grown in the Gulf Coast states. Adapted to humid regions with evenly distributed rainfall. Intolerant of either cold or drought. Withstands waterlogging.	Well-suited for grazing and hay.	
CARROT *Daucus carota*	Annual or biennial herb. Succulent thickened root ranges from 4 to 12 in. *(10 to 30 cm)* in length. Crown has compound leaves.	Originated in Europe. Longer maturing, longer rooted varieties are adapted to high elevations. Faster maturing, shorter rooted varieties are adapted to low elevations.	Crowns and roots are fed to livestock. Roots have high vitamin A activity and energy content on a moisture-free basis. Roots are sometimes dried and ground to be used as a vitamin supplement.	Raw carrots are sometimes fed to horses as a treat.

(Continued)

TABLE 30-1 (Continued)

Feedstuff	Description	Place of Origin / Geographical Adaptation / Cultural Characteristics	Importance/Use	Comments
CASSAVA (MANIOC, TAPIOCA, BRAZILIAN ARROWROOT, YUCA) *Manihot esculenta*	Herbaceous shrub or tree. Grows up to 13 ft (*4 m*) in height. Finger-like leaves. Tuberous, starchy roots form clusters at the base of the stem. Two types of cassava exist: (1) bitter varieties—contain high amounts of prussic acid and must be processed before use, and (2) sweet varieties—contain low amounts of prussic acid and can be fed raw.	Cultivated throughout the tropics and subtropics. Very little cultivated in the U.S.	Cassava roots, either cooked or raw, are fed to pigs, cattle, sheep, and goats in many areas. Molasses or water must be added to cassava-based poultry and swine rations in mash form to make them palatable. However, pelleting alleviates this need. Leaves are richer in protein and minerals than any other part of the plant.	Root meal is extremely deficient in methionine. Young leaves of the sweet varieties should be boiled for 15 minutes or dried for 3 weeks to inactivate the prussic acid.
CENTURY PLANT *Agave americana*	Has a large candelabralike stalk that can reach 30 ft (*9 m*) in height. Leaves grow next to the ground in a large rosette.	Native to Mexico. Found in hot, dry areas in the Mediterranean region and in Africa.	The skin is removed from the leaves, and the remaining parts are chopped and mixed into hay.	Saponin may cause illness if excessive amounts are fed. Sugar-rich sap is used for mescal and pulque—Mexican liquors.
CHESTNUTS *Castanea dentata*	Sweet, edible nuts of trees belonging to the genus *Castanea*. Fruits have spiny bur and become fibrous upon maturity.	Used to be abundant throughout the hardwood belt of the U.S., especially in the Appalachian region. Blight has killed most of the chestnuts in the U.S.	Shells and meats can be fed to livestock. Meats are low in fiber, medium in energy.	
CHUFA *Cyperus esculentus*	Tuber-producing plant. The tubers are small (about ¾ in. [*19 mm*] long), cylindrical, and hard. Grasslike top has simple leaves and a flower stalk that may grow to 3 ft (*0.9 m*) in height.	Cultivated primarily in southern Europe and Africa. Minor crop of the southern U.S. Grows best on sandy soils.	When properly supplemented with protein feeds and corn, chufas can be used to fatten swine.	Tubers may produce soft pork if improperly used.
CLOVER *Trifolium spp*	Perennial or annual legumes. Flowers of all species are borne in heads. Number of florets or individual flowers per head ranges from 5 to 200. Seeds per pod range from 1 to 8 depending on species. Leaves are trifoliate in most cases.	About 250 species of clover are recognized throughout the world. They are believed to have originated in southwestern Asia Minor and southeastern Europe. Wild species are found on all continents except Australia. In general, clovers thrive in cool, moist climates on soils having an available supply of calcium, phosphorus, and potassium. Photoperiodism (length of daylight) is critical in many species and varieties. Most are long-day plants.	Second only to alfalfa in importance as a legume forage crop.	The major species in the U.S. are ALSIKE CLOVER, ARROWLEAF CLOVER, CRIMSON CLOVER, HOP CLOVER, LADINO CLOVER (WHITE CLOVER), PERSIAN CLOVER, RED CLOVER, ROSE CLOVER, SEASIDE CLOVER, STRAWBERRY CLOVER, SUBTERRANEAN CLOVER, and WHITETIP CLOVER. Species of local importance are BALL CLOVER, BIGFLOWERED CLOVER, CLUSTER CLOVER, LAPPA CLOVER, STRIATE CLOVER, and ZIGZAG CLOVER.
COTTONTOP (ARIZONA COTTONGRASS) *Trichachne californica*	Fast-growing, coarse, leafy perennial grass. Slender, erect stems grow to 40 in. (*102 cm*) in height. Strong, wooly, knotted rootstocks. Silky-cottony panicle with paired lance-shaped spikelets covered with long white hairs.	Common on the deserts and foothills of southern New Mexico and Arizona and in the woodlands and semidesert areas of the Southwest, except California. Responds quickly to spring and summer rains.	Highly palatable forage when immature. Upon maturity, it becomes tough. Valuable grass when grown in combination with slower-growing grasses (*e.g.*, grama). The animals will graze the cottontop first, allowing the other grasses time to become established.	Cures well on the ground and makes good winter forage.
CRABGRASS *Digitaria spp*	Erect or prostrate annuals and perennials.	Adapted to a wide variety of conditions.	Good forage grasses.	CRABGRASS is a common weed in cultivated fields.
CRAMBE (ABYSSINIAN KALE, ABYSSINIAN CABBAGE) *Crambe abyssinica*	Erect annual herb. Grows to 3 ft (*0.9 m*) in height. Upon maturity, the leaves drop off and the pods and stems turn tan in color.		Oil meal must be treated with ammonia or sodium carbonate or heated to neutralize sulfur compounds that make the feed unpalatable. Treated oil meal can be fed to ruminants at levels up to 50% of the supplement.	Oil is extracted from the seeds. The oil meal is high in protein.
CROWN VETCH *Coronilla varia*	Perennial legume. Vetchlike leaves. Arrangement of florets resemble a crown. Deep taproot with numerous lateral roots. Flowers are variegated white to purple. Upon drying, the pods break in segments, each of which contains a rod-shaped seed.	Found predominantly in the Mediterranean region, central and southern Europe, southwestern Asia, and northern Africa. Grown in northern ⅔ of the U.S. Prefers fertile, well-drained soils of pH 6 or above. However, once established, it is relatively tolerant of soil acidity and infertility.	Palatable forage, especially well-suited to the improvement of permanent bluegrass pasture. Used extensively as a ground cover for the stabilization of roadbanks and along waterways.	The three most widely used cultivars in the U.S. are CHEMUNG, PENNGIFT, and EMERALD. Crown vetch is nonbloating.

(Continued)

Glossary of Feedstuffs

TABLE 30-1 (Continued)

Feedstuff	Description	Place of Origin Geographical Adaptation Cultural Characteristics	Importance/Use	Comments
CURLY MESQUITE *Hilaria belangeri*	Stolon-forming grass. Grows in tufts. Slender stems up to 1 ft (30.5 cm) tall. Short, narrow leaves.	Native to central Texas, Arizona, and throughout Mexico. Grows on dry soils that vary from clay to gravelly in texture. Highly drought-resistant. Responds readily to summer rains.	Highly regarded for its forage value. Very palatable to all classes of range livestock.	Produces a fair amount of forage for its size.
D				
DALLISGRASS *Paspalum dilatatum*	Foot growing, stout perennial. Smooth leaves; deep root system. Grows in clumps. Slender stems usually droop with weight of the seed.	Native to Argentina, Brazil, and Uruguay. Widely used throughout the Cotton Belt of the U.S. Prefers moist, fertile, clay, and loam bottomland. Drought- and moisture-tolerant.	Highly palatable. Dallisgrass and legume combination used primarily for pasture. Produces good yields of hay.	In southeastern U.S., growth occurs from early spring to late fall; but it is dormant during the coldest winter months.
DATE PALM *Phoenix dactylifera*	Strong palm tree of up to 100 ft (30.5 m) in height. Long, stiff leaves. Cylindrical fruits of 1 to 2 in. (2.5 to 5 cm) grow profusely on long strands. Fruits lose most of their moisture before harvest.	Found throughout tropical regions of the world. Most dates produced in the U.S. are grown in California and Arizona.	Whole fruits, ground seeds, and dried pulp can be fed to livestock. Ground seeds are good ruminant feeds when supplemented with a protein feed. Dried pulp can be fed to all classes of livestock.	
E				
EGGSHELL MEAL	Dried mixture of eggshells, shell membranes, and egg contents.	Product of egg-breaking plants.	Mineral supplement. About 94% of the shell is calcium carbonate. Other 6% contains calcium and magnesium phosphates, boron, copper, chromium, and iodine.	
F				
FESCUE *Festuca* spp	Annual and perennial grasses. Classified by leaf types: (1) broad-leaved, or (2) fine-leaved.	About 100 species are found throughout temperate or cool zones. Vary widely in growth habitat.	Perennial species provide excellent forage and turf. Annual species may be troublesome weeds.	TALL FESCUE and MEADOW FESCUE, both broad-leaved species, are the most widely cultivated fescues. Some of the important fine-leaved species are IDAHO FESCUE, RED FESCUE, SHEEP FESCUE, and CHEWING'S FESCUE.
FLAT PEA *Lathyrus sylvestris*	Perennial legume.	Native to Europe. Used on logged-off land in the Pacific Coast states.	Used for forage and green manure. Contains high amount of highly digestible protein.	Poisonous when fed in large amounts after the seed has formed.
FOXTAIL *Alopecurus* spp	Low or moderately tall perennial grasses or some annuals with flat blades and soft, dense, spikelike panicles.	Native to Europe and Asia. Has been cultivated in the U.S. since about 1750. Adapted to temperate climates, especially to low, overflow areas.	All species are highly palatable and nutritious, but are not found in abundance in the U.S. Meadow foxtail is used as a meadow grass in the eastern U.S. and the Pacific Northwest.	CREEPING FOXTAIL and MEADOW FOXTAIL are the most important species found in the U.S. The awns of ripe foxtail may pierce animal tissue.
G				
GAMAGRASS *Tripsacum* spp	Robust perennial grass. Broad, flat blades.	Eastern U.S., Mexico, and Central America.	Good forage grasses but not of importance in the U.S.	Related to corn; some hybrids with corn have been successful.
GRAMA *Bouteloua* spp	Vary greatly according to species.	Some 18 species are native to the U.S.; found primarily throughout the Great Plains and West. Summer growers. Most species cure naturally; thus, stands from previous year's growth are palatable and nutritious.	Valuable forage for range and pasture.	SIDE-OATS GRAMA and BLUE GRAMA are the two most widely grown species in the U.S. BLACK GRAMA is of local importance.
H				
HAIRY INDIGO *Indigofera hirsuta*	Climbing annual legume with heavy foliage on fine stems. Leaves are covered with short hairs.	Adapted to the Coastal Plains of the U.S. Prefers sandy loam soils.	Produce large amounts of high-quality forage. Used for hay, pasture, cover crops, and green manure.	
HARDINGGRASS (See CANARYGRASS.)				

(Continued)

TABLE 30-1 *(Continued)*

Feedstuff	Description	Place of Origin Geographical Adaptation Cultural Characteristics	Importance/Use	Comments
HEMP (MARIJUANA) *Cannabis sativa*	Erect dioecious annual; only the female plant yields seeds. May grow as high as 16 ft (5 m).	Adapted to most tropical and temperate regions. Requires rich bottomland.	In tropical countries, hemp is cultivated for fiber and its oil-rich seed. The resultant oil cake is of high fiber and low digestibility. Hence, it is used primarily as a ruminant feed. Whole seeds are used for poultry feeds.	Illegal to grow in the U.S. because of its narcotic properties upon smoking or ingestion, but can be found growing wild in pastures and ditches and along roadsides. Was grown extensively in the U.S. for fiber during World War II.
HORSE CHESTNUT *Aesculus hippocastanum*	Tree; up to 65 ft (20 m) tall. Yields spherical or elongated spiny fruit.	Native to northern Greece. Thrives in moist temperate areas.	Fruit is sometimes used for livestock feed.	

I

Feedstuff	Description	Place of Origin Geographical Adaptation Cultural Characteristics	Importance/Use	Comments
INDIANGRASS *Sorghastrum nutans*	Tall, erect perennial, warm season grass. Short, scaly rhizomes.	Found east of the Rocky Mountains from Canada to the Gulf of Mexico. Thrives on fertile bottom soils but will grow on sandy soils and dry slopes.	Fresh Indiangrass is highly palatable. As it dries, palatability decreases.	Common constituent of prairie hay in the Great Plains.
INDIAN RICEGRASS *Oryzopsis hymenoides*	Densely tufted perennial grass. Grows to 1 to 2 ft (30 to 61 cm) in height. Long, slender leaves.	Widely distributed throughout the western U.S. Tolerant of drought and alkali soils. Found primarily on dry, sandy soils.	Considered to be an outstanding winter forage.	It was once commonly found throughout the western ranges, but was almost wiped out by overgrazing. At one time, ricegrass seed was harvested by Indians for use as meal and flour when the corn crop failed.
INDIAN WHEAT, WOOLY (WOOLY PLANTAIN) *Plantago purshii*	Small, silvery annual weed. Small flowers are found in a dense cylindrical spike, resembling a spike of wheat.	Distributed throughout western U.S., except California. Grows in pastures, waste areas, and along roadsides. Prefers dry, open loamy soils.	Considered to be fair- to good-quality forage. Outstanding feed on desert lambing grounds.	

J

Feedstuff	Description	Place of Origin Geographical Adaptation Cultural Characteristics	Importance/Use	Comments
JACK BEAN (HORSE BEAN) *Canavalia ensiformis*	Fast-growing, erect annual legume. Deep-rooted.	Grown primarily in the southern U.S. Drought-resistant.	Forage. Palatable only when dried. Can be toxic in large amounts. Maximum safe levels in cattle feed is 30% of total feed.	Heat treatment renders the seeds and pods harmless. Sometimes used for food or green manure.
JERUSALEM ARTICHOKE *Helianthus tuberosus*	Herbaceous perennial of the sunflower family. Produces fleshy rootstocks with tubers.	Native to North America. Cultivated in tropical and subtropical countries.	Fresh immature tops can be used for forage. Tubers are often fed to swine. Also used for human food.	Tubers contain inulin rather than starch; do not store well.
JOHNSONGRASS (See SORGHUM.)				
JUNEGRASS *Koeleria cristata*	Tufted, perennial bunchgrass.	Native to western North America. It also occurs in Europe and Asia. It is one of the most common of the western grasses, thriving on a wide variety of soils.	Fair to good forage. Relished early in the growing season by all types of livestock, but sheep do not graze the stalks after seed maturity. Not a heavy yielding grass because its leaves are short and basal.	Closely resembles bluegrass. Tends to grow in scattered clumps rather than in solid stands.

K

Feedstuff	Description	Place of Origin Geographical Adaptation Cultural Characteristics	Importance/Use	Comments
KALE *Brassica oleracea*	Hardy biennial plant. Cultivated as an annual for the production of food.	Native to the eastern Mediterranean region. In the northern U.S., it is planted in the spring for summer production. In the southern U.S., it is planted in the fall and harvested throughout the winter.	Green stems and leaves rich in carotene. Dehydrated leaf meal is used for protein and vitamin supplementation—carotene and riboflavin. In poultry rations, it can be substituted for alfalfa meal.	Grown in the northern Pacific Coast region for forage; particularly for sheep.
KELP (SEAWEED) *Fucaceae* and *Laminariaceae*	Generally fed in dry meal form. High in certain vitamins and minerals, notably iodine.	Digestibility of protein is low.	Used primarily as a source of vitamins and minerals. Can constitute up to 10% of cattle feeds. Sheep can be fed up to 35 g/head/day.	Feeding large amounts may result in iodine toxicity. Can produce a fishy odor in pork. Has a positive effect on egg yolk color in poultry.
KLEINGRASS *Panicum coloratum*	Bunch- and sod-forming panicgrass. Slender stems can grow to 4 ft (1.2 m). Abundant dark green leaves. Fibrous root system. Spreads by tillers.	Introduced from Africa, primarily to South Texas. Adapted to moist, heavy soils.	Used for pasture, hay, and silage. Suitable for cattle only.	Causes *photosensitization* in lambs and some adult sheep. Cattle are not affected.

(Continued)

TABLE 30-1 *(Continued)*

Feedstuff	Description	Place of Origin Geographical Adaptation Cultural Characteristics	Importance/Use	Comments
KUDZU *Pueraria lobata*	Coarse, woody, perennial, leguminous vine. Forms long runners that root and form crowns at the nodes. Long, broad leaves.	Introduced into the U.S. from Japan. In the U.S., it is commonly found in the Southeast. Adapted to subtropical and humid, warm temperate climates. Drought-resistant. Aboveground parts killed by frost. Cannot withstand trampling.	Equal to alfalfa in nutritive value and palatability. Valuable cover for rough land, providing pasture and erosion control. Makes excellent silage. Hay is rather coarse.	Difficult to establish, but once established, it is long-lived and hardy. A troublesome weed in woodlands. Close grazing will generally kill the stand. Difficult to harvest mechanically because of the viny growth of the plant.
L				
LAMB'S QUARTER *Chenopodium album*	Annual weed of 1 to 10 ft *(30 cm to 3 m)* in height.	Introduced from Eurasia. Range plant.	Range forage.	Young leaves are sometimes used in salads or as a cooked vegetable.
LOCUSTS *Schistocerea gregaria*	Migratory grasshoppers that travel in swarms.	Used in Africa for food and feed.	Dried locust meal is readily consumed by swine and poultry. However, it may impart an off-flavor to meat.	Freshly killed locusts have an objectionable odor; sun-curing alleviates this problem.
LOVEGRASS *Eragrostis* spp	Annual or perennial grasses of various habits. Many lovegrasses produce an abundance of seed.	Over 250 species of lovegrass are known—about 40 are native to the U.S. Produces abundant growth on soils of low fertility.	In India and Australia, it is cultivated as a forage crop. Minimal use as a forage in the U.S. One species, Teff, is grown as a cereal in Ethiopia. Widely used for erosion control.	The most important species in the U.S. are BOER LOVEGRASS, WEEPING LOVEGRASS, LEHMANN LOVEGRASS, and SAND LOVEGRASS.
M				
MANGEL (See BEET).				
MANGO *Mangifera indica*	Large, spreading evergreen tree. Some varieties produce leaves that smell like turpentine when they are crushed. Produces round or oval fruits weighing up to 1 lb *(454 g)*. In the center of the fruit, there is a large, fibrous, flat seed containing a kernel.	Native to India but cultivated throughout the tropics of the world.	Fruit and kernels can be fed to livestock. Kernels can constitute up to 20% of poultry rations.	Fruits are commonly used for human consumption and are seldom fed to animals.
MESQUITE *Prosopis* spp	Leguminous shrub.	Characteristic of warm, dry subtropical and tropical climates. Abundant in the southern U.S.	Pods provide good forage. Leaves are rarely eaten. Also used for fence posts and fuel.	HONEY MESQUITE, VELVET MESQUITE, and COMMON MESQUITE are the three primary species in the U.S.
MILLET, JAPANESE (BARNYARD GRASS, CHIWAGA) *Echinochloa crusgalli*	Tall, robust, coarse annual grass. Grows to 4 ft *(1.2 m)* or more. Large, broad leaves. Dense, drooping seedheads.	Native to Asia. Grown to a limited extent in the northeastern U.S. Requires moist, well-drained fertile soils. Tolerant of temporary flooding.	High yields of coarse forage. Fair feeding value. Should be strip-grazed.	Superior in cool summers to SUDANGRASS or FOXTAIL MILLET. Troublesome weed in corn and soybeans, and sometimes in alfalfa and clover fields.
MILLET, PEARL (CATTAIL MILLET, BULRUSH MILLET, INDIAN MILLET, HORSE MILLET) *Pennisetum typhoides*	Tall, erect annual grass. Grows to 15 ft *(4.6 m)*. Coarse, pithy stems grow in dense clumps. Long, pointed leaves. Plants tiller profusely.	Native to India. Adapted to the southern U.S. Cultivated extensively in India and Africa for grain. Rapid rate of growth and maturity.	Used for pasture and silage for livestock.	May cause a depression in milk fat test in dairy cows.
MINT *Mentha* spp	Aromatic perennial herbs.	About 15 species can be found in the U.S. Typically, they grow in wet or moist places, most commonly at medium elevations in the West.	Palatability ranges from fair to fairly good for cattle and fair to good for sheep.	SPEARMINT and PEPPERMINT are grown for their aromatic oils. After steam distillation, plant residues are fed to animals.
N				
NAPIERGRASS (ELEPHANTGRASS, UGANDA GRASS) *Pennisetum purpureum*	High-yielding, tall, erect perennial grass with thick stems.	Native to Africa. Grown in the warmest regions of the U.S. Prefers deep soils of moderate to fairly heavy texture. Tolerant of short drought but not waterlogging.	Commonly used for hay, silage, and rotational grazing.	The strain, MERKER GRASS, is fine-stemmed and is more resistant to eyespot disease and drought than the other varieties.
NEEDLEGRASS *Stipa* spp	Feathery-awned grass. Each spikelet has one flower terminating in a needlelike awn.	About 30 species are found in the western U.S. Adapted to temperate climates.	Good source of forage because of its abundance, long-growing period and ability to cure well on the ground. Provides good early spring and winter grazing. However, the needlelike awns can cause sore mouths in livestock.	Some of the most important species are NEEDLE-AND-THREAD GRASS, PORCUPINEGRASS, GREEN NEEDLEGRASS, and CALIFORNIA NEEDLEGRASS.

(Continued)

TABLE 30-1 (Continued)

Feedstuff	Description	Place of Origin Geographical Adaptation Cultural Characteristics	Importance/Use	Comments
	**O**			
OAK *Quercus* spp	Very large genus of trees (about 500 species). LIVE OAKS have thick evergreen, persistent leaves. CHESTNUT OAKS have chestnutlike leaves. The third class of oaks is dwarf, shrubby, shinnery oaks (SHIN OAK).	Restricted to the northern hemisphere; primarily in the temperate regions and tropical mountain areas.	Leaves of deciduous oaks are generally more tender and nutritious than those of live oaks. Sheep and goats browse oaks more heavily than do cattle or horses. See ACORNS.	Eastern oaks have too much tannin to be used as browse. Several species of oak are considered to be poisonous, and care should be exercised when allowing livestock to graze in areas where these species abound.
OATGRASS *Dathonia* spp	Perennial bunchgrass.	Widely distributed in warm and temperate regions of the world, especially in South Africa and Australia. Seven species occur in the U.S.— six of them in the western U.S.	All of the western species, except for poverty oatgrass, are fairly palatable. Poverty oatgrass is of no value to livestock.	Some of the important species are CALIFORNIA OATGRASS, PARRY OATGRASS, TIMBER OATGRASS, ONE-SPIKE OATGRASS, and FLAT-STEM OATGRASS.
OLIVE *Olea europaea*	Low, spreading evergreens. Fruits are thin and smooth-skinned, changing from green to black at maturity.	Cultivated in the Mediterranean region, North Africa, and in the Americas. Very long-lived and drought-resistant.	Olive leaves can be fed fresh or ensiled. The press cake from the oil extraction of the fruit can be fed to livestock, but since it is susceptible to spoilage, it must be dried or ensiled.	Pulp is bitter because of tannin in the raw fruit and contains up to 20% oil. Kernels are of low feeding value and can cause digestive disorders.
ORCHARDGRASS *Dactylis glomerata*	Long-lived, bunch-type perennial grass. Folded leaf blades and compressed sheaths.	Native to Europe. Cultivated in the U.S. since the colonial period. In the U.S., it is found primarily in the Northeast, eastern Great Plains, western mountain regions, and throughout the irrigated areas of the West. Grows rapidly in cool temperatures. Shade and drought-tolerant. Does not tolerate wet soils or waterlogging.	Well-suited for spring pasture and rotational grazing, often in combination with a legume. Also used extensively for hay and silage. Palatability and nutritive value decreases rapidly as the grass approaches maturity. One of the earliest growing grasses in the spring.	Grown alone, it will yield 1 to 2 tons of hay per acre (2,240 to 4,480 kg/ha). With clover or alfalfa, yields can be increased to 2 to 3 tons per acre (4,480 to 6,720 kg/ha).
	**P**			
PANGOLAGRASS *Digitaria decumbens*	Creeping perennial grass. Grows to a height of 4 ft (1.2 m). Produces many semidecumbent surface runners that form roots at the joints or nodes.	Native to South Africa. Adapted to subtropical and tropical areas. In the U.S., it is grown in the Gulf Coast states and California. Adapted to fertile, moist, well-drained soils. Somewhat drought-resistant. Withstands heavy grazing.	One of the most important forage grasses in regions to which it is adapted. Highly palatable. Superior in dry matter and protein yields to Coastal Bermudagrass, bahiagrass, and guineagrass.	Should be grazed rotationally.
PANICGRASS, BLUE *Panicum antidotale*	Deep rooted, erect glabrous perennial grass. Forms tough crowns via short, thick rhizomes.	Native to India. Used in the southwestern U.S. for dry-land and irrigated pastures. Adapted to heavy loam or dark, clay soils.	Useful for either hay or pasture. Palatability decreases upon maturing.	Intolerant of close, intensive grazing or cutting.
PARAGRASS (MALOJILLO) *Panicum purpurascens*	Coarse, creeping, sod-forming perennial grass. Roots readily at the nodes. The nodes, leaves, and leaf sheaths are hairy. Sparse seed producer.	Native to Africa. Commonly cultivated in most tropical countries. Grown along the Gulf Coast region of the U.S. Adapted to moist soils and can withstand waterlogging.	Commonly used for hay and pasture. Although coarse in texture, paragrass hay is of good quality if cut before the stems become woody. Should be used for pasture only after the grass is well-sodded and is at least 1½ ft (46 cm) high.	Should be grazed rotationally. Occasional disking may stimulate growth.
PEAVINE *Lathyrus* spp	Smooth, weak stemmed, trailing or climbing plants.	Peavines are well represented in the western U.S. Found on moist, rich soils and on dry scablands, in open exposures, and in the shade of coniferous and broadleaf timber—being most typical of open aspen areas.	Palatability varies considerably. Trailing or climbing species with tendrils range from fair to good forage for cattle, sheep, and goats. The erect species range from poor to fair forage. All species are used for forage, primarily in the summer and fall. After the first heavy frosts, most species dry up and disappear. Can also be used for silage.	The best known peavine is the SWEET PEA. Can withstand heavy grazing.
PINE *Pinus sylvestris*	Tree having needlelike, long leaves and dark bark.	Widely grown in temperate climates and the cooler areas of subtropical countries for timber and pulpwood.	Excessive browsing of yellow pine (*P. ponderosa*) can cause abortion. In times of famine, leaves can be fed to ruminants, if intake is carefully monitored. One lb (454 g) of leaves per day can be fed safely to sheep.	

(Continued)

Glossary of Feedstuffs

TABLE 30-1 (Continued)

Feedstuff	Description	Place of Origin Geographical Adaptation Cultural Characteristics	Importance/Use	Comments
POTATO, IRISH *Solanum tuberosum*	Tuberous annual with branched stems having slightly hairy leaves. Tubers are extremely rich in starch.	Widely cultivated tuber crop that thrives in cool climates. Tuber production is retarded at soil temperatures above 68° F *(20° C)*, but freezing will injure the tubers. Adapted to a number of soil types—from sandy loam to peat. Prefers well-drained, moist soils.	Cull potatoes provide excellent source of energy. They may be fed raw to ruminants but must be cooked prior to being fed to swine and poultry. Potatoes can be ensiled effectively. If they are to be ensiled in a raw state, ground corn (2 to 3% by weight) should be added.	Tubers are very low in protein, fiber, fat, carotene, and vitamin D. They contain about 10% protein and 85% carbohydrate. Sprouted potatoes are poisonous to livestock.
POTATO, SWEET *Ipomoea batatas*	Perennial vine that is cultivated as an annual. Leaves are commonly heart shaped. Adventitious roots produce swollen tubers.	Widely cultivated in the tropics and warm temperate regions. Requires at least 4 months of warm weather. Intolerant of frost.	Highly digestible source of energy. Culls can be fed fresh or dehydrated. Young leaves are rich in protein and highly palatable. Difficult to cure into hay but make good silage.	Cooking increases the feeding value. Sweet potatoes produce hard pork.
PRICKLY PEAR (See CACTUS, PRICKLY PEAR.)				
PUMPKIN (CALABAZA) *Cucurbita* spp	Edible fruits of trailing annual plants having large 5-pointed leaves. Fruits are highly variable in size, color, shape and weight. They have a moderately hard rind with a thick, edible flesh surrounding a central seed cavity.	Native to tropical and subtropical climates.	Seeds are very high in oil and can be fed to livestock, especially swine. Hulls and oil meal can be fed to livestock.	Livestock fed a ration too high in pumpkin seeds will develop indigestion.
Q				
QUACKGRASS *Agropyron repens*	Troublesome weed grass. Spreads by seeds and creeping rhizomes. Stems grow up to 3 ft *(0.9 m)*. Thin, flat leaf blades.	Native to Europe. Widely distributed throughout most temperate regions of the world.	Good forage qualities. Used for pasture, hay, and silage.	Good soil binder for embankments.
R				
RATANY, RANGE (PURPLE HEATHER) *Krameria glandulosa*	Low, bushy, diffusely branched shrub, up to 2 ft *(61 cm)* high. Bluish-green twigs and bluish leaves. Sweet-scented flowers. Spiny burlike fruit.	Found from western Texas to southern Utah and southern California. Common at elevations between 2,000 and 4,500 ft *(610 and 1,373 m)*. Grows on dry, hot foothills.	Ranks as fair to fairly good forage.	Good emergency forage. Goats browse it freely.
REDTOP *Agrostis alba*	Perennial grass. Stems are both upright and creeping. Flat, sharp-pointed leaves. Forms a less dense sod than Kentucky bluegrass.	Believed to be native to Europe, but some forms of it are native to North America. Distribution in the U.S. is very similar to that of Kentucky bluegrass in the East. However, it can be grown in most areas of the U.S., except in the drier regions and the extreme South. Thrives on very acid soils, poor clay soils of low fertility, and on poorly drained land. Matures from mid-June to mid-August, depending on the latitude.	Slightly less palatable than Kentucky bluegrass. Prior to 1940, redtop was the second most important pasture grass in the U.S. It has now declined to minor significance. If grown on good soil, it is of satisfactory quality as a forage. However, of the common northern pasture grasses, redtop is among the lowest in palatability.	Used mainly on poor or wet land for hay or pasture, usually mixed with alsike clover or annual lespedeza.
RESCUEGRASS *Bromus catharticus*	Short-lived perennial bromegrass.	Native to Argentina. Adapted to humid areas with mild winters.	On good soils, rescuegrass provides a good amount of palatable forage.	Furnishes good winter pasture in the southern U.S.
RHODESGRASS *Chloris gayana*	Fine-stemmed, leafy, perennial grass. Grows up to 3 ft *(0.9 m)*. Produces abundant seeds and can also spread by running stolons that can reach 6 ft *(2 m)* in length.	Native to South Africa. Adapted in the U.S. to the Gulf Coast and southwestern states. Relatively drought-resistant, but ample moisture is required for good production. Grows well on sandy soils, well-drained peat soils, and alkaline soils.	In tropical countries, it is noted as one of the best hay grasses. If grown under favorable conditions, high yields of palatable forage can be obtained.	With irrigation, it has succeeded on soil too alkaline for other crops. Withstands heavy trampling.
ROYAL PALM *Roystonea regia*	Tall palm.	Grows in Central America and Cuba.	Fruits are sometimes fed to swine. When fresh fruits are fed to swine, they will eat the outer fleshy part and leave the kernel. Once the kernel has dried, swine will also eat it.	

(Continued)

TABLE 30-1 (Continued)

Feedstuff	Description	Place of Origin / Geographical Adaptation / Cultural Characteristics	Importance/Use	Comments
RUSSIAN-THISTLE *Salsola kali*	Annual weed that can grow to 4 ft (1.2 m) in height and form a dense bushlike plant ranging from 2 to 6 ft (61 cm to 2 m) in diameter.	Native to the U.S.S.R. Widely distributed over the western U.S. and Canada. Salt-resistant and drought-resistant.	On early spring ranges, Russian-thistle rates as fair forage for all range livestock. However, once the plant matures and forms thorns, it is useless. If harvested before maturity, it can be ensiled.	Russian-thistle is a host of the sugar beet leafhopper which carries a virus that causes curly-top of sugar beets and "blight" of beans, tomatoes, and other plants.
RUTABAGA (SWEDE, SWEDISH TURNIP, TURNIP-ROOTED CABBAGE, RUSSIAN TURNIP, LAURENTIAN TURNIP) *Brassica napus*	Biennial plant cultivated as an annual. Glabrous leaves. Yellow-fleshed roots.	Grown primarily in northern Eurasia and northern North America. Extensively grown in Europe. Prefers cool climate.	Used as a human food and livestock feed. Sheep prefer it to all other roots.	More nutritious than turnips. Can produce off-flavors in milk.
RYEGRASS *Lolium* spp	Bunchgrasses. No creeping habit of growth. Perennial ryegrasses grow to 3 ft (0.9 m) in height with erect culms. Awns are usually absent. Annual ryegrasses behave like short-lived perennials. Taller than perennial ryegrasses, sometimes growing to 4¼ ft (1.3 m). Typically, awns are present.	Commonly used for forage in Australia, New Zealand, the British Isles, and the temperate areas of western Europe and the U.S. Can be grown on a wide variety of soils. Tolerant of short periods of flooding. Less winter-hardy than other grasses; e.g., timothy and orchardgrass.	Commonly used for hay, pasture, silage, soil conservation, and turf. Can be seeded alone or with other grasses or legumes. Hay is of excellent quality for horses. In the southeastern states, ryegrass is used for fall, winter, and spring pasture. Used for overseeding lawns in warm areas to produce green growth in the winter.	The common name ryegrass is generally applied to the species ITALIAN RYEGRASS and PERENNIAL RYEGRASS.

S

Feedstuff	Description	Place of Origin / Geographical Adaptation / Cultural Characteristics	Importance/Use	Comments
SACATON *Sporobolus wrightii*	Extremely robust perennial bunchgrass. Leaf stalks can grow to 6 (2 m) and sometimes 8 ft (2.4 m) tall. Should not be confused with its relative, alkali sacaton, which is smaller and less coarse.	Occurs from Arizona to western Texas and southward into Mexico. Found primarily on low alluvial flats, bottomlands, and channels subject to flooding. Will not tolerate alkali soils. Drought-resistant.	Young shoots are relished by cattle and horses. Much of the herbage cures well and provides fairly good winter forage. When properly prepared, sacaton makes good, nutritious hay, especially for horses.	Ranchers sometimes burn off the dead growth in late winter to promote new growth.
SAGEBRUSH *Artemisia* spp	Woody shrubs of the Mayweed tribe.	Sagebrushes occur in a wide variety of climates on a diverse range of soils. They compete with grasses for moisture and nutrients and, consequently, lower the forage value of range areas.	In the U.S., sagebrushes are of the highest forage value in the southwest. Palatability ranges from worthless to very good. In tall stands, sagebrush is worthless as a forage.	Abundant sagebrush growth is often indicative of overgrazing. FRINGED SAGEBRUSH (western North America, northern Asia, and Europe) and BIG SAGEBRUSH (western North America) are species most commonly used for browse.
SAINFOIN *Onobrychio viriaefolia*	Erect, deep-rooted, perennial legume. Stout stems originate from a crown. Leaves are oddly pinnate with 13 to 21 leaflets per leaf. Flowers are generally pink, though sometimes white. A single kidney-shaped bean is produced in a bean-shaped, bilaterally compressed pod.	Found in most of southern Europe eastward to Lake Baykal in Siberia. Very little is grown in the U.S. However, it is well adapted to the dry calcareous soils of the northern Rocky Mountain region. Long-lived on dry land but short-lived on irrigated land. Grows well on low phosphorus soils.	Shows promise as a pasture and hay crop in the northern plains of the U.S. About equal to alfalfa in feeding value.	Sainfoin is nonbloating.
SAINT AUGUSTINE GRASS *Stenotaphrum secundatum*	Broad-leaved, creeping, perennial, sod-forming grass. Stolons with long intervals. Short, leafy, and flat branches.	Native to the West Indies, Australia, and southern Mexico. In the U.S., it is found along the southern Atlantic and Gulf Coast regions from South Carolina to Texas. Adapted to almost all soil types, especially muck soils. Thrives in partial shade and can withstand salt spray.	Generally not considered to be an important pasture grass but it can be of some local importance. Considered to be the most dependable pasture grass for the organic soils of southern Florida.	Well suited and primarily used for lawns. Forms a dense sod that can withstand heavy trampling.
SALTBUSH *Atriplex* spp	Some species are shrubs, others are herbs.	Well adapted to the arid regions of the western U.S. Generally tolerant of drought and alkali conditions.	Some species are unsurpassed as browse plants with regards to volume of browse and usefulness as winter pasture.	FOURWING SALTBUSH and SHADSCALE SALTBUSH are the two most important species in the western U.S. Several species are considered to be weeds. Plants of this genus serve as a host for the sugar beet leafhopper.
SALTGRASS *Distichlis* spp	Dioecious, low perennial grass. Extensive creeping, scaly rhizomes.	Four species are found in the U.S. Occupy alkaline or saline areas.	Generally considered to be of little forage value, except where better grasses are not available.	High in salt content.

(Continued)

TABLE 30-1 (Continued)

Feedstuff	Description	Place of Origin Geographical Adaptation Cultural Characteristics	Importance/Use	Comments
SEDGE *Carex* spp	One of the world's largest genuses of flowering plants. Perennials propagated by rootstocks. Solid, unjointed, usually three-angled stems. Mostly basal, closed sheaths. Flat, long, thin grasslike leaves. Most species range from 8 to 20 in. *(20 to 51 cm)* in height, but some species can grow to 3 ft *(0.9 m)*.	**A**dapted to a wide variety of soils that are moderately fertile and that receive favorable moisture. **P**refer neutral or acid soils. Most sedges are adapted to full sunlight.	**T**he small group of slope- and timber-inhabiting species are usually most valuable for forage in the spring and fall. **S**mall-leaved, low, dry-land species are often an important range forage, being most palatable in the spring. **M**oist-meadow species are the most palatable sedges providing fine, green, tender foliage. **W**et-site species are good forage for cattle but grow on land too wet for sheep. **R**obust, large-leaved, wet-site sedges are of low palatablility for sheep, fair for cattle.	**O**VALHEAD SEDGE (western U.S.) and THREADLEAF SEDGE (Great Plains and western U.S.) are the two most important sedges.
SORGHUM *Sorghum* spp	The sorghum genus can be broken down into 4 broad classifications according to use: (1) grain sorghums, (2) grass sorghums for hay, silage, and pasture, (3) syrup sorghums or sorgos, and (4) broomcorns for fiber production. *Grain sorghums.* Annual grass. Most types in the U.S. are less than 5 ft. *(1.5 m)* tall. Broad, waxy-loomed leaves. Flowers are borne in dense panicles. Small, near round to broad-coned kernels weigh about ⅔ that of wheat. *Grass sorghums.* Annual grass. Primarily summer annuals. However, Johnsongrass is a perennial. Broad, coarse leaves. *Syrup sorghums.* Annual grass. Produce sweet, juicy stalks. *Broomcorn sorghums.* Annual grass. Produce tough stems and panicles suitable for use in brooms.	**S**orghum has been cultivated in Africa and Asia for 4,000 years. **M**ost sorghums cultivated in the U.S. are grown in the south central states. They thrive where temperatures are uniformly high throughout the growing season. Valuable in areas of uncertain rainfall due to their resistance to drought and wilting. They grow best in deep, fertile, sandy loams but can do well on well-drained heavy clays. More tolerant of alkali soils than most plants.	*Grain sorghums.* Nearly all of the grain sorghum grown in the U.S. is fed to livestock. In other countries, they are commonly used for human consumption (in some areas for the production of native beer.) **F**eeding value is about equal to corn. **S**tarch is sometimes extracted and used in the same manner as cornstarch. Also, processed for the production of grain and butyl alcohol. Green grain sorghum plants are poisonous due to the presence of prussic acid. Hence, they are not suited for pasture. *Grass sorghums.* Valuable for hay and silage, especially in the central and southern plains of the U.S. **G**reen grass sorghums may contain the toxin prussic acid. Curing renders this toxin harmless. *Syrup sorghums.* Primarily used for the production of sugar and syrup but can be used as hay or silage. *Broomcorn sorghums.* Very little fed to livestock. Most broomcorn sorghums are grown for fiber.	*Grain sorghums.* Grain sorghums varieties are classed in 7 agronomic groups: (1) KAFIR, (2) MILO, (3) FETERITA, (4) DURRA, (5) SHALLU, (6) KOALIANG, and (7) HEGARI. **M**ost grain varieties grown in the U.S. are crosses involving Kafir and Milo. **P**rocessing generally improves the utilization of grain sorghums. *Grass sorghums.* JOHNSONGRASS, SORGHUM-SUDANGRASS HYBRIDS, and SUDANGRASS are the primary grass sorghums. *Syrup sorghums.* SORGO is the primary syrup sorghum.
STARGRASS *Cynodon plectostachyus*	**S**preading perennial grass. **S**tout, fast-growing stolons form a dense turf.	**A**dapted to a more tropical climate than Bermudagrass and is more productive than Bermudagrass in dry areas. **C**ommonly found on dry lake beds.	**V**aluable pasture for dry areas.	
SUDANGRASS (See SORGHUM.)				
SWEET CLOVER *Melilotus* spp	**M**ost cultivated species of the legume are biennial but a few are annual. The first season's growth consists of one multibranched stem. Deep-set taproot. Trifoliate, toothed leaves. White or yellow flowers. Pods usually contain one seed, sometimes two, and are loosely attached so that seeds shatter as they mature.	**N**ative to temperate Europe and Asia. **G**rown primarily in the Corn Belt and central third of the U.S. **A**dapted to a wide range of soils and climatic conditions, but will not tolerate acid soils.	**C**ommonly used for pasture and silage. **F**irst-year growth provides substantial pasture in the southern areas. **F**airly good hay can be made if cut in the bud stage. However, it is hard to cure and is seldom harvested as hay. **B**est silage is made when sweet clover is harvested in the prebloom stage.	**R**ated as one of the best soil-improving crops. **C**ontains the antimetabolite coumarin, a substance that inhibits proper blood clotting and consequently causes hemorrhaging, commonly called *sweet clover disease*. Low-coumarin cultivars are available.
SWITCHGRASS *Panicum virgatum*	**E**rect, often purplish perennial panicgrass. Numerous scaly, creeping rootstocks.	**F**ound in all states east of the Mississippi River, in the southwestern U.S., south through Mexico into Central America.	**O**ccasionally cultivated for pasture and hay. It is more valuable as a hay crop.	**E**xcellent soil-binder. **B**ecomes woody and unpalatable at maturity.

(Continued)

TABLE 30-1 (Continued)

Feedstuff	Description	Place of Origin Geographical Adaptation Cultural Characteristics	Importance/Use	Comments
T				
TEFF *Eragrostis tef*	Leafy, quick-maturing annual. Stems can grow to 4 ft (*1.2 m*) in height.	Grows in dry areas with a short rainy season. Prefers heavy soils. Highly resistant to disease and pests but not to weed competition.	Cultivated primarily for its grain. Used also as an annual hay grass for arid areas. Elsewhere, it is too stemmy and unproductive for cultivation.	An economically important cereal grass in Africa.
THREE-AWN (NEEDLEGRASS, WIREGRASS, POVERTY GRASS) *Aristida* spp	Annual and perennial grasses. Depend on seeds for reproduction. Seeds characterized by 3-branched beard at the tip.	Widely distributed throughout the western U.S., especially the Southwest. Found primarily on dry, sandy soils. Common grasses on semidesert areas, plains, and lower elevations in the mountains.	In the Southwest, three-awns are considered to be good spring and summer forage before the seeds mature. The small annuals and a few perennials produce little leafage and are considered to be of little forage value. A large number of perennials are leafy and provide considerable forage. Once mature, three-awn is unpalatable.	Barbed seeds may cause trouble in grazing animals, resulting in sore mouths and inflaming the eyes, ears, and nostrils of livestock. Generally classified as an undesirable range species which appears when overgrazing is practiced.
TOBOSA *Hilaria mutica*	Erect, perennial very much resembling galleta. Spikelets are bearded at the base. Culms form a tough rhizomatous base.	Grows from western Texas to Arizona and into Mexico. Prefers fine, somewhat compact soils on open flats, swales, depressions, and, to a certain extent, foothills. Commonly found in areas subject to flooding in the rainy season.	When green and succulent, tobosa rates as good palatable forage, especially for cattle and horses.	Withstands heavy grazing. Of little value as winter forage.
TREFOIL *Lotus* spp	Deep rooted, fine stemmed leafy perennial legumes. Somewhat decumbent as single plants but erect in thick stands. Leaves contain 5 leaflets. Yellow, showy flowers. Spreading seed pods resemble a bird's foot.	Found in a variety of conditions.	High feeding value as a forage, especially as a late summer feed. Primarily used for pasture, either alone or in mixtures. Good hay can also be obtained.	BIG TREFOIL and BIRDSFOOT TREFOIL are the most important species.
TUMBLEGRASS *Schedonnardus paniculatus*	Low, tufted perennial. Stiff, slender, divergent spikes. Leaves crowded at the base. At maturity, the plant breaks away and rolls in the wind as a tumbleweed.	Found primarily in the prairies and plains of the U.S.	Represents a small portion of the forage of the Great Plains.	
V				
VELVET BEAN *Stizolobium* spp	Vigorous growing, summer annual legume. Vines can grow more than 25 ft (*7.6 m*) in length. Trifoliate leaves. Two types of pods: (1) dense, black, velvety pubescence, and (2) white or gray hairs. Pods generally contain 3 to 6 seeds per pod, depending on species. Numerous fleshy surface roots.	Probably originated in India. Adapted to warm climates. Most velvet beans are grown in the South Atlantic and Gulf Coast states. Intolerant of cold, wet soils.	Excellent source of winter pasture. Many dairymen prefer corn-velvetbean silage to corn silage alone. Meal can be fed to livestock effectively. Seldom used for hay because of handling difficulties.	Excellent green manure crop.
VETCH *Vicia* spp	Most species are viny annual legumes. Large flowers, spherical seeds, and elongated, somewhat compressed pods.	Of the approximately 150 species of vetch, about 15 are native to the U.S. Cultivated vetches are native to Europe and adjacent parts of Asia and Africa. Widely distributed throughout the temperate regions of the world. 75% of the vetch acreage in the U.S. is in Oklahoma, Arkansas, Texas, and Louisiana. Adapted to most types of soils. More tolerant of acid soils than the other legumes. Require cool temperature but mild winters. No species are drought-resistant.	All commercial species make good pasturage, hay, silage, and green manure. Can also be used for green feed and as cover crops.	The most important cultivated vetches of the U.S. are BIGFLOWER VETCH, COMMON VETCH, HAIRY VETCH, HUNGARIAN VETCH, and PURPLE VETCH. Reseeding (hard-seeded) common vetches have been developed through interspecific hybridization.

(Continued)

TABLE 30-1 (Continued)

Feedstuff	Description	Place of Origin Geographical Adaptation Cultural Characteristics	Importance/Use	Comments
W				
WATER HYACINTH (MILLION DOLLAR WEED) *Eichhornia crassipas*	Unattached, free floating water plant. Leaves above the water surface. Roots below the water surface.	Found in tropical regions where it multiplies rapidly and clogs lakes, rivers, and ponds. Difficult to eradicate.	Fresh water hyacinth is unpalatable due to prickly crystals. Boiled water hyacinth is commonly fed to swine in Asia. Hay is unpalatable to cattle unless 20% molasses is added (urea may also be added). For ensilage, it should be allowed to wilt in the shade for 48 hours before it is chopped and ensiled. Molasses should be added, and it is advisable to add urea and salt. Also, a combination of 4 parts water hyacinth to 1 part straw makes good silage. Plant juice can be used for beef protein-concentrate.	Easily harvested by nets.
WATERMELON (COCORICO) *Citrullus vulgaris*	Trailing, tendriled, scabrous monoecious annual vine. Large, deeply incised leaves. Fleshy, spherical to oval fruits range in weight from 8 to 50 lb (3.6 to 23 kg) or more, depending on the variety. Fruit consists of a firm outer rind, a layer of white inner rind flesh, and an interior edible pulp surrounding the seeds.	Believed to be native to Africa. Grows in sandy soils. Persists for a long time into the dry season.	Highly succulent feed. Can be used as a water source in dry areas. Seeds and seed hulls are also used for feed.	A popular human food.
WHEATGRASS *Agropyron spp*	The important species are hardy, erect, cool season perennial grasses. Either sod-forming or have a bunch type of growth. Some species have rhizomes. Seed heads resemble wheat heads.	About 150 species are widely distributed throughout the temperate regions of the world. Over 30 species are native to North America. In the U.S., native species are found in vast areas of the central and northern Great Plains, the intermountain region, and higher elevations of the Rocky Mountains. Thirteen species are found in Alaska. QUACKGRASS, a weed, is the predominant wheatgrass in the eastern U.S.	Most important of the grasses in the western U.S. for the production of highly nutritious early season forage. Also used extensively to control wind and water erosion.	The most important native wheatgrasses of the U.S. are WESTERN WHEAT GRASS (central and northern Great Plains), BLUEBUNCH WHEATGRASS (northern intermountain region), and SLENDER WHEATGRASS (widespread over mountains, foothills, and high plains). CRESTED WHEATGRASS is the most important species introduced into the U.S.
WILD-RYE GRASS *Elymus spp*	Most species are perennial bunch growers, but some form sods. Foliage is tall, coarse, and harsh. Many species have twisted or bearded flower heads.	Fairly large genus which is found in the north temperate zone. The western U.S. is the area containing the most species. In the western U.S., wild-rye grasses occur from the lower semidesert areas to the aspen and spruce belts. Some typically grow in bottomlands and meadows while others are found in brush and woodland areas.	Wild-ryes are coarse at maturity, but are of moderate to high palatability if harvested early. They are sometimes useful in mixtures and for the quick establishment of cover.	Highly susceptible to ergot fungus, a very toxic organism to livestock. The most useful wild-rye species of the U.S. are CANADA WILD-RYE (Plains and western states), GIANT WILD-RYE (western states), BLUE WILD RYE (western states), RUSSIAN WILD-RYE (northern Great Plains and intermountain states), and BASIN WILD-RYE (western states.).
WINTERFAT (WHITE SAGE) *Eurotia lanata*	Low, silvery-white shrub. Grows up to 15 ft (4.6 m). New stems are produced each year from the basal part of the previous year's stems and from the woody crown. Covered with a mat of silvery-white hair. Deep taproot with numerous lateral roots.	Widely distributed throughout the western U.S. and Canada. Frequently found in lower foothills, plains, and valley on dry soils moderately impregnated with salt or white alkali material. Drought resistant.	Grazed by all types of livestock. Considered to be good forage for goats and horses.	Overgrazing has dramatically reduced its population.
X, Y, Z				
YUCCA *Yucca spp*	Shrubs with cup-shaped flowers and stiff, sword-shaped leaves.	Confined to North and Central America, Bermuda, and the West Indies. In the U.S., most are found in the desert areas of the Southwest.	When finely chopped or shredded, yucca can be used as forage for cattle.	

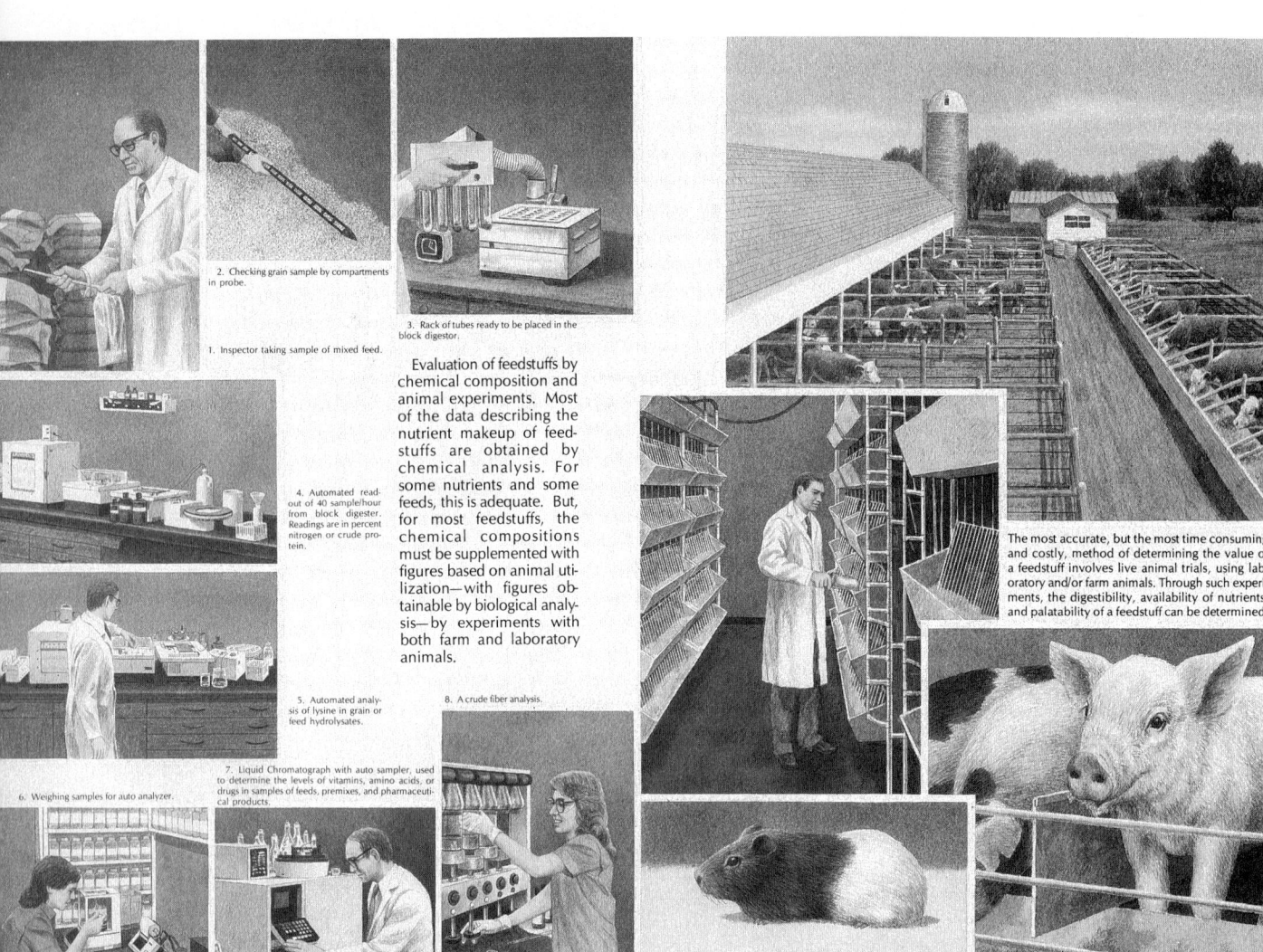

COMPOSITION OF FEEDS, above, is from an original by the noted artist, Tom Phillips (3333 17th Street, San Francisco, California 94110), prepared especially for this book. It portrays the artist's conception of the evaluation of feedstuffs (1) by chemical composition (left side), and (2) by animal experiments (right side).

SECTION V

COMPOSITION OF FEEDS

Both livestock producers and nutritionists have need for accurate and up-to-date composition of feedstuffs in order (1) to formulate rations for maximum production and net returns, and (2) to predict the production response of animals when they are fed rations of a given composition. In recognition of this need and its importance, compositions of a great array of feeds are presented in Section V.

COMPOSITION OF FEEDS

Contents	Page
Feed Names	708
Moisture Content of Feeds	708
Carotene	708
Pertinent Information About Data	708

In addition to the discussion that follows pertaining to composition of feeds, the reader is referred to Chapter 15, Feed Analysis/Feed Evaluation, and Chapter 18, Feeding Standards/Ration Formulation.

Both nutritionists and livestock producers should have access to accurate and up-to-date composition of feedstuffs in order to formulate rations for maximum production and net returns. The ultimate goal of feedstuff analysis, and the reason for feed composition tables, is to be able to predict the productive responses of animals when they are fed rations of a given composition. In recognition of this need and its importance, the authors spared no time or expense in compiling the feed composition tables presented in this section. At the outset, a survey of the industry was made in order to determine what kind of feed composition tables would be most useful, in both format and content. Secondly, it was decided to utilize, to the extent available, the feed compositions which, for many years, were compiled by Lorin Harris, Utah State University, now carried forward by the USDA, National Agricultural Library, Feed Composition Bank. These data were augmented by the authors with feed compositions from the National Academy of Sciences, NRC, and from experimental reports, industries, and other reliable sources.

To facilitate quick and easy use, the feeds in Table V are classified and separated into the following subtables:

1. **Table V-1A, Energy Feeds.** This includes feeds which are high in energy and low in fiber (under 18%), and which generally contain less than 20% protein.

2. **Table V-1B, Protein Feeds.** This includes feeds that contain more than 20% protein or protein equivalent.

3. **Table V-1C, Dry Forages.** This includes feeds which are bulky, low in weight per unit volume, and relatively low in energy, and which in the dry state contain more than 18% crude fiber.

4. **Table V-1D, Silages and Haylages.** Silage (ensilage) is fermentable high-moisture forage stored under anaerobic conditions in a silo, consisting of either green crops or crops to which moisture has been added, chopped when stored and containing 65 to 70% moisture.

Haylages are low-moisture silages, made from grasses and/or legumes that are wilted to 40 to 55% moisture content before ensiling.

5. **Table V-1E, Pasture and Range Plants.** This includes grass, browse, and other plants that are harvested by grazing animals.

6. **Table V-1F, Mineral Supplements.** This includes rich natural and synthetic sources of inorganic elements needed to perform certain essential body functions.

7. **Table V-1G, Vitamin Supplements.** This includes rich synthetic or natural feed sources of one or more complex compounds, called vitamins, that are required by animals in minute amounts for normal growth, production, reproduction, and/or health.

8. **Table V-2, Amino Acids.** This gives the known amino acid composition of certain feeds.

9. **Table V-3, Apparent Ileal Digestibility of Crude Protein and Essential Amino Acids in Feedstuffs for swine, and Digestible Amino Acid Recommendations for Swine Feed Formulation.** The protein requirement of the pig is primarily a requirement for essential amino acids. Currently, apparent ileal digestibility data give the best practical assessment of the availability of amino acids in various feed ingredients.

10. **Table V-4, True Digestibility of Essential Amino Acids for Poultry, and True Digestibility Amino Acid Recommendations for Poultry Feed Formulation.** Like all monogastric animals, poultry do not require protein as such; instead, they need well-defined amounts of available amino acids to perform at a desired level. Currently, digestibility assays give the best assessment of the availability of amino acids in various feed ingredients.

(Also see Chapter 24, Section on "Protein • True digestibility of amino acids.")

Some feeds fit the criteria of more than one of the above classes. For example, whole soybeans are used as both an energy feed and a protein feed; hence, they are listed in both Table V-1A and Table V-1B.

In Tables V-1A to V-1E, and V-1G, covering 6 of the respective feedstuff classifications indicated above, values for each feed are presented in tabular form on 4 pages (2 double-page spreads), with one page devoted to each of the following categories:

Left-hand, proximate analysis
Right-hand, energy
Left-hand, minerals
Right-hand, vitamins

...me should conjure up the same meaning
... e it, and it should provide helpful infor-
... the guiding philosophy of the authors
when choosing the names given in the Feed Composition
Tables. Genus and species—latin names—are also included.
To facilitate worldwide usage, the International Feed Number of each feed is given. To the extent possible, consideration was also given to source (or parent material), variety or kind, stage of maturity, processing, part eaten, and grade.

Where feeds are known by more than one name, cross-referencing was used.

MOISTURE CONTENT OF FEEDS

It is necessary to know the moisture content of feeds in ration formulation and buying. Usually, the composition of a feed is expressed according to one or more of the following bases:

1. **As-fed; A-F (wet, fresh).** This refers to feed as normally fed to animals. As-fed may range from near 0% to 100% dry matter.

2. **Air-dry (approximately 90% dry matter).** This refers to feed that is dried by means of natural air movement, usually in the open. It may either be an actual or an assumed dry matter content; the latter is approximately 90%. Most feeds are fed in an air-dry state.

3. **Moisture-free; M-F (oven-dry, 100% dry matter).** This refers to a sample of feed that has been dried in an oven at 221°F until all the moisture has been removed.

Where available, feed compositions are presented on both As-Fed (A-F) and Moisture-Free (M-F) bases. Formulas for adjusting moisture content from moisture-free to as-fed, or as-fed to moisture-free, are given in Chapter 18, in the section headed "Adjusting Moisture Content."

CAROTENE

Where carotene has been converted to vitamin A, the conversion rate of the rat has been used as the standard value, with 1 mg of β-carotene equal to 1,667 IU of vitamin A.

PERTINENT INFORMATION ABOUT DATA

The information which follows is pertinent to the feed composition tables presented in this section.

• **Variations in composition**—Feeds vary in their composition. Thus, actual analysis of a feedstuff should be obtained and used whenever possible, especially where a large lot of feed from one source is involved. Many times, however, it is either impossible to determine actual compositions or there is insufficient time to obtain such analysis. Under such circumstances, tabulated data may be the only information available.

• **Feed compositions change**—Feed compositions change over a period of time, primarily due to (1) the introduction of new varieties, and (2) modifications in the manufacturing process from which by-products evolve.

• **Biological value**—The response of animals when fed a feed is termed the biological value, which is a function of its chemical composition and the ability of the animal to derive useful nutrient value from the feed. The latter relates to the digestibility, or availability, of the nutrients in the feed. Thus, soft coal and shelled corn may have the same gross energy value in a bomb calorimeter but markedly different useful energy values (TDN, digestible energy, metabolizable energy, and net energy) when consumed by an animal. Biological tests of feeds are more laborious and costly than chemical analysis, but they are much more accurate in predicting the response of animals to a feed.

• **Where information is not available**—Where information is not available or reasonable estimates could not be made, no values are shown. Hopefully, such information will become available in the future.

• **Calculated on a dry matter (DM) basis**—All data were calculated on a 100% dry matter basis (moisture-free), then converted to an as-fed basis by multiplying the decimal equivalent of the DM content times the compositional value shown in the table.

• **Fiber**—Four values relating to dietary fiber are given in the feed composition tables—crude fiber, neutral detergent fiber (NDF), acid detergent fiber (ADF), and lignin.

Crude fiber, methods for the determination of which were developed more than 100 years ago, is declining as a measure of low digestible material in the more fibrous feeds. The newer method of forage analysis, developed by Van Soest and associates of the U.S. Department of Agriculture, separates feed dry matter into two fractions: a neutral detergent fibrous fraction; and an acid detergent fibrous fraction. Also the amount of lignin in the ADF may be determined. (See Chapter 15, Fig. 15–4.)

1. **Crude fiber (CF).** This fraction is an indicator of the relative indigestibility and bulkiness of the sample. It is the residue that remains after boiling a feed in a weak acid, and then in a weak alkali, in an attempt to imitate the process that occurs in the digestive tract. This procedure is based on the supposition that carbohydrates which are readily dissolved also will be readily digested by animals, and that those not soluble under such conditions are not readily digested. Unfortunately, the treatment dissolves much of the lignin, a nondigestible component. Hence, crude fiber is only an approximation of the indigestible material in feedstuffs. Nevertheless, it is a rough indicator of the energy value of feeds. Also, the crude fiber value is needed for the computation of TDN.

(Also see Chapter 15, section on "Crude Fiber [CF].")

2. **Neutral detergent fiber (NDF).** This is the fraction of the feed which is not soluble in neutral detergent. It consists of plant cell walls, including lignin, cellulose, and hemicellulose. NDF is closely related to feed intake because it contains all the fiber components that occupy space in the rumen and are slowly digested. The lower the NDF, the more forage the animal will eat; hence, a low percentage of NDF is desirable.

(Also see Chapter 15, section on "Neutral Detergent Fiber [NDF] and Acid Detergent Fiber [ADF].")

3. **Acid detergent fiber (ADF).** This is the fraction of the feed which is not soluble in acid detergent. It consists of cellulose (digestible) and lignin (indigestible). ADF is an indicator of forage digestibility because it contains a high proportion of lignin which is the indigestible fiber fraction. The lower the ADF, the more feed an animal can digest; hence, a low percentage of ADF is desirable.

(Also see Chapter 15, section on "Neutral Detergent Fiber [NDF] and Acid Detergent Fiber [ADF].")

4. **Lignin.** This fraction is essentially indigestible by all animals and is the substance that limits the availability of cellulose carbohydrates in the plant cell wall to rumen bacteria.

The acid detergent fiber procedure is used as a preparatory step in determining the lignin content of a forage sample. Hemicellulose is solubilized during this procedure, while the lignocellulose fraction of the feed remains insoluble. Cellulose is then separated from lignin by the addition of sulfuric acid. Only lignin and acid-insoluble ash remain upon completion of this step. This residue is then ashed, and the difference of the weights before and after ashing yields the amount of lignin present in the feed.

- **Nitrogen-free extract**—The nitrogen-free extract was calculated with mean data as: mean nitrogen-free extract (%) = 100 − % ash − % crude fiber − % ether extract − % protein.

- **Protein values**—Both crude protein and digestible protein values are given. Crude protein is determined by finding the nitrogen content and multiplying the result by 6.25. The nitrogen content of proteins averages about 16% (100 ÷ 16 = 6.25).

In addition to the compositions given in the Feed Composition Tables in this section, bypass protein values of selected feeds are given in Chapter 11, Table 11–1 of this book.

- **Ruminant values**—The ruminant values represent a pooling of cattle, sheep, and goat data.

- **Energy**—Many of the energy values given in the feed composition tables were derived from complex formulas developed by L. E. Harris and other animal scientists.

The following four measures of energy are shown:

1. **Total digestible nutrients (TDN).** This value is given because there are more of them, and because it has been the standard method of expressing the energy value of feeds for many years. However, the following disadvantages are inherent in the TDN system: (a) Only digestive losses are considered—it does not take into account other important losses, such as those in the urine, gases, and increased heat production; (b) there is a poor relationship between crude fiber and NFE digestibility in certain feeds; and (c) it overestimates roughages in relation to concentrates when animals are fed for high rates of production, due to the higher heat loss per pound of TDN in high-fiber feeds.

2. **Digestible energy (DE).** Digestible energy is that portion of the gross energy in a feed that is not excreted in the feces. It is roughly comparable to TDN.

For most animals, digestible energy is relatively easy to determine. With poultry, however, true digestibility is very difficult to measure because undigested residues and urinary wastes are excreted together.

3. **Metabolizable energy (ME).** Metabolizable energy represents that portion of the gross energy that is not lost in the feces, urine, and gas (mainly methane). It does not take into account the energy lost as heat, commonly called heat increment. As a result, it overevaluates roughages compared with concentrates, as do TDN and DE.

Metabolizable energy is considered to be the most accurate evaluation of the energy of feedstuffs for the scientific formulation of poultry feeds.

4. **Net energy (NE).** Net energy represents the energy fraction in a feed that is left after the fecal, urinary, gas, and heat losses are deducted from the GE. Because of its greater accuracy, net energy is being used increasingly in ration formulations, especially in computerized formulations for large operations. However, net energy is difficult and expensive to determine.

Two systems of net energy evaluation are presently used: (a) net energy for maintenance (NE_m) and net energy for gain (NE_g), and (b) net energy for lactation (NE_{lc}).

Note to dairy producers and nutritionists: In *Nutrient Requirements of Dairy Cattle*, Sixth Revised Edition, 1988, Table 7–1, Composition of Feeds Commonly Used in Dairy Cattle Diets on a 100% Dry Matter Basis, the NRC committee assumed an average decrease of 4% per unit of dry matter intake above maintenance in calculating NE_{lc} values for feed ingredients, or an average discount of 8% based on their assumption that lactating cows are fed at 3X maintenance. For the convenience of those dairy producers and dairy nutritionists who wish to use these values, the authors selected from *Nutrient Requirements of Dairy Cattle*, Sixth Revised Edition, 1988, Table 7–1, the feeds most commonly used in dairy cattle rations and reproduced them in Chapter 20, Feeding Dairy Cattle, pages 436, 437, and 438, Table 20–7, in a three-page spread.

- **Minerals**—The level of minerals in forages is largely determined by the mineral content of the soil on which the feeds are grown. Calcium, phosphorus, iodine, and selenium are well-known examples of soil nutrient—plant nutrient relationships.

- **Vitamins**—Generally speaking, it is unwise to rely on harvested feeds as a source of carotene (vitamin A value), unless the forage being fed is fresh (pasture or green chop) or of a good green color and not over a year old.

The authors are very grateful to Lorin E. Harris, Ph.D., and Clyde R. Richards, Ph.D., Utah State University, Logan, for their interest and invaluable assistance in preparing the Feed Composition Tables for this book. Also, in the preparation of Table V–1F, Composition of Mineral Supplements, the authors acknowledge with appreciation the review and added values provided by International Minerals and Chemical Corporation, Northbrook, Illinois; and R. F. Klay, Ph.D., Research Department, Moorman Mfg. Co., Quincy, Illinois.

TABLE V-1A ENERGY FEEDS, COMPOSITION OF FEEDS, DATA EXPRESSED AS-FED

Entry Number	Feed Name Description	International Feed Number	Proximate Analysis									Digestible Protein		
			Dry Matter	Ash	Crude Fiber	Neutral Det. Fib. (NDF)	Acid Det. Fib. (ADF)	Lignin	Ether Extract (Fat)	N-Free Extract	Crude Protein	Ruminant	Swine	Horse
			%	%	%	%	%	%	%	%	%	%	%	%
1	ALMOND, HULLS	4-00-359	90	6.1	13.5	28.8	25.2	—	2.9	63.8	4.1	0.0	2.0	2.4
	ANIMAL													
2	FAT, HYDROLYZED	4-00-376	99	—	—	—	—	—	98.4	—	—	—	—	—
3	TALLOW	4-08-127	97	0.1	—	—	—	—	96.8	—	1.5	-1.6	-0.4	0.3
4	ANIMAL–POULTRY, FAT	4-00-409	99	—	—	—	—	—	99.1	—	—	—	—	—
	BARLEY													
5	GRAIN, ALL ANALYSES	4-00-549	88	2.4	5.0	16.8	10.7	1.5	1.7	67.7	11.7	8.8	9.6	9.6
6	GRAIN, PACIFIC COAST	4-07-939	89	2.5	6.5	—	—	—	2.0	68.2	9.5	7.1	6.9	6.7
7	MALT SPROUTS, DEHY	4-00-545	93	5.6	14.2	—	—	—	1.4	48.8	22.9	19.2	21.5	17.2
	BEAN													
8	SEEDS, KIDNEY	5-00-600	89	3.6	4.1	—	—	—	1.2	57.8	21.8	14.6	20.6	16.5
9	SEEDS, NAVY	5-00-623	89	4.0	4.4	—	—	—	1.4	56.7	22.9	20.2	21.6	17.6
10	SEEDS, PINTO	5-00-624	90	4.3	4.0	—	—	—	1.3	57.9	22.7	17.2	21.4	17.3
11	BEAN, LIMA, SEEDS	5-00-613	90	4.1	4.5	—	—	—	1.3	58.9	20.7	16.7	19.4	15.1
12	BEAN, MUNG, SEEDS	5-08-185	90	3.8	3.9	—	—	—	1.3	57.1	23.9	19.6	22.6	18.7
13	BEAN, TEPARY, SEEDS	5-08-349	90	4.2	3.4	—	—	—	1.4	59.3	22.2	18.0	20.9	16.7
	BEET, SUGAR													
14	MOLASSES, MORE THAN 48% INVERT SUGAR, MORE THAN 79.5 DEGREES BRIX	4-00-668	78	8.9	—	—	—	—	0.2	62.2	6.6	3.6	4.5	4.5
15	PULP, DEHY	4-00-669	91	4.8	18.2	53.6	26.3	4.5	0.5	58.4	8.8	4.3	3.6	6.2
16	PULP WITH MOLASSES, DEHY	4-00-672	92	5.7	15.2	—	24.5	2.4	0.6	61.1	9.3	6.1	2.3	6.5
17	BREWERS GRAINS, DEHY	5-02-141	92	3.6	13.0	38.7	23.9	4.6	6.6	41.6	27.3	20.1	21.8	21.0
18	BUCKWHEAT, COMMON, GRAIN	4-00-994	88	2.1	10.6	—	14.9	—	2.4	61.5	11.1	8.0	8.5	7.2
19	CARROT, ROOTS, FRESH	4-01-145	11	1.0	1.1	1.0	0.9	—	0.2	8.1	1.2	0.8	0.8	0.8
20	CASSAVA, TUBERS, FRESH	4-01-150	32	1.3	1.5	—	—	—	0.3	28.2	1.2	-0.1	0.5	0.6
	CITRUS													
21	PULP WITHOUT FINES, DEHY (DRIED CITRUS PULP)	4-01-237	91	6.0	11.6	20.9	20.0	—	3.4	63.9	6.1	2.7	3.8	4.0
22	SYRUP (MOLASSES)	4-01-241	67	5.1	—	—	—	—	0.2	55.7	5.8	2.0	3.9	3.9
	CORN, DENT YELLOW													
23	GRAIN, ALL ANALYSES	4-02-935	88	1.3	2.3	—	3.8	—	3.6	71.0	9.9	8.1	7.3	7.0
24	GRAIN, GRADE 2, 54 lb/bu (69.5 kg/hl)	4-02-931	87	1.2	2.1	—	—	—	4.0	71.3	8.9	6.8	7.0	6.2
25	GRAIN, HIGH MOISTURE	4-20-770	77	1.2	2.1	17.5	3.8	1.3	3.3	61.9	8.1	4.9	5.9	5.7
26	DISTILLERS GRAINS, DEHY	5-02-842	93	2.2	11.5	40.0	15.8	—	8.9	43.1	27.8	20.1	19.4	22.7
27	DISTILLERS SOLUBLES, DEHY	5-28-237	93	7.2	4.6	21.4	6.5	—	8.6	45.2	27.4	21.5	20.0	22.3
28	EARS, GROUND (CORN AND COB MEAL)	4-28-238	87	1.7	8.2	24.4	9.6	—	3.2	65.7	7.8	4.7	5.6	5.4
29	GLUTEN, MEAL, 60% PROTEIN	5-28-242	90	1.7	1.8	12.6	4.5	—	2.1	23.7	60.8	54.5	-1.3	-0.4
30	GLUTEN FEED	5-28-243	90	6.6	8.7	40.5	10.8	—	2.1	49.4	23.0	19.8	18.2	17.7
31	GRITS (HOMINY GRITS)	4-03-011	90	2.8	4.8	49.5	11.7	—	6.5	65.8	10.3	7.3	7.6	7.3
32	OIL	4-16-450	99	—	—	—	—	—	99.0	—	—	—	—	—
33	CORN, OPAQUE 2, HIGH LYSINE, GRAIN	4-11-445	90	1.6	3.0	—	—	—	4.4	71.0	10.1	6.3	7.4	7.2
34	CORN, SWEET, CANNERY RESIDUE, FRESH	2-02-975	77	3.5	17.0	—	22.3	—	1.9	47.6	6.8	3.9	4.1	—
35	COTTON, SEEDS	5-13-749	91	4.7	19.4	40.0	30.9	—	20.4	24.5	21.8	17.7	20.5	16.2
	DISTILLERS PRODUCTS (ALSO SEE CORN; RYE; SORGHUM; WHEAT)													
36	GRAINS, DEHY	5-02-144	93	1.5	12.8	—	—	—	7.4	43.5	27.3	18.9	26.0	22.3
37	SOLUBLES, DEHY	5-02-147	92	6.2	3.4	—	—	—	8.9	44.8	28.8	24.0	27.4	24.0
	FATS AND OILS													
38	FAT, ANIMAL, HYDROLYZED	4-00-376	99	—	—	—	—	—	98.4	—	—	—	—	—
39	FAT, ANIMAL-POULTRY	4-00-409	99	—	—	—	—	—	99.1	—	—	—	—	—
40	FLAX, COMMON, SEEDS	5-02-042	96	5.0	6.3	—	—	—	35.9	30.7	17.8	13.9	16.5	11.2
41	GARBAGE, HOTEL AND RESTAURANT, BOILED, WET	4-07-865	23	1.3	0.7	—	—	—	5.4	11.9	3.6	2.5	2.8	2.6
42	HOMINY GRITS (CORN, DENT YELLOW, GRITS)	4-03-011	90	2.8	4.8	—	—	—	6.5	65.8	10.3	7.3	7.6	7.3
43	KAFIR SORGHUM, GRAIN	4-04-428	89	1.5	2.0	—	1.8	—	2.8	72.0	10.8	6.6	8.3	7.7
44	KELP (SEAWEED), WHOLE, DEHY	1-08-073	91	35.0	6.5	—	—	—	0.5	42.4	6.5	2.9	2.2	3.4
45	LARD (FAT), SWINE	4-04-790	99	—	—	—	—	—	99.3	—	—	—	—	—
46	MANGEL, BEET, ROOTS, FRESH	4-00-637	11	1.1	0.8	—	—	—	0.1	7.7	1.3	0.9	1.0	0.9
47	MANURE, CATTLE, WITHOUT BEDDING, DEHY	1-01-190	93	17.9	30.7	—	43.8	—	2.5	30.1	12.0	7.7	6.9	7.6
48	MILLET, GRAIN	4-03-098	90	2.7	5.8	—	—	—	4.0	65.2	12.1	6.8	8.8	8.7
49	MILLET, FOXTAIL, GRAIN	4-03-102	89	3.4	7.4	—	—	—	4.1	63.0	11.4	7.5	8.6	8.2
50	MILO SORGHUM, GRAIN	4-04-444	89	1.6	2.2	20.6	4.7	—	2.8	71.9	10.1	6.8	7.2	7.1
	MOLASSES AND SYRUP													
51	BEET, SUGAR, MOLASSES, MORE THAN 48% INVERT SUGAR, MORE THAN 79.5 DEGREES BRIX	4-00-668	78	8.9	—	—	—	—	0.2	62.2	6.6	3.6	4.5	4.5
52	CITRUS, SYRUP (CITRUS MOLASSES)	4-01-241	67	5.1	—	—	—	—	0.2	55.7	5.8	2.0	3.9	3.9
53	SUGAR CANE, MOLASSES, DEHY	4-04-695	94	12.5	6.3	—	—	—	0.9	65.0	9.7	5.3	7.0	6.8
54	SUGAR CANE, MOLASSES, MORE THAN 46% INVERT SUGAR, MORE THAN 79.5 DEGREES BRIX (BLACKSTRAP)	4-04-696	74	9.8	0.4	—	0.3	0.2	0.2	59.7	4.3	0.6	1.3	2.7
55	WOOD, MOLASSES	4-05-502	62	4.1	0.5	—	1.2	—	0.3	56.6	0.6	-1.3	-0.6	-0.1
	OATS													
56	GRAIN, ALL ANALYSES	4-03-309	89	3.1	10.7	26.4	14.2	2.7	4.7	58.9	11.9	9.2	9.7	9.1
57	CEREAL BY-PRODUCT (FEEDING OAT MEAL; OAT MIDDLINGS)	4-03-303	91	2.3	3.6	—	—	—	6.4	63.7	14.8	10.9	11.6	10.8
58	GROATS	4-03-331	90	2.1	2.5	—	—	—	6.2	63.0	15.8	11.1	13.7	11.6
59	PEA, SPLIT SEED BY-PRODUCT (PEA FEED; PEA MEAL)	1-08-478	90	3.5	23.7	—	—	—	1.4	43.7	17.7	14.5	12.1	12.0
60	PEA, FIELD, SEEDS	5-08-481	91	2.9	5.9	—	—	—	1.3	57.8	23.2	19.9	21.8	17.7
61	PEARL MILLET, GRAIN	4-03-118	90	2.2	3.7	—	—	—	4.3	67.2	13.0	8.9	7.9	9.4
62	PINEAPPLE, CANNERY RESIDUE, DEHY (PINEAPPLE BRAN)	4-03-722	87	3.0	18.2	63.5	32.2	—	1.3	60.5	4.0	0.7	2.0	2.4
	POTATO													
63	TUBERS, FRESH	4-03-787	24	1.1	0.6	—	—	—	0.1	19.5	2.2	1.4	0.7	1.5
64	TUBERS, BOILED	4-03-784	24	1.3	0.7	—	—	—	0.1	19.6	2.2	0.3	1.5	1.5
	RICE													
65	GRAIN, GROUND (GROUND ROUGH RICE; GROUND PADDY RICE)	4-03-938	89	5.3	8.6	—	—	—	1.6	65.9	7.5	4.0	5.1	5.1
66	BRAN WITH GERMS (RICE BRAN)	4-03-928	91	11.3	11.9	28.0	25.7	3.6	13.5	41.0	13.0	8.6	9.5	9.4
67	GROATS, POLISHED (RICE, POLISHED)	4-03-942	89	0.5	0.4	14.2	0.9	—	0.5	80.3	7.0	3.6	5.9	4.7
68	GROATS (RICE, BROWN)	4-03-936	88	1.0	0.8	—	—	—	1.8	77.2	7.4	3.2	5.1	5.1
69	POLISHINGS	4-03-943	90	7.6	3.2	—	3.6	—	12.6	54.9	12.0	8.6	10.1	8.6

Composition of Feeds

ENERGY FEEDS

Entry Number	TDN			Digestible Energy						Metabolizable Energy								Net Energy					
	Ruminant	Swine	Horse	Ruminant		Swine		Horse		Ruminant		Swine		Poultry ME_n		Horse		Ruminant NE_m		Ruminant NE_g		Lactating Cows NE_{lc}	
	%	%	%	Mcal		kcal		Mcal		Mcal		kcal		kcal		Mcal		Mcal		Mcal		Mcal	
				lb	kg	lb	kg	lb	kg	lb	kg	lb	kg	lb	kg	lb	kg	lb	kg	lb	kg	lb	kg
1	66	73	45	1.34	2.95	1457	3213	0.86	1.89	1.15	2.54	1375	3032	—	—	0.70	1.55	0.77	1.70	0.51	1.12	0.66	1.46
2	223	209	—	4.46	9.84	4144	9135	—	—	4.31	9.49	3834	8452	3685	8125	—	—	2.95	6.49	2.23	4.91	2.43	5.35
3	203	—	—	4.07	8.97	—	—	—	—	3.91	8.62	—	—	3304	7285	—	—	2.60	5.73	1.95	4.29	2.21	4.87
4	188	196	—	3.57	7.87	3750	8267	—	—	3.40	7.51	3617	7973	3482	7677	—	—	2.65	5.85	1.99	4.38	2.03	4.48
5	75	70	73	1.55	3.42	1396	3078	1.31	2.88	1.17	2.57	1331	2934	1180	2602	1.07	2.36	0.78	1.73	0.52	1.16	0.82	1.81
6	75	70	66	1.51	3.32	1405	3097	1.20	2.64	1.34	2.96	1324	2918	1171	2582	0.98	2.16	0.81	1.77	0.54	1.20	0.76	1.67
7	66	61	—	1.31	2.89	1219	2688	—	—	1.13	2.50	1046	2307	722	1592	—	—	0.69	1.52	0.43	0.95	0.67	1.48
8	73	79	—	1.46	3.22	1584	3493	—	—	1.29	2.85	1388	3059	—	—	—	—	—	—	—	—	—	—
9	76	78	—	1.53	3.37	1737	3830	—	—	1.36	3.00	1531	3376	1052	2320	—	—	0.83	1.83	0.56	1.24	0.78	1.72
10	74	78	—	1.47	3.25	1569	3460	—	—	1.30	2.87	1372	3025	—	—	—	—	0.81	1.79	0.54	1.20	0.77	1.69
11	78	78	—	1.55	3.42	1556	3430	—	—	1.39	3.05	1360	2999	—	—	—	—	—	—	—	—	—	—
12	79	81	—	1.53	3.37	1612	3554	—	—	1.36	3.00	1412	3113	—	—	—	—	0.82	1.80	0.55	1.21	0.77	1.70
13	69	80	—	1.37	3.02	1606	3542	—	—	1.20	2.65	1406	3100	—	—	—	—	—	—	—	—	—	—
14	61	57	—	1.20	2.64	1130	2491	—	—	1.04	2.29	1061	2338	875	1929	—	—	0.70	1.54	0.47	1.04	0.65	1.43
15	67	67	59	1.31	2.89	1305	2878	1.09	2.40	1.10	2.43	1225	2700	294	648	0.89	1.97	0.73	1.60	0.47	1.03	0.70	1.55
16	69	70	62	1.39	3.05	1390	3064	1.13	2.50	1.21	2.67	1307	2881	299	660	0.93	2.05	0.73	1.61	0.47	1.04	0.71	1.55
17	65	66	48	1.25	2.76	1045	2303	—	—	1.01	2.22	1038	2288	1047	2308	—	—	0.64	1.41	0.39	0.86	0.67	1.48
18	63	69	62	1.26	2.77	1377	3036	1.13	2.50	1.09	2.40	1297	2859	1200	2645	0.93	2.05	0.65	1.43	0.41	0.90	0.63	1.40
19	10	10	8	0.19	0.43	208	458	0.14	0.31	0.17	0.38	195	430	208	458	0.12	0.26	0.10	0.23	0.07	0.16	0.10	0.22
20	26	26	24	0.51	1.13	511	1127	0.43	0.95	0.45	0.99	482	1063	—	—	0.36	0.78	0.28	0.61	0.18	0.41	0.26	0.58
21	75	46	48	1.41	3.10	1386	3055	—	—	1.12	2.46	1097	2420	606	1336	—	—	0.74	1.63	0.48	1.06	0.77	1.71
22	51	54	—	1.01	2.22	1084	2390	—	—	0.86	1.89	1022	2253	—	—	—	—	0.57	1.26	0.38	0.83	0.52	1.15
23	80	79	64	1.63	3.59	1514	3338	1.17	2.57	1.26	2.78	1472	3246	1523	3359	0.96	2.11	0.86	1.90	0.59	1.31	0.82	1.81
24	80	81	61	1.57	3.47	1586	3498	1.11	2.45	1.45	3.19	1526	3365	1567	3456	0.91	2.01	0.89	1.96	0.61	1.35	0.84	1.85
25	71	68	54	1.41	3.11	1354	2984	0.99	2.18	1.27	2.80	1282	2826	—	—	0.81	1.79	0.79	1.75	0.55	1.22	0.74	1.63
26	81	65	—	1.54	3.41	1246	2746	—	—	1.28	2.82	1196	2636	894	1970	—	—	0.87	1.92	0.59	1.30	0.84	1.86
27	80	78	—	1.55	3.42	1474	3250	—	—	1.34	2.96	1419	3128	1324	2919	—	—	0.92	2.03	0.63	1.40	0.86	1.89
28	72	69	56	1.45	3.20	1410	3109	1.03	2.28	1.26	2.78	1260	2779	1238	2730	0.85	1.87	0.86	1.90	0.60	1.31	0.76	1.68
29	—	—	—	—	—	—	—	—	—	—	—	—	—	3370	7431	—	—	—	—	—	—	—	—
30	75	75	—	1.44	3.17	1375	3031	—	—	1.21	2.67	1037	2287	784	1729	—	—	0.82	1.80	0.55	1.21	0.78	1.72
31	84	82	49	1.69	3.73	1620	3571	0.91	2.01	1.29	2.85	1533	3381	1313	2894	0.75	1.65	0.88	1.95	0.61	1.34	0.91	2.00
32	203	208	—	3.57	7.86	3421	7543	—	—	3.40	7.50	3438	7579	3917	8635	—	—	2.65	5.84	1.99	4.38	2.20	4.86
33	80	79	61	1.53	3.38	1643	3623	1.12	2.47	1.37	3.01	1557	3432	1527	3367	0.92	2.03	0.81	1.78	0.54	1.20	0.76	1.68
34	54	—	42	1.05	2.31	—	—	0.79	1.74	0.90	1.99	—	—	—	—	0.65	1.43	0.57	1.26	0.36	0.79	0.56	1.23
35	87	71	—	1.57	3.46	1430	3152	—	—	1.39	3.07	1243	2739	—	—	—	—	0.97	2.13	0.67	1.49	0.91	2.01
36	78	82	—	1.61	3.54	1637	3609	—	—	1.43	3.16	1433	3160	1316	2901	—	—	0.93	2.06	0.65	1.43	0.87	1.92
37	68	75	31	1.39	3.07	1493	3291	0.62	1.37	1.22	2.69	1409	3107	—	—	0.51	1.12	0.77	1.70	0.51	1.12	0.74	1.63
38	223	209	—	4.46	9.84	4144	9135	—	—	4.31	9.49	3834	8452	3685	8125	—	—	2.95	6.49	2.23	4.91	2.43	5.35
39	188	196	—	3.57	7.87	3750	8267	—	—	3.40	7.51	3617	7973	3482	7677	—	—	2.65	5.85	1.99	4.38	2.03	4.48
40	—	—	—	—	—	—	—	—	—	—	—	—	—	—	—	—	—	—	—	—	—	—	—
41	20	25	—	0.42	0.92	501	1105	—	—	0.38	0.83	478	1053	—	—	—	—	0.24	0.53	0.17	0.37	0.22	0.49
42	84	82	49	1.69	3.73	1620	3571	0.91	2.01	1.29	2.85	1533	3381	1313	2894	0.75	1.65	0.88	1.95	0.81	1.34	0.91	2.00
43	75	81	70	1.47	3.24	1521	3354	1.26	2.77	1.30	2.87	1437	3169	1529	3372	1.03	2.27	0.79	1.74	0.53	1.16	0.75	1.64
44	29	—	—	0.58	1.27	—	—	—	—	0.40	0.87	—	—	—	—	—	—	—	—	—	—	—	—
45	—	—	—	—	—	3669	8089	—	—	—	—	3537	7798	3936	8677	—	—	—	—	—	—	—	—
46	9	9	8	0.18	0.39	181	399	0.14	0.32	0.15	0.34	171	377	—	—	0.12	0.26	0.09	0.21	0.06	0.14	0.09	0.20
47	44	—	25	0.88	1.94	1281	2825	0.53	1.17	0.70	1.54	1222	2695	—	—	0.43	0.96	0.36	0.79	0.13	0.29	0.43	0.96
48	61	66	61	1.54	3.39	1315	2900	1.11	2.46	1.17	2.58	1224	2697	1441	3178	0.91	2.01	0.86	1.90	0.59	1.29	0.81	1.79
49	76	72	57	1.45	3.19	1436	3167	1.04	2.30	1.28	2.82	1355	2988	—	—	0.86	1.89	0.75	1.65	0.49	1.09	0.71	1.58
50	76	77	68	1.41	3.12	1520	3350	1.23	2.71	1.13	2.49	1474	3250	1467	3234	1.01	2.23	0.75	1.66	0.50	1.09	0.74	1.62
51	61	57	—	1.20	2.64	1130	2491	—	—	1.04	2.29	1061	2338	875	1929	—	—	0.70	1.54	0.47	1.04	0.65	1.43
52	51	54	—	1.01	2.22	1084	2390	—	—	0.86	1.89	1022	2253	—	—	—	—	0.57	1.26	0.38	0.83	0.52	1.15
53	66	70	61	1.36	2.99	1208	2663	1.12	2.48	1.18	2.60	1127	2485	1227	2706	0.92	2.03	0.70	1.55	0.44	0.97	0.68	1.51
54	60	56	—	1.22	2.68	1135	2502	—	—	1.12	2.46	995	2194	870	1918	—	—	0.77	1.70	0.53	1.18	0.64	1.41
55	52	51	—	1.03	2.27	1033	2278	—	—	0.91	2.02	976	2151	—	—	—	—	—	—	—	—	—	—
56	69	65	65	1.36	3.00	1278	2818	1.19	2.63	1.19	2.62	1026	2263	1150	2536	0.98	2.15	0.80	1.77	0.54	1.19	0.70	1.55
57	86	77	57	1.72	3.79	1641	3618	1.04	2.30	1.55	3.42	1554	3426	1432	3158	0.86	1.89	0.95	2.10	0.67	1.47	0.89	1.95
58	87	84	59	1.72	3.80	1410	3108	1.08	2.39	1.46	3.21	1328	2928	1475	3251	0.89	1.96	1.02	2.24	0.72	1.58	0.88	1.94
59	78	55	62	1.34	2.96	—	—	1.13	2.49	1.17	2.59	—	—	—	—	0.93	2.04	0.57	1.26	0.33	0.74	0.58	1.28
60	76	82	—	1.51	3.34	1638	3611	—	—	1.34	2.96	1435	3163	1108	2442	—	—	0.80	1.77	0.54	1.18	0.76	1.68
61	65	65	64	1.36	3.01	1351	2977	1.16	2.57	0.99	2.19	1269	2799	1155	2546	0.96	2.11	0.63	1.39	0.39	0.85	0.67	1.48
62	64	61	50	1.27	2.81	1223	2696	0.93	2.04	1.11	2.45	1147	2530	—	—	0.76	1.67	0.70	1.55	0.46	1.00	0.67	1.49
63	19	20	19	0.38	0.84	398	878	0.35	0.77	0.34	0.74	377	830	—	—	0.29	0.63	0.20	0.45	0.14	0.30	0.19	0.43
64	17	21	19	0.33	0.73	416	918	0.35	0.77	0.29	0.63	394	869	—	—	0.29	0.63	0.16	0.35	0.10	0.21	0.16	0.35
65	68	72	59	1.35	2.98	1492	3290	1.09	2.39	1.18	2.61	1409	3107	1210	2668	0.89	1.96	0.71	1.57	0.46	1.01	0.68	1.51
66	64	69	—	1.10	2.42	1474	3250	—	—	0.99	2.18	1347	2971	920	2028	—	—	0.62	1.38	0.38	0.84	0.53	1.18
67	78	86	—	1.59	3.51	1697	3741	—	—	1.44	3.17	1658	3656	1399	3085	—	—	0.90	1.98	0.62	1.37	0.85	1.87
68	78	—	—	1.56	3.43	1666	3674	—	—	1.39	3.07	1642	3619	—	—	—	—	0.83	1.82	0.56	1.23	0.78	1.71
69	81	88	—	1.56	3.45	1684	3713	—	—	1.43	3.16	1555	3428	1367	3015	—	—	0.88	1.93	0.60	1.32	0.83	1.82

(Continued)

TABLE V-1A ENERGY FEEDS, MINERAL AND VITAMIN COMPOSITION OF FEEDS, DATA EXPRESSED **AS-FED**

Entry Number	Feed Name Description	Dry Matter	Macro Minerals							Micro Minerals						
			Calcium (Ca)	Phosphorus (P)	Sodium (Na)	Chlorine (Cl)	Magnesium (Mg)	Potassium (K)	Sulfur (S)	Cobalt (Co)	Copper (Cu)	Iodine (I)	Iron (Fe)	Manganese (Mn)	Selenium (Se)	Zinc (Zn)
		%	%	%	%	%	%	%	%	ppm or mg/kg	ppm or mg/kg	ppm or mg/kg	%	ppm or mg/kg	ppm or mg/kg	ppm or mg/kg
1	**ALMOND,** HULLS	90	0.19	0.09	—	—	—	0.48	0.10	—	—	—	—	—	—	—
	ANIMAL															
2	FAT, HYDROLYZED	99	—	—	—	—	—	—	—	—	—	—	—	—	—	—
3	TALLOW	97	—	—	—	—	—	—	—	—	—	—	—	—	—	—
4	**ANIMAL-POULTRY,** FAT	99	—	—	—	—	—	—	0.23	—	—	—	—	—	—	—
	BARLEY															
5	GRAIN, ALL ANALYSES	88	0.05	0.34	0.03	0.12	0.13	0.46	0.15	0.171	7.6	0.044	0.008	16.0	0.158	39.3
6	GRAIN, PACIFIC COAST	89	0.05	0.34	0.02	0.15	0.12	0.51	0.14	0.087	8.1	—	0.009	16.0	0.101	15.2
7	MALT SPROUTS, DEHY	93	0.18	0.63	0.88	0.36	0.17	0.25	0.79	—	5.9	—	0.018	29.4	0.416	56.4
	BEAN															
8	SEEDS, KIDNEY	89	0.11	0.43	0.01	—	0.10	0.93	—	0.508	6.1	—	0.007	17.0	0.355	24.3
9	SEEDS, NAVY	89	0.17	0.54	0.04	0.06	0.13	1.31	0.23	—	9.9	—	0.010	21.1	—	—
10	SEEDS, PINTO	90	0.18	0.48	0.01	—	0.24	2.59	—	0.551	13.0	—	0.014	17.3	—	40.0
11	**BEAN, LIMA,** SEEDS	90	0.08	0.38	0.02	0.03	0.18	1.61	0.20	—	8.2	—	0.009	16.1	—	—
12	**BEAN, MUNG,** SEEDS	90	0.13	0.35	0.01	—	—	1.04	—	—	—	—	0.009	—	—	—
13	**BEAN, TEPARY,** SEEDS	90	—	—	—	—	—	—	—	—	—	—	—	—	—	—
	BEET, SUGAR															
14	MOLASSES, MORE THAN 48% INVERT SUGAR, MORE THAN 79.5 DEGREES BRIX	78	0.12	0.03	1.16	1.28	0.23	4.73	0.46	0.362	16.8	—	0.007	4.5	—	14.0
15	PULP, DEHY	91	0.63	0.09	0.19	0.04	0.26	0.18	0.20	0.074	12.5	—	0.027	34.2	—	0.7
16	PULP WITH MOLASSES, DEHY	92	0.56	0.09	0.48	—	0.15	1.63	0.39	0.209	14.7	—	0.017	18.4	—	5.1
17	**BREWERS GRAINS,** DEHY	92	0.30	0.51	0.21	0.15	0.15	0.09	0.30	0.076	21.7	0.066	0.024	37.2	—	27.3
18	**BUCKWHEAT, COMMON,** GRAIN	88	0.10	0.33	0.05	0.04	0.10	0.45	0.14	0.049	9.5	—	0.005	33.7	—	8.8
19	**CARROT,** ROOTS, FRESH	11	0.05	0.04	0.06	0.06	0.02	0.32	0.02	—	1.2	—	0.002	3.6	—	—
20	**CASSAVA,** TUBERS, FRESH	32	0.05	0.05	—	—	—	0.33	—	—	—	—	—	—	—	—
	CITRUS															
21	PULP WITHOUT FINES, DEHY (DRIED CITRUS PULP)	91	1.69	0.12	0.07	—	0.15	0.71	0.17	0.169	5.0	—	0.033	6.6	—	13.7
22	SYRUP (MOLASSES)	67	1.18	0.09	0.28	0.07	0.14	0.09	0.16	0.109	72.8	—	0.035	40.9	—	92.4
	CORN, DENT YELLOW															
23	GRAIN, ALL ANALYSES	88	0.05	0.28	0.01	0.05	0.11	0.33	0.11	0.378	3.5	—	0.004	5.7	0.127	19.4
24	GRAIN, GRADE 2, 54 lb/bu *(69.5 kg/hl)*	87	0.02	0.29	0.02	0.04	0.11	0.31	0.12	0.029	3.8	—	0.003	5.3	—	13.7
25	GRAIN, HIGH MOISTURE	77	0.01	0.25	0.01	0.04	0.11	0.28	0.11	—	2.2	—	0.003	5.3	—	25.4
26	DISTILLERS GRAINS, DEHY	93	0.09	0.39	0.09	0.07	0.16	0.43	0.43	0.076	38.9	0.048	0.020	19.3	0.352	41.7
27	DISTILLERS SOLUBLES, DEHY	93	0.30	1.30	0.23	0.26	0.60	1.70	0.37	0.167	77.9	0.079	0.052	72.0	0.371	88.0
28	EARS, GROUND (CORN AND COB MEAL)	87	0.06	0.24	0.02	0.04	0.12	0.46	0.14	0.273	6.8	0.023	0.008	19.9	0.074	12.1
29	GLUTEN, MEAL, 60% PROTEIN	90	0.07	0.45	0.05	0.09	0.08	0.18	0.65	0.045	26.1	0.018	0.023	6.3	0.829	30.6
30	GLUTEN FEED	90	0.32	0.74	0.12	0.22	0.33	0.57	0.21	0.087	47.1	0.066	0.043	23.1	0.272	64.6
31	GRITS (HOMINY GRITS)	90	0.05	0.51	0.08	0.05	0.24	0.59	0.03	0.055	13.6	—	0.007	14.5	—	—
32	OIL	99	—	—	—	—	—	—	—	—	—	—	—	—	—	—
33	**CORN, OPAQUE 2, HIGH LYSINE,** GRAIN	90	0.03	0.20	—	—	0.13	0.35	0.10	—	—	—	—	—	—	—
34	**CORN, SWEET,** CANNERY RESIDUE, FRESH	77	0.25	0.54	0.02	—	0.18	0.88	0.10	—	5.4	—	0.016	—	—	—
35	**COTTON,** SEEDS	91	0.14	0.69	0.03	—	0.32	1.11	0.24	—	49.0	—	0.014	11.1	—	—
	DISTILLERS PRODUCTS (ALSO SEE CORN; RYE; SORGHUM; WHEAT)															
36	GRAINS, DEHY	93	0.12	0.54	0.05	0.05	0.09	0.20	0.46	0.092	47.9	—	0.027	35.0	—	—
37	SOLUBLES, DEHY	92	0.24	1.35	0.45	—	0.53	1.97	—	0.196	71.6	—	0.031	64.1	—	138.0
	FATS AND OILS															
38	FAT, ANIMAL, HYDROLYZED	99	—	—	—	—	—	—	—	—	—	—	—	—	—	—
39	FAT, ANIMAL-POULTRY	99	—	—	—	—	—	—	0.23	—	—	—	—	—	—	—
40	**FLAX, COMMON,** SEEDS	96	0.28	0.55	—	—	—	—	—	—	—	—	—	—	—	—
41	**GARBAGE,** HOTEL AND RESTAURANT, BOILED, WET	23	0.10	0.06	—	—	0.01	—	—	—	5.0	—	0.010	5.0	—	—
42	**HOMINY GRITS** (CORN, DENT YELLOW, GRITS)	90	0.05	0.51	0.08	0.05	0.24	0.59	0.03	0.055	13.6	—	0.007	14.5	—	—
43	**KAFIR SORGHUM,** GRAIN	89	0.03	0.31	0.05	0.10	0.15	0.34	0.16	0.387	7.0	—	0.007	15.8	0.797	13.5
44	**KELP (SEAWEED),** WHOLE, DEHY	91	2.47	0.28	—	—	0.85	—	—	—	—	—	—	—	—	—
45	**LARD (FAT),** SWINE	99	—	—	—	—	—	—	—	—	—	—	—	—	—	—
46	**MANGEL, BEET,** ROOTS, FRESH	11	0.02	0.02	0.07	0.16	0.02	0.25	0.02	—	0.6	—	0.002	—	—	—
47	**MANURE, CATTLE,** WITHOUT BEDDING, DEHY	93	1.35	1.08	—	—	—	0.47	—	—	—	—	—	—	—	—
48	**MILLET,** GRAIN	90	0.05	0.29	0.04	0.14	0.16	0.43	0.13	0.044	21.8	—	0.007	29.9	—	13.9
49	**MILLET, FOXTAIL,** GRAIN	89	—	0.41	—	—	—	0.31	—	—	—	—	0.010	—	—	—
50	**MILO SORGHUM,** GRAIN	89	0.04	0.30	0.04	0.08	0.13	0.31	0.11	0.471	4.3	0.061	0.005	15.8	0.201	16.9
	MOLASSES AND SYRUP															
51	BEET, SUGAR, MOLASSES, MORE THAN 48% INVERT SUGAR, MORE THAN 79.5 DEGREES BRIX	78	0.12	0.03	1.16	1.28	0.23	4.73	0.46	0.362	16.8	—	0.007	4.5	—	14.0
52	CITRUS, SYRUP (CITRUS MOLASSES)	67	1.18	0.09	0.28	0.07	0.14	0.09	0.16	0.109	72.8	—	0.035	40.9	—	92.4
53	SUGAR CANE, MOLASSES, DEHY	94	1.04	0.42	0.19	—	0.44	3.40	0.43	1.145	74.9	—	0.024	54.1	—	31.2
54	SUGAR CANE, MOLASSES, MORE THAN 46% INVERT SUGAR, MORE THAN 79.5 DEGREES BRIX (BLACKSTRAP)	74	0.74	0.08	0.16	2.26	0.31	2.98	0.35	1.180	48.9	1.564	0.020	43.7	—	15.6
55	WOOD, MOLASSES	62	1.17	0.05	0.03	0.12	0.07	0.04	0.03	—	—	—	—	12.6	—	—
	OATS															
56	GRAIN, ALL ANALYSES	89	0.08	0.34	0.05	0.09	0.14	0.40	0.21	0.056	6.0	0.112	0.007	35.8	0.215	34.9
57	CEREAL BY-PRODUCT (FEEDING OAT MEAL; OAT MIDDLINGS)	91	0.07	0.44	0.09	0.05	0.14	0.50	0.22	0.046	5.2	—	0.039	43.8	—	139.5
58	GROATS	90	0.08	0.43	0.05	0.08	0.11	0.35	0.20	—	6.0	0.108	0.008	27.8	—	0.0
59	**PEA,** SPLIT SEED BY-PRODUCT (PEA FEED; PEA MEAL)	90	—	—	—	—	—	—	—	—	—	—	—	—	—	—
60	**PEA, FIELD,** SEEDS	91	0.16	0.38	0.00	—	0.14	1.36	—	1.700	11.7	—	0.020	21.2	0.393	46.8
61	**PEARL MILLET,** GRAIN	90	0.05	0.31	0.04	0.14	0.16	0.43	0.13	0.045	22.1	—	0.006	31.0	—	13.3
62	**PINEAPPLE,** CANNERY RESIDUE, DEHY (PINEAPPLE BRAN)	87	0.20	0.11	—	—	—	—	—	—	—	—	0.049	—	—	—
	POTATO															
63	TUBERS, FRESH	24	0.01	0.06	0.02	0.07	0.03	0.51	0.02	—	6.7	—	0.002	9.8	—	—
64	TUBERS, BOILED	24	0.01	0.05	—	—	—	—	—	—	—	—	—	—	—	—
	RICE															
65	GRAIN, GROUND (GROUND ROUGH RICE; GROUND PADDY RICE)	89	0.07	0.32	0.06	0.07	0.13	0.47	0.05	—	—	—	—	18.0	—	15.0
66	BRAN WITH GERMS (RICE BRAN)	91	0.07	1.44	0.03	0.07	0.85	1.69	0.18	1.383	11.0	—	0.019	337.6	—	37.4
67	GROATS, POLISHED (RICE, POLISHED)	89	0.02	0.11	0.01	0.04	0.09	0.23	0.08	0.846	5.4	—	0.002	29.6	—	13.7
68	GROATS (RICE, BROWN)	88	0.03	0.20	0.02	0.07	0.08	0.30	0.04	0.727	3.7	—	0.003	20.3	—	14.3
69	POLISHINGS	90	0.05	1.34	0.04	0.11	0.60	1.28	0.17	3.890	8.0	—	0.009	126.8	—	63.2

Composition of Feeds

ENERGY FEEDS

Entry Number	Fat-Soluble Vitamins					Water-Soluble Vitamins								
	A (1 mg Carotene = 1667 IU Vit A)	Carotene (Provitamin A)	D	E	K	B-12	Biotin	Choline	Folacin (Folic Acid)	Niacin	Pantothenic Acid (B-3)	(Pyridoxine) B-6	Riboflavin (B-2)	Thiamin (B-1)
	IU/g	ppm or mg/kg	IU/kg	ppm or mg/kg	ppm or mg/kg	ppb or mcg/kg	ppm or mg/kg	ppm or mg/kg	ppm or mg/kg	ppm or mg/kg	ppm or mg/kg	ppm or mg/kg	ppm or mg/kg	ppm or mg/kg
1	–	–	–	–	–	–	–	–	–	–	–	–	–	–
2	–	–	–	–	–	–	–	–	–	–	–	–	–	–
3	–	–	–	–	–	–	–	–	–	–	–	–	–	–
4	–	–	–	7.9	–	–	–	–	–	–	–	–	–	–
5	3.4	2.0	–	23.2	0.22	–	0.15	1036	0.57	76	7.9	5.80	1.6	4.5
6	–	–	–	26.2	–	–	0.15	970	0.50	47	7.1	2.89	1.5	4.2
7	–	–	–	3.7	–	–	4.09	1591	0.20	55	9.0	8.62	2.8	8.3
8	–	–	–	–	–	–	–	–	–	24	–	–	1.8	5.7
9	–	–	–	1.0	–	–	0.11	1017	1.29	24	2.4	0.30	1.7	6.3
10	–	–	–	–	–	–	–	–	–	22	2.2	–	3.1	8.6
11	–	–	–	–	–	–	–	–	3.31	20	8.4	–	1.7	4.6
12	–	–	–	–	–	–	–	–	–	25	–	–	2.1	3.9
13	–	–	–	–	–	–	–	–	–	–	–	–	–	–
14	–	–	–	4.0	–	–	–	827	–	41	4.5	–	2.3	–
15	0.4	0.2	1	–	–	–	–	820	–	17	1.4	–	0.7	0.4
16	0.4	0.2	–	–	–	–	–	814	–	16	1.5	–	0.7	–
17	0.8	0.5	–	26.7	–	3.6	0.44	1651	0.22	44	8.2	1.03	1.5	0.6
18	–	–	–	–	–	–	–	439	–	18	11.5	–	4.7	3.7
19	129.9	77.9	–	6.9	–	–	0.01	–	0.14	7	3.5	1.39	0.6	0.7
20	–	–	–	–	–	–	–	–	–	–	–	–	–	–
21	0.4	0.2	–	–	–	–	–	789	–	22	14.0	–	2.1	1.5
22	–	–	–	–	–	–	–	–	–	27	17.2	–	6.2	–
23	9.5	5.7	–	20.9	0.22	–	0.07	504	0.31	23	5.1	6.16	1.1	3.7
24	2.9	1.7	–	21.6	–	–	0.06	569	0.35	24	3.9	6.88	1.3	3.5
25	–	–	–	–	–	–	–	–	–	–	–	–	–	–
26	5.2	3.1	–	–	–	0.3	0.41	1113	1.00	38	11.3	4.22	5.0	1.8
27	1.1	0.7	–	45.9	–	4.2	1.49	4751	1.34	124	23.3	9.41	15.1	6.8
28	5.3	3.2	–	17.5	–	–	0.03	357	0.24	17	4.2	5.97	0.9	2.9
29	–	–	–	14.6	–	–	–	–	–	–	–	6.39	–	–
30	9.8	5.9	–	12.1	–	–	0.33	1514	0.27	70	13.6	13.93	2.2	2.0
31	15.4	9.2	–	–	–	–	0.13	1154	0.31	47	8.2	10.95	2.1	8.1
32	–	–	–	–	–	–	–	–	–	–	–	–	–	–
33	7.8	4.7	–	–	–	–	–	518	–	19	4.7	–	1.1	–
34	17.3	10.4	–	–	–	–	–	–	–	–	–	–	–	–
35	–	–	–	–	–	–	–	–	–	–	–	–	–	–
36	13.0	7.8	–	30.5	–	–	–	2645	–	47	11.9	6.00	6.6	2.5
37	1.9	1.1	–	–	–	2.9	2.84	4992	–	143	25.3	8.66	11.3	6.9
38	–	–	–	–	–	–	–	–	–	–	–	–	–	–
39	–	–	–	7.9	–	–	–	–	–	–	–	–	–	–
40	–	–	–	–	–	–	–	–	–	–	–	–	–	–
41	–	–	–	–	–	–	–	–	–	–	–	–	–	–
42	15.4	9.2	–	–	–	–	0.13	1154	0.31	47	8.2	10.95	2.1	8.1
43	0.6	0.4	–	–	–	–	0.24	439	0.20	38	12.0	6.68	1.2	3.8
44	–	–	–	–	–	–	–	–	–	–	–	–	–	–
45	–	–	–	22.8	–	–	–	–	–	–	–	–	–	–
46	0.2	0.1	–	–	–	–	–	–	0.17	3	1.0	0.43	0.4	0.3
47	–	–	–	–	–	–	–	–	–	–	–	–	–	–
48	–	–	–	–	–	–	–	739	0.22	48	9.0	–	1.5	6.6
49	–	–	–	–	–	–	–	–	–	33	–	–	1.1	3.8
50	0.4	0.2	0	12.1	0.22	–	0.23	638	0.21	37	11.0	4.69	1.1	4.1
51	–	–	–	4.0	–	–	–	827	–	41	4.5	–	2.3	–
52	–	–	–	–	–	–	–	–	–	27	17.2	–	6.2	–
53	–	–	–	5.2	–	–	–	–	–	–	–	–	–	–
54	–	–	–	5.4	–	–	0.69	764	0.11	36	37.4	4.21	2.8	0.9
55	–	–	–	–	–	–	–	–	–	–	–	–	–	–
56	0.2	0.1	–	14.9	–	–	0.27	967	0.39	14	9.9	2.53	1.4	6.0
57	–	–	–	23.7	–	–	0.22	1157	0.46	24	17.6	–	1.7	7.0
58	–	–	–	14.8	–	–	–	1132	0.51	10	13.8	1.00	1.2	6.5
59	–	–	–	–	–	–	–	–	–	–	–	–	–	–
60	–	–	–	–	–	–	0.19	654	0.36	34	7.4	1.01	1.4	4.1
61	4.3	2.6	–	–	–	–	–	790	–	52	8.8	–	1.8	7.1
62	78.4	47.0	–	–	–	–	–	–	–	–	–	–	–	–
63	–	–	–	–	–	–	–	–	–	17	–	–	0.5	1.2
64	–	–	–	–	–	–	–	–	–	–	–	–	–	–
65	–	–	–	14.0	–	–	–	926	0.25	40	7.1	–	0.7	–
66	–	–	–	60.4	–	–	0.43	1230	2.20	299	22.8	13.24	2.6	22.4
67	–	–	–	3.5	–	–	–	901	0.15	15	3.5	0.39	0.6	0.7
68	–	–	–	10.3	–	–	0.09	–	0.19	43	10.7	7.00	0.6	2.9
69	–	–	–	90.2	–	–	0.62	1248	–	506	46.4	27.89	1.8	20.0

(Continued)

TABLE V-1A ENERGY FEEDS, COMPOSITION OF FEEDS, DATA EXPRESSED **AS-FED**—(Continued)

Entry Number	Feed Name Description	International Feed Number	Proximate Analysis									Digestible Protein		
			Dry Matter	Ash	Crude Fiber	Neutral Det. Fib. (NDF)	Acid Det. Fib. (ADF)	Lignin	Ether Extract (Fat)	N-Free Extract	Crude Protein	Ruminant	Swine	Horse
			%	%	%	%	%	%	%	%	%	%	%	%
	RYE													
70	GRAIN, ALL ANALYSES	4-04-047	87	1.6	2.2	—	—	—	1.5	70.0	12.0	8.4	9.1	8.7
71	DISTILLERS GRAINS, DEHY	5-04-023	92	2.3	12.3	—	—	—	6.0	48.3	23.0	13.8	21.7	17.5
72	DISTILLERS GRAINS WITH SOLUBLES, DEHY	5-04-024	90	6.4	8.1	—	—	—	4.1	44.7	27.2	22.6	25.9	22.3
73	**SAFFLOWER,** SEEDS	4-07-958	93	3.0	23.6	—	37.2	—	30.8	20.9	14.9	7.2	11.7	10.9
74	**SCREENINGS, GRAIN, CEREAL,** ALL ANALYSES (ALSO SEE BARLEY; WHEAT)	4-02-156	90	5.4	12.0	—	—	—	3.7	56.7	12.1	8.7	7.8	8.7
	SORGHUM													
75	GRAIN, ALL ANALYSES	4-04-383	90	1.8	2.6	16.2	8.1	1.2	2.7	71.6	11.5	7.1	8.7	8.2
76	KAFIR, GRAIN	4-04-428	89	1.5	2.0	—	1.8	—	2.8	72.0	10.8	6.6	8.3	7.7
77	MILO, GRAIN	4-04-444	89	1.6	2.2	20.6	4.7	—	2.8	71.9	10.1	6.8	7.2	7.1
78	SUDANGRASS, GRAIN	4-08-520	92	12.0	25.4	—	—	—	2.4	38.4	14.2	9.9	11.1	10.3
	SOYBEAN													
79	SEEDS	5-04-610	92	5.1	5.4	—	10.1	—	17.2	25.9	38.4	34.5	31.5	34.9
80	OIL	4-07-983	100	0.3	—	—	—	—	95.0	7.3	1.4	-1.8	-0.6	0.2
	SUGARCANE													
81	MOLASSES, DEHY	4-04-695	94	12.5	6.3	—	—	—	0.9	65.0	9.7	5.3	7.0	6.8
82	MOLASSES, MORE THAN 46% INVERT SUGAR, MORE THAN 79.5 DEGREES BRIX	4-04-696	74	9.8	0.4	—	0.3	0.2	0.2	59.7	4.3	0.6	1.3	2.7
83	**SUNFLOWER,** SEEDS	5-08-530	94	3.7	22.7	—	—	—	32.3	14.4	20.9	16.7	19.6	14.9
	SWEET POTATO													
84	TUBERS, DEHY, MEAL	4-08-536	89	3.8	3.7	—	—	—	1.0	74.5	6.4	0.9	4.9	4.3
85	TUBERS, FRESH	4-04-788	33	1.1	1.4	—	2.6	—	0.4	28.5	1.7	0.5	0.9	1.0
86	**TRITICALE,** GRAIN	4-20-362	89	1.8	3.0	11.9	—	—	1.5	67.3	15.4	11.1	12.2	11.3
87	**TURNIP,** ROOTS, FRESH	4-05-067	9	0.8	1.1	4.0	3.1	—	0.2	5.9	1.2	0.9	0.4	0.9
	WHEAT													
88	GRAIN, ALL ANALYSES	4-05-211	89	1.8	2.6	—	7.1	—	1.8	69.7	13.1	10.5	11.1	9.5
89	BRAN	4-05-190	89	5.9	10.0	40.9	12.0	2.6	4.0	53.6	15.5	12.0	11.8	13.1
90	MIDDLINGS, LESS THAN 9.5% FIBER	4-05-205	89	4.7	7.7	32.9	8.9	—	4.3	55.7	16.4	12.4	13.9	12.1
91	MILL RUN, LESS THAN 9.5% FIBER	4-05-206	90	5.1	8.2	—	9.9	—	4.1	57.4	15.1	11.4	12.0	11.0
92	RED DOG, LESS THAN 4% FIBER	4-05-203	88	2.4	2.9	—	—	—	3.4	64.0	15.6	12.9	13.6	11.5
93	SHORTS, LESS THAN 7% FIBER	4-05-201	88	4.4	6.4	—	—	—	4.6	56.5	16.5	12.9	13.2	12.2
94	**WOOD,** MOLASSES	4-05-502	62	4.1	0.5	—	1.2	—	0.3	56.6	0.6	-1.3	-0.6	-0.1

TABLE V-1B PROTEIN FEEDS, COMPOSITION OF FEEDS, DATA EXPRESSED **AS-FED**

Entry Number	Feed Name Description	International Feed Number	Proximate Analysis									Digestible Protein		
			Dry Matter	Ash	Crude Fiber	Neutral Det. Fib. (NDF)	Acid Det. Fib. (ADF)	Lignin	Ether Extract (Fat)	N-Free Extract	Crude Protein	Ruminant	Swine	Horse
			%	%	%	%	%	%	%	%	%	%	%	%
1	**AMMONIUM,** POLYPHOSPHATE SOLUTION	6-08-042	60	—	—	—	—	—	—	—	54.8	—	—	—
2	**BARLEY,** MALT SPROUTS, DEHY	5-00-545	93	5.6	14.2	43.7	16.7	—	1.4	48.8	22.9	19.2	21.5	17.2
	BEAN													
3	SEEDS, KIDNEY	5-00-600	89	3.6	4.1	—	—	—	1.2	57.8	21.8	14.6	20.6	16.5
4	SEEDS, NAVY	5-00-623	89	4.0	4.4	—	—	—	1.4	56.7	22.9	20.2	21.6	17.6
5	SEEDS, PINTO	5-00-624	90	4.3	4.0	—	—	—	1.3	57.9	22.7	17.2	21.4	17.3
6	**BEAN, LIMA,** SEEDS	5-00-613	90	4.1	4.5	—	—	—	1.3	58.9	20.7	16.7	19.4	15.1
7	**BEAN, MUNG,** SEEDS	5-08-185	90	3.8	3.9	—	—	—	1.3	57.1	23.9	19.6	22.6	18.7
8	**BEAN, TEPARY,** SEEDS	5-08-349	90	4.2	3.4	—	—	—	1.4	59.3	22.2	18.0	20.9	16.7
9	**BLOOD,** MEAL	5-00-380	91	5.3	1.0	—	—	—	1.3	3.1	80.5	57.2	60.1	82.5
	BREWERS (SEE SPECIFIC GRAINS)													
10	GRAINS, DEHY	5-02-141	92	3.6	13.0	38.7	23.9	4.6	6.6	41.6	27.3	20.1	21.8	21.0
11	GRAINS, WET	5-02-142	22	0.9	3.1	9.2	5.0	1.1	1.5	10.7	5.8	4.2	5.5	4.5
	BUTTERMILK, CATTLE													
12	CONDENSED	5-01-159	29	3.6	0.1	—	—	—	2.4	12.4	10.8	9.2	10.0	9.5
13	DEHY	5-01-160	92	9.1	0.3	—	—	—	5.2	45.9	31.7	28.7	29.5	27.3
14	**CASEIN,** ACID PRECIPITATED, DEHY	5-01-162	91	2.2	0.2	0	0	—	0.6	3.6	84.0	81.5	82.2	86.5
15	**CHEESE, CATTLE,** RIND	5-01-163	85	7.7	0.2	—	—	—	19.8	11.7	45.5	39.4	44.1	43.6
	CHICKEN													
16	BY-PRODUCT, FRESH (VISCERA WITH FEET, WITH HEADS)	5-07-951	39	4.9	—	—	0.5	—	14.7	3.0	17.6	15.1	17.0	16.2
17	BY-PRODUCT, WITHOUT FEET, FRESH (VISCERA WITHOUT FEET, WITH HEADS)	5-07-952	34	1.1	0.2	—	0.7	—	16.1	3.0	13.8	11.7	13.2	12.4
	CITRUS													
18	MOLASSES, AMMONIATED	5-01-240	61	4.7	—	—	—	—	2.2	32.5	21.4	18.0	20.4	18.5
19	PULP AMMONIATED, DEHY	4-01-238	87	4.6	13.2	—	—	—	5.6	51.9	12.1	8.2	9.3	8.7
20	**COCONUT,** KERNELS WITH COATS, MEAL MECH EXTD (COPRA MEAL)	5-01-572	92	6.4	12.1	—	18.3	—	6.8	45.0	21.2	17.6	15.5	15.4
	CORN													
21	DISTILLERS GRAINS, DEHY	5-02-842	93	2.2	11.5	—	—	—	8.9	43.1	27.8	20.1	19.4	22.7
22	DISTILLERS SOLUBLES, DEHY	5-02-844	93	7.2	4.6	—	—	—	8.6	45.2	27.4	21.5	20.0	22.3
23	GERM MEAL, WET MILLED, SOLV EXTD	5-02-898	92	3.9	12.2	—	—	—	1.5	53.5	20.7	16.6	19.4	14.9
24	GLUTEN FEED	5-02-903	90	6.6	8.7	—	—	—	2.1	49.4	23.0	19.8	18.2	17.7
25	GLUTEN MEAL	5-02-900	91	3.1	4.5	33.7	8.2	—	2.2	38.4	43.2	36.3	41.7	40.3
	COTTON													
26	SEEDS WITHOUT LINT	5-13-749	91	4.7	19.4	—	—	—	20.4	24.5	21.8	17.7	20.5	16.2

Composition of Feeds

ENERGY FEEDS

Entry Number	TDN Ruminant %	TDN Swine %	TDN Horse %	Digestible Energy Ruminant Mcal lb	Digestible Energy Ruminant Mcal kg	Digestible Energy Swine kcal lb	Digestible Energy Swine kcal kg	Digestible Energy Horse Mcal lb	Digestible Energy Horse Mcal kg	Metabolizable Energy Ruminant Mcal lb	Metabolizable Energy Ruminant Mcal kg	Metabolizable Energy Swine kcal lb	Metabolizable Energy Swine kcal kg	Metabolizable Energy Poultry ME$_n$ kcal lb	Metabolizable Energy Poultry ME$_n$ kcal kg	Metabolizable Energy Horse Mcal lb	Metabolizable Energy Horse Mcal kg	Net Energy Ruminant NE$_m$ Mcal lb	Net Energy Ruminant NE$_m$ Mcal kg	Net Energy Ruminant NE$_g$ Mcal lb	Net Energy Ruminant NE$_g$ Mcal kg	Net Energy Lactating Cows NE$_{lc}$ Mcal lb	Net Energy Lactating Cows NE$_{lc}$ Mcal kg
70	73	75	75	1.42	3.12	1474	3249	1.35	2.96	1.18	2.60	1319	2909	1202	2651	1.10	2.43	0.80	1.75	0.54	1.18	0.74	1.63
71	54	79	—	1.08	2.38	1586	3497	—	—	0.90	1.99	1387	3057	—	—	—	—	0.45	1.00	0.22	0.49	0.50	1.09
72	73	68	—	1.46	3.21	1354	2985	—	—	1.29	2.83	1173	2586	—	—	—	—	0.78	1.72	0.52	1.14	0.74	1.64
73	83	—	—	1.07	2.36	—	—	—	—	1.19	2.62	—	—	—	—	—	—	0.93	2.06	0.64	1.42	0.87	1.92
74	62	55	53	1.23	2.71	1493	3291	0.98	2.16	1.06	2.34	1409	3107	831	1833	0.80	1.77	0.63	1.39	0.39	0.85	0.62	1.37
75	67	—	71	1.34	2.96	1570	3462	1.28	2.82	1.17	2.58	1426	3143	—	—	1.05	2.31	0.53	1.18	0.30	0.66	0.55	1.21
76	75	81	70	1.47	3.24	1521	3354	1.26	2.77	1.30	2.87	1437	3169	1529	3372	1.03	2.27	0.79	1.74	0.53	1.16	0.75	1.64
77	76	77	68	1.41	3.12	1520	3350	1.23	2.71	1.13	2.49	1474	3250	1467	3234	1.01	2.23	0.75	1.66	0.50	1.09	0.74	1.62
78	47	—	—	0.94	2.08	—	—	—	—	0.76	1.68	—	—	—	—	—	—	—	—	—	—	—	—
79	84	93	—	1.69	3.72	1820	4012	—	—	1.52	3.34	1605	3539	1534	3382	—	—	0.93	2.04	0.64	1.41	0.86	1.90
80	193	—	—	3.86	8.51	3412	7523	—	—	3.70	8.15	3287	7247	4050	8929	—	—	2.40	5.30	1.79	3.94	2.09	4.61
81	66	70	61	1.36	2.99	1208	2663	1.12	2.48	1.18	2.60	1127	2485	1227	2706	0.92	2.03	0.70	1.55	0.44	0.97	0.68	1.51
82	60	56	—	1.22	2.68	1135	2502	—	—	1.12	2.46	995	2194	870	1918	—	—	0.77	1.70	0.53	1.18	0.64	1.41
83	78	84	—	1.56	3.44	1685	3715	—	—	1.38	3.05	1476	3254	—	—	—	—	0.89	1.95	0.60	1.33	0.83	1.83
84	72	64	68	1.42	3.12	1280	2823	1.23	2.72	1.25	2.75	1202	2651	—	—	1.01	2.23	0.76	1.67	0.50	1.11	0.72	1.60
85	27	27	25	0.52	1.16	530	1169	0.45	0.99	0.46	1.02	501	1104	—	—	0.37	0.81	0.28	0.63	0.19	0.42	0.27	0.60
86	75	—	79	1.44	3.17	1453	3203	1.42	3.13	1.27	2.80	1420	3130	1420	3130	1.16	2.57	0.75	1.65	0.49	1.09	0.71	1.57
87	8	7	6	0.16	0.34	146	322	0.11	0.24	0.14	0.31	138	304	—	—	0.09	0.20	0.09	0.19	0.06	0.13	0.08	0.18
88	77	79	76	1.54	3.40	1544	3404	1.36	2.99	1.28	2.82	1486	3274	1402	3092	1.11	2.45	0.88	1.93	0.60	1.33	0.81	1.79
89	63	57	44	1.26	2.78	1119	2466	0.84	1.84	1.09	2.40	1027	2265	556	1225	0.69	1.51	0.67	1.48	0.42	0.93	0.64	1.41
90	74	68	59	1.39	3.07	1321	2912	1.08	2.39	1.16	2.57	1225	2702	940	2072	0.89	1.96	0.78	1.72	0.52	1.15	0.79	1.73
91	71	72	58	1.46	3.21	1438	3170	1.07	2.35	1.14	2.52	1254	2765	803	1771	0.88	1.93	0.76	1.68	0.50	1.11	0.76	1.68
92	77	72	69	1.47	3.23	1429	3149	1.26	2.77	1.39	3.06	1305	2876	1164	2566	1.03	2.27	0.82	1.80	0.55	1.22	0.77	1.70
93	76	71	59	1.47	3.24	1413	3116	1.08	2.39	1.25	2.75	1327	2926	1001	2206	0.89	1.96	0.85	1.88	0.58	1.28	0.79	1.75
94	52	51	—	1.03	2.27	1033	2278	—	—	0.91	2.02	976	2151	—	—	—	—	—	—	—	—	—	—

PROTEIN FEEDS

Entry Number	TDN Ruminant %	TDN Swine %	TDN Horse %	Digestible Energy Ruminant Mcal lb	Digestible Energy Ruminant Mcal kg	Digestible Energy Swine kcal lb	Digestible Energy Swine kcal kg	Digestible Energy Horse Mcal lb	Digestible Energy Horse Mcal kg	Metabolizable Energy Ruminant Mcal lb	Metabolizable Energy Ruminant Mcal kg	Metabolizable Energy Swine kcal lb	Metabolizable Energy Swine kcal kg	Metabolizable Energy Poultry ME$_n$ kcal lb	Metabolizable Energy Poultry ME$_n$ kcal kg	Metabolizable Energy Horse Mcal lb	Metabolizable Energy Horse Mcal kg	Net Energy Ruminant NE$_m$ Mcal lb	Net Energy Ruminant NE$_m$ Mcal kg	Net Energy Ruminant NE$_g$ Mcal lb	Net Energy Ruminant NE$_g$ Mcal kg	Net Energy Lactating Cows NE$_{lc}$ Mcal lb	Net Energy Lactating Cows NE$_{lc}$ Mcal kg
1	—	—	—	—	—	—	—	—	—	—	—	—	—	—	—	—	—	—	—	—	—	—	—
2	66	61	—	1.31	2.89	1219	2688	—	—	1.13	2.50	1046	2307	722	1592	—	—	0.69	1.52	0.43	0.95	0.67	1.48
3	73	79	—	1.46	3.22	1584	3493	—	—	1.29	2.85	1388	3059	—	—	—	—	—	—	—	—	—	—
4	76	78	—	1.53	3.37	1737	3830	—	—	1.36	3.00	1531	3376	1052	2320	—	—	0.83	1.83	0.56	1.24	0.78	1.72
5	74	78	—	1.47	3.25	1569	3460	—	—	1.30	2.87	1372	3025	—	—	—	—	0.81	1.79	0.54	1.20	0.77	1.69
6	78	78	—	1.55	3.42	1556	3430	—	—	1.39	3.05	1360	2999	—	—	—	—	—	—	—	—	—	—
7	79	81	—	1.53	3.37	1612	3554	—	—	1.36	3.00	1412	3113	—	—	—	—	0.82	1.80	0.55	1.21	0.77	1.70
8	69	80	—	1.37	3.02	1606	3542	—	—	1.20	2.65	1406	3100	—	—	—	—	—	—	—	—	—	—
9	61	61	—	1.20	2.65	1220	2690	—	—	0.99	2.19	1012	2231	1282	2826	—	—	0.63	1.38	0.38	0.84	0.60	1.33
10	65	66	48	1.25	2.76	1045	2303	—	—	1.01	2.22	1038	2288	1047	2308	—	—	0.64	1.41	0.39	0.86	0.67	1.48
11	15	18	—	0.32	0.70	351	774	—	—	0.27	0.60	305	673	—	—	—	—	0.18	0.40	0.12	0.28	0.17	0.38
12	26	22	—	0.52	1.14	442	974	—	—	0.46	1.02	383	844	—	—	—	—	0.29	0.63	0.20	0.43	0.27	0.59
13	82	77	—	1.56	3.44	1578	3479	—	—	1.29	2.84	1381	3045	1248	2752	—	—	0.88	1.94	0.60	1.32	0.85	1.87
14	81	80	—	1.56	3.44	1590	3506	—	—	1.22	2.68	1391	3068	1867	4116	—	—	0.82	1.81	0.55	1.22	0.78	1.72
15	78	80	—	1.69	3.73	1609	3547	—	—	1.54	3.39	1371	3023	—	—	—	—	1.06	2.34	0.76	1.68	0.98	2.15
16	—	—	—	—	—	—	—	—	—	—	—	—	—	—	—	—	—	—	—	—	—	—	—
17	—	—	—	—	—	—	—	—	—	—	—	—	—	—	—	—	—	—	—	—	—	—	—
18	46	—	—	0.93	2.05	—	—	—	—	0.81	1.80	—	—	—	—	—	—	—	—	—	—	—	—
19	76	65	42	1.40	3.09	1293	2850	0.80	1.77	1.24	2.73	1216	2680	—	—	0.66	1.45	0.69	1.53	0.45	0.99	0.67	1.47
20	69	73	—	1.37	3.01	1460	3218	—	—	1.20	2.64	1386	3055	692	1525	—	—	0.74	1.63	0.48	1.06	0.70	1.55
21	81	65	—	1.54	3.41	1246	2746	—	—	1.28	2.82	1196	2636	894	1970	—	—	0.87	1.92	0.59	1.30	0.84	1.86
22	80	78	—	1.55	3.42	1474	3250	—	—	1.34	2.96	1419	3128	1324	2919	—	—	0.92	2.03	0.63	1.40	0.86	1.89
23	68	69	—	1.41	3.11	1568	3456	—	—	1.24	2.73	1372	3025	772	1702	—	—	0.73	1.61	0.47	1.04	0.71	1.55
24	75	75	—	1.44	3.17	1375	3031	—	—	1.21	2.67	1037	2287	784	1729	—	—	0.82	1.80	0.55	1.21	0.78	1.72
25	78	80	—	1.50	3.31	1637	3609	—	—	1.24	2.74	1437	3168	1367	3015	—	—	0.84	1.85	0.57	1.25	0.80	1.75
26	87	71	—	1.57	3.46	1430	3152	—	—	1.39	3.07	1243	2739	—	—	—	—	0.97	2.13	0.67	1.49	0.91	2.01

(Continued)

TABLE V-1A ENERGY FEEDS, MINERAL AND VITAMIN COMPOSITION OF FEEDS, DATA EXPRESSED AS-FED—(Continued)

Entry Number	Feed Name Description	Dry Matter	Macro Minerals							Micro Minerals						
			Calcium (Ca)	Phosphorus (P)	Sodium (Na)	Chlorine (Cl)	Magnesium (Mg)	Potassium (K)	Sulfur (S)	Cobalt (Co)	Copper (Cu)	Iodine (I)	Iron (Fe)	Manganese (Mn)	Selenium (Se)	Zinc (Zn)
		%	%	%	%	%	%	%	%	ppm or mg/kg	ppm or mg/kg	ppm or mg/kg	%	ppm or mg/kg	ppm or mg/kg	ppm or mg/kg
	RYE															
70	GRAIN, ALL ANALYSES	87	0.06	0.31	0.02	0.03	0.12	0.46	0.15	–	7.5	–	0.007	72.0	–	28.1
71	DISTILLERS GRAINS, DEHY	92	0.15	0.48	0.17	0.05	0.17	0.07	0.44	–	–	–	–	18.4	–	–
72	DISTILLERS GRAINS WITH SOLUBLES, DEHY	90	–	–	–	–	–	–	–	–	–	–	–	–	–	–
73	**SAFFLOWER,** SEEDS	93	0.24	0.57	0.06	–	0.34	0.74	0.06	–	10.0	–	0.032	1.1	–	30.0
74	**SCREENINGS, GRAIN, CEREAL,** ALL ANALYSES (ALSO SEE BARLEY; WHEAT)	90	0.33	0.35	0.40	–	0.12	0.30	–	–	–	–	–	44.4	–	–
	SORGHUM															
75	GRAIN, ALL ANALYSES	90	0.05	0.32	0.03	0.08	0.14	0.35	0.15	0.275	9.7	–	0.007	9.8	–	42.4
76	KAFIR, GRAIN	89	0.03	0.31	0.05	0.10	0.15	0.34	0.16	0.387	7.0	–	0.007	15.8	0.797	13.5
77	MILO, GRAIN	89	0.04	0.30	0.04	0.08	0.13	0.31	0.11	0.471	4.3	0.061	0.005	15.8	0.201	16.9
78	SUDANGRASS, GRAIN	92	–	–	–	–	–	–	–	–	–	–	–	–	–	–
	SOYBEAN															
79	SEEDS	92	0.25	0.60	0.00	0.03	0.27	1.66	0.22	–	18.2	–	0.009	36.4	0.111	56.9
80	OIL	100	–	–	–	–	–	–	–	–	–	–	–	–	–	–
	SUGARCANE															
81	MOLASSES, DEHY	94	1.04	0.42	0.19	–	0.44	3.40	0.43	1.145	74.9	–	0.024	54.1	–	31.2
82	MOLASSES, MORE THAN 46% INVERT SUGAR, MORE THAN 79.5 DEGREES BRIX	74	0.74	0.08	0.16	2.26	0.31	2.98	0.35	1.180	48.9	1.564	0.020	43.7	–	15.6
83	**SUNFLOWER,** SEEDS	94	0.16	0.67	0.02	–	0.37	0.68	0.28	–	23.5	–	0.006	21.9	–	68.6
	SWEET POTATO															
84	TUBERS, DEHY, MEAL	89	0.12	0.15	–	–	–	–	–	–	–	–	–	–	–	–
85	TUBERS, FRESH	33	0.03	0.05	0.02	0.02	0.05	0.35	0.04	–	1.4	–	0.002	3.7	–	–
86	**TRITICALE,** GRAIN	89	0.04	0.30	0.01	–	0.23	0.51	–	0.078	8.3	–	0.005	42.5	–	31.2
87	**TURNIP,** ROOTS, FRESH	9	0.06	0.03	0.01	0.06	0.02	0.26	0.04	–	2.0	–	0.002	3.9	–	2.7
	WHEAT															
88	GRAIN, ALL ANALYSES	89	0.05	0.35	0.06	0.08	0.14	0.41	0.18	0.442	5.8	0.090	0.006	41.5	0.256	31.4
89	BRAN	89	0.13	1.16	0.06	0.05	0.58	1.23	0.22	0.075	11.0	0.066	0.015	114.9	0.641	94.6
90	MIDDLINGS, LESS THAN 9.5% FIBER	89	0.13	0.89	0.01	0.04	0.34	0.98	0.17	0.502	15.9	0.109	0.009	114.0	0.736	96.9
91	MILL RUN, LESS THAN 9.5% FIBER	90	0.10	1.02	–	–	0.48	1.20	0.30	0.209	18.5	–	0.010	104.1	–	–
92	RED DOG, LESS THAN 4% FIBER	88	0.06	0.51	0.01	0.14	0.18	0.52	0.24	0.117	6.3	–	0.005	52.1	0.324	65.0
93	SHORTS, LESS THAN 7% FIBER	88	0.09	0.80	0.03	0.05	0.27	0.93	0.21	0.105	11.5	–	0.008	114.1	0.476	102.4
94	**WOOD,** MOLASSES	62	1.17	0.05	0.03	0.12	0.07	0.04	0.03	–	–	–	–	12.6	–	–

TABLE V-1B PROTEIN FEEDS, MINERAL AND VITAMIN COMPOSITION OF FEEDS, DATA EXPRESSED AS-FED

Entry Number	Feed Name Description	Dry Matter	Macro Minerals							Micro Minerals						
			Calcium (Ca)	Phosphorus (P)	Sodium (Na)	Chlorine (Cl)	Magnesium (Mg)	Potassium (K)	Sulfur (S)	Cobalt (Co)	Copper (Cu)	Iodine (I)	Iron (Fe)	Manganese (Mn)	Selenium (Se)	Zinc (Zn)
		%	%	%	%	%	%	%	%	ppm or mg/kg	ppm or mg/kg	ppm or mg/kg	%	ppm or mg/kg	ppm or mg/kg	ppm or mg/kg
1	**AMMONIUM,** POLYPHOSPHATE SOLUTION	60	0.10	13.44	–	–	–	–	0.50	–	–	0.505	–	–	–	–
2	**BARLEY,** MALT SPROUTS, DEHY	93	0.18	0.63	0.88	0.36	0.17	0.25	0.79	–	5.9	–	0.018	29.4	0.416	56.4
	BEAN															
3	SEEDS, KIDNEY	89	0.11	0.43	0.01	–	0.10	0.93	–	0.508	6.1	–	0.007	17.0	0.355	24.3
4	SEEDS, NAVY	89	0.17	0.54	0.04	0.06	0.13	1.31	0.23	–	9.9	–	0.010	21.1	–	–
5	SEEDS, PINTO	90	0.18	0.48	0.01	–	0.24	2.59	–	0.551	13.0	–	0.014	17.3	–	40.0
6	**BEAN, LIMA,** SEEDS	90	0.08	0.38	0.02	0.03	0.18	1.61	0.20	–	8.2	–	0.009	16.1	–	–
7	**BEAN, MUNG,** SEEDS	90	0.13	0.35	0.01	–	–	1.04	–	–	–	–	0.009	–	–	–
8	**BEAN, TEPARY,** SEEDS	90	–	–	–	–	–	–	–	–	–	–	–	–	–	–
9	**BLOOD,** MEAL	91	0.29	0.25	0.32	0.30	0.22	0.09	0.34	0.088	12.6	–	0.372	5.3	0.731	4.4
	BREWERS (ALSO SEE CORN; RYE; SORGHUM; WHEAT)															
10	GRAINS, DEHY	92	0.30	0.51	0.21	0.15	0.15	0.09	0.30	0.076	21.7	0.066	0.024	37.2	–	27.3
11	GRAINS, WET	22	0.06	0.12	0.06	0.03	0.03	0.02	0.07	0.022	4.9	–	0.006	9.0	–	23.2
	BUTTERMILK, CATTLE															
12	CONDENSED	29	0.44	0.26	0.31	0.12	0.19	0.23	0.03	–	–	–	–	–	–	–
13	DEHY	92	1.32	0.94	0.83	0.44	0.48	0.83	0.08	–	1.0	–	0.001	3.5	–	40.2
14	**CASEIN,** ACID PRECIPITATED, DEHY	91	0.61	0.82	0.01	–	0.01	0.01	–	–	4.1	–	0.002	3.5	–	31.8
15	**CHEESE, CATTLE,** RIND	85	0.98	0.56	0.81	0.60	0.02	0.28	–	–	–	–	–	–	–	–
	CHICKEN															
16	BY-PRODUCT, FRESH (VISCERA WITH FEET, WITH HEADS)	39	–	–	–	–	–	–	–	–	–	–	–	–	–	–
17	BY-PRODUCT, WITHOUT FEET, FRESH (VISCERA WITHOUT FEET, WITH HEADS)	34	0.34	0.24	–	–	–	–	–	–	–	–	–	–	–	–
	CITRUS															
18	MOLASSES, AMMONIATED	61	0.76	0.16	–	–	0.08	–	–	–	–	–	–	–	–	–
19	PULP, AMMONIATED, DEHY	87	1.66	0.12	–	–	0.07	–	–	–	–	–	–	–	–	–
20	**COCONUT,** KERNELS WITH COATS, MEAL MECH EXTD (COPRA MEAL)	92	0.19	0.60	0.04	–	0.30	1.65	0.34	0.127	16.7	–	0.068	70.1	–	48.5
	CORN															
21	DISTILLERS' GRAINS, DEHY	93	0.09	0.39	0.09	0.07	0.07	0.16	0.43	0.076	38.9	0.048	0.020	19.3	0.352	41.7
22	DISTILLERS' SOLUBLES, DEHY	93	0.30	1.30	0.23	0.26	0.60	1.70	0.37	0.167	77.9	0.079	0.052	72.0	0.371	88.0
23	GERM MEAL, WET MILLED, SOLV EXTD	92	0.04	0.51	0.04	0.04	0.16	0.35	0.31	–	4.5	–	9.034	3.8	0.340	104.8
24	GLUTEN FEED	90	0.32	0.74	0.12	0.22	0.33	0.57	0.21	0.087	47.1	0.066	0.043	23.1	0.272	64.6
25	GLUTEN MEAL	91	0.15	0.46	0.09	0.06	0.06	0.03	0.20	0.077	27.7	–	0.039	7.7	1.015	173.7
	COTTON															
26	SEEDS WITHOUT LINT	91	0.14	0.69	0.03	–	0.32	1.11	0.24	–	49.0	–	0.014	11.1	–	–

Composition of Feeds

ENERGY FEEDS

Entry Number	Fat-Soluble Vitamins					Water-Soluble Vitamins								
	A (1 mg Carotene = 1667 IU Vit A)	Carotene (Provitamin A)	D	E	K	B-12	Biotin	Choline	Folacin (Folic Acid)	Niacin	Pantothenic Acid (B-3)	(Pyridoxine) B-6	Riboflavin (B-2)	Thiamin (B-1)
	IU/g	ppm or mg/kg	IU/kg	ppm or mg/kg	ppm or mg/kg	ppb or mcg/kg	ppm or mg/kg	ppm or mg/kg	ppm or mg/kg	ppm or mg/kg	ppm or mg/kg	ppm or mg/kg	ppm or mg/kg	ppm or mg/kg
70	0.1	0.1	–	14.5	–	–	0.06	419	0.62	14	7.5	–	1.7	4.1
71	–	–	–	–	–	–	–	–	–	17	5.2	–	3.3	1.3
72	–	–	–	–	–	–	–	–	–	63	17.4	–	8.2	3.1
73	–	–	–	–	–	–	–	–	–	–	–	–	–	–
74	–	–	–	–	–	–	–	1044	1.06	10	12.8	–	1.8	–
75	2.0	1.2	–	–	–	–	0.26	880	0.22	47	10.2	5.41	1.2	4.5
76	0.6	0.4	–	–	–	–	0.24	439	0.20	38	12.0	6.68	1.2	3.8
77	0.4	0.2	0	12.1	0.22	–	0.23	638	0.21	37	11.0	4.69	1.1	4.1
78	–	–	–	–	–	–	–	–	–	–	–	–	–	–
79	1.5	0.9	–	33.7	–	–	0.38	2931	–	23	16.0	11.04	2.9	11.3
80	–	–	–	–	–	–	–	–	–	–	–	–	–	–
81	–	–	–	5.2	–	–	–	–	–	–	–	–	–	–
82	–	–	–	5.4	–	–	0.69	764	0.11	36	37.4	4.21	2.8	0.9
83	–	–	–	–	–	–	–	–	–	–	–	–	3.3	0.4
84	117.3	70.4	–	–	–	–	–	–	–	–	–	–	–	–
85	236.8	142.1	–	–	–	–	–	–	–	7	–	–	0.7	1.1
86	–	–	–	–	–	–	–	457	–	–	–	–	0.4	–
87	–	–	–	–	–	–	–	92	0.26	7	1.7	–	0.6	0.7
88	–	–	–	15.5	–	0.9	0.10	918	0.43	59	11.3	3.74	1.3	4.3
89	4.4	2.6	–	14.3	–	–	0.38	1232	1.77	197	28.0	10.34	3.6	8.4
90	5.1	3.1	–	23.8	–	–	0.24	1246	1.24	95	17.8	9.14	2.0	14.2
91	–	–	–	31.9	–	–	0.31	1005	1.08	116	13.7	11.09	2.1	15.2
92	–	–	–	37.4	–	–	0.11	1453	0.82	46	13.3	5.40	2.2	21.8
93	5.1	3.1	–	36.0	–	–	–	1697	1.51	105	21.9	–	4.1	19.5
94	–	–	–	–	–	–	–	–	–	–	–	–	–	–

PROTEIN FEEDS

Entry Number	Fat-Soluble Vitamins					Water-Soluble Vitamins								
	A (1 mg Carotene = 1667 IU Vit A)	Carotene (Provitamin A)	D	E	K	B-12	Biotin	Choline	Folacin (Folic Acid)	Niacin	Pantothenic Acid (B-3)	(Pyridoxine) B-6	Riboflavin (B-2)	Thiamin (B-1)
	IU/g	ppm or mg/kg	IU/kg	ppm or mg/kg	ppm or mg/kg	ppb or mcg/kg	ppm or mg/kg	ppm or mg/kg	ppm or mg/kg	ppm or mg/kg	ppm or mg/kg	ppm or mg/kg	ppm or mg/kg	ppm or mg/kg
1	–	–	–	–	–	–	–	–	–	–	–	–	–	–
2	–	–	–	3.7	–	–	4.09	1591	0.20	55	9.0	8.62	2.8	8.3
3	–	–	–	–	–	–	–	–	–	24	–	–	1.8	5.7
4	–	–	–	1.0	–	–	0.11	1017	1.29	24	2.4	0.30	1.7	6.3
5	–	–	–	–	–	–	–	–	–	22	2.2	–	3.1	8.6
6	–	–	–	–	–	–	–	–	3.31	20	8.4	–	1.7	4.6
7	–	–	–	–	–	–	–	–	–	25	–	–	2.1	3.9
8	–	–	–	–	–	–	–	–	–	–	–	–	–	–
9	–	–	–	–	–	44.3	0.09	780	0.10	31	2.3	4.41	2.0	0.3
10	0.8	0.5	–	26.7	–	3.6	0.44	1651	0.22	44	8.2	1.03	1.5	0.6
11	–	–	–	5.5	–	–	–	–	–	–	–	–	–	–
12	25.6	15.4	–	–	–	–	–	–	–	–	–	–	12.6	–
13	–	–	–	6.3	–	19.6	0.29	1746	0.39	9	37.0	2.47	30.6	3.4
14	–	–	–	–	–	–	0.04	208	0.47	1	2.7	0.42	1.5	0.4
15	–	–	–	–	–	–	–	–	–	–	–	–	–	–
16	–	–	–	–	–	–	–	–	–	–	–	–	–	–
17	–	–	–	–	–	–	–	–	–	–	–	–	–	–
18	–	–	–	–	–	–	–	–	–	–	–	–	–	–
19	–	–	–	–	–	–	–	–	–	–	–	–	–	–
20	–	–	–	–	–	–	–	1089	0.30	24	6.5	4.36	3.5	–
21	5.2	3.1	–	–	–	0.3	0.41	1113	1.00	38	11.3	4.22	5.0	1.8
22	1.1	0.7	–	45.9	–	4.2	1.49	4751	1.34	124	23.3	9.41	15.1	6.8
23	3.4	2.0	–	85.8	–	–	0.22	1586	0.20	39	4.2	–	3.8	4.5
24	9.8	5.9	–	12.1	–	–	0.33	1514	0.27	70	13.6	13.93	2.2	2.0
25	27.3	16.3	–	29.3	–	–	0.19	360	0.30	50	10.0	7.98	1.5	0.2
26	–	–	–	–	–	–	–	–	–	–	–	–	–	–

(Continued)

TABLE V-1B PROTEIN FEEDS, COMPOSITION OF FEEDS, DATA EXPRESSED AS-FED—(Continued)

Entry Number	Feed Name Description	International Feed Number	Proximate Analysis									Digestible Protein		
			Dry Matter %	Ash %	Crude Fiber %	Neutral Det. Fib. (NDF) %	Acid Det. Fib. (ADF) %	Lignin %	Ether Extract (Fat) %	N-Free Extract %	Crude Protein %	Ruminant %	Swine %	Horse %
	COTTON (Continued)													
27	SEEDS, MEAL MECH EXTD, 41% PROTEIN	5-01-617	93	6.1	11.9	25.9	18.5	5.6	4.7	28.9	41.0	35.1	33.8	37.7
28	SEEDS, MEAL SOLV EXTD, 41% PROTEIN	5-01-621	91	6.5	12.1	23.6	18.4	5.5	1.5	29.6	41.2	31.3	35.1	38.2
29	SEEDS, MEAL SOLV EXTD, 46% PROTEIN	5-26-100	92	7.2	8.9	25.8	19.3	—	1.6	26.8	47.6	41.1	46.1	45.2
30	SEEDS, MEAL SOLV EXTD, 48% PROTEIN	5-26-101	90	7.4	7.0	—	—	—	1.9	24.7	49.3	42.7	47.8	47.4
31	SEEDS WITHOUT HULLS, MEAL, PREPRESSED, SOLV EXTD, 50% PROTEIN	5-07-874	93	6.6	8.2	—	—	—	1.3	26.8	50.3	40.7	48.7	48.2
32	COTTON, GLANDLESS, SEEDS; MEAL SOLV EXTD	5-08-979	95	7.3	2.5	—	—	—	1.9	23.8	59.8	52.1	58.2	58.7
33	**COWPEA, COMMON,** SEEDS	5-01-661	89	3.5	5.0	—	—	—	1.5	55.5	23.8	19.6	22.4	18.7
34	**DIAMMONIUM PHOSPHATE,** (NH$_4$)$_2$HPO$_4$	6-00-370	98	35.5	—	—	—	—	—	—	112.9	—	—	—
	DISTILLERS PRODUCTS (ALSO SEE CORN; RYE; SORGHUM; WHEAT)													
35	GRAINS, DEHY	5-02-144	93	1.5	12.8	—	—	—	7.4	43.5	27.3	18.9	26.0	22.3
36	SOLUBLES, DEHY	5-02-147	92	6.2	3.4	—	—	—	8.9	44.8	28.8	24.0	27.4	24.0
37	**FEATHERS, POULTRY,** MEAL, HYDROLYZED	5-03-795	93	3.2	1.4	—	—	6.1	5.1	—	83.8	74.1	61.2	86.1
	FISH													
38	LIVER, MEAL MECH EXTD	5-01-968	93	6.1	1.2	—	—	—	17.3	5.4	62.8	54.9	61.2	62.3
39	SOLUBLES, CONDENSED	5-01-969	50	10.1	0.5	—	—	—	6.1	2.2	31.5	28.0	29.3	30.9
40	SOLUBLES, DEHY	5-01-971	93	12.7	2.0	—	—	—	9.0	8.7	60.4	52.7	58.8	59.7
41	FISH, ANCHOVY, MEAL MECH EXTD	5-01-985	92	14.7	1.0	—	—	—	4.1	6.7	65.4	57.3	55.0	65.4
42	FISH, MENHADEN, MEAL MECH EXTD	5-02-009	92	19.1	0.9	—	—	—	9.6	0.8	61.2	49.6	49.6	60.7
43	FISH, SARDINE, MEAL MECH EXTD	5-02-015	93	15.8	1.0	—	—	—	5.0	6.1	65.2	53.5	63.6	65.1
44	**FLAX, COMMON,** SEEDS, MEAL SOLV EXTD, 35% PROTEIN (LINSEED MEAL)	5-26-090	90	5.8	8.9	—	—	—	1.7	38.2	35.7	30.3	34.3	32.0
45	**LENTIL, COMMON,** SEEDS	5-02-506	88	2.6	3.4	—	—	—	1.0	57.1	24.4	19.3	23.1	19.4
46	**LINSEED,** SEEDS, MEAL SOLV EXTD, 35% PROTEIN (LINSEED MEAL)	5-26-090	90	5.8	8.9	—	—	—	1.7	38.2	35.7	30.3	34.3	32.0
47	**LIVER,** MEAL	5-00-389	93	6.3	1.4	—	—	—	15.7	3.2	66.1	57.9	64.4	66.1
	MANURE, POULTRY													
48	WITH LITTER, DEHY	5-05-587	86	17.8	15.0	—	—	8.0	2.6	25.7	24.6	14.3	23.3	19.8
49	WITHOUT LITTER, DEHY	5-14-015	90	29.2	12.0	36.3	14.4	2.1	2.0	21.6	25.4	21.0	24.1	20.4
	MEAT													
50	MEAL RENDERED	5-00-385	94	28.1	2.7	—	—	—	9.1	3.3	50.7	43.8	49.1	48.5
51	WITH BLOOD, MEAL RENDERED (TANKAGE)	5-00-386	92	21.1	1.8	—	—	—	8.7	0.1	60.5	52.8	58.9	59.8
52	WITH BLOOD, WITH BONE, MEAL RENDERED (TANKAGE)	5-00-387	93	28.2	2.2	—	—	—	12.7	3.1	46.6	40.2	45.1	44.1
53	WITH BONE, MEAL RENDERED	5-00-388	93	28.0	2.4	—	—	—	10.0	2.6	50.4	45.8	43.3	48.3
	MILK													
54	FRESH (CATTLE)	5-01-168	12	0.8	—	—	—	—	3.6	4.7	3.3	3.2	3.2	2.6
55	SKIMMED, FRESH (CATTLE)	5-01-170	10	0.7	—	—	—	—	0.1	5.8	3.0	2.5	3.0	2.5
56	SKIMMED, DEHY (CATTLE)	5-01-175	94	8.0	0.2	0.0	—	—	1.1	51.6	33.3	30.0	31.8	28.9
57	FRESH (GOAT)	5-02-128	13	0.8	—	—	—	—	4.2	4.8	3.4	2.8	3.2	2.6
58	**MOLASSES AND SYRUP,** SUGARCANE, MOLASSES, AMMONIATED	5-04-702	65	5.9	—	—	—	—	—	—	26.3	15.5	25.3	23.7
59	**MONOAMMONIUM PHOSPHATE,** NH$_4$H$_2$PO$_4$	6-09-338	98	53.0	—	—	—	—	—	—	69.4	—	—	—
	PEA													
60	SEEDS	5-03-600	89	2.9	5.5	—	—	—	1.1	56.7	23.2	19.0	21.9	18.0
61	SPLIT SEED BY-PRODUCT (PEA FEED; PEA MEAL)	1-08-478	90	3.5	23.7	—	—	—	1.4	43.7	17.7	14.5	12.1	12.0
62	**PEA, FIELD,** SEEDS	5-08-481	91	2.9	5.9	—	—	—	1.3	57.8	23.2	19.9	21.8	17.7
63	**PEANUT,** SEEDS WITHOUT HULLS, MEAL SOLV EXTD (PEANUT MEAL)	5-03-650	93	5.8	7.7	—	—	—	2.2	27.9	49.0	42.3	47.4	46.7
	POULTRY													
64	BY-PRODUCT, MEAL RENDERED	5-03-798	94	14.8	2.2	—	—	—	13.1	2.6	61.2	53.4	53.3	60.5
65	FEATHERS, MEAL, HYDROLYZED	5-03-795	93	3.2	1.4	—	—	6.1	5.1	—	83.8	74.1	61.2	86.1
66	**RAPE (CANOLA),** SEEDS, MEAL SOLV EXTD, 34% PROTEIN	5-26-092	90	7.0	13.0	—	—	—	2.5	33.5	34.0	—	—	—
67	**RICE,** HULLS, AMMONIATED	1-05-698	92	—	44.7	—	—	—	0.9	16.9	10.4	6.3	5.6	6.4
	RYE													
68	DISTILLERS GRAINS, DEHY	5-04-023	92	2.3	12.3	—	—	—	6.0	48.3	23.0	13.8	21.7	17.5
69	DISTILLERS GRAINS WITH SOLUBLES, DEHY	5-04-024	90	6.4	8.1	—	—	—	4.1	44.7	27.2	22.6	25.9	22.3
	SAFFLOWER													
70	SEEDS, MEAL SOLV EXTD, 20% PROTEIN	5-26-095	92	4.6	32.2	—	39.6	—	1.1	32.7	21.6	17.4	20.2	15.8
71	SEEDS WITHOUT HULLS, MEAL SOLV EXTD, 42% PROTEIN	5-26-094	92	6.5	14.6	—	19.2	—	1.3	26.3	42.7	36.7	41.3	39.8
72	**SESAME,** SEEDS, MEAL SOLV EXTD, 44% PROTEIN	5-26-096	92	13.1	6.8	—	—	—	1.4	25.8	45.0	39.4	44.4	42.3
73	**SHRIMP,** CANNERY RESIDUE, MEAL (SHRIMP MEAL)	5-04-226	90	26.7	14.1	—	16.6	—	3.9	6.7	38.7	33.0	37.3	35.4
	SORGHUM													
74	DISTILLERS GRAINS, DEHY	5-04-374	94	4.3	12.1	—	—	—	8.3	38.3	30.8	24.7	29.4	26.1
75	DISTILLERS GRAINS WITH SOLUBLES, DEHY	5-04-375	95	4.2	10.1	—	—	—	9.4	38.0	33.1	27.8	31.7	28.6
76	GLUTEN MEAL	5-04-388	90	1.7	4.9	—	—	—	4.4	34.9	44.4	38.2	43.0	41.8
	SOYBEAN													
77	SEEDS	5-04-610	92	5.1	5.4	—	9.2	—	17.2	25.9	38.4	34.5	31.5	34.9
78	SEEDS, MEAL SOLV EXTD, 44% PROTEIN	5-20-637	89	6.4	6.2	12.5	8.9	—	1.5	30.6	44.4	37.8	40.4	41.9
79	SEEDS WITHOUT HULLS, MEAL SOLV EXTD, 49% PROTEIN	5-20-638	90	6.1	3.7	6.6	6.2	—	1.2	29.8	49.0	42.4	44.6	47.1
80	**SUNFLOWER, COMMON,** SEEDS WITHOUT HULLS, MEAL SOLV EXTD, 44% PROTEIN	5-26-098	93	7.7	11.0	—	—	—	2.9	24.6	46.8	42.1	—	—
81	**UREA,** 45% NITROGEN, 281% PROTEIN EQUIVALENT	5-05-070	99	—	—	0	0	—	—	—	281.7	—	—	—
82	**UREA,** CONDITIONER ADDED, 42% NITROGEN, 262% PROTEIN EQUIVALENT	5-20-705	100	—	—	—	—	—	—	—	261.9	—	—	—
	WHEAT													
83	DISTILLERS GRAINS, DEHY	5-05-193	93	3.0	11.8	—	—	—	6.7	40.2	31.6	26.5	30.2	27.1
84	DISTILLERS SOLUBLES, DEHY	5-05-195	94	6.8	3.4	—	—	—	2.2	50.5	31.1	—	—	—
85	GERM MEAL	5-05-218	88	4.3	3.1	—	4.4	—	8.5	48.1	24.4	22.9	23.1	19.4
86	GLUTEN	5-05-221	91	0.9	0.4	—	—	—	0.8	9.8	79.0	71.1	77.3	80.9
	WHEY, CATTLE													
87	FRESH	4-08-134	7	0.7	—	0	0	—	0.3	5.1	0.9	0.6	0.7	0.7
88	DEHY	4-01-182	93	8.8	0.2	0.3	0.2	—	0.8	70.2	13.3	8.9	13.1	9.6
89	**YEAST, BREWERS,** DEHY	7-05-527	93	6.5	3.0	—	3.7	—	0.9	38.8	43.8	39.0	—	—
90	**YEAST, IRRADIATED,** DEHY	7-05-529	94	6.2	6.2	—	—	—	1.1	32.4	48.1	—	—	—
91	**YEAST, TORULA,** DEHY	7-05-534	93	8.0	2.5	—	3.7	—	1.6	31.5	49.6	45.1	40.6	—

Composition of Feeds

PROTEIN FEEDS

Entry Number	TDN			Digestible Energy						Metabolizable Energy								Net Energy					
	Ruminant	Swine	Horse	Ruminant		Swine		Horse		Ruminant		Swine		Poultry ME$_n$		Horse		Ruminant NE$_m$		Ruminant NE$_g$		Lactating Cows NE$_{lc}$	
	%	%	%	Mcal		kcal		Mcal		Mcal		kcal		kcal		Mcal		Mcal		Mcal		Mcal	
				lb	kg	lb	kg	lb	kg	lb	kg	lb	kg	lb	kg	lb	kg	lb	kg	lb	kg	lb	kg
27	72	69	—	1.49	3.29	1305	2878	—	—	1.12	2.47	1197	2638	1025	2261	—	—	0.72	1.59	0.64	1.41	0.76	1.67
28	68	61	—	1.48	3.27	1209	2666	—	—	1.17	2.57	1069	2356	889	1960	—	—	0.74	1.64	0.63	1.40	0.73	1.61
29	71	64	—	1.42	3.13	1282	2826	—	—	1.25	2.74	1105	2436	950	2095	—	—	0.77	1.70	0.51	1.12	0.74	1.63
30	73	65	—	1.44	3.17	1292	2847	—	—	1.27	2.79	1116	2459	983	2166	—	—	0.80	1.77	0.54	1.18	0.76	1.67
31	70	—	—	1.43	3.15	1295	2854	—	—	1.25	2.76	1185	2613	971	2141	—	—	0.76	1.67	0.49	1.08	0.73	1.60
32	81	78	—	1.62	3.58	1560	3438	—	—	1.45	3.19	1359	2996	948	2089	—	—	0.92	2.03	0.63	1.39	0.86	1.90
33	76	75	—	1.48	3.27	1500	3307	—	—	1.32	2.90	1309	2886	—	—	—	—	0.80	1.76	0.53	1.18	0.75	1.66
34	—	—	—	—	—	—	—	—	—	—	—	—	—	—	—	—	—	—	—	—	—	—	—
35	76	83	—	1.52	3.36	1669	3680	—	—	1.35	2.97	1463	3225	1138	2509	—	—	0.87	1.92	0.59	1.31	0.82	1.80
36	78	82	—	1.61	3.54	1637	3609	—	—	1.43	3.16	1433	3160	1316	2901	—	—	0.93	2.06	0.65	1.43	0.87	1.92
37	67	62	—	1.21	2.66	1238	2729	—	—	0.84	1.84	1004	2213	1104	2434	—	—	0.48	1.05	0.24	0.53	0.65	1.43
38	97	96	—	1.94	4.28	1914	4219	—	—	1.77	3.90	1688	3722	—	—	—	—	1.12	2.48	0.80	1.76	1.03	2.28
39	41	44	—	0.85	1.87	866	1909	—	—	0.78	1.73	736	1623	755	1665	—	—	0.54	1.20	0.38	0.84	0.44	0.97
40	77	66	—	1.50	3.30	1467	3234	—	—	1.21	2.66	1278	2818	1322	2915	—	—	0.81	1.78	0.54	1.19	0.77	1.70
41	72	69	—	1.45	3.19	1370	3020	—	—	1.27	2.81	1124	2478	1245	2745	—	—	0.79	1.73	0.52	1.14	0.75	1.65
42	67	61	—	1.33	2.94	1578	3479	—	—	1.16	2.55	1194	2633	1292	2848	—	—	0.75	1.65	0.49	1.08	0.72	1.58
43	70	67	—	1.40	3.09	1327	2925	—	—	1.23	2.71	1148	2531	1313	2896	—	—	0.76	1.67	0.49	1.08	0.73	1.60
44	70	67	—	1.41	3.10	1336	2945	—	—	1.24	2.73	1156	2549	—	—	—	—	0.75	1.65	0.49	1.08	0.72	1.58
45	74	84	—	1.48	3.26	1675	3692	—	—	1.31	2.89	1471	3244	—	—	—	—	0.81	1.79	0.55	1.21	0.77	1.69
46	70	67	—	1.41	3.10	1336	2945	—	—	1.24	2.73	1156	2549	—	—	—	—	0.75	1.65	0.49	1.08	0.72	1.58
47	89	93	—	1.79	3.94	1867	4116	—	—	1.62	3.57	1645	3627	1306	2878	—	—	1.00	2.21	0.70	1.55	0.93	2.04
48	47	9	—	0.99	2.19	175	386	—	—	0.83	1.83	88	194	—	—	—	—	0.52	1.16	0.30	0.66	0.54	1.18
49	46	—	—	0.92	2.02	—	—	—	—	0.74	1.63	—	—	—	—	—	—	0.54	1.18	0.30	0.66	0.55	1.22
50	67	64	—	1.20	2.64	936	2064	—	—	0.89	1.95	1009	2225	947	2088	—	—	0.52	1.15	0.28	0.62	0.69	1.53
51	67	67	—	1.34	2.96	1112	2451	—	—	1.17	2.58	951	2096	1212	2673	—	—	0.74	1.62	0.48	1.05	0.71	1.56
52	63	68	—	1.44	3.17	1382	3047	—	—	1.26	2.79	1199	2644	1188	2618	—	—	0.90	1.99	0.62	1.36	0.85	1.86
53	66	68	—	1.32	2.91	1028	2267	—	—	1.14	2.51	981	2162	946	2086	—	—	0.71	1.57	0.45	1.00	0.69	1.52
54	16	15	—	0.32	0.71	309	681	—	—	0.30	0.65	275	606	—	—	—	—	0.19	0.42	0.14	0.30	0.18	0.40
55	9	9	—	0.18	0.39	188	415	—	—	0.16	0.35	166	365	—	—	—	—	0.10	0.22	0.07	0.16	0.10	0.21
56	80	86	—	1.43	3.15	1758	3876	—	—	1.07	2.37	1630	3593	1152	2539	—	—	0.69	1.52	0.43	0.95	0.83	1.82
57	17	—	—	0.34	0.75	—	—	—	—	0.32	0.70	—	—	—	—	—	—	—	—	—	—	—	—
58	47	—	—	0.95	2.08	—	—	—	—	0.82	1.81	—	—	—	—	—	—	—	—	—	—	—	—
59	—	—	—	—	—	—	—	—	—	—	—	—	—	—	—	—	—	—	—	—	—	—	—
60	77	—	—	1.56	3.44	1483	3268	—	—	1.39	3.07	1293	2850	959	2115	—	—	0.87	1.91	0.59	1.31	0.81	1.79
61	78	55	62	1.34	2.96	—	—	1.13	2.49	1.17	2.59	—	—	—	—	0.93	2.04	0.57	1.26	0.33	0.74	0.58	1.28
62	76	82	—	1.51	3.34	1638	3611	—	—	1.34	2.96	1435	3163	1108	2442	—	—	0.80	1.77	0.54	1.18	0.76	1.68
63	73	74	—	1.43	3.15	1296	2857	—	—	1.29	2.84	1264	2787	1229	2709	—	—	0.78	1.72	0.51	1.13	0.74	1.63
64	74	76	—	1.44	3.17	1406	3101	—	—	1.21	2.66	1301	2869	1300	2865	—	—	0.81	1.78	0.53	1.18	0.76	1.68
65	67	62	—	1.21	2.66	1238	2729	—	—	0.84	1.84	1004	2213	1104	2434	—	—	0.48	1.05	0.24	0.53	0.65	1.43
66	—	—	—	—	—	—	—	—	—	—	—	—	—	—	—	—	—	—	—	—	—	—	—
67	—	—	—	—	—	—	—	—	—	—	—	—	—	—	—	—	—	—	—	—	—	—	—
68	54	79	—	1.08	2.38	1586	3497	—	—	0.90	1.99	1387	3057	—	—	—	—	0.45	1.00	0.22	0.49	0.50	1.09
69	73	68	—	1.46	3.21	1354	2985	—	—	1.29	2.83	1173	2586	—	—	—	—	0.78	1.72	0.52	1.14	0.74	1.64
70	46	—	—	0.87	1.92	1089	2401	—	—	0.86	1.90	929	2047	623	1374	—	—	0.51	1.12	0.27	0.59	0.40	0.89
71	66	56	—	1.15	2.53	1508	3325	—	—	0.97	2.15	1317	2904	885	1951	—	—	0.58	1.27	0.33	0.74	0.56	1.24
72	69	81	—	1.38	3.04	1610	3549	—	—	1.13	2.48	1389	3063	1178	2598	—	—	0.71	1.56	0.46	1.01	0.79	1.75
73	41	—	—	0.82	1.82	—	—	—	—	0.65	1.43	—	—	871	1920	—	—	0.29	0.65	0.08	0.17	0.38	0.84
74	78	77	—	1.56	3.45	1535	3385	—	—	1.39	3.06	1338	2949	—	—	—	—	0.85	1.88	0.57	1.27	0.80	1.77
75	81	82	—	1.63	3.58	1649	3635	—	—	1.45	3.19	1442	3178	—	—	—	—	0.88	1.94	0.60	1.32	0.83	1.83
76	80	91	—	1.60	3.52	1581	3486	—	—	1.43	3.15	1386	3056	1237	2727	—	—	0.85	1.88	0.58	1.28	0.80	1.76
77	84	93	—	1.69	3.72	1820	4012	—	—	1.52	3.34	1605	3539	1534	3382	—	—	0.93	2.04	0.64	1.41	0.86	1.90
78	76	75	—	1.45	3.19	1565	3450	—	—	1.17	2.59	1430	3153	1005	2216	—	—	0.79	1.74	0.53	1.16	0.75	1.66
79	78	77	—	1.51	3.32	1591	3508	—	—	1.10	2.42	1433	3160	1124	2478	—	—	0.72	1.59	0.47	1.03	0.79	1.75
80	65	69	—	1.22	2.70	1366	3012	—	—	1.01	2.23	1173	2585	919	2026	—	—	0.59	1.30	0.34	0.74	0.63	1.40
81	—	—	—	—	—	—	—	—	—	—	—	—	—	—	—	—	—	—	—	—	—	—	—
82	—	—	—	—	—	—	—	—	—	—	—	—	—	—	—	—	—	—	—	—	—	—	—
83	78	80	—	1.52	3.35	1600	3527	—	—	1.35	2.96	1398	3081	—	—	—	—	0.81	1.79	0.54	1.19	0.77	1.70
84	—	—	—	—	—	—	—	—	—	—	—	—	—	—	—	—	—	—	—	—	—	—	—
85	83	80	—	1.66	3.66	1727	3807	—	—	1.50	3.30	1522	3357	1223	2696	—	—	0.96	2.11	0.67	1.48	0.89	1.95
86	84	—	—	1.69	3.73	2002	4413	—	—	1.52	3.35	1771	3905	—	—	—	—	0.96	2.11	0.67	1.47	0.89	1.96
87	7	—	—	0.13	0.29	—	—	—	—	0.12	0.26	—	—	—	—	—	—	—	—	—	—	—	—
88	76	77	—	1.51	3.33	1444	3183	—	—	1.28	2.83	1411	3110	880	1939	—	—	0.87	1.92	0.59	1.30	0.78	1.71
89	73	70	—	1.46	3.21	—	—	—	—	1.28	2.83	1299	2865	928	2047	—	—	0.81	1.79	0.54	1.19	0.77	1.69
90	72	—	—	1.43	3.16	—	—	—	—	1.26	2.77	—	—	—	—	—	—	—	—	—	—	—	—
91	72	64	—	1.49	3.29	1287	2837	—	—	1.27	2.81	1096	2416	840	1851	—	—	0.82	1.81	0.55	1.21	0.78	1.71

TABLE V-1B PROTEIN FEEDS, MINERAL AND VITAMIN COMPOSITION OF FEEDS, DATA EXPRESSED AS-FED—(Continued)

Entry Number	Feed Name Description	Dry Matter	Macro Minerals							Micro Minerals						
			Calcium (Ca)	Phosphorus (P)	Sodium (Na)	Chlorine (Cl)	Magnesium (Mg)	Potassium (K)	Sulfur (S)	Cobalt (Co)	Copper (Cu)	Iodine (I)	Iron (Fe)	Manganese (Mn)	Selenium (Se)	Zinc (Zn)
		%	%	%	%	%	%	%	%	ppm or mg/kg	ppm or mg/kg	ppm or mg/kg	%	ppm or mg/kg	ppm or mg/kg	ppm or mg/kg
	COTTON (Continued)															
27	SEEDS, MEAL MECH EXTD, 41% PROTEIN	93	0.19	1.07	0.04	0.04	0.53	1.33	0.40	0.626	18.5	—	0.018	22.3	—	61.8
28	SEEDS, MEAL SOLV EXTD, 41% PROTEIN	91	0.17	1.11	0.04	0.04	0.54	1.37	0.25	0.483	19.5	—	0.019	20.6	—	60.7
29	SEEDS, MEAL SOLV EXTD, 46% PROTEIN	92	—	—	—	—	—	—	—	—	—	—	—	—	—	—
30	SEEDS, MEAL SOLV EXTD, 48% PROTEIN	90	0.20	1.20	—	—	—	—	—	—	—	—	—	—	—	—
31	SEEDS WITHOUT HULLS, MEAL, PREPRESSED, SOLV EXTD, 50% PROTEIN	93	0.18	1.16	0.05	0.05	0.46	1.45	0.52	0.042	14.5	—	0.012	23.0	—	73.8
32	COTTON, GLANDLESS, SEEDS, MEAL SOLV EXTD	95	—	—	—	—	—	—	—	—	—	—	—	—	—	—
33	**COWPEA, COMMON**, SEEDS	89	0.09	0.44	0.04	0.04	0.26	1.16	0.25	—	4.4	—	0.021	40.2	—	—
34	**DIAMMONIUM PHOSPHATE**, (NH₄)₂HPO₄	98	0.50	20.09	0.04	—	0.45	—	2.47	—	80.7	—	1.514	504.3	—	302.6
	DISTILLERS PRODUCTS (ALSO SEE CORN; RYE; SORGHUM; WHEAT)															
35	GRAINS, DEHY	93	0.12	0.54	0.05	0.05	0.09	0.20	0.46	0.092	47.9	—	0.027	35.0	—	—
36	SOLUBLES, DEHY	92	0.24	1.35	0.45	—	0.53	1.97	—	0.196	71.6	—	0.031	64.1	—	138.0
37	**FEATHERS, POULTRY**, MEAL, HYDROLYZED	93	0.30	0.62	0.63	0.28	0.18	0.27	1.50	0.116	7.3	0.044	0.023	11.9	0.913	71.9
	FISH															
38	LIVER, MEAL MECH EXTD	93	—	—	—	—	—	—	—	—	—	—	—	—	—	—
39	SOLUBLES, CONDENSED	50	0.16	0.57	2.45	2.93	0.03	1.64	0.12	0.069	46.6	1.111	0.028	13.2	—	43.2
40	SOLUBLES, DEHY	93	0.40	1.27	1.70	—	0.30	2.50	0.45	—	20.0	—	0.095	50.4	2.692	76.7
41	**FISH, ANCHOVY**, MEAL MECH EXTD	92	3.74	2.48	0.88	1.00	0.25	0.72	0.78	0.173	9.1	3.137	0.022	11.0	1.355	105.0
42	**FISH, MENHADEN**, MEAL MECH EXTD	92	5.19	2.88	0.41	0.55	0.15	0.70	0.56	0.153	10.3	1.091	0.055	37.0	2.147	144.2
43	**FISH, SARDINE**, MEAL MECH EXTD	93	4.61	2.68	0.18	0.41	0.10	0.32	—	0.183	20.2	—	0.030	23.2	1.772	—
44	**FLAX, COMMON**, SEEDS, MEAL SOLV EXTD, 35% PROTEIN (LINSEED MEAL)	90	0.40	0.82	0.14	—	0.60	1.37	0.39	—	—	—	—	—	—	—
45	**LENTIL, COMMON**, SEEDS	88	0.08	0.38	0.03	—	—	0.79	—	—	—	—	0.007	—	—	—
46	**LINSEED**, SEEDS, MEAL SOLV EXTD, 35% PROTEIN (LINSEED MEAL)	90	0.40	0.82	0.14	—	0.60	1.37	0.39	—	—	—	—	—	—	—
47	**LIVER**, MEAL	93	0.56	1.26	—	—	0.10	—	—	0.135	89.4	—	0.064	8.8	—	61.8
	MANURE, POULTRY															
48	WITH LITTER, DEHY	86	2.67	1.69	0.41	—	0.43	1.32	—	—	283.3	—	0.046	281.1	0.559	359.4
49	WITHOUT LITTER, DEHY	90	8.07	2.22	0.61	0.86	0.56	1.99	0.16	—	24.6	—	—	—	—	366.4
	MEAT															
50	MEAL RENDERED	94	8.61	4.58	1.05	1.11	0.25	0.55	0.46	2.250	9.6	—	0.050	11.8	0.505	74.3
51	WITH BLOOD, MEAL RENDERED (TANKAGE)	92	5.87	3.09	1.67	1.73	0.36	0.55	0.70	0.153	38.8	—	0.211	19.2	—	—
52	WITH BLOOD, WITH BONE, MEAL RENDERED (TANKAGE)	93	11.16	5.41	—	—	—	—	0.26	—	—	—	—	—	0.261	—
53	WITH BONE, MEAL RENDERED	93	10.00	4.94	0.72	0.75	1.02	1.33	0.25	0.181	1.5	1.317	0.066	13.3	0.263	94.3
	MILK															
54	FRESH (CATTLE)	12	0.12	0.09	0.05	0.11	0.01	0.14	0.04	0.001	0.1	—	0.002	—	—	2.3
55	SKIMMED, FRESH (CATTLE)	10	0.13	0.10	0.04	0.05	0.01	0.12	0.03	0.011	1.1	—	0.001	0.2	—	4.9
56	SKIMMED, DEHY (CATTLE)	94	1.28	1.02	0.51	0.90	0.12	1.60	0.32	0.113	11.7	—	0.001	2.1	0.124	38.5
57	FRESH (GOAT)	13	0.13	0.11	0.04	0.18	0.03	0.19	0.00	—	0.3	—	0.001	—	—	—
58	**MOLASSES AND SYRUP**, SUGARCANE, MOLASSES, AMMONIATED	65	0.79	0.13	—	—	—	—	—	—	—	—	—	—	—	—
59	**MONOAMMONIUM PHOSPHATE**, NH₄H₂PO₄	98	0.38	24.42	0.08	—	0.46	0.14	0.82	—	85.7	—	0.991	461.7	—	639.6
	PEA															
60	SEEDS	89	0.12	0.41	0.04	0.05	0.12	0.95	—	—	—	—	0.007	2.9	—	23.0
61	SPLIT SEED BY-PRODUCT (PEA FEED; PEA MEAL)	90	—	—	—	—	—	—	—	—	—	—	—	—	—	—
62	**PEA, FIELD**, SEEDS	91	0.16	0.38	0.00	—	0.14	1.36	—	1.700	11.7	—	0.020	21.2	0.393	46.8
63	**PEANUT**, SEEDS WITHOUT HULLS, MEAL SOLV EXTD (PEANUT MEAL)	93	0.36	0.61	0.03	0.03	0.27	1.16	0.31	—	—	—	—	—	—	—
	POULTRY															
64	BY-PRODUCT, MEAL RENDERED	94	3.97	2.06	0.78	0.54	0.14	0.51	0.53	4.926	19.9	3.101	0.064	16.5	0.920	193.5
65	FEATHERS, MEAL, HYDROLYZED	93	0.30	0.62	0.63	0.28	0.18	0.27	1.50	0.116	7.3	0.044	0.023	11.9	0.913	71.9
66	**RAPE**, SEEDS, MEAL SOLV EXTD, 34% PROTEIN	90	—	—	—	—	—	—	—	—	—	—	—	—	—	—
67	**RICE**, HULLS, AMMONIATED	92	0.15	0.19	—	—	—	—	—	—	—	—	—	—	—	—
	RYE															
68	DISTILLERS' GRAINS, DEHY	92	0.15	0.48	0.17	0.05	0.17	0.07	0.44	—	—	—	—	18.4	—	—
69	DISTILLERS' GRAINS WITH SOLUBLES, DEHY	90	—	—	—	—	—	—	—	—	—	—	—	—	—	—
	SAFFLOWER															
70	SEEDS, MEAL SOLV EXTD, 20% PROTEIN	92	0.31	0.61	—	—	0.32	0.74	0.20	—	9.6	—	0.043	17.7	—	39.6
71	SEEDS WITHOUT HULLS, MEAL SOLV EXTD, 42% PROTEIN	92	0.38	1.08	—	—	1.18	1.18	0.34	1.832	80.6	—	0.091	36.6	—	168.5
72	**SESAME**, SEEDS, MEAL SOLV EXTD, 44% PROTEIN	92	2.01	1.28	—	—	—	—	—	—	—	—	—	47.5	—	—
73	**SHRIMP**, CANNERY RESIDUE, MEAL (SHRIMP MEAL)	90	10.40	1.85	1.57	1.04	0.54	0.83	—	—	—	—	0.011	29.8	—	28.4
	SORGHUM															
74	DISTILLERS' GRAINS, DEHY	94	0.15	0.69	0.05	—	0.18	0.36	0.17	—	—	—	0.005	—	—	—
75	DISTILLERS' GRAINS WITH SOLUBLES, DEHY	95	0.17	0.92	—	—	—	—	—	—	—	—	—	104.5	—	—
76	GLUTEN MEAL	90	0.03	0.27	—	—	0.16	0.48	—	—	—	—	—	15.6	—	—
	SOYBEAN															
77	SEEDS	92	0.25	0.60	0.00	0.03	0.27	1.66	0.22	—	18.2	—	0.009	36.4	0.111	56.9
78	SEEDS, MEAL SOLV EXTD, 44% PROTEIN	89	0.35	0.64	0.03	—	0.27	1.98	0.41	1.381	19.9	—	0.017	31.6	0.486	50.5
79	SEEDS WITHOUT HULLS, MEAL SOLV EXTD, 49% PROTEIN	90	0.25	0.63	0.00	0.07	0.37	1.79	0.41	2.693	13.5	0.152	0.010	49.5	—	51.1
80	**SUNFLOWER, COMMON**, SEEDS WITHOUT HULLS, MEAL SOLV EXTD, 44% PROTEIN	93	—	—	—	—	—	—	—	—	—	—	—	—	—	—
81	**UREA**, 45% NITROGEN, 281% PROTEIN EQUIVALENT	99	—	—	—	0.00	—	—	—	—	6.9	—	0.018	—	—	6.9
82	**UREA, CONDITIONER ADDED**, 42% NITROGEN, 262% PROTEIN EQUIVALENT	100	—	—	—	—	—	—	—	—	—	—	—	—	—	—
	WHEAT															
83	DISTILLERS' GRAINS, DEHY	93	0.11	0.58	—	—	—	—	—	—	—	—	—	15.0	—	—
84	DISTILLERS' SOLUBLES, DEHY	94	—	—	—	—	—	—	—	—	—	—	—	—	—	—
85	GERM MEAL	88	0.06	0.95	0.02	0.06	0.25	0.94	0.27	0.120	9.2	—	0.006	132.5	0.463	119.4
86	GLUTEN	90	0.06	0.23	0.06	—	0.04	0.02	0.95	0.049	11.6	0.058	0.006	18.1	3.753	38.5
	WHEY, CATTLE															
87	FRESH	7	0.06	0.05	—	—	—	0.19	—	—	—	—	0.003	0.2	—	—
88	DEHY	93	0.86	0.76	0.62	0.07	0.13	1.11	1.04	0.111	46.5	—	0.017	5.9	—	3.2
89	**YEAST, BREWERS**, DEHY	93	0.14	1.36	0.07	0.07	0.24	1.69	0.43	0.506	38.4	0.358	0.009	6.7	0.911	39.0
90	**YEAST, IRRADIATED**, DEHY	94	0.78	1.42	—	—	—	—	2.14	—	—	—	—	—	—	—
91	**YEAST, TORULA**, DEHY	93	0.55	1.61	0.01	0.02	0.14	1.92	0.55	0.031	11.9	2.502	0.011	9.3	—	99.5

PROTEIN FEEDS

Composition of Feeds

ENERGY FEEDS

Entry Number	Fat-Soluble Vitamins A (1 mg Carotene = 1667 IU Vit A) IU/g	Carotene (Provitamin A) ppm or mg/kg	D IU/kg	E ppm or mg/kg	K ppm or mg/kg	Water-Soluble Vitamins B-12 ppb or mcg/kg	Biotin ppm or mg/kg	Choline ppm or mg/kg	Folacin (Folic Acid) ppm or mg/kg	Niacin ppm or mg/kg	Pantothenic Acid (B-3) ppm or mg/kg	(Pyridoxine) B-6 ppm or mg/kg	Riboflavin (B-2) ppm or mg/kg	Thiamin (B-1) ppm or mg/kg
27	0.4	0.2	—	32.3	—	—	0.91	2753	2.45	35	10.2	5.00	5.2	7.1
28	—	—	—	14.6	—	—	0.55	2780	2.55	41	13.7	5.41	4.7	7.3
29	—	—	—	—	—	—	—	—	—	—	—	—	—	—
30	—	—	—	—	—	—	—	3316	—	51	15.5	—	6.0	—
31	—	—	—	11.3	—	—	0.44	2962	0.93	45	14.3	6.29	4.9	8.2
32	—	—	—	—	—	—	—	—	—	—	—	—	—	—
33	0.4	0.2	—	—	—	—	—	—	—	24	15.5	—	2.3	9.3
34	—	—	—	—	—	—	—	—	—	—	—	—	4.0	—
35	13.0	7.8	—	30.5	—	—	—	2645	—	47	11.9	6.00	6.6	2.5
36	1.9	1.1	—	—	—	2.9	2.84	4992	—	143	25.3	8.66	11.3	6.9
37	—	—	—	—	—	80.4	0.04	894	0.22	21	8.9	4.39	2.0	0.1
38	—	—	—	—	—	—	—	—	—	—	—	—	—	—
39	2.2	1.3	—	—	—	506.6	0.14	3370	0.22	176	35.7	12.20	12.9	5.5
40	—	—	—	6.1	—	485.9	0.40	5525	0.57	256	50.4	19.71	13.5	7.4
41	—	—	—	3.7	—	214.5	0.20	3700	0.16	81	10.0	4.71	7.3	0.5
42	—	—	—	6.8	—	122.0	0.18	3112	0.15	55	8.6	3.80	4.8	0.6
43	—	—	—	—	—	238.0	0.10	3277	—	75	11.0	—	5.4	0.3
44	—	—	—	5.9	—	—	—	1216	2.85	30	—	9.93	2.9	9.4
45	—	—	—	—	—	—	—	—	—	20	—	—	2.2	3.7
46	—	—	—	5.9	—	—	—	1216	2.85	30	—	9.93	2.9	9.4
47	—	—	—	—	—	501.3	0.02	11370	5.56	205	29.2	—	36.2	0.2
48	—	—	—	—	—	—	—	—	—	—	—	—	—	—
49	—	—	—	—	—	21.1	—	—	—	19	—	—	11.7	—
50	—	—	—	0.9	—	75.2	0.12	1980	0.39	56	6.0	4.23	5.2	0.2
51	—	—	—	—	—	89.4	—	2203	1.54	38	3.2	—	2.2	0.3
52	—	—	—	0.8	—	104.4	0.07	2067	0.57	58	4.8	—	5.0	0.2
53	—	—	—	0.9	—	118.4	0.10	2049	0.37	51	5.5	5.86	4.7	0.2
54	—	—	—	—	—	—	—	904	—	1	8.4	—	1.7	0.3
55	—	—	—	—	—	—	—	—	—	1	3.5	—	2.0	0.4
56	—	—	0	9.1	—	50.9	0.33	1394	0.62	11	36.4	4.10	19.1	3.7
57	—	—	—	—	—	—	—	—	—	—	—	—	—	—
58	—	—	—	—	—	—	—	—	—	—	—	—	—	—
59	—	—	—	—	—	—	—	—	—	—	—	—	—	—
60	1.2	0.7	—	3.0	—	—	0.18	547	0.22	31	27.8	1.97	1.8	4.6
61	—	—	—	—	—	—	—	—	—	—	—	—	—	—
62	—	—	—	—	—	—	0.19	654	0.36	34	7.4	1.01	1.4	4.1
63	—	—	—	2.9	—	—	—	1896	—	178	36.8	5.95	5.3	—
64	—	—	—	2.2	—	304.1	0.09	6052	0.51	54	12.4	4.43	10.6	0.2
65	—	—	—	—	—	80.4	0.04	894	0.22	21	8.9	4.39	2.0	0.1
66	—	—	—	—	—	—	—	—	—	—	—	—	—	—
67	—	—	—	—	—	—	—	—	—	—	—	—	—	—
68	—	—	—	—	—	—	—	—	—	17	5.2	—	3.3	1.3
69	—	—	—	—	—	—	—	—	—	63	17.4	—	8.2	3.1
70	—	—	—	0.9	—	—	—	1541	—	12	36.2	474.43	2.2	—
71	—	—	—	0.6	—	—	1.56	3156	1.47	21	38.2	10.71	2.3	4.2
72	—	—	—	—	—	—	—	1517	—	—	6.3	—	3.7	—
73	—	—	—	—	—	—	—	5497	—	—	—	—	3.9	—
74	—	—	—	—	—	—	0.31	805	—	—	—	—	—	—
75	—	—	—	—	—	—	—	844	—	61	12.3	—	4.2	1.3
76	—	—	—	—	—	—	—	680	—	37	9.3	—	1.5	—
77	1.5	0.9	—	33.7	—	—	0.38	2931	—	23	16.0	11.04	2.9	11.3
78	—	—	—	3.0	0.22	2.0	0.36	2706	0.69	26	13.8	5.90	3.0	6.6
79	—	—	0	3.3	—	2.0	0.38	2772	0.59	24	14.1	5.59	2.9	3.5
80	—	—	—	—	—	—	—	—	—	—	—	—	—	—
81	—	—	—	—	—	—	—	—	—	—	—	—	—	—
82	—	—	—	—	—	—	—	—	—	—	—	—	—	—
83	1.8	1.1	—	—	—	—	—	—	—	56	8.2	—	3.7	2.0
84	—	—	—	—	—	—	—	—	—	—	—	—	—	—
85	—	—	—	141.2	—	—	0.22	3062	2.12	68	18.6	9.97	6.0	23.1
86	—	—	0	34.1	—	73.1	0.00	577	0.74	74	5.8	2.26	0.7	0.9
87	—	—	—	—	—	—	—	—	—	1	5.3	—	1.4	0.3
88	—	—	—	0.2	—	18.9	0.35	1790	0.85	11	46.2	3.21	27.4	4.0
89	—	—	—	2.1	—	1.1	1.04	3847	9.69	443	81.5	36.67	34.1	85.2
90	—	—	—	—	—	—	—	—	—	—	—	—	18.5	—
91	—	—	—	—	—	4.0	1.19	2981	25.66	512	107.5	34.48	47.7	6.8

PROTEIN FEEDS

TABLE V-1C DRY FORAGES, COMPOSITION OF FEEDS, DATA EXPRESSED AS-FED

Entry Number	Feed Name Description	International Feed Number	Proximate Analysis									Digestible Protein		
			Dry Matter %	Ash %	Crude Fiber %	Neutral Det. Fib. (NDF) %	Acid Det. Fib. (ADF) %	Lignin %	Ether Extract (Fat) %	N-Free Extract %	Crude Protein %	Ruminant %	Swine %	Horse %
	ALFALFA (LUCERNE)													
1	HAY, SUN-CURED, ALL ANALYSES	1-00-078	90	8.6	28.2	35.4	30.9	8.9	1.7	35.9	16.0	11.2	7.5	11.9
2	HAY, PREBLOOM, SUN-CURED	1-00-054	90	8.3	20.7	38.3	29.6	5.5	4.1	36.4	20.2	15.3	14.3	14.0
3	HAY, EARLY BLOOM, SUN-CURED	1-00-059	91	8.4	25.8	36.8	29.0	5.8	2.6	35.8	17.9	13.3	12.3	12.2
4	HAY, MIDBLOOM, SUN-CURED	1-00-063	91	7.8	25.5	43.2	33.4	6.7	3.3	37.4	17.1	12.0	11.5	11.6
5	HAY, FULL BLOOM, SUN-CURED	1-00-068	91	7.1	27.3	45.0	35.2	6.9	3.1	37.9	15.5	11.3	10.1	10.3
6	HAY, MATURE, SUN-CURED	1-00-071	91	6.7	29.3	50.1	40.1	11.3	2.9	37.0	15.2	11.3	9.8	10.1
7	HAY, RAINED ON, SUN-CURED	1-00-130	89	6.6	33.5	—	—	—	0.8	30.4	17.9	12.9	12.2	12.2
8	MEAL, DEHY, 17% PROTEIN	1-00-023	92	9.7	24.0	41.3	31.5	9.7	2.8	37.8	17.4	12.6	11.7	11.8
9	MEAL, DEHY, 20% PROTEIN	1-00-024	92	10.2	20.8	38.6	28.5	—	3.3	37.1	20.2	15.1	14.4	14.0
10	MEAL, DEHY, 22% PROTEIN	1-07-851	93	10.2	18.3	36.3	26.0	—	4.1	38.1	22.2	16.4	16.4	15.5
11	ALFALFA-BROMEGRASS, SMOOTH, HAY, SUN-CURED	1-00-255	91	6.1	31.0	—	—	—	2.1	37.7	14.1	9.6	8.9	9.2
12	ALFALFA-GRASS, HAY, SUN-CURED	1-08-331	91	6.7	30.3	—	36.6	—	2.1	37.8	14.5	9.9	9.2	9.5
13	ALFALFA-ORCHARDGRASS, HAY, SUN-CURED	1-00-322	89	7.3	28.7	—	—	—	1.9	36.8	14.8	10.2	9.5	9.8
14	ALMOND, HULLS	4-00-359	90	6.1	13.5	28.8	25.2	—	2.9	63.8	4.1	0.0	2.0	2.4
15	APPLE, POMACE, DEHY	4-00-423	89	3.0	16.2	—	23.1	—	4.3	61.2	4.4	-0.4	2.3	2.7
16	BAGASSE (SUGARCANE), PULP, DEHY	1-04-686	91	2.9	42.3	78.8	54.5	12.8	0.7	43.8	1.4	-1.5	-2.3	-0.5
	BARLEY													
17	HAY, SUN-CURED	1-00-495	88	6.6	23.6	—	—	—	1.9	48.5	7.8	4.3	3.4	4.4
18	STRAW	1-00-498	91	6.7	37.9	77.5	51.1	6.9	1.7	41.1	4.0	0.6	0.0	1.4
	BEAN													
19	HAY, SUN-CURED	1-00-583	90	6.3	24.1	—	—	—	1.7	44.4	13.9	9.4	8.7	9.1
20	STRAW	1-00-585	89	8.7	36.8	—	50.0	—	1.3	35.2	7.1	3.5	2.8	3.9
21	BERMUDAGRASS, HAY, SUN-CURED	1-00-703	91	8.0	28.4	—	—	—	1.8	43.7	9.2	5.5	4.5	5.4
22	BERMUDAGRASS, COASTAL, HAY, SUN-CURED	1-00-716	91	6.3	27.0	69.2	34.6	—	2.0	43.9	11.7	7.5	6.7	7.4
23	BIRDSFOOT TREFOIL, HAY, SUN-CURED	1-05-044	91	6.7	29.3	—	—	—	1.9	38.9	13.9	9.6	8.7	9.1
24	BLUEGRASS, CANADA, HAY, SUN-CURED	1-00-762	92	6.5	27.6	—	—	—	2.4	45.8	9.5	4.1	4.8	5.7
25	BLUEGRASS, KENTUCKY, HAY, SUN-CURED	1-00-776	89	5.9	26.8	—	—	—	3.0	44.3	9.1	5.3	4.6	5.4
26	BLUESTEM, HAY, SUN-CURED	1-00-819	90	6.3	30.8	—	—	—	2.2	45.6	4.9	1.5	0.8	2.1
27	BROMEGRASS, HAY, SUN-CURED	1-00-890	91	7.1	29.8	—	31.7	4.3	1.9	43.3	8.7	4.8	4.1	5.1
28	BUFFALOGRASS, HAY, SUN-CURED	1-01-003	90	11.9	24.9	—	—	—	1.5	45.1	6.9	3.7	2.5	3.7
29	BUR-CLOVER, TOOTHED, HAY, SUN-CURED	1-01-030	90	8.7	22.5	—	—	—	2.5	38.9	17.0	12.2	11.5	11.5
30	CANADA BLUEGRASS, HAY, SUN-CURED	1-00-762	92	6.5	27.6	—	—	—	2.4	45.8	9.5	4.1	4.8	5.7
31	CANARYGRASS, REED, HAY, SUN-CURED	1-01-104	89	7.3	30.2	62.9	32.7	—	2.7	40.0	9.1	5.8	4.6	5.5
32	CEREALS, IMMATURE, DEHY	1-26-069	92	14.4	16.0	—	—	—	4.8	32.5	24.4	19.0	—	—
33	CLOVER, ALSIKE, HAY, SUN-CURED	1-01-313	88	7.6	26.2	—	—	—	2.4	39.1	12.4	8.3	7.5	8.0
34	CLOVER, ALYCE, HAY, SUN-CURED	1-00-361	90	5.7	36.2	—	—	—	1.6	35.3	10.9	6.7	6.1	6.8
35	CLOVER, CRIMSON, HAY, SUN-CURED	1-01-328	88	7.8	28.1	—	—	—	2.0	35.2	14.7	10.2	9.5	9.8
36	CLOVER, LADINO, HAY, SUN-CURED	1-01-378	89	8.4	18.5	32.1	28.5	5.9	2.4	39.9	20.0	15.4	14.1	13.8
37	CLOVER, RED, HAY, SUN-CURED, ALL ANALYSES	1-01-415	88	6.7	27.1	49.5	36.2	8.8	2.5	39.2	13.0	7.7	8.0	8.4
38	CLOVER, RED-GRASS, HAY, FULL BLOOM, SUN-CURED	1-01-532	89	5.9	29.9	—	—	—	1.6	37.4	14.0	9.8	8.9	9.2
39	CLOVER, SUBTERRANEAN, HAY, SUN-CURED	1-20-278	90	10.0	9.1	—	—	—	3.3	40.1	27.5	21.7	—	20.5
40	CLOVER, SWEET, YELLOW, HAY, SUN-CURED, ALL ANALYSES	1-04-754	89	7.6	28.9	—	—	—	1.9	36.4	13.7	9.5	8.6	9.2
41	CLOVER, WHITE, HAY, SUN-CURED	1-01-464	90	7.6	21.9	—	—	—	2.4	40.1	16.9	12.4	11.4	11.5
42	CLOVER-TIMOTHY, HAY, FULL BLOOM, SUN-CURED	1-01-484	90	5.7	32.2	—	—	—	3.1	39.6	9.5	5.5	4.8	5.7
	CORN													
43	FODDER WITH EARS, WITH HUSKS, SUN-CURED	1-02-775	90	4.9	32.6	—	—	—	7.4	29.5	15.6	4.0	3.6	11.0
44	STOVER WITHOUT EARS, WITHOUT HUSKS, SUN-CURED	1-02-776	85	6.1	29.3	57.0	33.2	—	1.1	43.2	5.4	2.3	1.5	2.7
45	HUSKS, SUN-CURED	1-02-785	89	3.2	30.0	—	—	—	0.8	51.2	3.3	0.7	-0.5	0.9
46	COBS, GROUND	1-02-782	90	1.6	32.2	80.1	31.5	—	0.6	52.7	2.8	-0.4	-1.0	0.4
47	CORN, SWEET, FODDER WITH EARS, WITH HUSKS, SUN-CURED	1-08-407	88	9.0	26.4	—	—	—	1.8	41.3	9.2	5.9	4.7	5.5
48	COTTON, HULLS	1-01-599	90	2.6	43.2	81.0	65.7	21.0	1.5	39.3	3.8	-0.4	-0.2	1.3
49	COWPEA, COMMON, HAY, SUN-CURED	1-01-645	90	10.5	24.4	—	—	—	2.6	35.1	17.7	12.2	12.0	12.0
50	CRESTED WHEATGRASS, HAY, SUN-CURED	1-05-418	92	6.4	30.8	—	33.2	5.1	2.1	42.2	10.3	6.4	5.5	6.3
51	CRIMSON CLOVER, HAY, SUN-CURED	1-01-328	88	7.8	28.1	—	—	—	2.0	35.2	14.7	10.2	9.5	9.8
52	DALLISGRASS, HAY, SUN-CURED	1-01-737	91	7.9	30.8	—	—	—	1.9	41.1	9.2	5.3	4.6	5.5
53	FESCUE, TALL (ALTA), HAY, SUN-CURED	1-05-684	89	5.9	32.6	61.7	35.6	—	2.0	41.4	7.2	3.6	2.9	3.9
54	FLAX, COMMON, STRAW	1-02-038	93	5.5	48.1	—	—	—	2.2	32.4	5.0	2.2	0.8	2.2
55	FOXTAIL, MEADOW, HAY, SUN-CURED	1-02-072	90	8.4	25.5	—	—	—	2.0	41.1	12.8	8.5	7.7	8.2
56	GRAMA, HAY, SUN-CURED	1-02-162	89	8.4	29.1	—	—	—	1.5	44.5	5.6	2.2	1.5	2.7
57	GRASS, HAY, SUN-CURED, ALL ANALYSES	1-02-250	89	7.3	29.7	—	34.4	—	2.3	41.2	8.9	5.2	4.4	5.7
58	GRASS-LEGUME, HAY, SUN-CURED	1-02-301	89	5.4	31.6	—	33.9	—	2.3	39.5	10.3	6.0	5.6	6.3
59	JOHNSONGRASS SORGHUM, HAY, SUN-CURED	1-04-407	91	7.7	30.4	—	—	—	2.0	43.7	6.7	3.0	2.4	3.6
	KAFIR SORGHUM—SEE SORGHUM													
60	KELP (SEAWEED), WHOLE, DEHY	1-08-073	91	35.0	6.5	—	—	—	0.5	42.4	6.5	2.9	2.2	3.4
61	KENTUCKY BLUEGRASS, HAY, SUN-CURED	1-00-776	89	5.9	26.8	—	—	—	3.0	44.3	9.1	5.3	4.6	5.4
62	LESPEDEZA, COMMON, HAY, SUN-CURED, ALL ANALYSES	1-08-591	89	4.7	28.4	—	—	—	2.5	39.4	13.8	5.9	8.7	9.1
63	LESPEDEZA, SERICEA (CHINESE LESPEDEZA), HAY, SUN-CURED, ALL ANALYSES	1-02-607	90	4.8	29.9	—	—	—	2.0	42.9	10.7	3.5	5.9	6.6
64	MEADOW FESCUE, HAY, SUN-CURED	1-01-912	88	7.9	28.0	74.0	43.8	—	2.4	41.0	8.2	4.6	3.9	5.3
65	MEADOW FOXTAIL, HAY, SUN-CURED	1-02-072	90	8.4	25.5	—	—	—	2.0	41.1	12.8	8.5	7.7	8.2
66	MILLET, FOXTAIL, HAY, SUN-CURED	1-03-099	87	7.5	25.7	—	—	—	2.5	43.7	7.5	3.7	3.2	4.2
67	MILLET, JAPANESE (BARNYARD GRASS), HAY, SUN-CURED	1-03-105	87	8.5	26.3	—	—	—	1.7	41.8	8.8	5.4	4.4	5.3
	MILO SORGHUM—SEE SORGHUM													
68	NEEDLEGRASS, HAY, SUN-CURED	1-03-202	88	6.2	30.0	—	—	—	2.2	41.9	7.8	4.1	3.4	4.4
69	OATGRASS, TALL, HAY, SUN-CURED	1-03-259	88	6.7	29.6	—	—	—	2.0	42.5	6.7	3.1	2.5	3.6
	OATS													
70	HAY, SUN-CURED, ALL ANALYSES	1-03-280	91	7.2	29.1	—	34.8	—	2.2	43.6	8.6	4.5	4.1	5.0
71	STRAW	1-03-283	92	7.2	37.2	65.7	43.1	7.0	2.0	41.6	4.1	0.8	0.0	2.4
72	OATS-PEA, HAY, SUN-CURED	1-03-398	88	7.4	26.9	—	—	—	3.0	38.7	11.7	8.3	6.9	7.4
73	ORCHARDGRASS, HAY, SUN-CURED	1-03-438	89	6.5	31.0	64.1	36.0	—	2.8	39.7	9.4	5.7	4.8	5.6
74	PEA, HAY, SUN-CURED	1-03-572	87	7.4	23.7	—	—	—	2.6	41.7	11.8	7.7	7.0	7.6
75	PEA, FIELD, HAY (VINES WITHOUT SEEDS, WITH PODS), SUN-CURED	1-03-607	88	6.8	23.4	—	—	—	2.3	42.5	12.7	8.5	7.8	8.2

Composition of Feeds

DRY FORAGES

Entry Number	TDN Ruminant %	TDN Swine %	TDN Horse %	Digestible Energy Ruminant Mcal/lb	Digestible Energy Ruminant Mcal/kg	Digestible Energy Swine kcal/lb	Digestible Energy Swine kcal/kg	Digestible Energy Horse Mcal/lb	Digestible Energy Horse Mcal/kg	Metabolizable Energy Ruminant Mcal/lb	Metabolizable Energy Ruminant Mcal/kg	Metabolizable Energy Swine kcal/lb	Metabolizable Energy Swine kcal/kg	Poultry ME$_n$ kcal/lb	Poultry ME$_n$ kcal/kg	Metabolizable Energy Horse Mcal/lb	Metabolizable Energy Horse Mcal/kg	Net Energy Ruminant NE$_m$ Mcal/lb	Net Energy Ruminant NE$_m$ Mcal/kg	Net Energy Ruminant NE$_g$ Mcal/lb	Net Energy Ruminant NE$_g$ Mcal/kg	Lactating Cows NE$_{lc}$ Mcal/lb	Lactating Cows NE$_{lc}$ Mcal/kg
1	51	32	48	1.03	2.28	—	—	0.90	1.98	0.90	1.99	—	—	—	—	0.74	1.62	0.48	1.06	0.25	0.55	0.49	1.07
2	54	64	48	1.25	2.75	—	—	0.90	1.98	1.02	2.24	—	—	—	—	0.74	1.63	0.57	1.26	0.34	0.74	0.58	1.27
3	52	50	48	1.15	2.54	—	—	0.90	1.99	0.96	2.12	—	—	—	—	0.74	1.63	0.60	1.33	0.36	0.80	0.55	1.21
4	52	54	46	1.12	2.46	—	—	0.86	1.90	0.94	2.07	—	—	—	—	0.71	1.56	0.52	1.14	0.28	0.62	0.51	1.13
5	51	50	44	1.09	2.39	—	—	0.83	1.83	0.88	1.95	—	—	—	—	0.68	1.50	0.49	1.09	0.26	0.57	0.52	1.15
6	55	47	43	1.22	2.68	—	—	0.82	1.81	0.93	2.06	—	—	—	—	0.67	1.48	0.47	1.04	0.24	0.53	0.51	1.12
7	47	—	50	1.00	2.20	—	—	0.94	2.07	0.83	1.83	—	—	—	—	0.77	1.70	0.44	0.96	0.21	0.46	0.48	1.05
8	55	44	47	1.12	2.47	643	1418	0.89	1.96	0.96	2.12	601	1326	682	1504	0.73	1.61	0.60	1.32	0.36	0.79	0.57	1.25
9	57	48	50	1.07	2.36	943	2079	0.94	2.08	0.93	2.06	872	1923	737	1625	0.77	1.70	0.58	1.28	0.34	0.75	0.59	1.30
10	60	49	52	1.20	2.65	991	2186	0.98	2.15	1.02	2.26	841	1855	768	1692	0.80	1.76	0.63	1.39	0.38	0.84	0.63	1.38
11	58	41	45	1.11	2.44	—	—	0.85	1.88	0.93	2.06	—	—	—	—	0.70	1.54	0.59	1.30	0.35	0.76	0.59	1.31
12	51	42	45	1.01	2.23	—	—	0.86	1.89	0.83	1.84	—	—	—	—	0.70	1.55	0.50	1.10	0.26	0.58	0.53	1.16
13	52	42	47	1.03	2.27	—	—	0.88	1.93	0.86	1.89	—	—	—	—	0.72	1.58	0.49	1.08	0.26	0.57	0.52	1.14
14	66	73	45	1.34	2.95	1457	3213	0.86	1.89	1.15	2.54	1375	3032	—	—	0.70	1.55	0.77	1.70	0.51	1.12	0.66	1.46
15	60	69	40	1.21	2.66	1380	3043	0.77	1.71	1.04	2.28	1300	2867	—	—	0.64	1.40	0.66	1.46	0.42	0.91	0.64	1.42
16	43	—	—	0.88	1.93	—	—	—	—	0.69	1.53	—	—	—	—	—	—	0.37	0.82	0.15	0.32	0.40	0.88
17	50	40	42	0.99	2.17	—	—	0.81	1.77	0.79	1.73	—	—	—	—	0.66	1.46	0.45	0.99	0.22	0.49	0.50	1.10
18	43	—	28	0.84	1.86	—	—	0.57	1.26	0.64	1.40	—	—	—	—	0.47	1.03	0.29	0.64	0.07	0.15	0.42	0.91
19	59	40	52	1.13	2.48	—	—	0.97	2.13	0.95	2.10	—	—	—	—	0.79	1.74	0.52	1.15	0.29	0.64	0.54	1.20
20	43	—	30	0.87	1.91	—	—	0.60	1.32	0.69	1.53	—	—	—	—	0.49	1.08	0.36	0.79	0.14	0.30	0.42	0.93
21	43	35	40	0.88	1.93	—	—	0.77	1.70	0.64	1.41	—	—	—	—	0.63	1.40	0.29	0.65	0.07	0.16	0.40	0.88
22	49	43	46	1.04	2.30	—	—	0.86	1.90	0.66	1.44	—	—	—	—	0.71	1.56	0.31	0.68	0.09	0.20	0.53	1.16
23	54	41	46	0.91	2.01	—	—	0.87	1.91	0.83	1.84	—	—	—	—	0.71	1.56	0.55	1.22	0.32	0.69	0.57	1.25
24	57	42	41	1.14	2.52	—	—	0.79	1.75	0.97	2.13	—	—	—	—	0.65	1.43	0.61	1.35	0.37	0.81	0.61	1.35
25	54	44	37	1.06	2.34	—	—	0.73	1.60	0.89	1.96	—	—	—	—	0.60	1.31	0.51	1.12	0.28	0.62	0.53	1.17
26	41	31	33	0.88	1.95	—	—	0.65	1.43	0.71	1.56	—	—	—	—	0.53	1.17	0.44	0.96	0.21	0.47	0.48	1.06
27	48	35	39	1.12	2.48	—	—	0.75	1.65	0.95	2.09	—	—	—	—	0.61	1.35	0.50	1.11	0.27	0.60	0.50	1.11
28	48	30	35	0.93	2.05	—	—	0.69	1.53	0.76	1.66	—	—	—	—	0.57	1.25	0.40	0.88	0.18	0.39	0.45	1.00
29	54	52	49	1.08	2.39	—	—	0.92	2.03	0.91	2.01	—	—	—	—	0.75	1.66	0.54	1.18	0.30	0.66	0.55	1.21
30	57	42	41	1.14	2.52	—	—	0.79	1.75	0.97	2.13	—	—	—	—	0.65	1.43	0.61	1.35	0.37	0.81	0.61	1.35
31	44	36	43	0.93	2.06	—	—	0.82	1.80	0.76	1.68	—	—	—	—	0.67	1.47	0.46	1.02	0.24	0.52	0.50	1.10
32	56	—	—	1.13	2.48	—	—	—	—	0.92	2.02	—	—	—	—	—	—	—	—	—	—	—	—
33	51	42	41	1.01	2.24	—	—	0.78	1.72	0.85	1.86	—	—	—	—	0.64	1.41	0.49	1.08	0.26	0.58	0.51	1.13
34	44	—	38	0.89	1.96	—	—	0.74	1.63	0.71	1.57	—	—	—	—	0.61	1.34	0.30	0.66	0.08	0.18	0.38	0.85
35	50	41	43	1.00	2.21	—	—	0.82	1.81	0.84	1.84	—	—	—	—	0.67	1.49	0.54	1.19	0.31	0.68	0.55	1.21
36	58	60	57	1.16	2.55	—	—	1.05	2.31	0.99	2.18	—	—	—	—	0.86	1.89	0.58	1.27	0.34	0.75	0.58	1.28
37	52	44	42	1.21	2.67	—	—	0.81	1.78	0.95	2.10	—	—	—	—	0.66	1.46	0.50	1.11	0.28	0.61	0.53	1.16
38	51	39	47	1.02	2.25	—	—	0.88	1.93	0.85	1.87	—	—	—	—	0.72	1.59	0.48	1.06	0.25	0.56	0.51	1.12
39	—	—	—	—	—	—	—	—	—	—	—	—	—	—	—	—	—	—	—	—	—	—	—
40	49	39	43	0.98	2.16	—	—	0.81	1.79	0.81	1.78	—	—	—	—	0.67	1.47	0.49	1.07	0.26	0.57	0.51	1.13
41	57	52	50	1.11	2.45	—	—	0.94	2.07	0.94	2.07	—	—	—	—	0.77	1.70	0.53	1.18	0.30	0.66	0.55	1.22
42	51	39	34	1.01	2.23	—	—	0.67	1.47	0.84	1.85	—	—	—	—	0.55	1.21	0.49	1.08	0.26	0.57	0.52	1.14
43	45	45	44	0.91	2.00	—	—	0.84	1.85	0.74	1.64	—	—	—	—	0.69	1.52	0.44	0.97	0.12	0.26	0.46	1.01
44	51	26	36	1.01	2.23	—	—	0.70	1.54	0.85	1.87	—	—	—	—	0.57	1.26	0.51	1.12	0.29	0.63	0.52	1.15
45	55	29	41	1.11	2.44	—	—	0.79	1.74	0.94	2.06	—	—	—	—	0.65	1.43	0.69	1.51	0.44	0.97	0.66	1.46
46	44	28	28	0.91	2.00	—	—	0.57	1.25	0.74	1.62	—	—	—	—	0.46	1.02	0.39	0.87	0.17	0.37	0.44	0.96
47	56	34	38	1.03	2.28	—	—	0.73	1.61	0.86	1.90	—	—	—	—	0.60	1.32	0.43	0.95	0.21	0.46	0.47	1.04
48	42	—	29	0.86	1.90	—	—	0.59	1.30	0.60	1.31	—	—	—	—	0.48	1.07	0.25	0.55	0.03	0.08	0.39	0.86
49	54	48	46	1.14	2.52	—	—	0.86	1.90	0.92	2.02	—	—	—	—	0.71	1.56	0.58	1.28	0.34	0.75	0.56	1.24
50	49	38	41	0.99	2.19	—	—	0.78	1.73	0.82	1.80	—	—	—	—	0.64	1.42	0.48	1.06	0.25	0.55	0.52	1.14
51	50	41	43	1.00	2.21	—	—	0.82	1.81	0.84	1.84	—	—	—	—	0.67	1.49	0.54	1.19	0.31	0.68	0.55	1.21
52	49	33	37	0.97	2.14	—	—	0.73	1.60	0.80	1.76	—	—	—	—	0.60	1.32	0.44	0.98	0.22	0.47	0.49	1.07
53	48	31	41	0.95	2.10	—	—	0.78	1.71	0.78	1.72	—	—	—	—	0.64	1.40	0.44	0.96	0.21	0.47	0.48	1.06
54	38	—	—	0.81	1.78	—	—	—	—	0.66	1.46	—	—	—	—	—	—	0.37	0.81	0.14	0.30	0.44	0.96
55	57	42	44	1.08	2.37	—	—	0.84	1.86	0.90	1.99	—	—	—	—	0.69	1.52	0.48	1.07	0.25	0.56	0.51	1.13
56	39	27	35	0.84	1.86	—	—	0.68	1.49	0.67	1.47	—	—	—	—	0.55	1.22	0.40	0.89	0.18	0.40	0.46	1.00
57	51	35	40	1.08	2.38	—	—	0.76	1.68	0.91	2.00	—	—	—	—	0.63	1.38	0.53	1.17	0.30	0.66	0.52	1.15
58	52	37	38	1.04	2.30	—	—	0.74	1.64	0.87	1.92	—	—	—	—	0.61	1.34	0.47	1.04	0.24	0.54	0.50	1.11
59	51	31	35	1.01	2.23	—	—	0.68	1.51	0.84	1.85	—	—	—	—	0.56	1.24	0.48	1.06	0.25	0.56	0.51	1.13
60	29	—	—	0.58	1.27	—	—	—	—	0.40	0.87	—	—	—	—	—	—	—	—	—	—	—	—
61	54	44	37	1.06	2.34	—	—	0.73	1.60	0.89	1.96	—	—	—	—	0.60	1.31	0.51	1.12	0.28	0.62	0.53	1.17
62	44	46	45	0.98	2.15	—	—	0.86	1.89	0.80	1.77	—	—	—	—	0.70	1.55	0.53	1.16	0.29	0.65	0.54	1.19
63	42	40	44	0.93	2.06	—	—	0.83	1.84	0.76	1.67	—	—	—	—	0.68	1.51	0.50	1.09	0.26	0.58	0.52	1.15
64	53	35	37	1.06	2.35	—	—	0.72	1.58	0.90	1.98	—	—	—	—	0.59	1.30	0.55	1.20	0.32	0.70	0.56	1.22
65	57	42	44	1.08	2.37	—	—	0.84	1.86	0.90	1.99	—	—	—	—	0.69	1.52	0.48	1.07	0.25	0.56	0.51	1.13
66	51	38	35	1.00	2.19	—	—	0.68	1.49	0.83	1.83	—	—	—	—	0.56	1.22	0.46	1.01	0.24	0.52	0.49	1.08
67	47	34	38	0.94	2.07	—	—	0.73	1.62	0.77	1.70	—	—	—	—	0.60	1.32	0.43	0.95	0.21	0.46	0.47	1.04
68	42	34	35	0.90	1.98	—	—	0.69	1.52	0.73	1.60	—	—	—	—	0.57	1.25	0.45	0.99	0.23	0.50	0.49	1.07
69	46	32	34	0.93	2.04	—	—	0.67	1.48	0.76	1.67	—	—	—	—	0.55	1.21	0.43	0.95	0.21	0.47	0.47	1.04
70	52	36	38	1.04	2.29	—	—	0.74	1.62	0.93	2.04	—	—	—	—	0.60	1.33	0.57	1.26	0.33	0.73	0.55	1.22
71	46	—	44	1.15	2.53	—	—	0.84	1.85	0.78	1.72	—	—	—	—	0.69	1.52	0.43	0.94	0.20	0.43	0.48	1.06
72	52	43	37	1.03	2.26	—	—	0.72	1.58	0.86	1.89	—	—	—	—	0.59	1.30	0.49	1.08	0.27	0.59	0.52	1.14
73	51	37	40	1.23	2.71	—	—	0.77	1.69	0.96	2.11	—	—	—	—	0.63	1.38	0.50	1.09	0.27	0.59	0.52	1.15
74	51	45	42	1.02	2.24	—	—	0.79	1.75	0.85	1.87	—	—	—	—	0.65	1.43	0.50	1.10	0.27	0.60	0.53	1.15
75	52	47	46	1.04	2.29	—	—	0.86	1.89	0.87	1.91	—	—	—	—	0.70	1.55	0.51	1.12	0.28	0.62	0.53	1.17

(Continued)

TABLE V-1C DRY FORAGES, MINERAL AND VITAMIN COMPOSITION OF FEEDS, DATA EXPRESSED **AS-FED**

			Macro Minerals							Micro Minerals						
Entry Number	Feed Name Description	Dry Matter	Calcium (Ca)	Phosphorus (P)	Sodium (Na)	Chlorine (Cl)	Magnesium (Mg)	Potassium (K)	Sulfur (S)	Cobalt (Co)	Copper (Cu)	Iodine (I)	Iron (Fe)	Manganese (Mn)	Selenium (Se)	Zinc (Zn)
		%	%	%	%	%	%	%	%	ppm or mg/kg	ppm or mg/kg	ppm or mg/kg	%	ppm or mg/kg	ppm or mg/kg	ppm or mg/kg
	ALFALFA (LUCERNE)															
1	HAY, SUN-CURED, ALL ANALYSES	90	1.28	0.24	0.07	0.33	0.30	1.85	0.25	0.250	10.0	—	0.019	41.4	—	21.9
2	HAY, PREBLOOM, SUN-CURED	90	1.34	0.30	0.10	0.31	0.19	2.25	0.48	0.256	10.2	—	0.021	42.2	—	33.5
3	HAY, EARLY BLOOM, SUN-CURED	91	1.48	0.20	0.14	0.34	0.31	2.32	0.27	0.264	11.4	—	0.021	32.8	0.497	27.3
4	HAY, MIDBLOOM, SUN-CURED	91	1.27	0.22	0.11	0.34	0.32	1.42	0.26	0.359	16.1	—	0.021	55.1	—	28.1
5	HAY, FULL BLOOM, SUN-CURED	91	1.08	0.22	0.06	—	0.25	1.42	0.27	0.210	9.0	—	0.015	38.5	—	23.7
6	HAY, MATURE, SUN-CURED	91	1.07	0.19	0.07	—	0.20	1.88	0.23	0.370	12.5	—	0.015	35.1	—	20.1
7	HAY, RAINED ON, SUN-CURED	89	2.04	0.21	0.05	—	0.24	2.16	—	—	2.5	—	0.026	22.1	—	23.8
8	MEAL, DEHY, 17% PROTEIN	92	1.40	0.23	0.10	0.47	0.29	2.38	0.23	0.302	8.6	0.148	0.041	31.0	0.335	19.3
9	MEAL, DEHY, 20% PROTEIN	92	1.59	0.28	0.11	0.47	0.33	2.41	0.50	0.259	12.2	0.135	0.036	45.2	0.285	21.8
10	MEAL, DEHY, 22% PROTEIN	93	1.69	0.30	0.12	0.52	0.31	2.40	0.30	0.311	9.8	0.166	0.036	36.4	0.534	19.5
11	ALFALFA-BROMEGRASS, SMOOTH, HAY, SUN-CURED	91	1.02	0.25	0.21	0.43	0.52	1.77	0.21	0.079	15.5	—	0.012	36.2	—	—
12	ALFALFA-GRASS, HAY, SUN-CURED	91	1.33	0.23	0.11	—	0.28	2.31	0.22	—	11.4	—	0.020	60.7	—	20.1
13	ALFALFA-ORCHARDGRASS, HAY, SUN-CURED	89	—	—	—	—	—	—	—	—	—	—	—	—	—	—
14	ALMOND, HULLS	90	0.19	0.09	—	—	—	0.48	0.10	—	—	—	—	—	—	—
15	APPLE, POMACE, DEHY	89	0.11	0.10	0.12	—	0.06	0.43	0.02	—	—	—	0.027	7.2	—	—
16	BAGASSE (SUGARCANE), PULP, DEHY	91	0.47	0.26	0.04	—	0.08	0.34	0.09	—	—	—	0.019	—	—	—
	BARLEY															
17	HAY, SUN-CURED	88	0.21	0.25	0.12	—	0.14	1.30	0.15	0.059	3.9	—	0.027	34.8	—	—
18	STRAW	91	0.27	0.07	0.13	0.61	0.21	2.16	0.16	0.061	4.9	—	0.019	15.1	—	6.8
	BEAN															
19	HAY, SUN-CURED	90	—	—	—	—	—	—	—	—	—	—	—	—	—	—
20	STRAW	89	1.65	0.13	—	—	0.12	1.02	—	—	—	—	—	—	—	—
21	BERMUDAGRASS, HAY, SUN-CURED	91	0.43	0.16	0.07	—	0.16	1.40	0.19	0.111	24.3	0.105	0.027	99.4	—	53.0
22	BERMUDAGRASS, COASTAL, HAY, SUN-CURED	91	0.38	0.17	—	—	0.16	1.46	0.19	—	—	—	0.028	—	—	18.2
23	BIRDSFOOT TREFOIL, HAY, SUN-CURED	91	1.54	0.21	0.06	—	0.46	1.74	0.23	0.100	8.4	—	0.021	26.0	—	69.9
24	BLUEGRASS, CANADA, HAY, SUN-CURED	92	0.28	0.24	0.10	—	0.30	1.73	0.12	—	—	—	0.028	84.9	—	—
25	BLUEGRASS, KENTUCKY, HAY, SUN-CURED	89	0.40	0.27	0.10	0.55	0.19	1.66	0.12	—	8.8	—	0.025	76.2	—	—
26	BLUESTEM, HAY, SUN-CURED	90	—	—	0.01	0.04	—	—	—	—	—	—	—	—	—	—
27	BROMEGRASS, HAY, SUN-CURED	91	0.32	0.15	0.03	—	0.09	1.49	0.18	—	—	—	0.019	—	—	—
28	BUFFALOGRASS, HAY, SUN-CURED	90	0.56	0.12	—	—	—	1.38	—	—	—	—	—	—	—	—
29	BUR-CLOVER, TOOTHED, HAY, SUN-CURED	90	—	—	—	—	—	—	—	—	—	—	—	—	—	—
30	CANADA BLUEGRASS, HAY, SUN-CURED	92	0.28	0.24	0.10	—	0.30	1.73	0.12	—	—	—	0.028	84.9	—	—
31	CANARYGRASS, REED, HAY, SUN-CURED	89	0.32	0.21	0.01	—	0.19	2.60	—	—	10.6	—	0.014	82.5	—	—
32	CEREALS, IMMATURE, DEHY	92	0.65	0.46	—	—	—	—	—	—	—	—	—	—	—	—
33	CLOVER, ALSIKE, HAY, SUN-CURED	88	1.14	0.22	0.40	0.68	0.40	1.95	0.17	—	5.3	—	0.023	60.5	—	—
34	CLOVER, ALYCE, HAY, SUN-CURED	90	—	—	—	—	—	—	—	—	—	—	—	—	—	—
35	CLOVER, CRIMSON, HAY, SUN-CURED	88	1.23	0.19	0.34	0.55	0.25	2.10	0.25	—	—	0.059	0.062	183.3	—	—
36	CLOVER, LADINO, HAY, SUN-CURED	89	1.30	0.30	0.12	0.27	0.42	2.17	0.19	0.144	8.4	0.268	0.042	109.7	—	15.2
37	CLOVER, RED, HAY, SUN-CURED, ALL ANALYSES	88	1.22	0.22	0.16	0.28	0.34	1.60	0.15	0.138	18.8	0.217	0.022	95.2	—	32.5
38	CLOVER, RED-GRASS, HAY, SUN-CURED, FULL BLOOM	89	—	—	—	—	—	—	—	0.137	—	—	—	—	—	—
39	CLOVER, SUBTERRANEAN, HAY, SUN-CURED	90	—	—	—	—	—	—	—	—	—	—	—	—	—	—
40	CLOVER, SWEET, YELLOW, HAY, SUN-CURED, ALL ANALYSES	89	1.44	0.24	0.08	0.33	0.39	1.35	0.42	—	8.8	—	0.015	95.4	—	—
41	CLOVER, WHITE, HAY, SUN-CURED	90	1.71	0.29	—	—	—	—	—	—	—	—	—	—	—	—
42	CLOVER-TIMOTHY, HAY, FULL BLOOM, SUN-CURED	90	—	—	—	—	—	—	—	—	—	—	—	—	—	—
	CORN															
43	FODDER WITH EARS, WITH HUSKS, SUN-CURED	90	0.45	0.23	0.03	0.17	0.26	0.84	0.13	—	6.9	—	0.009	61.4	—	—
44	STOVER WITHOUT EARS, WITHOUT HUSKS, SUN-CURED	85	0.49	0.08	0.06	—	0.34	1.24	0.15	—	4.3	—	0.018	115.9	—	—
45	HUSKS, SUN-CURED	89	0.16	0.12	—	—	—	0.57	—	—	—	—	—	—	—	—
46	COBS, GROUND	90	0.11	0.04	—	0.06	0.78	0.42	—	0.117	6.6	—	0.021	5.6	—	—
47	CORN, SWEET, FODDER WITH EARS, WITH HUSKS, SUN-CURED	88	—	0.17	—	—	—	0.98	—	—	—	—	—	—	—	—
48	COTTON, HULLS	90	0.13	0.09	0.02	0.02	0.13	0.78	0.08	0.018	12.0	—	0.012	107.8	—	19.8
49	COWPEA, COMMON, HAY, SUN-CURED	90	1.26	0.31	0.24	0.15	0.41	2.04	0.32	0.064	—	—	0.055	438.4	—	—
50	CRESTED WHEATGRASS, HAY, SUN-CURED	92	0.24	0.14	—	—	—	—	—	0.219	—	—	—	—	—	—
51	CRIMSON CLOVER, HAY, SUN-CURED	88	1.23	0.19	0.34	0.55	0.25	2.10	0.25	—	—	0.059	0.062	183.3	—	—
52	DALLISGRASS, HAY, SUN-CURED	91	0.46	0.18	—	—	0.67	—	—	—	—	—	0.011	—	—	—
53	FESCUE, TALL (ALTA), HAY, SUN-CURED	89	0.35	0.21	0.05	—	0.20	2.12	—	—	—	—	—	—	—	—
54	FLAX, COMMON, STRAW	93	0.51	0.05	—	0.25	0.29	1.62	0.25	—	—	—	—	7.6	—	—
55	FOXTAIL, MEADOW, HAY, SUN-CURED	90	0.95	0.18	0.01	—	0.12	1.57	—	—	—	—	—	—	—	—
56	GRAMA, HAY, SUN-CURED	89	0.34	0.18	—	—	—	—	—	—	—	—	—	—	—	—
57	GRASS, HAY, SUN-CURED, ALL ANALYSES	89	0.44	0.18	0.01	—	0.20	1.38	0.19	0.133	6.5	—	0.013	75.4	—	15.1
58	GRASS-LEGUME, HAY, SUN-CURED	89	0.66	0.19	0.06	—	0.23	1.62	0.13	0.118	5.0	—	0.014	42.5	—	23.2
59	JOHNSONGRASS SORGHUM, HAY, SUN-CURED	91	0.80	0.27	0.01	—	0.31	1.22	0.09	—	—	—	0.054	—	—	—
	KAFIR SORGHUM—SEE SORGHUM															
60	KELP (SEAWEED), WHOLE, DEHY	91	2.47	0.28	—	—	0.85	—	—	—	—	—	—	—	—	—
61	KENTUCKY BLUEGRASS, HAY, SUN-CURED	89	0.40	0.27	0.10	0.55	0.19	1.66	0.12	—	8.8	—	0.025	76.2	—	—
62	LESPEDEZA, COMMON, HAY, SUN-CURED, ALL ANALYSES	89	0.78	0.25	—	—	0.20	1.21	0.21	—	7.1	—	0.019	99.6	—	21.3
63	LESPEDEZA, SERICEA (CHINESE LESPEDEZA), HAY, SUN-CURED, ALL ANALYSES	90	0.93	0.22	—	—	0.20	0.99	—	—	—	—	0.027	91.1	—	—
64	MEADOW FESCUE, HAY, SUN-CURED	88	0.33	0.25	—	—	0.44	1.61	—	0.119	—	—	—	21.4	—	—
65	MEADOW FOXTAIL, HAY, SUN-CURED	90	0.95	0.18	0.01	—	0.12	1.57	—	—	—	—	—	—	—	—
66	MILLET, FOXTAIL, HAY, SUN-CURED	87	0.29	0.16	0.09	0.11	0.20	1.69	0.14	—	—	—	—	120.1	—	—
67	MILLET, JAPANESE (BARNYARD GRASS), HAY, SUN-CURED	87	0.20	—	—	—	—	2.11	—	—	—	—	—	—	—	—
	MILO SORGHUM—SEE SORGHUM															
68	NEEDLEGRASS, HAY, SUN-CURED	88	—	—	—	—	—	—	—	—	—	—	—	—	—	—
69	OATGRASS, TALL, HAY, SUN-CURED	88	0.30	0.27	0.01	—	0.17	1.74	—	—	—	—	—	32.2	—	—
	OATS															
70	HAY, SUN-CURED, ALL ANALYSES	91	0.29	0.23	0.17	0.47	0.26	1.35	0.21	0.067	4.4	—	0.037	89.6	—	40.8
71	STRAW	92	0.22	0.06	0.39	0.72	0.16	2.35	0.21	—	9.5	—	0.016	29.0	—	5.5
72	OATS-PEA, HAY, SUN-CURED	88	0.71	0.22	—	—	—	1.02	—	—	—	—	—	—	—	—
73	ORCHARDGRASS, HAY, SUN-CURED	89	0.34	0.23	0.01	0.37	0.16	2.68	0.23	0.339	12.9	—	0.014	162.7	—	32.0
74	PEA, HAY, SUN-CURED	87	—	—	—	—	—	—	—	—	—	—	—	—	—	—
75	PEA, FIELD, HAY (VINES WITHOUT SEEDS, WITH PODS), SUN-CURED	88	—	—	—	—	—	—	—	—	—	—	—	—	—	—

Composition of Feeds

DRY FORAGES

	Fat-Soluble Vitamins					Water-Soluble Vitamins								
Entry Number	A (1 mg Carotene = 1667 IU Vit A)	Carotene (Provitamin A)	D	E	K	B-12	Biotin	Choline	Folacin (Folic Acid)	Niacin	Pantothenic Acid (B-3)	(Pyri-doxine) B-6	Ribo-flavin (B-2)	Thiamin (B-1)
	IU/g	ppm or mg/kg	IU/kg	ppm or mg/kg	ppm or mg/kg	ppb or mcg/kg	ppm or mg/kg	ppm or mg/kg	ppm or mg/kg	ppm or mg/kg	ppm or mg/kg	ppm or mg/kg	ppm or mg/kg	ppm or mg/kg
1	45.0	27.0	2	55.9	–	–	0.18	892	3.07	43	18.1	–	9.5	3.1
2	300.2	180.1	–	–	–	–	–	–	–	–	–	–	–	–
3	210.9	126.5	2	23.5	–	–	–	–	–	–	–	–	–	–
4	50.5	30.3	1	–	–	–	–	–	–	–	–	–	9.6	–
5	98.5	59.1	–	–	–	–	–	–	–	–	–	–	–	–
6	17.6	10.6	1	–	–	–	–	–	–	–	–	–	–	–
7	–	–	1	–	–	–	–	–	–	–	–	–	–	–
8	200.3	120.2	–	105.7	8.24	–	0.33	1369	4.37	37	29.7	7.18	12.9	3.4
9	265.4	159.2	–	143.3	14.19	–	0.35	1417	2.96	48	35.5	8.72	15.2	5.4
10	391.4	234.8	–	221.3	11.65	–	0.33	1605	5.15	50	39.0	8.28	17.6	5.9
11	39.4	23.7	–	–	–	–	–	–	–	25	21.4	–	6.1	–
12	28.9	17.3	–	–	–	–	–	–	–	–	–	–	–	–
13	–	–	–	–	–	–	–	–	–	–	–	–	–	–
14	–	–	–	–	–	–	–	–	–	–	–	–	–	–
15	–	–	–	–	–	–	–	–	–	–	–	–	–	–
16	–	–	–	–	–	–	–	–	–	–	–	–	–	–
17	77.4	46.4	1	–	–	–	–	–	–	–	–	–	–	–
18	3.5	2.1	1	–	–	–	–	–	–	–	–	–	–	–
19	–	–	–	–	–	–	–	–	–	–	–	–	–	–
20	–	–	–	–	–	–	–	–	–	–	–	–	–	–
21	87.5	52.5	–	–	–	–	–	–	–	–	–	–	–	–
22	123.7	74.2	–	–	–	–	–	–	–	–	–	–	–	–
23	217.8	130.6	1	–	–	–	–	–	–	–	–	–	14.6	6.2
24	378.4	227.0	–	–	–	–	–	–	–	–	–	–	–	–
25	–	–	–	–	–	–	–	–	–	–	–	–	9.9	–
26	62.4	37.4	–	–	–	–	–	–	–	–	–	–	–	–
27	50.3	30.2	–	–	–	–	–	–	–	–	–	–	–	–
28	–	–	–	–	–	–	–	–	–	–	–	–	–	–
29	–	–	–	–	–	–	–	–	–	–	–	–	–	–
30	378.4	227.0	–	–	–	–	–	–	–	–	–	–	–	–
31	28.2	16.9	–	–	–	–	–	–	–	–	–	–	8.5	3.6
32	–	–	–	–	–	–	–	–	–	–	–	–	–	–
33	272.1	163.2	–	–	–	–	–	–	–	–	–	–	15.1	4.2
34	–	–	–	–	–	–	–	–	–	–	–	–	–	–
35	32.9	19.8	–	–	–	–	–	–	–	–	–	–	–	–
36	239.5	143.7	–	–	–	–	–	–	–	10	1.0	–	15.2	3.7
37	40.5	24.3	–	–	–	–	0.09	–	–	38	9.9	–	15.7	2.0
38	16.6	10.0	–	–	–	–	–	–	–	–	–	–	–	–
39	–	–	–	–	–	–	–	–	–	–	–	–	–	–
40	145.8	87.4	2	–	–	–	–	–	–	–	–	–	–	–
41	92.1	55.3	–	115.5	–	–	–	–	–	–	–	–	–	–
42	–	–	–	–	–	–	–	–	–	–	–	–	–	–
43	6.6	4.0	1	–	–	–	–	–	–	–	–	–	–	–
44	6.3	3.8	1	–	–	–	–	–	–	–	–	–	–	–
45	–	–	–	–	–	–	–	–	–	–	–	–	–	–
46	1.0	0.6	–	–	–	–	–	–	–	7	3.8	–	1.0	0.9
47	–	–	–	–	–	–	–	–	–	–	–	–	–	–
48	–	–	–	–	–	–	–	–	–	–	–	–	3.7	–
49	52.7	31.6	–	–	–	–	–	–	–	–	–	–	–	–
50	34.2	20.5	–	–	–	–	–	–	–	–	–	–	–	–
51	32.9	19.8	–	–	–	–	–	–	–	–	–	–	–	–
52	–	–	–	–	–	–	–	–	–	–	–	–	–	–
53	30.8	18.5	–	–	–	–	–	–	–	–	–	–	–	–
54	–	–	–	–	–	–	–	–	–	–	–	–	–	–
55	–	–	–	–	–	–	–	–	–	–	–	–	–	–
56	–	–	–	–	–	–	–	–	–	–	–	–	–	–
57	40.8	24.5	–	–	–	–	–	–	–	–	–	–	–	–
58	18.0	10.8	2	–	–	–	–	–	–	–	–	–	–	–
59	58.8	35.3	–	–	–	–	–	–	–	–	–	–	–	–
60	–	–	–	–	–	–	–	–	–	–	–	–	–	–
61	–	–	–	–	–	–	–	–	–	–	–	–	9.9	–
62	73.9	44.3	–	–	–	–	–	–	–	–	–	–	8.7	–
63	59.3	35.6	–	–	–	–	–	–	–	–	–	–	8.7	–
64	105.8	63.4	–	118.6	–	–	–	–	–	–	–	–	–	–
65	–	–	–	–	–	–	–	–	–	–	–	–	–	–
66	86.9	52.1	–	–	–	–	–	–	–	–	–	–	–	–
67	–	–	–	–	–	–	–	–	–	–	–	–	–	–
68	–	–	–	–	–	–	–	–	–	–	–	–	–	–
69	–	–	–	–	–	–	–	–	–	–	–	–	–	–
70	45.0	27.0	1	–	–	–	–	–	–	–	–	–	–	–
71	5.8	3.5	1	–	–	–	–	–	–	–	–	–	–	–
72	–	–	–	–	–	–	–	–	–	–	–	–	–	–
73	28.9	17.3	–	170.7	–	–	–	–	–	–	–	–	6.1	2.6
74	–	–	–	–	–	–	–	–	–	–	–	–	–	–
75	–	–	–	–	–	–	–	–	–	–	–	–	–	–

(Continued)

TABLE V-1C DRY FORAGES, COMPOSITION OF FEEDS, DATA EXPRESSED AS-FED—(Continued)

Entry Number	Feed Name Description	International Feed Number	Proximate Analysis								Digestible Protein		
			Dry Matter %	Ash %	Crude Fiber %	Neutral Det. Fib. (NDF) %	Acid Det. Fib. (ADF) %	Lignin %	Ether Extract (Fat) %	N-Free Extract %	Crude Protein %	Ruminant %	Swine %	Horse %
76	PEANUT, HAY, SUN-CURED	1-03-619	91	8.2	30.3	—	37.2	—	3.3	39.1	9.9	5.7	5.2	6.0
77	PEARL MILLET, HAY, SUN-CURED	1-03-112	87	8.9	32.2	—	—	—	1.8	37.2	7.3	4.6	3.1	4.1
78	PEAVINE, HAY, SUN-CURED	1-03-666	91	6.8	24.9	—	—	—	2.6	36.6	19.8	15.4	13.9	13.6
79	PRAIRIE GRASS, MIDWEST (PRAIRIE HAY), HAY, SUN-CURED	1-03-191	91	7.2	30.7	—	—	—	2.1	45.2	5.8	2.9	1.6	2.9
80	REDTOP, HAY, SUN-CURED	1-03-885	92	6.0	28.4	—	—	—	2.8	47.4	7.4	3.7	3.0	4.1
81	REED CANARYGRASS, HAY, SUN-CURED	1-01-104	89	7.3	30.2	62.9	32.7	—	2.7	40.0	9.1	5.8	4.6	5.5
	RICE													
82	HULLS	1-08-075	92	19.0	38.9	71.9	62.3	9.6	1.0	30.3	3.0	0.1	-0.9	0.6
83	STRAW	1-03-925	91	15.4	31.9	64.4	50.1	4.4	1.3	38.2	3.9	0.7	-0.1	1.4
84	RYE, STRAW	1-04-007	91	3.8	38.3	—	—	—	1.4	44.7	2.8	-0.3	-1.1	0.5
85	RYEGRASS, HAY, SUN-CURED	1-04-057	88	7.1	25.3	56.3	37.0	—	1.8	46.2	7.5	3.8	3.2	4.2
86	SAWDUST, WOOD	1-07-714	90	0.7	71.5	—	—	—	0.8	16.7	0.3	-2.5	-3.2	-1.4
87	SEAWEED (KELP), WHOLE, DEHY	1-08-073	91	35.0	6.5	—	—	—	0.5	42.4	6.5	2.9	2.2	3.4
	SORGHUM													
88	FODDER WITH HEADS, SUN-CURED	1-07-960	90	8.9	25.6	—	—	—	2.0	47.4	6.2	2.4	2.0	3.2
89	STOVER WITHOUT HEADS, SUN-CURED	1-04-302	92	8.9	29.9	—	39.9	—	1.6	46.8	4.4	0.7	0.4	1.8
90	KAFIR, FODDER WITH HEADS, SUN-CURED	1-04-418	90	9.4	25.7	—	—	—	2.2	44.5	8.3	4.2	3.8	4.8
91	KAFIR, STOVER WITHOUT HEADS, SUN-CURED	1-04-419	82	8.1	26.9	—	—	—	1.5	40.7	4.6	1.6	0.9	2.1
92	MILO, FODDER WITH HEADS, SUN-CURED	1-04-433	89	8.5	21.8	—	—	—	2.9	49.4	6.5	2.5	2.3	3.4
93	MILO, STOVER WITHOUT HEADS, SUN-CURED	1-04-434	91	11.8	31.1	—	—	—	1.3	42.5	4.4	-0.8	0.4	1.3
94	SUDANGRASS, HAY, SUN-CURED	1-04-480	91	10.7	26.2	60.2	20.4	35.5	1.6	41.7	10.9	4.7	6.1	6.8
	SOYBEAN													
95	HAY, SUN-CURED	1-04-558	89	7.2	30.6	—	35.7	—	2.3	35.0	14.1	9.5	8.9	9.3
96	HULLS	1-04-560	91	4.6	36.2	59.4	42.4	1.8	2.0	37.0	10.8	6.5	3.4	6.7
97	STRAW	1-04-567	88	5.6	38.9	—	—	—	1.3	37.4	4.6	1.3	0.6	1.9
98	STARGRASS, HAY, SUN-CURED	1-13-407	90	8.6	28.8	—	—	—	1.5	44.4	7.7	2.0	—	2.8
99	SUGARCANE (BAGASSE), DEHY	1-04-686	91	2.9	42.3	78.8	54.5	12.8	0.7	43.8	1.4	-1.5	-2.3	-0.5
100	SWEET POTATO, HAY, SUN-CURED	1-04-779	90	10.2	21.3	—	23.4	—	3.2	43.5	11.9	4.6	7.0	7.6
101	TIMOTHY, HAY, SUN-CURED, ALL ANALYSES	1-04-893	91	4.6	30.3	63.7	36.4	—	2.4	47.3	6.8	2.8	2.4	3.3
102	TIMOTHY-CLOVER, HAY, SUN-CURED	1-04-973	89	5.1	31.7	—	—	—	2.0	42.7	7.8	4.1	3.4	4.4
103	TREFOIL, BIRDSFOOT, HAY, SUN-CURED	1-05-044	91	6.7	29.3	42.8	32.8	—	1.9	38.9	13.9	9.6	8.7	9.1
104	VELVETBEAN, HAY, SUN-CURED	1-05-080	93	7.4	27.5	—	—	—	3.1	38.4	16.4	9.5	10.8	11.0
105	VETCH, HAY, SUN-CURED	1-05-106	89	7.8	24.8	42.7	29.4	—	2.7	35.1	18.4	13.4	12.7	12.6
106	WATERGRASS, HAY, SUN-CURED	1-05-154	94	18.0	31.2	—	—	—	1.1	36.3	7.6	3.8	3.1	4.2
107	WHEAT, STRAW	1-05-175	90	6.9	37.4	70.3	47.7	8.4	1.8	40.4	3.2	0.3	-0.6	0.6
108	WHEATGRASS, CRESTED, HAY, SUN-CURED	1-05-418	92	6.4	30.8	—	33.2	5.1	2.1	42.2	10.3	6.4	5.5	6.3
109	WHITE CLOVER, HAY, SUN-CURED	1-01-464	90	9.0	21.9	—	—	—	2.4	40.1	16.9	12.4	11.4	11.5
110	WOOD, SAWDUST	1-07-714	90	0.7	71.5	—	—	—	0.8	16.7	0.3	-2.5	-3.2	-1.4

TABLE V-1D SILAGES AND HAYLAGES, COMPOSITION OF FEEDS, DATA EXPRESSED AS-FED

Entry Number	Feed Name Description	International Feed Number	Proximate Analysis								Digestible Protein		
			Dry Matter %	Ash %	Crude Fiber %	Neutral Det. Fib. (NDF) %	Acid Det. Fib. (ADF) %	Lignin %	Ether Extract (Fat) %	N-Free Extract %	Crude Protein %	Ruminant %	Swine %	Horse %
	ALFALFA (LUCERNE), SILAGE													
1	ALL ANALYSES	3-00-212	27	2.6	8.6	—	10.7	—	0.9	10.4	4.7	3.4	—	—
2	CORN GRAIN ADDED	3-00-226	23	2.3	7.0	—	—	—	0.9	9.0	4.1	3.0	—	—
3	MOLASSES ADDED	3-00-238	30	2.6	7.8	—	—	—	1.5	12.4	5.6	3.9	—	—
	ALFALFA-BROMEGRASS, SMOOTH, SILAGE													
4	ALL ANALYSES	3-00-268	26	2.0	9.4	—	—	—	1.1	10.4	3.6	2.3	—	—
5	WILTED	3-00-269	46	4.1	15.2	—	—	—	1.1	18.6	7.1	4.8	—	—
6	ALFALFA-ORCHARDGRASS, SILAGE, 30-50% DRY MATTER	3-08-144	37	4.0	11.2	—	16.8	3.7	1.5	13.7	6.7	4.7	—	—
7	BEET, MANGEL, SILAGE, TOPS WITH CROWNS	3-00-635	13	2.8	1.9	—	—	—	0.6	5.7	1.8	1.2	—	—
8	CITRUS, SILAGE, PULP	4-01-234	21	1.2	3.3	—	4.2	—	2.1	12.8	1.5	0.7	1.0	1.0
	CORN, SILAGE													
9	ALL ANALYSES	3-02-822	26	1.5	6.6	—	8.9	1.2	0.8	15.2	2.2	1.0	1.3	—
10	MOLASSES ADDED	3-02-834	29	1.9	6.8	—	—	—	0.8	17.2	2.3	1.1	—	—
11	EARS WITH HUSKS	4-02-839	43	1.5	4.8	—	—	—	1.6	31.5	4.0	2.2	2.8	2.7
12	HUSKS (HUSKLAGE)	3-26-074	78	2.7	26.8	—	—	—	0.7	45.0	2.9	—	—	—
13	CORN, SWEET, SILAGE, CANNERY RESIDUE	3-07-955	31	1.7	10.1	—	10.5	—	1.3	15.4	2.5	1.2	—	—
14	CORN-SORGHUM, SILAGE, MATURE	3-03-013	34	2.1	8.5	—	—	—	1.1	19.7	2.7	1.7	—	—
15	CORN-SOYBEAN, SILAGE	3-03-015	27	1.8	7.0	—	—	—	0.9	14.6	2.7	1.7	—	—
	GRASS, SILAGE													
16	EARLY BLOOM	3-02-218	23	2.1	7.3	—	—	—	0.6	10.6	2.8	1.7	—	—
17	MOLASSES ADDED	3-02-261	27	3.8	6.4	—	—	—	1.4	10.6	4.3	3.1	—	—
	GRASS-LEGUME, SILAGE													
18	ALL ANALYSES	3-02-303	31	2.5	10.1	18.1	12.4	3.1	1.1	13.6	3.7	1.9	—	—
19	BARLEY GRAIN ADDED	3-02-305	34	2.2	8.6	—	—	—	1.4	16.7	5.2	3.4	—	—
20	MOLASSES ADDED	3-02-309	28	2.0	9.1	—	—	—	1.1	12.8	3.4	2.0	—	—
21	KAFIR SORGHUM, SILAGE	3-04-425	30	2.2	8.1	—	—	—	1.0	16.1	2.1	0.9	—	—

Composition of Feeds

DRY FORAGES

Entry Number	TDN Ruminant %	TDN Swine %	TDN Horse %	Digestible Energy Ruminant Mcal/lb	Mcal/kg	Swine kcal/lb	kcal/kg	Horse Mcal/lb	Mcal/kg	Metabolizable Energy Ruminant Mcal/lb	Mcal/kg	Swine kcal/lb	kcal/kg	Poultry ME$_n$ kcal/lb	kcal/kg	Horse Mcal/lb	Mcal/kg	Net Energy Ruminant NE$_m$ Mcal/lb	Mcal/kg	Ruminant NE$_g$ Mcal/lb	Mcal/kg	Lactating Cows NE$_{lc}$ Mcal/lb	Mcal/kg
76	48	39	32	0.95	2.10	–	–	0.64	1.41	0.78	1.71	–	–	–	–	0.52	1.15	0.39	0.85	0.16	0.36	0.45	0.99
77	50	24	30	0.94	2.06	–	–	0.61	1.33	0.77	1.69	–	–	–	–	0.50	1.09	0.38	0.84	0.17	0.37	0.44	0.96
78	55	56	53	1.12	2.46	–	–	0.99	2.18	0.94	2.08	–	–	–	–	0.81	1.78	0.57	1.26	0.33	0.73	0.58	1.28
79	46	31	34	0.94	2.06	–	–	0.67	1.48	0.76	1.68	–	–	–	–	0.55	1.21	0.44	0.97	0.21	0.46	0.48	1.07
80	50	41	37	1.02	2.25	–	–	0.73	1.60	0.84	1.86	–	–	–	–	0.60	1.31	0.50	1.11	0.27	0.59	0.53	1.17
81	44	36	43	0.93	2.06	–	–	0.82	1.80	0.76	1.68	–	–	–	–	0.67	1.47	0.46	1.02	0.24	0.52	0.50	1.10
82	11	–	12	0.27	0.60	–	–	0.30	0.67	0.16	0.35	–	–	36	79	0.25	0.55	–0.26	–0.57	–0.47	–1.04	0.08	0.17
83	40	12	22	0.80	1.76	–	–	0.47	1.04	0.62	1.37	–	–	–	–	0.39	0.85	0.34	0.75	0.12	0.26	0.41	0.91
84	39	–	32	0.78	1.72	–	–	0.64	1.40	0.60	1.32	–	–	–	–	0.52	1.15	0.25	0.55	0.03	0.07	0.35	0.78
85	53	37	40	1.06	2.34	–	–	0.76	1.68	0.89	1.97	–	–	–	–	0.62	1.38	0.54	1.19	0.31	0.68	0.55	1.22
86	32	–	–	0.65	1.43	–	–	–	–	0.47	1.04	–	–	–	–	–	–	–	–	–	–	–	–
87	29	–	–	0.58	1.27	–	–	–	–	0.40	0.87	–	–	–	–	–	–	–	–	–	–	–	–
88	51	34	36	1.02	2.24	–	–	0.71	1.56	0.84	1.86	–	–	–	–	0.58	1.28	0.51	1.12	0.28	0.61	0.53	1.17
89	47	27	33	0.92	2.02	–	–	0.66	1.46	0.74	1.64	–	–	–	–	0.54	1.20	0.40	0.89	0.18	0.39	0.42	0.93
90	52	37	37	1.01	2.22	–	–	0.71	1.57	0.83	1.83	–	–	–	–	0.59	1.29	0.45	1.00	0.23	0.50	0.49	1.08
91	47	24	30	0.93	2.05	–	–	0.59	1.30	0.77	1.71	–	–	–	–	0.48	1.07	0.45	0.99	0.24	0.53	0.47	1.05
92	57	43	36	1.07	2.36	–	–	0.70	1.54	0.90	1.99	–	–	–	–	0.57	1.26	0.49	1.07	0.26	0.57	0.51	1.13
93	48	19	29	0.96	2.11	–	–	0.59	1.29	0.78	1.72	–	–	–	–	0.48	1.06	0.44	0.98	0.21	0.47	0.49	1.07
94	51	35	41	1.00	2.20	–	–	0.79	1.74	0.82	1.82	–	–	–	–	0.65	1.43	0.48	1.07	0.25	0.55	0.52	1.14
95	49	40	41	1.11	2.45	–	–	0.79	1.74	0.86	1.90	–	–	–	–	0.65	1.43	0.52	1.14	0.29	0.63	0.55	1.21
96	69	47	39	1.20	2.65	851	1877	0.75	1.65	1.03	2.26	305	671	301	665	0.61	1.35	0.75	1.66	0.50	1.09	0.72	1.59
97	36	–	29	0.71	1.57	–	–	0.58	1.27	0.54	1.19	–	–	–	–	0.47	1.04	0.21	0.46	0.00	0.00	0.32	0.71
98	49	–	35	0.90	1.98	–	–	0.65	1.44	0.74	1.62	–	–	–	–	0.53	1.17	0.41	0.90	0.08	0.18	0.41	0.90
99	43	–	–	0.88	1.93	–	–	–	–	0.69	1.53	–	–	–	–	–	–	0.37	0.82	0.15	0.32	0.40	0.88
100	49	49	39	1.02	2.25	–	–	0.76	1.67	0.85	1.86	–	–	–	–	0.62	1.37	0.51	1.13	0.28	0.62	0.54	1.18
101	53	38	43	1.06	2.35	–	–	0.81	1.79	0.85	1.88	–	–	–	–	0.67	1.47	0.50	1.10	0.27	0.58	0.51	1.13
102	50	34	38	0.99	2.18	–	–	0.73	1.60	0.82	1.80	–	–	–	–	0.60	1.31	0.46	1.01	0.23	0.49	0.49	1.09
103	54	41	46	0.91	2.01	–	–	0.87	1.91	0.83	1.84	–	–	–	–	0.71	1.56	0.55	1.22	0.32	0.69	0.57	1.25
104	56	51	46	1.12	2.47	–	–	0.87	1.91	0.94	2.08	–	–	–	–	0.71	1.57	0.55	1.22	0.31	0.69	0.57	1.26
105	55	52	48	1.09	2.41	–	–	0.91	2.00	0.92	2.04	–	–	–	–	0.74	1.64	0.55	1.22	0.32	0.71	0.56	1.24
106	39	14	27	0.79	1.75	–	–	0.55	1.22	0.61	1.34	–	–	–	–	0.45	1.00	0.30	0.66	0.07	0.16	0.39	0.87
107	40	–	36	0.86	1.90	–	–	0.71	1.56	0.70	1.54	–	–	–	–	0.58	1.28	0.36	0.79	0.14	0.30	0.39	0.86
108	49	38	41	0.99	2.19	–	–	0.78	1.73	0.82	1.80	–	–	–	–	0.64	1.42	0.48	1.06	0.25	0.55	0.52	1.14
109	57	52	50	1.11	2.45	–	–	0.94	2.07	0.94	2.07	–	–	–	–	0.77	1.70	0.53	1.18	0.30	0.66	0.55	1.22
110	32	–	–	0.65	1.43	–	–	–	–	0.47	1.04	–	–	–	–	–	–	–	–	–	–	–	–

SILAGES AND HAYLAGES

Entry Number	TDN Ruminant %	TDN Swine %	TDN Horse %	Digestible Energy Ruminant Mcal/lb	Mcal/kg	Swine kcal/lb	kcal/kg	Horse Mcal/lb	Mcal/kg	Metabolizable Energy Ruminant Mcal/lb	Mcal/kg	Swine kcal/lb	kcal/kg	Poultry ME$_n$ kcal/lb	kcal/kg	Horse Mcal/lb	Mcal/kg	Net Energy Ruminant NE$_m$ Mcal/lb	Mcal/kg	Ruminant NE$_g$ Mcal/lb	Mcal/kg	Lactating Cows NE$_{lc}$ Mcal/lb	Mcal/kg
1	16	–	–	0.28	0.61	–	–	–	–	0.27	0.59	–	–	–	–	–	–	0.16	0.36	0.09	0.20	0.16	0.35
2	13	–	–	0.27	0.59	–	–	–	–	0.22	0.49	–	–	–	–	–	–	0.14	0.31	0.08	0.17	0.14	0.32
3	18	–	–	0.40	0.88	–	–	–	–	0.33	0.72	–	–	–	–	–	–	0.21	0.46	0.13	0.28	0.20	0.44
4	16	–	–	0.31	0.69	–	–	–	–	0.26	0.57	–	–	–	–	–	–	0.15	0.34	0.09	0.19	0.16	0.36
5	26	–	–	0.53	1.16	–	–	–	–	0.44	0.97	–	–	–	–	–	–	0.24	0.54	0.13	0.28	0.26	0.58
6	20	–	–	0.42	0.93	–	–	–	–	0.35	0.77	–	–	–	–	–	–	0.19	0.41	0.09	0.20	0.20	0.45
7	7	–	–	0.14	0.32	–	–	–	–	0.12	0.26	–	–	–	–	–	–	0.07	0.15	0.04	0.08	0.07	0.16
8	18	18	5	0.37	0.80	354	781	0.11	0.24	0.33	0.72	335	739	–	–	0.09	0.19	0.21	0.45	0.14	0.31	0.19	0.42
9	18	19	–	0.34	0.75	–	–	–	–	0.30	0.66	–	–	–	–	–	–	0.19	0.42	0.12	0.26	0.17	0.37
10	19	–	–	0.38	0.85	–	–	–	–	0.33	0.73	–	–	–	–	–	–	0.19	0.42	0.11	0.25	0.19	0.42
11	31	35	27	0.66	1.45	692	1526	0.49	1.09	0.57	1.27	653	1439	–	–	0.41	0.89	0.33	0.74	0.21	0.47	0.32	0.71
12	47	–	–	0.92	2.03	–	–	–	–	0.78	1.72	–	–	–	–	–	–	–	–	–	–	–	–
13	22	–	–	0.42	0.92	–	–	–	–	0.36	0.79	–	–	–	–	–	–	0.24	0.53	0.15	0.34	0.23	0.51
14	24	–	–	0.45	1.00	–	–	–	–	0.39	0.86	–	–	–	–	–	–	0.22	0.49	0.13	0.29	0.22	0.49
15	19	–	–	0.37	0.82	–	–	–	–	0.32	0.70	–	–	–	–	–	–	0.21	0.46	0.13	0.29	0.20	0.44
16	14	–	–	0.27	0.61	–	–	–	–	0.23	0.51	–	–	–	–	–	–	0.13	0.28	0.07	0.15	0.14	0.30
17	18	–	–	0.34	0.75	–	–	–	–	0.29	0.64	–	–	–	–	–	–	0.16	0.36	0.09	0.20	0.17	0.37
18	18	–	–	0.38	0.84	–	–	–	–	0.31	0.68	–	–	–	–	–	–	0.19	0.42	0.11	0.24	0.19	0.41
19	23	–	–	0.45	1.00	–	–	–	–	0.39	0.86	–	–	–	–	–	–	0.23	0.51	0.14	0.31	0.23	0.50
20	16	–	–	0.33	0.72	–	–	–	–	0.28	0.61	–	–	–	–	–	–	0.17	0.37	0.10	0.22	0.17	0.37
21	17	–	–	0.35	0.77	–	–	–	–	0.29	0.64	–	–	–	–	–	–	0.18	0.40	0.10	0.23	0.19	0.41

(Continued)

TABLE V-1C DRY FORAGES, MINERAL AND VITAMIN COMPOSITION OF FEEDS, DATA EXPRESSED AS-FED—(Continued)

Entry Number	Feed Name Description	Dry Matter	Macro Minerals							Micro Minerals						
			Calcium (Ca)	Phosphorus (P)	Sodium (Na)	Chlorine (Cl)	Magnesium (Mg)	Potassium (K)	Sulfur (S)	Cobalt (Co)	Copper (Cu)	Iodine (I)	Iron (Fe)	Manganese (Mn)	Selenium (Se)	Zinc (Zn)
		%	%	%	%	%	%	%	%	ppm or mg/kg	ppm or mg/kg	ppm or mg/kg	%	ppm or mg/kg	ppm or mg/kg	ppm or mg/kg
76	PEANUT, HAY, SUN-CURED	91	1.12	0.14	—	—	0.44	1.25	0.21	0.072	—	—	—	—	—	—
77	PEARL MILLET, HAY, SUN-CURED	87	—	—	—	—	—	—	—	—	—	—	—	—	—	—
78	PEAVINE, HAY, SUN-CURED	91	—	—	—	—	—	—	—	—	—	—	—	—	—	—
79	PRAIRIE GRASS, MIDWEST (PRAIRIE HAY), HAY, SUN-CURED	91	0.32	0.13	—	—	0.24	0.98	—	—	—	—	0.008	—	—	—
80	REDTOP, HAY, SUN-CURED	92	0.39	0.20	0.06	0.06	0.20	1.74	0.23	0.134	3.6	0.092	0.015	207.7	—	—
81	REED CANARYGRASS, HAY, SUN-CURED	89	0.32	0.21	0.01	—	0.19	2.60	—	—	10.6	—	0.014	82.5	—	—
	RICE															
82	HULLS	92	0.11	0.10	0.02	0.07	0.41	0.64	0.08	2.046	3.1	—	0.010	295.0	—	22.0
83	STRAW	91	0.19	0.07	0.28	—	0.10	1.20	—	—	—	—	—	313.9	—	—
84	RYE, STRAW	91	0.22	0.08	0.12	0.22	0.07	0.88	0.10	—	3.6	—	—	6.0	—	—
85	RYEGRASS, HAY, SUN-CURED	88	—	—	—	—	—	—	—	—	—	—	—	—	—	—
86	SAWDUST, WOOD	90	—	—	—	—	—	—	—	—	—	—	—	—	—	—
87	SEAWEED (KELP), WHOLE, DEHY	91	2.47	0.28	—	—	0.85	—	—	—	—	—	—	—	—	—
	SORGHUM															
88	FODDER WITH HEADS, SUN-CURED	90	0.56	0.17	0.02	—	0.27	1.12	—	—	—	—	—	—	—	—
89	STOVER WITHOUT HEADS, SUN-CURED	92	0.37	0.10	—	—	—	1.10	—	—	—	—	—	—	—	—
90	KAFIR, FODDER WITH HEADS, SUN-CURED	90	0.35	0.18	—	—	0.26	1.53	—	—	—	—	—	—	—	—
91	KAFIR, STOVER WITHOUT HEADS, SUN-CURED	82	0.50	0.08	—	—	—	—	—	—	—	—	—	—	—	—
92	MILO, FODDER WITH HEADS, SUN-CURED	89	0.35	0.18	—	—	—	—	—	—	—	—	—	—	—	—
93	MILO, STOVER WITHOUT HEADS, SUN-CURED	91	0.51	0.15	0.10	—	0.23	0.41	—	—	—	—	—	—	—	—
94	SUDANGRASS, HAY, SUN-CURED	91	0.47	0.28	0.01	—	0.34	1.90	0.06	0.116	28.6	—	0.015	69.5	—	34.6
	SOYBEAN															
95	HAY, SUN-CURED	89	1.13	0.22	0.10	0.13	0.71	0.92	0.25	0.083	8.0	0.216	0.026	94.3	—	21.5
96	HULLS	91	0.45	0.19	0.03	—	—	1.15	0.08	0.109	16.1	—	0.030	9.9	—	21.8
97	STRAW	88	1.40	0.05	0.11	—	0.81	0.49	0.23	—	—	—	0.027	44.9	—	—
98	STARGRASS, HAY, SUN-CURED	90	—	—	—	—	—	—	—	—	—	—	—	—	—	—
99	SUGARCANE (BAGASSE), DEHY	91	0.47	0.26	0.04	—	0.08	0.34	0.09	—	—	—	0.019	—	—	—
100	SWEET POTATO, HAY, SUN-CURED	90	0.08	0.32	—	—	—	—	—	—	—	—	—	—	—	—
101	TIMOTHY, HAY, SUN-CURED, ALL ANALYSES	91	0.38	0.17	0.03	0.49	0.11	1.43	0.11	0.071	4.3	0.034	0.010	45.2	—	15.5
102	TIMOTHY-CLOVER, HAY, SUN-CURED	89	0.62	0.17	0.17	0.48	0.17	1.33	0.13	—	6.3	—	0.011	48.7	—	—
103	TREFOIL, BIRDSFOOT, HAY, SUN-CURED	91	1.54	0.21	0.06	—	0.46	1.74	0.23	0.100	8.4	—	0.021	26.0	—	69.9
104	VELVET BEAN, HAY, SUN-CURED	93	—	0.24	—	—	—	2.20	—	—	—	—	—	—	—	—
105	VETCH, HAY, SUN-CURED	89	1.21	0.30	0.46	—	0.24	1.88	0.13	0.315	8.8	0.437	0.044	53.9	—	—
106	WATERGRASS, HAY, SUN-CURED	94	—	—	—	—	—	—	—	—	—	—	—	—	—	—
107	WHEAT, STRAW	90	0.16	0.05	0.13	0.29	0.11	1.27	0.17	0.041	3.2	—	0.015	36.7	—	5.8
108	WHEATGRASS, CRESTED, HAY, SUN-CURED	92	0.24	0.14	—	—	—	—	—	0.219	—	—	—	—	—	—
109	WHITE CLOVER, HAY, SUN-CURED	90	1.71	0.29	—	—	—	—	—	—	—	—	—	—	—	—
110	WOOD, SAWDUST	90	—	—	—	—	—	—	—	—	—	—	—	—	—	—

TABLE V-1D SILAGES AND HAYLAGES, MINERAL AND VITAMIN COMPOSITION OF FEEDS, DATA EXPRESSED AS-FED

Entry Number	Feed Name Description	Dry Matter	Macro Minerals							Micro Minerals						
			Calcium (Ca)	Phosphorus (P)	Sodium (Na)	Chlorine (Cl)	Magnesium (Mg)	Potassium (K)	Sulfur (S)	Cobalt (Co)	Copper (Cu)	Iodine (I)	Iron (Fe)	Manganese (Mn)	Selenium (Se)	Zinc (Zn)
		%	%	%	%	%	%	%	%	ppm or mg/kg	ppm or mg/kg	ppm or mg/kg	%	ppm or mg/kg	ppm or mg/kg	ppm or mg/kg
	ALFALFA (LUCERNE), SILAGE															
1	ALL ANALYSES	27	0.48	0.07	0.04	0.11	0.09	0.64	0.09	—	3.0	—	0.008	13.5	—	11.1
2	CORN GRAIN ADDED	23	—	—	—	—	—	—	—	—	—	—	—	—	—	—
3	MOLASSES ADDED	30	0.49	0.09	0.05	—	0.10	0.78	0.11	—	—	—	0.009	—	—	—
	ALFALFA-BROMEGRASS, SMOOTH, SILAGE															
4	ALL ANALYSES	26	0.17	0.05	—	—	0.05	0.49	—	0.030	3.0	—	0.004	8.0	—	—
5	WILTED	46	0.72	0.12	—	—	—	—	—	—	—	—	—	—	—	—
6	ALFALFA-ORCHARDGRASS, SILAGE, 30-50% DRY MATTER	37	0.32	0.11	0.04	—	0.12	1.07	0.10	—	—	—	0.010	—	—	—
7	BEET, MANGEL, SILAGE, TOPS WITH CROWNS	13	—	—	—	—	—	—	—	—	—	—	—	—	—	—
8	CITRUS, SILAGE, PULP	21	0.42	0.03	0.02	—	0.03	0.13	0.00	—	—	—	0.004	—	—	3.3
	CORN, SILAGE															
9	ALL ANALYSES	26	0.08	0.07	0.01	0.05	0.06	0.32	0.03	0.026	2.4	—	0.005	10.8	—	5.5
10	MOLASSES ADDED	29	0.07	0.09	—	—	—	—	—	—	—	—	—	—	—	—
11	EARS WITH HUSKS	43	0.04	0.12	0.00	—	0.05	0.21	0.06	—	—	—	0.004	—	—	—
12	HUSKS (HUSKLAGE)	78	—	—	—	—	—	—	—	—	—	—	—	—	—	—
13	CORN, SWEET, SILAGE, CANNERY RESIDUE	31	0.10	0.24	0.01	—	0.07	0.36	0.03	—	—	—	0.007	—	—	—
14	CORN-SORGHUM, SILAGE, MATURE	34	0.15	0.08	—	—	—	0.35	—	—	—	—	—	—	—	—
15	CORN-SOYBEAN, SILAGE	27	0.19	0.09	0.01	—	0.08	0.29	0.05	—	2.2	—	0.012	12.4	—	7.6
	GRASS, SILAGE															
16	EARLY BLOOM	23	—	—	—	—	—	—	—	—	—	—	—	—	—	—
17	MOLASSES ADDED	27	0.28	0.07	—	—	—	—	—	—	—	—	—	—	—	—
	GRASS-LEGUME, SILAGE															
18	ALL ANALYSES	31	0.26	0.08	0.02	0.33	0.08	0.57	0.17	0.040	1.9	—	0.013	18.0	—	8.7
19	BARLEY GRAIN ADDED	34	0.26	0.12	—	—	—	—	—	—	—	—	—	—	—	—
20	MOLASSES ADDED	28	0.30	0.10	0.04	—	0.09	0.55	0.07	—	—	—	0.015	—	—	—
21	KAFIR SORGHUM, SILAGE,	30	0.07	0.05	—	—	0.08	0.50	—	—	—	—	—	—	—	—

Composition of Feeds

DRY FORAGES

Entry Number	Fat-Soluble Vitamins					Water-Soluble Vitamins								
	A (1 mg Carotene = 1667 IU Vit A)	Carotene (Provitamin A)	D	E	K	B-12	Biotin	Choline	Folacin (Folic Acid)	Niacin	Pantothenic Acid (B-3)	(Pyridoxine) B-6	Riboflavin (B-2)	Thiamin (B-1)
	IU/g	ppm or mg/kg	IU/kg	ppm or mg/kg	ppm or mg/kg	ppb or mcg/kg	ppm or mg/kg	ppm or mg/kg	ppm or mg/kg	ppm or mg/kg	ppm or mg/kg	ppm or mg/kg	ppm or mg/kg	ppm or mg/kg
76	52.6	31.5	–	–	–	–	–	–	–	–	–	–	8.8	–
77	–	–	–	–	–	–	–	–	–	–	–	–	–	–
78	48.0	28.8	–	–	–	–	–	–	–	–	–	–	8.4	–
79	–	–	1	–	–	–	–	–	–	–	–	–	–	–
80	6.1	3.7	–	–	–	–	–	–	–	–	–	–	–	–
81	28.2	16.9	–	–	–	–	–	–	–	–	–	–	8.5	3.6
82	–	–	–	7.5	–	–	–	–	–	28	7.9	0.07	0.5	2.2
83	–	–	–	–	–	–	–	–	–	–	–	–	–	–
84	–	–	–	–	–	–	–	–	–	–	–	–	–	–
85	175.8	105.5	–	–	–	–	–	–	–	–	–	–	–	–
86	–	–	–	–	–	–	–	–	–	–	–	–	–	–
87	–	–	–	–	–	–	–	–	–	–	–	–	–	–
88	–	–	–	–	–	–	–	–	–	–	–	–	–	–
89	–	–	–	–	–	–	–	–	–	–	–	–	–	–
90	–	–	–	–	–	–	–	–	–	–	–	–	–	–
91	–	–	–	–	–	–	–	–	–	–	–	–	–	–
92	2.8	1.8	–	–	–	–	–	–	–	–	–	–	–	–
93	–	–	–	–	–	–	–	–	–	–	–	–	–	–
94	–	–	–	–	–	–	–	–	–	–	–	–	–	–
95	53.1	31.8	1	26.3	–	–	–	–	–	–	–	–	–	–
96	–	–	–	6.6	–	–	–	500	–	25	13.4	1.70	3.6	1.6
97	–	–	–	–	–	–	–	–	–	–	–	–	–	–
98	–	–	–	–	–	–	–	–	–	–	–	–	–	–
99	–	–	–	–	–	–	–	–	–	–	–	–	–	–
100	–	–	–	–	–	–	–	–	–	–	–	–	–	–
101	39.8	23.8	2	57.6	–	–	0.06	741	2.09	31	7.2	–	9.2	1.5
102	39.7	23.8	–	–	–	–	–	–	–	–	–	–	–	–
103	217.8	130.6	1	–	–	–	–	–	–	–	–	–	14.6	6.2
104	–	–	–	–	–	–	–	–	–	–	–	–	–	–
105	–	–	–	–	–	–	–	–	–	–	–	–	–	–
106	–	–	–	–	–	–	–	–	–	–	–	–	–	–
107	3.3	2.0	1	–	–	–	–	–	–	–	–	–	2.2	–
108	34.2	20.5	–	–	–	–	–	–	–	–	–	–	–	–
109	92.1	55.3	–	115.5	–	–	–	–	–	–	–	–	–	–
110	–	–	–	–	–	–	–	–	–	–	–	–	–	–

SILAGES AND HAYLAGES

Entry Number	Fat-Soluble Vitamins					Water-Soluble Vitamins								
	A (1 mg Carotene = 1667 IU Vit A)	Carotene (Provitamin A)	D	E	K	B-12	Biotin	Choline	Folacin (Folic Acid)	Niacin	Pantothenic Acid (B-3)	(Pyridoxine) B-6	Riboflavin (B-2)	Thiamin (B-1)
	IU/g	ppm or mg/kg	IU/kg	ppm or mg/kg	ppm or mg/kg	ppb or mcg/kg	ppm or mg/kg	ppm or mg/kg	ppm or mg/kg	ppm or mg/kg	ppm or mg/kg	ppm or mg/kg	ppm or mg/kg	ppm or mg/kg
1	42.2	25.3	–	–	–	–	–	–	–	–	–	–	–	–
2	–	–	–	–	–	–	–	–	–	–	–	–	–	–
3	53.8	32.3	–	–	–	–	–	–	–	–	–	–	–	–
4	–	–	–	–	–	–	–	–	–	–	–	–	–	–
5	–	–	–	–	–	–	–	–	–	–	–	–	–	–
6	–	–	–	–	–	–	–	–	–	–	–	–	–	–
7	–	–	–	–	–	–	–	–	–	–	–	–	–	–
8	–	–	–	–	–	–	–	–	–	–	–	–	–	–
9	15.2	9.1	0	–	–	–	–	–	–	11	–	–	–	–
10	–	–	–	–	–	–	–	–	–	–	–	–	–	–
11	–	–	–	–	–	–	–	–	–	–	–	–	–	–
12	–	–	–	–	–	–	–	–	–	–	–	–	–	–
13	6.9	4.2	–	–	–	–	–	–	–	–	–	–	–	–
14	3.1	1.9	–	–	–	–	–	–	–	–	–	–	–	–
15	89.0	53.4	–	–	–	–	–	–	–	–	–	–	–	–
16	–	–	–	–	–	–	–	–	–	–	–	–	–	–
17	–	–	–	–	–	–	–	–	–	–	–	–	–	–
18	102.4	61.4	0	–	–	–	–	–	–	14	–	–	–	–
19	–	–	–	–	–	–	–	–	–	–	–	–	–	–
20	–	–	–	–	–	–	–	–	–	–	–	–	–	–
21	5.3	3.2	–	–	–	–	–	–	–	–	–	–	–	–

(Continued)

TABLE V-1D SILAGES AND HAYLAGES, COMPOSITION OF FEEDS, DATA EXPRESSED AS-FED—(Continued)

Entry Number	Feed Name Description	International Feed Number	Proximate Analysis									Digestible Protein		
			Dry Matter	Ash	Crude Fiber	Neutral Det. Fib. (NDF)	Acid Det. Fib. (ADF)	Lignin	Ether Extract (Fat)	N-Free Extract	Crude Protein	Ruminant	Swine	Horse
			%	%	%	%	%	%	%	%	%	%	%	%
22	LESPEDEZA, COMMON-KOREAN, SILAGE	3-08-455	30	1.8	9.5	—	—	—	0.8	13.8	4.3	2.8	—	—
23	LESPEDEZA, SERICEA (CHINESE), SILAGE	3-02-614	30	1.7	9.5	—	—	—	0.9	14.0	4.3	2.7	—	—
24	MILLET, JAPANESE (BARNYARD GRASS)-SOYBEAN, SILAGE	3-26-066	21	2.9	7.3	—	—	—	1.1	6.9	2.9	1.7	—	—
25	MILO SORGHUM, SILAGE	3-04-437	31	2.5	5.5	—	—	—	0.5	20.0	2.3	1.0	—	—
	OATS, SILAGE													
26	DOUGH STAGE	3-03-296	35	2.4	11.6	—	—	—	1.4	16.1	3.5	2.0	—	—
27	MOLASSES ADDED	3-03-300	33	2.5	10.4	—	—	—	1.2	15.8	2.9	1.5	—	—
28	OATS-PEA, SILAGE	3-03-402	26	2.5	8.7	—	11.6	1.2	0.9	10.9	2.8	1.5	—	—
29	OATS-VETCH, SILAGE, MILK STAGE	3-03-408	27	2.2	8.0	—	—	—	1.2	12.4	3.4	2.1	—	—
30	PEA, SILAGE, VINES (WITHOUT SEEDS, WITH PODS)	3-03-596	25	2.2	7.3	—	—	—	0.8	11.0	3.2	1.9	—	—
31	PEA, FIELD, SILAGE	3-03-609	27	2.5	7.5	—	—	—	1.2	12.2	3.8	2.5	—	—
	POTATO, SILAGE													
32	TUBERS, ALFALFA HAY ADDED	3-03-770	35	2.2	7.1	—	—	—	0.5	20.7	4.2	2.6	—	—
33	VINES	3-03-765	15	2.8	3.4	—	—	—	0.5	5.7	2.3	1.7	—	—
	SORGHUM, SILAGE													
34	DOUGH STAGE	3-04-321	29	2.5	8.3	19.2	10.9	1.9	0.9	15.1	2.3	0.9	—	—
35	STOVER WITHOUT HEADS	3-04-326	62	5.6	16.2	—	—	—	1.2	36.3	2.7	—	—	0.1
36	KAFIR	3-04-425	30	2.2	8.1	—	—	—	1.0	16.1	2.1	0.9	—	—
37	MILO	3-04-437	31	2.5	5.5	—	—	—	0.5	20.0	2.3	1.0	—	—
38	SOYBEAN, SILAGE	3-04-581	30	3.0	9.0	—	12.0	—	0.8	12.2	5.2	3.2	—	—
39	SUGAR BEET, SILAGE, TOPS WITH CROWNS	3-00-660	25	8.6	3.2	—	—	—	0.7	9.3	3.4	2.2	—	—
40	SUGARCANE, SILAGE, ALL ANALYSES	3-04-693	22	1.4	7.5	—	—	—	0.4	11.6	1.1	0.3	—	—
41	SUNFLOWER, SILAGE, MILK STAGE	3-04-733	21	2.1	6.2	—	—	—	1.3	9.5	2.1	1.0	—	—
42	SWEET CORN, SILAGE, CANNERY RESIDUE	3-07-955	31	1.7	10.1	—	10.5	—	1.3	15.4	2.5	1.2	—	—
43	SWEET POTATO, SILAGE, VINES	3-04-785	13	1.5	3.6	—	—	—	0.5	5.9	1.7	0.7	—	—
44	VETCH, SILAGE	3-05-112	30	2.4	9.8	—	—	—	1.0	13.4	3.5	2.0	—	—

TABLE V-1E PASTURE AND RANGE PLANTS, COMPOSITION OF FEEDS, DATA EXPRESSED AS-FED

Entry Number	Feed Name Description	International Feed Number	Proximate Analysis									Digestible Protein		
			Dry Matter	Ash	Crude Fiber	Neutral Det. Fib. (NDF)	Acid Det. Fib. (ADF)	Lignin	Ether Extract (Fat)	N-Free Extract	Crude Protein	Ruminant	Swine	Horse
			%	%	%	%	%	%	%	%	%	%	%	%
	ALFALFA (LUCERNE)													
1	FRESH, ALL ANALYSES	2-00-196	26	2.5	6.0	11.8	—	—	1.0	11.2	5.3	4.0	3.8	—
2	PREBLOOM, FRESH	2-00-181	20	2.1	4.9	6.4	5.2	1.5	0.6	8.5	4.3	3.3	3.0	—
3	EARLY BLOOM, FRESH	2-00-184	24	3.1	6.6	9.0	7.2	2.0	0.7	8.2	5.4	4.3	3.9	—
4	ALFALFA-BROMEGRASS, FRESH	2-08-328	23	2.2	5.3	—	—	—	0.8	9.4	4.8	3.3	3.5	—
5	ALFALFA-BROMEGRASS, SMOOTH, FRESH	2-00-262	22	2.1	5.5	—	—	—	0.8	9.0	4.2	3.2	3.0	—
6	ALFALFA-ORCHARDGRASS, FRESH	2-00-323	25	—	—	—	—	—	—	—	—	—	—	—
7	ALFILERIA, REDSTEM (FILAREE), FRESH	2-00-356	18	2.4	3.6	—	5.3	—	0.6	8.2	2.8	2.0	1.9	—
8	ALTA (TALL) FESCUE, FRESH	2-01-889	28	2.5	7.5	19.5	—	—	0.9	14.3	2.7	1.7	1.7	—
9	ALYCE CLOVER, FRESH	2-00-362	20	2.2	5.5	—	—	—	0.6	8.6	3.1	—	—	—
10	BAHIAGRASS, FRESH	2-00-464	30	3.3	9.0	20.4	11.4	—	0.5	14.2	2.6	1.5	1.6	—
11	BEARDGRASS (BLUESTEM), IMMATURE, FRESH	2-00-821	27	2.4	6.7	—	—	—	0.7	13.6	3.4	2.3	2.3	—
12	BEET, SUGAR, TOPS WITH CROWNS, FRESH	2-00-649	17	3.4	1.8	—	—	—	0.3	8.6	2.5	1.9	1.7	—
13	BERMUDAGRASS, FRESH	2-00-712	29	3.3	7.6	—	—	—	0.6	13.0	4.2	3.0	2.9	—
14	BERMUDAGRASS, COASTAL, FRESH	2-00-719	29	1.8	8.3	—	—	—	1.1	13.6	4.4	3.2	3.0	—
15	BLUEGRASS, CANADA, FRESH	2-00-764	31	2.8	8.3	—	—	—	1.2	13.8	5.3	3.9	3.7	—
	BLUEGRASS, KENTUCKY													
16	IMMATURE, FRESH	2-00-777	31	2.9	7.8	17.1	9.0	—	1.1	13.7	5.4	4.0	3.8	—
17	EARLY BLOOM, FRESH	2-00-779	35	2.5	9.6	22.8	11.2	—	1.4	15.7	5.8	4.2	4.1	—
18	BLUEGRASS, KENTUCKY-CLOVER, WHITE, FRESH	2-08-356	24	2.7	4.5	—	—	—	0.9	11.3	5.0	3.8	3.6	—
19	BLUESTEM (BEARDGRASS), IMMATURE, FRESH	2-00-821	27	2.4	6.7	—	—	—	0.7	13.6	3.4	2.3	2.3	—
20	BRISTLEGRASS, FRESH	2-00-876	26	3.2	8.3	—	—	—	0.5	10.9	3.1	2.0	2.0	—
21	BROMEGRASS, IMMATURE, FRESH	2-00-892	34	3.8	7.5	19.0	10.5	—	1.2	15.5	5.8	4.7	4.1	—
22	BROMEGRASS, SMOOTH, FRESH	2-00-963	27	—	7.7	—	—	—	0.8	13.1	3.1	2.0	2.0	—
23	BUFFALOGRASS, FRESH	2-01-010	46	5.6	12.7	33.9	16.7	2.9	0.9	22.0	4.7	2.5	3.0	—
24	BUFFELGRASS, PREBLOOM, FRESH	2-10-253	21	2.8	—	15.1	—	—	—	—	1.7	0.9	1.0	—
25	BUR-CLOVER, CALIFORNIA, FRESH	2-01-035	27	1.8	—	—	—	—	0.6	—	6.2	—	—	—
26	CABBAGE, HEADS, FRESH	2-01-046	9	0.9	1.0	—	—	—	0.2	5.3	1.9	1.6	1.6	1.4
27	CANADA BLUEGRASS, FRESH	2-00-764	31	2.8	8.3	—	—	—	1.2	13.8	5.3	3.9	3.7	—
28	CANARYGRASS, FRESH	2-01-093	26	2.4	6.9	—	—	—	1.0	12.1	3.4	2.3	2.3	—
29	CANARYGRASS, REED, FRESH	2-01-113	23	2.3	5.6	10.6	6.5	1.0	0.9	10.1	3.9	2.9	2.7	—
30	CLOVER, ALSIKE, FRESH	2-01-316	22	2.1	5.2	—	—	—	0.8	10.3	4.1	3.0	2.9	—
31	CLOVER, CRIMSON, FRESH	2-01-336	18	1.7	4.9	—	—	—	0.6	7.5	3.0	2.3	2.1	—
32	CLOVER, LADINO, FRESH	2-01-383	18	1.9	2.5	—	—	—	0.9	8.1	4.4	3.6	3.2	—
33	CLOVER, RED, EARLY BLOOM, FRESH	2-01-428	20	2.0	4.6	8.0	6.2	—	1.0	8.3	3.8	2.8	2.7	—
34	CLOVER, STRAWBERRY, FRESH	2-26-067	20	3.2	3.1	—	—	—	0.8	7.3	5.6	—	—	—
35	CLOVER, SUBTERRANEAN, FRESH	2-26-068	21	2.1	6.0	—	—	—	1.0	9.4	2.6	—	—	—
36	CLOVER, WHITE, FRESH	2-01-468	18	2.1	2.8	—	—	—	0.6	7.2	5.0	4.0	3.7	—

Composition of Feeds

SILAGES AND HAYLAGES

Entry Number	TDN Ruminant	TDN Swine	TDN Horse	Digestible Energy Ruminant		Digestible Energy Swine		Digestible Energy Horse		Metabolizable Energy Ruminant		Metabolizable Energy Swine		Metabolizable Energy Poultry ME_n		Metabolizable Energy Horse		Net Energy Ruminant NE_m		Net Energy Ruminant NE_g		Net Energy Lactating Cows NE_{lc}	
	%	%	%	Mcal lb	kg	kcal lb	kg	Mcal lb	kg	Mcal lb	kg	kcal lb	kg	kcal lb	kg	Mcal lb	kg	Mcal lb	kg	Mcal lb	kg	Mcal lb	kg
22	15	–	–	0.33	0.73	–	–	–	–	0.27	0.60	–	–	–	–	–	–	0.18	0.40	0.10	0.22	0.19	0.41
23	19	–	–	0.38	0.84	–	–	–	–	0.32	0.71	–	–	–	–	–	–	0.19	0.41	0.11	0.24	0.19	0.42
24	–	–	–	–	–	–	–	–	–	–	–	–	–	–	–	–	–	–	–	–	–	–	–
25	18	–	–	0.40	0.88	–	–	–	–	0.34	0.75	–	–	–	–	–	–	0.17	0.38	0.10	0.21	0.18	0.40
26	20	–	–	0.39	0.85	–	–	–	–	0.34	0.75	–	–	–	–	–	–	0.18	0.40	0.09	0.20	0.18	0.40
27	17	–	–	0.37	0.81	–	–	–	–	0.31	0.68	–	–	–	–	–	–	0.20	0.43	0.11	0.24	0.20	0.44
28	15	–	–	0.29	0.65	–	–	–	–	0.24	0.54	–	–	–	–	–	–	0.14	0.31	0.16	0.15	0.33	
29	17	–	–	0.34	0.75	–	–	–	–	0.29	0.64	–	–	–	–	–	–	0.17	0.38	0.10	0.22	0.17	0.38
30	14	–	–	0.28	0.61	–	–	–	–	0.23	0.51	–	–	–	–	–	–	0.13	0.29	0.07	0.15	0.14	0.31
31	18	–	–	0.35	0.76	–	–	–	–	0.29	0.65	–	–	–	–	–	–	0.17	0.38	0.10	0.22	0.17	0.38
32	20	–	–	0.42	0.93	–	–	–	–	0.35	0.78	–	–	–	–	–	–	0.22	0.48	0.13	0.28	0.22	0.49
33	8	–	–	0.17	0.36	–	–	–	–	0.14	0.30	–	–	–	–	–	–	0.07	0.16	0.04	0.08	0.08	0.18
34	16	–	–	0.31	0.68	–	–	–	–	0.24	0.53	–	–	–	–	–	–	0.16	0.35	0.08	0.18	0.16	0.35
35	29	–	–	0.69	1.53	–	–	–	–	0.57	1.25	–	–	–	–	–	–	0.34	0.74	0.14	0.31	0.35	0.78
36	17	–	–	0.35	0.77	–	–	–	–	0.29	0.64	–	–	–	–	–	–	0.18	0.40	0.10	0.23	0.19	0.41
37	10	–	–	0.40	0.88	–	–	–	–	0.34	0.75	–	–	–	–	–	–	0.17	0.38	0.10	0.21	0.18	0.40
38	16	–	–	0.32	0.71	–	–	–	–	0.28	0.58	–	–	–	–	–	–	0.15	0.33	0.07	0.16	0.16	0.36
39	13	–	–	0.26	0.58	–	–	–	–	0.21	0.47	–	–	–	–	–	–	0.13	0.28	0.06	0.14	0.14	0.30
40	13	–	–	0.26	0.57	–	–	–	–	0.22	0.48	–	–	–	–	–	–	0.12	0.26	0.06	0.14	0.13	0.28
41	11	–	–	0.24	0.53	–	–	–	–	0.20	0.44	–	–	–	–	–	–	0.10	0.21	0.04	0.09	0.11	0.24
42	22	–	–	0.42	0.92	–	–	–	–	0.36	0.79	–	–	–	–	–	–	0.24	0.53	0.15	0.34	0.23	0.51
43	6	–	–	0.14	0.31	–	–	–	–	0.12	0.26	–	–	–	–	–	–	0.08	0.17	0.04	0.10	0.08	0.18
44	19	–	–	0.38	0.83	–	–	–	–	0.32	0.71	–	–	–	–	–	–	0.19	0.42	0.11	0.25	0.19	0.43

PASTURE AND RANGE PLANTS

Entry Number	TDN Ruminant	TDN Swine	TDN Horse	Digestible Energy Ruminant		Digestible Energy Swine		Digestible Energy Horse		Metabolizable Energy Ruminant		Metabolizable Energy Swine		Metabolizable Energy Poultry ME_n		Metabolizable Energy Horse		Net Energy Ruminant NE_m		Net Energy Ruminant NE_g		Net Energy Lactating Cows NE_{lc}	
	%	%	%	Mcal lb	kg	kcal lb	kg	Mcal lb	kg	Mcal lb	kg	kcal lb	kg	kcal lb	kg	Mcal lb	kg	Mcal lb	kg	Mcal lb	kg	Mcal lb	kg
1	16	–	–	0.32	0.70	–	–	–	–	0.27	0.59	–	–	–	–	–	–	0.16	0.35	0.09	0.20	0.16	0.36
2	12	12	–	0.24	0.54	–	–	–	–	0.20	0.44	–	–	–	–	–	–	0.12	0.27	0.07	0.15	0.13	0.28
3	15	–	–	0.30	0.66	–	–	–	–	0.25	0.56	–	–	–	–	–	–	0.15	0.34	0.09	0.20	0.15	0.34
4	14	–	–	0.29	0.64	–	–	–	–	0.25	0.55	–	–	–	–	–	–	0.16	0.35	0.10	0.22	0.16	0.35
5	14	–	–	0.28	0.62	–	–	–	–	0.24	0.53	–	–	–	–	–	–	0.14	0.30	0.08	0.18	0.14	0.31
6	–	–	–	–	–	–	–	–	–	–	–	–	–	–	–	–	–	–	–	–	–	–
7	10	–	–	0.21	0.46	–	–	–	–	0.18	0.39	–	–	–	–	–	–	0.11	0.25	0.07	0.15	0.11	0.25
8	17	–	–	0.34	0.76	–	–	–	–	0.29	0.64	–	–	–	–	–	–	0.17	0.38	0.10	0.22	0.18	0.39
9	–	–	–	–	–	–	–	–	–	–	–	–	–	–	–	–	–	–	–	–	–	–
10	16	–	–	0.32	0.70	–	–	–	–	0.26	0.57	–	–	–	–	–	–	0.15	0.33	0.08	0.17	0.16	0.36
11	18	–	–	0.35	0.78	–	–	–	–	0.30	0.67	–	–	–	–	–	–	0.18	0.39	0.10	0.23	0.18	0.39
12	11	–	–	0.21	0.46	–	–	–	–	0.18	0.39	–	–	–	–	–	–	0.10	0.22	0.06	0.13	0.10	0.23
13	17	–	–	0.35	0.77	–	–	–	–	0.29	0.65	–	–	–	–	–	–	0.18	0.39	0.10	0.22	0.18	0.40
14	19	–	–	0.37	0.82	–	–	–	–	0.32	0.70	–	–	–	–	–	–	0.18	0.40	0.11	0.23	0.19	0.41
15	20	–	–	0.41	0.90	–	–	–	–	0.35	0.77	–	–	–	–	–	–	0.23	0.51	0.15	0.32	0.23	0.50
16	22	–	–	0.42	0.93	–	–	–	–	0.37	0.81	–	–	–	–	–	–	0.24	0.52	0.15	0.33	0.23	0.51
17	24	–	–	0.49	1.07	–	–	–	–	0.42	0.93	–	–	–	–	–	–	0.26	0.57	0.16	0.36	0.25	0.55
18	17	–	–	0.34	0.74	–	–	–	–	0.29	0.64	–	–	–	–	–	–	0.18	0.39	0.11	0.24	0.17	0.38
19	18	–	–	0.35	0.78	–	–	–	–	0.30	0.67	–	–	–	–	–	–	0.18	0.39	0.10	0.23	0.18	0.39
20	15	–	–	0.30	0.66	–	–	–	–	0.25	0.55	–	–	–	–	–	–	0.14	0.31	0.07	0.16	0.15	0.33
21	25	–	–	0.50	1.11	–	–	–	–	0.44	0.97	–	–	–	–	–	–	0.24	0.53	0.15	0.33	0.24	0.52
22	17	–	–	0.34	0.74	–	–	–	–	0.28	0.62	–	–	–	–	–	–	0.17	0.37	0.10	0.22	0.17	0.38
23	26	–	–	0.52	1.15	–	–	–	–	0.43	0.95	–	–	–	–	–	–	0.26	0.57	0.14	0.31	0.27	0.59
24	–	–	–	–	–	–	–	–	–	–	–	–	–	–	–	–	–	–	–	–	–	–
25	–	–	–	–	–	–	–	–	–	–	–	–	–	–	–	–	–	–	–	–	–	–
26	8	6	7	0.14	0.32	126	279	0.12	0.27	0.13	0.28	118	261	–	–	0.10	0.22	0.07	0.15	0.04	0.10	0.07	0.15
27	20	–	–	0.41	0.90	–	–	–	–	0.35	0.77	–	–	–	–	–	–	0.23	0.51	0.15	0.32	0.23	0.50
28	16	–	–	0.32	0.71	–	–	–	–	0.27	0.60	–	–	–	–	–	–	0.16	0.36	0.10	0.21	0.17	0.37
29	14	–	–	0.29	0.64	–	–	–	–	0.24	0.54	–	–	–	–	–	–	0.15	0.34	0.09	0.20	0.15	0.34
30	16	–	–	0.32	0.71	–	–	–	–	0.28	0.61	–	–	–	–	–	–	0.17	0.37	0.11	0.24	0.17	0.36
31	11	–	–	0.23	0.50	–	–	–	–	0.19	0.42	–	–	–	–	–	–	0.11	0.25	0.07	0.15	0.12	0.25
32	13	–	–	0.27	0.60	–	–	–	–	0.24	0.52	–	–	–	–	–	–	0.15	0.33	0.10	0.22	0.14	0.32
33	14	–	–	0.27	0.58	–	–	–	–	0.23	0.50	–	–	–	–	–	–	0.14	0.32	0.09	0.20	0.14	0.31
34	–	–	–	–	–	–	–	–	–	–	–	–	–	–	–	–	–	–	–	–	–	–
35	–	–	–	–	–	–	–	–	–	–	–	–	–	–	–	–	–	–	–	–	–	–
36	13	–	–	0.26	0.57	–	–	–	–	0.22	0.49	–	–	–	–	–	–	0.14	0.30	0.09	0.19	0.13	0.29

(Continued)

TABLE V-1D SILAGES AND HAYLAGES, MINERAL AND VITAMIN COMPOSITION OF FEEDS, DATA EXPRESSED AS-FED—(Continued)

Entry Number	Feed Name Description	Dry Matter	Macro Minerals							Micro Minerals						
			Calcium (Ca)	Phosphorus (P)	Sodium (Na)	Chlorine (Cl)	Magnesium (Mg)	Potassium (K)	Sulfur (S)	Cobalt (Co)	Copper (Cu)	Iodine (I)	Iron (Fe)	Manganese (Mn)	Selenium (Se)	Zinc (Zn)
		%	%	%	%	%	%	%	%	ppm or mg/kg	ppm or mg/kg	ppm or mg/kg	%	ppm or mg/kg	ppm or mg/kg	ppm or mg/kg
22	LESPEDEZA, COMMON-KOREAN, SILAGE	30	—	—	—	—	—	—	—	—	—	—	—	—	—	—
23	LESPEDEZA, SERICEA (CHINESE), SILAGE	30	—	—	—	—	—	—	—	—	—	—	—	—	—	—
24	MILLET, JAPANESE (BARNYARD GRASS)-SOYBEAN, SILAGE	21	—	—	—	—	—	—	—	—	—	—	—	—	—	—
25	MILO SORGHUM, SILAGE,	31	0.11	0.06	—	—	—	—	—	—	—	—	—	—	—	—
	OATS, SILAGE															
26	DOUGH STAGE	35	0.17	0.12	—	—	—	—	—	—	—	—	—	—	—	—
27	MOLASSES ADDED	33	0.10	0.09	—	—	—	0.31	—	—	—	—	—	—	—	—
28	OATS-PEA, SILAGE	26	0.16	0.08	—	0.31	0.11	0.48	—	0.063	4.8	—	0.017	8.5	—	—
29	OATS-VETCH, SILAGE, MILK STAGE	27	—	—	—	—	—	—	—	—	—	—	—	—	—	—
30	PEA, SILAGE, VINES (WITHOUT SEEDS, WITH PODS)	25	0.32	0.06	0.00	—	0.10	0.34	0.06	—	—	—	0.003	—	—	—
31	PEA, FIELD, SILAGE	27	0.37	0.08	—	—	0.11	0.38	0.07	—	—	—	—	—	—	—
	POTATO, SILAGE															
32	TUBERS, ALFALFA HAY ADDED	35	—	—	—	—	—	—	—	—	—	—	—	—	—	—
33	VINES	15	0.31	0.03	—	0.06	0.02	0.59	0.06	—	—	—	—	—	—	—
	SORGHUM, SILAGE															
34	DOUGH STAGE	29	—	—	—	—	—	—	—	—	—	—	—	—	—	—
35	STOVER WITHOUT HEADS	62	0.25	0.07	—	—	—	—	—	—	—	—	—	91.7	—	—
36	KAFIR	30	0.07	0.05	—	—	0.08	0.50	—	—	—	—	—	—	—	—
37	MILO	31	0.11	0.06	—	—	—	—	—	—	—	—	—	—	—	—
38	SOYBEAN, SILAGE	30	0.40	0.13	0.03	—	0.12	0.39	0.09	—	2.9	—	0.010	42.7	—	10.3
39	SUGAR BEET, SILAGE, TOPS WITH CROWNS	25	0.39	0.07	0.14	—	0.27	1.45	0.14	—	—	—	0.006	—	—	—
40	SUGARCANE, SILAGE, ALL ANALYSES	22	0.08	0.04	—	—	0.05	—	—	—	—	—	—	—	—	—
41	SUNFLOWER, SILAGE, MILK STAGE	21	—	—	—	—	—	—	—	—	—	—	—	—	—	—
42	SWEET CORN, SILAGE, CANNERY RESIDUE	31	0.10	0.24	0.01	—	0.07	0.36	0.03	—	—	—	0.007	—	—	—
43	SWEET POTATO, SILAGE, VINES	13	—	—	—	—	—	—	—	—	—	—	—	—	—	—
44	VETCH, SILAGE	30	—	—	—	—	—	—	—	—	—	—	—	—	—	—

TABLE V-1E PASTURE AND RANGE PLANTS, MINERAL AND VITAMIN COMPOSITION OF FEEDS, DATA EXPRESSED AS-FED

Entry Number	Feed Name Description	Dry Matter	Macro Minerals							Micro Minerals						
			Calcium (Ca)	Phosphorus (P)	Sodium (Na)	Chlorine (Cl)	Magnesium (Mg)	Potassium (K)	Sulfur (S)	Cobalt (Co)	Copper (Cu)	Iodine (I)	Iron (Fe)	Manganese (Mn)	Selenium (Se)	Zinc (Zn)
		%	%	%	%	%	%	%	%	ppm or mg/kg	ppm or mg/kg	ppm or mg/kg	%	ppm or mg/kg	ppm or mg/kg	ppm or mg/kg
	ALFALFA (LUCERNE)															
1	FRESH, ALL ANALYSES	26	0.40	0.07	0.05	0.12	0.09	0.83	0.10	0.092	3.2	—	0.009	24.1	—	9.4
2	PREBLOOM, FRESH	20	0.44	0.07	0.04	0.09	0.05	0.44	0.10	0.034	2.2	—	0.003	8.3	—	—
3	EARLY BLOOM, FRESH	24	0.39	0.07	0.04	—	0.12	0.88	—	0.107	4.4	—	0.008	33.2	—	9.6
4	ALFALFA-BROMEGRASS, FRESH	23	0.28	0.07	—	—	—	0.63	—	—	—	—	—	—	—	—
5	ALFALFA-BROMEGRASS, SMOOTH, FRESH	22	0.33	0.08	0.09	—	0.08	0.84	0.05	—	—	—	0.003	—	—	—
6	ALFALFA-ORCHARDGRASS, FRESH	25	0.10	0.13	—	—	0.06	—	—	—	—	—	—	—	—	—
7	AFILERIA REDSTEM (FILAREE), FRESH	18	0.35	0.08	—	—	—	0.59	—	—	—	—	—	—	—	—
8	ALTA (TALL) FESCUE, FRESH	28	0.13	0.05	0.03	—	0.07	0.70	—	0.113	1.0	—	0.003	18.0	—	5.9
9	ALYCE CLOVER, FRESH	20	—	—	—	—	—	—	—	0.018	—	—	—	—	—	—
10	BAHIAGRASS, FRESH	30	0.14	0.06	—	—	0.07	0.43	—	—	—	—	—	—	—	—
11	BEARDGRASS (BLUESTEM), IMMATURE, FRESH	27	0.17	0.05	—	—	—	0.46	—	—	12.6	—	0.024	28.5	—	—
12	BEET, SUGAR, TOPS WITH CROWNS, FRESH	17	0.17	0.04	0.09	0.09	0.18	0.96	0.09	—	2.3	—	0.003	9.0	—	—
13	BERMUDAGRASS, FRESH	29	0.16	0.06	0.13	—	0.07	0.55	—	0.022	1.6	—	0.033	28.6	—	—
14	BERMUDAGRASS, COASTAL, FRESH	29	0.14	0.08	—	—	—	—	—	—	—	—	—	—	—	—
15	BLUEGRASS, CANADA, FRESH	31	0.12	0.12	0.04	—	0.05	0.64	0.05	—	—	—	0.010	24.8	—	—
	BLUEGRASS, KENTUCKY															
16	IMMATURE, FRESH	31	0.15	0.14	0.04	—	0.05	0.70	0.05	—	—	—	0.010	—	—	—
17	EARLY BLOOM, FRESH	35	0.16	0.14	0.05	—	0.04	0.70	0.06	—	—	—	0.011	—	—	—
18	BLUEGRASS, KENTUCKY-CLOVER, WHITE, FRESH	24	0.31	0.11	—	—	—	—	—	—	—	—	—	—	—	—
19	BLUESTEM (BEARDGRASS), IMMATURE, FRESH	27	0.17	0.05	—	—	—	0.46	—	—	12.6	—	0.024	28.5	—	—
20	BRISTLEGRASS, FRESH	26	0.10	0.05	—	—	0.07	1.51	—	—	1.7	—	—	10.9	—	—
21	BROMEGRASS, IMMATURE, FRESH	34	0.20	0.13	0.01	—	0.06	1.46	0.07	—	—	—	0.007	—	—	—
22	BROMEGRASS, SMOOTH, FRESH	27	—	—	—	—	—	—	—	0.022	—	—	—	—	—	—
23	BUFFALOGRASS, FRESH	46	0.26	0.09	—	—	0.06	0.33	—	—	—	—	—	—	—	—
24	BUFFELGRASS, PREBLOOM, FRESH	21	0.19	0.03	0.03	—	0.12	0.89	—	0.063	2.0	—	0.014	27.7	—	9.5
25	BUR-CLOVER, CALIFORNIA, FRESH	27	—	—	—	—	—	—	—	—	—	—	—	—	—	—
26	CABBAGE, HEADS, FRESH	9	0.06	0.03	0.01	0.05	0.02	0.26	0.11	—	1.3	—	0.001	2.8	—	—
27	CANADA BLUEGRASS, FRESH	31	0.12	0.12	0.04	—	0.05	0.64	0.05	—	—	—	0.010	24.8	—	—
28	CANARYGRASS, FRESH	26	0.11	0.08	—	—	0.07	0.82	—	—	—	—	—	—	—	—
29	CANARYGRASS, REED, FRESH	23	0.08	0.08	—	—	—	0.83	—	—	—	—	—	—	—	—
30	CLOVER, ALSIKE, FRESH	22	0.31	0.06	0.10	0.17	0.07	0.61	0.05	—	1.3	—	0.010	26.3	—	—
31	CLOVER, CRIMSON, FRESH	18	0.24	0.05	0.07	0.11	0.05	0.55	0.05	—	—	—	43.1	—	—	—
32	CLOVER, LADINO, FRESH	18	0.22	0.07	0.02	—	0.09	0.33	0.02	—	—	—	0.007	12.7	—	—
33	CLOVER, RED, EARLY BLOOM, FRESH	20	0.45	0.08	0.04	—	0.10	0.49	0.03	—	—	—	0.006	—	—	—
34	CLOVER, STRAWBERRY, FRESH	20	0.37	0.09	—	—	—	—	—	—	—	—	—	—	—	—
35	CLOVER, SUBTERRANEAN, FRESH	21	0.31	0.07	—	—	—	—	—	—	—	—	—	—	—	—
36	CLOVER, WHITE, FRESH	18	0.25	0.09	0.07	0.11	0.08	0.38	0.06	—	—	—	0.006	54.4	—	—

Composition of Feeds

SILAGES AND HAYLAGES

Entry Number	Fat-Soluble Vitamins					Water-Soluble Vitamins								
	A (1 mg Carotene = 1667 IU Vit A)	Carotene (Provitamin A)	D	E	K	B-12	Biotin	Choline	Folacin (Folic Acid)	Niacin	Pantothenic Acid (B-3)	(Pyridoxine) B-6	Riboflavin (B-2)	Thiamin (B-1)
	IU/g	ppm or mg/kg	IU/kg	ppm or mg/kg	ppm or mg/kg	ppb or mcg/kg	ppm or mg/kg	ppm or mg/kg	ppm or mg/kg	ppm or mg/kg	ppm or mg/kg	ppm or mg/kg	ppm or mg/kg	ppm or mg/kg
22	–	–	–	–	–	–	–	–	–	–	–	–	–	–
23	–	–	–	–	–	–	–	–	–	–	–	–	–	–
24	–	–	–	–	–	–	–	–	–	–	–	–	–	–
25	–	–	–	–	–	–	–	–	–	–	–	–	–	–
26	35.1	21.1	–	–	–	–	–	–	–	–	–	–	–	–
27	–	–	–	–	–	–	–	–	–	–	–	–	–	–
28	33.5	20.1	–	–	–	–	–	–	–	–	–	–	–	–
29	–	–	–	–	–	–	–	–	–	–	–	–	–	–
30	77.2	46.3	–	–	–	–	–	–	–	–	–	–	–	–
31	–	–	–	–	–	–	–	–	–	–	–	–	–	–
32	–	–	–	–	–	–	–	–	–	–	–	–	–	–
33	–	–	–	–	–	–	–	–	–	–	–	–	–	–
34	–	–	–	–	–	–	–	–	–	–	–	–	–	–
35	7.1	4.2	–	–	–	–	–	–	–	–	–	–	–	–
36	5.3	3.2	–	–	–	–	–	–	–	–	–	–	–	–
37	–	–	–	–	–	–	–	–	–	–	–	–	–	–
38	52.2	31.3	–	–	–	–	–	–	–	–	–	–	–	–
39	–	–	–	–	–	–	–	–	–	–	–	–	–	–
40	–	–	–	–	–	–	–	–	–	–	–	–	–	–
41	–	–	–	–	–	–	–	–	–	–	–	–	–	–
42	–	–	–	–	–	–	–	–	–	–	–	–	–	–
43	6.9	4.2	–	–	–	–	–	–	–	–	–	–	–	–
44	–	–	–	–	–	–	–	–	–	–	–	–	–	–

PASTURE AND RANGE PLANTS

Entry Number	Fat-Soluble Vitamins					Water-Soluble Vitamins								
	A (1 mg Carotene = 1667 IU Vit A)	Carotene (Provitamin A)	D	E	K	B-12	Biotin	Choline	Folacin (Folic Acid)	Niacin	Pantothenic Acid (B-3)	(Pyridoxine) B-6	Riboflavin (B-2)	Thiamin (B-1)
	IU/g	ppm or mg/kg	IU/kg	ppm or mg/kg	ppm or mg/kg	ppb or mcg/kg	ppm or mg/kg	ppm or mg/kg	ppm or mg/kg	ppm or mg/kg	ppm or mg/kg	ppm or mg/kg	ppm or mg/kg	ppm or mg/kg
1	101.3	60.8	0	–	–	–	0.13	374	0.64	15	8.9	1.66	4.6	1.7
2	–	–	0	34.8	–	–	–	–	–	–	–	–	–	–
3	69.9	41.9	–	–	–	–	–	–	–	–	–	–	–	–
4	–	–	–	–	–	–	–	–	–	–	–	–	–	–
5	–	–	–	–	–	–	–	–	–	–	–	–	–	–
6	–	–	–	–	–	–	–	–	–	–	–	–	–	–
7	–	–	–	–	–	–	–	–	–	–	–	–	–	–
8	–	–	–	–	–	–	–	–	–	–	–	–	–	–
9	35.0	21.0	–	–	–	–	–	–	–	–	–	–	–	–
10	89.7	53.8	–	–	–	–	–	–	–	–	–	–	–	–
11	97.9	58.7	–	–	–	–	–	–	–	–	–	–	–	–
12	9.6	5.8	–	–	–	–	–	–	–	–	–	–	1.1	–
13	147.8	88.7	–	–	–	–	–	–	–	–	–	–	–	–
14	160.3	96.1	–	–	–	–	–	–	–	–	–	–	–	–
15	199.9	119.9	–	–	–	–	–	–	–	–	–	–	–	–
16	247.6	148.5	–	47.8	–	–	–	–	–	–	–	–	–	–
17	163.4	98.0	–	–	–	–	–	–	–	–	–	–	–	–
18	–	–	–	–	–	–	–	–	–	–	–	–	–	–
19	97.9	58.7	–	–	–	–	–	–	–	–	–	–	–	–
20	–	–	–	–	–	–	–	–	–	–	–	–	–	–
21	259.6	155.7	–	–	–	–	–	–	–	–	–	–	–	–
22	142.0	85.2	0	–	–	–	–	–	–	–	–	–	2.1	0.8
23	71.6	42.9	–	–	–	–	–	–	–	–	–	–	–	–
24	–	–	–	–	–	–	–	–	–	–	–	–	–	–
25	–	–	–	–	–	–	–	–	–	–	–	–	–	–
26	0.7	0.4	–	–	–	–	–	249	0.63	3	–	–	0.5	0.6
27	199.9	119.9	–	–	–	–	–	–	–	–	–	–	–	–
28	–	–	–	–	–	–	–	–	–	–	–	–	–	–
29	–	–	–	–	–	–	–	–	–	–	–	–	–	–
30	–	–	–	–	–	–	–	–	–	–	–	–	4.4	2.0
31	–	–	–	–	–	–	–	–	–	–	–	–	–	–
32	96.2	57.7	–	–	–	–	–	–	–	–	–	–	4.2	–
33	81.5	48.9	–	–	–	–	–	–	–	–	–	–	–	–
34	–	–	–	–	–	–	–	–	–	–	–	–	–	–
35	–	–	–	–	–	–	–	–	–	–	–	–	–	–
36	44.0	26.4	–	54.6	–	–	–	–	–	11	–	–	16.0	2.5

(Continued)

TABLE V-1E PASTURE AND RANGE PLANTS, COMPOSITION OF FEEDS, DATA EXPRESSED **AS-FED**—(Continued)

Entry Number	Feed Name Description	International Feed Number	Proximate Analysis								Digestible Protein		
			Dry Matter	Ash	Crude Fiber	Neutral Det. Fib. (NDF)	Acid Det. Fib. (ADF)	Lignin	Ether Extract (Fat)	N-Free Extract	Crude Protein	Ruminant	Swine	Horse
			%	%	%	%	%	%	%	%	%	%	%	%
37	CORN, FRESH, ALL ANALYSES	2-02-799	23	1.3	5.7	—	—	—	0.9	12.9	2.4	1.5	1.5	—
38	CORN, SWEET, STOVER WITHOUT EARS, WITHOUT HUSKS, FRESH	2-02-969	22	1.4	5.7	—	—	—	0.4	12.8	1.6	0.8	0.9	—
39	COWPEA, COMMON, FRESH	2-01-655	25	3.0	6.1	—	—	—	1.0	10.4	4.0	2.9	2.9	—
40	CRESTED WHEATGRASS, EARLY BLOOM, FRESH	2-05-422	41	—	8.9	—	—	—	1.7	—	4.8	3.1	3.1	—
	CURLY MESQUITE													
41	BROWSE, FRESH	2-01-728	35	5.1	9.8	—	—	—	0.7	16.4	3.0	1.7	1.8	—
42	BROWSE, MATURE, FRESH	2-01-729	50	7.6	14.3	32.0	—	—	1.1	23.9	2.8	1.1	1.4	—
43	DALLISGRASS, FRESH	2-01-741	25	3.0	7.3	—	—	—	0.6	11.1	3.0	2.0	2.0	—
44	DROPSEED, SAND, STEM CURED, FRESH	2-05-596	88	7.0	31.6	—	—	5.2	1.1	43.0	5.4	1.3	2.8	—
45	FESCUE, TALL (ALTA), FRESH	2-01-889	28	2.5	7.5	19.5	—	—	0.9	14.3	2.7	1.7	1.7	—
46	FILAREE (ALFILERIA, REDSTEM), FRESH	2-00-356	18	2.4	3.6	—	5.3	—	0.6	8.2	2.8	2.0	1.9	—
47	FOXTAIL, MEADOW, IMMATURE, FRESH	2-02-073	26	2.8	5.6	—	—	—	1.2	12.0	4.5	3.3	3.2	—
48	GALLETA, STEM CURED, FRESH	2-05-594	86	13.3	28.4	—	—	—	1.4	38.5	4.3	1.3	2.0	—
49	GRAMA, IMMATURE, FRESH	2-02-163	41	4.6	11.2	—	—	—	0.8	19.0	5.4	3.7	3.6	—
50	GRASS-LEGUME, FRESH	2-08-439	24	2.5	5.6	—	—	—	0.8	10.4	4.2	3.1	2.9	—
51	GREASEWOOD, BROWSE, FRESH	2-02-312	50	7.3	11.7	—	—	—	1.7	18.6	10.7	8.2	7.7	—
52	INDIANGRASS, FRESH	2-08-770	57	4.1	19.4	—	—	—	1.3	29.4	2.8	0.9	1.2	—
53	INDIAN RICEGRASS, FRESH	2-03-944	48	—	—	—	—	—	—	—	—	—	—	—
	JOHNSONGRASS SORGHUM—SEE SORGHUM													
54	JUNEGRASS, IMMATURE, FRESH	2-02-437	28	2.2	6.4	—	—	—	0.6	12.8	6.0	4.7	4.4	—
	KENTUCKY BLUEGRASS—SEE BLUEGRASS, KENTUCKY													
55	KOA HAOLE (LEAD TREE, WHITE POPINAC), BROWSE, FRESH	2-02-495	30	1.9	10.7	—	—	—	0.6	11.5	5.5	4.1	3.9	—
56	LESPEDEZA, COMMON, IMMATURE, FRESH	2-20-879	24	3.1	7.7	—	—	—	0.5	8.8	3.9	2.9	2.7	—
57	LESPEDEZA, COMMON-KOREAN, MATURE, FRESH	2-26-032	35	2.6	15.8	—	—	—	0.7	11.8	4.4	3.0	—	2.9
58	LESPEDEZA, SERICEA (CHINESE LESPEDEZA), FRESH	2-02-611	33	2.0	7.5	—	—	—	1.2	16.2	5.9	4.4	4.2	—
59	LOVEGRASS, IMMATURE, FRESH	2-02-647	43	2.8	13.1	—	—	—	1.3	20.2	5.4	3.6	3.6	—
60	MEADOW FOXTAIL, IMMATURE, FRESH	2-02-073	26	2.8	5.6	—	—	—	1.2	12.0	4.5	3.3	3.2	—
61	MEDIC, BLACK (YELLOW TREFOIL), FRESH	2-03-070	23	2.3	5.6	—	—	—	0.8	9.1	4.9	3.8	3.5	—
62	MESQUITE, COMMON, BROWSE, FRESH	2-03-081	35	2.1	9.6	—	—	—	1.2	14.8	7.4	5.7	5.3	—
63	MILLET, FOXTAIL, FRESH	2-03-101	29	2.5	9.2	—	—	—	0.9	13.4	2.8	1.7	1.7	—
64	MILLET, JAPANESE (BARNYARD GRASS), FRESH	2-03-108	22	1.6	6.8	—	—	—	0.6	11.0	1.7	1.0	1.0	—
65	MILLET, PEARL (PEARL MILLET), FRESH	2-03-115	21	1.9	6.5	—	—	—	0.6	9.7	2.1	1.3	1.3	—
66	MILLET, PROSO (BROOMCORN; HOG MILLET), FRESH	2-03-811	25	1.8	7.4	—	—	—	0.6	13.1	2.1	1.1	1.2	—
67	MILO SORGHUM, FRESH	2-04-436	30	1.8	7.6	19.5	12.0	—	0.5	17.2	2.6	1.5	1.5	—
68	MUHLY, BUSH, MIDBLOOM, FRESH	2-05-619	43	2.4	16.2	—	—	—	0.7	20.9	2.8	1.3	—	1.5
69	NEEDLEGRASS, MATURE, FRESH	2-03-205	42	2.9	17.1	—	—	—	0.5	18.4	3.2	1.7	1.8	—
70	OATGRASS, TALL, FRESH	2-03-267	30	2.0	10.5	—	—	—	0.9	14.3	2.6	1.5	1.6	—
71	OATS, IMMATURE, FRESH	2-03-286	16	1.7	4.0	—	—	—	0.4	7.4	2.5	1.8	1.7	—
72	OATS-VETCH, MILK STAGE, FRESH	2-03-407	33	2.5	9.1	—	—	—	1.0	16.3	3.5	2.2	2.3	—
73	ORCHARDGRASS, FRESH, ALL ANALYSES	2-03-451	26	2.6	6.4	13.9	—	—	1.6	11.3	3.9	2.7	2.7	—
74	PANGOLAGRASS, FRESH	2-03-493	20	1.5	6.6	—	7.5	1.0	0.5	9.8	1.8	0.9	0.9	—
75	PANICUM, FRESH	2-03-499	29	4.1	8.6	—	—	—	0.7	11.9	3.7	2.5	2.5	—
76	PARAGRASS, FRESH	2-03-525	26	3.0	9.1	—	—	—	0.4	11.9	2.0	1.2	1.1	—
77	PEA, FIELD, FRESH	2-03-603	18	1.7	4.6	—	—	—	0.6	7.6	3.6	3.0	2.6	—
78	PEA-OATS, FRESH	2-08-483	23	1.9	6.4	—	—	—	0.9	10.3	3.2	2.4	2.2	—
79	PEARL MILLET (MILLET, PEARL), FRESH	2-03-115	21	1.9	6.5	—	—	—	0.6	9.7	2.1	1.3	1.3	—
80	PEAVINE, FRESH	2-03-669	17	5.0	4.2	—	—	—	0.6	4.4	3.2	2.5	2.3	—
81	RAPE (Canola), FRESH	2-03-867	17	2.1	2.4	—	—	—	0.6	8.5	2.9	2.4	2.1	—
82	REDTOP, FULL BLOOM, FRESH	2-03-891	26	1.8	6.6	16.6	—	—	0.9	14.8	2.1	1.2	1.3	—
83	REED CANARYGRASS, FRESH	2-01-113	23	2.3	5.6	10.6	6.5	1.0	0.9	10.1	3.9	2.9	2.7	—
84	RESCUEGRASS (BROMEGRASS, RESCUE), FRESH	2-08-361	29	4.0	6.7	—	—	—	1.0	12.2	5.0	3.7	3.5	—
85	RHODESGRASS, FRESH	2-03-916	26	3.1	9.9	—	—	—	0.4	11.0	2.0	1.2	1.1	—
86	RYE, FRESH	2-04-018	20	1.9	5.9	—	—	—	0.8	8.1	3.6	2.8	2.5	—
87	RYEGRASS, FRESH	2-04-062	24	1.8	7.0	—	—	—	0.7	12.1	2.5	1.5	1.6	—
88	RYEGRASS, ITALIAN, FRESH	2-04-073	23	3.9	4.7	—	—	—	0.9	9.0	4.0	3.0	2.9	—
89	SAGEBRUSH, BIG, BROWSE, STEM CURED, FRESH	2-07-992	65	4.3	—	—	—	—	6.4	—	6.1	3.2	3.7	—
90	SALTGRASS, DESERT, FRESH	2-04-171	29	2.0	8.6	—	—	—	0.5	16.2	1.7	0.7	0.9	—
91	SEDGE, FRESH	2-04-195	25	2.2	—	15.4	—	1.0	—	—	3.0	2.0	2.0	—
	SORGHUM													
92	JOHNSONGRASS, IMMATURE, FRESH	2-04-409	20	2.1	5.6	—	—	—	0.6	8.4	3.1	2.2	2.1	—
93	JOHNSONGRASS, FULL BLOOM, FRESH	2-04-410	35	3.5	11.4	—	—	—	0.8	16.4	2.8	1.6	1.7	—
94	KAFIR, FRESH	2-04-424	24	1.9	6.6	—	—	—	0.7	12.0	2.4	1.5	1.5	—
95	MILO, FRESH	2-04-436	30	1.8	7.6	19.5	12.0	—	0.5	17.2	2.6	1.5	1.5	—
96	SUDANGRASS, MATURE, FRESH	2-04-487	30	2.4	10.6	—	—	—	0.5	14.5	1.6	0.6	0.8	—
97	SOYBEAN, FRESH	2-04-574	23	2.4	6.3	—	—	—	0.9	9.2	4.1	3.2	2.9	—
98	STARGRASS, FRESH	2-09-730	63	5.0	—	51.6	—	—	—	—	—	—	—	—
99	SUGARCANE, FRESH	2-04-689	28	1.6	8.8	—	—	—	0.6	15.2	1.4	0.8	0.7	—
100	SUNFLOWER, FRESH	2-04-723	15	1.7	4.7	—	—	—	0.4	7.2	1.4	0.8	0.8	—
101	SWEET CLOVER, YELLOW, FRESH	2-04-766	23	1.8	6.9	—	—	—	0.7	9.5	4.3	3.4	3.1	—
102	SWITCHGRASS, FRESH	2-04-800	55	3.5	19.2	41.7	—	5.3	1.3	27.8	3.5	1.6	1.9	—
103	TALL (ALTA) FESCUE, FRESH	2-01-889	28	2.5	7.5	19.5	—	—	0.9	14.3	2.7	1.7	1.7	—
104	THREE-AWN (WIREGRASS), FRESH	2-04-838	39	2.3	13.3	—	—	—	0.9	18.6	3.8	2.3	2.4	—
105	TIMOTHY, FRESH	2-04-912	28	2.3	7.5	19.4	—	—	1.1	13.4	3.4	2.1	2.3	—
106	TOBOSA, IMMATURE, FRESH	2-08-578	40	4.5	12.6	27.6	—	—	0.5	18.2	3.8	2.8	2.3	—
107	TREFOIL, BIRDSFOOT (DEERVETCH, BIRDSFOOT), FRESH	2-20-786	19	2.2	4.1	9.5	—	—	0.8	8.5	3.7	2.8	2.7	—
108	VELVETBEAN, DOUGH STAGE, FRESH	2-05-084	22	2.4	5.3	—	—	—	0.5	10.7	3.5	2.5	2.5	—
109	VETCH, FRESH	2-05-111	22	2.1	6.2	—	—	—	0.5	8.9	4.7	3.5	3.4	—
110	VETCH-OATS, FRESH	2-05-133	26	2.8	7.0	—	—	—	0.9	11.4	4.3	3.3	3.0	—
111	WHEAT, IMMATURE, FRESH	2-05-176	22	3.0	3.9	10.2	6.3	1.0	1.0	8.3	6.1	4.9	4.5	—
112	WHEATGRASS, BLUEBUNCH, PREBLOOM, FRESH	2-05-387	32	2.4	8.0	22.8	—	—	0.8	16.0	4.5	3.2	3.1	—
113	WHEATGRASS, CRESTED, IMMATURE, FRESH	2-05-420	28	2.9	6.2	—	—	—	0.6	12.9	6.0	5.1	4.3	—
114	WHEATGRASS, SLENDER, FRESH	2-05-439	32	3.0	10.4	23.1	—	—	1.4	14.5	3.1	1.8	1.9	—
115	WHITE CLOVER, FRESH	2-01-468	18	2.1	2.8	—	—	—	0.6	7.2	5.0	4.0	3.7	—
116	WILD-RYE, RUSSIAN, FRESH	2-05-469	35	3.1	7.8	25.6	—	—	0.9	18.9	4.2	2.8	2.8	—
117	WINTERFAT, COMMON, BROWSE, STEM CURED, FRESH	2-26-142	80	12.7	—	—	—	—	2.2	—	8.7	5.4	5.6	—

PASTURE AND RANGE PLANTS

Entry Number	TDN			Digestible Energy						Metabolizable Energy								Net Energy					
	Ruminant	Swine	Horse	Ruminant		Swine		Horse		Ruminant		Swine		Poultry ME$_n$		Horse		Ruminant NE$_m$		Ruminant NE$_g$		Lactating Cows NE$_{lc}$	
	%	%	%	Mcal		kcal		Mcal		Mcal		kcal		kcal		Mcal		Mcal		Mcal		Mcal	
				lb	kg	lb	kg	lb	kg	lb	kg	lb	kg	lb	kg	lb	kg	lb	kg	lb	kg	lb	kg
37	16	—	11	0.32	0.70	—	—	0.21	0.46	0.28	0.61	—	—	—	—	0.17	0.38	0.16	0.35	0.10	0.22	0.16	0.35
38	12	—	—	0.26	0.58	—	—	—	—	0.22	0.49	—	—	—	—	—	—	0.14	0.31	0.08	0.19	0.14	0.32
39	16	15	—	0.29	0.65	—	—	—	—	0.26	0.58	—	—	—	—	—	—	0.15	0.32	0.08	0.18	0.14	0.32
40	—	—	—	—	—	—	—	—	—	—	—	—	—	—	—	—	—	—	—	—	—	—	—
41	21	—	—	0.40	0.88	—	—	—	—	0.33	0.73	—	—	—	—	—	—	0.18	0.40	0.09	0.20	0.19	0.43
42	27	—	—	0.53	1.17	—	—	—	—	0.44	0.96	—	—	—	—	—	—	0.24	0.52	0.11	0.25	0.26	0.58
43	16	—	—	0.31	0.67	—	—	—	—	0.26	0.57	—	—	—	—	—	—	0.14	0.31	0.08	0.17	0.15	0.33
44	52	—	—	0.96	2.12	—	—	—	—	0.83	1.82	—	—	—	—	—	—	0.47	1.04	0.25	0.54	0.50	1.11
45	17	—	—	0.34	0.76	—	—	—	—	0.29	0.64	—	—	—	—	—	—	0.17	0.38	0.10	0.22	0.18	0.39
46	10	—	—	0.21	0.46	—	—	—	—	0.18	0.39	—	—	—	—	—	—	0.11	0.25	0.07	0.15	0.11	0.25
47	17	—	—	0.34	0.75	—	—	—	—	0.29	0.64	—	—	—	—	—	—	0.18	0.40	0.11	0.24	0.18	0.39
48	44	—	—	0.72	1.58	—	—	—	—	0.59	1.29	—	—	—	—	—	—	0.51	1.13	0.29	0.63	0.53	1.16
49	25	—	—	0.51	1.12	—	—	—	—	0.43	0.94	—	—	—	—	—	—	0.25	0.54	0.14	0.31	0.25	0.56
50	15	—	—	0.31	0.68	—	—	—	—	0.26	0.58	—	—	—	—	—	—	0.16	0.35	0.10	0.21	0.16	0.35
51	23	—	—	0.55	1.20	—	—	—	—	0.45	0.99	—	—	—	—	—	—	0.32	0.70	0.18	0.41	0.32	0.71
52	33	—	—	0.66	1.45	—	—	—	—	0.55	1.21	—	—	—	—	—	—	0.32	0.70	0.17	0.38	0.33	0.74
53	—	—	—	—	—	—	—	—	—	—	—	—	—	—	—	—	—	—	—	—	—	—	—
54	20	—	—	0.40	0.88	—	—	—	—	0.35	0.76	—	—	—	—	—	—	0.21	0.45	0.13	0.29	0.20	0.44
55	18	—	—	0.37	0.82	—	—	—	—	0.31	0.69	—	—	—	—	—	—	0.19	0.43	0.11	0.25	0.20	0.43
56	14	—	—	0.29	0.63	—	—	—	—	0.24	0.53	—	—	—	—	—	—	0.13	0.29	0.07	0.16	0.14	0.31
57	20	—	—	—	—	—	—	—	—	—	—	—	—	—	—	—	—	—	—	—	—	—	—
58	21	—	—	0.44	0.96	—	—	—	—	0.37	0.83	—	—	—	—	—	—	0.22	0.47	0.13	0.28	0.22	0.48
59	27	—	—	0.54	1.19	—	—	—	—	0.46	1.01	—	—	—	—	—	—	0.28	0.61	0.16	0.36	0.28	0.61
60	17	—	—	0.34	0.75	—	—	—	—	0.29	0.64	—	—	—	—	—	—	0.18	0.40	0.11	0.24	0.18	0.39
61	14	—	—	0.29	0.64	—	—	—	—	0.25	0.54	—	—	—	—	—	—	0.16	0.35	0.10	0.21	0.16	0.34
62	24	—	—	0.49	1.08	—	—	—	—	0.42	0.93	—	—	—	—	—	—	0.26	0.56	0.16	0.35	0.25	0.55
63	18	—	—	0.35	0.77	—	—	—	—	0.30	0.65	—	—	—	—	—	—	0.17	0.37	0.09	0.21	0.17	0.38
64	14	—	—	0.27	0.60	—	—	—	—	0.23	0.51	—	—	—	—	—	—	0.13	0.29	0.07	0.16	0.13	0.29
65	13	—	—	0.25	0.56	—	—	—	—	0.21	0.47	—	—	—	—	—	—	0.12	0.27	0.07	0.15	0.13	0.28
66	16	—	—	0.31	0.68	—	—	—	—	0.26	0.58	—	—	—	—	—	—	0.15	0.34	0.09	0.19	0.16	0.34
67	17	—	—	0.36	0.80	—	—	—	—	0.31	0.67	—	—	—	—	—	—	0.20	0.43	0.12	0.26	0.20	0.44
68	—	—	—	—	—	—	—	—	—	—	—	—	—	—	—	—	—	—	—	—	—	—	—
69	24	—	—	0.48	1.06	—	—	—	—	0.40	0.88	—	—	—	—	—	—	0.22	0.48	0.11	0.25	0.24	0.52
70	17	—	—	0.35	0.78	—	—	—	—	0.30	0.65	—	—	—	—	—	—	0.18	0.39	0.10	0.22	0.18	0.41
71	10	—	—	0.21	0.46	—	—	—	—	0.18	0.39	—	—	—	—	—	—	0.10	0.23	0.06	0.14	0.10	0.23
72	20	—	—	0.41	0.90	—	—	—	—	0.34	0.76	—	—	—	—	—	—	0.21	0.46	0.12	0.27	0.21	0.46
73	17	—	—	0.35	0.77	—	—	—	—	0.30	0.66	—	—	—	—	—	—	0.19	0.42	0.12	0.26	0.18	0.41
74	12	11	—	0.24	0.54	—	—	—	—	0.20	0.45	—	—	—	—	—	—	0.12	0.26	0.07	0.14	0.12	0.26
75	17	—	—	0.33	0.74	—	—	—	—	0.28	0.61	—	—	—	—	—	—	0.15	0.34	0.08	0.18	0.17	0.36
76	14	—	—	0.28	0.63	—	—	—	—	0.23	0.51	—	—	—	—	—	—	0.13	0.29	0.07	0.15	0.14	0.32
77	13	—	—	0.25	0.55	—	—	—	—	0.21	0.47	—	—	—	—	—	—	0.13	0.28	0.08	0.17	0.12	0.27
78	14	—	—	0.29	0.64	—	—	—	—	0.25	0.54	—	—	—	—	—	—	0.15	0.32	0.09	0.19	0.15	0.33
79	13	—	—	0.25	0.56	—	—	—	—	0.21	0.47	—	—	—	—	—	—	0.12	0.27	0.07	0.15	0.13	0.28
80	9	—	—	0.19	0.42	—	—	—	—	0.16	0.34	—	—	—	—	—	—	—	—	—	—	—	—
81	13	—	—	0.25	0.54	—	—	—	—	0.21	0.47	—	—	—	—	—	—	0.12	0.26	0.07	0.16	0.12	0.26
82	16	—	—	0.33	0.72	—	—	—	—	0.28	0.61	—	—	—	—	—	—	0.16	0.36	0.10	0.21	0.17	0.37
83	14	—	—	0.29	0.64	—	—	—	—	0.24	0.54	—	—	—	—	—	—	0.15	0.34	0.09	0.20	0.15	0.34
84	20	—	—	0.38	0.84	—	—	—	—	0.33	0.72	—	—	—	—	—	—	0.18	0.40	0.10	0.23	0.18	0.40
85	15	15	—	0.31	0.68	—	—	—	—	0.26	0.58	—	—	—	—	—	—	0.15	0.33	0.08	0.18	0.15	0.33
86	14	—	—	0.27	0.60	—	—	—	—	0.23	0.51	—	—	—	—	—	—	0.12	0.27	0.07	0.15	0.13	0.28
87	15	—	—	0.30	0.66	—	—	—	—	0.25	0.56	—	—	—	—	—	—	0.15	0.34	0.09	0.20	0.16	0.34
88	14	—	—	0.28	0.61	—	—	—	—	0.23	0.51	—	—	—	—	—	—	0.14	0.31	0.08	0.18	0.14	0.32
89	27	—	—	0.66	1.46	—	—	—	—	0.37	0.81	—	—	—	—	—	—	—	—	—	—	—	—
90	18	—	—	0.35	0.78	—	—	—	—	0.30	0.66	—	—	—	—	—	—	0.18	0.39	0.10	0.22	0.18	0.40
91	—	—	—	—	—	—	—	—	—	—	—	—	—	—	—	—	—	—	—	—	—	—	—
92	12	—	—	0.25	0.54	—	—	—	—	0.21	0.46	—	—	—	—	—	—	0.12	0.27	0.07	0.15	0.13	0.28
93	20	—	—	0.40	0.89	—	—	—	—	0.34	0.74	—	—	—	—	—	—	0.19	0.42	0.10	0.22	0.20	0.45
94	14	—	—	0.29	0.64	—	—	—	—	0.25	0.54	—	—	—	—	—	—	0.15	0.33	0.09	0.19	0.15	0.33
95	17	—	—	0.36	0.80	—	—	—	—	0.31	0.67	—	—	—	—	—	—	0.20	0.43	0.12	0.26	0.20	0.44
96	19	—	—	0.36	0.79	—	—	—	—	0.30	0.67	—	—	—	—	—	—	0.16	0.35	0.08	0.18	0.17	0.37
97	14	—	—	0.29	0.64	—	—	—	—	0.25	0.54	—	—	—	—	—	—	0.15	0.33	0.09	0.19	0.15	0.33
98	—	—	—	—	—	—	—	—	—	—	—	—	—	—	—	—	—	—	—	—	—	—	—
99	16	16	—	0.33	0.73	—	—	—	—	0.28	0.62	—	—	—	—	—	—	0.17	0.36	0.09	0.21	0.16	0.36
100	9	—	—	0.18	0.39	—	—	—	—	0.15	0.33	—	—	—	—	—	—	0.09	0.19	0.05	0.10	0.09	0.20
101	15	—	—	0.30	0.66	—	—	—	—	0.25	0.56	—	—	—	—	—	—	0.16	0.34	0.09	0.21	0.16	0.34
102	33	—	—	0.65	1.44	—	—	—	—	0.55	1.20	—	—	—	—	—	—	0.32	0.70	0.18	0.39	0.33	0.73
103	17	—	—	0.34	0.76	—	—	—	—	0.29	0.64	—	—	—	—	—	—	0.17	0.38	0.10	0.22	0.18	0.39
104	23	—	—	0.47	1.04	—	—	—	—	0.40	0.87	—	—	—	—	—	—	0.24	0.52	0.13	0.30	0.24	0.53
105	18	—	—	0.35	0.78	—	—	—	—	0.30	0.66	—	—	—	—	—	—	0.19	0.42	0.12	0.26	0.19	0.42
106	22	—	—	0.44	0.97	—	—	—	—	0.36	0.80	—	—	—	—	—	—	0.21	0.47	0.11	0.25	0.23	0.50
107	13	—	—	0.26	0.58	—	—	—	—	0.23	0.50	—	—	—	—	—	—	0.16	0.35	0.10	0.22	0.15	0.33
108	15	—	—	0.30	0.67	—	—	—	—	0.26	0.58	—	—	—	—	—	—	0.15	0.32	0.09	0.19	0.15	0.33
109	13	—	—	0.28	0.62	—	—	—	—	0.24	0.52	—	—	—	—	—	—	0.15	0.33	0.09	0.20	0.15	0.33
110	17	—	—	0.34	0.74	—	—	—	—	0.29	0.63	—	—	—	—	—	—	0.17	0.37	0.10	0.22	0.17	0.38
111	17	—	—	0.33	0.73	—	—	—	—	0.29	0.64	—	—	—	—	—	—	0.19	0.42	0.13	0.28	0.18	0.40
112	21	—	—	0.42	0.92	—	—	—	—	0.36	0.79	—	—	—	—	—	—	0.21	0.47	0.13	0.29	0.21	0.47
113	21	—	—	0.41	0.90	—	—	—	—	0.35	0.78	—	—	—	—	—	—	0.20	0.45	0.13	0.28	0.20	0.44
114	19	—	—	0.37	0.83	—	—	—	—	0.31	0.69	—	—	—	—	—	—	0.19	0.41	0.10	0.23	0.19	0.43
115	23	—	15	0.46	1.02	—	—	0.29	0.64	0.39	0.87	—	—	—	—	0.24	0.52	0.24	0.52	0.14	0.31	0.24	0.52
116	23	—	—	0.46	1.00	—	—	—	—	0.39	0.86	—	—	—	—	—	—	0.23	0.52	0.14	0.31	0.23	0.52
117	28	—	—	0.60	1.33	—	—	—	—	0.48	1.05	—	—	—	—	—	—	—	—	—	—	—	—

TABLE V-1E PASTURE AND RANGE PLANTS, COMPOSITION OF FEEDS, DATA EXPRESSED AS-FED—(Continued)

Entry Number	Feed Name Description	Dry Matter	Macro Minerals							Micro Minerals						
			Calcium (Ca)	Phosphorus (P)	Sodium (Na)	Chlorine (Cl)	Magnesium (Mg)	Potassium (K)	Sulfur (S)	Cobalt (Co)	Copper (Cu)	Iodine (I)	Iron (Fe)	Manganese (Mn)	Selenium (Se)	Zinc (Zn)
		%	%	%	%	%	%	%	%	ppm or mg/kg	ppm or mg/kg	ppm or mg/kg	%	ppm or mg/kg	ppm or mg/kg	ppm or mg/kg
37	CORN, FRESH, ALL ANALYSES	23	0.07	—	—	—	0.21	—	—	—	1.8	—	0.008	24.6	—	16.1
38	CORN, SWEET, STOVER WITHOUT EARS, WITHOUT HUSKS, FRESH	22	—	—	—	—	—	—	—	—	—	—	—	—	—	—
39	COWPEA, COMMON, FRESH	25	0.38	0.08	0.06	0.05	0.11	0.41	0.08	—	—	—	0.020	—	—	—
40	CRESTED WHEATGRASS, EARLY BLOOM, FRESH	41	0.09	0.07	—	—	—	—	—	—	—	—	—	—	—	—
	CURLY MESQUITE															
41	BROWSE, FRESH	35	0.18	0.05	—	—	0.06	0.23	—	—	3.5	—	—	16.4	—	—
42	BROWSE, MATURE, FRESH	50	0.27	0.04	—	—	0.08	0.19	—	—	—	—	—	—	—	—
43	DALLISGRASS, FRESH	25	0.14	0.05	0.09	—	0.10	0.43	—	0.019	—	—	0.005	—	—	—
44	DROPSEED, SAND, STEM CURED, FRESH	88	0.40	0.07	0.01	—	0.06	0.28	—	0.503	13.5	0.599	0.043	41.4	—	36.8
45	FESCUE, TALL (ALTA), FRESH	28	0.13	0.05	0.03	—	0.07	0.70	—	0.113	1.0	—	0.003	18.0	—	5.9
46	FILAREE (ALFILERIA, REDSTEM), FRESH	18	0.35	0.08	—	—	—	0.59	—	—	—	—	—	—	—	—
47	FOXTAIL, MEADOW, IMMATURE, FRESH	26	0.15	0.12	—	—	—	—	—	—	—	—	—	—	—	—
48	GALLETA, STEM CURED, FRESH	86	0.60	0.06	0.01	—	0.07	0.41	0.09	0.591	16.3	—	0.044	67.7	—	19.5
49	GRAMA, IMMATURE, FRESH	41	0.22	0.08	—	—	—	—	—	—	2.3	—	—	18.2	—	—
50	GRASS-LEGUME, FRESH	24	0.15	0.08	—	—	0.08	0.40	—	—	—	—	—	—	—	—
51	GREASEWOOD, BROWSE, FRESH	50	0.46	0.09	—	—	—	—	—	0.030	7.8	—	—	12.9	—	—
52	INDIANGRASS, FRESH	57	0.19	0.04	—	—	—	—	—	—	—	—	—	—	—	—
53	INDIAN RICEGRASS, FRESH	48	0.28	0.02	—	—	0.07	—	0.07	—	—	—	—	—	—	—
	JOHNSONGRASS SORGHUM—SEE SORGHUM															
54	JUNEGRASS, IMMATURE, FRESH	28	0.09	0.07	—	—	—	—	—	—	—	—	—	—	—	—
	KENTUCKY BLUEGRASS—SEE BLUEGRASS, KENTUCKY															
55	KOA HAOLE (LEAD TREE, WHITE POPINAC), BROWSE, FRESH	30	—	—	—	—	—	—	—	—	—	—	—	—	—	—
56	LESPEDEZA, COMMON, IMMATURE, FRESH	24	—	—	—	—	—	—	—	—	—	—	—	—	—	—
57	LESPEDEZA, COMMON-KOREAN, MATURE, FRESH	35	0.35	0.07	—	—	—	—	—	—	—	—	—	—	—	—
58	LESPEDEZA, SERICEA (CHINESE LESPEDEZA), FRESH	33	0.42	0.10	—	—	0.07	0.39	—	0.024	—	—	0.008	34.1	—	—
59	LOVEGRASS, IMMATURE, FRESH	43	0.20	0.10	—	—	—	—	—	—	—	—	—	—	—	—
60	MEADOW FOXTAIL, IMMATURE, FRESH	26	0.15	0.12	—	—	—	—	—	—	—	—	—	—	—	—
61	MEDIC, BLACK (YELLOW TREFOIL), FRESH	23	—	—	—	—	—	—	—	—	—	—	—	—	—	—
62	MESQUITE, COMMON, BROWSE, FRESH	35	0.68	0.07	—	—	0.08	0.49	—	—	—	—	—	—	—	—
63	MILLET, FOXTAIL, FRESH	29	0.09	0.05	—	—	—	0.56	—	—	—	—	—	—	—	—
64	MILLET, JAPANESE (BARNYARD GRASS), FRESH	22	0.11	0.07	—	—	—	0.52	—	—	—	—	—	—	—	—
65	MILLET, PEARL (PEARL MILLET), FRESH	21	—	—	—	—	—	—	—	—	—	—	—	—	—	—
66	MILLET, PROSO (BROOMCORN; HOG MILLET), FRESH	25	—	—	—	—	—	—	—	—	—	—	—	—	—	—
67	MILO SORGHUM, FRESH	30	0.09	0.05	—	—	—	0.81	—	—	—	—	—	—	—	—
68	MUHLY BUSH, MIDBLOOM, FRESH	43	0.13	0.03	0.00	—	0.02	0.22	—	0.215	0.2	—	0.006	5.2	—	5.2
69	NEEDLEGRASS, MATURE, FRESH	42	—	—	—	—	—	—	—	—	—	—	—	—	—	—
70	OATGRASS, TALL, FRESH	30	0.12	0.14	—	—	—	0.91	—	—	—	—	—	—	—	—
71	OATS, IMMATURE, FRESH	16	—	—	0.02	0.02	—	—	0.01	—	—	—	—	—	—	—
72	OATS-VETCH, MILK STAGE, FRESH	33	—	—	—	—	—	—	—	—	—	—	—	—	—	—
73	ORCHARDGRASS, FRESH, ALL ANALYSES	26	0.09	0.05	0.03	—	0.06	0.74	—	0.055	2.5	—	0.003	28.5	—	5.3
74	PANGOLAGRASS, FRESH	20	0.08	0.05	—	—	0.04	0.29	—	—	—	—	—	—	—	—
75	PANICUM, FRESH	29	0.14	0.05	—	—	0.10	0.93	—	—	—	—	—	—	—	—
76	PARAGRASS, FRESH	26	0.10	0.10	—	—	—	0.42	—	—	—	—	—	—	—	—
77	PEA, FIELD, FRESH	18	0.22	0.04	—	—	0.04	0.27	—	0.028	—	—	0.008	15.4	—	—
78	PEA-OATS, FRESH	23	0.17	0.07	—	—	—	0.38	—	—	—	—	—	—	—	—
79	PEARL MILLET (MILLET, PEARL), FRESH	21	—	—	—	—	—	—	—	—	—	—	—	—	—	—
80	PEAVINE, FRESH	17	—	—	—	—	—	—	—	—	—	—	—	—	—	—
81	RAPE (CANOLA), FRESH	17	0.25	0.07	—	—	0.01	0.56	0.11	—	1.4	—	0.004	7.7	—	—
82	REDTOP, FULL BLOOM, FRESH	26	0.16	0.10	0.01	—	0.07	0.62	0.04	—	—	—	0.006	—	—	—
83	REED CANARYGRASS, FRESH	23	0.08	0.08	—	—	—	0.83	—	—	—	—	—	—	—	—
84	RESCUEGRASS (BROMEGRASS, RESCUE), FRESH	29	0.15	0.08	—	—	—	—	—	—	—	—	—	—	—	—
85	RHODESGRASS, FRESH	26	0.13	0.10	—	—	—	0.61	—	—	—	—	—	—	—	—
86	RYE, FRESH	20	0.09	0.08	0.01	—	0.06	0.69	—	—	—	—	—	—	—	—
87	RYEGRASS, FRESH	24	—	—	—	—	—	—	—	—	—	—	—	—	—	—
88	RYEGRASS, ITALIAN, FRESH	23	0.15	0.09	0.00	—	0.08	0.45	0.02	—	—	—	0.023	—	—	—
89	SAGEBRUSH, BIG, BROWSE, STEM CURED, FRESH	65	0.46	0.12	—	—	—	—	—	—	—	—	—	—	—	—
90	SALTGRASS, DESERT, FRESH	29	0.05	0.03	—	—	—	—	—	—	—	—	—	—	—	—
91	SEDGE, FRESH	25	—	0.05	0.05	0.06	—	—	0.06	—	—	—	—	—	—	—
	SORGHUM															
92	JOHNSONGRASS, IMMATURE, FRESH	20	0.18	0.06	—	—	—	—	—	—	—	—	—	—	—	—
93	JOHNSONGRASS, FULL BLOOM, FRESH	35	0.29	0.06	—	—	—	—	—	—	—	—	—	—	—	—
94	KAFIR, FRESH	24	0.09	0.04	—	—	—	0.40	—	—	—	—	—	—	—	—
95	MILO, FRESH	30	0.09	0.05	—	—	—	0.81	—	—	—	—	—	—	—	—
96	SUDANGRASS, MATURE, FRESH	30	0.09	0.06	—	—	—	—	—	—	—	—	—	—	—	—
97	SOYBEAN, FRESH	23	0.25	0.07	—	—	0.12	0.21	—	—	2.1	—	0.005	27.3	—	—
98	STARGRASS, FRESH	63	0.39	0.19	0.02	—	0.20	2.07	—	0.133	6.7	—	0.012	50.6	—	38.1
99	SUGARCANE, FRESH	28	0.11	0.05	—	—	0.11	0.29	—	—	0.6	—	0.005	17.3	—	6.9
100	SUNFLOWER, FRESH	15	—	—	—	—	—	—	—	—	—	—	—	—	—	—
101	SWEET CLOVER, YELLOW, FRESH	23	0.31	0.06	0.02	0.09	0.08	0.38	0.11	—	2.3	—	0.004	29.0	—	—
102	SWITCHGRASS, FRESH	55	0.16	0.05	—	—	—	—	—	—	—	—	—	—	—	—
103	TALL (ALTA) FESCUE, FRESH	28	0.13	0.05	0.03	—	0.07	0.70	—	0.113	1.0	—	0.003	18.0	—	5.9
104	THREE-AWN (WIREGRASS), FRESH	39	—	—	—	—	—	—	—	—	—	—	—	—	—	—
105	TIMOTHY, FRESH	28	0.14	0.08	0.03	0.14	0.06	0.69	0.04	0.041	2.2	—	0.004	24.6	—	7.5
106	TOBOSA, IMMATURE, FRESH	40	0.18	0.05	0.01	—	0.04	0.21	—	0.277	5.6	—	0.023	32.3	—	12.3
107	TREFOIL, BIRDSFOOT (DEERVETCH, BIRDSFOOT), FRESH	19	0.34	0.05	0.02	—	0.08	0.63	0.05	0.094	2.5	—	0.006	16.0	—	6.0
108	VELVET BEAN, DOUGH STAGE, FRESH	22	—	—	—	—	—	—	—	—	—	—	—	—	—	—
109	VETCH, FRESH	22	—	—	0.11	0.42	—	—	0.03	0.068	—	—	—	—	—	—
110	VETCH-OATS, FRESH	26	0.18	0.08	—	—	0.06	0.45	—	—	—	—	—	—	—	—
111	WHEAT, FRESH	22	0.09	0.09	0.04	—	0.05	0.78	0.05	—	—	—	0.003	—	—	—
112	WHEATGRASS, BLUEBUNCH, PREBLOOM, FRESH	32	0.12	0.08	0.02	—	0.06	1.03	—	0.063	2.6	—	0.008	15.7	—	9.2
113	WHEATGRASS, CRESTED, IMMATURE, FRESH	28	0.13	0.10	—	—	0.08	—	—	—	—	—	—	—	—	—
114	WHEATGRASS, SLENDER, FRESH	32	0.16	0.05	0.03	—	0.08	1.04	—	0.067	1.5	—	0.003	19.8	—	7.4
115	WHITE CLOVER, FRESH	18	0.25	0.09	0.07	0.11	0.08	0.38	0.06	—	—	—	0.006	54.4	—	—
116	WILD-RYE, RUSSIAN, FRESH	35	0.11	0.06	0.08	—	0.05	1.06	—	0.097	1.1	—	0.004	9.5	—	5.7
117	WINTERFAT, COMMON, BROWSE, STEM CURED, FRESH	80	1.58	0.09	—	—	—	—	—	—	—	—	—	—	—	—

Composition of Feeds

PASTURE AND RANGE PLANTS

Entry Number	Fat-Soluble Vitamins					Water-Soluble Vitamins								
	A (1 mg Carotene = 1667 IU Vit A)	Carotene (Provitamin A)	D	E	K	B-12	Biotin	Choline	Folacin (Folic Acid)	Niacin	Pantothenic Acid (B-3)	(Pyridoxine) B-6	Riboflavin (B-2)	Thiamin (B-1)
	IU/g	ppm or mg/kg	IU/kg	ppm or mg/kg	ppm or mg/kg	ppb or mcg/kg	ppm or mg/kg	ppm or mg/kg	ppm or mg/kg	ppm or mg/kg	ppm or mg/kg	ppm or mg/kg	ppm or mg/kg	ppm or mg/kg
37	–	–	–	–	–	–	–	–	–	–	–	–	–	–
38	–	–	–	–	–	–	–	–	–	–	–	–	–	–
39	–	–	–	–	–	–	–	–	–	–	–	–	–	–
40	–	–	–	–	–	–	–	–	–	–	–	–	–	–
41	–	–	–	–	–	–	–	–	–	–	–	–	–	–
42	–	–	–	–	–	–	–	–	–	–	–	–	–	–
43	126.0	75.6	–	–	–	–	–	–	–	–	–	–	–	–
44	14.0	8.4	–	–	–	–	–	–	–	–	–	–	–	–
45	160.0	96.0	–	46.9	–	–	–	–	–	–	–	–	2.4	3.4
46	–	–	–	–	–	–	–	–	–	–	–	–	–	–
47	–	–	–	–	–	–	–	–	–	–	–	–	–	–
48	0.3	0.2	–	–	–	–	–	–	–	–	–	–	–	–
49	–	–	–	–	–	–	–	–	–	–	–	–	–	–
50	–	–	–	–	–	–	–	–	–	–	–	–	–	–
51	36.2	21.7	–	–	–	–	–	–	–	–	–	–	–	–
52	92.5	55.5	–	–	–	–	–	–	–	–	–	–	–	–
53	0.4	0.2	–	–	–	–	–	–	–	–	–	–	–	–
54	–	–	–	–	–	–	–	–	–	–	–	–	–	–
55	–	–	–	–	–	–	–	–	–	–	–	–	–	–
56	–	–	–	–	–	–	–	–	–	–	–	–	–	–
57	–	–	–	–	–	–	–	–	–	–	–	–	–	–
58	–	–	–	–	–	–	–	–	–	–	–	–	–	–
59	–	–	–	–	–	–	–	–	–	–	–	–	–	–
60	–	–	–	–	–	–	–	–	–	–	–	–	–	–
61	–	–	–	–	–	–	–	–	–	–	–	–	–	–
62	–	–	–	–	–	–	–	–	–	–	–	–	–	–
63	–	–	–	–	–	–	–	–	–	–	–	–	–	–
64	–	–	–	–	–	–	–	–	–	–	–	–	–	–
65	63.0	37.8	–	–	–	–	–	–	–	–	–	–	–	–
66	–	–	–	–	–	–	–	–	–	–	–	–	–	–
67	–	–	–	–	–	–	–	–	–	–	–	–	–	–
68	–	–	–	–	–	–	–	–	–	–	–	–	–	–
69	–	–	–	–	–	–	–	–	–	–	–	–	–	–
70	–	–	–	–	–	–	–	–	–	–	–	–	–	–
71	150.0	90.0	–	–	–	–	–	–	–	–	–	–	–	–
72	–	–	–	–	–	–	–	–	–	–	–	–	–	–
73	137.1	82.2	–	112.3	–	–	–	–	–	–	–	–	–	1.9
74	–	–	–	–	–	–	–	–	–	–	–	–	–	–
75	–	–	–	–	–	–	–	–	–	–	–	–	–	–
76	–	–	–	–	–	–	–	–	–	–	–	–	–	–
77	–	–	–	–	–	–	–	–	–	–	–	–	–	–
78	–	–	–	–	–	–	–	–	–	–	–	–	–	–
79	63.0	37.8	–	–	–	–	–	–	–	–	–	–	–	–
80	29.7	17.8	–	–	–	–	–	–	–	–	–	–	2.2	–
81	–	–	–	–	–	–	–	–	–	–	–	–	–	–
82	66.9	40.1	–	–	–	–	–	–	–	–	–	–	–	–
83	–	–	–	–	–	–	–	–	–	–	–	–	–	–
84	–	–	–	–	–	–	–	–	–	–	–	–	–	–
85	–	–	–	–	–	–	–	–	–	–	–	–	–	–
86	115.0	69.0	–	–	–	–	–	–	–	–	–	–	–	–
87	–	–	–	–	–	–	–	–	–	–	–	–	–	–
88	–	–	–	–	–	–	–	–	–	–	–	–	–	–
89	17.3	10.4	–	–	–	–	–	–	–	–	–	–	–	–
90	–	–	–	–	–	–	–	–	–	–	–	–	–	–
91	–	–	–	–	–	–	–	–	–	–	–	–	–	–
92	–	–	–	–	–	–	–	–	–	–	–	–	–	–
93	–	–	–	–	–	–	–	–	–	–	–	–	–	–
94	6.9	4.2	–	–	–	–	–	–	–	9	3.3	1.41	1.0	–
95	–	–	–	–	–	–	–	–	–	–	–	–	–	–
96	–	–	–	–	–	–	–	–	–	–	–	–	–	–
97	121.6	73.0	–	64.2	–	–	–	–	–	–	–	–	–	–
98	–	–	–	–	–	–	–	–	–	–	–	–	–	–
99	–	–	–	–	–	–	–	–	–	–	–	–	–	–
100	–	–	–	–	–	–	–	–	–	–	–	–	–	–
101	102.5	61.5	–	–	–	–	–	–	–	8	–	–	19.4	1.2
102	83.0	49.8	–	–	–	–	–	–	–	–	–	–	–	–
103	–	–	–	–	–	–	–	–	–	–	–	–	–	–
104	–	–	–	–	–	–	–	–	–	–	–	–	–	–
105	103.2	61.9	–	42.6	–	–	–	–	–	–	–	–	3.2	0.8
106	–	–	–	–	–	–	–	–	–	–	–	–	–	–
107	–	–	–	–	–	–	–	–	–	–	–	–	–	–
108	–	–	–	–	–	–	–	–	–	–	–	–	–	–
109	–	–	–	–	–	–	–	–	–	–	–	–	–	–
110	–	–	–	–	–	–	–	–	–	–	–	–	–	–
111	192.5	115.4	–	–	–	–	–	–	–	13	4.7	–	6.1	–
112	173.6	104.2	–	–	–	–	–	–	–	–	–	–	–	–
113	205.8	123.4	–	–	–	–	–	–	–	–	–	–	3.4	1.6
114	–	–	–	–	–	–	–	–	–	–	–	–	–	–
115	44.0	26.4	–	54.6	–	–	–	–	–	11	–	–	16.0	2.5
116	–	–	–	–	–	–	–	–	–	–	–	–	–	–
117	24.1	14.5	–	–	–	–	–	–	–	–	–	–	–	–

TABLE V-1F MINERAL SUPPLEMENTS, COMPOSITION, DATA EXPRESSED AS-FED

Entry Number	Feed Name Description	International Feed Number	Proximate Analysis						Digestible Protein		
			Dry Matter	Ash	Crude Fiber	Ether Extract (Fat)	N-Free Extract	Crude Protein (6.25 × N)	Ruminant	Non-Ruminant	Horse
			%	%	%	%	%	%	%	%	%
1	AMMONIUM PHOSPHATE, MONOBASIC	6-09-338	98	53.0	–	–	–	69.4	–	–	–
2	AMMONIUM PHOSPHATE, DIBASIC	6-00-370	98	35.5	–	–	–	112.9	–	–	–
3	AMMONIUM POLYPHOSPHATE SOLUTION	6-08-042	60	–	–	–	–	54.8	–	–	–
4	BONE, CHARCOAL	6-00-402	94	79.3	3.7	1.1	1.8	–	–	–	–
5	BONE MEAL	6-00-397	94	60.5	2.9	6.5	–	24.8	17.1	–	–
6	BONE MEAL, STEAMED*	6-00-400	95	67.3	1.9	3.6	3.8	18.6	–	–	–
7	CALCIUM CARBONATE*	6-01-069	100	97.1	–	–	–	–	–	–	–
8	CALCIUM OXIDE	6-14-003	97	–	–	–	–	–	–	–	–
9	CALCIUM PERIODATE*	6-09-335	–	–	–	–	–	–	–	–	–
10	CALCIUM PHOSPHATE, MONOBASIC, FROM DEFLUORINATED PHOSPHORIC ACID	6-01-082	99	87.1	–	–	–	–	–	–	–
11	CALCIUM PHOSPHATE, MONOBASIC, FROM FURNACED PHOSPHORIC ACID	6-26-334	96	–	–	–	–	–	–	–	–
12	CALCIUM PHOSPHATE, DIBASIC, FROM DEFLUORINATED PHOSPHORIC ACID*	6-01-080	97	89.7	–	–	–	–	–	–	–
13	CALCIUM PHOSPHATE, DIBASIC, FROM FURNACED PHOSPHORIC ACID*	6-26-335	97	85.6	–	–	–	–	–	–	–
14	CALCIUM PHOSPHATE, TRIBASIC, FROM FURNACED PHOSPHORIC ACID	6-01-084	98	92.1	–	–	–	–	–	–	–
15	CALCIUM SULFATE, ANHYDROUS	6-01-087	–	–	–	–	–	–	–	–	–
16	CALCIUM SULFATE (GYPSUM)	6-01-090	95	–	–	–	–	–	–	–	–
17	COBALT CARBONATE*	6-01-566	99	–	–	–	–	–	–	–	–
18	COBALT SULFATE*	6-01-564	–	–	–	–	–	–	–	–	–
19	COBALTOUS CHLORIDE	6-01-558	98	–	–	–	–	–	–	–	–
20	COBALTOUS OXIDE	6-01-560	99	–	–	–	–	–	–	–	–
21	COLLOIDAL CLAY (SOFT ROCK PHOSPHATE)	6-03-947	100	–	–	–	–	–	–	–	–
22	COPPER (CUPRIC) CARBONATE	6-01-703	98	–	–	–	–	–	–	–	–
23	COPPER (CUPRIC) CHLORIDE	6-01-705	99	–	–	–	–	–	–	–	–
24	COPPER (CUPRIC) GLUCONATE	6-01-707	99	–	–	–	–	–	–	–	–
25	COPPER (CUPRIC) HYDROXIDE	6-01-709	98	–	–	–	–	–	–	–	–
26	COPPER (CUPRIC) ORTHOPHOSPHATE	6-01-713	99	–	–	–	–	–	–	–	–
27	COPPER (CUPRIC) OXIDE	6-01-711	99	–	–	–	–	–	–	–	–
28	COPPER (CUPRIC) SULFATE, PENTAHYDRATE*	6-01-719	99	–	–	–	–	–	–	–	–
29	COPPER (CUPROUS) IODIDE	6-01-721	–	–	–	–	–	–	–	–	–
30	COPPER (CUPROUS) OXIDE*	6-28-224	99	–	–	–	–	–	–	–	–
31	CURACAO PHOSPHATE, GROUND	6-05-586	99	94.1	–	–	–	–	–	–	–
32	DIAMMONIUM PHOSPHATE*	6-00-370	98	35.5	–	–	–	112.9	–	–	–
33	DIIODOSALICYLIC ACID*	6-01-787	99	–	–	–	–	–	–	–	–
34	ETHYLENEDIAMINE DIHYDROIODIDE*	6-01-842	98	–	–	–	–	54.3	–	–	–
35	FERRIC (IRON) AMMONIUM CITRATE	6-01-857	99	–	–	–	–	42.1	–	–	–
36	FERRIC (IRON) CHLORIDE	6-01-865	98	–	–	–	–	–	–	–	–
37	FERRIC (IRON) OXIDE*	6-02-431	97	–	–	–	–	–	–	–	–
38	FERROUS (IRON) CARBONATE*	6-01-863	99	–	–	–	–	–	–	–	–
39	FERROUS (IRON) FUMARATE	6-08-097	99	–	–	–	–	–	–	–	–
40	FERROUS (IRON) GLUCONATE	6-01-867	99	–	–	–	–	–	–	–	–
41	FERROUS (IRON) OXIDE	6-20-728	97	–	–	–	–	–	–	–	–
42	FERROUS (IRON) SULFATE, MONOHYDRATE*	6-01-869	98	98.0	–	–	–	–	–	–	–
43	FERROUS (IRON) SULFATE, HEPTAHYDRATE	6-20-734	99	–	–	–	–	–	–	–	–
44	KELP (SEAWEED), WHOLE, DEHY	1-08-073	91	35.0	6.5	0.5	42.4	6.5	2.9	2.2	3.4
45	LIMESTONE, GROUND*	6-02-632	100	93.8	–	–	–	–	–	–	–
46	LIMESTONE, MAGNESIUM (DOLOMITE), GROUND*	6-02-633	100	–	–	–	–	–	–	–	–
47	MAGNESIUM CARBONATE	6-02-754	98	–	–	–	–	–	–	–	–
48	MAGNESIUM HYDROXIDE	6-26-012	98	–	–	–	–	–	–	–	–
49	MAGNESIUM OXIDE*	6-02-756	98	98.3	–	–	–	–	–	–	–
50	MAGNESIUM SULFATE (EPSOM SALTS)*	6-02-758	99	–	–	–	–	–	–	–	–
51	MANGANESE CHLORIDE	6-03-038	99	–	–	–	–	–	–	–	–
52	MANGANESE DIOXIDE	6-03-042	98	–	–	–	–	–	–	–	–
53	MANGANOUS (MANGANESE) OXIDE*	6-03-054	99	–	–	–	–	–	–	–	–
54	MANGANOUS (MANGANESE) SULFATE*	6-26-136	100	–	–	–	–	–	–	–	–
55	OYSTER SHELLS, GROUND (FLOUR)*	6-03-481	99	79.0	1.8	0.3	17.0	0.7	–	–	–
56	PHOSPHATE, DEFLUORINATED	6-01-780	100	99.3	–	–	–	–	–	–	–
57	PHOSPHATE ROCK, GROUND (RAW)	6-03-945	–	–	–	–	–	–	–	–	–
58	PHOSPHATE ROCK, LOW FLUORINE*	6-03-946	–	–	–	–	–	–	–	–	–
59	PHOSPHATE SOFT ROCK (COLLOIDAL CLAY)	6-03-947	100	–	–	–	–	–	–	–	–
60	PHOSPHORIC ACID, FEED GRADE (ORTHO)*	6-03-707	75	–	–	–	–	–	–	–	–
61	POTASSIUM CHLORIDE*	6-03-755	100	98.9	–	–	–	–	–	–	–
62	POTASSIUM IODIDE*	6-03-759	–	–	–	–	–	–	–	–	–
63	POTASSIUM MAGNESIUM SULFATE	6-06-177	98	–	–	–	–	–	–	–	–
64	POTASSIUM SULFATE	6-08-098	98	97.0	–	–	–	–	–	–	–
65	SEAWEED (KELP), WHOLE, DEHY	1-08-073	91	35.0	6.5	0.5	42.4	6.5	2.9	2.2	3.4
66	SODIUM BICARBONATE*	6-04-272	100	–	–	–	–	–	–	–	–
67	SODIUM CHLORIDE*	6-04-152	97	93.0	–	–	–	–	–	–	–
68	SODIUM IODIDE*	6-04-279	–	–	–	–	–	–	–	–	–
69	SODIUM PHOSPHATE, MONOBASIC*	6-04-288	97	96.9	–	–	–	–	–	–	–
70	SODIUM PHOSPHATE, DIBASIC	6-04-286	97	96.7	–	–	–	–	–	–	–
71	SODIUM SELENATE*	6-26-014	99	–	–	–	–	–	–	–	–
72	SODIUM SELENITE*	6-26-013	99	–	–	–	–	–	–	–	–
73	SODIUM SULFATE, DECAHYDRATE	6-04-291	97	–	–	–	–	–	–	–	–
74	SODIUM TRIPOLYPHOSPHATE*	6-08-076	97	89.7	–	–	–	–	–	–	–
75	SULFUR*	6-04-705	99	–	–	–	–	–	–	–	–
76	ZINC ACETATE	6-05-547	99	–	–	–	–	–	–	–	–
77	ZINC CARBONATE	6-05-549	99	–	–	–	–	–	–	–	–
78	ZINC CARBONATE, TETRAHYDRATE	6-29-585	98	–	–	–	–	–	–	–	–
79	ZINC CHLORIDE	6-05-551	98	–	–	–	–	–	–	–	–
80	ZINC OXIDE*	6-05-553	–	–	–	–	–	–	–	–	–
81	ZINC SULFATE, MONOHYDRATE*	6-05-555	99	–	–	–	–	–	–	–	–
82	ZINC SULFATE, HEPTAHYDRATE	6-20-729	98	–	–	–	–	–	–	–	–

*Sources most commonly used in commercial feeds.

Composition of Feeds

MINERAL SUPPLEMENTS

Entry Number	Macro Minerals							Micro Minerals							
	Calcium (Ca)	Phosphorus (P)	Sodium (Na)	Chlorine (Cl)	Magnesium (Mg)	Potassium (K)	Sulfur (S)	Cobalt (Co)	Copper (Cu)	Fluorine (F)	Iodine (I)	Iron (Fe)	Manganese (Mn)	Selenium (Se)	Zinc (Zn)
	%	%	%	%	%	%	%	ppm or mg/kg	ppm or mg/kg	ppm or mg/kg	ppm or mg/kg	%	ppm or mg/kg	ppm or mg/kg	ppm or mg/kg
1	0.38	24.42	0.08	—	0.46	0.14	0.82	—	86	1833	—	0.991	462	—	640
2	0.50	20.09	0.04	—	0.45	—	2.47	—	81	1548	—	1.514	504	—	303
3	0.10	13.44	—	—	—	—	0.50	—	—	1341	—	0.505	—	—	—
4	31.92	14.84	—	—	0.55	0.15	—	—	—	—	—	—	—	—	—
5	24.52	11.43	0.61	0.22	0.35	0.14	0.12	—	19	2014	—	0.057	9	—	377
6	25.98	11.80	0.40	0.01	0.78	0.18	0.34	0	162	637	29	0.085	37	—	362
7	37.97	0.04	0.07	0.04	0.41	0.04	0.08	—	14	0	—	0.059	159	0.07	17
8	69.33	—	—	—	—	—	—	—	—	—	—	—	—	—	—
9	—	—	—	—	—	—	—	—	—	—	—	—	—	—	—
10	18.55	20.98	0.06	—	0.81	0.40	0.81	5	5	1410	—	1.007	201	—	419
11	22.00	23.00	—	—	—	—	—	—	—	300	—	—	—	—	—
12	22.00	18.43	1.56	—	0.51	0.10	0.69	8	9	940	—	0.844	253	—	122
13	23.00	18.50	0.08	—	0.60	0.07	—	—	80	1150	—	1.000	300	0.60	220
14	30.90	17.04	0.17	—	—	—	—	—	—	501	—	—	—	—	—
15	—	—	—	—	—	—	—	—	—	—	—	—	—	—	—
16	21.86	—	—	—	0.46	—	16.20	—	—	27	—	—	—	—	—
17	—	—	0.25	0.01	—	—	0.03	465000	15	—	—	0.020	100	—	15
18	—	—	—	—	—	—	—	—	—	—	—	—	—	—	—
19	—	—	—	29.20	—	—	0.07	242648	20	—	—	0.003	—	—	196
20	—	—	—	0.01	—	—	0.20	703494	—	—	—	0.050	—	—	—
21	16.01	9.00	0.10	—	0.38	—	—	—	—	12061	—	1.911	995	—	—
22	—	—	—	—	—	—	0.17	—	530000	—	—	0.147	—	—	196
23	—	—	—	41.17	—	—	0.03	—	368973	—	—	0.006	—	—	—
24	—	—	—	—	—	—	—	—	133353	—	—	—	—	—	—
25	—	—	—	—	—	—	—	—	602994	—	—	—	—	—	—
26	—	14.11	—	—	—	—	—	—	434214	—	—	—	—	—	—
27	0.01	—	—	—	0.00	—	—	—	753827	—	—	0.020	10	—	800
28	—	—	—	—	—	—	13.25	—	250976	—	—	0.010	2	—	9
29	—	—	—	—	—	—	—	—	—	—	—	—	—	—	—
30	—	—	—	—	—	—	—	—	879318	—	—	—	—	—	—
31	35.10	14.24	0.20	—	0.80	—	—	—	—	5445	—	0.347	—	—	—
32	0.50	20.09	0.04	—	0.45	—	2.47	—	81	1548	—	1.514	504	—	303
33	—	—	—	—	—	—	—	—	—	—	644391	—	—	—	—
34	—	—	—	—	—	—	—	—	—	—	787234	—	—	—	—
35	—	—	—	—	—	—	—	—	—	—	—	15.840	—	—	—
36	—	—	—	64.27	—	—	—	—	—	—	—	33.742	—	—	—
37	0.36	0.10	—	—	0.66	—	—	—	—	—	—	58.800	3600	—	—
38	1.24	0.01	—	—	0.33	—	1.77	200	3000	—	—	40.667	9000	—	—
39	—	—	—	—	—	—	—	—	—	—	—	32.542	—	—	—
40	—	—	—	—	—	—	—	—	—	—	—	11.465	—	—	—
41	—	—	—	—	—	—	—	—	—	—	—	75.369	—	—	—
42	—	—	—	—	0.50	—	17.80	—	—	—	—	31.000	—	—	—
43	—	—	—	—	0.21	—	11.00	—	100	—	—	20.899	0	—	100
44	2.47	0.28	—	—	0.85	—	—	—	—	—	—	—	—	—	—
45	37.12	0.21	0.06	0.03	1.13	0.11	0.04	—	11	—	—	0.357	269	—	19
46	20.61	0.02	0.38	0.12	10.37	0.27	0.01	—	20	—	—	0.053	—	—	—
47	0.02	—	—	—	30.19	—	—	—	—	—	—	0.020	—	—	—
48	—	—	—	—	40.86	—	—	—	—	—	—	—	—	—	—
49	1.66	—	—	—	55.19	—	0.10	501	5	251	—	1.048	80	0.35	9
50	0.02	—	—	0.01	9.60	—	13.00	—	—	—	—	—	—	10.12	—
51	—	—	—	35.47	—	—	—	—	—	—	—	—	274824	—	—
52	—	—	—	—	—	—	—	—	—	—	—	—	619262	—	—
53	0.16	0.10	0.06	—	0.70	0.58	0.01	300	724	—	—	3.436	620217	—	1349
54	—	—	—	—	0.30	—	19.01	—	—	—	—	0.040	250000	—	—
55	35.85	0.10	0.21	0.01	0.24	0.10	—	—	15	—	—	0.254	178	—	7
56	31.99	17.07	3.26	—	0.29	0.10	0.13	10	40	1794	—	0.840	496	—	90
57	—	—	—	—	—	—	—	—	—	—	—	—	—	—	—
58	—	—	—	—	—	—	—	—	—	—	—	—	—	—	—
59	16.01	9.00	0.10	—	0.38	—	—	—	—	12061	—	1.911	995	—	—
60	0.14	20.88	0.18	—	0.40	0.06	1.50	—	17	1900	—	0.913	500	—	210
61	0.05	—	1.00	46.88	0.23	51.31	0.32	—	7	—	—	0.061	7	—	9
62	—	—	—	—	—	—	—	—	—	—	—	—	—	—	—
63	0.06	—	0.75	1.24	11.58	18.45	21.97	—	2	10	—	0.010	10	—	9
64	0.15	—	0.09	1.50	0.59	43.04	17.64	—	3	—	—	0.069	9	—	4
65	2.47	0.28	—	—	0.85	—	—	—	—	—	—	—	—	—	—
66	—	—	26.87	—	—	0.01	—	—	—	450138	—	0.001	—	—	—
67	—	—	38.17	58.46	—	—	—	—	—	—	—	—	—	—	—
68	—	—	—	—	—	—	—	—	—	—	—	—	—	—	—
69	0.04	24.84	18.65	—	—	0.14	—	—	7	—	—	—	—	—	5
70	—	21.65	31.04	—	—	—	—	—	—	300	—	—	—	—	—
71	—	—	24.18	—	—	—	—	—	—	—	—	—	—	415898.96	—
72	—	—	26.40	—	0.01	—	—	—	10	—	—	0.031	—	452927.78	—
73	—	—	31.33	—	—	—	9.66	—	—	—	—	0.001	—	—	—
74	—	24.53	30.18	—	—	—	—	—	—	247	—	0.004	—	—	—
75	—	—	—	—	—	—	99.00	—	—	—	—	—	—	—	291951
76	—	—	—	—	—	—	0.07	—	—	—	—	0.001	—	—	294822
77	—	—	—	—	—	—	—	—	—	—	—	—	—	—	516285
78	—	—	—	—	—	—	—	—	—	—	—	—	—	—	534100
79	—	—	—	50.99	—	—	0.07	—	—	—	—	0.001	—	—	470008
80	—	—	—	—	—	—	—	—	—	—	—	—	—	—	—
81	0.05	—	—	0.20	—	—	17.62	—	55	—	—	0.053	169	99.24	359073
82	—	—	—	—	—	—	10.93	—	—	—	—	—	—	—	222460

TABLE V-1G VITAMIN SUPPLEMENTS, COMPOSITION OF FEEDS, DATA EXPRESSED AS-FED

Entry Number	Feed Name Description	International Feed Number	Proximate Analysis									Digestible Protein		
			Dry Matter	Ash	Crude Fiber	Neutral Det. Fib. (NDF)	Acid Det. Fib. (ADF)	Lignin	Ether Extract (Fat)	N-Free Extract	Crude Protein	Ruminant	Swine	Horse
			%	%	%	%	%	%	%	%	%	%	%	%
	ALFALFA (LUCERNE)													
1	IMMATURE, FRESH	2-00-177	21	2.3	4.3	–	–	–	0.7	8.6	5.4	4.5	4.0	–
2	HAY, SUN-CURED	1-00-078	90	8.6	28.2	35.4	30.9	8.9	1.7	35.9	16.0	11.2	7.5	11.9
3	HAY, SUN-CURED, PELLETED	1-00-124	92	10.2	25.4	–	31.9	6.1	1.9	39.1	15.7	11.5	10.2	10.4
4	LEAVES, SUN-CURED, GROUND	1-00-146	88	9.2	15.0	–	–	–	2.7	40.7	20.1	15.8	14.3	13.9
5	MEAL, DEHY, 17% PROTEIN	1-00-023	92	9.7	24.0	41.3	31.5	9.7	2.8	37.8	17.4	12.6	11.7	11.8
6	MEAL, DEHY, 20% PROTEIN	1-00-024	92	10.2	20.8	–	27.0	–	3.3	37.1	20.2	15.1	14.4	14.0
7	MEAL, DEHY, 22% PROTEIN	1-07-851	93	10.2	18.3	–	25.3	–	4.1	38.1	22.2	16.4	16.4	15.5
	ANIMAL													
8	LIVER-GLANDS, GROUND	5-00-390	93	5.9	1.8	–	–	–	15.8	3.3	66.5	58.2	64.8	66.5
9	LIVER, MEAL	5-00-389	93	6.3	1.4	–	–	–	15.7	3.2	66.1	57.9	64.4	66.1
10	MEAT SOLUBLES, DEHY	5-00-393	90	5.7	–	–	–	–	–	–	80.0	70.6	78.3	82.0
11	BLUEGRASS, KENTUCKY, IMMATURE, FRESH	2-00-777	31	2.9	7.8	–	–	–	1.1	13.7	5.4	4.0	3.8	–
12	BREWERS GRAINS, DEHY	5-02-141	92	3.6	13.0	38.7	23.9	4.6	6.6	41.6	27.3	20.1	21.8	21.0
13	BUTTERMILK, CATTLE, CONDENSED	5-01-159	29	3.6	0.1	–	–	–	2.4	12.4	10.8	9.2	10.0	9.5
14	CARROT, ROOTS, FRESH	4-01-145	11	1.0	1.1	–	–	–	0.2	8.1	1.2	0.8	0.8	0.8
	CATTLE													
15	BUTTERMILK, CONDENSED	5-01-159	29	3.6	0.1	–	–	–	2.4	12.4	10.8	9.2	10.0	9.5
16	LIVER, FRESH	5-01-166	28	1.4	0.2	–	–	–	5.1	1.9	19.5	17.0	19.0	19.4
17	WHEY, DEHY	4-01-182	93	8.8	0.2	0.3	0.2	–	0.8	70.2	13.3	8.9	13.1	9.6
	COD, FISH													
18	LIVER, MEAL	5-08-423	93	2.9	0.7	–	–	–	28.9	9.6	50.4	43.6	48.9	48.4
19	LIVER OIL	7-01-993	100	–	–	–	–	–	99.5	–	–	–	–	–
20	CORN, DISTILLERS GRAINS WITH SOLUBLES, DEHY	5-02-843	92	4.5	9.1	–	–	–	9.2	41.9	27.1	17.2	25.7	22.1
21	CRAB, CANNERY RESIDUE, MEAL (CRAB MEAL)	5-01-663	92	41.1	10.7	–	–	–	2.2	5.9	32.2	27.1	30.9	27.9
	DISTILLERS PRODUCTS (ALSO SEE CORN)													
22	GRAINS, DEHY	5-02-144	93	1.5	12.8	–	–	–	7.4	43.5	27.3	18.9	26.0	22.3
23	SOLUBLES, DEHY	5-02-147	92	8.2	3.4	–	–	–	8.9	44.8	28.8	24.0	27.4	24.0
	FATS AND OILS													
24	GERM OIL (WHEAT)	7-05-207	100	–	–	–	–	–	99.5	–	–	–	–	–
25	LIVER OIL (COD)	7-01-993	100	–	–	–	–	–	99.5	–	–	–	–	–
	FISH													
26	MEAL MECH EXTD	5-01-977	92	21.4	0.7	–	–	–	6.0	–	64.3	57.1	59.1	64.1
27	SOLUBLES, CONDENSED	5-01-969	50	10.1	0.5	–	–	–	6.1	2.2	31.5	28.0	29.3	30.9
28	SOLUBLES, DEHY	5-01-971	93	12.7	2.0	–	–	–	9.0	8.7	60.4	52.7	58.8	59.7
	FISH, COD													
29	LIVER, MEAL	5-08-423	93	2.9	0.7	–	–	–	28.9	9.6	50.4	43.6	48.9	48.4
30	LIVER OIL	7-01-993	100	–	–	–	–	–	99.5	–	–	–	–	–
	FISH, SARDINE													
31	MEAL MECH EXTD	5-02-015	93	15.8	1.0	–	–	–	5.0	6.1	65.2	53.5	63.6	65.1
32	SOLUBLES, CONDENSED	5-02-014	50	10.2	–	–	–	–	9.4	0.6	29.5	25.6	28.7	28.7
33	KENTUCKY BLUEGRASS, IMMATURE, FRESH	2-00-777	31	2.9	7.8	–	–	–	1.1	13.7	5.4	4.0	3.8	–
	LIVER													
34	CATTLE, FRESH	5-01-166	28	1.4	0.2	–	–	–	5.1	1.9	19.5	17.0	19.0	19.4
35	SHEEP, FRESH	5-08-116	29	1.4	–	–	–	–	3.9	2.9	21.0	18.4	20.5	21.0
36	SWINE, FRESH	5-04-792	30	1.6	0.1	–	–	–	5.0	2.8	20.8	18.2	20.3	20.7
37	MEAL	5-00-389	93	6.3	1.4	–	–	–	15.7	3.2	66.1	57.9	64.4	66.1
38	LIVER-GLANDS, MEAL	5-00-390	93	5.9	1.8	–	–	–	15.8	3.3	66.5	58.2	64.8	66.5
	LUCERNE–SEE ALFALFA													
39	MAIZE (CORN), DISTILLERS GRAINS WITH SOLUBLES, DEHY	5-02-843	92	4.5	9.1	–	–	–	9.2	41.9	27.1	17.2	25.7	22.1
40	MEAT, SOLUBLES, DEHY	5-00-393	90	5.7	–	–	–	–	–	–	80.0	70.6	78.3	82.0
41	OATS, IMMATURE, FRESH	2-03-286	16	1.7	4.0	–	–	–	0.4	7.4	2.5	1.8	1.7	–
	RICE													
42	BRAN WITH GERMS, MEAL SOLV EXTD (RICE BRAN, SOLV EXTD)	4-03-930	91	14.5	12.9	–	–	–	1.5	48.1	14.0	9.8	9.1	10.2
43	POLISHINGS	4-03-943	90	7.6	3.2	–	3.6	–	12.6	54.9	12.0	8.6	10.1	8.6
44	SHEEP, LIVER, FRESH	5-08-116	29	1.4	–	–	–	–	3.9	2.9	21.0	18.4	20.5	21.0
45	SPINACH, LEAVES, FRESH	2-08-125	9	2.2	0.7	–	–	–	0.4	3.0	3.1	2.5	2.3	–
46	SWINE, LIVER, FRESH	5-04-792	30	1.6	0.1	–	–	–	5.0	2.8	20.8	18.2	20.3	20.7
47	TURNIP, FRESH	2-05-603	13	2.1	1.4	–	–	–	0.3	6.8	2.9	1.1	2.1	–
48	WHALE, LIVER, DEHY	5-05-157	93	–	–	–	–	–	–	–	–	–	–	–
	WHEAT													
49	GERM MEAL	5-05-218	88	4.3	3.1	–	4.4	–	8.5	48.1	24.4	22.9	23.1	19.4
50	GERM OIL	7-05-207	100	–	–	–	–	–	99.5	–	–	–	–	–
51	WHEY, CATTLE, DEHY	4-01-182	93	8.8	0.2	0.3	0.2	–	0.8	70.2	13.3	8.9	13.1	9.6
52	YEAST, BREWERS, DEHY	7-05-527	93	6.5	3.0	–	3.7	–	0.9	38.8	43.8	39.0	–	–
53	YEAST, PRIMARY, DEHY	7-05-533	93	8.0	3.1	–	–	–	1.0	32.5	48.0	–	–	–
54	YEAST, TORULA, DEHY	7-05-534	93	8.0	2.5	–	3.7	–	1.6	31.5	49.6	45.1	40.6	–

Composition of Feeds

VITAMIN SUPPLEMENTS

Entry Number	TDN Ruminant %	TDN Swine %	TDN Horse %	Digestible Energy Ruminant Mcal lb	Digestible Energy Ruminant Mcal kg	Digestible Energy Swine kcal lb	Digestible Energy Swine kcal kg	Digestible Energy Horse Mcal lb	Digestible Energy Horse Mcal kg	Metabolizable Energy Ruminant Mcal lb	Metabolizable Energy Ruminant Mcal kg	Metabolizable Energy Swine kcal lb	Metabolizable Energy Swine kcal kg	Metabolizable Energy Poultry ME$_n$ kcal lb	Metabolizable Energy Poultry ME$_n$ kcal kg	Metabolizable Energy Horse Mcal lb	Metabolizable Energy Horse Mcal kg	Net Energy Ruminant NE$_m$ Mcal lb	Net Energy Ruminant NE$_m$ Mcal kg	Net Energy Ruminant NE$_g$ Mcal lb	Net Energy Ruminant NE$_g$ Mcal kg	Net Energy Lactating Cows NE$_{lc}$ Mcal lb	Net Energy Lactating Cows NE$_{lc}$ Mcal kg
1	15	—	—	0.30	0.66	—	—	—	—	0.26	0.57	—	—	—	—	—	—	0.16	0.35	0.10	0.22	0.15	0.34
2	51	32	48	1.03	2.28	—	—	0.90	1.98	0.90	1.99	—	—	—	—	0.74	1.62	0.48	1.06	0.25	0.55	0.49	1.07
3	54	44	48	1.01	2.22	—	—	0.90	1.98	0.84	1.86	—	—	—	—	0.74	1.62	0.49	1.09	0.26	0.57	0.53	1.16
4	56	63	56	1.13	2.49	—	—	1.03	2.27	0.96	2.12	—	—	—	—	0.85	1.86	0.58	1.28	0.35	0.77	0.58	1.28
5	55	44	47	1.12	2.47	643	1418	0.89	1.96	0.96	2.12	601	1326	682	1504	0.73	1.61	0.60	1.32	0.36	0.79	0.57	1.25
6	57	48	50	1.07	2.36	943	2079	0.94	2.08	0.93	2.06	872	1923	737	1625	0.77	1.70	0.58	1.28	0.34	0.75	0.59	1.30
7	60	49	52	1.20	2.65	991	2186	0.98	2.15	1.02	2.26	841	1855	768	1692	0.80	1.76	0.63	1.39	0.38	0.84	0.63	1.38
8	90	95	—	1.87	4.11	1711	3771	—	—	1.69	3.73	1503	3314	1323	2917	—	—	1.11	2.44	0.79	1.73	1.02	2.25
9	89	93	—	1.79	3.94	1867	4116	—	—	1.62	3.57	1645	3627	1306	2878	—	—	1.00	2.21	0.70	1.55	0.93	2.04
10	—	—	—	—	—	—	—	—	—	—	—	—	—	—	—	—	—	—	—	—	—	—	—
11	22	—	—	0.42	0.93	—	—	—	—	0.37	0.81	—	—	—	—	—	—	0.24	0.52	0.15	0.33	0.23	0.51
12	65	66	48	1.25	2.76	1045	2303	—	—	1.01	2.22	1038	2288	1047	2308	—	—	0.64	1.41	0.39	0.86	0.67	1.48
13	26	22	—	0.52	1.14	442	974	—	—	0.46	1.02	383	844	—	—	—	—	0.29	0.63	0.20	0.43	0.27	0.59
14	10	10	8	0.19	0.43	208	458	0.14	0.31	0.17	0.38	195	430	208	458	0.12	0.26	0.10	0.23	0.07	0.16	0.10	0.22
15	26	22	—	0.52	1.14	442	974	—	—	0.46	1.02	383	844	—	—	—	—	0.29	0.63	0.20	0.43	0.27	0.59
16	29	31	—	0.58	1.28	615	1356	—	—	0.53	1.17	544	1200	—	—	—	—	0.34	0.76	0.25	0.54	0.32	0.70
17	76	77	—	1.51	3.33	1444	3183	—	—	1.28	2.83	1411	3110	880	1939	—	—	0.87	1.92	0.59	1.30	0.78	1.71
18	109	118	—	2.18	4.81	2357	5196	—	—	2.01	4.44	2098	4625	—	—	—	—	1.27	2.81	0.92	2.03	1.17	2.57
19	—	—	—	—	—	—	—	—	—	—	—	—	—	—	—	—	—	—	—	—	—	—	—
20	81	79	—	1.55	3.43	1460	3232	—	—	1.34	2.95	1278	2817	1149	2533	—	—	0.92	2.03	0.63	1.40	0.86	1.89
21	27	—	—	0.54	1.18	686	1511	—	—	0.35	0.78	555	1224	827	1823	—	—	0.07	0.16	—	—	0.25	0.54
22	76	83	—	1.52	3.36	1669	3680	—	—	1.35	2.97	1463	3225	1138	2509	—	—	0.87	1.92	0.59	1.31	0.82	1.80
23	78	82	—	1.61	3.54	1637	3609	—	—	1.43	3.16	1433	3160	1316	2901	—	—	0.93	2.06	0.65	1.43	0.87	1.92
24	—	—	—	—	—	—	—	—	—	—	—	—	—	—	—	—	—	—	—	—	—	—	—
25	—	—	—	—	—	—	—	—	—	—	—	—	—	—	—	—	—	—	—	—	—	—	—
26	67	66	—	1.34	2.95	1317	2903	—	—	1.17	2.57	1138	2508	1174	2587	—	—	0.64	1.40	0.39	0.86	0.63	1.39
27	41	44	—	0.85	1.87	866	1909	—	—	0.78	1.73	736	1623	755	1665	—	—	0.54	1.20	0.38	0.84	0.44	0.97
28	77	66	—	1.50	3.30	1467	3234	—	—	1.21	2.66	1278	2818	1322	2915	—	—	0.81	1.78	0.54	1.19	0.77	1.70
29	109	118	—	2.18	4.81	2357	5196	—	—	2.01	4.44	2098	4625	—	—	—	—	1.27	2.81	0.92	2.03	1.17	2.57
30	—	—	—	—	—	—	—	—	—	—	—	—	—	—	—	—	—	—	—	—	—	—	—
31	70	67	—	1.40	3.09	1327	2925	—	—	1.23	2.71	1148	2531	1313	2896	—	—	0.76	1.67	0.49	1.08	0.73	1.60
32	—	—	—	—	—	—	—	—	—	—	—	—	—	—	—	—	—	—	—	—	—	—	—
33	22	—	—	0.42	0.93	—	—	—	—	0.37	0.81	—	—	—	—	—	—	0.24	0.52	0.15	0.33	0.23	0.51
34	29	31	—	0.58	1.28	615	1356	—	—	0.53	1.17	544	1200	—	—	—	—	0.34	0.76	0.25	0.54	0.32	0.70
35	—	—	—	—	—	—	—	—	—	—	—	—	—	—	—	—	—	—	—	—	—	—	—
36	31	32	—	0.62	1.37	650	1433	—	—	0.56	1.24	574	1266	—	—	—	—	0.37	0.80	0.26	0.57	0.34	0.74
37	89	93	—	1.79	3.94	1867	4116	—	—	1.62	3.57	1645	3627	1306	2878	—	—	1.00	2.21	0.70	1.55	0.93	2.04
38	90	95	—	1.87	4.11	1711	3771	—	—	1.69	3.73	1503	3314	1323	2917	—	—	1.11	2.44	0.79	1.73	1.02	2.25
39	81	79	—	1.55	3.43	1466	3232	—	—	1.34	2.95	1278	2817	1149	2533	—	—	0.92	2.03	0.63	1.40	0.86	1.89
40	—	—	—	—	—	—	—	—	—	—	—	—	—	—	—	—	—	—	—	—	—	—	—
41	10	—	—	0.21	0.46	—	—	—	—	0.18	0.39	—	—	—	—	—	—	0.10	0.23	0.06	0.14	0.10	0.23
42	55	72	—	1.15	2.54	1481	3264	—	—	0.98	2.15	1199	2643	909	2003	—	—	0.62	1.36	0.37	0.82	0.61	1.35
43	81	88	—	1.56	3.45	1684	3713	—	—	1.43	3.16	1555	3428	1367	3015	—	—	0.88	1.93	0.60	1.32	0.83	1.82
44	—	—	—	—	—	—	—	—	—	—	—	—	—	—	—	—	—	—	—	—	—	—	—
45	—	—	—	—	—	—	—	—	—	—	—	—	—	—	—	—	—	—	—	—	—	—	—
46	31	32	—	0.62	1.37	650	1433	—	—	0.56	1.24	574	1266	—	—	—	—	0.37	0.80	0.26	0.57	0.34	0.74
47	10	—	—	0.19	0.41	—	—	—	—	0.16	0.36	—	—	—	—	—	—	0.10	0.21	0.06	0.13	0.10	0.21
48	—	—	—	—	—	—	—	—	—	—	—	—	—	—	—	—	—	—	—	—	—	—	—
49	83	80	—	1.66	3.66	1727	3807	—	—	1.50	3.30	1522	3357	1223	2696	—	—	0.96	2.11	0.67	1.48	0.89	1.95
50	—	—	—	—	—	—	—	—	—	—	—	—	—	—	—	—	—	—	—	—	—	—	—
51	76	77	—	1.51	3.33	1444	3183	—	—	1.28	2.83	1411	3110	880	1939	—	—	0.87	1.92	0.59	1.30	0.78	1.71
52	73	70	—	1.46	3.21	—	—	—	—	1.28	2.83	1299	2865	928	2047	—	—	0.81	1.79	0.54	1.19	0.77	1.69
53	—	—	—	—	—	—	—	—	—	—	—	—	—	—	—	—	—	—	—	—	—	—	—
54	72	64	—	1.49	3.29	1287	2837	—	—	1.27	2.81	1096	2416	840	1851	—	—	0.82	1.81	0.55	1.21	0.78	1.71

TABLE V-1G VITAMIN SUPPLEMENTS, MINERAL AND VITAMIN COMPOSITION OF FEEDS, DATA EXPRESSED AS-FED

			Macro Minerals							Micro Minerals						
Entry Number	Feed Name Description	Dry Matter	Calcium (Ca)	Phosphorus (P)	Sodium (Na)	Chlorine (Cl)	Magnesium (Mg)	Potassium (K)	Sulfur (S)	Cobalt (Co)	Copper (Cu)	Iodine (I)	Iron (Fe)	Manganese (Mn)	Selenium (Se)	Zinc (Zn)
		%	%	%	%	%	%	%	%	ppm or mg/kg	ppm or mg/kg	ppm or mg/kg	%	ppm or mg/kg	ppm or mg/kg	ppm or mg/kg
	ALFALFA (LUCERNE)															
1	IMMATURE, FRESH	21	0.50	0.09	0.04	0.08	0.05	0.48	0.13	—	—	—	0.006	6.7	—	—
2	HAY, SUN-CURED	90	1.28	0.24	0.07	0.33	0.30	1.85	0.25	0.250	10.0	—	0.019	41.4	—	21.9
3	HAY, SUN-CURED, PELLETED	92	1.48	0.20	—	—	0.25	—	0.24	—	0.6	—	—	—	—	—
4	LEAVES, SUN-CURED, GROUND	88	2.32	0.24	—	—	0.36	1.47	—	—	—	—	0.031	29.8	—	—
5	MEAL, DEHY, 17% PROTEIN	92	1.40	0.23	0.10	0.47	0.29	2.38	0.23	0.302	8.6	0.148	0.041	31.0	0.335	19.3
6	MEAL, DEHY, 20% PROTEIN	92	1.59	0.28	0.11	0.47	0.33	2.41	0.50	0.259	12.2	0.135	0.036	45.2	0.285	21.8
7	MEAL, DEHY, 22% PROTEIN	93	1.69	0.30	0.12	0.52	0.31	2.40	0.30	0.311	9.8	0.166	0.036	36.4	0.534	19.5
	ANIMAL															
8	LIVER-GLANDS, GROUND	93	0.63	1.18	0.10	—	—	0.40	—	0.170	94.2	—	0.056	8.1	—	50.1
9	LIVER, MEAL	93	0.56	1.26	—	—	0.10	—	—	0.135	89.4	—	0.064	8.8	—	61.8
10	MEAT SOLUBLES, DEHY	90	0.45	0.67	—	—	—	—	—	—	—	—	—	—	—	—
11	**BLUEGRASS, KENTUCKY,** IMMATURE, FRESH	31	0.15	0.14	0.04	—	0.05	0.70	0.05	—	—	—	0.010	—	—	—
12	**BREWERS' GRAINS,** DEHY	92	0.30	0.51	0.21	0.15	0.15	0.09	0.30	0.076	21.7	0.066	0.024	37.2	—	27.3
13	**BUTTERMILK, CATTLE,** CONDENSED	29	0.44	0.26	0.31	0.12	0.19	0.23	0.03	—	—	—	—	—	—	—
14	**CARROT,** ROOTS, FRESH	11	0.05	0.04	0.06	0.06	0.02	0.32	0.02	—	1.2	—	0.002	3.6	—	—
	CATTLE															
15	BUTTERMILK, CONDENSED	29	0.44	0.26	0.31	0.12	0.19	0.23	0.03	—	—	—	—	—	—	—
16	LIVER, FRESH	28	0.01	0.23	0.10	—	0.01	0.20	—	—	6.1	—	0.005	2.8	—	26.6
17	WHEY, DEHY	93	0.86	0.76	0.62	0.07	0.13	1.11	1.04	0.111	46.5	—	0.017	5.9	—	3.2
	COD, FISH															
18	LIVER, MEAL	93	0.16	0.69	—	—	—	—	—	—	—	—	—	—	—	—
19	LIVER OIL	100	—	—	—	—	—	—	—	—	—	—	—	—	—	—
20	**CORN,** DISTILLERS GRAINS WITH SOLUBLES, DEHY	92	0.16	0.69	0.47	0.17	0.18	0.47	0.31	0.152	52.6	0.051	0.024	24.0	0.331	80.7
21	**CRAB,** CANNERY RESIDUE, MEAL (CRAB MEAL)	92	14.46	1.58	0.88	1.51	0.94	0.45	0.25	—	32.7	0.557	0.435	132.8	—	—
	DISTILLERS PRODUCTS (ALSO SEE CORN)															
22	GRAINS, DEHY	93	0.12	0.54	0.05	0.05	0.09	0.20	0.46	0.092	47.9	—	0.027	35.0	—	—
23	SOLUBLES, DEHY	92	0.24	1.35	0.45	—	0.53	1.97	—	0.196	71.6	—	0.031	64.1	—	138.0
	FATS AND OILS															
24	GERM OIL (WHEAT)	100	—	—	—	—	—	—	—	—	—	—	—	—	—	—
25	LIVER OIL (COD)	100	—	—	—	—	—	—	—	—	—	—	—	—	—	—
	FISH															
26	MEAL MECH EXTD	92	6.63	3.61	1.11	1.25	0.21	0.40	0.25	0.110	15.1	—	0.038	23.6	—	99.1
27	SOLUBLES, CONDENSED	50	0.16	0.57	2.45	2.93	0.03	1.64	0.12	0.069	46.6	1.111	0.028	13.2	—	43.2
28	SOLUBLES, DEHY	93	0.40	1.27	1.70	—	0.30	2.50	0.45	—	20.0	—	0.095	50.4	2.692	76.7
	FISH, COD															
29	LIVER, MEAL	93	0.16	0.69	—	—	—	—	—	—	—	—	—	—	—	—
30	LIVER OIL	100	—	—	—	—	—	—	—	—	—	—	—	—	—	—
	FISH, SARDINE															
31	MEAL MECH EXTD	93	4.61	2.68	0.18	0.41	0.10	0.32	—	0.183	20.2	—	0.030	23.2	1.772	—
32	SOLUBLES, CONDENSED	50	0.14	0.83	0.18	0.28	—	0.18	0.11	—	25.8	4.934	0.002	24.9	—	—
33	**KENTUCKY BLUEGRASS,** IMMATURE, FRESH	31	0.15	0.14	0.04	—	0.05	0.70	0.05	—	—	—	0.010	—	—	—
	LIVER															
34	CATTLE, FRESH	28	0.01	0.23	0.10	—	0.01	0.20	—	—	6.1	—	0.005	2.8	—	26.6
35	SHEEP, FRESH	29	0.01	0.35	0.06	—	0.02	0.20	—	—	37.5	—	0.007	3.6	—	36.0
36	SWINE, FRESH	30	0.01	0.37	0.07	—	0.01	0.26	—	0.255	56.4	0.340	0.015	1.8	0.340	44.2
37	MEAL	93	0.56	1.26	—	—	0.10	—	—	0.135	89.4	—	0.064	8.8	—	61.8
38	LIVER-GLANDS, MEAL	93	0.63	1.18	0.10	—	—	0.40	—	0.170	94.2	—	0.056	8.1	—	50.1
	LUCERNE—SEE ALFALFA															
39	**MAIZE (CORN),** DISTILLERS GRAINS WITH SOLUBLES, DEHY	92	0.16	0.69	0.47	0.17	0.18	0.47	0.31	0.152	52.6	0.051	0.024	24.0	0.331	80.7
40	**MEAT,** SOLUBLES, DEHY	90	0.45	0.67	—	—	—	—	—	—	—	—	—	—	—	—
41	**OATS,** IMMATURE, FRESH	16	—	—	0.02	0.02	—	—	0.01	—	—	—	—	—	—	—
	RICE															
42	BRAN WITH GERMS, MEAL SOLV EXTD (RICE BRAN, SOLV EXTD)	91	0.11	1.37	—	—	1.48	—	0.18	0.111	13.0	0.045	0.019	232.2	—	30.0
43	POLISHINGS	90	0.05	1.34	0.04	0.11	0.60	1.28	0.17	3.890	8.0	—	0.009	126.8	—	63.2
44	**SHEEP,** LIVER, FRESH	29	0.01	0.35	0.06	—	0.02	0.20	—	—	37.5	—	0.007	3.6	—	36.0
45	**SPINACH,** LEAVES, FRESH	9	0.09	0.05	0.07	—	—	0.48	—	—	—	—	0.004	—	—	—
46	**SWINE,** LIVER, FRESH	30	0.01	0.37	0.07	—	0.01	0.26	—	0.255	56.4	0.340	0.015	1.8	0.340	44.2
47	**TURNIP,** FRESH	13	0.40	0.05	—	0.26	0.07	0.41	0.04	—	2.4	—	0.006	55.2	—	5.0
48	**WHALE,** LIVER, DEHY	93	—	—	—	1.99	—	—	—	—	—	—	—	—	—	—
	WHEAT															
49	GERM MEAL	88	0.06	0.95	0.02	0.06	0.25	0.94	0.27	0.120	9.2	—	0.006	132.5	0.463	119.4
50	GERM OIL	100	—	—	—	—	—	—	—	—	—	—	—	—	—	—
51	**WHEY, CATTLE,** DEHY	93	0.86	0.76	0.62	0.07	0.13	1.11	1.04	0.111	46.5	—	0.017	5.9	—	3.2
52	**YEAST, BREWERS,** DEHY	93	0.14	1.36	0.07	0.07	0.24	1.69	0.43	0.506	38.4	0.358	0.009	6.7	0.911	39.0
53	**YEAST, PRIMARY,** DEHY	93	0.36	1.72	—	0.02	0.36	—	0.57	—	—	—	0.030	3.7	—	—
54	**YEAST, TORULA,** DEHY	93	0.55	1.61	0.01	0.02	0.14	1.92	0.55	0.031	11.9	2.502	0.011	9.3	—	99.5

Composition of Feeds

VITAMIN SUPPLEMENTS

Entry Number	Fat-Soluble Vitamins					Water-Soluble Vitamins								
	A (1 mg Carotene = 1667 IU Vit A)	Carotene (Provitamin A)	D	E	K	B-12	Biotin	Choline	Folacin (Folic Acid)	Niacin	Pantothenic Acid (B-3)	(Pyridoxine) B-6	Riboflavin (B-2)	Thiamin (B-1)
	IU/g	ppm or mg/kg	IU/kg	ppm or mg/kg	ppm or mg/kg	ppb or mcg/kg	ppm or mg/kg	ppm or mg/kg	ppm or mg/kg	ppm or mg/kg	ppm or mg/kg	ppm or mg/kg	ppm or mg/kg	ppm or mg/kg
1	85.9	51.5	–	–	–	–	–	–	–	–	8.9	–	–	–
2	45.0	27.0	2	55.9	–	–	0.18	892	3.07	43	18.1	–	9.5	3.1
3	48.2	28.9	–	–	–	–	–	–	–	–	–	–	–	–
4	97.7	58.6	–	–	–	–	–	–	–	–	–	–	20.9	–
5	200.3	120.2	–	105.7	8.24	–	0.33	1369	4.37	37	29.7	7.18	12.9	3.4
6	265.4	159.2	–	143.3	14.19	–	0.35	1417	2.96	48	35.5	8.72	15.2	5.4
7	391.4	234.8	–	221.3	11.65	–	0.33	1605	5.15	50	39.0	8.28	17.6	5.9
8	–	–	–	–	–	440.5	0.41	10610	4.00	172	90.8	5.01	42.4	0.2
9	–	–	–	–	–	501.3	0.02	11370	5.56	205	29.2	–	36.2	0.2
10	–	–	–	–	–	881.6	–	–	–	–	–	–	–	–
11	247.6	148.5	–	47.8	–	–	–	–	–	–	–	–	–	–
12	0.8	0.5	–	26.7	–	3.6	0.44	1651	0.22	44	8.2	1.03	1.5	0.6
13	25.6	15.4	–	–	–	–	–	–	–	–	–	–	12.6	–
14	129.9	77.9	–	6.9	–	–	0.01	–	0.14	7	3.5	1.39	0.6	0.7
15	25.6	15.4	–	–	–	–	–	–	–	–	–	–	12.6	–
16	–	–	–	7.1	–	425.8	0.98	1424	2.33	75	46.1	5.03	25.8	1.8
17	–	–	–	0.2	–	18.9	0.35	1790	0.85	11	46.2	3.21	27.4	4.0
18	–	–	–	–	–	–	–	–	–	132	46.1	32.85	33.3	18.1
19	845.8	–	–	39.5	–	–	–	–	–	–	–	–	–	–
20	6.2	3.7	1	39.0	–	1.5	0.69	2582	0.91	73	13.8	4.74	8.5	3.0
21	–	–	–	–	–	437.6	0.07	2008	0.11	45	6.6	6.62	6.1	0.4
22	13.0	7.8	–	30.5	–	–	–	2645	–	47	11.9	6.00	6.6	2.5
23	1.9	1.1	–	–	–	2.9	2.84	4992	–	143	25.3	8.66	11.3	6.9
24	–	–	–	–	18.66	–	–	–	–	–	–	–	–	–
25	845.8	–	–	39.5	–	–	–	–	–	–	–	–	–	–
26	–	–	–	19.2	–	258.6	–	3644	–	75	15.0	14.68	5.6	0.8
27	2.2	1.3	–	–	–	506.6	0.14	3370	0.22	176	35.7	12.20	12.9	5.5
28	–	–	–	6.1	–	485.9	0.40	5525	0.57	256	50.4	19.71	13.5	7.4
29	–	–	–	–	–	–	–	–	–	132	46.1	32.85	33.3	18.1
30	845.8	–	–	39.5	–	–	–	–	–	–	–	–	–	–
31	–	–	–	–	–	238.0	0.10	3277	–	75	11.0	–	5.4	0.3
32	–	–	–	–	–	1041.0	0.13	3009	–	356	41.2	–	16.8	4.0
33	247.6	148.5	–	47.8	–	–	–	–	–	–	–	–	–	–
34	–	–	–	7.1	–	425.8	0.98	1424	2.33	75	46.1	5.03	25.8	1.8
35	–	–	–	–	–	–	–	–	–	169	–	–	32.8	4.0
36	–	–	–	–	–	282.7	0.75	–	2.07	165	23.6	3.02	27.3	2.3
37	–	–	–	–	–	501.3	0.02	11370	5.56	205	29.2	–	36.2	0.2
38	–	–	–	–	–	440.5	0.41	10610	4.00	172	90.8	5.01	42.4	0.2
39	6.2	3.7	1	39.8	–	1.5	0.69	2582	0.91	73	13.8	4.74	8.5	3.0
40	–	–	–	–	–	881.6	–	–	–	–	–	–	–	–
41	150.0	90.0	–	–	–	–	–	–	–	–	–	–	–	–
42	–	–	–	60.7	–	–	0.42	1128	2.21	284	23.0	29.11	2.9	22.6
43	–	–	–	90.2	–	–	0.62	1248	–	506	46.4	27.89	1.8	20.0
44	–	–	–	–	–	–	–	–	–	169	–	–	32.8	4.0
45	56.4	33.8	–	38.5	–	–	–	–	–	6	–	–	2.0	1.0
46	–	–	–	–	–	282.7	0.75	–	2.07	165	23.6	3.02	27.3	2.3
47	–	–	–	–	–	–	–	–	–	11	–	–	5.4	2.9
48	–	–	–	–	–	499.0	–	3351	–	200	36.4	9.04	79.1	2.6
49	–	–	–	141.2	–	–	0.22	3062	2.12	68	18.6	9.97	6.0	23.1
50	–	–	–	–	18.66	–	–	–	–	–	–	–	–	–
51	–	–	–	0.2	–	18.9	0.35	1790	0.85	11	46.2	3.21	27.4	4.0
52	–	–	–	2.1	–	1.1	1.04	3847	9.69	443	81.5	36.67	34.1	85.2
53	–	–	–	–	–	6.2	1.61	–	31.13	301	312.0	–	38.8	6.4
54	–	–	–	–	–	4.0	1.19	2981	25.66	512	107.5	34.48	47.7	6.8

TABLE V-2 AMINO ACID, COMPOSITION OF FEEDS, DATA EXPRESSED **AS-FED**

Entry Number	Feed Name Description	International Feed Number	Dry Matter %	Crude Protein %	Arginine %	Cystine %	Glycine %	Histidine %	Iso-leucine %	Leucine %	Lysine %	Methi-onine %	Phenyl-alanine %	Serine %	Threo-nine %	Trypto-phan %	Tyrosine %	Valine %
	ENERGY FEEDS																	
1	BAKERY, WASTE, DEHY (DRIED BAKERY PRODUCT)	4-00-466	91	10.1	0.47	0.17	0.69	0.16	0.45	0.77	0.31	0.17	0.45	0.65	0.46	0.10	0.36	0.47
	BARLEY																	
2	GRAIN	4-00-549	88	11.7	0.51	0.20	0.37	0.25	0.46	0.75	0.40	0.16	0.58	0.46	0.36	0.15	0.35	0.57
3	GRAIN, PACIFIC COAST	4-07-939	89	9.5	0.44	0.19	0.30	0.21	0.40	0.60	0.26	0.14	0.47	0.32	0.31	0.12	0.31	0.46
4	GRAIN SCREENINGS	4-00-542	89	11.5	—	—	—	—	—	—	—	—	—	—	0.36	—	—	—
5	MALT SPROUTS, DEHY	5-00-545	93	22.9	1.05	0.23	0.81	0.43	0.88	1.36	1.12	0.31	0.80	0.47	0.85	0.41	0.46	1.16
6	**BEAN, NAVY,** SEEDS	5-00-623	89	22.9	1.19	0.23	0.80	—	—	—	1.29	0.23	—	—	—	—	0.24	—
7	**BEET, SUGAR,** PULP, DEHY	4-00-669	91	8.8	0.30	0.01	—	0.20	0.30	0.60	0.60	0.01	0.30	—	0.40	0.10	0.40	0.40
8	**BROOMCORN (HOG MILLET; MILLET, PROSO),** GRAIN	4-03-120	90	11.6	0.34	0.20	0.25	0.20	0.44	1.13	0.22	0.26	0.54	0.63	0.37	0.16	0.23	0.55
9	**BUCKWHEAT, COMMON,** GRAIN	4-00-994	88	11.1	0.96	0.17	0.61	0.26	0.37	0.59	0.62	0.19	0.44	0.41	0.44	0.18	0.21	0.53
10	**CITRUS,** PULP WITHOUT FINES, DEHY (DRIED CITRUS PULP)	4-01-237	91	6.1	0.25	0.11	—	0.09	0.18	0.31	0.20	0.09	0.18	—	0.18	0.06	—	0.25
11	**CORN, DENT WHITE,** GRAIN	4-02-928	90	10.8	0.27	0.09	—	0.18	0.45	0.90	0.27	0.09	0.36	—	0.36	0.09	0.45	0.36
	CORN, DENT YELLOW																	
12	GRAIN, ALL ANALYSES	4-02-935	88	9.9	0.43	0.12	0.37	0.27	0.35	1.19	0.30	0.18	0.46	0.49	0.36	0.09	0.31	0.48
13	GRAIN, GRADE 2, 54 lb/bu (69.5 kg/hl)	4-02-831	87	8.9	0.45	0.11	0.45	0.20	0.40	1.00	0.19	0.11	0.45	—	0.35	0.09	0.43	0.35
14	GRAIN, FLAKED	4-28-244	89	9.9	0.44	0.25	0.36	0.28	0.34	1.24	0.25	0.15	0.44	0.48	0.35	—	0.39	0.47
15	DISTILLERS SOLUBLES, DEHY	5-28-237	93	27.4	0.99	0.44	1.12	0.67	1.32	2.38	0.92	0.55	1.47	1.22	1.01	0.25	0.88	1.53
16	EARS, GROUND (CORN-AND-COB MEAL)	4-28-238	87	7.8	0.36	0.12	0.31	0.16	0.36	0.86	0.17	0.14	0.39	—	0.28	0.07	0.32	0.31
17	GERM MEAL, WET MILLED, SOLV EXTD	5-28-240	92	20.7	1.31	0.40	0.31	0.70	0.70	1.81	0.90	0.58	0.90	1.00	1.09	0.20	0.70	1.20
18	GRITS (HOMINY GRITS)	4-03-011	90	10.3	0.47	0.15	0.35	0.20	0.39	0.85	0.38	0.16	0.33	—	0.39	0.11	0.50	0.49
19	**CORN, OPAQUE 2 (HIGH LYSINE),** GRAIN	4-11-445	90	10.1	0.64	0.19	0.48	0.35	0.33	0.98	0.42	0.16	0.43	0.46	0.37	0.12	0.40	0.48
20	**COWPEA, COMMON,** SEEDS	5-01-661	89	23.8	1.70	—	—	0.70	1.10	2.31	2.10	0.20	1.30	—	0.80	0.30	1.10	1.20
21	**DISTILLERS PRODUCTS** (ALSO SEE CORN; WHEAT, SOLUBLES, DEHY)	5-02-147	92	28.8	1.06	0.40	1.20	0.66	1.21	2.35	0.95	0.50	1.24	0.93	1.00	0.24	0.93	1.40
22	**EMMER,** GRAIN	4-01-830	91	11.7	0.46	—	—	0.20	0.42	0.67	0.29	0.16	0.48	—	0.38	0.12	—	0.47
23	**GOOSEFOOT, LAMB'S QUARTER,** SEEDS	5-08-424	90	18.8	0.08	—	0.05	0.02	0.03	0.05	0.04	0.02	0.03	0.03	0.03	0.07	0.02	0.03
24	**GRAPE,** POMACE, DEHY (MARC)	1-02-208	90	12.1	0.67	0.17	0.89	0.26	0.55	1.63	0.50	0.18	0.55	—	0.38	—	0.16	1.09
25	**HOG MILLET (BROOMCORN; MILLET, PROSO),** GRAIN	4-03-120	90	11.6	0.34	0.20	0.25	0.20	0.44	1.13	0.22	0.26	0.54	0.63	0.37	0.16	0.23	0.55
26	**HOMINY GRITS (CORN, DENT YELLOW, GRITS)**	4-03-011	90	10.3	0.47	0.15	0.35	0.20	0.39	0.85	0.38	0.16	0.33	—	0.39	0.11	0.50	0.49
	MAIZE—SEE CORN																	
27	**MILLET,** GRAIN,	4-03-098	90	12.1	0.35	0.12	0.40	0.23	0.49	1.23	0.26	0.30	0.59	—	0.44	0.12	—	0.62
28	**MILLET, PROSO (BROOMCORN; HOG MILLET),** GRAIN	4-03-120	90	11.6	0.34	0.20	0.25	0.20	0.44	1.13	0.22	0.26	0.54	0.63	0.37	0.16	0.23	0.55
	OATS																	
29	GRAIN	4-03-309	89	11.9	0.71	0.19	0.51	0.17	0.48	0.87	0.40	0.18	0.57	0.50	0.38	0.15	0.45	0.62
30	GRAIN, GRADE 1, 34 lb/bu (43.8 kg/hl)	4-03-313	88	11.2	0.79	0.22	0.49	0.19	0.52	0.89	0.49	0.18	0.59	—	0.39	0.16	0.52	0.69
31	GRAIN, PACIFIC COAST	4-07-999	91	9.1	0.58	0.17	0.40	0.17	0.38	0.70	0.33	0.13	0.43	0.40	0.30	0.12	0.70	0.49
32	MIDDLINGS, LESS THAN 4% FIBER (FEEDING OAT MEAL)	4-03-303	91	14.8	0.81	0.25	0.62	0.30	0.56	1.05	0.53	0.21	0.69	0.70	0.48	0.20	0.72	0.74
33	GROATS	4-03-331	90	15.8	0.89	0.21	0.61	0.27	0.54	1.04	0.54	0.20	0.70	0.62	0.44	0.19	0.51	0.74
34	**PEA, GARDEN,** SEEDS	5-08-482	89	23.8	1.43	—	—	0.63	1.03	1.61	1.47	0.34	1.20	—	0.80	0.23	—	1.18
35	**PUMPKIN,** SEEDS	5-03-817	94	38.3	4.93	0.93	1.96	0.75	1.29	2.29	1.45	0.77	1.67	2.03	0.95	0.47	1.38	1.57
	RICE																	
36	GRAIN, GROUND (GROUND ROUGH RICE; GROUND PADDY RICE)	4-03-938	89	7.5	0.54	0.12	0.62	0.16	0.27	0.54	0.25	0.14	0.30	0.50	0.23	0.10	0.63	0.40
37	BRAN WITH GERMS (RICE BRAN)	4-03-928	91	13.0	0.82	0.16	0.81	0.29	0.50	0.84	0.54	0.26	0.53	0.73	0.44	0.10	0.59	0.75
38	GROATS, POLISHED (RICE, POLISHED)	4-03-942	90	7.0	0.48	0.09	0.42	0.18	0.28	0.47	0.24	0.17	0.31	0.29	0.25	0.09	0.23	0.40
39	POLISHINGS	4-03-943	90	12.0	0.57	0.14	0.65	0.19	0.37	0.73	0.51	0.22	0.43	0.49	0.35	0.11	0.45	0.68
40	**RYE,** GRAIN	4-04-047	87	12.0	0.52	0.19	0.44	0.25	0.48	0.68	0.42	0.17	0.56	0.61	0.35	0.12	0.26	0.58
41	**SAFFLOWER,** SEEDS	4-07-958	93	14.9	1.60	0.35	1.00	0.48	0.80	1.20	0.60	0.33	1.00	—	0.64	0.28	—	1.00
	SCREENINGS, GRAIN (CEREAL) (ALSO SEE WHEAT)																	
42	REFUSE	4-02-151	91	12.6	0.68	—	0.59	0.30	0.52	0.98	0.48	0.15	0.64	0.57	0.46	—	0.32	0.63
43	UNCLEANED	4-02-153	92	13.7	0.67	—	0.61	0.30	0.45	0.90	0.42	0.19	0.58	0.67	0.44	—	0.58	0.58
	SORGHUM																	
44	GRAIN, ALL ANALYSES	4-04-383	90	11.5	0.39	0.21	0.34	0.24	0.42	1.47	0.26	0.14	0.56	0.49	0.36	0.09	0.40	0.50
45	GRAIN, LESS THAN 9% PROTEIN	4-08-138	89	8.9	0.28	0.14	0.27	0.19	0.46	1.40	0.19	0.12	0.47	—	0.36	0.12	0.60	0.53
46	DARSO, GRAIN	4-04-357	90	10.1	0.36	—	—	0.18	0.45	1.23	0.19	0.11	0.48	—	0.31	0.11	—	0.48
47	FETERITA, GRAIN	4-04-369	89	11.7	0.46	—	—	0.26	0.58	1.78	0.20	0.18	0.67	—	0.46	0.17	—	0.67

Composition of Feeds

TABLE V-2 AMINO ACID, COMPOSITION OF FEEDS, DATA EXPRESSED **AS-FED**—(Continued)

Entry Number	Feed Name Description	International Feed Number	Dry Matter %	Crude Protein %	Arginine %	Cystine %	Glycine %	Histidine %	Iso-leucine %	Leucine %	Lysine %	Methi-onine %	Phenyl-alanine %	Serine %	Threo-nine %	Trypto-phan %	Tyrosine %	Valine %
	SORGHUM (Continued)																	
48	HEGARI, GRAIN	4-04-398	89	10.4	0.29	—	—	0.18	0.47	1.48	0.17	0.12	0.54	—	0.36	0.11	—	0.55
49	KAFIR, GRAIN	4-04-428	89	10.8	0.38	0.17	0.30	0.27	0.55	1.62	0.26	0.19	0.64	—	0.45	0.15	—	0.62
50	MILO, GRAIN	4-04-444	89	10.1	0.37	0.13	0.35	0.24	0.44	1.32	0.23	0.16	0.49	0.49	0.35	0.10	0.37	0.53
51	MILO, GLUTEN WITH BRAN, MEAL	5-08-089	89	23.2	0.90	0.20	0.68	0.60	1.00	2.51	0.70	0.40	1.00	—	0.80	0.20	0.90	1.30
52	SHALLU, GRAIN	4-04-456	90	11.5	0.31	—	—	0.19	0.38	0.97	0.19	0.17	0.40	—	0.30	0.10	—	0.46
	SOYBEAN																	
53	SEEDS	5-04-610	92	38.4	2.63	0.42	1.42	0.92	1.62	2.72	2.32	0.48	1.76	1.59	1.46	0.56	1.29	1.61
54	SOYBEAN MILL FEED	4-04-594	90	12.6	0.70	0.13	0.47	0.18	0.41	0.58	0.59	0.12	0.37	—	0.30	0.13	0.23	0.37
55	**SPELT**, GRAIN	4-04-651	90	12.0	0.45	—	—	0.18	0.36	0.63	0.27	0.18	0.45	—	0.36	0.09	—	0.45
56	**TRITICALE**, GRAIN	4-20-362	89	15.4	0.85	0.27	0.68	0.38	0.58	1.11	0.52	0.22	0.77	0.73	0.53	0.18	0.49	0.78
	WHEAT																	
57	GRAIN, ALL ANALYSES	4-05-211	89	13.1	0.61	0.22	0.59	0.30	0.49	0.90	0.39	0.18	0.61	0.63	0.40	0.15	0.37	0.61
58	GRAIN, HARD RED SPRING	4-05-258	88	14.2	0.64	0.22	0.60	0.27	0.52	0.91	0.38	0.20	0.66	0.61	0.39	0.15	0.45	0.61
59	GRAIN, HARD RED WINTER	4-05-268	88	12.8	0.65	0.30	0.58	0.30	0.53	0.91	0.36	0.22	0.63	0.59	0.37	0.17	0.46	0.58
60	GRAIN, SOFT RED WINTER	4-05-294	88	11.4	0.65	0.36	0.55	0.32	0.45	0.60	0.36	0.22	0.64	0.65	0.39	0.27	0.37	0.58
61	GRAIN, SOFT WHITE WINTER	4-05-337	90	10.2	0.47	0.27	0.50	0.22	0.42	0.66	0.32	0.16	0.46	0.46	0.32	0.13	0.37	0.45
62	GRAIN, SOFT WHITE WINTER, PACIFIC COAST	4-08-555	89	10.0	0.45	0.24	0.50	0.20	0.40	0.75	0.30	0.14	0.48	0.49	0.31	0.12	0.36	0.46
63	BRAN	4-05-190	89	15.5	0.85	0.26	0.77	0.33	0.55	0.89	0.54	0.17	0.50	0.38	0.40	0.25	0.38	0.67
64	DISTILLERS GRAINS, DEHY	5-05-193	93	31.6	1.10	0.30	—	0.80	2.01	1.71	0.70	—	1.71	—	0.90	—	0.50	1.71
65	ENDOSPERM	4-05-197	88	11.1	0.60	0.30	—	0.30	1.10	1.70	0.40	0.20	0.60	—	0.40	0.30	—	0.60
66	FLOUR, LESS THAN 1.5% FIBER	4-05-199	88	13.7	0.42	0.25	0.46	0.28	0.56	0.89	0.27	0.13	0.62	0.51	0.30	0.10	0.25	0.51
67	GRAIN SCREENINGS	4-05-216	88	13.3	0.68	0.12	0.53	0.30	0.45	0.74	0.43	0.26	0.49	0.40	0.33	0.13	0.23	0.55
68	MIDDLINGS, LESS THAN 9.5% FIBER	4-05-205	89	16.4	0.98	0.22	0.96	0.40	0.68	1.11	0.68	0.19	0.66	0.80	0.57	0.19	0.43	0.80
69	MILL RUN, LESS THAN 9.5% FIBER	4-05-206	90	15.1	0.94	0.23	0.53	0.40	0.70	1.20	0.57	0.33	—	—	0.50	0.21	0.50	0.80
70	RED DOG, LESS THAN 4% FIBER	4-05-203	88	15.6	0.96	0.36	0.74	0.38	0.58	1.38	0.60	0.22	0.65	0.76	0.50	0.20	0.46	0.73
71	SHORTS, LESS THAN 7% FIBER	4-05-201	88	16.5	1.20	0.38	0.96	0.44	0.57	1.07	0.80	0.28	0.67	0.77	0.60	0.23	0.47	0.82
72	**WHEAT, DURUM**, GRAIN	4-05-224	88	13.8	0.58	0.13	0.46	0.27	0.48	1.40	1.05	0.14	0.53	0.45	0.37	0.26	0.29	0.54
	PROTEIN FEEDS																	
73	**ACACIA, SWEET**, SEEDS	5-09-110	87	47.9	4.40	—	1.63	1.10	1.67	3.58	2.25	0.44	1.67	1.97	1.20	—	1.34	1.86
	ANIMAL																	
74	BLOOD, MEAL	5-00-380	91	80.5	3.23	1.25	3.45	3.93	0.85	10.07	6.43	0.94	5.56	3.95	3.59	1.01	1.94	6.56
75	BLOOD, SPRAY DEHY	5-00-381	93	86.0	3.59	1.03	3.83	5.18	0.91	10.97	7.44	1.05	5.89	3.53	3.63	1.05	2.26	7.52
76	LIVER, MEAL	5-00-389	93	66.1	4.04	0.94	5.61	1.48	3.11	5.31	5.22	1.22	2.92	2.50	2.50	0.69	1.70	4.15
77	MEAT, MEAL RENDERED	5-00-385	94	50.7	3.58	0.60	7.23	0.87	1.63	3.11	3.00	0.69	1.74	2.31	1.67	0.35	1.09	2.42
78	MEAT WITH BONE, MEAL RENDERED	5-00-385	93	50.4	3.53	0.53	6.49	1.01	1.64	3.10	2.93	0.67	1.71	1.90	1.66	0.31	0.89	2.44
79	TANKAGE, MEAL RENDERED	5-00-386	92	60.5	3.60	0.48	6.45	2.06	1.82	5.10	3.89	0.75	2.56	2.81	2.34	0.65	1.38	3.83
80	TANKAGE WITH BONE, MEAL RENDERED	5-00-387	93	46.6	2.82	0.27	6.58	1.76	1.87	5.26	3.32	0.69	2.28	—	2.18	0.62	—	3.42
	BABASSU																	
81	KERNELS WITH COATS, MEAL MECH EXTD (BABASSU OIL MEAL)	5-00-454	92	22.3	2.87	—	—	0.40	1.04	1.34	0.89	0.30	0.89	—	0.60	0.20	0.40	1.09
82	KERNELS WITH COATS, MEAL SOLV EXTD (BABASSU OIL MEAL)	5-00-455	93	21.2	3.19	—	—	0.41	0.88	1.40	0.98	0.53	1.35	—	0.71	0.24	—	1.19
83	**BEAN, PINTO**, SEEDS	5-00-624	90	22.7	1.55	—	—	0.64	1.14	1.11	1.60	0.26	1.20	—	1.09	0.32	—	1.23
84	BLOOD, MEAL	5-00-380	91	80.5	3.23	1.25	3.45	3.93	0.85	13.07	6.43	0.94	5.56	3.95	3.59	1.01	1.94	6.56
85	**BUTTERMILK (CATTLE)**, DEHY	5-01-160	92	31.7	1.08	0.39	0.47	0.85	2.42	5.21	6.43	0.71	1.46	1.50	1.52	0.49	1.00	2.58
86	**CASEIN**, ACID PRECIPITATED, DEHY	5-01-162	91	84.0	3.49	0.31	1.61	2.59	5.72	8.80	7.14	2.81	4.81	5.46	3.91	1.08	4.90	6.71
87	**CASTOR BEAN**, SEEDS WITHOUT TOXIN, MEAL	5-01-155	87	26.0	2.77	—	0.86	0.48	1.01	1.40	0.76	0.36	0.82	1.13	0.72	—	0.68	1.27
	CATTLE																	
88	BUTTERMILK, DEHY	5-01-160	92	31.7	1.08	0.39	0.47	0.85	2.42	3.21	2.28	0.71	1.46	1.50	1.52	0.49	1.00	2.58
89	MILK, FRESH	5-01-168	12	3.3	—	—	—	—	0.32	0.25	0.28	0.18	0.07	—	0.16	0.05	—	0.25
90	MILK, DEHY	5-01-167	95	25.3	0.92	—	—	0.71	1.32	2.54	2.24	0.61	1.32	—	1.02	0.41	1.32	1.73
91	SKIM MILK, DEHY	5-01-175	94	33.3	1.16	0.45	0.29	0.86	2.18	3.33	2.54	0.90	1.57	1.67	1.57	0.43	1.14	2.29
92	WHEY, DEHY	5-01-182	93	13.3	0.33	0.30	0.44	0.17	0.78	1.18	0.94	0.19	0.35	0.47	0.90	0.20	0.25	0.67
93	WHEY, LOW LACTOSE, DEHY	4-01-186	93	16.7	0.60	0.43	0.72	0.27	0.96	1.54	1.40	0.41	0.55	0.59	0.95	0.27	0.46	0.87
94	**CHICKPEA (GARBANZO; GRAM PEA)**, SEEDS	5-01-218	89	19.1	1.52	—	0.69	0.40	0.76	1.32	1.25	0.24	1.14	0.90	0.61	—	0.57	0.80
95	COCONUT, KERNELS WITH COATS, MEAL MECH EXTD, (COPRA MEAL)	5-01-572	92	21.2	2.30	0.21	1.05	0.33	0.90	1.35	0.55	0.31	0.81	—	0.60	0.20	0.58	0.98

(Continued)

TABLE V-2 AMINO ACID, COMPOSITION OF FEEDS, DATA EXPRESSED **AS-FED**—(Continued)

Entry Number	Feed Name Description	International Feed Number	Dry Matter %	Crude Protein %	Arginine %	Cystine %	Glycine %	Histidine %	Iso- leucine %	Leucine %	Lysine %	Methi- onine %	Phenyl- alanine %	Serine %	Threo- nine %	Trypto- phan %	Tyrosine %	Valine %
	CORN																	
96	DISTILLERS GRAINS, DEHY	5-02-842	93	27.8	0.99	0.23	0.75	0.62	1.00	3.01	0.76	0.42	0.99	1.01	0.56	0.20	0.84	1.21
97	DISTILLERS SOLUBLES, DEHY	5-02-844	93	27.4	0.99	0.44	1.12	0.67	1.32	2.38	0.92	0.55	1.47	1.22	1.01	0.25	0.88	1.53
98	GLUTEN FEED	5-02-903	90	23.0	0.78	0.42	0.85	0.61	0.88	2.20	0.64	0.37	0.81	0.85	0.78	0.15	0.72	1.10
99	GLUTEN MEAL	5-02-900	91	43.2	1.40	0.67	1.51	0.97	2.25	7.38	0.80	1.03	2.85	1.70	1.43	0.21	1.01	2.23
	COTTON																	
100	SEEDS, MEAL MECH EXTD, 36% PROTEIN	5-01-625	92	37.2	3.55	0.79	1.83	0.91	1.32	—	1.22	0.55	1.88	—	1.12	0.48	—	2.84
101	SEEDS, MEAL MECH EXTD, 41% PROTEIN	5-01-617	93	41.0	4.20	0.71	1.87	1.07	1.42	2.30	1.60	0.57	2.19	1.70	1.33	0.52	0.97	1.89
102	SEEDS, MEAL, PREPRESSED, SOLV EXTD, 41% PROTEIN	5-07-872	90	41.3	4.32	0.78	1.89	1.14	1.42	2.42	1.80	0.58	2.05	1.80	1.34	0.50	1.14	1.97
103	SEEDS, MEAL, SOLV EXTD, 41% PROTEIN	5-01-621	91	41.2	4.24	0.76	1.95	1.10	1.50	2.46	1.69	0.58	2.23	1.78	1.37	0.55	1.04	1.97
104	SEEDS WITHOUT HULLS, MEAL, PREPRESSED, SOLV EXTD, 50% PROTEIN	5-07-874	91	50.3	4.83	1.05	2.82	1.21	1.86	2.82	1.93	0.76	2.82	1.38	1.66	0.62	0.81	2.16
105	CRAB, CANNERY RESIDUE, MEAL (CRAB MEAL)	5-01-663	92	32.2	1.66	0.24	1.74	0.48	1.16	1.54	1.38	0.52	1.16	1.38	1.00	0.29	1.17	1.47
106	**CRAMBE, ABYSSINIAN,** SEEDS WITHOUT HULLS, MEAL MECH EXTD	5-18-453	92	45.8	—	—	—	—	—	—	—	—	—	—	—	—	—	—
	DISTILLERS PRODUCTS (ALSO SEE CORN, RYE)																	
107	GRAINS, DEHY	5-02-144	93	27.3	1.04	0.42	0.56	0.53	1.16	2.66	0.81	0.46	1.03	0.70	0.81	0.21	0.73	1.22
108	SOLUBLES, DEHY	5-02-147	92	28.8	1.06	0.40	1.20	0.66	1.21	2.35	0.95	0.50	1.24	0.93	1.00	0.24	0.93	1.40
	FISH																	
109	MEAL MECH EXTD	5-01-977	92	64.3	31.28	0.62	3.99	1.46	3.27	4.90	5.26	1.63	2.60	2.42	2.59	0.75	1.79	3.14
110	SOLUBLES, CONDENSED	5-01-969	50	31.5	1.63	0.39	3.87	1.54	1.09	1.94	1.85	0.70	1.07	1.05	0.90	0.33	0.50	1.26
111	**FISH, ANCHOVY,** MEAL MECH EXTD	5-01-985	92	65.4	3.78	0.60	3.69	1.60	3.11	4.99	5.02	1.99	2.78	2.42	2.76	0.75	2.24	3.50
112	**FISH, MENHADEN,** MEAL MECH EXTD	5-02-009	92	61.2	3.74	0.58	4.19	1.44	2.85	4.48	4.74	1.75	2.46	2.25	2.51	0.65	1.93	3.19
113	**FISH, TUNA,** MEAL MECH EXTD	5-02-023	93	59.0	3.43	0.47	4.09	1.75	2.45	3.79	4.06	1.47	2.15	2.08	2.31	0.57	1.69	2.77
114	**FISH, WHITE,** MEAL MECH EXTD	5-02-025	91	62.8	4.28	0.77	5.15	1.38	2.85	4.85	4.70	1.79	2.44	3.44	2.58	0.67	2.27	3.25
	FLAX, COMMON																	
115	SEEDS, MEAL MECH EXTD (LINSEED MEAL)	5-02-048	90	34.6	2.94	0.61	1.74	0.69	1.68	2.02	1.16	0.54	1.46	1.93	1.22	0.51	1.09	1.74
116	SEEDS, MEAL SOLV EXTD (LINSEED MEAL)	5-02-045	91	34.3	2.81	0.61	1.64	0.85	1.69	1.92	1.18	0.58	1.38	1.90	1.14	0.51	0.96	1.61
117	**GARBANZO (CHICKPEA; GRAM PEA),** SEEDS	5-01-218	89	19.1	1.52	—	0.69	0.40	0.76	1.32	1.25	0.24	1.14	0.90	0.61	—	0.57	0.80
118	**HORSE BEAN,** SEEDS	5-02-407	88	25.5	—	—	—	—	—	—	—	—	—	—	—	—	—	—
119	**LIVER,** MEAL	5-00-389	93	66.1	4.04	0.94	5.61	1.48	3.11	5.31	5.22	1.22	2.92	2.50	2.50	0.69	1.70	4.15
120	**LOCUST, NEW MEXICO,** SEEDS	5-09-055	89	36.5	3.01	—	1.32	0.73	0.87	1.75	1.32	0.26	1.02	1.28	0.87	—	0.84	1.17
	MAIZE—SEE CORN																	
	MEAT																	
121	MEAL RENDERED	5-00-385	94	50.7	3.58	0.60	7.23	0.87	1.63	3.11	3.00	0.69	1.74	2.31	1.67	0.35	1.09	2.42
122	WITH BLOOD, MEAL RENDERED (TANKAGE)	5-00-386	92	60.5	3.60	0.48	6.45	2.06	1.82	5.10	3.89	0.75	2.56	2.81	2.34	0.65	1.38	3.83
123	WITH BLOOD, WITH BONE, MEAL RENDERED (TANKAGE)	5-00-387	93	46.6	2.82	0.27	6.58	1.76	1.87	5.28	3.32	0.69	2.28	—	2.18	0.62	—	3.42
124	WITH BONE, MEAL RENDERED	5-00-388	93	50.4	3.53	0.53	6.49	1.01	1.84	3.10	2.93	0.67	1.71	1.90	1.66	0.31	0.89	2.44
	MILK																	
125	FRESH (CATTLE)	5-01-168	12	3.3	—	—	—	—	0.32	0.25	0.28	0.18	0.07	—	0.16	0.05	—	0.25
126	DEHY (CATTLE)	5-01-167	95	25.3	0.92	—	—	0.71	1.32	2.54	2.24	0.61	1.32	—	1.02	0.41	1.32	1.73
127	SKIMMED, DEHY (CATTLE)	5-01-175	94	33.3	1.16	0.45	0.29	0.86	2.18	3.33	2.54	0.90	1.57	1.67	1.57	0.43	1.14	2.29
128	FRESH (HORSE)	5-02-401	17	4.2	—	—	—	0.11	0.25	0.34	0.25	0.07	0.18	—	0.16	0.05	—	0.29
129	FRESH (SHEEP)	5-08-510	19	4.6	—	—	—	0.18	0.39	0.60	0.51	0.17	0.32	—	0.30	0.09	—	0.48
130	FRESH (SWINE)	5-08-537	20	7.3	—	—	—	0.20	0.42	0.59	0.50	0.14	0.34	—	0.37	0.09	—	0.45
131	**MILKWEED, COMMON,** SEEDS	5-09-137	86	31.8	3.05	—	1.65	0.73	1.12	1.97	1.56	0.45	1.53	1.31	0.86	0.20	1.08	1.37
132	**PALM,** KERNELS WITH COATS, MEAL SOLV EXTD	5-03-486	90	18.2	2.52	—	—	0.31	0.76	1.22	0.66	0.41	0.81	—	0.60	0.21	—	1.02
133	**PEA,** SEEDS	5-03-600	89	23.2	1.40	0.21	1.09	0.60	1.20	1.81	1.53	0.27	1.25	—	0.94	0.21	—	1.25
134	**PEA, FIELD,** SEEDS	5-08-481	91	23.2	1.86	0.26	1.05	0.51	0.91	1.59	1.44	0.23	1.00	1.05	0.82	0.22	0.77	1.00
	PEANUT																	
135	PODS WITH SEEDS, MEAL SOLV EXTD	5-03-656	92	47.4	5.19	0.70	2.39	1.10	1.92	3.20	1.75	0.43	2.49	3.05	1.38	0.49	1.88	2.48
136	SEEDS WITHOUT HULLS, MEAL MECH EXTD (PEANUT MEAL)	5-03-649	93	49.2	5.08	0.96	2.49	1.03	1.78	3.13	1.69	0.50	2.38	1.44	1.27	0.46	1.59	2.29
137	SEEDS WITHOUT HULLS, MEAL SOLV EXTD (PEANUT MEAL)	5-03-650	93	49.0	5.82	0.54	2.88	1.46	1.84	3.27	1.45	0.44	2.12	3.12	1.37	0.48	—	2.16
	POULTRY																	
138	BY-PRODUCT, MEAL RENDERED	5-03-798	94	61.2	4.01	0.85	6.09	1.13	2.35	4.10	3.12	1.14	2.04	2.88	2.10	0.47	1.84	2.94
139	FEATHERS, HYDROLYZED, MEAL	5-03-795	93	83.8	5.33	3.21	6.32	0.47	3.51	6.42	1.55	0.54	3.59	9.16	3.63	0.52	2.35	5.85
140	**RAPE (CANOLA), SUMMER,** SEEDS, MEAL, PREPRESSED, SOLV EXTD	5-08-135	92	40.5	2.23	—	1.94	1.09	1.46	2.71	2.15	0.77	1.54	1.70	1.70	0.49	0.85	1.94

Composition of Feeds

TABLE V-2 AMINO ACID, COMPOSITION OF FEEDS, DATA EXPRESSED **AS-FED**—(Continued)

Entry Number	Feed Name Description	International Feed Number	Dry Matter %	Crude Protein %	AMINO ACIDS													
					Arginine %	Cystine %	Glycine %	Histidine %	Iso-leucine %	Leucine %	Lysine %	Methionine %	Phenyl-alanine %	Serine %	Threonine %	Trypto-phan %	Tyrosine %	Valine %
	RYE																	
141	DISTILLERS GRAINS, DEHY	5-04-023	92	23.0	—	—	—	—	—	—	—	—	—	—	—	—	—	—
142	DISTILLERS GRAINS WITH SOLUBLES, DEHY	5-04-024	90	27.2	1.00	—	—	0.70	1.50	2.10	1.00	0.40	1.30	1.20	1.10	0.30	0.50	1.60
	SAFFLOWER																	
143	SEEDS WITHOUT HULLS, MEAL MECH EXTD	5-08-499	91	42.0	5.44	—	2.52	—	—	—	1.31	0.71	—	—	0.81	—	—	—
144	SEEDS WITHOUT HULLS, MEAL SOLV EXTD	5-07-959	91	42.8	3.87	0.71	2.36	0.97	1.58	2.42	1.26	0.67	1.73	—	1.30	0.59	1.01	2.17
145	SESAME, SEEDS, MEAL MECH EXTD	5-04-220	93	45.0	4.55	0.59	3.96	1.07	1.96	3.20	1.26	1.37	2.14	2.94	1.60	0.71	1.87	2.32
146	SHRIMP, CANNERY RESIDUE, MEAL (SHRIMP MEAL)	5-04-226	90	38.7	2.33	0.47	1.31	0.87	1.51	2.37	2.05	0.84	1.55	1.25	1.28	0.36	1.10	1.71
147	SORGHUM, GLUTEN MEAL	5-04-388	90	44.4	1.26	0.73	0.95	1.07	2.39	7.35	0.74	0.71	2.70	—	1.45	0.44	—	2.50
	SOYBEAN																	
148	FLOUR, SOLV EXTD	5-04-593	93	51.6	4.27	0.64	1.65	1.26	1.90	3.33	4.48	0.57	2.00	2.09	1.58	0.79	1.44	1.86
149	MEAL, SOLV EXTD	5-04-612	90	49.7	3.67	0.70	2.27	1.20	2.13	3.63	3.12	0.71	2.36	2.49	1.90	0.69	1.71	2.47
150	MEAL, SOLV EXTD, 44% PROTEIN	5-20-637	89	44.4	3.26	0.67	2.10	1.13	2.12	3.49	2.85	0.59	2.23	2.37	1.81	0.62	1.60	2.37
151	MEAL, SOLV EXTD, 49% PROTEIN	5-20-638	90	49.0	3.82	0.75	2.39	1.28	2.34	3.77	3.08	0.66	2.47	2.76	2.00	0.70	1.98	2.49
152	WHALE, MEAT, MEAL RENDERED	5-05-160	91	71.4	2.49	0.63	6.31	1.19	2.72	4.27	3.48	1.01	2.06	—	1.63	0.82	—	2.81
	WHEAT																	
153	GERM MEAL	5-05-218	88	24.4	1.83	0.47	1.46	0.62	0.95	1.47	1.53	0.41	0.93	1.12	0.94	0.30	0.74	1.16
154	GLUTEN	5-05-221	90	63.4	2.97	1.74	2.77	1.64	3.39	5.54	1.54	1.23	4.21	4.10	2.15	0.72	2.36	3.90
	WHEY—SEE CATTLE																	
155	YEAST, IRRADIATED, DEHY	7-05-529	94	48.1	2.46	—	—	1.00	2.94	3.56	3.70	1.00	2.77	—	2.41	0.73	—	3.06
156	YEAST, TORULA, DEHY	7-05-534	93	49.6	2.52	0.59	2.54	1.34	2.69	3.39	3.65	0.76	2.63	2.75	2.67	0.52	1.94	2.88

DRY FORAGES

	ALFALFA (LUCERNE)																	
1157	HAY, SUN-CURED	1-00-078	90	16.0	0.81	—	—	0.28	0.87	1.12	1.00	0.12	0.71	—	0.62	0.18	0.50	0.69
158	HAY, SUN-CURED, EARLY BLOOM, MEAL	1-00-108	92	22.5	—	—	—	—	—	—	—	—	—	—	—	—	—	—
159	LEAVES, SUN-CURED, MEAL	1-00-146	88	20.1	—	—	—	—	—	—	—	—	—	—	—	—	—	—
160	LEAVES, MEAL, DEHY	1-00-137	92	20.6	0.97	0.27	0.99	0.39	0.92	1.45	0.95	0.32	0.94	0.89	0.86	0.43	0.60	1.05
161	MEAL, DEHY, 15% PROTEIN	1-00-022	90	15.6	0.59	0.17	0.70	0.27	0.64	1.02	0.59	0.22	0.62	0.60	0.56	0.38	0.41	0.75
162	MEAL, DEHY, 17% PROTEIN	1-00-023	92	17.4	0.77	0.29	0.84	0.33	0.81	1.28	0.85	0.27	0.80	0.71	0.71	0.34	0.54	0.88
163	MEAL, DEHY, 20% PROTEIN	1-00-024	92	20.2	0.95	0.32	0.99	0.38	0.89	1.43	0.89	0.32	0.94	0.90	0.82	0.41	0.60	1.05
164	MEAL, DEHY, 22% PROTEIN	1-07-851	93	22.2	0.96	0.30	1.09	0.44	1.06	1.63	0.97	0.34	1.13	0.97	0.97	0.49	0.64	1.29
165	ALFALFA-GRASS, HAY, SUN-CURED	1-08-331	91	14.5	—	—	—	—	—	—	—	—	—	—	—	—	—	—
166	BEET, SUGAR, LEAVES, SUN-CURED	1-00-641	91	23.2	1.00	—	—	0.27	1.00	1.55	1.27	0.36	0.91	—	0.91	0.27	—	1.18
167	CLOVER, LADINO, HAY, SUN-CURED	1-01-378	89	20.0	—	—	—	—	—	—	—	—	—	—	—	—	—	—
168	COWPEA, COMMON, HAY, SUN-CURED	1-01-645	90	17.7	1.11	—	—	0.45	1.27	2.01	1.08	0.51	1.26	—	1.06	0.52	—	1.44
169	LESPEDEZA, COMMON, HAY, SUN-CURED	1-08-591	89	13.8	—	—	—	—	—	—	—	—	—	—	—	—	—	—
170	OATS, HULLS	1-03-281	92	3.7	0.15	0.06	0.15	0.08	0.15	0.25	0.17	0.08	0.15	—	0.16	0.09	0.14	0.19
171	SOYBEAN, HAY, SUN-CURED	1-04-558	89	14.1	—	—	—	—	—	—	—	—	—	—	—	—	—	—
172	VETCH, HAY, SUN-CURED	1-05-106	89	18.4	—	—	—	—	—	—	—	—	—	—	—	—	—	—

PASTURE AND RANGE PLANTS

173	COWPEA, COMMON, FRESH	2-01-855	25	4.0	—	—	—	—	—	—	—	—	—	—	—	—	—	—
174	SPINACH, LEAVES, FRESH	2-08-125	9	3.1	0.11	—	—	0.04	0.09	0.18	0.12	0.06	0.12	—	0.10	0.03	—	0.13

VITAMIN SUPPLEMENTS

	ALFALFA (LUCERNE)																	
175	MEAL, DEHY, 20% PROTEIN	1-00-024	92	20.2	0.95	0.32	0.99	0.38	0.89	1.43	0.89	0.32	0.94	0.90	0.82	0.41	0.60	1.05
176	MEAL, DEHY, 22% PROTEIN	1-07-851	93	22.2	0.96	0.30	1.09	0.44	1.06	1.63	0.97	0.34	1.13	0.97	0.97	0.49	0.64	1.29
177	BREWERS GRAINS, DEHY	5-02-141	92	27.3	1.27	0.35	1.09	0.53	1.57	2.53	0.88	0.46	1.46	1.30	0.93	0.37	1.16	1.58
178	CORN, DISTILLERS GRAINS WITH SOLUBLES, DEHY	5-02-843	92	27.1	0.97	0.31	0.59	0.64	1.33	2.31	0.70	0.50	1.47	1.21	0.93	0.18	0.72	1.47
179	FISH, SOLUBLES, DEHY	5-01-971	93	60.4	3.06	0.62	5.75	2.10	2.05	2.98	3.52	1.18	1.53	2.03	1.35	0.60	0.85	2.10
	FISH, SARDINE																	
180	MEAL MECH EXTD	5-02-015	93	65.2	2.70	0.80	4.50	1.80	3.34	—	5.91	2.01	2.00	—	2.60	0.50	2.79	4.10
181	SOLUBLES, CONDENSED	5-02-014	50	29.5	1.50	0.20	—	2.00	0.90	1.60	1.60	0.90	0.80	—	0.80	0.10	—	1.00
182	RICE, BRAN WITH GERM, MEAL SOLV EXTD (RICE BRAN, SOLV EXTD)	4-03-930	91	14.0	0.98	0.21	0.91	0.33	0.52	1.02	0.61	0.26	0.57	0.70	0.53	0.21	0.55	0.76
183	YEAST, BREWERS, DEHY	7-05-527	93	43.8	2.26	0.52	1.77	1.13	2.03	2.86	2.98	0.66	1.60	—	2.06	0.51	1.47	2.25
184	YEAST, PRIMARY, DEHY	7-05-533	93	48.0	2.60	0.50	—	5.60	3.60	3.70	3.80	1.00	2.50	—	2.50	0.40	—	3.20

AMINO ACIDS

TABLE V-3 APPARENT ILEAL DIGESTIBILITY OF CRUDE PROTEIN AND ESSENTIAL AMINO ACIDS IN FEEDSTUFFS FOR SWINE, AND

FEEDSTUFF (AS FED BASIS)	**Obs	DM %	Crude Protein Total %	Crude Protein Coefficient %	Crude Protein Digestible %	Arginine Total %	Arginine Coefficient %	Arginine Digestible %	Cystine Total %	Cystine Coefficient %	Cystine Digestible %	Histidine Total %	Histidine Coefficient %	Histidine Digestible %	Isoleucine Total %	Isoleucine Coefficient %	Isoleucine Digestible %
ALFALFA	3	89.85	16.95	39	6.61	0.66	59	0.39	0.18	15	0.03	0.32	46	0.15	0.66	55	0.36
BARLEY	33	87.74	10.59	70	7.41	0.52	75	0.39	0.22	74	0.16	0.24	71	0.17	0.37	73	0.27
*BARLEY, NAKED	1	87.20	11.70	78	9.13	0.64	85	0.54	—	—	—	0.30	90	0.27	0.45	87	0.39
BLOOD MEAL	3	90.57	87.87	87	76.45	3.68	94	3.46	—	—	—	5.44	94	5.11	0.72	70	0.50
BONE MEAL	2	94.11	28.10	72	20.23	1.78	79	1.41	0.20	38	0.08	0.26	71	0.18	0.54	71	0.38
BREWERS GRAINS	3	91.60	26.03	70	18.22	1.82	—	—	—	—	—	0.67	—	—	1.11	79	0.88
CANOLA MEAL	4	90.43	37.28	69	25.72	2.21	82	1.81	0.42	89	0.37	1.20	78	0.94	1.57	75	1.18
CASEIN	4	90.69	86.19	89	76.71	2.88	93	2.68	—	62	—	2.00	94	1.88	4.22	90	3.80
COCONUT OIL MEAL	1	94.70	21.58	52	11.22	2.18	—	—	0.29	—	—	0.42	—	—	0.60	65	0.39
CORN	10	88.34	9.19	77	7.08	0.41	82	0.34	0.17	74	0.13	0.25	82	0.21	0.33	80	0.26
CORN GLUTEN FEED	4	90.30	19.58	54	10.57	1.07	72	0.77	0.47	—	—	0.66	58	0.38	0.66	62	0.41
CORN GLUTEN MEAL	2	89.35	59.00	80	47.20	2.00	86	1.72	1.06	73	0.77	1.31	81	1.06	2.61	84	2.19
COTTONSEED MEAL GLANDED	9	91.60	41.14	72	29.62	4.42	88	3.89	—	—	—	1.10	77	0.85	1.28	67	0.86
COTTONSEED MEAL GLANDLESS	2	94.82	42.40	83	35.19	5.56	95	5.28	—	—	—	1.36	88	1.20	1.45	85	1.23
DRIED SKIM MILK	4	95.78	33.96	85	28.87	1.24	85	1.05	0.28	86	0.24	1.47	93	1.37	1.73	85	1.47
FEATHER MEAL	3	94.30	83.20	71	59.07	5.09	79	4.02	3.85	72	2.77	0.57	45	0.26	3.69	78	2.88
FISH MEAL	9	94.49	61.83	79	48.85	3.91	88	3.44	0.66	62	0.41	1.37	81	1.11	2.60	86	2.24
GROUNDNUT MEAL	2	89.30	41.70	85	35.45	4.92	95	4.67	0.49	78	0.38	0.88	85	0.75	1.38	85	1.17
LUPIN MEAL	2	88.25	33.07	80	26.46	3.50	93	3.26	0.43	70	0.30	0.77	85	0.65	1.47	83	1.22
L-LYSINE HCl	2	98.50	94.40	100	94.40	—	—	—	—	—	—	—	—	—	—	—	—
MEAT & BONE MEAL	29	93.38	55.07	67	36.90	3.44	77	2.65	0.63	51	0.32	0.93	71	0.66	1.53	68	1.04
OATS	1	87.60	13.00	61	7.93	0.65	—	—	0.29	—	—	0.21	—	—	0.39	—	—
OAT GROATS	1	91.62	16.32	84	13.71	1.15	90	1.04	1.15	90	1.04	0.46	86	0.40	0.52	86	0.45
PEANUT MEAL	4	93.87	44.23	73	32.29	5.20	90	4.68	—	—	—	1.04	73	0.76	1.46	76	1.11
PEAS	3	89.03	20.87	73	15.24	1.83	86	1.57	0.37	61	0.23	0.57	81	0.46	0.90	74	0.67
POULTRY-BY-PRODUCT MEAL	5	92.76	63.71	72	45.87	4.34	88	3.82	2.04	—	—	1.23	81	1.00	2.52	79	1.99
RAPESEED MEAL	7	91.45	35.32	70	24.72	2.05	83	1.70	0.83	73	0.61	0.99	83	0.82	1.48	75	1.11
*RICE	1	85.50	8.80	72	6.34	0.98	90	0.88	—	—	—	0.18	90	0.16	0.32	86	0.28
RYE	3	88.27	10.03	68	6.82	0.48	74	0.36	0.20	72	0.14	0.21	70	0.15	0.33	70	0.23
SAND EEL	1	89.42	72.56	79	57.32	3.45	—	—	0.70	—	—	1.63	54	0.88	2.93	88	2.58
SESAME MEAL	1	92.70	44.04	76	33.47	5.39	—	—	0.86	—	—	1.01	36	0.36	1.40	79	1.11
SORGHUM	2	89.99	9.83	81	7.96	0.45	85	0.38	—	—	—	0.24	81	0.19	0.44	88	0.39
SOY FLOUR	1	94.44	50.68	82	41.56	4.02	91	3.66	0.74	78	0.58	1.39	88	1.22	2.42	83	2.01
SOYBEANS EXTRUDED	2	92.68	35.11	74	25.98	2.72	84	2.23	0.57	64	0.36	0.99	80	0.79	1.69	74	1.25
SOYBEAN MEAL 44%	9	89.62	44.27	80	35.42	3.29	90	2.96	0.69	74	0.51	1.18	86	1.01	2.03	83	1.68
SOYBEAN MEAL 48%	18	89.51	47.29	80	37.83	3.56	90	3.20	0.66	79	0.52	1.28	86	1.10	2.22	84	1.86
SOYFLAKES RAW	1	90.72	48.72	35	17.05	3.91	56	2.19	0.64	35	0.22	1.41	48	0.68	2.42	43	1.04
SOYFLAKES HEATED	1	91.96	49.29	79	38.94	3.91	88	3.44	0.63	74	0.47	1.40	82	1.15	2.48	78	1.93
SUNFLOWER MEAL	10	91.90	35.29	75	26.47	2.93	90	2.64	0.57	74	0.42	0.89	76	0.68	1.45	78	1.13
TRITICALE	5	89.14	13.31	78	10.38	0.55	87	0.48	—	—	—	0.25	84	0.21	0.36	83	0.30
WHEAT	24	87.93	12.68	82	10.40	0.57	83	0.47	0.20	80	0.16	0.25	80	0.20	0.43	83	0.36
WHEAT BRAN 7% FIBER	3	88.30	13.50	67	9.05	0.88	85	0.75	0.37	71	0.26	0.35	79	0.28	0.47	74	0.35
WHEAT FLOUR	1	88.00	13.46	91	12.25	0.43	91	0.39	—	—	—	0.26	91	0.24	0.50	94	0.47
WHEAT MIDDS 9% FIBER	5	89.14	16.63	67	11.14	1.18	84	0.99	—	—	—	0.45	79	0.36	0.55	70	0.39
WHEAT OFFAL 9% FIBER	1	88.00	15.84	70	11.09	0.97	95	0.92	—	—	—	0.39	79	0.31	0.51	73	0.37

*True Ileal Digestibility Values

**The number of observations for tryptophan may differ substantially from the value indicated.

DIGESTIBLE AMINO ACID RECOMMENDATIONS

Period Weight, lb.	Starting 10 - 22	Starting 22 - 55	Growing-Finishing[1] 55 - 110	Growing-Finishing[1] 110 - 220
DIGESTIBLE MINIMUMS, %				
CRUDE PROTEIN	16.50	14.50	12.50	11.00
LYSINE/CALORIE (g/Mcal ME)	3.60	3.30	2.50	2.00
LYSINE[2]	1.25	1.00	0.75	0.61
THREONINE	0.81	0.65	0.49	0.40
TRYPTOPHAN	0.23	0.18	0.14	0.11
ISOLEUCINE	0.75	0.60	0.45	0.37
METHIONINE & CYSTINE	0.69	0.55	0.41	0.34
HISTIDINE	0.41	0.33	0.25	0.20
LEUCINE	1.35	1.08	0.81	0.66
PHENYLALANINE & TYROSINE	1.50	1.20	0.90	0.73
VALINE	0.91	0.73	0.55	0.45

[1] Boar requirements are 10 to 15% higher than gilts or barrows

Heartland Lysine would like to express its appreciation to each of the authors who contributed data to this work. In particular our thanks are extended to Dr. Darrell Knabe (Texas A&M University), Dr. Malcolm Fuller (Rowett Research Institute), and Dr. Michael Taverner (Animal Research Institute) for their contributions.

This chart reflects our interpretation of published literature on digestible amino acids for swine nutrition. It is the responsibility of the purchaser of our products to determine the best application of our products for their needs. Information and recommendations regarding our products and/or nutrient levels for swine feeding are to the best of our knowledge accurate. We do not warrant the accuracy or completeness of this information. Our making this information available does not relieve the purchaser or user of his obligation to verify the suitability of our products and recommendations for their intended application.

Composition of Feeds

DIGESTIBLE AMINO ACID RECOMMENDATIONS FOR SWINE FEED FORMULATION[1]

Leucine			Lysine			Methionine			Phenylalanine			Threonine			Tryptophan			Valine		
1.10	60	0.66	0.69	48	0.33	0.21	64	0.13	0.68	59	0.40	0.62	49	0.30	0.16	—	—	0.83	52	0.43
0.71	75	0.53	0.39	67	0.26	0.18	79	0.14	0.52	78	0.41	0.38	63	0.24	0.12	73	0.09	0.53	71	0.38
0.95	85	0.81	0.54	89	0.48	0.25	87	0.22	0.76	90	0.68	0.47	87	0.41	0.17	—	—	0.62	87	0.54
11.89	93	11.06	9.05	93	8.42	—	—	—	6.65	92	6.12	4.68	87	4.07	0.99	89	0.88	8.61	92	7.92
1.20	74	0.89	1.05	72	0.76	0.23	78	0.18	0.71	75	0.53	0.67	68	0.46	—	50	—	1.00	73	0.73
2.23	—	—	1.05	68	0.71	—	—	—	1.49	—	—	0.99	67	0.66	0.28	71	0.20	1.53	—	—
2.77	78	2.16	2.19	73	1.60	0.98	82	0.80	1.57	77	1.21	1.78	68	1.21	0.43	71	0.31	2.03	67	1.36
7.81	94	7.34	6.09	95	5.79	2.61	96	2.51	4.34	95	4.12	3.22	86	2.77	0.98	91	0.89	5.62	92	5.17
1.20	70	0.84	1.98	53	1.05	0.38	—	—	0.89	76	0.68	0.68	49	0.33	0.22	—	—	0.99	70	0.69
1.12	88	0.99	0.27	68	0.18	0.18	86	0.15	0.43	84	0.36	0.32	71	0.23	0.06	72	0.04	0.44	78	0.34
2.01	77	1.55	0.67	48	0.32	0.37	—	—	0.75	75	0.56	0.74	50	0.37	0.07	33	0.02	1.00	68	0.68
10.71	90	9.64	1.10	74	0.81	1.42	86	1.22	3.76	88	3.31	2.10	80	1.68	0.27	72	0.19	2.85	81	2.31
2.29	70	1.60	1.68	59	0.99	0.64	72	0.46	2.16	80	1.73	1.28	61	0.78	0.47	72	0.34	2.57	71	1.82
2.52	84	2.12	2.04	82	1.67	0.77	84	0.65	2.62	91	2.38	1.47	78	1.15	0.57	81	0.46	2.83	85	2.41
3.21	95	3.05	2.79	95	2.65	0.78	96	0.75	1.72	96	1.65	1.49	89	1.33	0.56	—	—	2.15	88	1.89
6.63	77	5.11	1.59	51	0.81	0.39	71	0.28	4.29	81	3.47	3.49	74	2.58	0.42	60	0.25	6.75	78	5.27
4.51	87	3.92	4.69	87	4.08	1.87	91	1.70	2.43	84	2.04	2.71	81	2.20	0.55	74	0.41	2.95	83	2.45
2.46	87	2.14	1.46	82	1.20	0.38	84	0.32	1.88	89	1.67	1.04	77	0.80	—	—	—	1.58	83	1.31
2.35	83	1.95	1.54	79	1.22	0.21	65	0.14	1.30	82	1.07	1.19	77	0.92	—	—	—	1.41	79	1.11
			78.80	100	78.80															
3.32	71	2.36	2.75	70	1.93	0.75	77	0.58	1.79	72	1.29	1.75	63	1.10	0.26	54	0.14	2.51	70	1.76
0.75	—	—	0.40	70	0.28	0.17	79	0.13	0.48	—	—	0.34	55	0.19	0.10	—	—	0.52	—	—
1.04	85	0.88	0.64	82	0.52	0.23	89	0.20	0.68	90	0.61	0.53	78	0.41	0.17	81	0.14	0.78	85	0.66
2.65	78	2.07	1.55	72	1.12	—	—	—	2.12	85	1.80	1.16	68	0.79	0.35	68	0.24	1.75	76	1.33
1.50	75	1.13	1.54	82	1.26	0.23	74	0.17	0.98	75	0.74	0.79	75	0.59	—	—	—	0.96	71	0.68
4.61	80	3.69	3.57	81	2.89	1.07	—	—	2.37	82	1.94	2.53	71	1.80	0.43	79	0.34	3.39	77	2.61
2.49	79	1.97	2.00	73	1.46	0.71	85	0.60	1.46	80	1.17	1.53	68	1.04	—	—	—	1.91	70	1.34
0.67	86	0.58	0.33	81	0.27	0.19	87	0.17	0.43	88	0.38	0.31	84	0.26	0.11	—	—	0.48	87	0.42
0.59	71	0.42	0.36	65	0.23	0.15	77	0.12	0.44	78	0.34	0.32	59	0.19	—	—	—	0.45	69	0.31
5.40	89	4.81	5.58	91	5.08	2.01	—	—	2.77	87	2.41	3.06	80	2.45	0.91	—	—	3.54	86	3.04
2.75	82	2.26	1.06	71	0.75	1.32	—	—	1.92	85	1.63	1.38	66	0.91	0.65	—	—	1.97	80	1.58
1.47	91	1.34	0.24	75	0.18	0.18	87	0.16	0.65	92	0.60	0.40	78	0.31	0.09	80	0.07	0.53	86	0.46
4.23	81	3.43	3.36	88	2.96	0.79	91	0.72	2.66	87	2.31	2.03	76	1.54	0.64	79	0.51	4.12	81	3.34
2.89	73	2.11	2.39	80	1.91	0.57	78	0.44	1.86	80	1.49	1.44	69	0.99	0.47	70	0.33	2.36	73	1.72
3.40	83	2.82	2.81	86	2.42	0.67	85	0.57	2.22	85	1.89	1.74	76	1.32	0.53	80	0.42	2.83	81	2.29
3.61	83	3.00	3.05	85	2.59	0.67	87	0.58	2.39	85	2.03	1.89	77	1.46	0.58	78	0.45	2.45	80	1.96
4.04	37	1.49	3.35	44	1.47	0.79	47	0.37	2.69	45	1.21	2.04	32	0.65	0.57	25	0.14	4.13	35	1.45
4.15	80	3.32	3.35	85	2.85	0.79	82	0.65	2.67	84	2.24	2.03	72	1.46	0.56	77	0.43	4.15	78	3.24
2.11	77	1.62	1.30	74	0.96	0.78	87	0.68	1.56	80	1.25	1.26	71	0.89	0.42	76	0.32	1.71	75	1.28
0.67	84	0.56	0.40	73	0.29	0.16	84	0.13	0.58	83	0.48	0.34	65	0.22	0.14	—	—	0.47	83	0.39
0.83	83	0.69	0.33	71	0.23	0.20	84	0.17	0.55	86	0.47	0.36	69	0.25	0.15	81	0.12	0.53	77	0.41
0.88	76	0.67	0.52	72	0.37	0.21	79	0.17	0.57	79	0.45	0.44	66	0.29	—	—	—	0.65	72	0.47
0.92	95	0.87	0.23	84	0.19	0.16	94	0.15	0.70	96	0.67	0.33	85	0.28	—	—	—	0.56	93	0.52
1.06	72	0.76	0.68	72	0.49	0.23	79	0.18	0.68	82	0.56	0.55	60	0.33	0.19	72	0.14	0.78	73	0.57
0.95	74	0.70	0.54	66	0.36	0.23	78	0.18	0.60	76	0.46	0.41	54	0.22	—	—	—	0.71	72	0.51

Column labels (per amino acid block, left to right): Total %, Coefficient %, Digestible %

FOR SWINE FEED FORMULATION

	Sows Gestation	Sows Lactation	Mature Boar
DIGESTIBLE MINIMUMS, %			
CRUDE PROTEIN	9.50	12.00	12.00
LYSINE/CALORIE (g/Mcal ME)	1.60	2.00	2.00
LYSINE[2]	0.51	0.67	0.67
THREONINE	0.33	0.44	0.44
TRYPTOPHAN	0.09	0.12	0.12
ISOLEUCINE	0.30	0.40	0.40
METHIONINE & CYSTINE	0.28	0.37	0.37
HISTIDINE	0.17	0.22	0.22
LEUCINE	0.54	0.73	0.73
PHENYLALANINE & TYROSINE	0.60	0.80	0.80
VALINE	0.37	0.49	0.49

[2] For each 1% added fat digestible lysine should increase by .02%.

[1] Table V-3 data was assembled by, and is presented through the courtesy of, Heartland Lysine, Inc., 8430 West Bryn Mawr Avenue, Suite 650, Chicago, IL 60631.

TABLE V-4 TRUE DIGESTIBILITY OF ESSENTIAL AMINO ACIDS FOR POULTRY, AND TRUE DIGESTIBLE AMINO ACID RECOMMENDATIONS

FEEDSTUFF (AS FED BASIS)	Obs	DM%**	CP	Arginine			Cystine			Histidine			Isoleucine		
				Total %	Coefficient %	Digestible %	Total %	Coefficient %	Digestible %	Total %	Coefficient %	Digestible %	Total %	Coefficient %	Digestible %
ALFALFA MEAL	9	89.86	16.23	0.70	81.3 (6.1)	0.57	0.27	41.2 (10.1)	0.11	0.34	74.5 (5.8)	0.25	0.60	76.6 (7.0)	0.46
ALGAE	2	90.00	33.64	2.10	87.1 (5.0)	1.83	1.06	83.2 (0.6)	0.88	0.53	82.4 (4.5)	0.44	1.33	81.2 (5.0)	1.08
BARLEY	30	89.43	11.43	0.56	83.5 (4.8)	0.47	0.28	81.5 (8.5)	0.22	0.26	86.0 (4.5)	0.22	0.37	80.7 (5.9)	0.30
BLOOD MEAL	22	90.00	87.56	4.05	86.5 (4.2)	3.51	1.27	75.8 (5.3)	0.96	5.57	87.1 (3.4)	4.85	0.91	79.8 (6.2)	0.73
BONE MEAL	1	91.50	39.80	2.38	70.5 (0.0)	1.68	0.20	48.7 (0.0)	0.10	0.43	60.5 (0.0)	0.26	0.83	76.2 (0.0)	0.63
BREWERS GRAINS	1	90.00	22.70	1.04	80.9 (0.0)	0.84	—	—	—	0.42	73.8 (0.0)	0.31	0.78	80.7 (0.0)	0.63
CANOLA MEAL	25	90.38	35.72	2.21	89.8 (1.6)	1.98	0.93	72.4 (8.4)	0.68	0.97	87.0 (2.2)	0.84	1.29	83.5 (2.6)	1.08
CASEIN	2	90.00	61.00	3.94	98.3 (1.0)	3.87	—	—	—	1.86	88.9 (9.4)	1.65	4.36	98.4 (0.9)	4.28
COCONUT MEAL	1	90.00	20.93	2.58	86.5 (0.0)	2.23	0.41	52.9 (0.0)	0.22	0.35	74.8 (0.0)	0.26	0.60	83.6 (0.0)	0.50
CORN	12	88.75	8.45	0.37	94.6 (4.8)	0.35	0.17	83.7 (7.8)	0.14	0.23	89.2 (7.3)	0.20	0.30	91.3 (3.0)	0.27
CORN GERM MEAL	1	90.00	21.43	1.47	90.3 (0.0)	1.32	0.40	57.8 (0.0)	0.23	0.67	93.7 (0.0)	0.56	0.67	86.2 (0.0)	0.57
CORN GLUTEN FEED	10	90.00	21.88	1.12	88.8 (2.7)	0.99	0.53	64.4 (6.4)	0.34	0.66	84.6 (2.7)	0.56	0.62	92.9 (4.8)	0.51
CORN GLUTEN MEAL	13	90.03	61.75	2.07	95.8 (2.1)	1.98	1.17	87.5 (3.7)	1.02	1.24	94.7 (1.3)	1.18	2.24	95.5 (1.0)	2.14
COTTONSEED MEAL	1	87.40	34.70	3.75	80.0 (0.0)	3.00	—	—	—	0.81	82.2 (0.0)	0.67	0.90	68.0 (0.0)	0.61
D L METHIONINE	2	99.00	58.10	—	—	—	—	—	—	—	—	—	—	—	—
D L MHA CALCIUM	2	99.00	50.47	—	—	—	—	—	—	—	—	—	—	—	—
FEATHER MEAL	14	90.27	79.58	6.05	82.1 (4.8)	4.06	4.2	54.9 (6.9)	2.33	0.80	69.5 (15.3)	0.56	3.50	93.6 (5.0)	2.92
FISH MEAL	34	90.25	62.87	4.01	92.5 (2.4)	3.71	0.57	77.9 (5.9)	0.44	1.46	89.1 (3.1)	1.30	2.40	93.3 (2.9)	2.24
FISH MEAL ANALOG	16	89.00	62.73	4.02	88.4 (2.2)	3.55	—	—	—	1.59	75.5 (6.6)	1.20	1.79	87.5 (2.2)	1.57
GROUNDNUT MEAL	4	89.26	43.80	4.90	90.7 (3.7)	4.44	0.50	79.5 (3.0)	0.40	0.86	85.4 (5.0)	0.73	1.37	88.7 (2.4)	1.22
HAIR (HYDROLYZED)	2	90.00	66.07	5.77	62.5 (2.6)	3.60	—	—	—	0.65	51.9 (1.0)	0.34	2.11	64.0 (5.4)	1.35
L-LYSINE HCl	1	98.50	94.40	—	—	—	—	—	—	—	—	—	—	—	—
LIVER MEAL	1	90.00	73.80	5.51	76.3 (0.0)	4.20	—	—	—	2.04	73.1 (0.0)	1.49	3.82	73.8 (0.0)	2.82
LUCERNE MEAL	2	90.70	13.90	0.78	82.5 (1.2)	0.64	0.15	24.8 (0.0)	0.04	0.34	65.6 (19.9)	0.22	0.68	74.7 (8.5)	0.51
LUPINSEED MEAL	3	88.90	32.80	2.86	92.7 (3.5)	2.65	0.43	93.7 (0.0)	0.40	0.75	98.4 (3.9)	0.66	1.49	90.0 (5.0)	1.34
MEAT MEAL	20	90.83	47.70	3.27	86.0 (6.2)	2.81	0.48	55.1 (11.0)	0.27	0.91	84.9 (4.3)	0.77	1.14	84.7 (6.7)	0.97
MEAT AND BONE MEAL	22	90.51	54.02	3.79	85.6 (6.9)	3.25	0.70	62.5 (14.1)	0.44	1.04	79.3 (8.1)	0.83	1.54	84.0 (6.3)	1.29
OATS	14	90.00	10.78	0.72	93.2 (3.9)	0.67	0.47	84.2 (9.5)	0.40	0.24	92.4 (4.3)	0.22	0.38	87.9 (4.3)	0.33
POULTRY BYPRODUCT MEAL	7	90.19	55.10	3.96	87.0 (4.9)	3.44	1.35	60.5 (8.5)	0.82	1.15	70.5 (7.4)	0.81	2.22	83.1 (6.7)	1.84
POULTRY OFFAL MEAL	2	90.00	56.40	4.14	88.9 (0.0)	3.67	—	—	—	0.90	82.3 (3.8)	0.75	2.46	86.1 (1.0)	2.11
RICE (ROUGH)	1	86.60	6.60	0.56	90.6 (0.0)	0.51	—	—	—	0.14	91.9 (0.0)	0.13	0.22	78.9 (0.0)	0.17
RICE BRAN (DEFATTED)	6	89.68	14.56	1.15	85.7 (2.9)	0.98	0.32	63.4 (7.0)	0.20	0.39	81.7 (3.9)	0.32	0.47	74.1 (5.4)	0.35
SESAMESEED MEAL	3	90.00	43.37	3.82	77.6 (21.0)	2.97	0.90	81.7 (3.5)	0.74	0.87	71.0 (25.9)	0.62	1.31	72.1 (28.1)	0.94
SHRIMP MEAL	1	90.00	36.28	2.25	93.5 (0.0)	2.10	0.43	78.6 (0.0)	0.34	0.90	91.0 (0.0)	0.82	1.40	95.1 (0.0)	1.33
SINGLE CELL PROTEIN	2	90.00	68.18	2.51	88.6 (6.0)	2.22	0.48	66.9 (0.0)	0.32	1.22	88.7 (1.4)	1.08	2.27	87.8 (4.6)	1.99
SORGHUM	2	89.30	9.30	0.43	88.4 (0.9)	0.38	0.17	76.5 (0.0)	0.13	0.24	92.6 (3.4)	0.22	0.39	87.5 (2.8)	0.34
SOYBEAN MEAL (48%)	30	89.84	48.47	3.77	92.1 (2.1)	3.47	0.76	82.5 (4.2)	0.63	1.31	91.4 (4.5)	1.20	2.05	91.9 (1.7)	1.88
SOYBEAN MEAL (FULL FAT)	1	90.00	34.35	2.62	94.6 (0.0)	2.48	—	—	—	0.88	94.0 (0.0)	0.83	1.59	93.4 (0.0)	1.49
SUNFLOWER MEAL	5	91.51	33.12	2.79	95.3 (2.4)	2.66	0.53	80.9 (6.8)	0.43	0.86	89.1 (4.3)	0.77	1.41	91.2 (1.2)	1.29
TRITICALE	1	90.00	19.00	0.83	93.2 (0.0)	0.77	—	—	—	0.51	94.8 (0.0)	0.48	0.56	94.6 (0.0)	0.53
WHEAT	30	89.58	15.77	0.69	87.1 (4.3)	0.61	0.37	87.3 (6.3)	0.32	0.35	90.7 (3.5)	0.32	0.48	87.9 (3.9)	0.42
WHEAT BRAN	3	89.27	14.80	1.03	79.7 (1.0)	0.82	0.29	71.8 (0.4)	0.21	0.38	79.4 (1.1)	0.30	0.47	76.2 (2.5)	0.36
WHEAT MIDDS	2	89.75	17.00	0.58	84.3 (2.7)	0.49	0.37	77.2 (0.0)	0.29	0.44	83.1 (0.2)	0.36	0.46	67.4 (10.3)	0.31
WHEAT POLLARD	2	90.00	16.50	0.95	79.3 (5.0)	0.75	—	—	—	0.39	84.5 (4.0)	0.33	0.49	75.3 (1.0)	0.37
WHEAT SCREENINGS	5	90.00	13.65	0.89	89.4 (5.0)	0.80	0.35	74.3 (14.5)	0.26	0.31	85.5 (8.1)	0.26	0.47	84.4 (5.9)	0.39
WHEAT SHORTS	15	90.00	17.44	1.29	86.4 (4.0)	1.11	0.41	69.1 (7.5)	0.28	0.48	83.8 (3.5)	0.40	0.50	82.3 (4.0)	0.41

*Combination of data from conventional and caecectomized precision-fed rooster assays. Due to lack of data, tryptophan is not included

**90% dry matter assumed where data unavailable

***Values in parentheses represent standard deviations of digestibility coefficients

TRUE DIGESTIBLE AMINO ACID RECOMMENDATIONS

	BROILER CHICKENS			TURKEYS							DUCKS			
Period	Starting	Growing	Finishing						Holding	Breeding Hens	Starting	Growing	Rearing	Breeding
Age (weeks)	0-21	22-42[2]	43+[2]	0-4	4-8	8-12[2]	12-16[2]	16+[2]			0-3	3-8[2]	8-breeding[2]	
Metabolizable Energy (Kcal/kg)	3200	3250	3300	2900	3000	3100	3200	3300	2750	3000	2900	2950	2900	2400
Crude Protein[1] %	21.00	19.00	17.00	26.00	24.00	21.50	18.50	15.00	12.00	13.50	20.00	16.00	14.00	14.00
Methionine + Cystine %	.86	.75	.65	.97	.88	.79	.70	.62	.44	.53	.83	.66	.51	.49
Methionine + Cystine %/Mcal	.269	.230	.197	.334	.293	.255	.219	.188	.160	.177	.286	.224	.176	.169
Methionine %	.50	.43	.36	.56	.51	.46	.41	.36	.23	.29	.48	.38	.28	.27
Methionine %/Mcal	.156	.132	.109	.193	.170	.148	.128	.109	.084	.097	.166	.129	.097	.093
Lysine %	1.14	1.00	.90	1.65	1.49	1.29	1.05	.90	.57	.65	1.10	.94	.72	.71
Lysine %/Mcal	.356	.308	.273	.569	.497	.416	.328	.273	.207	.217	.379	.319	.248	.245
Threonine %	.69	.63	.59	.88	.79	.70	.61	.52	.35	.40	.62	.53	.42	.41
Threonine %/Mcal	.216	.194	.179	.303	.263	.226	.191	.158	.127	.133	.214	.180	.145	.141
Arginine %	1.30	1.12	.96	1.49	1.34	1.19	1.03	.90	.60	.65	1.12	1.01	.77	.75
Arginine %/Mcal	.406	.345	.291	.514	.447	.384	.322	.273	.218	.217	.386	.342	.266	.259
Tryptophan[1] %	.22	.18	.17	.25	.225	.20	.175	.15	.09	.12	.20	.17	.13	.13
Tryptophan %/Mcal	.069	.055	.052	.086	.075	.065	.547	.045	.033	.040	.069	.058	.045	.045
Histidine %	.34	.29	.26	.54	.49	.43	.38	.32	.24	.28	.33	.29	.22	.22
Histidine %/Mcal	.106	.089	.079	.186	.163	.139	.119	.097	.087	.093	.114	.098	.076	.076
Isoleucine %	.74	.66	.58	.99	.89	.79	.68	.58	.41	.46	.72	.63	.50	.49
Isoleucine %/Mcal	.231	.203	.176	.341	.297	.255	.213	.176	.149	.153	.248	.214	.172	.169
Leucine %	1.35	1.18	1.02	1.74	1.56	1.39	1.21	1.05	.71	.90	1.31	1.17	.90	.88
Leucine %/Mcal	.422	.363	.309	.600	.520	.448	.378	.318	.258	.300	.452	.397	.310	.303
Phenylalanine %	.67	.60	.52	.93	.84	.74	.65	.56	.37	.53	.65	.59	.45	.44
Phenylalanine %/Mcal	.209	.185	.158	.321	.280	.239	.203	.170	.135	.177	.224	.200	.155	.152
Valine %	.76	.73	.60	1.07	.98	.86	.75	.64	.48	.54	.74	.72	.55	.54
Valine %/Mcal	.238	.225	.182	.369	.327	.277	.234	.194	.175	.180	.255	.244	.190	.186

[1] Minimum Total Content

[2] 70 F Temperature assumed during growing and finishing periods. For each 10 F increase in temperature, increase amino acid levels by 3% of value

Composition of Feeds

FOR POULTRY FEED FORMULATION[1]

Leucine				Lysine				Methionine				Phenylalanine				Threonine				Valine			
1.11	79.5	(6.1)	0.88	0.78	60.1	(9.1)	0.47	0.19	76.3	(9.1)	0.14	0.72	78.7	(5.6)	0.57	0.68	72.5	(7.0)	0.49	0.77	76.4	(6.0)	0.59
2.98	82.9	(5.4)	2.47	1.83	82.5	(4.3)	1.51	0.77	83.4	(5.3)	0.65	1.76	83.4	(5.0)	1.47	1.89	78.0	(4.3)	1.47	2.01	81.1	(4.8)	1.63
0.77	84.8	(4.3)	0.65	0.42	77.8	(5.5)	0.33	0.18	78.3	(10.4)	0.14	0.55	85.3	(5.5)	0.47	0.38	75.8	(6.0)	0.29	0.50	80.1	(5.1)	0.40
11.19	90.8	(4.9)	10.16	7.97	87.1	(4.8)	6.94	1.10	92.5	(3.1)	1.02	6.07	92.3	(3.5)	5.61	4.24	89.1	(4.4)	3.78	6.99	90.1	(4.8)	6.30
1.85	77.9	(0.0)	1.44	1.54	69.2	(0.0)	1.07	0.34	79.8	(0.0)	0.27	1.09	77.5	(0.0)	0.84	1.01	75.4	(0.0)	0.76	1.38	75.6	(0.0)	1.04
1.39	81.6	(0.0)	1.13	0.73	72.7	(0.0)	0.53	0.36	80.4	(0.0)	0.29	0.99	75.1	(0.0)	0.74	0.69	72.2	(0.0)	0.50	1.02	78.2	(0.0)	0.80
2.49	87.3	(2.4)	2.17	1.97	78.3	(4.6)	1.54	0.77	89.7	(2.8)	0.69	1.39	86.8	(2.9)	1.20	1.56	78.7	(4.1)	1.23	1.65	81.8	(3.1)	1.35
8.19	99.1	(0.8)	8.11	5.59	97.1	(2.5)	5.43	2.49	99.2	(0.4)	2.46	4.51	99.1	(0.7)	4.47	3.63	97.7	(1.6)	3.54	5.63	98.8	(0.8)	5.55
1.12	80.0	(0.0)	0.89	0.64	79.7	(0.0)	0.51	0.44	88.1	(0.0)	0.39	0.87	86.2	(0.0)	0.75	0.57	65.6	(0.0)	0.37	0.93	83.8	(0.0)	0.78
1.03	95.6	(1.8)	0.99	0.25	84.8	(6.3)	0.21	0.15	94.5	(2.6)	0.15	0.39	93.6	(2.9)	0.36	0.29	86.3	(3.5)	0.25	0.41	91.0	(3.8)	0.37
1.63	87.2	(0.0)	1.42	0.91	83.0	(0.0)	0.75	0.37	85.6	(0.0)	0.32	0.91	88.0	(0.0)	0.80	0.78	78.6	(0.0)	0.62	1.13	85.6	(0.0)	0.96
2.12	89.8	(2.5)	1.90	0.71	71.9	(4.5)	0.51	0.40	84.8	(3.4)	0.34	3.71	93.1	(0.8)	0.81	0.79	76.9	(4.2)	0.61	0.96	84.0	(4.3)	0.81
10.13	98.0	(0.5)	9.93	1.07	89.1	(3.1)	0.95	1.59	97.2	(0.7)	1.54	3.95	97.4	(0.6)	3.85	2.12	92.9	(1.1)	1.97	2.64	95.5	(1.1)	2.52
1.87	74.8	(0.0)	1.40	1.36	61.0	(0.0)	0.83	0.44	80.8	(0.0)	0.36	1.68	83.2	(0.0)	1.40	1.09	70.9	(0.0)	0.77	1.23	74.1	(0.0)	0.91
								99.0	99.7	(0.0)	98.70												
								86.0	97.4	(1.4)	83.76												
6.60	81.5	(5.2)	5.38	2.24	62.5	(9.0)	1.40	0.57	73.9	(7.1)	0.42	3.83	84.4	(4.7)	3.23	3.77	70.5	(4.8)	2.66	5.40	79.6	(4.4)	4.30
4.47	94.0	(2.2)	4.20	4.66	90.1	(2.9)	4.20	1.75	92.7	(2.6)	1.62	2.35	92.5	(2.4)	2.18	2.62	91.4	(2.4)	2.40	2.85	92.5	(2.9)	2.64
5.81	89.6	(2.5)	5.20	4.00	87.5	(2.6)	3.50	1.49	94.2	(1.6)	1.41	3.11	90.7	(2.2)	2.82	2.84	84.3	(3.0)	2.39	4.32	89.8	(2.0)	3.88
2.53	90.4	(1.0)	2.29	1.34	75.4	(6.4)	1.01	0.40	87.1	(0.9)	0.35	1.93	91.8	(0.6)	1.78	1.09	84.7	(1.4)	0.92	1.65	89.2	(1.1)	1.47
4.53	76.1	(6.6)	3.45	1.86	61.2	(11.1)	1.14	0.36	73.2	(10.8)	0.26	1.47	70.1	(8.7)	1.03	4.21	33.3	(3.9)	1.40	3.41	54.6	(6.6)	1.86
				78.80	100.0	(0.0)	78.80																
7.31	73.8	(0.0)	5.40	5.40	69.5	(0.0)	3.75	2.02	75.0	(0.0)	1.51	3.84	73.3	(0.0)	2.82	3.84	71.3	(0.0)	2.74	4.90	75.7	(0.0)	3.71
1.16	77.3	(6.2)	0.90	0.78	64.6	(10.5)	0.50	0.23	81.2	(3.8)	0.18	0.73	76.2	(4.6)	0.56	0.69	67.1	(9.1)	0.46	0.85	71.6	(8.3)	0.61
2.61	91.2	(4.5)	2.38	1.76	87.3	(2.9)	1.54	0.45	85.0	(7.4)	0.38	1.56	90.8	(5.4)	1.41	1.35	87.3	(5.2)	1.18	1.64	88.1	(5.1)	1.44
2.78	86.7	(6.4)	2.41	2.38	81.8	(9.3)	1.95	0.63	87.6	(5.1)	0.55	1.50	87.1	(6.3)	1.31	1.54	82.8	(8.6)	1.27	1.84	84.3	(8.2)	1.55
3.44	84.6	(7.6)	2.91	2.84	82.6	(7.6)	2.34	0.72	86.6	(6.6)	0.62	1.85	83.0	(7.5)	1.54	1.88	80.6	(7.5)	1.52	2.42	82.6	(8.0)	2.00
0.81	90.2	(4.5)	0.73	0.47	86.5	(4.1)	0.40	0.17	85.5	(4.7)	0.15	0.54	92.2	(3.8)	0.49	0.37	83.0	(6.4)	0.31	0.53	86.6	(4.9)	0.46
4.16	82.4	(6.6)	3.43	3.03	80.0	(6.7)	2.42	0.93	83.6	(7.0)	0.78	2.26	83.0	(7.0)	1.87	2.27	78.9	(6.7)	1.71	3.02	82.0	(6.9)	2.48
4.44	85.4	(1.3)	3.79	2.46	84.3	(3.4)	2.07	0.80	90.3	(1.2)	0.72	2.50	86.2	(0.2)	2.15	2.57	81.7	(1.9)	2.10	3.52	84.6	(1.1)	2.97
0.48	85.7	(0.0)	0.41	0.23	62.7	(0.0)	0.14	0.14	89.5	(0.0)	0.13	0.30	82.2	(0.0)	0.25	0.22	79.4	(0.0)	0.17	0.33	87.4	(0.0)	0.29
0.98	73.1	(4.8)	0.72	0.66	73.4	(5.2)	0.48	0.30	76.1	(3.1)	0.23	0.64	74.4	(4.8)	0.47	0.56	68.6	(4.9)	0.37	0.74	75.2	(4.1)	0.56
2.62	73.1	(25.4)	1.92	0.79	58.4	(41.4)	0.46	1.15	76.6	(24.8)	0.88	1.81	74.9	(25.9)	1.35	1.16	60.7	(37.2)	0.70	1.74	73.0	(25.5)	1.27
2.18	95.2	(0.0)	2.07	2.09	90.0	(0.0)	1.88	0.80	96.2	(0.0)	0.77	12.02	99.1	(0.0)	11.92	1.64	91.1	(0.0)	1.49	1.79	93.5	(0.0)	1.67
3.99	88.0	(3.8)	3.51	3.15	86.6	(2.4)	2.73	1.14	88.8	(0.0)	1.01	2.19	80.9	(12.0)	1.77	2.57	84.5	(4.1)	2.17	2.98	85.8	(4.2)	2.56
1.31	93.6	(0.6)	1.22	0.24	81.3	(0.0)	0.20	0.18	86.0	(3.2)	0.15	0.53	89.5	(0.7)	0.47	0.36	81.0	(1.8)	0.29	0.51	88.5	(0.2)	0.45
3.75	92.1	(1.6)	3.45	3.09	89.8	(2.0)	2.77	0.70	92.2	(2.2)	0.65	2.39	92.9	(1.8)	2.23	1.92	89.3	(2.0)	1.71	2.13	90.8	(2.4)	1.93
2.68	93.6	(0.0)	2.51	2.19	92.2	(0.0)	2.02	0.39	88.9	(0.0)	0.35	1.68	88.8	(0.0)	1.49	1.38	89.5	(0.0)	1.23	1.72	91.9	(0.0)	1.58
2.05	90.9	(1.5)	1.86	1.23	83.6	(7.8)	1.03	0.78	93.8	(1.3)	0.73	1.50	92.7	(0.9)	1.39	1.21	86.4	(4.3)	1.04	1.65	88.6	(2.8)	1.46
1.13	95.5	(0.0)	1.08	0.57	90.7	(0.0)	0.52	0.24	95.7	(0.0)	0.23	0.81	95.7	(0.0)	0.78	0.52	91.9	(0.0)	0.48	0.69	91.7	(0.0)	0.63
1.00	90.0	(3.7)	0.90	0.42	81.4	(6.4)	0.34	0.23	86.7	(3.6)	0.20	0.71	91.4	(3.1)	0.65	0.41	81.9	(5.7)	0.34	0.58	85.9	(4.1)	0.50
0.89	77.1	(1.2)	0.69	0.61	74.1	(2.1)	0.45	0.21	77.3	(9.7)	0.16	0.56	86.5	(9.6)	0.48	0.49	71.4	(1.0)	0.35	0.70	74.9	(2.1)	0.52
0.99	80.9	(0.6)	0.80	0.63	77.8	(4.3)	0.49	0.24	82.4	(0.5)	0.20	0.61	82.5	(0.5)	0.50	0.50	73.0	(0.6)	0.37	0.64	75.1	(4.0)	0.48
0.98	78.0	(0.2)	0.76	0.62	75.7	(0.5)	0.47	0.22	74.9	(1.5)	0.16	0.60	74.0	(4.5)	0.44	0.54	69.7	(1.6)	0.38	0.74	74.5	(0.1)	0.55
0.96	87.1	(5.4)	0.84	0.56	79.3	(6.7)	0.44	0.24	82.4	(4.2)	0.20	0.59	87.3	(5.4)	0.52	0.48	79.8	(6.5)	0.38	0.59	83.6	(6.2)	0.50
1.09	84.2	(2.7)	0.92	0.75	81.0	(5.9)	0.60	0.27	79.5	(2.5)	0.22	0.68	85.3	(2.7)	0.58	0.57	78.6	(4.1)	0.45	0.74	81.7	(4.1)	0.60

Columns under each amino acid: Total %, Coefficient %, (...), Digestible %

FOR POULTRY FEED FORMULATION

GEESE		EGG-TYPE CHICKENS				Laying[3]
Growing	Breeding	Growing				
0-5	5+[2]	0-6	6-14[2]	14-20[2]		
2900	2950	2900	2900	2900	2900	Metabolizable Energy (Kcal/kg)	2900
20.00	15.00	14.00	18.00	12.00	12.00	Crude Protein[1] %	14
						Crude Protein mg/hen/day	15,400
70	66	55	62	53	40	Methionine + Cystine %	54
241	224	190	214	183	138	Methionine + Cystine mg/hen/day	594
40	38	32	33	28	23	Methionine %	32
138	129	110	114	097	079	Methionine mg/hen/day	352
1.10	82	71	87	70	60	Lysine %	69
379	278	245	300	241	207	Lysine mg/hen/day	759
62	52	45	53	44	33	Threonine %	40
214	176	155	183	152	114	Threonine mg/hen/day	440
1.1	87	72	93	75	62	Arginine %	68
379	295	248	321	259	214	Arginine mg/hen/day	748
16	15	12	16	12	11	Tryptophan %[1]	13
055	051	041	055	041	038	Tryptophan mg/hen/day	143
27	25	21	27	21	18	Histidine %	20
093	085	072	093	072	062	Histidine mg/hen/day	220
59	54	45	57	46	38	Isoleucine %	46
203	183	155	197	159	131	Isoleucine mg/hen/day	506
1.08	1.02	85	1.08	85	71	Leucine %	89
372	346	293	372	293	245	Leucine mg/hen/day	979
53	50	42	51	42	33	Phenylalanine %	37
183	169	145	176	145	114	Phenylalanine mg/hen/day	407
61	.58	.48	58	48	38	Valine %	51
210	.197	.166	200	166	131	Valine mg/hen/day	561

[1] Assumes an average daily intake of 110 g of feed/hen daily. Protein and Amino Acid levels should be adjusted according to feed intake to result in a constant nutrient intake in mg/hen/day.

Heartland Lysine would like to express its appreciation to each of the authors who contributed data to this work. In particular our thanks are extended to Dr. Ian Sibbald (Agriculture Canada) and Dr. Carl Parsons (University of Illinois) for their contributions to the digestibility of feedstuff data.

This chart reflects our interpretation of published literature on digestible amino acids for poultry nutrition. It is the responsibility of the purchaser of our products to determine the best application of our products for their needs. Information and recommendations regarding our products and/or nutrient levels for poultry feeding are to the best of our knowledge accurate. We do not warrant the accuracy or completeness of this information. Our making this information available does not relieve the purchaser or user of his obligation to verify the suitability of our products and recommendations for their intended application.

[1] Table V-4 data was assembled by, and is presented through the courtesy of, Heartland Lysine, Inc., 8430 West Bryn Mawr Avenue, Suite 650, Chicago, IL 60631.

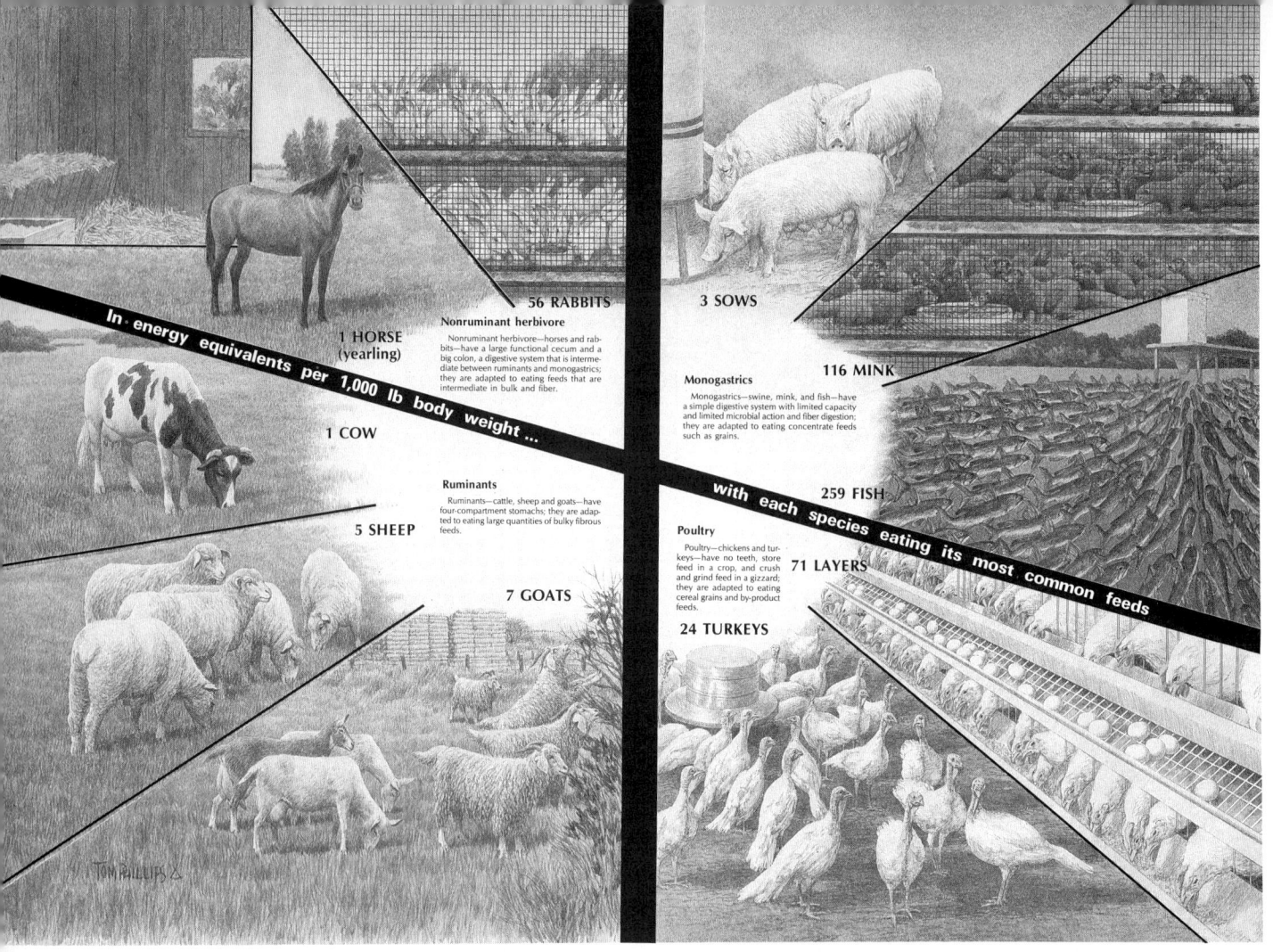

ANIMAL UNITS, above, is from an original painting by the noted artist, Tom Phillips (3333 17th Street, San Francisco, California 94110), prepared specially for this book. It portrays the artist's conception of animal units—the energy equivalents per 1,000 lb body weight, with each species eating its most common feeds.

SECTION VI

APPENDIX

The Appendix is essential to the completeness of *Feeds & Nutrition*. It provides useful information on weights and measures, animal units, and poison information centers.

Original painting by Tom Phillips

APPENDIX

Contents	Page
Weights and Measures	753
Metric System	753
Temperature	757
Weights and Measures of Common Feeds	758
Storage Space Requirements for Feed and Bedding	758
Grain Weight in a Bin	759
Hay Weight in a Barn or Stack	759
Animal Weights	760
Beef Cattle Weights	760
Dairy Heifer Weights	760
Sheep and Goat Weights	761
Swine Weights	761
Horse Weights	761
Animal Units	762
Poison Information Centers	762
Careers in Animal Industries	763
Career Opportunities in the Animal Industries	763
Magnitude of U.S. Animal Industries	763
Agriculture—the Nation's Largest Industry	764
New Frontiers in the 21st Century	765
World Without End—With Animals	766
Consumers Benefit From Animal Industries	766
The Functions of Animals	766
The Food Industry	768
A Galaxy of Career Opportunities	768
Animal Production and Management	769
Agribusiness Serving Animal Production	770
Food Processing	770
Marketing, Merchandizing, and Sales	771
Education and Research	771
Communications and Service Specialists	774
Public Service	775
Sustainable Agriculture—Conservation, Wildlife, Ecology	775
Why Go To School?	777

Useful supplementary information to all that has gone before is contained in this section.

WEIGHTS AND MEASURES

Weights and measures are the standards employed in arriving at weights, quantities, and volumes. Even among primitive people, such standards were necessary; and with the growing complexity of life, they become of greater and greater importance.

Weights and measures form one of the most important parts of modern agriculture. This section contains pertinent information relative to the most common standards used in the U.S. livestock industry.

Metric System[1]

The United States and a few other countries use standards that belong to the *customary,* or English, system of measurement. This system evolved in England from older measurement standards, beginning about the year 1200. All other countries—including England—now use a system of measurements called the *metric system,* which was created in France in the 1790s. Increasingly, the metric system is being used in the United States. Hence, everyone should have a working knowledge of it.

The basic metric units are the *meter* (length/distance), the *gram* (weight), and the *liter* (capacity). The units are then expanded in multiples of 10 or made smaller by $\frac{1}{10}$. The prefixes, which are used in the same way with all basic metric units, follow:

"milli-"	=	$\frac{1}{1000}$	"deca-"	=	10
"centi-"	=	$\frac{1}{100}$	"hecto-"	=	100
"deci-"	=	$\frac{1}{10}$	"kilo-"	=	1,000

The following tables will facilitate conversion from metric units to U.S. customary, and vice versa:

  Table VI-1 Weights and Measures—
    Weight
    Length
    Surface/Area
    Volume

  Table VI-2 Temperature

---

[1]For additional conversion factors, or for greater accuracy, see *Misc. Pub. 223,* the National Bureau of Standards.

TABLE VI-1
WEIGHTS AND MEASURES

Weight

Unit	Is Equal To	
Metric system:	(metric)	(U.S. customary)
1 microgram (mcg)	.001 mg	
1 milligram (mg)	.001 g	.015432356 grain
1 centigram (cg)	.01 g	.15432356 grain
1 decigram (dg)	.1 g	1.5432 grains
1 gram (g)	1,000 mg	.03527396 oz
1 decagram (dkg)	10 g	5.643833 dr
1 hectogram (hg)	100 g	3.527396 oz
1 kilogram (kg)	1,000 g	35.274 oz; 2.2046223 lb
1 ton	1,000 kg	2,204.6 lb; 1.102 tons (short or 0.984 ton (long)
U.S. customary:	(U.S. customary)	(metric)
1 grain	.037 dr	64.798918 mg; .064798918 g
1 dram (dr)	.063 oz	1.771845 g
1 ounce (oz)	16 dr	28.349527 g
1 pound (lb)	16 oz	453.5924 g or 0.4536 kg
1 hundredweight (cwt)	100 lb	
1 ton (short)	2,000 lb	907.18486 kg or 0.907 (metric) ton
1 ton (long)	2,200 lb	1,016.05 kg or 1.016 (metric) ton
1 part per million (ppm)	1 microgram/gram; 1 mg/l; 1 mg/kg	.4535924 mg/lb; .907 g/ton; .0001%; .00013 oz/gal
1 percent (%) (1 part in 100 parts)	10,000 ppm; 10 g/l	1.28 oz/gal; 8.34 lb/100 gal

Weight Conversions

U.S. Customary to Metric		Metric to U.S. Customary	
To Change	Multiply By	To Change	Multiply By
grains to milligrams	64.799	grams to ounces	0.035
ounces to grams	28.35		
pounds to grams	453.6		
pounds to kg	0.454	kg to pounds	2.205
tons to metric tons	0.9	metric tons to tons	1.102

Weight—Unit Conversion Factors

To Change	Multiply By	To Change	Multiply By
mg/lb to g/ton	2	mg/g to mg/lb	453.6
g/lb to g/ton	2,000	mg/kg to mg/lb	0.4536
lb/ton to g/ton	453.6	mcg/kg to g/lb	0.4536
ppm to mg/lb	0.4536	g/ton to g/lb	0.0005
ppm to %	move decimal 4 places to left	g/ton to lb/ton	0.0022
mg/lb to ppm	2.2046	g/ton to %	0.00011
		% to g/ton	9,072.00
ppm to g/ton	0.907	g/ton to ppm	1.1

(Continued)

Appendix

TABLE VI-1 *(Continued)*

Length

Unit	Is Equal To	
	(metric)	(U.S. customary)
Metric system:		
1 millimicron (m)	.000000001 m	.000000039 in.
1 micron ( )	.000001 m	.000039 in.
1 millimeter (mm)	.001 m	.0394 in.
1 centimeter (cm)	.01 m	.3937 in.
1 decimeter (dm)	.1 m	3.937 in.
1 meter (m)	1 m	39.37 in.; 3.281 ft; 1.094 yd
1 hectometer (hm)	100 m	328.08 ft; 19.8338 rd
1 kilometer (km)	1,000 m	3,280.8 ft; 0.621 mi
U.S. customary:	(U.S. customary)	(metric)
1 inch (in.)	1 in.	25 mm; 2.54 cm
1 hand*	4 in.	10.16 cm
1 foot (ft)	12 in.	30.48 cm; .305 m
1 yard (yd)	3 ft	.914 m
1 fathom** (fath)	6.08 ft	1.829 m
1 rod (rd), pole, or perch	16½ ft; 5½ yd	5.029 m
1 chain	792 in.; 66 ft; 22 yd	20.116 m
1 furlong (fur.)	220 yd; 40 rd	201.168 m
1 mile (mi)	5,280 ft; 1,760 yd; 320 rd; 8 fur.	1,609.35 m; 1.609 km
1 knot or nautical mile	6,080 ft; 1.15 land miles	1.85 km
1 league (land)	3 mi (land)	4.827 km
1 league (nautical)	3 mi (nautical)	4.827 km

Length Conversions

U.S. Customary to Metric		Metric to U.S. Customary	
To Change	Multiply By	To Change	Multiply By
inches to millimeters	25.4	millimeters to inches	0.04
inches to centimeters	2.54	centimeters to inches	0.4
feet to centimeters	30.5	centimeters to feet	0.033
feet to meters	0.305	meters to feet	3.3
yards to meters	0.914	meters to yards	1.1
miles to kilometers	1.609	kilometers to miles	0.6

*Used in measuring height of horses.
**Used in measuring depth at sea.

(Continued)

TABLE VI-1 (Continued)

Surface/Area

Unit	Is Equal To	
Metric system:	(metric)	(U.S. customary)
1 square millimeter (mm²)	.000001 m²	.00155 in.²
1 square centimeter (cm²)	.0001 m²	.155 in.²
1 square decimeter (dm²)	.01 m²	15.50 in.²
1 square meter (m²)	1 centare (ca)	1,550 in.²; 10.76 ft²; 1.196 yd²
1 are (a)	100 m²	119.6 yd²
1 hectare (ha)	10,000 m²	2.47 acres
1 square kilometer (km²)	1,000,000 m²	247.1 acres; .386 mi²
U.S. customary:	(U.S. customary)	(metric)
1 square inch (in.²)	1 in. × 1 in.	6.452 cm²
1 square foot (ft²)	144 in.²; 0.111 yd²	.093 m²
1 square yard (yd²)	1,296 in.²; 9 ft²	.836 m²
1 square rod (rd²)	272.25 ft²; 30.25 yd²	25.29 m²
1 rood	40 rd²	10.117 a
1 acre	43,560 ft²; 4,840 yd²; 160 rd²; 4 roods	4,046.87 m²; 0.405 ha
1 square mile (mi²)	640 acres; 1 section	2.59 km² or 259 ha
1 township	36 sections; 6 miles square	

Surface/Area Conversions

U.S. Customary to Metric		Metric to U.S. Customary	
To Change	Multiply By	To Change	Multiply By
sq in. to cm²	6.452	cm² to sq in.	0.155
sq ft to m²	0.09	m² to sq ft	10.764
sq yd to m²	0.836	m² to sq yd	1.196
sq mi to km²	2.6	km² to sq mi	0.4
acres to ha	0.4	ha to acres	2.5

Weights/Measures/Unit Area

Unit	Is Equal To
Volume per unit area:	
1 liter/hectare	0.107 gal/acre
1 gal/acre	9.354 liter/ha
Weight per unit area:	
1 kilogram/cm²	14.22 lb/in²
1 kilogram/hectare	0.892 lb/acre
1 lb/sq in.	0.0703 kg/cm²
1 lb/acre	1.121 kg/ha
Area per unit weight:	
1 square centimeter/kilogram	0.0703 in.²/lb
1 sq in./lb	14.22 cm²/kg

(Continued)

TABLE VI-1 *(Continued)*

Volume

Unit	Is Equal To		
Metric system		(U.S. customary) (liquid)	(U.S. customary) (dry)
liquid and dry:			
1 milliliter (ml) .001 liter		.271 dram (fl)	.061 in.³
1 centiliter (cl) .01 liter		.338 oz (fl)	.610 in.³
1 deciliter (dl) .1 liter		3.38 oz (fl)	
1 liter 1,000 cc		1.057 qt or 0.2642 gal (fl)	.908 qt
1 hectoliter (hl) 100 liters		26.418 gal	2.838 bu
1 kiloliter (kl) 1,000 liters		264.18 gal	1,308 yd³
U.S. customary			
liquid:	(ounces)	(cubic inches)	(metric)
1 teaspoon (t) 60 drops	⅛		5 ml
1 dessert spoon 2 t			
1 tablespoon (T) 3 t	½		15 ml
1 fl oz	1	1.805	29.57 ml
1 gill (gi) ½ c	4	7.22	118.29 ml
1 cup (c) 16 T	8	14.44	236.58 ml or 0.24 liter
1 pint (pt) 2 c	16	28.88	.47 liter
1 quart (qt) 2 pt	32	57.75	.95 liter
1 gallon (gal) 4 qt	8.34 lb	231	3.79 liters
1 barrel (bbl) 31½ gal			
1 hogshead (hhd) 2 bbl			
Dry:			
1 pint (pt) ½ qt		33.6	.55 liter
1 quart (qt) 2 pt		67.20	1.10 liters
1 peck (pk) 8 qt		537.61	8.81 liters
1 bushel (bu) 4 pk		2,150.42	35.24 liters

Unit	Is Equal To	
Solid		
metric system:	(metric)	(U.S. customary)
1 cubic millimeter (mm³)	.001 cc	.061 cu in.
1 cubic centimeter (cc)	1,000 mm³	61.023 cu in.
1 cubic decimeter (dm³)	1,000 cc	35.315 ft³; 1.308 yd³
1 cubic meter (m³)	1,000 dm³	
U.S. customary:	(U.S.customary)	(metric)
1 cubic inch (in.³)		16.387 cc
1 board foot (fbm)	144 in.³	2,359.8 cc
1 cubic foot (ft³)	1,728 in.³	.028 m³
1 cubic yard (yd³)	27 ft³	.765 m³
1 cord	128 ft³	3.625 m³

Volume Conversions

U.S. Customary to Metric		Metric to U.S. Customary	
To Change	Multiply By	To Change	Multiply By
ounces (fluid) to cc	29.57	cc to oz (fluid)	0.034
ounces to ml	29.57	ml to oz	0.034
qt to liters	0.946	liters to qt	1.057
cu in. to cc	16.387	cc to cu in	0.061
cu yd to cm	0.765	cm to cu yd	1.308

TABLE VI-2
TEMPERATURE

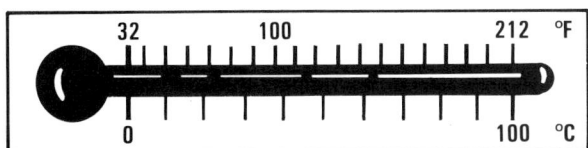

Fig. VI-1. Fahrenheit-Centigrade scale for direct conversion and reading.

One Fahrenheit (F) degree is 1/180 of the difference between the temperature of melting ice and that of water boiling at standard atmospheric pressure. One Fahrenheit degree equals 0.556°C.

One Centigrade (C) degree is 1/100 the difference between the temperature of melting ice and that of water boiling at standard atmospheric pressure. One Centigrade degree equals 1.8°F.

To Change	To	Do This
Degrees Fahrenheit	Degrees Centigrade	Subtract 32, then multiply by .556 (5/9)
Degrees Centigrade	Degrees Fahrenheit	Multiply by 1.8 (9/5) and add 32

Weights and Measures of Common Feeds

In calculating rations and mixing concentrates, it is usually necessary to use weights rather than measures. However, in practical feeding operations it is often more convenient for the farmer or rancher to measure the concentrates by volume. Table VI-3 will serve as a guide in feeding by measure.

TABLE VI-3
WEIGHTS AND MEASURES OF COMMON FEEDS

Feed	Approximate Weight	
	Lb per Quart[1]	Lb per Bushel[1]
Alfalfa meal	0.6	19
Barley	1.5	48
Beet pulp (dried)	0.6	19
Brewers' grain (dried)	0.6	19
Buckwheat	1.6	51
Buckwheat bran	1.0	32
Corn, husked ear	—	70
Corn, cracked	1.6	51
Corn, shelled	1.8	58
Corn meal	1.6	51
Corn-and-cob meal	1.4	45
Cottonseed	0.9–1.0	29–32
Cottonseed meal	1.5	48
Cowpeas	1.9	61
Distillers' grain (dried)	0.6	19
Fish meal	1.0	32
Flax	1.7	54
Gluten feed	1.3	42
Linseed meal (old process)	1.1	35
Linseed meal (new process)	0.9	29
Meat scrap	1.3	42
Milo (grain sorghum)	1.7	54
Molasses feed	0.8	26
Oat middlings	1.5	48
Oats	1.0	32
Oats, ground	0.7	22
Peanut meal	1.0	32
Peas	1.9	61
Rice	1.4	45
Rice bran	0.8	26
Rye	1.7	54
Sorghum (grain)	1.7	54
Soybeans	1.8	58
Sunflower	0.7	22
Tankage	1.6	51
Velvet beans, shelled	1.8	58
Wheat	1.9	61
Wheat bran	0.5	16
Wheat middlings, standard	0.8	26
Wheat screenings	1.0	32

[1]32 qts per bushel.

Storage Space Requirements for Feed and Bedding

The space requirements for feed storage for the livestock enterprise—whether it be for cattle, sheep, hogs, or horses, or, as is more frequently the case, a combination of these—vary so widely that it is difficult to provide a standard method of calculating space requirements applicable to such diverse conditions. The amount of feed to be stored depends primarily upon (1) length of pasture season, (2) method of feeding and management, (3) kind of feed, (4) climate, and (5) the proportion of feeds produced on the farm or ranch in comparison with those purchased. Normally, the storage capacity should be sufficient to handle all feed grain and silage grown on the farm and to hold purchased supplies. Forage and bedding may or may not be stored under cover. In those areas where weather conditions permit, hay and straw are frequently stacked in the fields or near the barns in loose, baled, or chopped form. Sometimes poled, framed sheds or a cheap cover of waterproof paper, grass, or cereal straw grass are used for protection. Other forms of low-cost storage include temporary upright silos, trench silos, temporary grain bins, and open-walled buildings for hay.

Table VI-4 gives the storage space requirements for feed and bedding. This information may be helpful to the individual operator who desires to compute the barn storage space required for a specific livestock enterprise. This table provides a convenient means of estimating the amount of feed or bedding in storage.

TABLE VI-4
STORAGE SPACE REQUIREMENTS FOR FEED AND BEDDING

Kind of Feed or Bedding	Pounds per Cubic Foot	Cubic Feet per Ton	Pounds per Bushel of Grain
Hay-Straw:[1]			
1. Loose			
Alfalfa	4.4–4.0	450– 500	
Nonlegume	4.4–3.3	450– 600	
Straw	3.0–2.0	670–1,000	
2. Baled			
Alfalfa	10.0–6.0	200– 330	
Nonlegume	8.0–6.0	250– 330	
Straw	5.0–4.0	400– 500	
3. Chopped			
Alfalfa	7.0–5.5	285– 360	
Nonlegume	6.7–5.0	300– 400	
Straw	8.0–5.7	250– 350	
Corn:			
15½% moisture:			
Shelled	44.8		56.0
Ear	28.0		70.0
Shelled, ground	38.0		48.0
Ear, ground	36.0		45.0
30% moisture:			
Shelled	54.0		67.5
Ear, ground	35.8		89.6
Barley, 15% moisture	38.4		48.0
Ground	28.0		37.0
Flax, 11% moisture	44.8		56.0
Oats, 16% moisture	25.6		32.0
Ground	18.0		23.0
Rye, 16% moisture	44.8		56.0
Ground	38.0		48.0
Sorghum grain, 15% moisture	44.8		56.0
Soybeans, 14% moisture	48.0		60.0
Wheat, 14% moisture	48.0		60.0
Ground	43.0		50.0

[1]Many factors—other than kind of hay-straw, form (loose, baled, chopped), and period of settling—affect the density of hay-straw in a stack or in a barn, including (a) moisture content at haying time, and (b) texture and foreign material.

Appendix

Grain Weight in a Bin

Sometimes farmers need to estimate the weight of grain in storage. Such estimates are difficult to make because of differences in moisture content, depth of material stored, and other factors. However, the following procedure will enable one to figure feed quantities fairly closely.

1. **Corn (shelled) or small grain in rectangular cribs or bins.** Multiply the width by the length by the average depth (all in feet) and multiply by 0.8 to get the number of bushels (multiplying by 0.8 is the same as dividing by 1¼, the number of cubic feet in a bushel).

2. **Ear corn in rectangular cribs or bins.** Multiply the width by the length by the average depth (all in feet) and multiply by 0.4 to get the number of bushels (multiplying by 0.4 is the same as dividing by 2½, the number of cubic feet in a bushel of ear corn).

3. **Round bins or cribs.** To find the cubic feet in a cylindrical bin, multiply the squared radius by 3.1416 by the depth.

Thus, the volume of a round bin 20 ft in diameter and 10 ft deep is determined as follows:

  a. The radius is half the diameter, or 10 ft

  b. 10 × 10 = 100

  c. 100 × 3.1416 = 314.16

  d. 314.16 × 10 = 3,141.6 cu ft

  e. Where shelled corn or small grain is involved, one would multiply 3,141.6 × 0.8, which equals 2,513.28 bu of grain that it would hold if full.

  f. Where ear corn is involved, one would multiply 3,141.6 × 0.4 which equals 1,256.64 bu of ear corn that it would hold if full.

Hay Weight in a Barn or Stack

Livestock producers and hay dealers frequently buy and sell large quantities of hay in the stack or in the barn. This practice is especially prevalent in the western and Great Plains states where cattle and sheep are brought into the valleys to be wintered on hay bought from valley hay producers. Under such circumstances, the weight of hay is usually estimated, because (1) no scales are available, and/or (2) it is impractical to weigh the hay due to the time, labor, and wastage involved. In many such instances, the hay is fed directly from the stack or barn, in racks arranged about it. Under these and other circumstances, there is need for a simple and reasonably accurate method of estimating the weight of hay in a stack or in a barn.

In order to estimate the tonnage of hay in a stack or in a barn, it is necessary (1) to compute the volume of hay, and (2) to know the number of cubic feet per ton of hay. Table VI-5 gives the latter information.

TABLE VI-5
CUBIC FEET PER TON OF HAY

Feed	Settled 1–2 Months	Settled Over 3 Months
	(cu ft)	(cu ft)
Alfalfa	485	470
Clover	512	500
Hay, baled (closely stacked)	150–300	150–200
Hay, chopped	225	210
Straw, baled	200	200
Straw, loose	1,000	600–1,000
Timothy	640	625
Wild hay	600	450

In using Table VI-5, it should be recognized that many factors—other than kind of hay, form (loose, chopped, or baled), and period of settling—affect the density of hay in a barn or in a stack, including (1) moisture content at haying time, and (2) texture and foreign material.

It is relatively simple to compute the volume of hay in a mow, but it is more difficult to determine the volume of a stack. Although different rules or formulas may be and are used, the following are recommended by the U.S. Department of Agriculture.[2]

1. **Volume of hay in barns.** Multiply the width by the length by the height, all in feet, and divide by the cubic feet per ton as given in Table VI-6.

2. **Volume of hay in oblong stacks.** Three types of oblong stacks are common, as shown in Fig. VI-2.

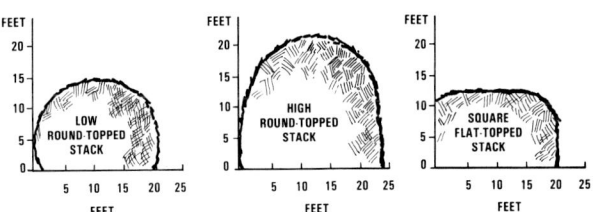

Fig. VI-2. Three common types of oblong stacks.

The volume of each type of oblong stack may be determined as follows:

  a. For low, round-topped stacks—

    (0.52 × O) − (0.44 × W) × W × L

  b. For high, round-topped stacks—

    (0.52 × O) − (0.46 × W) × W × L

  c. For square, flat-topped stacks—

    (0.56 × O) − (0.55 × W) × W × L

In these formulas "O" is the "over" or "overthrow," which is the distance in feet from the ground on one side

---

[2]*Measuring Hay in Stacks*, USDA Leaflet No. 72.

of the stack, up and over the stack and down to the ground on the other side; W is the width; and L is the length.

The application of this formula is illustrated as follows:

Example. *It is desired to estimate the amount of alfalfa hay in a low, round-topped type of oblong stack that has settled for 4 months. The stack is 20 ft wide, 30 ft long, and has an over of 40 ft.*

The answer is secured as follows:

a. Volume = (0.52 × 40) − (0.44 × 20) × 20 × 30 = 7,200 cu ft.

b. Table VI-6 shows that there are 470 cu ft per ton of settled alfalfa.

c. 7,200 ÷ 470 = 15 tons of hay.

3. **Volume of hay in round stacks.** The rules or formulas used for oblong stacks do not apply to round stacks. The volume of round stacks can be calculated by using the following formula:

Volume = (0.04 × O) − (0.012 × C) × C²

In this formula, C equals the circumference or distance around the stack at the ground, and O equals the over or distance from the ground on one side over the peak to the ground on the other side (usually it is best to take 2 over measurements at right angles to each other, and to average them).

Thus, the computation of the volume of a large round stack may be illustrated by the following example:

Example. *It is desired to determine the amount of alfalfa hay in a round stack that is 100 ft in circumference and has an average over of 60 ft.*

The answer is secured as follows:

a. Volume = (0.04 × 60) − (0.012 × 100) × (100)² = 12,000 cu ft.

b. Table VI-6 shows that there are 470 cu ft per ton of settled alfalfa.

c. 12,000 ÷ 470 = 25.5 tons of hay.

Animal Weights

Feeders who finish large numbers of animals have scales in their feedyards for use in determining in-weights, out-weights, and interim weight gains of animals while they are on feed. Likewise, both purebred and commercial breeders usually have scales. However, those with only one animal, or a few head—such as 4-H Club and FFA members, and part-time farmers—may not have scales. As a result, rations cannot be accurately evaluated, rate of gain cannot be calculated, and an animal's "weight readiness" for a livestock show or for market cannot be determined. Under such circumstances, a simple but reasonably accurate method of estimating body weight is very useful. Fortunately, animal weights may be determined with reasonable accuracy by taking two body measurements (body length and heart girth), then applying an appropriate formula.

BEEF CATTLE WEIGHTS

Here is how to do it:

Step 1. Measure the circumference (heart girth), from a point slightly behind the shoulder blade, thence down over the foreribs and under the body, behind the elbow (distance C of Fig. VI-3).

Step 2. Measure the length of body, from the point of the shoulder to the point of the rump (pinbone), in inches (distance A-B of Fig. VI-3).

Step 3. Take the values obtained in Steps 1 and 2 and apply the following formula to calculate body weight:

Heart girth × heart girth × body length ÷ 300 = weight in pounds

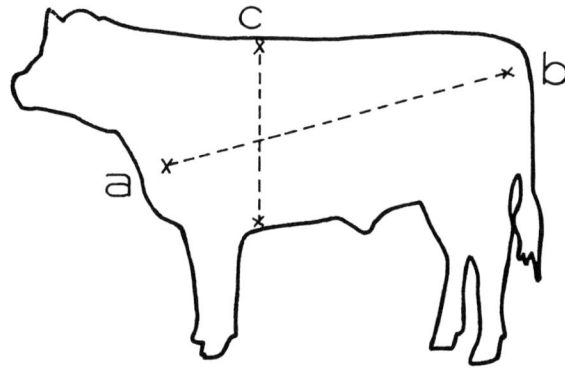

Fig. VI-3. How and where to measure beef cattle.

Example of a beef animal. *Assume that the heart girth measures 76 in. and the body length, 66 in. How much does the animal weigh?*

76 × 76 = 5,776
5,776 × 66 = 381,216
381,216 ÷ 300 = 1,270 lb

DAIRY HEIFER WEIGHTS

Weight for age is important in dairy heifers from the standpoint of determining the growth progress made by herd replacements.

Table 20-24 of Chapter 20, Feeding Dairy Cattle, shows the weight and heart girth measurements of dairy calves or heifers at monthly intervals up to 22 months of age. If producers do not have scales, they can measure the heart girth with a tape (see Fig. VI-4) and use Table 20-24 to estimate weight within 95% accuracy.

Appendix

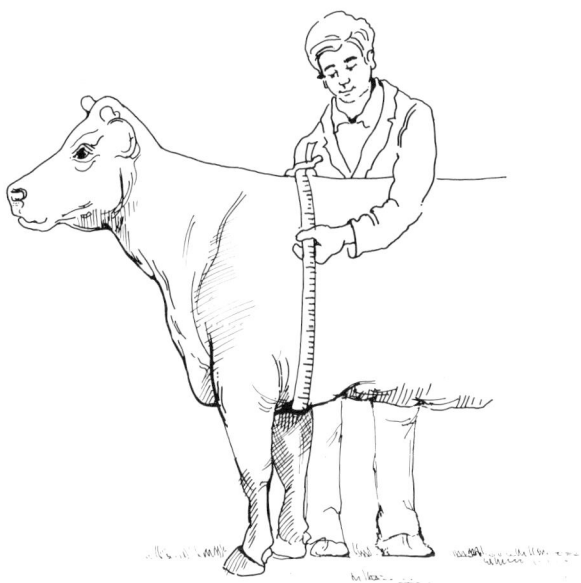

Fig. VI-4. How to tape measure a dairy heifer.

SHEEP AND GOAT WEIGHTS

The weight of sheep and goats is estimated in the same way as for beef cattle; hence, it involves making the measurements and applying the formula given for beef cattle. There is one important precaution, however; with unshorn sheep, be sure to part, or compress, the wool to ensure an accurate heart girth measurement.

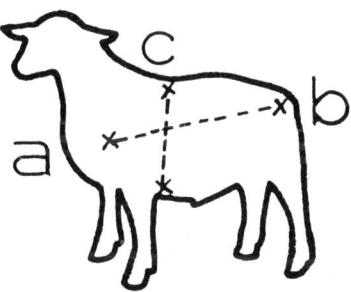

Fig. VI-5. How and where to measure sheep.

SWINE WEIGHTS

Hog weights can be calculated from body measurements, similar to beef cattle, but a different formula must be used. Here is how to estimate the weight of hogs:

Step 1. Measure the circumference (heart girth) of the animal (C in diagram).

Step 2. Measure the length of body (A-B in Fig. VI-6). With the animal standing or restrained in the position shown in Fig. VI-6, measure the distance from the poll (between the ears), over the backbone, to the base of the tail.

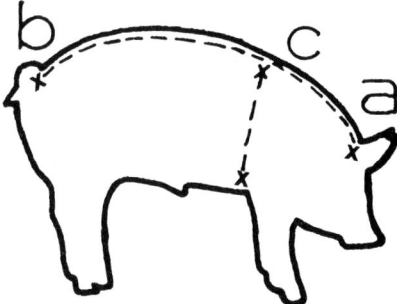

Fig. VI-6. How and where to measure hogs.

Step 3. Apply the following formula:

Heart girth × heart girth × length ÷ 400 = weight in pounds.

Note: For hogs weighing less than 150 lb, add 7 lb to the weight figure obtained from the formula. For animals weighing 151 to 400 lb, no adjustment is necessary.

HORSE WEIGHTS

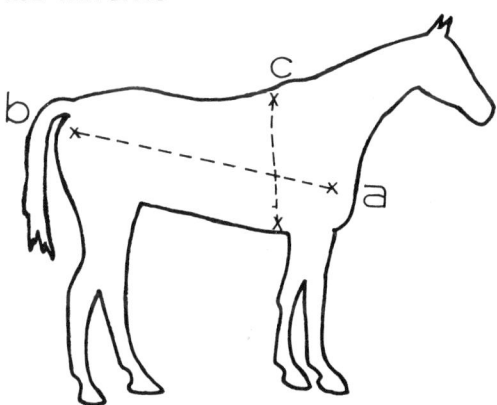

Fig. VI-7. How and where to measure horses.

It is easy to estimate the weight of a horse; and tests have shown that the results obtained this way are accurate within 3% of actual scale weight. This procedure is as follows:

Step 1. Measure the circumference (heart girth) of the body in inches (C in diagram).

Step 2. Measure the length of body from the point of the shoulder to the point of croup (A-B in the diagram).

Step 3. Apply the following formula to calculate the weight of the horse:

Heart girth × heart girth × length ÷ 300 + 50 lb = weight of horse.

Example. *Assume that the heart girth is 70 in. and the body length is 65 in. How much does the horse weigh?*

70 × 70 × 65 ÷ 300 + 50 lb = weight
4,900 × 65 = 318,500
318,500 ÷ 300 = 1,061 lb
1,061 + 50 = 1,111 lb body weight

ANIMAL UNITS

An animal unit is a common animal denominator, based on feed consumption. It is assumed that 1 mature cow represents an animal unit. Then, the comparative (to a mature cow) feed consumption of other age groups or classes of animals determines the proportion of an animal unit which they represent. For example, it is generally estimated that the ration of one mature cow will feed 5 mature ewes, or that 5 mature ewes equal 1.0 animal unit.

The original concept of an animal unit included a weight stipulation—an animal unit referred to a 1,000-lb cow, with or without a calf at side. Unfortunately, in recent years, the 1,000-lb qualification has been dropped. Certainly, there is a wide difference in the daily feed requirements of a 900-lb cow and of a 1,500-lb cow. Both will consume dry matter on a daily basis at a level equivalent to about 2% of their body weight.

Hence, a 1,500-lb cow will consume 50% more feed than a 1,000-lb cow.

Also, the period of time to be grazed has an effect on the total carrying capacity. For example, if an animal is carried for 1 month only, it will take $\frac{1}{12}$ of the total feed required to carry the same animal 1 year. For this reason, the term *animal unit months* is becoming increasingly important. So, in addition to the weight factor, the time factor has a distinct bearing on the ultimate carrying capacity of a tract of land.

Table VI-6 gives the animal units of different classes and ages of livestock.

TABLE VI-6
ANIMAL UNITS

Type of Livestock	Animal Units
Cattle:	
Cow, with or without unweaned calf at side, or heifer 2 years old or older	1.0
Bull, 2 years old or older	1.3
Young cattle, 1 to 2 years	0.8
Weaned calves to yearlings	0.6
Horses:	
Horse, mature	1.3
Horse, yearling	1.0
Weanling colt or filly	0.75
Sheep:	
5 mature ewes, with or without unweaned lambs at side	1.0
5 rams, 2 years old or over	1.3
5 yearlings	0.8
5 weaned lambs to yearlings	0.6
Swine:	
Sow	0.4
Boar	0.5
Pigs to 200 lb	0.2
Chickens:	
75 layers or breeders	1.0
325 replacement pullets to 6 months of age	1.0
650 8-week-old broilers	1.0
Turkeys:	
35 breeders	1.0
40 turkeys raised to maturity	1.0
75 turkeys to 6 months of age	1.0

Fig. VI-8. White turkeys on the range. Forty turkeys raised to maturity equal one animal unit. (Courtesy, J. C. Allen & Son, West Lafayette, Ind.)

POISON INFORMATION CENTERS

With the large number of chemical sprays, dusts, and gases now on the market for use in agriculture, accidents may arise because of operators being careless in their use. Also, there is always the hazard that a child may eat or drink something that may be harmful. Centers have been established in various parts of the country where doctors can obtain prompt and up-to-date information on treatment of such cases, if desired.

Local medical doctors have information relative to the Poison Information Centers of their area, along with some of the names of their directors, telephone numbers, and street numbers. When calling any of these centers, one should ask for the "Poison Information Center." If this information cannot be obtained locally, call the U.S. Public Health Service at Atlanta, Ga., or Wenatchee, Wash.

Appendix

CAREERS IN ANIMAL INDUSTRIES

Animal Industries embraces all the technical, professional, and business aspects connected with America's largest and most important single industry—that of producing, processing, and distributing meat, milk, eggs, and fiber for all the people. It extends from the nation's farms and ranches to its kitchens, with many steps between. It also includes animals kept for clothing, recreation, sport, and companionship.

Career Opportunities in the Animal Industries

Of course, there are many good and interesting professions. In fact, one of the most cherished heritages of American youth is the opportunity to choose a profession—then build it into a life career. In reaching this decision, two primary considerations are important: (1) a knowledge of various fields, and (2) an evaluation of one's ability, aptitude, and interests. Let's pursue the first one from the standpoint of Animal Industries.

Animal Industries are scientific, specialized, mechanized, and industrialized. They provide job opportunities in the areas described on the pages that follow. Actually, there are countless distinct occupations in Animal Industries, most of which are in businesses and industries closely related to agriculture rather than on the farm.

Magnitude of U.S. Animal Industries

The far-flung livestock business comprises one of the largest industries in the United States. The enormity of providing America's meat, milk, eggs, fiber, and horses is evidenced by the following:

1. **Land area and number of farms and ranches devoted to animal production.** Animals are the largest users of the nation's land. More than 29% of the U.S. land area is devoted to pasture and rangeland, exclusive of forested grazing land—more than any other land use; and 37% of U.S. land is devoted to the production of livestock feed (pasture and rangeland, hay and other forage crops, and grain).

2. **Comparative cash income derived from different agricultural pursuits.** Consistently, livestock and their products account for more cash income than crops. Fig.VI-9 shows the cash receipts derived from U.S. livestock and crops in 1987.

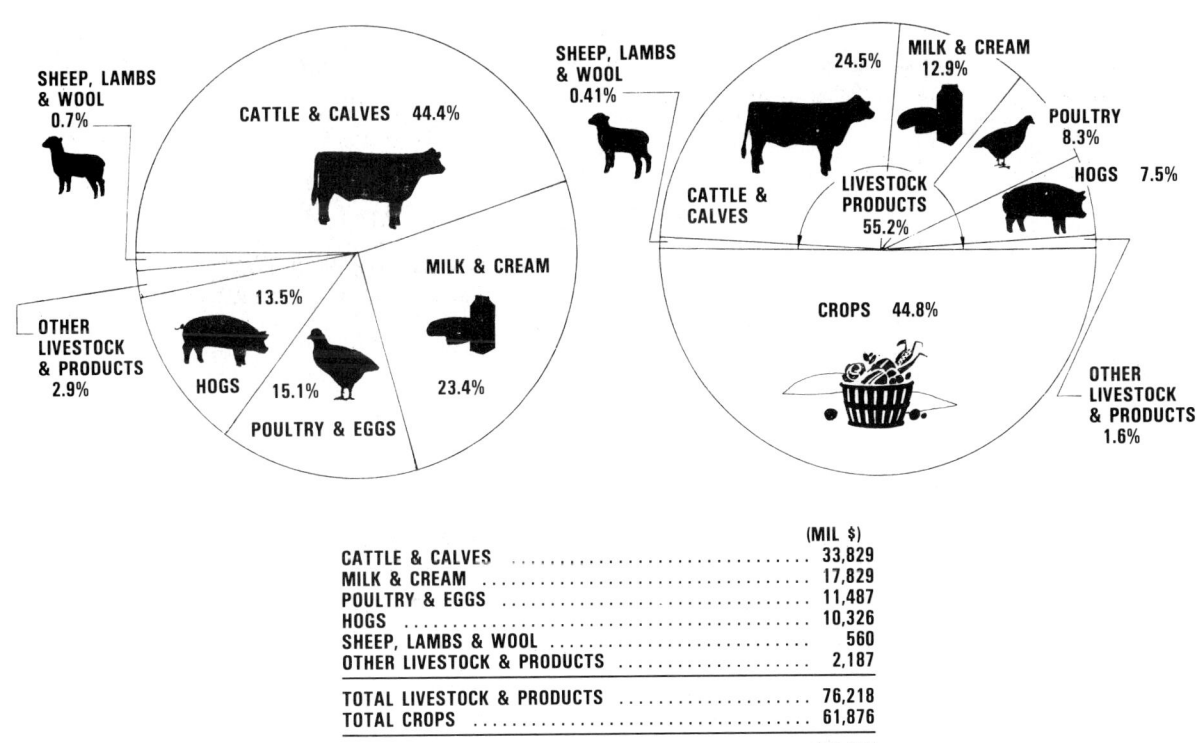

Fig. VI-9. Comparative cash income derived from different agricultural pursuits in 1987. (From *Agricultural Statistics 1988*, USDA, p. 409, Table 581)

The pie diagram on the left shows the proportion of livestock income accounted for by each class of livestock. The dominant position of cattle and calves is quite obvious; they accounted for nearly half of all income derived from livestock and livestock products in 1987.

The chart on the right shows the relative importance of U.S. livestock and crops in terms of cash receipts. It is noteworthy that livestock and their products accounted for 55.2% of the cash income received by U.S. farmers and ranchers, while crops accounted for 44.8%.

3. **Number and value of livestock on farms and ranches.** In 1988, there were 166 million head of cattle, hogs, sheep, and goats on U.S. farms, with an aggregate value of $57 billion (exclusive of the value of chickens, turkeys, and horses). Thus, animals are a big investment.

4. **Size and diversity of the meat team.** Getting meats from producers to consumers involves (1) 1.8 million cattle, sheep, and hog farms and ranches; (2) 42,000 feedlots; (3) 1,780 livestock marketing points; (4) 6,753 meat packers; (5) 181,000 meat retailers; and (6) 727,000 food service operations. Thus, the meat team involves many people doing many different things. (See Fig. VI-10)

HOW MEAT REACHES THE TABLE

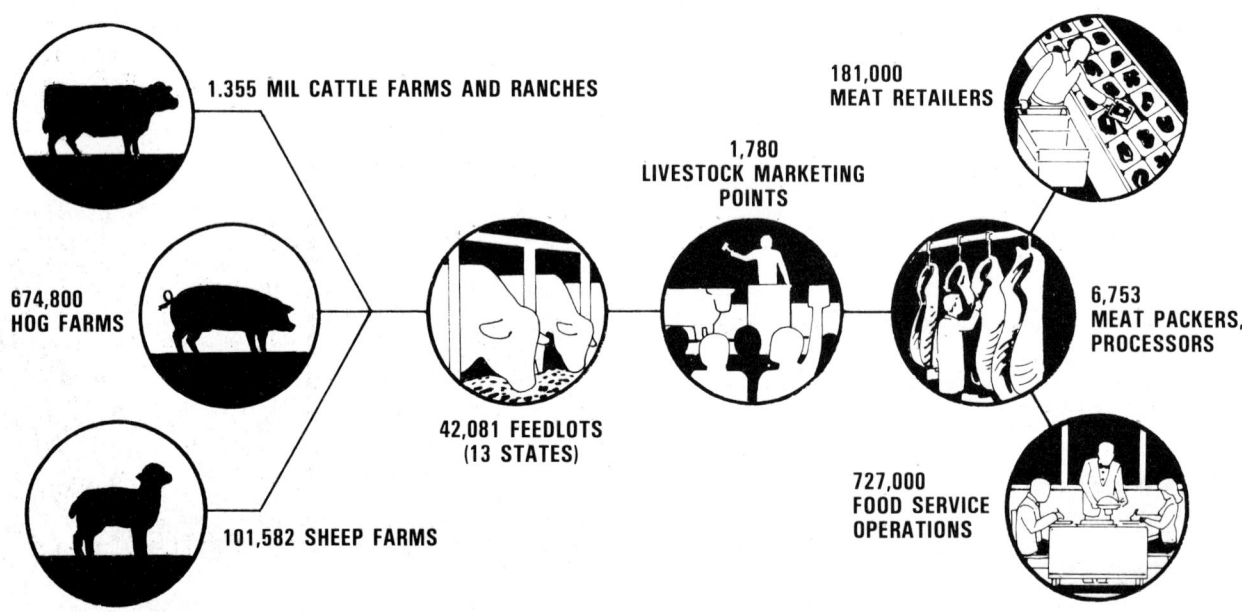

Fig. VI-10. From whence meat comes. Cattle, hogs, and sheep are produced on more than 1.8 million farms and ranches, thence animals and meats progress through many intermediate steps from farms to consumers. (Data provided by the Economic Research Service, USDA, in 1989)

Agriculture—The Nation's Largest Industry

Agriculturalists can stand high and be proud. Daniel Webster put it this way:

> "Let us never forget that the cultivation of the earth is the most important labor of man. When tillage begins, other arts follow. The farmers, therefore, are the founders of civilization." (Daniel Webster, *On Agriculture*, Jan. 13, 1840)

Although Daniel Webster made this statement a century and a half ago, it is still true today. Agriculture is the nation's largest industry. The following facts and figures point up the enormity and importance of American agriculture:

1. Total farm assets exceed $800 billion, which is equal to about 40% of the total capital assets of all manufacturing corporations in the United States.

2. More than 21 million people are employed in some phase of agriculture; they are either producing, processing, or marketing farm commodities, or they are supplying goods used on the farm. About one job out of every five in the U.S. is farm/food related.

3. A hundred years ago, half of the nation's population was required on farms to feed themselves and the other half. In 1820, each farm worker supplied farm products for 4 people (self and 3 others). By 1940, one farm worker produced enough for 11 people (self and 10 others). Today, one farm worker supplies enough agricultural products for 95 people (self and 94 others), more than 8 times the 1940 productivity (see Fig. VI-11)—thereby freeing 94 people to produce such luxuries as autos, refrigerators, TV sets, and a host of other goods and services for modern American living.

Appendix

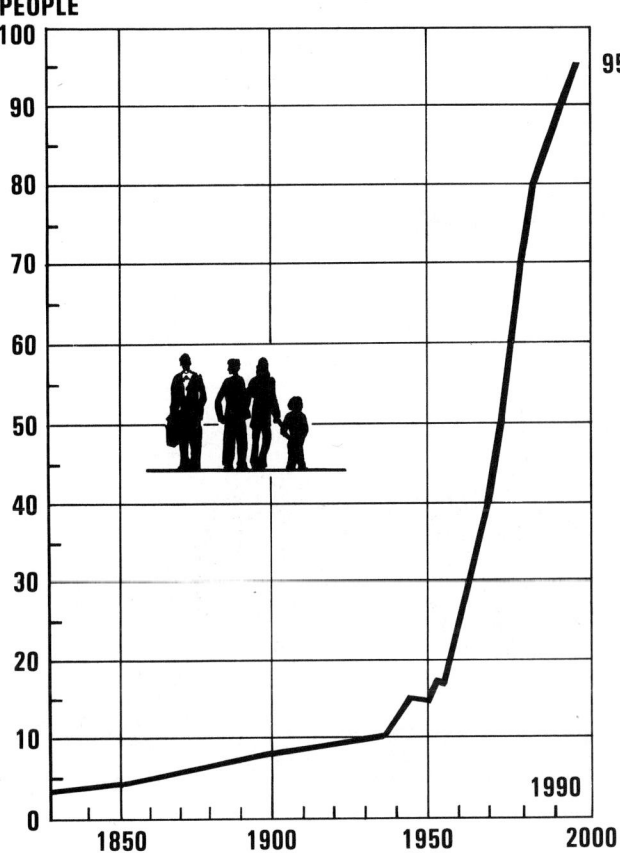

Fig. VI-11. Productivity of farm worker; the number of persons supplied farm products by one farm worker. In 1820, each farm worker supplied farm products for 4 people (self and 3 others). Today, one farm worker supplies enough agricultural products for 95 people (self and 94 others). (*Source:* USDA)

ducts produced; along with the production of large quantities of drugs and chemicals. While some aspects of biotechnology are decades away from commercial production, others are near, and still others are here now.

But advanced technology calls for advanced animal adaptation, welfare, and environmental control. We need to breed and select animals adapted to an artificially-made environment—animals that not only survive, but thrive, under the conditions in which they are kept. We need to heed the warnings of endangered animals, endangered people, and an endangered planet—presaged by increased pollution, the greenhouse effect, acid rain, depletion of the ozone layer, and destruction of rain forests.

Biotechnology will be the key to unlocking vast improvement per animal, per feed unit, and per dollar investment. The speed with which innovations are developed will be limited only by people.

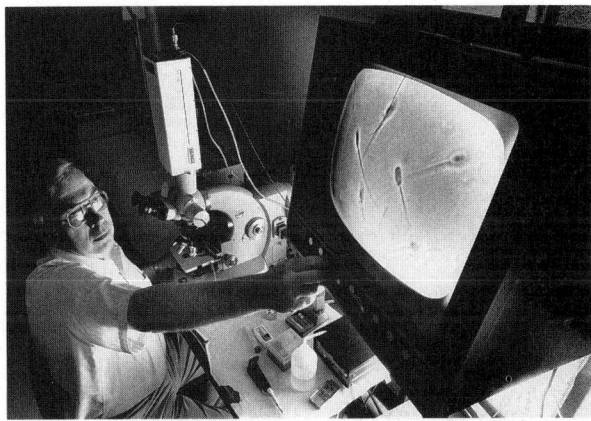

Fig. VI-12. Sex control research. USDA Animal Physiologist Dr. Lawrence Johnson, watches male sperm cells to evaluate their motility, a procedure that precedes laser separation of x-chromosome (female) sperm from y-chromosome (male) sperm. This research is still in the experimental stage. If perfected, farmers could select the sex of their animals. For example, a dairy producer could produce all heifer calves for herd replacements, and eliminate unwanted bull calves. (Courtesy, USDA)

4. In recent years, the production from about one-fourth of U.S. cropland has moved into export markets. So, farm exports make a most important contribution to the U.S. balance of trade.

5. Farmers are big buyers of goods produced by other industries. Annually, they normally spend over $6 billion for tractors, trucks, cars, and machinery. Each year, they spend another $9.5 billion to fuel, lubricate, and maintain this fleet. Additionally, farmers make annual expenditures of about $19 billion for feed and seed and nearly $6 billion for fertilizers and lime.

New Frontiers in the 21st Century

The past is prologue! Genetic wizardry by gene splicing is giving rise to a major scientific revolution called biotechnology and spawning many new developments exceeding our fondest dreams. Biotechnology will involve every facet of animal production from breeding and feeding to the finished product, including the genetic makeup of animals and the feeds they eat; the digestion, physiology, stress tolerance, disease resistance, and efficiency of production of animals; the composition, quality, and quantity of pro-

Fig. VI-13. USDA researcher Dr. Robert A. Bellows flushes 7-day-old embryos from a cow treated with a hormone that induces multiple ovulation. Recovered embryos are implanted into surrogate mothers or frozen for later use. (Courtesy, USDA)

World Without End—With Animals

Global food demand is rising significantly. World population will grow from over 5 billion now to more than 6 billion soon after the year 2,000. Thus, early in the 21st century, there will be another billion mouths to feed.

World food consumption is determined by population and the amount eaten per person. It is expected to double over the next three decades, led by greater per capita consumption linked to rising incomes, changing tastes, and improved food supplies in the developing countries.

Practicality dictates that in the years ahead a hungry world will meet its increased food needs through having plants and animals play complementary roles—with animal products complementing the deficiencies of plant products. Grazing land is highly efficient in the capture of solar energy—requiring little energy for a high return. However, grass does not store the energy in a form available to humans. It follows that ruminants which can utilize grass not suitable for human consumption and convert it into meat and milk are essential.

For a world without end, the developing countries need a massive infusion of research, technology, and education, with emphasis on plant-animal relationships. Other approaches serve only to prolong and aggravate the current disparities.

Fig. VI-14. An Ecuadoran farmer, high in the Andes Mountains, tills with a primitive plough which has a wooden share and is hard to pull. (Courtesy, FAO, United Nations, Rome, Italy)

Consumers Benefit From Animal Industries

Animal Industries are essential to a well-nourished and happy people.

Back of animals are feeds; and back of the feeds are soil resources, spring rains, and the energy of the sun. With the aid of science, technology, and animals, farmers and ranchers combine these to produce a tasty platter of meat and eggs for the table, cream for the peaches, butter for the biscuits, and cheese for the macaroni—all derived from the sun through the process known as photosynthesis.

But animal products are far more than just tempting and delicious foods! From a nutrition standpoint, foods of animal origin contribute certain essentials to the American diet; they supply 35% of the energy, 70% of the protein—along with the essential amino acids, 80% of the calcium, 60% of the phosphorus, and significant amounts of the other minerals and vitamins needed in the human diet. It is noteworthy, too, that animal products contain vitamin B-12, which does not occur in plant foods, and that they are a rich source of iron, the availability of which is twice as high as in plants.

In addition to eating extremely well, Americans eat at a very low cost. Recently, the average U.S. family has been spending only about 10% of its income for food. Each year, this provides the average person with about 135 pounds of red meats, 82 pounds of poultry meats, 90 pounds of fruits, 72 pounds of potatoes, 80 pounds of fresh vegetables, and 263 pounds of milk and milk products. In comparison, consumers in other countries spend far more of their income for food. In France and Japan, they spend about 16% of their income to feed themselves; in Poland, Spain, the U.S.S.R., and South Korea, they spend 30 to 38% of their income for food; and in the Philippines, 49%. In 1988, one pound of sirloin steak cost $5.21 in Washington, D.C.; in Tokyo, Japan, it cost $23.27.

Fig. VI-15. The Ron Davis family prepares to eat a noon meal in their home near Visalia, Calif. (Courtesy, USDA)

The Functions of Animals

The primary utility functions of animals are the production of food, clothing, and power—upon which people become more and more dependent with each succeeding step in their advancing civilization. There will always be an animal agriculture because of these and other functions, which follow:

1. **Food.** Today, nearly half of the food supply is contributed by mammalian, avian, and aquatic life. But foods of animal origin are more costly than foods of vegetable origin; so, more than half of the current annual grocery bill of more than $283 billion goes for animal products.

Fig. VI-16. Animals contribute meat for the table. This shows a pork loin roast. (Courtesy, National Live Stock and Meat Board, Chicago, Ill.)

Appendix

2. **Clothing.** The chief contributions of animal life to clothing are in the form of wool, leather, and hair.

Fig. VI-17. Animals contribute clothing. Basic black and vivid fuchsia team up in this attractive wool jersey combination. (Courtesy, The Wool Bureau, Inc., New York, N.Y.)

3. **Power.** In the developing countries, cattle, water buffalo, donkeys, mules, horses, and camels still provide much of the agricultural power. Work animals are part of the agricultural scene of Asia, Africa, the Near East, Latin America, and parts of Europe—areas characterized by small farms, low incomes, an abundance of human labor, and lack of capital. But animals have certain advantages! They can be fueled on roughages to produce power, a most important consideration in time of energy shortage; and both cattle and water buffalo can be used for work, milk, and meat. Although the general trend in the world is toward more and more mechanization, animals will continue to provide most of the agricultural power in many of the developing countries.

Fig. VI-18. Animals contribute power. A team of mules pulling a wooden plow.

4. **Recreation and companionship.** No mechanical device has such wide appeal as horses for recreation and sport. Also, animals provide companionship; people need animals, and animals need people.

Fig VI-19. Animals contribute recreation and sport. This shows instruction in Western Equitation at Mississippi State University, Mississippi State, Miss.)

5. **Animals serve as an important companion of grain production.** Livestock feeding provides a large and flexible outlet for grain supplies. Thus, animals give stability to grain farming. Also, meat animals serve as a "storehouse" of food in times of scarcity.

6. **Ruminants utilize grass and roughage.** Ruminants provide practically the only means of marketing millions of acres of grass and other roughages.

Fig. VI-20. Dairy cows and other ruminants can utilize roughages such as this pearl millet. (Courtesy, Northrup King Co., Minneapolis, Minn.)

7. **Animals can utilize land that is not cultivatable.** Vast acreages throughout the world—including arid and semiarid grazing lands; and brush, forest, cutover, and swamp lands—are unsuited to the production of bread grains or any other type of farming. Their highest and best use is, and will remain, for grazing and forest.

Fig. VI-21. Sheep on the range, utilizing land that is not suited to cultivation. (Courtesy, *The Sheepman's Production Handbook*, Denver, Colo.)

8. **Animals utilize by-product feeds.** Animals provide a practical outlet for a host of by-product feeds derived from plants and animals, which are not suited for human consumption.

9. **Animals provide medicinal and other products.** Animals are not processed for meat alone. They are the source of hundreds of important by-products, including medicines, and everything from candles to cosmetics; without which the health and life-style of many people would never be the same.

10. **Animals maintain soil fertility.** Animals provide manure for the fields, a fact which is often forgotten when chemical fertilizers are relatively abundant and cheap.

The Food Industry

Through the years, consumers have demanded, and gotten, more and more processing and packaging of foods—more built-in services. Today, few consumers are interested in buying a live hog—or even a whole carcass. Instead, they want a pound of pork chops—all trimmed, packaged, and either ready for cooking or pre-cooked. Likewise, few housewives are interested in buying flour and baking bread.

The increased demand for processed food has resulted from the rising standard of living, the desire for a more diversified diet the year round, an expanding urbanization, an increase in the total population, more women working outside the home, and the desire for more leisure time.

Scientific knowledge of food processing and human inventiveness have successfully extended the period of availability of foods in forms that retain their nutritive and esthetic values. This has improved human health, added variety to the diet, reduced the drudgery of food preservation, and increased the mobility of consumers. Also, the technology of processing has greatly expanded the markets for agricultural products, both at home and abroad. Today, food is big business. Annually, the U.S. food system contributes about 20% to our gross national product.

Fig. VI-23. Meat packers process about one-fourth of the nation's manufactured food products. This shows a refrigerated truck at the loading dock of Monfort of Colorado, a major meat packer. (Courtesy, Union Pacific Railroad Co., Omaha, Nebr.)

A Galaxy of Career Opportunities

During the last three decades, there have been tremendous technological innovations in the livestock industry. Computers have revolutionized animal research, production, and marketing. Alternative feed sources have been developed, new growth promoting compounds have been discovered, estrous synchronization has been perfected, and embryo transfer has been commercialized. While these developments were occurring, molecular geneticists unraveled the complexities of mammalian genes and the field of biotechnology was born. Hand in hand with this progress, knowledge of immunology and animal behavior made for greater animal efficiency.

To meet the challenges in Animal Science and related industries, career opportunities abound. The United States Department of Agriculture reports that more than 48,000 jobs are being created annually for college graduates with expertise in agriculture and related industries. The livestock

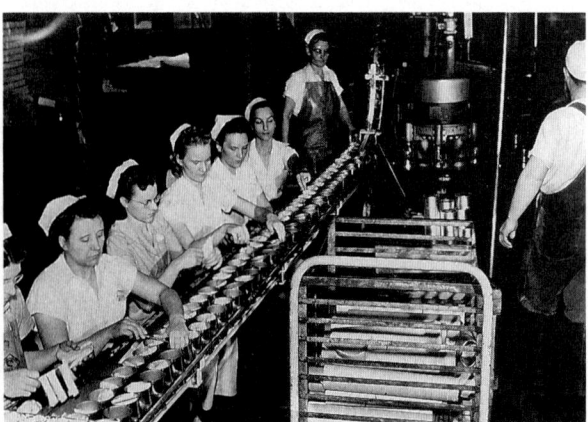

Fig. VI-22. Canning meat. This method of meat processing and preservation was first initiated by Arthur Libby in 1874. (Courtesy, *Meat Magazine*)

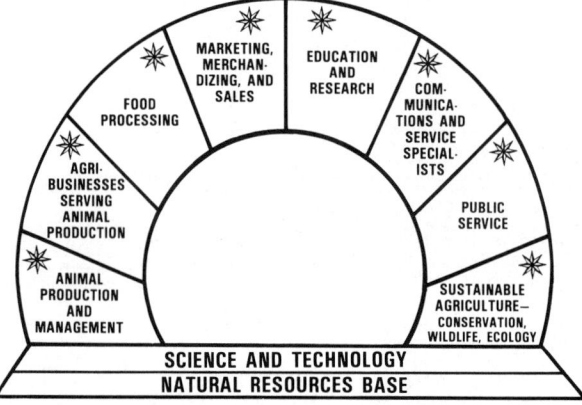

Fig. VI-24. The U.S. Animal Industries are like an arch. The arch rests on two vital stones: (1) natural resources—animals, feed, and water; and (2) science and technology. The other critical components include eight segments which form the arch.

Appendix

industry and related fields are major components of the United States economy; hence, career opportunities therein are excellent for both men and women trained in Animal Science.

Rather than attempt the impossible task of listing all the types of jobs available in Animal Science and related industries, the authors have categorized the career opportunities into a galaxy of eight career opportunities. These are portrayed in Fig. VI-24, followed by a discussion of each of them.

✷ ANIMAL PRODUCTION AND MANAGEMENT

Livestock producers may be concerned with one or more species: beef cattle, dairy cattle, sheep, goats, swine, poultry, and/or horses. They should be knowledgeable relative to the breeding, feeding, care, management, health, behavior, and marketing of their animals and their products. For success, they should also possess experience, industry, and judgment Additionally, if they have a great love for animals, the job will be an exciting career instead of a task.

Fig. VI-27. Swine production. Grower-finishing hogs in an air conditioned building. (Courtesy, *National Hog Farmer*, St. Paul, Minn.)

Fig. VI-25. Beef production. Cowboy and cutting horse sorting Santa Gertrudis cattle on world-famed King Ranch of Texas.

Fig. VI-28. Egg production. Layers in an air conditioned house and fed by automation. (Courtesy, De Kalb Genetics, Corp., De Kalb, Ill.)

Fig. VI-26. Dairy production. Open-air, free-stall housing for dairy cows, with milking parlor and maternity barn in the background. (Courtesy, Babson Bros. Co., Oak Brook, Ill.)

Fig. VI-29. Dairy goat in milking parlor, being prepped before attaching the milking unit. (Courtesy, Babson Bros. Col, Oak Brook, Ill.)

✳ AGRIBUSINESSES SERVING ANIMAL PRODUCTION

This refers to those related businesses that serve animal production. It includes the commercial feed and animal health industries, credit sources, manufacturers of livestock buildings and equipment, and consultants. For men and women with the necessary training, experience, judgment, and integrity, there are unlimited opportunities in agribusinesses.

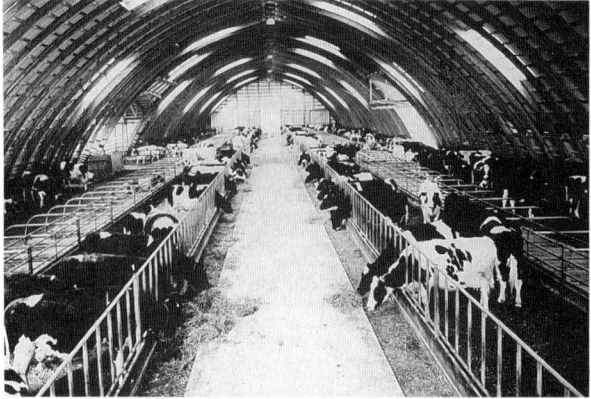

Fig. VI-30. Buildings manufacture, an animal-related business. This shows an environmentally-controlled dairy barn—completely enclosed, with air, temperature, humidity, and light control. (Courtesy, Sperry New Holland, New Holland, Penn.)

Fig. VI-31. Equipment manufacture, an animal-related business. This shows farrowing stalls in an environmentally controlled swine farrowing unit. (Courtesy, J. C. Allen and Son, West Lafayette, Ind.)

Fig. VI-32. Hay-making equipment manufacture, an animal-related business. This machine makes loaf-like stacks of hay, of one-ton size. (Courtesy, Deere & Company, Moline, Ill.)

✳ FOOD PROCESSING

Food processors are involved in the fabrication of food from the time it leaves the farms, ranches, and feedlots until it reaches the nation's kitchens. They make it possible for consumers to choose between quick-frozen, dry-frozen, quick-cooking, ready-to-heat, ready-to-eat, and many other conveniences. Also, they develop new products, specialty food items, and products that appeal to a health-conscious public. Today, the food processing industry is a major employer—more than 1.6 million people are engaged therein.

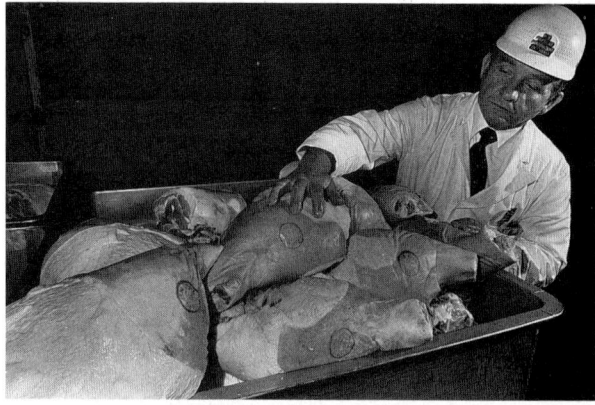

Fig. VI-33. Hams of the famed Smithfield brand, processed in the packing plant at Smithfield, Virginia. (Courtesy, USDA)

Fig. VI-34. Fabricated and boxed beef. Each cut is trimmed to rigid specifications, placed in a vacuum sealed protective bag, and boxed for shipment. (Courtesy, Iowa Beef Processors, Inc., Dakota City, Nebr.)

Fig. VI-35. Manufactured dairy products utilize about 60% of the U.S. milk supply. This shows Delton Parks (left), a graduate from Michigan State University, now President of County Fresh Dairy, Grand Rapids, Mich., a food processor with annual sales of $230 million. (Courtesy, Michigan State University, East Lansing, Mich.)

Appendix

Fig. VI-36. Harvesting catfish on a 150-acre fish farm, near Tunico, Miss. Proper harvesting is the first requisite in processing fish. (Courtesy, USDA)

 MARKETING, MERCHANDIZING, AND SALES

The marketing, merchandizing, and sales of agricultural products is big business, with food in the lead. In 1987, U.S. consumers spent $193.9 billion for the meat, eggs, dairy products, and poultry produced by farmers. Economic activity of this magnitude requires a continuing infusion of educated professionals.

Fig. VI-37. Successful marketers know their products. This shows Dr. Floyd McKeith, University of Illinois, giving instructions on various meat cuts. (Courtesy, National Live Stock and Meat Board, Chicago, Ill.)

Fig. VI-38. Hectic futures trading action at the Chicago Mercantile Exchange, Chicago, Ill. (Courtesy, Chicago Mercantile Exchange, Chicago, Ill.)

Fig. VI-39. A modern meat market. (Courtesy, Iowa Beef Processors, Inc., Dakota City, Nebr.)

 EDUCATION AND RESEARCH

• **Education**—No more noble or rewarding profession exists than that of imparting knowledge, and of motivating and guiding people toward a better understanding of their problems and their solution. Also, it is recognized that the custodians of today's knowledge and the producers of tomorrow's are the products of good teaching.

Animal science teachers may find rewarding work in college teaching, as extension specialists or county extension agents, in adult education or youth work, or as vocational agriculture, biology, or chemistry teachers.

• **Research**—Research is the tomorrow mind instead of the yesterday mind. Animal scientists are involved in every phase of animal research—breeding and genetics, feeding and nutrition, physiology, bacteriology, management, animal behavior, statistics, zoology, computer science, and biotechnology. A degree in Animal Science will help to extend the boundaries of knowledge about animal functions which are important in production, such as growth, reproduction, milk secretion, and wool and egg production. Students with inquisitive minds and the desire to explore the unknown will find an interesting and challenging career in animal research.

Fig. VI-40. Teaching equitation at Texas A & M University. (Courtesy, Texas A & M University, College Station, Texas)

Fig. VI-41. Instruction in beef cattle judging at the University of Missouri. (Courtesy, University of Missouri—Columbia)

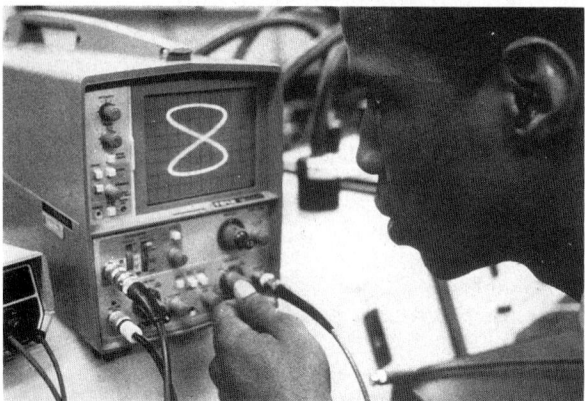

Fig. VI-42. Computers and other diagnostic technology give students the opportunity to experience the most up-to-date problem solving methods. (Courtesy, Michigan State University, East Lansing, Mich.)

Fig. VI-43. Tissue culturing and micropropagation are just two areas of biotechnology that students have the opportunity to investigate in high-tech laboratories. (Courtesy, Michigan State University, East Lansing, Mich.)

Fig. VI-44. An extension worker giving instruction on various beef cookery methods. (Courtesy, National Live Stock and Meat Board, Chicago, Ill.)

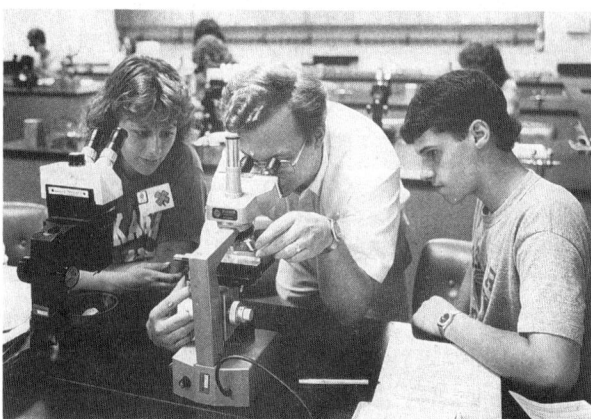

Fig. VI-45. Two 4-H Club members learn about using a microscope from Dr. Michael Smith, Department of Animal Science, University of Missouri. (Courtesy, University of Missouri—Columbia)

Appendix

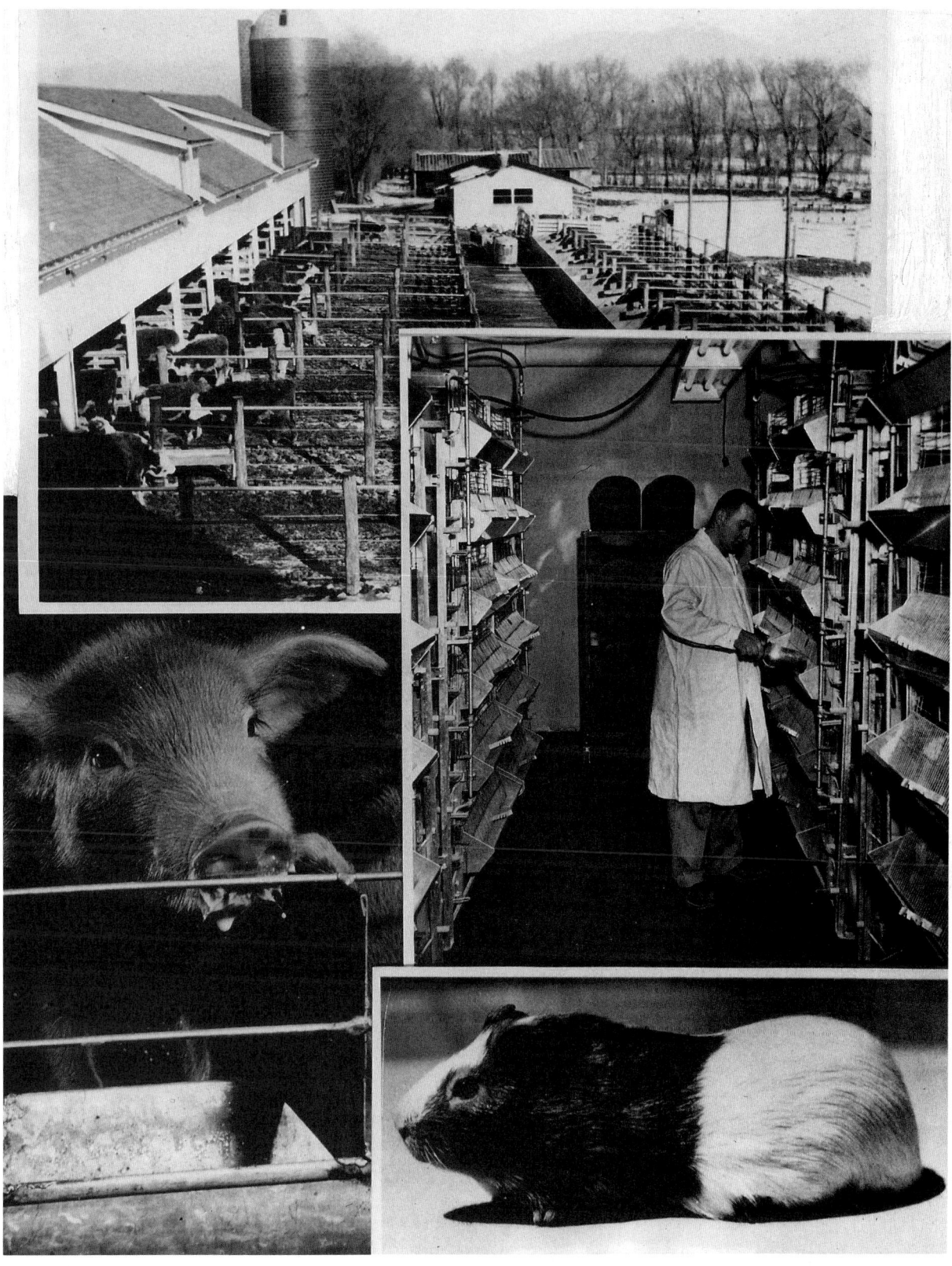

Fig. VI-46. Research—a montage.

Fig. VI-47. Researcher studying a rib eye of beef. (Courtesy, Cal Poly State University, San Luis Obispo, Calif.)

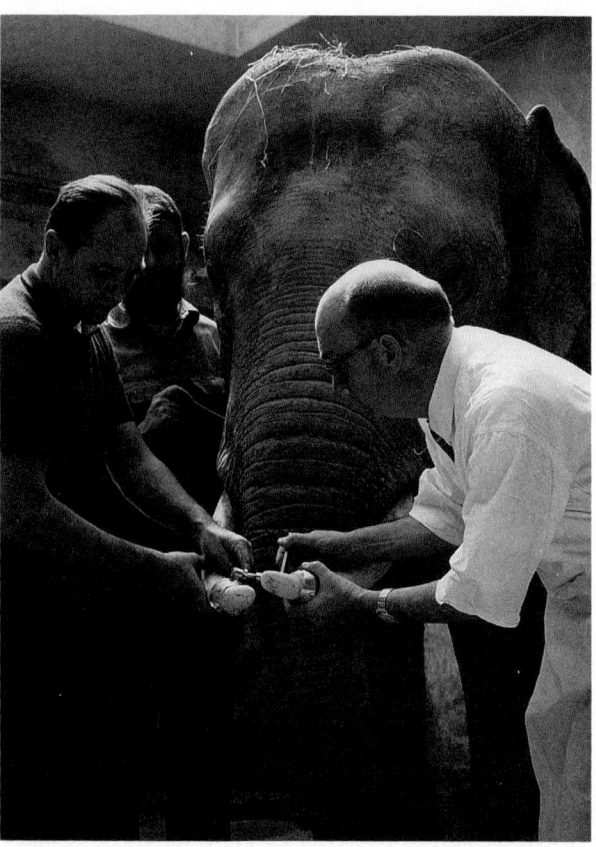

Fig. VI-49. Zoos require animal scientists and veterinarians, too. (Courtesy, National Institute of Health, U.S. Department of Health, Education, and Welfare, Bethesda, Maryland)

Fig. VI-48. Sheep in metabolism stalls, used in digestion studies. (Courtesy, The University of Nebraska, Lincoln, Nebr.)

 COMMUNICATIONS AND SERVICE SPECIALISTS

Jobs in communications offer excitement, adventure, and opportunity to those who combine knowledge in Animal Science with skill in writing, speaking, and/or illustrating.

Livestock publications, newspapers, television stations, advertising agencies, and public relations firms need talented people who can communicate. Also, good communication skills are sought by breed associations, livestock commodity groups, artificial insemination services, banks and other financial institutions, insurance companies, real estate agencies, and power companies.

Students who wish to pursue a career in communications should combine training in Animal Science with journalism and/or speech. Tools of their trade may also include computers, satellite uplinks, laser disks and videotapes, radio and television, and magazines and newspapers.

Appendix

Fig. VI-50. Communication is very important—and very difficult. This shows Jim Bremer, a graduate of Michigan State University, communicating over the telephone. Mr. Bremer is President of Farm Credit Services of Mid-Michigan. (Courtesy, Michigan State University, East Lansing, Mich.)

✹ SUSTAINABLE AGRICULTURE— CONSERVATION, WILDLIFE, ECOLOGY

Endangered planet, endangered people and animals, and endangered agriculture! Among the deluge of warnings of environmental catastrophes are: the greenhouse effect, depleting the atmosphere's protective ozone layer; toxic and radioactive wastes; acid rain; and destruction of tropical rain forests.

Now the situation is being righted. Worldwide, environmental quality and economic sufficiency are in vogue. In the United States, the movement is called *Sustainable Agriculture*.

Men and women with interest in this type of work would do well to take course work better to prepare them for a career in this new and exciting field.

✹ PUBLIC SERVICE

Federal, state, and local governments and agencies employ a large number of animal science graduates in research, marketing, and regulatory work. Opportunities are available to those with bachelor, master, and doctorate degrees in animal science to serve with civil service rank.

Fig. VI-51. Government agencies employ many animal science graduates. This shows an inspector taking a sample of mixed feed. (Courtesy, College of Agriculture, University of Missouri—Columbia)

Fig. VI-52. Highly erodible lands can provide substantial wildlife habitat for a number of species including upland game birds like this pheasant. (Courtesy, USDA—Soil Conservation Service, Washington, D.C.)

Fig. VI-53 (above). This seriously eroded condition resulted from over-cropping and over-grazing steep lands that were first denuded of forests.

Fig. VI-54 (below). Contour strips of alfalfa and corn help control soil erosion on this Wisconsin Farm. (Courtesy, USDA—Soil Conservation Service, Washington, D.C.)

Appendix

Fig. VI-55. Dr. John L. Merrill, XXX Ranch, Crowley, Texas; and Director of Range Management, Texas Christian University, Ft. Worth, Texas, shown checking range vegetation. (Courtesy, John Merrill)

Fig. VI-56. Soil Conservationist Judy Hill measures the electrical conductivity of poultry waste as part of a national water quality monitoring program. (Courtesy, USDA, Soil Conservation Service, Washington, D.C.)

Why Go To School?

Practical young people graduate from high school and college for four primary reasons:

1. Because it takes trained people to operate high-tech; which is already here in the form of such things as computer-controlled growth environments, robots that harvest crops, gene transfers that make plants and animals resistant to disease, and laser-guided machinery. Remember, too, that futurists predict that during the next quarter century the most important scientific discoveries in genetic engineering will be made by food and agricultural scientists.

2. Because in many cases a diploma is a requisite to getting a position. There are fewer unskilled agricultural jobs; more positions for managers, technicians, and scientists.

3. Because it increases earning power. As a result, they can enjoy a higher standard of living and some of the finer things of life. Indeed, a college education is a big key to financial success (see Fig. VI-57). But remember that a job is more than a paycheck! It reflects who you are, your success in life, and your position in the community. It also affects your well-being. Remember, too, that when work is play and play is work, the job changes from an 8-hour per day *task* to an *exciting career*.

4. Because most people like to do some good in the world. Certainly everyone wants to make money; and there's no doubt that an education is most helpful from this standpoint. But an education also enhances those things of the spirit and intellect that do not wear dollar marks—the pure enjoyment from living and the enlarged contribution to society. Generally speaking, those who command respect as leaders, and whose lives are fullest and most productive, are college graduates.

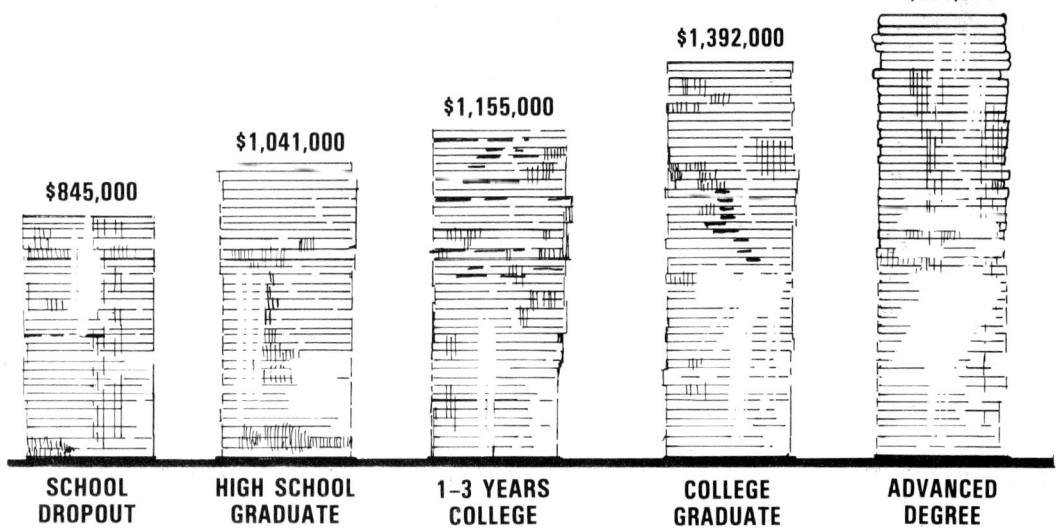

Fig. VI-57. In a lifetime, and on the average, this shows what you may expect to earn. Prepared by the authors from the most current data available: *Lifetime Earning Estimates for Men and Women in the United States*: 1979, U.S. Bureau of the Census (*Current Population Reports*, Series P-60, Number 139, page 3). The data are for 18-year-old, year-round, full-time, male workers, for 1978-80 and are expressed in 1981 dollars. In future years, lifetime earnings will be affected by (1) supply of, and demand for, the educated labor force, and (2) inflation/value of the dollar; in the decades to come, it is expected that lifetime earnings of each group will be higher, and that the gap between the least and most educated will grow wider.

Success stories like those that follow are being repeated by graduates all over the world.

Fig. VI-58. Lee Eaton, who attended the University of Kentucky; now a noted Thoroughbred breeder and agent. Lee Eaton operates a 400-acre horse farm, near Lexington, Kentucky. Also, he is the senior member of the Eaton-Williams, Agent, which during each of the past three years has sold an average of 278 Thoroughbred yearlings and breeding stock at auctions throughout North America for an average annual total of $47 million. (Courtesy, The Blood-Horse, Lexington, Ky.)

Fig. VI-60. Ron Baker, a graduate in Animal Science from Washington State University, owns and operates C & B Livestock, Inc., Hermiston, Oregon, and Western Meat Producers, Inc., (Meat Packing), Pasco, Washington. Ron Baker farms 4,000 acres, has 1,200 cows, operates a 20,000 head cattle feedlot, and markets his own brand of beef—C & B Natural Light. (Courtesy, Ron Baker)

Fig. VI-59. Guillermo Osuna, Jr., a graduate of Southern Methodist University, Dallas, treating a calf which he has just roped on the family-owned ranch—Infante Ranch, Muzquiz, Coahuila, Mexico.

Fig. VI-61. John A. Hannah (seated), received a degree in Agriculture from Michigan State University. He began his distinguished career as an Extension Poultry Specialist. He went on to become President of Michigan State University from 1941 to 1969. From 1969 to 1974, he served as Administrator of the U.S. Agency for International Development; and in 1975, he was named as Executive Director of the United Nations World Food Council.)

For More Information, Contact:

1. The Animal Science Department at the Land Grant or State University.
2. The American Society of Animal Science, 309 West Clark St., Champaign, Illinois 61820.
3. The local County Extension Agent or FFA Instructor.
4. The career offices in local colleges.
5. The state employment service.
6. The public libraries. Among other references, see (1) *Employment Opportunities for College Graduates in the Food and Agricultural Sciences*, 1986, by Jane Coulter, Marge Stanton, and Allan D. Goecker; and (2) *Technical Addendum to Employment Opportunities for College Graduates in the Food and Agricultural Sciences*, by Kyle Jane Coulter and Marge Stanton. Both publications prepared and distributed by Higher Education Programs, U.S. Department of Agriculture, Washington, D.C.

INDEX

A

Ablactation 675
Abomasum 35
   displaced 91
Abortifacients 253
Abortion 502
Absorption 27, 31, 36, 38
   small intestine 36
Acacia 693
Acclimation 325
Acclimatization 325
Acetone 675
Acetonemia 66, 79
Acetyl-CoA 675
Achromotrichia 675
Acid(s) 180
   detergent fiber 152, 288
   detergent lignin 288
Acidosis 39, 66, 462
Acorns 95, 693
Actin 675
Acute tympany 39
Adaptation of animals 325
Additive(s), feed .. 124, 179, 247, 253, 675
   beef cattle, for 385
   dairy cattle, for 446, 447
   definition 124, 247
   goats, for 500
   mink, for 650
   nonnutritive 275
   poultry, for 569-571
   rabbits, for 632
   safe use 14, 248
   sheep, for 480, 481
   swine, for 522-524
Adipose tissue 675
Ad libitum 675
Adrenal gland 675
Aerial part 675
Aerobe 675
African marigold 693
Aftermath 675
Agar 675
Age, effect on nutritive needs 21
Agglomerated feed 675
Agonistic behavior 309, 315
Agricultural
   chemicals 94
   drugs 94
Agriculture
   animal 6
   sustainable 332
Agroforestry 675
Air dry feed 675
AIV preservative 675
Alfalfa meal 207
Alfileria 693
Algae 244, 675
Alkali
   disease 67, 107
   sacaton 693
Allelomimetic behavior 308, 310-311
Allergy 675
Almond 694
   hulls 236
Alopecia 675

Alpha-tocopherol equivalent 675
Alyce clover 694
Ambient temperature 675
American Feed Industry Association
   (AFIA) 675
Amino Acid(s) 51, 52, 203
   analysis 289
   era 14
   essential 51
   nonessential 51
   supplements 223, 249
   synthesis of 53
   transamination 53
Ammonia toxicity 108
Ammoniated 676
   feeds 50
   products 219
Ammoniation 277, 279
Amphoteric 676
Amylopectin 676
Anabolism, definition of 41, 676
Anaerobe 676
Analogue (analog) 676
Anamalous (abnormal) behavior 315
   cattle 315
   chicken 317
   horse 316
   sheep 315
   swine 315
   turkeys 318
Anaphylaxis 676
Anasarca (dropsy) 676
Anatomy, digestive 28
Anemia 67
Aneurism 676
Angora goats 505
Animal(s)
   agriculture 6
   factors favoring 6-8
   behavior 307, 308
      abnormal 308
   by-product
      consumption 7
      feeds 122, 226-228
   environment 318, 319
      effect on 319
   fat 227
   nutritionist 676
   protein(s) 6, 211, 676
   units 762
   wastes 239, 277
      litter 237-242
      manure 237-242
      processing 277
   weights estimated 760
   welfare/rights 333
Anion 676
Anise 694
Anorexia 676
Anoxia 676
Antacid 676
Antagonist 676
Anthelmintic(s) 254, 676
Antibiotics 14, 254-256
   dairy cattle, for 447
   definition 254
   fish, for 665

   goats, for 500
   horse, for 610
   mink, for 650
   mode of action 254
   poultry, for 569
   rabbits, for 632
   safety 255
   sheep, for 481
   swine, for 523
Antibody 676
   production 327
Anticoccidial drugs 632
Antifungals 569
Antigen 676
Antihistamines 676
Antimetabolite 676
Antioxidant(s) 256, 569, 650, 676
   rabbit feeds 633
Antivitamin factors 676
Appetite 27
   gastric influences 28
   hypothalmic control of 28
   thermostatic control of 28
Appetizer 676
Apple(s) pomace 229
Aquarium fish culture 672
Aquatic plants 243, 244
Arrowroot 694
Arsenicals 256, 570
Arsenic poisoning 95
Artichoke 269, 694
Ascorbic acid 58, 253
As-fed (feed) 676
Ash 676
   analysis 286
   leaves 233
Aspen 694
   wood as feed 233
Aspirating 676
Assay 676
Assimilation 676
Ataxia 676
Atom 676
Atrophy 676
Available nutrient 676
Average daily gain (ADG) 676
Avian digestive system 30-31, 35
Avidin 676
Avoirdupois weights and measures 753
Azoturia 68

B

Baby pig shakes 68, 77
Backgrounding 407
Bacteria 676
Bactericide 676
Bagasse 228, 236
Bakery by-products 123, 243
Balancing rations 338
   methods
      computer 340-343
      net energy 344
      square 338
      trial and error 339
Bamboo 694

Barley
   malting 192
   milling 192
   straw 236
Barn sour **316**
Basal diet **676**
Basal metabolic rate (BMR) **677**
Base **677**
Battery **677**
B complex vitamins **252**
Bedding storage space requirements ... **758**
Beechnuts **694**
Beef cattle
   backgrounding 407
   bulls, feeding 402
   calves, feeding 403–405
      early weaned 405
   calving
      control/daytime calving 398
      2-year-olds 407
   compensatory growth 410
   concentrates 387
   confinement cows 401
   confinement feeding partial/semi . 402
   creep feeding 403–405
   crop residues 388–389
   drylot 401
   economic importance of feed 348
   energy needs 372
   fats/cottonseed feed 199
   feed(s) 348, 386–388
      allowance 393–395
      by-products 387
      consumption 7
      corn residues 398
      cottonseed 198
      crop residues 398
      preparation 393
      substitution tables 388–392
   feeders 407
   feeding 347, 396
      breeding cattle 396–407
      brood cows 396
         nutritional requirements ... 396
      winter 397
   finishing (fattening) 411–421
      age 412
      amount to feed 417
      boxed beef 414
      breeding 414
      bulls 413
      cold confinement 416
      confinement feeding 415
      crossbreds 414
      cull out 418
      custom-contract 419
      dairy beef 414
      definition 25
      facilities 415–416
      feed(s) 416
         bunk management 418
         preparation 416
      gain rate factor 417
      grade of cattle 413
      kinds of cattle 412
      managing 416
      mud problem 418
      net energy ration formulation . 416
      new cattle arrivals 417
      nutritional feedlot diseases .. 421
      open pen 415
      overfinishing 418
      pasture 420–421
      rations 416
      records 419
      sex to feed 413
      slotted floors 415
      spayed heifers 413
      test weights 419
      tests 419
      top out 418
      warm confinement 416
      weather affects 418
      weight 412
   growth stimulants 384–385
   hay(s) 386
   implants 384–385
   mineral(s) 374–380
      chart 374–379
      requirements 372
      supplements 387
      toxic levels 372
   NRC requirements 349–372
   nutritional
      diseases 396
      reproductive failure 397
   nutritive needs 348
   orphan calves, feeding 403
   pasture(s) 129–130, 386
      feeding 399
      winter 398–399
   preconditioning 405–407
   protein needs 373
   range
      feeding 399
      nutrient deficiencies 400
      supplementing 400
   ration(s) 393–395
      cows with calves 398
      dry cows 397, 401
      fitting 395–396
      pregnant cows 397
      show 395–396
   replacement heifers 405–407
   root crops 386
   roughage(s) 386
   sale fitting 394–396, 402
   show fitting 394–396
   silage 386
   stockers 408–411
      contracts 411
      definition 407
      programs, types 408
      rations 408–410
      wintering level 410
   urea for beef cattle 387
   vitamin(s) 380–383
      chart 382–383
      requirements 381
      supplements 388
   water 384
      daily intake 384
   weaning 405
   weights estimated 760
Beet **694**
   molasses 200
   pulp 231
   tops 231
Behavior
   abnormal 308, 315–318
      cattle 315
      chicken 317
      horse 316
      sheep 315
      swine 315
      turkey 318
   agonistic 309
   allelomimetic 308, 310, 311
   anamalous (abnormal) 315
   animal 307, 308
   care-giving 309
   care-seeking 309
   dominance 313
   eliminative 309–311
   feeding 28
   gregarious 309–311
   ingestive 308, 310, 311
   interspecies 315
   investigative 309
   poultry 317–318
   sexual 309
   shelter-seeking 309
   social 312
   systems 308–309
Bentgrass **694**
Beriberi **57**
Bermudagrass **694**
Berzelius **13, 55**
Big neck **78**
Bile **35–36**
Bioassay **677**
Biochemistry **677**
Biological **677**
   activity 677
   analysis 289
   era 13
   value of protein 677
Biosynthesis **677**
Biotechnology **44**
   definition 15, 44
   era 14–15
Biotin **58**
Biting **316**
Black walnut toxicosis **96**
Blackwater **68**
Bleeders, equine **330**
Blended **677**
Blind staggers **107**
Bloat **39, 69, 462**
   control 69, 70, 256
Block(s) **677**
   feed 273
   protein 221
Blood **32**
   meal 212, 227
Bloom (animal) **677**
   feeds imparting 612
Blueback **318**
Bluegrass **694**
Bluestem **695**
Body
   composition 16–17
   measurements 21
   size 20
   weight 21
Bolting feed **316, 677**
Bolus definition **32, 677**
Bomb calorimeter **288, 677**
Botulism **97, 652**
Boussingault, J. B. **13**
Bovatec (lasalocid) **261**
Bovine
   bonkers disease 108
   pulmonary emphysema 84
   somatatropin (BST) 260, 447
Boxed beef **414**

Index

Brand name 677
Bran mash 272, 612
Breeders (chickens) 553, 578
Breeding diary
  age 23
  size 23
Brewers'
  by-products 231
  dried grains 207, 232
  yeast, dried 233
Brewing by-products 231
British Thermal Unit (BTU) 48, 677
Brix (molasses) 200, 677
Brody, Samuel 22
Broilers 552, 553, 580–581
  definition 677
Bromegrass 694
Browse 129, 500
Brush control 142
Buckwheat 695
Buffalograss 695
Buffelgrass 695
Buffers 257, 447, 677
Bur-clover 695
Bureau of Indian Affairs 144
Bureau of Land Management 143
Burroughs 14
Buttermilk 214, 226
  condensed 214
  dried 214
Bypass protein 204, 439, 471
By-product feeds 122, 225, 677
  animal 122, 226
  use 7
  bakery 123
  beef cattle 387
  brewing 231–233
  definition 122, 225
  distilling 231, 232
  economy of 226
  fermentation 122
  fruit 229–230
  grain 229
  high-protein 223
  industrial 122
  marine 214, 216, 226
  meat processing 212
  mill 120
  milling 191
  molasses 231
  nut 229
  oilseed 228
  paper 123, 233–234
  plant 122, 226–234
  poultry 227
  sugar 230–231
  vegetable 229
  waste 237
  wood 123, 233–234

C

Cachexia 677
Cactus, prickly pear 695
Cake (presscake) 677
Calcification 677
Calcitonin 677
Calcium 56
  supplements 251
Calculi 677
Calf scours 460
Caloric 677

Calorie 48
  source 4
  system 49, 50
Calorimeter 677
Calorimetric feed evaluation .. 293
  direct 293
  indirect 293
Calves
  dairy 458–460
  early weaned 405
  preconditioning 405
  starter rations 459
  weaning 405
Canarygrass 695
Cane molasses 200
Cannery residue 228
Cannibalism
  definition 677
  poultry 317, 318
Cannula 677
Cannulation, feed evaluation .. 290
Canola meal 209, 228
Cantaloupe 695
Caprine arthritis 502
Carbohydrate(s) 36, 44, 45, 193
  absorption 36
  digestion 36
Carbonaceous feeds 120
Carcass weight 677
Carcinogen 677
Carcinogenic 677
Care-giving behavior 309
Care-seeking behavior 309
Carib grass 695
Carnivore(s) 28, 678
Carotene 251
  feed content 388
Carotinoids 253
Carp 671
Carrier 678
Carrot(s) 230, 695
  energy value 201
Carrying capacity 678
Cassava 696
  energy value 201
Catabolism 41, 678
Catalyst 678
Catfish diets 668
Cation 678
Cattle
  age 412
  behavior, abnormal 315
  breeding 414
  feedlots 415
  grade 413
  kinds to feed 412
  managing 416
  manure (feed) 237
  sex of feedlot cattle 413
  type 414
  waste 239
  weight 412
Cecum 37
  colon region 37
Cell(s) 41, 42
Cellulose feeds 279
  treatment 279
Centigrade 757
Century plant 696
Cereal diet, advantages 4
Cerebrocortical nechrosis 83
Chaff 678

Charging 316
Chastek paralysis 650
Cheese rind 214
Chelate(s) 39, 56, 251, 678
Chemical
  agriculture 94
  energy era 12
Chemostatic appetite control .. 678
Chemotherapeutics 258
Chestnuts 696
Chicken (see Poultry)
Chlorine 56
Chlorophyll 678
Choking 91
Cholesterol 253, 263, 571, 678
Choline 58
Chopping forage 276
Chossat, C. J. 13
Chromatography 288
Chufa 696
  energy value 201
Chyme 678
Cirrhosis of the liver 678
Citrus
  molasses 200, 230
  pulp, dehydrated 230
Clover 696, 706
Coagulated 678
Cobalt 56
  deficiency 70, 86
Coconut meal 209
Coefficient of digestibility .. 678
Coenzyme(s) 678
Cold confinement 415
Colic 40, 71
Collagen 678
Colloid 678
Colon 36, 37
Colorimetry 289
Colostral defense 328
Colostrum, definition 458
Combat behavior 309
Combustion 678
Comfort zone 19, 324, 678
Commercial feeds
  .... 206, 300–301, 447, 501, 615, 678
  tag 303
  types of 302
Compaction 678
Comparative slaughter method .. 294
Compensatory growth 22, 320, 678
  beef cattle 410
Complete rations 279, 678
  feeds 122
Composite sample 678
Composition of feeds 707–751
Compudose 261
Computer ration formulation .. 340, 451
Concentrate(s) 119, 678
  beef cattle 387
  consumption 119
  dairy cattle 446
  definition 119, 187
  -hay proportion 165
  goats 501
  horses 611
  percent feed for livestock .. 117
  ration formulation 336
  sheep 481
  swine 525
  tonnage fed 119
Condition 678

Conditioning 678
Confinement beef cattle 401, 415
Congenital 678
Continuous grazing 134, 140
Contracts
  finishing cattle 419
  hog production 545
  stocker-grower 411
Cooking feed 271
Cooperative feed companies 303
Copper 56, 258
Copra meal 209
Coprophagy 678
Corn 525
  cobs 237
  energy value 193
  gluten feed 210, 229
  gluten meal 210, 229
  high-lysine 194, 526
  high-moisture 526
  -hog ratio 545
  husk 237
  milling 191
  nutrient composition 192-193
  opaque-2 194
  residues 398
  silage 176, 178
  stalk poisoning 102-103
  stover 237
  structure (seed) 188, 189
  types 120
  unprocessed (whole) 274
Cornstalk poisoning 102
Cotton fur 651
Cottonseed 197, 228
  gin trash 237
  hulls 237
  meal 209, 228
    feeding value 210
    quality 210
  whole seed 228
Cottontop 696
Cow asthma 84
Crabgrass 696
Cracking feed 270
Crambe 696
Crayfish culture 672
Crazy cattle disease 108
Creatinine 678
Creep feeding
  calves 403-405
  lambs 486
  pigs 541
Creep grazing 135
Cribbing 316
Crimping feed 270
Crop residues 122, 225, 234-237, 398
  definition 122, 225
  economy 226, 235
  feeding systems for 236
  harvesting 235
  kinds 234
  nutrient value 235
  quantity 234
  storing 236
  transporting, packaging 236
  treating 235
Crown vetch 696
Crude
  fat 678
  fiber analysis 287, 678
  protein analysis 286, 678
Crumbling feed 272
Crushing feed 270

Cubing (feed) 276, 679
Cud 679
Curd 679
Curly mesquite 697
Custom
  cattle feeding 419
  cattle grazing 129
Cyanocobalamin 679
Cystitis 679

D

Dairy (cattle) 425
  acidosis 462
  additives 446-447
    antibiotics 447
    bovine somatatropin 447
    buffers 447
    ionophores 447
    isoacids 447
  beef (from dairy cattle) 414, 461
  bloat 462
  butterfat records 462
  by-product feeds 213-214, 226, 446
  calf
    amount/method/frequency 459
    hay 459
    production 458-460
    scours 460
    starters 459
    water 459
  colostrum 458
  concentrates 446, 449
  displaced abomasum 462
  dry matter intake 430
  drugs/pesticides 463
  energy needs 430
  fat cow syndrome 462
  fats and oils 199, 446
  feed(s) 445-448
    commercial 447
    composition 436-438
    considerations 448
    cottonseed 198
    economic importance 426
    preparation 448
    substitutions 448
  feeding 445-448
    calves 458-460
    challenge 455
    dairy
      beef 461
      bulls 461
    dry cows 458
    frequency 461
    group 455-457
    lactating cows 448-458
      concentrates 449
      forages 449
      rations 450-451
    phase 454-455
    show and sale 461
    systems 454
  fiber 447
  forages 445
  grass tetany 462
  green chop 445
  hardware disease 462
  hay(s) 445
  haylage 446
  health disorders 462
  heifer growth 21, 460

ketosis 462
management 461
mastitis 463
milk
  fat records 462
  -feed price ratio 450
  fever 463
  testing programs 462
  milking frequency 462
mineral needs 440-443
NRC requirements 426-434
nutritive needs 426
pastures 129-130, 445
  feeding on 458
products 213-214
protein
  degraded 439
  needs 439
  undegraded 439
ration(s) 448
  balancing 450
  computer formulated 451-453
replacement heifers 460-461
  breeding weights 460
  normal growth 460
retained placenta 463
root crops 446
silages 445
toxic levels of elements 435
udder edema 463
urea/NPN products 440
veal calves 460
vitamin needs 443-445
water needs 445
weights estimated 760
Dairy goats 503
Dallisgrass 697
Database 679
Date palm 697
Davy 55
Daytime calving 398
Debeaking 679
Decortification 679
Defecation 679
  behavior 309, 310
Deferred grazing 140
Deficiency disease 66-89, 679
Defluorinated phosphate 679
Deglutition definition 32
Dehydrate 679
Dehydroascorbic acid 253
Delaney Clause 14, 248
Deoxyribonucleic acid 42
Depraved appetite 679
DES 14
Desiccate 679
Diahrrea 40
Dialysis 679
Dicoumarol 679
Digestible
  dry matter 679
  dry matter intake 679
  energy 49, 50
  nutrient 679
  protein 679
Digestion 27, 36, 38
  coefficient 679
  factors affecting 38
  physiology of 32
  process of 31
  small intestine 36
  trials 291

Index

Digestive system
  anatomy 28
  avian 30–31, 35
  capacity 31
  chicken 31
  dysfunction 39
  fish 29
  functional cecum 29
  horse 29
  mink 29
  monogastric 29
  nonruminant 28
  processes 31–32
  ruminant 29–30
  swine 29
  tract capacity 31
Diluent 679
Disaccharide(s) 679
Disease defense 328
Diseases, nutritional 66
Displaced abomasum 91, 462
Distillers'
  by-products 232
  dried grains 207, 233
  dried grains with solubles 233
Diuresis 679
Diuretic 679
DNA 14, 42, 43
Docking tails (pigs) 543
Dominance 313
Dormant 679
Dough stage 679
Doves 587
Dressed 680
Drought area feeds 123
Drugs 94, 463, 679
Dry
  matter (DM) 679
    basis 679
    intake (DMI) 679
  -rendered 679
Drying
  feed 272
  forage 276
Drylot 679
  rations 401
Ducks
  feeding 584
  NRC requirements 557
  rations 585
Dung 679
Dystrophy 679

E

Early
  leaf 679
  weaning 679
Easy keeper 680
Eating bedding 317
Eating with relish 680
E. coli 43
Eczema 680
Edema 680
Eel culture 672
Efficiency of feed conversion 680
Egg production 24
Eggshell meal 213, 697
Eijkman 13, 57
Electrolyte(s) 259, 680
Element 680
Eliminative behavior 309–311
Emaciated 680
Embden-Meyerhof pathway 680

Emulsification of fats 47
Emulsifier 680
Endocrine 680
Endogenous 680
Endosperm 680
Enema 680
Energy 2–4, 15, 44, 64
  balance trials 292
  beef cattle, for 372
  by-product sources 202
  conserve 2–3
  conversion 48
  dairy cattle, for 430
  deficiency 44, 64
  definition 18
  digestible 50
  efficiency 8
  feeds 633
  fish, for 659
  food efficiency 3
  goats, for 494
  gross 50
  horses, for 597
  measuring 48
  metabolizable 50
  mink, for 643
  net 50
  output of lands 8
  poultry, for 559
  rabbits, for 625, 633
  sheep, for 470
  swine, for 512
  systems 48
  utilization 49
  values 48
Ensiled 680
Ensiling 241, 276
  definition 170
  process 170
Enteritis 680
Enteric disease 502, 638
Enterotoxemia 71, 638
Environment 15, 307, 546
  animal 318
  breeds/adaptation 325–326
  comfort zone 324
  control 326–327
    buildings 19
  definition 318, 327
  /facilities interactions 326
  /feed interactions 319–322
  effect 319, 546
  habituation 325
  /health interactions 327–328
    pollution of 331
  interactions 319
  /stress interactions 330–331
  thermoneutral comfort zone 324
  /water interactions 322–323
  /weather interactions 324–326
Enzymatic 680
Enzymes 31
Epiphysitis 599
Ergosterol 252, 263, 680
Ergot 97, 680
Ergotism 97
Eructation 30, 680
Esophagus(eal) 33
  groove 30
  region 33
Essential fatty acids 47
Ester 680
Estrogen 680
Ether extract 287, 680

Excreta 680
Exercise 18
Exogenous 680
Expeller process 680
Experiment 680
Exploding feed 271
Extracellular fluid 680
Extrinsic factor 680
Extruded 680
Extruding feed 269
Exudate 680

F

Facilities/environmental interactions ... 326
Facultative anaerobes 680
Fahrenheit 757
Fat(s) 45–46, 124, 198,
      278, 446, 526, 598, 644, 660
  added to feeds 277
  carbon chain length 47
  cow syndrome 462
  emulsification 47
  hydrogenation 47
  iodine number 46
  metabolism 47–48
  monounsaturated 46
  polyunsaturated 46
  rancidity 47
  saponification 47
  saturated 46
  soluble 680
Fattening 25
Fatty acids 46–47
  essential 47, 48
FDA 14, 248, 680
Feather
  meal 213, 227
  pulling 318
Febrile 680
Fecal energy 49
Feces 680
Federal lands 137, 143
Feed (feedstuff) 114, 680
  additive era 14
  additives (see Additives) ... 124, 247, 253
  analysis 283, 284, 285
    biological 289
    physical 284
    proximate 285
  beef cattle 386
  bunk management 418
  buying (best) 297, 298, 300
  by-products 120, 122, 225
  carbonaceous 120
  classification 116
  commercial 297, 300–303, 447
  complete 122
  compositions 39
  contaminants 656
  cooperatives 303
  costs 298
  dairy cattle 445–448
  definition 115, 124
  economic importance 116, 348
  efficiency 4, 5, 21, 680
  /environmental interactions 319–322
  evaluation 283, 284, 290
    digestion trials 291
    metabolism trials 291
    nutrient balance trials 291
    techniques 290
  expenditure for 298
  exports and value 116

fish, for 665
/food efficiency 4, 5
futures trading 299
goats 500
grades 296, 680
grain 680
high-energy 120, 187
   definition 120
high-moisture 120
homegrown 298
home-mixed 300
horses 611
implant era 14
implants 124, 247
intake 38
kinds consumed 116, 297
labeling 302
laws 297, 303
liquids 121
marine 215
medicated 262
microscopic examination 285
mink 652
mixed 284
moisture content 286, 300
passage of 38
percentage consumption 117
poultry 570
preparation 38, 416, 448, 528
   fish 665
   goats 501
   horse 613
   mink 653
   poultry 571
   rabbits 633
   sheep 481
   swine 528
principal 115
processing 38, 265
   concentrates 268
   farm 280
   forages 275
   heat treatment 270
   mechanical alterations 269
   method of choice 280
   moisture alterations 272
   nonnutritional considerations 268
   nutritional considerations 268
   pelleting 271
   purposes 266–268
protein 207
purchased 298
quality 283, 303
rabbits 633
rate of passage 38
requisites 299
sampling 284
semiliquid 121
specialty 123
storage space requirements 758
substitutions (tables for) 299
   beef cattle 388–392
   dairy cattle 388–392, 448
   goats 502
   horses 615–616
   poultry 573
   rabbits 636
   sheep 481–483
   swine 539–540
supplements 247, 248
   definition 124, 247

swine 525
tag 303
tonnage fed 116
trials 291, 295
watered feeds 273
weights and measures of feeds 758
wet 125
Feeder
   calves 407
   pig production 543
Feeder's margin **681**
Feeding
   beef cattle 347, 396
   behavior 28
   custom 419
   dairy cattle 425
   fish 657
   goats 491, 503–505
   horses 591
   mink 641
   poultry 549
   rabbits 623, 637
   sheep 465
   standard era 13
   standards 335–336
      limitations 336
      NRC 336
   swine 507
   trials 295
Feedlot cattle
   diseases 421
   facilities 415–416
   managing 416
   tests 295, 419, 420
Feedstuff(s) (feed) **681**
   classification 116
   definition 115
   importance 115, 116
   physical evaluation 284
   roles 115
   storage effect 275
   types 115
Fermentation **681**
   aids 180
   by-products 122
Fermented **681**
Fermenting 273
Ferritin **681**
Fescue **697**
   foot 72
   toxicosis 72
Fiber 45, 447
   content of feed 681
   fish 665
   rabbits 632
Fibrous **681**
Fighting 315
Fill **681**
Finaplex 261
Fines **681**
Finish **681**
Finishing (fattening) **25, 411**
   animals 681
   beef cattle (See Beef cattle,
     finishing [fattening], pp. 411–421.)
   definition 25
   sheep (See Sheep, feeding,
     growing-finishing lambs, pp. 486, 487.)
   swine (See Swine, growing-finishing, p. 541.)
Fish 657
   antibiotics 665

   antioxidants 665
   aquarium feeding 672
   carp 671
   catfish 668, 669, 670
   coldwater fish 669
   crayfish 672
   digestive system 29
   eel culture 672
   energy 659–660
   fats (lipids) 660
   feed(s) 665, 666
      artificial wet 665
      economic importance 658
      formulated 665
      handling 666
      natural 665
      quality affecting 672
      storage 666
   feeding
      habits 658
      systems 671
   fiber for 665
   herring 216
   lobster culture 672
   manure for fertilizing ponds 671
   meal (fish) 206, 215, 227
   menhaden 216
   minerals 661–663
      chart 662
   nutritive needs 658
   pigment producing 665
   prawn culture 672
   predacious fish 658, 673
   processing 215
   protein 660
      as energy 660
   rations 666
   salmon 216, 666
   sardine 216
   shrimp 673
   sturgeon culture 673
   tilapia culture 673
   toxicants 673
   trout 666
   types of eaters 658
   vitamins 662–664
      chart 663
   warmwater fish 670
   water (in fish diet) 664
   white fish meal 216
Fistula **681**
Fistulation feed evaluation 290
Fitting
   definition 25, 681
   rations for beef cattle 396
Flaking feed **271**
Flat pea **697**
Flatulence **681**
Flavoring agent(s) **259, 681**
Flocking behavior **309, 310**
Flora **681**
Fluorine **56**
   poisoning 98
   toxicity, cattle 98
Fluorosis **98**
Flushing
   definition 23, 321, 681
   ewes 485
   sows 541
Fodder **681**
Fogfever **74**

Index

Folacin 58
Folic acid 58
Food 681
   and Drug Administration (FDA) . 14, 248, 681
   energy efficiency 3
   improving world situation 8
   nutrients from animal production 6
Forage(s) 116
   characteristics of 117
   consumption 7, 117
   definition 117, 681
   growing hay 153
   pasture 117, 127
   processing methods 275-277
      chopping 276
      cubing 276
      drying 276
      ensiling 276
      grinding 276
      pelleting 277
      shredding 276
      wafering 276
   range 117, 127
Forbs 129, 500
Formula feed 681
Fortity 681
Fossil fuels 3
Founder 73
Fowler's solution 681
Foxtail 697
Fraps 14
Free-choice 681
Freeze drying 681
Freezing feeds 277
Fresh 681
Fruit(s) 120
   by-products 229
Full-feed 681
Fungi 290, 682
   inhibitors 262
   toxin producing 100
Fungicide 681
Funk, Casimir 13, 57
Fur-eating rabbits 638
Futures trading 299, 681
     feed, kinds of 299
     hedging 299

G

Gain, composition of 16-17
Gamagrass 697
Game
   birds 587, 588
   population 140, 145
Garbage 123, 242-243, 528
Gastric 681
   juice 681
   region 33
Gastritis 681
Gastroenteritis 681
Gastrointestinal 681
   hormones 37
   neurological control 37-38
Gavage 682
Geese
   feeding 586
   NRC requirements 588
   rations 586
Gelatinization 269
Gelatinized 682

Gene-splicing 14
Genetic engineering 43
Germ (seed) 682
Gestation 682
Glossitis 682
Gluconeogenesis 682
Gluten 682
   feed and meal 207
Glyceride 682
Glycerol 682
Glycogenesis 682
Glycogenolysis 682
Glycolysis 682
Goats 491
   abortion 502
   additives 500
   antibiotics 500
   browse 500
   caprine arthritis 502
   concentrates 501
   energy 494
   enterotoxemia 502
   feed(s) 492, 500-502
     commercial 501
     economic importance 492
     preparation 501
     substitution table 502
   feeding 491
     Angora 505
     dairy goats 503
     does 503
     lactating 503
     kids 504, 505
     show, for 504
     Spanish (meat) 505
     yearlings 505
   forbs 500
   hay 501
   haylages 501
   milk 501
     replacers 501
   minerals 495-498
     chart 494
   mohair 25
   nonnutritive 500
   NRC requirements 492-494
   nutrition disorders 502
   nutritive
     needs 492
     requirements 492
   pasture 500
   protein 495
   rations 503-505
   root crops 501
   roughage 501
   silages 501
   vitamins 498-499
     chart 498-499
   water 500
   weight, estimated 761
Goiter 74, 78
Goitrogenic 682
Gossypol 682
Grain(s) 116, 120, 187, 188, 284
   alleurone 190
   by-products 191, 229
   characteristics 285
   composition 192-193
   consumption by humans 4
   definition 120, 188
   diet 4

feed 7
food 7
high-moisture 120, 184, 185, 272
   acid preservation of 185
   buying and selling 186
   definition 120
   feeding value 185, 359
   harvesting 184
   storage 184
milling 191
   by-products 191
nutrient composition 192-194
   carbohydrate 193
   high-lysine corn 194
   lipids 193
   minerals 194
   protein 193
   stage of maturity effect 194
   vitamins 194
proteins 116
quantity fed 188
reconstituted 186, 272
sampling 284
sprouted 197
standards, U.S. 194-195
storage 195
   biological damage 196
   insects and mites 196
   microbial contamination 196
   rodents 196
   structure 188, 189
Grama 697
Gras 682
Grass(es) 1, 127, 129
   adapted 129, 130
   definition 129
   legume silage 177, 178
   staggers 74
   tetany 74, 462
Gravel 88
Gray underfur 651
Graze 682
Grazing
   continuous 134, 140
   controlled 133
   creep 135
   deferred 133, 140
   distribution of animals 142
   extending season 133
   fee 682
   improvement methods 141
   intensive 134
   multispecies 144
   public lands 143
   reseeding 142
   rotation 134, 140
   rotation-deferred 140
   Savory system 141
   selective 322
   short duration 141
   stocking 141
   strip 135
   systems 134
   zero 118
Green chop 118, 135, 445
   definition of 118, 135
Green revolution 682
Gregarious behavior 309-311
Grinding
   feed 269
   forage 276

Grit(s) 254, 570, 682
Groats 682
Grouse 587
Grouven 13
Grow out 682
Growth 20-22
  compensatory 22
  dairy heifers 21
  definition 20
  hormone 259
  implants 763
  measures of 21
  nutritive needs for 21-22
  Quarter Horses 22
  rate 22
  species effect 21-22
  stimulants 384
Growthy 682
Gruel 682
Guinea fowl feeding 588

H

Habituation 325
Haecker 14
Hair
  analysis 110
  growth 321
Hairy indigo 697
Halter pulling 317
Handler aversion 317
Hardinggrass 697
Hard keeper 682
Hardware disease 92, 462
Hatchery by-product 213
Haustral contractions 37
Hay 118, 147, 284
  acid detergent fiber (ADF) of 152, 163
  additives 162
  advantages of 118
  alfalfa cuttings value .. 158
  analysis: NDF, ADF, NIRS .. 152
  balers 154
  bales 159, 165
  bailing wire danger 165
  beef cattle, for 386
  belly 682
  buying 162
  characteristics, quality .. 285
  chemical
    analysis 151
    composition 151
    conditioning 157
  chopped 159
  cocking 156
  comparative value 149
  -concentrate ratio 165
  conditioning, chemical .. 157
  crop silage 178
  cubes (wafers) 160
  curing 156-158
  cutting 156
    value of 158
  dairy cattle, for 446
  definition 118, 147
  dehydrators, artificial .. 161
  disadvantages of 118
  drying, artificial 160
  economic importance 148
  energy source 148
  equipment 153-154
  equivalent 48
  evaluating 153, 162
  feeding 164
    schedule 165
    systems 165
  goats, for 501
  grades 163
  grain replacement 149
  growing forage 153
  harvesting stage 154-155
  hidden costs 164
  history 148
  horse, for 611
  importance 148
  infrared analysis 152
  kinds 150
  legume vs grass 164
  long 159
  loose 159
  losses 148, 164
  magnitude 148
  making systems 158-161
  market grades 163
  moisture 156
  mow curing 161
  neutral detergent fiber (NDF) of 152
  nitrogenation 158
  packaged 159
  pellets 160
  preparation 164
  preservatives 157, 278
  pricing 162-163
  quality 150, 153-162, 682
  rain damaged 157
  raking 156
  refusal 166
  ruminant needs 166
  sampling 152, 284
  scorecard 151
  selling 162
  shattering 156
  sheep, for 481
  shrinkage 164
  sources 162
  spontaneous combustion .. 162
  stacks 160
  stage of maturity ... 150-151, 154-155
  storing 161-162
  stretching supply 166
  supplementing 166
  systems of hay making ... 158
  tests 153
  toxic free residue 164
  value 149, 158
  visual inspection 151
  wafers 160
  wagon dryers 161
  waste 166
  weeds for 150
  weight in barn or stack .. 759
Haylage(s) 118, 119, 169, 178
  definition 119
HCN poisoning 105
Health, animal 19, 22, 327
  definnition 682
  /environmental interactions 327-328
Head picking 317
Heat
  increment 682
  labile 682
  of fermentation 682
  moist processing 271
  -processed feed 682
  -rendered feed 682
  treatments 270
Heaves 77
Heifers
  feeding spayed 413
  replacements 405
Heme 682
Hemicellulose(s) 682
Hemoglobin 682
Hemoglobinuria 68
Hemorrhage 682
Hemp 698
Henry, W. A. 14
Hepatic region 35-36
Herb 682
Herbivore(s) 28, 682
Herringbone fish meal 216
High-energy feeds 120, 187
High-lysine corn 194, 526
High-moisture feeds 125
High-moisture grain .. 120, 169, 184, 272, 526
  acid preservation 185
  buying/selling 186
  harvesting 184
  preservatives 185
  reconstituted 186
  storage 184
  value 185
High-nutrient density 682
Hippocrates 12, 57
Histamine 652
Hives 682
Hogging down crops 528
Home-mixed feeds 300
Home range 682
Hominy feed 229
Homogenized 682
Hormone(s) 683
  additives 259
  compounds 259
  gastrointestinal 37
Horse(s) 591
  antibiotics 610
  behavior, abnormal 316
  bloom feeds 612
  broodmare feeding 617
  bran mash 612
  chestnut 698
  concentrates 611
  digestive system 29
  energy 597
  epiphysitis 599
  fats 598
  feed(s) 611-613
    allowance 596, 613
    bloom 612
    commercial 615
    economic importance .. 592
    fat 199, 598
    forage 611
    home-mixed 615
    milk replacer 612
    palatability 612
    preparation 613
    substitutions 615-616
    sweet 615
  feeding 591, 617
    art of 620
    rules 620
    sale, for 619

Index

show, for 619
starting 620
systems 619
fitting for show/sale 619
foal feeding 618
orphan 618
shaker 619
hay 611
health signs 621
minerals 599-605
chart 600
feeding 605
imbalances 605
NRC requirements 593-596
nutrient(s)
allowances 596
requirements 596
nutritional diseases 641
nutritive needs 592
osteochondritis 599
overfeeding 615
pastures 129-130, 611
pleasure horse feeding 617
protein 598
poisoning 599
supplements 612
racehorse feeding 617
rations 613, 614
silage 611
stallion feeding 618
training horse feeding 617
treats (feeds) 612
2- and 3-year-old feeding 619
unidentified factors 610
vitamins 605-610
chart 606
imbalances 610
water 610
weanling feeding 619
weights estimated 761
yearling feeding 619
Horsepower 683
Hunger 4
definition 27
Hydraulic process 683
Hydrocyanic acid 105
Hydrogenation (fats) 47, 683
Hydrolysis 683
Hydroponics 274
Hygroscopic 683
Hyperthyroidism 683
Hypertrophied 683
Hypervitaminosis 683
Hypocalcemia 81, 683
Hypoglycemia 77, 683
Hypokalemia 683
Hypomagnesemia 683
Hypomagnesemic tetany 74
Hyponatremia 683
Hypothalamus 28, 320
Hysteria, poultry 318

I

ICU (international chick unit) 683
immune suppression 328
Immunoglobulins 92, 683
Impaction (horses) 93, 683
Implants 124, 247, 253, 261
beef cattle 384
definition 124, 248, 683
guidelines 248
safe use 14, 248

Indiangrass 698
Indian ricegrass 698
Indian wheat, wooly 698
Industrial by-products 122
Inert 683
Ingest 683
Ingesta 683
Ingestion 683
Ingestive behavior 308, 310, 311
Ingredient 683
Injections 253
Inoculation, legume 683
Inositol 58
Insulin 683
Intensive grazing 134
Interspecies relationships 315
Intersucking 316
Intestine, small 36
Intestinal juice 683
Intolerance 683
Intracellular 683
Intradermal 683
Intramuscular 683
Intraperitoneal 683
Intravenous 683
Intrinsic factor 683
Investigative behavior 308, 310, 311
In vitro 295, 683
In vivo 683
Iodine 56
deficiency 74, 78
number 46, 683
Ion 683
Ionophores 261, 447
Irish potato 701
Iron 56
Irradiated yeast 683
Irradiation 683
feedstuffs 277
Irrigated pastures 131, 135
Isoacids (IsoPlus) 262, 447
Isolated soybean protein 683
Isomer 683
Isotope 683
IU 683

J

Jack bean 698
Jerusalem artichoke 698
Johnsongrass 698
Joule 48, 684
Jukes 14
Junegrass 698

K

Kale 698
Kellner 14
Kelp 244, 262, 698
Keratin(s) 684
Kernel 684
Ketogenic 684
Ketosis 79, 462, 684
Kibbled 684
Kilocalorie 48
Kjeldahl method 684
Kleingrass 698
Krebs citric acid cycle 684
Kudzu 699
Kwashiorkor 685

L

Labile 684
Lachrymation 684
Lactation (nutrient needs) 20, 24
Lactic acid 684
Lactic acidosis 66
Lactose 684
Lamb's-quarter 699
Laminitis 73, 684
Land
agencies 143-144
conservation 144
farm and ranches, in 128
federal 143
government 144
Indian 143
ownership 128, 143
private 143
public 143
railroad 144
state 143-144
uses 7, 128, 144
Lavoisier 12
Laws, feed 297, 303
Layer(s) 551, 552, 574-577, 684
Leader-follower behavior 314
Lead poisoning 99
Leaf protein concentrate 211
Lecithin 684
Legume(s) 129
adapted 129, 130
inoculated seed 131
Lehmann 14
Lignification 684
Lignin 684
Limited feeding 684
Limiting amino acids 684
Lind 13, 57
Linear programming 340
Linseed meal 206, 209, 228
Lipase 684
Lipid(s) 36, 45, 47, 193
absorption 36
Lipolysis 684
Liquid(s) 121
feeds 121
protein supplements 221, 684
supplements 219, 273, 278
Litter 237
Liver
abscesses 80, 68
Livestock
percent feed from concentrates 117
percent feed from roughages 117
Liveweight 684
Lobster culture 672
Locusts 699
Lovegrass 699
Lunin 13, 57
Lupine 210
Lymph 31, 684
Lyophilization 684

M

Maintenance 18
factors affecting 18-20
requirement—definition 18, 684
Maize 684
Malignant 684
Malnutrition 4, 684

Malt sprouts 233
Malting 192
Malthus, Thomas Robert 4
Manganese 56
Mangel 699
Mango 699
Manure 237–242, 684
   cattle (feed) 239
   composition 239
   feed, as 239, 277
   gasses 238
   poultry (feed) 240, 243
   processing for feed 241, 277
   production 238–239
   regulatory aspects 242
   swine (feed) 240
   uses 239
Marasmus 684
Marine by-products 214–216, 226
Mash 685
Mastication, definition 32, 685
Mastitis 463
McCollum and Davis 13
Meal 685
Measures 753
Meat
   analogs 685
   by-product feeds 212
   meal 212, 227
   packing by-products 212
Mechanically extracted 685
Meconium 685
Medicated feed 262, 685
Medicinal products, animals 8
Megacalorie 48
Melengestrol acetate (MGA) 261
Mendel, Gregor Johann 42
Mesquite 699
Metabolism 41, 44
   definition 41, 685
   trials 291
Metabolite 685
Metabolizable energy 49, 50
Metric system 753
Microbe 685
Microbial enhancers 263
Microbiological assay 289
Microflora 685
Microingredient 685
Micronizing feed 270
Microorganisms, rumen 29–30, 685
Microscopic examinations (feeds) .. 285
Middlings, energy value 193
Milestones, nutrition 15
Milk
   condensed buttermilk 214
   dried 206, 214
   ejection 685
   fever 81, 463
   products 213–214
   protein products 214
   replacers 123, 612, 501
   skimmed 214, 226
   stage (of seed) 685
   whey 214
   whole dried 214
Mill by-product 120, 685
Millet
   Japanese 699
   pearl 699
Mill feeds 120

Milling grain 191
   energy value 193
Mill run 685
Mineral(s) 37, 55, 65, 194
   absorption 37
   beef cattle, for 374–380
   chart 374
   classification of 56
   dairy cattle, for 440–443
   deficiencies 65
      areas 111
   definition 56
   discovery 55
   era 12
   fish, for 661, 662
   functions 56
   goats, for 495–498
   history 55
   horse, for 599–605
   interrelationship 55
   macro- or major 56, 250
   micro- or trace 56, 250
   mink, for 644–647
   poultry, for 561–563
   rabbits, for 627–629
   role 55
   sheep, for 473–477
   supplementation 250
   supplements 124, 249–251, 387, 685
      definition of 124
   swine, for 514–519
   trace or micro- 56
Mink 641
   additives 650
   antibiotics 650
   antioxidants 650
   botulism 652
   Chastek paralysis 650
   cotton fur 651
   energy 643
   fats 644
   feed(s) 652
      allowance 654
      contaminants 656
      economic importance 641
      preparation 653
   feeding 641
      breeding stock 654
      kits 655
      methods 655
   gray underfur 651
   histamine 652
   minerals 644–647
      chart 644
   mycotoxins 652
   NRC requirements 642–643
   nursing sickness 651
   nutritive needs 642
   protein 645
   rations 654
   salmonella 652
   thyroid glands 652
   urinary calculi 651
   vitamins 647–650
      chart 647
   water 650
      methods 655
   wet-belly disease 652
   yellow fat disease 652
Mint 699

Mixing 685
Mohair 25
Moisture 195, 337, 685
   adjusting 252
   analysis 286
   feeds 286, 300
Moisture-free feed 685
Molasses 200, 231, 277
   beet 200, 231
   cane 200, 231
   citrus 200
   wood 200, 234
Mold(s) 290, 685
   inhibitors 262
   toxin producing 100
Mollgaard 14
Molting 576
Molybdenosis 100
Molybdenum 56
   toxicity 100
Monday morning disease 68
Monensin 262
Monogastric digestive system ... 29
Morbidity 685
Mucoid enteropathy 638
Mucus 685
Muddy lots 328
Mud problems—cattle 418
Municipal waste 123, 242
Muscular dystrophy 89
Mycotoxins 100, 197, 652, 685
Myosin 685

N

Napiergrass 699
National Research Council (NRC) .. 336, 685
Naturalistic era 12
Near infrared reflectance
   spectroscopy 152, 289
Necrosis 685
Needlegrass 699
Nephritis 685
Net energy 49, 50, 416
   feeding 344
Neurological control 37–38
Neutral detergent fiber ... 152, 288, 685
Niacin 58
Nicotinamide 58
Nicotinic acid 58
Nitrate poisoning 102
Nitrite poisoning 102
Nitrogen
   balance 685
   balance trials 292
   definition 685
   fixation 685
   free-extract (NFE) 287, 685
Nitrogenous feeds 50
Nonprotein nitrogen . 6, 50, 121, 203, 216–219
   ammoniated feeds 219
   definition 216, 686
   slow-release 219
   sources 216
   urea 217–219
Nonruminant(s) 33–34
   digestive system 28
Nourish 686
Noxious 686

Index

NRC requirements
  beef cattle 349-372
  dairy cattle 426-434
    compositionn of common feeds .. 436-438
  goats 492-494
  horses 593-596
  mink 642-643
  poultry 550-559
  rabbits 626
  sheep 466-480
  swine 508-511
Nucleic acids 42
Nursing sickness 651
Nutrient(s) 41, 44, 686
  allowances 686
  balance trials 291
  carriers 31, 32
  classification of 17
  deficiency, range 137-138
  -deficient animals 289
  definition 17, 41, 686
  functions 17
  requirements 686
Nutrition
  definition 11, 686
  /disease interaction 90
  milestones 12, 15
  parasite interaction 90
  perspective 11
  principles 11
  related disorders 91
  reproductive failure 23-24
Nutritional
  ailments 66-89, 396
  anemia 67, 543
  deficiencies 63, 65
  diseases 66-89, 396, 621
  disorders 63, 91-93
  feedlot diseases 421
  imbalances 63
  toxins 63
Nutritionist, definition 11
Nutritive ratio (NR) 686
Nutriture 686
Nuts 120
Nylon bag technique 290

O

Oak 700
Oat(s)
  milling 192
  straw 237
Oatgrass 700
Oat hay poisoning 102
Obese 686
Offal 686
Off feed 686
Oil(s) 120, 198
  crops 686
Oilseed meals 116, 208
  by-products 228
  feeding value 210
  protein, quality 210
Old process 686
Olfactory 686
Olive 700
Omasum 35
Omnivores 28
Opaque-2 corn 194
Open pen feedlot 415

Oral
  definition 686
  region 32
Orchardgrass 700
Organic
  acids 278
  farming 111
Orphan
  calves 403
  foals 618
  lambs 487
  pigs 541
Orts 686
Osborn-Mendel 13
Osmosis 31
Osmotic pressure 686
Ossification 686
Osteitis 686
Osteochondritis 599
Osteomalacia 82, 686
Osteoporosis 686
Ostrich feeding 588
Overeating disease 71, 502
Overfeeding 24, 321, 686
Overfinishing cattle 418, 686
Overripe 686
Oxalates 39
Oxidation 686
Oxytocin 686
Oyster shell 686

P

Palatability 322, 686
Pancreatic region 35
Pangolagrass 700
Panicgrass, blue 700
Pantothenic acid 58
Paper
  by-products 123, 233
  waste 234
Papillae 686
Para-aminobenzoic acid 58
Paragrass 700
Parakeratosis 39
Parasites 90
Parathyroid glands 686
Partridges 587
Parts per billion (ppb) 686
Parts per million (ppm) 686
Parturient paresis 81
Pasture 117, 127, 128, 284
  adapted grasses and legumes 129-130
  advantages 117
  areas, U.S. 128-129
  beef cattle 129-130, 386
  classes 128
  consumption 7
  cultivated 117
  custom grazing 129
  dairy cattle 129-130, 445
  definition of 117, 128
  disadvantages 118
  establishing 131
  extending season 133
  horses 129-130, 611
  importance 128
  irrigated 131, 135
  leasing 129
  location 128
  management 131-134
  maturity changes 132

  native 117, 128
  permanent 117, 128
  renovating 131
  rotation 117, 128
  sampling 284
  seeded 128
  seeding 131
  semipermanent 128
  sheep 129-130, 481
  supplemental 117, 129
  supplementing 400
  swine 131, 527
  temporary 117, 129
  types 128-129
  value, factors affecting .. 132
  winter 399
Pawing 317
Pea(s), cull 230
Peafowl feeding 589
Peanut meal 206, 210, 228
Pearled 687
Pear cannery residue 230
Peavine 700
Pectin(s) 687
Pellagra 57
Pellet binders 263, 687
Pelleting 271
  forage 277
Pellets 221
People-animal relationships 315
Peptide linkage 687
Pericarp 190, 370
Peristalsis 37
Per oral 687
Per os 687
Pests and pesticides 463
Pesticide poisoning 103
pH 687
Pharyngeal region 33
Phase feeding 687
  dairy 454
  poultry 576
Pheasant feeding 587-588
Phospholipids 687
Phosphorus 56
  supplements 251
Photosensitization 82
Photosynthesis 1, 2, 45, 687
Physiological
  fuel values 687
  saline solution 687
Phytate 687
Phytic acid 39
Pica 65, 315, 317, 687
Pigeon(s), feeding 587
  rations 587
Pigment producing factors 665
Pine 700
  needle abortion 104
Pineapple(s) bran 230
Plant(s)
  by-products 122
  feeds 228-233
  poisonous 93
    preventing losses from ... 94
  proteins 121, 207, 687
Plasma 687
Poison information centers 109
Poisonous
  plants 93-94, 139
    common 93
    diagnosis 109
    treatment 93-94

Poisons **94-109, 546**
   definition of 94
   diagnosing 109
   treating 109
Polar **687**
Polio **83**
Polioencephalomalacia **83**
Pollution
   control 16, 328
   laws 331
Poly- **687**
Polydipsia **318, 688**
Polyneuritis **687**
Polyunsaturated fatty acids **687**
Popping **270**
Population, world **8**
Porcine
   somatatropin 261, 524
   stress syndrome (PSS) 329
Potato(es)
   cull 201, 230
   dehydrated 201, 230
   Irish 201, 701
   sweet 201, 230, 701
Pot-bellied **687**
Poult **687**
Poultry **549**
   additives 569-570
   antibiotics 569
   behavior, abnormal 317-318
   breeder rations 577-578
   broiler rations 580-581
   by-products 213, 227
      meal 213, 228
   cholesterol in eggs 571
   digestive system 30-31
   ducks (see Ducks)
   egg production 24
   energy 559, 570
      apparent metabolizable 559
      definition 559
      true metabolizable 559
   fat 199, 228
   feed(s) 570-571
      economic importance 550
      efficiency 550
      energy 570
      preparation 571
      substitution table 573
   feeding 549
      chickens 574
      phase 576
      programs 574
   game birds (see Game birds)
   geese (see Geese)
   grit 570
   guinea fowl (see Guinea fowl)
   hatchery by-products 213
   layer(s) rations 574-577
   minerals 561-563, 571
      chart 560
      supplements 571
   molting 576
   NRC requirements 550-559
   nutrient needs 550
   ostriches (see Ostriches)
   peafowl (see Peafowl)
   phase feeding 576
   pheasants (see Pheasants)
   pigeons (see Pigeons)
   protein 560, 570
      apparent digestibility of amino acids .. 561
      supplements 570
      true digestibility of amino acids 561

   pullet rations 578
   quail (see Quail)
   rations ... 575, 578, 579, 581, 583, 585-588
   salmonella 589
   substitution table 573
   toxic elements 589
   turkeys (see Turkeys)
   unidentified factors 569
   vitamins 564-568, 571
      chart 564
      supplements 571
   wastes 212-213, 240
   water 569
   xanthophylls 570
Prawn culture **672**
Preconditioning
   beef cattle 405, 407
   definition 322, 687
Precursor **687**
Predacious fish culture **673**
Predators **140**
Pregnancy disease **79, 84**
Prehension, definition **32, 687**
Premix **122, 687**
   definition 122
Preservatives, feeds **278, 687**
Preservatives, silage **179, 180, 278**
   mineral inorganic acids 180
   organic acids 180
Pressed **687**
Pressure
   cooker 687
   flaking 271
Prickly pear **701**
Probiotics **263**
Processing feeds **265**
Production level **20**
Product(s), definition **687**
Prolactin **687**
Prostaglandin **687**
Protein(s) **36-37, 51, 64, 193, 203**
   absorption 36
   amino acid requirements
      nonruminants, for 204
      ruminants, for 204
   analysis 289
   animal 6, 116, 121, 211
   assessment system 223
   beef cattle, for 373
   biological value 53-54
   blocks 221
   -bound iodine test 687
   bumps 687
   bypass (protected, escaped)
      .... 204-207, 278, 439, 471
      feeds, in 205
   by-product feeds 223
   consumption/person 6
   crude 286
   dairy cattle, for 439
   deficiency 64
   definition 203
   degradability 34, 204
      protecting methods 206-207
   dietary protection 204
   digestion 36
   equivalent 204, 687
   escaped 204
   feeds 116, 207
   fish, for 660
   goats, for 495
   high-protein by-product feeds 223
   horses, for 598
   importance of feeds 207

   microbial 206
   mink, for 645
   nonprotein nitrogen 121, 203
   plant 121, 207
   poisoning 599
   poultry, for 560
   protected 204
   pulse 211
   quality 203
   rabbits, for 626
   requirements 204
   salt-feed mixtures 222
   sheep, for 471
   single-cell 122, 220
   solubility 206
   sources of 54
   sparing 688
   supplements .. 121, 203, 221, 249, 612, 633
      amino acid 223
      blocks 221
      definition 121, 688
      hand-feeding 221
      importance 121
      liquid 221
      pasture 221
      pellets 221
      range cubes 221
      self feeding salt mixtures 222
      types and methods of feeding 221
   swine, for 513
Prothrombin **688**
Protoplasm **688**
Provitamin **688**
Provitamin A **688**
Proximate analysis **285**
Prussic acid poisoning **105**
Public lands **143**
   agencies 143
Pullet (replacement) . **552, 554, 578, 579, 688**
Pulmonary emphysema **84**
Pulp **688**
Pulpy kidney disease **71**
Pulse proteins **211**
Pumpkin **701**
Purgative **688**
Purified diet **688**
Putrefaction **688**

Q

Quackgrass **701**
Quail
   Bobwhite NRC requirements 558
   Japanese NRC requirements 559
Quality **688**
   protein 688
Quarter Horse growth **22**

R

Rabbits **623**
   additives 632
   antibiotics 632
   anticoccidial drugs 632
   antioxidants 633
   coprophagy 626
   energy 625
   enteric diseases 638
   enterotoxemia 638
   feed(s) 633
      allowance 634
      economic importance 624
      energy 633

Index

preparation 633
protein 633
roots 633
substitution table 636
tubers 633
feeding 637
hand 637
methods 637
self- 637
fiber 632
forages 633
fur-eating 638
grasses 633
minerals 627–629
chart 628
mucoid enteropathy 638
nonnutritive factors 632
NRC requirements 625, 626
nutritive needs 625
orphans 637
protein 626
rations 635
Tyzzer's disease 638
vitamins 630–632
chart 630
water 632
Rachitic 688
Radiant heat 688
Radiation 688
Radioactive 688
active ^{131}I uptake tests 688
isotope 688
tracer 688
Railroad-owned lands 144
Ralgro 261
Rancid fats 47, 688
Range 136
area 137
brush control 142
cubes 221, 688
distribution of animals 142
feeding range sheep 485
forages 117, 127, 688
grasses 127
grazing systems 140
continuous 140
deferred rotation 140
rotation 140
Savory system 141
short duration 141
improvement 141–142
kind of livestock 139
management 138–139
nutrient deficiencies 137–138, 400
pellets 221
poisonous plants 139
protein supplements 221
reseeding 142
season of use 139
sheep 485
stocking rate 139, 141
supplementing 138, 400
weed control 142
western 136
Rangeland 688
Rapeseed meal 209, 228
Ratany, range 701
Rate of passage (feed) 688
Ration(s)
balancing 337
beef cattle rations 394–395
complete 279
dairy cattle rations 449–457
definition 688
fish rations 666–667, 669
formulation 337–343
balancing 337
flexible formula 343
health considerations 337
ingredients 336
methods 338
computer 340–343
square 338
trial and error 339
moisture content 337
net energy method 344
steps 337, 338–340
worksheets 343
goat rations 503, 504, 505
health considerations in rations .. 337
horse rations 614, 618
mink rations 654
moisture content adjusting 337
poultry rations
..... 575, 578, 579, 581, 583, 585–588
rabbit rations 635
roughages in rations 336
sheep rations 484, 486, 488, 489
supplements in rations 336
swine rations 530–538
Ratios 688
Rearing 317
Recombinant DNA 14, 43
Reconstituted grain 272
Records 419
Red meat 688
Redtop 701
Regurgitation 688
Relationships 315
interspecies 315
people/animal 315
Relative feed value 688
Rennin 688
Replacement heifers (beef) 405
rations 406
Reproduction 22–23
nutritional effects 23
nutritional failure 23–24
Rescuegrass 701
Residue 689
corn 237
crop 122, 234
economy 235
feeding 236
harvesting 235
nutrient value 235
storing 236
transporting 236
treating 235
Respiratory quotient (RQ) 689
Restricted feeding 689
Retained placenta 463
Retarded growth 689
Reticulitis 39
Reticular groove 30
Reticulum 34
Retinol equivalent 689
Retort-charred 689
Rhodesgrass 701
Riboflavin 58
Ribonucleic 42
Rice
milling 192
polishings 192, 229
Rickets 57, 63, 85
RNA 42, 43
Roasting feed 270
Rodents 196
Rolling feed 270
steam 220
Root(s) 120, 200, 446, 481, 501, 633
crops 386
definition 690
Rose, William C. 14
Rotation grazing 134, 140
Roughage(s)
consumption 7
definition 689
percent feed for livestock 117
substitutes 254
Royal palm 701
Rumen 34
bypass 278
culture 689
fill 689
flora 689
microorganisms 29, 30
Rumensin (monensin) 262
Ruminal acidosis 39
Ruminant(s) 1, 3, 6, 7, 10, 34
digestive system 29
need hay 156
newborn 35
nonprotein nitrogen conversion ... 6
roughage use 7
stomach 30
Rumination 30, 689
Running 25
Russian-thistle 702
Rutabaga 702
Rye
milling 192
nutrient composition 193
Ryegrass 702

S

Sacaton 702
Saccharides 689
Safflower meal 210, 328
Sagebrush 702
Sainfoin 702
Saint Augustine grass 702
Saline 689
Saliva 32
Salivary glands 32–33
Salmonella 106–107, 589, 652, 689
Salmonellosis 106, 689
Salmon
feeding 666
meal 216
Salt 251
-feed mixtures 222, 278
sick 86
Saltbush 702
Saltgrass 702
Saponification 47, 689
Satiety 689
Saturated fats 46, 689
Savage sow syndrome 316
Savory grazing system 141
Sawdust 234
Scalping 689
Schwartz 56
Scouring 544
Scours (calf) 40
Scratch poultry feed 689
Screened (feed) 689

Screenings	229
Scurvy	689
Seaweed	244
Sedge	703
Seed, parts of	190
Selenium	
poisoning	67
toxicity	107
Self-fed	689
Self-feeder	689
Self-feeding governors	278
Semipermeable membrane	689
Serum	689
Sesame meal	210, 229
Sewage sludge	242
Sex difference in nutrition needs	22
Sexual behavior	309
Shaker foal	619
Sheep	465
additives	480, 481
antibiotics	481
behavior, abnormal	315
concentrates	481
creep feeding	486
energy needs	470
feed(s)	481
economic importance	466
preparation	481
substitution table	481–483
feeding	465, 484
breeding ewes	484, 485
creep	486
drylot	485, 488
early weaned	487
finishing lambs	487–489
areas and types of	487
field	487
rations	488–489
flushing	485
growing-finishing lambs	486, 487
lactating ewes	485
lambing time	485
pregnant ewes	485
rams	484
range sheep	485, 486
show	489
hay(s)	481
implants	481
minerals	473–477
chart	473
NRC requirements	466–480
nutritional disorders/toxins	489
nutritive needs	466
orphan lambs	487
pastures	129–130, 481
protein needs	471
range	
nutrient deficiencies	485
supplements	485
rations	484, 485, 489
fitting	488
growing-finishing	488
root crops	481
silages	481
vitamins	478–480
chart	478–480
water	480
weight estimated	761
wool	25
pullling	315
Shelter-seeking behavior	309
Short feed	689
Shredded	689
Shredding forage	276
Shrimp	
culture	673
meal	227
Shrinkage	689
Shying	317
Sifted	689
Silage(s)	116, 118, 119, 284
additives	179, 180
definition of	119, 169
advantages	170, 171
beef cattle, for	386
characteristics	183, 285
combining crops	179
composition	176
corn	176–178
cut length	181
dairy cattle, for	445
definition	118, 169, 170
direct cut	177
disadvantages	170
distribute forage evenly	182
drought-stricken crop	179
economy of	170, 183
ensiling process	170
fill rapidly	182
frosted crop	179
gasses, danger	171, 183
goats, for	501
grass/hay crop	177, 178
grass/legume	177, 178
harvesting	
methods	181
stage of maturity	181
haylage	778
high-moisture grain	184–186
horses, for	611
kinds	176–179
losses, storage	176
low-moisture	178
machinery	181
milk	
flavor effect	183
odor effect	183
moisture	
control	182
determination	182
moldy	183
NPN additives	179
potato	178
preservatives	179, 180, 278
rain-damaged hay	179
sampling	284
sealing	183
sheep, for	481
sorghum	176, 178
residue	176
sunflower	178
swine, for	527
urea-mix	217–219
value, feeding	183
weight	175
wilting	178
Silicon	56
Silo(s)	171–176
bunker (self-feeder)	172
conventional upright	171
gastight	171
horizontal	172
kinds	171
losses	176
pit	172
plastic	172–173
bag	173
bale, round	173
seal	183
size determination	173–175
stack	
enclosed	172
modified trench	172
open	172
upright	172
storage losses	176
temporary	172
tower	171, 174
trench	172, 175
upright	171
weight in	175
Single-cell protein	122, 220
definition	220
problems	220
types	220
Skimmed milk	214, 226
condensed cultured	214
dried cultured	214
Sleek coat	689
Slotted floors	415, 689
Slow release treatments	278
Small grain refuse	689
Soap	689
Social	
behavior, definition	312
order, dominance	313
organization	312–313
relationships	312
cattle	312
goats	312
horses	312
interspecies	315
leader-follower	314
people-animal	315
poultry	313
sheep	312
swine	312
Sodium	56
bentonite	690
bicaronate	690
chloride	56
Soft pork	544, 690
Soil	
erosion control	145
fertility	8
good	110
/livestock	110
nutrients	110
testing	111
Soiling	118
Solubles	690
Solution	690
Solvent-extracted	690
Somatatropin	259, 260
Sorghum	690, 703
milling	192
nutrient composition	192–193
residue	176
silage	176, 178
stover	237
Soybean(s)	
cooked	526
full-fat	527

Index

hulls 237
meal 208, 229, 526
   expeller extraction 208
   feeding value 210
   hydraulic extraction 208
   processing 208
   quality 210
Spanish goats 505
Spayed heifers 413
Specific
   dynamic action (SDA) 690
   gravity 690
   heat 690
Spectrophotometry 289
SPF (pigs) 690
Spontaneous combustion of hay ... 162
Spray-dehydrated 690
Sprouted grain 197
Square method feed formulation .. 338
Stabilized 690
Stalk (plant) 690
Stall walking 317
Standards, feeding 335-336
Stargrass 703
Steam
   flaking 271
   rolling feed 270
Steamed 690
Stearin 690
Steroids 263
Sterol 690
Stiff-lamb disease 87, 89
Stilbestrol 14
Stocker(s), cattle (feeder) 407
   backgrounding 407
   compensatory growth 410
   contracts 411
   feeders, definition 407
   growth stimulants 410
   implants 410
   preconditioning 407
   programs 408, 411
   rations 408-410
   replacement heifers 407
   wintering 410
Stocking rate 139, 690
Stomach
   nonruminant 29
   ruminant 30
Stomatitis 690
Stones 88
Stool (fecal) 690
Storage of feedstuffs, grain 195
Stover 228, 690
Straw(s) 690
Stray voltage 329
Stress 19, 322, 330, 690
Striking 317
Strip grazing 135
Stubble 690
Stump sucker 316
Sturgeon culture 673
Subcutaneous 690
Succulence 690
Suckle 690
Sucrase 690
Sucrose 690
Sudangrass 703
Sugar
   beets 231
      energy value 201
   by-products 230

Sugarcane 231, 236
Sun-cured 690
Sunflower meal 210, 229
Supplements 124, 247, 692
   amino acids 249
   calcium-phosphorus 251
   definition 124, 247, 690
   feed 247, 248
   liquid 273
   mineral 124, 249-251
   protein 121, 203, 249
   vitamin 124, 251-253
Sustainable agriculture 332
Sweet clover 703
   disease 87
Sweet corn cannery refuse 237
Sweet feed 615, 690
Sweet potato(es) 201, 230, 703
Swine 507
   additives 522-524
   anemia 543
   antibiotics 523
   behavior, abnormal 315
   concentrates 525
   contract production 545
   corn 525
      high-lysine 526
      high moisture 526
      -hog ratio 545
   creep feeding 541
   digestive system 29
   docking tails 543
   energy 512
   environmental effects 546
   fats and oils 199, 526
   feeder pig production 543
   feed(s) 525-528
      allowances 511, 529-539
      economic importance 507
      energy 525
      /market pig 544
      preparation 528
      protein 526
      soybean meal 526
      soybean, full-fat 527
      substitution table 539-540
   feeding 507
      baby pigs 541
      boars 541
      breeding gilts 541
      brood sows 541
      growing-finishing 541
      orphans 541
      programs 528, 529
      show and sale 542
      systems 542
   flushing sows 541
   forages, dry 527
   garbage 528
   hogging down crops 528
   leasing hogs 545
   manure, feed 239
   minerals 514-519
      chart 514-517
   NRC requirements 508-511
   nutrient
      allowances 511-512
      needs 508
   nutritional anemia 543
   pastures 131, 527
   poisons/toxins 546
   porcine somatatropin 524

   protein 513
   rations 529-539
   scouring 544
   sex effect on performance 544
   silages 527
   soft pork 544
   sulfonamides 523
   unidentified factors 519
   vitamins 519-522
      chart 520
   water 522
   waste 240
   weaning 544
   weights estimated 761
Switchgrass 703
Symbiosis 690
Syndrome 690
Synovex 261
Synthesis 690
Synthetics 691

T

Tail biting 316, 691
Tankage 212, 227
Taylor Grazing Act 136, 143
TDN 48, 49, 691
   converting to Mcal 691
Teart 691
Teeth 32
Teff 704
Temperament, animal 20
Temperature 195
   Centigrade/Fahrenheit 757
   comfort zone 19
   critical 18-19
      lower 324
      upper 324
   optimum 19, 324
Tempering 691
Tetany 691
Tether 691
Thaer 13
Therm 691
Thermal 691
Thermoneutral zone 324
Thermostatic appetite control 691
Thiamin 58
   -stimulating hormone 691
Thiaminase 691
Three-awn 704
Thrifty 691
Thrombosis 691
Thyroid gland as mink feed 652
Thyroxin 691
Tilapia culture 673
Toasted 691
Tobosa 704
Tocopherol(s) 691
Toe picking 317
Tomato(es) pomace 230
Tongue 32
Tonic 691
Torula yeast 234
Total digestible nutrients (TDN) ... 691
Total milk solids 691
Toxemia 691
Toxic elements 589, 673, 691
Toxins 63, 100-102, 546
Trace minerals 56, 691
   element 691

Tracer isotope 691
   digestibility 290
Tranquilizers 264
Transamination 53
Transferrin 691
Treats (horses) 612, 691
Trefoil 704
Trial and error feed formulation 339-340
   computer 340
Triglycerides 691
Trout diets 666
True protein 691
Trypsin 691
Tuber(s) 200, 691
   energy value 201
Tumblegrass 704
Turkey(s)
   behavior, abnormal 318
   body weight/feed consumption 556
   feeding 582
   growth rate/feed consumption 556
   NRC requirements 555
   rations 583-584
Turnip(s) 201, 230
   grazing 134
Twenty-eight hour rail law 691
Tympany, acute 39
Tyzzer's disease 638

U

Udder edema 463
Ulcers 330
Ultra-violet 691
Underfeeding 24, 322, 692
Unidentified factors ... 253, 519, 569, 610
Unsaturated fat 692
Unsaturated fatty acid 692
Unthriftiness 692
Urea 7, 50, 205, 217-219
   beef cattle 387
   dairy cattle 440
   feeding facts 218-219
   toxicity 108, 218
   utilization 218
Urease 692
Uremia 692
Urinary calculi 88, 651
Urine 692
Urolithiasis 88
U.S. Forest Service 143
USP (unit) 692

V

Vacuum-dehydrated 692
Valence 692
Veal calf production 460
Vealer 692
Vegetable
   by-products 229
   protein 692
Velvet bean 704
Vent picking 317
Vermifuges 254
Vetch 704
VFA 692

Villi 692
Vitamin(s) 13, 57, 65, 194, 380
   A 58, 251
   absorption 37
   beef cattle, for 380-383
   B-12 58
   B complex 58-59, 252
   biological era 13
   C 58, 253
   classification 58
   D 252
   dairy cattle, for 443-445
   deficiencies 65
   definition of 58
   digestion 37
   discovery 52-58
   E 58, 252
   fat-soluble 58, 382
   fish, for 662-664
   functions 58, 59
   goats, for 498-499
   history 57-58
   horses, for 605-610
   imbalances 253
   K 58, 252
   mink, for 647-650
   poultry, for 564-568
   rabbits, for 630-632
   sheep, for 478-480
   supplements 124, 251-253, 692
      definition 124
   swine, for 519-522
   water-soluble 58, 382
Void 692
Volatile fatty acids (VFA) 692
Vomiting 692

W

Wafering forage 276
Wafers 692
Warm confinement 416, 692
Waste
   by-product feeds 237
   poultry 212, 240
Water 59-61, 145
   balance 59
   beef cattle, for 384
   belly disease 88, 652
   consumption by species 60
   dairy cattle, for 445
   drinking 60
   /environmental interactions 322-323
   excretion 61
   feeds, in 60
   fish, for 664
   goats, for 500
   horse, for 610
   hyacinth 244, 705
   metabolic 61
   mink, for 650
   poultry, for 569
   rabbits, for 632
   sheep, for 480
   sources 60
   swine, for 522
Watered feeds 273

Water hyacinth 705
Watermelon 705
Weaning 405, 544, 692
Weather 18
   /environmental interactions 324-326
Weed(s) 142
   as feed 124
   control 124, 142
Weight ratio 692
Weights and measures 753
   animal estimates 760
Western range 136
Wet
   feeds 125
   milled 692
   rendered 692
Wheat
   bran 229
   chaff 237
   energy value 193
   middlings 229
   mill feeds 207
   milling 192
   nutrient composition 192
   straw 237
   structure (seed) 188
Wheatgrass 705
Whey products 214, 226
Whitefish meal 216
White muscle disease 87, 89
Wildlife 145
Wild-rye grass 705
Wilted 692
Wind sucker 316
Winterfat 705
Winter pastures 398
Winter tetany 74
Wolff 13
Woll 14
Wood
   by-products 123, 233-234
   chewing 317, 692
   molasses 200, 234
Wool 25
Work (running) 25
World population 8
   food improving 8
Wormers 254

X

Xanthophylls 253, 570, 692
Xerosis 692

Y

Yeast 692
   brewers', dried 233
   torula 234
Yellow fat disease 652
Yucca 705

Z

Zero grazing 118, 692
Zero tolerance 14
Zinc 56